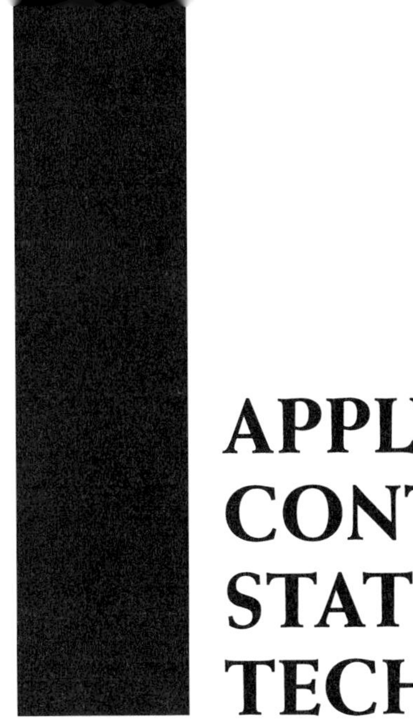

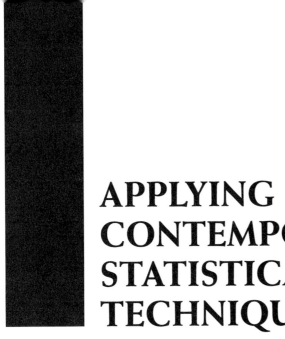

APPLYING CONTEMPORARY STATISTICAL TECHNIQUES

APPLYING CONTEMPORARY STATISTICAL TECHNIQUES

Rand R. Wilcox

University of Southern California
Los Angeles, CA

ACADEMIC PRESS

An imprint of Elsevier Science

Amsterdam Boston London New York Oxford Paris
San Diego San Francisco Singapore Sydney Tokyo

Senior Editor, Mathematics	Barbara A. Holland
Senior Project Manager	Angela Dooley
Editorial Coordinator	Tom Singer
Product Manager	Anne O'Mara
Cover Design	Shawn Girsberger
Copyeditor	Elliot Simon
Composition	CEPHA
Printer	Maple-Vail

This book is printed on acid-free paper. ♾

Academic Press
An imprint of Elsevier Science
525 B Street, Suite 1900, San Diego, California 92101-4495, USA
http://www.academicpress.com

Academic Press
An imprint of Elsevier Science
84 Theobald's Road, London WC1X 8RR, UK
http://www.academicpress.com

Academic Press
An imprint of Elsevier Science
200 Wheeler Road, Burlington, Massachusetts 01803, USA
http://www.academicpressbooks.com

Library of Congress Catalog Card Number: 2002111027

International Standard Book Number: 0-12-751541-0

PRINTED IN THE UNITED STATES OF AMERICA
02 03 04 05 06 9 8 7 6 5 4 3 2 1

CONTENTS

PREFACE

Overview

The goals in this book are: (1) to describe fundamental principles in a manner that takes into account many new insights and advances that are often ignored in an introductory course, (2) to summarize basic methods covered in a graduate level, applied statistics course dealing with ANOVA, regression, and rank-based methods, (3) to describe how and why conventional methods can be unsatisfactory, and (4) to describe recently developed methods for dealing with the practical problems associated with standard techniques. Another goal is to help make contemporary techniques more accessible by supplying and describing easy-to-use S-PLUS functions. Many of the S-PLUS functions included here have not appeared in any other book. (Chapter 1 provides a brief introduction to S-PLUS so that readers unfamiliar with S-PLUS can employ the methods covered in the book.) Problems with standard statistical methods are well known among quantitative experts but are rarely explained to students and applied researchers. The many details are simplified and elaborated upon in a manner that is not available in any other book. No prior training in statistics is assumed.

Features

The book contains many methods beyond those in any other book and provides a much more up-to-date look at the strategies used to address nonnormality and heteroscedasticity. The material on regression includes several estimators that have recently been found to have practical value. Included is the deepest regression line estimator recently proposed by Rousseeuw and his colleagues. The last chapter covers rank-based methods, but unlike any other book, the latest information on handling tied values is described. (Brunner and Cliff describe different strategies for dealing with ties and both are considered.) Recent results on two-way designs are covered, including repeated measures designs.

Chapter 7 provides a simple introduction to bootstrap methods, and chapters 8–14 include the latest information on the relative merits of different bootstrap techniques when dealing with ANOVA and regression. The best non-bootstrap methods are covered as well. Again, methods and advances not available in any other book are described.

Chapters 13–14 include many new insights about robust regression that are not available in any other book. For example, many estimators often provide substantial improvements over ordinary least squares, but recently it has been found that some of these estimators do not always correct commonly occurring problems. Improved methods are covered in this book. Smoothers are described and recent results on checking for linearity are included.

Acknowledgments

The author is grateful to Sam Green, Philip Ramsey, Jay Devore, E. D. McCune, Xuming He, and Christine Anderson-Cook for their helpful comments on how to improve this book. I am also grateful to Pat Goeters and Matt Carlton for their checks on accuracy, but of course I am responsible for any remaining errors. I'm especially grateful to Harvey Keselman for many stimulating conversations regarding this book as well as inferential methods in general.

Rand R. Wilcox
Los Angeles, California

INTRODUCTION

The goals of this book are to describe the basics of applied statistical methods in a manner that takes into account the many insights from the last half century, to describe contemporary approaches to commonly encountered statistical problems, to provide an intuitive sense of why modern methods — developed after the year 1960 — have substantial advantages over conventional techniques, and to make these new methods practical and accessible. Once basic concepts are covered, the main goal will be to address two general types of statistical problems that play a prominent role in applied research. The first is finding methods for comparing groups of individuals or things, and the second has to do with studying how two or more variables are related.

To elaborate on the first general problem to be considered, imagine that you give one group of 20 individuals a drug for lowering their cholesterol level and that a second group of 20 gets a placebo. Suppose the average decrease for the first group is 9.5 and for the second group is 7.2. What can we say about the population of all individuals who might take this drug? A natural guess is that if all individuals of interest took the drug, the average drop in cholesterol will be lower versus using the placebo. But obviously this conclusion might be wrong, because for each group we are attempting to generalize from a sample of 20 individuals to the millions of people who might take the drug. A general goal in statistics is to describe conditions and methods where the precision of generalizations can be assessed.

The most common approach to the problem just posed is based on a general strategy developed by Pierre-Simon Laplace about two centuries ago. In the drug example, 9.5 is the average based on the 20 available participants. But as an estimate of the average we would get if millions of people took the drug, chances are that 9.5 is inaccurate. That is, it differs from the *population* average we would get if all potential individuals took the drug. So a natural question is whether we can find some way of measuring the precision of the estimate, 9.5. That is, can we rule out certain values for the population average based on the data at hand, and can we specify a range of values that is likely to contain it?

Laplace actually developed two general approaches to the problem of assessing precision. His first approach was based on what we now call a Bayesian method.[1] His second approach is now called the *frequentist* approach to statistical problems. It is described in Chapter 4, it is covered in almost all introductory statistics books, and currently it forms the backbone of statistical methods routinely used in applied research. Laplace's method is based in part on assuming that if all individuals of interest could be measured and the results were plotted, we would get a particular bell-shaped curve called a *normal distribution*. Laplace realized that there was no particular reason to assume normality, and he dealt with this issue by using his central limit theorem, which he publicly announced in 1810. Simply put, the *central limit theorem* says that if a sufficiently large number of observations is randomly sampled, then normality can be assumed when using Laplace's method for making inferences about a population of people (or things) based on the data available to us. (Details about the central limit theorem are covered in Chapter 4.)

One obvious concern about the central limit theorem is the phrase *sufficiently large*. Just how many observations do we require so that normality can be assumed? Some books claim that the answer is 40, and others state that even 25 observations suffice. These statements are not wild speculations; they stem from results discussed in Chapter 4. But we now know that this view can be highly misleading and inaccurate. For some of the simplest problems to be covered, practical situations arise where hundreds of observations are needed. For other routinely used techniques, inaccurate inferences are possible no matter how many observations happen to be available. Yet it seems fair to say that despite the insights made during the last 40 years, conventional wisdom still holds that the most frequently used techniques perform in a satisfactory manner for the majority of situations that arise in practice. Consequently, it is important to understand why the methods typically taught in an introductory statistics course can be highly unsatisfactory and how modern technology can be used to address this problem.

In our earlier illustration, two groups of individuals are being compared; but in many situations multiple groups are compared instead. For example, there might be interest in three experimental drugs for lowering cholesterol and how they compare to a placebo. So now a total of four experimental groups might take part in an experiment. Another common view is that the more groups of individuals we compare, the more certain we can be that conventional methods (methods developed prior to 1960 and routinely used today) perform in a satisfactory manner. Unfortunately, this speculation is incorrect as well, and again it is important to understand why in order to appreciate the modern techniques described in this book.

The other general problem covered in this book has to do with discovering and describing associations among variables of interest. Two examples will help clarify what this means. The first has to do with a classic problem in astronomy: Is the universe expanding? At one point Albert Einstein assumed that the answer is no — all stars are fixed in space. This view was based on a collective intuition regarding the nature of space and time built up through everyday experiences over thousands of years. But one implication of Einstein's general theory of relativity is that the

1 The Reverend Thomas Bayes was the first to propose what we now call the Bayesian approach to statistics. But it appears that Laplace invented this approach independent of Bayes, and it was certainly Laplace who developed and extended the method so that it could be used for a wide range of problems.

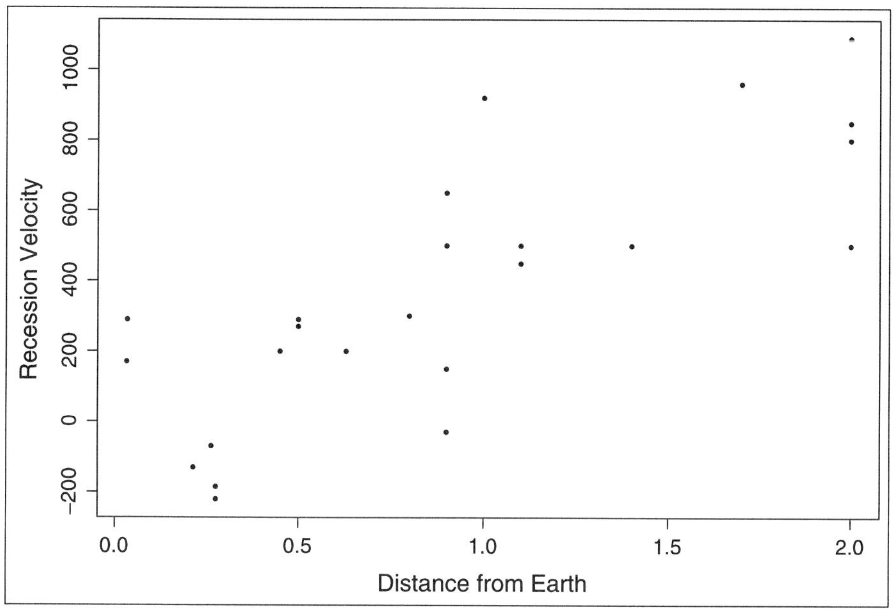

FIGURE 1.1 A scatterplot of Hubble's data.

universe cannot be static. In fact, during the early 1920s, the Russian meteorologist Alexander Friedmann provided the details showing that Einstein's theory implied an expanding universe. But during the early years of the twentieth century, the notion of a never-changing universe was so ingrained that even Einstein could not accept this implication of his theory. For this reason, he revisited his equations and introduced what is known as the cosmological constant, a term that avoids the prediction of a changing universe.

But 12 years later, Edwin Hubble made some astronomical measurements indicating that galaxies are either approaching or receding from our own Milky Way Galaxy. Moreover, Hubble concluded that typically, the further away a galaxy happens to be from our own, the faster it is moving away. A scatterplot of his observations is shown in Figure 1.1 and shows the rate (in kilometers per second) at which some galaxies are receding from our own galaxy versus its distance (in megaparsecs) from us. (The data are given in Table 6.1.) Hubble's empirical evidence convinced Einstein that the universe is generally expanding, and there has been considerable confirmation during the ensuing years (but alternative views cannot be completely ruled out, for reasons reviewed by Clark, 1999). Based on Hubble's data, is the conclusion of an expanding universe reasonable? After all, there are billions of galaxies, and his observations reflect only a very small proportion of the potential measurements he might make. In what sense can we use the data available to us to generalize to all the galaxies in our universe?

Here is another example where we would like to understand how two variables are related: Is there an association between breast cancer rates (per 100,000 women) and solar radiation (in calories per square centimeter)? Figure 1.2 shows a scatterplot, based on 24 cities in the United States, of the breast cancer rate among

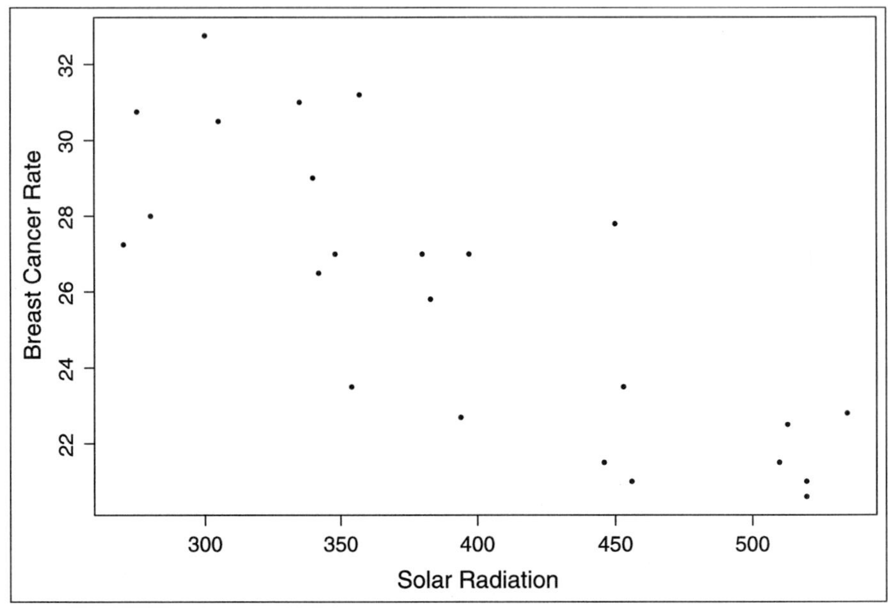

FIGURE 1.2 Breast cancer rates versus solar radiation.

100,000 women versus the average daily amount of solar radiation in calories per square centimeter. Can we make reasonable inferences about the association between these two variables regarding all geographical regions we might measure? What must be assumed to make such inferences? To what extent can we violate these assumptions and still arrive at reasonably accurate conclusions? Again it was Laplace who laid down the basic tools and assumptions that are used today. The great mathematician Carl Gauss extended and refined Laplace's techniques in ways that will be described in subsequent chapters. More refinements would come about a century later that are routinely used today—methods that are in some sense dictated by a lack of access to high-speed computers. But two fundamental assumptions routinely made in the applied work of both Laplace and Gauss are at the heart of the conventional methods that play a dominant role in modern research. Now that we are about two centuries beyond Laplace's great insight, what can be said about the accuracy of his approach and the conventional modifications routinely used today? They are, after all, mere approximations of reality. How does access to high-speed computers help us analyze data? Do modern methods and computers open the door to new ways of analyzing data that have practical value?

The answer to the last question is an unequivocal yes. Nearly a half century ago it became obvious from a theoretical point of view that conventional methods have an inherent problem with potentially devastating implications for applied researchers. And more recently, new insights have raised additional concerns of great practical importance. In simple terms, if groups differ or variables are related in some manner, conventional methods might be poorly designed to discover this. Moreover, the precision and accuracy of conventional methods can be relatively poor unless sample sizes are fairly large. One strategy for dealing with these problems is simply to hope they never arise in practice. But all indications are that such situations are rather common. Interestingly, even Laplace had derived theoretical results hinting of serious problems associated with techniques routinely used today. But because of both technical and computational difficulties, finding practical alternatives proved to be extremely difficult until very recently.

In addition to theoretical concerns regarding standard statistical methods are empirical studies indicating practical difficulties. The first such study was conducted by Bessell in 1818 with the goal to determine whether the normal curve provides a good approximation of what we find in nature. Bessell's data reflected a property that is frequently encountered and poorly handled by conventional techniques. But unfortunately, Bessell did not have the mathematical tools needed to understand and appreciate the possible importance of what he saw in his data. Indeed, it would be nearly 150 years before the importance of Bessell's observation would be appreciated. Today, a variety of empirical studies support concerns about traditional techniques used to analyze data, as will be illustrated in subsequent chapters.

1.1 Software

One goal in this book is to provide easy access to many of the modern statistical methods that have not yet appeared in popular commercial software. This is done by supplying S-PLUS[2] functions that are very easy to use and can be downloaded, as described in Section 1.2. For most situations, you simply input your data, and a single call to some function will perform the computations described in subsequent chapters. S-PLUS is a powerful and vast software package that is described in various books (e.g., Krause and Olson, 2000) and manuals.[3] Included are a wide range of built-in functions not described in this book.[4]

An alternative to S-PLUS is R, which is nearly identical to S-PLUS and can be downloaded for free from www.R-project.org. Both zipped and unzipped files containing R are available. (Files ending in .tgz are zipped.) The zipped file can be downloaded more quickly, but it requires special software to unzip it so that it can be used. Also available from this Web site is a free manual explaining how to use R that can serve as a guide to using S-PLUS as well. Unfortunately, S-PLUS has a few built-in functions that are not standard in R but that are used in subsequent chapters.

The goal in the remainder of this section is to describe the basic features of S-PLUS that are needed to apply the statistical methods covered in subsequent chapters. An exhaustive description of the many features and nuances of S-PLUS go well beyond the scope of this book.

Once you start S-PLUS you will see this prompt:

```
>
```

---

2 S-PLUS is a registered trademark of Insightful Corporation, which can be contacted at www.insightful.com

3 See, in particular, the *S-PLUS User's Guide* as well as *S-PLUS 4 Guide to Statistics*, Data Analysis Products Division, Mathsoft, Seattle, WA.

4 For software that links the S-PLUS functions in this book to SPSS, see the Web site zumastat.com

It means that S-PLUS is waiting for a command. To quit S-PLUS, use the command

```
> q()
```

1.1.1 Entering Data

To begin with the simplest case, imagine you want to store the value 5 in an S-PLUS variable called *dat*. This can be done with the command

```
> dat<-5,
```

where < − is a "less than" sign followed by a minus sign. Typing *dat* and hitting Return will produce the value 5 on the computer screen.

To store the values 2, 4, 6, 8, 12 in the S-PLUS variable dat, use the c command, which stands for "combine." That is, the command

```
> dat<-c(2,4,6,8,12)
```

will store these values in the S-PLUS variable dat.

To read data stored in a file into an S-PLUS variable, use the scan command. The simplest method assumes that values are separated by one or more spaces. *Missing values* are recorded as NA, for "not available." For example, imagine that a file called *ice.dat* contains

```
6 3 12 8 9
```

Then the command

```
> dat<-scan(file="ice.dat")
```

will read these values from the file and store them in the S-PLUS variable dat. When using the scan command, the file name must be in quotes. If instead you have a file called *dis.data* that contains

```
12   6  4
 7  NA  8
 1  18  2
```

then the command

```
> dat2<-scan(file="dis.data")
```

will store the data in the S-PLUS variable dat2. Typing dat2 and hitting Enter returns

```
12 6 4 7 NA 8 1 18 2
```

Values stored in S-PLUS variables stay there until they are removed. (On some systems, enabling this feature might require the command !SPLUS CHAPTER.) So in this last example, if you turn off your computer and then turn it back on, typing dat2 will again return the values just displayed. To remove data, use the rm command. For example,

```
> rm(dat)
```

would remove the data stored in dat.

S-PLUS variables are case sensitive. So, for example, the command

```
> Dat2<-5
```

would store the value 5 in Dat2, but the S-PLUS variable dat2 would still contain the nine values listed previously, unless of course they had been removed.

S-PLUS has many built-in functions, and generally it is advisable not to store data in an S-PLUS variable having the same name as a built-in function. For instance, S-PLUS has a built-in function called *mean* that computes the average of the values stored in some S-PLUS variable. For example, the command

```
> mean(x)
```

will compute the average of the values stored in x and print it on the screen. In some situations S-PLUS will tell you that a certain variable name is reserved for special purposes and will not allow you to use it for your own data. In other situations it is allowed even when the variable name also corresponds to a built-in function. For example, the command

```
> mean<-2
```

will store the value 2 in an S-PLUS variable called *mean*, but mean(x) will still compute the average values stored in x. However, to avoid problems, particularly when using the functions written for this book, it is suggested that you do not use a built-in function name as an S-PLUS variable for storing data. A simple way to find out whether something is a built-in function is to type the name and hit Return. For instance, typing

```
> mean
```

will return

```
function(x, trim = 0, na.rm = F)
```

That is, mean is a built-in function with three arguments. The latter two arguments are optional, with default values if not specified. For example, na.rm indicates whether missing values are to be removed. By default, na.rm=F (for false), meaning that missing values are not removed. So if there are any missing values stored in x, mean(x) will result in the value NA. If, for example, you use the command mean(z,na.rm=T), any missing values will be removed and the average of the remaining values is computed. (Some details about built-in functions are provided by the help command. For example, help(mean) provides details about the function mean.)

If you type

```
> blob
```

and hit Enter, S-PLUS returns

```
Object "blob" not found
```

because data were never stored in the S-PLUS variable blob and there is no built-in function with this name.

One of the optional arguments associated with the scan command is called *skip*. It allows you to skip one or more lines in a file before beginning to read your data. For example, if a file called *dis1.dat* contains

```
This is Phase One
for my Dissertation
12   6          4
 7   3          8
 1  18          2
```

the command

```
> datl<-scan(file="disl.dat",skip=2)
```

will skip the first two lines in the file dis1.dat before beginning to read the data.

1.1.2 Storing Data in a Matrix

For many purposes it is convenient to store data in a matrix. Imagine, for example, that for each of five individuals you have measures taken at three different times. For instance, you might be interested in how blood pressure changes during the day, so you measure diastolic blood pressure in the morning, in the afternoon, and in the evening. One convenient way of storing these data is in a matrix having five rows and three columns. If the data are stored in the file bp.dat in the form

```
140  120  115
 95  100  100
110  120  115
 90   85   80
 85   90   85
```

then the command

```
> m<-matrix(scan(file="bp.dat"),ncol=3,byrow=T)
```

will read the data from the file into a matrix called *m* having three columns. Here the argument ncol indicates how many columns the matrix is to have. (The number of rows can be specified as well with the argument nrow.) Typing m and hitting Return outputs

```
      [,1] [,2] [,3]
[1,]   140  120  115
[2,]    95  100  100
[3,]   110  120  115
[4,]    90   85   80
[5,]    85   90   85
```

on the computer screen.

The argument byrow=T (where T is for "true") means that data will be read by rows. That is, the first row of the matrix will contain 140, 120, and 115, the second row will contain 95, 100, and 100, and so forth. If not specified, byrow defaults to F (for "false"), meaning that the data will be read by columns instead. In the example, the first row of data would now contain 140, 95, 110 (the first three values stored in column 1), the second row would contain 90, 85, and 120, and so on.

Once stored in a matrix, it is a simple matter to access a subset of the data. For example, m[1,1] contains the value in the first row and first column, m[1,3] contains the value in the first row and third column, and m[2,4] contains the value in row 2 and column 4. The symbol [,1] refers to the first column and [2,] is the second row. So typing m[,2] and hitting Enter returns

```
[1] 120 100 120 85 90
```

which is the data in the second column.

As before, when reading data from a file, you can skip lines using the skip command. For example, if the data in your file were

```
My  data  on snakes
21  45
67  81
32  72
```

then the command

```
> fdat<-matrix(scan("data.dat",skip=1),ncol=2,byrow=T)
```

would skip the first line and begin reading data.

1.1.3 Storing Data in List Mode

For certain purposes it is convenient to store data in what is called *list mode*. As a simple example, imagine you have three groups of individuals who are treated for anorexia via different methods. For illustrative purposes, suppose a rating method has been devised and that the observations are

```
G1:  36  24  82  12  90  33  14  19
G2:   9  17   8  22  15
G3:  43  56  23  10
```

In some situations it is convenient to have the data stored under one variable name, and this can be done using list mode. One way of storing data in list mode is as follows. First create a variable having list mode. If you want the variable to be called *gdat*, use the command

```
> gdat<-list()
```

Then the data for group 1 can be stored via the command

```
> gdat[[1]]<-c(36, 24, 82, 12, 90, 33, 14, 19),
```

the group 2 data would be stored via the command

```
> gdat[[2]]<-c(9, 17, 8, 22, 15),
```

and group 3 data would be stored by using the command

```
> gdat[[3]]<-c(43, 56, 23, 10)
```

Typing the command gdat and hitting Enter returns

```
[[1]]:
[1] 36 24 82 12 90 33 14 19

[[2]]:
[1] 9 17 8 22 15

[[3]]:
[1] 43 56 23 10
```

That is, gdat contains three vectors of numbers corresponding to the three groups under study.

Another way to store data in list mode is with a variation of the scan command. Suppose the data are stored in a file called *mydata.dat* and are arranged as follows:

```
36   9  43
24  17  56
82   8  23
12  22  10
90  15  NA
33  NA  NA
14  NA  NA
19  NA  NA
```

Then the command

```
> gdat<-scan("mydata.dat",list(g1=0,g2=0,g3=0))
```

will store the data in gdat in list mode. Typing gdat and hitting Enter returns

```
$g1:
 [1]  36  24  82  12  90  33  14  19

$g2:
 [1]   9  17   8  22  15  NA  NA  NA

$g3:
 [1]  43  56  23  10  NA  NA  NA  NA
```

So the data for group 1 are stored in gdat$g1, for group 2 they are in gdat$g2, and for group 3 they are in gdat$g3. An alternative way of accessing the data in group 1 is with gdat[[1]]. Note that as used, scan assumes that the data for group 1 are stored in column 1, group 2 data are stored in column 2, and group 3 data are in column 3.

1.1.4 Arithmetic Operations

In the simplest case, arithmetic operations can be performed on numbers using the operators +, −, * (multiplication), / (division), and ^ (exponentiation). For example, to compute 1 plus 5 squared, use the command

```
> 1+5^2,
```

TABLE 1.1 Some Basic S-PLUS Functions.

Function	Description
exp	exponential
log	natural logarithm
sqrt	square root
cor	correlation
mean	arithmetic mean (with a trimming option)
median	median
min	smallest value
max	largest value
quantile	quantiles
range	max value minus the min value
sum	arithmetic sum
var	variance and covariance

which returns

```
[1] 26.
```

To store the answer in an S-PLUS variable — say, ans — use the command

```
> ano<- 1+5^2.
```

If a vector of observations is stored in an S-PLUS variable, arithmetic operations applied to the variable name will be performed on all the values. For example, if the values 2, 5, 8, 12, and 25 are stored in the S-PLUS variable vdat, then the command

```
> vinv<-1/vdat
```

will compute 1/2, 1/5, 1/8, 1/12, and 1/25 and store the results in the S-PLUS variable vinv.

Most S-PLUS commands consist of a name of some function followed by one or more arguments enclosed in parentheses. There are hundreds of functions that come with S-PLUS, and Section 1.2 describes how to obtain the library of functions written for this book and described in subsequent chapters. For convenience, some of the more basic functions are listed in Table 1.1.

EXAMPLE. If the values 2, 7, 9, and 14 are stored in the S-PLUS variable x, the command

```
> min(x)
```

returns 2, the smallest of the four values stored in x. The average of the numbers is computed with the command mean(x) and is 8. The command range(x) returns the difference between the largest and smallest values stored in x and is $14 - 2 = 12$, and sum(x) returns the value $2 + 7 + 9 + 14 = 32$.

Continued

EXAMPLE. (*Continued*) Suppose you want to subtract the average from each value stored in the S-PLUS variable blob. The command

```
> blob-mean(blob)
```

accomplishes this goal. If in addition you want to square each of these differences and then sum the results, use the command

```
> sum((blob-mean(blob))^2).
```

You can apply arithmetic operations to specific rows or columns of a matrix. For example, to compute the average of all values in column 1 of the matrix m, use the command

```
> mean(m[,1]).
```

The command

```
> mean(m[2,])
```

will compute the average of all values in row 2. In contrast, the command mean(m) will average all of the values in m. In a similar manner, if x has list mode, then

```
> mean(x[[2]])
```

will average the values in x[[2]]. ■

1.1.5 Data Management

There are many ways to manipulate data in S-PLUS. Here attention is focused on those methods that are particularly useful in subsequent chapters.

For certain purposes it is common to want to split data into two groups. For example, situations might arise where you want to focus on those values stored in x that are less than or equal to 6. One way to do this is with the command

```
> z<-x[x<=6],
```

which will take all values stored in x that are less than or equal to 6 and store them in z. More generally, S-PLUS will evaluate any logical expression inside the brackets and operate only on those for which the condition is true. The basic conditions are: == (equality), != (not equal to), < (less than), <= (less than or equal to) > (greater than), >= (greater than or equal to), && (and), || (or). So the command

```
> z<-x[x<=6 || x > 32]
```

will take all values in x that are less than or equal to 6 or greater than 32 and store them in z. The command

```
> z<-x[x>=4 && x<=40]
```

will store all values between 4 and 40, inclusive, in z.

Now suppose you have two measures for each of 10 individuals that are stored in the variables x and y. To be concrete, it is assumed the values are:

x	y
9	23
14	19
23	36
29	24
36	32
42	45
49	39
50	60
63	71
88	92

Situations arise where there is interest in those y values for which the x values satisfy some condition. If you want to operate on only those y values for which x is less than 42, say, use y[x<42]. So the command

```
mean(y[x<42])
```

would average all of the y values for which the corresponding x value is less than 42. In the example, this command would compute

```
(23+19+36+24+32)/5.
```

To compute the average of the y values for which the corresponding x value is less than or equal to the average of the x values, use the command

```
mean(y[x<=mean(x)]).
```

To compute the average of the y values for which the corresponding x value is less than or equal to 14 or greater than or equal to 50, use

```
mean(y[x<=14 || x>=50]).
```

Situations also arise where you might need to change the storage mode used. For example, Chapter 13 describes methods for detecting outliers (points that are unusually far from the majority of points) in multivariate data. Some of the functions for accomplishing this important goal assume data are stored in a matrix. For the data in the example, the values in x and y can be stored in a 10×2 matrix called *m* via the command

```
> m<-cbind(x,y)
```

That is, cbind combines columns of data. (The command rbind combines rows of data instead.)

1.1.6 S-PLUS Function selby

A common situation is where one column of data indicates group membership. For example, imagine a file called *dis.dat* with the following values:

G	OUTCOME
1	34
1	23
1	56
2	19
2	32
1	41
3	29
3	62

There are three groups of individuals corresponding to the values stored under G. The first group, for example, has four individuals with the values 34, 23, 56, and 41. The problem is storing the data in a manner that can be used by the functions described in subsequent chapters. To facilitate matters, the function

```
selby(m,grpc,coln)
```

has been supplied for separating the data by groups and storing it in list mode. (This function is part of the library of functions written for this book.) The first argument, m, can be any S-PLUS variable containing data stored in a matrix. The second argument (grpc) indicates which column indicates group membership, and coln indicates which column contains the measures to be analyzed. In the example, if the data are stored in the S-PLUS matrix dis, the command

```
> selby(dis,1,2)
```

will return

```
$x:
$x[[1]]:
[1] 34 23 56 41

$x[[2]]:
[1] 19 32

$x[[3]]:
[1] 29 62

$grpn:
[1] 1 2 3
```

If the command

```
> ddat<-selby(dis,1,2)
```

is used, ddat$x[[1]] contains the data for group 1, ddat$x[[2]] contains the data for group 2, and so forth. More generally, the data are now stored in list mode in a variable called *ddat$x*—not *ddat*, as might be thought. The command

```
> tryit<-selby(dis,1,2)
```

would store the data in a variable called *tryit$x* instead.

1.2 R and S-PLUS Functions Written for This Book

A rather large library of S-PLUS functions has been written for this book. They can be obtained via anonymous ftp at ftp.usc.edu. That is, use the login name anonymous and use your e-mail address as the password. Once connected, change directories to pub/wilcox; on a UNIX system you can use the command

```
cd pub/wilcox
```

The functions are stored in two files called *allfunv1* and *allfunv2*. Alternatively, these files can be downloaded from www-rcf.usc.edu/~rwilcox/ using the Save As command. When using this Web site, on some systems the file allfunv1 will be downloaded into a file called allfunv1.txt rather than just allfunv1, and of course the same will be true with allfunv2. On other systems, allfunv1 will be downloaded into the file allfunv1.html. When using R, download the files Rallfunv1 and Rallfunv2 instead. (They are nearly identical to the S-PLUS functions, but a few changes were needed to make them run under R.) When using ftp on a Unix machine, use the get command to download them to your computer. For example, the command

```
get allfunv1
```

will download the first file.

The files allfunv1 and allfunv2 should be stored in the same directory where you are using S-PLUS. To make these functions part of your version of S-PLUS, use the command

```
> source("allfunv1").
```

When running under a UNIX system, this command assumes that the file allfunv1 is stored in the directory from which S-PLUS was invoked. When using a PC, the easiest method is to store allfunv1 in the directory being used by S-PLUS. For example, when running the Windows 2000 version, the top of the window indicates that S-PLUS is using the directory

```
C: Program Files\sp2000\users\default
```

Storing allfunv1 in the subdirectory default, the source command given earlier will cause the library of functions stored in allfunv1 to become a part of your version of S-PLUS, until you remove them. Of course, for the remaining functions in allfunv2, use the command

```
source("allfunv2")
```

The arguments used by any of these functions can be checked with the args command. For example, there is a function called *yuen*, and the command

```
> args(yuen)
```

returns

```
function(x, y, tr = 0.2, alpha = 0.05).
```

The first two arguments are mandatory and are assumed to contain data. Arguments with an = are optional and default to the value shown. Here, the argument tr defaults to .2 and alpha defaults to .05. The command

```
yuen(x,y,tr=0,alpha=.1)
```

would use tr=0 and alpha=.1 (the meaning of which is described in Chapter 8).

Each function also contains a brief description of itself that can be read by typing the function name only (with no parentheses) and hitting Enter. For example, the first few lines returned by the command

```
> yuen
```

are

```
#
# Perform Yuen's test for trimmed means on the data in
  x and y.
# The default amount of trimming is 20%.
# Missing values (values stored as NA) are
  automatically removed.
#
# A confidence interval for the trimmed mean of x
  minus the
# the trimmed mean of y is computed and returned in
  yuen$ci.
# The significance level is returned in yuen$siglevel.
#
# For an omnibus test with more than two independent
  groups,
# use t1way.
#
```

The remaining lines are the S-PLUS commands used to perform the analysis, which presumably are not of interest to most readers.

Many of the data sets used in this book can be downloaded as well. You would proceed as was described when downloading allfunv1 and allfunv2, only download the files ending in .dat. For example, read.dat contains data from a reading study that is used in various chapters.

PROBABILITY AND RELATED CONCEPTS

This chapter covers the fundamentals of probability and some related concepts that will be needed in this book. Some ideas are basic and in all likelihood familiar to most readers. But some concepts are not always covered or stressed in an introductory statistics course, whereas other features are rarely if ever discussed, so it is suggested that even if the reader has had some training in basic statistics and probability, the information in this chapter should be scrutinized carefully, particularly Section 2.7.

2.1 Basic Probability

The term *probability* is of course routinely used; all of us have some vague notion of what it means. Yet there is disagreement about the philosophy and interpretation of probability. Devising a satisfactory definition of the term is, from a technical point of view, a nontrivial issue that has received a great deal of scrutiny from stellar mathematicians. Here, however, consideration of these issues is not directly relevant to the topics covered. For present purposes it suffices to think about probabilities in terms of proportions associated with some population of people or things that are of interest. For example, imagine you are a psychologist interested in mental health and one of your goals is to assess feelings of loneliness among college students. Further assume that a measure of loneliness has been developed where an individual can get one of five scores consisting of the integers 1 through 5. A score of 1 indicates relatively no feelings of loneliness and a score of 5 indicates extreme feelings of loneliness. Among the entire population of college students, imagine that 15% would get a loneliness score of 1. Then we say that the probability of the score 1 is .15. Again, when dealing with the mathematical foundations of probability, this view is not completely satisfactory, but attaching a probabilistic interpretation to proportions is all that is required in this book.

In statistics, an uppercase roman letter is typically used to represent whatever measure happens to be of interest, the most common letter being X. For the loneliness study, X represents a measure of loneliness, and the possible values of X are the integers 1 through 5. But X could just as well represent how tall someone is, how much she weighs, her IQ, and so on. That is, X represents whatever happens to be of interest in a given situation. In the illustration, we write $X = 1$ to indicate the event that a college student receives a score of 1 for loneliness, $X = 2$ means a student got a score of 2, and so on.

In the illustration there are five possible events: $X = 1, X = 2, \ldots, X = 5$, and the notation

$$p(x) \tag{2.1}$$

is used to indicate the probability assigned to the value x. So $p(1)$ is the probability that a college student will have a loneliness score of 1, $p(2)$ is the probability of a score of 2, and so forth. Generally, $p(x)$ is called the *probability function* associated with the variable X.

Unless stated otherwise, it is assumed that the possible responses we might observe are mutually exclusive and exhaustive. In the illustration, describing the five possible ratings of loneliness as being *mutually exclusive* means that a student can get one and only one rating. By assumption, it is impossible, for example, to have ratings of both 2 and 3. *Exhaustive* means that a complete list of the possible values we might observe has been specified. If we consider only those students who get a rating between 1 and 5, meaning, for example, that we exclude the possibility of no response, then the ratings 1–5 are exhaustive. If instead we let 0 represent no response, then an exhaustive list of the possible responses would be 0, 1, 2, 3, 4, and 5.

The set of all possible responses is called a *sample space*. If in our ratings illustration the only possible responses are the integers 1–5, then the sample space consists of the numbers 1, 2, 3, 4, and 5. If instead we let 0 represent no response, then the sample space is 0, 1, 2, 3, 4, and 5. If our goal is to study birth weight among humans, the sample space can be viewed as all numbers greater than or equal to zero. Obviously some birth weights are impossible — there seems to be no record of someone weighing 100 pounds at birth — but for convenience the sample space might contain outcomes that have zero probability of occurring.

It is assumed that the reader is familiar with the most basic principles of probability. But as a brief reminder, and to help establish notation, these basic principles are illustrated with the ratings example assuming that the outcomes 1, 2, 3, 4, and 5 are mutually exclusive and exhaustive. The basic principle is that in order for $p(x)$ to qualify as a *probability function*, it must be the case that

- $p(x) \geq 0$ for any x.
- For any two mutually exclusive outcomes — say, x and y — $p(x \text{ or } y) = p(x) + p(y)$.
- $\sum p(x) = 1$, where the notation $\sum p(x)$ means that $p(x)$ is evaluated for all possible values of x and the results are summed. In the loneliness example where the sample space is x: 1, 2, 3, 4, 5, $\sum p(x) = p(1) + p(2) + p(3) + p(4) + p(5) = 1$.

In words, the first criterion is that any probability must be greater than or equal to zero. The second criterion says, for example, that if the responses 1 and 2 are mutually exclusive, then the probability that a student gets a rating of 1 or 2 is equal to the probability of a 1 plus the probability of a 2. Notice that this criterion makes perfect sense when probabilities are viewed as relative proportions. If, for example, 15% of students have a rating of 1, and 20% have a rating of 2, then the probability of a rating of 1 or 2 is just the sum of the proportions: $.15 + .20 = .35$. The third criterion is that if we sum the probabilities of all possible events that are mutually exclusive, we get 1. (In more formal terms, the probability that an observation belongs to the sample space is 1.)

2.2 Expected Values

A fundamental tool in statistics is the notion of *expected values*. Most of the concepts and issues in this book can be understood without employing expected values, but *expected values* is a fairly simple idea that might provide a deeper understanding of some important results to be covered. Also, having an intuitive understanding of expected values facilitates communications with statistical experts, so this topic is covered here.

To convey the basic principle, it helps to start with a simple but unrealistic situation. Still using our loneliness illustration, imagine that the entire population of college students consists of 10 people; that is, we are interested in these 10 individuals only. So in particular we have no desire to generalize to a larger group of college students. Further assume that two students have a loneliness rating of 1, three a rating of 2, two a rating of 3, one a rating of 4, and two a rating of 5. So for this particular population of individuals, the probability of the rating 1 is 2/10, the proportion of individuals who have a rating of 1. Written in a more formal manner, $p(1) = 2/10$. Similarly, the probability of the rating 2 is $p(2) = 3/10$. As is evident, the average of these 10 ratings is

$$\frac{1+1+2+2+2+3+3+4+5+5}{10} = 2.8.$$

Notice that the left side of this last equation can be written as

$$\frac{1(2)+2(3)+3(2)+4(1)+5(2)}{10} = 1\frac{2}{10} + 2\frac{3}{10} + 3\frac{2}{10} + 4\frac{1}{10} + 5\frac{2}{10}.$$

But the fractions in this last equation are just the probabilities associated with the possible outcomes. That is, the average rating for all college students, which is given by the right side of this last equation, can be written as

$$1p(1) + 2p(2) + 3p(3) + 4p(4) + 5p(5).$$

EXAMPLE. If there are a million college students, and the proportion of students associated with the five possible ratings 1, 2, 3, 4, and 5 are .1, .15, .25, .3, and .2, respectively, then the average rating for all 1 million students is

$$1(.1) + 2(.15) + 3(.25) + 4(.3) + 5(.2) = 3.35$$

■

EXAMPLE. If there are a billion college students, and the probabilities associated with the five possible ratings are .15, .2, .25, .3, and .1, respectively, then the average rating of all 1 billion students is

$$1(.15) + 2(.2) + 3(.25) + 4(.3) + 5(.1) = 3.$$

■

Next we introduce some general notation for computing an average based on the view just illustrated. Again let a lowercase x represent a particular value you might observe associated with the variable X. The *expected value* of X, written $E(X)$, is

$$E(X) = \sum xp(x), \tag{2.2}$$

where the notation $\sum xp(x)$ means that you compute $xp(x)$ for every possible value of x and sum the results. So if, for example, the possible values for X are the integers 0, 1, 2, 3, 4, and 5, then

$$\sum xp(x) = 0p(0) + 1p(1) + 2p(2) + 3p(3) + 4p(4) + 5p(5).$$

The expected value of X is so fundamental it has been given a special name: the *population mean*. Typically the population mean is represented by μ. So

$$\mu = E(X)$$

is the average value for all individuals in the population of interest.

EXAMPLE. Imagine that an auto manufacturer wants to evaluate how potential customers will rate handling for a new car being considered for production. So here, X represents ratings of how well the car handles, and the population of individuals who are of interest consists of all individuals who might purchase it. If all potential customers were to rate handling on a four-point scale, 1 being poor and 4 being excellent, and if the corresponding probabilities associated with these ratings are $p(1) = .2$, $p(2) = .4$, $p(3) = .3$, and $p(4) = .1$, then the population mean is

$$\mu = E(X) = 1(.2) + 2(.4) + 3(.3) + 4(.1) = 2.3.$$

That is, the average rating is 2.3. ■

2.3 Conditional Probability and Independence

Conditional probability refers to the probability of some event given that some other event has occurred; it plays a fundamental role in statistics. The notion of conditional probability is illustrated in two ways. The first is based on what is called a *contingency table*,

TABLE 2.1 Hypothetical Probabilities for Sex and Political Affiliation

Sex	Democrat (D)	Republican (R)	
M	0.25	0.20	0.45
F	0.28	0.27	0.55
	0.53	0.47	1.00

an example of which is shown in Table 2.1. In the contingency table are the probabilities associated with four mutually exclusive groups: individuals who are (1) both male and belong to the Republican party, (2) male and belong to the Democratic party, (3) female and belong to the Republican party, and (4) female and belong to the Democratic party. So according to Table 2.1, the proportion of people who are both female and Republican is 0.27. The last column shows what are called the *marginal probabilities*. For example, the probability of being male is $0.20+0.25 = 0.45$, which is just the proportion of males who are a Democrat plus the proportion who are Republican. The last line of Table 2.1 shows the marginal probabilities associated with party affiliation. For example, the probability of being a Democrat is $0.25 + 0.28 = 0.53$.

Now consider the probability of being a Democrat given that the individual is male. According to Table 2.1, the proportion of people who are male is 0.45. So among the people who are male, the proportion who belong to the Democratic party is $0.25/0.45 = 0.56$. Put another way, the probability of being a Democrat, given that the individual is male, is 0.56.

Notice that a conditional probability is determined by altering the sample space. In the illustration, the proportion of all people who belong to the Democratic party is 0.53. But restricting attention to males, meaning that the sample space has been altered to include males only, the proportion is $0.25/0.45 = 0.56$. In a more general notation, if A and B are any two events, and if we let $P(A)$ represent the probability of event A and $P(A \text{ and } B)$ represent the probability that events A and B occur simultaneously, then the conditional probability of A, given that B has occurred, is

$$P(A|B) = \frac{P(A \text{ and } B)}{P(B)}. \tag{2.3}$$

In the illustration, A is the event of being a Democrat, B is the event that a person is male. According to Table 2.1, $P(A \text{ and } B) = 0.25$, $P(B) = 0.45$, so $P(A|B) = 0.25/0.45$, as previously indicated.

EXAMPLE. From Table 2.1, the probability that someone is a female, given that she is Republican, is

$$0.27/0.47 = 0.5745.$$

■

Roughly, two events are *independent* if the probability associated with the first event is not altered when the second event is known. If the probability is altered, the events are *dependent*.

EXAMPLE. According to Table 2.1, the probability that someone is a Democrat is 0.53. The event that someone is a Democrat is independent of the event someone is male if when we are told that someone is male, the probability of being a Democrat remains 0.53. We have seen, however, that the probability of being a Democrat, given that the person is male, is 0.56, so these two events are dependent. ■

Consider any two variables—say, X and Y—and let x and y be any two possible values corresponding to these variables. We say that the variables X and Y are independent if for any x and y we might pick

$$P(Y = y|X = x) = P(Y = y). \tag{2.4}$$

Otherwise they are said to be dependent.

EXAMPLE. Imagine that married couples are asked to rate the effectiveness of the President of the United States. To keep things simple, assume that both husbands and wives rate effectiveness with the values 1, 2, and 3, where the values stand for fair, good, and excellent, respectively. Further assume that the probabilities associated with the possible outcomes are as shown in Table 2.2. We see that the probability a wife (Y) gives a rating of 1 is 0.2. In symbols, $P(Y = 1) = 0.2$. Furthermore, $P(Y = 1|X = 1) = .02/.1 = .2$, where $X = 1$ indicates that the wife's husband gave a rating of 1. So the event $Y = 1$ is independent of the event $X = 1$. If the probability had changed, we could stop and say that X and Y are dependent. But to say that they are independent requires that we check all possible outcomes. For example, another possible outcome is $Y = 1$ and $X = 2$. We see that $P(Y = 1|X = 2) = .1/.5 = .2$, which again is equal to $P(Y = 1)$. Continuing in this manner, it can be seen that for any possible values for Y and X, the corresponding events are independent, so we say that X and Y are independent. That is, they are independent regardless of what their respective values might be. ■

Now, the notion of *dependence* is described and illustrated in another manner. A common and fundamental question in applied research is whether information about one variable influences the probabilities associated with another variable. For example, in a study dealing with diabetes in children, one issue of interest was the association between a child's age and the level of serum C-peptide at diagnosis. For convenience, let X represent age and Y represent C-peptide concentration. For any child we

TABLE 2.2 Hypothetical Probabilities for Presidential Effectiveness

	Husband (X)			
Wife (Y)	1	2	3	
1	.02	.10	.08	0.2
2	.07	.35	.28	0.7
3	.01	.05	.04	0.1
	0.1	0.5	0.4	

might observe, there is some probability that her C-peptide concentration is less than 3, or less than 4, or less than c, where c is any constant we might pick. The issue at hand is whether information about X (a child's age) alters the probabilities associated with Y (a child's C-peptide level). That is, does the conditional probability of Y, given X, differ from the probabilities associated with Y when X is not known or ignored. If knowing X does not alter the probabilities associated with Y, we say that X and Y are independent. Equation (2.4) is one way of providing a formal definition. An alternative way is to say that X and Y are independent if

$$P(Y < y|X = x) = P(Y \leq y) \tag{2.5}$$

for any x and y values we might pick. Equation (2.5) implies Equation (2.4). Yet another way of describing independence is that for any x and y values we might pick,

$$\frac{P(Y = y \text{ and } X = x)}{P(X = x)} = P(Y = y), \tag{2.6}$$

which follows from Equation (2.4). From this last equation it can be seen that if X and Y are independent, then

$$P(X = x \text{ and } Y = y) = P(X = x)P(Y = y). \tag{2.7}$$

Equation (2.7) is called the *product rule* and says that if two events are independent, the probability that they occur simultaneously is equal to the product of their individual probabilities.

EXAMPLE. If two wives rate presidential effectiveness according to the probabilities in Table 2.2, and if their responses are independent, then the probability that both give a response of 2 is $.7 \times .7 = .49$. ■

EXAMPLE. Suppose that for all children we might measure, the probability of having a C-peptide concentration less than or equal to 3 is $P(Y \leq 3) = .4$.

Continued

EXAMPLE. (*Continued*) Now consider only children who are 7 years old and imagine that for this subpopulation of children, the probability of having a C-peptide concentration less than 3 is 0.2. In symbols, $P(Y \leq 3|X = 7) = 0.2$. Then C-peptide concentrations and age are said to be dependent, because knowing that the child's age is 7 alters the probability that the child's C-peptide concentration is less than 3. If instead $P(Y \leq 3|X = 7) = 0.4$, the events $Y \leq 3$ and $X = 7$ are independent. More generally, if, for any x and y we pick, $P(Y \leq y|X = x) = P(Y = y)$, then C-peptide concentration and age are independent. ■

Attaining a graphical intuition of independence will be helpful in subsequent chapters. To be concrete, imagine a study where the goal is to study the association between a person's general feeling of well-being (Y) and the amount of chocolate they consume (X). Assume that an appropriate measure for these two variables has been devised and that the two variables are independent. If we were to measure these two variables for a very large sample of individuals, what would a plot of the results look like? Figure 2.1 shows a scatterplot of observations where values were generated on a computer with X and Y independent. As is evident, there is no visible pattern.

If X and Y are dependent, generally—but not always—there is some discernible pattern. But it is important to keep in mind that there are many types of patterns that can and do arise. (Section 6.5 describes situations where patterns are not evident based on a scatterplot, yet X and Y are dependent.) Figure 2.2 shows four types of patterns where feelings of well-being and chocolate consumption are dependent.

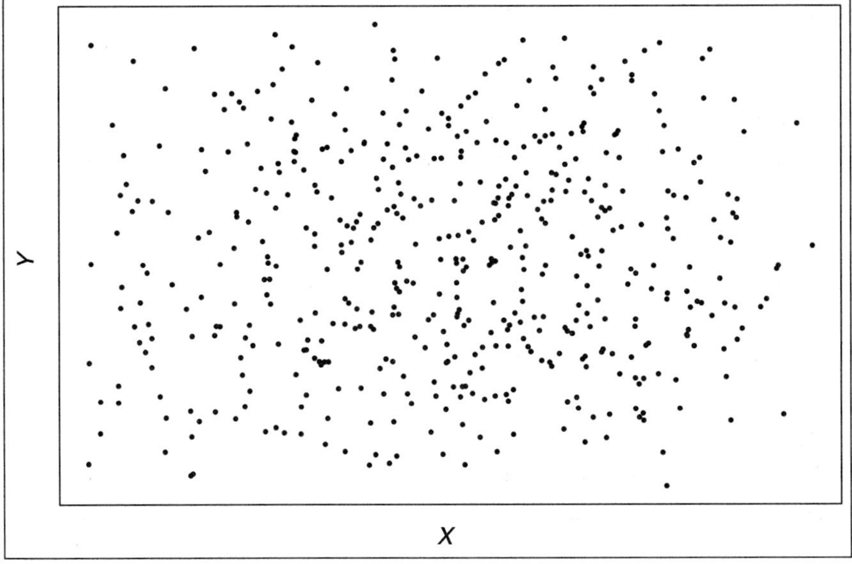

FIGURE 2.1 A scatterplot of two independent variables.

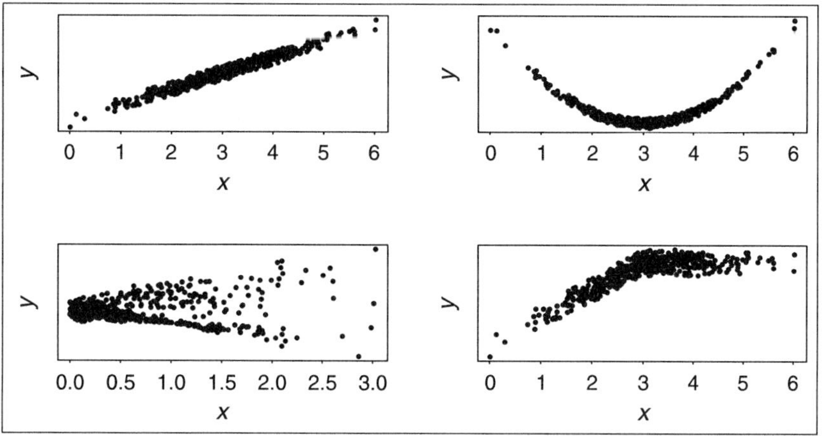

FIGURE 2.2 Different types of associations that might be encountered.

The two upper scatterplots show some rather obvious types of dependence that might arise. The upper left scatterplot, for example, shows a linear association where feelings of well-being increase with chocolate consumption. The upper right scatterplot shows a curved, nonlinear association. The type of dependence shown in the lower two scatterplots are, perhaps, less commonly considered when describing dependence, but in recent years both have been found to be relevant and very important in applied work, as we shall see. In the lower left scatterplot we see that the variation in feelings of well-being differs depending on how much chocolate is consumed. The points in the left portion of this scatterplot are more tightly clustered together. For the left portion of this scatterplot there is, for example, virtually no possibility that someone's feeling of well-being exceeds 1. But for the right portion of this scatterplot, the data were generated so that among individuals with a chocolate consumption of 3, there is a .2 probability that the corresponding value of well-being exceeds 1. That is, $P(Y \leq 1|X)$ decreases as X gets large, so X and Y are dependent. Generally, any situation where the variation among the Y values changes with X implies that X and Y are dependent. Finally, the lower right scatterplot shows a situation where feelings of well-being tend to increase for consumption less than 3, but for $X > 3$ this is no longer the case. Considered as whole, X and Y are dependent, but in this case, if attention is restricted to $X > 3$, X and Y are independent.

The lower left scatterplot of Figure 2.2 illustrates a general principle that is worth stressing: If knowing the value of X alters the range of possible values for Y, then X and Y are dependent. In the illustration, the range of possible values for well-being increases as chocolate consumption increases, so they must be dependent.

2.4 Population Variance

Associated with every probability function is a quantity called the *population variance*. The *population variance* reflects the average squared difference between the population mean and an observation you might make.

Consider, for example, the following probability function:

x:	0	1	2	3
$p(x)$:	.1	.3	.4	.2

The population mean is $\mu = 1.7$. If, for instance, we observe the value 0, its squared distance from the population mean is $(0 - 1.7)^2 = 2.89$ and reflects how far away the value 0 is from the population mean. Moreover, the probability associated with this squared difference is .1, the probability of observing the value 0. In a similar manner, the squared difference between 1 and the population mean is .49, and the probability associated with this squared difference is .3, the same probability associated with the value 1. More generally, for any value x, it has some squared difference between it and the population mean, namely, $(x - \mu)^2$, and the probability associated with this squared difference is $p(x)$. So if we know the probability function, we know the probabilities associated with all squared differences from the population mean. For the probability function considered here, we see that the probability function associated with all possible values of $(x - \mu)^2$ is

$(x - \mu)^2$:	2.89	0.49	0.09	1.69
$p(x)$:	.1	.3	.4	.2

Because we know the probability function associated with all possible squared differences from the population mean, we can determine the average squared difference as well. This average squared difference, called the *population variance*, is typically labeled σ^2. More succinctly, the population variance is

$$\sigma^2 = E[(X - \mu)^2], \tag{2.8}$$

the expected value of $(X - \mu)^2$. Said another way,

$$\sigma^2 = \sum (x - \mu)^2 p(x).$$

The *population standard deviation* is σ, the (positive) square root of the population variance. (Often it is σ, rather than σ^2, that is of interest in applied work.)

EXAMPLE. Suppose that for a five-point scale of anxiety, the probability function for all adults living in New York City is

x:	1	2	3	4	5
$p(x)$:	.05	.1	.7	.1	.05

Continued

EXAMPLE. (*Continued*) The population mean is

$$\mu = 1(.05) + 2(.1) + 3(.7) + 4(.1) + 5(.05) = 3,$$

so the population variance is

$$\sigma^2 = (1-3)^2(.05)+(2-3)^2(.1)+(3-3)^2(.7)+(4-3)^2(.1)+(5-3)^2(.05) = .6,$$

and the population standard deviation is $\sigma = \sqrt{.6} = .775$. ■

Understanding the practical implications associated with the magnitude of the population variance is a complex task that is addressed at various points in this book. There are circumstances where knowing σ is very useful, but there are common situations where it can mislead and give a highly distorted view of what a variable is like. For the moment, complete details must be postponed. But to begin to provide some sense of what σ tells us, consider the following probability function:

x:	1	2	3	4	5
$p(x)$:	.2	.2	.2	.2	.2

It can be seen that $\mu = 3$, the same population mean associated with the probability function in the last example, but the population variance is

$$\sigma^2 = (1-3)^2(.2) + (2-3)^2(.2) + (3-3)^2(.2) + (4-3)^2(.2) + (5-3)^2(.2) = 2.$$

Notice that this variance is larger than the variance in the previous example, where $\sigma^2 = .6$. The reason is that in the former example, it is much less likely for a value to be far from the mean than is the case for the probability function considered here. Here, for example, there is a .4 probability of getting the value 1 or 5. In the previous example, this probability is only .1. Here the probability that an observation differs from the population mean is .8, but in the previous example it was only .3. This illustrates the crude rule of thumb that larger values for the population variance reflect situations where observed values are likely to be far from the mean, and small population variances indicate the opposite.

For discrete data, it is common to represent probabilities graphically with the height of spikes. Figure 2.3 illustrates this approach with the last two probability functions used to illustrate the variance. The left panel shows the probability function

x:	1	2	3	4	5
$p(x)$:	.05	.1	.7	.1	.05

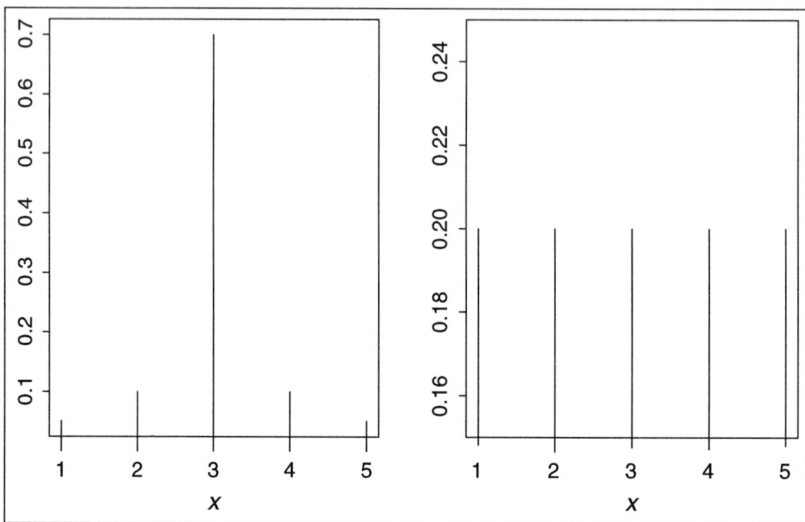

FIGURE 2.3 Examples of how probabilities associated with discrete variables are graphed.

The right panel graphically shows the probability function

x:	1	2	3	4	5
$p(x)$:	.2	.2	.2	.2	.2

Look at the graphed probabilities in Figure 2.3 and notice that the graphed probabilities in the left panel indicate that an observed value is more likely to be close to the mean; in the right panel they are more likely to be further from the mean. That is, the graphs suggest that the variance is smaller in the left panel because, probabilistically, observations are more tightly clustered around the mean.

2.5 The Binomial Probability Function

The most important discrete distribution is the binomial. It arises in situations where only two possible outcomes are possible when making a single observation. The outcomes might be yes and no, success and failure, agree and disagree. Such random variables are called *binary*. Typically the number 1 is used to represent a success, and a failure is represented by 0. A common convention is to let p represent the probability of success and to let $q = 1 - p$ be the probability of a failure.

Before continuing, a comment about notation might help. Consistent with Section 2.1, we follow the common convention of letting X be a variable that represents the number of successes among n observations. The notation $X = 2$, for example, means we observed two successes; more generally, $X = x$ means we observed x successes, where the possible values for x are $0, 1, \dots, n$.

In applied work, often the goal is to estimate p, the probability of success, given some data. But before taking up this problem, we must first consider how to compute probabilities given p. For example, suppose you ask five people whether they approve of a certain political leader. If these five people are allowed to respond only yes or no, and if the probability of a yes response is $p = .6$, what is the probability that exactly three of five randomly sampled people will say yes? There is a convenient formula for solving this problem based on the *binomial probability function*. It says that among n observations, the probability of exactly x successes, $P(X = x)$, is given by

$$p(x) = \binom{n}{x} p^x q^{n-x}. \tag{2.9}$$

The first term on the right side of this equation, called the *binomial coefficient*, is defined to be

$$\binom{n}{x} = \frac{n!}{x!(n-x)!},$$

where $n!$ represents n factorial. That is,

$$n! = 1 \times 2 \times 3 \times \cdots \times (n-1) \times n.$$

For example, $1! = 1$, $2! = 2$, and $3! = 6$. By convention, $0! = 1$.

In the illustration, you have $n = 5$ randomly sampled people and you want to know the probability that exactly $x = 3$ people will respond yes when the probability of a yes is $p = .6$. To solve this problem, compute

$$n! = 1 \times 2 \times 3 \times 4 \times 5 = 120,$$

$$x! = 1 \times 2 \times 3 = 6,$$

$$(n-x)! = 2! = 2,$$

in which case

$$p(3) = \frac{120}{6 \times 2}\left(.6^3\right)\left(.4^2\right) = .3456.$$

As another illustration, suppose you randomly sample 10 couples who recently got married, and your experiment consists of assessing whether they are happily married at the end of 1 year. If the probability of success is $p = .3$, the probability that exactly $x = 4$ couples will report that they are happily married is

$$p(4) = \frac{10!}{4! \times 6!}\left(.3^4\right)\left(.7^6\right) = .2001.$$

Often attention is focused on the probability of *at least* x successes in n trials or *at most* x successes, rather than the probability of getting *exactly* x successes. In the last illustration, you might want to know the probability that four couples or fewer are happily married as opposed to exactly four. The former probability consists of five mutually exclusive events, namely, $x = 0$, $x = 1$, $x = 2$, $x = 3$, and $x = 4$. Thus, the probability that four couples or fewer are happily married is

$$P(X \leq 4) = p(0) + p(1) + p(2) + p(3) + p(4).$$

In summation notation,

$$P(X \leq 4) = \sum_{x=0}^{4} p(x).$$

More generally, the probability of k successes or less in n trials is

$$P(X \leq k) = \sum_{x=0}^{k} p(x)$$

$$= \sum_{x=0}^{k} \binom{n}{x} p^x q^{n-x}.$$

Table 2 in Appendix B gives the values of $P(X \leq k)$ for various values of n and p. Returning to the illustration where $p = .3$ and $n = 10$, Table 2 reports that the probability of four successes or less is .85. Notice that the probability of five successes or more is just the complement of getting four successes or less, so

$$P(X \geq 5) = 1 - P(X \leq 4) = 1 - .85$$

$$= .15.$$

In general,

$$P(X \geq k) = 1 - P(X \leq k - 1),$$

so $P(X \geq k)$ is easily evaluated with Table 2.

Expressions like

$$P(2 \leq x \leq 8),$$

meaning you want to know the probability that the number of successes is between 2 and 8, inclusive, can also be evaluated with Table 2 by noting that

$$P(2 \leq x \leq 8) = P(x \leq 8) - P(x \leq 1).$$

In words, the event of eight successes or less can be broken down into the sum of two mutually exclusive events: the event that the number of successes is less than or equal to 1 and the event that the number of successes is between 2 and 8, inclusive. Rearranging terms yields the last equation. The point is that $P(2 \leq x \leq 8)$ can be written in terms of two expressions that are easily evaluated with Table 2 in Appendix B.

EXAMPLE. Assume $n = 10$ and $p = .5$. From Table 2 in Appendix B, $P(X \leq 1) = .011$ and $P(X \leq 8) = .989$, so

$$P(2 \leq X \leq 8) = .989 - .011 = .978.$$

■

A related problem is determining the probability of one success or less or nine successes or more. The first part is simply read from Table 2 and can be seen to be .011. The probability of nine successes or more is the complement of eight successes or less, so

$$P(X \geq 9) = 1 - P(X \leq 8) = 1 - .989 = .011,$$

again assuming that $n = 10$ and $p = .5$. Thus, the probability of one success or less or nine successes or more is $.011 + .011 = .022$. In symbols,

$$P(X \leq 1 \text{ or } X \geq 9) = .022.$$

There are times when you will need to compute the mean and variance of a binomial probability function once you are given n and p. It can be shown that the mean and variance are given by

$$\mu = E(X)$$
$$= np,$$

and

$$\sigma^2 = npq.$$

For example, if $n = 16$ and $p = .5$, the mean of the binomial probability function is $\mu = np = 16(.5) = 8$. That is, on average, 8 of the 16 observations in a random sample will be a success, while the other 8 will not. The variance is $\sigma^2 = npq = 16(.5)(.5) = 4$, so the standard deviation is $\sigma = \sqrt{4} = 2$. If, instead, $p = .3$, then $\mu = 16(.3) = 4.8$. That is, the average number of successes is 4.8.

In most situations, p, the probability of a success, is not known and must be estimated based on x, the observed number of successes. The result, $E(X) = np$, suggests that x/n be used as an estimator of p; and indeed this is the estimator that is typically used. Often this estimator is written as

$$\hat{p} = \frac{x}{n}.$$

Note that $\hat{p}$ is just the proportion of successes in n trials. It can be shown (using the rules of expected values covered in Section 2.9) that

$$E(\hat{p}) = p.$$

That is, if you were to repeat an experiment infinitely many times, each time randomly sampling n observations, the average of these infinitely many $\hat{p}$ values is p. It can also be shown that the variance of $\hat{p}$ is

$$\sigma^2_{\hat{p}} = \frac{pq}{n}.$$

EXAMPLE. If you sample 25 people and the probability of success is .4, then the variance of $\hat{p}$ is

$$\sigma^2_{\hat{p}} = \frac{.4 \times .6}{25} = .098.$$

■

The characteristics and properties of the binomial probability function can be summarized as follows:

- The experiment consists of exactly n independent trials.
- Only two possible outcomes are possible on each trial, usually called *success* and *failure*.
- Each trial has the same probability of success, p.
- $q = 1 - p$ is the probability of a failure.
- There are x successes among the n trials.
- $p(x) = \binom{n}{x} p^x q^{n-x}$ is the probability of x successes in n trials, $x = 0, 1, \ldots, n$.
- $\binom{n}{x} = (n!/x!(n-x)!)$.
- You estimate p with $\hat{p} = x/n$, where x is the total number of successes.
- $E(\hat{p}) = p$.
- The variance of $\hat{p}$ is $\sigma^2 = pq/n$.
- The average or expected number of successes in n trials is $\mu = E(X) = np$.
- The variance of X is $\sigma^2 = npq$.

2.6 Continuous Variables and the Normal Curve

For various reasons (described in subsequent chapters), *continuous* variables, meaning that the variables can have any value over some range of values, play a fundamental and useful role in statistics. In contrast to discrete variables, probabilities associated with continuous variables are given by the area under a curve. The equation for this curve is called a *probability density function*. If, for instance, we wanted to know the probability that a variable has a value between 2 and 5, say, this is represented by the area under the curve and between 2 and 5.

EXAMPLE. Suppose X represents the proportion of time someone spends on pleasant tasks at their job. So, of course, for any individual we observe, X has some value between 0 and 1. Assume that for the population of all working adults, the probability density function is as shown in Figure 2.4. Further assume that we want to know the probability that the proportion of time spent working on pleasant tasks is less than or equal to .4. In symbols,

Continued

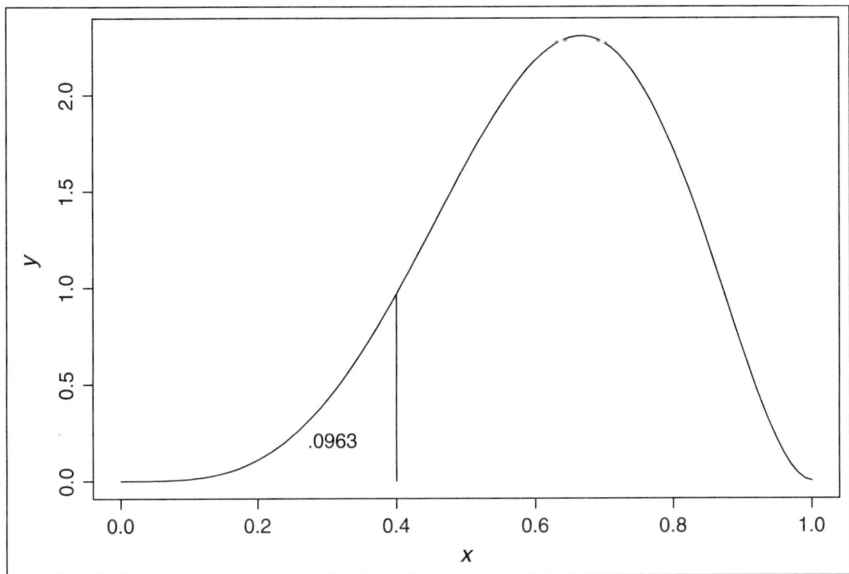

FIGURE 2.4 Probabilities associated with continuous variables are represented by the area under a curve. Here the area under the curve and to the left of 0.4 is .0963. That is, according to this probability curve, $P(X \le .4) = .0963$.

EXAMPLE. (*Continued*) we want to know $P(X \le .4)$. In Figure 2.4, the area under the curve and to the left of .4 is .096. That is, the probability we seek is .096. In symbols, $P(X \le .4) = .096$. ■

If $P(X \le 5) = .8$ and X is a continuous variable, then the value 5 is called the .8 quantile. If $P(X \le 3) = .4$, then 3 is the .4 quantile. In general, if $P(X \le c) = q$, then c is called the qth *quantile*. In Figure 2.4, for example, .4 is the .096 quantile. *Percentiles* are just quantiles multiplied by 100. So in Figure 2.4, .4 is the 9.6 percentile. There are some mathematical difficulties when defining quantiles for discrete data; there is a standard method for dealing with this issue (e.g., Serfling, 1980, p. 3), but the details are not important here.

The .5 quantile is called the *population median*. If $P(X \le 6) = .5$, then 6 is the population median. The median is centrally located in a probabilistic sense, because there is a .5 probability that a value is less than the median and there is a .5 probability that a value is greater than the median instead.

2.6.1 The Normal Curve

The best-known and most important probability density function is the normal curve, an example of which is shown in Figure 2.5. Normal curves have the following important properties:

1. The total area under the curve is 1. (This is a requirement of any probability density function.)

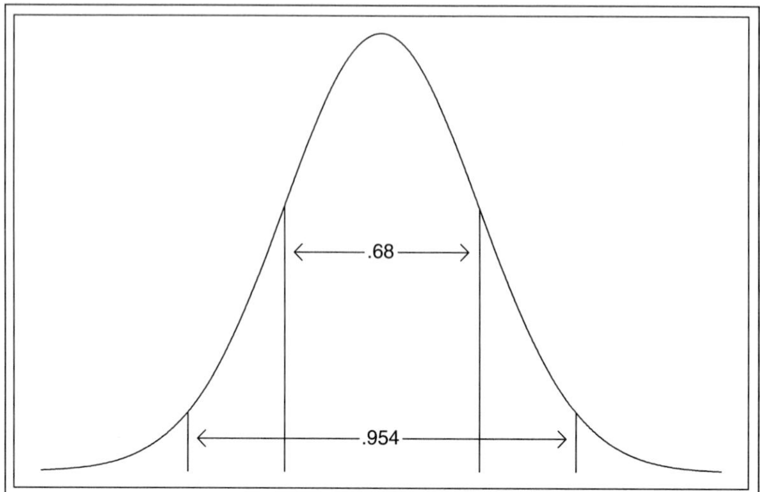

FIGURE 2.5 For any normal curve, the probability that an observation is within one standard deviation of the mean is always .68. The probability of being within two standard deviations is always .954.

2. All normal distributions are bell shaped and symmetric about their mean, μ. It follows that the population mean and median are identical.
3. Although not indicated in Figure 2.5, all normal curves extend from $-\infty$ to ∞ along the x-axis.
4. If the variable X has a normal distribution, the probability that X has a value within one standard deviation of the mean is .68, as indicated in Figure 2.5. In symbols, if X has a normal distribution, then

   $$P(\mu - \sigma < X < \mu + \sigma) = .68$$

   regardless of what the population mean and variance happen to be. The probability of being within two standard deviations is approximately .954. In symbols,

   $$P(\mu - 2\sigma < X < \mu + 2\sigma) = .954.$$

   The probability of being within three standard deviations is

   $$P(\mu - 3\sigma < X < \mu + 3\sigma) = .9975.$$

5. The probability density function of a normal distribution is

   $$f(x) = \frac{1}{\sigma\sqrt{2\pi}} \exp\left[-\frac{(x-\mu)^2}{2\sigma^2}\right], \tag{2.10}$$

   where, as usual, μ and σ^2 are the mean and variance. This rather complicated-looking equation does not play a direct role in applied work, so no illustrations are given on how it is evaluated. Be sure to notice, however, that the probability density function is determined by the mean and variance. If, for example, we want to determine the probability that a variable is less than 25, this probability is completely determined by the mean and variance *if* we assume normality.

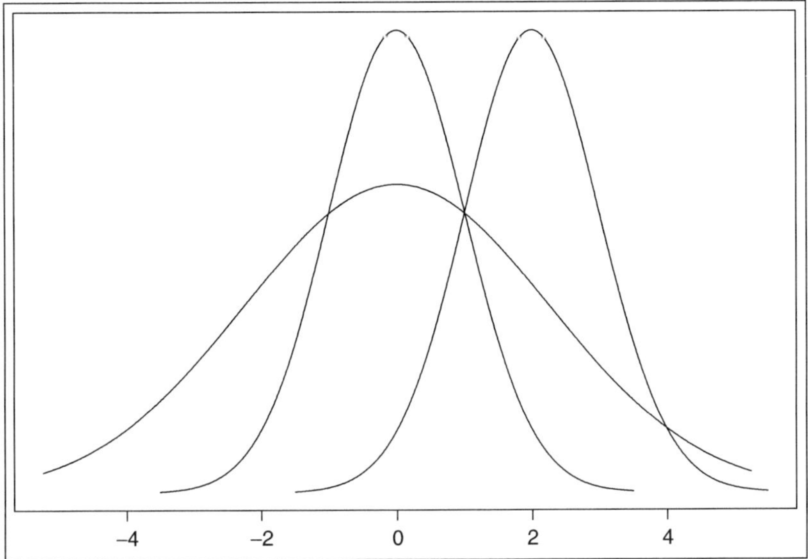

FIGURE 2.6 Two of these normal distributions have equal means and two have equal variances. Note that for normal distributions, increasing the standard deviation from 1 to 1.5 results in a substantial change in the probability curve. (Compare this with Figure 2.8.)

Figure 2.6 shows three normal distributions, two of which have equal means of zero but standard deviations $\sigma = 1$ and $\sigma = 1.5$, and the third again has standard deviation $\sigma = 1$ but with a mean of $\mu = 2$. There are two things to notice. First, if two normal distributions have equal variances but unequal means, the two probability curves are centered around different values but otherwise are identical. Second, for *normal* distributions, there is a distinct and rather noticeable difference between the two curves when the standard deviation increases from 1 to 1.5.

2.6.2 Computing Probabilities Associated with Normal Curves

Assume that human infants have birth weights that are normally distributed with a mean of 3700 grams and a standard deviation of 200 grams. What is the probability that a baby's birth weight will be less than or equal to 3000 grams? As previously explained, this probability is given by the area under the normal curve, but simple methods for computing this area are required. Today the answer is easily obtained on a computer. (For example, the S-PLUS function pnorm can be used.) But for pedagogical reasons a more traditional method is covered here. We begin by considering the special case where the mean is zero and the standard deviation is 1 ($\mu = 0$, $\sigma = 1$), after which we illustrate how to compute probabilities for any mean and standard deviation.

Standard Normal

The *standard normal distribution* is a normal distribution with mean $\mu = 0$ and standard deviation $\sigma = 1$; it plays a central role in many areas of statistics. As is typically done,

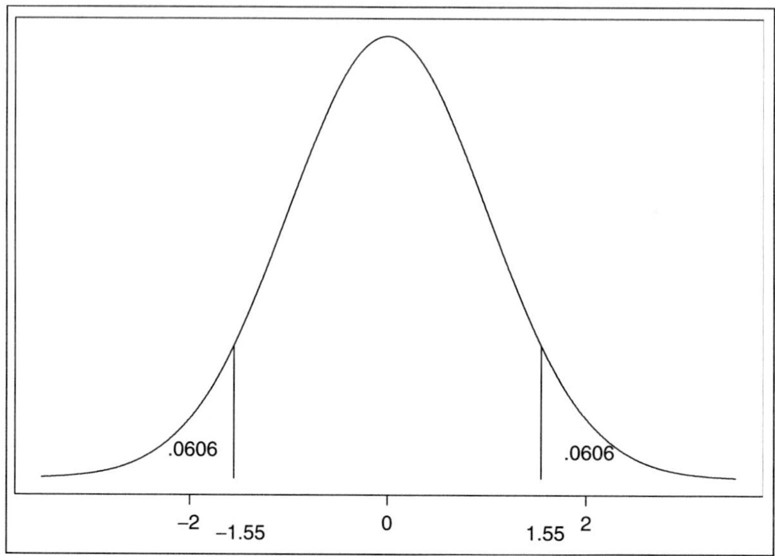

FIGURE 2.7 The standard normal probability curve. The probability that an observation is less than or equal to -1.55 is .0606, and the probability that an observation is greater than or equal to 1.55 is .0606.

Z is used to represent a variable that has a standard normal distribution. Our immediate goal is to describe how to determine the probability that an observation randomly sampled from a standard normal distribution is less than any constant c we might choose.

These probabilities are easily determined using Table 1 in Appendix B, which reports the probability that a standard normal random variable has probability less than or equal to c for $c = -3.00, -2.99, -2.98, \ldots, -0.01, 0, .01, \ldots, 3.00$. The first entry in the first column shows -3. The column next to it gives the corresponding probability, .0013. That is, the probability that a standard normal random variable is less than or equal to -3 is $P(Z \leq -3) = .0013$. Put another way, -3 is the .0013 quantile of the standard normal distribution. Going down the first column we see the entry -2.08; the column next to it indicates that the probability that a standard normal variable is less than or equal to -2.08 is .0188. This says that -2.08 is the .0188 quantile. Looking at the last entry in the third column, we see -1.55; the entry just to the right, in the fourth column, is .0606, so $P(Z \leq -1.55) = .0606$. This probability corresponds to the area in the left portion of Figure 2.7. Because the standard normal curve is symmetric about zero, the probability that X is greater than 1.55 is also .0606, which is shown in the right portion of Figure 2.7. Again looking at the first column of Table 1 in Appendix B, we see the value $z = 1.53$; next to it is the value .9370, meaning that $P(Z \leq 1.53) = .9370$.

In applied work, there are three types of probabilities that need to be determined:

1. $P(Z \leq c)$, the probability that a standard normal random variable is less than or equal to c
2. $P(Z \geq c)$, the probability that a standard normal random variable is greater than or equal to c

3. $P(a \le Z \le b)$, the probability that a standard normal random variable is between the values a and b

The first of these is determined from Table 1 in Appendix B, as already indicated. Because the area under the curve is 1, the second is given by

$$P(Z \ge c) = 1 - P(Z \le c).$$

The third is given by

$$P(a \le Z \le b) = P(Z \le b) - P(Z \le a).$$

EXAMPLE. Determine $P(Z \ge 1.5)$, the probability that a standard normal random variable is greater than 1.5. From Table 1 in Appendix B, $P(Z \le 1.5) = .9332$. Therefore, $P(Z \ge 1.5) = 1 - .9332 = .0668$. ■

EXAMPLE. Next we determine $P(-1.96 \le Z \le 1.96)$, the probability that a standard normal random variable is between -1.96 and 1.96. From Table 1 in Appendix B, $P(Z \le 1.96) = .975$. Also, $P(Z \le -1.96) = .025$, so

$$P(\ 1.96 \le Z \le 1.96) = .975 \quad .025 = .95.$$

■

In some situations it is necessary to use Table 1 (in Appendix B) backwards. That is, we are given a probability and the goal is to determine c. For example, if we are told that $P(Z \le c) = .99$, what is c? We simply find where .99 happens to be in Table 1 under the columns headed by $P(Z \le z)$ and then read the number to the left, under the column headed by z. The answer is 2.33.

Two related problems also arise. The first is determining c given the value of

$$P(Z \ge c).$$

A solution is obtained by noting that the area under the curve is 1, so $P(Z \ge c) = 1 - P(Z \le c)$, which involves a quantity we can determine from Table 1. That is, you compute $d = 1 - P(Z \ge c)$ and then determine c such that

$$P(Z \le c) = d.$$

EXAMPLE. To determine c if $P(Z \ge c) = .9$, first compute $d = 1 - P(Z \le c) = 1 - .9 = .1$. Then c is given by $P(Z \le c) = .1$. Referring to Table 1 in Appendix B, $c = -1.28$. ■

The other type of problem is determining c given

$$P(-c \le Z \le c).$$

Letting $d = P(-c \le Z \le c)$, the answer is given by

$$P(Z \le c) = \frac{1 + d}{2}.$$

EXAMPLE. To determine c if $P(-c \le Z \le c) = .9$, let $d = P(-c \le Z \le c) = .9$ and then compute $(1 + d)/2 = (1 + .9)/2 = .95$. Then c is given by $P(Z \le c) = .95$. Referring to Table 1 in Appendix B, $c = 1.645$. ■

Solution for Any Normal Distribution

Now consider any normal random variable having mean μ and standard deviation σ. The next goal is to describe how to determine the probability that an observation is less than c, where, as usual, c is any constant that might be of interest. The solution is based on *standardizing* a normal random variable, which means that we subtract the population mean μ and divide by the standard deviation, σ. In symbols, we standardize a normal random variable X by transforming it to

$$Z = \frac{X - \mu}{\sigma}. \tag{2.11}$$

It can be shown that if X has a normal distribution, then the distribution of Z is standard normal. In particular, the probability that a normal random variable X is less than or equal to c is

$$P(X \le c) = P\left(Z \le \frac{c - \mu}{\sigma}\right). \tag{2.12}$$

EXAMPLE. Suppose it is claimed that the cholesterol levels in adults have a normal distribution with mean $\mu = 230$ and standard deviation $\sigma = 20$. If this is true, what is the probability that an adult will have a cholesterol level less than or equal to $c = 200$? Referring to Equation (2.12), the answer is

$$P(X \le 200) = P\left(Z \le \frac{200 - 230}{20}\right) = P(Z < -1.5) = .0668,$$

where .0668 is read from Table 1 in Appendix B. This means that the probability that an adult has a cholesterol level less than 200 is .0668. ■

In a similar manner, we can determine the probability that an observation is greater than or equal to 240 or between 210 and 250. More generally, for any constant c that is of interest, we can determine the probability that an observation is greater than c with the equation

$$P(X \ge c) = 1 - P(X \le c),$$

the point being that the right side of this equation can be determined with Equation (2.12). In a similar manner, for any two constants a and b,

$$P(a \le X \le b) = P(X \le b) - P(X \le a).$$

EXAMPLE. Continuing the last example, determine the probability of observing an adult with a cholesterol level greater than or equal to 240. We have that

$$P(X \ge 240) = 1 - P(X \le 240).$$

Referring to Equation (2.12),

$$P(X \le 240) = P\left(Z \le \frac{240 - 230}{20}\right) = P(Z < .5) = .6915,$$

so

$$P(X \ge 240) = 1 - .6915 = .3085.$$

In words, the probability that an adult has a cholesterol level greater than or equal to 240 is .3085. ■

EXAMPLE. Continuing the cholesterol example, we determine

$$P(210 \le X \le 250).$$

We have that

$$P(210 \le X \le 250) = P(X \le 250) - P(X \le 210).$$

Now,

$$P(X \le 250) = P\left(Z < \frac{250 - 230}{20}\right) = P(Z \le 1) = .8413$$

and

$$P(X \le 210) = P\left(Z < \frac{210 - 230}{20}\right) = P(Z \le -1) = .1587,$$

so

$$P(210 \le X \le 250) = .8413 - .1587 = .6826,$$

meaning that the probability of observing a cholesterol level between 210 and 250 is .6826. ■

2.7 Understanding the Effects of Nonnormality

Conventional statistical methods are based on the assumption that observations follow a normal curve. It was once thought that violating the normality assumption rarely had a detrimental impact on these methods, but theoretical and empirical advances have made it clear that two general types of nonnormality cause serious practical problems in a wide range of commonly occurring situations. Indeed, even very slight departures from normality can be a source of concern. To appreciate the practical utility of modern statistical techniques, it helps to build an intuitive sense of how nonnormality influences the population mean and variance and how this effect is related to determining probabilities.

The so-called contaminated, or mixed normal, distribution is a classic way of illustrating some of the more important effects of nonnormality. Consider a situation where we have two subpopulations of individuals or things. Assume each subpopulation has a normal distribution but that they differ in terms of their means or variances or both. When we mix the two populations together we get what is called a *mixed*, or *contaminated*, normal. Generally, mixed normals fall outside the class of normal distributions. That is, for a distribution to qualify as normal, the equation for its curve must have the form given by Equation (2.10), and the mixed normal does not satisfy this requirement. When the two normals mixed together have a common mean but unequal variances, the resulting probability curve is again symmetric about the mean, but even then the mixed normal is not a normal curve.

To provide a more concrete description of the mixed normal, consider the entire population of adults living around the world and let X represent the amount of weight they have gained or lost during the last year. Let's divide the population of adults into two groups: those who have tried some form of dieting to lose weight and those that have not. For illustrative purposes, assume that for adults who have not tried to lose weight, the distribution of their weight loss is standard normal (so $\mu = 0$ and $\sigma = 1$). As for adults who have dieted to lose weight, assume that their weight loss is normally distributed, again with mean $\mu = 0$ but with standard deviation $\sigma = 10$. Finally, suppose that 10% of all adults went on a diet last year to lose weight. So if we were to randomly pick an adult, there is a 10% chance of selecting someone who has dieted. That is, there is a 10% chance of selecting an observation from a normal distribution having standard deviation 10, so there is a 90% chance of selecting an observation from a normal curve having a standard deviation of 1.

Now, if we mix these two populations of adults together, the exact distribution of X (the weight loss for a randomly sampled adult) can be derived and is shown in Figure 2.8. Also shown is the standard normal distribution, and as is evident there is little separating the two curves. Let $P(X \leq c)$ be the probability that an observation is less than c when sampling from the mixed normal, and let $P(Z \leq c)$ be the probability when sampling from the standard normal instead. For any constant c we might pick, it can be shown that $P(X \leq c)$ does not differ from $P(Z \leq c)$ by more than .04. For example, for a standard normal curve, we see from Table 1 in Appendix B that $P(Z \leq 1) = .8413$. If X has the mixed normal distribution considered here, then the probability that X has a value less than or equal to 1 will not differ from .8413 by more than .04; it will be between .8013 and .8813. The exact value happens to be .81.

Here is the point: Very small departures from normality can greatly influence the value of the population variance. For the standard normal in Figure 2.8 the variance is 1, but for the mixed normal it is 10.9. The full implications of this result are impossible to appreciate at this point, but they will become clear in subsequent chapters.

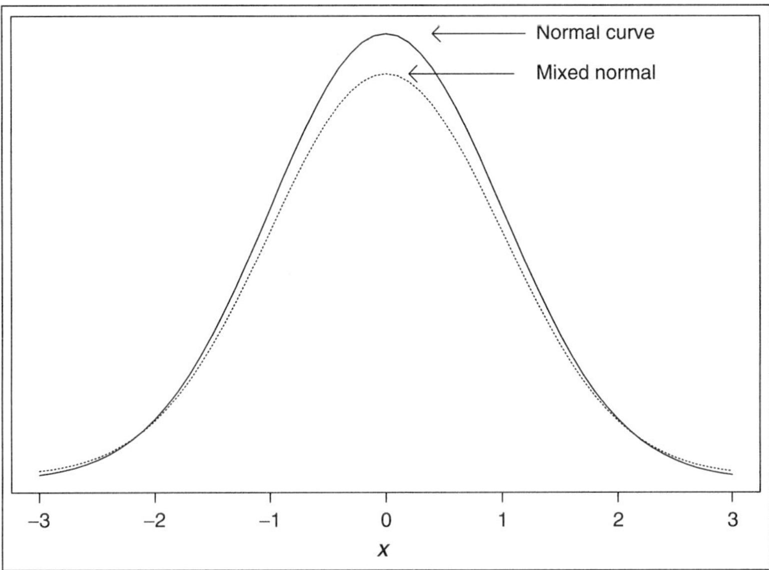

FIGURE 2.8 A standard normal curve having variance $\sigma^2 = 1$ and a mixed normal curve having variance $\sigma^2 = 10.9$. Figure 2.6 illustrated that slight increases in σ result in a substantial change in a normal curve. But the contaminated normal illustrates that two distributions can have substantially different variances even though their probability curves are very similar.

The main goal now is to lay the foundation for understanding some of the problems associated with conventional methods to be described.

To illustrate one of the many implications associated with the mixed normal, consider the following problem: Given the population mean and variance, how can we determine the probability that an observation is less than c. If, for example, $\mu = 0$ and $\sigma^2 = 10.9$, and if we want to know the probability that an observation is less than 1, we get an answer if we assume normality and use the method described in Section 2.6.2. The answer is .619. But for the mixed normal having the same mean and variance, the answer is .81, as previously indicated. So determining probabilities assuming normality when in fact a distribution is slightly nonnormal can lead to a fairly inaccurate result. Figure 2.9 graphically illustrates the problem. Both curves have equal means and variances, yet there is a very distinct difference.

Figure 2.9 illustrates another closely related point. As previously pointed out, normal curves are completely determined by their mean and variance, and Figure 2.6 illustrated that under normality, increasing the variance from 1 to 1.5 results in a very noticeable difference in the graphs of the probability curves. If we assume that curves are normal, or at least approximately normal, this might suggest that in general, if two distributions have equal variances, surely they will appear very similar in shape. But this is not necessarily true even when the two curves are symmetric about the population mean and are bell-shaped. Again, knowing σ is useful in some situations to be covered, but there are many situations where it can mislead.

Figure 2.10 provides another illustration that two curves can have equal means and variances yet differ substantially.

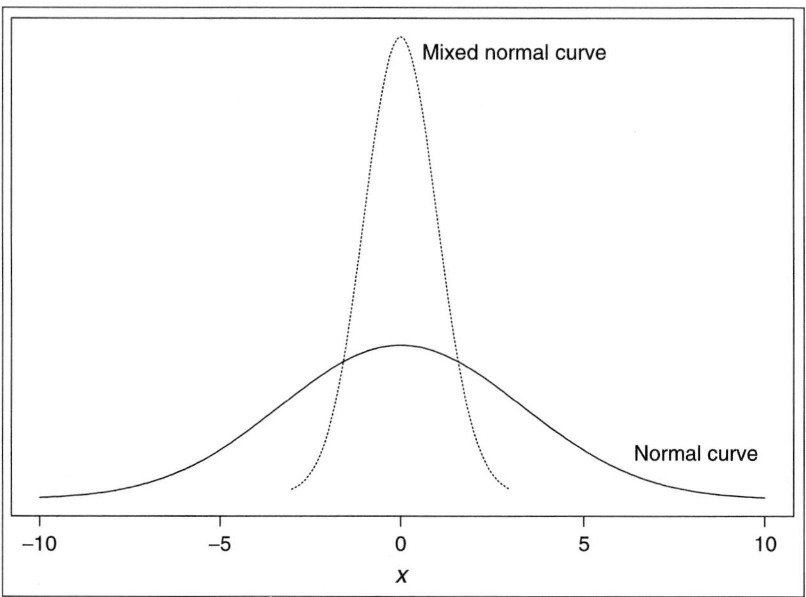

FIGURE 2.9 Two probability curves having equal means and variances.

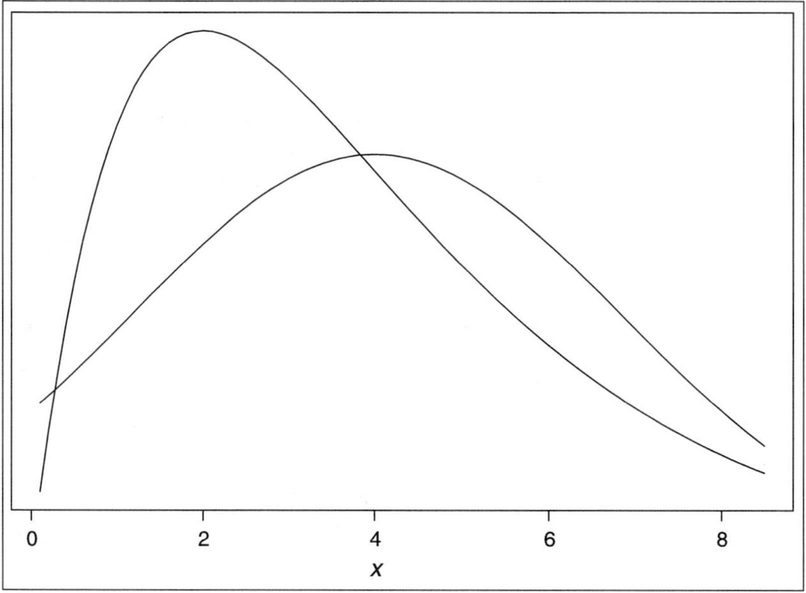

FIGURE 2.10 Two probability curves having equal means and variances.

Here is another way σ might mislead. We saw that for a normal distribution, there is a .68 probability that an observation is within one standard deviation of the mean. It is incorrect to conclude, however, that for nonnormal distributions, this rule always applies. The mixed normal is approximately normal in a sense already described, yet the probability of being within one standard deviation of the mean now exceeds .999.

One reason this last point is important in applied work is related to the notion of outliers. *Outliers* are values that are unusually large or small. For a variety of reasons to be described in subsequent chapters, detecting outliers is important. Assuming normality, a common rule is to declare a value an outlier if it is more than two standard deviations from the mean. In symbols, declare X an outlier if

$$|X - \mu| > 2\sigma. \tag{2.13}$$

So, for example, if $\mu = 4$ and $\sigma = 3$, the value $X = 5$ would not be declared an outlier because $|5 - 4|$ is less than $2 \times 3 = 6$. In contrast the value 12 would be labeled an outlier. The idea is that if a value lies more than two standard deviations from the mean, then probabilistically it is unusual. For normal curves, the probability that an observation is more than two standard deviations from the mean is .046.

To illustrate a concern about this rule, consider what happens when the probability density function is the mixed normal in Figure 2.8. Because the variance is 10.9, we would declare X an outlier if

$$|X - \mu| > 2\sqrt{10.9} = 6.6.$$

But $\mu = 0$, so we declare X to be an outlier if $|X| > 6.6$. It can be seen that now the probability of declaring a value an outlier is 4×10^{-11}—it is virtually impossible. (The method used to derive this probability is not important here.) The value 6, for example, would not be declared an outlier, even though the probability of getting a value greater than or equal to 6 is 9.87×10^{-10}. That is, from a probabilistic point of view, 6 is unusually large, because the probability of getting this value or larger is less than 1 in a billion, yet Equation (2.13) does not flag it as being unusual.

Note that in Figure 2.8, the tails of the mixed normal lie above the tails of the normal. For this reason, the mixed normal is often described as being *heavy-tailed*. Because the area under the extreme portions of a heavy-tailed distribution is larger than the area under a normal curve, extreme values or outliers are more likely when sampling from the mixed normal. Generally, very slight changes in the tail of any probability density function can inflate the variance tremendously, which in turn can make it difficult and even virtually impossible to detect outliers using the rule given by Equation (2.13), even though outliers are relatively common. There are very effective methods for dealing with this problem, but the details are postponed until Chapter 3.

2.7.1 Skewness

Heavy-tailed distributions are one source of concern when employing conventional statistical techniques. Another is *skewness*, which generally refers to distributions that are not exactly symmetric. It is too soon to discuss all the practical problems associated with skewed distributions, but one of the more fundamental issues can be described here.

Consider how we might choose a single number to represent the typical individual or thing under study. A seemingly natural approach is to use the population mean. If a distribution is symmetric about its mean, as is the case when a distribution is normal, there is general agreement that the population mean is indeed a reasonable reflection of what is typical. But when distributions are skewed, at some point doubt begins to

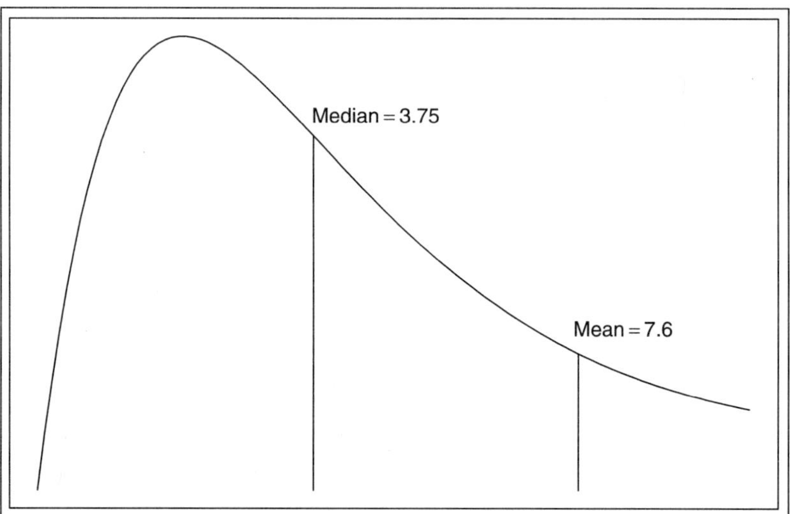

FIGURE 2.11 The population mean can be located in the extreme portion of the tail of a probability curve.

arise as to whether the mean is a good choice. Consider, for example, the distribution shown in Figure 2.11, which is skewed to the right. In this particular case the population mean is located in the extreme right portion of the curve. In fact, the probability that an observation is less than the population mean is 0.74. So from a probabilistic point of view, the population mean is rather atypical. In contrast, the median is located near the more likely outcomes and would seem to better reflect what is typical.

One strategy is to routinely use means and hope that you never encounter a situation where extreme skewness occurs. We will see empirical evidence, however, that such situations arise in practice. Another strategy is simply to switch to the median. If a distribution is symmetric, the population mean and median are identical; but if a distribution is skewed, the median can be argued to be a better indication of what is typical. In some applied settings, the median is a good choice, but unfortunately the routine use of the median can be rather unsatisfactory as well. The reasons are rather involved, but they will be made clear in subsequent chapters. For the moment it is merely remarked that dealing with skewness is a complex issue that has received a great deal of attention. In addition to the concern illustrated by Figure 2.11, there are a variety of other problems, which become evident in Chapter 5. Yet another strategy is to use some simple transformation of the data in an attempt to deal with skewness. A common method is to take logarithms, but this can fail as well, for reasons described in Chapters 3 and 4.

2.8 Pearson's Correlation

This section introduces Pearson's correlation and some of its properties. Pearson's correlation is covered in detail in Chapter 6, but to address certain technical issues it is convenient to introduce it here. The goal is to provide a slightly deeper understanding of why some seemingly natural strategies for analyzing data are theoretically unsound and cannot be recommended. Verbal and graphical explanations are provided as well, and so this section can be skipped by readers who are willing to accept certain results.

Imagine any situation where we have two measures. For example, for each individual among a population of individuals, the two measures might be height and weight or measures of gregariousness and severity of heart disease. Or another situation might be where we sample married couples and the two measures are the cholesterol levels of the wife and husband. For convenience we label the two measures X and Y. Associated with these two variables are two quantities that play a major role in statistics: their covariance and Pearson's correlation.

The *covariance* between X and Y is

$$\sigma_{xy} = E[(X - \mu_x)(Y - \mu_y)].$$

In words, if for some population of individuals we subtract the mean of X from every possible value for X and do the same for Y, then the covariance between X and Y is defined to be the average of the products of these differences. It might help to note that the covariance of X with itself is just its variance, and the same is true for Y. That is, the idea of covariance generalizes the notion of variance to two variables. Pearson's correlation is the covariance divided by the product of the standard deviations and is typically labeled ρ. That is,

$$\rho = \frac{\sigma_{xy}}{\sigma_x \sigma_y}. \tag{2.14}$$

Here are the properties that will be important in some of the chapters to follow:

- $-1 \le \rho \le 1$ (Pearson's correlation always has a value between -1 and 1.)
- If X and Y are independent, then $\rho = \sigma_{xy} = 0$.
- If $\rho \ne 0$, then X and Y are dependent.
- For any two variables, the variance of their sum is

$$\text{VAR}(X + Y) = \sigma_x^2 + \sigma_y^2 + 2\rho\sigma_x\sigma_y. \tag{2.15}$$

This last property is important when explaining why some seemingly reasonable strategies for analyzing data (covered in Chapter 4) are technically incorrect. Note that when we add any two measures together, their sum will have some average value. Using the rules of expected values (covered in Section 2.9), the average of this sum is simply $\mu_x + \mu_y$, the sum of the means. Equation (2.15) says that if we add two measures together, the average squared difference between any sum we might observe, and the mean of the sum, is completely determined by the individual variances plus the correlation.

EXAMPLE. There are two variables in Table 2.2: a wife's rating and a husband's rating. The variance associated with the wives can be seen to be $\sigma_y^2 = 0.29$. As for the husband's, $\sigma_x^2 = .41$. It was already pointed out that the ratings for

Continued

EXAMPLE. (*Continued*) husbands and wives are independent, so $\rho = 0$. Consequently, the variance of the sum of the ratings is $.29 + .41 = .7$. That is, without even determining the probability function associated with this sum, its variance can be determined. ■

A cautionary note should be added. Although independence implies $\rho = 0$, it is *not* necessarily the case that $\rho = 0$ implies independence. In the lower left scatterplot of Figure 2.2, for example, points were generated on a computer with $\rho = 0$, yet there is dependence because, as already indicated, the variation in the Y values increases with X. (This issue is discussed in more detail in Chapter 6.)

2.8.1 Computing the Population Covariance

This subsection is added for readers interested in understanding how the population covariance is computed when probabilities are known. These details are not crucial in what follows, but they might provide a better sense of how the covariance is defined, which in turn provides some details about how the population correlation is defined as well. But readers not interested in technical issues can skip this section.

Consider the probabilities shown in Table 2.3. It can be seen that the expected value of X is $\mu_x = 2.08$ and that the expected value of Y is $\mu_y = 2.02$. Let $p(x, y)$ be the probability of observing the values $X = x$ and $Y = y$ simultaneously. So, according to Table 2.3, the probability that $Y = 1$ and $X = 1$ is $p(1, 1) = .13$; and $p(2, 3) = .18$. To compute the population covariance, you simply perform the calculations shown in Table 2.4. That is, for every combination of values for X and Y, you subtract the corresponding means, yielding the values in columns three and four of Table 2.4. The probabilities associated with all possible pairs of values are shown in column five. Column six shows the product of the values in columns three, four, and five. The population covariance is the sum of the values in column six, which is .0748. Under independence, X and Y must have a covariance of zero, so we have established that the variables considered here are dependent, because the covariance differs from zero.

TABLE 2.3 Hypothetical Probabilities for Pearson's Correlation

		X		
Y	1	2	3	
1	.13	.15	.06	0.34
2	.04	.08	.18	0.30
3	.10	.15	.11	0.36
	.27	.38	.35	

TABLE 2.4 How to Compute the Covariance for the Probabilities in Table 2.3

x	y	$x - \mu_x$	$y - \mu_y$	$p(x,y)$	$(x - \mu_x)(y - \mu_y)p(x,y)$
1	1	−1.08	−1.02	.13	0.143208
1	2	−1.08	−0.02	.04	0.000864
1	3	−1.08	0.98	.10	−0.063504
2	1	−0.08	−1.02	.15	0.003264
2	2	−0.08	−0.02	.08	0.000128
2	3	−0.08	0.98	.15	−0.014112
3	1	0.92	−1.02	.06	−0.093840
3	2	0.92	−0.02	.18	−0.002760
3	3	0.92	0.98	.11	0.099176
					0.0748

EXAMPLE. To compute the population correlation for the values and probabilities shown in Table 2.3, first compute the covariance, which we just saw is $\sigma_{xy} = 0.0748$. It is left as an exercise to show that the variances are $\sigma_x^2 = 0.6136$ and $\sigma_y^2 = .6996$. Consequently,

$$\rho = \frac{.0748}{\sqrt{0.6136} \times \sqrt{.6996}} = .11.$$

So according to Equation (2.15), the variance of the sum, $X + Y$, is

$$0.6136 + .6996 + 2 \times 0.0748 = 1.46.$$

■

2.9 Some Rules About Expected Values

This section summarizes some basic rules about expected values that will be useful in subsequent chapters. The first rule is that if we multiply a variable by some constant c, its expected value is multiplied by c as well. In symbols,

$$E(cX) = cE(X). \tag{2.16}$$

This is just a fancy way of saying, for example, that if the average height of some population of children is five feet ($\mu = 5$), then the average in inches is 60 (the average in feet multiplied by 12). More formally, if $E(X) = \mu = 5$, then $E(12X) = 12 \times 5 = 60$.

The second rule is that if we add c to every possible value, the expected value increases by c as well. That is,

$$E(X + c) = E(X) + c. \tag{2.17}$$

So if $\mu = 6$ and 4 is added to every possible value for X, the average becomes 10. Or, in terms of Equation (2.17), $E(X + 4) = E(X) + 4 = 6 + 4 = 10$.

EXAMPLE. Because μ is a constant, $E(X - \mu) = E(X) - \mu = \mu - \mu = 0$. That is, if we subtract the population mean from every possible value we might observe, the average value of this difference is always zero. If, for instance, the average height of all adult men is 5.9 feet and we subtract 5.9 from everyone's height, the average will be zero. ■

To provide some intuition about the next rule, imagine that for the population of all married women, if they were to rate their marital satisfaction, the probability function would be

x:	1	2	3	4	5
$p(x)$:	.2	.1	.4	.2	.1

Now consider two individuals, Mary and Jane, and suppose they are asked about how they would rate their level of satisfaction regarding their married life. For convenience, label Mary's response X_1 and Jane's response X_2. So before Mary rates her marriage, there is a .2 probability that she will rate her marriage satisfaction as 1, a .1 probability she will rate it as 2, and so on. The same is assumed to be true for Jane. Now consider the sum of their two ratings, which we label $X = X_1 + X_2$. What is the expected value of this sum? That is, on average, what is the value of X?

One way of solving this problem is to attempt to derive the probability function of the sum, X. The possible values for X are $1 + 1 = 2$, $1 + 2 = 3, \ldots, 6 + 6 = 12$, and if we could derive the probabilities associated with these values, we could determine the expected value of X. But there is a much simpler method, because it can be shown that the expected value of a sum is just the sum of the expected values. That is,

$$E(X_1 + X_2) = E(X_1) + E(X_2), \tag{2.18}$$

so the expected value of the sum can be determined if we know the probability function associated with each of the observations we make. But given the probability function, we can do just that. We see that $E(X_1) = 2.9$; in a similar manner $E(X_2) = 2.9$. So the expected value of their sum is 5.8. That is, if two women are asked to rate their marital satisfaction, the average sum of their ratings, over all pairs of women we might interview, is 5.8.

This last illustration demonstrates a more general principle that will be helpful. If X_1 and X_2 have identical probability functions, so in particular the variables have a common mean, μ, then the expected value of their sum is 2μ. So using our rule for constants, we see that the average of these two ratings is μ. That is,

$$E\left[\frac{1}{2}(X_1 + X_2)\right] = \frac{1}{2}(\mu + \mu) = \mu.$$

Here is a summary of the rules for expected values, where c is any constant:

- $E(cX) = cE(X) = c\mu$.

- $E(X + c) = E(X) + c = \mu + c$.
- $E(X_1 + X_2) = E(X_1) + E(X_2)$. For the special case where X_1 and X_2 have a common mean μ, which occurs when they have identical probability functions, $E(X_1 + X_2) = 2\mu$.

2.10 Chi-Squared Distributions

There is an important family of distributions related to the standard normal distribution that will play a role in subsequent chapters. It is called the family of chi-squared distributions and arises as follows. Suppose Z has a standard normal distribution and let $Y = Z^2$. The distribution of Y is so important, it has been given a special name: a chi-squared distribution with one degree of freedom. Next, suppose two independent observations are made, Z_1 and Z_2, both of which have standard normal distributions. Then the distribution of

$$Y = Z_1^2 + Z_2^2$$

is called chi-square distribution with two degrees of freedom. More generally, for n independent standard normal variables, $Z_1, \ldots, Z_n$,

$$Y = Z_1^2 + \cdots + Z_n^2$$

is said to have a chi-square distribution with n degrees of freedom. There are many statistical methods that utilize the family of chi-squared distributions. The only goal now is to introduce the distribution and to emphasize that any use of a chi-squared distribution is intimately connected to normality.

2.11 Exercises

1. For the probability function

x:	0,	1
$p(x)$:	.7,	.3

verify that the mean and variance are .3 and .21. What is the probability of getting a value less than the mean?

2. Standardizing the possible values in Exercise 1 means that we transform the possible values (0 and 1) by subtracting the population mean and dividing by the population standard deviation. Here this yields $(1 - .3)/\sqrt{.21} = .7/\sqrt{.21}$ and $(0 - .3)/\sqrt{.21} = -.3/\sqrt{.21}$, respectively. The probabilities associated with these two values are .3 and .7. Verify that the expected value of the standardized values is zero and the variance is 1.
3. For the probability function

x:	1,	2,	3,	4,	5
$p(x)$:	.15,	.2,	.3,	.2,	.15

determine the mean, the variance, and $P(X \le \mu)$.

4. For the probability function

x:	1,	2,	3,	4,	5
$p(x)$:	.1,	.25,	.3,	.25,	.1

would you expect the variance to be larger or smaller than the variance associated with the probability function used in the previous exercise? Verify your answer by computing the variance for the probability function given here.
5. For the probability function

x:	1,	2,	3,	4,	5
$p(x)$:	.2,	.2,	.2,	.2,	.2

would you expect the variance to be larger or smaller than the variance associated with the probability function used in the previous exercise? Verify your answer by computing the variance.
6. Verify that if we standardize the possible values in Exercise 5, the resulting mean is zero and the variance is 1.
7. For the following probabilities, determine (a) the probability that someone is under 30, (b) the probability that someone has a high income given that he or she is under 30, (c) the probability that someone has a low income given that he or she is under 30, and (d) the probability that someone has a medium income given that he or she is over 50.

	Income		
Age	High	Medium	Low
<30	.030	.180	.090
30–50	.052	.312	.156
>50	.018	.108	.054

8. For Exercise 7, are income and age independent?
9. Coleman (1964) interviewed 3,398 schoolboys and asked them about their self-perceived membership in the "leading crowd." Their response was either yes, they were a member, or no, they were not. The same boys were also asked about their attitude concerning the leading crowd. In particular, they were asked whether membership meant that it does not require going against one's principles sometimes or whether they think it does. Here, the first response will be indicated by a 1 and the second will be indicated by a 0. The results were as follows:

	Attitude	
Member?	1	0
Yes	757	496
No	1071	1074

These values, divided by the sample size, 3,398, are called *relative frequencies*. For example, the relative frequency of the event (Yes, 1) is 757/3398. Treat the relative frequencies as probabilities and determine (a) the probability that an arbitrarily chosen boy responds Yes, (b) $P(\text{Yes}|1)$, (c) $P(1|\text{Yes})$, (d) whether the response Yes is independent of the attitude 0, (e) the probability of a (Yes and 1) or a (No and 0) response, (f) the probability of not responding (Yes and 1), (g) the probability of responding Yes or 1.

10. The probability density function associated with a so-called *uniform distribution* is given by $f(x) = 1/(b-a)$, where a and b are given constants and $a \le x \le b$. That is, the possible values you might observe range between the constants a and b, with every value between a and b equally likely. If, for example, $a = 0$ and $b = 1$, the possible values you might observe lie between 0 and 1. For a uniform distribution over the interval $a = 1$ and $b = 4$, draw the probability density function and determine the median and the .1 and .9 quantiles.
11. For the uniform distribution over the interval -3 to 2, determine (a) $P(X \le 1)$, (b) $P(X < -1.5)$, (c) $P(X > 0)$, (d) $P(-1.2 \le X \le 1)$, (e) $P(X = 1)$.
12. For the uniform distribution in Exercise 11, determine the median and the .25 and .9 quantiles.
13. For the uniform distribution with $a = -1$ and $b = 1$, determine c such that (a) $P(X \le c) = .9$, (b) $P(X \le c) = .95$, (c) $P(X > c) = .99$.
14. For the uniform distribution with $a = -1$ and $b = 1$, determine c such that (a) $P(-c \le X \le c) = .9$, (b) $P(-c \le X \le c) = .95$, (c) $P(-c \le X \le c) = .99$.
15. Suppose the waiting time at a traffic light has a uniform distribution from 0 to 20 seconds. Determine the probability of waiting (a) exactly 12 seconds, (b) less than 5 seconds, (c) more than 10 seconds.
16. When you look at a clock, the number of minutes past the hour—say, X—is some number between 0 and 60. Assume the number of minutes past the hour has a uniform distribution. Determine (a) $P(X = 30)$, (b) $P(X \le 10)$, (c) $P(X \ge 20)$, (d) $P(10 \le X < 20)$.
17. For Exercise 16, determine the .8 quantile.
18. Given that Z has a standard normal distribution, use Table 1 in Appendix B to determine (a) $P(Z \ge 1.5)$, (b) $P(Z \le -2.5)$, (c) $P(Z < -2.5)$, (d) $P(-1 \le Z \le 1)$.
19. If Z has a standard normal distribution, determine (a) $P(Z \le .5)$, (b) $P(Z > -1.25)$, (c) $P(-1.2 < Z < 1.2)$, (d) $P(-1.8 \le Z < 1.8)$.
20. If Z has a standard normal distribution, determine (a) $P(Z < -.5)$, (b) $P(Z < 1.2)$, (c) $P(Z > 2.1)$, (d) $P(-.28 < Z < .28)$.
21. If Z has a standard normal distribution, find c such that (a) $P(Z \le c) = .0099$, (b) $P(Z < c) = .9732$, (c) $P(Z > c) = .5691$, (d) $P(-c \le Z \le c) = .2358$.
22. If Z has a standard normal distribution, find c such that (a) $P(Z > c) = .0764$, (b) $P(Z > c) = .5040$, (c) $P(-c \le Z < c) = .9108$, (d) $P(-c \le Z \le c) = .8$.
23. If X has a normal distribution with mean $\mu = 50$ and standard deviation $\sigma = 9$, determine (a) $P(X \le 40)$, (b) $P(X < 55)$, (c) $P(X > 60)$, (d) $P(40 \le X \le 60)$.
24. If X has a normal distribution with mean $\mu = 20$ and standard deviation $\sigma = 9$, determine (a) $P(X < 22)$, (b) $P(X > 17)$, (c) $P(X > 15)$, (d) $P(2 < X < 38)$.

25. If X has a normal distribution with mean $\mu = .75$ and standard deviation $\sigma = .5$, determine (a) $P(X < .25)$, (b) $P(X > .9)$, (c) $P(.5 < X < 1)$, (d) $P(.25 < X < 1.25)$.
26. If X has a normal distribution, determine c such that

$$P(\mu - c\sigma < X < \mu + c\sigma) = .95.$$

27. If X has a normal distribution, determine c such that

$$P(\mu - c\sigma < X < \mu + c\sigma) = .8.$$

28. Assuming that the scores on a math achievement test are normally distributed with mean $\mu = 68$ and standard deviation $\sigma = 10$, what is the probability of getting a score greater than 78?
29. In Exercise 28, how high must someone score to be in the top 5%? That is, determine c such that $P(X > c) = .05$.
30. A manufacturer of car batteries claims that the life of their batteries is normally distributed with mean $\mu = 58$ months and standard deviation $\sigma = 3$. Determine the probability that a randomly selected battery will last at least 62 months.
31. Assume that the income of pediatricians is normally distributed with mean $\mu = \$100{,}000$ and standard deviation $\sigma = 10{,}000$. Determine the probability of observing an income between \$85,000 and \$115,000.
32. Suppose the winnings of gamblers at Las Vegas are normally distributed with mean $\mu = -300$ (the typical person loses \$300) and standard deviation $\sigma = 100$. Determine the probability that a gambler does not lose any money.
33. A large computer company claims that their salaries are normally distributed with mean \$50,000 and standard deviation 10,000. What is the probability of observing an income between \$40,000 and \$60,000?
34. Suppose the daily amount of solar radiation in Los Angeles is normally distributed with mean 450 calories and standard deviation 50. Determine the probability that for a randomly chosen day, the amount of solar radiation is between 350 and 550.
35. If the cholesterol levels of adults are normally distributed with mean 230 and standard deviation 25, what is the probability that a randomly sampled adult has a cholesterol level greater than 260?
36. If after one year, the annual mileage of privately owned cars is normally distributed with mean 14,000 miles and standard deviation 3,500, what is the probability that a car has mileage greater than 20,000 miles?
37. Can small changes in the tails of a distribution result in large changes in the population mean, μ, relative to changes in the median?
38. Explain in what sense the population variance is sensitive to small changes in a distribution.
39. For normal random variables, the probability of being within one standard deviation of the mean is .68. That is, $P(\mu - \sigma \le X \le \mu + \sigma) = .68$ if X has a normal distribution. For nonnormal distributions, is it safe to assume that this probability is again .68? Explain your answer.

40. If a distribution appears to be bell-shaped and symmetric about its mean, can we assume that the probability of being within one standard deviation of the mean is .68?
41. Can two distributions differ by a large amount yet have equal means and variances?
42. If a distribution is skewed, is it possible that the mean exceeds the .85 quantile?
43. Determine $P(\mu - \sigma \le X \le \mu + \sigma)$ for the probability function

x:	1,	2,	3,	4
$p(x)$:	.2,	.4,	.3,	.1

44. The U.S. Department of Agriculture reports that 75% of people who invest in the futures market lose money. Based on the binomial probability function with $n = 5$, determine:

    (a) The probability that all five lose money
    (b) The probability that all five make money
    (c) The probability that at least two lose money

45. If for a binomial, $p = .4$ and $n = 25$, determine (a) $P(X < 11)$, (b) $P(X \le 11)$, (c) $P(X > 9)$, (d) $P(X \ge 9)$.
46. In Exercise 45, determine $E(X)$, the variance of X, $E(\hat{p})$, and the variance of $\hat{p}$.

3

SUMMARIZING DATA

Chapter 2 covered some ways of describing and summarizing a population of individuals (or things) when we know the probabilities associated with some variable of interest. For example, the population mean and median can be used to reflect the typical individual, and σ provides some indication of the variation among the individuals under study. But of course in most situations we do not know the probabilities, and often we have little or no information about the probability density function, so the population mean and median are not known. If we could measure every individual of interest, then the probabilities would be known, but obviously measuring every individual can be difficult or impossible to do. However, suppose we are able to obtain a sample of individuals, meaning a subset of the population of individuals under study. One of our main goals is to find ways of making inferences about the entire population of individuals based on this sample. Simultaneously, we need to describe conditions under which accurate inferences can be made. But before addressing these important problems, we first describe some methods for summarizing a sample of observations. We begin with standard methods typically covered in an introductory course, and then we introduce some nonstandard techniques that play an important role in this book.

3.1 Basic Summation Notation

To make this book as self-contained as possible, basic summation is briefly described for the benefit of any readers not familiar with it. Imagine that 15 college students are asked to rate their feelings of optimism about their future on a six-point scale. If the first student gives a rating of 6, this result is typically written $X_1 = 6$, where the subscript 1 indicates that this is the first student interviewed. If you sample a second student, who gets a score of 4, you write this as $X_2 = 4$, where now the subscript 2 indicates that this is the second student you measure. Here we assume 15 students are interviewed, and their ratings are represented by $X_1, \ldots, X_{15}$. The notation X_i is used to represent the ith subject. In the example with a total of 15 subjects, the possible values for i are the integers $1, 2, \ldots, 15$. Typically, the sample size is represented by n. In the illustration, there are 15 subjects, and this is written as $n = 15$. Table 3.1 illustrates the notation, along with the ratings you might get. The first subject ($i = 1$) got a score of 3, so $X_1 = 3$. The next subject ($i = 2$) got a score of 7, so $X_2 = 7$.

TABLE 3.1 Hypothetical Data Illustrating Commonly Used Notation.

Subject's name	i	X_i
Tom	1	3
Alice	2	7
Dick	3	6
Harry	4	4
Quinn	5	8
Bryce	6	9
Bruce	7	10
Nancy	8	4
Linda	9	5
Karen	10	4
George	11	5
Peter	12	6
Adrian	13	5
Marsha	14	7
Jean	15	6

The notation $\sum X_i$ is a shorthand way of indicating that the observations are to be summed. That is,

$$\sum X_i = X_1 + X_2 + \cdots + X_n. \tag{3.1}$$

For the data in Table 3.1,

$$\sum X_i = 3 + 7 + 6 + \cdots + 7 + 6 = 89.$$

3.2 Measures of Location

One of the most common approaches to summarizing a sample of subjects or a batch of numbers, is to use a so-called measure of location. Roughly, a *measure of location* is a number intended to reflect the typical individual or thing under study. Measures of location are also called *measures of central tendency*, the idea being that they are intended to reflect the middle portion of a set of observations. Examples of population measures of location are the population mean (μ) and the population median. Here attention is focused on sample analogs of these measures plus some additional measures of location that play a prominent role in this book.

3.2.1 The Sample Mean

A natural and very common way of summarizing a batch of numbers is to compute their average. In symbols, the average of n numbers, $X_1, \ldots, X_n$, is

$$\bar{X} = \frac{1}{n} \sum X_i, \tag{3.2}$$

where the notation $\bar{X}$ is read "X bar." In statistics, $\bar{X}$ is called the *sample mean*. As is probably evident, the sample mean is intended as an estimate of the population mean, μ. Of course, a fundamental problem is determining how well the sample mean estimates the population mean, and we begin to discuss this issue in Chapter 4. For now, attention is restricted to other important properties of the sample mean.

EXAMPLE. You sample 10 married couples and determine the number of children they have. The results are 0, 4, 3, 2, 2, 3, 2, 1, 0, 8, and the sample mean is $\bar{X} = 2.5$. Based on this result, it is estimated that if we could measure all married couples, the average number of children would be 2.5. In more formal terms, $\bar{X} = 2.5$ is an estimate of μ. In all likelihood the population mean is not 2.5, so there is the issue of how close the sample mean is likely to be to the population mean. Again, we get to this topic in due course. ■

To elaborate on how the population mean and the sample mean are related, it helps to describe how the sample mean can be computed based on the frequencies of the observations, particularly when the number of observations is large. Here we let f_x represent the number of times the value x was observed among a sample of n observations. That is, f_x represents the *frequency* associated with x. In the last example, the frequencies associated with the number of children are $f_0 = 2$, $f_1 = 1$, $f_2 = 3$, $f_3 = 2$, $f_4 = 1$, and $f_8 = 1$. So there were two couples with 0 children, one couple had 1 child, three had 2 children, and so forth.

The summation notation introduced in Section 3.1 is used almost exclusively in this book. But in this subsection it helps to introduce a variation of this notation:

$$\sum_x$$

This indicates that a sum is to be computed over all possible values of x. For example,

$$\sum_x f_x.$$

means that we sum the frequencies for all the x values available. Continuing the illustration in the last paragraph, the observed values for x are 0, 1, 2, 3, 4, and 8, so

$$\sum_x f_x = f_0 + f_1 + f_2 + f_3 + f_4 + f_8 = 10.$$

The sum of the observations is just the sum of every possible value multiplied by its frequency. In the present notation, the sum of the observations is

$$\sum x f_x = 0f_0 + 1f_1 + 2f_2 + 3f_3 + 4f_4 + 8f_8$$

$$= 0(2) + 1(1) + 2(3) + 3(2) + 4(1) + 8(1)$$

$$= 25.$$

Dividing this sum by the sample size, n, gives the sample mean. In symbols, another way of writing the sample mean is

$$\frac{1}{n}\sum_x x f_x = \sum_x x\frac{f_x}{n}. \tag{3.3}$$

In words, the sample mean can be computed by multiplying every observed value x by its frequency, f_x, dividing by the sample size, n, and then summing the results.

Note that a natural way of estimating the probability associated with the value x is with the proportion of times it is observed among a sample of observations. In more formal terms, f_x/n, the relative frequency of the value x, is used to estimate $p(x)$. This reveals a close connection between the description of the sample mean just given and the description of the population mean given in Chapter 2. The main difference is that the population mean, $\mu = \sum x\,p(x)$, is defined in terms of $p(x)$, the proportion of all individuals among the entire population of individuals having the response x, whereas the sample mean uses f_x/n in place of $p(x)$.

EXAMPLE. Consider a sample of $n = 1000$ couples where the proportions of couples having 0, 1, 2, 3, 4, or 5 children are .12, .18, .29, .24, .14, and .02, respectively. In symbols, the relative frequencies for the number of children are $f_0/n = .12$, $f_1/n = .18$, $f_2/n = .29$, $f_3/n = .24$, $f_4/n = .14$, and $f_5/n = .02$. Then the sample mean is easily determined by substituting the appropriate values into Equation (3.3). This yields

$$\bar{X} = 0(.12) + 1(.18) + 2(.29) + 3(.24) + 4(.14) + 5(.02) = 2.14.$$

That is, based on these 1000 couples, the estimate of the population mean, the average number of children among all couples, is 2.14. ■

Chapter 2 demonstrated that the population mean can lie in the extreme tails of a distribution and can be argued to provide a misleading reflection of what is typical in some situations. That is, the population mean can in fact be an extreme value that is relatively atypical. A similar argument can be made about the sample mean, as demonstrated by the following example, based on data from an actual study.

EXAMPLE. Why is it that so many marriages in the United States end in divorce? One proposed explanation is that humans, especially men, seek multiple sexual partners and that this propensity is rooted in our evolutionary past. In support of this view, some researchers have pointed out that when young

Continued

TABLE 3.2 Desired Number of Sexual Partners for 105 Males

x:	0	1	2	3	4	5	6	7	8	9
f_x:	5	49	4	5	9	4	4	1	1	2
x:	10	11	12	13	15	18	19	30	40	45
f_x:	3	2	3	1	2	1	2	2	1	1
x:	150	6000								
f_x:	2	1								

EXAMPLE. (*Continued*) males are asked how many sexual partners they desire over their lifetime, the average number has been found to be substantially higher than the corresponding responses given by females. Pedersen, Miller, Putcha-Bhagavatula, and Yang (2002) point out, however, that the data are typically skewed. In one portion of the study by Pedersen et al., the responses given by 105 males, regarding the desired number of sexual partners over the next 30 years, were as shown in Table 3.2. The sample mean is 64.9. This is, however, a dubious indication of the desired number of sexual partners, because 97% of the observations fall below the sample mean. Notice, for example, that 49 of the males said they wanted one sexual partner, and more than half gave a response of zero or one. In fact 5 gave a response of zero. ■

A criticism of the sample mean is that a single outlier can greatly influence its value. In the last example, one individual responded that he wanted 6000 sexual partners over the next 30 years. This response is unusually large and has an inordinate influence on the sample mean. If, for example, it is removed, the mean of the remaining observations is 7.9. But even 7.9 is rather misleading, because over 77% of the remaining observations fall below 7.9.

One way of quantifying the sensitivity of the sample mean to outliers is with the so-called finite-sample breakdown point. The *finite-sample breakdown point* of the sample mean is the smallest proportion of observations that can make it arbitrarily large or small. Said another way, the finite-sample breakdown point of the sample mean is the smallest proportion of n observations that can render it meaningless. A single observation can make the sample mean arbitrarily large or small, regardless of what the other values might be, so its finite-sample breakdown point is $1/n$.

3.2.2 The Sample Median

Another important measure of location is the *sample median*, which is intended as an estimate of the population median. Simply put, if the sample size is odd, the sample median is the middle value after putting the observations in ascending order. If the sample size is even, the sample median is the average of the two middle values.

Chapter 2 noted that for symmetric distributions, the population mean and median are identical, so for this special case the sample median provides another way of estimating the population mean. But for skewed distributions the population mean and median differ, so generally the sample mean and median are attempting to estimate different quantities.

It helps to describe the sample median in a more formal manner in order to illustrate a commonly used notation. For the observations $X_1, \ldots, X_n$, let $X_{(1)}$ represent the smallest number, $X_{(2)}$ the next smallest, and $X_{(n)}$ the largest. More generally,

$$X_{(1)} \leq X_{(2)} \leq X_{(3)} \leq \cdots \leq X_{(n)}$$

is the notation used to indicate that n values are to be put in ascending order. The sample median is computed as follows:

1. If the number of observations, n, is odd, compute $m = (n + 1)/2$. Then the sample median is

$$M = X_{(m)},$$

the mth value after the observations are put in order.

2. If the number of observations, n, is even, compute $m = n/2$. Then the sample median is

$$M = (X_{(m)} + X_{(m+1)})/2,$$

the average of the mth and $(m + 1)$th observations after putting the observed values in ascending order.

EXAMPLE. Consider the values 1.1, 2.3, 1.7, 0.9, and 3.1. The smallest of the five observations is 0.9, so $X_{(1)} = 0.9$. The smallest of the remaining four observations is 1.1, and this is written as $X_{(2)} = 1.1$. The smallest of the remaining three observations is 1.7, so $X_{(3)} = 1.7$; the largest of the five values is 3.1, and this is written as $X_{(5)} = 3.1$. ■

EXAMPLE. Seven subjects are given a test that measures depression. The observed scores are

34, 29, 55, 45, 21, 32, 39.

Because the number of observations is $n = 7$, which is odd, $m = (7 + 1)/2 = 4$. Putting the observations in order yields

21, 29, 32, **34**, 39, 45, 55.

The fourth observation is $X_{(4)} = 34$, so the sample median is $M = 34$. ■

EXAMPLE. We repeat the last example, only with six subjects having test scores

$$29, 55, 45, 21, 32, 39.$$

Because the number of observations is $n = 6$, which is even, $m = 6/2 = 3$. Putting the observations in order yields

$$21, 29, \mathbf{32}, \mathbf{39}, 45, 55.$$

The third and fourth observations are $X_{(3)} = 32$ and $X_{(4)} = 39$, so the sample median is $M = (32 + 39)/2 = 35.5$. ■

Notice that nearly half of any n values can be made arbitrarily large without making the value of the sample median arbitrarily large as well. Consequently, the finite-sample breakdown point is approximately .5, the highest possible value. So the mean and median lie at two extremes in terms of their sensitivity to outliers. The sample mean can be affected by a single outlier, but nearly half of the observations can be outliers without affecting the median. For the data in Table 3.2, the sample median is $M = 1$, which gives a decidedly different picture of what is typical as compared to the mean, which is 64.9.

Based on the single criterion of having a high breakdown point, the median beats the mean. But it is stressed that this is not a compelling reason to routinely use the median over the mean. Chapter 4 describes other criteria for judging measures of location, and situations will be described where both the median and mean are unsatisfactory.

Although we will see several practical problems with the mean, it is not being argued that the mean is always inappropriate. Imagine that someone invests $200,000 and reports that the median amount earned per year, over a 10-year period, is $100,000. This sounds good, but now imagine that the earnings for each year are:

```
$100,000, $200,000, $200,000, $200,000, $200,000,
$200,000, $200,000, $300,000, $300,000, $-1,800,000.
```

So at the end of 10 years this individual has earned nothing and in fact has lost the initial $200,000 investment. Certainly the long-term total amount earned is relevant, in which case the sample mean provides a useful summary of the investment strategy that was followed.

3.2.3 A Weighted Mean

A general goal of this book is to build an understanding of and appreciation for the practical benefits of contemporary statistical methods. To do this requires some understanding of the circumstances under which more conventional methods are optimal. Many introductory books make it clear that the sample mean is optimal when sampling from a normal distribution, but there are some additional details that will be important in this book, which we begin to discuss in Chapter 4. With this goal in mind, this subsection introduces the weighted mean.

Let $w_1, \ldots, w_n$ be any n constants. A *weighted mean* is just

$$\sum w_i X_i = w_1 X_1 + \cdots + w_n X_n. \tag{3.4}$$

An important special case of a weighted mean is where the constants $w_1, \ldots, w_n$ sum to 1. That is,

$$\sum w_i = 1.$$

In this case, any weighted mean provides a reasonable estimate of the population mean. So an important issue is determining which weights should be used in applied work; we begin to examine this problem in Chapter 4. Weighted means where the weights do not sum to 1 are in common use (as we shall see in connection with least squares regression). But for now attention is focused on situations where the weights sum to 1. The sample mean is a special case where

$$w_1 = w_2 = \cdots = w_n = \frac{1}{n}.$$

EXAMPLE. For the weights $w_1 = .2$, $w_2 = .1$, $w_3 = .4$, $w_4 = .3$, the weighted mean corresponding to $X_1 = 6$, $X_2 = 12$, $X_3 = 14$, and $X_4 = 10$ is

$$.2(6) + .1(12) + .4(14) + .3(10) = 11.$$

■

3.2.4 A Trimmed Mean

A *trimmed mean* refers to a situation where a certain proportion of the largest and smallest observations are removed and the remaining observations are averaged. Trimmed means contain as special cases the sample mean, where no observations are trimmed, and the median, where the maximum possible amount of trimming is used.

EXAMPLE. Consider the values

37, 14, 26, 17, 21, 43, 25, 6, 9, 11.

When computing a trimmed mean it is convenient first to put the observations in order. Here this yields

6, 9, 11, 14, 17, 21, 25, 26, 37, 43.

A 10% trimmed mean indicates that 10% of the smallest observations are removed, as are 10% of the largest, and the remaining values are averaged.

Continued

EXAMPLE. (*Continued*) Here there are 10 observations, so the 10% trimmed mean removes the smallest and largest observations and is given by

$$\bar{X}_t = \frac{9+11+14+17+21+25+26+37}{8} = 20.$$

The 20% trimmed mean removes 20% of the largest and smallest values and is

$$\bar{X}_t = \frac{11+14+17+21+25+26}{6} = 19.$$

■

A more general notation for describing a trimmed mean is useful, to avoid any ambiguity about how it is computed. Let γ be some constant, $0 \le \gamma < .5$, and set $g = [\gamma n]$, where the notation $[\gamma n]$ means that γn is rounded down to the nearest integer. For example, if $\gamma = .1$ and $n = 99$, then $g = [\gamma n] = [9.9] = 9$. The γ-trimmed mean is just the average of the values after the g smallest and g largest observations are removed. In symbols, the γ-trimmed mean is

$$\bar{X}_t = \frac{1}{n-2g}(X_{(g+1)} + X_{(g+2)} + \cdots + X_{(n-g)}). \tag{3.5}$$

Setting $\gamma = .1$ yields the 10% trimmed mean, and $\gamma = .2$ is the 20% trimmed mean.

The finite-sample breakdown point of the γ-trimmed mean is γ. So, in particular, the 10% trimmed mean has a breakdown point of .1, and the 20% trimmed mean has a breakdown point of .2. This says that when using the 20% trimmed mean, for example, more than 20% of the values must be altered to make the 20% trimmed mean arbitrarily large or small.

A fundamental issue is deciding how much to trim. At some level it might seem that no trimming should be done in most cases; otherwise, information will be lost somehow. We will see, however, that when addressing a variety of practical goals, 20% trimming often offers a considerable advantage over no trimming and the median. Moreover, Huber (1993) has argued that any estimator with a breakdown point less than or equal to .1 is dangerous and should be avoided. Eventually some of the reasons for this remark will become clear, but for now the details must be postponed.

3.2.5 S-PLUS Function for the Trimmed Mean

S-PLUS has a built-in function for computing a trimmed mean that has the general form

```
mean(x,tr=0),
```

where tr indicates the amount of trimming and x is now an S-PLUS variable containing a batch of numbers. Following the standard conventions used by S-PLUS functions, the notation tr=0 indicates that the amount of trimming defaults to 0, meaning that the S-PLUS function mean will compute the sample mean if a value for tr is not specified. For example, if the values 2, 6, 8, 12, 23, 45, 56, 65, 72 are stored in the S-PLUS variable x, then mean(x) will return the value 32.111, which is the sample mean. The command mean(x,.2) returns the 20% trimmed mean, which is 30.71. [For convenience the S-PLUS function tmean(x,tr=.2) has been supplied, which by default computes a 20% trimmed mean.]

3.2.6 A Winsorized Mean

In order to deal with some technical issues described in Chapter 4, we will need the so-called *Winsorized mean*. The Winsorized mean is similar to the trimmed mean, only the smallest and largest observations are not removed, but rather transformed by "pulling them in." To explain, we first describe what it means to Winsorize a batch of numbers.

Recall that when computing the 10% trimmed mean, you remove the smallest 10% of the observations. Winsorizing the observations by 10% simply means that rather than remove the smallest 10%, their values are set equal to the smallest value not trimmed when computing the 10% trimmed mean. Simultaneously, the largest 10% are reset to the largest value not trimmed. In a similar manner, 20% Winsorizing means that the smallest 20% of the observations are pulled up to the smallest value not trimmed when computing the 20% trimmed mean, and the largest 20% are pulled down to the largest value not trimmed. The Winsorized mean is just the average of the Winsorized values, which is labeled $\bar{X}_w$. The finite-sample breakdown point of the 20% Winsorized mean is .2, the same as the 20% trimmed mean. More generally, the finite-sample breakdown point of the γ-Winsorized mean is γ.

EXAMPLE. Consider again the 10 values

$$37, 14, 26, 17, 21, 43, 25, 6, 9, 11.$$

Because $n = 10$, with 10% trimming $g = [.1(10)] = 1$, meaning that the smallest and largest observations are removed when computing the 10% trimmed mean. The smallest value is 6, and the smallest value not removed when computing the 10% trimmed mean is 9. So 10% Winsorization of these values means that the value 6 is reset to the value 9. In a similar manner, the largest observation, 43, is pulled down to the next largest value, 37. So 10% Winsorization of the values yields

$$37, 14, 26, 17, 21, 37, 25, 9, 9, 11.$$

The 10% Winsorized mean is just the average of the Winsorized values:

$$\bar{X}_w = \frac{37 + 14 + 26 + 17 + 21 + 37 + 25 + 9 + 9 + 11}{10} = 20.6.$$

■

EXAMPLE. To compute a 20% Winsorized mean using the values in the previous example, first note that when computing the 20% trimmed mean, $g = [.2(10)] = 2$, so the two smallest values, 6 and 9, would be removed and the smallest value not trimmed is 11. Thus, Winsorizing the values means that the values 6 and 9 become 11. Similarly, when computing the 20% trimmed mean, the largest value not trimmed is 26, so Winsorizing means that the two largest values become 26. Consequently, 20% Winsorization of the data yields

26, 14, 26, 17, 21, 26, 25, 11, 11, 11.

The 20% Winsorized mean is $\bar{X}_w = 18.8$, the average of the Winsorized values. ■

Here is a general description of the Winsorized mean using a common notation. To compute a γ-Winsorized mean, first compute g as was done when computing a γ-trimmed mean. That is, g is γn rounded down to the nearest integer. The γ-Winsorized mean is

$$\bar{X}_w = \frac{1}{n}\{(g+1)X_{(g+1)} + X_{(g+2)} + \cdots + X_{(n-g-1)} + (g+1)X_{(n-g)}\}. \tag{3.6}$$

3.2.7 S-PLUS Function winmean

S-PLUS does not have a built-in function for computing the Winsorized mean, so one has been provided in the library of S-PLUS functions written especially for this book. (These functions can be obtained as described in the Section 1.2.) The function has the form

```
winmean(x,tr=.2),
```

where tr now indicates the amount of Winsorizing, which defaults to .2. So if the values 2, 6, 8, 12, 23, 45, 56, 65, 72 are stored in the S-PLUS variable x, the command winmean(x) returns 31.78, which is the 20% Winsorized mean. The command winmean(x,0) returns the sample mean, 32.11.

3.2.8 M-Estimators

So-called *M-estimators* provide yet another class of measures of location that have practical value. To provide some intuitive sense of M-estimators, imagine a game where someone has written down five numbers and two contestants are asked to pick a number that is close to all five without knowing what the five numbers happen to be. Further, suppose that the first contestant picks the number 22 and that the five numbers written down are

46, 18, 36, 23, 9.

We can measure how close 22 is to the first value simply by taking the absolute value of their difference: $|46 - 22| = 24$. In a similar manner, the accuracy of the first contestant's guess compared to the second number is $|18 - 22| = 4$. To get an overall measure of accuracy for the first contestant's guess, we might use the sum of the absolute errors associated with each of the five values:

$$|46 - 22| + |18 - 22| + |36 - 22| + |23 - 22| + |9 - 22| = 56.$$

If the second contestant guessed 28, we can measure her overall accuracy in a similar manner:

$$|46 - 28| + |18 - 28| + |36 - 28| + |23 - 28| + |9 - 28| = 60.$$

So based on our criterion of the sum of the absolute differences, contestant 1 is generally more accurate.

Now consider what happens if instead of absolute values, we use squared differences to measure accuracy. So for the first contestant, the accuracy relative to the first number written down is $(46 - 22)^2 = 576$, and the overall accuracy is

$$(46 - 22)^2 + (18 - 22)^2 + (36 - 22)^2 + (23 - 22)^2 + (9 - 22)^2 = 958.$$

As for the second contestant, we get

$$(46 - 28)^2 + (18 - 28)^2 + (36 - 28)^2 + (23 - 28)^2 + (9 - 28)^2 = 874.$$

Now the overall accuracy of the second contestant is judged to be better, in contrast to the situation where absolute differences were used. That is, the choice of a winner can change depending on whether we use squared error or absolute error.

Generalizing, let c be any number and suppose we measure its closeness to the five values under consideration with

$$\sum |X_i - c| = |46 - c| + |18 - c| + |36 - c| + |23 - c| + |9 - c|.$$

A value for c that is closest based on this criterion is the median, $M = 23$. In our contest, if one of the contestants had picked the value 23, they could not be beat by the other contestant, provided we use the sum of the absolute differences to measure closeness. But if we use squared differences instead,

$$\sum (X_i - c)^2 = (46 - c)^2 + (18 - c)^2 + (36 - c)^2 + (23 - c)^2 + (9 - c)^2,$$

the optimal choice for c is the sample mean, $\bar{X} = 26.4$.

Generalizing even further, if for any n values $X_1, \ldots, X_n$ we want to choose c so that it minimizes the sum of squared errors,

$$\sum (X_i - c)^2 = (X_1 - c)^2 + (X_2 - c)^2 + \cdots + (X_n - c)^2, \tag{3.7}$$

it can be shown that it must be the case that

$$(X_1 - c) + (X_2 - c) + \cdots + (X_n - c) = 0. \tag{3.8}$$

From this last equation it can be seen that $c = \bar{X}$. That is, when we choose a measure of location based on minimizing the sum of the squared errors given by Equation (3.7), which is an example of what is called the *least squares principle*, this leads to using the sample mean. But if we measure how close c is to the n values using the sum of the absolute differences, the sample median M minimizes this sum.

We can state this result in a more formal manner as follows. The sign of a number is -1, 0, or 1, depending on whether the number is less than, equal to, or greater than zero. So the sign of -6 is -1 and 10 has a sign of 1. A common abbreviation for the sign of a number is simply sign(X). So for any constant c, sign($X - c$) is equal to -1, 0, or 1, depending on whether the value X is less than c, equal to c, or greater than c, respectively. If, for instance, $c = 10$, $\text{sign}(12 - c) = 1$. It can be shown that if we want to choose c so as to minimize

$$\sum |X_i - c| = |X_1 - c| + |X_2 - c| + \cdots + |X_n - c|, \tag{3.9}$$

then it must be that

$$\text{sign}(X_1 - c) + \text{sign}(X_2 - c) + \cdots + \text{sign}(X_n - c) = 0, \tag{3.10}$$

and the sample median satisfies this last equation.

Here is the point: There are infinitely many ways of measuring closeness that lead to reasonable measures of location. For example, one might measure closeness using the absolute difference between an observation and c raised to some power, say, a. (That is, use $|X - c|^a$.) Setting $a = 1$ leads to the median, as just explained, and $a = 2$ results in the mean. In 1844, the Cambridge mathematician R. L. Ellis pointed out that an even broader class of functions might be used. (See Hald, 1998, p. 496.) Ellis noted that for any function Ψ having the property $\Psi(-x) = -\Psi(x)$, we get a reasonable measure of location, provided the probability curve is symmetric, if we choose c so that it satisfies

$$\Psi(X_1 - c) + \Psi(X_2 - c) + \cdots + \Psi(X_n - c) = 0. \tag{3.11}$$

(For some choices for Ψ, reasonable measures of location require special treatment when distributions are skewed, as is explained in Section 3.5.) Measures of location based on this last equation are called *M-estimators*, a modern treatment of which was first made by Huber (1964). For example, if we take $\Psi(x) = x$, in which case $\Psi(X - c) = X - c$, we get the mean, and $\Psi(x) = \text{sign}(x)$ leads to the median. Said another way, it is arbitrary how we measure closeness, and because different measures of closeness lead to different measures of location, there is the obvious dilemma of deciding which measure of closeness to use. Chapter 4 will begin to address this problem. For now it is merely remarked that two additional choices for Ψ will be seen to have practical value. The first is *Huber's* Ψ, where $\Psi(x) = x$, provided $|x| < K$, with K some constant to be determined. (For reasons explained in Chapter 4, a good choice for general use is $K = 1.28$.) If $x < -K$, then $\Psi(x) = -K$; if $X > K$, then $\Psi(x) = K$. (A technical detail is being ignored at this point but is addressed in Section 3.5.) Said another way, Huber's Ψ is like the Ψ corresponding to least squares, provided an observation is not too far from zero. The constant K is chosen so as to deal with certain practical issues, but the details are best postponed for now.

Another well-studied choice for Ψ is the so-called *biweight*, where $\Psi(x) = x(1 - x^2)$ if $|x| \leq 1$, otherwise $\Psi(x) = 0$. In Section 3.5 we see that the biweight has a serious practical problem when estimating location, but it has practical value for some other goals that will be discussed. Figure 3.1 shows a graph of the four choices for Ψ just discussed.

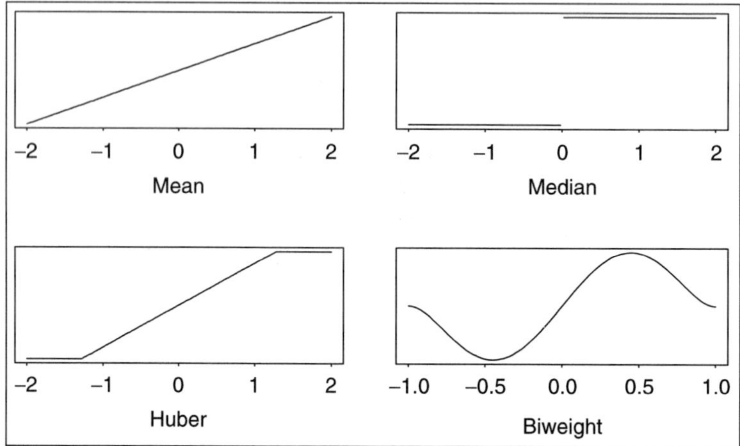

FIGURE 3.1 Examples of Ψ functions that have been considered in connection with M-estimators of location.

There remains the problem of how to compute an M-estimator of location once we have data. But before addressing this issue, we need some results in Section 3.3.

3.2.9 Other Measures of Location

The list of measures of location already covered is far from exhaustive, but it covers the measures that form the heart of this book. However, for completeness, some other classes of location estimators are mentioned. One consists of what are called *R-estimators*, which includes the so-called *Hodges–Lehmann estimator*, as a special case. To compute the Hodges–Lehmann estimator, first average every pair of observations. The median of all such averages is the Hodges–Lehmann estimator. This measure of location, as well as R-estimators in general, often have good properties when sampling from a perfectly symmetric distribution. But for asymmetric distributions, they can be quite unsatisfactory (e.g., Bickel & Lehmann, 1975; Huber, 1981, p. 65). Consequently, further details are not given here. (For more details about R-estimators, see Hettmansperger, 1984; Hettmansperger & McKean, 1998.) For a situation where the Hodges–Lehmann estimator can be unsatisfactory even when sampling from a symmetric distribution, see Morgenthaler and Tukey (1991, p. 15.)

L-estimators form yet another class of estimators that include the sample mean and trimmed means as special cases. L-estimators are like weighted means, only the weights are applied to the observed values after putting them in order. In formal terms, an L-estimator is

$$w_1X_{(1)} + w_2X_{(2)} + \cdots + w_nX_{(n)},$$

where, as in Section 3.2.3, $w_1, \ldots, w_n$ are constants chosen to achieve some desired goal.

For an extensive list of location estimators, and how they compare, see Andrews et al. (1972). Morgenthaler and Tukey (1991) describe yet another interesting class of estimators, but currently they can be difficult to compute.

3.3 Measures of Variation or Scale

As is evident, not all people are alike — they respond differently to similar conditions. If we ask 15 people to rate the job the president is doing on a 10-point scale, some might give a rating of 1 or 2, and others might give a rating of 9 or 10. There is *variation* among their ratings. In many ways, it is variation that provides the impetus for sophisticated statistical techniques.

For example, suppose all adults would give the same rating for the president — say, 7 — if they were polled. In statistical terms, the population mean is $\mu = 7$ because, as is evident, if all ratings are equal to 7, the average rating will be 7. This implies that we need sample only one adult in order to determine that the population mean is 7. That is, with $n = 1$ adult, the sample mean, $\bar{X}$, will be exactly equal to the population mean. Because in reality there is variation, in general the sample mean will not be equal to the population mean. If the population mean is $\mu = 7$ and we poll 15 adults, we might get a sample mean of $\bar{X} = 6.2$. If we poll another 15 adults, we might get $\bar{X} = 7.3$. We get different sample means because of the variation among the population of adults we want to study. One common goal in statistics is finding ways of taking into account how variation affects our ability to estimate the population mean with the sample mean. In a similar manner, there is the issue of how well the sample trimmed mean estimates the population trimmed mean.

To make progress, we need appropriate measures of variation, which are also called *measures of scale*. Like measures of location, many measures of scale have been proposed and studied.

3.3.1 Sample Variance

Imagine you sample 10 adults ($n = 10$) and ask each to rate the president; the ratings are:

3, 9, 10, 4, 7, 8, 9, 5, 7, 8.

The sample mean is $\bar{X} = 7$, and this is your estimate of the population mean, μ. The *sample variance* is

$$s^2 = \frac{\sum (X_i - \bar{X})^2}{n - 1}. \tag{3.12}$$

In words, the sample variance is computed by subtracting the sample mean from each observation and squaring. Then you add the results and divide by $n - 1$, the number of observations minus 1. For the data at hand, the calculations can be summarized as follows:

i	X_i	$X_i - \bar{X}$	$(X_i - \bar{X})^2$
1	3	−4	16
2	9	2	4
3	10	3	9

Continued

i	X_i	$X_i - \bar{X}$	$(X_i - \bar{X})^2$
4	4	−3	9
5	7	0	0
6	8	1	1
7	9	2	4
8	5	−2	4
9	7	0	0
10	8	1	1
$\sum$		0	48

The sum of the observations in the last column is $\sum (X_i - \bar{X})^2 = 48$. Then, because there are $n = 10$ subjects, the sample variance is

$$s^2 = \frac{48}{10 - 1} = 5.33.$$

The sample variance, s^2, is used to estimate the population variance σ^2, the variance we would get if only we could poll all adults. Under random sampling (which is formally described in Section 4.2), the sample variance gives us an increasingly more accurate estimate of the population variance as the sample size gets large. The square root of s^2, s, is called the *sample standard deviation* and estimates the population standard deviation, σ.

It is important to realize that a single unusual value can dominate the sample variance. This is one of several facts that wreaks havoc with standard statistical techniques. To provide a glimpse of problems to come, consider the values

$$8, 8, 8, 8, 8, 8, 8, 8, 8, 8, 8.$$

The sample variance is $s^2 = 0$, meaning there is no variation. If we increase the last value to 10, the sample variance is $s^2 = .36$. Increasing the last observation to 12, $s^2 = 1.45$, and increasing it to 14, $s^2 = 3.3$. The point is, even though there is no variation among the bulk of the observations, a single value can make the sample variance arbitrarily large. In modern terminology, the sample variance is not *resistant*, meaning roughly that a single unusual value can inflate the sample variance and give a misleading indication of how much variation there is among the bulk of the observations. Said more formally, the sample variance has a finite-sample breakdown point of only $1/n$. In some cases, this sensitivity to extreme values is desirable, but for many applied problems it is not, as will be seen.

3.3.2 The Interquartile Range

Another measure of scale or dispersion that is frequently used in applied work, particularly when the goal is to detect outliers, is called the *interquartile range*. For a population of individuals, let Q_u and Q_ℓ be the .75 and .25 quartiles, respectively. Q_u and Q_ℓ are called the *upper* and *lower quartiles*. So the probability that an observation is less than Q_ℓ is .25, the probability that an observation is less than Q_u is .75, and the probability that an observation is between Q_ℓ and Q_u is .5. The difference between the upper and lower quartiles, $Q_u - Q_\ell$, is the *population interquartile range* and reflects the variation of the middle portion of a distribution.

How should the quartiles be estimated based on data available to us? Many methods have been proposed and compared. (See Harrell & Davis, 1982; Dielman, Lowry, & Pfaffenberger, 1994; Parrish, 1990.) The choice of estimation method can depend in part on how the quartiles are to be used. In this book their main use is to detect unusually small or large values among a batch of numbers called *outliers*, in which case results in Cleveland (1985), Hoaglin and Iglewicz (1987), Hyndman and Fan (1996), Frigge, Hoaglin, and Iglewicz (1989) as well as Carling (2000) are relevant.

As usual, let $X_{(1)} \leq \cdots \leq X_{(n)}$ be the observations written in ascending order. Then estimates of the lower quartile typically have the form

$$q_1 = (1 - h)X_{(j)} + hX_{(j+1)}, \tag{3.13}$$

and the problem is determining appropriate choices for j and h. Among the eight choices considered by Frigge, Hoaglin, and Iglewicz (1989) when trying to detect outliers, the method based on the so-called *ideal fourth*, also known as the *machine fourth*, was found to be best, where j is the integer portion of $(n/4) + (5/12)$, meaning that j is $(n/4) + (5/12)$ rounded down to the nearest integer, and

$$h = \frac{n}{4} + \frac{5}{12} - j.$$

The estimate of the upper quartile is taken to be

$$q_2 = (1 - h)X_{(k)} + hX_{(k-1)}, \tag{3.14}$$

where $k = n - j + 1$, in which case the interquartile range is estimated with

$$\text{IQR} = q_2 - q_1. \tag{3.15}$$

EXAMPLE. Consider the values

$$-29.6, -20.9, -19.7, -15.4, -12.3, -8.0, -4.3, 0.8, 2.0, 6.2, 11.2, 25.0.$$

There are 12 observations ($n = 12$), so

$$\frac{n}{4} + \frac{5}{12} = 3.41667.$$

Rounding this last quantity down to the nearest integer gives $j = 3$, so $h = 3.416667 - 3 = .41667$. Because $X_{(3)} = -19.7$, the resulting estimate of the lower quartile is

$$q_1 = (1 - .41667)(-19.7) + .41667(-15.4) = -17.9.$$

In a similar manner, an estimate of the upper quartile is

$$q_2 = (1 - .41667)(6.2) + .41667(2) = 4.45,$$

so the estimate of the interquartile range is

$$\text{IQR} = 4.45 - (-17.9) = 22.35.$$

■

3.3.3 Winsorized Variance

When working with the trimmed mean, we will see that the so-called Winsorized variance plays an important role. To compute the Winsorized variance, simply Winsorize the observations as was done when computing the Winsorized mean in Section 3.2.6. The Winsorized variance is just the sample variance of the Winsorized values. Its finite-sample breakdown point is γ. So, for example, when computing a 20% Winsorized sample variance, more than 20% of the observations must be changed in order to make the sample Winsorized variance arbitrarily large.

3.3.4 S-PLUS Function winvar

Among the library of S-PLUS functions written for this book is

winvar(x,tr = .2),

which computes the Winsorized variance, where again tr represents the amount of Winsorizing and defaults to .2. So if the values

12, 45, 23, 79, 19, 92, 30, 58, 132

are stored in the S-PLUS variable x, winvar(x) returns the value 937.9, which is the 20% Winsorized variance. The command winvar(x,0) returns the sample variance, s^2, which is 1596.8. Typically the Winsorized variance will be smaller than the sample variance s^2 because Winsorizing pulls in extreme values.

3.3.5 Median Absolute Deviation

Another measure of dispersion, which plays an important role when trying to detect outliers (using a method described in Section 3.4.2) is the *median absolute deviation* (MAD) statistic. To compute it, first compute the sample median, M, subtract it from every observed value, and then take absolute values. In symbols, compute

$$|X_1 - M|, \ldots, |X_n - M|.$$

The median of the n values just computed is the MAD. Its finite sample breakdown point is .5.

EXAMPLE. Again using the values

12, 45, 23, 79, 19, 92, 30, 58, 132,

the median is $M = 45$, so $|X_1 - M| = |12 - 45| = 33$ and $|X_2 - M| = 0$. Continuing in this manner for all nine values yields

33, 0, 22, 34, 26, 47, 15, 13, 87.

The MAD is the median of the nine values just computed: 26. ■

There is a useful and commonly employed connection between the sample standard deviation, s, and the MAD. Recall that s is intended as an estimate of the population standard deviation, σ. In general, MAD does not estimate σ, but it can be shown that when sampling from a normal distribution, MAD/.6745 estimates σ as well. In Section 3.4.2 we will see that this suggests an approach to detecting outliers that plays an important role in data analysis. For convenience we set

$$\text{MADN} = \frac{\text{MAD}}{.6745}. \tag{3.16}$$

Statisticians define MAD in the manner just described. S-PLUS has a built-in function called mad, but it computes MADN, not MAD. So if you wanted to compute MAD using S-PLUS, you would use the command .6745 * mad(x). In most applications, MADN is employed, so typically you would use the S-PLUS function mad without multiplying it by .6745.

3.3.6 Average Absolute Distance from the Median

There is a measure of dispersion closely related to MAD that is frequently employed. Rather than take the median of the values $|X_1 - M|, \ldots, |X_n - M|$, take the average of these values instead. That is, use

$$D = \frac{1}{n} \sum |X_i - M|. \tag{3.17}$$

Despite using the median, D has a finite-sample breakdown point of only $1/n$. If, for example, we increase $X_{(n)}$, the largest of the X_i values, M does not change, but the difference between the largest value and the median becomes increasingly large, which in turn can make D arbitrarily large as well.

3.3.7 Biweight Midvariance and Percentage Bend Midvariance

There are many methods for measuring the variation among a batch of numbers, over 150 of which were compared by Lax (1985). There is little reason to list all of them here, but some additional methods should be mentioned: the *biweight midvariance* and the *percentage midvariance*. Recall that the Winsorized variance and MAD measure the variation of the middle portion of your data. In contrast, both the biweight and percentage bend midvariances make adjustments according to whether a value is flagged as being unusually large or small. The biweight midvariance empirically determines whether a value is unusually large or small using a slight modification of the outlier detection method described in Section 3.4.2. These values are discarded and the variation among the remaining values is computed. But the motivation for the remaining computational details is not remotely obvious without delving deeper into the theory of M-estimators, so further details regarding the derivation of these scale estimators are omitted. The percentage bend midvariance uses a different outlier detection rule and treats outliers in a different manner. Computational details are described in Box 3.1. Presumably readers will use S-PLUS to perform the computations, so detailed illustrations are omitted.

BOX 3.1 How to Compute the Percentage Bend Midvariance and the Biweight Midvariance

Computing the Percentage Bend Midvariance

As usual, let $X_1, \ldots, X_n$ represent the observed values. Choose a value for the finite-sample breakdown point and call it β. A good choice for general use is .2. Set

$$m = [(1-\beta)n + .5],$$

the value of $(1-\beta)n + .5$ rounded down to the nearest integer. Let $W_i = |X_i - M|$, $i = 1, \ldots, n$, and let $W_{(1)} \leq \cdots \leq W_{(n)}$ be the W_i values written in ascending order. Set

$$\hat{\omega}_\beta = W_{(m)},$$

$$Y_i = \frac{X_i - M}{\hat{\omega}_\beta},$$

$$a_i = \begin{cases} 1, & \text{if } |Y_i| < 1 \\ 0, & \text{if } |Y_i| \geq 1, \end{cases}$$

in which case the estimated percentage bend midvariance is

$$\hat{\zeta}^2_{pb} = \frac{n\hat{\omega}^2_\beta \sum \{\Psi(Y_i)\}^2}{\left(\sum a_i\right)^2}, \tag{3.18}$$

where

$$\Psi(x) = \max[-1, \min(1, x)].$$

Computing the Biweight Midvariance

Set

$$Y_i = \frac{X_i - M}{9 \times MAD},$$

$$a_i = \begin{cases} 1, & \text{if } |Y_i| < 1 \\ 0, & \text{if } |Y_i| \geq 1, \end{cases} \tag{3.19}$$

$$\hat{\zeta}_{\text{bimid}} = \frac{\sqrt{n}\sqrt{\sum a_i (X_i - M)^2 \left(1 - Y_i^2\right)^4}}{\left|\sum a_i \left(1 - Y_i^2\right)\left(1 - 5Y_i^2\right)\right|}.$$

The biweight midvariance is $\hat{\zeta}^2_{\text{bimid}}$.

The percentage bend midvariance plays a role when searching for robust analogs of Pearson's correlation. Such measures can be useful when trying to detect an association between two variables, as we see in Chapter 13. Also, the most common strategy for comparing two groups of individuals is in terms of some measure of location, as indicated in Chapter 1. However, comparing measures of variation can be of interest, in which case both the biweight and percentage midvariances can be useful, for reasons best postponed until Chapter 8.

The finite-sample breakdown point of the percentage bend midvariance can be controlled through the choice of a constant labeled β in Box 3.1. Setting $\beta = .1$, the finite-sample breakdown point is approximately .1, and for $\beta = .2$ it is approximately .2. Currently it seems that $\beta = .2$ is a good choice for general use, based on criteria to be described. The finite-sample breakdown point of the biweight midvariance is .5.

3.3.8 S-PLUS Functions bivar and pbvar

The S-PLUS functions

pbvar(x, beta = .2) and bivar(x)

(written for this book) compute the percentage bend midvariance and the biweight midvariance, respectively. Storing the values

12, 45, 23, 79, 19, 92, 30, 58, 132

in the S-PLUS variable x, pbvar(x) returns the value 1527.75. If the largest value, 132, is increased to 1000, pbvar still returns the value 1527.75. If the two largest values (92 and 132) are increased to 1000, again pbvar returns the value 1527.75. With beta equal to .2, it essentially ignores the two largest and two smallest values for the observations used here.

A point that cannot be stressed too strongly is that when we discard outliers, this is not to say that they are uninteresting or uninformative. Outliers can be very interesting, but for some goals they do more harm than good. Again, this issue is discussed in detail after some more basic principles are covered.

For the original values used to illustrate pbvar, the S-PLUS function bivar returns 1489.4, a value very close to the percentage bend midvariance. But increasing the largest value (132) to 1000 means bivar now returns the value 904.7. Its value decreases because it did not consider the value 132 to be an outlier, but increasing it to 1000 means pbvar considers 1000 to be an outlier and subsequently ignores it. Increasing the value 92 to 1000 means bivar returns 739. Now bivar ignores the two largest values because it flags both as outliers.

It is not the magnitude of the biweight midvariance that will interest us in future chapters. Rather, the biweight midvariance plays a role when comparing groups of individuals, and it plays a role when studying associations among variables, as we see in Chapter 13.

3.4 Detecting Outliers

For reasons that will become clear, detecting outliers — unusually large or small values among a batch of numbers — can be very important. This section describes several strategies for accomplishing this goal. The first, which is based on the sample mean and sample variance, is frequently employed and is a rather natural strategy based on properties of the normal curve described in Chapter 2. Unfortunately, it can be highly unsatisfactory, for reasons illustrated in the next subsection. The other methods are designed to correct the problem associated with the first.

3.4.1 A Natural but Unsatisfactory Method for Detecting Outliers

Equation (2.13) described a method for detecting outliers assuming normality: Declare a value an outlier if it is more than two standard deviations from the mean. Here, however, consistent with a general approach to outlier detection suggested by Rousseeuw and van Zomeren (1990), a very slight modification of this rule is used: Declare a value an outlier if it is more than 2.24 standard deviations from the mean. So the value X is flagged an outlier if

$$\frac{|X - \mu|}{\sigma} > 2.24. \tag{3.20}$$

Under normality, the probability of declaring a value an outlier using Equation (3.20) is .025.

Section 2.7 described a problem with this outlier detection rule, but another aspect of this rule should be described and emphasized. Generally we do not know μ and σ, but they can be estimated from data using the sample mean and sample variance. This suggests the commonly used strategy of declaring X an outlier if

$$\frac{|X - \bar{X}|}{s} > 2.24. \tag{3.21}$$

EXAMPLE. Consider the values

2, 2, 2, 2, 2, 3, 3, 3, 3, 3, 4, 4, 4, 4, 4, 1000.

The sample mean is $\bar{X} = 65.94$, the sample variance is $s = 249.1$,

$$\frac{|1000 - 65.94|}{249.1} = 3.75,$$

3.75 is greater than 2.24, so the value 1000 is declared an outlier. As is evident, the value 1000 is certainly unusual, and in this case our outlier detection rule gives a reasonable result. ■

EXAMPLE. Consider

$$2, 2, 2, 2, 2, 3, 3, 3, 3, 3, 4, 4, 4, 4, 4, 1000, 10{,}000.$$

These are the same values as in the last example, but with another outlier added. The value 10,000 is declared an outlier using Equation (3.21). Surely 1000 is unusual compared to the bulk of the observations, but it is not declared an outlier. The reason is that the two outliers inflate the sample mean and especially the sample standard deviation. Moreover, the influence of the outliers on s is so large, the value 1000 is not declared an outlier. In particular, $\bar{X} = 650.3$, $s = 2421.4$, so

$$\frac{|1000 - \bar{X}|}{s} = .14.$$

■

EXAMPLE. Now consider the values

$$2, 2, 3, 3, 3, 4, 4, 4, 100{,}000, 100{,}000.$$

It is left as an exercise to verify that the value 100,000 is not declared an outlier using Equation (3.21), yet surely it is unusually large. ■

The last two examples illustrate the problem known as *masking*. Outliers inflate both the sample mean and the sample variance, which in turn can mask their presence when using Equation (3.21). What is needed is a rule for detecting outliers that is not itself affected by outliers. One way of accomplishing this goal is to switch to measures of location and scale that have a reasonably high breakdown point.

3.4.2 A Better Outlier Detection Rule

Here is a simple outlier detection rule that has received a great deal of attention. Declare X to be an outlier if

$$\frac{|X - M|}{\text{MAD}/.6745} > 2.24. \tag{3.22}$$

When sampling from a normal curve, Equation (3.22) mimics our rule based on Equation (3.21) because for this special case the sample median M estimates the population mean μ, and MAD/.6745 estimates the population standard deviation σ. An important advantage of the method is that it addresses the problem of masking by employing measures of location and scale both of which have a breakdown point of .5. That is, the method can handle a large number of outliers without making the problem of masking an issue.

The use of the value 2.24 in Equation (3.22) stems from Rousseeuw and van Zomeren (1990). It should be noted that Equation (3.22) is known as the

Hampel identifier, but Hampel used the value 3.5 rather than 2.24. (For some refinements, see Davies & Gather, 1993.)

EXAMPLE. Consider again the values

$$2, 2, 3, 3, 3, 4, 4, 4, 100{,}000, 100{,}000.$$

Using our rule based on the sample mean and sample variance, we saw that the two values equal to 100,000 are not declared outliers. It can be seen that $M = 3.5$, MADN = MAD/.6745 = .7413, and

$$\frac{100{,}000 - 3.5}{.7413} = 134{,}893.4.$$

So in contrast to the rule based on the mean and variance, 100,000 would now be declared an outlier. ■

3.4.3 S-PLUS Function out

An S-PLUS function

out(x)

has been supplied that detects outliers using Equation (3.22). If the values from the last example are stored in the S-PLUS variable data, then part of the output from the command out(data) is

```
$out.val:
[1] 100000 100000

$out.id:
[1] 9 10
```

That is, there are two values declared outliers, both equal to 100,000, and they are the ninth and tenth observations stored in the S-PLUS variable data. (That is, the outliers are stored in data[9] and data[10].)

3.4.4 The Boxplot

Proposed by Tukey (1977), a *boxplot* is a commonly used graphical summary of data that provides yet another method for detecting outliers. The example boxplot shown in Figure 3.2 was created with the built-in S-PLUS command boxplot. As indicated, the ends of the rectangular box mark the lower and upper quartiles. That is, the box indicates where the middle half of the data lie. The horizontal line inside the box indicates the position of the median. The lines extending out from the box are called *whiskers*.

The boxplot declares the value X to be an outlier if

$$X < q_1 - 1.5(\text{IQR}) \tag{3.23}$$

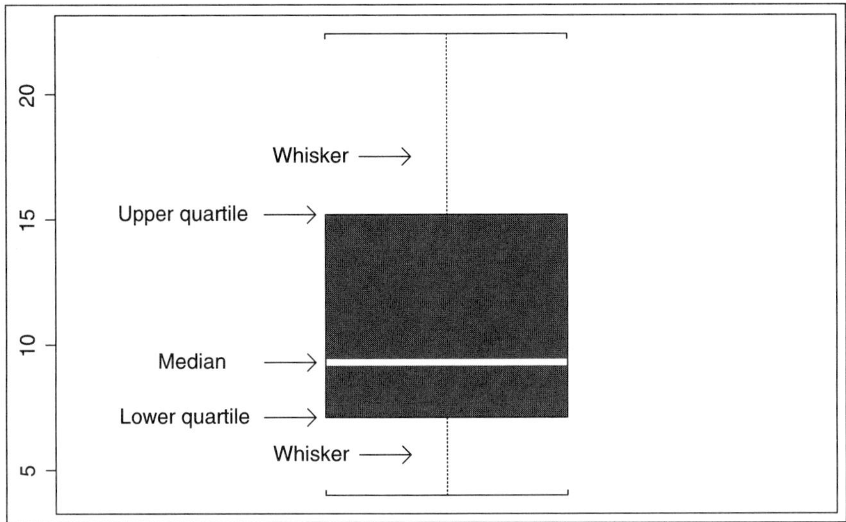

FIGURE 3.2 Example of a boxplot with no outliers.

or

$$X > q_2 + 1.5(\text{IQR}), \tag{3.24}$$

where IQR is the interquartile range defined in Section 3.3.2 and q_1 and q_2 are estimates of the lower and upper quartiles. To complicate matters, different software packages use different estimates of the quartiles. S-PLUS, for example, uses $j = n/4$ in Equation (3.13). If $n/4$ is an integer, S-PLUS sets $g = .5$, otherwise it uses $g = 0$. (A numerical summary of which points are declared outliers with the S-PLUS function boxplot can be determined with the S-PLUS command

```
print(boxplot(x,plot=F)$out).
```

The S-PLUS command summary uses yet another method for estimating quartiles.)

Figure 3.3 shows a boxplot with two outliers. The ends of the whiskers are called *adjacent values*. They are the smallest and largest values not declared outliers. Because the interquartile range has a finite sample breakdown point of .25, it takes more than 25% of the data to be outliers before the problem of masking occurs. A breakdown point of .25 seems to suffice in most situations, but exceptions can occur.

EXAMPLE. For the data in Table 3.2, a boxplot declares all values greater than 13.5 to be outliers. This represents 11.4% of the data. In contrast, the rule in Section 3.4.2, based on the median and MAD, declares all values greater than or equal to 6 to be outliers which is 27.6% of the data. This suggests that masking might be a problem for the boxplot because the proportion of outliers using the rule in Section 3.4.2 exceeds .25, the breakdown point of the boxplot. If we use the sample mean and standard deviation to detect outliers, only the value 6,000 is declared an outlier. ■

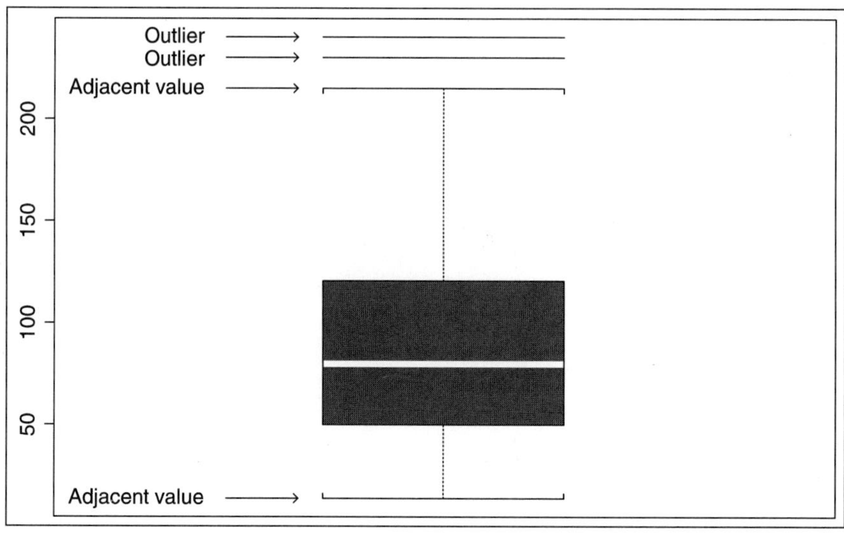

FIGURE 3.3 Boxplot with two values flagged as outliers.

3.4.5 A Modified Boxplot Rule for Detecting Outliers

A criticism of the traditional boxplot rule for detecting outliers is that the expected proportion of numbers that are declared outliers depends on the sample size; it is higher in situations where the sample size is small. To correct this problem, Carling (2000) uses the following rule: Declare the value X an outlier if

$$X > M + k\,\text{IQR} \tag{3.25}$$

or if

$$X < M - k\,\text{IQR}, \tag{3.26}$$

where

$$k = \frac{17.63n - 23.64}{7.74n - 3.71},$$

IQR is estimated with the ideal fourths, as described in Section 3.3.2, and, as usual, M is the sample median. So unlike standard boxplot rules, the median plays a role in determining whether a value is an outlier.

The choice between the outlier detection rule in Section 3.4.2 over the method just described is not completely straightforward. The rule used here is designed so that for normal distributions, the expected proportion of numbers declared outliers is .04. The choice .04 is arbitrary at some level and is clouded by the problem that the notion of outliers is vague. The extent to which this rate differs when using the method in Section 3.4.2 has not been studied. The method used here has a breakdown point of .25, in contrast to a breakdown point of .5 using the method in Section 3.4.2. As previously noted, generally a breakdown point of .25 suffices, but exceptions can occur. Moreover, for many purposes the method in Section 3.4.2, or some slight modification of it, has proven to have practical value in situations to be covered.

Finally, it is noted that there is a vast literature on detecting outliers. A few additional issues will be covered in subsequent chapters. Readers interested in a book-length treatment of this topic are referred to Barnett and Lewis (1994).

3.4.6 S-PLUS Function outbox

The S-PLUS function

outbox(x,mbox=F,gval=NA)

checks for outliers using one of two methods. With mbox=F, it uses the rule given by Equations (3.23) and (3.24), but unlike S-PLUS, quartiles are estimated with the ideal fourths. The argument gval corresponds to the constant 1.5 in Equations (3.23) and (3.24). Using gval=2, for example, causes this constant to be changed to 2. Setting mbox=T, outliers are detected using the method in the previous subsection.

EXAMPLE. For the data in Table 3.2, values greater than or equal to 13 are declared outliers with the S-PLUS function outbox with mbox=T. So in this particular case, the values declared outliers are the same as those using the built-in S-PLUS function boxplot. But with mbox=F, 13 is no longer declared an outlier. Situations also arise where a value is declared an outlier with mbox=F but not when mbox=T. ■

3.5 Computing an M-Estimator of Location

Section 3.2.8 introduced the notion of an M-estimator based on Huber's Ψ, but no details were given on how it is computed. This is because we needed first to describe some measures of scale plus some outlier detection rules in order to address an important technical issue.

For any measure of location, it should be the case that if we multiply all observed values by some constant b, the measure of location should be multiplied by b as well. Such measures of location are said to be *scale equivariant*. For example, if the weights of five children are measured in pounds and found to be 65, 55, 72, 80, and 70, the sample mean is 68.4. So if we convert to kilograms by dividing each child's weight by 2.2, the sample mean becomes 68.4/2.2 = 31.09. That is, the sample mean is scale equivariant. In a similar manner, the median in pounds is 70, and converting to kilograms the median becomes 70/2.2 = 31.8. When defining an M-estimator as described by Equation (3.11), we do not get this property automatically when using Huber's Ψ or the biweight. However, there is a simple method for addressing this problem: Include a measure of scale in Equation (3.11). It can be shown that if we use Huber's Ψ plus a measure of scale that has a high finite-sample breakdown point, the resulting M-estimator will have the same breakdown point as the measure of scale. So if we use MAD as our measure of scale, which has a breakdown point of .5, the resulting M-estimator, still using Huber's Ψ, will be .5 as well. In this case, Equation (3.11) becomes

$$\Psi\left(\frac{X_1 - c}{\text{MAD}}\right) + \cdots + \Psi\left(\frac{X_n - c}{\text{MAD}}\right) = 0, \tag{3.27}$$

where, as before, c is our measure of location.

An explicit equation for computing c once we have observations cannot be derived, but there are two simple and effective methods for dealing with this problem. One of these is to use a particular iterative technique for determining c that is easily applied on a computer. The second is to use a single step in this iterative method that inherits the positive features of M-estimators. The latter strategy is called a one-step M-estimator, and the resulting estimate of the measure of location is computed as follows. Let i_1 be the number of observations X_i for which $(X_i - M)/\text{MADN} < -K$, and let i_2 be the number of observations such that $(X_i - M)/\text{MADN} > K$, where typically $K = 1.28$ is used (for reasons that are difficult to explain until concepts in Chapter 4 are covered). The one-step M-estimator of location (based on Huber's Ψ) is

$$\hat{\mu}_{os} = \frac{K(\text{MADN})(i_2 - i_1) + \sum_{i=i_1+1}^{n-i_2} X_{(i)}}{n - i_1 - i_2}. \tag{3.28}$$

This one-step M-estimator almost uses the following strategy: Determine which values are outliers using the method in Section 3.4.2, except that Equation (3.22) is replaced by

$$\frac{|X - M|}{\text{MAD}/.6745} > K. \tag{3.29}$$

Next, remove the values flagged as outliers and average the values that remain. But for technical reasons, the one-step M-estimator makes an adjustment based on MADN, a measure of scale plus the number of outliers above and below the median.

EXAMPLE. Computing a one-step M-estimator (with $K = 1.28$) is illustrated with the following ($n = 19$) observations:

77 87 88 114 151 210 219 246 253 262

296 299 306 376 428 515 666 1310 2611.

It can be seen that $M = 262$ and that $\text{MADN} = \text{MAD}/.6745 = 169$. If for each observed value we subtract the median and divide by MADN we get

−1.09 − 1.04 −1.035 −0.88 −0.66 −0.31 −0.25 −0.095 −0.05

0.00 0.20 0.22 0.26 0.67 0.98 1.50 2.39 6.2 13.90

Continued

EXAMPLE. (*Continued*) So there are four values larger than the median that are declared outliers: 515, 666, 1310, 2611. That is, $i_2 = 4$. No values less than the median are declared outliers, so $i_1 = 0$. The sum of the values not declared outliers is

$$77 + 87 + \cdots + 428 = 3411.$$

So the value of the one-step M-estimator is

$$\frac{1.28(169)(4 - 0) + 3411}{19 - 0 - 4} = 285.$$

■

3.5.1 S-PLUS Function onestep

The S-PLUS function

onestep(x,bend=1.28)

computes a one-step M-estimator with Huber's Ψ. The second argument, bend, corresponds to the constant K in Equation (3.28) and defaults to 1.28. If, for example, the data in Table 3.2 are stored in the S-PLUS variable sexm, onestep(sexm) returns the value 2.52.

3.5.2 A Modified One-Step M-Estimator

The one-step M-estimator has desirable theoretical properties, but when sample sizes are small, problems can arise when comparing groups using the methods covered in subsequent chapters. This section describes a simple modification of the one-step M-estimator given by Equation (3.28). The modification consists of dropping the term containing MADN. That is, use

$$\hat{\mu}_{\text{mom}} = \frac{\sum_{i=i_1+1}^{n-i_2} X_{(i)}}{n - i_1 - i_2} \tag{3.30}$$

as a measure of location, where now $K = 2.24$ is used to determine i_1 and i_2. In effect, use Equation (3.22) to detect outliers, discard any outliers that are found, and then use the mean of the remaining values. The finite-sample breakdown point is .5. This modified one-step M-estimator, MOM, is very similar to what are known as *skipped estimators*, which were studied by Andrews et al. (1972). (Skipped estimators use a boxplot rule to detect outliers rather than the median and MAD.) Initially, technical problems precluded skipped estimators from being used to compare groups of individuals, but recent advances have made MOM a viable measure of location. Note that MOM introduces a certain amount of flexibility versus using trimmed means. For example, MOM might discard zero observations. Moreover, if a distribution is heavy-tailed and highly skewed to the right, it might be desirable to trim more observations from the right tail versus the left, and MOM contains the possibility of doing this. (The relative merits of MOM versus M-estimators are discussed in subsequent chapters.)

It might help to summarize a fundamental difference among trimmed means, M-estimators, and MOM (or more generally the class of skipped estimators). Each represents a different approach to measuring location. Trimmed means discard a fixed proportion of large and small observations. MOM, and skipped estimators in general, empirically determine how many observations are to be trimmed and includes the possibility of different amounts of trimming in the tails as well as no trimming at all. M-estimators are based on how we measure the overall distance between some measure of location and the observations. Huber's measure of distance leads to the one-step M-estimator given by Equation (3.28), which has certain similarities to MOM, but unlike MOM, an adjustment is made based on a measure of scale when the amount of trimming in the left tail differs from the amount in the right.

3.5.3 S-PLUS Function mom

The S-PLUS function

mom(x,bend=2.24)

computes the modified one-step M-estimator just described. As a brief illustration, consider again the data used in the last example. It can be seen that according to Equation (3.22), the three largest values (666, 1310, and 2611) are outliers. Discarding them and averaging the remaining values yields 245.4, which is the value returned by mom. So in contrast to the 20% trimmed mean, none of the lower values are discarded.

3.6 Histograms

Two additional graphical tools for summarizing data should be mentioned. One of these is the histogram; the other is a so-called *kernel density estimator*, which is described in the next section.

A histogram is illustrated with data from a heart transplant study conducted at Stanford University between October 1, 1967, and April 1, 1974. Of primary concern is whether a transplanted heart will be rejected by the recipient. With the goal of trying to address this issue, a so-called T5 mismatch score was developed by Dr. C. Bieber. It measures the degree of dissimilarity between the donor and the recipient tissue with respect to HL-A antigens. Scores less than 1 represent a good match, and scores greater than 1 a poor match. Of course, of particular interest is how well a T5 score predicts rejection, but this must wait for now. The T5 scores, written in ascending order, are shown in Table 3.3 and are taken from R. G. Miller (1976).

A histogram simply groups the data into categories and plots the corresponding frequencies. To illustrate the basic idea, we group the T5 values into eight categories: (1) values between −0.5 and 0.0, (2) values greater than 0.0 but less than or equal

TABLE 3.3 T5 Mismatch Scores from a Heart Transplant Study

0.00	0.12	0.16	0.19	0.33	0.36	0.38	0.46	0.47	0.60	0.61	0.61	0.66
0.67	0.68	0.69	0.75	0.77	0.81	0.81	0.82	0.87	0.87	0.87	0.91	0.96
0.97	0.98	0.98	1.02	1.06	1.08	1.08	1.11	1.12	1.12	1.13	1.20	1.20
1.32	1.33	1.35	1.38	1.38	1.41	1.44	1.46	1.51	1.58	1.62	1.66	1.68
1.68	1.70	1.78	1.82	1.89	1.93	1.94	2.05	2.09	2.16	2.25	2.76	3.05

TABLE 3.4 Frequencies and Relative Frequencies for Grouped T5 scores, $n = 65$

Test score (x)	Frequency	Relative frequency
−0.5–0.0	1	1/65 = .015
0.0–0.5	8	8/65 = .123
0.5–1.0	20	20/65 = .308
1.0–1.5	18	18/65 = .277
1.5–2.0	12	12/65 = .138
2.0–2.5	4	4/65 = .062
2.5–3.0	1	1/65 = .015
3.0–3.5	1	1/65 = .015

to 0.5, (3) values greater than 0.5 but less than or equal to 1.0, and so on. The frequency and relative frequency associated with each of these intervals is shown in Table 3.4. For example, there are eight T5 mismatch scores in the interval extending from 0.0 to 0.5, and the proportion of all scores belonging to this interval is .123.

Figure 3.4 shows the histogram for the T5 scores that was created with the built-in S-PLUS function hist. Notice that the base of the leftmost shaded rectangle extends from 0 to 0.5 and has a height of 9. This means that there are nine cases where a T5 score has a value between 0 and 0.5. The base of the next shaded rectangle extends from 0.5 to 1 and has a height of 20. This means that there are 20 T5 scores having a value between 0.5 and 1. The base of the next rectangle extends from 1 to 1.5 and has a height of 18, so there are 18 T5 scores between 1 and 1.5.

3.7 Kernel Density Estimators

As indicated in Section 2.6, probabilities associated with continuous variables are determined by the area under a curve called a *probability density function*. The equation for this curve is typically labeled $f(x)$. An example is Equation (2.10), which gives the equation for the probability density function of the normal distribution. For some purposes to be covered, it is useful to have an estimate of $f(x)$ (the equation for the probability density function) based on observations we make. A histogram provides a crude estimate, but for various reasons it can be unsatisfactory (e.g., Silverman, 1986, pp. 9–11).

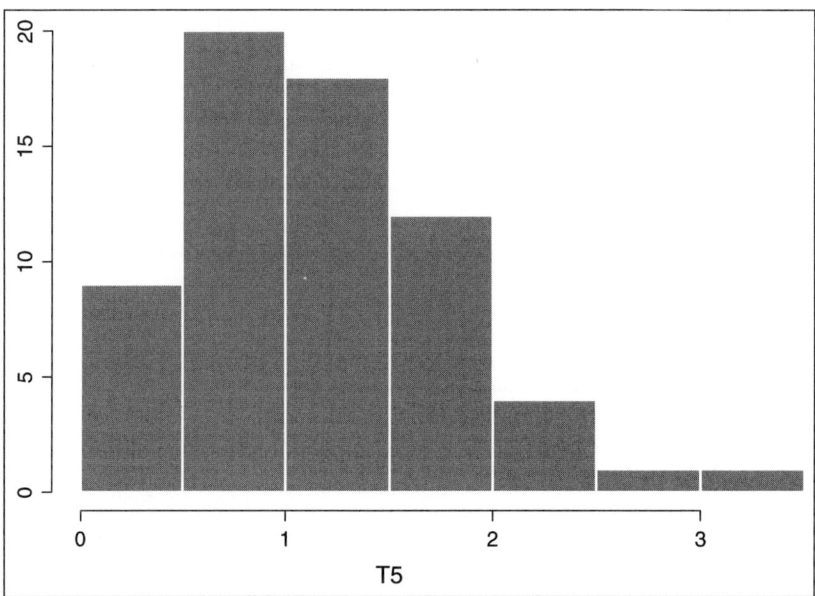

FIGURE 3.4 Example of a histogram.

Another seemingly natural but unsatisfactory approach to estimating the probability density function is to assume observations are sampled from a normal curve and replace the population mean and variance (μ and σ^2) in Equation (2.10) with the sample mean and variance ($\bar{X}$ and s^2). But this approach can be highly unsatisfactory, even when sampling from a perfectly symmetric distribution. To understand why, assume we are sampling from a mixed normal and that we happen to get an exact estimate of the population mean and variance. The dotted curve in Figure 2.9 is the probability density function being estimated, and the normal curve in Figure 2.9 is the estimate of the mixed normal following the strategy just indicated. As is evident, there is a substantial difference between the two curves, indicating that we get a poor estimate.

For some purposes we get a much more effective estimate using what is called a *kernel density estimator*. There are many variations of kernel density estimators (Silverman, 1986), but only one is used here. It is based on what is called *Rosenblatt's shifted histogram* and employs results derived by D. W. Scott (1979) as well as Freedman and Diaconis (1981). In particular, to estimate $f(x)$ for any x, set

$$h = \frac{1.2(\text{IQR})}{n^{1/5}},$$

where IQR is the interquartile range. Let A be the number of observations less than or equal to $x + h$. Let B be the number of observations strictly less than $x - h$. Then the estimate of $f(x)$ is

$$\hat{f}(x) = \frac{A - B}{2nh}.$$

A numerical illustration is not given because presumably readers will use the S-PLUS function described in the next subsection.

3.7.1 S-PLUS Function kdplot

The S-PLUS function kdplot estimates the probability density function $f(x)$ for a range of x values, using the kernel density estimator just described, and then plots the results. (The interquartile range is estimated with the S-PLUS built-in function summary.) It has the form

kdplot(data,rval=15),

where data is any S-PLUS variable containing data. By default, the function begins by estimating $f(x)$ at 15 values spread over the range of observed values. For example, if the data consist of values ranging between 2 and 36, kdplot picks 15 values between 2 and 36, estimates $f(x)$ at these points, and then plots the results. For very large sample sizes ($n \geq 500$) it might help to plot $f(x)$ at 20 or 25 values instead. This can be accomplished with the second argument. For example, rval=25 will plot an estimate of $f(x)$ at 25 x values evenly spaced between the smallest and largest of the observed values. For small and even moderate sample sizes, rval=15 or smaller gives a better (less ragged) estimate of the probability density function in most cases. For small sample sizes ($n < 40$), kdplot can be rather unrevealing. Figure 3.5 shows an estimate of the probability density function associated with the T5 mismatch scores in Table 3.3 based on the S-PLUS function kdplot (with the argument rval set equal to 10).

For another approach to estimating $f(x)$, readers might consider the built-in S-PLUS function density. The S-PLUS command plot(density(x)) plots an estimate of $f(x)$ using the data in the S-PLUS variable x.

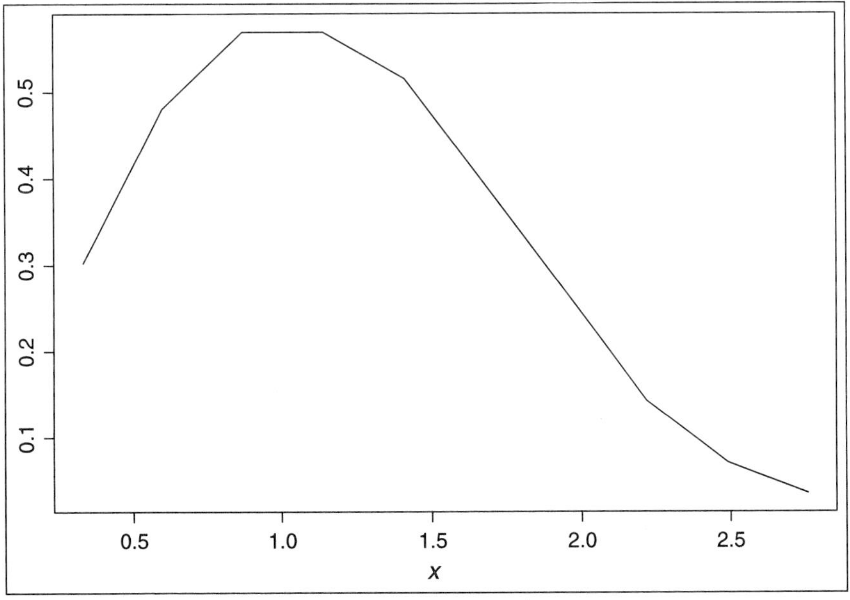

FIGURE 3.5 A kernel density estimate of the distribution of T5 mismatch scores.

TABLE 3.5 Word Identification Scores

58	58	58	58	58	64	64	68	72	72	72	75	75	77	77	79	80
82	82	82	82	82	84	84	85	85	90	91	91	92	93	93	93	95
95	95	95	95	95	95	95	98	98	99	101	101	101	102	102	102	102
102	103	104	104	104	104	104	105	105	105	105	105	107	108	108	110	111
112	114	119	122	122	125	125	125	127	129	129	132	134				

3.8 Stem-and-Leaf Displays

A stem-and-leaf display is another method of gaining some overall sense of what data are like. The method is illustrated with measures taken from a study aimed at understanding how children acquire reading skills. A portion of the study was based on a measure that reflects the ability of children to identify words.[1] Table 3.5 lists the observed scores in ascending order.

The construction of a stem-and-leaf display begins by separating each value into two components. The first is the *leaf*, which in this example is the number in the 1s position (the single digit just to the left of the decimal place). For example, the leaf corresponding to the value 58 is 8. The leaf for the value 64 is 4, and the leaf for 125 is 5. The digits to the left of the leaf are called the *stem*. Here the stem of 58 is 5, the number to the left of 8. Similarly, 64 has a stem of 6 and 125 has a stem of 12. We can display the results for all 81 children as follows:

Stems	Leaves
5	88888
6	448
7	22255779
8	0222224455
9	011233355555555889
10	1112222223444445555788
11	01249
12	22555799
13	24

There are five children who have the score 58, so there are five scores with a leaf of 8, and this is reflected by the five 8s displayed to the right of the stem 5, in the Leaves column. Two children got the score 64, and one child got the score 68. That is, for the stem 6, there are two leaves equal to 4 and one equal to 8, as indicated by the list of leaves in the display. Now look at the third row of numbers, where the stem is 7. The leaves listed are 2, 2, 2, 5, 5, 7, 7, and 9. This indicates that the value 72 occurred three times, the value 75 occurred two times, as did the value 77, and the value 79 occurred once. Notice that the display of the leaves gives us some indication of the

1 These data were supplied by L. Doi.

values that occur most frequently and which are relatively rare. Like the histogram, the stem-and-leaf display gives us an overall sense of what the values are like.

The choice of which digit is to be used as the leaf depends in part on which digit provides a useful graphical summary of the data. But details about how to address this problem are not covered here. Suffice it to say that algorithms have been proposed for deciding which digit should be used as the leaf and determining how many lines a stem-and-leaf display should have (e.g., Emerson & Hoaglin, 1983). Here we merely note that S-PLUS has a built-in function for computing a stem-and-leaf display that has the form

stem(x),

where, as usual, x now represents any S-PLUS variable containing data. For the T5 mismatch scores, the stem-and-leaf display created by S-PLUS is

```
Decimal point is at the colon

0 : z122344
0 : 556667777788889999
1 : 00000111111223334444
1 : 5566777788999
2 : 0122
2 : 8
3 : 0
```

The z in the first row stands for zero. This function also reports the median and the quartiles.

3.9 Exercises

1. For the observations

   21, 36, 42, 24, 25, 36, 35, 49, 32

   verify that the sample mean, trimmed mean, and median are $\bar{X} = 33.33$, $\bar{X}_t = 32.9$, and $M = 35$.
2. The largest observation in Exercise 1 is 49. If 49 is replaced by the value 200, verify that the sample mean is now $\bar{X} = 50.1$ but the trimmed mean and median are not changed. What does this illustrate about the resistance of the sample mean?
3. For the data in Exercise 1, what is the minimum number of observations that must be altered so that the 20% trimmed mean is greater than 1000?
4. Repeat the previous problem but use the median instead. What does this illustrate about the resistance of the mean, median, and trimmed mean?
5. For the observations

   6, 3, 2, 7, 6, 5, 8, 9, 8, 11

   verify that the sample mean, trimmed mean, and median are $\bar{X} = 6.5$, $\bar{X}_t = 6.7$, and $M = 6.5$.

6. A class of fourth-graders was asked to bring a pumpkin to school. Each of the 29 students counted the number of seeds in his or her pumpkin and the results were

```
250, 220, 281, 247, 230, 209, 240, 160, 370, 274,
210, 204, 243, 251, 190, 200, 130, 150, 177, 475,
221, 350, 224, 163, 272, 236, 200, 171, 98.
```

   (These data were supplied by Mrs. Capps at the La Cañada Elementary School, La Cañada, CA.) Verify that the sample mean, trimmed mean, and median are $\bar{X} = 229.2$, $\bar{X}_t = 220.8$, and $M = 221$.
7. Suppose health inspectors rate sanitation conditions at restaurants on a five-point scale, where a 1 indicates poor conditions and a 5 is excellent. Based on a sample of restaurants in a large city, the frequencies are found to be $f_1 = 5$, $f_2 = 8$, $f_3 = 20$, $f_4 = 32$, and $f_5 = 23$. What is the sample size, n? Verify that the sample mean is $\bar{X} = 3.7$.
8. For the frequencies $f_1 = 12$, $f_2 = 18$, $f_3 = 15$, $f_4 = 10$, $f_5 = 8$, and $f_6 = 5$, verify that the sample mean is 3.
9. For the observations

   21, 36, 42, 24, 25, 36, 35, 49, 32

   verify that the sample variance and the sample Winsorized variance are $s^2 = 81$ and $s_w^2 = 51.4$, respectively.
10. In Exercise 9, what is your estimate of the standard deviation, σ?
11. In general, will the Winsorized sample variance, s_w^2, be less than the sample variance, s^2?
12. Among a sample of 25 subjects, what is the minimum number of subjects that must be altered to make the sample variance arbitrarily large?
13. Repeat Exercise 12 but for s_w^2 instead. Assume 20% Winsorization.
14. For the observations

   6, 3, 2, 7, 6, 5, 8, 9, 8, 11

   verify that the sample variance and Winsorized variance are 7.4 and 1.8, respectively.
15. For the data in Exercise 6, verify that the sample variance is 5584.9 and the Winsorized sample variance is 1375.6.
16. For the data in Exercise 14, determine the ideal fourths and the corresponding interquartile range.
17. For the data in Exercise 6, which values would be declared outliers using the rule based on MAD?
18. Referring to the description of the population variance, σ^2, devise a method of estimating σ^2 based on the frequencies, f_x.
19. Referring to Exercise 18, and using the data in Exercise 7, estimate σ, the population standard deviation.
20. For a sample of n subjects, the relative frequencies are $f_0/n = .2$, $f_1/n = .4$, $f_2/n = .2$, $f_3/n = .15$, and $f_4/n = .05$. The sample mean is 1.45. Using your answer to Exercise 18, verify that the sample variance is 1.25.

21. Snedecor and Cochran (1967) report results from an experiment on weight gain in rats as a function of source of protein and levels of protein. One of the groups was fed beef with a low amount of protein. The weight gains were

$$90, 76, 90, 64, 86, 51, 72, 90, 95, 78.$$

Verify that there are no outliers among these values when using a boxplot but that there is an outlier using Equation (3.22).

22. For the values 1, 2, 3, 4, 5, 6, 7, 8, 9, and 10 there are no outliers. If you increase 10 to 20, then 20 would be declared an outlier by a boxplot. If the value 9 is also increased to 20, would the boxplot find two outliers? If the value 8 is also increased to 20, would the boxplot find all three outliers?
23. Use the results of the last problem to come up with a general rule about how many outliers a boxplot can detect.
24. For the data in Table 3.3, verify that the value 3.05 is an outlier, based on a boxplot.
25. Figure 3.6 shows a boxplot of data from a reading study where the general goal was to find predictors of reading ability in children. What, approximately, are the lower and upper quartiles and the interquartile range? How large or small would an observation have to be in order to be declared an outlier?
26. Under normality, it can be shown that $(n - 1)s^2/\sigma^2$ has a chi-squared distribution with $n - 1$ degrees of freedom. If $n = 11$ and $\sigma^2 = 100$, use Table 3 in Appendix B to determine $P(s^2 > 159.8)$.
27. A researcher claims that for a certain population of adults, cholesterol levels have a normal distribution with $\sigma^2 = 10$. If you randomly sample 21 individuals, verify that $P(s^2 \leq 14.2) = .9$.

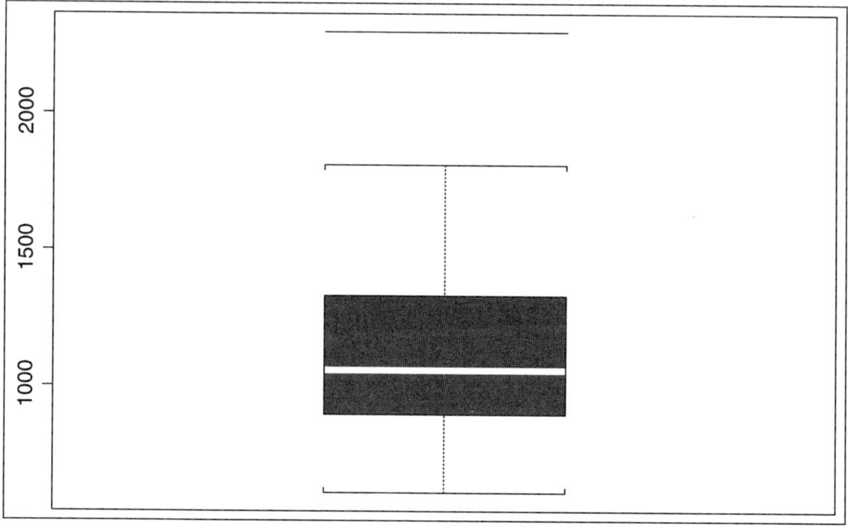

FIGURE 3.6 Boxplot of reading study data.

SAMPLING DISTRIBUTIONS AND CONFIDENCE INTERVALS

Understanding and appreciating modern statistical methods requires, among other things, a close look at the notion of a *sampling distribution*. We begin with the basics typically covered in an introductory course and then take up issues that explain why contemporary statistical techniques have practical value. This chapter also introduces the notion of a *confidence interval*, but important practical issues not typically covered in an introductory course are described.

4.1 Basics

To illustrate the notion of a *sampling distribution* in a concrete manner, imagine you are interested in the health risks associated with ozone in the atmosphere and that one of your concerns is how ozone affects weight gain in infants. Obviously you cannot experiment on human infants, so suppose you conduct an experiment where 22 rats are exposed to an ozone environment and their average weight gain is $\bar{X} = 11$ grams. Now imagine that another team of researchers repeats your experiment with a new sample of 22 rats. Of course their sample mean will probably differ from yours; they might get $\bar{X} = 16$. In a similar manner, a third team of researchers might get $\bar{X} = 9$. If infinitely many teams of researchers could repeat your experiment, then we would know what is called the *sampling distribution of the sample mean*. In particular, we would know the probability that the sample mean is less than 1, less than 10, or less than c for any constant c we might pick.

Being able to approximate the sampling distribution of the sample mean allows us to address a problem of fundamental importance. There are millions of rats that could have been used in your experiment. How well does the sample mean ($\bar{X}$) estimate the population mean (μ), the average weight gain you would observe if all living rats took part in this experiment? Based on your data, can you be reasonably certain that the population mean has a value close to the sample mean? Can you be reasonably certain that the population mean is less than 30 or less than 20? Can you be reasonably certain that it is greater than 1? If we can get a good approximation of the sampling distribution of the sample mean, we can answer these questions.

Sampling distributions also provide a perspective on how different location estimators compare to one another. That is, they provide both a graphical and a numerical summary of how the accuracy of the sample mean, for example, compares to the accuracy of the sample median, trimmed mean, or MOM estimate of location. Section 4.3 covers this important topic and provides results that begin to reveal the relative merits of the location estimators described in Section 3.2.

The conventional method for approximating the sampling distribution of the sample mean was derived by Laplace nearly 200 years ago. Laplace's strategy, which forms the foundation of many methods routinely used today, consists of several components, each of which must be understood if we want to appreciate the advantages of modern techniques. The immediate goal is to describe each of these components and then to indicate how they come together in an attempt to make inferences about the population mean. Then we will scrutinize these components and consider how we might get better results with modern technology.

Before proceeding, it might help to outline in a bit more detail where we are going. Continuing our illustration concerning weight gain in rats, imagine that we repeat an experiment infinitely many times and that for each experiment we sample 22 rats and compute the sample mean. Then there will be some average value for these infinitely many sample means. In formal terms, this average is written as $E(\bar{X})$, the expected value of the sample mean. Simultaneously, there will be variation among the sample means, and we can measure this variation in terms of the expected squared difference between the sample mean and the population mean. That is, we use a strategy similar to how the population variance was defined, only rather than use the average squared distance of a *single* observation from the mean, we now focus on the average squared distance between the sample mean and the population mean. In formal terms, we use

$$E(\bar{X} - \mu)^2,$$

which is called the *expected squared error* of the sample mean.

Notice that the expected squared error of the sample mean reflects how close the sample mean tends to be to the population mean. That is, it provides a crude indication of whether the sample mean tends to be an accurate estimate of μ. If the sample mean is always identical to the population mean, then the expected squared error of the sample mean is zero. But if situations arise where the sample mean provides a poor estimate of the population mean, the expected squared error of the sample mean will be large.

Now, our immediate goal is to approximate the distribution of the sample means without actually repeating our experiment infinitely many times. Recall from Chapter 2 that if we assume normality, probabilities are determined exactly once we know the mean and variance. So if we assume that a plot of infinitely many sample means would be a normal curve, then we would know the distribution of the sample mean if we could determine its mean ($E(\bar{X})$) and variance ($\text{VAR}(\bar{X})$). Laplace realized that these quantities can be determined if we assume random sampling, which is formally described in the next section.

A comment should be made about the normality assumption. We have seen indications that nonnormality is a practical concern. Under what conditions can normality be assumed when trying to approximate the plot of infinitely many sample means? There are two that are important here. The first is that if we (randomly) sample observations from a normal curve, then the sample mean will have a normal distribution as well. In the event sampling is not from a normal curve, Laplace appealed to his central limit theorem, which is formally introduced in Section 4.6. There are indeed situations where this latter strategy provides reasonably accurate results, but it can fail miserably, as we shall see.

4.2 Random Sampling

We need to be more formal about the notion of random sampling in order to understand why some strategies for dealing with nonnormality are theoretically unsound. *Random sampling* means that the observations available to us satisfy two conditions: (1) They are identically distributed, and (2) they are independent. (This is sometimes called *simple random sampling* to distinguish it from other types of sampling strategies not covered in this book.) Two measures are identically distributed if they have the same probability function. For continuous variables this means that for any constant c we might pick, the probability that an observation is less than c is the same regardless of which measure we use. So, for example, if X_1 is the first rat in our experiment on the effects of ozone on weight gain, there is a certain probability that this rat will gain 14 grams or less during the course of the experiment. Of course, there is some corresponding probability for the second rat, X_2. If these two probabilities are the same and in fact $P(X_1 < c) = P(X_2 < c)$ for any c we might pick, then X_1 and X_2 are said to be identically distributed. More generally, n observations, which we label $X_1, \ldots, X_n$, are said to be identically distributed if any two of them are identically distributed. Consequently, the expected value of each observation is the same, namely, μ, the population mean. That is,

$$E(X_1) = E(X_2) = \cdots = E(X_n) = \mu.$$

So a slight extension of results in Section 2.9 shows that if the observations are identically distributed, then the average value of their sum is just the sum of their average values. That is,

$$E(X_1 + \cdots + X_n) = \mu + \cdots + \mu = n\mu.$$

Using our rule for expected values when we multiply by a constant, if we multiply the sum in this last equation by $1/n$, we have that

$$E(\bar{X}) = \frac{n\mu}{n} = \mu. \tag{4.1}$$

In words, on average, the sample mean estimates the population mean when observations are identically distributed. When Equation (4.1) is satisfied, we say that the sample mean is an *unbiased* estimate of the population mean.

The variance of the sample mean refers to the variance of all sample means if we were to repeat a study infinitely many times. Said another way, it is the variance associated with the sampling distribution. The variance of any measure is defined as the expected squared distance from its mean (or average) value. For example, if we have a single observation X, its expected value is labeled μ, and the variance of X is the expected or average value of the squared distance between X and μ. In symbols, the variance of X is $E(X-\mu)^2$, as explained in Chapter 2. In a similar manner, the variance of the sample mean is the average squared difference between it and its mean. We just saw that the expected value of the sample mean is μ, so by definition its variance is $E(\bar{X}-\mu)^2$, which happens to be its expected squared error as well. It is common to write the variance of the sample mean as $\text{VAR}(\bar{X})$ or $\sigma^2_{\bar{X}}$. So we have that

$$\text{VAR}(\bar{X}) = E(\bar{X}-\mu)^2 = \sigma^2_{\bar{X}}.$$

To make a clear distinction between the variance of a single observation and the variance of the sample mean, the variance of the sample mean is often called the *squared standard error of the sample mean.*

Now consider the problem of getting an expression for the variance of the sample mean. The assumption that observations are identically distributed implies that they have identical variances. That is,

$$\text{VAR}(X_1) = \text{VAR}(X_2) = \cdots = \text{VAR}(X_n) = \sigma^2.$$

As indicated by Equation (2.15), the variance of a sum of observations requires that we take into account their correlation. But if we assume independence, which is implied when we assume (simple) random sampling, each pair of observations has zero correlation, so the variance of the sum is just the sum of the variances. That is, under random sampling,

$$\text{VAR}(X_1 + X_2 + \cdots + X_n) = n\sigma^2. \tag{4.2}$$

Finally, the rules for expected values can be used to show that as a result,

$$\text{VAR}(\bar{X}) = \frac{\sigma^2}{n}. \tag{4.3}$$

Said another way, σ^2/n is the squared standard error of the sample mean, and $\sigma/\sqrt{n}$ is its standard error.

The results just given are so important, they are illustrated in another manner to make sure they are clear. To be concrete, imagine you have been asked to determine the attitude of the typical adult regarding the death penalty. You randomly sample $n = 100$ adults, and each gives one of three responses: the death penalty is not a deterrent, it makes a slight difference in some cases, or it is a strong deterrent. Suppose the responses are recorded as 1, 2, 3, respectively and that *unknown* to you, among the millions of adults you might interview, the probability function is

x:	1	2	3
$p(x)$:	.3	.5	.2

It can be seen that the population mean and variance are $\mu = 1.9$ and $\sigma^2 = .49$, respectively. If in your particular study 10 adults respond that it is not a deterrent,

TABLE 4.1 Illustration of a Sampling Distribution

Sample	X_1	X_2	X_3	...	X_{100}	$\bar{X}$	s^2
1	2	3	2	...	2	2.4	.43
2	1	3	2	...	1	2.2	.63
3	1	2	1	...	3	2.1	.57
4	3	1	3	...	1	2.3	.62
⋮	⋮	⋮	⋮	⋮	⋮	⋮	⋮
Average of $\bar{X}$ values is $\mu = 1.9$; i.e., $E(\bar{X}) = \mu$							
Average of s^2 values is $\sigma^2 = .49$; i.e., $E(s^2) = \sigma^2$							

$n = 100$ for each sample.

25 say it makes a slight difference, and the remaining 65 say it is a strong deterrent, then the sample mean is

$$\bar{X} = 1\frac{10}{100} + 2\frac{25}{100} + 3\frac{65}{100} = 2.55.$$

So there is a discrepancy between the sample mean and the population mean, as we would expect.

Suppose we repeat the survey of adults millions of times, and each time we interview $n = 100$ subjects. Again assume that the possible outcomes we observe are 1, 2, and 3. Table 4.1 summarizes what might be observed. Each row represents a replication of the study, and there are 100 columns, corresponding to the 100 subjects observed each time the study is replicated. The 100 subjects in the first row differ from the 100 subjects in the second row, which differ from the 100 subjects in third row, and so on. Then for the first column, among the millions of times we repeat the experiment, there is a certain proportion of times we would observe a 1 or a 2 or a 3, and we can think of these proportions as the corresponding probabilities. If the probability of a 1 is .3, then the proportion of 1's in the column headed by X_1 will be .3. Similarly, the column headed by X_2 has a certain proportion of 1's, 2's, and 3's, and again they can be thought of as probabilities. If the probability of a 2 is .5, then 50% of the values in this column will be 2. When we say that X_1 and X_2 are identically distributed, we mean that the proportion of 1's, 2's, and 3's in columns 1 and 2 are exactly the same. That is, it does not matter whether a subject is first or second when determining the probability of a 1 or a 2 or a 3. More generally, if all 100 observations are identically distributed, then any column we pick in Table 4.1 will have the same proportions of 1's, 2's, and 3's over the millions of subjects.

Now, an implication of having identically distributed random variables is that if we average all the sample means in the next to last column of Table 4.1, we would get μ, the population mean. That is, the average or expected value of all sample means we might observe based on n subjects is equal to μ, which in this case is 1.9. In symbols, $E(\bar{X}) = \mu$, as previously indicated. Moreover, the variance of these sample means is σ^2/n, which in this particular case is $.49/100 = .0049$, so the standard error of the sample mean is $\sqrt{.0049} = .07$.

But how do we determine the standard error of the sample mean in the more realistic situation where the probability function is not known? A natural guess is to estimate σ^2 with the sample variance, s^2, in which case VAR($\bar{X}$) is estimated with s^2/n, and this is exactly what is done in practice.

To provide a bit more justification for estimating σ^2 with s^2, again consider the survey of adults regarding the death penalty. Imagine that for a sample of 100 adults we compute the sample variance, s^2. Repeating this process infinitely many times, each time sampling 100 adults and computing s^2, the sample variances we might get are shown in the last column of Table 4.1. It can be shown that under random sampling, the average of the resulting sample variances is σ^2, the population variance from which the observations were sampled. In symbols, $E(s^2) = \sigma^2$ for any sample size n we might use, which means that the sample variance is an *unbiased estimator* of σ^2, the population variance.[1] Because s^2 estimates σ^2, we estimate the standard error of the sample mean with $s/\sqrt{n}$.

EXAMPLE. Imagine you are a health professional interested in the effects of medication on the diastolic blood pressure of adult women. For a particular drug being investigated, you find that for $n = 9$ women, the sample mean is $\bar{X} = 85$ and the sample variance is $s^2 = 160.78$. An estimate of the squared standard error of the sample mean, assuming random sampling, is $s^2/n = 160.78/9 = 17.9$. An estimate of the standard error of the sample mean is $\sqrt{17.9} = 4.2$, the square root of s^2/n. ■

4.3 Approximating the Sampling Distribution of $\bar{X}$

Consider again the ozone experiment described at the beginning of this chapter. Suppose a claim is made that if all rats could be measured, we would find that weight gain is normally distributed, with $\mu = 14$ grams and a standard deviation of $\sigma = 6$. Under random sampling, this claim implies that the distribution of the sample mean has a mean of 14 as well; and because there are $n = 22$ rats, the variance of the sample mean is $6^2/22$. In symbols, $E(\bar{X}) = 14$ and VAR($\bar{X}$) $= \sigma^2/22 = 6^2/22 = 36/22$. The sample mean based on the 22 rats in the experiment is $\bar{X} = 11$. Is it reasonable to expect a sample mean this low if the claimed values for the population mean (14) and standard deviation (6) are correct? In particular, what is the probability of observing $\bar{X} \le 11$?

Recall from Chapter 2 that if X has a normal distribution, $P(X \le c)$ can be determined by standardizing X. That is, subtract the mean and divide by the standard deviation, yielding $Z = (X-\mu)/\sigma$. If X is normal, Z has a standard normal distribution. That is, we solve the problem of determining $P(X \le c)$ by transforming it into a problem involving the standard normal distribution. *Standardizing a random variable* is a recurrent theme in statistics, and it is useful when determining $P(\bar{X} \le c)$ for any constant c.

1 Recall that the sample variance is given by $s^2 = \sum (X_i - \bar{X})^2/(n-1)$. If we divide by n rather than $n-1$, it is no longer true that $E(s^2) = \sigma^2$.

Again consider the ozone experiment, where $n = 22$. Still assuming that the mean is $\mu = 14$ and the standard deviation is $\sigma = 6$, suppose we want to determine the probability that the sample mean will be less than 11. In symbols, we want to determine $P(\bar{X} \leq 11)$. When sampling from a normal distribution, $\bar{X}$ also has a normal distribution, so essentially the same technique described in Chapter 2 can be used to determine $P(\bar{X} \leq 11)$: We convert the problem into one involving a standard normal. This means that we *standardize the sample mean* by subtracting its mean and dividing by its standard deviation which is $\sigma/\sqrt{n}$. That is, we compute

$$Z = \frac{\bar{X} - \mu}{\sigma/\sqrt{n}}.$$

When sampling from a normal distribution, it can be mathematically verified that Z has a standard normal distribution. This means that

$$P(\bar{X} \leq 11) = P\left(Z \leq \frac{11 - \mu}{\sigma/\sqrt{n}}\right).$$

For the problem at hand,

$$\frac{11 - \mu}{\sigma/\sqrt{n}} = \frac{11 - 14}{6/\sqrt{22}} = -2.35,$$

and Table 1 in Appendix B tells us that the probability that a standard normal variable is less than -2.35 is .0094. More succinctly,

$$P(Z \leq -2.35) = .0094,$$

so the probability of getting a sample mean less than or equal to 11 is .0094. That is, if simultaneously the assumption of random sampling from a normal distribution is true, the population mean is $\mu = 14$, and the population standard deviation is $\sigma = 6$, then the probability of getting a sample mean less than or equal to 11 is .0094.

More generally, given c, some constant of interest, the probability that the sample mean is less than c is

$$P(\bar{X} < c) = P\left(\frac{\bar{X} - \mu}{\sigma/\sqrt{n}} < \frac{c - \mu}{\sigma/\sqrt{n}}\right)$$

$$= P\left(Z < \frac{c - \mu}{\sigma/\sqrt{n}}\right). \qquad (4.4)$$

In other words, $P(\bar{X} < c)$ is equal to the probability that a standard normal random variable is less than $(c - \mu)/(\sigma/\sqrt{n})$.

EXAMPLE. To be sure Equation (4.4) is clear, the ozone example is repeated in a more concise manner. The claim is that $\mu = 14$ and $\sigma = 6$. With $n = 22$ rats, what is the probability that the sample mean will be less than or equal to $c = 11$ if this claim is true and it is assumed that we are randomly sampling from a normal distribution? To answer this question, compute

$$\frac{c - \mu}{\sigma/\sqrt{n}} = \frac{11 - 14}{6/\sqrt{22}} = -2.35.$$

Then according to Equation (4.4),

$$P(\bar{X} \leq 11) = P(Z \leq -2.35).$$

Referring to Table 1 in Appendix B, the probability that a standard normal random variable is less than or equal to -2.35 is .0094. That is, if the claim is correct,

$$P(\bar{X} \leq 11) = .0094.$$

■

For the 22 rats in the ozone experiment, the sample mean is $\bar{X} = 11$. As just indicated, getting a sample mean less than or equal to 11 is unlikely if the claims $\mu = 14$ and $\sigma = 6$ are true. That is, the data suggest that perhaps these claims are false.

Analogous to results in Chapter 2, we can determine the probability that the sample mean is greater than some constant c, and we can determine the probability that the sample mean is between two numbers, say, a and b. The probability of getting a sample mean greater than c is

$$P(\bar{X} > c) = 1 - P\left(Z < \frac{c - \mu}{\sigma/\sqrt{n}}\right), \tag{4.5}$$

and the probability that the sample mean is between the numbers a and b is

$$P(a < \bar{X} < b) = P\left(Z < \frac{b - \mu}{\sigma/\sqrt{n}}\right) - P\left(Z < \frac{a - \mu}{\sigma/\sqrt{n}}\right). \tag{4.6}$$

EXAMPLE. A researcher claims that for college students taking a particular test of spatial ability, the scores have a normal distribution with mean 27 and variance 49. If this claim is correct, and you randomly sample 36 subjects, what is the probability that the sample mean will be greater than $c = 28$? Referring to Equation (4.5), first compute

$$\frac{c - \mu}{\sigma/\sqrt{n}} = \frac{28 - 27}{\sqrt{49/36}} = .857.$$

Continued

EXAMPLE. (*Continued*) Because $P(Z \leq .857) = .20$, Equation (4.5) tells us that

$$P(\bar{X} > 28) = 1 - P(Z \leq .857) = 1 - .20 = .80.$$

This means that if we randomly sample $n = 25$ subjects and the claims of the researcher are true, the probability of getting a sample mean greater than 28 is .8. ■

EXAMPLE. Suppose observations are randomly sampled from a normal distribution with $\mu = 5$ and $\sigma = 3$. If $n = 36$, what is the probability that the sample mean is between $a = 4$ and $b = 6$? To find out, compute

$$\frac{b - \mu}{\sigma/\sqrt{n}} = \frac{6 - 5}{3/\sqrt{36}} = 2.$$

Referring to Table 1 in Appendix B, $P(\bar{X} < 4) = P(Z < 2) = .9772$. Similarly,

$$\frac{a - \mu}{\sigma/\sqrt{n}} = \frac{4 - 5}{3/\sqrt{36}} = -2,$$

and $P(Z < -2) = .0228$. So according to Equation (4.6),

$$P(2 < \bar{X} < 4) = .9772 - .0228 = .9544.$$

This means that if $n = 36$ observations are randomly sampled from a normal distribution with mean $\mu = 5$ and standard deviation $\sigma = 3$, there is a .9544 probability of getting a sample mean between 4 and 6. ■

An important point is that as the number of observations increases, the sample mean will provide a better estimate of the population mean, μ. The sampling distribution of $\bar{X}$, under normality, provides an illustration of the sense in which this is true. Suppose we randomly sample a single observation ($n = 1$) from a standard normal distribution. This single observation provides an estimate of the population mean, and the standard normal distribution in Figure 4.1 graphically illustrates how close this observation will be to the mean, μ. Figure 4.1 also shows the distribution of $\bar{X}$ when $n = 16$. Notice that this distribution is more tightly centered around the mean. That is, $\bar{X}$ is more likely to be close to μ when $n = 16$ rather than 1. If we increase n to 25, the distribution of $\bar{X}$ would be even more tightly centered around the mean.

4.4 The Sample Mean versus MOM, the Median, Trimmed Mean, and M-Estimator

The notion of a sampling distribution generalizes to any of the location estimators considered in Chapter 3. For example, if we conduct a study and compute the median based on 20 observations and repeat the study billions of times, each time computing

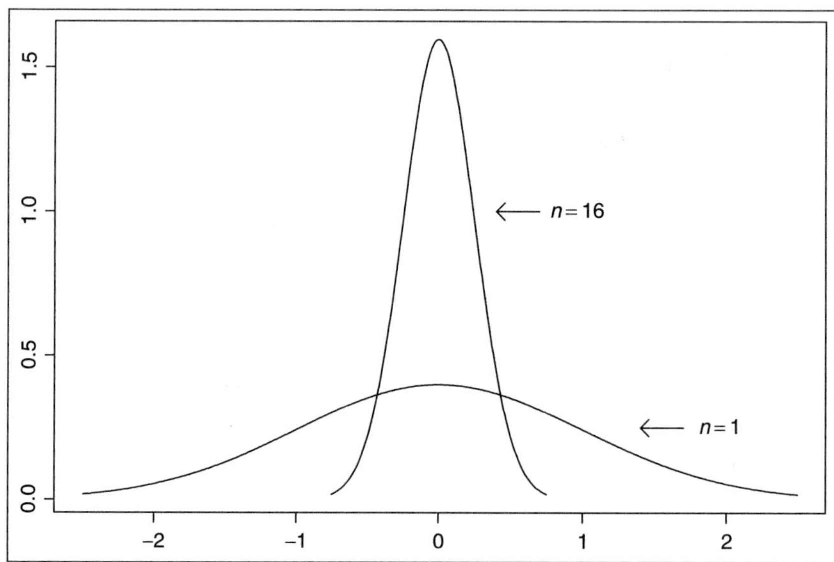

FIGURE 4.1 The distribution of $\bar{X}$, under normality, when $n = 1$ and $n = 16$.

a median based on 20 observations, we would know the sampling distribution of the median. That is, we would know the probability that the median is less than 4, less than 10, or less than any constant c that might be of interest. Temporarily assume sampling is from a normal distribution, and, for convenience only, suppose the population mean is zero and the variance is 1. Because sampling is from a symmetric distribution, the sample median, M, is a reasonable estimate of the population mean, μ. But of course the sample median will be in error, as was the case when using the sample mean.

Now, when sampling from a normal distribution, theory tells us that, on average, the sample mean will be a more accurate estimate of the population mean versus the sample median. To illustrate the extent to which this is true, 20 observations were generated on a computer from a standard normal distribution, the sample mean and median were computed, and this process was repeated 5000 times. Figure 4.2 shows a plot of the resulting means and medians. That is, Figure 4.2 shows an approximation of the sampling distributions of both the mean and median. Note that the plot of the means is more tightly centered around the value zero, the value both the mean and median are trying to estimate. This indicates that in general, the sample mean is more accurate.

But what about nonnormal distributions? How does the median compare to the mean in accuracy? To illustrate an important point, rather than sample from the standard normal distribution, we repeat the computer experiment used to create Figure 4.2, only now sampling is from the mixed normal shown in Figure 2.8. Recall that the mixed normal represents a small departure from normality in the sense described in Section 2.7.

Figure 4.3 shows the results of our computer experiment, and, in contrast to Figure 4.2, now the sample median is much more accurate, on average, relative to the mean. That is, the sample median is likely to be substantially closer to the true population mean than is the sample mean, $\bar{X}$. This illustrates a general result of considerable

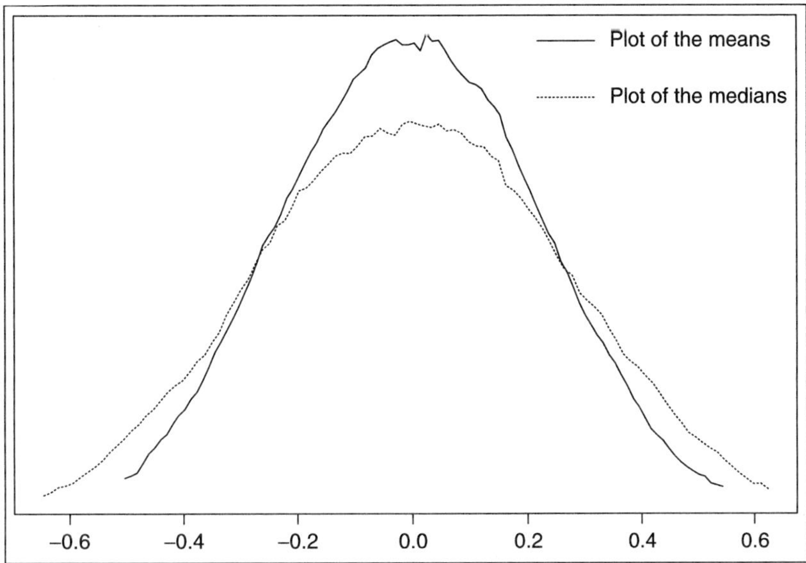

FIGURE 4.2 A plot of 5000 means and 5000 medians when sampling from a normal distribution.

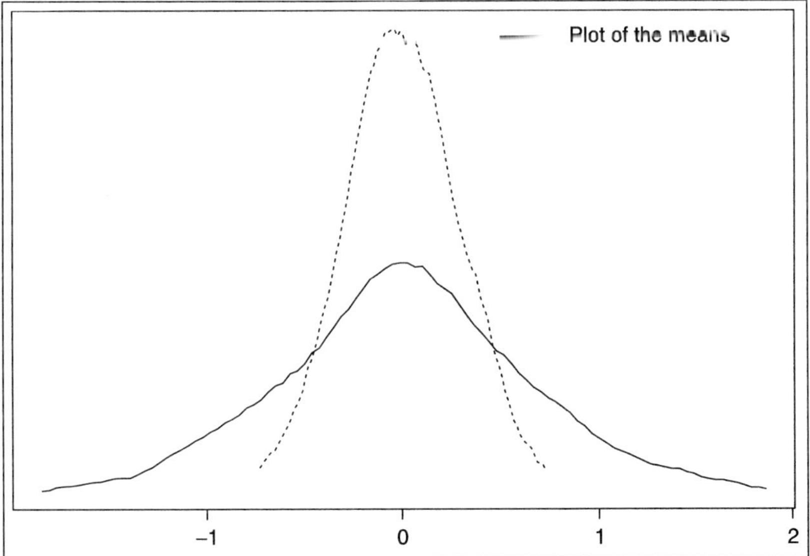

FIGURE 4.3 A plot of 5000 means and 5000 medians when sampling from a mixed normal distribution.

importance. Even for an extremely small departure from normality, the median can be a much more accurate estimate of the center of a symmetric distribution than the mean. But this is not a very compelling reason to routinely use the median, because Figure 4.2 illustrates that it can be substantially less accurate than the mean as well.

Now we consider the accuracy of the 20% trimmed mean and the M-estimator (based on Huber's Ψ) versus the mean. We repeat the process used to create

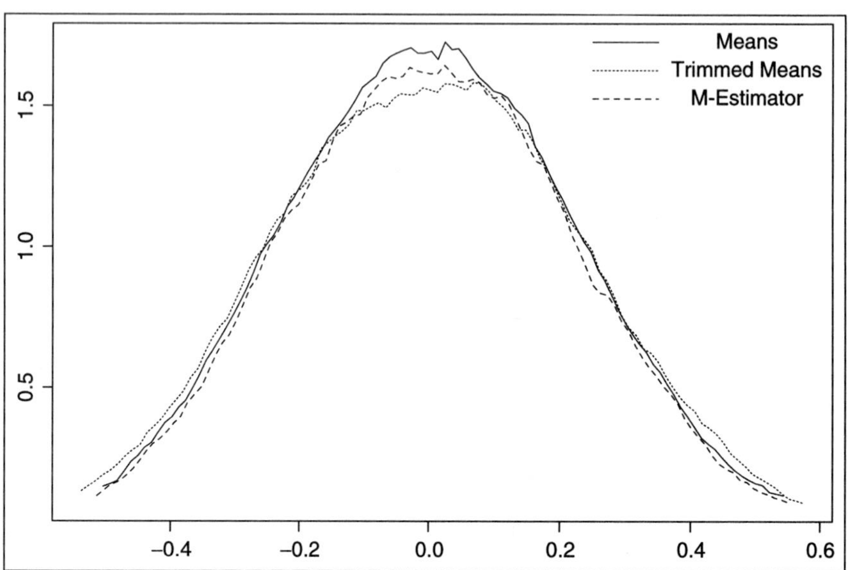

FIGURE 4.4 Plots of 5000 means, 20% trimmed means and M-estimators when sampling from a normal distribution.

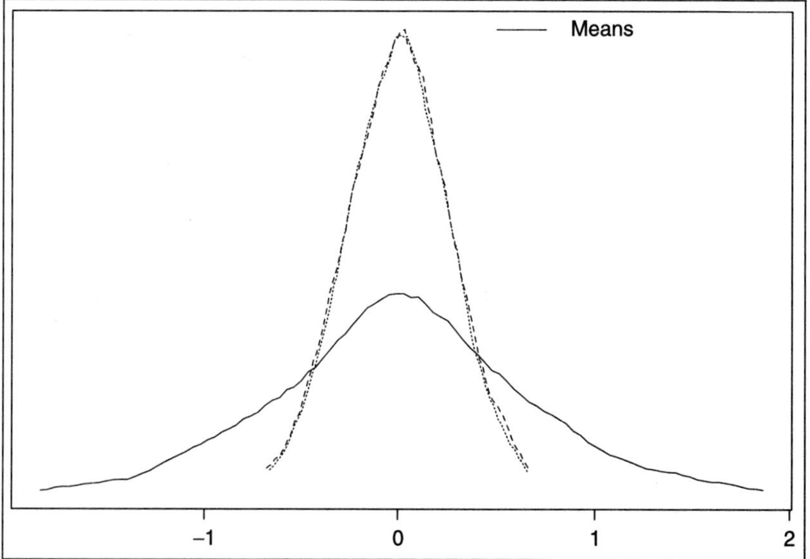

FIGURE 4.5 Plots of 5000 means, 20% trimmed means and M-estimators when sampling from a mixed normal distribution.

Figures 4.2 and 4.3, only the median is replaced by $\bar{X}_t$ and $\hat{\mu}_{os}$, where $\hat{\mu}_{os}$ is the one-step M-estimator given by Equation (3.28). Figure 4.4 shows the resulting plots when sampling is from a normal distribution. The mean is most accurate, but its improvement over the 20% trimmed mean and M-estimator is not very striking. If we use the MOM estimator given by Equation (3.30), the plot of the values reveals that it is slightly less accurate than the M-estimator used here. Figure 4.5 shows the sampling distributions when observations are sampled from a mixed normal instead. As is evident, the sample mean performs poorly relative to the other two estimators considered. Here, MOM gives nearly the same results as the 20% trimmed mean and M-estimator.

The results illustrated in Figures 4.4 and 4.5 are not surprising, in the sense that MOM, the 20% trimmed mean, and the M-estimator used here were designed to give nearly the same accuracy as the mean when sampling from a normal distribution. That is, theoretical results tell us how much we can trim without sacrificing too much accuracy under normality, and theory also suggests how to design an M-estimator so that again relatively good accuracy is obtained. Simultaneously, theory tells us that these estimators will continue to perform relatively well when sampling from a heavy-tailed distribution, such as the mixed normal.

To broaden our perspective on the relative merits of the mean, we now describe a general situation where the sample mean is optimal among all weighted means, described in Section 3.2.3. Without assuming normality, Gauss showed that under random sampling, among all the weighted means we might consider, the sample mean will have a smaller variance (or squared standard error) than any other weighted mean we might consider. (This result is a special case of the Gauss–Markov theorem, which is discussed in more detail when we take up regression.) So if we were to plot the sampling distribution of any weighted mean versus the sample mean, the sample mean would be more tightly centered around the population mean.

The result just described might seem to contradict our finding that the median, 20% trimmed mean, and one-step M-estimator can be substantially more accurate than the mean. There is no contradiction, however, because all three of these estimators do not belong to the class of weighted means. The median and trimmed mean, for example, involve more than weighting the observations — they require putting the observations in order.

It is easy to see why Gauss's result might suggest using the mean rather than the median or the trimmed mean. The 20% trimmed mean, for example, removes 40% of the observations. How could this possibly improve accuracy as opposed to giving some weight to all of the values? It is important to develop some intuition about this issue because despite graphical and mathematical arguments demonstrating that a 20% trimmed mean can be much more accurate than the sample mean, often there is reluctance to use any trimming at all because it seems counterintuitive. One way of understanding this issue is covered in Section 4.9. Here, an alternate but less technical explanation is given.

From the perspective about to be described, it is not surprising that a trimmed mean beats the mean, but it is rather amazing that the sample mean is optimal in any situation at all. To explain, suppose we sample 20 observations from a standard normal distribution. So the population mean and population 20% trimmed mean are zero. Now consider the smallest of the 20 values. It can be shown that, with probability .983, this value will be less than −0.9, and with probability .25 it will be less than −1.5. That is, with fairly high certainty, it will not be close to zero, the value we are trying to estimate. In a similar manner, the largest of the 20 observations has probability .983 of being greater than 0.9 and probability .25 of being greater than 1.5. Simultaneously, if we put the observations in ascending order, the probability that the two middle values do not differ by more than .5 from zero is .95. So a natural reaction is that extreme values should be given less weight in comparison to the observations in the middle. A criticism of this simple argument is that the smallest value will tend to be less than zero, the largest will be greater than zero, and their average value will be exactly zero, so it might seem that there is no harm in using them to estimate the population mean. However, the issue is how much these extreme values contribute to the variance of our estimator. When sampling from a normal distribution, we are better off on average using the sample mean, despite the fact that the extreme values are highly likely to be inaccurate. But as we move away from a normal distribution toward a heavier-tailed distribution, the sample mean becomes extremely inaccurate relative to the 20% trimmed mean.

There is, however, a practical concern about 20% trimming that should be stressed: Its standard error might be substantially higher versus the standard error of MOM. This can happen even when the number of outliers in both tails does not exceed 20%.

EXAMPLE. Consider the data

```
77 87 87 114 151 210 219 246 253 262
296 299 306 376 428 515 666 1310 2611,
```

which are from a study on self-awareness conducted by E. Dana (1990). The estimated standard error of the 20% trimmed mean is 56.1 (using the method described in Section 4.9.2). But the estimated standard error of MOM (using a method described in Chapter 7) is 37, which is substantially smaller. The outlier detection rule based on M and MAD [given by Equation (3.22)] flags the three largest values as outliers, and these are removed by MOM as well as the 20% trimmed mean. But the 20% trimmed also removes the three smallest values, which are not flagged as outliers. ■

4.5 A Confidence Interval for the Population Mean

If you were a psychologist, you might be interested in a new method for treating depression and how it compares to a standard, commonly used technique. Assume that based on a widely used measure of effectiveness, the standard method has been applied thousands of times and found to have a mean effectiveness of $\mu = 48$. That is, the standard method has been used so many times, for all practical purposes we know that the population mean is 48. Suppose we estimate the effectiveness of the new method by trying it on $n = 25$ subjects and computing the sample mean. A crucial issue is whether the resulting sample mean is close to the population mean being estimated, the mean we would obtain if all depressed individuals were treated with the new method. Assume that for the experimental method we obtain a sample mean of $\bar{X} = 54$. This means that based on our experiment, the average effectiveness of the new method is estimated to be 54, which is larger than the average effectiveness of the standard method, suggesting that the new method is better for the typical individual.

But suppose that unknown to us, the average effectiveness of the new method is actually 46, meaning that on average the standard technique is better for treating depression. By chance we might get a sample mean of 54 and incorrectly conclude that the experimental method is best. What is needed is some way of determining whether $\bar{X} = 54$ makes it unlikely that the mean is 46 or some other value less than 48. If we can be reasonably certain that 54 is close to the actual population mean, and in particular we can rule out the possibility that the population mean is less than 48, we have evidence that, on average, the new treatment is more effective.

A major advantage of being able to determine the sampling distribution of $\bar{X}$ is that it allows us to address the issue of how well the sample mean $\bar{X}$ estimates the population mean μ. We can address this issue with what is called a confidence interval for μ. A *confidence interval* for μ is just a range of numbers that contains μ with some specified probability. If for the experimental method for treating depression you can be reasonably certain that the population mean is between 50 and 55, then there is evidence that its mean is greater than 48, the average effectiveness of the standard technique. That is, the new method appears to be more effective than the standard technique.

Suppose we want to use the observations to determine a range of values that contains μ with probability .95. From the previous section, if sampling is from a normal distribution, then $(\bar{X} - \mu)/(\sigma/\sqrt{n})$ has a standard normal distribution. Recall from Chapter 2 that the probability of a standard normal random variable being between -1.96 and 1.96 is .95. In symbols,

$$P\left(-1.96 \leq \frac{\bar{X} - \mu}{\sigma/\sqrt{n}} \leq 1.96\right) = .95.$$

We can rearrange terms in this last equation to show that

$$P\left(\bar{X} - 1.96\frac{\sigma}{\sqrt{n}} \leq \mu \leq \bar{X} + 1.96\frac{\sigma}{\sqrt{n}}\right) = .95.$$

This says that although the population mean, μ, is not known, there is a .95 probability that its value is between

$$\bar{X} - 1.96\frac{\sigma}{\sqrt{n}} \quad \text{and} \quad \bar{X} + 1.96\frac{\sigma}{\sqrt{n}}.$$

When sampling from a normal distribution,

$$\left(\bar{X} - 1.96\frac{\sigma}{\sqrt{n}}, \bar{X} + 1.96\frac{\sigma}{\sqrt{n}}\right) \tag{4.7}$$

is called a .95 confidence interval for μ. This means that if the experiment were repeated billions of times, and each time a confidence interval is computed using Equation (4.7), 95% of the resulting confidence intervals will contain μ if observations are randomly sampled from a normal distribution.

EXAMPLE. In the example for treating depression, assume for illustrative purposes that the standard deviation is $\sigma = 9$ and sampling is from a normal distribution. Because the sample mean is $\bar{X} = 54$ and the sample size is $n = 25$, the .95 confidence interval for μ is

$$\left(54 - 1.96\frac{9}{\sqrt{25}}, 54 + 1.96\frac{9}{\sqrt{25}}\right) = (50.5, 57.5).$$

That is, based on the 25 subjects available to us, we can be reasonably certain that μ is somewhere between 50.5 and 57.5. ■

DEFINITION. The *probability coverage* of a confidence interval is the probability that the interval contains the parameter being estimated. The previous example described a confidence interval for the mean that has probability coverage .95.

A standard notation for the probability that a confidence interval does *not* contain the population mean, μ, is α. When computing a .95 confidence interval, $\alpha = 1 - .95 = .05$. For a .99 confidence interval, $\alpha = .01$. The quantity α is the probability of making a mistake. That is, if we perform an experiment with the goal of computing a .95 confidence interval, there is a .95 probability that the resulting interval contains the mean, but there is an $\alpha = 1 - .95 = .05$ probability that it does not.

The method of computing a confidence interval can be extended to any value of $1 - \alpha$ you might choose. The first step is to determine c such that the probability that a standard normal random variable lies between $-c$ and c is $1 - \alpha$. In symbols, determine c such that

$$P(-c \le Z \le c) = 1 - \alpha.$$

From Chapter 2, this means that you determine c such that

$$P(Z \le c) = \frac{1 + (1 - \alpha)}{2}$$

$$= 1 - \frac{\alpha}{2}.$$

Put another way, c is the $1 - \alpha/2$ quantile of a standard normal distribution. For example, if you want to compute a $1 - \alpha = .95$ confidence interval, then

$$\frac{1 + (1 - \alpha)}{2} = \frac{1 + .95}{2} = .975,$$

and from Table 1 in Appendix B we know that

$$P(Z \le 1.96) = .975,$$

so $c = 1.96$. For convenience, the values of c for $1 - \alpha = .9$, .95, and .99 are listed in Table 4.2.

TABLE 4.2 Common Choices for $1-\alpha$ and c

$1-\alpha$	c
.90	1.645
.95	1.96
.99	2.58

Once c is determined, a $1-\alpha$ confidence interval for μ is

$$\left(\bar{X} - c\frac{\sigma}{\sqrt{n}}, \bar{X} + c\frac{\sigma}{\sqrt{n}}\right). \tag{4.8}$$

Equation (4.8) is a special case of a general technique developed by Laplace.

EXAMPLE. A college president claims that IQ scores at her institution are normally distributed with a mean of $\mu = 123$ and a standard deviation of $\sigma = 14$. Suppose you randomly sample $n = 20$ students and find that $\bar{X} = 110$. Does the $1-\alpha = .95$ confidence interval for the mean support the claim that the average of all IQ scores at the college is $\mu = 123$? Because $1-\alpha = .95$, $c = 1.96$, as just explained, so the .95 confidence interval is

$$\left(110 - 1.96\frac{14}{\sqrt{20}}, 110 + 1.96\frac{14}{\sqrt{20}}\right) = (103.9, 116.1).$$

The interval (103.9,116.1) does not contain the value 123, suggesting that the president's claim might be false. Note that there is a .05 probability that the confidence interval will not contain the true population mean, so there is some possibility that the president's claim is correct. ■

EXAMPLE. For 16 observations randomly sampled from a normal distribution, $\bar{X} = 32$ and $\sigma = 4$. To compute a .9 confidence interval (meaning that $1-\alpha = .9$), first note from Table 4.2 that $c = 1.645$. So a .9 confidence interval for μ is

$$\left(32 - 1.645\frac{4}{\sqrt{16}}, 32 + 1.645\frac{4}{\sqrt{16}}\right) = (30.355, 33.645).$$

Although $\bar{X}$ is not, in general, equal to μ, the confidence interval provides some sense of how well $\bar{X}$ estimates the population mean. ■

4.6 An Approach to Nonnormality: The Central Limit Theorem

We have seen how to compute a confidence interval for the mean when the standard deviation is known and sampling is from a normal distribution. Assuming observations

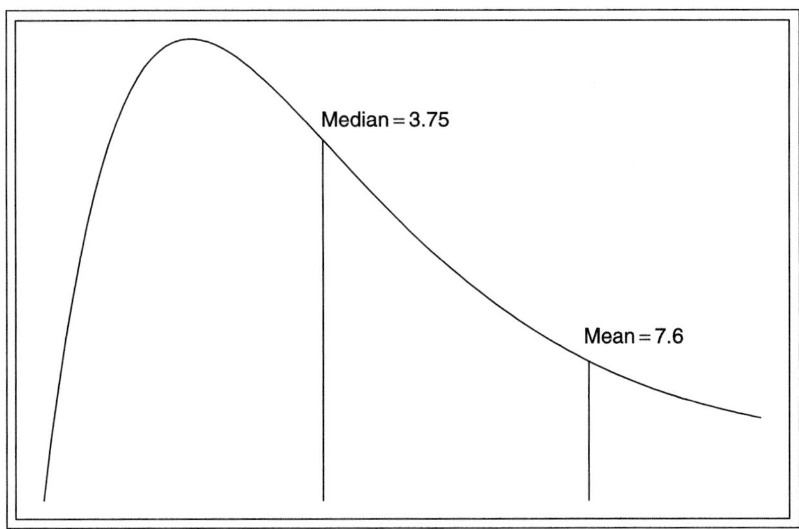

FIGURE 4.6 Example of a skewed, heavy-tailed distribution.

are randomly sampled from a normal distribution is convenient because the sampling distribution of $\bar{X}$ turns out to be a normal distribution as well. But how do we deal with nonnormal distributions? In the ozone experiment, how do we compute a confidence interval for μ if, unknown to us, observations are randomly sampled from the distribution shown in Figure 4.6, which is a reproduction of Figure 2.11.

Laplace's solution was to appeal to his central limit theorem, which says that under very general conditions, even if observations are randomly sampled from a nonnormal distribution, the sampling distribution of the sample mean will approach a normal distribution as the sample size gets large. In more practical terms, if n is sufficiently large, we can pretend that $\bar{X}$ has a normal distribution with mean μ and variance σ^2/n, in which case the method in the previous section can be employed. Of course this last statement is rather vague, in an important sense. How large must n be? Many books claim that $n = 25$ suffices and others claim that $n = 40$ is more than sufficient. These are not wild speculations, but now we know that much larger sample sizes are needed for many practical problems. The immediate goal is to provide some sense of why larger sample sizes are needed than once thought.

To begin, there is no theorem telling us when n is sufficiently large. The answer depends in part on the skewness of the distribution from which observations are sampled. (Boos & Hughes-Oliver, 2000, provide a recent summary of relevant details.) Generally, we must rely on empirical investigations, at least to some extent, in our quest to address this problem.

To illustrate how such empirical studies are done and why it might seem that $n = 40$ is sufficiently large for most purposes, suppose $n = 20$ observations are randomly sampled from the distribution in Figure 4.7, which is an example of a uniform distribution, and then we compute the sample mean. As is evident, the uniform distribution in Figure 4.7 does not remotely resemble a normal curve. If we generate 5000 sample means in this manner and plot the results, we get the curve shown in Figure 4.8. Also shown is the normal curve implied by the central limit

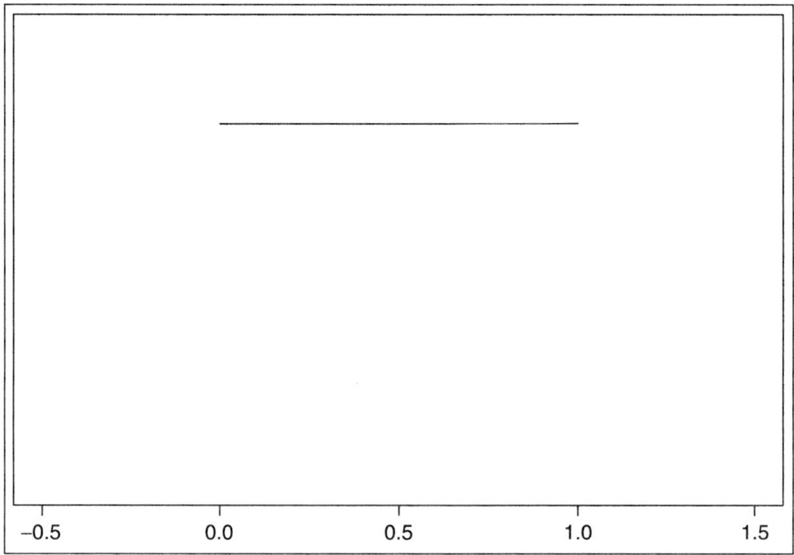

FIGURE 4.7 A uniform distribution.

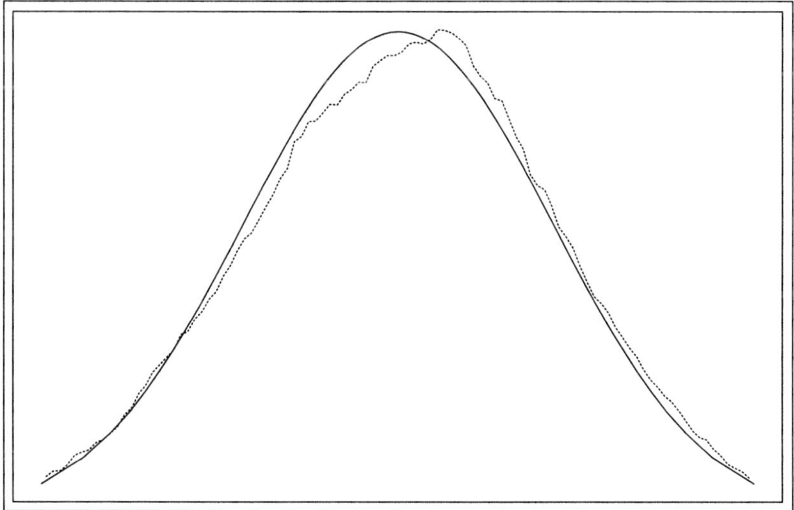

FIGURE 4.8 A plot of 5000 means based on observations sampled from a uniform distribution. The solid, symmetric curve is the plot of the means based on the central limit theorem.

theorem. As we see, the two curves are very similar, indicating that the central limit theorem is performing rather well in this particular case. That is, if we sample 20 observations from a uniform distribution, for all practical purposes we can assume the sample mean has a normal distribution.

Now consider the probability curve in Figure 4.9 (which is called an *exponential distribution*). Again this curve is nonnormal in an obvious way. If we repeat our computer experiment used to create Figure 4.8, but sampling from the distribution in Figure 4.9, then the plot of the resulting sample means appears as shown in Figure 4.10.

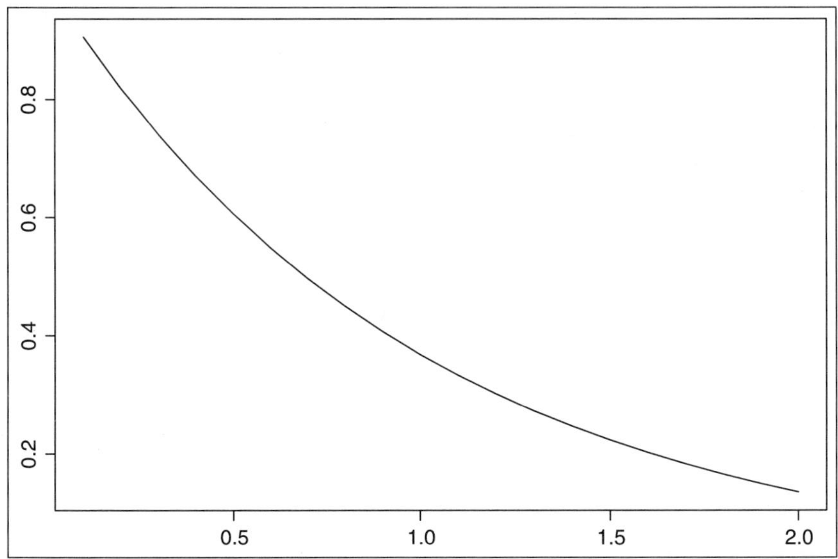

FIGURE 4.9 An exponential distribution which is relatively light-tailed.

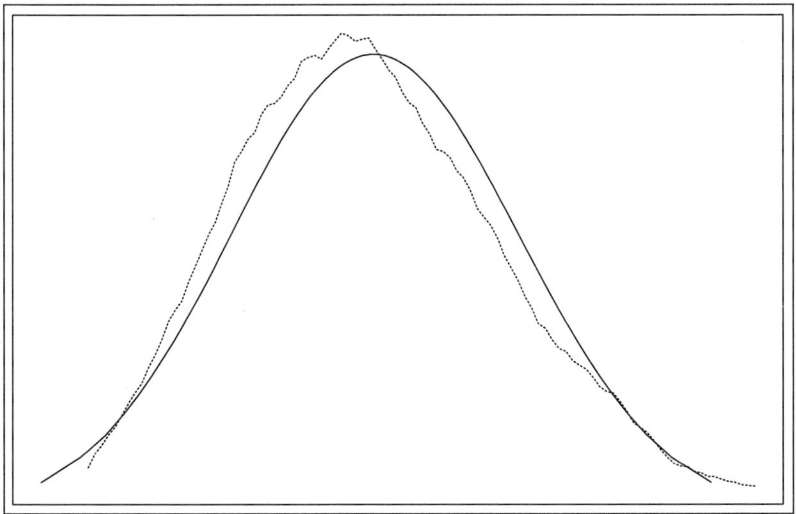

FIGURE 4.10 A plot of 5000 means when sampling from an exponential distribution.

Again we get fairly close agreement between the empirical distribution for the sample means and the theoretical distribution implied by the central limit theorem. These two illustrations are classic ways of demonstrating the central limit theorem, and the obvious speculation based on these results is that in general, with $n \geq 25$, we can assume the sample mean has a normal distribution. There are, however, two fundamental problems that have been overlooked. The first of these is illustrated here, and the second and even more serious problem is described in Section 4.7.

We repeat our computer experiment one more time, only now we sample 25 observations from the distribution shown in Figure 4.6. Figure 4.11 shows the resulting

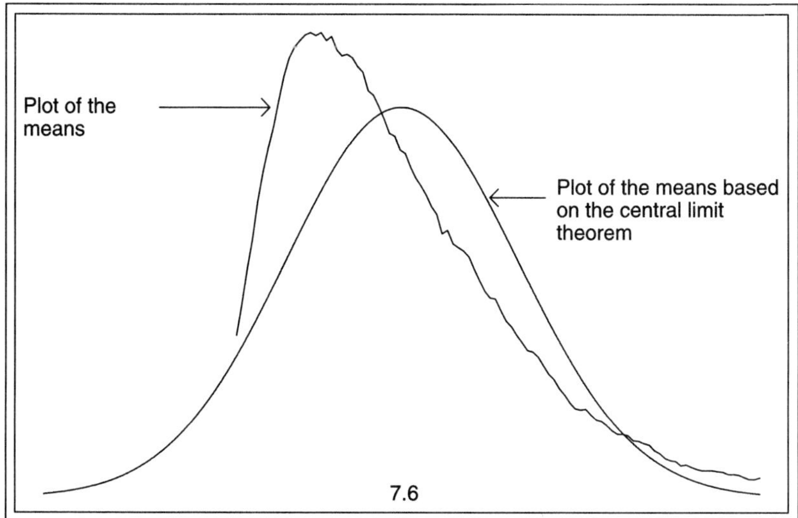

FIGURE 4.11 A plot of 5000 means when sampling from the distribution in Figure 4.6. The sample size for each mean is 25.

plot of the sample means. Now there is an obvious discrepancy between the plot of the sample means and the curve implied by the central limit theorem. Why do we get a different result from the two previous situations? The reason is that the curves in Figures 4.7 and 4.9 are distributions with relatively light tails. That is, outliers are fairly rare when sampling observations. In contrast, Figure 4.11 is obtained by sampling from a heavy-tailed distribution, where outliers are more common.

To provide some sense of how quickly matters improve as the sample size increases, we again sample observations from the distribution in Figure 4.6, but with a sample size of $n = 50$. Figure 4.12 shows the results. Note that the left tail of the plot of the means is lighter than what we would expect via the central limit theorem. Increasing n to 100, the central limit theorem gives a reasonable approximation of the actual distribution of the sample mean.

The illustrations just given might seem to suggest that as long as a distribution is not too heavy-tailed or if $n = 100$, an accurate confidence interval for μ can be computed using Equation (4.8). So a seemingly reasonable speculation is that if a boxplot indicates that there are no outliers, the actual probability coverage is reasonably close to the nominal level. Unfortunately, this strategy can fail, for reasons covered in Section 4.8. Moreover, even with $n = 100$, practical problems might occur. Briefly, when we take up the more realistic situation where σ is unknown and estimated with s, problems arise even when sampling from light-tailed distributions and $n = 100$.

Notice that when sampling from the distribution shown in Figure 4.6, although the population mean is in the extreme right portion of the distribution under consideration, the sample means become more tightly centered around the population mean, $\mu = 7.6$, as the sample size increases. So the sample mean is fulfilling its intended goal: As the sample size increases, it provides a better estimate of the population mean, which in this case is a quantity that happens to be in the tail of the distribution.

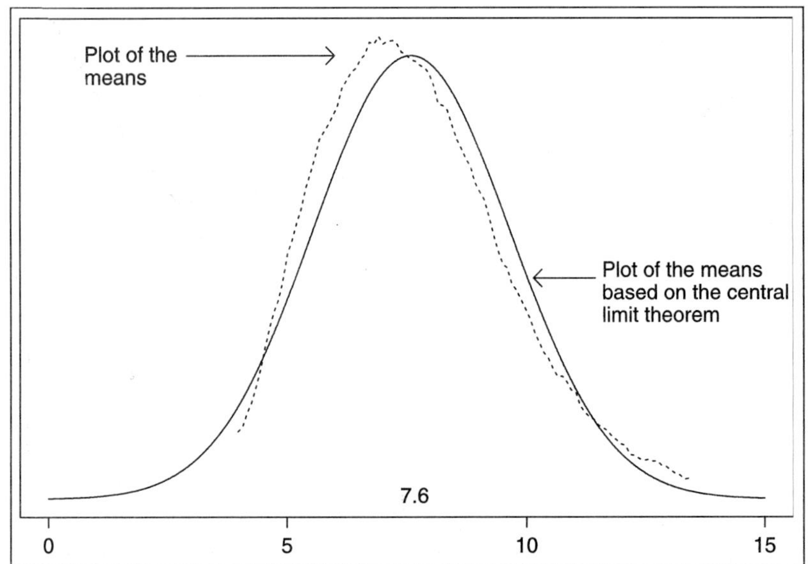

FIGURE 4.12 A plot of 5000 means when sampling from the distribution in Figure 4.6, only now the sample size for each mean is 50.

A criticism of the illustrations based on Figure 4.6 is that we are considering a hypothetical distribution. Perhaps in practice the central limit theorem will give satisfactory results even with $n = 25$. There are several reasons, however, for concluding that, in practice, problems can arise even when $n = 100$. Details are covered in Section 4.8.

One point should be stressed, because it plays an important role in subsequent chapters. Again consider the ozone experiment, and first assume that weight gain in rats has a normal distribution with variance $\sigma^2 = 1$. Then because the sample mean is $\bar{X} = 11$ and $n = 22$, the .95 confidence interval for μ is

$$\left(11 - 1.96\frac{1}{\sqrt{22}}, 11 + 1.96\frac{1}{\sqrt{22}}\right) = (10.58, 11.42).$$

The length of the confidence interval is just the difference between these two numbers, namely, $11.42 - 10.58 = .84$. When computing a .95 confidence interval, we want the length of the confidence interval to be as short as possible because a short confidence interval means we can be reasonably certain about what the value of μ happens to be. In the ozone experiment, we can be reasonably certain that the population mean is somewhere between 10.58 and 11.42, and this reflects how well $\bar{X}$ estimates μ.

Now suppose that sampling is from a mixed normal instead, which was discussed in Chapter 2. Although the mixed normal differs only slightly from the standard normal, the mixed normal has variance 10.9. Consequently, the .95 confidence interval for μ is

$$\left(11 - 1.96\frac{\sqrt{10.9}}{\sqrt{22}}, 11 + 1.96\frac{\sqrt{10.9}}{\sqrt{22}}\right) = (9.62, 12.38).$$

The length of this interval is $12.38 - 9.62 = 2.76$, more than three times longer than the situation where we sampled from a standard normal distribution instead. This illustrates that *small shifts away from normality, toward a heavy-tailed distribution, can drastically increase the length of a confidence interval*. Modern methods have been found for getting much shorter confidence intervals in situations where the length of the confidence interval based on Equation (4.8) is relatively long, some of which will be described.

4.7 Confidence Intervals when σ Is Unknown

The previous section described how to compute a confidence interval for μ when the standard deviation, σ, is known. However, typically σ is not known, so a practical concern is finding a reasonably satisfactory method for dealing with this issue. This section describes the classic method for addressing this problem, which was derived by William Gosset about a century ago. It is used routinely today, but unfortunately problems with nonnormality are exacerbated relative to the situation in Section 4.6, where σ is known.

Consider again the study of women's blood pressure, where the goal is to determine the average diastolic blood pressure of adult women taking a certain drug. Based on $n = 9$ women, we know that $\bar{X} = 85$, and $s^2 = 160.78$, but, as is usually the case, we do not know the population variance, σ^2. Although σ is not known, it can be estimated with s, the sample standard deviation, which in turn yields an estimate of the standard error of the sample mean, namely, $s/\sqrt{n} = 4.2$, as previously explained. If we assume that 4.2 is indeed an accurate estimate of $\sigma/\sqrt{n}$, then a reasonable suggestion is to assume that $\sigma/\sqrt{n} = 4.2$ when computing a confidence interval. In particular, a .95 confidence interval for the mean would be

$$(85 - 1.96(4.2), 85 + 1.96(4.2)) = (76.8, 93.2).$$

Prior to the year 1900, this was the strategy used, and based on a version of the central limit theorem, it turns out that this approach is reasonable if the sample size is sufficiently large, assuming random sampling. However, even when sampling from a normal distribution, a concern is that when the sample size is small, the population standard deviation, σ, might differ enough from its estimated value, s, to cause practical problems. Gosset realized that problems can arise and derived a solution assuming random sampling from a normal distribution. Gosset worked for a brewery and was not immediately allowed to publish his results, but eventually he was permitted to publish under the pseudonym Student.

Let

$$T = \frac{\bar{X} - \mu}{s/\sqrt{n}}. \tag{4.9}$$

The random variable T in Equation (4.9) is the same as Z used in Section 4.3, except σ has been replaced by s. Note that like $\bar{X}$ and Z, T has a distribution. That is, for any constant c we might pick, there is a certain probability that $T < c$ based on a random

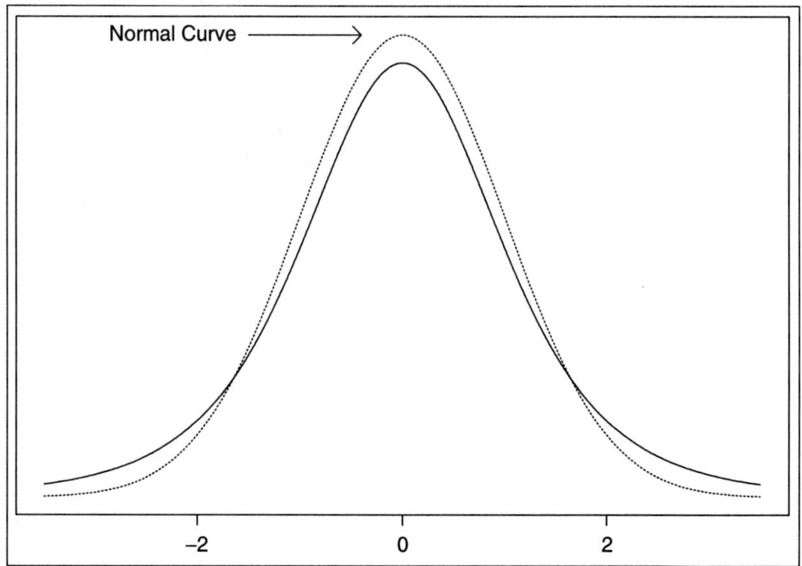

FIGURE 4.13 Student's T distribution with 4 degrees of freedom.

sample of n subjects. If the distribution of T can be determined, then a confidence interval for μ could be computed without knowing σ.

If we assume that observations are randomly sampled from a normal distribution, then the distribution of T, called *Student's T Distribution*, can be determined exactly. It turns out that the distribution depends on the sample size, n. By convention, the quantiles of the distribution are reported in terms of *degrees of freedom*: $\nu = n - 1$. Figure 4.13 shows Student's T distribution with $\nu = 4$ degrees of freedom. Note that the distribution is similar to a standard normal. In particular, it is symmetric about zero, so $E(T) = 0$. With infinite degrees of freedom, Student's T and the standard normal are identical.

Table 4 in Appendix B reports some quantiles of Student's T distribution. The first column gives the degrees of freedom. The next column, headed by $t_{.9}$, reports the .9 quantiles. For example, with $\nu = 1$, we see 3.078 under the column $t_{.9}$. This means that $P(T < 3.078) = .9$. That is, if we randomly sample two observations from a normal distribution, in which case $\nu = n - 1 = 1$, there is a .9 probability that the resulting value for T is less than 3.078. Similarly, if $\nu = 24$, then $P(T < 1.318) = .9$. The column headed by $t_{.99}$ lists the .99 quantiles. For example, if $\nu = 3$, we see 4.541 under the column headed $t_{.99}$, so the probability that T is less than 4.541 is .99. If $\nu = 40$, Table 2 indicates that $P(T < 2.423) = .99$. Many modern computer programs, such as Minitab and S-PLUS, contain functions that compute Student's T distribution for any $\nu \geq 1$. [In S-PLUS you can use the built-in function pt. For example, pt(1,5) will return the probability that T is less than 1 with $\nu = 5$ degrees of freedom.]

Similar to the situation when working with normal distributions,

$$P(T \geq c) = 1 - P(T \leq c), \tag{4.10}$$

where c is any constant that might be of interest. For example, with $\nu = 4$, $P(T \leq 2.132) = .95$, as previously indicated, so $P(T \geq 2.132) = 1 - P(T \leq 2.132) = .05$.

To assist in learning how to use Table 4 in Appendix B, the following examples are presented.

EXAMPLE. Suppose you are involved in a study on the effects of alcohol on reaction times. Assuming normality, you randomly sample $n = 13$ observations and compute the sample mean and variance. To determine the probability that $T = (\bar{X} - \mu)/(s/\sqrt{n})$ is less than 2.179, first note that the degrees of freedom are $\nu = n - 1 = 13 - 1 = 12$. From Table 4 in Appendix B, looking at the row with $\nu = 12$, we see 2.179 in the column headed by $t_{.975}$, so $P(T < 2.179) = .975$. ■

EXAMPLE. If $\nu = 30$ and $P(T > c) = .005$, what is c? Because Table 4 gives the probability that T is less than or equal to some constant, we must convert the present problem into one where Table 4 can be used. Based on Equation (4.10), if $P(T > c) = .005$, then

$$P(T \leq c) = 1 - P(T > c) = 1 - .005 = .995.$$

Looking at the column headed by $t_{.995}$ in Table 4, we see that with $\nu = 30$, $P(T < 2.75) = .995$, so $c = 2.75$. ■

With Student's T distribution, we can compute a confidence interval for μ when σ is not known, assuming that observations are randomly sampled from a normal distribution. Recall that when σ is known, the $1 - \alpha$ confidence interval for μ is

$$\bar{X} \pm c\frac{\sigma}{\sqrt{n}} = \left(\bar{X} - c\frac{\sigma}{\sqrt{n}}, \bar{X} + c\frac{\sigma}{\sqrt{n}}\right),$$

where c is the $1 - \alpha/2$ quantile of a standard normal distribution and read from Table 1 in Appendix B. When σ is not known, this last equation becomes

$$\bar{X} \pm c\frac{s}{\sqrt{n}}, \tag{4.11}$$

where now c is the $1 - \alpha/2$ quantile of Student's T distribution with $n - 1$ degrees of freedom and read from Table 4 of Appendix B. If observations are randomly sampled from a normal distribution, then the probability coverage is exactly $1 - \alpha$. [The S-PLUS built-in function t.test computes a confidence interval using Equation (4.11).]

EXAMPLE. Returning to the ozone experiment, we compute a $1 - \alpha = .95$ confidence interval for μ. Because there are $n = 22$ rats, the degrees of freedom

Continued

EXAMPLE. (*Continued*) are $n-1 = 22-1 = 21$. Because $1-\alpha = .95$, $\alpha = .05$; so $\alpha/2 = .025$, and $1-\alpha/2 = .975$. Referring to Table 4 in Appendix B, we see that the .975 quantile of Student's T distribution with 22 degrees of freedom is approximately $c = 2.08$. Because $\bar{X} = 11$ and $s = 19$, a .95 confidence interval is

$$11 \pm 2.08 \frac{19}{\sqrt{22}} = (2.6, 19.4).$$

That is, although both the population mean and variance are not known, we can be reasonably certain that the population mean, μ, is between 2.6 and 19.4, if the assumption of sampling from a normal distribution is true. ■

EXAMPLE. Suppose you are interested in the reading abilities of fourth-graders. A new method for enhancing reading is being considered. You try the new method on 11 students and then administer a reading test yielding the scores

12, 20, 34, 45, 34, 36, 37, 50, 11, 32, 29.

For illustrative purposes, imagine that after years of using a standard method for teaching reading, the average scores on the reading test have been found to be $\mu = 25$. Someone claims that if the new teaching method is used, the population mean will remain 25. Assuming normality, we determine whether this claim is consistent with the .99 confidence interval for μ. That is, does the .99 confidence interval contain the value 25? It can be seen that the sample mean is $\bar{X} = 30.9$ and $s/\sqrt{11} = 3.7$. Because $n = 11$, the degrees of freedom are $\nu = 11-1 = 10$. Because $1-\alpha = .99$, it can be seen that $1-\alpha/2 = .995$, so, from Table 4 in Appendix B, $c = 3.169$. Consequently, the .99 confidence interval is

$$30.9 \pm 3.169(3.7) = (19.2, 42.6).$$

This interval contains the value 25, so the claim that $\mu = 25$ cannot be refuted based on the available data. Note, however, that the confidence interval also contains 35 and even 40. Although we cannot rule out the possibility that the mean is 25, there is some possibility that the new teaching method enhances reading by a substantial amount, but with only 11 subjects, the confidence interval is too long to resolve how effective the new method happens to be. ■

4.8 Student's *T* and Nonnormality

In the next to last example dealing with weight gain in an ozone environment, it was shown that the .95 confidence interval for the mean is (2.6,19.4) *if* observations are randomly sampled from a normal distribution. But can we be reasonably certain that it contains μ if sampling is from a nonnormal distribution instead?

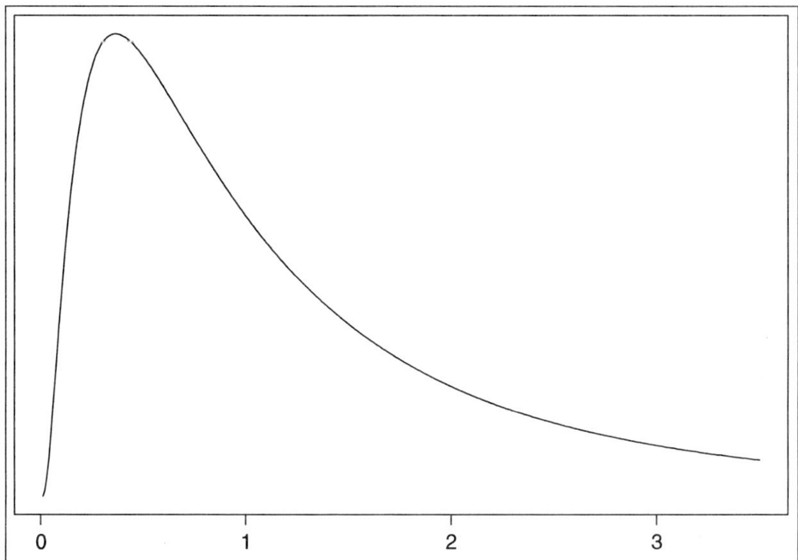

FIGURE 4.14 Example of what is called a lognormal distribution.

Unfortunately, if observations are randomly sampled from a skewed distribution, the actual probability coverage can be less than .7 (e.g., Wilcox, 1997a, p. 74). That is, the confidence interval is too short and there is a $1 - .7 = .3$ probability that it does not contain μ. To get .95 probability coverage, we need a longer interval. If we increase the sample size, the actual probability coverage will be closer to .95, as desired, but probability coverage might be unsatisfactory even with $n = 160$ (e.g., Westfall & Young, 1993, p. 40).

To elaborate a little, imagine that, unknown to us, observations have the distribution shown in Figure 4.14. This is an example of a lognormal distribution, which is skewed to the right and has relatively light tails, meaning that outliers can occur but the number of outliers is relatively low on average. (With $n = 20$, the expected number of outliers using the method in Section 3.4.5 is about 1.4. The median number of outliers is approximately 1.) The symmetric smooth curve in the left plot of Figure 4.15 shows Student's T distribution when $n = 20$ and sampling is from a normal distribution. The other curve shows a close approximation of the actual distribution of T when sampling from the distribution in Figure 4.14 instead. (The approximation is based on 5000 T-values generated on a computer.) Note that the actual distribution is skewed, not symmetric. Moreover, its mean is not zero but $-.5$, approximately. The right plot of Figure 4.15 shows the distribution of T when $n = 100$. There is closer agreement between the two distributions, but the tails of the distributions differ enough that practical problems arise.

It was noted that in the left plot of Figure 4.15, T has a mean of $-.5$. This might seem to be impossible because the numerator of T is $\bar{X} - \mu$, which has an expected value of zero. Under normality, T does indeed have a mean of zero, the proof of which is based on the result that under normality, $\bar{X}$ and s are independent. But for nonnormal distributions, $\bar{X}$ and s are dependent, and this makes it possible for T to

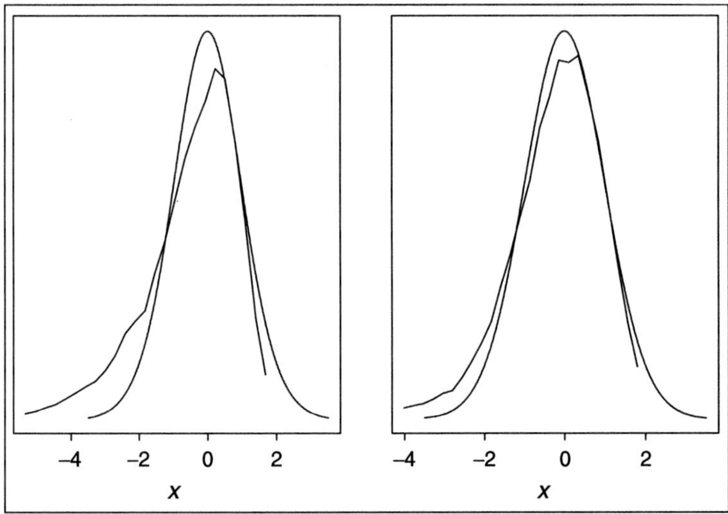

FIGURE 4.15 A plot of 5000 T values when sampling from a lognormal distribution. The left plot is based on $n = 20$ and the right plot is based on $n = 100$.

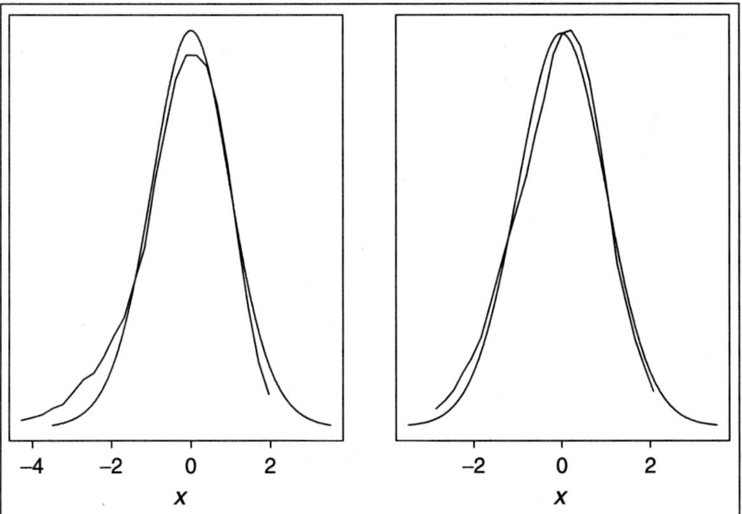

FIGURE 4.16 A plot of 5000 T values when sampling from an exponential distribution. The left plot is with $n = 20$ and the right plot is with $n = 100$.

have a mean that differs from zero. (Gosset was aware of this problem but did not have the tools and technology to study it to the degree he desired.)

It might be thought that the lognormal distribution (shown in Figure 4.14) represents an extreme case and therefore denigrates Student's T in an unfair manner. However, when working with Student's T distribution, other skewed, light-tailed distributions—where outliers are rare—also cause serious practical problems. Consider the (exponential) distribution shown in Figure 4.9. The left panel of Figure 4.16 shows the distribution of Student's T when $n = 20$. Again the left tails of the actual distribution differ noticeably from the distribution we get under normality. For example, under normality, the probability that T is less than -2 is $P(T \leq -2) = .03$, but when sampling from the exponential distribution, it is .08. The right panel shows the distribution when $n = 50$. Now, $P(T \leq -2) = .026$ when sampling from a normal distribution, but for the exponential distribution it is .053. In some situations this discrepancy is unsatisfactory. Increasing n to 100 gives good results.

Figure 4.16 illustrates another point worth stressing. We saw that when sampling observations from the distribution in Figure 4.9, the sampling distribution of the sample mean is in fairly close agreement with the distribution implied by the central limit theorem. We have just illustrated, however, that this does not imply that the actual distribution of T will be in close agreement with the distribution we get under normality.

When sampling from a skewed heavy-tailed distribution, the discrepancy between the actual distribution of T and the distribution obtained under normality becomes even more striking. If, for example, we sample $n = 20$ observations from the distribution shown in Figure 4.6, the distribution of T is as shown in Figure 4.17 and differs substantially from the distribution we get under normality, particularly in the left tail.

The illustrations so far are based on hypothetical distributions. Experience with actual data suggests that the problems just illustrated are real and, in at least some situations, these theoretical illustrations appear to underestimate problems with Student's T. To describe one reason for this remark, we consider data from a study on hangover symptoms reported by the sons of alcoholics. The 20 observed values were

1 0 3 0 3 0 15 0 6 10 1 1 0 2 24 42 0 0 0 2.

(These data were supplied by M. Earleywine.) Figure 4.18 shows an approximation of the sampling distribution of T (which was obtained using methods covered

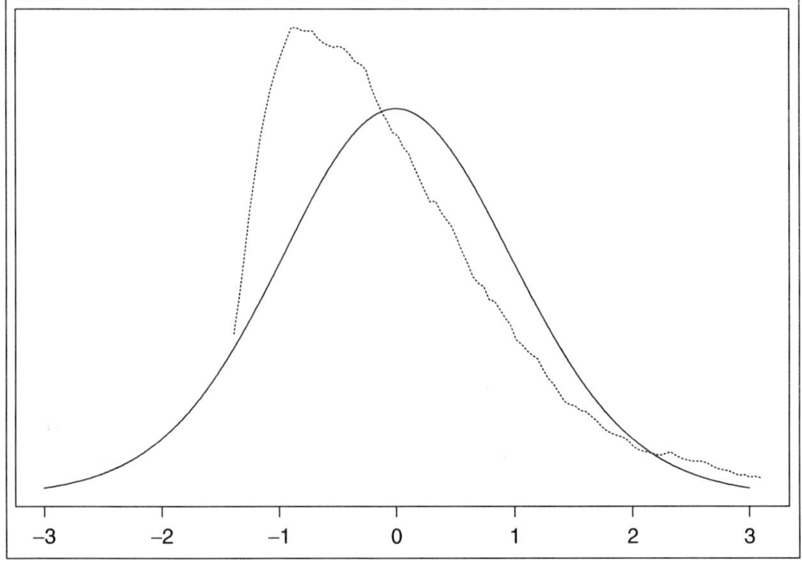

FIGURE 4.17 A plot of 5000 T values when sampling from a skewed, heavy-tailed distribution, $n = 20$.

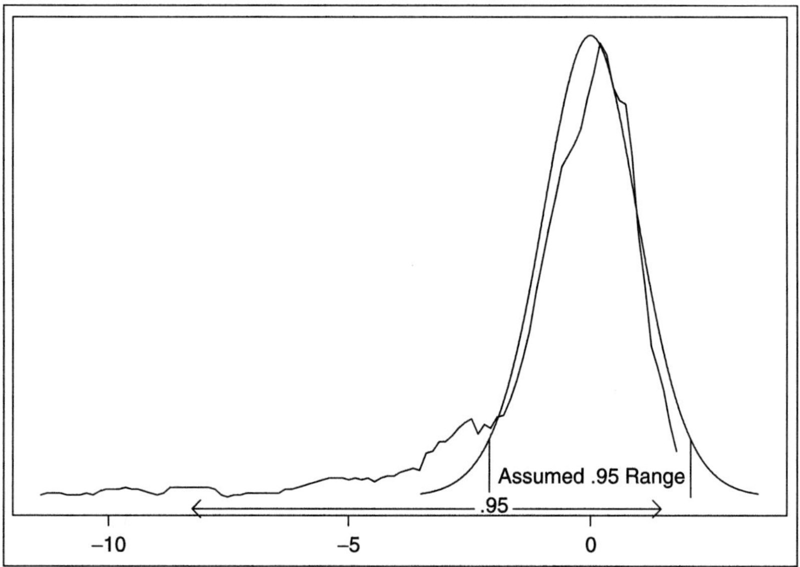

FIGURE 4.18 A plot of 5000 T values when resampling with replacement using data from a study dealing with sons of alcoholics. The smooth symmetric curve is the plot we get when sampling from a normal distribution.

in Chapter 7). Also shown is the distribution of T when sampling from a normal distribution instead. As is evident, there is a considerable discrepancy between the two distributions, particularly in the left tail. The practical implication is that when using T, the actual probability coverage, when computing a confidence interval based on Equation (4.11), might differ substantially from the nominal level.

It might be argued that this last example is somehow unusual or that with a slightly larger sample size, satisfactory probability coverage will be obtained. First note that if we were to repeat an experiment 5000 times, each time computing Student's T based on observations sampled from a normal distribution, a boxplot of the T values would be symmetric about the value zero. Now consider again the data in Table 3.2, where $n = 105$. We can use these observations to approximate the boxplot of T values we would get if this particular study were repeated 5,000 times. (The approximation is based on a method covered in Chapter 7.) Figure 4.19 shows the result. As is evident, there is extreme skewness, indicating that any confidence interval based on Student's T will be highly inaccurate.

An objection to this last illustration is that there is an extreme outlier among the data. Although the data are from an actual study, it might be argued that having such an extreme outlier is a highly rare event. So we repeat the last illustration, but with this extreme outlier removed. Figure 4.20 shows an approximation of the distribution of T. Again we see that there is a substantial difference as compared to the distribution implied by the central limit theorem.

Yet another possible objection to illustrations based on data from actual studies is that although we are trying to determine empirically the correct distribution for T, nevertheless we are approximating the correct distribution, and perhaps the method used to approximate the distribution is itself in error. That is, if were to take millions of

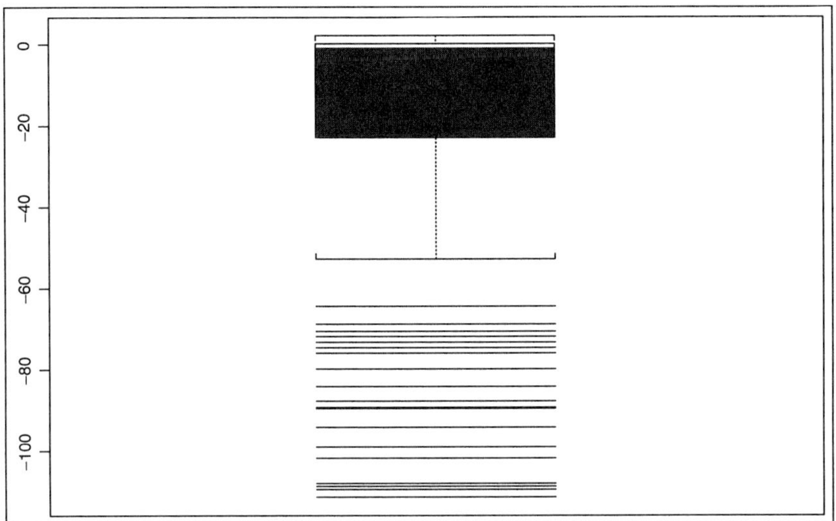

FIGURE 4.19 A boxplot of 5000 T values when resampling with replacement from the data in Table 3.2.

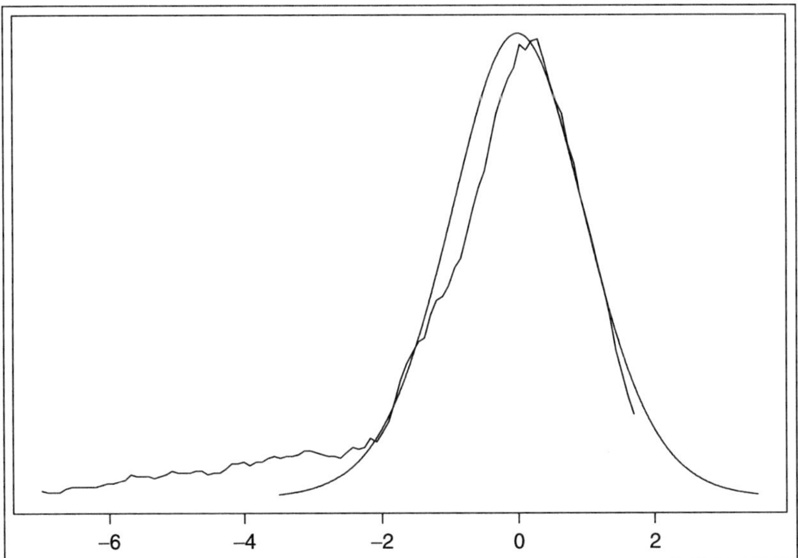

FIGURE 4.20 A plot of 5000 T values when resampling with replacement from the data in Table 3.2, but with the extreme outlier removed.

samples from the population under study, each time computing T, and if we were then to plot the results, perhaps this plot would better resemble a Student's T distribution. There is in fact reason to suspect that the approximation of the distribution of T used here is in error, but unfortunately all indications are that problems with T are being *underestimated*. For example, in Figure 4.20, it is highly likely that the actual distribution of T is more skewed and that the left tail extends even farther to the left than is indicated.

4.9 Confidence Intervals for the Trimmed Mean

There are at least three practical concerns with computing confidence intervals for the mean with Student's T. First, the probability coverage can be unsatisfactory, for reasons explained in the last section. Second, as noted in Chapter 2, there is the concern that when distributions are skewed, the population mean might provide a poor reflection of the typical subject under study. Third, slight departures from normality can greatly inflate the length of the confidence interval, regardless of whether sampling is from a skewed or a symmetric distribution. Theoretical results (e.g., Huber, 1981; Staudte and Sheather, 1990; Wilcox, 1993a) suggest a strategy that addresses all of these concerns: Switch to the Tukey–McLaughlin confidence interval for the population trimmed mean, μ_t (which is described in Section 4.9.3). But before describing this method, we first consider the more fundamental problem of how the standard error of the trimmed mean might be estimated.

4.9.1 Estimating VAR($\bar{X}_t$): A Natural but Incorrect Method

A seemingly natural method for computing a confidence when using a trimmed mean is to apply Student's T method to the values left after trimming. But this strategy is unsatisfactory, because the remaining observations are no longer identically distributed and they are not independent. Consequently, we are using an incorrect estimate of the standard error. We need to use a technique that addresses this problem, otherwise we run the risk of getting an increasingly inaccurate confidence interval as the sample size gets large.

When first encountered, the statement just made might seem counterintuitive. To be sure the problem just described is appreciated, we now elaborate. First we illustrate that if we trim observations, the remaining observations are dependent. Suppose $n = 5$ observations are randomly sampled from a standard normal distribution. We might get the values $X_1 = 1.5$, $X_2 = -1.2$, $X = 3.89$, $X_4 = .4$, and $X_5 = -.6$. Random sampling means that the observations are independent, as explained in Section 4.2. Now suppose we repeat this process 500 times, each time generating five observations but only recording the fourth and fifth values we observe. Figure 4.21 shows a scatterplot of the 500 pairs of points; this is the type of scatterplot we should observe if the observations are independent.

Next, suppose we generate five values from a normal distribution as was done before, but now we put the values in ascending order. As was done in Chapter 3, we label the ordered values $X_{(1)} \leq X_{(2)} \leq X_{(3)} \leq X_{(4)} \leq X_{(5)}$. If we observe $X_1 = 1.5$, $X_2 = -1.2$, $X_3 = .89$, $X_4 = .4$, and $X_5 = -.6$, then $X_{(1)} = -1.2$ is the smallest of the five values, $X_{(2)} = -.6$ is the second smallest, and so on. Now we repeat this process 500 times, each time randomly sampling five observations, but this time we record the two largest values. Figure 4.22 shows the resulting scatterplot for $X_{(4)}$ (the x-axis) versus $X_{(5)}$. There is a discernible pattern because they are dependent.

An important point is that we get dependence when sampling from any distribution, including normal distributions as a special case. If, for example, you are told that

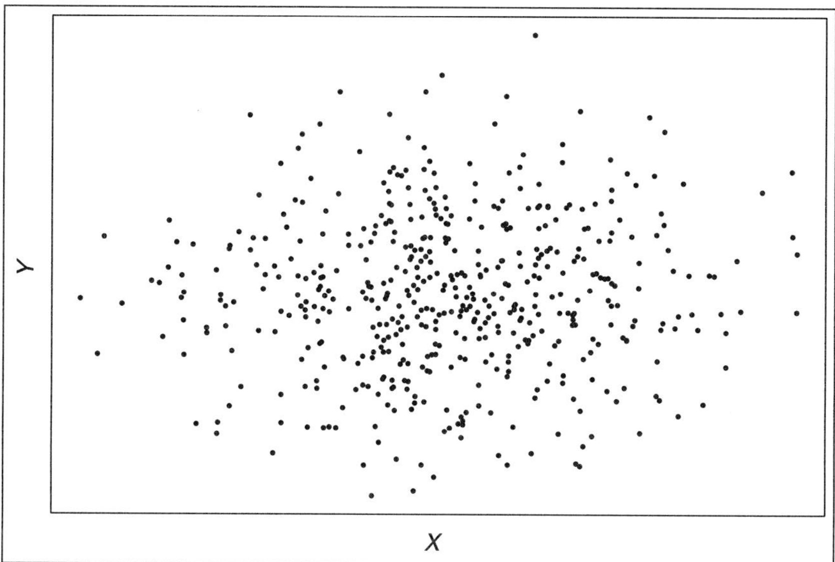

FIGURE 4.21 Five values were randomly sampled from a normal distribution and the fourth and fifth observations were recorded. Repeating this process 500 times yielded the pairs of points shown, which are independent.

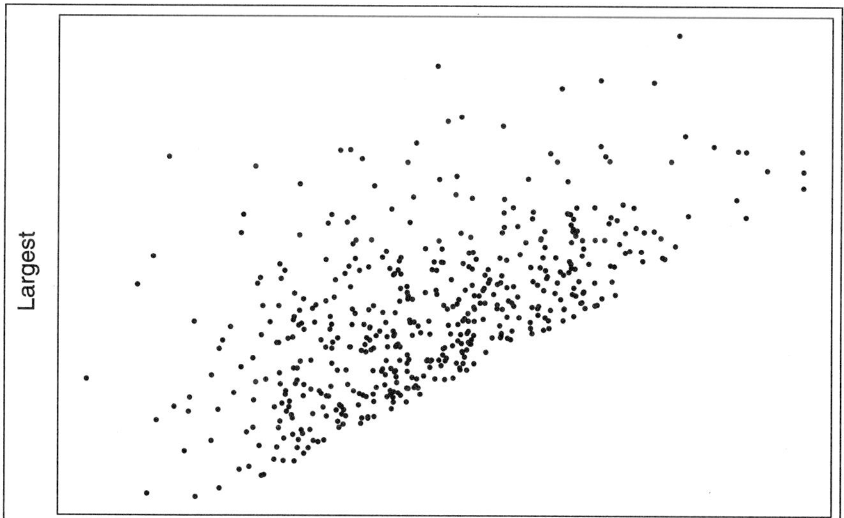

FIGURE 4.22 Five values were randomly sampled from a normal distribution, but now the two largest observations were recorded. Repeating this process 500 times yielded the pairs of points shown, illustrating that the two largest observations are dependent.

$X_{(4)}$ is .89, it follows that the largest value ($X_{(5)}$, still assuming $n = 5$) cannot be .8, .2, 0, or -1; it must be as large or larger than .89. Put another way, if we focus attention on the largest value, there is some probability—greater than zero—that its value is less than .89. In symbols, $P(X_{(5)} < .89) > 0$. But this probability is altered if you are told that $X_{(4)} = .89$; now it is exactly equal to zero. That is, it is impossible for the largest value to be less than .89 if the second-largest value is equal to .89. Said another way, if knowing the value of $X_{(4)}$ alters the range of possible values for $X_{(5)}$, then $X_{(4)}$ and $X_{(5)}$ are dependent. (See the end of Section 2.3.) This argument generalizes: Any two ordered values, say, $X_{(i)}$ and $X_{(j)}$, $i \neq j$, are dependent. Moreover, if we discard even one unusually large or small observation, the remaining observations are dependent.

The point of all this is that the method for determining the variance of the sample mean cannot be used to determine the variance of the trimmed mean. Recall that the derivation of the variance of the sample mean made use of the fact that under random sampling, all pairs of observations have zero correlation. But when we discard extreme observations, the remaining observations are dependent and have nonzero correlations, and this needs to be taken into account when trying to derive an expression for the variance of the trimmed mean. In symbols, we could determine the variance of the trimmed mean if we could determine

$$\text{VAR}(X_{(g+1)} + \cdots + X_{(n-g)}),$$

where $X_{(g+1)} + \cdots + X_{(n-g)}$ is the numerator of the sample trimmed mean as given by Equation (3.5). The difficulty is that the variables in this last equation have nonzero correlations, so this variance is *not* equal to $\text{VAR}(X_{(g+1)}) + \cdots + \text{VAR}(X_{(n-g)})$, the sum of the individual variances. That is, after trimming, it is no longer true that the variance of the sum of the remaining observations is equal to the sum of the variances of the individual observations. (There is also the problem that the observations left after trimming do not have the same variance as the distribution from which they were sampled.)

We conclude this subsection with the following remark. Various studies suggest that outliers and heavy-tailed distributions are common in applied research. A tempting method for dealing with outliers is simply to discard any that are found and then to compute a confidence interval with Student's T using only the data that remain. This strategy is theoretically unsound, however, because, as just indicated, the observations not discarded are dependent, in which case the mathematical derivation of Student's T is no longer valid.

4.9.2 A Theoretically Correct Approach

During the 1960s, some mathematical techniques were developed that provide a convenient and useful method for estimating the variance of a trimmed mean. The theoretical details are described in Huber (1981), but here we merely describe the resulting method. In particular, the variance of the 20% trimmed mean can be estimated with

$$\frac{s_w^2}{.6^2 n}, \tag{4.12}$$

where s_w^2 is the Winsorized sample variance introduced in Chapter 3. The .6 in the denominator is related to the amount of trimming, which is assumed to be 20%.

More generally, for a γ-trimmed mean, the squared standard error is estimated with

$$\frac{s_w^2}{(1-2\gamma)^2 n}, \tag{4.13}$$

where now s_w^2 is the γ-Winsorized variance. For example, if 10% trimming is used instead, the .6 in Equation (4.12) is replaced by .8. The standard error of the trimmed mean is estimated with the square root of this last equation.

EXAMPLE. Again consider the weight-gain data for rats and assume 20% trimming is to be used. The sample Winsorized standard deviation can be computed as described in Chapter 3 and is $s_w = 3.927$. The trimmed mean is $\bar{X}_t = 23.27$. Because there are $n = 23$ rats, the estimated standard error of the sample trimmed is

$$\frac{3.927}{.6\sqrt{23}} = 1.4.$$

In contrast, the estimated standard error of the sample mean is $s/\sqrt{n} = 2.25$. Note that the ratio of these two values is $1.4/2.25 = .62$, and this is substantially less than 1 when viewed in light of techniques to be covered. Put another way, it might seem that the trimmed mean would have a larger standard error because only six of the ten observations are used to compute the trimmed mean. In fact the exact opposite is true, and this turns out to be important. ■

EXAMPLE. The data in Table 4.3, from a study on self-awareness, reflect how long an individual could keep a portion of an apparatus in contact with a specified target. The trimmed mean is $\bar{X}_t = 283$ and its estimated standard error is 56.1. In contrast, the standard error of the sample mean is $s/\sqrt{n} = 136$, a value approximately 2.4 times larger than the sample standard error of the trimmed mean. This difference will be seen to be substantial. ■

A practical issue is whether using a correct estimate of the standard error of the trimmed mean can make a difference in applied work versus the strategy of applying methods for means to the data left after trimming. That is, if we trim and ignore the

TABLE 4.3 Self-Awareness Data

77	87	88	114	151	210	219	246	253	262
296	299	306	376	428	515	666	1310	2611	

dependence among the remaining values, can this have any practical consequences? The answer is an unequivocal yes, as illustrated in the next example.

EXAMPLE. For the data in Table 3.2, the estimated standard error of the 20% trimmed mean is .532 using Equation (4.12). If we trim 20% and simply use the method for the sample mean on the remaining 63 values (meaning that we compute s using these 63 values only and then compute $s/\sqrt{63}$), we get 0.28, which is less than half of the value based on Equation (4.12). So we see that using a theoretically motivated estimate of the standard error of the trimmed mean, rather than using methods for the sample mean based on data not trimmed, is not an academic matter. The incorrect estimate can differ substantially from the estimate based on theory. ■

4.9.3 A Confidence Interval for the Population Trimmed Mean

As was the case when working with the population mean, we want to know how well the sample trimmed mean, $\bar{X}_t$, estimates the population trimmed mean, μ_t. What is needed is a method for computing a $1-\alpha$ confidence interval for μ_t. A solution was derived by Tukey and McLaughlin (1963) and is computed as follows. Let h be the number of observations left after trimming, as described in Chapter 3. Let c be the $1-\alpha/2$ quantile of the Student's T distribution with $h-1$ degrees of freedom and let s_w be the Winsorized sample standard deviation, which is also described in Chapter 3. A confidence interval for the γ-trimmed mean is

$$\left(\bar{X}_t - c\frac{s_w}{(1-2\gamma)\sqrt{n}}, \bar{X}_t + c\frac{s_w}{(1-2\gamma)\sqrt{n}}\right). \qquad (4.14)$$

So for the special case of 20% trimming, a $1-\alpha$ confidence interval is given by

$$\left(\bar{X}_t - c\frac{s_w}{.6\sqrt{n}}, \bar{X}_t + c\frac{s_w}{.6\sqrt{n}}\right). \qquad (4.15)$$

In terms of probability coverage, we get reasonably accurate confidence intervals for a much broader range of nonnormal distributions versus confidence intervals for μ based on Student's T.

Section 4.4 provided one reason why a trimmed mean can be a more accurate estimate of the population mean when sampling from a symmetric distribution. The method for computing a confidence interval just described provides another perspective and explanation. We saw in Chapter 3 that generally the Winsorized standard deviation, s_w, can be substantially smaller than the standard deviation s. Consequently, a confidence interval based on a trimmed mean can be substantially shorter. However, when computing a confidence interval based on 20% trimming, for example, the estimate of the standard error of the trimmed mean is $s_w/(.6\sqrt{n})$. Because $s_w/.6$ can be greater than s, such as when sampling from a normal distribution, it is possible to get a shorter confidence interval using means. Generally, however, any improvement achieved with the mean is small, but substantial improvements based on a trimmed mean are often possible.

EXAMPLE. Suppose a test of open-mindedness is administered to $n = 10$ subjects, yielding the observations

$$5, 60, 43, 56, 32, 43, 47, 79, 39, 41.$$

We compute a .95 confidence interval for the 20% trimmed mean and compare the results to the confidence interval for the mean. With $n = 10$, the number of trimmed observations is four, as explained in Chapter 3. That is, the two largest and two smallest observations are removed, leaving $h = 6$ observations, and the average of the remaining observations is the trimmed mean, $\bar{X}_t = 44.8$. The mean using all 10 observations is $\bar{X} = 44.5$. This suggests that there might be little difference between the population mean, μ, and the population trimmed mean, μ_t. With $\nu = 6 - 1 = 5$ degrees of freedom, Table 4 in Appendix B indicates that the .975 quantile of Student's T distribution is $c = 2.57$. It can be seen that the Winsorized sample variance is $s_w^2 = 54.54$, so $s_w = \sqrt{54.54} = 7.385$, and the resulting confidence interval for the trimmed mean is

$$44.8 \pm 2.57 \frac{7.385}{.6\sqrt{10}} = (34.8, 54.8).$$

In contrast, the .95 confidence interval for the mean is (30.7, 58.3). The ratio of the lengths of the confidence intervals is

$$\frac{54.8 - 34.8}{58.3 - 30.7} = .72.$$

That is, the length of the confidence interval based on the trimmed mean is substantially shorter. ■

In the previous example, a boxplot of the data reveals that there is an outlier. This explains why the confidence interval for the mean is longer than the confidence interval for the trimmed mean: The outlier inflates the sample variance, s^2, but has no effect on the Winsorized sample variance, s_w^2. Yet another method for trying to salvage means is to check for outliers, and if none are found, compute a confidence interval for the mean. Recall, however, that even when sampling from a skewed light-tailed distribution, the distribution of T can differ substantially from the case where observations are normal. This means that even though no outliers are detected, when computing a .95 confidence interval for μ, the actual probability coverage could be substantially smaller than intended unless the sample size is reasonably large. When attention is turned to comparing multiple groups of subjects, this problem becomes exacerbated, as will be seen. Modern theoretical results tell us that trimmed means reduce this problem substantially.

TABLE 4.4 Average LSAT Scores for 15 Law Schools

545	555	558	572	575	576	578	580
594	605	635	651	653	661	666	

EXAMPLE. Table 4.4 shows the average LSAT scores for the 1973 entering classes of 15 American law schools. (LSAT is a national test for prospective lawyers.) The sample mean is $\bar{X} = 600.3$ with an estimated standard error of 10.8. The 20% trimmed mean is $\bar{X}_t = 596.2$ with an estimated standard error of 14.92. The .95 confidence interval for μ_t is (561.8, 630.6). In contrast, the .95 confidence interval for μ is (577.1, 623.4), assuming T does indeed have a Student's T distribution. Note that the length of the confidence interval for μ is smaller and, in fact, is a subset of the confidence interval for μ_t. This might suggest that the sample mean is preferable to the trimmed mean for this particular set of data, but closer examination suggests that this might not be true. The concern here is the claim that the confidence interval for the mean has probability coverage .95. If sampling is from a light-tailed, skewed distribution, the actual probability coverage for the sample mean can be substantially smaller than the nominal level. Figure 4.23 shows a boxplot of the data, indicating that the central portion of the data is skewed to the right. Moreover, there are no outliers, suggesting the possibility that sampling is from a relatively light-tailed distribution. Thus, the actual probability coverage of the confidence interval for the mean might be too low—a longer confidence interval might be needed to achieve .95 probability coverage. That is, an unfair comparison of the two confidence intervals has probably been made because they do not have the same probability coverage. If we were able to compute a .95 confidence interval for the mean, there is some possibility that it would be longer than the confidence interval for the trimmed mean. When sampling from a skewed, heavy-tailed distribution, problems with the mean can be exacerbated (as illustrated in Chapter 7). ■

The confidence interval for the 20% trimmed mean given by Equation (4.15) assumes that

$$T_t = \frac{.6(\bar{X}_t - \mu_t)}{s_w/\sqrt{n}}$$

has a Student's T distribution with $h - 1 = n - 2g - 1$ degrees of freedom, where $g = [.2n]$ and $[.2n]$ is $.2n$ rounded down to the nearest integer. To graphically illustrate how nonnormality affects this assumption, we repeat the method used to create the left panel of Figure 4.15. That is, we sample $n = 20$ observations from the (lognormal) distribution in Figure 4.14, compute T_t, and repeat this 5000 times, yielding 5000 T_t values. The left panel of Figure 4.24 shows a plot of these T_t values versus Student's T with 11 degrees of freedom. To facilitate comparisons, the right panel shows a plot of 5000 T values (based on the mean and variance) versus Student's T distribution

FIGURE 4.23 A boxplot of the data in Table 4.4.

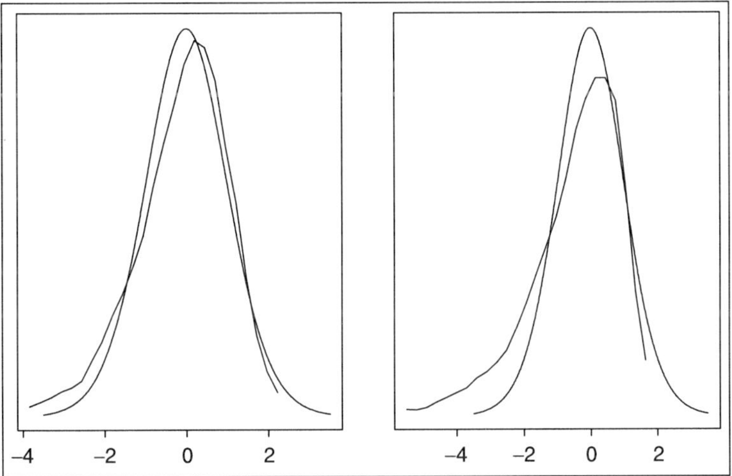

FIGURE 4.24 The left plot is based on 5000 T_t values (with 20% trimming) when sampling from a lognormal distribution. The right plot is based on 5000 T values (no trimming).

with 19 degrees of freedom. Student's T distribution gives a better approximation of the actual distribution of T_t versus T. Of particular importance is that the tails of the distribution are better approximated when using a trimmed mean. This indicates that more accurate probability coverage will be achieved using a 20% trimmed mean versus using the mean with no trimming. Generally, as the sample size increases, problems with nonnormality diminish more rapidly when using a trimmed mean versus using a mean. But switching to a trimmed mean does not eliminate all practical problems when sample sizes are small. Fortunately there are methods for getting even more accurate results, as we shall see.

4.9.4 S-PLUS Functions trimse and trimci

The S-PLUS function

trimse(x,tr=.2,alpha=.05)

has been supplied that computes an estimate of the standard error of the trimmed mean for the data stored in any S-PLUS variable x. The default amount of trimming (tr) is .2. The function

trimci(x,tr=.2,alpha=.05)

computes a $1 - \alpha$ confidence interval for μ_t. If the argument alpha is unspecified, $\alpha = .05$ is used.

EXAMPLE. If the data in Table 4.4 are stored in the S-PLUS variable blob, the command trimci(blob) returns a .95 confidence interval for the 20% trimmed mean: (561.8, 630.6). The command trimci(blob,tr=0,alpha=.01) returns a .99 confidence interval for the population mean using Student's T method described in Section 4.7. ■

4.10 Transforming Data

For completeness, it should be remarked that simple transformations of data are often recommended for dealing with problems due to nonnormality. For example, a common recommendation is to replace all the values in a study with their logarithm and then to use methods based on means. Another common strategy when all observations are positive is to replace each value by its square root. These simple transformations can be useful for certain purposes, but for the situations covered in this book they are not always satisfactory. One problem is that they do not eliminate the deleterious effect of outliers (e.g., Rasmussen, 1989). Transformations can make a distribution appear to be reasonably normal, at least in some cases, but when using simple transformations it has been found that trimming is still beneficial (Doksum & Wong, 1983). For this reason, simple transformations of data are not discussed.

4.11 Confidence Interval for the Population Median

Although the median is a member of the class of trimmed means, the method for computing a confidence interval for the trimmed mean gives absurd results for the extreme case of the median. One way of dealing with this problem is to use a result derived by Laplace, which gives an expression for the variance of the median that depends in part on $f(x)$, the equation for the probability density function from which observations are randomly sampled. Details of this approach are covered in Wilcox (1997a) but not here. Instead we rely on the method outlined in Box 4.1, which was suggested by Hettmansperger and Sheather (1986). Results supporting the use of this method are reported by Sheather and McKean (1987) as well as Hall and Sheather (1988).

BOX 4.1 How to Compute a Confidence Interval for the Median

As indicated by Equation (2.9), for some p, $0 \le p \le 1$, the binomial probability function is

$$f(x) = \binom{n}{x} p^x (1-p)^{n-x}, x = 0, \ldots, n.$$

For any integer k between 0 and $n/2$, let

$$\gamma_k = f(k) + f(k+1) + \cdots + f(n-k)$$

when $p = .5$.

Then

$$(X_{(k)}, X_{(n-k+1)}) \tag{4.16}$$

is a confidence interval for the median that has probability coverage exactly equal to γ_k.

Because the binomial distribution is discrete, it is not possible, in general, to choose k so that the probability coverage is exactly equal to $1-\alpha$. To get a $1-\alpha$ confidence interval, first determine k such that $\gamma_{k+1} < 1-\alpha < \gamma_k$. Next, compute

$$I = \frac{\gamma_k - 1 - \alpha}{\gamma_k - \gamma_{k+1}} \quad \text{and} \quad \lambda = \frac{(n-k)I}{k + (n-2k)I}.$$

Then an approximate $1-\alpha$ confidence interval is

$$(\lambda X_{(k+1)} + (1-\lambda)X_{(k)}, \lambda X_{(n-k)} + (1-\lambda)X_{(n-k+1)}). \tag{4.17}$$

4.11.1 S-PLUS Function sint

The S-PLUS function

sint(x,alpha=.05)

has been supplied for applying the method in Box 4.1. As usual, x is any S-PLUS variable containing data.

EXAMPLE. Staudte and Sheather (1990) illustrate the use of Equation (4.17) with data from a study on the lifetimes of EMT6 cells. The values are 10.4, 10.9, 8.8, 7.8, 9.5, 10.4, 8.4, 9.0, 22.2, 8.5, 9.1, 8.9, 10.5, 8.7, 10.4, 9.8, 7.7, 8.2, 10.3, 9.1. Storing these values in the S-PLUS variable blob, the command sint(blob) returns a .95 confidence interval of (8.72, 10.38). The command sint(blob,.01) returns a .99 confidence interval of (8.5, 10.4). ■

4.11.2 Estimating the Standard Error of the Sample Median

The method just described for computing a confidence interval for the median does not require an estimate of the standard error of M, the sample median. However, for situations to be covered, an explicit estimate will be needed. Many strategies have been proposed, comparisons of which can be found in Price and Bonett (2001). Here, the method derived by McKean and Schrader (1984) is described because it is very simple and currently appears to have practical value for problems addressed in subsequent chapters.

Compute

$$k = \frac{n+1}{2} - z_{.995}\sqrt{\frac{n}{4}},$$

where k is rounded to the nearest integer and $z_{.995}$ is the .995 quantile of a standard normal distribution. Put the observed values in ascending order, yielding $X_{(1)} \leq \cdots \leq X_{(n)}$. Then the McKean–Schrader estimate of the squared standard error of M is

$$\left(\frac{X_{(n-k+1)} - X_{(k)}}{2z_{.995}}\right)^2.$$

4.11.3 S-PLUS Function msmedse

The S-PLUS function

```
msmedse(x)
```

computes the estimated standard error of M given by the square root of the last equation.

4.12 A Remark About MOM and M-Estimators

How do we compute a confidence interval based on an M-estimator of location or the MOM estimator in Section 3.5.2? An expression for the standard error of an M-estimator has been derived and can be used to derive a confidence interval using methods similar to those described in this chapter. But for $n < 100$ it yields a reasonably accurate confidence interval only when sampling from a perfectly symmetric distribution. With n sufficiently large, an accurate confidence interval can be computed when sampling from a skewed distribution, but it remains unclear just how large n must be. As for MOM, no expression for the standard error has been derived. However, Chapter 7 describes an effective method for estimating its standard and computing accurate confidence intervals.

4.13 Confidence Intervals for the Probability of Success

Section 2.5 introduced the binomial probability function, where p represents the probability of success and x represents the number of successes among n randomly sampled observations. As was noted, the usual estimate of p is simply

$$\hat{p} = \frac{x}{n},$$

the proportion of successes among the n observations.

Results in Chapter 2 plus the central limit theorem suggest a simple method for computing a $1 - \alpha$ confidence interval for p:

$$\hat{p} \pm c\sqrt{\frac{p(1-p)}{n}},$$

where c is the $1 - \alpha/2$ quantile of a standard normal distribution. We do not know the value of the quantity under the radical, but it can be estimated with $\hat{p}$, in which case a simple $1 - \alpha$ confidence interval for p is

$$\hat{p} \pm c\sqrt{\frac{\hat{p}(1-\hat{p})}{n}}. \tag{4.18}$$

The resulting probability coverage will be reasonably close to $1 - \alpha$ if n is not too small and p is not too close to zero or 1. Just how large n must be depends on how close p is to zero or 1. An obvious concern is that we do not know p, so there is some difficulty in deciding whether n is sufficiently large. Numerous methods have been proposed for dealing with this issue. For the special cases $x = 0$, 1, $n - 1$, and n, Blyth (1986) suggests proceeding as follows:

- If $x = 0$, use

$$(0, 1 - \alpha^{1/n}).$$

- If $x = 1$, use

$$\left(1 - \left(1 - \frac{\alpha}{2}\right)^{1/n}, 1 - \left(\frac{\alpha}{2}\right)^{1/n}\right).$$

- If $x = n - 1$, use

$$\left(\left(\frac{\alpha}{2}\right)^{1/n}, \left(1 - \frac{\alpha}{2}\right)^{1/n}\right).$$

- If $x = n$, use

$$(\alpha^{1/n}, 1).$$

For all other situations, Blyth's comparisons of various methods suggest using Pratt's (1968) approximate confidence interval, which is computed as shown in Box 4.2.

BOX 4.2 Computing a $1-\alpha$ Confidence Interval for p Based on x Successes Among n Trials

Let c be the $1-\alpha/2$ quantile of a standard normal distribution read from Table 1 in Appendix B. That is, if Z is a standard normal random variable, then $P(Z \le c) = 1-\alpha/2$. To determine c_U, the upper end of the confidence interval, compute

$$A = \left(\frac{x+1}{n-x}\right)^2$$

$$B = 81(x+1)(n-x) - 9n - 8$$

$$C = -3c\sqrt{9(x+1)(n-x)(9n+5-c^2)+n+1}$$

$$D = 81(x+1)^2 - 9(x+1)(2+c^2)+1$$

$$E = 1 + A\left(\frac{B+C}{D}\right)^3$$

$$c_U = \frac{1}{E}.$$

To get the lower end of the confidence interval, c_L, compute

$$A = \left(\frac{x}{n-x-1}\right)^2$$

$$B = 81(x)(n-x-1) - 9n - 8$$

$$C = 3c\sqrt{9x(n-x-1)(9n+5-c^2)+n+1}$$

$$D = 81x^2 - 9x(2+c^2)+1$$

$$E = 1 + A\left(\frac{B+C}{D}\right)^3$$

$$c_L = \frac{1}{E}.$$

An approximate $1-\alpha$ confidence interval for p is

$$(c_L, c_U).$$

4.13.1 S-PLUS Function binomci

The S-PLUS function

```
binomci(x=sum(y), nn=length(y), y=NA, n=NA, alpha=0.05)
```

has been supplied to compute Pratt's approximate confidence interval for p. In the event x $= 0, 1, n-1$, or n, Blyth's method is used instead. The first argument, x, is the number of successes and the second argument, nn, indicates the value of n, the number of observations. If the data are stored as a vector of 1's and 0's in some S-PLUS variable, where a 1 indicates a success and 0 a failure, use the third argument, y. (Generally, the fourth argument can be ignored.)

EXAMPLE. The command

binomci(5,25)

returns (0.07, 0.41) as a .95 confidence interval for p based on five successes among 25 observations. If the values 1, 1, 1, 0, 0, 1, 1, 0, 0, 0, 0, 1 are stored in the S-PLUS variable obs, the command

binomci(y=obs)

returns (0.25, 0.79) as a .95 confidence interval for p. ■

4.14 Exercises

1. Explain the meaning of a .95 confidence interval.
2. If you want to compute a .80, .92, or .98 confidence interval for μ when σ is known, and sampling is from a normal distribution, what values for c should you use in Equation (4.8)?
3. Assuming random sampling is from a normal distribution with standard deviation $\sigma = 5$, if you get a sample mean of $\bar{X} = 45$ based on $n = 25$ subjects, what is the .95 confidence interval for μ?
4. Repeat the previous example, but compute a .99 confidence interval instead.
5. A manufacturer claims that their light bulbs have an average life span of $\mu = 1200$ hours with a standard deviation of $\sigma = 25$. If you randomly test 36 light bulbs and find that their average lifespan is $\bar{X} = 1150$, does a .95 confidence interval for μ suggest that the claim $\mu = 1200$ is unreasonable?
6. Compute a .95 confidence interval for the mean in the following situations: (a) $n = 12$, $\sigma = 22$, $\bar{X} = 65$, (b) $n = 22$, $\sigma = 10$, $\bar{X} = 185$, (c) $n = 50$, $\sigma = 30$, $\bar{X} = 19$.
7. Describe the two components of a random sample.
8. If $n = 10$ observations are randomly sampled from a distribution with mean $\mu = 9$ and variance $\sigma^2 = 8$, what is the mean and variance of the sample mean?
9. Determine $E(\bar{X})$ and $\sigma^2_{\bar{X}}$ for a random sample of $n = 12$ observations from a discrete distribution with the following probability function:

x:	1	2	3	4
$p(x)$:	.2	.1	.5	.2

10. In Exercise 9, again suppose you sample $n = 12$ subjects and compute the sample mean. If you repeat this process 1000 times, each time using $n = 12$ subjects, and if you averaged the resulting 1000 sample means, approximately what would be the result? That is, approximate the average of the 1000 sample means.
11. Answer the same question posed in Exercise 10, except replace means with sample variances.
12. Estimate the variance and standard error of the sample mean for a random sample of $n = 8$ subjects from whom you get

$$2, 6, 10, 1, 15, 22, 11, 29.$$

13. If you randomly sample a single observation and get 32, what is the estimate of the population mean, μ? Can you get an estimate of the squared standard error? Explain, in terms of the squared standard error, why only a single observation is likely to be a less accurate estimate of μ versus a sample mean based on $n = 15$ subjects.
14. As part of a health study, a researcher wants to know the average daily intake of vitamin E for the typical adult. Suppose that for $n = 12$ adults, the intake is found to be

$$450, 12, 52, 80, 600, 93, 43, 59, 1000, 102, 98, 43.$$

Estimate the squared standard error of the sample mean.
15. In Exercise 14, verify that there are outliers. Based on results in Chapter 2, what are the effects of these outliers on the estimated squared standard error?
16. Estimate the variance and standard error of the sample mean when you randomly sample $n = 8$ subjects and get

$$2, 6, 10, 1, 15, 22, 11, 29.$$

17. In Exercise 16, if the observations are dependent, can you still estimate the standard error of the sample mean?
18. Section 2.7 described a mixed normal distribution that differs only slightly from a standard normal. Suppose we randomly sample $n = 25$ observations from a standard normal distribution. Then the squared standard error of the sample mean is 1/25. Referring back to Section 2.7, what is the squared standard error if sampling is from the mixed normal instead? What does this indicate about what might happen under slight departures from normality?
19. Explain why knowing the mean and squared standard error is not enough to determine the distribution of the sample mean. Relate your answer to results on nonnormality described in Section 2.7.
20. Suppose $n = 16$, $\sigma = 2$, and $\mu = 30$. Assume normality and determine (a) $P(\bar{X} < 29)$, (b) $P(\bar{X} > 30.5)$, (c) $P(29 < \bar{X} < 31)$.
21. Suppose $n = 25$, $\sigma = 5$, and $\mu = 5$. Assume normality and determine (a) $P(\bar{X} < 4)$, (b) $P(\bar{X} > 7)$, (c) $P(3 < \bar{X} < 7)$.
22. Someone claims that within a certain neighborhood, the average cost of a house is $\mu = \$100,000$ with a standard deviation of $\sigma = \$10,000$. Suppose that based on $n = 16$ homes, you find that the average cost of a house is

$\bar{X} = \$95{,}000$. Assuming normality, what is the probability of getting a sample mean this low or lower if the claims about the mean and standard deviation are true?

23. In Exercise 22, what is the probability of getting a sample mean between \$97,500 and \$102,500?
24. A company claims that the premiums paid by its clients for auto insurance have a normal distribution with mean $\mu = \$750$ and standard deviation $\sigma = \$100$. Assuming normality, what is the probability that for $n = 9$ randomly sampled clients, the sample mean will have a value between \$700 and \$800?
25. You sample 16 observations from a discrete distribution with mean $\mu = 36$ and variance $\sigma^2 = 25$. Use the central limit theorem to determine (a) $P(\bar{X} < 34)$, (b) $P(\bar{X} < 37)$, (c) $P(\bar{X} > 33)$, (d) $P(34 < \bar{X} < 37)$.
26. You sample 25 observations from a nonnormal distribution with mean $\mu = 25$ and variance $\sigma^2 = 9$. Use the central limit theorem to determine (a) $P(\bar{X} < 24)$, (b) $P(\bar{X} < 26)$, (c) $P(\bar{X} > 24)$, (d) $P(24 < \bar{X} < 26)$.
27. Describe a situation where Equation (4.11), used in conjunction with the central limit theorem, might yield a relatively long confidence interval.
28. Describe a type of continuous distribution where the central limit theorem gives good results with small sample sizes.
29. Compute a .95 confidence interval if (a) $n = 10$, $\bar{X} = 26$, $s = 9$, (b) $n = 18$, $\bar{X} = 132$, $s = 20$, (c) $n = 25$, $\bar{X} = 52$, $s = 12$.
30. Repeat Exercise 29, but compute a .99 confidence interval instead.
31. Table 4.3 reports data from a study on self-awareness. Compute a .95 confidence interval for the mean.
32. Rats are subjected to a drug that might affect aggression. Suppose that for a random sample of rats, measures of aggression are found to be

    5, 12, 23, 24, 18, 9, 18, 11, 36, 15.

    Compute a .95 confidence for the mean assuming the scores are from a normal distribution.
33. Describe in general terms how nonnormality can affect Student's T distribution.
34. When sampling from a light-tailed, skewed distribution, where outliers are rare, a small sample size is needed to get good probability coverage, via the central limit theorem, when the variance is known. How does this contrast with the situation where the variance is not known and confidence intervals are computed using Student's T distribution?
35. Compute a .95 confidence for the trimmed mean if (a) $n = 24$, $s_w^2 = 12$, $\bar{X}_t = 52$, (b) $n = 36$, $s_w^2 = 30$, $\bar{X}_t = 10$, (c) $n = 12$, $s_w^2 = 9$, $\bar{X}_t = 16$.
36. Repeat Exercise 35, but compute a .99 confidence interval instead.
37. Compute a .95 confidence interval for the 20% trimmed mean using the data in Table 4.3.
38. Compare the length of the confidence interval in Exercise 37 to the length of the confidence interval for the mean you got in Exercise 31. Comment on why they differ.

39. In a portion of a study of self-awareness, Dana observed the values

$$59, 106, 174, 207, 219, 237, 313, 365, 458, 497, 515,$$
$$529, 557, 615, 625, 645, 973, 1065, 3215.$$

Compare the lengths of the confidence intervals based on the mean and 20% trimmed mean. Why is the latter confidence interval shorter?

40. The ideal estimator of location would have a smaller standard error than any other estimator we might use. Explain why such an estimator does not exist.

41. Under normality, the sample mean has a smaller standard error than the trimmed mean or median. If observations are sampled from a distribution that appears to be normal, does this suggest that the mean should be preferred over the trimmed mean and median?

42. Chapter 3 reported data on the number of seeds in 29 pumpkins. The results were

$$250, 220, 281, 247, 230, 209, 240, 160, 370, 274, 210, 204, 243, 251, 190,$$
$$200, 130, 150, 177, 475, 221, 350, 224, 163, 272, 236, 200, 171, 98.$$

The 20% trimmed mean is $\bar{X}_t = 220.8$ and the mean is $\bar{X} = 229.2$. Verify that the .95 confidence interval for μ is (200.7, 257.6) and that for the trimmed mean, μ_t, it is (196.7, 244.9).

43. In Exercise 42, the length of the confidence interval for μ is $257.6 - 200.7 = 56.9$ and the length based on the trimmed mean is $244.9 - 196.7 = 48.2$. Comment on why the length of the confidence interval for the trimmed mean is shorter.

44. If the mean and trimmed mean are nearly identical, it might be thought that it makes little difference which measure of location is used. Based on your answer to Exercise 43, why might it make a difference?

45. For the past 16 presidential elections in the United States, the incumbent party won or lost the election depending on whether the Washington Redskins, American football team, won their last game just prior to the election. That is, there has been perfect agreement between the two events during the last 16 elections. Verify that according to Blyth's method, a .99 confidence for the probability of agreement is (.75, 1).

46. An ABC news program reported that a standard method for rendering patients unconscious led patients to wake up during surgery. These individuals were not only aware of their plight, they suffered from nightmares later on. Some physicians tried monitoring brain function during surgery to avoid this problem, the strategy being to give patients more medication if they showed signs of regaining consciousness, and they found that among 200,000 trials, no patients woke during surgery. However, administrators concerned about cost argued that with only 200,000 trials, the probability of waking up using the new method could not be accurately estimated. Verify that a .95 confidence interval for p, the probability of waking up, is (0, .000015).

HYPOTHESIS TESTING

In applied research, it is common to make some speculation about the population mean and then to try to assess whether this speculation is reasonable based on the data that are available. Roughly, if the likelihood of an observed value for the sample mean is small based on an assumed value for the population mean, then perhaps the assumed value of the population mean is incorrect. Another possibility is that the sample mean does not accurately reflect the population mean. That is, the speculation about the population mean is correct, but by chance the sample mean differs substantially from the population mean.

As an example, imagine a researcher who claims that on a test of open-mindedness, the population mean (μ) for adult men is 50. Suppose you randomly sample $n = 10$ adult males, give them the test for open-mindedness, and get the scores

$$25, 60, 43, 56, 32, 43, 47, 59, 39, 41.$$

The sample mean is $\bar{X} = 44.5$. Does this make the claim $\mu = 50$ unreasonable? Do the data support the claim? If in reality $\mu = 50$, what is the probability that you will get a sample mean less than 45?

Chapter 4 touched on how you might decide whether the claim $\mu = 50$ is reasonably consistent with the 10 open-mindedness scores just given. If the .95 confidence interval for μ happens to be (40, 48), then this interval does not contain 50, suggesting that the claim $\mu = 50$ is not reasonable. If the .95 confidence interval is (46, 52), this interval contains 50, which suggests that the claim $\mu = 50$ should not be ruled out. The purpose of this chapter is to expand on the topic of making decisions about whether some claim about the population mean or some other parameter of interest is consistent with data. As usual, we begin by describing basic concepts and techniques typically covered in an introductory statistics course. Then we describe modern insights into when and why the standard method based on the mean might be highly unsatisfactory and how these practical problems might be addressed.

5.1 The Basics of Hypothesis Testing

We continue the illustration described in the introduction to this chapter, but for convenience we first consider a situation where it is claimed that the population mean is greater than or equal to 50. A typical way of writing this claim more succinctly is

$$H_0 : \mu \geq 50,$$

where the notation H_0 is read "H naught." This last expression is an example of what is called a *null hypothesis*. A null hypothesis is just a statement — some speculation — about some characteristic of a distribution. In the example, the null hypothesis is a speculation about the population mean, but it could just as easily be some speculation about the population median or trimmed mean. If someone claims that the mean is greater than or equal to 60, then our null hypothesis would be written as $H_0 : \mu \geq 60$. If there is some reason to speculate that $\mu \leq 20$, and the goal is to see whether this speculation is consistent with observations we make, then the null hypothesis is $H_0 : \mu \leq 20$.

The goal of statistical hypothesis testing is to find a decision rule about whether the null hypothesis is true, or should be ruled out, based on observations we make. When the null hypothesis is rejected, this means you decide that the corresponding alternative hypothesis is accepted. For example, if the null hypothesis is $H_0 : \mu \geq 50$, the *alternative hypothesis* is typically written as

$$H_1 : \mu < 50,$$

and if you reject H_0, you in effect accept H_1. That is, you conclude that the mean is less than 50 based on the data available in your study.

Suppose we sample some adult men, measure their open-mindedness, and find that the resulting sample mean is $\bar{X} = 61$. Thus, our estimate of the population mean μ is 61, which is consistent with the null hypothesis $H_0 : \mu \geq 50$, so there is no empirical evidence to doubt the claim that the population mean is greater than or equal to 50. For the data given at the beginning of this chapter, $\bar{X} = 44.5$. That is, μ is estimated to be less than 50, which suggests that the null hypothesis is false and should be rejected. But if it were true that $\mu = 50$, then there is some possibility of observing a sample mean less than or equal to 44.5. That is, if we reject the null hypothesis and conclude that μ is less than 50 based on this observed sample mean, there is some possibility that our decision is in error.

DEFINITION. A *Type I error* refers to a particular type of mistake, namely, rejecting the null hypothesis when in fact it is correct. A common notation for the probability of a Type I error is α, which is often referred to as the *level of significance*.

We can avoid a Type I error by never rejecting the null hypothesis. In this case, $\alpha = 0$, meaning that the probability of erroneously rejecting the null hypothesis is zero. But a problem with this rule is that it is impossible to discover situations where indeed the null hypothesis is false. If in our illustration $\mu = 46$, then the null hypothesis is false and we want a method that will detect this. That is, we need a rule that allows the possibility of rejecting, but simultaneously we want to control the probability of a Type I error.

A natural strategy is to try to determine how small the sample mean must be before we reject the hypothesis that μ is greater than or equal to 50. But rather than work with $\bar{X}$, it is more convenient to work with

$$Z = \frac{\bar{X} - 50}{\sigma/\sqrt{n}},$$

where for the moment we assume the population standard deviation (σ) is known. Using Z is convenient because when sampling from a normal distribution, it provides a simple method for controlling the probability of a Type I error.

For illustrative purposes, temporarily assume $\sigma = 12$ and $n = 10$. So if $\bar{X} = 50$, then $Z = 0$. If $\bar{X} = 49$, then $Z = -0.26$; and if $\bar{X} = 48$, then $Z = -0.53$. That is, as the sample mean decreases and moves further away from the null hypothesis, Z decreases as well. But *if* the null hypothesis is true and in fact $\mu = 50$, then the assumption that observations are sampled from a normal distribution implies that Z has a standard normal distribution. This in turn provides a simple way of controlling the probability of making a Type I error. For example, if the null hypothesis is true and $\mu = 50$, then from Table 1 in Appendix B we see that $P(Z \leq -1.645) = .05$. So if we reject when $Z \leq -1.645$, the probability of a Type I error is $\alpha = .05$.

EXAMPLE. For the hypothesis $H_0 : \mu \geq 50$, imagine that, based on $n = 10$ subjects, with $\sigma = 12$, you find that the sample mean is $\bar{X} = 48$. Then, as previously indicated, $Z = -0.53$. If you want the probability of a Type I error to be .05, then you should reject the null hypothesis only if Z is less than or equal to -1.645. Because -0.53 is greater than -1.645, you fail to reject. That is, the sample mean is less than 50, but you do not have convincing evidence for ruling out the possibility that the population mean is greater than or equal to 50. This does *not* mean, however, that it is reasonable to accept H_0 and conclude that $\mu \geq 50$. (This issue is elaborated upon in Section 5.2.) ■

Figure 5.1 illustrates the decision rule just described. If the null hypothesis is true, and in particular $\mu = 50$, then Z has a standard normal distribution as shown in Figure 5.1, in which case the probability that Z is less than -1.645 is the area of the shaded region, which is .05. In summary, if you assume the null hypothesis is true, are willing to have a Type I error probability of .05, and $Z \leq -1.645$, then this suggests that the assumption $\mu \geq 50$ is not reasonably consistent with empirical evidence and should be rejected.

Recall from Chapter 2 that $P(Z \leq -1.96) = .025$. This means that if we reject $H_0 : \mu \geq 50$ when $Z \leq -1.96$, then the probability of a Type I error is .025 if it happens to be the case that $\mu = 50$. In a similar manner, $P(Z \leq -2.58) = .005$, and if we reject when $Z \leq -2.58$, then the probability of a Type I error is .005. A *critical value* is the value used to determine whether the null hypothesis should be rejected. If it is desired to have a Type I error probability of .05 when testing $H_0 : \mu \geq 50$, the critical value is -1.645, meaning that you reject if $Z \leq -1.645$ (assuming normality and that σ is known). The set of all Z values such that $Z \leq -1.645$ is called the *critical region*; it corresponds to the shaded region in Figure 5.1. If you want the probability of a Type I error to be $\alpha = .025$, then the critical value is -1.96 and the critical region

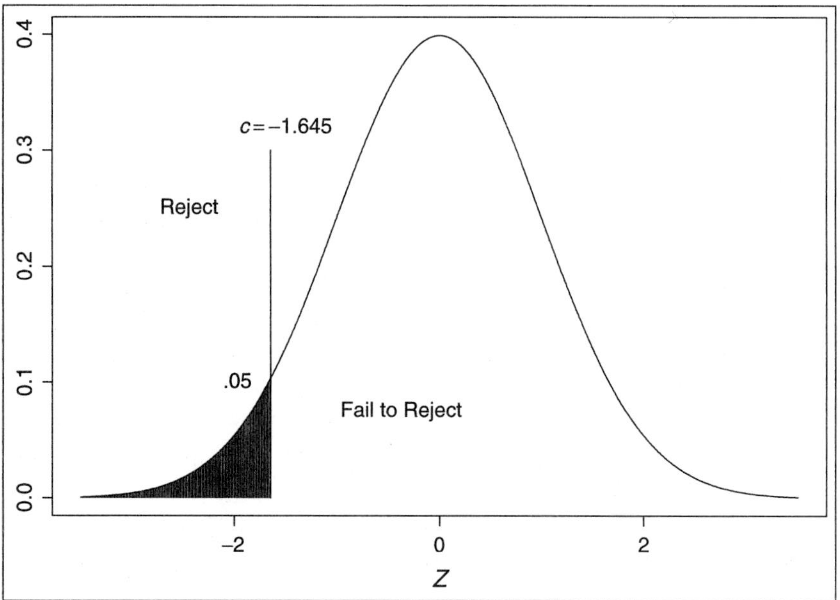

FIGURE 5.1 Graphical depiction of a decision rule. Here you reject when $Z \leq -1.645$ and fail to reject otherwise. The shaded region corresponds to the probability of a Type I error, which is .05.

consists of all Z values less than or equal to -1.96. If $\alpha = .005$, then the critical value is -2.58 and the critical region is the set of all Z values less than -2.58.

EXAMPLE. Continuing the illustration regarding open-mindedness, where the goal is to test $H_0: \mu \geq 50$, suppose we want the probability of a Type I error to be .025, in which case the critical value is -1.96, the .025 quantile of a standard normal distribution. By assumption, $\sigma = 12$, there are $n = 10$ subjects, and, for the data reported at the beginning of this chapter, $\bar{X} = 44.5$. Therefore,

$$Z = \frac{44.5 - 50}{12/\sqrt{10}} = -1.45.$$

Because -1.45 is greater than the critical value, -1.96, you fail to reject the null hypothesis. If instead you are willing to have a Type I error probability of .1, the critical value is -1.28, and because $Z = -1.45$ is less than -1.28, you reject. That is, you conclude that the mean is less than 50, and the probability of making a mistake, rejecting when you should not have rejected, is .1. ■

EXAMPLE. Rather than test $H_0: \mu \geq 50$, imagine that the goal is to test $H_0: \mu \geq 60$. The calculations are exactly the same as before, except that 50 is

Continued

EXAMPLE. (*Continued*) rcplaccd by 60 whcn computing Z. So now

$$Z = \frac{44.5 - 60}{12/\sqrt{10}} = -4.08.$$

If you test at the .005 level, the critical value is −2.58, −4.08 is less than −2.58, so you reject. That is, you conclude that the sample mean is not consistent with the assumption that $\mu \geq 60$ and that μ is less than 60. Again, there is the possibility of incorrectly rejecting when in fact $\mu \geq 60$ is true, and by design the probability of making this mistake is $\alpha = .005$ when $\mu = 60$. ■

There are two variations of the hypothesis-testing method just described. To illustrate the first, imagine that you work in the research and development department of a company that helps students train for the SAT examination. After years of experience, it is found that the typical student attending the training course gets an SAT mathematics score of $\mu = 580$ and the standard deviation is $\sigma = 50$. You suspect that the training course could be improved and you want to empirically determine whether this is true. You try the new method on $n = 20$ students and get a sample mean of $\bar{X} = 610$. For illustrative purposes, assume that the standard deviation is again 50. We need to consider carefully how the null hypothesis should be stated. You have evidence that the new training method is better for the typical student because the estimate of the population mean is 610, which is greater than 580. But you need to convince management, so you assume that the new method is actually worse, with the goal of determining whether this assumption can be ruled out based on $\bar{X} = 610$. That is, you decide to test the hypothesis $H_0: \mu \leq 580$, so if you reject, there is empirical evidence suggesting that it is unreasonable to believe that the new method is not beneficial for the typical student.

Testing $H_0: \mu \leq 580$ is like the mirror image of testing $H_0: \mu \geq 580$. If you get a sample mean of $\bar{X} = 550$, your estimate of μ is 550, and this is consistent with the hypothesis that μ is less than or equal to 580. That is, you would not reject. In the illustration, $\bar{X} = 610$, which suggests that the null hypothesis might be false. But if $\mu = 580$, getting a mean of 610 or larger could happen by chance. The issue is whether 610 is large enough to rule out the possibility that, beyond a reasonable doubt, $\mu \leq 580$. To find out, you compute Z as before, only now you reject if Z is sufficiently large. If you reject when $Z \geq 1.645$, the probability of a Type I error is $\alpha = P(Z \geq 1.645) = .05$. In the illustration,

$$Z = \frac{\bar{X} - \mu}{\sigma/\sqrt{n}} = \frac{610 - 580}{50/\sqrt{20}} = 2.68.$$

Because 2.68 is greater than 1.645, you reject and conclude that the mean is greater than 580. That is, you have empirical evidence to present to management that the new training method offers an advantage over the conventional approach, and there is a .05 probability that you made a mistake.

EXAMPLE. We repeat the illustration just given, only now imagine you want the probability of a Type I error to be .005 instead. From Table 1 in Appendix B, we see that $P(Z \geq 2.58) = .005$. This means that if you reject $H_0 : \mu \leq 580$ when Z is greater than or equal to 2.58, the probability of a Type I error is $\alpha = .005$. As already indicated, $Z = 2.68$, and because this exceeds 2.58, you again reject and conclude that the mean is greater than 580. ■

5.1.1 *p*-Value (Significance Level)

There is an alternative way of describing hypothesis testing that is frequently employed and therefore important to understand. It is based on what is called the *significance level*, or *p-value*, an idea that appears to have been proposed first by Deming (1943), which is just the probability of a Type I error if the observed value of Z is used as a critical value.[1] If you reject when the p-value is less than or equal to .05, then the probability of a Type I error is .05, assuming normality. If you reject when the p-value is less than or equal to .01, then the probability of a Type I error is .01.

EXAMPLE. Again consider the open-mindedness example where we want to test $H_0 : \mu \geq 50$ and $\sigma = 12$. Imagine that you randomly sample $n = 10$ subjects and compute the sample mean, $\bar{X}$. If, for example, you get $\bar{X} = 48$, then

$$Z = \frac{48 - 50}{12/\sqrt{10}} = -0.53.$$

The p-value is just the probability of a Type I error if you reject when Z is less than or equal to -0.53. This probability is

$$P(Z \leq -0.53) = .298.$$

If you want the probability of a Type I error to be no greater than .05, then you would not reject, because .298 is greater than .05. Put another way, if you reject when your test statistic, Z, is less than or equal to -0.53, the probability of a Type I error is .298. ■

The idea behind the p-value (or significance level) is that it gives you more information about the α level at which the null hypothesis would be rejected. If you are told that you reject with $\alpha = .05$, and nothing else, this leaves open the issue of whether you would also reject with $\alpha = .01$. If you are told that the p-value is .024, say, then you know that you would reject with $\alpha = .05$ but not $\alpha = .01$. If the p-value is .003, then in particular you would reject with $\alpha = .05$, $\alpha = .01$, and even $\alpha = .005$.

1 Level of significance refers to α, the Type I error probability specified by the investigator. Consequently, some authorities prefer the term *p-value* over the expression *significance level* as it is used here.

In case it helps, the p-value can be described in a slightly different fashion. Again consider the null hypothesis $H_0 : \mu \geq 50$ with $n = 10$ and $\sigma = 10$. *Given* $\bar{X}$, let

$$p = P\left(Z \leq \frac{\bar{X} - 50}{\sigma/\sqrt{n}}\right). \tag{5.1}$$

The quantity p is called the *significance level*, or *p-value*, associated with the null hypothesis $H_0 : \mu \geq 50$. In the example, $\bar{X} = 48$,

$$\begin{aligned} p &= P\left(Z \leq \frac{48 - 50}{12/\sqrt{10}}\right) \\ &= P(Z \leq -0.53) \\ &= .298, \end{aligned}$$

so, as in the preceding example, the p-value is .298.

Next, consider a situation where the null hypothesis is that the mean is less than or equal to some specified value. To be concrete, consider $H_0 : \mu \leq 580$. Then given $\bar{X}$, the significance level is now

$$p = P\left(Z \geq \frac{\bar{X} - 580}{\sigma/\sqrt{n}}\right).$$

If $\bar{X} = 590$, $\sigma = 60$, and $n = 20$, then the significance level is

$$\begin{aligned} p &= P\left(Z \geq \frac{590 - 580}{60/\sqrt{20}}\right) \\ &= P(Z \geq 0.745) \\ &= .228 \end{aligned}$$

This means that when $\bar{X} = 590$, $Z = 0.745$ and that if you reject when $Z \geq 0.745$, the probability of a Type I error is .228.

5.1.2 A Two-Sided Test: Testing for Exact Equality

One other variation of hypothesis testing needs to be described: testing the hypothesis that the mean is exactly equal to some specified value. Returning to the example regarding open-mindedness, suppose it is claimed that the average score of all adult men is exactly 50, as opposed to being greater than or equal to 50. Then the null hypothesis is

$$H_0 : \mu = 50.$$

If the sample mean is exactly equal to 50, you would not reject, because this is consistent with H_0. If $\bar{X} > 50$, then the larger the sample mean happens to be, the more doubt there is that $\mu = 50$. Similarly, if $\bar{X} < 50$, then the smaller the sample mean,

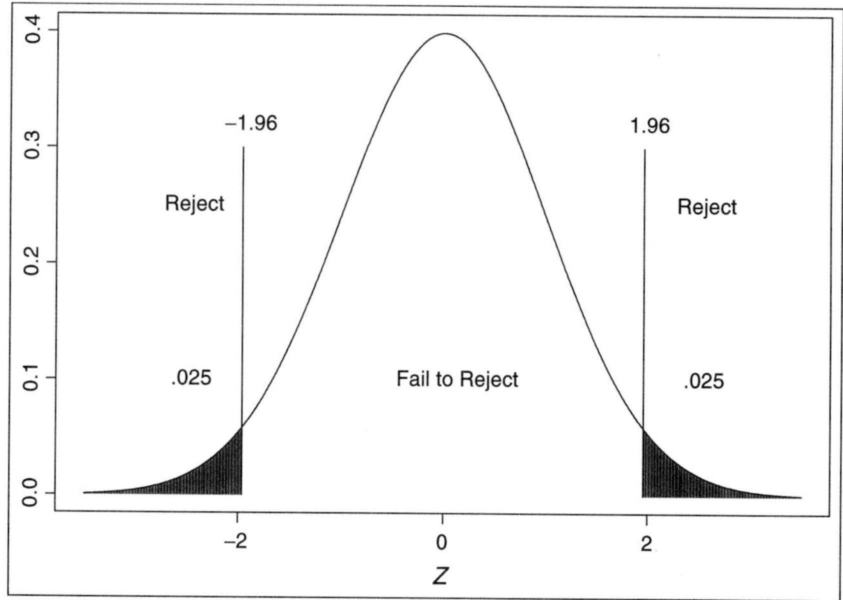

FIGURE 5.2 Critical region for a two-sided test such that P(Type I error) = .05.

the more doubt there is that $\mu = 50$. That is, now it is reasonable to reject H_0 if $\bar{X}$ is either too large or too small. An equivalent way of saying this is that you should reject if

$$Z = \frac{\bar{X} - 50}{\sigma/\sqrt{n}}$$

is too large or too small.

Suppose you reject H_0 if either $Z \leq -1.96$ or $Z \geq 1.96$. A more succinct way of describing this decision rule is that you reject if the absolute value of Z is greater than or equal to 1.96. In symbols, reject if $|Z| \geq 1.96$. If the null hypothesis is true and sampling is from a normal distribution, then Z has a standard normal distribution, so the probability of rejecting is

$$P(Z \leq -1.96) + P(Z \geq 1.96) = .025 + .025 = .05,$$

which is the total area of the two shaded regions in Figure 5.2.

EXAMPLE. Imagine a list of 55 minor malformations babies might have at birth. For illustrative purposes, it is assumed that the average number of malformations is 15 and the population standard deviation is $\sigma = 6$. For babies born to diabetic women, is the average number different from 15? That is, can you reject the hypothesis $H_0 : \mu = 15$? To find out, you sample $n = 16$ babies having diabetic mothers, count the number of malformations for each, and find

Continued

EXAMPLE. (*Continued*) that the average number of malformations is $\bar{X} = 19$. Then

$$Z = \frac{19 - 15}{6/\sqrt{16}} = 2.67.$$

If the goal is to have the probability of a Type I error equal to .05, then the critical values are -1.96 and 1.96, for the reasons just given. Because 2.67 is greater than 1.96, you reject the null hypothesis and conclude that the average number of malformations is greater than 15. ■

The significance level, or p-value, can be determined when testing for exact equality, but you must take into account that the critical region consists of both tails of the standard normal distribution.

EXAMPLE. Continuing the last example, where $Z = 2.67$, if you had decided to reject the null hypothesis if $Z \leq -2.67$ or if $Z \geq 2.67$, then the probability of a Type I error is

$$P(Z \leq -2.67) + P(Z \geq 2.67) = .0038 + .0038 = .0076.$$

This means that the significance level is 0.0076. ■

5.1.3 Criticisms of Two-Sided Hypothesis Testing and *p*-Values

Testing for exact equality has met with some criticism, on the grounds that exact equality is impossible. If, for example, one tests $H_0: \mu = 50$, the argument is that surely μ differs from 50 at some decimal place, meaning that the null hypothesis is false and will be rejected with a sufficiently large sample size. A related criticism is that because H_0 is surely false, the p-value (or significance level) is meaningless.

Assuming that these criticisms have merit, there are at least two ways one might address them. One is to reformulate the goal (cf. Shaffer, 1974). Rather than test $H_0: \mu = 50$, for example, suppose the goal is to determine whether μ is less than or greater than 50. Further assume that when addressing this goal we make one of three decisions: (1) if $Z \leq -1.96$, then decide $\mu < 50$, (2) if $Z \geq 1.96$, then decide $\mu > 50$, and (3) if $-1.96 < Z < 1.96$, then make no decision about whether μ is less than or greater than 50. Then there is some probability of making an incorrect decision, and, still assuming normality, the maximum probability of deciding $\mu < 50$ when in fact $\mu \geq 50$ is .025. Similarly, the maximum probability of deciding $\mu > 50$ when in fact $\mu \leq 50$ is .025. So in this context, α, or the p-value, tells us something about how certain we can be that μ is less than or greater than 50. However, the p-value tells us nothing about the degree to which μ differs from the hypothesized value. (This last issue is discussed in more detail in Chapter 8.)

A second approach is to rely exclusively on a confidence interval for μ. A confidence interval not only tells us whether we should reject H_0 and conclude that μ is less than or greater than some hypothesized value, it also provides information about the degree to which the population mean differs from the hypothesized value. For example, if the .95 confidence interval is $(12, 19)$, we would reject $H_0 : \mu = 10$ because this interval does not contain 10. In general, if we compute a $1 - \alpha$ confidence interval for μ, and if we reject whenever this confidence interval does not contain the hypothesized value for μ, the probability of a Type I error is α. In the example, the confidence interval tells us more. In addition to rejecting H_0, we can be reasonably certain that the population mean exceeds 10 (the hypothesized value) by at least 2.

5.1.4 Summary and Generalization

The basics of hypothesis testing, assuming normality and that σ is known, can be summarized in the following manner. Let μ_0 (read "mu naught") be some specified constant. The goal is to make some inference about how the population mean, μ, compares to μ_0. For the hypothesis $H_0 : \mu \leq 50$, $\mu_0 = 50$. For $H_0 : \mu = 15$, $\mu_0 = 15$; and for $H_0 : \mu \geq 580$, $\mu_0 = 580$. Furthermore, you want the probability of a Type I error to be α. Once the sample mean has been determined, compute

$$Z = \frac{\bar{X} - \mu_0}{\sigma/\sqrt{n}}.$$

Case 1. $H_0 : \mu \geq \mu_0$. Reject H_0 if $Z \leq c$, the α quantile of a standard normal distribution.

Case 2. $H_0 : \mu \leq \mu_0$. Reject H_0 if $Z \geq c$, the $1 - \alpha$ quantile of a standard normal distribution.

Case 3. $H_0 : \mu = \mu_0$. Reject H_0 if $Z \geq c$ or if $Z \leq -c$, where now c is the $1 - \alpha/2$ quantile of a standard normal distribution. Equivalently, reject if $|Z| \geq c$.

The hypotheses $H_0 : \mu \geq \mu_0$ and $H_0 : \mu \leq \mu_0$ are called *one-sided hypotheses*. In contrast, $H_0 : \mu = \mu_0$ is called a *two-sided hypothesis*.

5.1.5 A Property of *p*-Values

p-Values have a property that is rarely discussed in applied statistics books, but it might provide some sense of why certain modern techniques (described in subsequent chapters) are reasonable. Imagine that a the null hypothesis is true and that an experiment is repeated infinitely many times. Here it is assumed that the probability of a Type I error can be controlled exactly for any α we might pick. This will be true, for example, when using Z as described in Section 5.1.4 and sampling is from a normal distribution. Further imagine that each time the experiment is performed, the *p*-value (or significance level) is computed. If the infinitely many *p*-values were plotted, we would get the uniform distribution (described in Exercise 10 of Chapter 2). That is, all *p*-values are equally likely and are centered around .5 (e.g., Sackrowitz & Samuel-Cahn, 1999).

In particular, the probability of getting a p-value less than or equal to .025 or greater than or equal to .975 is exactly .05. More generally, when sampling from a nonnormal distribution, if a method for testing some hypothesis controls the probability of a Type I error for a sufficiently large sample size (meaning that the central limit theorem applies), then the distribution of p converges to a uniform distribution as the sample size increases.

5.2 Power and Type II Errors

After years of production, a manufacturer of batteries for automobiles finds that on average, their batteries last 42.3 months with a standard deviation of $\sigma = 4$. A new manufacturing process is being contemplated and one goal is to determine whether the batteries have a longer life on average. Ten batteries are produced by the new method and their average life is found to be 43.4 months. For illustrative purposes, assume that the standard deviation is again $\sigma = 4$. Based on these $n = 10$ test batteries, it is estimated that the average life of all the batteries produced using the new manufacturing method is greater than 42.3 (the average associated with the standard manufacturing method), in which case the new manufacturing process has practical value. To add support to this speculation, it is decided to test $H_0: \mu \leq 42.3$ versus $H_1: \mu > 42.3$, where μ is the population mean using the new method.

The idea is to determine whether $\bar{X}$ is sufficiently larger than 42.3 to rule out the possibility that $\mu \leq 42.3$. That is, the goal is to determine whether the new method is no better and possibly worse on average. If H_0 is rejected, there is empirical evidence that the new method should be adopted. As explained in the previous section, you test this hypothesis by computing $Z = (43.4 - 42.3)/(4/\sqrt{10}) = .87$. If you want the probability of a Type I error to be $\alpha = .01$, the critical value is 2.33, because $P(Z \leq 2.33) = .99$. In the present context, a Type I error is concluding that the new method is better on average when in reality it is not. Because .87 is less than 2.33, you fail to reject. Does this imply that you should accept the alternative hypothesis that μ is less than 42.3? In other words, should you conclude that the average battery lasts less than 42.3 months under the new manufacturing method?

Suppose that if the null hypothesis is not rejected, you conclude that the null hypothesis is true and that the population mean is less than 42.3. Then there are four possible outcomes, which are summarized in Table 5.1. The first possible outcome is

TABLE 5.1 Four Possible Outcomes When Testing Hypotheses

	Reality	
Decision	H_0 **true**	H_0 **false**
H_0 true	Correct decision	Type II error (probability β)
H_0 false	Type I error (probability α)	Correct decision (power)

that the null hypothesis is true and you correctly decide not to reject. The second possible outcome is that the null hypothesis is false but you fail to reject and therefore make a mistake. That is, your decision that $\mu \leq 42.3$ is incorrect—in reality the mean is greater than 42.3. The third possible outcome is that the null hypothesis is true but you make a mistake and reject. This is a Type I error, already discussed in Section 5.1. The fourth possible outcome is that in reality $\mu > 42.3$ and you correctly detect this by rejecting H_0.

This section is concerned with the error depicted by the upper right portion of Table 5.1. That is, the null hypothesis is false but you failed to reject. If, for example, the actual average life of a battery under the new manufacturing method is $\mu = 44$, the correct conclusion is that $\mu > 42.3$. The practical problem is that even if in reality $\mu = 44$, by chance you might get $\bar{X} = 41$, suggesting that the hypothesis $H_0: \mu \leq 42.3$ should be accepted. And even if $\bar{X} > 42.3$, it might be that the sample mean is not large enough to reject even though in reality H_0 is false. Failing to reject when you should reject is called a *Type II error*.

DEFINITION. A *Type II error* is failing to reject a null hypothesis when it should be rejected. The probability of a Type II error is often labeled β.

DEFINITION. *Power* is the probability of rejecting H_0 when in fact it is false. In symbols, power is $1 - \beta$, which is 1 minus the probability of a Type II error. In the illustration, if the new manufacturing method is actually better, meaning that μ is greater than 42.3, and the probability of rejecting $H_0: \mu \leq 42.3$ is .8, say, this means that power is $1 - \beta = .8$, and the probability of a Type II error is $\beta = .2$.

Power and the probability of making a Type II error are of great practical concern. In the illustration, if $\mu = 44$, the manufacturer has found a better manufacturing method, and clearly it is in their interest to discover this. What is needed is a method for ascertaining power, meaning the probability of correctly determining that the new method is better when in fact $\mu > 42.3$. If power is high but the company fails to detect an improvement over the standard method of production, the new method can be discarded. That is, there is empirical evidence that H_0 is true and the new method has no practical value. However, if power is low, meaning that there is a low probability of discovering that the new method produces longer-lasting batteries even when the new method is in fact better, then simply failing to reject does not provide convincing empirical evidence that H_0 is true.

In the present context, power depends on four quantities: σ, α, n, and the value of $\mu - \mu_0$, where μ is the unknown mean of the new manufacturing method. Although μ is not known, you can address power by considering values of μ that are judged to be interesting and important in a given situation. In the illustration, suppose you want to adopt the new manufacturing method if $\mu = 44$. That is, the average life of a battery using the standard method is 42.3 and you want to be reasonably certain of adopting the new method if the average life is now 44. In the more formal terminology of hypothesis testing, you want to test $H_0: \mu \leq 42.3$, and if $\mu = 44$, you want power to be reasonably close to 1. What is needed is a convenient way of assessing power given α, σ, μ, μ_0, and the sample size, n, you plan to use.

BOX 5.1 How to Compute Power, σ Known

Goal

Assuming normality, compute power when testing $H_0: \mu < \mu_0$ or $H_0: \mu > \mu_0$ or $H_0: \mu = \mu_0$ given:

1. n, the sample size
2. σ, the standard deviation
3. α, the probability of a Type I error
4. some specified value for μ
5. μ_0, the hypothesized value.

Case 1

$H_0: \mu < \mu_0$. Determine the critical value c as described in Section 5.1. (The critical value is the $1-\alpha$ quantile of a standard normal distribution.) Then power, the probability of rejecting the null hypothesis, is

$$1-\beta = P\left(Z \ge c - \frac{\sqrt{n}(\mu-\mu_0)}{\sigma}\right).$$

In words, power is equal to the probability that a standard normal random variable is greater than or equal to

$$c - \frac{\sqrt{n}(\mu-\mu_0)}{\sigma}.$$

Case 2

$H_0: \mu > \mu_0$. Determine the critical value c, which is now the α quantile of a standard normal distribution. Then power is

$$1-\beta = P\left(Z \le c - \frac{\sqrt{n}(\mu-\mu_0)}{\sigma}\right).$$

Case 3

$H_0: \mu = \mu_0$. Now c is the $1-\alpha/2$ quantile of a standard normal distribution. Power is

$$1-\beta = P\left(Z \le -c - \frac{\sqrt{n}(\mu-\mu_0)}{\sigma}\right) + P\left(Z \ge c - \frac{\sqrt{n}(\mu-\mu_0)}{\sigma}\right).$$

Box 5.1 summarizes how to compute power given n, σ, μ, μ_0, and α. Continuing the illustration where $H_0: \mu \le 42.3$, suppose $\alpha = .05$, $n = 10$, and you want to determine how much power there is when $\mu = 44$. Because $\alpha = .05$, the critical value is $c = 1.645$. Referring to Box 5.1, power is

$$1 - \beta = P\left(Z \geq c - \frac{\sqrt{n}(\mu - \mu_0)}{\sigma}\right)$$

$$= P\left(Z \geq 1.645 - \frac{\sqrt{10}(44 - 42.3)}{4}\right)$$

$$= P(Z \geq .30)$$

$$= .38.$$

This says that if battery life has a normal distribution and, unknown to us, the actual average life of a battery under the new manufacturing method is $\mu = 44$, then the probability of rejecting the hypothesis that the mean is less than 42.3 is .38. That is, for this situation where we should reject and conclude that the new manufacturing method is better on average, there is a .38 probability of making the correct decision that the null hypothesis is false. Consequently, the probability of committing a Type II error and failing to reject even though the null hypothesis is false is $1 - .38 = .62$.

Figure 5.3 graphically illustrates power when testing $H_0 : \mu \leq 42.3$ with $\alpha = .05$ and $\mu = 46$. It can be seen that power is $1 - \beta = .9$, so the probability of a Type II error is $\beta = .1$. The left normal distribution is the distribution of Z when the null hypothesis is true; it is standard normal and you reject if $Z \geq 1.645$, as already discussed. When the null hypothesis is false and in fact $\mu = 46$, Z still has a normal distribution, but its mean is no longer zero—it is larger, as indicated by Figure 5.3. That is, the right distribution reflects the actual distribution of Z when $\mu = 46$. Power is the area under the right (nonnull) curve and to the right of the critical value ($c = 1.645$). The area of the shaded region represents the probability of a Type II error, which is .1.

Notice that we do not know the actual value of μ, the average life of batteries manufactured with the new method. To deal with this issue, we must ask ourselves a series of questions: What if $\mu = 44$ or 45 or 46, and so on? By computing power for each of these situations, we get some idea about the probability of rejecting when in fact the null hypothesis is false. Figure 5.4 graphs power as μ increases. Notice that the more the mean μ exceeds the hypothesized value of 42.3, the higher the power. This is, of course, a property we want. The larger the difference between the hypothesized value and the actual value of μ, the more likely we are to reject and correctly conclude that μ is greater than 42.3.

5.2.1 Understanding How n, α, and σ Are Related to Power

Power is a function of three fundamental components of any study: the sample size, n, the Type I error probability you pick, α, and the population standard deviation, σ. As already explained, power plays a crucial role in applied work, so it is important to understand how each of these quantities is related to power.

First consider how the sample size, n, affects power. If the null hypothesis is false, we want the probability of rejecting to go up as the sample size increases. That is,

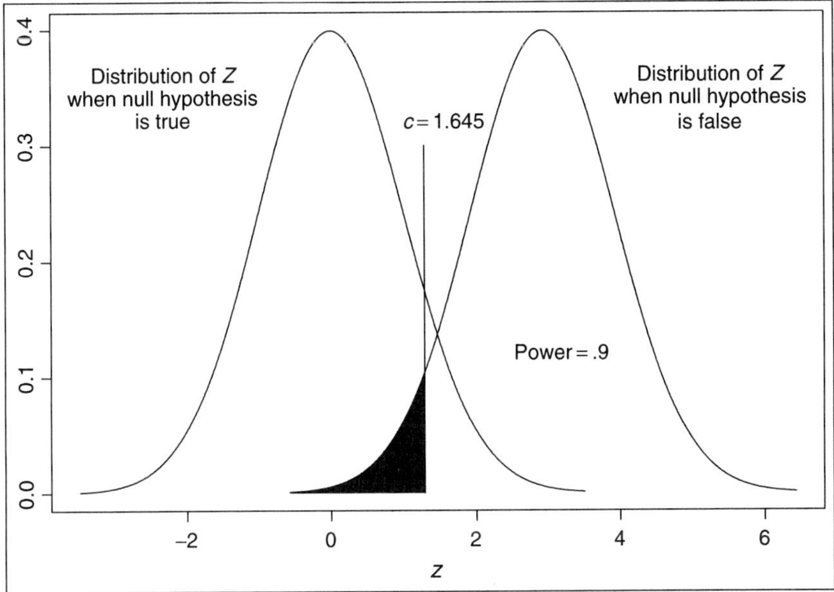

FIGURE 5.3 Illustration of power. The left distribution is the distribution of Z when the null hypothesis is true. The right distribution is the distribution of Z when the null hypothesis is false. The area of the shaded region, which is the area under the right (nonnull) distribution and to the left of 1.645 is .1; it is equal to the probability of a Type II error. So power is the area under the nonnull distribution and to the right of 1.645.

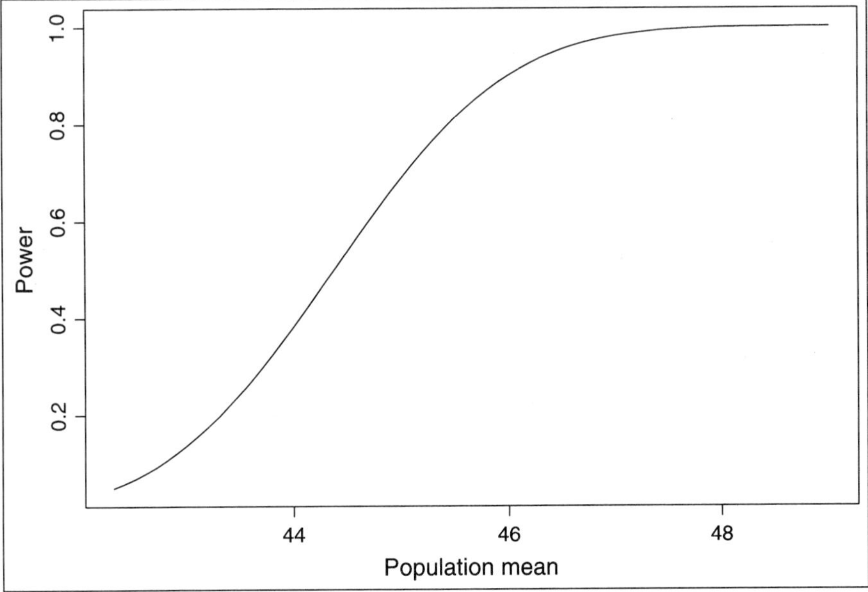

FIGURE 5.4 Power curve for Z when the hypothesized value for the population mean (μ_0) is 42.3. As μ increases, meaning that we are moving away from the hypothesized value for μ, power increases.

as the sample size, n, gets large, we should have an increasingly higher probability of making a correct decision about H_0 when it is false. Examining the expressions for power in Box 5.1 reveals that is exactly what happens.

EXAMPLE. Consider the battery example once again, where we want to test $H_0: \mu \leq 42.3$, and suppose we want to know how much the power is when $\mu = 44$, but this time we consider three sample sizes: 10, 20, and 30. For $n = 10$ we have already seen that power is .38. If $n = 20$, then power is

$$
\begin{aligned}
1 - \beta &= P\left(Z \geq c - \frac{\sqrt{n}(\mu - \mu_0)}{\sigma}\right) \\
&= P\left(Z \geq 1.645 - \frac{\sqrt{20}(44 - 42.3)}{4}\right) \\
&= P(Z \geq -0.256) \\
&= .60
\end{aligned}
$$

Increasing n to 30, it can be seen that power is now $1 - \beta = .75$, meaning that your probability of making a correct decision and rejecting H_0 when it is false is now .75. ■

This example illustrates how the sample size might be determined in applied work. First determine the difference between μ and μ_0 that is important in a given situation. In the battery illustration, it might be decided that if $\mu - \mu_0 = 44 - 42.3 = 1.7$, we want to be reasonably certain of rejecting the null hypothesis and deciding that the new manufacturing method is better on average. Next, compute power for some value of n. If the power is judged to be sufficiently large, use this sample size in your study. If not, increase the sample size.

Your choice for α, the probability of a Type I error you are willing to allow, also affects power.

EXAMPLE. For the battery example with $n = 30$, consider three choices for α: .05, .025, and .01. For $\alpha = .05$ we have already seen that power is .75 when testing $H_0: \mu < 42.3$ and $\mu = 44$. For $\alpha = .025$, the critical value is now $c = 1.96$, so power is

$$
\begin{aligned}
1 - \beta &= P\left(Z \geq c - \frac{\sqrt{n}(\mu - \mu_0)}{\sigma}\right) \\
&= P\left(Z \geq 1.96 - \frac{\sqrt{20}(44 - 42.3)}{4}\right) \\
&= P(Z \geq .059) \\
&= .47.
\end{aligned}
$$

Continued

EXAMPLE. (*Continued*) If instead you use $\alpha = .01$, the critical value is now $c = 2.33$ and power can be seen to be .33. This illustrates that if we adjust the critical value so that the probability of a Type I error goes down, power goes down as well. Put another way, the more careful you are not to commit a Type I error by choosing α close to zero, the more likely you are to commit a Type II error if the null hypothesis happens to be false. ■

Finally, the population standard deviation, σ, is also related to power. The larger σ happens to be, the lower the power will be given α, n, and a value for $\mu - \mu_0$.

EXAMPLE. For the battery example, consider $\alpha = .05$, $\mu - \mu_0 = 1.7$, and $n = 30$, with $\sigma = 4$. Then power is .75, as previously explained. But if $\sigma = 8$, power is now .31. If $\sigma = 12$, power is only .19. ■

The results just described on how n, α, and σ are related to power can be summarized as follows:

- As the sample size, n, gets large, power goes up, so the probability of a Type II error goes down.
- As α goes down, in which case the probability of a Type I error goes down, power goes down and the probability of a Type II error goes up.
- As the standard deviation, σ, goes up, with n, α, and $\mu - \mu_0$ fixed, power goes down.

Notice that once you have chosen an outcome variable of interest (X) and the population of individuals you want to study, there are two types of factors that affect power. The first type consists of factors that are under your control: n, the sample size, and α, the probability of a Type I error you are willing to allow. By increasing n or α, you increase power. The population standard deviation also affects power, but it is not under your control, it merely reflects a state of nature. (In some situations the variance of X can be influenced based on how an outcome variable is designed or constructed.) However, understanding how σ affects power is important in applied work, because it plays a role in choosing an accurate hypothesis-testing method, as will be seen.

5.3 Testing Hypotheses About the Mean When σ Is Not Known

Next we describe the classic method for testing hypotheses about the population mean when the population standard deviation is not known. Then we describe recent insights into why this technique has several practical problems.

When σ is known, we can test hypotheses about the population mean if we can determine the distribution of

$$Z = \frac{\bar{X} - \mu_0}{\sigma/\sqrt{n}}.$$

When σ is not known, we estimate σ with s, the sample standard deviation, and we can test hypotheses if the distribution of

$$T = \frac{\bar{X} - \mu_0}{s/\sqrt{n}} \tag{5.2}$$

can be determined. As indicated in Chapter 4, the distribution of T can be determined when sampling from a normal distribution. For this special case, hypotheses can be tested as described in Section 5.1.4, except that critical values are read from Table 4 in Appendix B with the degrees of freedom set to $\nu = n - 1$. The details can be summarized as follows.

GOAL: Test hypotheses regarding how the population mean, μ, compares to a specified constant, μ_0. The probability of a Type I error is to be α.

ASSUMPTIONS: Random sampling and normality.

DECISION RULES:

- For $H_0 : \mu \geq \mu_0$, reject if $T \leq c$, where c is the α quantile of Student's T distribution with $\nu = n - 1$ degrees of freedom and T is given by Equation (5.2).
- For $H_0 : \mu \leq \mu_0$, reject if $T \geq c$, where c is now the $1 - \alpha$ quantile of Student's T distribution with $\nu = n - 1$ degrees of freedom.
- For $H_0 : \mu = \mu_0$, reject if $T \geq c$ or $T \leq -c$, where c is now the $1 - \alpha/2$ quantile of Student's T distribution with $\nu = n - 1$ degrees of freedom. Equivalently, reject if $|T| \geq c$.

EXAMPLE. For the measures of open-mindedness given at the beginning of this chapter, test the hypothesis $H_0 : \mu \geq 50$ with $\alpha = .05$. The sample standard deviation is $s = 11.4$ and the sample mean is $\bar{X} = 44.5$. Because $n = 10$, the degrees of freedom are $\nu = n - 1 = 9$ and

$$T = \frac{\bar{X} - \mu_0}{s/\sqrt{n}} = \frac{44.5 - 50}{11.4/\sqrt{10}} = -1.5.$$

Referring to Table 4 in Appendix B, $P(T \leq -1.83) = .05$, so the critical value is -1.83. This means that if we reject when T is less than or equal to -1.83, the probability of a Type I error will be .05, assuming normality. Because the observed value of T is -1.5, which is greater than the critical value, you fail to reject. In other words, the sample mean is not sufficiently smaller than 50 to be reasonably certain that the speculation $\mu \geq 50$ is false. ■

As you can see, the steps you follow when σ is not known mirror the steps you use to test hypotheses when σ is known.

EXAMPLE. Suppose you observe the values

12, 20, 34, 45, 34, 36, 37, 50, 11, 32, 29

and the goal is to test $H_0: \mu = 25$ such that the probability of a Type I error is $\alpha = .05$. Here, $n = 11$, $\mu_0 = 25$, and it can be seen that $\bar{X} = 33.24$, $s/\sqrt{11} = 3.7$, so

$$T = \frac{\bar{X} - \mu}{s/\sqrt{n}} = \frac{33.24 - 25}{3.7} = 2.23.$$

The null hypothesis is that the population mean is exactly equal to 25. So the critical value is the $1 - \alpha/2 = .975$ quantile of Student's T distribution with degrees of freedom $\nu = 11 - 1 = 10$. Table 4 in Appendix B indicates that

$$P(T \leq 2.28) = .975,$$

so our decision rule is to reject H_0 if the value of T is greater than or equal to 2.28 or less than or equal to −2.28. Because the absolute value of T is less than 2.28, you fail to reject. ■

5.4 Controlling Power and Determining *n*

Problems of fundamental importance are determining what sample size to use and finding methods that ensure power will be reasonably close to 1. Two approaches are described in this section, both of which assume random sampling from a normal distribution. The first is based on choosing n prior to collecting any data. The second is used after data are available and is aimed at determining whether n was sufficiently large to ensure that power is reasonably high. One fundamental difference between the two methods is how they measure the extent to which the null hypothesis is false.

5.4.1 Choosing *n* Prior to Collecting Data

First consider how one might choose n prior to collecting data so that power is reasonably close to 1. To begin, we need a measure of the difference between the hypothesized value for the mean (μ_0) and its true value (μ). One possibility is

$$\delta = \mu - \mu_0, \tag{5.3}$$

which is consistent with how we discussed power in Section 5.2 when σ is known. However, when using T, it is impossible to control power given some value for δ without first obtaining data, because when using T, power depends on the unknown variance (Dantzig, 1940). The standard method for dealing with this problem is to replace δ with

$$\Delta = \frac{\mu - \mu_0}{\sigma}. \tag{5.4}$$

So if $\Delta = 1$, for example, the difference between the means is one standard deviation. That is, $\mu - \mu_0 = \sigma$. If $\Delta = .5$, the difference between the mean is half a standard deviation. (That is, $\mu - \mu_0 = .5\sigma$.) We saw in Chapter 2 that for normal distributions, σ has a convenient probabilistic interpretation, but Section 2.7 illustrated that for even a small departure from normality, this interpretation breaks down. This causes practical problems when using Δ, but we temporarily ignore this issue. (These practical problems are discussed in detail in Chapter 8.) Here it is merely remarked that under normality, power can be determined for any choice of n, Δ, and α. Rather than describe the details, we merely provide an S-PLUS function that performs the computations.

5.4.2 S-PLUS Function pow1

The S-PLUS function

```
pow1(n,Del,alpha)
```

(written for this book) computes power when performing a one-sided test, where the argument Del is Δ and alpha is α. For example, if you want to determine how much power you have when testing $H_0: \mu \geq 15$ with $n = 10$, $\Delta = -.3$, and $\alpha = .05$, the S-PLUS command pow1(10, −.3, .05) returns the value .219. Increasing n to 30, power is now .479. With $n = 100$, power is .9. So in this particular case, $n = 10$ is inadequate if $\Delta = -.3$ is judged to be a difference that is important to detect. To ensure high power requires a sample size of around 100. In a similar manner, if the goal is to test $H_0: \mu \leq 15$ and now $\Delta = .3$, then pow1(10, .3, .05) again returns the value .219. (The function assumes that if Δ is positive, you are testing $H_0: \mu \leq \mu_0$, and that if Δ is negative, you are testing $H_0: \mu \geq \mu_0$.)

This S-PLUS function can handle two-sided tests in a simple manner. You simply divide α by 2. In the previous illustration, if instead you want to test $H_0: \mu = 15$ at the .05 level, the command pow1(30, .3, .025) returns the value .35, indicating that power is .35. (The same result is returned if the argument Del is −.3.) If this amount of power is judged to be too small, simply increase n until a more satisfactory power level is obtained.

5.4.3 Stein's Method

Assuming normality, Stein (1945) derived a method that indicates whether the sample size n is sufficiently large to achieve some specified amount of power. In contrast to the method in Section 5.4.1, it is used after data are collected and it is based on $\delta = \mu - \mu_0$ rather than Δ. Said another way, if you fail to reject some null hypothesis, Stein's method helps you decide whether this is because power is low due to too small an n. Stein's method does even more; it indicates how many additional observations are needed to achieve some specified amount of power.

For convenience, assume that a one-sided test is to be performed. Also assume that n observations have been randomly sampled from some normal distribution, yielding a sample variance s^2. If the goal is to ensure that power is at least $1-\beta$, then compute

$$d = \left(\frac{\delta}{t_{1-\beta} - t_\alpha}\right)^2,$$

where $t_{1-\beta}$ and t_α are, respectively, the $1-\beta$ and α quantiles of Student's T distribution with $\nu = n-1$ degrees of freedom. For example, if $n = 10$ and you want power to be $1-\beta = .9$, then $t_{1-\beta} = t_{.9} = 1.383$, which can be read from Table 4 in Appendix B. Then the number of required observations is

$$N = \max\left(n, \left[\frac{s^2}{d}\right] + 1\right), \tag{5.5}$$

where the notation $[s^2/d]$ means you compute s^2/d and round down to the nearest integer and max refers to the larger of the two numbers inside the parentheses. Continuing the example where $n = 10$ and $1-\beta = .9$, if $s = 21.4$, $\delta = 20$, and $\alpha = .01$, then $\nu = 9$,

$$d = \left(\frac{20}{1.383 - (-2.82)}\right)^2 = 22.6,$$

so

$$N = \max\left(10, \left[\frac{(21.4)^2}{22.6}\right] + 1\right) = \max(10, 21) = 21.$$

If $N = n$, the sample size is adequate, but in the illustration $N - n = 21 - 10 = 11$. That is, 11 additional observations are needed to achieve the desired amount of power.

A two-sided test ($H_0: \mu = \mu_0$) is handled in a similar manner. The only difference is that α is replaced by $\alpha/2$. So if in the last example we wanted the Type I error probability to be .02 when testing a two-sided test, then again $N = 21$. If we want the Type I error probability to be .05, then $t_{\alpha/2} = t_{.025} = -2.26$, so if again we want power to be .9 when $\delta = \mu - \mu_0 = 20$,

$$d = \left(\frac{20}{1.383 - (-2.26)}\right)^2 = 30.14,$$

so we need a total of

$$N = \max\left(10, \left[\frac{(21.4)^2}{30.14}\right] + 1\right) = 16$$

observations.

Stein also indicated how to test the null hypothesis if the additional $(N - n)$ observations can be obtained. But rather than simply perform Student's T on all N values, Stein used instead

$$T_s = \frac{\sqrt{n}(\hat{\mu} - \mu_0)}{s}, \tag{5.6}$$

where $\hat{\mu}$ is the mean of all N observations. You test hypotheses by treating T_s as having a Student's T distribution with $\nu = n - 1$ degrees of freedom. That is, you test hypotheses as described in Section 5.3 but with T replaced by T_s. For example, you reject $H_0 : \mu \leq \mu_0$ if $T_s \geq c$, where c is the $1 - \alpha$ quantile of Student's T distribution with $\nu = n - 1$ degrees of freedom. A two-sided confidence interval for the population mean is given by

$$\hat{\mu} \pm c\frac{s}{\sqrt{n}}, \tag{5.7}$$

where c is now the $1 - \alpha/2$ quantile of Student's T distribution with $n - 1$ degrees of freedom. For the special case where $N = n$ (meaning that the original sample size was sufficient for your power needs), $T_s = T$ and you are simply using Student's T test, but with the added knowledge that the sample size meets your power requirements. What is unusual about Stein's method is that if $N > n$, it uses the sample variance of the original n observations—not the sample variance of all N observations. Also, the degrees of freedom remain $n - 1$ rather than the seemingly more natural $N - 1$. By proceeding in this manner, Stein showed that power will be at least $1 - \beta$ for whatever value of $1 - \beta$ you pick. (Simply performing Student's T on all N values, when $N > n$, results in certain technical problems that are described by Stein, 1945, but not here. For a survey of related methods, see Hewett & Spurrier, 1983.)

A popular alternative to Stein's method when trying to assess power when a nonsignificant result is obtained is based on what it called *observed power*. The approach assumes that the observed difference between the sample mean and its hypothesized value is indeed equal to the true difference $(\mu - \mu_0)$; it also assumes that $s^2 = \sigma^2$, and then based on these assumptions one computes power. Hoenig and Heisey (2001) illustrate that this approach is generally unsatisfactory.

5.4.4 S-PLUS Functions stein1 and stein2

The S-PLUS function

```
stein1(x,del,alpha=.05,pow=.8,oneside=F)
```

returns N, the sample size needed to achieve power given by the argument pow (which defaults to .8), given some value for δ (which is the argument del) and α. The function assumes that a two-sided test is to be performed. For a one-sided test, set the argument oneside to T for true.

The S-PLUS function

stein2(x1,x2,mu0=0,alpha=.05)

tests the hypothesis $H_0: \mu = \mu_0$ using Stein's method, assuming that the initial n observations are stored in x1 and that the additional $N - n_1$ observations are stored in x2. The argument mu0 is the hypothesized value, μ_0, which defaults to 0.

EXAMPLE. The last example in Section 5.3 used Student's T to test $H_0: \mu = 25$ with $\alpha = .05$, based on the data

12, 20, 34, 45, 34, 36, 37, 50, 11, 32, 29.

A nonsignificant result was obtained. If we want power to be .9 when $\mu = 28$, in which case $\delta = 28 - 25 = 3$, was the sample size sufficiently large? Storing these data in the S-PLUS variable y, the command

stein1(y,3,pow=.9)

returns the value 220. That is, we need $N = 220$ observations to achieve this much power. Since we have only 11 observations, $220 - 11 = 209$ additional observations are needed. For a one-sided test, $N = 94$. ■

5.5 Practical Problems with Student's T

Student's T deals with the common situation where σ is not known, but it assumes observations are randomly sampled from a normal distribution. Because distributions are never exactly normal, it is important to understand how nonnormality affects conclusions based on T. A version of the central limit theorem tells us that as n gets large, the distribution of T becomes more like Student's T distribution, and in fact its distribution approaches a standard normal. That is, if the sample size is large enough and observations are randomly sampled, then violating the normality assumption is not a serious concern. Conventional wisdom is that assuming T has a Student's T distribution with $\nu = n - 1$ degrees of freedom provides reasonably accurate results with n fairly small, and surely accurate results are obtained with $n = 100$. However, in recent years, much more sophisticated methods have been derived for understanding how nonnormality affects T, and serious concerns have been discovered, two of which are described here.

The first is that *very* small departures from normality can drastically reduce power. The main reason is that even small departures from normality can inflate the population variance; this in turn inflates the standard error of the sample mean, so power can be relatively low. As indicated in Section 5.2.1, as σ gets large, power goes down when using Z to test hypotheses, and the same is true when using T. (One of the earliest results indicating theoretical concerns about this problem can be found in Bahadur & Savage, 1956.)

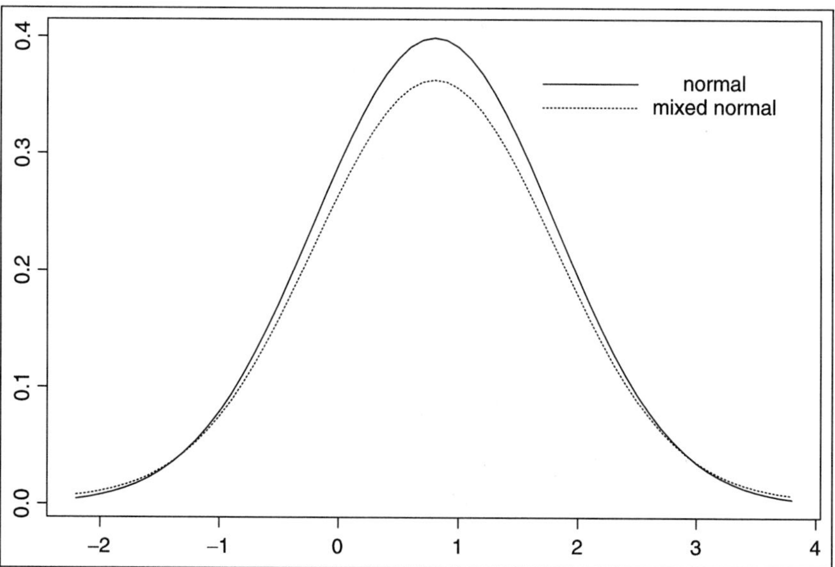

FIGURE 5.5 Small departures from normality can greatly reduce power when using Student's T. If we sample 20 observations from the normal distribution shown here, which has mean .8, and we test $H_0 : \mu = 0$ at the .05 level, power is .93. But if we sample from the mixed normal instead, power is only .39.

EXAMPLE. As an illustration, first consider using Student's T to test $H_0 : \mu = 0$ with $\alpha = .05$ when sampling from the normal distribution shown in Figure 5.5, which has mean .8 and standard deviation $\sigma = 1$. It can be shown that with $n = 20$, power is .93. That is, there is a 93% chance of correctly rejecting the null hypothesis that the mean is $\mu = 0$. Now suppose that sampling is from the other distribution shown in Figure 5.5. This is the mixed normal distribution described in Chapter 2, but with mean .8. This distribution is very similar to the normal distribution, in the sense described in Chapter 2, yet power is now .39. This demonstrates that if you test hypotheses with Student's T or any method based on the sample mean, small departures from normality can result in a substantial decrease in your ability to detect situations where the null hypothesis is false. ■

The second problem is that nonnormality can affect your ability to control the probability of a Type I error or control the probability coverage when computing a confidence interval. First consider situations where sampling is from a perfectly symmetric distribution and imagine you want the probability of a Type I error to be .05. When sampling is from a normal distribution, you can accomplish your goal with Student's T test, as already demonstrated. However, if you happen to be sampling observations from a mixed normal instead, the actual probability of a Type I error is only .022 with $n = 20$. This might seem desirable because the probability of incorrectly rejecting when the null hypothesis is true is less than the nominal level of .05. However, recall that the smaller α happens to be, the lower your power. This means that the probability of correctly rejecting the null hypothesis when it is false might be low, contrary to what you want, because you are inadvertently testing at the .022 level. Increasing the sample size to 100, now the actual probability of a Type I error is .042, but low power remains a possible concern because sampling is from a heavy-tailed distribution.

If observations are sampled from a skewed distribution, the actual probability of a Type I error can be substantially higher or lower than .05. With $n = 12$, there are situations where the actual probability of a Type I error is .42 when testing at the .05 level, and it can be as low as .001 when testing a one-sided test at the .025 level (e.g., Wilcox, 1997a, p. 74). As another illustration, imagine you are interested in how response times are affected by alcohol and that, *unknown to you*, response times have the skewed (lognormal) distribution shown in Figure 4.14, which has mean $\mu = 1.65$. Further imagine that you want to test the hypothesis $H_0 : \mu \geq 1.65$ with $\alpha = .05$. That is, unknown to you, the null hypothesis happens to be true, so you should not reject. With $n = 20$ observations, the actual probability of a Type I error is .14. That is, your intention was to have a 5% chance of rejecting in the event the null hypothesis is true, but in reality there is a 14% chance of rejecting by mistake. Increasing the sample size to $n = 160$, the actual probability of Type I error is now .11. That is, control over the probability of a Type I error improves as the sample size gets large, in accordance with the central limit theorem, but at a rather slow rate. Even with $n = 160$, the actual probability of rejecting might be more than twice as large as intended. The seriousness of a Type I error will depend on the situation, but at least in some circumstances the discrepancy just described would be deemed unsatisfactory.

The lognormal distribution used in the previous paragraph is relatively light-tailed. When sampling from a skewed, heavy-tailed distribution, Student's T can deteriorate even more. Consider the distribution in Figure 5.6, which has a mean of .0833. With $n = 20$ and $\alpha = .05$, the actual probability of a Type I error is .20. Increasing n to 100, the actual probability of a Type I error drops to only .19. It is getting closer to the nominal level, in accordance with the central limit theorem, but at a very slow rate. (For theoretical results indicating drastic sensitivity to nonnormality, see Basu & DasGupta, 1995.)

When you use Student's T test under the assumption that sampling is from a normal distribution, you are assuming that T has a symmetric distribution about zero. But as pointed out in Section 4.8, the actual distribution of T can be asymmetric with a mean that differs from zero. For example, when sampling from the (lognormal) distribution in Figure 4.14, with $n = 20$, the distribution of T is skewed, with a mean of -0.5, the result being that the probability of a Type I error is not equal to the value you want. In fact, Student's T test is *biased*, meaning that the probability of rejecting is not minimized when the null hypothesis is true. That is, situations arise where there is a higher probability of rejecting when the null hypothesis is true than in a situation where the null hypothesis is false. (Generally, any hypothesis-testing method is said to be unbiased if the probability of rejecting is minimized when the null hypothesis is true. Otherwise it is biased.)

To illustrate the possible effect of skewness on the power curve of Student's T, suppose we sample 20 observations from the (lognormal) distribution shown in

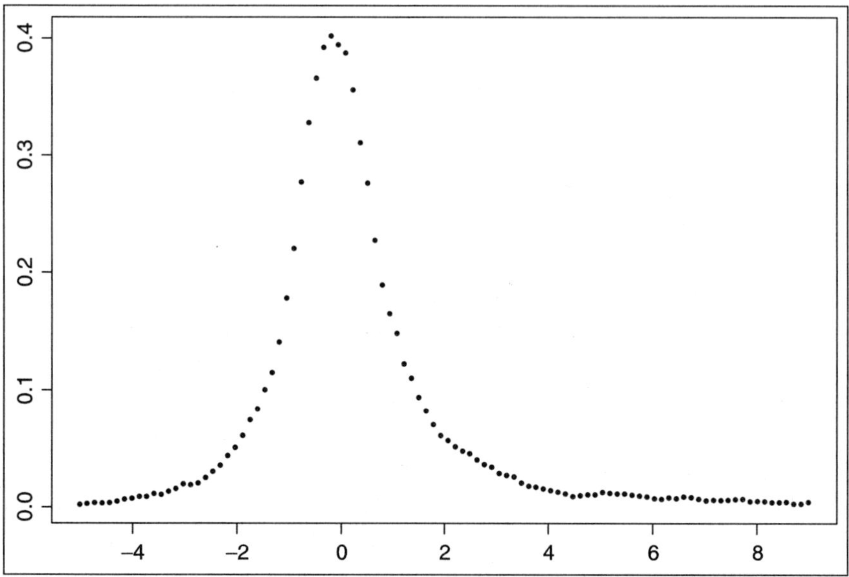

FIGURE 5.6 Example of a skewed, heavy-tailed distribution used to illustrate the effects of nonnormality on Student's T. This distribution has a mean of .8033.

Figure 4.14, which has a population mean approximately equal to 1.649. So if we test $H_0 : \mu = 1.649$ with $\alpha = .05$, the intention is to have the probability of a Type I error equal to .05, but in reality it is approximately .14. Now suppose we add δ to every observation. In effect, we shift the distribution in Figure 4.14 so that its mean is now $1.649 + \delta$. If, for example, we use $\delta = .3$, we have in effect increased the population mean from 1.649 to 1.949. So now we should reject $H_0 : \mu = 1.649$, but the actual probability of rejecting is .049. That is, we have described a situation where the null hypothesis is false, yet we are less likely to reject than in the situation where the null hypothesis is true. If we set $\delta = .6$ so that now $\mu = 2.249$ and we again test $H_0 : \mu = 1.649$, the probability of rejecting is .131, approximately the same probability of rejecting when the null hypothesis is true. As we increase δ even more, power continues to increase as well. Figure 5.7 shows the power curve of Student's T for δ ranging between 0 and 1.

Section 5.2.1 summarized factors that are related to power when using Z. Now we summarize factors that influence power for the more common situation where T is used to make inferences about the population mean. These features include the same features listed in Section 5.2.1 plus some additional features related to nonnormality.

- As the sample size, n, gets large, power goes up, so the probability of a Type II error goes down.
- As α goes down, in which case the probability of a Type I error goes down, power goes down and the probability of a Type II error goes up.
- As the standard deviation, σ, goes up, with n, α, and $\mu - \mu_0$ fixed, power goes down.

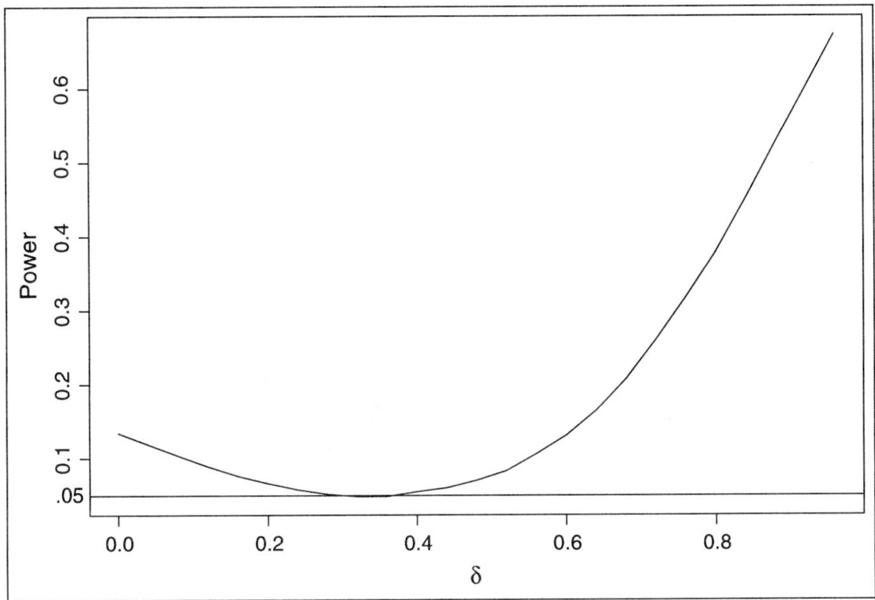

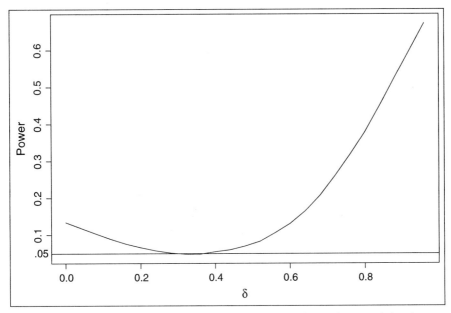

FIGURE 5.7 Power curve of Student's T when sampling from a lognormal distribution, $n = 20$, $\alpha = .05$. The null hypothesis corresponds to $\delta = 0$. Ideally the power curve should be strictly increasing as δ gets large.

- Small departures from normality can inflate the standard error of the sample mean ($\sigma/\sqrt{n}$), which in turn can substantially reduce power.
- Student's T can be biased due to skewness. That is, power might be low (relative to other inferential methods you might use) because as μ moves away from the hypothesized value, the probability of rejecting can actually decrease. Practical problems arise even when sampling from a distribution where outliers are rare.

5.6 Hypothesis Testing Based on a Trimmed Mean

An argument for testing hypotheses based on the mean is that under normality, the sample mean has a smaller standard error than any other measure of location we might use. This means that no other hypothesis-testing method will have more power than the method based on Student's T. However, this argument is not very compelling, because arbitrarily small departures from normality can result in extremely low power relative to other methods you might use, and of course there is the concern about getting accurate confidence intervals and good control over the probability of a Type I error. Currently, there seem to be two general strategies for dealing with these problems that are relatively effective. The first is to switch to a robust measure of location, and the trimmed mean is particularly appealing based on recently published studies. The second is to switch to a rank-based method, some of which are described in Chapter 15.

Here, attention is focused on the 20% trimmed mean. The method for computing a confidence interval for the trimmed mean described in Chapter 4 is easily extended to the problem of testing some hypothesis about μ_t, the population trimmed mean. The process is the same as Student's T test, only you adjust the degrees of freedom, you replace the sample mean with the trimmed mean, you replace the sample variance with the Winsorized variance, and for technical reasons you multiply by .6. In symbols, your test statistic is now

$$T_t = \frac{.6(\bar{X}_t - \mu_0)}{s_w/\sqrt{n}}, \tag{5.8}$$

where again μ_0 is some specified value of interest and s_w is the 20% Winsorized standard deviation. Then reject $H_0: \mu_t = \mu_0$ if $T_t \le -c$ or $T_t \ge c$, where c is now the $1-\alpha/2$ quantile of Student's T distribution with $n-2g-1$ degrees of freedom and g is the number of observations trimmed from each tail, as described and illustrated in Chapter 3. (The total number of trimmed observations is $2g$, so $n-2g$ is the number of observations left after trimming.)

More generally, when using a γ-trimmed mean,

$$T_t = \frac{(1-2\gamma)(\bar{X}_t - \mu_0)}{s_w/\sqrt{n}}, \tag{5.9}$$

where s_w is now the γ-Winsorized standard deviation. (In the previous paragraph, $\gamma = .2$.) The degrees of freedom are $\nu = n - 2g - 1$, where $g = [\gamma n]$ and $[\gamma n]$ is γn rounded down to the nearest integer.

As for the one-sided hypothesis $H_0: \mu \ge \mu_0$, reject if $T_t \le c$, where c is now the α quantile of Student's T distribution with $n-2g-1$ degrees of freedom. The hypothesis $H_0: \mu \le \mu_0$ is rejected if $T_t \ge c$, where c is now the $1-\alpha$ quantile of Student's T distribution with $n-2g-1$ degrees of freedom.

5.6.1 S-PLUS Function trimci

The S-PLUS function

trimci(x,tr = .2,alpha = .05,nv = 0),

introduced in Chapter 4, also tests hypotheses about the population trimmed mean. By default it tests $H_0: \mu_t = 0$. To test $H_0: \mu_t = 2$, set the argument nv equal to 2. In addition to a confidence interval, the function returns the p-value.

EXAMPLE. Doksum and Sievers (1976) report data on weight gain among rats. One group was the control and the other was exposed to an ozone environment. (The data are given in Table 8.6.) For illustrative purposes, attention is focused on the control group, and we consider the claim that the typical weight

Continued

EXAMPLE. (*Continued*) gain is 26.4. If we test the hypothesis $H_0: \mu = 26.4$ with Student's T, we get

$$T = \frac{\bar{X} - \mu_0}{s/\sqrt{n}} = \frac{22.4 - 26.4}{10.77/\sqrt{23}} = -1.8.$$

With $\nu = 23 - 1 = 22$ degrees of freedom, and $\alpha = .05$, the critical value is $c = 2.07$. Because $|T| = 1.8$ is less than 2.07, we fail to reject. In contrast, the 20% trimmed mean is $\bar{X}_t = 23.3$, $s_w = 3.9$, and for $H_0: \mu_t = 26.4$ we see that

$$T_t = \frac{.6(\bar{X}_t - \mu_0)}{s_w/\sqrt{n}} = \frac{.6(23.3 - 26.4)}{3.9/\sqrt{23}} = -2.3.$$

Because there are 23 rats, $g = 4$, so the number of trimmed observations is $2g = 8$, the degrees of freedom are $\nu = 23 - 8 - 1 = 14$, and the critical value is $c = 2.14$. Because $|T| = |-2.3| = 2.3$ is greater than the critical value, we reject the hypothesis that the trimmed mean is 26.4. Thus, although you cannot rule out the possibility that the population mean for all rats is 26.4 if our Type I error probability is to be .05, it is unlikely that the population trimmed mean has this value. ■

In the last example, the sample mean exceeds the hypothesized value by more than the sample trimmed mean. The difference between the hypothesized value of 26.4 and the mean is $22.4 - 26.4 = -4$. The difference between the hypothesized value and the trimmed mean is $23.3 - 26.4 = -3.1$, yet you reject with the trimmed mean but not with the mean. The reason is that the standard error of the trimmed mean is smaller than the standard error of the mean. This illustrates one of the practical advantages of using a trimmed mean. Situations often arise where the trimmed mean has a substantially smaller standard error, and this can translate into a substantial gain in power.

Once again it is stressed that for skewed distributions, population means and trimmed means are generally not equal. So, for example, the null hypothesis $H_0: \mu = 26.4$ is not necessarily the same as $H_0: \mu_t = 26.4$. In the context of hypothesis testing, an argument for the trimmed mean is that good control over the probability of a Type I error can be achieved in situations where Student's T gives poor results. Trimmed means often have a smaller standard error than the mean which can result in substantially higher power. If there is some reason for preferring the mean to the trimmed mean in a particular study, Student's T might be unsatisfactory unless the sample size is very large. Just how large n must be depends on the *unknown* distribution from which observations were sampled. In some cases even a sample size of 300 is unsatisfactory. A small sample size will suffice in some instances, but an effective method for establishing whether this is the case, simply by examining your data, has not been found.

In this book, only two-sided trimming is considered. If a distribution is skewed to the right, for example, a natural reaction is to trim large observations but not small ones. An explanation can now be given as to why one-sided trimming is not recommended. In terms of Type I errors and probability coverage, you get more accurate results if two-sided trimming is used. There is nothing obvious or intuitive about this result, but all of the studies cited by Wilcox (1997a) support this view.

Thanks to the central limit theorem, we know that when working with means, problems with Student's T diminish as the sample size increases. Theory and simulations indicate that when using a 20% trimmed mean instead, problems diminish much more rapidly. That is, smaller sample sizes are required to get good control over the probability of a Type I error; but with very small sample sizes, practical problems persist. (Methods covered in Chapter 7 provide a basis for dealing with very small sample sizes in an effective manner.)

Finally, no mention has been made about how to determine whether power is adequate based on the sample size used when making inferences about a trimmed mean. Theoretical results suggest how an analog of Stein's method might be derived; a reasonable speculation is that the method should perform well when sample sizes are small. But this issue has not yet been investigated. An alternative solution has been found, but we will need some tools covered in Chapter 7 before describing it.

5.7 Exercises

1. Given that $\bar{X} = 78$, $\sigma^2 = 25$, $n = 10$, and $\alpha = .05$, test $H_0: \mu > 80$, assuming observations are randomly sampled from a normal distribution. Also, draw the standard normal distribution indicating where Z and the critical value are located.
2. Repeat Exercise 1, but test $H_0: \mu = 80$.
3. For Exercise 2, compute a .95 confidence interval and verify that this interval is consistent with your decision about whether to reject the null hypothesis.
4. For Exercise 1, determine the p-value.
5. For Exercise 2, determine the p-value.
6. Given that $\bar{X} = 120$, $\sigma = 5$, $n = 49$, and $\alpha = .05$, test $H_0: \mu > 130$, assuming observations are randomly sampled from a normal distribution.
7. Repeat Exercise 6, but test $H_0: \mu = 130$.
8. For Exercise 7, compute a .95 confidence interval and compare the result with your decision about whether to reject H_0.
9. If $\bar{X} = 23$ and $\alpha = .025$, can you make a decision about whether to reject $H_0: \mu < 25$ without knowing σ?
10. An electronics firm mass-produces a component for which there is a standard measure of quality. Based on testing vast numbers of these components, the company has found that the average quality is $\mu = 232$ with $\sigma = 4$. However, in recent years the quality has not been checked, so management asks you to check their claim with the goal of being reasonably certain that an average quality of less than 232 can be ruled out. That is, assume the quality is poor and in fact less than 232, with the goal of empirically establishing that this assumption is unlikely. You get $\bar{X} = 240$ based on a sample $n = 25$ components, and you want the probability of a Type I error to be .01. State the null hypothesis, and perform the appropriate test assuming normality and $\sigma = 4$.

11. An antipollution device for cars is claimed to have an average effectiveness of exactly 546. Based on a test of 20 such devices you find that $\bar{X} = 565$. Assuming normality and that $\sigma = 40$, would you rule out the claim with a Type I error probability of .05?
12. Comment on the relative merits of using a .95 confidence interval for addressing the effectiveness of the antipollution device in Exercise 11.
13. For $n = 25$, $\alpha = .01$, $\sigma = 5$, and $H_0: \mu \geq 60$, verify that power is .95 when $\mu = 56$.
14. For $n = 36$, $\alpha = .025$, $\sigma = 8$, and $H_0: \mu \leq 100$, verify that power is .61 when $\mu = 103$.
15. For $n = 49$, $\alpha = .05$, $\sigma = 10$, and $H_0: \mu = 50$, verify that power is approximately .56 when $\mu = 47$.
16. A manufacturer of medication for migraine headaches knows that their product can cause liver damage if taken too often. Imagine that by a standard measuring process, the average liver damage is $\mu = 48$. A modification of their product is being contemplated, and, based on $n = 10$ trials, it is found that $\bar{X} = 46$. Assuming $\sigma = 5$, they test $H_0: \mu \geq 48$, the idea being that if they reject, there is convincing evidence that the average amount of liver damage is less than 48. Then

$$Z = \frac{46 - 48}{5/\sqrt{10}} = -1.3.$$

With $\alpha = .05$, the critical value is -1.645, so they do not reject, because Z is not less than the critical value. What might be wrong with accepting H_0 and concluding that the modification results in an average amount of liver damage greater than or equal to 48?
17. For Exercise 16, verify that power is .35 if $\mu = 46$.
18. Exercise 17 indicates that power is relatively low with only $n = 10$ observations. Imagine that you want power to be at least .8. One way of getting more power is to increase the sample size, n. Verify that for sample sizes of 20, 30, and 40, power is .56, .71, and .81, respectively.
19. For Exercise 18, rather than increase the sample size, what else might you do to increase power? What is a negative consequence of using this strategy?
20. Test the hypothesis $H_0: \mu = 42$ with $\alpha = .05$ and $n = 25$ given the following values for $\bar{X}$ and s: (a) $\bar{X} = 44$, $s = 10$, (b) $\bar{X} = 43$, $s = 10$, (c) $\bar{X} = 43$, $s = 2$.
21. For part b of Exercise 20, you fail to reject, but you reject for the situation in part c. What does this illustrate about power?
22. Test the hypothesis $H_0: \mu < 42$ with $\alpha = .05$ and $n = 16$ given the following values for $\bar{X}$ and s: (a) $\bar{X} = 44$, $s = 10$, (b) $\bar{X} = 43$, $s = 10$, (c) $\bar{X} = 43$, $s = 2$.
23. Repeat Exercise 22, except test $H_0: \mu > 42$.
24. A company claims that on average, when exposed to their toothpaste, 45% of all bacteria related to gingivitis are killed. You run ten tests and find that the percentages of bacteria killed among these tests are 38, 44, 62, 72, 43, 40, 43, 42, 39, 41. The mean and standard deviation of these values are $\bar{X} = 46.4$ and $s = 11.27$. Assuming normality, test the hypothesis that the average percentage is 45 with $\alpha = .05$.

25. A portion of a study by Wechsler (1958) reports that for 100 males taking the Wechsler Adult Intelligence Scale (WAIS), the sample mean and variance on picture completion are $\bar{X} = 9.79$ and $s = 2.72$. Test the hypothesis $H_0 : \mu \geq 10.5$ with $\alpha = .025$.
26. Assuming 20% trimming, test the hypothesis $H_0 : \mu_t = 42$ with $\alpha = .05$ and $n = 20$ given the following values for $\bar{X}_t$ and s_w: (a) $\bar{X}_t = 44$, $s_w = 9$, (b) $\bar{X}_t = 43$, $s_w = 9$, (c) $\bar{X}_t = 43$, $s_w = 3$.
27. Repeat Exercise 26, except test the hypothesis $H_0 : \mu_t < 42$ with $\alpha = .05$ and $n = 16$.
28. For the data in Exercise 24, the trimmed mean is $\bar{X}_t = 42.17$ with a Winsorized standard deviation of $s_w = 1.73$. Test the hypothesis that the population trimmed mean is 45 with $\alpha = .05$.
29. A standard measure of aggression in 7-year-old children has been found to have a 20% trimmed mean of 4.8 based on years of experience. A psychologist wants to know whether the trimmed mean for children with divorced parents differs from 4.8. Suppose $\bar{X}_t = 5.1$ with $s_w = 7$ based on $n = 25$. Test the hypothesis that the population trimmed mean is exactly 4.8 with $\alpha = .01$.

LEAST SQUARES REGRESSION AND PEARSON'S CORRELATION

Two common goals are determining whether and how two variables are related. This chapter describes the basics of the two most frequently used approaches to these problems. The first is based on what is called *least squares regression*. The second employs an estimate of Pearson's correlation (ρ). It will be seen that these methods are satisfactory for some purposes, but they are completely inadequate for others. In terms of hypothesis testing, the techniques in this chapter inherit the problems associated with means, and new problems are introduced. There are methods for dealing with these difficulties, but the details are given in subsequent chapters. The main goal here is to summarize some basic principles and to motivate the use of more modern techniques.

6.1 Fitting a Straight Line to Data: The Least Squares Principle

When describing the association between two variables, certainly the most common strategy is to assume that the association is linear. There are practical situations where the assumption of a linear association is unreasonable, but this issue is ignored for the moment.

We illustrate the basic strategy for fitting a straight line to a scatterplot of points using data from a classic study in astronomy. Is there some pattern to how the galaxies in the universe are moving relative to one another? Edwin Hubble collected data on two measures relevant to this issue in the hope of gaining some insight into how the universe was formed. He measured the distance of 24 galaxies from earth plus their recession velocity. His measurements, published in 1929, are shown in Table 6.1, where X is a galaxy's distance from earth in megaparsecs and Y is its speed in kilometers per second. (One parsec is 3.26 light-years.) For example, the first galaxy is .032 megaparsecs from earth and moving away from earth at the rate of 170 kilometers per second. The third galaxy is .214 megaparsecs from earth and approaching earth at the rate of 130 kilometers per second.

TABLE 6.1 Hubble's Data on the Distance and Recession Velocity of 24 Galaxies

Distance (X):	0.032	0.034	0.214	0.263	0.275	0.275	0.450	0.500	0.500	0.630	0.800	0.900
	0.900	0.900	0.900	1.000	1.100	1.100	1.400	1.700	2.000	2.000	2.000	2.000
Velocity (Y):	170	290	−130	−70	−185	−220	200	290	270	200	300	−30
	650	150	500	920	450	500	500	960	500	850	800	1090

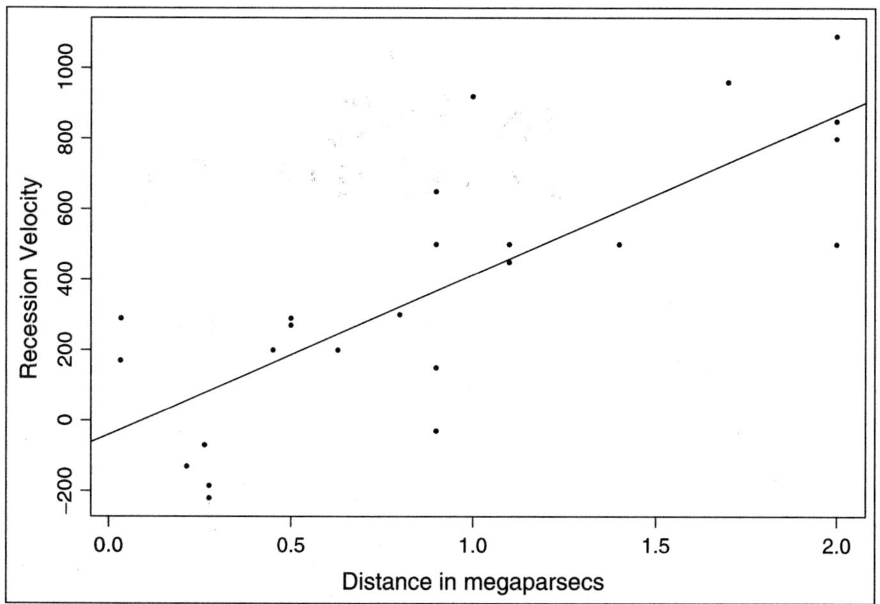

FIGURE 6.1 Scatterplot of Hubble's data on the recession velocity of galaxies. The straight line has slope 454 and intercept −40.8.

A common notation for representing n pairs of points is

$$(X_1, Y_1), \ldots, (X_n, Y_n).$$

So in Table 6.1, $(X_1, Y_1) = (.032, 170)$, where the subscript 1 indicates the first pair of observations. Here $n = 24$ and $(X_{24}, Y_{24}) = (2, 1090)$.

To fit a straight line to Hubble's data, we need a criterion for judging how well any line fits a scatterplot of the points. Given a criterion, we can then search for the line that is optimal. More precisely, consider lines having the form

$$\hat{Y} = b_0 + b_1 X, \tag{6.1}$$

where b_1 and b_0 are, respectively, the unknown slope and intercept that are to be determined from the data and the notation $\hat{Y}$ is used to make a distinction between the predicted value of Y based on X, which is $\hat{Y}$, and the Y we observe. Figure 6.1 shows a scatterplot of Hubble's data together with the line $\hat{Y} = 454X - 40.8$ (so $b_1 = 454$ and $b_0 = -40.8$). For the first galaxy in Table 6.1, which is $X_1 = .032$ megaparsecs from earth, its predicted recession velocity is $\hat{Y}_1 = 454(.032) - 40.8 = -26.3$ kilometers per second. But from Table 6.1 we see that its actual recession velocity is $Y_1 = 170$,

TABLE 6.2 Fitted Values and Residuals for the Data in Table 6.1

Observation number	Y_i	$\hat{Y}_i$	r_i
1	170	−26.3	196.3
2	290	−25.3	315.3
3	−130	56.4	−186.4
4	−70	78.7	−148.7
5	−185	84.1	−269.1
6	−220	84.1	−304.1
7	200	163.6	36.4
8	290	186.3	103.7
9	270	186.3	83.7
10	200	245.3	−45.3
11	300	322.5	−22.5
12	−30	368.0	−398.0
13	650	282	367.0
14	150	368.0	−217.0
15	500	368.0	132.0
16	920	413.4	506.6
17	450	458.8	−8.8
18	500	458.8	41.2
19	500	595.0	−95.0
20	960	731.3	228.7
21	500	867.5	−367.5
22	850	867.5	−17.5
23	800	867.5	−67.5
24	1090	867.5	222.5

so for the first galaxy having $(X_1, Y_1) = (.032, 170)$, there is a discrepancy between the actual and predicted velocity of $r_1 = Y_1 - \hat{Y}_1 = 170 + 26.3 = 196.3$. More generally, for the ith pair of observations, there is a discrepancy between the observed and predicted Y values, given by

$$r_i = Y_i - \hat{Y}_i, \tag{6.2}$$

where r_i $(i = 1, \ldots, n)$ is called the ith *residual*. Table 6.2 shows the $\hat{Y}$ values and residuals for the data in Table 6.1.

For any slope (b_1) and intercept (b_0) we might choose, one way of judging the overall fit of the resulting line to a scatterplot of points is to use the sum of the squared residuals:

$$\sum r_i^2 = \sum (Y_i - b_1 X_i - b_0)^2. \tag{6.3}$$

If we choose the slope and intercept so as to minimize the sum of squared residuals, we are using what is called the *least squares principle*. Without making any assumptions about the distribution of X or Y, least squares leads to taking the slope to be

$$b_1 = \frac{\sum (X_i - \bar{X})(Y_i - \bar{Y})}{\sum (X_i - \bar{X})^2} \tag{6.4}$$

and the intercept is

$$b_0 = \bar{Y} - b_1\bar{X}. \tag{6.5}$$

That is, if we use the regression equation

$$\hat{Y} = b_1 X + b_0,$$

with b_1 and b_0 given by Equations (6.4) and (6.5), we minimize the sum of squared residuals. The straight line in Figure 6.1 is the least squares regression line for Hubble's data.

Chapter 3 noted that when summarizing a single measure, the least squares principle leads to the sample mean, which has a finite-sample breakdown point of only $1/n$. That is, only one unusual point can render the sample mean meaningless. And even when sampling from a perfectly symmetric distribution, in which case the population mean provides a reasonable measure of location, the sample mean can be a relatively inaccurate estimate of μ. These problems extend to the situation at hand, and there are new ways practical problems can arise.

To begin to explain the practical problems associated with least squares regression, we first note that the estimate of the slope can be written as a weighted mean of the Y values, with the weights depending on the X values. In particular, the least squares estimate of the slope can be written as

$$b_1 = \sum w_i Y_i,$$

where

$$w_i = \frac{X_i - \bar{X}}{(n-1)s_x^2}$$

and s_x^2 is the sample variance of the X values. But we saw in Chapter 3 that a weighted mean has a finite-sample breakdown point of only $1/n$, so the finite-sample breakdown point of the least squares estimate of the slope is $1/n$ as well. In particular, a single unusual X value can have an inordinate influence on the estimate of the slope, as can a single unusual Y value. This suggests checking whether any of the X values are outliers, doing the same for the Y values; if no outliers are found, assume that there are no influential points that result in a poor fit to the bulk of the points. Unfortunately, this relatively simple strategy can be inadequate, for at least two reasons. First, there are situations where an outlying X value is beneficial. (Details are covered in Section 6.3.1.) Second, a small number of unusual points can greatly influence the least squares estimate, giving a distorted view of how the bulk of the points are associated, even though none of the corresponding X or Y values is declared outliers using the methods of Section 3.4. When addressing this latter problem, what is needed is some method that takes into account the overall structure of the points, but for now we merely illustrate the problem.

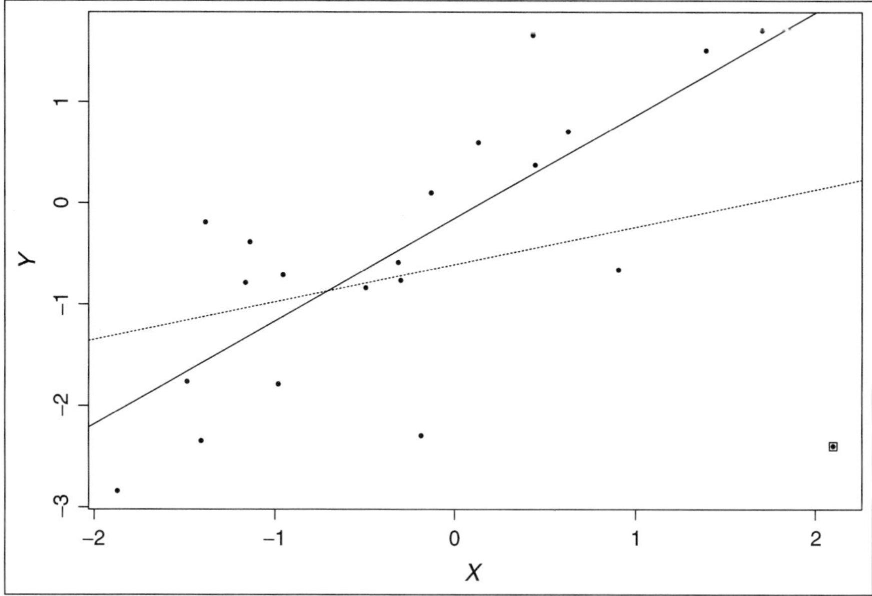

FIGURE 6.2 The two points marked by the square in the lower right corner have a substantial impact of the least squares regression line. Ignoring these two points, the least squares regression is given by the solid line. Including them, the least squares regression line is given by the dashed line. Moreover, none of the X or Y values are declared outliers using the methods in Chapter 3.

EXAMPLE. To illustrate the last problem, Figure 6.2 shows 20 points that were generated on a computer, where both X and Y are normal and the points are centered around the line $Y = X$. So the true slope is 1. The solid straight line passing through the bulk of the points in Figure 6.2 is this least squares estimate of the regression line and has slope $b_1 = 1.01$. Then two additional points were added at $X = 2.1$ and $Y = -2.4$ and are marked by the square in the lower right corner of Figure 6.2. Among all 22 X values, none is declared an outlier by any of the methods in Section 3.4, and the same is true for the Y values. Yet these two additional points are clearly unusual relative to the other 20, and they have a substantial influence on the least squares estimate of the slope. Now the estimated slope is .37, and the resulting least squares regression line is represented by the dashed line in Figure 6.2. Note that the estimated intercept changes substantially as well. ■

EXAMPLE. Figure 6.3 shows the surface temperature (X) of 47 stars versus their light intensity. The solid line is the least squares regression line. As is evident, the regression line does a poor job of summarizing the association

Continued

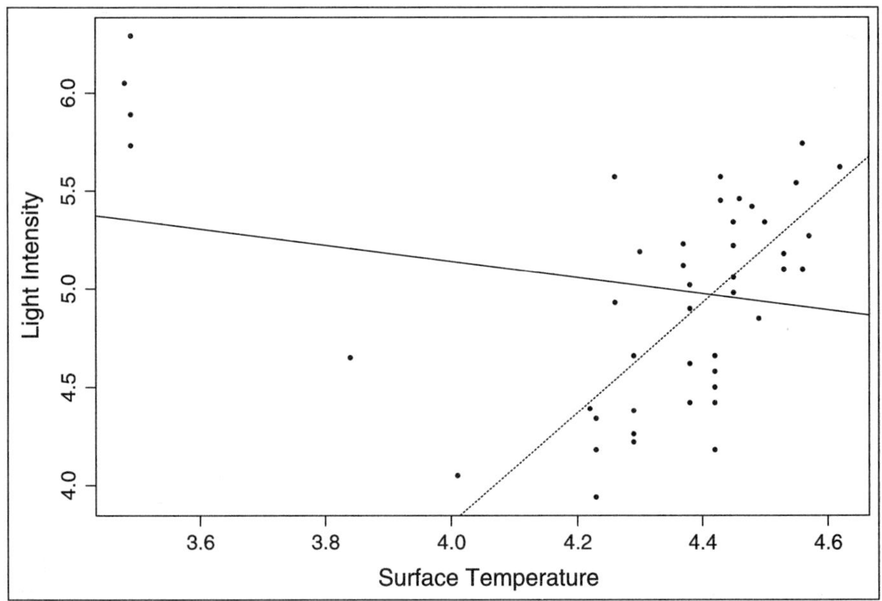

FIGURE 6.3 The solid line is the least squares regression line using all of the star data. When ignoring outliers among the X values, the least squares regression line is given by the dashed line.

EXAMPLE. (*Continued*) between the two variables under study. In this particular case it is clear that the four points in the upper left portion of Figure 6.3 are unusual. Moreover, the S-PLUS function outbox (described in Section 3.4.6) indicates that X values less than or equal to 3.84 are outliers. If we simply exclude all points with X values declared outliers and apply least squares regression to the points that remain, we get the dashed line shown in Figure 6.3. This provides a better summary for the bulk of the points, but even then the least squares regression line does not seem quite satisfactory. ■

One point should be stressed. In the last example we simply restricted the range of X values to get a better fit to the bulk of the points under study. A similar strategy might be used when dealing with unusual Y values. But when restricting the range of Y values, caution must be used. The reason is that when testing hypotheses about the slope and intercept, simply eliminating points with unusual Y values leads to technical problems that are described in Section 6.3.3. Special techniques for dealing with unusual Y values are required, and they are described in Chapter 13.

6.2 The Standard Least Squares Model

This section describes the standard least squares model, which is important to understand for at least two general reasons. First, it describes conditions under which the least squares estimate of the slope and intercept [given by Equations (6.4) and (6.5)] have the smallest standard error among a large class of estimators. Said another way, understanding this model is important when trying to understand when and why the least squares estimate of the slope and intercept can be unsatisfactory. Second, a special case of this model is used when computing confidence intervals and testing hypotheses, and understanding the standard model helps explain some of the reasons conventional hypothesis-testing methods can be highly inaccurate.

To be concrete, we describe the model using a study conducted by G. Margolin and A. Medina, where the goal was to examine how children's information processing is related to a history of exposure to marital aggression. Results for two of their measures are shown in Table 6.3. The first, labeled X, is a measure of marital aggression that reflects physical, verbal, and emotional aggression during the last year; Y is a child's score on a recall test. If aggression in the home (X) has a relatively low value, what would we expect a child to score on the recall-test (Y)? If the measure of aggression is high, now what would we expect the recall-test score to be?

TABLE 6.3 Measures of Marital Aggression and Recall-Test Scores

Family i	Aggression X_i	Test score Y_i	Family i	Aggression X_i	Test score Y_i
1	3	0	25	34	2
2	104	5	26	14	0
3	50	0	27	9	4
4	9	0	28	28	0
5	68	0	29	7	4
6	29	6	30	11	6
7	74	0	31	21	4
8	11	1	32	30	4
9	18	1	33	26	1
10	39	2	34	2	6
11	0	17	35	11	6
12	56	0	36	12	13
13	54	3	37	6	3
14	77	6	38	3	1
15	14	4	39	3	0
16	32	2	40	47	3
17	34	4	41	19	1
18	13	2	42	2	6
19	96	0	43	25	1
20	84	0	44	37	0
21	5	13	57	11	2
22	4	9	46	14	11
23	18	1	47	0	3
24	76	4			

Among the millions of homes in the world, temporarily consider all homes that have an aggression measure of 50 ($X = 50$). Among these homes there will be some average value on the recall test (Y). In more formal terms, this average is $E(Y|X = 50)$, the mean of Y given that $X = 50$. The standard regression model assumes that the population mean of Y, given X, is $\beta_0 + \beta_1 X$, where β_1 and β_0 are unknown parameters that we want to estimate based on observations available to us. [Least squares regression estimates β_1 and β_0 with b_1 and b_0, respectively, given by Equations (6.4) and (6.5).] One problem is that in general, still restricting attention to those homes with $X = 50$, the Y values will differ from one another. In particular, they will not always be equal to the population mean of the Y values. Of course typically this will be true for any value of X we might pick. That is, for a randomly sampled pair of observations, (X, Y), ordinarily there will be a discrepancy between Y and its (conditional) mean given X. In formal terms, this discrepancy is $e = Y - \beta_1 X - \beta_0$. Rearranging the terms of this last equation we get the *standard regression model*:

$$Y = \beta_0 + \beta_1 X + e, \tag{6.6}$$

where e is the so-called error term, which is assumed to have a mean of zero. That is, $E(e) = 0$ is assumed, which implies that given X, $E(Y) = \beta_0 + \beta_1 X$.

This regression model is said to be *homoscedastic* if the variance of the error term does not depend on X. In our example, this means, for example, that the variance of recall-test scores (Y), given that aggression in the home (X) is 50, is equal to the variance of the recall-test scores given that aggression in the home is 75. More generally, for any value of X we might pick, homoscedasticity means that the conditional variance of Y, given X, does not change with X—it is some constant value, as illustrated in Figure 6.4.

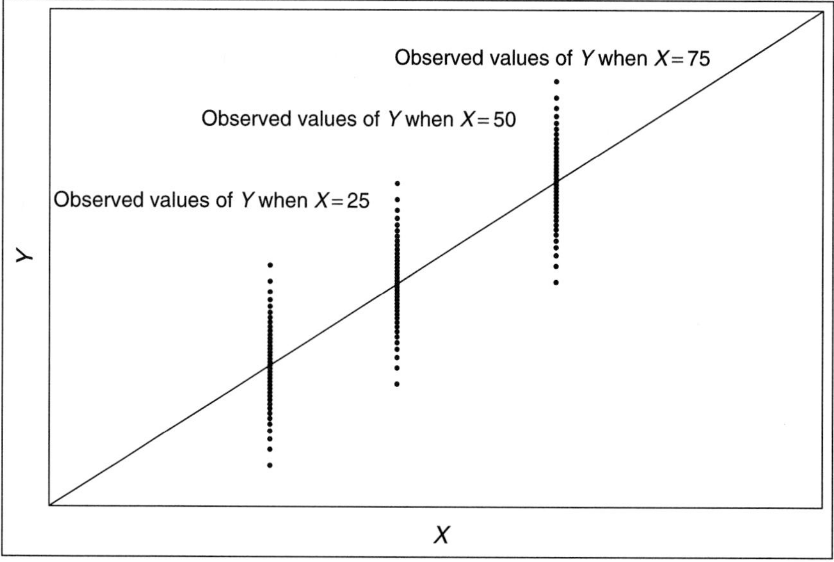

FIGURE 6.4 Example of homoscedasticity. The conditional variance of Y, given X, does not change with X.

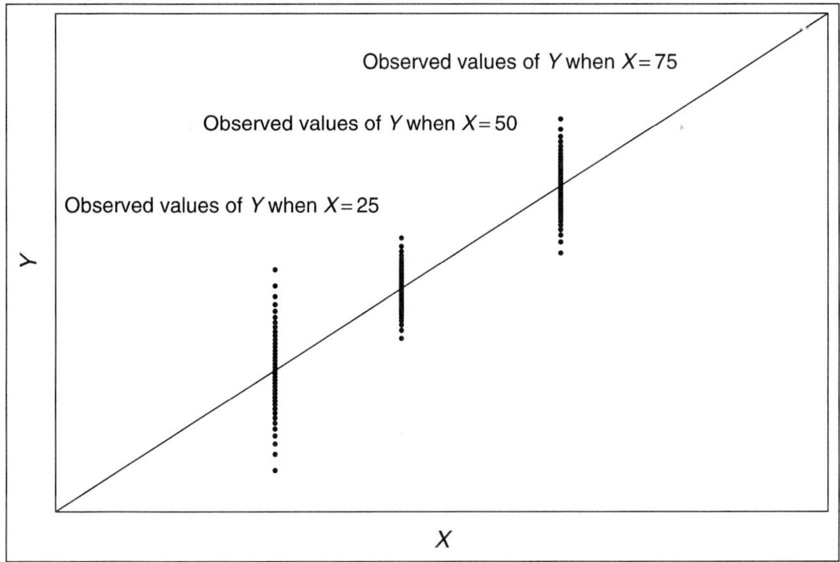

FIGURE 6.5 Example of heteroscedasticity. The conditional variance of Y, given X, changes with X.

A *heteroscedastic* regression model refers to a situation where this conditional variance changes with X, such as illustrated in Figure 6.5, where the variance of Y is smaller when $X = 50$ versus the other two X values shown.

Notice that b_1 has a sampling distribution. That is, if we were to repeat a study infinitely many times with each replication based on n randomly sampled pair of observations, we would get a collection of estimated slopes that generally differ from one another, and of course the same is true for the intercept. For the regression model given by Equation (6.6), it can be shown that

$$E(b_1) = \beta_1 \qquad \text{and} \qquad E(b_0) = \beta_0.$$

That is, b_1 and b_0 are unbiased estimates of the slope and intercept, respectively, roughly meaning that on average, they give a correct estimate of the true slope and intercept. The squared standard error of b_1 refers to the variance of the b_1 values obtained when repeating an experiment infinitely many times. If we can find an unbiased estimator with a smaller standard error, then generally it will be a more accurate estimate of the true slope, β_1. Our immediate goal is to describe conditions under which the least squares estimates of the slope and intercept have the smallest possible standard error.

For the homoscedastic regression model just described, among all the weighted means of the Y values we might consider for estimating the slope and intercept, the least squares estimate given by Equations (6.4) and (6.5) has the smallest standard error. This result is a special case of what is known as the *Gauss–Markov* theorem. This theorem also yields the weighted mean with the smallest standard error in the heteroscedastic case. In particular, Gauss showed that if the conditional variance of the Y values, given that $X = X_i$, is σ_i^2, say, the optimal estimates of the slope and intercept are the values b_1 and b_0 that minimize

$$\sum w_i(Y_i - b_1X_i - b_0)^2, \tag{6.7}$$

where $w_i = 1/\sigma_i^2$. In our illustration, for example, we see from Table 6.3 that the first aggression score is $X_1 = 3$, the second is $X_2 = 104$, and so on. If we knew σ_1^2, the variance of recall-test scores (Y) given that the aggression measure is $X_1 = 3$, and more generally if we knew σ_i^2, the conditional variance of Y given that $X = X_i$, we could compute the optimal weighted mean estimate of the slope and intercept, which are the values b_1 and b_0 that minimize Equation (6.7). In the homoscedastic case, all of the w_i values in Equation (6.7) have a common value and the b_1 and b_0 values that minimize this equation are the same values that minimize Equation (6.3). Determining the slope and intercept with Equation (6.7) is an example of what is called *weighted least squares*, and using Equation (6.3) is called *ordinary least squares*.

A problem is that typically the σ_i^2 are not known, so a common strategy is to assume homoscedasticity and simply estimate the slope and intercept with ordinary least squares [using Equations (6.4) and (6.5)], the optimal least squares estimates in the homoscedastic case. A practical issue is whether knowing the σ_i^2 would result in a substantially more accurate estimate of the slope and intercept. If the answer is no, heteroscedasticity is not a concern; but if such situations arise, methods for dealing with heteroscedasticity become important.

EXAMPLE. Consider a situation where both X and the error term (ϵ) are standard normal and the slope is $\beta_1 = 1$. Further assume that the standard deviation of Y, given X, is 1 if $|X| < 1$, otherwise it is $|X| + 1$. So, for example, if $X = .5$, Y has variance 1; but if $X = 1.5$, Y has standard deviation 2.5. Figure 6.6 shows a scatterplot of points generated in this fashion. Figure 6.7 shows the sampling distribution of the weighted and ordinary least squares estimates of the slope. The standard error of the ordinary least squares estimator is .52, versus .37 using the weighted least squares estimator instead. As is evident, the weighted least squares estimate tends to be much closer to the correct value for the slope. ■

EXAMPLE. Consider the same situation as in the last example, except that the conditional standard deviation of Y, given X, is X^2. Now the standard error of the ordinary least squares estimator is more than 10 times larger than the standard error of the optimal weighted least squares estimator. So we see that even under normality, the ordinary least squares estimate of the slope and intercept can be highly inaccurate relative to the optimal weighted least squares approach. Nonnormality can make the least squares estimator perform even more poorly. ■

One way of trying to improve upon ordinary least squares is to attempt to estimate the variance of Y given X. If we could do this in a reasonably accurate manner for

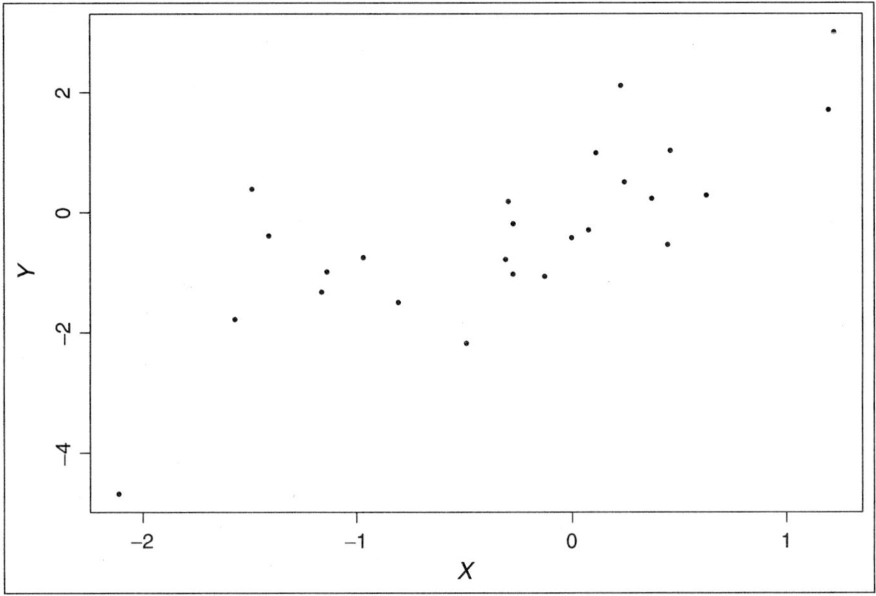

FIGURE 6.6 Example of what points look like when both X and the error term are standard normal and the variance of Y is 1 if $|X| < 1$ and $|X| + 1$ if not.

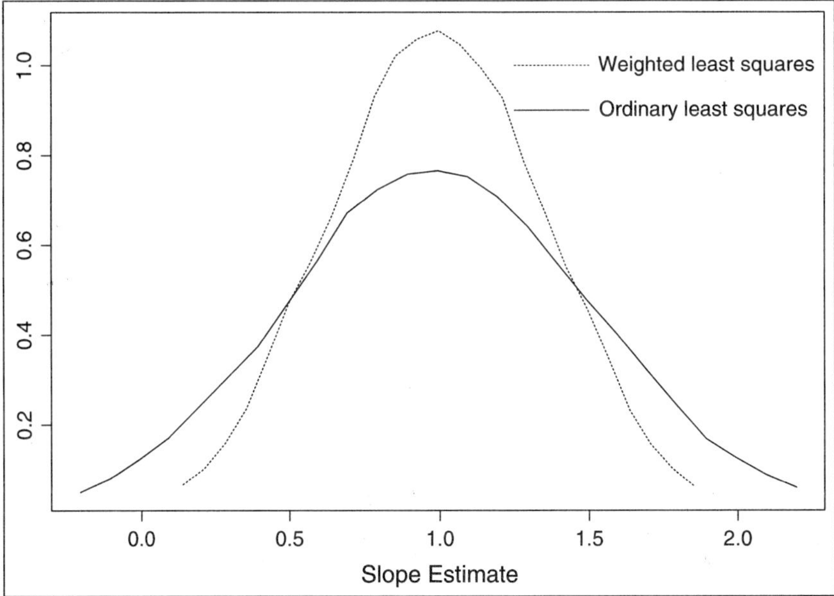

FIGURE 6.7 Comparison of the sampling distributions of ordinary least squares estimator versus the optimal weighted least squares estimator. There is heteroscedasticity, which is why the weighted least squares estimator tends to be closer to the correct value of 1.

every X value we observe, weighted least squares could be employed. In the aggression study, for example, the first home had an aggression score of 3, so we need to estimate the variance of the recall-test scores for all homes having an aggression score of 3. In a similar manner, the second home had an aggression score of 104, so we need to estimate the variance of the recall-test scores for all homes having an aggression score of 104. In more formal terms, $X_2 = 104$, and to apply weighted least squares we need to know σ_2^2, the variance of Y given that $X = X_2$. The third home has an aggression score of $X_3 = 50$, so we need to know σ_3^2 as well, which is the variance of the recall-test scores when the aggression measure is 50. If we had a reasonably large number of homes with an aggression score of 50, we could simply compute the sample variance of the corresponding recall-test scores to get an estimate of σ_3^2. The practical problem is that we have only one home with an aggression score of 50, so this method cannot be used.

There are at least two general strategies for dealing with heteroscedasticity when the goal is to find an estimator of the slope and intercept that has a relatively small standard error. The first uses what are called *smoothers* to estimate the σ_i^2 values (e.g., Müller, 1988, p. 153; M. Cohen, Dalal, & Tukey, 1993; Wilcox, 1996a). We do not describe these methods here because either they ignore certain problems caused by outliers, or there is no known method that performs reasonably well when testing hypotheses or computing confidence intervals and the sample sizes are small or even moderately large. Instead we rely on the second general strategy, which uses one of the robust regression methods covered in Chapter 13. So the main point here is that heteroscedasticity matters in terms of getting an estimator of the slope and intercept that has a relatively small standard error, but we postpone how to deal with this problem for now.

6.2.1 Comments About Linearity and Homoscedasticity

Caution must be exercised when assuming that a regression line is straight. Consider, for example, the aggression data in Table 6.3, where Y is a recall-test score. If we fit a straight line using the least squares principle, we find that $b_1 = -0.0405$ and $b_0 = 4.581$. Figure 6.8 shows a scatterplot of the 47 pairs of observations along with the least squares regression line used to predict test scores. If, for instance, the measure of marital aggression is $X = 20$, the fitted value for the score on the recall test is $\hat{Y} = -0.0405(20) + 4.581 = 3.77$. Similarly, if $X = 40$, the estimate is $\hat{Y} = 2.961$. The estimated recall-test score corresponding to $X = 40$ is less than it is when $X = 20$ because the slope is negative. Generally, the regression equation $Y = -0.0405X + 4.581$ suggests that the higher the measure of aggression, the lower the score on the recall test. Is there some possibility that this is a misleading representation of the data?

Now consider the right portion of Figure 6.8, consisting of the 21 aggression scores greater than or equal to 25. For these 21 values, if we estimate the regression line, ignoring the other 26 values, we find that

$$\hat{Y} = 0.002285X + 1.93.$$

Note that the slope is slightly larger than zero. Now the fitted value for $X = 25$ is $\hat{Y} = 1.98$, and for $X = 40$ it is $\hat{Y} = 2.02$. In contrast, for the aggression scores

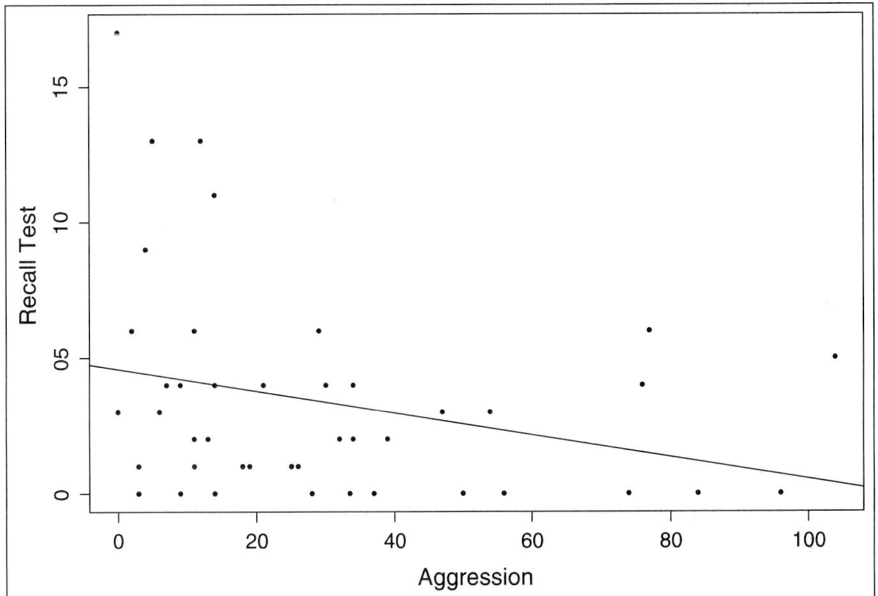

FIGURE 6.8 Scatterplot and least squares regression line for the aggression data in Table 6.3.

less than 25, ignoring the scores greater than 25, the regression equation is

$$\hat{Y} = -0.194X + 6.33.$$

The left panel of Figure 6.9 shows a scatterplot of the aggression data, with squares around the points having aggression scores greater than 25. Also shown is the least squares regression line based on these rightmost points, ignoring the points not marked with a square. The right panel shows the least squares regression line based on the remaining 26 pairs of observations, which are now marked with squares. As is evident, these two plots suggest something quite different from the regression line based on all of the data. In particular, one possibility is that there is a negative association for low aggression scores, but as the aggression scores increase, there seems to be little or no association at all. This means that, for the range of aggression scores available, perhaps there is a nonlinear association between aggression and test scores. Another possibility is that with more data, a straight regression line would prove to be adequate.

In recent years, better and more sophisticated methods have been developed for studying, describing, and detecting nonlinear associations. The only goal here is to illustrate a situation where nonlinearity might be an issue. Some very useful exploratory methods for studying nonlinearity can be found in Hastie and Tibshirani (1990). (A portion of these methods will be covered in Chapter 13.) There are also methods that can be used to test the hypothesis that an association between two random variables is linear.

A natural strategy when dealing with curvature is to add a quadratic term. That is, use a prediction rule having the form $\hat{Y} = \beta_0 + \beta_1 X + \beta_2 X^2$, or, more generally, include a term with X raised to some power. This might suffice in some situations,

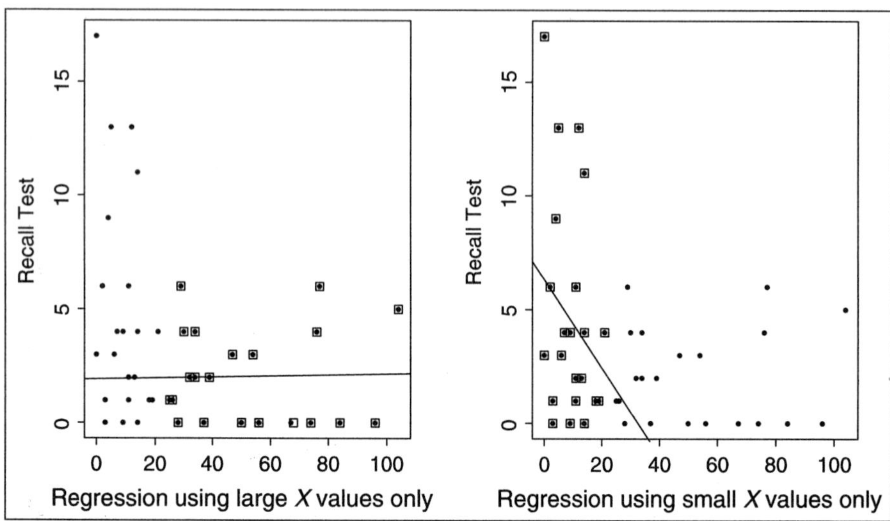

FIGURE 6.9 The left panel shows the least squares regression line for the data in Figure 6.8 using only the points marked with a square. The right panel shows the least squares regression line using only the data ignored in the right panel. As is evident, the left portion of the data gives a substantially different impression of how aggression and recall test scores are associated.

but experience with modern methods suggests that another type of nonlinearity is common: A straight line with a nonzero slope gives a reasonable prediction rule over some range of X values, but for X values outside this range there is little or no association at all. That is, the slope is zero. The aggression data in Table 6.3 are one example where this might be true.

Yet another interesting feature of the aggression data is the possibility is that there is heteroscedasticity. Perhaps for fairly low X values, the variance of Y is large relative to higher X values. It might even be the case that X and Y are dependent but that this dependence is due primarily to how the variance of Y, as opposed to the mean, is related to X. In any event, the three largest Y values, which appear in the upper left portion of Figure 6.8, have a substantial impact on the least squares regression line. If we ignore these three points, the least squares slope goes from $-.041$ to $-.017$.

6.3 Hypothesis Testing and Confidence Intervals

So far, attention has been focused on fitting a straight line to data without making any inferences about what the slope and intercept might be if you could sample all subjects of interest. That is, β_1 and β_0 represent the slope and intercept if all subjects could be sampled, and there is interest in determining which values for these parameters can be ruled out based on the data at hand. One of the most common goals is to test $H_0: \beta_1 = 0$. If $\beta_1 = 0$, knowing X provides no help in our attempts to estimate Y using a straight regression line. In the aggression study, for example, the regression equation becomes $\hat{Y} = 3.4$, the sample mean of all the Y values.

TABLE 6.4 Selling Price of Homes (divided by 1000) versus Size (in square feet)

Home i	Size (X_i)	Price (Y_i)	Home i	Size (X_i)	Price (Y_i)
1	2359	510	15	3883	859
2	3397	690	16	1937	435
3	1232	365	17	2565	555
4	2608	592	18	2722	525
5	4870	1125	19	4231	805
6	4225	850	20	1488	369
7	1390	363	21	4261	930
8	2028	559	22	1613	375
9	3700	860	23	2746	670
10	2949	695	24	1550	290
11	688	182	25	3000	715
12	3147	860	26	1743	365
13	4000	1050	27	2388	610
14	4180	675	28	4522	1290

This says that regardless of what the aggression score X might be, you estimate the score on the recall test to be 3.4. In general, if a regression equation is assumed to have the form given by Equation (6.1) and if $\beta_1 = 0$, then $\hat{Y} = \bar{Y}$. Thus, rejecting $H_0: \beta_1 = 0$ says that knowing X provides some help in estimating Y, assuming a linear equation having the form $\hat{Y} = \beta_0 + \beta_1 X$ is to be used.

To begin to understand the standard method for testing hypotheses, consider the data in Table 6.4, which shows the selling price (divided by 1000) of homes in a particular suburb of Los Angeles during the month of May 1998. Also shown is the size of the home, in square feet. Given that you are interested in buying a home with 2000 square feet, what would you expect to pay? What would you expect to pay for a house having 1500 square feet?

If we assume that the mean selling price of a home (Y), given its square feet (X), is

$$E(Y) = \beta_0 + \beta_1 X,$$

then the least squares estimates of the slope and intercept are, respectively,

$$b_1 = .215 \qquad \text{and} \qquad b_0 = 38.192.$$

So, for example, the estimated cost of a house with 2000 square feet (in thousands of dollars) is

$$.00215(2000) + .38192 = 468.7921,$$

or \$468,792. But the data in Table 6.4 do not represent all homes sold in this area. If all homes were included, we would know β_1 and β_0. How can we compute confidence intervals or test hypotheses about these two parameters?

The standard solution is to assume random sampling, homoscedasticity, and that the distribution of the error term (e) in the regression model [given by Equation (6.6)] has a normal distribution. This last assumption means that the selling price of homes, given that they have 1400 square feet, for example, is normally distributed. Similarly, the selling price of homes, among all homes having 2000 square feet, is normally distributed. More generally, the distribution of Y, given X, is assumed to be normal.

As previously explained, homoscedasticity means that the conditional variance of Y, given X, does not change with X. Let σ^2 represent this common variance. In formal terms, $\sigma^2 = \text{VAR}(Y|X) = \text{VAR}(e)$, which does not depend on X. The standard estimate of σ^2 is

$$\hat{\sigma}^2 = \frac{1}{n-2}\sum r_i^2,$$

the sum of the squared residuals divided by $n - 2$.

EXAMPLE. For the housing data in Table 6.4, the sum of squared residuals is $\sum r_i^2 = 26.99$. There are $n = 14$ homes, so, assuming homoscedasticity, the estimate of σ^2 is

$$\hat{\sigma}^2 = \frac{26.99}{14-2} = 2.25.$$

■

6.3.1 Conventional Hypothesis Testing and Confidence Intervals

Under the assumptions that the regression model is true and that there is random sampling, homoscedasticity, and normality, confidence intervals for the slope and intercept can be computed and hypotheses can be tested. In particular, a $1 - \alpha$ confidence interval for the slope, β_1, is

$$b_1 \pm t\sqrt{\frac{\hat{\sigma}^2}{\sum(X_i - \bar{X})^2}}, \qquad (6.8)$$

where t is the $1 - \alpha/2$ quantile of Student's T distribution with $\nu = n - 2$ degrees of freedom. (The value of t is read from Table 4 in Appendix B.) The quantity

$$\sqrt{\frac{\hat{\sigma}^2}{\sum(X_i - \bar{X})^2}}$$

is the *estimated standard error* of b_1. As for the intercept, β_0, a $1-\alpha$ confidence interval is given by

$$b_0 \pm t\sqrt{\frac{\hat{\sigma}^2 \sum X_i^2}{n\sum (X_i - \bar{X})^2}}. \tag{6.9}$$

The quantity

$$\sqrt{\frac{\hat{\sigma}^2 \sum X_i^2}{n\sum (X_i - \bar{X})^2}}$$

is the *estimated standard error* of b_0. For the common goal of testing $H_0: \beta_1 = 0$, the hypothesis that the slope is zero, you reject if the confidence interval does not contain zero. Alternatively, you reject if

$$|T| \geq t,$$

where again t is the $1-\alpha/2$ quantile of Student's T distribution with $\nu = n-2$ degrees of freedom and

$$T = b_1\sqrt{\frac{\sum (X_i - \bar{X})^2}{\hat{\sigma}^2}}. \tag{6.10}$$

For convenience, it is noted that the hypothesis-testing methods just described can be applied in S-PLUS with the command

```
summary(lm(y~x))
```

where x is any S-PLUS variable containing the predictor values and y is any S-PLUS variable containing the outcome values.

EXAMPLE. Using the aggression data in Table 6.3, we test the hypothesis $H_0: \beta_1 = 0$ with the goal that the probability of a Type I error be .05, assuming normality and that the error term is homoscedastic. Because $\alpha = .05$, $1-\alpha/2 = .975$. There are $n = 47$ pairs of observations, so the degrees of freedom are $\nu = 47 - 2 = 45$, and the critical value is $c = 2.01$. The least squares estimate of the slope is $b_1 = -0.0405$, and it can be seen that $\sum (X_1 - \bar{X})^2 = 34659.74$ and $\hat{\sigma}^2 = 14.15$, so the test statistic [given by Equation (6.10)] is

$$T = -0.0405\sqrt{\frac{34659.74}{14.5}} = -1.98.$$

Because $|T| = 1.98 < 2.01$, fail to reject. ■

In regression, any outlier among the X values is called a *leverage point*. Notice that a single leverage point can inflate $\sum (X_i - \bar{X})^2$, which is just the numerator of the sample variance for the X values. But $\sum(X_i - \bar{X})^2$ appears in the denominator of the expression for the standard error of b_1, so a single leverage point can cause the standard error of b_1 to be smaller compared to a situation where no leverage points occurred. In practical terms, we can get shorter confidence intervals and more power when there are leverage points, but caution must be exercised because leverage points can result in a poor fit to the bulk of the data.

6.3.2 Violating Assumptions

Currently, the hypothesis-testing method described in the previous subsection is routinely employed. Unfortunately, violating the assumptions of the method can cause serious problems.

Consider a situation where the normality assumption is valid but there is heteroscedasticity. For the housing data, for example, imagine that the variation among the selling prices differs depending on how many square feet a house happens to have. For instance, the variation among houses having 1500 square feet might differ from the variation among homes having 2000 square feet. Then the standard method for testing hypotheses about the slope, given by Equation (6.10), might provide poor control over the probability of a Type I error and poor probability coverage (e.g., Long & Ervin, 2000; Wilcox, 1996b). If the distributions are not normal, the situation gets worse. In some cases, the actual probability of a Type I error can exceed .5 when testing at the $\alpha = .05$ level! Perhaps an even more serious concern is that violating the homoscedasticity assumption might result in a substantial loss in power.

The homoscedasticity assumption is valid when X and Y are independent. (Independence implies homoscedasticity, but $\beta_1 = 0$, for example, does not necessarily mean that there is homoscedasticity.) Practical problems arise when X and Y are dependent because now there is no particular reason to assume homoscedasticity; and if there is heteroscedasticity, the wrong standard error is being used to compute confidence intervals and test hypotheses. If we could determine how VAR($Y|X$) changes with X, a correct estimate of the standard error could be employed, but currently it seems that alternate strategies for dealing with heteroscedasticity (covered in subsequent chapters) are more effective.

There are methods for testing the assumption that there is homoscedasticity (see, for example, Lyon & Tsai, 1996). But given some data, it is unknown how to tell whether any of these tests have enough power to detect situations where there is enough heteroscedasticity to cause practical problems with standard inferential methods. Even if there is homoscedasticity, nonnormality remains a serious concern. Currently, a more effective approach appears to be to switch to some method that allows heteroscedasticity.

Long and Ervin (2000) compare three simple methods for computing confidence intervals when there is heteroscedasticity. One of these they recommend for general use. However, in situations where leverage points are likely and simultaneously the error term has a normal or light-tailed distribution, their recommended method can be unsatisfactory. A more effective method is described in Chapter 7, so no details are given here about the method recommended by Long and Ervin.

6.3.3 Restricting the Range of Y

The derivation of the inferential methods in Section 6.3.1 treats the X values as constants and the Y values as random variables. That is, the methods are derived by conditioning on the X values, which makes it a fairly simple matter to derive an estimate of the standard error of the least squares estimate of the slope and intercept. For example, we saw that

$$b_1 = \sum w_i Y_i,$$

where

$$w_i = \frac{X_i - \bar{X}}{(n-1)s_x^2}.$$

Treating the X values as constants, and using the rules of expected values in Chapter 2, it can be shown that the squared standard error of b_1 is given by

$$\frac{\sigma^2}{\sum (X_i - \bar{X})^2}, \tag{6.11}$$

as noted in Section 6.3.1. A practical implication is that if we restrict the range of X values, no technical problems arise when trying to estimate the standard error of b_1; we simply use Equation (6.11) on the points that remain (with n reduced to the number of points remaining). But if we restrict the range of Y values by eliminating outliers, the methods in Section 6.3.1 are no longer valid, even under normality and homoscedasticity. We saw in Section 4.9.1 that if we eliminate extreme values and compute the mean using the data that remain, the standard error of this mean should not be estimated with the sample variance based on the data that remain. A similar problem arises here. If we eliminate extreme Y values, the remaining Y values are no longer independent. So if we use least squares to estimate the slope based on the pairs of points not eliminated, estimating the standard error of the slope becomes a nontrivial problem—the dependence among the Y values must be taken into account.

6.3.4 Standardized Regression

Popular statistical software reports what is called a *standardized regression coefficient*. This simply means that rather than compute the least squares estimator using the raw data, the observations are first converted to Z scores. For the aggression data in Table 6.3, for example, it can be seen that the test scores (Y) have mean $\bar{Y} = 3.4$ and standard deviation $s_y = 3.88$. The first test score is $Y_1 = 0$, and its Z-score equivalent is

$$Z = \frac{0 - 3.4}{3.88} = -0.88.$$

Of course, the remaining Y values can be converted to a Z score in a similar manner. In symbols, for the ith observation, you compute

$$Z_{yi} = \frac{Y_i - \bar{Y}}{s_y}.$$

Next, convert all of the aggression values to Z scores. That is, for the ith pair of observations, compute

$$Z_{xi} = \frac{X_i - \bar{X}}{s_x}.$$

For the aggression data, $\bar{X} = 28.5$ and $s_x = 27.45$. For example, the first entry in Table 6.3 has $X_1 = 3$, so $Z_{x1} = -0.93$. Next, you determine the least squares estimate of the slope using the transformed X and Y values just computed. The resulting estimate of the slope will be labeled b_z. The resulting estimate of the intercept is always zero, so the regression equation takes the form

$$\hat{Z}_y = b_z Z_x.$$

For the aggression data, it can be seen that $b_z = -0.29$, so $\hat{Z}_y = -0.29(Z_x)$.

The standardized regression coefficient, b_z, can be computed in another manner. First compute the least squares estimate of the slope using the original data yielding b_1. Then

$$b_z = b_1 \frac{s_x}{s_y}.$$

EXAMPLE. For the aggression data in Table 6.3, the sample standard deviations of the X and Y values are $s_y = 3.88$ and $s_x = 27.45$. As previously indicated, the least squares estimate of the slope is $b_1 = -0.0405$. The standardized slope is just

$$b_z = -0.0405 \frac{27.45}{3.88} = -0.29.$$

■

One reason standardized regression has some appeal is that it attempts to provide perspective on the magnitude of a predicted value for Y. Recall from Chapter 2 that for normal distributions, the value of $Z = (X - \mu)/\sigma$ has a convenient probabilistic interpretation under normality. For example, half the observations fall below a Z score of zero. A Z score of 1 indicates we are 1 standard deviation above the mean, and about 84% of all observations are to below this point when observations have a normal distribution. (From Table 1 in Appendix B, $Z = 1$ is the .84

quantile.) Similarly, $Z = 2$ refers to a point 2 standard deviations above the mean, and approximately 98% of all observations are below this value. Thus, for normal distributions, Z scores give you some sense of how large or small a value happens to be. Standardized regression attempts to tell us, for example, how a change of 1 standard deviation in X is related to changes in Y, again measured in standard deviations.

EXAMPLE. For the aggression data, assume normality and suppose we want to interpret the standardized regression estimate of the recall test when the measure of aggression is 1 standard deviation above or below the mean. One standard deviation above the mean of the aggression scores, X, corresponds to $Z_x = 1$, so, as previously indicated,

$$\hat{Z}_y = (-0.29)1 = -0.29.$$

For a standard normal distribution, the probability of being less than -0.29 is approximately .39, and this provides a perspective on how the recall test is related to the measure of aggression. In a similar manner, 1 standard deviation below the mean of the aggression scores corresponds to $Z_x = -1$, so now

$$\hat{Z}_y = (-0.29)(-1) = 0.29,$$

and there is approximately a .61 probability that a standard normal random variable is less than .29. ■

For nonnormal distributions, situations arise where Z scores can be interpreted in much the same way as when distributions are normal. But based on results in Chapters 2 and 3, there are two points to keep in mind when using Z scores. The first is that they can give a misleading representation of the sample of observations being studied. The second is that interpretation problems can arise even with an arbitrarily large sample size and a very small departure from normality.

EXAMPLE. The last example illustrated how to interpret $\hat{Z}_y$ assuming normality, but now we take a closer look at the data to see whether this interpretation might be misleading. Table 6.5 shows all 47 Z_y scores for the recall-test values written in ascending order. We see that 24 of the 47 values are below $-.29$, so the proportion below -0.29 is $24/47 = .51$. This means that based on the available data, your estimate is that there is .51 probability of having a Z_y score less than -0.29. Put another way, a Z score of -0.29 corresponds, approximately, to the median. In contrast, for normal distributions, $Z_y = 0$ is the median and the probability of getting a Z_y score less than -0.29 is approximately .39.

Continued

TABLE 6.5 Z Scores for the Recall-Test Scores in Table 6.3 (in ascending order)

−0.88	−0.88	−0.88	−0.88	−0.88	−0.88	−0.88	−0.88	−0.88	−0.88	−0.88	−0.88
−0.62	−0.62	−0.62	−0.62	−0.62	−0.62	−0.62	−0.36	−0.36	−0.36	−0.36	−0.36
−0.10	−0.10	−0.10	−0.10	0.15	0.15	0.15	0.15	0.15	0.15	0.15	0.41
0.67	0.67	0.67	0.67	0.67	0.67	1.44	1.96	2.47	2.47	3.51	

EXAMPLE. (*Continued*) Thus, there is some discrepancy between the empirical estimate of the probability of getting a Z_y score less than −0.29 versus the probability you get assuming normality. The main point here is that switching to standardized regression does not necessarily provide a perspective that is readily interpretable. ■

A criticism of this last example is that the estimated probability of getting a Z score less than −0.29 is based on only 47 observations. Perhaps with a larger sample size, the estimated probability would be reasonably close to 0.39, the value associated with a normal distribution. However, results in Chapter 2 indicate that even with a large sample size, there can be a considerable difference, so caution is recommended when interpreting standardized regression equations.

6.4 Pearson's Correlation

Chapter 2 introduced Pearson's correlation, ρ. This section takes up the problem of estimating ρ based on data available to us, plus the issue of interpreting what this estimate tells us about the association between the two measures under study.

Recall from Chapter 2 that

$$\rho = \frac{\sigma_{xy}}{\sigma_x \sigma_y},$$

where σ_{xy} is the (population) covariance between X and Y (as defined in Section 2.8). Given n pairs of observations

$$(X_1, Y_1), \ldots, (X_n, Y_n),$$

the covariance is typically estimated with

$$s_{xy} = \frac{1}{n-1} \sum (X_i - \bar{X})(Y_i - \bar{Y}).$$

The usual estimate of ρ is

$$r = \frac{s_{xy}}{s_x s_y}, \tag{6.12}$$

where s_x and s_y are the standard deviations of the X and Y values, respectively.

It can be shown that $-1 \leq r \leq 1$. That is, the estimate of Pearson's correlation, based on r, always lies between -1 and 1, as does ρ. If all of the points lie on a straight line with a positive slope, $r = 1$. If all of the points lie on a straight line with a negative slope, $r = -1$.

The estimate of the slope of the least squares regression line (b_1) is related to the estimate of Pearson's correlation in the following manner:

$$b_1 = r\frac{s_y}{s_x}. \tag{6.13}$$

So if $r > 0$, the least squares regression line has a positive slope; if $r < 0$, the slope is negative. In standardized regression (as discussed in Section 6.3.4), the slope is equal to the correlation between X and Y. That is, the least squares regression line between

$$Z_x = \frac{X - \bar{X}}{s_x} \quad \text{and} \quad Z_y = \frac{Y - \bar{Y}}{s_y}$$

is

$$\hat{Z}_y = rZ_x.$$

6.4.1 Five Features of Data That Affect the Magnitude of *r*

Interpreting r is complicated by the fact that various features of the data under study affect its magnitude. Five such features are described here, and a sixth is briefly indicated at the end of this section.

Assuming that there is a linear association between X and Y, the first feature is the distance of the points from the line around that they are centered. That is, the magnitude of the residuals is associated with the magnitude of r. The left panel of Figure 6.10 shows a scatterplot of points with $r = .92$. The right panel shows another scatterplot of points that are centered around the same line as in the left panel, only they are farther from the line around which they are centered. Now $r = .42$.

A second feature that affects the magnitude of r is the magnitude of the slope around which the points are centered (e.g., Barrett, 1974; Loh, 1987). Figure 6.11 shows the same points as in the left panel of Figure 6.10, only rotated so that the slope around which they are centered has been decreased from 1 to .5. This causes the correlation to drop from .92 to .83. If we continue to rotate the points until they are centered around the x-axis, $r = 0$.

A third feature of data that affects r is outliers. This is not surprising, because we already know that the least squares regression line has a breakdown point of only $1/n$, and we have seen how r is related to the least squares estimate of the slope [as indicated by Equation (6.13)]. For the star data in Figure 6.3, $r = -.21$, which is consistent with the negative slope associated with the least squares regression line. But we have already seen that for the bulk of the points, there is a positive association. Generally, a single unusual value can cause r to be close to zero even when the remaining points

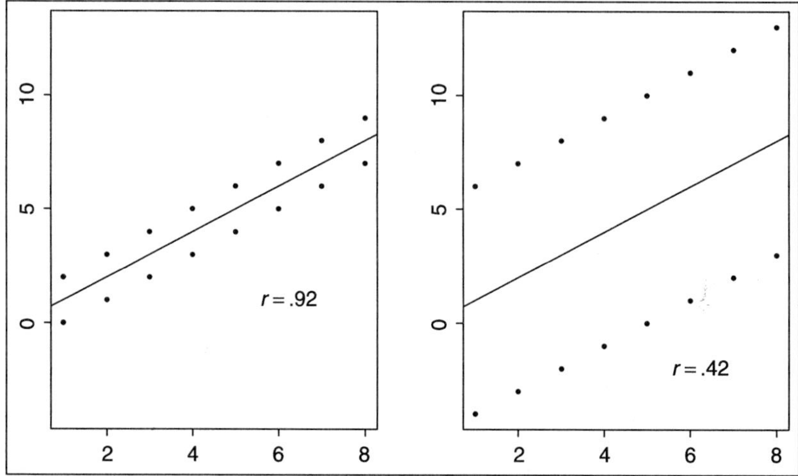

FIGURE 6.10 Illustration showing that the magnitude of the residuals affects Pearson's correlation.

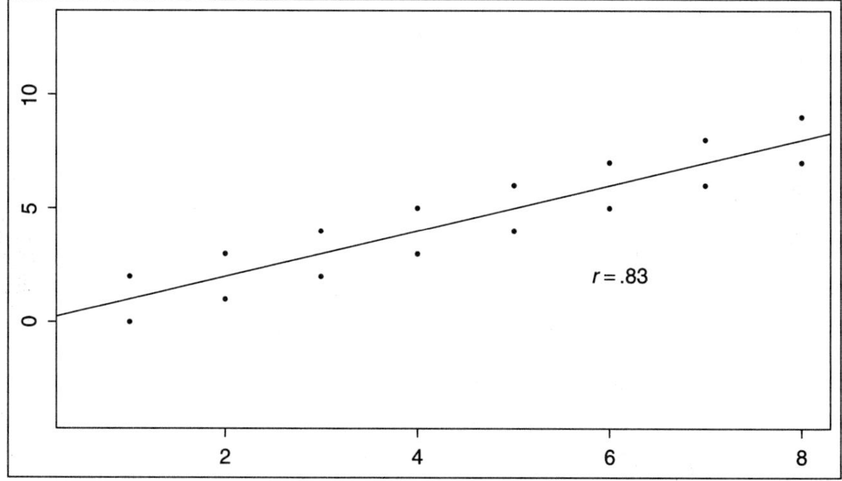

FIGURE 6.11 Illustration showing that Pearson's correlation is related to the magnitude of the slope of the line around which points are clustered.

are centered around a line having a nonzero slope, and one outlier can cause $|r|$ to be fairly large even when there is no association among the remaining points.

Moreover, when sampling points in situations where outliers are likely to occur, even small departures from normality can greatly affect the population correlation ρ, and r can be affected as well no matter how large the sample size might be. To provide some indication of why, the left panel of Figure 6.12 shows the distribution between X and Y when both X and Y are normal and $\rho = .8$. In the right panel, again X and Y are normal, but now $\rho = .2$. So under normality, decreasing ρ from .8 to .2 has a very noticeable effect on the joint distribution of X and Y. Now look at Figure 6.13. It looks similar to the left panel of Figure 6.12, where $\rho = .8$, but now $\rho = .2$. In Figure 6.13 X is again normal, but Y has the mixed normal distribution (described

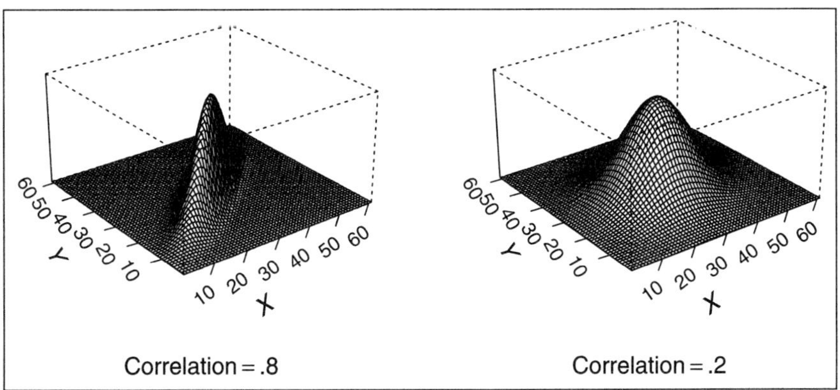

FIGURE 6.12 When both X and Y are normal, increasing ρ from .2 to .8 has a noticeable effect on the bivariate distribution of X and Y.

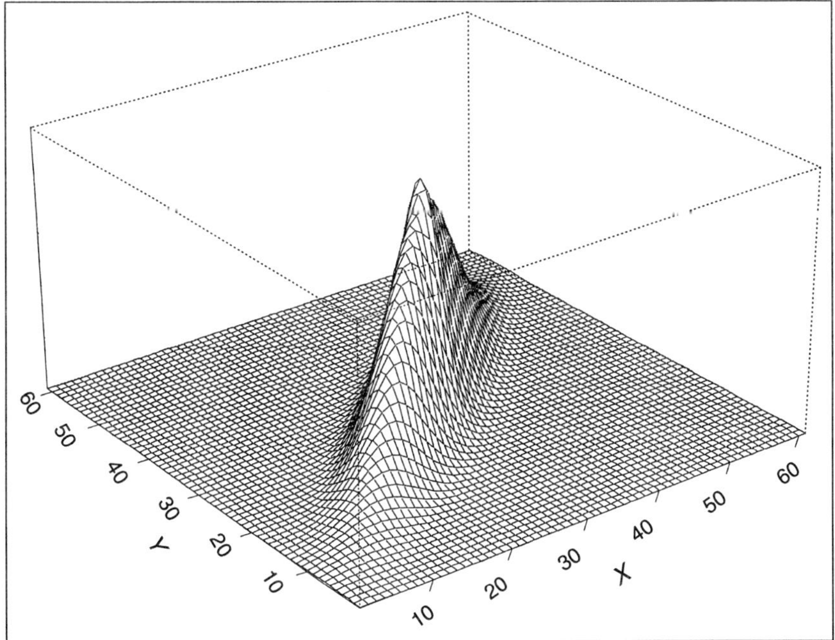

FIGURE 6.13 Two bivariate distributions can appear to be very similar yet have substantially different correlations. Shown is a bivariate distribution with $\rho = .2$, but the graph is very similar to the left panel of Figure 6.12, where $\rho = .8$.

in Section 2.7). This demonstrates that a very small change in any distribution can have a very large impact on ρ. Also, no matter how large the sample size might be, a slight departure from normality can drastically affect r.

A fourth feature that affects the magnitude of r is any restriction in range among the X (or Y) values. To complicate matters, restricting the range of X can increase or decrease r. For example, the left panel of Figure 6.14 shows a scatterplot of points for which $r = .98$. When we eliminate the points with $|X| > 1$, leaving the points shown in the right panel of Figure 6.14, $r = .79$.

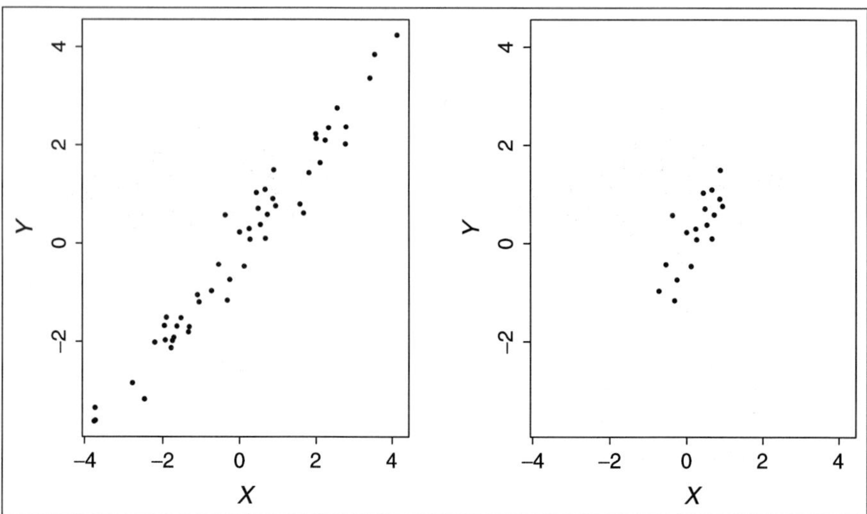

FIGURE 6.14 Restricting the range of X can reduce Pearson's correlation.

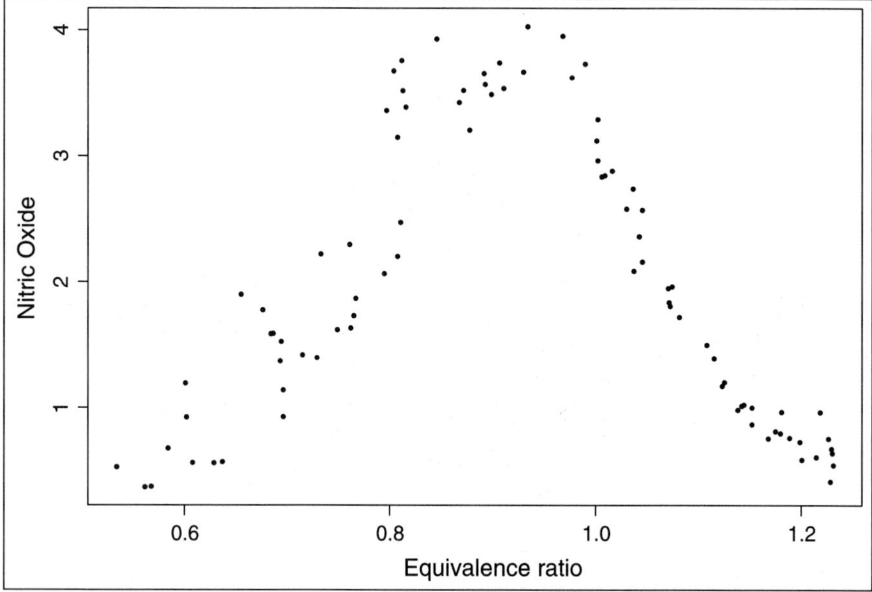

FIGURE 6.15 Curvature occurs in applied work, as illustrated here with data from a study of how concentrations of nitric oxides in engine exhaust are related to its equivalence ratio. Here, Pearson's correlation is only −.1 despite a rather strong association.

The star data in Figure 6.3 illustrate that restricting the range of X (or Y) can increase r as well. If we eliminate all points having $X \leq 4.1$, r increases from −.21 to .65.

A fifth feature that affects r is curvature. Figure 6.15 shows a scatterplot of points relating the concentration of nitric oxides in engine exhaust versus its equivalence ratio, a measure of the richness of the air–ethanol mix. There is a rather obvious association, but the correlation is $r = -.1$, a value relatively close to zero. As another example, if X is standard normal and $Y = X^2$, there is an exact association between X and Y, but $\rho = 0$.

In summary, the following features of data influence the magnitude of Pearson's correlation:

- The slope of the line around which points are clustered
- The magnitude of the residuals
- Outliers
- Restricting the range of the X values, which can cause r to go up or down
- Curvature

A point worth stressing is that although independence implies that $\rho = 0$, $\rho = 0$ does not necessarily imply independence. In fact there are various ways in which X and Y can be dependent, yet ρ is exactly zero. For example, if X and e are independent and $Y = |X|e$, then X and Y are dependent because there is heteroscedasticity, yet $\rho = 0$. More generally, if there is heteroscedasticity and the least squares slope is zero, then $\rho = 0$ as well. As another example, suppose U, V, and W are independent standard normal random variables. Then it can be shown that $X = U/W^2$ and $Y = V/W^2$ are dependent (roughly because both X and Y have the same denominator), yet they have correlation $\rho = 0$.

We conclude this section by noting that the foregoing list of factors that affect the magnitude of r is not exhaustive. Yet another feature of data that affects the magnitude of r is the reliability of the measures under study (e.g., Lord & Novick, 1968), but the details go beyond the scope of this book.

6.5 Testing $H_0 : \rho = 0$

Next we describe the classic test of

$$H_0 : \rho = 0. \tag{6.14}$$

If we can reject this hypothesis, then by implication X and Y are dependent.

If we assume that X and Y are independent and if at least one of these two variables is normal, then

$$T = r\sqrt{\frac{n-2}{1-r^2}} \tag{6.15}$$

has a Student's T distribution with $\nu = n - 2$ degrees of freedom (Muirhead, 1982, p. 146; also see Hogg & Craig, 1970, pp. 339–341). So the decision rule is to reject H_0 if $|T| \geq t$, where t is the $1 - \alpha/2$ quantile of Student's T distribution with $n - 2$ degrees of freedom.

EXAMPLE. For the data in Table 6.3, $n = 47$, $r = -0.286$, so $\nu = 45$ and

$$T = -0.286\sqrt{\frac{45}{1-(-0.286)^2}} = -2.$$

Continued

EXAMPLE. (*Continued*) With $\alpha = .05$, the critical value is $t = 2.01$; because $|-2| < 2.01$, we fail to reject. That is, we are unable to conclude that the aggression scores and recall-test scores are dependent with $\alpha = .05$. ■

Caution must be exercised when interpreting the implications associated with rejecting the hypothesis of a zero correlation with T. Although it is clear that T given by Equation (6.15) is designed to be sensitive to r, homoscedasticity plays a crucial role in the derivation of this test. When in fact there is heteroscedasticity, the derivation of T is no longer valid and can result in some unexpected properties. For instance, it is possible to have $\rho = 0$, yet the probability of rejecting $H_0 : \rho = 0$ *increases* as the sample size gets large.

EXAMPLE. Figure 6.16 shows a scatterplot of 40 points generated on a computer, where both X and Y have normal distributions and $\mu_x = 0$. In this particular case, Y has variance 1 unless $|X| > .5$, in which case Y has standard deviation $|X|$. So X and Y are dependent, but $\rho = 0$. For this situation, when testing at the $\alpha = .05$ level, the actual probability of rejecting H_0 with T is .098 with $n = 20$. For $n = 40$ it is .125, and for $n = 200$ it is .159. The probability of rejecting is *increasing* with n even though $\rho = 0$. When we reject, a correct conclusion is that X and Y are dependent, but it would be incorrect to conclude that $\rho \neq 0$. ■

Some experts might criticize this last example on the grounds that it would be highly unusual to encounter a situation where $\rho = 0$ and there is heteroscedasticity. That is, perhaps we are describing a problem that is theoretically possible but unlikely ever to be encountered. Even if we agree with this argument, the more salient issue is that employing the wrong standard error can lead to highly erroneous results. Generally, conventional methods use correct estimates of standard errors when variables are independent but incorrect estimates when they are dependent, and in the latter case this can lead to poor power and highly inaccurate confidence intervals.

A common alternative to T when making inferences about ρ is to employ what is known as Fisher's *r-to-z transformation*, but we provide no details here because of results in Duncan and Layard (1973). Briefly, Fisher's method requires normality. For nonnormal distributions there are general conditions where the method does not converge to the correct answer as the sample size increases. That is, the method violates the basic principle that the accuracy of our results should increase as n gets large.

6.5.1 The Coefficient of Determination

A positive feature of r is that it provides a useful characterization of how well the least squares regression line summarizes the association between two variables. To explain,

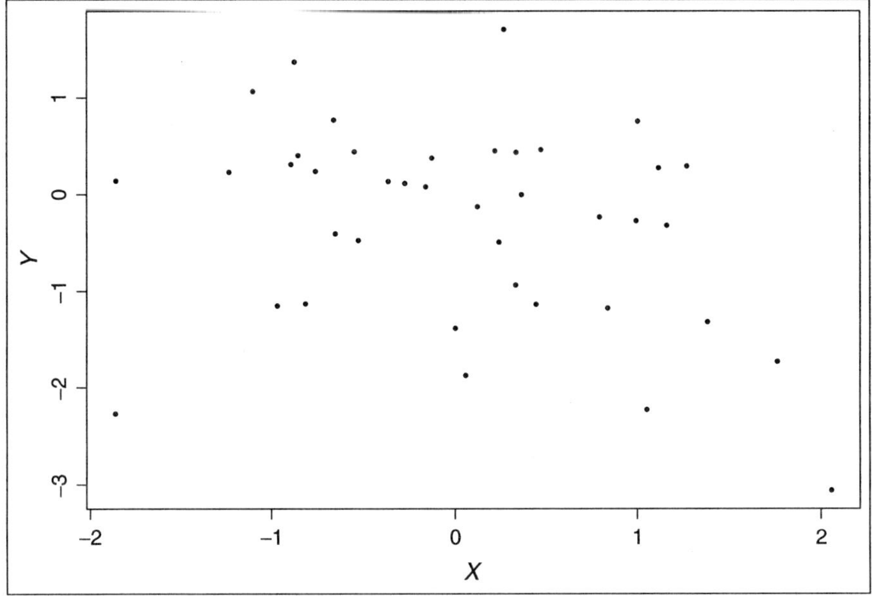

FIGURE 6.16 Example of 40 points that were generated from distributions where $\rho = 0$ and there is heteroscedasticity. Student's T test of $H_0 : \rho = 0$ is unsatisfactory in terms of making inferences about ρ, but it does detect the dependence between X and Y.

imagine that we ignore X in our attempts to predict Y and simply use $\hat{Y} = \bar{Y}$. Then we can measure the accuracy of our prediction rule with

$$\sum (Y_i - \bar{Y})^2,$$

the sum of the squared discrepancies between the Y values we observe and the predicted values, $\bar{Y}$. (This is the method for measuring accuracy already discussed in Section 3.2.8.) Notice that this sum is the numerator of the sample variance of the Y values. If instead we use the least squares regression line $\hat{Y} = \beta_0 + \beta_1 X_i$ to predict Y, then an overall measure of the accuracy of our prediction rule is

$$\sum (Y_i - \hat{Y}_i)^2,$$

as already explained. The difference between these two sums measures the extent to which using $\hat{Y}$ improves upon using $\bar{Y}$. In symbols, this difference is

$$\sum (Y_i - \bar{Y})^2 - \sum (Y_i - \hat{Y}_i)^2.$$

Finally, if we divide this difference by $\sum (Y_i - \bar{Y})^2$, we get a measure of how much $\hat{Y}$ improves upon $\bar{Y}$ relative to $\bar{Y}$. It can be seen that this measure is just r^2, which is called the *coefficient of determination*. In symbols, the coefficient of determination is,

$$r^2 = \frac{\sum (Y_i - \bar{Y})^2 - \sum (Y_i - \hat{Y}_i)^2}{\sum (Y_i - \bar{Y})^2}. \qquad (6.16)$$

We have already seen that even a single outlier can have a substantial impact on both r and the least squares regression line. So when the coefficient of determination (r^2) is fairly large, this does not necessarily mean that the least squares regression line accurately reflects the association among the bulk of the points under study. It simply reflects the extent to which the least squares regression line improves upon the prediction rule $\hat{Y} = \bar{Y}$, the point being that both rules might be very ineffective.

6.5.2 Establishing Independence

Establishing that two measures are independent is a much more difficult goal than showing that they are dependent. If we reject the hypothesis that Pearson's correlation is zero, we can be reasonably certain that the two measures are dependent even though this tells us virtually nothing about what the dependence is like. But if we fail to reject, this is not remotely convincing evidence that we have independence. The basic problem is that the test of $H_0: \rho = 0$ given by Equation (6.15) may not be sensitive to the type of association that exists between the variables under study. There is a rather lengthy list of alternative methods for detecting dependence that attempt to address this problem. (See, for example, Kallenberg & Ledwina, 1999, plus the references they cite.) A few alternative techniques are described in subsequent chapters.

6.6 Concluding Remarks

The purpose of this chapter was to introduce basic concepts and to describe standard hypothesis-testing methods associated with least squares regression and Pearson's correlation. Another goal was to provide some indication of what might go wrong with these standard methods. Some contemporary techniques for addressing these problems are covered in subsequent chapters. It is stressed, however, that regression is a vast topic and that not all methods and issues are discussed in this book. For more about regression, see Li (1985), Montgomery and Peck (1992), Staudte and Sheather (1990), Hampel, Ronchetti, Rousseeuw, and Stahel (1986), Huber (1981), Rousseeuw and Leroy (1987), Belsley, Kuh, and Welsch (1980), Cook and Weisberg (1992), Carroll and Ruppert (1988), Hettmansperger (1984), Hettmansperger and McKean (1998), and Wilcox (1997a).

6.7 Exercises

1. For the following pairs of points, verify that the least squares regression line is $\hat{Y} = 1.8X - 8.5$.

X:	5,	8,	9,	7,	14
Y:	3,	1,	6,	7,	19

2. Compute the residuals using the results from Exercise 1. Verify that if you square and sum the residuals, you get 47, rounding to the nearest integer.
3. Verify that for the data in Exercise 1, if you use $\hat{Y} = 2X - 9$, the sum of the squared residuals is larger than 47. Why would you expect a value greater than 47?
4. Suppose that based on $n = 25$ values, $s_x^2 = 12$, $s_y^2 = 25$, and $r = .6$. What is the slope of least squares regression?
5. Verify that for the data in Table 6.3, the least squares regression line is $\hat{Y} = -0.0405X + 4.581$.
6. The following table reports breast cancer rates plus levels of solar radiation (in calories per day) for various cities in the United States. Fit a least squares regression to the data with the goal of predicting cancer rates and comment on what this line suggests.

City	Rate	Daily calories	City	Rate	Daily calories
New York	32.75	300	Chicago	30.75	275
Pittsburgh	28.00	280	Seattle	27.25	270
Boston	30.75	305	Cleveland	31.00	335
Columbus	29.00	340	Indianapolis	26.50	342
New Orleans	27.00	348	Nashville	23.50	354
Washington, DC	31.20	357	Salt Lake City	22.70	394
Omaha	27.00	380	San Diego	25.80	383
Atlanta	27.00	397	Los Angeles	27.80	450
Miami	23.50	453	Fort Worth	21.50	446
Tampa	21.00	456	Albuquerque	22.50	513
Las Vegas	21.50	510	Honolulu	20.60	520
El Paso	22.80	535	Phoenix	21.00	520

7. For the following data, compute the least squares regression line for predicting gpa given SAT.

SAT:	500	530	590	660	610	700	570	640
gpa:	2.3	3.1	2.6	3.0	2.4	3.3	2.6	3.5

8. For the data in the Exercise 7, verify that the coefficient of determination is .36 and interpret what this tells you.
9. For the following data, compute the least squares regression line for predicting Y from X.

X:	40	41	42	43	44	45	46
Y:	1.62	1.63	1.90	2.64	2.05	2.13	1.94

10. In Exercise 6, what would be the least squares estimate of the cancer rate given a solar radiation of 600? Indicate why this estimate might be unreasonable.
11. Maximal oxygen uptake (mou) is a measure of an individual's physical fitness. You want to know how mou is related to how fast someone can run a mile. Suppose you randomly sample six athletes and get

mou (milliliters/kilogram):	63.3	60.1	53.6	58.8	67.5	62.5
time (seconds):	241.5	249.8	246.1	232.4	237.2	238.4

Compute the correlation. Can you be reasonably certain about whether it is positive or negative with $\alpha = .05$?
12. Verify that for the following pairs of points, the least squares regression line has a slope of zero. Plot the points and comment on the assumption that the regression line is straight.

X:	1	2	3	4	5	6
Y:	1	4	7	7	4	1

13. Repeat Exercise 12; but for the points

X:	1	2	3	4	5	6
Y:	4	5	6	7	8	2

14. Vitamin A is required for good health. You conduct a study and find that as vitamin A intake decreases, there is a linear association with bad health. However, one bite of polar bear liver results in death because it contains a high concentration of vitamin A. Comment on what this illustrates in the context of regression.
15. Sockett et al. (1987) report data related to patterns of residual insulin secretion in children. A portion of the study was concerned with whether age can be used to predict the logarithm of C-peptide concentrations at diagnosis. The observed values are

Age (X):	5.2	8.8	10.5	10.6	10.4	1.8	12.7	15.6	5.8	1.9
	2.2	4.8	7.9	5.2	0.9	11.8	7.9	1.5	10.6	8.5
	11.1	12.8	11.3	1.0	14.5	11.9	8.1	13.8	15.5	9.8
	11.0	12.4	11.1	5.1	4.8	4.2	6.9	13.2	9.9	12.5
	13.2	8.9	10.8							

Continued

C-peptide (Y):	4.8	4.1	5.2	5.5	5.0	3.4	3.4	4.9	5.6
	3.7	3.9	4.5	4.8	4.9	3.0	4.6	4.8	5.5
	4.5	5.3	4.7	6.6	5.1	3.9	5.7	5.1	5.2
	3.7	4.9	4.8	4.4	5.2	5.1	4.6	3.9	5.1
	5.1	6.0	4.9	4.1	4.6	4.9	5.1		

Replace the C-peptide values with their (natural) logarithms. For example, the value 4.8 would be replaced by $\log(4.8) = 1.5686$. Create a scatterplot for these data and consider whether a linear rule for predicting Y with X is reasonable. Also verify that $r = .4$ and that you reject $H_0 : \rho = 0$ with $\alpha = .05$.

16. For the data in Exercise 15, verify that a least squares regression line using only X values (age) less than 7 yields $b_1 = 0.247$ and $b_0 = 3.51$. Verify that when using only the X values greater than 7 you get $b_1 = .009$ and $b_0 = 4.8$. What does this suggest about using a linear rule for all of the data?
17. The housing data in Table 6.4 are from a suburb of Los Angeles where even a small empty lot would cost at least \$200,000 (and probably much more) at the time the data were collected. Verify that based on the least squares regression line for these data, if we estimate the cost of an empty lot by setting the square feet of a house to $X = 0$, we get 38,192. What does this suggest about estimating Y using an X value outside the range of observed X values?
18. For the data in Table 6.1, the sizes of the corresponding lots are:

```
18,200 12,900 10,060 14,500 76,670 22,800 10,880
10,880 23,090 10,875 3,498 42,689 17,790 38,330
18,460 17,000 15,710 14,180 19,840 9,150 40,511
9,060 15,038 5,807 16,000 3,173 24,000 16,600.
```

Verify that the least squares regression line for estimating the selling price, based on the size of the lot, is $\hat{Y} = 11X + 436{,}834$.

19. Imagine two scatterplots where in each scatterplot the points are clustered around a line having slope .3. If for the first scatterplot $r = .8$, does this mean that points are more tightly clustered around the line versus the other scatterplot, where $r = .6$?
20. You measure stress (X) and performance (Y) on some task and get

X:	18	20	35	16	12
Y:	36	29	48	64	18

Verify that you do not reject $H_0 : \beta_1 = 0$ using $\alpha = .05$. Is this result consistent with what you get when testing $H_0 : \rho = 0$? Why would it be incorrect to conclude that X and Y are independent?

21. Suppose you observe

X:	12.2,	41,	5.4,	13,	22.6,	35.9,	7.2,	5.2,	55,	2.4,	6.8,	29.6,	58.7,
Y:	1.8,	7.8,	0.9,	2.6,	4.1,	6.4,	1.3,	0.9,	9.1,	0.7,	1.5,	4.7,	8.2

TABLE 6.6 Reading Data

X:	34	49	49	44	66	48	49	39	54	57	39	65	43	43	44	42
	71	40	41	38	42	77	40	38	43	42	36	55	57	57	41	66
	69	38	49	51	45	141	133	76	44	40	56	50	75	44	181	45
	61	15	23	42	61	146	144	89	71	83	49	43	68	57	60	56
	63	136	49	57	64	43	71	38	74	84	75	64	48			
Y:	129	107	91	110	104	101	105	125	82	92	104	134	105	95	101	104
	105	122	98	104	95	93	105	132	98	112	95	102	72	103	102	102
	80	125	93	105	79	125	102	91	58	104	58	129	58	90	108	95
	85	84	77	85	82	82	111	58	99	77	102	82	95	95	82	72
	93	114	108	95	72	95	68	119	84	75	75	122	127			

Verify that the .95 confidence interval for the slope is (0.14, 0.17). Would you reject $H_0: \beta_1 = 0$? Based on this confidence interval only, can you be reasonably certain that, generally, as X increases, Y increases as well?

22. The data in Table 6.6 are from a study, conducted by L. Doi, where the goal is to understand how well certain measures predict reading ability in children. Verify that the .95 confidence interval for the slope is (−0.16, .12) based on Equation (6.8).

BASIC BOOTSTRAP METHODS

This chapter covers the basics of a modern statistical tool called the *bootstrap*. There is a rather large collection of bootstrap methods, most of which will not be described here. For book-length descriptions of these techniques, the reader is referred to Efron and Tibshirani (1993), Chernick (1999), Davison and Hinkley (1997), Hall and Hall (1995), Lunneborg (2000), Mooney and Duval (1993), and Shao and Tu (1995). The goal here is to introduce and illustrate some of the more basic versions that have considerable practical value in applied work when computing confidence intervals or testing hypotheses.

7.1 The Percentile Method

We begin with what is called the *percentile bootstrap method*. It is stressed that this technique does not perform well when the goal is to make inferences about the population mean based on the sample mean, unless the sample size is very large. However, it has considerable practical value for a wide range of other problems, even with very small sample sizes, and modifications of the method have practical value as well. Although the percentile bootstrap is not recommended when working with the sample mean, it is perhaps easiest to explain in terms of the sample mean, so we start with this special case.

Imagine we want to compute a .95 confidence interval for the population mean, μ. The strategy behind the percentile bootstrap method is to estimate the .025 and .975 quantiles of the sampling distribution of the sample mean and then to use this estimate as a .95 confidence interval for μ. For example, if we estimated that $P(\bar{X} \leq 5) = .025$ and $P(\bar{X} \leq 26) = .975$, then 5 and 26 are the .025 and .975 quantiles, respectively, and (5, 26) would be a .95 confidence interval for μ. Of course, the practical problem is that we do not know the sampling distribution of $\bar{X}$, let alone what the .025 and .975 quantiles might be. If we assume that $\bar{X}$ has a normal distribution, which is reasonable with a sufficiently large sample size, the .025 and .975 quantiles are easily estimated: You simply estimate μ and σ^2 with the sample mean and sample variance, and, assuming these estimates are fairly accurate, the method in Section 4.3 can be used.

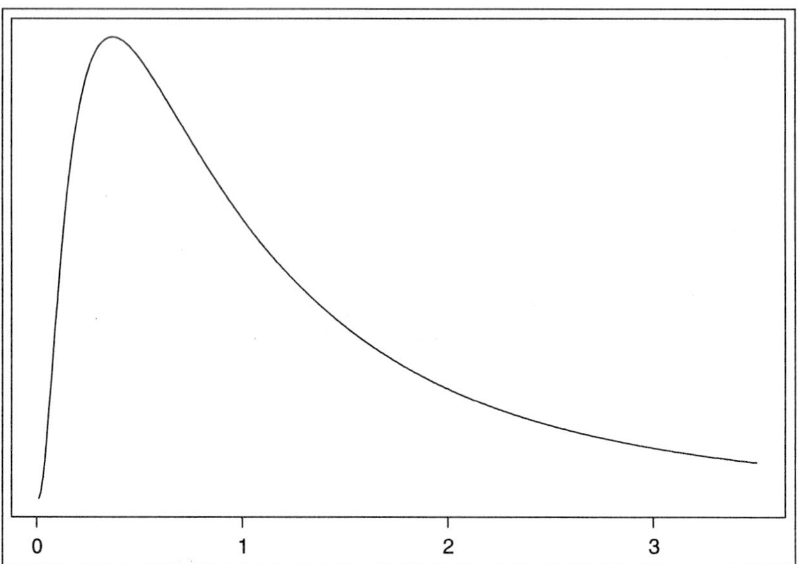

FIGURE 7.1 Example of a skewed, light-tailed distribution. This particular curve belongs to the family of lognormal distributions.

The percentile bootstrap method uses another estimate of the sampling distribution of the sample mean that makes no assumptions about what this distribution might be. In particular, it does not assume the distribution is normal. To understand the basic strategy, first we review a description of the sampling distribution given in Chapter 4. If we were to repeat an experiment infinitely many times, each time computing $\bar{X}$ based on n observations, we would know $P(\bar{X} \le c)$, the probability that the sample mean is less than or equal to c, for any constant c we might choose. Said another way, if we knew the distribution from which observations were sampled, we could use a computer to get a very accurate estimate of the sampling distribution of $\bar{X}$.

For example, imagine we want to determine the sampling distribution of $\bar{X}$ when n observations are randomly sampled from the distribution shown in Figure 7.1. This is an example of a lognormal distribution, and observations can be generated from it using standard software. (In S-PLUS, the function rlnorm accomplishes this goal.) So if we want to know the sampling distribution of $\bar{X}$ when $n = 20$, say, simply generate 20 observations from the lognormal distribution and compute $\bar{X}$. If we repeat this process many times we will have an excellent approximation of what the sampling distribution happens to be. Figure 7.2 shows a plot of 2000 sample means generated in this manner.

Of course, the practical problem is that we do not know the distribution from which observations are sampled. However, we can use the data available to us to estimate this distribution, and this is the basic idea behind all bootstrap techniques. In the simplest case, the strategy is to use the observed relative frequencies as estimates of the probabilities associated with possible values we might observe, and then we simply use a computer to generate observations based on these probabilities. By repeatedly generating many samples of size n, the resulting sample means provide an estimate of the sampling distribution of $\bar{X}$, and the middle 95% of these generated sample means provide an approximate .95 confidence interval for μ.

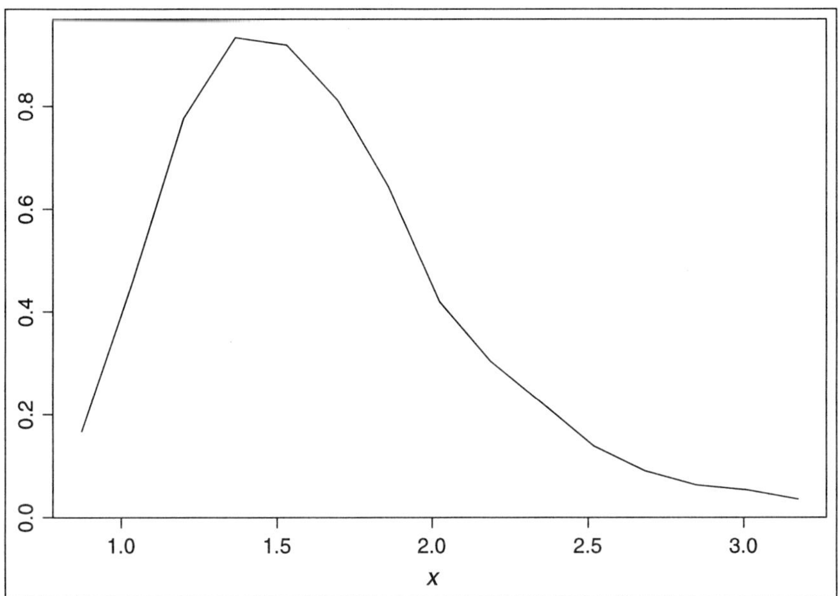

FIGURE 7.2 Plot of 2000 sample means, each mean based on $n = 20$ observations generated from the distribution in Figure 7.1.

To illustrate the idea, imagine we observe the following 10 values:

$$1, 4, 2, 19, 4, 12, 29, 4, 9, 16.$$

The value 1 occurred only once, so we estimate that the value 1 has probability 1/10. Similarly, the value 4 occurred three times, so we estimate its probability to be 3/10. The value 3 occurred zero times, so we estimate its probability to be 0. In the notation of Chapter 3, if the number of times we observe the value x is f_x, $p(x)$ is estimated with f_x/n.

A *bootstrap sample* is obtained by randomly sampling, *with replacement*, observations from the observed values. In our illustration, if we randomly sample a single observation from the values listed in the previous paragraph, we might get the value 12. The probability of getting 12 is 1/10. Or we might get the value 4, and the probability of getting a 4 is 3/10. If we randomly sample a second value from among all 10 values, we might again get 12, or we might get 9 or any of the 10 values from our original sample. If we randomly sample n observations in this manner, we get what is called a *bootstrap sample of size n*. In our example we might get

$$2, 9, 16, 2, 4, 12, 4, 29, 16, 19.$$

The mean of this bootstrap sample is $\bar{X}^* = 11.3$. This is in contrast to the sample mean of our original observations, which is $\bar{X} = 10$. If we were to obtain a second bootstrap sample, the mean of these bootstrap values will typically differ from the first bootstrap sample mean; it might be $\bar{X}^* = 9.6$.

Now imagine that we repeat this process B times, yielding B bootstrap sample means. If B is reasonably large, we get a collection of bootstrap sample means that yields an approximation of the sampling distribution of the sample mean. As an

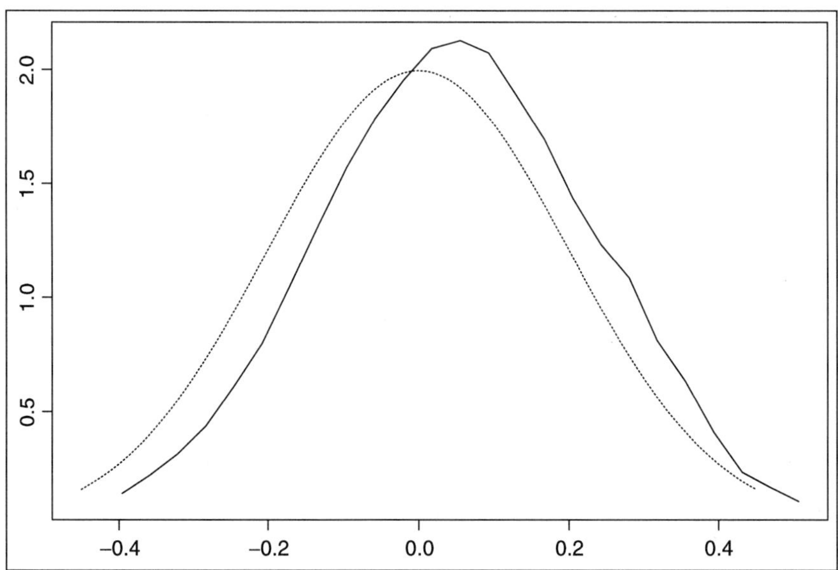

FIGURE 7.3 The solid line is a bootstrap approximation of the sampling distribution of $\bar{X}$ based on 25 values generated from a normal curve. The dashed line shows the exact sampling distribution of $\bar{X}$ under random sampling.

illustration, 25 observations were randomly sampled from a standard normal distribution and then 1000 bootstrap samples were generated in the manner just described. For each of these 1000 bootstrap samples, the sample mean was computed; a plot of these bootstrap means is shown in Figure 7.3. Also shown is the exact sampling distribution of $\bar{X}$. In this particular case, the middle 95% of the 1000 bootstrap means extend from −.32 to .41, and this interval contains 0. That is, (−.32, .41) is an approximate .95 confidence interval for μ and it happens to contain the population mean. This is an example of what is called a *percentile bootstrap confidence interval,* because the strategy is to estimate percentiles of the sampling distribution. However, for the situation at hand, the actual probability that the bootstrap confidence interval will contain μ is less than .95. And for nonnormal distributions this method for computing a confidence interval for μ can be quite unsatisfactory. On the positive side, there are some nonnormal distributions for which the resulting confidence interval is more accurate than Student's T method given in Chapters 4 and 5. So progress has been made, but more needs to be done.

A more general and more formal description of the percentile bootstrap method will be helpful. Let $X_1, \dots, X_n$ represent a random sample of observations, and let $X_1^*, \dots, X_n^*$ represent a bootstrap sample of size n that is obtained by randomly sampling, with replacement, n values from $X_1, \dots, X_n$. The sample mean of this bootstrap sample is just

$$\bar{X}^* = \frac{1}{n} \sum X_i^*.$$

Now suppose we repeat the process of generating a bootstrap sample mean B times, and we label these sample means $\bar{X}_1^*, \dots, \bar{X}_B^*$. Then an approximate $1 - \alpha$ confidence interval for μ is

$$\left(\bar{X}^*_{(\ell+1)}, \bar{X}^*_{(u)}\right), \tag{7.1}$$

where $\bar{X}^*_{(1)} \leq \cdots \leq \bar{X}^*_{(B)}$ are the B bootstrap means written in ascending order, $\ell = \alpha B/2$, rounded to the nearest integer, and $u = B - \ell$. So if $B = 20$ and $1 - \alpha = .8$, ℓ is 2 and u is 18. For the special case $\alpha = .05$, $\ell = .025B$ (still rounding to the nearest integer), and $\bar{X}^*_{(\ell+1)}$ and $\bar{X}^*_{(u)}$ estimate the .025 and .975 quantiles of the distribution of $\bar{X}$, respectively. In general, $\bar{X}^*_{(\ell+1)}$ and $\bar{X}^*_{(u)}$ contain the middle $(1 - \alpha)$ percent of the B bootstrap sample means.

One final point might help before ending this section. An attempt has been made to provide some intuitive sense of how the percentile bootstrap is applied. But from a theoretical point of view, more needs to be done to justify this technique. Here it is merely noted that a formal justification has been derived (e.g., Hall, 1988a, 1988b; Liu and Singh, 1997). Although complete theoretical details cannot be given here, a crude description of the theoretical underpinnings might help. To this end, imagine you want to test the hypothesis $H_0: \mu = 12$ and that the null hypothesis is true. When you generate a bootstrap sample, there will be some probability that the bootstrap sample mean will be less than 12. For convenience, label this probability p^*. That is, $p^* = P(\bar{X}^* < 12)$. Of course, p^* is not known, but it can be estimated with $\hat{p}^*$, the proportion of the B bootstrap sample means that is less than 12. Note that associated with every random sample is some p^* value. That is, repeating a study infinitely many times will produce infinitely many p^* values. With a sufficiently large sample size, p^* will have a uniform distribution when the null hypothesis is true and in fact is like a significance level or p-value. (Under random sampling, if the null hypothesis is true, the significance level of Student's T test also converges to a uniform distribution as the sample size increases.) Moreover, with B sufficiently large, $\hat{p}^*$ will provide a reasonably accurate estimate of p^*, and $2\hat{p}^*$ is the estimated significance level (or p-value) when testing a two-sided hypothesis. This also leads to the confidence interval given by Equation (7.1).

7.1.1 A Bootstrap Estimate of Standard Errors

Situations arise where expressions for the standard error of some estimator is not known or it takes on a rather complicated form. Examples are M-estimators and MOM. So if it is desired to estimate their standard errors, it would be convenient to have a relatively simple method for accomplishing this goal. The percentile bootstrap method, just described, is one way of tackling this problem.

Let $\hat{\mu}$ be any measure of location and let $\hat{\mu}^*$ be its value based on a bootstrap sample. Let $\hat{\mu}^*_b$ $(b = 1, \ldots, B)$ be B bootstrap estimates of the measure of location. Then an estimate of the squared standard error of $\hat{\mu}$ is

$$S^2 = \frac{1}{B-1} \sum_{b=1}^{B} (\hat{\mu}^*_b - \bar{\mu}^*)^2,$$

where $\bar{\mu}^* = \sum_{b=1}^{B} \hat{\mu}^*_b / B$.

7.1.2 S-PLUS Function bootse

The S-PLUS function

```
bootse(x,nboot=1000,est=median)
```

computes S, the bootstrap estimate of the standard error of the measure of location specified by the argument est. By default, the median is used, and the argument nboot corresponds to B, which defaults to 1000.

7.2 The Bootstrap-*t* Interval

This section describes what is called the bootstrap-t (or the percentile-t) method. The basic idea is that if we knew the distribution of

$$T = \frac{\bar{X} - \mu}{s/\sqrt{n}}, \tag{7.2}$$

a confidence interval for the population mean could be computed. As explained in Chapter 4, the conventional strategy is to assume normality or to assume that the sample size is sufficiently large, in which case T has a Student's T distribution. But we have already seen that confidence intervals and control over the probability of a Type I error can be unsatisfactory with $n = 160$ when sampling from a skewed, light-tailed distribution. And sample sizes greater than 300 can be required when sampling from a skewed, heavy-tailed distribution instead. A better approximation of the distribution of T is needed.

The bootstrap strategy for estimating the distribution of T begins in the same manner used in the percentile method: Obtain a bootstrap sample of size n. As in the previous section, we let $X_1, \ldots, X_n$ represent the original observations and $X_1^*, \ldots, X_n^*$ represent a bootstrap sample of size n that is obtained by randomly sampling, with replacement, n values from $X_1, \ldots, X_n$. Let $\bar{X}^*$ and s^* be the mean and standard deviation based on this bootstrap sample. That is,

$$\bar{X}^* = \frac{1}{n}\sum X_i^*$$

and

$$s^* = \sqrt{\frac{1}{n-1}\sum (X_i^* - \bar{X}^*)^2}.$$

Also let

$$T^* = \frac{\bar{X}^* - \bar{X}}{s^*/\sqrt{n}}. \tag{7.3}$$

Notice that when obtaining a bootstrap sample, we know the mean of the distribution from which the bootstrap sample was obtained. It is $\bar{X}$. So in the bootstrap world, $\bar{X}$ plays the role of μ, and $\bar{X}^*$ plays the role of $\bar{X}$.

If we repeat the foregoing process B times, yielding B T^* values, we obtain an approximation of the sampling distribution of T, and in particular we have an estimate of its .025 and .975 quantiles. The estimate of these quantiles is based on the middle 95% of the T^* values. In more formal terms, if we let $T^*_{(1)} \le T^*_{(2)} \le \cdots \le T^*_{(B)}$ be the B bootstrap T^* values written in ascending order, and we let $\ell = .025B$, rounded to the nearest integer, and $u = B - \ell$, an estimate of the .025 and .975 quantiles of the distribution of T is $T^*_{(\ell+1)}$ and $T^*_{(u)}$. The resulting .95 confidence interval for μ is

$$\left(\bar{X} - T^*_{(u)}\frac{s}{\sqrt{n}}, \bar{X} - T^*_{(\ell+1)}\frac{s}{\sqrt{n}}\right). \tag{7.4}$$

(In this last equation, $T^*_{(\ell+1)}$ is negative, which is why it is subtracted, not added, from $\bar{X}$. Also, it might seem that $T^*_{(u)}$ should be used to compute the upper end of the confidence interval, not the lower end, but it can be shown that this is not the case.)

HYPOTHESIS TESTING. To test $H_0: \mu = \mu_0$, compute

$$T = \frac{\bar{X} - \mu_0}{s/\sqrt{n}}$$

and reject if

$$T \le T^*_{(\ell+1)},$$

or if

$$T \ge T^*_{(u)}.$$

EXAMPLE. Forty observations were generated from a standard normal distribution, and then the bootstrap-t method was used to approximate the distribution of T with $B = 1000$. A plot of the 1000 bootstrap T^* values is shown in Figure 7.4. The smooth symmetric curve is the correct distribution (a Student's T distribution with $\nu = 39$). In this particular case, the bootstrap estimate of the distribution of T is fairly accurate. The bootstrap estimates of the .025 and .975 quantiles are $-T^*_{(u)} = -2.059$ and $-T^*_{(\ell)} = 2.116$. The correct answers are -2.022 and 2.022, respectively. ■

Both theoretical and simulation studies indicate that generally, the bootstrap-t performs better than the percentile bootstrap or Student's T when computing a confidence interval or testing some hypothesis about μ. There are exceptions, such as when sampling from a normal distribution, but to avoid poor probability coverage, the bootstrap-t method is preferable to Student's T or the percentile bootstrap. (However, when working with robust measures of location, we will see that typically the percentile bootstrap is preferable to the bootstrap-t.)

From a theoretical point of view, the improvements achieved by the bootstrap-t method over Student's T are not surprising. To roughly explain why, note that when computing a $1 - \alpha$ confidence interval with Student's T, there will be some discrepancy between the actual probability coverage and the value for $1 - \alpha$ that you

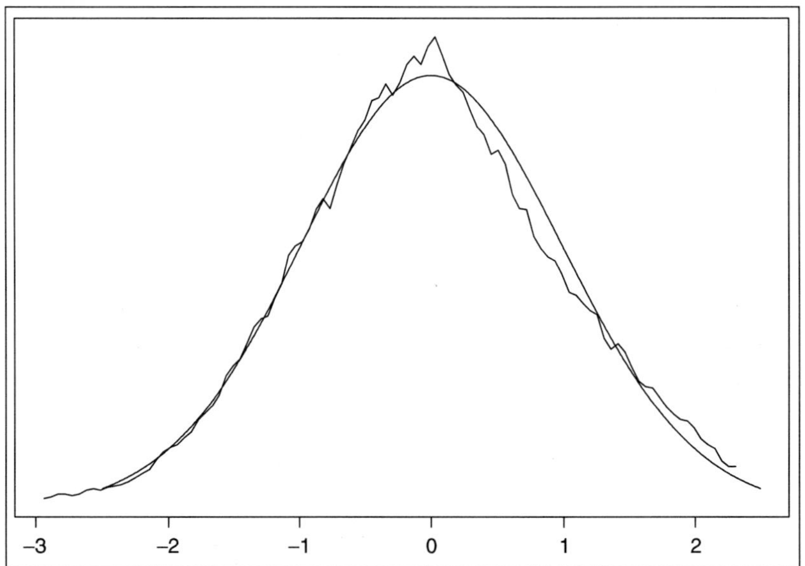

FIGURE 7.4 A plot of 1000 bootstrap T^* values. These T^* values are attempting to approximate the smooth, symmetric curve.

have picked. *When the sample size is large*, mathematicians are able to characterize the rate at which this discrepancy goes to zero; it is $1/\sqrt{n}$. When using the bootstrap-t interval instead, the rate this discrepancy goes to zero is now $1/n$. The discrepancy goes to zero faster using the bootstrap-t, suggesting that it will have better probability coverage and better control over the probability of a Type I error. This mathematical result is encouraging, but the theoretical tools being used tell us only what happens when sample sizes are large. There are known situations where these tools are highly misleading when sample sizes are small — say, less than 150 — but simulation studies aimed at assessing performance when sample sizes are small again indicate that the bootstrap-t is preferable to the percentile bootstrap or Student's T (e.g., Westfall & Young, 1993).

But despite the theoretical appeal of the bootstrap-t method when trying to find an accurate confidence interval for the mean, and even though it improves upon Student's T in certain situations, the method can be unsatisfactory. For example, if we sample 20 observations from the mixed normal shown in Figure 2.8, and we compute a .95 bootstrap-t confidence interval with $B = 1000$, the actual probability coverage is only .9. Put another way, if we reject $H_0 : \mu = \mu_0$ if the .95 bootstrap-t confidence interval does not contain μ_0, the actual probability of a Type I error will not be .05 as intended, but close to .1. Theory tells us that as both n and B get large, if we compute a $1 - \alpha$ confidence interval with the bootstrap-t method, the actual probability coverage will converge to $1 - \alpha$. For the situation at hand, simply increasing B, with n fixed, does not improve matters very much. Increasing n to 100, the actual probability of a Type I error (still testing at the .05 level) is .09. The seriousness of a Type I error will vary from one situation to the next, but some authorities would argue that when testing some hypothesis with $\alpha = .05$, usually the actual probability of a Type I error should not exceed .075 and should not drop below .025 (e.g., Bradley, 1978). One argument for being dissatisfied with an actual Type I error probability of .075 is that if a researcher believes that a Type I error probability of .075 is acceptable, she would have set $\alpha = .075$ in the first place to achieve higher power.

If we sample observations from a skewed heavy-tailed distribution, such as the one shown in Figure 5.6, and then we apply the bootstrap-*t* method at the $\alpha = .05$ level with $n = 20$, the actual probability of a Type I error is .198. This is not much better than using Student's T, where the actual Type I error probability is .202. Increasing n to 100 it drops to .168 using the bootstrap-*t* method. Student's T is even less satisfactory: The actual Type I error probability drops to only .190. So both methods are improving as the sample size gets large, but at a rather slow rate. Even with $n = 300$ the actual Type I error probability remains above .15 when using the bootstrap-*t*, and it is worse using Student's T.

We saw in Chapter 5 that Student's T is biased: When testing $H_0 : \mu = \mu_0$, the probability of rejecting is not minimized when $\mu = \mu_0$. (In practical terms, the probability of rejecting might be higher when H_0 is true versus certain situations where it is false.) The bootstrap-*t* method reduces this problem but does not eliminate it.

Chapter 5 pointed out that arbitrarily small departures from normality can destroy power when using Student's T to make inferences about the population mean. Switching to the bootstrap-*t* method, or any other bootstrap method, does not address this problem.

7.2.1 Symmetric Confidence Intervals

A variation of the bootstrap-*t* method should be mentioned that can be used when testing a two-sided hypothesis only. Rather than use T^* as defined by Equation (7.3), use

$$T^* = \frac{|\bar{X}^* - \bar{X}|}{s^*/\sqrt{n}}, \tag{7.5}$$

and reject $H_0 : \mu = \mu_0$ if $|T| \geq T^*_{(c)}$, where $c = (1 - \alpha)B$ rounded to the nearest integer and again $T^*_{(1)} \leq \cdots \leq T^*_{(B)}$ are the B bootstrap T^* values written in ascending order. An approximate $1 - \alpha$ confidence interval for μ is now given by

$$\bar{X} \pm T^*_{(c)} \frac{s}{\sqrt{n}}. \tag{7.6}$$

This is called a *symmetric* two-sided confidence interval, meaning that the same quantity $(T^*_{(c)} s/\sqrt{n})$ is added and subtracted from the mean when computing a confidence interval. In contrast is the confidence interval given by Equation (7.4), which is called an *equal-tailed* confidence interval. With large sample sizes, the symmetric two-sided confidence interval enjoys some theoretical advantages over the equal-tailed confidence interval (Hall, 1988a, 1988b). The main point here is that when sample sizes are small, probability coverage and control over the probability of a Type I error can again be unsatisfactory. In some cases the actual probability coverage of these two methods differs very little, but exceptions arise. For example, when sampling from the mixed normal ($n = 20$) and testing at the .05 level, the actual Type I error probability using the symmetric confidence interval [given by Equation (7.6)] has probability coverage .014, compared to .10 when using the equal-tailed method [given by equation (7.4)]. So in this particular case, the symmetric confidence interval does a better job of avoiding a Type I error that is substantially higher than the nominal level. But there are situations where the symmetric confidence interval is less satisfactory than the equal-tailed method. Moreover, even when the equal-tailed method has a Type I error probability substantially higher than the nominal α level, switching to the symmetric confidence interval can make matters worse. In practical terms, given some data, it is difficult knowing which of these two methods should be preferred. With a large sample size, currently it seems that it makes little practical difference.

7.3 A Modified Percentile Method for Least Squares Regression and Pearson's Correlation

Both the percentile and bootstrap-t methods have been considered when computing a .95 confidence interval for the slope of a least squares regression line—and both have been found to be unsatisfactory when sample sizes are small or even moderately large. For example, there are known situations where the percentile method requires $n = 250$ to get reasonably accurate probability coverage. However, in a comparison of several methods, Wilcox (1996b) found a slight modification of the percentile method that performs reasonably well over a relatively broad class of nonnormal distributions, even when $n = 20$. Moreover, unlike the conventional method for computing a confidence interval for the slope, the method performs well when there is heteroscedasticity.

To begin, we first note that in regression, two general methods have been considered for generating a bootstrap sample. The first is based on resampling values from the residuals, but no details are given because the method deals poorly with heteroscedasticity (e.g., Wu, 1986). The second method is much more flexible (for general reasons detailed by Efron and Tibshirani, 1993, pp. 113–115). It allows heteroscedasticity, so we focus on this method here. In particular, we obtain a bootstrap sample simply by resampling, *with replacement*, n pairs of values from the original n pairs of values used to compute the least squares estimate of the slope and intercept. In symbols, if we observe $(X_1, Y_1), \ldots, (X_n, Y_n)$, a bootstrap sample is obtained by resampling n pairs of these points, with each pair of points having probability $1/n$ of being resampled. If, for example, we observe

$$(6, 2), (12, 22), (10, 18), (18, 24), (16, 29),$$

a bootstrap sample might be

$$(10, 18), (16, 29), (10, 18), (6, 2), (6, 2).$$

That is, there are $n = 5$ pairs of points, so with each resample there is a 1/5 probability that the first pair of points selected will be (6, 2), and this is true for the other four pairs of values as well. A common notation for a bootstrap sample obtained in this manner is $(X_1^*, Y_1^*), \ldots, (X_n^*, Y_n^*)$. In our illustration, $(X_1^*, Y_1^*) = (10, 18)$ and $(X_2^*, Y_2^*) = (16, 29)$.

The least squares estimate of the slope and intercept based on this bootstrap sample is represented by b_1^* and b_0^*, respectively.

The basic percentile bootstrap method described in Section 7.1 extends to the situation at hand in a simple manner. To compute a .95 confidence interval for the slope, first repeat the process of generating a bootstrap sample of size n B times, yielding B bootstrap estimates of the slope, which we label $b_{11}^*, \ldots, b_{1B}^*$. Then an approximate .95 confidence interval for the slope is given by the middle 95% of these bootstrap estimates. In symbols, we write these B bootstrap estimates of the slope in ascending order as $b_{1(1)}^* \leq b_{1(2)}^* \leq \cdots \leq b_{1(B)}^*$. Letting $\ell = .025B$ and setting $u = B - \ell$ then rounding ℓ and u to the nearest integer, an approximate .95 confidence interval for the slope is

$$\left(b_{1(\ell+1)}^*, b_{1(u)}^*\right).$$

Although the probability coverage of the confidence interval just given can differ substantially from .95 when n is less than 250, it has a property of considerable practical value: Given n, the actual probability coverage is fairly stable over a relatively wide range of distributions, even when there is a fairly large degree of heteroscedasticity and the sample size is small. This suggests a method for getting a reasonably accurate confidence interval: Adjust the confidence interval so that the actual probability coverage is close to .95 when sampling from a normal distribution and there is homoscedasticity. Then use this adjusted confidence interval for nonnormal distributions or when there is heteroscedasticity. So we adjust the percentile bootstrap method when computing a .95 confidence interval, based on the least squares regression estimator, in the following manner. Take $B = 599$, and for each bootstrap sample compute the least squares estimate of the slope. Next, put these 599 values in ascending order yielding $b_{1(1)}^* \leq \cdots \leq b_{1(599)}^*$. The .95 confidence interval is

$$\left(b_{1(a)}^*, b_{1(c)}^*\right), \tag{7.7}$$

where for $n < 40$, $a = 7$ and $c = 593$; for $40 \leq n < 80$, $a = 8$ and $c = 592$; for $80 \leq n < 180$, $a = 11$ and $c = 588$; for $180 \leq n < 250$, $a = 14$ and $c = 585$; while for $n \geq 250$, $a = 15$ and $c = 584$. Said another way, these choices for a and c stem from Gosset's strategy for dealing with small sample sizes: Assume normality for a given sample size determine the (critical) value so that the probability of a Type I error is α, and then hope that these values continue to give good results under nonnormality. This strategy performs relatively well here (Wilcox, 1996b), but it does not perform very well for other problems, such as when computing a confidence interval for the mean. Confidence intervals based on Equation (7.7) will be called the *modified percentile* bootstrap method.

HYPOTHESIS TESTING. Reject $H_0: \beta_1 = 0$ if the confidence interval for the slope, given by Equation (7.7), does not contain zero.

A confidence interval for the intercept can be computed in a similar manner. You simply replace b_1, the least squares estimate of the slope, with b_0, the estimate of the intercept in the description of the modified bootstrap method just given.

Although situations arise where we get more accurate probability coverage than the conventional method covered in Chapter 6, practical problems still occur. That is, we get accurate probability coverage when computing a confidence interval for the slope under a relatively broad range of situations, but the modified percentile bootstrap method is less successful when dealing with the intercept. How to improve upon the modified bootstrap method when dealing with the intercept remains unknown.

There are two practical points to keep in mind when comparing the bootstrap confidence interval for the slope just described to the conventional method in Chapter 6. First, often the bootstrap method yields a longer confidence interval because its probability coverage is generally much closer to the nominal .95 level — the actual probability coverage of the conventional method is often much smaller than .95. In some cases the actual probability coverage drops below .5! That is, the conventional method often gives a shorter confidence interval because it is not nearly as accurate as the modified percentile bootstrap method. Second, despite having longer confidence intervals, situations arise where the bootstrap method rejects $H_0: \beta_1 = 0$ and the conventional method does not.

EXAMPLE. Using the aggression data in Table 6.3, it was already illustrated that the hypothesis $H_0: \beta_1 = 0$ is not rejected with $\alpha = .05$ using the conventional Student's T test given by Equation (6.10). Using the modified percentile bootstrap method, the .95 confidence interval for the slope is (−0.105, −0.002), this interval does not contain zero, so you reject. The .95 confidence interval based on Student's T [using Equation (6.8)] is (−0.08, 0.0002). ■

EXAMPLE. For the selling price of homes in Table 6.4, the .95 confidence interval using the bootstrap method is (.166, .265) versus (.180, .250) using Student's T. Student's T gives a shorter confidence interval, but it might be substantially less accurate because it is sensitive to violations of the assumptions of normality and homoscedasticity. ■

7.3.1 S-PLUS Function lsfitci

The S-PLUS function

```
lsfitci(x,y)
```

computes a modified .95 percentile bootstrap confidence interval for the slope and intercept of a least squares regression line. Here x is a vector of predictor values and y is a corresponding vector of outcome values.

7.3.2 Testing for Zero Correlation

The modified percentile bootstrap method just described performs relatively well when the goal is to test the hypothesis of a zero correlation (Wilcox & Muska, 2001).

You proceed exactly as already described in this section, except for every bootstrap sample you compute Pearson's correlation r rather than the least squares estimate of the slope. So now we have B bootstrap values for r, which, when written in ascending order, we label $r^*_{(1)} \leq \cdots \leq r^*_{(B)}$. Then a .95 confidence interval for ρ is

$$\left(r^*_{(a)}, r^*_{(c)}\right),$$

where again for $n < 40$, $a = 7$ and $c = 593$; for $40 \leq n < 80$, $a = 8$ and $c = 592$; for $80 \leq n < 180$, $a = 11$ and $c = 588$; for $180 \leq n < 250$, $a = 14$ and $c = 585$; while for $n \geq 250$, $a = 15$ and $c = 584$. As usual, if this interval does not contain zero, reject $H_0: \rho = 0$.

We saw in Chapter 6 that heteroscedasticity causes Student's T test of $H_0: \rho = 0$ to have undesirable properties. All indications are that the modified percentile bootstrap eliminates these problems. When $\rho \neq 0$, the actual probability coverage remains fairly close to the .95 level provided ρ is not too large. But if, for example, $\rho = .8$, the actual probability coverage of the modified percentile bootstrap method can be unsatisfactory in some situations (Wilcox & Muska, 2001). There is no known method for correcting this problem.

7.3.3 S-PLUS Function corb

The S-PLUS function

corb(x,y)

computes a .95 confidence interval for ρ using the modified percentile bootstrap method. Again, x and y are S-PLUS variables containing vectors of observations.

EXAMPLE. For the aggression data in Table 6.3, Student's T test fails to reject the hypothesis that the correlation is zero. Using the modified percentile bootstrap method instead, the S-PLUS function corb returns a .95 confidence interval of (−0.54, −0.01). So now we reject $H_0: \rho = 0$ (because the confidence interval does not contain zero), and we conclude that these two variables are dependent. ■

7.4 More About the Population Mean

For many situations encountered in statistics, it is now possible to compute reasonably accurate confidence intervals even under fairly extreme departures from standard assumptions. But making accurate inferences about the population mean remains one of the more difficult problems. In terms of avoiding Type I errors greater than the nominal level, Student's T is satisfactory when sampling from a perfectly symmetric distribution. But for skewed distributions, it can be quite unsatisfactory, even with a sample size of 300. Yet another strategy is to use the modified bootstrap method introduced in Section 7.3. To provide some sense of how the modified bootstrap

TABLE 7.1 Actual Type I Error Probabilities for Four Methods Based on the Mean, $\alpha = .05$

	Dist.	Method BT	SB	MP	T
$n = 20$	N	.054	.051	.041	.050
	LN	.078	.093	.096	.140
	MN	.100	.014	.050	.022
	SH	.198	.171	.190	.202
$n = 100$	N	.048	.038	.049	.050
	LN	.058	.058	.063	.072
	MN	.092	.018	.054	.041
	SH	.168	.173	.177	.190

N = normal; LN = lognormal; MN = mixed normal; SH = skewed, heavy-tailed; BT = equal-tailed, bootstrap-*t*; SB = symmetric bootstrap-*t*; MP = modified percentile bootstrap; T = Student's T.

performs, Table 7.1 shows the actual probability of a Type I error when testing $H_0: \mu = \mu_0$ with $\alpha = .05$ and $n = 20$ and 100. In Table 7.1, BT indicates the equal-tailed bootstrap-*t* method [given by Equation (7.4)], SB is the symmetric bootstrap-*t* method [given by Equation (7.6)], MP is the modified bootstrap method described in Section 7.3 in conjunction with least squares regression, and T indicates Student's T. The distributions considered here are normal (N), lognormal (LN), which is shown in Figure 7.1 and represents a distribution that is skewed with relatively light tails, mixed normal (MN), and a skewed, heavy-tailed distribution (SH), which is shown in Figure 5.6. So, for example, when sampling from a lognormal distribution and testing at the .05 level with $n = 20$, the actual probability of a Type I error with Student's T is .14.

Notice that for distribution LN, the equal-tailed bootstrap-*t* method is substantially better than Student's T; this has been one of the reasons the equal-tailed bootstrap-*t* method has been recommended. In this particular case, the equal-tailed bootstrap-*t* also beats the symmetric bootstrap-*t* as well as the modified percentile method, which is more accurate than the percentile method described in Section 7.1. However, for the mixed normal, the equal-tailed bootstrap-*t* method is the least satisfactory. In this particular case, if we switch to the symmetric bootstrap-*t* method, the Type I error probability is .014 with $n = 20$ and .018 with $n = 100$. But when sampling from a lognormal distribution, the actual Type I error probability is .093, and with $n = 100$ there is little difference between the two bootstrap-*t* methods for this special case. For the mixed normal, the symmetric bootstrap-*t* has an actual Type I error probability well below the nominal .05 level, suggesting that in this particular case its power might be low relative to the modified percentile bootstrap method. Unfortunately, all of these methods are highly unsatisfactory when sampling from a skewed, heavy-tailed distribution, and the situation improves very slowly as the sample size increases. So we see that in situations where Student's T is unsatisfactory in terms of Type I errors, we can get improved results with some type of bootstrap method. But the choice of which bootstrap method to use depends on the situation, and all four methods considered here can be unsatisfactory, even with $n = 100$, if sampling happens to be from a skewed, heavy-tailed distribution. One could check whether a distribution appears to be skewed and heavy-tailed, but an effective diagnostic tool that detects situations where these four methods fail to control the probability of a Type I error has not been established.

The lognormal distribution is a relatively light-tailed distribution. We have just seen that as we move toward a skewed distribution, where outliers are more common, all four methods in Table 7.1 begin to break down. Sutton (1993) proposed a bootstrap method that improves upon a method for handling skewed distributions proposed by Johnson (1978), but Sutton's method deals with skewed distributions for a certain special case only. In particular, if it is known that a distribution is skewed to the right and the goal is to test $H_0: \mu \leq \mu_0$, the method can be employed, but the method is not designed to handle $H_0: \mu \geq \mu_0$. If the distribution is skewed to the left, now you can test $H_0: \mu \geq \mu_0$ but not the other. More recently, Chen (1995) proposed a modification that avoids the bootstrap, but it too is based on the same restrictions. That is, if a distribution is skewed to the right, you can test $H_0: \mu \leq \mu_0$, but not $H_0: \mu \geq \mu_0$. Chen's method appears to perform well, provided the distribution is not too heavy-tailed. If the distribution is heavy-tailed, its control over the probability of a Type I error becomes unsatisfactory. Currently, no method has been found that provides accurate inferences about μ when sampling from a skewed, heavy-tailed distribution unless the sample size is very large. The only certainty is that in some situations, even $n = 300$ is not large enough.

7.5 Inferences About a Trimmed Mean

The Tukey–McLaughlin method for making inferences about the trimmed mean (covered in Sections 4.9 and 5.6) reduces the problems associated with Student's T. Generally, as we increase the amount of trimming, the problem of low power under very small departures from normality is reduced, and we get improved control over the probability of a Type I error. But problems with controlling the Type I error probability, or probability coverage when computing a confidence interval, persist. Combining trimmed means with an appropriate bootstrap method reduces these problems considerably. In fact, with 20% trimming, good control over the probability of a Type I error can be achieved under fairly extreme departures from normality, even with $n = 11$.

7.5.1 Using the Percentile Method

The percentile bootstrap is applied using a simple modification of the method described in Section 7.1: Simply replace the sample mean by the trimmed mean. So if we generate a bootstrap sample of size n and compute the trimmed mean, $\bar{X}_t^*$, and if we repeat this B times, yielding $\bar{X}_{t1}^*, \ldots, \bar{X}_{tB}^*$, then an approximate $1 - \alpha$ confidence interval for the population trimmed mean is

$$\left(\bar{X}_{t(\ell+1)}^*, \bar{X}_{t(u)}^*\right), \tag{7.8}$$

where $\bar{X}^*_{t(1)} \leq \cdots \leq \bar{X}^*_{t(B)}$ are the B bootstrap trimmed means written in ascending order, $\ell = \alpha B/2$, rounded to the nearest integer, and $u = B - \ell$. As usual, reject $H_0 : \mu_t = \mu_0$ if the confidence interval for the trimmed mean [Equation (7.8) in this particular case] does not contain the hypothesized value, μ_0.

The performance of the percentile bootstrap improves in terms of Type I errors and probability coverage as we increase the amount of trimming and becomes fairly accurate with at least 20% trimming. Moreover, with 20% trimming, the modified percentile bootstrap considered in Section 7.3 is no longer needed and performs in an unsatisfactory manner. But the minimum amount of trimming needed to justify the percentile bootstrap is not known. The only rule currently available is that with a minimum of 20% trimming, accurate results can be obtained even with very small sample sizes. So in particular, the percentile bootstrap method performs well when working with the median. Perhaps the percentile bootstrap continues to perform well with 15% trimming, or even 10% trimming, but this has not been established.

7.5.2 Singh's Modification

Imagine 10 observations with two extreme outliers. For example, suppose we observe

2, 3, 6, 3, 9, 12, 15, 7, 200, 300.

With 20% trimming, the two outliers have no influence on $\bar{X}_t$. Notice, however, that when we generate a bootstrap sample, by chance we might get three outliers. That is, the number of outliers in a bootstrap sample might exceed the finite-sample breakdown point of the trimmed mean even though this is not the case for the original observations. The result is that the bootstrap trimmed mean becomes inflated, and this can lead to a relatively long confidence interval when using the percentile bootstrap method.

Singh (1998) showed that from a theoretical point of view, we can address this problem by first Winsorizing the data before we take bootstrap samples. The only restriction imposed by theory is that the amount of Winsorizing must be less than or equal to the amount of trimming. So in our example, if we plan to use a 20% trimmed mean, we are allowed to Winsorize our values by 20%, in which case the 10 values in our example become

6, 6, 6, 3, 9, 12, 15, 7, 15, 15.

Now we generate bootstrap samples as before, but we resample with replacement from the Winsorized values rather than from the original observations. After generating B bootstrap trimmed means, we apply the percentile bootstrap method in the usual manner. For example, the middle 95% of the bootstrap trimmed means provide a .95 confidence interval.

Unfortunately, if we Winsorize as much as we trim, it seems that probability coverage based on a percentile bootstrap method can be unsatisfactory, at least when the sample size is small (Wilcox, 2001d). However, if when we use 20% trimming we Winsorize by 10%, good probability coverage is obtained.

7.5.3 Using the Bootstrap-*t* Method

The bootstrap-*t* method can be applied with a trimmed mean in the following manner. Generate a bootstrap sample of size n and compute the trimmed mean and Winsorized standard deviation, which we label $\bar{X}_t^*$ and s_w^*, respectively. Next, compute

$$T_t^* = \frac{(1-2\gamma)(\bar{X}_t^* - \bar{X}_t)}{s_w^*/\sqrt{n}}, \tag{7.9}$$

where, as usual, γ is the amount of trimming, which we usually take to be .2. Repeating this process B times yields B T_t^* values. Writing these B values in ascending order we get $T_{t(1)}^* \le T_{t(2)}^* \le \cdots \le T_{t(B)}^*$. Letting $\ell = .025B$, rounded to the nearest integer, and $u = B - \ell$, an estimate of the .025 and .975 quantiles of the distribution of T_t is $T_{(\ell+1)}^*$ and $T_{(u)}^*$. The resulting .95 confidence interval for μ_t (the population trimmed mean) is

$$\left(\bar{X}_t - T_{t(u)}^* \frac{s_w}{(1-2\gamma)\sqrt{n}}, \bar{X}_t - T_{t(\ell+1)}^* \frac{s_w}{(1-2\gamma)\sqrt{n}}\right). \tag{7.10}$$

HYPOTHESIS TESTING. As for testing $H_0: \mu_t = \mu_0$, compute

$$T_t = \frac{(1-2\gamma)(\bar{X}_t - \mu_0)}{s_w/\sqrt{n}}$$

and reject if

$$T_t \le T_{t(\ell+1)}^*,$$

or if

$$T_t \ge T_{t(u)}^*.$$

That is, we use the same method employed when making inferences about the mean, except we replace the sample mean with the trimmed mean and we replace the sample standard deviation s with $s_w/(1-2\gamma)$.

The symmetric bootstrap-*t* method can be used as well when testing a two-sided hypothesis. Now we use

$$T_t^* = \frac{|(1-2\gamma)(\bar{X}_t^* - \bar{X}_t)|}{s_w^*/\sqrt{n}}. \tag{7.11}$$

and reject H_0 if $|T_t| > T_{t(c)}^*$, where $c = (1-\alpha)B$ rounded to the nearest integer. An approximate $1-\alpha$ confidence interval for μ_t is

$$\bar{X}_t \pm T_{t(c)}^* \frac{s_w}{(1-2\gamma)\sqrt{n}}. \tag{7.12}$$

Table 7.1 reported the actual probability of a Type I error when using means with one of four methods. None of the methods was satisfactory for all four distributions

TABLE 7.2 Actual Type I Error Probabilities Using 20% Trimmed Means, $\alpha = .05$

		Method			
	Dist.	BT	SB	P	TM
$n = 20$	N	.067	.052	.063	.042
	LN	.049	.050	.066	.068
	MN	.022	.019	.053	.015
	SH	.014	.018	.066	.020

N = normal; LN = lognormal; MN = mixed normal; SH = skewed, heavy-tailed; BT = equal-tailed, bootstrap-*t*; SB = symmetric bootstrap-*t*; P = percentile bootstrap; TM = Tukey–McLaughlin.

considered, even after increasing the sample size to 300. Table 7.2 shows the actual probability of a Type I error when using 20% trimmed means instead. Notice that the percentile bootstrap method is the most stable; the actual probability of a Type I error ranges between .053 and .066. The other three methods do a reasonable job of avoiding Type I error probabilities above the nominal .05 level. But they can have actual Type I error probabilities well below the nominal level, which is an indication that their power might be less than when using the percentile method instead. With the caveat that no method is best in all situations, the percentile bootstrap with a 20% trimmed mean is a good candidate for general use.

In the previous subsection we noted that we can Winsorize our data before we take bootstrap samples. Theory allows us to do the same when working with the bootstrap-*t* method. But when sample sizes are small, probability coverage can be poor. Apparently with a sufficiently large sample size this problem becomes negligible, but just how large the sample size must be remains unknown.

7.5.4 S-PLUS Functions trimpb and trimcibt

The S-PLUS function

```
trimpb(x,tr=.2,alpha=.05,nboot=2000,WIN=F,win=.1)
```

(written for this book) computes a confidence interval for a trimmed mean using the percentile bootstrap method. The argument tr indicates the amount of trimming and defaults to .2 if not specified. As usual, alpha is α and defaults to .05. It appears that $B = 500$ suffices, in terms of achieving accurate probability coverage with 20% trimming. But to be safe, B (nboot) defaults to 2000. (An argument for using $B = 2000$ can be made along the lines used by Booth & Sarker, 1998.) The argument WIN indicates whether the values should be Winsorized before bootstrap samples are taken. By default WIN=F, for false, meaning that Winsorizing will not be done. If WIN=T, the amount of Winsorizing is given by the argument win, which defaults to .1.

The S-PLUS function

```
trimcibt(x, tr = 0.2, alpha = 0.05, nboot = 599, side = F)
```

computes a bootstrap-t confidence interval. The argument side indicates whether an equal-tailed or a symmetric confidence interval is to be computed. As indicated, side defaults to F, meaning that an equal-tailed confidence interval [given by Equation (7.10)] will be computed. Using side=T results in a symmetric confidence interval [given by Equation (7.12)].

EXAMPLE. Table 3.2 reported data on the desired number of sexual partners among 105 college males. As previously indicated, these data are highly skewed, with a relatively large number of outliers, and this can have a deleterious effect on many methods for computing a confidence interval and testing hypotheses. If we compute the Tukey–McLaughlin .95 confidence interval for the 20% trimmed mean [using Equation (4.41)], we get (1.62, 3.75). Using the S-PLUS function trimcibt with side=F yields an equal-tailed .95 confidence interval of (1.28, 3.61). With side=T it is (1.51, 3.61). Using the percentile bootstrap method, the S-PLUS function trimpb returns (1.86, 3.95). So in this particular case, the lengths of the confidence intervals do not vary that much among the methods used, but the intervals are centered around different values, which in some cases might affect any conclusions made. ■

In summary, all indications are that the percentile bootstrap is more stable (with at least 20% trimming) than the bootstrap-t method. That is, the actual Type I error probability tends to be closer to the nominal level. And it has the added advantage of more power, at least in some situations, compared to any other method we might choose. However, subsequent chapters will describe situations where the bootstrap-t method outperforms the percentile method. And there are additional situations where the percentile bootstrap is best. So both methods are important to know.

7.6 Estimating Power When Testing Hypotheses About a Trimmed Mean

As when working with means, if we test some hypothesis about a trimmed mean and fail to reject, this might be because the null hypothesis is true, or perhaps power is too low to detect a meaningful difference. If we can estimate how much power we have based on the same data used to test some hypothesis, we are better able to discern which reason accounts for a nonsignificant result. We saw in Section 5.4.3 how, given some data, power can be controlled using Stein's method. A natural strategy is to use some analog of this method when working with trimmed means, but this approach has not been investigated as yet. There are several alternative methods one might employ, but most have proven to be unsatisfactory when sample sizes are small or even moderately large (Wilcox & Keselman, 2002). This section outlines the method that currently performs best when estimating power. It is a special case of another method covered in Chapter 8. So for brevity, we merely describe what the method attempts to do and then provide some software for implementing it. Readers interested in computational details can refer to Chapter 8.

The method in this section is designed specifically for the case where the percentile bootstrap method is used to test some hypothesis about a 20% trimmed mean with $\alpha = .05$. There are two goals: (1) Compute an (unbiased) estimate of power for some given value of $\delta = \mu_t - \mu_0$, and (2) provide a conservative estimate of power meaning a (one-sided) confidence interval for how much power we have. Roughly, the method estimates the standard error of the trimmed mean and then, given δ, provides an estimate of how much power we have. A possible concern, however, is that this estimate might underestimate power. Based on data, we might estimate power to be .7, but in reality it might be .5 or it might be as low as .4. So the method also computes a (lower) .95 confidence interval for the actual amount of power using a percentile bootstrap technique. Briefly, for every bootstrap sample, the standard error of the trimmed mean is estimated, which yields an estimate of power corresponding to whatever δ value is of interest. Repeating this process B times yields B estimates of power, which, when put into ascending order, we label $\hat{\xi}_{(1)} \le \cdots \le \hat{\xi}_{(B)}$. Then a conservative estimate of power is $\hat{\xi}_{(a)}$, where $a = .05B$ rounded to the nearest integer. So if $\hat{\xi}_{(a)} = .4$, say, we estimate that with probability .95, power is at least .4. If $\hat{\xi}_{(a)} = .6$, we estimate that with probability .95, power is at least .6.

7.6.1 S-PLUS Functions powt1est and powt1an

The S-PLUS function

powt1est(x, delta=0, ci=F, nboot=800)

returns an estimate of how much power there is for some value of δ. As usual, x now represents any S-PLUS variable containing data. The argument ci defaults to F (for false), meaning that no confidence interval for power is computed. If ci=T is used, a percentile bootstrap is used to get a conservative estimate of power. As usual, nboot indicates how many bootstrap samples are used. (That is, nboot corresponds to B.)

EXAMPLE. Consider the values

12, 20, 34, 45, 34, 36, 37, 50, 11, 32, 29.

Using the S-PLUS function trimpb, we get a .95 confidence interval for the 20% trimmed mean of (22.86, 38.71). So we would not reject $H_0 : \mu_t = 32$. To gain perspective on why, we estimate how much power we have when $\delta = \mu_t - \mu_0 = 2$. The S-PLUS command powt1est(x,2,T) returns

```
$est.power:
[1] 0.09033971

$ci:
[1] 0.06400682
```

meaning that the estimated power is only .09 and that with probability .95 power is at least .06. So power is estimated to be inadequate, suggesting that H_0 should not be accepted. (Using Stein's method for means, we see that 368 observations are needed to get power equal to .8 with $\delta = 2$.) ■

The S-PLUS function

powt1an(x, ci = F, plotit = T, nboot = 800)

provides a power analysis without having to specify a value for δ. Rather, the function chooses a range of δ values so that power will be between .05 and .9, approximately. Then it estimates how much power there is for each δ value that it selects and plots the results. That is, the function estimates the power curve associated with the percentile bootstrap method of testing hypotheses with a 20% trimmed mean. If the argument plotit is set to F (for false), no plot is generated; the function merely reports the δ values it generates and the corresponding power. The function also reports a lower estimate of power if ci=T is used. That is, with probability .95, power is at least as high as this lower estimate.

EXAMPLE. Figure 7.5 shows the plot created by powt1an(x) using the data from the last example. This plot indicates that power is reasonably high with $\delta = 12$, but for $\delta = 6$ power is only about .3.

Figure 7.6 shows the plot generated by the command powt1an(x,ci=T). The upper, solid line is the same as in Figure 7.5 and represents the estimated power. The lower, dashed line indicates a conservative estimate of power. So we see that for $\delta = 12$, it is estimated that power exceeds .8, and we can be approximately 95% certain that power exceeds .4. That is, it appears that power is adequate for $\delta = 12$, but there is some possibility that it is not — perhaps power is adequate only for δ values much greater than 12. ■

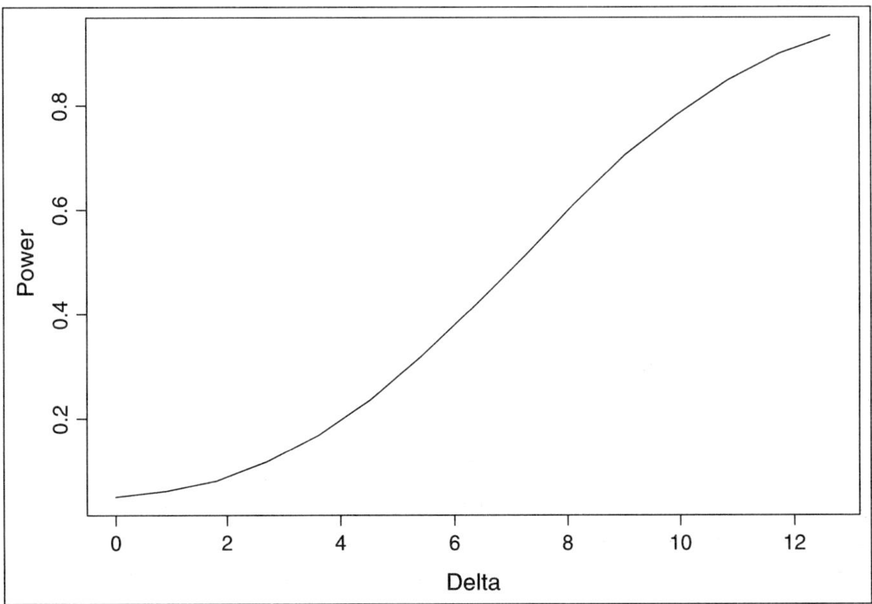

FIGURE 7.5 Estimate of the power curve returned by the S-PLUS function powt1an based on the data in Section 7.6.1.

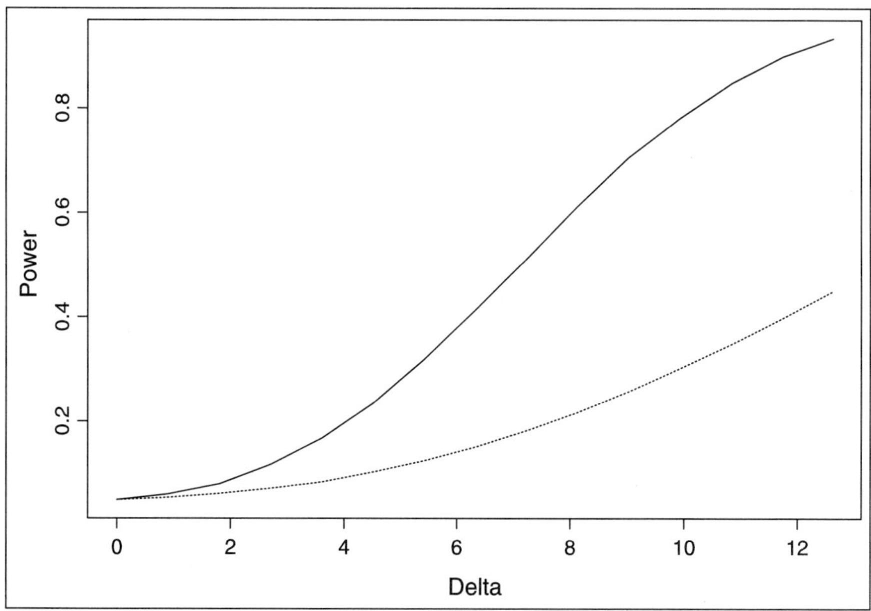

FIGURE 7.6 Same as Figure 7.5 but with lower confidence interval for power, indicated by the dashed lines.

7.7 Inferences Based on MOM and M-Estimators

Currently it appears that the best method for making inferences about an M-estimator of location is the percentile bootstrap technique. For very small sample sizes, probability coverage and control over the probability of a Type I error are a bit more stable using a 20% trimmed mean, but with a sample size of at least 20 the percentile bootstrap appears to perform reasonably well with an M-estimator. So to compute a confidence interval for μ_{os}, the measure of location being estimated by $\hat{\mu}_{os}$ [the one-step M-estimator given by Equation (3.28)], generate bootstrap samples as before until you have B values for $\hat{\mu}_{os}$. As usual, put these B values in ascending order, yielding $\hat{\mu}^*_{os(1)} \leq \cdots \leq \hat{\mu}^*_{os(B)}$. Then a $1 - \alpha$ confidence interval for μ_{os} is

$$\left(\hat{\mu}^*_{os(\ell+1)}, \hat{\mu}^*_{os(u)}\right) \tag{7.13}$$

where $\ell = \alpha B/2$, rounded to the nearest integer, and $u = B - \ell$. For $\alpha = .05$, all indications are that in terms of probability coverage, $B = 399$ suffices and that using $B = 599$ offers no practical advantage.

As for the MOM estimator, simply replace $\hat{\mu}_{os}$ with $\hat{\mu}_{mom}$, which is given by Equation (3.30). All indications are that confidence intervals based on MOM (with $B = 500$) provide accurate probability coverage and good control over the probability of a Type I error, even in situations where control over the probability of a Type I error using an M-estimator is unsatisfactory. In particular, currently it seems that good results are obtained even with $n = 11$.

HYPOTHESIS TESTING. Reject $H_0 : \mu_{os} = \mu_0$ if the bootstrap confidence interval given by Equation (7.13) does not contain μ_0. A similar strategy is used when the one-step M-estimator is replaced by MOM.

7.7.1 S-PLUS Function mestci

The S-PLUS function

mestci(x,alpha=.05,nboot=399)

computes a confidence interval based on an M-estimator, where x is an S-PLUS variable containing data, alpha is α, which defaults to .05, and nboot is B, the number of bootstrap samples to be used, which defaults to 399. (This function contains two additional arguments, the details of which can be found in Wilcox, 1997a, p. 85.)

7.7.2 S-PLUS Functions momci and onesampb

The S-PLUS function

momci(x,alpha=.05,nboot=500)

computes a confidence interval based on the measure of location MOM described in Section 3.5.2. Unlike the S-PLUS function mestci, the default number of bootstrap samples (nboot) is 500. The S-PLUS function

onesampb(x,est=mom,alpha=.05,nboot=500,...)

can also be used to make inferences based on MOM; but unlike momci, any other measure of location or scale can be used through the argument est. For example, using est=pbvar would compute a confidence interval for the percentage bend midvariance.

EXAMPLE. For 15 law schools, the undergraduate GPA of entering students, in 1973, was

3.39 3.30 2.81 3.03 3.44 3.07 3.00 3.43 3.36 3.13 3.12 2.74 2.76 2.88 2.96.

The .95 confidence interval returned by mestci is (2.95, 3.28). So among all law schools, it is estimated that the typical GPA of entering students is between 2.95 and 3.28. Using the MOM estimator instead (the S-PLUS function momci), the .95 confidence interval is (2.88, 3.35). ■

7.8 Detecting Nonlinear Associations

A possible concern about any method for testing the hypothesis that Pearson's correlation (ρ) is zero is that it might routinely miss an association that is not linear. The method described in this section, which allows heteroscedasticity, is one possible way of dealing with this problem. The theoretical justification for the method stems from Stute, Manteiga, and Quindimil (1998). Here, a very slight modification of their method is used that performs better in terms of Type I errors (in simulations) when the sample size is small.

The basic idea is to test the hypothesis that the regression line between two variables is both flat and straight. So the method is designed to be sensitive to any curved association between two variables; in the event the regression line is straight, the method is designed to detect situations where the slope differs from zero. Said another way, in the context of the regression model in Section 6.2, the method can be used to detect situations where the expected value of Y, given X, is not a constant. That is, the mean of Y, given X, changes with X in some unspecified manner that may be nonlinear. This is in contrast to standard regression methods, where it is assumed that the mean of Y, given X, is given by some straight line: $\beta_1 X + \beta_0$. Because the method in this section is designed to be sensitive to a broader range of associations than any test of $H_0 : \rho = 0$, a natural speculation is that it is more likely to detect an association, and experience suggests that this is indeed the case.

As in Section 7.3, let $(X_1, Y_1), \ldots, (X_n, Y_n)$ be a random sample of n pairs of points. The test statistic is computed as follows. Consider the jth pair of points: (X_j, Y_j). For any j we might pick $(1 \le j \le n)$, some of the X values will be less than or equal to X_j. For these X values, compute the sum of the difference between the corresponding Y values and $\bar{Y}$. We represent this sum, divided by the square root of the sample size, by R_j. Said more formally, if we fix j and set $I_i = 1$ if $X_i \le X_j$, otherwise $I_i = 0$, then

$$\begin{aligned} R_j &= \frac{1}{\sqrt{n}} \sum I_i (Y_i - \bar{Y}) \\ &= \frac{1}{\sqrt{n}} \sum I_i r_i, \end{aligned} \tag{7.14}$$

where now

$$r_i = Y_i - \bar{Y}.$$

The test statistic is the maximum absolute value of all the R_j values. That is, the test statistic is

$$D = \max |R_j|, \tag{7.15}$$

where max means that D is equal to the largest of the $|R_j|$ values.

It is noted that the method just described is heteroscedastic. That is, unlike Student's T test of $H_0 : \rho = 0$, this wild bootstrap method is not sensitive to changes in the variation of Y (the conditional variance of Y given X) as X increases.

EXAMPLE. We illustrate the computation of D with the data in Table 7.3. These data are from a study conducted about 200 years ago with the goal of determining whether the earth bulges at the equator, as predicted by Newton, or whether it bulges at the poles. The issue was addressed by measuring latitude at

Continued

TABLE 7.3 Data on Meridian Arcs

Place	Transformed latitude (X)	Arc length (Y)	$r_i = Y_i - \bar{Y}$	R
Quito	0.0000	56,751	−301.6	−134.88
Cape of Good Hope	0.2987	57,037	−15.6	−141.86
Rome	0.4648	56,979	−73.6	−174.77
Paris	0.5762	57,074	21.4	−165.20
Lapland	0.8386	57,422	369.4	0.00

EXAMPLE. (*Continued*) various points on the earth and trying to determine how latitude is related to a measure of arc length. To compute R_1, we note that $X_1 = 0.0$ and there are no other X values less than or equal to 0.0, so

$$R_1 = \frac{r_1}{\sqrt{5}} = \frac{-301.6}{\sqrt{5}} = -134.88.$$

As for R_2, $X_2 = 0.2987$, there are two X values less than or equal to 0.2987 (namely, X_1 and X_2); the corresponding r_i values are −301.6 and −15.6. So

$$R_2 = \frac{1}{\sqrt{5}}(-301.6 - 15.6) = -141.86.$$

The remaining R values are computed in a similar manner and are shown in the last column of Table 7.3. The largest absolute value in this column is 174.77, so $D = 174.77$. ■

To determine how large D must be to reject the null hypothesis, a so-called *wild bootstrap* is used. (The other types of bootstrap methods already covered are known to be theoretically unsound for the problem at hand.) Let

$$r_i = Y_i - \bar{Y}$$

be the ith residual corresponding to the regression line $\hat{Y} = \bar{Y}$. Generate n observations from a uniform distribution (which is shown in Figure 4.7) and label the results $U_1, \ldots, U_n$. (This can be done in S-PLUS with the built-in function runif.) So each U has a value between 0 and 1, and all values between 0 and 1 are equally likely. Next, for every value of i ($i = 1, \ldots, n$) set

$$V_i = \sqrt{12}(U_i - .5),$$

$$r_i^* = r_i V_i,$$

$$Y_i^* = \bar{Y} + r_i^*.$$

Then based on the n pairs of points $(X_1, Y_1^*), \ldots, (X_n, Y_n^*)$, compute the test statistic as described in the previous paragraph and label it D^*. Repeat this process B times, and label the resulting (bootstrap) test statistics $D_1^*, \ldots, D_B^*$. Finally, put these B values in ascending order, which we label $D^*_{(1)} \leq \cdots \leq D^*_{(B)}$. Then the critical value is $D^*_{(u)}$, where $u = (1-\alpha)B$ rounded to the nearest integer. That is, reject if

$$D \geq D^*_{(u)}.$$

There is a variation of the method that should be mentioned where the test statistic D is replaced by

$$W = \frac{1}{n}\left(R_1^2 + \cdots + R_n^2\right). \tag{7.16}$$

The critical value is determined in a similar manner as before. First, generate a wild bootstrap sample and compute W, yielding W^*. Repeating this B times, you reject if

$$W \geq W^*_{(u)},$$

where again $u = (1-\alpha)B$ rounded to the nearest integer, and $W^*_{(1)} \leq \cdots \leq W^*_{(B)}$ are the B W^* values written in ascending order. The test statistic D is called the *Kolmogorov–Smirnov* test statistic, and W is called the *Cramér–von Mises* test statistic. The choice between these two test statistics is not clear-cut. For the situation at hand, currently it seems that there is little separating them in terms of controlling Type I errors. But for situations where there are multiple predictors, as described in Chapter 14, the Cramér–von Mises test statistic seems to have an advantage (Wilcox & Muska, 2001).

The method just described can be used with $\bar{Y}$ replaced by a trimmed mean, $\bar{Y}_t$. So now $r_i = Y_i - \bar{Y}_t$ and $Y_i^* = \bar{Y}_t + r_i^*$. Otherwise the computations are exactly the same as before. We have already seen that replacing the sample mean with a 20% trimmed mean can make a substantial difference in power and control over the probability of a Type I error. Here, however, all indications are that the improvement in Type I error control is negligible. As for power, often it makes little difference whether a mean or trimmed mean is used, but situations do arise where you reject with a mean but not a trimmed mean, and the reverse happens as well.

By design, the method in this section is not sensitive to heteroscedasticity. That is, if the regression line is horizontal and straight, the probability of a Type I error will be approximately α, even when there is heteroscedasticity. In contrast, Student's T test of $H_0: \rho = 0$, covered in Section 6.5, is sensitive to heteroscedasticity. This would seem to suggest that in some instances, Student's T test might have more power, but experience suggests that in practice the method in this section is more likely to detect an association.

7.8.1 S-PLUS Function indt

The S-PLUS function

indt(x, y, tr = 0.2, nboot = 500, alpha = 0.05, flag = 1)

(written for this book) performs the test of independence just described. As usual, x and y are S-PLUS variables containing data, tr indicates the amount of trimming, and nboot is B. The argument flag indicates which test statistic will be used:

- flag=1 means the Kolmogorov–Smirnov test statistic, D, is used.
- flag=2 means the Cramér–von Mises test statistic, W, is used.
- flag=3 both test statistics are computed.

EXAMPLE. For the aggression data in Table 6.3, Pearson's correlation is $r = -0.286$ and we fail to reject $H_0: \rho = 0$ at the $\alpha = .05$ level. So we fail to conclude that there is an association between marital aggression and recall-test scores among children. (The significance level is .051, so we nearly reject.) The output from indt(x,y,flag=3) is

```
$dstat:
[1] 8.002455

$wstat:
[1] 37.31641

$critd:
[1] 6.639198

$critw:
[1] 24.36772
```

So $D = 8.002455$, the critical value is 6.639198, and because D exceeds the critical value, reject and conclude that marital aggression and recall-test scores are dependent. Similarly, the Cramér–von Mises test statistic is $W = 37.31641$; it exceeds the critical value 24.36772, so again we reject. So we have empirical evidence that aggression in the home is associated with recall-test scores among children living in the home, but the function indt tells us nothing about what this association might be like. ■

7.9 Exercises

1. For the following 10 bootstrap sample means, what would be an appropriate .8 confidence interval for the population mean?

   7.6, 8.1, 9.6, 10.2, 10.7, 12.3, 13.4, 13.9, 14.6, 15.2.

2. Rats are subjected to a drug that might affect aggression. Suppose that for a random sample of rats, measures of aggression are found to be

   5, 12, 23, 24, 18, 9, 18, 11, 36, 15.

   Verify that the equal-tailed .95 confidence interval for the mean returned by the S-PLUS function trimcibt is (11.8, 25.9). Compare this confidence interval to the confidence interval you got for Exercise 32 in Chapter 4.

3. For the data in Exercise 2, verify that the .95 confidence interval for the mean returned by the S-PLUS function trimpb is (12.3, 22.4).
4. Referring to the previous two Exercises, which confidence interval is more likely to have probability coverage at least .95?
5. For the data in Exercise 2, verify that the equal-tailed .95 confidence interval for the population 20% trimmed mean using trimcibt is (9.1, 25.7).
6. For the data in Exercise 2, verify that the .95 confidence interval for the population 20% trimmed mean using trimpb is (11.2, 22.0).
7. Which of the two confidence intervals given in the last two exercises is likely to have probability coverage closer to .95?
8. For the following observations, verify that the .95 confidence interval based on a one-step M-estimator returned by the S-PLUS function mestci is (7, 21.8).

   2, 4, 6, 7, 8, 9, 7, 10, 12, 15, 8, 9, 13, 19, 5, 2, 100, 200, 300, 400

9. For the data in the previous exercise, verify that the .95 confidence interval for the 20% trimmed mean returned by trimpb is (7.25, 64). Why would you expect this confidence interval to be substantially longer than the confidence interval based on a one-step M-estimator?
10. Use trimpb on the data used in the previous two exercises, but this time Winsorize the data first by setting the argument WIN to T. Verify that the .95 confidence interval for the trimmed mean is now (7.33, 57.7). Why do you think this confidence interval is shorter than the confidence interval in the last exercise?
11. Repeat the last exercise, only now Winsorize 20% by setting the argument win to .2. Verify that the .95 confidence interval is now (7.67, 14). Why is this confidence interval so much shorter than the confidence interval in the last exercise?
12. For the confidence interval obtained in the last exercise, what practical problem might have occurred regarding probability coverage?
13. For the data in Exercise 8, compute a .95 confidence interval for the median using the S-PLUS function sint, described in Section 4.11.1. What does this suggest about using a median versus a one-step M-estimator?
14. Exercise 22 in Chapter 6 reports that the .95 confidence interval for the slope of the least squares regression line, based on the data in Table 6.6, is (−0.16, .12). Using the S-PLUS function lsfitci, described in Section 7.3.1, verify that the .95 confidence interval based on the modified bootstrap method is (−0.27, 0.11).
15. Again using the data in Table 6.6, verify that you do not reject $H_0: \rho = 0$ using the S-PLUS function corb.
16. Using the data in Table 6.6, verify that when using the method in Section 7.8 based on a mean (meaning that you set tr=0 when using the S-PLUS function indt), both the Kolmogorov–Smirnov Cramér–von Mises test statistics reject the hypothesis that these two variables are independent.
17. The previous exercise indicates that the variables in Table 6.6 are dependent, but the results in Exercises 14 and 15 failed to detect any association. Describe a possible reason for the discrepant results.
18. Create a scatterplot of the data in Table 6.6, and note that six points having X values greater than 125 are visibly separated from the bulk of the observations.

Now compute a .95 confidence interval for the slope of the least squares regression line with the six points having X values greater than 125 eliminated. The resulting confidence interval is (−0.84, −0.17), so you reject the hypothesis of a zero slope and conclude that the two variables are dependent. Comment on this result in the context of factors that affect the magnitude of Pearson's correlation, which are described in Section 6.4.

19. Restricting the range as was done in the previous exercise, verify that the S-PLUS function corb returns a .95 confidence interval of (−0.63, −0.19). Compare this to the result in Exercise 15.
20. Using the data in Exercise 6 of Chapter 6, verify that the S-PLUS function lsfitci returns a .95 confidence interval for the slope (when predicting cancer rates given solar radiation) of (−0.049, −0.25). The conventional .95 confidence interval based on Equation (6.8) is (−0.047, −0.24). What does this suggest about the conventional method?
21. For Hubble's data on the recession velocity of galaxies, shown in Table 6.1, verify that the modified bootstrap method yields a .95 confidence interval of (310.1, 630.1) for the slope.

COMPARING TWO INDEPENDENT GROUPS

One of the most common goals in applied research is comparing two independent variables or groups. For example if one group of individuals receives an experimental drug for treating migraine headaches and a different, independent group of individuals receives a placebo, and we measure the effectiveness of a drug using some standard technique, how might we compare the outcomes corresponding to the two groups? How does the reading ability of children who watch 30 hours or more of television per week compare to children who watch 10 hours or less? How does the birth weight of newborns among mothers who smoke compare to the birth weight among mothers who do not smoke? In general terms, if we have two independent variables, how might we compare these two measures?

In this book, attention is focused on four interrelated and overlapping methods one might use to compare two independent groups or variables:

- Compare the groups using some measure of location, such as the mean, trimmed mean, or median. In particular, we might test the hypothesis that the measures of location are identical, or we might compute a confidence interval to get some sense of how much they differ.
- Test the hypothesis that the two groups have identical distributions. Identical distributions means that for any constant c we might pick, the probability that a randomly sampled observation is less than c is the same for both groups.
- Determine the probability that a randomly sampled observation from the first group will be less than a randomly sampled observation from the second. If the groups do not differ, this probability will be .5.
- Compare variances or some other measure of scale.

Each approach has its advantages and disadvantages. Each provides a different and useful perspective, and no single approach is optimal in all situations. A general goal is to explain the relative merits of each of the four strategies just listed, plus the practical advantages associated with the methods based on a specific strategy, so that applied researchers can make an informed decision as to which approach and which method might be used in a given situation.

The emphasis in this chapter is on the first strategy, but some methods covered here, including the best-known method for comparing means, are related to the second strategy as well. The fourth strategy is taken up in Section 8.10 (and the third strategy is discussed in Chapter 15). Section 8.12 describes how Pearson's correlation and regression slopes can be compared.

8.1 Student's *T*

We begin with the classic and most commonly used method for comparing two independent groups: Student's T test. The goal is to test

$$H_0 : \mu_1 = \mu_2, \tag{8.1}$$

the hypothesis that the two groups have identical means. That is, the goal is to determine whether the typical individual in the first group differs from the typical individual in the second. If we can find a method for computing a confidence interval for $\mu_1 - \mu_2$, we can get some sense of the degree to which the typical individual in the first group differs from the typical individual in the second. This assumes, of course, that the population mean provides a reasonable measure of what is typical. We have already seen that this assumption is dubious in some situations, but we ignore this issue for the moment.

Exact control over the probability of a Type I error can be had under the following assumptions:

- Sampling is random.
- Sampling is from normal distributions.
- The two groups have equal variances; that is, $\sigma_1^2 = \sigma_2^2$, where σ_1^2 and σ_2^2 are the variances corresponding to the groups having means μ_1 and μ_2, respectively.

The last assumption is called *homoscedasticity*. If the variances differ ($\sigma_1^2 \neq \sigma_2^2$), we say that there is *heteroscedasticity*.

Before describing how to test the hypothesis of equal means, first consider how we might estimate the assumed common variance. For convenience, let σ_p^2 represent the common variance and let s_1^2 and s_2^2 be the sample variances corresponding to the two groups. Also let n_1 and n_2 represent the corresponding sample sizes. The typical estimate of the assumed common variance is

$$s_p^2 = \frac{(n_1 - 1)s_1^2 + (n_2 - 1)s_2^2}{n_1 + n_2 - 2}, \tag{8.2}$$

where the subscript p is used to indicate that the sample variances are being pooled. Because s_1^2 and s_2^2 are assumed to estimate the same quantity, σ_p^2, a natural strategy for combining them into a single estimate is to average them, and this is exactly what is done when the sample sizes are equal. But with unequal sample sizes a slightly different strategy is used, as indicated by Equation (8.2). (A weighted average is used instead.)

Now consider the problem of testing the null hypothesis of equal means. Simultaneously, we want a confidence interval for the difference between the population means, $\mu_1 - \mu_2$. Under the assumptions already stated, the probability of a Type I error will be exactly α if we reject H_0 when

$$|T| \geq t, \tag{8.3}$$

where

$$T = \frac{\bar{X}_1 - \bar{X}_2}{\sqrt{s_p^2\left(\frac{1}{n_1} + \frac{1}{n_2}\right)}}, \tag{8.4}$$

and t is the $1 - \alpha/2$ quantile of Student's T distribution with $\nu = n_1 + n_2 - 2$ degrees of freedom. An exact $1 - \alpha$ confidence interval for the difference between the population means is

$$(\bar{X}_1 - \bar{X}_2) \pm t\sqrt{s_p^2\left(\frac{1}{n_1} + \frac{1}{n_2}\right)}. \tag{8.5}$$

EXAMPLE. Salk (1973) conducted a study where the general goal was to examine the soothing effects of a mother's heartbeat on her newborn infant. Infants were placed in a nursery immediately after birth, and they remained there for four days except when being fed by their mothers. The infants were divided into two groups. One group was continuously exposed to the sound of an adult's heartbeat; the other group was not. Salk measured, among other things, the weight change of the babies from birth to the fourth day. Table 8.1 reports the weight change for the babies weighing at least 3,510 grams at birth. The estimate of the assumed common variance is

$$s_p^2 = \frac{(20-1)(60.1^2) + (36-1)(88.4^2)}{20 + 36 - 2} = 6335.9.$$

So

$$T = \frac{18 - (-52.1)}{\sqrt{6335.9\left(\frac{1}{20} + \frac{1}{36}\right)}} = \frac{70.1}{22.2} = 3.2.$$

The degrees of freedom are $\nu = 20 + 36 - 2 = 54$. If we want the Type I error probability to be $\alpha = .05$, then, from Table 4 in Appendix B, $t = 2.01$. Because $|T| \geq 2.01$, reject H_0 and conclude that the means differ. That is, we conclude that among all newborns we might measure, the average weight gain would be higher among babies exposed to the sound of a heartbeat compared to those who are not exposed. By design, the probability that our conclusion is in error is .05, assuming normality and homoscedasticity. The .95 confidence interval

Continued

TABLE 8.1 Weight Gain (in grams) for Large Babies

Group 1 (heartbeat)*				Group 2 (no heartbeat)†							
Subject	Gain	Subject	Gain	Subject	Gain	Subject	Gain	Subject	Gain	Subject	Gain
1	190	11	10	1	140	11	−25	21	−50	31	−130
2	80	12	10	2	100	12	−25	22	−50	32	−155
3	80	13	0	3	100	13	−25	23	−60	33	−155
4	75	14	0	4	70	14	−30	24	−75	34	−180
5	50	15	−10	5	25	15	−30	25	−75	35	−240
6	40	16	−25	6	20	16	−30	26	−85	36	−290
7	30	17	−30	7	10	17	−45	27	−85		
8	20	18	−45	8	0	18	−45	28	−100		
9	20	19	−60	9	−10	19	−45	29	−110		
10	10	20	−85	10	−10	20	−50	30	−130		

*$n_1 = 20$, $\bar{X}_1 = 18.0$, $s_1 = 60.1$, $s_1/\sqrt{n_1} = 13$.
†$n_2 = 36$, $\bar{X}_2 = -52.1$, $s_2 = 88.4$, $s_2/\sqrt{n_2} = 15$.

EXAMPLE. (*Continued*) for $\mu_1 - \mu_2$, the difference between the means, is

$$[18 - (-52.1)] \pm 2.01\sqrt{6335.9\left(\frac{1}{20} + \frac{1}{36}\right)} = (25.5, 114.7).$$

This interval does not contain zero, and it indicates that the difference between the means is likely to be at least 25.5, so again you would reject the hypothesis of equal means. ■

8.2 Relative Merits of Student's T

We begin by describing some positive features of Student's T. If distributions are nonnormal, but otherwise *identical*, Student's T performs reasonably well in terms of controlling Type I errors. This result is somewhat expected based on features of the one-sample Student's T covered in Chapters 4 and 5. To get a rough idea of why, we first note that for any two independent variables having identical distributions, their difference will have a perfectly symmetric distribution about zero, even when the distributions are skewed. For example, suppose that in Salk's study, the first group of infants has weight gains that follow the (lognormal) distribution shown in Figure 4.14, and the second group has the same distribution. If we randomly sample an observation from the first group and do the same for the second group, then the distribution of the difference will be symmetric about zero. (That is, repeating this process billions of times, a plot of the resulting differences will be symmetric about zero.) More generally, if we sample n observations from each of two identical distributions, the difference between the sample means will have a symmetric distribution.

Note that when two distributions are identical, then not only are their means equal, but their variances are also equal.

Chapters 4 and 5 noted that in the one-sample case, problems with controlling Type I errors — ensuring that the actual Type I error probability does not exceed the nominal level — arise when sampling from a skewed distribution. For symmetric distributions, this problem is of little concern. So for the two-sample problem considered here, the expectation is that if there is absolutely no difference between the two groups, implying that they have not only equal means but equal variances and the same skewness, then the actual Type I error probability will not exceed by very much the α value you specify. All indications are that this argument is correct, but some simple extensions of this argument lead to erroneous conclusions. In particular, it might seem that generally, if we sample from perfectly symmetric distributions, probability coverage and control of Type I error probabilities will be satisfactory, but we will see that this is not always the case — even under normality. In particular, perfectly symmetric distributions with unequal variances can create practical problems. Nevertheless, all indications are that generally, when comparing identical distributions (so that in particular the hypothesis of equal means is true and the variances are equal), Type I error probabilities will not exceed the nominal level by very much, and for this special case power is not an issue.

Student's T begins having practical problems when distributions differ in some manner. If sampling is from normal distributions with equal sample sizes but unequal variances, Student's T continues to perform reasonably well in terms of Type I errors (Ramsey, 1980). But when sampling from nonnormal distributions, this is no longer the case. And even under normality there are problems when sample sizes are unequal. Basically, Type I error control, power and probability coverage can be very poor. In fact, when sample sizes are unequal, Cressie and Whitford (1986) describe general conditions under which Student's T does not even converge to the correct answer as the sample sizes get large.

A reasonable suggestion for salvaging the assumption that the variances are equal is to test it. That is, test $H_0 : \sigma_1^2 = \sigma_2^2$; if a nonsignificant result is obtained, proceed with Student's T. But this strategy has been found to be unsatisfactory, even under normality (Markowski & Markowski, 1990; Moser, Stevens & Watts, 1989; Wilcox, Charlin & Thompson, 1986). Nonnormality makes this strategy even less satisfactory. A basic problem is that tests of the hypothesis of equal variances do not always have enough power to detect situations where the assumption should be discarded.

Another general problem is that Student's T is designed to be sensitive to the differences between the means, but in reality it is sensitive to the myriad ways the distributions might differ, such as unequal skewnesses. Said in a more formal manner, if $F_1(x)$ is the probability that an observation from the first group is less than or equal to x, and $F_2(x)$ is the probability that an observation from the second group is less than or equal to x, then Student's T provides a reasonably satisfactory test of

$$H_0 : F_1(x) = F_2(x), \qquad \text{for any } x \tag{8.6}$$

in terms of controlling the probability of a Type I error. That is, the hypothesis is that for any constant x we pick, the probability of getting an observation less than or equal to x is the same for both groups being compared.

As a method for testing the hypothesis of equal population means, one might defend Student's T in the following manner. If we reject, we can be reasonably certain that the distributions differ. And if the distributions differ, some authorities would argue that by implication, the means are not equal. That is, they would argue that in practice we never encounter situations where distributions differ in shape but have equal means. But as soon as we agree that the distributions differ, there is the possibility that the actual probability coverage of the confidence interval for the difference between the means, given by Equation (8.5), differs substantially from the $1-\alpha$ value you have specified. A crude rule is that the more the distributions differ, particularly in terms of skewness, the more inaccurate the confidence interval [given by Equation (8.5)] might be. Outliers and heteroscedasticity can contribute to this problem. Said another way, Student's T can be used to establish that the means differ by implication, but it can be very inaccurate in terms of indicating the magnitude of this difference.

One more problem with Student's T is that situations arise where it is biased. That is, there is a higher probability of rejecting when the means are equal compared to some situations where the means differ. If the goal is to find a method that is sensitive to differences between the means, surely this property is undesirable.

Yet another concern about Student's T is that for a variety of reasons, its power can be very low relative to many other methods one might use. One reason is that very slight departures from normality can inflate the variances. Consider, for example, the two normal distributions shown in the left panel of Figure 8.1. Both have a variance of 1, and the means differ by 1. Using Student's T with $\alpha = .05$, power is .96 with $n_1 = n_2 = 25$. But if we sample from the two distributions shown in the right panel of Figure 8.1, power is only .28. One reason power is low is that when sampling from a heavy-tailed distribution, the actual probability of a Type I error can be substantially lower than the specified α value. For example, if you use Student's T with $\alpha = .05$, the actual probability of a Type I error can drop below .01. If an adjustment could be made so that the actual probability of a Type I error is indeed .05, power would be better, but it would still be low relative to alternative methods that are less sensitive to outliers. And even when outliers are not a concern, having unequal variances or even different degrees of skewness can result in relatively poor power as well.

Finally, we note that when using Student's T, even a single outlier in only one group can result in a rather undesirable property. The following example illustrates the problem.

EXAMPLE. Consider the following values:

Group 1:	4	5	6	7	8	9	10	11	12	13
Group 2:	1	2	3	4	5	6	7	8	9	10

The corresponding sample means are $\bar{X}_1 = 8.5$ and $\bar{X}_2 = 5.5$ and $T = 2.22$. With $\alpha = .05$, the critical value is $T = 2.1$, so Student's T would reject the hypothesis of equal means and conclude that the first group has a larger

Continued

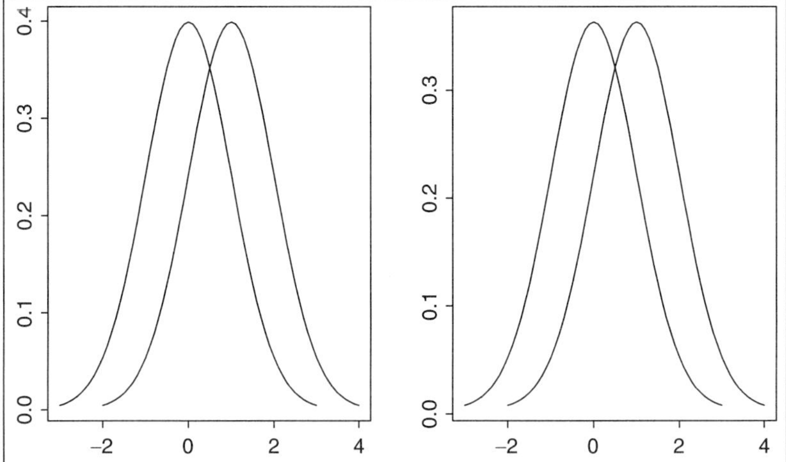

FIGURE 8.1 When sampling from the distributions on the left, power is .96 when using Student's T with $\alpha = .05$ and $n_1 = n_2 = 25$. But for the distributions on the right, power is only .28. This illustrates that very small departures from normality can destroy power when using means.

EXAMPLE. (*Continued*) population mean than the second (because the first group has the larger sample mean). Now, if we increase the largest observation in the first group from 13 to 23, the sample mean increases to $\bar{X}_1 = 9.5$. So the difference between $\bar{X}_1$ and $\bar{X}_2$ has increased from 3 to 4, and this would seem to suggest that we have stronger evidence that the population means differ and in fact the first group has the larger population mean. However, increasing the largest observation in the first group also inflates the corresponding sample variance, s_1^2. In particular, s_1^2 increases from 9.17 to 29.17. The result is that T *decreases* to $T = 2.04$ and we no longer reject. That is, increasing the largest observation has more of an effect on the sample variance than on the sample mean, in the sense that now we are no longer able to conclude that the population means differ. Increasing the largest observation in the first group to 33, the sample mean increases to 10.5, the difference between the two sample means increases to 5, and now $T = 1.79$. So again we do not reject and in fact our test statistic is getting smaller! This illustration provides another perspective on how outliers can mask differences between population means. ■

8.3 Welch's Heteroscedastic Method for Means

A step toward addressing the negative features of Student's T, still assuming that the goal is to compare means, is to switch to a method that allows unequal variances. Many techniques have been proposed, none of which is completely satisfactory. We describe only one such method here. One reason it was chosen is because it is a special case of a more general technique that gives more satisfactory results. That is, our goal is to build our way up to a method that performs relatively well over a broad range of situations.

Proposed by Welch (1938), we test the hypothesis of equal means ($H_0: \mu_1 = \mu_2$) as follows. For the jth group ($j = 1, 2$) let

$$q_j = \frac{s_j^2}{n_j}, \tag{8.7}$$

where, as usual, s_1^2 and s_2^2 are the sample variances corresponding to the two groups being compared. That is, q_1 and q_2 are the usual estimates of the squared standard errors of the sample means, $\bar{X}_1$ and $\bar{X}_2$, respectively. We can get a confidence interval having probability coverage exactly equal to $1 - \alpha$ if we can determine the distribution of

$$W = \frac{(\bar{X}_1 - \bar{X}_2) - (\mu_1 - \mu_2)}{\sqrt{\frac{s_1^2}{n_1} + \frac{s_2^2}{n_2}}}. \tag{8.8}$$

Welch's strategy was to approximate the distribution of W using a Student's T distribution with degrees of freedom determined by the sample variances and sample sizes. In particular, the degrees of freedom are estimated with

$$\hat{\nu} = \frac{(q_1 + q_2)^2}{\frac{q_1^2}{n_1 - 1} + \frac{q_2^2}{n_2 - 1}}. \tag{8.9}$$

The $1 - \alpha$ confidence interval for $\mu_1 - \mu_2$ is

$$(\bar{X}_1 - \bar{X}_2) \pm t\sqrt{\frac{s_1^2}{n_1} + \frac{s_2^2}{n_2}}, \tag{8.10}$$

where t is the $1 - \alpha/2$ quantile of Student's T distribution and is again read from Table 4 in Appendix B.

HYPOTHESIS TESTING. When the hypothesis of equal means is true (so $\mu_1 - \mu_2 = 0$), Equation (8.8) reduces to

$$W = \frac{\bar{X}_1 - \bar{X}_2}{\sqrt{\frac{s_1^2}{n_1} + \frac{s_2^2}{n_2}}} \tag{8.11}$$

and you reject $H_0: \mu_1 = \mu_2$ if

$$|W| \geq t, \tag{8.12}$$

where again t is the $1 - \alpha/2$ quantile of Student's T distribution with $\hat{\nu}$ degrees of freedom.

One-sided tests are performed in a similar manner. To test $H_0 : \mu_1 \le \mu_2$, reject if $W \ge t$, where now t is the $1-\alpha$ quantile of Student's T distribution with $\hat{\nu}$ degrees of freedom. As for $H_0 : \mu_1 \ge \mu_2$, reject if $W \le t$, where t is the α quantile of Student's T.

EXAMPLE. For Salk's data in Table 8.1, the value of the test statistic given by Equation (8.11) is

$$W = \frac{18 - (-52.1)}{\sqrt{\frac{60.1^2}{20} + \frac{88.4^2}{36}}} = \frac{70.1}{19.9} = 3.52.$$

To compute the estimated degrees of freedom, first compute

$$q_1 = \frac{60.1^2}{20} = 180.6,$$
$$q_2 = \frac{88.4^2}{36} = 217,$$

in which case

$$\hat{\nu} = \frac{(180.6 + 217)^2}{\frac{180.6^2}{19} + \frac{217^2}{35}} \approx 52.$$

So if the Type I error probability is to be .05, then for a two-sided test ($H_0 : \mu_1 = \mu_2$), $t = 2.01$, approximately, and we reject because $|W| = 3.52 > 2.01$. That is, the empirical evidence indicates that infants exposed to the sounds of a mother's heart beat gain more weight, on average. The .95 confidence interval for the difference between the means can be seen to be (30, 110). In contrast, Student's T method yields a .95 confidence interval of (25.5, 114.7). So Welch's procedure indicates that the difference between the means of the two groups is at least 30, but Student's T leads to the conclusion that the difference between the means is at least 25.5. Generally, Welch's method provides more accurate confidence intervals than Student's T, so for the problem at hand, results based on Welch's method should be used. Also, Welch's method can provide a shorter confidence interval. There are situations where Student's T is more accurate than Welch's method, but in general the improvement is modest at best, while the improvement of Welch's method over Student's T can be substantial. ■

8.3.1 Nonnormality and Welch's Method

One reason Welch's method improves upon Student's T is that under random sampling, Welch's method satisfies the basic requirement of converging to the correct answer as the sample sizes get large when randomly sampling from nonnormal distributions, even when the sample sizes are unequal. That is, as the sample sizes increase, the actual probability coverage for the difference between the means will converge to $1 - \alpha$ when using Equation (8.10). For Student's T, this is true only when the sample sizes are equal. When sampling from normal distributions, Welch's method does a better job of handling unequal variances. This can translate into more power as well as shorter and more accurate confidence intervals. But unfortunately, nonnormality can be devastating in terms of probability coverage (e.g., Algina, Oshima & Lin, 1994) and especially power. Moreover, Welch's method can be biased: The probability of rejecting can be higher when the means are equal compared to situations where they are not. When the two groups being compared have identical distributions, Welch's method performs well in terms of controlling the probability of a Type I error, and because for this special case the population means are equal, power is not an issue. But when distributions differ in some manner, such as having different skewnesses or even unequal variances, Welch's method can have poor properties under nonnormality. So although Welch's method is an improvement over Student's T, more needs to be done.

8.4 Comparing Groups with Individual Confidence Intervals: An Example of What Not to Do

It might seem that one could compare population means simply by computing a confidence interval for μ_1 using Equation (4.8), computing a confidence interval for μ_2, and then checking whether these two confidence intervals overlap. If they do not overlap, it might seem reasonable to reject the hypothesis of equal population means. So, for example, if we compute a .95 confidence interval for the first mean, yielding (5, 8), and for the second mean we get a .95 confidence interval of (9, 13), it might appear reasonable to reject $H_0 : \mu_1 = \mu_2$. However, this strategy violates a basic principle. Because there seems to be increasing interest in this approach, particularly when using error bars (defined shortly) to summarize data, some comments seem in order. (Also see Schenker & Gentleman, 2001; cf. Tryon, 2001.)

First we note that a general way of standardizing any variable, call it Y, is to subtract its mean and then divide this difference by the standard error of Y. In symbols, we standardize Y by transforming it to

$$\frac{Y - E(Y)}{\text{SE}(Y)}, \tag{8.13}$$

where SE(Y) indicates the standard error of Y. That is, SE(Y) is the square root of the variance of Y. For example, when working with a single sample mean, in which case $Y = \bar{X}$, we know that the standard error of the sample mean is $\sigma/\sqrt{n}$, $E(\bar{X}) = \mu$, so we standardize the sample mean by transforming it to

$$\frac{\bar{X} - \mu}{\sigma/\sqrt{n}}.$$

In practice we do not know σ, so we estimate it with s, the standard deviation. Then this last equation becomes

$$T = \frac{\bar{X} - \mu}{s/\sqrt{n}},$$

and a version of the central limit theorem tells us that with n sufficiently large, T is approximately standard normal. That is, inferences about the population mean can be made assuming T has a normal distribution, and this satisfies the basic requirement of converging to the correct answer as the sample size gets large.

When working with two independent groups, the standard error of the difference between the sample means is

$$\sqrt{\frac{\sigma_1^2}{n_1} + \frac{\sigma_2^2}{n_2}}. \tag{8.14}$$

This expression for the standard error is being estimated by Welch's method, which is part of the reason why Welch's method converges to the correct answer as the sample sizes get large. That is, a correct expression for the standard error is being used to derive Welch's method, and this is a fundamental component of any method when standardizing is being done. In contrast, under general conditions, Student's T does not converge to the correct answer when the sample sizes differ, and the reason is roughly because it does not use an estimate of the correct standard error.

To keep the discussion simple, we focus on what are called error bars. *Error bars*, often used in graphical summaries of data, are simply vertical lines plotted above and below the sample means. In symbols, error bars for the first means correspond to the values

$$\bar{X}_1 - \frac{s_1}{\sqrt{n_1}} \quad \text{and} \quad \bar{X}_1 + \frac{s_1}{\sqrt{n_1}},$$

so they are simply a type of confidence interval. As for the second group, error bars correspond to the values

$$\bar{X}_2 - \frac{s_2}{\sqrt{n_2}} \quad \text{and} \quad \bar{X}_2 + \frac{s_2}{\sqrt{n_2}}.$$

Now consider the strategy of deciding that the population means differ if the intervals based on these error bars do not overlap. In symbols, reject the null hypothesis of equal means if

$$\bar{X}_1 + \frac{s_1}{\sqrt{n_1}} < \bar{X}_2 - \frac{s_2}{\sqrt{n_2}}$$

or if

$$\bar{X}_1 - \frac{s_1}{\sqrt{n_1}} > \bar{X}_2 + \frac{s_2}{\sqrt{n_2}}.$$

Here is a fundamental problem with this strategy. Rearranging terms, we are rejecting if

$$\frac{|\bar{X}_1 - \bar{X}_2|}{\frac{s_1}{\sqrt{n_1}} + \frac{s_2}{\sqrt{n_2}}} \geq 1. \tag{8.15}$$

The denominator,

$$\frac{s_1}{\sqrt{n_1}} + \frac{s_2}{\sqrt{n_2}},$$

violates a basic principle — it does not estimate a correct expression for the standard error of the difference between the sample means, which is given by Equation (8.14). Consequently, the left side of Equation (8.15) does not converge to a standard normal distribution and statements about Type I error probabilities are at best difficult to make. A correct estimate of the standard error is

$$\sqrt{\frac{s_1^2}{n_1} + \frac{s_2^2}{n_2}},$$

which is used by Welch's method and which can differ substantially from the incorrect estimate used in Equation (8.15). So regardless of how large the sample sizes might be, using error bars to make decisions about whether groups differ can be highly inaccurate, even under normality. More generally, if we reject the hypothesis of equal means when confidence intervals for the individual means do not overlap, the wrong standard error is being used, so any statements about the probability of a Type I error are difficult to make. Yet another problem, for reasons explained in Chapter 4, is that under nonnormality the probability coverage for the individual population means can differ substantially from the nominal level, depending on the type of distributions from which we sample. (There is a variation of the method just described where error bars are computed assuming homoscedasticity, but again basic principles are being violated.)

8.5 A Bootstrap Method for Comparing Means

This section describes how we might improve upon Welch's method for comparing means. Similar to Chapter 7, we can compute accurate confidence intervals for the difference between means or test hypotheses about the equality of means if we can determine the distribution of W given by Equation (8.8). A bootstrap approximation of this distribution is obtained as follows. Generate a bootstrap sample of size n_1 from the first group and label the resulting sample mean and standard deviation $\bar{X}_1^*$ and s_1^*, respectively. Do the same for the second group and label the bootstrap sample mean

TABLE 8.2 Data Generated from a Mixed Normal (Group 1) and a Standard Normal (Group 2)

Group 1:	3.73624506	2.10039320	−3.56878819	−0.26418493	−0.27892175	
	0.87825842	−0.70582571	−1.26678127	−0.30248530	0.02255344	
	14.76303893	−0.78143390	−0.60139147	−4.46978177	1.56778991	
	−1.14150660	−0.20423655	−1.87554928	−1.62752834	0.26619836	
Group 2:	−1.1404168	−0.2123789	−1.7810069	−1.2613917	−0.3241972	1.4550603
	−0.5686717	−1.7919242	−0.6138459	−0.1386593	−1.5451134	−0.8853377
	0.3590016	0.4739528	−0.2557869			

and standard deviation $\bar{X}_2^*$ and s_2^*. Let

$$W^* = \frac{(\bar{X}_1^* - \bar{X}_2^*) - (\bar{X}_1 - \bar{X}_2)}{\sqrt{\frac{(s_1^*)^2}{n_1} + \frac{(s_2^*)^2}{n_2}}}. \tag{8.16}$$

Repeating this process B times, yields $W_1^*, \ldots, W_B^*$. Next, put these B values in ascending order, which we label $W_{(1)}^* \le \cdots \le W_{(B)}^*$ in our usual way. Let $\ell = \alpha B/2$, rounded to the nearest integer, and $u = B - \ell$. Then an approximate $1 - \alpha$ confidence interval for the difference between the means $(\mu_1 - \mu_2)$ is

$$\left((\bar{X}_1 - \bar{X}_2) - W_{(u)}^* \sqrt{\frac{(s_1^*)^2}{n_1} + \frac{(s_2^*)^2}{n_2}},\ (\bar{X}_1 - \bar{X}_2) - W_{(\ell+1)}^* \sqrt{\frac{(s_1^*)^2}{n_1} + \frac{(s_2^*)^2}{n_2}} \right). \tag{8.17}$$

EXAMPLE. We illustrate that the confidence interval based on Welch's method can differ substantially from the confidence interval based on the bootstrap-*t* method just described. The data for group 1 in Table 8.2 were generated from a mixed normal distribution (using S-PLUS) and the data for group 2 were generated from a standard normal distribution. So both groups have population means equal to 0. Applying Welch's method, the .95 confidence interval for the difference between the means is (− 0.988, 2.710). Using the bootstrap-*t* method instead, it is (− 2.21, 2.24). ■

8.6 A Permutation Test Based on Means

The so-called *permutation test*, introduced by R. A. Fisher in the 1930s, is sometimes recommended for comparing means. The method is somewhat similar in spirit to the bootstrap, but a fundamental difference between it and the bootstrap is that the bootstrap resamples with replacement and the permutation test does not. We first outline the method in formal terms, then we illustrate the steps, and finally we indicate what this test tells us.

The permutation test based on means is applied as follows:

1. Compute the sample means for each group and label the difference $d = \bar{X}_1 - \bar{X}_2$.
2. Pool the data.
3. Randomly permute the pooled data. That is, rearrange the order of the pooled data in a random fashion.
4. Compute the sample mean for the first n_1 observations resulting from step 3, compute the sample mean of the remaining n_2 observations, and note the difference.
5. Repeat steps 3–4 B times and label the differences between the resulting sample means $d_1, \ldots, d_B$.

EXAMPLE. Imagine we have two groups with the following observations:

Group 1:	6,	19,	34,	15	
Group 2:	9,	21,	8,	53,	25

Pooling the observations yields

$$6, 19, 34, 15, 9, 21, 8, 53, 25.$$

The sample mean for group 1 is $\bar{X}_1 = 18.5$; for group 2 it is $\bar{X}_2 = 23.2$; and the difference between these means is

$$d = 18.5 - 23.2 = -4.7.$$

Next we permute the pooled data in a random fashion and for illustrative purposes we assume this yields

$$34, 21, 8, 25, 6, 19, 15, 9, 53.$$

The sample mean for the first $n_1 = 4$ observations is 22, the remaining observations have a sample mean of 20.4, and the difference between these means is $d_1 = 22 - 20.4 = 1.6$. Repeating this process of randomly permuting the pooled data B times yields B differences between the resulting sample means, which we label $d_1, \ldots, d_B$. If we want the Type I error probability to be .05, we conclude that the groups differ if the middle 95% of the values $d_1, \ldots, d_B$ do not contain d. ■

Although the method just described is sometimes recommended for making inferences about the population means, in reality it is testing the hypothesis that the two groups being compared have identical distributions. Even under normality but unequal variances, the method fails to control the probability of a Type I error when testing the hypothesis of equal means (e.g., Boik, 1987). On the positive side, when testing the hypothesis of identical distributions, the probability of a Type I error is controlled exactly if all possible permutations of the data are used rather than just B randomly sampled permutations as is done here. An argument in favor of using the permutation test to compare means is that if distributions differ, surely the population means differ. But even if we accept this argument, the permutation test gives us little or no information about how the groups differ, let alone the magnitude of the difference between the population means, and it tells us nothing about the precision of the estimated difference between the population means based on the sample means. That is, it does not provide a confidence interval for $\mu_1 - \mu_2$. (For yet another argument in favor of the permutation test, see Ludbrook & Dudley, 1998.)

It is noted that the permutation test can be applied with any measure of location or scale, but again the method is testing the hypothesis of equal distributions. If, for example, we use variances, examples can be constructed where we are likely to reject because the distributions differ, even though the population variances are equal.

8.6.1 S-PLUS Function permg

The S-PLUS function

permg(x, y, alpha = 0.05, est = mean, nboot = 1000)

performs the permutation test just described. By default it uses means, but any measure of location or scale can be used by setting the argument est to an appropriate expression. For example, est=var would use variances rather than means.

8.7 Yuen's Method for Comparing Trimmed Means

Yuen (1974) derived a method for comparing the population γ-trimmed means of two independent groups that reduces to Welch's method for means when there is no trimming. As usual, 20% trimming ($\gamma = .2$) is a good choice for general use, but situations arise where more than 20% trimming might be beneficial (such as when the proportion of outliers in either tail of an empirical distribution exceeds 20%).

Generalizing slightly the notation in Chapter 3, let $g_j = [\gamma n_j]$, where again n_j is the sample size associated with the jth group ($j = 1, 2$) and let $h_j = n_j - 2g_j$. That is, h_j is the number of observations left in the jth group after trimming. Let

$$d_j = \frac{(n_j - 1)s_{wj}^2}{h_j(h_j - 1)}, \tag{8.18}$$

where s_{wj}^2 is the γ-Winsorized variance for the jth group. Yuen's test statistic is

$$T_y = \frac{\bar{X}_{t1} - \bar{X}_{t2}}{\sqrt{d_1 + d_2}}. \tag{8.19}$$

The degrees of freedom are

$$\hat{\nu}_y = \frac{(d_1 + d_2)^2}{\frac{d_1^2}{h_1-1} + \frac{d_2^2}{h_2-1}}.$$

CONFIDENCE INTERVAL. The $1-\alpha$ confidence interval for $\mu_{t1} - \mu_{t2}$, the difference between the population trimmed means, is

$$(\bar{X}_{t1} - \bar{X}_{t2}) \pm t\sqrt{d_1 + d_2}, \tag{8.20}$$

where t is the $1 - \alpha/2$ quantile of Student's T distribution with $\hat{\nu}_y$ degrees of freedom.

HYPOTHESIS TESTING. The hypothesis of equal trimmed means ($H_0 : \mu_{t1} = \mu_{t2}$) is rejected if

$$|T_y| \geq t.$$

As before, t is the $1 - \alpha/2$ quantile of Student's T distribution with $\hat{\nu}_y$ degrees of freedom.

The improvement in power, probability coverage, and control over Type I errors can be substantial when using Yuen's method with 20% trimming rather than Welch. For example, Wilcox (1997a, p. 111) describes a situation where when testing at the .025 level, the actual probability of rejecting with Welch's test is .092, nearly four times as large as the nominal level. Switching to Yuen's test, the actual probability of a Type I error is .042. So control over the Type I error probability is much better, but more needs to be done.

8.7.1 Comparing Medians

Although the median can be viewed as belonging to the class of trimmed means, special methods are required for comparing groups based on medians. An approach that currently seems to have practical value is as follows. Let M_j be the sample median corresponding to the jth group ($j = 1, 2$) and let S_j^2 be the McKean–Schrader estimate of the squared standard error of M_j, which is described in Section 4.11.2. Then an approximate $1 - \alpha$ confidence interval for the difference between the population medians is

$$(M_1 - M_2) \pm z_{1-\alpha/2}\sqrt{S_1^2 + S_2^2},$$

where $z_{1-\alpha/2}$ is the $1 - \alpha/2$ quantile of a standard normal distribution. Alternatively, reject the hypothesis of equal population medians if

$$\frac{|M_1 - M_2|}{\sqrt{S_1^2 + S_2^2}} \geq z_{1-\frac{\alpha}{2}}.$$

8.7.2 S-PLUS Function msmed

The S-PLUS function

msmed(x,y,alpha=.05)

has been supplied for comparing medians using the McKean–Schrader estimate of the standard error. (This function contains some additional parameters that are explained in Chapter 12.)

EXAMPLE. For the data in Table 8.1, the .95 confidence interval for the difference between the medians is (18.5, 91.5). ■

8.8 Bootstrap Methods for Comparing Trimmed Means

The bootstrap methods for trimmed means, described in Chapter 7, can be extended to the two-sample case. Again there are three versions of the bootstrap method that should be described and discussed.

8.8.1 The Percentile Method

Generalizing the notation in Section 8.6, generate a bootstrap sample of size n_1 from the first group, generate a bootstrap sample of size n_2 from the second group, let $\bar{X}^*_{t1}$ and $\bar{X}^*_{t2}$ be the bootstrap trimmed means corresponding to groups 1 and 2, respectively, and let

$$D^* = \bar{X}^*_{t1} - \bar{X}^*_{t2}$$

be the difference between the bootstrap trimmed means. Now suppose we repeat this process B times, yielding $D^*_1, \ldots, D^*_B$. Then an approximate $1 - \alpha$ confidence interval for the difference between the population trimmed means, $\mu_{t1} - \mu_{t2}$, is

$$\left(D^*_{(\ell+1)}, D^*_{(u)}\right), \tag{8.21}$$

where, as usual, $\ell = \alpha B/2$, rounded to the nearest integer, and $u = B - \ell$. So for a .95 confidence interval, $\ell = .025B$.

HYPOTHESIS TESTING. Reject the hypothesis of equal population trimmed means if the confidence interval given by Equation (8.21) does not contain zero.

If the amount of trimming is at least .2, the percentile bootstrap method just described is one of the most effective methods for obtaining accurate probability coverage, minimizing bias, and achieving relatively high power. But with no trimming the method performs poorly. (As noted in Chapter 7, a modification of the percentile bootstrap method performs well when working with the least squares regression estimator, even though this estimator has a finite-sample breakdown point of only $1/n$. But this modification does not perform particularly well when comparing the means of two independent groups.) The minimum amount of trimming needed to justify using a percentile bootstrap method, rather than some competing technique, has not been determined.

In subsequent chapters we will take up the problem of comparing multiple groups. To lay the foundation for one of the more effective methods, we describe the percentile bootstrap method for comparing trimmed means in another manner. Let

$$p^* = P\left(\bar{X}^*_{t1} > \bar{X}^*_{t2}\right). \tag{8.22}$$

That is, p^* is the probability that a bootstrap trimmed mean from the first group is greater than a bootstrap trimmed mean from the second. The value of p^* reflects the degree of separation between the two groups being compared, in the following sense. If the trimmed means based on the observed data are identical, meaning that $\bar{X}_{t1} = \bar{X}_{t2}$, then p^* will have a value approximately equal to .5. In fact, as the sample sizes increase, the value of p^* will converge to .5 for this special case. Moreover, if the population trimmed means are equal, then p^* will have, approximately, a uniform distribution, provided the sample sizes are not too small. That is, if $H_0: \mu_{t1} = \mu_{t2}$ is true, p^* will have a value between 0 and 1, with all possible values between 0 and 1 equally likely if the sample sizes are not too small. (Hall, 1988a, provides relevant theoretical details and results in Hall, 1988b, are readily extended to trimmed means.) This suggests the following decision rule: Reject the hypothesis of equal trimmed means if p^* is less than or equal to $\alpha/2$ or greater than or equal to $1 - \alpha/2$. Said another way, if we let p^*_m be equal to p^* or $1 - p^*$, whichever is smaller, then reject if

$$p^*_m \leq \frac{\alpha}{2}. \tag{8.23}$$

We do not know p^*, but it can be estimated with the proportion of times a bootstrap trimmed mean from the first group is greater than a bootstrap trimmed mean from the second. That is, if A represents the number of values among $D^*_1, \ldots, D^*_B$ that are greater than zero, then we estimate p^* with

$$\hat{p}^* = \frac{A}{B}. \tag{8.24}$$

Finally, we reject the hypothesis of equal population trimmed means if $\hat{p}^*$ is less than or equal to $\alpha/2$ or greater than or equal to $1 - \alpha/2$. Or setting $\hat{p}^*_m$ to $\hat{p}^*$ or $1 - \hat{p}^*$, whichever is smaller, reject if

$$\hat{p}^*_m \leq \frac{\alpha}{2}. \tag{8.25}$$

The quantity $2\hat{p}^*_m$ is the estimated p-value.

8.8.2 Bootstrap-*t* Methods

Bootstrap-*t* methods for comparing trimmed means are preferable to the percentile bootstrap when the amount of trimming is close to zero. An educated guess is that the bootstrap-t is preferable if the amount of trimming is less than or equal to 10%, but it is stressed that this issue is in need of more research. The only certainty is that with no trimming, all indications are that the bootstrap-t outperforms the percentile bootstrap.

Bootstrap-t methods for comparing trimmed means are performed as follows:

1. Compute the sample trimmed means, $\bar{X}_{t1}$ and $\bar{X}_{t2}$, and Yuen's estimate of the squared standard errors, d_1 and d_2, given by Equation (8.18).
2. For each group, generate a bootstrap sample and compute the trimmed means, which we label $\bar{X}^*_{t1}$ and $\bar{X}^*_{t2}$. Also, compute Yuen's estimate of the squared standard error, again using Equation (8.18), which we label d^*_1 and d^*_2.
3. Compute

$$T^*_y = \frac{(\bar{X}^*_{t1} - \bar{X}^*_{t2}) - (\bar{X}_{t1} - \bar{X}_{t2})}{\sqrt{d^*_1 + d^*_2}}.$$

4. Repeat steps 2 and 3 B times, yielding $T^*_{y1}, \ldots, T^*_{yB}$. In terms of probability coverage, $B = 599$ appears to suffice in most situations when $\alpha = .05$.
5. Put the $T^*_{y1}, \ldots, T^*_{yB}$ values in ascending order, yielding $T^*_{y(1)} \leq \cdots \leq T^*_{y(B)}$. The T^*_{yb} values ($b = 1, \ldots, B$) provide an estimate of the distribution of

$$\frac{(\bar{X}_{t1} - \bar{X}_{t2}) - (\mu_{t1} - \mu_{t2})}{\sqrt{d_1 + d_2}}.$$

6. Set $\ell = \alpha B/2$ and $u = B - \ell$, where ℓ is rounded to the nearest integer.

The equal-tailed $1 - \alpha$ confidence interval for the difference between the population trimmed means ($\mu_{t1} - \mu_{t2}$) is

$$\left(\bar{X}_{t1} - \bar{X}_{t2} - T^*_{y(u)}\sqrt{d_1 + d_2}, \bar{X}_{t1} - \bar{X}_{t2} - T^*_{y(\ell+1)}\sqrt{d_1 + d_2}\right). \tag{8.26}$$

To get a symmetric two-sided confidence interval, replace step 3 with

$$T^*_y = \frac{|(\bar{X}^*_{t1} - \bar{X}^*_{t2}) - (\bar{X}_{t1} - \bar{X}_{t2})|}{\sqrt{d^*_1 + d^*_2}},$$

set $a = (1 - \alpha)B$, rounding to the nearest integer, in which case a $1 - \alpha$ confidence interval for $\mu_{t1} - \mu_{t2}$ is

$$(\bar{X}_{t1} - \bar{X}_{t2}) \pm T^*_{y(a)}\sqrt{d_1 + d_2}. \tag{8.27}$$

HYPOTHESIS TESTING. As usual, reject the hypothesis of equal population trimmed means ($H_0: \mu_{t1} = \mu_{t2}$) if the $1 - \alpha$ confidence interval for the difference between the trimmed means does not contain zero. Alternatively, compute Yuen's test statistic

$$T_y = \frac{\bar{X}_{t1} - \bar{X}_{t2}}{\sqrt{d_1 + d_2}},$$

and reject if

$$T_y \le T^*_{y(\ell+1)}$$

or if

$$T_y \ge T^*_{y(u)}.$$

When using the symmetric, two-sided confidence interval method, reject if

$$|T_y| \ge T^*_{y(a)}.$$

8.8.3 Winsorizing

Section 7.5.2 indicated that theory allows us to Winsorize the observations before taking bootstrap samples provided the amount of Winsorizing does not exceed the amount of trimming. A possible advantage of Winsorizing is shorter confidence intervals. However, we saw in Chapter 7 that if we Winsorize as much as we trim, probability coverage can be poor, at least with small to moderate sample sizes. This continues to be the case when comparing groups. But when using the percentile bootstrap method, if, for example, we trim 20% and Winsorize 10% and if the smallest sample size is at least 15, it seems that probability coverage is reasonably close to the nominal level, at least when $\alpha = .05$. Winsorizing is not recommended when using a bootstrap-t method when sample sizes are small. Perhaps this strategy provides good probability coverage with moderately large sample sizes, but this has not been determined as yet.

If a situation arises where Winsorizing makes a practical difference in terms of power and length of confidence intervals, a competing strategy is not to Winsorize but instead simply to increase the amount of trimming. The relative merits of these two strategies have not been determined.

8.8.4 S-PLUS Functions trimpb2 and yuenbt

The S-PLUS functions trimpb2 and yuenbt are supplied for applying the bootstrap methods just described. The function

trimpb2(x, y, tr = 0.2, alpha = 0.05, nboot = 2000, WIN = F, win = 0.1)

performs the percentile bootstrap method, where x is any S-PLUS variable containing the data for group 1 and y contains the data for group 2. The amount of trimming, tr, defaults to 20%, α defaults to .05, and nboot (B) defaults to 2000. The argument WIN defaults to F, for false, meaning that Winsorizing will not be done prior to generating bootstrap samples. Setting WIN equal to T, for true, Winsorizing will be done with the amount of Winsorizing determined by the argument win, which defaults to .1 (10%). This function returns the estimated significance level (or p-value), labeled sig.level, plus a $1 - \alpha$ confidence interval for the difference between the trimmed means.

The function

yuenbt(x, y, tr = 0.2, alpha = 0.05, nboot = 599, side = F)

performs the bootstrap-t method, which is based on Yuen's procedure for comparing trimmed means. The arguments are the same as before, except for the argument labeled side, which indicates whether a symmetric or equal-tailed confidence interval will be used. Side defaults to F, for false, meaning that the equal-tailed confidence interval [given by Equation (8.26)] will be computed. Setting side equal to T yields the symmetric confidence interval given by Equation (8.27).

EXAMPLE. Table 8.3 shows data from a study dealing with the effects of consuming alcohol. (The data were generously supplied by M. Earleywine.) Group 1, a control group, reflects hangover symptoms after consuming a specific amount of alcohol in a laboratory setting. Group 2 consisted of sons of alcoholic fathers. Storing the group 1 data in the S-PLUS variable A1, and the group 2 data in A2, the command trimpb2(A1,A2) returns the following output:

```
$sig.level:
[1] 0.038

$ci:
[1] 0.1666667 8.3333333
```

This says that a .95 confidence interval for the difference between the population trimmed means is (.17, 8.3). The significance level is .038, so in particular you would reject $H_0 : \mu_{t1} = \mu_{t2}$ at the .05 level. If we set the argument WIN to T, so that Winsorizing is done, then the .95 confidence interval is (0.58, 7.17), which is shorter than the confidence interval without Winsorizing. In contrast, if we use Welch's method for means, the .95 confidence interval is (−1.6, 10.7). This interval contains zero, so we no longer reject, the only point being that it can make a difference which method is used. Notice that the Winsorized confidence interval using trimpb2 is substantially shorter than Welch's confidence interval, the ratio of the lengths being

$$\frac{10.7 + 1.6}{7.17 - .58} = 1.87.$$

Yet all indications are that the percentile bootstrap confidence interval generally has more accurate probability coverage. ■

Notice that the default value for nboot (B) when using yuenbt is only 599, compared to 2000 when using trimpb2. Despite this, trimpb2 tends to have faster execution time, because it is merely computing trimmed means; yuenbt requires estimating the standard error for each bootstrap sample, which increases the execution

TABLE 8.3 Effect of Alcohol

Group 1:	0	32	9	0	2	0	41	0	0	0
	6	18	3	3	0	11	11	2	0	11
Group 2:	0	0	0	0	0	0	0	0	1	8
	0	3	0	0	32	12	2	0	0	0

TABLE 8.4 Self-Awareness Data

Group 1:	77	87	88	114	151	210	219	246	253		
	262	296	299	306	376	428	515	666	1310	2611	
Group 2:	59	106	174	207	219	237	313	365	458	497	515
	529	557	615	625	645	973	1065	3215			

time considerably. When using the bootstrap-*t* method (the S-PLUS function yuenbt), published papers indicate that increasing B from 599 to 999, say, does not improve probability coverage by very much, if at all, when $\alpha = .05$. This means that if we were to repeat the experiment billions of times, each time computing a .95 confidence interval, the proportion of times the resulting confidence interval contains the true difference between the population trimmed means will not be appreciably closer to .95 if we increase B from 599 to 999.

There is, however, a practical matter that should be mentioned. Consider the data in Table 8.4 and focus on group 1. Notice that the values are in ascending order. The S-PLUS function yuenbt begins by generating a bootstrap sample. The first value that it chooses might be the third value listed, which is 88. But suppose you store the data in S-PLUS in descending order instead. Then if yuenbt chooses the third value to be in the bootstrap sample, it is no longer 88 but rather 666. This means that the bootstrap sample will be altered, resulting in a different bootstrap sample trimmed mean. With B large enough, this will not change the resulting confidence interval and significance level by much. But with $B = 599$ the results might be altered enough to change your conclusion about whether to reject the null hypothesis. That is, a nearly significant result might become significant if the order of the observations is altered before invoking the bootstrap method or if we simply increase B. To reduce the likelihood of this possibility, consider using $B = 1999$ instead.

EXAMPLE. In an unpublished study by Dana (1990), the general goal was to investigate issues related to self-awareness and self-evaluation. In one portion of the study, he recorded the times individuals could keep an apparatus in contact with a specified target. The results, in hundredths of seconds, are shown in Table 8.4. Storing the data for group 1 in the S-PLUS variable G1 and storing

Continued

EXAMPLE. (*Continued*) the data for group 2 in G2, the command

yuenbt(G1,G2)

returns a .95 confidence interval of $(-312.5, 16.46)$. This interval contains zero, so we would not reject. If we increase the number of bootstrap samples (B) by setting the argument nboot to 999, now the confidence interval is $(-305.7, 10.7)$. We still do not reject, but increasing B alters the confidence interval slightly. In contrast, comparing medians via the method in Section 8.7.1, the .95 confidence interval is $(-441.4, -28.6)$, so we reject, the only point being that even among robust estimators, the choice of method can alter the conclusions reached. ■

8.8.5 Estimating Power and Judging the Sample Sizes

Imagine we compare two groups using the percentile bootstrap method with 20% trimming, as described in Section 8.8.1. If we fail to reject, this might be because there is little or no difference between the groups. Another possibility is that the population trimmed means differ by a substantively important amount but power is low. To help differentiate between these two possibilities, you can estimate how much power there was based on the data available to you.

The basic strategy is to estimate the standard errors associated with the 20% trimmed means and then to use these estimates to estimate power for a given value of the difference between the population trimmed means $(\mu_{t1} - \mu_{t2})$. The computational details, which stem from Wilcox and Keselman (in press), are shown in Box 8.1 and apply to 20% trimmed means only. Adjustments of the method when the amount of trimming is altered have not been studied. Another strategy for estimating power is to use a nested bootstrap similar to the one studied by Boos and Zhang (2000). A concern, however, is that the precision of the estimate cannot be easily assessed. That is, there is no known way of computing a reasonably accurate confidence for the actual amount of power if a nested bootstrap is used. In contrast, a confidence interval can be computed using the method in Box 8.1.

BOX 8.1 Power When Comparing 20% Trimmed Means with a Percentile Bootstrap Method

Goal:

Given data, estimate power associated with some specified value of $\delta = \mu_{t1} - \mu_{t2}$. Alternatively, estimate the power curve based on the data at hand.

Continued

BOX 8.1 (*Continued*)

Computations:

For $i = 1, 2, \ldots, 35$, let h_i be given by

500, 540, 607, 706, 804, 981, 1176, 1402, 1681, 2008, 2353, 2769,

3191, 3646, 4124, 4617, 5101, 5630, 6117, 6602, 7058, 7459,

7812, 8150, 8479, 8743, 8984, 9168, 9332, 9490, 9607,

9700, 9782, 9839, 9868.

For example, $h_1 = 500$ and $h_4 = 706$. For any two distributions, given $\delta = \mu_{t1} - \mu_{t2}$ and an estimate of the standard error of $\bar{X}_{t1} - \bar{X}_{t2}$, namely, $S = \sqrt{d_1 + d_2}$, where d_1 and d_2 are given by Equation (8.18), power is estimated as follows. Let $v = [8\delta/S] + 1$, where $[8\delta/S]$ indicates that $8\delta/S$ is rounded down to the nearest integer, and let

$$a = 8\left(\frac{\delta}{S} - \frac{v-1}{8}\right).$$

Then power is estimated to be

$$\hat{\gamma} = \frac{h_v}{10{,}000} + a\left(\frac{h_{v+1}}{10{,}000} - \frac{h_v}{10{,}000}\right).$$

In the event $v = 36$, h_{v+1} is taken to be 10,000 in the previous equation. If $v > 36$, power is estimated to be 1.

8.8.6 S-PLUS Functions powest and pow2an

The S-PLUS function

powest(x,y,delta)

estimates how much power there is when the difference between the population 20% trimmed means is delta. This is done by computing the standard errors of the sample trimmed means using the data in the S-PLUS variable x (Group 1) and the S-PLUS variable y (Group 2) and then performing the calculations in Box 8.1.

The S-PLUS function

pow2an(x,y,ci=F,plotit=T,nboot=800)

computes a power curve using the data in the S-PLUS variables x and y. That is, the function chooses a range of values for the difference between the population means, and for each difference it computes power using the S-PLUS function powest. By default, the power curve is plotted. To avoid the plot and get the numerical results only, set the argument plotit to F, for false. Setting the argument ci to T will result in a lower .95 confidence interval for the power curve to be computed using a bootstrap method based on nboot (B) bootstrap samples.

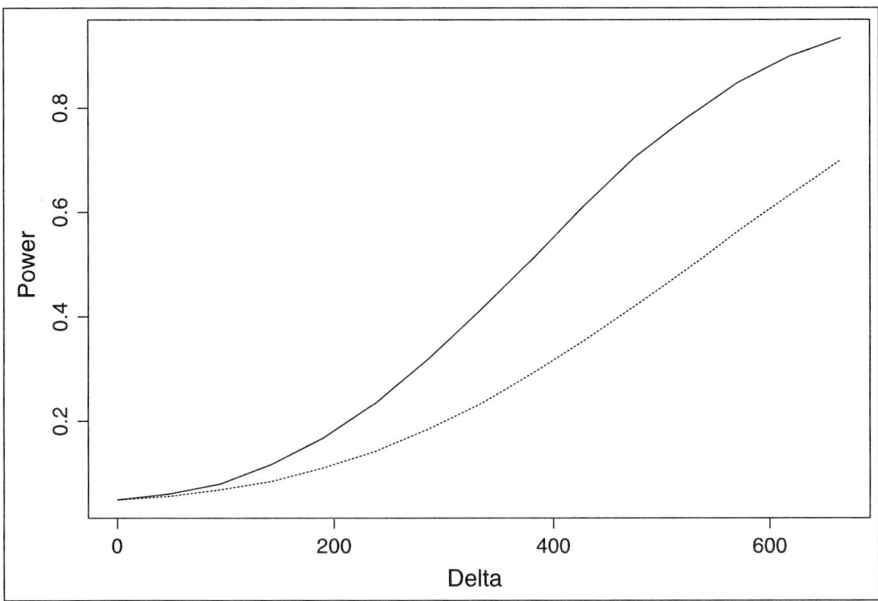

FIGURE 8.2 Estimate of the power returned by the S-PLUS function pow2an.

EXAMPLE. The S-PLUS functions powest and pow2an are illustrated with data from a reading study. (These data were generously supplied by Frank Manis.) Theoretical arguments suggest that groups should differ, but a nonsignificant result was obtained when comparing 20% trimmed means (or when comparing means with any of the previously described techniques). One possible explanation is that there is little or no difference between the groups, but another possibility is that power is low due to relatively large standard errors, meaning that detecting a substantively interesting difference is unlikely based on the sample sizes used. For a difference of 600 between the population 20% trimmed means, powest estimates that power is .8. Figure 8.2 shows the estimated power curve returned by pow2an. The lower, dashed line is a lower .95 confidence interval for the actual amount of power. That is, the solid line provides an approximately unbiased estimate of power, but a possibility is that power is as low as indicated by the dashed line. Based on this analysis it was concluded that power is low and that accepting the hypothesis of equal trimmed means is not warranted. ■

8.9 Comparing MOM-Estimators, M-Estimators, and Other Measures of Location

As is probably evident, when comparing two groups, any measure of location, and indeed virtually any parameter that characterizes a distribution (such as measures of scale), can be used with a percentile bootstrap method. Basically, generate B bootstrap samples from the first group; do the same for the second, in which case the middle 95% of the differences provides an approximate .95 confidence interval. That is, proceed as was described in Section 8.8.1 when comparing trimmed means, except that the sample trimmed mean is replaced with whatever estimator you care to use. Currently, all indications are that as a general rule, if the finite-sample breakdown point of the estimator is at least .2, relatively accurate probability coverage will be obtained. So, for example, if the goal is to compare the medians of two groups, or the quartiles, the percentile bootstrap is a relatively effective method.

It is possible to use a bootstrap-*t* method with M-estimators. (The method requires estimating the standard error of the M-estimator, which can be done as described in Wilcox, 1997a.) However, all indications are that the resulting confidence interval is not as accurate as the confidence interval associated with the percentile bootstrap method, particularly when sampling from skewed distributions. Consequently, the rather involved computations are not described.

A negative feature of using M-estimators is that for very small sample sizes — say, less than 20 — the probability coverage may not be as accurate as the probability coverage obtained with 20% trimmed means. Also, using a bootstrap with an M-estimator is not always possible, for reasons described in Section 7.7. (Bootstrap samples can have MAD = 0.) On the positive side, the M-estimator described in Section 3.5 has a high finite-sample breakdown point, which could translate into more power, compared to using a 20% trimmed mean, when sampling is from distributions where the proportion of outliers exceeds 20%. Of course one could increase the amount of trimming to deal with this problem. And when many outliers are common, the median might be an excellent choice. An advantage of an M-estimator is that it empirically adjusts to the number of outliers, a consequence being that, compared to the median, it performs better in terms of power when sampling is from a normal distribution.

8.9.1 S-PLUS Function pb2gen

The S-PLUS function

pb2gen(x, y, alpha = 0.05, nboot = 2000, est = mom, ...)

computes a percentile bootstrap confidence interval for the difference between any two measures of location. As usual, x and y are any S-PLUS variables containing data and nboot is B, the number of bootstrap samples to be used. By default, $B = 2000$ is used. The argument est indicates which measure of location is to be employed. It can be any S-PLUS function that computes a measure of location and defaults to the S-PLUS function mom (written for this book), which is the MOM-estimator described in Chapter 3. The argument ... can be used to reset certain default settings associated with the argument est. For example, if est=mean is used, means are compared. In contrast, the command

pb2gen(x, y, alpha = 0.05, nboot = 2000, est = mean, tr = .2)

would compare 20% trimmed means instead. (In this case, pb2gen and trimpb2, described in Section 8.8.4, give the same results.) The command

pb2gen(x, y, alpha = 0.05, nboot = 2000, est = median)

would compare medians.

EXAMPLE. A study was conducted comparing the EEG (electroencephalogram) readings of convicted murderers to the EEG readings of a control group with measures taken at various sites in the brain. For one of these sites the results were

Control	−0.15	−0.22	0.07	−0.07	0.02	0.24	−0.60
group:	−0.17	−0.33	0.23	−0.69	0.70	1.13	0.38
Murderers:	−0.26	0.25	0.61	0.38	0.87	−0.12	0.15
	0.93	0.26	0.83	0.35	1.33	0.89	0.58

(These data were generously supplied by A. Raine.) The sample medians are −0.025 and 0.48, respectively. Storing the data in the S-PLUS variables x1 and x2, the command

pb2gen(x1, x2, est = median)

returns a .95 confidence interval for the difference between the population medians of (−0.97, −0.085). So the hypothesis of equal population medians is rejected because this interval does not contain 0, and the data indicate that the typical measure for the control group is less than the typical measure among convicted murderers. Using the nonbootstrap method in Section 8.7.1 instead, the .95 confidence interval is (−0.89, −0.119). ■

EXAMPLE. Table 3.2 contains data on the desired number of sexual partners over the next 30 years reported by male undergraduates. The responses by 156 females are shown in Table 8.5. Does the typical response among males differ from the typical response among females? If we simply apply Student's T, we fail to reject, which is not surprising because there is an extreme outlier among the responses for males. (See the last example in Section 8.2.) But if we trim only 1% of the data, Yuen's method rejects, suggesting that the two distributions differ. However, with so little trimming, accurate confidence intervals might be difficult to obtain. Moreover, the median response among both males and females is 1, suggesting that in some sense the typical male and typical female are similar. To add perspective, we compare the .75 quantiles of the distributions, which can be estimated with the built-in S-PLUS function quantile. For example, if the responses for the males are stored in the S-PLUS variable sexm, the S-PLUS command quantile(sexm,probs=.75) estimates the .75 quantile to be 6; for females the estimate is 3. The command

pb2gen(sexm, sexf, est = quantile, probs = .75)

compares the .75 quantiles of the two groups and returns a .95 confidence interval of (1, 8). So we reject the hypothesis of equal .75 quantiles, indicating

Continued

TABLE 8.5 Desired Number of Sexual Partners for 156 Females

x:	0	1	2	3	4	5	6	7	8	10
f_x:	2	101	11	10	5	11	1	1	3	4
x:	11	12	15	20	30					
f_x:	1	1	2	1	2					

EXAMPLE. (*Continued*) that the groups differ among the higher responses. That is, in some sense the groups appear to be similar because they have identical medians. But if we take the .75 quantiles to be the typical response among the higher responses we might observe, the typical male appears to respond higher than the typical female. ■

8.10 Comparing Variances or Other Measures of Scale

Although the most common approach to comparing two independent groups is to use some measure of location, situations arise where there is interest in comparing variances or some other measure of scale. For example, in agriculture, one goal when comparing two crop varieties might be to assess their relative stability. One approach is to declare the variety with the smaller variance as being more stable (e.g., Piepho, 1997). As another example, consider two methods for training raters of some human characteristic. For example, raters might judge athletic ability or they might be asked to rate aggression among children in a classroom. Then one issue is whether the variances of the ratings differ depending on how the raters were trained. Also, in some situations, two groups might differ primarily in terms of the variances rather than their means or some other measure of location. To take a simple example, consider two normal distributions both having means zero with the first having variance one and the second having variance three. Then a plot of these distributions would show that they differ substantially, yet the hypotheses of equal means, equal trimmed means, equal M-estimators, and equal medians are all true. That is, to say the first group is comparable to the second is inaccurate, and it is of interest to characterize how they differ.

There is a vast literature on comparing variances and as usual not all methods are covered here. For studies comparing various methods, the reader is referred to Conover, Johnson, and Johnson (1981), Brown and Forsythe (1974b), Wilcox (1992), plus the references they cite.

8.10.1 Comparing Variances

We begin with testing

$$H_0: \sigma_1^2 = \sigma_2^2, \tag{8.28}$$

the hypothesis that the two groups have equal variances. Many methods have been proposed. The classic technique assumes normality and is based on the ratio of the largest sample variance to the smallest. So if $s_1^2 > s_2^2$, the test statistic is $F = s_1^2/s_2^2$; otherwise you use $F = s_2^2/s_1^2$. When the null hypothesis is true, F has a so-called F distribution, which is described in Chapter 9. But this approach has long been known to be highly unsatisfactory when distributions are nonnormal (e.g., Box, 1953), so additional details are omitted.

Currently, the most successful method in terms of maintaining control over the probability of a Type I error and achieving relatively high power is to use a slight modification of the percentile bootstrap method. In particular, set $n_m = \min(n_1, n_2)$, and, for the jth group ($j = 1, 2$), take a bootstrap sample of size n_m. Ordinarily we take a bootstrap sample of size n_j from the jth group, but when sampling from heavy-tailed distributions, and when the sample sizes are unequal, control over the probability of a Type I error can be extremely poor for the situation at hand. Next, for each group, compute the sample variance based on the bootstrap sample and set D^* equal to the difference between these two values. Repeat this $B = 599$ times, yielding 599 bootstrap values for D, which we label $D_1^*, \ldots, D_{599}^*$. As usual, when writing these values in ascending order, we denote this by $D_{(1)}^* \leq \cdots \leq D_{(B)}^*$. Then an approximate .95 confidence interval for the difference between the population variances is

$$\left(D_{(\ell)}^*, D_{(u)}^*\right), \tag{8.29}$$

where for $n_m < 40$, $\ell = 7$ and $u = 593$; for $40 \leq n_m < 80$, $\ell = 8$ and $u = 592$; for $80 \leq n_m < 180$, $\ell = 11$ and $u = 588$; for $180 \leq n_m < 250$, $\ell = 14$ and $u = 585$; and for $n_m \geq 250$, $\ell = 15$ and $u = 584$. (For results on the small-sample properties of this method, see Wilcox, in press.) Notice that these choices for ℓ and u are the same as those used in Section 7.3 when making inferences about the least squares regression slope and Pearson's correlation. The hypothesis of equal variances is rejected if the confidence interval given by Equation (8.29) does not contain zero.

Using the confidence interval given by Equation (8.29) has two practical advantages over the many alternative methods one might use to compare variances. First, compared to many methods, it provides higher power. Second, among situations where distributions differ in shape, extant simulations indicate that probability coverage remains relatively accurate, in contrast to many other methods one might use. If the standard percentile bootstrap method is used instead, then with sample sizes of 20 for both groups, the Type I error probability can exceed .1 when testing at the .05 level, and with unequal sample sizes it can exceed .15.

8.10.2 S-PLUS Function comvar2

The S-PLUS function

```
comvar2(x,y)
```

compares variances using the bootstrap method described in the previous subsection. The method can only be applied with $\alpha = .05$; modifications that allow other α values have not been derived. The arguments x and y are S-PLUS variables containing data for group 1 and group 2, respectively. The function returns a .95 confidence interval for $\sigma_1^2 - \sigma_2^2$ plus an estimate of $\sigma_1^2 - \sigma_2^2$ based on the difference between the sample variances, $s_1^2 - s_2^2$, which is labeled difsig.

8.10.3 Brown–Forsythe Method

Section 3.3.6 described a measure of scale based on the average absolute distance of observations from the median. In the notation used here, if M_1 is the median of the first group, the measure of scale for the first group is

$$\hat{\tau}_1 = \frac{1}{n_1} \sum |X_{i1} - M_1|, \tag{8.30}$$

where again $X_{11}, \ldots, X_{n_1 1}$ are the observations randomly sampled from the first group. For the second group this measure of scale is

$$\hat{\tau}_2 = \frac{1}{n_2} \sum |X_{i2} - M_2|, \tag{8.31}$$

where M_2 is the median for the second group. Notice that these measures of scale do *not* estimate the population variance (σ^2) or the population standard deviation (σ). There is a commonly recommended method for comparing groups based on these measures of scale, so it is important to comment on its relative merits.

For convenience, let

$$Y_{ij} = |X_{ij} - M_j|,$$

$i = 1, \ldots, n_j$; $j = 1, 2$. That is, the ith observation in the jth group (X_{ij}) is transformed to $|X_{ij} - M_j|$, its absolute distance from the median of the jth group. So the sample mean of the Y values for the jth group is

$$\bar{Y}_j = \frac{1}{n_j} \sum Y_{ij},$$

which is just the measure of scale described in the previous paragraph. Now let τ_j be the population value corresponding to $\bar{Y}_j$. That is, τ_j is the value of $\bar{Y}_j$ we would get if all individuals in the jth group could be measured. The goal is to test

$$H_0 : \tau_1 = \tau_2. \tag{8.32}$$

If we reject, we conclude that the groups differ based on this measure of dispersion.

The Brown and Forsythe (1974b) test of the hypothesis given by Equation (8.32) consists of applying Student's T to the Y_{ij} values. We have already seen, however, that when distributions differ in shape, Student's T performs rather poorly, and there are general conditions under which it does not converge to the correct answer as the sample sizes get large. We can correct this latter problem by switching to Welch's test, but problems remain when distributions differ in shape. For example, suppose we sample $n_1 = 20$ observations from a normal distribution and $n_2 = 15$ observations from the observations shown in Figure 5.6. Then when testing at the $\alpha = .05$ level, the actual probability of a Type I error is approximately .21. Like Student's T or Welch's method, the Brown–Forsythe test provides a test of the hypothesis that distributions are identical. Although it is designed to be sensitive to a reasonable measure of scale, it can be sensitive to other ways the distributions might differ. So if the goal is to compute a confidence interval for $\tau_1 - \tau_2$, the Brown–Forsythe method can be unsatisfactory if $\tau_1 \neq \tau_2$. Presumably some type of bootstrap method could improve matters, but this has not been investigated and indirect evidence suggests that practical problems will remain. Moreover, if there is explicit interest in comparing variances (σ_1^2 and σ_2^2), the Brown–Forsythe test is unsatisfactory, because $\hat{\tau}_1$ and $\hat{\tau}_2$ do not estimate the population variances, σ_1^2 and σ_2^2, respectively.

8.10.4 Comparing Robust Measures of Scale

There are at least 150 measures of scale that have been proposed. One criterion for choosing from among them is that an estimator have a relatively small standard error when sampling from any of a range of distributions. Lax (1985) compared many measures of scale in this manner where the distributions ranged between a normal and symmetric distributions with very heavy tails. In the context of hypothesis testing, having a relatively small standard error can help increase power. Two scale estimators that performed well were the percentage bend midvariance and biweight midvariance described in Section 3.3.7. These measures of scale can be compared with the S-PLUS function pb2gen, described in Section 8.9.1. (As noted in Section 3.3.8, the S-PLUS functions pbvar and bivar have been supplied for computing these two measures of scale, respectively.)

EXAMPLE. Twenty-five observations were generated from the mixed normal distribution shown in Figure 2.8, and another 25 observations were sampled from a standard normal. As explained in Section 2.7, the corresponding population variances differ considerably even though the corresponding probability curves are very similar. Storing the data in the S-PLUS variables x and y, comvar2(x, y) returned a .95 confidence interval of (0.57, 32.2), so we correctly reject the hypothesis of equal population variances. The sample variances were $s_x^2 = 14.08$ and $s_y^2 = 1.23$. In contrast, the S-PLUS command

```
pb2gen(x,y,est=pbvar)
```

returns a .95 confidence interval of (− 1.59, 11.6) for the difference between the percentage bend midvariances. The values of the percentage bend midvariances were 2.3 and 1.7. In reality, the population values of the percentage bend midvariances differ slightly, and we failed to detect this due to low power. ■

EXAMPLE. We repeat the last example, only now we sample from the two distributions shown in Figure 8.3. (These two distributions are the normal and

Continued

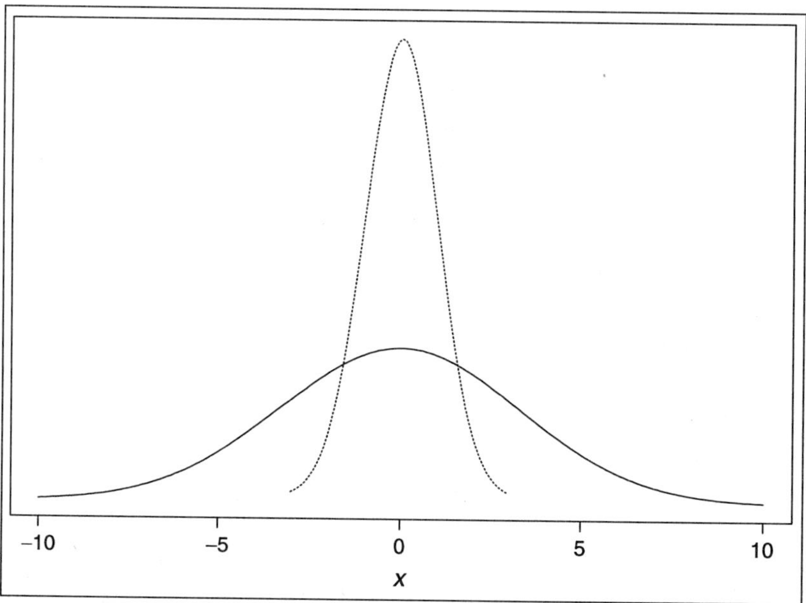

FIGURE 8.3 Two distributions that have equal variances. Testing the hypothesis of equal variances, we should not reject, even though there is a striking difference between the two distributions.

EXAMPLE. (*Continued*) mixed normal shown in Figure 2.9.) Although there is a clear and rather striking difference between these two distributions, the population means and variances are equal. So we should not reject the hypothesis of equal population variances, and with data generated by the author (with both sample sizes equal to 25) we indeed fail to reject; the .95 confidence interval returned by comvar2 is $(-13.3, 20.8)$. In contrast, comparing the percentage bend midvariances, the .95 confidence interval is $(-28.7, -1.8)$; this interval does not contain zero, so we reject. If we compare the biweight midvariances with the command

pb2gen(x,y,est=bivar),

the resulting .95 confidence interval is $(-18.8, -5.6)$ and again we reject. ■

These two examples merely illustrate that different methods can lead to different conclusions. In the first example, it is certainly true that the two distributions being compared are very similar, in the sense described in Section 2.7. But the tails of the mixed normal differ from the tails of the normal, and in the context of measuring stability or reliability, there is a difference that might have practical importance. This difference happens to be detected in the first example of this subsection by comparing variances but not when comparing the percentage bend midvariances or the biweight midvariances. In fairness, however, for this particular situation the power associated with comparing variances is not very high. However, in the second example, clearly the distributions differ considerably in terms of scale, as indicated in Figure 8.3. Comparing the percentage bend midvariances or biweight midvariances happens to detect this, but comparing the variances does not.

8.11 Measuring Effect Size

It has long been recognized that merely rejecting the hypothesis of equal means (or any other measure of location) tells us virtually nothing about the magnitude of the difference between the two groups (e.g., Cohen, 1994). If we reject at the .001 level and the first group has a larger sample mean than the second, then we conclude that the first group has the larger population mean. But this tells us nothing about the magnitude of the difference. An article in *Nutrition Today* (19, 1984, 22–29) illustrates the importance of this issue. A study was conducted on whether a particular drug lowers the risk of heart attacks. Those in favor of using the drug pointed out that the number of heart attacks in the group receiving the drug was significantly lower than in the group receiving a placebo when testing at the $\alpha = .001$ level. However, critics of the drug argued that the difference between the number of heart attacks was trivially small. They concluded that because of the expense and side effects of using the drug, there is no compelling evidence that patients with high cholesterol levels should be put on this medication. A closer examination of the data revealed that the standard errors corresponding to the two groups were very small, so it was possible to get a statistically significant result that was clinically unimportant.

Generally, how might we measure the difference between two groups? Three approaches are considered in this section:

- Compute a confidence interval for the difference between some measure of location.
- Use a so-called standardized difference.
- Use a global comparison of the distributions.

The first approach has already been discussed, so no additional comments are given here. The second approach, which is commonly used, is typically implemented by assuming the two groups have a common variance, which we label σ^2. That is, $\sigma_1^2 = \sigma_2^2 = \sigma^2$ is assumed. Then the so-called standardized difference between the groups is

$$\Delta = \frac{\mu_1 - \mu_2}{\sigma}. \tag{8.33}$$

Assuming normality, Δ can be interpreted using results in Chapter 2. For example, if $\Delta = 2$, then the difference between the means is 2 standard deviations, and for normal distributions we have some probabilistic sense of what this means. We estimate Δ with

$$\hat{\Delta} = \frac{\bar{X}_1 - \bar{X}_2}{s_p},$$

where s_p is the pooled standard deviation given by Equation (8.2).

Unfortunately, Δ suffers from some fundamental problems. First, if groups differ, there is no reason to assume that the variances are equal. Indeed, some authorities would argue that surely they must be unequal. We could test the hypothesis of equal variances, but how much power is needed to justify the conclusion that variances are equal if we fail to reject? Another possibility is to replace σ with the standard deviation from one of the groups. That is, we might use

$$\Delta_1 = \frac{\mu_1 - \mu_2}{\sigma_1} \quad \text{or} \quad \Delta_2 = \frac{\mu_1 - \mu_2}{\sigma_2},$$

which we would estimate, respectively, with

$$\hat{\Delta}_1 = \frac{\bar{X}_1 - \bar{X}_2}{s_1} \quad \text{and} \quad \hat{\Delta}_2 = \frac{\bar{X}_1 - \bar{X}_2}{s_2}.$$

But an even more serious problem is nonnormality.

The left panel of Figure 8.1 shows two normal distributions, where the difference between the means is 1 ($\mu_1 - \mu_2 = 1$) and both standard deviations are 1. So

$$\Delta = 1.$$

Cohen (1977) defines a large effect size as one that is visible to the naked eye, and he concludes (p. 40) that for normal distributions, $\Delta = .8$ is large, $\Delta = .5$ is a medium effect size, and $\Delta = .2$ is small. Now look at the right panel of Figure 8.1. As is evident, the difference between the two distributions appears to be very similar to the difference shown in the left panel, so according to Cohen we again have a large effect size. However, in the right panel, $\Delta = .3$ because these two distributions are mixed normals with variances 10.9. This illustrates the general principle that arbitrarily small departures from normality can render the magnitude of Δ meaningless. In practical terms, if we rely exclusively on Δ to judge whether there is a substantial difference between two groups, situations will arise where we will grossly underestimate the degree to which groups differ.

Here is another concern about Δ when trying to characterize how groups differ. Look at Figure 8.3. These two distributions have equal means and equal variances, but they differ in an obvious way that might have practical importance. Although the difference between measures of location provides a useful measure of effect size, we need additional ways of gaining perspective on the extent to which groups differ.

To describe one way of measuring the degree of separation between two groups, imagine that we randomly sample an observation from one of the two distributions in Figure 8.3 and that we get the value -5. Then Figure 8.3 indicates that this observation probably came from the first group (the one with the probability density function given by the solid line) because the probability of getting a value as low as -5 from the second group is virtually zero. If we had gotten the value 0, it is more likely that the observation came from the second group because the probability of getting a value near zero is higher for the second group. More formally let $f_1(x)$ be the equation for the solid line in Figure 8.3 (which is its probability density function) and let $f_2(x)$ be the equation for the dashed line. The likelihood that the value x came from the first group is $f_1(x)$, and the likelihood that the value x came from the second group is $f_2(x)$. So if $f_1(x) > f_2(x)$, a natural rule is to decide that x came from group 1, and if $f_1(x) < f_2(x)$ to decide that x came from group 2. We do not know $f_1(x)$ and $f_2(x)$, but they can be estimated, as indicated in Section 3.7. The result is a relatively effective method for deciding whether the value x came from group 1 or 2. (e.g., Silverman, 1986). There are many other strategies one might use to decide from which group the observation came, but the method just described has been found to be relatively effective for the problem at hand.

Let Q be the probability of correctly deciding whether a randomly sampled observation came from group 1 using the strategy just outlined. Then Q provides a measure of the separation between the two groups. If the distributions are identical, then $Q = .5$. If they are completely distinct, then $Q = 1$. To add perspective, if for two normal distributions $\Delta = .8$, which is typically labeled a large effect size, then $Q = .66$. If $\Delta = .2$, then $Q = .55$. But unlike Δ, Q does not change drastically with small shifts away from a normal distribution. For example, for the left panel of Figure 8.1, $Q = .66$ and for the right panel $Q = .69$, so in both cases a large effect size is indicated. This is in contrast to Δ, which drops from 1 to .3.

It is *not* being suggested that Q be used as a measure of effect size to the exclusion of all other measures. Measuring the difference between two distributions is a complex issue that often requires several perspectives. Also, rigid adherence to the idea that $Q = .66$ is large and that $Q = .55$ is small is not being recommended. What constitutes a large difference can vary from one situation to the next. We are comparing Q to Δ merely to add perspective.

Currently, the most accurate estimate of Q is based on a rather complex bootstrap method (called the .632 estimator). A description of this estimator in a much more general context is given in Efron and Tibshirani (1993). The use of this bootstrap method when estimating Q has been investigated by Wilcox and Muska (1999). Here the computational details are relegated to Box 8.2. It is noted that the estimate of Q, $\hat{Q}_{.632}$, can be less than .5 even though we know $Q \geq .5$. So $\hat{Q}_{.632} < .5$ suggests that there is little or no difference between the groups.

BOX 8.2 A Bootstrap Estimate of Q

For each group, compute the kernel density estimator as described in Section 3.7, and label the results $\hat{f}_1(x)$ and $\hat{f}_2(x)$, respectively. Set $\hat{\eta}(X_{i1}) = 1$ if $\hat{f}_1(X_{i1}) > \hat{f}_2(X_{i1})$, otherwise $\hat{\eta}(X_{i1}) = 0$. (That is, decide X_{i1} came from group 1 if $\hat{f}_1(X_{i1}) > \hat{f}_2(X_{i1})$.) In contrast, set $\hat{\eta}(X_{i2}) = 1$ if $\hat{f}_1(X_{i2}) < \hat{f}_2(X_{i2})$, otherwise $\hat{\eta}(X_{i2}) = 0$. Generate a bootstrap sample from each group and let $\hat{\eta}^*$ be the resulting estimate of η. Repeat this process B times, yielding $\hat{\eta}^*_b$,

Continued

BOX 8.2 (*Continued*) $b = 1, \ldots, B$. Let

$$\hat{\epsilon}_1 = \frac{1}{n_1}\sum_{i=1}^{n_1}\frac{1}{B_{1i}}\sum_{b\in C_{1i}}\hat{\eta}_b^*(X_{1i}),$$

$$\hat{\epsilon}_2 = \frac{1}{n_2}\sum_{i=1}^{n_2}\frac{1}{B_{2i}}\sum_{b\in C_{2i}}\hat{\eta}_b^*(X_{2i}).$$

For the bootstrap samples obtained from the first group, C_{1i} is the set of indices of the bth bootstrap sample not containing X_{1i}, and B_{1i} is the number of such bootstrap samples. The notation $b \in C_{1i}$ means that b is an element of C_{1i}. That is, the second sum in the definition of $\hat{\epsilon}_1$ is over all bootstrap samples not containing X_{1i}. Similarly, for the bootstrap samples from the second group, C_{2i} is the set of indices of the bth bootstrap sample not containing X_{2i}, and B_{2i} is the number of such bootstrap samples. Let

$$\hat{Q}_{.632,1} = .368\hat{Q}_{ap1} + .632\hat{\epsilon}_1,$$

where

$$\hat{Q}_{ap1} = \frac{1}{n_1}\sum \hat{\eta}(X_{1i}),$$

and define $\hat{Q}_{.632,2}$ in an analogous fashion. The .632 estimator, which is computed by the S-PLUS function qhat in Section 8.11.1, is taken to be

$$\hat{Q}_{.632} = \frac{1}{n_1+n_2}\left(n_1\hat{Q}_{.632,1} + n_2\hat{Q}_{.632,2}\right).$$

EXAMPLE. To illustrate a portion of the computations in Box 8.2, consider five bootstrap samples from the first group with the following observation numbers:

Bootstrap sample				
1	2	3	4	5
1	16	25	1	14
5	5	4	7	10
23	16	12	12	2
11	24	16	7	8
11	11	14	14	13
17	15	24	1	1
8	21	3	21	17

Continued

EXAMPLE. (*Continued*) So the first bootstrap sample (column 1) contains observation numbers 1, 5, 23, 11, 11, 17, and 8. That is, it contains the first observation followed by the fifth observation, followed by the twenty-third observation, and so on. Note that observation 1 appears in bootstrap samples 1, 4, and 5, but not in samples 2 and 3. That is, $C_{11} = (2, 3)$, and $B_{11} = 2$, the number of elements in $C_{11} = (2, 3)$. So when $i = 1$, the second sum when computing $\hat{\epsilon}$ is over the bootstrap samples $b = 2$ and 3. That is, when $i = 1$, $\hat{\eta}^*_b(X_{1i})$ is being computed using X_{i1} values that do not appear in the bootstrap sample used to determine $\hat{\eta}^*_b$. Similarly, the second observation appears in only one bootstrap sample, the fifth. So $C_{12} = (1, 2, 3, 4)$. That is, when $i = 2$, the second sum is over $b = 1, 2, 3$, and 4. ■

8.11.1 S-PLUS Function qhat

The S-PLUS function

qhat(x,y)

estimates Q using the data stored in the S-PLUS variables x and y. (Execution time can be quite high.)

EXAMPLE. For the alcohol data in Table 8.3 we rejected the hypothesis of equal 20% trimmed means. If we use a standardized difference between the two groups based on the means and the standard deviation of the first group, we get $\hat{\Delta}_1 = .4$. Using the standard deviation of the second group yields $\hat{\Delta}_2 = .6$. So taken together, and assuming normality, these results suggest a medium effect size. The S-PLUS function qhat returns

```
qhat.632 = .61,
```

supporting the view that there is a medium difference between the two groups. ■

8.11.2 The Shift Function

Again, it currently seems that no single method for characterizing how groups differ is satisfactory in all situations. A criticism of the methods covered so far is that they do not capture some of the global details of how groups differ. For example, imagine that two methods for treating depression are being compared and that the higher X happens to be the more effective the method. Further assume that the distribution associated with the first method is given by the solid line in Figure 8.3 and that the distribution for the second group is given by the dashed line. Then in terms of the population means or any other measure of location, it makes no difference which method is used. However, for the first method, there is a good chance that an individual will have a level of effectiveness less than -3; but under the second method, the probability of an effectiveness level less than -3 is virtually zero. That is, we can avoid a poor effectiveness rating by using method 2. Similarly, there is a good chance of an effectiveness rating greater than or equal to 3 using method 1, but not with method 2. That is, the relative merits of the methods change depending on where we look.

In this last example we could simply compare some measure of scale. But consider the distributions in Figure 2.10, which have equal means and variances. How might we summarize the extent to which these distributions differ? Doksum and Sievers (1976) suggest the following strategy. Rather than just compare a single measure of location, notice that we could, for example, compare the lower quartiles corresponding to these two groups. That is, we compare low values in the first group to the corresponding low values in the second. Of course, we can compare the upper quartiles as well. So, for example, if the upper quartile of the first group is 12 and the upper quartile of the second group is 8, we can say that the typical individual in the high end of the first group is better off than the comparable person in the second group. Of course, there is nothing sacred about the quartiles; we can use any quantile and get a more detailed sense about how the groups differ in contrast to using a single measure of location. If we compare the .25, .5, and .75 quantiles, we are comparing the quartiles in addition to the median. But we might also compare the .1, .2, .3, .4, .6, .7, .8, and .9 quantiles or any other set of quantiles we choose. To help convey the differences between all of the quantiles, Doksum and Sievers (1976) suggest plotting the differences between all quantiles versus the quantiles in the first group. So if x_q is the qth quantile of the first group, meaning that $P(X \leq x_q) = q$, the suggestion is to estimate all possible quantiles for each group and plot x_q versus $y_q - x_q$, where y_q is the qth quantile of the second group. If we do this for the two distributions shown in Figure 2.10, we get Figure 8.4. This says that low-scoring individuals in the second group score higher

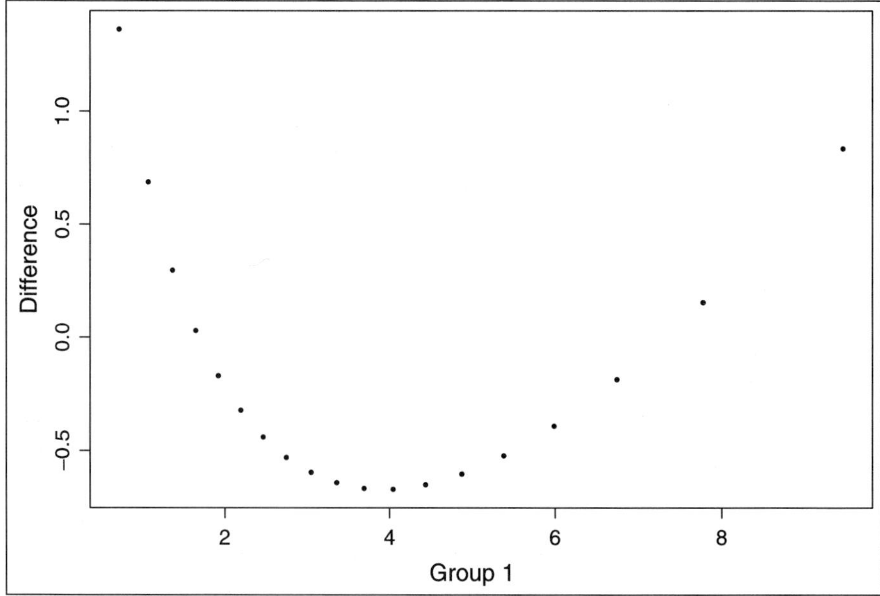

FIGURE 8.4 Plot of the quantiles of the first group versus the difference between the quantiles.

than the comparable low-scoring individuals in the first group, up to about the value 1.9. For the middle range of values in the first group, namely, between 2 and 7, the reverse is true; but for values greater than 7, again individuals in the second group score higher than comparable individuals in the first.

Doksum and Sievers also indicate how to compute a confidence band for the difference between all quantiles. That is, for each possible quantile, a confidence interval for the difference between the quantiles is computed with the property that with probability $1-\alpha$, it will be simultaneously true that all such confidence intervals contain the true difference $(y_q - x_q)$. Said another way, their method computes a confidence interval for $y_q - x_q$ for all values of q between 0 and 1, and if we reject the hypothesis of equal quantiles when this interval does not contain zero, then the probability of *at least one* Type I error is α. The computational details of their method are not given here, but an S-PLUS function for applying the method is supplied.

8.11.3 S-PLUS Function sband

The S-PLUS function

sband(x,y)

computes an estimate of the difference between the quantiles using the data stored in the S-PLUS variables x and y and plots these differences as a function of the estimated quantiles associated with the first group, the first group being the data stored in the S-PLUS variable x. (For more details about this function plus variations of the method, see Wilcox, 1997a.)

EXAMPLE. Table 8.6 contains data from a study designed to assess the effects of ozone on weight gain in rats. (These data were taken from Doksum & Sievers, 1976.) The experimental group consisted of 22 70-day-old rats kept in an ozone environment for 7 days. A control group of 23 rats of the same age was kept in an ozone-free environment. Storing the data for the control group in the S-PLUS variable x, and storing the data for the ozone group in y, sband produces the graph shown in Figure 8.5. The + indicates the location of the median for the control group (the data stored in the first argument, x), and the lower and upper quartiles are marked with an o to the left and right of the +. For example, the median of the control group is $M = 22.7$, as indicated by the +, and the difference between the median of the ozone group (which is 11.1) and the control group is given by the solid line and is equal to -11.6. So based on the medians it is estimated that the typical rat in the ozone group gains less weight than the typical rat in the control group. However, the difference between the upper quartiles is $26.95 - 17.35 = 9.6$. Now there is more weight gain among rats in the ozone group. Looking at the graph as a whole suggests that the effect of ozone becomes more pronounced as we move along the x-axis, up to about 19, but then the trend reverses, and in fact in the upper end we see more weight

Continued

TABLE 8.6 Weight Gain (in grams) of Rats in Ozone Experiment

Control:	41.0	38.4	24.4	25.9	21.9	18.3	13.1	27.3	28.5	−16.9
Ozone:	10.1	6.1	20.4	7.3	14.3	15.5	−9.9	6.8	28.2	17.9
Control:	26.0	17.4	21.8	15.4	27.4	19.2	22.4	17.7	26.0	29.4
Ozone:	−9.0	−12.9	14.0	6.6	12.1	15.7	39.9	−15.9	54.6	−14.7
Control:	21.4	26.6	22.7							
Ozone:	44.1	−9.0								

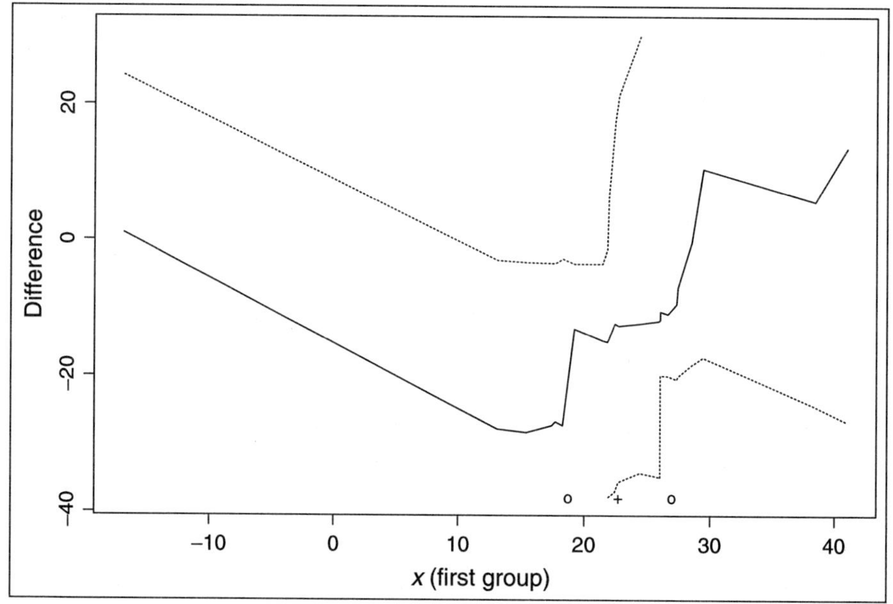

FIGURE 8.5 Example of a shift function. The dashed lines are confidence bands. The + along the x-axis indicates the location of the median of the first group; and the o's indicate the lower and upper quartiles.

EXAMPLE. (*Continued*) gain in the ozone group. If there are no differences between the quantiles, the shift function should be a straight horizontal line at 0. The dashed lines in Figure 8.5 mark the confidence band for the difference between the quantiles. The hypothesis of equal quantiles is rejected if the lower (upper) dashed line is above (below) zero. ■

Notice that the left end of the lower dashed line in Figure 8.5 begins at approximately $x = 22$. This is because for $x < 22$, the lower confidence band extends down to $-\infty$. That is, the precision of the estimated differences between the quantiles might be poor in this region based on the sample sizes used. Similarly, the upper dashed line terminates around $x = 27$. This is because for $x > 27$, the upper confidence band extends up to ∞.

The methods listed in this section are not exhaustive. Another approach is to examine a so-called quantile–quantile plot. If the groups have identical quantiles, a plot of the quantiles should be close to a line having slope 1 and intercept zero. Another approach to measuring effect size is the so-called *overlapping coefficient*. You estimate the distributions associated with both groups and then compute the area under the intersection of these two curves; see Clemons and Bradley (2000) for recent results on how this might be done. An area of zero corresponds to no overlap, and an area of 1 occurs when the distributions are identical and the groups do not differ in any manner whatsoever. Another useful approach is to create a boxplot for both groups on the same scale. This is easily done in S-PLUS. For example, the command boxplot(x,y) will create a boxplot for the data in both x and y. Yet another strategy is to estimate the probability that a randomly sampled observation from the first group is less than a randomly sampled observation from the second. Details about this approach will be covered in Chapter 15.

8.12 Comparing Correlations and Regression Slopes

Rather than compare measures of location or scale, situations arise where the goal is to compare correlations or regression parameters instead. That is, for every individual we have two measures and the goal is to determine whether the association for the first group differs from the association for the second. For example, in a study of schizophrenia, Dawson, Schell, Hazlett, Nuechterlein, and Filion (2000) were interested in, among other things, the association between prepulse inhibition and measures of schizophrenic symptoms. A portion of their study dealt with comparing correlations of individuals with positive symptoms to the correlation of those with negative symptoms. Also, comparing correlations or regression slopes is one strategy for determining whether a third variable modifies the association between two other variables. (See the example given at the end of this section.)

Methods for comparing correlations have been studied by Yu and Dunn (1982) and Duncan and Layard (1973), but all of these methods are known to be rather unsatisfactory, and using Fisher's r-to-z transformation is unsatisfactory, for reasons indicated in Section 6.5. Currently, the most effective procedure is to use the modified percentile bootstrap method, but adjusted to take into account the total number of observations. If we have n_1 pairs of observations for the first group, yielding a correlation of r_1, and n_2 pairs of observations for the second group, yielding a correlation of r_2, the goal is to test

$$H_0: \rho_1 = \rho_2,$$

the hypothesis that the two groups have equal population correlation coefficients. If we reject, this indicates that the association differs for each group, but for reasons outlined in Chapter 6, how the associations differ is vague and unclear.

To apply the modified percentile bootstrap method to the present problem, let $N = n_1 + n_2$ be the total number of pairs of observations available. For the jth group, generate a bootstrap sample of n_j pairs of observations as described in Section 7.3. Let r_1^* and r_2^* represent the resulting correlation coefficients and set

$$D^* = r_1^* - r_2^*.$$

Repeat this process 599 times, yielding $D_1^*, \ldots, D_{599}^*$. Then a .95 confidence interval for the difference between the population correlation coefficients ($\rho_1 - \rho_2$) is

$$\left(D^*_{(\ell)}, D^*_{(u)}\right),$$

where $\ell = 7$ and $u = 593$ if $N < 40$; $\ell = 8$ and $u = 592$ if $40 \le N < 80$; $\ell = 11$ and $u = 588$ if $80 \le N < 180$; $\ell = 14$ and $u = 585$ if $180 \le N < 250$; $\ell = 15$ and $u = 584$ if $N \ge 250$. If the resulting confidence interval does not contain zero, reject the hypothesis of equal correlations. Note that this is just a simple modification of the method used to compute a confidence interval for the slope of a regression line that was described in Chapter 7.

When using least squares regression, the slopes can be compared in a similar manner. To test

$$H_0 : \beta_1 = \beta_2,$$

where β_1 and β_2 are the slopes corresponding to the two groups, simply proceed as was done when working with Pearson's correlation, except replace r with the least squares estimate of the slope.

8.12.1 S-PLUS Functions twopcor and twolsreg

The S-PLUS function

twopcor(x1,y1,x2,y2)

computes a confidence interval for the difference between two Pearson correlations corresponding to two independent groups using the modified bootstrap method just described. The data for group 1 are stored in the S-PLUS variables x1 and y1, and the data for group 2 are stored in x2 and y2. The S-PLUS function

twolsreg(x1,y1,x2,y2)

computes a confidence interval for the difference between the slopes based on the least squares estimator described in Chapter 6.

EXAMPLE. In an unpublished study by L. Doi, there was interest in whether a measure of orthographic ability (Y) is associated with a measure of sound blending (X). Here we consider whether an auditory analysis variable (Z) *modifies* the association between X and Y. This was done by partitioning the pairs of points (X, Y) according to whether $Z \le 14$ or $Z > 14$, and then entering the resulting pairs of points into the S-PLUS function twopcor. The .95 confidence interval for $\rho_1 - \rho_2$, the difference between the correlations, is $(-0.64, 0.14)$. This interval contains zero, so we would not reject the hypothesis of equal correlations. If we compare regression slopes instead, the .95 confidence interval is $(-0.55, 0.18)$ and again we fail to reject. It is stressed, however, that this analysis does not establish that the association does not differ for the two groups under study. A concern is that power might be low when attention is

Continued

EXAMPLE. (*Continued*) restricted to Pearson's correlation or least squares regression. (Methods covered in subsequent chapters indicate that the measure of auditory analysis does modify the association between orthographic ability and sound blending.) ■

8.13 Comparing Two Binomials

This section considers the problem of comparing the probability of success associated with two independent binomials. For example, if the probability of surviving an operation using method 1 is p_1, and if the probability of surviving using method 2 is p_2, do p_1 and p_2 differ, and if they do differ, by how much? As another example, to what degree do men and women differ in whether they believe the President of the United States is an effective leader?

Many methods have been proposed for comparing binomials, two of which are described here. These two methods were chosen based on results in Storer and Kim (1990) and Beal (1987), where comparisons of several methods were made. It is noted, however, that competing methods have been proposed that apparently have not been compared directly to the methods covered here (e.g., Berger, 1996; Coe & Tamhane, 1993). The Storer–Kim method tests $H_0: p_1 = p_2$ using the calculations shown in Box 8.3, and Beal's method computes a $1-\alpha$ confidence interval for $p_1 - p_2$ using the calculations in Box 8.4. The choice between these two methods is not completely clear. An appeal of Beal's method is that it provides a confidence interval and the Storer–Kim method does not. Situations arise in subsequent chapters where the Storer–Kim method has less power than Beal's method when comparing multiple groups of individuals, but when comparing two groups only, we find situations where the Storer–Kim method rejects and Beal's method does not.

BOX 8.3 Storer–Kim Methods for Comparing Two Independent Binomials

You observe r_1 successes among n_1 trials in the first group and r_2 successes among n_2 trials in the second. The goal is to test $H_0: p_1 = p_2$. Note that the possible number of successes in the first group is any integer, x, between 0 and n_1, and for the second group it is any integer, y, between 0 and n_2. For any x between 0 and n_1 and any y between 0 and n_2, set

$$a_{xy} = 1$$

if

$$\left|\frac{x}{n_1} - \frac{y}{n_2}\right| \geq \left|\frac{r_1}{n_1} - \frac{r_2}{n_2}\right|;$$

Continued

BOX 8.3 (*Continued*)
otherwise

$$a_{xy} = 0.$$

Let

$$\hat{p} = \frac{r_1 + r_2}{n_1 + n_2}.$$

The test statistic is

$$T = \sum_{x=0}^{n_1} \sum_{y=0}^{n_2} a_{xy} b(x, n_1, \hat{p}) b(y, n_2, \hat{p}),$$

where

$$b(x, n_1, \hat{p}) = \binom{n_1}{x} \hat{p}^x (1 - \hat{p})^{n_1 - x}$$

and $b(y, n_2, \hat{p})$ is defined in an analogous fashion. You reject if

$$T \leq \alpha.$$

That is, T is the significance level.

BOX 8.4 Beal's Method for Computing a Confidence Interval for $p_1 - p_2$

Following the notation in Box 8.3, let $\hat{p}_1 = r_1/n_1$ and $\hat{p}_2 = r_2/n_2$ and let $c = z^2_{1-\alpha/2}$, where $z_{1-\alpha/2}$ is the $1 - \alpha$ quantile of a standard normal distribution. (So c is the $1 - \alpha$ quantile of a chi-squared distribution with one degree of freedom.) Compute

$$a = \hat{p}_1 + \hat{p}_2$$

$$b = \hat{p}_1 - \hat{p}_2$$

$$u = \frac{1}{4}\left(\frac{1}{n_1} + \frac{1}{n_2}\right)$$

$$v = \frac{1}{4}\left(\frac{1}{n_1} - \frac{1}{n_2}\right)$$

$$V = u\{(2 - a)a - b^2\} + 2v(1 - a)b$$

Continued

BOX 8.4 (*Continued*)

$$A = \sqrt{c\{V + cu^2(2 - a)a + cv^2(1 - a)^2\}}$$

$$B = \frac{b + cv(1 - a)}{1 + cu}.$$

The $1 - \alpha$ confidence interval for $p_1 - p_2$ is

$$B \pm \frac{A}{1 + cu}.$$

8.13.1 S-PLUS Functions twobinom and twobici

The S-PLUS function

twobinom(r1 = sum(x), n1 = length(x), r2 = sum(y), n2 = length(y),
x = NA, y = NA)

has been supplied to test $H_0 : p_1 = p_2$ using the Storer–Kim method in Box 8.3. The function can be used by specifying the number of successes in each group (arguments r1 and r2) and the sample sizes (arguments n1 and n2), or the data can be in the form of two vectors containing 1's and 0's, in which case you use the arguments x and y. Beal's method can be applied with the S-PLUS function

twobici(r1 = sum(x), n1 = length(x), r2 = sum(y), n2 = length(y), x = NA,
y = NA, alpha = 0.05)

EXAMPLE. If for the first group we have 7 successes among 12 observations, for the second group we have 22 successes among 25 observations, the command

twobinom(7,12,22,25)

returns a significance level of .044; this is less than .05, so we would reject with $\alpha = .05$. The .95 confidence interval for $p_1 - p_2$ returned by the command

twobici(7,12,22,25)

is (−0.61, 0.048); this interval contains zero, so in contrast to the Storer–Kim method we do not reject the hypothesis $H_0 : p_1 = p_2$, the only point being that different conclusions might be reached depending on which method is used. ■

EXAMPLE. In Table 8.5 we see that 101 of the 156 females responded that they want one sexual partner during the next 30 years. As for the 105 males in this study, 49 gave the response 1. Does the probability of a 1 among males differ from the probability among females? The S-PLUS function twobinom returns a significance level of .0037, indicating that the probabilities differ even with $\alpha = .0037$. The command

twobici(49, 105, 101, 156)

returns a .95 confidence interval of $(-0.33, -0.04)$, so again we reject, but there is some possibility that the difference between the two probabilities is fairly small. ■

8.14 Exercises

1. Suppose that the sample means and variances are $\bar{X}_1 = 15$, $\bar{X}_2 = 12$, $s_1^2 = 8$, $s_2^2 = 24$ with sample sizes $n_1 = 20$ and $n_2 = 10$. Verify that $s_p^2 = 13.14$ and $T = 2.14$ and that Student's T test rejects the hypothesis of equal means with $\alpha = .05$.
2. For two independent groups of subjects, you get $\bar{X}_1 = 45$, $\bar{X}_2 = 36$, $s_1^2 = 4$, $s_2^2 = 16$ with sample sizes $n_1 = 20$ and $n_2 = 30$. Assume the population variances of the two groups are equal and verify that the estimate of this common variance is 11.25.
3. Still assuming equal variances, test the hypothesis of equal means using the data in Exercise 2 assuming random sampling from normal distributions. Use $\alpha = .05$.
4. Repeat the previous exercise, but use Welch's test for comparing means.
5. Comparing the test statistics for the preceding two exercises, what do they suggest regarding the power of Welch's test versus Student's T test for the data being examined?
6. For two independent groups of subjects, you get $\bar{X}_1 = 86$, $\bar{X}_2 = 80$, $s_1^2 = s_2^2 = 25$, with sample sizes $n_1 = n_2 = 20$. Assume the population variances of the two groups are equal and verify that Student's T rejects with $\alpha = .01$.
7. Repeat Exercise 6 using Welch's method.
8. Comparing the results of Exercises 6 and 7, what do they suggest about using Student's T versus Welch's method when the sample variances are approximately equal?
9. If for two independent groups, you get $\bar{X}_{t1} = 42$, $\bar{X}_{t2} = 36$, $s_{w1}^2 = 25$, $s_{w2}^2 = 36$, $n_1 = 24$, and $n_2 = 16$, test the hypothesis of equal trimmed means with $\alpha = .05$.
10. Referring to Exercise 9, compute a .99 confidence interval for the difference between the trimmed means.
11. For $\bar{X}_1 = 10$, $\bar{X}_2 = 5$, $s_1^2 = 21$, $s_2^2 = 29$, and $n_1 = n_2 = 16$, compute a .95 confidence interval for the difference between the means using Welch's method, and state whether you would reject the hypothesis of equal means.

12. Repeat Exercise 11, but use Student's T instead.
13. Two methods for training accountants are to be compared. Students are randomly assigned to one of the two methods. At the end of the course, each student is asked to prepare a tax return for the same individual. The returns reported by the students are

Returns								
Method 1:	132	204	603	50	125	90	185	134
Method 2:	92	−42	121	63	182	101	294	36

Using Welch's test, would you conclude that the methods differ in terms of the average return? Use $\alpha = .05$.
14. Repeat Exercise 13, but compare 20% trimmed means instead.
15. You compare lawyers to professors in terms of job satisfaction and fail to reject the hypothesis of equal means or equal trimmed means. Does this mean it is safe to conclude that the typical lawyer has about the same amount of job satisfaction as the typical professor?
16. Responses to stress are governed by the hypothalamus. Imagine you have two groups of subjects. The first shows signs of heart disease and the other does not. You want to determine whether the groups differ in terms of the weight of the hypothalamus. For the first group of subjects, with no heart disease, the weights are

11.1, 12.2, 15.5, 17.6, 13.0, 7.5, 9.1, 6.6, 9.5, 18.0, 12.6.

For the other group, with heart disease, the weights are

18.2, 14.1, 13.8, 12.1, 34.1, 12.0, 14.1, 14.5, 12.6, 12.5, 19.8,
13.4, 16.8, 14.1, 12.9.

Determine whether the groups differ based on Welch's test. Use $\alpha = .05$.
17. Repeat Exercise 16, but use Yuen's test with 20% trimmed means.
18. Use Δ and Q to measure effect size using the data in the previous two exercises.
19. Published studies indicate that generalized brain dysfunction may predispose someone to violent behavior. Of interest is determining which brain areas may be dysfunctional in violent offenders. In a portion of such a study conducted by Raine, Buchsbaum, and LaCasse (1997), glucose metabolism rates of 41 murderers were compared to the rates for 41 control subjects. Results for the left hemisphere, lateral prefrontal region of the brain yielded a sample mean of 1.12 for the controls and 1.09 for the murderers. The corresponding standard deviations were 0.05 and 0.06. Verify that Student's $T = 2.45$ and that you reject with $\alpha = .05$.
20. In the previous exercise, you rejected the hypothesis of equal means. What does this imply about the accuracy of the confidence interval for the difference between the population means based on Student's T?
21. For the data in Table 8.6, if we assume that the groups have a common variance, verify that the estimate of this common variance is $s_p^2 = 236$.

22. The sample means for the data in Table 8.6 are 22.4 and 11. If we test the hypothesis of equal means using Student's T, verify that $T = 2.5$ and that you would reject with $\alpha = .05$.
23. Verify that the .95 confidence interval for the difference between the means, based on the data in Table 8.6 and Student's T, is (2.2, 20.5). What are the practical problems with this confidence interval?
24. Student's T rejects the hypothesis of equal means based on the data in Table 8.6. Interpret what this means.
25. For the data in Table 8.6, the sample variances are 116.04 and 361.65, respectively. Verify that the .95 confidence interval for the difference between the means based on Welch's method is (1.96, 20.83). Check this result with the S-PLUS function yuen.
26. In the previous exercise you do not reject the hypothesis of equal variances. Why is this *not* convincing evidence that the assumption of equal variances, when using Student's T, is justified?
27. The 20% Winsorized standard deviation (s_w) for the first group in Table 8.6 is 1.365 and for the second group it is 4.118. Verify that the .95 confidence interval for the difference between the 20% trimmed means, using Yuen's method, is (5.3, 22.85).
28. Create a boxplot of the data in Table 8.6, and comment on why the probability coverage, based on Student's T or Welch's method, might differ from the nominal α level.
29. For the self-awareness data in Table 8.4, verify that the S-PLUS function yuenbt, with the argument tr set to 0, returns (−571.4, 302.5) as a .95 confidence interval for the difference between the means.
30. For the data in Table 8.4, use the S-PLUS function comvar2 to verify that the .95 confidence interval for the difference between the variances is (−1165766.8, 759099.7).
31. Describe a general situation where comparing medians will have more power than comparing means or 20% trimmed means.
32. For the data in Table 8.4, verify that the .95 confidence interval for the difference between the biweight midvariances is (−159234, 60733).
33. The last example in Section 8.9 dealt with comparing males to females regarding the desired number of sexual partners over the next 30 years. Using Student's T, we fail to reject, which is not surprising because there is an extreme outlier among the responses given by males. If we simply discard this one outlier and compare groups using Student's T or Welch's method, what criticism might be made even if we could ignore problems with nonnormality?

ONE-WAY ANOVA

This chapter (and the three chapters that follow) addresses the common situation where more than two groups are to be compared based on some measure of location. We begin with a classic and commonly used method for comparing the means of independent groups under the assumption of normality and homoscedasticity. We have already seen serious practical problems with this approach when comparing two groups only. For various reasons, problems are exacerbated when comparing more than two groups, and indeed new problems are introduced. However, these problems are not completely obvious and should be described in order to motivate more modern methods. As usual, it is not assumed that the reader has any prior knowledge about the classic approach to comparing groups based on means.

To help fix ideas, we begin with a concrete example. Clinical psychologists have long tried to understand schizophrenia. One issue of interest to some researchers is whether various groups of individuals differ in terms of measures of skin resistance. In such a study, four groups of individuals were identified: (1) no schizophrenic spectrum disorder, (2) schizotypal or paranoid personality disorder, (3) schizophrenia, predominantly negative symptoms, (4) schizophrenia, predominantly positive symptoms. Table 9.1 presents the first 10 observations for each group, where the entries are measures of skin resistance (in ohms) following presentation of a generalization stimulus. (These data were supplied by S. Mednick, Dept. of Psychology, University of Southern California.) Note that the actual sample sizes in the study were larger; only the first 10 observations for each group are listed here.

You could, of course, compare all pairs of groups in terms of some measure of location. If, for example, you decide to use means, then you might simply compare the mean of the first group to the mean of the second, then compare the mean of first group to the mean of third, and so on. In symbols, you could test

$$H_0 : \mu_1 = \mu_2,$$

$$H_0 : \mu_1 = \mu_3,$$

$$H_0 : \mu_1 = \mu_4,$$

TABLE 9.1 Measures of Skin Resistance for Four Groups

No schiz.	Schizotypal	Schiz. neg.	Schiz. pos.
0.49959	0.24792	0.25089	0.37667
0.23457	0.00000	0.00000	0.43561
0.26505	0.00000	0.00000	0.72968
0.27910	0.39062	0.00000	0.26285
0.00000	0.34841	0.11459	0.22526
0.00000	0.00000	0.79480	0.34903
0.00000	0.20690	0.17655	0.24482
0.14109	0.44428	0.00000	0.41096
0.00000	0.00000	0.15860	0.08679
1.34099	0.31802	0.00000	0.87532
$\bar{X}_1 = 0.276039$	$\bar{X}_2 = 0.195615$	$\bar{X}_3 = 0.149543$	$\bar{X}_4 = 0.399699$
$s_1^2 = 0.1676608$	$s_2^2 = 0.032679$	$s_3^2 = 0.0600529$	$s_4^2 = 0.0567414$
$n_1 = n_2 = n_3 = n_4 = 10$			

$$H_0 : \mu_2 = \mu_3,$$

$$H_0 : \mu_2 = \mu_4,$$

$$H_0 : \mu_3 = \mu_4,$$

using the methods in Chapter 8, or you might compare trimmed means or MOMs or M-estimators instead. There is, however, a technical issue that arises if you do this. Suppose there are no differences among the groups, in which case none of the six null hypotheses just listed should be rejected. To keep things simple for the moment, assume all four groups have normal distributions with equal variances, in which case Student's T test in Chapter 8 provides exact control over the probability of a Type I error when testing any single hypothesis. Further assume that each of the six hypotheses just listed are tested with $\alpha = .05$. So for *each* hypothesis, the probability of a Type I error is .05. But what is the probability of *at least one* Type I error when you perform all six tests? That is, what is the probability of making one or more mistakes and rejecting when in fact all pairs of groups being compared have equal means?

If you perform each of the tests with $\alpha = .05$, the probability of at least one Type I error will be larger than .05. The more tests you perform, the more likely you are to reject the hypothesis of equal means when in fact the groups do not differ. A common goal, then, is to test all pairs of means so that the probability of one or more Type I errors is α, where, as usual, α is some value you pick. Put another way, when comparing all pairs of groups, the goal is to have the probability of making no Type I errors equal to $1 - \alpha$.

There are two general approaches to the problem of controlling the probability of committing one or more Type I errors. The first and most popular begins by testing the hypothesis that all the groups being compared have equal means. In the illustration, the null hypothesis is written as

$$H_0 : \mu_1 = \mu_2 = \mu_3 = \mu_4.$$

More generally, when comparing J groups, the null hypothesis is

$$H_0 : \mu_1 = \mu_2 = \cdots = \mu_J. \tag{9.1}$$

If this hypothesis of equal means is rejected, you then make decisions about which pairs of groups differ (using a method described in Chapter 12.) The second general approach is to skip the methods in this chapter and use one of the appropriate techniques in Chapter 12. There are circumstances under which this latter strategy has practical advantages, but the details must be postponed for now.

9.1 Analysis of Variance (ANOVA) for Independent Groups

Assuming normality, Student's T test of the hypothesis that two independent groups have equal means can be extended to the problem of testing Equation (9.1), the hypothesis that J independent groups have equal means. Even though the goal is to test the hypothesis of equal means, the method is called *analysis of variance*, or ANOVA; it was derived by Sir Ronald Fisher. Like Student's T test, homoscedasticity is assumed, meaning that all groups have a common variance. That is, if $\sigma_1^2, \ldots, \sigma_J^2$ are the population variances of the J groups, homoscedasticity means that

$$\sigma_1^2 = \sigma_2^2 = \cdots = \sigma_J^2. \tag{9.2}$$

As was done with Student's T test, this common variance will be labeled σ_p^2. As in previous chapters, heteroscedasticity refers to a situation where not all the variances are equal.

To help convey the strategy of the traditional ANOVA method, we temporarily restrict attention to the situation where all of the sample sizes are equal. Box 9.1 summarizes the computations for this special case. (Box 9.2 covers the more general case where samples sizes might be unequal.) Here the common sample size is labeled n. In symbols, it is temporarily assumed that $n_1 = n_2 = \cdots = n_J = n$. For the schizophrenia data in Table 9.1, $n = 10$. You begin by computing the sample mean and sample variance for each group. The average of the J sample means, called the *grand mean*, is then computed and labeled $\bar{X}_G$. As indicated in Box 9.1,

$$\bar{X}_G = \frac{1}{J}(\bar{X}_1 + \cdots + \bar{X}_J).$$

Next, you compute four quantities called the *sum of squares between groups*, the *mean squares between groups*, the *sum of squares within groups*, and the *mean squares within groups*, as described in Box 9.1. For convenience, these four quantities are labeled SSBG, MSBG, SSWG, and MSWG, respectively. Finally you compute the test statistic, F, which is just MSBG divided by MSWG. If F is sufficiently large, reject the hypothesis of equal means.

BOX 9.1 Summary of How to Compute the ANOVA F-Test with Equal Sample Sizes

Goal

Test $H_0 : \mu_1 = \cdots = \mu_J$, the hypothesis of equal means among J independent groups

Assumptions

- Random sampling
- Normality
- Equal variances

Computations

Compute the sample means, $\bar{X}_1, \ldots, \bar{X}_J$, and sample variances, $s_1^2, \ldots, s_J^2$. Then compute

$$\bar{X}_G = \frac{1}{J}\sum \bar{X}_j$$

(the grand mean),

$$N = nJ$$

(the total number of observations), and

$$\text{SSBG} = n\sum_{j=1}^{J}(\bar{X}_j - \bar{X}_G)^2,$$

$$\text{MSBG} = \frac{\text{SSBG}}{J-1},$$

$$\text{SSWG} = (n-1)\sum_{j=1}^{J} s_j^2,$$

$$\text{MSWG} = \frac{\text{SSWG}}{N-J} = \frac{1}{J}\sum_{j=1}^{J} s_j^2.$$

Test Statistic

$$F = \frac{\text{MSBG}}{\text{MSWG}}$$

Decision Rule

Reject H_0 if $F \geq f$, the $1-\alpha$ quantile of an F-distribution with $\nu_1 = J-1$ and $\nu_2 = N-J$ degrees of freedom.

When the null hypothesis is true, the exact distribution of F has been derived under the assumptions of normality and equal variances. This means that you are able to

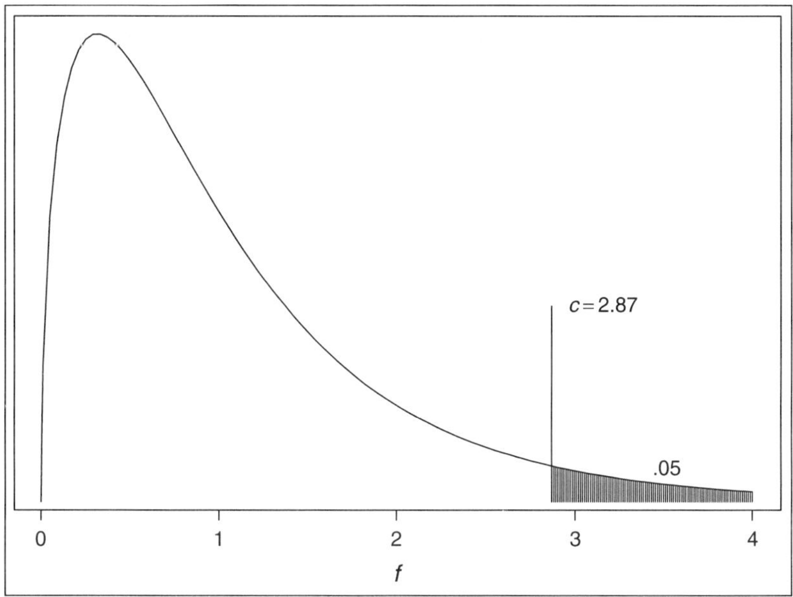

FIGURE 9.1 An F-distribution with 3 and 36 degrees of freedom. The probability that F is greater than 2.87 is .05, it is depicted by the area of the shaded region.

determine how large F must be in order to reject H_0. The distribution of F, when the null hypothesis is true, is called an *F-distribution* with degrees of freedom

$$\nu_1 = J - 1 \qquad \text{and} \qquad \nu_2 = N - J.$$

That is, the distribution depends on two quantities: the number of groups being compared, J, and the total number of observations in all of the groups, N.

For the schizophrenia illustration, there are $J = 4$ groups with a total of $N = 40$ observations, so the degrees of freedom are $\nu_1 = 4 - 1 = 3$ and $\nu_2 = 40 - 4 = 36$. Figure 9.1 shows the distribution of F with these degrees of freedom. The shaded region indicates the critical region when $\alpha = .05$; it extends from 2.87 to infinity. That is, if you want the probability of a Type I error to be .05, then reject the null hypothesis of equal means if $F \geq 2.87$, which is the .95 quantile of the F-distribution.

More generally, reject if $F \geq f$, where f is the $1 - \alpha$ quantile of an F-distribution with $\nu_1 = J - 1$ and $\nu_2 = N - J$ degrees of freedom. Tables 5–8 in Appendix B report critical values, f, for $\alpha = .1$, .05, .025, and .01 and various degrees of freedom. For example, with $\alpha = .05$, $\nu_1 = 6$, $\nu_2 = 8$, Table 6 indicates that the .95 quantile is $f = 3.58$. That is, there is a .05 probability of getting a value for F that exceeds 3.58 when in fact the population means are equal. For $\alpha = .01$, Table 8 says that the .99 quantile is 4.38. This means that if you reject when $F \geq 4.38$, the probability of a Type I error will be .01, assuming normality and that the groups have equal variances.

Now we illustrate the computations by computing F for the data in Table 9.1 and determining whether the hypothesis of equal means should be rejected with $\alpha = .05$. From Table 9.1, the sample means are 0.276039, 0.195615, 0.149543, 0.399699. The grand mean is just the average of these four sample means and is given by

$$\bar{X}_G = \frac{1}{4}(0.276039 + 0.195615 + 0.149543 + 0.399699) = 0.255224.$$

There are $n = 10$ subjects for each group, so the sum of squares between groups is

$$\begin{aligned} \text{SSBG} &= n\sum_{j=1}^{J}(\bar{X}_j - \bar{X}_G)^2 \\ &= 10\{(0.276039 - 0.255224)^2 + (0.195615 - 0.255224)^2 \\ &\quad + (0.149543 - 0.255224)^2 + (0.399699 - 0.255224)^2\} \\ &= 10(0.036) \\ &= 0.36. \end{aligned}$$

Therefore, the mean squares between groups is

$$\begin{aligned} \text{MSBG} &= \frac{\text{SSBG}}{J-1} \\ &= \frac{0.36}{4-1} \\ &= 0.12. \end{aligned}$$

With equal sample sizes, the mean squares within groups is just the average of the sample variances:

$$\begin{aligned} \text{MSWG} &= \frac{1}{J}\left(s_1^2 + \cdots + s_J^2\right) \\ &= \frac{1}{4}(0.1676608 + 0.032679 + 0.0600529 + 0.0567414) \\ &= .0793. \end{aligned}$$

Therefore,

$$\begin{aligned} F &= \frac{\text{MSBG}}{\text{MSWG}} \\ &= \frac{0.12}{0.0793} \\ &= 1.51. \end{aligned}$$

As already indicated, the degrees of freedom are $\nu_1 = 3$ and $\nu_2 = 36$ with a critical value $f = 2.87$. Because 1.51 is less than 2.87, you do not reject the hypothesis of equal means. This means that you do not have convincing empirical evidence that the hypothesis of equal means is unreasonable. As was the case in previous chapters, failing to reject the null hypothesis can be due to one of two reasons: The null hypothesis is true, or the null hypothesis is false but you failed to detect this because your sample size is too small to achieve reasonably high power. Methods for assessing power prior to collecting data, assuming normality, are nicely summarized by Cohen (1977). Section 9.3 describes a method for addressing power once data are available.

TABLE 9.2 ANOVA Summary Table

Source of variation	Degrees of freedom	Sum of squares	Mean square	F
Between groups	$J-1$	SSBG	MSBG	$F = \frac{\text{MSBG}}{\text{MSWG}}$
Within groups	$N-J$	SSWG	MSWG	
Totals	$N-1$	SSBG + SSWG		

TABLE 9.3 ANOVA Summary Table for the Data in Table 9.1

Source of variation	Degrees of freedom	Sum of squares	Mean square	F
Between groups	3	0.36	0.12	1.51
Within groups	36	2.8542	0.0793	
Totals	39	3.2142		

Table 9.2 outlines what is called an *analysis of variance summary table*, a common way of summarizing the computations associated with ANOVA. Table 9.3 illustrates the summary table using the data in Table 9.1.

The computations outlined in Box 9.1 are convenient for describing certain conceptual details covered in Section 9.1.1. However, an alternative method for computing F is a bit faster and easier and allows unequal sample sizes. The details are summarized in Box 9.2.

EXAMPLE. The computations in Box 9.2 are illustrated with the following data.

```
Group 1:  7, 9, 8, 12, 8, 7, 4, 10, 9, 6
Group 2:  10, 13, 9, 11, 5, 9, 8, 10, 8, 7
Group 3:  12, 11, 15, 7, 14, 10, 12, 12, 13, 14
```

We see that

$$A = 7^2 + 9^2 + \cdots + 14^2 = 3026,$$

$$B = 7 + 9 + \cdots + 14 = 290,$$

$$C = \frac{(7+9+\cdots+6)^2}{10} + \frac{(10+13+\cdots+7)^2}{10} + \frac{(13+11+\cdots+14)^2}{10} = 2890,$$

Continued

EXAMPLE. *(Continued)*

$$N = 10 + 10 + 10 = 30,$$

$$\text{SST} = 3026 - \frac{290^2}{30} = 222.67,$$

$$\text{SSBG} = 2890 - \frac{290^2}{30} = 86.67,$$

$$\text{SSWG} = 3026 - 2890 = 136,$$

$$\text{MSBG} = \frac{86.67}{3-1} = 43.335,$$

$$\text{MSWG} = \frac{136}{30-3} = 5.03.$$

So

$$F = \frac{43.335}{5.03} = 8.615.$$

■

BOX 9.2 Summary of the ANOVA F-Test With or Without Equal Sample Sizes

Notation

X_{ij} refers to the ith observation from the jth group, $i = 1, \ldots, n_j$; $j = 1, \ldots, J$. (There are n_j observations randomly sampled from the jth group.)

Computations

$$A = \sum\sum X_{ij}^2$$

(In words, square each value, add the results, and call it A.)

$$B = \sum\sum X_{ij}$$

(In words, sum all the observations and call it B.)

$$C = \sum_{j=1}^{J} \frac{1}{n_j}\left(\sum_{i=1}^{n_j} X_{ij}\right)^2$$

Continued

BOX 9.2 (*Continued*) (Sum the observations for each group, square the result, divide by the sample size, and add the results corresponding to each group.)

$$N = \sum n_j$$

$$\text{SST} = A - \frac{B^2}{N}$$

$$\text{SSBG} = C - \frac{B^2}{N}$$

$$\text{SSWG} = \text{SST} - \text{SSBG} = A - C$$

$$\nu_1 = J - 1$$

$$\nu_2 = N - J$$

$$\text{MSBG} = \frac{\text{SSBG}}{\nu_1}$$

$$\text{MSWG} = \frac{\text{SSWG}}{\nu_2}$$

Test Statistic

$$F = \frac{\text{MSBG}}{\text{MSWG}}.$$

Decision Rule

Reject H_0 if $F \geq f$, the $1 - \alpha$ quantile of an F-distribution with $\nu_1 = J - 1$ and $\nu_2 = N - J$ degrees of freedom.

The degrees of freedom are $\nu_1 = 3 - 1 = 2$ and $\nu_2 = 30 - 3 = 27$. With $\alpha = .01$ we see, from Table 8 in Appendix B, that the critical value is $f = 4.6$. Because $22.13 > 4.6$, reject the hypothesis of equal means.

9.1.1 Some Conceptual Details

Some of the technical and conceptual details of the ANOVA F-test are important because they play a role in commonly used or recommended methods for summarizing how groups compare. It is also important to gain some sense of the strategy behind the ANOVA F-test so that its performance, relative to other methods you might use, is understood. Again, primarily for convenience, equal sample sizes are assumed. That is, $n_1 = n_2 = \cdots n_J = n$.

First look at the expression for SSBG in Box 9.1 and notice that the sum

$$\sum_{j=1}^{J} (\bar{X}_j - \bar{X}_G)^2,$$

is similar to the expression for the sample variance s^2 given in Chapter 3. Also recall that the variance of the jth sample mean is σ_j^2/n_j and because we assume all J groups have equal variances, each sample mean has variance σ_p^2/n. Moreover, if the null hypothesis of equal means is true, then

$$\frac{1}{J-1}\sum_{j=1}^{J}(\bar{X}_j-\bar{X}_G)^2,$$

is an unbiased estimate of the variance of the sample means. That is, by assumption, $\bar{X}_1,\ldots,\bar{X}_J$ each have variance σ_p^2/n and so the sample variance of these sample means estimates σ_p^2/n. In symbols,

$$E\left(\frac{1}{J-1}\sum_{j=1}^{J}(\bar{X}_j-\bar{X}_G)^2\right)=\frac{\sigma_p^2}{n}.$$

Using our rules of expected values covered in Chapter 2, it follows that

$$E(\text{MSBG})=\sigma_p^2.$$

That is, if the null hypothesis of equal means is true, MSBG is an unbiased estimate of the assumed common variance.

Now consider the average of the sample variances:

$$\frac{1}{J}\sum_{j=1}^{J}s_j^2.$$

Assuming homoscedasticity, each of these sample variances is an unbiased estimate of σ_p^2, so in particular the average of these sample variances is an unbiased estimate of the assumed common variance. That is,

$$E(\text{MSWG})=\sigma_p^2,$$

regardless of whether the null hypothesis of equal means is true or false. So when the null hypothesis is true, MSBG and MSWG should have similar values. In terms of the test statistic, $F=\text{MSBG}/\text{MSWG}$, if the null hypothesis is true, then F will tend to be close to 1. However, when the null hypothesis is false, MSBG will tend to be larger than MSWG. In fact, the more unequal the population means, the larger MSBG will be on average. In particular, it can be shown that

$$E(\text{MSBG})=\sigma_p^2+\frac{n\sum(\mu_j-\bar{\mu})^2}{J-1},$$

where $\bar{\mu}$ is the average of the population means being compared. That is,

$$\bar{\mu}=\frac{1}{J}(\mu_1+\mu_2+\cdots+\mu_J),$$

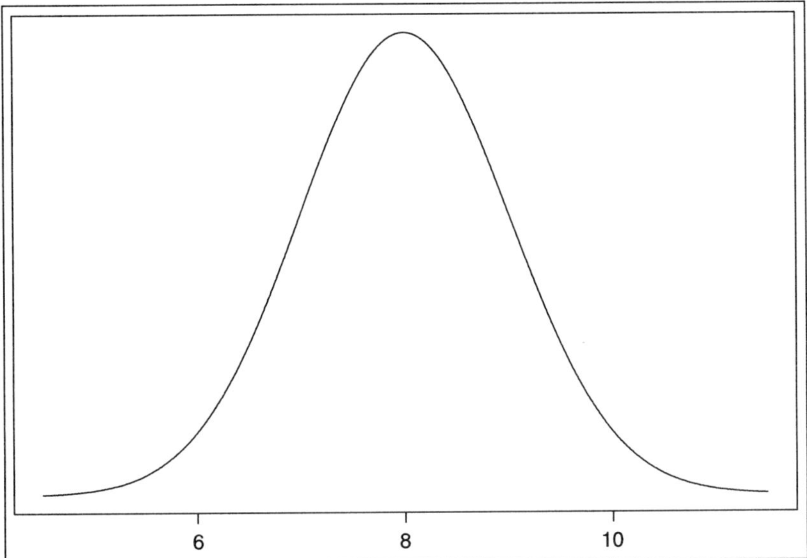

FIGURE 9.2 If three groups have normal distributions with a common variance and a common mean of 8, then the distributions will be identical, as shown. In this case, the sample means ($\bar{X}_1$, $\bar{X}_2$, and $\bar{X}_3$) will tend to be close to 8.

which is called the population *grand mean*. Consequently, if F is larger than what we would expect when the null hypothesis is true, the hypothesis of equal means should be rejected.

Now we can understand why the method in this section is called *analysis of variance* even though the goal is to test the hypothesis that the means have a common value. In essence, the ANOVA F-test compares the variation among the sample means to the variation within the groups. The variation among the sample means is measured by MSBG, and the variance for each of the groups is estimated with MSWG.

A graphical description of the strategy behind the ANOVA F-test might help. First consider the situation where the null hypothesis is true. Because normality and equal variances are assumed, all three groups will have identical distributions, as shown in Figure 9.2. In contrast, Figure 9.3 shows what the distributions might look like when the null hypothesis is false. The population means are 2, 4, and 8. The distributions are spread out, versus the situation in Figure 9.2, meaning that the sample means associated with the three groups are likely to be more spread out. In Figure 9.2, all of the sample means will tend to be close to 8, but in Figure 9.3 the sample mean for the first group will tend to be close to 2, the sample mean of the second group will tend to be close to 8, and the sample mean of the third group will tend to be close to 4. This means that if MSBG is large enough, which reflects the variation among the sample means, the hypothesis of equal means should be rejected.

Box 9.3 summarizes some properties of the ANOVA F-test that are sometimes used in applied work, particularly when measuring effect size (the extent to which the groups differ).

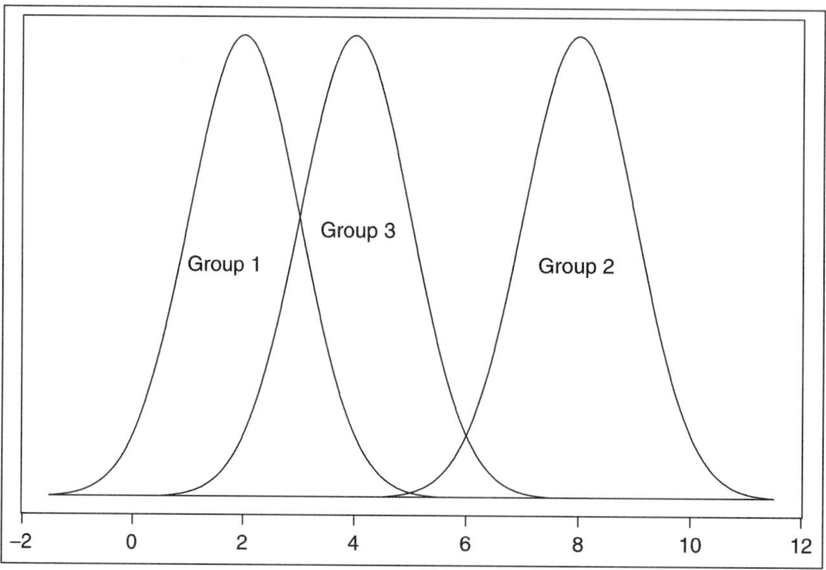

FIGURE 9.3 If the null hypothesis of equal means is false, the distributions of the three groups, still assuming normality and a common variance, might appear as shown. Compared to Figure 9.2, the sample means will tend to be more spread out.

BOX 9.3 Properties of the ANOVA Method When There Are Equal Sample Sizes

Let $\bar{\mu}$ be the average of the population means being compared. That is,

$$\bar{\mu} = \frac{1}{J}(\mu_1 + \mu_2 + \cdots + \mu_J).$$

Then

$$E(\text{MSBG}) = \sigma_p^2 + \frac{n\sum(\mu_j - \bar{\mu})^2}{J-1}.$$

When the null hypothesis of equal means is true,

$$\sum(\mu_j - \bar{\mu})^2 = 0,$$

so

$$E(\text{MSBG}) = \sigma_p^2.$$

In words, if the null hypothesis is true, MSBG estimates the common variance, σ_p^2; but if the null hypothesis is false, MSBG estimates a quantity that is larger than σ_p^2.

Continued

BOX 9.3 *(Continued)*

Regardless of whether the null hypothesis is true,

$$E(\text{MSWG}) = \sigma_p^2.$$

This means that if the null hypothesis is true, MSBG and MSWG estimate the same quantity and therefore will tend to have similar values. If, however, the null hypothesis is false, MSBG estimates a larger quantity than MSWG, so if $F = \text{MSBG}/\text{MSWG}$ is sufficiently large, reject the hypothesis of equal means.

EXAMPLE. *If* the null hypothesis is true, there are two ways of estimating the assumed common variance based on the results in Table 9.3. The first is with MSBG. So for the data in Table 9.1, the estimate is MSBG = 0.12. That is, if the hypothesis of equal means is true and all J groups have variances equal to σ_p^2, then an estimate of σ_p^2 is 0.12. MSWG provides a second estimate; it estimates the assumed common variance regardless of whether the means are equal, as indicated in Box 9.3. In the illustration, MSWG = 0.0793. ■

EXAMPLE. For the data in Table 9.1, is it reasonable to estimate the common variance with MSBG, assuming that the groups do indeed have equal variances? MSBG is a reasonable estimate of the assumed common variance if the null hypothesis of equal means is true. As indicated in Table 9.3, F is not large enough to reject the hypothesis of equal means, but this does not necessarily imply that it is reasonable to assume the means are equal. It might be that the means are not equal but that power was not high enough to detect this. Consequently, there is not convincing evidence that MSBG provides a reasonable estimate of the common variance. ■

We conclude this section by describing the conventional ANOVA model. Many books use the notation

$$\alpha_j = \mu_j - \bar{\mu}$$

to represent the difference between the mean of the jth group and the *grand mean*,

$$\bar{\mu} = \frac{1}{J}\sum \mu_j.$$

Another common notation is

$$\begin{aligned}\epsilon_{ij} &= X_{ij} - \bar{\mu} - \alpha_j \\ &= X_{ij} - \mu_j.\end{aligned}$$

In words, ϵ_{ij} is the difference between the ith observation in the jth group and the corresponding mean, μ_j. That is, ϵ_{ij} is an error term: It measures the extent to which X_{ij} differs from the population mean of the jth group. Rearranging terms, this last equation becomes

$$X_{ij} = \bar{\mu} + \alpha_j + \epsilon_{ij}. \tag{9.3}$$

The ANOVA F-test is obtained by assuming that ϵ_{ij} has a normal distribution with mean 0 and variance σ_p^2.

9.2 Dealing with Unequal Variances

Improvements on the ANOVA F-test have been phenomenal, particularly in recent years. These improvements deal with problems associated with sampling from non-normal distributions and problems due to having unequal variances. Violating the equal variance assumption associated with the F-test can result in poor power and undesirable power properties, even when sampling from a normal distribution. We saw in Chapter 8 that even if the population variances are unequal but the sample sizes are equal, Student's T controls Type I errors fairly well when sampling from normal distributions except when sample sizes are very small. In contrast, problems arise when using the ANOVA F-statistic even when the sample sizes are equal. That is, in a very real sense, as the number of groups increases, practical problems with unequal variances increase, even under normality.

Another serious problem is that even if distributions are normal but have unequal variances, the power of the F-test can be low relative to more modern methods. Moreover, it can be biased, meaning that the probability of rejecting can actually drop as the population means become unequal, and there are concerns about its ability to control the probability of a Type I error. For example, imagine you want to compare four groups, the null hypothesis of equal means is true, and you want the probability of a Type I error to be $\alpha = .05$. Situations arise where the actual probability of rejecting exceeds .27, due to comparing normal distributions that have unequal variances. When comparing six groups, the probability of a Type I error can exceed .3. That is, the actual probability of Type I error is substantially higher than the stated level of .05.

A reasonable suggestion for trying to salvage the F-test is first to test the hypothesis of equal variances and, if not significant, to assume equal variances and use F. This strategy is known to fail (Markowski & Markowski, 1990; Moser, Stevens, & Watts, 1989; Wilcox, Charlin, & Thompson, 1986). As in Chapter 8, the basic problem is that tests for equal variances do not have enough power to detect situations where violating the assumption of equal variances causes practical problems. All indications are that the F-test should be abandoned in favor of some more modern technique. The two main reasons for including the F-test here are: (1) it is commonly used because its practical problems are relatively unknown, and (2) it provides a relatively simple first step toward describing effective methods for comparing groups.

As in Chapter 8, some authorities would argue that it is virtually impossible for the null hypothesis of equal means to be true and simultaneously to have unequal variances. If we accept this argument, the probability of a Type I error is no longer an issue. But this does not salvage the F-test, because problems controlling the probability of a Type I error when variances differ reflects problems with bias.

9.2.1 Welch's Test

Many methods have been proposed for testing the equality of J means without assuming equal variances (e.g., S. Chen & Chen, 1998; Mehrotra, 1997; James, 1951; Krutchkoff, 1988; Alexander & McGovern, 1994; Fisher, 1935, 1941; Cochran & Cox, 1950; Wald, 1955; Asiribo & Gurland, 1989; Scariano & Davenport, 1986; Matuszewski & Sotres, 1986; Pagurova, 1986; Weerahandi, 1995). Unfortunately, all of these methods, plus many others, have been found to have serious practical problems (e.g., Keselman, Wilcox, Taylor, & Kowalchuk, 2000; Keselman & Wilcox, 1999). One of these problems is poor control over the probability of a Type I error and another is low power under nonnormality, a problem that cannot be escaped when using sample means. The method described here performs reasonably well under normality and heteroscedasticity and it forms the basis of a technique that deals with nonnormality. The method is due to Welch (1951) and it generally outperforms the F-test. The computational details are described in Box 9.4. (The S-PLUS function t1way, described in Section 9.4.1, contains Welch's test as a special case.)

BOX 9.4 Computations for Welch's Method

Goal

Without assuming equal variances, test $H_0: \mu_1 = \mu_2 = \cdots = \mu_J$, the hypothesis that J independent groups have equal means.

Computations

Let

$$w_1 = \frac{n_1}{s_1^2},\ w_2 = \frac{n_2}{s_2^2}, \ldots, w_J = \frac{n_J}{s_J^2}.$$

Next, compute

$$U = \sum w_j$$

$$\tilde{X} = \frac{1}{U} \sum w_j \bar{X}_j$$

$$A = \frac{1}{J-1} \sum w_j (\bar{X}_j - \tilde{X})^2$$

$$B = \frac{2(J-2)}{J^2 - 1} \sum \frac{(1 - \frac{w_j}{U})^2}{n_j - 1}$$

Continued

BOX 9.4 *(Continued)*

$$F_w = \frac{A}{1+B}.$$

When the null hypothesis is true, F_w has, approximately, an F-distribution with

$$\nu_1 = J - 1$$

and

$$\nu_2 = \left[\frac{3}{J^2 - 1}\sum\frac{(1 - w_j/U)^2}{n_j - 1}\right]^{-1}$$

degrees of freedom.

Decision Rule
Reject H_0 if $F_w \geq f$, where f is the $1 - \alpha$ quantile of the F-distribution with ν_1 and ν_2 degrees of freedom.

EXAMPLE. Welch's test is illustrated with the schizophrenia data in Table 9.1, which lists the sample means and variances. Referring to Table 9.1 and Box 9.4, we see that

$$w_1 = \frac{10}{0.1676608} = 59.6, \qquad w_2 = \frac{10}{0.032679} = 306.0,$$

$$w_3 = \frac{10}{0.0600529} = 166.5, \qquad w_4 = \frac{10}{0.0567414} = 176.2.$$

Therefore,

$$U = 59.6 + 306.0 + 166.5 + 176.2 = 708.3,$$

$$\tilde{X} = \frac{1}{708.3}\{59.6(0.276039) + 306(0.195615) + 166.5(0.149543)$$
$$+ 176.2(0.399699\}$$
$$= .242,$$

$$A = \frac{1}{4-1}\{59.6(0.276039 - 0.242)^2 + 306(0.195615 - 0.242)^2$$
$$+ 166.5(0.149543 - 0.242)^2 + 176.2(0.399699 - 0.242)^2\}$$
$$= 2.18,$$

Continued

EXAMPLE. (*Continued*)

$$B = \frac{2(4-2)}{4^2-1}\left\{\frac{(1-59.6/708.3)^2}{9} + \frac{(1-306.0/708.3)^2}{9} + \frac{(1-166.5/708.3)^2}{9} + \frac{(1-176.2/708.3)^2}{9}\right\}$$

$$= 0.0685,$$

$$F_w = \frac{2.18}{1+0.0685} = 2.04.$$

The degrees of freedom are

$$\nu_1 = 4 - 1 = 3$$

and

$$\nu_2 = \left[\frac{3}{4^2-1}(0.256)\right]^{-1}$$

$$= \frac{1}{.0512}$$

$$= 19.5$$

If you want the probability of a Type I error to be $\alpha = .05$, then, referring to Table 6 in Appendix B, the critical value is approximately 3.1. Because $F_w = 2.04$, which is less than 3.1, you fail to reject the hypothesis of equal means. ■

For the data in Table 9.1, both Welch's test and the ANOVA F-test fail to reject the hypothesis of equal means. It is stressed, however, that in applied work, situations arise where Welch's test rejects and the F-test does not. That is, in applied work, it can matter which test you use. The following example illustrates this point.

EXAMPLE. Consider the following data.

```
Group 1:  53 2 34 6 7 89 9 12
Group 2:  7 34 5 12 32 36 21 22
Group 3:  5 3 7 6 5 8 4 3
```

The ANOVA F-test yields $F = 2.7$, with a critical value of 3.24, so you do not reject. (The significance level is .09.) In contrast, Welch's test yields $W = 8$, with a critical value of 4.2, so now you reject. (The significance level is .009.) ■

9.3 Judging Sample Sizes and Controlling Power When Comparing Means

When you fail to reject the hypothesis of equal means, this might be because there are small or no differences among the groups being compared. Another possibility is that there is an important difference but that you failed to detect this due to low power. Power might be low because the sample sizes are small relative to the variances. Section 5.4.3 describes Stein's method, which might be used to help distinguish between these two possibilities. That is, given some data, you can determine how large the sample sizes must be to achieve a desired amount of power. If the sample sizes you used are small compared to what is needed to achieve high power, you have empirical evidence that the null hypothesis should not be accepted. This section describes an extension of Stein's method, called the Bishop–Dudewicz ANOVA, for judging the sample sizes when testing the hypothesis of equal means. Normality is assumed, but unlike the ANOVA F-test, homoscedasticity is not required. In fact, under normality, the method provides exact control over both Type I error probabilities and power.

Imagine that you want power to be $1 - \beta$ for some given value of

$$\delta = \sum (\mu_j - \bar{\mu})^2.$$

In case it helps, it is noted that if

$$\mu_1 = \cdots = \mu_{J-1} \quad \text{but} \quad \mu_J - \mu_{J-1} = a,$$

then

$$\delta = \frac{a^2(J-1)}{J}.$$

That is, if $J - 1$ of the population means are equal but the other population mean exceeds all of the other means by a, then we have a simple method for determining δ that might help when trying to specify what δ should be. For example, if for three groups you want power to be high when $\mu_1 = \mu_2 = 5$ but $\mu_3 = 6$, then $a = 1$ and $\delta = 2/3$. Given δ, α, $1 - \beta$, and n_j observations randomly sampled from the jth group, Box 9.5 shows how to determine N_j, the number of observations needed to achieve the desired amount of power.

BOX 9.5 Judging Sample Sizes

Given α, δ, and n_j observations randomly sampled from the jth group, determine N_j, the number of observations for the jth group needed to achieve power $1 - \beta$.

Continued

BOX 9.5 (*Continued*)

Let z be the $1 - \beta$ quantile of the standard normal random distribution. For the jth group, let $\nu_j = n_j - 1$. Compute

$$\nu = \frac{J}{\sum \frac{1}{\nu_j - 2}} + 2,$$

$$A = \frac{(J-1)\nu}{\nu - 2}, \qquad B = \frac{\nu^2}{J} \times \frac{J-1}{\nu - 2},$$

$$C = \frac{3(J-1)}{\nu - 4}, \qquad D = \frac{J^2 - 2J + 3}{\nu - 2},$$

$$E = B(C + D),$$

$$M = \frac{4E - 2A^2}{E - A^2 - 2A},$$

$$L = \frac{A(M-2)}{M},$$

$$c = Lf,$$

where f is the $1 - \alpha$ quantile of an F-distribution with L and M degrees of freedom,

$$b = \frac{(\nu - 2)c}{\nu},$$

$$A_1 = \frac{1}{2}\left\{\sqrt{2}z + \sqrt{2z^2 + 4(2b - J + 2)}\right\},$$

$$B_1 = A_1^2 - b,$$

$$d = \frac{\nu - 2}{\nu} \times \frac{\delta}{B_1}.$$

Then

$$N_j = \max\left\{n_j + 1, \left[\frac{s_j^2}{d}\right] + 1\right\}. \tag{9.4}$$

For technical reasons, the number of observations needed for the jth group, N_j, cannot be smaller than $n_j + 1$. (The notation $[s_j^2/d]$ means you compute s_j^2/d and then round down to the nearest integer.)

EXAMPLE. Suppose you have three groups and you want power to be $1 - \beta = .8$ if one of the three groups has a population mean 2.74 larger than

Continued

EXAMPLE. (*Continued*) the other two. That is, $a = 2.74$, so $\delta = 5$. For illustrative purposes, assume $\alpha = .05$ and that the sample sizes are $n_1 = n_2 = n_3 = 10$, so $\nu_1 = \nu_2 = \nu_3 = 9$, and

$$\nu = \frac{3}{\frac{1}{7} + \frac{1}{7} + \frac{1}{7}} + 2 = 9.$$

The tedious calculations eventually yield a critical value of $c = 8.15$ and $d = .382$. If the sample variance for the first group is $s_1^2 = 5.43$, then N_1 is equal to either $n_1 + 1 = 11$ or $[s_1^2/d] + 1 = [5.43/.382] + 1 = 15$, whichever is larger. In this particular case $N_1 = 15$, suggesting that the original sample size of 10 is not quite satisfactory. If $N_1 = n_1 + 1$, this suggests that the available sample sizes are reasonably adequate. If the second group has a sample variance of $s_2^2 = 10$, then $N_2 = 27$, and if $s_3^2 = 20$, $N_3 = 53$. So for group 3, you have only 10 observations and about five times as many observations are required for the specified amount of power. ■

The method just described indicates that an additional $N_j - n_j$ observations are needed for the jth group to achieve the desired amount of power. It is noted that if these additional observations can be obtained, the hypothesis of equal means can be tested without assuming homoscedasticity. More precisely, assuming normality, the probability of a Type I error will be exactly α, and power will be at least $1 - \beta$ using the Bishop–Dudewicz ANOVA method outlined in Box 9.6.

BOX 9.6 Bishop–Dudewicz ANOVA

Goal

Test $H_0 : \mu_1 = \cdots = \mu_J$ such that power is $1 - \beta$ and the probability of a Type I error is α.

Assumptions

Normality and random sampling are assumed. It is further assumed that initially you have n_j observations randomly sampled from the jth group, labeled X_{ij}, $i = 1, \ldots, n_j$, you have computed N_j as described in Box 9.5, and that you have then randomly sampled $N_j - n_j$ additional observations from the jth group, which are labeled X_{ij}, $i = n_j + 1, \ldots, N_j$. For the jth group, compute

$$T_j = \sum_{i=1}^{n_j} X_{ij},$$

$$U_j = \sum_{i=n_j+1}^{N_j} X_{ij}$$

Continued

BOX 9.6 (*Continued*)

$$b_j = \frac{1}{N_j}\left(1 + \sqrt{\frac{n_j(N_j d - s_j^2)}{(N_j - n_j)s_j^2}}\right),$$

$$\tilde{X}_j = \frac{T_j\{1 - (N_j - n_j)b_j\}}{n_j} + b_j U_j.$$

The test statistic is

$$\tilde{F} = \frac{1}{d}\sum(\tilde{X}_j - \tilde{X})^2,$$

where

$$\tilde{X} = \frac{1}{J}\sum \tilde{X}_j.$$

Decision Rule

Reject H_0 if $\tilde{F} \geq c$, where c is the critical value given in Box 9.5.

9.3.1 S-PLUS Functions bdanova1 and bdanova2

The S-PLUS function

bdanova1(x, alpha = 0.05, power = 0.9, delta = NA)

performs the calculations in Box 9.5 and returns the number of observations required to achieve the specified amount of power. The argument power indicates how much power you want and defaults to .9. The argument delta corresponds to δ, and the data are stored in the S-PLUS variable x, which can be an n-by-J matrix or it can have list mode. (In the latter case it is assumed that x[[1]] contains the data for group 1, x[[2]] contains the data for group 2, and so on.)

The S-PLUS function

bdanova2(x1, x2, alpha = 0.05, power = 0.9, delta = NA)

performs the second-stage analysis once the additional observations are obtained, as described in Box 9.6. Here x1 and x2 contain the first-stage and second-stage data, respectively.

9.4 Trimmed Means

Welch's heteroscedastic method for comparing means can be extended to trimmed means. That is, the goal is to test

$$H_0 : \mu_{t1} = \mu_{t2} = \cdots = \mu_{tJ},$$

the hypothesis that J independent groups have a common population trimmed mean. As in Chapter 8, trimming can greatly reduce practical problems (low power, poor control over the probability of a Type I error, and bias) associated with methods for comparing means.

Compute

$$d_j = \frac{(n_j - 1)s_{wj}^2}{h_j \times (h_j - 1)},$$

$$w_j = \frac{1}{d_j},$$

$$U = \sum w_j,$$

$$\tilde{X} = \frac{1}{U} \sum w_j \bar{X}_{tj},$$

$$A = \frac{1}{J-1} \sum w_j (\bar{X}_{tj} - \tilde{X})^2,$$

$$B = \frac{2(J-2)}{J^2 - 1} \sum \frac{(1 - w_j/U)^2}{h_j - 1},$$

$$F_t = \frac{A}{1+B}. \tag{9.5}$$

When the null hypothesis is true, F_t has, approximately, an F-distribution with

$$\nu_1 = J - 1$$

$$\nu_2 = \left[\frac{3}{J^2 - 1} \sum \frac{(1 - w_j/U)^2}{h_j - 1} \right]^{-1}$$

degrees of freedom. (For $J > 2$, the expression for ν_2 reduces to $2(J-2)/3B$.)

DECISION RULE: Reject the hypothesis of equal population trimmed means if $F_t \geq f$, the $1 - \alpha$ quantile of an F-distribution with ν_1 and ν_2 degrees of freedom.

9.4.1 S-PLUS Function t1way

The S-PLUS function

t1way(x,tr=.2,grp=NA)

tests the hypothesis of equal trimmed means using the method just described. The argument x can have list mode or it can be a matrix. In the former case, the data for group 1 are stored in the S-PLUS variable x[[1]], group 2 is stored in x[[2]], and so on. In the latter case, x is an n-by-J matrix, where column 1 contains the data for group 1, column 2 contains the data for group 2, and so forth. The argument tr indicates the amount of trimming; and when tr=0, this function performs

Welch's method for means described in Section 9.2.1. The argument grp allows you to compare a selected subset of the groups. By default all groups are used. If you set grp=c(1,3,4), then the trimmed means for groups 1, 3, and 4 will be compared, with the remaining data ignored. The function returns the value of the test statistic and the corresponding significance level (so specifying a value for α is not necessary).

If you used the S-PLUS function selby (described in Section 1.1.6) to store your data in list mode in the S-PLUS variable dat, note that you would use the S-PLUS variable data$x to analyze your data. For example, to compare the groups based on 20% trimming, use the command t1way(dat$x).

EXAMPLE. For the data in Table 9.1 and using the default amount of trimming, a portion of the output from t1way is

```
$TEST:
[1] 5.059361

$nu1:
[1] 3

$nu2:
[1] 10.82531

$siglevel:
[1] 0.01963949
```

This says that the test statistic F_t has a value of 5.06 and the significance level is .0194. So in particular you would reject the hypothesis of equal trimmed means with $\alpha = .05$ or even .02. In contrast, as previously indicated, if we compare means with Welch's method or the ANOVA F-test, we fail to reject with $\alpha = .05$, illustrating that the choice of method can alter the conclusions reached. Setting the argument tr to zero, t1way reports the results of Welch's test to be

```
$TEST:
[1] 2.038348

$nu1:
[1] 3

$nu2:
[1] 19.47356

$siglevel:
[1] 0.1417441
```

So switching from means to 20% trimmed means, the significance level drops from .14 to about .02. ■

EXAMPLE. It is common to find situations where we get a lower significance level using trimmed means than using means. However, the reverse situation can and does occur. This point is illustrated with data taken from Le (1994), where the goal is to compare the testosterone levels of four groups of male smokers: heavy smokers (group 1), light smokers (group 2), former smokers (group 3), and nonsmokers (group 4). The data are:

G1	G2	G3	G4
.29	.82	.36	.32
.53	.37	.93	.43
.33	.77	.40	.99
.34	.42	.86	.95
.52	.74	.85	.92
.50	.44	.51	.56
.49	.48	.76	.87
.47	.51	.58	.64
.40	.61	.73	.78
.45	.60	.65	.72

The significance level using Welch's method is .0017. Using 20% trimmed means instead, the significance level is .029. One reason this is not surprising is that a boxplot for each group reveals no outliers, and it can be seen that the estimated standard errors for the means are smaller than the corresponding estimated standard errors when 20% trimming is used instead. However, a boxplot for the first group suggests that the data are skewed, which might be having an effect on the significance level of Welch's test beyond any differences among the means. ■

9.4.2 Comparing Groups Based on Medians

The median has a relatively large standard error under normality and more generally when sampling from a light-tailed distribution, but with a sufficiently heavy-tailed distribution its standard error can be relatively small. If there is a specific interest in comparing medians, the general method for trimmed means described in this section is not recommended. That is, it is not recommended that you set tr=.5 when using the S-PLUS function t1way in Section 9.4.1 because this results in using a relatively poor estimate of the standard error of the median. A better approach is to estimate the standard error of each median with the McKean–Schrader method in Chapter 4 and then to use a slight modification of the method for trimmed means. In particular, let S_j^2 be the McKean–Schrader estimate of the squared standard error of M_j, the sample median corresponding to the jth group ($j = 1, \ldots, J$). Let

$$w_j = \frac{1}{S_j^2},$$

$$U = \sum w_j,$$

$$\tilde{M} = \frac{1}{U}\sum w_j M_j,$$

$$A = \frac{1}{J-1}\sum w_j (M_j - \tilde{M})^2,$$

$$B = \frac{2(J-2)}{J^2-1}\sum \frac{(1-w_j/U)^2}{n_j - 1},$$

$$F_m = \frac{A}{1+B}. \tag{9.6}$$

DECISION RULE: Reject the hypothesis of equal population medians if $F_m \geq f$, the $1-\alpha$ quantile of an F distribution with $\nu_1 = J-1$ and $\nu_2 = \infty$ degrees of freedom.

9.4.3 S-PLUS Function med1way

The hypothesis of equal population medians can be tested with the S-PLUS function

med1way(x,grp),

where the argument grp can be used to analyze a subset of the groups if desired. (See Section 9.4.1.) The function returns the value of the test statistic, F_m, and the significance level.

9.5 Bootstrap Methods

This section describes how the percentile and bootstrap-t methods described in Chapter 8 can be extended to testing hypotheses based on some robust measure of location, such as MOM or trimmed means. With small sample sizes, all indications are that some type of bootstrap method has a practical advantage over the method for trimmed means in Section 9.4, assuming that we want a method that is sensitive to differences among the trimmed means only. This is particularly true for the special case where means are to be compared. With sufficiently large sample sizes, the method in Section 9.4 can be used in place of the bootstrap when comparing trimmed means, but it remains unclear how large the sample sizes must be. An argument for the method in Section 9.4 over any bootstrap techniques we might use is that if the former method rejects, we can be reasonably certain that the groups have different distributions. A possible concern, however, is that nonbootstrap methods are more sensitive to bias problems.

9.5.1 A Bootstrap-*t* Method

The bootstrap-t method in Chapter 8 can be extended to the problem of testing $H_0: \mu_{t1} = \mu_{t2} = \cdots = \mu_{tJ}$, the hypothesis of equal trimmed means. The strategy is to use the available data to estimate an appropriate critical value for the test statistic,

F_t, described in Section 9.4. First, for the jth group, set

$$Y_{ij} = X_{ij} - \bar{X}_{tj}.$$

That is, for the jth group, subtract the sample trimmed from each of the observed values. Next, for the jth group, generate a bootstrap sample of size n_j from the Y_{ij} values, which we denote by Y^*_{ij}, $i = 1, \ldots, n_j$; $j = 1, \ldots, J$. So, in effect, the Y^*_{ij} values represent a random sample from distributions all of which have zero trimmed means. That is, in the bootstrap world, when working with the Y_{ij} values, the null hypothesis of equal trimmed means is true. Said another way, the observations are shifted so that each has a trimmed mean of zero, with the goal of empirically determining an appropriate critical value. The value of the test statistic F_t given by Equation (9.5) and based on the Y^*_{ij} is labeled F^*_t. The strategy is to use a collection of F^*_t values to estimate the distribution of F_t, the test statistic based on the original observations, when the null hypothesis is true. If we can do this reasonably well, then in particular we can determine an appropriate critical value.

To estimate the critical value we repeatedly generate bootstrap samples in the manner just described, each time computing the test statistic F^*_t based on the Y^*_{ij} values. Doing this B times yields $F^*_{t1}, \ldots, F^*_{tB}$. Next, put these B values in ascending order, yielding $F^*_{t(1)} \leq \cdots \leq F^*_{t(B)}$, and let u be the value of $(1 - \alpha)B$ rounded to the nearest integer. Then the hypothesis of equal trimmed means is rejected if

$$F_t \geq F^*_{t(u)}. \tag{9.7}$$

9.5.2 S-PLUS Function t1waybt

The S-PLUS function

(x, tr = 0.2, alpha = 0.05, grp = NA, nboot = 599)

performs the percentile-t bootstrap method for trimmed means that was just described. The argument x is any S-PLUS variable containing data that are stored in list mode or in a matrix. In the first case x[[1]] contains the data for group 1, x[[2]] contains the data for group 2, and so on. If x is a matrix, column 1 contains the data for group 1, column 2 the data for group 2, and so forth. The argument grp can be used to analyze a subset of the groups. For example, grp=c(2,4,5) would compare groups 2, 4, and 5 only. As usual, alpha is α and nboot is B, the number of bootstrap samples to be used.

EXAMPLE. If the data in Table 9.1 are stored in the S-PLUS variable skin, the command

t1waybt(skin,tr=0)

tests the hypothesis of equal means and returns

```
$test:
[1] 2.04
```

Continued

EXAMPLE. (*Continued*)

```
$crit:
[1] 4.65336
```

Because the test statistic $F_t = 2.04$ is less than the bootstrap estimate of the critical value, 4.65, you fail to reject. Note that in Section 9.2.1, the estimated critical value, assuming normality, was 3.1. ■

9.5.3 Two Percentile Bootstrap Methods

There are many variations of the percentile bootstrap method that can be used to test the hypothesis that J groups have a common measure of location, but only two are described here. The first is related to a test statistic mentioned by Schrader and Hettmansperger (1980) and studied by He, Simpson, and Portnoy (1990). Let θ be any population measure of location, such as the 20% trimmed mean (μ_t) or a median, and let $\hat{\theta}_j$ be the estimate of θ based on data from the jth group ($j = 1, \dots, J$). The test statistic is

$$H = \frac{1}{N} \sum n_j (\hat{\theta}_j - \bar{\theta})^2,$$

where $N = \sum n_j$ and

$$\bar{\theta} = \frac{1}{J} \sum \hat{\theta}_j.$$

To determine a critical value, shift the empirical distributions of each group so that the measure of location being used has a value of zero. That is, set $Y_{ij} = X_{ij} - \hat{\theta}_j$, as was done in Section 9.5.1. Then generate bootstrap samples from each group in the usual way from the Y_{ij} values and compute the test statistic based on the bootstrap samples, yielding H^*. Repeat this B times, resulting in $H_1^*, \dots, H_B^*$, and put these B values in order, yielding $H_{(1)}^* \le \dots \le H_{(B)}^*$. Then an estimate of an appropriate critical value is $H_{(u)}^*$, where $u = (1 - \alpha)B$, rounded to the nearest integer, and H_0 is rejected if $H \ge H_{(u)}^*$. (For simulation results on how this method performs when comparing M-estimators, see Wilcox, 1993b.)

The second method stems from general results derived by Liu and Singh (1997). Let

$$\delta_{jk} = \theta_j - \theta_k,$$

where for convenience it is assumed that $j < k$. That is, the δ_{jk} values represent all pairwise differences among the J groups. When working with means, for example, δ_{12} is the difference between the means of groups 1 and 2, and δ_{35} is the difference for groups 3 and 5. If all J groups have a common measure of location (i.e., $\theta_1 = \dots = \theta_J$), then in particular

$$H_0 : \delta_{12} = \delta_{13} = \dots = \delta_{J-1,J} = 0 \tag{9.8}$$

is true. It can be seen that the total number of δ's in Equation (9.8) is $L = (J^2 - J)/2$. For example, if $J = 3$, there are $L = 3$ values: δ_{12}, δ_{13}, and δ_{23}.

For each group, generate bootstrap samples from the *original* values and compute the measure of location of interest for each group. That is, the observations are *not* centered as was done in the previous method. Said another way, bootstrap samples are *not* generated from the Y_{ij} values but rather from the X_{ij} values. Repeat this B times. The resulting estimates of location are represented by

$$\hat{\theta}^*_{jb} (j = 1, \ldots, J;\, b = 1, \ldots, B)$$

and the corresponding estimates of δ are denoted by $\hat{\delta}^*_{jkb}$. (That is, $\hat{\delta}^*_{jkb} = \hat{\theta}^*_{jb} - \hat{\theta}^*_{kb}$.) The general strategy is to determine how deeply $\mathbf{0} = (0, \ldots, 0)$ is nested within the bootstrap values $\hat{\delta}^*_{jkb}$ (where $\mathbf{0}$ is a vector having length L). For the special case where only two groups are being compared, this is tantamount to determining the proportion of times $\hat{\theta}^*_{1b} > \hat{\theta}^*_{2b}$, among all B bootstrap samples, which is how we proceeded in Section 8.8.1. But here we need special techniques for comparing more than two groups.

There remains the problem of measuring how deeply $\mathbf{0}$ is nested within the bootstrap values. Several strategies have been proposed for dealing with this problem (e.g., Liu & Singh, 1997). But in terms of Type I error probabilities and power, it remains unclear whether the choice among these methods is relevant for the problem at hand. Accordingly, only one method is described, based on a very slight modification of what is called the *Mahalanobis distance*. The details are relegated to Box 9.7, assuming familiarity with basic matrix algebra. (Appendix C summarizes the matrix algebra used in this book.)

BOX 9.7 Details About How to Test H_0 Given by Equation (9.8)

Let $\hat{\delta}_{jk} = \hat{\theta}_j - \hat{\theta}_k$ be the estimate of δ_{jk} based on the original data and let $\hat{\delta}^*_{jkb} = \hat{\theta}^*_{jb} - \hat{\theta}^*_{kb}$ based on the bth bootstrap sample ($b = 1, \ldots, B$). (It is assumed that $j < k$.) For notational convenience, we rewrite the $L = (J^2 - J)/2$ differences $\hat{\delta}_{jk}$ as $\hat{\Delta}_1, \ldots, \hat{\Delta}_L$ and the corresponding bootstrap values are denoted by $\hat{\Delta}^*_{\ell b}$ ($\ell = 1, \ldots, L$). Let

$$\bar{\Delta}^*_\ell = \frac{1}{B} \sum_{b=1}^{B} \hat{\Delta}^*_{\ell b},$$

$$Y_{\ell b} = \hat{\Delta}^*_{\ell b} - \bar{\Delta}^*_\ell + \hat{\Delta}_\ell,$$

(so the $Y_{\ell b}$ values are the bootstrap values shifted to have mean $\hat{\Delta}_\ell$), and let

$$S_{\ell m} = \frac{1}{B-1} \sum_{b=1}^{B} (Y_{\ell b} - \bar{Y}_\ell)(Y_{mb} - \bar{Y}_m),$$

Continued

BOX 9.7 (*Continued*)
where

$$\bar{Y}_\ell = \frac{1}{B}\sum_{b=1}^{B} Y_{\ell b}.$$

(Note that in the bootstrap world, the bootstrap population mean of $\bar{\Delta}^*_\ell$ is known and is equal to $\hat{\Delta}_\ell$.) Next, compute

$$D_b = \left(\hat{\mathbf{\Delta}}^*_b - \hat{\mathbf{\Delta}}\right) \mathbf{S}^{-1} \left(\hat{\mathbf{\Delta}}^*_b - \hat{\mathbf{\Delta}}\right)',$$

where $\hat{\mathbf{\Delta}}^*_b = (\hat{\Delta}^*_{1b}, \ldots, \hat{\Delta}^*_{Lb})$ and $\hat{\mathbf{\Delta}} = (\hat{\Delta}_1, \ldots, \hat{\Delta}_L)$. D_b measures how closely $\hat{\mathbf{\Delta}}_b$ is located to $\hat{\mathbf{\Delta}}$. If $\mathbf{0}$ (the null vector) is relatively far from $\hat{\mathbf{\Delta}}$, reject. In particular, put the D_b values in ascending order, yielding $D_{(1)} \le \cdots \le D_{(B)}$, and let $u = (1-\alpha)B$, rounded to the nearest integer.

Decision Rule
Reject H_0 if

$$T \ge D_{(u)},$$

where

$$T = (\mathbf{0} - \hat{\mathbf{\Delta}})\mathbf{S}^{-1}(\mathbf{0} - \hat{\mathbf{\Delta}})'.$$

Notice that with three groups ($J = 3$), $\theta_1 = \theta_2 = \theta_3$ can be true if and only if $\theta_1 = \theta_2$ and $\theta_2 = \theta_3$. So in terms of Type I errors, it suffices to test

$$H_0: \theta_1 - \theta_2 = \theta_2 - \theta_3 = 0$$

as opposed to testing

$$H_0: \theta_1 - \theta_2 = \theta_2 - \theta_3 = \theta_1 - \theta_3 = 0,$$

the hypothesis that all pairwise differences are zero. However, if groups differ, then rearranging the groups could alter the conclusions reached if the first of these hypotheses is tested. For example, if the groups have means 6, 4, and 2, then the difference between groups 1 and 2, as well as between 2 and 3, is 2. But the difference between groups 1 and 3 is 4, so comparing groups 1 and 3 could mean more power. That is, we might not reject when comparing group 1 to group 2 and group 2 to group 3, but we might reject if instead we compare group 1 to group 3 and group 2 to group 3. To help avoid different conclusions depending on how the groups are arranged, all pairwise differences among the groups were used in Box 9.7.

Between the two methods described in this section, it currently seems that the method in Box 9.7 is better in terms of Type I error probabilities when comparing groups based on MOM. How these two methods compare when comparing M-estimators has not been investigated as yet. The method in Box 9.7 can be used to compare trimmed means; but with a relatively small amount of trimming, it seems that the bootstrap-t method is preferable.

For the special case where the goal is to compare medians, again the method in Box 9.7 can be used. Whether it offers any practical advantages versus the method for medians in Section 9.4.2 has not been investigated.

One last comment might be helpful. Before using the percentile bootstrap methods described in this section, it is strongly recommended that the reader take into account results described in Chapter 12. There is a common convention dictating how the methods in this chapter are to be used in conjunction with those in Chapter 12, but modern insights reveal that this convention can be detrimental in some situations.

9.5.4 S-PLUS Functions b1way and pbadepth

The S-PLUS function

b1way(x,est=onestep,alpha=.05,nboot=599)

performs the first percentile bootstrap method described in the previous subsection. By default it uses an M-estimator (with Huber's Ψ). The function

pbadepth(x,est=mom,con=0,alpha=.05,nboot=500,op=F,allp=T,...)

performs the other percentile bootstrap method and uses the MOM estimator by default. As usual, the argument ... can be used to reset default settings associated with the estimator being used. The argument op determines how depth is measured. By default, a Mahalanobis-type depth, outlined in Box 9.7, is used. (Setting op=T results in the minimum covariance determinant method for measuring depth, which is described in Chapter 13.) The argument allp indicates how the null hypothesis is defined. Setting allp=T, all pairwise differences are used. Setting allp=F, the function tests

$$H_0: \theta_1 - \theta_2 = \theta_2 - \theta_3 = \cdots = \theta_{J-1} - \theta_J = 0.$$

A negative consequence of using allp=T is that in some situations, $\mathbf{S}^{-1}$ in Box 9.7 cannot be computed. This problem appears to be rare with $J \leq 4$, but it can occur otherwise. This problem might be avoided by setting allp=F, but perhaps a better way of dealing with this problem is to use the method described in Section 12.7.3.

9.6 Random Effects Model

The ANOVA methods covered so far deal with what is called a *fixed effect* design, roughly meaning that we are interested in comparing J specific (fixed) groups. In contrast is a random effects design, where the goal is to generalize to a larger population of groups. For example, consider a study where it is suspected that the personality of the experimenter has an effect on the results. Among all the experimenters we might use, do the results vary depending on who conducts the experiment? Here, the notion of J groups corresponds to a sample of J experimenters, and for the jth experimenter we have results on n_j participants. The goal is not only to compare results among the J experimenters but to generalize to the entire population of experimenters we might use.

TABLE 9.4 Estrone Assay Measurements of a Single Blood Sample from Each of Five Postmenopausal Women

	Individuals				
Vial	P1	P2	P3	P4	P5
1	23	25	38	14	46
2	23	33	38	16	36
3	22	27	41	15	30
4	20	27	38	19	29
5	25	30	38	20	36
6	22	28	32	22	31
7	27	24	38	16	30
8	25	22	42	19	32
9	22	26	35	17	32
10	22	30	40	18	31
11	23	30	41	20	30
12	23	29	37	18	32
13	27	29	28	12	25
14	19	37	36	17	29
15	23	24	30	15	31
16	18	28	37	13	32

A study reported by Fears, Benichou, and Gail (1996) provides another illustration where 16 estrone measures (in pg/mL) from each of five postmenopausal women were taken and found to be as shown in Table 9.4. Of interest was whether the estrone levels vary among women. That is, we envision the possibility of taking many measures from each woman, but the goal is not to simply compare the five women in the study but rather to generalize to all women who might have taken part in the study.

A study by Cronbach, Gleser, Nanda, and Rajaratnam (1972, Chap. 6) provides yet another example. The Porch index of communicative ability (PICA) is a test designed for use by speech pathologists. It is intended for initial diagnosis of patients with aphasic symptoms and for measuring the change during treatment. The oral portion of the test consists of several subtests; but to keep the illustration simple, only one subtest is considered here. This is the subtest where a patient is shown an object (such as a comb) and asked how the object is used. The response by the patient is scored by a rater on a 16-point scale. A score of 6, for example, signifies a response that is "intelligible but incorrect," and a score of 11 indicates a response that is "accurate but delayed and incomplete." A concern is that one set of objects might lead to a different rating, compared to another set of objects one might use. Indeed, we can imagine a large number of potential sets of objects that might be used. To what extent do ratings differ among all of the potential sets of objects we might employ?

Let $\mu_G = E(\mu_j)$, where the expected value of μ_j is taken with respect to the process of randomly sampling a group. If all groups have a common mean, then of course no matter which J groups you happen to pick, it will be the case that

$$\mu_1 = \mu_2 = \cdots = \mu_J.$$

A more convenient way of describing the situation is to say that there is no variation among all the population means. A way of saying this in symbols is that $\sigma_\mu^2 = 0$, where

$$\sigma_\mu^2 = E(\mu_j - \mu_G)^2,$$

and again the expectation is taken with respect to the process of randomly selecting μ_j. That is, among all groups of interest, σ_μ^2 is the variance of the population means. Testing the hypothesis that all groups have the same mean is equivalent to testing

$$H_0 : \sigma_\mu^2 = 0.$$

To test H_0, the following assumptions are typically made:

1. Regardless of which group you choose, the observations within that group have a normal distribution with a common variance, σ_p^2. That is, a homogeneity of variance assumption is imposed.
2. The difference $\mu_j - \mu_G$ has a normal distribution with mean 0 and variance σ_μ^2.
3. The difference $X_{ij} - \mu_j$ is independent of the difference $\mu_j - \mu_G$.

Let MSBG and MSWG be as defined in Section 9.1, and, primarily for notational convenience, temporarily assume equal sample sizes. That is,

$$n = n_1 = \cdots = n_J.$$

Based on the assumptions just described, it can be shown that

$$E(\text{MSBG}) = n\sigma_\mu^2 + \sigma_p^2$$

and that

$$E(\text{MSWG}) = \sigma_p^2.$$

When the null hypothesis is true, $\sigma_\mu^2 = 0$ and

$$E(\text{MSBG}) = \sigma_p^2.$$

That is, when the null hypothesis is true, MSBG and MSWG estimate the same quantity, so the ratio

$$F = \frac{\text{MSBG}}{\text{MSWG}}$$

should have a value reasonably close to 1. If the null hypothesis is false, MSBG will tend to be larger than MSWG; so if F is sufficiently large, reject. It can be shown that F has an F-distribution with $J - 1$ and $N - J$ degrees of freedom when the null hypothesis is true, so reject if $F \geq f_{1-\alpha}$, where $f_{1-\alpha}$ is the $1 - \alpha$ quantile of an F-distribution with $\nu_1 = J - 1$ and $\nu_2 = N - J$ degrees of freedom. Put more simply, the computations are exactly the same as they are for the fixed effects ANOVA F-test

TABLE 9.5 Hypothetical Data Used to Illustrate a Random Effects Model

Dosage 1	Dosage 2	Dosage 3
7	3	9
0	0	2
4	7	2
4	5	7
4	5	1
7	4	8
6	5	4
2	2	4
3	1	6
7	2	1

in Section 9.1. The only difference is how the experiment is performed. Here the levels are chosen at random, whereas in Section 9.1 they are fixed.

As mentioned in Section 9.1, the fixed effects ANOVA model is often written as

$$X_{ij} = \bar{\mu} + \alpha_j + \epsilon_{ij},$$

where ϵ_{ij} has a normal distribution with mean zero and variance σ_p^2. In contrast, the random effects model is

$$X_{ij} = \mu_G + a_j + \epsilon_{ij},$$

where $a_j = \mu_j - \mu_G$. The main difference between these two models is that in the fixed effects model, α_j is an unknown *parameter*, but in the random effects model, a_j is a *random variable* that is assumed to have a normal distribution.

EXAMPLE. Suppose that for three randomly sampled dosage levels of a drug, you get the results shown in Table 9.5. To test the null hypothesis of equal means among all dosage levels you might use, compute the degrees of freedom and the F-statistic as described in Section 9.1. This yields $\nu_1 = 2$, $\nu_2 = 27$, and $F = .53$, which is not significant at the $\alpha = .05$ level. That is, among all dosage levels you might have used, you fail to detect a difference among the corresponding means. ■

9.6.1 A Measure of Effect Size

As pointed out in Chapter 8, if you test and reject the hypothesis of equal means, there remains the issue of measuring the extent to which two groups differ. As already illustrated, the significance level can be unsatisfactory. From Chapter 8, it is evident that finding an appropriate measure of effect size is a complex issue. When dealing with more than two groups, the situation is even more difficult. Measures have been proposed under the assumption of equal variances; they are far from satisfactory, but few alternative measures are available. However, measures derived under the assumption of equal variances are in common use, so it is important to discuss them here.

Suppose you randomly sample a group from among all the groups you are interested in, and then you randomly sample an individual and observe the outcome X. Let σ_X^2 be the variance of X. It can be shown that

$$\sigma_X^2 = \sigma_\mu^2 + \sigma_p^2$$

when the assumptions of the random effects model are true and where σ_p^2 is the assumed common variance among all groups we might compare. A common measure of effect size is

$$\rho_I = \frac{\sigma_\mu^2}{\sigma_\mu^2 + \sigma_p^2},$$

which is called the *intraclass correlation coefficient*. The value of ρ_I is between 0 and 1 and measures the variation among the means relative to the variation among the observations. If there is no variation among the means, in which case they have identical values, $\rho_I = 0$.

To estimate ρ_I, compute

$$n_0 = \frac{1}{J-1}\left(N - \sum \frac{n_j^2}{N}\right),$$

where $N = \sum n_j$ is the total sample size. The usual estimate of σ_μ^2 is

$$s_u^2 = \frac{\text{MSBG} - \text{MSWG}}{n_0},$$

in which case the estimate of ρ_I is

$$\begin{aligned} r_I &= \frac{s_u^2}{s_u^2 + \text{MSWG}} \\ &= \frac{\text{MSBG} - \text{MSWG}}{\text{MSBG} + (n_0 - 1)\text{MSWG}} \\ &= \frac{F - 1}{F + n_0 - 1} \end{aligned}$$

For the data in Table 9.1 it was found that $F = 6.05$, $n_0 = 8$, so

$$r_I = \frac{6.05 - 1}{6.05 + 8 - 1} = .387.$$

That is, about 39% of the variation among the observations is due to the variation among the means.

Donner and Wells (1986) compared several methods for computing an approximate confidence interval for ρ_I, and their results suggest using a method derived by

Smith (1956). Smith's confidence interval is given by

$$r_I \pm z_{1-\alpha/2}V,$$

where $z_{1-\alpha/2}$ is the $1-\alpha/2$ quantile of the standard normal distribution, read from Table 1 in Appendix B, and

$$V = \sqrt{A(B+C+D)},$$

where

$$A = \frac{2(1-r_I)^2}{n_0^2}$$

$$B = \frac{[1+r_I(n_0-1)]^2}{N-J}$$

$$C = \frac{(1-r_I)[1+r_I(2n_0-1)]}{(J-1)}$$

$$D = \frac{r_I^2}{(J-1)^2}\left(\sum n_j^2 - \frac{2}{N}\sum n_j^3 + \frac{1}{N^2}\left(\sum n_j^2\right)^2\right).$$

For equal sample sizes an exact confidence interval is available, still assuming that sampling is from normal distributions with equal variances (Searle, 1971). Let $f_{1-\alpha/2}$ be the $1-\alpha/2$ quantile of the F-distribution with $\nu_1 = J-1$ and $\nu_2 = N-J$ degrees of freedom. Similarly, $f_{\alpha/2}$ is the $\alpha/2$ quantile. Then an exact confidence interval for ρ_I is

$$\left(\frac{F/f_{1-\alpha/2}-1}{n+F/f_{1-\alpha/2}-1}, \frac{F/f_{\alpha/2}-1}{n+F/f_{\alpha/2}-1}\right),$$

where n is the common sample size. The tables in Appendix B give only the upper quantiles of an F-distribution, but you need the lower quantiles when computing a confidence interval for ρ_I. To determine $f_{\alpha/2,\nu_1,\nu_2}$, you reverse the degrees of freedom and look up $f_{1-\alpha/2,\nu_2,\nu_1}$, in which case

$$f_{\alpha/2,\nu_1,\nu_2} = \frac{1}{f_{1-\alpha/2,\nu_2,\nu_1}}.$$

For example, if $\alpha = .05$ and you want to determine $f_{.025}$ with $\nu_1 = 2$ and $\nu_2 = 21$ degrees of freedom, you first look up $f_{.975}$ with $\nu_1 = 21$ and $\nu_2 = 2$ degrees of freedom. The answer is 39.45. Then $f_{.025}$ with 2 and 21 degrees of freedom is the reciprocal of 39.45. That is,

$$f_{.025,2,21} = \frac{1}{39.45} = .025.$$

EXAMPLE. Assume that professors are rated on their level of extroversion and you want to investigate how their level of extroversion is related to student

Continued

TABLE 9.6 Students' Ratings

Group 1	Group 2	Group 3
3	4	6
5	4	7
2	3	8
4	8	6
8	7	7
4	4	9
3	2	10
9	5	9
$\bar{X}_1 = 4.75$	$\bar{X}_2 = 4.62$	$\bar{X}_3 = 7.75$

EXAMPLE. (*Continued*) evaluations of a course. Suppose you randomly sample three professors, and their student evaluations are as shown in Table 9.6. (In reality one would of course want to sample more than three professors, but the goal here is to keep the illustration simple.) To illustrate how a confidence interval for ρ_I is computed, suppose you choose $\alpha = .05$. Then $n = 8$, $f_{.025} = .025$, $f_{.975} = 4.42$, $F = 6.05$, and the .95 confidence interval for ρ_I is

$$\left(\frac{\frac{6.05}{4.42} - 1}{8 + \frac{6.05}{4.42} - 1}, \frac{\frac{6.05}{.025} - 1}{8 + \frac{6.05}{.025} - 1}\right) = (0.047, 0.967).$$

Hence, you can be reasonably certain that ρ_I has a value somewhere between .047 and .967. Notice that the length of the confidence interval is relatively large, since ρ_I has a value between 0 and 1. Thus, in this case, the data might be providing a relatively inaccurate estimate of the intraclass correlation. ■

In some situations you might also want a confidence interval for σ_μ^2. Methods for accomplishing this goal are available, but no details are given here. For a recent discussion of this problem, see C. Brown and Mosteller (1991).

9.6.2 A Heteroscedastic Method

One serious concern about the conventional random effects model just described is the assumption of equal variances. We have seen that violating this assumption can result in poor power and undesirable power properties in the fixed effects design, and this problem continues for the situation at hand. This section describes a method derived by Jeyaratnam and Othman (1985) for handling unequal variances. (For an alternative approach, see Westfall, 1988.) As usual, let s_j^2 be the sample variance for the jth group, let $\bar{X}_j$ be the sample mean, and let $\bar{X} = \sum \bar{X}_j/J$ be the average of the J sample means. To test $H_0 : \sigma_\mu^2 = 0$, compute

$$q_j = \frac{s_j^2}{n_j},$$

$$\text{BSS} = \frac{1}{J-1} \sum (\bar{X}_j - \bar{X})^2,$$

$$\text{WSS} = \frac{1}{J} \sum q_j,$$

in which case the test statistics is

$$F_{jo} = \frac{\text{BSS}}{\text{WSS}}$$

with

$$\nu_1 = \frac{\left(\frac{J-1}{J} \sum q_j\right)^2}{\left(\sum \frac{q_j}{J}\right)^2 + \frac{J-2}{J} \sum q_j^2}$$

and

$$\nu_2 = \frac{\left(\sum q_j\right)^2}{\sum \frac{q_j^2}{n_j - 1}}$$

degrees of freedom. In the illustration regarding students' ratings,

$$\text{BSS} = 3.13$$

$$\text{WSS} = .517$$

$$F_{jo} = 6.05.$$

(The numerical details are left as an exercise.) The degrees of freedom are $\nu_1 = 1.85$ and $\nu_2 = 18.16$, and the critical value is 3.63. Because $6.05 > 3.63$, reject and conclude there is a difference among students' ratings.

When there are unequal variances, a variety of methods have been suggested for estimating σ_μ^2, several of which were compared by P. Rao, Kaplan, and Cochran (1981). Their recommendation is that when $\sigma_\mu^2 > 0$, σ_μ^2 be estimated with

$$\hat{\sigma}_\mu^2 = \frac{1}{J} \sum \ell_j^2 (\bar{X}_j - \tilde{X})^2,$$

where

$$\ell_j = \frac{n_j}{n_j + 1}$$

$$\tilde{X} = \frac{\sum \ell_j \bar{X}_j}{\sum \ell_j}.$$

Evidently there are no results on how this estimate performs under nonnormality.

9.6.3 A Method Based on Trimmed Means

Under normality with unequal variances, the F-test can have a Type I error probability as high as .179 when testing at the $\alpha = .05$ level with equal sample sizes of 20 in each group (Wilcox, 1994a). The Jeyaratnam–Othman test statistic, F_{jo}, has a probability of a Type I error close to .05 in the same situation. However, when the normality assumption is violated, the probability of a Type I error using both F and F_{jo} can exceed .3. Another concern with both F and the Jeyaratnam–Othman method is that there are situations where power decreases even when the difference among the means increases. This last problem appears to be reduced considerably when using trimmed means. Trimmed means provide better control over the probability of a Type I error and can yield substantially higher power when there are outliers. Of course there are exceptions. Generally no method is best in all situations. But if the goal is to reduce the problems just described, an extension of the Jeyaratnam–Othman method to trimmed means has considerable practical value. (Extensions of the random effects model based on MOM have not been investigated as yet.) The computational details are relegated to Box 9.8. Readers interested in the derivation and technical details of the method are referred to Wilcox (1997a, Sec. 6.3).

BOX 9.8 Comparing Trimmed Means in a Random Effects Model

For each of the J groups, Winsorize the observations as described in Section 3.2.6 and label the results Y_{ij}. For example, if the observations for group 3 are:

$$37, 14, 26, 17, 21, 43, 25, 6, 9, 11,$$

then the 20% Winsorized values are $Y_{13} = 26$, $Y_{23} = 14$, $Y_{33} = 26$, and so forth. To test the hypothesis of no differences among the trimmed means, let h_j be the effective sample size of the jth group (the number of observations left after trimming), and compute

$$\bar{Y}_j = \frac{1}{n_j}\sum_{i=1}^{n_j} Y_{ij},$$

$$s_{wj}^2 = \frac{1}{n_j - 1}\sum (Y_{ij} - \bar{Y}_j)^2,$$

$$\bar{X}_t = \frac{1}{J}\sum \bar{X}_{tj},$$

$$\text{BSST} = \frac{1}{J-1}\sum_{j=1}^{J} (\bar{X}_{tj} - \bar{X}_t)^2,$$

$$\text{WSSW} = \frac{1}{J}\sum_{j=1}^{J}\sum_{i=1}^{n_j} \frac{(Y_{ij} - \bar{Y}_j)^2}{h_j(h_j - 1)},$$

Continued

BOX 9.8 (*Continued*)

$$D = \frac{\text{BSST}}{\text{WSSW}}.$$

Let

$$q_j = \frac{(n_j - 1)s_{wj}^2}{J(h_j)(h_j - 1)}.$$

The degrees of freedom are estimated to be

$$\hat{\nu}_1 = \frac{\left((J-1)\sum q_j\right)^2}{\left(\sum q_j\right)^2 + (J-2)J\sum q_j^2},$$

$$\hat{\nu}_2 = \frac{\left(\sum q_j\right)^2}{\sum q_j^2/(h_j - 1)}.$$

Reject if $D \geq f$, the $1-\alpha$ quantile of an F-distribution with $\hat{\nu}_1$ and $\hat{\nu}_2$ degrees of freedom.

9.6.4 S-PLUS Function rananova

The S-PLUS function

rananova(x,tr=.2,grp=NA)

performs the calculations in Box 9.8. As usual, x is any S-PLUS variable that has list mode or is a matrix (with columns corresponding to groups), tr is the amount of trimming, which defaults to .2, and grp can be used to specify some subset of the groups if desired. If grp is not specified, all groups stored in x are used. If the data are not stored in a matrix or in list mode, the function terminates and prints an error message. The function returns the value of the test statistic, D, which is stored in rananova$teststat, the significance level is stored in rananova$siglevel, and an estimate of a Winsorized intraclass correlation is returned in the S-PLUS variable rananova$rho. This last quantity is like the intraclass correlation ρ_I, but with the variance of the means and observations replaced by a Winsorized variance.

EXAMPLE. Assuming the data in Table 9.5 are stored in the S-PLUS variable data, the command rananova(data) returns.

```
$teststat:
[1] 140.0983

$df:
[1] 3.417265 36.663787
```

Continued

EXAMPLE. *(Continued)*

```
$siglevel:
[1] 0

$rho:
[1] 0.9453473
```

So we reject the hypothesis of equal trimmed means. The value for rho indicates that about 95% of the Winsorized variance among the observations is accounted for by the Winsorized variance of the *Winsorized* means. (For technical reasons, Winsorized means are used rather than trimmed means when deriving a robust analog of the intraclass correlation coefficient.) The command rananova(data,tr=0) compares means instead and returns

```
$teststat:
[1] 76.23691

$df:
[1] 3.535804 63.827284

$siglevel:
[1] 0

$rho:
[1] 0.782022
```

So again we reject. In the latter case the intraclass correlation is estimated to be .78, meaning that the variation among the means is estimated to account for 78% of the variation among all possible observations. ■

9.7 Exercises

1. For the following data, assume that the three groups have a common population variance, σ_p^2. Estimate σ_p^2.

Group 1	Group 2	Group 3
3	4	6
5	4	7
2	3	8
4	8	6
8	7	7
4	4	9
3	2	10
9	5	9
$\bar{X}_1 = 4.75$	$\bar{X}_2 = 4.62$	$\bar{X}_3 = 7.75$
$s_1^2 = 6.214$	$s_2^2 = 3.982$	$s_3^2 = 2.214$

2. For the data in the previous exercise, test the hypothesis of equal means using the ANOVA F. Use $\alpha = .05$.
3. For the data in Exercise 1, verify that Welch's test statistic is $F_w = 7.7$ with degrees of freedom $\nu_1 = 2$ and $\nu_2 = 13.4$. Then verify that you would reject the hypothesis of equal means with $\alpha = .01$. Check your results with the S-PLUS function t1way in Section 9.4.1.
4. Construct an ANOVA summary table using the following data, as described in Section 9.1, and then test the hypothesis of equal means with $\alpha = .05$.

Group 1	Group 2	Group 3	Group 4
15	9	17	13
17	12	20	12
22	15	23	17

5. In the previous exercise, what is your estimate of the assumed common variance?
6. For the data used in the preceding two exercises, verify that for Welch's test, $F_w = 3.38$ with $\nu_1 = 3$ and $\nu_2 = 4.42$.
7. Based on the results of the previous exercise, would you reject the hypothesis of equal means with $\alpha = .1$?
8. Why would you not recommend the strategy of testing for equal variances and, if not significant, using the ANOVA F-test rather than Welch's method?
9. For the data in Table 9.1, assume normality and that the groups have equal variances. As already illustrated, the hypothesis of equal means is not rejected. If the hypothesis of equal means is true, an estimate of the assumed common variance is MSBG = .12, as already explained. Describe a reason why you would prefer to estimate the common variance with MSWG rather than MSBG.
10. Five independent groups are compared, with $n = 15$ observations for each group. Fill in the missing values in the following summary table.

Source of variation	Degrees of freedom	Sum of squares	Mean square	F
Between groups	______	50	______	______
Within groups	______	150	______	

11. Referring to Box 9.2, verify that for the following data, MSBG = 14.4 and MSWG = 12.59.

G1	G2	G3
9	16	7
10	8	6
15	13	9
	6	

12. Consider $J = 5$ groups with population means 3, 4, 5, 6, and 7 and a common variance $\sigma_p^2 = 2$. If $n = 10$ observations are sampled from each group, determine the value estimated by MSBG, and comment on how this differs from the value estimated by MSWG.
13. For the following data, verify that you do not reject with the ANOVA F-test with $\alpha = .05$ but that you do reject with Welch's test. What might explain the discrepancy between the two methods?

Group 1:	10	11	12	9	8	7
Group 2:	10	66	15	32	22	51
Group 3:	1	12	42	31	55	19

14. Given the following ANOVA summary table; verify that the number of groups is $J = 4$, that the total number of observations is $N = 12$, and that with $\alpha = .025$ the critical value is 5.42.

Source of variation	Degrees of freedom	Sum of squares	Mean square	F
Between groups	3	300	100	10
Within groups	8	80	10	
Total	11	428		

15. For the data in Table 9.1, the ANOVA F-test and Welch's test were not significant with $\alpha = .05$. Imagine that you want power to be .9 if the mean of one of the groups differs from the others by .2. (In the notation of Section 9.3, $a = .2$.) Verify that according to the S-PLUS function bdanova1, the required sample sizes for each group are 110, 22, 40, and 38.
16. For the data in Table 9.1, use the S-PLUS function pbadepth to compare the groups based on MOM. Verify that the significance level is .17.
17. Compare the groups in Table 9.4 with the S-PLUS function pbadepth. You should find that the function terminates with an error if allp=T is used. How might you deal with this problem given the goal of comparing groups based on MOM?
18. Store the data in Table 9.7 in an S-PLUS variable having matrix mode with two columns corresponding to the two columns shown. There are three groups, with the first column indicating to which group the value in the second column belongs. For example, the first row indicates that the value 12 belongs to group 1 and the fourth row indicates that the value 42 belongs to group 3. Use the S-PLUS function selby in Section 1.1.6 to separate the data into three groups and then store it in list mode. Then compare the three groups with Welch's test, the method in Section 9.4 with 20% trimming, and then use the

S-PLUS function pbadepth in Section 9.5.4 to compare the groups based on MOM.

19. Based on the properties summarized in Box 9.3, how would you estimate

$$\sum (\mu_j - \bar{\mu})^2?$$

20. For the data in Table 9.8, verify that the significance level, based on Welch's test, is .98. Use the S-PLUS Function t1way.
21. For the data in Table 9.8, use t1way to compare 20% trimmed means and verify that the significance level is less than .01.

TABLE 9.7 Data for Exercise 18

G	X
1	12
1	8
1	22
3	42
3	8
3	12
3	9
3	21
2	19
2	24
2	53
2	17
2	10
2	9
2	28
2	21
1	19
1	21
1	56
1	18
1	16
1	29
1	20
3	32
3	10
3	12
3	39
3	28
3	35
2	10
2	12

TABLE 9.8 Hypothetical Data for Exercises 20–22

Group 1	Group 2	Group 3	Group 4	Group 5
10.1	10.7	11.6	12.0	13.6
9.9	9.5	10.4	13.1	11.9
9.0	11.2	11.9	13.2	13.6
10.7	9.9	11.7	11.0	12.3
10.0	10.2	11.8	13.3	12.3
9.3	9.1	11.6	10.5	11.3
10.6	8.0	11.6	14.4	12.4
11.5	9.9	13.7	10.5	11.8
11.4	10.7	13.3	12.2	10.4
10.9	9.7	11.8	11.0	13.1
9.5	10.6	12.3	11.9	14.1
11.0	10.8	15.5	11.9	10.5
11.1	11.0	11.4	12.4	11.2
8.9	9.6	13.1	10.9	11.7
12.6	8.8	10.6	14.0	10.3
10.7	10.2	13.1	13.2	12.0
10.3	9.2	12.5	10.3	11.4
10.8	9.8	13.9	11.6	12.1
9.2	9.8	12.2	11.7	13.9
8.3	10.9	11.9	12.1	12.7
93.0	110.6	119.6	112.8	112.8
96.6	98.8	113.6	108.0	129.2
94.8	107.0	107.5	113.9	124.8

22. For the data in the Table 9.8, use pbadepth with the argument allp set equal to *F* and verify that the significance level is zero. The text mentioned a possible concern with using allp=F. Why is this not an issue here?

TWO-WAY ANOVA

This chapter takes up an extension of the analysis of variance method described in Chapter 9. As usual, we begin by describing basic concepts and summarizing the standard approach based on means. Then robust methods are described.

10.1 The Basics of a Two-Way ANOVA Design

The basic concepts are illustrated with a study where the goal is to understand the effect of diet on weight gains in rats. Specifically, four diets are considered that differ in: (1) amounts of protein (high and low) and (2) the source of the protein (beef versus cereal). The results for these four groups, reported in Table 10.1, are taken from Snedecor and Cochran (1967). Different rats were used in the four groups, so the groups are independent. The first column gives the weight gains of rats fed a low-protein diet, with beef the source of protein. The next column gives the weight gains for rats on a high-protein diet, again with beef the source of protein, and the next two columns report results when cereal is substituted for beef.

It is convenient to depict the population means as shown in Table 10.2. Table 10.2 indicates, for example, that μ_1 is the population mean associated with rats receiving a low-protein diet from beef. That is, μ_1 is the average weight gain if all of the millions of rats we might study are fed this diet. Similarly, μ_4 is the population mean for rats receiving a high-protein diet from cereal.

The study just described is an example of what is called a *two-way design*, meaning that there are two *independent variables*, or *factors*, being studied. Here, the first factor is *source of protein*, which has two levels: beef and cereal. The second factor is *amount of protein*, which also has two levels: low and high. A more precise description is that you have a 2-by-2 design, meaning you have two factors, both of which have two levels. If you compare three methods for increasing endurance and simultaneously take into account three different diets, you have a two-way design with both factors having three levels. More succinctly, this is called a 3-by-3 design. The first factor is method and the second is diet. As a final example, imagine you want to compare four methods for teaching statistics and simultaneously take into account

TABLE 10.1 Weight Gains (in grams) of Rats on One of Four Diets

Beef, low	Beef, high	Cereal, low	Cereal, high
90	73	107	98
76	102	95	75
90	118	97	56
64	104	80	111
86	81	98	95
51	107	74	88
72	100	74	82
90	87	67	77
95	117	89	86
78	111	58	92
$\bar{X}_1 = 79.2$	$\bar{X}_2 = 100$	$\bar{X}_3 = 83.9$	$\bar{X}_4 = 85.9$

TABLE 10.2 Depiction of the Population Means for Four Diets

	Source	
Amount	Beef	Cereal
High	μ_1	μ_2
Low	μ_3	μ_4

the amount of previous training in mathematics. If you categorize students as having poor or good training in mathematics, you have a 4-by-2 design. That is, you have a two-way design where the first factor is method of teaching and the second is previous training. If you ignore previous training and consider only the four methods, you have what is called a *one-way design* with four levels. This means you have one factor of interest (method of training) and four different methods are to be compared.

Returning to Table 10.2, you could compare these four groups by testing

$$H_0 : \mu_1 = \mu_2 = \mu_3 = \mu_4,$$

the hypothesis that all of the means are equal. However, there are other comparisons you might want to make. For example, you might want to compare the rats receiving a high-protein versus low-protein diet, *ignoring* the source of the protein. To illustrate how this might be done, imagine that the values of the population means are as shown in Table 10.3. For rats on a high-protein diet, the mean is 45 when consuming beef versus 60 when consuming cereal instead. If you want to characterize the typical weight gain for a high-protein diet while ignoring source, a natural strategy is to average the two population means, yielding $(45 + 60)/2 = 52.5$. That is, the typical rat on a protein diet gains 52.5 grams. For the more general situation depicted by Table 10.2, the typical weight gain on a high-protein diet would be $(\mu_1 + \mu_2)/2$.

TABLE 10.3 Hypothetical Population Means for Illustrating Main Effects and Interactions

	Source	
Amount	Beef	Cereal
High	$\mu_1 = 45$	$\mu_2 = 60$
Low	$\mu_3 = 80$	$\mu_4 = 90$

Similarly, the typical weight gain for a rat on a low-protein diet would be $(\mu_3 + \mu_4)/2$, which for Table 10.3 is $(80 + 90)/2 = 85$ grams. Of course, you can do the same when characterizing source of protein while ignoring amount. The typical weight gain for a rat eating beef, ignoring amount of protein, is $(45 + 80)/2 = 62.5$, and for cereal it is $(60 + 90)/2 = 75$.

As usual, you do not know the population means. What is needed is some way of testing the hypothesis that weight gain is different for a high-protein diet versus a low-protein diet, ignoring source of protein. One way of doing this is to test

$$H_0: \frac{\mu_1 + \mu_2}{2} = \frac{\mu_3 + \mu_4}{2},$$

the hypothesis that the average of the populations means in the first row of Table 10.2 is equal to the average for the second row. If this hypothesis is rejected, then there is said to be a *main effect* for the amount of protein. More generally, a main effect for the first factor (amount) is said to exist if

$$\frac{\mu_1 + \mu_2}{2} \neq \frac{\mu_3 + \mu_4}{2}.$$

Similarly, you might want to compare source of protein while ignoring amount. One way of doing this is to test

$$H_0: \frac{\mu_1 + \mu_3}{2} = \frac{\mu_2 + \mu_4}{2},$$

the hypothesis that the average of the means in the column for beef in Table 10.2 is equal to the average for the cereal column. If this hypothesis is rejected, then there is said to be a *main effect* for the source of protein. More generally, a main effect for the second factor is said to exist if

$$\frac{\mu_1 + \mu_3}{2} \neq \frac{\mu_2 + \mu_4}{2}.$$

Consider again the 4-by-2 design where you want to compare four methods for teaching statistics and simultaneously take into account the amount of previous training in mathematics. The means corresponding to the eight groups can be written as shown in Table 10.4. Main effects for the first factor (method) can be addressed by testing the hypothesis that the averages of the means in each row have equal values. That is, test

$$H_0: \frac{\mu_1 + \mu_2}{2} = \frac{\mu_3 + \mu_4}{2} = \frac{\mu_5 + \mu_6}{2} = \frac{\mu_7 + \mu_8}{2}.$$

TABLE 10.4 Depiction of the Population Means for a 4-by-2 Design

	Previous Training	
Method	High	Low
1	μ_1	μ_2
2	μ_3	μ_4
3	μ_5	μ_6
4	μ_7	μ_8

Typically this is referred to as the hypothesis of *no main effects for factor A*, where *factor A* is a generic term for the first of the two factors under study. The hypothesis of no main effects for the second factor, previous training, is

$$H_0: \frac{\mu_1 + \mu_3 + \mu_5 + \mu_7}{4} = \frac{\mu_2 + \mu_4 + \mu_6 + \mu_8}{4}.$$

That is, the average of the means in column 1 is hypothesized to be equal to the average of the means in column 2. This is called a hypothesis of no main effects for the second factor, previous training. A generic term for the second factor is *factor B*, and the hypothesis of no main effects for factor B refers to the hypothesis that the averages of the means in each column have a common value.

10.1.1 Interactions

There is one other important feature of a two-way design. Consider again a 2-by-2 design where the goal is to compare high- and low-protein diets in conjunction with two protein sources. Suppose the *population* means associated with the four groups are as depicted in Table 10.3. Now look at the first row (high amount of protein) and notice that the weight gain for a beef diet is 45 grams versus a weight gain of 60 for cereal. As is evident, there is an increase of 15 grams. In contrast, with a low-protein diet, switching from beef to cereal results in an increase of 10 grams on average. That is, in general, switching from beef to cereal results in an increase for the average amount of weight gained, but the increase differs depending on whether we look at high or low protein. This is an example of what is called an *interaction*.

In a 2-by-2 design with means as shown in Table 10.2, an *interaction* is said to exist if

$$\mu_1 - \mu_2 \neq \mu_3 - \mu_4.$$

In words, an interaction exists if for the first level of factor A the difference between the means is not equal to the difference between the means associated with the second level of factor A. *No interaction* means that

$$\mu_1 - \mu_2 = \mu_3 - \mu_4.$$

Various types of interactions arise and can be important when considering how groups differ. Notice that in the beef-versus-cereal illustration as depicted in Table 10.3, there is an increase in the average weight gain when switching from beef to cereal for both high- and low-protein diets. For high protein there is an increase in the population mean from 45 to 60, and for low protein there is an increase from 80 to 90. In both cases, though, the largest gain in weight is associated with cereal. This is an example of what is called an *ordinal interaction*. In a 2-by-2 design as depicted in Table 10.2, an *ordinal interaction* is said to exist if

$$\mu_1 > \mu_2 \quad \text{and} \quad \mu_3 > \mu_4$$

or if

$$\mu_1 < \mu_2 \quad \text{and} \quad \mu_3 < \mu_4.$$

In words, an interaction is said to be ordinal if the relative rankings remain the same; in the illustration, cereal always results in the largest weight gain, regardless of whether a low- or high-protein diet is used.

If there is a change in the relative rankings of the means, a *disordinal interaction* is said to exist. As an illustration, imagine that the population means are as follows:

	Source	
Amount	Beef	Cereal
High	80	110
Low	50	30

Observe that for the first row (a high-protein diet), the average weight gain increases from 80 to 110 as we move from beef to cereal. In contrast, for the low-protein diet, the average weight gain decreases from 50 to 30. Moreover, when comparing beef to cereal, the relative rankings change depending on whether a high- or low-protein diet is used. For a high-protein diet, cereal results in the largest gain; for a low-protein diet, beef results in a larger gain than cereal. This is an example of a disordinal interaction. In general, for the population means in Table 10.2, a disordinal interaction is said to exist if

$$\mu_1 > \mu_2 \quad \text{and} \quad \mu_3 < \mu_4$$

or if

$$\mu_1 < \mu_2 \quad \text{and} \quad \mu_3 > \mu_4.$$

Research articles often present graphical displays reflecting ordinal and disordinal interactions. Figure 10.1 is an example based on the sample means just used to illustrate a disordinal interaction. Along the x-axis we see the levels of the first factor (high and low protein). The line extending from 110 down to 30 reflects the change in means when the source of protein is beef. The other line reflects the change in means associated with cereal. As is evident, the lines cross, which reflects a disordinal interaction.

Now suppose the population means are as shown in Table 10.3. Regardless of whether rats are given a high- or low-protein diet, cereal always results in the largest average weight gain. The left panel of Figure 10.2 graphs the means. Observe that the lines are not parallel, but they do not cross, indicating an ordinal interaction.

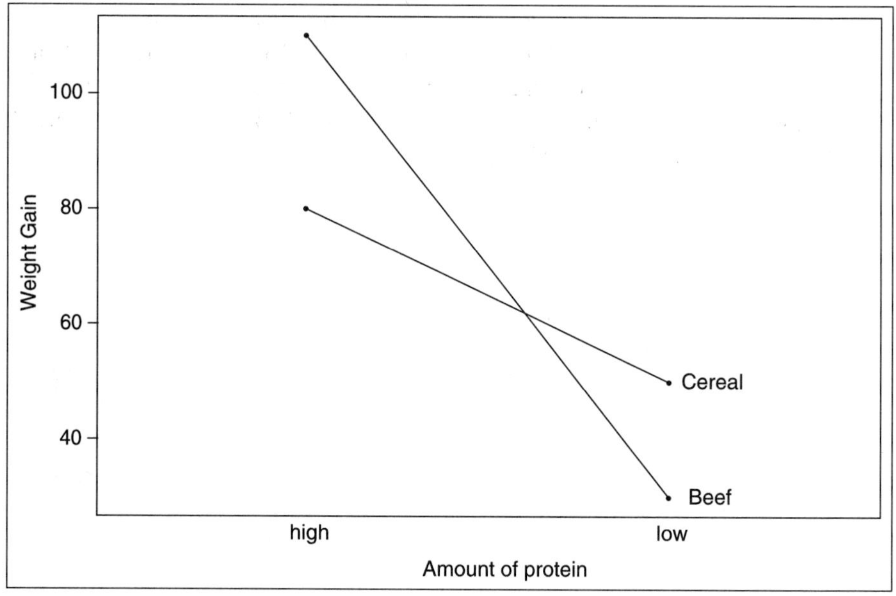

FIGURE 10.1 Example of a disordinal interaction.

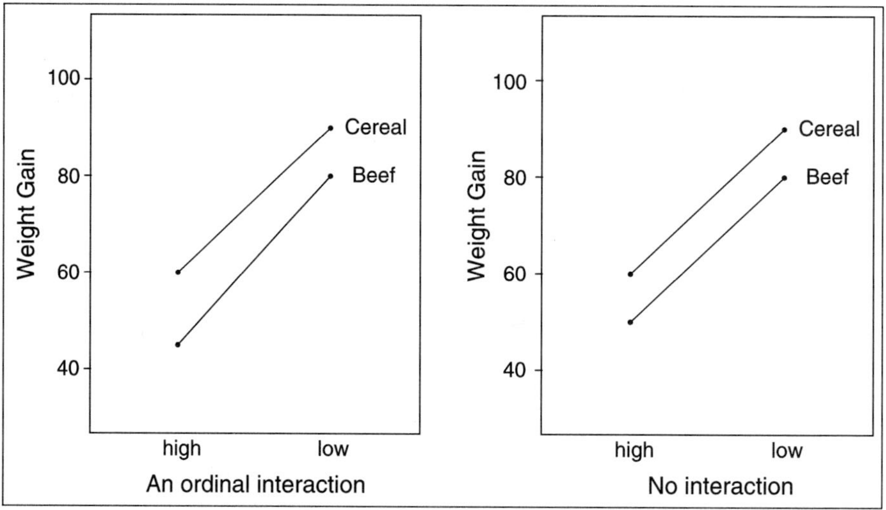

FIGURE 10.2 Two types of interactions.

Now imagine that the population means are

	Source	
Amount	Beef	Cereal
High	50	60
Low	80	90

There is no interaction because for the first row, the means increase by 10 (from 50 to 60), and the increase is again 10 for the second row. The right panel of

Figure 10.2 graphs the means. Notice that the lines are parallel. That is, when there is no interaction, a graph of the means results in parallel lines.

Notice that in the discussion of ordinal versus disordinal interactions, attention was focused on comparing means within rows. Not surprisingly, ordinal and disordinal interactions can also be defined in terms of columns. Consider again the example where the means are

	Source	
Amount	Beef	Cereal
High	80	110
Low	50	30

As previously explained, there is a disordinal interaction for rows. Now, however, look at the population means in the first column and notice that 80 is greater than 50, and for the second column 110 is greater than 30. There is an interaction because $80 - 50 \neq 110 - 30$. Moreover, the interaction is ordinal because for beef, average weight gain is largest on a high-protein diet, and the same is true for a cereal diet.

When graphing the means to illustrate a disordinal or ordinal interaction for rows, the levels of factor A (high and low amounts of protein) were indicated by the x-axis, as illustrated by Figures 10.1 and 10.2. When describing ordinal or disordinal interactions by columns, now the x-axis contains the levels of the second factor, which in this example is source of protein (beef versus cereal). Figure 10.3 illustrates what the graph looks like for the means considered here.

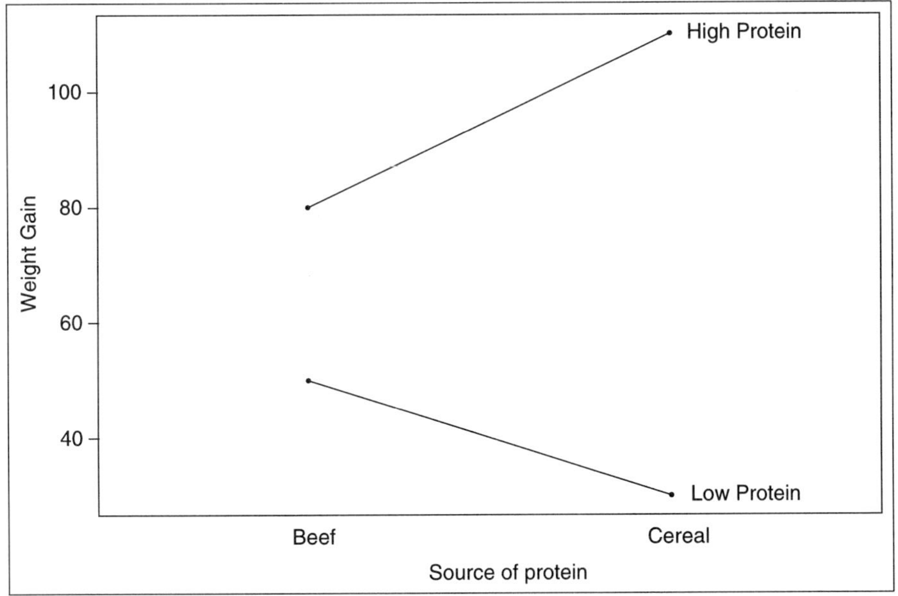

FIGURE 10.3 Graph of an ordinal interaction based on columns.

Situations also arise where there is a disordinal interaction for both rows and columns. For example, if the population means happen to be

	Source	
Amount	Beef	Cereal
High	80	110
Low	100	95

then there is a disordinal interaction for rows, because for the first row, cereal results in the largest gain, but for the second row (low amount of protein) the largest gain is for beef. Simultaneously, there is a disordinal interaction for columns, because for the first column (beef), a low-protein diet results in the largest average gain, but for the second column (cereal) a high-protein diet has the largest mean.

10.1.2 Interactions When There Are More Than Two Levels

So far, attention has been focused on a 2-by-2 design. What does no interaction mean when there are more than two levels for one or both factors? Basically, *no interaction* means that for any two levels of factor A and any two levels of factor B, there is no interaction for the corresponding cells. As an illustration, consider the population means in Table 10.5. Pick any two rows — say, the first and third. Then pick any two columns — say, the first and second. The population means for the first row and the first and second columns are 10 and 20. For the third row, the means for these two columns are 30 and 40; these four means are in boldface in Table 10.5. Notice that for these four means, there is no interaction. The reason is that for the first row, the means increase from 10 to 20, and for the third row the means increase from 30 to 40. That is, for both rows there is an increase of 10 when switching from column 1 to column 2. In a similar manner, again looking at rows 1 and 2, we see that there is an increase of 20 as we move from column 1 to column 3. That is, there is no interaction for these four means either. An interaction is said to exist among the JK means if any two rows and two columns have an interaction.

EXAMPLE. Suppose the population means for a 3-by-4 design are

	Factor B			
Factor A	Level 1	Level 2	Level 3	Level 4
Level 1	40	40	40	40
Level 2	40	40	40	40
Level 3	40	40	40	40

Is there an interaction? The answer is no, because regardless of which two rows you pick, there is an increase of 0 as you move from any one column to another. ■

TABLE 10.5 Hypothetical Population Means Illustrating Interactions

	Factor B		
Factor A	Level 1	Level 2	Level 3
Level 1	10	20	30
Level 2	20	30	40
Level 3	30	40	50

EXAMPLE. For the population means used in this last example, is there a main effect for factor A or factor B? First consider factor A. As is evident, for any row we pick, the average of the four means is 40. That is, the average of the means is the same for all three rows; this means there is no main effect for factor A. In a similar manner, there is no main effect for factor B, because the average of the means in any column is again 40. ■

EXAMPLE. Suppose the population means for a 3-by-4 design are

	Factor B			
Factor A	Level 1	Level 2	Level 3	Level 4
Level 1	40	40	50	60
Level 2	20	20	50	80
Level 3	20	30	10	40

Is there an interaction? Looking at level 1 of factor A, we see that the means increase by 0 as we move from level 1 of factor B to level 2. The increase for level 2 of factor A is again 0, so there is no interaction for these four means. However, looking at level 1 of factor A, we see that the means increase by 10 as we move from level 1 to level 3 of factor B. In contrast, there is an increase of 30 for level 2 of factor A, which means that there is an interaction. ■

10.2 Testing Hypotheses About Main Effects and Interactions

So far, attention has been focused on explaining the meaning of main effects and interactions in terms of the population means. As usual, we do not know the population means, and this raises the issue of how we might test the hypotheses of no main effects and no interactions. As was the case in Chapter 9, the most commonly used method is based on the assumption that all groups have normal distributions with a common variance, which we again label σ_p^2.

TABLE 10.6 Sample Means for Illustrating Main Effects and Interactions

	Source	
Amount	Beef	Cereal
Low	$\bar{X}_{11} = 79.2$	$\bar{X}_{12} = 83.9$
High	$\bar{X}_{21} = 100$	$\bar{X}_{22} = 85.9$

The computations begin by computing the sample mean corresponding to the *j*th level of factor A and the *k*th level of factor B. The computations are easier to describe if we switch notation and represent this sample mean with $\bar{X}_{jk}$. That is, $\bar{X}_{jk}$ is the sample mean corresponding to the *j*th level of factor A and the *k*th level of factor B. For example, for the 2-by-2 study comparing high versus low protein and beef versus cereal, $\bar{X}_{11}$ is the sample mean for level 1 of factor A and level 1 of factor B. Referring to Table 10.1, $\bar{X}_{11}$ is the sample mean of the rats on a low-protein diet (level 1 of factor A) that consume beef (level 1 of factor B). From the first column in Table 10.1, the average weight gain for the 10 rats on this diet can be seen to be $\bar{X}_{11} = 79.2$. Level 1 of factor A and level 2 of factor B correspond to low protein from a cereal diet. The data are given in the third column of Table 10.1, and the sample mean is $\bar{X}_{12} = 83.9$. For level 2 of factor A and level 1 of factor B, $\bar{X}_{21} = 100$, the sample mean of the values in column 2 of Table 10.1. Finally, $\bar{X}_{22}$ is the sample mean for level 2 of factor A and level 2 of factor B and is 85.9. These sample means are summarized in Table 10.6.

Under normality and equal variances, a test of the hypothesis of no main effects for factor A can be performed that provides exact control over the probability of a Type I error. Box 10.1 summarizes the bulk of the calculations when all groups have a common sample size, *n*. (The method is not robust to violations of assumptions, particularly when the sample sizes are unequal, so the computational details for handling unequal samples are omitted.) In Box 10.1, s_{jk}^2 is the sample variance corresponding to the data used to compute $\bar{X}_{jk}$. As in the one-way design, MSWG estimates the assumed common variance, σ_p^2.

BOX 10.1 Computations for a Two-Way Design

Assumptions

1. Random sampling
2. Normality

Continued

BOX 10.1 *(Continued)*

3. Equal variances
4. Equal sample sizes

Notation

J is the number of levels for factor A, and K is the number of levels for factor B.

Computations

$$\bar{X}_G = \frac{1}{JK} \sum_{j=1}^{J} \sum_{k=1}^{K} \bar{X}_{jk}$$

($\bar{X}_G$ is the average of all JK sample means.)

$$\bar{X}_{j.} = \frac{1}{K} \sum_{k=1}^{K} \bar{X}_{jk}$$

($\bar{X}_{j.}$ is the average of the sample means for the jth level of factor A.)

$$\bar{X}_{.k} = \frac{1}{J} \sum_{j=1}^{J} \bar{X}_{jk}$$

($\bar{X}_{.k}$ is the average of the sample means for the kth level of factor B.)

$$A_j = \bar{X}_{j.} - \bar{X}_G, \qquad B_k = \bar{X}_{.k} - \bar{X}_G$$

$$C_{jk} = \bar{X}_{jk} - \bar{X}_{j.} - \bar{X}_{.k} + \bar{X}_G$$

$$\text{SSA} = nK \sum A_j^2, \qquad \text{SSB} = nJ \sum B_k^2$$

$$\text{SSINTER} = n \sum \sum C_{jk}^2$$

$$\text{MSWG} = \frac{1}{JK} \sum \sum s_{jk}^2, \qquad \text{SSWG} = (n-1)JK(\text{MSWG})$$

$$\text{MSA} = \frac{\text{SSA}}{J-1}, \qquad \text{MSB} = \frac{\text{SSB}}{K-1}$$

$$\text{MSINTER} = \frac{\text{SSINTER}}{(J-1)(K-1)}.$$

(Some books write MSINTER and SSINTER as MSAB and SSAB, respectively, where AB denotes the interaction of factors A and B.)

DECISION RULES: Once you complete the computations in Box 10.1, the relevant hypotheses are tested as follows:

- **Factor A**. The hypothesis of no main effects for factor A is tested with

$$F = \frac{\text{MSA}}{\text{MSWG}},$$

and you reject if $F \geq f$, the $1 - \alpha$ quantile of an F-distribution with $\nu_1 = J - 1$ and $\nu_2 = N - JK$ degrees of freedom.

- **Factor B**. The hypothesis of no main effects for factor B is tested with

$$F = \frac{\text{MSB}}{\text{MSWG}},$$

and you reject if $F \geq f$, the $1 - \alpha$ quantile of an F-distribution with $\nu_1 = K - 1$ and $\nu_2 = N - JK$ degrees of freedom.

- **Interactions**. The hypothesis of no interactions is tested with

$$F = \frac{\text{MSINTER}}{\text{MSWG}},$$

and you reject if $F \geq f$, the $1 - \alpha$ quantile of an F-distribution with $\nu_1 = (J - 1)(K - 1)$ and $\nu_2 = N - JK$ degrees of freedom.

In Box 10.1, one of the assumptions is normality. In terms of Type I errors, this assumption can be violated if all groups have identically shaped distributions and the sample sizes are reasonably large. This means, in particular, that the equal variance assumption is true. Put another way, if you reject with the ANOVA F-test, this indicates that the distributions differ, but it remains unclear how they differ and by how much. One possibility is that you reject because the populations means differ. Another possibility is that you reject primarily because the variances differ. As for power, again practical problems arise under very slight departures from normality, for reasons discussed in previous chapters. (See in particular Section 5.5.) Generally, the more groups you compare, the more likely the ANOVA F-tests described in this section will be unsatisfactory when indeed groups differ.

Table 10.7 outlines a typical ANOVA summary table for a two-way design. The notation *SS* in the first row stands for sum of squares, *DF* is degrees of freedom, and *MS* is mean squares.

TABLE 10.7 Typical ANOVA Summary Table for a Two-Way Design

Source	SS	DF	MS	F
A	SSA	$J-1$	$\text{MSA} = \frac{\text{SSA}}{J-1}$	$F = \frac{\text{MSA}}{\text{MSWG}}$
B	SSB	$K-1$	$\text{MSB} = \frac{\text{SSB}}{K-1}$	$F = \frac{\text{MSB}}{\text{MSWG}}$
INTER	SSINTER	$(J-1)(K-1)$	$\text{MSINTER} = \frac{\text{SSINTER}}{(J-1)(K-1)}$	$F = \frac{\text{MSINTER}}{\text{MSWG}}$
WITHIN	SSWG	$N-JK$	$\text{MSWG} = \frac{\text{SSWG}}{N-JK}$	

EXAMPLE. Consider the following ANOVA summary table.

Source	SS	DF	MS	F
A	200	1	200	1.94
B	300	2	150	1.46
INTER	500	2	250	2.42
WITHIN	620	6	103	

Referring to Table 10.7, we see that $J-1$ corresponds to the value 1 in the example, indicating that $J-1=1$, so the first factor has $J=2$ levels. Similarly, the second factor has $K=3$ levels. Table 10.7 indicates that $N-JK=6$, but $JK=6$, so $N-JK=N-6=6$, and therefore $N=12$. That is, the total number of observations among the 6 groups is 12. The estimate of the common variance is MSWG$=103$. If we use $\alpha=.05$, then from Table 6 in Appendix B, the critical values for the three hypotheses are 5.98, 5.14, and 5.14. The F-values are less than their corresponding critical values, so you do not reject the hypotheses of no main effects for factor A, as well as for factor B and the hypothesis of no interactions. ■

10.2.1 Inferences About Disordinal Interactions

It should be mentioned that if you reject the hypothesis of no interactions, simply looking at the means is not enough to determine whether the interaction is ordinal or disordinal. Consider again the study on weight gain in rats described at the beginning of this section. Suppose that unknown to you, the population means are

	Source	
Amount	Beef	Cereal
Low	$\mu_1 = 60$	$\mu_2 = 60$
High	$\mu_3 = 50$	$\mu_4 = 70$

There is an interaction, because $60 - 60 \neq 50 - 70$, but the interaction is not disordinal. Further assume that you reject the hypothesis of no interactions based on the following sample means:

	Source	
Amount	Beef	Cereal
Low	$\bar{X}_1 = 55$	$\bar{X}_2 = 45$
High	$\bar{X}_3 = 49$	$\bar{X}_4 = 65$

That is, you have correctly concluded that there is an interaction. Notice that the sample means suggest that the interaction is disordinal, because 55 is greater than 45 but 49 is less than 65. By chance, rats on a low-protein diet with beef as the source got a smaller sample mean versus rats on a low-protein diet with cereal as the source. The important point is that to establish that a disordinal interaction exists for the rows, you must also reject the hypotheses $H_0: \mu_1 = \mu_2$ and $H_0: \mu_3 = \mu_4$. If, for example, $\mu_1 = \mu_2$, you do not have a disordinal interaction. Moreover, simply rejecting the hypothesis of no interaction does not tell you whether you should reject $H_0: \mu_1 = \mu_2$ or $H_0: \mu_3 = \mu_4$. Under normality, these hypotheses can be tested with Student's T test or Welch's test, described in Chapter 8. If both of these hypotheses are rejected and if

$$\bar{X}_1 > \bar{X}_2 \quad \text{and} \quad \bar{X}_3 < \bar{X}_4$$

or if

$$\bar{X}_1 < \bar{X}_2 \quad \text{and} \quad \bar{X}_3 > \bar{X}_4,$$

then you have empirical evidence that there is a disordinal interaction.

A similar strategy is used when checking for a disordinal interaction for columns. That is, to establish that a disordinal interaction exists, you must reject both $H_0: \mu_1 = \mu_3$ and $H_0: \mu_2 = \mu_4$. If both hypotheses are rejected and the sample means satisfy

$$\bar{X}_1 > \bar{X}_3 \quad \text{and} \quad \bar{X}_2 < \bar{X}_4$$

or if

$$\bar{X}_1 < \bar{X}_3 \quad \text{and} \quad \bar{X}_2 > \bar{X}_4,$$

you conclude that there is a disordinal interaction for the columns.

An illustration of how to detect a disordinal interaction is postponed until Chapter 12 because there is yet another technical issue that must be addressed: The more tests you perform, the more likely you are to reject even if none of the means differs. There is a large collection of methods for dealing with this issue, many of which will be described in Chapter 12.

10.2.2 The Two-Way ANOVA Model

There is a classic model associated with the two-way ANOVA method that generalizes the ANOVA model in Chapter 9. Although the traditional ANOVA method described in this section is relatively ineffective by modern standards, variations of the model turn out to have value (as will be seen in Chapter 15), so it is briefly described here.

The population grand mean associated with the *JK* groups is

$$\bar{\mu} = \frac{1}{JK}\sum_{j=1}^{J}\sum_{k=1}^{K}\mu_{jk},$$

the average of the population means. Let

$$\mu_{j.} = \frac{1}{K}\sum_{k=1}^{K}\mu_{jk},$$

be the average of the K means among the levels of factor B associated with the jth level of factor A. Similarly,

$$\mu_{.k} = \frac{1}{J}\sum_{j=1}^{J}\mu_{jk}$$

is the average of the J means among the levels of factor A associated with the kth level of factor B. The main effects associated with factor A are

$$\alpha_1 = \mu_{1.} - \bar{\mu}, \ldots, \alpha_J = \mu_{J.} - \bar{\mu}.$$

So the main effect associated with the jth level is the difference between the grand mean and the average of the means associated with the jth level, namely, $\alpha_j = \mu_{j.} - \bar{\mu}$. There are no main effects for factor A if

$$\alpha_1 = \cdots = \alpha_J = 0.$$

As for factor B, main effects are defined by

$$\beta_1 = \mu_{.1} - \bar{\mu}, \ldots, \beta_K = \mu_{.K} - \bar{\mu}.$$

The hypothesis of no main effects for factor B can be expressed as

$$H_0: \beta_1 = \cdots = \beta_K = 0.$$

As for interactions, let

$$\begin{aligned}\gamma_{jk} &= \mu_{jk} - \alpha_j - \beta_k - \bar{\mu} \\ &= \mu_{jk} - \mu_{j.} - \mu_{.k} + \bar{\mu}.\end{aligned}$$

Then no interactions means that

$$\gamma_{11} = \gamma_{12} = \cdots = \gamma_{JK} = 0.$$

Although the γ_{jk} terms are not very intuitive, they provide a convenient framework for deriving an appropriate test of the hypothesis that there are no interactions among any two levels of factor A and factor B.

The discrepancy between the ith observation in the jth level of factor A and kth level of factor B, versus the terms just described, is

$$\epsilon_{ijk} = X_{ijk} - \bar{\mu} - \alpha_j - \beta_k - \gamma_{jk}.$$

Rearranging terms yields

$$X_{ijk} = \bar{\mu} + \alpha_j + \beta_k + \gamma_{jk} + \epsilon_{ijk}.$$

Assuming that the error term, ϵ_{ijk}, has a normal distribution with a common variance among the groups results in the standard two-way ANOVA model, which forms the basis of the hypothesis-testing methods covered in this section.

Based on the model just described, it can be shown that

$$E(\text{MSWG}) = \sigma_p^2,$$

$$E(\text{MSA}) = \sigma_p^2 + \frac{nK}{J-1}\sum \alpha_j^2,$$

$$E(\text{MSB}) = \sigma_p^2 + \frac{nJ}{K-1}\sum \beta_k^2,$$

$$E(\text{MSINTER}) = \sigma_p^2 + \frac{n\sum\sum \gamma_{jk}^2}{(J-1)(K-1)}.$$

10.3 Heteroscedastic Methods for Trimmed Means

The tests for main effects and interactions just described assume sampling is from normal distributions with a common variance. As was the case in Chapters 8 and 9, violating the assumption of equal variances causes serious practical problems. That is, unequal variances can result in poor power properties (power can go down as the means become unequal), unsatisfactory control over the probability of a Type I error, and relatively low power versus other methods that might be used. The main reason for describing the extension of the F-test to a two-way design is that it is commonly used, but its practical problems are relatively unknown and many popular computer programs have not yet added more modern methods to their library of techniques. Another reason for describing the conventional method is to be sure readers understand the basic concepts and goals associated with two-way designs.

As in the previous two chapters, our goal is to work up to a method that gives more satisfactory results when groups differ in some manner. In previous chapters we addressed this problem by first describing a heteroscedastic method for means and then indicating how to extend the method to trimmed means. Here, for brevity, we merely describe a heteroscedastic method for trimmed means that includes the problem of comparing means as a special case. The method for dealing with main effects stems from Welch's method, described in Chapter 9. The test statistics for the main effects are computed as shown in Box 10.2. For interactions, it currently seems that an extension of a method derived by Johansen (1980), which is summarized in Box 10.3, gives more satisfactory results.

BOX 10.2 A Heteroscedastic Test for Main Effects Based on Trimmed Means

Let n_{jk} represent the number of observations for the jth level of factor A and the kth level of B, and let h_{jk} be the number of observations left after trimming. Compute

$$R_j = \sum_{k=1}^{K} \bar{X}_{tjk}, \qquad W_k = \sum_{j=1}^{J} \bar{X}_{tjk}$$

$$d_{jk} = \frac{(n_{jk} - 1)s^2_{wjk}}{h_{jk}(h_{jk} - 1)}$$

$$\hat{\nu}_j = \frac{(\sum_k d_{jk})^2}{\sum_k d^2_{jk}/(h_{jk} - 1)}, \qquad \hat{\omega}_k = \frac{(\sum_j d_{jk})^2}{\sum_j d^2_{jk}/(h_{jk} - 1)}$$

$$r_j = \frac{1}{\sum_k d_{jk}}, \qquad w_k = \frac{1}{\sum_j d_{jk}}$$

$$\hat{R} = \frac{\sum r_j R_j}{\sum r_j}, \qquad \hat{W} = \frac{\sum w_k W_k}{\sum w_k}$$

$$B_a = \sum_{j=1}^{J} \frac{1}{\hat{\nu}_j}\left(1 - \frac{r_j}{\sum r_j}\right)^2, \qquad B_b = \sum_{k=1}^{K} \frac{1}{\hat{\omega}_k}\left(1 - \frac{w_k}{\sum w_k}\right)^2$$

$$V_a = \frac{1}{(J-1)\left(1 + \frac{2(J-2)B_a}{J^2-1}\right)} \sum_{j=1}^{J} r_j (R_j - \hat{R})^2,$$

$$V_b = \frac{1}{(K-1)\left(1 + \frac{2(K-2)B_b}{K^2-1}\right)} \sum_{k=1}^{K} w_k (W_k - \hat{W})^2.$$

Continued

BOX 10.2 (*Continued*) The degrees of freedom for factor A are

$$\nu_1 = J - 1, \qquad \nu_2 = \frac{J^2 - 1}{3B_a}.$$

The degrees of freedom for factor B are

$$\nu_1 = K - 1, \qquad \nu_2 = \frac{K^2 - 1}{3B_b}.$$

Decision Rule

Reject the hypothesis of no main effect for factor A if $V_a \geq f_{1-\alpha}$, the $1 - \alpha$ quantile of an F-distribution with the degrees of freedom for factor A. Similarly, reject for factor B if $V_b \geq f_{1-\alpha}$, where now the degrees of freedom are for factor B.

BOX 10.3 A Heteroscedastic Test of the Hypothesis of No Interactions Based on Trimmed Means

Let

$$d_{jk} = \frac{(n_{jk} - 1)s_{wjk}^2}{h_{jk}(h_{jk} - 1)}$$

$$D_{jk} = \frac{1}{d_{jk}}$$

$$D_{.k} = \sum_{j=1}^{J} D_{jk}, \qquad D_{j.} = \sum_{k=1}^{K} D_{jk}$$

$$D_{..} = \sum D_{jk}$$

$$\tilde{X}_{tjk} = \sum_{\ell=1}^{J} \frac{D_{\ell k}\bar{X}_{t\ell k}}{D_{.k}} + \sum_{m=1}^{K} \frac{D_{jm}\bar{X}_{tjm}}{D_{j.}} - \sum_{\ell=1}^{J}\sum_{m=1}^{K} \frac{D_{\ell m}\bar{X}_{t\ell m}}{D_{..}}.$$

The test statistic is

$$V_{ab} = \sum_{j=1}^{J}\sum_{k=1}^{K} D_{jk}(\bar{X}_{tjk} - \tilde{X}_{tjk})^2.$$

Continued

BOX 10.3 (*Continued*) Let c be the $1-\alpha$ quantile of a chi-squared distribution with $\nu = (J-1)(K-1)$ degrees of freedom. Reject if $V_{ab} \geq c + h(c)$, where

$$h(c) = \frac{c}{2(J-1)(K-1)}\left\{1 + \frac{3c}{(J-1)(K-1)+2}\right\}A,$$

$$A = \sum_j \sum_k \frac{1}{f_{jk}}\left\{1 - D_{jk}\left(\frac{1}{D_{j.}} + \frac{1}{D_{.k}} - \frac{1}{D_{..}}\right)\right\}^2$$

$$f_{jk} = h_{jk} - 3.$$

(From Johansen, 1980, it might appear that this last expression should be $h_{jk} - 1$, but $h_{jk} - 3$ gives better control over the probability of a Type I error.)

10.3.1 S-PLUS Function t2way

The S-PLUS function

```
t2way(J, K, x, tr=0.2, grp=c(1:p), alpha=0.05, p=J*K)
```

tests the hypotheses of no main effects and no interaction, as described in Boxes 10.2 and 10.3. Here J and K denote the number of levels associated with factors A and B, respectively. Like t1way, the data are assumed to be stored in x, which can be any S-PLUS variable that is a matrix or has list mode. If stored in list mode, the first K groups are assumed to be the data for the first level of factor A, the next K groups are assumed to be data for the second level of factor A, and so on. In S-PLUS notation, x[[1]] is assumed to contain the data for level 1 of factors A and B, x[[2]] is assumed to contain the data for level 1 of factor A and level 2 of factor B, and so forth. If, for example, a 2-by-4 design is being used, the data are stored as follows:

```
                       Factor B
             x[[1]]  x[[2]]  x[[3]]  x[[4]]
Factor A
             x[[5]]  x[[6]]  x[[7]]  x[[8]]
```

For instance, x[[5]] contains the data for the second level of factor A and the first level of factor B.

If the data are stored in a matrix, the first K columns are assumed to be the data for the first level of factor A, the next K columns are assumed to be data for the second level of factor A, and so on.

If the data are not stored in the assumed order, the argument grp can be used to correct this problem. As an illustration, suppose the data are stored as follows:

```
                          Factor B
Factor A    x[[2]]  x[[3]]  x[[5]]  x[[8]]
            x[[4]]  x[[1]]  x[[6]]  x[[7]]
```

That is, the data for level 1 of factors A and B are stored in the S-PLUS variable x[[2]], the data for level 1 of A and level 2 of B are stored in x[[3]], and so forth. To use t2way, first enter the S-PLUS command

grp<-c(2,3,5,8,4,1,6,7).

Then the command t2way(2,4,x,grp=grp) tells the function how the data are ordered. In the example, the first value stored in grp is 2, indicating that x[[2]] contains the data for level 1 of both factors A and B, the next value is 3, indicating that x[[3]] contains the data for level 1 of A and level 2 of B, and the fifth value is 4, meaning that x[[4]] contains the data for level 2 of factor A and level 1 of B. As usual, tr indicates the amount of trimming, which defaults to .2, and alpha is α, which defaults to .05. The function returns the test statistic for factor A, V_a, in the S-PLUS variable t2way$test.A, and the significance level is returned in t2way$sig.A. Similarly, the test statistics for factor B, V_b, and interaction, V_{ab}, are stored in t2way$test.B and t2way$test.AB, respectively, with the corresponding significance levels stored in t2way$sig.B and t2way$sig.AB.

As a more general example, the command

t2way(2,3,z,tr=.1,grp=c(1,3,4,2,5,6),alpha=.1)

would perform the tests for no main effects and no interactions for a 2-by-3 design for the data stored in the S-PLUS variable z, assuming the data for level 1 of factors A and B are stored in z[[1]], the data for level 1 of A and level 2 of B are stored in z[[3]], and so on. The analysis would be based on 10% trimmed means and $\alpha = .1$.

Note that t2way contains an argument p. Generally this argument can be ignored; it is used by t2way to check whether the total number of groups being passed to the function is equal to JK. If JK is not equal to the number of groups in x, the function prints a warning message. If, however, you want to perform an analysis using some subset of the groups stored in x, this can be done simply by ignoring the warning message. For example, suppose x contains data for 10 groups but that you want to use groups 3, 5, 1, and 9 in a 2-by-2 design. That is, groups 3 and 5 correspond to level 1 of the first factor and levels 1 and 2 of the second. The command

t2way(2,2,x,grp=c(3,5,1,9))

accomplishes this goal.

EXAMPLE. A total of $N = 50$ male Sprague–Dawley rats were assigned to one of six conditions, corresponding to a 2-by-3 ANOVA. (The data in this example were supplied by U. Hayes.) The two levels of the first factor have to do with whether an animal was placed on a fluid-restriction schedule one week prior to the initiation of the experiment. The other factor had to do with the injection of one of three drugs. One of the outcome measures was sucrose consumption shortly after acquisition of a LiCl-induced conditioned taste avoidance. The output from t2way appears as follows:

```
$test.A:
[1] 11.0931

$sig.A:
[1] 0.001969578

$test.B:
[1] 3.764621

$sig.B:
[1] 0.03687472

$test.AB:
[1] 2.082398

$critinter:
[1] 7.385763
```

So based on 20% trimmed means, there is a main effect for both factors A and B, but no interaction is detected. ■

10.3.2 S-PLUS Function selby2

Chapter 1 mentioned an S-PLUS function called selby that is aimed at assisting with data manipulation when data are stored in a matrix, with some particular column indicating group membership. The function separates the data into groups and stores it in list mode, which in turn can be used by the S-PLUS functions in Chapter 9. For a two-way design, situations arise where one column of a matrix indicates the levels of factor A and another column indicates the levels of factor B. For example, suppose the data are stored in a file as follows:

```
A  B  X
1  3  46
1  2  23
2  1  21
1  1  35
   ⋮
```

That is, the first column indicates the level of factor A, the second column indicates the level of factor B, and the third column is the outcome of interest. So here, the first row of data indicates that for level 1 of factor A and level 3 of factor B, the outcome is 46. Suppose these data have been stored in an S-PLUS matrix m. The problem is storing the data so that they can be fed into the function t2way. The S-PLUS function

```
selby2(m,grpc,coln=NA)
```

is designed to sort the data in the matrix m into groups and store them in list mode, which in turn can be used in t2way. The argument grpc is a vector containing two values indicating which columns of the matrix m reflect the levels of factors A and B. For example,

```
selby2(m,grpc=c(1,2),coln=3)
```

indicates that there are two factors with the levels of the first factor stored in column 1, the levels of factor B stored in column 2, and the outcome variable to be analyzed is stored in column 3. So the command

```
> dat<-selby2(m,grpc=c(1,2),coln=3)
```

will determine how many levels there are for each factor and store the data in the S-PLUS variable dat$x. The variable dat$x[[1]] will contain the data for level 1 of both factors, dat$x[[2]] will contain the data for level 1 of factor A and level 2 of factor B, dat$x[[3]] will contain the data for level 1 of factor A and level 3 of factor B, and so on. That is, the data are automatically stored as described in Section 10.3.1.

10.4 Bootstrap Methods

The bootstrap methods described in previous chapters are readily extended to a two-way design. To apply the bootstrap-*t* method with a trimmed mean, you proceed in a manner similar to that in Chapter 9. That is, for the *j*th level of factor A and the *k*th level of factor B you subtract the trimmed mean ($\bar{X}_{tjk}$) from each of the n_{jk} observations; this is done for all *JK* groups. In symbols, for the *j*th level of factor A and the *k*th level of factor B, bootstrap samples are generated from $C_{1jk}, \ldots, C_{n_{jk}jk}$, where $C_{ijk} = X_{ijk} - \bar{X}_{tjk}$. Said yet another way, center the data for each of the *JK* groups by subtracting out the corresponding trimmed mean, in which case the empirical distributions of all *JK* groups have a trimmed mean of zero. That is, the distributions are shifted so that the null hypothesis is true, with the goal of empirically determining an appropriate critical value. Then you generate bootstrap samples from each of these *JK* groups and compute the test statistics as described in Boxes 10.2 and 10.3. For the main effect associated with factor A, we label these *B* bootstrap test statistics as $V^*_{a1}, \ldots, V^*_{aB}$; these *B* values provide an approximation of the distribution of V_a when the null hypothesis is true. Put these values in ascending order and label the results $V^*_{a(1)} \leq \cdots \leq V^*_{a(B)}$. If $V_a \geq V^*_{a(c)}$, where $c = (1 - \alpha)B$, rounded to the nearest integer, reject. The hypotheses of no main effect for factor B and no interaction are tested in an analogous manner.

Other robust measures of location can be compared with the percentile bootstrap method. Again there are many variations that might be used (which include important techniques covered in Chapter 12). Here, only one of these methods is described.

Let θ be any measure of location and let

$$\Upsilon_1 = \frac{1}{K}(\theta_{11} + \theta_{12} + \cdots + \theta_{1K}),$$

$$\Upsilon_2 = \frac{1}{K}(\theta_{21} + \theta_{22} + \cdots + \theta_{2K}),$$

$$\vdots$$

$$\Upsilon_J = \frac{1}{K}(\theta_{J1} + \theta_{J2} + \cdots + \theta_{JK}).$$

So Υ_j is the average of the K measures of location associated with the jth level of factor A. The hypothesis of no main effects for factor A is

$$H_0 : \Upsilon_1 = \Upsilon_2 = \cdots = \Upsilon_J,$$

and one variation of the percentile bootstrap method is to test this hypothesis using a slight modification of the method in Box 9.7. For example, one possibility is to test

$$H_0 : \Delta_1 = \cdots = \Delta_{J-1} = 0, \tag{10.1}$$

where

$$\Delta_j = \Upsilon_j - \Upsilon_{j+1},$$

$j = 1, \ldots, J - 1$. Briefly, generate bootstrap samples in the usual manner, yielding $\hat{\Delta}^*_j$, a bootstrap estimate of Δ_j. Then proceed as described in Box 9.7. That is, determine how deeply $\mathbf{0} = (0, \ldots, 0)$ is nested within the bootstrap samples. If $\mathbf{0}$ is relatively far from the center of the bootstrap samples, reject. (Chapter 12 describes another approach, where all pairwise comparisons of the rows are done instead. That is, for every j and ℓ, $j < \ell$, test $H_0 : \Upsilon_j = \Upsilon_\ell$.)

For reasons described in Section 9.5.3, the method just described is satisfactory when dealing with the probability of a Type I error; but when the groups differ, this approach might be unsatisfactory in terms of power, depending on the pattern of differences among the Υ_j values. One way of dealing with this issue is to compare all pairs of the Υ_j instead. That is, for every $j < j'$, let

$$\Delta_{jj'} = \Upsilon_j - \Upsilon_{j'},$$

and then test

$$H_0 : \Delta_{12} = \Delta_{13} = \cdots = \Delta_{J-1,J} = 0. \tag{10.2}$$

Of course, a similar method can be used when dealing with factor B.

The percentile bootstrap method just described for main effects can be extended to the problem of testing the hypothesis of no interactions. Box 10.4 outlines how to proceed.

BOX 10.4 How to Test for No Interactions Using the Percentile Bootstrap Method

For convenience, label the JK measures of location as follows:

$$
\begin{array}{lcccc}
 & & \text{Factor B} & & \\
 & \theta_1 & \theta_2 & \cdots & \theta_K \\
 & \theta_{K+1} & \theta_{K+2} & \cdots & \theta_{2K} \\
\text{Factor A} & \vdots & \vdots & \cdots & \vdots \\
 & \theta_{(J-1)K+1} & \theta_{(J-1)K+2} & \cdots & \theta_{JK}
\end{array}
$$

If **A** is any r-by-s matrix and **B** is any t-by-u matrix, the Kronecker product of **A** and **B**, written as $\mathbf{A} \otimes \mathbf{B}$, is

$$
\begin{pmatrix}
a_{11}B & a_{12}B & \ldots & a_{1s}B \\
 & & \vdots & \\
a_{r1}B & a_{r2}B & \ldots & a_{rs}B
\end{pmatrix}.
$$

Let $\mathbf{C}_J$ be a $(J-1)$-by-J matrix having the form

$$
\begin{pmatrix}
1 & -1 & 0 & 0 & \ldots & 0 \\
0 & 1 & -1 & 0 & \ldots & 0 \\
 & & & \vdots & & \\
0 & 0 & \ldots & 0 & 1 & -1
\end{pmatrix}.
$$

That is, $c_{ii} = 1$ and $c_{i,i+1} = -1$; $i = 1, \ldots, J-1$, and $\mathbf{C}_K$ is defined in a similar fashion. A test of no interactions corresponds to testing

$$H_0: \Psi_1 = \cdots = \Psi_{(J-1)(K-1)} = 0,$$

where

$$\Psi_L = \sum c_{L\ell}\theta_\ell,$$

$L = 1, \ldots, (J-1)(K-1)$, $\ell = 1, \ldots, J(K-1)$ and $c_{L\ell}$ be the entry in the Lth row and ℓth column of $\mathbf{C}_J \otimes \mathbf{C}_K$. So in effect we have a situation similar to that in Box 9.7. That is, generate bootstrap samples yielding $\hat{\Psi}^*_L$ values, do this B times, and then determine how deeply $\mathbf{0} = (0, \ldots, 0)$ is nested within these bootstrap samples.

Continued

BOX 10.4 (*Continued*) A criticism of this approach is that when groups differ, not all possible tetrad differences are being tested, which might affect power. One way of dealing with this problem is, for every $j < j'$ and $k < k'$, set

$$\Psi_{jj'kk'} = \theta_{jk} - \theta_{jk'} + \theta_{j'k} - \theta_{j'k'},$$

and then test

$$H_0 : \Psi_{1212} = \cdots = \Psi_{J-1,J,K-1,K} = 0. \qquad (10.3)$$

10.4.1 S-PLUS Function pbad2way

The S-PLUS function

pbad2way(J, K, x, est = mom, conall= T, alpha = 0.05, nboot = 2000,

grp = NA, op = F,...)

performs the percentile bootstrap method just described, where J and K indicate the number of levels associated with factors A and B. The argument conall=T indicates that all possible pairs are to be tested [as described by Equation (10.2)], and conall=F means that the hypotheses given by Equation (10.1) will be used instead. The remaining arguments are the same as those used in the S-PLUS function pbadepth described in Section 9.5.4.

EXAMPLE. The data in Table 10.1 are used to illustrate the S-PLUS function pbad2way. Storing the data in the S-PLUS variable weight, the command

pbad2way(2,2,weight,est=median)

tests all relevant hypotheses using medians. It is left as an exercise to verify that the significance levels for factors A and B are .39 and .056, respectively. The test for no interaction has a significance level of .16. ■

10.5 Testing Hypotheses Based on Medians

As was the case in Chapter 9, the heteroscedastic method for trimmed means in Boxes 10.2 and 10.3 should be modified for the special case where the goal is to compare medians. Currently, the methods described in Boxes 10.5 and 10.6 appear to perform relatively well and are based on simple modifications of the method for trimmed means plus the McKean–Schrader estimate of the standard error of the median (which was described in Section 4.11.2). Medians can also be compared using the bootstrap method described in the previous section, but the advantages of using this bootstrap method, versus the methods in Boxes 10.5 and 10.6, have not been investigated.

BOX 10.5 A Heteroscedastic Test for Main Effects Based on Medians

Let M_{jk} be the sample median for the jth level of factor A and the kth level of B, and let n_{jk} and S_{jk}^2 be the corresponding sample size and estimate of the squared standard error of M_{jk}. Here S_{jk}^2 is the McKean–Schrader estimate. Compute

$$R_j = \sum_{k=1}^{K} M_{jk}, \qquad W_k = \sum_{j=1}^{J} M_{jk},$$

$$d_{jk} = S_{jk}^2,$$

$$\hat{\nu}_j = \frac{(\sum_k d_{jk})^2}{\sum_k d_{jk}^2/(n_{jk}-1)}, \qquad \hat{\omega}_k = \frac{(\sum_j d_{jk})^2}{\sum_j d_{jk}^2/(n_{jk}-1)}$$

$$r_j = \frac{1}{\sum_k d_{jk}}, \qquad w_k = \frac{1}{\sum_j d_{jk}}$$

$$r_s = \sum_{j=1}^{J} r_j, \qquad w_s = \sum_{k=1}^{K} w_k,$$

$$\hat{R} = \frac{\sum_j r_j R_j}{r_s}, \qquad \hat{W} = \frac{\sum_k w_k W_k}{w_s}$$

$$B_a = \sum_{j=1}^{J} \frac{1}{\hat{\nu}_j}\left(1 - \frac{r_j}{\sum r_j}\right)^2, \qquad B_b = \sum_{k=1}^{K} \frac{1}{\hat{\omega}_k}\left(1 - \frac{w_k}{\sum w_k}\right)^2$$

$$V_a = \frac{\sum_j r_j (R_j - \hat{R})^2}{(J-1)\left(1 + \frac{2(J-2)B_a}{J^2-1}\right)}, \qquad V_b = \frac{\sum_k w_k (W_k - \hat{W})^2}{(K-1)\left(1 + \frac{2(K-2)B_b}{K^2-1}\right)}.$$

The degrees of freedom for factor A are $\nu_1 = J - 1$ and $\nu_2 = \infty$. For factor B the degrees of freedom are $\nu_1 = K - 1$ and $\nu_2 = \infty$.

Decision Rule

Reject the hypothesis of no main effect for factor A if $V_a \geq f_{1-\alpha}$, the $1 - \alpha$ quantile of an F-distribution with the degrees of freedom for factor A. Similarly, reject for factor B if $V_b \geq f_{1-\alpha}$, with the degrees of freedom for factor B.

BOX 10.6 A Heteroscedastic Test of the Hypothesis of No Interactions Based On Medians

Again let $d_{jk} = S_{jk}^2$ be the McKean–Schrader estimate of the squared standard error of M_{jk}. Let

$$D_{jk} = \frac{1}{d_{jk}}$$

$$D_{.k} = \sum_{j=1}^{J} D_{jk}, \quad D_{j.} = \sum_{k=1}^{K} D_{jk}$$

$$D_{..} = \sum_{j=1}^{J}\sum_{k=1}^{K} D_{jk}$$

$$\tilde{M}_{jk} = \sum_{\ell=1}^{J} \frac{D_{\ell k} M_{\ell k}}{D_{.k}} + \sum_{m=1}^{K} \frac{D_{jm} M_{jm}}{D_{j.}} - \sum_{\ell=1}^{J}\sum_{m=1}^{K} \frac{D_{\ell m} M_{\ell m}}{D_{..}}.$$

The test statistic is

$$V_{ab} = \sum_{j=1}^{J}\sum_{k=1}^{K} D_{jk}(\bar{X}_{jk} - \tilde{M}_{jk})^2.$$

Let c be the $1-\alpha$ quantile of a chi-squared distribution with $\nu = (J-1)(K-1)$ degrees of freedom.

Decision Rule
Reject if $V_{ab} \geq c$.

10.5.1 S-PLUS Function med2way

The computations for comparing medians, described in Boxes 10.5 and 10.6, are performed by the S-PLUS function

med2way(J,K,x,alpha=.05)

EXAMPLE. The example in Section 10.3.1 is repeated, only now medians are compared instead. The output from med2way is as follows:

```
$test.A:
[1] 8.124937
```

Continued

EXAMPLE. *(Continued)*

```
$sig.A:
[1] 0.004366059

$test.B:
[1] 1.805773

$sig.B:
[1] 0.1643474

$test.AB:
[1] 2.417318

$sig.AB:
[1] 0.2985974
```

So again there is a main effect for factor A and no interaction is found. But unlike before, the main effect for factor B is not significant at the .05 level. ■

10.6 Exercises

1. State the hypotheses of no main effects and no interactions for a 2-by-4 design with the following population means.

	Factor B			
Factor A	Level 1	Level 2	Level 3	Level 4
Level 1	μ_1	μ_2	μ_3	μ_4
Level 2	μ_5	μ_6	μ_7	μ_8

2. For the following 2-by-2 design with population means, state whether there is a main effect for factor A or for factor B and whether there is an interaction.

	Factor B	
Factor A	Level 1	Level 2
Level 1	$\mu_1 = 110$	$\mu_2 = 70$
Level 2	$\mu_3 = 80$	$\mu_4 = 40$

3. For the following 2-by-2 design with population means, determine whether there is an interaction. If there is an interaction, determine whether there is an ordinal interaction for rows.

	Factor B	
Factor A	Level 1	Level 2
Level 1	$\mu_1 = 10$	$\mu_2 = 20$
Level 2	$\mu_3 = 40$	$\mu_4 = 10$

4. Make up an example where the population means in a 3-by-3 design have no interaction effect but main effects for both factors exist.
5. For the following ANOVA summary table, fill in the missing values and then determine the number of levels, the total number of observations used, the estimate of the common variance, and whether the hypotheses of no main effects or no interaction should be rejected with $\alpha = .025$.

Source	SS	DF	MS	F
A	800	2	____	____
B	600	3	____	____
INTER	1200	____	____	____
WITHIN	4800	36	____	

6. For the following ANOVA summary table, fill in the missing values and then determine the number of levels, the total number of observations used, the estimate of the common variance, and whether the hypotheses of no main effects or no interaction should be rejected with $\alpha = .025$.

Source	SS	DF	MS	F
A	667	1	____	____
B	212.4	5	____	____
INTER	884	____	____	____
WITHIN	3900	48	____	

7. Imagine a study where two methods are compared for treating depression. A measure of effectiveness has been developed; the higher the measure, the more effective the treatment. Further assume there is reason to believe that males might respond differently to the methods than females. If the sample means are

	Factor B	
Factor A	Males	Females
Method 1	$\bar{X}_1 = 50$	$\bar{X}_2 = 70$
Method 2	$\bar{X}_3 = 80$	$\bar{X}_4 = 60$

and if the hypothesis of no main effects for factor A is rejected but the hypothesis of no interactions is not rejected, what does this suggest about which method should be used?

8. In the previous exercise, suppose the hypothesis of no interactions is rejected and in fact there is a disordinal interaction. What does this suggest about when you might use method 1 versus method 2?
9. Use your answer to Exercise 8 to make a general comment on interpreting main effects when there is a disordinal interaction.
10. Referring to Exercise 7, imagine the hypothesis of no interactions is rejected. Is it reasonable to conclude that the interaction is disordinal?
11. This exercise is based on a study, where the general goal is to study people's reactions to unprovoked verbal abuse. In the study, 40 subjects were asked to sit alone in a cubicle and answer a brief questionnaire. After the subjects had waited far longer than it took to fill out the form, a research assistant returned to collect the responses. Half the subjects received an apology for the delay and the other half were told, among other things, that they could not even fill out the form properly. Each of these 20 subjects were divided into two groups: Half got to retaliate against the research assistant by giving her a bad grade, and the other half did not get a chance to retaliate. All subjects were given a standardized test of hostility. Imagine that the sample means are

	Abuse	
Retaliation	Insult	Apology
Yes	$\bar{X}_1 = 65$	$\bar{X}_2 = 54$
No	$\bar{X}_3 = 61$	$\bar{X}_4 = 57$

Further assume that the hypotheses of no main effects are rejected and that the hypothesis of no interactions was also rejected. Interpret this result.

12. Verify that when comparing the groups in Table 10.1 based on medians and with the S-PLUS function pbad2way in Section 10.4.1, the significance levels for factors A and B are .39 and .056, respectively, and that for the test of no interaction the significance level is .16.
13. Read the data in Table 10.8 into an S-PLUS variable having matrix mode, and use the function selby2 in Section 10.3.2 to store the data in list mode. Store the output from selby2 in the S-PLUS variable dat, and then use the function t2way in Section 10.3.1 to test the hypotheses of no main effects or interactions based on means. Verify that the significance levels for factors A and B are .835 and .951, respectively.
14. For the data in the previous exercise, verify that if t2way is used to compare 20% trimmed means, the significance level when testing the hypothesis of no interaction is reported to be Inf, meaning that it is infinitely large due to division by zero. Explain why this happens based on the computations outlined in Box 10.3.
15. For the data in Exercise 13, verify that if t2way is used to compare 20% trimmed means instead, the significance levels for factors A and B are .15 and

TABLE 10.8 Data for Exercise 13

A	B	Outcome
1	1	32
1	1	21
1	3	19
1	2	21
1	3	46
1	3	33
1	3	10
1	2	11
1	2	13
1	2	12
1	2	59
1	3	28
1	1	19
1	1	72
1	1	35
2	1	33
2	1	45
2	1	31
2	2	42
2	2	67
2	2	51
2	1	19
2	2	18
2	2	21
2	3	39
2	3	63
2	3	41
2	3	10
2	3	11
2	3	34
2	3	47
2	1	21
2	1	29
2	3	26

.93, respectively. Why does the significance level for factor A drop from .835 when comparing means to .15 with 20% trimmed means instead?

16. The data used to illustrate the S-PLUS function med2way can be downloaded via anonymous ftp, as described in Section 1.2; it is stored in the file hayes.dat. Use these data to compare medians with the S-PLUS function pbad2way in Section 10.4.1.

17. The data used in the example at the end of Section 10.3.1 are stored in the file hayes.dat. (See Section 1.2 on how to download this file.) Compare the groups with med2way. Verify that you reject at the .05 level for factor A but not for factor B or when testing the hypothesis of no interaction. Note that in contrast, when comparing 20% trimmed means, you reject when dealing with factor B.

COMPARING DEPENDENT GROUPS

Chapters 8–10 described methods for comparing independent groups. This chapter describes methods for dealing with dependent groups or variables. For example, imagine you sample 10 married couples and want to compare husbands and wives on some measure of open-mindedness. That is, the goal is to characterize how, for a typical couple, the wife compares to her husband. If open-mindedness scores are independent among the population of married couples, we can compare husbands to their wives using the methods in Chapter 8. But there is no particular reason to assume that scores among women are independent of their spouse's score, so special methods are needed to take this into account. As another example, imagine that the endurance of athletes is measured before a particular training program is begun and again four weeks after training under some experimental method. This is an example of what is called a *repeated measures* or a *within-subjects* design, simply meaning that we repeatedly measure the same individuals over time. Of interest is whether endurance levels have changed. But because there is no particular reason to assume that endurance levels before undergoing the new training method are independent of the scores after training, the methods in Chapter 8 may not be valid. As a final example, C. R. Rao (1948) discussed a study where there was interest in the weight of cork borings from trees. Of specific interest was the difference in weight for the north, east, south, and west sides of the trees. Here we focus on the north versus east sides. Table 11.1 reports the data for 28 trees. Do the typical weights differ in some sense, and, if so, how do they differ and by how much? Because samples taken from the same tree may be dependent, again special methods are required that take this dependence into account.

When attention is restricted to comparing means, it is stressed that a plethora of methods have been proposed that are not covered in this chapter. For book-length descriptions of these techniques, which include procedures especially designed for handling longitudinal data, see Crowder and Hand (1990), Jones (1993), and Diggle, Liang, and Zeger (1994). This chapter covers the more basic methods for means that are typically used in applied research; methods that address the practical problems associated with these techniques are then described.

TABLE 11.1 Cork Boring Data

i	X_{i1} (North)	X_{i2} (East)	$D_i = X_{i1} - X_{i2}$
1	72	66	6
2	60	53	7
3	56	57	−1
4	41	29	12
5	32	32	0
6	30	35	−5
7	39	39	0
8	42	43	−1
9	37	40	−3
10	33	29	4
11	32	30	2
12	63	45	18
13	54	46	8
14	47	51	−4
15	91	79	12
16	56	68	−12
17	79	65	14
18	81	80	1
19	78	55	23
20	46	38	8
21	39	35	4
22	32	30	2
23	60	50	10
24	35	37	−2
25	39	36	3
26	50	34	16
27	43	37	6
28	48	54	−6

Before continuing, a comment about notation should be made. Notice that the second column in Table 11.1 is headed by the symbol X_{i1}. The notation X_{11} refers to the weight of the cork boring for the north side of the first tree. Similarly, X_{12} is the weight for the east side for the first tree. More generally, X_{i1} is the measurement for the north side of the ith tree, and X_{i2} is the east side of the ith tree. The sample mean for the first group (the north side) is $\bar{X}_1$, and for the second group, or east side, the average is denoted by $\bar{X}_2$. More succinctly, the mean of the jth group is

$$\bar{X}_j = \frac{1}{n}\sum_{i=1}^{n} X_{ij}.$$

The distribution associated with the jth group, ignoring all other groups, called the jth *marginal distribution*, has a population mean labeled μ_j. As was the case when working with independent groups, $\bar{X}_j$ is an unbiased estimate of μ_j under random sampling. That is, $E(\bar{X}_j) = \mu_j$.

11.1 The Paired *T*-Test for Means

Now we describe the most common method for comparing the means of two dependent groups. To add perspective, it is noted that the method in Chapter 8 for comparing independent groups is based on the result that the squared standard error of the difference between the sample means is just the sum of the individual squared standard errors. In symbols,

$$\text{VAR}\left(\bar{X}_1 - \bar{X}_2\right) = \frac{\sigma_1^2}{n_1} + \frac{\sigma_2^2}{n_2},$$

the point being that the difference between the sample means can be standardized, and this leads to a convenient method for comparing the means using the Laplace–Gosset strategy covered in Chapter 4. However, when the groups are dependent, $\text{VAR}(\bar{X}_1 - \bar{X}_2)$ takes on a more complicated form that is inconvenient from a technical point of view. The usual method for dealing with this technical problem is to use the differences between the pairs of observations to make inferences about the means.

Let D_i be the difference between the ith pair of observations. That is,

$$D_i = X_{i1} - X_{i2},$$

$i = 1, \ldots, n$. For example, for the first tree ($i = 1$), the weights are $X_{11} = 72$ and $X_{12} = 66$ and the difference is $D_1 = X_{11} - X_{12} = 72 - 66 = 6$. For the second tree ($i = 2$), $D_2 = X_{21} - X_{22} = 60 - 53 = 7$. It can be shown that the population mean associated with the difference scores is just the difference between the population means of the two groups or variables under study. In the illustration, if μ_1 represents the population mean for the north side of a tree and μ_2 the mean for the east side, then $\mu_1 - \mu_2$ is equal to μ_D, the population mean associated with the difference scores $D_1, \ldots, D_n$. More succinctly,

$$\mu_D = E(D) = \mu_1 - \mu_2.$$

This means that if we randomly sampled millions of trees and computed the sample mean of the difference scores, we would get the same result if instead we averaged the results for the east side and then subtracted this from the average for the north side of the trees. For the special case where the population means μ_1 and μ_2 are equal, $\mu_D = 0$. That is, testing the hypothesis $H_0: \mu_1 - \mu_2 = 0$ is the same as testing the hypothesis $H_0: \mu_D = 0$. Assuming normality, $H_0: \mu_D = 0$ can be tested using Student's T-test described in Chapter 5. The calculations are summarized in Box 11.1 and this approach to comparing the means of two dependent groups is called the *paired T-test*.

EXAMPLE. For the data in Table 11.1, the sample mean and standard deviation of the D_i values are $\bar{D} = 4.36$ and $s_D = 7.93$, respectively. There are $n = 28$ pairs of observations (or D_i values), so, referring to Box 11.1, the test statistic is

$$T_D = \frac{4.36}{7.93/\sqrt{28}} = 2.9.$$

With $\nu = 28 - 1 = 27$ degrees of freedom and $\alpha = .01$, the critical value is the $1 - \alpha/2 = 1 - .01/2 = .995$ quantile of Student's T distribution, which is $c = 2.77$. Because $|T| = 2.9 \geq 2.77$, reject the hypothesis of equal means and conclude that the average weight for the north side of a tree is greater than the average weight of the east side. The .99 confidence interval for the difference between the means is

$$4.36 \pm 2.77\frac{7.93}{\sqrt{28}} = (.21, 8.5).$$

This says that, assuming normality, we can be reasonably certain that the difference between the means is at least .21 but not larger than 8.5. ■

BOX 11.1 How to Perform the Paired T-Test

Goal:
For two dependent groups, test $H_0: \mu_1 = \mu_2$, the hypothesis that they have equal means.

We observe n pairs observations: $(X_{11}, X_{12}), \ldots, (X_{n1}, X_{n2})$.

Assumptions:
Random sampling from normal distributions

Computations:
Begin by forming the differences between the paired observations:

$$D_1 = X_{11} - X_{12}$$
$$D_2 = X_{21} - X_{22}$$
$$D_3 = X_{31} - X_{32}$$
$$\vdots$$
$$D_n = X_{n1} - X_{n2}$$

Testing the hypothesis of equal means is accomplished by testing the hypothesis that the D_i values have a population mean of zero. That is, test $H_0: \mu_D = 0$. To do this, compute the mean and variance of the

Continued

BOX 11.1 *(Continued)* D_i values:

$$\bar{D} = \frac{1}{n}\sum_{i=1}^{n} D_i$$

and

$$s_D^2 = \frac{1}{n-1}\sum_{i=1}^{n}(D_i - \bar{D})^2.$$

Next, compute

$$T_D = \frac{\bar{D}}{s_D/\sqrt{n}}.$$

The critical value is c, the $1 - \alpha/2$ quantile of Student's T distribution with $\nu = n - 1$ degrees of freedom.

Decision Rule:
The hypothesis of equal means is rejected if $|T_D| \geq c$.

Confidence Interval:
A $1 - \alpha$ confidence interval for $\mu_1 - \mu_2$, the difference between the means, is

$$\bar{D} \pm c\frac{s_D}{\sqrt{n}}.$$

As indicated in Box 11.1, the paired T-test has $n - 1$ degrees of freedom. If the groups are independent, Student's T-test in Section 8.1 has $2(n - 1)$ degrees of freedom, twice as many as the paired T-test. This means that if we compare independent groups (having equal sample sizes) using the method in Box 11.1, power will be lower than when using Student's T in Section 8.1. However, if the correlation between the observations is sufficiently high, the paired T-test will have more power instead. To provide an indication of why, it is noted that when pairs of observations are dependent, then

$$\mathrm{VAR}\left(\bar{X}_1 - \bar{X}_2\right) = \frac{\sigma_1^2 + \sigma_2^2 - 2\rho\sigma_1\sigma_1}{n},$$

where σ_1^2 and σ_2^2 are the variances associated with groups 1 and 2, respectively. From this last equation we see that as the correlation, ρ, increases, the variance of the difference between the sample means (the squared standard error of $\bar{X}_1 - \bar{X}_2$) goes down. As noted in Chapters 5 and 8, as the standard error goes down, power goes up. In practical terms, the loss of degrees of freedom when using the paired T-test will be more than compensated for if the correlation is reasonably high. Just how high it must be in order to get more power with the paired T-test is unclear. For normal distributions, a rough guideline is that when $\rho > .25$, the paired T-test will have more power (Vonesh, 1983).

TABLE 11.2 Hypothetical Cholesterol Levels Before and After Training

X:	250,	320,	180,	240,	210,	255,	175,	280,	250,	200
Y:	230,	340,	185,	200,	190,	225,	185,	285,	210,	190

11.1.1 Assessing Power

When you fail to reject the hypothesis of equal means with the paired T-test, this might be because there is little or no difference between the groups, or perhaps there is a difference but you failed to detect it due to low power. Stein's method, described in Section 5.4.3, can be used to help distinguish between these two possibilities. You simply perform the computations in Section 5.4.3 on the D_i values. In particular, you can use the S-PLUS functions stein1 and stein2 described in Section 5.4.4.

EXAMPLE. Table 11.2 shows some hypothetical data on cholesterol levels before individuals undergo an experimental exercise program (X) and the levels after four weeks of training (Y). Applying the paired T-test, we fail to reject the hypothesis of equal means using $\alpha = .05$. But suppose we want power to be at least .8 when the difference between the means is 10. If we store the X values in the S-PLUS variable x and the Y values in the S-PLUS variable y, then the S-PLUS command stein1(x-y,10) returns the value 46. That is, a total of 46 pairs of observations are required to achieve power equal to .8 when the difference between the means is 10. Only 10 observations were used, so an additional $46 - 10 = 36$ observations are required. This analysis suggests that we should not accept the null hypothesis of equal means, because the sample size is too small to achieve a reasonable amount of power. ■

11.2 Comparing Trimmed Means

The good news about the paired T-test described in Section 11.1 is that if the observations in the first group (the X_{i1} values) have the same distribution as the observations in the second group (the X_{i2} values), so in particular they have equal means and variances, then Type I error probabilities substantially higher than the nominal level can generally be avoided. The reason is that for this special case, the difference scores (the D_i values) have a symmetric distribution, in which case methods based on means perform reasonably well in terms of avoiding Type I error probabilities substantially higher than the nominal level. However, as was the case in Chapter 8, practical problems arise when the two groups (or the two dependent variables) differ in some manner. Again, arbitrarily small departures from normality can destroy power, even when comparing groups having symmetric distributions. And if groups differ in terms of skewness, the paired T-test can be severely biased, meaning that power can actually decrease as the difference between the means gets large. Yet another problem is poor probability coverage when computing a confidence interval for the difference between the means as described in Box 11.1. If the goal is to test the hypothesis that the two variables under study have identical distributions, the paired T-test is satisfactory in terms of Type I errors. But if we reject, there is doubt as to whether this is due primarily to the difference between the means or to some other way the distributions differ.

A general strategy for addressing problems with a paired T-test is to switch to some robust measure of location, such as a trimmed mean. One possibility is simply to apply the methods for trimmed means described in Section 5.6 or the bootstrap methods in Section 7.5 to the difference scores (the D_i values). This is a reasonable approach; in some situations it offers an advantage over other strategies that might be used. But there is a technical issue that should be made clear. First it is noted that when working with means,

$$\bar{D} = \bar{X}_1 - \bar{X}_2.$$

That is, the average of the difference scores is just the difference between the averages associated with the two groups. Moreover, $\mu_D = \mu_1 - \mu_2$, which makes it possible to test $H_0: \mu_1 = \mu_2$ simply by testing $H_0: \mu_D = 0$. However, when using trimmed means, typically, but not always, the trimmed mean of the difference scores is not equal to the difference between the trimmed means of the marginal distributions. That is, usually, $\bar{D}_t \neq \bar{X}_{t1} - \bar{X}_{t2}$. An exception occurs when the pairs of observations are identical.

EXAMPLE. Consider the following pairs of observations:

G1:	1,	2,	3,	4,	5,	6,	7,	8,	9,	10
G2:	1,	2,	3,	4,	5,	6,	7,	8,	9,	10

Here both groups have a 20% trimmed mean of 5.5, so of course $\bar{X}_{t1} - \bar{X}_{t2} = 0$. As is evident, the difference scores (the D_i values) are all equal to zero, so the 20% trimmed mean of the difference scores is zero as well. That is, $\bar{D}_t = \bar{X}_{t1} - \bar{X}_{t2}$. However, if for the second group we rearrange the values so that now

G2: 1, 8, 2, 3, 4, 5, 6, 7, 9, 10

and the observations for G1 remain the same, then the difference scores are

0, −6, 1, 1, 1, 1, 1, 10, 0,

which have a 20% trimmed mean of 0.67. That is, $\bar{D}_t \neq \bar{X}_{t1} - \bar{X}_{t2}$. ■

More generally, if two dependent groups have identical distributions, then the population trimmed mean of the difference scores is equal to the difference between the individual trimmed means, which is zero. In symbols, $\mu_{tD} = \mu_{t1} - \mu_{t2} = 0$.

However, if the distributions differ, this equality is not necessarily true and in general it will be the case that $\mu_{tD} \neq \mu_{t1} - \mu_{t2}$. In practical terms, computing a confidence interval for μ_{tD} is not necessarily the same as computing a confidence interval for $\mu_{t1} - \mu_{t2}$. So an issue is whether one should test

$$H_0 : \mu_{tD} = 0, \tag{11.1}$$

or

$$H_0 : \mu_{t1} = \mu_{t2}. \tag{11.2}$$

The latter case will be called comparing the *marginal* trimmed means. In terms of Type I errors, currently it seems that there is no reason to prefer one approach over the other for the situation at hand. But for more than two groups, differences between these two approaches will be described later in this chapter. As for power, the optimal choice depends on how the groups differ, which of course is unknown. In some situations it will make little difference which method is used. But for the general problem of comparing groups based on some measure of location, we will see situations where the choice between these two approaches alters our conclusions and can provide different perspectives on how groups differ.

For the situation at hand, inferences about the trimmed mean of difference scores can be made with the method in Section 4.9.3, and the S-PLUS function trimci in Section 4.9.4 can be used to perform the calculations. Another option is to use the bootstrap method in Section 7.5. The remainder of this section describes how to test the hypothesis of equal trimmed means, which corresponds to Equation (11.2). In our illustration, the goal is to test the hypothesis that the north and east sides of a tree have equal population trimmed means.

To test the hypothesis of equal trimmed means using Laplace's general strategy, modified along the lines used by Gosset, we first Winsorize the observations. Here, however, it is important to keep observations paired together when Winsorizing. In symbols, fix j and let $X_{(1)j} \leq X_{(2)j} \leq \cdots \leq X_{(n)j}$ be the n values in the jth group, written in ascending order. Winsorizing the observations means computing

$$Y_{ij} = \begin{cases} X_{(g+1)j} & \text{if } X_{ij} \leq X_{(g+1)j} \\ X_{ij} & \text{if } X_{(g+1)j} < X_{ij} < X_{(n-g)j} \\ X_{(n-g)j} & \text{if } X_{ij} \geq X_{(n-g)j}, \end{cases}$$

where again g is the number of observations trimmed or Winsorized from each end of the distribution corresponding to the jth group. As usual, 20% trimming is assumed unless stated otherwise, in which case $g = [.2n]$, where $[.2n]$ means to round $.2n$ down to the nearest integer. The expression for Y_{ij} says that $Y_{ij} = X_{ij}$ if X_{ij} has a value between $X_{(g+1)j}$ and $X_{(n-g)j}$. If X_{ij} is less than or equal to $X_{(g+1)j}$, set $Y_{ij} = X_{(g+1)j}$; if X_{ij} is greater than or equal to $X_{(n-g)j}$, set $Y_{ij} = X_{(n-g)j}$.

EXAMPLE. As a simple illustration, consider these eight pairs of observations:

X_{i1}:	18	6	2	12	14	12	8	9
X_{i2}:	11	15	9	12	9	6	7	10

With 20% Winsorization, $g = 1$, and we see that $X_{(1)1} = 2$ and $X_{(2)1} = 6$. That is, the smallest value in the first row of data is 2, and Winsorizing changes its value to 6. If it had been the case that $g = 2$, then the values 6 and 2 in the first row (the two smallest values) would be increased to 8, the third smallest value in the first row. In this latter case, the first row of data would look like this:

$$Y_{i1}: \quad 18\ 8\ 8\ 12\ 14\ 12\ 8\ 9.$$

Notice that the order of the values remains the same—we simply Winsorize the g smallest values. If $g = 3$, then we would have

$$Y_{i1}: \quad 18\ 9\ 9\ 12\ 14\ 12\ 9\ 9.$$

Returning to $g = 1$, Winsorizing means that the largest observation is pulled down to the next largest. In our example, the 18 in the first row would become 14. Winsorizing the second row of data using this same process, 6 becomes 7 and 15 becomes 12. So after both rows are Winsorized, this yields

Y_{i1}:	14	6	6	12	14	12	8	9
Y_{i2}:	11	12	9	12	9	7	7	10

Yuen's method in Chapter 8 can be extended to the problem at hand in the following manner. Let $h = n - 2g$ be the effective sample size (the number of observations in each group after trimming). Let

$$d_j = \frac{1}{h(h-1)} \sum_{i=1}^{n} (Y_{ij} - \bar{Y}_j)^2$$

and

$$d_{12} = \frac{1}{h(h-1)} \sum_{i=1}^{n} (Y_{i1} - \bar{Y}_1)(Y_{i2} - \bar{Y}_2).$$

(The term d_{12} plays a role in adjusting for the dependence between the groups being compared.) Then the hypothesis of equal trimmed means can be tested with

$$T_y = \frac{\bar{X}_{t1} - \bar{X}_{t2}}{\sqrt{d_1 + d_2 - 2d_{12}}}, \tag{11.3}$$

which is rejected if $|T_y| \geq t$, where t is the $1 - \alpha/2$ quantile of Student's T distribution with $h - 1$ degrees of freedom. A $1 - \alpha$ confidence interval for the difference between the trimmed means ($\mu_{t1} - \mu_{t2}$) is

$$(\bar{X}_{t1} - \bar{X}_{t2}) \pm t\sqrt{d_1 + d_2 - 2d_{12}}.$$

■

11.2.1 S-PLUS Function yuend

The S-PLUS function

yuend(x,y,tr=.2,alpha=.05)

has been supplied to perform the calculations just described, where x and y are any S-PLUS variables containing the data for groups 1 and 2. As usual, the argument tr indicates the amount of trimming and defaults to .2 (20%), and alpha is α, which defaults to .05.

EXAMPLE. If the data in Table 11.1 for the north side of a tree are stored in the S-PLUS variable cork1 and the data for the east side are stored in cork2, then the S-PLUS command

yuend(cork1,cork2)

returns

```
$ci:
[1] 0.3658136 7.4119642

$siglevel:
[1] 0.03245657

$dif:
[1] 3.888889

$se:
[1] 1.66985

$teststat:
[1] 2.328885

$df:
[1] 17
```

So the .95 confidence interval for the difference between the 20% trimmed means is (0.37, 7.41); this interval does not contain zero, so you reject the hypothesis of equal trimmed means. As indicated, the significance level is .03. The estimated standard error of the difference between the sample trimmed means is 1.67. To test Equation (11.1), the hypothesis that the difference scores have a trimmed mean of zero, simply use the command

trimci(cork1-cork2),

where trimci is the S-PLUS function described in Section 4.9.4. It returns a .95 confidence interval of (0.254, 7.41), which is fairly similar to the confidence interval given by yuend. ■

11.2.2 Comparing Medians

For the special case where the goal is to test the hypothesis that difference scores have a median of zero, simply use the method in Chapter 4. In particular, the S-PLUS function sint can be used to compute a confidence interval for the median of the difference scores. (If the data are stored in x and y, use the command sint(x-y).) As for testing $H_0: \theta_1 = \theta_2$, the hypothesis that the population medians are equal, it currently seems that the best approach is to use a bootstrap method described in the next section of this chapter.

11.3 Bootstrap Methods

The bootstrap methods covered in previous chapters are readily extended to the problem of comparing dependent groups. However, based on what is currently known, the recommended type of bootstrap method differs in some cases. That is, the choice between percentile versus bootstrap-*t* is not always the same as described in Chapters 9 and 10, and there is the additional issue of whether to use difference scores. Currently, when comparing trimmed means, it seems that with 20% trimming or less, the bootstrap-*t* is preferable to the percentile bootstrap, particularly when comparing more than two dependent groups. (In contrast, when comparing independent groups as described in Section 9.5.3, a percentile bootstrap with 20% trimmed means appears to perform quite well in terms of controlling the probability of a Type I error.) But for 25% trimming or more, a percentile bootstrap seems better. (This is a very rough comparison and more detailed studies might alter this conclusion.) More generally, when using any location estimator with a breakdown point of at least .25 (such as MOM or the median), a percentile bootstrap method appears to perform relatively well. For the special case where the goal is to compare medians, inferences based on difference scores can be made with the method in Section 4.11. The extent to which a bootstrap method offers an advantage for this special case has not been investigated. As for comparing the medians of the marginal distributions, method RMPB2, described in the next section, can be used.

11.3.1 The Percentile Bootstrap

We begin with the percentile bootstrap. The methods in this section can be used with MOM, medians, M-estimators, and trimmed means (when the amount of trimming is sufficiently high or the amount of trimming is close to zero and n is large). As usual, we let θ be the population value for any of these measures of location, and we let $\hat{\theta}$ be the estimate of θ based on data.

But before continuing, we expand upon a point made in Section 11.2. It was noted that there are two closely related but different methods for comparing trimmed means. The first uses difference scores and the second compares marginal measures of location. This distinction carries over to the other robust measures of location considered here. For example, we could test the hypothesis that difference scores have a population value for MOM that is equal to zero, or we could test

$$H_0: \theta_1 = \theta_2.$$

At the moment, using MOM with difference scores seems to be one of the best methods for controlling Type I error probabilities when the sample size is very small. It seems that testing $H_0: \theta_1 = \theta_2$ also performs well in terms of Type I error probabilities, at least when the estimator being used has a reasonably high breakdown point. But choosing between these methods is not an academic matter, because they can reach different conclusions about whether groups differ. Also, using difference scores is known to yield higher power in some situations, and presumably situations arise where the reverse is true.

Method RMPB1

First consider how to compare groups based on difference scores. That is, the goal is to test

$$H_0: \theta_D = 0,$$

the hypothesis that the typical difference score (the typical D_i value) is zero. In the context of the bootstrap, this means that we obtain a bootstrap sample by resampling with replacement n D_i values. Let $\hat{\theta}^*_D$ be the estimate of θ_D based on this bootstrap sample. Next, repeat this process B times, yielding, $\hat{\theta}^*_{D1}, \ldots, \hat{\theta}^*_{DB}$ and let $\hat{p}^*$ be the proportion of these bootstrap values that are greater than zero. So if A is the number of times $\hat{\theta}^*_{Db} > 0$, then

$$\hat{p}^* = \frac{A}{B}.$$

Then reject H_0 if

$$\hat{p}^* \le \frac{\alpha}{2}$$

or if

$$\hat{p}^* \ge 1 - \frac{\alpha}{2}.$$

The S-PLUS function onesampb in Section 7.7.2 can be used to perform the calculations just described.

EXAMPLE. Again assume the data in Table 11.1 are stored in cork1 and cork2. Then the command

```
onesampb(cork1-cork2)
```

computes a confidence interval for the difference scores based on MOM and returns a .95 confidence interval of (0.27, 7.82). So the typical difference between the north and east sides of a tree tends to be greater than zero. That is, the north side tends to have a higher cork boring weight. ■

Method RMPB2

If instead the goal is to test

$$H_0: \theta_1 = \theta_2,$$

bootstrap samples are generated by resampling pairs of observations. (This is in contrast to independent groups, where a bootstrap sample is obtained by resampling observations from the first group and then resampling observations from the second.) For example, suppose you observe the following five pairs of values:

6	12
19	2
23	5
32	21
15	18

Then a bootstrap sample consists of randomly resampling with replacement five pairs of values from the five pairs just listed. So the first pair, (6, 12), has probability 1/5 of being chosen each time a bootstrap pair is generated, and the same is true for the other four pairs. A bootstrap sample from these five pairs of values might look like this.

23	5
32	21
6	12
32	21
6	12

More generally, if you observe n pairs of values — say, $(X_{11}, X_{12}), \ldots, (X_{n1}, X_{n2})$ — generating a bootstrap sample means that each pair has probability $1/n$ of being selected each time a pair is chosen. That is, you randomly resample with replacement pairs of observations. The bootstrap sample of paired observations generated in this manner will be labeled $(X^*_{11}, X^*_{12}), \ldots, (X^*_{n1}, X^*_{n2})$.

After generating a bootstrap sample of n pairs of observations, label the resulting estimates of location as $\hat{\theta}^*_1$ and $\hat{\theta}^*_2$. Repeat this B times, yielding $(\hat{\theta}^*_{1b}, \hat{\theta}^*_{2b})$, $b = 1, \ldots, B$, and let $\hat{p}^*$ be the proportion of times the bootstrap estimate of location for the first group is greater than the estimate for the second. In symbols, $\hat{p}^*$ is an estimate of

$$P\left(\hat{\theta}^*_1 > \hat{\theta}^*_2\right)$$

for a random bootstrap sample. As in Chapter 8, it is convenient to set $\hat{p}^*_m$ equal to $\hat{p}^*$ or $1 - \hat{p}^*$, whichever is smaller, in which case $2p^*_m$ is like a significance level, and you reject the hypothesis of equal measures of location if

$$\hat{p}^*_m \leq \frac{\alpha}{2}. \tag{11.4}$$

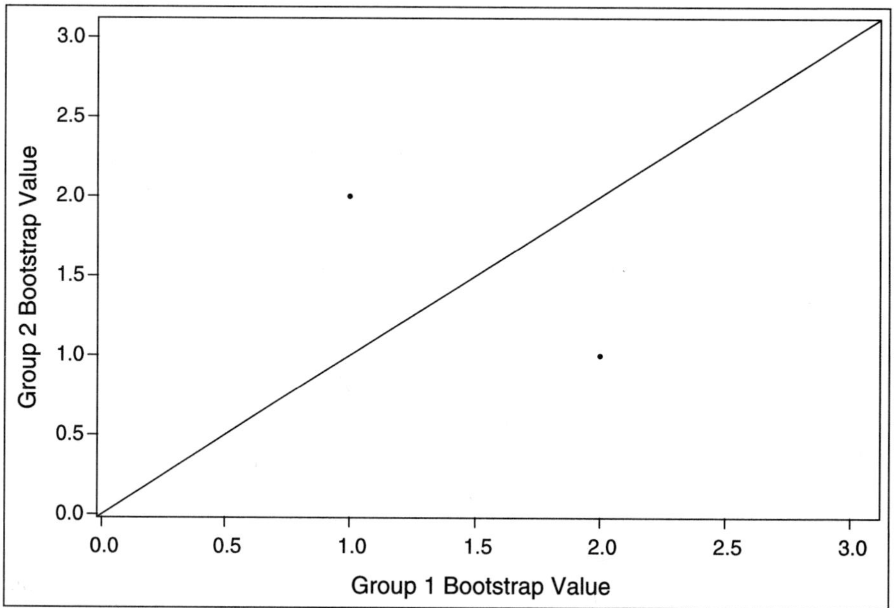

FIGURE 11.1 The lower right point corresponds to a situation where for a pair of bootstrap estimates, Group 1 has the larger value. For a point above the line, the reverse is true.

It helps to graphically illustrate the bootstrap method just described. First imagine that we generate a single bootstrap sample, compute some measure of location $\hat{\theta}$ for each group (or variable), yielding the pair $(\hat{\theta}_1^*, \hat{\theta}_2^*)$, and then plot the point corresponding to this pair of values. Then from basic principles, if $\hat{\theta}_1^* > \hat{\theta}_2^*$, $(\hat{\theta}_1^*, \hat{\theta}_2^*)$ corresponds to a point below (or to the right of) a line having a slope of 1 and an intercept of zero, as indicated in Figure 11.1. If $\hat{\theta}_1^* < \hat{\theta}_2^*$, the point $(\hat{\theta}_1^*, \hat{\theta}_2^*)$ will appear above (or to the left) of this line. And if $\hat{\theta}_1^* = \hat{\theta}_2^*$, the point $(\hat{\theta}_1^*, \hat{\theta}_2^*)$ is somewhere on the line shown in Figure 11.1.

More generally, if the null hypothesis is true and B bootstrap pairs are generated and plotted, then it should be the case that approximately half of the plotted points will be below the line having slope 1 and intercept zero, and about half should be above the line instead, as illustrated in Figure 11.2. However, if the first group has a larger population measure of location than the second ($\theta_1 > \theta_2$), then typically a majority of the bootstrap values, when plotted, will be to the right of the line having a slope of 1 and an intercept of zero, as shown in the left panel of Figure 11.3. If the reverse is true and the typical measure for group 1 is less than the typical measure for group 2, then a scatterplot of the points will appear as in the right panel of Figure 11.3. In particular, if we set $\alpha = .05$ and if 97.5% of the bootstrap values are to the right of the line in Figure 11.3, reject and conclude that the first group typically has a larger value than the second. If 97.5% of the bootstrap values are to the left of the line in Figure 11.3, reject and conclude that the first group typically has a smaller value. A $1 - \alpha$ confidence interval for $\theta_1 - \theta_2$ is

$$\left(V_{(\ell+1)}, V_{(u)}\right), \tag{11.5}$$

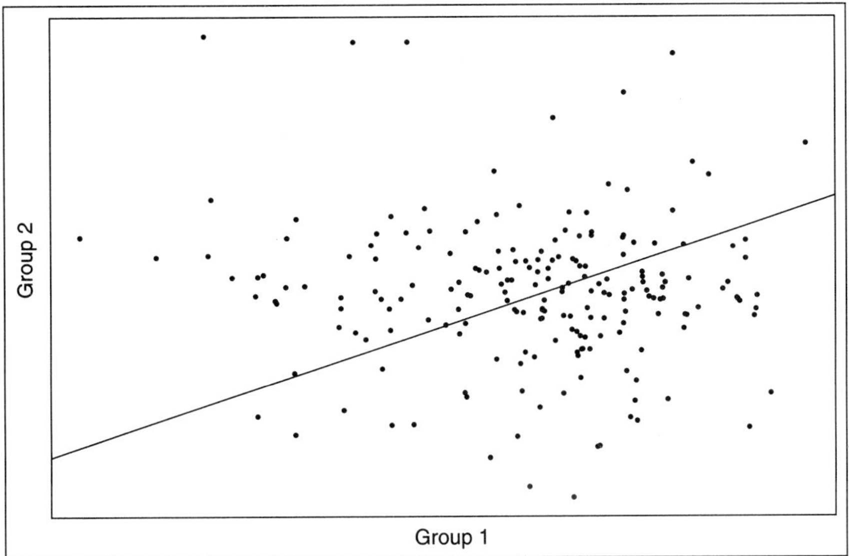

FIGURE 11.2 A plot of bootstrap values where the null hypothesis is true.

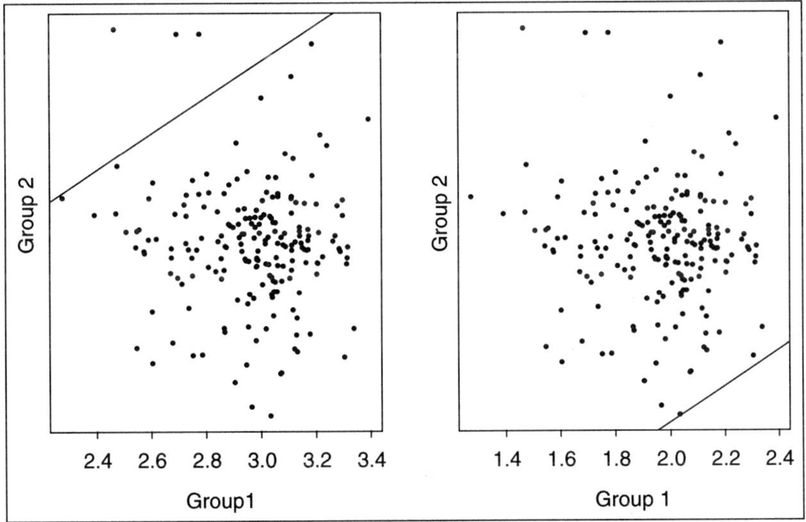

FIGURE 11.3 Bootstrap values where the null hypothesis is false. In the left panel, Group 1 has the larger measure of location. In the left panel, the reverse is true.

where $\ell = \alpha B/2$, rounded to the nearest integer, $u = B - \ell$, $V_b = \hat{\theta}^*_{1b} - \hat{\theta}^*_{2b}$, $b = 1, \ldots, B$, and $V_{(1)} \leq \cdots \leq V_{(B)}$ are the V_b values written in ascending order.

The estimated significance level when using the method just described is $2\hat{p}^*_m$, where $\hat{p}^*_m$ is given by Equation (11.4). A concern is that as the correlation between the two variables under study increases, the method can become too conservative in terms of the probability of a Type I error when using MOM or 20% trimmed means. For example, when testing at the .05 level, the actual probability of a Type I error can drop below .025 and even go as low as .01, suggesting that power might be relatively low. Currently, a method that appears to give better results in some situations is applied as follows. Set $C_{ij} = X_{ij} - \hat{\theta}_j$. That is, shift the data so that both groups have measures of location equal to zero, so in particular the null hypothesis is true. Generate bootstrap samples from the C_{ij} values and let $\hat{q}^*$ be the proportion of times, among the B bootstrap samples, that the measure of location from the first group is larger than the measure of location from the second. Because the null hypothesis is true, $\hat{q}^*$ should be approximately equal to .5. A so-called bias-adjusted significance level is

$$2\min\left(\hat{p}_a^*, 1 - \hat{p}_a^*\right),$$

where $\hat{p}_a^* = \hat{p}^* - .1(\hat{q}^* - .5)$.

11.3.2 S-PLUS Function rmmcppb

The S-PLUS function

rmmcppb(x,y=NA,alpha=.05,est=mom,dif=T,plotit=T,nboot=NA,BA=F,...)

has been supplied for comparing dependent groups based on the percentile bootstrap methods RMPB1 and RMPB2 described in the previous section. (This function contains another argument, con, which is explained in Chapter 12.) You can enter data through x, which can be a matrix or can have list mode. When x is a matrix, columns correspond to groups. (That is, column 1 is group 1, column 2 is group 2, and so forth.) Alternatively, when comparing two groups only, the data for the first group can be stored in x and the data for the second stored in y. (When comparing more than two groups, this function uses the method outlined in Section 12.8.4.) The argument est indicates which measure of location is used and defaults to MOM. The default value for the argument dif (T for true) indicates that difference scores will be used. That is, method RMPB1 in Section 11.3.1 is applied; dif=F results in using method RMPB2. When comparing two groups, this function also plots the bootstrap values if plotit=T is used. For method RMPB2, pairs of bootstrap values are plotted and the central $1 - \alpha$ percent of these values is indicated by a polygon. These centrally located values provide an approximate $1 - \alpha$ confidence region for the values of the parameters being estimated. (The following example will help clarify what this means.) Included in the plot is a line having a slope of 1 and an intercept of zero to help provide perspective on how much the groups differ. When dif=F is used, setting BA=T will cause bias-adjusted significance levels to be computed.

EXAMPLE. Figure 11.4 shows the plot created by rmmcppb, with dif=F, based on the cork data in Table 11.1. Each point represents a pair of bootstrap estimates of the measures of location, which here is taken to be MOM. The polygon contains the central $1 - \alpha$ percent of these values and provides a two-dimensional analog of a confidence interval. That is, it is estimated

Continued

EXAMPLE. (*Continued*) that (θ_1, θ_2) is somewhere inside this polygon. For example, the value (50, 45) lies inside the polygon, indicating that having the typical weight for the north side of a tree equal to 50 and simultaneously having the typical weight for the east side equal to 45, is a reasonable possibility. The point (55, 40) does not lie inside the polygon indicating that it is unlikely that simultaneously, $\theta_1 = 55$ and $\theta_2 = 40$. Note, however, that $\theta_1 = 55$, taken by itself, is a reasonable estimate of the typical weight for the north side of a tree. The value 55 is within the .95 confidence interval based on the S-PLUS function momci. That is, ignoring the data on the east side, the typical weight on the north side could be as high as 55. But if the typical weight on the north side is 55, Figure 11.4 suggests that it is unlikely that the typical weight on the east side is 40. In contrast, if we ignore the north side data, now 40 is a reasonable estimate for the east side. To add perspective, Figure 11.4 includes the line having slope 1 and intercept zero. If the null hypothesis is true, that is, $\theta_1 = \theta_2$, then the true values for θ_1 and θ_2 must be somewhere on this line.

Figure 11.5 shows the plot created when dif=T is used. When the typical difference score is zero, about half of the bootstrap values should be above the horizontal line having intercept zero. In this particular case, the proportion of points below this horizontal line is only .014, indicating a significance level of $2 \times .014 = .028$.

A portion of the printed output from rmmcppb (with dif=F) looks like this:

```
$output:
     psihat  sig.level  crit.sig   ci.lower  ci.upper
   5.421652      0.174      0.05  -2.178571  10.85165
```

The value under psihat, 5.42, is $\hat{\theta}_1 - \hat{\theta}_2$, the estimated difference between the typical weight for group 1 minus the typical weight for group 2. Under sig.level we see 0.174, which is the significance level. That is, the proportion of points in Figure 11.4 above the straight line is $.174/2 = .087$. If only 2.5% were above the line, you would reject at the .05 level. Under crit.sig we see the α value used and to the right is a $1-\alpha$ confidence interval for $\theta_1 - \theta_2$: (−2.18, 10.85). When using difference scores, we reject at the .05 level, but here we fail to reject, the only point being that the choice of method can make a practical difference. ■

EXAMPLE. To test the hypothesis that the two variables in Table 11.1 have equal population medians, use the command

rmmcppb(cork,est=median,dif=F),

assuming the data are stored in cork. The resulting significance level is .176. To compare 20% trimmed means, use the command

rmmcppb(cork,est=mean,tr=.2).

■

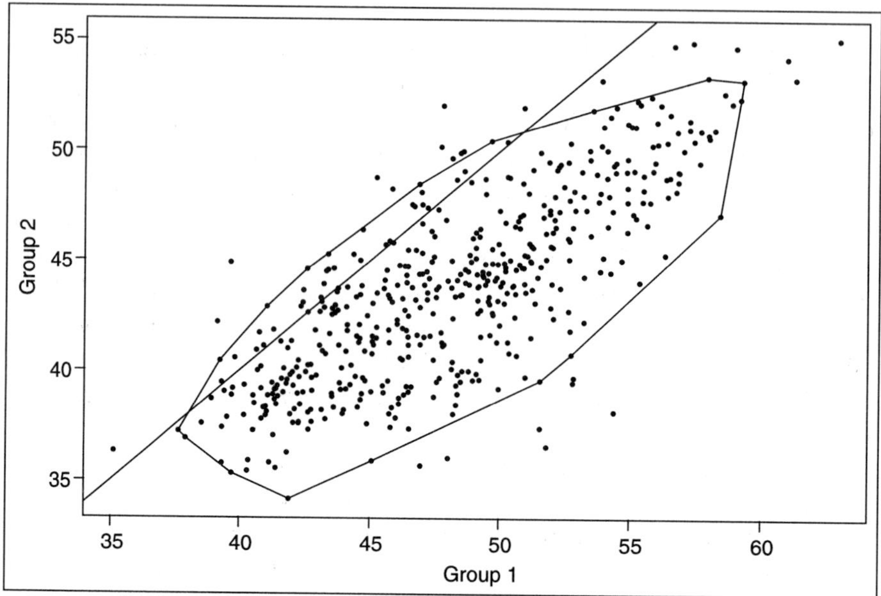

FIGURE 11.4 A plot created by the S-PLUS function rmmcppb with the argument dif=F.

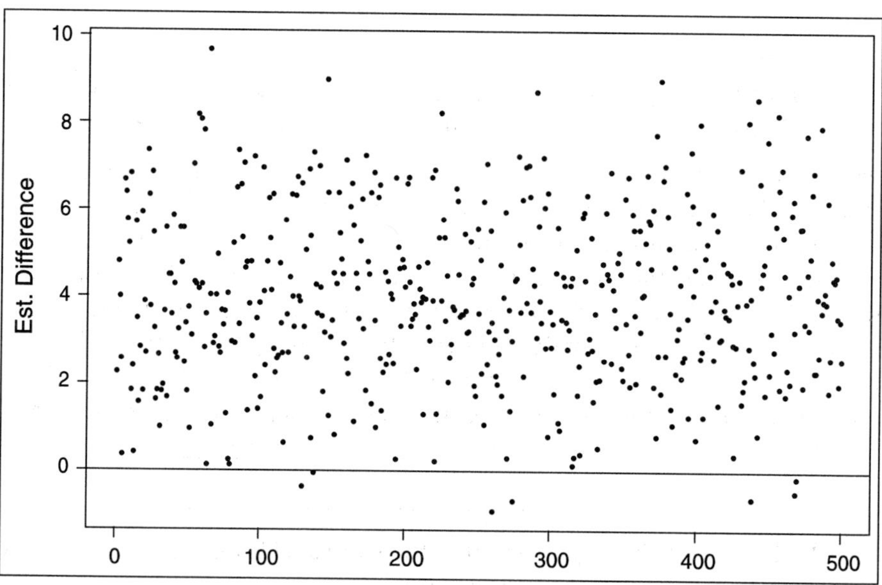

FIGURE 11.5 A plot created by the S-PLUS function rmmcppb with the argument dif=T.

11.3.3 Bootstrap-*t*

Unless the sample size is fairly large, in terms of controlling the probability of a Type I error or achieving accurate probability coverage when computing a confidence interval, the percentile bootstrap method performs poorly when working with means or when using a trimmed mean and the amount of trimming is close to zero. All indications are that when comparing dependent groups, the amount of trimming should be at least .25 when using a percentile bootstrap method to compare the marginal trimmed means. Indeed, with the amount of trimming sufficiently high, the percentile bootstrap seems preferable to a bootstrap-t in terms of Type I errors. (Some adjustments can be made to percentile bootstrap methods when the amount of trimming is between .2 and .25 that can render it relatively competitive with the bootstrap-t. The adjustments depend on the sample size, but currently, from a practical point of view, there seems to be no reason to describe them here.) Also, when comparing more than two groups, extant studies have focused on testing for equal population-trimmed means, as opposed to making inferences based on the trimmed mean of the difference scores. For this reason, the bootstrap-t for comparing dependent groups focuses on comparing trimmed means with the goal of testing $H_0: \mu_{t1} = \mu_{t2}$. (There might be practical reasons for considering a bootstrap-t on difference scores, but this has not been established as yet.)

The goal is to estimate an appropriate critical value when the null hypothesis of equal trimmed means is true and when using the test statistic T_y given by Equation (11.3). This is done by centering the data, computing T_y based on a bootstrap sample generated from the centered data, and then repeating this process B times to get an estimate of the distribution of T_y when the null hypothesis of equal trimmed means is true. More formally, set $C_{ij} = X_{ij} - \bar{X}_{tj}$ and let C^*_{ij} be a bootstrap sample obtained by resampling with replacement n pairs of values from the C_{ij} values. Let T^*_y be the value of T_y based on the C^*_{ij} values. Repeat this process B times, yielding T^*_{yb}, $b = 1, \dots, B$. Let $T^*_{y(1)} \leq \cdots \leq T^*_{y(B)}$ be the T^*_{yb} values written in ascending order. Set $\ell = \alpha B/2$, rounding to the nearest integer, and $u = B - \ell$. Then an estimate of the lower and upper critical values is $T^*_{y(\ell+1)}$ and $T^*_{y(u)}$, respectively. An equal-tailed $1 - \alpha$ confidence interval for $\mu_{t1} - \mu_{t2}$ is

$$\left(\bar{X}_{t1} - \bar{X}_{t2} + T^*_{y(\ell+1)}\sqrt{d_1 + d_2 - 2d_{12}}, \bar{X}_{t1} - \bar{X}_{t2} + T^*_{y(u)}\sqrt{d_1 + d_2 - 2d_{12}}\right). \quad (11.6)$$

To get a symmetric confidence interval, replace T^*_{yb} by its absolute value and set $a = (1 - \alpha)B$, rounding to the nearest integer, in which case the $(1 - \alpha)$ confidence interval for $(\mu_{t1} - \mu_{t2})$ is

$$(\bar{X}_{t1} - \bar{X}_{t2}) \pm T^*_{y(a)}\sqrt{d_1 + d_2 - 2d_{12}}.$$

(The S-PLUS function rmanovb described in Section 11.6.4 performs the calculations.)

11.4 Measuring Effect Size

Section 8.11 describes some measures of effect size and indicates why they are important. It is noted that analogs of these methods are available when comparing dependent groups. For example, a simple measure of effect size is

$$\Delta = \frac{\mu_1 - \mu_2}{\sigma_d}, \tag{11.7}$$

where σ_d is the population standard deviation of the difference scores (the D_i values). But for reasons already described, this approach can grossly underestimate the extent to which two groups differ, and it does not capture some of the global differences between the groups. Another approach is to compute a confidence interval for $\mu_1 - \mu_2$ or some robust measure of location, such as a trimmed mean or MOM.

A simple but useful way of assessing effect size is with a boxplot of the difference scores (the D_i values). When comparing dependent groups with identical distributions, the D values will tend to be symmetrically distributed about zero. So in particular, a boxplot will tend to have a median of zero. (Also see the discussion of the sign test in Chapter 15.)

Another simple measure of effect size is $\hat{p}^*$ or $2\hat{p}^*_m$, the bootstrap estimate of the significance level. These measures of effect size reflect the separation of the distributions but not the magnitude of the differences between the measures of location.

Another approach is to use an analog of Q described in Section 8.11. That is, Q represents the likelihood of correctly determining whether an observation came from group 1 or group 2. If we make an arbitrary decision, meaning that we randomly decide whether an observation came from group 1, then $Q = .5$. But suppose we decide that an observation came from group 1 using the method outlined in Box 8.2. Roughly, you first estimate the likelihood that an observation came from group 1 and then do the same for group 2. If the estimated likelihood is higher that it came from group 1, then simply decide that it indeed came from group 1. If the groups are identical, meaning they have identical (marginal) distributions, then the probability of a correct decision is $Q = .5$. The more separated the groups happen to be, the closer Q will be to 1. For dependent groups, Q can be estimated with the so-called .632 bootstrap estimator, $\hat{Q}_{.632}$, which is computed as summarized in Box 11.2. As noted in Chapter 8, a rough guideline is that $\hat{Q}_{.632} = .55$ reflects a small difference between the groups and $\hat{Q}_{.632} = .66$ is large. It is noted that $\hat{Q}_{.632} < .5$ is possible even though we know $Q \geq .5$. So $\hat{Q}_{.632} < .5$ suggests that there is little or no difference between the groups.

BOX 11.2 A Bootstrap Estimate of Q for Dependent Groups

For each group, compute the kernel density estimator as described in Section 3.7 and label the results $\hat{f}_1(x)$ and $\hat{f}_2(x)$, respectively. Set $\hat{\eta}(X_{i1}) = 1$ if $\hat{f}_1(X_{i1}) > \hat{f}_2(X_{i1})$; otherwise $\hat{\eta}(X_{i1}) = 0$. Generate a bootstrap sample by resampling n pairs of observations and let $\hat{\eta}^*$ be the bootstrap analog of $\hat{\eta}$. Repeat this process B times, yielding $\hat{\eta}^*_b$, $b = 1, \ldots, B$. Let

$$\hat{\epsilon} = \frac{1}{n}\sum_{i=1}^{n}\frac{1}{B_i}\sum_{b\in C_i}\hat{\eta}^*_b(X_{1i}).$$

Continued

BOX 11.2 *(Continued)*
C_i is the set of indices of the bth bootstrap sample not containing the pair (X_{1i}, X_{2i}), and B_i is the number of such bootstrap samples. (The illustration in conjunction with Box 8.2 provides some details about how $\hat{\epsilon}$ is computed.) Let

$$\hat{Q}_{ap} = \frac{1}{n}\sum \hat{\eta}(X_{1i}).$$

The estimate of Q is

$$\hat{Q}_{.632} = .368\hat{Q}_{ap} + .632\hat{\epsilon},$$

11.4.1 S-PLUS Functions qhatd

The S-PLUS function

qhatd(x,y,nboot=50)

computes the measure of effect size, $\hat{Q}_{.632}$, as described in Box 11.2. As usual, x and y can be any S-PLUS variables containing data.

EXAMPLE. For the data in Table 11.1, qhatd returns $\hat{Q}_{.632} = .42$, suggesting that if there is a difference between the groups, it is small. However, some caution is required because the precision of the estimate of Q, based on $\hat{Q}_{.632}$, is not known. The S-PLUS command

onesampb(x-y,est=median)

returns (0, 8) as a .95 confidence interval for the median of the difference scores. This again suggests that if the groups differ, in some sense the difference might be small, because we cannot rule out the possibility that the median difference is zero. ■

11.5 Comparing Variances

As noted in Chapter 8, situations arise where the goal is to compare variances rather than some measure of location. When comparing robust measures of scale, all indications are that the percentile bootstrap method is the most effective approach. But for the special case where there is specific interest in the variances, a variation of the percentile bootstrap method is required. The method described here is based on a bootstrap analog of the so-called *Morgan–Pitman test* and currently appears to be the most satisfactory method among the many that have been proposed.

As usual, we have n pairs of randomly sampled observations: $(X_{11}, X_{12}), \ldots, (X_{n1}, X_{n2})$. Set

$$U_i = X_{i1} - X_{i2} \quad \text{and} \quad V_i = X_{i1} + X_{i2},$$

$(i = 1, \ldots, n)$. That is, for the ith pair of observations, U_i is the difference and V_i is the sum. Let σ_1^2 be the variance associated with the first group (the X_{i1} values) and let σ_2^2 be the variance associated with the second. The goal is to test the hypothesis that these variances are equal. It can be shown that if

$$H_0 : \sigma_1^2 = \sigma_2^2$$

is true, then Pearson's correlation between the U and V values is zero. That is, we can test the hypothesis of equal variances by testing

$$H_0 : \rho_{uv} = 0,$$

where ρ_{uv} is Pearson's correlation between U and V. Section 6.5 describes the conventional method for testing the hypothesis of a zero correlation based on Student's T distribution. Applying this method to the U and V values yields what is known as the *Morgan–Pitman test* for equal variances. But we have already seen that the conventional test of a zero correlation has undesirable properties when variables are dependent; these problems render the Morgan–Pitman test unsatisfactory when distributions differ. To deal with this problem, we simply apply the modified percentile bootstrap as was done in Section 7.3. That is, take a bootstrap sample of n pairs of the U and V values, compute the correlation between these values, repeat this 599 times, and label the results $r_1^*, \ldots, r_{599}^*$, in which case a .95 confidence interval for ρ_{uv} is

$$\left(r_{(a)}^*, r_{(c)}^*\right),$$

where for $n < 40$, $a = 7$ and $c = 593$; for $40 \leq n < 80$, $a = 8$ and $c = 592$; for $80 \leq n < 180$, $a = 11$ and $c = 588$; for $180 \leq n < 250$, $a = 14$ and $c = 585$; while for $n \geq 250$, $a = 15$ and $c = 584$. To apply the method, you can use the S-PLUS function pcorb (described in Section 7.3.3), as illustrated in the next example.

EXAMPLE. Imagine that the data in Table 11.1 corresponding to the north side of a tree are stored in the S-PLUS variable x and that the east side data are stored in the S-PLUS variable y. Then the command pcorb(x-y,x+y) tests the hypothesis that these two dependent groups have equal variances. The function returns a .95 confidence interval of (−0.154, 0.677); this interval contains zero, so you fail to reject. (The S-PLUS function comvar2d can be used as well. That is, the command comvar2d(x,y) will test the hypothesis of equal variances.) ■

11.5.1 Comparing Robust Measures of Scale

As for comparing dependent groups using a robust measure of scale, a basic percentile bootstrap method can be used. For example, if the goal is to test the hypothesis that two dependent groups have equal population percentage bend midvariances, simply proceed as described in Section 11.3, where a bootstrap sample is obtained by resampling with replacement n pairs of observations. For each group, compute the percentage bend midvariance as described by Equation (3.18) (in Box 3.1) or whichever measure of scale is of interest, and let d^* be the difference between the two estimates. Repeat this process B times, yielding $d_1^*, \ldots, d_B^*$, and then put these B values in ascending order, and label them in the usual way, namely, $d_{(1)}^* \leq \cdots \leq d_{(B)}^*$. Then a $1-\alpha$ confidence interval for the difference between the measures of scale is $(d_{(\ell+1)}^*, d_{(u)}^*)$, where as usual $\ell = \alpha(B)/2$, rounded to the nearest integer, and $u = B - \ell$. In essence, use method RMPB2 (described in Section 11.3.1), except use a robust measure of scale rather than a robust measure of location. The calculations can be done with the S-PLUS function rmmcppb described in Section 11.3.2, with the argument dif set equal to F.

EXAMPLE. Chapter 8 describes a study where EEG measures of convicted murderers were compared to a control group. In fact measures for both groups were taken at four sites in the brain. For illustrative purposes, the first two sites for the control group are compared using the percentage bend midvariance described in Chapter 3. The data are

Site 1:	−0.15	−0.22	0.07	−0.07	0.02	0.24	−0.60	−0.17	−0.33
	0.23	−0.69	0.70	1.13	0.38				
Site 2:	−0.05	−1.68	−0.44	−1.15	−0.16	−1.29	−2.49	−1.07	−0.84
	−0.37	0.01	−1.24	−0.33	0.78				

If we store the data in the S-PLUS variables eeg1 and eeg2, the command

rmmcppb(eeg1,eeg2,est=pbvar,dif=F)

returns a significance level of .13. So we are not able to detect a difference in the variation between these two sites when testing at the .05 level. If the argument est=pbvar is replaced with est=mad so that the MAD measure of scale is used, the significance level is .25. ■

11.6 Comparing More Than Two Groups

This section describes how to compare measures of location corresponding to multiple dependent groups. One strategy is to test

$$H_0 : \theta_1 = \cdots = \theta_J. \tag{11.8}$$

Another approach is to test some appropriate hypothesis based on difference scores. A simple possibility is to test

$$H_0 : \theta_{d1} = \cdots = \theta_{d,J-1} = 0, \tag{11.9}$$

where θ is any measure of location, such as a median or MOM, and θ_{dj} is the value of θ based on the difference scores between groups j and $j+1$ ($j = 1, \ldots, J-1$).

That is, θ_{dj} is the population value of θ corresponding to $Y_{ij} = X_{ij} - X_{i,j+1}$. We begin with a nonbootstrap method based on trimmed means and then consider bootstrap methods that can be used with trimmed means and other robust estimators. Section 11.8 discusses the relative merits of the methods about to be described.

11.6.1 A Method for Trimmed Means

This section generalizes the methods in Sections 11.1 and 11.2 to test

$$H_0 : \mu_{t1} = \cdots = \mu_{tJ}, \tag{11.10}$$

the hypothesis that all J groups have equal trimmed means. The test statistic, labeled F, is computed as summarized in Box 11.3. You reject if $F \geq f$, where f is the $1 - \alpha$ quantile of an F-distribution with ν_1 and ν_2 degrees for freedom, where ν_1 and ν_2 are computed as described in Box 11.4.

For the special case where there is no trimming, F reduces to a standard test statistic for comparing means (which is typically referred to as an ANOVA F-test). An early approach to determining an appropriate critical value for this special case was based on assuming that the variances and correlations among the J groups follow a rather restrictive pattern called *sphericity* or *circularity* (see Huynh & Feldt, 1976). However, violating this assumption is known to cause practical problems. Methods for testing the assumption that sphericity holds have been proposed, but all indications are that such tests do not have enough power to detect situations where the assumption should be discarded (e.g., Boik, 1981; Keselman, Rogan, Mendoza, & Breen, 1980), so no details are given here. Rather, we simply rely on the method in Box 11.4 for determining a critical value that was derived without assuming sphericity. (Readers interested in more details about sphericity are referred to Kirk, 1995, as well as Rogan, Keselman, & Mendoza, 1979.) For simulation results on how the method in Box 11.3 performs with 20% trimmed means, see Wilcox, Keselman, Muska, and Cribbie, 2000.

BOX 11.3 Test Statistic for Comparing the Trimmed Means of Dependent Groups

Winsorize the observations by computing

$$Y_{ij} = \begin{cases} X_{(g+1)j}, & X_{ij} \leq X_{(g+1)j} \\ X_{ij}, & X_{(g+1)j} < X_{ij} < X_{(n-g)j} \\ X_{(n-g)j}, & X_{ij} \geq X_{(n-g)j}. \end{cases}$$

Let $h = n - 2g$ be the effective sample size, where $g = [\gamma n]$, $[\gamma n]$ is γn rounded down to the nearest integer, and γ is the amount of trimming. Compute

$$\bar{X}_t = \frac{1}{J} \sum \bar{X}_{tj}$$

Continued

BOX 11.3 *(Continued)*

$$Q_c = (n - 2g) \sum_{j=1}^{J} (\bar{X}_{tj} - \bar{X}_t)^2$$

$$Q_e = \sum_{j=1}^{J} \sum_{i=1}^{n} (Y_{ij} - \bar{Y}_{.j} - \bar{Y}_{i.} + \bar{Y}_{..})^2,$$

where

$$\bar{Y}_{.j} = \frac{1}{n} \sum_{i=1}^{n} Y_{ij}$$

$$\bar{Y}_{i.} = \frac{1}{J} \sum_{j=1}^{J} Y_{ij}$$

$$\bar{Y}_{..} = \frac{1}{nJ} \sum_{j=1}^{J} \sum_{i=1}^{n} Y_{ij}.$$

The test statistic is

$$F = \frac{R_c}{R_e},$$

where

$$R_c = \frac{Q_c}{J - 1}$$

$$R_e = \frac{Q_e}{(h - 1)(J - 1)}.$$

Decision Rule:
Reject if $F \geq f$, the $1 - \alpha$ quantile of an F-distribution with degrees of freedom computed as described in Box 11.4.

BOX 11.4 How to Compute Degrees of Freedom When Comparing Trimmed Means

Let

$$v_{jk} = \frac{1}{n - 1} \sum_{i=1}^{n} (Y_{ij} - \bar{Y}_{.j})(Y_{ik} - \bar{Y}_{.k})$$

Continued

BOX 11.4 (*Continued*)
for $j = 1, \ldots, J$ and $k = 1, \ldots, J$, where, as in Box 11.3, Y_{ij} is the Winsorized observation corresponding to X_{ij}. When $j = k$, $v_{jk} = s_{wj}^2$, the Winsorized sample variance for the jth group; and when $j \neq k$, v_{jk} is a Winsorized analog of the sample covariance.

Let

$$\bar{v}_{..} = \frac{1}{J^2}\sum_{j=1}^{J}\sum_{k=1}^{J} v_{jk}$$

$$\bar{v}_d = \frac{1}{J}\sum_{j=1}^{J} v_{jj}$$

$$\bar{v}_{j.} = \frac{1}{J}\sum_{k=1}^{J} v_{jk}$$

$$A = \frac{J^2(\bar{v}_d - \bar{v}_{..})^2}{J-1}$$

$$B = \sum_{j=1}^{J}\sum_{k=1}^{J} v_{jk}^2 - 2J\sum_{j=1}^{J}\bar{v}_{j.}^2 + J^2\bar{v}_{..}^2$$

$$\hat{\epsilon} = \frac{A}{B}$$

$$\tilde{\epsilon} = \frac{n(J-1)\hat{\epsilon} - 2}{(J-1)\{n - 1 - (J-1)\hat{\epsilon}\}}.$$

The degrees of freedom are

$$\nu_1 = (J-1)\tilde{\epsilon}$$

$$\nu_2 = (J-1)(h-1)\tilde{\epsilon},$$

where h is the effective sample size (the number of observations left in each group after trimming).

11.6.2 S-PLUS Function rmanova

The S-PLUS function rmanova, which comes with this book, compares the trimmed means of J dependent groups using the calculations in Boxes 11.3 and 11.4. The function has the general form

```
rmanova(x,tr=.2,grp=c(1:length(x))).
```

The data are stored in x, which can be either an n-by-J matrix, with the jth column containing the data for jth group, or an S-PLUS variable having list mode. In the latter case, x[[1]] contains the data for group 1, x[[2]] contains the data for group 2, and so on. As usual, tr indicates the amount of trimming, which defaults to .2, and grp can be used to compare a subset of the groups. By default, the trimmed means of all J groups are compared. If, for example, there are five groups but the goal is to test $H_0: \mu_{t2} = \mu_{t4} = \mu_{t5}$, then the command

rmanova(x,grp=c(2,4,5))

accomplishes this goal using 20% trimming.

EXAMPLE. The S-PLUS function rmanova is illustrated with the weight of cork borings described in the introduction to this chapter, only now all four sides of each tree are used. The data are reproduced in Table 11.3. If we store the data in the S-PLUS variable cork in a matrix having four columns, then the output from the command rmanova(cork) is

```
$test:
[1] 2.311757

$df:
[1] 2.636236 44.816010

$siglevel:
[1] 0.09624488
```

So with $\alpha = .05$, we are unable to reject the hypothesis that the typical weights of cork borings differ among the four sides of the trees. ■

11.6.3 A Bootstrap-*t* Method for Trimmed Means

When comparing trimmed means, a bootstrap-*t* method can be applied in basically the same way as described in previous chapters, only again we generate bootstrap samples in a manner consistent with dependent groups. As usual, when working with a bootstrap-*t* method, we begin by centering the data. That is, we set

$$C_{ij} = X_{ij} - \bar{X}_{tj}$$

with the goal of estimating an appropriate critical value, based on the test statistic F in Box 11.3, when the null hypothesis is true. The remaining steps are as follows:

1. Generate a bootstrap sample by randomly sampling, with replacement, n rows of data from the matrix

$$\begin{pmatrix} C_{11}, \ldots, C_{1J} \\ \vdots \\ C_{n1}, \ldots, C_{nJ} \end{pmatrix},$$

TABLE 11.3 Cork Boring Weights for the North, East, South, and West Sides of Trees

N	E	S	W
72	66	76	77
60	53	66	63
56	57	64	58
41	29	36	38
32	32	35	36
30	35	34	26
39	39	31	27
42	43	31	25
37	40	31	25
33	29	27	36
32	30	34	28
63	45	74	63
54	46	60	52
47	51	52	53
91	79	100	75
56	68	47	50
79	65	70	61
81	80	68	58
78	55	67	60
46	38	37	38
39	35	34	37
32	30	30	32
60	50	67	54
35	37	48	39
39	36	39	31
50	34	37	40
43	37	39	50
48	54	57	43

yielding

$$\begin{pmatrix} C^*_{11}, \ldots, C^*_{1J} \\ \vdots \\ C^*_{n1}, \ldots, C^*_{nJ} \end{pmatrix}.$$

2. Compute the test statistic F in Box 11.3 based on the C^*_{ij} values generated in step 1, and label the result F^*.
3. Repeat steps 1 and 2 B times and label the results $F^*_1, \ldots, F^*_B$.
4. Put these B values in ascending order and label the results $F^*_{(1)} \leq \cdots \leq F^*_{(B)}$.

The critical value is estimated to be $F^*_{(c)}$, where $c = (1 - \alpha)B$ rounded to the nearest integer. That is, reject the hypothesis of equal trimmed means if

$$F \geq F^*_{(c)},$$

where F is the statistic given in Box 11.3 based on the X_{ij} values.

11.6.4 S-PLUS Function rmanovab

The S-PLUS function

rmanovab(x, tr = 0.2, alpha = 0.05, grp = 0, nboot = 599)

performs the bootstrap-t method just described. The arguments have their usual meaning; see, for example, Section 9.5.2.

11.7 Percentile Bootstrap Methods for Other Robust Measures of Location

When comparing independent groups, some version of the percentile bootstrap method generally performs better than the bootstrap-t, in terms of controlling the probability of a Type I error, when the amount of trimming is at least 20%. For the situation at hand, however, no percentile bootstrap method has been shown to perform better than the bootstrap-t when the amount of trimming is 20%. But in fairness, some promising percentile bootstrap methods have not been examined. When comparing 20% trimmed means, all indications are that the bootstrap-t method in Section 11.6.3 performs well, in terms of Type I errors, with sample sizes as small as 21 (Wilcox, Keselman, Muska, & Cribbie, 2000). But when comparing groups based on MOMs, medians, and M-estimators or when the amount of trimming is at least 25%, certain variations of the percentile bootstrap method provide excellent control over the probability of a Type I error.

There are in fact many variations of the percentile bootstrap method that might be used to compare dependent groups. Most are not described here, either because they are known to be relatively unsatisfactory or because little is known about how well they perform when sample sizes are small. To complicate matters, there is the issue of whether to use difference scores. So the goal in the remainder of this section is to describe the methods that currently seem to have practical value and then in Section 11.8 to comment on their relative merits.

11.7.1 Methods Based on Marginal Measures of Location

Again let θ_j be any measure of location associated with the jth group and let $\hat{\theta}_j$ be an estimate of θ_j based on the available data (the X_{ij} values). The goal is to test

$$H_0 : \theta_1 = \cdots = \theta_J,$$

the hypothesis that all J dependent groups have identical measures of location. That is, measures of location associated with marginal distributions are being compared, as opposed to using difference scores.

Method RMPB3

The first method uses the test statistic

$$Q = \sum \left(\hat{\theta}_j - \bar{\theta}\right)^2,$$

where $\bar{\theta} = \sum \hat{\theta}_j / J$ is the average of the $\hat{\theta}_j$ values. An appropriate critical value is estimated using an approach similar to the bootstrap-t technique. First, set $C_{ij} = X_{ij} - \hat{\theta}_j$. That is, shift the empirical distributions so that the null hypothesis is true. Next a bootstrap sample is obtained by resampling, with replacement, as described in step 1 of Section 11.6.3. Again we label the results

$$\begin{pmatrix} C^*_{11}, \ldots, C^*_{1J} \\ \vdots \\ C^*_{n1}, \ldots, C^*_{nJ} \end{pmatrix}.$$

For the jth column of the bootstrap data just generated, compute the measure location that is of interest and label it $\hat{\theta}^*_j$. Compute

$$Q^* = \sum \left(\hat{\theta}^*_j - \bar{\theta}^*\right)^2,$$

where $\bar{\theta}^* = \sum \hat{\theta}^*_j / J$, and repeat this process B times, yielding $Q^*_1, \ldots, Q^*_B$. Put these B values in ascending order, yielding $Q^*_{(1)} \le \cdots \le Q^*_{(B)}$. Then reject the hypothesis of equal measures of location if $Q > Q^*_{(c)}$, where again $c = (1 - \alpha)B$. (The S-PLUS function bd1way described in Section 11.7.2 performs these calculations.)

Method RMPB4

If the null hypothesis is true, then all J groups have a common measure of location, θ. The next method estimates this common measure of location and then checks to see how deeply it is nested within the bootstrap values obtained when resampling from the original values. That is, in contrast to method RMPB3, the data are not centered, and bootstrap samples are obtained by resampling rows of data from

$$\begin{pmatrix} X_{11}, \ldots, X_{1J} \\ \vdots \\ X_{n1}, \ldots, X_{nJ} \end{pmatrix},$$

yielding

$$\begin{pmatrix} X^*_{11}, \ldots, X^*_{1J} \\ \vdots \\ X^*_{n1}, \ldots, X^*_{nJ} \end{pmatrix}.$$

For the jth group (or column of bootstrap values) compute $\hat{\theta}_j^*$. Repeating this pro cess B times yields $\hat{\theta}_{jb}^*$ ($j = 1, \ldots, J$; $b = 1, \ldots, B$). The remaining calculations are performed as outlined in Box 11.5 (and are done by the S-PLUS function ddep in Section 11.7.2). Notice that for $J = 2$, this method does not reduce to method RMPB2 described in Section 11.3.1.

BOX 11.5 Repeated-Measures ANOVA Based on the Depth of the Grand Mean

Goal:
Test the hypothesis

$$H_0 : \theta_1 = \cdots = \theta_J.$$

1. Compute

$$S_{jk} = \frac{1}{B-1} \sum_{b=1}^{B} \left(\hat{\theta}_{jb}^* - \bar{\theta}_j^*\right)\left(\hat{\theta}_{kb}^* - \bar{\theta}_k^*\right),$$

where

$$\bar{\theta}_j^* = \frac{1}{B} \sum_{b=1}^{B} \hat{\theta}_{jb}^*.$$

(The quantity S_{jk} is the sample covariance of the bootstrap values corresponding to the jth and kth groups.)

2. Let

$$\hat{\theta}_b^* = \left(\hat{\theta}_{1b}^*, \ldots, \hat{\theta}_{Jb}^*\right)$$

and compute

$$d_b = \left(\hat{\theta}_b^* - \hat{\theta}\right) \mathbf{S}^{-1} \left(\hat{\theta}_b^* - \hat{\theta}\right)',$$

where $\mathbf{S}$ is the matrix corresponding to S_{jk}, $\hat{\theta} = (\hat{\theta}_1, \ldots, \hat{\theta}_J)$, $\hat{\theta}_j$ is the estimate of θ based on the original data for the jth group (the X_{ij} values, $i = 1, \ldots, n$), and $\hat{\theta}_b = (\hat{\theta}_{1b}, \ldots, \hat{\theta}_{Jb})$. The value of d_b measures how far away the bth bootstrap vector of location estimators is from $\hat{\theta}$, which is roughly the center of all B bootstrap values.

3. Put the d_b values in ascending order: $d_{(1)} \leq \cdots \leq d_{(B)}$.

4. Let $\hat{\theta}_G = (\bar{\theta}, \ldots, \bar{\theta})$, where $\bar{\theta} = \sum \hat{\theta}_j / J$, and compute

$$D = \left(\hat{\theta}_G - \hat{\theta}\right) \mathbf{S}^{-1} \left(\hat{\theta}_G - \hat{\theta}\right)'.$$

D measures how far away the estimated common value is from the observed measures of location (based on the original data).

5. Reject if $D \geq d_{(u)}$, where $u = (1 - \alpha)B$, rounded to the nearest integer.

11.7.2 S-PLUS Functions bd1way and ddep

The S-PLUS functions

bd1way(x, est = onestep, nboot = 599, alpha = 0.05)

and

ddep(x, alpha = 0.05, est = mom, grp = NA, nboot = 500, ...)

perform the percentile bootstrap methods just described. The first function performs method RMPB3. By default it uses the one-step M-estimator of location (based on Huber's Ψ), but any other estimator can be used via the argument est. As usual, x is any S-PLUS variable that is a matrix or has list mode, nboot is B, the number of bootstrap samples to be used, and grp can be used to analyze a subset of the groups, with the other groups ignored. (That is, grp is used as illustrated in Section 10.3.1.) The function ddep performs method RMPB4, described in Box 11.5.

EXAMPLE. We reanalyze the data in Table 11.3 using the S-PLUS functions just described. Assuming the data are stored in the S-PLUS matrix cork, the command bd1way(cork) returns

```
$test:
17.08

$crit:
34.09
```

So comparing one-step M-estimators, we fail to reject the hypothesis that the typical weight of a cork boring is the same for all four sides of a tree. If we compare groups using MOM in conjunction with method RMPB4 in Box 11.5, the significance level is .385. ■

11.7.3 Percentile Bootstrap Methods Based on Difference Scores

The following method, based on difference scores, has been found to have practical value, particularly in terms of controlling Type I error probabilities when sample sizes are very small. Rather than test

$$H_0 : \theta_1 = \cdots = \theta_J, \qquad (11.11)$$

first consider the goal of testing the hypothesis that a measure of location associated with the difference scores $D_{ij} = X_{ij} - X_{i,j+1}$ has the value zero. That is, use the difference between the ith observation in group j and the ith observation in group $j+1, j = 1, \ldots, J-1$. Let θ_j be any measure of location associated with the D_{ij} values. So, for example, θ_1 might be the population value associated with MOM based on the difference scores between groups 1 and 2, and θ_2 the population MOM value associated with difference scores between groups 2 and 3. A simple alternative to

Equation (11.11) is to test

$$H_0 : \theta_1 = \cdots = \theta_{J-1} = 0, \tag{11.12}$$

the hypothesis that the typical difference scores do not differ and are all equal to zero. However, a criticism of this approach is that the outcome can depend on how we order the groups. That is, rather than take differences between groups 1 and 2, we could just as easily take differences between groups 1 and 3 instead, which might alter our conclusions about whether to reject. We can avoid this problem by instead taking differences among all pairs of groups. There are a total of

$$L = \frac{J^2 - J}{2}$$

such differences, which are labeled $D_{i\ell}$, $i = 1, \ldots, n$; $\ell = 1, \ldots, L$. In particular,

$$D_{i1} = X_{i1} - X_{i2},$$
$$D_{i2} = X_{i1} - X_{i3},$$
$$\vdots$$
$$D_{iL} = X_{i,J-1} - X_{iJ}.$$

EXAMPLE. For four groups ($J = 4$), there are $L = 6$ differences, given by

$$D_{i1} = X_{i1} - X_{i2},$$
$$D_{i2} = X_{i1} - X_{i3},$$
$$D_{i3} = X_{i1} - X_{i4},$$
$$D_{i4} = X_{i2} - X_{i3},$$
$$D_{i5} = X_{i2} - X_{i4},$$
$$D_{i6} = X_{i3} - X_{i4}.$$

■

The goal is to test

$$H_0 : \theta_1 = \cdots = \theta_L = 0, \tag{11.13}$$

where θ_ℓ is the population measure of location associated with the ℓth set of difference scores, $D_{i\ell}$ ($i = 1, \ldots, n$). To test H_0 given by Equation (11.13), resample vectors of D values; but unlike the bootstrap-t, observations are not centered. That is, a bootstrap sample now consists of resampling with replacement n rows from the matrix

$$\begin{pmatrix} D_{11}, \ldots, D_{1L} \\ \vdots \\ D_{n1}, \ldots, D_{nL} \end{pmatrix},$$

yielding

$$\begin{pmatrix} D^*_{11}, \ldots, D^*_{1L} \\ \vdots \\ D^*_{n1}, \ldots, D^*_{nL} \end{pmatrix}.$$

For each of the L columns of the D^* matrix, compute MOM or whatever measure of location is of interest, and for the ℓth column label the result $\hat{\theta}^*_\ell$ ($\ell = 1, \ldots, L$). Next, repeat this B times, yielding $\hat{\theta}^*_{\ell b}$, $b = 1, \ldots, B$, and then determine how deeply the vector $\mathbf{0} = (0, \ldots, 0)$, having length L, is nested within the bootstrap values $\hat{\theta}^*_{\ell b}$. For two groups, this is tantamount to determining how many bootstrap values are greater than zero. If most are greater than (or less than) zero, we reject, as indicated in Section 11.3.1 in conjunction with method RMPB1. For more than two groups, you use a method similar to the approach in Box 11.5. The details are relegated to Box 11.6.

BOX 11.6 Repeated Measures ANOVA Based on Difference Scores and the Depth of Zero

Goal:
Test the hypothesis, given by Equation (11.13), that all difference scores have a typical value of zero.

1. Let $\hat{\theta}_\ell$ be the estimate of θ_ℓ. Compute bootstrap estimates as described in Section 11.7.3 and label them $\hat{\theta}^*_{\ell b}$, $\ell = 1, \ldots, L$; $b = 1, \ldots, B$.
2. Compute the L-by-L matrix

$$S_{\ell\ell'} = \frac{1}{B-1} \sum_{b=1}^{B} \left(\hat{\theta}^*_{\ell b} - \hat{\theta}_\ell\right)\left(\hat{\theta}^*_{\ell' b} - \hat{\theta}_{\ell'}\right).$$

Readers familiar with multivariate statistical methods might notice that $S_{\ell\ell'}$ uses $\hat{\theta}_\ell$ (the estimate of θ_ℓ based on the original difference values) rather than the seemingly more natural $\bar{\theta}^*_\ell$, where

$$\bar{\theta}^*_\ell = \frac{1}{B} \sum_{b=1}^{B} \hat{\theta}^*_{\ell b}.$$

If $\bar{\theta}^*_\ell$ is used, unsatisfactory control over the probability of a Type I error can result.

Continued

BOX 11.6 *(Continued)*

3. Let $\hat{\theta} = (\hat{\theta}_1, \ldots, \hat{\theta}_L)$, $\hat{\theta}^*_b = (\hat{\theta}^*_{1b}, \ldots, \hat{\theta}^*_{Lb})$, and compute
$$d_b = \left(\hat{\theta}^*_b - \hat{\theta}\right) \mathbf{S}^{-1} \left(\hat{\theta}^*_b - \hat{\theta}\right)',$$
where S is the matrix corresponding to $S_{\ell\ell'}$.
4. Put the d_b values in ascending order: $d_{(1)} \leq \cdots \leq d_{(B)}$.
5. Let
$$\mathbf{0} = (0, \ldots, 0),$$
having length L.
6. Compute
$$D = (\mathbf{0} - \hat{\theta})\mathbf{S}^{-1}(\mathbf{0} - \hat{\theta})'.$$
D measures how far away the null hypothesis is from the observed measures of location (based on the original data). In effect, D measures how deeply $\mathbf{0}$ is nested within the cloud of bootstrap values.
7. Reject if $D \geq d_{(u)}$, where $u = (1 - \alpha)B$, rounded to the nearest integer.

11.7.4 S-PLUS Function rmdzero

The S-PLUS function

rmdzero(x,est = mom, grp = NA, nboot = NA,...)

performs the test on difference scores outlined in Box 11.6.

EXAMPLE. For the cork data in Table 11.3, rmdzero returns a significance level of .044, so in particular reject with $\alpha = .05$. That is, conclude that the typical difference score is not equal to zero for all pairs of groups. This result is in sharp contrast to comparing marginal measures of location based on MOM and the method in Box 11.5, which has a significance level of .385. Currently, it seems that the method in Box 11.5 does an excellent job of avoiding Type I error probabilities larger than the nominal level, but that in many situations it is too conservative. That is, the actual probability of a Type I error can be substantially smaller than the nominal level, suggesting that it might have relatively poor power. Switching to difference scores appears to correct this problem. ■

11.8 Comments on Which Method to Use

Several reasonable methods for comparing groups have been described, so there is the issue of which one to use. As usual, no method is perfect in all situations.

The expectation is that in many situations where groups differ, all methods based on means perform poorly relative to approaches based on some robust measure of location, such as MOM or a 20% trimmed mean. Currently, with a sample size as small as 21, the bootstrap-*t* method in Section 11.6.3, used in conjunction with 20% trimmed means, appears to provide excellent control over the probability of a Type I error. Its power compares reasonably well to most other methods that might be used. But as noted in previous chapters, different methods are sensitive to different features of your data, and arguments for some other measure of location, such as MOM, have been made.

The percentile bootstrap methods in Section 11.7 also do an excellent job of avoiding Type I errors greater than the nominal level, but there are indications that method RMPB3 in Section 11.7.1 can be too conservative when sample sizes are small. That is, the actual probability of a Type I error can be substantially less than α, suggesting that some other method might provide better power. Nevertheless, if there is specific interest in comparing marginal distributions with M-estimators, it is suggested that method RMPB3 in Section 11.7.1 be used and that the sample size be greater than 20. There is some indirect evidence that a larger sample size might be needed when using this method. (This is in contrast to comparing independent groups, where sample sizes greater than 20 seem to suffice.) Also, it can be used to compare groups based on MOM. But with very small sample sizes there are some indications that its power might be inadequate, at least in some situations, relative to other techniques that might be used.

Currently, among the techniques covered in this chapter, it seems that the two best methods for controlling Type I error probabilities and simultaneously providing reasonably high power are the bootstrap-*t* method based on 20% trimmed means and the percentile bootstrap method in Box 11.6 used in conjunction with MOM, which uses difference scores. (Other excellent options are covered in Chapter 12.) With near certainty, situations arise where some other technique is more optimal, but typically the improvement is small. However, comparing groups with MOM is not the same as comparing means, trimmed means, or M-estimators, and certainly there will be situations where some other estimator has higher power than any method based on MOM or a 20% trimmed mean. If the goal is to maximize power, several methods are contenders for routine use, but as usual, standard methods based on means are generally the least satisfactory. With sufficiently large sample sizes, trimmed means can be compared without resorting to the bootstrap-*t* method, but it remains unclear just how large the sample size must be. When using MOM, currently a bootstrap method is required regardless of how large the sample size might be.

As for the issue of whether to use difference scores versus robust measures of location based on the marginal distributions, each approach provides a different perspective on how groups differ, and they can give different results regarding whether groups are significantly different. There is some evidence that difference scores typically provide more power and better control over the probability of a Type I error, but more detailed study is needed to resolve this issue.

As previously mentioned, method RMPB4, outlined in Box 11.5 (and performed by the S-PLUS function ddep), is very conservative in terms of Type I errors, meaning that when testing at the .05 level, say, often the actual probability of a Type I error will be less than or equal to α and typically smaller than any other method described in this chapter. But a concern is that the actual Type I error probability can be substantially smaller than the nominal level, resulting in low power relative to many other methods you might use.

11.9 Between-by-Within, or Split-Plot, Designs

Chapter 10 covered a two-way ANOVA design involving JK independent groups, with J levels associated with the first factor and K levels with the second. A between-by-within, or split-plot, design refers to a situation where for a given level of the first factor, the measures associated with the K levels of the second factor are dependent instead. As a simple illustration, again consider the situation where endurance is measured before and after training, but now we have two training methods. Moreover, a sample of n_1 athletes undergo training method 1 and a different, independent sample of athletes undergo training method 2. So between methods, observations are independent, but they are possibly dependent between occasions. The population means are:

	Time (B)	
Method (A)	1	2
1	μ_{11}	μ_{12}
2	μ_{21}	μ_{22}

Moreover, the notions of main effects and interactions (described in Chapter 10) extend immediately to the situation at hand. For example, the hypothesis of no main effect for factor A (method) is

$$H_0: \frac{\mu_{11} + \mu_{12}}{2} = \frac{\mu_{21} + \mu_{22}}{2},$$

and the hypothesis of no interaction is

$$H_0: \mu_{11} - \mu_{12} = \mu_{21} - \mu_{22}.$$

11.9.1 Method for Trimmed Means

The computational details on how to compare trimmed means (including means as a special case) are tedious at best. Presumably most readers are more interested in applying the method versus understanding the computational details, so the bulk of the computations are relegated to Box 11.7 (assuming familiarity with basic matrix algebra). In Box 11.7, X_{ijk} represents the ith observation in level j of factor A and level k of factor B, $i = 1, \dots, n_j$; $j = 1, \dots, J$ and $k = 1, \dots, K$. Once the quantities Q and A are computed as described in Box 11.7, let

$$c = k + 2A - \frac{6A}{k+2}.$$

When the null hypothesis is true, Q/c has, approximately, an F-distribution with $\nu_1 = k$ and $\nu_2 = k(k+2)/3A$ degrees of freedom (where k represents the number of rows corresponding to the matrix $\mathbf{C}$ in Box 11.7). For factor A, $k = J - 1$; for factor B, $k = K - 1$; and for interactions, $k = (J-1)(K-1)$.

DECISION RULE: Reject if $Q/c \geq f_{1-\alpha}$, the $1-\alpha$ quantile of an F-distribution with ν_1 and ν_2 degrees of freedom. (Also see Keselman, Algina, Boik, & Wilcox, 1999).

BOX 11.7 Computations for a Split-Plot Design When Using Trimmed Means

The hypotheses of no main effects and no interactions can be written in the form

$$H_0 : \mathbf{C}\mu_t = \mathbf{0},$$

where $\mathbf{C}$ is a k-by-JK matrix (having rank k) that reflects the null hypothesis of interest. (Here, μ_t is a column vector of population trimmed means having length JK.) Let $\mathbf{C}_J$ be defined as in Box 10.4 and let $\mathbf{j}'_J$ be a $1 \times J$ matrix of 1's. Then for factor A, $\mathbf{C} = \mathbf{C}_J \otimes \mathbf{j}'_K$ and $k = J - 1$. For factor B, $\mathbf{C} = \mathbf{j}'_J \otimes \mathbf{C}_K$, $k = K - 1$, and the test for no interactions uses $\mathbf{C} = \mathbf{C}_J \otimes \mathbf{C}_K$, where now $k = (J-1)(K-1)$.

For every level of factor A, there are K dependent random variables, and each pair of these dependent random variables has a Winsorized covariance that must be estimated. For fixed j, let g_j be the number of observations trimmed from both tails. (If the amount of trimming is γ, $g_j = [\gamma n_j]$, as in Chapter 3.) The Winsorized covariance between the mth and ℓth levels of factor B is estimated with

$$s_{jm\ell} = \frac{1}{n_j - 1} \sum_{i=1}^{n_j} \left(Y_{ijm} - \bar{Y}_{.jm}\right)\left(Y_{ij\ell} - \bar{Y}_{.j\ell}\right),$$

where $\bar{Y}_{.jm} = \sum_{i=1}^{n_j} Y_{ijm}/n_j$,

$$Y_{ijk} = \begin{cases} X_{(g_j+1),jk} & \text{if } X_{ijk} \leq X_{(g_j+1),jk} \\ X_{ijk} & \text{if } X_{(g_j+1),jk} < X_{ij} < X_{(n-g_j),jk} \\ X_{(n-g_j),jk} & \text{if } X_{ijk} \geq X_{(n_j-g),jk}. \end{cases}$$

For fixed j, let $\mathbf{S}_j = (s_{jm\ell})$, which is the matrix of Winsorized variances and covariances for level j of factor A. Let

$$\mathbf{V}_j = \frac{(n_j - 1)\mathbf{S}_j}{h_j(h_j - 1)}, \qquad j = 1, \ldots, J,$$

Continued

BOX 11.7 (*Continued*) and let $\mathbf{V} = \text{diag}(\mathbf{V}_1, \ldots, \mathbf{V}_J)$ be a block diagonal matrix. The test statistic is

$$Q = \bar{\mathbf{X}}'\mathbf{C}'(\mathbf{CVC}')^{-1}\mathbf{C}\bar{\mathbf{X}}, \tag{11.14}$$

where $\bar{\mathbf{X}}' = (\bar{X}_{t11}, \ldots, \bar{X}_{tJK})$. Let $\mathbf{I}_{K\times K}$ be a K-by-K identity matrix, let $\mathbf{Q}_j$ be a JK-by-JK block diagonal matrix (consisting of J blocks, each block being a K-by-K matrix), where the tth block ($t = 1, \ldots, J$) along the diagonal of $\mathbf{Q}_j$ is $\mathbf{I}_{K\times K}$ if $t = j$, and all other elements are zero. (For example, if $J = 3$ and $K = 4$, then $\mathbf{Q}_1$ is a 12-by-12 matrix block diagonal matrix, where the first block is a 4-by-4 identity matrix and all other elements are zero. As for $\mathbf{Q}_2$, the second block is an identity matrix, and all other elements are zero.) Compute

$$A = \frac{1}{2}\sum_{j}^{J} \frac{\text{tr}(\{\mathbf{VC}'(\mathbf{CVC}')^{-1}\mathbf{CQ}_j\}^2) + \{\text{tr}(\mathbf{VC}'(\mathbf{CVC}')^{-1}\mathbf{CQ}_j)\}^2}{h_j - 1}.$$

The remaining calculations are described in the text.

11.9.2 S-PLUS Function tsplit

The S-PLUS function

tsplit(J, K, x, tr = 0.2, grp = c(1:p), p = J * K)

performs the calculations in Box 11.4. Here, J is the number of independent groups, K is the number of dependent groups, x is any S-PLUS variable that is a matrix or has list mode, and, as usual, tr indicates the amount of trimming, which defaults to .2 if unspecified. If the data are stored in list mode, it is assumed that x[[1]] contains the data for level 1 of both factors, x[[2]] contains the data for level 1 of the first factor and level 2 of the second, and so on. If the data are not stored in the proper order, grp can be used to indicate how they are stored. For example, if a 2-by-2 design is being used, the S-PLUS command

tsplit(2,2,x,grp=c(3,1,2,4))

indicates that the data for the first level of both factors are stored in x[[3]], the data for level 1 of factor A and level 2 of factor B are in x[[1]], and so forth. The last argument, p, can be ignored. It is needed only to satisfy certain requirements of S-PLUS. If the data are stored in a matrix, it is assumed that the first column contains the data for level 1 of both factors, the second column contains the data for level 1 of the first factor and level 2 of the second, and so forth.

11.9.3 A Bootstrap-*t* Method

To apply a bootstrap-*t* method when working with trimmed means, you first center the data in the usual way. In the present context, this means you compute

$$C_{ijk} = X_{ijk} - \bar{X}_{tjk},$$

$i = 1, \ldots, n_j$; $j = 1, \ldots, J$; and $k = 1, \ldots, K$. That is, for the group corresponding to the jth level of factor A and the kth level of factor B, subtract the corresponding trimmed mean from each of the observations. Next, for the jth level of factor A, generate a bootstrap sample by resampling with replacement n_j vectors of observations from the data in level j of factor A. That is, for each level of factor A, you have an n_j-by-K matrix of data, and you generate a bootstrap sample from this matrix of data as described in Section 11.6.3. Label the resulting bootstrap samples C^*_{ijk}. Compute the test statistic F, based on the C^*_{ijk} values as described in Box 11.3, and label the result F^*. Repeat this B times, yielding $F^*_1, \ldots, F^*_B$, and then put these B values in ascending order, yielding $F^*_{(1)} \leq \cdots \leq F^*_{(B)}$. Next, compute F using the original data (the X_{ijk} values) as described in Box 11.3 and reject if $F \geq F^*_{(c)}$, where $c = (1 - \alpha)$ rounded to the nearest integer.

A crude rule that seems to apply to a wide variety of situations is: The more distributions associated with groups differ, the more beneficial it is to use some type of bootstrap method, at least when sample sizes are small. Keselman, Algina, Wilcox, and Kowalchuk (2000) compared the bootstrap-t method just described to the nonbootstrap method for a split-plot design, covered in Section 11.9.1. For the situations they examined, this rule did not apply; it was found that the bootstrap-t offered little or no advantage. Their study included situations where the correlations (or covariances) among the dependent groups differ across the independent groups being compared. However, the more complicated the design, the more difficult it becomes to consider all the factors that might influence operating characteristics of a particular method. One limitation of their study was that the differences among the covariances were taken to be relatively small. Another issue that has not been addressed is how the bootstrap-t performs when distributions differ in skewness. Having differences in skewness is known to be important when dealing with the simple problem of comparing two groups only. There is no reason to assume that this problem diminishes as the number of groups increases, and indeed there are reasons to suspect that it becomes a more serious problem. So currently, it seems that if groups do not differ in any manner or the distributions differ slightly, it makes little difference whether you use a bootstrap-t versus a nonbootstrap method for comparing trimmed means. However, if distributions differ in shape, there is indirect evidence that the bootstrap-t might offer an advantage when using a split-plot design, but the extent to which this is true is not well understood.

11.9.4 S-PLUS Function tsplitbt

The S-PLUS function

```
tsplitbt(J,K,x,tr=.2,alpha=.05,JK=J*K,grp=c(1 : JK),
    nboot=599,monitor=F)
```

performs a bootstrap-t method for a split-plot design, as just described. The data are assumed to be arranged as indicated in conjunction with the S-PLUS function tsplit (as described in Section 11.9.2), and the arguments J, K, tr, and alpha have the same meaning as before. The argument JK can be ignored, and grp can be used to rearrange the data if they are not stored as expected by the function. (See Section 10.3.1 for an illustration on how to use grp.) The argument monitor can be used to monitor the progress of the function. If we set monitor=T, the function prints a message each time it completes a bootstrap iteration. This way you can get some sense of how long it will take before the computations are complete.

11.9.5 Using MOMs, Medians, and M-Estimators

Comparing groups based on MOMs, medians, and M-estimators in a split-plot design is possible using extensions of the bootstrap methods considered in this chapter as well as Chapters 9 and 10. Generally, bootstrap samples must be generated in a manner that reflects the dependence among the levels of factor B, and then some appropriate hypothesis is tested using some slight variation of one of the methods already described. In fact, the problem is *not* deriving a method, but rather deciding which method should be used among the many that are available. This section summarizes some approaches that are motivated by published papers, but it is noted that alternative methods might be more optimal and that for the specific situation at hand more research is needed to better understand the relative merits of different techniques. (Methods in Chapter 12 provide yet another perspective and should be considered before analyzing data.)

Again consider a two-way design where factor A consists of J independent groups and factor B corresponds to K dependent groups. First consider the dependent groups. One approach to comparing these K groups, ignoring factor A, is simply to form difference scores and then to apply the method in Box 11.6. More precisely, imagine you observe X_{ijk} ($i = 1, \dots, n_j$; $j = 1, \dots, J$; $k = 1, \dots, K$). That is, X_{ijk} is the ith observation in level j of factor A and level k of factor B. Note that if we ignore the levels of factor A, we can write the data as Y_{ik}, $i = 1, \dots, N$; $k = 1, \dots, K$, where $N = \sum n_j$. Now consider levels k and k' of factor B ($k < k'$) and set

$$D_{ikk'} = Y_{ik} - Y_{ik'};$$

let $\theta_{kk'}$ be some measure of location associated with $D_{ikk'}$. Then the levels of factor B can be compared, ignoring factor A, by testing

$$\theta_{12} = \cdots = \theta_{k-1,k} = 0 \tag{11.15}$$

using the method in Section 11.7.3. In words, the null hypothesis is that the typical difference score between any two levels of factor B, ignoring factor A, is zero.

As for factor A, ignoring factor B, one approach is as follows. Momentarily focus on the first level of factor B and note that the levels of factor A can be compared using the method in Box 9.7. That is, the null hypothesis of no differences among the levels of factor A is

$$H_0 : \theta_{11} = \theta_{21} = \cdots = \theta_{J1},$$

where of course these J groups are independent, and a percentile bootstrap method can be used as described in Chapter 9. More generally, for any level of factor B — say, the kth — no main effects is

$$H_0 : \theta_{1k} = \theta_{2k} = \cdots = \theta_{Jk},$$

($k = 1, \ldots, K$), and the goal is to test the hypothesis that these K hypotheses are simultaneously true. Here we take this to mean that we want to test

$$H_0: \theta_{11} - \theta_{21} = \cdots \theta_{J-1,1} - \theta_{J1} = \cdots = \theta_{J-1,K} - \theta_{JK} = 0 \tag{11.16}$$

In this last equation, there are $C = K(J^2 - J)/2$ differences, all of which are hypothesized to be equal to zero. Generalizing the method in Box 9.7, for each level of factor A, generate bootstrap samples as is appropriate for K dependent groups (see Section 11.7.1), and then test Equation (11.16). To briefly outline the computations, label the C differences as $\delta_1, \ldots, \delta_C$ and then denote bootstrap estimates by $\hat{\delta}_c^*$ ($c = 1, \ldots C$). For example, $\hat{\delta}_1^* = \theta_{11}^* - \theta_{21}^*$. Then we test Equation (11.16) by determining how deeply the vector $(0, \ldots, 0)$, having length C, is nested within B bootstrap values, which is done in the manner described in Box 11.6.

For factor A an alternative approach is to average the measures of location across the K levels of factor B and then to proceed in the manner described in Box 9.7. In symbols, let

$$\bar{\theta}_{j.} = \frac{1}{K}\sum_{k=1}^{K} \theta_{jk},$$

in which case the goal is to test

$$H_0: \bar{\theta}_{1.} = \cdots = \bar{\theta}_{J.}.$$

Again for each level of factor A, you generate B samples for the K dependent groups as described in Section 11.7.1 in conjunction with method RMPB4. Let $\bar{\theta}_{j.}^*$ be the bootstrap estimate for the jth level of factor A. For levels j and j' of factor A, $j < j'$, set $\delta_{jj'}^* = \bar{\theta}_{j.}^* - \bar{\theta}_{j'.}^*$. Then you determine how deeply $\mathbf{0}$, having length $(J^2 - J)/2$, is nested within the B bootstrap values for $\delta_{jj'}^*$ using the method described in Box 11.6.

As for interactions, again there are several approaches one might adopt. Here an approach based on difference scores among the dependent groups is used. To explain, first consider a 2-by-2 design, and for the first level of factor A let $D_{i1} = X_{i11} - X_{i12}$, $i = 1, \ldots, n_1$. Similarly, for level 2 of factor A let $D_{i2} = X_{i21} - X_{i22}$, $i = 1, \ldots, n_2$, and let θ_{d1} and θ_{d2} be the population measure of location corresponding to the D_{i1} and D_{i2} values, respectively. Then the hypothesis of no interaction is taken to be

$$H_0: \theta_{d1} = \theta_{d2},$$

which of course is the same as

$$H_0: \theta_{d1} - \theta_{d2} = 0. \tag{11.17}$$

Again the basic strategy for testing hypotheses is generating bootstrap estimates and determining how deeply $\mathbf{0}$ is embedded in the B values that result. For the more general case of a J-by-K design, there are a total of

$$C = \frac{J^2 - J}{2} \times \frac{K^2 - K}{2}$$

equalities, one for each pairwise difference among the levels of factor B and any two levels of factor A.

11.9.6 S-PLUS Functions sppba, sppbb, and sppbi

The S-PLUS function

sppba(J,K,x,est=mom,grp=c(1:JK),avg=F,nboot=500,...)

tests the hypothesis of no main effects for factor A in the manner just described. Setting the argument avg to T (for true) indicates that the averages of the measures of location (the $\bar{\theta}_{j.}$ values) will be used. That is, $H_0: \bar{\theta}_{1.} = \cdots = \bar{\theta}_{J.}$ is tested. Otherwise, the hypothesis given by Equation (11.16) is tested. The remaining arguments have their usual meaning. The S-PLUS function

sppbb(J,K,x,est=mom,grp=c(1:JK),nboot=500,...)

tests the hypothesis of no main effects for factor B (as described in the previous section), and

sppbi(J,K,x,est=mom,grp=c(1:JK),nboot=500,...)

tests the hypothesis of no interactions.

EXAMPLE. We examine once more the EEG measures for murderers versus a control group, only now we use the data for all four sites in the brain where measures were taken. If we label the typical measures for the control group as $\theta_{11},\ldots,\theta_{14}$ and the typical measures for the murderers as $\theta_{21},\ldots,\theta_{24}$, we have a 2-by-4 between-by-within design, and a possible approach to comparing the groups is testing

$$H_0: \theta_{11} - \theta_{21} = \theta_{12} - \theta_{22} = \theta_{13} - \theta_{23} = \theta_{14} - \theta_{24} = 0.$$

This can be done with the S-PLUS function sppba with the argument avg set to F. If the data are stored in a matrix called eeg having eight columns, with the first four corresponding to the control group, then the command sppba(2,4,eeg) performs the calculations based on the MOM measure of location and returns a significance level of .098. An alternative approach is to average the value of MOM over the four brain sites for each group and then to compare these averages. That is, test $H_0: \bar{\theta}_{1.} = \bar{\theta}_{2.}$, where $\bar{\theta}_{j.} = \sum \theta_{jk}/4$. This can be done with the command

sppba(2,4,eeg,avg=T).

Now the significance level is .5, so we see that the significance level can vary tremendously depending on how we compare the groups. ■

11.9.7 The S-PLUS Function selby

Section 1.1.6 describes an S-PLUS function that is convenient when data are stored in a matrix, with one of the columns indicating the group to which an observation belongs. Basically, the function takes the data in a matrix and sorts it into groups in a manner that can be used by the S-PLUS functions written for this book. It is noted that this function can be used when there are multiple columns of data rather than just one, as illustrated in Section 1.1.6. That is, the third argument, coln, can be a vector.

EXAMPLE. Imagine that you have four dependent variables stored in columns 1–4 of the S-PLUS matrix mat and that column 5 indicates the group to which a vector of observations belongs. So, for example, the data might look like this:

v1	v2	v3	v4	G
34	42	63	19	1
26	99	45	29	1
33	42	18	32	2
		⋮		

The problem is to sort the data into groups, based on the values listed in column 5 (under G), and to store the data in a manner that can be used by, for example, the functions in Section 11.9.6. The command

```
dat <- selby(mat,5,c(1:4))
```

accomplishes this goal. (This command stores the data in dat$x, having list mode.) So if there are five groups corresponding to the values in column 5 of mat, interactions could be investigated with the command

```
sppbi(5,4,dat$x).
```

■

11.10 Exercises

1. For the data in Table 11.3, perform the paired T-test for means using the weights for the east and south sides of the trees. Verify that the significance level is .09.
2. Repeat the previous exercise, but use 20% trimmed means instead, using the difference scores in conjunction with the S-PLUS function trimci in Section 4.9.4. Note that the significance level is .049.
3. If in Exercise 1 you compare the marginal 20% trimmed means with the S-PLUS function yuend in Section 11.2.1, verify that now the significance level is .121.
4. Generally, why is it possible to get a different significance level comparing the marginal trimmed means than when making inferences about the trimmed mean of the difference scores?
5. Based on what is currently known, would you expect more power when comparing the marginal trimmed means or when making inferences about the trimmed mean of the difference scores?

6. Repeat Exercise 1, but now use a percentile bootstrap method based on MOM and the difference scores. That is, use with the S-PLUS function rmmcppb. Verify that the significance level is .092.
7. Repeat the last exercise, except now compare the marginal measures of location based on MOM. So in the S-PLUS function rmmcppb, set dif=F. Verify that the significance level is .578.
8. Based on the S-PLUS output from rmmcppb, if you were to repeat the last exercise, except adding 6.75 to every observation in the second group, would you reject with $\alpha = .05$? Verify your answer by adding 6.75 to every observation in the second group and invoking rmmcpp.
9. For the data in Exercise 1, verify that the .95 confidence interval for the median of the difference scores, based on the method in Chapter 4, is $(-7, .55)$.
10. Compare the marginal medians of the data in Exercise 1 using the S-PLUS function rmmcppb in Section 11.3.2. Verify that the .95 confidence interval for the difference between the population medians is $(-14, 4.5)$.
11. In this chapter and in Chapter 8 reference was made to a study dealing with EEG measures for murderers versus a control group. In another portion of this study, the measures for four sites in the brain were found to be as reported in Table 11.4. (These observations differ from those used in previous illustrations.) Using difference scores, compare site 1 and site 3 for murderers using both means and 20% trimmed means. (Use the S-PLUS function trimci.) Verify that the significance levels are .61 and .27, respectively.
12. For the data in Table 11.4, compare all four sites for murderers with the S-PLUS function rmanova in Section 11.6.2. Verify that you reject with both means and 20% trimmed means with $\alpha = .01$.

TABLE 11.4 EEG Measures

Murderers				Controls			
Site 1	Site 2	Site 3	Site 4	Site 1	Site 2	Site 3	Site 4
−0.19	−1.05	−0.04	−1.48	0.45	−1.57	0.96	−0.76
0.39	−1.39	1.08	−0.95	0.24	−1.12	1.25	−0.31
0.09	−0.49	0.19	−2.14	0.33	−1.53	−0.64	−1.57
0.54	−0.76	1.06	−0.33	0.81	−0.28	0.29	−1.25
0.78	−0.36	−0.36	−1.09	1.30	−0.53	−0.05	−0.98
0.59	−1.17	0.80	−1.06	0.46	−1.09	0.09	−2.35
0.04	−1.75	0.11	−1.41	−0.01	−1.98	1.07	−0.94
0.38	−0.83	1.05	−0.29	1.11	−0.84	0.88	−1.62
0.25	−0.40	−0.07	−1.90	0.16	−1.25	0.28	−0.55
0.01	−1.06	0.50	−0.07	1.02	−1.07	0.00	−1.31
0.40	−1.36	0.54	−0.63	0.67	−0.92	−0.08	−2.18
0.52	−1.30	1.69	−0.22	1.37	−0.69	0.62	−0.86
1.35	−0.45	0.01	−1.22	0.59	−0.64	−0.02	−0.06
0.02	−0.86	−0.07	−1.65	0.66	−1.43	−0.48	−0.93

13. For the data in Table 11.4, compare all groups using the S-PLUS function tsplit in Section 11.9.2. Verify that when comparing murderers to the control group, the main effect has a significance level of .9955.
14. For the data in Table 11.4, compare murderers versus controls using MOM and difference scores as outlined in Section 11.9.5. That is, use the S-PLUS function sppba in Section 11.9.6. Verify that the significance level is .138, and compare this result with the previous exercise.
15. In the previous exercise, you could have used averages rather than difference scores when comparing the murderers to the controls using the approach described in Section 11.9.6. Using the S-PLUS function sppba with the argument avg set equal to T, verify that the significance level is .89.
16. Compare the results from the previous three exercises and comment on finding the optimal method for detecting a true difference among groups.
17. For the data in Table 11.4, use the S-PLUS function sppbi to test the hypothesis of no interactions based on MOM. Verify that the significance level is .1.

MULTIPLE COMPARISONS

Chapters 9–11 describe how to test the hypothesis that two or more groups have a common measure of location. When working with means, for example, one goal was to test

$$H_0: \mu_1 = \cdots = \mu_J, \tag{12.1}$$

the hypothesis that J groups have equal population means. It is common, however, to want to know more about how the groups compare: Which groups differ? How do they differ, and by how much? When addressing these questions, what role should the methods in Chapters 9–11 play? A very common strategy is first to test Equation (12.1) and, if a nonsignificant result is obtained, to stop and fail to declare any of the groups to be different. One goal in this chapter is to cover modern insights into the relative merits of this approach.

The other general goal is to describe methods for controlling what is called the *familywise error rate* (FWE) (sometimes called the *experimentwise error rate*) that is, the probability of making at least one Type I error when performing multiple tests. To elaborate, imagine you have four independent groups and for the moment assume normality and homoscedasticity. Suppose that for each pair of means we test the hypothesis of equal means using Student's T. That is, the goal is to test

$$H_0: \mu_1 = \mu_2,$$
$$H_0: \mu_1 = \mu_3,$$
$$H_0: \mu_1 = \mu_4,$$
$$H_0: \mu_2 = \mu_3,$$
$$H_0: \mu_2 = \mu_4,$$
$$H_0: \mu_3 = \mu_4.$$

If we test each of these hypotheses at the .05 level ($\alpha = .05$), then of course the probability of a Type I error will be .05 for each test. But what is the probability of *at least one* Type I error among the six hypotheses of interest? That is, what is the probability of erroneously concluding that one or more pairs of means differ when in fact none differ at all? Determining this probability is complicated by the fact that the individual tests are not all independent. For example, when testing the first two hypotheses, the corresponding Student's T-tests will be dependent because both have the first sample mean in the numerator of the test statistic. That is, $\bar{X}_1$ is used both times. If all six tests were independent, then we could use the binomial probability function to determine the probability of at least one Type I error. But because there is dependence, special methods are required. Bernhardson (1975) describes a situation where the probability of at least one Type I error (FWE) can be as high as .29 for the situation at hand. When comparing all pairs of 10 groups, this probability can be as high as .59.

12.1 Homoscedastic Methods for the Means of Independent Groups

We begin by describing classic methods for comparing means that assume normality and equal variances. Problems with these classic methods have long been established (e.g., Wilcox, 1996c), but they are commonly used; their pitfalls remain relatively unknown, so they are included here merely for completeness and future reference.

12.1.1 Fisher's Least Significant Difference (LSD) Method

One of the earliest strategies for comparing multiple groups is the so-called *least significant difference* (LSD) method due to Sir Ronald Fisher. Assuming normality and homoscedasticity, first perform the ANOVA F-test in Section 9.1. If a significant result is obtained, apply Student's T to all pairs of means, but unlike the approach in Chapter 8, typically the assumption of equal variances is taken advantage of by using the estimate of the assumed common variance when performing Student's T. Under normality and homoscedasticity, this has the advantage of increasing the degrees of freedom, which in turn can mean more power.

To be more concrete, suppose the ANOVA F-test in Section 9.1 is significant for some specified value of α and let MSWG (described in Box 9.1) be the estimate of the assumed common variance. To test

$$H_0 : \mu_j = \mu_k, \tag{12.2}$$

the hypothesis that the mean of the jth group is equal to the mean of the kth group, compute

$$T = \frac{\bar{X}_j - \bar{X}_k}{\sqrt{\text{MSWG}\left(\frac{1}{n_j} + \frac{1}{n_k}\right)}}. \tag{12.3}$$

When the assumptions of normality and homoscedasticity are met, T has a Student's T-distribution with $\nu = N - J$ degrees of freedom, where J is the number of groups being compared and $N = \sum n_j$ is the total number of observations in all J groups. So when comparing the jth group to the kth group, you reject the hypothesis

TABLE 12.1 Hypothetical Data for Three Groups

G1	G2	G3
3	4	6
5	4	7
2	3	8
4	8	6
8	7	7
4	4	9
3	2	10
9	5	9

of equal means if

$$|T| \geq t_{1-\alpha/2},$$

where $t_{1-\alpha/2}$ is the $1 - \alpha/2$ quantile of Student's T-distribution with $N - J$ degrees of freedom.

EXAMPLE. For the data in Table 12.1, it can be shown that MSWG = 4.14, the sample means are $\bar{X}_1 = 4.75$, $\bar{X}_2 = 4.62$, and $\bar{X}_3 = 7.75$, and the F-test is significant, so according to Fisher's LSD procedure, you would proceed by comparing each pair of groups with Student's T-test. For the first and second groups,

$$T = \frac{4.75 - 4.62}{\sqrt{4.14(\frac{1}{8} + \frac{1}{8})}} = .128.$$

The degrees of freedom are $\nu = 21$, and with $\alpha = .05$, Table 4 in Appendix B says that the critical value is 2.08. Therefore, you fail to reject. That is, the F-test indicates that there is a difference among the three groups, but Student's T suggests that the difference does not correspond to groups 1 and 2. For groups 1 and 3,

$$T = \frac{4.75 - 7.75}{\sqrt{4.14(\frac{1}{8} + \frac{1}{8})}} = -2.94,$$

and because 2.94 is greater than the critical value, 2.08, reject. That is, conclude that groups 1 and 3 differ. In a similar manner, you conclude that groups 2 and 3 differ as well, because $T = 3.08$. ■

When the assumptions of normality and homoscedasticity are true, Fisher's method controls FWE when $J = 3$. That is, the probability of at least one Type I error will be less than or equal to α. However, when there are more than three groups ($J > 3$), this is no longer true (Hayter, 1986). To gain some intuition as to why, suppose four groups are to be compared, the first three have equal means, but the mean of the fourth group is so much larger than the other three means that power is close to 1. That is, with near certainty, you will reject with the ANOVA F-test and proceed to compare all pairs of means with Student's T at the α level. So in particular you will test

$$H_0: \mu_1 = \mu_2,$$

$$H_0: \mu_1 = \mu_3,$$

$$H_0: \mu_2 = \mu_3,$$

each at the α level, and the probability of at least one Type I error among these three tests will be greater than α.

12.1.2 The Tukey–Kramer Method

Tukey was the first to propose a method that controls FWE. He assumed normality and homoscedasticity and obtained an exact solution when all J groups have equal sample sizes. Kramer (1956) proposed a generalization that provides an approximate solution when the sample sizes are unequal, and Hayter (1984) showed that when there is homoscedasticity and sampling is from normal distributions, Kramer's method is conservative. That is, it guarantees that FWE will be less than or equal to α.

When comparing the jth group to the kth group, the Tukey–Kramer $1 - \alpha$ confidence interval for $\mu_j - \mu_k$ is

$$(\bar{X}_j - \bar{X}_k) \pm q\sqrt{\frac{\text{MSWG}}{2}\left(\frac{1}{n_j} + \frac{1}{n_k}\right)}, \quad (12.4)$$

where n_j is the sample size of the jth group, MSWG is the mean square within groups, which estimates the assumed common variance (see Box 9.1), and q is a constant read from Table 9 in Appendix B, which depends on the values of α, J (the number of groups being compared), and the degrees of freedom,

$$\nu = N - J,$$

where again N is the total number of observations in all J groups. Under normality, equal variances and equal sample sizes, the *simultaneous probability coverage* is exactly $1-\alpha$. That is, with probability $1-\alpha$, it will be simultaneously true that the confidence interval for $\mu_1 - \mu_2$ will indeed contain $\mu_1 - \mu_2$, the confidence interval for $\mu_1 - \mu_3$

TABLE 12.2 Ratings of Three Types of Cookies

Method 1	Method 2	Method 3
5	6	8
4	6	7
3	7	6
3	8	8
4	4	7
5	5	
3	8	
4	5	
8		
2		

will indeed contain $\mu_1 - \mu_3$, and so on. You reject $H_0: \mu_j = \mu_k$ if

$$\frac{|\bar{X}_j - \bar{X}_k|}{\sqrt{\frac{\text{MSWG}}{2}\left(\frac{1}{n_j} + \frac{1}{n_k}\right)}} \geq q.$$

EXAMPLE. Table 12.2 shows some hypothetical data on the ratings of three brands of cookies. Each brand is rated by a different sample of individuals. There are a total of $N = 23$ observations, so the degrees of freedom are $\nu = 23 - 3 = 20$, the sample means are $\bar{X}_1 = 4.1$, $\bar{X}_2 = 6.125$, and $\bar{X}_3 = 7.2$, the estimate of the common variance is MSWG $= 2.13$, and with $\alpha = .05$, Table 9 in Appendix B indicates that $q = 3.58$. The confidence interval for $\mu_1 - \mu_3$ is

$$(4.1 - 7.2) \pm 3.58\sqrt{\frac{2.13}{2}\left(\frac{1}{10} + \frac{1}{5}\right)} = (-5.12, -1.1).$$

This interval does not contain zero, so you reject the hypothesis that typical ratings of brands 1 and 3 are the same. You can compare brand 1 to brand 2 and brand 2 to brand 3 in a similar manner, but the details are left as an exercise. ■

12.1.3 A Step-Down Method

All-pairs power refers to the probability of detecting all true differences among all pairwise differences among the means. For example, suppose you want to compare four groups where $\mu_1 = \mu_2 = \mu_3 = 10$ but $\mu_4 = 15$. In this case, all-pairs power refers to the probability of rejecting $H_0: \mu_1 = \mu_4$ and $H_0: \mu_2 = \mu_4$ and $H_0: \mu_3 = \mu_4$. Still assuming normality and homoscedasticity, it is possible to achieve higher all-pairs power than with the Tukey–Kramer method in Section 12.1.2 using what is called a *step-down* technique. One price for this increased power is that you can no longer compute confidence intervals for the differences among the pairs of means. Box 12.1 summarizes the details.

BOX 12.1 Summary of the Step-Down Procedure

The goal is to perform all pairwise comparisons of the means of J independent groups such that the familywise Type I error probability is α.

1. Test $H_0 : \mu_1 = \cdots = \mu_J$ at the $\alpha_J = \alpha$ level of significance. Assuming normality and homoscedasticity, this is done with the ANOVA F-test in Chapter 9. If you fail to reject, stop; otherwise continue to the next step.
2. For each subset of $J - 1$ means, test the hypothesis that these means are equal at the $\alpha_{J-1} = \alpha$ level of significance. If all such tests are nonsignificant, stop. Otherwise continue to the next step.
3. For each subset of $J - 2$ means, test the hypothesis that they are equal at the $\alpha_{J-2} = 1 - (1 - \alpha)^{(J-2)/J}$ level of significance. If all of these tests are nonsignificant, stop; otherwise continue to the next step.
4. In general, test the hypothesis of equal means, for all subsets of p means, at the $\alpha_p = 1 - (1 - \alpha)^{p/J}$ level of significance, when $p \le J - 2$. If all of these tests are nonsignificant, stop and fail to detect any differences among the means; otherwise continue to the next step.
5. The final step consists of testing all pairwise comparisons of the means at the $\alpha_2 = 1 - (1 - \alpha)^{2/J}$ level of significance. In this final step, when comparing the jth group to the kth group, you either fail to reject, you fail to reject by implication from one of the previous steps, or you reject.

EXAMPLE. Consider $J = 5$ methods designed to increase the value of a client's stock portfolio, which we label methods A, B, C, D, and E. Further assume that when comparing these five methods, you are willing to sacrifice confidence intervals to enhance your all-pairs power. Assume that you want the familywise error rate to be $\alpha = .05$. The first step is to test

$$H_0 : \mu_1 = \mu_2 = \mu_3 = \mu_4 = \mu_5$$

at the $\alpha_5 = \alpha = .05$ level of significance, where the subscript 5 on α_5 indicates that in the first step, all $J = 5$ means are being compared. If you fail to reject H_0, stop and decide that there are no pairwise differences among the five methods. If you reject, proceed to the next step, which consists of testing the equality of the means for all subsets of four groups. In the illustration, suppose the F-test for equal means is applied as described in Chapter 9, yielding

Continued

EXAMPLE. (*Continued*) $F = 10.5$. Assuming the critical value is 2.6, you would reject and proceed to the next step. That is, you test

$$H_0: \mu_1 = \mu_2 = \mu_3 = \mu_4$$

$$H_0: \mu_1 = \mu_2 = \mu_3 = \mu_5$$

$$H_0: \mu_1 = \mu_2 = \mu_4 = \mu_5$$

$$H_0: \mu_1 = \mu_3 = \mu_4 = \mu_5$$

$$H_0: \mu_2 = \mu_3 = \mu_4 = \mu_5.$$

In this step you test each of these hypotheses at the $\alpha_4 = \alpha = .05$ level of significance, where the subscript 4 indicates that each test is comparing the means of four groups. Note that both the first and second steps use the same significance level, α. If in the second step, all five tests are nonsignificant, you stop and fail to detect any pairwise differences among the five methods; otherwise you proceed to the next step. In the illustration, suppose the values of your test statistic, F, are 9.7, 10.2, 10.8, 11.6, and 9.8, with a critical value of 2.8. So you reject in every case, but even if you reject in only one case, you proceed to the next step.

The third step consists of testing all subsets of exactly three groups, but this time you test at the

$$\alpha_3 = 1 - (1 - \alpha)^{3/5}$$

level of significance, where the subscript 3 is used to indicate that subsets of three groups are being compared. In the illustration, this means you test

$$H_0: \mu_1 = \mu_2 = \mu_3$$

$$H_0: \mu_1 = \mu_2 = \mu_4$$

$$H_0: \mu_1 = \mu_3 = \mu_4$$

$$H_0: \mu_1 = \mu_2 = \mu_5$$

$$H_0: \mu_1 = \mu_3 = \mu_5$$

$$H_0: \mu_1 = \mu_4 = \mu_5$$

$$H_0: \mu_2 = \mu_3 = \mu_4$$

$$H_0: \mu_2 = \mu_3 = \mu_5$$

$$H_0: \mu_2 = \mu_4 = \mu_5$$

$$H_0: \mu_3 = \mu_4 = \mu_5$$

Continued

EXAMPLE. (*Continued*) using $\alpha_3 = 1 - (1 - .05)^{3/5} = .030307$. If none of these hypotheses is rejected, you stop and fail to detect any pairwise difference among all pairs of methods; otherwise you continue to the next step.

The final step is to compare the jth group to the kth group by testing

$$H_0 : \mu_j = \mu_k.$$

This time you test at the

$$\alpha_2 = 1 - (1 - \alpha)^{2/5}$$

level. In the illustration, $\alpha_2 = .020308$. In this final stage, you make one of three decisions: You fail to reject H_0, you fail to reject H_0 due to the results from a previous step, or you reject. To clarify the second decision, suppose you fail to reject $H_0 : \mu_1 = \mu_3 = \mu_4$. Then by implication, you would conclude that $\mu_1 = \mu_3$, $\mu_1 = \mu_4$, and $\mu_3 = \mu_4$, *regardless* of what you got in the final step. That is, $H_0 : \mu_1 = \mu_3$, $H_0 : \mu_1 = \mu_4$, and $H_0 : \mu_3 = \mu_4$ would be declared not significant by implication, even if they were rejected in the final step. This might seem counterintuitive, but it is necessary if you want to control the familywise Type I error probability. Table 12.3 summarizes the results. ■

As stressed in Chapter 9, the ANOVA F-test performs rather poorly when the normality or homoscedasticity assumption is violated. One particular problem is low power under arbitrarily small departures from normality. When comparing means with a step-down procedure, there is a sense in which this problem is exacerbated. To illustrate why, imagine you are comparing four groups; all of the groups have unequal means, the first three groups have normal distributions, but the fourth has the mixed normal described in Section 2.7. Then the ANOVA F-test, applied to all four groups, can have low power, which in turn can mean that the step-down method described here has low power as well. In fact, even a single outlier in one group can mask substantial differences among the other groups being compared. (Dunnett & Tamhane, 1992, describe results on a step-up method that has practical advantages under normality and homoscedasticity, but it suffers from the same problem just described.)

12.2 ANOVA *F* Versus Multiple Comparisons

It is common practice to use Tukey's method only if the ANOVA F-test is significant. More generally, when using any multiple comparison procedure to compare groups based on some measure of location θ, it is common first to test

$$H_0 : \theta_1 = \cdots = \theta_J \tag{12.5}$$

and, if a nonsignificant result is obtained, to fail to detect any differences among the groups. Testing this omnibus test obviously plays a central role in a step-down method, but there is an important negative consequence of this strategy: Many modern multiple comparison procedures are designed to control FWE without first testing Equation (12.5). One example is Tukey's procedure. Under normality, homoscedasticity, and

TABLE 12.3 Illustration of the Step-Down Procedure

Groups	F	α	ν_1	Critical value	Decision
ABCDE	11.5	$\alpha_5 = .05$	4	2.61	Significant
ABCD	9.7	$\alpha_4 = .05$	3	2.84	Significant
ABCE	10.2				Significant
ABDE	10.8				Significant
ACDE	11.6				Significant
BCDE	9.8				Significant
ABC	2.5	$\alpha_3 = .0303$	2	3.69	Not significant
ABD	7.4				Significant
ACD	8.1				Significant
ABE	8.3				Significant
ACE	12.3				Significant
ADE	18.2				Significant
BCD	2.5				Not significant
BCE	9.2				Significant
BDE	8.1				Significant
CDE	12.4				Significant
AB	5.1	$\alpha_2 = .0203$	1	5.85	Not significant by implication
AC	6.0				Not significant by implication
AD	19.2				Significant
AE	21.3				Significant
BC	1.4				Not significant by implication
BD	6.0				Not significant by implication
BE	15.8				Significant
CD	4.9				Not significant by implication
CE	13.2				Significant
DE	3.1				Not significant

equal sample sizes, it guarantees that FWE will be exactly equal to α. But if Tukey's method is used contingent on rejecting with the ANOVA F, this is no longer true — FWE will be less than α, indicating that power will be reduced (Bernhardson, 1975). Generally, most modern multiple comparison procedures are designed so that FWE will be approximately equal to α. That is, they do not require that you first apply one of the methods in Chapters 9–11 and reject the hypothesis that groups have a common measure of location. Indeed, if you use modern methods only when you first reject, you run the risk of lowering the actual Type I error probability by an unknown amount, which in turn might mask true differences among the groups being compared. Robust and heteroscedastic ANOVA methods remain relevant, however, because they might be useful in some type of step-down or step-up technique.

12.3 Heteroscedastic Methods for the Means of Independent Groups

Consistent with earlier chapters, when groups have identical distributions, the methods in the previous section appear to provide reasonably good control over the probability of a Type I error. But when distributions differ in some manner, they suffer from the same problems associated with the homoscedastic methods described in previous chapters: Poor power, undesirable power properties (bias, meaning that power can decrease as the difference among the means increases), and poor probability coverage. This section describes heteroscedastic methods that reduce these problems, but serious practical concerns remain, some of which are not addressed even with very large sample sizes. However, the methods in this section set the stage for effective techniques.

12.3.1 Dunnett's T3

For multiple comparison procedures based on means, Dunnett (1980a, 1980b) documented practical problems with methods that assume homoscedasticity and then compared several heteroscedastic methods, two of which stood out when sampling from normal distributions. Although nonnormality can ruin these methods, they are important because they provide a basis for deriving substantially improved techniques.

Dunnett's so-called T3 procedure is just Welch's method described in Section 8.3, but with the critical value adjusted so that FWE is approximately equal to α when sampling from normal distributions. Let s_j^2 be the sample variance for the jth group, again let n_j be the sample size and set

$$q_j = \frac{s_j^2}{n_j}, j = 1, \ldots, J.$$

When comparing group j to group k, the degrees of freedom are

$$\hat{\nu}_{jk} = \frac{(q_j + q_k)^2}{\frac{q_j^2}{n_j - 1} + \frac{q_k^2}{n_k - 1}}.$$

The test statistic is

$$W = \frac{\bar{X}_j - \bar{X}_k}{\sqrt{q_j + q_k}},$$

and you reject $H_0: \mu_j = \mu_k$ if $|W| \geq c$, where the critical value, c, is read from Table 10 in Appendix B. (This table provides the .05 and .01 quantiles of what is called the *Studentized maximum modulus distribution*.) When using Table 10, you need to know the total number of comparisons you plan to perform. When performing all pairwise comparisons, the total number of comparisons is

$$C = \frac{J^2 - J}{2}.$$

In the illustration, there are $J = 3$ groups, so the total number of comparisons is

$$C = \frac{3^2 - 3}{2} = 3.$$

If you have $J = 4$ groups *and* you plan to perform all pairwise comparisons, $C = (4^2 - 4)/2 = 6$.

EXAMPLE. Suppose the goal is to compare five groups to a control group and that only these five comparisons are to be done. That is, the goal is to test $H_0: \mu_j = \mu_6$, $j = 1, \ldots, 5$, so $C = 5$. If $\alpha = .05$ and the degrees of freedom are 30, the critical value is $c = 2.73$. If you have five groups and plan to do all pairwise comparisons, $C = 10$; and with $\alpha = .01$ and $\nu = 20$, the critical value is 3.83. ■

A confidence interval for $\mu_j - \mu_k$, the difference between the means of groups j and k, is given by

$$(\bar{X}_j - \bar{X}_k) \pm c\sqrt{\frac{s_j^2}{n_j} + \frac{s_k^2}{n_k}}.$$

By design, the simultaneous probability coverage will be approximately $1 - \alpha$, under normality, when computing C confidence intervals and c is read from Table 10 in Appendix B.

EXAMPLE. Table 9.1 reports skin resistance for four groups of individuals. If the goal is to compare all pairs of groups with $\alpha = .05$, then $C = 6$, and the confidence interval for the difference between the means of the first two groups is $(-0.35, 0.52)$; this interval contains zero, so you fail to detect a difference. The degrees of freedom are 12.3, the critical value is $c = 3.07$, the test statistic is $W = 0.56$, and again you fail to reject, because $|W| < 3.07$. It is left as an exercise to show that for the remaining pairs of means, you again fail to reject. ■

12.3.2 Games–Howell Method

An alternative to Dunnett's T3 is the Games and Howell (1976) method. When comparing the jth group to the kth group, you compute the degrees of freedom, $\hat{\nu}_{jk}$, exactly as in Dunnett's T3 procedure, and then you read the critical value, q, from Table 9 in Appendix B. (Table 9 reports some quantiles of what is called the Studentized range distribution.) The $1-\alpha$ confidence interval for $\mu_j - \mu_k$ is

$$(\bar{X}_j - \bar{X}_k) \pm q\sqrt{\frac{1}{2}\left(\frac{s_j^2}{n_j} + \frac{s_k^2}{n_k}\right)}.$$

You reject $H_0 : \mu_j = \mu_k$ if this interval does not contain zero, which is the same as rejecting if

$$\frac{|\bar{X}_j - \bar{X}_k|}{\sqrt{\frac{1}{2}\left(\frac{s_j^2}{n_j} + \frac{s_k^2}{n_k}\right)}} \geq q.$$

Under normality, the Games–Howell method appears to provide more accurate probability coverage than Dunnett's T3 method when all groups have a sample size of at least 50. A close competitor under normality is Dunnett's (1980b) C method, but no details are given here.

EXAMPLE. Imagine you have three groups, with $\bar{X}_1 = 10.4$, $\bar{X}_2 = 10.75$,

$$\frac{s_1^2}{n_1} = .11556,$$

$$\frac{s_2^2}{n_2} = .156.$$

Then $\hat{\nu} = 19$ and with $\alpha = .05$, $q = 3.59$, so the confidence interval for $\mu_1 - \mu_2$ is

$$(10.4 - 10.75) \pm 3.59\sqrt{\frac{1}{2}(.11556 + .156)} = (-.167, 0.97).$$

This interval contains 0, so you do not reject the hypothesis of equal means. ■

12.3.3 Alternative Methods Based on Adjustments of α

Dunnett's T3 and the Games–Howell method use adjusted critical values to attempt to control FWE. There is a collection of methods for controlling FWE that adjust the p-values in a more direct fashion that should be mentioned. The easiest to use is based on the Bonferroni inequality, which, if C hypotheses are to be tested, test each hypothesis at the α/C level. Provided the probability of a Type I error can be controlled for each of the individual tests, FWE will be at most α.

Other approaches are based on what are called *sequentially rejective* methods. For example, Hochberg's (1988) method is applied as follows. Let $P_1, \ldots, P_C$ be the p-values associated with the C tests, put these p-values in descending order, and label the results $P_{[1]} \geq P_{[2]} \geq \cdots \geq P_{[C]}$. The method is applied in steps, where the total number of steps is at most C. Beginning with $k = 1$ (step 1), reject all hypotheses if

$$P_{[k]} \leq \frac{\alpha}{k}.$$

That is, reject all hypotheses if the largest p-value is less than or equal to α. If $P_{[1]} > \alpha$, proceed as follows:

1. Increment k by 1. If

$$P_{[k]} \leq \frac{\alpha}{k},$$

   stop and reject all hypotheses having a p-value less than or equal to $P_{[k]}$.
2. If $P_{[k]} > \alpha/k$, repeat step 1.
3. Repeat steps 1 and 2 until you reject or all C hypotheses have been tested.

Benjamini and Hochberg (1995) proposed a similar method, only in step 1 of Hochberg's method, $P_{[k]} \leq \alpha/k$ is replaced by

$$P_{[k]} \leq \frac{(C - k + 1)\alpha}{C}.$$

Results in Williams, Jones, and Tukey (1999) support the use of the Benjamini–Hochberg method over Hochberg. Both of these procedures have power greater than or equal to the Bonferroni method. However, the Bonferroni method can be used to compute confidence intervals, and currently there is uncertainty about how one should compute confidence intervals when using either of the two sequentially rejective methods just described.

EXAMPLE. Suppose six hypotheses are tested with the Benjamini–Hochberg method based on the following results:

Number	Test	p-value	
1	$H_0: \mu_1 = \mu_2$	$P_1 = .010$	$P_{[5]}$
2	$H_0: \mu_1 = \mu_3$	$P_2 = .015$	$P_{[3]}$
3	$H_0: \mu_1 = \mu_4$	$P_3 = .005$	$P_{[6]}$
4	$H_0: \mu_2 = \mu_3$	$P_4 = .620$	$P_{[1]}$
5	$H_0: \mu_2 = \mu_4$	$P_5 = .130$	$P_{[2]}$
6	$H_0: \mu_3 = \mu_4$	$P_6 = .014$	$P_{[4]}$

Because $P_{[1]} > .05$, fail to reject the fourth hypothesis. Had it been the case that $P_{[1]} \leq .05$, you would stop and reject all six hypotheses. Because you did not reject, set $k = 2$; and because $C = 6$, we see that

$$\frac{(C - k + 1)\alpha}{C} = \frac{5(.05)}{6} = .0417.$$

Continued

EXAMPLE. (*Continued*) Because $P_{[2]} = .130 > .0417$, fail to reject the fifth hypothesis and proceed to the next step. Incrementing k to 3,

$$\frac{(C - k + 1)\alpha}{C} = \frac{4(.05)}{6} = .0333,$$

and because $P_{[3]} = .015 \leq .0333$, reject this hypothesis and the remaining hypotheses having p-values less than or equal to .0333. That is, reject hypotheses 1, 2, 3, and 6 and fail to reject hypotheses 4 and 5. If the Bonferroni inequality had been used instead, we see that $.05/6 = .00833$, so only hypothesis 3 would be rejected. ■

A criticism of the Benjamini–Hochberg method is that situations can be found where some hypotheses are true, some are false, and the probability of at least one Type I error will exceed α among the hypotheses that are true (Hommel, 1988; cf. Keselman, Cribbie, & Holland, 1999). In contrast, Hochberg's method does not suffer from this problem. However, the Benjamini–Hochberg method does have the following property. When C hypotheses are tested, let Q be the proportion of hypotheses that are true and rejected. That is, Q is the proportion of Type I errors among the null hypotheses that are correct. If all hypotheses are false, then of course $Q = 0$, but otherwise Q can vary from one experiment to the next. That is, if we repeat a study many times, the proportion of erroneous rejections will vary. The *false-discovery rate* is the expected value of Q. That is, if a study is repeated infinitely many times, the false-discovery rate is the average proportion of Type I errors among the hypotheses that are true. Benjamini and Hochberg (1995) show that their method ensures that the false-discovery rate is less than or equal to α. The Benjamini–Hochberg method can be improved if the number of true hypotheses is known. Of course it is not known how many null hypotheses are in fact correct, but Benjamini and Hochberg (2000) suggest how this number might be estimated, which can result in higher power. (For related results, see Finner and Roters, 2002; Sarkar, 2002.)

12.4 Linear Contrasts

When dealing with a two-way design, as described in Chapter 10, it is convenient to describe relevant multiple comparison procedures in the context of what are called *linear contrasts*. For a J-by-K design, let $L = JK$ represent the total number of groups. (So in a one-way design, $L = J$.) By definition, a *linear contrast* is any linear combination of the means among L groups having the form

$$\Psi = \sum_{\ell=1}^{L} c_\ell \mu_\ell, \tag{12.6}$$

where $c_1, \ldots, c_L$, called *contrast coefficients*, are constants that sum to 0. In symbols, Ψ is a linear contrast if

$$\sum c_\ell = 0.$$

EXAMPLE. Section 10.1 describes a two-by-two design dealing with weight gain in rats. The two factors are source (beef versus cereal) and amounts of protein (low versus high), and here the population means are represented as follows:

	Source	
Amount	Beef	Cereal
Low	μ_1	μ_2
High	μ_3	μ_4

Consider the hypothesis of no main effect for the first factor, amount of protein. As explained in Chapter 10, the null hypothesis is

$$H_0 : \frac{\mu_1 + \mu_2}{2} = \frac{\mu_3 + \mu_4}{2}.$$

Rearranging terms in this last equation, the null hypothesis can be written as a linear contrast, namely,

$$H_0 : \mu_1 + \mu_2 - \mu_3 - \mu_4 = 0.$$

That is $\Psi = \mu_1 + \mu_2 - \mu_3 - \mu_4$, the contrast coefficients are $c_1 = c_2 = 1$, $c_3 = c_4 = -1$, and the null hypothesis is $H_0 : \Psi = 0$. ■

EXAMPLE. Now consider a three-by-two design:

	Factor B	
Factor A	1	2
1	μ_1	μ_2
2	μ_3	μ_4
3	μ_5	μ_6

An issue that is often of interest is not just whether there is a main effect for factor A, but which levels of factor A differ and by how much. That is, the goal is to compare level 1 of factor A to level 2, level 1 to level 3, and level 2 to level 3. In symbols, the three hypotheses of interest are

$$H_0 : \frac{\mu_1 + \mu_2}{2} = \frac{\mu_3 + \mu_4}{2},$$

Continued

EXAMPLE. (*Continued*)

$$H_0: \frac{\mu_1 + \mu_2}{2} = \frac{\mu_5 + \mu_6}{2},$$

$$H_0: \frac{\mu_3 + \mu_4}{2} = \frac{\mu_5 + \mu_6}{2}.$$

In terms of linear contrasts, the goal is to test

$$H_0: \Psi_1 = 0,$$
$$H_0: \Psi_2 = 0,$$
$$H_0: \Psi_3 = 0,$$

where

$$\Psi_1 = \mu_1 + \mu_2 - \mu_3 - \mu_4,$$
$$\Psi_2 = \mu_1 + \mu_2 - \mu_5 - \mu_6,$$
$$\Psi_3 = \mu_3 + \mu_4 - \mu_5 - \mu_6.$$

The interactions can be written as linear contrasts as well. For example, the hypothesis of no interaction for the first two rows corresponds to

$$H_0: \mu_1 - \mu_2 = \mu_3 - \mu_4,$$

which is the same as testing

$$H_0: \Psi_4 = 0,$$

where

$$\Psi_4 = \mu_1 - \mu_2 - \mu_3 + \mu_4.$$

Similarly, for rows 1 and 3, the hypothesis of no interaction is

$$H_0: \Psi_5 = 0,$$

where

$$\Psi_5 = \mu_1 - \mu_2 - \mu_5 + \mu_6.$$

In general, there are a collection of C linear contrasts that one might want to test; the goal is to devise a method for performing these C tests in a manner that controls FWE. ■

12.4.1 Scheffé's Homoscedastic Method

Assuming normality and homoscedasticity, Scheffé's classic method can be used to test C hypotheses about C linear contrasts such that FWE is less than or equal to

α — regardless of how large C might be! Let Ψ be any specific hypothesis and let

$$\hat{\Psi} = \sum_{\ell=1}^{L} c_\ell \bar{X}_\ell. \tag{12.7}$$

That is, estimate the mean, μ_ℓ, of the ℓth group with $\bar{X}_\ell$, the sample mean of the ℓth group, and then plug this estimate into Equation (12.6) to get an estimate of Ψ. Then the confidence interval for Ψ is

$$(\hat{\Psi} - S, \hat{\Psi} + S), \tag{12.8}$$

where

$$S = \sqrt{(L-1)f_{1-\alpha}(\text{MSWG}) \sum \frac{c_\ell^2}{n_\ell}},$$

MSWG is the mean square within groups (described in Chapters 9 and 10) that estimates the assumed common variance, and $f_{1-\alpha}$ is the $1-\alpha$ quantile of an F-distribution with $\nu_1 = L - 1$ and $\nu_2 = N - L$ degrees of freedom, where $N = \sum n_j$ is the total number of observations in all L groups. (For a one-way design with J levels, $L = J$; for a J-by-K design, $L = JK$.) In particular, $H_0: \Psi = 0$ is rejected if the confidence interval given by Equation (12.8) does not contain zero.

EXAMPLE. For the special case where all pairwise comparisons of J independent groups are to be performed, Scheffé's confidence interval for the difference between the means of the jth and kth groups, $\mu_j - \mu_k$, is

$$(\bar{X}_j - \bar{X}_k) \pm S,$$

where

$$S = \sqrt{(J-1)f_{1-\alpha}(\text{MSWG})\left(\frac{1}{n_j} + \frac{1}{n_k}\right)},$$

and $f_{1-\alpha}$ is the ANOVA F critical value based on $\nu_1 = J - 1$ and $\nu_2 = N - J$ degrees of freedom. ■

Scheffé's method remains one of the more popular multiple comparison procedures in applied work, but it suffers from the same problems associated with other homoscedastic methods already covered. Even under normality but heteroscedasticity, problems arise (Kaiser & Bowden, 1983). Also, under normality and homoscedasticity, the Tukey–Kramer method should give shorter confidence intervals than the Scheffé method, but for certain collections of linear contrasts the reverse is true (Scheffé, 1959, p. 76).

12.4.2 Heteroscedastic Methods

Two heteroscedastic methods for linear contrasts are covered in this section.

Welch–Šidák Method

The first contains Dunnett's T3 method as a special case and is called the Welch–Šidák method. (The computations are performed by the S-PLUS function lincon described in Section 12.6.1.) Again let L represent the total number of groups being compared, let Ψ be any linear contrast of interest, and let C represent the total number of contrasts to be tested. An expression for the squared standard error of $\hat{\Psi}$ is

$$\sigma^2_{\hat{\Psi}} = \sum \frac{c^2_\ell \sigma^2_\ell}{n_\ell},$$

where σ^2_ℓ and n_ℓ are the variance and sample size of the ℓth group, respectively. An estimate of this quantity is obtained simply by replacing σ^2_ℓ with s^2_ℓ, the sample variance associated with the ℓth group. That is, estimate $\sigma^2_{\hat{\Psi}}$ with

$$\hat{\sigma}^2_{\hat{\Psi}} = \sum \frac{c^2_\ell s^2_\ell}{n_\ell}.$$

Let

$$q_\ell = \frac{c^2_\ell s^2_\ell}{n_\ell}.$$

The degrees of freedom are estimated to be

$$\hat{\nu} = \frac{(\sum q_\ell)^2}{\sum \frac{q^2_\ell}{n_\ell - 1}}.$$

The test statistic is

$$T = \frac{\hat{\Psi}}{\hat{\sigma}_{\hat{\Psi}}}.$$

The critical value, c, is a function of $\hat{\nu}$ and C (the total number of hypotheses you plan to perform) and is read from Table 10 in Appendix B. Reject if $|T| \geq c$, and a confidence interval for Ψ is

$$\hat{\Psi} \pm c\hat{\sigma}_{\hat{\Psi}}. \tag{12.9}$$

Kaiser–Bowden Method

A heteroscedastic analog of Scheffé's method was derived by Kaiser and Bowden (1983). The computations are exactly the same as in the Welch–Šidák method, except the squared critical value is now

$$A = (L-1)\left(1 + \frac{L-2}{\hat{\nu}_2}\right)f,$$

where f is the $1-\alpha$ quantile of an F-distribution with $\nu_1 = L-1$ and

$$\hat{\nu}_2 = \frac{(\sum q_\ell)^2}{\sum \frac{q_\ell^2}{n_\ell - 1}}.$$

That is, $\hat{\nu}_2 = \hat{\nu}$, the estimated degrees of freedom used by the Welch–Šidák method. The confidence interval for Ψ is

$$\hat{\Psi} \pm \sqrt{A \sum q_\ell},$$

where again $q_\ell = c_\ell^2 s_\ell^2 / n_\ell$. Generally, the Welch–Šidák method will provide shorter confidence intervals and more power than the Kaiser–Bowden method. But the critical values in Table 10 in Appendix B for the Welch–Šidák method are limited to $\alpha = .05$ and .01 and at most 28 hypotheses.

EXAMPLE. For the special case where all pairwise comparisons among J independent groups are to be performed, the Kaiser–Bowden method for computing a confidence interval for $\mu_j - \mu_k$ is

$$(\bar{X}_j - \bar{X}_k) \pm \sqrt{A\left(\frac{s_j^2}{n_j} + \frac{s_k^2}{n_k}\right)},$$

where now

$$A = (J-1)\left(1 + \frac{J-2}{\hat{\nu}_{jk}}\right)f,$$

$\hat{\nu}_{jk}$ is as given in Section 12.3.1 (in conjunction with Dunnett's T3 procedure), and f is the $1-\alpha$ quantile of an F-distribution with $\nu_1 = J-1$ and $\nu_2 = \hat{\nu}_{jk}$ degrees of freedom. ■

EXAMPLE. Consider a 2-by-2 design and suppose the sample means are $\bar{X}_1 = 10$, $\bar{X}_2 = 14$, $\bar{X}_3 = 18$, $\bar{X}_4 = 12$; the sample variances are $s_1^2 = 20$, $s_2^2 = 8$, $s_3^2 = 12$, $s_4^2 = 4$; and the sample sizes are $n_1 = n_2 = n_3 = n_4 = 4$. Further assume you want to test three hypotheses with the Welch–Šidák method: no

Continued

EXAMPLE. (*Continued*) main effects for factor A, no main effects for factor B, and no interaction. In terms of linear contrasts, the goal is to test

$$H_0 : \Psi_1 = 0,$$
$$H_0 : \Psi_2 = 0,$$
$$H_0 : \Psi_3 = 0,$$

where

$$\Psi_1 = \mu_1 + \mu_2 - \mu_3 - \mu_4,$$
$$\Psi_2 = \mu_1 + \mu_3 - \mu_2 - \mu_4,$$
$$\Psi_3 = \mu_1 - \mu_2 - \mu_3 + \mu_4.$$

Moreover, the probability of at least one Type I error is to be .05 among these three tests. For the first hypothesis, the estimate of Ψ_1 is

$$\hat{\Psi}_1 = 10 + 14 - 18 - 12 = -6.$$

The estimate of the squared standard error is

$$\hat{\sigma}^2_{\hat{\Psi}_1} = \frac{1^2(20)}{4} + \frac{1^2(8)}{4} + \frac{(-1)^2(12)}{4} + \frac{(-1)^2(4)}{4}$$
$$= 11,$$

the degrees of freedom for the Welch–Šidák method can be shown to be 9.3; and with $C = 3$, the critical value is approximately $c = 2.87$. The test statistic is

$$T = \frac{-6}{\sqrt{11}} = -1.8,$$

and because $|T| < 2.87$, fail to reject. If instead the Kaiser–Bowden method is used, because there is a total of $L = 4$ groups, the critical value is

$$c = \sqrt{(4-1)\left(1 + \frac{4-2}{9.3}\right) 3.81} = 3.7$$

which is considerably larger than the critical value based on the Welch–Šidák method. The other two hypotheses can be tested in a similar manner, but the details are left as an exercise. ■

EXAMPLE. In Chapter 10 it was noted that when investigating the possibility of a disordinal interaction, it can become necessary to test a collection of relevant hypotheses. In the previous example, note that $\bar{X}_1 < \bar{X}_2$ but that $\bar{X}_3 > \bar{X}_4$. To establish that there is a disordinal interaction, you must be able to reject both $H_0 : \mu_1 = \mu_2$ and $H_0 : \mu_3 = \mu_4$. One way to control FWE among these two hypotheses is with the Welch–Šidák method. ■

12.5 Judging Sample Sizes

As noted in Chapter 8, failing to reject some hypothesis might be because there is little or no difference between the groups, or there might be an important difference that was missed due to low power. So issues of general importance are determining whether the sample size used was sufficiently large to ensure adequate power and how large the sample size should have been if power is judged to be inadequate; there is the related problem of controlling the length of confidence intervals. This section describes two methods for accomplishing the last of these goals when working with linear contrasts.

12.5.1 Tamhane's Procedure

The first of the two methods is due to Tamhane (1977). A general form of his results can be used to deal with linear contrasts, but here attention is restricted to all pairwise comparisons among J independent groups. The goal is achieve confidence intervals having some specified length $2m$, given s_j^2, the sample variance from the jth group based on n_j observations ($j = 1, \ldots, J$). The computational details are summarized in Box 12.2.

BOX 12.2 Summary of Tamhane's Method

Goal

Compute confidence intervals for all pairwise differences among J independent groups such that the simultaneous probability coverage is equal to $1-\alpha$ and the length of each confidence interval is $2m$. Normality is assumed but unequal variances are allowed.

Compute

$$A = \sum \frac{1}{n_j - 1}$$

$$\nu = \left[\frac{J}{A}\right],$$

where the notation $[J/A]$ means you compute J/A and round down to the nearest integer. Next, determine h from Table 11 in Appendix B with ν degrees of freedom. (Table 11 gives the quantiles of the range of independent Student T variates.) Note that Table 11 assumes $\nu \leq 59$. For larger degrees of freedom, use Table 9 in Appendix B instead. Let

$$d = \left(\frac{m}{h}\right)^2.$$

Continued

BOX 12.2 (*Continued*) Letting s_j^2 be the sample variance for the jth group, the total number of observations required from the jth group is

$$N_j = \max\left\{n_j + 1, \left[\frac{s_j^2}{d}\right] + 1\right\}.$$

Once the additional observations are available, compute the generalized sample mean, $\tilde{X}_j$, for the jth group as described and illustrated in Box 9.6 in connection with the Bishop–Dudewicz ANOVA. The confidence interval for $\mu_j - \mu_k$ is

$$\left(\tilde{X}_j - \tilde{X}_k - m, \tilde{X}_j - \tilde{X}_k + m\right).$$

EXAMPLE. Assume that for three groups, $n_1 = 11, n_2 = 21, n_3 = 41$, and the goal is to compute confidence intervals having length 4 and having simultaneous probability coverage $1 - \alpha = .95$. So

$$A = \frac{1}{10} + \frac{1}{20} + \frac{1}{40} = .175,$$

$$\nu = \left[\frac{3}{.175}\right] = [17.14] = 17,$$

and $h = 3.6$. Confidence intervals with length 4 means that $m = 2$, so

$$d = \left(\frac{2}{3.6}\right)^2 = .3086.$$

If the sample variance for the first group is $s_1^2 = 2$, then

$$\begin{aligned} N_1 &= \max\left\{11 + 1, \left[\frac{2}{.3086}\right] + 1\right\} \\ &= \max(12, [6.4] + 1) \\ &= 12. \end{aligned}$$

You already have $n_1 = 11$ observations, so you need $12 - 11 = 1$ more. If you get $\tilde{X}_1 = 14$ and $\tilde{X}_2 = 17$, the confidence interval for the difference between the means is

$$(14 - 17 - 2, 14 - 17 + 2) = (-5, -1).$$

■

12.5.2 S-PLUS Function tamhane

Tamhane's two-stage multiple comparison procedure can be applied with the S-PLUS function

tamhane(x, x2 = NA, cil = NA, crit = NA).

The first-stage data are stored in x (in a matrix or in list mode) and the second-stage data in x2. If x2 contains no data, the function prints the degrees of freedom, which can be used to determine h as described in Box 12.2. Once h has been determined, store it in the argument crit. If no critical value is specified, the function terminates with an error message; otherwise it prints the total number of observations needed to achieve confidence intervals having length cil (given by the third argument). (So in the notation of Section 12.5.1, the value of cil divided by 2 corresponds to m.) If it finds data in the argument x2, confidence intervals are computed as described in Box 12.2 and returned by the function tamhane in the S-PLUS variable ci.mat; otherwise, the function returns ci.mat with the value NA.

EXAMPLE. For the data in Table 9.1, if the goal is to perform all pairwise comparisons such that the confidence intervals have length 1 (so $m = .5$) and FWE is to be .05, then the degrees of freedom are 9 and, from Table 11 in Appendix B, $h = 4.3$. If the data are stored in the S-PLUS variable skin, the command

tamhane(skin,cil=1,crit=4.3)

returns

```
$n.vec:
[1] 13 11 11 11
```

indicating that the required sample sizes are 13, 11, 11, and 11, respectively. ■

12.5.3 Hochberg's Procedure

This section describes another two-stage procedure derived by Hochberg (1975). In contrast to Tamhane's procedure, Hochberg's method allows the possibility of no additional observations being required in the second stage, uses the usual sample mean (rather than the generalized sample mean), and ensures that the lengths of the confidence intervals are at most $2m$. When using Tamhane's approach, the confidence intervals have length exactly equal to $2m$. (As before, m is some constant chosen by the investigator.) Box 12.3 summarizes the computations for the more general case, where C linear contrasts are to be tested, n_j represents the sample size for the jth group in the first stage, and s_j^2 is the sample variance for the jth group based on these n_j observations. (Hochberg's original derivation assumed equal sample sizes in the first stage, but the adjustment for unequal sample sizes in Box 12.3 appears to perform well.)

BOX 12.3 Summary of Hochberg's Method

As in Tamhane's procedure, read the critical value h from Table 11 in Appendix B, with the degrees of freedom, ν, computed as in Box 12.2. If $\nu > 59$, use Table 9 instead. Compute

$$d = \left(\frac{m}{h}\right)^2.$$

The total number of observations you need to sample from the jth group is

$$N_j = \max\left(n_j, \left[\frac{s_j^2}{d}\right] + 1\right). \tag{12.10}$$

Sample an additional $N_j - n_j$ observations from the jth group, and compute the sample mean, $\bar{X}_j$, based on all N_j values. For all pairwise comparisons, the confidence interval for $\mu_j - \mu_k$ is

$$(\bar{X}_j - \bar{X}_k) \pm hb,$$

where

$$b = \max\left(\frac{s_j}{\sqrt{N_j}}, \frac{s_k}{\sqrt{N_k}}\right).$$

Be sure to notice that the sample variances used to compute b are not recomputed once the additional observations are available. For technical reasons, you use the sample variances based on the initial n observations.

As for the linear contrast $\Psi = \sum c_j\mu_j$, sum the positive c_j values and label the result a. Again N_j is given by Equation (12.10), except

$$d = \left(\frac{m}{ha}\right)^2.$$

Let

$$b_j = \frac{c_j s_j}{\sqrt{N_j}}.$$

Let A be the sum of the positive b_j values and C be the sum of the negative b_j values. Compute

$$D = \max(A, -C),$$

$$\hat{\Psi} = \sum c_j\bar{X}_j,$$

in which case the confidence interval for Ψ is

$$\hat{\Psi} \pm hD.$$

EXAMPLE. To illustrate Hochberg's method, imagine that with four groups, each having a sample size of $n = 25$, one of your goals is to compute a confidence interval for $H_0: \Psi = 0$ with $\alpha = .05$, where

$$\Psi = \mu_1 + \mu_2 - \mu_3 - \mu_4,$$

and the length of the confidence interval is to be at most $2m = 8$. So $\nu = 24$ and, from Table 11 in Appendix B, $b = 3.85$. We see that the sum of the positive contrast coefficients is $a = 2$, so

$$d = \left(\frac{4}{3.85(2)}\right)^2 = .2699.$$

If $s_1^2 = 4$, $s_2^2 = 12$, $s_3^2 = 16$, and $s_4^2 = 20$, then

$$N_1 = \max\left(25, \left[\frac{4}{.2699}\right] + 1\right) = 25.$$

Hence, no additional observations are required. The sample sizes for the other three groups are $N_2 = 45$, $N_3 = 60$, and $N_4 = 75$.

Once the additional observations are available, compute

$$b_1 = \frac{1 \times 2}{\sqrt{25}} = .4.$$

Similarly, $b_2 = .5164$, $b_3 = -.5164$, and $b_4 = -.5164$. The sum of the positive b_j values is $A = b_1 + b_2 = .4 + .5164 = 0.9164$. The sum of the negative b_j values is $C = -1.0328$, so $-C = 1.0328$,

$$D = \max(.9164, 1.0328) = 1.0328,$$

$$bD = 3.85 \times 1.0328 = 3.976.$$

If, after the additional observations are sampled, the sample means are $\bar{X}_1 = 10$, $\bar{X}_2 = 12$, $\bar{X}_3 = 8$, and $\bar{X}_4 = 18$, then

$$\hat{\Psi} = 10 + 12 - 8 - 18 = -4,$$

and the confidence interval is

$$-4 \pm 3.976 = (-7.976, \ -0.024).$$

Thus, you would reject H_0. ■

12.5.4 S-PLUS Function hochberg

The S-PLUS function

hochberg(x, x2 = NA, cil = NA, crit = NA)

performs Hochberg's two-stage procedure. The arguments are used in the same manner as in Section 12.5.2.

12.6 Methods for Trimmed Means

The serious practical problems with methods based on means, described in previous chapters, extend to the situation at hand. One way of addressing these problems is to switch to generalizations of the Welch–Šidák and Kaiser–Bowden methods to trimmed means. Letting L represent the total number of groups, details are summarized in Box 12.4. (For a J-by-K design, $L = JK$.)

The method in Box 12.4 can be used to compute confidence intervals for all pairwise differences of the trimmed means for the special case of J independent groups. That is, the goal is to compute a confidence interval for $\mu_{tj} - \mu_{tk}$, for all $j < k$, such that the simultaneous probability coverage is approximately $1 - \alpha$. The confidence interval for $\mu_{tj} - \mu_{tk}$ is

$$(\bar{X}_{tj} - \bar{X}_{tk}) \pm c\sqrt{d_j + d_k},$$

where

$$d_j = \frac{(n_j - 1)s_{wj}^2}{h_j(h_j - 1)},$$

h_j is the number of observations left in group j after trimming, and c is read from Table 10 in Appendix B with

$$\hat{\nu} = \frac{(d_j + d_k)^2}{\frac{d_j}{h_j - 1} + \frac{d_k}{h_k - 1}}$$

degrees of freedom.

BOX 12.4 Heteroscedastic Tests of Linear Contrasts Based on Trimmed Means

Let $\mu_{t1}, \ldots, \mu_{tL}$ be the trimmed means corresponding to L independent groups, and let

$$\Psi = \sum_{\ell=1}^{L} c_\ell \mu_{t\ell}$$

be some linear contrast of interest. It is assumed that there is a total of C such linear contrasts; for each the goal is to test $H_0: \Psi = 0$ with FWE equal to α.

Continued

BOX 12.4 (*Continued*) Compute

$$\hat{\Psi} = \sum_{\ell=1}^{L} c_\ell \bar{X}_{t\ell},$$

$$d_\ell = \frac{c_\ell^2 (n_\ell - 1) s_{w\ell}^2}{h_\ell (h_\ell - 1)},$$

where $s_{w\ell}^2$ and h_ℓ are the Winsorized variance and effective sample size (the number of observations left after trimming) of the ℓth group, respectively. An estimate of the squared standard error of $\hat{\Psi}$ is

$$S_e = \sum d_\ell.$$

Letting

$$G = \sum \frac{d_\ell^2}{h_\ell - 1},$$

the estimated degrees of freedom are

$$\hat{\nu} = \frac{S_e^2}{G}.$$

Then a confidence interval for Ψ, based on a trimmed analog of the Welch–Šidák method, is

$$\hat{\Psi} \pm c\sqrt{S_e},$$

where c is read from Table 10 in Appendix B (and is again a function of C and $\hat{\nu}$).

As for a trimmed analog of the Kaiser–Bowden method, the confidence interval is

$$\hat{\Psi} \pm \sqrt{AS_e},$$

where

$$A = (L-1)\left(1 + \frac{L-2}{\hat{\nu}}\right) f,$$

and f is the $1-\alpha$ quantile of an F-distribution with $\nu_1 = L - 1$ and $\nu_2 = \hat{\nu}$ degrees of freedom.

12.6.1 S-PLUS Function lincon

The S-PLUS function

```
lincon(x,con=0,tr=.2,alpha=.05, KB=F)
```

tests hypotheses for a collection of linear contrasts based on trimmed means. The default value for con is zero, indicating that all pairwise comparisons are to be done.

Other linear contrasts can be specified by storing a matrix of linear contrast coefficients in con. By assumption, the rows of con correspond to groups, and the columns correspond to the linear contrasts. That is, column 1 contains the contrast coefficients for the first linear contrast of interest, column 2 has the contrast coefficients for the second linear contrast, and so forth. So in general, con must be a matrix having J rows and C columns, where C is the number of linear contrasts to be tested. If $\alpha = .05$ or .01, and simultaneously the number of contrasts is less than or equal to 28, lincon uses the generalization of the Welch–Šidák method to trimmed means as described in Box 12.4; otherwise it uses the extension of the Kaiser–Bowden method. Setting the argument KB to T forces the function to use Kaiser–Bowden.

EXAMPLE. If the data in Table 9.1 are stored in the S-PLUS variable skin, the command lincon(skin) returns:

```
Group Group      test     crit         se       df
    1     2 0.2909503 3.217890 0.11539200 9.627086
    1     3 0.8991534 3.306229 0.08713196 8.568838
    1     4 2.2214991 3.307845 0.08703807 8.550559
    2     3 1.1074715 3.406642 0.10105753 7.609956
    2     4 1.5823636 3.408688 0.10097658 7.593710
    3     4 4.0624997 3.190010 0.06688001 9.999866

Group Group       psihat   ci.lower    ci.upper
    1     2 -0.03357333 -0.4048921  0.33774542
    1     3  0.07834500 -0.2097332  0.36642319
    1     4 -0.19335500 -0.4812635  0.09455346
    2     3  0.11191833 -0.2323484  0.45618510
    2     4 -0.15978167 -0.5039794  0.18441602
    3     4 -0.27170000 -0.4850479 -0.05835214
```

So, for example, the test statistic for comparing group 1 to group 2 is 0.29, the critical value is 3.2, the estimate of $\Psi_1 = \mu_{t1} - \mu_{t2}$ is -0.03, and the confidence interval for this difference is $(-0.40, 0.34)$. ■

EXAMPLE. A classic problem is comparing treatment groups to a control group. That is, rather than perform all pairwise comparisons, the goal is to compare group 1 to group J, group 2 to group J, and so on. In symbols, the goal is to test

$$H_0: \mu_{t1} = \mu_{tJ},$$
$$H_0: \mu_{t2} = \mu_{tJ},$$
$$\vdots$$
$$H_0: \mu_{t,J-1} = \mu_{tJ}$$

Continued

EXAMPLE. (*Continued*) with FWE equal to α. For illustrative purposes, assume $J = 4$. Then to compare the first three groups to the control (group 4) with the S-PLUS function lincon, store the matrix

$$\begin{pmatrix} 1 & 0 & 0 \\ 0 & 1 & 0 \\ 0 & 0 & 1 \\ -1 & -1 & -1 \end{pmatrix}$$

in some S-PLUS variable—say, mmat. This can be done with the S-PLUS command

```
mmat<-matrix(c(1,0,0,-1,0,1,0,-1,0,0,1,-1),ncol=3).
```

Then the command

```
lincon(x,con=mmat)
```

will perform the relevant comparisons. For the schizophrenia data in Table 9.1, the results are

```
$test:
        con.num       test       crit          se       df
              1  -2.221499  2.919435  0.08703807  8.550559
              2  -1.582364  2.997452  0.10097658  7.593710
              3  -4.062500  2.830007  0.06688001  9.999866
$psihat:
           con.num       psihat    ci.lower    ci.upper
                 1  -0.1933550  -0.4474570   0.0607470
                 2  -0.1597817  -0.4624542   0.1428908
                 3  -0.2717000  -0.4609709  -0.0824291
```

So again the conclusion is that groups 3 and 4 differ. ■

EXAMPLE. Consider a 3-by-3 ANOVA design with population trimmed means labeled in the usual way:

	Factor B		
Factor A	1	2	3
1	μ_{t1}	μ_{t2}	μ_{t3}
2	μ_{t4}	μ_{t5}	μ_{t6}
3	μ_{t7}	μ_{t8}	μ_{t9}

All pairwise comparisons for the main effects of factor A are given by the three

Continued

EXAMPLE. (*Continued*) linear contrasts

$$\Psi_1 = \mu_{t1} + \mu_{t2} + \mu_{t3} - \mu_{t4} - \mu_{t5} - \mu_{t6},$$

$$\Psi_2 = \mu_{t1} + \mu_{t2} + \mu_{t3} - \mu_{t7} - \mu_{t8} - \mu_{t9},$$

$$\Psi_3 = \mu_{t4} + \mu_{t5} + \mu_{t6} - \mu_{t7} - \mu_{t8} - \mu_{t9}.$$

So the matrix of contrast coefficients is

$$\begin{pmatrix} 1 & 1 & 0 \\ 1 & 1 & 0 \\ 1 & 1 & 0 \\ -1 & 0 & 1 \\ -1 & 0 & 1 \\ -1 & 0 & 1 \\ 0 & -1 & -1 \\ 0 & -1 & -1 \\ 0 & -1 & -1 \end{pmatrix}.$$

■

12.6.2 S-PLUS Function mcp2atm for Two-Way Designs

As indicated in the previous section, multiple comparisons for a two-way ANOVA design can be performed by specifying an appropriate set of linear contrasts. For convenience the S-PLUS function

mcp2atm(J,K,x,tr = 0.2,con = 0, alpha = 0.05, grp = NA, op = F)

is provided for performing all pairwise comparisons for the main effects as well as for all interactions. This function creates the appropriate linear contrast coefficients for you and calls the function lincon. By default, FWE is set at α when performing the $M_a = (J^2 - J)/2$ pairwise comparisons for factor A; the same is done for the $M_b = (K^2 - K)/2$ pairwise comparisons for factor B. As for interactions, FWE is based on all $M_i = M_a M_b$ interactions. If op=T is used, FWE is controlled for all $M = M_a + M_b + M_i$ tests to be performed.

12.6.3 Linear Contrasts Based on Medians

The method for testing linear contrasts based on trimmed means should not be used for the special case where the goal is to compare medians, but a modification of the method for trimmed means appears to perform relatively well when the goal is to compare medians. Again let L indicate the number of groups to be compared and let M_ℓ represent the sample median of the ℓth group. Let $\theta_1, \ldots, \theta_L$ be the population medians, and now let

$$\Psi = \sum_{\ell=1}^{L} c_\ell \theta_\ell$$

be some linear contrast of interest. Compute

$$\hat{\Psi} = \sum_{\ell=1}^{L} c_\ell M_\ell,$$

and let

$$S_e^2 = \sum c_\ell^2 S_\ell^2,$$

where S_ℓ^2 is the McKean–Schrader estimate of the squared standard error of M_ℓ. Then an approximate $1 - \alpha$ confidence interval for Ψ is

$$\hat{\Psi} \pm c S_e,$$

where c is read from Table 10 in Appendix B with degrees of freedom $\nu = \infty$ and where C is again the number of hypotheses to be tested. As usual, this method is designed so that FWE will be approximately $1 - \alpha$.

12.6.4 S-PLUS Functions msmed and mcp2med

The S-PLUS function

msmed(x, y = NA, con = 0, alpha = 0.05)

can be used just like the function lincon, ignoring the second argument y. So if data are stored in a matrix called mat having six columns corresponding to six groups, msmed(mat) will perform all pairwise comparisons using medians. (If only two groups are being compared, the second argument, y, can be used as indicated in Section 8.7.2.) The S-PLUS function

mcp2med(J, K, x, con = 0, alpha = 0.05, grp = NA, op = F)

performs multiple comparisons among main effects and interactions and is used exactly like the function mcp2atm in Section 12.6.2; the only difference is that now medians are compared rather than trimmed means.

12.7 Bootstrap Methods

As was the case in Chapter 8, when comparing means, all indications are that the bootstrap-*t* method performs better than the percentile bootstrap. However, with at least 20% trimming or when using some other robust measure of location, extant results support the use of some type of percentile bootstrap method instead. When comparing groups with trimmed means, the minimum amount of trimming required to justify switching from the bootstrap-*t* to the percentile bootstrap is unknown. That is, with 15% or perhaps 10% trimming it might be preferable to use some variation of the percentile bootstrap, but this has not been established. But with 20% trimming, all indications are that a percentile bootstrap has practical value. When comparing groups using MOMs or M-estimators, currently some type of percentile bootstrap method is recommended.

12.7.1 Bootstrap-*t*

As usual, let $\Psi_1, \ldots, \Psi_C$ indicate C linear contrasts of interest, which include all pairwise comparisons of the groups as a special case. A bootstrap-*t* method for testing the hypothesis that each of these C linear contrasts is zero is outlined in Box 12.5.

BOX 12.5 Bootstrap-*t* Method for Trimmed Means

Goal

Test $H_0 : \Psi = 0$, for each of C linear contrasts such that FWE is α.

1. For each of the L groups, generate a bootstrap sample, $X^*_{i\ell}$, $i = 1, \ldots, n_\ell$; $\ell = 1, \ldots, L$. For each of the L bootstrap samples, compute the trimmed mean, $\bar{X}^*_{t\ell}$,
$$\hat{\Psi}^* = \sum c_\ell \bar{X}^*_{t\ell},$$
$$d^*_\ell = \frac{c^2_\ell (n_\ell - 1)(s^*_{w\ell})^2}{h_\ell (h_\ell - 1)},$$
where $(s^*_{w\ell})^2$ is the Winsorized variance based on the bootstrap sample taken from the ℓth group and h_ℓ is the effective sample size of the ℓth group (the number of observations left after trimming).
2. Compute
$$T^* = \frac{|\hat{\Psi}^* - \hat{\Psi}|}{\sqrt{A^*}},$$
where $\hat{\Psi} = \sum c_\ell \bar{X}_{t\ell}$ and $A^* = \sum d^*_\ell$. The results for each of the C linear contrasts are labeled $T^*_1, \ldots, T^*_C$.
3. Let
$$T^*_q = \max\{T^*_1, \ldots, T^*_C\}.$$
In words, T^*_q is the maximum of the C values $T^*_1, \ldots, T^*_C$.
4. Repeat steps 1–3 B times, yielding T^*_{qb}, $b = 1, \ldots, B$.

Let $T^*_{q(1)} \le \cdots \le T^*_{q(B)}$ be the T^*_{qb} values written in ascending order, and let $u = (1 - \alpha)B$, rounded to the nearest integer. Then the confidence interval for Ψ is
$$\hat{\Psi} \pm T^*_{q(u)}\sqrt{A},$$
where $A = \sum d_\ell$, $d_\ell = \frac{c^2_\ell (n_\ell - 1)s^2_{w\ell}}{h_\ell (h_\ell - 1)}$, and the simultaneous probability coverage is approximately $1 - \alpha$.

12.7.2 S-PLUS Function linconb

The S-PLUS function

linconb(x, con = 0, tr = 0.2, alpha = 0.05, nboot = 599)

has been supplied to perform the bootstrap-*t* method for trimmed means just described. As usual, the argument con can be used to specify the contrast coefficients. If con is not passed to the function, all pairwise comparisons are performed.

12.7.3 Sequentially Rejective Methods for MOMs and M-Estimators

When comparing groups using MOMs or M-estimators, the bootstrap method in Section 8.8.1 can be generalized to control FWE in a variety of ways. Currently, a slight modification of a sequentially rejective method derived by Rom (1990) appears to perform well when sample sizes are less than or equal to 100. With larger sample sizes, perhaps a direct application of Rom's method is preferable, but this issue needs further research.

Consider the problem of all pairwise comparisons among J groups based on any robust measure of location. As usual the measure of location associated with the jth group is labeled θ_j; here the goal is to test for every $j < k$ the hypothesis

$$H_0 : \theta_j = \theta_k \tag{12.11}$$

with FWE equal to α. The total number of hypotheses to be tested is

$$C = \frac{J^2 - J}{2}.$$

Compute the test statistic $\hat{p}^*$ described in Section 8.8.1 for comparing the jth group to the kth group. That is, among the B bootstrap samples, $\hat{p}^*$ is the proportion of times the bootstrap estimate for the jth group is greater than the bootstrap estimate of the kth. There are a total of C $\hat{p}^*$ values and for convenience they are labeled $\hat{p}^*_c$, ($c = 1, \ldots, C$). So $\hat{p}^*_1$, for example, is the proportion of times among B bootstrap samples that a bootstrap estimate of θ_1 is larger than a bootstrap estimate of θ_2, $\hat{p}^*_2$ is the estimate when comparing group 1 to group 3, and $\hat{p}^*_C$ is the estimate when comparing group $J - 1$ to group J. Let

$$\hat{p}^*_{mc} = \min\left(\hat{p}^*_c, 1 - \hat{p}^*_c\right).$$

The value $2\hat{p}^*_{mc}$ represents an estimated p-value for the corresponding hypothesis, as was explained in Chapter 8.

The modification of Rom's method is applied as follows. Put the $\hat{p}^*_{mc}$ values in *descending* order, yielding $\hat{p}^*_{m[1]} \geq \hat{p}^*_{m[2]} \geq \cdots \geq \hat{p}^*_{m[C]}$. So, for example, $\hat{p}^*_{m[1]}$ is the largest of the C values just computed and $\hat{p}^*_{m[C]}$ is the smallest. Decisions about the individual hypotheses are made as follows. If $\hat{p}^*_{m[1]} \leq \alpha_1$, where α_1 is read from Table 12.4, reject all C of the hypotheses. Put another way, if the largest

TABLE 12.4 Values of α_c for $\alpha = .05$ and .01

c	$\alpha = .05$	$\alpha = .01$
1	.02500	.00500
2	.02500	.00500
3	.01690	.00334
4	.01270	.00251
5	.01020	.00201
6	.00851	.00167
7	.00730	.00143
8	.00639	.00126
9	.00568	.00112
10	.00511	.00101

estimated p-value, $2\hat{p}^*_{m[1]}$, is less than or equal to α, reject all C hypotheses. If $\hat{p}^*_{m[1]} > \alpha_1$ but $\hat{p}^*_{m[2]} \leq \alpha_2$, fail to reject the hypothesis associated with $\hat{p}^*_{m[1]}$, but the remaining hypotheses are rejected. If $\hat{p}^*_{m[1]} > \alpha_1$ and $\hat{p}^*_{m[2]} > \alpha_2$ but $\hat{p}^*_{m[3]} \leq \alpha_3$, fail to reject the hypotheses associated with $\hat{p}^*_{m[1]}$ and $\hat{p}^*_{m[2]}$, but reject the remaining hypotheses. In general, if $\hat{p}^*_{m[c]} \leq \alpha_c$, reject the corresponding hypothesis and all other hypotheses having smaller $\hat{p}^*_m$ values. For other values of α (assuming $c > 1$) or for $c > 10$, use

$$\alpha_c = \frac{\alpha}{c}$$

(which corresponds to a slight modification of a sequentially rejective method derived by Hochberg, 1988.) This will be called *method SR*.

Method SR is unusual, in the sense that familiarity with many multiple comparison procedures suggests a slightly different approach. In particular, a natural guess at how to proceed is to compute the estimated significance level for the cth hypothesis, $2\hat{p}^*_{mc}$, and then to use Rom's method outlined in Section 12.8.2. But a practical concern is that with small to moderate sample sizes, now FWE will be substantially smaller than intended. In fact, if the goal is to have FWE equal to α, typically the actual FWE will be less than $\alpha/2$, which can affect power. A better approach would be to use the Benjamini–Hochberg method in Section 12.3.3; but again, with small sample sizes the actual FWE level can be too small compared to the nominal level. For large sample sizes, the Benjamini–Hochberg method might be preferable, but this issue has received little attention.

Simulation studies indicate that method SR performs well in terms of controlling FWE when sample sizes are less than 100. However, a criticism of method SR is that it is unknown what happens to FWE as all sample sizes get large. If all pairwise comparisons among four groups are to be performed and all groups have equal sample sizes of 100, then FWE is approximately .06. With sample sizes of 200, FWE is approximately .074, so for large sample sizes, perhaps Rom's method is preferable.

EXAMPLE. To illustrate method SR, imagine you test five hypotheses corresponding to the linear contrasts $\Psi_1, \ldots, \Psi_5$ and get $\hat{p}^*_{m1} = .02$, $\hat{p}^*_{m2} = .005$, $\hat{p}^*_{m3} = .23$, $\hat{p}^*_{m4} = .002$, $\hat{p}^*_{m5} = .013$, respectively. Further assume that you want FWE to be .05. The largest of these five values is .23, which corresponds to $H_0: \Psi_3 = 0$. That is, $\hat{p}^*_{m[1]} = .23$; with $\alpha = .05$, $\alpha_1 = .025$; this is less than .23, so you fail to reject $H_0: \Psi_3 = 0$. Had it been the case that $\hat{p}^*_{m[1]}$ was less than or equal to .025, you would stop and reject all five hypotheses. The next largest $\hat{p}^*$ value is .02, which corresponds to $H_0: \Psi_1 = 0$; this is less than $\alpha_2 = .025$, so $H_0: \Psi_1 = 0$ is rejected. Moreover, the remaining hypotheses are rejected regardless of what their $\hat{p}^*$ value happens to be. ■

12.7.4 S-PLUS Functions pbmcp and mcp2a

The S-PLUS function

$$\text{pbmcp(x, alpha} = 0.05\text{, nboot} = \text{NA, grp} = \text{NA, est} = \text{mom,}$$
$$\text{con} = 0\text{, bhop} = \text{F,} \ldots)$$

performs multiple comparisons using method SR described in the previous section. By default, all pairwise comparisons are performed, but a collection of linear contrasts can be specified instead via the argument con, which is used as illustrated in Section 12.6.1. The function computes $2\,\hat{p}^*_m$, the estimated significance level for each hypothesis, and lists it in the column headed sig.test. The appropriate critical value based on method SR, which is taken from Table 12.4, is listed under sig.crit. (The value listed is $2\alpha_c$, which is compared to $2\,\hat{p}^*_m$ as previously described.) At the end of the output is a value for sig.num, the number of significant results. When all groups have sample sizes of at least 100, it might be preferable to set the argument bhop to T. This causes the significant critical levels to be computed via the Benjamini–Hochberg method in Section 12.3.3.

EXAMPLE. For the data in Table 9.1, the S-PLUS function pbmcp returns

```
$output:
     con.num      psihat sig.test sig.crit ci.lower    ci.upper
[1,]       1 -0.03790389   0.6495  0.05000 -0.359385  0.43023143
[2,]       2  0.10149361   0.4945  0.05000 -0.184108  0.55663000
[3,]       3 -0.14128764   0.0770  0.02040 -0.549206  0.27054725
[4,]       4  0.13939750   0.3520  0.03380 -0.167386  0.37823232
[5,]       5 -0.10338375   0.2390  0.02540 -0.532423  0.12117667
[6,]       6 -0.24278125   0.0130  0.01702 -0.587199 -0.03574371
```

Continued

EXAMPLE. (*Continued*)

```
$con:
        [,1]  [,2]  [,3]  [,4]  [,5]  [,6]
[1,]       1     1     1     0     0     0
[2,]      -1     0     0     1     1     0
[3,]       0    -1     0    -1     0     1
[4,]       0     0    -1     0    -1    -1
$num.sig:
[1] 1
```

So one significant result was obtained and corresponds to the sixth hypothesis where group 3 is compared to group 4. That is, among individuals with schizophrenia, the typical measure of skin resistance among those with predominantly negative symptoms is lower than among those with predominantly positive symptoms instead. In this particular case, similar results are obtained when comparing 20% trimmed means with the function linconb (described in Section 12.7.2), but no significant results are obtained when comparing medians with the S-PLUS msmed in Section 12.6.4.

For convenience when dealing with a two-way ANOVA design, the S-PLUS function

mcp2a(J,K,x,est=mom,con=0,alpha=.05,nboot=NA,grp=NA,...)

performs all pairwise multiple comparisons among the rows and the columns and then does all tetrad differences relevant to interactions. ■

12.7.5 A Percentile Bootstrap Method for 20% Trimmed Means

The 20% trimmed mean has received considerable attention in recent years, because it performs nearly as well as the mean when distributions are normal and it can handle a fair degree of heavy-tailedness (situations where outliers are likely to appear). In particular, in contrast to medians, power remains relatively high under normality. Although arguments for preferring MOM can be made (e.g., MOM can handle more outliers), the 20% trimmed mean remains one of the better measures of location. When performing all pairwise comparisons based on 20% trimmed means, a special variation of the percentile bootstrap method currently seems best. The method is basically the same as the percentile bootstrap method in Section 12.7.3, except that a single critical value is used for testing all C hypotheses; this critical value is designed specifically for comparing 20% trimmed means. An approximation of the critical value for $\alpha = .05$ is

$$p_{\text{crit}} = \frac{0.0268660714}{C} - 0.0003321429,$$

which is based on results in Wilcox (2001d). Now you reject $H_0 : \Psi_c = 0$ if $2\hat{p}^*_{mc} \leq 2p_{\text{crit}}$.

It is possible that, in terms of Type I errors, the method in Section 12.7.3 is as good as and perhaps slightly better than the approach mentioned here when comparing 20% trimmed means. That is, you might be able to use the S-PLUS function pbmcp with est = tmean (a 20% trimmed mean), but this issue needs further investigation before any recommendations can be made.

12.7.6 S-PLUS Function mcppb20

The S-PLUS function

mcppb20(x, crit = NA, con = 0, tr = 0.2, alpha = 0.05, nboot = 2000, grp = NA)

performs the multiple comparison procedure just described. If no value for crit, the critical value, is specified, the function chooses a critical value based on results in Wilcox (2001d). For situations where a critical value has not been determined, the function approximates the critical value with p_{crit} given above if $\alpha = .05$. Otherwise it uses α/C as the critical value, a strategy that stems from the Bonferroni method covered in Section 12.8.1.

12.8 Methods for Dependent Groups

Multiple comparison methods for independent groups typically take advantage of the independence among the groups in some manner. When comparing dependent groups instead, generally some modification of the methods for independent groups must be made.

12.8.1 Bonferroni Method

One of the simplest and earliest methods for comparing dependent groups is based on what is known as the *Bonferroni inequality*. The strategy is simple: If you plan to test C hypotheses and want FWE to be α, test each of the individual hypotheses at the α/C level of significance. So, for example, when comparing groups having normal distributions, if you plan to test five hypotheses and want FWE to be at most .05, perform the five paired T-tests at the .01 level. If each of the resulting confidence intervals has probability coverage .99, then the simultaneous probability coverage will be greater than .95. That is, with probability at least .95, all of the confidence intervals will contain the true value of the parameter being estimated. Moreover, if the individual tests are able to control the probability of a Type I error, then the Bonferroni method controls FWE.

EXAMPLE. Consider J dependent groups and imagine that all pairwise comparisons are to be performed. Then the number of hypotheses to be tested

Continued

EXAMPLE. (*Continued*) is $C = (J^2 - J)/2$. So if, for each pair of groups, the hypothesis of equal (population) trimmed means is tested as described in Section 11.2, and if the goal is to have FWE equal to .05, perform each test at the $.05/C$ level. ■

12.8.2 Rom's Method

Several improvements on the Bonferroni method have been published, and one that stands out is a so-called sequentially rejective method derived by Rom (1990), which has been found to have good power relative to several competing methods (e.g., Olejnik, Li, Supattathum, & Huberty, 1997). To apply it, compute significance levels for each of the C tests to be performed and label them $P_1, \ldots, P_C$. Next, put the significance levels in descending order, which are now labeled $P_{[1]} \geq P_{[2]} \geq \cdots \geq P_{[C]}$. Proceed as follows:

1. Set $k = 1$.
2. If $P_{[k]} \leq d_k$, where d_k is read from Table 12.5, stop and reject all C hypotheses; otherwise, go to step 3.
3. Increment k by 1. If $P_{[k]} \leq d_k$, stop and reject all hypotheses having a significance level less than or equal to d_k.
4. If $P_{[k]} > d_k$, repeat step 3.
5. Continue until you reject or all C hypotheses have been tested.

An advantage of Rom's method is that its power is greater than or equal to that of the Bonferroni approach. In fact, Rom's method always rejects as many or more hypotheses. A negative feature is that confidence intervals are not readily computed.

A closely related method was derived by Hochberg (1988) where, rather than use Table 12.5, use $d_k = \alpha/(C - k + 1)$. For $k = 1$ and 2, d_k is the same as in

TABLE 12.5 Critical Values, d_k, for Rom's Method

k	$\alpha = .05$	$\alpha = .01$
1	.05000	.01000
2	.02500	.00500
3	.01690	.00334
4	.01270	.00251
5	.01020	.00201
6	.00851	.00167
7	.00730	.00143
8	.00639	.00126
9	.00568	.00112
10	.00511	.00101

TABLE 12.6 Illustration of Rom's method

Number	Test	Significance level	
1	$H_0: \mu_1 = \mu_2$	$P_1 = .010$	$P_{[5]}$
2	$H_0: \mu_1 = \mu_3$	$P_2 = .015$	$P_{[3]}$
3	$H_0: \mu_1 = \mu_4$	$P_3 = .005$	$P_{[6]}$
4	$H_0: \mu_2 = \mu_3$	$P_4 = .620$	$P_{[1]}$
5	$H_0: \mu_2 = \mu_4$	$P_5 = .130$	$P_{[2]}$
6	$H_0: \mu_3 = \mu_4$	$P_6 = .014$	$P_{[4]}$

Rom's method. An advantage of Hochberg's method is that it does not require special tables and can be used with $k > 10$.

EXAMPLE. Imagine you want to perform all pairwise comparisons among four dependent groups and you apply some method for means and get the significance levels shown in Table 12.6. Further assume that you want FWE to be .05. The largest significance level is .62; this is greater than .05, so you fail to reject the corresponding hypothesis, $H_0: \mu_2 = \mu_3$. The next largest significance level is .130; this is greater than $d_2 = .025$, so you fail to reject $H_0: \mu_2 = \mu_4$. The next largest significance level is .015; this is less than $d_3 = .0167$, so you stop and reject the corresponding hypothesis as well as those having smaller significance levels. ■

12.8.3 Linear Contrasts Based on Trimmed Means

More generally, a collection of linear contrasts can be tested when working with trimmed means corresponding to dependent groups. First consider a single linear contrast based on the marginal trimmed means of L groups:

$$\Psi = \sum_{\ell=1}^{L} c_\ell \mu_\ell.$$

Then $H_0: \Psi = 0$ can be tested as outlined in Box 12.6. Alternatively, a generalization of difference scores can be used instead. That is, set

$$D_i = \sum_{\ell=1}^{L} c_\ell X_{i\ell}$$

and test the hypothesis that the population trimmed mean of the D_i values is zero. When testing C such hypotheses, FWE can be controlled with Rom's method.

BOX 12.6 How to Test a Linear Contrast Based on the Marginal Trimmed Means of Dependent Groups

Goal

Test $H_0 : \Psi = 0$ for each of C linear contrasts such that FWE is α. (There are L groups.)

Let Y_{ij} ($i = 1, \ldots, n$; $j = 1, \ldots, L$) be the Winsorized values, which are computed as described in Box 11.3. Let

$$A = \sum_{j=1}^{L} \sum_{\ell=1}^{L} c_j c_\ell d_{j\ell},$$

where

$$d_{j\ell} = \frac{1}{h(h-1)} \sum_{i=1}^{n} (Y_{ij} - \bar{Y}_j)(Y_{i\ell} - \bar{Y}_\ell)$$

and h is the number of observations left in each group after trimming. Let

$$\hat{\Psi} = \sum_{\ell=1}^{L} c_\ell \bar{X}_{t\ell}.$$

Test statistic:

$$T = \frac{\hat{\Psi}}{\sqrt{A}}.$$

Decision Rule

Reject if $|T| \geq t$, where t is the $1 - \alpha/2$ quantile of a Student's T-distribution with $\nu = h - 1$ degrees of freedom. When testing more than one linear contrast, FWE can be controlled with Rom's method.

12.8.4 S-PLUS Function rmmcp

The S-PLUS function

rmmcp(x, con = 0, tr = 0.2, alpha = 0.05, dif = T)

performs multiple comparisons among dependent groups using trimmed means and Rom's method for controlling FWE. By default, difference scores are used. Setting dif=F results in comparing marginal trimmed means. (When α differs from both .05 and .01, FWE is controlled with Hochberg's method as described in Section 12.8.2.)

EXAMPLE. Imaginc a two-way ANOVA design where husbands and wives are measured at two different times. Then $L = 4$, all four groups are dependent, and this is an example of a two-way ANOVA with a within-subjects design on both factors. That is, the levels of factor A are possibly dependent, as are the levels of factor B. Assume that for wives, the trimmed means are μ_{t1} and μ_{t2} at times 1 and 2, respectively and that the trimmed means at times 1 and 2 for the husbands are μ_{t3} and μ_{t4}, respectively. If the goal is to detect an interaction by testing

$$H_0 : \mu_{t1} - \mu_{t2} = \mu_{t3} - \mu_{t4},$$

the linear contrast is $\Psi = \mu_{t1} - \mu_{t2} - \mu_{t3} + \mu_{t4}$. To use rmmcp, first store the contrast coefficients in some S-PLUS variable. For example, use the command

```
mat<-matrix(c(1,-1,-1,1)).
```

Then if the data are stored in the S-PLUS matrix m1, the command

```
rmmcp(m1,con=mat,dif=F)
```

will perform the computations. ■

12.8.5 Percentile Bootstrap Methods

A percentile bootstrap method for multiple comparisons among dependent groups can be performed using a simple combination of techniques already described. First consider all pairwise comparisons where the goal is to compare group j to k by testing

$$H_0 : \theta_j = \theta_k \tag{12.12}$$

for all $j < k$. As usual, the goal is to have FWE equal to α. Generate bootstrap samples as described in Section 11.7.1 in conjunction with method RMPB4. For any two specific groups, compute a significance level (or p-value) as indicated in Section 11.3.1 in connection with method RMPB2. That is, compute the proportion of bootstrap values from the first group that are greater than the bootstrap values from second group, label the result $\hat{p}^*$, set

$$\hat{p}^*_m = \min(\hat{p}^*, 1 - \hat{p}^*),$$

in which case the estimated significance level is $2\hat{p}^*_m$. Then FWE is controlled using a modification of Rom's method. In particular, let $\hat{p}^*_{mc}$ be the value of $\hat{p}^*_m$ when performing the cth comparison, $c = 1, \dots, C$. Again, if all pairwise comparisons are being performed, $C = (J^2 - J)/2$. Now proceed as in Section 12.7.3. That is, put the $\hat{p}^*_{mc}$ in descending order, yielding $\hat{p}^*_{m[1]} \geq \cdots \geq \hat{p}^*_{m[C]}$. Then use Table 12.4 as illustrated in Section 12.7.3.

Section 11.3.1 mentioned an adjustment of $\hat{p}^*$ used in conjunction with method RMPB2. It was called a bias-adjusted critical value and labeled $\hat{p}^*_a$. Here, when testing the cth hypothesis, $\hat{p}^*_a$ is labeled $\hat{p}^*_{ca}$. If this adjustment is used, the estimated significance level is $2\min(\hat{p}^*_{ca}, 1 - \hat{p}^*_{ca})$. Now it seems that a direct application of Rom's method can be used to control FWE. When using the method in the previous paragraph, there are some indications that it might not be quite satisfactory with large sample sizes, but switching to $\hat{p}^*_{ca}$ and using Rom's method seems to correct this.

12.8.6 Using Difference Scores

Another approach is to use difference scores. That is, when comparing group j to group k, set

$$D_{ijk} = X_{ij} - X_{ik},$$

the difference between the ith pair of observations. (Now bootstrap samples, D^*_{ijk}, are obtained by resampling n vectors of observations from the n-by-C matrix of difference scores, where $C = (J^2 - J)/2$.) Then, when testing the hypothesis that the typical difference is zero, you control FWE using either the Bonferroni method or Rom's technique.

To elaborate, it is again convenient to relabel the D^*_{ijk} values as D^*_{ic}, $c = 1, \ldots, C$. So here, $c = 1$ corresponds to comparing group 1 to group 2, $c = 2$ is comparing group 1 to group 3, and so on. For the cth comparison, let $\hat{p}^*_c$ be the proportion of times among B bootstrap resamples that $D^*_{ic} > 0$. As usual, let

$$\hat{p}^*_{mc} = \min(\hat{p}^*_c, 1 - \hat{p}^*_c),$$

in which case $2\hat{p}^*_{mc}$ is the estimated significance level for the cth comparison. Then put the p-values in descending order and make decisions about which hypotheses are to be rejected using method SR outlined in Section 12.7.3. That is, once the $\hat{p}^*_{mc}$ is computed, reject the hypothesis corresponding to $\hat{p}^*_{mc}$ if $\hat{p}^*_{mc} \le \alpha_c$, where α_c is read from Table 12.4. Alternatively, reject if the estimated significance level $2\hat{p}^*_{mc} \le 2\alpha_c$.

As for linear contrasts, consider any specific linear contrast with contrast coefficients $c_1, \ldots, c_J$, set

$$D_i = \sum c_j X_{ij},$$

and let θ_d be the typical (population) value of this sum. Then $H_0: \theta_d = 0$ can be tested by generating a bootstrap sample from the D_i values, repeating this B times, computing $\hat{p}^*$, the proportion of bootstrap estimates that are greater than zero, in which case $2\min(\hat{p}^*, 1 - \hat{p}^*)$ is the estimated significance level. Then FWE can be controlled in the manner just outlined.

When comparing groups using MOM, at the moment it seems that the method based on difference scores often provides the best power versus testing Equation (12.12). Both approaches do an excellent job of avoiding Type I error probabilities greater than the nominal α level, still using MOM. But when testing Equation (12.12), the actual Type I error probability can drop well below the nominal level in situations where the method based on difference scores avoids this problem. This suggests that the method based on difference scores will have more power; and indeed, there are situations where this is the case even when the two methods have comparable Type I error probabilities. It is stressed, however, that a comparison of these methods, in terms of power, needs further study, and perhaps the bias-adjusted critical value mentioned in Section 12.8.5 helps increase power when testing Equation (12.12), but this issue has not yet been investigated.

12.8.7 S-PLUS Function rmmcppb

The S-PLUS function

rmmcppb(x,y = NA, alpha = 0.05, con = 0, est = mom, plotit = T,
dif = T, grp = NA, nboot = NA, BA = F,...)

performs multiple comparisons among dependent groups using the percentile bootstrap methods just described. The argument dif defaults to T (for true), indicating that difference scores will be used. If dif=F, difference scores are not used. For example, when comparing all pairs of groups, hypotheses given by Equation (12.12) will be tested instead. If dif=F and BA=T, the significance levels are computed as described in Section 11.3.1 and Rom's method is used to control FWE. (With BA=F, a slight modification of Rom's method is used instead.) If no value for con is specified, then all pairwise differences will be tested as given by Equation (12.12). As usual, if the goal is test hypotheses other than all pairwise comparisons, con can be used to specify the linear contrast coefficients. (See Section 12.6.1 for an illustration of how to use con.)

12.9 Analyzing Between-by-Within Designs

There are various ways of performing multiple comparisons when dealing with a between-by-within (or split-plot) design using a combination of methods already described. A few specific possibilities are summarized here in the hope that one of them will match the needs of the reader. We begin with nonbootstrap methods for trimmed means and then consider bootstrap methods for other measures of location.

As in Chapter 11, it is assumed that levels of factor B correspond to dependent groups and that the levels of factor A are independent. First consider factor A. A simple approach is simply to average over the levels of factor B. So X_{ijk} becomes $Y_{ij} = \sum_k X_{ijk}/K$. Then all pairs of levels of factor A can be compared as based on the Y_{ij} values described in Section 12.6, and the computations can be performed with the S-PLUS function lincon in Section 12.6.1.

An alternative approach, one that provides more detail about how groups differ, is to perform all pairwise comparisons among the levels of factor A for each level of factor B. So, for any k, the goal is to test

$$H_0 : \mu_{tjk} = \mu_{tj'k},$$

for all $j < j'$ and $k = 1, \ldots, K$. The total number of hypotheses is $K(J^2 - J)/2$, and it is desired to control FWE among all of these tests. For fixed k, because independent groups are being compared, one approach is simply to create the appropriate linear contrasts and use the S-PLUS function lincon in Section 12.6.1. (For convenience, the S-PLUS function bwamcp, described in the next section, creates these linear contrasts for you.)

As for factor B, one approach is to ignore the levels of factor A and to test hypotheses based on the trimmed means corresponding to the difference scores associated with any two levels of factor B. So now there are $(K^2 - K)/2$ hypotheses to be tested, where each hypothesis is that the difference scores have a trimmed mean of zero. One way of controlling FWE is with the method in Section 12.8.2. From Chapter 11, another approach is not to take difference scores, but rather to use the marginal trimmed means.

A more detailed approach is as follows. Consider the jth level of factor A. Then there are $(K^2 - K)/2$ pairs of groups that can be compared. If, for each of the J levels of factor A, all pairwise comparisons are performed, the total number of comparisons is $J(K^2 - K)/2$. In symbols, the goal is to test

$$H_0 : \mu_{tjk} = \mu_{tjk'},$$

for all $k < k'$ and $j = 1, \ldots, J$. And of course an alternative approach is to use difference scores instead.

As for interactions, take any two levels of factor A—say, j and j'—and do the same for factor B—say, levels k and k'. Form the difference scores

$$D_{ij} = X_{ijk} - X_{ijk'}, \qquad \text{and} \qquad D_{ij'} = X_{ij'k} - X_{ij'k'},$$

and let μ_{tj} and $\mu_{tj'}$ be the population trimmed means associated with these difference scores. Then one way of stating the hypothesis of no interaction for these specific levels of factors A and B is with

$$H_0 : \mu_{tj} = \mu_{tj'},$$

and of course this can be done for any two levels of factors A and B. The goal is to test this hypothesis for all $j < j'$ and $k < k'$ in a manner that controls FWE, and this might be done as described in Section 12.8.2.

12.9.1 S-PLUS Functions bwamcp, bwbmcp, and bwimcp

Three S-PLUS functions are supplied for applying the methods just described. The first is

```
bwamcp(J, K, x, tr = 0.2, JK = J * K, grp = c(1 : JK),
        alpha = 0.05, KB = F, op = T).
```

The default value for the argument op is T, meaning that the hypotheses $H_0 : \mu_{tjk} = \mu_{tj'k}$ for all $j < j'$ and $k = 1, \ldots, K$ are tested. In essence, the function creates the appropriate set of linear contrasts and calls the S-PLUS function lincon. (The argument KB is used as described in Section 12.6.1.) Setting op=F results in averaging the data over the levels of factor B and performing all pairwise comparisons corresponding to the levels of factor A.

The S-PLUS function

bwbmcp(J, K, x, tr = 0.2, JK = J * K, grp = c(1 : JK), con = 0,
alpha = 0.05, dif = T, pool = F)

compares the levels of factor B instead. If pool=T is used, it simply pools the data for you and then calls the function rmmcp. If dif=F is used, the marginal trimmed means are compared instead. By default, pool=F, meaning that $H_0 : \mu_{tjk} = \mu_{tjk'}$ is tested for all $k < k'$ and $j = 1, \ldots, J$. For each level of factor A, the function simply selects data associated with the levels of factor B and sends them to the S-PLUS function rmmcp described in Section 12.8.3.

As for interactions, the S-PLUS function

bwimcp(J, K, x, tr = 0.2, JK = J * K, grp = c(1 : JK), alpha = 0.05)

can be used.

12.9.2 Bootstrap Methods

This section describes methods aimed at testing a collection of linear contrasts based on some bootstrap method. One advantage is that it is easy to test some variations of the methods described at the beginning of this section. Consider, for example, factor A and let $\bar{\theta}_{j.} = \sum \theta_{jk}/K$ be the average measure of location for the jth level. That is, for a fixed level of factor A, $\bar{\theta}_{j.}$ is the average of the measures of location across the levels of factor B. Then an approach to comparing the levels of factor A is to test

$$H_0 : \bar{\theta}_{j.} = \bar{\theta}_{j'.} \qquad (12.13)$$

for every $j < j'$. That is, for all pairs of rows, compare the average measure of location among the dependent groups. There are $C = (J^2 - J)/2$ such comparisons, the individual tests can be performed as described in Chapter 11, and the resulting significance levels can be used in conjunction with the modified Rom's method to control FWE. (That is, use method SR in Section 12.7.3.) Of course, another possibility is to focus on the kth level of factor B and test

$$H_0 : \theta_{jk} = \theta_{j'k} \qquad (12.14)$$

for all $j < j'$ and then to do this for all K levels for factor B (i.e., $k = 1, \ldots, K$). So now there is a total of $C = K(J^2 - J)/2$ tests to be performed.

As for factor B, again a simple approach is to ignore the levels of factor A, simply view the data as coming from K dependent groups, and use the methods in Section 12.8.1 or 12.8.2. As before, difference scores can be used or marginal measures of location can be compared. In the latter case, again ignore factor A and let θ_k be the population measure of location associated with the kth level of factor B. Then, for every $k < k'$, test

$$H_0 : \theta_k = \theta_{k'}$$

and control FWE using method SR in Section 12.7.3.

As for Interactions, first consider a two-by-two design — say, two independent groups measured at two different times. For the first group, let θ_{d1} be the population value for some measure of location based on the time 1 measure minus the time 2 measure. For example, θ_{d1} might be the population M-estimators corresponding to the difference scores between times 1 and 2. Similarly, for group 2, let θ_{d2} be the difference for group 2. Then an approach to interactions is to test

$$H_0 : \theta_{d1} = \theta_{d2}.$$

This can be done by first taking a bootstrap sample based on the difference scores associated with group 1, repeating this for group 2, and labeling the difference between these two bootstrap estimates $\hat{\delta}^*$. Then repeat this B times and let $\hat{p}^*$ be the proportion of times $\hat{\delta}^*$ is greater than zero. Then $2\hat{p}^*$ is the estimated significance level. For the general case of J independent groups and K dependent groups, this process can be repeated for any two levels of factor A and any two levels of factor B, and FWE can again be controlled by adopting method SR in Section 12.7.3.

12.9.3 S-PLUS Functions spmcpa, spmcpb, and spmcpi

The S-PLUS function

```
spmcpa(J, K, x, est = mom, JK = J * K, grp = c(1 : JK),
       avg = F, nboot = NA, ...)
```

performs pairwise comparisons for factor A of a split-plot design as described in the previous section. Setting est=tmean results in using 20% trimmed means. The argument avg indicates whether measures of location are to be averaged. That is, the goal is to test the hypotheses given by Equation (12.13) if avg=T is used. If avg is not specified (in which case it defaults to F, for false), the hypotheses given by Equation (12.14) will be tested. The function determines B if the argument nboot is omitted. Otherwise, the arguments have their usual meaning.

The S-PLUS function

```
spmcpb(J, K, x, est = mom, JK = J * K, grp = c(1 : JK),
       dif=T, nboot = NA, ...)
```

performs pairwise comparisons among the levels of factor B. Setting est=tmean results in using 20% trimmed means. The argument dif=T indicates that difference scores will be used. Setting dif=F results in marginal measures of location being compared. The S-PLUS function

```
spmcpi(J,K,x,est=mom,JK=J*K,grp=c(1:JK),nboot=NA, ...)
```

tests hypotheses related to no interactions.

When using difference scores, the following might help when reading the output. For every two levels of factor B, the function creates difference scores. In effect the number of levels for factor B becomes $(K^2 - K)/2$ and the contrast coefficients reported correspond to the total number of parameters, which is $J(K^2 - K)/2$.

EXAMPLE. Imagine that for each of two independent groups, the goal is to compare K measures in a manner that controls FWE. This problem is a special case of a split-plot or a between-by-within subjects design. For example, one group might receive an experimental drug at four different times, a control group receives a placebo, and the goal is to compare the groups at time 1, at time 2, at time 3, and at time 4. That is, four hypotheses are to be performed and the goal is to have FWE equal to α. If the data are stored in the S-PLUS variable x, the command bwamcp(2,4,x) will perform the analysis. ■

EXAMPLE. Another example stems from an illustration in Chapter 11 where EEG for murderers was measured at four sites in the brain and the same was done for a control group, in which case the goal might be, among other things, to determine which sites differ between the two groups. When working with means under the assumption of normality, the problem is simple: Compare each site with Welch's method and control FWE with Rom's procedure. To compare groups with outliers removed (using MOM), use the S-PLUS function spmcpa described earlier. If the EEG data are stored in the S-PLUS function eeg, the command spmcpa(2, 4, eeg) returns:

```
$output:
     con.num       psihat sig.test crit.sig   ci.lower    ci.upper
[1,]    1    -0.54895604    0.020   0.0254 -0.9635714 -0.03111111
[2,]    2    -0.02000000    0.802   0.0500 -0.7363889  0.60989011
[3,]    3     0.16000000    0.550   0.0338 -0.5131319  0.91923077
[4,]    4    -0.01275641    0.902   0.0500 -0.7785714  0.65279221

$con:
        [,1]  [,2]  [,3]  [,4]
[1,]       1     0     0     0
[2,]       0     1     0     0
[3,]       0     0     1     0
[4,]       0     0     0     1
[5,]      -1     0     0     0
[6,]       0    -1     0     0
[7,]       0     0    -1     0
[8,]       0     0     0    -1

$num.sig:
[1] 1
```

The contrast coefficients are reported in $con. For example, among the eight groups, the first column indicates that measures of location corresponding to the first and fifth groups are compared. (That is, for level 1 of factor B, compare

Continued

EXAMPLE. (*Continued*) levels 1 and 2 of factor A.) The value of num.sig is 1, meaning that one difference was found based on whether the value listed under sig.test is less than or equal to the corresponding value listed under crit.sig. The data indicate that the typical EEG for murderers differs from the control group at the first site where measures were taken. ■

12.10 Exercises

1. Assuming normality and homoscedasticity, what problem occurs when comparing multiple groups with Student's T-test?
2. For five independent groups, assume that you plan to do all pairwise comparisons of the means and you want FWE to be .05. Further assume that $n_1 = n_2 = n_3 = n_4 = n_5 = 20$, $\bar{X}_1 = 15$, $\bar{X}_2 = 10$, $s_1^2 = 4$, $s_2^2 = 9$, $s_3^2 = s_4^2 = s_5^2 = 15$; test $H_0: \mu_1 = \mu_2$ using (a) Fisher's method (assuming the ANOVA F-test rejects), (b) Tukey–Kramer, (c) Dunnett's T3, (d) Games–Howell, (e) Scheffé's method, (f) Kaiser–Bowden.
3. Repeat the previous exercise, but now with $n_1 = n_2 = n_3 = n_4 = n_5 = 10$, $\bar{X}_1 = 20$, $\bar{X}_2 = 12$, $s_1^2 = 5$, $s_2^2 = 6$, $s_3^2 = 4$, $s_4^2 = 10$, and $s_5^2 = 15$.
4. You perform six tests and get significance levels .07, .01, .40, .001, .1, and .15. Based on the Bonferroni inequality, which would be rejected with FWE equal to .05?
5. For the previous exercise, if you use Rom's method, which tests would be rejected?
6. You perform five tests and get significance levels .049, .048, .045, .047, and .042. Based on the Bonferroni inequality, which would be rejected with FWE equal to .05?
7. Referring to the previous exercise, which would be rejected with Rom's procedure?
8. Imagine you compare four groups with Fisher's method and you reject the hypothesis of equal means for the first two groups. If the largest observation in the fourth group is increased, what happens to MSWG? What does this suggest about power when comparing groups 1 and 2 with Fisher's method?
9. Repeat the previous exercise, but with the Tukey–Kramer and Scheffé's methods instead.
10. For the data in Table 9.1, each group has 10 observations, so when using Tamhane's method to compute confidence intervals for all pairwise differences, the degrees of freedom are 9 and the value of h in Table 11 of Appendix B is 4.3. Verify that if the goal is to have confidence intervals with FWE equal to .05 and lengths .5, the required sample sizes for each group are 50, 11, 18, and 17.
11. Repeat the previous exercise using the S-PLUS function hochberg. Verify that the required sample sizes are 50, 10, 18, and 17.
12. Use the S-PLUS function lincon to verify the results in the first example of Section 12.6.1.

13. Use the S-PLUS function msmed to compare all pairs of groups based on medians and the data in Table 9.1. Verify that the output from msmed is

```
$test:
      Group Group      test crit        se
[1,]      1     2 0.1443384 2.63 0.2742167
[2,]      1     3 0.4313943 2.63 0.3025886
[3,]      1     4 0.5795942 2.63 0.3019699
[4,]      2     3 0.9624726 2.63 0.1767479
[5,]      2     4 0.7709186 2.63 0.1756865
[6,]      3     4 1.4059718 2.63 0.2173265

$psihat:
      Group Group    psihat   ci.lower  ci.upper
[1,]      1     2 -0.039580 -0.7607699 0.6816099
[2,]      1     3  0.130535 -0.6652731 0.9263431
[3,]      1     4 -0.175020 -0.9692008 0.6191608
[4,]      2     3  0.170115 -0.2947319 0.6349619
[5,]      2     4 -0.135440 -0.5974955 0.3266155
[6,]      3     4 -0.305555 -0.8771238 0.2660138
```

14. Use the S-PLUS function msmedse to compute the standard errors for the median corresponding to the four groups in Table 9.1. Compare the results to the estimated standard error for the 20% trimmed means returned by trimse. Now note that when comparing groups 3 and 4, you reject with 20% trimmed means, as was illustrated in Section 12.6.1, but not with medians, as illustrated in Exercise 13.
15. Perform all pairwise comparisons of the groups in Table 11.3 using the S-PLUS function rmmcppb in Section 12.8.4. Use the MOM estimate of location and use difference scores. Verify that a difference between groups 2 and 3 is found with FWE set at .05.
16. A. Thompson and Randall–Maciver (1905) report four measurements of male Egyptian skulls from five different time periods. The first was maximal breadth of skull, and the five time periods were 4000 bc, 3300 bc, 1850 bc, 200 bc, and 150 ad. A portion of the output from lincon, when comparing means (with the argument tr set equal to zero), is

```
$psihat:
     Group Group     psihat  ci.lower    ci.upper
[1,]     1     2 -1.0000000 -4.728058  2.72805781
[2,]     1     3 -3.1000000 -6.402069  0.20206887
[3,]     1     4 -4.1333333 -7.563683 -0.70298384
[4,]     1     5 -4.8000000 -8.729236 -0.87076449
[5,]     2     3 -2.1000000 -5.258496  1.05849612
[6,]     2     4 -3.1333333 -6.427215  0.16054796
```

```
       Group Group     psihat  ci.lower   ci.upper
 [7,]     2     5 -3.8000000 -7.615383 0.01538268
 [8,]     3     4 -1.0333333 -3.813622 1.74695545
 [9,]     3     5 -1.7000000 -5.103517 1.70351741
[10,]     4     5 -0.6666667 -4.193848 2.86051459
```

So when comparing means, significant results are obtained when comparing group 1 to group 4 and group 1 to group 5. The output from pbmcp (when comparing groups based on MOM) is

```
$output:
      con.num      psihat sig.test sig.crit  ci.lower    ci.upper
 [1,]       1 -0.65134100   0.5910  0.05000 -4.358466  2.75714286
 [2,]       2 -3.62997347   0.0110  0.01460 -7.172174 -0.05952381
 [3,]       3 -3.70689655   0.0050  0.01278 -7.266667 -0.42142857
 [4,]       4 -4.37356322   0.0005  0.01022 -8.733333 -1.23333333
 [5,]       5 -2.97863248   0.0145  0.01702 -6.040134  0.19333333
 [6,]       6 -3.05555556   0.0215  0.02040 -6.347619  0.42450142
 [7,]       7 -3.72222222   0.0030  0.01136 -7.540404 -0.56410256
 [8,]       8 -0.07692308   0.8545  0.05000 -3.433333  3.26638177
 [9,]       9 -0.74358974   0.3580  0.02540 -4.817949  2.26819923
[10,]      10 -0.66666667   0.4475  0.03380 -4.869565  2.50000000

$con:
      [,1] [,2] [,3] [,4] [,5] [,6] [,7] [,8] [,9] [,10]
[1,]     1    1    1    1    0    0    0    0    0     0
[2,]    -1    0    0    0    1    1    1    0    0     0
[3,]     0   -1    0    0   -1    0    0    1    1     0
[4,]     0    0   -1    0    0   -1    0   -1    0     1
[5,]     0    0    0   -1    0    0   -1    0   -1    -1

$num.sig:
[1] 5
```

Interpret the results and contrast them with the analysis based on means.

17. Among the five groups in the previous exercise, only one group was found to have an outlier based on any of the boxplot rules in Section 3.4. What might explain why more significant differences are found when comparing groups based on MOM versus the mean?

ROBUST AND EXPLORATORY REGRESSION

Chapter 7 indicates how to make inferences about regression parameters when using the least squares estimator if standard assumptions (normality and homoscedasticity) are violated. A comparable method for making inferences about Pearson's correlation is also described there, but there are other fundamental problems with least squares regression and Pearson's correlation that need to be addressed. One basic concern is that outliers can greatly distort both of these methods. A second concern is that heteroscedasticity can grossly inflate the standard error of the ordinary least squares estimator, relative to other estimators one might use, even under normality. This means that in terms of power (the probability of detecting an association), using least squares regression can be relatively ineffective. A related problem is getting an accurate and relatively short confidence interval for the slope and intercept. This chapter describes some of the tools one might use to address these problems. Some issues related to multiple predictors are discussed, but the emphasis in this chapter is on simple regression, meaning that there is only one predictor. (Chapter 14 expands upon strategies for dealing with multiple predictors.)

13.1 Detecting Outliers in Multivariate Data

First consider the problem of detecting outliers in bivariate data. As in Chapter 6, imagine we have n pairs of observations, which we label $(X_1, Y_1), \ldots, (X_n, Y_n)$. At first glance the problem might appear to be trivial: Simply apply one of the outlier detection methods in Chapter 3 to the X values and do the same to the Y values. Comrey (1985) gives a rough indication of why this approach can be unsatisfactory with the following example. It is not unusual for an individual to be young or for an individual to have hardening of the arteries. But it is unusual for someone to be both young and have hardening of the arteries. More formally, this approach suffers from a fundamental problem: If the points are rotated, values that were declared outliers might no longer be declared outliers, and points that were not declared outliers might

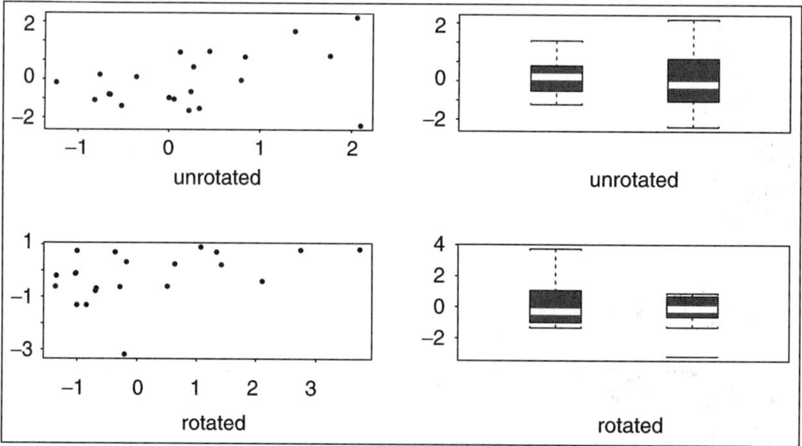

FIGURE 13.1 For multivariate data, simply checking for outliers among the marginal distributions can lead to different conclusions depending on how the points are rotated.

become outliers. That is, if you look at a scatterplot of points on a sheet of paper, declaring a point an outlier should not depend on whether you hold your head up straight versus tilting your head slightly to the right.

To illustrate this important point, look at the upper left panel of Figure 13.1. The points in this plot were created by generating 20 values from a standard normal curve for both X and the error term (ϵ) and setting

$$Y_i = X_i + \epsilon_i.$$

That is, points were generated according to the standard regression model discussed in Chapter 6, where the regression line has a slope of 1 and an intercept of zero. Then an additional point was added at $X = 2.1$ and $Y = -2.4$; it appears in the lower right corner of the scatterplot. To the right of the scatterplot are boxplots for the X and Y values. As is evident, no outliers are found, and none are found using any of the other outlier detection methods in Chapter 3. Note, however, that the point $(X, Y) = (2.1, -2.4)$ is unusual by construction. The reason is that when $X = 2.1$, Y should have a value that is reasonably close to 2.1 as well (because the regression line is simply $Y = X$). But given that $X = 2.1$, $Y = -2.4$ is located 4.5 standard deviations away from the regression line, which makes it unusual. Indeed, a casual glance at the scatterplot suggests that it is somehow removed from the bulk of the points.

The bottom left portion of Figure 13.1 shows the same points rotated by 45 degrees. So in effect the points are rotated so that now they are centered around a regression line having a slope of zero. The lower right panel of Figure 13.1 shows the resulting boxplots of the rotated points. Now an outlier is found among the Y values, which corresponds to the unusual point in the scatterplot of the unrotated points. What we need is an outlier detection method that takes into account the overall structure of the scatterplot. In particular, outliers should remain outliers under any rotation of the points we might make.

In the context of regression, the problem just illustrated is important, because when fitting a straight line to data, even when no outliers are found among the X values and none are found among the Y values, it is possible for a few points to be separated from the bulk of the observations in a way that has an inordinate effect on the least squares regression line. That is, in a very real way, points can be outliers in a scatterplot even though they are not deemed outliers when attention is restricted to the X values or the Y values. As an illustration, again consider the points in the scatterplot in the upper left panel of Figure 13.1, only now we add two points at $(X, Y) = (2.1, -2.4)$. For the original 20 points, the least squares slope of this line is $b_1 = 1.063$. So in this particular case, the least squares regression line provides a fairly accurate estimate of the true slope, which is 1. What is particularly important is that the two points added at $(X, Y) = (2.1, -2.4)$ have a tremendous influence on the least squares regression line — the estimate of the slope drops from 1.063 to 0.316.

A criticism of the illustration just given might be that the two points added to the scatterplot should have some influence on the estimated slope. However, another point of view is that a few unusual values should not mask a true association. If we test the hypothesis that the slope is zero using the conventional method covered in Section 6.3.1, then, based on the original 20 values, we reject with $\alpha = .001$. But when the two unusual values are added to the data, the significance level increases to .343. In this particular case, simply restricting the range of X to values less than 2.1 corrects this problem. But the simple strategy of restricting the range of X is not always effective when trying to detect associations that might be masked by outliers. Indeed, conventional wisdom is that restricting the range of X can actually mask an association, and this is in fact a realistic concern, as will be illustrated in Section 13.4.

13.1.1 A Relplot

An outlier detection method that satisfies our goal of dealing with the rotation of points and taking into account the overall structure of the data is the so-called *relplot* proposed by Goldberg and Iglewicz (1992); it is a bivariate analog of the boxplot. The somewhat involved computations are not particularly important for present purposes and therefore not given. (Computational details can be found in Goldberg & Iglewicz, 1992; and Wilcox, 1997a, Section 7.6.) However, familiarity with a relplot helps convey other concepts and strategies used in this chapter. The basic idea is first to compute a measure of location that has a reasonably high breakdown point. That is, any reasonable measure of location should be embedded in the central portion of a scatterplot of the data, and we want to avoid having a measure of location that is not near the center due to a few points that are unusually separated from the bulk of the observations. (Recall from Chapter 3 that we need estimators with a reasonably high breakdown point when searching for outliers, which continues to be the case here.) The particular measure of location used by Goldberg and Iglewicz is related to the M-estimator described in Chapter 3. They also compute a measure of covariance that is based primarily on the centrally located points. Based on these measures of location and covariance, a relplot creates two ellipses. The inner ellipse contains the central half of the points; points outside the outer ellipse are declared outliers.

13.1.2 S-PLUS Function relplot

The S-PLUS function relplot, written for this book, has the form

relplot(x,y,plotit=T)

and computes a relplot. As usual, x and y are any S-PLUS variables containing data. The function also returns a correlation coefficient based on the centrally located data (labeled mrho), but it plays no important role in this book and in fact suffers from a practical problem illustrated momentarily.

EXAMPLE. To illustrate how a relplot can alter our perceptions about the association between two variables, first look at Figure 13.2, which shows a scatterplot of data from a study where the goal is to predict reading ability. (These data were generously supplied by L. Doi. The X values are stored in column 4 of the file read.dat, and the Y values are in column 8.) Also shown is the least squares regression line, which has an estimated slope of −0.032. Testing the hypothesis of a zero slope using the conventional method in Section 6.3.1, the significance level is .764 (which is the same significance level obtained when testing $H_0 : \rho = 0$ with Student's T) and the .95 confidence interval for the slope is (−0.074, 0.138). Using the modified bootstrap method in Section 7.3, the .95 confidence interval is (−0.27, 0.11). So again no association is detected between the two variables under study. ■

Figure 13.3 shows a relplot of the same data. The inner ellipse contains the central half of the data; points outside the outer ellipse are declared outliers. As is evident,

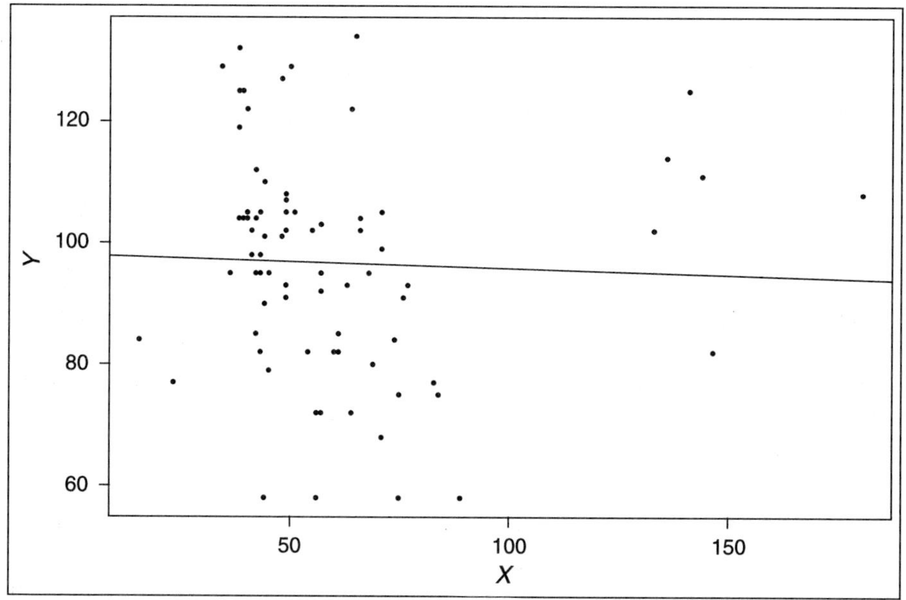

FIGURE 13.2 Scatterplot and least squares regression line for the reading data.

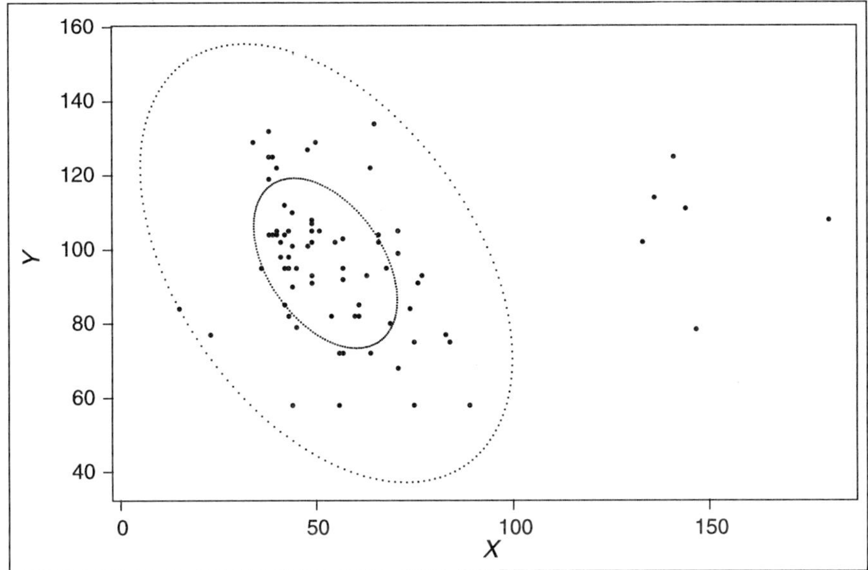

FIGURE 13.3 Relplot for the reading data in Figure 13.2.

we get a rather different impression about the association between these two variables versus the scatterplot in Figure 13.2. Figure 13.3 suggests that for the bulk of the points there might be a negative association that is masked when using least squares due to the outliers in the right portion of the scatterplot. (Methods covered later in this chapter support this conclusion.)

EXAMPLE. A criticism of the relplot is that outliers might affect our overall sense of how points are associated, depending on where the outliers happen to be located. Consider again the data in the upper portion of Figure 13.1 — only momentarily we ignore the unusual point that was added at $(X, Y) = (2.1, -2.4)$. Figure 13.4 shows a relplot of the data (and it reports a correlation of .66, which is reasonably close to Pearson's correlation, $r = .68$). ■

Now we add two points at $(X, Y) = (2.1, -2.4)$. Figure 13.5 shows the relplot. It correctly identifies the two outliers, but the association among the points not declared outliers is less pronounced. In particular, the relplot correlation drops from .66 to .32 and Pearson's correlation becomes $r = .21$.

13.1.3 MVE and MCD Estimators

The relplot is certainly an improvement on the simple strategy of checking for outliers among the X values only (ignoring the Y values) and then doing the same for the Y values. However, in addition to the concern just illustrated, a limitation is that it has not been extended to situations where we have more than two measures for each individual. There is an interesting alternative called a *bagplot* (Rousseeuw, Ruts,

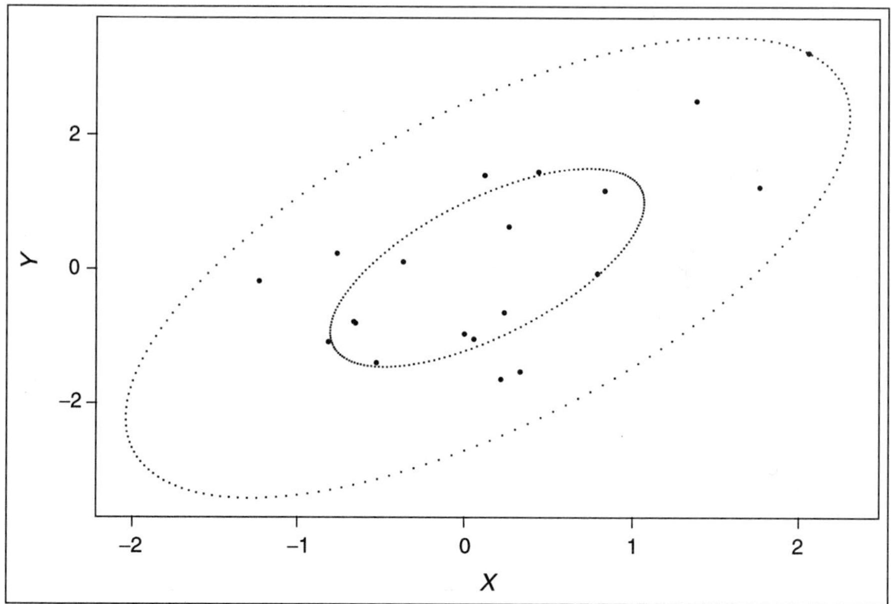

FIGURE 13.4 Relplot where both X and Y have normal distributions.

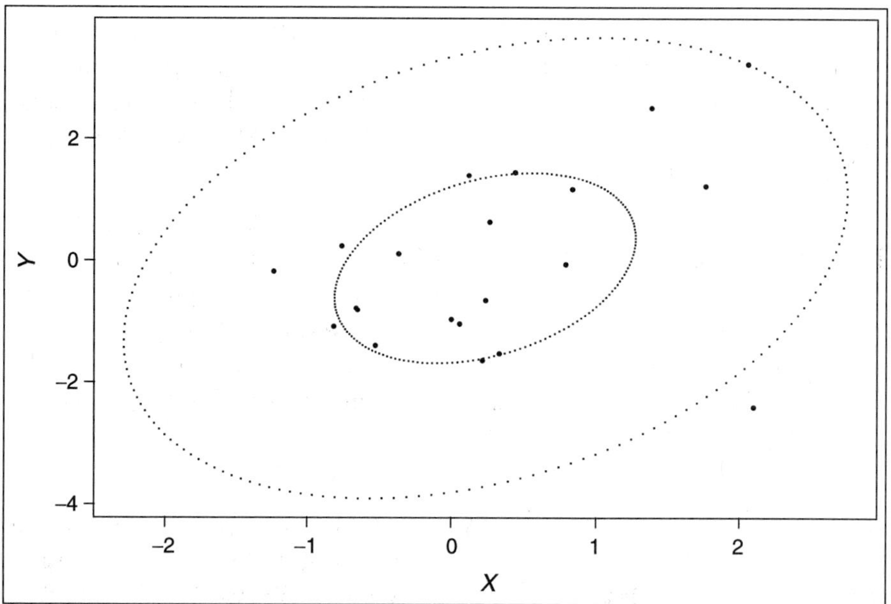

FIGURE 13.5 Adding two outliers can substantially alter the relplot in Figure 13.4.

& Tukey, 1999), but it too is restricted to the bivariate case. This is not to say that the bagplot has no practical value, but to conserve space attention is now restricted to methods that have been generalized to more than two variables. (The bagplot might soon meet this requirement if certain technical difficulties can be addressed.) But before describing multivariate alternatives to the relplot, we first need to consider robust analogs of the covariance between X and Y.

Recall from Chapter 6 that the sample covariance between X and Y is

$$s_{xy} = \frac{1}{n-1}\sum_{i=1}^{n}(X_i - \bar{X})(Y_i - \bar{Y}).$$

This quantity is the numerator of Pearson's correlation, and it has a finite-sample breakdown point of only $1/n$. That is, a single unusual point can have an inordinate influence on its value. One general strategy for dealing with outliers in bivariate data hinges on analogs of s_{xy} that can handle a large number of outliers. One such strategy arises as follows. Consider all ellipses that contain half of the data points. The inner ellipse in Figure 13.3 is one example for the reading data. The strategy behind the *minimum-volume ellipsoid* estimator, typically called the *MVE* estimator, is to search among all of the ellipses containing half of the data and identify the one having the smallest area. Once this ellipse is identified, a robust analog of the usual covariance (s_{xy}) is the covariance based only on the points inside this particular ellipse, and the mean of these points provides a measure of location with a high breakdown point. (The finite-sample breakdown point is .5.) Although the details as to how to find the ellipse with the smallest area are not straightforward, S-PLUS has a built-in function that performs the calculations for you. (It has the form cov.mve(m), where m is a matrix having n rows.) Moreover, the estimate is automatically rescaled so that when sampling from normal distributions, it estimates the variance and Pearson's correlation. (A comparable function, called MVE, can be found in SAS.)

Given the covariance and measure of location just described, one can measure the distance of each point from the center of the scatterplot using an analog of what is called the *Mahalanobis distance*. This distance can be used to judge how far away a point happens to be from the centrally located portion of a scatterplot. Rousseeuw and van Zomeren (1990) proposed a rule for deciding if a point is sufficiently far from the center to be declared an outlier. For readers familiar with matrix algebra, the details are relegated to Box 13.1 for the more general case where we have p measures for each individual. (Basic matrix algebra is summarized in Appendix C.)

BOX 13.1 How to Detect Outliers Using the MVE or MCD Measures of Covariance

We have a sample of p measures for each of n individuals, which is denoted by $\mathbf{X}_i' = (X_{i1}, \ldots, X_{ip})$, $i = 1, \ldots, n$. Let **C** and **M** be the center and covariance matrix, respectively, of the data determined by the MVE estimator (which is computed by the built-in S-PLUS function cov.mve) or by the MCD estimator (which is computed by the built-in S-PLUS function cov.mcd). Let

$$D_i = \sqrt{(\mathbf{X}_i - \mathbf{C})'\mathbf{M}^{-1}(\mathbf{X}_i - \mathbf{C})}.$$

Continued

BOX 13.1 (*Continued*)
The value D_i measures how far the point $\mathbf{X}_i'$ is from the center of the data and is a generalization of what is called the *Mahalanobis distance*. (See for example Mardia, Kent, & Bibby, 1979, for details about the Mahalanobis distance.) The point $\mathbf{X}_i'$ is declared an outlier if $D_i > \sqrt{\chi^2_{.975,p}}$, the square root of the .975 quantile of a chi-squared distribution with p degrees of freedom.

An alternative to the MVE estimator is the so-called *minimum-covariance determinant* (MCD) estimator. To convey the basic strategy, we first must describe the notion of a generalized variance (introduced by S. Wilks in 1934), which is intended to measure the extent to which a scatterplot of points is tightly clustered together. For bivariate data, this measure of dispersion is given by

$$s_g^2 = s_x^2 s_y^2 (1 - r^2), \tag{13.1}$$

where s_x^2 and s_y^2 are the sample variances associated with the X and Y values, respectively, and r is Pearson's correlation between X and Y. (In the multivariate case, s_g^2 is the determinant of the covariance matrix.) Recall from Chapter 3 that the smaller the variance of the X values, the more tightly clustered together are the X values, and of course a similar result applies to the Y values. We have seen that a single outlier can inflate the sample variance tremendously, but what is more important here is that the sample variance can be small only if the X values are tightly clustered together with no outliers. And in Chapter 6 we saw that the correlation is sensitive to how far points happen to be from the regression line around which they are centered. When the sample variances are small and Pearson's correlation is large, the generalized variance will be small.

To provide some perspective, the left panel of Figure 13.6 shows a scatterplot of 100 points for which $s_g^2 = 0.82$. The right panel shows a scatterplot of another 100 points, except that they are more tightly clustered around the line $Y = X$, and the generalized variance has decreased to $s_g^2 = 0.23$. The left panel of Figure 13.7 shows another 100 points that were generated in the same manner as those shown in the left panel of Figure 13.6, except that the variance of the X values was reduced from 1 to 0.5. Now $s_g^2 = 0.20$. In the right panel of Figure 13.7, the points are more tightly clustered together, and the generalized variance has decreased to $s_g^2 = 0.06$.

Now consider any subset of the data containing half of the points. The strategy behind the MCD estimator is to search among all such subsets and identify the one with the smallest generalized variance. Then the MCD measure of location and covariance is just the mean and covariance of these points. (For results supporting the use of the MCD estimator over the MVE estimator, see Woodruff & Rocke, 1994.) As with the MVE estimator, computing the MCD measure of location and covariance is a nontrivial, computer-intensive problem, but S-PLUS has a built-in function that performs the calculations for you (called mcd.cov); SAS has an analog of this function called MCD. (For a description of the algorithm used, see Rousseeuw & van Driesen, 1999.) Once these measures of location and scale are available, you can measure the relative distance of a point from the center using the method in Box 13.1, and

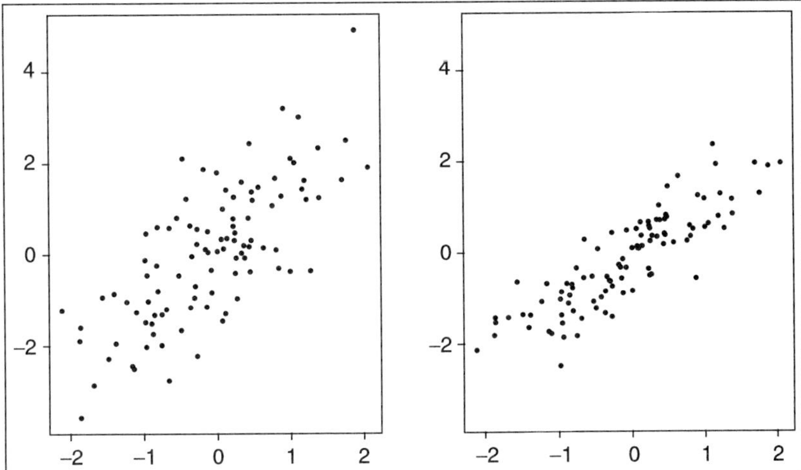

FIGURE 13.6 The more tightly points are clustered around a line, the smaller the generalized variance. The right panel has a smaller generalized variance than the left panel.

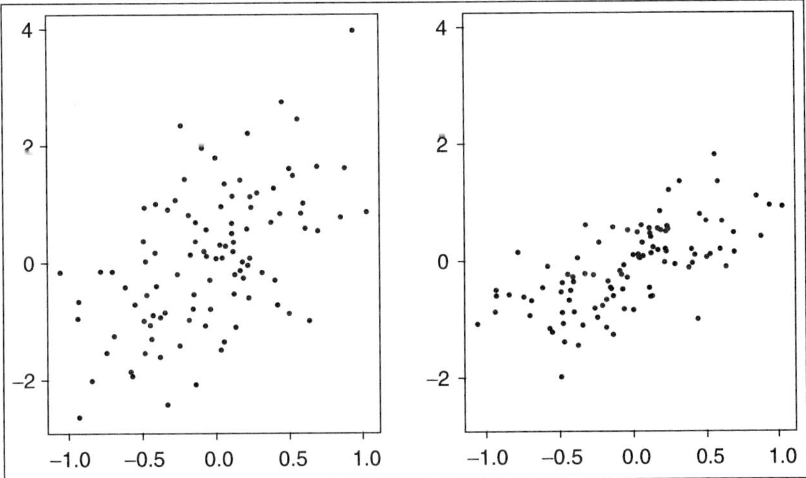

FIGURE 13.7 The generalized variance is also related to the variance of X and Y. In the left panel, $s_g^2 = .2$; in the right panel, $s_g^2 = .06$.

these distances can be used to detect outliers. (For an extension of this method, see Rocke & Woodruff, 1996.)

13.1.4 S-PLUS Function out

The S-PLUS function

```
out(m,mcd=F,plotit=T)
```

detects outliers using the MVE method when mcd=F is used; otherwise, the MCD method is used. It is common for the MCD method to find more outliers than the method based on MVE.

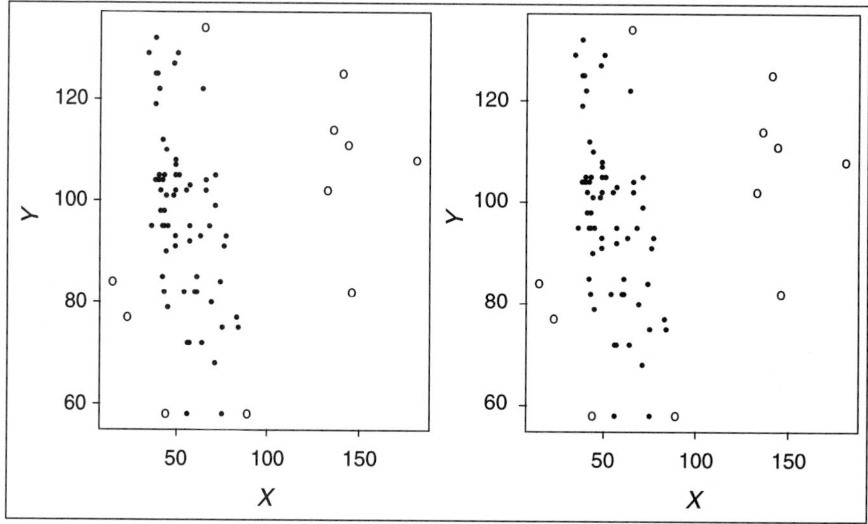

FIGURE 13.8 Output from the S-PLUS function out. The left panel is with mcd=F, meaning that the MVE method is used; the right panel is with mcd=T, meaning that the MCD method is used. Both methods return the same results in this particular case.

EXAMPLE. If the reading data used to create Figure 13.2 are stored in the S-PLUS variable blob, the command out(blob) creates the plot shown in the left panel of Figure 13.8. Points marked with a circle are declared outliers. The right panel shows the plot created by the function out with MCD=T. So in this particular instance, both methods flag the same points as outliers, and more points are declared outliers than with the relplot. ■

13.1.5 The Minimum Generalized Variance Method

It might seem that we could simply discard any outliers detected by the MVE or MCD methods and estimate the regression line with the data that remain. That is, use a method similar in spirit to the MOM estimator in Section 3.5.2. However, many variations of this approach are known to be unsatisfactory — they can mask the overall association (cf. Fung, 1993). This is somewhat expected based on properties of Pearson's correlation coefficient, r, covered in Chapter 6. In particular, we saw that restricting the range of X or Y can greatly influence r, so it is not too surprising that if we focus on the middle 50% of the data only, we might be misled regarding the association between X and Y.

EXAMPLE. As an illustration, the MVE and MCD methods are applied to the original 20 points in Figure 13.1. Both X and Y were generated from normal distributions, with the regression line between X and Y having a slope

Continued

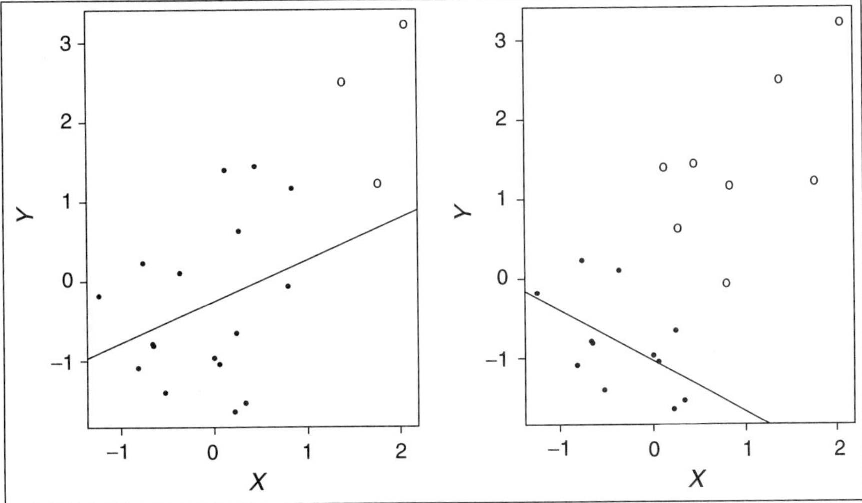

FIGURE 13.9 Eliminating outliers using the MVE or MCD method and fitting a least squares regression line to the data that remain can result in a poor estimate of the true slope, which is 1 in this particular case.

EXAMPLE. (*Continued*) equal to 1. The left panel of Figure 13.9 shows the plot created by out plus the least squares regression line based on the points not flagged as outliers. This line has a slope of .55, so it poorly estimates the true slope. The right panel shows the results when the MCD method is applied instead. As is evident, MCD finds more outliers. But what is perhaps more important, if the outliers found by MCD are discarded, the true association between X and Y is completely lost, as indicated by the least squares regression line based on the points not declared outliers. ■

EXAMPLE. To add perspective, the process used to generate Figure 13.9 was repeated 500 times, and each time the least squares estimate of the slope was computed using the points not flagged as outliers. So again the true slope is 1. The first boxplot in the left panel of Figure 13.10 shows the estimated slopes when using MVE to detect outliers; the second boxplot is based on the least squares estimate of the slope using all of the data instead. (That is, outliers are not discarded.) The second panel shows a boxplot of the least squares estimate when outliers detected by the MCD are removed; again, the other boxplot is based on the estimated slopes when outliers are not removed. As is evident, discarding outliers and applying least squares is a relatively unsatisfactory strategy in this particular case, because the least squares estimate based on all of the data tends to be closer to the true slope being estimated. More generally, where

Continued

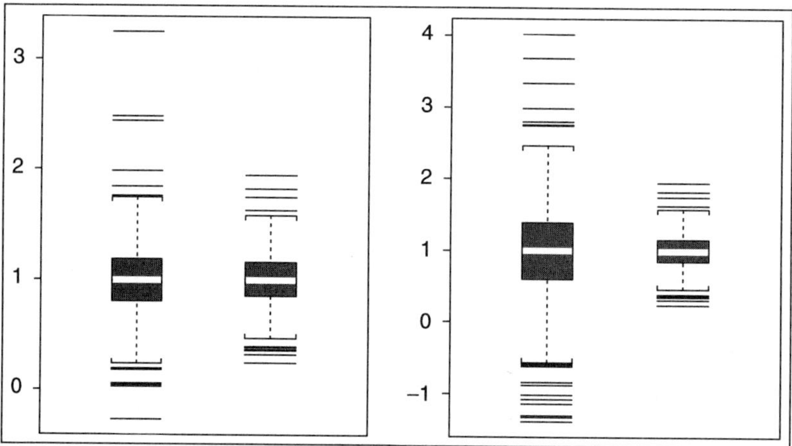

FIGURE 13.10 In the left panel, the first boxplot shows estimated slopes, based on least squares, with outliers detected by the MVE method removed. The other boxplot shows the estimated slopes when retaining all points. The right panel is the same as the left, but with the MVE method replaced by MCD.

EXAMPLE. (*Continued*) observations do not have a normal distribution, discarding outliers and fitting a least squares regression line to the remaining data can be highly unsatisfactory. ■

There are many strategies for discarding outliers and fitting a line to the data that remain. Currently, most seem to offer little or no improvement over other regression estimators described in this chapter. However, a variation of this strategy, where outliers are discarded and then a line is fit to the remaining points, does have practical value, at least in some situations, and is based on a different approach to detecting outliers. When searching for outliers, of particular importance is detecting so-called *bad leverage points*. A *leverage point* is an outlier among the X values. A *regression outlier* is a point with a relatively large residual. A *bad leverage point* is a leverage point that is also a regression outlier. A *good leverage point* is a leverage point that is not a regression outlier. That is, a good leverage point is a point that is reasonably close to the regression line, as illustrated in Figure 13.11. Good leverage points lower the standard error of the least squares estimate of the slope without giving a distorted indication of the association among the bulk of the observations. Bad leverage points can result in a poor fit to the majority of the data, even when using various robust estimators. So the hope is to be able to eliminate the effects of bad leverage points yet achieve a relatively accurate estimate of the slope and intercept, even under normality and homoscedasticity. In some situations this means that we want to avoid identifying and discarding so many points that the true association is lost. There are direct methods for detecting regression outliers (Rousseeuw & van Zomeren, 1990), but currently it seems that an indirect method performs best, based on the criterion of achieving a relatively low standard error, when estimating the slope.

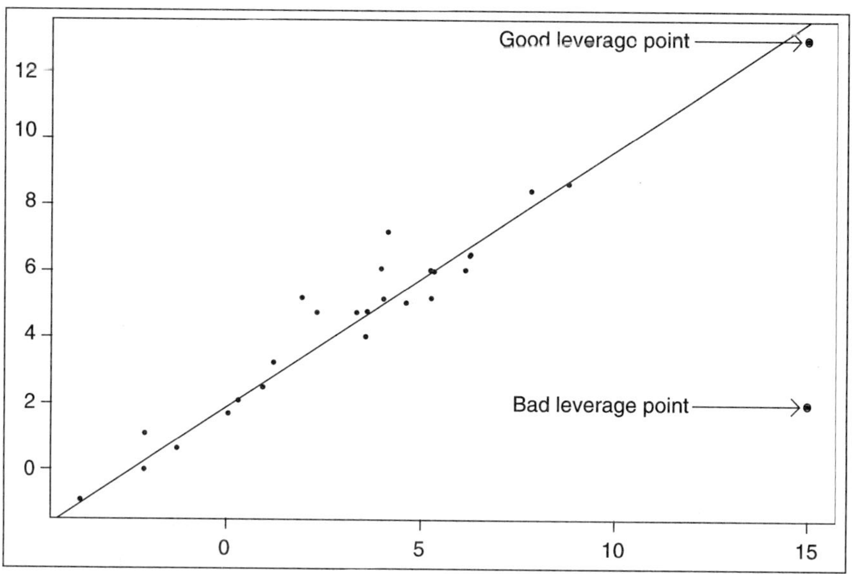

FIGURE 13.11 Illustration of a good versus bad leverage point.

For p-variate data, an outlier detection method that has been found to have practical value, in the context just described, begins by determining the p centrally located points and then determining how close each point is to the center using the notion of a generalized variance. There are many ways of finding the center of a cloud of points, some of which use some multivariate analog of the median (e.g., Small, 1990), and perhaps for the situation at hand the method used here can be improved upon; this issue is in need of further investigation.

The steps for applying this alternative outlier detection method, called the *MGV method*, are as follows.

1. Initially, all n points are described as belonging to set A.
2. Find the p points that are most centrally located. In the bivariate case, for the ith pair of points, compute

$$d_i = \sum_{j=1}^{n} \sqrt{\frac{(X_j - X_i)^2}{\text{MAD}_x^2} + \frac{(Y_j - Y_i)^2}{\text{MAD}_y^2}}, \tag{13.2}$$

where MAD_x and MAD_y are the values of MAD for the X and Y values, respectively. The two most centrally located points are taken to be the two points having the smallest d_i values. For the more general case where we have p measures for each individual, $(X_{i1}, \ldots, X_{ip})$, $i = 1, \ldots, n$,

$$d_i = \sum_{j=1}^{n} \sqrt{\sum_{\ell=1}^{p} \frac{(X_{j\ell} - X_{i\ell})^2}{\text{MAD}_\ell^2}}, \tag{13.3}$$

where MAD_ℓ is the value of MAD based on $X_{1\ell}, \ldots, X_{n\ell}$.

3. Remove the centrally located points from set A and put them into set B. At this step, the generalized variance of the points in set B is zero. (When dealing with p measures, any p distinct points will have a generalized variance of zero.)
4. If the ith point in set A is put in set B, the generalized variance of the points in set B will be changed to some value that is labeled s^2_{gi}. That is, associated with every point in A is the value s^2_{gi}, which is the resulting generalized variance when it, and it only, is placed in set B. Compute s^2_{gi} for every point in A.
5. Among the s^2_{gi} values computed in the previous step, permanently remove the point associated with the smallest s^2_{gi} value from set A and put it in set B. That is, find the point in set A that is most tightly clustered together with the points in set B. Once this point is identified, permanently remove it from A and leave it in B henceforth.
6. Repeat steps 4 and 5 until all points are now in set B.

The first p points removed from set A have a generalized variance of zero, which is labeled $s^2_{g(1)} = \cdots = s^2_{g(p)} = 0$. When the next point is removed from A and put into B (using steps 4 and 5), the resulting generalized variance of set B is labeled $s^2_{g(p+1)}$; continuing this process, each point has associated with it some generalized variance when it is put into set B. Note that by construction, $s^2_{g(1)} \le s^2_{g(2)} \le \cdots \le s^2_{g(n)}$.

Based on the process just described, the ith point has associated with it one of the ordered generalized variances just computed. For example, in the bivariate case, associated with the ith point (X_i, Y_i) is some value $s^2_{g(j)}$ indicating that the ith point was removed in the jth step of the process used to compute the values $s^2_{g(1)} \le s^2_{g(2)} \le \cdots \le s^2_{g(n)}$. For convenience, the generalized variance associated with the ith point, $s^2_{g(j)}$, is labeled D_i. The p deepest points have D values of zero. Points located at the edges of a scatterplot have the highest D values, meaning that they are relatively far from the center of the cloud of points. Moreover, we can detect outliers simply by applying one of the outlier detection rules in Chapter 3 to the D_i values. Note, however, that we would not declare a point an outlier if D_i is small, only if D_i is large. If we use the rule based on the median and MAD, for example, then according to Equation (3.22), the point (X_i, Y_i) is declared an outlier if

$$\frac{|D_i - M_D|}{\text{MAD}_D/.6745} > 2.24, \tag{13.4}$$

where M_D and MAD_D are the median and the value of MAD, respectively, based on the D values.

Of course, an alternative to Equation (13.4) is some type of boxplot rule. Currently, in the context of regression, the boxplot rule described in Section 3.4.4 has received the most attention and will be used henceforth. (When trying to estimate regression parameters, the effect of using Equation (13.4) has not been studied.) In particular, declare the ith point an outlier if

$$D_i > q_2 - 1.5(\text{IQR}), \tag{13.5}$$

where IQR $= q_2 - q_1$ and where q_1 and q_2 are the ideal fourths based on the D_i values. For the more general case where there are p variables, replace Equation (13.5) with

$$D_i > M_D + \sqrt{\chi^2_{.975,p}}(\text{IQR}),$$

where $\sqrt{\chi^2_{.975,p}}$ is the square root of the .975 quantile of a chi-squared distribution with p degrees of freedom.

13.1.6 S-PLUS Function outmgv

The S-PLUS function

outmgv(x,y=NA, plotit=T, outfun=outbox,...)

applies the MGV outlier detection method just described. If the second argument is not specified, it is assumed that x is a matrix with p columns corresponding to the p variables under study. So, for example, in the bivariate case, x could be a matrix having n rows and two columns. If the second argument, y, is specified, the function combines the data in x with the data in y and checks for outliers among these $p + 1$ variables. In particular, the data do not have to be stored in a matrix; they can be stored in two vectors (x and y) and the function combines them into a single matrix for you. If plotit=T is used and bivariate data are being studied, a plot of the data will be produced, with outliers marked by a circle. The argument outfun can be used to change the outlier detection rule applied to the depths of the points (the D_i values in the previous section). By default, the boxplot rule based on Equation (13.5) is used.

EXAMPLE. Consider the following five pairs of points:

X:	6	22	19	29	33
Y:	11	7	42	22	26

To find the two centrally located points, first note that for the X values MAD is 7 and for the Y values MAD is 11. For the first pair of points, $(X, Y) = (6, 11)$,

$$d_1 = \sqrt{\frac{(6-6)^2}{7^2} + \frac{(11-11)^2}{11^2} + \cdots + \frac{(6-33)^2}{7^2} + \frac{(11-26)^2}{11^2}} = 6.73.$$

In a similar manner, $d_2 = 4.896$, $d_3 = 5.76$, $d_4 = 4.523$, and $d_5 = 5.36$. The two smallest d_i values are 4.896 and 4.523, which correspond to the points (29, 22) and (22, 7). ■

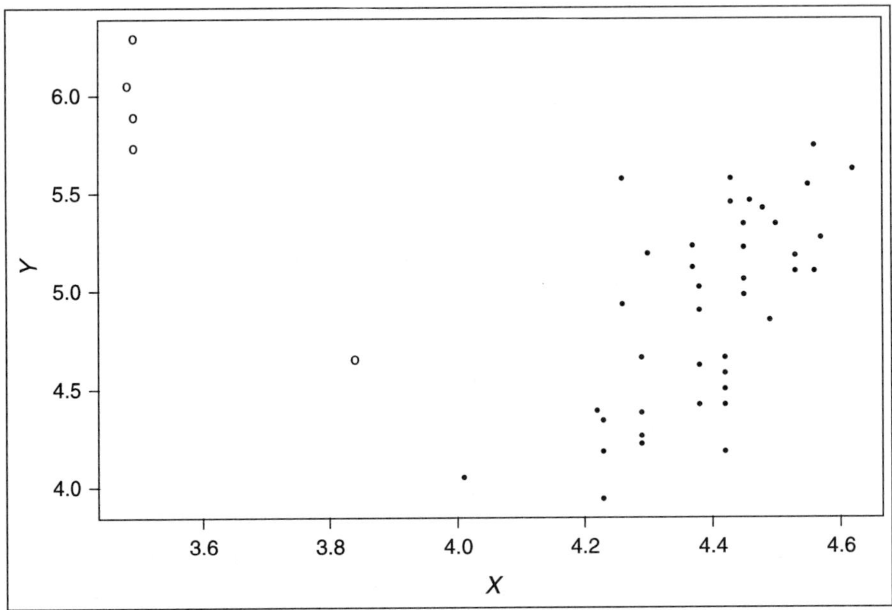

FIGURE 13.12 For the star data, the MGV outlier detection method flags the points indicated by a circle as outliers.

EXAMPLE. If the star data in Figure 6.3 are stored in the S-PLUS variables starx and stary, the command

outmgv(starx,stary)

produces the plot shown in Figure 13.12. If the data are stored in the n-by-2 matrix mstar, the command outmgv(mstar) will again produce Figure 13.12. ■

13.1.7 A Variation of the MGV Method for Large Sample Sizes

The MGV outlier detection method has practical value when fitting a straight line to data. However, a criticism is that as n gets large, execution time increases substantially using the software provided with this book. (Much faster software could be written, but this has not been done as yet.) When fitting a line to data, a regression method based in part on a variation of the MGV method performs well and has faster execution time. This alternative method begins by determining for each point the generalized variance when it is removed. That is, we remove the ith point, (X_i, Y_i), and compute the generalized variance for the remaining $n - 1$ points, which we label as D_i. So, for example, D_1 is the generalized variance if (X_1, Y_1) is eliminated from the data, and D_2 is the generalized variance when (X_1, Y_1) is put back and (X_2, Y_2) is removed instead. Extreme points will tend to have smaller D values versus points near the center of the data. So points with unusually small D values are declared outliers, and this will be called method *MGVF* or the *inward depth* method. Here, the boxplot rule for detecting

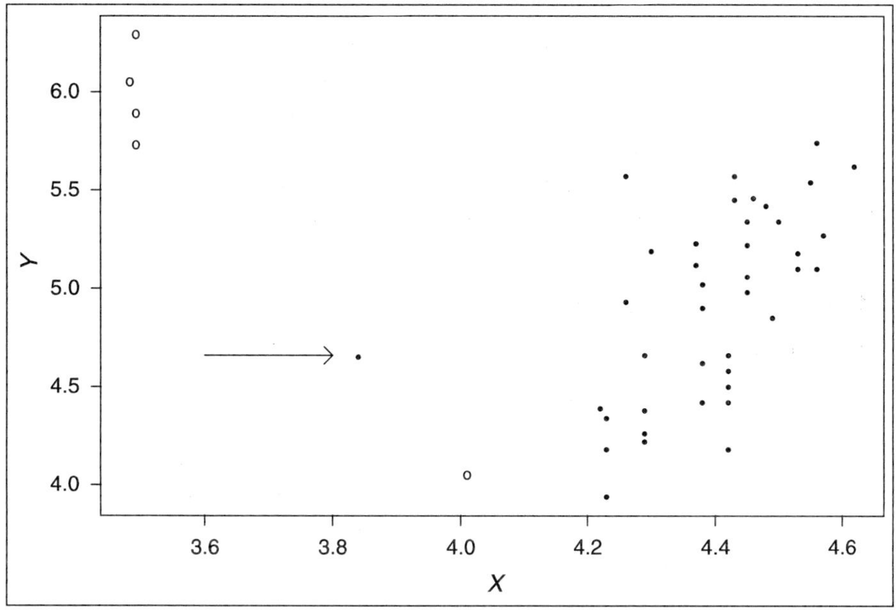

FIGURE 13.13 Output from the MGVF outlier detection method applied to the star data. The MGVF method misses what appears to be an obvious outlier, but despite this it has practical value when estimating the slope.

outliers (described in Section 3.4.4) will be applied to the D values unless stated otherwise.

It is stressed that if the main goal is to detect outliers, the method just described can be unsatisfactory. For example, if we apply the MGVF method to the data in Figure 13.12, we get the plot shown in Figure 13.13. Notice the point indicated by the arrow. It is declared an outlier in Figure 13.12 and it certainly seems to be relatively far from the majority of points. However, it is not declared an outlier by method MGVF. But despite this shortcoming, method MGVF will be seen to have practical value when we take up regression.

For completeness, there are several other approaches to measuring how deeply a point is embedded in a scatterplot (see, for example, Liu & Singh, 1997). Perhaps some of these techniques will be found to have practical value versus the measures of depth described here, but this remains to be seen. Also, there are other methods for detecting outliers that might have more practical value when fitting a straight line to data. For example, Fung (1993) begins by declaring points an outlier with the MVE method in Section 13.1.3. Then an iterative method is used to determine whether any of these outliers should be put back into the set of points not declared an outlier. (Some preliminary checks do not support the use of this method, but a more detailed study is needed.) Another possibility is to modify the MVE or MCD methods so that rather than use the central 50% of the data to determine the center and correlations, some higher proportion is used instead. This lowers the breakdown point, but in terms of estimating the association between two variables, perhaps this approach has practical value. So far, variations of this approach have been found to be relatively unsatisfactory. Another possibility is to use method MGVF but rather than use the boxplot rule on the D values, use instead the method in Section 3.4.2 which is based on the median and MAD. This strategy now flags the point marked by the arrow in Figure 13.12 as an outlier, but an additional point is marked an outlier beyond the five outliers found in Figure 13.11. Other approaches to detecting outliers in multivariate data were recently proposed by Rocke and Woodruff (1996); Poon, Lew, and Poon (2000); and Peña and Prieto (2001). Perhaps they can be used effectively in regression, relative to the other methods considered here, but this remains to be determined.

13.1.8 S-PLUS Function outmgvf

The S-PLUS function

outmgvf(x,y = NA, plotit = T, outfun = outbox,...)

checks for outliers using the MGVF method. The arguments x and y are used as described in Section 13.1.6. By default a boxplot rule is applied to the measures of depth, but this can be altered with the argument outfun. For example, outfun=out would use MAD and the median instead.

13.1.9 A Projection Method for Detecting Outliers

If the main goal is to detect outliers, all of the methods described so far are open to criticism. One issue has to do with the so-called *outside rate per observation*. This is just the expected or average proportion of outliers among n randomly sampled vectors. When working with a single random variable and when sampling from a normal distribution, a goal has been to achieve an outside rate per observation roughly equal to .05. The basic boxplot rule in Section 3.4.4 achieves this goal reasonably well (Hoaglin, Iglewicz, & Tukey, 1986), but the outside rate per observation is a bit unstable as the sample size increases from small to moderately large values. The method in Section 3.4.5 was developed to help correct this problem.

It seems that for most outlier detection methods aimed at multivariate data, little or nothing is known about their outside rate per observation. Checks on this rate when sampling from bivariate normal data indicate that the rate can be well above .05 when using the MVE method, and it is even higher when using MCD when variables are correlated (cf. Fung, 1993). These methods are well known and now easy to apply, but for some purposes an alternative strategy might be in order. This section outlines one approach for which the outside rate per observation is roughly equal to .05. The method reflects a blend of techniques that have been proposed, and it appears to have practical value for a wide range of situations.

The method begins by computing the Donoho and Gasko (1992) estimate of the median of the data (which can be done with the S-PLUS function dmean). The computational details are too involved to give here, but an outline of the strategy might help. For simplicity, attention is restricted to the bivariate case, but the method can be extended to more than two variables. The Donoho–Gasko median is based on something called *halfspace depth*, which is a method for measuring how deeply a point is nested within a scatterplot of all the data. For any point in a scatterplot, consider any line going through this point. As is evident, a certain proportion of the points in a scatterplot will be on or above this line, and a certain proportion will be on or below it. For convenience, the smaller of these two proportions is labeled P_m. Now, among the infinitely many lines going through some specific point there will be a minimum P_m value; this is called the *halfspace depth* of the point. (For bivariate data, halfspace depth can be computed exactly using the algorithm in Rousseeuw & Ruts, 1996. For more than two variables, an approximation has been derived by Rousseeuw & Struyf, 1998.) A high halfspace depth indicates that a point is deeply nested within the cloud of data. The Donoho–Gasko median is the average of all points having the largest depth. (For a single variable it reduces to the usual sample median.) An alternative and much simpler approach when trying to find the center of a scatterplot is to compute the median for each of the variables under study, but there are theoretical concerns about this strategy that go well beyond the scope of this book.

Before continuing, the notion of a projection of a point onto a line is needed. Consider any line through a scatterplot of data. For convenience, call this line $\mathcal{L}$. Now consider any point in the scatterplot, say, (X, Y). If we draw a line through this point that is perpendicular to the line $\mathcal{L}$, it will intersect with $\mathcal{L}$ at some point, say, (X_p, Y_p). The point (X_p, Y_p) is the (orthogonal) projection of (X, Y) onto the line $\mathcal{L}$.

Consider any point among the scatterplot of the data, and form the line connecting this point with the Donoho–Gasko median. Then project all points onto this line. (For computational details, see for example, Graybill, 1983, Section 4.4.) The distance between the projected points can be used to check for outliers. One possibility is simply to apply a boxplot rule, but a slight modification is needed to achieve an outside rate per observation reasonably close to .05. In particular, use Equation (3.25) on the distances of the projected points, but with $k = 2.4$ when dealing with bivariate data. (For the general case of p-variate data, use $k = \sqrt{\chi^2_{.95,p}}$.) Equation (3.26) is not used, because only points with large distances are declared outliers. As in Section 3.4.5, use the interquartile range based on the ideal fourths. The process just described is repeated for every point in the scatterplot. That is, n projections are considered. Any point is declared an outlier if it is found to be an outlier for any of the projections. (Complete computational details can be found in Wilcox, 2002.)

13.1.10 S-PLUS Function outpro

The S-PLUS function

outpro(m, gval = NA, plotit = T, op=T)

checks for outliers using the projection method just described. Here m is assumed to be a matrix having two or more columns. The argument gval is k and defaults to $\sqrt{\chi^2_{.95,p}}$ if not specified. If op=T is used with bivariate data and plotit=T, the function creates a scatterplot of the data and draws a polygon containing the centrally located points and ignoring the outliers. (This polygon is the .5 depth contour as described by Liu, Parelius, & Singh, 1999. It encompasses approximately half of the data, corresponding to the points having the highest halfspace depths.) Setting op=F, the function does not ignore the outliers when drawing the .5 depth contour. (An illustration will be given in Figure 13.21.)

13.2 Some Robust Regression Methods

This section summarizes some robust regression estimators that appear to have considerable practical value. All of the regression methods in this chapter can be used when there are multiple (p) predictors. However, explaining the basic strategy behind some of these methods is easier if we first focus on the single predictor case. As in Chapter 6, if there is a single predictor, it is assumed that

$$Y_i = \beta_0 + \beta_1 X_i + \epsilon \tag{13.6}$$

and X and ϵ are independent and that for p predictors the standard regression model is that

$$Y_i = \beta_0 + \beta_p X_{ip} + \cdots + \beta_1 X_{i1} + \epsilon,$$

$i = 1, \ldots, n$. Typically it is assumed that $E(\epsilon) = 0$, but here it is assumed that ϵ has a median of zero instead. The goal is to estimate the unknown slope and intercept (β_1 and β_0). With p predictors there are $p + 1$ parameters to be estimated: $\beta_0, \ldots, \beta_p$.

Although it is clear that the blind use of least squares regression is highly unsatisfactory, all indications are that no single regression method is always optimal among the many situations encountered in practice. That is, regression method A might have substantial advantages relative to method B in some situations, but situations arise where the reverse can happen as well. To complicate matters, several criteria are used to compare different regression methods, making it difficult and seemingly impossible to single out one method for routine use — several methods need to be considered. So for the moment we merely describe some regression methods and then try to convey their relative merits. At a minimum it is suggested that the methods in this section be given serious consideration, and it is recommended that several of the estimators in Section 13.3 be considered as well.

Another important point is that not all regression estimators are covered in this chapter. The omission of some methods was not arbitrary, but some experts might argue passionately that additional methods should have been included in this chapter. There might be merit to these arguments. Moreover, views about regression estimators continue to evolve. Simultaneously, some would argue that too many estimators are covered in this chapter and that providing such a seemingly bewildering array of methods will only confuse individuals learning about modern techniques. Currently, it seems that familiarity with multiple methods is a must. As will be illustrated, even among the better robust estimators, different results can be obtained with different methods, and choosing a method that provides relatively short confidence intervals is a nontrivial task, as will become evident. During the exploratory phases of an investigation, it seems that several estimators should be considered. An educated guess is that some of the estimators in this section can be ignored in most applications, but there is no compelling evidence that this can be done safely, so the goal is to cover a reasonable number of estimators in the hope that at least one of them will be valuable to the reader. Some authorities would argue that simply looking at a scatterplot enables us to tell which points are influential and that this should tell us which estimator should be used. But some of the illustrations later in this chapter suggest that dealing with regression is not always that simple.

In case it helps, a brief list of some of the estimators omitted from this chapter is given here. Maronna and Morgenthaler (1986) discuss how one might approach regression via robust covariances. A variation and extension of this method is discussed by Wilcox (1997a). This chapter describes two M-estimators, but certain variations of this approach are not covered. In particular, GM-based estimators (with Mallows weights) are not discussed. Recent results on this estimator are reported by Bianco, Boente, and Rienzo (2000), but work reviewed by Wilcox (1997a) suggests that it does not compete well with other estimators when there is heteroscedasticity. Despite this, perhaps arguments can be made for using this estimator in applied work, but this remains to be established. For a survey of results related to M-estimators, see Maronna, Yohai, and Zamar (1993). Some methods approach heteroscedasticity assuming that it can be modeled using some *known* function (e.g., Carroll & Ruppert, 1982; Giltinan, Carroll, & Ruppert, 1986). Such situations are not discussed here. For methods that deal with heteroscedasticity by attempting to estimate the optimal weights in weighted least squares, see Cohen, Dalal, and Tukey (1993) as well as Wilcox (1996a). (Rank-based approaches are briefly discussed in Chapter 15.)

13.2.1 The Theil–Sen (TS) Estimator

As is evident, any two distinct points determine a line. A method proposed by Theil (1950) and Sen (1968) estimates the slope of a regression line by computing the slope for all pairs of points having distinct X values and then computing the median of these slopes; the result will be labeled $b_{1\text{ts}}$. More formally, let X_i and $X_{i'}$ be any two X values such that $X_i > X_{i'}$. The slope corresponding to the two points (X_i, Y_i) and $(X_{i'}, Y_{i'})$ is

$$b_{1ii'} = \frac{Y_i - Y_{i'}}{X_i - X_{i'}}. \tag{13.7}$$

Computing the slope for all pairs of points having $X_i > X_{i'}$, the median of these slopes is the Theil–Sen estimate of β_1 and is labeled $b_{1\text{ts}}$. The intercept is estimated with

$$b_{0\text{ts}} = M_y - b_{1\text{ts}} M_x,$$

where M_y and M_x are the sample medians corresponding to the Y and X values, respectively.

When there is one predictor, the finite-sample breakdown point of the Theil–Sen estimator is approximately .29 (Dietz, 1989), meaning that about 29% of the data must be altered to make the resulting estimate of the slope and intercept arbitrarily large or small. A negative feature is that the finite-sample breakdown point has not been established when there are two or more predictors, but it appears to decrease as p gets large. An advantage is that its standard error can be tens, even hundreds, of times

TABLE 13.1 Boscovich's Data on Meridian Arcs

X:	0.0000	0.2987	0.4648	0.5762	0.8386
Y:	56,751	57,037	56,979	57,074	57,422

smaller than the ordinary least squares estimator when the error term is heteroscedastic. Even when the error term has a normal distribution but is heteroscedastic, the Theil–Sen estimator can be substantially more accurate.

EXAMPLE. The computation of the Theil–Sen estimator is illustrated with the data in Table 13.1, which were collected about 200 years ago and analyzed by Roger Boscovich. The data deal with determining whether the earth bulges at the center, as predicted by Newton, or whether it bulges at the poles as suggested by Cassini. Here X is a transformed measure of latitude and Y is a measure of arc length. (Newton's prediction implied that $\beta_1/(3\beta_0) \approx 1/230$.) For the first two pairs of points, the estimated slope is

$$\frac{57{,}037 - 56{,}751}{0.2987 - 0} = 1560.1.$$

Computing the slope for the remaining nine pairs of points and taking the median yields 756.6. It is left as an exercise to verify that the intercept is estimated to be 56,685. Interestingly, $b_{1\text{ts}}/(3b_{0\text{ts}}) = 0.0044$, which is fairly close to Newton's prediction: $1/230 = 0.0043$. (Least squares gives a very similar result.) ■

13.2.2 S-PLUS Function tsreg

The S-PLUS function

tsreg(x,y)

(written for this book) computes the Theil–Sen estimate of the slope and intercept. Here, x can be any n-by-p matrix of predictors and y is any S-PLUS variable containing the Y values. For the data in Table 13.1 this function returns

```
$coef:
[1] 56685.3235  756.6191

$residuals:
[1]  65.67654  125.67443  -58.00000  -47.28736  102.17580
```

where $residuals are the residuals given by $r_i = Y_i - b_{1\text{ts}}X_i - b_{0\text{ts}}$.

It is noted that generalizations of the Theil–Sen estimator to multiple predictors are available. One approach, which is used by the S-PLUS function tsreg, is based on the so-called *Gauss–Seidel algorithm*, which is described in a general context by Hastie and Tibshirani (1990), but no details are given here. Other extensions have been considered (e.g., Hussain & Sprent, 1983; Wilcox, 1998), but currently there is no known reason for preferring them to the method used here.

13.2.3 M-Estimators

Section 3.2.8 introduced M-estimators of location, which stem from the realization that different ways of measuring closeness lead to different measures of location. To quickly review, if the goal is to choose a constant c that is close to all of the values $X_1, \ldots, X_n$, and if we measure the overall closeness of c to these n values with

$$\sum (X_i - c)^2,$$

the sum of all squared differences, then the value of c that minimizes this sum is $c = \bar{X}$, the sample mean. If we use

$$\sum |X_i - c|$$

to measure closeness, this leads to taking c to be the median. M-estimators of location consider a broad class of measures that might be used to measure closeness, and members of this class have been identified that have desirable properties. In particular, if we use Huber's Ψ (described in Section 3.2.8), we get an estimator that has a finite-sample breakdown point of .5, and the accuracy of the estimator (its standard error) compares well to the sample mean in the event sampling is from a normal distribution. But unlike the sample mean, this particular M-estimator can maintain a relatively small standard error when sampling is from a heavy-tailed distribution instead, where outliers are fairly common.

The basic idea behind M-estimators of location is readily extended to regression. Rather than measure the fit of a regression line with the sum of squared residuals, M-estimators are based on a broad class of functions that includes least squares as a special case. This approach leads to choosing the slope and intercept to be the values satisfying

$$\sum \Psi\left(\frac{r_i}{\hat{\tau}}\right) = 0, \tag{13.8}$$

where Ψ is some function to be determined, $r_1, \ldots, r_n$ are the residuals corresponding to some choice for the slope and intercept, and $\hat{\tau}$ is some measure of scale based on the residuals. [Equation (13.8) is a generalization of Equation (3.11).]

In the context of regression, many M-estimators (choices for Ψ) have been proposed and studied. Several have excellent theoretical properties, but no attempt is made to list all of these methods here. (Readers interested in more details are referred to Coakley & Hettmansperger, 1993; Staudte & Sheather, 1990; Rousseeuw & Leroy, 1987; Wilcox, 1997a.) Rather, two methods are described in Box 13.2. The first is based on Huber's Ψ in conjunction with what are called *Schweppe weights*. Roughly, the strategy is to look for outliers among the X values (using the h_i values given in Box 13.2) and to make adjustments if any outliers are found. The method also makes an adjustment if any unusually large residuals are found. The second method described in Box 13.2 was derived by Coakley and Hettmansperger (1993).

BOX 13.2 Two Regression M-Estimators

The first method begins by setting $k = 0$ and

$$h_i = \frac{1}{n} + \frac{(X_i - \bar{X})^2}{\sum (X_i - \bar{X})^2}.$$

Compute the least squares estimate of the intercept and slope as described in Chapter 6. Denote them by b_{0k} and b_{1k}, respectively. Proceed as follows:

1. Compute the residuals, $r_{i,k} = y_i - b_{0k} - b_{1k}X_{i1}$, let M_k be equal to the median of the largest $n - 1$ of the $|r_{i,k}|$, $\hat{\tau}_k = 1.48M_k$, and let $e_{i,k} = r_{i,k}/\hat{\tau}_k$.
2. Form weights,

$$w_{ik} = \frac{\sqrt{1 - h_i}}{e_{ik}} \Psi\left(\frac{e_{ik}}{\sqrt{1 - h_i}}\right),$$

   where

$$\Psi(x) = \max[-K, \min(K, x)]$$

   is Huber's Ψ with $K = 2\sqrt{2/n}$.
3. Use the w_{ik} values to obtain a weighted least squares estimate of the slope and intercept. That is, find the values $b_{0,k+1}$ and $b_{1,k+1}$ that minimize

$$\sum w_{ik} r_i^2.$$

   (S-PLUS and other software have built-in functions for computing weighted least squares estimates of the slope and intercept.) Increase k by 1.
4. Repeat steps 1–3 until convergence. That is, iterate until the change in the estimates of the slope and intercept are small.

The second estimator, called the *Coakley–Hettmansperger estimator*, is described for the general case of p predictors. Begin by computing the LTS estimator in Section 13.3.7, a vector having length $p + 1$. Compute the residuals, r_i $(i = 1, \ldots, n)$, and let

$$\hat{\tau} = 1.4826\left(1 + \frac{5}{n - p}\right) \times \text{med}\{|r_i|\}$$

and

$$w_i = \min\left\{1, \left[\frac{b}{(\mathbf{x}_i - \mathbf{m}_x)'\mathbf{C}^{-1}(\mathbf{x}_i - \mathbf{m}_x)}\right]^{a/2}\right\},$$

BOX 13.2 (*Continued*)
where the quantities $\mathbf{m_x}$ and $\mathbf{C}$ are the minimum volume ellipsoid (MVE) estimators of location and covariance associated with the predictors. (The notation med$\{|r_i|\}$ refers to the median of the values $|r_1|, \ldots, |r_n|$.) Let $\Psi'(r_i/\hat{\tau}w_i) = 1$ if $|r_i/\hat{\tau}w_i| \leq K$; otherwise $\Psi'(r_i/\hat{\tau}w_i) = 0$. Coakley and Hettmansperger suggest using $K = 1.345$. Let $\mathbf{W} = \text{diag}(w_i)$ and $\mathbf{B} = \text{diag}(\Psi'(r_i/\hat{\tau}w_i))$. The Coakley–Hettmansperger estimator is

$$b_{\text{ch}} = b_{\text{lts}} + (\mathbf{X}'\mathbf{B}\mathbf{X})^{-1}\mathbf{X}'\mathbf{W}\Psi\left(\frac{r_i}{w_i\hat{\tau}}\right)\hat{\tau}.$$

A serious criticism of the first M-estimator in Box 13.2 is that the finite-sample breakdown point is only $2/n$. That is, it can handle one outlier, but two might cause practical problems. One way of addressing this concern is to switch to the Coakley–Hettmansperger estimator. It begins by computing the LTS estimate of the slope and intercept as described in Section 13.3.7. Then it forms weights based on how deeply each X_i is nested within all of the X values. It then uses these weights, in conjunction with Huber's Ψ, to adjust the initial estimate of the slope and intercept. The result is an estimator with the highest possible breakdown point of .5. It can be substantially more accurate than the first method outlined in Box 13.2, and it enjoys excellent theoretical properties. For the special case of a single predictor, often it seems that some other estimator is a bit more satisfactory, but it remains one of the many estimators that should be given serious consideration. A possible argument for the Coakley–Hettmansperger estimator is that software is available that has relatively fast execution time, even when the sample size is fairly large and there are multiple predictors.

13.2.4 S-PLUS Functions bmrg and chreg

The S-PLUS functions

bmreg(x,y) and chreg(x,y)

compute the two M-estimators just described. The function bmreg performs the calculations for the first method in Box 13.2; chreg computes the Coakley–Hettmansperger estimator.

13.2.5 MGV and MGVF Regression

The so-called *MGV regression estimator* first checks for outliers using Equation (13.5). If any outliers are found, they are discarded, and the Theil–Sen estimator is applied to the data that remain. (So this approach is similar in spirit to the class of skipped estimators mentioned in Section 3.5.2.) All indications are that this does a relatively good job of eliminating any points that cause contamination bias when using Theil–Sen, where, roughly, *contamination bias* refers to the ability of a very small number of unusual values to result in a poor fit to the bulk of the observations. (More details are given in Section 13.4.) If instead you apply the least squares estimator after eliminating outliers, the accuracy of your estimate can be relatively poor, particularly when there is heteroscedasticity. That is, it might seem that least squares might be salvaged by first eliminating outliers; but based on the standard error of this method, it cannot be recommended (Wilcox, 2001b). Of course, some alternative to both Theil–Sen and least squares might be used after outliers are removed, but so far Theil–Sen seems to be a very effective choice.

As will be illustrated, the MGV estimator appears to have practical value when dealing with contamination bias, and it seems to compete reasonably well in terms of achieving a relatively small standard error. But a practical concern is that as the sample size increases, execution time can become unacceptably high when using the software written for this book, particularly when computing confidence intervals using the bootstrap method in Section 13.5. An alternative approach is to replace the MGV outlier detection method with the inward method described in Section 13.1.7. That is, discard any outliers found by the method in Section 13.1.7 and again apply Theil–Sen to the data that remain. This will be called the *MGVF regression estimator*.

13.2.6 S-PLUS Functions mgvreg and mgvfreg

The S-PLUS function

mgvreg(x,y,regfun=tsreg,outfun=outbox)

computes the MGV estimate of the slope and intercept as just described. The argument regfun indicates which estimator is applied after outliers are removed and defaults to Theil–Sen. The argument outfun indicates which boxplot rule will be applied to the depths of the points (the D values described in Section 13.1.5). By default, a boxplot rule is used. The function

mgvfreg(x,y,regfun=tsreg,outfun=outbox)

is like mgvreg, except that the inward outlier detection method (described in Section 13.1.7) is used instead.

13.3 More Regression Estimators

When there is a single predictor, it seems that often the Theil–Sen estimator is a good choice, but there are indications that with multiple predictors it might not compete well with other methods. (This issue is in need of more study.) Generally, the methods in the previous section seem to provide reasonably good alternatives to least squares, but currently the only certainty is that exceptions can occur. Accordingly, some additional estimators are described here that might prove to be useful.

13.3.1 S-Estimators and a Modification of Theil–Sen

S-estimators of regression parameters search for the slope and intercept values that minimize some measure of scale associated with the residuals. Least squares, for example, minimizes the variance of the residuals and is a special case of S-estimators. The hope is that by replacing the variance with some measure of scale that is relatively insensitive to outliers, we will obtain estimates of the slope and intercept that are relatively insensitive to outliers as well. As noted in Chapter 3, there are many measures of scale. The main point is that if, for example, we use the percentage bend midvariance (described in Section 3.3.7), situations arise where the resulting estimate of the slope and intercept has advantages over other regression estimators we might use. This is not to say that other measures of scale never provide a more satisfactory estimate of the regression parameters. But for general use, it currently seems that the percentage bend midvariance is a good choice.

Here a simple approximation of the S-estimator is used. (There are other ways of computing S-estimators, e.g., Croux, Rousseeuw, & Hössjer, 1994; Ferretti et al., 1999. Perhaps they have practical advantages, but it seems that this possibility has not been explored.) As with the Theil–Sen estimator, consider any two points such that $X_i < X_{i'}$ and again let $b_{1ii'}$ be the corresponding slope given by Equation (13.7). For convenience we let K represent the total number of slopes that can be computed in this manner. In the event all n of the X values are distinct, there are a total of $K = n(n-1)/2$ such slopes. Next, let

$$v_j = Y_j - b_{1ii'}X_j,$$

$j = 1, \ldots, n$. That is, for each of the n points and the slope corresponding to X_i and $X_{i'}$ ($b_{1ii'}$), compute the difference between Y_j and $b_{1ii'}X_j$, ($j = 1, \ldots, n$) and label the results $v_1, \ldots, v_n$. Let $S_{ii'}$ be some measure of scale based on the v values just computed, repeat this process for all K slopes, and let $S_{\min}$ be the smallest of the corresponding $S_{ii'}$ values. The final estimate of the slope is the value of $b_{1ii'}$ corresponding to $S_{\min}$, which we label b_1. The intercept is taken to be

$$b_0 = M_y - b_1M_x,$$

where M_x and M_y are the medians of the X and Y values, respectively. This will be called *method STS*. (As with the Theil–Sen estimator, the Gauss–Seidel method is used to handle multiple predictors.)

13.3.2 S-PLUS Function stsreg

The S-PLUS function

```
stsreg(x,y,sc=pbvar,...)
```

computes the STS estimator (the S-type modification of the Theil–Sen estimator) just described. The arguments x and y are used as described in Section 13.2.2. The argument sc indicates which measure of scale will be applied to the residuals and defaults to the percentage bend midvariance. The argument ... can be replaced by arguments related to the chosen measure of scale, sc.

EXAMPLE. For the data in Table 13.1, the S-PLUS function stsreg returns $b_1 = 490.5$ and $b_0 = 56{,}809$. In contrast, the Theil–Sen estimate of the slope is 756.6, the only point being that these two methods can give substantially different results. ■

13.3.3 An Extension of S-Type Estimators

An appeal of the STS estimator in Section 13.2.3 is that it can provide a reasonable fit to the majority of the points in situations where many other estimators provide a poor fit instead. There are exceptions, but there are situations where it has practical value. However, even when this estimator gives a reasonably good fit to the majority of points, it can provide a relatively poor estimate of the true slope when the error term is heteroscedastic. That is, its standard error can be relatively high compared to other estimators that might be used.

One approach toward this problem is to use the STS estimator in Section 13.2.3 as a preliminary fit to data, with the goal of detecting points with unusually large residuals. Such points can result in a poor fit to the bulk of the observations and a highly inaccurate estimate of the true slope (the slope we would get if all individuals could be measured). So one strategy is first to check for points that have unusually large residuals, called *regression outliers*, to remove them, and then to fit a line to the data that remain. (For an extensive comparison of methods for detecting regression outliers when there is homoscedasticity, see Wisnowski, Montgomery, & Simpson, 2001.) One specific strategy for fitting a line to data after outliers are removed is simply to use the least squares estimator covered in Chapter 6. But this approach can perform poorly when the error term is heteroscedastic and should be used with caution. The Theil–Sen estimator performs relatively well when there is heteroscedasticity, little accuracy is lost, versus least squares, in the event that the error term is homoscedastic, so here the Theil–Sen estimator is used to fit a line to data once regression outliers are removed.

The specific method used to detect regression outliers is applied as follows. First, fit a line to the data using the S-type modification of the Theil–Sen estimator (the STS estimator) described in Section 13.3.1. Let b_1 and b_0 be the resulting estimates of the slope and intercept and let

$$r_i = Y_i - b_1 X_1 - b_0,$$

($i = 1, \ldots, n$) be the usual residuals. Let M_r be the median of the residuals and let MAD_r be the median of the values $|r_1 - M_r|, \ldots, |r_n - M_r|$. Then the ith point (X_i, Y_i) is declared a regression outlier if

$$|r_i - M_r| > \frac{2(\text{MAD}_r).}{.6745}. \tag{13.9}$$

The final estimate of the slope and intercept is obtained by applying the Theil–Sen estimator to those points not declared regression outliers. When there are

p predictors, again compute the residuals based on STS and use Equation (13.9) to eliminate any points with large residuals. This will be called *method TSTS*.

13.3.4 S-PLUS Function tstsreg

The estimator just described is computed by the S-PLUS function

tstsreg(x,y,sc=pbvar,...),

which was written for this book. The argument sc indicates which measure of scale will be used when method STS is employed to detect regression outliers, and the default measure of scale is the percentage bend midvariance.

13.3.5 Least Median of Squares (LMS) Estimator

For the general case of p predictors, the *least median of squares* (LMS) estimator simply chooses the values $b_0, b_1, \ldots, b_p$ so as to minimize the median of the squared residuals. That is, choose $b_0, b_1, \ldots, b_p$ so as to minimize

$$\text{MED}\left(r_1^2, \ldots, r_n^2\right),$$

the median of the values $r_1^2, \ldots, r_n^2$, where

$$r_i = Y_i - b_0 - b_p X_{ip} \ldots, b_1 X_{i1},$$

$(i = 1, \ldots, n)$ are the residuals.

The LMS estimator has the highest possible breakdown point, .5. Its main use has been as an exploratory tool, such as when trying to detect regression outliers. Its standard error generally compares poorly to many other estimators, and it can give a poor fit to the majority of points, so this method should be used cautiously. However, despite its many shortcomings, situations do arise where in the preliminary stages of analysis it gives a reasonable fit to data when many other methods do not.

13.3.6 S-PLUS Function lmsreg

The LMS (least median of squares) estimator can be computed with the built-in S-PLUS function

lmsreg(x,y).

13.3.7 Least Trimmed Squares (LTS) Estimator

Rather than minimize the sum of the squared residuals or the median of the squared residuals, another approach is to minimize the sum of the trimmed squared residuals instead. That is, now the slope and intercept are taken to be the values that minimize

$$\sum_{i=1}^{h} r_{(i)}^2, \tag{13.10}$$

where $r_{(1)}^2 \leq r_{(2)}^2 \leq \ldots \leq r_{(n)}^2$ are the squared residuals written in ascending order. This is called the *least trimmed squares* (LTS) estimator. Typically, $h = [n/2] + 1$ is used to achieve the highest possible breakdown point, which is approximately .5. (The notation $[n/2]$ means that $n/2$ is rounded down to the nearest integer.)

Consideration has been given to increasing the efficiency of the LTS estimator by lowering its breakdown point. That is, investigations have been conducted to see whether increasing h in the previous paragraph results in situations where the standard error of the LTS estimator competes more favorably with other estimators that might be used. In particular, consideration has been given to using $h = [(1 - \gamma)n]$ with $\gamma = .2$ and .25 (Wilcox, 2001b). Generally, LTS has a smaller standard error than least squares when the error term is sufficiently heteroscedastic and $h = [n/2] + 1$ is used. Switching to $\gamma = .2$ or .25 does not appear to improve its performance appreciably, so this variation of the LTS estimator is not considered henceforth.

13.3.8 S-PLUS Function ltsreg

The built-in S-PLUS function

ltsreg(x,y)

computes the LTS (least trimmed squares) estimator just described with $h = [n/2] + 1$, the highest amount of trimming.

13.3.9 Least Trimmed Absolute (LTA) Value Estimator

A close variation of the LTS estimator is the least trimmed absolute (LTA) value estimator. Rather than choose the intercept and slope so as to minimize Equation (13.10), the goal is to minimize

$$\sum_{i=1}^{h} |r|_{(i)}, \tag{13.11}$$

where $|r|_{(i)}$ is the ith smallest absolute residual and h is as defined as in Section 13.3.7. (For recent results on the LTA estimator, see Hawkins & Olive, 1999.) Like LTS, the LTA estimator can have a much smaller standard error than the least squares estimator, but its improvement over the LTS estimator seems to be marginal at best, at least based on what is currently known. (For some comparisons of LTA and LTS, see Wilcox, 2001b.)

13.3.10 S-PLUS Function ltareg

The S-PLUS function

ltareg(x,y)

computes the LTA estimate of the slope and intercept.

13.3.11 Deepest Regression Line

The outlier detection methods covered in Section 13.1 are based in part on measuring how deeply a point is embedded in a scatterplot. Today there are comparable (numerical) methods for measuring how deeply a line is embedded in a scatterplot. This leads to yet another method of fitting a line to data: Search for the line that is most deeply nested within a cloud of points. Rousseeuw and Hubert (1999) have examined how this might be done and provided an algorithm for implementing the technique. Box 13.3 provides a brief outline of how this approach is applied, but complete details go well beyond the scope of this book. A negative feature is that the software provided with this book allows one predictor only.

BOX 13.3 A Brief Outline of the Deepest Regression Line Estimator

Let b_1 and b_0 be any choice for the slope and intercept, respectively, and let r_i ($i = 1, \ldots, n$) be the corresponding residuals. Roughly, this candidate fit is called a nonfit if you can find some partition or splitting of the X values such that all of the residuals for the lower X values are negative (positive) but that for all of the higher X values the residuals are positive (negative). So, for example, if all of the points lie above a particular straight line, in which case all of the residuals are positive, this line is called a nonfit. More formally, a candidate fit is called a nonfit if and only if a value for v can be found such that

$$r_i < 0 \quad \text{for all } X_i < v$$

and

$$r_i > 0 \quad \text{for all } X_i > v$$

or

$$r_i > 0 \quad \text{for all } X_i < v$$

and

$$r_i < 0 \quad \text{for all } X_i > v.$$

The regression depth of a fit (b_1, b_0) relative to $(X_1, Y_1), \ldots, (X_n, Y_n)$ is the smallest number of observations that need to be removed to make (b_1, b_0) a nonfit. The deepest regression estimator corresponds to the values of b_1 and b_0 that maximize regression depth.

13.3.12 S-PLUS Function depreg

The S-PLUS function

```
depreg(x,y)
```

computes the deepest regression line. A negative feature of this function is that execution time can be relatively high compared to many other estimators you might use.

13.4 Comments on Choosing a Regression Estimator

Again it is stressed that no single regression estimator is optimal in all situations. However, this does not mean that choosing an estimator is an academic matter. The standard least squares estimator covered in Chapter 6 is satisfactory when variables are independent. But when there is an association, different methods can lead to drastically different conclusions, and the routine use of least squares, to the exclusion of all other estimators that might be used, can be disastrous. Although authorities are not in agreement as to which method is best, familiarity with their relative merits might help the reader choose a method. For some purposes, particularly in the exploratory phases of an investigation, multiple methods might be considered. However, when testing hypotheses, there is the problem of controlling the probability of a Type I error. If multiple tests are performed using different regression estimators, the familywise error rate can be controlled along the lines covered in Chapter 12. A simple strategy would be to use the Bonferroni method in Section 12.8.1; or a sequentially rejective method might be used as described in Section 12.8.2. However, a consequence of such adjustments is lower power versus testing a single hypothesis.

Given the goal of finding a straight line that gives a good fit to the bulk of the points, a minimum requirement is that an estimator have a reasonably high finite-sample breakdown point. This eliminates least squares plus the first of the two M-estimators covered in Section 13.2.3. However, having a high finite-sample breakdown point is not sufficient to guarantee a good fit to the majority of points. That is, some estimators can be highly influenced by a few unusual points even though they have a high breakdown point. Said yet another way, some estimators guard against complete disaster, meaning that a few unusual values cannot result in estimates that are arbitrarily large (or small). But some of these estimators might poorly reflect the association among the vast majority of the data, depending on where the outliers are located. So as a general rule, regardless of which estimator is used, it seems prudent always to graphically check how well a line fits the data.

EXAMPLE. For the data in Figure 13.2, the least squares estimate of the slope is -0.02. In contrast, both TS and LTA estimate the slope to be -0.28, the MGV estimate is -0.69, and the STS estimate is -0.8, the only point being that the choice of an estimator can make a practical difference. In this particular case, LTS and the Coakley–Hettmansperger estimate give results similar to least squares. As was indicated in Section 13.1.2 (see Figure 13.3), the rightmost six points are outliers. If we eliminate these outliers by restricting the range of X to values less than 100, the least squares estimate is now -0.48, with a significance level less than .001, and the Coakley–Hettmansperger estimate is -0.64.

Continued

EXAMPLE. (*Continued*) However, simply restricting the range of the X values does not necessarily deal with problems due to heteroscedasticity and outliers among the Y values. (Restricting the range of the Y values and applying standard hypothesis-testing methods results in using the wrong standard error.) Even without restricting the range, the MGV and STS estimates suggest that the association might be more negative for the bulk of the observations than indicated by least squares when the range of X is restricted. ■

Next it is illustrated that although robust estimators can be substantially more accurate than least squares, situations do arise where some robust estimators are greatly influenced by a few outliers, despite having a high breakdown point. That is, it can make a difference which robust method is used, and switching to some robust method does not necessarily eliminate the need to check the fit to data graphically.

EXAMPLE. Twenty points were generated as in Figure 13.4, so the true slope is 1. Then two aberrant points were added to the data at $(X, Y) = (2.1, -2.4)$. Figure 13.14 shows a scatterplot of the points plus the LTS regression line, which has an estimated slope of −0.94. So in this case, LTS is a complete disaster in terms of detecting how the majority of the points were generated. The least squares estimate is −0.63. The Coakley–Hettmansperger estimator relies on LTS as an initial estimate of the slope, and despite its high breakdown point the estimate of the slope is −0.65. In contrast, the MGV estimate of the slope is .97, nearly equal to the true slope, 1. The deepest regression line estimate is .66, and the STS estimate performs poorly in this particular case, the estimate being −0.98. The LMS estimate of the slope is 1.7, so it performs poorly as well in this particular instance. Again, this is not to suggest that LMS, LTS, STS, and the Coakley–Hettmansperger estimators be excluded from consideration. Rather, the point is that despite the robust properties they enjoy, they can perform poorly in some situations where other methods do well. Also, although MGV does very well here, this is not to suggest that it be used to the exclusion of all other methods. ■

In Figure 13.14, it is evident that the rightmost point is an outlier, and eliminating it by restricting the range of the X values improves matters considerably for the estimators that performed poorly in this particular case. But it cannot be stressed too strongly that there is more to robust regression than simply restricting the range of the X values, as will be illustrated.

To add perspective, the process used to generate the data in Figure 13.14 was repeated 500 times, and estimates of the slope were computed using least squares, the M-estimator with Schweppe weights (estimator bmreg), the Coakley–Hettmansperger estimator (chreg), the Theil–Sen estimator (tsreg), and the deepest regression line. Boxplots of the results are shown in Figure 13.15. Notice that the

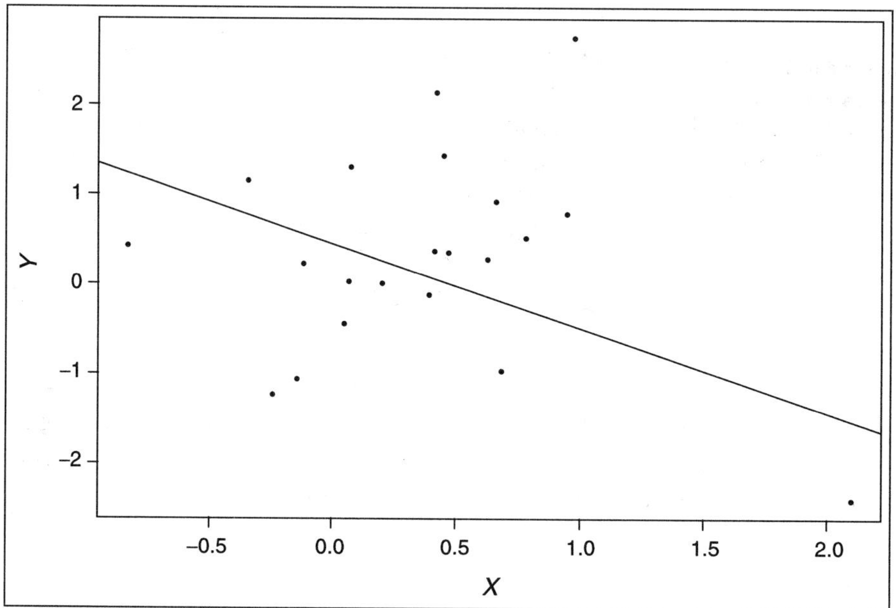

FIGURE 13.14 Twenty points where the true slope is 1, plus two outliers. The solid line is the LTS regression line, which performs poorly in this particular case.

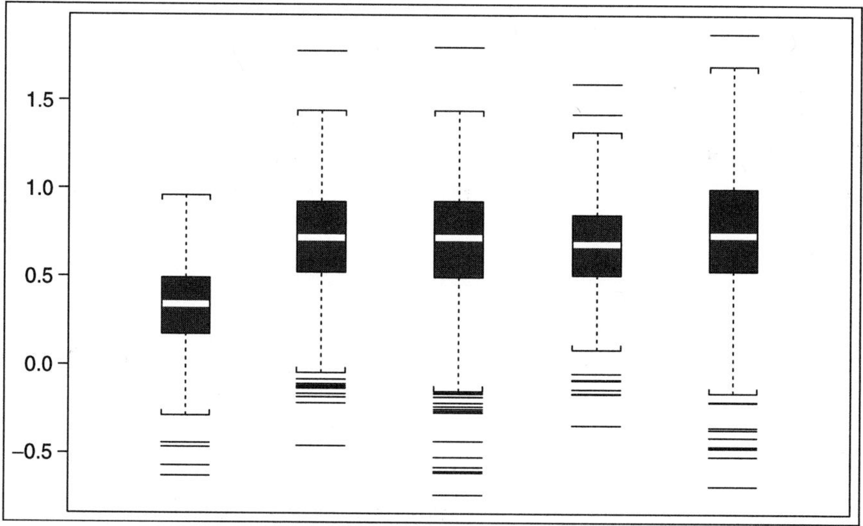

FIGURE 13.15 Boxplots of the estimated slope using (from left to right) least squares, M-estimator (with Schweppe weights), the Coakley–Hettmansperger estimate, Theil–Sen, and the deepest regression line. The true slope is 1, but all five estimators are influenced by two outliers.

medians of all these estimators differ from 1, the value being estimated. This illustrates that these estimators can be sensitive to a type of *contamination bias*. That is, despite having a reasonably high finite-sample breakdown point, it is possible for a few unusual points to result in a poor fit to the bulk of the observations. So these

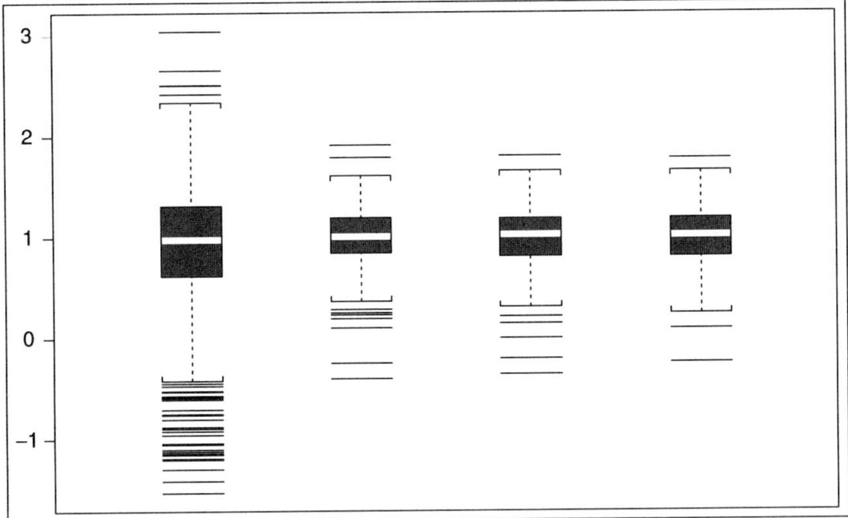

FIGURE 13.16 As in Figure 13.15, the only estimators used here are LTS with a breakdown point of .5, LTS with a breakdown point of .2, TSTS, and MGV. All four estimators have a median value approximately equal to the true slope, despite the outliers.

estimators can provide substantial advantages versus least squares, but they do not eliminate all practical concerns.

Figure 13.16 shows the results when using LTS with a breakdown point of .5, LTS with a breakdown point of .2, TSTS, and the MGV estimator in Section 13.2.5. (The LTA estimator gives results similar to LTS.) In contrast to the estimators in Figure 13.15, the median of all the estimators is approximately 1. So in this particular situation, these estimators do a better job of avoiding contamination bias. Note that there is considerably more variation among the LTS estimates based on a breakdown point of .5. This illustrates that under normality, this estimator is substantially less accurate than the others, at least on average. Again, this is not to suggest that LTS has no practical value.

There is some evidence that the STS estimator generally gives a better fit to the majority of points versus LTS and LMS. In particular, it seems common to encounter situations where STS is less affected by a few aberrant points. However, exceptions occur, as is illustrated next, so again it seems that multiple methods should be considered.

EXAMPLE. Figure 13.17 shows 20 points that were generated in the same manner as in Figure 13.14. So the two aberrant points located at $(X, Y) = (2.1, -2.4)$ are positioned relatively far from the true regression line, which again has a slope of 1 and an intercept of zero. Also shown in Figure 13.17 are the STS, LMS, and MGV estimates of the regression line. In this particular case, STS performs poorly; the estimated slope is $-.23$. The estimated slope

Continued

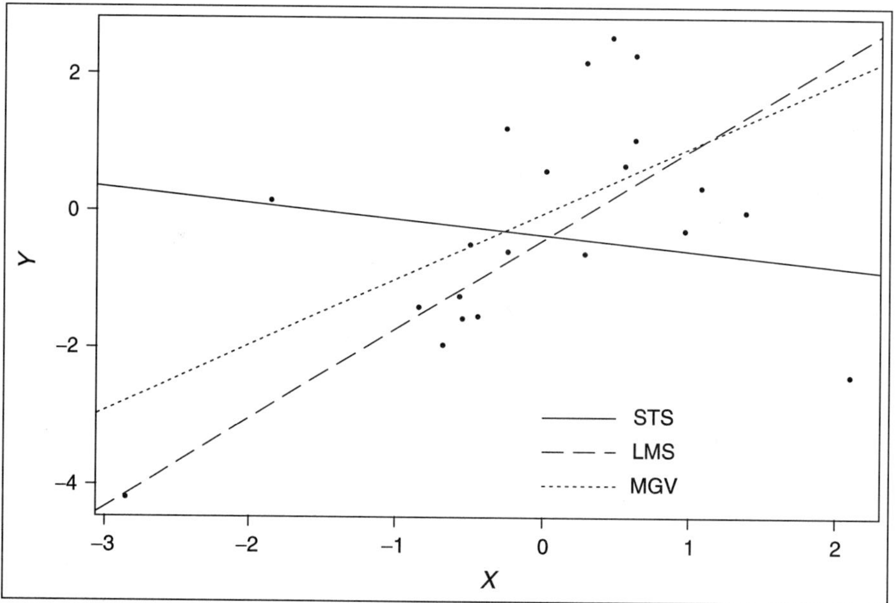

FIGURE 13.17 Illustration that some robust estimators can give substantially different results from others, depending on where the outliers happen to be.

EXAMPLE. (*Continued*) and intercept based on the MGV estimator are 0.96 and −0.03, respectively, which are closer to the true values than are the other estimates considered here. The estimated slope based on the TSTS estimator is .58, and least squares returns an estimate of .4. So once again we see that the choice of which robust estimator to use can make a substantial difference in how the association between X and Y is summarized. ■

To provide more perspective about the relative merits of various robust estimators, some comparisons are made of their small-sample accuracy versus the least squares estimator, including situations where there is heteroscedasticity. This is done in terms of the standard error of the least squares estimate of the slope divided by the standard error of some competing estimator. So if, for example, this ratio is less than 1, least squares tends to be more accurate. A ratio of .5, for instance, means that least squares has a standard error that is only half as large, and a ratio of 3 indicates that it is three times as large instead. Here it is assumed that $Y = X + \lambda(X)\epsilon$, where the function $\lambda(X)$ reflects heteroscedasticity. Setting $\lambda(X) = 1$ corresponds to the homoscedastic case. Table 13.2 shows estimates of these ratios for the estimators TS, MGVF, MGV, the deepest regression line estimator (T^*), and TSTS, where VP 1 corresponds to $\lambda(X) = 1$, VP 2 is where $\lambda(X) = X^2$, and VP 3 is

$$\lambda(X) = \left(1 + \frac{2}{(|X| + 1)}\right)\epsilon.$$[1]

1 The estimated ratios of the standard errors in Table 13.2 are based on simulations with 5000 replications.

TABLE 13.2 Estimated Ratios of Standard Errors, X Distribution Symmetric, $n = 20$

X	ϵ	VP	TS	MGVF	MGV	T^*	TSTS
N	N	1	0.91	0.91	0.91	0.76	0.88
		2	2.64	2.64	2.62	3.11	2.36
		3	202.22	201.18	196.31	187.89	135.70
N	SH	1	4.28	4.28	4.27	4.42	3.51
		2	10.67	10.66	10.94	11.03	8.66
		3	220.81	220.29	214.31	228.59	121.35
N	AL	1	1.13	1.13	1.13	0.92	1.05
		2	3.21	3.21	3.21	3.69	2.84
		3	183.74	182.69	177.53	146.89	106.70
N	AH	1	8.89	8.84	8.85	16.41	7.05
		2	26.66	26.59	27.07	25.81	20.89
		3	210.37	209.85	204.25	182.20	103.04
SH	N	1	0.81	0.80	0.72	0.61	0.76
		2	40.57	40.55	42.30	55.47	27.91
		3	41.70	40.54	34.44	40.08	22.57
SH	SH	1	3.09	3.08	2.78	2.88	2.41
		2	78.43	78.41	83.56	90.84	47.64
		3	38.70	38.03	31.93	45.29	17.80
SH	AL	1	0.99	0.98	0.87	0.73	0.90
		2	46.77	46.74	49.18	63.60	31.46
		3	39.32	38.68	32.70	31.70	19.76
SH	AH	1	6.34	6.31	5.64	6.75	4.62
		2	138.53	138.49	146.76	108.86	78.35
		3	43.63	43.22	37.34	39.38	18.40

N = normal; SH = symmetric, heavy-tailed; AL = asymmetric, light-tailed; AH = asymmetric, heavy-tailed.

So for VP 2, the error term has more variance, corresponding to extreme X values, and VP 3 is a situation where the error term has more variance when the value of X is near its median. The results are limited to situations where the distribution for X is symmetric; but very similar results are obtained when X has an asymmetric distribution instead. In Table 13.2, the distributions for ϵ are taken to be normal (N), symmetric and heavy-tailed (SH), asymmetric and relatively light-tailed (AL), and asymmetric and relatively heavy-tailed (AH). (For precise information about these four distributions, see Wilcox, 2001b.)

The results in Table 13.2 can be roughly summarized as follows. If the error term is homoscedastic and simultaneously has a light-tailed distribution, the ordinary least squares estimator covered in Chapter 6 competes well against the alternative estimators considered here. However, as the distribution of the error term becomes more heavy-tailed (meaning outliers become more common), the least squares estimator becomes unsatisfactory. Moreover, if the error term is sufficiently heavy-tailed, the least squares estimator is disastrous, even when there is homoscedasticity. In fact, even when the error term is normal but heteroscedastic, the least squares estimator can be highly unsatisfactory. Note that for VP 3, situations arise where the least squares estimator has a standard error more than 100 times larger than competing estimators.

Said another way, if two variables are independent, then power is not an issue when testing the hypothesis that the slope is zero (because the null hypothesis is true). However, if the variables are dependent, then the results in Table 13.2 suggest that the choice of estimator can make a substantial difference in terms of the likelihood of detecting a true association.

For the situations in Table 13.2, the MGV and MGVF estimators do not seem to have a striking advantage over the other robust estimators considered. It is stressed, however, that while several estimators compete well with least squares, it seems to be easy to find fault with any estimator that has been proposed. For example, the execution time required for MGV and MGVF increases substantially as the sample gets large, which makes them impractical for many situations based on the software and computer hardware currently available. Also, MGVF tends to have faster execution time than MGV, so the results in Table 13.2 might seem to suggest using MGVF over MGV. However, situations can be constructed where MGV gives a good fit to the bulk of the points when all of the other estimators give a unsatisfactory fit instead. Again, it seems that no single estimator is always ideal.

It is noted that the M-estimators in Section 13.2.3 also offer an advantage over least squares when there is heteroscedasticity. If there is reasonable certainty that the Coakley–Hettmansperger estimator is not being affected by contamination bias, it is a possible option and is fairly fast in terms of execution time.

So which estimator should be used? A rough strategy might be as follows. First plot the points and then plot the estimated regression line using several of the estimators considered in this book. (The S-PLUS command

```
abline(mgvreg(x,y)$coef),
```

for example, will plot the line corresponding to the MGV estimator.) The goal at this point is merely to make sure that a reasonable fit to the data is obtained. (At this stage, it is strongly recommended that a smooth of the data, described in Section 14.1, be checked as well.) In some cases, simply restricting the range of the X values might improve the fit and provide relatively short confidence intervals for the slope and intercept. If several estimators appear to give an adequate fit, then one strategy is to choose the estimator that will have relatively high power and relatively short confidence intervals, assuming of course that probability coverage will be adequate. The results in Table 13.2 illustrate that the choice of estimator might make a practical difference, and it will be illustrated that in applied work this is indeed the case. Although no single estimator is optimal in all situations, there seems to be general agreement on the worst possible strategy: Apply standard least squares regression and assume all is well. At a minimum, use an estimator with a reasonably high breakdown point. Also, it is strongly recommended that an estimator perform reasonably well (in terms of achieving a relatively low standard error) when there is heteroscedasticity.

13.5 Hypothesis Testing and Confidence Intervals

Confidence intervals for the slope and intercept can be computed with a percentile bootstrap method with bootstrap samples generated as described in Section 7.3. When there is one predictor, if we observe $(X_1, Y_1), \ldots, (X_n, Y_n)$, a bootstrap sample is obtained by resampling with replacement n pairs of these points, with each pair of points having probability $1/n$ of being resampled. The resulting bootstrap sample will be labeled $(X_1^*, Y_1^*), \ldots, (X_n^*, Y_n^*)$. When there are p predictors, resample n vectors of observations from

$$(X_{11}, \ldots, X_{1p}, Y_1), \ldots, (X_{n1}, \ldots, X_{np}, Y_n).$$

Next, compute the slope and intercept based on one of the robust estimators in this chapter, applied to the bootstrap sample just obtained, and repeat this process B times. If we label the resulting estimates of the slope as $b_{11}^*, \ldots, b_{1B}^*$, then an approximate $1-\alpha$ confidence interval for the slope is given by the middle 95% of these bootstrap estimates:

$$\left(b_{1(\ell+1)}^*, b_{1(u)}^*\right),$$

where $\ell = \alpha B/2$, rounded to the nearest integer, and $u = B - \ell$. If $\hat{p}^*$ is the proportion of bootstrap estimates greater than zero, the significance level (when testing $H_0: \beta_1 = 0$) is

$$2 \min\left(\hat{p}^*, 1 - \hat{p}^*\right).$$

A confidence interval and significance level for the intercept, and the other slope coefficients when $p > 1$, can be computed in the same manner.

13.5.1 S-PLUS Function regci

The S-PLUS function

```
regci(x,y,regfun=tsreg,nboot=599,alpha=.05)
```

computes percentile bootstrap confidence intervals using the method just described. As indicated, the default regression estimator is tsreg (the Theil–Sen estimator). This default estimator was chosen based on its ability to improve substantially upon the least squares estimator and because its execution time is fairly fast. But it is stressed that situations arise where some other estimator might offer a better fit to the majority of the data and yield substantially shorter confidence intervals.

EXAMPLE. The diabetes data in Exercise 15 of Chapter 6 serve to illustrate the output from regci when the least trimmed squares estimator is used. Assuming the data are stored in the S-PLUS variables x and y, the command regci(x,y,regfun=ltsreg) returns

Continued

EXAMPLE. (*Continued*)

```
$regci:
               [,1]        [,2]
[1,]    1.22786964  1.64758295
[2,]   -0.00727429  0.03702609

$sig.level:
[1]  0.0000000  0.1866667

$se:
[1]  0.11215894  0.01120903
```

So the .95 confidence interval for the intercept is (1.23, 1.65) with a significance level of 0.0, and the confidence interval for the slope is (−0.007, .04) with a significance level of 0.187. The estimated standard errors for the intercept and slope are 0.11 and 0.011, respectively. ■

EXAMPLE. Sometimes restricting the range of X values eliminates outliers that mask an association or create contamination bias, but exceptions are encountered. Consider again the diabetes data. One goal was to understand how age is related to C-peptide concentrations, and here we focus on children under the age of seven (for reasons motivated by results covered in Chapter 14). The MGV and MVE outlier detection methods in Sections 13.1.5 and 13.1.3 find one outlier. However, to eliminate it by restricting the range of the X values would entail eliminating points that are not flagged as outliers. (The MCD outlier detection method finds an additional outlier, but to eliminate both outliers by restricting the range of X values would again entail eliminating points that are not flagged as outliers.) If we simply apply the S-PLUS function regci using the default Theil–Sen estimator, the .95 confidence interval is (0.0, .11), with a significance level of .023. So with $\alpha = .05$ we reject the hypothesis of a zero slope. Using the Coakley–Hettmansperger estimator, the significance level is .063, so now we fail to reject. As for the S-estimator in Section 13.3.1, the significance level is .07. Using least squares in conjunction with the method in Chapter 7, the .95 confidence interval is (−0.001, .106), so we fail to reject, although we come close when testing at the .05 level. With the MGV estimator, the .95 confidence interval is (.004, .14) with a significance level of .013, about half the significance level based on Theil–Sen, which merely illustrates that the choice of method can make a practical difference. ■

EXAMPLE. Figure 6.3 shows some star data where the least squares regression line provides a poor fit to the bulk of the observations. It was noted that if we

Continued

EXAMPLE. (*Continued*) eliminate the five points having the smallest X values (values less than 4), which are flagged as outliers among the X values using the boxplot methods in Chapter 3, a much better fit to the data is obtained. If we compute a .95 confidence interval for the slope using least squares and the percentile bootstrap method in Section 7.3, we get (1.85, 3.95). Using the Theil–Sen estimator (tsreg) or the MGVFREG estimator, the .95 confidence interval based on the S-PLUS function regci is (2.11, 4.00). So in this particular case we get a slightly longer confidence interval using least squares. The least squares estimate of the slope is 2.93. The Theil–Sen estimate, as well as the estimate returned by MGVFREG, is 3.07, and MGVREG estimates the slope to be 3.22. However, if we use the outlier detection rule in Section 3.4.2 (based on MAD and M), an additional outlier among the X values is found: 4.01. If we restrict X to those values greater than 4.01, the least squares estimate of the slope increases slightly to 2.98 and Theil–Sen increases to 3.22, the same value returned by MGVREG. ■

EXAMPLE. Here is another illustration that restricting the range of the X values can decrease the strength of an association. Fifty values for the error term (ϵ) were generated on a computer from a standard normal distribution, the same was done for X, and $Y = .35X + \epsilon$ was computed, yielding 50 values for Y. Pearson's correlation between X and Y was found to be .26 with a significance level of .026 using Student's T-test of $H_0: \rho = 0$ described in Chapter 6. Restricting the range of X to values less than 1.3 results in eliminating five values, and the significance level increases to .094. Restricting the range of X so that $|X| \leq 1.5$ results in the elimination of six points with the significance level now equal to .14. ■

EXAMPLE. Table 6.3 reports data on aggression in the home (X) versus a measure of cognitive functioning among children (Y). In Figure 6.9, it was illustrated that if we split the data into two groups according to whether X has a value greater than or equal to 25, we get two strikingly different impressions about the association between X and Y. For $X \geq 25$ the least squares estimate of the slope is close to zero, but for $X < 25$ the estimated slope is -0.19. In Section 7.3 we saw that the least squares slope is significantly different from zero (based on the modified percentile bootstrap method) when using all of the data. So a case might be made that there is a negative association between X and Y up to about $X = 25$ and that for $X \geq 25$ there appears to be little or no association at all. But is it adequate to describe the association as linear and negative for $X < 25$, or is it more accurate to say that the association is primarily due to heteroscedasticity? Figure 13.18 shows the plot created by the

Continued

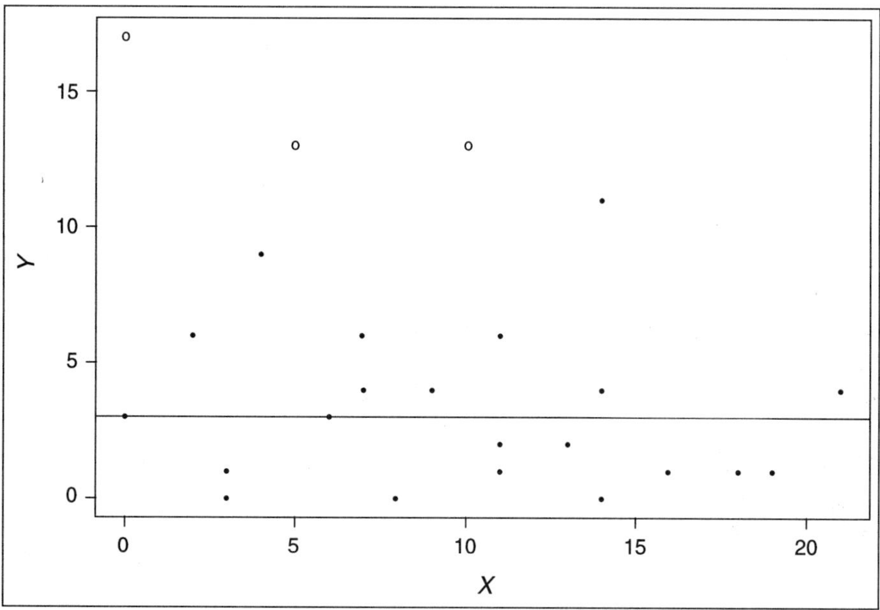

FIGURE 13.18 For the aggression data, with $X < 25$, the horizontal line is the MGV regression line, which suggests that the typical value of Y changes very little with X and that the apparent dependence between X and Y is perhaps better explained by other perspectives (such as heteroscedasticity).

EXAMPLE. (*Continued*) S-PLUS function outmgv when attention is restricted to $X < 25$. The horizontal line is the regression line based on the MGV estimator. (Theil–Sen, the Coakley–Hettmansperger estimator, and MGVF return estimated slopes of -0.11, -0.13, and -0.11, respectively.) If we use the method in Section 8.10.1 to compare the variance of the Y values corresponding to $X < 25$ versus $X \geq 25$, the .95 confidence interval for the difference between the variances is (2.49, 32.34), so we reject at the .05 level of significance, indicating heteroscedasticity. In this particular case, our ability to reject hinges on using a measure of scale that is sensitive to outliers. If, for example, we use the percentage bend midvariance instead (a robust measure of scale described in Chapter 3), now we fail to reject. Also, with $X < 25$, the least squares regression line is no longer significantly different from zero; the .95 confidence interval (based on the modified bootstrap method in Chapter 7) is $(-0.51, 0.17)$. Using Theil–Sen, in conjunction with the S-PLUS function regci in Section 13.5.1, we get a .95 confidence interval of $(-0.43, 0.11)$ and again we fail to reject. Of course, failing to reject the hypothesis of a zero slope might be due to low power resulting from a reduction of the sample size. ■

In summary, there is evidence that recall scores are associated with aggression in the home, but it seems that some care must be taken when describing what this association might be. In particular, there is evidence of decreasing variability among the Y values as X gets large, in the sense that relatively large Y values become less likely as X increases. In terms of typical Y values as X increases, again there are some indications of a negative association. But it seems that if the apparent negative association is real, it is weaker in some sense compared to the decrease in variation and number of outliers among the Y values as X gets large. For example, the estimated slope based on the MGV estimator is exactly equal to zero for the data in Figure 13.18. That is, *typical* recall scores are estimated to have no association with aggression in the home; but despite this, relatively high recall test scores appear to become less likely as X gets large.

EXAMPLE. Figure 13.19 shows a scatterplot of the same points shown in Figures 13.2 and 13.3. As previously noted, the least squares estimate of the slope is highly nonsignificant, but it is fairly evident the six rightmost points are outliers. If we ignore these outliers by restricting attention to points having $X < 100$, the least squares regression line is the dashed line in Figure 13.19, which has an estimated slope of -0.48, and the .95 confidence interval for the slope (using the modified percentile bootstrap method in Section 7.3) is $(-0.87, -0.17)$. The Theil–Sen estimate of the slope is -0.6, and the bootstrap confidence interval for the slope (using the S-PLUS function regci) is $(-0.90, -0.27)$. In this particular case, both methods yield confidence intervals having approximately the same length. ■

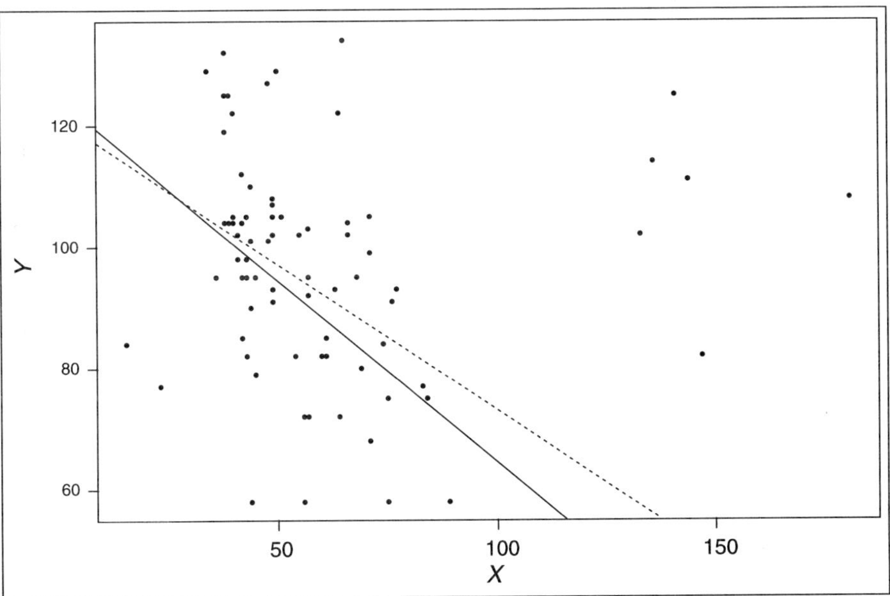

FIGURE 13.19 For the reading data in Figure 13.2, outliers can be eliminated simply by restricting the range of X values. In this particular case, various methods give similar confidence intervals for the slope based on the remaining data.

We have seen that in some situations, restricting the range of the X values can do a reasonable job of salvaging least squares. But it is important to realize that simply restricting the range of the X values does not always salvage least squares, especially when computing a confidence interval or testing hypotheses. In particular, situations arise where least squares yields an estimated slope similar to other robust estimators, but the length of the confidence interval can be substantially larger.

EXAMPLE. Consider the data in Figure 13.20, which is from the reading study mentioned in connection with Figure 13.2. The solid, nearly horizontal line is the least squares regression line, and the dashed line is based on the TSTS estimator in Section 13.3.3. It is evident that the upper six points are outliers, but if we simply eliminate them and apply the conventional methods in Chapter 6 to test the hypothesis of a zero slope, we are using the wrong standard error. Still using least squares, if a .95 bootstrap confidence interval for the slope is computed as described in Chapter 7, we get (−0.417, 0.414). As is evident, this confidence interval is not remotely close to rejecting the hypothesis of a zero slope. In contrast, the .95 confidence interval based on the Theil–Sen estimator (using the S-PLUS function regci) is (−0.428, 0) — about half the length of the confidence interval based on least squares — with a significance level of .056; the estimated slope is −0.23. The MGV estimator estimates the slope to be −0.35 and the TSTS estimate is −0.426. The .95 confidence interval based on

Continued

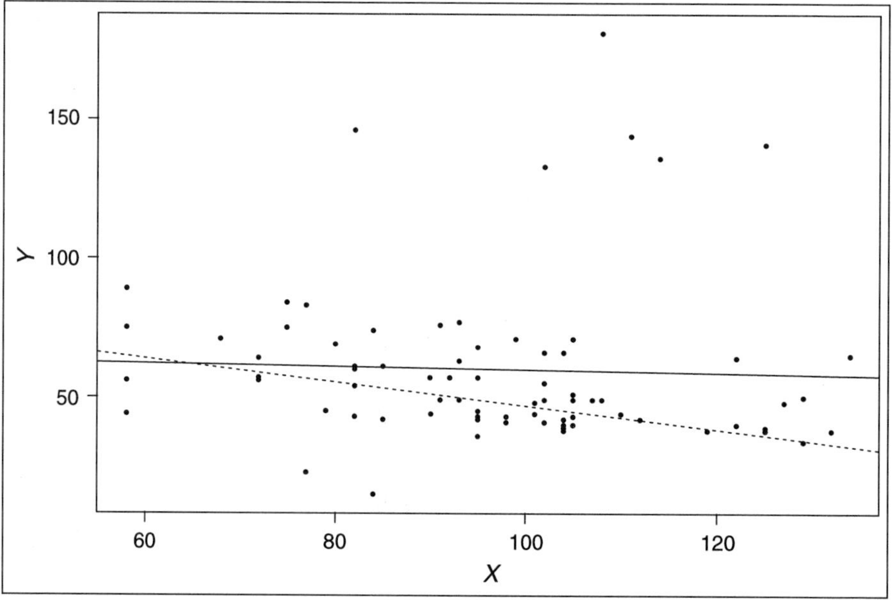

FIGURE 13.20 Example where the length of the confidence interval depends in a crucial way on the regression estimator used.

EXAMPLE. (*Continued*) this last estimator is (−0.6, −0.125) with a significance level of .003. The .95 confidence interval based on the MGV estimator is (−0.56, −0.125) with a significance level less than .001. ■

13.6 Robust Measures of Correlation

This section summarizes some robust analogs of Pearson's correlation, r. We begin with two relatively simple measures of association and then consider how they might be improved.

13.6.1 Winsorized Correlation

The Winsorized correlation coefficient is obtained by Winsorizing the n pairs of observations as described in Section 11.2 and in Box 11.3. (For technical reasons, trimming is a less satisfactory approach to defining a robust correlation.) The Winsorized correlation between X and Y is just Pearson's correlation applied to the Winsorized values. The resulting correlation coefficient will be labeled r_w.

Letting ρ_w be the population analog of r_w, it can be shown that when X and Y are independent, $\rho_w = 0$. If we assume independence — implying homoscedasticity — a simple test of

$$H_0 : \rho_w = 0$$

is to compute

$$T_w = r_w \sqrt{\frac{n-2}{1-r_w^2}}. \tag{13.12}$$

Let

$$\nu = n - 2g - 2,$$

where, as in Chapter 3, $g = [\gamma n]$ and γ is the amount of Winsorizing. (Remember that $[\gamma n]$ means to compute γn and then to round down to the nearest integer.) Then reject if $|T_w| \geq t_{1-\alpha/2}$, the $1-\alpha/2$ quantile of Student's T-distribution with ν degrees of freedom. Setting the amount of Winsorizing to zero (i.e., using $\gamma = 0$) reduces T_w to the test statistic covered in Section 6.5.

13.6.2 Percentage Bend Correlation

A criticism of the Winsorized correlation is that the amount of Winsorizing is fixed in advance rather than determined by your data. One solution is to use what is called the *percentage bend correlation*, r_{pb}, which is computed as described in Box 13.4. Under independence, the population value of r_{pb}, ρ_{pb}, is zero. To test $H_0 : \rho_{pb} = 0$, compute

$$T_{pb} = r_{pb}\sqrt{\frac{n-2}{1-r_{pb}^2}} \tag{13.13}$$

and reject if $|T_{pb}| \geq t_{1-\alpha/2}$, where $t_{1-\alpha/2}$ is the $1 - \alpha/2$ quantile of Student's T-distribution with $n - 2$ degrees of freedom.

BOX 13.4 How to Compute the Percentage Bend Correlation

For the observations $X_1, \ldots, X_n$, let $\hat{\theta}$ be the sample median. Choose a value for β between 0 and 1 and compute

$$W_i = |X_i - \hat{\theta}|,$$

$$m = [(1-\beta)n],$$

where the notation $[(1-\beta)n]$ is $(1-\beta)n$ rounded down to the nearest integer. Using $\beta = .2$ appears to be a good choice in most situations. (The value of β determines the finite-sample breakdown point of a measure of scale used to detect outliers.) Let $W_{(1)} \leq \cdots \leq W_{(n)}$ be the W_i values written in ascending order and let

$$\hat{\omega}_x = W_{(m)}.$$

Let i_1 be the number of X_i values such that $(X_i - \hat{\theta})/\hat{\omega}_x < -1$ and let i_2 be the number of X_i values such that $(X_i - \hat{\theta})/\hat{\omega}_x > 1$. Compute

$$S_x = \sum_{i=i_1+1}^{n-i_2} X_{(i)}$$

$$\hat{\phi}_x = \frac{\hat{\omega}_x(i_2 - i_1) + S_x}{n - i_1 - i_2}.$$

Set $U_i = (X_i - \hat{\phi}_x)/\hat{\omega}_x$. Repeat these computations for the Y_i values, yielding $V_i = (Y_i - \hat{\phi}_y)/\hat{\omega}_y$. Let

$$\Psi(x) = \max[-1, \min(1, x)].$$

Set $A_i = \Psi(U_i)$ and $B_i = \Psi(V_i)$. The percentage bend correlation is estimated to be

$$r_{pb} = \frac{\sum A_i B_i}{\sqrt{(\sum A_i^2)(\sum B_i^2)}}.$$

13.6.3 S-PLUS Functions wincor and pbcor

The S-PLUS functions

```
wincor(x,y,tr=.2)    and    pbcor(x,y,beta=.2)
```

compute the Winsorized and percentage bend correlations. The argument tr determines how much Winsorizing is done and defaults to .2. This function also tests $H_0 : \rho_w = 0$ using T_w given by Equation (13.12). Setting tr=0 results in the conventional test based on Pearson's correlation, which is described in Section 6.5. The function pbcor tests the hypothesis of independence using Equation (13.13).

The practical advantages of the Winsorized and percentage bend correlation are that they limit the influence of a few unusual X values, and they do the same for the Y values. However, a concern is that neither one takes into account the overall structure of the data, as is illustrated next.

EXAMPLE. Look again at the scatterplot in Figure 13.4 and recall that Pearson's correlation is $r = .68$. Figure 13.5 shows the same data but with two unusual values added at $(X, Y) = (2.1, -2.4)$, which causes r to drop to .21. For the data in Figure 13.4, the Winsorized correlation is $r_w = .59$ with a significance level of .011 when testing $\rho_w = 0$ with Equation (13.12). But in Figure 13.5, the Winsorized correlation is only $r_w = .33$ with a significance level of .14. As for the percentage bend correlation, in Figure 13.4, $r_{pb} = .37$ with a significance level of .008; but in Figure 13.5, $r_{pb} = .25$ with a significance level of .26. Of course, one might argue that the association between X and Y is weaker in Figure 13.5. Certainly this is true in some sense, but simultaneously it is erroneous to conclude that there is no association. ■

13.6.4 Heteroscedastic Bootstrap Confidence Intervals for Robust Correlations

Chapter 6 noted that heteroscedasticity can result in undesirable properties when using the conventional method for testing the hypothesis that Pearson's correlation is zero. These same concerns apply when using the methods in Sections 13.6.1 and 13.6.2. When these methods reject, it is reasonable to conclude that the variables under study are dependent, but the reason for rejecting remains a bit unclear. One strategy is to replace the methods in Sections 13.6.1 and 13.6.2 with techniques that allow heteroscedasticity. This can be done by using a bootstrap to compute a confidence interval for some robust correlation coefficient. Essentially, proceed as described in Section 13.5 but with the estimated slope replaced by the correlation of interest.

13.6.5 S-PLUS Function corb

The S-PLUS function

```
corb(x,y,corfun=pbcor,nboot=599)
```

tests the hypothesis of a zero correlation using a heteroscedastic percentile bootstrap method. By default, the percentage bend correlation is used.

EXAMPLE. For the reading data in Figure 13.2, the test of a zero percentage bend correlation, based on Equation (13.12), has a significance level of .014. Is heteroscedasticity playing a role here? Using a percentile bootstrap method instead, a .95 confidence interval for the percentage bend correlation is (−0.63, −0.19); this interval does not contain zero, so again we reject. ■

13.6.6 Correlation with Outliers Removed

We saw in Section 13.1.5 that simply discarding outliers using the MVE or MCD methods and then applying Pearson's correlation to the remaining data can mask an association. We have also seen that the MGVF outlier detection method can miss outliers. However, it generally detects any blatant outliers, so another strategy is first to check for outliers using MGVF, to remove any that are found, and then to apply the percentage bend correlation to the data that remain. By using the percentage bend correlation, you guard against any outliers that were not found by MGVF but that might have an inordinate influence on the correlation. However, a criticism of this approach is that there are no published papers on how one might test the hypothesis of independence based on the resulting correlation coefficient. A bootstrap method might give good control over the probability of a Type I error, but more research is needed before this approach can be recommended.

Yet another strategy is to remove outliers with the projection method described in Section 13.1.9 and to compute Pearson's correlation on the data that remain. One appeal of this method is that under normality, the resulting estimate of the population correlation coefficient, ρ, is relatively accurate. That is, if our criterion is to get a good estimate of ρ in the event data follow a bivariate normal distribution, this is one of the few methods that performs reasonably well. Also, there is a simple test of a zero correlation that performs well under homoscedasticity.

Let r_p be Pearson's correlation based on the data not flagged as outliers using the projection method. Let

$$T_p = r_p\sqrt{\frac{n-2}{1-r_p^2}}.$$

Then, when testing at the .05 level, reject the hypothesis of zero correlation when $|T_p| \geq c$, where

$$c = \frac{6.947}{n} + 2.3197.$$

The value of r_p can be computed quickly with $n = 100$, but as the sample size increases, execution time will get high with the software currently available. There are types of heteroscedasticity where the method does not control the probability of a Type I error. However, if the only goal is to control the probability of a Type I error when there is independence, all indications are that T_p is satisfactory (Wilcox, 2002).

13.6.7 S-PLUS Functions ocor and scor

The S-PLUS function

ocor(x, y, corfun = pbcor, outfun = outmgvf, pcor = F, plotit = F)

computes a correlation coefficient by removing outliers detected by the function outfun, which defaults to the MGVF method, and then applying the function indicated by the argument corfun. By default, corfun is the percentage bend correlation. To use the Winsorized correlation, set corfun=wincor. To apply Pearson's correlation instead, set the argument pcor to T, or you can use corfun=pcor. To plot the data, set plotit to T.

The S-PLUS function

scor(x,y=NA, plotit = T)

computes Pearson's correlation after outliers, identified with the S-PLUS function outpro, are removed. If no values for y are specified, it is assumed that x is a matrix with p columns of data. The function tests the hypothesis of a zero correlation at the .05 level. If plotit=T is used, the same plot created by outpro is produced. (Methods for controlling FWE, the familywise error rate, have not been investigated as yet.)

EXAMPLE. If the star data in Figure 13.12 are stored in the S-PLUS variables starx and stary, the command

ocor(starx,stary)

returns a correlation of .63. This is in contrast to Pearson's correlation, $r = -.21$. If we eliminate points where starx is less than 4.1, now Pearson's correlation is $r = .65$. Often, restricting the range of X values gives similar results to robust methods, but this is not always the case. Figure 13.21 shows the plot created by scor (which is the same plot created by outpro). Here, $r_p = .68$, $|T_p| = 6.26$, and the .05 critical value is 2.47, so reject and conclude there is dependence. The command

ocor(starx,stary,outfun=out)

removes outliers with the MVE method and returns a value of .71, but now no simple test of a zero correlation has been derived. (A bootstrap method might be used, but this has not been investigated as yet.) So in some instances using MVE (or MCD) can lower the correlation, but situations arise where it increases the correlation as well. ■

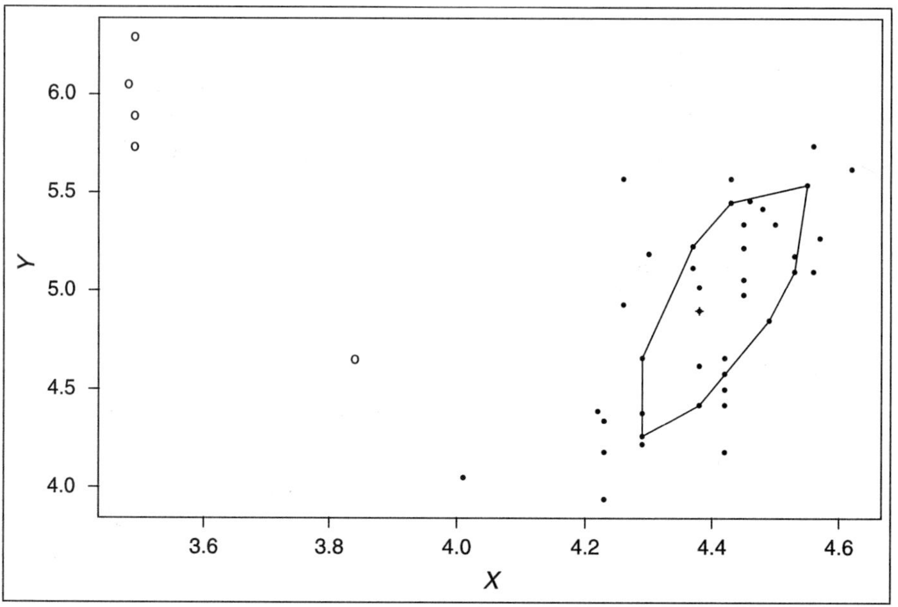

FIGURE 13.21 Plot created by the S-PLUS function scor based on the star data.

13.6.8 Explanatory Power

This section describes a general approach to measuring association based on what is called *explanatory power*. Roughly, it measures the variation among the predicted Y values (based on X) relative to the variation of Y when the predictor X is ignored. Let $S(Y)$ indicate some (population) measure of variability. For example, $S(Y)$ might be the variance of Y (σ_y^2), or it could be the Winsorized variance, the percentage bend midvariance, or even MAD. Let $\hat{Y}$ be any method for predicting Y given some value for X. For example, $\hat{Y}$ might be the least squares estimate of Y given X, or it could be any of the robust regression estimators described in this chapter. In its most general form *explanatory power* is defined to be

$$\eta^2 = \frac{S(\hat{Y})}{S(Y)}. \tag{13.14}$$

If we use the sample variance to measure variation and least squares regression to predict Y given X, then η^2 is just ρ^2, the square of Pearson's correlation, which is the population value of the coefficient of determination covered in Chapter 6. Note that η^2 can be turned into a correlation coefficient by taking the square root and multiplying by the sign of the estimated slope. That is, if the regression line has a negative slope, make η negative as well; otherwise, η is taken to be positive.

To provide some graphical intuition about explanatory power, look at the left panel of Figure 13.22, which shows 10 points all falling on the straight line having slope 1 and intercept zero. As is evident, $\hat{Y} = Y = X$, and the variance of both the Y and the $\hat{Y}$ values is 115.2. That is, the explanatory power is 1. In the right panel of

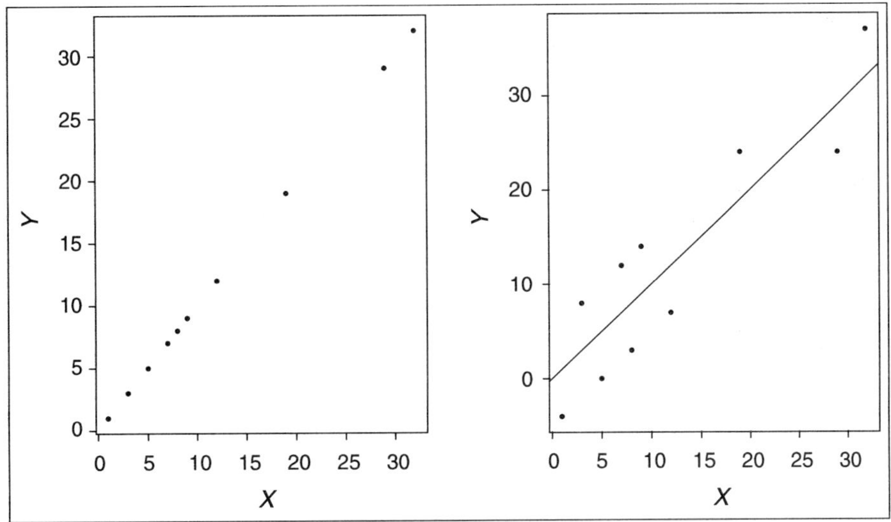

FIGURE 13.22 Graphical explanation of explanatory power.

Figure 13.21, 10 points are centered around the straight line having slope 1 and intercept zero, but they are not on the line. The predicted Y values (the $\hat{Y}_i$ values) fall on the straight line and again have variance 115.2. In contrast, the observed Y values have variance 159.6, so the explanatory power is 115.2/159.6 = .72.

A natural estimate of η^2, based on the least squares regression line and the usual sample variance, is

$$\hat{\eta}_1^2 = \frac{\sum (\hat{Y}_i - \tilde{Y})^2}{\sum (Y_i - \bar{Y})^2}, \tag{13.15}$$

where $\tilde{Y} = \sum \hat{Y}_i/n$, the average of the predicted Y values. A little algebra shows that this estimate is just r^2, the coefficient of determination. However, results in Doksum and Samarov (1995) as well as Wilcox (2000a) suggest estimating explanatory power by computing the squared correlation between the values $(Y_i, \hat{Y}_i)$. That is, estimate η^2 with

$$\hat{\eta}_2^2 = (\text{COR}(\hat{Y}, Y))^2. \tag{13.16}$$

If the goal is to use the variance as a measure of explanatory power, then use Pearson's correlation. If instead you want to use a Winsorized variance, use the Winsorized correlation, and use the percentage bend correlation for the percentage bend midvariance.

Before ending this section, it is remarked that yet another approach to measuring association stems from the following connection between Pearson's correlation and the least squares regression line:

$$r = b_1 \frac{s_x}{s_y}. \tag{13.17}$$

So a reasonable generalization is

$$r_g = b_g \frac{\hat{\tau}_x}{\hat{\tau}_y}, \tag{13.18}$$

where now b_g is any estimate of the slope and $\hat{\tau}_x^2$ is an estimate of some measure of variation among the X values, such as the Winsorized variance or the percentage bend midvariance described in Chapter 3.

13.6.9 S-PLUS Functions epow and ecor

The S-PLUS function

epow(x,y, pcor = F, regfun = tsreg, corfun = pbcor, outkeep = F,

outfun = outmgvf, varfun = pbvar, op = T)

estimates explanatory power (η^2) as just described. (When there are p predictors, x is assumed to be an n-by-p matrix.) If you want to use Pearson's correlation between $\hat{Y}$ and Y to estimate explanatory power, set the argument pcor to T; otherwise the function uses the measure of correlation specified by corfun, which defaults to the percentage bend correlation. The argument regfun indicates which regression estimator will be used; the default method is the Theil–Sen regression line. The measure of variation (varfun) defaults to the percentage bend midvariance. The argument op controls how η^2 is estimated. By default (op=T), the squared correlation between the $\hat{Y}$ and Y values is used. Should a situation arise where there is some reason to use Equation (13.15), use op=F. (The argument varfun is ignored if op=T.) The argument outkeep indicates whether outliers are to be retained when estimating explanatory power. If this argument is F, the data are checked for outliers using the function outfun (which defaults to the method in Section 13.1.7) and outliers are removed.

The S-PLUS function

ecor(x,y,pcor = F, regfun = tsreg, corfun = pbcor, outkeep = F,

outfun = outmgvf)

computes the explanatory power correlation coefficient. The arguments are the same as those used by epow. The function ecor assumes that x is a vector or a matrix having a single column of numbers. This is in contrast to epow, which allows x to be a matrix with multiple predictors.

EXAMPLE. For the data in Figure 13.19, first consider the situation where outliers are kept. The command

epow(x,y,outkeep=T)

Continued

EXAMPLE. (*Continued*)
returns a value of only .078. However, restricting the range of X to values less than 100 increases the estimate substantially to .23. If you use instead the command epow(x,y) so that again the range of X values is not restricted, outliers are removed and now the estimate is .28. That is, by removing outliers, in effect the range of X values is restricted, except that now we get a slightly larger measure of explanatory power because even when attention is restricted to $X < 100$, a few outliers are detected. So roughly, there is a fair amount of explanatory power among the points clustered together, but there are outliers that, when included, reduce the explanatory power considerably. Because the slope of the regression line is negative, if we were to convert the estimate .078 into a correlation, we would get $-1\sqrt{.078} = -.28$, which is the value returned by the S-PLUS function ecor. In this particular case, the percentage bend correlation is also equal to −.28. ■

EXAMPLE. Again the aggression data in Table 6.3 are analyzed, except that now all of the data are used and consideration is given to fitting the data with $Y = \beta_0 + \beta_1 X + \beta_2 X^2$. Figure 13.23 shows a scatterplot of the data plus the least squares regression line (the dashed line in the figure) and the regression line based on the Coakley–Hettmansperger estimator. For both methods,

Continued

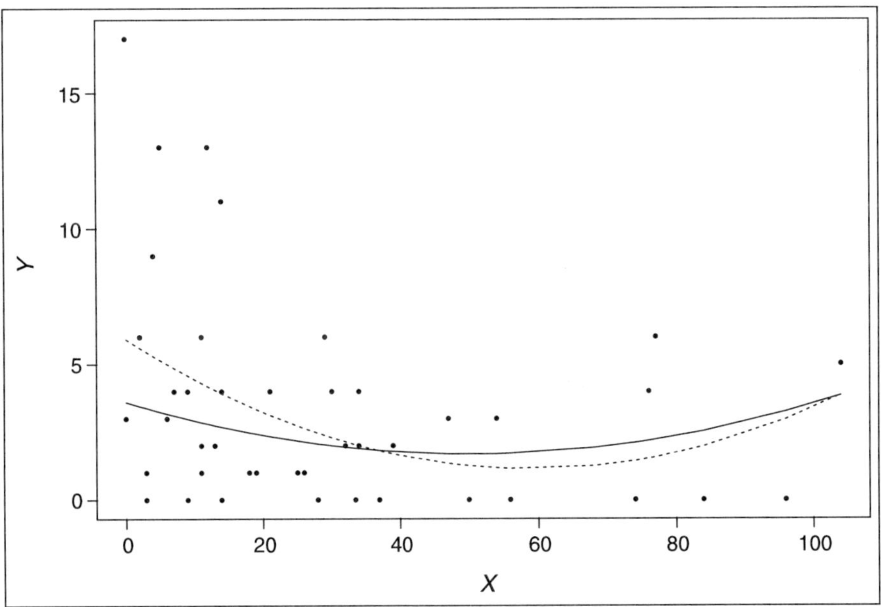

FIGURE 13.23 Quadratic regression line fit to the aggression data.

EXAMPLE. (*Continued*) the hypothesis $H_0 : \beta_2 = 0$ is not rejected. The explanatory power based on least squares is .098; for the Coakley–Hettmansperger estimator, in conjunction with the percentage bend variance, it is .051. If the quadratic term is ignored, the estimates are now .088 and .040, respectively, suggesting a modest improvement in the fit when the quadratic term is included. ■

13.6.10 Detecting Associations in Contingency Tables

Section 2.3 introduced the notion of a contingency table. Special methods for analyzing contingency tables have been developed, and there are many books devoted exclusively to this topic. These books are typically classified under the analysis of categorical data, recent examples of which are Agresti (1990, 1996), Andersen (1997), and Powers and Xie (1999). Although it is impossible to cover the many important issues here, a few comments about detecting associations might help.

Consider a two-by-two contingency table. To be concrete, look at the hypothetical data in Table 13.3, which deals with blood pressure and personality types. For example, there were eight individuals who were classified as having a type A personality and simultaneously having high blood pressure. A fundamental issue is determining whether personality type and blood pressure are dependent. An approach to this problem is to test the assumption that they are independent. If you are able to reject, you conclude that they are dependent, and this means that knowing whether a person has a type A personality provides you with some information about that person's probability of having high blood pressure.

For the four cells in Table 13.3, let n_{11} be the number of individuals falling in the first row and first column. In Table 13.3, $n_{11} = 8$. More generally, n_{ij} represents the entry in the ith row and jth column. So, for example, $n_{21} = 5$. Let

$$n_{i+} = n_{i1} + n_{i2} \quad \text{and} \quad n_{+j} = n_{1j} + n_{2j}.$$

In the example, $n_{1+} = 8 + 67 = 75$ and $n_{+2} = 87$. The total number of observations is represented by n, which is 100 in Table 13.3.

A classic test of the hypothesis of independence is based on

$$X^2 = \frac{n(n_{11}n_{22} - n_{12}n_{21})^2}{n_{1+}n_{2+}n_{+1}n_{+2}}.$$

TABLE 13.3 Hypothetical Results on Personality versus Blood Pressure

	Blood pressure		
Personality	High	Not high	Total
A	8	67	75
B	5	20	25
Total	13	87	100

When the null hypothesis of independence is true, X^2 has, approximately, a chi-squared distribution with 1 degree of freedom. For the more general case where there are R rows and C columns,

$$X^2 = \sum_{i=1}^{R} \sum_{j=1}^{C} \frac{n\left(n_{ij} - \frac{n_{i+}n_{+j}}{n}\right)^2}{n_{i+}n_{+j}}$$

and the degrees of freedom are

$$\nu = (R - 1)(C - 1).$$

Now

$$n_{i+} = \sum_j n_{ij} \quad \text{and} \quad n_{+j} = \sum_i n_{ij}.$$

If X^2 equals or exceeds the $1-\alpha$ quantile of a chi-squared distribution with ν degrees of freedom, which is read from Table 3 in Appendix B, you reject. In the illustration

$$X^2 = \frac{100[8(20) - 67(5)]^2}{75(25)(13)(87)} = 1.4.$$

With $\nu = 1$ degree of freedom and $\alpha = .05$, the critical value is 3.84; because $1.4 < 3.84$, you fail to reject. This means that you are unable to detect any dependence between personality type and blood pressure. Generally, the chi-squared test of independence performs reasonably well in terms of Type I errors (e.g., Hosmane, 1986), but difficulties can arise, particularly when the number of observations in any of the cells is relatively small. For instance, if any of the n_{ij} values is less than or equal to 5, problems might occur in terms of Type I errors. There are a variety of methods for improving upon the chi-squared test, but details are not given here. Interested readers can refer to Agresti (1990).

It is noted that measuring the strength of the association in a contingency table is a nontrivial matter. An early approach was to use the so-called *phi coefficient*:

$$\phi = \frac{X}{\sqrt{n}}.$$

But this measure, plus all other functions of X^2, have been found to have little value as measures of association (e.g., Fleiss, 1981). Effective methods have been devised, but the details go beyond the scope of this book.

13.7 Exercises

1. The following data were collected from 29 lakes in Florida by the U.S. Environmental Protection Agency (and are taken from Stromberg, 1993).

NIN:	5.548	4.896	1.964	3.586	3.824	3.111	3.607	3.557	2.989	18.053
3.773	1.253	2.094	2.726	1.758	5.011	2.455	0.913	0.890	2.468	4.168
4.810	34.319	1.531	1.481	2.239	4.204	3.463	1.727			
TW:	0.137	2.499	0.419	1.699	0.605	0.677	0.159	1.699	0.340	2.899
0.082	0.425	0.444	0.225	0.241	0.099	0.644	0.266	0.351	0.027	0.030
3.400	1.499	0.351	0.082	0.518	0.471	0.036	0.721			

NIN is the average influent nitrogen concentration and TW is the water retention time. Plot the points, and verify that six points are declared outliers by the MVE and MCD methods but that the MGV outlier detection method in Section 13.1.5 detects only two outliers, namely, points 10 and 23. (In this particular case, the MGV method will flag the same points as the MVE and MCD methods if, in the the argument list of outmgv, the option se=T is used. This option standardizes the data.)

2. In the study mentioned in the previous exercise, there was a third variable of interest, Y: the mean annual nitrogen concentration. One particular goal was to understand how Y is related to NIN and TW. The Y values are

   2.590 3.770 1.270 1.445 3.290 0.930 1.600 1.250 3.450 1.096 1.745
   1.060 0.890 2.755 1.515 4.770 2.220 0.590 0.530 1.910 4.010
   1.745 1.965 2.550 0.770 0.720 1.730 2.860 0.760

   Create a scatterplot of Y versus NIN. Verify that the least squares regression line has a slope of .012, and plot this line with the S-PLUS command

   ```
   abline(lsfit(NIN,Y)$coef),
   ```

   assuming that the data are stored in the S-PLUS variables NIN and Y. As will be evident, the two largest NIN values are outliers. Verify that if they are eliminated, then now the least squares regression line has a slope of .56. Plot this line.
3. Repeat the previous exercise, except use TW to predict Y and verify that the six rightmost points are outliers.
4. The results in the previous exercise suggest restricting the range of TW values to those less than 1. Verify that if we eliminate these values and the corresponding Y values, and if the S-PLUS function out is used to check for outliers among the scatterplot of the remaining Y and TW values, there are outliers among the Y values. Also verify that the least squares estimate of the slope is -1.975 versus -1.88 using Theil–Sen.
5. In the previous exercise, why would it be improper to eliminate the outliers among the Y values by restricting the range of Y values and computing a confidence interval for the slope using the conventional method in Section 6.3.1?
6. In the previous two exercises, why might some robust estimator provide a substantially different confidence interval for the slope than the conventional method in Section 6.3.1? What must be done to determine whether a robust

method does indeed give a shorter confidence interval? Apply this method using Theil–Sen and the MGV estimators.

7. The star data in Figures 6.3 and 13.12 are examined again, only now we consider the problem of predicting the surface temperature given the light intensity. The data are

Y: 4.37 4.56 4.26 4.56 4.30 4.46 3.84 4.57 4.26 4.37 3.49 4.43 4.48 4.01 4.29 4.42 4.23 4.42 4.23 3.49 4.29 4.29 4.42 4.49 4.38 4.42 4.29 4.38 4.22 3.48 4.38 4.56 4.45 3.49 4.23 4.62 4.53 4.45 4.53 4.43 4.38 4.45 4.50 4.45 4.55 4.45 4.42

X: 5.23 5.74 4.93 5.74 5.19 5.46 4.65 5.27 5.57 5.12 5.73 5.45 5.42 4.05 4.26 4.58 3.94 4.18 4.18 5.89 4.38 4.22 4.42 4.85 5.02 4.66 4.66 4.90 4.39 6.05 4.42 5.10 5.22 6.29 4.34 5.62 5.10 5.22 5.18 5.57 4.62 5.06 5.34 5.34 5.54 4.98 4.50

Plot the data, and note that not all outliers can be eliminated simply by restricting the range of the X values. Using all of the data, verify that the .95 confidence interval for the slope, using least squares and the modified percentile bootstrap method in Chapter 7, is $(-0.33, 0.14)$ and so you fail to reject. Then verify that when using the Theil–Sen estimate and the S-PLUS function regci, the .95 confidence interval is $(0.0, 0.18)$ with a significance level of .017. Finally, verify that when using the MGV estimator, the .95 confidence interval is $(0.05, 0.22)$ with a significance level of .013.

8. For the data in the previous exercise, the least trimmed squares estimate of the slope is .169, the STS estimator in Section 13.2.3 estimates the slope to be .149, and the MGV estimate is .161. It might be argued that these estimates are similar enough that it makes no practical difference which estimator is used. What is wrong with this argument?

9. Verify the results in Section 13.2.2.

10. The MGV outlier detection method can fail to declare points outliers that are clearly unusual and flagged as outliers by other methods. Why is this desirable?

11. For the data in Exercise 7, the MGVF estimate of the slope is only .107 and the Coakley–Hettmansperger estimate is .13. The latter estimator has the highest possible breakdown point, .5. Both of these estimates are smaller than the least trimmed squares and the MGV estimates reported in Exercise 8. What might explain this?

12. For the data in Table 13.1, the percentage bend correlation is $r_{pb} = .9367$. Verify that you reject $H_0: \rho_{pb} = 0$ with $\alpha = .05$.

13. The earthquake data in Table 13.4 were taken from a brochure published by the Southern California Earthquake Center. For the first two variables (magnitude and fault length) check for outliers among the data using the MVE, MCD, projection, and MGV methods. (Use the S-PLUS functions out, outpro, and outmgv, and be sure to examine the plots created by these functions.) Verify that the MCD method identifies five points as outliers, the MVE method identifies four, and the MGV and projection methods identify

TABLE 13.4 Earthquake Data

Magnitude	Fault length (kilometers)	Duration (seconds)
7.8	360	130
7.7	400	110
7.5	75	27
7.3	70	24
7.0	40	7
6.9	50	15
6.7	16	8
6.7	14	7
6.6	23	15
6.5	25	6
6.4	30	13
6.4	15	5
6.1	15	5
5.9	20	4
5.9	6	3
5.8	5	2

only two. Imagine we want to predict fault length based on magnitude. If we used the strategy of removing outliers and fitting a line to the remaining data, why might the MGV method be preferable when fitting a regression line to the bulk of the observations?

14. Repeat the previous exercise, but now use magnitude versus duration.
15. For the data in Table 13.4, remove the two points with the largest magnitude and check for outliers when examining magnitude and fault length with duration ignored. Why might it be a poor idea to remove outliers and use the MGV regression estimator on the data that remain?
16. For the data in Table 13.4, based on a plot of magnitude versus fault length, would you expect Pearson's correlation to be larger or smaller than the correlation returned by the S-PLUS function ocor in Section 13.6.7? Check your response by actually computing the correlations.
17. The percentage bend correlation between magnitude and fault length is .897. Verify that you reject the hypothesis of a zero correlation with $\alpha = .05$.
18. The average LSAT scores (X) for the 1973 entering classes of 15 American law schools, and the corresponding grade point averages (Y), are as follows.

X:	576	635	558	578	666	580	555	661	651	605	653	575	545	572	594
Y:	3.39	3.30	2.81	3.03	3.44	3.07	3.00	3.43	3.36	3.13	3.12	2.74	2.76	2.88	2.96

Create a boxplot for the X values and note that no outliers are found. Do the same for the Y values. Now verify that the function relplot also finds no outliers but that outliers are detected when using MGVF, MVE, and MCD.

19. For the data in the previous exercise, verify that the .95 confidence interval for the slope, using least squares and the bootstrap method in Section 7.3.1, is (0.0022, 0.0062) and that the .95 confidence interval for the slope using the S-PLUS function regci in Section 13.5.1 (with the Theil–Sen estimator) is (0.0031, 0.0061). Note that the ratio of the lengths of the confidence interval is

$$\frac{.0062 - .0022}{.0061 - .0031} = 1.33.$$

Based on the results in the previous exercise, what might explain why the Theil–Sen estimator provides a substantially shorter confidence interval?

MORE REGRESSION METHODS

Regression is an extremely vast topic that is difficult to cover in a single book let alone a few chapters. The goal in this chapter is to cover some modern methods that are particularly relevant in applied work, with the understanding that many issues and techniques are not discussed.

14.1 Smoothers

One way of describing a fundamental goal in regression is to say that for any given value of X, we want to estimate some (conditional) measure of location associated with Y. In Chapter 6, for example, where the standard least squares regression model was introduced [given by Equation (6.6)], the assumption is that the conditional mean of Y, given X, is $\beta_1 X + \beta_0$. More formally,

$$E(Y|X) = \beta_0 + \beta_1 X,$$

a result that is implied by assuming that the error term (ϵ) has a mean of zero. That is, the mean of Y is assumed to have a linear association with X. Often this assumption provides a reasonable summary of the data, but it is common to encounter situations where it is highly unsatisfactory. This section describes some exploratory methods, called *smoothers*, that attempt to estimate a regression line without forcing it to have a particular shape, such as a straight line. There are many smoothers, two of which are described in this section. (S-PLUS has at least six built-in smoothers.) Readers interested in more details about smoothers are referred to Hastie and Tibshirani (1990).

14.1.1 Cleveland's Smoother

Suppose we want to estimate the mean of Y given that $X = 6$ based on n pairs of observations: $(X_1, Y_1), \ldots, (X_n, Y_n)$. One strategy is to focus on the observed X values close to 6 and use the corresponding Y values to estimate the mean of Y.

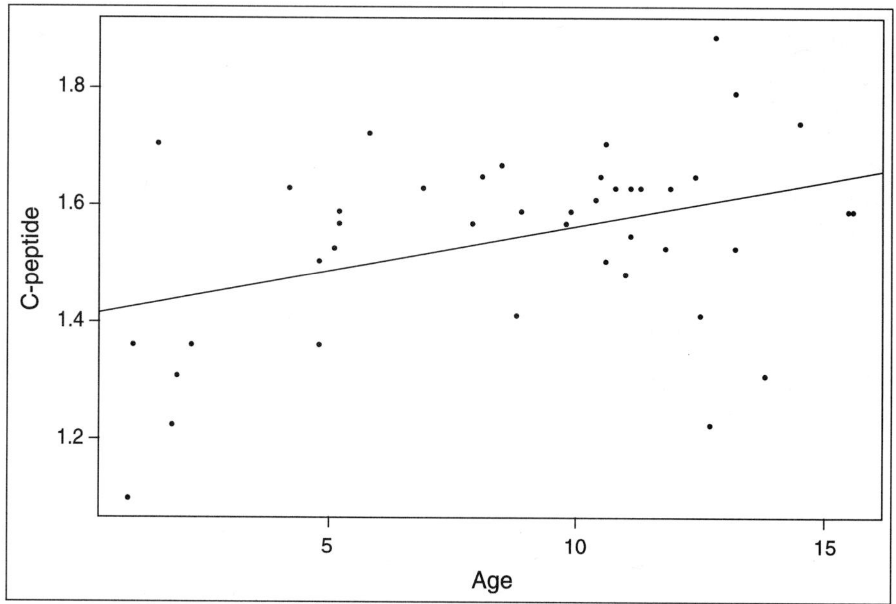

FIGURE 14.1 Least squares regression line for the diabetes data.

A specific technique for implementing this strategy was derived by Cleveland (1979) and is known as a *locally weighted running-line smoother*. The idea is that, although a regression line might not be linear over the entire range of X values, it will be approximately linear over small intervals of X. That is, for X values close to 6, say, a linear regression line might perform reasonably well and can be used to estimate Y given that $X = 6$. But for X values far from 6, some other regression estimate of Y should be used instead.

To be more concrete, look at Figure 14.1, which shows the logarithm of C-peptide concentrations in children versus their age. (The data are given in Exercise 15 of Chapter 6 and are taken from a study on diabetes in children.) The solid straight line is the least squares regression line. Cleveland's strategy begins by measuring how far away each X_i value is from the point of interest. If you want to predict or estimate the typical C-peptide concentration for a 10-year-old you measure how far away each X_i is from 10, while for an 11-year-old you measure how far away each X_i is from 11. The idea is to give less credence to those who are relatively far away from the age that is of interest. In fact, those individuals furthest away are ignored altogether. In particular, given the goal of estimating the mean of Y corresponding to some specific value for X, you measure the distance of X from each of the observed X_i values with

$$\delta_i = |X_i - X|.$$

For example, if the goal is to predict C-peptide concentrations for 10-year-old children, then $X = 10$; because $X_1 = 5.2$, $X_2 = 8.8$, and $X_3 = 10.5$, it follows that $\delta_1 = |5.2 - 10| = 4.8$, $\delta_2 = |8.8 - 10| = 1.2$, $\delta_3 = |10.5 - 10| = 0.5$, and so forth. When predicting C-peptide levels for 11-year-old children, $X = 11$, $\delta_1 = |5.2 - 11| = 5.8$, $\delta_2 = |8.8 - 11| = 2.2$, and so on.

Next, sort the δ_i values and retain the pn pairs of points that have the smallest δ_i values, where p is a number between 0 and 1. The value of p represents the proportion of points used to predict Y and is generally referred to as the *span*. For the moment, suppose $p = 1/2$. In the illustration, $n = 43$, so this means that you retain 22 of the pairs of points that have X_i values closest to $X = 10$. These 22 points are the *nearest neighbors* to $X = 10$. If you want to predict C-peptide concentrations for 11-year-old children, you retain the 22 pairs of points with X_i values closest to $X = 11$. Let δ_m be the maximum value of the δ_i values that are retained. For $X = 10$ and $p = 0.5$, $\delta_m = 2.7$. (The details are left as an exercise.) Set

$$Q_i = \frac{|X - X_i|}{\delta_m};$$

if $0 \le Q_i < 1$, set

$$w_i = \left(1 - Q_i^3\right)^3,$$

otherwise set

$$w_i = 0.$$

Finally, use *weighted least squares* to predict Y using w_i as weights (cf. Fan, 1992). That is, determine the values b_1 and b_0 that minimize

$$\sum w_i(Y_i - b_0 - b_1X_i)^2$$

and estimate the mean of Y corresponding to X to be $\hat{Y} = b_0 + b_1X$. Because the weights (the w_i values) change with X, generally a different regression estimate of Y is used when X is altered. Finally, let $\hat{Y}_i$ be the estimated mean of Y given that $X = X_i$ based on the method just described. Then an estimate of the regression line is obtained by the line connecting the points $(X_i, \hat{Y}_i)$ $(i = 1, \ldots, n)$ and is called a *smooth*.

The span, p, controls the raggedness of the smooth. If p is close to 1, we get a straight line stronger even when there is curvation. If p is too close to zero, an extremely ragged line is obtained instead. By choosing a value for p between .2 and .8, curvature can usually be detected and a relatively smooth regression line is obtained.

14.1.2 S-PLUS Function lowess

The built-in S-PLUS function

```
lowess(x,y,p=2/3)
```

computes Cleveland's smoother, just described. The value for p, the span, defaults to 2/3. You can create a scatterplot of points that contains this smooth with the S-PLUS commands

```
plot(x,y)
lines(lowess(x,y))
```

If the line appears to be rather ragged, try increasing p to see what happens. If the line appears to be approximately horizontal, indicating no association, check to see what happens when p is lowered.

EXAMPLE. For the diabetes data in Figure 14.1, testing the hypothesis that the slope is zero with the conventional method in Chapter 6 returns a significance level of .008. The heteroscedastic, .95 confidence interval for the slope, based on the modified bootstrap method in Section 7.3, is (−0.001, .029), and the bootstrap test of independence covered in Section 7.8 rejects at the .05 level. So although the confidence interval based on the modified bootstrap method is not quite able to reject at the .05 level (because the .95 confidence interval contains 0), there are indications that age and C-peptide concentrations are dependent, and based on Figure 14.1 it might seem that C-peptide concentrations tend to increase with age.

Now look at the solid line in Figure 14.2, which is the smooth created by lowess. Notice that it increases up to about the age of 7 and then flattens out. That is, this smooth suggests that there is a positive association up to about age seven, and then the association seems to disappear. However, some caution is warranted in this particular case. The dashed line shows the smooth when X values less than 4 are ignored; it is nearly horizontal. Because there are only six points with $X < 4$, it is difficult to know how age is related to C-peptide concentrations for very young children. The smooth suggests that generally, there is little or no association; it hints that this might not be true for children under the age of 4, but more data in this age range are needed to understand the association when $X < 4$. Generally, the ends of a smooth must be viewed with caution, because data are often sparse in these regions and there is an inherent

Continued

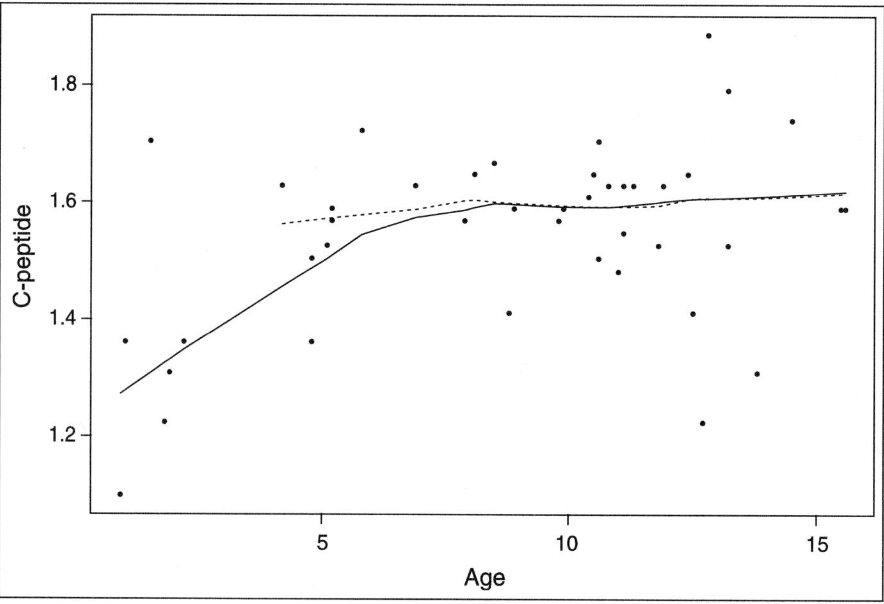

FIGURE 14.2 Two smooths for the diabetes data based on the S-PLUS function lowess. The dashed line is the smooth ignoring points with $X < 4$.

EXAMPLE. (*Continued*) bias associated with these points. For example, if $X = 3$ is the smallest observed X value, estimating $E(Y|X = 3)$ is hampered by the fact that the estimate is based on X values that are greater than or equal to 3. Ideally, a smooth would be based on X values that are both less than and greater than 3. ■

14.2 Smooths Based on Robust Measures of Location

A natural guess at how to extend Cleveland's method to robust measures of location is to replace the weighted least squares estimator with one of the robust regression estimators covered in Chapter 13. However, this strategy is known to be highly unsatisfactory. The reason is that robust regression estimators can be insensitive to curvature, so the resulting smooth often misses curvature when in fact it exists. A better strategy is to use what is called the *running-interval smoother*. To estimate some measure of location for Y, given some value for X, a running interval smoother searches for *all* points close to the value of X that are of interest and then simply computes the measure of location based on the corresponding Y values. Note that the number of X_i values close to X will depend on the value of X. In contrast, Cleveland's method uses the k nearest points, with k fixed and chosen in advance.

To elaborate, compute MAD based on the X values and label it MAD_X. Let f be some constant that is chosen in a manner to be described and illustrated. Then the value X is said to be close to X_i if

$$|X_i - X| \le f\left(\frac{\text{MAD}_X}{.6745}\right).$$

So for normal distributions, X is close to X_i if X is within f standard deviations of X_i. Now consider all of the Y_i values corresponding to the X_i values that are close to X. Then an estimate of the typical value of Y, given X, is the estimated measure of location based on the Y values just identified. For example, if six X_i values are identified as being close to $X = 22$ and the corresponding Y values are 2, 55, 3, 12, 19, and 21, then the estimated mean of Y, given that $X = 22$, would be the average of these six numbers: 18.7. The estimated 20% trimmed mean of Y, given that $X = 22$, would be the trimmed mean of these six values, which is 13.75.

A running-interval smoother is created as follows. For each X_i, determine which of the X_j values are close to X_i, compute a measure of location associated with the corresponding Y_j values, and label this result $\hat{Y}_i$. So we now have the following n pairs of numbers: $(X_1, \hat{Y}_1), \dots, (X_n, \hat{Y}_n)$. The running-interval smooth is the line formed by connecting these points. The span, f, controls how ragged the line will be. As with Cleveland's method, if f is too close to 1, the smooth will be a straight line, even when there is curvature; if f is too close to zero, the result is a very ragged line.

14.2.1 S-PLUS Functions rungen and runmean

The S-PLUS function

```
rungen(x,y,est=mom,fr=.8,plotit=T,scat=T,pyhat=F,...)
```

computes a running interval smooth assuming that there is only one predictor. (For multiple predictors, see Sections 14.2.3 and 14.2.4.) The argument est determines the measure of location that is used and defaults to MOM. The argument fr corresponds to the span (f) and defaults to .8. The function returns the $\hat{Y}_i$ values if the argument pyhat is set to T. By default, a scatterplot of the points is created with a plot of the smooth. To avoid the scatterplot, set scat=F. If there is specific interest in using a trimmed mean, the function

```
runmean(x,y,fr=.8,tr=.2,pyhat=F)
```

is a bit more convenient to use. As usual, the argument tr controls the amount of trimming and defaults to 20%.

EXAMPLE. The left panel of Figure 14.3 shows the smooth created by rungen for the reading data shown in Figures 13.2 and 13.8. The dashed line is the smooth created by lowess. In this particular case both methods give similar results, but strategically placed outliers can cause lowess to give a distinctly different impression about the association between two variables. To illustrate that curvature can be detected correctly, the right panel shows the smooth returned by runmean for 40 points generated on a computer where $Y = X^2 + \epsilon$, with both X and ϵ having standard normal distributions. ■

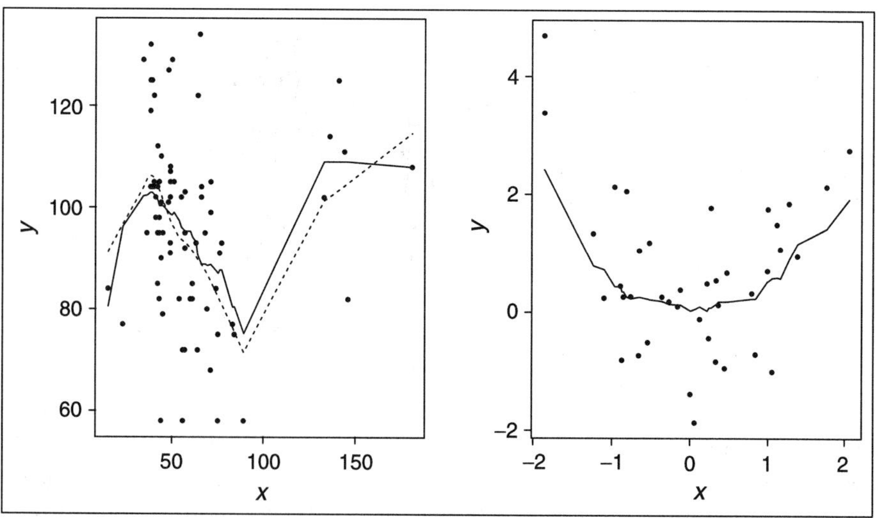

FIGURE 14.3 The left panel shows the smooths for the reading data based on lowess (the dashed line) and rungen. The right panel shows the smooths for data where $Y = X^2 + \epsilon$.

14.2.2 Prediction When X is Discrete: The S-PLUS Function rundis

In some situations the predictor X has very few values and there are multiple Y values for each X value. In this case it might be of interest to compute a measure of location for Y corresponding to each X value and to plot the results. For example, for all Y values corresponding to $X = 1$, say, compute some measure of location, do the same for $X = 2$, and so on. This is in contrast to the smooths previously described, where you search for all X values close to 2, for example, and compute a measure of location based on the corresponding Y values. For convenience, the S-PLUS function

```
rundis(x,y,est=mom,plotit=T,pyhat=F,...)
```

has been supplied to perform this task.

14.2.3 Multiple Predictors

The basic idea behind the running-interval smoother can be extended to multiple predictors. First consider the case with two predictors, so we have n pairs of X values: $(X_{11}, X_{12}), \ldots, (X_{n1}, X_{n2})$. For the ith pair of predictors, (X_{i1}, X_{i2}), imagine you want to estimate the corresponding typical value of Y. For instance, if $X_{i1} = 6$ and $X_{i2} = 12$, the goal is to estimate the typical value of Y given that $(X_{i1}, X_{i2}) = (6, 12)$. Like the running-interval smoother, the strategy is to apply a measure of location to those Y values for which the corresponding (X_{j1}, X_{j2}) values are close to (X_{i1}, X_{i2}). To determine whether (X_{j1}, X_{j2}) is close to (X_{i1}, X_{i2}), we proceed in a manner that has certain similarities to the outlier detection method described in Box 13.1. For the more general case of p predictors, let $\mathbf{X}_i = (X_{i1}, \ldots, X_{ip})$ and let $\mathbf{M}$ be the MVE or MCD covariance matrix based on $\mathbf{X}_1, \ldots, \mathbf{X}_n$. Then a measure of the distance between the two points $\mathbf{X}_i$ and $\mathbf{X}_j$ is

$$D_{ij} = \sqrt{(\mathbf{X}_i - \mathbf{X}_j)\mathbf{M}^{-1}(\mathbf{X}_i - \mathbf{X}_j)'}.$$

We say that $\mathbf{X}_j$ is close to $\mathbf{X}_i$ if $D_{ij} \le f$, where f again plays the role of the span. Generally, $f = 1$ seems to give good results; but, as usual, exceptions arise. Now, for all those $\mathbf{X}_j$ that are close to $\mathbf{X}_i$, compute a measure of location associated with the corresponding Y values and label it $\hat{Y}_i$. That is, the typical value of Y, given that $\mathbf{X} = \mathbf{X}_i$, is estimated to be $\hat{Y}_i$.

14.2.4 S-PLUS Functions runm3d and rung3d

The S-PLUS function

```
rung3d(x,y,est = mom, fr = 1, plotit = T, pyhat = F,...)
```

creates a running-interval smooth of the data based on the measure of location specified by the argument est, which defaults to MOM. Here, x is assumed to be a matrix with p columns corresponding to the p predictors. The argument fr is the span and

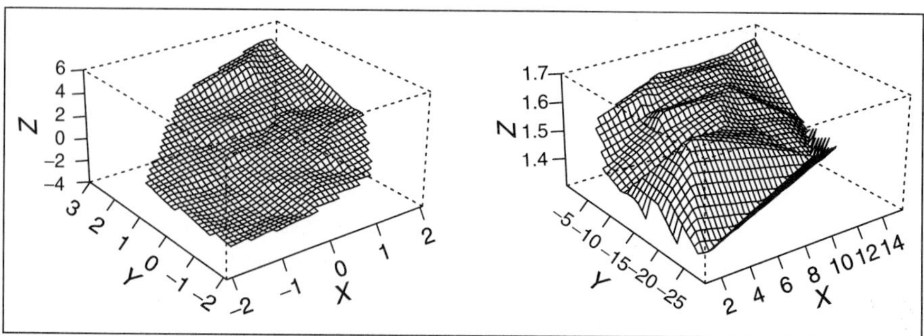

FIGURE 14.4 The left panel is a smooth for data where the true regression line is a plane. The right panel is a smooth based on data from a diabetes study.

pyhat=T will return the $\hat{Y}_i$ values. To get 20% trimmed means, use est=mean and tr=.2. The function

runm3d(x, y, fr = 1, tr = 0.2, plotit = T, pyhat = F)

is designed for trimmed means only.

EXAMPLE. The left panel of Figure 14.4 shows a smooth for 100 points generated according to $Y = X_1 + X_2 + \epsilon$, where X_1, X_2, and ϵ are independent standard normal variables. Notice that the smooth is well approximated by a plane, which is consistent with how the data were generated. ■

EXAMPLE. The diabetes data mentioned in Section 14.1 are reconsidered, only now two predictors of C-peptide concentration are used: age and base deficit. The right panel of Figure 14.4 shows the smooth created by rung3d. In this case, it seems that a plane does not provide an adequate summary of the data. ■

14.2.5 Seeing Curvature with More Than Two Predictors

The running-interval smoother can be used with more than two predictors, but, as is evident, visualizing curvature cannot be done in a simple manner. A variety of techniques have been proposed for dealing with this problem, and a comparison of several methods was made by Berk and Booth (1995). (They focused on predicting means, but the strategies they considered are readily extended to robust measures of location.) Generally, these methods can help, but situations can be created where any one method fails.

A simple strategy, sometimes called the *partial response plot*, is to check a smooth for each individual predictor while ignoring the others. An alternative approach is to plot the residuals versus the predicted values. Experience suggests that this strategy often can be unsatisfactory. Another strategy is based on what is called a *partial residual plot*. The idea dates back to Ezekiel (1924, p. 443) and was named a partial residual plot by Larsen and McLeary (1972). To explain it, imagine there are p predictors and that you want to check for curvature associated with predictor j. Assuming that the other predictors have a linear association with Y, fit a regression plane to the data, ignoring the jth predictor. The *partial residual plot* simply plots the resulting residuals versus X_j. A smooth applied to this plot can be used to check for curvature.

14.2.6 S-PLUS Function prplot

The S-PLUS function

prplot(x,y,pvec=ncol(x),regfun=tsreg,fr=.8,est=mom,...)

creates a partial residual plot assuming that curvature is to be checked for the predictor indicated by the argument pvec. The argument x is assumed to be an n-by-p matrix. By default, it is assumed that curvature is to be checked using the data stored in the last column of the matrix x. The argument regfun indicates which regression method will be used, fr is the span used to create the smooth, and est indicates which measure of location is used with the smooth.

EXAMPLE. The model $Y = \beta_0 + \beta_1 X_1 + \beta_2 X_2 + \beta_3 X_1 X_2 + \epsilon$ is often used to investigate interactions in regression and appears to have been first suggested by Saunders (1955, 1956). One hundred vectors of observations were generated according to the model $Y = X_1 + X_2 + X_1 X_2 + \epsilon$, where X_1, X_2, and ϵ are independent standard normal random variables. The partial residual plot provides a partial check on the adequacy of this model. If the model is correct, then, in particular, a partial residual plot based on the term $X_1 X_2$ should produce a reasonably straight line. Applying the S-PLUS function prplot to the data creates the smooth shown in Figure 14.5. We see that for the bulk of the centrally located points, we do indeed get a straight line, which has a slope approximately equal to the true slope, 1. ■

EXAMPLE. For another portion of the reading study (previously mentioned in Chapter 13), there was interest in how a measure of orthographic ability (Y) is related to a measure of sound blending (X_1) and a measure of auditory analysis (X_2). A smooth between X_1 and Y strongly indicates a linear association. Assuming that the association between X_1 and Y is linear, is it reasonable to assume that the association between Y, X_1, and X_2 is linear? That is, does the model $Y = \beta_0 + \beta_1 X_1 + \beta_2 X_2 + \epsilon$ provide an adequate summary of the data for some choice for β_0, β_1, and β_2? Figure 14.6 shows the plot created by prplot. As is evident, there seems to be a bend at approximately $X_2 = 6$, which is near the center of the X_2 values. That is, it appears that something might be wrong

Continued

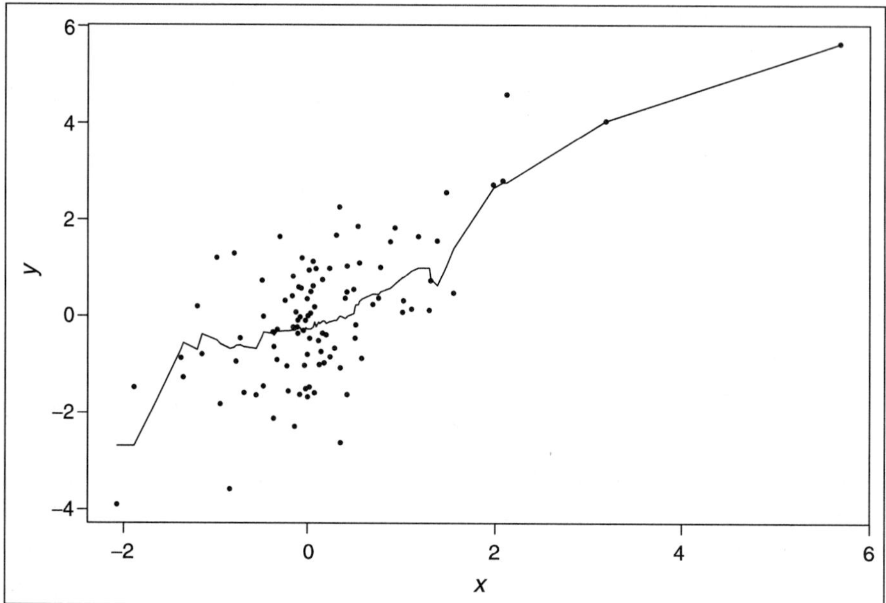

FIGURE 14.5 Output from the S-PLUS function prplot, where the goal is to check a commonly used method for modeling interactions.

EXAMPLE. (*Continued*) with this particular model. Because a smooth indicates that Y and X_1 have a linear association, it seems that there might be curvature associated with X_2. A smooth (or partial response plot) between Y and X_2 again indicates a curvilinear association. One possible way of dealing with this apparent curvature is simply to divide the data into two groups according to whether $X_2 \leq 6$. ■

14.2.7 Some Alternate Methods

The methods already covered for detecting and describing curvature when there are multiple predictors are far from exhaustive. Although complete details about other methods are not provided, it might help to mention some of the alternate strategies that have been proposed.

One approach is to assume that for some collection of functions $f_1, \ldots, f_p$,

$$Y = \beta_0 + f_1(X_1) + \cdots + f_p(X_p) + \epsilon,$$

and then to try to approximate these functions using some type of smoother. This is called a *generalized additive model*; details can be found in Hastie and Tibshirani (1990), who described an algorithm for applying it to data. (S-PLUS has built-in functions for applying this technique; see the methods listed under generalized additive models in the S-PLUS *Guide to Statistics*.)

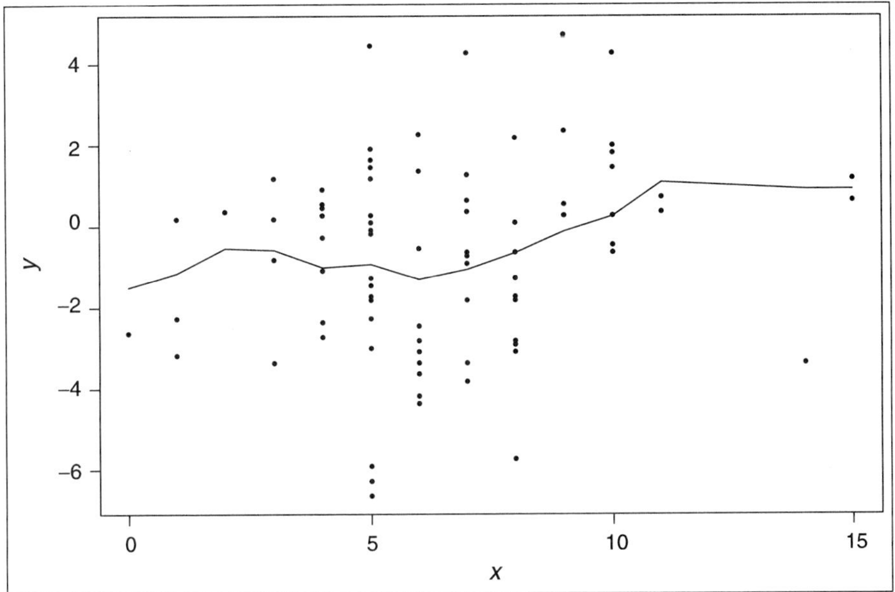

FIGURE 14.6 Output from the S-PLUS function prplot based on the reading data.

The *augmented partial residual plot* (Mallows, 1986) is like the partial residual plot, except that it includes a quadratic term for the predictor being investigated. For a generalization of this method, see Cook (1993).

Another approach is to search for nonlinear transformations of both Y and the predictors that results in an additive model. That is, search for functions $f_y, f_1, \ldots, f_p$ such that

$$f_y(Y) = f_1(X_1) + \cdots + f_p(X_p) + \epsilon$$

provides a good fit to data. An algorithm for implementing the method was derived by Breiman and Friedman (1985) and is called *alternating conditional expectations*, or *ace*. (S-PLUS has built-in functions for applying the technique.) For some refinements and extensions, see Tibshirani (1988).

14.3 Comparing the Slopes of Two Independent Groups

Consider two independent groups and imagine that for each group you have an outcome variable Y and a predictor X. For the jth group ($j = 1, 2$), assume

$$Y_j = \beta_{0j} + \beta_{1j}X_j + \epsilon_j,$$

where ϵ_j is independent of X_j and $E(\epsilon_j) = 0$. That is, for each group, the standard regression model in Section 6.2 holds. Then a common goal is to test

$$H_0 : \beta_{11} = \beta_{12}, \tag{14.1}$$

the hypothesis that the two groups have equal slopes.

A variety of methods have been proposed for testing Equation (14.1), many of which are summarized by Conerly and Mansfield (1988), but the bulk of the methods are known to be relatively unsatisfactory, including a popular technique derived by Chow (1960). Here attention is focused on comparing slopes with a percentile bootstrap method plus one of the robust estimators covered in Chapter 13. As in Chapter 13, the error terms can be heteroscedastic, and it is not assumed that the distribution for the two error terms are similar in any manner. (Methods that assume homoscedasticity or that groups have error terms with identical distributions can be highly unsatisfactory, for the basic reasons mentioned in previous chapters.) To test Equation (14.1), generate bootstrap samples from each group as described in Chapter 7, compute the slope for each group based on these bootstrap samples, and label them b^*_{11} and b^*_{12}. Next, repeat this process B times; let d^*_b be the difference between the bootstrap estimate of the slopes based on the bth bootstrap sample ($b = 1, \ldots, B$) and let $\hat{p}^*$ be the proportion of times the bootstrap estimate for the first group is less than the bootstrap estimate from the second. That is, $\hat{p}^*$ is the proportion of d^*_b values less than zero. Setting

$$\hat{p}^*_m = \min(\hat{p}^*, 1 - \hat{p}^*),$$

the estimated significance level for the hypothesis given by Equation (14.1) is $2\hat{p}^*_m$. Then reject the null hypothesis if $2\hat{p}^*_m \le \alpha$. Putting the d^*_b values in ascending order, the $1 - \alpha$ confidence interval for the difference between the slopes, $\beta_{11} - \beta_{12}$, is $(d^*_{(\ell+1)}, d^*_{(u)})$, where, as usual, $\ell = \alpha B/2$ and $u = B - \ell$. In essence, use the method in Section 8.8.1 (designed for measures of location) adapted to the problem at hand.

14.3.1 S-PLUS Functions reg2ci and runmean2g

The S-PLUS function

```
reg2ci(x1, y1, x2, y2, regfun=tsreg, nboot=599, alpha=0.05, plotit=T)
```

compares the slopes of two groups using the method just described. The data for group 1 are stored in x1 and y1, and for group 2 they are stored in x2 and y2. As usual, nboot is B, the number of bootstrap samples, regfun indicates which regression estimator is to be used and defaults to the Theil–Sen estimator, and plotit=T creates a plot of the bootstrap estimates.

To provide some visual sense of how the regression lines differ and to provide an informal check on whether both regression lines are reasonably straight, the S-PLUS function

```
runmean2g(x1, y1, x2, y2, fr=1, est=mom, ...)
```

has been supplied. It creates a scatterplot for both groups (with a + used to indicate points that correspond to the second group) and it plots a smooth for both groups. The smooth for the first group is indicated by a solid line, and a dashed line is used for the other.

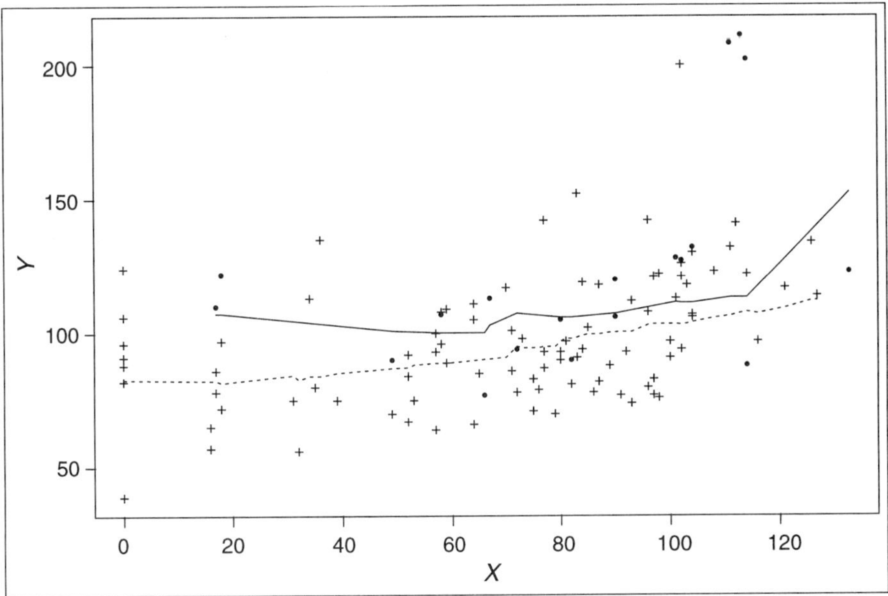

FIGURE 14.7 A smooth for two groups of students.

EXAMPLE. A controversial issue is whether teachers' expectancies influence intellectual functioning. A generic title for studies that address this issue is Pygmalion in the classroom. Rosenthal and Jacobson (1968) argue that teachers' expectancies influence intellectual functioning, and others argue that it does not. A brief summary of some of the counterarguments can be found in Snow (1995). Snow illustrates his concerns with data collected by Rosenthal, where children in grades 1 and 2 were used. Here, other issues are examined using robust regression methods. One of the analyses performed by Rosenthal involved comparing experimental children, for whom positive expectancies had been suggested to teachers, with control children, for whom no expectancies had been suggested. (The data used here are taken from Elashoff and Snow, 1970.) One measure was a reasoning IQ pretest score, and a second was a reasoning IQ posttest score. Here we consider whether the slopes of the regression lines differ for the two groups when predicting posttest scores based on pretest scores. Figure 14.7 shows the output from the S-PLUS function runmean2g. The .95 confidence interval for the difference between the slopes, returned by reg2ci and based on the Theil–Sen estimator, is (−0.72, 0.18) with a significance level of .22. ■

EXAMPLE. For the example in connection with Figure 14.6, a partial residual plot suggested that there is a curvilinear relationship between a measure of

Continued

EXAMPLE. (*Continued*) orthographic ability Y and X_2, a measure of auditory analysis. In particular, there seems to be a distinct bend near $X_2 = 6$. If we split the data into two groups according to whether $X_2 \leq 6$, does this alter the association between Y and X_1, a measure of sound blending? One way of checking this possibility is to test the hypothesis that these two groups have identical slopes. The .95 confidence interval for the difference between the slopes is $(-1.5, 0.0)$ with a significance level of .09, so we fail to detect a change in the association at the .05 level. ■

14.3.2 Comparing Correlations

The method for comparing slopes just described can be used to compare correlation coefficients simply by replacing estimates of the slopes with a measure of association that is of interest. When working with Pearson's correlation, however, a modified percentile bootstrap method is required. In particular, if there are n_1 pairs of observations for the first group and n_2 for the second, let $n = n_1 + n_2$ be the total number of observations, and set $B = 599$. Now generate bootstrap samples from both groups and for the bth bootstrap sample from each group ($b = 1, \dots, 599$), let d_b^* be the difference between the bootstrap estimates of the correlations. Then a .95 confidence interval for $\rho_1 - \rho_2$, the difference between the population (Pearson) correlation coefficients, is

$$\left(d^*_{(a)}, d^*_{(c)}\right),$$

where for $n < 40$, $a = 7$ and $c = 593$; for $40 \leq n < 80$, $a = 8$ and $c = 592$; for $80 \leq n < 180$, $a = 11$ and $c = 588$; for $180 \leq n < 250$, $a = 14$ and $c = 585$; while for $n \geq 250$, $a = 15$ and $c = 584$. If this interval does not contain zero, reject the hypothesis of equal Pearson correlations. Adjustments (choices for a and c) for other values of α have not been determined.

14.3.3 S-PLUS Functions twocor and twopcor

The S-PLUS function

twocor(x1, y1, x2, y2, corfun = pbcor, nboot = 599, alpha = .05, plotit = T, ...)

compares the correlations corresponding to two independent groups. The argument corfun indicates which correlation is to be used and defaults to the percentage bend correlation. The argument plotit indicates whether the bootstrap estimates are to be plotted. To compare Pearson correlations, use

twopcor(x1, y1, x2, y2).

14.4 Tests for Linearity

Smooths provide an informal check on curvature. This section describes two methods that can be used to establish curvature in a more formal manner.

14.4.1 A Method Based on a Split of the Data

The first strategy is to split the data into two groups based on the X values. If there is no curvature, then these two sets of observations should have identical slopes, which can be tested as described in Section 14.3.

EXAMPLE. Consider the diabetes data in Figure 14.2. As previously noted, a smooth suggests that there might be a positive association between age and C-peptide concentrations up to about the age of 4, or possibly 7, but then the association seems to disappear. For illustrative purposes, split the data into two groups according to whether a child is less than or greater than 7 years old. If there is no curvature, then the regression lines for these two groups should have identical slopes. Applying the S-PLUS function reg2ci to these two groups and using the default regression method (the Theil–Sen estimator), the .95 confidence interval for the difference between the slopes is (0, .125). This interval contains zero, so we fail to reject, although we come very close to detecting curvature. ■

14.4.2 An Alternate Method

This section describes an alternate approach to detecting curvature by testing the hypothesis that there is a linear association between Y and some set of predictors. That is, the goal is to test the hypothesis that for some $\beta_0, \ldots, \beta_p$, $Y = \beta_0 + \beta_1 X_1 + \cdots + \beta_p X_p + \epsilon$. Here, it is *not* assumed that the error term is homoscedastic. The theoretical justification for the method in this section is due to Stute, Manteiga, and Quindimil (1998) and is essentially a generalization of the test of independence covered in Section 7.8. For simplicity, the method is described for the case of a single predictor only, but multiple predictors can be handled as well. Basically, proceed as was done in Section 7.8, only with $\bar{Y}$ replaced by $\hat{Y}$, where $\hat{Y}$ is the estimate of Y based on some regression estimator that assumes there is a linear association. The details are summarized in Box 14.1.

BOX 14.1 Test the Hypothesis That a Regression Line is Straight

Let $\hat{Y}$ be some regression estimate of Y. Least squares could be used, but it has been shown that this can lead to problems in terms of controlling the probability of a Type I error. So it is suggested that some robust estimator be used instead. For fixed j $(1 \le j \le n)$, set $I_i = 1$ if $X_i \le X_j$; otherwise $I_i = 0$,

Continued

BOX 14.1 (*Continued*) and let

$$R_j = \frac{1}{\sqrt{n}} \sum I_i(Y_i - \hat{Y}_i)$$
$$= \frac{1}{\sqrt{n}} \sum I_i r_i, \tag{14.2}$$

where $r_i = Y_i - \hat{Y}_i$ are the usual residuals. The (Kolmogorov) test statistic is the maximum absolute value of all the R_j values. That is, the test statistic is

$$D = \max|R_j|, \tag{14.3}$$

where max means that D is equal to the largest of the $|R_j|$ values. As in Chapter 7, a Cramér–von Mises test statistic can be used instead, where now

$$D = \frac{1}{n} \sum R_j^2. \tag{14.4}$$

A critical value is determined with the bootstrap method described in Section 7.8. Generate n observations from a uniform distribution and label the results $U_1, \ldots, U_n$. Next, for $i = 1, \ldots, n$, set

$$V_i = \sqrt{12}(U_i - .5),$$
$$r_i^* = r_i V_i,$$
$$Y_i^* = \hat{Y}_i + r_i^*.$$

Then based on the n pairs of points $(X_1, Y_1^*), \ldots, (X_n, Y_n^*)$, compute the test statistic and label it D^*. Repeat this process B times and label the resulting test statistics $D_1^*, \ldots, D_B^*$. Finally, put these B values in ascending order, yielding $D_{(1)}^* \le \cdots \le D_{(B)}^*$. The critical value is $D_{(u)}^*$, where $u = (1 - \alpha)B$ rounded to the nearest integer. That is, reject if

$$D \ge D_{(u)}^*.$$

14.4.3 S-PLUS Functions lintest and linchk

The S-PLUS function

```
lintest(x,y,regfun=tsreg,nboot=500,alpha=.05)
```

tests the hypothesis that a regression surface is a plane using the method outlined in Box 14.1. (Execution time is fairly fast with one predictor, but it might be slow when there are multiple predictors instead.) When reading the output, the Kolmogorov test statistic is labeled dstat and its critical value is labeled critd. The Cramér–von Mises test statistic is labeled wstat. The default regression method (indicated by the argument regfun) is Theil–Sen.

For convenience, the function

linchk(x,y,sp,pv=1,regfun=tsreg,nboot=599,alpha=.05)

is supplied, which splits the data into two groups according to whether predictor pv has a value less than the value stored in the argument sp. For example,

linchk(x,y,sp=10,pv=3)

would split the data into two groups based on whether predictor 3 has a value less than 10. Then it compares the regression parameters for these two groups with the function reg2ci.

EXAMPLE. For the diabetes data shown in Figure 14.1, the Kolmogorov test statistic returned by lintest is $D = .179$; it reports a .05 critical value of .269, so you fail to reject. Note that based on results in Section 14.1, a smooth suggests that for the bulk of the data, the association is linear, with a nearly horizontal regression line. It was previously remarked, however, that this association might change for children under the age of 4. But even if a change exists, with only six children under the age of 4, detecting it is difficult at best. ■

14.5 Inferential Methods with Multiple Predictors

This section takes up the problem of making inferences about regression parameters when there are multiple predictors. We begin with the classic approach, which assumes that there is a linear relationship between the p predictors $X_1,\ldots,X_p$ and some variable Y, that Y has a normal distribution, and that there is homoscedasticity. That is,

$$Y = \beta_0 + \beta_1 X_1 + \beta_2 X_2 + \cdots + \beta_p X_p + \epsilon \tag{14.5}$$

is assumed, where ϵ has a normal distribution with variance σ^2, and σ^2 does not depend on the values of the predictors. With one predictor ($p = 1$), Equation (14.5) reduces to the regression model given by Equation (6.6). The goal is to test

$$H_0: \beta_1 = \cdots = \beta_p = 0, \tag{14.6}$$

the hypothesis that all p slope parameters are zero.

Let $\hat{Y} = b_0 + b_1 X_1 + \cdots + b_p X_p$ be the least squares regression line. That is, the values $b_0,\ldots,b_p$ minimize $\sum (Y_i - \hat{Y}_i)^2$, the sum of the squared residuals. The *squared multiple correlation coefficient* is

$$R^2 = 1 - \frac{\sum \left(Y_i - \hat{Y}_i\right)^2}{\sum \left(Y_i - \bar{Y}\right)^2} \tag{14.7}$$

and can be seen to be the squared (Pearson) correlation between Y_i and $\hat{Y}_i$. Note that when using least squares regression, R^2 is an estimate of explanatory power, as discussed in Section 13.6.8. The classic method for testing the hypothesis given by

Equation (14.6) is based on the test statistic

$$F = \left(\frac{n-p-1}{p}\right)\left(\frac{R^2}{1-R^2}\right). \tag{14.8}$$

Under normality and homoscedasticity, F has an F-distribution with $\nu_1 = p$ and $\nu_2 = n - p - 1$ degrees of freedom. So the null hypothesis is rejected if $F \geq f_{1-\alpha}$, the $1 - \alpha$ quantile of an F-distribution with ν_1 and ν_2 degrees of freedom.

If you have three predictors with the data stored in the S-PLUS variables x1, x2, and x3 and the outcome predictor stored in y, then the built-in S-PLUS command

```
summary(lm(y~x1+x2+x3))
```

will perform the F-test just described (and it performs a Student's T-test of $H_0: \beta_j = 0$, $j = 0, 1, \ldots, p$, for each of the $p + 1$ regression parameters).

14.5.1 A Bootstrap Method

As was the case in Chapter 6, the conventional hypothesis-testing method just described performs well in terms of controlling Type I error probabilities when Y is independent of all p predictors, which implies that there is homoscedasticity. When Y and the p predictors are dependent, the conventional F-test can be very unsatisfactory—even under normality.

A straightforward application of a particular bootstrap method can be used to test Equation (14.6), the hypothesis that the parameters in a linear regression model have a common value of zero. The method is similar in spirit to the method for comparing measures of location among dependent groups that was covered in Box 11.6 and is described in Box 14.2. Basically, generate B bootstrap estimates of the p slope parameters and then check to see how deeply the vector $\mathbf{0} = (0, \ldots, 0)$, having length p, is nested within the bootstrap values. When $p = 1$, the method in Box 14.2 is essentially a bootstrap-t method with the standard error of the regression estimators estimated by the bootstrap values. This is in contrast to the method in Section 13.5 for computing a confidence interval for the individual parameters, which uses a percentile bootstrap method instead.

BOX 14.2 Test the Hypothesis That All Slope Parameters Are Zero

First, generate a bootstrap sample as was described in Chapter 7. For the p predictors case, this means that among the n vectors of observations

$$(Y_1, X_{11}, \ldots, X_{1p})$$
$$\vdots$$
$$(Y_n, X_{n1}, \ldots, X_{np}),$$

Continued

BOX 14.2 (*Continued*)
randomly resample with replacement n vectors, yielding

$$\left(Y_1^*, X_{11}^*, \quad \ldots, \quad X_{1p}^*\right)$$
$$\vdots$$
$$\left(Y_n^*, X_{n1}^*, \quad \ldots, \quad X_{np}^*\right).$$

Let $\hat{\beta}_j^*$ be the estimate of the jth regression parameter, $j = 1, \ldots, p$. Next, repeat this process B times, yielding $\hat{\beta}_{jb}^*$, $j = 1, \ldots, p$; $b = 1, \ldots, B$, and let

$$s_{jk} = \frac{1}{B-1} \sum_{b=1}^{B} \left(\hat{\beta}_{jb}^* - \hat{\beta}_j\right)\left(\hat{\beta}_{kb}^* - \hat{\beta}_k\right),$$

where $\hat{\beta}_j$ is the estimate of β_j based on the original data. Compute

$$d_b = \left(\hat{\beta}_b^* - \hat{\beta}\right) \mathbf{S}^{-1} \left(\hat{\beta}_b^* - \hat{\beta}\right)',$$

where $\mathbf{S}$ is the matrix corresponding to s_{jk}, $\hat{\beta} = (\hat{\beta}_1, \ldots, \hat{\beta}_p)$, and $\hat{\beta}_b^* = (\hat{\beta}_{1b}^*, \ldots, \hat{\beta}_{pb}^*)$. The value of d_b measures how far away the bth bootstrap vector of estimated slope parameters is from the center of all B bootstrap values. Put the d_b values in ascending order, yielding $d_{(1)} \leq \cdots \leq d_{(B)}$. The test statistic is

$$D = (0 - \hat{\beta})\mathbf{S}^{-1}(0 - \hat{\beta})'$$

and measures how far away the null hypothesis is from the estimated slope parameters. Reject if $D \geq d_{(u)}$, where $u = (1 - \alpha)B$, rounded to the nearest integer.

The bootstrap method in Box 14.2 has been found to perform relatively well when using robust regression estimators such as those covered in Chapter 13. Limited studies suggest that it even performs reasonably well when using least squares regression, provided $n \geq 40$. This means that when testing at the .05 level, the actual Type I error probability will be less than .05. However, a criticism is that when using least squares, the probability of a Type I error can be substantially less than .05, suggesting that power might be relatively low. For smaller sample sizes, the function lsfitci (described in Section 7.3), which uses a modified percentile bootstrap method, provides more accurate results, but it is limited to testing for a zero slope coefficient for each parameter at the $\alpha = .05$ level. That is, for each j, it tests $H_0 : \beta_j = 0$, but it does not test Equation (14.6) and it does not control the probability of at least one Type I error among all the tests that are performed. The method in Box 14.2 can be seen to differ in a crucial way from the method in Section 7.3. For the special case of a single predictor, both of these methods can be used to test $H_0 : \beta_1 = 0$. Based on the goal of controlling Type I error probabilities, little is known about their relative merits for this special case.

14.5.2 S-PLUS Function regtest

The S-PLUS function

regtest(x,y,regfun=tsreg,nboot=600,alpha=.05,grp,nullvec)

tests the hypothesis that the regression coefficients are equal to some specified set of constants using the method in Box 14.2. By default, the hypothesis given by Equation (14.6) is tested. The argument nullvec can be used to set the hypothesized values to something other than zero. For example, if there are two predictors, the S-PLUS command

regtest(x,y,nullvec=c(2,4))

will test the hypothesis that $\beta_1 = 2$ and $\beta_2 = 4$. The argument grp can be used to indicate that a subset of the parameters is to be tested, which can include the intercept term. For example, when calling the function, setting grp=c(0,3) will test $H_0: \beta_0 = \beta_3 = 0$, assuming the argument nullvec is not specified. The command

regtest(x,y,grp=c(2,4,7))

will test $H_0: \beta_2 = \beta_4 = \beta_7 = 0$.

EXAMPLE. For the error term (ϵ) and each of two predictors, 30 observations were generated from a standard normal distribution and Y was determined by $Y = .33X_1 + 0X_2 + \epsilon$. Applying the standard F-test [given by Equation (14.8)] for $H_0: \beta_1 = \beta_2 = 0$, the significance level is .23. Now suppose the same data are used, except that when $X_1 > .8$, the error term is taken to be $X_1^2\epsilon$ rather than just ϵ. So in general the data follow the standard regression model, except when the first predictor is somewhat large, in which case there is heteroscedasticity. Now the significance level of the standard F-test is .90. If we use the S-PLUS function regtest with the default regression estimator (the Theil–Sen estimator), the significance level is .79. Situations arise where the Theil–Sen estimator substantially increases power when there is heteroscedasticity, but here it does not do much better than least squares. ■

EXAMPLE. This example illustrates that it is possible to reject the hypothesis that the slope parameters are all equal to zero [the hypothesis given by Equation (14.6)] but fail to reject for any of the individual slope parameters. Table 14.1 shows data from a study by Hald (1952) concerning the heat evolved, in calories per gram (Y), versus the amount of each of four ingredients in the mix: tricalcium aluminate (X_1), tricalcium silicate (X_2), tetracalcium alumino ferrite (X_3), and dicalcium silicate (X_4). Consider the first and third predictors and suppose we test $H_0: \beta_1 = \beta_3 = 0$ with the S-PLUS function regtest. Figure 14.8 shows the bootstrap estimates returned by the function regtest when using least squares.

Continued

TABLE 14.1 Hald's Cement Data

Y	X_1	X_2	X_3	X_4
78.5	7	26	6	60
74.3	1	29	15	52
104.3	11	56	8	20
87.6	11	31	8	47
95.9	7	52	6	33
109.2	11	55	9	22
102.7	3	71	17	6
72.5	1	31	22	44
93.1	2	54	18	22
115.9	21	47	4	26
83.8	1	40	23	34
113.3	11	66	9	12
109.4	10	68	8	12

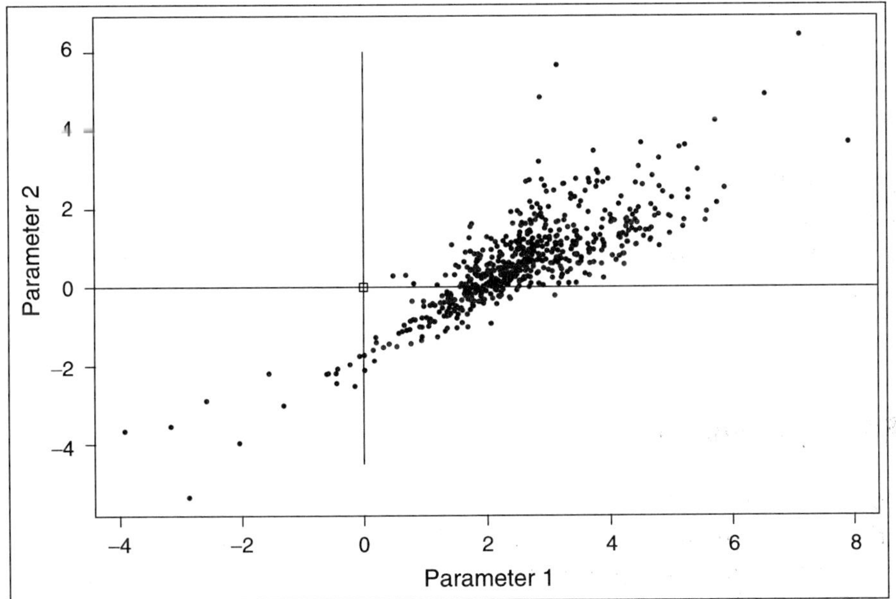

FIGURE 14.8 Bootstrap estimates of β_1 and β_3 using the cement data in Table 14.1.

EXAMPLE. (*Continued*) (That is, regfun=lsfit was used.) The significance level is .047, so in particular we would reject with $\alpha = .05$. However, if we test the individual slope parameters with the S-PLUS function lsfitci (see Section 7.3.1), the .95 confidence intervals for β_1 and β_3 are $(-0.28, 5.93)$ and $(-2.3, 3.9)$, respectively, so we fail to reject for either of the predictor variables.

Continued

EXAMPLE. (*Continued*) This phenomenon, where the omnibus test is significant but the individual tests are not, is known to occur when using the conventional F-test [given by Equation (14.8)] as well (e.g., Fairley, 1986). The reason is that when two estimators have a reasonably strong association, the resulting confidence region for the two parameters is a relatively narrow ellipse. Figure 14.8 shows a plot of the bootstrap estimates for the data under consideration and provides some indication of why this phenomenon occurs. The square, where the horizontal and vertical lines intersect, corresponds to the hypothesized values and, as is evident, is relatively far from the bulk of the bootstrap estimates. However, in order to reject $H_0: \beta_1 = 0$ at the .05 level, which corresponds to parameter 1 in Figure 14.8, 97.5% of the bootstrap estimates would need to be either above or below the horizontal line. To reject $H_0: \beta_3 = 0$, 97.5% of the bootstrap estimates would need to be to the right or to the left of the vertical line. Said another way, computing separate confidence intervals is essentially computing a rectangular confidence region for the two parameters under investigation. When the two estimators are approximately independent, this tends to give similar results to those obtained with the confidence region used by the S-PLUS function regtest, but otherwise it is possible for one method to reject when the other does not. ■

EXAMPLE. Repeating the previous example using the conventional F-test, we again reject $H_0: \beta_1 = \beta_3 = 0$, only now Student's T-test of $H_0: \beta_1 = 0$ rejects as well. Using regtest with the Theil–Sen estimator, we reject $H_0: \beta_1 = \beta_3 = 0$ once more and the S-PLUS function regci rejects $H_0: \beta_1 = 0$. ■

14.6 Identifying the Best Predictors

A problem that has received considerable attention is identifying a subset of predictors that might be used in place of the p predictors that are available. If p is large, the variance of the regression equation can be relatively large. If a subset of the p predictors can be identified that performs relatively well in some sense, not only do we get a simpler model, but we can get a regression equation with a lower variance. (For example, the variance of a sum of two variables—say, X_1 and X_2—is $\sigma_1^2 + \sigma_2^2 + 2\rho\sigma_1\sigma_2$, where σ_1 and σ_2 are the standard deviations associated with X_1 and X_2 and ρ is Pearson's correlation. So if $\rho > 0$, the variance of the sum is larger than the variance of the individual variables.) If we have 40 predictors, surely it would be convenient if a subset of, say, five predictors could be found that could be used instead. Of particular concern in this book is subset selection when using a robust regression estimator and the number of predictors is relatively small. This is an extremely complex problem that has received relatively little attention. Based on what is known, some type of bootstrap estimate of prediction error (which is formally defined later) appears to be relatively effective, and so this approach is described here. It is stressed, however, that this area is in need of more research and perhaps some alternative strategy will be found to have practical advantages over the approach used here.

Perhaps the best-known method for selecting a subset of the predictors is *stepwise regression*, but it is known that the method can be rather unsatisfactory (e.g., Montgomery & Peck, 1992, Section 7.2.3; Derksen & Keselman, 1992), and the same is true when using a related (forward selection) method, so for brevity these techniques are not covered here. (Also see Kuo & Mallick, 1998; Huberty, 1989; Chatterjee & Hadi, 1988; cf. A.J. Miller, 1990.) Generally, methods based on R^2 (given by Equation (14.7)), F (given by Equation (14.8)) and a homoscedastic approach based on

$$C_p = \frac{1}{\hat{\sigma}^2} \sum (Y_i - \hat{Y}_i)^2 - n + 2p,$$

called Mallows' (1973) C_p criterion, cannot be recommended either (A.J. Miller, 1990).[1] Another approach is based on what is called *ridge regression*, but it suffers from problems listed by Breiman (1995). Three alternative approaches are cross-validation, bootstrap methods (such as the .632 estimator described in Box 14.3), and the so-called *nonnegative garrote technique* derived by Breiman (1995). Efron and Tibshirani (1993, Chapter 17) discuss cross-validation, but currently it seems that some type of bootstrap method is preferable, so no details are given here. (Breiman's method is appealing when the number of predictors is large. For an interesting variation of Breiman's method, see Tibshirani, 1996.) Here, henceforth, attention is restricted to methods that allow heteroscedasticity.

Imagine you observe n pairs of values $(X_1,Y_1), \ldots, (X_n,Y_n)$, you estimate the regression line to be $\hat{Y} = b_0 + b_1X$, and now you observe a new X value, which will be labeled X_0. Based on this new X value you can, of course, estimate Y with $\hat{Y}_0 = b_0 + b_1X_0$. That is, you do not observe the value Y_0 corresponding to X_0, but you can estimate it based on past observations. *Prediction error* refers to the discrepancy between the predicted value of Y, $\hat{Y}_0$, and the actual value of Y, Y_0, if only you could observe it. One way of measuring the typical amount of prediction error is with

$$E[(Y_0 - \hat{Y}_0)^2],$$

the expected squared difference between the observed and predicted values of Y. Of course squared error might be replaced with some other measure, but for now this issue is ignored. As is evident, the notion of prediction error is easily generalized to multiple predictors. The basic idea is that via some method we get a predicted value for Y, which we label $\hat{Y}$, and the goal is to measure the discrepancy between $\hat{Y}_0$ (the predicted value of Y based on a future collection of X values) and the actual value of Y, Y_0, if it could be observed.

A simple estimate of prediction error is the so-called *apparent error rate*, meaning you simply average the error when predicting the observed Y values with $\hat{Y}$. To elaborate,

---

1 When using C_p, $\hat{\sigma}^2$ is usually taken to be $\sum r_i^2/(n-2)$, where the residuals are based on all of the predictors under consideration.

let $Q(Y,\hat{Y})$ be some measure of the discrepancy between an observation, Y, and its predicted value, $\hat{Y}$. So squared error corresponds to

$$Q(Y,\hat{Y}) = (Y - \hat{Y})^2.$$

The goal is to estimate the typical amount of error for future observations. In symbols, the goal is to estimate

$$\eta = E[Q(Y_0, \hat{Y}_0)],$$

the expected error between a predicted value for Y, based on a future value of X, and the actual value of Y, Y_0, if it could be observed. A simple estimate of η is the *apparent error*:

$$\hat{\eta}_{ap} = \frac{1}{n} \sum Q(Y_i, \hat{Y}_i).$$

So for squared error, the apparent error is

$$\hat{\eta}_{ap} = \frac{1}{n} \sum (Y_i - \hat{Y}_i)^2,$$

the average of the squared residuals.

A practical concern is that the apparent error is biased downward because the data used to come up with a prediction rule ($\hat{Y}$) are also being used to estimate error (Efron & Tibshirani, 1993). That is, it tends to underestimate the true error rate, η. The so-called .632 bootstrap estimator, described in Box 14.3, is designed to address this problem and currently seems to be a relatively good choice for identifying the best predictors. It is stressed, however, that more research is needed when dealing with this very difficult problem, particularly when using robust methods.

BOX 14.3 How to Compute the .632 Bootstrap Estimate of η

Generate a bootstrap sample as described in Box 14.2, except rather than resample n vectors of observations, as is typically done, resample $m < n$ vectors of observations instead. (Setting $m = n$, Shao, 1995, shows that the probability of selecting the correct model may not converge to 1 as n gets large.) Here, $m = 5\log(n)$ is used, which was derived from results reported by Shao (1995). Let $\hat{Y}_i^*$ be the estimate of Y_i based on the bootstrap sample, $i = 1, \ldots, n$. Repeat this process B times, yielding $\hat{Y}_{ib}^*$, $b = 1, \ldots, B$. Then an

Continued

BOX 14.3 *(Continued)* estimate of η is

$$\hat{\eta}_{\text{Boot}} = \frac{1}{nB}\sum_{b=1}^{B}\sum_{i=1}^{n} Q\left(Y_i, \hat{Y}_{ib}^*\right).$$

A refinement of $\hat{\eta}_{\text{Boot}}$ is to take into account whether a Y_i value is contained in the bootstrap sample used to compute $\hat{Y}_{ib}^*$. Let

$$\hat{\epsilon}_0 = \frac{1}{n}\sum_{i=1}^{n}\frac{1}{B_i}\sum_{b\in C_i} Q\left(Y_i, \hat{Y}_{ib}^*\right),$$

where C_i is the set of indices of the bth bootstrap sample not containing Y_i and B_i is the number of such bootstrap samples. Then the .632 estimate of the prediction error is

$$\hat{\eta}_{.632} = .368\hat{\eta}_{\text{ap}} + .632\hat{\epsilon}_0. \tag{14.9}$$

This estimator arises in part from a theoretical argument showing that .632 is approximately the probability that a given observation appears in a bootstrap sample of size n. [For a refinement of the .632 estimator given by Equation (14.9), see Efron & Tibshirani, 1997.]

14.6.1 S-PLUS function regpre

The S-PLUS function

```
regpre(x,y,regfun = lsfit, error = sqfun, nboot = 100,
    mval = round(5 log(length(y))),model = NA)
```

estimates prediction error for a collection of models specified by the argument model, which is assumed to have list mode. For example, imagine you have three predictors and you want to consider the following models:

$$Y = \beta_0 + \beta_1 X_1 + \epsilon,$$
$$Y = \beta_0 + \beta_1 X_1 + \beta_2 X_2 + \epsilon,$$
$$Y = \beta_0 + \beta_1 X_1 + \beta_3 X_3 + \epsilon,$$
$$Y = \beta_0 + \beta_1 X_1 + \beta_2 X_2 + \beta_3 X_3 + \epsilon.$$

Then the commands

```
model< - list( )
model[[1]]< - 1
model[[2]]< - c(1,2)
```

```
model[[3]]< − c(1,3)
model[[4]]< − c(1,2,3)
regpre(x,y,model=model)
```

result in estimating prediction error for the four models. For example, the values in model[[3]], namely, 1 and 3, indicate that predictors 1 and 3 will be used and predictor 2 will be ignored. The argument error determines how error is measured; it defaults to squared error. Setting error=absfun will result in using absolute error.

EXAMPLE. For the Hald data in Table 14.1, if we test the hypothesis given by Equation (14.5) with the conventional F-test in Section 14.5 [given by Equation (14.8)], the significance level is less than .001, indicating that there is some association between the outcome variable and the four predictors. However, for each of the four predictors, Student's T-tests of $H_0 : \beta_j = 0$ ($j = 1, 2, 3, 4$) have significance levels .07, .5, .9, and .84, respectively. That is, we fail to reject for any specific predictor at the .05 level, yet there is evidence of some association. Now consider the eight models

$$Y = \beta_0 + \beta_1 X_1 + \epsilon,$$
$$Y = \beta_0 + \beta_2 X_2 + \epsilon,$$
$$Y = \beta_0 + \beta_3 X_3 + \epsilon,$$
$$Y = \beta_0 + \beta_4 X_4 + \epsilon,$$
$$Y = \beta_0 + \beta_1 X_1 + \beta_2 X_2 + \epsilon,$$
$$Y = \beta_0 + \beta_1 X_1 + \beta_3 X_3 + \epsilon,$$
$$Y = \beta_0 + \beta_1 X_1 + \beta_4 X_4 + \epsilon,$$
$$Y = \beta_0 + \beta_1 X_1 + \beta_2 X_2 + \beta_3 X_3 + \beta_4 X_4 + \epsilon.$$

The estimated prediction errors for these models, based on least squares regression, are 142, 94.7, 224, 94, 7.6, 219, 9.6, and 638, respectively. Notice that the full model (containing all of the predictors) has the highest prediction error, suggesting that it is the least satisfactory model considered. Model 5 has the lowest prediction error, indicating that $Y = \beta_0 + \beta_1 X_1 + \beta_2 X_2 + \epsilon$ provides the best summary of the data among the models considered. ■

14.7 Detecting Interactions

This section illustrates the methods previously covered in this chapter in the context of detecting interactions. Consider some outcome variable, Y, and two predictors, X_1 and X_2. Roughly, the issue is whether knowing the value of X_2 modifies the association between Y and X_1. For example, for the reading study introduced in Section 13.1, there was interest in how a measure of orthographic ability (Y) is related to a measure of auditory analysis, X_1. A third variable in this study was a measure of sound blending (X_2). Does knowing the value of this third variable alter the association between Y and X_1; and if it does, how? More generally, there is interest in knowing whether a particular factor affects the magnitude of some effect size. Such factors are called *moderators* (e.g., Judd, Kenny, & McClelland, 2001).

Graphically, an interaction, in the context of regression, can be roughly described as follows. Let x_2 be any value of the second predictor variable, X_2. For example, x_2 could be the median of the X_2 values, but any other value could be used in what follows. Now consider the outcome variable, Y, and the first predictor (X_1), and imagine that we split the n pairs of points $(Y_1, X_{11}), \ldots, (Y_n, X_{n1})$ into two groups: those pairs for which the corresponding X_2 value is less than x_2 and those for which the reverse is true. No interaction means that the regression lines corresponding to these two groups are parallel. For example, if for the first group $Y = X_1^2 + \epsilon$ and for the second $Y = X_1^2 + 6 + \epsilon$, these regression lines are parallel and there is no interaction. But if for the second group $Y = X_1^2 + 8X_1 + 3 + \epsilon$, say, then the regression lines intersect (at $X_1 = -3/8$) and we say that X_2 modifies the association between Y and X_1.

A very popular method for checking and modeling interaction is with

$$Y = \beta_0 + \beta_1 X_1 + \beta_2 X_2 + \beta_3 X_1 X_2 + \epsilon. \tag{14.10}$$

That is, use the product of the two predictors to model interaction and conclude that an interaction exists if $H_0: \beta_3 = 0$ can be rejected. (Saunders, 1955, 1956, appears to be the first to suggest this approach to detecting interactions in regression; cf. Cronbach, 1987; Baron & Kenny, 1986.) This hypothesis can be tested using methods already covered. That is, for the observations $(Y_1, X_{11}, X_{12}), \ldots, (Y_n, X_{n1}, X_{n2})$, set $X_{i3} = X_{i1}X_{i2}$ and for the model $Y = \beta_0 + \beta_1 X_1 + \beta_2 X_2 + \beta_3 X_3 + \epsilon$, test $H_0: \beta_3 = 0$. However, it currently seems that a collection of tools is needed to address the issue of interactions in an adequate manner.

To add perspective on the product term just described, suppose we fix (or condition on) X_2. That is, we treat it as a constant. Then a little algebra shows that Equation (14.10) can be written as

$$Y = (\beta_0 + \beta_2 X_2) + (\beta_1 + \beta_3 X_2)X_1 + \epsilon.$$

So the intercept term becomes $(\beta_0 + \beta_2 X_2)$ and the slope for X_1 changes as a linear function of X_2. If $\beta_3 = 0$, then

$$Y = (\beta_0 + \beta_2 X_2) + \beta_1 X_1 + \epsilon.$$

That is, knowing X_2 alters the intercept term but not the slope. Said another way, if we split the data into two groups according to whether X_2 is less than or greater than some constant x_2, the corresponding regression lines will be parallel, consistent with the description given earlier.

There are various ways one might model interaction in a more general fashion. For example, one could replace Equation (14.10) with

$$Y = \beta_0 + \beta_1 X_1 + \beta_2 X_2 + g(X_1 X_2) + \epsilon \tag{14.11}$$

for some function g of the product. Equation (14.10) corresponds to the special case $g(X_1X_2) = \beta_3X_1X_2$. A simple method for examining whether some function of the product might be useful is to create a partial residual plot as described in Section 14.2.5. This assumes, of course, that Y has a linear association with X_1 and X_2, and for the reading data, for example, we have already seen that this assumption seems dubious.

Before continuing, an exploratory tool is described that is useful when checking for a modifier variable; the method is based on a slight extension of the running-interval smoother. The basic idea is to plot the strength of the association between Y and X_1 as a function of X_2. Consider the ith observed value for X_2, X_{i2}. Briefly, identify all of the X_{j2} values that are close to X_{i2} in the manner outlined in Section 14.2. Then for the corresponding pairs of points, (X_{1j}, Y_j), compute some correlation coefficient and label it $\hat{\theta}_i$. Repeat this process for $i = 1, \ldots, n$ and then plot the pairs $(X_{i2}, \hat{\theta}_i)$. The computations are performed by the S-PLUS function runcor described in Section 14.7.1 (cf. Doksum, Blyth, Bradlow, Meng, & Zhao, 1994).

EXAMPLE. Some of the strategies that might be used to detect modifier variables are illustrated with data from a reading study previously mentioned. Again, we take as the outcome variable a measure of orhographic ability (Y), only this time the first predictor (X_1) is a measure of sound blending and the issue is whether a measure of phonological awareness (X_2) modifies the association between Y and X_1. First consider the model given by Equation (14.10), where $X_3 = X_1X_2$. If we apply the conventional F-test to $H_0 : \beta_1 = \beta_2 = \beta_3 = 0$, the significance level is .044, but the significance levels for the three individual slope parameters are .46, .67, and .92, respectively. Estimating prediction error as described in Section 14.6, among the models

$$Y = \beta_0 + \beta_1X_1 + \epsilon,$$
$$Y = \beta_0 + \beta_2X_2 + \epsilon,$$
$$Y = \beta_0 + \beta_3X_3 + \epsilon,$$
$$Y = \beta_0 + \beta_1X_1 + \beta_2X_2 + \epsilon,$$
$$Y = \beta_0 + \beta_1X_1 + \beta_2X_2 + \beta_3X_3 + \epsilon,$$

the first has the lowest prediction error (using least squares regression and squared error). This suggests that the first predictor is important but the others are not. But before making any final decisions about the associations under study, it is important to see whether other methods paint a different picture. ■

To get some feel for the data, Figure 14.9 shows a smooth created by the S-PLUS function runm3d, described in Section 14.2.4, when using X_1 and X_2 to predict Y. Notice that the regression surface does not appear to be well approximated by a plane. That is, assuming a linear association might provide an inadequate representation of the data, contrary to what is typically assumed when investigating interactions.

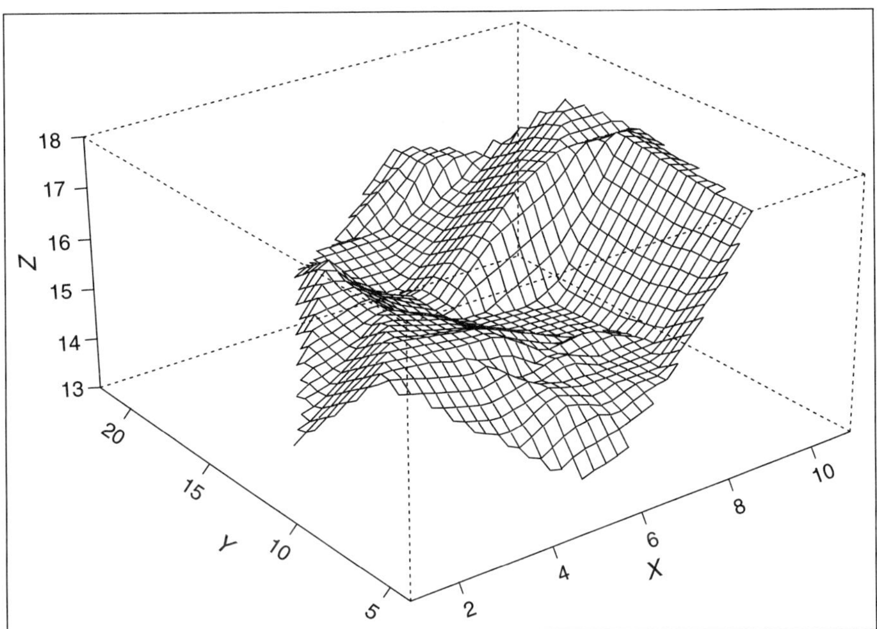

FIGURE 14.9 A smooth for two predictors used in a reading study.

Next we examine the association between X_1 and Y, ignoring X_2 for the moment. Typically it is assumed that this association is linear and so one goal is to get some sense of whether this assumption is reasonable for the problem at hand. Figure 14.10 shows a smooth of these two variables that suggests a nonlinear association. Using the S-PLUS function linchk (described in Section 14.4.3) to check for linearity, with the data split at $X_1 = 7$, the .95 confidence interval for the difference between the slopes is $(-1.5, 0)$ with a significance level of .027. The estimated slopes, using Theil–Sen, are 0 (for $X_1 < 7$) and .5. Checking for linearity with the method in Box 14.1, the test statistic reported by the S-PLUS function lintest is 7.76 and the .05 critical value is 4.27, again suggesting that the hypothesis of a linear association is false. One way to proceed from here is to incorporate a measure of association that is sensitive to monotonic relationships that are not necessarily linear. Two classic approaches are available, but the details are postponed until Chapter 15. (For an analysis of these data based on one of these approaches, see the example in Section 15.12.) It is instructive, however, to proceed under the assumption that problems with nonlinearity can ignored.

Next we plot the smooth relating the (percentage bend) correlation between Y and X_1 as a function of X_2. Figure 14.11 shows the results based on the S-PLUS function runcor (described in Section 14.7.1). The + indicates the location of the median of X_2, and the quartiles are indicated by a |. Note that the smooth is fairly horizontal on the left but that there is some indication that it begins to increase around $X_2 = 12$ or perhaps $X_2 = 15$. To provide perspective, the S-PLUS function runmean2g is used to simultaneously plot a smooth between Y and X_1 based on points corresponding to $X_2 \leq 12$ as well as for $X_2 > 12$. That is, runmean2g creates

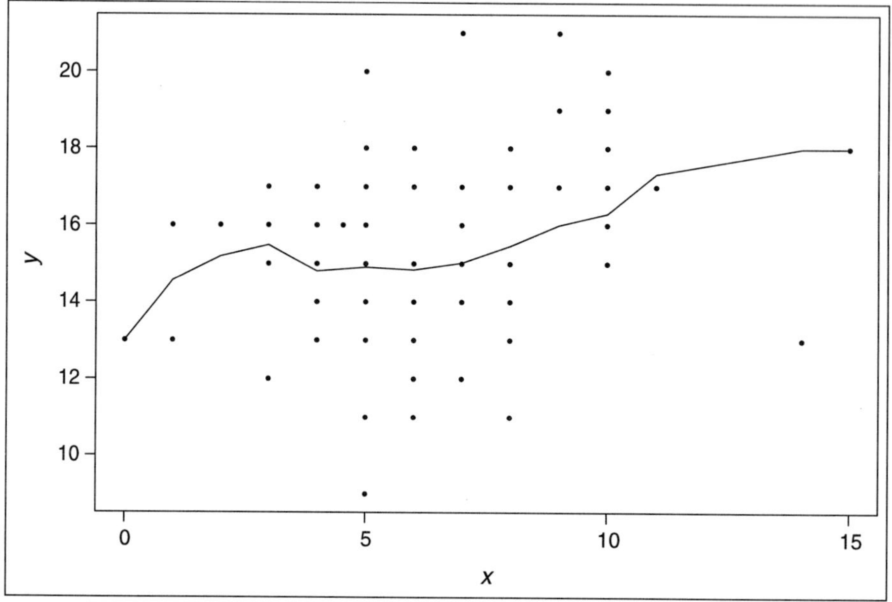

FIGURE 14.10 A smooth based on the first of the two predictors used in the reading example.

a smooth when predicting Y based on X_1 for two groups. The first group corresponds to $X_2 \leq 12$, and the second is simply $X_2 > 12$. (To facilitate this part of the analysis, the S-PLUS function regi, described in Section 14.7.1, is provided.) Figure 14.12 shows the results. Notice that for the second group (with points indicated by a +), there is only one point with $X_1 < 4$. This means that the left portion of the smooth for the second group might be relatively inaccurate, for reasons mentioned in Section 14.1. Focusing on the X values in Figure 14.12 that are greater than or equal to 4, the two smooths appear to be fairly parallel, which suggests that there is no interaction. Also, curvature between Y and X_1 now seems to be minimal. Testing the hypothesis of equal slopes with the S-PLUS function reg2ci using the Theil–Sen estimator, the significance level is .47, so we fail to reject and it might seem that it is safe to conclude that there is no interaction. The estimates of the slopes are 0 and .33.

There are, however, several concerns. The first is that for convenience, Theil–Sen was used. Checking for outliers using the MGV method and comparing the MGV regression estimate of the slope to Theil–Sen suggests that the slopes for the two groups are even more similar than indicated: The estimated slope for the second group drops slightly to .29. This might seem to support the conclusion that there is no interaction, but the MGV estimator can have a much smaller standard error, which could result in rejecting the hypothesis of equal slopes. Another concern is that the data were split according to whether $X_2 \leq 12$. Is this the optimal split for detecting an interaction? Empirical results previously described suggest splitting the data in this manner, but perhaps a slightly different split would make a practical difference. A third concern, which is always an issue when failing to reject, is whether power is sufficiently high to accept the null hypothesis of equal slopes. Addressing this problem for the

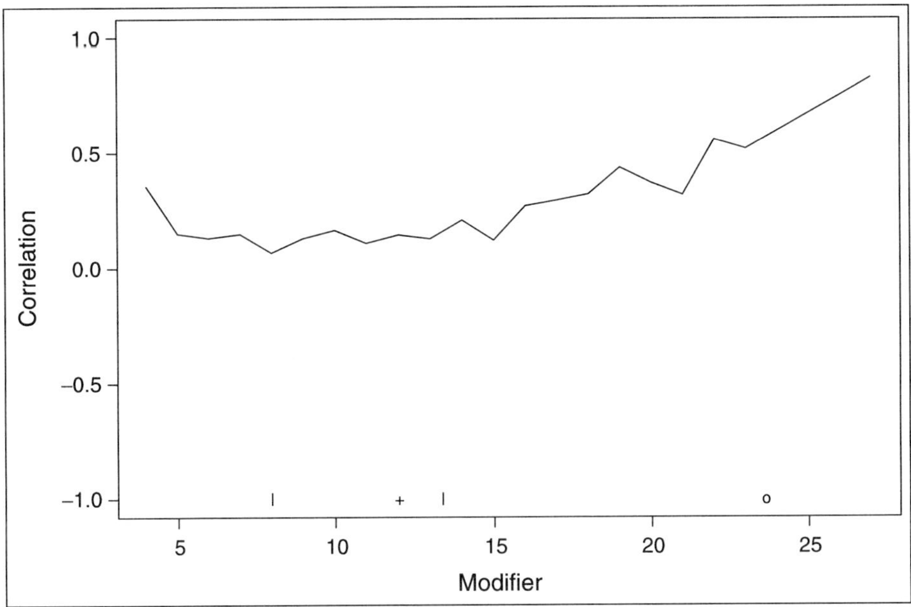

FIGURE 14.11 A smooth based on the reading data that shows the percentage bend correlation between Y and X_1 as a function of X_2.

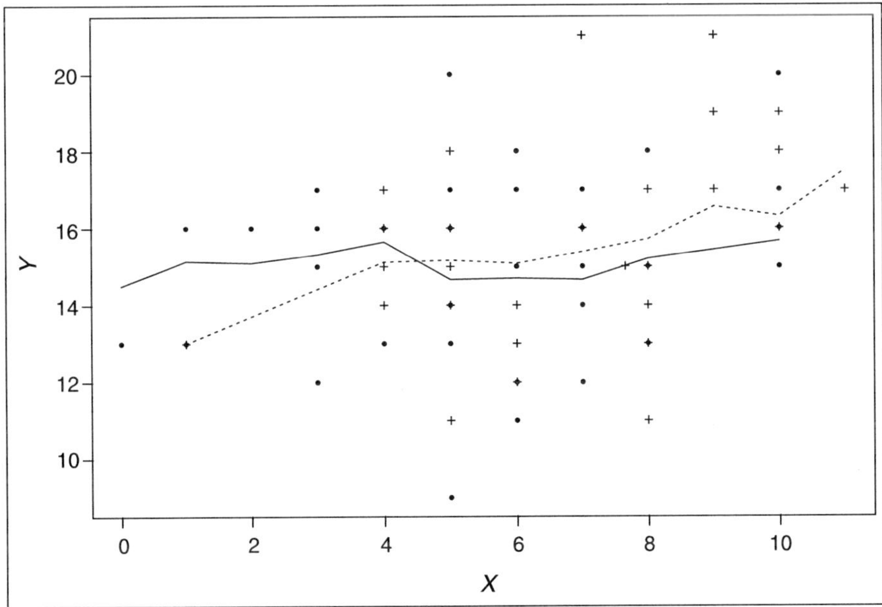

FIGURE 14.12 A smooth, based on the reading data, between Y and X_1. The solid line is based on points with $X_2 \leq 12$ and the dashed line is for $X_2 > 12$.

problem at hand is difficult at best. And finally, it is unclear whether the apparent curvilinear association between X_1 and Y can be ignored. (In Section 15.12, we will see that using a measure of association that is sensitive to this curvilinearity makes a practical difference.)

EXAMPLE. The same two predictors in the previous example are used again, but now the outcome variable of interest (Y) is a measure of word accuracy and we consider whether phonological awareness modifies the association between sound blending and Y. Here we merely split the data into two groups according to whether X_2 is less than or equal to its median. Checks on the second group indicate that a single unusual X_1 value is seriously affecting the estimated slope based on both the Theil–Sen estimator and least squares. By restricting the range in this second group so as to eliminate this seemingly aberrant point, the .95 confidence interval for the difference between the slopes, using Theil–Sen, is (0, 4) with a significance level of .057 when testing the hypothesis that the slopes are equal. That is, for this particular split of the data, we are unable to reject at the .05 level the hypothesis that the two groups have unequal slopes. But we come fairly close to rejecting and concluding that phonological awareness modifies the association between sound blending and word accuracy. Testing the hypothesis that in Equation (14.10), $\beta_3 = 0$, the significance level is .14 using least squares and the conventional Student's T-test. ■

14.7.1 S-PLUS Functions runcor, regi, and cori

The S-PLUS function

```
runcor(x, y, z, fr = 1, corflag = F, corfun = pbcor, plotit = T, rhat = F)
```

plots the correlation between x and y as a function of the data stored in the argument z. Setting the argument corflag to T (for true) results in using Pearson's correlation; otherwise the function uses the correlation specified by the argument corfun, which defaults to the percentage bend correlation. Setting the argument rhat to T, the function returns the values of the estimated correlations corresponding to each value stored in z. For example, if the first value in z is 6, then the first value returned in rhat is the correlation between x and y for points for which the corresponding z values are close to 6. As usual, fr is the span and plotit=F will suppress the plot.

The S-PLUS function

```
regi(x, y, z, pt=median(z),fr=.8,est=mom,regfun=tsreg,testit=F,...)
```

creates two smooths. The first is based on the x and y values for which the corresponding value for z is less than the value stored in the argument pt. By default, pt is taken to be the median of the z values. The other smooth is based on the x and y values for which the corresponding value for z is greater than pt. The smooth is created with the measure of location given by the argument est, which defaults to MOM. If testit=T is used, the slopes (and intercepts) of the two regression lines, based on the function regfun (which defaults to the Theil–Sen estimator), are compared. (This is done by splitting the data for you and calling the function reg2ci.)

The S-PLUS function

```
cori(x, y, z, pt=median(z),fr=.8,est=mom,corfun=pbcor,testit=F,...)
```

is like regi, except that setting testit to T causes the correlations to be compared rather than the regression slopes. By default, the percentage bend correlation is used, but the argument corfun can be used to specify some other measure of association. (For example, corfun=spear results in using Spearman's correlation, a measure of association covered in Chapter 15.)

14.8 ANCOVA

This section takes up a topic known as the *analysis of covariance*, or ANCOVA. As was the case in Chapter 8, the goal is to compare two independent groups in terms of some measure of location, but here an additional goal is to take into account the information provided by some predictor variable called a *covariate*. As a simple illustration, imagine that men and women are compared in terms of their typical score on some mathematics aptitude test and it is found that the typical male scores higher than the typical woman. However, test scores might be related to previous training in mathematics, and if we compare men and women having comparable training, now women might score higher than men. Here the covariate is previous training.

There is a standard ANCOVA method that is based on least squares regression. (See Huitema, 1980; Rutherford, 1992.) Not only does the method assume normality and homoscedasticity, but it assumes that the regression lines for the two groups being compared are parallel. As previously mentioned, there are many methods for testing the assumption that regression lines are parallel. But it is unknown how to determine whether they have enough power to detect nonparallel lines in situations where violating this assumption has practical consequences. Some least squares methods that allow nonparallel regression lines are available, a classic example being the Johnson–Neyman method (P. Johnson & Neyman, 1936). Here, however, attention is focused on modern robust methods. Unlike conventional approaches, it is not assumed that the regression line is straight. Rather, a smooth is used to approximate the regression lines and then typical values for Y, given some value for the covariate, are compared using methods covered in previous chapters.

For the jth group, let $m_j(X)$ be some population measure of location associated with Y given X. So for the first group, $m_1(6)$ might represent the population mean of Y given that $X = 6$, or it could be the population value for MOM or a 20% trimmed mean. Given X, the problem is determining how the typical value of Y in the first group compares to the typical value in the second. In the Pygmalion study introduced in Section 14.3.1, the goal might be to determine how the 20% trimmed mean of the experimental group compares to the trimmed mean of the control group, given that a student's IQ reasoning pretest score is $X = 90$. Of course, a more general goal is to determine how the trimmed means compare as X varies. A common strategy is to assume that a straight regression line can be used to predict Y from X. In the present notation, it is assumed that for the jth group,

$$m_j(X) = \beta_{0j} + \beta_{1j}X_{1j},$$

$j = 1, 2$. However, when working with robust regression methods, currently this approach to ANCOVA has been found to be relatively unsatisfactory when testing hypotheses.

A more satisfactory approach is based in part on the running-interval smooth described in Section 14.1. So in particular, it is *not* assumed that a straight line provides an adequate summary of the data; in the event it does, all indications are that the method described here continues to perform relatively well in terms of both Type I errors and power (Wilcox, 1997b). Even under normality, the conventional ANCOVA method appears to have only a minor advantage.

To elaborate on how the method is applied, first assume that an X value has been chosen with the goal of computing a confidence interval for $m_1(X) - m_2(X)$. For the jth group, let X_{ij}, $i = 1, \ldots, n_j$ be values of the predictors that are available. The value $m_j(X)$ is estimated as described in Section 14.2. That is, for fixed j, estimate $m_j(X)$ using the Y_{ij} values corresponding to the X_{ij} values that are close to X. Let $N_j(X)$ be the number of observations used to compute the estimate of $m_j(X)$. That is, $N_j(X)$ is the number of points in the jth group that are close to X, which in turn is the number of Y_{ij} values used to estimate $m_j(X)$. Provided that both $N_1(X)$ and $N_2(X)$ are not too small, a reasonably accurate confidence interval for $m_1(X) - m_2(X)$ can be computed using methods already covered. When comparing the regression lines at more than one design point, confidence intervals for $m_1(X) - m_2(X)$, having simultaneous probability coverage approximately equal to $1 - \alpha$, can be computed as described in Chapter 12.

14.8.1 S-PLUS Functions ancova, ancpb, and ancboot

The S-PLUS function

```
ancova(x1,y1,x2,y2,fr1=1,fr2=1,tr=0.2,alpha=0.05,plotit=T,pts = NA)
```

performs ANCOVA with trimmed means as just described. The arguments x1, y1, x2, y2, tr, and alpha have their usual meaning. The arguments fr1 and fr2 are the spans used for groups 1 and 2, respectively. The argument pts can be used to specify the X values at which the two groups are to be compared. For example, pts=12 will result in comparing the trimmed mean for group 1 (based on the y1 values) to the trimmed mean of group 2 given that $X = 12$. If there is no trimming, the null hypothesis is $H_0 : E(Y_1|X = 12) = E(Y_2|X = 12)$, where Y_1 and Y_2 are the outcome variables of interest corresponding to the two groups. Using pts=c(22,36) will result in testing two hypotheses. The first is $H_0 : m_1(22) = m_2(22)$ and the second is $H_0 : m_1(36) = m_2(36)$. If no values for pts are specified, then the function picks five X values and performs the appropriate tests. The values that it picks are reported in the output illustrated later. Generally, this function controls FWE using the method in Section 12.6. If plotit=T is used, the function also creates a scatterplot and smooth for both groups, with a + and a dashed line indicating the points and the smooth, respectively, for group 2.

The function

```
ancpb(x1,y1,x2,y2,est=mom,pts=NA,fr1=1,fr2=1,nboot=599,plotit=T,...)
```

is like the S-PLUS function ancova, except that only a percentile bootstrap method is used to test hypotheses and by default the measure of location is MOM. Now FWE is controlled as described in Section 12.7.3. In essence, the function creates groups based on the values in pts; in conjunction with the strategy behind the smooth, it creates the appropriate set of linear contrasts and then calls the function pbmcp, described in Section 12.7.4.

Finally, the function

ancboot(x1,y1,x2,y2,fr1=1,fr2=1,tr=0.2,nboot=599,pts=NA,plotit=T)

compares trimmed means using a bootstrap-t method. Now FWE is controlled as described in Section 12.7.1.

EXAMPLE. The ANCOVA methods described in this section are illustrated with the Pygmalion data described in Section 14.3.1. The goal is to compare posttest scores for the two groups, taking into account the pretest scores. If the data for the experimental group are stored in the S-PLUS matrix pyge, with the pretest scores in column 1, and the data for the control are stored in pygc, the command

ancova(pyge[,1],pyge[,2],pygc[,1],pygc[,2])

returns

```
  X  n1  n2       DIF     TEST        se     ci.low    ci.hi
 72  12  63  13.39103 1.848819  7.243016  -9.015851 35.79790
 82  16  68  14.79524 1.926801  7.678655  -8.211174 37.80165
101  14  59  22.43243 1.431114 15.674806 -26.244186 71.10905
111  12  47  23.78879 1.321946 17.995286 -35.644021 83.22161
114  12  43  21.59722 1.198906 18.014112 -37.832791 81.02724
```

The first column, headed by X, says that posttest scores are being compared given that pretest scores (X) have the values 72, 82, 101, 111, and 114. The sample sizes used to make the comparisons are given in the next two columns. For example, when $X = 72$, there are 12 observations being used from the experimental group and 63 from the control. That is, there are 12 pretest scores in the experimental group and 63 values in the control group that are close to $X = 72$. The column headed DIF contains the estimated difference between the trimmed means. For example, the estimated difference between the posttest scores, given that $X = 72$, is 13.39. The last two columns indicate the ends of the confidence intervals. These confidence intervals are designed so that FWE is approximately α. The critical value is also reported and is 3.33 for the situation here. All of the confidence intervals contain zero, and none of the test statistics exceeds the critical value, so we fail to detect any differences between posttest scores taking into account the pretest scores of these individuals. ■

Figure 14.13 shows the plot created by the S-PLUS function ancova. Note that $X = 72$ appears near the center of the plot, yet this is the smallest X value used.

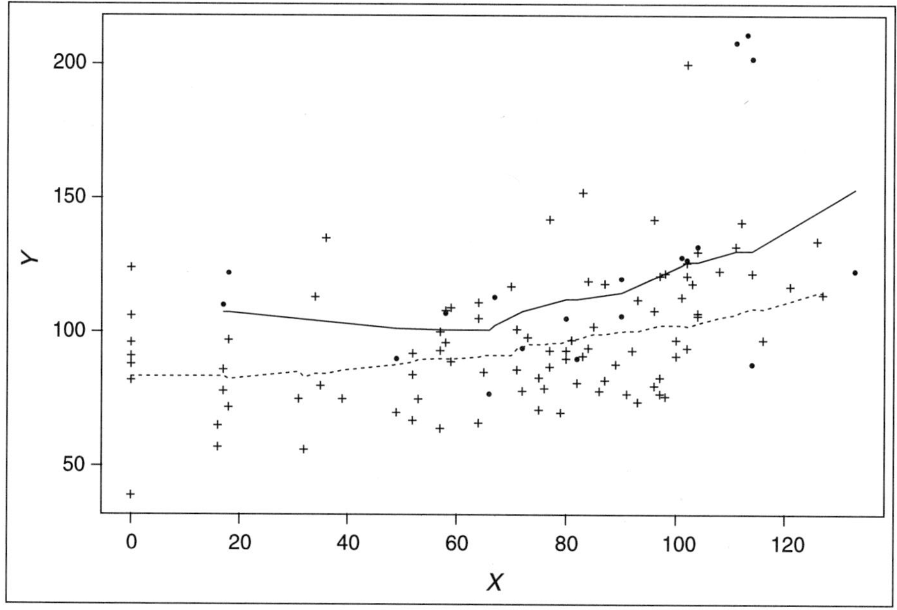

FIGURE 14.13 A plot created by the S-PLUS function ANCOVA.

The reason is that for the experimental group, there are only six cases where $X < 72$. If for example we try to compare the posttest scores given that the pretest scores are $X = 70$, there are too few individuals in the experimental group with X close to 70 to make meaningful comparisons.

If we apply the function ancpb instead, a portion of the output is

```
$mat:
        X  n1  n2
[1,]   72  12  63
[2,]   82  16  68
[3,]  101  14  59
[4,]  111  12  47
[5,]  114  12  43


      con.num     psihat  sig.level  sig.crit   ci.lower  ci.upper
[1,]        1  12.334699 0.05008347    0.0102  -4.507937  35.28625
[2,]        2   7.907925 0.10350584    0.0127  -7.200000  57.44683
[3,]        3   8.092476 0.12020033    0.0169  -5.282468  62.18519
[4,]        4   6.917874 0.18697830    0.0250  -7.025000  63.63889
[5,]        5   5.388889 0.23706177    0.0500  -5.887805  55.28488
```

Again we fail to find any differences, but note that the length of the confidence intervals are generally substantially shorter than the confidence intervals from the function ancova. For example, with $X = 114$, the length of the confidence interval here is $(55.3+5.9) = 61.2$. In contrast, the length of the confidence interval reported by ancova is 118.8, and the ratio of the lengths is .51. Moreover, MOM, used in conjunction with the percentile bootstrap, can provide accurate probability coverage in situations where trimmed means (with trimming less than 20%) and nonbootstrap methods are unsatisfactory.

14.9 Exercises

1. For the predictor X_2 in Table 14.1, MAD/.6745=20.756. If in Section 14.2 you take the span to be $f = 1$, what would be the estimate of Y given that $X_2 = 250$, based on a 20% trimmed mean? What would be the estimate based on MOM?
2. Exercises 1 and 2 in Chapter 13 report data on 29 lakes in Florida. Assuming that you want to predict Y (the mean annual nitrogen concentration) given TW (water retention time), plot a smooth, and comment on whether a straight regression line is advisable.
3. Repeat the previous exercise, only now use NIN (the average influent nitrogen concentration) to predict Y.
4. Again referring to Exercises 1 and 2 in Chapter 13, check for any outliers among the TW and NIN values using the MVE method, eliminate any outliers that are found, and plot a smooth for predicting Y using the S-PLUS functions rungen and runmean in Section 14.2.1. Compare the results to smooths you get when the outliers are not eliminated. Comment on why retaining the outliers among the TW and NIN values might have an adverse effect on fitting a regression plane to the data.
5. For the data in Exercises 1 and 2 of Chapter 13, check for curvature using a partial residual plot.
6. For the data in Exercises 1 and 2 of Chapter 13, eliminate any outliers among the predictor values found by the MVE method, and for the remaining data, test $H_0: \beta_1 = \beta_2 = 0$ using the methods in Section 14.5.
7. Table 14.2 contains a portion of the data reported by Thompson and Randall-Maciver (1905) dealing with skull measurements of male Egyptians from different time periods. Here, only the data from 4000 bc and 150 ad are reported. Pool the data from both periods and create a smooth using runmean, rungen, and lowess. What do these smooths suggest about the association between X and Y? Check this possibility with the S-PLUS function lintest in Section 14.4.3.
8. For the data in Table 14.2, create a smooth using the 4000 bc data only and compare it to the smooth for the 150 ad data.
9. Compare the regression slopes in the previous exercise using the S-PLUS function reg2ci. Verify that the significance level is .04 when using the default settings.
10. Repeat the previous exercise, only now use the S-PLUS function ancova to compare the regression lines. At which X values do you get a significant difference with $\alpha = .05$?

TABLE 14.2 Skull Measurements

X (4000 BC)	138 131 132 132 143 137 130 136 134 134 138 121 129 136 140 134 137 133 136 131 133 135 124 134 130 135 132 129 136 138
X (150 AD)	123 131 126 134 127 138 138 126 132 135 120 136 135 134 135 134 125 135 125 129 136 129 126 124 127 125 128 135 129 133
Y (4000 BC)	89 92 99 96 100 89 108 93 102 99 95 95 109 100 100 97 103 93 96 101 102 103 93 103 104 100 93 106 114 101
Y (150 AD)	91 95 91 92 86 101 97 92 99 92 95 101 95 93 96 95 99 96 92 89 92 97 88 91 97 85 81 103 87 97

X is basibregmatic height of skull; Y is basialveolar length of skull.

11. Data were generated from normal distributions for X_1, X_2, and ϵ. Setting $X_3 = X_1X_2$, consider the following models:

$$Y = \beta_0 + \beta_1 X_1 + \epsilon,$$
$$Y = \beta_0 + \beta_1 X_2 + \epsilon,$$
$$Y = \beta_0 + \beta_1 X_3 + \epsilon,$$
$$Y = \beta_0 + \beta_1 X_1 + \beta_2 X_2 + \epsilon,$$
$$Y = \beta_0 + \beta_1 X_1 + \beta_2 X_2 + \beta_3 X_3 + \epsilon.$$

The output from the S-PLUS function regpre (in Section 14.6.1) is

```
$estimates:
     apparent.error boot.est   err.632
[1,]      1.4623288 1.639636 1.625111
[2,]      1.2262127 1.364102 1.353811
[3,]      1.0169318 1.136070 1.124203
[4,]      1.2100394 1.440440 1.422938
[5,]      0.8807802 1.121439 1.095643
```

Based on this output, which model appears to be the best summary of the data?

12. In the previous exercise, imagine that you fail to reject $H_0: \beta_3 = 0$. Describe some reasons why it might be erroneous to conclude that there is no interaction.

13. An exploratory method for dealing with interactions is to assume that the model $Y = \beta_0 + \beta_1 X_1 + \beta_2 X_2 + \beta_3 X_1 X_2 + \epsilon$ is true and to use a partial residual plot to check the adequacy of the third term ($\beta_3 X_1 X_2$), assuming that Y has a linear association between $\beta_1 X_1$ and $\beta_2 X_2$. Describe some reasons why this approach might be unsatisfactory.

RANK-BASED AND NONPARAMETRIC METHODS

This chapter covers basic nonparametric and so-called rank-based methods. Generally, the techniques covered here provide a different and interesting perspective on how groups differ and how variables are related versus the methods covered in previous chapters. Often the methods in this chapter are recommended for dealing with problems that arise when sampling from nonnormal distributions, and so one goal is to try to convey their relative merits versus techniques covered in previous chapters. Many conventional rank-based methods suffer from serious practical problems when comparing groups with different distributions, but substantial progress has been made regarding how to overcome these difficulties.

15.1 Comparing Two Independent Groups

This section describes methods for comparing two independent groups. We begin with a classic technique, outline its practical problems, and then cover modern methods for dealing with these issues.

15.1.1 Wilcoxon–Mann–Whitney Test

The standard rank-based method for comparing two independent groups is called the Wilcoxon–Mann–Whitney (WMW) test. It was originally derived by Wilcoxon (1945), and later it was realized that Wilcoxon's method was the same as a procedure proposed by Mann and Whitney (1947). To describe the basic goal, imagine you randomly sample an observation from the first group and do the same for the second. Now temporarily assume that these two observations cannot have equal values. For example, if the observation from the first group is 6, it is assumed that there is zero probability that you will get 6 when sampling from the other group. Let p be the probability that a randomly sampled observation from the first group is less than a randomly sampled observation from the second. If the groups do not differ in any way, meaning that they have identical distributions, then $p = .5$. Also, p provides a perspective (a measure of effect size) on how groups differ not provided by any of the techniques covered in previous chapters. So a general goal of interest is making inferences about p based on observations we make. This includes estimating p, computing a confidence interval for p, and testing the hypothesis

$$H_0 : p = .5. \tag{15.1}$$

First consider the problem of estimating p. For illustrative purposes suppose we observe

Group 1:	30, 60, 28, 38, 42, 54
Group 2:	19, 21, 27, 73, 71, 25, 59, 61

Now focus on the first value in the first group, 30, and notice that it is less than four of the eight observations in the second group. So a reasonable estimate of p is 4/8. In a similar manner, the second observation in the first group is 60; it is less than three of the values in the second group, so a reasonable estimate of p is 3/8. These two estimates of p differ, and a natural way of combining them into a single estimate of p is to average them. More generally, if we have n_1 observations in group 1 and n_2 observations in group 2, focus on the ith observation in the first group and suppose this value is less than V_i of the observation in group 2. So based on the ith observation in group 1, an estimate of p is V_i/n_2, and we have n_1 estimates of p: $V_1/n_2, \ldots, V_{n_1}/n_2$. To combine these n_1 estimates of p into a single estimate, average them, yielding

$$\hat{p} = \frac{1}{n_1 n_2} \sum V_i. \tag{15.2}$$

As is usually done, let

$$U = n_1 n_2 \hat{p}. \tag{15.3}$$

The quantity U is called the Mann–Whitney U statistic; typically a test of $H_0 : p = .5$ is described in terms of U. If $p = .5$, it can be shown that $E(U) = n_1 n_2/2$. More generally,

$$E\left(\frac{U}{n_1 n_2}\right) = p.$$

Next, consider the problem of estimating VAR(U), the squared standard error of U. If we assume there are no tied values and that both groups have identical distributions, the classic estimate of the standard error can be derived. (By "no tied values" it is meant that each observed value occurs only once. So if we observe the value 6, for example, it never occurs again among the remaining observations.) The expression for VAR(U) is

$$\sigma_u^2 = \frac{n_1 n_2 (n_1 + n_2 + 1)}{12}.$$

This means that the null hypothesis given by Equation (15.1) can be tested with

$$Z = \frac{U - n_1 n_2/2}{\sigma_u}, \tag{15.4}$$

which has, approximately, a standard normal distribution when the assumptions are met and H_0 is true. In particular, reject if

$$|Z| \geq Z_{1-\alpha/2},$$

where $Z_{1-\alpha/2}$ is the $1-\alpha/2$ quantile of a standard normal distribution. Hodges, Ramsey, and Wechsler (1990) suggest estimating the p-value as outlined in Box 15.1.

BOX 15.1 Computing the p-Value of the Wilcoxon–Mann–Whitney Test

Let

$$y = \frac{1}{\sigma_u}\left(U + 0.5 - \frac{n_1 n_2}{2}\right),$$

$$k = \frac{20 n_1 n_2 (n_1 + n_2 + 1)}{n_1^2 + n_2^2 + n_1 n_2 + n_1 + n_2},$$

$$S = y^2,$$

$$T_1 = S - 3,$$

$$T_2 = \frac{155S^2 - 416S - 195}{42},$$

$$c = 1 + \frac{T_1}{k} + \frac{T_2}{k^2}.$$

If cy is negative, the one-sided p-value is

$$P(Z \leq cy),$$

where Z is a standard normal random variable. So the p-value can be determined from Table 1 in Appendix B. If cy is positive, the one-sided p-value is

$$1 - P(Z \leq cy).$$

The two-sided p-value is

$$2[1 - P(Z \leq |cy|)].$$

EXAMPLE. Continuing the illustration using the data just following Equation (15.1), it can be seen that $\hat{p} = .479$, so $U = 23$ and

$$Z = \frac{23 - 24}{7.75} = -0.129.$$

With $\alpha = .05$, the critical value is 1.96; $|Z|$ is less than 1.96, so fail to reject. ■

Sometimes the Wilcoxon–Mann–Whitney test is described as a method for comparing medians. However, it is relatively unsatisfactory for this purpose because it is not based on a direct estimate of the population medians. For example, there are situations where power decreases as the difference between the population medians increases, and confidence intervals for the difference cannot be computed (Kendall & Stuart, 1973; Hettmansperger, 1984).

Another characterization of the Wilcoxon–Mann–Whitney method is that it tests the hypothesis that two groups have identical distributions. In symbols, if F_1 and F_2 are the distributions of the two groups being compared, the goal is to test

$$H_0 : F_1(x) = F_2(x), \tag{15.5}$$

which says that for any x, the probability that a randomly sampled observation is less than or equal to x is the same for both groups. A more accurate description of the method in Box 15.1 is that it approximates the p-value of the Wilcoxon–Mann–Whitney test when the goal is to test Equation (15.5) as opposed to testing $H_0 : p = .5$.

When tied values occur with probability zero and the goal is to test $H_0 : F_1(x) = F_2(x)$, the probability of a Type I error can be controlled exactly by computing a critical value as described, for example, in Hogg and Craig (1970, p. 373). Let

$$W = U + \frac{n_2(n_2 + 1)}{2} \tag{15.6}$$

and suppose H_0 [given by Equation (15.5)] is rejected if W is sufficiently large or small. If H_0 is rejected when

$$W \le c_L$$

or when

$$W \ge c_U,$$

where c_L is read from Table 13 in Appendix B and

$$c_U = n_2(n_2 + n_1 + 1) - c_L,$$

then the actual probability of a Type I error will not exceed .05 under random sampling.

15.1.2 S-PLUS Function wmw

Using the data stored in the S-PLUS variables x and y, the S-PLUS function

wmw(x,y)

computes the significance level of the Wilcoxon–Mann–Whitney test as described in Box 15.1.

15.1.3 Handling Ties and Heteroscedasticity

A practical concern is that if groups differ, then under general circumstances the wrong standard error is being used by the Wilcoxon–Mann–Whitney test in

Equation (15.4), which can result in relatively poor power and an unsatisfactory confidence interval for p. Said another way, if groups have different distributions, generally σ_u^2 is the wrong standard error for U. Another problem is how to handle tied values. Currently there are two general approaches to both of these problems that appear to be relatively effective.

First consider the problem of tied values and note that if we randomly sample a single observation from both groups, there are three possible outcomes: the observation from the first group is greater than the observation from the second, the observations have identical values, and the observation from the first group is less than the observation from the second. The probabilities associated with these three mutually exclusive outcomes are labeled p_1, p_2, and p_3. In symbols, if X_{ij} represents the ith observation from the jth group, then

$$p_1 = P(X_{i1} > X_{i2}),$$
$$p_2 = P(X_{i1} = X_{i2}),$$
$$p_3 = P(X_{i1} < X_{i2}).$$

So in the notation of Section 15.1.1, $p_3 = p$. Cliff (1996) focuses on testing

$$H_0: \delta = p_1 - p_3 = 0. \tag{15.7}$$

In the event tied values occur with probability zero, in which case $p_2 = 0$, Equation (15.7) becomes $H_0: p_1 = p_3 = .5$, which is the same as Equation (15.1). It can be shown that another way of expressing Equation (15.7) is with

$$H_0: p_3 + .5p_2 = .5.$$

For convenience, let $P = p_3 + .5p_2$, in which case this last equation becomes

$$H_0: P = .5. \tag{15.8}$$

Of course, when there are no tied values, $P = p_3$. The parameter δ [in Equation (15.7)] is related to P in the following manner:

$$\delta = 1 - 2P, \tag{15.9}$$

so

$$P = \frac{1 - \delta}{2}. \tag{15.10}$$

Cliff derived a heteroscedastic confidence interval for δ, which is computed as summarized in Box 15.2. If the confidence interval for δ does not contain zero, reject $H_0: \delta = 0$. When ties occur with probability zero, this is tantamount to rejecting $H_0: p = .5$.

BOX 15.2 Cliff's Method for Two Independent Groups

As usual, let X_{ij} be the ith observation from the jth group, $j = 1, 2$. For the ith observation in group 1 and the hth observation in group 2, let

$$d_{ih} = \begin{cases} -1 & \text{if } X_{i1} < X_{h2}, \\ 0 & \text{if } X_{i1} = X_{h2}, \\ 1 & \text{if } X_{i1} > X_{h2}. \end{cases}$$

An estimate of $\delta = P(X_{i1} > X_{i2}) - P(X_{i1} < X_{i2})$ is

$$\hat{\delta} = \frac{1}{n_1 n_2} \sum_{i=1}^{n_1} \sum_{h=1}^{n_2} d_{ih}, \tag{15.11}$$

the average of the d_{ih} values. Let

$$\bar{d}_{i.} = \frac{1}{n_2} \sum_h d_{ih},$$

$$\bar{d}_{.h} = \frac{1}{n_2} \sum_i d_{ih},$$

$$s_1^2 = \frac{1}{n_1 - 1} \sum_{i=1}^{n_1} (\bar{d}_{i.} - \hat{\delta})^2,$$

$$s_2^2 = \frac{1}{n_2 - 1} \sum_{h=1}^{n_2} (\bar{d}_{.h} - \hat{\delta})^2,$$

$$\tilde{\sigma}^2 = \frac{1}{n_1 n_2} \sum \sum (d_{ih} - \hat{\delta})^2.$$

Then

$$\hat{\sigma}^2 = \frac{(n_1 - 1)s_1^2 + (n_2 - 1)s_2^2 + \tilde{\sigma}^2}{n_1 n_2}$$

estimates the squared standard error of $\hat{\delta}$. Let z be the $1 - \alpha/2$ quantile of a standard normal distribution. Rather than use the more obvious confidence interval for δ, Cliff (1996, p. 140) recommends

$$\frac{\hat{\delta} - \hat{\delta}^3 \pm z\hat{\sigma}\sqrt{(1 - \hat{\delta}^2)^2 + z^2\hat{\sigma}^2}}{1 - \hat{\delta}^2 + z^2\hat{\sigma}^2}.$$

If there are no tied values, alternative heteroscedastic methods have been proposed by Mee (1990) as well as Fligner and Policello (1981). Currently it seems that for this special case, these methods offer no practical advantage over the method in Box 15.2.

15.1.4 S-PLUS function cid

The S-PLUS function

cid(x,y,alpha=.05),

written for this book, performs the calculations in Box 15.2.

15.1.5 The Brunner–Munzel Procedure

An alternative approach to both tied values and heteroscedasticity stems from Brunner and Munzel (2000). Their approach is based in part on what are called the *midranks* for handling tied values. To explain, first consider the values 45, 12, 32, 64, 13, and 25. There are no tied values and the smallest value is said to have *rank* 1, the next smallest has rank 2, and so on. A common notation for the rank corresponding to the ith observation is R_i. So in the example, the first observation is $X_1 = 45$ and its rank is $R_1 = 5$. Similarly, $X_2 = 12$ and its rank is $R_2 = 1$.

Now consider a situation where there are tied values: 45, 12, 13, 64, 13, and 25. Putting these values in ascending order yields 12, 13, 13, 25, 45, 64. So the value 12 gets a rank of 1, but there are two identical values having a rank of 2 and 3. The *midrank* is simply the average of the ranks among the tied values. Here, this means that the rank assigned to the two values equal to 13 would be $(2 + 3)/2 = 2.5$, the average of their corresponding ranks. So the ranks for all six values would be 1, 2.5, 2.5, 4, 5, 6.

Generalizing, consider

7, 7.5, 7.5, 8, 8, 8.5, 9, 11, 11, 11.

There are 10 values, so if there were no tied values, their ranks would be 1, 2, 3, 4, 5, 6, 7, 8, 9, and 10. But because there are two values equal to 7.5, their ranks are averaged, yielding a rank of 2.5 for each. There are two values equal to 8; their original ranks were 4 and 5, so their final ranks (their midranks) are both 4.5. There are three values equal to 11; their original ranks are 8, 9, and 10, the average of these ranks is 9, so their midranks are all equal to 9. So the ranks for the 10 observations are

1, 2.5, 2.5, 4.5, 4.5, 6, 7, 9, 9, 9.

Now consider testing $H_0: P = .5$, where P is as defined in Section 15.1.3. As usual, let X_{ij} be the ith observation from the jth group ($i = 1, \ldots, n_j$; $j = 1,2$). To apply the Brunner–Munzel method, first pool all $N = n_1 + n_2$ observations and assign ranks. In the event there are tied values, ranks are averaged as just illustrated. The results for the jth group are labeled R_{ij}, $i = 1, \ldots, n_j$. That is, R_{ij} is the rank corresponding to X_{ij} among the pooled values. Let $\bar{R}_1$ be the average of the ranks corresponding to group 1 and $\bar{R}_2$ be the average for group 2. So for the jth group,

$$\bar{R}_j = \frac{1}{n_j} \sum_{i=1}^{n_j} R_{ij}.$$

Next, for the first group, rank the observations, ignoring group 2, and label the results $V_{11}, \ldots, V_{n_1 1}$. Do the same for group 2 (ignoring group 1), and label the ranks $V_{12}, \ldots, V_{n_2 2}$. The remaining calculations are shown in Box 15.3.

BOX 15.3 The Brunner–Munzel Method for Two Independent Groups

Compute

$$S_j^2 = \frac{1}{n_j - 1} \sum_{i=1}^{n_j} \left(R_{ij} - V_{ij} - \bar{R}_j + \frac{n_j + 1}{2} \right)^2,$$

$$s_j^2 = \frac{S_j^2}{(N - n_j)^2},$$

$$s_e = \sqrt{N} \sqrt{\frac{s_1^2}{n_1} + \frac{s_2^2}{n_2}},$$

$$U_1 = \left(\frac{S_1^2}{N - n_1} + \frac{S_2^2}{N - n_2} \right)^2,$$

$$U_2 = \frac{1}{n_1 - 1} \left(\frac{S_1^2}{N - n_1} \right)^2 + \frac{1}{n_2 - 1} \left(\frac{S_2^2}{N - n_2} \right)^2.$$

The test statistic is

$$W = \frac{\bar{R}_2 - \bar{R}_1}{\sqrt{N} s_e},$$

and the degrees of freedom are

$$\hat{\nu} = \frac{U_1}{U_2}.$$

Decision Rule

Reject $H_0 : P = .5$ if $|W| \geq t$, where t is the $1 - \alpha/2$ quantile of a Student's T-distribution with $\hat{\nu}$ degrees of freedom. An estimate of P is

$$\hat{P} = \frac{1}{n_1} \left(R_2 - \frac{n_2 + 1}{2} \right) = \frac{1}{N} (\bar{R}_2 - \bar{R}_1) + \frac{1}{2}.$$

The estimate of δ is

$$\hat{\delta} = 1 - 2\hat{P}.$$

An approximate $1 - \alpha$ confidence interval for P is

$$\hat{P} \pm t s_e.$$

There is a connection between the method just described and the Wilcoxon–Mann–Whitney test that is worth mentioning:

$$U = n_2\bar{R}_2 - \frac{n_2(n_2+1)}{2}.$$

That is, if you sum the ranks of the second group (which is equal to $n_2\bar{R}_2$) and subtract $n_2(n_2+1)/2$, you get the Wilcoxon–Mann–Whitney U statistic given by Equation (15.3). Many books describe the Wilcoxon–Mann–Whitney method in terms of U rather than the approach used here.

Note that both the Cliff and Brunner–Munzel rank-based methods offer protection against low power due to outliers. If, for example, the largest observation among a batch of numbers is increased from 12 to 1 million, its rank does not change. But how should one choose between rank-based methods covered here versus the robust methods in Chapter 8? If our only criterion is high power, both perform well, with weak evidence that in practice, robust methods are a bit better. But the more important point is that they provide different information about how groups compare. Some authorities argue passionately that as a measure of effect size, P and δ, as defined in this section, reflect what is most important and what we want to know. Others argue that measures of location also provide useful information; they reflect what is typical and provide a sense of the magnitude of the difference between groups that is useful and not provided by rank-based methods. The only certainty is that at present, there is no agreement about which approach should be preferred or even if it makes any sense to ask the question of which is better.

Often Cliff's method gives similar results to the Brunner–Munzel technique. But when the probability of a tied value is high and there are relatively few outcomes possible (i.e., there are few possible X values that can be observed), the Brunner–Munzel procedure can have a higher Type I error probability than Cliff. Based on a very limited comparison of the two methods, the author has found that generally there seems to be little separating Cliff's approach from Brunner–Munzel. However, situations can be constructed where, with many tied values, Cliff's approach seems to be better at guaranteeing an actual Type I error probability less than the nominal α level; and when testing at the .05 level, Cliff's method seems to do an excellent job of avoiding actual Type I error probabilities less than .04. In contrast, the Brunner–Munzel method can have an actual Type I error rate close to .07 when tied values are common and sample sizes are small. This suggests the possibility that the Brunner–Munzel method might have more power and reject when Cliff's method does not, but the issue of how these two methods compare needs closer scrutiny.

15.1.6 S-PLUS function bmp

The S-PLUS function

```
bmp(x,y,alpha=.05)
```

performs the Brunner–Munzel method. It returns the p-value (or significance level) when testing $H_0: P = .5$ plus an estimate of P labeled phat and a confidence interval for P labeled ci.p (an estimate of δ, labeled d.hat, is returned as well).

EXAMPLE. Table 8.3 reports data from a study of hangover symptoms among sons of alcoholics versus a control. Note that there are many tied values among these data. In the second group, for example, 14 of the 20 values are zero. Welch's test for means has a significance level of .14, Yuen's test has a significance level of .076, the Brunner–Munzel method has a significance level of .042, and its .95 confidence interval for P is (.167, .494). Cliff's method also rejects at the .05 level, the .95 confidence interval for δ being (0.002, 0.60). ■

15.1.7 The Kolmogorov–Smirnov Test

Yet another way of testing the hypothesis that two independent groups have identical distributions is with the so-called Kolmogorov–Smirnov test. This test forms the basis of the shift function in Section 8.11.2. Like the WMW test in Section 15.1.1, exact control over the probability of a Type I error can be had by assuming random sampling only. When there are no tied values, the method in Kim and Jennrich (1973) can be used to compute the exact significance level. With tied values, the exact significance level can be computed with a method derived by Schroër and Trenkler (1995). The S-PLUS function supplied to perform the Kolmogorov–Smirnov test has an option for computing the exact significance level, but the details of the method are not given here. (Details can be found in Wilcox, 1997a.)

To apply the Kolmogorov–Smirnov test, let $\hat{F}_1(x)$ be the proportion of observations in group 1 that are less than or equal to x, and let $\hat{F}_2(x)$ be the corresponding proportion for group 2. Let

$$U_i = |\hat{F}_1(X_{i1}) - \hat{F}_2(X_{i1})|,$$

$i = 1, \ldots, n_1$. In other words, for X_{i1}, the ith observation in group 1, compute the proportion of observations in group 1 that are less than or equal to X_{i1}, do the same for group 2, take the absolute value of the difference, and label the result U_i. Repeat this process for the observations in group 2 and label the results

$$V_i = |\hat{F}_1(X_{i2}) - \hat{F}_2(X_{i2})|,$$

$i = 1, \ldots, n_2$. The Kolmogorov–Smirnov test statistic is

$$\text{KS} = \max\{U_1, \ldots, U_{n_1}, V_1, \ldots, V_{n_2}\}, \tag{15.12}$$

the largest of the pooled U and V values. For large sample sizes, an approximate critical value when $\alpha = .05$ is

$$1.36\sqrt{\frac{n_1 + n_2}{n_1 n_2}}.$$

Reject when KS is greater than or equal to the critical value. When there are no tied values, the Kolmogorov–Smirnov test can have relatively high power; but with ties, its power can be relatively low.

15.1.8 S-PLUS Function ks

The S-PLUS function

ks(x,y,w=F,sig=T)

performs the Kolmogorov–Smirnov test. The argument w can be used to invoke a weighted version of the Kolmogorov–Smirnov test not covered here. (See, for example, Wilcox, 1997a.) By default, w=F, meaning that the version described here will be used. With sig=T, the exact critical value will be used. With large sample sizes, computing the exact critical value can result in high execution time. Setting sig=F avoids this problem, but now only the approximate critical value with $\alpha = .05$ can be used.

EXAMPLE. For the data in Table 8.3, the function ks returns

```
$test:
[1] 0.35

$critval:
[1] 0.4300698

$siglevel:
[1] 0.03942698
```

This says that the Kolmogorov–Smirnov test statistic is KS = 0.35, the approximate .05 critical value is 0.43, which is greater than KS, yet the exact significance level, assuming only random sampling, is .039. ■

15.2 Comparing More Than Two Groups

15.2.1 The Kruskall–Wallis Test

The best-known rank-based method for comparing multiple groups is the Kruskall–Wallis test. The goal is to test

$$H_0 : F_1(x) = F_2(x) = \cdots = F_J(x), \tag{15.13}$$

the hypothesis that J independent groups have identical distributions. The method begins by pooling all $N = \sum n_j$ observations and assigning ranks. In symbols, if X_{ij} is the ith observation in the jth group, let R_{ij} be its rank among the pooled data. When there are tied values, use midranks, as described in connection with the Brunner–Munzel method. Next, sum the ranks for each group. In symbols, compute

$$R_j = \sum_{i=1}^{n_j} R_{ij},$$

$(j = 1, \ldots, J)$. Letting

$$S^2 = \frac{1}{N-1}\left(\sum_{j=1}^{J}\sum_{i=1}^{n_j} R_{ij}^2 - \frac{N(N+1)^2}{4}\right),$$

the test statistic is

$$T = \frac{1}{S^2}\left(-\frac{N(N+1)^2}{4} + \sum \frac{R_j^2}{n_j}\right).$$

If there are no ties, S^2 simplifies to

$$S^2 = \frac{N(N+1)}{12},$$

and T becomes

$$T = -3(N+1) + \frac{12}{N(N+1)}\sum \frac{R_j^2}{n_j}.$$

The hypothesis of identical distributions is rejected if $T \geq c$, where c is some appropriate critical value. For small sample sizes, exact critical values are available from Iman, Quade, and Alexander (1975). For large sample sizes, the critical value is approximately equal to the $1 - \alpha$ quantile of a chi-squared distribution with $J - 1$ degrees of freedom.

EXAMPLE. Table 15.1 shows data for three groups and the corresponding ranks. For example, after pooling all $N = 10$ values, $X_{11} = 40$ has a rank of $R_{11} = 1$, the value 56 has a rank of 6, and so forth. The sum of the ranks corresponding to each group are $R_1 = 1 + 6 + 2 = 9$, $R_2 = 3 + 7 + 8 = 18$, and $R_3 = 9 + 10 + 5 + 4 = 28$. The number of groups is $J = 3$, so the degrees of freedom are $\nu = 2$; from Table 3 in Appendix B, the critical value is approximately $c = 5.99$ with $\alpha = .05$. Because there are no ties among the N observations,

$$T = -3(10+1) + \frac{12}{10 \times 11}\left(\frac{9^2}{3} + \frac{18^2}{3} + \frac{28^2}{4}\right) = 3.109.$$

Because $3.109 < 5.99$, fail to reject. That is, you are unable to detect a difference among the distributions. ■

15.2.2 The BDM Method

The Kruskall–Wallis test performs relatively well when the null hypothesis of identical distributions is true, but concerns arise when the null hypothesis is false. In particular, the Kruskall–Wallis test is homoscedastic, which might affect power.

TABLE 15.1 Hypothetical Data Illustrating the Kruskall–Wallis Test

Group 1		Group 2		Group 3	
X_{i1}	R_{i1}	X_{i2}	R_{i2}	X_{i3}	R_{i3}
40	1	45	3	61	9
56	6	58	7	65	10
42	2	60	8	55	5
				47	4

A heteroscedastic analog of the Kruskall–Wallis test that allows tied values was derived by Brunner, Dette, and Munk (1997), which will be called the BDM method. Again the goal is to test the hypothesis that all J groups have identical distributions. The basic idea is that if J independent groups have identical distributions and we assign ranks to the pooled data as was done in the Kruskall–Wallis test, then for each group the average of the ranks should be approximately equal. (This greatly oversimplifies the technical issues.) To apply it, compute the ranks of the pooled data as was done in connection with the Kruskall–Wallis test. As before, let $N = \sum n_j$ be the total number of observations. The remaining calculations are relegated to Box 15.4. The values in the vector **Q** in Box 15.4 are called the *relative effects* and reflect the average ranks among the groups, which provide some sense of how the groups compare.

BOX 15.4 BDM Heteroscedastic Rank-Based ANOVA Method

Let $\bar{R}_j$ be the average of the pooled ranks corresponding to the jth group. So if R_{ij} is the rank of X_{ij} after the data are pooled, then

$$\bar{R}_j = \frac{1}{n_j} \sum_{i=1}^{n_j} R_{ij}.$$

Let

$$\mathbf{Q} = \frac{1}{N}\left(\bar{R}_1 - \frac{1}{2}, \ldots, \bar{R}_J - \frac{1}{2}\right).$$

For the jth group, compute

$$s_j^2 = \frac{1}{N^2(n_j - 1)} \sum_{i=1}^{n_j} (R_{ij} - \bar{R}_j)^2,$$

and let

$$\mathbf{V} = N \operatorname{diag}\left\{\frac{s_1^2}{n_1}, \ldots, \frac{s_J^2}{n_J}\right\}.$$

Continued

BOX 15.4 (*Continued*)

Let **I** be a J-by-J identity matrix, let **J** be a J-by-J matrix of 1's, and set

$$\mathbf{M} = \mathbf{I} - \frac{1}{J}\mathbf{J}.$$

(The diagonal entries in **M** have a common value, a property required to satisfy certain theoretical restrictions.) The test statistic is

$$F = \frac{N}{\mathrm{tr}(\mathbf{M}_{11}\mathbf{V})}\mathbf{Q}\mathbf{M}\mathbf{Q}', \quad (15.14)$$

where tr indicates trace and $\mathbf{Q}'$ is the transpose of the matrix **Q**. (See Appendix C for how the trace and transpose of a matrix are defined.)

Decision Rule

Reject if $F \geq f$, where f is the $1 - \alpha$ quantile of an F-distribution with

$$\nu_1 = \frac{M_{11}[\mathrm{tr}(\mathbf{V})]^2}{\mathrm{tr}(\mathbf{MVMV})} \qquad \nu_2 = \frac{[\mathrm{tr}(\mathbf{V})]^2}{\mathrm{tr}(\mathbf{V}^2\mathbf{\Lambda})}$$

degrees of freedom and

$$\mathbf{\Lambda} = \mathrm{diag}\{(n_1 - 1)^{-1}, \ldots, (n_J - 1)^{-1}\}.$$

An alternative heteroscedastic method, one that assumes there are no tied values, was derived by Rust and Fligner (1984). In the event there are no tied values, it is unknown how the Rust–Fligner and BDM methods compare. Perhaps the use of midranks in conjunction with the Rust–Fligner procedure performs reasonably well when tied values occur, but it seems that this issue has not been investigated. (The S-PLUS function rfanova(x), written for this book, performs the Rust–Fligner technique but is not described because currently it seems that the BDM method suffices.)

15.2.3 S-PLUS Function bdm

The S-PLUS function

bdm(x)

has been supplied to perform the BDM rank-based ANOVA described in Box 15.4. Here, x can have list mode or it can be a matrix with columns corresponding to groups. The function returns the value of the test statistic, the degrees of freedom, the vector of relative effects, which is labeled q.hat, and the significance level.

EXAMPLE. For the schizophrenia data in Table 9.1, the S-PLUS function bmd returns a significance level of .040. The relative effect sizes (the **Q** values) are

Continued

EXAMPLE. (*Continued*) reported as

```
$ output$q.hat:
           [,1]
[1,]     0.4725
[2,]     0.4725
[3,]     0.3550
[4,]     0.7000
```

So the conclusion is that the distributions associated with these four groups differ, and we see that the average of the ranks among the pooled data is smallest for group 3 and highest for group 4. This is consistent with the means. That is, group 3 has the smallest mean and group 4 the largest. The same is true when using a 20% trimmed mean or MOM. ■

15.3 Multiple Comparisons Among Independent Groups

One way of extending the Cliff and the Brunner–Munzel methods when comparing all pairs of J groups, $J > 2$, is to proceed in the manner used to derive Dunnett's T3 (described in Section 12.3.1). In particular, use a critical value from Table 10 in Appendix B (which reports some quantiles of the Studentized maximum modulus distribution). Here, assuming all pairwise comparisons are to be made, there are a total of $C = (J^2 - J)/2$ hypotheses to be tested. First consider an extension of the Brunner–Munzel method. When comparing group j to group k, simply perform the calculations in Section 15.1.5, ignoring the other groups. Let $\hat{\nu}_{jk}$ be the resulting degrees of freedom, let s_{ejk} be the corresponding value of s_e, and let $\hat{P}_{jk}$ be the estimate of P. So when there are no ties, P_{jk} is the probability that a randomly sampled observation from group j is less than a sampled observation from group k. The confidence interval for P_{jk} is

$$\hat{P}_{jk} \pm cs_{ejk},$$

where c is the critical value read from Table 10 in Appendix B, which depends on C and $\hat{\nu}_{jk}$.

As for extending Cliff's method, the same strategy can be used. That is, proceed as described in Box 15.2, but when computing a confidence interval for δ, replace z with a critical value read from Table 10 with degrees of freedom $\nu = \infty$. For example, if there are four groups and all pairwise comparisons are to be performed, the total number of hypotheses to be tested is $C = 6$, so if $\alpha = .05$, the critical value is 2.63. That is, replace z with 2.63 in Box 15.2.

15.3.1 S-PLUS Functions cidmul and bmpmul

The S-PLUS function

bmpmul(x,alpha=.05)

performs the extension of the bmp method in Box 15.3 when the goal is to perform all pairwise comparisons among the groups stored in x, where x has list mode or is a matrix with columns corresponding to groups. The S-PLUS function

cidmul(x,alpha=.05)

is like bmpmul, except that Cliff's method for making inferences about δ is used instead. Both of these functions are limited to $\alpha = .05$ and .01. If the argument alpha has any value other than .05, $\alpha = .01$ is assumed.

15.3.2 Multiple Comparisons Based on BDM

Rather than perform multiple comparisons based on estimates of P, another approach is to compare each pair of groups in terms of their distributions. That is, for the jth and kth groups, test

$$H_0 : F_j(x) = F_k(x),$$

with the goal of controlling FWE among all the hypotheses to be tested. For each pair of groups, the significance level associated with this hypothesis can be computed as described in Box 15.4. Here, FWE is controlled using either Rom's method, which is described in Section 12.8.2, or the Benjamini–Hochberg technique in Section 12.3.3.

15.3.3 S-PLUS r1mcp

The S-PLUS function

r1mcp(x,alpha=.05,bhop=F)

performs all pairwise comparisons of J independent groups, each comparison being based on the BDM method in Section 15.2.2; if bhop=F, then FWE is controlled via Rom's procedure. Here, the argument alpha corresponds to the desired value for FWE. Setting bhop=T, the Benjamini and Hochberg method (described in Section 12.3.3) is used to control FWE.

15.4 Two-Way Designs

The BDM method described in Section 15.3.2 can be extended to a two-way design. (For an extension to a mixed design where one factor is fixed and the other is random, see Brunner & Dette, 1992.) Following the notation in Chapter 10, now we have J levels associated with the first factor and K levels for the other, for a total of JK independent groups. The observations are represented by X_{ijk}, $i = 1, \ldots, n_{jk}$; $j = 1, \ldots, J$; and $k = 1, \ldots, K$. Let $F_{jk}(x)$ be the distribution associated with the jth and kth levels. So, for example, $F_{23}(6)$ is the probability that for the second level of the first factor and the third level of the second factor, a randomly sampled observation has a value less than or equal to 6.

Pool all of the observations and assign ranks. In the event of tied values, ranks are averaged as in Section 15.1.5. For convenience, let $L = JK$ represent the total number of groups and assume the first K groups correspond to level 1 of factor A, the next K correspond to level 2 of factor A, and so on. So the groups are arranged as described in Section 10.3.1. Furthermore, the sample size for the ℓth group is n_ℓ. For each of the L groups, compute the average of the corresponding ranks, label the results $\bar{R}_1, \ldots, \bar{R}_L$, and let

$$s_\ell^2 = \frac{1}{N^2(n_\ell - 1)} \sum_{i=1}^{n_\ell} (R_{i\ell} - \bar{R}_\ell)^2,$$

where $N = \sum n_\ell$ is the total sample size and $R_{i\ell}$ is the rank of the ith observation in the ℓth group. Main effects and interactions are tested as described in Box 15.5.

BOX 15.5 BDM Two-Way, Heteroscedastic Rank-Based ANOVA Method

Set

$$\mathbf{V} = N \operatorname{diag}\left\{\frac{s_1^2}{n_1}, \ldots, \frac{s_L^2}{n_L}\right\}.$$

Let $\mathbf{I}_J$ be a J-by-J identity matrix, let $\mathbf{H}_J$ be a J-by-J matrix of 1's, and let

$$\mathbf{P}_J = \mathbf{I}_J - \frac{1}{J}\mathbf{H}_J, \qquad \mathbf{M}_A = \mathbf{P}_J \otimes \frac{1}{K}\mathbf{H}_K,$$

$$\mathbf{M}_B = \frac{1}{J}\mathbf{H}_J \otimes \mathbf{P}_K, \qquad \mathbf{M}_{AB} = \mathbf{P}_J \otimes \mathbf{P}_K.$$

(The notation $\otimes$ refers to the right Kronecker product, which is described in Appendix C.) As in Box 15.4, let

$$\mathbf{Q} = \frac{1}{N}\left(\bar{R}_1 - \frac{1}{2}, \ldots, \bar{R}_L - \frac{1}{2}\right)$$

be the relative effects. The test statistics are:

$$F_A = \frac{N}{\operatorname{tr}(M_{A11}\mathbf{V})}\mathbf{Q}\mathbf{M}_A\mathbf{Q}', \qquad F_B = \frac{N}{\operatorname{tr}(M_{B11}\mathbf{V})}\mathbf{Q}\mathbf{M}_B\mathbf{Q}',$$

$$F_{AB} = \frac{N}{\operatorname{tr}(M_{AB11}\mathbf{V})}\mathbf{Q}\mathbf{M}_{AB}\mathbf{Q}'.$$

Decision rules

For factor A, reject if $F_A \geq f$, where f is the $1 - \alpha$ quantile of an F-distribution with degrees of freedom

$$\nu_1 = \frac{M_{A11}[\operatorname{tr}(\mathbf{V})]^2}{\operatorname{tr}(\mathbf{M}_A\mathbf{V}\mathbf{M}_A\mathbf{V})}, \qquad \nu_2 = \frac{[\operatorname{tr}(\mathbf{V})]^2}{\operatorname{tr}(\mathbf{V}^2\mathbf{\Lambda})},$$

Continued

BOX 15.5 (*Continued*)

where

$$\Lambda = \mathrm{diag}\{(n_1 - 1)^{-1}, \ldots, (n_L - 1)^{-1}\}.$$

Here M_{A11} is the first diagonal element of the matrix $\mathbf{M}_A$. (By design, all of the diagonal elements of $\mathbf{M}_A$ have a common value.) For factor B, reject if $F_B \geq f$, where

$$\nu_1 = \frac{M_{B11}[\mathrm{tr}(\mathbf{V})]^2}{\mathrm{tr}(\mathbf{M}_B\mathbf{V}\mathbf{M}_B\mathbf{V})}.$$

(The value for ν_2 remains the same.) As for the hypothesis of no interactions, reject if $F_{AB} \geq f$, where now

$$\nu_1 = \frac{M_{AB11}[\mathrm{tr}(\mathbf{V})]^2}{\mathrm{tr}(\mathbf{M}_{AB}\mathbf{V}\mathbf{M}_{AB}\mathbf{V})}$$

and ν_2 is the same value used to test for main effects.

Before continuing it might help to be more precise about how the null hypotheses are formulated. The basic idea stems from Akritas and Arnold (1994) and the particular case of a two-way design was addressed by Akritas, Arnold, and Brunner (1997). For any value x, let

$$\bar{F}_{j.}(x) = \frac{1}{K}\sum_{k=1}^{K} F_{jk}(x)$$

be the average of the distributions among the K levels of factor B corresponding to the jth level of factor A.

EXAMPLE. Consider a 2-by-2 design and suppose that $F_{11}(6) = .5$ and $F_{12}(6) = .3$. That is, for the first level of factor A and the first level of factor B, the probability that an observation is less than 6 is .5. For the first level of factor A and the second level of factor B, this probability is .3. So $\bar{F}_{1.}(6) = (.5 + .3)/2 = .4$. More generally, for any x,

$$\bar{F}_{1.}(x) = \frac{F_{11}(x) + F_{12}(x)}{2}.$$

■

The hypothesis of no main effects for factor A is

$$H_0 : \bar{F}_{1.}(x) = \bar{F}_{2.}(x) = \cdots = \bar{F}_{J.}(x).$$

for any x. Letting

$$\bar{F}_{.k}(x) = \frac{1}{J}\sum_{j=1}^{J} F_{jk}(x)$$

be the average of the distributions for the kth level of factor B, the hypothesis of no main effects for factor B is

$$H_0 : \bar{F}_{.1}(x) = \bar{F}_{.2}(x) = \cdots = \bar{F}_{.K}(x).$$

As for interactions, first consider a 2-by-2 design. Then no interaction is taken to mean that for any x,

$$F_{11}(x) - F_{12}(x) = F_{21}(x) - F_{22}(x),$$

which has a certain similarity to how no interaction based on means was defined in Chapter 10. Here, no interaction in a J-by-K design means that for any two rows and any two columns, there is no interaction as just described. From a technical point of view, a convenient way of stating the hypothesis of no interactions among all JK groups is with

$$H_0 : F_{jk}(x) - \bar{F}_{j.}(x) - \bar{F}_{.k}(x) + \bar{F}_{..}(x) = 0,$$

for any x, all j ($j = 1, \dots, J$) and all k ($k = 1, \dots, K$), where

$$\bar{F}_{..}(x) = \frac{1}{JK} \sum_{j=1}^{J} \sum_{k=1}^{K} F_{jk}(x).$$

15.4.1 S-PLUS Function bdm2way

The S-PLUS function

bdm2way(J, K, x)

performs the two-way ANOVA method described in Box 15.5.

15.5 Multiple Comparisons in a Two-Way Design

One approach when dealing with multiple comparisons among the levels of each factor in a two-way design is to perform each comparison of interest using the BDM method and control FWE using Rom's method or perhaps the Benjamini–Hochberg technique described in Chapter 12.

To elaborate, first consider factor A and imagine that all pairwise comparisons among the J levels are to be performed. This means that for any two levels — say, j and j' — the goal is to test

$$H_0 : \bar{F}_{j.} = \bar{F}_{j'.}$$

and simultaneously to control FWE among all pairwise comparisons. As noted in Chapter 12, if all pairwise comparisons are to be made among the J levels of factor A, there are a total of $C = (J^2 - J)/2$ hypotheses to be tested. To compare level j to level j', compute F and ν_2 as indicated in Box 15.4, ignoring all other groups. (That is, perform the calculations in Box 15.4 as if you were testing the hypothesis of no main effect for factor A in a 2-by-K design, in which case $\nu_1 = 1$.) Then for each of the individual hypotheses, make a decision about whether to reject using the significance levels as described in Section 12.3.3. (That is, use the Rom or Benjamini–Hochberg method.) The levels of factor B can be compared in a similar manner using the methods in Section 12.8 to control FWE.

As for interactions, a basic approach is to test the hypothesis of no interaction for any two levels of factor A and factor B and to repeat this for all pairs of rows and all pairs of columns. That is, for every $j < j'$ and $k < k'$, test

$$F_{jk}(x) - F_{jk'}(x) = F_{j'k}(x) - F_{j'k'}(x).$$

Again Rom's method or the Benjamini–Hochberg technique can be used to control FWE (the familywise Type I error rate).

15.5.1 S-PLUS Function r2mcp

The S-PLUS function

```
r2mcp(J,K,x,grp=NA,alpha=.05,bhop=F)
```

performs the multiple comparisons method just described. The groups are assumed to be arranged as in Section 10.3.1; if not arranged in this manner, the argument grp can be used to address this problem as illustrated in Section 10.3.1. The default value for bhop causes critical significance levels to be computed using Rom's method; bhop=T results in using the Benjamini and Hochberg method instead.

15.5.2 The Patel–Hoel Approach to Interactions

Patel and Hoel (1973) proposed an alternative approach to interactions in a 2-by-2 design that can be extended to a multiple comparison method for a J-by-K design, even when there are tied values. To describe the basic idea, first consider a 2-by-2 design where X_{ijk} is the ith observation randomly sampled from the jth level of factor A and the kth level of factor B. Temporarily assume ties never occur and let

$$p_{11,12} = P(X_{i11} < X_{i12}).$$

In words, X_{i11} represents a randomly sampled observation from level 1 of factors A and B, X_{i12} is a randomly sampled observation from level 1 of factor A and level 2 of factor B, and $p_{11,12}$ is the probability that X_{i11} is less than X_{i12}. Note that ignoring level 2 of factor A, levels 1 and 2 of factor B can be compared by testing $H_0 : p_{11,12} = 0$, as described in Sections 15.1.3 and 15.1.5. The Patel–Hoel definition of no interaction is that $p_{11,12} = p_{21,22}$. That is, the probability that an observation is smaller under level 1 of factor B than under level 2 is the same for both levels of factor A. In the event ties can occur, then define

$$p_{11,12} = P(X_{i11} \leq X_{i12}) + \frac{1}{2}P(X_{i11} = X_{i12}),$$

$$p_{21,22} = P(X_{i21} \leq X_{i22}) + \frac{1}{2}P(X_{i21} = X_{i22}),$$

and the hypothesis of no interaction is

$$H_0 : p_{11,12} = p_{21,22}.$$

Again, temporarily ignore level 2 of factor A and note that the 2 independent groups corresponding to the 2 levels of factor B can be compared in terms of δ, as defined in Section 15.1.3. Let δ_1 represent δ when focusing on level 1 of factor A with level 2 ignored, and let $\hat{\delta}_1$ be the estimate of δ as given by Equation (15.11). An estimate of its squared standard error is computed as indicated in Box 15.2 and will be labeled $\hat{\sigma}_1^2$. Similarly, let δ_2 be δ when focusing on level 2 of factor A with level 1 ignored, and denote its estimate with $\hat{\delta}_2$, which has an estimated squared standard error $\hat{\sigma}_2^2$. It can be seen that the null hypothesis of no interaction just defined corresponds to

$$H_0 : \Delta = \frac{\delta_2 - \delta_1}{2} = 0.$$

Moreover, the results in Box 15.2 can be used to estimate $p_{11,12} - p_{21,22}$ and compute an appropriate $1 - \alpha$ confidence interval. The estimate is

$$\hat{\Delta} = \frac{\hat{\delta}_2 - \hat{\delta}_1}{2},$$

the estimated squared standard error of $\hat{\Delta}$ is

$$S^2 = \frac{1}{4}(\hat{\sigma}_1^2 + \hat{\sigma}_2^2),$$

where $\hat{\sigma}_j^2$ is the value of $\hat{\sigma}^2$ in Box 15.2 corresponding to the jth level of factor A, and a $1 - \alpha$ confidence interval for Δ is

$$\hat{\Delta} \pm z_{1-\alpha/2} S,$$

where $z_{1-\alpha/2}$ is the $1 - \alpha/2$ quantile of a standard normal distribution. The hypothesis of no interaction is rejected if this confidence interval does not contain zero.

There remains the problem of controlling FWE for the more general case of a J-by-K design. Here, an analog of Dunnett's T3 method is used, but Rom's method or the Benjamini–Hochberg approach are other possibilities. When working with levels j and j' of factor A and levels k and k' of factor B, we represent the parameter Δ by $\Delta_{jj'kk'}$, its estimate is labeled $\hat{\Delta}_{jj'kk'}$, and the estimated squared standard error is denoted by $S^2_{jj'kk'}$. For every $j < j'$ and $k < k'$, the goal is to test

$$H_0 : \Delta_{jj'kk'} = 0.$$

The total number of hypotheses to be tested is

$$C = \frac{J^2 - J}{2} \times \frac{K^2 - K}{2}.$$

The critical value, c, is read from Table 10 in Appendix B with degrees of freedom $\nu = \infty$. The confidence interval for $\Delta_{jj'kk'}$ is

$$\hat{\Delta}_{jj'kk'} \pm c S_{jj'kk'},$$

and the hypothesis of no interaction, corresponding to levels j and j' of factor A and levels k and k' of factor B, is rejected if this confidence interval does not contain zero.

Yet another approach to testing the hypothesis of no interactions can be based on the results in Section 15.1.5. The method just described seems to perform well in simulations (Wilcox, 2000b), so currently there seems to be little motivation for extending the method in Section 15.1.5 to the problem at hand.

15.5.3 S-PLUS Function rimul

The S-PLUS function

rimul(J,K,x,p=J*K,grp=c(1:p))

performs the test for interactions just described. (The fourth argument, p=J*K, is not important in applied work; it is used to deal with certain conventions in S-PLUS.) The groups are assumed to be arranged as in Section 10.3.1, and the argument grp is used as illustrated in Section 10.3.1.

EXAMPLE. Table 10.1 shows data on weight gains in rats under four conditions: source of protein (beef versus cereal), which is factor A, and amount of protein (factor B), which has two levels: low and high. Assume that for beef, the data for low protein is stored in x[[1]], and for high protein it is stored in x[[2]]. As for cereal, low-and high-protein data are stored in x[[3]] and x[[4]], respectively. Then the S-PLUS command rimul(2,2,x) returns:

```
$test:
          Factor A  Factor A Factor B Factor B delta
[1,]             1         2        1        2  0.31

ci.lower             ci.upper
-0.02708888         0.6470889
```

Let $p_{11,12}$ represent the probability that for beef, the weight gain for a randomly sampled rat will be smaller under a low-versus high-protein diet. Similarly, for cereal, let $p_{21,22}$ represent the probability that weight gain is smaller under a low-protein diet. The estimate of $\Delta = p_{11,12} - p_{21,22}$ is listed under delta and is 0.31, and a .95 confidence interval for Δ is $(-0.027, 0.647)$. So the hypothesis of no interaction is not rejected at the $\alpha = .05$ level. ■

15.6 Comparing Two Dependent Groups

15.6.1 The Sign Test

A simple method for comparing dependent groups is the so-called *sign test*. In essence it is based on making inferences about the probability of success associated with a binomial distribution, which is discussed in Section 4.13, but here we elaborate a bit on its properties.

As in Chapter 11, let $(X_{11}, X_{12}), \ldots, (X_{n1}, X_{n2})$ be a random sample of n pairs of observations. That is, X_{ij} is the ith observation from the jth group. Primarily for convenience, it is temporarily assumed that tied values never occur. Let p be the probability that for a randomly sampled pair of observations, the observation from group 1 is less than the observation from group 2. In symbols,

$$p = P(X_{i1} < X_{i2}).$$

Letting $D_i = X_{i1} - X_{i2}$, an estimate of p is simply the proportion of D_i values that are less than zero. More formally, let $V_i = 1$ if $D_i < 0$; otherwise $V_i = 0$. Then an estimate of p is

$$\hat{p} = \frac{1}{n} \sum V_i. \tag{15.15}$$

Because $X = \sum V_i$ has a binomial probability function, results in Section 4.13 provide a confidence interval for p. If this interval does not contain zero, reject

$$H_0 : p = .5$$

and conclude that the groups differ. If $p > .5$, group 1 is more likely to have a lower observed value than group 2; if $p < .5$, the reverse is true.

Now consider a situation where ties can occur. Given the goal of making inferences about p, one strategy is simply to ignore or discard cases where $D_i = 0$. So if among n pairs of observations, there are N D_i values not equal to zero, then an estimate of p is

$$\hat{p} = \frac{1}{N} \sum V_i, \tag{15.16}$$

where V_i is defined as before.

To add perspective, look at Figure 15.1, which shows a scatterplot of the cork boring data given in Table 11.1. Also shown is a diagonal line through the origin. Points to the right of this line correspond to D_i values that are greater than zero, and points to the left are D_i values for which the reverse is true. So graphically, the sign test is based on the proportion of points to the right of this line. Points falling exactly on this line are ignored. In this particular case, $n = 28$, two points fall exactly on the line, so $N = 26$, $\hat{p} = .31$, and the .95 confidence interval for p is (.15, .52); this interval contains .5, so fail to reject.

Notice that some of the methods in Chapter 11 found a difference between these two groups, in contrast to the sign test. One reason this can happen is that the sign test does not take into account how far a point is from the line in Figure 15.1 Also, it might help to note the graphical similarity between the sign test as depicted in Figure 15.1 and the graphical description of the bootstrap in Figure 11.2. The sign test is based on whether the observations are to the left or the right of the diagonal line. In contrast, the bootstrap is based on whether bootstrap estimates of the measures of location are to the left or the right of this line.

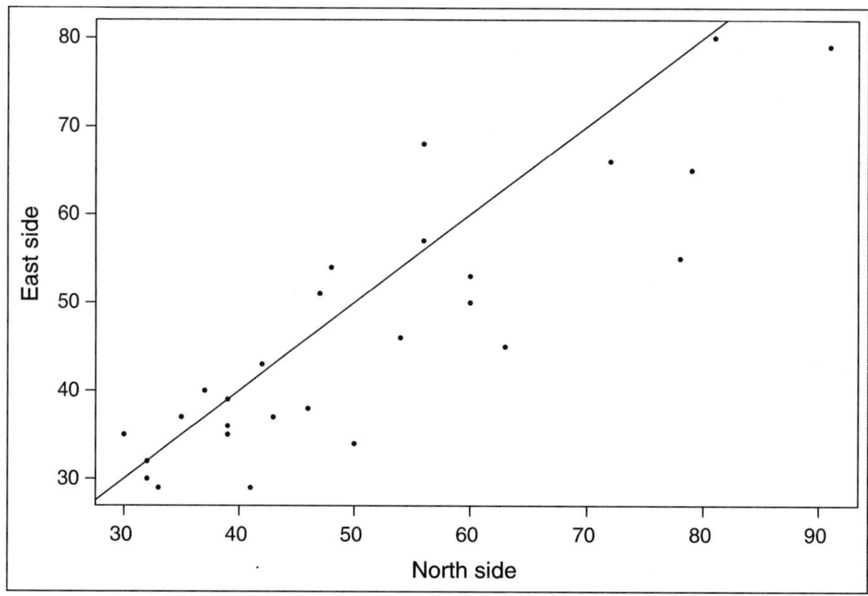

FIGURE 15.1 Scatterplot of the data in Table 11.1. In order for the sign test to reject, most of the points need to be above or below the diagonal line.

15.6.2 S-PLUS Function signt

The S-PLUS function

signt(x, y = NA, alpha = 0.05)

tests $H_0 : p = .5$ with the sign test as just described. If the argument y is not specified, it is assumed either that x is a matrix with two columns corresponding to the two dependent groups or that x has list mode. The function computes the differences $X_{i1} - X_{i2}$, eliminates all differences that are zero, leaving N values, determines the number of pairs for which $X_{i1} < X_{i2}$, $i = 1, \ldots, N$, and then calls the function binomci (see Section 4.13.1).

EXAMPLE. The output from signt based on the cork data in Figure 15.1 is

```
$phat:
[1] 0.3076923

$ci:
[1] 0.1530612 0.5179361

$n:
[1] 28

$N:
[1] 26
```

■

15.6.3 Wilcoxon Signed Rank Test

The sign test provides an interesting, useful, and reasonable perspective on how two groups differ. However, a common criticism is that its power can be low relative to other techniques that might be used. One alternative approach is the Wilcoxon signed rank test, which tests

$$H_0 : F_1(x) = F_2(x),$$

the hypothesis that two dependent groups have identical distributions. To apply it, first form difference scores as was done in conjunction with the paired T-test in Chapter 11 and discard any difference scores that are equal to zero. It is assumed that there are n difference scores not equal to zero. That is, for the ith pair of observations, compute

$$D_i = X_{i1} - X_{i2},$$

$i = 1, \ldots, n$ and each D_i value is either less than or greater than zero. Next, rank the $|D_i|$ values and let U_i denote the result for $|D_i|$. So, for example, if the D_i values are 6, −2, 12, 23, −8, then $U_1 = 2$, because after taking absolute values, 6 has a rank of 2. Similarly, $U_2 = 1$, because after taking absolute values, the second value, −2, has a rank of 1. Next set

$$R_i = U_i,$$

if $D_i > 0$; otherwise

$$R_i = -U_i.$$

Positive numbers are said to have a sign of 1 and negative numbers a sign of −1, so R_i is the value of the rank corresponding to $|D_i|$ multiplied by the sign of D_i.

If the sample size (n) is less than or equal to 40 and there are no ties among the $|D_i|$ values, the test statistic is W, the sum of the positive R_i values. For example, if $R_1 = 4$, $R_2 = -3$, $R_3 = 5$, $R_4 = 2$, and $R_5 = -1$, then

$$W = 4 + 5 + 2 = 11.$$

A lower critical value, c_L, is read from Table 12 in Appendix B. So for $\alpha = .05$ and $n = 5$, the critical value corresponds to $\alpha/2 = .025$ and is 0, so reject if $W \leq 0$. The upper critical value is

$$c_U = \frac{n(n+1)}{2} - c_L.$$

In the illustration, because $c_L = 0$,

$$c_U = \frac{5(6)}{2} - 0 = 15,$$

meaning that you reject if $W \geq 15$. Because $W = 11$ is between 1 and 15, fail to reject.

If there are ties among the $|D_i|$ values or the sample size exceeds 40, the test statistic is

$$W = \frac{\sum R_i}{\sqrt{\sum R_i^2}}.$$

If there are no ties, this last equation simplifies to

$$W = \frac{\sqrt{6}\sum R_i}{\sqrt{n(n+1)(2n+1)}}.$$

For a two-sided test, reject if $|W|$ equals or exceeds $z_{1-\alpha/2}$, the $1-\alpha/2$ quantile of a standard normal distribution.

Rejecting with the signed rank test indicates that two dependent groups have different distributions. Although the signed rank test can have more power than the sign test, a criticism is that it does not provide certain details about how the groups differ. For instance, in the cork boring example, rejecting indicates that the distribution of weights differs for the north versus east side of a tree, but how might we elaborate on what this difference is? One possibility is to estimate p, the probability that the weight from the north side is less than the weight from the east side. So despite lower power, one might argue that the sign test provides a useful perspective on how groups compare.

15.6.4 S-PLUS Function wilcox.test

The built-in S-PLUS function

wilcox.test(x, y, paired = F, exact = T)

performs the Wilcoxon signed rank test just described by setting the argument paired to T, for true. (With paired=F, a one-sample version of the test is used.)

15.7 Comparing Multiple Dependent Groups

There are a variety of rank-based methods for comparing multiple dependent groups. Recent advances and techniques can be found in Agresti and Pendergast (1986); Akritas and Arnold (1994); Brunner and Denker (1994); Brunner, Munzel, and Puri (1999); and Munzel and Brunner (2000b). Here, only portions of these methods are described, plus some classic techniques that are typically covered in an introductory course.

15.7.1 Friedman's Test

The classic rank-based method for comparing multiple dependent groups is Friedman's test. The goal is to test

$$H_0 : F_1(x) = \cdots = F_J(x),$$

the hypothesis that all J dependent groups have identical distributions. The method begins by assigning ranks within rows. For example, imagine that for each individual, measures are taken at three different times, yielding

Time 1	Time 2	Time 3
9	7	12
1	10	4
8	2	1
	⋮	

The ranks corresponding to the first row (the values 9, 7, and 12) are 2, 1, and 3. For the second row the ranks are 1, 3, and 2, and continuing in this manner the data become

Time 1	Time 2	Time 3
2	1	3
1	3	2
3	2	1
	⋮	

Let R_{ij} be the resulting rank corresponding to X_{ij} ($i = 1, \ldots, n$; $j = 1, \ldots, J$). Compute

$$A = \sum_{j=1}^{J}\sum_{i=1}^{n} R_{ij}^2$$

$$R_j = \sum_{i=1}^{n} R_{ij}$$

$$B = \frac{1}{n}\sum_{j=1}^{J} R_j^2$$

$$C = \frac{1}{4}nJ(J+1)^2.$$

If there are no ties, the equation for A simplifies to

$$A = \frac{nJ(J+1)(2J+1)}{6}.$$

The test statistic is

$$F = \frac{(n-1)(B-C)}{A-B}. \tag{15.17}$$

Reject if $F \geq f_{1-\alpha}$, or if $A = B$, where $f_{1-\alpha}$ is the $1-\alpha$ of an F-distribution with $\nu_1 = J-1$ and $\nu_2 = (n-1)(J-1)$ degrees of freedom.

15.7.2 Agresti–Pendergast Test

A variety of improvements on Friedman's test have been proposed (e.g., Iman, 1974; Quade, 1979). One such method that currently stands out was proposed by Agresti and Pendergast (1986); it provides better control over the probability of a Type I error than does Iman's method and it can have higher power than Friedman's test. (For theoretical results on the Agresti–Pendergast test, see Kepner & Robinson, 1988.) Basically, the method tests the hypothesis of equal distributions based on the average ranks among the groups. Box 15.6 describes the computations.

BOX 15.6 The Agresti–Pendergast Method for Dependent Groups

Pool all the observations and assign ranks. Let R_{ij} be the resulting rank of the ith observation in the jth group. Compute

$$\bar{R}_j = \frac{1}{n}\sum_{i=1}^{n} R_{ij}$$

$$s_{jk} = \frac{1}{n-J+1}\sum_{i=1}^{n}(R_{ij}-\bar{R}_j)(R_{ik}-\bar{R}_k).$$

Let the vector $\mathbf{R}'$ be defined by

$$\mathbf{R}' = (\bar{R}_1, \ldots, \bar{R}_J),$$

and let $\mathbf{C}$ be the $(J-1)$-by-J matrix given by

$$\begin{pmatrix} 1 & -1 & 0 & \ldots & 0 & 0 \\ 0 & 1 & -1 & \ldots & 0 & 0 \\ \cdot & \cdot & \cdot & \cdot & \cdot & \cdot \\ 0 & 0 & 0 & \ldots & 1 & -1 \end{pmatrix}.$$

The test statistic is

$$F = \frac{n}{J-1}(\mathbf{CR})'(\mathbf{CSC}')^{-1}\mathbf{CR},$$

where

$$\mathbf{S} = (s_{jk}).$$

Decision Rule

Reject if $F \geq f_{1-\alpha}$, the $1-\alpha$ quantile of an F-distribution with $\nu_1 = J-1$ and $\nu_2 = (J-1)(n-1)$ degrees of freedom.

15.7.3 S-PLUS Function apanova

The S-PLUS function

apanova(x)

performs the Agresti–Pendergast ANOVA method based on ranks.

15.8 One-Way Multivariate Methods

There are several rank-based methods for dealing with situations where there are J independent groups with K measures for each individual (e.g., Choi & Marden, 1997; Liu & Singh, 1993; Hettmansperger, Möttönen, & Oja, 1997; Munzel & Brunner, 2000a). The theoretical details of the method developed by Hettmansperger, Möttönen, and Oja assume that tied values occur with probability zero and that distributions are symmetric, so the details of their method are not given here.

15.8.1 The Munzel–Brunner Method

This section describes a one-way multivariate method derived by Munzel and Brunner (2000a). (A variation of the Munzel–Brunner method can be used in place of the Agresti–Pendergast procedure in Section 15.7.2, but the relative merits of these two techniques have not been explored.) For the jth group, there are n_j randomly sampled vectors of observations, with each vector containing K measures. Let $F_{jk}(x)$ be the distribution associated with the jth group and kth measure. So, for example, $F_{32}(6)$ is the probability that for the third group, the second variable will be less than or equal to 6 for a randomly sampled individual. For the kth measure, the goal is to test the hypothesis that all J groups have identical distributions. And the more general goal is to test the hypothesis that simultaneously, all groups have identical distributions for each of the K measures under consideration. That is, the goal is to test

$$H_0 : F_{1k}(x) = \cdots = F_{Jk}(x) \qquad \text{for all } k = 1, \ldots, K. \tag{15.18}$$

To apply the method, begin with the first of the K measures, pool all the observations among the J groups, and assign ranks. Ties are handled in the manner described in Section 15.1.3. Repeat this process for all K measures and label the results R_{ijk}. That is, R_{ijk} is the rank of the ith observation in the jth group and for the kth measure. Let

$$\bar{R}_{jk} = \frac{1}{n_j}\sum_{i=1}^{n_j} R_{ijk},$$

be the average rank for the jth group corresponding to the kth measure. Set

$$\hat{Q}_{jk} = \frac{\bar{R}_{jk} - .5}{n},$$

where $n = \sum n_j$ is the total number of randomly sampled vectors among the J groups. The remaining calculations are summarized in Box 15.7. The $\hat{Q}$ values are called the *relative effects* and reflect the ordering of the average ranks. If, for example, $\hat{Q}_{11} < \hat{Q}_{21}$, the typical rank for variable 1 in group 1 is less than the typical rank for variable 1 in group 2. More generally, if $\hat{Q}_{jk} < \hat{Q}_{j'k}$, then based on the kth measure, the typical rank (or observed value) for group j is less than the typical rank for group j'.

BOX 15.7 The Munzel–Brunner One-Way Multivariate Method

Let

$$\hat{\mathbf{Q}} = (\hat{Q}_{11}, \hat{Q}_{12}, \ldots, \hat{Q}_{1K}, \hat{Q}_{21}, \ldots, \hat{Q}_{JK})',$$

$$\mathbf{R}_{ij} = (R_{ij1}, \ldots, R_{ijK})',$$

$$\bar{\mathbf{R}}_j = (\bar{R}_{j1}, \ldots, \bar{R}_{jK})',$$

$$\mathbf{V}_j = \frac{1}{nn_j(n_j - 1)} = \sum_{i=1}^{n_j} (\mathbf{R}_{ij} - \bar{\mathbf{R}}_j)(\mathbf{R}_{ij} - \bar{\mathbf{R}}_j)',$$

where $n = \sum n_j$, and let

$$\mathbf{V} = \text{diag}\{\mathbf{V}_1, \ldots, \mathbf{V}_J\}.$$

Compute the matrix $\mathbf{M}_A$ as described in Section 15.4. The test statistic is

$$F = \frac{n}{\text{tr}(\mathbf{M}_A\mathbf{V})} \hat{\mathbf{Q}}'\mathbf{M}_A\hat{\mathbf{Q}}.$$

Decision Rule

Reject if $F \geq f$, where f is the $1 - \alpha$ quantile of an F-distribution with

$$\nu_1 = \frac{(\text{tr}(\mathbf{M}_A\mathbf{V}))^2}{\text{tr}(\mathbf{M}_A\mathbf{V}\mathbf{M}_A\mathbf{V})}$$

and $\nu_2 = \infty$ degrees of freedom.

15.8.2 S-PLUS Function mulrank

The S-PLUS function

mulrank(J, K, x)

performs the one-way multivariate method in Box 15.7. The data are stored in x, which can be a matrix or have list mode. If x is a matrix, the first K columns correspond to the K measures for group 1, the second K correspond to group 2, and so forth. If stored in list mode, x[[1]], ..., x[[K]] contain the data for group 1, x[[K+1]], ..., x[[2K]] contain the data for group 2, and so on.

TABLE 15.2 CGI and PGI Scores After Four Weeks of Treatment

Exercise		Clomipramine		Placebo	
CGI	PGI	CGI	PGI	CGI	PGI
4	3	1	2	5	4
1	1	1	1	5	5
2	2	2	0	5	6
2	3	2	1	5	4
2	3	2	3	2	6
1	2	2	3	4	6
3	3	3	4	1	1
2	3	1	4	4	5
5	5	1	1	2	1
2	2	2	0	4	4
5	5	2	3	5	5
2	4	1	0	4	4
2	1	1	1	5	4
2	4	1	1	5	4
6	5	2	1	3	4

EXAMPLE. Table 15.2 summarizes data (reported by Munzel & Brunner, 2000a) from a psychiatric clinical trial where three methods are compared for treating individuals with panic disorder. The three methods are exercise, clomipramine, and placebo. The two measures of effectiveness were a clinical global impression (CGI) and the patient's global impression (PGI). The test statistic is $F = 12.7$ with $\nu_1 = 2.83$ and a significance level less than .001. The relative effects are:

```
$q.hat:
           [,1]        [,2]
[1,]  0.5074074   0.5096296
[2,]  0.2859259   0.2837037
[3,]  0.7066667   0.7066667
```

So among the three groups, the second group, clomipramine, has the lowest relative effects. That is, the typical ranks were lowest for this group, and the placebo group had the highest ranks on average. ■

15.8.3 The Choi–Marden Multivariate Rank Test

This section describes a multivariate analog of the Kruskal–Wallis test derived by Choi and Marden (1997). There are actually many variations of the approach they considered, but here attention is restricted to the version they focused on. As with the method in Section 15.8.1, we have K measures for each individual and there are J

independent groups. For the jth group and any vector of constants $\mathbf{x} = (x_1, \ldots, x_K)$, let

$$F_j(\mathbf{x}) = P(X_{j1} \le x_1, \ldots, X_{jK} \le x_K).$$

So, for example, $F_1(\mathbf{x})$ is the probability that for the first group, the first of the K measures is less than or equal to x_1, the second of the K measures is less than or equal to x_2, and so forth. The null hypothesis is that for any $\mathbf{x}$,

$$H_0 : F_1(\mathbf{x}) = \cdots = F_J(\mathbf{x}), \tag{15.19}$$

which is sometimes called the *multivariate hypothesis*, to distinguish it from Equation (15.18), which is called the *marginal hypothesis*. The multivariate hypothesis is a stronger hypothesis, in the sense that if it is true, then by implication the marginal hypothesis is true as well. For example, if the marginal distributions for both groups are standard normal distributions, the marginal hypothesis is true; but if the groups have different correlations, the multivariate hypothesis is false.

The Choi–Marden method represents an extension of a technique derived by Möttönen and Oja (1995) and is based on a generalization of the notion of a rank to multivariate data, which was also used by Chaudhuri (1996, Section 4). First consider a random sample of n observations with K measures for each individual or thing and denote the ith vector of observations by

$$\mathbf{X}_i = (X_{i1}, \ldots, X_{iK}).$$

Let

$$A_{ii'} = \sqrt{\sum_{k=1}^{K} (X_{ik} - X_{i',k})^2},$$

Here, the "rank" of the ith vector is itself a vector (having length K), given by

$$\mathbf{R}_i = \frac{1}{n} \sum_{i=1}^{n} \frac{\mathbf{X}_i - \mathbf{X}_{i'}}{A_{ii'}},$$

where

$$\mathbf{X}_i - \mathbf{X}_{i'} = (X_{i1} - X_{i'1}, \ldots, X_{iK} - X_{i'K}).$$

The remaining calculations are summarized in Box 15.8. All indications are that this method provides good control over the probability of a Type I error when ties never occur. There are no known problems when there are tied values, but this issue is in need of more research.

BOX 15.8 The Choi–Marden Method

Pool the data from all J groups and compute rank vectors as described in the text. The resulting rank vectors are denoted by $\mathbf{R}_1, \ldots, \mathbf{R}_n$, where $n = \sum n_j$

Continued

BOX 15.8 (*Continued*) is the total number of vectors among the J groups. For each of the J groups, average the rank vectors and denote the average of these vectors for the jth group by $\bar{\mathbf{R}}_j$.

Next, assign ranks to the vectors in the jth group, ignoring all other groups. We let $\mathbf{V}_{ij}$ (a column vector of length K) represent the rank vector corresponding to the ith vector of the jth group ($i = 1, \ldots, n_j$; $j = 1, \ldots, J$) to make a clear distinction with the ranks based on the pooled data. Compute

$$S = \frac{1}{n - J} \sum_{j=1}^{J} \sum_{i=1}^{n_j} \mathbf{V}_{ij}\mathbf{V}'_{ij},$$

where $\mathbf{V}'_{ij}$ is the transpose of $\mathbf{V}_{ij}$ (so S is a K-by-K matrix). The test statistic is

$$H = \sum_{j=1}^{J} n_j \bar{\mathbf{R}}'_j \mathbf{S}^{-1} \bar{\mathbf{R}}_j. \tag{15.20}$$

(For $K = 1$, H does not quite reduce to the Kruskall–Wallis test statistic. In fact, H avoids a certain technical problem that is not addressed by the Kruskall–Wallis method.)

Decisions Rule

Reject if $H \geq c$, where c is the $1 - \alpha$ quantile of a chi-squared distribution with degrees of freedom $K(J - 1)$.

15.8.4 S-PLUS Function cmanova

The S-PLUS function

cmanova(J,K,x)

performs the Choi–Marden method just described. The data are assumed to be stored in x as described in Section 15.8.2.

15.9 Between-by-Within Designs

There are a variety of rank-based methods one might use in a between-by-within-subjects design, or what is called a *split-plot design*. That is, as in Section 11.9, we have a two-way design where the J levels of the first factor are independent and the K levels of the other factor are possibly dependent. For comparing the independent groups, one approach is to use the methods in Section 15.8. As for the dependent groups, one possibility is to ignore the levels of factor A and use the methods in Section 15.6. For example, if the two independent groups are males and females, and measures for every individual are taken at three different times, you could simply pool the males and females and test the hypothesis that the distributions are identical at times 1, 2, and 3.

Yet another approach is to proceed along the lines in Section 11.9, only rather than compare measures of location, hypotheses are formulated in terms of distributions, as was done in Section 15.4. So as in Section 15.4, main effects for factor A are given in terms of

$$\bar{F}_{j.}(x) = \frac{1}{K}\sum_{k=1}^{K} F_{jk}(x),$$

the average of the distributions among the K levels of factor B corresponding to the jth level of factor A. The only difference between the present situation and Section 15.4 is that the average of dependent groups is being used. The hypothesis of no main effects for factor A is

$$H_0 : \bar{F}_{1.}(x) = \bar{F}_{2.}(x) = \cdots = \bar{F}_{J.}(x)$$

for any x. Letting

$$\bar{F}_{.k}(x) = \frac{1}{J}\sum_{j=1}^{J} F_{jk}(x)$$

be the average of the distributions for the kth level of factor B, the hypothesis of no main effects for factor B is

$$H_0 : \bar{F}_{.1}(x) = \bar{F}_{.2}(x) = \cdots = \bar{F}_{.K}(x).$$

As for interactions, again proceed as before. So for a 2-by-2 design, no interaction is taken to mean that for any x,

$$F_{11}(x) - F_{12}(x) = F_{21}(x) - F_{22}(x).$$

More generally, the hypothesis of no interactions among all JK groups is

$$H_0 : F_{jk}(x) - \bar{F}_{j.}(x) - \bar{F}_{.k}(x) + \bar{F}_{..}(x) = 0,$$

for any x, all j ($j = 1, \ldots, J$) and all k ($k = 1, \ldots, K$), where

$$\bar{F}_{..}(x) = \frac{1}{JK}\sum_{j=1}^{J}\sum_{k=1}^{K} F_{jk}(x).$$

A technical difficulty is taking into account the dependence among the levels of factor B, and here the methods covered in Brunner, Domhof, and Langer (2002, Chapter 8) are used. (Beasley, 2000, suggests another approach to interactions; perhaps it has a practical advantage over the approach described here, but this issue has not been investigated as yet.)

As usual, let X_{ijk} represent the ith observation for level j of factor A and level k of factor B. Here, $i = 1, \ldots, n_j$. That is, the jth level of factor A has n_j vectors of observations, each vector containing K values. So for the jth level of factor A there are a total of n_jK observations; and among all the groups, the total number of observations is denoted by N. So the total number of vectors among the J groups is $n = \sum n_j$, and the total number of observations is $N = K \sum n_j = Kn$.

Pool all N observations and assign ranks. As usual, midranks are used if there are tied values. Let R_{ijk} represent the rank associated with X_{ijk}. The remaining calculations for factor A are relegated to Box 15.9; factor B and interactions are tested as described in Box 15.10.

BOX 15.9 Main Effects for Factor A in a Between-by-Within Design

Let

$$\bar{R}_{.jk} = \frac{1}{n_j} \sum_{i=1}^{n_j} R_{ijk},$$

$$\bar{R}_{.j.} = \frac{1}{K} \sum_{k=1}^{K} \bar{R}_{.jk},$$

$$\bar{R}_{ij.} = \frac{1}{K} \sum_{k=1}^{K} R_{ijk},$$

$$\hat{\sigma}_j^2 = \frac{1}{n_j - 1} \sum_{i=1}^{n_j} (\bar{R}_{ij.} - \bar{R}_{.j.})^2,$$

$$S = \sum_{j=1}^{J} \frac{\hat{\sigma}_j^2}{n_j},$$

$$U = \sum_{j=1}^{J} \left(\frac{\hat{\sigma}_j^2}{n_j}\right)^2,$$

$$D = \sum_{j=1}^{J} \frac{1}{n_j - 1} \left(\frac{\hat{\sigma}_j^2}{n_j}\right)^2.$$

Factor A

The test statistic is

$$F_A = \frac{J}{(J-1)S} \sum_{j=1}^{J} (\bar{R}_{.j.} - \bar{R}_{...})^2,$$

Continued

BOX 15.9 (*Continued*)
where $\bar{R}_{...} = \sum \bar{R}_{.j.}/J$. The degrees of freedom are

$$\nu_1 = \frac{(J-1)^2}{1 + J(J-2)U/S^2},$$

$$\nu_2 = \frac{S^2}{D}.$$

Decision Rule
Reject if $F_A \geq f$, where f is the $1-\alpha$ quantile of an F-distribution with ν_1 and ν_2 degrees of freedom.

BOX 15.10 Interactions and Main Effects for Factor B in a Between-by-Within Design

Factor B
Following the notation in Box 15.9, let

$$\mathbf{R}_{ij} = (R_{ij1}, \ldots, R_{ijK})',$$

$$\bar{\mathbf{R}}_j = \frac{1}{n_j}\sum_{i=1}^{n_j} \mathbf{R}_{ij}, \bar{\mathbf{R}}_{..} = \frac{1}{J}\sum_{j=1}^{J} \bar{\mathbf{R}}_j, \qquad n = \sum n_j \quad (\text{so } N = nK),$$

$$\mathbf{V}_j = \frac{n}{N^2 n_j(n_j - 1)} \sum_{i=1}^{n_j} (\mathbf{R}_{ij} - \bar{\mathbf{R}}_j)(\mathbf{R}_{ij} - \bar{\mathbf{R}}_j)'.$$

So $\mathbf{V}_j$ is a K-by-K matrix of covariances based on the ranks. Let

$$\mathbf{S} = \frac{1}{J^2}\sum_{j=1}^{J} \mathbf{V}_j$$

and let $\mathbf{P}_K$ be defined as in Box 15.5.
The test statistic is

$$F_B = \frac{n}{N^2 \mathrm{tr}(\mathbf{P}_K \mathbf{S})} \sum_{k=1}^{K} (\bar{R}_{..k} - \bar{R}_{...})^2.$$

The degrees of freedom are

$$\nu_1 = \frac{(\mathrm{tr}(\mathbf{P}_K \mathbf{S}))^2}{\mathrm{tr}(\mathbf{P}_K \mathbf{S} \mathbf{P}_K \mathbf{S})}, \qquad \nu_2 = \infty.$$

Continued

BOX 15.10 (*Continued*)

Interactions

Let $\mathbf{V}$ be the block diagonal matrix based on the matrices $\mathbf{V}_j$, $j = 1, \ldots, J$. (See Appendix C for a definition of a block diagonal matrix.) Letting $\mathbf{M}_{AB}$ be defined as in Box 15.5, the test statistic is

$$F_{AB} = \frac{n}{N^2 \text{tr}(\mathbf{M}_{AB}\mathbf{V})} \sum_{j=1}^{J} \sum_{k=1}^{K} (\bar{R}_{.jk} - \bar{R}_{.j.} - \bar{R}_{..k} + \bar{R}_{...})^2.$$

The degrees of freedom are

$$\nu_1 = \frac{(\text{tr}(\mathbf{M}_{AB}\mathbf{V}))^2}{\text{tr}(\mathbf{M}_{AB}\mathbf{V}\mathbf{M}_{AB}\mathbf{V})}, \qquad \nu_2 = \infty.$$

Decision Rule

Reject if $F_A \geq f$ (or if $F_{AB} \geq f$), where f is the $1 - \alpha$ quantile of an F-distribution with ν_1 and ν_2 degrees of freedom.

15.9.1 S-PLUS Function bwrank

The S-PLUS function

bwrank(J,K,x)

performs a between-by-within ANOVA based on ranks using the method just described. In addition to testing hypotheses as indicated in Boxes 15.9 and 15.10, the function returns the average ranks ($\bar{R}_{.jk}$) associated with all JK groups as well as the relative effects, $(\bar{R}_{.jk} - .5)/N$.

EXAMPLE. Lumley (1996) reports data on shoulder pain after surgery; the data are from a study by Jorgensen et al. (1995). Table 15.3 shows a portion of the results where two treatment methods are used and measures of pain are taken at three different times. The output from bwrank is

```
$test.A:
[1] 12.87017

$sig.A:
[1] 0.001043705

$test.B:
[1] 0.4604075

$sig.B:
[1] 0.5759393
```

Continued

EXAMPLE. *(Continued)*

```
$test.AB:
[1] 8.621151

$sig.AB:
[1] 0.0007548441

$avg.ranks:
          [,1]      [,2]      [,3]
[1,]  58.29545  48.40909  39.45455
[2,]  66.70455  82.36364  83.04545

$rel.effects:
          [,1]      [,2]      [,3]
[1,] 0.4698817 0.3895048 0.3167036
[2,] 0.5382483 0.6655580 0.6711013
```

Continued

TABLE 15.3 Shoulder Pain Data

Active treatment			No active treatment		
Time 1	Time 2	Time 3	Time 1	Time 2	Time 3
1	1	1	5	2	3
3	2	1	1	5	3
3	2	2	4	4	4
1	1	1	4	4	4
1	1	1	2	3	4
1	2	1	3	4	3
3	2	1	3	3	4
2	2	1	1	1	1
1	1	1	1	1	1
3	1	1	1	5	5
1	1	1	1	3	2
2	1	1	2	2	3
1	2	2	2	2	1
3	1	1	1	1	1
2	1	1	1	1	1
1	1	1	5	5	5
1	1	1	3	3	3
2	1	1	5	4	4
4	4	2	1	3	3
4	4	4			
1	1	1			
1	1	1			

1 = low, 5 = high.

EXAMPLE. (*Continued*)
So treatment methods are significantly different and there is a significant interaction, but no significant difference is found over time. Note that the average ranks and relative effects suggest that a disordinal interaction might exist. In particular, for group 1 (the active treatment group), time 1 has higher average ranks than time 2, and the reverse is true for the second group. However, the Wilcoxon signed rank test fails to reject at the .05 level when comparing time 1 to time 2 for both groups. When comparing time 1 to time 3 for the first group, again using the Wilcoxon signed rank test, you reject at the .05 level, but a nonsignificant result is obtained for group 2. So again a disordinal interaction appears to be a possibility, but the empirical evidence is not compelling. ■

15.9.2 Multiple Comparisons Based on Comparing Distributions

Multiple comparisons for a between-by-within design can be performed in essentially the same way used to compare all independent groups as described in Section 15.5. For example, when dealing with factor A, simply compare level j to level j', ignoring the other levels. For all pairwise comparisons among the J levels of factor A, FWE is controlled using Rom's method or the Benjamini–Hochberg technique, and the same is done for factor B and the collection of all interactions corresponding to any two levels of factor A and any two levels of factor B.

15.9.3 S-PLUS Function bwrmcp

The S-PLUS function

```
bwrmcp(J,K,x,grp=NA,alpha=.05,bhop=F)
```

performs all pairwise multiple comparisons using the method for a between-by-within-subjects design described in the previous section. The value for alpha indicates the FWE for all pairwise comparisons among the levels of factor A, as well as all pairwise comparisons among the levels of factor B and all interactions. So if $J = 3$ and $K = 4$, the default value for FWE when performing all three comparisons among any two levels of factor A will be .05. There are a total of 18 interactions that would be tested, and again the default FWE will be .05 among these 18 hypotheses.

15.9.4 Multiple Comparisons When Using a Patel–Hoel Approach to Interactions

Rather than compare distributions when dealing with a between-by-within design, one could use a simple analog of the Patel–Hoel approach instead. First consider a two-by-two design and focus on level 1 of factor A. Then the two levels of factor B are dependent and can be compared with the sign test described in Section 15.6.1. In essence, inferences are being made about p_1, the probability that for a randomly sampled pair of observations, the observation from level 1 of factor B is less than the corresponding observation from level 2. Of course, for level 2 of factor A, we can again compare levels 1 and 2 of factor B with the sign test. Now we let p_2 be the probability that for a randomly sampled pair of observations, the observation from level 1 of factor B is less than the corresponding observation from level 2. Then as in Section 15.5.2, no interaction can be defined as $p_1 = p_2$.

The hypothesis of no interaction,

$$H_0 : p_1 = p_2,$$

is just the hypothesis that two independent binomials have equal probabilities of success, which can be tested using one of the methods described in Section 8.13. Here, Beal's method (described in Box 8.4) is used rather than the Storer–Kim method in Box 8.3, because it currently seems that Beal's method provides more accurate control over FWE for the problem at hand, execution time can be much lower when sample sizes are large, and, unlike the Storer–Kim procedure, Beal's method provides confidence intervals. There are various ways FWE might be controlled using the methods in Chapter 12. However, it seems that they result in an actual FWE that can be substantially smaller than the nominal level, so the following modification is used.

Among a collection of techniques considered by Wilcox (2001c) for controlling FWE, the following method was found to be relatively effective. Let q be the critical value read from Table 10 in Appendix B with degrees of freedom $\nu = \infty$. Recall that in Table 10, C is the total number of hypotheses to be tested. Assuming that all pairs of rows and columns are to be considered when checking for interactions,

$$C = \frac{J^2 - J}{2} \times \frac{K^2 - K}{2}.$$

Let Z be a standard normal random variable (so $P(Z \le z)$ can be read from Table 1 in Appendix B). Then if FWE is to be α, test each of the C hypotheses at the α_a level, where

- If $(J, K) = (5, 2)$, then $\alpha_a = 2(1 - P(Z \le q))$.
- If $(J, K) = (3,2)$, $(4, 2)$, or $(2, 3)$, then $\alpha_a = 3(1 - P(Z \le q))$.
- For all other J and K values, $\alpha_a = 4(1 - P(Z \le q))$.

These adjusted α values appear to work well when the goal is to achieve FWE less than or equal to .05. Whether this remains the case with FWE equal to .01 is unknown. For $C > 28$ and FWE equal to .05, use

$$q = 2.383904C^{1/10} - .202.$$

(Of course, for $C = 1$, no adjustment is necessary; simply use Beal's method as described in Chapter 8.)

Tied values are handled in the same manner as with the sign test in Section 15.6.1: Pairs of observations with identical values are simply discarded. So among the remaining observations, for every pair of observations, the observation from level 1 of factor B, for example, is either less than or greater than the corresponding value from level 2.

A criticism of the method in this section is that compared to the test of no interaction in Section 15.9.2, power is generally lower. However, the method in Section 15.9.2 tests hypotheses about entire distributions based on average ranks.

The method in this section adds information about how the groups differ that is not supplied merely by comparing distributions. The situation is similar to how the sign test compares to the Wilcoxon signed rank test. The sign test generally has lower power, but it provides a direct estimate of the probability that an observation in the first group is less than an observation in the second. This probability adds perspective that is not available when attention is restricted to the Wilcoxon signed rank test. The method in this section deals directly with how p_1 compares to p_2, which provides a characterization how groups differ that we do not get when comparing groups based on the average ranks.

A variation of the approach in this section is where, for level 1 of factor B, p_1 is the probability that an observation from level 1 of factor A is less than an observation from level 2. Similarly, p_2 is now defined in terms of the two levels of factor A when working with level 2 of factor B. However, the details of how to implement this approach have not been studied.

15.9.5 S-PLUS Function sisplit

The method just described for testing hypotheses of no interaction can be applied with the S-PLUS function

sisplit(J,K,x)

This function assumes $\alpha = .05$; other values are not allowed.

15.10 Rank-Based Correlations

As in previous chapters, imagine that we have n randomly sampled pairs of observations, which are labeled $(X_1, Y_1), \ldots, (X_n, Y_n)$. Two goals are determining whether these two measures are dependent and, if they are, characterizing what this dependence is like. There are two well-known rank-based measures of association aimed at accomplishing these goals: Kendalls' tau and Spearman's rho.

15.10.1 Kendall's tau

Kendall's tau is based on the following idea. Consider two pairs of observations, which are labeled in the usual way as (X_1, Y_1) and (X_2, Y_2). For convenience, assume that $X_1 < X_2$. If $Y_1 < Y_2$, then these two pairs of numbers are said to be concordant. That is, if Y increases as X increases or if Y decreases as X decreases, we have concordant pairs of observations. If two pairs of observations are not concordant, they are said to be discordant.

Roughly, among all pairs of points, Kendall's tau is just the average number that are concordant minus the average number that are discordant. If the measures X and Y are independent, then this difference should be approximately equal to zero. To describe its computation in a more formal manner, let $K_{ij} = 1$ if the ith and jth pairs of observations are concordant; otherwise $K_{ij} = -1$. Then Kendall's tau is given by

$$\hat{\tau} = \frac{2\sum_{i<j} K_{ij}}{n(n-1)} \tag{15.21}$$

and has a value between −1 and 1. If $\hat{\tau}$ is positive, there is a tendency for Y to increase with X—possibly in a nonlinear fashion—and if $\hat{\tau}$ is negative, the reverse is true.

The population analog of $\hat{\tau}$ is labeled τ and can be shown to be zero when X and Y are independent. To test

$$H_0 : \tau = 0,$$

compute

$$\sigma_\tau^2 = \frac{2(2n+5)}{9n(n-1)},$$

$$Z = \frac{\hat{\tau}}{\sigma_\tau},$$

and reject if

$$|Z| \geq z_{1-\frac{\alpha}{2}},$$

where $z_{1-\alpha/2}$ is the $1-\alpha/2$ quantile of a standard normal distribution (which can be read from Table 1).

Cliff (1996) suggests computing a confidence interval for τ by estimating the variance of $\hat{\tau}$ and applying Laplace's strategy. There are various ways this might be done, and a method that seems to be relatively effective is outlined in Box 15.11. (Also see J. D. Long & Cliff, 1997.) Another approach is to use the heteroscedastic bootstrap method in Section 13.6.4, which can be done with the S-PLUS function corb in Section 13.6.5. Direct comparisons between the heteroscedastic bootstrap method and the method in Box 15.11 have not been made. An educated guess is that the bootstrap method is better for general use, but this issue is in need of further study.

BOX 15.11 Confidence Interval for Kendall's tau

For the ith and hth pair of observations, set

$$U_{ih} = \text{sign}(X_i - X_h),$$

where sign(X) is 1, 0, or −1 according to whether X is greater than, equal to, or less than zero. Let

$$W_{ih} = \text{sign}(Y_i - Y_h),$$

$$t_{ih} = U_{ih} W_{ih},$$

$$t_{i.} = \frac{1}{n-1} \sum_h t_{ih},$$

$$V_1 = \frac{1}{n-1} \sum_i (t_{i.} - \hat{\tau})^2,$$

Continued

BOX 15.11 (*Continued*)

$$V_2 = \frac{1}{[n(n-1)]-1}\left[\left(\sum\sum t_{ib}^2\right) - n(n-1)\hat{\tau}^2\right],$$

$$s_t^2 = \frac{1}{n(n-1)}(4(n-2)V_1 + 2V_2).$$

The $1-\alpha$ confidence interval for τ is

$$\hat{\tau} \pm z_{1-\alpha/2}s_t,$$

where z is the $1-\alpha/2$ quantile of a standard normal distribution (read from Table 1 in Appendix B).

For a general discussion about handling tied values when dealing with Kendall's tau, readers are referred to Cliff (1996), who provides an excellent summary of various approaches. Here it is merely remarked that different strategies lead to different measures of association, which include the Goodman–Kruskal γ, Somers' d, and Yule's Q.

EXAMPLE. Imagine we observe the 10 values $X = .1, .2, \ldots, 1$ and that $Y = X^2$. Then there is a perfect monotonic increasing relationship between X and Y, $\hat{\tau} = 1$, and Pearson's correlation is $r = .975$. So in this particular case there is little separating the two coefficients. However, Kendall's tau provides some protection against missing an association due to one or more outliers. For example, if the largest Y value is increased from 1 to 10, again $\hat{\tau} = 1$ but now $r = .59$ with a significance level (based on Student's T) equal to .07. Increasing the largest Y value to 50, $r = .54$.

As just indicated, Kendall's tau provides protection against outliers among the X values; it does the same among the Y values, but it does not take into account the overall structure of the data. That is, it does not address the concerns raised in Section 13.3. ■

EXAMPLE. Following the second illustration in Section 13.3, 20 points were generated according to the model $Y = X + \epsilon$, where both X and ϵ have standard normal distributions. The points are shown in Figure 13.4. Kendall's tau is estimated to be .368 and when testing $H_0: \tau = 0$, the significance level is .023 based on the method in Box 15.11. Now we add two aberrant points at $(X, Y) = (2.1, -2.4)$; a scatterplot of the points now appears as shown in Figure 13.5. The estimate of Kendall's tau drops to .13 and the significance level increases to .398 — these two unusual points have a substantial impact on Kendall's tau. ■

15.10.2 Kendall's tau and the Theil–Sen Estimator

There is a connection between Kendall's tau and the Theil–Sen regression estimator in Section 13.2.1 that is worth mentioning. As was done in Chapter 13, let b_1 be some estimate of the slope. A general approach for determining b_1 from data is to take it to be the value that results in a zero correlation between the n pairs of points Y_i and $Y_i - b_1X_i$. If the correlation used is Kendall's tau, b_1 is the Theil–Sen estimate in Chapter 13.

15.10.3 S-PLUS Function tau

The S-PLUS function

tau(x,y)

has been supplied for computing Kendall's tau and testing $H_0: \tau = 0$. The function returns an estimate of τ plus the significance level. To compute a confidence interval for τ, it is suggested that the S-PLUS function corb in Section 13.6.5 be used.

15.10.4 Spearman's rho

Spearman's rho, labeled r_s, is just Pearson's correlation based on the ranks associated with X versus the ranks associated with Y. Under independence, the population analog of r_s, ρ_s, is zero. Also, like Kendall's tau, Spearman's rho is exactly equal to 1 if there is a monotonic increasing relationship between X and Y. That is, Y is a strictly increasing function of X. And $\rho_s = -1$ if the association is monotonic decreasing instead.

The usual approach to testing

$$H_0: \rho_s = 0$$

is based on

$$T = \frac{r_s\sqrt{n-2}}{\sqrt{1-r_s^2}}.$$

When there is independence, T has, approximately, a Student's T-distribution with $\nu = n - 2$ degrees of freedom. So reject and conclude there is an association if $|T| \geq t$, where t is the $1 - \alpha/2$ quantile of a Student's T-distribution with $n - 2$ degrees of freedom.

Like Kendall's tau, Spearman's rho provides protection against outliers among the X values or among the Y values, but it does not take into account the overall structure of the data. That is, a few unusual points, properly placed, can have a substantial influence on its value.

EXAMPLE. In the last example it was illustrated that Kendall's tau can be influenced by a few unusual values. We repeat this illustration with Spearman's rho. For the original 20 values, $r_s = .54$ with a significance level of .014. But when the two outliers are added at $(X, Y) = (2.1, -2.4)$, now $r_s = .16$ with a significance level of .48. ■

15.10.5 S-PLUS Function spear

The S-PLUS function

spear(x,y)

computes Spearman's rho and returns the significance level when testing $H_0: \rho_s = 0$ as described in Section 15.10.3. A confidence interval for ρ_s can be computed with the S-PLUS function corb in Section 13.6.5.

15.11 Comparing Rank-Based Correlations

Section 14.3.2 describes a bootstrap method for comparing the correlations associated with two independent groups that allows heteroscedasticity. The computations are performed with the S-PLUS function twocor in Section 14.3.3. It is noted that this method can be used to compare rank-based correlations as well. That is, we have two independent groups with two measures for each participant. Letting ρ_{s1} and ρ_{s2} represent the population Spearman correlations, the S-PLUS function computes a confidence interval for $\rho_{s1} - \rho_{s2}$, so, in particular, the hypothesis

$$H_0: \rho_{s1} = \rho_{s2}$$

can be tested.

EXAMPLE. In Section 14.7, one of the examples was aimed at investigating whether the association between two variables from a reading study (orthographic ability, Y, and a measure of sound blending, X_1) is modified by a third variable, phonological awareness (X_2). A smooth of the data, using both X_1 and X_2 as predictors, is shown in Figure 14.9. Attempts at establishing that X_2 modifies the association failed, but it was noted that there seems to be a nonlinear association between Y and X_1. Rank-based correlations are sensitive to monotonic associations that are not necessarily linear, and it was suggested that using a rank-based correlation with these data might make a difference in the conclusions reached. If we use the S-PLUS function runcor (in Section 14.7.1) with the argument corfun set equal to spear (so that Spearman's correlation is used), we see that the association between X_1 and Y appears to change around $X_2 = 14$. Using the S-PLUS function twocor to compare Spearman's correlation between X_1 and Y when $X_2 < 14$, versus $X_2 \geq 14$, the .95 confidence interval for the difference between these two correlations is $(-0.775, -0.012)$ with a significance level of .043. So there is empirical evidence that X_2 modifies the association between X_1 and Y. ■

15.12 Rank-Based Regression

Yet another approach to regression, beyond those covered in Chapters 6 and 13, is to minimize sum function of the ranks of the residuals. In simple regression, for example, with n pairs of values, $(X_1, Y_1), \ldots, (X_n, Y_n)$, consider any choice for the slope and intercept, say, b_1 and b_0. As in previous chapters, $\hat{Y}_i = b_1X_i + b_0$ is the predicted value of Y based on X_i. Least squares chooses the slope and intercept so as to minimize the sum of squared residuals, $\sum r_i^2$, where as usual $r_i = Y_i - \hat{Y}_i$. A rank-based approach simply replaces the residuals with their ranks, and often some function of the ranks is used instead (e.g., Hössjer, 1994). Here attention is focused on a slight variation of this method, which is called the Wilcoxon R estimate. The method begins by choosing the slope (b_1) so as to minimize

$$\sum_{i=1}^{n} = a(R(Y_i - b_1X_i))(Y_i - b_1X_i), \qquad (15.22)$$

where $R(Y_i - b_1X_i)$ is the rank of $Y_i - b_1X_i$ among $Y_1 - b_1X_1, \ldots, Y_n - b_1X_n$,

$$a(i) = \phi\left(\frac{i}{n+1}\right),$$

and

$$\phi(u) = \sqrt{12}\left(u - \frac{1}{2}\right).$$

Once b_1 is determined, one way of estimating the intercept is with

$$b_0 = \text{med}\{Y_i - b_1X_i\},$$

the median of the values $Y_1 - b_1X_1, \ldots, Y_n - b_1X_n$. (Readers interested in a theoretical treatment of this estimator are referred to Hettmansperger & McKean, 1998.)

15.12.1 S-PLUS Function wreg

The S-PLUS function

```
wreg(x,y)
```

computes an estimate of the slope and intercept using the Wilcoxon R estimate just described. A negative feature of this function is that execution time can be somewhat high compared to some of the other estimators in Chapter 13.

EXAMPLE. For the reading data in Figure 13.18, wreg estimates the slope to be zero, which is close to the least squares estimate of −0.02 as well as the Coakley–Hettmansperger M-estimate of −0.04. It is fairly evident, however, that the six most right points are outliers. If these six points are ignored, wreg estimates the slope to be −0.53. The Theil–Sen estimate of the slope is −0.6, again ignoring the outliers; but even with the outliers, the Theil–Sen estimate is −0.28. So switching to the rank-based estimate does not eliminate the possibility that a few points can dominate the estimate of the slope and give a relatively poor fit to the bulk of the points. ■

EXAMPLE. For the star data in Figure 13.12, wreg estimates the slope to be −0.477 versus the Theil–Sen estimate of 1.73, which gives a better fit to the bulk of the observations. Again, ignoring the outliers by restricting X to be greater than 4.01, a good fit to the data is obtained with the rank-based regression method. ■

EXAMPLE. As with many of the other estimators covered in Chapter 13, restricting the range of X values by checking for outliers among the X values and eliminating any that are found is no guarantee that a good fit to the data will be obtained with wreg. For example, in Figure 13.5, the data were generated having a slope of 1 and then two unusual points were added, but no outliers are found among the X values using the methods in Chapter 3. The function wreg estimates the slope to be .498. ■

As stressed in Chapter 13, it currently seems that no single regression estimator is ideal under all circumstances, and in applied work it seems that multiple methods are required, at least in the preliminary stages of data analysis. The rank-based regression estimator described here does not perform well in some situations where other estimators give excellent results, but it remains among the group of estimators that seem to have practical value.

One appeal of the Wilcoxon R estimator is that it appears to perform reasonably well when there are multiple predictors and the goal is to detect curvature using a partial residual plot (as described in Section 14.2.5). McKean and Sheather (2000) provide results on this issue and compare the use of the Wilcoxon R estimator to partial residual plots based on least squares.

Yet another issue is testing hypotheses when using the Wilcoxon R estimator. One possibility is the bootstrap method in Section 14.5. It is possible to avoid the bootstrap using results in Hettmansperger and McKean (1998), but it seems the relative merits of this approach, versus the bootstrap method have not been investigated.

15.12.2 Other Rank-Based Estimators

For completeness, there are other rank-based estimators that are briefly mentioned here. Jaeckel (1972) suggests estimating the slope by minimizing

$$\sum_{i<j} |(Y_i - b_1X_i) - (Y_j - b_1X_j)|,$$

which turns out to be tantamount to minimizing a function of the ranks of the residuals. A generalization of this method was derived by Naranjo and Hettmansperger (1994). An appeal of their method is that it appears to control Type I error probabilities relatively well without resorting to the bootstrap. A negative feature is that when there are leverage points (outliers among the X values), power can be low versus least squares (Wilcox, 1995). For yet another rank-based approach, see Cliff (1996).

15.13 The Rank-Transform Method

It should be noted that a simple approach when dealing with rank-based methods is to replace the observations by their ranks and apply a method for means. There are situations where this strategy gives reasonable results, but there are general conditions where it performs poorly. For criticisms of the method, see Blair, Sawilowski, and Higgens (1987); Akritas (1990); G. Thompson and Ammann (1990); and G. Thompson (1991). For more details, see McKean and Vidmar (1994); Hettmansperger and McKean (1998, Section 4.7); and Brunner, Domhof, and Langer (2002, Section 5.8). Headrick and Rotou (2001) studied this approach when dealing with multiple regression and found it to be unsatisfactory.

15.14 Exercises

1. Two methods for reducing shoulder pain after laparoscopic surgery were compared by Jorgensen et al. (1995). The data were

Group 1:	1 2 1 1 1 1 1 1 1 1 2 4 1 1
Group 2:	3 3 4 3 1 2 3 1 1 5 4

   Verify that all of the methods in Section 15.1 reject at the .05 level. Although the Kolmogorov–Smirnov test rejects with $\alpha = .05$, why might you suspect that the Kolmogorov–Smirnov test will have relatively low power in this particular situation? Check your results using the S-PLUS functions provided.

2. Imagine two groups of cancer patients are compared, the first group having a rapidly progressing form of the disease and the other having a slowly progressing form instead. At issue is whether psychological factors are related to the progression of cancer. The outcome measure is one where highly negative scores indicated a tendency to present the appearance of serenity in the presence of stress. The results are

Group 1:	−25 −24 −22 −22 −21 −18 −18 −18 −18 −17 −16 −14 −14 −13 −13 −13 −13 −9 −8 −7 −5 1 3 7 7
Group 2:	−21 −18 −16 −16 −16 −14 −13 −13 −12 −11 −11 −11 −9 −9 −9 −9 −7 −6 −3 −2 3 10

   Verify that the Wilcoxon–Mann–Whitney test rejects at the .05 level but that none of the other methods in Section 15.1 reject. Check your results using the S-PLUS functions provided. What might explain this?

3. Chapter 14 mentions data from a study regarding four skull measurements from five different time periods. If we compare the five groups based on these four measures with the S-PLUS function mulrank in Section 15.8.2, the output is as follows.

```
$test.stat:
[1] 4.197179

$nul:
[1] 14.13007

$sig.level:
[1] 1.717545e-07

$N:
[1] 150

$q.hat:
          [,1]      [,2]      [,3]      [,4]
[1,] 0.3605556 0.5682222 0.6316667 0.4626667
[2,] 0.3881111 0.4988889 0.6518889 0.4422222
[3,] 0.5380000 0.5635556 0.4756667 0.4568889
[4,] 0.5898889 0.4855556 0.3880000 0.5961111
[5,] 0.6234444 0.3837778 0.3527778 0.5421111
```

Interpret the results.

4. If a significant result is obtained with the S-PLUS function cmanova in Section 15.8.4, is this evidence that we should also reject the hypothesis tested by mulrank in Section 15.8.2?
5. Imagine that you get a significant result when using the ANOVA F-test in Chapter 9. If we increase the largest observation among the J groups, it is generally the case that eventually we will no longer reject. Is the same thing true when using the rank-based methods in Section 15.2?
6. Two independent groups are given different cold medicines and the goal is to compare reaction times. Suppose that the decreases in reaction times when taking drug A versus B are as follows.

A:	1.96,	2.24,	1.71,	2.41,	1.62,	1.93	
B:	2.11,	2.43,	2.07,	2.71,	2.50,	2.84,	2.88

Compare these two groups with the Mann–Whitney–Wilcoxon test using Equation (15.4). What is your estimate of the probability that a randomly sampled participant receiving drug A will have less of a reduction in reaction time than a randomly sampled participant receiving drug B?

7. Repeat the previous exercise, only now use the Cliff as well as the Brunner–Munzel methods.
8. For two dependent groups you get

Group 1:	10	14	15	18	20	29	30	40
Group 2:	40	8	15	20	10	8	2	3

Compare the two groups with the sign test and the Wilcoxon signed rank test with $\alpha = .05$. Verify that according to the sign test, $\hat{p} = .36$, that the .95

confidence interval for p is (.13, .69), and that the Wilcoxon signed rank test has an approximate significance level of .46.

9. For two dependent groups you get

Group 1:	86	71	77	68	91	72	77	91	70	71	88	87
Group 2:	88	77	76	64	96	72	65	90	65	80	81	72

Apply the Wilcoxon signed rank test with $\alpha = .05$. Verify that $W = .7565$ and that you fail to reject.

10. A developing nation is trying to improve its ability to grow its own crops. Four methods of growing corn are being considered, and you have been asked to help determine whether it makes a difference which method is used. To find out, four adjacent plots of land are used to grow the corn, and the yield is measured at the end of the season. This process is repeated in 12 locations located throughout the country, and the results are as shown in Table 15.4. Because results for adjacent plots of land might not be independent, you decide to use Friedman's test. Verify that a significant result among the four methods is obtained with $\alpha = .05$.
11. Repeat the previous exercise using the Agresti–Pendergast procedure. (Use the S-PLUS function apanova.)
12. Use the Agresti–Pendergast procedure to perform all pairwise comparisons of the groups in the previous two exercises. Use Rom's method so that the familywise error rate does not exceed 0.05. Verify that a significant result is obtained only when comparing groups 1 and 3.
13. For the following pairs of observations, test $H_0 : \tau = 0$ with $\alpha = .05$.

Time 1:	10	16	15	20
Time 2:	25	8	18	9

Verify that $\hat{\tau} = -0.667$, $Z = -1.36$, so you fail to reject.

TABLE 15.4 Data for Exercise 10

A	B	C	D
10	7	6	4
11	6	9	5
9	5	5	10
8	2	7	9
10	5	4	8
6	6	6	7
4	11	5	12
6	9	2	8
9	4	5	9
10	3	5	4
9	4	8	3
6	2	3	6

14. Repeat the previous exercise, except use Spearman's rho instead. Verify that $r_s = -0.8$ and that again you fail to reject.
15. Section 15.8.3 makes a distinction between the multivariate versus the marginal hypothesis. Describe a situation where the marginal hypothesis is true but the multivariate hypothesis is false.
16. Apply the S-PLUS function mulrank to the data in Table 15.2, and verify the results in the example of Section 15.8.2.
17. Using the S-PLUS function cmanova, apply the Choi–Marden method to the data in Table 15.2 and verify that the significance level is less than .01.
18. Apply the S-PLUS function bwrank to the data in Table 15.2. Verify that the significance level associated with the main effect for factor A is less than .001 but that you get a highly nonsignificant result for factor B and the hypothesis of no interaction.
19. Using the S-PLUS function bwrmcp, verify that for the data in Table 15.2, all pairwise comparisons among the levels of factor A are significant when the goal is to have FWE equal to .05.
20. Verify that no interaction is found using the method in Section 15.9.4 with the data in Table 15.2.
21. In the previous three exercises, beyond random sampling, what is being assumed about the variables when making inferences about the levels of factor B and interactions?
22. Verify that for the data in Table 15.3, the S-PLUS function bwrmcp returns

```
$Factor.A:
      Level  Level   test.stat     sig.level  sig.crit
[1,]      1      2    12.87017   0.001043705      0.05

$Factor.B:
      Level  Level   test.stat     sig.level  sig.crit
[1,]      1      2   0.3048713     0.5808447    0.0250
[2,]      1      3   0.2224258     0.6371979    0.0500
[3,]      2      3   0.5858703     0.4440207    0.0169

$Factor.AB:
        Lev.A     Lev.A     Lev.B     Lev.B    test.stat
[1,]        1         2         1         2     6.520209
[2,]        1         2         1         3    11.170399
[3,]        1         2         2         3     3.728463

    sig.level   sig.crit
0.0106656917     0.0250
0.0008311581     0.0169
0.0534928817     0.0500
```

23. Based on the results of the previous exercise, can you conclude that based on the average ranks among the groups, there is a disordinal interaction?
24. Look at the data in Table 15.3 and comment on why the method in Section 15.9.4 would be expected to have relatively low power.

25. Verify that for the data in Table 15.3, time 1 and time 2 among the active treatment group have a Kendall's tau equal to .31 with a significance level of .04.
26. For the data in Table 15.3, compute a .95 confidence interval for Spearman's rho using the data for time 1 and time 2 among the active treatment group. Use the S-PLUS function corb in Section 13.6.5. Verify that the .95 confidence interval is (0.036, 0.84), so again you reject $H_0 : \rho_s = 0$ at the .05 level.
27. For the data used in the last exercise, plot the points and comment on the association and the robustness of Spearman's rho and Kendall's tau.
28. Verify that for the data in Table 15.3, the Choi–Marden method has a significance level less than .001.
29. Verify that the hypothesis given by Equation (15.7) is the same as the hypothesis given by Equation (15.8).

APPENDIX A

Solutions to Selected Exercises

Chapter 2

3. $\mu = 3$, $\sigma^2 = 1.6$. 4. Smaller, $\sigma^2 = 1.3$. 5. Larger. 7. (a) .3, (b) .03/.3, (c) .09/.3, (d) .108/.18. 8. Yes. For example, probability of a high income given that they are under 30 is $.03/.3 = .1$, which is equal to the probability of a high income. 9. (a) 1253/3398, (b) 757/1828, (c) 757/1253, (d) no, (e) 1831/3398. 10. Median = 2.5. The .1 quantile—say, y—is given by $(y - 1) * (1/3) = .1$, so the .1 quantile is 1.3. The .9 quantile is 3.7. 11. (a) $4 \times (1/5)$, (b) $1.5 \times (1/5)$, (c) $2 \times (1/5)$, (d) $2.2 \times (1/5)$, (e) 0. 12. Median $= -.5$. The .25 quantile, y, is given by $(y - (-3)) \times 1/5 = .25$, so $y = -1.75$. 13. (a) $(c - (-1)) \times (1/2) = .9$, so $c = .8$, (b) $(c - (-1)) \times (1/2) = .95$, $c = .9$, (c) $(c - (-1)) \times (1/2) = .01$, $c = -.98$. 14. (a) .9, (b) $c = .95$, (c) $c = .99$. 15. (a) 0, (b) .25, (c) .5. 16. (a) 0, (b) 1/6, (c) 2/3, (d) 1/6. 17. $y \times (1/60) = .8$, $y = 48$. 18. (a) 0.0668, (b) 0.0062, (c) 0.0062, (d) .683. 19. (a) 0.691, (b) 0.894, (c) .77. 20. (a) .31, (b) .885, (c) 0.018, (d).221. 21. (a) −2.33, (b) 1.93, (c) −0.174, (d) .3. 22. (a) 1.43, (b) −0.01, (c) 1.7, (d) 1.28. 23. (a) .133, (b) .71, (c) .133, (d) .733. 24. (a) .588, (b) .63, (c) .71, (d) .95. 26. $c = 1.96$. 27. 1.28. 28. .16. 29. 84.45. 30. 1 − .91. 31. .87. 32. .001. 33. .68. 34. .95. 35. .115. 36. .043. 37. Yes. 39. No, for small departures from normality this probability can be close to 1. 40. No, for reasons similar to those in the previous exercise. 41. Yes. 42. Yes. 43. $\mu = 2.3$, $\sigma = .9$, and $P(\mu - \sigma \le X \le \mu + \sigma) = .7$. 44. (a) $.75^5$, (b) $.25^5$, (c) $1 - .25^5 - 5(.75)(.25)^4$. 45. (a) .586, (b) .732, (c) 1 − .425, (d) 1 − .274. 46. .4(25), .4(.6)(25), .4, .4(.6)/25.

Chapter 3

3. Two. 7. $n = 88$. 10. 9. 11. Yes. 12. One. 13. 20%. 17. 98, 350, 370, and 475. 18. $\sum (x - \bar{X})^2 (f_x/n)$. 25. The lower and upper quartiles are approximately 900 and 1300, respectively. So the standard boxplot rule would declare a value an outlier if it is less than 900 − 1.5(1300 − 900) or greater than 1300 + 1.5(1300 − 900). 26. .1.

Chapter 4

2. 1.28, 1.75, 2.326. **3**. 45±1.96. **4**. 45±2.58. **5**. Yes, upper end of confidence interval is 1158. **6**. (a) $65 \pm 1.96(22)/\sqrt{12}$, (b) $185 \pm 1.96(10)/\sqrt{22}$, (c) $19 \pm 1.96(30)/\sqrt{50}$. **8**. 9 and 8/10. **9**. 2.7 and 1.01/12. **10**. 2.7. **11**. 1.01. **12**. 94.3/8 and $\sqrt{94.3/8}$. **13**. 32. No. **14**. 93,663.52/12. **15**. They inflate the standard error. **16**. 94.3/8, $\sqrt{94.3/8}$. **17**. No. **18**. 10.9/25; small departures from normality can inflate the standard error. **20**. (a) .0228, (b) .159. **21**. (a) .16, (b) .023, (c) .977 − .028. **22**. .023. **24**. .933 − .067. **25**. (a) .055, (b) .788, (c) .992, (d) .788 − .055. **26**. (a) .047, (b) .952, (c) 1 − .047, (d) .952 − .047. **27**. Sample from a heavy-tailed distribution. **28**. Sampling from a light-tailed distribution, the distribution of the sample mean will be well approximated by the central limit theorem. **29**. (a) $26 \pm 2.26(9)/\sqrt{10}$, (b) $132 \pm 2.09(20)/\sqrt{18}$, (c) $52 \pm 2.06(12)/\sqrt{25}$. **31**. (161.4, 734.7). **32**. (10.7, 22.4). **35**. (a) $52 \pm 2.13\sqrt{12}/(.6\sqrt{24})$, (b) $10 \pm 2.07\sqrt{30}/(.6\sqrt{36})$. **37**. (160.4, 404.99). **38**. Outliers. **39**. Outliers. **41**. No.

Chapter 5

1. $Z = -1.265$, fail to reject. **2**. Fail to reject. **3**. (74.9, 81.1). **4**. .103. **5**. .206. **6**. $Z = -14$, reject. **7**. Reject. **8**. (118.6, 121.4). **9**. Yes, because $\bar{X}$ is consistent with H_0. **10**. $Z = 10$, reject. **11**. $Z = 2.12$, reject. **19**. Increase α. **20**. (a) $T = 1$, fail to reject. (b) $T = .5$, fail to reject. (c) $T = 2.5$, reject. **22**. (a) $T = .8$, fail to reject. (b) $T = .4$, fail to reject. (c) $T = 2$, fail to reject. **24**. $T = .39$, fail to reject. **25**. $T = -2.61$, reject. **26**. (a) $T_t = .596$, fail to reject. (b) $T_t = .298$, fail to reject. (c) $T_t = .894$, fail to reject. **28**. $T_t = -3.1$, reject. **29**. $T_t = .129$, fail to reject.

Chapter 6

4. .87. **6**. $b_1 = -0.0355$, $b_0 = 39.93$. **7**. $b_1 = .0039$, $b_0 = .485$. **10**. One concern is that $X = 600$ lies outside the range of X values used to compute the least squares regression line. **9**. $b_1 = -0.0754$, $b_0 = -1.253$. **11**. $r = -.366$. Not significantly different from zero, so can't be reasonably certain about whether ρ is positive or negative. **14**. Health improves as vitamin intake increases, but health deteriorates with too much vitamin A. That is, there is a nonlinear association. **17**. Extrapolation can be misleading.

Chapter 7

1. (8.1, 14.6), the middle 80%. **4**. The one based on trimcibt, which uses the bootstrap-t method. **7**. The one based on trimpb, which uses the percentile bootstrap method. **10**. In a bootstrap sample, outliers are more likely to inflate the 20% trimmed mean if no Winsorizing is done. **12**. The amount of Winsorizing equals the amount of trimming, which might mean inaccurate probability coverage. **13**. (7, 14.5). Even though the one-step M-estimator removes outliers, situations arise where using a median might yield a substantially shorter confidence interval. **17**. There might be a nonlinear association. **18**. Restricting the range of the X values can reveal an association that is not otherwise detected. **19**. Now you reject, before you did not reject, so again restricting the range of X can reveal an association that is otherwise missed. **20**. The conventional method appears to have accurate probability coverage in this particular situation.

Chapter 8

3. $T = (45 - 36)/\sqrt{11.25(.083333)} = 9.3$, $\nu = 48$, reject. **4**. $W = (45 - 36)/\sqrt{8/20 + 24/10} = 10.49$, reject. **5**. Welch's test might have more power. **6**. $T = 3.79$, $\nu = 38$, reject. **7**. $W = 3.8$, $\nu = 38$, reject. **8**. With equal sample sizes and equal sample variances, T and W give exactly the same result. This suggests that if the sample variances are approximately equal, it makes little difference which method is used. But if the sample variances differ enough, Welch's method is generally more accurate. **9**. $h_1 = 16$, $h_2 = 10$, $d_1 = 2.4$, $d_2 = 6$, $\nu = 16.1$, so $t = 2.12$. $T_y = 2.07$, fail to reject. **10**. .99 confidence interval is $(-2.5, 14.5)$. **11**. $\nu = 29$, $t = 2.045$, CI is (1.38, 8.62), reject. **12**. CI is (1.39, 8.6), reject. **13**. $W = 1.2$, $\hat{\nu} = 11.1$, $t = 2.2$, fail to reject. **14**. $T_y = 1.28$, $\hat{\nu} = 9.6$, fail to reject. **15**. No, power might be low. **16**. .95 CI is $(-7.4, .2)$, fail to reject. **17**. Fail to reject. **20**. The data indicate that the distributions differ, so the confidence interval based on Student's T might be inaccurate. **24**. The distributions differ, so some would argue that by implication the means in particular differ. **26**. Power. **28**. One concern is that the second group appears to be more skewed than the first, suggesting that probability coverage, when using means, might be poor. Another concern is that the first group has an outlier. **31**. If the tails of the distributions are sufficiently heavy, medians have smaller standard errors and hence can have more power. **33**. An improper estimate of the standard error is being used if extreme observations are discarded and methods for means are applied to the data that remain. See Section 4.9.1.

Chapter 9

17. Set allp=F, but this might lower power. A better approach would be to use the method in Section 12.7.3. **19**. $(\text{MSBG} - \text{MSWG})(J - 1)/n$. **22**. Using allp=F might lower power, but this is not an issue here because you reject.

Chapter 10

2. There are main effects for both factors A and B but no interaction. **3**. There is a disordinal interaction. **4**. Row 1: 10, 20, 30; Row 2: 20, 30, 40; Row 3: 30, 40, 50. **5**.

Source	SS	DF	MS	F
A	800	2	400	3.00
B	600	3	200	1.5
INTER	1200	6	200	1.5
WITHIN	4800	36	133.3	

This is a 3-by-4 design. The total number of observations is $N = 48$. None of the hypotheses is rejected.

7. The means suggest that there is a disordinal interaction by rows. That is, method 1 is best for females, but method 2 is best for males. Although no interaction was detected, accepting the null hypothesis is not warranted, because power might be too low to detect a true interaction.
9. Disordinal interactions means that interpreting main effects may not be straightforward. **10**. No, more needs to be done to establish that a disordinal interaction exists.

Chapter 11

4. One possible reason is that these measures of location differ. That is, they are testing different hypotheses; depending on how the groups differ, one approach could have more power than the other.
5. Currently it seems that trimmed means based on difference scores usually provide more power.

Chapter 12

1. Does not control the familywise error rate (the probability of at least one Type I error).
2. MSWG $= 11.6$. (a) $T = |15 - 10|/\sqrt{11.6(1/20 + 1/20)} = 4.64$, $\nu = 100 - 5 = 95$, reject. (b) $T = |15 - 10|/\sqrt{11.6(1/20 + 1/20)/2} = 6.565$, $q = 3.9$, reject. (c) $W = (15 - 10)/\sqrt{4/20 + 9/20} = 6.2$, $\hat{\nu} = 33$, $c = 2.99$, reject. (d) $(15 - 10)/\sqrt{.5(4/20 + 9/20)} = 8.77$, $q = 4.1$, reject. (e) $f = 2.47$, $S = \sqrt{4(2.47)(11.6)(1/20 + 1/20)} = 3.39$, reject. (f) $f = 2.66$, $A = 4(1 + 3/33)2.66 = 11.61$, reject.
3. MSWG $= 8$. (a) $T = |20 - 12|/\sqrt{8(1/10 + 1/10)} = 6.325$, $\nu = 50 - 5 = 45$, reject. (b) $T = |20 - 12|/\sqrt{8(1/10 + 1/10)/2} = 8.94$, $q = 4.01$, reject. (c) $W = (20 - 12)/\sqrt{5/10 + 6/10} = 7.63$, $\hat{\nu} = 37.7$, $c = 2.96$, reject. (d) $(20 - 12)/\sqrt{.5(5/10 + 6/10)} = 10.79$, $q = 4.06$, reject. (e) $f = 2.58$, $S = \sqrt{4(2.58)(8)(1/10 + 1/10)} = 4.06$, reject. (f) $f = 2.62$, $A = 11.3$, reject.
4. Reject if the significance level is less than or equal to $.05/6 = .0083$. So the fourth test is significant.
5. Fourth test, having significance level .001. **6**. None is rejected. **7**. All are rejected. **8**. MSWG goes up and eventually you will no longer reject when comparing groups one and two. **9**. The same problem occurs as in Exercise 8. **16**. Significant results are obtained when comparing groups 1 to 3, 1 to 4, 1 to 5, as well as 2 to 5. **17**. One possibility is that the boxplot rule is missing outliers. Using the rule based on MAD and M in Section 3.4.2, group 3 is found to have four outliers, but none of the boxplot rules in Chapter 3 finds an outlier in this group. Comparing the standard errors of the means versus the standard errors when using MOM, sometimes MOM has a smaller standard error, but sometimes the reverse is true. Skewness can affect power when comparing means, even when there are no outliers, and in general the differences among the MOM population values might be larger than those among the population means.

Chapter 13

6. Both outliers among the Y values and heteroscedasticity can affect the confidence intervals. The only known method that effectively determines whether robust methods make a difference is simply to try the methods and compare the results. The conventional .95 confidence interval for the slope, using the method in Section 6.3.1, is $(-4.23, 0.285)$ with a significance level of .083. Using the heteroscedastic bootstrap method in Chapter 7 it is $(-4.62, 0.41)$. Using Theil–Sen it is $(-4.26, 0.26)$, and for MGV it is $(-4.55, 0.37)$. So all four methods fail to reject. **8**. When computing a confidence interval, the length of the confidence interval can change substantially, depending on which estimator is used. That is, power can depend to a large extent on which regression method is employed. **10**. Eliminating too many points can mask the true association among the bulk of the observations.

Chapter 14

1. All but three of the X_3 values are close to 50. The three that are not are observations 1, 2, and 7. Excluding the corresponding Y values, the 20% trimmed mean of the remaining Y values is 99.9 and the value of MOM is 98.5. **5**. It appears that the association might be nonlinear. However, lintest fails to reject the hypothesis of a linear association. **8**. $X = 131$ and 133.

Chapter 15

6. $\hat{p} = .9$. Significance level is .015. 7. Using Cliff's method, the .95 confidence interval for δ is $(-0.966, -0.21)$, so reject with $\alpha = .05$. The Brunner–Munzel method returns a significance level of .00066. **15**. For the first group, suppose there are two variables, each having standard normal distributions with a correlation of zero. Imagine for the second group that, again, the two variables each have standard normal distributions, only now $\rho = .5$. Then the marginal hypothesis is true but the multivariate hypothesis is false. **21**. A basic assumption is that it is meaningful to compare the variables. For example, if for every individual we measure height and weight, we could analyze the data with the methods in Section 15.9, but the analysis is meaningless when dealing with factor B or interactions. We could, however, compare groups, based on these measures, using the methods in Section 15.8. **23**. No. The average ranks need to be compared for the appropriate groups. **24**. There are many tied values. **27**. The significant results associated with Spearman's rho and Kendall's tau are due to the two points where both time 1 and time 2 scores are equal to 4. These two points appear in the upper right corner of a scatterplot. Ignoring these two points, both Kendall's tau and Spearman's rho are no longer significant.

APPENDIX B

Tables

TABLE 1 Standard Normal Distribution

z	$P(Z \leq z)$	z	$P(Z \leq z)$	z	$P(Z \leq z)$	z	$P(Z \leq z)$
−3.00	0.0013	−2.99	0.0014	−2.98	0.0014	−2.97	0.0015
−2.96	0.0015	−2.95	0.0016	−2.94	0.0016	−2.93	0.0017
−2.92	0.0018	−2.91	0.0018	−2.90	0.0019	−2.89	0.0019
−2.88	0.0020	−2.87	0.0021	−2.86	0.0021	−2.85	0.0022
−2.84	0.0023	−2.83	0.0023	−2.82	0.0024	−2.81	0.0025
−2.80	0.0026	−2.79	0.0026	−2.78	0.0027	−2.77	0.0028
−2.76	0.0029	−2.75	0.0030	−2.74	0.0031	−2.73	0.0032
−2.72	0.0033	−2.71	0.0034	−2.70	0.0035	−2.69	0.0036
−2.68	0.0037	−2.67	0.0038	−2.66	0.0039	−2.65	0.0040
−2.64	0.0041	−2.63	0.0043	−2.62	0.0044	−2.61	0.0045
−2.60	0.0047	−2.59	0.0048	−2.58	0.0049	−2.57	0.0051
−2.56	0.0052	−2.55	0.0054	−2.54	0.0055	−2.53	0.0057
−2.52	0.0059	−2.51	0.0060	−2.50	0.0062	−2.49	0.0064
−2.48	0.0066	−2.47	0.0068	−2.46	0.0069	−2.45	0.0071
−2.44	0.0073	−2.43	0.0075	−2.42	0.0078	−2.41	0.0080
−2.40	0.0082	−2.39	0.0084	−2.38	0.0087	−2.37	0.0089
−2.36	0.0091	−2.35	0.0094	−2.34	0.0096	−2.33	0.0099
−2.32	0.0102	−2.31	0.0104	−2.30	0.0107	−2.29	0.0110
−2.28	0.0113	−2.27	0.0116	−2.26	0.0119	−2.25	0.0122
−2.24	0.0125	−2.23	0.0129	−2.22	0.0132	−2.21	0.0136
−2.20	0.0139	−2.19	0.0143	−2.18	0.0146	−2.17	0.0150
−2.16	0.0154	−2.15	0.0158	−2.14	0.0162	−2.13	0.0166
−2.12	0.0170	−2.11	0.0174	−2.10	0.0179	−2.09	0.0183
−2.08	0.0188	−2.07	0.0192	−2.06	0.0197	−2.05	0.0202
−2.04	0.0207	−2.03	0.0212	−2.02	0.0217	−2.01	0.0222
−2.00	0.0228	−1.99	0.0233	−1.98	0.0239	−1.97	0.0244
−1.96	0.0250	−1.95	0.0256	−1.94	0.0262	−1.93	0.0268
−1.92	0.0274	−1.91	0.0281	−1.90	0.0287	−1.89	0.0294
−1.88	0.0301	−1.87	0.0307	−1.86	0.0314	−1.85	0.0322
−1.84	0.0329	−1.83	0.0336	−1.82	0.0344	−1.81	0.0351
−1.80	0.0359	−1.79	0.0367	−1.78	0.0375	−1.77	0.0384
−1.76	0.0392	−1.75	0.0401	−1.74	0.0409	−1.73	0.0418
−1.72	0.0427	−1.71	0.0436	−1.70	0.0446	−1.69	0.0455
−1.68	0.0465	−1.67	0.0475	−1.66	0.0485	−1.65	0.0495
−1.64	0.0505	−1.63	0.0516	−1.62	0.0526	−1.61	0.0537
−1.60	0.0548	−1.59	0.0559	−1.58	0.0571	−1.57	0.0582
−1.56	0.0594	−1.55	0.0606	−1.54	0.0618	−1.53	0.0630
−1.52	0.0643	−1.51	0.0655	−1.50	0.0668	−1.49	0.0681
−1.48	0.0694	−1.47	0.0708	−1.46	0.0721	−1.45	0.0735
−1.44	0.0749	−1.43	0.0764	−1.42	0.0778	−1.41	0.0793
−1.40	0.0808	−1.39	0.0823	−1.38	0.0838	−1.37	0.0853

continued

TABLE 1 *continued*

z	$P(Z \leq z)$	z	$P(Z \leq z)$	z	$P(Z \leq z)$	z	$P(Z \leq z)$
−1.36	0.0869	−1.35	0.0885	−1.34	0.0901	−1.33	0.0918
−1.32	0.0934	−1.31	0.0951	−1.30	0.0968	−1.29	0.0985
−1.28	0.1003	−1.27	0.1020	−1.26	0.1038	−1.25	0.1056
−1.24	0.1075	−1.23	0.1093	−1.22	0.1112	−1.21	0.1131
−1.20	0.1151	−1.19	0.1170	−1.18	0.1190	−1.17	0.1210
−1.16	0.1230	−1.15	0.1251	−1.14	0.1271	−1.13	0.1292
−1.12	0.1314	−1.11	0.1335	−1.10	0.1357	−1.09	0.1379
−1.08	0.1401	−1.07	0.1423	−1.06	0.1446	−1.05	0.1469
−1.04	0.1492	−1.03	0.1515	−1.02	0.1539	−1.01	0.1562
−1.00	0.1587	−0.99	0.1611	−0.98	0.1635	−0.97	0.1662
−0.96	0.1685	−0.95	0.1711	−0.94	0.1736	−0.93	0.1762
−0.92	0.1788	−0.91	0.1814	−0.90	0.1841	−0.89	0.1867
−0.88	0.1894	−0.87	0.1922	−0.86	0.1949	−0.85	0.1977
−0.84	0.2005	−0.83	0.2033	−0.82	0.2061	−0.81	0.2090
−0.80	0.2119	−0.79	0.2148	−0.78	0.2177	−0.77	0.2207
−0.76	0.2236	−0.75	0.2266	−0.74	0.2297	−0.73	0.2327
−0.72	0.2358	−0.71	0.2389	−0.70	0.2420	−0.69	0.2451
−0.68	0.2483	−0.67	0.2514	−0.66	0.2546	−0.65	0.2578
−0.64	0.2611	−0.63	0.2643	−0.62	0.2676	−0.61	0.2709
−0.60	0.2743	−0.59	0.2776	−0.58	0.2810	−0.57	0.2843
−0.56	0.2877	−0.55	0.2912	−0.54	0.2946	−0.53	0.2981
−0.52	0.3015	−0.51	0.3050	−0.50	0.3085	−0.49	0.3121
−0.48	0.3156	−0.47	0.3192	−0.46	0.3228	−0.45	0.3264
−0.44	0.3300	−0.43	0.3336	−0.42	0.3372	−0.41	0.3409
−0.40	0.3446	−0.39	0.3483	−0.38	0.3520	−0.37	0.3557
−0.36	0.3594	−0.35	0.3632	−0.34	0.3669	−0.33	0.3707
−0.32	0.3745	−0.31	0.3783	−0.30	0.3821	−0.29	0.3859
−0.28	0.3897	−0.27	0.3936	−0.26	0.3974	−0.25	0.4013
−0.24	0.4052	−0.23	0.4090	−0.22	0.4129	−0.21	0.4168
−0.20	0.4207	−0.19	0.4247	−0.18	0.4286	−0.17	0.4325
−0.16	0.4364	−0.15	0.4404	−0.14	0.4443	−0.13	0.4483
−0.12	0.4522	−0.11	0.4562	−0.10	0.4602	−0.09	0.4641
−0.08	0.4681	−0.07	0.4721	−0.06	0.4761	−0.05	0.4801
−0.04	0.4840	−0.03	0.4880	−0.02	0.4920	−0.01	0.4960
0.01	0.5040	0.02	0.5080	0.03	0.5120	0.04	0.5160
0.05	0.5199	0.06	0.5239	0.07	0.5279	0.08	0.5319
0.09	0.5359	0.10	0.5398	0.11	0.5438	0.12	0.5478
0.13	0.5517	0.14	0.5557	0.15	0.5596	0.16	0.5636
0.17	0.5675	0.18	0.5714	0.19	0.5753	0.20	0.5793
0.21	0.5832	0.22	0.5871	0.23	0.5910	0.24	0.5948
0.25	0.5987	0.26	0.6026	0.27	0.6064	0.28	0.6103

continued

TABLE 1 *continued*

z	$P(Z \leq z)$	z	$P(Z \leq z)$	z	$P(Z \leq z)$	z	$P(Z \leq z)$
0.29	0.6141	0.30	0.6179	0.31	0.6217	0.32	0.6255
0.33	0.6293	0.34	0.6331	0.35	0.6368	0.36	0.6406
0.37	0.6443	0.38	0.6480	0.39	0.6517	0.40	0.6554
0.41	0.6591	0.42	0.6628	0.43	0.6664	0.44	0.6700
0.45	0.6736	0.46	0.6772	0.47	0.6808	0.48	0.6844
0.49	0.6879	0.50	0.6915	0.51	0.6950	0.52	0.6985
0.53	0.7019	0.54	0.7054	0.55	0.7088	0.56	0.7123
0.57	0.7157	0.58	0.7190	0.59	0.7224	0.60	0.7257
0.61	0.7291	0.62	0.7324	0.63	0.7357	0.64	0.7389
0.65	0.7422	0.66	0.7454	0.67	0.7486	0.68	0.7517
0.69	0.7549	0.70	0.7580	0.71	0.7611	0.72	0.7642
0.73	0.7673	0.74	0.7703	0.75	0.7734	0.76	0.7764
0.77	0.7793	0.78	0.7823	0.79	0.7852	0.80	0.7881
0.81	0.7910	0.82	0.7939	0.83	0.7967	0.84	0.7995
0.85	0.8023	0.86	0.8051	0.87	0.8078	0.88	0.8106
0.89	0.8133	0.90	0.8159	0.91	0.8186	0.92	0.8212
0.93	0.8238	0.94	0.8264	0.95	0.8289	0.96	0.8315
0.97	0.8340	0.98	0.8365	0.99	0.8389	1.00	0.8413
1.01	0.8438	1.02	0.8461	1.03	0.8485	1.04	0.8508
1.05	0.8531	1.06	0.8554	1.07	0.8577	1.08	0.8599
1.09	0.8621	1.10	0.8643	1.11	0.8665	1.12	0.8686
1.13	0.8708	1.14	0.8729	1.15	0.8749	1.16	0.8770
1.17	0.8790	1.18	0.8810	1.19	0.8830	1.20	0.8849
1.21	0.8869	1.22	0.8888	1.23	0.8907	1.24	0.8925
1.25	0.8944	1.26	0.8962	1.27	0.8980	1.28	0.8997
1.29	0.9015	1.30	0.9032	1.31	0.9049	1.32	0.9066
1.33	0.9082	1.34	0.9099	1.35	0.9115	1.36	0.9131
1.37	0.9147	1.38	0.9162	1.39	0.9177	1.40	0.9192
1.41	0.9207	1.42	0.9222	1.43	0.9236	1.44	0.9251
1.45	0.9265	1.46	0.9279	1.47	0.9292	1.48	0.9306
1.49	0.9319	1.50	0.9332	1.51	0.9345	1.52	0.9357
1.53	0.9370	1.54	0.9382	1.55	0.9394	1.56	0.9406
1.57	0.9418	1.58	0.9429	1.59	0.9441	1.60	0.9452
1.61	0.9463	1.62	0.9474	1.63	0.9484	1.64	0.9495
1.65	0.9505	1.66	0.9515	1.67	0.9525	1.68	0.9535
1.69	0.9545	1.70	0.9554	1.71	0.9564	1.72	0.9573
1.73	0.9582	1.74	0.9591	1.75	0.9599	1.76	0.9608
1.77	0.9616	1.78	0.9625	1.79	0.9633	1.80	0.9641
1.81	0.9649	1.82	0.9656	1.83	0.9664	1.84	0.9671
1.85	0.9678	1.86	0.9686	1.87	0.9693	1.88	0.9699
1.89	0.9706	1.90	0.9713	1.91	0.9719	1.92	0.9726

continued

TABLE 1 *continued*

z	P(Z ≤ z)	z	P(Z ≤ z)	z	P(Z ≤ z)	z	P(Z ≤ z)
1.93	0.9732	1.94	0.9738	1.95	0.9744	1.96	0.9750
1.97	0.9756	1.98	0.9761	1.99	0.9767	2.00	0.9772
2.01	0.9778	2.02	0.9783	2.03	0.9788	2.04	0.9793
2.05	0.9798	2.06	0.9803	2.07	0.9808	2.08	0.9812
2.09	0.9817	2.10	0.9821	2.11	0.9826	2.12	0.9830
2.13	0.9834	2.14	0.9838	2.15	0.9842	2.16	0.9846
2.17	0.9850	2.18	0.9854	2.19	0.9857	2.20	0.9861
2.21	0.9864	2.22	0.9868	2.23	0.9871	2.24	0.9875
2.25	0.9878	2.26	0.9881	2.27	0.9884	2.28	0.9887
2.29	0.9890	2.30	0.9893	2.31	0.9896	2.32	0.9898
2.33	0.9901	2.34	0.9904	2.35	0.9906	2.36	0.9909
2.37	0.9911	2.38	0.9913	2.39	0.9916	2.40	0.9918
2.41	0.9920	2.42	0.9922	2.43	0.9925	2.44	0.9927
2.45	0.9929	2.46	0.9931	2.47	0.9932	2.48	0.9934
2.49	0.9936	2.50	0.9938	2.51	0.9940	2.52	0.9941
2.53	0.9943	2.54	0.9945	2.55	0.9946	2.56	0.9948
2.57	0.9949	2.58	0.9951	2.59	0.9952	2.60	0.9953
2.61	0.9955	2.62	0.9956	2.63	0.9957	2.64	0.9959
2.65	0.9960	2.66	0.9961	2.67	0.9962	2.68	0.9963
2.69	0.9964	2.70	0.9965	2.71	0.9966	2.72	0.9967
2.73	0.9968	2.74	0.9969	2.75	0.9970	2.76	0.9971
2.77	0.9972	2.78	0.9973	2.79	0.9974	2.80	0.9974
2.81	0.9975	2.82	0.9976	2.83	0.9977	2.84	0.9977
2.85	0.9978	2.86	0.9979	2.87	0.9979	2.88	0.9980
2.89	0.9981	2.90	0.9981	2.91	0.9982	2.92	0.9982
2.93	0.9983	2.94	0.9984	2.95	0.9984	2.96	0.9985
2.97	0.9985	2.98	0.9986	2.99	0.9986	3.00	0.9987

Note: This table was computed with IMSL subroutine ANORIN.

TABLE 2 Binomial Probability Function (Values of Entries are $P(X \leq k)$)

$n = 5$

					p					
k	.05	.1	.2	.3	.4	.5	.6	.7	.8	.9	.95
0	0.774	0.590	0.328	0.168	0.078	0.031	0.010	0.002	0.000	0.000	0.000
1	0.977	0.919	0.737	0.528	0.337	0.188	0.087	0.031	0.007	0.000	0.000
2	0.999	0.991	0.942	0.837	0.683	0.500	0.317	0.163	0.058	0.009	0.001
3	1.000	1.000	0.993	0.969	0.913	0.813	0.663	0.472	0.263	0.081	0.023
4	1.000	1.000	1.000	0.998	0.990	0.969	0.922	0.832	0.672	0.410	0.226

continued

TABLE 2 *continued*

$n = 6$

k	.05	.1	.2	.3	.4	p .5	.6	.7	.8	.9	.95
0	0.735	0.531	0.262	0.118	0.047	0.016	0.004	0.001	0.000	0.000	0.000
1	0.967	0.886	0.655	0.420	0.233	0.109	0.041	0.011	0.002	0.000	0.000
2	0.998	0.984	0.901	0.744	0.544	0.344	0.179	0.070	0.017	0.001	0.000
3	1.000	0.999	0.983	0.930	0.821	0.656	0.456	0.256	0.099	0.016	0.002
4	1.000	1.000	0.998	0.989	0.959	0.891	0.767	0.580	0.345	0.114	0.033
5	1.000	1.000	1.000	0.999	0.996	0.984	0.953	0.882	0.738	0.469	0.265

$n = 7$

k	.05	.1	.2	.3	.4	p .5	.6	.7	.8	.9	.95
0	0.698	0.478	0.210	0.082	0.028	0.008	0.002	0.000	0.000	0.000	0.000
1	0.956	0.850	0.577	0.329	0.159	0.062	0.019	0.004	0.000	0.000	0.000
2	0.996	0.974	0.852	0.647	0.420	0.227	0.096	0.029	0.005	0.000	0.000
3	1.000	0.997	0.967	0.874	0.710	0.500	0.290	0.126	0.033	0.003	0.000
4	1.000	1.000	0.995	0.971	0.904	0.773	0.580	0.353	0.148	0.026	0.004
5	1.000	1.000	1.000	0.996	0.981	0.938	0.841	0.671	0.423	0.150	0.044
6	1.000	1.000	1.000	1.000	0.998	0.992	0.972	0.918	0.790	0.522	0.302

$n = 8$

k	.05	.1	.2	.3	.4	p .5	.6	.7	.8	.9	.95
0	0.663	0.430	0.168	0.058	0.017	0.004	0.001	0.000	0.000	0.000	0.000
1	0.943	0.813	0.503	0.255	0.106	0.035	0.009	0.001	0.000	0.000	0.000
2	0.994	0.962	0.797	0.552	0.315	0.145	0.050	0.011	0.001	0.000	0.000
3	1.000	0.995	0.944	0.806	0.594	0.363	0.174	0.058	0.010	0.000	0.000
4	1.000	1.000	0.990	0.942	0.826	0.637	0.406	0.194	0.056	0.005	0.000
5	1.000	1.000	0.999	0.989	0.950	0.855	0.685	0.448	0.203	0.038	0.006
6	1.000	1.000	1.000	0.999	0.991	0.965	0.894	0.745	0.497	0.187	0.057
7	1.000	1.000	1.000	1.000	0.999	0.996	0.983	0.942	0.832	0.570	0.337

$n = 9$

k	.05	.1	.2	.3	.4	p .5	.6	.7	.8	.9	.95
0	0.630	0.387	0.134	0.040	0.010	0.002	0.000	0.000	0.000	0.000	0.000
1	0.929	0.775	0.436	0.196	0.071	0.020	0.004	0.000	0.000	0.000	0.000
2	0.992	0.947	0.738	0.463	0.232	0.090	0.025	0.004	0.000	0.000	0.000
3	0.999	0.992	0.914	0.730	0.483	0.254	0.099	0.025	0.003	0.000	0.000
4	1.000	0.999	0.980	0.901	0.733	0.500	0.267	0.099	0.020	0.001	0.000
5	1.000	1.000	0.997	0.975	0.901	0.746	0.517	0.270	0.086	0.008	0.001
6	1.000	1.000	1.000	0.996	0.975	0.910	0.768	0.537	0.262	0.053	0.008
7	1.000	1.000	1.000	1.000	0.996	0.980	0.929	0.804	0.564	0.225	0.071
8	1.000	1.000	1.000	1.000	1.000	0.998	0.990	0.960	0.866	0.613	0.370

continued

TABLE 2 *continued*

$n = 10$

k	p .05	.1	.2	.3	.4	.5	.6	.7	.8	.9	.95
0	0.599	0.349	0.107	0.028	0.006	0.001	0.000	0.000	0.000	0.000	0.000
1	0.914	0.736	0.376	0.149	0.046	0.011	0.002	0.000	0.000	0.000	0.000
2	0.988	0.930	0.678	0.383	0.167	0.055	0.012	0.002	0.000	0.000	0.000
3	0.999	0.987	0.879	0.650	0.382	0.172	0.055	0.011	0.001	0.000	0.000
4	1.000	0.998	0.967	0.850	0.633	0.377	0.166	0.047	0.006	0.000	0.000
5	1.000	1.000	0.994	0.953	0.834	0.623	0.367	0.150	0.033	0.002	0.000
6	1.000	1.000	0.999	0.989	0.945	0.828	0.618	0.350	0.121	0.013	0.001
7	1.000	1.000	1.000	0.998	0.988	0.945	0.833	0.617	0.322	0.070	0.012
8	1.000	1.000	1.000	1.000	0.998	0.989	0.954	0.851	0.624	0.264	0.086
9	1.000	1.000	1.000	1.000	1.000	0.999	0.994	0.972	0.893	0.651	0.401

$n = 15$

k	p .05	.1	.2	.3	.4	.5	.6	.7	.8	.9	.95
0	0.463	0.206	0.035	0.005	0.000	0.000	0.000	0.000	0.000	0.000	0.000
1	0.829	0.549	0.167	0.035	0.005	0.000	0.000	0.000	0.000	0.000	0.000
2	0.964	0.816	0.398	0.127	0.027	0.004	0.000	0.000	0.000	0.000	0.000
3	0.995	0.944	0.648	0.297	0.091	0.018	0.002	0.000	0.000	0.000	0.000
4	0.999	0.987	0.836	0.515	0.217	0.059	0.009	0.001	0.000	0.000	0.000
5	1.000	0.998	0.939	0.722	0.403	0.151	0.034	0.004	0.000	0.000	0.000
6	1.000	1.000	0.982	0.869	0.610	0.304	0.095	0.015	0.001	0.000	0.000
7	1.000	1.000	0.996	0.950	0.787	0.500	0.213	0.050	0.004	0.000	0.000
8	1.000	1.000	0.999	0.985	0.905	0.696	0.390	0.131	0.018	0.000	0.000
9	1.000	1.000	1.000	0.996	0.966	0.849	0.597	0.278	0.061	0.002	0.000
10	1.000	1.000	1.000	0.999	0.991	0.941	0.783	0.485	0.164	0.013	0.001
11	1.000	1.000	1.000	1.000	0.998	0.982	0.909	0.703	0.352	0.056	0.005
12	1.000	1.000	1.000	1.000	1.000	0.996	0.973	0.873	0.602	0.184	0.036
13	1.000	1.000	1.000	1.000	1.000	1.000	0.995	0.965	0.833	0.451	0.171
14	1.000	1.000	1.000	1.000	1.000	1.000	1.000	0.995	0.965	0.794	0.537

$n = 20$

k	p .05	.1	.2	.3	.4	.5	.6	.7	.8	.9	.95
0	0.358	0.122	0.012	0.001	0.000	0.000	0.000	0.000	0.000	0.000	0.000
1	0.736	0.392	0.069	0.008	0.001	0.000	0.000	0.000	0.000	0.000	0.000
2	0.925	0.677	0.206	0.035	0.004	0.000	0.000	0.000	0.000	0.000	0.000
3	0.984	0.867	0.411	0.107	0.016	0.001	0.000	0.000	0.000	0.000	0.000
4	0.997	0.957	0.630	0.238	0.051	0.006	0.000	0.000	0.000	0.000	0.000
5	1.000	0.989	0.804	0.416	0.126	0.021	0.002	0.000	0.000	0.000	0.000
6	1.000	0.998	0.913	0.608	0.250	0.058	0.006	0.000	0.000	0.000	0.000
7	1.000	1.000	0.968	0.772	0.416	0.132	0.021	0.001	0.000	0.000	0.000
8	1.000	1.000	0.990	0.887	0.596	0.252	0.057	0.005	0.000	0.000	0.000

continued

TABLE 2 *continued*

$n = 20$

k	.05	.1	.2	.3	.4	p .5	.6	.7	.8	.9	.95
9	1.000	1.000	0.997	0.952	0.755	0.412	0.128	0.017	0.001	0.000	0.000
10	1.000	1.000	0.999	0.983	0.872	0.588	0.245	0.048	0.003	0.000	0.000
11	1.000	1.000	1.000	0.995	0.943	0.748	0.404	0.113	0.010	0.000	0.000
12	1.000	1.000	1.000	0.999	0.979	0.868	0.584	0.228	0.032	0.000	0.000
13	1.000	1.000	1.000	1.000	0.994	0.942	0.750	0.392	0.087	0.002	0.000
14	1.000	1.000	1.000	1.000	0.998	0.979	0.874	0.584	0.196	0.011	0.000
15	1.000	1.000	1.000	1.000	1.000	0.994	0.949	0.762	0.370	0.043	0.003
16	1.000	1.000	1.000	1.000	1.000	0.999	0.984	0.893	0.589	0.133	0.016
17	1.000	1.000	1.000	1.000	1.000	1.000	0.996	0.965	0.794	0.323	0.075
18	1.000	1.000	1.000	1.000	1.000	1.000	0.999	0.992	0.931	0.608	0.264
19	1.000	1.000	1.000	1.000	1.000	1.000	1.000	0.999	0.988	0.878	0.642

$n = 25$

k	.05	.1	.2	.3	.4	p .5	.6	.7	.8	.9	.95
0	0.277	0.072	0.004	0.000	0.000	0.000	0.000	0.000	0.000	0.000	0.000
1	0.642	0.271	0.027	0.002	0.000	0.000	0.000	0.000	0.000	0.000	0.000
2	0.873	0.537	0.098	0.009	0.000	0.000	0.000	0.000	0.000	0.000	0.000
3	0.966	0.764	0.234	0.033	0.002	0.000	0.000	0.000	0.000	0.000	0.000
4	0.993	0.902	0.421	0.090	0.009	0.000	0.000	0.000	0.000	0.000	0.000
5	0.999	0.967	0.617	0.193	0.029	0.002	0.000	0.000	0.000	0.000	0.000
6	1.000	0.991	0.780	0.341	0.074	0.007	0.000	0.000	0.000	0.000	0.000
7	1.000	0.998	0.891	0.512	0.154	0.022	0.001	0.000	0.000	0.000	0.000
8	1.000	1.000	0.953	0.677	0.274	0.054	0.004	0.000	0.000	0.000	0.000
9	1.000	1.000	0.983	0.811	0.425	0.115	0.013	0.000	0.000	0.000	0.000
10	1.000	1.000	0.994	0.902	0.586	0.212	0.034	0.002	0.000	0.000	0.000
11	1.000	1.000	0.998	0.956	0.732	0.345	0.078	0.006	0.000	0.000	0.000
12	1.000	1.000	1.000	0.983	0.846	0.500	0.154	0.017	0.000	0.000	0.000
13	1.000	1.000	1.000	0.994	0.922	0.655	0.268	0.044	0.002	0.000	0.000
14	1.000	1.000	1.000	0.998	0.966	0.788	0.414	0.098	0.006	0.000	0.000
15	1.000	1.000	1.000	1.000	0.987	0.885	0.575	0.189	0.017	0.000	0.000
16	1.000	1.000	1.000	1.000	0.996	0.946	0.726	0.323	0.047	0.000	0.000
17	1.000	1.000	1.000	1.000	0.999	0.978	0.846	0.488	0.109	0.002	0.000
18	1.000	1.000	1.000	1.000	1.000	0.993	0.926	0.659	0.220	0.009	0.000
19	1.000	1.000	1.000	1.000	1.000	0.998	0.971	0.807	0.383	0.033	0.001
20	1.000	1.000	1.000	1.000	1.000	1.000	0.991	0.910	0.579	0.098	0.007
21	1.000	1.000	1.000	1.000	1.000	1.000	0.998	0.967	0.766	0.236	0.034
22	1.000	1.000	1.000	1.000	1.000	1.000	1.000	0.991	0.902	0.463	0.127
23	1.000	1.000	1.000	1.000	1.000	1.000	1.000	0.998	0.973	0.729	0.358
24	1.000	1.000	1.000	1.000	1.000	1.000	1.000	1.000	0.996	0.928	0.723

TABLE 3 Percentage Points of the Chi-Squared Distribution

ν	$\chi^2_{.005}$	$\chi^2_{.01}$	$\chi^2_{.025}$	$\chi^2_{.05}$	$\chi^2_{.10}$
1	0.0000393	0.0001571	0.0009821	0.0039321	0.0157908
2	0.0100251	0.0201007	0.0506357	0.1025866	0.2107213
3	0.0717217	0.1148317	0.2157952	0.3518462	0.5843744
4	0.2069889	0.2971095	0.4844186	0.7107224	1.0636234
5	0.4117419	0.5542979	0.8312111	1.1454763	1.6103077
6	0.6757274	0.8720903	1.2373447	1.6353836	2.2041321
7	0.9892554	1.2390423	1.6898699	2.1673594	2.8331099
8	1.3444128	1.6464968	2.1797333	2.7326374	3.4895401
9	1.7349329	2.0879011	2.7003908	3.3251143	4.1681604
10	2.1558590	2.5582132	3.2469759	3.9403019	4.8651857
11	2.6032248	3.0534868	3.8157606	4.5748196	5.5777788
12	3.0738316	3.5705872	4.4037895	5.2260313	6.3037949
13	3.5650368	4.1069279	5.0087538	5.8918715	7.0415068
14	4.0746784	4.6604300	5.6287327	6.5706167	7.7895403
15	4.6009169	5.2293501	6.2621403	7.2609539	8.5467529
16	5.1422071	5.8122101	6.9076681	7.9616566	9.3122330
17	5.6972256	6.4077673	7.5641880	8.6717682	10.0851974
18	6.2648115	7.0149183	8.2307510	9.3904572	10.8649368
19	6.8439512	7.6327391	8.9065247	10.1170273	11.6509628
20	7.4338474	8.2603989	9.5907822	10.8508148	12.4426041
21	8.0336685	8.8972015	10.2829285	11.5913391	13.2396393
22	8.6427155	9.5425110	10.9823456	12.3380432	14.0414886
23	9.2604370	10.1957169	11.6885223	13.0905151	14.8479385
24	9.8862610	10.8563690	12.4011765	13.8484344	15.6587067
25	10.5196533	11.5239716	13.1197433	14.6114349	16.4734497
26	11.1602631	12.1981506	13.8439331	15.3792038	17.2919159
27	11.8076019	12.8785095	14.5734024	16.1513977	18.1138763
28	12.4613495	13.5647125	15.3078613	16.9278717	18.9392395
29	13.1211624	14.2564697	16.0470886	17.7083893	19.7678223
30	13.7867584	14.9534760	16.7907562	18.4926147	20.5992126
40	20.7065582	22.1642761	24.4330750	26.5083008	29.0503540
50	27.9775238	29.7001038	32.3561096	34.7638702	37.6881561
60	35.5294037	37.4848328	40.4817810	43.1865082	46.4583282
70	43.2462311	45.4230499	48.7503967	51.7388763	55.3331146
80	51.1447754	53.5226593	57.1465912	60.3912201	64.2818604
90	59.1706543	61.7376862	65.6405029	69.1258850	73.2949219
100	67.3031921	70.0493622	74.2162018	77.9293976	82.3618469

continued

TABLE 3 *continued*

ν	$\chi^2_{.900}$	$\chi^2_{.95}$	$\chi^2_{.975}$	$\chi^2_{.99}$	$\chi^2_{.995}$
1	2.7056	3.8415	5.0240	6.6353	7.8818
2	4.6052	5.9916	7.3779	9.2117	10.5987
3	6.2514	7.8148	9.3486	11.3465	12.8409
4	7.7795	9.4879	11.1435	13.2786	14.8643
5	9.2365	11.0707	12.8328	15.0870	16.7534
6	10.6448	12.5919	14.4499	16.8127	18.5490
7	12.0171	14.0676	16.0136	18.4765	20.2803
8	13.3617	15.5075	17.5355	20.0924	21.9579
9	14.6838	16.9191	19.0232	21.6686	23.5938
10	15.9874	18.3075	20.4837	23.2101	25.1898
11	17.2750	19.6754	21.9211	24.7265	26.7568
12	18.5494	21.0263	23.3370	26.2170	28.2995
13	19.8122	22.3627	24.7371	27.6882	29.8194
14	21.0646	23.6862	26.1189	29.1412	31.3193
15	22.3077	24.9970	27.4883	30.5779	32.8013
16	23.5421	26.2961	28.8453	31.9999	34.2672
17	24.7696	27.5871	30.1909	33.4087	35.7184
18	25.9903	28.8692	31.5264	34.8054	37.1564
19	27.2035	30.1434	32.8523	36.1909	38.5823
20	28.4120	31.4104	34.1696	37.5662	39.9968
21	29.6150	32.6705	35.4787	38.9323	41.4012
22	30.8133	33.9244	36.7806	40.2893	42.7958
23	32.0069	35.1725	38.0757	41.6384	44.1812
24	33.1962	36.4151	39.3639	42.9799	45.5587
25	34.3815	37.6525	40.6463	44.3142	46.9280
26	35.5631	38.8852	41.9229	45.6418	48.2899
27	36.7412	40.1134	43.1943	46.9629	49.6449
28	37.9159	41.3371	44.4608	48.2784	50.9933
29	39.0874	42.5571	45.7223	49.5879	52.3357
30	40.2561	43.7730	46.9792	50.8922	53.6721
40	51.8050	55.7586	59.3417	63.6909	66.7660
50	63.1670	67.5047	71.4201	76.1538	79.4899
60	74.3970	79.0820	83.2977	88.3794	91.9516
70	85.5211	90.5283	95.0263	100.4409	104.2434
80	96.5723	101.8770	106.6315	112.3434	116.3484
90	107.5600	113.1425	118.1392	124.1304	128.3245
100	118.4932	124.3395	129.5638	135.8203	140.1940

Note: This table was computed with IMSL subroutine CHIIN.

TABLE 4 Percentage Points of Student's T-Distribution

ν	$t_{.9}$	$t_{.95}$	$t_{.975}$	$t_{.99}$	$t_{.995}$	$t_{.999}$
1	3.078	6.314	12.706	31.821	63.6567	318.313
2	1.886	2.920	4.303	6.965	9.925	22.327
3	1.638	2.353	3.183	4.541	5.841	10.215
4	1.533	2.132	2.776	3.747	4.604	7.173
5	1.476	2.015	2.571	3.365	4.032	5.893
6	1.440	1.943	2.447	3.143	3.707	5.208
7	1.415	1.895	2.365	2.998	3.499	4.785
8	1.397	1.856	2.306	2.897	3.355	4.501
9	1.383	1.833	2.262	2.821	3.245	4.297
10	1.372	1.812	2.228	2.764	3.169	4.144
12	1.356	1.782	2.179	2.681	3.055	3.930
15	1.341	1.753	2.131	2.603	2.947	3.733
20	1.325	1.725	2.086	2.528	2.845	3.552
24	1.318	1.711	2.064	2.492	2.797	3.467
30	1.310	1.697	2.042	2.457	2.750	3.385
40	1.303	1.684	2.021	2.423	2.704	3.307
60	1.296	1.671	2.000	2.390	2.660	3.232
120	1.289	1.658	1.980	2.358	2.617	3.160
∞	1.282	1.645	1.960	2.326	2.576	3.090

Entries were computed with IMSL subroutine TIN.

TABLE 5 Percentage Points of the F-Distribution, $\alpha = .10$

	ν_1								
ν_2	1	2	3	4	5	6	7	8	9
1	39.86	49.50	53.59	55.83	57.24	58.20	58.91	59.44	59.86
2	8.53	9.00	9.16	9.24	9.29	9.33	9.35	9.37	9.38
3	5.54	5.46	5.39	5.34	5.31	5.28	5.27	5.25	5.24
4	4.54	4.32	4.19	4.11	4.05	4.01	3.98	3.95	3.94
5	4.06	3.78	3.62	3.52	3.45	3.40	3.37	3.34	3.32
6	3.78	3.46	3.29	3.18	3.11	3.05	3.01	2.98	2.96
7	3.59	3.26	3.07	2.96	2.88	2.83	2.79	2.75	2.72
8	3.46	3.11	2.92	2.81	2.73	2.67	2.62	2.59	2.56
9	3.36	3.01	2.81	2.69	2.61	2.55	2.51	2.47	2.44
10	3.29	2.92	2.73	2.61	2.52	2.46	2.41	2.38	2.35

continued

TABLE 5 *continued*

ν_2	ν_1 1	2	3	4	5	6	7	8	9
11	3.23	2.86	2.66	2.54	2.45	2.39	2.34	2.30	2.27
12	3.18	2.81	2.61	2.48	2.39	2.33	2.28	2.24	2.21
13	3.14	2.76	2.56	2.43	2.35	2.28	2.23	2.20	2.16
14	3.10	2.73	2.52	2.39	2.31	2.24	2.19	2.15	2.12
15	3.07	2.70	2.49	2.36	2.27	2.21	2.16	2.12	2.09
16	3.05	2.67	2.46	2.33	2.24	2.18	2.13	2.09	2.06
17	3.03	2.64	2.44	2.31	2.22	2.15	2.10	2.06	2.03
18	3.01	2.62	2.42	2.29	2.20	2.13	2.08	2.04	2.00
19	2.99	2.61	2.40	2.27	2.18	2.11	2.06	2.02	1.98
20	2.97	2.59	2.38	2.25	2.16	2.09	2.04	2.00	1.96
21	2.96	2.57	2.36	2.23	2.14	2.08	2.02	1.98	1.95
22	2.95	2.56	2.35	2.22	2.13	2.06	2.01	1.97	1.93
23	2.94	2.55	2.34	2.21	2.11	2.05	1.99	1.95	1.92
24	2.93	2.54	2.33	2.19	2.10	2.04	1.98	1.94	1.91
25	2.92	2.53	2.32	2.18	2.09	2.02	1.97	1.93	1.89
26	2.91	2.52	2.31	2.17	2.08	2.01	1.96	1.92	1.88
27	2.90	2.51	2.30	2.17	2.07	2.00	1.95	1.91	1.87
28	2.89	2.50	2.29	2.16	2.06	2.00	1.94	1.90	1.87
29	2.89	2.50	2.28	2.15	2.06	1.99	1.93	1.89	1.86
30	2.88	2.49	2.28	2.14	2.05	1.98	1.93	1.88	1.85
40	2.84	2.44	2.23	2.09	2.00	1.93	1.87	1.83	1.79
60	2.79	2.39	2.18	2.04	1.95	1.87	1.82	1.77	1.74
120	2.75	2.35	2.13	1.99	1.90	1.82	1.77	1.72	1.68
∞	2.71	2.30	2.08	1.94	1.85	1.77	1.72	.167	1.63

ν_2	ν_1 10	12	15	20	24	30	40	60	120	∞
1	60.19	60.70	61.22	61.74	62.00	62.26	62.53	62.79	63.06	63.33
2	9.39	9.41	9.42	9.44	9.45	9.46	9.47	9.47	9.48	9.49
3	5.23	5.22	5.20	5.19	5.18	5.17	5.16	5.15	5.14	5.13
4	3.92	3.90	3.87	3.84	3.83	3.82	3.80	3.79	3.78	3.76
5	3.30	3.27	3.24	3.21	3.19	3.17	3.16	3.14	3.12	3.10
6	2.94	2.90	2.87	2.84	2.82	2.80	2.78	2.76	2.74	2.72
7	2.70	2.67	2.63	2.59	2.58	2.56	2.54	2.51	2.49	2.47
8	2.54	2.50	2.46	2.42	2.40	2.38	2.36	2.34	2.32	2.29
9	2.42	2.38	2.34	2.30	2.28	2.25	2.23	2.21	2.18	2.16
10	2.32	2.28	2.24	2.20	2.18	2.16	2.13	2.11	2.08	2.06

continued

TABLE 5 *continued*

ν_2	ν_1 10	12	15	20	24	30	40	60	120	∞
11	2.25	2.21	2.17	2.12	2.10	2.08	2.05	2.03	2.00	1.97
12	2.19	2.15	2.10	2.06	2.04	2.01	1.99	1.96	1.93	1.90
13	2.14	2.10	2.05	2.01	1.98	1.96	1.93	1.90	1.88	1.85
14	2.10	2.05	2.01	1.96	1.94	1.91	1.89	1.86	1.83	1.80
15	2.06	2.02	1.97	1.92	1.90	1.87	1.85	1.82	1.79	1.76
16	2.03	1.99	1.94	1.89	1.87	1.84	1.81	1.78	1.75	1.72
17	2.00	1.96	1.91	1.86	1.84	1.81	1.78	1.75	1.72	1.69
18	1.98	1.93	1.89	1.84	1.81	1.78	1.75	1.72	1.69	1.66
19	1.96	1.91	1.86	1.81	1.79	1.76	1.73	1.70	1.67	1.63
20	1.94	1.89	1.84	1.79	1.77	1.74	1.71	1.68	1.64	1.61
21	1.92	1.87	1.83	1.78	1.75	1.72	1.69	1.66	1.62	1.59
22	1.90	1.86	1.81	1.76	1.73	1.70	1.67	1.64	1.60	1.57
23	1.89	1.84	1.80	1.74	1.72	1.69	1.66	1.62	1.59	1.55
24	1.88	1.83	1.78	1.73	1.70	1.67	1.64	1.61	1.57	1.53
25	1.87	1.82	1.77	1.72	1.69	1.66	1.63	1.59	1.56	1.52
26	1.86	1.81	1.76	1.71	1.68	1.65	1.61	1.58	1.54	1.50
27	1.85	1.80	1.75	1.70	1.67	1.64	1.60	1.57	1.53	1.49
28	1.84	1.79	1.74	1.69	1.66	1.63	1.59	1.56	1.52	1.48
29	1.83	1.78	1.73	1.68	1.65	1.62	1.58	1.55	1.51	1.47
30	1.82	1.77	1.72	1.67	1.64	1.61	1.57	1.54	1.50	1.46
40	1.76	1.71	1.66	1.61	1.57	1.54	1.51	1.47	1.42	1.38
60	1.71	1.66	1.60	1.54	1.51	1.48	1.44	1.40	1.35	1.29
120	1.65	1.60	1.55	1.48	1.45	1.41	1.37	1.32	1.26	1.19
∞	1.60	1.55	1.49	1.42	1.38	1.34	1.30	1.24	1.17	1.00

Note: Entries in this table were computed with IMSL subroutine FIN.

TABLE 6 Percentage Points of the F-Distribution, $\alpha = .05$

ν_2	ν_1 1	2	3	4	5	6	7	8	9
1	161.45	199.50	215.71	224.58	230.16	233.99	236.77	238.88	240.54
2	18.51	19.00	19.16	19.25	19.30	19.33	19.35	19.37	19.38
3	10.13	9.55	9.28	9.12	9.01	8.94	8.89	8.85	8.81
4	7.71	6.94	6.59	6.39	6.26	6.16	6.09	6.04	6.00
5	6.61	5.79	5.41	5.19	5.05	4.95	4.88	4.82	4.77
6	5.99	5.14	4.76	4.53	4.39	4.28	4.21	4.15	4.10
7	5.59	4.74	4.35	4.12	3.97	3.87	3.79	3.73	3.68
8	5.32	4.46	4.07	3.84	3.69	3.58	3.50	3.44	3.39
9	5.12	4.26	3.86	3.63	3.48	3.37	3.29	3.23	3.18
10	4.96	4.10	3.71	3.48	3.33	3.22	3.14	3.07	3.02

continued

TABLE 6 *continued*

ν_2	ν_1: 1	2	3	4	5	6	7	8	9
11	4.84	3.98	3.59	3.36	3.20	3.09	3.01	2.95	2.90
12	4.75	3.89	3.49	3.26	3.11	3.00	2.91	2.85	2.80
13	4.67	3.81	3.41	3.18	3.03	2.92	2.83	2.77	2.71
14	4.60	3.74	3.34	3.11	2.96	2.85	2.76	2.70	2.65
15	4.54	3.68	3.29	3.06	2.90	2.79	2.71	2.64	2.59
16	4.49	3.63	3.24	3.01	2.85	2.74	2.66	2.59	2.54
17	4.45	3.59	3.20	2.96	2.81	2.70	2.61	2.55	2.49
18	4.41	3.55	3.16	2.93	2.77	2.66	2.58	2.51	2.46
19	4.38	3.52	3.13	2.90	2.74	2.63	2.54	2.48	2.42
20	4.35	3.49	3.10	2.87	2.71	2.60	2.51	2.45	2.39
21	4.32	3.47	3.07	2.84	2.68	2.57	2.49	2.42	2.37
22	4.30	3.44	3.05	2.82	2.66	2.55	2.46	2.40	2.34
23	4.28	3.42	3.03	2.80	2.64	2.53	2.44	2.37	2.32
24	4.26	3.40	3.01	2.78	2.62	2.51	2.42	2.36	2.30
25	4.24	3.39	2.99	2.76	2.60	2.49	2.40	2.34	2.28
26	4.23	3.37	2.98	2.74	2.59	2.47	2.39	2.32	2.27
27	4.21	3.35	2.96	2.73	2.57	2.46	2.37	2.31	2.25
28	4.20	3.34	2.95	2.71	2.56	2.45	2.36	2.29	2.24
29	4.18	3.33	2.93	2.70	2.55	2.43	2.35	2.28	2.22
30	4.17	3.32	2.92	2.69	2.53	2.42	2.33	2.27	2.21
40	4.08	3.23	2.84	2.61	2.45	2.34	2.25	2.18	2.12
60	4.00	3.15	2.76	2.53	2.37	2.25	2.17	2.10	2.04
120	3.92	3.07	2.68	2.45	2.29	2.17	2.09	2.02	1.96
∞	3.84	3.00	2.60	2.37	2.21	2.10	2.01	1.94	1.88

ν_2	ν_1: 10	12	15	20	24	30	40	60	120	∞
1	241.88	243.91	245.96	248.00	249.04	250.08	251.14	252.19	253.24	254.3
2	19.40	19.41	19.43	19.45	19.45	19.46	19.47	19.48	19.49	19.50
3	8.79	8.74	8.70	8.66	8.64	8.62	8.59	8.57	8.55	8.53
4	5.97	5.91	5.86	5.80	5.77	5.74	5.72	5.69	5.66	5.63
5	4.73	4.68	4.62	4.56	4.53	4.50	4.46	4.43	4.40	4.36
6	4.06	4.00	3.94	3.87	3.84	3.81	3.77	3.74	3.70	3.67
7	3.64	3.57	3.51	3.44	3.41	3.38	3.34	3.30	3.27	3.23
8	3.35	3.28	3.22	3.15	3.12	3.08	3.04	3.00	2.97	2.93
9	3.14	3.07	3.01	2.94	2.90	2.86	2.83	2.79	2.75	2.71
10	2.98	2.91	2.85	2.77	2.74	2.70	2.66	2.62	2.58	2.54

continued

TABLE 6 *continued*

ν_2	ν_1 10	12	15	20	24	30	40	60	120	∞
11	2.85	2.79	2.72	2.65	2.61	2.57	2.53	2.49	2.45	2.40
12	2.75	2.69	2.62	2.54	2.51	2.47	2.43	2.38	2.34	2.30
13	2.67	2.60	2.53	2.46	2.42	2.38	2.34	2.30	2.25	2.21
14	2.60	2.53	2.46	2.39	2.35	2.31	2.27	2.22	2.18	2.13
15	2.54	2.48	2.40	2.33	2.29	2.25	2.20	2.16	2.11	2.07
16	2.49	2.42	2.35	2.28	2.24	2.19	2.15	2.11	2.06	2.01
17	2.45	2.38	2.31	2.23	2.19	2.15	2.10	2.06	2.01	1.96
18	2.41	2.34	2.27	2.19	2.15	2.11	2.06	2.02	1.97	1.92
19	2.38	2.31	2.23	2.16	2.11	2.07	2.03	1.98	1.93	1.88
20	2.35	2.28	2.20	2.12	2.08	2.04	1.99	1.95	1.90	1.84
21	2.32	2.25	2.18	2.10	2.05	2.01	1.96	1.92	1.87	1.81
22	2.30	2.23	2.15	2.07	2.03	1.98	1.94	1.89	1.84	1.78
23	2.27	2.20	2.13	2.05	2.00	1.96	1.91	1.86	1.81	1.76
24	2.25	2.18	2.11	2.03	1.98	1.94	1.89	1.84	1.79	1.73
25	2.24	2.16	2.09	2.01	1.96	1.92	1.87	1.82	1.77	1.71
26	2.22	2.15	2.07	1.99	1.95	1.90	1.85	1.80	1.75	1.69
27	2.20	2.13	2.06	1.97	1.93	1.88	1.84	1.79	1.73	1.67
28	2.19	2.12	2.04	1.96	1.91	1.87	1.82	1.77	1.71	1.65
29	2.18	2.10	2.03	1.94	1.90	1.85	1.81	1.75	1.70	1.64
30	2.16	2.09	2.01	1.93	1.89	1.84	1.79	1.74	1.68	1.62
40	2.08	2.00	1.92	1.84	1.79	1.74	1.69	1.64	1.58	1.51
60	1.99	1.92	1.84	1.75	1.70	1.65	1.59	1.53	1.47	1.39
120	1.91	1.83	1.75	1.66	1.61	1.55	1.50	1.43	1.35	1.25
∞	1.83	1.75	1.67	1.57	1.52	1.46	1.39	1.32	1.22	1.00

Note: Entries in this table were computed with IMSL subroutine FIN.

TABLE 7 Percentage Points of the F-Distribution, $\alpha = .025$

ν_2	ν_1 1	2	3	4	5	6	7	8	9
1	647.79	799.50	864.16	899.59	921.85	937.11	948.22	956.66	963.28
2	38.51	39.00	39.17	39.25	39.30	39.33	39.36	39.37	39.39
3	17.44	16.04	15.44	15.10	14.88	14.74	14.63	14.54	14.47
4	12.22	10.65	9.98	9.61	9.36	9.20	9.07	8.98	8.90
5	10.01	8.43	7.76	7.39	7.15	6.98	6.85	6.76	6.68

continued

TABLE 7 *continued*

ν_2	ν_1 1	2	3	4	5	6	7	8	9
6	8.81	7.26	6.60	6.23	5.99	5.82	5.70	5.60	5.52
7	8.07	6.54	5.89	5.52	5.29	5.12	5.00	4.90	4.82
8	7.57	6.06	5.42	5.05	4.82	4.65	4.53	4.43	4.36
9	7.21	5.71	5.08	4.72	4.48	4.32	4.20	4.10	4.03
10	6.94	5.46	4.83	4.47	4.24	4.07	3.95	3.85	3.78
11	6.72	5.26	4.63	4.28	4.04	3.88	3.76	3.66	3.59
12	6.55	5.10	4.47	4.12	3.89	3.73	3.61	3.51	3.44
13	6.41	4.97	4.35	4.00	3.77	3.60	3.48	3.39	3.31
14	6.30	4.86	4.24	3.89	3.66	3.50	3.38	3.29	3.21
15	6.20	4.77	4.15	3.80	3.58	3.41	3.29	3.20	3.12
16	6.12	4.69	4.08	3.73	3.50	3.34	3.22	3.12	3.05
17	6.04	4.62	4.01	3.66	3.44	3.28	3.16	3.06	2.98
18	5.98	4.56	3.95	3.61	3.38	3.22	3.10	3.01	2.93
19	5.92	4.51	3.90	3.56	3.33	3.17	3.05	2.96	2.88
20	5.87	4.46	3.86	3.51	3.29	3.13	3.01	2.91	2.84
21	5.83	4.42	3.82	3.48	3.25	3.09	2.97	2.87	2.80
22	5.79	4.38	3.78	3.44	3.22	3.05	2.93	2.84	2.76
23	5.75	4.35	3.75	3.41	3.18	3.02	2.90	2.81	2.73
24	5.72	4.32	3.72	3.38	3.15	2.99	2.87	2.78	2.70
25	5.69	4.29	3.69	3.35	3.13	2.97	2.85	2.75	2.68
26	5.66	4.27	3.67	3.33	3.10	2.94	2.82	2.73	2.65
27	5.63	4.24	3.65	3.31	3.08	2.92	2.80	2.71	2.63
28	5.61	4.22	3.63	3.29	3.06	2.90	2.78	2.69	2.61
29	5.59	4.20	3.61	3.27	3.04	2.88	2.76	2.67	2.59
30	5.57	4.18	3.59	3.25	3.03	2.87	2.75	2.65	2.57
40	5.42	4.05	3.46	3.13	2.90	2.74	2.62	2.53	2.45
60	5.29	3.93	3.34	3.01	2.79	2.63	2.51	2.41	2.33
120	5.15	3.80	3.23	2.89	2.67	2.52	2.39	2.30	2.22
∞	5.02	3.69	3.12	2.79	2.57	2.41	2.29	2.19	2.11

ν_2	ν_1 10	12	15	20	24	30	40	60	120	∞
1	968.62	976.71	984.89	993.04	997.20	1001	1006	1010	1014	1018
2	39.40	39.41	39.43	39.45	39.46	39.46	39.47	39.48	39.49	39.50
3	14.42	14.33	14.26	14.17	14.13	14.08	14.04	13.99	13.95	13.90
4	8.85	8.75	8.66	8.56	8.51	8.46	8.41	8.36	8.31	8.26
5	6.62	6.53	6.43	6.33	6.28	6.23	6.17	6.12	6.07	6.02
6	5.46	5.37	5.27	5.17	5.12	5.06	5.01	4.96	4.90	4.85
7	4.76	4.67	4.57	4.47	4.41	4.36	4.31	4.25	4.20	4.14
8	4.30	4.20	4.10	4.00	3.95	3.89	3.84	3.78	3.73	3.67
9	3.96	3.87	3.77	3.67	3.61	3.56	3.51	3.45	3.39	3.33
10	3.72	3.62	3.52	3.42	3.37	3.31	3.26	3.20	3.14	3.08

continued

TABLE 7 *continued*

ν_2	ν_1: 10	12	15	20	24	30	40	60	120	∞
11	3.53	3.43	3.33	3.23	3.17	3.12	3.06	3.00	2.94	2.88
12	3.37	3.28	3.18	3.07	3.02	2.96	2.91	2.85	2.79	2.72
13	3.25	3.15	3.05	2.95	2.89	2.84	2.78	2.72	2.66	2.60
14	3.15	3.05	2.95	2.84	2.79	2.73	2.67	2.61	2.55	2.49
15	3.06	2.96	2.86	2.76	2.70	2.64	2.59	2.52	2.46	2.40
16	2.99	2.89	2.79	2.68	2.63	2.57	2.51	2.45	2.38	2.32
17	2.92	2.82	2.72	2.62	2.56	2.50	2.44	2.38	2.32	2.25
18	2.87	2.77	2.67	2.56	2.50	2.44	2.38	2.32	2.26	2.19
19	2.82	2.72	2.62	2.51	2.45	2.39	2.33	2.27	2.20	2.13
20	2.77	2.68	2.57	2.46	2.41	2.35	2.29	2.22	2.16	2.09
21	2.73	2.64	2.53	2.42	2.37	2.31	2.25	2.18	2.11	2.04
22	2.70	2.60	2.50	2.39	2.33	2.27	2.21	2.14	2.08	2.00
23	2.67	2.57	2.47	2.36	2.30	2.24	2.18	2.11	2.04	1.97
24	2.64	2.54	2.44	2.33	2.27	2.21	2.15	2.08	2.01	1.94
25	2.61	2.51	2.41	2.30	2.24	2.18	2.12	2.05	1.98	1.91
26	2.59	2.49	2.39	2.28	2.22	2.16	2.09	2.03	1.95	1.88
27	2.57	2.47	2.36	2.25	2.19	2.13	2.07	2.00	1.93	1.85
28	2.55	2.45	2.34	2.23	2.17	2.11	2.05	1.98	1.91	1.83
29	2.53	2.43	2.32	2.21	2.15	2.09	2.03	1.96	1.89	1.81
30	2.51	2.41	2.31	2.20	2.14	2.07	2.01	1.94	1.87	1.79
40	2.39	2.29	2.18	2.07	2.01	1.94	1.88	1.80	1.72	1.64
60	2.27	2.17	2.06	1.94	1.88	1.82	1.74	1.67	1.58	1.48
120	2.16	2.05	1.95	1.82	1.76	1.69	1.61	1.53	1.43	1.31
∞	2.05	1.94	1.83	1.71	1.64	1.57	1.48	1.39	1.27	1.00

Note: Entries in this table were computed with IMSL subroutine FIN.

TABLE 8 Percentage Points of the F-Distribution, $\alpha = .01$

ν_2	ν_1: 1	2	3	4	5	6	7	8	9
1	4052	4999	5403	5625	5764	5859	5928	5982	6022
2	98.50	99.00	99.17	99.25	99.30	99.33	99.36	99.37	99.39
3	34.12	30.82	29.46	28.71	28.24	27.91	27.67	27.50	27.34
4	21.20	18.00	16.69	15.98	15.52	15.21	14.98	14.80	14.66
5	16.26	13.27	12.06	11.39	10.97	10.67	10.46	10.29	10.16
6	13.75	10.92	9.78	9.15	8.75	8.47	8.26	8.10	7.98
7	12.25	9.55	8.45	7.85	7.46	7.19	6.99	6.84	6.72
8	11.26	8.65	7.59	7.01	6.63	6.37	6.18	6.03	5.91
9	10.56	8.02	6.99	6.42	6.06	5.80	5.61	5.47	5.35
10	10.04	7.56	6.55	5.99	5.64	5.39	5.20	5.06	4.94

continued

TABLE 8 *continued*

ν_2	ν_1 1	2	3	4	5	6	7	8	9
11	9.65	7.21	6.22	5.67	5.32	5.07	4.89	4.74	4.63
12	9.33	6.93	5.95	5.41	5.06	4.82	4.64	4.50	4.39
13	9.07	6.70	5.74	5.21	4.86	4.62	4.44	4.30	4.19
14	8.86	6.51	5.56	5.04	4.69	4.46	4.28	4.14	4.03
15	8.68	6.36	5.42	4.89	4.56	4.32	4.14	4.00	3.89
16	8.53	6.23	5.29	4.77	4.44	4.20	4.03	3.89	3.78
17	8.40	6.11	5.18	4.67	4.34	4.10	3.93	3.79	3.68
18	8.29	6.01	5.09	4.58	4.25	4.01	3.84	3.71	3.60
19	8.18	5.93	5.01	4.50	4.17	3.94	3.77	3.63	3.52
20	8.10	5.85	4.94	4.43	4.10	3.87	3.70	3.56	3.46
21	8.02	5.78	4.87	4.37	4.04	3.81	3.64	3.51	3.40
22	7.95	5.72	4.82	4.31	3.99	3.76	3.59	3.45	3.35
23	7.88	5.66	4.76	4.26	3.94	3.71	3.54	3.41	3.30
24	7.82	5.61	4.72	4.22	3.90	3.67	3.50	3.36	3.26
25	7.77	5.57	4.68	4.18	3.85	3.63	3.46	3.32	3.22
26	7.72	5.53	4.64	4.14	3.82	3.59	3.42	3.29	3.18
27	7.68	5.49	4.60	4.11	3.78	3.56	3.39	3.26	3.15
28	7.64	5.45	4.57	4.07	3.75	3.53	3.36	3.23	3.12
29	7.60	5.42	4.54	4.04	3.73	3.50	3.33	3.20	3.09
30	7.56	5.39	4.51	4.02	3.70	3.47	3.30	3.17	3.07
40	7.31	5.18	4.31	3.83	3.51	3.29	3.12	2.99	2.89
60	7.08	4.98	4.13	3.65	3.34	3.12	2.95	2.82	2.72
120	6.85	4.79	3.95	3.48	3.17	2.96	2.79	2.66	2.56
∞	6.63	4.61	3.78	3.32	3.02	2.80	2.64	2.51	2.41

ν_2	ν_1 10	12	15	20	24	30	40	60	120	∞
1	6056	6106	6157	6209	6235	6261	6287	6313	6339	6366
2	99.40	99.42	99.43	99.45	99.46	99.46	99.47	99.48	99.49	99.50
3	27.22	27.03	26.85	26.67	26.60	26.50	26.41	26.32	26.22	26.13
4	14.55	14.37	14.19	14.02	13.94	13.84	13.75	13.65	13.56	13.46
5	10.05	9.89	9.72	9.55	9.46	9.38	9.30	9.20	9.11	9.02
6	7.87	7.72	7.56	7.40	7.31	7.23	7.15	7.06	6.97	6.88
7	6.62	6.47	6.31	6.16	6.07	5.99	5.91	5.82	5.74	5.65
8	5.81	5.67	5.52	5.36	5.28	5.20	5.12	5.03	4.95	4.86
9	5.26	5.11	4.96	4.81	4.73	4.65	4.57	4.48	4.40	4.31
10	4.85	4.71	4.56	4.41	4.33	4.25	4.17	4.08	4.00	3.91

continued

TABLE 8 *continued*

ν_2	ν_1 10	12	15	20	24	30	40	60	120	∞
11	4.54	4.40	4.25	4.10	4.02	3.94	3.86	3.78	3.69	3.60
12	4.30	4.16	4.01	3.86	3.78	3.70	3.62	3.54	3.45	3.36
13	4.10	3.96	3.82	3.66	3.59	3.51	3.43	3.34	3.25	3.17
14	3.94	3.80	3.66	3.51	3.43	3.35	3.27	3.18	3.09	3.00
15	3.80	3.67	3.52	3.37	3.29	3.21	3.13	3.05	2.96	2.87
16	3.69	3.55	3.41	3.26	3.18	3.10	3.02	2.93	2.84	2.75
17	3.59	3.46	3.31	3.16	3.08	3.00	2.92	2.83	2.75	2.65
18	3.51	3.37	3.23	3.08	3.00	2.92	2.84	2.75	2.66	2.57
19	3.43	3.30	3.15	3.00	2.92	2.84	2.76	2.67	2.58	2.49
20	3.37	3.23	3.09	2.94	2.86	2.78	2.69	2.61	2.52	2.42
21	3.31	3.17	3.03	2.88	2.80	2.72	2.64	2.55	2.46	2.36
22	3.26	3.12	2.98	2.83	2.75	2.67	2.58	2.50	2.40	2.31
23	3.21	3.07	2.93	2.78	2.70	2.62	2.54	2.45	2.35	2.26
24	3.17	3.03	2.89	2.74	2.66	2.58	2.49	2.40	2.31	2.21
25	3.13	2.99	2.85	2.70	2.62	2.54	2.45	2.36	2.27	2.17
26	3.09	2.96	2.81	2.66	2.58	2.50	2.42	2.33	2.23	2.13
27	3.06	2.93	2.78	2.63	2.55	2.47	2.38	2.29	2.20	2.10
28	3.03	2.90	2.75	2.60	2.52	2.44	2.35	2.26	2.17	2.06
29	3.00	2.87	2.73	2.57	2.49	2.41	2.33	2.23	2.14	2.03
30	2.98	2.84	2.70	2.55	2.47	2.39	2.30	2.21	2.11	2.01
40	2.80	2.66	2.52	2.37	2.29	2.20	2.11	2.02	1.92	1.80
60	2.63	2.50	2.35	2.20	2.12	2.03	1.94	1.84	1.73	1.60
120	2.47	2.34	2.19	2.03	1.95	1.86	1.76	1.66	1.53	1.38
∞	2.32	2.18	2.04	1.88	1.79	1.70	1.59	1.47	1.32	1.00

Note: Entries in this table were computed with IMSL subroutine FIN.

TABLE 9 Studentized Range Statistic, q, for $\alpha = .05$

ν	J (number of groups) 2	3	4	5	6	7	8	9	10	11
3	4.50	5.91	6.82	7.50	8.04	8.48	8.85	9.18	9.46	9.72
4	3.93	5.04	5.76	6.29	6.71	7.05	7.35	7.60	7.83	8.03
5	3.64	4.60	5.22	5.68	6.04	6.33	6.59	6.81	6.99	7.17
6	3.47	4.34	4.89	5.31	5.63	5.89	6.13	6.32	6.49	6.65
7	3.35	4.17	4.69	5.07	5.36	5.61	5.82	5.99	6.16	6.30
8	3.27	4.05	4.53	4.89	5.17	5.39	5.59	5.77	5.92	6.06
9	3.19	3.95	4.42	4.76	5.03	5.25	5.44	5.59	5.74	5.87
10	3.16	3.88	4.33	4.66	4.92	5.13	5.31	5.47	5.59	5.73

continued

TABLE 9 *continued*

ν	J (number of groups) 2	3	4	5	6	7	8	9	10	11
11	3.12	3.82	4.26	4.58	4.83	5.03	5.21	5.36	5.49	5.61
12	3.09	3.78	4.19	4.51	4.76	4.95	5.12	5.27	5.39	5.52
13	3.06	3.73	4.15	4.45	4.69	4.88	5.05	5.19	5.32	5.43
14	3.03	3.70	4.11	4.41	4.64	4.83	4.99	5.13	5.25	5.36
15	3.01	3.67	4.08	4.37	4.59	4.78	4.94	5.08	5.20	5.31
16	3.00	3.65	4.05	4.33	4.56	4.74	4.90	5.03	5.15	5.26
17	2.98	3.63	4.02	4.30	4.52	4.70	4.86	4.99	5.11	5.21
18	2.97	3.61	4.00	4.28	4.49	4.67	4.83	4.96	5.07	5.17
19	2.96	3.59	3.98	4.25	4.47	4.65	4.79	4.93	5.04	5.14
20	2.95	3.58	3.96	4.23	4.45	4.62	4.77	4.90	5.01	5.11
24	2.92	3.53	3.90	4.17	4.37	4.54	4.68	4.81	4.92	5.01
30	2.89	3.49	3.85	4.10	4.30	4.46	4.60	4.72	4.82	4.92
40	2.86	3.44	3.79	4.04	4.23	4.39	4.52	4.63	4.73	4.82
60	2.83	3.40	3.74	3.98	4.16	4.31	4.44	4.55	4.65	4.73
120	2.80	3.36	3.68	3.92	4.10	4.24	4.36	4.47	4.56	4.64
∞	2.77	3.31	3.63	3.86	4.03	4.17	4.29	4.39	4.47	4.55
					$\alpha = .01$					
2	14.0	19.0	22.3	24.7	26.6	28.2	29.5	30.7	31.7	32.6
3	8.26	10.6	12.2	13.3	14.2	15.0	15.6	16.2	16.7	17.8
4	6.51	8.12	9.17	9.96	10.6	11.1	11.5	11.9	12.3	12.6
5	5.71	6.98	7.81	8.43	8.92	9.33	9.67	9.98	10.24	10.48
6	5.25	6.34	7.04	7.56	7.98	8.32	8.62	8.87	9.09	9.30
7	4.95	5.92	6.55	7.01	7.38	7.68	7.94	8.17	8.37	8.55
8	4.75	5.64	6.21	6.63	6.96	7.24	7.48	7.69	7.87	8.03
9	4.59	5.43	5.96	6.35	6.66	6.92	7.14	7.33	7.49	7.65
10	4.49	5.28	5.77	6.14	6.43	6.67	6.88	7.06	7.22	7.36
11	4.39	5.15	5.63	5.98	6.25	6.48	6.68	6.85	6.99	7.13
12	4.32	5.05	5.51	5.84	6.11	6.33	6.51	6.67	6.82	6.94
13	4.26	4.97	5.41	5.73	5.99	6.19	6.38	6.53	6.67	6.79
14	4.21	4.89	5.32	5.63	5.88	6.08	6.26	6.41	6.54	6.66
15	4.17	4.84	5.25	5.56	5.80	5.99	6.16	6.31	6.44	6.55
16	4.13	4.79	5.19	5.49	5.72	5.92	6.08	6.22	6.35	6.46
17	4.10	4.74	5.14	5.43	5.66	5.85	6.01	6.15	6.27	6.38
18	4.07	4.70	5.09	5.38	5.60	5.79	5.94	6.08	6.20	6.31
19	4.05	4.67	5.05	5.33	5.55	5.73	5.89	6.02	6.14	6.25
20	4.02	4.64	5.02	5.29	5.51	5.69	5.84	5.97	6.09	6.19
24	3.96	4.55	4.91	5.17	5.37	5.54	5.69	5.81	5.92	6.02
30	3.89	4.45	4.80	5.05	5.24	5.40	5.54	5.65	5.76	5.85
40	3.82	4.37	4.69	4.93	5.10	5.26	5.39	5.49	5.60	5.69
60	3.76	4.28	4.59	4.82	4.99	5.13	5.25	5.36	5.45	5.53
120	3.70	4.20	4.50	4.71	4.87	5.01	5.12	5.21	5.30	5.37
∞	3.64	4.12	4.40	4.60	4.76	4.88	4.99	5.08	5.16	5.23

Note: The values in this table were computed with the IBM SSP subroutines DQH32 and DQG32.

TABLE 10 Studentized Maximum Modulus Distribution

		C (Number of Tests Performed)								
ν	α	2	3	4	5	6	7	8	9	10
2	.05	5.57	6.34	6.89	7.31	7.65	7.93	8.17	8.83	8.57
	.01	12.73	14.44	15.65	16.59	17.35	17.99	18.53	19.01	19.43
3	.05	3.96	4.43	4.76	5.02	5.23	5.41	5.56	5.69	5.81
	.01	7.13	7.91	8.48	8.92	9.28	9.58	9.84	10.06	10.27
4	.05	3.38	3.74	4.01	4.20	4.37	4.50	4.62	4.72	4.82
	.01	5.46	5.99	6.36	6.66	6.89	7.09	7.27	7.43	7.57
5	.05	3.09	3.39	3.62	3.79	3.93	4.04	4.14	4.23	4.31
	.01	4.70	5.11	5.39	5.63	5.81	5.97	6.11	6.23	6.33
6	.05	2.92	3.19	3.39	3.54	3.66	3.77	3.86	3.94	4.01
	.01	4.27	4.61	4.85	5.05	5.20	5.33	5.45	5.55	5.64
7	.05	2.80	3.06	3.24	3.38	3.49	3.59	3.67	3.74	3.80
	.01	3.99	4.29	4.51	4.68	4.81	4.93	5.03	5.12	5.19
8	.05	2.72	2.96	3.13	3.26	3.36	3.45	3.53	3.60	3.66
	.01	3.81	4.08	4.27	4.42	4.55	4.65	4.74	4.82	4.89
9	.05	2.66	2.89	3.05	3.17	3.27	3.36	3.43	3.49	3.55
	.01	3.67	3.92	4.10	4.24	4.35	4.45	4.53	4.61	4.67
10	.05	2.61	2.83	2.98	3.10	3.19	3.28	3.35	3.41	3.47
	.01	3.57	3.80	3.97	4.09	4.20	4.29	4.37	4.44	4.50
11	.05	2.57	2.78	2.93	3.05	3.14	3.22	3.29	3.35	3.40
	.01	3.48	3.71	3.87	3.99	4.09	4.17	4.25	4.31	4.37
12	.05	2.54	2.75	2.89	3.01	3.09	3.17	3.24	3.29	3.35
	.01	3.42	3.63	3.78	3.89	3.99	4.08	4.15	4.21	4.26
14	.05	2.49	2.69	2.83	2.94	3.02	3.09	3.16	3.21	3.26
	.01	3.32	3.52	3.66	3.77	3.85	3.93	3.99	4.05	4.10
16	.05	2.46	2.65	2.78	2.89	2.97	3.04	3.09	3.15	3.19
	.01	3.25	3.43	3.57	3.67	3.75	3.82	3.88	3.94	3.99
18	.05	2.43	2.62	2.75	2.85	2.93	2.99	3.05	3.11	3.15
	.01	3.19	3.37	3.49	3.59	3.68	3.74	3.80	3.85	3.89
20	.05	2.41	2.59	2.72	2.82	2.89	2.96	3.02	3.07	3.11
	.01	3.15	3.32	3.45	3.54	3.62	3.68	3.74	3.79	3.83
24	.05	2.38	2.56	2.68	2.77	2.85	2.91	2.97	3.02	3.06
	.01	3.09	3.25	3.37	3.46	3.53	3.59	3.64	3.69	3.73
30	.05	2.35	2.52	2.64	2.73	2.80	2.87	2.92	2.96	3.01
	.01	3.03	3.18	3.29	3.38	3.45	3.50	3.55	3.59	3.64
40	.05	2.32	2.49	2.60	2.69	2.76	2.82	2.87	2.91	2.95
	.01	2.97	3.12	3.22	3.30	3.37	3.42	3.47	3.51	3.55
60	.05	2.29	2.45	2.56	2.65	2.72	2.77	2.82	2.86	2.90
	.01	2.91	3.06	3.15	3.23	3.29	3.34	3.38	3.42	3.46
∞	.05	2.24	2.39	2.49	2.57	2.63	2.68	2.73	2.77	2.79
	.01	2.81	2.93	3.02	3.09	3.14	3.19	3.23	3.26	3.29

continued

TABLE 10 *continued*

		C								
ν	α	11	12	13	14	15	16	17	18	19
2	.05	8.74	8.89	9.03	9.16	9.28	9.39	9.49	9.59	9.68
	.01	19.81	20.15	20.46	20.75	20.99	20.99	20.99	20.99	20.99
3	.05	5.92	6.01	6.10	6.18	6.26	6.33	6.39	6.45	6.51
	.01	10.45	10.61	10.76	10.90	11.03	11.15	11.26	11.37	11.47
4	.05	4.89	4.97	5.04	5.11	5.17	5.22	5.27	5.32	5.37
	.01	7.69	7.80	7.91	8.01	8.09	8.17	8.25	8.32	8.39
5	.05	4.38	4.45	4.51	4.56	4.61	4.66	4.70	4.74	4.78
	.01	6.43	6.52	6.59	6.67	6.74	6.81	6.87	6.93	6.98
6	.05	4.07	4.13	4.18	4.23	4.28	4.32	4.36	4.39	4.43
	.01	5.72	5.79	5.86	5.93	5.99	6.04	6.09	6.14	6.18
7	.05	3.86	3.92	3.96	4.01	4.05	4.09	4.13	4.16	4.19
	.01	5.27	5.33	5.39	5.45	5.50	5.55	5.59	5.64	5.68
8	.05	3.71	3.76	3.81	3.85	3.89	3.93	3.96	3.99	4.02
	.01	4.96	5.02	5.07	5.12	5.17	5.21	5.25	5.29	5.33
9	.05	3.60	3.65	3.69	3.73	3.77	3.80	3.84	3.87	3.89
	.01	4.73	4.79	4.84	4.88	4.92	4.96	5.01	5.04	5.07
10	.05	3.52	3.56	3.60	3.64	3.68	3.71	3.74	3.77	3.79
	.01	4.56	4.61	4.66	4.69	4.74	4.78	4.81	4.84	4.88
11	.05	3.45	3.49	3.53	3.57	3.60	3.63	3.66	3.69	3.72
	.01	4.42	4.47	4.51	4.55	4.59	4.63	4.66	4.69	4.72
12	.05	3.39	3.43	3.47	3.51	3.54	3.57	3.60	3.63	3.65
	.01	4.31	4.36	4.40	4.44	4.48	4.51	4.54	4.57	4.59
14	.05	3.30	3.34	3.38	3.41	3.45	3.48	3.50	3.53	3.55
	.01	4.15	4.19	4.23	4.26	4.29	4.33	4.36	4.39	4.41
16	.05	3.24	3.28	3.31	3.35	3.38	3.40	3.43	3.46	3.48
	.01	4.03	4.07	4.11	4.14	4.17	4.19	4.23	4.25	4.28
18	.05	3.19	3.23	3.26	3.29	3.32	3.35	3.38	3.40	3.42
	.01	3.94	3.98	4.01	4.04	4.07	4.10	4.13	4.15	4.18
20	.05	3.15	3.19	3.22	3.25	3.28	3.31	3.33	3.36	3.38
	.01	3.87	3.91	3.94	3.97	3.99	4.03	4.05	4.07	4.09
24	.05	3.09	3.13	3.16	3.19	3.22	3.25	3.27	3.29	3.31
	.01	3.77	3.80	3.83	3.86	3.89	3.91	3.94	3.96	3.98
30	.05	3.04	3.07	3.11	3.13	3.16	3.18	3.21	3.23	3.25
	.01	3.67	3.70	3.73	3.76	3.78	3.81	3.83	3.85	3.87
40	.05	2.99	3.02	3.05	3.08	3.09	3.12	3.14	3.17	3.18
	.01	3.58	3.61	3.64	3.66	3.68	3.71	3.73	3.75	3.76
60	.05	2.93	2.96	2.99	3.02	3.04	3.06	3.08	3.10	3.12
	.01	3.49	3.51	3.54	3.56	3.59	3.61	3.63	3.64	3.66
∞	.05	2.83	2.86	2.88	2.91	2.93	2.95	2.97	2.98	3.01
	.01	3.32	3.34	3.36	3.38	3.40	3.42	3.44	3.45	3.47

continued

TABLE 10 *continued*

ν	α	C 20	21	22	23	24	25	26	27	28
2	.05	9.77	9.85	9.92	10.00	10.07	10.13	10.20	10.26	10.32
	.01	22.11	22.29	22.46	22.63	22.78	22.93	23.08	23.21	23.35
3	.05	6.57	6.62	6.67	6.71	6.76	6.80	6.84	6.88	6.92
	.01	11.56	11.65	11.74	11.82	11.89	11.97	12.07	12.11	12.17
4	.05	5.41	5.45	5.49	5.52	5.56	5.59	5.63	5.66	5.69
	.01	8.45	8.51	8.57	8.63	8.68	8.73	8.78	8.83	8.87
5	.05	4.82	4.85	4.89	4.92	4.95	4.98	5.00	5.03	5.06
	.01	7.03	7.08	7.13	7.17	7.21	7.25	7.29	7.33	7.36
6	.05	4.46	4.49	4.52	4.55	4.58	4.60	4.63	4.65	4.68
	.01	6.23	6.27	6.31	6.34	6.38	6.41	6.45	6.48	6.51
7	.05	4.22	4.25	4.28	4.31	4.33	4.35	4.38	4.39	4.42
	.01	5.72	5.75	5.79	5.82	5.85	5.88	5.91	5.94	5.96
8	.05	4.05	4.08	4.10	4.13	4.15	4.18	4.19	4.22	4.24
	.01	5.36	5.39	5.43	5.45	5.48	5.51	5.54	5.56	5.59
9	.05	3.92	3.95	3.97	3.99	4.02	4.04	4.06	4.08	4.09
	.01	5.10	5.13	5.16	5.19	5.21	5.24	5.26	5.29	5.31
10	.05	3.82	3.85	3.87	3.89	3.91	3.94	3.95	3.97	3.99
	.01	4.91	4.93	4.96	4.99	5.01	5.03	5.06	5.08	5.09
11	.05	3.74	3.77	3.79	3.81	3.83	3.85	3.87	3.89	3.91
	.01	4.75	4.78	4.80	4.83	4.85	4.87	4.89	4.91	4.93
12	.05	3.68	3.70	3.72	3.74	3.76	3.78	3.80	3.82	3.83
	.01	4.62	4.65	4.67	4.69	4.72	4.74	4.76	4.78	4.79
14	.05	3.58	3.59	3.62	3.64	3.66	3.68	3.69	3.71	3.73
	.01	4.44	4.46	4.48	4.50	4.52	4.54	4.56	4.58	4.59
16	.05	3.50	3.52	3.54	3.56	3.58	3.59	3.61	3.63	3.64
	.01	4.29	4.32	4.34	4.36	4.38	4.39	4.42	4.43	4.45
18	.05	3.44	3.46	3.48	3.50	3.52	3.54	3.55	3.57	3.58
	.01	4.19	4.22	4.24	4.26	4.28	4.29	4.31	4.33	4.34
20	.05	3.39	3.42	3.44	3.46	3.47	3.49	3.50	3.52	3.53
	.01	4.12	4.14	4.16	4.17	4.19	4.21	4.22	4.24	4.25
24	.05	3.33	3.35	3.37	3.39	3.40	3.42	3.43	3.45	3.46
	.01	4.00	4.02	4.04	4.05	4.07	4.09	4.10	4.12	4.13
30	.05	3.27	3.29	3.30	3.32	3.33	3.35	3.36	3.37	3.39
	.01	3.89	3.91	3.92	3.94	3.95	3.97	3.98	4.00	4.01
40	.05	3.20	3.22	3.24	3.25	3.27	3.28	3.29	3.31	3.32
	.01	3.78	3.80	3.81	3.83	3.84	3.85	3.87	3.88	3.89
60	.05	3.14	3.16	3.17	3.19	3.20	3.21	3.23	3.24	3.25
	.01	3.68	3.69	3.71	3.72	3.73	3.75	3.76	3.77	3.78
∞	.05	3.02	3.03	3.04	3.06	3.07	3.08	3.09	3.11	3.12
	.01	3.48	3.49	3.50	3.52	3.53	3.54	3.55	3.56	3.57

Note: This table was computed using the FORTRAN program described in Wilcox (1986b).

TABLE 11 Percentage Points, h, of the Range of J Independent t Variates

α	$\nu = 5$	$\nu = 6$	$\nu = 7$	$\nu = 8$	$\nu = 9$	$\nu = 14$	$\nu = 19$	$\nu = 24$	$\nu = 29$	$\nu = 39$	$\nu = 59$
						$J = 2$ groups					
.05	3.63	3.45	3.33	3.24	3.18	3.01	2.94	2.91	2.89	2.85	2.82
.01	5.37	4.96	4.73	4.51	4.38	4.11	3.98	3.86	3.83	3.78	3.73
						$J = 3$ groups					
.05	4.49	4.23	4.07	3.95	3.87	3.65	3.55	3.50	3.46	3.42	3.39
.01	6.32	5.84	5.48	5.23	5.07	4.69	5.54	4.43	4.36	4.29	4.23
						$J = 4$ groups					
.05	5.05	4.74	4.54	4.40	4.30	4.03	3.92	3.85	3.81	3.76	3.72
.01	7.06	6.40	6.01	5.73	5.56	5.05	4.89	4.74	4.71	4.61	4.54
						$J = 5$ groups					
.05	5.47	5.12	4.89	4.73	4.61	4.31	4.18	4.11	4.06	4.01	3.95
.01	7.58	6.76	6.35	6.05	5.87	5.33	5.12	5.01	4.93	4.82	4.74
						$J = 6$ groups					
.05	5.82	5.42	5.17	4.99	4.86	4.52	4.38	4.30	4.25	4.19	4.14
.01	8.00	7.14	6.70	6.39	6.09	5.53	5.32	5.20	5.12	4.99	4.91
						$J = 7$ groups					
.05	6.12	5.68	5.40	5.21	5.07	4.70	4.55	4.46	4.41	4.34	4.28
.01	8.27	7.50	6.92	6.60	6.30	5.72	5.46	5.33	5.25	5.16	5.05
						$J = 8$ groups					
.05	6.37	5.90	5.60	5.40	5.25	4.86	4.69	4.60	4.54	4.47	4.41
.01	8.52	7.73	7.14	6.81	6.49	5.89	5.62	5.45	5.36	5.28	5.16
						$J = 9$ groups					
.05	6.60	6.09	5.78	5.56	5.40	4.99	4.81	4.72	4.66	4.58	4.51
.01	8.92	7.96	7.35	6.95	6.68	6.01	5.74	5.56	5.47	5.37	5.28
						$J = 10$ groups					
.05	6.81	6.28	5.94	5.71	5.54	5.10	4.92	4.82	4.76	4.68	4.61
.01	9.13	8.14	7.51	7.11	6.83	6.10	5.82	5.68	5.59	5.46	5.37

Source: Reprinted, with permission, from R. Wilcox, "A table of percentage points of the range of independent t variables," *Technometrics* 25: 201–204, 1983.

TABLE 12 Lower Critical Values for the One-Sided Wilcoxon Signed Rank Test

n	$\alpha = .005$	$\alpha = .01$	$\alpha = .025$	$\alpha = .05$
4	0	0	0	0
5	0	0	0	1
6	0	0	1	3
7	0	1	3	4
8	1	2	4	6
9	2	4	6	9
10	4	6	9	11
11	6	8	11	14
12	8	10	14	18
13	10	13	18	22

continued

TABLE 12 *continued*

n	$\alpha = .005$	$\alpha = .01$	$\alpha = .025$	$\alpha = .05$
14	13	16	22	26
15	16	20	26	31
16	20	24	30	36
17	24	28	35	42
18	28	33	41	48
19	33	38	47	54
20	38	44	53	61
21	44	50	59	68
22	49	56	67	76
23	55	63	74	84
24	62	70	82	92
25	69	77	90	101
26	76	85	111	125
27	84	94	108	120
28	92	102	117	131
29	101	111	127	141
30	110	121	138	152
31	119	131	148	164
32	129	141	160	176
33	139	152	171	188
34	149	163	183	201
35	160	175	196	214
36	172	187	209	228
37	184	199	222	242
38	196	212	236	257
39	208	225	250	272
40	221	239	265	287

Entries were computed as described in Hogg and Craig, 1970, p. 361.

TABLE 13 Lower Critical Values, c_L, for the One-Sided Wilcoxon–Mann–Whitney Test

	$\alpha = .025$							
n_2	$n_1 = 3$	$n_1 = 4$	$n_1 = 5$	$n_1 = 6$	$n_1 = 7$	$n_1 = 8$	$n_1 = 9$	$n_1 = 10$
3	6	6	7	8	8	9	9	10
4	10	11	12	13	14	15	15	16
5	16	17	18	19	21	22	23	24
6	23	24	25	27	28	30	32	33
7	30	32	34	35	37	39	41	43
8	39	41	43	45	47	50	52	54
9	48	50	53	56	58	61	63	66
10	59	61	64	67	70	73	76	79

continued

TABLE 13 *continued*

	$\alpha = .005$							
n_2	$n_1 = 3$	$n_1 = 4$	$n_1 = 5$	$n_1 = 6$	$n_1 = 7$	$n_1 = 8$	$n_1 = 9$	$n_1 = 10$
3	10	10	10	11	11	12	12	13
4	15	15	16	17	17	18	19	20
5	21	22	23	24	25	26	27	28
6	28	29	30	32	33	35	36	38
7	36	38	39	41	43	44	46	48
8	46	47	49	51	53	55	57	59
9	56	58	60	62	65	67	69	72
10	67	69	72	74	77	80	83	85

Entries were determined with the algorithm in Hogg and Craig, 1970, p. 373.

APPENDIX C

Basic Matrix Algebra

A matrix is a two dimensional array of numbers or variables having r rows and c columns.

EXAMPLE.

$$\begin{pmatrix} 32 & 19 & 67 \\ 11 & 21 & 99 \\ 25 & 56 & 10 \\ 76 & 39 & 43 \end{pmatrix}$$

is a matrix having four rows and three columns. ■

The matrix is said to be square if $r = c$ (the number of rows equals the number of columns). A matrix with $r = 1$ ($c = 1$) is called a row (column) vector.

A common notation for a matrix is $\mathbf{X} = (x_{ij})$, meaning that $\mathbf{X}$ is a matrix where x_{ij} is the value in the ith row and jth column. For the matrix just shown, the value in the first row and first column is $x_{11} = 32$ and the value in the third row and second column is $x_{32} = 56$.

EXAMPLE. A commonly encountered square matrix is the correlation matrix. That is, for every individual, we have p measures with r_{ij} being Pearson's correlation between the ith and jth measures. Then the correlation matrix is $\mathbf{R} = (r_{ij})$. If $p = 3$, $r_{12} = .2$, $r_{13} = .4$, and $r_{23} = .3$, then

$$\mathbf{R} = \begin{pmatrix} 1 & .2 & .4 \\ .2 & 1 & .3 \\ .4 & .3 & 1 \end{pmatrix}.$$

(The correlation of a variable with itself is 1.) ■

The transpose of a matrix is just the matrix obtained when the rth row becomes the rth column. More formally, the transpose of the matrix $\mathbf{X} = (x_{ij})$ is

$$\mathbf{X}' = (x_{ji}),$$

which has c rows and r columns.

EXAMPLE. The transpose of the matrix

$$\mathbf{X} = \begin{pmatrix} 23 & 91 \\ 51 & 29 \\ 63 & 76 \\ 11 & 49 \end{pmatrix}$$

is

$$\mathbf{X}' = \begin{pmatrix} 23 & 51 & 63 & 11 \\ 91 & 29 & 76 & 49 \end{pmatrix}.$$

■

The matrix $\mathbf{X}$ is said to be symmetric if $\mathbf{X} = \mathbf{X}'$. That is, $x_{ij} = x_{ji}$.

The diagonal of an r-by-r (square) matrix refers to x_{ii}, $i = 1, \ldots, r$. A diagonal matrix is an r-by-r matrix where the off-diagonal elements (the x_{ij}, $i \neq j$) are zero. An important special case is the identity matrix, which has 1's along the diagonal and zeros elsewhere. For example,

$$\begin{pmatrix} 1 & 0 & 0 \\ 0 & 1 & 0 \\ 0 & 0 & 1 \end{pmatrix}$$

is the identity matrix when $r = c = 3$. A common notation for the identity matrix is $\mathbf{I}$.

Two $r \times c$ matrices, $\mathbf{X}$ and $\mathbf{Y}$, are said to be equal if for every i and j, $x_{ij} = y_{ij}$. That is, every element in $\mathbf{X}$ is equal to the corresponding element in $\mathbf{Y}$.

The sum of two matrices having the same number of rows and columns is

$$z_{ij} = x_{ij} + y_{ij}.$$

EXAMPLE.

$$\begin{pmatrix} 1 & 3 \\ 4 & -1 \\ 9 & 2 \end{pmatrix} + \begin{pmatrix} 8 & 2 \\ 4 & 9 \\ 1 & 6 \end{pmatrix} = \begin{pmatrix} 9 & 5 \\ 8 & 8 \\ 10 & 8 \end{pmatrix}.$$

■

Multiplication of a matrix by a scalar — say, a — is

$$aX = (ax_{ij}).$$

That is, every element of the matrix **X** is multiplied by a.

EXAMPLE.

$$2\begin{pmatrix} 8 & 2 \\ 4 & 9 \\ 1 & 6 \end{pmatrix} = \begin{pmatrix} 16 & 4 \\ 8 & 18 \\ 2 & 12 \end{pmatrix}.$$

■

For an n-by-p matrix (meaning we have p measures for each of n individuals), the sample mean is

$$\bar{X} = (\bar{X}_1, \ldots, \bar{X}_p),$$

the vector of the sample means corresponding to the p measures. That is,

$$\bar{X}_j = \sum_{i-1}^{n} X_{ij}, \quad j = 1, \ldots, p.$$

If **X** is an r-by-c matrix and **Y** is a c-by-t matrix (the number of columns for **X** is the same as the number of rows for **Y**), the product of **X** and **Y** is the r-by-t matrix **Z** = **XY**, where

$$z_{ij} = \sum_{k=1}^{c} x_{ik} y_{kj}.$$

EXAMPLE.

$$\begin{pmatrix} 8 & 2 \\ 4 & 9 \\ 1 & 6 \end{pmatrix} \begin{pmatrix} 5 & 3 \\ 2 & 1 \end{pmatrix} = \begin{pmatrix} 44 & 26 \\ 38 & 21 \\ 17 & 9 \end{pmatrix}$$

■

EXAMPLE. Consider a random sample of n observations, $X_1, \ldots, X_n$, and let **J** be a row matrix of 1's. That is, **J** = $(1, 1, \ldots, 1)$. Letting **X** be a column matrix containing $X_1, \ldots, X_n$, then

$$\sum X_i = \mathbf{JX}.$$

Continued

EXAMPLE. (*Continued*)
The sample mean is

$$\bar{X} = \frac{1}{n}\mathbf{J}\mathbf{X}.$$

The sum of the squared observations is

$$\sum X_i^2 = \mathbf{X}'\mathbf{X}.$$

■

Let X be an n-by-p matrix of p measures taken on n individuals. Then $\mathbf{X}_i$ is the ith row (vector) in the matrix X and $(\mathbf{X}_i - \bar{\mathbf{X}})'$ is a p-by-1 matrix consisting of the ith row of X minus the sample mean. Moreover, $(\mathbf{X}_i - \bar{\mathbf{X}})'(\mathbf{X}_i - \bar{\mathbf{X}})$ is a p-by-p matrix. The (sample) covariance matrix is

$$\mathbf{S} = \frac{1}{n-1}\sum_{i=1}^{n}(\mathbf{X}_i - \bar{\mathbf{X}})'(\mathbf{X}_i - \bar{\mathbf{X}}).$$

That is, $\mathbf{S} = (s_{jk})$, where s_{jk} is the covariance between the jth and kth measures. When $j = k$, s_{jk} is the sample variance corresponding to the jth variable under study.

For any square matrix X, the matrix $\mathbf{X}^{-1}$ is said to be the inverse of X if

$$\mathbf{X}\mathbf{X}^{-1} = \mathbf{I},$$

the identity matrix. If an inverse exists, X is said to be *nonsingular*; otherwise it is *singular*. The inverse of a nonsingular matrix can be computed with the S-PLUS built-in function

solve(m),

where m is any S-PLUS variable having matrix mode, with the number of rows equal to the number of columns.

EXAMPLE. Consider the matrix

$$\begin{pmatrix} 5 & 3 \\ 2 & 1 \end{pmatrix}.$$

Storing it in the S-PLUS variable m, the command solve(m) returns

$$\begin{pmatrix} -1 & 3 \\ 2 & -5 \end{pmatrix}.$$

It is left as an exercise to verify that multiplying these two matrices together yields I. ■

EXAMPLE. It can be shown that the matrix

$$\begin{pmatrix} 2 & 5 \\ 2 & 5 \end{pmatrix}$$

does not have an inverse. The S-PLUS function solve, applied to this matrix, reports that the matrix appears to be singular. ■

Consider any r-by-c matrix X, and let k indicate any square submatrix. That is, consider the matrix consisting of any k rows and any k columns taken from X. The *rank* of X is equal to the largest k for which a k-by-k submatrix is nonsingular.

The *trace* of a square matrix is just the sum of the diagonal elements and is often denoted by tr. For example, if

$$A = \begin{pmatrix} 5 & 3 \\ 2 & 1 \end{pmatrix},$$

then

$$\text{tr}(A) = 5 + 1 = 6.$$

The notation

$$\text{diag}\{x_1, \ldots, x_n\}$$

refers to a *diagonal* matrix with the values $x_1, \ldots, x_n$ along the diagonal. For example,

$$\text{diag}\{4,5,2\} = \begin{pmatrix} 4 & 0 & 0 \\ 0 & 5 & 0 \\ 0 & 0 & 2 \end{pmatrix}.$$

A *block diagonal* matrix refers to a matrix where the diagonal elements are themselves matrices.

EXAMPLE. If

$$V_1 = \begin{pmatrix} 9 & 2 \\ 4 & 15 \end{pmatrix} \quad \text{and} \quad V_2 = \begin{pmatrix} 11 & 32 \\ 14 & 29 \end{pmatrix},$$

then

$$\text{diag}(V_1, V_2) = \begin{pmatrix} 9 & 2 & 0 & 0 \\ 4 & 15 & 0 & 0 \\ 0 & 0 & 11 & 32 \\ 0 & 0 & 14 & 29 \end{pmatrix}.$$

■

Let A be an $m_1 \times n_1$ matrix, and let B be an $m_2 \times n_2$ matrix. The (right) Kronecker product of A and B is the $m_1 m_2 \times n_1 n_2$ matrix

$$\mathrm{A} \otimes \mathrm{B} = \begin{pmatrix} a_{11}\mathrm{B} & a_{12}\mathrm{B} & \cdots & a_{1n_1}\mathrm{B} \\ a_{21}\mathrm{B} & a_{22}\mathrm{B} & \cdots & a_{2n_1}\mathrm{B} \\ \vdots & \vdots & \vdots & \vdots \\ a_{m_1 1}\mathrm{B} & a_{m_1 2}\mathrm{B} & \cdots & a_{m_1 n_1}\mathrm{B} \end{pmatrix}$$

A matrix X^- is said to be a *generalized inverse* of the matrix X if

1. XX^- is symmetric.
2. $\mathrm{X}^-\mathrm{X}$ is symmetric.
3. $\mathrm{XX}^-\mathrm{X} = \mathrm{X}$.
4. $\mathrm{X}^-\mathrm{XX}^- = \mathrm{X}^-$.

A method for computing a generalized inverse can be found in Graybill (1983).

REFERENCES

Agresti, A. (1990). *Categorical Data Analysis*. New York: Wiley.

Agresti, A. (1996). *An Introduction to Categorical Data Analysis*. New York: Wiley.

Agresti, A., & Pendergast, J. (1986). Comparing mean ranks for repeated measures data. *Communications in Statistics—Theory and Methods, 15*, 1417–1433.

Akritas, M. G. (1990). The rank transform method in some two-factor designs. *Journal of the American Statistical Association, 85*, 73–78.

Akritas, M. G., & Arnold, S. F. (1994). Fully nonparametric hypotheses for factorial designs I: Multivariate repeated measures designs. *Journal of the American Statistical Association, 89*, 336–343.

Akritas, M. G., Arnold, S. F., & Brunner, E. (1997). Nonparametric hypotheses and rank statistics for unbalanced factorial designs. *Journal of the American Statistical Association, 92*, 258–265.

Alexander, R. A., & Govern, D. M. (1994). A new and simpler approximation for ANOVA under variance heterogeneity. *Journal of Educational Statistics, 19*, 91–101.

Algina, J., Oshima, T. C., & Lin, W.-Y. (1994). Type I error rates for Welch's test and James's second-order test under nonnormality and inequality of variance when there are two groups. *Journal of Educational and Behavioral Statistics, 19*, 275–291.

Andersen, E. B. (1997). *Introduction to the Statistical Analysis of Categorical Data*. New York: Springer.

Andrews, D. F., Bickel, P. J., Hampel, F. R., Huber, P. J., Rogers, W. H., & Tukey, J. W. (1972). *Robust Estimates of Location: Survey and Advances*. Princeton University Press, Princeton, NJ.

Asiribo, O., & Gurland, J. (1989). Some simple approximate solutions to the Behrens–Fisher problem. *Communications in Statistics—Theory and Methods, 18*, 1201–1216.

Bahadur, R., & Savage, L. (1956). The nonexistence of certain statistical procedures in nonparametric problems. *Annals of Statistics, 25*, 1115–1122.

Barnett, V., & Lewis, T. (1994). *Outliers in Statistical Data*. New York: Wiley.

Baron, R. M., & Kenny, D. A. (1986). The moderator–mediator variable distinction in social psychological research: Conceptual, strategic, and statistical considerations. *Journal of Personality and Social Psychology 51*, 1173–1182.

Barrett, J. P. (1974). The coefficient of determination—Some limitations. *Annals of Statistics, 28*, 19–20.

Basu, S., & DasGupta, A. (1995). Robustness of standard confidence intervals for location parameters under departure from normality. *Annals of Statistics, 23*, 1433–1442.

Beal, S. L. (1987). Asymptotic confidence intervals for the difference between two binomial parameters for use with small samples. *Biometrics, 43*, 941–950.

Beasley, T. M. (2000). Nonparametric tests for analyzing interactions among intrablock ranks in multiple group repeated measures designs. *Journal of Educational and Behavioral Statistics, 25*, 20–59.

Belsley, D. A., Kuh, E., & Welsch, R. E. (1980). *Regression Diagnostics: Identifying Influential Data and Sources of Collinearity*. New York: Wiley.

Benjamini, Y., & Hochberg, Y. (1995). Controlling the false discovery rate: A practical and powerful approach to multiple testing. *Journal of the Royal Statistical Society, B, 57*, 289–300.

Benjamini, Y., & Hochberg, Y. (2000). On the adaptive control of the false discovery rate in multiple testing with independent statistics. *Journal of Educational and Behavioral Statistics, 25*, 60–83.

Berger, R. L. (1993). More powerful tests from confidence interval p values. *American Statistician, 50*, 314–318.

Berk, K. N., & Booth, D. E. (1995). Seeing a curve in multiple regression. *Technometrics, 37*, 385–398.

Bernhardson, C. (1975). Type I error rates when multiple comparison procedures follow a significant *F* test of ANOVA. *Biometrics, 31*, 719–724.

Bianco, A., Boente, G., & di Rienzo, J. (2000). Some results for robust GM-estimators in heteroscedastic regression models. *Journal of Statistical Planning and Inference, 89*, 215–242.

Bickel, P. J., & Lehmann, E. L. (1975). Descriptive statistics for nonparametric models II. Location. *Annals of Statistics, 3*, 1045–1069.

Bishop, T., & Dudewicz, E. J. (1978). Exact analysis of variance with unequal variances: Test procedures and tables. *Technometrics, 20*, 419–430.

Blair, R. C., Sawilowski, S. S., & Higgens, J. J. (1987). Limitations of the rank transform statistic in tests for interactions. *Communications in Statistics—Simulation and Computation, 16*, 1133–1145.

Blyth, C. R. (1986). Approximate binomial confidence limits. *Journal of the American Statistical Association, 81*, 843–855.

Boik, R. J. (1981). A priori tests in repeated measures designs: Effects of nonsphericity. *Psychometrika, 46*, 241–255.

Boik, R. J. (1987). The Fisher–Pitman permutation test: A nonrobust alternative to the normal theory *F* test when variances are heterogeneous. *British Journal of Mathematical and Statistical Psychology, 40*, 26–42.

Boos, D. B., & Hughes-Oliver, J. M. (2000). How large does *n* have to be for *Z* and *t* intervals? *American Statistician, 54*, 121–128.

Boos, D. B., & Zhang, J. (2000). Monte Carlo evaluation of resampling-based hypothesis tests. *Journal of the American Statistical Association, 95*, 486–492.

Booth, J. G., & Sarkar, S. (1998). Monte Carlo approximation of bootstrap variances. *American Statistician, 52*, 354–357.

Box, G. E. P. (1953). Non-normality and tests on variances. *Biometrika, 40*, 318–335.

Bradley, J. V. (1978) Robustness? *British Journal of Mathematical and Statistical Psychology, 31*, 144–152.

Breiman, L. (1995). Better subset regression using the nonnegative garrote. *Technometrics, 37*, 373–384.

Breiman, L., & Friedman, J. H. (1985). Estimating optimal transformations for multiple regression and correlation (with discussion). *Journal of the American Statistical Association, 80*, 580–619.

Brown, C., & Mosteller, F. (1991). Components of variance. In D. Hoaglin, F. Mosteller, & J. Tukey (Eds.), *Fundamentals of Exploratory Analysis of Variance*, pp. 193–251. New York: Wiley.

Brown, M. B., & Forsythe, A. (1974a). The small-sample behavior of some statistics which test the equality of several means. *Technometrics, 16*, 129–132.

Brown, M. B., & Forsythe, A. (1974b). Robust tests for the equality of variances. *Journal of the American Statistical Association, 69*, 364–367.

Brunner, E., & Denker, M. (1994). Rank statistics under dependent observations and applications to factorial designs. *Journal of Statistical Planning and Inference, 42*, 353–378.

Brunner, E., & Dette, H. (1992). Rank procedures for the two-factor mixed model. *Journal of the American Statistical Association, 87*, 884–888.

Brunner, E., Dette, H., & Munk, A. (1997). Box-type approximations in nonparametric factorial designs. *Journal of the American Statistical Association, 92*, 1494–1502.

Brunner, E., Domhof, S., & Langer, F. (2002). *Nonparametric Analysis of Longitudinal Data in Factorial Experiments*. New York: Wiley.

Brunner, E., & Munzel, U. (2000). The nonparametric Behrens–Fisher problem: Asymptotic theory and small-sample approximation. *Biometrical Journal 42*, 17–25.

Brunner, E., Munzel, U., & Puri, M. L. (1999). Rank-score tests in factorial designs with repeated measures. *Journal of Multivariate Analysis, 70*, 286–317.

Carling, K. (2000). Resistant outlier rules and the non-Gaussian case. *Computational Statistics & Data Analysis, 33*, 249–258.

Carroll, R. J., & Ruppert, D. (1982). Robust estimation in heteroscedastic linear models. *Annals of Statistics, 10*, 429–441.

Carroll, R. J., & Ruppert, D. (1988). *Transformation and Weighting in Regression*. New York: Chapman and Hall.

Chatterjee, S., & Hadi, A. S. (1988). *Sensitivity Analysis in Linear Regression Analysis*. New York: Wiley.

Chaudhuri, P. (1996). On a geometric notion of quantiles for multivariate data. *Journal of the American Statistical Association, 91*, 862–872.

Chen, L. (1995). Testing the mean of skewed distributions. *Journal of the American Statistical Association, 90*, 767–772.

Chen, S., & Chen, H. J. (1998). Single-stage analysis of variance under heteroscedasticity. *Communications in Statistics—Simulation and Computation 27*, 641–666.

Chernick, M. R. (1999). *Bootstrap Methods: A Practitioner's Guide*. New York: Wiley.

Choi, K., & Marden, J. (1997). An approach to multivariate rank tests in multivariate analysis of variance. *Journal of the American Statistical Association, 92*, 1581–1590.

Chow, G. C. (1960). Tests of equality between sets of coefficients in two linear regressions. *Econometrika, 28*, 591–606.

Clark, S. (1999). *Towards the Edge of the Universe: A Review of Modern Cosmology*. New York: Springer.

Clemons, T. E., & Bradley Jr., E. L. (2000). A nonparametric measure of the overlapping coefficient. *Computational Statistics & Data Analysis, 34*, 51–61.

Cleveland, W. S. (1979). Robust locally weighted regression and smoothing scatterplots. *Journal of the American Statistical Association, 74*, 829–836.

Cleveland, W. S. (1985). *The Elements of Graphing Data*. Summit, NJ: Hobart Press.

Cliff, N. (1996). *Ordinal Methods for Behavioral Data Analysis*. Mahwah, NJ: Erlbaum.

Coakley, C. W., & Hettmansperger, T. P. (1993). A bounded influence, high breakdown, efficient regression estimator. *Journal of the American Statistical Association, 88*, 872–880.

Cochran, W. G., & Cox, G. M. (1950). *Experimental Design*. New York: Wiley.

Coe, P. R., & Tamhane, A. C. (1993). Small-sample confidence intervals for the difference, ratio, and odds ratio of two success probabilities. *Communications in Statistics—Simulation and Computation, 22*, 925–938.

Cohen, J. (1977). *Statistical Power Analysis for the Behavioral Sciences*. New York: Academic Press.

Cohen, J. (1994). The earth is round ($p < .05$). *American Psychologist, 49*, 997–1003.

Cohen, M., Dalal, S. R., & Tukey, J. W. (1993). Robust, smoothly heterogeneous variance regression. *Applied Statistics, 42*, 339–353.

Coleman, J. S. (1964). *Introduction to Mathematical Sociology*. New York: Free Press.

Comrey, A. L. (1985). A method for removing outliers to improve factor analytic results. *Multivariate Behavioral Research, 20*, 273–281.

Conerly, M. D., & Mansfield, E. R. (1988). An approximate test for comparing heteroscedastic regression models. *Journal of the American Statistical Association, 83*, 811–817.

Conover, W., Johnson, M., & Johnson, M. (1981). A comparative study of tests for homogeneity of variances, with applications to the outer continental shelf bidding data. *Technometrics, 23*, 351–361.

Cook, R. D. (1993). Exploring partial residuals plots. *Technometrics, 35*, 351–362.

Cook, R. D., & Weisberg, S. (1992). *Residuals and Influence in Regression*. New York: Chapman and Hall.

Cressie, N. A. C., & Whitford, H. J. (1986). How to use the two sample *t*-test. *Biometrical Journal, 28*, 131–148.

Cronbach, L. J. (1987). Statistical tests for moderator variables: Flaws in analyses recently proposed. *Psychological Bulletin, 102*, 414–417.

Cronbach, L. J., Gleser, G. C., Nanda, H., & Rajaratnam, N. (1972). *The Dependability of Behavioral Measurements*. New York: Wiley.

Crowder, M. J., & Hand, D. J. (1990). *Analysis of Repeated Measures*. London: Chapman.

Croux, C., Rousseeuw, P. J., & Hössjer, O. (1994). Generalized S-estimators. *Journal of the American Statistical Association, 89*, 1271–1281.

Dana, E. (1990). Salience of the self and salience of standards: Attempts to match self to standard. Unpublished Ph.D. dissertation, University of Southern California.

Dantzig, G. (1940). On the non-existance of tests of "Student's" hypothesis having power functions independent of σ. *Annals of Mathematical Statistics, 11*, 186.

Davies, L., & Gather, U. (1993). The identification of multiple outliers (with discussion). *Journal of the American Statistical Association, 88*, 782–792.

Davison, A. C., & Hinkley, D. V. (1997). *Bootstrap Methods and Their Application*. Cambridge: Cambridge University Press.

Dawson, M. E., Schell, A. M., Hazlett, E. A., Nuechterlein, K. H., & Filion, D. L. (2000). On the clinical and cognitive meaning of impaired sensorimotor gating in schizophrenia. *Psychiatry Research, 96*, 187–197.

Deming, W. E. (1943). *Statistical Adjustment of Data*. New York: Wiley.

Derksen, S., & Keselman, H. J. (1992). Backward, forward and stepwise automated subset selection algorithms: Frequency of obtaining authentic and noise variables. *British Journal of Mathematical and Statistical Psychology, 45*, 265–282.

Dielman, T., Lowry, C., & Pfaffenberger, R. (1994). A comparison of quantile estimators. *Communications in Statistics—Simulation and Computation, 23*, 355–371.

Dietz, E. J. (1989). Teaching regression in a nonparametric statistics course. *American Statistician, 43*, 35–40.

Diggle, P. J., Liang, K.-Y., & Zeger, S. L. (1994). *Analysis of Longitudinal Data*. Oxford: Oxford University Press.

Doksum, K. A. (1974). Empirical probability plots and statistical inference for nonlinear models in the two-sample case. *Annals of Statistics, 2*, 267–277.

Doksum, K. A. (1977). Some graphical methods in statistics. A review and some extensions. *Statistica Neerlandica, 31*, 53–68.

Doksum, K. A., Blyth, S., Bradlow, E., Meng, X., & Zhao, H. (1994). Correlation curves as local measures of variance explained by regression. *Journal of the American Statistical Association, 89*, 571–582.

Doksum, K. A., & Samarov, A. (1995). Nonparametric estimation of global functionals and a measure of the explanatory power of covariates in regression. *Annals of Statistics, 23*, 1443–1473.

Doksum, K. A., & Sievers, G. L. (1976). Plotting with confidence: Graphical comparisons of two populations. *Biometrika, 63*, 421–434.

Doksum, K. A., & Wong, C.-W. (1983). Statistical tests based on transformed data. *Journal of the American Statistical Association, 78*, 411–417.

Donner, A., & Wells, G. (1986). A comparison of confidence interval methods for the intraclass correlation coefficient. *Biometrics, 42*, 401–412.

Donoho, D. L., & Gasko, M. (1992). Breakdown properties of location estimates based on halfspace depth and projected outlyingness. *Annals of Statistics, 20*, 1803–1827.

Drouet, D., & Kotz, S. (2001). *Correlation and Dependence*. London: Imperial College Press.

Duncan, G. T., & Layard, M. W. (1973). A Monte-Carlo study of asymptotically robust tests for correlation. *Biometrika, 60*, 551–558.

Dunnett, C. W. (1980a). Pairwise multiple comparisons in the unequal variance case. *Journal of the American Statistical Association, 75*, 796–800.

Dunnett, C. W. (1980b). Pairwise multiple comparisons in the homogeneous variance, unequal sample size case. *Journal of the American Statistical Association, 75*, 796–800.

Dunnett, C. W., & Tamhane, A. C. (1992). A step-up multiple test procedure. *Journal of the American Statistical Association, 87*, 162–170.

Efron, B., & Tibshirani, R. J. (1993). *An Introduction to the Bootstrap*. New York: Chapman and Hall.

Efron, B., & Tibshirani, R. J. (1997). Improvements on cross-validation: The .632+ bootstrap method. *Journal of the American Statistical Association, 92*, 548–560.

Elashoff, J. D., & Snow, R. E. (1970). A case study in statistical inference: Reconsideration of the Rosenthal–Jacobson data on teacher expectancy. Technical report no. 15, School of Education, Stanford University, Stanford, CA.

Emerson, J. D., & Hoaglin, D. C. (1983). Stem-and-leaf displays. In D. C. Hoaglin, F. Mosteller, & J. W. Tukey (Eds.), *Understanding Robust and Exploratory Data Analysis*, pp. 7–32. New York: Wiley.

Ezekiel, M. (1924). A method for handling curvilinear correlation for any number of variables. *Journal of the American Statistical Association, 19*, 431–453.

Fairly, D. (1986). Cherry trees with cones? *American Statistician, 40*, 138–139.

Fan, J. (1992). Design-adaptive nonparametric regression. *Journal of the American Statistical Association, 87*, 998–1004.

Fears, T. R., Benichou, J., & Gail, M. H. (1996). A reminder of the fallibility of the Wald Statistic. *Journal of the American Statistical Association, 94*, 226–227.

Ferretti, N., Kelmansky, D., Yohai, V. J., & Zamar, R. (1999). A class of locally and globally robust regression estimates. *Journal of the American Statistical Association, 94*, 174–188.

Finner, H., & Roters, M. (2002). Multiple hypotheses testing and expected number of type I errors. *Annals of Statistics, 30*, 220–238.

Fisher, R. A. (1935). The fiducial argument in statistical inference. *Annals of Eugenics, 6*, 391–398.

Fisher, R. A. (1941). The asymptotic approach to Behren's integral, with further tables for the *d* test of significance. *Annals of Eugenics, 11*, 141–172.

Fleiss, J. L. (1981). *Statistical Methods for Rates and Proportions*, 2nd ed. New York: Wiley.

Fligner, M. A., & Policello II, G. E. (1981). Robust rank procedures for the Behrens–Fisher problem. *Journal of the American Statistical Association, 76*, 162–168.

Freedman, D., & Diaconis, P. (1981). On the histogram as density estimator: L_2 theory. *Z. Wahrsche. verw. Ge., 57*, 453–476.

Freedman, D., & Diaconis, P. (1982). On inconsistent *M*-estimators. *Annals of Statistics, 10*, 454–461.

Frigge, M., Hoaglin, D. C., & Iglewicz, B. (1989). Some implementations of the Boxplot. *American Statistician, 43*, 50–54.

Fung, W.-K. (1993). Unmasking outliers and leverage points: A confirmation. *Journal of the American Statistical Association, 88*, 515–519.

Games, P. A., & Howell, J. (1976). Pairwise multiple comparison procedures with unequal *n*'s and/or variances: A Monte Carlo study. *Journal of Educational Statistics, 1*, 113–125.

Giltinan, D. M., Carroll, R. J., & Ruppert, D. (1986). Some new estimation methods for weighted regression when there are possible outliers. *Technometrics, 28*, 219–230.

Goldberg, K. M., & Iglewicz, B. (1992). Bivariate extensions of the boxplot. *Technometrics, 34*, 307–320.

Graybill, F. A. (1983). *Matrices with Applications in Statistics*. Belmont, CA: Wadsworth.

Hald, A. (1952). *Statistical Theory with Engineering Applications*. New York: Wiley.

Hald, A. (1998). *A History of Mathematical Statistics from 1750 to 1930*. New York: Wiley.

Hall, P. (1988a). On symmetric bootstrap confidence intervals. *Journal of the Royal Statistical Society, Series B, 50*, 35–45.

Hall, P. (1988b). Theoretical comparison of bootstrap confidence intervals. *Annals of Statistics, 16*, 927–953.

Hall, P., & Hall, D. (1995). *The Bootstrap and Edgeworth Expansion*. New York: Springer Verlag.

Hall, P., & Sheather, S. J. (1988). On the distribution of a Studentized quantile. *Journal of the Royal Statistical Society, Series B, 50*, 380–391.

Hampel, F. R., Ronchetti, E. M., Rousseeuw, P. J., & Stahel, W. A. (1986). *Robust Statistics: The Approach Based on Influence Functions*. New York: Wiley.

Hand, D. J., & Crowder, M. J. (1996). *Practical Longitudinal Data Analysis*. London: Chapman and Hall.

Harrell, F. E., & Davis, C. E. (1982). A new distribution-free quantile estimator. *Biometrika, 69*, 635–640.

Hastie, T. J., & Tibshirani, R. J. (1990). *Generalized Additive Models*. New York: Chapman and Hall.

Hawkins, D. M., & Olive, D. (1999). Applications and algorithms for least trimmed sum of absolute deviations regression. *Computational Statistics & Data Analysis, 28*, 119–134.

Hayter, A. (1984). A proof of the conjecture that the Tukey–Kramer multiple comparison procedure is conservative. *Annals of Statistics, 12*, 61–75.

Hayter, A. (1986). The maximum familywise error rate of Fisher's least significant difference test. *Journal of the American Statistical Association, 81*, 1000–1004.

He, X., & Portnoy, S. (1992). Reweighted LS estimators converge at the same rate as the initial estimator. *Annals of Statistics, 20*, 2161– 2167.

He, X., Simpson, D. G., & Portnoy, S. L. (1990). Breakdown robustness of tests. *Journal of the American Statistical Association, 85*, 446–452.

Headrick, T. C., & Rotou, O. (2001). An investigation of the rank transformation in multiple regression. *Computational Statistics & Data Analysis, 38*, 203–215.

Hettmansperger, T. P. (1984). *Statistical Inference Based on Ranks*. New York: Wiley.

Hettmansperger, T. P., & McKean, J. W. (1977). A robust alternative based on ranks to least squares in analyzing linear models. *Technometrics, 19*, 275–284.

Hettmansperger, T. P., & McKean, J. W. (1998). *Robust Nonparametric Statistical Methods*. London: Arnold.

Hettmansperger, T. P., Möttönen, J., & Oja, H. (1997). Affine-invariant one-sample signed-rank tests. *Journal of the American Statistical Association, 92*, 1591–1600.

Hettmansperger, T. P., & Sheather, S. J. (1986). Confidence intervals based on interpolated order statistics. *Statistics and Probability Letters 4*, 75–79.

Hewett, J. E., & Spurrier, J. D. (1983). A survey of two-stage tests of hypotheses: Theory and application. *Communications in Statistics—Theory and Methods, 12*, 2307–2425.

Hoaglin, D. C., & Iglewicz, B. (1987). Fine-tuning some resistant rules for outlier labeling. *Journal of the American Statistical Association, 82*, 1147–1149.

Hoaglin, D. C., Iglewicz, B., & Tukey, J. W. (1986). Performance of some resistant rules for outlier labeling. *Journal of the American Statistical Association, 81*, 991–999.

Hochberg, Y. (1975). Simultaneous inference under Behrens–Fisher conditions: A two sample approach. *Communications in Statistics, 4*, 1109–1119.

Hochberg, Y. (1988). A sharper Bonferroni procedure for multiple tests of significance. *Biometrika, 75*, 800–802.

Hodges, J. L., Ramsey, P. H., & Wechsler, S. (1990). Improved significance probabilities of the Wilcoxon test. *Journal of Educational Statistics, 15*, 249–265.

Hoenig, J. M., & Heisey, D. M. (2001). The abuse of power: The pervasive fallacy of power calculations for data analysis. *American Statistician, 55*, 19–24.

Hogg, R. V., & Craig, A. T. (1970). *Introduction to Mathematical Statistics*. New York: Macmillan.

Hollander, M., & Sethuraman, J. (1978). Testing for agreement between two groups of judges. *Biometrika, 65*, 403–411.

Hommel, G. (1988). A stagewise rejective multiple test procedure based on a modified Bonferroni test. *Biometrika, 75*, 383–386.

Hosmane, B. S. (1986). Improved likelihood ratio tests and Pearson chi-square tests for independence in two-dimensional tables. *Communications Statistics—Theory and Methods, 15*, 1875–1888.

Hössjer, O. (1994). Rank-based estimates in the linear model with high breakdown point. *Journal of the American Statistical Association, 89*, 149–158.

Huber, P. J. (1964). Robust estimation of location parameters. *Annals of Mathematical Statistics, 35*, 73–101.

Huber, P. J. (1981). *Robust Statistics*. New York: Wiley.

Huber, P. J. (1993). Projection pursuit and robustness. In S. Morgenthaler, E. Ronchetti, & W. Stahel (Eds.), *New Directions in Statistical Data Analysis and Robustness*. Boston: Birkhäuser Verlag.

Huberty, C. J. (1989). Problems with stepwise methods—Better alternatives. *Advances in Social Science Methodology, 1*, 43–70.

Huitema, B. E. (1980). *The Analysis of Covariance and Alternatives*. New York: Wiley.

Hussain, S. S., & Sprent, P. (1983). Nonparametric regression. *Journal of the Royal Statistical Society, 146*, 182–191.

Huynh, H., & Feldt, L. S. (1976). Estimation of the Box correction for degrees of freedom from sample data in randomized block and split-plot designs. *Journal of Educational Statistics, 1*, 69–82.

Hyndman, R. B., & Fan, Y. (1996). Sample quantiles in social packages. *American Statistician, 50*, 361–365.

Iman, R. L. (1974). A power study of a rank transform for the two-way classification model when interactions may be present. *Canadian Journal of Statistics, 2*, 227–239.

Iman, R. L., Quade, D., & Alexander, D. A. (1975). Exact probability levels for the Kruskal–Wallis test. *Selected Tables in Mathematical Statistics, 3*, 329–384.

Jaeckel, L. A. (1972). Estimating regression coefficients by minimizing the dispersion of residuals. *Annals of Mathematical Statistics, 43*, 1449–1458.

James, G. S. (1951). The comparison of several groups of observations when the ratios of the population variances are unknown. *Biometrika, 38*, 324–329.

Jeyaratnam, S., & Othman, A. R. (1985). Test of hypothesis in one-way random effects model with unequal error variances. *Journal of Statistical Computation and Simulation, 21*, 51–57.

Johansen, S. (1980). The Welch–James approximation to the distribution of the residual sum of squares in a weighted linear regression. *Biometrika, 67*, 85–93.

Johnson, N. J. (1978). Modifed *t*-tests and confidence intervals for asymmetrical populations. *Journal of the American Statistical Association, 73*, 536–576.

Johnson, P., & Neyman, J. (1936). Tests of certain linear hypotheses and their application to some educational problems. *Statistical Research Memoirs, 1*, 57–93.

Jones, R. H. (1993). *Longitudinal Data with Serial Correlation: A State-Space Approach*. London: Chapman & Hall.

Jorgensen, J., Gilles, R. B., Hunt, D. R., Caplehorn, J. R. M., & Lumley, T. (1995). A simple and effective way to reduce postoperative pain after laparoscopic cholecystectomy. *Australia and New Zealand Journal of Surgery, 65*, 466–469.

Judd, C. M., Kenny, D. A., & McClelland, G. H. (2001). Estimating and testing mediation and moderation in within-subjects designs. *Psychological Methods, 6*, 115–134.

Kaiser, L., & Bowden, D. (1983). Simultaneous confidence intervals for all linear contrasts of means with heterogeneous variances. *Communications in Statistics—Theory and Methods, 12*, 73–88.

Kallenberg, W. C. M., & Ledwina, T. (1999). Data-driven rank tests for independence. *Journal of the American Statistical Association, 94*, 285–310.

Kendall, M. G., & Stuart, A. (1973). *The Advanced Theory of Statistics*, Vol. 2. New York: Hafner.

Kepner, J. L., & Robinson, D. H. (1988). Nonparametric methods for detecting treatment effects in repeated-measures designs. *Journal of the American Statistical Association, 83*, 456–461.

Keselman, H. J., Algina, J., Boik, R. J., & Wilcox, R. R. (1999). New approaches to the analysis of repeated measurements. In B. Thompson (Ed.), *Advances in Social Science Methodology, 5*, 251–268. Greenwich, CT: JAI Press.

Keselman, H. J., Algina, J., Wilcox, R. R., & Kowalchuk, R. K. (2000). Testing repeated measures hypotheses when covariance matrices are heterogeneous: Revisiting the robustness of the Welch–James test again. *Educational and Psychological Measurement, 60*, 925–938.

Keselman, H. J., & Wilcox, R. R. (1999). The "improved" Brown and Forsythe test for mean equality: Some things can't be fixed. *Communications in Statistics —Simulation and Computation, 28*, 687–698.

Keselman, H. J., Wilcox, R. R., Taylor, J., & Kowalchuk, R. K. (2000). Tests for mean equality that do not require homogeneity of variances: Do they really work? *Communications in Statistics—Simulation and Computation, 29*, 875–895.

Keselman, J. C., Cribbie, R., & Holland, B. (1999). The pairwise multiple comparison multiplicity problem: An alternative approach to familywise and comparisonwise Type I error control. *Psychological Methods, 4*, 58–69.

Keselman, J. C., Rogan, J. C., Mendoza, J. L., & Breen, L. J. (1980). Testing the validity conditions of repeated measures *F*-tests. *Psychological Bulletin, 87*, 479–481.

Kim, P. J., & Jennrich, R. I. (1973). Tables of the exact sampling distribution of the two-sample Kolmogorov–Smirnov criterion, D_{mn}, $m \leq n$. In H. L. Harter, & D. B. Owen (Eds.), *Selected Tables in Mathematical Statistics*, Vol. I. Providence, RI: American Mathematical Society.

Kirk, R. E. (1995). *Experimental Design*. Monterey, CA: Brooks/Cole.

Kramer, C. (1956). Extension of multiple range test to group means with unequal number of replications. *Biometrics, 12*, 307–310.

Krause, A., & Olson, M. (2000). *The Basics of S and S-PLUS*. New York: Springer.

Krutchkoff, R. G. (1988). One-way fixed-effects analysis of variance when the error variances may be unequal. *Journal of Statistical Computation and Simulation, 30*, 259–271.

Kuo, L., & Mallick, B. (1998). Variable selection for regression models. *Sankhya, Series B, 60*, 65–81.

Larsen, W. A., & McCleary, S. J. (1972). The use of partial residual plots in regression analysis. *Technometrics, 14*, 781–790.

Lax, D. A. (1985). Robust estimators of scale: Finite sample performance in long-tailed symmetric distributions. *Journal of the American Statistical Association, 80*, 736–741.

Le, C. T. (1994). Some tests of linear trend of variances. *Communications in Statistics—Theory and Methods, 23*, 2269–2282.

Li, G. (1985). Robust regression. In D. Hoaglin, F. Mosteller, & J. Tukey (Eds.), *Exploring Data Tables, Trends, and Shapes*, pp. 281–343. New York: Wiley.

Liu, R. G., Parelius, J. M., & Singh, K. (1999). Multivariate analysis by data depth. *Annals of Statistics, 27*, 783–840.

Liu, R. G., & Singh, K. (1993). A quality index based on data depth and multivariate rank tests. *Journal of the American Statistical Association, 88*, 257–262.

Liu, R. G., & Singh, K. (1997). Notions of limiting *P* values based on data depth and bootstrap. *Journal of the American Statistical Association, 92*, 266–277.

Lloyd, C. J. (1999). *Statistical Analysis of Categorical Data*. New York: Wiley.

Loh, W.-Y. (1987). Does the correlation coefficient really measure the degree of clustering around a line? *Journal of Educational Statistics, 12*, 235–239.

Long, J. D., & Cliff, N. (1997). Confidence intervals for Kendall's tau. *British Journal of Mathematical and Statistical Psychology, 50*, 31–42.

Long, J. S., & Ervin, L. H. (2000). Using heteroscedasticity consistent standard errors in the linear regression model. *American Statistician, 54*, 217–224.

Lord, F. M., & Novick, M. R. (1968). *Statistical Theories of Mental Test Scores*. Reading, MA: Addison-Wesley.

Ludbrook, J., & Dudley, H. (1998). Why permutation tests are superior to *t*- and *F*-tests in biomedical research. *American Statistician, 52*, 127–132.

Lumley, T. (1996). Generalized estimating equations for ordinal data: A note on working correlation structures. *Biometrics, 52*, 354–361.

Lunneborg, C. E. (2000). *Data Analysis by Resampling: Concepts and Applications*. Pacific Grove, CA: Duxbury.

Lyon, J. D., & Tsai, C.-L. (1996). A comparison of tests for homogeneity. *Statistician, 45*, 337–350.

Mallows, C. L. (1973). Some comments on C_p. *Technometrics, 15*, 661–675.

Mallows, C. L. (1986). Augmented partial residuals. *Technometrics, 28*, 313–319.

Mann, H. B., & Whitney, D. R. (1947). On a test of whether one of two random variables is stochastically larger than the other. *Annals of Mathematical Statistics, 18*, 50–60.

Mardia, K. V., Kent, J. T., & Bibby, J. M. (1979). *Multivariate Analysis*. San Diego, CA: Academic Press.

Markowski, C. A., & Markowski, E. P. (1990). Conditions for the effectiveness of a preliminary test of variance. *American Statistician, 44*, 322–326.

Maronna, R., & Morgenthaler, S. (1986). Robust regression through robust covariances. *Communications in Statistics—Theory and Methods, 15*, 1347–1365.

Maronna, R., Yohai, V. J., & Zamar, R. (1993). Bias-robust regression estimation: A partial survey. In S. Morgenthaler, E. Ronchetti, & W. A. Stahel (Eds.), *New Directions in Statistical Data Analysis and Robustness*. Boston: Fuller Verlag.

Matuszewski, A., & Sotres, D. (1986). A simple test for the Behrens–Fisher problem. *Computational Statistics and Data Analysis, 3*, 241–249.

McKean, J. W., & Schrader, R. M. (1984). A comparison of methods for studentizing the sample median. *Communications in Statistics—Simulation and Computation, 13*, 751–773.

McKean, J. W., & Sheather, S. J. (2000). Partial residual plots based on robust fits. *Technometrics, 42*, 249–261.

McKean, J. W., & Vidmar, T. J. (1994). A comparison of two rank-based methods for the analysis of linear models. *American Statistician, 48*, 220–229.

Mee, R. W. (1990). Confidence intervals for probabilities and tolerance regions based on a generalization of the Mann–Whitney statistic. *Journal of the American Statistical Association, 85*, 793–800.

Mehrotra, D. V. (1997). Improving the Brown–Forsythe solution to the generalized Behrens–Fisher problem. *Communications in Statistics—Simulation and Computation, 26*, 1139–1145.

Miller, A. J. (1990). *Subset Selection in Regression*. London: Chapman and Hall.

Miller, R. G. (1974). The jackknife—A review. *Biometrika, 61*.

Miller, R. G. (1976). Least squares regression with censored data. *Biometrika, 63*, 449–464.

Montgomery, D. C., & Peck, E. A. (1992). *Introduction to Linear Regression Analysis*. New York: Wiley.

Mooney, C. Z., & Duval, R. D. (1993). *Bootstrapping: A Nonparametric Approach to Statistical Inference*. Newbury Park, CA: Sage.

Morgenthaler, S., & Tukey, J. W. (1991). *Configural Polysampling*. New York: Wiley.

Moser, B. K., Stevens, G. R., & Watts, C. L. (1989). The two-sample *t*-test versus Satterthwaite's approximate *F*-test. *Communications in Statistics—Theory and Methods, 18*, 3963–3975.

Möttönen, J., & Oja, H. (1995). Multivariate spatial sign and rank methods. *Nonparametric Statistics, 5*, 201–213.

Muirhead, R. J. (1982). *Aspects of Multivariate Statistical Theory*. New York: Wiley.

Müller, H.-G. (1988). *Nonparametric Regression Analysis*. New York: Springer-Verlag.

Munzel, U. (1999). Linear rank score statistics when ties are present. *Statistics and Probability Letters, 41*, 389–395.

Munzel, U., & Brunner, E. (2000a). Nonparametric test in the unbalanced multivariate one-way design. *Biometrical Journal, 42*, 837–854.

Munzel, U., & Brunner, E. (2000b). Nonparametric methods in multivariate factorial designs. *Journal of Statistical Planning and Inference, 88*, 117–132.

Naranjo, J. D., & Hettmansperger, T. P. (1994). Bounded influence rank regression. *Journal of the Royal Statistical Society, B, 56*, 209–220.

Olejnik, S., Li, J., Supattathum, S., & Huberty, C. J. (1997). Multiple testing and statistical power with modified Bonferroni procedures. *Journal of Educational and Behavioral Statistics, 22*, 389–406.

Pagurova, V. I. (1968). On a comparison of means of two normal samples. *Theory of Probability and Its Applications, 13*, 527–534.

Parrish, R. S. (1990). Comparison of quantile estimators in normal sampling. *Biometrics, 46*, 247–257.

Patel, K. M., & Hoel, D. G. (1973). A nonparametric test for interactions in factorial experiments. *Journal of the American Statistical Association, 68*, 615–620.

Pedersen, W. C., Miller, L. C., Putcha-Bhagavatula, A. D., & Yang, Y. (2002). Evolved sex differences in sexual strategies: The long and the short of it. *Psychological Science, 13*, 157–161.

Peña, D., & Prieto, F. J. (2001). Multivariate outlier detection and robust covariance matrix estimation. *Technometrics, 43*, 286–299.

Piepho, H.-P. (1997). Tests for equality of dispersion in bivariate samples—Review and empirical comparison. *Journal of Statistical Computation and Simulation, 56*, 353–372.

Poon, W. Y., Lew, S. F., & Poon, Y. S. (2000). A local-influence approach to identifying multiple multivariate outliers. *British Journal of Mathematical and Statistical Psychology, 53*, 255–273.

Powers, D. A., & Xie, Y. (1999). *Statistical Methods for Categorical Data Analysis*. San Diego, CA: Academic Press.

Pratt, J. W. (1968). A normal approximation for binomial, *F*, beta, and other common, related tail probabilities, I. *Journal of the American Statistical Association, 63*, 1457–1483.

Price, R. M., & Bonett, D. G. (2001). Estimating the variance of the median. *Journal of Statistical Computation and Simulation, 68*, 295–305.

Quade, D. (1979). Using weighted rankings in the analysis of complete blocks with additive block effects. *Journal of the American Statistical Association, 74*, 680–683.

Raine, A., Buchsbaum, M., & LaCasse, L. (1997). Brain abnormalities in murderers indicated by positron emission tomography. *Biological Psychiatry, 42*, 495–508.

Ramsey, P. H. (1980). Exact Type I error rates for robustness of Student's *t*-test with unequal variances. *Journal of Educational Statistics, 5*, 337–349.

Rao, C. R. (1948). Tests of significance in multivariate analysis. *Biometrika, 35*, 58–79.

Rao, P. S., Kaplan, J., & Cochran, W. G. (1981). Estimators for the one-way random effects model with unequal error variances. *Journal of the American Statistical Association, 76*, 89–97.

Rasmussen, J. L. (1989). Data transformation, Type I error rate and power. *British Journal of Mathematical and Statistical Psychology, 42*, 203–211.

Rocke, D. M. (1996). Robustness properties of *S*-estimators of multivariate location and shape in high dimensions. *Annals of Statistics, 24*, 1327–1345.

Rocke, D. M., & Woodruff, D. L. (1996). Identification of outliers in multivariate data. *Journal of the American Statistical Association, 91*, 1047–1061.

Rogan, J. C., Keselman, H. J., & Mendoza, J. L. (1979). Analysis of repeated measurements. *British Journal of Mathematical and Statistical Psychology, 32*, 269–286.

Rom, D. M. (1990). A sequentially rejective test procedure based on a modified Bonferroni inequality. *Biometrika, 77*, 663–666.

Rosenthal, R., & Jacobson, L. (1968). *Pygmalion in the Classroom: Teacher Expectations and Pupil's Intellectual Development.* New York: Holt, Rinehart and Winston.

Rousseeuw, P. J., & Hubert, M. (1999). Regression depth. *Journal of the American Statistical Association, 94*, 388–402.

Rousseeuw, P. J., & Leroy, A. M. (1987). *Robust Regression & Outlier Detection*. New York: Wiley.

Rousseeuw, P. J., & Ruts, I. (1996). AS 307: Bivariate location depth. *Applied Statistics, 45*, 516–526.

Rousseeuw, P. J., Ruts, I., & Tukey, J. W. (1999). The bagplot: A bivariate boxplot. *American Statistician, 53*, 382–387.

Rousseeuw, P. J., & Struyf, A. (1998). Computing location depth and regression depth in higher dimensions. *Statistical Computations, 8*, 193–203.

Rousseeuw, P. J., & van Driesen, K. (1999). A fast algorithm for the minimum covariance determinant estimator. *Technometrics, 41*, 212–223.

Rousseeuw, P. J., & van Zomeren, B. C. (1990). Unmasking multivariate outliers and leverage points (with discussion). *Journal of the American Statistical Association, 85*, 633–639.

Rust, S. W., & Fligner, M. A. (1984). A modification of the Kruskal–Wallis statistic for the generalized Behrens–Fisher problem. *Communications in Statistics—Theory and Methods, 13*, 2013–2027.

Rutherford, A. (1992). Alternatives to traditional analysis of covariance. *British Journal of Mathematical and Statistical Psychology, 45*, 197–223.

Salk, L. (1973). The role of the heartbeat in the relations between mother and infant. *Scientific American, 235*, 26–29.

Sackrowitz, H., & Samuel-Cahn, E. (1999). *P* values as random variables—Expected *P* values. *American Statistician, 53*, 326–331.

Sarkar, S. K. (2002). Some results on false discovery rate in stepwise multiple testing procedures. *Annals of Statistics, 30*, 239–257.

Saunders, D. R. (1955). The "moderator variable" as a useful tool in prediction. In Proceedings of the 1954 Invitational Conference on Testing Problems (pp. 54–58). Princeton, NJ: Educational Testing Service.

Saunders, D. R. (1956). Moderator variables in prediction. *Educational and Psychological Measurement, 16*, 209–222.

Scariano, S. M., & Davenport, J. M. (1986). A four-moment approach and other practical solutions to the Behrens–Fisher problem. *Communications in Statistics—Theory and Methods, 15*, 1467–1501.

Scheffé, H. (1959). *The Analysis of Variance*. New York: Wiley.

Schenker, N., & Gentleman, J. F. (2001). On judging the significance of differences by examining the overlap between confidence intervals. *American Statistician, 55*, 182–186.

Schrader, R. M., & Hettmansperger, T. P. (1980). Robust analysis of variance. *Biometrika, 67*, 93–101.

Schroër, G., & Trenkler, D. (1995). Exact and randomization distributions of Kolmogorov–Smirnov tests two or three samples. *Computational Statistics and Data Analysis, 20*, 185–202.

Scott, D. W. (1979). On optimal and data-based histograms. *Biometrika, 66*, 605–610.

Scott, W. A. (1955). Reliability of content analysis: The case of nominal scale coding. *Public Opinion Quarterly, 19*, 321–325.

Searle, S. R. (1971). *Linear Models*. New York: Wiley.

Sen, P. K. (1968). Estimate of the regression coefficient based on Kendall's tau. *Journal of the American Statistical Association*, *63*, 1379–1389.

Serfling, R. J. (1980). *Approximation Theorems of Mathematical Statistics*. New York: Wiley.

Shaffer, J. P. (1974). Bidirectional unbiased procedures. *Journal of the American Statistical Association*, *69*, 437–439.

Shao, J. (1995). Bootstrap model selection. *Journal of the American Statistical Association*, *91*, 655–665.

Shao, J., & Tu, D. (1995). *The Jackknife and the Bootstrap*. New York: Springer-Verlag.

Sheather, S. J., & McKean, J. W. (1987). A comparison of testing and confidence intervals for the median. *Statistical Probability Letters*, *6*, 31–36.

Silverman, B. W. (1986). *Density Estimation for Statistics and Data Analysis*. New York: Chapman and Hall.

Singh, K. (1998). Breakdown theory for bootstrap quantiles. *Annals of Statistics*, *26*, 1719–1732.

Small, C. G. (1990). A survey of multidimensional medians. *International Statistical Review*, *58*, 263–277.

Smith, C. A. B. (1956). Estimating genetic correlations. *Annals of Human Genetics*, *44*, 265–284.

Snedecor, G. W., & Cochran, W. (1967). *Statistical Methods*, 6th ed. Ames, IA: University Press.

Snow, R. E. (1995). Pygmalion and Intelligence? *Current Directions in Psychological Science*, *4*, 169–172.

Sockett, E. B., Daneman, D., Clarson, C., & Ehrich, R. M. (1987). Factors affecting and patterns of residual insulin secretion during the first year of type I (insulin-dependent) diabetes mellitus in children. *Diabetes*, *30*, 453–459.

Staudte, R. G., & Sheather, S. J. (1990). *Robust Estimation and Testing*. New York: Wiley.

Stein, C. (1945). A two-sample test for a linear hypothesis whose power is independent of the variance. *Annals of Statistics*, *16*, 243–258.

Storer, B. E., & Kim, C. (1990). Exact properties of some exact test statistics for comparing two binomial proportions. *Journal of the American Statistical Association*, *85*, 146–155.

Stromberg, A. J. (1993). Computation of high-breakdown nonlinear regression parameters. *Journal of the American Statistical Association*, *88*, 237–244.

Stute, W., Manteiga, W. G., & Quindimil, M. P. (1998). Bootstrap approximations in model checks for regression. *Journal of the American Statistical Association*, *93*, 141–149.

Sutton, C. D. (1993). Computer-intensive methods for tests about the mean of an asymmetrical distribution. *Journal of the American Statistical Association*, *88*, 802–810.

Tamhane, A. C. (1977). Multiple comparisons in model I one-way ANOVA with unequal variances. *Communications in Statistics—Theory and Methods*, *6*, 15–32.

Theil, H. (1950). A rank-invariant method of linear and polynomial regression analysis. *Indagationes Mathematicae*, *12*, 85–91.

Thompson, A., & Randall-Maciver, R. (1905). *Ancient Races of the Thebaid*. Oxford: Oxford University Press.

Thompson, G. L. (1991). A unified approach to rank tests for multivariate and repeated measures designs. *Journal of the American Statistical Association*, *86*, 410–419.

Thompson, G. L., & Ammann, L. P. (1990). Efficiencies of interblock rank statistics for repeated measures designs. *Journal of the American Statistical Association*, *85*, 519–528.

Tibshirani, R. (1988). Estimating transformations for regression via additivity and variance stabilization. *Journal of the American Statistical Association*, *83*, 394–405.

Tibshirani, R. (1996). Regression shrinkage and selection via the lasso. *Journal of the Royal Statistical Society, B*, *58*, 267–288.

Tryon, W. W. (2001). Evaluating statistical difference, equivalence, and indeterminacy using inferential confidence intervals: An integrated alternative method of conducting null hypothesis statistical tests. *Psychological Methods*, *6*, 371–386.

Tukey, J. W. (1960). A survey of sampling from contaminated normal distributions. In I. Olkin et al. (Eds.), *Contributions to Probability and Statistics*. Stanford, CA: Stanford University Press.

Tukey, J. W. (1977). *Exploratory Data Analysis*. Reading, MA: Addison-Wesley.

Tukey, J. W., & McLaughlin, D. H. (1963). Less vulnerable confidence and significance procedures for location based on a single sample: Trimming/Winsorization 1. *Sankhya A*, *25*, 331–352.

Vonesh, E. (1983). Efficiency of repeated measures designs versus completely randomized designs based on multiple comparisons. *Communications in Statistics—Theory and Methods*, *12*, 289–302.

Wald, A. (1955). Testing the difference between the means of two normal populations with unknown standard deviations. In T. W. Anderson et al. (Eds.), *Selected Papers in Statistics and Probability by Abraham Wald*. New York: McGraw-Hill.

Wechsler, D. (1958). *The Measurement and Appraisal of Adult Intelligence*. Baltimore: Williams and Wilkins.

Weerahandi, S. (1995). ANOVA under unequal error variances. *Biometrics, 51,* 589–599.

Welch, B. L. (1938). The significance of the difference between two means when the population variances are unequal. *Biometrika, 29,* 350–362.

Welch, B. L. (1951). On the comparison of several mean values: An alternative approach. *Biometrika, 38,* 330–336.

Westfall, P. (1988). Robustness and power of tests for a null variance ratio. *Biometrika, 75,* 207–214.

Westfall, P. H., & Young, S. S. (1993). *Resampling-Based Multiple Testing.* New York: Wiley.

Wilcox, R. R. (1983). A table of percentage points of the range of independent t variables. *Technometrics, 25,* 201–204.

Wilcox, R. R. (1992). An improved method for comparing variances when distributions have nonidentical shapes. *Computational Statistics & Data Analysis, 13,* 163–172.

Wilcox, R. R. (1993a). Some results on the Tukey–McLaughlin and Yuen methods for trimmed means when distributions are skewed. *Biometrical Journal, 36,* 259–273.

Wilcox, R. R. (1993b). Comparing one-step M-estimators of location when there are more than two groups. *Psychometrika, 58,* 71–78.

Wilcox, R. R. (1994a). A one-way random-effects model for trimmed means. *Psychometrika, 59,* 289–306.

Wilcox, R. R. (1994b). The percentage bend correlation coefficient. *Psychometrika, 59,* 601–616.

Wilcox, R. R. (1995). Some small-sample results on a bounded influence rank regression method. *Communications in Statistics—Theory and Methods, 24,* 881–888.

Wilcox, R. R. (1996a). Estimation in the simple linear regression model when there is heteroscedasticity of unknown form. *Communications in Statistics—Theory and Methods, 25,* 1305–1324.

Wilcox, R. R. (1996b). Confidence intervals for the slope of a regression line when the error term has nonconstant variance. *Computational Statistics & Data Analysis, 22,* 89–98.

Wilcox, R. R. (1996c). *Statistics for the Social Sciences.* San Diego, CA: Academic Press.

Wilcox, R. R. (1997a). *Introduction to Robust Estimation and Hypothesis Testing.* San Diego, CA: Academic Press.

Wilcox, R. R. (1997b). ANOVA based on comparing a robust measure of location at empirically determined design points. *British Journal of Mathematical and Statistical Psychology, 50,* 93–103.

Wilcox, R. R. (1998). Simulation results on extensions of the Theil–Sen regression estimator. *Communications in Statistics—Simulation and Computation, 27,* 1117–1126.

Wilcox, R. R. (2000a). Some exploratory methods for studying curvature in robust regression. *Biometrical Journal, 42,* 335–347.

Wilcox, R. R. (2000b). Rank-based tests for interactions in a two-way design when there are ties. *British Journal of Mathematical and Statistical Psychology, 53,* 145–153.

Wilcox, R. R. (2001a). *Fundamentals of Modern Statistical Methods: Substantially Increasing Power and Accuracy.* New York: Springer.

Wilcox, R. R. (2001b). Robust regression estimators that reduce contamination bias and have high efficiency when there is heteroscedasticity. Unpublished technical report, Dept. of Psychology, University of Southern California.

Wilcox, R. R. (2001c). Rank-based multiple comparisons for interactions in a split-plot design. Unpublished technical report, Dept. of Psychology, University of Southern California.

Wilcox, R. R. (2001d). Pairwise comparisons of trimmed means for two or more groups. *Psychometrika, 66,* 343–256.

Wilcox, R. R. (2002). Inferences based on a skipped correlation coefficient. Unpublished technical report, Dept. of Psychology, University of Southern California.

Wilcox, R. R. (2002). Comparing the variances of independent groups. *British Journal of Mathematical and Statistical Psychology, 55,* 169–176.

Wilcox, R. R., & Charlin, V. (1986). Comparing medians: A Monte Carlo study. *Journal of Educational Statistics, 11,* 263–274.

Wilcox, R. R., Charlin, V., & Thompson, K. L. (1986). New Monte Carlo results on the robustness of the ANOVA F, W, and F^* statistics. *Communications in Statistics—Simulation and Computation, 15,* 933–944.

Wilcox, R. R., & Keselman, H. J. (2002). Power analyses when comparing trimmed means. *Journal of Modern Statistical Methods, 1,* 24–31.

Wilcox, R. R., Keselman, H. J., Muska, J., & Cribbie, R. (2000). Repeated measures ANOVA: Some new results on comparing trimmed means and means. *British Journal of Mathematical and Statistical Psychology, 53,* 69–82.

Wilcox, R. R., & Muska, J. (1999). Measuring effect size: A nonparametric analogue of ω^2. *British Journal of Mathematical and Statistical Psychology, 52,* 93–110.

Wilcox, R. R., & Muska, J. (2001). Inferences about correlations when there is heteroscedasticity. *British Journal of Mathematical and Statistical Psychology, 54*, 39–47.

Wilcox, R. R., & Muska, J. (2002). Comparing correlation coefficients. *Communications in Statistics—Simulation and Computation, 31*, 49–59.

Wilcoxon, F. (1945). Individual comparisons by ranking methods. *Biometrics, 1*, 80–83.

Williams, V. S. L., Jones, L. V., & Tukey, J. W. (1999). Controlling error in multiple comparisons, with examples from state-to-state differences in educational achievement. *Journal of Educational and Behavioral Statistics, 24*, 42–69.

Wisnowski, J. W., Montgomery, D. C., & Simpson, J. R. (2001). A comparative analysis of multiple outlier detection procedures in the linear regression model. *Computational Statistics & Data Analysis, 36*, 351–382.

Woodruff, D. L., & Rocke, D. M. (1994). Computable robust estimation of multivariate location and shape in high dimension using compound estimators. *Journal of the American Statistical Association, 89*, 888–896.

Wu, C. F. J. (1986). Jackknife, bootstrap, and other resampling methods in regression analysis. *The Annals of Statistics, 14*, 1261–1295.

Yu, M. C., & Dunn, O. J. (1986). Robust test for the equality of two correlations: A Monte Carlo study. *Educational and Psychological Measurement, 42*, 987–1004.

Yuen, K. K. (1974). The two-sample trimmed *t* for unequal population variances. *Biometrika, 61*, 165–170.

INDEX

T

U

V

W

Y

Introduction to Programming

A Problem Solving Approach

John Dean
Park University

Raymond Dean
University of Kansas

The McGraw·Hill Companies

INTRODUCTION TO PROGRAMMING WITH JAVA: A PROBLEM SOLVING APPROACH: SECOND EDITION

Published by McGraw-Hill, a business unit of The McGraw-Hill Companies, Inc., 1221 Avenue of the Americas, New York, NY 10020.

This book is printed on acid-free paper.

6 7 8 9 LCR 21 20 19 18 17 16

ISBN 978–0–07–337606–6
MHID 0–07–337606–x

Vice President, General Manager: *Marty Lange*
Editorial Director: *Michael Lange*
Publisher: *Raghothaman Srinivasan*
Marketing Manager: *Curt Reynolds*
Development Editor: *Katie Neubauer*
Project Manager: *Melissa M. Leick*
Buyer: *Sandy Ludovissy*
Media Project Manager: *Prashanthi Nadipalli*
Cover Design: *Studio Montage, St. Louis, MO*
Cover image: *Chris Johnson*
Compositor: *S4Carlisle Publishing Services*
Typeface: *10/12 Times Roman*
Printer: *LSC Communications Crawfordsville, IN*

Figure 1.1b: © PhotoDisc/Getty Images; Figure 1.1c: © BigStock Photos; Figure 1.1d: © PhotoDisc/Getty Images; Figure 1.1e: © BrandX/Punchstock; Figure 1.1f: © Ryan McVay/Getty Images; Figure 1.2a: © BigStock Photos; Figure 1.2b: © BigStock Photos; Figure 1.2c: © BigStock Photos; Figure 1.4a: © BigStock Photos; Figure 1.4b: © Getty Royalty Free; Figure 1.4c: © Oleksiy Mark | Dreamstime.com; Figure 1.4d: © BigStock Photos; Figure ta2.1: Courtesy of the Naval Surface Warfare Center, Dahlgren, VA, 1988. US Naval History and Heritage Command Photograph; Figure ta4.2: © Sergey Mostovoy | Dreamstime.com; Figure 5.13: Photo courtesy of the authors; Figure 5.16: Photo courtesy of the authors; Figure 11.1: Courtesy of Rafael de Sant'Anna Neri; Figure 14.5a: © Michael Pettigrew | Dreamstime.com

All credits appearing on page or at the end of the book are considered to be an extension of the copyright page.

Library of Congress Cataloging-in-Publication Data

Dean, John, 1962-
Introduction to programming Java : with a problem solving approach / John S. Dean, Raymond H. Dean.
p. cm.
Includes bibliographical references and index.
ISBN 978–0–07–337606–6 (alk. paper) ISBN 0–07–337606–X (alk. paper) 1. Java (Computer program language) I. Dean, Ray, 1936- II. Title.

QA76.73.J38D4264 2012
005.2'762—dc23

2012036870

www.mhhe.com

Dedication

—To Jordan and Caiden

About the Authors

John Dean is an Associate Professor in the Computer Science and Mathematics Department at Park University. He earned a Ph.D. degree in computer science from Nova Southeastern University and an M.S. degree in computer science from the University of Kansas. He is Sun Java–certified and has worked in industry as a software engineer and project manager, specializing in Java and various Web technologies—JavaScript, JavaServer Pages, and servlets. He has taught a full range of computer science courses, including Java programming and Java-based Web programming.

Raymond Dean is a Professor Emeritus, Electrical Engineering and Computer Science, University of Kansas. He earned an M.S. degree from MIT and a Ph.D. degree from Princeton University, and he is a senior member of IEEE. He has published numerous scientific papers and has 21 U.S. patents. He has industry experience in manufacturing HVAC equipment and energy-management controls, as well as in system energy analysis. At the University of Kansas, he taught a wide range of courses in electrical engineering and computer science.

Contents

CHAPTER 4

Control Statements 115

CHAPTER 5

Using Prebuilt Methods 162

Interlude 209

CHAPTER 6

Object-Oriented Programming 213

CHAPTER 7

Object-Oriented Programming—Additional Details 262

CHAPTER 8

Software Engineering 317

CHAPTER 13

Aggregation, Composition, and Inheritance 565

CHAPTER 14

Inheritance and Polymorphism 612

CHAPTER 15

Exception Handling 661

CHAPTER 16

Files, Buffers, Channels, and Paths 709

CHAPTER 17

GUI Programming Basics 761

CHAPTER 18

GUI Programming—Component Layout, Additional GUI Components 810

Appendices

On the Book's Website

Preface

In this book, we lead you on a journey into the fun and exciting world of computer programming. Throughout your journey, we'll provide you with lots of problem-solving practice. After all, good programmers need to be good problem solvers. We'll show you how to implement your problem solutions with Java programs. We provide a plethora of examples, some short and focused on a single concept, some longer and more "real world." We present the material in a conversational, easy-to-follow manner aimed at making your journey a pleasant one. When you're done with the book, you should be a proficient Java programmer.

Our textbook targets a wide range of readers. Primarily, it targets students in a standard college-level "Introduction to Programming" course or course sequence where no prerequisite programming experience is assumed. We have tried to include all the topics recommended by the College Board for students studying for advanced placement (AP) in computer science. So this text should be good for these students as well.

In addition to targeting students with no prerequisite programming experience, our textbook targets industry practitioners and college-level students who have some programming experience and want to learn Java. This second set of readers can skip the early chapters on general programming concepts and focus on the features of Java that differ from the languages that they already know. In particular, because C++ and Java are similar, readers with a C++ background should be able to cover the textbook in a single three-credit-hour course. (But let us reiterate for those of you with no programming experience: You should be fine. No prerequisite programming experience is required.)

Finally, our textbook targets high school students and readers outside of academia with no programming experience. This third set of readers should read the entire textbook at a pace determined on a case-by-case basis.

What's New in This Edition?

The changes in this edition are big and small. Big changes include new chapters, reorganized chapter sections, new programming constructs, and new programs. Smaller changes include updating descriptions, anecdotes, examples, exercises, and projects. We've combed the entire book for opportunities to improve the book's clarity and readability. The following list highlights the more significant changes we've made for this edition.

- **Language Enhancements**

  We wrote the first edition when the standard Java compiler was Java 6. Back then, we described most of Java 6's language features—but not all. With this edition, we present additional language features that were previously omitted, including the `assert` keyword, enumerated types (`enum`s), and variable-length argument lists (varargs).

  We describe several Java language enhancements that were introduced in Java 7—numeric literals with underscores, binary literals, `switch` statements with strings, and type inferences with the diamond operator.

  We present closures, a language feature introduced in Java 8, as an appendix on the book's website. A *closure* is a block of code that can be passed to a method and executed later. Among other things, closures are intended to clean up the messiness of anonymous inner classes.

- **Introductory Chapter**

  To keep up with the computer industry's continued growth, we've made many changes to Chapter 1, such as updating the information in the computer hardware and Java history sections. In addition to the many updates in the preexisting sections, we've added a new short section on computer ethics.

- **Accommodation for Late Objects Advocates**

  Some of our first edition readers suggested that before launching into a full discussion of object-oriented programming (OOP) in Chapter 6, we should describe multiple-method programs in a non-OOP environment. To accommodate that suggestion, we include a short Interlude "mini-chapter" named "Multiple-Method Programs in a Non-Object-Oriented Environment" between Chapters 5 and 6. We anticipate that most readers will benefit from and be satisfied by that brief discussion. However, we realize that some readers will want more. Some of the readers of our first edition advocated for a "late objects" approach, with complete details on multiple-method programs and complete details on arrays, all before starting OOP in Chapter 6. To support that approach, we provide those details in two supplemental chapters on the book's website. The Interlude encourages late objects advocates to read those supplemental chapters before returning to Chapter 6.

- **Reduced Emphasis on Class Variables and Class Methods**

  Based on feedback from readers of the first edition, we've eliminated the "Classes with Class Members" chapter and moved a rewritten subset of that material to the end of Chapter 7.

- **Software Engineering**

  To provide more emphasis on documentation, we've moved a subset of the javadoc material forward, from the Javadoc appendix up to Chapter 8. In Chapter 8, we introduce the notion of pre- and post-conditions and include pre- and post-condition comments in code examples.

- **New Chapter—`ArrayLists` and an Introduction to the Java Collections Framework**

  We've written a new chapter that describes various classes and interfaces in the Java Collections Framework, with an emphasis on the `ArrayList` class, the `List` interface, queues, and stacks. Within this context, we describe Java generics and multiclass interfaces. The chapter provides an easily accessible glimpse of data structures and their applications.

- **New Chapter—Recursion**

  We've written a new chapter on recursion that combines the content from the first edition's "Recursion" appendix with new content. The Recursion chapter explains how to analyze and write recursive methods, and it presents a significant number of complete recursive programs, including a merge sort program and a graphical user interface (GUI) fractal program that draws an animated picture of a grove of growing trees.

- **Performance Analysis**

  In our chapters on arrays, collections, and recursion, we've added some performance analysis discussion to our presentation of various algorithms and data structures. For the sequential search, binary search, merge sort, and several other types of algorithms, we include brief discussions of how execution times depend on the number of items being processed. For Java's `ArrayList`, `LinkedList`, and `ArrayDeque` classes, we compare measured execution times for various operations. In a separate section at the end of the Recursion chapter, we introduce the Big O notation and use it to quantify the performance of various algorithms.

- **New AP Computer Science Case Study**

  In sections at the ends of Chapters 12 and 13, we describe and provide enhancements to the GridWorld program, a legacy case study from the College Board's Advanced Placement (AP) Computer Science curriculum. The GridWorld program implements simple animation. It shows bugs crawling around, bumping into things, changing direction as a result, eating plants or other critters, planting new flowers, changing color based on events or passage of time, and so on. We provide keyed exercises to put GridWorld features in the context of our book's presentation.

- **Rewritten Chapter—Exception Handling**

  We've rewritten about half of Chapter 15, the exception handling chapter, taking advantage of new Java 7 exception handling constructs—automatic resource management within a `try` block heading and multiple parameters within a `catch` block heading.

- **Rewritten Chapter—File Handling**

  We've rewritten almost the entire file handling chapter, Chapter 16, taking advantage of NIO.2, which is Java 7's new file system application programming interface (API). The chapter puts greater emphasis on file handling that relates to Internet communication by including alternate character sets, alternate input/output (IO) options, and random access to buffered and channeled IO. Additionally, the chapter includes memory mapping and file system traversal techniques.

  For readers who want to use files early, we've introduced in Chapter 3 an optional "quick and dirty" file input technique that appends a `throws Exception` clause to the main method.

- **Better Visuals and More GUI**

  We describe the Nimbus look and feel and explain how users can configure their computers to use it. Oracle rolled out Nimbus in a later release of Java 6 and started recommending it with the advent of Java 7. We use the Nimbus look and feel for all window screenshots throughout the book.

  Also, we have added a description of image decoration that uses semitransparent layers.

- **New Supplemental GUI Chapters Using JavaFX**

  With the release of Java 8, JavaFX replaced Swing as Oracle's preferred GUI toolkit. Thus, we've written two supplemental GUI chapters, S17 and S18, found on the book's website, that are dedicated to describing JavaFX. Because Swing will remain popular for quite some time, we explain and use Swing in the early optional GUI tracks and in Chapters 17 and 18. Our parallel descriptions of Swing and JavaFX give readers a chance to see how the old compares with the new.

- **New Appendix—Number Systems and Conversions Between Them**

  We've written a new appendix that describes how numbers are represented in different bases—decimal, binary, octal, and hexadecimal. The appendix provides algorithms for converting between the number bases.

Compliant with the College Board's AP Computer Science Curriculum

As noted by teachers using the first edition in their AP Computer Science courses, the first edition covered most of the College Board's AP Computer Science curriculum content. We have put a great deal of effort into ensuring that this second edition follows all the AP Computer Science guidelines, as put forth by the College Board.

This second edition has been reviewed by a former member of the AP Computer Science Development Committee, Judy Hromcik. She states that at the time of its printing, this edition is fully compliant with the College Board's AP Computer Science curriculum.

Textbook Cornerstone #1: Problem Solving

Being able to solve problems is a critical skill that all programmers must possess. We teach programmatic problem solving by emphasizing two of its key elements—algorithm development and program design.

Emphasis on Algorithm Development

In Chapter 2, we immerse readers into algorithm development by using pseudocode for the algorithm examples instead of Java. In using pseudocode, students are able to work through non-trivial problems on their own without getting bogged down in Java syntax—no need to worry about class headings, semicolons, braces, and so on.[1] Working through non-trivial problems enables students to gain an early appreciation for creativity, logic, and organization. Without that appreciation, Java students tend to learn Java syntax with a rote-memory attitude. But with that appreciation, students tend to learn Java syntax more quickly and effectively because they have a motivational basis for learning it. In addition, they are able to handle non-trivial Java homework assignments fairly early because they have prior experience with similarly non-trivial pseudocode homework assignments.

In Chapter 3 and in later chapters, we rely primarily on Java for algorithm-development examples. But for the more involved problems, we sometimes use high-level pseudocode to describe first-cut proposed solutions. Using pseudocode enables readers to bypass syntax details and focus on the algorithm portion of the solution.

Emphasis on Program Design

Problem solving is more than just developing an algorithm. It also involves figuring out the best implementation for the algorithm. That's program design. Program design is extremely important, and that's why we spend so much time on it. Frequently, we explain the thought processes that a person might go through when coming up with a solution. For example, we explain how to choose between different loop types, how to split up a method into multiple methods, how to decide on appropriate classes, how to choose between instance and class members, and how to determine class relationships using inheritance and composition. We challenge students to find the most elegant implementations for a particular task.

We devote a whole chapter to program design—Chapter 8, "Software Engineering." In that chapter, we provide an in-depth look at coding-style conventions and documentation for programmers and users. We discuss design strategies like separation of concerns, modularization, and encapsulation. Also in the chapter, we describe alternative design strategies—top-down, bottom-up, case-based, and iterative enhancement.

[1] Inevitably, we use a particular style for our pseudocode, but we repeatedly emphasize that other pseudocode styles are fine so long as they convey the intended meaning. Our pseudocode style is a combination of free-form description for high-level tasks and more specific commands for low-level tasks. We've chosen a pseudocode style that is intuitive, to welcome new programmers, and structured, to accommodate program logic.

Problem-Solving Sections

We often address problem solving (algorithm development and program design) in the natural flow of explaining concepts. But we also cover problem solving in sections that are wholly devoted to it. In each problem-solving section, we present a situation that contains an unresolved problem. In coming up with a solution for the problem, we try to mimic the real-world problem-solving experience by using an iterative design strategy. We present a first-cut solution, analyze the solution, and then discuss possible improvements to it. We use a conversational trial-and-error format (e.g., "What type of layout manager should we use? We first tried the `GridLayout` manager. That works OK, but not great. Let's now try the `BorderLayout` manager."). This casual tone sets the student at ease by conveying the message that it is normal, and in fact expected, that a programmer will need to work through a problem multiple times before finding the best solution.

Additional Problem-Solving Mechanisms

We include problem-solving examples and problem-solving advice throughout the text (not just in Chapter 2, Chapter 8, and the problem-solving sections). As a point of emphasis, we insert a problem-solving box, with an icon and a succinct tip, next to the text that contains the problem-solving example and/or advice.

We are strong believers in learning by example. As such, our textbook contains a multitude of complete program examples. Readers are encouraged to use our programs as recipes for solving similar programs on their own.

Textbook Cornerstone #2: Fundamentals First

Postpone Concepts That Require Complex Syntax

We feel that many introductory programming textbooks jump too quickly into concepts that require complex syntax. In using complex syntax early, students get in the habit of entering code without fully understanding it or, worse yet, copying and pasting from example code without fully understanding the example code. That can lead to less-than-ideal programs and students who are limited in their ability to solve a wide variety of problems. Thus, we prefer to postpone concepts that require complex syntax. We prefer to introduce such concepts later on, when students are better able to understand them fully.

As a prime example of that philosophy, we cover the simpler forms of GUI programming early (in an optional graphics track), but we cover the more complicated forms of GUI programming later in the book. Specifically, we postpone event-driven GUI programming until the end of the book. This is different from some other Java textbooks, which favor early full immersion into event-driven GUI programming. We feel that strategy is a mistake because proper event-driven GUI programming requires a great deal of programming maturity. When they learn it at the end of the book, our readers are better able to understand it fully.

Tracing Examples

To write code effectively, it's imperative to understand code thoroughly. We've found that step-by-step tracing of program code is an effective way to ensure thorough understanding. Thus, in the earlier parts of the textbook, when we introduce a new programming structure, we often illustrate it with a meticulous trace. The detailed tracing technique we use illustrates the thought process programmers employ while debugging. It's a printed alternative to the sequence of screen displays generated by debuggers in integrated development environment (IDE) software.

Input and Output

In the optional GUI-track sections and in the GUI chapters at the end of the book, we use GUI commands for input and output (I/O). But because of our emphasis on fundamentals, we use console commands for I/O for the rest of the book.[2] For console input, we use the `Scanner` class. For console output, we use the standard `System.out.print`, `System.out.println`, and `System.out.printf` methods.

Textbook Cornerstone #3: Real World

More often than not, today's classroom students and industry practitioners prefer to learn with a hands-on, real-world approach. To meet this need, our textbook and its associated website include:

- compiler tools
- complete program examples
- practical guidance in program design
- coding-style guidelines based on industry standards
- Unified Modeling Language (UML) notation for class relationship diagrams
- practical homework-project assignments

Compiler Tools

We do not tie the textbook to any particular compiler tool—you are free to use any compiler tool(s) that you like. If you do not have a preferred compiler in mind, then you might want to try out one or more of these:

- Java Standard Edition Development Kit (JDK), by Oracle
- TextPad, by Helios
- Eclipse, by the Eclipse Foundation
- Netbeans, backed by Oracle
- BlueJ, by the University of Kent and Deaken University

To obtain the above compilers, visit our textbook website at http://www.mhhe.com/dean2e, find the appropriate compiler link(s), and download away for free.

Complete Program Examples

In addition to providing code fragments to illustrate specific concepts, our textbook contains lots of complete program examples. With complete programs, students are able to (1) see how the analyzed code ties in with the rest of a program, and (2) test the code by running it.

Coding-Style Conventions

We include coding-style tips throughout the textbook. The coding-style tips are based on Oracle's coding conventions (http://www.oracle.com/technetwork/java/codeconventions-150003.pdf) and industry practice. In Appendix 5, we provide a complete reference for the book's coding-style conventions and an associated example program that illustrates these conventions.

[2]We introduce GUI I/O early on with the `JOptionPane` class. That opens up an optional door for GUI fans. If readers are so inclined, they can use `JOptionPane` to implement all our programs with GUI I/O rather than console I/O. To do so, they replace all console I/O method calls with `JOptionPane` method calls.

UML Notation

UML has become a standard for describing the entities in large software projects. Rather than overwhelm beginning programmers with syntax for the entire UML (which is quite extensive), we present a subset of UML. Throughout the textbook, we incorporate UML notation to represent classes and class relationships pictorially. For those interested in more details, we provide additional UML notation in Appendix 7.

Homework Problems

We provide homework problems that are illustrative, practical, and clearly worded. The problems range from easy to challenging. They are grouped into three categories—review questions, exercises, and projects. We include review questions and exercises at the end of each chapter, and we provide projects on our textbook's website.

The review questions tend to have short answers, and the answers are in the textbook. The review questions use these formats: short-answer, multiple-choice, true/false, fill-in-the-blank, tracing, debugging, and write a code fragment. Each review question is based on a relatively small part of the chapter.

The exercises tend to have short to moderate-length answers, and the answers are not in the textbook. The exercises use these formats: short-answer, tracing, debugging, and write a code fragment. Exercises are keyed to the highest prerequisite section number in the chapter, but they sometimes integrate concepts from several parts of the chapter.

The projects consist of problem descriptions whose solutions are complete programs. Project solutions are not in the textbook. Projects require students to employ creativity and problem-solving skills and apply what they've learned in the chapter. These projects often include optional parts, which provide challenges for the more talented students. Projects are keyed to the highest prerequisite section number in the chapter, but they often integrate concepts from several preceding parts of the chapter.

An important special feature of this book is the way that it specifies problems. "Sample sessions" show the precise output generated for a particular set of input values. These sample sessions include inputs that represent typical situations and sometimes also extreme or boundary situations.

Academic-Area Projects

To enhance the appeal of projects and to show how the current chapter's programming techniques might apply to different areas of interest, we take project content from several academic areas:

- computer science and numerical methods
- business and accounting
- social sciences and statistics
- math and physics
- engineering and architecture
- biology and ecology

The academic-area projects do not require prerequisite knowledge in a particular area. Thus, instructors are free to assign any of the projects to any of their students. To provide a general reader with enough specialized knowledge to work a problem in a particular academic area, we sometimes expand the problem statement to explain a few special concepts in that academic area.

Most of the academic-area projects do not require students to have completed projects from earlier chapters; that is, the projects do not build on each other. Thus, instructors are free to assign projects without worrying about prerequisite projects. In some cases, a project repeats a previous chapter's project with a different approach. The teacher may elect to take advantage of this repetition to dramatize the availability of alternatives, but this is not necessary.

Project assignments can be tailored to fit readers' needs. For example:

- For readers outside of academia—
  Readers can choose projects that match their interests.
- When a course has students from one academic area—
  Instructors can assign projects from the relevant academic area.
- When a course has students with diverse backgrounds—
  Instructors can ask students to choose projects from their own academic areas, or instructors can ignore the academic-area delineations and simply assign projects that are most appealing.

To help you decide which projects to work on, we've included a "Project Summary" section after the preface. It lists all the projects by chapter, and for each project, it specifies:

- the associated section within the chapter
- the academic area
- the length and difficulty
- a brief description

After using the "Project Summary" section to get an idea of which projects you might like to work on, see the textbook's website for the full project descriptions.

Organization

In writing this book, we lead readers through three important programming methodologies: structured programming, OOP, and event-driven programming. For our structured programming coverage, we introduce basic concepts such as variables and operators, `if` statements, and loops. Then we show readers how to call prebuilt methods from Oracle's Java API library. Many of these methods, like those in the `Math` class, are non-OOP methods that can be called directly. Others, like those in the `String` class, are OOP methods that must be called by a previously created object. After an "interlude" that gives readers a brief taste of what it's like to write methods in a non-OOP environment, we move into OOP programming, and introduce basic OOP concepts such as classes, objects, instance variables, instance methods, and constructors. We also introduce class variables and class methods, which are useful in certain situations. However, we note that they should be used less often than instance variables and instance methods. Next, we move on to more advanced OOP concepts—arrays, collections, interfaces, and inheritance. Chapters on exception handling and files provide a transition into event-driven GUI programming. We describe event-driven GUI programming in the final two chapters.

The content and sequence we promote enable students to develop their skills from a solid foundation of programming fundamentals. To foster this fundamentals-first approach, our book starts with a minimum set of concepts and details. It then gradually broadens concepts and adds detail later. We avoid overloading early chapters by deferring certain less-important details to later chapters.

GUI Track

Many programmers find GUI programming to be fun. As such, GUI programming can be a great motivational tool for keeping readers interested and engaged. That's why we include graphics sections throughout the book, starting in Chapter 1. We call those sections our "GUI track." For readers who do not have time for the GUI track, no problem. Any or all of the GUI track sections may be skipped as they cover material that is independent of later material.

Chapter 1

In Chapter 1, we first explain basic computer terms—what are the hardware components, what is source code, what is object code, and so on. We then narrow our focus and describe the programming language we'll be using for the remainder of the book—Java. Finally, we give students a quick view of the classic bare-bones "Hello World" program. We explain how to create and run the program using minimalist software—Microsoft's Notepad text editor and Oracle's command-line JDK tools.

Chapter 2

In Chapter 2, we present problem-solving techniques with an emphasis on algorithmic design. In implementing algorithm solutions, we use generic tools—flowcharts and pseudocode—with pseudocode given greater weight. As part of our algorithm-design explanation, we describe structured programming techniques. In order to give students an appreciation for semantic details, we show how to trace algorithms.

Chapters 3–5

We present structured programming techniques using Java in Chapters 3–5. Chapter 3 describes sequential programming basics—variables, input/output, assignment statements, and simple method calls. Chapter 4 describes non-sequential program flow—`if` statements, `switch` statements, and loops. In Chapter 5, we explain methods in more detail and show readers how to use prebuilt methods in the Java API library. In all three chapters, we teach algorithm design by solving problems and writing programs with the newly introduced Java syntax.

Interlude

This "mini-chapter" contains a program that shows how to write multiple methods without using OOP. The Interlude presents a fork in the road between two study sequences. For the standard study sequence, read the chapters in the standard order (Chapters 1 through 18). For the "objects later" study sequence, after reading Chapter 5, read the supplemental chapters S6 and S9 on the book's website before returning to Chapter 6, where you'll begin your study of OOP in earnest.

Chapters 6–7

Chapter 6 introduces the basic elements of OOP in Java. This includes implementing classes and implementing methods and variables within those classes. We use UML class diagrams and object-oriented tracing techniques to illustrate these concepts.

Chapter 7 provides additional OOP details. It explains how reference variables are assigned, tested for equality, and passed as arguments to a method. It explains overloaded methods and constructors. It also explains the use of class variables and methods and named constants.

Chapter 8

While the art of program design and the science of computerized problem-solving are developed throughout the textbook, in Chapter 8, we focus on these aspects in the context of OOP. This chapter begins with an organized treatment of programming style. It introduces `javadoc`, the Java application that automatically generates documentation for user-programmers. It describes ways to communicate with users who are not programmers. It describes organizational strategies like separation of concerns, modularization, encapsulation, and provision of general-purpose utilities. Coded examples show how to implement these strategies. It describes the major programming paradigms—top-down design, bottom-up design, using pre-written software for low-level modules, and prototyping.

Chapters 9–10

Chapter 9 describes arrays, including arrays of primitives, arrays of objects, and multidimensional arrays. It illustrates array use with complete programs that sort, search, and construct histograms. Chapter 10 describes Java's powerful array alternative, `ArrayList`. This provides a simple example of generic-element specification. It also introduces the Java Collections Framework, which in turn, provides natural illustrations of Java interfaces. The prewritten classes in the Java Collections Framework provide a simple introduction of sets, maps, and queues. A relatively short but complete program shows how the pre-written Java implementations of these data structures can be used to create and traverse a multiconnected random network.

Chapter 11

Chapter 11 describes the other way to process a collection of data—recursion. This chapter includes a discussion of various recursive strategies. It introduces recursion with a real-life example and a familiar problem that one can solve easily with either looping or recursion. Then it moves gradually to problems that are harder to solve with looping and more easily solved with recursion. Although this chapter appears after the chapter on `ArrayLists` and the Java Collections Framework, it does not depend on these concepts—it uses just ordinary arrays.

Chapter 12

Early on, students need to be immersed in problem-solving activities. Covering too much syntax detail early can detract from that objective. Thus, we initially gloss over some less-important syntax details and come back to those details later in Chapter 12. This chapter provides more details on items such as these:

- the `byte` and `short` primitive types
- the Unicode character set
- type promotions
- postfix versus prefix modes for the increment and decrement operators
- the conditional operator
- short-circuit evaluation
- the `enum` data type

The chapter ends with a friendly introduction to a relatively large public-domain program called GridWorld, which the College Board used for many years as part of its recommended course of study for advanced placement in computer science. This gives students a glimpse of how larger programs are organized.

Chapters 13–14

We describe class relationships in Chapters 13 and 14. We spend two full chapters on class relationships because the subject matter is so important. We take the time to explain class relationship details in depth and provide numerous examples. Chapter 13 describes aggregation, composition, and inheritance. Chapter 14 describes more advanced inheritance-related details such as the `Object` class, polymorphism, `abstract` classes, and the finer points of interfaces. A section at the end of Chapter 13 extends our discussion of the GridWorld program. And numerous exercises in these two chapters relate chapter material to corresponding GridWorld features.

Chapters 15–16

We cover exception handling in Chapter 15 and files in Chapter 16. We present exception handling before files because file-handling code utilizes exception handling; for example, opening a file requires that you check for an exception. Our treatment of exception handling includes multiple-`catch` parameters and try-with-resources. In addition to simple text I/O, our treatment of files includes buffering, random access, channeling, and memory mapping.

Chapters 17–18

We describe event-driven GUI programming at the end of the book in Chapters 17 and 18. By learning event-driven GUI programming late, students are better able to grasp its inherent complexities.

Chapters S17–S18

Chapters 17 and 18 present GUI concepts using the Swing toolkit because Swing is very popular and will remain so for years to come. However, with the advent of Java 8, Oracle started supporting JavaFX as the default GUI toolkit. In Chapters S17 and S18 (the *S*'s stands for supplemental), we present GUI concepts using the JavaFX toolkit.

Appendices

Most of the appendices cover reference material, such as the ASCII character set and the operator precedence table. But the last two appendices introduce advanced Java material—multithreading and closures.

Subject-Matter Dependencies and Sequence-Changing Opportunities

We've positioned the textbook's material in a natural order for someone who wants fundamentals and also wants an early introduction to OOP. We feel that our order is the most efficient and effective one for learning how to become a proficient OOP programmer. Nonetheless, we realize that different readers have different content-ordering preferences. To accommodate those different preferences, we've provided some built-in flexibility. Figure 0.1 illustrates that flexibility by showing chapter dependencies and, more importantly, chapter non-dependencies. For example, the arrow between Chapter 3 and Chapter 4 means that Chapter 3 must be read prior to Chapter 4. Because there are no arrows going out of Chapters 1, 11, and 16 that point to other complete chapters, you may skip those chapters without losing prerequisite material that later chapters need. We use rectangles with rounded corners to indicate chapter sections that you may want to read in advance. If you choose that option, you'll want to return to the normal chapter sequence after completing the advanced sections.

Here are some sequence-changing opportunities revealed by Figure 0.1:

- Readers can skip Chapter 1, "Introduction to Computers and Programming."
- For an earlier introduction to OOP, readers can read the OOP overview section in Chapter 6 after reading Chapter 1.
- Readers can learn OOP syntax and semantics in Chapter 6 after finishing Java basics in Chapter 3.

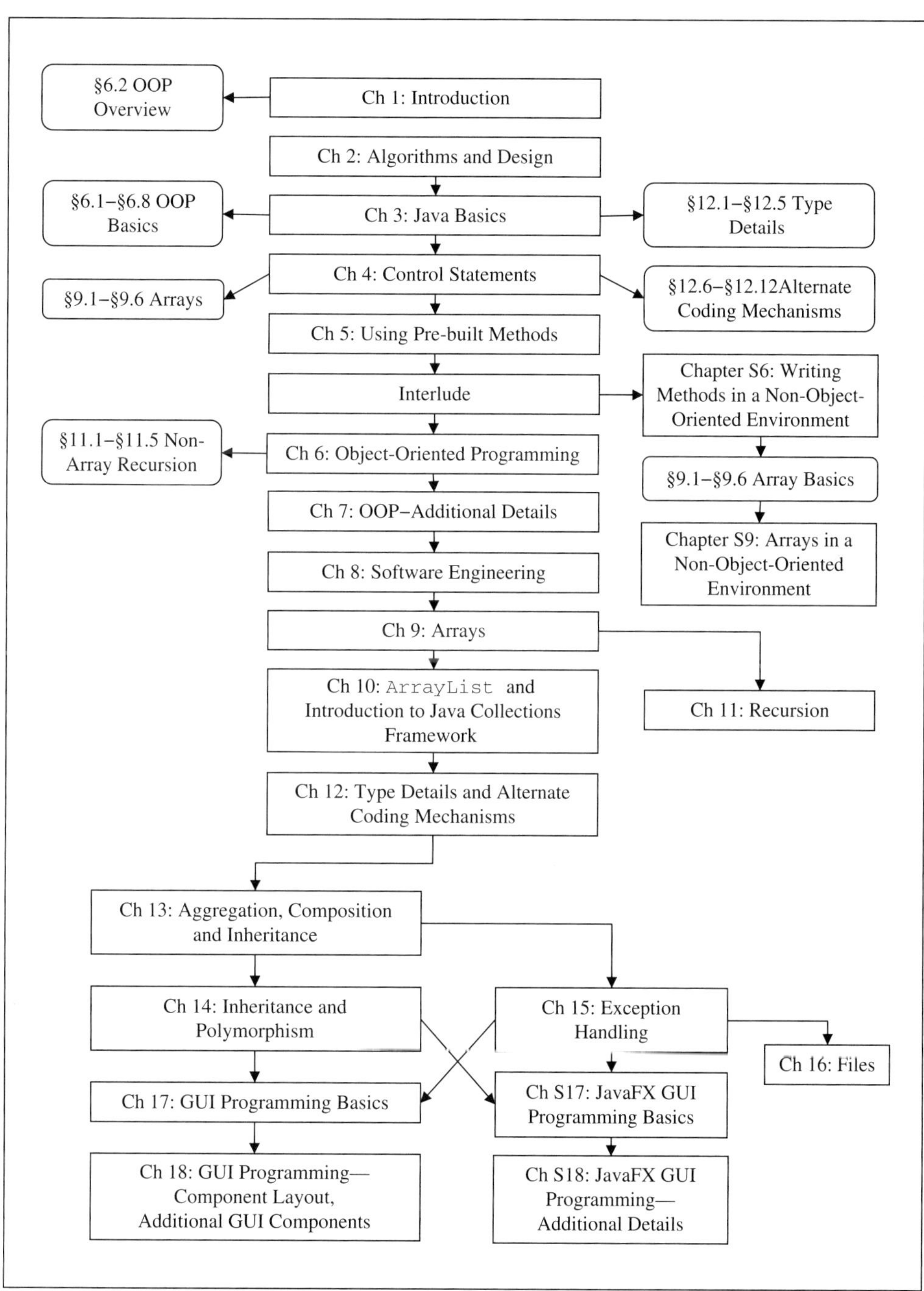

Figure 0.1 Chapter dependencies

- For additional looping practice, readers can learn about arrays in Chapter 9 after finishing loops in Chapter 4.
- Readers can skip Chapter 11, "Recursion," and Chapter 16, "Files."
- Readers who prefer a late objects approach can postpone reading Chapter 6, "Object-Oriented Programming," by first reading Chapter S6, "Writing Methods in a Non-Object-Oriented Environment," Sections 9.1–9.6, "Array Basics," and Chapter S9, "Arrays in a Non-Object-Oriented Environment."
- For GUI programming, readers who prefer the Swing toolkit should read Chapters 17 and 18, whereas readers who prefer the JavaFX toolkit should read Chapters S17 and S18.

To support content-ordering flexibility, the book contains "hyperlinks." A hyperlink is an optional jump forward from one place in the book to another place. The jumps are legal in terms of prerequisite knowledge, meaning that the jumped-over (skipped) material is unnecessary for an understanding of the later material. We supply hyperlinks for each of the non-sequential arrows in Figure 0.1. For example, we supply hyperlinks that go from Chapter 1 to Chapter 6 and from Chapter 3 to Chapter 12. For each hyperlink tail end (in the earlier chapter), we tell the reader where they may optionally jump to. For each hyperlink target end (in the later chapter), we provide an icon at the side of the target text that helps readers find the place where they are to begin reading.

Pedagogy

Icons

Program elegance.
Indicates that the associated text deals with a program's coding style, readability, maintainability, robustness, and scalability. Those qualities comprise a program's elegance.

Problem solving.
Indicates that the associated text deals with problem-solving issues. Comments associated with this icon attempt to generalize highlighted material in the adjacent text.

Common errors.
Indicates that the associated text deals with common errors.

Hyperlink target.
Indicates the target end of a hyperlink.

Program efficiency.
Indicates that the associated text refers to program-efficiency issues.

Student Resources

At the textbook website, http://www.mhhe.com/dean2e, students (and also teachers) can view and download these resources:

- Links to compiler software—for Oracle's JDK, Helios's TextPad, Eclipse, NetBeans, and BlueJ
- TextPad tutorial

- Eclipse tutorials
- Textbook errata
- Student-version Microsoft PowerPoint lecture slides without hidden notes
  - The student-version slides are identical to the teacher-version slides except that the hidden notes, hidden slides, and quizzes are omitted.
  - Omitting the hidden notes forces the students to go to lecture to hear the sage on the stage fill in the blanks. ☺
- GridWorld code
- Project assignments
- All textbook example programs and associated resource files
- Supplemental chapters
- Supplemental appendices

Instructor Resources

At the textbook website, http://www.mhhe.com/dean2e, instructors can view and download these resources:

- Teacher-version PowerPoint lecture slides with hidden notes
  - Hidden notes provide comments that supplement the displayed text in the lecture slides.
  - For example, if the displayed text asks a question, the hidden notes provide the answer.
- Exercise solutions
- Project solutions
- Test bank materials

Acknowledgments

Anyone who has written a textbook can attest to what a large and well-orchestrated team effort it requires. Such a book can never be the work of only one person, or even a few people. We are deeply indebted to the team at McGraw-Hill Higher Education who have shown continued faith in our writing and invested generously in it.

It was a pleasure to work with Alan Apt during the original book's two-year review period. He provided excellent guidance on several large design issues. Helping us through the various stages of this edition's production was Project Manager Melissa Leick. We would also like to thank the rest of the editorial and marketing team, who helped in the final stages: Raghu Srinivasan, Publisher; Katie Neubauer, Developmental Editor; and Curt Reynolds, Marketing Manager.

All the professionals we have encountered throughout the McGraw-Hill organization have been wonderful to work with, and we sincerely appreciate their efforts.

We would like to acknowledge with appreciation the numerous and valuable comments, suggestions, and constructive criticisms and praise from the many instructors who have reviewed the book. In particular,

Robert Burton, *Brigham Young University*

William Duncan, *Louisiana State University*

Frantisek Franek, *McMaster University*

Junilda Spirollari, *New Jersey Institute of Technology*

Geoffrey Decker, *Northern Illinois University*
Patricia Roth Pierce, *Southern Polytechnic State University*
Jeffrey A. Meunier, *University of Connecticut*
Chris Johnson, *University of Wisconsin, Eau Claire*
Mark Pauley, *University of Nebraska at Omaha*
Christopher Taylor, *Milwaukee School of Engineering*

We would also like to thank Park University colleagues Guillermo Tonsmann and Ken Dewey for their help over the past several years, making suggestions for improvements to the first edition. We appreciate Daniel Green from the Oracle Corporation, who provided advice about new developments in the JavaFX platform and in the Java language itself. A special shout-out goes to Judy Hromcik, a former member of the Advanced Placement (AP) Computer Science Development Committee. In reviewing the second edition, she provided many helpful suggestions, with an eye towards making the book work well in both a college setting and also in an AP Computer Science high school setting.

Thanks to Bruce Jenkins, who worked on the JavaFX chapters. Thanks to Rafael Neri and Miki Katuwal, who created the "Woman seen yesterday" recursive picture in Chapter 11. Finally, thanks to the students who provided feedback and searched diligently for mistakes in order to earn bonus points on their homework. In particular, thank you Komi Labitoko, Nodirbek Hojimatov, Lance Delaney, Ryan Todd, Rob Liebsch, Nirendra Shakya, and Anna Luo.

Sincerely,
John and Ray

Project Summary

One of the special features of this text is the diversity of its projects. Project subject matter spans six broad academic areas, as this short table shows:

Abbreviation	Description	Easy	Moderate	Difficult	Total
CS	computer science and numerical methods	15	14	6	35
Business	business and accounting	11	12	3	26
Sociology	social sciences and statistics	6	8	5	19
Math & Phys	math and physics	10	6	3	19
Engineering	engineering and architecture	2	8	6	16
Biol & Ecol	biology and ecology	0	3	4	7
	Totals	**44**	**51**	**27**	**122**

The abbreviation in the first column above will be used in a larger table below as a brief identification of a particular academic area. The four right-side columns in the above table indicate the number of projects in various categories. Of course, the highest number of projects (35) occurs in the area of computer science and numerical methods. The 29 easy and moderate CS projects are typical CS introductory programming problems. The 6 difficult CS projects provide gentle introductions to some advanced topics like linked list operations, database operations, and simulated annealing.

In addition, there are 26 projects in business and accounting, which include miscellaneous financial calculations, simple bookkeeping problems, and cost-accounting applications. There are 19 projects in social sciences and statistics, which include applications in sociology and political science, as well as general experience. There are 19 projects in math and physics, which include applications in both classical and chaotic mechanics. There are 16 projects in engineering and architecture, which include applications in heating ventilating and air conditioning (HVAC), electrical engineering, and civil engineering. Finally, there are 7 projects in biology and ecology, which include realistic growth and predator-prey simulations. Although we've associated each project with one primary academic area, many of these projects can fit into other academic areas as well.

Because many of these projects apply to disciplines outside the field of computer science, we do not expect that the average reader will already know about all of these "other" topics. Therefore, in our problem statements, we usually take considerable time to explain the topic as well as the problem. And we often explain how to go about solving the problem—in layman's terms. Therefore, working many of these projects will be like implementing computer solutions for customers who are not programmers themselves but understand their subject matter and know what they want you (the programmer) to do for them. They will explain their problem and how to go about solving it. But then they will expect you to create the program that actually solves that problem.

Because our project explanations frequently take considerable printed space, instead of putting them in the book itself, we put them on our website:

http://www.mhhe.com/dean2e

The following table provides a summary of the projects on the book's website. This table lists all of the book's projects in a sequence that matches the book's sequence. The first column identifies the first point at which you should be able to do the project, by chapter and section, in the form: ChapterNumber.SectionNumber. The second column is a unique project number for the chapter in question. The third column identifies the project's primary academic area with an abbreviation that's explained in the shorter table above. The fourth column indicates the approximate number of pages of code that our solution contains. The fifth column indicates the difficulty relative to where you are in your study of Java. For example, you can see that what we call "easy" involves progressively more pages of code as you progress through the book. The last two columns provide a title and brief description of each project.

Project Summary

Ch./Sec	Proj.	Academic Area	Sol. Pages	Difficulty	Title	Brief Description
2.7	1	Business	0.6	Easy	Annual Bonus–(Flowchart)	Draw a flowchart for an algorithm that computes an annual bonus.
2.7	2	Business	0.3	Easy	Annual Bonus—(Pseudocode)	Write pseudocode for an algorithm that computes an annual bonus.
2.7	3	Business	0.6	Easy	Number of Stamps—(Flowchart)	Draw a flowchart for an algorithm that calculates the number of stamps needed for an envelope. Use one stamp for every five sheets of paper.
2.7	4	Business	0.3	Easy	Number of Stamps—(Pseudocode)	Write pseudocode for an algorithm that calculates the number of stamps needed for an envelope. Use one stamp for every five sheets of paper.
2.7	5	Biol & Ecol	0.5	Moderate	Five Kingdoms—(Pseudocode)	Write pseudocode for an algorithm that identifies a biological kingdom from a set of characteristics.
2.7	6	Math & Phys	0.6	Easy	Speed of Sound—(Flowchart)	Draw a flowchart for an algorithm that provides the speed of sound in a particular medium.
2.7	7	Math & Phys	0.4	Easy	Speed of Sound—(Pseudocode)	Write pseudocode for an algorithm that provides the speed of sound in a particular medium.
2.7	8	Business	0.6	Moderate	Stock Market Return—(Flowchart)	Draw a flowchart for an algorithm that prints the type of market and its probability given a particular rate of return.
2.7	9	Business	0.4	Moderate	Stock Market Return—(Pseudocode)	Write pseudocode for an algorithm that prints the type of market and its probability given a particular rate of return.
2.8	10	Business	0.3	Moderate	Bank Balance—(Pseudocode)	Write pseudocode for an algorithm that determines the number of years until a growing bank balance reaches a million dollars.
2.9	11	Engineering	1.0	Moderate	Loop Termination by User Query—(Flowchart)	Draw a flowchart for an algorithm that calculates the overall miles per gallon for a series of miles and gallons user inputs.

Project Summary

Ch./Sec	Proj.	Academic Area	Sol. Pages	Difficulty	Title	Brief Description
2.9	12	Engineering	0.5	Easy	Loop Termination by User Query—(Pseudocode)	Write pseudocode for an algorithm that calculates the overall miles per gallon for a series of miles and gallons user inputs.
2.9	13	Engineering	0.4	Moderate	Loop Termination by Sentinal Value—(Pseudocode)	Write pseudocode for an algorithm that calculates the overall miles per gallon for a series of miles and gallons user inputs.
2.9	14	Engineering	0.3	Easy	Loop Termination by Counter—(Pseudocode)	Write pseudocode for an algorithm that calculates the overall miles per gallon for a series of miles and gallons user inputs.
2.10	15	CS	0.4	Moderate	Average Weight—(Pseudocode)	Write pseudocode for an algorithm that determines average weight for a group of items.
3.2	1	CS	NA	Easy	Hello World Experimentation	Experiment with the `Hello.java` program to learn the meanings of typical compile-time and runtime error messages.
3.3	2	CS	NA	Moderate	Research	Study Oracle's Java Coding Conventions.
3.3	3	CS	NA	Moderate	Research	Study Appendix 5, "Java Coding-Style Conventions."
3.16 3.23	4	Engineering	2.5	Difficult	Truss Analysis	Given the load in the center of a bridge and the weights of all truss members, compute the compression or tension force in each truss member.
3.17	5	CS	1.0	Easy	Sequence of Commands	Trace a sequence of commands and write a program that executes those commands.
3.17 3.23	6	CS	1.7	Moderate	Computer Speed	Given a simple set of hardware and software characteristics, write a program that estimates the total time to run a computer program.
3.17 3.23	7	Engineering	2.7	Moderate	HVAC Load	Calculate the heating and cooling loads for a typical residence.
3.17 3.23	8	Sociology	3.5	Difficult	Campaign Planning	Write a program to help organize estimates of votes, money, and labor.
3.22	9	CS	1.0	Easy	String Processing	Trace a set of string processing operations and write a program that implements them.
3.23	10	CS	1.2	Easy	Swapping	Trace an algorithm that swaps the values in two variables, and write a program that implements that algorithm.

(continued)

Project Summary

Ch./Sec	Proj.	Academic Area	Sol. Pages	Difficulty	Title	Brief Description
3.23	11	Math & Phys	1.0	Easy	Circle Parameters	Write a program that generates and prints circle-related values.
3.23	12	Sociology	0.4	Easy	One-Hundredth Birthday	Write a program that prompts the user for his/her birthday month, day, and year and prints the date of the user's one-hundredth birthday.
4.3	1	Math & Phys	1.7	Easy	Stopping Distance	Write a program which determines whether a vehicle's tailgating distance is safe, given the speed of the vehicle, the vehicle's tailgating distance, and a formula that gives the distance required to stop the vehicle.
4.3 4.8	2	Engineering	1.9	Moderate	Column Safety	Write a program that determines whether a structural column is thick enough to support the column's expected load.
4.3	3	Business	1.1	Easy	Economic Policy	Write a program that reads in growth rate and inflation values and outputs a recommended economic policy.
4.8	4	Business	2.0	Moderate	Bank Balance	Write a program that determines the number of years until a growing bank balance reaches a million dollars.
4.9 4.12	5	CS	2.6	Difficult	Game of NIM	Implement the game of NIM. Start the game with a user-specified number of stones in a pile. The user and the computer take turns removing either one or two stones from the pile. The player who takes the last stone loses.
4.12	6	Math & Phys	1.0	Easy	Triangle	Write a program that generates an isosceles triangle made of asterisks, given user input for triangle size.
4.12	7	Sociology	0.8	Easy	Mayan Calendar	Implement an algorithm that determines the number of Tzolkins and the number of Haabs in one Calendar Round.
4.12	8	CS	0.9	Easy	Input Validation	Implement an algorithm that repeatedly prompts for inputs until they fall within an acceptable range and computes the average of valid inputs.
4.14	9	Business	2.6	Moderate	Tax Preparation	Write a program that calculates customers' income taxes using the following rules: • The amount of taxes owed equals the taxable income times the tax rate. • Taxable income equals gross income minus $1,000 for each exemption. • The taxable income cannot be less than zero.

Project Summary

Ch./Sec	Proj.	Academic Area	Sol. Pages	Difficulty	Title	Brief Description
4.14	10	CS	1.7	Moderate	Text Parsing	Write a program that converts words to pig Latin.
5.3	1	Math & Phys	1.2	Easy	Trigonometric Functions	Write a demonstration program that asks the user to select one of three possible inverse functions, arcsin, arccos, or arctan, and input a trigonometric ratio. It should generate appropriate output, with diagnostics.
5.3	2	Math & Phys	0.7	Easy	Combining Decibels	Determine the acoustical power level produced by the combination of two sound sources.
5.3	3	Business	1.0	Moderate	Net Present Value Calculation	Write a program that computes the net present value of a proposed investment, given a discount rate and an arbitrary set of future cash flows.
5.5	4	CS	1.5	Moderate	Variable Name Checker	Write a program that checks the correctness of a user-entered variable name, i.e., whether it is: (1) illegal, (2) legal, but poor style, or (3) good style. Assume that "good style" variable names use letters and digits only, and use a lowercase letter for the first character.
5.6	5	CS	1.0	Moderate	Phone Number Dissector	Implement a program that reads phone numbers, and for each phone number, it displays the phone number's three components—country code, area code, and local number.
5.6	6	CS	1.1	Difficult	Phone Number Dissector—robust version	Implement a more robust version of the above phone number program. Allow for shortened phone numbers—phone numbers that have just a local digit group and nothing else, and phone numbers that have just a local digit group and an area code and nothing else.
6.4	1	Biol & Ecol	1.5	Moderate	Plant Germination Observation	Write a program that (1) creates an object called `tree` from the `MapleTree` class; (2) calls a `plant` method to record the planting of the seed; (3) calls a `germinate` method to record the first observation of a seedling and record its height; and (4) calls a `dumpData` method to display the current values of all instance variables.

(continued)

Project Summary

Ch./Sec	Proj.	Academic Area	Sol. Pages	Difficulty	Title	Brief Description
6.4	2	Business	0.5	Easy	Bank Account	Given the code for a `BankAccount` class, provide a driver that tests that class by instantiating an object and calling its methods—`setCustomer`, `setAccountNum`, and `printAccountInfo`.
6.8	3	Math & Phys	1.5	Moderate	Logistic Equation	Exercise the logistic equation: nextX = presentX + r × presentX × (1 − presentX), where presentX = (present x)/(maximum x), and r is a growth factor.
6.9	4	Math & Phys	0.9	Easy	Circle	Given the code for a `CircleDriver` class, write a `Circle` class that defines a `radius` instance variable, a `setRadius` method, and a `printAndCalculateCircleData` method that uses the circle's radius to calculate and print the circle's diameter, circumference, and area.
6.10	5	Engineering	2.0	Moderate	Digital Filter	Given a formula for a "Chebyshev second-order low-pass" filter or a "Butterworth second-order low-pass" filter, with appropriate parameter values, write a program that asks the user to supply a sequence of raw input values and generates the corresponding filtered output.
6.10	6	Sociology	3.1	Difficult	Vending Machine	Write a program that mimics the operations of a vending machine. The program should read amounts of money inserted into the vending machine, ask the user to select an item, and then print the change that's returned to the user.
6.12	7	Math & Phys	1.1	Easy	Rectangle	Implement a `Rectangle` class that defines a rectangle with length and width instance variables, mutator and accessor methods, and a `boolean isSquare` method.
6.12	8	Biol & Ecol	4.0	Difficult	Predator-Prey Dynamics	Write a program that models a species that could be either predator or prey or both. Run a simulation that includes predators, prey, and limited renewable sustenance for the prey.
6.13	9	Math & Phys	2.1	Moderate	Guitar Mechanics	Write a program that simulates the motion of a plucked guitar string.

Project Summary

Ch./Sec	Proj.	Academic Area	Sol. Pages	Difficulty	Title	Brief Description
7.5 7.9	1	CS	3.5	Difficult	Linked List	Given the code for a driver, implement a `Recipe` class that creates and maintains a linked list of recipes. The problem assignment specifies all instance variables and methods in UML class diagrams.
7.7	2	CS	2.5	Easy	Automobile Description	Use method-call chaining to help display properties of automobiles.
7.7 7.9	3	Biol & Ecol	4.6	Difficult	Carbon Cycle	Given the code for a driver, write a pair of classes for a program that models the carbon cycle in an ecosystem. Use two generic classes. One class, `Entity`, defines things. The other class, `Relationship`, defines interactions.
7.8	4	CS	1.4	Easy	IP Address	Implement an `IpAddress` class that stores an Internet Protocol (IP) address as a dotted-decimal string and as four octet `ints`.
7.9	5	Math & Phys	4.5	Moderate	Fraction Handler	Given the `main` method of a driver class, write a `Fraction` class. Include the following instance methods: `add`, `multiply`, `print`, `printAsDouble`, and a separate accessor method for each instance variable.
7.10	6	Math & Phys	1.1	Easy	Rectangles	Write a class that processes rectangular objects. Include a variable that holds the total number of objects and a method that gets that number.
7.11	7	Sociology	2.7	Easy	Person Class	Define a class that simulates the creation and display of `Person` objects.
7.11	8	Sociology	3.9	Difficult	Political Approval Rating	Write a program that determines the mean and standard deviation of statistical samples.
7.12	9	Sociology	2.7	Moderate	Homework Scores	Write a program that handles homework scores. Use instance variables for actual and maximum points on a particular homework, and use class variables for actual total and maximum total points of all homeworks combined.
7.12	10	Engineering	5.7	Difficult	Solar Input for HVAC and Solar Collectors	Write a program that tracks the sun and determines how much solar energy penetrates a glass window of any orientation, at any place and time.

(continued)

Project Summary

Ch./Sec	Proj.	Academic Area	Sol. Pages	Difficulty	Title	Brief Description
7.13	11	Engineering	2.8	Moderate	Electric Circuit	Write branch and node classes for lumped-circuit elements. A branch carries current through a resistor in series with an inductor. A node holds voltage on a capacitor connected to a common ground. Driver code is provided in the problem assignment.
7.13	12	Business	5.1	Difficult	Cost Accounting	Write an object-oriented program that demonstrates cost accounting in a manufacturing plant.
7.13	13	Sociology	6.4	Difficult	Political Campaign	Write a program to help organize estimates of votes, money, and labor. This is an object-oriented version of Project 8 in Chapter 3.
7.13	14	Business	2.7	Moderate	Net Present Value Calculation	Write a program that computes the net present value of a proposed investment, given a discount rate and an arbitrary set of future cash flows. This is an OOP version of Project 3 in Chapter 5.
7.13	15	Math & Phys	7.0	Difficult	Three-Body Problem	Write a program to model the three-body problem in which two equally sized moons circle the Earth in different orbits. This illustrates chaotic dynamic motion.
8.5	1	CS	1.6	Easy	Input Validation	Implement an algorithm that repeatedly prompts for inputs until they fall within an acceptable range and computes the average of valid inputs. This is an object-oriented version of Project 8 in Chapter 4.
8.5	2	Engineering	4.0	Difficult	HVAC Load	Calculate the heating and cooling loads for a typical residence. This is an object-oriented version of Project 7 in Chapter 3.
8.8	3	Sociology	2.6	Moderate	Elevator Control	Write a program that mimics the operations of the inside of an elevator. The program should simulate what happens when the user chooses to go to a particular floor and when the user pulls the fire alarm.
8.11	4	CS	2.0	Easy	Prototype Restructuring	Consider the NestedLoopRectangle program in Figure 4.17 in Section 4.12 to be a prototype. Using top-down methodology, restructure it into OOP format.

Project Summary

Ch./Sec	Proj.	Academic Area	Sol. Pages	Difficulty	Title	Brief Description
9.4	1	Biol & Ecol	5.0	Difficult	Demographic Projections	Write a program that projects future world population and average individual wealth as a function of fertility rates and resource extraction rates, and includes the effects of governmental taxation and spending.
9.6	2	CS	3.3	Moderate	Dice-Throwing Simulator	Write a program that simulates the rolling of a pair of dice and prints a histogram showing the frequencies of possible results.
9.6	3	CS	5.1	Difficult	Simulated Annealing—the Traveling Salesman Problem	Write a program that uses simulated annealing to find the shortest itinerary that visits all of the world's major cities exactly one time.
9.7	4	Sociology	2.1	Easy	Party Guest List	Write a program that creates a `Party` object, adds guests to the party, and prints party information.
9.9	5	Sociology	2.7	Easy	Vowel Counter	Write a program that counts the number of uppercase and lowercase vowels in user-entered lines of text and prints a summary report of vowel counts.
9.9	6	Math & Phys	7.6	Difficult	Solution of Simultaneous Algebraic Equations	Write a program that loads a set of simultaneous algebraic equations into two-dimensional arrays and solves the equations by Lower-Upper Decomposition.
9.9	7	Math & Phys	2.5	Moderate	Linear Regression	Write a program that computes a linear regression by fitting a straight line to a series of random data.
9.10	8	Business	3.4	Moderate	Purchase Vouchers	Write a program that creates business vouchers that record purchases, displays current voucher information, and records payments for those purchases.
10.2	1	Sociology	1.1	Easy	Deck of Cards	Write a class that uses an `ArrayList` to hold a deck of cards.
10.4	2	Business	1.9	Easy	Bookstore	Write a program that models the storing and retrieving of books based on title.
10.9	3	Business	0.9	Easy	LIFO Inventory	Write a program that uses a stack to model last-in-first-out inventory costing.
10.10	4	CS	0.7	Easy	Queue Behavior	Write a program that illustrates the behavior of ordinary and priority queues.

(continued)

Project Summary

Ch./Sec	Proj.	Academic Area	Sol. Pages	Difficulty	Title	Brief Description
10.10	5	CS	2.1	Moderate	Survey with a two-variable key	Add hash codes to form the key of a map that relates a person and an event to an assessment of that event.
10.10	6	Engineering	4.8	Difficult	Discrete Event Simulation of Queuing Systems	Using iterative enhancement, write code for a discrete-event simulator for: (1) single-server queue with constant service time; (2) priority queue with random service time; (3) multiple servers and queue length limit.
11.5	1	Business	1.6	Easy	Loan Payments and Balances	Write a program that uses recursion to determine the level payment amount needed to pay off a loan in a given number of equal installments. Display balance after each payment.
11.5	2	Math & Phys	2.0	Moderate	Graphical Display of Hénon Map	Convert a given GUI applet into a GUI application. Then modify the application to zoom in to see the fine structure in a classical fractal. Convert from recursion to iteration and zoom in more.
11.6	3	Sociology	2.0	Moderate	Traversing a Maze	Use recursion to traverse a maze by following one wall. Modify the program to back up at dead ends, and modify again to find an interior object.
11.9	4	CS	2.0	Moderate	Enhanced Tree Simulation	Enhance the program in Section 11.9 by adding colored leaves, giving tree trunks and branches color and varying thickness, and randomizing branch lengths and branching angles.
12.3	1	CS	0.7	Easy	ASCII Table	Write a program that prints the 128-character ASCII table in eight tab-separated columns.
12.7	2	CS	0.8	Easy	Circular Queue	A given program implements a circular-array queue. Rewrite the `isFull`, `remove`, and `showQueue` methods by replacing conditional operators, embedded assignments, and embedded increment operators with simpler, more understandable code.
12.7	3	Math & Phys	4.1	Moderate	Polynomial Interpolation	Fit a polynomial to points on either side of a pair of points in an array of data and use that to estimate the value at a position between the pair of points.
12.9	4	CS	1.4	Moderate	Bitwise Operations	Use arithmetic and logical shifting to display the binary values of numbers.

Project Summary

Ch./Sec	Proj.	Academic Area	Sol. Pages	Difficulty	Title	Brief Description
12.11	5	CS	3.5	Moderate	Heap Sort	Use the heap-sort algorithm to sort data.
10.9 12.14	6	Biol & Ecol	5.5	Difficult	Game of Spawn	This "game" simulates reproduction and growth in a rectangular grid of cells. An X indicates life. A dead cell comes to life when it has exactly three living neighbor cells. A living cell remains alive only when surrounded by two or three living neighbor cells.
13.2	1	Business	1.7	Easy	Savings Accounts	Compute and display savings account balances that accumulate with compound interest.
13.4	2	Math & Phys	13.4	Difficult	Statistics Functions	Write a program that generates values for the Gamma, Incomplete Gamma, Beta, Incomplete Beta, and Binomial statistical functions.
13.5	3	Business	3.3	Easy	Car Program	Using inheritance, write a program that keeps track of information about new and used cars.
13.10	4	Sociology	16.4	Difficult	Game of Hearts	Write a program that simulates a basic game of hearts with an arbitrary number of players. Give all players an identical set of good strategies that optimize the chances of winning.
14.7	1	Business	9.0	Difficult	Grocery Store Inventory	Write an inventory program that keeps track of various kinds of food items. Use different methods in an `Inventory` class to process heterogeneous objects representing generic and branded food items. Store the objects together in a common `ArrayList`.
14.7	2	Engineering	8.7	Difficult	Electric Circuit Analysis	Write a program that calculates the steady-state currents in a two-loop electric circuit with resistors, inductors, capacitors, and voltage sources. Include methods to perform addition, subtraction, multiplication, and division of complex numbers.
14.8	3	Business	5.4	Moderate	Payroll	Use polymorphism to write an employee payroll program that calculates and prints the weekly payroll for a company with hourly, salaried, and salaried plus commission employees, where each type of employee gets paid using a different formula. Use an abstract base class.

(continued)

Project Summary

Ch./Sec	Proj.	Academic Area	Sol. Pages	Difficulty	Title	Brief Description
14.8	4	Business	2.9	Moderate	Bank Accounts	Write a bank account program that handles bank account balances for an array of bank accounts. Use two types of bank accounts, checking and savings, derived from an abstract class named `BankAccount`.
15.4	1	Sociology	4.0	Moderate	Body Mass Index	Write a program that prompts the user for height and weight values and displays the associated body mass index (BMI).
15.5	2	CS	6.4	Difficult	Storage and Retrieval of Objects in an Array	Search for a match with the key value in a relational table, using two different search algorithms, a sequential search, and a hashed search.
15.8	3	Biol & Ecol	3.8	Moderate	Whale Watching	Estimate whale length from observed fluke span. Use exceptions to help user correct input format errors.
15.9	4	CS	2.5	Moderate	Date Formatting	Create a class named `Date` that stores date values and prints out the date in either a numeric format or an alphabetic format.
15.9	5	CS	5.5	Difficult	Input Utility	Write a utility class that reads and parses keyboard inputs for the following: `String`, `char`, `double`, `float`, `long`, `int`.
16.2	1	Engineering	3.7	Moderate	Road Use Survey	Model traffic flowing on a highway past a particular place, store observations, and read file later for analysis.
16.2	2	Business	2.9	Moderate	Mail Merge	Write a program that reads a form letter from a text file and modifies custom fields.
16.4	3	CS	1.5	Easy	Appending Data to an Object File	Implement code needed to append data to an object file.
16.2 16.9	4	CS	5.0	Moderate	File Converter	Write a program that changes whitespace in text files.
17.12	1	Engineering	4.1	Moderate	Animated Garage Door	Write a program that simulates the operation of an automatic garage door and its controls and visually display its position as it operates.
17.14	2	Sociology	3.0	Moderate	Color Memorization	Write a program that tests the user's ability to memorize a sequence of colors.
17.14	3	Business	8.7	Difficult	Grocery Inventory GUI	Write a GUI version of the Grocery Store Inventory project in Chapter 14.
17.15	4	Sociology	4.2	Moderate	Word Order Game	Create a simple interactive game that helps kids practice their alphabetic skills.

Project Summary

Ch./Sec	Proj.	Academic Area	Sol. Pages	Difficulty	Title	Brief Description
17.16	5	Business	3.8	Moderate	Airline Reservations	Write a GUI program that assigns seats on airline flights.
18.3	1	CS	1.7	Easy	Changing Color and Alignment	Write an interactive program that modifies the color and position of buttons in a GUI window.
18.7	2	CS	1.9	Easy	Click Tracker	Write an interactive program that modifies the borders and labels of buttons in a GUI window.
18.11	3	Sociology	3.4	Moderate	Tic-Tac-Toe	Create an interactive Tic-Tac-Toe game.
18.11	4	Sociology	4.3	Moderate	Word Order Game, revisited	Modify Chapter 17's Word Order Game program so that it uses embedded layout managers.
18.11	5	Engineering	7.5	Difficult	Thermal Diffusion in a Ground-Source Heat Pump's Well	Write a program to compute and display temperature around a ground-source heat pump well as a function of distance from the well center and the time of year.

CHAPTER 1

Introduction to Computers and Programming

Objectives

- Describe the various components that make up a computer.
- List the steps involved in program development.
- Know what it means to write algorithms using pseudocode.
- Know what it means to write programs with programming language code.
- Understand source code, object code, and the compilation process.
- Describe how bytecode makes Java portable.
- Become familiar with Java's history—why it was initially developed, how it got its name, and so forth.
- Enter, compile, and run a simple Java program.

Outline

1.1 Introduction

This book is about problem solving. Specifically, it is about creating solutions to problems through a set of precisely stated instructions. We call such a set of instructions (when in a format that can be entered into and executed on a computer) a *program*. To understand what a program is, think about the following situation.

Suppose you manage a department store, and you don't know when to restock the shelves because you have difficulty keeping track of inventory. The solution to the problem is to write a set of instructions that keeps track of items as they arrive at your store and as they are purchased. If the instructions are correct and in a format that is understood by a computer, you can enter the instructions as a program, run the program, and enter item-arrival and item-purchase data as they occur. You can then retrieve inventory information from the computer any time you need it. That accurate and easily accessible knowledge enables you to restock your shelves effectively, and you are more likely to turn a profit.

The first step to learning how to write programs is to learn the background concepts. In this chapter, we teach background concepts. In subsequent chapters, we use these background concepts while explaining the really good stuff—how to program.

We start this chapter by describing the various parts of a computer. We then describe the steps involved in writing a program and in running a program. Next, we narrow our focus and describe the programming language we'll be using for the remainder of the book—Java. We present step-by-step instructions on how to enter and run a real Java program, so that you'll be able to gain some hands-on experience early on. We finish the chapter with an optional GUI-track section that describes how to enter and run a graphical user interface (GUI) program.

1.2 Hardware Terminology

A *computer system* is made up of all the components that are necessary for a computer to operate and the connections between those components. There are two basic categories of components—*hardware* and *software*. Hardware consists of the physical components associated with a computer. Software consists of the programs that tell a computer what to do. For now, let's focus on hardware.

Our description of a computer's hardware provides you with the information you'll need as a beginning programmer. (A *programmer* is a person who writes programs.) After you master the material here, if you decide you want more, go to Webopedia's hardware web page at http://www.webopedia.com/hardware.

The Big Picture

Figure 1.1 shows the basic hardware components in a computer system. It shows input devices at the left (keyboard, mouse, and scanner), output devices at the right (monitor and printer), storage devices at the bottom, and the central processing unit (CPU) and main memory in the center. The arrows in Figure 1.1 represent connections between the components. For example, the arrow from the keyboard to the CPU-main memory represents a cable (a connecting wire) that transmits information from the keyboard to the CPU and main memory. Throughout this section, we explain the CPU, main memory, and all the devices in Figure 1.1.

Input and Output Devices

Input and output devices are collectively called *I/O devices.* There are different definitions of an *input device,* but usually the term refers to a device that transfers information into a computer. Remember—information going into a computer is input. For example, a keyboard is an input device because when a person presses a key, the keyboard sends information into the computer (it tells the computer which key was pressed).

There are different definitions of an *output device,* but usually the term refers to a device that transfers information out of a computer. Remember—information going out of a computer is output. For example, a *monitor* (also called a *display* or a *screen*) is an output device because it displays information going out of the computer.

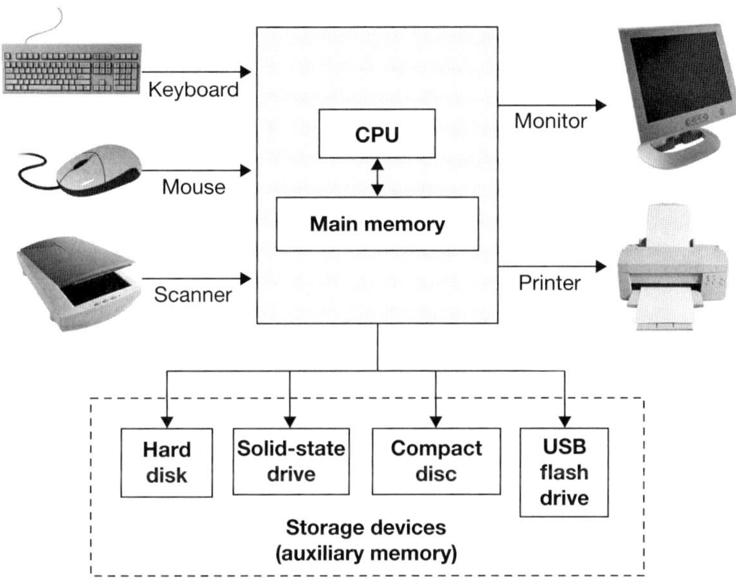

Figure 1.1 A simplified view of a computer

Central Processing Unit

The *central processing unit,* often referred to as the *processor* or *microprocessor,* can be considered the computer's brain. As with a biological brain, the CPU splits its time between two basic activities— thinking and managing the rest of its system. The "thinking" activities occur when the CPU reads a program's instructions and executes them. The "managing its system" activities occur when the CPU transfers information to and from the computer system's other devices.

Here's an example of a CPU's thinking activities. Suppose you have a program that keeps track of a satellite's position in its orbit around the Earth. Such a program contains quite a few mathematical calculations. The CPU performs those mathematical calculations.

Here's an example of a CPU's managing-its-system activities. Suppose you have a job application program. The program displays boxes in which a person enters his/her name, phone number, and so on. After entering information, the person uses his/her mouse and clicks a Done button. For such a program, the CPU manages its system as follows. To display the initial job application form, the CPU sends information to the monitor. To gather the person's data, the CPU reads information from the keyboard and mouse.

If you're thinking about buying a computer, you'll need to judge the quality of its components. To judge the quality of its components, you need to know certain component details. For CPUs, you should know the popular CPUs and the range of typical CPU speeds. We present the following CPUs and CPU speeds with hesitation because such things change in the computer world at a precipitous rate. By presenting such details, we're dating our book mercilessly. Nonetheless, we forge ahead. . . .

At the time of this second edition's writing:

- Popular CPUs—Core i7 (manufactured by Intel), Phenom II (manufactured by AMD).
- Current CPU speeds—anywhere from 2.5 GHz up to 3.8 GHz.

What is *GHz,* you ask? GHz stands for *gigahertz* and is pronounced with hard *g*'s, as in *giggle. Giga* means billion, and *hertz* is a unit of measure that deals with the number of times that something occurs per second. A 2.5 GHz CPU uses a clock that ticks 2.5 billion times per second. That's fast, but a 3.8 GHz CPU is even faster—it uses a clock that ticks 3.8 billion times per second. A CPU's clock speed provides a rough measure for how fast the CPU gets things done. Clock ticks are the initiators for computer tasks. With more clock ticks per second, there are more opportunities for getting tasks done.

Main Memory

When a computer executes instructions, it often needs to save intermediate results. For example, in calculating the average speed for 100 speed measurements, the CPU needs to calculate the sum of all the speed values prior to dividing by the number of measurements. The CPU calculates the sum by creating a storage area for it. For each speed value, the CPU adds the value to the sum storage area. Think of memory as a collection of storage boxes. The sum is stored in one of memory's storage boxes.

There are two categories of memory— *main memory* and *auxiliary memory.* The CPU works more closely with main memory. Think of main memory as a storage room next to the boss's office. The boss is the CPU, and he/she stores things in the storage room's storage boxes whenever the need arises. Think of auxiliary memory as a warehouse that's across the street from the boss's building. The boss uses the warehouse to store things, but doesn't go there all that often. Since auxiliary memory is considered secondary to main memory, auxiliary memory is sometimes referred to as *secondary storage.* We'll consider auxiliary memory details in the next subsection. For now, we'll focus on main memory details.

The CPU relies on main memory a lot. It's constantly storing data in main memory and reading data from main memory. With this constant interaction, it's important that the CPU and main memory are able to communicate quickly. To ensure quick communication, the CPU and main memory are physically close together. They are both constructed on *chips,* and they both plug into the computer's main circuit board, the *motherboard.* See Figure 1.2 for a picture of a motherboard, a CPU chip, and main memory chips.

Figure 1.2 Motherboard, CPU chip, and main memory chips

Main memory contains storage boxes, and each storage box contains a piece of information. For example, if a program stores our last name, Dean, it uses eight storage boxes: one for the first half of D, one for the second half of D, one for the first half of e, one for the second half of e, and so on. After storing the four letters, the program will probably need to retrieve them at some point later on. For information to be retrievable, it must have an address. An *address* is a specifiable location. A postal address uses street, city, and ZIP code values to specify a location. A computer address uses the information's position within main memory to specify a location. Main memory's first storage box is at the zero position, so we say it's at address 0. The second storage box is at the one position, so we say it's at address 1. See Figure 1.3. It shows Dean stored in memory starting at address 50,000.

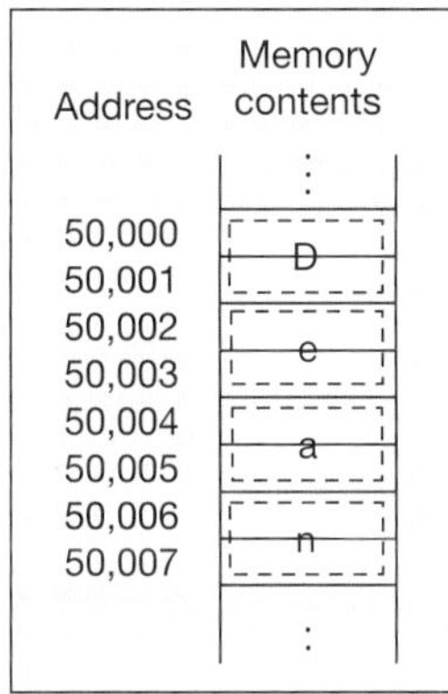

Figure 1.3 The characters D, e, a, n stored in memory starting at address 50,000

It's important to understand the formal terminology when talking about the size of main memory. Suppose you're buying a computer and you want to know how big the computer's main memory is. If you ask a salesperson how many "storage boxes" it contains, you'll probably get a perplexed look. What you need to do is ask about its *capacity*—that's the formal term for its size. If you ask for the main memory's capacity, the salesperson will say something like, "It's 1 *gigabyte*." You already know that *giga* means billion. A *byte* refers to the size of one storage box. So a 1 gigabyte capacity main memory holds 1 billion storage boxes.

Let's describe storage boxes in more detail. You know that storage boxes can hold characters, like the letter D. But computers aren't very smart—they don't understand the alphabet. They only understand 0's and 1's. So computers map each alphabet character to a series of sixteen 0's and 1's. For example, the letter D is 00000000 01000100. So in storing the letter D, main memory actually stores 00000000 01000100. Each of the 0's and 1's is called a *bit*. And each of the eight-bit groupings is called a *byte*.

Are you wondering why computers use 0's and 1's? Computers understand only high-energy signals versus low-energy signals. When a computer generates a low-energy signal, that's a 0. When a computer generates a high-energy signal, that's a 1.

You know that computers store characters as 0's and 1's, but did you know that computers also store numbers as 0's and 1's? Formally, we say that computers use the *binary number system*. The binary number system uses just two digits, 0 and 1, to represent all numbers. For example, computers store the number 19 as 32 bits, 00000000 00000000 00000000 00010011. The reason those 32 bits represent 19 is that each 1-value bit represents a power of 2. Note that there are three 1-value bits. As shown here, the 1-value bits are at positions 0, 1, and 4, where the positions start at 0 from the right side. A bit's position determines its power of 2. Thus, the rightmost bit, at position 0, represents 2 raised to the power 0, which is 1 ($2^0 = 1$). The bit at position 1 represents

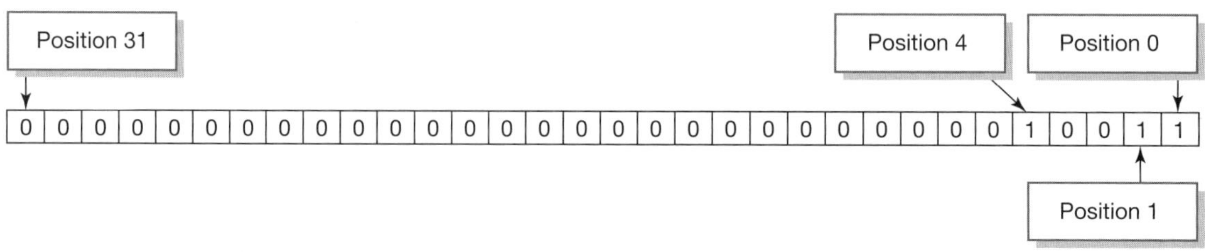

2 raised to the power 1, which is 2 ($2^1 = 2$). And the bit at position 4 represents 2 raised to the power 4, which is 16 ($2^4 = 16$). Add the three powers and you get 19 ($1 + 2 + 16 = 19$). Voila! Appendix 8 contains additional information about the binary number system. Feel free to peruse the appendix now, or wait until after you've read Chapter 12, which introduces another number system—hexadecimal. The appendix provides an in-depth discussion of the binary and hexadecimal number systems, as well as a third number system—octal.

Be aware that main memory is often referred to as *RAM.* RAM stands for *random access memory.* Main memory is considered "random access" because data can be directly accessed at any address; that is, at a "random" address. That's in contrast to some storage devices where data is accessed by starting at the very beginning and stepping through all the data until the target data is reached.

Once again, if you're buying a computer, you'll need to judge the quality of its components. For the main memory/RAM component, you'll need to know whether its capacity is adequate. At the time of this book's writing typical main memory capacities range from 2 GB up to 8 GB, where *GB* stands for *gigabyte.*

Auxiliary Memory

Main memory is *volatile,* which means that data is lost when power to the computer goes off. You might ask: If data is lost when power goes off, how can anyone save anything permanently on a computer? The answer is something you do (or should do) frequently. When you perform a save command, the computer makes a copy of the main memory data you're working on and stores the copy in auxiliary memory. Auxiliary memory is *nonvolatile,* which means that data is not lost when power to the computer goes off.

One advantage of auxiliary memory over main memory is that it's nonvolatile. Another advantage is that its cost per unit of storage is much less than main memory's cost per unit of storage. A third advantage is that it is more *portable* than main memory (i.e., it can be moved from one computer to another more easily).

The disadvantage of auxiliary memory is that its *access time* is quite a bit slower than main memory's access time. Access time is the time it takes to locate a single piece of data and make it available to the computer for processing.

Auxiliary memory comes in many different forms, the most common of which are hard disks, solid-state drives (SSDs), universal serial bus (USB) flash drives, and compact discs. All these devices are called *storage media,* or simply *storage devices.* Figure 1.4 shows pictures of them.

Hard Disks and Solid-State Drives

Hard disks and *solid-state drives* serve the same basic purpose: They provide the primary permanent storage for a computer. They have different advantages and disadvantages, which make them attractive to different types of computers. Most *desktop* computers (computers that remain stationary on or next to a desk) use hard disks. On the other hand, some *laptop* computers (computers that are portable enough to sit on

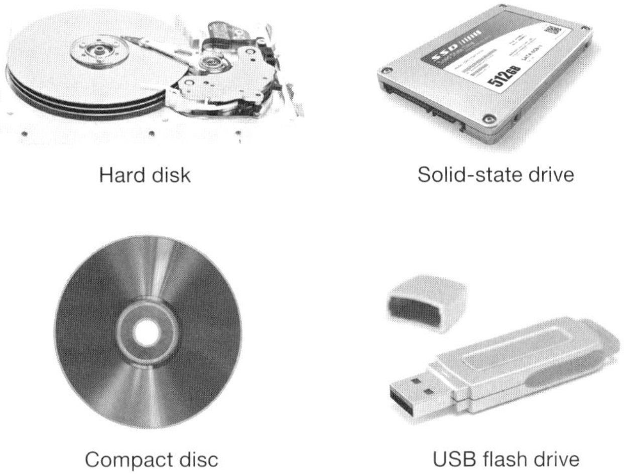

Figure 1.4 Hard-disk drive, solid-state drive, compact disc, and USB flash drive

someone's lap) and many *tablet* computers (computers that use a touch screen for their primary input device, rather than a keyboard and mouse) use solid-state drives instead of hard disks.

Solid-state drives are particularly suitable for laptop and tablet computers because they have no moving mechanical parts. As such, they are more resistant to damage when subject to travel. Also, they are smaller and lighter than hard disks, which once again makes them particularly suitable for laptop and tablet computers. A disadvantage of solid-state drives is their cost: Given a solid-state drive and a hard disk device with the same capacity, the solid-state drive will be quite a bit more expensive. Thus, solid-state drives are usually cost prohibitive for computers with large capacity needs. Desktop computers fall into the large-capacity camp, so they usually use hard disks. The trend is for portable computers to rely on *cloud storage,* where computers transmit their data over the Internet for storage on pools of computers hosted by third-party data centers. With cloud storage, portable computers' local storage needs are reduced and solid-state drives become more affordable.

Although some high-end desktop computers use SSDs, most use hard disks. Hard disks are slower than SSDs because for a computer to access a particular piece of data on a disk, the computer must wait for the disk to spin the data to the place where the data can be read. The spinning and reading mechanisms are part of the disk's *drive.* As a disk drive rotates its disks, its *heads* (electronic sensors) access the disks' data as it spins past.

Access to solid-state-drive storage is faster than access to hard disk storage because solid-state drives don't have to wait for mechanical parts to move. Instead, they simply have to wait for electronic signals to arrive. Electronic signals, traveling at the speed of light, move much faster than spinning disks.

Off-line Storage

Although hard disk and solid-state drives are sometimes located outside a computer's metal case and connected to the computer with a cable (such drives are called *external drives*), in the interest of speed, most hard disk and solid-state drives are located inside a computer's metal case. Their internal location makes it difficult for them to be transferred from one computer to another. On the other hand, *off-line storage devices,* such as USB flash drives and compact discs, can be easily transferred from one computer to another because they are designed to connect and disconnect to and from a computer easily.

USB flash drives, also called *thumb drives,* are particularly portable because they are the size of a person's thumb and they can be *hot swapped* into virtually any computer. (Hot swapping is when you plug a device into a computer while the computer is on.) The "USB" in USB flash drive stands for "universal serial bus," and it refers to a particular type of connection to the computer. More specifically, it refers to a particular type of connection wire and connection socket. A flash drive uses that type of connection, and we therefore call it a *USB flash drive.* USB flash drives plug into USB ports, where *port* is the formal term for a connection socket. USB ports are ubiquitous on computers, and that is another reason that USB flash drives are particularly portable.

USB flash drives are built with *flash memory,* which is a popular form of nonvolatile storage with no moving mechanical parts. Solid-state drives also use flash memory. However, USB flash drives are much slower than solid-state drives (and slightly slower than hard disks) because USB flash drives are connected to the computer with a narrow, relatively slow USB interface, whereas solid-state drives are connected to the rest of the computer with a wide, relatively fast interface.

Compact discs provide a less expensive and slower form of off-line storage. The most popular types of compact discs can be grouped as follows:

- CD-Audio—for storing recorded music, usually referred to as just "CD" (for compact disc)
- CD-ROM, CD-R, CD-RW—for storing computer data and recorded music
- DVD, DVD-R, DVD-RW—for storing video, computer data, and recorded music
- Blu-ray—for storing video and computer data, designed to replace DVDs

The "ROM" in CD-ROM stands for "read-only memory." *Read-only* memory refers to memory that can be read from, but not written to. Thus, you can read a CD-ROM, but you can't change its contents. With CD-Rs, you can write once and read as many times as you like. With CD-RWs, you can write and read as often as you like.

DVD stands for "digital versatile disc" or "digital video disc." DVDs parallel CD-ROMs in that you can read from them, but you can't write to them. Likewise, DVD-Rs and DVD-RWs parallel CD-Rs and CD-RWs in terms of their reading and writing capabilities.

Blu-ray, also known as Blu-ray Disc (BD), is the name of an optical disc format that was invented to store high-definition videos and large amounts of data. The technology is called Blu-ray because unlike DVDs, which use a red laser, Blu-ray discs are accessed using a blue-violet laser. Eventually, Blu-ray discs will replace DVDs in terms of popularity, but not for several years, because (1) Blu-ray technology is more expensive, (2) Blu-ray technology has slower access speeds, and (3) Blu-ray players are backward compatible (so DVDs can run on Blu-ray players).

Storage Capacity Comparison

Different storage devices have different storage capacities. At the time of this book's writing:

- Typical hard disks have a capacity range from 250 GB up to 3 TB (*TB* stands for *terabyte,* where *tera* is 1 trillion).
- Typical solid-state drives have a capacity range from 120 GB up to 512 GB.
- Typical USB flash drives have a capacity range from 8 GB up to 64 GB.
- Typical CD-ROMs, CD-Rs, and CD-RWs have a capacity of 700 MB (*MB* stands for *megabyte,* where *mega* is 1 million).
- Typical DVDs, DVD-Rs, and DVD-RWs have a capacity range from 4.7 GB up to 8.5 GB.
- Typical Blu-ray discs have a capacity range from 25 GB up to 50 GB.

File Access

To access data on your computer, you'll need to access the file that contains the data. A *file* is a group of related instructions or a group of related data. For example, (1) a program is a file that holds a set of instructions, and (2) a Word document is a file that holds text data created by Microsoft Word.

Files are stored on auxiliary memory storage devices. In order to retrieve a file (for the purpose of viewing, copying, etc.), you need to specify the storage device on which the file is stored. On computers that use Microsoft Windows, the different storage devices are specified using a drive letter followed by a colon. If your computer has a hard disk drive or a solid-state drive, your computer will refer to one of the drives using drive letter C (`C:`). If your computer has additional hard disk drives or solid-state drives, it will refer to them using subsequent drive letters (`D:`, `E:`, etc.). If your computer has compact-disc drives, it will refer to them using the first unused drive letters starting no earlier than `D:`. If your computer has additional storage devices, such as external hard drives and USB flash drives, it will refer to them using the next unused drive letters, starting no earlier than `D:`.

You might have noticed that drive letters A and B were not mentioned in this discussion so far. In the past, `A:` and `B:` were used for floppy disk drives. *Floppy disks* (also called *diskettes*) are off-line storage devices that were very popular from the mid-1970s through about 2005. They are called "floppy" because the original forms of these devices would bend, in contrast to hard disks, which do not bend. Computer manufacturers no longer provide floppy disk drives because floppy disks have been superseded by more durable, greater capacity off-line storage devices, such as USB flash drives and compact discs.

Even though floppy disks are no longer used, their legacy lives on. Because Windows-based computers reserved drive letters A and B for floppy disk drives in the past, Windows-based computers continue to start with drive letter C for hard-disk and solid-state drives. Because floppy disks became synonymous with file storage in the 1980s and 1990s, software manufacturers introduced floppy disk icons that, when clicked, would save the user's current file. Using a floppy disk icon for a file-save operation is still the norm today. This standard floppy disk icon should look familiar:

Common Computer-Hardware Vocabulary

When buying a computer or when talking about computers with your computer friends, you'll want to make sure to understand the vernacular—the terms that people use in everyday speech as opposed to the terms found in textbooks—so that you will be able to understand what's going on. When a computer-savvy person refers to a computer's memory by itself, the person typically means main memory—the computer's RAM. When someone refers to a computer's *disk space,* the person typically means the capacity of the computer's hard disk.

Pace of Computer Improvements

For as long as memory and CPU components have been around, manufacturers of these devices have been able to improve their products' performances at a consistently high rate. For example, RAM and hard disk capacities double approximately every two years. CPU speeds also double approximately every two years.

An *urban legend* is a story that spreads spontaneously in various forms and is popularly believed to be true. The following exchange is a classic Internet urban legend that comments on the rapid pace of computer improvements.[1] Although the exchange never took place, the comments, particularly the first one, are relevant.

[1]Snopes.com, *Rumor Has It,* on the Internet at http://www.snopes.com/humor/jokes/autos.asp (visited in September, 2012).

At a COMDEX computer exposition in the 1990's, Bill Gates reportedly compared the computer industry with the auto industry and stated, "If GM had kept up with the technology like the computer industry has, we would all be driving $25.00 cars that got 1,000 miles to the gallon."

In response to Bill's comments, General Motors issued a press release stating:

If GM had developed technology like Microsoft, we would all be driving cars with the following characteristics:

1. For no reason whatsoever, your car would crash twice a day.
2. Every time they repainted the lines in the road, you would have to buy a new car.
3. Occasionally your car would die on the freeway for no reason. You would just accept this, restart, and drive on.
4. Apple would make a car powered by the sun, reliable, five times as fast, and twice as easy to drive—but it would run on only 5 percent of the roads.
5. The oil, water temperature, and alternator warning lights would be replaced by a single "This Car Has Performed an Illegal Operation" warning light.
6. Occasionally, for no reason whatsoever, your car would lock you out and refuse to let you in until you simultaneously lifted the door handle, turned the key, and grabbed the radio antenna.
7. The airbag system would ask "Are you sure?" before deploying.

1.3 Program Development

As mentioned earlier, a program is a set of instructions that can be used to solve a problem. Often, a program contains many instructions, and the instructions are rather complicated. Therefore, developing a successful program requires some effort. It requires careful planning, careful implementation, and ongoing maintenance. Here is a list of typical steps involved in the program development process:

- Requirements analysis
- Design
- Implementation
- Testing
- Documentation
- Maintenance

Requirements analysis is determining the program's needs and goals. *Design* is writing a rough outline of the program. *Implementation* is writing the program itself. *Testing* is verifying that the program works. *Documentation* is writing a description of what the program does. *Maintenance* is making improvements and fixing errors later on. The steps are ordered in a reasonable sequence in that you'll normally perform requirements analysis first, design second, and so on. But some of the steps should be performed throughout the development process rather than at one particular time. For example, you should work on the documentation step throughout the development process, and you should work on the testing step during and after the implementation step and also after the maintenance step. Be aware that you'll often need to repeat the sequence of steps as needs arise. For example, if one of the program's goals changes, you'll need to repeat all of the steps in varying degrees.

We discuss the requirements analysis step and the design step in this section. We discuss the design step in detail in Chapter 2, and we illustrate it with examples throughout the book. We discuss the implementation step in this chapter's "Source Code" section, and we illustrate it with examples throughout the book. We discuss the testing step in Chapter 8. We discuss the documentation step starting in Chapter 3 and illustrate it with examples throughout the book. We discuss the maintenance step in Chapter 8 and illustrate it with examples throughout the book.

Requirements Analysis

The first step in the program development process is a requirements analysis, where you determine the needs and goals of your program. It's important that the programmer thoroughly understands the customer's wishes. Unfortunately, it's all too common for a programmer to produce a program, only to find out later that the customer wanted something different. This unfortunate circumstance can often be blamed on imprecise communication between the customer and the programmer at the beginning of the project. If a customer and programmer rely solely on a verbal description of the proposed solution, it's easy to omit important details. Later on, those omitted details can be a problem when the customer and programmer realize that they had different assumptions about how the details would be implemented.

To aid the up-front communication process, the customer and programmer should create *screen shots* of data-entry screens and output reports. A screen shot is a picture of what the computer screen looks like. To create screen shots, you can write short programs that print data-entry screens with hypothetical input, and you can write short programs that print reports with hypothetical results. As a quicker alternative, you can create screen shots with the help of drawing software or, if you're a decent artist, with pencil and paper.

Program Design

After the requirements analysis step, the second step is program design, where you write a draft of your program and focus on the basic logic, not the wording details. More specifically, you write instructions that are coherent and logically correct, but you don't worry about missing minor steps or misspelling words. That sort of program is referred to as an *algorithm*. For example, a cake recipe is an algorithm. It contains instructions for solving the problem of baking a cake. The instructions are coherent and logically correct, but they don't contain every minor step, like covering your hands with pot holders prior to removing the cake from the oven.

Pseudocode

In writing an algorithm, you should focus on organizing the flow of the instructions, and you should try to avoid getting bogged down in details. To facilitate that focus, programmers often write an algorithm's instructions using *pseudocode*. Pseudocode is an informal language that uses regular English terms to describe a program's steps. With pseudocode, precise computer *syntax* is not required. Syntax refers to the words, grammar, and punctuation that make up a language. Pseudocode syntax is more forgiving: Pseudocode must be clear enough so that humans can understand it, but the words, grammar, and punctuation don't have to be perfect. We mention this leniency in order to contrast it with the precision required for the next phase in a program's development. In the next section, we'll cover the next phase, and you'll see that it requires perfect words, grammar, and punctuation.

Example—Using Pseudocode to Find Average Miles Per Hour

Suppose you are asked to write an algorithm that finds the average miles per hour value for a given car trip. Let's step through the solution for this problem. To determine the average miles per hour, you'll need to divide the total distance traveled by the total time. Let's assume that you have to calculate the total distance from two given locations. To determine the total distance, you'll need to take the ending-point location, called "ending location," and subtract the starting-point location, called "starting location," from it. Let's assume that you have to calculate the total time in the same manner, subtracting the starting time from the ending time. Putting it all together, the pseudocode for calculating average miles per hour looks like this:

Calculate ending location minus starting location.

Put the result in total distance.

Calculate ending time minus starting time.

Put the result in total time.

Divide total distance by total time.

At this point, some readers might want to learn about a relatively advanced form of program development—object-oriented programming, or OOP as it's commonly called. OOP is the idea that when you're designing a program, you should first think about the program's components (objects) rather than the program's tasks. You don't need to learn about OOP just yet, and you're not properly prepared to learn about OOP implementation details, but if you're interested in a high-level overview, you can find it in Chapter 6, Section 2.

1.4 Source Code

In the early stages of a program's development, you write an algorithm using pseudocode. Later, you translate the pseudocode to *source code*. Source code is a set of instructions written in a programming language.

Programming Languages

A *programming language* is a language that uses specially defined words, grammar, and punctuation that a computer understands. If you try to run pseudocode instructions on a computer, the computer won't understand them. On the other hand, if you try to run programming language instructions (i.e., source code) on a computer, the computer will understand them.

Just as there are many spoken languages in the world (English, Chinese, Hindi, etc.), there are many programming languages as well. Some of the more popular programming languages are Visual Basic, C++, and Java. Each programming language has its own set of syntax rules. In this book, we'll focus on the Java programming language. If you write your program in Java, you must follow Java's syntax rules precisely in terms of words, grammar, and punctuation. If you write Java source code using incorrect syntax (e.g., you misspell a word or forget necessary punctuation, like a semicolon), and you try to run such source code on a computer, the computer won't be able to understand it.

Example—Using Java to Find Average Miles Per Hour

Continuing with the earlier example where you wrote pseudocode to find the average miles per hour value for a given car trip, let's now translate the pseudocode into Java source code. In the following table, the pseudocode at the left translates into the Java source code at the right.

Programmers normally refer to Java source code instructions as Java *statements*. For Java statements to work, they must use precise syntax. For example, as shown above, Java statements must (1) use a - for

Pseudocode	Java Source Code
Calculate ending location minus starting location. Put the result in total distance.	`distanceTotal = locationEnd - locationStart;`
Calculate ending time minus starting time. Put the result in total time.	`timeTotal = timeEnd - timeStart;`
Divide total distance by total time.	`averageMPH = distanceTotal / timeTotal;`

subtraction, (2) use a / for division, and (3) end with a semicolon. The precision required by Java statements contrasts with the flexibility of pseudocode. Pseudocode allows any syntax, so long as it is understandable by a person. For example, in pseudocode, it would be acceptable to represent subtraction with a - or the word "subtract." Likewise, it would be acceptable to represent division with a / or a ÷ or the word "divide."

Skipping the Pseudocode Step

Initially, programming language code will be harder for you to understand than pseudocode. But after gaining experience with a programming language, you may become so comfortable with it that you're able to skip the pseudocode step entirely and go right to the second step, where you write the program using programming language code.

For larger programs, we recommend that you do not skip the pseudocode step. Why? Because with larger programs, it's important to first focus on the big picture because if you don't get that right, then nothing else matters. And it's easier to focus on the big picture if you use pseudocode where you're not required to worry about syntax details. After implementing a pseudocode solution, it's relatively easy to convert the pseudocode to source code.

1.5 Compiling Source Code into Object Code

After writing a program, you'll want to have a computer perform the tasks specified by the program. Getting that to work is normally a two-step process: (1) Perform a compile command. (2) Perform a run command. When you perform a *compile* command, you tell the computer to translate the program's source code to code that the computer can run. When you perform a *run* command, you tell the computer to run the translated code and perform the tasks specified by the code. In this section, we describe the translation process.

The computer contains a special program called a *compiler* that's in charge of the translation process. If you submit source code to a compiler, the compiler translates it to code that the computer can run. More formally, the compiler compiles the source code and produces *object code* as the result.[2] Object code is a set of binary-format instructions that can be directly run by a computer to solve a problem. An object-code instruction is made up of all 0's and 1's because computers understand only 0's and 1's. Here's an example of an object-code instruction:

```
0100001111101010
```

[2]Most compilers produce object code, but not all. As you'll see in the next section, Java compilers produce an intermediate form of instructions. At a later time, that intermediate form of instructions is translated into object code.

This particular object-code instruction is referred to as a *16-bit instruction* because each of the 0's and 1's is called a *bit,* and there are 16 of them. Each object-code instruction is in charge of only a simple computer task. For example, one object-code instruction might be in charge of copying a single number from some place in main memory to some place in the CPU. There's no need for general-purpose computer programmers to understand the details of how object code works. That's the computer's job, not the programmer's job.

Programmers sometimes refer to object code as *machine code.* Object code is called machine code because it's written in binary, and that's what a computer "machine" understands.

1.6 Portability

In Section 1.2's "Auxiliary Memory" subsection, we said that auxiliary memory is more portable than main memory because it can be moved from one computer to another more easily. In that context, portability referred to hardware. Portability can also refer to software. A piece of software is *portable* if it can be used on many different types of computers.

Portability Problem with Object Code

Object code is not very portable. As you now know, object code is comprised of binary-format instructions. Those binary-format instructions are intimately tied to a particular type of computer. If you have object code that was created on a type X computer, then that object code can run only on a type X computer. Likewise, if you have object code that was created on a type Y computer, then that object code can run only on a type Y computer.[3]

So what's all the fuss about portability? Who cares that object code is not very portable? Software manufacturers care. If they want to sell a program that runs on different computer types, they typically have to compile their program on the different computer types. That produces different object-code files, and they then sell those files. Wouldn't it be easier if software manufacturers could provide one form of their program that runs on all types of computers?

Java's Solution to the Portability Problem

The inventors of Java attempted to address the inherent lack of portability in object code by introducing the *bytecode* level between the source code and object code levels. Java compilers don't compile all the way down to object code. Instead, they compile down to bytecode, which possesses the best features of both object code and source code:

- Like object code, bytecode uses a format that works closely with computer hardware, so it runs fast.
- Like source code, bytecode is generic, so it can be run on any type of computer.

How can bytecode be run on any type of computer? As a Java program's bytecode runs, the bytecode is translated into object code by the computer's bytecode interpreter program. The bytecode interpreter program is known as the *Java virtual machine,* or *JVM* for short. Figure 1.5 shows how the JVM translates bytecode to object code. It also shows how a Java compiler translates source code to bytecode.

[3]There are about 15 or so different computer types that are in common use today. Those 15 computer types correspond to 15 categories of CPUs. Each CPU category has its own distinct *instruction set.* An instruction set defines the format and meanings of all the object-code instructions that work on a particular type of CPU. A full discussion of instruction sets is beyond the scope of this book. If you'd like to learn more, see Wikipedia's website at http:// www.wikipedia.org/ and enter "instruction set" in the search box.

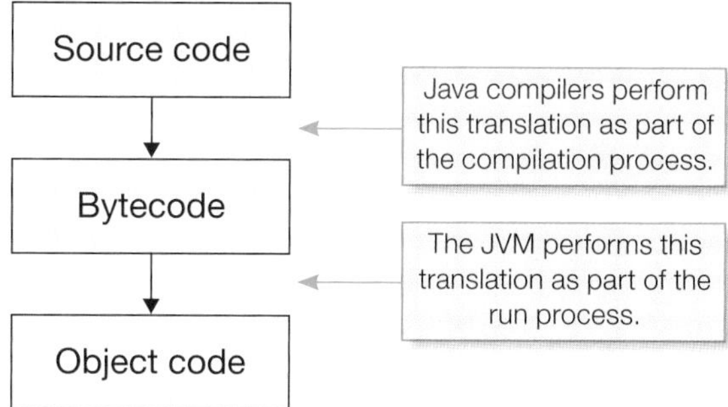

Figure 1.5 How a Java program is converted from source code to object code

To run Java bytecode, a computer must have a JVM installed on it. Fortunately, installing a JVM is straightforward. It's a small program, so it doesn't take up much space in memory. And it's easy to obtain—anyone can download a JVM for free from the Internet. In Section 1.9, we explain how to download a JVM and install it on your own computer.

Why Is the Bytecode Interpreter Program Called a "Java virtual machine"?

We'll now explain the origin of the name "Java virtual machine." For programs written with most programming languages, the CPU "machine" runs the program's compiled code. For programs written in Java, the bytecode interpreter program runs the program's compiled code. So with Java, the bytecode interpreter program acts like a CPU machine. But the bytecode interpreter is just a piece of software, not a piece of hardware like a real CPU. Thus, it's a virtual machine. And that's why Java designers decided to call the bytecode interpreter program a Java virtual machine.

1.7 Emergence of Java

Home-Appliance Software

In the early 1990s, putting intelligence into home appliances was thought to be the next "hot" technology. Examples of intelligent home appliances include coffee pots controlled by a computer and televisions controlled by an interactive programmable device. Anticipating a strong market for such items, Sun Microsystems in 1991 funded a team of researchers to work on the secretive "Green Project," whose mission was to develop software for intelligent home appliances.

An intelligent home appliance's intelligence comes from its embedded processor chips and the software that runs on those processor chips. Appliance processor chips change often because engineers continually find ways to make them smaller, less expensive, and more powerful. To accommodate the frequent turnover of new chips, the software that runs on them should be extremely flexible.

Originally, Sun planned to use C++ for its home-appliance software, but it soon realized that C++ wasn't sufficiently portable. Rather than write C++ software and fight C++'s inherent portability problems, Sun decided to develop a whole new programming language for its home-appliance software.

Sun's new language was originally named Oak (for the tree that was outside project leader James Gosling's window), but it turned out that Oak was already being used as the name of another programming language. As the story goes, while a group of Sun employees was on break at a local coffee shop, they came up with the name "Java." They liked the name "java" because of the significant role caffeine plays in the lives of software developers. ☺

World Wide Web

When the market for intelligent home-appliance software proved to be less fertile than anticipated, Sun almost pulled the plug on its Java project during the prerelease development phase. Fortunately for Sun (and for all of today's Java lovers), the World Wide Web exploded in popularity. Sun realized that the Web's growth could fuel demand for a language like Java, so Sun decided to continue with its Java development efforts. Those efforts bore fruit when they presented Java's first release at the May 1995 SunWorld Conference. Soon thereafter, Netscape, the world's most popular browser manufacturer at the time, announced its intention to use Java in its browser software. With support from Netscape, Java started with a bang and it's been going strong ever since.

The Web relies on web pages being downloaded and run on many different types of computers. To work in such a diverse environment, web page software must be extremely portable. You're probably thinking, Java to the rescue! Actually, that would be a bit of an exaggeration. The Web didn't need rescuing—the Web was doing reasonably well even before Java came into the picture, thank you very much. But Java was able to add some much-needed functionality to plain old web pages.

Prior to Java, web pages were limited to one-way communication with their users. web pages sent information to users, but users did not send information to web pages. More specifically, web pages displayed information for users to read, but users did not enter data for web pages to process. When the Web community figured out how to embed Java programs inside web pages, that opened the door to more exciting web pages. Java-embedded web pages were able to read and process user input, and that provided users with a more enjoyable, interactive experience.

Java Today

Today, programmers use Java in many different environments. Java programs in web pages, called *applets*, helped Java grow into one of the leading programming languages in the world. Although applets are fast and versatile, their versatility creates security problems and some of the other types of web-page programs are beginning to exhibit comparable performance and versatility without concomitant security issues.

To help with the small talk at your next Java social event, we'll provide brief descriptions of some of the more popular uses for Java. An applet is a Java program that's embedded in a web page. A *servlet* is a Java program that supports a web page, but it runs on a different computer than the web page. A *Java Server Page* (JSP) is a web page that has fragments of a Java program (as opposed to a complete Java program, like an applet) embedded in it. An advantage of servlets and JSPs over applets is that servlets and JSPs lead to web pages that display more quickly. Servlets and JSPs are written with support from the *Java Enterprise Edition (Java EE)* platform. Java EE provides an underlying framework of software that enables programmers to write secure, large-scale programs that rely on various technologies such as servlets, JSPs, and databases. (A *database* is a collection of data organized in a manner such that its data can be retrieved, updated, or sorted easily.) A *Java Micro Edition (Java ME)* application is a Java program that runs on a limited-resource device. The most prominent example of a limited-resource device is a mobile phone, which is constrained in terms of memory capacity and power. A *Java Standard Edition (Java SE)* application is a Java program that runs on a standard computer—a desktop or a laptop. In this book, we focus on Java SE applications, as opposed to the other types of Java programs, because Java SE applications are the most general purpose and they provide the best environment for learning programming concepts.

In January 2010, Oracle Corporation acquired Sun Microsystems for a purchase price of $7.4 billion. With the acquisition, Oracle now oversees the development of the Java programming language. According to chief executive officer Larry Ellison, owning Java helps Oracle with the fastest growing part of its business—Java middleware for its Oracle Database software.[4] *Middleware* is software that connects two other software components that would normally have a hard time communicating with each other. In Oracle's case, its middleware connects its database software with user-interface software. That leads to database software that's easier to use, and ease of use leads to happier customers. With Oracle's Fusion middleware written in Java, Oracle is firmly committed to Java's success.

1.8 Computer Ethics

Before you learn how to write your first Java program in the next section, you might want to take a moment to ponder the ethical and social implications of programming and using computer software. Such ponderings are part of a large field of study—computer ethics. We don't pretend that our book will teach you all about computer ethics, but you should at least be aware that the field exists. As you read through this book, if something sparks your interest about a computer ethics issue, we encourage you to do a Google search and see what you can find. Or check out the Computer Ethics Institute's website at http://computerethicsinstitute.org or the Computer Professionals for Social Responsibility's website at http://cpsr.org.

In 1992, the Computer Ethics Institute created the "Ten Commandments of Computer Ethics." In a nod to the original Ten Commandments (from the Bible and the Torah), the Computer Ethics Institute's commandments use the archaic "Thou shalt" and "Thou shalt not" format. Here are the commandments.[5]

1. Thou shalt not use a computer to harm other people.
2. Thou shalt not interfere with other people's computer work.
3. Thou shalt not snoop around in other people's computer files.
4. Thou shalt not use a computer to steal.
5. Thou shalt not use a computer to bear false witness.
6. Thou shalt not copy or use proprietary software for which you have not paid.
7. Thou shalt not use other people's computer resources without authorization or proper compensation.
8. Thou shalt not appropriate other people's intellectual output.
9. Thou shalt think about the social consequences of the program you are writing or the system you are designing.
10. Thou shalt always use a computer in ways that ensure consideration and respect for your fellow humans.

Although certainly not as entrenched as the Golden Rule ("He who has the gold makes the rules"), these commandments are generally well regarded. Note that the commandments suggest behaviors for individual people. Another way to look at computer ethics is to focus less on individual behaviors and more on holistic goals. For example, the goal of *system reliability* is to ensure that software does what it's supposed to do. The goal of *privacy* is to enable a person or a group of people to hide personal information from the outside world. The goal of *intellectual property protection* is to enable a person or a group of people to protect creations of the mind (e.g., inventions, literary works, or software) so that outsiders are unable to claim them as their own.

[4] Patrick Thibodeau, "Oracle's Sun buy: Ellison praises Solaris, thumbs nose at IBM," *ComputerWorld*, April 20, 2009.

[5] Barquin, Roman, In Pursuit of a *"Ten Commandments" for Computer Ethics*, Computer Ethics Institute, 1992.

1.9 First Program—Hello World

Earlier in this chapter, you learned what it means to compile and run a Java program. But learning by reading only goes so far. It's now time to learn by doing. In this section, you'll enter a Java program into a computer, compile the program, and run it. What fun!

Development Environments

There are different ways to enter a Java program into a computer. You can use an integrated development environment, or you can use a plain text editor. We'll briefly describe the two options.

An *integrated development environment* (IDE) is a rather large piece of software that allows you to enter, compile, and run programs. The entering, compiling, and running are all part of a program's development, and those three functions are integrated together into one environment. Thus, the name "integrated development environment." Some IDEs are free and some are quite expensive. We provide tutorials for several popular IDEs on the book's website.

A *plain text editor* is a piece of software that allows you to enter text and save your text as a file. Plain text editors know nothing about compiling or running a program. If you use a plain text editor to enter a program, you'll need to use separate software tools to compile and run your program. Note that *word processors,* like Microsoft Word, can be called text editors, but they're not plain text editors. A word processor allows you to enter text and save your text as a file. But the saved text is not "plain." When a word processor saves text to a file, it adds hidden characters that provide formatting for the text like line height, color, etc. And those hidden characters create problems for Java programs. If you attempt to enter a program into a computer using a word processor, your program won't compile successfully, and it certainly won't run.

Different types of computers have different plain text editors. For example, computers that use Windows have a plain text editor called Notepad. Computers that use UNIX or Linux have a plain text editor called vi. Computers that use Mac OS X have a plain text editor called TextEdit. Note: Windows, UNIX, Linux, and Mac OS X are *operating systems.* An operating system is a collection of programs whose purpose is to help run the computer system. In running the computer system, the operating system manages the transfer of information between computer components, like a traffic cop.

For the rest of this section, we'll describe how you can enter, compile, and run a program using free, bare-bones tools. You'll use a plain text editor for entering your program, and you'll use simple software tools from Oracle for compiling and running your program. If you have no interest in using such bare-bones tools, and you prefer instead to stick exclusively with an IDE, then refer to the IDE tutorials on the book's website and feel free to skip the rest of this section. If you're unsure what to do, we encourage you to try out the bare-bones tools. They're free and they don't require as much memory as the IDEs. They serve as a standard baseline that you should be able to use on almost all computers.

Entering a Program into a Computer

We'll now describe how you can enter a program into a computer using Notepad, the plain text editor that comes with all versions of Microsoft Windows.

Move your mouse cursor on top of the Start button at the bottom-left corner of your Windows desktop. Click the Start button. (When we ask you to "click" an item, we want you to move your mouse on top of the item and press the left mouse button.) That should cause a menu to appear. On the menu, move your mouse on top of the Programs option. That should cause another menu to appear. On that menu, move your mouse on top of the Accessories option. That should cause another menu to appear. On that menu, click on the Notepad option. That should cause the Notepad text editor to appear.

```
public class Hello
{
  public static void main(String[] args)
  {
    System.out.println("Hello, world!");
  }
}
```

Figure 1.6 The Notepad text editor with the Hello World program entered into it

In the newly opened Notepad text editor, enter the source code for your first program. More specifically, click somewhere in the middle of the Notepad window and then enter the seven lines of text that are shown in Figure 1.6. When you enter the text, be sure to type the letters with uppercase and lowercase exactly as shown. Specifically, use uppercase for the `H`'s in `Hello` and the `S`'s in `String` and `System`. Use lowercase for all the other letters. Use spaces, not tabs, for indentations. Your entered text comprises the source code for what is known as the Hello World program. The Hello World program is the traditional first program for all programming students. It simply prints a hello message. In Chapter 3, we'll describe the meaning behind the words in the Hello World source code. In this chapter, we're more interested in hands-on experience, and we show you how to enter, compile, and run the Hello World program.

After entering the source code into the Notepad window, you'll need to save your work by storing it in a file. To save your source code in a file, click the File menu in the top-left corner of the Notepad window. That should cause a menu to appear. On the menu, select the Save As option. That should cause a Save As *dialog box* to appear. A dialog box is a small window that performs one task. For this dialog box, the task is to save a file.

Note the File name: box at the bottom of the dialog box. That's where you'll enter the name of your file. But first, you should create a *directory* to store your file in. A directory, also called a *folder,* is an organizational entity that contains a group of files and other directories.[6]

At the left side of your Save As dialog box, you should see a directory tree. In the directory tree, scroll down until you see a right-facing triangle (▷) with Computer on its right. Click on the triangle to display your computer's contents. Within the now-opened computer container, you should see a right-facing triangle with `C:` on its right. If you'd like to save on your `C:` drive, click on `C:`. If, instead, you'd like to save on a USB flash drive, make sure you have a USB flash drive plugged into your computer, search within your computer container for the drive letter associated with your USB flash drive, and click on that drive letter.

[6]In the Windows and Macintosh worlds, people tend to use the term *folder.* In the UNIX and Linux worlds, people tend to use the term *directory.* As you'll see in Chapter 16, Oracle uses the term *directory* as part of the Java programming language. We like to follow Oracle, and we therefore use the term *directory* rather than *folder.*

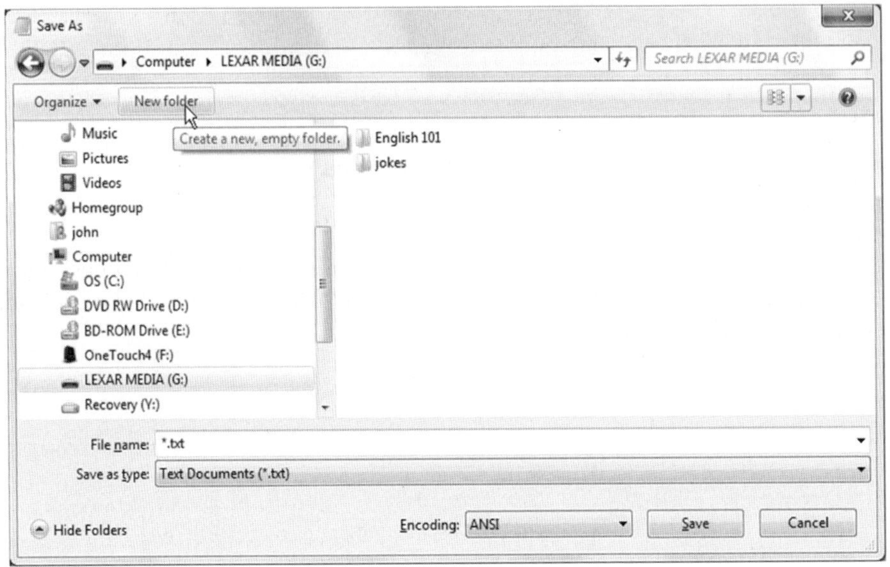

Figure 1.7 Notepad's **Save As** dialog box, with the user about to create a new folder

Clicking the drive letter should cause the drive letter to appear at the top of your **Save As** dialog box. Verify that your **Save As** dialog box now looks similar to the **Save As** dialog box in Figure 1.7. In particular, note the G: drive at the top of the **Save As** dialog box. Your drive may be different, depending on what drive letter you clicked.

As shown in Figure 1.7, move your mouse cursor over the **New Folder** button near the top-left corner of the **Save As** dialog box. Click the button. That should cause a new directory to appear in the directory tree. The name of the new directory is `New Folder` by default. The `New Folder` name should be selected and highlighted. Enter `myJavaPgms`, and it should overlay the `New Folder` name. Click the **Open** button in the bottom-right corner of the dialog box. That should cause your new `myJavaPgms` directory to appear at the top of your **Save As** dialog box.

Enter `"Hello.java"` in the **File name:** box at the bottom of the dialog box. You must enter `"Hello.java"` exactly as shown here:

Don't forget the quotes, the uppercase H, and the lowercase subsequent letters. Click the **Save** button in the bottom-right corner of the dialog box. That should cause the **Save As** dialog box to disappear, and the top of the Notepad window should now say **Hello.java**. Shut down Notepad by clicking on the X in the top-right corner of the Notepad window.

Installing a Java Compiler and the JVM

In the previous subsection, you entered the Hello World program and saved it to a file. Normally, the next step would be to compile the file. Remember what compiling is? That's when a compiler translates a source code file into a bytecode file. For our Hello World program, the compiler will translate your `Hello.java` source code file into a `Hello.class` bytecode file. If you're working in a school's computer lab, chances are pretty good that your computer already has a Java compiler installed on it. If your computer does not have a Java compiler installed on it, you'll need to install it now in order to complete the hands-on portion of this section.

Normally, if someone is interested in installing the Java compiler (to compile Java programs), they are also interested in installing the JVM (to run Java programs). To make the installation easier, Oracle bundles the Java compiler together with the JVM. Oracle calls the bundled software the *Java Development Kit,* or *JDK* for short.

To install the JDK on your computer, go to http://www.mhhe.com/dean2e and click on the Links to Compiler Software link. Then click on the Oracle's Java Download Site link. On Oracle's site, download Java SE and install the JDK as instructed.

Compiling a Java Program

We'll next describe how you can compile a program using a *command prompt window* (also called a *console*). A command prompt window allows you to enter operating system instructions where the instructions are in the form of words. The words are referred to as *commands.* For example, on a computer that runs the Windows operating system, the command for deleting a file is `del` (for delete). On a computer that runs the UNIX or Linux operating system, the command for deleting a file is `rm` (for remove).

To open a command-prompt window on a computer that runs the Windows 7 operating system, click the Start button at the bottom-left corner of your Windows desktop. That should cause a menu to appear. On the menu, click the All Programs option. That should cause another menu to appear. In the new menu, click the Accessories directory. That should cause the Accessories directory to expand and a Command Prompt option to appear. Click the Command Prompt option. That should cause a command-prompt window to appear. Figure 1.8 shows the newly opened command-prompt window.

In Figure 1.8, note this line:

```
C:\Users\john
```

That's a *prompt.* In general, a prompt tells you to do something. For a command-prompt window, the prompt tells you to enter a command. Very soon, you'll enter commands in your actual command-prompt window. But first, note the text at the left of the > symbol. The text `C:\Users\john` forms the *path* to the current directory. A path specifies the location of a directory. More specifically, a path starts with a drive letter and contains a series of one or more slash-separated directory names. In our example, `C:` refers to the hard

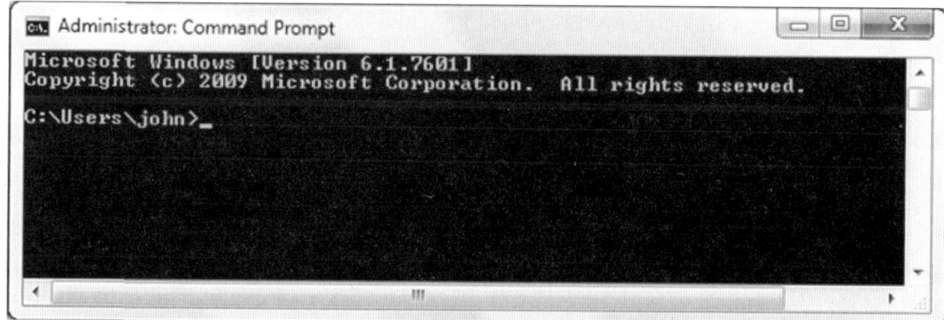

Figure 1.8 A command-prompt window when it first opens up

drive, `Users` refers to the `Users` directory that's on the hard drive, and `john` refers to the `john` directory that's contained within the `Users` directory.

To compile your Hello World program, you'll need to go first to the drive and directory where it resides. Suppose your command-prompt window's prompt indicates that your current drive is `C:`, and you saved `Hello.java` on F:. Then you'll need to change your drive to `F:`. To do so, enter `f:` in your command-prompt window.

To change to the Hello World program's directory, enter this `cd` command (*cd* stands for change directory):

```
cd \myJavaPgms
```

Now you're ready to compile your program. Enter this `javac` command (*javac* stands for java compile):

```
javac Hello.java
```

In entering that command, if your command-prompt window displays an error message, refer to Figure 1.9 for possible solutions. If your command-prompt window displays no error messages, that indicates success.

The compilation error message says something like this:	**Explanation:**
`'javac' is not recognized`	All three error messages indicate that the computer doesn't understand the `javac` command because it can't find the javac compiler program. The error is probably due to the PATH variable being set improperly. Review the JDK installation instructions and reset the PATH variable accordingly.
`javac: command not found`	
`bad command or filename`	
`Hello.java:` *number*: *text*	There is a syntax error in the `Hello.java` source code. The specified *number* provides the approximate line number in `Hello.java` where the error occurs. The specified *text* provides an explanation for the error. Review the contents of the `Hello.java` file and make sure that every character is correct and uses the proper case (lowercase, uppercase).

Figure 1.9 Compilation errors and explanations

More specifically, it indicates that the compiler created a bytecode file named `Hello.class.` To run the `Hello.class` file, enter this `java` command:

```
java Hello
```

Your command-prompt window should now display your program's output—`Hello, world!` See Figure 1.10. It shows the command-prompt window after completing the steps described previously.

1.10 GUI Track: Hello World and Look and Feel (Optional)

This section is the first installment of our optional GUI track. In each GUI-track section, we provide a short introduction to a GUI concept. For example, in this section, we describe how to display a message in a GUI window. In another GUI track section, we describe how to draw lines and shapes. For readers who do not have time for the GUI track, no problem. Any or all of the GUI track sections may be skipped, as they cover material that is independent of later material. Note that we cover hard-core GUI material in earnest in Chapters 17 and 18 at the end of the book. The GUI material in Chapters 17 and 18 is independent of the

```
Administrator: Command Prompt
Microsoft Windows [Version 6.1.7601]
Copyright (c) 2009 Microsoft Corporation.  All rights reserved.

C:\Users\john>g:

G:\>cd \myJavaPgms

G:\myJavaPgms>javac Hello.java

G:\myJavaPgms>java Hello
Hello, world!

G:\myJavaPgms>_
```

Figure 1.10 Compiling and running the Hello World program

GUI material in the GUI track, so, once again, it's OK to skip the GUI track. But why skip it? GUI programming is sooooo much fun!

GUI Hello World

In this sub-section, we present a GUI version of the Hello World program. Before examining the program's code, we'll examine its user interface. GUI programs such as the GUI Hello World program use graphical tools for their interfaces. Figure 1.11 shows alternate outputs of the GUI Hello World program. In either case, the displayed window includes these graphical tools: a title bar (the bar at the top of the window), a close-window button (the "X" in the top-right corner), an OK button, and an i icon. Here's how the tools work: If you drag the title bar with your mouse, the window moves. If you click the close-window button or the OK button, the window closes. The i icon is a visual cue that indicates the nature of the window—the i stands for "information" since the window simply displays information.

The dashed boxes in Figure 1.12 indicate code that differs from the code in the previous section's `Hello` program. For now, don't worry about the meaning of the program's code. We'll explain it later on. For now, the goal is to give you some fun and valuable hands-on experience.

Go ahead and enter the program code into a text editor. If you need a refresher on how to do that, see the previous section. This time, save your source code file with the name `HelloGUI.java` instead of

Figure 1.11 Alternative windows produced by the GUI Hello World program

```
import javax.swing.JOptionPane;

public class HelloGUI
{
  public static void main (String[] args)
  {
    JOptionPane.showMessageDialog(null, "Hello, world!");
  }
}
```

The dashed boxes indicate code that differs from the code in the previous section's `Hello` program.

Figure 1.12 GUI version of the Hello World program

`Hello.java`. When saving `HelloGUI.java`, make sure you spell the filename with capitals for H, G, U, and I since that's how `HelloGUI` is spelled in your program's third line. Next, you'll want to compile and run the program. Once again, if you need a refresher, see the previous section.

Look and Feel

Look and feel refers to the default appearance and behavior of the components within a GUI program's window. In Figure 1.11, for each of the two windows, note the background shading, the position of the OK button, the size of the i icon, and the font used for "Hello, world!" Those are examples of appearance features controlled by the program's look and feel. Imagine a user moving his or her mouse over the OK button. With some Java programs, that will cause the OK button's color to become lighter. That is an example of a behavior feature controlled by the program's look and feel.

There are many different look and feels to choose from. With the advent of Java 7, Oracle recommends its new Nimbus look and feel because of its ability to display GUI components consistently and clearly across different platforms. In the interest of maintaining compatibility with older versions of Java, Java continues to use Metal for its default look and feel. Most of the look and feel features are the same for Nimbus and Metal, but in examining the two different look and feel windows in Figure 1.11, you'll see differences in the background shading, in the position of the OK button, and in the size of the i icon. Another difference between the two windows is the behavior when the mouse moves over the OK button. With Nimbus, the OK button's color becomes lighter. With Metal, the OK button remains unchanged.

Throughout the remainder of the book, we use the Nimbus look and feel. If you'd like to change your default look and feel to Nimbus, use a simple text editor to create a file named `swing.properties` and store this line of text in it:

```
swing.defaultlaf=javax.swing.plaf.nimbus.NimbusLookAndFeel
```

For this line to work, you must enter it exactly as shown, with no spaces before swing and no spaces at the end.

Have your computer administrator (you?) save the `swing.properties` file in your *<java-home>*`\lib` directory, where *<java-home>* will be different on different types of computers. On a typical Windows computer, with the initial version of Java7, *<java-home>* will be `C:\Program Files\Java\jdk1.#.#\jre`, where `#.#` will be replaced with your computer's Java compiler version number.

Summary

- A computer system is all the components that are necessary for a computer to operate and the connections between those components. More specifically, a computer system consists of the CPU, main memory, auxiliary memory, and I/O devices.
- Programmers write algorithms as first attempt solutions for programming problems.
- Algorithms are written with pseudocode—similar to programming language code except that precise syntax (words, grammar, and punctuation) isn't required.
- Source code is the formal term for programming language instructions.
- Object code is a set of binary-encoded instructions that can be executed directly by a computer.
- Most non-Java compilers compile from source code to object code.
- Java compilers compile from source code to bytecode.
- As a Java program runs, the Java virtual machine translates the program's bytecode to object code.
- Originally, Sun developed Java for use in home appliance software.
- To expedite development, Java programmers often use integrated development environments, but you can use a plain text editor and command-prompt window.

Review Questions

§1.2 Hardware Terminology

1. What do the following abbreviations mean?
   a) I/O
   b) CPU
   c) RAM
   d) GHz
   e) MB
2. Identify two important computer input devices.
3. Identify two important computer output devices.
4. Assertions:
   a) Main memory is faster than auxiliary memory. (T / F)
   b) Auxiliary memory is volatile. (T / F)
   c) The first position in main memory is at address 1. (T / F)
   d) The CPU is considered to be a peripheral device. (T / F)
   e) Hot swapping is when you plug a device into a computer while the computer is on. (T / F)
5. Which type of storage device is normally preferred for tablet computers—hard disk or SSD? Why?

§1.3 Writing Algorithms Using Pseudocode

6. What is an algorithm?
7. What is pseudocode?

§1.4 Translating Pseudocode into Programming Language Code

8. Syntax rules are more lenient for which type of code—pseudocode or programming language code?

§1.5 Compiling Source Code into Object Code

9. What happens when you compile a program?
10. What is object code?

§1.6 Portability

11. What is a Java virtual machine?

§1.7 Emergence of Java

12. List five different types of Java programs.

Exercises

1. [after §1.2] For each of the following items, determine whether it is associated with main memory or auxiliary memory.
   a) solid-state drive main or auxiliary?
   b) RAM main or auxiliary?
   c) hard disk main or auxiliary?
   d) CD-RW main or auxiliary?

2. [after §1.2] What is a bit?

3. [after §1.2] What is a byte?

4. [after §1.2] What type of computer component does `C:` usually refer to?

5. [after §1.2] For each of the following computer system components, identify parallel components in a bear's biological system.
   a) CPU
   b) input devices
   c) output devices

6. [after §1.2] What is "Moore's Law"? You won't find the answer to the question in the book, but you can find it on the Internet. (*Hint:* Gordon Moore was one of the founders of Intel.)

7. [after §1.3] This question is not very specific, so don't worry about whether your solution conforms to some prescribed answer. Just do whatever seems reasonable to you.
   Using pseudocode in the form of short statements, provide an algorithm for a bear that describes the steps involved in gathering honey. If a certain step or a group of steps is to be repeated, use an if statement and an arrow to show the repetition. For example, your algorithm might include something like this:

```
<statement>   <------------+
<statement>                |
<statement>                |
.                          |
.                          |
<statement>                |
If still hungry, repeat ---+
```

8. [after §1.5] Humans generally prefer to work with source code rather than object code because source code is easier to understand than object code. So why is object code necessary?

9. [after §1.6] Most programming languages compile down to object code. Java compiles down to bytecode. What is the primary benefit of bytecode over object code?

10. [after §1.6] What does the Java virtual machine do?

11. [after §1.7] What was the original name for the Java programming language?

12. [after §1.8] According to BBC News, Wikileaks is a whistle-blowing website "which has established a reputation for publishing sensitive material from governments and other high-profile organizations."

Provide at least one argument in favor of the work done by WikiLeaks. Provide at least one argument in opposition to the work done by WikiLeaks.

13. [after §1.9] On a computer whose operating system is a recent version of Microsoft Windows, invoke Start > Programs > Accessories > Command Prompt. Navigate to the directory that has the `Hello.java` source code. Enter `dir Hello.*` to list all files starting with "Hello". If this list includes `Hello.class`, delete that file by entering `del Hello.class`. Enter `javac Hello.java` to compile the source code. Again enter `dir Hello.*` and verify that the bytecode file, `Hello.class`, has been created. Now you can enter `java Hello` to execute the compiled program. Enter `type Hello.java` and type `Hello.class` to get a feeling for how bytecode differs from source code.

14. [after §1.9] Experiment with the `Hello.java` program to learn the meanings of typical compilation and runtime error messages:
- **a)** Omit the final / from the header block.
- **b)** Omit any part of the argument in the parentheses after `main`.
- **c)** Omit the semicolon from the end of the output statement.
- **d)** One at a time, omit the braces—{ and }.
- **e)** Try using lowercase, $, _, or a number for the first character in the class name.
- **f)** Make the program filename different from the class name.
- **g)** Change `main` to `Main`.
- **h)** One at a time, try omitting `public`, `static`, and `void` from before `main`.

15. [after §1.9] Learn how to use TextPad by working your way through the "Getting Started with TextPad" tutorial on the book's website. Submit hardcopy of the source code for your Countdown program (i.e., print your program from within TextPad). Note that you're not required to submit source code for your Hello World program or submit output for either program.

Review Question Solutions

1. What do the following abbreviations mean?
- **a)** I/O: input/output devices
- **b)** CPU: central processing unit or processor
- **c)** RAM: random access memory or main memory
- **d)** GHz: Gigahertz = billions of cycles per second
- **e)** MB: Megabytes = millions of bytes, where 1 byte is 8 bits, and 1 bit is the answer to a single yes/no question

2. The keyboard and a mouse are the two most obvious examples of input devices. Another possible input device is a telephone modem.

3. The display screen and a printer are the two most obvious examples of important output devices. Other examples are a telephone modem and speakers.

4. Assertions:
- **a)** True. Main memory is physically closer to the processor, and the bus that connects the main memory to the processor is faster than the bus that connects the auxiliary memory to the processor. Main memory is also more expensive and therefore usually smaller.
- **b)** False. When power goes off, main memory loses its information, while auxiliary memory does not. An unexpected power failure might corrupt information in auxiliary memory, however.
- **c)** False. The first position in main memory is at address 0.
- **d)** False. The CPU is considered to be part of the computer itself; it's not a peripheral device.
- **e)** True. Hot swapping is when you plug a device into a computer while the computer is on.

5. An SSD is normally preferred for tablet computers, because it is smaller and lighter and more resistant to damage.
6. An algorithm is a step-by-step procedure for solving a problem.
7. Pseudocode is an informal language that uses regular English terms to describe a program's steps.
8. Syntax rules are more lenient for pseudocode (as opposed to programming language code).
9. Most compilers convert source code to object code. Java compilers convert source code to bytecode.
10. Object code is the formal term for binary-format instructions that a processor can read and understand.
11. A Java virtual machine (JVM) is an interpreter that translates Java bytecode into object code.
12. Five different types of Java programs are applets, servlets, JSP pages, micro edition applications, and standard edition applications.

CHAPTER 2

Algorithms and Design

Objectives

- Learn how to write an informal text description of what you want a computer program to do.
- Understand how a flowchart describes what a computer program does.
- Become familiar with the standard well-structured control patterns.
- Learn how to structure conditional executions.
- Learn how to structure and terminate looping operations, including nested loops.
- Learn how to "trace through" a program's sequence of operation.
- See how you can describe program operation at different levels of detail.

Outline

2.1 Introduction

As indicated in Chapter 1, writing a computer program involves two basic activities: (1) figuring out what you want to do and (2) writing code to do it. You might be tempted to skip the first step and jump immediately to the second step—writing code. Try to resist that urge. Jumping immediately into the code often results in bad programs that work poorly and are hard to fix because poor organization makes them hard to understand. Therefore, for all but the very simplest problems, it's best to start by thinking about what you want to do and then organize your thoughts.

As part of the organization process, you'll want to write an *algorithm*.[1] An algorithm is a sequence of instructions for solving a problem. It's a recipe. When specifying an algorithm, two formats are common:

1. The first format is a natural-language outline called *pseudocode*, where the prefix pseudo means fictitious or pretended, so it's not "real" code. Pseudocode, like real code, is composed of one or more *statements*. A statement is the equivalent of a "sentence" in a natural language. If the sentence is simple, the corresponding statement usually appears on one line, but if the sentence is complex, the statement may be spread out over several lines. Statements can be nested inside each other, as in an outline. We'll use the term statement a lot, and you'll get a better appreciation for it as we go along.
2. The second format is an arrangement of boxes and arrows that help you visually step through the algorithm. The most detailed form of boxes and arrows is called a *flowchart*. The boxes in a flowchart typically contain short statements that are similar to pseudocode statements.

This chapter shows you how to apply pseudocode and flowcharts to a fundamental set of standard programming problems—problems that appear in almost all large programs. The chapter also shows you how to *trace* an algorithm—step through it one statement at a time—to see what it's actually doing. Our goal is to give you a basic set of informal tools which you can use to describe what you want a program to do. The tools help you organize your thinking before you start writing the actual program. Tracing helps you figure out how an algorithm (or completed program) actually works. It helps you verify correctness and identify problems when things are not right.

2.2 Output

The first problem to consider is the problem of displaying a program's final result—its output. This may sound like something to consider last, so why consider it first? The output is what the *end user*—the client, the person who eventually uses the program—wants. It's the goal. Thinking about the output first keeps you from wasting time solving the wrong problem.

Put yourself in user's place.

Hello World Algorithm

In Chapter 1, we showed you a Java program that generated "Hello, world!" output on the computer screen. Now we'll revisit that problem, but focus on the algorithm, not the program. You may recall that Chapter 1's Hello World program was seven lines long. Figure 2.1 shows the Hello World algorithm—it contains just one line, a pseudocode print statement. The point of an algorithm is to show the steps necessary to solve a problem without getting bogged down in syntax details. The Hello World algorithm does just that. It shows a simple print statement, which is the only step needed to solve the Hello World problem.

```
print "Hello, world!"
```

Figure 2.1 Hello World algorithm that prints the message "Hello, world!"

[1] Ninth-century Persian mathematician Muhammad ibn Musa al-Khwarizmi is considered to be the father of algebra. The term *algorithm* comes from Algoritmi, which is the Latin form of his shortened name, al-Khwarizmi.

Figure 2.1's "Hello, world!" message is a string literal. A *string* is a generic term for a sequence of characters. A *string literal* is a string whose characters are written out explicitly and enclosed in quotation marks. If you print a string literal, you print the characters literally as they appear in the command. So Figure 2.1's algorithm prints the characters H, e, l, l, o, comma, space, w, o, r, l, d, and !.

Rectangle Algorithm

For the next example, suppose you want to display the area of a particular rectangle. First, consider what you want the algorithm to do. In Figure 2.2, look at the `area = 40` line under Output. That shows what the algorithm's output looks like.

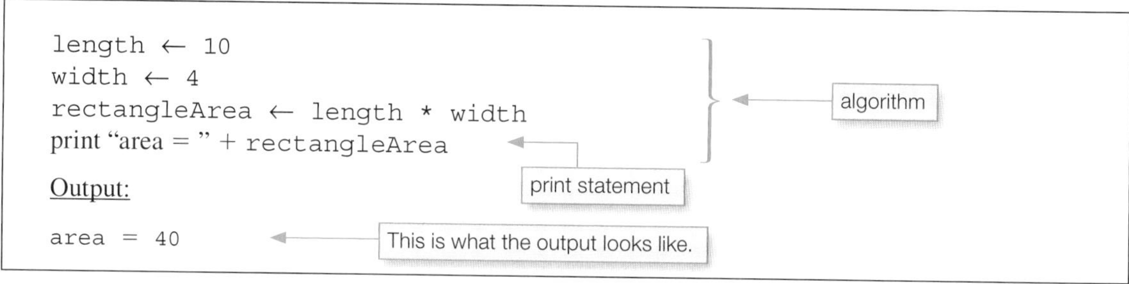

Figure 2.2 Rectangle algorithm that prints the area of a rectangle

The top part of Figure 2.2 is the algorithm for calculating a rectangle's area. Note that some of the words, like `length` and `width`, appear with monospace font. *Monospace font* is when each character's width is uniform. We use monospace font to indicate that something is a variable. A *variable* is a container that holds a value. To indicate that you want a value to be assigned into a variable, use the backwards arrow symbol, ←, with the variable at the left and the value at the right. For example, the first two lines in Figure 2.2's algorithm assign 10 to the `length` variable and 4 to the `width` variable, respectively. Those statements, and all statements that use the backwards arrow, are pseudocode assignment statements. The Rectangle algorithm's third line is a more complicated assignment statement. The third line describes two operations: First, compute the area by multiplying `length` times `width`. (The `*` is the multiplication "times" symbol.) Then assign the result (the product) to the variable, `rectangleArea`. The fourth line prints two items – the string literal "area =" and the value of the `rectangleArea` variable. When a variable appears in a print statement, the print statement prints the value stored inside the variable. `retangleArea` contains 40, so the print statement prints the value 40. In the print statement, note the + symbol separating "area =" and `rectangleArea`. We do this to help you develop good habits for later Java coding. Java requires that whenever you use a print statement to print a string plus something else, you insert a + sign between each of the things that you're printing. This chapter is about algorithms and pseudocode, and pseudocode is very forgiving, so if you forget to use the + sign in this chapter to separate a string from another thing that is to be printed, it's not a big deal.

2.3 Variables

Now let's consider variables in more detail. Figure 2.2's Rectangle algorithm has three variables—`length`, `width`, and `rectangleArea`. In `rectangleArea`, notice how we run together the two words, "rectangle" and "area", and notice how we start the second word with a capital letter. We do this to help you develop

good habits for later Java coding (that's the same rationale that we used for inserting a + sign between things that you want to print). Java does not permit any spaces in a variable name. Therefore, if you'd like to use a variable name with more than one word, don't use spaces to separate the words. Instead, to distinguish the words, we suggest that you capitalize the first letter of all words after the first word. All other letters, including the very first letter, should be lowercase. This technique for writing multiple-word variable names is called *camelCase,* because of the bump(s) in the middle. Here are two examples that show how to name variables with camelCase:

Description	A Good Variable Name
sports team name	`teamName`
weight in grams	`weightInGrams`

Variables can hold different *types* of data. Which type of data would the `teamName` variable probably hold—a number or a string? It would probably be used to hold a string (e.g., "Jayhawks" or "Pirates"). Which type of data would the `weightInGrams` variable probably hold—a number or a string? It would probably be used to hold a number (e.g., 12.5). It's relatively easy for a human to determine the type of a named variable by just thinking about the name, but this kind of thinking is very difficult for a computer. So in a real Java program, we must tell the computer the type of each data item.

However, since pseudocode is designed strictly for humans and not computers, in pseudocode we don't bother with type specification. Notice that Figure 2.2's pseudocode representation of a Rectangle program does not contain any mention of data type. Pseudocode ignores data type so that focus can be kept on the algorithm's essence—its instructions.

2.4 Operators and Assignment Statements

The previous section described variables by themselves. Now let's consider relationships between variables by looking at operators and assignments.

Here is the third statement from Figure 2.2's Rectangle algorithm:

```
rectangleArea ← length * width
```

As indicated earlier, the `*` symbol is the multiplication operator. The other common arithmetic operators are + for addition, - for subtraction, and / for division. These should be familiar to everyone. The `length` and `width` variables are *operands.* In mathematics, and again in programming, an operand is an entity (e.g., a variable or a value) that is operated on by an operator. The `length` and `width` variables are operands because they are operated on by the `*` operator.

When we say "`variableA` ← `x`," we mean "put the value of `x` into `variableA`" or "assign the value of `x` to `variableA`." So the `rectangleArea ← length * width` statement puts the product of `length` times `width` into the `rectangleArea` variable. A picture is worth a thousand words. See Figure 2.3—it visually describes what the statement does.

Figure 2.3 includes a pair of parentheses not shown in the pseudocode statement. As you might remember from grade school math, parentheses have the highest *precedence* of all the math operators. That means that any operations inside parentheses (like multiplication or addition) should be performed before

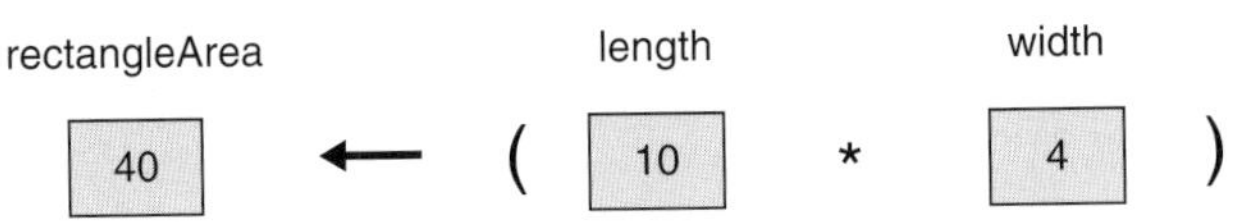

Figure 2.3 Assignment operation represented by left-pointing arrow

operations that appear outside parentheses. So Figure 2.3 indicates that everything on the right side of the backwards arrow (←) should be evaluated completely before the result is assigned to the variable on the left side of the backwards arrow. That's the way it is in pseudocode, and that's also the way it is in Java. You can include the parentheses in your code if you like, but most people don't bother because they know that the parentheses are implied. The backwards arrow's direction—right to left—indicates that the assignment statement flows right to left, and that helps to make the implied parentheses more obvious. In other words, with a right-to-left flow, the right side of the backwards arrow must be evaluated before its value can be transferred to the left side of the backwards arrow.

2.5 Input

In the preceding Rectangle algorithm, the algorithm itself supplied the values for the `length` and `width` variables. We did it that way to make the introductory discussion as simple as possible. Sometimes this is an appropriate strategy, but in this particular case, it's silly, because the algorithm solves the problem only for one particular set of values. To make the algorithm more general, instead of having the algorithm supply the values for `length` and `width`, you should have the *user* (the person who runs the program) supply the values. When a user supplies a value(s) for a program, that's called *user input,* or just *input.* Figure 2.4 presents an improved Rectangle algorithm, where input `length` and input `width` perform user input operations.

Note the first two `print` statements in Figure 2.4—they're called *prompts* because they tell (or prompt) the user what to enter. Without prompts, most users would be left with an unpleasant sensation and the puzzling question, "What do I do now?"

Throughout the book, we provide *sample sessions* as a means of showing what happens when an algorithm or program is run with a typical set of inputs. When there is space, we include the sample session in the figure with the algorithm or program that generates it. Can you identify the user-input values in the sample session in Figure 2.4? Our convention is to italicize sample session input values to distinguish them from output. Thus, *10* and *4* are user-input values.

The combination of a pseudocode algorithm and a sample session represents a convenient and efficient way to specify a simple algorithm or program. The sample session shows the format of desired inputs and outputs. It also shows representative input and output numerical values, which allow a programmer to verify that his/her completed program actually behaves as required. In many of the book's projects (projects are on the book's website), we provide some combination of pseudocode and sample session to specify the problem we are asking you to solve.

Write what you'll do and how you'll do it.

```
print "Enter a length in meters: "
input length
print "Enter a width in meters: "
input width
rectangleArea ← length * width
print "The area is" + rectangleArea + "square meters."
```

Sample session:

User inputs are italicized.

```
Enter a length in meters: 10
Enter a width in meters: 4
The area is 40 square meters.
```

Figure 2.4 Rectangle algorithm that gets length and width values from a user

2.6 Flow of Control and Flowcharts

In the preceding sections, we described various statements—print statements, assignment statements, and input statements—and we focused on the mechanics of how each statement works. Now it's time to focus on the relationships between statements. More specifically, we'll focus on *flow of control*. Flow of control is the order in which program statements are executed. In our discussion of flow of control, we'll refer to both algorithms and programs. Flow of control concepts apply equally to both.

Flow of control is best explained with the help of flowcharts. Flowcharts are helpful because they are pictures. As such, they help you to "see" an algorithm's logic. A flowchart uses two basic symbols: (1) rectangles, which contain commands like print, assign, and input, and (2) diamonds, which contain yes/no questions. At each diamond, the flow of control splits. If the answer is "yes," flow goes one way. If the answer is "no," flow goes another way.

The dashed boxes in Figure 2.5 show three standard structures for flow-of-control—a sequential structure, a conditional structure, and a looping structure. The flowchart on the left—the sequential structure—is a picture of the Rectangle algorithm described in Figure 2.2. *Sequential structures* contain statements that are executed in the sequence/order in which they are written. In other words, after executing a statement, the computer executes the statement immediately below it. *Conditional structures* contain a yes/no question, and the answer to the question determines whether to execute the subsequent block of statements or skip it. *Looping structures* also contain a yes/no question, and the answer to the question determines whether to repeat the loop's block of statements or move on to the statements after the loop.

Structured programming is a discipline that requires programs to limit their flow of control to sequential, conditional, and looping structures. A program is considered to be well structured if it can be decomposed into the patterns in Figure 2.5. You should strive for well-structured programs because they tend to be easier to understand and work with. To give you an idea of what not to do, see Figure 2.6. Its flow of control is bad because there are two points of entry into the loop, and when you're inside the loop, it's hard to know what's happened in the past. When a program is hard to understand, it's error-prone and hard to fix. Code that implements an algorithm like this is sometimes called *spaghetti* code because when you draw a flowchart of the code, the flowchart looks like spaghetti. When you see spaghetti, untangle it!

Don't write spaghetti code.

In addition to standardizing sequential, conditional, and looping control structures, structured programming also splits up large problems into smaller subproblems. In Java, we put the solution to each subproblem in a separate block of code called a *method*. We'll discuss methods in Chapter 5, but for now, we'll focus on the three control structures shown in Figure 2.5.

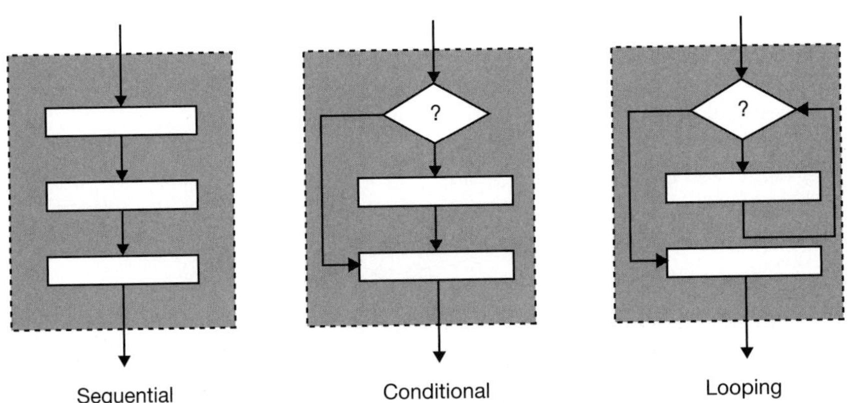

Figure 2.5 Well-structured flow of control

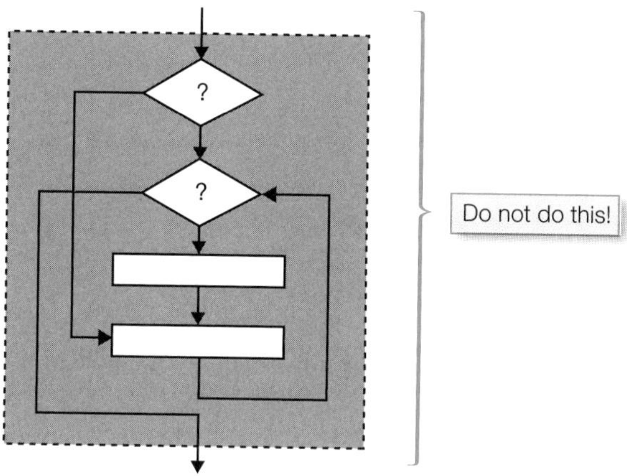

Figure 2.6 Poorly structured flow of control

2.7 if Statements

In previous sections describing print, assignment, and input statements, you saw examples of the sequential control structure on the left side of Figure 2.5. Now let's consider the conditional control structure in the center of Figure 2.5. In going through a sequence of steps, sometimes you get to a "fork in the road," at which point you must choose which way to go. The choice you make depends on the situation. More specifically, it depends on the answer to a question. When a program has a fork in the road, programmers use an *if statement* to implement the fork. The if statement asks a question and the answer to the question tells the algorithm which way to go. More formally, the if statement contains a *condition*. A condition is a question whose answer is either yes or no. The answer to the condition's question determines which statement executes next. Here are three forms for the if statement:

"if"
"if, else"
"if, else if"

Now let's look at each of these three forms separately.

"if"

First, suppose you want to do either one thing or nothing at all. In that case, you should use the simple "if" form of the if statement. Here is its format:

if *<condition>* ← if statement's heading
    <statement(s)>

↑ Indent subordinate statements

Note the angled brackets "<>" that surround "condition" and "statement(s)." Throughout the book, we use the italics and angled bracket notation for items that require a description. Thus, when you see "*<condition>*," it tells you that an actual condition, not the word "condition," is supposed to follow the word "if." Likewise, when you see "*<statement(s)>*," it tells you that one or more actual statements, not the word "statements," is supposed to go underneath the if statement's heading.

In the above if statement illustration, note how *<statement(s)>* is indented. Pseudocode emulates a natural-language outline by using indentation to show encapsulation or subordination. The statements under an if statement's heading are subordinate to the if statement because they are considered to be part of the larger, encompassing if statement. Since they are subordinate, they should be indented.

Here's how the simple "if" form of the if statement works:

- If the condition is true, execute all subordinate statements; that is, execute all indented statements immediately below the "if."
- If the condition is false, jump to the line after the last subordinate statement; that is, jump to the first un-indented statement below the "if."

Let's put these concepts into practice by showing an if statement in the context of a complete algorithm. Figure 2.7's Lightning algorithm determines how far the user is from a lightning strike. The algorithm prompts the user for the number of seconds between seeing a lightning flash and hearing the lightning's associated thunderclap. The standard way to calculate the distance from a lightning strike is to divide the number of seconds by a "lightning factor." The lightning factor divisor is different depending on whether the user is on the ground or in an airplane. The algorithm uses an if statement to distinguish between these two cases. Specifically, in Figure 2.7, note how the algorithm initializes the lightning factor to 3 and then, using an if statement, it changes the lightning factor to 3.4 if the user is in an airplane. After the if statement (i.e., after there are no more indented statements inside the if statement's heading), note how the algorithm divides the entered seconds value by the lightning factor and then prints the resulting distance.[2]

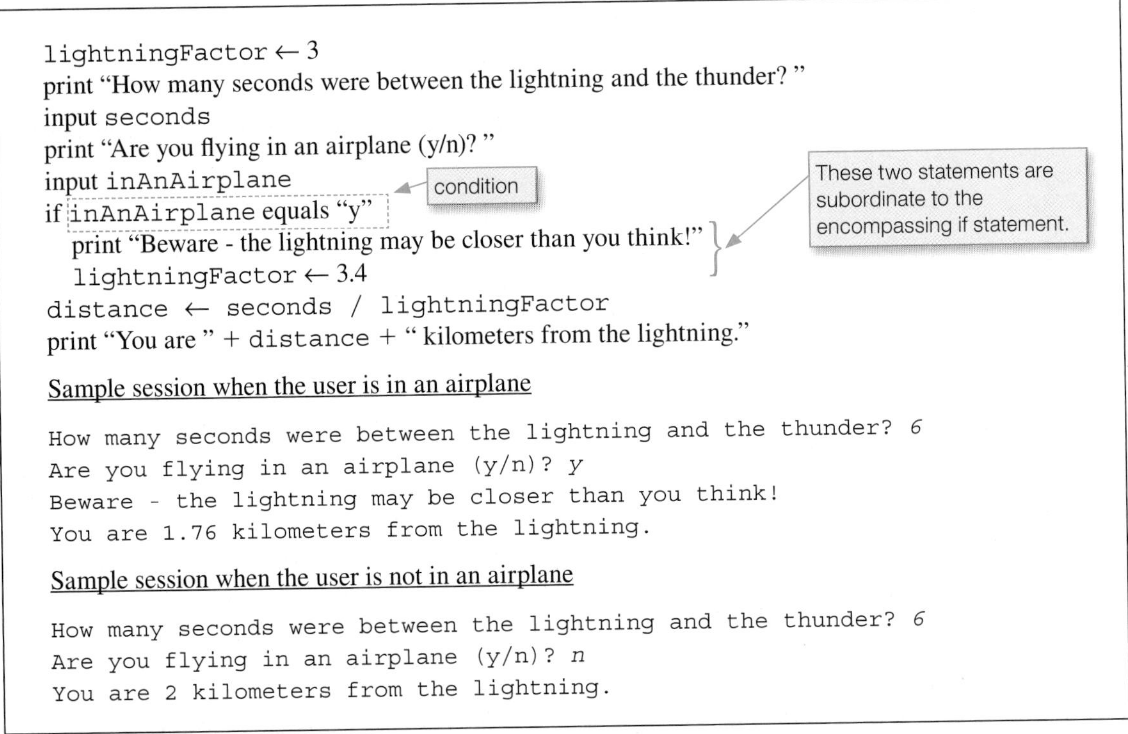

Figure 2.7 Lightning algorithm that calculates the user's distance from a lightning strike

In the Lightning algorithm, note that the if statement's condition is `inAnAirplane` equals "y". The `inAnAirplane` variable holds the user's input, "y" or "n," which indicates whether the user is in an airplane or not. The condition compares the contents of the `inAnAirplane` variable to the "y" value to see if they're equal. If they are equal, the if statement's two indented statements are executed.

There's one more thing to learn from the Lightning algorithm. Note how the `distance` assignment command and the subsequent print command are separate statements. That's perfectly acceptable and quite common, but you should be aware of an alternative implementation where the two commands are merged into one statement:

print "You are " + (`seconds` / `lightningFactor`) + " kilometers from the lightning."

In this case, we put parentheses around the mathematical calculation to emphasize that we want the computer to print the result of the calculation, rather than the values of the individual variables.

"if, else"

Now for the second form of the if statement—the "if, else" form. Use the "if, else" form if you want to do either one thing or another thing. Here is its format:

if <*condition*>
  <*statement(s)*>
else
  <*statement(s)*>

And here's how the "if, else" form of the if statement works:

- If the condition is true, execute all statements subordinate to the "if," and skip all statements subordinate to the "else."
- If the condition is false, skip all statement(s) subordinate to the "if," and execute all statements subordinate to the "else."

Here's an example that uses the "if, else" form of the if statement:

```
if grade ≥ 60
   print "Pass"
else
   print "Fail"
```

Note how we indent the print "Pass" statement because it is subordinate to the if condition, and we indent the print "Fail" statement because it is subordinate to the "else."

"if, else if"

The "if, else" form of the if statement addresses situations in which there are exactly two possibilities. But what if there are more than two possibilities? For example, suppose that you want to print one of five

[2]For the case of being in an airplane, the lightning-factor divisor is greater, so the calculated distance is less. Why is the calculated distance less when you're in an airplane? When flying in an airplane at a typical cruising altitude of 10,000 meters (= 33,000 feet), the air temperature is around −57°C (= −70°F). Sound travels more slowly in cold air, so in an airplane, when you hear thunder, it's closer than the delay would normally indicate.

possible letter grades for a particular numerical score. You can do it by using the "if, else if" form of the if statement to establish parallel paths:

```
if grade ≥ 90
   print "A"
else if grade ≥ 80
   print "B"
else if grade ≥ 70
   print "C"
else if grade ≥ 60
   print "D"
else
   print "F"
```

What happens if the grade is 85? The print "A" statement is skipped, and the print "B" statement is executed. Once one of the conditions is found to be true, then the rest of the entire if statement is skipped. So the third, fourth, and fifth print statements are not executed.

What happens if all the conditions are false? If all the conditions are false, then the subordinate statement under "else" is executed. So if the grade is 55, print "F" is executed. Note that you're not required to have an "else" with the "if, else if" statement. If you don't have an "else" and all the conditions are false, then no statements are executed.

if Statement Summary

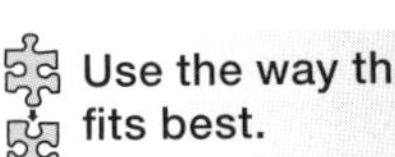

Use the way that fits best.

Use the first form ("if") for problems where you want to do one thing or nothing. Use the second form ("if, else") for problems where you want to do either one thing or another thing. Use the third form ("if, else if") for problems where there are three or more possibilities.

Practice Problem with Flowchart and Pseudocode

Let's practice what you've learned about if statements by presenting a flowchart and having you write the corresponding pseudocode for an algorithm that cuts a CEO's excessively large salary in half. Figure 2.8 shows the flowchart.

In flowcharts, we omit the word "if" from the condition in diamonds and add a question mark to turn the condition into a question. The question format fits well with the "yes" and "no" on the exiting arrows. If the condition is true, the answer to the question is "yes." If the condition is false, the answer to the question is "no." Given the flowchart in Figure 2.8, try to write a pseudocode version of the cut-CEO-salary-in-half algorithm. When you're done, compare your answer to our answer:

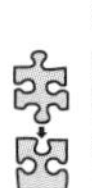

Practice writing a pseudocode algorithm.

```
print "Enter CEO Salary: "
input ceoSalary
if ceoSalary > 500000
   ceoSalary ← ceoSalary * 0.5
   print "Reduced CEO Salary is $" + ceoSalary
```

In Figure 2.8's flowchart, note how the diamond's left arrow and the bottom rectangle's down arrow are connected. That's because every flowchart is supposed to have one starting point and one ending point. Don't forget to connect the bottom components of your flowcharts.

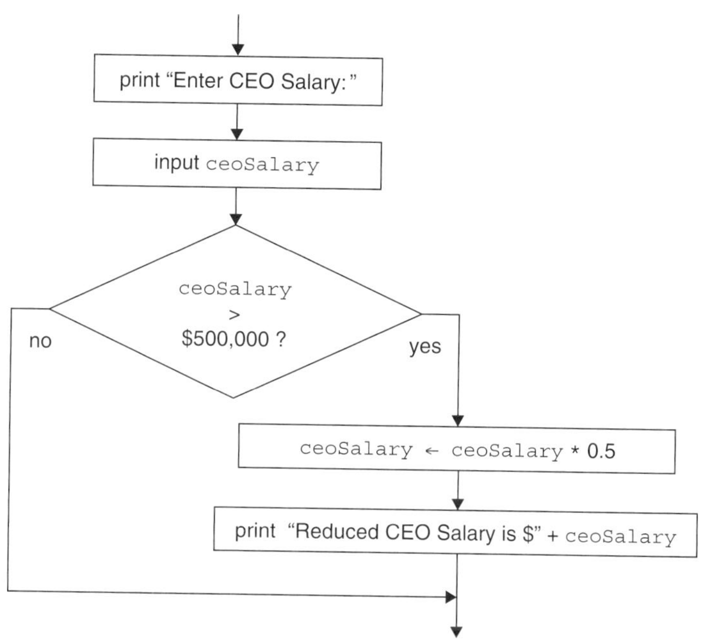

Figure 2.8 Flowchart for reducing CEO salaries

Practice Problems with Pseudocode Only

Everybody knows the saying, "A picture is worth a thousand words." This may be true, but compare the space consumed by and the effort to construct Figure 2.8's flowchart with the space consumed by and the effort to write the corresponding pseudocode. Pictures help you get started, but text is more efficient once you know what you're doing. So now let's try skipping the flowchart and going immediately to pseudocode.

First, let's write an algorithm that prints "No school!" if the temperature is below 0 degrees Fahrenheit. Which if statement form should you use for this problem? Because the problem description says to do either something or nothing, you should use the simple "if" form:

```
print "Enter a temperature: "
input temperature
if temperature < 0
   print "No school!"
```

Next, let's write an algorithm that prints "warm" if the temperature is above 50 degrees and prints "cold" otherwise. Which if statement form should we use? Because the problem description says to do one thing or another thing, you should use the "if, else" form:

```
print "Enter a temperature: "
input temperature
if temperature > 50
   print "warm"
else
   print "cold"
```

Finally, let's write an algorithm that prints "hot" if the temperature is above 80 degrees, prints "OK" if it's between 50 and 80 degrees, and prints "cold" if it's less than 50 degrees. For this problem, it's appropriate to use the "if, else if" form, like this:

```
print "Enter a temperature: "
input temperature
if temperature > 80
   print "hot"
else if temperature ≥ 50
   print "OK"
else
   print "cold"
```

2.8 Loops

We've now discussed two of the three structures in Figure 2.5—sequential structures and conditional structures. Let's now discuss the third structure—*looping structures*. Looping structures repeat the execution of a particular sequence of statements. If you need to execute a block of code many times, you could, of course, repeatedly write the code wherever you need it. However, that leads to redundancy, which is something you want to avoid in a computer program because it opens the door to inconsistency. It's better to write the code once and then reuse it. The simplest way to reuse a block of code is to go back up to before where that block starts, and run through it again. That's called a *loop*. Every loop has a condition that determines how many times to repeat the loop. Think of driving through western Kansas and seeing a sign for "Prairie Dog Town." Your kids demand that you take the prairie dog drive-through tour. The decision about how many times to repeat the tour parallels the condition in a loop statement.

A Simple Example

Suppose you want to print "Happy birthday!" 100 times. Rather than writing 100 print "Happy birthday!" statements, wouldn't it be better to use a loop? Figure 2.9 presents a solution to this problem in the form of a flowchart with a loop. The flowchart implements the looping logic with an arrow that goes from "count ← count + 1" back up to the "count ≤ 100?" condition.

In a loop, you'll often use a count variable that keeps track of the number of times the loop has repeated. You can either count up or count down. The Happy Birthday flowchart counts up.

To count up, you'll need to add 1 to the count variable. In the flowchart, here's the statement that counts up:

```
count ← count + 1
```

A common mistake by beginning programmers is to try to count up by using the following statement and nothing else:

```
count + 1
```

That code does indeed add 1 to the current value of count, but then what happens to the sum? It just dies. It isn't saved anywhere. If you ever need to increment a variable, you must do more than just add 1 to the variable. You must use an assignment operator to assign the new value to the variable itself. In pseudocode, we use the backwards arrow for assignment operations, so here, once again, is how you implement the counting part of an algorithm that needs to count (yes, we know we presented the same code 30 seconds ago, but this is so important that we decided to show it twice):

```
count ← count + 1
```

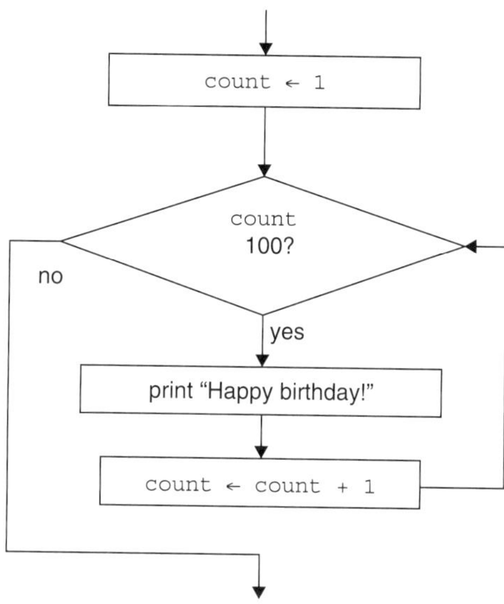

Figure 2.9 Flowchart for the Happy Birthday algorithm

It might look weird to have the same variable on both sides of the backwards arrow, but trust us—it's necessary. It reflects the fact that the statement uses the `count` variable twice, for different purposes. As you know, assignment statements work right to left, so the `count` + `1` part of the statement is performed first. Next, the result (the sum of `count` + `1`) is assigned into the `count` variable, replacing the original `count` value with the new value.

In practice, all loops should have some kind of termination. That is, they should stop executing at some point. A loop that counts up normally uses a maximum value as a termination condition. For example, Figure 2.9's loop continues so long as `count` is less than or equal to 100, and it terminates (stops looping) when `count` reaches 101. A loop that counts down normally uses a minimum value as a termination condition. For example, a loop might start with `count` equal to 100 and continue so long as `count` is greater than zero. Then the loop would terminate when `count` reached zero.

When a loop's condition compares a counter variable to a maximum value, the question often arises about whether to use "less than or equal to" or just "less than." Likewise, when a loop's condition compares a counter variable to a minimum value, the question often arises about whether to use "greater than or equal to" or just "greater than." There are no absolute answers to those questions. Sometimes you'll need to do it one way, and sometimes you'll need to do it the other way—it depends on the situation. For example, look again at the decision condition in Figure 2.9's Happy Birthday algorithm. Suppose you used "less than." Then, when `count` equaled 100, you would quit before printing the last (100th) "Happy birthday!" Therefore, in this case you should use "less than or equal to." If you mistakenly used "less than," that would be an *off-by-one error*. Such errors are called "off by one" because they occur when you execute a loop one more time than you should or one less time than you should. To avoid off-by-one errors, you should always double-check the borderline cases for your algorithms' loops.

The while Loop

Most popular programming languages have several different types of loops. Although it may be awkward, theoretically, there's always a way to convert any one type of loop to any other type of loop. So, for simplicity and brevity, in this discussion of algorithms, we'll consider only one type of loop and look at the other types when we get into the details of the Java language. The type of loop we'll consider now is a very popular one, the while loop, which has this format:

```
while <condition>
{
  <statement(s)>
}
```

This format should look familiar because it's similar to the if statement's format. The condition is at the top, and the subordinate statements are indented. But for the while loop, the subordinate statements are enclosed by braces. The subordinate statements are called the loop's *body*. The number of times that a loop executes its body is called the number of *iterations*. It's possible for a loop to repeat forever, which is called an *infinite loop*. It's also possible for a loop to repeat zero times. There's no special name for the zero-iteration occurrence, but it's important to be aware that this sometimes happens. For an example, let's see how Figure 2.9's Happy Birthday flowchart looks when it's presented as pseudocode with a while loop. This is shown in Figure 2.10.

```
count ← 1
while count ≤ 100
{
  print "Happy birthday!"
  count ← count + 1
}
```

Figure 2.10 Pseudocode for another Happy Birthday algorithm

Here's how the while loop works:

- If the condition is true, execute all of the loop's subordinate statements, and then jump back to the top of the loop to check the loop's condition again.
- When the loop's condition finally becomes false, jump to below the loop (that is, the first statement after the loop's last subordinate statement) and continue execution there.

2.9 Loop Termination Techniques

In this section, we describe three common ways to terminate loops:

- Counter

  Use a counter variable to keep track of the number of iterations.

- User query

  Ask the user if he or she wants to continue. If the user responds yes, then execute the body of the loop. After each pass through the subordinate statements in the loop, ask the user again if he or she wants to continue.

- Sentinel value

  When a loop includes a data-input statement, identify a special value (a sentinel value) that is outside the normal range of input, and use it to indicate that looping should terminate. For example, if the normal range of input is positive numbers, the sentinel value could be a negative number like –1. Here's how you do it: Continue to read in values and execute the loop until the entered value equals the sentinel value, and then stop the looping. In the real world, a sentinel is a guard who lets people continue to pass until the enemy arrives. So a program's sentinel value is like a human sentinel—it allows the loop to continue or not.

Counter Termination

Figure 2.10's Happy Birthday algorithm is a good example of using a counter to terminate a looping operation. We should point out, however, that the normal place for a computer to start counting is 0, rather than one. If we use the standard start-at-zero convention, Figure 2.10's pseudocode changes to this:

```
count ← 0
while count < 100
{
   print "Happy birthday!"
   count ← count + 1
}
```

Notice that as we change the initial count value from 1 to 0, we also change condition comparison from ≤ to <. This will produce the same 100 iterations, but this time, the count values will be 0, 1, 2, ...98, 99. Each time you create a counter loop, it's important to assure yourself that the number of iterations will be exactly what you want. Because you can start with numbers different than 1, and because the termination condition can employ different comparison operators, it's sometimes hard to be sure about the total number of iterations you'll get. Here's a handy trick to give you more confidence:

Simplify the problem to check its essence.

To check a loop's terminal condition, temporarily change the terminal condition to produce what you think will be exactly one iteration. For example, in this most recent pseudocode version of the Happy Birthday algorithm (where the initial count is 0), change the final count from 100 to 1. Then ask yourself, "How many print operations will occur?" In this case, the initial count is 0. The first time the condition is tested, the condition is "0 < 1," which is true. So the condition is satisfied and the loop's subordinate statements execute. Since the final statement in the loop increments the count to 1, the next time the condition is tested, the condition is "1 < 1," which is false. So the condition is not satisfied, and looping terminates. Because using 1 in the loop condition produces one iteration, you can have confidence that using 100 in the loop condition will produce 100 iterations.

User Query Termination

To understand user query termination, consider an algorithm that repeatedly asks a user for numbers and calculates and prints the squares of the input values. This activity should continue so long as the user answers "y" to a "Continue?" prompt.

Figure 2.11 displays this algorithm as pseudocode. Within the while loop body, the first statement prompts the user to enter a number, the third statement does the computation, and the fourth statement prints the result. The query "Continue? (y/n)" and the corresponding input come just before the end of the body. This loop always executes at least one time, because we assign "y" to the continue variable before the loop begins.

Suppose you want to give the user the opportunity to quit before entering even one number to square. You can do that by replacing the first statement:

```
continue ← "y"
```

```
continue ← "y"
while continue equals "y"
{
  print "Enter a number: "
  input num
  square ← num * num
  print num + " squared is " + square
  print "Continue? (y/n): "
  input continue
}
```

Figure 2.11 Print Squares algorithm that uses a query loop

with these two statements:

```
print "Do you want to print a square? (y/n): "
input continue
```

This provides the user the option to enter "n" so that no squares will be computed.

Sentinel-Value Termination

To understand sentinel-value termination, consider an algorithm that reads in bowling scores repeatedly until a sentinel value of –1 is entered. Then the algorithm prints the average score.

Mull it over. Often, you should spend time just thinking about a problem's solution before writing anything down. And you should think first about the solution at a high level, without worrying about all the details. With that said, we encourage you to set the book aside now and think about the steps needed in the Bowling Score algorithm.

Are you done thinking? If so, compare your thoughts to this high-level description:

> Read in scores repeatedly and find the sum of all the scores.
> Then, when –1 is entered, divide the sum by the number of scores entered.

There are two details in this high-level description that you now need to address. First, you need to think about how to find the sum of all the scores. Before asking for any input, and before any looping, assign an initial value of zero to a `totalScore` variable. In other words, *initialize* it to zero. Then, in the same loop which repeatedly asks the user for the next score, right after inputting that score, add it to the `totalScore` variable to accumulate the scores as they come in. This way, after all the scores are in, the `totalScore` variable will already contain the sum of all scores.

The sum of all scores is useful because the goal is to determine the average score, and to compute an average you need the sum. But to compute an average, you also need the total number of items, and that's not known ahead of time. How can you keep track of the number of scores entered so far? Initialize and accumulate a `count` variable while you initialize and update the `totalScore` variable. Note that just one loop does all three activities (inputting, updating `totalScore`, and updating `count`). We chose –1 as a sentinel value for a Bowling Score algorithm because it's a value that would never be a valid bowling-score entry. But any negative number would work as the sentinel value.

Figure 2.12 illustrates the algorithm solution for this problem. Note how the prompt messages say "(–1 to quit)." That is necessary because without it, the user wouldn't know how to quit. In general, always provide enough prompting information so that the user knows what to do next and knows how to quit. 

```
totalScore ← 0
count ← 0
print "Enter score (−1 to quit): "
input score
while score ≠ −1
{
   totalScore ← totalScore + score
   count ← count + 1
   print "Enter score (−1 to quit): "
   input score
}
avg ← totalScore / count
print "Average score is " + avg
```

Figure 2.12 Bowling Score algorithm using a sentinel-value loop

What would you expect to happen if the user enters –1 as the very first input? That causes the loop body to be skipped, and the count variable never gets updated from its original initialized value, 0. When the average assignment statement attempts to calculate the average score, it divides totalScore by count. Because count is 0, it divides by 0. As you may recall from your math courses, division by zero creates problems. If an algorithm divides by zero, the result is undefined. If a Java program divides by zero, the computer prints a cryptic error message and then immediately shuts down the program. Because the Bowling Score algorithm allows for the possibility of division by zero, it is not very *robust*. To be robust, it should behave in a way that a typical user would consider to be both sensible and courteous, even when the input is unreasonable. To make it more robust, replace the last two statements in Figure 2.12's algorithm with an if statement like this:

```
if count ≠ 0
   avg ← totalScore / count
   print "Average score is " + avg
else
   print "No entries were made."
```

Using this if statement enables the program to tell the user why a normal output was not produced, and it avoids the problems inherent with division by zero.

2.10 Nested Looping

In the preceding two sections, we presented algorithms where each algorithm contained one loop. As you proceed through the book and as you proceed through your programming career, you'll find that most programs contain more than one loop. If a program has loops that are independent (i.e., the first loop ends before the second loop begins), then the program's flow should be reasonably straightforward. On the other hand, if a program has a loop inside a loop, then the program's flow can be harder to understand. In this section, we'll try to make you comfortable with a *nested loop*, which is the formal term for an inner loop that's inside an outer loop.

Suppose you're asked to write an algorithm that finds the largest prime number from a list of user-entered numbers. The user indicates that he or she is done by entering a negative number. For example, suppose the user enters this:

5 80 13 21 1 –3

The number 5 is prime because it has only two factors, 1 and 5. The number 13 is also prime because it has only two factors, 1 and 13. The numbers 80 and 21 are not prime because they each have more than two factors. The number 1 is not prime by the definition of prime numbers. Thus, the largest prime number in the list is 13.

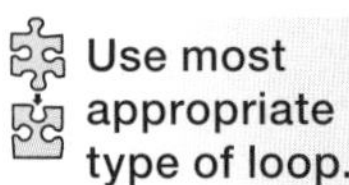
Use most appropriate type of loop.

When working on a nontrivial algorithm, before writing anything down, you should think about a very important question: What types of loops should be used? For the find-the-largest-prime-number algorithm, you'll need an outer loop that terminates when the user enters a negative number. What type of loop will that be—a counter loop, user-query loop, or sentinel-value loop? It will be a sentinel-value loop, with a negative number as the sentinel value.

Figuring out the inner loop is not as easy. Actually, you might not even realize at first that an inner loop is necessary. To determine whether a number x is prime, try to find a number greater than 1 and less than x that divides evenly into x. If you cannot find such a number, then x is prime. So where's the loop? To answer that question, you need to break down your thought process into steps that are simple enough to be part of an algorithm. Specifically, what steps are necessary to find a number greater than 1 and less than x that divides evenly into x? Before you read the answer to that question in the next paragraph, stop and try to answer the question on your own. Hint: You'll need to use a loop.

To find a number greater than 1 and less than x that divides evenly into x, you can use a counter loop that counts from 2 to x – 1.[3] For each iteration of the loop, determine whether the `count` variable's value divides evenly into x. To make that determination, divide x by the `count` variable and if the remainder is 0, then we say that the `count` variable's value divides evenly into x. If the `count` variable's value divides evenly into x, then the `count` variable's value is a factor of x, and x is not prime. If the `count` variable gets all the way up to x – 1 without finding a factor of x, then x is prime.

Now look at the complete algorithm solution in Figure 2.13. Note that the algorithm does indeed use a sentinel-value outer loop—the loop reads values into a variable named x and terminates when the user enters a negative number. Note that the algorithm does indeed use a counter inner loop – the algorithm initializes a `count` variable to 2, increments `count` by 1 inside the loop, and terminates the loop when `count` reaches the value of x. Now study Figure 2.13's inner loop pseudocode carefully. It does the usual counter loop things, but it includes some additional logic that terminates the loop immediately if it finds a factor of x before `count` increments all the way up to x. Specifically, the code uses a `prime` variable to keep track of whether x is a prime number. For each user-entered x value, the algorithm initializes `prime` to "yes" and then, within the inner loop, changes `prime` to "no" if it finds that x has a factor. The change to "no" causes the inner loop to terminate.

[3]Actually, you need to check for factors only from 2 up to the square root of x. If you find a factor greater than the square root of x, it will have a "partner factor" that is less than the square root of x, such that the product of the two factors equals x. Thus, if numbers less than the square root of x are checked before numbers greater than the square root of x, then finding a factor greater than the square root of x will reveal nothing new.

```
largestPrime ← 1
print "Enter a number (negative to quit): "
input x
while x ≥ 0                                          ← outer loop
{
    count ← 2
    prime ← "yes"
    while count < x and prime equals "yes"           ← inner loop
    {
        if x / count has no remainder
            prime ← "no"
        else
            count ← count + 1
    }
    if prime equals "yes" and x > largestPrime
        largestPrime ← x
    print "Enter a number (negative to quit): "
    input x
}
if largestPrime equals 1
    print "No prime numbers were entered."
else
    print "The largest prime number entered was " + largestPrime + "."
```

Figure 2.13 Algorithm that finds the largest prime number from a user-entered list of numbers

There's one more important thing to learn from the find-the-largest-prime-number algorithm. The algorithm's goal isn't just to find prime numbers. It's to find the largest prime number from a list of numbers. Finding the largest value from a list of values is a common problem, and you should be familiar with its solution. Let's think about the solution at a high level. After each new prime number is found, the algorithm should ask the question: Is the new prime number larger than the previous largest prime number? If the new prime number is larger, then it becomes the new "champion"; that is, the new largest prime number. Note that the preceding sentence starts with the word *if.* That's a good indication that you can implement that logic with an if statement. Find the if statement immediately after Figure 2.13's inner loop and verify that it implements the aforementioned logic. You'll see that the if statement checks the new number to see if (1) the new number is prime and (2) the new number is larger than the previous largest prime number. If both conditions are met, the algorithm assigns the new number into the `largestPrime` variable. That assignment crowns the new number as the new champion.

Initialize with a terrible value.

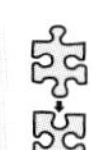

Note the `largestPrime` ← 1 initialization at the top of the algorithm. What's the point of initializing `largestPrime` to 1? You should initialize the champion variable (`largestPrime`) with a terrible starting value so that it will lose automatically the first time a new prime number is compared to it. You know that 1 will lose to the first entered prime number in a find-the-largest-prime-number contest because prime numbers are greater than 1 by definition. After the first prime number replaces `largestPrime's` 1 initial value, subsequent prime numbers might or might not replace `largestPrime's` value, depending on the value of the new prime number and the value of `largestPrime`. If the user enters no prime numbers, then `largestPrime` will retain its original 1 value. The bottom of the algorithm checks for this possibility and prints "No prime numbers were entered" if `largestPrime` equals 1.

2.11 Tracing

Dig into details.

Up to now, we have focused on design. Now let's look at *analysis*—the breaking up of a whole into its parts. In the present context, that means going through the details of an already-existing algorithm. The analysis technique we'll use is called *tracing,* where you essentially pretend that you're the computer. You step through an algorithm (or a program) line by line and carefully record everything that happens. In the early parts of this book, we'll use tracing to illustrate programming details that we're trying to explain. Tracing gives you a way to make sure that you really understand newly learned programming mechanisms. Tracing also gives you a way to verify whether an existing algorithm or Java code is correct, or whether it has *bugs*.

What are bugs, and why are they called that? One of the early digital computers, the Harvard Mark II, used mechanical relays rather than transistors, and programmers programmed by changing electrical connections. As the story goes,[4] even though all the electrical connections were right, the computer kept making a mistake. Finally the programmer discovered a moth squeezed between the contacts of one of the relays. Apparently, the moth had been squashed when the relay contacts closed, and the moth's dead body was interrupting the proper flow of electricity between those contacts. After the programmer pulled the moth out—"debugged" the computer program—the computer gave the right answer. When you're tracing an algorithm or program to find software bugs, you may sometimes feel like one of these old-timers crawling around inside the CPU, looking for moths.

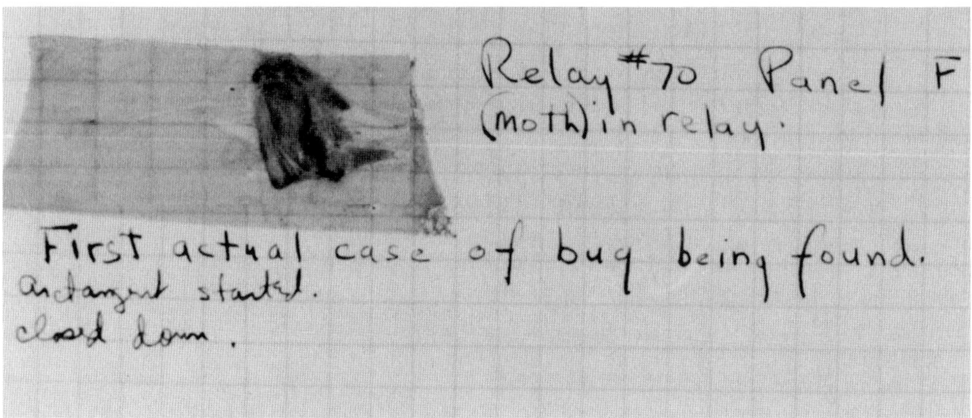

Courtesy of the Naval Surface Warfare Center, Dahlgren, VA., 1988. US Naval Historical Center Photograph.

[4]http://www.faqs.org/docs/jargon/B/bug.html

Short-Form Tracing

We present two tracing forms—a short form, described in this subsection, and a long form, described in the next subsection. The short-form tracing procedure is commonly used in industry and in classrooms. It works well in a dynamic environment, where you can move back and forth between pseudocode (or Java code, later) and a trace listing, and fill information in as you go. You may see your teacher go through this dynamic operation on a whiteboard. For example, here's an algorithm that prints the "Happy Birthday" song:

```
print "What is your name?"
input name
count ← 0
while count < 2
{
   print "Happy birthday to you."
   count ← count + 1
}
print "Happy birthday, dear " + name + "."
print "Happy birthday to you."
```

Here's what the short-form trace looks like after the trace is complete:

input	name	count	output
~~Arjun~~	Arjun	~~0~~	What is your name?
		~~1~~	Happy birthday to you.
		2	Happy birthday to you.
			Happy birthday, dear Arjun.
			Happy birthday to you.

The above trace listing has four columns—input, name, count, and output. The input column shows hypothetical input for the algorithm. The output column shows what the algorithm produces when the algorithm runs with the given input. The name and count columns show the values stored in the `name` and `count` variables. In this example, we started with the input value "Arjun." Then we stepped through the code, one line at a time. In stepping through the code, we added values under the name, count, and output columns, and we crossed out old `count` values as they were overwritten by new `count` values. Figure 2.14 describes the general procedure.

Short-form tracing works well in a live interactive context, but it does not work as well in a static context like the pages of a printed book. That's because in a book, the short-form tracing does not portray the dynamics of the updating process very well. With our simple Happy Birthday algorithm, you may have been able to visualize the dynamics. But for more involved algorithms, a short-form trace listing on the page of a book just "blows through" the details it needs to highlight. Therefore, in this book, we'll use a long-form tracing procedure that keeps better track of each step as the process unfolds.

Long-Form Tracing

With the long-form tracing procedure, there's an added emphasis on keeping track of where you are in the algorithm. To implement that emphasis, (1) you need to have a separate row in the tracing table for each step that's executed in the algorithm, and (2) for each row in the tracing table, you need to provide a line number that tells you the row's associated line in the algorithm. For an example, see the long-form happy birthday trace in Figure 2.15.

Trace setup:

- If there is input, provide a column heading labeled <u>input</u>.
- Provide a column heading for each variable.
- Provide a column heading labeled <u>output</u>.

Trace the program by executing the algorithm one line at a time, and for each line, do this:

- For an `input` statement, cross off the next input value under the input column heading.
- For an assignment statement, update a variable's value by writing the new value under the variable's column heading. If there are already values under the column heading, insert the new value below the bottom value and cross off the old value.
- For a `print` statement, write the printed value under the output column heading. If there are already values under the output column heading, insert the new printed value below the bottom of the output column.

Figure 2.14 Short-form tracing procedure

```
1  print "What is your name? "
2  input name
3  count ← 0
4  while count is < 2
5  {
6     print "Happy birthday to you."
7     count ← count + 1
8  }
9  print "Happy birthday, dear " + name + "."
10 print "Happy birthday to you."
```

<u>input</u>
Arjun

line#	name	count	output
1			What is your name?
2	Arjun		
3		0	
6			Happy birthday to you.
7		1	
6			Happy birthday to you.
7		2	
9			Happy birthday, dear Arjun.
10			Happy birthday to you.

Figure 2.15 Happy Birthday trace—long form

Figure 2.15's long-form trace looks somewhat like the previous short-form trace, with a few notable exceptions. The input column has been moved above the main part of the tracing table. In its place is the line# column, which holds line numbers in the algorithm that correspond to rows in the tracing table. Notice how lines 6 and 7 are executed once, and then executed again. That shows how the trace "unrolls" the loop and repeats the sequence of statements within the loop for each loop iteration.

Using a Trace To Find a Bug

It's time for you to get your money's worth from all this tracing talk. We'll provide you with an algorithm, and it's up to you to determine whether it works properly. More specifically, trace the algorithm to determine whether each step produces reasonable output. If it produces faulty output, find the algorithm's bug and fix the algorithm.

Check each step.

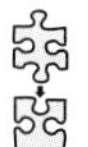

Suppose that Park University's Student Housing office wrote the algorithm shown in Figure 2.16. The algorithm is supposed to read in the names of freshmen and assign each freshman to one of two dormitories. Freshmen with names that begin with A through M are assigned to Chestnut Hall, and freshmen with names that begin with N through Z are assigned to Herr House. Using the trace setup provided in Figure 2.16, try to either complete the trace or get to a point in the trace where you've identified a problem.

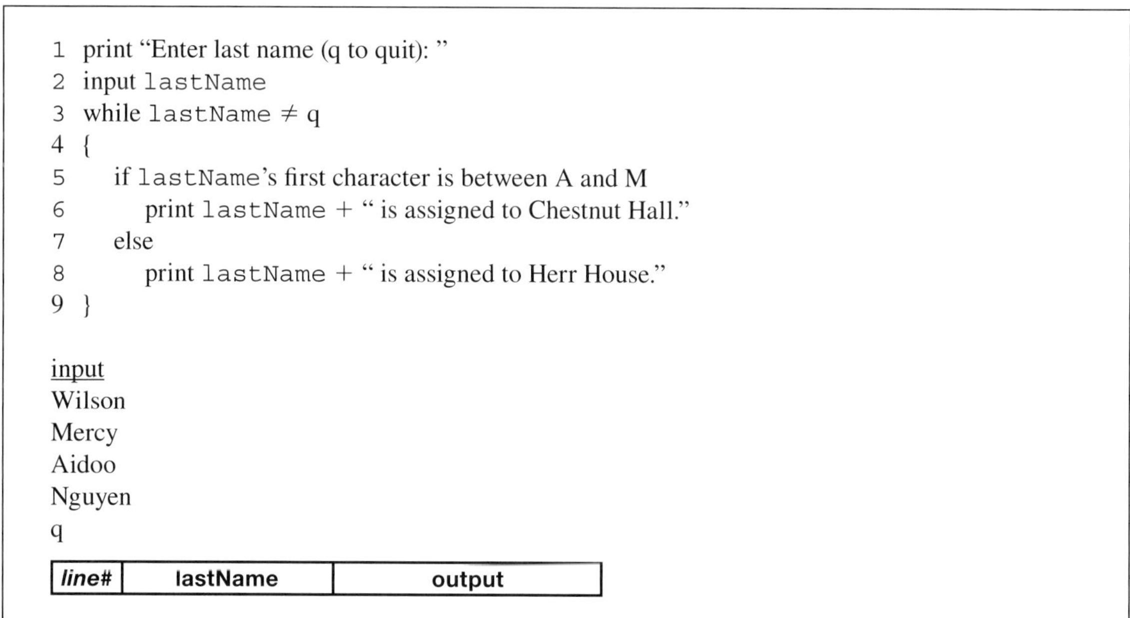

```
1 print "Enter last name (q to quit): "
2 input lastName
3 while lastName ≠ q
4 {
5    if lastName's first character is between A and M
6       print lastName + " is assigned to Chestnut Hall."
7    else
8       print lastName + " is assigned to Herr House."
9 }
```

input
Wilson
Mercy
Aidoo
Nguyen
q

line#	lastName	output

Figure 2.16 Freshmen dormitory assignment algorithm and trace setup

Have you finished working on the trace? If so, compare your answer to this:

line#	lastName	output
1		Enter last name (q to quit):
2	Wilson	
8		Wilson is assigned to Herr House.
8		Wilson is assigned to Herr House.
8		Wilson is assigned to Herr House.
⋮		⋮

The trace points out a problem—the algorithm repeatedly prints Wilson's dorm assignment, but no one else's. There appears to be an infinite loop. Can you identify the bug? The trace shows that `lastName` gets the first input value, Wilson, but it never gets any other input values. Referring back to Figure 2.16, you can see that the algorithm prompts for the last name above the loop, but not inside the loop. Therefore, the first input value is read in, but no others. The solution is to add another last name prompt inside the while loop, at its bottom. Here is the corrected algorithm:

```
print "Enter last name (q to quit): "
input lastName
while lastName ≠ q
{
   if lastName's first character is between A and M
      print lastName + "is assigned to Chestnut Hall."
   else
      print lastName + "is assigned to Herr House."
   print "Enter last name (q to quit): "
   input lastName
}
```

We encourage you to trace the corrected algorithm on your own, and you'll find that all four freshmen are assigned to appropriate dorms. Yeah!

Software Development Tools

Most software development tools temporarily label each line of code with a line number to help identify the locations of programming errors. Those line numbers are not actually part of the code, but when they are available, you can use them as identifiers in the *line#* column of a long-form trace. Many software development tools also include a *debugger* that enables you to step through a program one line at a time as it executes. The debugger enables you to look at variable values as you go. Our tracing procedure emulates a debugger's step-by-step type of evaluation. Experience with the tracing used in this book will make it easier for you to understand what an automated debugger is telling you.

2.12 Problem Solving: Other Pseudocode Formats and an Asset Management Example

Pseudocode comes in many different varieties. In this section, we start by describing several alternative pseudocode variations. We then focus on high-level pseudocode, which is particularly helpful for large-scale projects. Finally, we show an example of high-level pseudocode being used for the initial design phase of a large-scale water system asset management project.

Alternative Pseudocode Formats

Up to this point, we've provided one set of constructs for the basic pseudocode commands: the backwards arrow (←) for assignment, the word "equals" for testing for equality, the word "while" for repetition, and so on. The pseudocode constructs in this book are fairly common, but we don't want to give you the impression that they form a standard that is universally followed. On the contrary, pseudocode by its very nature is supposed to be flexible. Its purpose is to provide a mechanism for describing the steps necessary to solve a problem without getting bogged down in syntax details. So if you misspell "while" with two *i*'s (i.e., "whiile"), no problem—the pseudocode's meaning is still clear. On the other hand, it is important to follow rules that help with clarity—like the rule about indenting when you're inside an if heading or a while loop heading.

In the interest of consistency, we will stick with the pseudocode constructs that we've presented thus far. However, you should be aware of a few other common pseudocode constructs. For pseudocode assignment, some programmers use the equals sign (=), and some use the words *set* and *to*. For example:

```
x = y
set x to y
```

Those constructs are acceptable (and you should use them if that's what your teacher likes), but we prefer the backwards arrow for assignment because it emphasizes that the assignment operation flows right to left, with the evaluation on the right side occurring before the result gets transferred to the left.

In pseudocode, to check to see if two entities are equal, some programmers use the equals sign (=) and some use two equals signs (==). For example:

```
if (answer = "yes")
if (answer == "yes")
```

Those constructs are acceptable (and you should use them if that's what your teacher likes), but we prefer the word *equals* to check to see if two entities are equal because we feel that it's clearer. Also, if you get in the habit of using = or == for equality testing in your pseudocode algorithms, your habit might carry over to Java, and that can be problematic. As you'll learn later, using = to test for equality in Java can lead to difficult-to-detect bugs. And using == to test for equality in Java can lead to difficult-to-detect bugs if you are comparing strings.

High-Level Pseudocode

Because pseudocode is so flexible, you can also use it to describe algorithms at a higher, more macroscopic level—with more abstraction. The trick is to ignore the details of subordinate operations and just describe and keep track of inputs to and outputs from those subordinate operations. This strategy presents the "big picture" as seen by the outside world. It looks at the "forest" rather than the "trees." It helps keep you on the right track—so you don't solve the wrong problem!

For example, the following Bowling Scores algorithm uses a more high-level pseudocode than what you've seen in the past:

```
Input all scores.
Compute average score.
Print the average score.
```

This high-level description presents only the major features, not all the details. It indicates what the program is supposed to do, but not how to do it.

An Asset Management Example

In this subsection, we ask you to think about a real-world managerial problem at a fairly abstract level. Imagine that you are the information technology (IT) specialist working in the government of a small city. The head of that city's water department respects your organizational skills and has asked you to come to a city council meeting and lead a discussion of how you might set up a computer program to help the council manage the assets of that city's water system.

First, you suggest that the city council members help you come up with an overall sequence of steps. On a blackboard, you'll write high-level pseudocode for the "program." To avoid jargon, you'll just call this high-level pseudocode a "to-do list."

After some discussion, the council members agree on—and you list—the following overall steps:[5]

1. Make an inventory of all water system assets.
2. Prioritize those assets.
3. Schedule future changes, replacements, and additions to those assets.
4. Prepare a long-range budget.

The council thanks you for your help, and for the next meeting, they ask you to flesh out this list with enough detail to show how you plan to implement each of the four steps. They don't want to see a bunch of Java source code. They just want to see how you'd proceed—to get a feeling for the difficulty of the project.

Back in your office, you get out your scratch pad and begin to work. The four steps that you presented in the meeting constitute high-level pseudocode. To "flesh out the list," you decide to write more detailed pseudocode for each list item. For step 1, you identify seven variables: `assetName`, `expectedLife`, `condition`, `serviceHistory`, `adjustedLife`, `age`, and `remainingLife`. For each asset, you'll have to ask someone in the water department to provide appropriate input for each of the first six variables. Then your program will calculate a value for the last variable. You'll have to repeat this for each significant asset. So here's an abbreviated pseudocode description of the implementation of step 1:

```
more ← 'y'
while more equals 'y'
{
   input assetName
   input expectedLife
   input condition
   input serviceHistory
   input adjustedLife
   input age
   remainingLife ← adjustedLife − age
   print "Another asset? (y/n): "
   input more
}
```

This algorithm does not include prompts for the individual variables. Some of these variables may have multiple components, and you may wish to establish and enforce certain conventions for what input values will be acceptable. For example, `condition` and `serviceHistory` may each have several subordinate components. You'll deal with all those details later.

For step 2, you have five variables: `assetName`, `remainingLife`, `importance`, `redundancy`, and `priority`. The `assetName` and `remainingLife` variables are the same as two of the variables used for step 1, so you won't need to input those again. But wait! If this is a separate loop, you'll still have to identify each asset to make sure the new values are being associated with the right asset. You could do this by asking the user to re-enter the `assetName`, or you could do it by looping through all the existing assets and printing out each name just before asking for the required additional information for that asset.

[5] These four steps and their subsequent elaboration are based on recommendations in *Asset Management: A Handbook for Small Water Systems,* Office of Water (4606M) EPA 816-R-03-016, www.epa.gov/safewater, September, 2003.

The second strategy is easier for the user, so you pick it. Here's an abbreviated pseudocode description of the implementation of step 2:

```
while another asset exists
{
   print assetName
   input importance
   input redundancy
   input priority
}
```

Again, the algorithm does not include prompts, and it does not establish and enforce input conventions. You'll deal with those details later.

For step 3, you identify five variables: `assetName`, `activity`, `yearsAhead`, `dollarCost`, and `annualReserve`. Again, `assetName` is already in the system, so again, you can identify it by printing it out. But in scheduling things, the council members will want to deal with the most important things first, so before you start going through the assets, you'll want the program to sort them by priority. The sorting operation might be a little tricky. But if you're lucky, someone else already will have written code for that popular computer task, and you'll be able to use it instead of "reinventing the wheel."

The `activity`, `yearsAhead`, and `dollarCost` are inputs, and you'll want the program to compute `annualReserve` as `dollarCost / yearsAhead`. After computing the annual reserve for each individual asset, you'll want the program to add it to a `totalAnnualReserve` variable, and after the loop, you'll want it to print the final value of `totalAnnualReserve`. Here's an abbreviated pseudocode description of the implementation of step 3:

```
sort assets by priority
totalAnnualReserve ← 0
while another asset exists
{
   print assetName
   input activity
   input yearsAhead
   input dollarCost
   annualReserve ← dollarCost / yearsAhead
   totalAnnualReserve ← totalAnnualReserve + annualReserve
}
print totalAnnualReserve
```

Again, the algorithm does not include prompts. You'll deal with all those details later.

For step 4, you identify the three variables, `totalAnnualReserve`, `currentNetIncome`, and `additionalIncome`. For this, you need to get someone in the accounting department to provide a value for `currentNetIncome`. Then have the program subtract it from the `totalAnnualReserve` computed in step 3 to obtain the `additionalIncome` required to make the plan work. If the answer comes out negative, you'll want it to just print zero to indicate that your city won't have to come up with any additional income. Here's a pseudocode description of the implementation of step 4:

```
input currentNetIncome
additionalIncome ← currentNetIncome − totalAnnualReserve
```

```
if additionalIncome < 0
   additionalIncome ← 0
print "Additional income needed = " + additionalIncome
```

OK, that's probably enough preparation for next week's city council meeting. At least you'll be able to give the council members a reasonable sense of the amount of work required.

Summary

- Use pseudocode to write informal descriptions of algorithms. Use understandable names for variables. Indent subordinate statements.
- When your program needs an input, provide an informative prompt to tell the user what kind of information to supply.
- A flowchart provides a visual picture of how the elements of a program are related and how control flows through those elements as the program executes.
- There are three basic well-structured flow-of-control patterns—sequential, conditional, and looping.
- You can implement conditional execution using the three forms of the if statement: "if," "if, else," and "if, else if."
- Provide all loops with some kind of terminating condition such as counter, user query, or sentinel value.
- Use a nested loop if there's a need to repeat something during each iteration of an outer loop.
- Use tracing to (1) obtain an intimate understanding of what an algorithm does and (2) debug programs that have logical errors.
- Use more abstract language to describe larger and more complex programming operations succinctly.

Review Questions

§2.2 Output

1. Describe what this statement does:

   ```
   print "user name = " + userName
   ```

§2.3 Variables

2. Provide an appropriate variable name for a variable that holds the total number of students.

§2.4 Operators and Assignment Statements

3. Write a line of pseudocode that tells the computer to assign `distance` divided by `time` into a `speed` variable.

§2.5 Input

4. Write a line of pseudocode that tells the computer to put a user entry into a variable called `height`.

§2.6 Flow of Control and Flowcharts

5. What are the three types of control flow described in this chapter?
6. Looping is appropriate whenever the next thing done is something previously done. (T / F)

§2.7 if Statements

7. Consider the following pseudocode:

> if it is night, set `speedLimit` to 55;
> otherwise, set `speedLimit` to 65.

Suppose the value of the variable, `night`, is "false." After this code runs, what should be the value of the variable, `speedLimit`?

8. The above pseudocode does not have the exact form suggested in the text. Is that OK?
9. Draw a flowchart that implements this logic:
   If the temperature is greater than 10°C and it's not raining, print "walk." Otherwise, print "drive."
10. Provide a solution to the previous problem in the form of pseudocode.

§2.8 Loops

11. Where is a `while` loop's terminating decision made?
12. When a `while` loop terminates, what executes next?
13. Is it possible for a `while` loop to have an infinite number of iterations?
14. Is it possible for a `while` loop to have zero iterations?

§2.9 Loop Termination Techniques

15. What are the three loop termination techniques described in this chapter?
16. A *sentinel value* is used to do which of the following?
    a) Specify the first value printed.
    b) Print an error message.
    c) Signal the end of input.

§2.10 Nested Looping

17. How does the form of pseudocode we use in most of this chapter differentiate an inner loop from an outer loop?

§2.11 Tracing

18. Which of the following is true?
    a) Tracing shows sequence of execution.
    b) Tracing helps you debug a program.
    c) Tracing highlights errors in loop initialization and termination.
    d) All of the above.
19. Trace the following Bowling Score algorithm (taken from Section 2.9). Use the setup shown below the algorithm.

```
totalScore ← 0
count ← 0
print "Enter score (−1 to quit): "
input score
while score ≠ −1
{
   totalScore ← totalScore + score
   count ← count + 1
   print "Enter score (−1 to quit): "
   input score
}
avg ← totalScore / count
print "Average score is " + avg
```

Trace setup:

input
94
104
114
–1

line#	score	totalScore	count	avg	output

Exercises

1. [after §2.5] Write pseudocode for an algorithm that (1) asks the user to input the length of the side of a square, (2) computes the square's area, and (3) prints the square's area. Use the following sample session.

   Sample session:

   The italics signify user input.

   ```
   Enter length of side of square in meters: 15
   The area of the square is 225 square meters.
   ```

2. [after §2.8] What is an infinite loop?

3. [after §2.8] Given the following pseudocode, circle the statements that are considered to be within the body of the while loop:

   ```
   input time
   while time < 8
   {
      print time
      time ← time + 1
   }
   ```

4. [after §2.9] In exercise 3, suppose the user's input for `time` is 3. How many lines of output will the algorithm generate?

5. [after §2.11] Trace the following algorithm. The book presents two ways to do tracing—a short form and a long form. To give you a head start, the setup for the short form and also the long form are given below. For your answer, pick one setup and use it. Skip the other setup.

   ```
   y ← 0
   input x
   while x ≠ y
   {
      y ← x
      input x
      x ← x + y
      print "x = " + x
      print "y = " + y
   }
   ```

   Short-form setup:

input	x	y	output
2			
3			

4
0

Long-form setup:

<u>input</u>
2
3
4
0

line#	x	y	output

6. [after §2.11] Trace the following algorithm. The book presents two ways to do tracing—a short form and a long form. To give you a head start, the setup for the short form and also the long form are given below. For your answer, pick one setup and use it. Skip the other setup.

```
num ← 2
count ← 1
while count < 5
{
   count ← count * num
   if count / 2 < 2
      print "Hello"
   else
      while count < 7
      {
         count ← count + 1
      }
   print "The count is " + count + "."
}
```

Short-form setup:

<u>num</u>   <u>count</u>   <u>output</u>

Long-form setup:

line#	num	count	output

Review Question Solutions

1. The statement prints what is in quotation marks literally, and then prints the current value of the variable `userName`.
2. `totalNumberOfStudents`
3. Pseudocode that tells the computer to assign distance divided by time into a speed variable:
   `speed ← distance / time`
4. Pseudocode statement:
   input `height`

5. The three types of control flow discussed in Chapter 2 are sequential, conditional, and looping.
6. True. Looping is appropriate whenever the next thing done is something previously done.
7. After the code executes, the value of the variable, `speedLimit`, should be 65.
8. Yes. It's OK because it's only pseudocode, and it conveys the meaning unambiguously. However, if it were supposed to be code the computer could compile, the syntax would have to conform exactly to prescribed rules for a particular programming language like Java.
9. Flowchart that implements walk/drive logic:

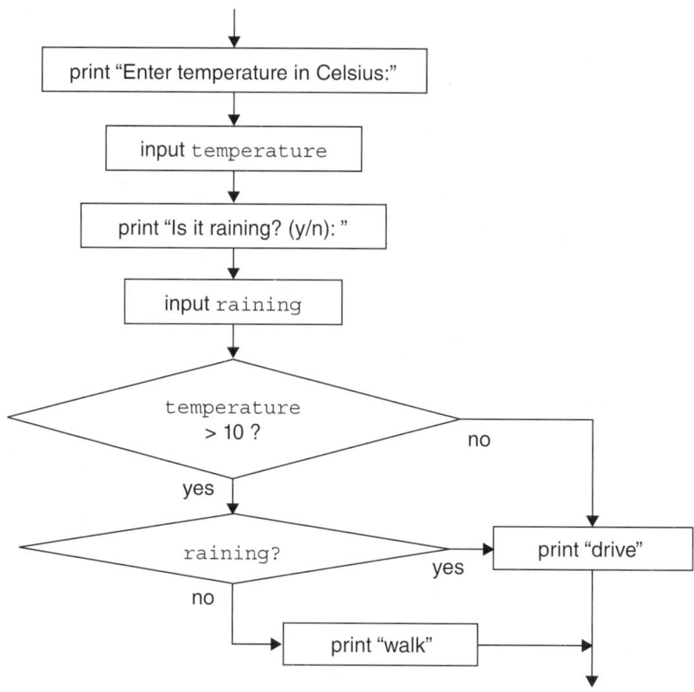

10. Provide a solution to the previous problem in the form of pseudocode.

```
print "Enter temperature in Celsius: "
input temperature
print "Is it raining? (y/n): "
input raining
if temperature > 10 and raining equals "n"
  print "walk"
else
  print "drive"
```

11. A while loop's terminating decision is made at the beginning of the loop.
12. After a while loop terminates, the next thing to execute is the first statement after the end of the loop.
13. Yes.
14. Yes.
15. The three loop termination techniques described in this chapter are: counter, user query, and sentinel value.

16. A sentinel value is used to: c) signal the end of input.

17. The inner loop is entirely inside the outer loop. The entire inner loop is shifted to the right compared to the outer loop.

18. d) All of above. Tracing shows sequence of execution, helps debug, and highlights initialization and termination errors.

19. Bowling Score algorithm trace:

<u>input</u>
94
104
114
–1

line#	Score	totalscore	count	avg	output
1		0			
2			0		
3					Enter score (–1 to quit):
4	94				
7		94			
8			1		
9					Enter score (–1 to quit):
10	104				
7		198			
8			2		
9					Enter score (–1 to quit):
10	114				
7		312			
8			3		
9					Enter score (–1 to quit):
10	–1				
12				104	
13					Average score is 104

CHAPTER 3

Java Basics

Objectives

- Write simple Java programs.
- Learn about style issues such as comments and readability.
- Declare, assign, and initialize variables.
- Understand primitive data types—integer, floating point, and character.
- Understand reference variables.
- Use the `String` class's methods for string manipulation.
- Use the `Scanner` class for user input.
- Optionally, learn about GUI input and output with the `JOptionPane` class.

Outline

3.1 Introduction

In solving a problem, it's best to spend time first thinking about what you want to do and organizing your thoughts. In Chapter 2, you focused on the thinking and organizing by writing pseudocode algorithm solutions for given problem descriptions. In this chapter, you'll take the next step—you'll focus on writing solutions using a real programming language, Java. By using a real programming language, you'll be able to run your program on a computer and produce results on a computer screen.

As you progress through this chapter, you'll find that much of Java's code parallels pseudocode. The primary difference is the precise syntax required for Java. Pseudocode syntax is lenient: Pseudocode must be clear enough so that humans can understand it, but the spelling and grammar need not be perfect. Programming-code syntax is stringent: It must be perfect in terms of spelling and grammar. Why? Because regular programming code is read by computers, and computers are not able to understand instructions unless they're perfect.

Because this chapter is your first real taste of Java, we'll stick to the basics. We'll present Java syntax that's needed for simple *sequential-execution* programs. A sequential-execution program is one in which all the program's statements are executed in the order in which they are written. As we write such programs, we'll show you output, assignment, and input statements. In addition, we'll describe data types and arithmetic operations. Toward the end of the chapter, we'll present a few slightly more advanced topics—type casting and string methods—that will add important functionality without adding much complexity. Let us begin the Java journey.

3.2 "I Have a Dream" Program

In this section, we present a simple program that prints a single line of text. In the next several sections, we'll analyze the different components of the program. The analysis may be a bit dry, but bear with us. It's important to understand the program's components because all future programs will use those same components. In the rest of the chapter, we'll introduce new concepts that enable us to present more substantial programs.

See Figure 3.1. It shows a program that prints "I have a dream!"[1] In the upcoming sections, we'll refer to it as the Dream program. The program contains comments for human readers and instructions for the computer to execute. We'll analyze the comments first, and then we'll move on to the instructions. You can use this tiny program as a common starting point for all other Java programs. Enter it, run it, and see what it does. Modify it, run it again, and so on, until you have what you need.

Start every program with this code's structure.

[1]Dr. Martin Luther King, Jr., presented his famous "I have a dream" speech on the steps of the Lincoln Memorial as part of an August 28, 1963 civil rights march on Washington, D.C. The speech supported desegregation and helped spur passage of the 1964 Civil Rights Act.

```
/*****************************************
 * Dream.java
 * Dean & Dean
 *
 * This program prints "I have a dream."
 *****************************************/

public class Dream
{
  public static void main(String[] args)
  {
    System.out.println("I have a dream!");
  }
} // end class Dream
```


Figure 3.1 Dream program

3.3 Comments and Readability

In the real world, you'll spend a lot of your time looking at and fixing other people's code. And other people will spend a lot of their time looking at and fixing your code after you've moved on to something else. With all this looking at other people's code going on, everyone's code needs to be understandable. One key to understanding is good comments. *Comments* are words that humans read but the compiler[2] ignores.

One-Line-Comment Syntax

There are two types of comments—one-line comments and block comments. If your comment text is short enough to fit on one line, use a one-line comment. One-line comments start with two slashes. Here's an example:

```
} // end class Dream
```

The compiler ignores everything from the first slash to the end of the line. So in the above line, the compiler pays attention only to the right brace (}) and ignores the rest of the line. Why is the comment helpful? If you're viewing a long piece of code on a computer screen and you've scrolled to the bottom of the code, it's nice to see a description of the code (e.g., `end class Dream`) without having to scroll all the way back up to the beginning of the code.

Block-Comment Syntax

If your comment text is too long to fit on one line, you can use multiple one-line comments, but it's a bit of a pain to retype the `//`'s for every line. As an alternative, you can use a block comment. Block comments start with an opening `/*` and end with a closing `*/`. Here's an example:

[2]A *compiler*, defined in Chapter 1, is a special program that converts a source-code program into an executable program. An *executable program* is a program that the computer can execute directly.

```
/*
The following code displays the androids in a high-speed chase,
wreaking havoc on nearby vehicles.
*/
```

The compiler ignores everything between the first slash and the last slash.

Prologue

We use the term *prologue* to describe a special example of a block comment. It provides information about the program so that a programmer can glance at it quickly and get an idea of what the program is all about. You should put a prologue at the top of every one of your programs. To make it stand out, it's common to enclose the prologue in a box of asterisks. Here's the `Dream` program's prologue:

start of the block comment

```
/****************************************
* Dream.java
* Dean & Dean
*
* This program prints "I have a dream."
****************************************/
```

end of the block comment

Note that the opening `/*` and the closing `*/` blend in with the other asterisks. That's OK. The compiler still recognizes the `/*` and `*/` as the start and end points of the block comment.

Include these items in your program's prologue section:

- a line of `*`'s
- filename
- programmer's name
- a line with a single `*` at its left
- program description
- a line of `*`'s

Readability and Blank Lines

We say that a program is *readable* if a programmer can easily understand what the program does. Comments are one way to improve a program's readability. Another way to improve a program's readability is to use blank lines. How are blank lines helpful? Isn't it easier to understand several short, simple recipes rather than a single long, complicated recipe? Likewise, it's easier to understand small chunks of code rather than one large chunk of code. Using blank lines allows you to split up large chunks of code into smaller chunks of code. In a prologue, we insert a blank line to separate the filename-author section from the description section. Also, we insert a blank line below the prologue to separate it from the rest of the program.

By the way, computers don't care about readability; they just care about whether a program works. More specifically, computers skip all comments, blank lines, and contiguous space characters. Because computers don't care about readability, your computer would be perfectly happy to compile and execute this Dream program:

```
public class Dream{public static void
main(String[]args){System.out.println("I have a dream!");}}
```

But a person trying to read the program would probably be annoyed because of the program's poor readability. 

3.4 The Class Heading

So far, we've focused on code that the computer ignores—comments. Now let's talk about code that the computer pays attention to. Here's the first non-comment line in the Dream program:

```
public class Dream
```

That line is called a *class heading* because it's the heading for the definition of the program's *class*. What's a class? For now, think of a class simply as a container for your program's code.

Let's examine the three words in the class heading. First, the last word—`Dream`. `Dream` is the name of the class. The compiler allows the programmer to choose any name for the class, but in the interest of making your code readable, you should choose a word or words that describe the program. Because the Dream program prints "I have a dream," `Dream` is a reasonable class name.

The first two words in the class heading, `public` and `class`, are *reserved words*. Reserved words, also called *keywords,*[3] are words that are defined by the Java language for a particular purpose. They cannot be redefined by a programmer to mean something else. That means programmers cannot use reserved words when choosing names in their programs. For example, we were able to choose `Dream` for the class name because Dream is not a reserved word. We would not have been allowed to choose `public` or `class` for the class name.

So what are the meanings of the `public` and `class` reserved words? The word `class` is a marker that signifies the beginning of the class. For now, with our simple one-class programs, the word `class` also signifies the beginning of the program.

The word `public` is an *access* modifier—it modifies the class's permissions so that the class is accessible by the "public." Making the class publicly accessible is crucial so that when a user attempts to run it, the user's run command will be able to find it.

There are certain coding conventions that most programmers follow. We list such conventions in our "Java Coding-Style Conventions" appendix. Throughout the book, when we refer to "standard coding conventions," we're referring to the coding conventions found in the appendix. Standard coding conventions dictate that class names start with an uppercase first letter; thus, the `D` in the `Dream` class name is uppercase. Java is *case-sensitive,* which means that the Java compiler distinguishes between lowercase and uppercase letters. Because Java is case-sensitive, the filename should also start with an uppercase first letter.

3.5 The `main` Method's Heading

We've talked about the class heading. Now it's time to talk about the heading that goes below the class heading—the `main` method heading. In starting a program, the computer looks for a `main` method heading, and execution begins with the first statement after the `main` method heading. The `main` method heading must have this form:

```
public static void main(String[] args)
```

Let's start our analysis of the `main` method heading by explaining the word `main` itself. So far, all you know about `main` is that in starting a program, the computer looks for it. But `main` is more than that; it's a Java *method.* A Java method is similar to a mathematical function. A mathematical function takes arguments, performs a calculation, and returns an answer. For example, the sin(x) mathematical function takes

[3] In Java, reserved words and keywords are the same. But in some programming languages, there is a subtle difference. In those languages, both terms refer to words that are defined by the programming language, but keywords can be redefined by the programmer, while reserved words cannot be redefined by the programmer.

the x argument, calculates the sine of the given x angle, and returns the calculated sine of x. Likewise, a Java method may take arguments, will perform a calculation, and may return an answer.

The rest of the `main` heading contains quite a few mysterious words whose explanations may be confusing at this point. In later chapters, when you're better prepared, we'll explain the words in detail. For now, it's OK to treat the `main` method heading as a line of text that you simply copy and paste under the class heading. We realize that some of you may be uncomfortable with that. For you folks, the rest of this section explains `main` method heading details.

Explanation of `main` Method Heading Details

We'll now explain the three reserved words at the left of the `main` method heading—`public static void`. As previously mentioned, the word `public` is an access modifier—it grants permissions so that `main` is accessible by the "public." Because `main` is the starting point for all Java programs, it must be publicly accessible.

While `public` specifies <u>who</u> can access the `main` method (everyone), the word `static` specifies <u>how</u> to access the `main` method. With a non-`static` method, you must do some extra work prior to accessing it.[4] On the other hand, a `static` method can be accessed immediately, without doing the extra work. Because `main` is the starting point for all Java programs, it must be immediately accessible, and therefore it requires the word `static`.

Now for the third reserved word in the `main` heading—`void`. Remember that a method is like a mathematical function—it calculates something and returns the calculated value. Well actually, a Java method sometimes returns a value and sometimes returns nothing. `void` indicates that a method returns nothing. Because the `main` method returns nothing, we use `void` in the `main` method's heading.

Now for the `(String[] args)` portion of the `main` heading. Remember that a mathematical function takes arguments. Likewise, the `main` method takes arguments.[5] Those arguments are represented by the word `args`. In Java, if you ever have an argument, you need to tell the computer what type of value the argument can hold. In this case, the argument's type is defined to be `String[]`, which tells the computer that the `args` argument can hold an array of strings. The square brackets, `[]`, indicate an array. An *array* is a structure that holds a collection of elements of the same type. In this case `String[]` is an array that holds a collection of strings. A *string* is a sequence of characters. You'll learn more about strings later in this chapter in Section 3.22, and you'll learn about arrays in Chapter 9.

3.6 Braces

In the Dream program, we inserted opening braces, {, below the class heading and below the `main` heading, and we inserted closing braces, }, at the bottom of the program. Braces identify groupings for humans and for the computer. They must come in pairs—whenever you have an opening brace, you'll need an associated closing brace. In the Dream program, the top and bottom braces group the contents of the entire class, and the interior braces group the contents of the `main` method. For readability's sake, you should put an opening brace on a line by itself in the same column as the first character of the previous line. Look at the following code fragment and note how the opening braces are positioned correctly.

[4]To access a non-`static` method (more formally called an *instance method*), you must first instantiate an object. We describe object instantiation in Chapter 6.

[5]Although the `main` method takes arguments, it's rare for the `main` method to use those arguments. The book's programs do not use the `main` method's arguments.

```
public class Dream
{
  public static void main(String[] args)
  {
    System.out.println("I have a dream!");
  }
} // end class Dream
```

The first brace is positioned immediately below the first character in the class heading, and the second brace is positioned immediately below the first character in the `main` heading. For readability's sake, you should put a closing brace on a line by itself in the same column as its partner opening brace. Look at the above code fragment and note how the closing braces are positioned correctly.

In pseudocode, remember how you indented statements that were logically inside something else? You should do the same thing in Java. You can rely on the braces to remind you when something is inside something else. In the Dream program, the `Dream` class's braces surround the `main` method, so you should indent the entire `main` method. Likewise, the `main` method's braces surround the `System.out.println` statement, so you should indent that statement further. Speaking of the `System.out.println` statement. . . .

3.7 System.out.println

In the Dream program, the `main` method contains this one statement:

```
System.out.println("I have a dream!");
```

The `System.out.println` statement tells the computer to print something. The word `System` refers to the computer. `System.out` refers to the output part of the computer system—the computer's monitor. The word `println` (pronounced "print line") refers to the Java `println` method that's in charge of printing a message to the computer screen. The above statement would normally be referred to as a `println` method call. You *call* a method when you want to execute it.

The parentheses after `println` contain the message that is to be printed. The above statement prints this message on a computer screen:

```
I have a dream!
```

Note the double quotes in `System.out.println("I have a dream!");` To print a group of characters (e.g., I, space, h, a, v, e, . . .), you need to group them together. As you learned in Chapter 2, the double quotes are in charge of grouping together characters to form a string literal.

Note the semicolon at the end of `System.out.println("I have a dream!");` A semicolon in the Java language is like a period in natural language. It indicates the end of a statement. You'll need to put a semicolon at the end of every `System.out.println` statement.

You'll be calling the `System.out.println` method a lot, so you might want to try to memorize its wording. To help with your memorization, think of it as an acronym—"Sop" for `System`, `out`, and `println`. Don't forget that the *S* is uppercase and the rest of the command is lowercase.

The `System.out.println` method prints a message and then moves to the beginning of the next line. That means that if there is another `System.out.println` method call, it starts its printing on the next line. The upcoming example illustrates what we're talking about.

An Example

In our Dream program, we print just one short line—"I have a dream!" In our next example, we print multiple lines of varying lengths. See Figure 3.2's Sayings program and its associated output. Note how each of the three `println` method calls produces a separate line of output. Note how the second `println` method call is too long to fit on one line, so we split it just to the right of the left parenthesis. The third `println` method call is longer than the second `println` method call, and as such, it could not fit on two lines if it were split after the left parenthesis. In other words, this does not work:

```
/*********************************************************************
* Sayings.java
* Dean & Dean
*
* This program prints several sayings.
*********************************************************************/

public class Sayings
{
  public static void main(String[] args)
  {
    System.out.println("The future ain't what it used to be.");
    System.out.println(
      "Always remember you're unique, just like everyone else.");
    System.out.println("If you are not part of the solution," +
      " you are part of the precipitate.");
  } // end main
} // end class Sayings
```

This connects/concatenates the split-apart strings.

Output:

```
The future ain't what it used to be.
Always remember you're unique, just like everyone else.
If you are not part of the solution, you are part of the precipitate.
```

Figure 3.2 Sayings program and its associated output

```
System.out.println(
  "If you are not part of the solution, you are part of the pr
```

Not enough room.

Thus, we split the third `println` method call in the middle of the string that is to be printed. To split a string literal, you need to put opening and closing quotes around each of the two split-apart substrings, and you need to insert a + between the substrings. See the quotes and the + in Figure 3.2's third `println` method call.

3.8 Compilation and Execution

Up to this point in the chapter, you've been exposed only to the theory behind Java code (the theory behind the Dream program's code and the theory behind the Sayings program's code). To gain a more complete appreciation for code, you need to enter it on a computer, compile it, and run it. After all, learning how to program requires lots of hands-on practice. It's a "contact sport"! We've provided several tutorials on the book's website that step you through the compilation and execution of a few simple Java programs. We recommend that you now take the time to work your way through one or more of those tutorials. The rest of this section covers some basic concepts related to compilation and execution. Be aware that we cover these concepts plus additional details in the tutorials.

After entering a program's source code on a computer, save it in a file whose name is comprised of the class name plus a `.java` extension. For example, because the Dream program's class name is `Dream`, its source-code filename must be `Dream.java`.

After saving a program's source code in an appropriately named file, create Java bytecode[6] by submitting the source code file to a Java compiler. In compiling the source code, the compiler generates a bytecode program file whose name is comprised of the class name plus a `.class` extension. For example, because the Dream program's class name is `Dream`, its bytecode filename will be `Dream.class`.

The next step after creating the bytecode program file is to run it. To run a Java program, submit the bytecode program file to the Java virtual machine (JVM).

3.9 Identifiers

So far in this chapter, you've learned Java by looking at code. Eventually, you'll need to learn it by writing your own code. When you do so, you'll need to pick out names for your program components. Java has certain rules for naming your program components. We'll look at those rules now.

An *identifier* is the technical term for a program component's name—the name of a class, the name of a method, and so on. In our Dream program, `Dream` was the identifier for the class name, and `main` was the identifier for the method name.

Identifiers must consist entirely of letters, digits, dollar signs ($), and/or underscore (_) characters. The first character must not be a digit. If an identifier does not follow these rules, your program won't compile.

Coding-convention rules are narrower than compiler rules when it comes to identifiers. Coding conventions suggest that you limit identifiers to just letters and digits. Do not use dollar signs, and (except for named constants—to be described later) do not use underscores. They also suggest that you use lowercase for all your identifier letters except:

- Start class names with an uppercase letter. For example, our `Dream` class starts with an uppercase `D`.
- Run together the words in a multiple-word identifier, using an uppercase letter for the first letter in the second word, third word, and so on. For example, if a method prints a favorite color, an appropriate method name would be `printFavoriteColor`.

Perhaps the most important coding-convention identifier rule is the one that says identifiers must be descriptive. Returning to the example of a method that prints a favorite color, `printFavoriteColor` is plenty descriptive. But how about `favColor`? Nope, not good enough. Some programmers like to use abbreviations (like "fav") in their identifiers. That works OK sometimes, but not all that often. We recommend

[6] *Bytecode*, defined in Chapter 1, is a binary-encoded version of the source code. The computer cannot execute source code, but it can execute bytecode.

staying away from abbreviations unless they're standard. Using complete and meaningful words in identifiers promotes self documentation. A program is *self-documenting* if the code itself explains the meaning, without needing a manual or lots of comments.

If you break a coding-conventions rule, it won't affect your program's ability to compile, but it will detract from your program's readability. Suppose you have a `sngs` method that prints a list of the week's top 40 songs. Even though `sngs` might work, you should rename it to something like `printTop40Songs` to improve your program's readability.

3.10 Variables

To this point, our programs haven't done a whole lot; they've just printed a message. If you want to do more than that, you'll need to be able to store values in variables. A Java variable can hold only one type of value. For example, an integer variable can hold only integers and a string variable can hold only strings.

Variable Declarations

How does the computer know which type of data a particular variable can hold? Before a variable is used, its type must be declared in a *declaration statement.*

Declaration statement syntax:

<type> <list-of-variables-separated-by-commas>;

Example declarations:

```
int row, col;
String firstName;    // student's first name
String lastName;     // student's last name
int studentId;
```

In each declaration statement, the word on the left specifies the type for the variable or variables on the right. For example, in the first declaration statement, `int` is the type for the `row` and `col` variables. Having an `int` type means that the `row` and `col` variables can hold only integers (`int` stands for integer). In the second declaration statement, `String` is the type for the `firstName` variable. Having a `String` type means that the `firstName` variable can hold only strings.

Have you noticed that we sometimes spell string with an uppercase S and we sometimes spell it with a lowercase s? When we use "string" in the general sense, to refer to a sequence of characters, we use a lowercase s. In Java, `String` is a data type that happens to be a class name also. As you now know, coding conventions dictate that class names begin with an uppercase letter. Thus, the `String` class/data type begins with an uppercase `S`. So when we refer to `String` as a data type, in code and in conversational text, we use an uppercase `S`.

When you declare a variable or variables, don't forget to put a semicolon at the end of each declaration statement. When you declare more than one variable with one declaration statement, don't forget to separate the variables with commas.

Style Issues

The compiler will accept a variable declaration anywhere in a block of code, as long as it's above where the variable is used. However, in the interest of readability, you should normally put your declarations at the top of the `main` method. That makes them easy to find.

Although it may waste some space, you should normally declare only one variable per declaration statement. That way, you'll be able to provide a comment for each variable (and you should normally provide a comment for each variable).

We do make exceptions to these recommendations. Note how these `row` and `col` variables are declared together with one declaration statement:

```
int row, col;
```

That's acceptable because they are intimately related. Note that the `row` and `col` variables are declared without a comment. That's acceptable because `row` and `col` are standard names that all programmers should understand. It would be overkill to include a comment like this:

```
int row, col;       // row and col hold row and column index numbers
```

Note how this `studentId` variable is declared without a comment:

```
int studentId;
```

That's acceptable because the `studentId` name is so completely descriptive that everyone should be able to understand it. It would be overkill to include a comment like this:

```
String studentId;  // a student's ID value
```

Variable names are identifiers. Thus, when you name your variables, you should follow the identifier rules covered earlier. The `studentId` variable is well named—it uses all lowercase letters except for the first letter in its second word, `Id`.

One final recommendation for your variable declarations: Try to align your comments so that they all begin in the same column. For example, note how the `//`'s are in the same column here:

```
String lastName;       // student's last name
String firstName;      // student's first name
```

3.11 Assignment Statements

You now know how to declare a variable in Java. After declaring a variable, you'll want to use it, and the first step in using a variable is to put a value inside it. We'll now consider the assignment statement, which allows you to assign/put a value into a variable.

Java Assignment Statements

Java uses the single equals sign (=) for assignment statements. See Figure 3.3's BonusCalculator program. In particular, note the `salary = 50000;` line. That's an example of a Java assignment statement. It assigns the value 50000 to the variable `salary`.

In the BonusCalculator program, note the blank line below the declaration statements. In accordance with the principles of good style, you should insert blank lines between logical chunks of code. A group of declaration statements is usually considered to be a logical chunk of code, so you should normally insert a blank line below your bottom declaration statement.

Let's analyze the program's `bonusMessage` assignment statement. Note the `*` operator. The `*` operator performs multiplication. Note the + operator. If a + operator appears between a string and something else (e.g., a number or another string), then the + operator performs *string concatenation*. That means that the JVM appends the item at the right of the + to the item at the left of the +, forming a new string. In our

```
/****************************************************************
* BonusCalculator.java
* Dean & Dean
*
* This program calculates and prints a person's work bonus.
****************************************************************/

public class BonusCalculator
{
  public static void main(String[] args)
  {
    int salary;             // person's salary
    String bonusMessage;    // specifies work bonus

    salary = 50000;
    bonusMessage = "Bonus = $" + (.02 * salary);
    System.out.println(bonusMessage);
  } // end main
} // end class BonusCalculator
```

(Callout pointing to `+`: string concatenation operator)

Figure 3.3 BonusCalculator program

example, the mathematical expression, `.02 * salary`, is evaluated first since it's inside parentheses. The JVM then appends the result, 1000, to "Bonus = $", forming the new string "Bonus = $1000".

In the `bonusMessage` assignment statement, note the parentheses around `.02 * salary`. Although the parentheses are not required by the compiler, we prefer to include them here because they improve the code's readability. They improve readability by making it clear that the math operation (.02 × salary) is separate from the string concatenation operation. Use of discretionary parentheses to enhance clarity is an art. Sometimes it's helpful, but don't get carried away. If you use parentheses too often, your code can look cluttered.

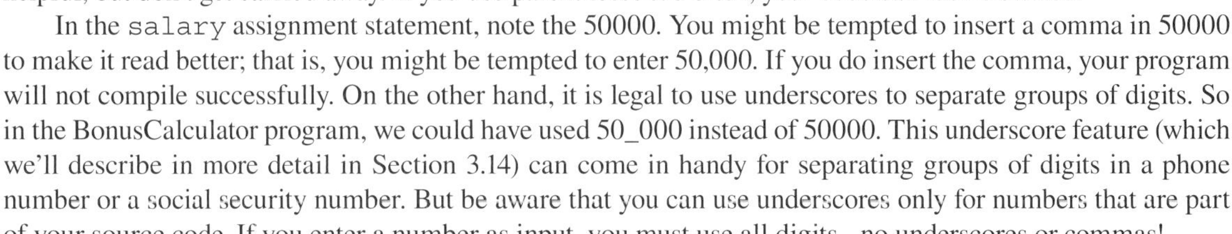

In the `salary` assignment statement, note the 50000. You might be tempted to insert a comma in 50000 to make it read better; that is, you might be tempted to enter 50,000. If you do insert the comma, your program will not compile successfully. On the other hand, it is legal to use underscores to separate groups of digits. So in the BonusCalculator program, we could have used 50_000 instead of 50000. This underscore feature (which we'll describe in more detail in Section 3.14) can come in handy for separating groups of digits in a phone number or a social security number. But be aware that you can use underscores only for numbers that are part of your source code. If you enter a number as input, you must use all digits - no underscores or commas!

Tracing

As part of a program's presentation, we'll sometimes ask you to trace the program. Tracing forces you to understand program details thoroughly. And understanding program details thoroughly is important for writing good programs. To set up a trace, provide a column heading for each variable and for output. Then execute each statement, starting with the first statement in `main`. For declaration statements, write a ? in the declared variable's column, indicating that the variable exists, but it doesn't have a value yet. For assignment statements, write the assigned value in the variable's column. For a print statement, write the printed value in the output column.[7]

[7] If you'd like a more detailed discussion of tracing, see Chapter 2, Section 2.11.

For your first Java trace, we'll make things easy. Rather than asking you to do a trace on your own, we just ask you to study the completed trace in Figure 3.4. But please do study it. Make sure you understand how all the column values get filled in.[8]

```
1  int salary;
2  String bonusMessage;
3
4  salary = 50_000;
5  bonusMessage = "Bonus = $" + (.02 * salary);
6  System.out.println(bonusMessage);
```

line#	salary	bonusMessage	output
1	?		
2		?	
4	50000		
5		Bonus = $1000.0	
6			Bonus = $1000.0

Figure 3.4 Calculating a bonus—code fragment and its associated trace

3.12 Initialization Statements

A declaration statement specifies a data type for a particular variable. An assignment statement puts a value into a variable. An initialization statement is a combination of the declaration and assignment statements—it specifies a data type for a variable, and it puts a value into that variable.

The Java language is *strongly typed,* meaning that all variable types are fixed. Once a variable is declared, it cannot be redeclared. Therefore, you can have only one declaration statement for a particular variable. Likewise, since an initialization statement is a specialized form of a declaration statement, you can have only one initialization statement for a particular variable.

Here's the syntax for an initialization statement:

```
<type> <variable> = <value>;
```

And here are some initialization examples:

```
String name = "John Doe"; // student's name
int creditHours = 0;       // student's total credit hours
```

The `name` variable is declared to be a `String` type, and it's given the initial value of "John Doe."[9] The `creditHours` variable is declared to be an `int`, and it's given the initial value of 0.

[8] If you run the code fragment on a computer, you'll see a .0 at the end of the output (`Bonus = 1000.0`). The .0 should make sense when you learn about mixed expressions and promotion later in this chapter.

[9] John Doe is commonly used as a filler in the United States and Great Britain when a person's real name is unknown. We use it here as a default value for a student's name. It serves as an indication that the student's real name has not yet been filled in.

Here's an alternative way to do the same thing using declaration and assignment statements (instead of using initialization statements):

```
String name;          // student's name
int creditHours;      // student's total credit hours

name = "John Doe";
creditHours = 0;
```

It's OK to use either technique—initialization or declaration/assignment. You'll see it done both ways in the real world. Initialization has the benefit of compactness. Declaration/assignment has the benefit of leaving more room in the declaration for a comment.

3.13 Numeric Data Types—`int`, `long`, `float`, `double`

Integers

We've already mentioned one Java numeric data type—`int`. We'll now discuss numeric types in more detail. Variables that hold whole numbers (e.g., 1000, −22) should normally be declared with the `int` data type or the `long` data type. A whole number is a number with no decimal point and no fractional component.

An `int` uses 32 bits of memory. A `long` uses 64 bits of memory (twice as many bits as an `int`). The range of values that can be stored in an `int` variable is approximately −2 billion to +2 billion. The range of values that can be stored in a `long` variable is approximately -9×10^{18} to $+9 \times 10^{18}$. Here's an example that declares `studentId` to be an `int` variable and `satelliteDistanceTraveled` to be a `long` variable:

```
int studentId;
long satelliteDistanceTraveled;
```

If you attempt to store a really big number (a number over 2 billion) in an `int` variable, you'll get an "integer number too large" error when you compile your program. So to be safe, why shouldn't you just always declare your integer variables as type `long` rather than type `int`? An `int` takes up less storage in memory. And using less storage means your computer will run faster because there's more free space. So in the interest of speed/efficiency, use an `int` rather than a `long` for a variable that holds values less than 2 billion.[10] If you're not sure whether a variable will hold values greater than 2 billion, play it safe and use a `long`. If you want the greatest possible precision in financial calculations, convert everything to cents, and use `long` variables to hold all values.

Floating-Point Numbers

Most mathematics books refer to numbers that contain a decimal point (like 66. and −1234.5) as real numbers. In Java, such numbers are called *floating-point* numbers. Why? Because a floating-point number can be written with different forms by shifting (floating) its decimal point. For example, the number −1234.5 can be written equivalently as -1.2345×10^3. See how the decimal point has "floated" to the left in the second version of the number?

There are two types for floating-point numbers—`float` and `double`. A `float` uses 32 bits of memory. A `double` uses 64 bits of memory. A `double` is called a "double" because it uses twice as many bits as a `float`.

[10] The suggestion to use an `int` for efficiency reasons is valid, but be aware that the speed difference is only occasionally noticeable. It's noticeable only if you have lots of `long` numbers and you have got a small amount of available memory, such as when you're running only a program on a personal digital assistant (PDA).

Here's an example that declares gpa as a float variable and cost as a double variable:

```
float gpa;
double cost;
```

The double data type is used much more often than the float data type. You should normally declare your floating-point variables to be double rather than float because (1) double variables can hold a wider range of numbers[11] and (2) double variables can store numbers with greater precision. Greater precision means more significant digits. You can rely on 15 significant digits for a double variable but only 6 significant digits for a float variable.

Six significant digits may seem like a lot, but for many cases, six significant digits are not enough. With only six significant digits, accuracy errors can creep into float-based programs whenever there's a mathematical operation (addition, multiplication, etc.). If such a program performs a significant number of mathematical operations, then the accuracy errors become nontrivial. So as a general rule, use double rather than float for programs that perform a significant number of floating-point mathematical operations. And because accuracy is particularly important with money, scientific measurements, and engineering measurements, use double rather than float for calculations that involve those items.

The 15 significant digits in a double variable should be sufficient for all your floating-point programming needs, but if you come across a situation where you need a floating-point variable with more than 15 significant digits, you can declare your variable using the word BigDecimal rather than the word double. BigDecimal variables handle an unlimited number of significant digits. For a thorough understanding of BigDecimal, you'll need to learn a few advanced concepts that are better off postponed. Nonetheless, if you're champing at the bit,[12] feel free to jump ahead to Chapter 12 for a sneak preview. Chapter 12 describes not only BigDecimal, but also BigInteger. BigInteger variables store integers with an unlimited number of significant digits.

Assignments Between Different Types

You've learned about assigning integer values into integer variables and floating-point values into floating-point variables, but you haven't learned about assignments where the types are different.

Assigning an integer value into a floating-point variable works just fine. Note this example:

```
double bankAccountBalance = 1000;
```

Assigning an integer value into a floating-point variable is like putting a small item into a large box. The int type goes up to approximately 2 billion. It's easy to fit 2 billion into a double "box" because a double goes all the way up to 1.8×10^{308}.

On the other hand, assigning a floating-point value into an integer variable is like putting a large item into a small box. You can't do it. For example, this generates an error:

```
int temperature = 26.7;
```

Because 26.7 is a floating-point value, it cannot be assigned to the int variable, temperature. That should make sense when you realize that it's impossible to store .7, the fractional portion of 26.7, in an int. After all, int variables don't store fractions; they store only whole numbers.

---

[11] A float variable can store positive values between 1.2×10^{-38} and $3.4 \times 10^{+38}$ and negative values between $-3.4 \times 10^{+38}$ and -1.2×10^{-38}. A double variable can store positive values between 2.2×10^{-308} and $1.8 \times 10^{+308}$ and negative values between $-1.8 \times 10^{+308}$ and -2.2×10^{-308}.

[12] "Champing at the bit" is an idiom referring to the habit of horses chewing their bits when they're (supposedly) impatient. A bit is a small metal rod placed in a horse's mouth which, when attached to the reins, enables the rider to guide the horse. To our equine readers: we apologize in advance for bringing up what might be an uncomfortable subject.

This statement also generates an error:

```
int count = 0.0;
```

The rule says that it's illegal to assign a floating-point value to an integer variable. 0.0 is a floating-point value. It doesn't matter that the fractional portion of 0.0 is insignificant (it's .0); 0.0 is still a floating-point value, and it's always illegal to assign a floating-point value into an integer variable. That type of error is known as a *compile-time error* or *compilation error* because the error is identified by the compiler during the compilation process.

Later in the book, we provide additional details about integer and floating-point data types. You don't need those details now, but if you can't wait, you can find the details in Chapter 12, Section 12.2.

3.14 Constants

We've used numeric and string values in our examples, but we haven't given you the formal name for them. Numeric and string values are called *constants*. They're called constants because their values are fixed—they don't change. Here are some examples:

Integer Constants	Floating-Point Constants	String Constants
`8`	`-34.6`	`"Hi, Bob"`
`-45`	`.009`	`"yo"`
`2000000`	`8.`	`"dog"`
`2_000_000`	`0.577_215`	

In the bottom line, note the underscores in the integer and floating-point constants. For large numbers, underscores can help to improve readability. For example, it's easy to identify the bottom integer as 2 million. Likewise, it's easy to see that the bottom floating-point number contains 6 digits of precision. In evaluating an underscored number, the JVM discards the underscores and just pays attention to the number itself. Thus, in Java, 2_000_000 equals 2_00_00_00. It's legal to place underscores between any two digits within a number (thus precluding underscores at the number's left side, at the number's right side, or next to the decimal point). But because most countries group digits by the thousands, you should normally use underscores only between every third digit. Although many countries use a comma, point, or space for the digit group separator, those symbols won't work in Java. For example, if you try to use 2,000,000 (for 2 million) in a Java program, you'll get a compilation error.

For a constant to be a floating-point constant, it must contain a decimal point, but numbers to the right of the decimal point are optional. Thus, 8. and 8.0 represent the same floating-point constant.

What is the default type for integer constants—`int` or `long`? You'd probably guess `int` because *integer* sounds like `int`. And that guess is correct—the default type for an integer constant is `int`. So the above integer examples (8, −45, and 2000000) are all `int` constants.

What is the default type for floating-point constants—`float` or `double`? Although you might be tempted to say `float` because "floating point" sounds like "float", the correct answer is `double`. Because this is so easily forgotten, let us repeat: The default type for floating-point constants is `double`.

Try to identify the compile-time errors in this code fragment:

```
float gpa = 2.30;
float mpg;
mpg = 28.6;
```

The 2.30 and 28.6 constants both default to type `double`, which uses 64 bits. The 64 bits can't squeeze into the 32-bit `gpa` and `mpg` variables so this code generates "possible loss of precision" error messages.

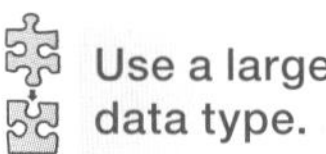
Use a larger data type.

There are two possible solutions for these types of errors. The easiest solution is to use `double` variables instead of `float` variables all the time. Here's another solution: Explicitly force the floating-point constants to be `float` by using an `f` or `F` suffix, like this:

```
float gpa = 2.30f;
float mpg;
mpg = 28.6F;
```

Two Categories of Constants

Constants can be split into two categories—hard-coded constants and named constants. The constants we've covered so far can be referred to as hard-coded constants. A *hard-coded constant* is an explicitly specified value. Hard-coded constants are also called *literals*. "Literal" is a good, descriptive term because literals refer to items that are interpreted literally; for example, 5 means 5, "hello" means "hello." In the following statement, the forward slash (/) is the division operator, and 299_792_458.0 is a hard-coded constant:

```
propagationDelay = distance / 299_792_458.0;
```

Assume that this code fragment is part of a program that calculates delays in messages carried through space. What's the meaning behind the value 299_792_458.0? Not very obvious, eh? Read on.

In space, message signals travel at the speed of light. Because time = distance / velocity, the time it takes a message signal to travel from a satellite equals the satellite's distance divided by the speed of light. Thus, in the code fragment, the number 299_792_458.0 represents the speed of light.

The above code fragment is somewhat confusing. The meaning behind the hard-coded constant 299_792_458.0 may be clear to science techies, but it isn't very clear to the rest of us. For a better solution, use a named constant.

Named Constants

A *named constant* is a constant that has a name associated with it. For example, in this code fragment, `SPEED_OF_LIGHT` is a named constant:

```
final double SPEED_OF_LIGHT = 299_792_458.0; // in meters/sec
. . .
propagationDelay = distance / SPEED_OF_LIGHT;
```

As you should be able to discern from this code fragment, a named constant is really a variable. Now there's an oxymoron—a constant is a variable. Note how `SPEED_OF_LIGHT` is declared to be a `double` variable, and it's initialized to the value 299_792_458.0. How is the `SPEED_OF_LIGHT` initialization different from initializations that you've seen in the past? The word `final` appears at the left. Also, notice that we inserted optional underscores to help identify the proper number of digits.

The reserved word `final` is a *modifier*—it modifies `SPEED_OF_LIGHT` so that its value is fixed or "final." And being fixed is the whole point of a named constant. Thus, all named constants use the `final` modifier. The `final` modifier tells the compiler to generate an error if your program ever tries to change the `final` variable's value at a later time.

Standard coding conventions suggest that you capitalize all characters in a named constant and use an underscore to separate the words in a multiple-word named constant. Example: `SPEED_OF_LIGHT`. The rationale for the uppercase is that uppercase makes things stand out. And you want named constants to stand out because they represent special values.

Named Constants Versus Hard-Coded Constants

Not all constants should be named constants. For example, if you need to initialize a `count` variable to 0, use a hard-coded 0 like this:

```
int count = 0;
```

So how do you know when to use a hard-coded constant versus a named constant? Use a named constant if it makes the code easier to understand. The above count initialization is clear the way it is now. If you replace the 0 with a named constant (e.g., `int count = COUNT_STARTING_VALUE`), it does not improve the clarity, so stick with the hard-coded constant. On the other hand, this code is unclear:

```
propagationDelay = distance / 299_792_458.0;
```

By replacing 299_792_458.0 with a `SPEED_OF_LIGHT` named constant, it does improve the clarity, so switch to the named constant.

There are two main benefits of using named constants:

1. Named constants make code more self-documenting, and therefore more understandable.
2. If a programmer ever needs to change a named constant's value, the change is easy—find the named constant initialization at the top of the method and change the initialization value. That implements the change automatically everywhere within the program. There is no danger of forgetting to change one of many occurrences of some constant value. There is consistency.

Make it easy to change.

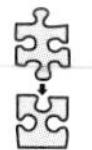

An Example

Let's put what you've learned about constants into practice by using them within a complete program. In Figure 3.5's TemperatureConverter program, we convert a Fahrenheit temperature value to a Celsius temperature value. Note the two named constant initializations at the top of the program: (1) the `FREEZING_POINT` named constant gets initialized to 32.0 and (2) the `CONVERSION_FACTOR` named constant gets initialized to 5.0 / 9.0. Usually, you'll want to initialize each named constant to a single hard-coded constant. For example, `FREEZING_POINT`'s initialization value is 32.0. But be aware that it's legal to use a constant expression for a named constant initialization value. For example, `CONVERSION_FACTOR`'s initialization value is 5.0 / 9.0. That expression is considered to be a constant expression because constant values are used, not variables.

In the TemperatureConverter program, this statement performs the conversion:

```
celsius = CONVERSION_FACTOR * (fahrenheit - FREEZING_POINT);
```

By using named constants, `CONVERSION_FACTOR` and `FREEZING_POINT`, we're able to embed some meaning into the conversion code. Without named constants, the statement would look like this:

```
celsius = 5.0 / 9.0 * (fahrenheit - 32.0);
```

The 5.0 / 9.0 may be distracting to some readers. They may spend time wondering about the significance of the 5.0 and the 9.0. By using a `CONVERSION _ FACTOR` named constant, we tell the reader "Don't worry about it; it's just a conversion factor that some scientist came up with." If someone who is unfamiliar with the Fahrenheit scale reads the above statement, they won't know the significance of the 32.0. Using a `FREEZING_POINT` named constant makes things clearer.

```
/*********************************************************************
* TemperatureConverter.java
* Dean & Dean
*
* This program converts a Fahrenheit temperature to Celsius
*********************************************************************/

public class TemperatureConverter
{
  public static void main(String[] args)
  {
    final double FREEZING_POINT = 32.0;
    final double CONVERSION_FACTOR = 5.0 / 9.0;
    double fahrenheit = 50;    // temperature in Fahrenheit
    double celsius;            // temperature in Celsius

    celsius = CONVERSION_FACTOR * (fahrenheit - FREEZING_POINT);
    System.out.println(fahrenheit + " degrees Fahrenheit = " +
      celsius + " degrees Celsius.");
  } // end main
} // end class TemperatureConverter
```

Output:

```
50.0 degrees Fahrenheit = 10.0 degrees Celsius.
```

Figure 3.5 TemperatureConverter program and its output

3.15 Arithmetic Operators

We've talked about numbers for a while now—how to declare numeric variables, how to assign numbers, and how to work with numeric constants. In addition, we've shown a few examples of using numbers in mathematical expressions. In this section and the next two sections, we study expressions in more depth. An *expression* is a combination of operands and operators that performs a calculation. Operands are variables and constants. An operator is a symbol, like + or -, that performs an operation. In this section, we'll look at arithmetic operators for numeric data types. Later, we'll look at operators for other data types.

Addition, Subtraction, and Multiplication

Java's +, -, and * arithmetic operators should be familiar to you. They perform addition, subtraction, and multiplication, respectively.

Floating-Point Division

Java performs division differently depending on whether the numbers/operands being divided are integers or whether they're floating-point numbers. Let's first discuss floating-point division.

When the Java virtual machine (JVM) performs division on floating-point numbers, it performs "calculator division." We call it "calculator division" because Java's floating-point division works the same as division performed by a standard calculator. For example, if you enter this on your calculator, what is the result?

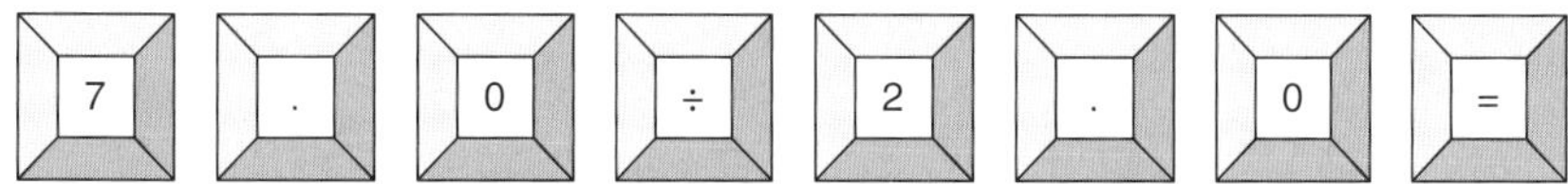

The result is 3.5. Likewise, this line of Java code prints 3.5:

```
System.out.println (7.0 / 2.0);
```

Although many calculators display a ÷ symbol on the key for division, Java uses the / character.

To explain arithmetic operators, we'll need to evaluate lots of expressions. To simplify that discussion, we'll use the ⇒ symbol. It means "evaluates to." Thus, this next line says that 7.0 / 2.0 evaluates to 3.5:

7.0 / 2.0 ⇒ 3.5

This next line asks you to determine what 5 / 4. evaluates to:

5 / 4. ⇒ ?

5 is an `int` and 4. is a `double`. This is an example of a *mixed expression.* A mixed expression is an expression that contains operands with different data types. Because they contain a fractional component, `double` values are considered to be more complex than `int` values. Whenever there's a mixed expression, the JVM temporarily *promotes* the less complex operand's type so that it matches the more complex operand's type, and then the JVM applies the operator. In the 5 / 4. expression, the JVM promotes 5 to a `double` and then performs floating-point division on the two floating-point values. The expression evaluates to 1.25.

Integer Division

When the JVM performs division on integers, it performs "grade school division." We call it grade school division because Java's integer division works the same as the division you did by hand in grade school. Remember how you calculated two values for each division operation? You calculated a quotient and also a remainder. Likewise, Java has the ability to calculate both a quotient and a remainder when integer division is called for. But Java doesn't calculate both values simultaneously. If Java's / operator is used, then the quotient is calculated. If Java's `%` operator is used, then the remainder is calculated. The `%` operator is more formally called the *modulus* operator. Note these examples:

7 / 2 ⇒ 3

7 % 2 ⇒ 1

These correspond to the equivalent grade school arithmetic notation:

```
   3   <---- quotient
2 |7
  -6
   1   <---- remainder
```

We'll give you many expression evaluation problems like this. As a sanity check, we recommend that you verify at least some of the calculated results by executing the expressions on a computer. To execute the expressions, embed the expressions into print statements, embed the print statements into a test program, and run the test program. For example, to execute the above expressions, use the TestExpressions program in Figure 3.6.

Print details to see what the computer does.

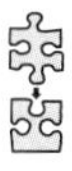

```
public class TestExpressions
{
  public static void main(String[] args)
  {
    System.out.println("7 / 2 = " + (7 / 2));
    System.out.println("7 % 2 = " + (7 % 2));
    System.out.println("8 / 12 = " + (8 / 12));
    System.out.println("8 % 12 = " + (8 % 12));
  } // end main
} // end class TestExpressions
```

Output:

```
7 / 2 = 3
7 % 2 = 1
8 / 12 = 0
8 % 12 = 8
```

Figure 3.6 TestExpressions program and its output

Figure 3.6 also illustrates these additional examples:

8 / 12 ⇒ 0

8 % 12 ⇒ 8

Here is the corresponding grade school arithmetic notation:

```
    0 <---- quotient
12 |8
   -0
    8 <---- remainder
```

At this point, you might be thinking "Gosh, that integer division stuff sure is fun, but can I use it for something practical?" Yes, indeed! You can use it to split apart numbers into their constituent parts. For example, if you're writing a program for a candy vending machine, each time the customer inserts a dollar, your program can use the integer division operators to split apart the customer's change into the proper number of quarters, dimes, and nickels. You are asked to do that for one of the projects in Chapter 6. As another example, the following code fragment shows how to split apart a number that represents the current time into its constituent parts—hours and minutes

```
int time;
System.out.print("Enter the current time as an integer (no colon): ");
time = stdIn.nextInt();
System.out.println(
  "hours = " + (time / 100) + ", minutes = " + (time % 100));
```

If the user enters 1208, then:

1208 / 100 ⇒ 12
1208 % 100 ⇒ 8

So the output would be:

```
hours = 12, minutes = 8
```

3.16 Expression Evaluation and Operator Precedence

In the above examples, the expressions were pretty basic—they each contained only one operator—so they were fairly easy to evaluate. Expressions are sometimes fairly complicated. In this section, we discuss how to evaluate those more complicated expressions.

Average Bowling Score Example

Suppose you'd like to calculate the average bowling score for three bowling games. Would this statement work?

```
bowlingAverage = game1 + game2 + game3 / 3;
```

The code looks reasonable. But it's not good enough to rely on your sense of what looks reasonable. To be a good programmer, you need to be sure. The code you should be focusing on is the expression on the right side: `game1 + game2 + game3 / 3`. More specifically, you should be asking yourself, "Which operator executes first—the left addition operator or the division operator?" To answer that question, we turn to the operator precedence table.

Operator Precedence Table

The key to understanding complicated expressions is to understand the operator precedence shown in Figure 3.7. Please study Figure 3.7's operator precedence table now.

The operator precedence table might need some clarification. The groups at the top have higher precedence than the groups at the bottom. That means that if one of the top operators appears in an expression along with one of the bottom operators, then the top operator executes first. For example, if * and + both appear in the same expression, then the * operator executes before the + operator (because the * operator's group is higher in the table than the + operator's group). If parentheses appear within an expression, then the items inside the parentheses execute before the items that are outside the parentheses (because parentheses are at the very top of the table).

If an expression has two or more operators in the same group (from Figure 3.7's groups), then apply the operators from left to right. In mathematics, that's referred to as *left-to-right associativity*. In Java, that means that operators appearing on the left should be executed before operators appearing on the right. For example, because the * and / operator are in the same group, if * and / both appear in the same expression and / appears further to the left than * within that expression, division is performed before multiplication

The operators in the second-from-the-top group are *unary operators*. A unary operator is an operator that applies to just one operand. The unary + operator is cosmetic; it does nothing. The unary - operator (negation) reverses the sign of its operand. For example, if the variable `x` contains a 6, then `-x` evaluates to a negative 6. The (<*type*>) operator represents the cast operators. We'll get to cast operators later in this chapter.

1. grouping with parentheses:
   (<*expression*>)
2. unary operators:
   ```
   +x
   -x
   (<type>) x
   ```
3. multiplication and division operators:
   ```
   x * y
   x / y
   x % y
   ```
4. addition and subtraction operators:
   ```
   x + y
   x - y
   ```

Figure 3.7 Abbreviated operator precedence table (see Appendix 2 for the complete table)
Operator groups at the top of the table have higher precedence than operator groups at the bottom of the table. All operators within a particular group have equal precedence, and they evaluate left to right.

Average Bowling Score Example Revisited

Let's return to the average bowling score example and apply what you've learned about operator precedence. Does the following statement correctly calculate the average bowling score for three bowling games?

```
bowlingAverage = game1 + game2 + game3 / 3;
```

No. The operator precedence table says that the / operator has higher priority than the + operator, so division is performed first. After the JVM divides `game3` by 3, the JVM adds `game1` and `game2`. The correct way to calculate the average is to add the three game scores first and then divide the sum by 3. In other words, you need to force the + operators to execute first. The solution is to use parentheses like this:

```
bowlingAverage = (game1 + game2 + game3) / 3;
```

Expression Evaluation Practice

Hand calculation helps your understanding.

Let's do some expression evaluation practice problems to ensure that you really understand this operator precedence material. Given these initializations:

```
int a = 5, b = 2;
double c = 3.0;
```

What does the following expression evaluate to?

```
(c + a / b) / 10 * 5
```

Here's the solution:

1. `(c + a / b) / 10 * 5` ⇒
2. `(3.0 + 5 / 2) / 10 * 5` ⇒

3. `(3.0 + 2) / 10 * 5` ⇒
4. `5.0 / 10 * 5` ⇒
5. `0.5 * 5` ⇒
6. `2.5`

In solving expression evaluation problems, we recommend that you show each step of the evaluation process so your solution is easy to follow. In the above solution, we show each step, and we also show line numbers. There's normally no need to show line numbers, but we do it here to help with our explanation. From line 1 to line 2, we replace variables with their values. From line 2 to line 3, we evaluate the highest priority operator, the / inside the parentheses. From line 3 to line 4, we evaluate the next highest priority operator, the + inside the parentheses. Study the remaining lines on your own.

Let's do one more expression evaluation practice problem. Given these initializations:

```
int x = 5;
double y = 3.0;
```

What does the following expression evaluate to?

```
(0 % x) + y + (0 / x)
```

Here's the solution:

```
(0 % x) + y + (0 / x) ⇒
(0 % 5) + 3.0 + (0 / 5) ⇒
0 + 3.0 + (0 / 5) ⇒
0 + 3.0 + 0 ⇒
3.0
```

Perhaps the trickiest part of the above solution is evaluating 0 % 5 and 0 / 5. They both evaluate to 0. This grade school arithmetic notation shows why:

```
   0 ◄──── quotient
5 |0
  -0
   0 ◄──── remainder
```

3.17 More Operators: Increment, Decrement, and Compound Assignment

So far, we've covered Java math operators that correspond to operations found in math books—addition, subtraction, multiplication, and division. Java provides additional math operators that have no counterparts in math books. In this section, we'll talk about the increment, decrement, and compound assignment operators.

Increment and Decrement Operators

It's fairly common for a computer program to count the number of times something occurs. For example, have you ever seen a web page that displays the number of "visitors"? The number of visitors is tracked by a program that counts the number of times the web page has been loaded on someone's web browser. Since counting is such a common task for programs, there are special operators for counting. The increment operator (++) counts up by 1. The decrement operator (--) counts down by 1.

Here's one way to increment the variable *x*:

```
x = x + 1;
```

And here's how to do it using the increment operator:

```
x++;
```

The two techniques are equivalent in terms of their functionality. Experienced Java programmers almost always use the second form rather than the first form, and proper style suggests using the second form. So use the second form.

Here's one way to decrement the variable *x*:

```
x = x - 1;
```

And here's how to do it using the decrement operator:

```
x--;
```

Once again, you should use the second form.

Compound Assignment Operators

Let's now discuss five of Java's *compound assignment* operators: +=, -=, *=, /=, and %=.

The += operator updates a variable by adding a specified value to the variable. Here's one way to increment x by 3:

```
x = x + 3;
```

And here's how to do it using the += operator:

```
x += 3;
```

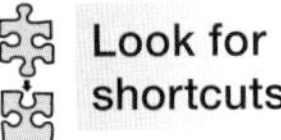

Look for shortcuts.

The two techniques are equivalent in terms of their functionality. Experienced Java programmers almost always use the shorter second form rather than the longer first form. And proper style suggests using the second form. So use the second form.

The -= operator updates a variable by subtracting a specified value from the variable. Here's one way to decrement x by 3:

```
x = x - 3;
```

And here's how to do it using the -= operator:

```
x -= 3;
```

Once again, you should use the second form.

The *=, /=, and %= operators parallel the += and -= operators, so we won't bore you with detailed explanations for those remaining three operators. But we do encourage you to study the *=, /=, and %= examples shown below:

```
x += 3;         ≡     x = x + 3;
x -= 4;         ≡     x = x - 4;
x *= y;         ≡     x = x * y;
x /= 4;         ≡     x = x / 4;
x %= 16;        ≡     x = x % 16;
x *= y + 1;     ≡     x = x * (y + 1);
```

The examples show assignment operator statements on the left and their equivalent long-form statements on the right. The ≡ symbol means "is equivalent to." It's better style to use the forms on the left rather than the forms on the right, but don't ignore the forms on the right. They show how the assignment operators work.

The bottom example is the only one in which the compound assignment operator uses an expression rather than a single value; that is, the expression to the right of the *= assignment operator is y + 1, rather than just 1. For cases like these, the compound assignment form is somewhat confusing. Therefore, for these cases, it's acceptable style-wise to use the equivalent long form rather than the compound assignment form.

Why are the +=, -=, *=, /=, and %= operators called compound assignment operators? Because they compound/combine a math operation with the assignment operation. For example, the += operator performs addition and assignment. The addition part is obvious, but what about the assignment part? The += does indeed perform assignment because the variable at the left of the += is assigned a new value.

3.18 Tracing

To make sure that you really understand the increment, decrement, and compound assignment operators, let's trace a program that contains those operators. Earlier in the chapter, we showed a trace, but the trace was for a very limited code fragment—the code fragment contained two assignment statements and that was it. In this section, we present a more complicated trace.

See the TestOperators program and associated trace table in Figure 3.8. In particular, look at the first three lines under the heading in the trace table. They contain the variables' initial values. For variables declared as part of an initialization, their initial value is the initialization value. For variables declared without an initialization, we say their initial value is *garbage* because its actual value is unknown. Use a question mark to indicate a garbage value.

We suggest you cover up the bottom part of the trace, and try to complete the trace on your own. When you're done, compare your answer to Figure 3.8's trace table.

Put yourself in the computer's place.

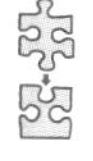

There are different modes for the increment and decrement operators—prefix mode and postfix mode. Later in the book, we explain the modes and provide details on how they work within the context of a trace. You don't need those details now, but if you can't wait, you can find the details in Chapter 12, Section 12.5.

3.19 Type Casting

We've now described simple arithmetic operators (+, -, *, /, %), increment and decrement operators (++, --), and compound assignment operators (+=, -=, *=, /=, %=). In this section, we'll discuss yet another operator, the cast operator.

Cast Operator

In writing a program, you'll sometimes need to convert a value to a different data type. The cast operator can be used to perform that sort of conversion. Here's the syntax:

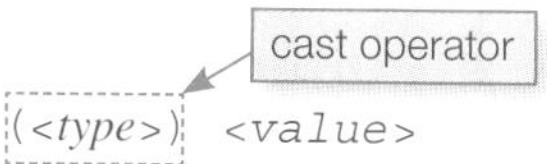

As shown above, a cast operator consists of a data type inside parentheses. You should place a cast operator at the left of the value that you'd like to convert.

```
1   public class TestOperators
2   {
3     public static void main(String[] args)
4     {
5       int x;
6       int y = 2;
7       double z = 3.0;
8
9       x = 5;
10      System.out.println("x + y + z = " + (x + y + z));
11      x += y;
12      y++;
13      z--;
14      z *= x;
15      System.out.println("x + y + z = " + (x + y + z));
16    } // end main
17  } // end class TestOperators
```

Trace:

line#	x	y	z	output
5	?			
6		2		
7			3.0	
9	5			
10				x + y + z = 10.0
11	7			
12		3		
13			2.0	
14			14.0	
15				x + y + z = 24.0

Figure 3.8 TestOperators program and its trace

Suppose you have a variable named `interest` that stores a bank account's interest as a `double`. You'd like to extract the dollars portion of the interest and store it in a variable of type `int` that is named `interestInDollars`. To do that, use the `int` cast operator like this:

```
interestInDollars = (int) interest;
```

The `int` cast operator returns the whole number portion of the casted value, truncating the fractional portion. Thus, if `interest` contains the value 56.96, after the assignment, `interestInDollars` contains the value 56. Note that the cast operation does not change the value of `interest`. After the assignment, `interest` still contains 56.96.

Use Parentheses to Cast an Expression

If you ever need to cast more than just a single value or variable, then make sure to put parentheses around the entire expression that you want to cast. Note this example:

```
double interestRate;
double balance;
int interestInDollars;
. . .
interestInDollars = (int) (balance * interestRate);
```

Parentheses are necessary here

In the `interestInDollars` assignment, `balance * interestRate` is the formula for calculating interest. This code fragment performs basically the same operation as the previous one-line code fragment. It extracts the dollars portion of the interest and stores it in an `int` variable named `interestInDollars`. The difference is that the interest this time is in the form of an expression, `balance * interestRate`, rather than in the form of a simple variable, `interest`. Because we want the cast operator to apply to the entire expression, we need to put parentheses around `balance * interestRate`.

In the above code fragment, what would happen if there were no parentheses around the expression, `balance * interestRate`? The cast would then apply only to the first thing at its right, `balance`, rather than the entire expression. That should make sense when you look at the operator precedence table. The operator precedence table shows that the cast operator has very high precedence. So without the parentheses, the cast operator would execute prior to the multiplication operator, and the cast would thus apply only to `balance`. And that leads to an incorrect calculation for interest in dollars.

Use a Floating-Point Cast to Force Floating-Point Division

Suppose you have a variable named `earnedPoints` that stores a student's earned grade points for a semester's worth of classes. Suppose you have a variable named `numOfClasses` that stores the number of classes taken by the student. The student's grade point average (GPA) is calculated by dividing earned points by number of classes. In the following statement, `earnedPoints` and `numOfClasses` are `int`s and `gpa` is a `double`. Does the statement correctly calculate the student's GPA?

```
gpa = earnedPoints / numOfClasses;
```

Compare output with what you expect.

Suppose `earnedPoints` holds 14 and `numOfClasses` holds 4. You'd like `gpa` to get a value of 3.5 (because $14 \div 4 = 3.5$). But alas, `gpa` gets a value of 3. Why? Because the / operator performs integer division on its two `int` operands. Integer division means the quotient is returned. The quotient of $14 \div 4$ is 3. The solution is to force floating-point division by introducing the cast operator. Here's the corrected code:

```
gpa = (double) earnedPoints / numOfClasses;
```

After casting `earnedPoints` to a `double`, the JVM sees a mixed expression and promotes `numOfClasses` to a `double`. Then floating-point division takes place.

For this example, you should not put parentheses around the `earnedPoints / numOfClasses` expression. If you did so, the / operator would have higher precedence than the cast operator, and the JVM would perform division (integer division) prior to performing the cast operation.

Later in the book, we provide additional details about type conversions. You don't need those details now, but if you can't wait, you can find the details in Chapter 12, Section 12.4.

3.20 char Type and Escape Sequences

In the past, when we've stored or printed text, we've always worked with groups of text characters (strings), not with individual characters. In this section, we'll use the char type to work with individual characters.

char Type

If you know that you'll need to store a single character in a variable, use a char variable. Here's an example that declares a char variable named ch and assigns the letter *A* into it.

```
char ch;
ch = 'A';
```

Note the 'A'. That's a char literal. char literals must be surrounded by single quotes. That syntax parallels the syntax for string literals—string literals must be surrounded by double quotes.

What's the point of having a char type? Why not just use one-character strings for all character processing? Because for applications that manipulate lots of individual characters, it's more efficient (faster) to use char variables, which are simple, rather than string variables, which are more complex. For example, the software that allows you to view web pages has to read and process individual characters as they're downloaded onto your computer. In processing the individual characters, it's more efficient if they're stored as separate char variables rather than as string variables.

String Concatenation with char

Remember how you can use the + symbol to concatenate two strings together? You can also use the + symbol to concatenate a char and a string. What do you think this code fragment prints?

```
char first, middle, last;       // a person's initials

first = 'J';
middle = 'S';
last = 'D';
System.out.println("Hello, " + first + middle + last + '!');
```

Here's the output:

```
Hello, JSD!
```

Escape Sequences

Usually, it's easy to print characters. Just stick them inside a System.out.println statement. But some characters are hard to print. We use *escape sequences* to print hard-to-print characters such as the tab character. An escape sequence is comprised of a backslash (\) and another character. See Java's most popular escape sequences in Figure 3.9.

If you print the tab character (\t), the computer screen's cursor moves to the next tab stop. The computer screen's cursor is the position on the screen where the computer prints next. If you print the newline character (\n), the computer screen's cursor moves to the beginning of the next line.

Here's an example of how you could print two column headings, BALANCE and INTEREST, separated by a tab, and followed by a blank line:

```
System.out.println("BALANCE" + '\t' + "INTEREST" + '\n');
```

```
\t   move the cursor to the next tab stop
\n   newline—go to first column in next line
\r   return to first column in current line

\"   print a literal double quote
\'   print a literal single quote
\\   print a literal backslash
```

Figure 3.9 Common escape sequences

Note that escape sequences are indeed characters, so to print the tab and newline characters, we've surrounded them with single quotes.

Normally the compiler interprets a double quote, a single quote, or a backslash as a *control character*. A control character is in charge of providing special meaning to the character(s) that follows it. The double quote control character tells the computer that the subsequent characters are part of a string literal. Likewise, the single quote control character tells the computer that the subsequent character is a `char` literal. The backslash control character tells the computer that the next character is to be interpreted as an escape sequence character.

But what if you'd like to print one of those three characters as is and bypass the character's control functionality? To do that, preface the control character (double quote, single quote, backslash) with a backslash. The initial backslash turns off the subsequent character's control functionality and thus allows the subsequent character to be printed as is. If that doesn't make sense, all you really have to know is this:

To print a double quote, use `\"`.

To print a single quote, use `\'`.

To print a backslash, use `\\`.

Suppose you'd like to print this message:

```
"Hello.java" is stored in the c:\javaPgms folder.
```

Here's how to do it:

```
System.out.println('\"' + "Hello.java" + '\"' +
  " is stored in the c:" + '\\' + "javaPgms folder.");
```

Embedding an Escape Sequence Within a String

Write a print statement that generates this heading for a computer-specifications report:

```
HARD DISK SIZE       RAM SIZE ("MEMORY")
```

Specifically, your print statement should generate a tab, a HARD DISK SIZE column heading, two more tabs, a RAM SIZE ("MEMORY") column heading, and then two blank lines. Here's one solution:

```
System.out.println('\t' + "HARD DISK SIZE" + '\t' + '\t' +
  "RAM SIZE (" + '\"' + "MEMORY" + '\"' + ")" + '\n' + '\n');
```

That's pretty cluttered. Fortunately, there's a better way. An escape sequence is designed to be used like any other character within a string of text, so it's perfectly acceptable to embed escape sequences within strings and omit the +'s and the single quotes. For example, here's an alternative solution for the PC specifications report heading problem where the +'s and single quotes have been removed:

```
System.out.println("\tHARD DISK SIZE\t\tRAM SIZE (\"MEMORY\")\n\n");
```

Everything is now all within just one string literal. By omitting the +'s and single quotes, the clutter is reduced and that makes everyone happy. (Exception—author John's kids love clutter and would thus abhor this second solution.)

Origin of the Word "Escape" for Escape Sequences

Why is the word "escape" used for escape sequences? The backslash forces an "escape" from the normal behavior of a specified character. For example, if `t` is in a print statement, the computer normally prints `t`. If `\t` is in a print statement, the computer escapes from printing `t`, and instead, it prints the tab character. If the double quote character (") is in a print statement, the computer normally treats it as the start or end of a string literal. If `\"` is in a print statement, the computer escapes from the start/end string behavior; instead the computer prints the double quote character.

Later in the book, we present relatively advanced syntax details that pertain to the `char` type. You don't need those details now, but if you can't wait, you can find the details in Chapter 12, Section 12.3.

3.21 Primitive Variables Versus Reference Variables

Throughout the chapter, we've defined and discussed various types of variables—`String`, `int`, `long`, `float`, `double`, and `char` variables. It's now time to step back and get a big-picture view of the two different categories of variables in Java—primitive variables and reference variables.

Primitive Variables

A *primitive variable* stores a single piece of data. It's helpful to think of a primitive variable's data item as being inherently indivisible. More formally, we say that it's "atomic" because, like an atom, it's a basic "building block" and it cannot be broken apart.[13] Primitive variables are declared with a *primitive type,* and those types include:

`int, long`	(integer types)
`float, double`	(floating-point types)
`char`	(character type)

There are additional primitive types (`boolean`, `byte`, `short`), which we'll get to in Chapters 4 and 12, but for most situations, these five primitive types are sufficient.

[13]The word "atom" comes from the Greek *a-tomos* and means indivisible. In 1897, J. J. Thomson discovered one of the atom's components—the electron—and thus dispelled the notion of an atom's indivisibility. Nonetheless, as a holdover from the original definition of atom, the term "atomic" still refers to something that is inherently indivisible.

Reference Variables

Whereas a primitive variable stores a single piece of data, a *reference variable* stores a memory location that points to a collection of data. This memory location is not a literal memory address, like a street address. It's a coded abbreviation, like a post-office box number. However, for everything you can do in Java, the value in a reference variable acts exactly like a literal memory address, so we'll pretend it is one. We said a reference variable's "address" points to a collection of data. More formally, it points to an *object*. You'll learn about object details in Chapter 6, but for now, just realize that an object is a collection of related data wrapped in a protective shell. To access an object's data, you need to use a reference variable (or *reference* for short) that points to the object.

String variables are examples of reference variables. A string variable holds a memory address that points to a string object. The string object holds the data—the string's characters.

Reference variables are declared with a *reference type*. A reference type is a type that provides for the storage of a collection of data. `String` is a reference type, and it provides for the storage of a collection of characters. So in the following example, declaring `name` with a `String` reference type means that `name` points to the collection of characters T, h, a, n, h, space, N, g, u, y, e, n.

```
String name = "Thanh Nguyen";
```

`String` is just one reference type from among a multitude of reference types. Classes, arrays, and interfaces are all considered to be reference types. You'll learn about arrays in Chapter 9 and interfaces in Chapter 10. You'll learn about class details in Chapter 6, but for now, it's good enough to know that a class is a generic description of the data in a particular type of object. For example, the `String` class describes the nature of the data in string objects. More specifically, the `String` class says that each string object can store zero or more characters and the characters are stored in a sequence.

An Example

Let's look at an example that uses primitive variables and reference variables. In this code fragment, we declare variables that keep track of a person's basic data:

```
int ssn;        // Social Security number
String name;    // person's name
Calendar bday;  // person's birthday
```

As you can tell by the `int` and `String` data types, `ssn` is a primitive variable and `name` is a reference variable. In the third line, `Calendar` is a class. That tells us that `bday` is a reference variable. The `Calendar` class allows you to store date information such as year, month, and day.[14] Since `bday` is declared with the `Calendar` class, `bday` is able to store year, month, and day data items.

3.22 Strings

We've used strings for quite a while now, but we've stored them and printed them and that's it. Many programs need to do more with strings than just store and print. For example, Microsoft Office programs (such as Word, Excel, and PowerPoint) all include text search and text replace capabilities. In this section, we describe how Java provides that sort of string-manipulation functionality in the `String` class.

[14]Explaining the `Calendar` class in depth is beyond the scope of this chapter. If you want an in-depth explanation, go to Oracle's Java documentation website http://docs.oracle.com/javase/7/docs/api/ and search for `Calendar`.

String Concatenation

As you know, strings are normally concatenated with the + operator. Note that strings can also be concatenated with the += compound assignment operator. In the following example, if the `animal` string references "dog" originally, it references "dogfish" after the statement is executed:

```
animal += "fish";
```

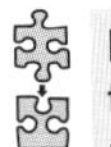

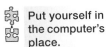

Put yourself in the computer's place. We recommend that you now go through a trace to make sure you thoroughly understand string concatenation. See the code fragment in Figure 3.10. Try to trace the code fragment on your own prior to looking at the solution.

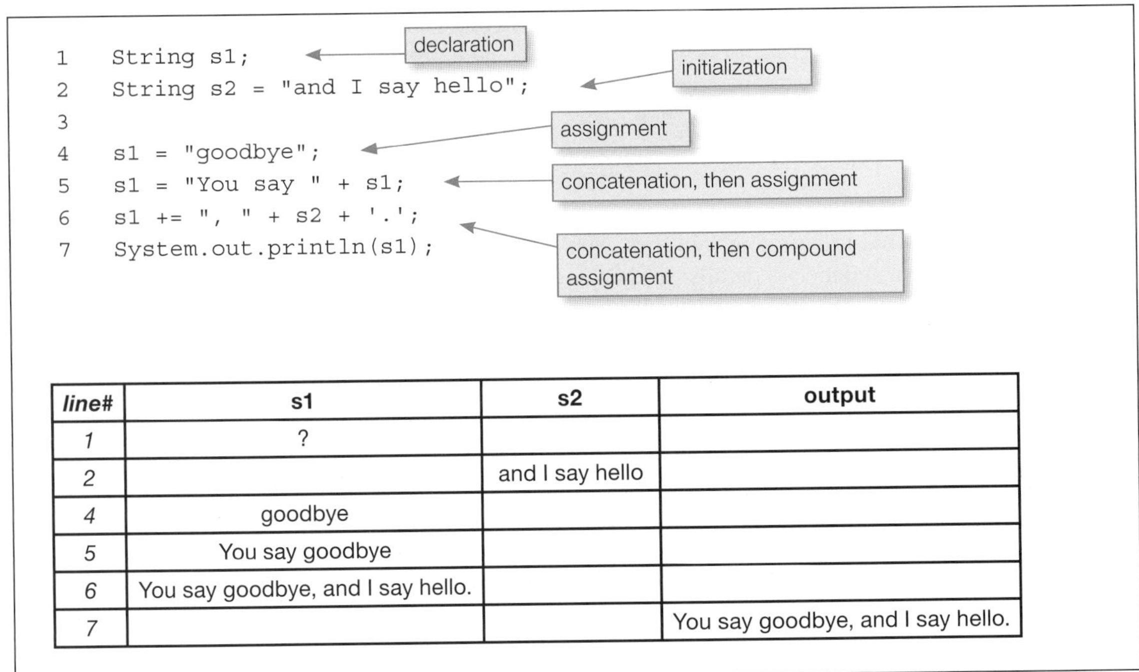

```
1   String s1;
2   String s2 = "and I say hello";
3
4   s1 = "goodbye";
5   s1 = "You say " + s1;
6   s1 += ", " + s2 + '.';
7   System.out.println(s1);
```


line#	s1	s2	output
1	?		
2		and I say hello	
4	goodbye		
5	You say goodbye		
6	You say goodbye, and I say hello.		
7			You say goodbye, and I say hello.

Figure 3.10 Code fragment and associated trace for string concatenation illustration

String Methods

In the previous section, we defined an object to be a collection of data. An object's data is normally protected, and, as such, it can be accessed only through special channels. Normally, it can be accessed only through the object's methods. A string object stores a collection of characters, and a string object's characters can be accessed only through its `charAt` method. In the remainder of this section, we'll describe the `charAt` method and three other popular string methods—`length`, `equals`, and `equalsIgnoreCase`. These methods, as well as many about other string methods, are defined in the `String` class.

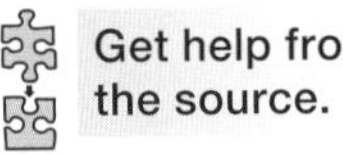

Get help from the source. If you'd like to learn more about the `String` class and all of its methods, visit Oracle's Java documentation website http://docs.oracle.com/javase/7/docs/api/ and follow links that take you to the `String` class.

The `charAt` Method

Suppose you initialize a string variable, `animal`, with the value "cow". The `animal` variable then points to a string object that contains three data items—the three characters 'c', 'o', and 'w'. To retrieve a data item (i.e., a character), call the `charAt` method. `charAt` stands for character at. The `charAt` method returns a character at a specified position. For example, if `animal` calls `charAt` and specifies the third position, then `charAt` returns 'w' because 'w' is the third character in "cow".

So how do you call the `charAt` method? Let us answer that question by comparing a `charAt` method call to a method call that you're already comfortable with—the `println` method call. See Figure 3.11.

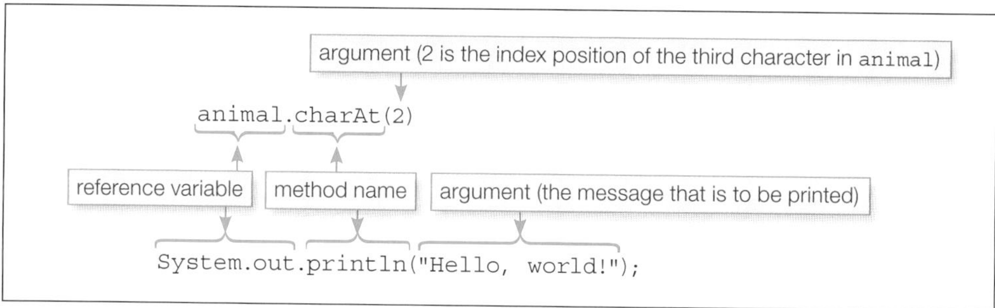

Figure 3.11 Comparison of `charAt` method call to `println` method call

In Figure 3.11, note how the `charAt` method call and the `println` method call both use this syntax:

<reference-variable>.*<method-name>*(*<argument>*)

In the `charAt` call, `animal` is the reference variable, `charAt` is the method name and 2 is the argument. The argument is the tricky part. The argument specifies the *index* of the character that is to be returned. The positions of characters within a string are numbered starting with index zero, not index one. For emphasis, we say again! The positions in a string start with index zero. So if `animal` contains "cow," what does `animal.charAt(2)` return? As the following table indicates, the 'w' character is at index 2, so `animal.charAt(2)` returns 'w.'

index:	0	1	2
"cow" string's characters:	c	o	w

If you call `charAt` with an argument that's negative or that's equal to or greater than the string's length, your code will compile OK, but it won't run properly. For example, suppose you run this program:

```
public class Test
{
  public static void main(String[] args)
  {
    String animal = "sloth";
    System.out.println("Last character: " + animal.charAt(5));
  }
}
```

Because sloth's last index is 4, not 5, the JVM prints an error message. More specifically, it prints this:

```
Exception in thread "main"
java.lang.StringIndexOutOfBoundsException:
  String index out of range: 5
  at java.lang.String.charAt(String.java:558)
  at Test.main(Test.java:6)
```

The 5 refers to the specified index; it is "out of range."

The 6 refers to the line number in the program where the error occurred.

Ask: What is the computer trying to tell me?

At first, such error messages are intimidating and depressing, but eventually you'll learn to love them. Well, maybe not quite love them, but you'll learn to appreciate the information they provide. They provide information about the type of error and where the error occurred. Try to view each error message as a learning opportunity! At this point, don't worry about understanding all the details in the above error message. Just focus on the two callouts and the lines that they refer to.

The above error is an example of a *runtime error*. A runtime error is an error that occurs while a program is running, and it causes the program to terminate abnormally. Said another way, it causes the program to *crash*.

The `length` Method

The `length` method returns the number of characters in a particular string. What does this code fragment print?

```
String s1 = "hi";
String s2 = "";
System.out.println("number of characters in s1 = " + s1.length());
System.out.println("number of characters in s2 = " + s2.length());
```

Because `s1`'s string contains two characters ('h' and 'i'), the first print statement prints this:

```
number of characters in s1 = 2
```

In the code fragment, `s2` is initialized with the `""` value. The `""` value is commonly known as the *empty string*. An empty string is a string that contains no characters. Its length is zero. The second print statement prints this:

```
number of characters in s2 = 0
```

In calling the `charAt` method, you need to insert an argument (an index value) in the method call's parentheses. For example, `animal.charAt(2)`. On the other hand, in calling the `length` method, there's no need to insert an argument in the method call's parentheses. For example, `s1.length()`. You may be thinking "With no argument, why bother with the parentheses?" In calling a method, you always need parentheses, even if they're empty. Without the parentheses, the compiler won't know that the method call is a method call.

The `equals` Method

To compare two strings for equality, it's necessary to step through the characters in both strings and compare same-positioned characters, one at a time. Fortunately, you don't have to write code to do that rather tedious comparison operation every time you want to see if two strings are equal. You just have to call the `equals` method, and it does the tedious comparison operation automatically, behind the scenes. More succinctly, the `equals` method returns `true` if two strings contain the exact same sequence of characters. It returns `false` otherwise.

We recommend that you now go through a trace to make sure you thoroughly understand the `equals` method. See the code fragment in Figure 3.12. Try to trace the code fragment on your own prior to looking at the solution.

Put yourself in the computer's place.

```
1   String animal1 = "Horse";
2   String animal2 = "Fly";
3   String newCreature;
4
5   newCreature = animal1 + animal2;
6   System.out.println(newCreature.equals("HorseFly"));
7   System.out.println(newCreature.equals("horsefly"));
```

line#	animal 1	animal 2	newCreature	output
1	Horse			
2		Fly		
3			?	
5			HorseFly	
6				true
7				false

Figure 3.12 Code fragment that illustrates the `equals` method and its associated trace

Because `newCreature` contains the value "HorseFly", the `equals` method returns a value of `true` when `newCreature` is compared to "HorseFly". On the other hand, when `newCreature` is compared to lowercase "horsefly", the `equals` method returns a value of `false`.

The `equalsIgnoreCase` Method

Sometimes you might want to disregard uppercase versus lowercase when comparing strings. In other words, you might want "HorseFly" and "horsefly" to be considered equal. To test for case-insensitive equality, call the `equalsIgnoreCase` method.

What does this code fragment print?

```
System.out.println("HorseFly".equalsIgnoreCase("horsefly"));
```

Because `equalsIgnoreCase` considers "HorseFly" and "horsefly" to be equal, the code fragment prints `true`.

3.23 Input—the `Scanner` Class

Programs are normally a two-way street. They produce output by displaying something on the computer screen, and they read input from the user. Up to this point, all our Java programs and code fragments have gone just one way—they've displayed something on the screen, but they haven't read any input. With no input, our programs have been rather limited. In this section, we'll discuss how to get input from a user. With input, we'll be able to write programs that are much more flexible and useful.

Ask: What if? Suppose you're asked to write a program that calculates earnings for a retirement fund. If there's no input, your program must make assumptions about contribution amounts, years before retirement, and so on. Your program then calculates earnings based on those assumptions. Bottom line: Your no-input program calculates earnings for one specific retirement-fund plan. If input is used, your program asks the user to supply contribution amounts, years before retirement, and so forth. Your program then calculates earnings based on those user inputs. So which version of the program is better—the no-input version or the input version? The input version is better because it allows the user to plug in what-if scenarios. What happens if I contribute more money? What happens if I postpone retirement until I'm 90?

Input Basics

The Java API library provides a prebuilt class named `Scanner`, which allows you to get input from either a keyboard or a file. We describe file input in Chapter 16. Prior to that, when we talk about input, you should assume that we're talking about keyboard input.

The `Scanner` class is not part of the core Java language. So if you use the `Scanner` class, you need to tell the compiler where to find it. You do that by importing the `Scanner` class into your program. More specifically, you need to include this `import` statement at the top of your program (right after your prologue section):

```
import java.util.Scanner;
```

We describe `import` details (like what is `java.util`?) in Chapter 5. For now, suffice it to say that you need to import the `Scanner` class in order to prepare your program for input.

There's one more thing you need to do to prepare your program for input. Insert this statement at the top of your `main` method:

```
Scanner stdIn = new Scanner(System.in);
```

The `new Scanner(System.in)` expression creates an object. As you now know, an object stores a collection of data. In this case, the object stores characters entered by a user at a keyboard. The `stdIn` variable is a reference variable, and it gets initialized to the address of the newly created `Scanner` object. After the initialization, the `stdIn` variable allows you to perform input operations.

With the above overhead in place, you can read and store a line of input by calling the `nextLine` method like this:

```
<variable> = stdIn.nextLine();
```

Let's put what you've learned into practice by using the `Scanner` class and the `nextLine` method call in a complete program. See the FriendlyHello program in Figure 3.13. The program prompts the user to enter his or her name, saves the user's name in a `name` variable, and then prints a greeting with the user's name embedded in the greeting.

In the FriendlyHello program, note the "Enter your name: " print statement. It uses a `System.out.print` statement rather than a `System.out.println` statement. Remember what the "ln" in `println` stands for? It stands for "line." The `System.out.println` statement prints a message and then moves the screen's cursor to the next line. On the other hand, the `System.out.print` statement prints a message and that's it. The cursor ends up on the same line as the printed message (just to the right of the last printed character).

So why did we bother to use a `print` statement instead of a `println` statement for the "Enter your name:" prompt? Because users are used to entering input just to the right of a prompt message. If we used `println`, then the user would have to enter input on the next line. One additional item: We inserted a colon and a blank space at the end of the prompt. Once again, the rationale is that that's what users are used to.

```
/*************************************************************
* FriendlyHello.java
* Dean & Dean
*
* This program displays a personalized Hello greeting.
*************************************************************/

import java.util.Scanner;

public class FriendlyHello
{
  public static void main(String[] args)
  {
    Scanner stdIn = new Scanner(System.in);
    String name;
    System.out.print("Enter your name: ");
    name = stdIn.nextLine();
    System.out.println("Hello " + name + "!");
  } // end main
} // end class FriendlyHello
```

These two statements create a keyboard-input connection.

This gets a line of input.

Figure 3.13 FriendlyHello program

Input Methods

In the FriendlyHello program, we called the `Scanner` class's `nextLine` method to get a line of input. The `Scanner` class contains quite a few other methods that get different forms of input. Here are some of those methods:

`next()`	Skip leading whitespace until a token is found. Return the token as a `String` value.
`nextInt()`	Skip leading whitespace until a token is found. Return the token as an `int` value.
`nextLong()`	Skip leading whitespace until a token is found. Return the token as a `long` value.
`nextFloat()`	Skip leading whitespace until a token is found. Return the token as a `float` value.
`nextDouble()`	Skip leading whitespace until a token is found. Return the token as a `double` value.

The above descriptions need some clarification:

1. What is leading whitespace?
   Whitespace refers to all characters that appear as blanks on a display screen or printer. This includes the space character, the tab character, and the newline character. The newline character is generated with the enter key. *Leading whitespace* refers to whitespace characters that are at the left side of the input.
2. The `next` method looks for a token. What is a token?
   Think of a *token* as a word since the `next` method is usually used for reading in a single word. But more formally, a *token* is a sequence of non-whitespace characters. For example, "gecko" and "53B@a!" are

tokens. But "Gila monster" forms two tokens, not one, because the space between "Gila" and "monster" signals the end of a token.

3. What happens if the user provides invalid input for `nextInt()`, `nextLong()`, `nextFloat()`, or `nextDouble()`?
   The JVM prints an error message and stops the program. For example, if a user enters `hedgehog`, `45g`, or `45.0` in response to a `nextInt()` call, the JVM prints an error message and stops the program.
4. What is a Boolean value?
   You'll learn about Boolean values at the start of the next chapter, in Section 4.2.

Examples

To make sure you understand `Scanner` methods, study the programs in Figures 3.14 and 3.15. They illustrate how to use the `nextDouble`, `nextInt`, and `next` methods. Pay particular attention to the sample sessions. The sample sessions show what happens when the programs run with typical sets of input. In

```
/****************************************************************
* PrintPO.java
* Dean & Dean
*
* This program calculates and prints a purchase order amount.
****************************************************************/

import java.util.Scanner;

public class PrintPO
{
  public static void main(String[] args)
  {
    Scanner stdIn = new Scanner(System.in);
    double price;  // price of purchase item
    int qty;       // number of items purchased

    System.out.print("Price of purchase item: ");
    price = stdIn.nextDouble();
    System.out.print("Quantity: ");
    qty = stdIn.nextInt();
    System.out.println("Total purchase order = $" + price * qty);
  } // end main
} // end class PrintPO
```

Sample session:

```
Price of purchase item: 34.14
Quantity: 2
Total purchase order = $68.28
```

Figure 3.14 PrintPO program that illustrates `nextDouble()` and `nextInt()`

```
/*******************************************************************
* PrintInitials.java
* Dean & Dean
*
* This program prints the initials for a user-entered name.
*******************************************************************/

import java.util.Scanner;

public class PrintInitials
{
  public static void main(String[] args)
  {
    Scanner stdIn = new Scanner(System.in);
    String first;  // first name
    String last;   // last name

    System.out.print(
      "Enter your first and last name separated by a space: ");
    first = stdIn.next();
    last = stdIn.next();
    System.out.println("Your initials are " +
      first.charAt(0) + last.charAt(0) + ".");
  } // end main
} // end class PrintInitials
```

Sample session:

```
Enter first and last name separated by a space: Ada Lovelace
Your initials are AL.
```

Figure 3.15 PrintInitials program that illustrates `next()`

Figure 3.14, note the italics for `34.14` and `2`. In Figure 3.15, note the italics for `Ada Lovelace`.[15] We italicize input values in order to distinguish them from the rest of the program. Be aware that the italicization is a pedagogical technique that we use for clarification purposes in the book. Input values are not really italicized when they appear on a computer screen.

A Problem with the `nextLine` Method

The `nextLine` method and the other `Scanner` methods don't play well together. It's OK to use a series of `nextLine` method calls in a program. It's also OK to use a series of `nextInt`, `nextLong`, `nextFloat`, `nextDouble`, and `next` method calls in a program. But if you use the `nextLine` method and the other `Scanner` methods in the same program, be careful. Here's why.

[15]Ada Lovelace, 1815–1852, is widely credited as the world's first computer programmer. She wrote a program to calculate Bernoulli numbers as a means to show the usefulness of her friend Charles Babbage's Analytical Engine, which is considered to be the world's first design of a programmable computer. Although Babbage never completed the construction of his Analytical Engine, Lovelace recognized its enormous potential to solve problems. In her honor, the U.S. Department of Defense created Ada, a popular programming language that has been praised for its flexibility but criticized for its relative slowness.

The `nextLine` method is the only method that processes leading whitespace. The other methods skip it. Suppose you have a `nextInt` method call and the user types `25` and then presses the enter key. The `nextInt` method call reads the 25 and returns it. The `nextInt` method call does not read in the enter key's newline character. Suppose that after the `nextInt` method call, you have a `nextLine` method call. The `nextLine` method call does not skip leading whitespace, so it's stuck with reading whatever is left over from the previous input call. In our example, the previous input call left the enter key's newline character. Thus, the `nextLine` call is stuck with reading it. Uh oh.

What happens if the `nextLine` method reads a newline character? It quits because it's done reading a line (the newline character marks the end of a line, albeit a very short line). So the `nextLine` method call doesn't get around to reading the next line, which is probably the line that the programmer intended it to read.

One solution to this `nextLine` problem is to include an extra `nextLine` method call whose sole purpose is to read in the leftover newline character. Another solution is to use one `Scanner` reference variable for `nextLine` input (e.g., `stdIn1`) and another `Scanner` reference variable for other input (e.g., `stdIn2`). But for the most part, we'll try to steer clear of the problem altogether. We'll try to avoid `nextLine` method calls that follow one of the other `Scanner` method calls.

3.24 Simple File Input for Repetitive Testing During Program Development

As you progress through the book, you'll see that input from the computer keyboard and output to the computer screen is all the I/O you need to solve a vast array of complex problems. But if you have a large amount of input, it might be easier and safer to use a simple text processor to write that input once into a file and then reread it from that file each time you rerun the program.

Figure 3.16 shows an updated version of Figure 3.15's `PrintInitials` program. This new version shows the original keyboard-input code commented out and replaced with code that reads the initials from the first name in a `names.txt` file. Note this line from the `PrintInitials2` program:

```
Scanner stdIn = new Scanner(new File("names.txt"));
```

The `new File("names.txt")` code causes input to come from a file named `names.txt`. For comparison purposes, look at the commented-out `Scanner` object code, where `System.in` is used for keyboard input.

In the `PrintInitials2` program's `new File("names.txt")` code, note the word `File`. The `File` class is not part of the core Java language, so if you use it, you need to import it at the top of your program. More specifically, you need to do this:

```
import java.io.File;
```

The attempt to create a `Scanner` object with `new Scanner(new File("names.txt"))` might not work. For example, it won't work if the operating system is unable to create the specified file. The Java compiler is a pessimist by nature, so, by default, it will generate a compilation error if you use the command `new Scanner(new File("names.txt"))` without some kind of "protective gear." The simplest way to avoid the compilation error is to append `throws Exception` to the `main` method heading, like this:

```
public static void main(String[] args) throws Exception
```

```
/*******************************************************************
* PrintInitials2.java
* Dean & Dean
*
* This prints leading character of first two words in a file.
*******************************************************************/

import java.util.Scanner;
import java.io.File;

public class PrintInitials2
{
  public static void main(String[] args) throws Exception
  {
//    Scanner stdIn = new Scanner(System.in);
    Scanner stdIn = new Scanner(new File("names.txt"));
    String first; // first name
    String last;  // last name

//    System.out.print (
//      "Enter your first and last name separated by a space: ");
    first = stdIn.next();
    last = stdIn.next();
    System.out.println("The initials are " +
      first.charAt(0) + last.charAt(0) + ".");
  } // end main
} // end PrintInitials2 class
```


Figure 3.16 Updated version of the PrintInitials program that reads input from a file

The `throws Exception` clause tells the Java compiler that you're aware that the JVM might not be able to do what you want. More specifically, you're telling the compiler that an error (exception) might be generated (thrown). The `throws Exception` clause won't magically guarantee that the file will be created successfully, but if you don't append it to the `main` method heading, your program won't even be able to try. Why? Because it won't compile. Appending `throws Exception` to the `main` method header is a "simple and dirty" way to force successful compilation.

Chapter 15 describes a more responsible treatment of exceptions. If you want to use this more responsible treatment when you write to or read from a file, look at the HTMLGenerator program in Section 16.2. It contains code that you can use as a template. That code is more complicated, however, than just appending `throws Exception` to `main`.

At this point, some readers might want to apply what they've learned to an object-oriented programming (OOP) context. OOP is the idea that programs should be organized into objects. You're not required to learn about OOP just yet, but if you can't wait, you can find such details in Chapter 6, Sections 6.1 through 6.8.

3.25 GUI Track: Input and Output with `JOptionPane` (Optional)

This section is the second installment of our optional graphical user interface (GUI) track. In each GUI track section, we provide an introduction to a GUI concept. In this section, we describe how to implement rudimentary input/output (I/O) in a GUI window.

Up to this point in the chapter, we've used *console windows* for displaying input and output. In this section, we use *GUI windows*. What's the difference? A console window is a window that can display text only. A GUI window is a window that can display not only text, but also graphical items like buttons, text boxes, and pictures. For an example, see Figure 3.17's GUI window. It displays text (an installation message), a button (an OK button), and a picture (a circled i icon).

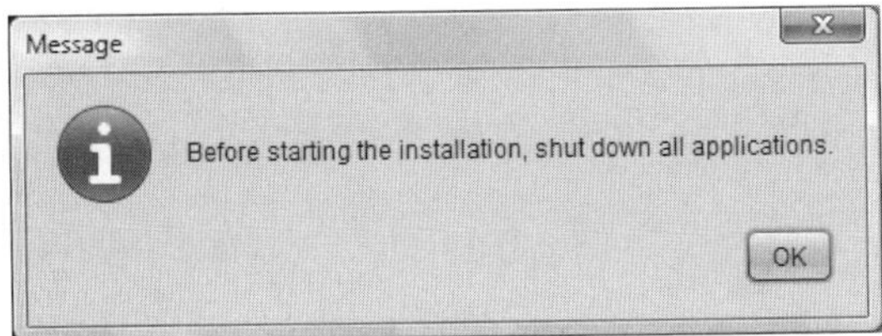

Figure 3.17 A dialog box that displays installation information

Figure 3.17's window is a specialized type of window. It's called a *dialog box*. A dialog box performs just one specific task. Figure 3.17's dialog box performs the task of displaying information (the *i* in the i icon stands for "information"). In later GUI track sections and again in Chapters 17 and 18, we'll use general-purpose standard windows. But for now, we'll stick with dialog boxes.

The `JOptionPane` Class and Its `showMessageDialog` Method

In order to display a dialog box, you need to use the `JOptionPane` class. The `JOptionPane` class is not part of the core Java language. So if you use the `JOptionPane` class, you need to tell the compiler where to find it. You do that by importing the `JOptionPane` class into your program. More specifically, you need to include this import statement at the top of your program:

```
import javax.swing.JOptionPane;
```

See the InstallationDialog program in Figure 3.18. It produces the dialog box shown in Figure 3.17. The InstallationDialog program's code should look familiar. At the top, it has the `JOptionPane` import statement. Then it has the standard class heading, standard `main` method heading, and braces. What's new is the `JOptionPane.showMessageDialog` method call.

The `showMessageDialog` method displays a message in a dialog box. Here's the syntax for calling the `showMessageDialog` method:

```
JOptionPane.showMessageDialog(null, <message>)
```

The `showMessageDialog` method takes two arguments. The first argument specifies the position of the dialog box on the computer screen. We'll keep things simple and go with the default position, which is the center of the screen. To go with the default position, specify `null` for the first argument. The second argument specifies the message that appears in the dialog box.

```
/****************************************************
* InstallationDialog.java
* Dean & Dean
*
* This program illustrates JOptionPane's message dialog.
****************************************************/

import javax.swing.JOptionPane;

public class InstallationDialog
{
  public static void main(String[] args)
  {
    JOptionPane.showMessageDialog(null,
      "Before starting the installation, " +
      "shut down all applications.");
  }
} // end class InstallationDialog
```

Figure 3.18 InstallationDialog program

Input Dialog Box

Note that a dialog box is often referred to simply as a *dialog*. Both terms are acceptable. There are several types of `JOptionPane` dialogs. We consider two of them—the message dialog for output and the input dialog for input.[16] We've already described the message dialog. That's what `showMessageDialog` produces. We'll now look at the input dialog.

The input dialog displays a prompt message and an input box. The first four dialogs in Figure 3.19 are input dialogs. These display a question mark icon as a visual cue that the dialog is asking a question and waiting for user input. Clicking OK processses the value entered. Clicking Cancel closes the dialog box without processing.

The purpose of an input dialog is to read a user-entered value and store it in a variable. To read text values, call `showInputDialog` like this:

```
<string-variable> = JOptionPane.showInputDialog(<prompt-message>);
```

To read in a number, you need to call `showInputDialog` and then convert the read-in string to a number. More specifically, to read an `int` value, do this:

```
<int-variable> = Integer.parseInt(JOptionPane.showInputDialog(<prompt-message>));
```

And to read a `double` value, do this:

```
<double-variable> = Double.parseDouble(JOptionPane.showInputDialog
                      (<prompt-message>));
```

[16] Other `JOptionPane` dialogs are a confirmation, invoked by `showConfirmDialog` and a combination, invoked by `showOptionDialog`. For more information, see http://docs.oracle.com/javase/7/docs/api/index.html.

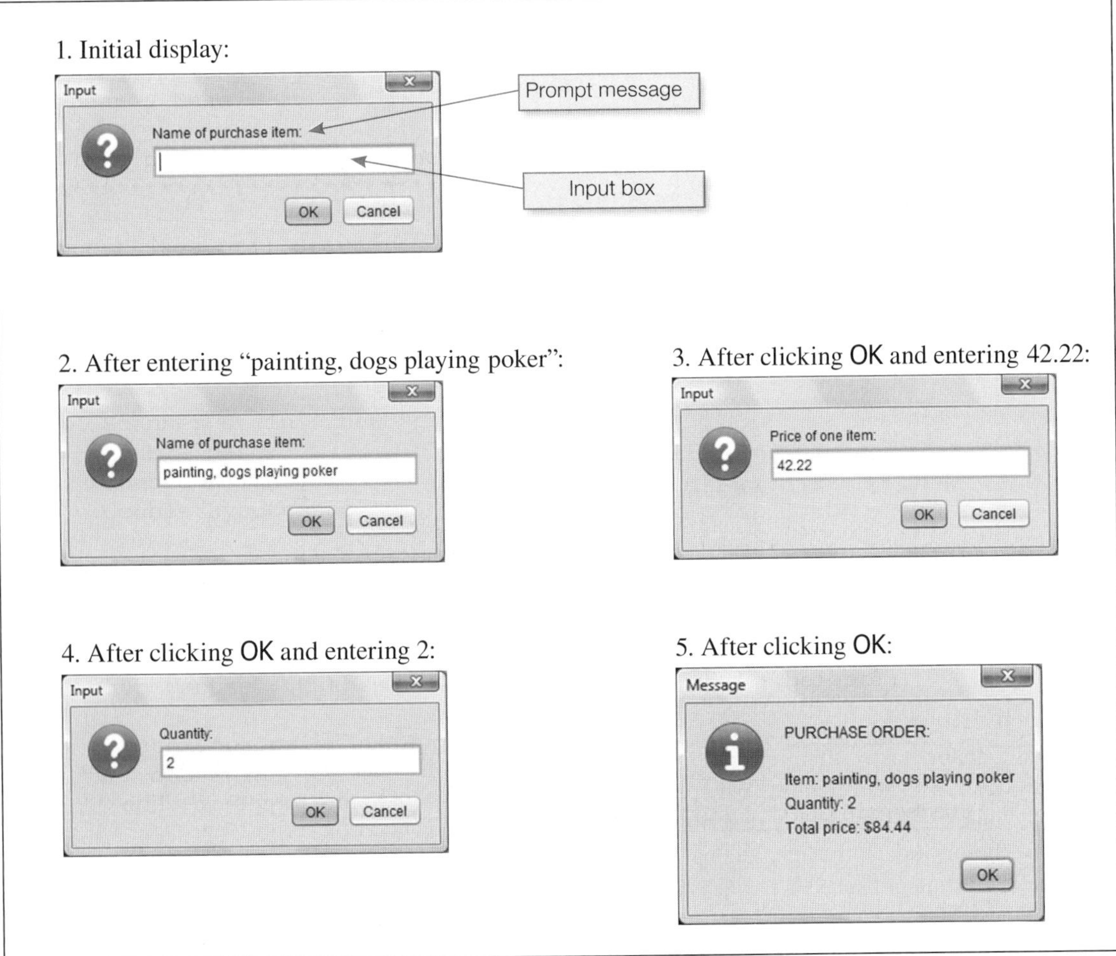

Figure 3.19 Sample session for `PrintPOGUI` program

Now look at Figure 3.20. It shows how to use these new statements to produce Figure 3.19's displays. `Integer.parseInt` converts the read-in string to an `int` value, and `Double.parseDouble` converts the read-in string to a `double` value. `Integer` and `Double` are wrapper classes. `parseInt` and `parseDouble` are wrapper class methods. We'll describe wrapper classes and their methods in Chapter 5.

I/O for the Remainder of the Book

For the GUI track sections and for the GUI chapters at the end of the book, we'll use GUI windows for I/O. But for the remainder of the book, we'll use console windows because that leads to simpler programs. Simpler programs are important so we can cut through clutter and focus on newly introduced material. But if you've decided that you love all things GUI and you can't get enough of it, feel free to convert all our console-window programs to GUI-window programs. To do so, replace all of our output code with `showMessageDialog` calls, and replace all of our input code with `showInputDialog` calls.

```
/*********************************************************************
* PrintPOGUI.java
* Dean & Dean
*
* This program calculates and prints a purchase order report.
*********************************************************************/

import javax.swing.JOptionPane;

public class PrintPOGUI
{
  public static void main(String[] args)
  {
    String itemName; // name of purchase item
    double price;    // price of purchase item
    int qty;         // number of items purchased

    itemName = JOptionPane.showInputDialog("Name of purchase item:");
    price = Double.parseDouble(
      JOptionPane.showInputDialog("Price of one item:"));
    qty = Integer.parseInt(
      JOptionPane.showInputDialog("Quantity:"));
    JOptionPane.showMessageDialog(null,
      "PURCHASE ORDER:\n\n" +
      "Item: " + itemName + "\nQuantity: " + qty +
      "\nTotal price: $" + price * qty);
  } // end main
} // end class PrintPOGUI
```

Figure 3.20 PrintPOGUI program

Summary

- Comments are used for improving a program's readability/understandability.
- The `System.out.println` method prints a message and then moves the screen's cursor to the next line. The `System.out.print` method prints a message and leaves the cursor on the same line as the printed message.
- Variables can hold only one type of data item, and that type is defined with a variable declaration statement.
- An assignment statement uses the = operator, and it puts a value into a variable.
- An initialization statement is a combination of a declaration statement and an assignment statement. It declares a variable's type and also gives the variable an initial value.
- Variables that hold whole numbers should normally be declared with the `int` data type or the `long` data type.

- Variables that hold floating-point numbers should normally be declared with the `double` data type. If you're sure that a variable is limited to small floating-point numbers, it's OK to use the `float` data type.
- Named constants use the `final` modifier.
- There are two types of integer division. One type finds the quotient (using the / operator). The other type finds the remainder (using the % operator).
- Expressions are evaluated using a set of well-defined operator precedence rules.
- The cast operator allows you to return a different-data-type version of a given value.
- Use an escape sequence (with a backslash) to print hard-to-print characters such as the tab character.
- A reference variable stores a memory address that points to an object. An object is a collection of related data wrapped in a protective shell.
- The `String` class provides methods that can be used for string processing.
- The `Scanner` class provides methods that can be used for input.

Review Questions

§3.2 "I Have a Dream" Program

1. What does this chapter's `Dream.java` program do?
2. What are the filename extensions for Java source code and bytecode, respectively?

§3.3 Comments and Readability

3. Why does source code have comments?

§3.4 The Class Heading

4. For a file with a `public` class, the program's filename must match the program's class name except that the filename has a `.java` extension added to it. (T / F)
5. Standard coding conventions dictate that class names start with a lowercase first letter. (T / F)
6. In Java, the case of a character does matter. (T / F)

§3.5 The `main` Method's Heading

7. A program's start-up method, `main`, should be in a class that is `public`. (T / F)
8. The `main` method itself must be `public`. (T / F)
9. From your memory alone (don't look for the answer in the book), write the `main` method heading.

§3.6 Braces

10. Identify two types of groupings that must be enclosed in braces.

§3.7 `System.out.println`

11. From your memory alone (don't look for the answer in the book), write the statement that tells the computer to display this string of text:
    `Here is an example`

§3.9 Identifiers

12. List all of the types of characters that may be used to form an identifier.
13. List all of the types of characters that may be used as the first character of an identifier.

§3.10 Variables

14. You should abbreviate variable names by omitting vowels, in order to save space. (T / F)
15. Why is it good practice to use a separate line to declare each separate variable?

§3.11 Assignment Statements

16. There must be a semicolon after every assignment statement. (T / F)

§3.12 Initialization Statements

17. Initialization "kills two birds with one stone." What are the "two birds"?

§3.13 Numeric Data Types—`int, long, float, double`

18. The most appropriate type to use for financial accounting is ____________.

19. For each statement, specify true or false:
  a) `1234.5` is a floating-point number. (T / F)
  b) `1234` is a floating-point number. (T / F)
  c) `1234.` is a floating-point number. (T / F)

20. If you try to assign an `int` value into a `double` variable, the computer automatically makes the conversion without complaining, but if you try to assign a `double` value into an `int` variable, the compiler generates an error. Why?

§3.14 Constants

21. For each statement, specify true or false:
  a) `0.1234` is a `float`. (T / F)
  b) `0.1234f` is a `float`. (T / F)
  c) `0.1234` is a `double`. (T / F)
  d) `1234.0` is a `double`. (T / F)

22. What modifier specifies that a variable's value is fixed/constant?

§3.15 Arithmetic Operators

23. What is the remainder operator?

24. Write the following mathematical expressions as legal Java expressions:
  a) $\frac{3x-1}{x^2}$
  b) $\frac{1}{2}+\frac{1}{xy}$

§3.16 Expression Evaluation and Operator Precedence

25. Assume this:

```
int m = 3, n = 2;
double x = 7.5;
```

Evaluate the following expressions:
  a) `(7 - n) % 2 * 7.5 + 9`
  b) `(4 + n / m ) / 6.0 * x`

§3.17 More Operators: Increment, Decrement, .and Compound Assignment

26. Write the shortest Java statement that increments `count` by 1.

27. Write the shortest Java statement that decrements `count` by 3.

28. Write the shortest Java statement that multiplies `number` by `(number - 1)` and leaves the product in `number`.

§3.18 Tracing

29. What does it mean if a variable contains garbage?

30. In a trace listing, what are line numbers for?

§3.19 Type Casting

31. Write a Java statement that assigns the `double` variable, `myDouble`, to the `int` variable, `myInteger`.

§3.20 `char` Type and Escape Sequences

32. What's wrong with the following initialization?

```
char letter = "y";
```

33. If we try to put a quotation mark (") somewhere inside a string literal to be printed, the computer interprets the quotation mark as the end of the string literal. How can we overcome this problem and force the computer to recognize the quotation mark as something we want to print?

34. When describing the location of a file or directory, computers use directory paths. In Windows environments, use the backslash character (\) to separate directories and files within a directory path. If you need to print a directory path within a Java program, how should you write the backslash character?

§3.21 Primitive Variables Versus Reference Variables

35. The type name for a primitive type is not capitalized, but the type name for a reference type is usually capitalized. (T / F)

36. List the primitive types this chapter describes, in the following categories:

a) Integer numbers
b) Floating-point numbers
c) Individual text characters and special symbols

§3.22 Strings

37. What two operators perform string concatenation, and what's the difference between these operators?

38. What method can be used to retrieve a character at a specified position within a string?

39. What two methods can be used to compare strings for equality?

§3.23 Input—the `Scanner` class

40. What is whitespace?

41. Write the statement that you must put before any other code to tell the compiler that you will be using the `Scanner` class.

42. Write the statement that creates a connection between your program and the computer's keyboard.

43. Write a statement that inputs a line of text from the keyboard and puts it into a variable named `line`.

44. Write a statement that inputs a `double` number from the keyboard and puts it into a variable named `number`.

Exercises

1. [after §3.3] Illustrate the two ways to provide comments in a Java program by writing the following as a comment in both formats:

```
This a very long comment with lots of useless and unnecessary words that
force us to use multiple lines to include it all.
```

When using the block syntax, minimize your use of asterisks.

2. [after §3.5] Why does `public static void Main(String[] args)` generate an error?

3. [after §3.6] What are braces used for?

4. [after §3.8] What program is in charge of

a) Reading Java source code and creating bytecode?
b) Executing bytecode?

5. [after §3.9] To enhance readability of an identifier that's comprised of several words, use periods between the words. (T / F)

6. [after §3.10] For each of the below variable names, indicate (with y or n) whether it's legal and whether it uses proper style. Note: You may skip the style question for illegal variable names because style is irrelevant in that case.

	legal (y/n)?	proper style (y/n)?
a) `_isReady`		
b) `3rdName`		
c) `num of wheels`		
d) `money#on#hand`		
e) `taxRate`		
f) `SeatNumber`		

7. [after §3.10] You don't need a semicolon after a variable declaration. (T / F)

8. [after §3.13] If we just write a floating-point number without specifying its type, what type does the computer assume it is?

9. [after §3.14] How would you specify the square root of two as a named constant? Use 1.41421356237309 for your named constant's value.

10. [after §3.15] Write the following mathematical expressions as legal Java expressions:

    a) $\left(\frac{3-k}{4}\right)^2$

    b) $\frac{9x-(4.5+y)}{2x}$

11. [after §3.16] Assume this:

```
int a = 9;
double b = 0.5;
```

Evaluate each of the following expressions by hand. Show your work, using a separate line for each evaluation step. Check your work by writing and executing a program that evaluates these expressions and outputs the results.

    a) `a + 3 / a`
    b) `25 / ((a - 4) * b)`
    c) `a / b * a`
    d) `a % 2 - 2 % a`

12. [after §3.19] Type Casting:
Assume the following declarations:

```
int integer;
double preciseReal;
float sloppyReal;
long bigInteger;
```

Rewrite those of the following statements which would generate a compile-time error using an appropriate cast that makes the error go away. Do not provide a cast for any statement which the compiler automatically promotes.

    a) `integer = preciseReal;`
    b) `bigInteger = sloppyReal;`
    c) `preciseReal = integer;`
    d) `sloppyReal = bigInteger;`

e) `integer = sloppyReal;`
f) `bigInteger = preciseReal;`
g) `sloppyReal = integer;`
h) `preciseReal = bigInteger;`
i) `integer = bigInteger;`
j) `sloppyReal = preciseReal;`
k) `preciseReal = sloppyReal;`
l) `bigInteger = integer;`

13. [after §3.20] Assuming that tab stops are 4 columns apart, what output does the following statement generate?

```
System.out.println("\"pathName:\"\n\tD:\\myJava\\Hello.java");
```

14. [after §3.21] Reference types begin with an uppercase letter. (T / F)

15. [after §3.22] Assume that you have a string variable named `myName`. Provide a code fragment that prints `myName`'s third character.

16. [after §3.22] What does this code fragment print?

```
String s = "hedge";
s += "hog";
System.out.println(s.equals("hedgehog"));
System.out.println((s.length() - 6) + " " + s.charAt(0) + "\'s");
```

17. [after §3.23] Modify the `PrintInitials` program so that it takes input from the keyboard but prints output to a file called "`Output.txt`".

Review Question Solutions

1. It generates the output:
   ```
   I have a dream!
   ```
2. The Java source code extension is `.java`. The bytecode extension is `.class`.
3. Source code has comments to help Java programmers recall or determine how a program works. (Comments are ignored by the computer, and they are not accessible to ordinary users.) The initial comment block includes the file name as a continuous reminder to the programmer. It contains program authors, for help and reference. It may include date and version number to identify context. It includes a short description to facilitate rapid understanding. Punctuation comments like `// end class` *<class-name>* help keep a reader oriented. Special comments identify variables and annotate obscure formulas.
4. True. If a file has a `public` class, the filename must equal this class name.
5. False. Class names should start with an uppercase first letter.
6. True. Java is case sensitive. Changing the case of any letter creates a completely different identifier.
7. True.
8. True. Otherwise, the startup procedure cannot be accessed.
9. `public static void main(String[] args)`
10. One must use braces for (1) all the contents of a class and (2) all the contents of a method.
11. `System.out.println("Here is an example");`

12. Upper-case characters, lower-case characters, numbers, underscore, and dollar sign.
13. Upper-case characters, lower-case characters, underscore, and dollar sign. No numbers.
14. False: In source code, saving space is not as important as good communication. Weird abbreviations are hard to say and not as easy to remember as real words.
15. If each variable is on a separate line, each variable has space at the right for an elaborating comment.
16. True.
17. Variable declaration and assigning a value into the variable.
18. Type `double`, or type `long`, with value in cents.
19. **a)** True; **b)** False; **c)** True
20. Assigning an integer value into a floating-point variable is like putting a small object into a large box. The `int` type goes up to approximately 2 billion. It's easy to fit 2 billion into a `double` "box" because a `double` goes all the way up to 1.8×10^{308}. On the other hand, assigning a floating-point value into an integer variable is like putting a large object into a small box. By default, that's illegal.
21. **a)** False; **b)** True; **c)** True; **d)** True
22. The `final` modifier specifies that a variable's value is fixed/constant.
23. The remainder operator is a percent sign: `%`.
24. Write the following mathematical expressions as legal Java expressions:
    a) (3 * x − 1) / (x * x)
    b) 1.0 / 2 + 1.0 / (x * y)
    or
    .5 + 1.0 / (x * y)
25. Expression evaluation:

    a)
```
(7 - n) % 2 * 7.5 + 9 ⇒
5 % 2 * 7.5 + 9 ⇒
1 * 7.5 + 9 ⇒
7.5 + 9 ⇒
16.5
```
    b)
```
(4 + n / m) / 6.0 * x ⇒
(4 + 2 / 3) / 6.0 * 7.5 ⇒
(4 + 0) / 6.0 * 7.5 =>
4 / 6.0 * 7.5 ⇒
0.6666666666666667 * 7.5 ⇒
5.0
```
26. `count++;`
27. `count -= 3;`
28. `number *= (number - 1);`
29. For variables declared without an initialization, the initial value is referred to as *garbage* because its actual value is unknown. Use a question mark to indicate a garbage value.
30. Line numbers tell you which statement in the code generates the current trace results.
31. `myInteger = (int) myDouble;`

32. The variable, `letter`, is of type `char`, but the double quotes in `"y"` specify that the initial value has type `String`, so the types are incompatible. It should be written:

```
char letter = 'y';
```

33. To print a double quotation mark, put a backslash in front of it, that is, use \".

34. To print a backslash, use two backslashes, that is, use \\.

35. True.

36. List the primitive types this chapter describes, in the following categories:
  a) Integer numbers: `int`, `long`
  b) Floating-point numbers: `float`, `double`
  c) Individual text characters and special symbols: `char`

37. The + and += operators perform *concatenation*. The + operator does not update the operand at its left. The += operator does update the operand at its left.

38. The `charAt` method can be used to retrieve a character at a specified position within a string.

39. The `equals` and `equalsIgnoreCase` methods can be used to compare strings for equality.

40. Whitespace = the characters associated with the spacebar, the tab key, and the enter key.

41. `import java.util.Scanner;`

42. `Scanner stdIn = new Scanner(System.in);`

43. `line = stdIn.nextLine();`

44. `number = stdIn.nextDouble();`

CHAPTER 4

Control Statements

Objectives

- Learn how to use `if` statements to alter a program's sequence of execution.
- Become familiar with Java's comparison and logical operators, and learn how to use them to describe complex conditions.
- Learn how to use the `switch` statement to alter a program's sequence of execution.
- Recognize repetitive operations, understand the various kinds of looping that Java supports, and learn how to select the most appropriate type of loop for each problem that requires repetitive evaluation.
- Be able to trace a looping operation.
- Learn how and when to nest a loop inside another loop.
- Learn how to use `boolean` variables to make code more elegant.
- Learn how to validate input data.
- Optionally, learn how to simplify complicated logical expressions.

Outline

4.1 Introduction

In Chapter 3, we kept things simple and wrote pure sequential programs. In a pure sequential program, statements execute in the sequence/order in which they are written; that is, after executing a statement, the computer executes the statement that immediately follows it. Pure sequential programming works well for trivial problems, but for anything substantial, you'll need the ability to execute in a nonsequential fashion. For example, if you're writing a recipe-retrieval program, you probably don't want to print all the program's recipes, one after another. You'll want to execute the chocolate chip cookie print statements if the user indicates an affinity for chocolate chip cookies, and you'll want to execute the crab quiche print statements if the user indicates an affinity for crab quiche. That sort of functionality requires the use of *control statements*. A control statement controls the order of execution of other statements. In Chapter 2, you used pseudocode if and while statements to control the order of execution within an algorithm. In this chapter, you'll use Java `if` and `while` statements, plus a few additional Java statements, to control the order of execution within a program.

In controlling the order of execution, a control statement uses a condition (a question) to decide which way to go. We start Chapter 4 with an overview of Java conditions. We then describe Java's control statements—the `if` statement, the `switch` statement, the `while` loop, the `do` loop, and the `for` loop. Along the way, we describe Java's logical operators `&&`, `||`, and `!`, which are needed when dealing with more complicated conditions. We conclude the chapter with several loop-related concepts—nested loops, input validation, and `boolean` variables. Good stuff!

4.2 Conditions and Boolean Values

In Chapter 2's flowcharts, we used diamond shapes to represent logical decision points—points where control flow went either one way or another. Into those diamonds we inserted various abbreviated questions like "`ceoSalary` greater than $500,000?" and "`count` less than or equal to 100?" Then we labeled alternate paths away from those diamonds with "yes" or "no" answers to those questions. In Chapter 2's pseudocode, we used "if" and "while" clauses to describe logical conditions. Examples included "if `shape` is a circle," "if `grade` is greater than or equal to 60," and "while `score` is not equal to —1." We considered pseudocode conditions to be either "true" or "false."

Informal condition expressions like these are fine for flowcharts and pseudocode, but when you start writing real Java code, you must be precise. The computer interprets each "if" condition or loop condition as a two-way choice. What are the two possible values recognized by Java? They are the Java values `true` and `false`. These values, `true` and `false`, are called *Boolean* values, after George Boole, a famous 19th-century logician. Throughout the rest of this chapter, you'll see `if` statements and loop statements where *conditions* appear as little fragments of code within a pair of parentheses, like this:

```
if (<condition>)
{
    . . .
}

while (<condition>)
{
    . . .
}
```

Whatever is in the places marked by <*condition*> always evaluates to either `true` or `false`.

Typically, each condition involves some type of comparison. With pseudocode, you may use words to describe comparisons, but in real Java code, you must use special *comparison operators*. Comparison operators (also called *equality* and *relational operators*) are like mathematical operators in that they link adjacent operands, but instead of combining the operands in some way, they compare them. When a mathematical operator combines two numbers, the combination is a number like the operands being combined. But when a comparison operator compares two numbers, the result is a different type. It is not a number like the operands being compared. It is a Boolean truth value—either `true` or `false`.

Here are Java's comparison operators:

```
==, !=, <, >, <=, >=
```

The == operator tests whether two values are equal. Notice that this symbol uses <u>two</u> equals signs! This is different from the single equals sign that we all use instinctively to represent equality. Why does Java use two equals signs for equality in a comparison? It's because Java already uses the single equals sign for assignment, and context is not enough to distinguish assignment from equality. Do not try to use a single = for comparison! The Java compiler will not like it. 

The `!=` operator tests whether two values are unequal. As you'd expect, the < operator tests whether a value on the left is less than a value on the right. The > operator tests whether a value on the left is greater than a value on the right. The <= operator tests whether a value on the left is less than or equal to a value on the right. The >= operator tests whether a value on the left is greater than or equal to a value on the right. The result of any one of these tests is always either `true` or `false`.

4.3 `if` Statements

Now, let's look at a simple example of the condition in an `if` statement. Here's a simple `if` statement that checks a car's temperature gauge value and prints a warning if the temperature is above 215 degrees:

condition

```
if (temperature > 215)
{
  System.out.println("Warning! Engine coolant is too hot.");
  System.out.println("Stop driving and allow engine to cool.");
}
```

The condition uses the > operator to generate a `true` value if the temperature is above 215 degrees or a `false` value if it is not. The subordinate statements (the statements in the braces) execute only if the condition generates a `true` value.

Syntax

In the above example, note the parentheses around the condition. Parentheses are required whenever you have a condition, regardless of whether it's for an `if` statement, a `while` loop, or some other control structure. Note the braces around the two subordinate print statements. Use braces to surround statements that are logically inside something else. For example, braces are required below the `main` method's heading and at the bottom of the `main` method because the statements inside the braces are logically inside the `main` method. Likewise, you should use braces to surround the statements that are logically inside an `if` statement.

To emphasize the point that statements inside braces are logically inside something else, you should always indent statements that are inside braces. Because this is so important, we'll say it again: Always indent when you're inside braces!

When an `if` statement includes two or more subordinate statements, you <u>must</u> enclose the subordinate statements in braces. Said another way, you must use a *block*. A block, also called a *compound statement*, is a set of zero or more statements surrounded by braces. A block can be used anywhere a standard statement can be used. If you don't use braces for the `if` statement's two subordinate statements, the computer considers only the first statement to be subordinate to the `if` statement. When there is supposed to be just one subordinate statement, you're not required to enclose it in braces, but we recommend that you do so anyway.

That way, you won't get into trouble if you come back later and want to insert additional subordinate statements at that point in your program.

Three Forms of the `if` Statement

There are three basic forms for an `if` statement:

- "if"—use when you want to do one thing or nothing.
- "if, else"—use when you want to do one thing or another thing.
- "if, else if"—use when there are three or more possibilities.

Chapter 2 presented pseudocode versions of these forms. Figures 4.1, 4.2, and 4.3 show Java forms.

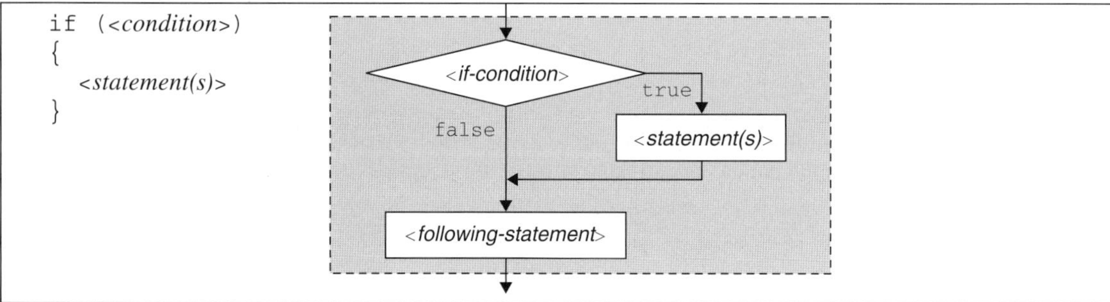

Figure 4.1 Syntax and semantics for the simple "if" form of the `if` statement

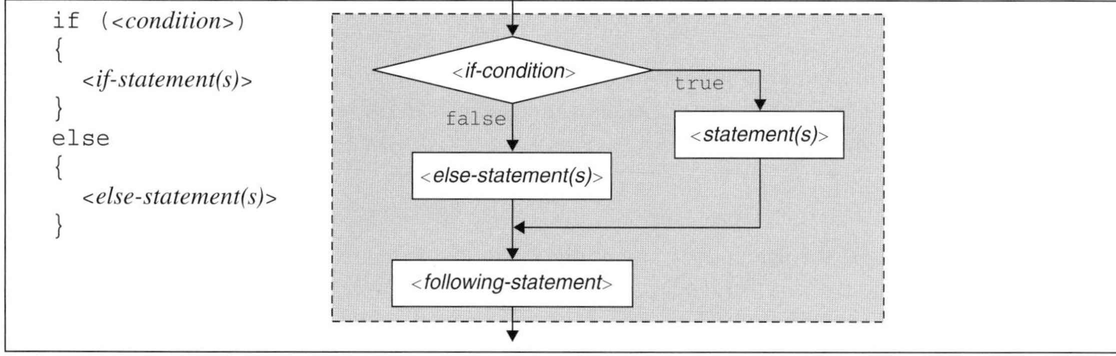

Figure 4.2 Syntax and semantics for the "if, else" form of the `if` statement

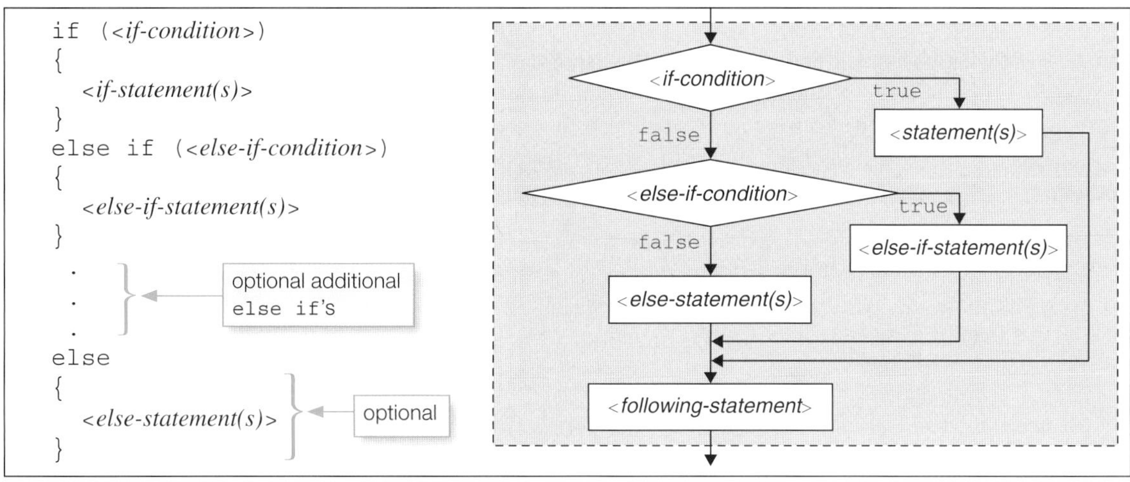

Figure 4.3 Syntax and semantics for the "if, else if" form of the `if` statement

Take several minutes and examine Figures 4.1, 4.2, and 4.3. The figures show the syntax and semantics for the three forms of the Java `if` statement. The *semantics* of a statement is a description of how the statement works. For example, Figure 4.1's flowchart illustrates the semantics of the "if" form of the `if` statement by showing the flow of control for different values of the `if` statement's condition.

Most of what you see in the `if` statement figures should look familiar because it parallels what you learned in Chapter 2. But the "if, else if" form of the `if` statement deserves some extra attention. You may include as many "else if" blocks as you like—more "else if" blocks for more choices. Note that the "else" block is optional. If all the conditions are `false` and there's no "else" block, none of the statement blocks is executed. Here's a code fragment that uses the "if, else if" form of the `if` statement to troubleshoot iPod[1] problems:

```
if (iPodProblem.equals("no response"))
{
  System.out.println("Unlock iPod's Hold switch.");
}
else if (iPodProblem.equals("songs don't play"))
{
  System.out.println("Use iPod Updater to update your software.");
}
else
{
  System.out.println("Visit http://www.apple.com/support.");
}
```

[1]The iPod is a portable media player designed and marketed by Apple Computer.

Practice Problem

Now let's put what you've learned into practice by using the `if` statement within a complete program. Suppose you're asked to write a sentence-tester program that checks whether a user-entered line ends with a period. Your program should print an error message if the last character in the line is not a period. In writing the program, use a sample session as a guide. Note that the italicized Mahatma Gandhi quote is a user-entered input value.

Use design output to specify the problem.

Sample session:

```
Enter a sentence:
Permanent good can never be the outcome of violence.
```

Another sample session:

```
Enter a sentence:
Permanent good can never be the outcome of
Invalid entry - your sentence is not complete!
```

As your first step in implementing a solution, use pseudocode to generate an informal outline of the basic logic:

```
print "Enter a sentence: "
input sentence
if sentence's last character is not equal to '.'
  print "Invalid entry – your sentence is not complete!"
```

Note the simple "if" form of the `if` statement. That's appropriate because there's a need to do something (print an invalid entry message) or nothing. Why nothing? Because the problem description does not say to print anything for user entries that are legitimate sentences. In other words, the program should skip what's in the `if` statement if you finish the sentence properly. Now we suggest that you try writing the Java code to implement this algorithm. You'll need to use a couple of the `String` methods described near the end of Chapter 3. When you're ready, look at the `SentenceTester` solution in Figure 4.4.

How does the SentenceTester program determine whether the last character is a period? Suppose the user enters "Hello." In that case, what value would be assigned to the `lastCharPosition` variable? `String`'s `length` method returns the number of characters in a string. The number of characters in "Hello." is six. Because the first position is zero, `lastCharPosition` would get assigned a value of (6 − 1) or 5. The purpose of finding the last character's index position value is to see if the last character in the user-entered value is a period. To see if it is, we use `lastCharPosition` as the argument in a `charAt` method call. `String`'s `charAt` method returns the character at a specified index position within a string. The index position of the period in "Hello." is 5, and the `if` condition checks whether the user-entered value's last character is a period.

4.4 && Logical Operator

Up to this point, all of our `if` statement examples have used simple conditions. A simple condition evaluates directly to either `true` or `false`. In the next three sections, we introduce you to logical operators, like the "and" operator (`&&`) and the "or" operator (`||`), which make it possible to construct compound conditions. A *compound condition* is a conjunction (either an "anding" or an "oring") of two or more conditions. When

```
/*********************************************************************
* SentenceTester.java
* Dean & Dean
*
* This program checks for a period at the end of a line of input.
*********************************************************************/

import java.util.Scanner;

public class SentenceTester
{
  public static void main(String[] args)
  {
    Scanner stdIn = new Scanner(System.in);
    String sentence;       // user-entered sentence
    int lastCharPosition; // index of sentence's last character

    System.out.println("Enter a sentence:");
    sentence = stdIn.nextLine();
    lastCharPosition = sentence.length() - 1;
    if (sentence.charAt(lastCharPosition) != '.')
    {
      System.out.println(
        "Invalid entry - your sentence needs a period!");
    }
  } // end main
} // end class SentenceTester
```

This condition checks for proper termination.

Figure 4.4 SentenceTester program

you have a compound condition, each part of the compound condition evaluates to either `true` or `false`, and then the parts combine to produce a composite `true` or `false` for the whole compound condition. The combining rules are what you might expect: When you "and" two conditions together, the combination is `true` only if the first condition is true and the second condition is true. When you "or" two conditions together, the combination is `true` if the first condition is `true` or the second condition is `true`. You'll see plenty of examples as the chapter progresses.

&& Operator Example

Let's begin our discussion of logical operators with an example that uses the && operator. (Note: && is pronounced "and"). Suppose you want to print "OK" if the temperature is between 50 degrees and 90 degrees and print "not OK" otherwise:

Here's a pseudocode description of the problem:

```
if temp ≥ 50 and ≤ 90
  print "OK"
else
  print "not OK"
```

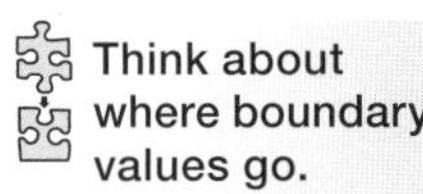

Think about where boundary values go.

Notice that the pseudocode condition uses ≥ and ≤ rather than > and <. The original problem specification says to print "OK" if the temperature is between 50 degrees and 90 degrees. When people say "between," they usually, but not always, mean to include the end points. Thus, we assumed that the 50 and 90 end points were supposed to be included in the OK range, and we chose to use ≥ and ≤ accordingly. But in general, if you're writing a program and you're unsure about the end points for a particular range, you should not assume. Instead, you should ask the customer what he or she wants. The end points are important.

See Figure 4.5. It shows the Java implementation for the temperature-between-50-and-90 problem. In Java, if both of two criteria must be met for a condition to be satisfied (e.g., `temp >= 50` and `temp <= 90`), then separate the two criteria with the && operator. As indicated by Figure 4.5's first callout, if both criteria use the same variable (e.g., `temp`), you must include the variable on both sides of the &&. Note the use of >= and <=. In pseudocode, it's OK to use ≥, ≤, or even the words "greater than or equal to," and "less than or equal to." But in Java, you must use >= and <=.

```
if ((temp >= 50) && (temp <= 90))
{
  System.out.println("OK");
}
else
{
  System.out.println("not OK");
}
```

`temp` must be repeated

Use <=, not ≤.

Figure 4.5 Java implementation of the temperature-between-50-and-90 problem

Operator Precedence

In Figure 4.5, note the parentheses around each of the two temperature comparisons. They force evaluation of the comparisons before evaluation of the `&&`. What would happen if we omitted those inner parentheses? To answer that sort of question, you need to refer to an operator precedence table. Appendix 2 provides a complete operator precedence table, but most of the cases you'll encounter are covered by the abbreviated precedence table in Figure 4.6. All operators within a particular numbered group have equal precedence, but operators at the top of the figure (in groups 1, 2, . . .) have higher precedence than operators at the bottom of the figure (in groups . . . 7, 8).

Figure 4.6 shows that the comparison operators `>=` and `<=` have higher precedence than the logical operator `&&`. Thus, the `>=` and `<=` operations execute before the `&&` operation—even if the inner parentheses in the condition in Figure 4.5 are omitted. In other words, we could have written Figure 4.5's condition more simply, like this:

1. grouping with parentheses:
   (*<expression>*)

2. unary operators:
   ```
   +x
   -x
   (<type>) x
   x++
   x--
   !x
   ```

3. multiplication and division operators:
   ```
   x * y
   x / y
   x % y
   ```

4. addition and subtraction operators:
   ```
   x + y
   x - y
   ```

5. less than and greater than relational operators:
   ```
   x < y
   x > y
   x <= y
   x >= y
   ```

6. equality operators:
   ```
   x == y
   x != y
   ```

7. "and" logical operator:
   ```
   x && y
   ```

8. "or" logical operator:
   ```
   x || y
   ```

Figure 4.6 Abbreviated operator precedence table (see Appendix 2 for complete table)
The operator groups at the top of the table have higher precedence than the operator groups at the bottom of the table. All operators within a particular group have equal precedence. If an expression has two or more same-precedence operators, then within that expression, operators on the left execute before those on the right.

```
if (temp >= 50 && temp <= 90)
```

You may include these extra parentheses or not, as you wish. We included them in Figure 4.5 to emphasize the order of evaluation in this initial presentation, but in the future, we will often omit them to minimize clutter. 

Another Example

For another example, consider commercial promotions at sports events. Suppose the local Yummy Burgers restaurant is willing to provide free French fries to all fans at a basketball game whenever the home team wins and scores at least 100 points. The problem is to write a program that prints the following message whenever that condition is satisfied:

"Fans: Redeem your ticket stub for a free order of French fries at Yummy Burgers."

Figure 4.7 shows the framework. Within the figure, note where it says *<insert code here>*. Before looking ahead at the answer, see if you can provide the inserted code on your own.

```
/*******************************************************************
 * FreeFries.java
 * Dean & Dean
 *
 * This program reads points scored by the home team and the
 * opposing team and determines whether the fans win free
 * french fries.
*******************************************************************/

import java.util.Scanner;

public class FreeFries
{
  public static void main(String[] args)
  {
    Scanner stdIn = new Scanner(System.in);
    int homePts;       // points scored by home team
    int opponentPts;   // points scored by opponents

    System.out.print("Home team points scored: ");
    homePts = stdIn.nextInt();
    System.out.print("Opposing team points scored: ");
    opponentPts = stdIn.nextInt();

    <insert-code-here>

  } // end main
} // end class FreeFries
```

Sample session:

```
Home team points scored: 103
Opposing team points scored: 87
Fans: Redeem your ticket stub for a free order of French fries at Yummy
Burgers.
```

Figure 4.7 FreeFries program with "and" condition

Here's what you should insert:

homePts must be repeated

```
if (homePts > opponentPts && homePts >= 100)
{
  System.out.println("Fans: Redeem your ticket stub for" +
    " a free order of French fries at Yummy Burgers.");
}
```

4.5 || Logical Operator

Now let's look at the complement to the "and" operator—the "or" operator. Assume that you have a variable named `response` that contains (1) a lowercase or uppercase "q" if the user wants to quit or (2) some other character if the user wants to continue. Write a code fragment that prints "Bye" if the user enters either a lowercase or uppercase "q." Using pseudocode, you'd probably come up with something like this for the critical part of the algorithm:

```
if response equals "q" or "Q"
  print "Bye"
```

Note the "or" in the `if` statement's condition. That works fine for pseudocode, where syntax rules are lenient, but for Java, you must use || for the "or" operation, not "or." (Note: || is pronounced "or.") To enter the || operator on your computer, look for the vertical bar key on your keyboard and press it twice. Here's a tentative Java implementation of the desired code fragment:

```
Scanner stdIn = new Scanner(System.in);
String response;

System.out.print("Enter q or Q: ");
response = stdIn.nextLine();
if (response == "q" || response == "Q")
{
  System.out.println("Bye");
}
```

When inserted in a `main` method, this compiles, but it does not "work"!

Note that the `response` variable appears twice in the `if` statement's condition. That's necessary because if both sides of an || condition involve the same variable, you must repeat the variable.

The callout indicates that something is wrong. What is it? If you insert this code fragment into a valid program shell, the program compiles and runs. But when a user responds to the prompt by dutifully entering either "q" or "Q," nothing happens. The program does not print "Bye." Why not? Should we have used interior parentheses in the "if" condition? Figure 4.6 shows that the == operator has a higher precedence than the || operator, so what we did was OK. The problem is something else.

Don't Use == to Compare Strings

The problem is with the `response == "q"` and `response == "Q"` expressions. We'll focus on the `response == "q"` expression. The `response` string variable and the "q" string literal both hold memory addresses that point to string objects; they don't hold string objects themselves. So when you use ==, you're comparing the memory addresses stored in the `response` string variable and the "q" string literal. If the `response` string variable and the "q" string literal contain different memory addresses (i.e., they point to different string objects), then the comparison evaluates to **false**, even if both string objects contain the same sequence of characters. The following picture shows what we're talking about. The arrows represent memory addresses. Because they point to two different objects, `response == "q"` evaluates to `false`.

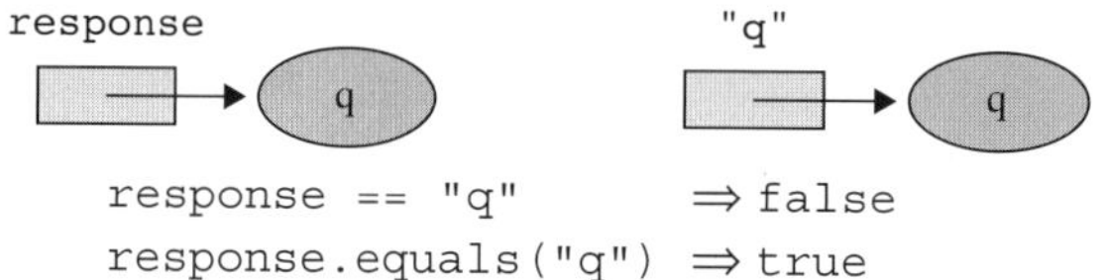

So what can you do to solve this problem? In Chapter 3, you learned to use the `equals` method to test strings for equality. The `equals` method compares the string objects pointed to by the memory addresses. In the above picture, the string objects hold the same sequence of characters, q and q, so the method call, `response.equals("q")`, returns `true`, which is what you want. Here's the corrected code fragment:

```
if (response.equals("q") || response.equals("Q"))
{
  System.out.println("Bye");
}
```

Or, as a more compact alternative, use the `equalsIgnoreCase` method like this:

```
if (response.equalsIgnoreCase("q"))
{
  System.out.println("Bye");
}
```

A third alternative is to use the `String` class's `charAt` method to convert the string input into a character and then use the == operator to compare that character with the character literals, 'q' and 'Q':

```
char resp = response.charAt(0);
if (resp == 'q' || resp == 'Q')
{
  System.out.println("Bye");
}
```

The devil is in the details.

These implementations are not trivial translations from the pseudocode that specified the algorithm. It's important to organize your thoughts before you start writing Java code. But even very good preparation does not eliminate the need to keep thinking as you proceed. Details matter also!

Errors

We made a big deal about not using == to compare strings because it's a very easy mistake to make and it's a hard mistake to catch. It's easy to make this mistake because you use == all the time when comparing primitive values. It's hard to catch this mistake because programs that use == for string comparison compile and run with no reported errors. No reported errors? Then why worry? Because although there are no reported errors, there are still errors—they're called logic errors.

Be careful. Test every aspect.

A *logic error* manifests itself when your program runs to completion without an error message, and the output is wrong. Logic errors are the hardest errors to find and fix because there's no error message glaring at you, telling you what you did wrong. To make matters worse, using == for string comparison generates a logic error only some of the time, not all the time. Because the logic error occurs only some of the time, programmers can be lulled into a false sense of confidence that their code is OK, when in reality it's not OK.

There are three main categories of errors—compile-time errors, runtime errors, and logic errors. A compile-time error is an error that is identified by the compiler during the compilation process. A runtime error is an error that occurs while a program is running and it causes the program to terminate abnormally. The compiler generates an error message for a compile-time error, and the Java virtual machine (JVM) generates an error message for a runtime error. Unfortunately, there are no error messages for a logic error. It's up to the programmer to fix logic errors by analyzing the output and thinking carefully about the code.

4.6 ! Logical Operator

Now it's time to consider the logical "not" operator (!). Assume that you have a `char` variable named `resp` that contains (1) a lowercase or uppercase 'q' if the user wants to quit or (2) some other character if the user wants to continue. This time, the goal is to print "Let's get started. . . ." if `resp` contains anything other than a lowercase or uppercase "q". You could use an "if, else" statement with an empty "if" block like this:

```
if (resp == 'q' || resp == 'Q')
{ }
else
{
  System.out.println("Let's get started. . . .");
  . . .
```

But this is not very elegant. Programmers often use the term *elegant* to describe code that is well written and has "beauty." More specifically, elegant code is easy to understand, easy to update, robust, reasonably compact, and efficient. The above code's empty "if" block is inelegant because it's not compact. If you ever have an empty "if" block with a nonempty "else" block, you should try to rewrite it as just an "if" block with no "else" block. The trick is to invert the `if` statement's condition. In the above example, that means testing for the absence of lowercase or uppercase 'q' rather than the presence of lowercase or uppercase 'q.' To test for the absence of lowercase or uppercase 'q,' use the `!` operator.

The `!` operator changes `true` values into `false` values and vice versa. This true-to-false, false-to-true toggling functionality is referred to as a "not" operation, and that's why the `!` operator is called the "not" operator. Because we want to print "Let's get started. . . ." if the above `if` statement's condition is <u>not</u> true, we insert `!` at the left of the condition like this:

```
if (!(resp == 'q' || resp == 'Q'))
{
  System.out.println("Let's get started. . . .");
  . . .
```

Note that the `!` is inside one set of parentheses and outside another set. Both sets of parentheses are required. The outer parentheses are necessary because the compiler requires parentheses around the entire condition. The inner parentheses are also necessary because without them, the `!` operator would operate on the `resp` variable instead of on the entire condition. Why? Because the operator precedence table (Figure 4.6) shows that the `!` operator has higher precedence than the == and || operators. The way to force the == and || operators to be executed first is to put them inside parentheses.

Don't confuse the `!` (not) operator with the != (inequality) operator. The `!` operator returns the opposite value of the given expression (a true expression returns `false` and a `false` expression returns `true`). The != operator asks a question—are the two expressions unequal?

4.7 `switch` Statement

The `switch` statement works similarly to the "if, else if" form of the `if` statement in that it allows you to follow one of several paths. But a key difference between the `switch` statement and the `if` statement is that the `switch` statement's determination of which path to take is based on a single value. (With an `if` statement, the determination of which path to take is based on multiple conditions, one for each path.) Having the determination based on a single value can lead to a more compact, more understandable implementation. Think of driving on Route 1 along the California coastline and coming to a junction with alternate routes through and around a city. The different routes are better at certain times of the day. If it's 8 AM or 5 PM, you should take the outer business loop to avoid rush-hour traffic. If it's 8 PM, you should take the coastal bluffs route to appreciate the scenic sunset view. If it's any other time, you should take the through-the-city route because it is the most direct and fastest. Using a single value, time of day, to determine the route parallels the decision-making process in a `switch` statement.

Syntax and Semantics

Study the `switch` statement's syntax in Figure 4.8. When executing a `switch` statement, control jumps to the `case` constant that matches the controlling expression's value, and the computer executes all subsequent statements up to a `break` statement. The `break` statement causes control to exit from the `switch` statement (to below the closing brace). If there are no `case` constants that match the controlling expression's value, then control jumps to the `default` label (if there is a `default` label) or out of the `switch` statement if there is no `default` label.

Usually, `break` statements are placed at the end of every `case` block. That's because you normally want to execute just one `case` block's subordinate statement(s) and then exit the `switch` statement. However, `break` statements are not required. Sometimes you want to omit them, and being able to omit them is a special feature of the `switch` construct. But accidentally forgetting to include a `break` statement that should be included is a common error. If there's no `break` at the bottom of a particular `case` block, control flows through subsequent `case` constants and executes all subordinate statements until a `break` statement is reached. If there's no `break` at the bottom of the last `case` block, control flows through to the (indented) subordinate statements in the `default` block (if there is a `default` block).

```
switch (<controlling expression>)
{
  case <constant>:
    <statement(s)>;  }
    break;           }  <- optional
  case < constant>:
    <statement(s)>;  }
    break;           }  <- optional
  . . .
  default:           }
    <statement(s)>;  }  <- optional
} // end switch
```

Figure 4.8 switch statement's syntax

Referring to Figure 4.8, take note of these details:

- There must be parentheses around the controlling expression.
- The controlling expression may evaluate to either an `int`, a `char`, or a `String`.[2] It's illegal to use a `boolean`, a `long`, or a floating-point value.
- Although it's common for the controlling expression to consist of a single variable, it can consist of a more complicated expression as well—multiple variables, operators, and even method calls are allowed—provided the expression evaluates to one of the permitted types.
- There must be braces around the `switch` statement's body.
- There must be a colon after each `case` constant.
- Even though statements following the `case` constants are indented, braces ({ }) are unnecessary. That's unusual in Java—it's the only time where you don't need braces around statements that are logically inside something else.
- It's good style to include `// end switch` after the `switch` statement's closing brace.

ZIP Code Example

To exercise your understanding of the `switch` statement, write a program that reads in a ZIP Code and uses the first digit to print the associated geographic area. Here's what we're talking about:

If ZIP Code begins with	Print this message
0, 2, 3	*<zip>* `is on the East Coast.`
4–6	*<zip>* `is in the Central Plains area.`
7	*<zip>* `is in the South`
8–9	*<zip>* `is in the West.`
other	*<zip>* `is an invalid ZIP Code.`

[2]Other permitted types are `byte` and `short`; the wrapped types, `Byte`, `Short`, `Integer`, and `Character`; and the special Java type, `enum`. We describe wrapped types in Chapter 5, and we describe `byte`, `short`, and `enum` in Chapter 12.

The first digit of a U.S. postal ZIP Code identifies a particular geographic area within the United States. ZIP Codes that start with 0, 2, or 3 are in the east, ZIP Codes that start with 4, 5, or 6 are in the central region, and so on.[3] Your program should prompt the user for his or her ZIP Code and use the first character of the entered value to print the user's geographical region. In addition to printing the geographical region, your program should *echo print* the user's ZIP Code. (Echo print means print out an input exactly as it was read in.) Here's an example of what the program should do:

Sample session:

```
Enter a ZIP Code: 56044
56044 is in the Central Plains area.
```

Now let's look at the problem solution in Figure 4.9. Look at the controlling expression, (zip.charAt(0)). This evaluates to the first character in zip. As an alternative, you could have started by reading the first character into a separate variable (for example, firstChar), and then inserted that variable into the controlling expression. But because the first character was needed only at one point, the code is made more compact by embedding the zip.charAt(0) method call directly in the controlling expression's parentheses.

The switch statement compares the character in its controlling expression with each of the case constants until it finds a match. Because the controlling expression's charAt method returns a char value, the case constants must all be chars. Therefore, the case constants must have single quotes around them. If you don't use single quotes—if you use double quotes or no quotes—you'll get a compile-time error. The switch statement is not very flexible!

As previously mentioned, it's a common error to omit a break statement at the end of a switch statement's case block. For example, suppose you did this in the ZipCode program:

```
case '4': case '5': case '6':
  System.out.println(
    zip + " is in the Central Plains area.");
case '7':
  System.out.println(zip + " is in the South.");
  break;
```

Note that there's no longer a break statement at the end of the case 4, 5, 6 block. The following sample session illustrates what happens. With an input of 56044, the switch statement searches for a '5' and stops when it reaches the case '5': label. Execution begins there and continues until it reaches a break statement. So it flows through the case '6': label and prints the Central Plains message. Then, in a logic error, the flow continues into the case '7': block and inappropriately prints the South message.

Sample session:

```
Enter a ZIP Code: 56044
56044 is in the Central Plains area.
56044 is in the South.  <------  Error!
```

[3]http://www.nass.usda.gov/census/census97/zipcode/zipcode.htm.

```
/*******************************************************************
* ZipCode.java
* Dean & Dean
*
* This program identifies geographical region from ZIP Code.
*******************************************************************/

import java.util.Scanner;

public class ZipCode
{
  public static void main(String[] args)
  {
    Scanner stdIn = new Scanner(System.in);
    String zip;  // user-entered ZIP code

    System.out.print("Enter a ZIP Code: ");
    zip = stdIn.nextLine();

    switch (zip.charAt(0))
    {
      case '0': case '2': case '3':
        System.out.println(zip + " is on the East Coast.");
        break;
      case '4': case '5': case '6':
        System.out.println(
          zip + " is in the Central Plains area.");
        break;
      case '7':
        System.out.println(zip + " is in the South.");
        break;
      case '8': case '9':
        System.out.println(zip + " is in the West.");
        break;
      default:
        System.out.println(zip + " is an invalid ZIP Code.");
    } // end switch
  } // end main
} // end class ZipCode
```

Figure 4.9 Using a `switch` statement to find a geographical region from a ZIP Code

`switch` Statement Versus "if, else if" Form of the `if` Statement

Now you know that the `switch` statement allows you to do one or more things from a list of multiple possibilities. But so does the "if, else if" form of the `if` statement, so why would you ever use a `switch` statement? Because the `switch` statement provides a more elegant solution (cleaner, better-looking organization) for certain kinds of problems.

Now for the opposite question: Why would you ever use the "if, else if" form of the `if` statement rather than the `switch` statement? Because `if` statements are more flexible. With a `switch` statement, each test (i.e., each `case` label) is limited to an exact match with one of the permitted types. With an `if` statement, each test can be a full-blooded expression, complete with operators, variables, and method calls.

In a nutshell, when you need to do one thing from a list of multiple possibilities:

- Use a `switch` statement if you need to match an `int`, `char`, or `String` value.
- Use an `if` statement if you need more flexibility.

4.8 `while` Loop

There are two basic categories of control statements—forward branching statements and looping statements. The `if` statement and `switch` statement implement *forward branching* functionality (so named because the decisions cause control to "branch" to a statement that is ahead of the current statement). The `while` loop, `do` loop, and `for` loop implement looping functionality. We describe the `while` loop in this section and the `do` and `for` loops in the next two sections. But first an overview of loops in general.

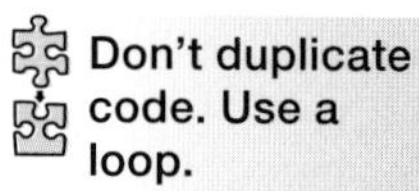

In solving a particular problem, one of the first and most important things to think about is whether there are any repetitive tasks. Repetitive tasks should normally be implemented with the help of a loop. For some problems, you can avoid a loop by implementing the repetitive tasks with consecutive sequential statements. For example, if you are asked to print "Happy Birthday!" 10 times, you could implement a solution with 10 consecutive print statements. But such a solution would be a poor one. A better solution is to insert a single print statement inside a loop that repeats 10 times. The loop implementation is better because it's more compact. Also, updating is easier and safer because the updated code appears in only one place. For example, if you need to change "Happy Birthday!" to "Bon Anniversaire!" ("Happy Birthday!" in French), then it's only a matter of changing one print statement inside a loop rather than updating 10 separate print statements.

`while` Loop Syntax and Semantics

Now let's look at the simplest kind of loop, the `while` loop. Figure 4.10 shows the syntax and semantics of the `while` loop. The syntax for the `while` loop looks like the syntax for the `if` statement except that the word `while` is used instead of the word `if`. Don't forget the parentheses around the condition. Don't forget the braces, and don't forget to indent the subordinate statements they enclose.

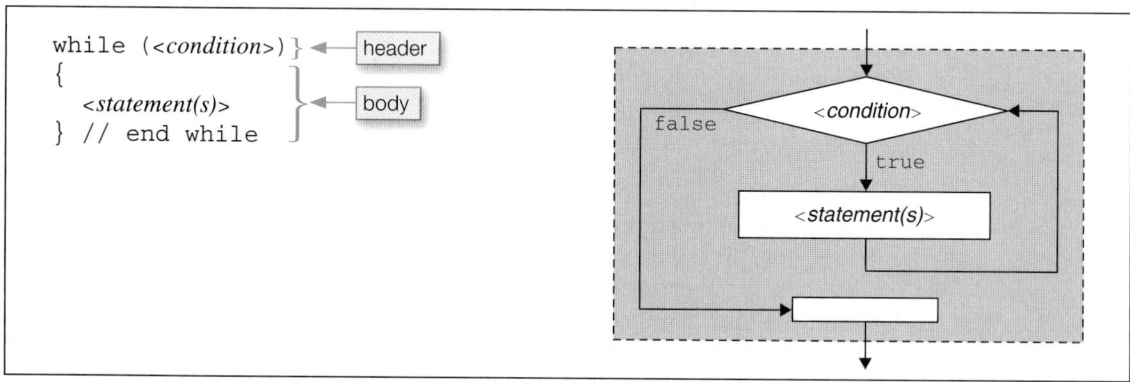

Figure 4.10 Syntax and semantics for the `while` loop

A `while` loop's condition is the same as an `if` statement's condition. It typically employs comparison and logical operators, and it evaluates to `true` or `false`. Here's how the `while` loop works:

1. Check the `while` loop's condition.
2. If the condition is `true`, execute the `while` loop's body (the statements that are inside the braces), jump back to the `while` loop's condition, and repeat step 1.
3. If the condition is `false`, jump to below the `while` loop's body and continue with the next statement.

Example

Now let's consider an example—a program that creates a bridal gift registry. More specifically, the program repeatedly prompts the user for two things—a gift item and the store where the gift can be purchased. When the user is done entering gift and store values, the program prints the bridal registry list. Study this sample session:

Sample session:

```
Do you wish to create a bridal registry list? (y/n): y
Enter item: candle holder
Enter store: Sears
Any more items? (y/n): y
Enter item: lawn mower
Enter store: Home Depot
Any more items? (y/n): n

Bridal Registry:
candle holder - Sears
lawn mower - Home Depot
```

That's the problem specification. Our solution appears in Figure 4.11. As you can tell by the `while` loop's `more == 'y'` condition and the query at the bottom of the loop, the program employs a user-query loop. The initial query above the `while` loop makes it possible to quit without making any passes through the loop. If you want to force at least one pass through the loop, you should delete the initial query and initialize `more` like this:

Use I/O sample to specify the problem.

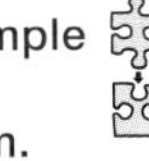

```
char more = 'y';
```

The BridalRegistry program illustrates several peripheral concepts that you'll want to remember for future programs. Within the `while` loop, note the += assignment statements, repeated here for your convenience:

```
registry += stdIn.nextLine() + " - ";
registry += stdIn.nextLine() + "\n";
```

The += operator comes in handy when you need to add incrementally to a string variable. The BridalRegistry program stores all the gift and store values in a single `String` variable named `registry`. Each new gift and store entry gets concatenated to the `registry` variable with the += operator.

At the top and bottom of the BridalRegistry program's `while` loop, note the `nextLine` and `charAt` method calls, repeated here for your convenience:

```
more = stdIn.nextLine().charAt(0);
```

The method calls are *chained* together by inserting a dot between them. The `nextLine()` method call reads a line of input from the user and returns the input as a string. That string then calls the `charAt(0)`, which returns the string's first character. Note that it's acceptable and fairly common to chain multiple method calls together like this.

```
/***********************************************************
* BridalRegistry.java
* Dean & Dean
*
* This makes entries in a bridal registry.
***********************************************************/

import java.util.Scanner;

public class BridalRegistry
{
  public static void main(String[] args)
  {
    Scanner stdIn = new Scanner(System.in);
    String registry = "";
    char more;

    System.out.print(
      "Do you wish to create a bridal registry list? (y/n): ");
    more = stdIn.nextLine().charAt(0);

    while (more == 'y')
    {
      System.out.print("Enter item: ");
      registry += stdIn.nextLine() + " - ";
      System.out.print("Enter store: ");
      registry += stdIn.nextLine() + "\n";
      System.out.print("Any more items? (y/n): ");
      more = stdIn.nextLine().charAt(0);
    } // end while

    if (!registry.equals(""))
    {
      System.out.println("\nBridal Registry:\n" + registry);
    }
  } // end main
} // end BridalRegistry class
```

Figure 4.11 BridalRegistry program with `while` loop and user-query termination

Infinite Loops

Suppose you're trying to print the numbers 1 through 10. Will the following code fragment work?

```
int x = 0;
while (x < 10)
{
  System.out.println(x + 1);
}
```

The `while` loop body does just one thing it prints 1 (because 0 + 1 is 1). It does not update `x`'s value (because there's no assignment or increment statement for `x`). With no update for `x`, the `while` loop's condition (`x < 10`) always evaluates to `true`. That's an example of an *infinite loop*. The computer executes the statements in the loop body over and over—forever. When you have an infinite loop, the computer seems to freeze or "hang up."

Sometimes what seems to be an infinite loop is just an extremely inefficient algorithm that takes a long time to finish. In either of these cases, you can figure out what's happening by inserting into the loop a diagnostic statement that prints a value you think should be changing in a certain way. Then run the program and watch what happens to that value.

Insert temporary print statements to see details.

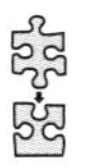

4.9 do Loop

Now let's consider a second type of Java loop—the `do` loop. A `do` loop is appropriate when you're sure that you want the loop body to be executed at least one time. Because the `do` loop matches the way most computer hardware performs looping operations, it is slightly more efficient than the other types of loops. Unfortunately, its awkwardness makes it prone to programming error, and therefore some programmers don't like to use it. But at the very least, you need to be aware of it.

Syntax and Semantics

Figure 4.12 shows the `do` loop's syntax and semantics. Note that the `do` loop's condition is at the bottom. This contrasts with the `while` loop, where the condition is at the top. Having the condition tested at the bottom is how the `do` loop guarantees that the loop executes at least one time. Note the semicolon at the right of the condition. That's required by the compiler, and omitting it is a common error. Finally, note that the `while` part is on the same line as the closing brace—that's good style. It's possible to put `while (`*<condition>*`);` on the line after the closing brace, but that would be bad style because it would look like you're trying to start a new `while` loop.

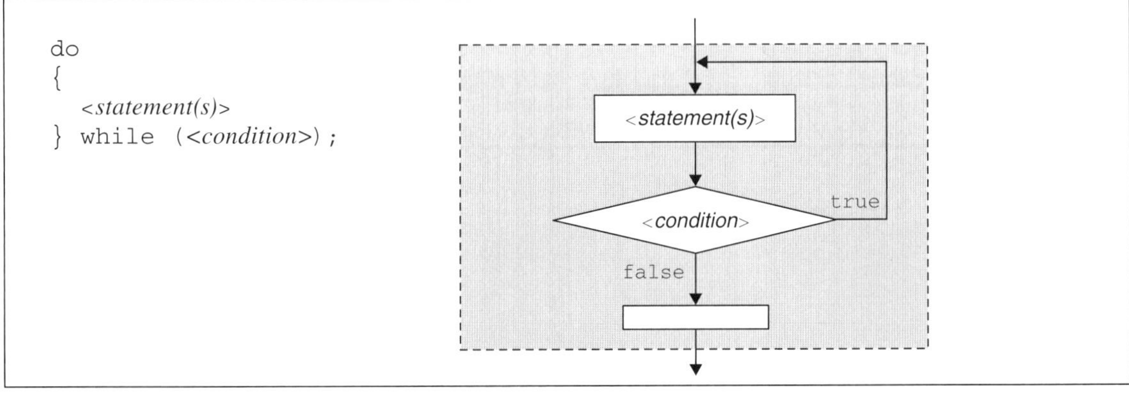

Figure 4.12 Syntax and semantics for the do loop

Here's how the do loop works:

1. Execute the do loop's body.
2. Check the final condition.
3. If the condition is `true`, jump back to the top of the do loop and repeat step 1.
4. If the condition is `false`, continue with the statement immediately below the loop.

Practice Problem

Now let's illustrate the do loop with an example problem. Suppose you're asked to write a program that prompts the user to enter length and width dimensions for each room in a proposed house so that total floor space can be calculated for the entire house. After each length/width entry, ask the user if there are any more rooms. When there are no more rooms, print the total floor space.

How many repeats?

To solve this problem, first ask whether a loop is appropriate. Does anything need to be repeated? Yes, you'll want to read in dimensions repeatedly, so a loop is appropriate. To determine the type of loop, ask yourself: Will you always need to execute the read-in-the—dimensions loop body at least once? Yes, every house must have at least one room, so you'll need to read in at least one set of dimensions. Thus, it's appropriate to use a do loop for this problem. Now that you've thought through the looping issues, you're ready to put pencil to paper and write down your solution. Go for it.

When you're done working out a solution on your own, look at our solution in Figure 4.13. Did you prompt for length and width values within your do loop and then add the length times width product to a total floor space variable? Did you then prompt the user for a continue decision?

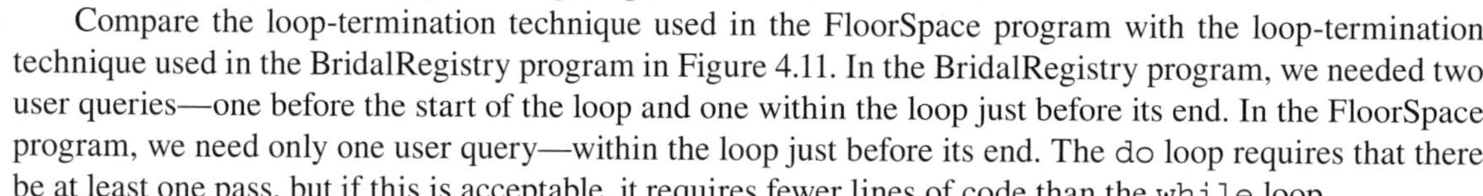

Compare the loop-termination technique used in the FloorSpace program with the loop-termination technique used in the BridalRegistry program in Figure 4.11. In the BridalRegistry program, we needed two user queries—one before the start of the loop and one within the loop just before its end. In the FloorSpace program, we need only one user query—within the loop just before its end. The do loop requires that there be at least one pass, but if this is acceptable, it requires fewer lines of code than the `while` loop.

Before leaving the FloorSpace program, take note of a style feature. Do you see the blank lines above and below the do loop? It's good style to separate logical chunks of code with blank lines. Because a loop is a logical chunk of code, it's nice to surround loops with blank lines unless the loop is very short (that is, less than about four lines).

4.10 `for` Loop

Now let's consider a third type of loop—the `for` loop. A `for` loop is appropriate when you know the exact number of loop iterations before the loop begins. For example, suppose you want to perform a countdown from 10, like this:

<u>Sample session:</u>

```
10  9  8  7  6  5  4  3  2  1  Liftoff!
```

In your program, you'll need to print 10 numbers, and you should print each number with the help of a print statement inside a loop. Because the print statement should execute 10 times, you know the exact number of iterations for the loop, 10. Therefore, you should use a `for` loop.

```
/*******************************************************************
* FloorSpace.java
* Dean & Dean
*
* This program calculates total floor space in a house.
*******************************************************************/

import java.util.Scanner;

public class FloorSpace
{
  public static void main(String[] args)
  {
    Scanner stdIn = new Scanner(System.in);
    double length, width;        // room dimensions
    double floorSpace = 0;       // house's total floor space
    char response;               // user's y/n response

    do
    {
      System.out.print("Enter the length: ");
      length = stdIn.nextDouble();
      System.out.print("Enter the width: ");
      width = stdIn.nextDouble();
      floorSpace += length * width;
      System.out.print("Any more rooms? (y/n): ");
      response = stdIn.next().charAt(0);
    } while (response == 'y' || response == 'Y');

    System.out.println("Total floor space is " + floorSpace);
  } // end main
} // end class FloorSpace
```

Figure 4.13 Using a do loop to calculate total floor space

For another example, suppose you want to find the factorial of a user-entered number, like this:

Sample session:

```
Enter a whole number: 4
4! = 24
```

For 4 factorial, you need to multiply the values 1 through 4: $1 \times 2 \times 3 \times 4 = 24$. The three $\times$'s indicate that three multiplications are necessary. So 4 factorial requires three loop iterations. For the general case, where you need to find the factorial for a user-entered number, store the user-entered number in a `count` variable. Then multiply the values 1 through `count` like this:

```
1 * 2 * 3 *...* count
```

`count - 1` number of `*`'s

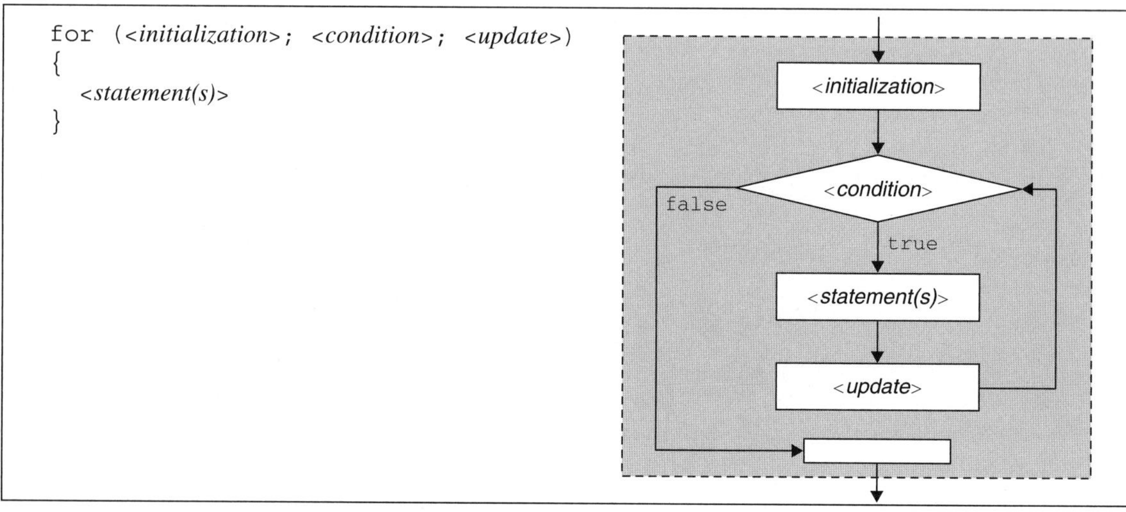

Figure 4.14 Syntax and semantics for the `for` loop

The `*`'s indicate that `count - 1` multiplications are necessary. So `count` factorial requires `count - 1` loop iterations. Because you know the number of iterations for the loop (`count - 1`), use a `for` loop.

Syntax and Semantics

Figure 4.14 shows the `for` loop's syntax and semantics. The `for` loop header does a lot of work. So much work that it's split into three components—the *initialization, condition,* and *update* components. The following list explains how the `for` loop uses the three components. As you read the list, refer to Figure 4.14's flowchart to get a better idea of what's going on.

1. Initialization component
   Before the first pass through the body of the loop, execute the initialization component.
2. Condition component
   Before each loop iteration, evaluate the condition component:
   - If the condition is `true`, execute the body of the loop.
   - If the condition is `false`, terminate the loop (exit to the statement below the loop's closing brace).
3. Update component
   After each pass through the body of the loop, return to the loop header and execute the update component. Then, recheck the continuation condition in the second component, and if it's satisfied, go through the body of the loop again.

The header of a `for` loop is quite flexible. We discuss alternative configurations later in this book, in Section 11.12 of Chapter 11.

Countdown Example

Here is a code fragment for the countdown example mentioned at the start of this section:

```
for (int i=10; i>0; i--)
{
  System.out.print(i + " ");
}
System.out.println("Liftoff!");
```

Note that the same variable, `i`, appears in all three components of the `for` loop header. That variable is given a special name. It's called an *index variable*. Index variables in `for` loops are often, but not always, named `i` for "index." Even when it is not the letter `i`, a `for`-loop index is usually abbreviated as one letter or just a few letters. As a rule, we discourage highly abbreviated variable names. It's OK in this instance, however, because the context of the `for`-loop header clearly explains the index variable's meaning.

Index variables often start at a low value, increment up, and then stop when they reach a threshold set by the condition component. But in the above example, the index variable does just the opposite. It starts at a high value (10), decrements down, and then stops when it reaches the threshold of 0. Let's informally trace the example:

The initialization component assigns 10 to the index, `i`.

The condition component asks "Is `i` > 0?" The answer is yes, so execute the body of the loop.

Print 10 (because `i` is 10), and append a space.

Because you're at the bottom of the loop, the update component decrements `i` from 10 to 9.

The condition component asks "Is `i` > 0?" The answer is yes, so execute the body of the loop.

Print 9 (because `i` is 9) and append a space.

Because you're at the bottom of the loop, the update component decrements `i` from 9 to 8.

The condition component asks "Is `i` > 0?" The answer is yes, so execute the body of the loop.

Repeat the previous printing and decrementing until you print 1.

. . .

After printing 1, because you're at the bottom of the loop, decrement `i` from 1 to 0.

The condition component asks "Is `i` > 0?" The answer is no, so quit the loop, drop down to the first statement after the closing brace, and print "Liftoff!"

Alternatively, we could have implemented the solution with a `while` loop or a `do` loop. Why is the `for` loop preferable? With a `while` loop or a `do` loop, you'd need two extra statements to initialize and update the count variable. That would work OK, but using a `for` loop is more elegant.

Factorial Example

Now, let's make sure you really understand how the `for` loop works by studying a formal trace of the second example mentioned at the start of this section—the calculation of a factorial. Figure 4.15 shows the factorial-calculation code listing and its associated trace. Note the input column in the top-left corner of the trace. You didn't have input in Chapter 3's trace examples, so input is worth mentioning now. When the program reads an input value, you copy the next input from the input column into the next row under the variable to which the input is assigned. In this case, when you get to `number = stdIn.nextInt()`, you copy the 4 from the input column to the next row in the `number` column.

This trace shows that the 8, 10 sequence repeats three times, so there are indeed three iterations, as expected. Suppose you entered `number` = 0. Does the program work for that extreme case? The loop header initializes `int i=2` and then immediately tests to see if `i<=number`. Because this condition is `false`, the loop terminates before it starts, and the code prints the initial value of `factorial`, which is 1.0. That's correct, because 0 factorial does indeed equal 1.

What about the other extreme case—when the input value is very large? The factorial of a number increases much more rapidly than the number itself increases. If we had declared `factorial` to be of type `int`, then input values greater than 12 would cause the `factorial` variable to overflow, and the output value would be horribly wrong! That's why we declared `factorial` to be of type `double`.

```
1  Scanner stdIn = new Scanner(System.in);
2  int number;
3  double factorial = 1.0;
4
5  System.out.print("Enter a whole number: ");
6  number = stdIn.nextInt();
7
8  for (int i=2; i<=number; i++)
9  {
10   factorial *= i;
11 }
12
13 System.out.println(number + "! = " + factorial);
```

Declare for loop index variables within the for loop header.

input

4

line#	number	factorial	i	output
2	?			
3		1.0		
5				Enter a whole number:
6	4			
8			2	
10		2.0		
8			3	
10		6.0		
8			4	
10		24.0		
8			5	
13				4! = 24.0

Figure 4.15 Code fragment that illustrates factorial calculation plus its associated trace

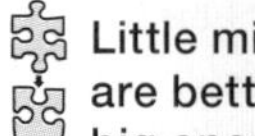

Little mistakes are better than big ones.

A `double` has more precision than an `int`, and it gives approximately correct answers even when its precision is inadequate. This makes the program more robust because it fails more *gracefully*. That is, when it fails, it fails just a little bit, not a lot.

Scope of `for` Loop Index

In the `for` loop examples presented so far, the loop's index variable (`i`) is initialized (declared and given an initial value) in the `for` loop header. This limits the *scope* or recognizable range of the index variable to the `for` loop itself. In other words, whenever a variable is declared in the `for` loop header, it exists and can be recognized and used only by code that is within the body of that `for` loop. For example, if you tried to use the value of the `i` index variable in the print statement that followed the final brace of the `for` loop in Figure 4.15, the compiler would say "cannot find symbol. . . variable i."

Sometimes variables used in a loop need to have a scope beyond the loop's scope. The above Factorial program illustrates what we're talking about. The `factorial` variable must be available for the print statement after the end of the loop, so it must be declared outside the loop. Because it is also needed in the loop, it must be declared before the loop, so we declare it at the beginning of the method with the other variables whose scopes extend throughout the method.

4.11 Solving the Problem of Which Loop to Use

It's helpful to keep in mind that all three types of loops have an initial condition, a test of that condition, and an update to the tested condition. In while loops and do loops, initialization occurs before the loop starts, and updating occurs within the body of the loop. In for loops, initialization, testing, and updating all occur in the header. The `do` loop's decision point is at the bottom of the loop. That's in contrast to the `while` and `for` loops, where the decision point is at the top of the loop. When the decision point is at the top of the loop, the decision stands out more and the code is therefore less prone to programming error.

With programming, as in life, there are usually many different ways to accomplish the same thing. For example, for a problem that requires repetition, you can actually use any of the three loops to solve any repetition problem. Even though that's the case, you should strive to make your programs elegant, and that means choosing the most appropriate loop even though any loop could be made to work.

A toolkit needs more than one tool.

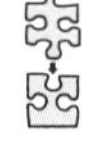

Flexibility makes programming fun if you like to be creative. But if you're just starting out, that flexibility can lead to confusion. In Figure 4.16, we provide a table that attempts to alleviate some of that confusion. It suggests a way to choose an appropriate type of loop and how to get started with that loop's code. We use angled brackets around text to indicate that the enclosed text is a description of code, not actual code. Thus, in using Figure 4.16's `do` loop and `while` loop templates, you'll need to replace *<prompt—do it again (y/n?)>* with actual code. For example, for a game program, you might use this actual code:

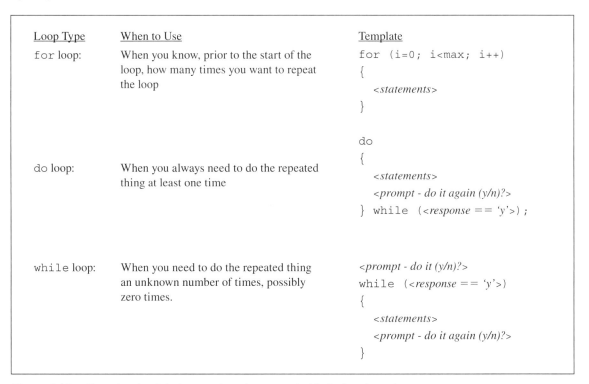

Loop Type	When to Use	Template

`for` loop:	When you know, prior to the start of the loop, how many times you want to repeat the loop	`for (i=0; i<max; i++)`
		{
		  <statements>
		}

`do` loop:	When you always need to do the repeated thing at least one time	`do`
		{
		  <statements>
		  <prompt - do it again (y/n)?>
		} `while` (*<response* == 'y'>);

`while` loop:	When you need to do the repeated thing an unknown number of times, possibly zero times.	*<prompt - do it (y/n)?>*
		`while` (*<response* == 'y'>)
		{
		  <statements>
		  <prompt - do it again (y/n)?>
		}

Figure 4.16 Choosing the right loop and getting started with the loop's code

```
System.out.print("Do you want to play another game (y/n)? ");
response = stdIn.nextLine().charAt(0);
```

When figuring out which loop to use, it's best to think about the loops in the order of appearance in Figure 4.16. Why? Note how the `for` loop uses the fewest lines, the `do` loop uses the next fewest lines, and the `while` loop uses the most lines. Thus, the `for` loop is the most compact, and the `do` loop is the next most compact. But the `while` loop is more popular than the `do` loop because its condition is at the beginning of the loop, which makes it more flexible and easier to find. Although you may wish to avoid the `do` loop because of its relatively awkward structure, in general, you should use the loop that's most appropriate for your particular problem.

When deciding how to write loop code, you can use the templates shown in Figure 4.16 as starting points. Be aware that in writing loop code, you have to do more than just copy code from Figure 4.16. You need to adapt the code to your particular problem. For example, in writing a `for` loop, it's common to use `i=0` for the initialization component, and that's why the `for` loop template's initialization component shows `i=0`. However, if some other initialization component is more appropriate, like `count=10`, then use the more appropriate code.

4.12 Nested Loops

A *nested loop* is a loop that's inside another loop. You'll see nested loops quite often in real-world programs. In this section, we discuss some of the common characteristics inherent to nested loops.

Suppose you're asked to write a program that prints a rectangle of characters where the user specifies the rectangle's height, the rectangle's width, and the character's value.

Sample session:

```
Enter height: 4
Enter width: 3
Enter character: <
<<<
<<<
<<<
<<<
```

Select the best tool for the job.

To figure out the loops, you first need to think about what needs to be repeated. So, . . . what needs to be repeated? You need to print rows of characters repeatedly. What type of loop should you use to print the rows repeatedly? First, try to use a `for` loop. The test for a `for` loop is whether you know the number of times you'll need to repeat the loop. Do you know the number of times you'll need to repeat this loop? Yes, the user enters the height, you can use that entered value to determine the number of rows, and that tells you the number of times to repeat the loop. Therefore, you should use a `for` loop to print successive rows.

Now that you know how to print multiple rows, you need to know how to print an individual row. Do you need to repeat anything when printing an individual row? Yes, you need to print characters repeatedly. So what type of loop should you use for that? Use another `for` loop because you can use the user's width entry to determine the number of characters to be printed.

So there you go—you need two `for` loops. Should you put one loop right after the other? No! You need to nest the second loop, the one that prints an individual row, inside the first loop. That should make sense if you word the goal carefully—"Print multiple rows, and within each row, print a sequence of characters." The key word is "within." That tells you to insert the second `for` loop inside the first `for` loop's braces.

Using this discussion as a guideline, now write a complete program solution. When you're done, compare your answer to the NestedLoopRectangle program in Figure 4.17.

```
/*************************************************************
* NestedLoopRectangle.java
* Dean & Dean
*
* This program uses nested looping to draw a rectangle.
*************************************************************/

import java.util.Scanner;

public class NestedLoopRectangle
{
  public static void main(String[] args)
  {
    Scanner stdIn = new Scanner(System.in);
    int height, width;         // rectangle's dimensions
    char printCharacter;

    System.out.print("Enter height: ");
    height = stdIn.nextInt();
    System.out.print("Enter width: ");
    width = stdIn.nextInt();
    System.out.print("Enter character: ");
    printCharacter = stdIn.next().charAt(0);

    for (int row=1; row<=height; row++)
    {
      for (int col=1; col<=width; col++)
      {
        System.out.print(printCharacter);
      }
      System.out.println();
    }
  } // end main
} // end class NestedLoopRectangle
```

Use `print` here, to stay on same line.

Use `println` here, to move to new line.

Figure 4.17 Program that uses nested loops to draw a rectangle

Note how we use the `print` method for the print statement inside the inner loop to keep subsequent printed characters on the same line. Then after the inner loop finishes, we use a separate `println` method to go to the next line.

For most problems where you're dealing with a two-dimensional picture like this rectangle example, you'll want to use nested `for` loops with index variables named `row` and `col` (`col` is short for column). Why? It makes code more understandable. For example, in the first `for` loop header, the `row` variable goes from 1 to 2 to 3, and so on, and that corresponds perfectly with the actual rows printed by the program. However, be aware that it's also common for nested `for` loops to use index variables named `i` and `j`. Why `i` and `j`? Because `i` stands for "index," and `j` comes after `i`.

In the NestedLoopRectangle program, there are two levels of nesting, but in general, there may be any number of nesting levels. Each level adds another dimension to the problem. Our NestedLoopRectangle

program is quite symmetrical. Both loops are the same type (they're both `for` loops), and both loops do the same kind of thing (they both help to count the number of times something is printed). In general, however, nested loops do not have to be the same type, and they do not have to do the same kinds of things.

Determining the Number of Iterations in a Nested `for` Loop

As you develop and debug code in the real world, you'll very often need to analyze loops to determine how many times they iterate. If you have a basic `for` loop that increments its index variable by 1, calculating its iterations is easy: index variable's maximum value minus index variable's minimum value plus 1. For example, in the NestedLoopRectangle program, the outer loop repeats `height` − 1 + 1 times or, more simply, `height` times (where `height` is `row`'s maximum value, and 1 is `row`'s minimum value). Similarly, you can determine that the NestedLoopRectangle program's inner loop repeats `width` times. To calculate the number of times the inner loop will execute its print statement, you multiply the outer loop's iterations by the inner loop's iterations. The result indicates that the print statement will be executed `height * width` times.

Now for a more complex problem. Consider this code fragment:

```
for (int j=0; j<n; j++)
{
  System.out.print(" ");
  for (int k=1; k<n-1; k++)
  {
    System.out.print("*");
  }
  System.out.println();
}
```

To determine how many times the code fragment prints an asterisk, first calculate the outer loop's iterations. Because the outer loop uses < rather than <=, the outer loop's maximum index variable value is actually `n` – 1, not `n`. Thus, the outer loop repeats (`n`−1) − 0 + 1 times, which simplifies to `n`. We next calculate the inner loop's iterations: (`n`−2) − 1 + 1, which simplifies to `n` − 2. Multiplying the outer loop's iterations by the inner loop's iterations yields the formula `n *` (`n` – 2).

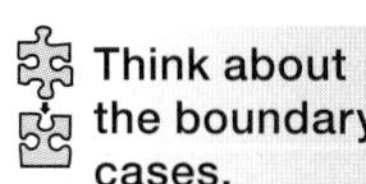

Think about the boundary cases.

Calculating the number of loop iterations can be tricky, so after you come up with what you think is the final formula, step back and think about the boundary cases. In the example above, `n *` (`n` – 2) correctly predicts the number of times the code fragment prints an asterisk, but only when `n` ≥ 2. What happens when `n` equals 1? The formula tells you that the code fragment prints –1 asterisks (because 1 * (1 – 2) equals –1), but that is impossible. By examining the inner loop's condition carefully, you should be able to determine that the inner loop iterates 0 times when `n` equals 1. Likewise, you should be able to determine that the inner loop iterates 0 times when `n` equals 0. Thus, when `n` < 2, the code fragment prints 0 asterisks.

4.13 `boolean` Variables

The conditions that appear in `if` statements and loops all evaluate to either `true` or `false`. We described these Boolean values in Section 4.2. Java also allows us to define a `boolean` variable, which is a variable that can hold a Boolean value. To declare a `boolean` variable, specify `boolean` for the variable's type, like this:

```
boolean upDirection;
```

In this section, we describe when to use boolean variables in general, and we provide a program that uses boolean variables, including the upDirection variable shown above.

When to Use a boolean Variable

Programs often need to keep track of the state of some condition. You can use a boolean variable to keep track of any two-way *state*—a yes/no, up/down, on/off attribute of some entity. For example, if you're writing a program that simulates the operations of an electronic garage door opener, you'll need to keep track of the state of the garage door's direction—is the direction up or down? You need to keep track of the direction "state" because the direction determines what happens when the garage door opener's button is pressed. If the direction state is up, then pressing the garage door button causes the direction to switch to down. If the direction state is down, then pressing the garage door button causes the direction to switch to up.

A boolean variable is good at keeping track of the state of some condition when the state has one of two values. For example:

Values for the state of a garage door opener's direction	**Comparable values for a boolean variable named upDirection**
up	true
down	false

Garage Door Opener Example

The following code skeleton illustrates how the upDirection variable works:

```
boolean upDirection = true;
do
{
  . . .
  upDirection = !upDirection;
  . . .
} while (<user-presses-the-garage-door-opener-button>);
```

The boolean upDirection = true; statement tells the program to start in the down/closed position and go up when the garage door opener is first pressed. Each iteration of the loop represents what happens when the user presses the garage door opener button. The upDirection = !upDirection statement implements the garage door opener's toggling operation. If upDirection holds the value true, this statement changes it to false, and vice versa.

Now let's look at the upDirection variable in the context of a complete GarageDoor program. In the program, each push of the Enter key on the computer keyboard simulates a push of the garage door opener button. The first push makes the door move upward. The second push makes the door stop. The third push makes the door move downward. The fourth push makes the door stop. And so forth, until the user enters 'q' to make the program quit. Here's an example of user interaction with the GarageDoor program:

Sample session:

```
GARAGE DOOR OPENER SIMULATOR

Press Enter, or enter 'q' to quit:
moving up
```

```
Press Enter, or enter 'q' to quit:
stopped
Press Enter, or enter 'q' to quit:
moving down
Press Enter, or enter 'q' to quit:
stopped
Press Enter, or enter 'q' to quit: q
```

Figure 4.18 contains the program's code. In the program, verify that `up-Direction` is used as previously discussed. Note that there's a second `boolean` variable, `inMotion`. The `upDirection boolean` variable keeps track of the state of going up or down. That one state variable would be good enough if pressing a garage door opener button always generated an up or down motion. But as shown in the sample session, that's not the case. Half the time, pressing the garage door opener causes the garage door to stop moving. Here's the key point: If the door is moving, the door stops, and if the door is stopped, the door starts moving. We keep track of whether the garage door is currently moving with the help of a second state variable, `inMotion`. The `inMotion` state variable toggles (goes from `false` to `true` or vice versa) at each button push, whereas the `upDirection` state variable toggles only when the door is stopped—at every other button push.

The GarageDoor program is *user-friendly* because it requires a minimum amount of user input. A given user entry serves one of two purposes. The simplest kind of entry (pressing the **Enter** key) simulates pushing the button on a garage door opener. Any other entry (not just a 'q' entry) terminates the looping process. Whenever a special data value (in this case anything except a plain **Enter**) tells a program to stop looping, we say we're using a *sentinel value* to terminate the looping process. Because the program imposes a minimum burden on the user in terms of input, and because the code is relatively concise and efficient, it's appropriate to call this an elegant implementation.

Beware of Comparing a `boolean` Variable Within a Condition

In the GarageDoor program, note how we use the `inMotion` and `upDirection boolean` variables by themselves as conditions in `if` statements:

```
if (inMotion)
{                          }  <-- good
  if (upDirection)
  {
    . . .
```

In the past, you used relational operators in your conditions (<. <=, ==, etc.). But the only rule for a condition is that it needs to evaluate to `true` or `false`. A `boolean` variable <u>is</u> either `true` or `false`, so using a `boolean` variable by itself for a condition is legal. Actually, using a `boolean` variable by itself for a condition is considered to be elegant. Why? Consider the alternative. The following `if` conditions are functionally equivalent to the prior `if` conditions:

```
if (inMotion == true)
{                               }  <-- not so good
  if (upDirection == true)
  {
    . . .
```

```
/*********************************************************************
* GarageDoor.java
* Dean & Dean
*
* This simulates the operation of a garage door opener.
*********************************************************************/

import java.util.Scanner;

public class GarageDoor
{
  public static void main(String[] args)
  {
    Scanner stdIn = new Scanner(System.in);
    String entry;                    // user's entry - enter key or q
    boolean upDirection = true;      // Is the current direction up?
    boolean inMotion = false;        // Is garage door currently moving?

    System.out.println("GARAGE DOOR OPENER SIMULATOR\n");

    do
    {
      System.out.print("Press Enter, or enter 'q' to quit: ");
      entry = stdIn.nextLine();

      if (entry.equals(""))          // pressing Enter generates ""
      {
        inMotion = !inMotion;        // button toggles run state
        if (inMotion)
        {
          if (upDirection)
          {
            System.out.println("moving up");
          }
          else
          {
            System.out.println("moving down");
          }
        }
        else
        {
          System.out.println("stopped");
          upDirection = !upDirection;  // direction reverses at stop
        }
      } // end if entry = ""
    } while (entry.equals(""));
  } // end main
} // end GarageDoor class
```

! operator toggles motion every time
! operator toggles direction when stopped

Figure 4.18 The GarageDoor program

What would happen if instead of using the double equals (==), you accidentally used the single equals (=)? The expressions in the two conditions would assign `true` to `inMotion` and `upDirection`, respectively. Then both conditions would evaluate to `true` (because an assignment operation evaluates to the assigned value), and code in the `if` statements would execute. That would give the correct result only when `inMotion` and `upDirection` were both true beforehand. But it would give an incorrect result if either `inMotion` or `upDirection` was `false` beforehand. Moreover, the unintended alteration of the values of `inMotion` and `upDirection` might upset other conditional evaluations. This kind of logic error is insidious because the compiler would not see it, and it would be hard for you to see it too, because = looks so much like ==.

Thus, to steer clear of accidentally using the assignment operator, we recommend that you never compare `boolean` variables to `true` or `false`. Instead, if you need to use a `boolean` variable within a condition, just use the `boolean` variable by itself.

By the way, for non-Boolean variables (like `int` variables and `double` variables), accidentally using the assignment operator within a condition isn't as much of a problem. What happens if you enter the following code?

```
if (score = -1)
```

Because the code assigns the integer –1 into `score`, the condition evaluates to an integer. The compiler requires conditions to evaluate to Boolean values, so the code generates a compilation error. Such a compilation error should be relatively easy to identify and fix. Certainly much easier to deal with than the nasty logic error due to assigning `true` or `false` to a `boolean` variable within a condition.

4.14 Input Validation

In the previous section, you learned to use a `boolean` variable to keep track of a two-way state. In this section, you'll learn to use a `boolean` variable for a particularly common two-way state—the state of a user's input in terms of whether it's valid or invalid.

Input validation is when a program checks a user's input to make sure it's valid (that is, correct and reasonable). If it's valid, the program continues. If it's invalid, the program enters a loop that warns the user about the erroneous input and then prompts the user to re-enter.

In the GarageDoor program, note how the program checks for an empty string (which indicates the user wants to continue). If the string is not empty, it assumes that the user entered a 'q', but it doesn't check specifically for a 'q'. Consequently, it does not deal well with the possibility that the user accidentally hits another key before pressing the `Enter` key. It interprets that input as a quit command instead of a mistake.

To make the program more robust, you should provide input validation. There are several possible ways to do this. One of the simplest ways is to insert a `while` loop whose condition and's together all bad possibilities and whose body warns the user about the erroneous input and then prompts the user to re-enter. For the GarageDoor program in Figure 4.18, input validation is provided by the code fragment in Figure 4.19.

```
while (!entry.equals("") && !entry.equalsIgnoreCase("q"))
{
  System.out.println("Invalid entry.");
  System.out.print("Press Enter, or enter 'q': ");
  entry = stdIn.nextLine();
}
```

Figure 4.19 Input validation loop to insert after the input statement in Figure 4.18

Where should you insert this code fragment? You want to validate the input right after the input is entered. So to make the GarageDoor program more robust, you should insert the above code fragment into Figure 4.18 immediately after this statement:

```
entry = stdIn.nextLine();
```

Running the modified program produces the following sample session:

<u>Sample session:</u>

```
GARAGE DOOR OPENER SIMULATOR

Press Enter, or enter 'q' to quit:
moving up
Press Enter, or enter 'q' to quit: stop  <---- invalid entry
Invalid entry.
Press Enter, or enter 'q':  <---- corrected entry
stopped
Press Enter, or enter 'q' to quit: q
```

Optional Forward References

At this point, some readers might want to learn about arrays. An *array* is a collection of related items of the same type. Array manipulations require the use of loops. As such, arrays provide a means for readers to gain further practice with the material presented in Chapter 4, specifically the loop material. You're not required to learn about arrays just yet, but if you can't wait, you can read about arrays in Chapter 9, Sections 9.1 through 9.6.

Later in the book, we present relatively advanced syntax details that pertain to control statements. For example, embedding an assignment expression in a loop header or using a `break` statement to break out of a loop. You're not required to learn those details just yet, but if you can't wait, you can read about them in Chapter 12, Sections 12.6 through 12.12.

4.15 Problem Solving with Boolean Logic (Optional)

Make the logic as clean as possible.

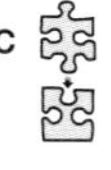

The conditions for `if` statements and loops can sometimes get complicated. For a better understanding of complicated conditions, we'll now look at the logic that comprises a condition. Learning how to manipulate logic should help you to (1) simplify condition code and (2) debug logical problems. You've seen how logical operators work when applied to comparison-operator conditions. For example, this code (which uses the `&&` operator in conjunction with the >= and <= comparison operators) probably already makes sense to you:

```
(temp >= 50.0 && temp <= 90.0)
```

Now you'll see how logical operators work with `boolean` variables. This is called *Boolean logic* or *Boolean algebra*. The most primitive and most general way to describe Boolean logic is with a *truth table*. A truth table is a list of all input combinations and their corresponding outputs. Figure 4.20 presents truth tables for the three basic building blocks of Boolean logic, the logical operators `!` for NOT, `&&` for AND, and `||` for OR.

NOT	
x	!x
false	true
true	false

AND		
x	y	x && y
false	false	false
false	true	false
true	false	false
true	true	true

OR		
x	y	x || y
false	false	false
false	true	true
true	false	true
true	true	true

Figure 4.20 Truth tables for the three basic logical operations

Boolean Algebra Basic Identities

Sometimes, however, a logical expression is harder to understand. This is particularly true when it includes several "not" (!) operators. To gain a better understanding of what the code means and is supposed to do, it's sometimes helpful to transform the logical expression to another form. Boolean algebra provides a special set of formulas called *basic identities,* which anyone can use to make transformations. These basic identities are listed in Figure 4.21. The precedence of the various operators is the precedence given in Figure 4.6. That

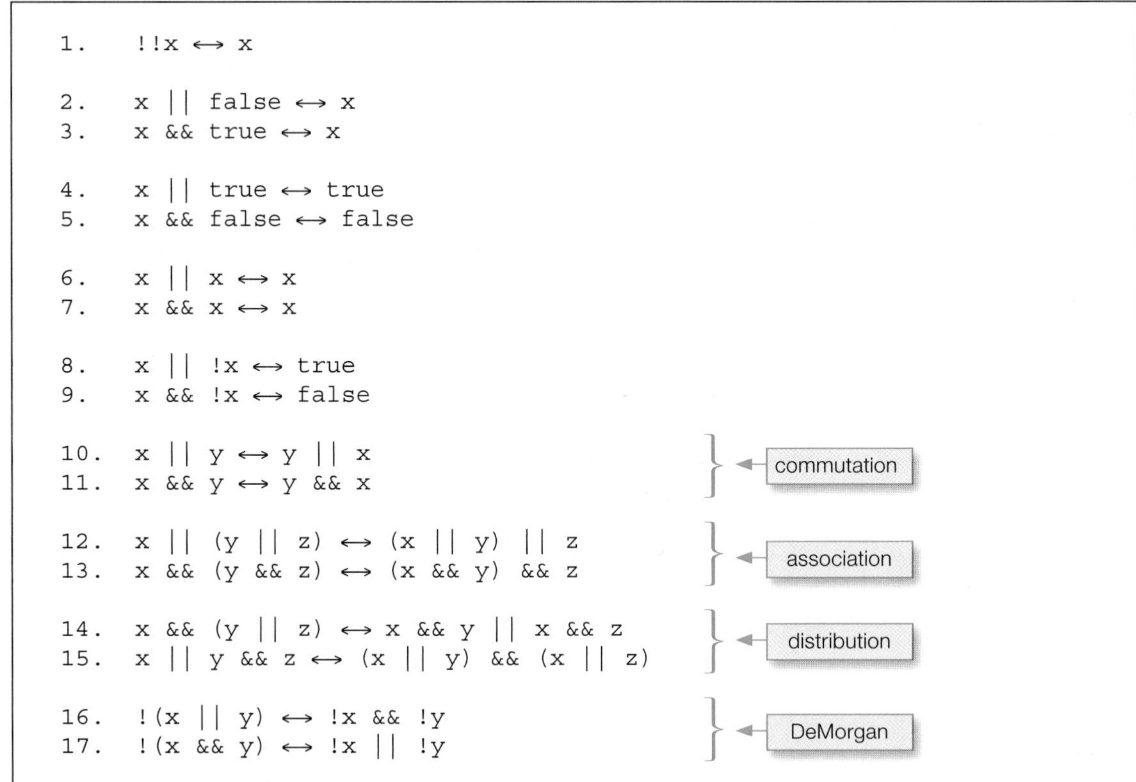

```
1.   !!x ↔ x

2.   x || false ↔ x
3.   x && true ↔ x

4.   x || true ↔ true
5.   x && false ↔ false

6.   x || x ↔ x
7.   x && x ↔ x

8.   x || !x ↔ true
9.   x && !x ↔ false

10.  x || y ↔ y || x
11.  x && y ↔ y && x

12.  x || (y || z) ↔ (x || y) || z
13.  x && (y && z) ↔ (x && y) && z

14.  x && (y || z) ↔ x && y || x && z
15.  x || y && z ↔ (x || y) && (x || z)

16.  !(x || y) ↔ !x && !y
17.  !(x && y) ↔ !x || !y
```


Figure 4.21 Basic identities of Boolean algebra
You can use these identities in any combination to change the form of any conditional expression.

is, `!` has the highest precedence, `&&` has the next highest precedence, and `||` has the lowest precedence. The ⟷ symbol means equivalence; that is, whatever is on the left side of the double arrow can be replaced by whatever is on the right side, and vice versa.

The first 13 identities are relatively straightforward, and you should be able to satisfy yourself of their validity by just thinking about them. Likewise, you shouldn't have to memorize them. You should be able to use them instinctively. For example, *commutation* means you can switch the order without changing anything, and *association* means you can move the parentheses without changing anything. The last four identities are more mysterious, and some of them might even seem unreasonable at first. For example, *distribution* is a kind of shuffling, and *DeMorgan's laws* say you can negate everything and exchange all and's and or's.

Proving the Boolean Identities

Now that you've seen the basic identities, let's see how to prove them. The proof technique is to write a program that compares two arbitrary logical expressions for all possible values of the `boolean` variables they contain. If the two expressions evaluate to the same truth values for all possible variable values, they are logically equivalent. Figure 4.22 contains a program that does just that for the special case of the expressions on either side of basic identity 16 in Figure 4.21.

It's straightforward to modify the TruthTable program in Figure 4.22 to test any of the other basic identities in Figure 4.21. In fact, you can modify the program to test any prospective logical equivalence. To test a different equivalence, substitute the left and right sides of the prospective equivalence for the expressions assigned to `result1` and `result2`, respectively.

Applications

There are many ways you can use Boolean identities. For example, consider the condition in the `if` statement in Figure 4.5, which looked like this:

```
((temp >= 50) && (temp <= 90))
```

If you apply the `!` operator to a comparison operator expression, that causes the expression's comparison operator to change. For example, `!(temp >= 50)` is equivalent to `(temp < 50)`. With that reasoning in mind plus basic identity 1, you can apply `!!` to each of the above comparison expressions and generate this equivalent condition:

```
(!(temp < 50) && !(temp > 90))
```

You can apply basic identity 16 to the above condition and come up with this equivalent condition:

```
!((temp < 50) || (temp > 90))
```

You can use the above condition as part of a replacement for Figure 4.5's original `if` statement where the `if` and `else` subordinate statements are swapped. Here's the resulting functionally equivalent `if` statement:

```
if ((temp < 50) || (temp > 90))
{
  System.out.println("not OK");
}
else
{
  System.out.println("OK");
}
```

```
/*************************************************************
* TruthTable.java
* Dean & Dean
*
* This proves equivalence of two boolean expressions
*************************************************************/

public class TruthTable
{
  public static void main(String[] args)
  {
    boolean x = false;
    boolean y = false;
    boolean result1, result2;

    System.out.println("x\ty\tresult1\tresult2");
    for (int i=0; i<2; i++)
    {
      for (int j=0; j<2; j++)
      {
        result1 = !(x || y);
        result2 = !x && !y;
        System.out.println(x + "\t" + y +
          "\t" + result1 + "\t" + result2);
        y = !y;
      } // end for j
      x = !x;
    } // end for i
  } // end main
} // end TruthTable class
```

To test the equivalence of any two `boolean` expressions, substitute them for these two (shaded) expressions.

Sample output:

```
x       y       result1 result2
false   false   true    true
false   true    false   false
true    false   false   false
true    true    false   false
```

Figure 4.22 Program that generates a truth table for two logical expressions
If `result1` and `result2` values are the same in all rows, the expressions are equivalent.

For another example, consider the condition in the `while` loop in Figure 4.19, which looks like this:

```
(!entry.equals("") && !entry.equalsIgnoreCase("q"))
```

You can apply basic identity 16 to the above condition and come up with this equivalent condition:

```
!(entry.equals("") || entry.equalsIgnoreCase("q"))
```

Chapter Summary

- You can alter a program's sequence of execution by using an `if` statement. The choice of which of two alternative paths to take is determined by the truth of the `if` statement's condition.
- Use the "if, else if" form of the `if` statement to choose among three or more alternatives.
- You must use braces around two or more subordinate statements within any part of an `if` statement, and it's advisable to use them even when there is only one subordinate statement.
- A condition's comparison operators (<, >, <=, >=, ==, and !=) have higher priority than its "and" (&&) and "or" (||) logical operators.
- To negate the result of && and/or || operations, enclose them in parentheses and precede them with a ! operator.
- Use a `switch` statement to choose among several alternatives on the basis of integer or character identifiers.
- Use `case` <*number*>: or `case` <*character*>: and a following `break;` to delimit each alternative in a `switch` statement.
- If the condition in a `while` loop's header is `true`, whatever is in the subsequent block executes, and then if the condition is still `true`, that execution repeats.
- A `do` loop executes its block at least once, and it repeats that execution as long as the condition after the final `while` remains `true`.
- A `for` loop executes its block as long as the condition in the second component of its header remains `true`. The first component in the header initializes a count variable before the first execution, and the third component in the header updates that count variable after each execution and before the next evaluation of the second component's condition.
- You can perform multidimensional iteration by putting loops inside other loops.
- To avoid duplication and/or clutter, assign complicated logical expressions to `boolean` variables, and use those variables in `if` statement or looping conditions.
- Use input validation to avoid bringing bad data into your programs.
- Optionally, use Boolean logic to simplify the expressions in `if` statement and looping conditions, and use truth tables to verify the equivalence of alternative logical expressions.

Review Questions

§4.2 Conditions and `boolean` Values

1. What are Java's two Boolean values?
2. Provide a list of Java's comparison operators.

§4.3 `if` Statements

3. Provide an `if` statement that implements this logic:
   When the water temperature is less than 120 °F, turn the heater on by assigning the value "on" to a `heater` string variable. When the water temperature is greater than 140 °F, turn the heater off by assigning the value "off" to a `heater` string variable. Don't do anything when the water temperature is between these two temperatures.
4. What is the maximum number of "else if" blocks allowed in an `if` statement that uses the "if, else if" form?

§4.4 && Logical Operator

5. The relational and equality operators have higher precedence than the arithmetic operators. (T / F)

§4.5 || Logical Operator

6. Correct the following code fragment so that it executes and outputs OK if a, an int variable, is equal to either 2 or 3:

```
if (a = 2 || 3)
{
  print("OK\n");
}
```

§4.6 ! Logical Operator

7. What Java operator reverses the truth or falsity of a condition?

§4.7 switch Statement

8. What happens if you forget to include `break;` at the end of a block of statements after a particular `case:` label?
9. If you are trying to substitute a `switch` statement for an "if, else" statement, you can use the `if` condition as the controlling expression in the `switch` statement. (T / F)
10. Suppose the controlling expression in a `switch` statement is `(stdIn.next().charAt(0))`, and you want to allow either `'Q'` or `'q'` to produce the same result, which is:
    `System.out.println("quitting");`
    Write the code fragment for the `case` that produces this result.

§4.8 while Loop

11. What must a `while` loop condition evaluate to?
12. Suppose you want to use the user-query technique to terminate a simple `while` loop. Where should you put the user query?

§4.9 do Loop

13. What's wrong with this code fragment?

```
int x = 3;
do
{
  x -= 2;1
} while (x >= 0)
```

§4.10 for Loop

14. If you know ahead of time the exact number of iterations through a loop, what type of loop should you use?
15. Implement the following as a `for` loop:

```
int age = 0;
while (age < 5)
{
  System.out.println("Happy Birthday# " + age);
  age = age + 1;
} // end while
```

What output will your equivalent `for` loop generate?

§4.11 Solving the Problem of Which Loop to Use

16. If you know that a loop should be executed at least one time, what type of loop is most appropriate?

§4.12 Nested Loops

17. Construct a template for a `for` loop inside a `for` loop. Use `i` for the outer `for` loop's index variable and use `j` for the inner `for` loop's index variable.

§4.13 Boolean Variables

18. Assume that the variable `OK` has been declared to be of type `boolean`. Replace the following code with an equivalent `for` loop:

```
OK = false;
while (!OK)
{
  <statement(s)>
}
```

§4.15 Problem Solving with Boolean Logic (Optional)

19. Given the logical expression:

```
!(!a || !b)
```

Replace it with an equivalent logical expression that is completely devoid of "not" operations.

Exercises

1. [after §4.3] Whenever you mail a letter, you must decide how much postage to put on the envelope. You like to use this rule of thumb—use one stamp for every five sheets of paper or fraction thereof. For example, if you have 11 sheets of paper, then you use three stamps. To save money, you simply don't mail the letter if an envelope requires more than three stamps.

Given that the number of sheets is stored in a variable named `numSheets`, write a code fragment that prompts the user and inputs the number of sheets, calculates the number of stamps required, and prints "Use *<number-of-stamps>* stamps" or "Don't mail," where *<number-of-stamps>* is an appropriate integer value.

2. [after §4.8] Given this code fragment:

```
double x = 2.1;

while (x * x <= 50)
{
  switch ((int) x)
  {
    case 6:
      x--;
      System.out.println("case 6, x= " + x);
    case 5:
      System.out.println("case 5, x= " + x);
    case 4:
      System.out.println("case 4, x= " + x);
      break;
    default:
      System.out.println("something else, x= " + x);
  } // end switch
  x +=2;
} // end while
```

Trace the code using either the short form or the long form. To help you get started, here's the trace setup. For the short form, you won't need the line# column.

line#	x	output

3. [after §4.9] The following `main` method is supposed to print the sum of the numbers 1 through 5 and the product of the numbers 1 through 5. Find all the bugs in the program and fix them. Do not add or delete statements. Just fix existing statements. We encourage you to check your work by running test code on a computer.

```
public static void main(String[] args)
{
  int count = 0;
  int sum = 0;
  int product = 0;
  do
  {
    count++;
    sum += count;
    product *= count;
    if (count == 5)
      System.out.println("Sum = " + sum);
      System.out.println("Product = " + product);
  } while (count < 5)
} // end main
```

Intended output:

```
Sum = 15
Product = 120
```

4. [after §4.10] Given this `main` method:

```
public static void main(String[] args)
{
  int i;
  String debug;
  for (i=0; i<3; i++)
  {
    switch (i * i)
    {
      case 0:
        debug = "first";
        break;
      case 1: case 2:
        debug = "second";
      case 3:
        debug = "third";
      default:
```

```
17          System.out.println("In default");
18        } // end switch
19      } // end for
20      System.out.println("i = " + i);
21    } // end main
```

Trace the code using either the short form or the long form. To help you get started, here's the trace setup. For the short form, you won't need the line# column.

line#	i	debug	output

5. [after §4.10] Given the below program skeleton, insert code in the *<insert-code-here>* section such that the program prints the product of even integers from `2` to `num`. You are not required to perform input validation.

```
public class ProductEvenInts
{
  public static void main(String[] args)
  {
    Scanner stdIn = new Scanner(System.in);
    int i, num, product;

    System.out.print("Enter a positive even number: ");
    num = stdIn.nextInt();

    <insert-code-here>

    System.out.println("Product = " + product);
  } // end main
} // end class ProductEvenInts
```

Sample session:

```
Enter a positive even number: 8
Product = 384
```

6. [after §4.12] Given this `main` method:

```
1     public static void main(String[] args)
2     {
3       for (int start=1; start<=5; start+=2)
4       {
5         for (int count=start; count>=1; count--)
6         {
7           System.out.println(count);
8         }
9         System.out.println("Liftoff!");
10      }
11    } // end main
```

Trace the code using either the short form or the long form. To help you get started, here's the trace setup. For the short form, you won't need the line# column.

line#	start	count	output

7. [after §4.12] Given the following code fragment, and assuming that *n* is an `int`, write an expression for the total number of lines of output as a function of *n*.

```
for (int j=n-1; j>=0; j--)
{
  for (int k=1; k<n; k++)
  {
    System.out.println("***");
  }
}
```

What values of *n* minimize the number of lines of output, and what is that minimum number of lines of output?

8. [after §4.13] Given this `main` method:

```
1   public static void main(String[] args)
2   {
3     boolean sheLovesMe = true;
4
5     for (int num=0; num<4; num++)
6     {
7       sheLovesMe = !sheLovesMe;
8     }
9     if (sheLovesMe)
10    {
11      System.out.println("She loves me!");
12    }
13    else
14    {
15      System.out.println("She loves me not!");
16    }
17  } // end main
```

Trace the code using either the short form or the long form. To help you get started, here's the trace setup. For the short form, you won't need the line# column.

line#	sheLovesMe	num	output

9. [after §4.13] Consider the BowlingScores program below.

```
/*************************************************************
* BowlingScores.java
* Dean & Dean
*
* This implements a bowling scores algorithm.
*************************************************************/

import java.util.Scanner;

public class BowlingScores
{
  public static void main(String[] args)
```

```
    {
      Scanner stdIn = new Scanner(System.in);
      int score;
      int totalScore = 0;
      int count = 0;
      double average;

      System.out.print("Enter score (-1 to quit): ");
      score = stdIn.nextInt();

      while (score >= 0)
      {
        totalScore += score;
        count++;
        System.out.print("Enter score (-1 to quit): ");
        score = stdIn.nextInt();
      }

      average = (double) totalScore / count;
      System.out.println("Average score is " + average);
    } // end main
  } // end BowlingScores class
```

Modify this program to avoid division by zero. Initialize a `boolean` variable called `more` with `true`, and use it as the `while` loop condition. Eliminate the prompt and input before the loop and move the prompt and input inside the loop to the top of the loop. Use an "if, else" structure in the loop to set `more` to `false` and bypass the normal calculation if the input is negative.

10. [after §4.13] Consider the following code fragment. Without changing the loop type, modify the code as follows. Incorporate an `if` statement in the loop body to prevent printout when the input equals the sentinel value of zero.

```
int x;
do
{
  x = stdIn.nextInt();
  System.out.println("square = " + (x * x));
} while (x != 0);
```

11. [after §4.15] Here's a brain teaser that uses Boolean logic:
You're traveling on a road, and you come to a fork in the road. You know that one path leads to a pot of gold and the other path leads to a dragon. There are two elves at the fork, both of whom know the way to the pot of gold. You know that one elf always tells the truth and the other elf always lies, but you don't know which elf is which. What single question should you ask to figure out the proper path to the pot of gold?

Review Question Solutions

1. Java's Boolean values are `true` and `false`.

2. Java's comparison operators are:

```
==, !=, <, >, <=, >=
```

3. Use an "if, else if" statement, like this:

```
if (temp < 120)
{
  heater = "on";
}
else if (temp > 140)
{
  heater = "off";
}
```

Do not include a final `else`.

4. There is no limit on the number of "else if" blocks that are allowed.

5. False. The arithmetic operators have higher precedence than the comparison operators.

6. The corrections are underlined:

```
(a == 2 || a == 3)
{
  System.out.print("OK\n");
}
```

7. The ! operator reverses the truth or falsity of a condition.

8. If you omit the `break`, control flows into the next `case` block, and that `case` block's statements execute as well.

9. False. An "if, else" condition evaluates to either `true` or `false`. The controlling expression in a `switch` statement must evaluate to either `int` or `char` (or `byte` or `short`).

10. When more than one identifier produces the same result, concatenate on the same line, if possible, using separate `case` *<identifier>*: for each identifier:

```
case 'Q': case 'q':
  System.out.println("quitting");
```

11. A `while` condition evaluates to either `true` or `false`.

12. The user query should occur just prior to where the termination condition is tested. A `while` loop tests the termination condition at the beginning of the loop. Therefore, the user query should occur just above the top of the loop and also just above the bottom of the loop. If you want the loop to always execute at least once, then omit the user query above the loop and replace it with an assignment that forces the termination condition to be `true`.

13. There is no semicolon after the `while` condition.

14. If you know ahead of time the exact number of iterations through a loop, use a `for` loop.

15. Happy birthday as a `for` loop:

```
for (int age=0; age < 5; age++)
{
  System.out.println("Happy Birthday# " + age);
} // end for
```

Output:

```
Happy Birthday# 0
Happy Birthday# 1
Happy Birthday# 2
Happy Birthday# 3
Happy Birthday# 4
```

16. A `do` loop is most appropriate in simple situations where there will always be at least one pass.

17. Template for a pair of nested for loops:

```
for (int i=0; i<imax; i++)
{
  for (int j=0; j<jmax; j++)
  {
    <statement(s)>
  } // end for j
} // end for i
```

18. A `for` loop representation of a `while` loop:

```
for (boolean OK=false; !OK;)
{
  <statement(s)>
}
```

19. Given the expression:

```
!(!a || !b)
```

Starting on the left side of basic identity 16 and going to the right side gives this:

```
!!a && !!b
```

Then using basic identity 1 gives this:

```
a && b
```

CHAPTER 5

Using Prebuilt Methods

Objectives

- See what it takes to incorporate Java's prebuilt application programming interface (API) software into your programs, and become acquainted with Oracle's documentation of the API software.
- Use the methods and named constants defined in Java's `Math` class.
- Use the parsing methods in wrapper classes to convert text representations of numbers into numerical format, and learn to use the `toString` methods to go the other way.
- Use methods in the `Character` class to identify and alter character types and formats.
- Use methods in the `String` class to find the first index of a particular character, extract or replace substrings, convert case, and trim leading and trailing whitespaces.
- Format output with the `System.out.printf` method.
- Optionally use the `Random` class to generate non-uniform random-number distributions.
- Optionally see how to draw geometric shapes, modify and display pictures, and display text on a graphics display window.

Outline

5.1 Introduction

In Chapters 3 and 4, we focused on basic Java programming language constructs—variables, assignments, operators, `if` statements, loops, and so on. We also introduced a more advanced programming technique—calling a method. Method calls provide a lot of "bang for your buck." In other words, they do a lot and require very little work on your part. For example, you get great benefit for little effort when you call the `print` and `println` methods for output, the `next`, `nextLine`, `nextInt`, and `nextDouble` methods for input, and the `charAt`, `length`, `equals`, and `equalsIgnoreCase` methods for string manipulation. In this chapter, we want to expose you to other methods that are already written, already tested, and are readily accessible to all Java programmers.

While this chapter raises your awareness of valuable already written methods, it also gives you a better feeling for what methods can do in general. And learning what methods can do is an important first step in learning about *object-oriented programming* (OOP). We describe OOP in all its glory in the next chapter, but for now, here's a pared-down explanation: OOP is the idea that programs should be broken down into objects. An *object* is a set of related data plus a set of behaviors. For example, a string is an object: A string's "set of related data" is its characters, and its "set of behaviors" is its methods (the `length` method, the `charAt` method, etc.). Each object is an instance of a class. For example, a single string object, "hello," is an instance of the `String` class. This chapter serves as a transition from Java basics in Chapters 3 and 4 to OOP in the remainder of the book. We carry out this transition by showing you how to use prebuilt OOP code without having to implement it yourself. More specifically, in this chapter, you learn how to use methods, and in the next chapter, you'll learn how to write your own classes and the methods that go inside those classes.

There are two basic types of methods, *instance methods* and *class methods,* and we provide examples of both in this chapter. Instance methods are methods that are associated with a particular instance of a class. For example, to call the `String` class's `length` method, you have to associate it with a particular string. So in the example below, note how the `firstName` string is associated with the `length` method:

```
firstNameSize = firstName.length();
```

The `firstName` string is an example of a *calling object.* As the name implies, a calling object is an object that calls a method. Whenever you call an instance method, you have to prefix the method name with a calling object and then a dot.

Class methods are methods that are associated with an entire class, not with a particular instance of a class. For example, there's a `Math` class that contains many class methods. Its methods are associated with math in general, not with a particular instance of math (referring to a particular instance of math doesn't even make sense). To call a class method, you prefix the method name with the name of the class that defines it. For example, the `Math` class contains a `round` method that returns the rounded version of a given value. To call the `round` method, you prefix it with `Math` like this:

```
paymentInDollars = Math.round(calculatedEarnings);
```

We start the chapter with an overview of the API library, which is Oracle's collection of prebuilt classes. We then examine the `Math` class, which provides methods for mathematical calculations. We next turn our attention to the wrapper classes, which encapsulate (wrap up) primitive data types. We then expand on our previous discussion of the `String` class by providing additional string methods. After that, we describe the `printf` method, which provides formatted output functionality. We then discuss the `Random` class, which provides methods for generating random numbers. We end the chapter with an optional GUI track section. In it, we discuss methods provided by the `Graphics` class and describe how to manipulate and display images and graphics. Very cool stuff!

5.2 The API Library

When working on a programming problem, you should normally check to see if there are prebuilt classes that meet your program's needs. If there are such prebuilt classes, then use those classes—don't "reinvent the wheel." For example, user input is a rather complicated task. Java's `Scanner` class handles user input. Whenever you need user input in a program, use the `Scanner` class rather than writing and using your own input class.

There are two primary advantages of using prebuilt classes. Using prebuilt classes can save you time because you don't have to write the classes yourself. Using prebuilt classes can also improve the quality of your programs because the classes have been thoroughly tested, debugged, and scrutinized for efficiency, and other programmers will find it easier to understand the programs you write.

Searching API Class Library Documentation

Java's prebuilt classes are stored in the *Application Programming Interface (API) class library,* which is more simply known as the API library. You should be able to find documentation for the API library at Oracle's Java API website:

http://download.oracle.com/javase/7/docs/api/index.html

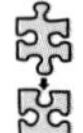

Use available resources.

The API library contains tens of thousands of prebuilt methods defined in thousands of classes. The classes are organized in about 200 groups called *packages* (a package is a group of classes). It's unlikely that you'll be able to memorize the names of all those methods, where they are, and what they do. So how do you locate the particular piece of prebuilt software that might be just what you need for your current programming project?

Use a textbook (like this one ☺) to get you started with selected sample classes and methods. Then go to Oracle's Java API website and browse. See Figure 5.1. It shows that the website's window is partitioned into three window panes. The top-left pane displays a list of all of Java's packages. The bottom-left pane displays a list of all of Java's classes. The right pane displays a variety of different content, where the type of content depends on what the user specifies.

The website provides several ways to look things up:

1. If you hope that the API library contains a method or class that might help with your current programming project, but you're unsure, you'll have to do some browsing. Start by making sure the Overview link at the top of the website is selected. When the Overview link is selected, the right window pane displays a list of all the packages and a brief description of each package. If you find a package that looks promising, click the package's name. That causes the right pane to display all the classes within the selected package and a brief description of each class. If you find a class that looks promising, click the class's name. That causes the right pane to display all the methods within the selected class and a brief description of each method. If you find a method that looks promising, click the method's name. That causes the right pane to display the method's complete details.
2. If you know the name of a particular class that you want details on, click anywhere in the bottom-left window pane and press Ctrl+f (hold the control key down and tap the f key). The f stands for "find," and pressing Ctrl+f causes a Find dialog box to appear. Enter the name of the class in the Find dialog box and click the Find Next button. That should cause the class to be found and highlighted in the bottom-left window pane. Click the class, which causes the right window pane to display the class's details.
3. If you know the name of a particular method but not its class, click the Index link at the top of the window. That causes the right window pane to display the letters of the alphabet. Click the letter that matches the first letter of the method you're interested in. That causes the right window pane to display methods (and other entities, like classes) that begin with the clicked letter. Find the method you're interested in and click it to display all its details.

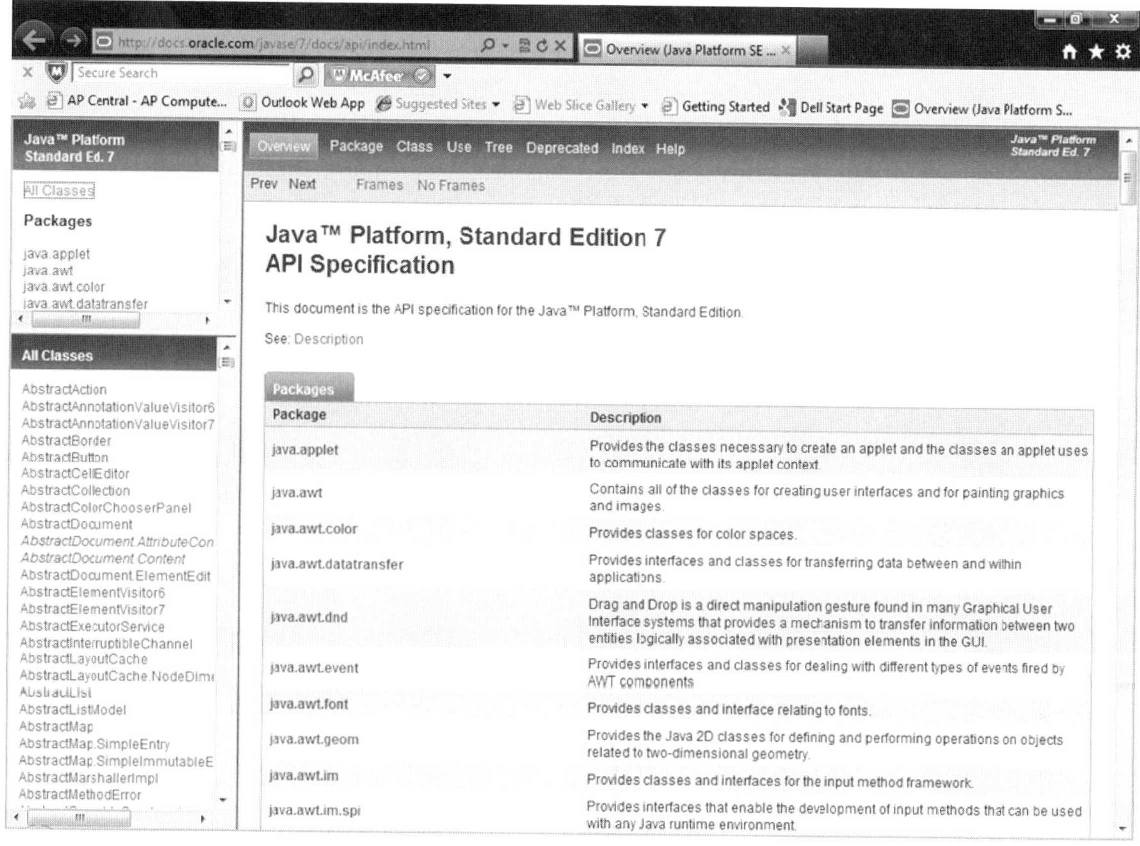

Figure 5.1 Oracle's Java API website

Using Oracle's Java API website is like surfing the web, but you're not surfing the whole world. You're just surfing the Java class library. You can do it, and we encourage you to give it a try whenever you're curious.

Using the API Class Library

To use an API class in your program, you must first import it (i.e., load it) into your program. For example, to use the `Scanner` class, you must include this statement at the top of your program:

```
import java.util.Scanner;
```

Note the `java.util` part of `java.util.Scanner`. The `java.util` part is the name of a package. The "`util`" stands for "utility," and the `java.util` package contains general-purpose utility classes. The only `java.util` class you'll need right now is the `Scanner` class. But there are many other useful classes in the `java.util` package. Examples are:

- The `Random` class, for helping you work with random numbers—discussed in an optional section at the end of this chapter.
- The `Calendar` class, for helping you work with times and dates—discussed in an optional section at the end of Chapter 8.
- The `Arrays`, `ArrayList`, `LinkedList`, and `Collections` classes, for helping you work with lists or collections of similar data—`ArrayLists` are discussed in Chapter 10.

If you have a program that needs to use more than one of the classes in a particular package, like two or more of the util package classes just mentioned, you can import them all at once using a statement like this:

```
import java.util.*;
```

The asterisk is a *wildcard*. In the above statement, the asterisk causes all classes in the java.util package to be imported—not just the Scanner class. There's no inefficiency in using the wildcard notation. The compiler includes only as much as it needs in the compiled program.

Several classes are so important that the Java compiler automatically imports them for you. These automatically imported classes are in the java.lang package, where lang stands for "language." In effect, the Java compiler automatically inserts this statement at the top of every Java program:

```
import java.lang.*;
```

Because this is automatic and understood, there's no need to write it explicitly.

The Math class is in the java.lang package, so there's no need for you to import the Math class if you want to perform math operations. Likewise, the System class is in the java.lang package, so there's no need for you to import the System class if you want to perform a System.out.println command.

Headings for API Methods

To use an API class, you don't need to know the internals of the class; you just need to know how to "interface" with it. To interface with a class, you need to know how to use the methods within the class. For example, to perform input, you need to know how to use the Scanner class's methods—next, nextLine, nextInt, nextDouble, and so on. To use a method, you need to know what type of *arguments* to pass to it and what type of value it *returns*. Arguments are the input you supply to a method when you call it, or ask it to do something for you, and the value it returns is the answer it gives you back.

The standard way to present method-interface information is to show the method's source code heading. For example, here's the source code heading for the Scanner class's nextInt method:

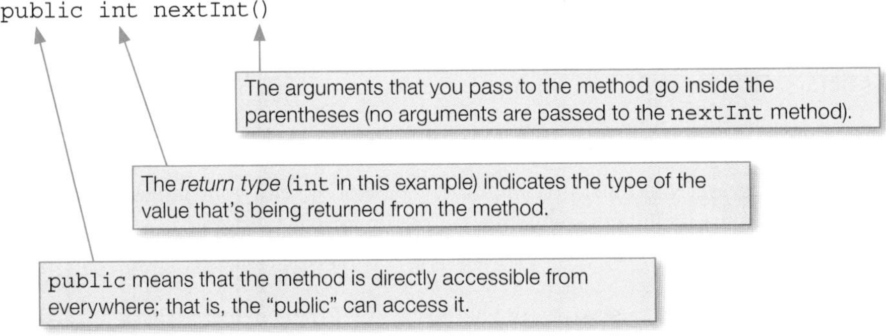

In the above nextInt heading, the public access modifier should look familiar because your main method headings all use public. We'll discuss private methods in Chapter 8. They're accessible only from within the class that defines them. Note that the nextInt method returns an int value and that it has no arguments inside the parentheses. Here's an example of a Java statement that shows how you might call the nextInt method:

```
int days = stdIn.nextInt();
```

5.3 Math Class

The Math class is one of the prebuilt classes in the always available java.lang package. This class contains methods that implement standard mathematical *functions*. A mathematical function generates a numeric value based on one or more other numeric values. For example, a square root function generates the square root of a given number. Likewise, the Math class's sqrt method returns the square root of a given number. In addition to providing mathematical methods, the Math class provides two mathematical constants—π (the ratio of a circle's circumference to its diameter) and e (the base of natural logarithms).

Basic Math Methods

Let's now look at some of the Math class's methods. Throughout the book, when there's a need to present a group of methods from the API library, we'll introduce the methods by showing a list of method headings and associated brief descriptions. Headings for API methods are commonly referred to as *API headings*. Figure 5.2 contains API headings for some of the more popular methods in the Math class, with associated brief descriptions.

As you read through Figure 5.2, we hope that you'll find most of the methods to be straightforward. But some items may need clarification. Note the static modifier at the left of all the Math methods. All the methods in the Math class are static. The static modifier means they are class methods and must be called by prefacing the method's name with the name of the class in which they are defined. For example, here's how you'd call the abs method:

Call Math methods by prefacing them with Math dot.

```
num = Math.abs(num);
```

The above statement updates num's value, so num gets the absolute value of its original value. For example, if num starts out with −15, it ends up with 15.

Note that the following statement does not work properly:

```
Math.abs(num);
```

It finds the absolute value of num, but it does not update the content stored inside num. Math methods return a value. They do not update a value. So if you want to update a value, you must use an assignment operator.

In Figure 5.2, note that there's only one pow method—one with double parameters. There's no pow method with int parameters. But that's no big deal because you can pass an int value to the pow method. More generally, it's legal to pass an integer value to a method that accepts a floating-point argument. It's like assigning an integer value into a floating-point variable, discussed in Chapter 3. Let's see how this works within a code fragment. There is an empirical rule called "Horton's Law," which says that the length of a river scales with the area drained by the river in accordance with this formula:

$$\text{length} \approx 1.4\,(\text{area})^{0.6}$$

Here's how you might implement Horton's Law in Java code:

OK to pass an int (area), into pow, which accepts double arguments.

```
int area = 10000;      // square miles drained
System.out.println("river length = " + 1.4 * Math.pow(area, 0.6));
```

Output:

```
river length = 351.66410041134117
```

```
public static double abs(double num)
```
Returns the absolute value of a `double num`.

```
public static int abs(int num)
```
Returns the absolute value of an `int num`.

```
public static double ceil(double num)
```
Returns the smallest whole number greater than or equal to `num`. `ceil` stands for "ceiling."

```
public static double exp(double power)
```
Returns `E` (base of natural logarithms) raised to the specified `power`.

```
public static double floor(double num)
```
Returns the largest whole number that is less than or equal to `num`.

```
public static double log(double num)
```
Returns the natural logarithm (base `E`) of `num`.

```
public static double log10(double num)
```
Returns the base 10 logarithm of `num`.

```
public static double max(double x, double y)
```
Returns the more positive of the two `double` values, `x` and `y`.

```
public static int max(int x, int y)
```
Returns the more positive of the two `int` values, `x` and `y`.

```
public static double min(double x, double y)
```
Returns the less positive of the two `double` values, `x` and `y`.

```
public static int min(int x, int y)
```
Returns the less positive of the two `int` values, `x` and `y`.

```
public static double pow(double num, double power)
```
Returns `num` raised to the specified `power`.

```
public static double random()
```
Returns a uniformly distributed value between 0.0 and 1.0, but not including 1.0.

```
public static long round(double num)
```
Returns the whole number that is closest to `num`.

```
public static double sqrt(double num)
```
Returns the square root of `num`.

Figure 5.2 API headings and brief descriptions of some of the methods in the `java.lang.Math` class

Note the round method in Figure 5.2. How is it different from using an (int) type cast operator on a double value? The (int) operator truncates the fraction, whereas the round method rounds up if the fraction is ≥0.5.

As shown in Figure 5.2, Math's random method returns a uniformly distributed value between 0.0 and 1.0, not including 1.0. "Uniformly distributed" means that there's the same chance of getting any value within the specified range. In other words, if you have a program that calls random, the chances are the same for random returning 0.317, 0.87, 0.02, or any value between 0.0 and 1.0, not including 1.0.

Why would you want to call the random method? If you need to analyze a real-world situation that involves random events, you should consider writing a program that uses the random method to model the random events. For example, if you work for a city transportation department, and you're in charge of improving traffic flow at traffic light intersections, you could write a program that uses the random method to model the arrival of automobiles at the traffic lights. For each traffic light that you're interested in, you'd set the traffic light's cycle time (e.g., two minutes between each new green signal) and then simulate automobiles arriving at the traffic light at random intervals. You'd run the program so that it simulates one week of traffic flow, and you'd keep track of average wait time for all vehicles. You'd then adjust the traffic light's cycle time (e.g., one minute and forty-five seconds between each new green signal), run the simulation again, and determine which traffic light cycle time produces shorter average wait times.

The Math.random method is handy, and we'll use it at various points throughout the chapter. If you need to do serious work with numbers, however, you'll want to use the API's Random class. We describe this class in this chapter's final section.

Let's wrap up the discussion of Figure 5.2's Math methods with a complete program example. Suppose you want to calculate the length of the hypotenuse of a right triangle, given the lengths of its base and height, as shown in this picture:

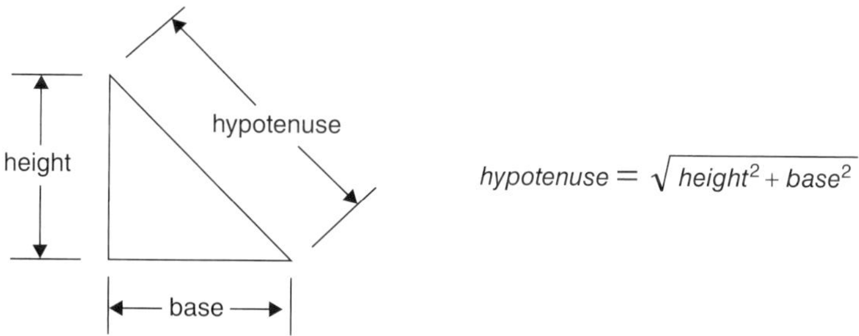

Figure 5.3 contains a simple program that asks the user to provide base and height values. Then it uses Math's sqrt method to calculate and print the square root of the sum of the squares. Notice that we did not use the Math.pow method to square the base and square the height. For small powers, it's more efficient just to multiply them out. 

Trigonometric Math Methods

Figure 5.4 contains API headings and descriptions for some of the methods in the Math class that can help you solve problems in trigonometry. The sin, cos, and tan methods implement the sine, cosine, and tangent functions, respectively. The asin, acos, and atan methods implement the arcsine, arccosine, and arctangent functions, respectively. The trigonometric and inverse trigonometric functions all use or return angle values as radians, not degrees. Using or assuming degrees is a common programming error. Be careful! 

```
/*******************************************************************
* FindHypotenuse.java
* Dean & Dean
*
* This program computes the hypotenuse of a right triangle.
*******************************************************************/

import java.util.Scanner;

public class FindHypotenuse
{
  public static void main(String[] args)
  {
    Scanner stdIn = new Scanner(System.in);
    double base;
    double height;
    double hypotenuse;

    System.out.print("Enter right triangle base: ");
    base = stdIn.nextDouble();
    System.out.print("Enter right triangle height: ");
    height = stdIn.nextDouble();
    hypotenuse = Math.sqrt(base * base + height * height);

    System.out.println("Hypotenuse length = " + hypotenuse);
  } // end main
} // end FindHypotenuse
```

call to `Math` class's `sqrt` method

Sample session:

```
Enter right triangle base: 3.0
Enter right triangle height: 4.0
Hypotenuse length = 5.0
```

Figure 5.3 FindHypotenuse program demonstrating the use of one of Java's prebuilt math functions

Named Constants

The `Math` class also contains `double` values for two important named constants, π and e:

```
PI = 3.14159265358979323846
E = 2.7182818284590452354
```

`PI` and `E` are standard mathematical constants. `PI` is the ratio of a circle's perimeter to its diameter. `E` is Euler's number, the base for natural logarithm calculations. The names `PI` and `E` are in all uppercase characters, because that's standard style for named constants. Constants have fixed values, and if you attempt to assign a value to them, you'll get a compilation error. Just as `Math`'s methods are called class methods, these constants are called *class constants,* and you access them through the `Math` class name. In other words, if you need π, specify `Math.PI`.

```
public static double acos(double ratio)
    Returns the angle in radians between 0.0 and π whose cosine equals the given value.

public static double asin(double ratio)
    Returns the angle in radians between −π/2 and +π/2 whose sine equals the given value.

public static double atan(double ratio)
    Returns the angle in radians between −π/2 and +π/2 whose tangent equals the given value.

public static double cos(double radians)
    Returns the cosine of an angle expressed in radians.

public static double sin(double radians)
    Returns the sine of an angle expressed in radians.

public static double tan(double radians)
    Returns the tangent of an angle expressed in radians.

public static double toDegrees(double radians)
    Converts an angle measured in radians to an angle measured in degrees.

public static double toRadians(double degrees)
    Converts an angle measured in degrees to an angle measured in radians.
```

Figure 5.4 API headings and brief descriptions of some trigonometric methods in the `java.lang.Math` class

Suppose you want to compute the water needed for a 10-centimeter-diameter water balloon. Here's the formula for the volume of a sphere:

$$\frac{\pi}{6}\,\mathit{diameter}^3$$

And here's the code and resulting output for computing the volume of water for the water balloon:

```
double diameter = 10.0;
double volume = Math.PI / 6.0 * diameter * diameter * diameter;
System.out.print("Balloon volume in cubic cm = " + volume);
```

Output:

```
Balloon volume in cubic cm = 523.5987755982989
```

Some of Java's `Math` class methods are extremely helpful when you need to evaluate a nontrivial mathematical function, like raising a floating-point number to a fractional power. Others do simple things you could do yourself. For example, can you think of a primitive way to do the same thing that `Math.round` does? It's pretty easy. Just add `0.5` to your original `double` number and then use a `long` cast operator on that `double` value to end up with a rounded version of the original number. (That's what was done in days of yore.) If it's that easy, why bother to use `Math.round`? Because it makes code more readable! The expression, `Math.round(number)`, is self-documenting. It's more informative than the odd-looking expression, `((long) (0.5 + number))`.

5.4 Wrapper Classes for Primitive Types

A *wrapper* is a construct that wraps (contains) a primitive data type and converts it to an object with a similar name, so it can be used in a situation where only objects are allowed. Wrapper classes do more than wrapping, however. They also provide some useful class methods and class constants. The `java.lang` package provides wrapper classes for all of the Java primitive types. Because this package is always available, you don't need to use `import` to access these classes. Here are the wrapper classes that we'll consider, along with the primitive types they encapsulate:

Wrapper Class	Primitive Type
`Integer`	`int`
`Long`	`long`
`Float`	`float`
`Double`	`double`
`Character`	`char`

For most wrapper classes, the wrapper class's name is the same as its associated primitive type except that it uses an uppercase first letter. There are two exceptions. The wrapper class for `int` is `Integer`, and the wrapper class for `char` is `Character`.

Methods

Like the `Math` class, wrapper classes contain methods and constants. We start with methods. We limit our coverage to just two sets of methods—methods that convert strings to primitives and methods that convert primitives to strings. So when would you need to convert a string to a primitive? For example, when would you need to convert the string "4" to the `int` 4? If you need to read a value in as a string and then later manipulate the value as a number, you'll need to perform a string-to-number conversion. Later in this section, we'll show a program that reads a value that could be either a number (for a lottery-number choice) or a "q" (for quitting). The program reads the user entry as a string, and if the value is not a "q", then the program converts the user entry to a number.

Now for the other direction—when would you need to convert a primitive to a string? If you need to call a method that takes a string argument and what you've got is a number argument, then you'll need to perform a number-to-string conversion. With graphical user interface (GUI) programs, all numeric output is string based. So to display a number, you need to convert the number to a string prior to calling the GUI display method. With GUI programs, all numeric input is string based, too. So to read a number, you first read the input as a string and then convert the string to a number. You'll see many examples of these processes later, in Chapters 17 and 18.

Here's the syntax for converting strings to primitives and primitives to strings:

Wrapper Class	String → Primitive	Primitive → String
`Integer`	`Integer.parseInt(`<*string*>`)`	`Integer.toString(`<#>`)`
`Long`	`Long.parseLong(`<*string*>`)`	`Long.toString(`<#>`)`
`Float`	`Float.parseFloat(`<*string*>`)`	`Float.toString(`<#>`)`
`Double`	`Double.parseDouble(`<*string*>`)`	`Double.toString(`<#>`)`

All the number wrapper classes work similarly. So if you understand how to convert from a string to an `int`, then you'll also understand how to convert from a string to another primitive type. To convert from a string to an `int`, use `int`'s wrapper class, `Integer`, to call `parseInt`. In other words, call `Integer.parseInt(`<*string*>`)` and the string's corresponding `int` is returned. Likewise, to convert from a string to a `double`, use `double`'s wrapper class, `Double`, to call `parseDouble`. In other words, call `Double.parseDouble(`<*string*>`)` and the string's corresponding `double` is returned. Later in this section, we'll show a non-trivial example that uses the wrapper class conversion methods. But first, we'll show some trivial examples to get you used to the method-call syntax. Here we use `parseInt` and `parseDouble` to convert from strings to primitives:

```
String yearStr = "2002";
String scoreStr = "78.5";
int year = Integer.parseInt(yearStr);
double score = Double.parseDouble(scoreStr);
```

To remember the syntax for the string-to-number method calls, think of <*type*>`.parse`<*type*> for `Integer.parseInt`, `Long.parseLong`, and so on.

To convert from an `int` to a string, use `int`'s wrapper class, `Integer`, to call `toString`. In other words, call `Integer.toString(`<*int-value*>`)` and the `int` value's corresponding string is returned. Likewise, to convert from a `double` to a string, use `double`'s wrapper class, `Double`, to call `toString`. In other words, call `Double.toString(`<*double-value*>`)` and the `double` value's corresponding string is returned. Note this example:

```
int year = 2002;
float score = 78.5;
String yearStr = Integer.toString(year);
String scoreStr = Float.toString(score);
```

About half the numerical wrapper-class methods are class methods. We're focusing on those methods. Because they're class methods, you call them by prefacing the method call with the wrapper class's name, just as we have done.

Named Constants

The wrapper classes contain more than just methods; they also contain named constants. All the number wrappers provide named constants for minimum and maximum values. The floating-point wrappers also provide named constants for plus and minus infinity and "Not a Number," which is the indeterminate value you get if you try to divide zero by zero. Here's how you access the most important named constants defined in the `Integer` and `Double` wrapper classes:

```
Integer.MAX_VALUE
Integer.MIN_VALUE
Double.MAX_VALUE
Double.POSITIVE_INFINITY
Double.NEGATIVE_INFINITY
Double.NaN
```

NaN stands for "not a number."

There are comparable named constants for the `Long` and `Float` wrappers.

An Example

Let's put the wrapper and `Math.random` material into practice by showing it in the context of a complete program. Figure 5.5's Lottery program prompts the user to guess a randomly generated number between 1 and the maximum `int` value. The user pays $1 for each guess and wins $1,000,000 if the guess is correct. The user enters a "q" to quit.

In the initialization of `winningNumber`, note how the program generates a random winning-number value:

```
winningNumber = (int) (Math.random() * Integer.MAX_VALUE + 1);
```

The starting point in the above equation is the call to `Math.random()`, which returns a random number between 0.0 and 1.0, not including 1.0. The goal is to return a value between 1 and the maximum `int value`, so the range is expanded by multiplying `Math.random()`'s value by `Integer.MAX_VALUE`. However, `Math.random()` returns a value less than 1. Therefore, the product of `Math.random()` and `Integer.MAX_VALUE` generates a number less than `Integer.MAX_VALUE`.

```
/*********************************************************************
* Lottery.java
* Dean & Dean
*
* This program asks the user to guess a randomly selected number.
*********************************************************************/

import java.util.Scanner;

public class Lottery
{
  public static void main(String[] args)
  {
    Scanner stdIn = new Scanner(System.in);
    String input;     // user's number guess, or q to quit
    int winningNumber =
      (int) (Math.random() * Integer.MAX_VALUE + 1);

    System.out.println("Want to win a million dollars?");
    System.out.println("If so, guess the winning number (a" +
      " number between 1 and " + Integer.MAX_VALUE + ").");
    do
    {
      System.out.print(
        "Insert $1.00 and enter your number or 'q' to quit: ");
      input = stdIn.nextLine();
      if (input.equals("give me a hint"))    //  a backdoor
      {
        System.out.println("try: " + winningNumber);
      }
```

Initialize with scaled random number, plus 1.

Figure 5.5a Lottery program illustrating use of the `Integer` wrapper class—part A

```
      else if (!input.equals("q"))
      {
        if (Integer.parseInt(input) == winningNumber)
        {
          System.out.println("YOU WIN!");
          input = "q"; // if someone wins, they're forced to quit
        }
        else
        {
          System.out.println(
            "Sorry, good guess, but not quite right.");
        }
      } // end else if
    } while (!input.equals("q"));
    System.out.println("Thanks for playing. Come again!");
  } // end main
} // end Lottery class
```

The `Integer.parseInt` method converts type from `String` to `int`.

Figure 5.5b Lottery program—part B

Adapt existing software to your needs

To ensure that the range of generated numbers includes `Integer.MAX_VALUE`, we add 1 to the generated product. For example, suppose the generated product is 2,147,483,646.33 (a number very near `Integer.MAX_VALUE`'s 2,147,483,647, which is the large end of the possible range of numbers). Adding 1 to it produces 2,147,483,647.33. As another example, suppose the generated product is 0.26 (a number very near 0, which is the small end of the possible range of numbers). Adding 1 to it produces 1.26. The resulting sum is a floating-point number in the range from 1.0 to slightly greater than `Integer.MAX_VALUE`. However, the result needs to be an integer in the range from 1 to `Integer.MAX_VALUE`. For that result, the fractional part of the sum needs to be removed. And how can you remove the fractional part? Use the `(int)` cast operator (as shown in the code fragment).

Note how the program reads in the user's number guess as a string:

```
input = stdIn.nextLine();
```

By reading the number guess as a string rather than a number, the program can handle the user entering a nonnumerical input, such as "q" for quit or "give me a hint" for a hint. If the user enters "q" the program quits. If the user enters "give me a hint," the program prints the winning number. Great hint, eh? In this case, the hint is really a *backdoor.* A backdoor is a secret technique for gaining access to a program. The Lottery program's backdoor can be used for testing purposes.

If the user does not enter "q" or "give me a hint," the program attempts to convert the user entry to a number by calling `Integer.parseInt`. The program then compares the converted number to the winning number and responds accordingly.

The Lottery program might produce the following output:

Sample session:

```
Want to win a million dollars?
If so, guess the winning number (a number between 0 and 2147483646).
Insert $1.00 and enter your number or 'q' to quit: 66761
Sorry, good guess, but not quite right.
```

```
Insert $1.00 and enter your number or 'q' to quit: 1234567890
Sorry, good guess, but not quite right.
Insert $1.00 and enter your number or 'q' to quit: give me a hint
try 1661533855
Insert $1.00 and enter your number or 'q' to quit: 1661533855
YOU WIN!
Thanks for playing. Come again!
```

5.5 Character Class

In the previous section, we mentioned the `Character` wrapper class, but we didn't explain it. It's time to explain it. Often, you'll need to write programs that manipulate individual characters in a string of text. For example, you might need to read in a phone number and store just the digits, skipping the other characters (dashes, spaces, etc.). To check for digits, use the `Character` class's `isDigit` method. Figure 5.6 shows some of the more popular methods in the `Character` class, including the `isDigit` method.

`public static boolean isDigit(char ch)`
Returns `true` if the specified character is a numerical digit.

`public static boolean isLetter(char ch)`
Returns `true` if the specified character is a letter of the alphabet.

`public static boolean isUpperCase(char ch)`
Returns `true` if the specified character is an uppercase letter.

`public static boolean isLowerCase(char ch)`
Returns `true` if the specified character is a lowercase letter.

`public static boolean isLetterOrDigit(char ch)`
Returns `true` if the specified character is a letter or a digit.

`public static boolean isWhitespace(char ch)`
Returns `true` if the specified character is any kind of whitespace (blank, tab, newline).

`public static char toUpperCase(char ch)`
Returns the specified character as an uppercase character.

`public static char toLowerCase(char ch)`
Returns the specified character as a lowercase character.

Figure 5.6 API headings and brief descriptions of some of the methods in the `Character` class

The IdentifierChecker program in Figure 5.7 illustrates the `Character` class in the context of a complete program. It uses the `Character` class's `isLetter` and `isLetterOrDigit` methods to check whether the user entry is a legal identifier.

Most of Figure 5.6's methods are straightforward, but the `toUpperCase` and `toLowerCase` methods may need some clarification. Because the two methods are so similar, we'll clarify only one of them, `toUpperCase`. If you call `toUpperCase` and pass in a lowercase letter, the method returns the uppercase version of the lowercase letter. But what if you call `toUpperCase` and pass in an uppercase letter or a

```
/****************************************************************
* IdentifierChecker.java
* Dean & Dean
*
* Check a user entry to see if it's a legal identifier.
****************************************************************/

import java.util.Scanner;

public class IdentifierChecker
{
  public static void main(String[] args)
  {
    Scanner stdIn = new Scanner(System.in);
    String line;            // user entry
    char ch;
    boolean legal = true; // Is entered line a legal identifier?

    System.out.println("This program checks the validity of a" +
      " proposed Java identifier.");
    System.out.print("Enter a proposed identifier: ");
    line = stdIn.nextLine();
    ch = line.charAt(0);
    if (!(Character.isLetter(ch) || ch == '$' || ch == '_'))
    {
      legal = false;
    }
    for (int i=1; i<line.length() && legal; i++)
    {
      ch = line.charAt(i);
      if (!(Character.isLetterOrDigit(ch) || ch == '$' || ch == '_'))
      {
        legal = false;
      }
    }
    if (legal)
    {
      System.out.println(
        "Congratulations, " + line + " is a legal Java identifier.");
    }
    else
    {
      System.out.println(
        "Sorry, " + line + " is not a legal Java identifier.");
    }
  } // end main
} // end class IdentifierChecker
```

Character method calls

Figure 5.7 IdentifierChecker program

nonletter? The method returns the passed-in character, unchanged. And what if you pass in a `char` variable to `toUpperCase` instead of a `char` constant? The method returns the uppercase version of the passed-in `char` variable, but it does not change the passed-in variable's value.

As evidenced by the `static` modifiers in Figure 5.6, most of the `Character` methods are class methods. Because they're class methods, you call them by prefacing the method call with the wrapper class's name. Let's look at an example. Suppose you have a `char` variable named `middleInitial` and you'd like to have its content be converted to an uppercase letter. Here's a first attempt at changing `middle-Initial`'s content to an uppercase letter:

```
Character.toUpperCase(middleInitial);
```

That statement compiles and runs, but it does not change `middleInitial`'s content. Here's the proper way to do it:

```
middleInitial = Character.toUpperCase(middleInitial);
```

5.6 `String` Methods

The `String` class is another one of the classes in the always-available `java.lang` package. In Chapter 3, you saw several examples of useful methods associated with objects of the `String` class, such as the `charAt` method, the `length` method, the `equals` method, and the `equalsIgnoreCase` method. In this section, we describe some additional `String` methods—the `String` methods shown in Figure 5.8. These `String` methods do not have the `static` access modifier, so they are not class methods, and you cannot access them with the class name. They are instance methods, and you must access them with a particular string instance. Or, said another way, you must access them with a calling-object string.

Lexicographical Ordering of Strings

You know that numbers can be compared to determine which number is greater. Strings can also be compared. When computers compare strings to determine which string is greater, they use *lexicographical ordering.* For the most part, lexicographical ordering is the same as dictionary order. The string "hyena" is greater than the string "hedgehog" because *hyena* comes after *hedgehog* in the dictionary.

The `String` class's `compareTo` method compares two strings to determine which is greater. As explained in Figure 5.8, `compareTo` returns a positive number if the calling string is greater than the argument string, a negative number if the calling string is less than the argument string, and zero if the calling string and argument string are the same. The following code fragment illustrates what we're talking about. It compares YouTube[1] video titles and prints the results of the comparisons. If you run this code fragment, don't be surprised if your first two output values are different from −10 and 10. According to the Java specification, the first two output values can be any negative number and any positive number, respectively.

```
String youTubeVideo = "Colbert Super Pac";
System.out.println(
  youTubeVideo.compareTo("Makana We Are the Many") + " " +
  youTubeVideo.compareTo("Colbert Immigration Testimony") + " " +
  youTubeVideo.compareTo("Colbert Super Pac"));
```

<u>Output:</u>

−10 10 0

[1]YouTube is a popular free video sharing website, acquired by Google in October 2006, which lets users upload, view, and share video clips.

```
public int compareTo(String str)
```
Returns an integer that indicates the lexicographical ordering of the calling string when compared to the argument string. If the calling string is "greater than" the argument string, a positive number is returned. If the calling string is "less than" the argument string, a negative number is returned. If the calling string equals the argument string, zero is returned.

```
public static String format(String format, Object... args)
```
Returns a formatted string, using the `printf` format specification and arguments as described in Section 5.7. The `...` notation is called *varargs*. It means there may be any number of arguments, including none.

```
public int indexOf(int ch)
```
Returns the position of the first occurrence of the specified character.
Returns −1 if not found.

```
public int indexOf(int ch, int fromIndex)
```
Returns the position of the first occurrence of the specified character at or after `fromIndex`.
Returns −1 if the specified character is not found.

```
public int indexOf(String str)
```
Returns the start position of the first occurrence of the specified string. Returns -1 if not found.

```
public int indexOf(String str, int fromIndex)
```
Returns the start position of the first occurrence of the specified string at or after `fromIndex`.
Returns −1 if the specified string is not found.

```
public boolean isEmpty()
```
Returns `true` if the calling string is the empty string (""). Otherwise, returns `false`.

```
public String replaceAll(String target, String replacement)
```
Returns a new string with all occurrences of the calling string's `target` replaced by `replacement`.

```
public String replaceFirst(String target, String replacement)
```
Returns a new string with the first occurrence of the calling string's `target` replaced by `replacement`.

```
public String substring(int beginIndex)
```
Returns the portion of the calling string from `beginIndex` to the end.

```
public String substring(int beginIndex, int afterEndIndex)
```
Returns the portion of the calling string from `beginIndex` to just before `afterEndIndex`.

```
public String toLowerCase()
```
Returns a new string with all characters in the calling string converted to lowercase.

```
public String toUpperCase()
```
Returns a new string with all characters in the calling string converted to uppercase.

```
public String trim()
```
Returns a new string with all whitespace removed from the start and end of the calling string.

Figure 5.8 API headings and brief descriptions of some of the methods in the `String` class

Checking for the Empty String

Previously, you learned that the empty string is a string that contains no characters, and it's represented by two quotes with nothing between them—"". Sometimes you'll need to check a string variable to see whether it contains the empty string. For example, when reading an input string from a user, you might want to check for the empty string as part of input validation. The following `if` condition illustrates:

```
if (userInput.equals(""))
```

Because checking for the empty string is such a common need, the Java API provides a method to handle that need. The `isEmpty` method returns `true` if the calling string contains the empty string and `false` otherwise. Figure 5.9's program uses the `isEmpty` method as part of an input validation `while` loop. The `while` loop forces the user to enter a non-empty name.

```
/*********************************************************************
* StringMethodDemo.java
* Dean & Dean
*
* This program exercises the String class's isEmpty method.
*********************************************************************/

import java.util.Scanner;

public class StringMethodDemo
{
  public static void main(String[] args)
  {
    Scanner stdIn = new Scanner(System.in);
    String name;

    System.out.print("Enter your name: ");
    name = stdIn.nextLine();

    while (name.isEmpty())
    {
      System.out.print("Invalid entry. You must enter your name: ");
      name = stdIn.nextLine();
    }
    System.out.println("Hello, " + name + "!");
  } // end main
} // end StringMethodDemo
```

This checks for the empty string.

Sample session:

```
Enter your name:
Invalid entry. You must enter your name: Virginia Maikweki
Hello, Virginia Maikweki!
```

The user immediately presses Enter here.

Figure 5.9 StringMethodDemo program exercising the `String` class's `isEmpty` method

Substring Retrieval

Note the two substring methods in Figure 5.8. The one-parameter substring method returns a string that is a subset of the calling-object string, starting at the beginIndex parameter's position and extending to the end of the calling-object string. The two-parameter substring method also returns a string that is a subset of the calling-object string. The returned substring spans from the position of its first parameter, beginIndex, to the position just to the left of its second parameter, afterEndIndex.

The following code fragment processes a quote from Candide.[2] In its candide.substring(8) method call, candide is the calling object, and 8 is the beginIndex parameter value. Because string indices start at 0, the 8 refers to candide's ninth character, which is 'c'. Thus, the first println statement prints "cultivate our garden." In the candide.substring(3,17) method call, the 3 and 17 refer to candide's 4th and 18th characters, which are the '*m*' in "*must*" and the blank space character after "*cultivate.*" Remember that substring's second argument indicates the position just to the right of the extracted string. Thus, the substring(3, 17) method call returns a string that spans from the '*m*' in "*must*" to the '*e*' in "*cultivate.*"

```
String candide = "we must cultivate our garden";
System.out.println(candide.substring(8));
System.out.println(candide.substring(3,17));
```

Output:

```
cultivate our garden
must cultivate
```

If you want to test the above code fragment or any of the following String method code fragments, use Figure 5.9's program as a template. More specifically, replace Figure 5.9's main method body with the new code fragment. Then compile and run the resulting program.

Position Determination

Note the one-parameter indexOf methods in Figure 5.8. They return the position of the first occurrence of a given character or substring within the calling-object string. If the given character or substring does not appear within the calling-object string, indexOf returns –1.

Note the two-parameter indexOf methods in Figure 5.8. They return the position of the first occurrence of a given character or substring within the calling-object string, starting the search at the position specified by indexOf's second parameter. If the given character or substring is not found, indexOf returns –1.

It's common to use one of the indexOf methods to locate a character or substring of interest and then use one of the substring methods to extract a nearby string. For example, consider this code fragment:[3]

Here's the beginning of the printed substring.

```
String hamlet = "To be, or not to be: that is the question;";
int index = hamlet.indexOf(':');
if (index != -1)
{
  System.out.println(hamlet.substring(index + 2));
}
```

Output:

```
that is the question;
```

[2]Voltaire, *Candide,* translated by Lowell Bair, Bantam Books, 1959, final sentence.

[3]Shakespeare, *Hamlet,* Act III, Sc. 1.

Text Replacement

Note the `replaceAll` and `replaceFirst` methods in Figure 5.8. The `replaceAll` method searches its calling-object string for `target`, `replaceAll`'s first parameter. It returns a new string, in which all occurrences of `target` are replaced with `replacement`, `replaceAll`'s second parameter. The `replaceFirst` method works the same as `replaceAll` except that only the first occurrence of the searched-for target string is replaced. Here's an example that illustrates both methods:[4]

```
String ladyMacbeth = "Out, damned spot! Out, I say!";
System.out.println(ladyMacbeth.replaceAll("Out", "Expunge"));
ladyMacbeth = ladyMacbeth.replaceFirst(", damned spot", "");
System.out.println(ladyMacbeth);
```

Update the content of the `ladyMacbeth` string variable.

Output:

```
Expunge, damned spot! Expunge, I say!
Out! Out, I say!
```

Note how the second statement prints the Lady Macbeth quote with both occurrences of "Out" replaced by "Expunge", but it does not change the content of the `ladyMacbeth` string object. You can tell that it doesn't change the content of the `ladyMacbeth` string object because the next two statements generate `Out! Out, I say!`, where "Out" is used, not "Expunge". The reason that the second statement's `replace-All` method does not change the content of the `ladyMacbeth` string object is that string objects are *immutable*, which means unchangeable. `String` methods such as `replaceAll` and `replaceFirst` return a new string, not an updated version of the calling-object string. If you really want to change the content of a string variable, you need to assign a new string object to it. That's what happens in the third statement, where the JVM assigns the result of the `replaceFirst` method call to the `ladyMacbeth` variable.

In the Lady Macbeth example, the `replaceFirst` method call deletes the "damned spot" by replacing it with an empty string. Because there is only one occurrence of "damned spot", `replaceAll` would yield the same result as `replaceFirst`. But `replaceFirst` is slightly more efficient, and that's why we use it here.

Whitespace Removal and Case Conversion

Note the `trim`, `toLowerCase`, and `toUpperCase` methods in Figure 5.8. The `trim` method removes all whitespace from before and after a calling-object string. The `toLowerCase` method returns a string identical to the calling-object string except that all the characters are lowercase. The `toUpperCase` method returns an uppercase version of the calling-object string. To see how these methods work, suppose we change the previous Hamlet code to this:

```
String hamlet = "To be, or not to be: that is the question;";
int index = hamlet.indexOf(':');
```

[4]Shakespeare, *Macbeth,* Act V, Sc. I.

```
String hamlet2 = hamlet.substring(index + 1);
System.out.println(hamlet2);
hamlet2 = hamlet2.trim();
hamlet2 = hamlet2.toUpperCase();
System.out.println(hamlet2);
```

Now the output looks like this:

Output:

```
 that is the question;
THAT IS THE QUESTION;
```

Note how the `trim` method strips the leading space from `hamlet2`'s string. Also, note how the `toUpperCase` method returns an all-uppercase version of `hamlet2`.

Insertion

To make an insertion, you must know where you want to make it. If you don't already know the index of where you want the insertion to start, you can find it by using the `indexOf` method with a unique substring argument. Then extract the substring up to that index, concatenate the desired insertion, and concatenate the substring after that index. The following code fragment performs two insertions within a string. More specifically, the code fragment starts with a philosophy espoused by 17th century French mathematician and philosopher René Descartes: "All nature will do as I wish it." It then inserts two strings and transforms the message into a starkly contrasting quote from Charles Darwin: "All nature is perverse & will not do as I wish it."[5]

```
String descartes = "All nature will do as I wish it.";
String darwin;
int index;
index = descartes.indexOf("will");
darwin = descartes.substring(0, index) +
  "is perverse & " +
  descartes.substring(index);
index = darwin.indexOf("do");
darwin = darwin.substring(0, index) +
  "not " +
  darwin.substring(index);
System.out.println(darwin);
```

Output:

```
All nature is perverse & will not do as I wish it.
```

[5]*Charles Darwin's Letters,* edited by Frederick Burkhardt, Cambridge (1996). Charles Darwin started college at the University of Edinburgh in 1825, studying to be a medical doctor like his father. A medical career didn't appeal to him, however, so he transferred to Cambridge University, where he earned a B.A. in preparation for a career as a country parson. But what he really enjoyed was searching for bugs in the family barn. Right after his graduation, and before he began his first job as a country parson, family connections, a good reference from a college professor, and a pleasant personality gave him the chance to travel around the world as the companion of a brilliant sea captain named Robert FitzRoy (who later invented weather forecasting). This trip launched Darwin's career as one of the most influential scientists of the modern world.

5.7 Formatted Output with the `printf` Method

You've used the `System.out.print` and `System.out.println` methods for quite a while now. They work fine most of the time, but there's a third `System.out` method that you'll want to use every now and then for formatted output. It's the `printf` method, where the "f" stands for "formatted." We describe the `printf` method in this section.

Formatted Output

For most programs, the goal is to calculate something and then display the result. It's important that the displayed result is understandable. If it's not understandable, then no one will bother to use the program, even if it calculates flawlessly. One way to make your displayed results understandable is to format your output. By that, we mean having data columns align properly, having floating-point numbers show the same number of digits after the decimal point, and so on. Note the formatting in the budget report below. The left column is left-aligned. The other columns are right aligned. The numbers show two digits at the right of the decimal point. The numbers show commas between every third digit at the left of the decimal point. And finally, the numbers show parentheses to indicate negativeness.

```
Account                    Actual       Budget      Remaining
-------                    ------       ------      ---------
Office Supplies          1,150.00     1,400.00         250.00
Photocopying             2,100.11     2,000.00       (100.11)

Total remaining: $149.89
```

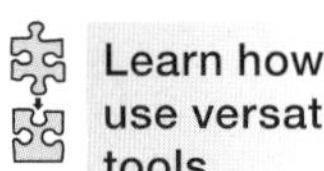

Learn how to use versatile tools.

The `System.out.printf` method is in charge of generating formatted output. The `printf` method has lots of formatting features. We'll keep things simple and explain only a few of the more popular features. We begin our explanation of the `printf` method by showing you how to generate the "Total remaining" line in the above budget report. Here's the code:

format specifier

```
System.out.printf(
  "\nTotal remaining: $%.2f\n", remaining1 + remaining2);
```

The `printf` method's first argument is known as the *format string*. It contains text that prints as is, plus format specifiers that handle formatted printing. In the above example, "\nTotal remaining: $...\n" is the text that prints as is. And `%.2f` is the format specifier. Think of a *format specifier* as a hole where you plug in a data item. In the above example, `remaining1 + remaining2` is the data item that gets plugged in. If `remaining1` holds 250 and `remaining2` holds −100.11, the sum is 149.89 and 149.89 gets plugged into the format specifier hole. The format specifier starts with `%` because all format specifiers must start with `%`. The format specifier's `.2` causes two digits to be displayed after the decimal point. The format specifier's `f` indicates that the data item is a floating-point number. The example shows only one format specifier. You can have as many format specifiers as you like in a given format string. For each format specifier, you should have a corresponding data item/argument. Here's an illustration of what we're talking about:

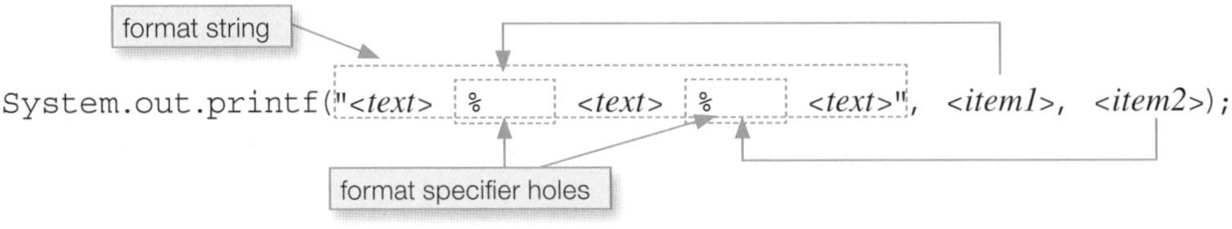

Format Specifier Details

Format specifiers are powerful little critters. We won't try to describe all of their power, but we'll provide enough details to get you up and running. If you come across a formatting issue that you can't resolve with our limited coverage, look up the `Formatter` class on Oracle's Java API website and search for format string details. But be prepared for lots of details. The Java API provides a tremendous number of options with the `printf` method.

Here's the syntax for a format specifier:

`%`[*flags*] [*width*] [*.precision*] *conversion-character*

You've already seen the `%` symbol. It indicates the start of a format specifier. The flags, width, precision, and conversion character represent the different parts of a format specifier. Each of them specifies a different formatting trait. We'll cover them in right-to-left order. Thus, we'll describe the conversion character first. But before jumping into conversion character details, note the square brackets. They indicate that something is optional. So the flags, width, and precision parts are optional. Only the `%` and the conversion character are required.

Conversion Character

The conversion character tells the JVM the type of data that is to be printed. For example, it might tell the JVM to print a string, or it might tell the JVM to print a floating-point number. Here is a partial list of conversion characters:

- `s` This displays a string.
- `d` This displays a decimal integer (an `int` or a `long`).
- `f` This displays a floating-point number (a `float` or a `double`) with a decimal point and at least one digit to the left of the decimal point.
- `e` This displays a floating-point number (`float` or `double`) in scientific notation.

In explaining each part of a format specifier (conversion character, precision, width, and flags), we'll provide short examples that illustrate the syntax and semantics. After we're done with all the explanations, we'll show a complete program example. Note this code fragment and its associated output:

```
System.out.printf("Planet: %s\n", "Neptune");
System.out.printf("Number of moons: %d\n", 13);
System.out.printf("Orbital period (in earth years): %f\n", 164.79);
System.out.printf(
"Average distance from the sun (in km): %e\n", 4498252900.0);
```

<u>Ouput:</u>

```
Planet: Neptune
Number of moons: 13
Orbital period (in earth years): 164.790000
Average distance from the sun (in km): 4.498253e+09
```

The f and e conversion characters print six digits by default.

Note that by default, the `f` and `e` conversion characters generate six digits to the right of the decimal point. Also, note that if you try to print an integer value with an `f` or `e` conversion character, you'll get a runtime error.

Precision and Width

The precision part of a format specifier works in conjunction with the `f` and `e` conversion characters; that is, it works with floating-point data items. It specifies the number of digits that are to be printed to the right of the decimal point. We'll refer to those digits as the *fractional digits.* If the data item has more fractional digits than the precision's value, then rounding occurs. If the data item has fewer fractional digits than the precision's value, then zeros are added to the right so that the printed value has the specified number of fractional digits.

The width part of a format specifier specifies the minimum number of characters that are to be printed. If the data item contains more than the specified number of characters, then all the characters are printed. If the data item contains fewer than the specified number of characters, then spaces are added. By default, output values are right-aligned, so when spaces are added, they go on the left side.

Note this code fragment and its associated output:

```
System.out.printf("Cows are %6s\n", "cool");
System.out.printf("But dogs %2s\n", "rule");
System.out.printf("PI = %7.4f\n", Math.PI);
```

<u>Ouput</u>

```
Cows are   cool
But dogs rule
PI =  3.1416
```

6 characters

7 characters

In the third statement above, note the `%7.4f` specifier. It's easy to get fooled by the 7.4. It looks like it might be saying "seven places to the left of the decimal point and four places to the right of the decimal point," but it's actually saying "seven total spaces, with four places to the right of the decimal point." And don't forget that the decimal point is counted as one of those seven total spaces. `Math.PI`'s value is 3.141592653589793, and when it gets printed with four places to the right of the decimal point, it gets rounded to 3.1416.

Flags

As a refresher, here's the syntax for a format specifier:

`%`[*flags*] [*width*] [.*precision*] *conversion-character*

We've described the conversion, precision, and width parts of a format specifier. It's now time to discuss flags. Flags allow you to add supplemental formatting features, one flag character for each formatting feature. Here's a partial list of flag characters:

`-`	Display the printed value using left justification.
`0`	If a numeric data item contains fewer characters than the width specifier's value, then pad the printed value with leading zeros (i.e., display zeros at the left of the number).
`,`	Display a numeric data item with locale-specific grouping separators. In the United States, that means commas are inserted between every third digit at the left of the decimal point.
`(`	Display a negative numeric data item using parentheses, rather than using a minus sign. Using parentheses for negative numbers is a common practice in the field of accounting.

Let's see how format specifiers work in the context of a complete program. See Figure 5.10's BudgetReport program. Note that we use the same format string for printing the column headers and the column underlines,

left justification

parenthesis for negatives, comma for group separators

```
/*************************************************************
* BudgetReport.java
* Dean & Dean
*
* This program generates a budget report.
*************************************************************/

public class BudgetReport
{
  public static void main(String[] args)
  {
    final String HEADING_FMT_STR = "%-25s%13s%13s%15s\n";
    final String DATA_FMT_STR = "%-25s%,13.2f%,13.2f%(,15.2f\n";
    double actual1 = 1149.999; // amount spent on 1st account
    double budget1 = 1400;     // budgeted for 1st account
    double actual2 = 2100.111; // amount spent on 2nd account
    double budget2 = 2000;     // budgeted for 2nd account
    double remaining1, remaining2; // unspent amounts

    System.out.printf(HEADING_FMT_STR,
      "Account", "Actual", "Budget", "Remaining");
    System.out.printf(HEADING_FMT_STR,
      "-------", "------", "------", "---------");

    remaining1 = budget1 - actual1;
    System.out.printf(DATA_FMT_STR,
      "Office Supplies", actual1, budget1, remaining1);
    remaining2 = budget2 - actual2;
    System.out.printf(DATA_FMT_STR,
      "Photocopying", actual2, budget2, remaining2);

    System.out.printf(
      "\nTotal remaining: $%(,.2f\n", remaining1 + remaining2);
  } // end main
} // end class BudgetReport
```

Output:

```
Account                         Actual       Budget      Remaining
-------                         ------       ------      ---------
Office Supplies               1,150.00     1,400.00         250.00
Photocopying                  2,100.11     2,000.00        (100.11)

Total remaining: $149.89
```

Figure 5.10 BudgetReport program and its ouput

and the format string is stored in a HEADING_FMT_STR named constant. If you use a format string in more than one place, it's a good idea to save the format string in a named constant and use the named constant in the printf statements. By storing the format string in one common place (in a named constant), you ensure consistency and make it easier to update the format string in the future.

In the BudgetReport program, note the minus sign in the HEADING_FMT_STR and DATA_FMT_STR format strings. That left justifies the first column's data. Note the commas in the DATA_FMT_STR format string. That causes locale-specific characters (commas in the United States) to appear between every third digit at the left of the decimal point. Note the left parenthesis in the DATA_FMT_STR format string. That causes negative numbers to use parentheses instead of a minus sign.

5.8 Problem Solving with Random Numbers (Optional)

This section will show you how to generate random variables with probability distributions that are different from the 0.0 to 1.0 uniform distribution assumed in a simple Math.random method call.

Using Math.random to Generate Random Numbers with Other Probability Distributions

As indicated in Figure 5.2, in Section 5.3, when you need a random number, you can use the Math.random method to generate one. Suppose you want a random number from a range that's different from the range 0.0 to 1.0. As we did in the initialization of winningNumber in Figure 5.5, you can expand the range to any maximum value by multiplying the random number generated by Math.random() by your desired maximum value. You can also offset the range by adding or subtracting a constant. For example, suppose you want to pick a random number that's uniformly distributed in the range between −5.0 and +15.0. Instead of using just plain old Math.random(), use this:

```
(20.0 * Math.random()) - 5.0.
```

It's possible to manipulate numbers produced by Math.random to get any kind of distribution you want. For example, you can generate any of the distributions shown in Figure 5.11.

Now, let's look at how to generate these five types of random numbers from Math.random.

1. The first type (a continuous uniform distribution) is easy. To get a value for a random number, x, uniformly distributed in the interval between zero and unity ($0.0 \le x < 1.0$), use a statement like this:

   ```
   double r1 = Math.random();
   ```

   This first type of random number is the basis of all other types of random numbers.

2. For the second type (an offset and expanded continuous uniform distribution), you must have some minimum and maximum values; for example:

   ```
   double minReal = 1.07;   // meters for shortest adult human
   double maxReal = 2.28;   // meters for tallest adult human
   ```

   Then you shift and expand the basic random number by using a statement like this:

   ```
   double r2 = minReal + Math.random() * (maxReal - minReal) ;
   ```

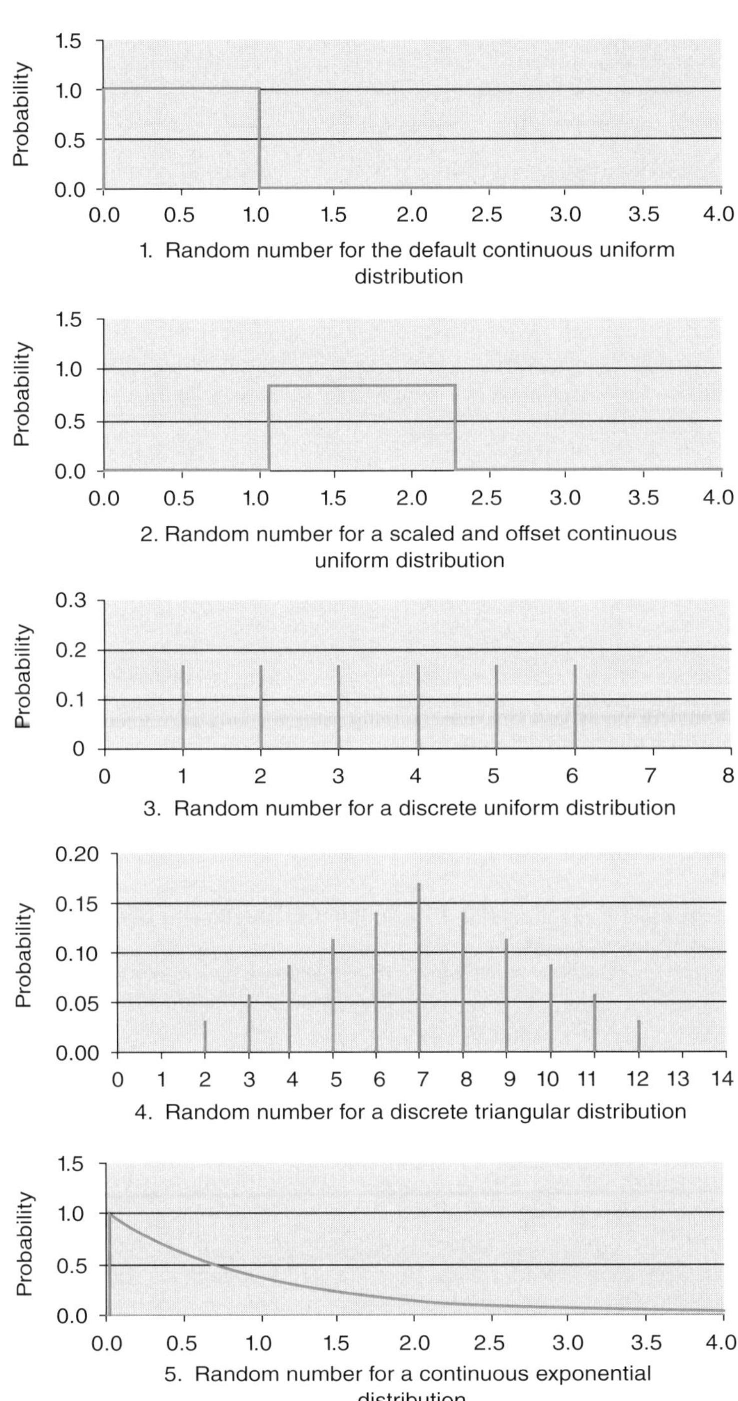

Figure 5.11 Important types of random number distributions

3. For the third type (a discrete uniform distribution), you create integer versions of the limits; for example:

```
int min = 1;              // fewest dots on one die
int max = 6;              // most dots on one die
```

Then you shift and expand the basic random number, sort of like you did for the second type:

```
double r3 = min + (int) (Math.random() * (max - min + 1)) ;
```

This time, you must remember that integer subtraction produces a distance that is one less than the number of integers in the range (6 minus 1 equals 5, not 6), so you have to add 1 to the difference like this (`max - min + 1`). The `double` returned by `Math.random` automatically promotes everything to `double`, so the shifted and expanded range is from 1.0 to 6.99999. The random selection gives equal weight to each of the six intervals above the integers of interest (1, 2, 3, 4, 5, and 6). The final `(int)` cast drops fractions.

4. For the fourth type (a discrete triangular distribution), at first you might think you could just use the third type with min = 2 and max = 12, but that would be wrong. It would generate just as many 2's and 12's as 7's, but the chance of getting a 7 is actually six times higher than getting either a 2 or a 12! The most straightforward way to get the right answer is to call `Math.random` twice, and add the results:

```
int twoDice = r3 + r3;
```

5. The fifth type of distribution (a continuous exponential distribution) has been included because it's used in models of many important real-world phenomena, such as:
   - Inter-arrival time of automobiles at an isolated traffic light
   - Time between infrequent telephone calls
   - Time between radioactive emissions from an unstable atom
   - Time to breakdown of a piece of machinery
   - Time to failure of a semiconductor device

   To generate a random variable with a continuous exponential distribution, use a statement like this:

```
double r5 = -Math.log(1.0 - Math.random()) * averageTimeBetweenEvents;
```

The logarithm of zero is −infinity, but that never occurs, because `Math.random` never generates a number as high as `1.0`, so `(1.0 - Math.random())` is never as low as zero.

Using the `Random` class

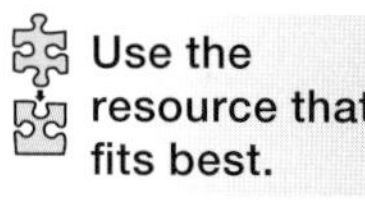
Use the resource that fits best.

Although it is possible to get any kind of distribution from `Math.random`, it's not always easy. For example, the algorithm that converts `Math.random`'s uniform distribution to a Gaussian (bell-curve) distribution is rather convoluted. So it would be nice to have some prebuilt methods that immediately generate random numbers from this and other distributions. The `Random` class in the `java.util` package provides help. Here are API headings for some of the `Random` class methods:

```
public double nextDouble()
public int nextInt()
public int nextInt(int n)
public boolean nextBoolean()
public double nextGaussian()
```

The `nextDouble` method does essentially the same thing as `Math.random` does. This distribution appears in the top graph in Figure 5.11. The zero-parameter `nextInt` method generates random integers uniformly from the entire range of integers, that is, from −2147483648 to +2147483647, inclusive. The one-parameter `nextInt` method generates random integers uniformly from zero to one less than the parameter value. This distribution is almost like what appears in Figure 5.11's third graph for the special case of n = 7, except zero is allowed also. The `nextBoolean` method generates random values of `true` or `false`. The `nextGaussian` method generates a `double` value from a distribution having a mean value of 0.0 and a standard deviation of 1.0.

Notice that the `Random` class methods do not have the `static` modifier, so they are not class methods, and you cannot use the `Random` class name to access these methods. As with `Scanner` (described in Section 3.23), you must create an object first, and then use that object's name to access these methods. The code in Figure 5.12 creates the object with the statement:

```
Random random = new Random();
```

This statement gives the variable called `random` a reference to this object. In the next two statements, this object generates two random numbers by calling the `nextInt` and `nextGausssian` methods. Due to the `Integer.MAX_VALUE` argument, `nextInt` generates a random number between 0 and one less than the maximum integer value. The `nextGaussian` method generates a random number drawn from a Gaussian distribution having a mean of 5.0 and a standard deviation of 0.8.

```
/*****************************************************************
* RandomTest.java
* Dean & Dean
*
* This program demonstrates methods of the Random class.
*****************************************************************/

import java.util.Random;

public class RandomTest
{
  public static void main(String[] args)
  {
    Random random = new Random();

    System.out.println(random.nextInt(Integer.MAX_VALUE));
    System.out.println(5.0 + 0.8 * random.nextGaussian());
  } // end main
} // end class RandomTest
```

Use new to create an object

Sample session:

```
1842579217
4.242694469045554
```

Figure 5.12 RandomTest program using `Random` class methods to generate random numbers from different distributions

When a program uses `Math.random()` or `new Random()` to generate random numbers, what appears is always a surprise because it's random! This unpredictability can be quite frustrating when you are trying to develop and test a program that uses random numbers, because every test run produces different numerical values. During development and testing, what you'd like is a repeatable sequence of "random" numbers, which turn out to be exactly the same every time you rerun a program you're testing.

Using a Fixed Seed to Freeze a Random Number Sequence

For a repeatable random sequence, you can create a `Random` object with a seed. A seed provides a starting point for the internal state of the random number generator. Suppose you change the body of the `main` method in Figure 5.12 to this:

```
Random random = new Random(123);

System.out.println(5.0 + 0.8 * random.nextGaussian()) ;
System.out.println(5.0 + 0.8 * random.nextGaussian()) ;
System.out.println(5.0 + 0.8 * random.nextGaussian()) ;
```

Now, if you run the program, you'll get this:

Sample session:

```
3.8495605526872745
5.507356060142144
5.1808496102657315
```

If you run the program again, and again, and again, you'll get exactly the same three "random" numbers every time! The 123 seed establishes a starting point, and this determines the "random" sequence precisely. If you pick a different seed, you'll get a different sequence, but that sequence will always be the same, as long as you stick with that particular seed. Now you know why the methods in the `Random` class are not class methods. They need an object to call the methods because the methods need to know some information the object contains—the seed and the current position in the random-number sequence.

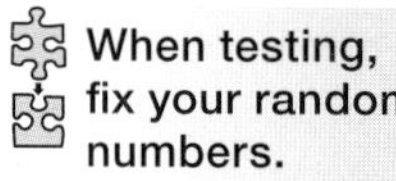
When testing, fix your random numbers.

You can use the deterministic nature of the seeded random-number generator to make your life a lot easier when you are developing and debugging programs that use random numbers. To establish a fixed random-number test set, you could write a simple program that prints a particular set of random numbers. You could copy those particular numbers into assignment statements in your program—that is, hard-code them in your program for development and testing. Then, after your program has been tested and verified, you could replace each hard-coded "random number" by a random-number generator that produces a different number every time it's invoked.

But the `Random` class provides a more elegant way to develop programs that have random variables. During development, create the Random object with a seed to produce exactly the same sequence of randomly distributed numbers every time that you run the program. Then, when all your bugs are fixed, simply delete the seed number to create the Random object, as in Figure 5.12, and—*voila*—your random-number generator produces completely different numbers from then on.

5.9 GUI Track: Drawing Images, Lines, Rectangles, and Ovals (Optional)

This section shows you how to display images, lines, rectangles, and ovals in a GUI window. The simplest way to do this is to call methods in Java's `Graphics` class from within a Java applet. As you might recall from Chapter 1, an applet is a Java program that's embedded in a web page. You execute a Java applet by calling it from the web page's HTML code (HTML is the base language for most web pages). You can run the web page by loading it within an Internet browser.

Image Files

Java can handle many kinds of images. Some images are stylized icons. Others are digitized photographs. For example, suppose you have a digitized photograph that includes a family member, like author John's older daughter, Jordan, the player closest to the ball in Figure 5.13.

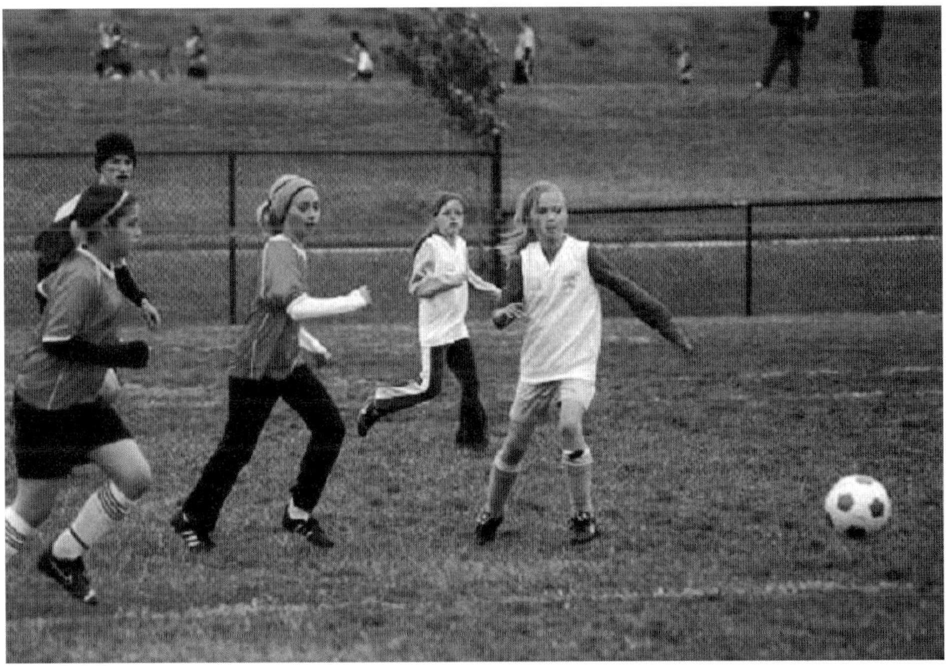

Figure 5.13 A typical image stored in a `.jpg` file

Suppose this photo is stored in a file named `dolphins.jpg`. ("Dolphins" is the name of Jordan's soccer team.) The extension, ".jpg," is short for JPEG, which stands for Joint Photographic Experts Group. Files with this extension should conform to JPEG standards for digital compression of photographic images. Exact representations of simple drawn images have been stored in files with a ".gif" extension and more recently are stored in files with a ".png" extension. GIF stands for Graphics Interchange Format, while PNG stands for Portable Network Graphics. For simplicity, we'll assume that the image file is in the current directory—the same directory that holds the Java program that will display it.

Before you can use a picture in a program, you need to know how big it is—its width and height in numbers of *pixels*. Pixels are the tiny dots of color a computer uses to display something on its screen. Figure 5.14 contains a program you can use to determine the pixel width and pixel height of the contents of an image file. The program imports Java's `Scanner` class to retrieve a keyboard entry of the filename, and it imports Java's `ImageIcon` class to read the image file and determine image properties. After prompting the user for a filename, the program uses that name to create an object we call `icon`, which manages information transfer from the image file—like our `stdIn` object manages information transfer from the keyboard. The `getWidth` and `getHeight` methods return the image's width and height in pixels. This gives the default size of the area required to display the image on a computer screen.

```
/*************************************************************
* ImageInfo.java
* Dean & Dean
*
* This supplies width and height of an image.
*************************************************************/

import java.util.Scanner;
import javax.swing.ImageIcon;

public class ImageInfo
{
  public static void main(String[] args)
  {
    Scanner stdIn = new Scanner(System.in);
    ImageIcon icon;

    System.out.print("Enter image filename: ");
    icon = new ImageIcon(stdIn.nextLine());
    System.out.println("image width = " + icon.getIconWidth());
    System.out.println("image height = " + icon.getIconHeight());
  }
} // end ImageInfo
```

Sample session:

```
Enter image filename: dolphins.jpg
image width = 864
image height = 576
```

Figure 5.14 ImageInfo program for finding the width and height of an image in an image file

Graphics Class Methods

The Java API class called `Graphics` contains several methods for displaying images and geometric shapes. Figure 5.15 presents API headings and descriptions for some of these methods. Notice that these headings do not have `static` modifiers. This means that you must use an object of the `Graphics` class to call all these methods, just as you must use a `String` object to call most of the methods of the `String` class. This `Graphics` object contains a reference to the window on which things are drawn, and it contains other necessary information, like the current position of that window on the computer screen, current color, and current font type.

Most of Figure 5.15's methods have pairs of parameters (like `int x, int y`), which indicate x and y coordinates in an image or window. These coordinates are always measured in pixels. The x coordinate is the number of pixels in from the left side of the image or window. The y coordinate is the number of pixels down from the top of the image or window. The x coordinate is like what you probably expect, but the y coordinate might seem funny, because normally we think of y as increasing upward. However, in a computer display, y increases downward because when a computer paints something on a screen, it paints the top line first, the line below second, and so on, in a top-to-bottom sequence.

In a method heading, x- and y-coordinate parameter names sometimes employ numerical suffixes to distinguish one point from another. In the `drawImage` method's parameters, there is also a character prefix before each coordinate identifier. The d prefix stands for "destination," which means position in the display window. The s prefix stands for "source," which means position in the original image.

Sometimes numbers in a sequence of parameters are not x and y coordinates. Instead, they are width and height, which are coordinate differences. This is the case for the `drawRect` and `fillOval` methods. 

```
public boolean drawImage(Image img,
  int dx1, int dy1, int dx2, int dy2,
  int sx1, int sy1, int sx2, int sy2,
  ImageObserver observer)
```
Selects whatever is between `sx1` and `sx2` pixels to the right of the left edge of the source image and between `sy1` and `sy2` pixels below the top of the source image. Scales this selection as required to fit between `dx1` and `dx2` pixels to the right of the left edge of the destination window and between `dy1` and `dy2` below the top of the destination window.

```
public void setColor(Color c)
```
Establishes a specified painting color.

```
public void drawRect(int x, int y, int width, int height)
```
Draws the outline of a rectangle whose upper left corner is `x` pixels to the right of the left side of the window and `y` pixels below the top of the window. Uses most recently set color.

```
public void drawLine(int x1, int y1, int x2, int y2)
```
Draws a straight line from a point `x1` pixels to the right of the left side of the window and `y1` pixels below the top of the window to a point `x2` pixels to the right of the left side of the window and `y2` pixels below the top of the window. Uses most recently set color.

```
public void fillOval(int x, int y, int width, int height)
```
Fills an ellipse bounded by the specified rectangle with the most recently set color.

```
public void drawString(String text, int x, int y)
```
Prints the specified text on a line that starts `x` pixels to the right of the left side of the window and `y` pixels down from the top of the window. Uses the most recently set color.

Figure 5.15 Selected methods from the Java API `Graphics` class

It's easy to forget which technique a particular method uses to specify width or height. Does it specify positions of upper-left and lower-right corners, or does it specify position of upper-left corner and then width and height? Be careful. This is a common source of GUI programming errors.

The `drawImage` method copies a rectangular portion of the source image and pastes an expanded or contracted version of it into a specified rectangle in the destination window. The first parameter is a reference to the source image. The second and third parameters are the destination (display window) coordinates of the top-left corner of the copied part of the image. The fourth and fifth parameters are the destination coordinates of the bottom-right corner of the copied part of the image. The sixth and seventh parameters are the coordinates of the top-left corner of the part of the source image to be copied. The eighth and ninth parameters are the coordinates of the bottom-right corner of the part of the source image to be copied. The last parameter enables the method to send out current-status information.

The `setColor` method establishes a color to be used in subsequent operations that draw lines or geometric figures or write text. You pass this method an argument like `Color.GREEN`, which identifies one of several named constants defined in the Java API `Color` class. You can find the names of the other named colors in the documentation for the Java API `Color` class. Most of those names are pretty obvious, so as a practical matter, you can just guess and see what happens.

The `drawRect` method draws the border of a rectangle using the most recently set color. The first and second parameters are the coordinates of the top left corner of the rectangle in the display window. The third and fourth parameters are the width and height of the rectangle.

The `drawLine` method draws a straight line between two specified points using the most recently set color. The first and second parameters are the coordinates of the starting point. The third and fourth parameters are the coordinates of the ending point.

The `fillOval` method draws an ellipse in the specified rectangle and fills it with the most recently set color. The parameters are like those for `drawRect`: The first and second parameters are the coordinates of the top left corner of the enclosing rectangle in the display window. The third and fourth parameters are the width and height of the enclosing rectangle, and the width and height of the oval itself.

The `drawString` method prints text at the specified position using the most recently set color. The first parameter is the string to be printed. The second and third parameters are the coordinates of the upper left corner of the string.

Using `Graphics` Methods

Figure 5.16 provides an example of how you can use the graphics methods in Figure 5.15 to manipulate a photographic image and add your own lines, shapes, and text to it. The overall window in Figure 5.16 is 640 pixels wide and 640 pixels high.

Figure 5.17 shows the Java code needed to render the display in Figure 5.16. The code is all in the body of a method called `paint`. The `paint` method's parameter `g` is used as a prefix on each of the method calls in `paint`. It refers to the `Graphics` object that manages the painting operation. Within `paint`, the first statement retrieves a reference to the source image. The next statement's `drawImage` method employs the source image reference to access the source image.

The `drawImage` method shrinks the image in Figure 5.13 to two thirds of its original size and pastes it into the upper left corner of the display window. The `setColor` method sets the current color to green. The `drawRect` method draws a square around an area of interest. Then four calls to the `drawLine` method draw four straight lines to the corners of where a three-times enlargement of the area of interest will go. Another call to the `drawImage` method pastes an enlarged version of the area of interest at this location. Another call to `drawRect` puts a rectangle around this enlargement. Then the `fillOval` method paints a green oval in the enlargement, and finally the `drawString` method prints the name

"JORDAN" in the center of this green oval. Notice how each subsequent operation overwrites or covers all previous operations.

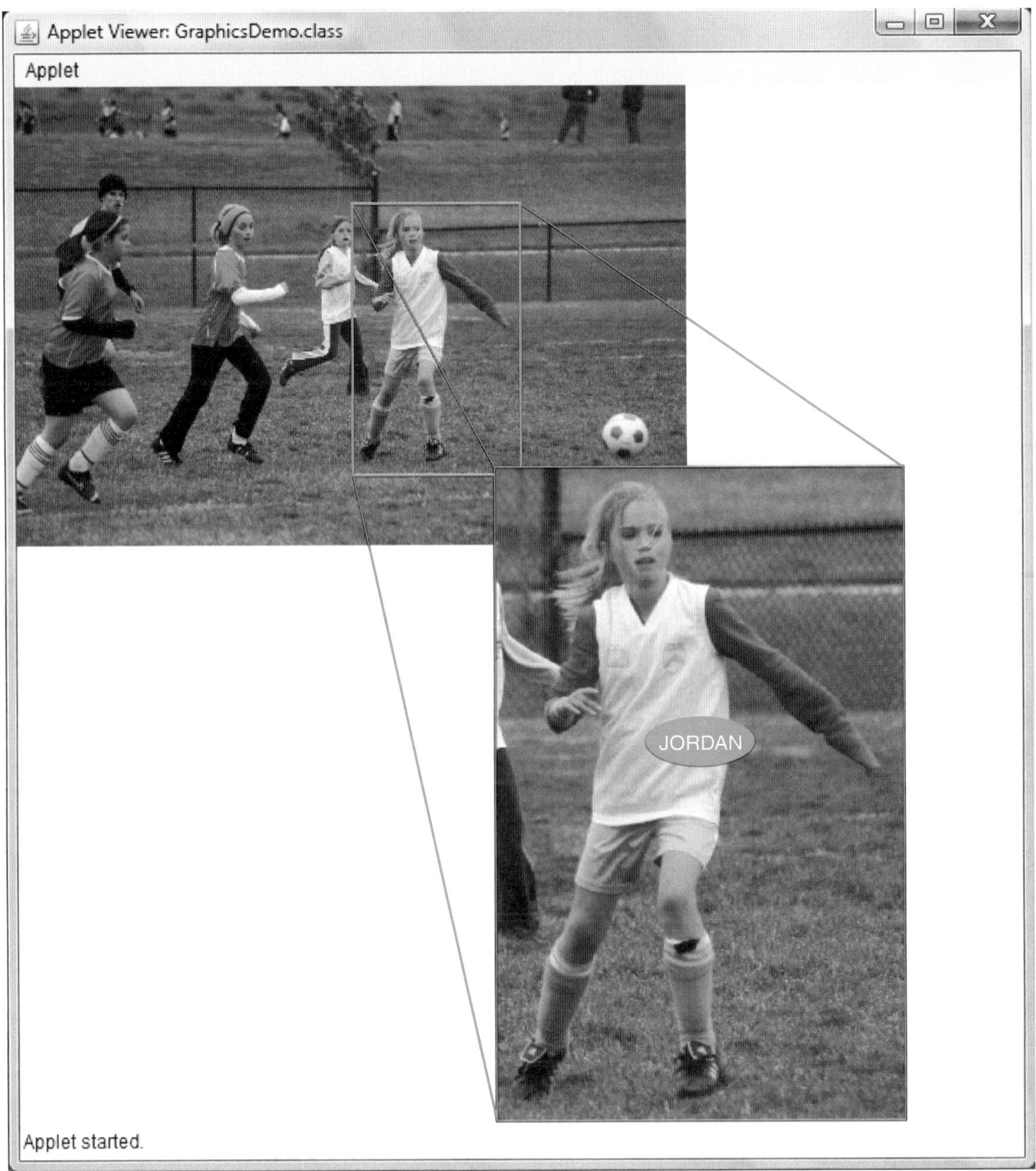

Figure 5.16 Output produced by the program in Figure 5.17

The code in Figure 5.17 is relatively sparse. For example, it does not include a definition of the `getImage` method. So how can it call that method? As you'll discover in Chapter 13, Java allows any class you define to borrow the methods of another class that's already been defined. In particular, the `GraphicsDemo` class

```
/*********************************************************************
* GraphicsDemo.java
* Dean & Dean
*
* This defines a Java applet that displays an image and graphics.
*********************************************************************/

import java.awt.*;       // for Graphics, Image, and Color classes
import javax.swing.JApplet;

public class GraphicsDemo extends JApplet
{
  public void paint (Graphics g)
  {
    Image image =
      this.getImage(getDocumentBase(),"dolphins.jpg");

    // display smaller complete image in upper left corner of window
    g.drawImage(image, 0, 0, 427, 284,     // destination topL, botR
      0, 0, 864, 576, this);               // source topL, botR

    // establish color of all lines to be drawn
    g.setColor(Color.GREEN);

    // draw rectangle around region to be expanded
    g.drawRect(214, 72, 107, 168);         // topL, width & height

    // draw lines between corners of rectangles
    g.drawLine(214, 72, 304, 235);         // upper left
    g.drawLine(321, 72, 565, 235);         // upper right
    g.drawLine(214, 240, 304, 646);        // lower left
    g.drawLine(321, 240, 566, 646);        // lower right

    // display expanded part of original image
    g.drawImage(image, 304, 235, 565, 646, // destination topL, botR
      432, 145, 650, 487, this);           // source topL, botR

    // draw rectangle around expanded part of image
    g.drawRect(304, 235, 261, 411);        // topL, width & height

    // create GREEN colored oval and write name on it
    g.fillOval(400, 390, 70, 30);          // topL, width & height
    g.setColor(Color.WHITE);               // change color for text
    g.drawString("JORDAN", 410, 410);      // string & start position
  } // end paint
} // end GraphicsDemo class
```

The `extends JApplet` appended to the class heading allows this class to "borrow" the `getImage` method from the already-defined Java API `JApplet` class.

Figure 5.17 GraphicsDemo Java applet that illustrates graphics methods listed in Figure 5.15
This applet produces the output shown in Figure 5.16.

defined in Figure 5.17 borrows the getImage method from the already-defined Java API JApplet class. It does this by appending the clause extends JApplet to its class heading. Of course, the compiler must know where to find the JApplet class, so the program must import it. It also imports the java.awt package to provide access to the Graphics, Image, and Color classes used by statements in the paint method.

Notice that the getImage method call has the word this as a prefix, and the word this also appears in the last argument in the two calls to the drawImage method. The next chapter will explain that the special Java term this refers to whatever object happened to call the currently executing method. In our GraphicsDemo program, this refers to the object that calls the paint method, and that object is an instance of the GraphicsDemo class defined by the code in Figure 5.17. But where is the code that creates a GraphicsDemo object, where is the code that retrieves a reference to the associated Graphics object, and where is the code that calls the paint method? It's in a separate file. . . .

Applet Execution

Did you notice that there is no main method in Figure 5.17? A Java applet is different from a Java application. Because a Java applet does not have a main method, it cannot be executed in the normal manner. Typically, it's embedded in another program, such that its code executes when that other program calls it. The primary purpose of a Java applet is to liven up a web page. So we typically call Java applets from HTML (HyperText Markup Language) programs that define web pages. Figure 5.18 contains a minimal HTML program that calls the GraphicsDemo applet defined in Figure 5.17.

Notice that the part of this HTML code that's specific to our particular applet is all in the seventh line. This identifies the compiled version of the applet, and it specifies the pixel width and pixel height of the window that will hold whatever the applet will display. You can write this HTML code with any primitive text editor, like Microsoft notepad or UNIX vi. Then save it in the same directory as the code that has the compiled version of the applet it drives. For example, in the directory that contains the GraphicsDemo.class file. When you save it, give it a name which has the html extension, like graphicsDemo.html.

You have three alternate ways to run the HTML file (and its associated Java applet):

1. Open a browser like Mozilla's Firefox, navigate to the directory that contains the HTML file, and double click on the HTML filename.
2. Open a Command Prompt window, navigate to the directory that contains the HTML file, and enter: `appletviewer graphicsDemo.html`
3. Select "Run a Java Applet" in your local integrated development environment (IDE), and select the desired HTML filename.

```
<!doctype html>
<html>
<head>
<title>Graphics Demo</title>
</head>
<body>
<applet code="GraphicsDemo.class" width="640" height="650">
</applet>
</body>
</html>
```


Figure 5.18 Code for an HTML file that runs the GraphicsDemo code in Figure 5.17

5.10 GUI Track: Decorating an Image by Covering It with a Tinted Pane (Optional)

This section continues the discussion of the preceding section by showing how you can form a composite image by covering a standard image with a semitransparent tinted pane, whose coloration varies from one place to another. This uses some other classes in the Java API, so it provides additional exposure to that large body of already written Java code that is always available for you to use.

The preceding section's program imported the `java.awt` package to use its `Graphics`, `Image`, and `Color` classes. It also used the individual `javax.swing.JApplet` class. This section's program uses the `JPanel` class instead of the `JApplet` class, and it uses several other classes in the `java.awt` package: `ImageIcon`, `Graphics2D`, `RadialGradientPaint`, `AlphaComposite`, and `Font`.

See Figure 5.19. Unlike the GraphicsDemo program, the GraphicsDemo2 program appends `extends JPanel` to its class heading and defines a customized `paint` method. This program's `paint` method starts with two declarations that retrieve a `.jpg` image.

A `Graphics2D` variable called `g2` provides access to `setPaint`, `setComposite`, and `fillRect` methods; `centerX` and `centerY` variables identify the center of a circular area; and a `radius` variable specifies that area's radius. The two values in the `gradientRange` variable identify the boundaries of a range through which the opacity of the covering tinted pane varies. In this example, from the center out to a radius equal to `0.7 * radius`, the covering pane is transparent. Then, from a radius equal to `0.7 * radius` out to a radius equal to `1.0 * radius`, the covering pane's opacity varies up to a maximum. Outside a radius equal to `1.0 * radius`, the opacity stays at this maximum. The two color values in the `colors` array are the minimum-opacity and maximum-opacity colors, respectively.

As in the previous program, the `drawImage` method call draws the underlying image. The `setPaint` method call establishes the pattern and amount of tinting in the covering tinted pane. The `setComposite` method call computes what the image would look like when seen through the tinted pane. The opaqueness of this tinted pane is modified by the final argument in the `setComposite` method call. Because the opaqueness is only `0.5f` rather than `1.0f`, you can still see through the pane in areas outside the circle. The `fillRect` method call repaints the pane in the displayed window with the computed composite image.

The `setColor` and `setFont` method calls set the color and the font type, style, and size for a text label. As in the previous section's example, the `drawString` method writes the text label. (This label contains the name of author John's younger daughter, Caiden.)

Instead of Figure 5.17's separate html file, the GraphicsDemo2 program started in Figure 5.19 also gets a `main` method, which contains the following statements:

```
JFrame frame = new JFrame("Graphics Demo 2");
frame.setSize(696, 520);
frame.add(new GraphicsDemo2());
frame.setVisible(true);
frame.setDefaultCloseOperation(JFrame.EXIT_ON_CLOSE);
```

Of course, the current directory also needs to contain the `.jpg` file that has the underlying image. Assuming that's true, when we run the program, we get what you see in Figure 5.20.

Notice that the text label extends into a tinted part of the composite image. Yet the text itself is not tinted. That's because the program writes the text after it paints the composite. If the `g.drawString` method call were before the `g2.fillRect` method call, but after the `g.drawImage` method call, the right side of the text especially would be darkened, as though seen through the tinted covering. If the `g.drawString` method call were before the `g.drawImage` method call, the text would be completely covered by the image and not seen at all.

```
/*************************************************************
* GraphicsDemo2.java
* Dean & Dean
*
* This darkens image around circular area of interest.
*************************************************************/

// Graphics, ImageIcon, Image, Graphics2D, Color,
// RadialGradientPaint, AlphaComposite, Font
import java.awt.*;
import javax.swing.*; // for JFrame, JPanel

public class GraphicsDemo2 extends JPanel
{
  public void paint(Graphics g)
  {
    ImageIcon icon = new ImageIcon("dolphins2.jpg");
    Image image = icon.getImage();
    Graphics2D g2 = (Graphics2D) g.create();
    float centerX = 340;
    float centerY = 240;
    float radius = 300;
    float[] gradientRange = {0.7f, 1.0f};    // times radius
    // Color components are: red, green, blue, & opaqueness
    // in the range between 0.0f and 1.0f
    Color insideColor = new Color(0.0f, 0.0f, 0.0f, 0.0f);
    Color outsideColor = Color.BLACK;
    Color[] colors = {insideColor, outsideColor};

    super.paint(g);                          // preserve background

    g.drawImage(image, 0, 0, 680, 480,       // destination topL, botR
      0, 0, 2306, 1625, this);               // source topL, botR
    g2.setPaint(new RadialGradientPaint(
      centerX, centerY, radius, gradientRange, colors));
    g2.setComposite(AlphaComposite.getInstance(
      AlphaComposite.SRC_OVER, 0.5f));
    g2.fillRect(0, 0, getWidth(), getHeight());
    g.setColor(Color.WHITE);
    g.setFont(new Font(Font.SERIF, Font.ITALIC, 20));
    g.drawString("CAIDEN", 435, 470);
  } // end paint
```

for tinted pane

for text

Figure 5.19 GraphicsDemo2 program's `paint` method, which forms composite image in Figure 5.20. This program also needs a `main` method whose contents are described in the text.

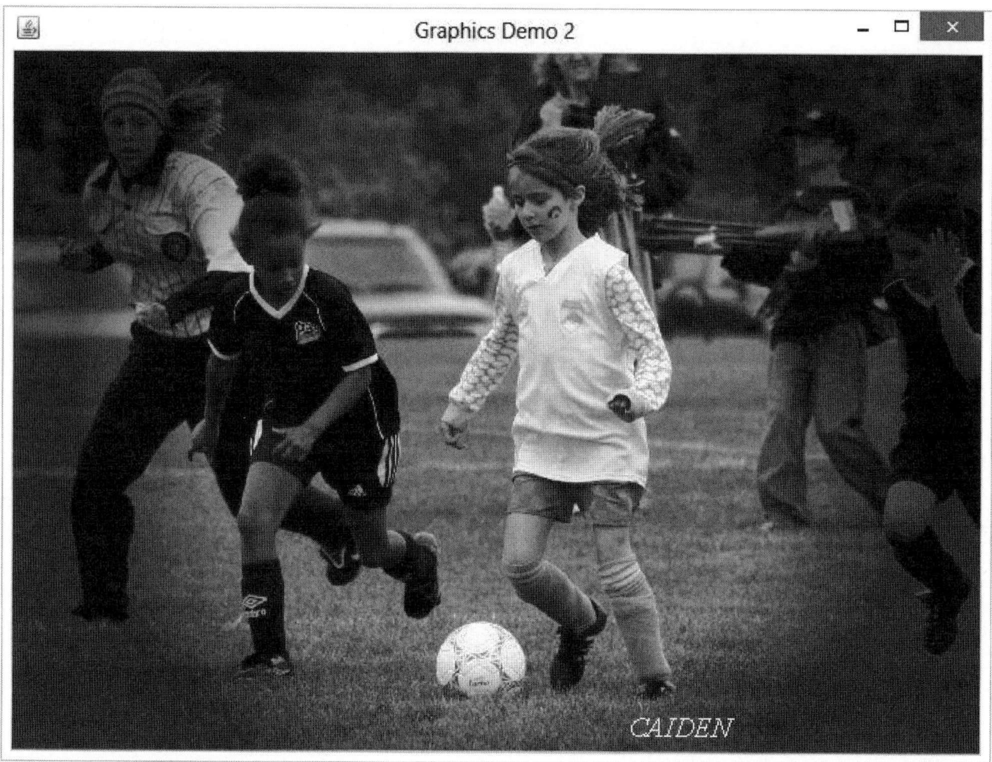

Figure 5.20 Output produced by the program whose `paint` method appears in Figure 5.19.

Substituting `JApplet` for `JPanel`

The Graphics Demo2 program in this section produces exactly the same results if you replace every `JPanel` with `JApplet`. This allows you to develop and test Java applets on a local computer with a `main` methods and defer the web-page hassles mentioned in Section 1.7

Summary

- Oracle's Java documentation identifies the public interface of all Java API software. It also provides a brief description of what it does and how to use it. The `java.lang` package is always available.
- The `Math` class provides methods that enable you to compute powers and roots, maximums or minimums, angle conversions, and many trigonometric functions. The `random` function generates a random number whose distribution is uniform in the range 0.0 to slightly less than 1.0. This class also provides named constant values for PI and E.
- Numerical wrapper classes like `Integer`, `Long`, `Float`, and `Double` contain parsing methods like `parseInt`, which enable you to convert `String` representations of numbers into numerical format. MIN_VALUE and MAX_VALUE named constants give minimum and maximum allowable values for the various numerical data types.

- The Character class provides methods that tell you whether a character is whitespace, a digit, or letter, and, if it's a letter, whether it's lowercase or uppercase. Other methods allow you to change case.
- The String class's indexOf method helps you find the position of a particular character in a string of text. The substring method allows you to extract any part of a given string of text. The replace-All and replaceFirst methods make substitutions within a string of text. You can make case conversions with the toLowerCase and toUpperCase methods, and you can use the trim method to remove whitespace from either end of a string of text.
- The first argument in the System.out.printf method is a format string which enables you to use a special code to specify the output format of text and numbers. For example, to display a double number called price as dollars and cents with commas between groups of three digits and a zero to the left of the decimal for values less than $1.00, you would write:

```
System.out.printf("$%,04.2f\n", price);
```

- Use the Random class in the java.util package to get various random number distributions or obtain exactly the same list of random numbers every time you run a particular program.
- To display photographic images, geometric figures, and text in graphics windows, call methods in Java's Graphics class from within a paint method in a new class that extends Java's JPanel class. Then in a separate main method, add your extension of Java's JPanel class to an instance of Java's JFrame class.

Review Questions

§5.3 Math Class

1. Given these declarations:

```
double diameter = 3.0;
double perimeter;
```

   Provide a statement that assigns the length of a circle's perimeter to the perimeter variable. Use the diameter variable.
2. What is the name of the class that contains the abs, min, and round methods?
   a) Arithmetic
   b) Math
   c) Number

§5.4 Wrapper Classes for Primitive Types

3. Provide a statement that assigns positive infinity into a double variable named num.
4. Provide a statement that converts a string variable named s to a long and assigns the result to a long variable named num.
5. Provide a statement that converts an int variable named num to a string and assigns the result to a String variable named numStr. Use a wrapper class method.

§5.5 Character Class

6. What does the following code fragment print?

```
System.out.println(Character.isDigit('#'));
System.out.println(Character.isWhitespace('\t'));
System.out.println(Character.toLowerCase('B'));
```

§5.6 String Methods

7. Given this declaration:[6]

```
String snyder = "Stick together.\nLearn the flowers.\nGo light.";
```

Write a Java statement that finds the index of the letter 'G' and prints everything in `snyder` from that point on. In other words, it prints `Go light.`

§5.7 Formatted Output with the printf Method

8. Write a format string that handles the display of three data items in three columns. The first column should be 20 spaces wide, and it should print a left-aligned string. The second column should be 10 spaces wide, and it should print a right-aligned integer. The third column should be 16 spaces wide, and it should print a right-aligned floating-point number in scientific format with 6 decimal places. Your format string should cause the screen's cursor to move to the next line after printing the third data item.
9. Provide a format specifier that handles the display of a floating-point data item. It should print a rounded version of the data item with no decimal places. It should insert grouping separators, and it should use parentheses if the number is negative.

§5.8 Problem Solving with Random Numbers (Optional)

10. Write a Java statement that prints a random number for the total number of dots on a roll of a pair of dice.
11. Write a program that prints five random `boolean` values with the seed, `123L`. Then display those values.

Exercises

1. [after §5.3] Write a statement that computes and prints the cube root of a `double` variable named `number`. [*Hint:* look for an appropriate method in the Java API documentation of the `Math` class.]
2. [after §5.3] In probability calculations, we frequently need to compute the value of the factorial of some number `n`. The factorial of a number `n` (designated `n!`) is given by the formula

   $n! \leftarrow n * (n-1) * (n-2) * ... * 3 * 2 * 1.$

   When n is a very large number, this is a time-consuming calculation. Fortunately, there is a handy formula, called *Stirling's Formula,* which gives a very good approximation of n! whenever n is large. Stirling's formula says:

   $n! \approx (1 + 1/(12n -1)) * sqrt(2n\pi) * (n/E)^n$

   The symbol π is the ratio of a circle's perimeter to its diameter, and the symbol E is the base of natural logarithms. The actual value of n! is always slightly smaller than the value given by this formula. For this exercise, write a Java code fragment that implements Stirling's formula.
3. [after §5.3] Write a `main` method that asks the user for an angle, θ, in degrees, and prints out the values of sin(θ), cos(θ), and tan(θ).

   Sample session:

   ```
   Enter an angle in degrees: 30
   sin(deg) = 0.49999999999999994
   cos(deg) = 0.8660254037844387
   tan(deg) = 0.5773502691896257
   ```

[6] Gary Snyder, "For the Children," in *Turtle Island,* New Directions (1974).

4. [after §5.3] Provide a statement that prints the length of the hypotenuse of a right triangle whose base is given by the variable `base`, and whose height is given by the variable `height`. In your statement, you must use the `Math` class's `hypot` method. To learn about the `hypot` method, see the Java API website.

5. [after §5.3] Given the base-e log, you can always find the log to any other base, with the formula: $\log_{base}(x) = \log_e(x) / \log_e(base)$. For example, a computer scientist might be interested in how many bits are needed to express a given positive integer x in binary. In that case, the total number of bits required is $\log_2(x)$, rounded up to the next higher integer. Write a Java statement that (1) calculates the number of bits required to store variable x's value and (2) assigns that calculated value into an `int` variable named `bits`.

6. [after §5.6] In the following program skeleton, replace <*Insert code here.*> with your own code. *Hint:* Use the variables that are already declared for you (`songs`, `searchText`, `foundIndex`, and `count`). The resulting program should prompt the user for a search string and then display the number of occurrences of the search string in a given list of songs. Study the sample session.

```
import java.util.Scanner;

public class CountSubstringOccurrences
{
  public static void main(String[] args)
  {
    Scanner stdIn = new Scanner(System.in);
    String songs =
      "1. Green Day - American Idiot\n" +
      "2. Jesus Jones - Right Here, Right Now\n" +
      "3. Indigo Girls - Closer to Fine\n" +
      "4. Peter Tosh - Equal Rights\n";

    String searchText; // text that is searched for
    int foundIndex;    // position of where text is found
    int count = 0;     // number of occurrences of search text

    System.out.print("Enter search text: ");
    searchText = stdIn.nextLine();

    <Insert code here.>

    System.out.println("Number of occurrences of \"" +
      searchText + "\": " + count);
  } // end main
} // end class CountSubstringOccurrences
```

Sample session:

```
Enter search text: Right
Number of occurrences of "Right": 3
```

7. [after §5.6] In the following program skeleton, replace <*Insert code here.*> with your own code. *Hint:* Use the variables that are already declared for you (`songs`, `songNum`, `songIndex`, `eolIndex`, and `song`). The resulting program should prompt the user for a song number and then extract the song number, plus the rest of that string's line from a given list of songs. Study the sample session. You may assume that the user enters a valid song number (no need for input validation).

```
import java.util.Scanner;

public class ExtractLine
{
  public static void main(String[] args)
  {
    Scanner stdIn = new Scanner(System.in);
    String songs =
      "1. Bow Wow - Fresh Azimiz\n" +
      "2. Weezer - Beverly Hills\n" +
      "3. Dave Matthews Band - Crash Into Me\n" +
      "4. Sheryl Crow - Leaving Las Vegas\n";

    String songNum; // song number that is searched for
    int songIndex;  // position of where song number is found
    int eolIndex;   // position of end of line character
    String song;    // the specified line

    System.out.print("Enter song number: ");
    songNum = stdIn.nextLine();

    <Insert code here.>

    System.out.println(song);
  } // end main
} // end class ExtractLine
```

Sample session:

```
Enter song number: 3
3. Dave Matthews Band - Crash Into Me
```

8. [after §5.7] Given the below program skeleton. Replace the four *<add code here>* items so that the program produces the below output. Try to mimic the output's format precisely, but it's OK if your column widths vary slightly from the shown column widths.

```
public class CarInventoryReport
{
  public static void main(String[] args)
  {
    final String HEADING_FMT_STR = <add code here>;
    final String DATA_FMT_STR = <add code here>;
    String item1 = "Mazda RX-8";
    int qty1 = 10;
    double price1 = 27999.99;
    String item2 = "MINI Cooper";
    int qty2 = 100;
    double price2 = 23000.25;

    System.out.printf(HEADING_FMT_STR,
      "Item", "Quantity", "Price", "Value");
    System.out.printf(HEADING_FMT_STR,
      "----", "--------", "-----", "-----");
```

```
    System.out.printf(DATA_FMT_STR, <add code here>);
    System.out.printf(DATA_FMT_STR, <add code here>);
  } // end main
} // end class CarInventoryReport
```

Output:

```
Item                 Quantity      Price       Value
----                 --------      -----       -----
Mazda RX-8                 10     28,000     280,000
MINI Cooper               100     23,000   2,300,025
```

9. [after §5.8] Provide a statement that uses `Math.Random` to generate the total number of dots on a rolled pair of dice.

Review Question Solutions

1. `perimeter = Math.PI * diameter;`
2. The class that contains the `abs`, `min`, and `round` methods is: b) `Math`
3. `num = Double.POSITIVE _ INFINITY;`
4. `num = Long.parseLong(s);`
5. `numStr = Integer.toString(num);`
6. Here is the code fragment's output.

```
false
true
b
```

7. `System.out.println(snyder.substring(snyder.indexOf('G')));`
8. `"%-20s%10d%16.6e\n"`
   or
   `"%-20s%10d%16e\n"`
   (It's OK to omit the .6 because the e conversion character prints 6 decimal places by default.)
9. `"%(,.0f"`
   or
   `"%,(.0f"`
   (The order of flag specifier characters is irrelevant.)
10. Statement that prints the total number of dots on a thrown pair of dice:

```
System.out.println(2 + (int) (6 * (Math.random())) +
  (int) (6 * (Math.random())));
```

11. Program that prints five random `boolean` values with seed 123L:

```
import java.util.Random;

public class RandomBoolean
{
  public static void main(String[] args)
  {
```

```
    Random random = new Random(123L);

    for (int i=0; i≤5; i++)
    {
      System.out.println(random.nextBoolean());
    }
  } // end main
} // end RandomBoolean
```

The values are:

```
true
false
true
false
false
```

Interlude

Multiple-Method Programs in a Non-Object-Oriented Environment

In Chapter 5, you learned how to call prebuilt methods in Oracle's Java API library. Calling prebuilt methods is an important step in your journey along the path to programming proficiency. Calling your own methods and implementing those called methods is the next logical step.

For a while now, you've been writing programs with one method—the `main` method. In this Interlude "mini-chapter," we describe how to implement multiple methods within a program and how to call the non-`main` methods from `main`. Following the description, we provide a complete program that illustrates the concepts. After that, there's a fork in the road. You'll need to learn more details about writing and calling methods, but there are two different strategies for learning those details. Many textbooks take the approach that because object-oriented programming (OOP) is such an integral part of Java programming, the additional method details should be taught in conjunction with learning OOP. On the other hand, some textbooks take a "late objects" approach and postpone a full discussion of OOP until after readers have learned the additional method details.

Each of the two approaches has its merits. The late objects approach has the advantage of being able to focus on writing methods without getting bogged down in simultaneously having to learn OOP. The other approach has the advantage of teaching good OOP habits early. Most (but not all) real-world Java programs are written using the OOP paradigm. Folks who learn with the late objects approach sometimes become enamored with non-OOP programming, to a point where it becomes harder for them to write OOP programs later on. To avoid that problem, we have a slight preference for OOP being taught in conjunction with readers learning additional method details, and that's why we do it that way in Chapter 6. As an alternative, we provide a late objects approach, where you can learn additional method details before you learn the nuts and bolts of OOP. Most late objects advocates like to learn not only about methods before OOP, but also how to program with arrays before OOP. If you like the late objects approach, here's what we suggest you do:

1. Read the supplemental chapter "Writing Methods in a Non-Object-Oriented Environment," which can be found at our book's resource center at http://www.mhhe.com/dean2e.
2. Read Sections 1–6 from Chapter 9 and then read the supplemental chapter "Arrays in a Non-Object-Oriented Environment," also at our book's resource center.
3. Read the rest of the book in the standard order, starting from the next chapter, Chapter 6. As part of its OOP presentation, Chapter 6 describes how to write methods that implement an object's behaviors. You should be able to cover that material quickly because it relies on many of the same concepts introduced in the "Writing Methods in a Non-Object-Oriented Environment" supplemental chapter. When you get to Chapter 9, you'll want to skip Sections 1–6 because you will have read them earlier.

Regardless of which approach you decide to take, we encourage you to read the rest of this interlude. If you decide to go with the late objects approach, the upcoming RollDice program will serve as a teaser for what you'll see in the first supplemental chapter. If you decide to go with the standard approach, and dive into Chapter 6 right after this interlude, the RollDice program will be particularly helpful. Many real-world programs (mostly programs written in other languages, but sometimes Java programs as well) do not use the OOP paradigm. Instead, they use the *procedural programming* paradigm, in which the emphasis is on the procedures or tasks that make up a program, not the objects. The following RollDice program will serve as an example for those types of programs.

Take a look at the RollDice program shown in Figures I.1a and I.1b (the I's in the figure numbers stand for "Interlude"). The RollDice program simulates rolling two dice and determines whether "doubles" are

```
/*******************************************************************
* RollDice.java
* Dean & Dean
*
* This program simulates rolling dice until the user rolls doubles.
*******************************************************************/

import java.util.Scanner;

public class RollDice
{
  public static void main(String args[])
  {
    Scanner stdIn = new Scanner(System.in);
    int die1, die2; // values of two dice

    System.out.println("Can you roll doubles?");

    do
    {
      System.out.print("Press enter to roll the dice:");
      stdIn.nextLine();
      die1 = rollDie();
      die2 = rollDie();
      printResult(die1, die2);
    } while (die1 != die2);
  } // end main

  //***************************************************************

  // This method returns the value of a randomly rolled die.

  public static int rollDie()
  {
    return (int) (Math.random() * 6 + 1);
  } // end rollDie
```


Figure I.1a RollDice program–part A

```
  //***************************************************************

  // Print the dice value and whether doubles were rolled.

  public static void printResult(int die1, int die2)
  {
    if (die1 == die2)
    {
      System.out.printf("Doubles! You rolled two %d's." +
        " Thank you for playing!\n", die1);
    }
    else
    {
      System.out.printf("No doubles. You rolled a %d and a %d." +
        " Try again.\n", die1, die2);
    }
  } // end printResult
} // end class RollDice
```


Figure I.1b RollDice program–part B

rolled. Doubles means that the two die values are the same. So long as the die values are different, the program prompts the user to roll again.

Within the RollDice program, three methods are defined—`main`, `rollDie`, and `printResult`. The entire program could have been written with just a `main` method and no other methods. That would work OK, but, as explained more fully in the upcoming chapters, whenever you have a nontrivial method, you should look for well-defined subtasks within the method. If you find such subtasks, you should consider implementing them as separate methods. In writing the RollDice program, we realized that rolling a die was a well-defined subtask, so we created a `rollDie` method. Likewise, we realized that printing the result of two dice being rolled was another well-defined subtask, so we implemented a `printResult` method. By implementing the subtasks with their own methods, we were able to keep the `main` method fairly short, and, generally speaking, short code is easier to understand than long code. Study RollDice's `main` method in Figure I.1a. It's fairly short, and we hope that it's relatively easy to understand. Inside the loop, the program rolls each die by calling the `rollDie` method twice. The program then prints the results by calling the `printResult` method.

In the upcoming chapters, you'll learn lots of details about how to implement methods, but for now, as an introduction, let's examine the `rollDie` method briefly. Here is the single statement that forms `rollDie`'s entire body:

```
return (int) (Math.random() * 6 + 1);
```

`Math.random()` generates a random floating-point number between 0.0 and 1.0, not including 1.0. Multiplying the random number by 6 generates a number between 0.0 and slightly less than 6.0. Adding 1 generates a number between 1.0 and slightly less than 7.0. Applying the `(int)` cast operator generates an integer between 1 and 6. Finally, at the left of the statement, the Java reserved word `return` causes the generated value (an integer between 1 and 6) to be passed back to the place where the method was called from. Looking at the RollDice program, you can see that `main` calls `rollDie` with these two statements:

```
die1 = rollDie();
die2 = rollDie();
```

When those two statements are executed, the JVM assigns `rollDie`'s returned values to `die1` and `die2`.

Again briefly, let's examine the `printResult` method. As you can see in the RollDice program and copied below for your convenience, the `printResult` method heading declares two variables—`die1` and `die2`:

```
public static void printResult(int die1, int die2)
```

Such method heading variables are called *parameters.* They store the arguments that are passed to them from their associated method calls. In the RollDice program, you can see that `main` calls `printResult` with this statement:

```
printResult(die1, die2);
```

That method call passes its `die1` and `die2` arguments to the `die1` and `die2` parameters in the `printResult` heading. The `printResult` method then compares the `die1` and `die2` values for equality and prints a "Doubles!" or "No doubles." message, accordingly.

There's something else noteworthy about the `printResult` method. Do you see the word `static` in the method's heading? As explained in Chapter 3, if you don't use static, then you'll be required to do some extra work before you can call the `printResult` method. Specifically, you'll be required to first instantiate an object. We'll describe instantiation in Chapter 6. On the other hand, if you're writing a non-OOP program, then you won't be instantiating objects, and you'll need to use `static` for each method heading. Thus, in the non-OOP RollDice program, `static` appears in the `printResult` method heading and also in the `rollDie` method heading.

That's it for your brief introduction to writing multiple methods in a non-object-oriented environment. If you prefer the late objects approach, you should now jump to the first supplemental chapter on the book's resource center and learn more about writing methods in a non-object-oriented environment. On the other hand, if you prefer the standard approach, you should move on to the next chapter, Chapter 6, and learn how to write methods in the context of an object-oriented programming environment. Either way, you'll get to the same final place. Now go forth and enjoy the journey!

CHAPTER 6

Object-Oriented Programming

Objectives

- Learn what an object is and how it relates to a class.
- Learn how to encapsulate and access data inside an object.
- Learn how to partition your programs into "driver" and "driven" classes, to create an object of the driven class, and to give the driver a reference to that object.
- Learn the differences between an object's data and data that is local to a method, and learn how to distinguish between those pieces of data when both have the same name.
- Understand implicit initialization (default values) of various kinds of variables.
- Learn how to trace an object-oriented program.
- Learn how to use a UML class diagram.
- Learn how to make a method return a suitable value.
- Learn how values are passed to methods.
- Write methods that get, set, and test the values of an object's data.
- Optionally learn how to improve the speed and accuracy of a simulation.

Outline

6.1 Introduction

As discussed in the Preface, we've written the book with some built-in flexibility in terms of content ordering. Readers who want an early introduction to object-oriented programming (OOP) have the option of reading Sections 6.1 through 6.8 after completing Chapter 3.

Chapter 5 served as a bridge from basic programming language constructs (variables, assignments, operators, `if` statements, loops, etc.) to OOP concepts. We focused primarily on one important aspect of OOP—learning how to use prebuilt methods. You used methods associated with an object, like `substring` and `indexOf` for string objects, and you used methods associated with a class, like `abs` and `pow` from the `Math` class. In this chapter, you'll learn how to do more than just use prebuilt classes and methods; you'll learn how to write your own classes and methods.

As you'll come to see, OOP makes large programs easier to work with. And making large programs easier to work with is very important because today's computers use lots of very large programs! The tension in learning OOP is that the first OOP programs a student can understand are necessarily small, and they can't show the power of OOP very well. But hang in there. Think of your study of this chapter and most of the next chapter as an investment. By the end of the next chapter, you'll be getting some return on that investment.

In this chapter, we start with an overview of basic OOP terms and concepts. We then step through the design and implementation of a simple OOP program. Typically, OOP design starts with a simple Unified Modeling Language (UML) class diagram, which provides a high-level, pictorial description of what you want the program to model. Then OOP design proceeds to the program's details. We'll show you how to adapt the previously described tracing technique to an OOP environment. We'll show you how to specify method details. In the previous chapter you looked at methods from the outside. Now you'll be looking at methods from the inside.

We end the chapter with an optional problem-solving section that introduces you to an important computer application—computer simulation. Computer simulation allows humans to solve problems that are difficult or impossible to solve by hand. We describe a special strategy that enables you to improve substantially both the accuracy and efficiency of computer simulations.

6.2 Object-Oriented Programming Overview

Readers who want a very early OOP overview have the option of reading this section after completing Chapter 1, Section 1.3 (Program Development).

Before OOP, the standard programming technique was *procedural programming.* Procedural programming is so named because the emphasis is on the procedures or tasks that make up a problem solution. You think first about what you want to do—your procedures. In contrast, the OOP programming paradigm invites you to think about what you want the program to represent. You typically respond to this invitation by identifying some things in the world that you want your program to model. Those things might be physical entities or conceptual entities. Once you have identified the things you want to model, you identify their basic properties/attributes. Then you determine what the things can do (their behaviors) or what the things can have done to them. You group each thing's properties and behaviors together into a coherent structure called an object. In writing an OOP program, you define objects, create them, and have them interact with each other.

Objects

An object is:

  a set of related data which identifies the current *state* of the object
\+ a set of *behaviors*

An object's state refers to the characteristics that currently define the object. For example, if you're writing a program that keeps track of employee salaries, you'd probably want to have employee objects, where an employee object's state consists of the employee's name and current salary.

An object's behaviors refer to the activities associated with the object. Once again, if you're writing a program that keeps track of employee salaries, you'd probably want to define a behavior that adjusts an employee's salary. That type of behavior parallels a real-world behavior—a pay raise or a pay cut. In Java, you implement an object's behaviors as methods. For example, you'd implement the salary adjustment behavior as an `adjustSalary` method. We'll describe method implementation details shortly. But it's important to complete our OOP overview first.

Here are some entities that would make good candidates for objects in an object-oriented program:

Physical Objects	Human Objects	Mathematical Objects
cars in a traffic-flow simulation	employees	points in a coordinate system
aircraft in an air-traffic control system	customers	complex numbers
electrical components in a circuit-design program	students	time

Let's think about the first example object. If a car is considered to be an object in a traffic-flow-simulation program, what is the data stored in each car object? In order to analyze traffic flow, each car's position and speed should be monitored. Therefore, those two pieces of data should be stored as part of a car object's state. And what behaviors are associated with the car objects? You'd need to be able to start the car, stop the car, slow down, and so on. So you'd probably want to implement these methods:

```
start, stop, slowDown
```

An object's behaviors can change an object's state. For example, a car object's `start` method causes the car's position and speed data items to change.

Encapsulation

Objects provide *encapsulation*. In general terms, encapsulation is when something is wrapped up inside a protective covering. When applied to objects, encapsulation means that an object's data are protected by being "hidden" inside the object. With hidden data, how can the rest of the program access an object's data? (*Accessing* an object's data refers to either reading the data or modifying it.) The rest of the program cannot access an object's data directly, but it can access the data with the help of the object's methods. Assuming that an object's methods are well written, the methods ensure that data is accessed in an appropriate manner. Returning to the employee-salaries program example, an employee object's salary should be modified only by calling the `adjustSalary` method. The `adjustSalary` method ensures that an employee object's salary is modified appropriately. For example, the `adjustSalary` method prevents an employee object's salary from becoming negative.

See Figure 6.1. It illustrates how an object's methods form the interface between an object's data and the rest of the program.

Benefits of OOP

Now that you have a basic idea of what OOP is, you may be asking yourself what all the hype is about. Why is OOP preferred over procedural programming for most of today's new programs? Here are some benefits of OOP:

- OOP programs have a more natural organization.
  Since people tend to think about real-world problems in terms of real-world objects, it's easier for people to understand a program that's organized around objects.

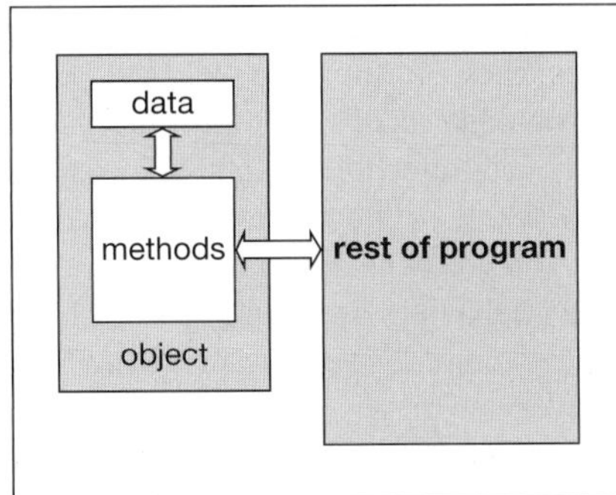

Figure 6.1 To access an object's data, you should use the object's methods as an interface

- OOP makes it easier to develop and maintain large programs.
  Although switching to OOP programming typically makes a small program more complicated, it naturally partitions things so that the program grows gracefully and does not evolve into a giant mess. Since objects provide encapsulation, bugs (errors) and bug repairs tend to be localized.

The second bullet item needs some clarification. When an object's data can be modified only by using one of that object's methods, it's hard for a programmer to mess up an object's data accidentally. Returning again to the employee-salaries program example, assume the only way to change an employee object's salary is to use its `adjustSalary` method. Then, if there's a bug relating to an employee's salary, the programmer immediately knows where to look for the problem—in the `adjustSalary` method or in one of the calls to the `adjustSalary` method.

Classes

Having discussed objects, it's now time to talk about an intimately related entity—a *class*. We'll start with a broad definition of a class, and we'll refine it later. Broadly speaking, a class is a description of all the objects it defines. As such, it is an *abstraction*—a concept apart from any particular instances. In Figure 6.2, note the three computers on a conveyor belt in a manufacturing plant. The three computers represent objects. The specifications document that hovers above the computers is a blueprint that describes the computers: it lists the computers' components and describes the computers' features. The computer-specification document represents a class. Each object is an instance of its class. Thus, for practical purposes, "object" and "instance" are synonyms.

One class can have any number of objects associated with it. A class can even have zero objects associated with it. This should make sense if you think about the computer-manufacturing example. Isn't it possible to have a blueprint for a computer, but not yet have any computers manufactured from that blueprint?

We'll now present a more complete description of a class. Above, we said that a class is a description for a set of objects. The description consists of:

    a list of variables
+   a list of methods

Classes can define two types of variables—*class variables* and *instance variables*. And classes can define two types of methods—*class methods* and *instance methods*. Chapter 5 showed you how to use the `Math` class's class methods, and you have been defining a class method called `main` since the beginning. In Chapter 7, we'll show you when it's appropriate to define other class methods and define and use class variables. But it's easy to fall into the trap of defining and using class methods and class variables improperly.

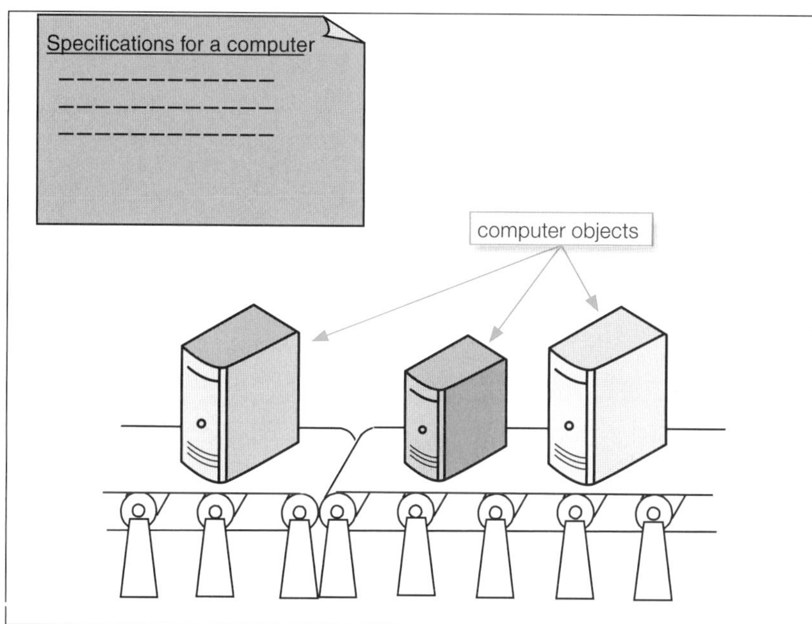

Figure 6.2 Conveyor belt portrayal of the class-objects relationship

We want to keep you away from that trap until after you have developed good OOP habits. Therefore, we focus on instance variables and instance methods throughout this chapter and the next several chapters.

A class's instance variables specify the type of data that an object can store. For example, if you have a class for computer objects, and the `Computer` class contains a `hardDiskSize` instance variable, then each computer object stores a value for the size of the computer's hard disk. A class's instance methods specify the behavior that an object can exhibit. For example, if you have a class for computer objects, and the `Computer` class contains a `printSpecifications` instance method, then each computer object can print a specifications report (the specifications report shows the computer's hard disk size, CPU speed, cost, etc.).

Note the use of the term "instance" in "instance variable" and "instance method." That reinforces the fact that instance variables and instance methods are associated with a particular object instance. For example, each `employee` object would have its own value for a `salary` instance variable, which would be accessed through its `adjustSalary` instance method. That contrasts with class methods. Class methods are associated with an entire class. For example, the `Math` class contains the `round` class method, which is not associated with a particular instance of the `Math` class.

6.3 First OOP Class

In the next several sections, we put what you've learned into practice by implementing a complete OOP program. The program will contain a `Mouse` class, and it will simulate the growth of two `Mouse` objects (we're talking about rodents here, not computer pointing devices). As is customary with OOP programs, we start the implementation process by describing the solution pictorially with a *UML class diagram*. A UML class diagram is a diagrammatic technique for describing classes, objects, and the relationships between them. It is widely accepted in the software industry as a standard for modeling OOP designs. After describing our mouse-simulation solution with a UML class diagram, we will present the Mouse program's source code and walk you through it.

Use UML to specify OOP. 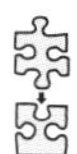

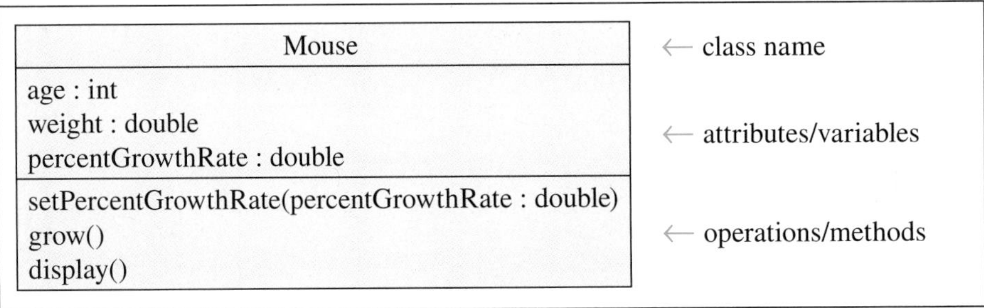

Figure 6.3 Abbreviated UML class diagram for a `Mouse` class

UML Class Diagram

See Figure 6.3. It contains an abbreviated UML class diagram for a `Mouse` class. A UML class diagram box is divided into three parts—class name at the top, *attributes* in the middle, and *operations* at the bottom. With Java programs, attributes equate to variables and operations equate to methods. Henceforth, we'll use the Java terms, *variables* and *methods,* rather than the formal UML terms, *attributes* and *operations*. Collectively, we refer to a class's variables and methods as the class's *members*. Let's now describe each `Mouse` member.

The `Mouse` class has three instance variables—`age`, `weight`, and `percentGrowthRate`. The `age` instance variable keeps track of how old a `Mouse` object is, in days. The `weight` instance variable keeps track of a `Mouse` object's weight, in grams. The `percentGrowthRate` instance variable is the percentage of its current weight that gets added to its weight each day. If the `percentGrowthRate` is 10 percent and the mouse's current weight is 10 grams, then the mouse gains 1 gram by the next day.

The `Mouse` class has three instance methods—`setPercentGrowthRate`, `grow`, and `display`. The `setPercentGrowthRate` method assigns a specified value to the `percentGrowthRate` instance variable. The `grow` method simulates one day of weight gain for a mouse. The `display` method prints a mouse's age and weight.

Referring to Figure 6.3, note how we specify variable types in a class diagram. The type appears at the right of the variable (e.g., age : int). That's opposite from Java declarations, where we write the type at the left of the variable (e.g., `int age;`)

Start documenting early.

Some programmers use UML class diagrams as a means to document programs after they've already been written. That's OK, but it's not how class diagrams were originally intended to be used. We encourage you to start drawing class diagrams as a first step in your solution implementation. The class diagram details provide an outline for your program. Depending on the complexity of the program and your affinity for pseudocode, you may want to code the methods directly with Java or you may want to code the methods first, with pseudocode as an intermediate step. For our `Mouse` example, the `Mouse` class's methods are straightforward, so we'll code them directly with Java. Let's now take a look at the `Mouse` class's Java source code.

Mouse Class Source Code

Figure 6.4 shows the `Mouse` class implemented with Java. Note the `Mouse` class's three instance variable declarations for `age`, `weight`, and `percentGrowthRate`. Instance variables must be declared outside all methods, and to make your code more self-documenting, you should declare them all at the beginning of the class definition. Instance variable declarations are very similar to variable declarations you've seen in the past: The variable's type goes at the left of the variable, and you can optionally assign an initial value to the variable. Do you remember what it's called when you assign a value to a variable as part of a declaration?

```
/*****************************************************************
* Mouse.java
* Dean & Dean
*
* This class models a mouse for a growth simulation program.
*****************************************************************/

public class Mouse
{
  private int age = 0;                    // age of mouse in days
  private double weight = 1.0;            // mouse weight in grams
  private double percentGrowthRate;       // increase per day

  //*************************************************************

  // This method assigns the mouse's percent growth rate.

  public void setPercentGrowthRate(double percentGrowthRate)
  {
    this.percentGrowthRate = percentGrowthRate;
  } // end setPercentGrowthRate

  //*************************************************************

  // This method simulates one day of growth for the mouse.

  public void grow()
  {
    this.weight +=
      (.01 * this.percentGrowthRate * this.weight);
    this.age++;
  } // end grow

  //*************************************************************

  // This method prints the mouse's age and weight.

  public void display()
  {
    System.out.printf("Age = %d, weight = %.3f\n",
      this.age, this.weight);
  } // end display
} // end class Mouse
```


Figure 6.4 Mouse class

That's called an *initialization*. Note the initializations for age and weight. We initialize age to 0 because newborn mice are zero days old. We initialize weight to 1 because newborn mice weigh approximately 1 gram.

The primary difference between instance variable declarations and variable declarations you've seen in the past is the private access modifier. If you declare a member to be private, then the member can be accessed only from within the member's class and not from the "outside world" (i.e., by code that's outside of the class in which the member resides). Instance variables are almost always declared with the private access modifier because you almost always want an object's data to be hidden. Making an instance variable private gives you control over how its value can be changed. For example, you could assure that a weight is never made negative. Constraining data access is what encapsulation is all about, and it's one of the cornerstones of OOP.

In addition to the private access modifier, there's also a public access modifier. Given the standard definitions of the words "public" and "private," you can probably surmise that public members are easier to access than private members. If you declare a member to be public, then the member can be accessed from anywhere (from within the member's class, and also from outside the member's class). You should declare a method to be public when you want it to be a portal through which the outside world accesses your objects' data. Go back and verify that all three methods in the Mouse class use the public access modifier. When you want a method to help perform a local task only, you should declare it to be private, but we'll delay that consideration until Chapter 8.

Look once again at the Mouse class's instance variable declarations. Note that we initialize age and weight to 0 and 1.0, respectively, but we don't initialize percentGrowthRate. That's because we're comfortable with age = 0 and weight = 1.0 for all newborn Mouse objects, but we're not comfortable with a predefined initial value for percentGrowthRate. Presumably, we'll want to use different percentGrowthRate values for different Mouse objects (mice in a doughnut-eating study might have higher percentGrowthRate values than mice in a cigarette-smoking study).

With no initialization for the percentGrowthRate instance variable, how can you set the growth rate for a Mouse object? You can have the Mouse object call the setPercentGrowthRate method with a growth rate value as an argument. For example, here's how a Mouse object can set its growth rate to 10 (percent):

```
setPercentGrowthRate(10);
```

As you may recall from Chapter 5, a method call's parenthetical values are referred to as *arguments*. Thus, in this example, 10 is an argument. The 10 gets passed into the percentGrowthRate variable in setPercentGrowthRate's heading. A method heading's parenthetical variables are referred to as *parameters*. Thus, in the example shown in Figure 6.4, percentGrowthRate is a parameter. Within the setPercentGrowthRate *method body* (the code between the method's opening and closing braces), the percentGrowthRate parameter is assigned into the percentGrowthRate instance variable. Here's the relevant assignment statement:

```
this.percentGrowthRate = percentGrowthRate;
```

Note the "this dot" in this.percentGrowthRate. The this dot is how you tell the Java compiler that the variable you're referring to is an instance variable. Because the percentGrowthRate variable at the right does not have this dot, the Java compiler knows that that percentGrowthRate refers to the percentGrowthRate parameter, not the percentGrowthRate instance variable. In Figure 6.4's setPercentGrowthRate method, the instance variable and the parameter have the same name. That's a common practice. There's no problem distinguishing between the two variables because the instance variable uses this dot and the parameter does not.

Now, take a look at the Mouse class's display and grow methods. The display method is straightforward; it prints a mouse's age and weight. The grow method simulates one day of weight gain for a mouse. The weight-gain formula adds a certain percentage of the current weight to the current weight. That means that the mouse will continue to grow every day of its life. That's a simple, but not very accurate, portrayal of normal weight gain. We've intentionally kept the weight-gain formula simple in order to avoid getting bogged down in complicated math. In the final section of this chapter, we provide more realistic growth models.

Finally, take a look at the Mouse class's comments. Note the descriptions above each method. Proper style suggests that, above each method, you should have a blank line, a line of asterisks, a blank line, a description of the method, and another blank line. The blank lines and asterisks serve to separate the methods. The method descriptions allow someone who's reading your program to quickly get an idea of what's going on.

6.4 Driver Class

What Is a Driver?

Driver is a common computer term that applies to a piece of software that runs or "drives" something else. For example, a printer driver is a program that is in charge of running a printer. Likewise, a *driver class* is a class that is in charge of running another class.

In Figure 6.5, we present a MouseDriver class. We name the class MouseDriver because it is in charge of driving the Mouse class. We say that the MouseDriver class drives the Mouse class because it creates Mouse objects and then manipulates them. For example, note the gus = new Mouse() and the jaq = new Mouse() statements. That code creates Mouse objects gus and jaq.[1] In addition, note the gus.setPercentGrowthRate(growthRate) code. That code manipulates the gus object by updating gus's percentGrowthRate value.

Normally, a driver class consists entirely of a main method and nothing else. The driver class, with its main method, is the starting point for the program. It calls upon the driven class to create objects and manipulate them. The driven class dutifully carries out the object creation and object manipulation requests. Normally, carrying out those tasks is the primary focus of the program, and their implementation requires the majority of the program's code. Thus, driven classes are typically (but not always) longer than driver classes.

Driver classes, such as the MouseDriver class, are in separate files from the classes that they drive. To make them accessible from the outside world, driver classes must be public. Each public class must be stored in a separate file whose name is the same as the class name, so the MouseDriver class must be stored in a file named MouseDriver.java. For MouseDriver's code to find the Mouse class, both classes should be in the same directory.[2]

Reference Variables

In the MouseDriver class, we create Mouse objects, and we refer to those Mouse objects using gus and jaq, where gus and jaq are *reference variables*. The value contained in a reference variable is a "reference" to an object (thus the name reference variable). More precisely, a reference variable holds the address of where an object is stored in memory. For a pictorial explanation, see Figure 6.6. In the figure, the little

[1]Gus and Jaq are mice in the Disney classic *Cinderella*.

[2]We're keeping things simple by telling you to put both classes in the same directory. Actually, the files may be in different directories, but then you'd need to use a package to group together your classes. Appendix 4 describes how to group classes into a package.

```
/*******************************************
* MouseDriver.java
* Dean & Dean
*
* This is a driver for the Mouse class.
*******************************************/

import java.util.Scanner;

public class MouseDriver
{
  public static void main(String[] args)
  {
    Scanner stdIn = new Scanner(System.in);
    double growthRate;
    Mouse gus = new Mouse();
    Mouse jaq = new Mouse();

    System.out.print("Enter % growth rate: ");
    growthRate = stdIn.nextDouble();
    gus.setPercentGrowthRate(growthRate);
    jaq.setPercentGrowthRate(growthRate);
    gus.grow();
    jaq.grow();
    gus.grow();
    gus.display();
    jaq.display();
  } // end main
} // end class MouseDriver
```


Figure 6.5 `MouseDriver` class that drives the `Mouse` class in Figure 6.4

boxes immediately to the right of `gus` and `jaq` represent addresses. So `gus`'s little box holds the address of the first object.

Industry OOP Vernacular

Most Java programmers in industry don't use the term reference variable. Instead, they just use the term object. This blurs the distinction between reference variables and objects. For example, in the `MouseDriver` class in Figure 6.5, this statement initializes the `gus` reference variable:

```
Mouse gus = new Mouse();
```

Even though it's a reference variable, most industry Java programmers would refer to `gus` as an object. Despite the common practice of using the word "object" as a substitute for "reference variable," it's important to know the difference—an object holds a group of data, and a reference variable holds the location where that group of data is stored in memory. Understanding the difference between an object and a reference variable will help you to understand the behavior of Java code.

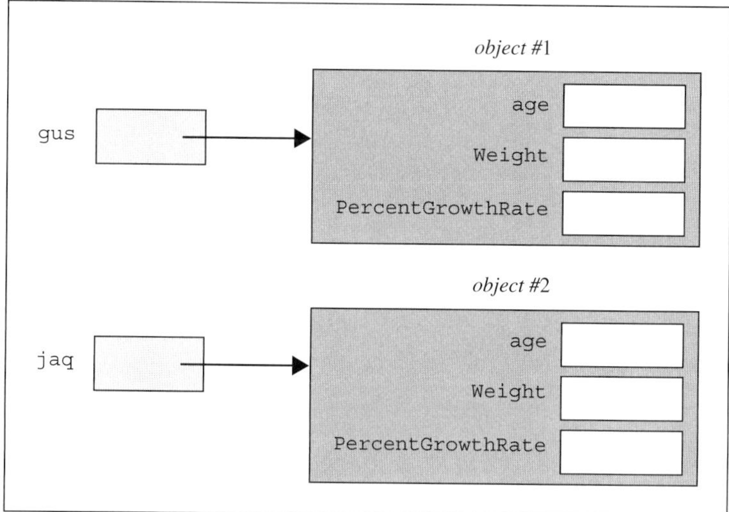

Figure 6.6 Reference variables and objects for the Mouse program in Figures 6.4 and 6.5
The two reference variables on the left, `gus` and `jaq`, contain references that point to the two objects on the right.

Declaring a Reference Variable

You must always declare a variable before you can use it. For example, in order to use an `int` variable named `count`, you must first declare `count` like this:

```
int count;
```

Likewise, in order to use a `gus` reference variable, you must first declare `gus` like this:

```
Mouse gus;
```

As you can see, the process for declaring reference variables mirrors the process for declaring primitive variables. The only difference is that instead of writing a primitive type on the left (e.g., `int`), for reference variables, you write a class name on the left (e.g., `Mouse`).

Instantiation and Assigning a Value to a Reference Variable

As you know, the point of a reference variable is to store a reference to an object. But before you can store a reference to an object, you have to have an object. So let's look at object creation.

To create an object, use the `new` operator. For example, to create a `Mouse` object, specify `new Mouse()`. The `new` operator should make sense when you realize that `new Mouse()` creates a new object. The formal term for creating an object is *instantiating* an object. So `new Mouse()` instantiates an object. The term *instantiate* is a verbalized form of the noun "instance." It is computer jargon for "make an instance of a class" or "create an object."

After instantiating an object, you'll normally assign it to a reference variable. For example, to assign a `Mouse` object to the `gus` reference variable, do this:

```
gus = new Mouse();
```

After the assignment, `gus` holds a reference to the newly created `Mouse` object.

Let's review. Here's how we declared a gus reference variable, instantiated a Mouse object, and assigned the object's address to gus:

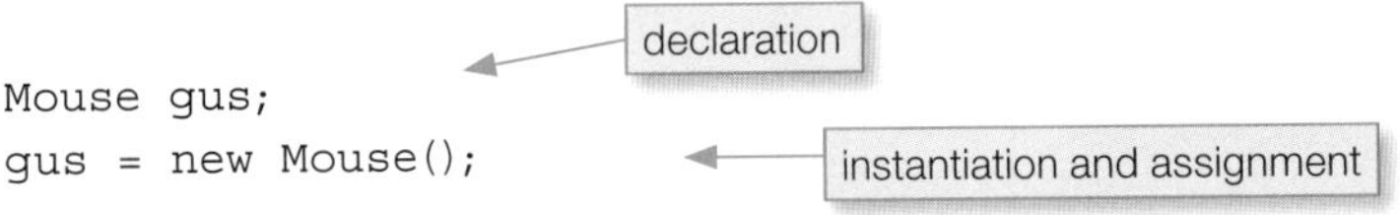


```
Mouse gus;
gus = new Mouse();
```

Now here's how to do the same thing with only one statement:

```
Mouse gus = new Mouse();
```

initialization

The above statement is what appears in Figure 6.5's MouseDriver class. It's an initialization. As mentioned previously, an initialization is when you declare a variable and assign it a value, all in one statement.

Calling a Method

After you instantiate an object and assign its reference to a reference variable, you can call/invoke an instance method using this syntax:

<reference-variable>.<method-name>(<comma-separated-arguments>);

Here are three example instance method calls from the MouseDriver class:

```
gus.setPercentGrowthRate(growthRate);
gus.grow();
gus.display();
```

Note how the three method calls mimic the syntax template. The first method call has one argument and the next two method calls have zero arguments. If we had a method with two parameters, we'd call it with two arguments separated by a comma.

When a program calls a method, it passes control from the calling statement to the first executable statement in the called method. For example, when the MouseDriver's main method calls the setPercentGrowthRate method with gus.setPercentGrowthRate(growthRate), control passes to this statement in the Mouse class's setPercentGrowthRate method:

```
this.percentGrowthRate = percentGrowthRate;
```

Go back to Figure 6.4's Mouse class and verify that the setPercentGrowthRate method contains the above statement.

After the last statement in any called method executes, control returns to the calling method at the point just after where the call was made. For a pictorial explanation, see Figure 6.7.

6.5 Calling Object, this Reference

Suppose you have two objects that are instances of the same class. For example, gus and jaq refer to two objects that are instances of the Mouse class. And suppose you want the two objects to call the same instance method. For example, you want both gus and jaq to call setPercentGrowthRate. For each method call, the Java virtual machine (JVM) needs to know which object to update (if gus calls setPercentGrowthRate, then the JVM should update gus's percentGrowthRate; if jaq calls

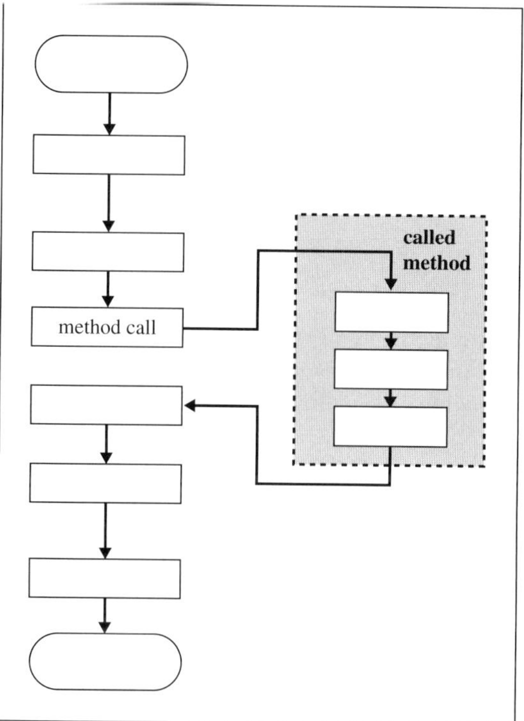

Figure 6.7 Calling a method

setPercentGrowthRate, then the JVM should update jaq's percentGrowthRate). This section describes how the JVM knows which object to update

Calling Object

As mentioned in Chapter 5, whenever an instance method is called, it is associated with a calling object. You can identify the calling object by looking to the left of the dot in an instance method call statement. Can you identify the calling objects in the following main method?

```
public static void main(String[] args)
{
  Scanner stdIn = new Scanner(System.in);
  double growthRate;
  Mouse gus = new Mouse();

  System.out.print("Enter % growth rate: ");
  growthRate = stdIn.nextDouble();
  gus.setPercentGrowthRate(growthRate);
  gus.grow();
  gus.display();
} // end main
```

The `gus` object is the calling object for these statements:

```
gus.setPercentGrowthRate(growthRate);
gus.grow();
gus.display();
```

Are there any other calling objects? Yes. The `stdIn` object is a calling object in this statement:

```
growthRate = stdIn.nextDouble();
```

The `this` Reference

It's easy to identify the calling object when you're looking at a method call statement. But what if you're inside the called method—how can you tell what object called the method? For example, when you're looking at the definition of the `Mouse` class in Figure 6.4, can you identify the calling object that called its `grow` method? Here is that method again:

```
public void grow()
{
  this.weight +=
    (0.01 * this.percentGrowthRate * this.weight);
  this.age++;
} // end grow
```

The pronoun `this` (called the *`this` reference*) stands for the calling object, but it doesn't tell you which object that is. Thus, you can't tell what the calling object is just by looking at the method that was called. You must look at what called that method. If the statement that called `grow` was `gus.grow()`, then `gus` is the calling object. Alternately, if the statement that called `grow` was `jaq.grow()`, then `jaq` is the calling object. As you'll see when we do the upcoming trace, you must know which object, `gus` or `jaq`, is the current calling object so that you update the proper object. Within the above `grow` method, note `this.weight` and `this.age`. The `this` reference reminds you that `weight` and `age` are instance variables. Instance variables in which object? In the calling object!

The `setPercentGrowthRate` method in Figure 6.4 provides another example. Here is that method again:

```
public void setPercentGrowthRate(double percentGrowthRate)
{
  this.percentGrowthRate = percentGrowthRate;
} // end setPercentGrowthRate
```

The `this` reference tells you the variable on the left side of this method's lone statement is an instance variable in the calling object. As indicated earlier, the `this` reference in this statement also helps the compiler and a human distinguish the variable on the left side from the variable on the right side. Before the advent of OOP, computer languages did not include `this` dot functionality. Then, the only way the compiler and a human could distinguish between variables in different places that referred to essentially the same thing was to give them similar but slightly different names.

The *ad hoc* (special case) nature of how old-time programmers devised slightly different names made programs confusing and increased programming errors. Java's `this` reference provides a standard way to make the distinction and show the relationship at the same time. You can use exactly the same name to show the relationship and then use `this` dot to make the distinction. So it is no longer necessary to use slightly different names for that purpose, and we recommend against that archaic practice.

To emphasize the meaning and utility of Java's `this` reference, we will use it with all examples of instance variables up to the end of the next chapter—even when it is not necessary to draw a distinction between an instance variable and a parameter. There is no performance penalty in using `this` dot, and it provides an immediate indicator to everyone that the variable is an instance variable. Thus, it helps to explain the program; that is, it provides useful self documentation.

6.6 Instance Variables

You've been exposed to instance variables for a while now. You know that an object stores its data in instance variables. You know that an instance method accesses its instance variables by prefacing them with the `this` reference (e.g., `this.weight`). In this section, we consider a few more instance variable details. Specifically, we consider default values and persistence.

Default Values for Instance Variables

As implied by the common definition of "default," a variable's *default value* is the variable's value when there's no explicitly assigned initial value; that is, when there's no initialization. Different types of variables have different default values.

There are two integer types that we've covered so far—`int` and `long`. Integer-type instance variables are assigned 0 by default. But in the `Mouse` class, notice that we initialize the `age` instance variable to 0:

```
private int age = 0; // age of mouse in days
```

Why bother with the explicit initialization? Wouldn't `age` be assigned 0 by default even if "`= 0`" was omitted? Yes, the program would work the same either way. But it's poor practice to depend on hidden default values. By explicitly assigning values to variables, we show our intent. That's a form of self-documenting code.

There are two floating-point types—`float` and `double`. Floating-point-type instance variables are assigned 0.0 by default. The `Mouse` class declares two floating-point instance variables—`weight` and `percentGrowthRate`:

```
private double weight = 1.0;       // weight of mouse in grams
private double percentGrowthRate; // % weight increase per day
```

In this case, we initialize the `weight` instance variable to 1.0, so the default value doesn't come into play. We do not initialize the `percentGrowthRate` value, so `percentGrowthRate` is initialized to 0.0 by default. Didn't we just say that it's poor practice to depend on hidden default values? Yes, but in this case, we're not depending on the default value. In the `MouseDriver` class, we overlay the `percentGrowthRate` default value with a custom value by calling `setPercentGrowthRate` like this:

```
gus.setPercentGrowthRate(growthRate);
```

By default, `boolean` instance variables are assigned `false`. For example, if you added a `boolean` instance variable named `vaccinated` to the `Mouse` class, `vaccinated` would be assigned `false` by default.

Reference-type instance variables are assigned `null` by default. For example, if you added a `String` instance variable named `breed` to the `Mouse` class, `breed` would be assigned `null` by default. Normally, a reference variable holds the address of an object and that address points to an object. The Java designers added `null` to the language as a way to indicate that a reference variable points to nothing. So the default for a reference-type instance variable is to point to nothing.

Here's a summary of default values for instance variables:

Instance Variable's Type	Default Value
integer	0
floating point	0.0
`boolean`	`false`
reference	`null`

Instance Variable Persistence

Now consider variable *persistence*. Persistence refers to how long a variable's value survives before it's wiped out. Instance variables persist for the duration of a particular object. Thus, if an object makes two method calls, the second called method does not reset the calling object's instance variables to their initialized values. Instead, the object's instance variables retain their values from one method call to the next. For example, in the `MouseDriver` class, `gus` calls `grow` twice. In the first call to `grow`, `gus`'s `age` increments from 0 to 1. In the second call to `grow`, `gus`'s `age` starts out as 1 and increments to 2. `gus`'s `age` retains its value from one `grow` call to the next because `age` is an instance variable.

6.7 Tracing an OOP Program

To reinforce what you've learned so far in this chapter, we'll trace the Mouse program. Remember the tracing procedure we used in prior chapters? It worked fine for programs with only one method—the `main` method. But for OOP programs with multiple classes and multiple methods, you'll need to keep track of which class and which method you're in and which object called that method. In addition, you'll need to keep track of parameters and instance variables. This requires a more elaborate trace table.

In tracing the Mouse program, we'll use a slightly different driver, the `MouseDriver2` class, shown in Figure 6.8. In `MouseDriver2`, we delay the instantiation of the individual mice and assign their growth rates (by calling `setPercentGrowthRate`) immediately after each instantiation. This is better style, because it more closely associates each object's instantiation with its growth rate assignment. However, in changing the driver, we "accidentally" forget to call `setPercentGrowthRate` for `jaq`, the second mouse. You can see the effect of this logic error in the output—`jaq` doesn't grow (after the first day, `jaq` still weighs 1 gram). But let's pretend that you don't know why this error occurs and use the trace to help find its cause. Remember—tracing is an effective tool when you need help debugging a program.

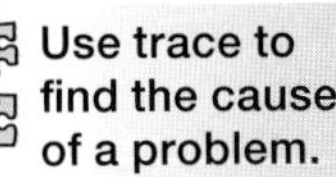
Use trace to find the cause of a problem.

To perform the trace, in addition to the driver, you'll also need the code for the driven class. For your convenience, we repeat the original driven `Mouse` class in Figure 6.9.

Trace Setup

Figure 6.10 shows the setup. As with the traces in the previous chapters, the input goes in the top-left corner. Unlike the traces in the previous chapters, the headings under the input now require more than one line. The first line of headings shows the class names—`MouseDriver2` and `Mouse`. Under each class name heading is a heading for each of the class's methods. In the trace setup, find the `setPercentGrowthRate`, `grow`, and `display` method headings (to save space, we abbreviated `setPercentGrowthRate` and `display` to setPGR and disp, respectively). And under each method-name heading, there's a heading for each of the method's local variables and parameters.

```
/*******************************************************
* MouseDriver2.java
* Dean & Dean
*
* This is a driver for the Mouse class.
*******************************************************/

import java.util.Scanner;

public class MouseDriver2
{
  public static void main(String[] args)
  {
    Scanner stdIn = new Scanner(System.in);
    double growthRate;
    Mouse gus, jaq;

    System.out.print("Enter % growth rate: ");
    growthRate = stdIn.nextDouble();
    gus = new Mouse();
    gus.setPercentGrowthRate(growthRate);
    gus.grow();
    gus.display();
    jaq = new Mouse();
    jaq.grow();
    jaq.display();
  } // end main
} // end class MouseDriver2
```

This declares reference variables but does not initialize them.

Try to group initialization activities.

There's a logic error here. We "accidentally" forget to initialize the growth rate in `jaq`.

Sample session:

```
Enter % growth rate: 10
Age = 1, weight = 1.100
Age = 1, weight = 1.000
```

`jaq` doesn't grow. A bug!

Figure 6.8 `MouseDriver2` class that drives the `Mouse` class in Figure 6.9

We'll discuss *local variables* in detail later, but for now, just realize that `growthRate` (abbreviated to rate in the trace setup), `gus`, and `jaq` are considered to be local variables because they're declared and used "locally" within one particular method, the `main` method. That's different from the `age`, `weight`, and `percentGrowthRate` instance variables, which are declared outside of all methods, at the top of the class. Note that `stdIn` is another local variable within `main`, but there's no need to trace it because it's instantiated from an application programming interface (API) class, `Scanner`. There's no need to trace API classes because they've already been traced and tested thoroughly by the developers of the Java language. You can assume that they work properly.

Now let's examine the trace setup's parameters. The `setPercentGrowthRate` method has two parameters—`percentGrowthRate`, abbreviated to rate in the trace setup, and the `this` reference,

```
/*************************************************************
* Mouse.java
* Dean & Dean
*
* This class models a mouse for a growth simulation program.
*************************************************************/

public class Mouse
{
  private int age = 0;                 // age of mouse in days
  private double weight = 1.0;         // mouse weight in grams
  private double percentGrowthRate;    // increase per day

  //**********************************************************

  // This method assigns the mouse's percent growth rate.

  public void setPercentGrowthRate(double percentGrowthRate)
  {
    this.percentGrowthRate = percentGrowthRate;
  } // end setPercentGrowthRate

  //**********************************************************

  // This method simulates one day of growth for the mouse.

  public void grow()
  {
    this.weight +=
      (.01 * this.percentGrowthRate * this.weight);
    this.age++;
  } // end grow

  //**********************************************************

  // This method prints the mouse's age and weight.

  public void display()
  {
    System.out.printf(
      "Age = %d, weight = %.3f\n", this.age, this.weight);
  } // end display
} // end class Mouse
```

Figure 6.9 Mouse class, repeated from Figure 6.4

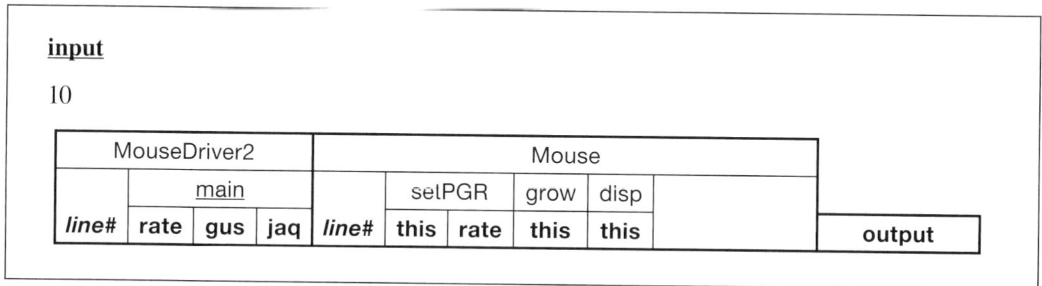

Figure 6.10 Trace setup for the Mouse program

an implicit parameter. As you may recall, the `this` reference points to the calling object. For the `setPercentGrowthRate`, `grow`, and `display` methods, we include a column for `this` so that the trace can keep track of which object called the method.

Note the vacant area under the `Mouse` heading. We'll fill in more headings there as we execute the trace.

Trace Execution

Using Figure 6.10's trace setup as a starting point, we'll walk you through the key sections of the trace shown in Figure 6.11. We'll focus on the OOP parts of the trace since those are the parts that are new to you. When starting a method, under the method's local variable headings, write initial values for each of the local variables. Use a question mark for local variables that are uninitialized. In the first three lines of Figure 6.11's trace, note the ?'s for the uninitialized `growthRate` (abbreviated to rate), `gus`, and `jaq` local variables.

When an object is instantiated, under the object's class-name heading, provide a column heading named "obj#", where # is a unique number. Under the obj# heading, provide an underlined column heading for each of the object's instance variables. Under the instance variable headings, write initial values for each of the instance variables. In Figure 6.11's trace, note the obj1 and obj2 column headings and their `age`, `weight`, and `percentGrowthRate` (abbreviated to rate) subheadings. Also note the initial values for the `age`, `weight`, and `percentGrowthRate` instance variables.

When there's an assignment into a reference variable, write obj# under the reference variable's column heading, where obj# matches up with the associated obj# in the object portion of the trace. For example, in Figure 6.11's trace, we created obj1 while tracing the `gus = new Mouse();` statement. Subsequently, we put obj1 under the `gus` column heading.

When there's a method call, under the called method's `this` column heading, write the calling object's obj#. In Figure 6.11's trace, note obj1 under `setPercentGrowthRate`'s `this` heading. If the method call contains an argument, write the argument's value under the called method's associated parameter. In the trace, note the passed-in 10 under the `setPercentGrowthRate`'s `percentGrowthRate` heading. Inside the method, if there's a `this` reference, find the obj# under the method's `this` column heading. Then go to the found obj#'s heading and read or update the obj#'s value accordingly. In Figure 6.9's `Mouse` class, note `this.percentGrowthRate` in the `setPercentGrowthRate` method body. In the trace, note that `setPercentGrowthRate`'s `this` reference refers to obj1, so obj1's `percentGrowthRate` is updated accordingly.

When you finish tracing a method, draw a horizontal line under the method's variable values to indicate the end of the method trace and to signify that the values in the method's local variables are wiped out. For example, in the trace, the heavy horizontal line in Mouse line #20 under setPGR indicates the end of the `setPercentGrowthRate` method, and it signifies that `percentGrowthRate`'s value is wiped out.

input

10

MouseDriver2				Mouse											
	main				setPGR		grow	disp	obj1			obj2			
line#	**rate**	**gus**	**jaq**	*line#*	**this**	**rate**	**this**	**this**	**age**	**wt**	**rate**	**age**	**wt**	**rate**	**output**
15	?														
16		?	?												
18															Enter % growth rate:
19	*10.0*														
20															
				10					0						
				11						1.000					
				12							0.0				
20		obj1													
21					obj1	10.0									
				20							10.0				
22							obj1								
				29						1.100					
				31					1						
23								obj1							
				40											Age = 1, weight = 1.100
24															
				10								0			
				11									1.000		
				12										0.0	
24			obj2												
25							obj2								
				29									1.000		
				31								1			
26								obj2							
				40											Age = 1, weight = 1.000

Figure 6.11 Completed trace for the Mouse program

Practice. Now that we've walked you through the new techniques for tracing an OOP program, we encourage you to go back to the trace setup in Figure 6.10 and do the entire trace on your own. Pay particular attention to what happens when `gus` and `jaq` call the `grow` method. Verify that `gus`'s weight increases (as it should) and `jaq`'s weight fails to increase (a bug). When you're done with the trace, compare your answer to Figure 6.11.

Experience with the long-form tracing used in this book will make it easier for you to understand what an automated debugger in an Integrated Development Environment (IDE) is telling you. As you step through a program that's running in debug mode under the control of an IDE debugger, when you get to a method call, you have two choices. You can "step into" and go through all the statements in the called method, like we do in Figure 6.11, or you can "step over" and just see what happens after the method returns. In a typical debugging activity, you will use a combination of stepping over and stepping in. For the example problem we have been considering, the sample session in Figure 6.8 tells you that the simulation is OK for the first object. The problem is with the second object. So, the appropriate thing to do is step over the method calls down through line 23 in the `MouseDriver2` class. Then, starting at line 24 in the `MouseDriver2` class, step into the methods calls to zero in on what caused the problem.

Paper trace emulates IDE debugger.

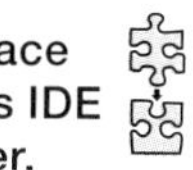

6.8 UML Class Diagrams

The `Mouse` class's `grow` method is not very flexible—it forces the driver to call the `grow` method separately for each day or to provide a `for` loop for each multiple-day simulation. It isn't good style to include such things in a driver. It's better to include multiple-day functionality within the driven class. In this section, we do just that. We present a revised mouse class with a `grow` method that handles any number of days, not just one day.

To specify a second-generation mouse class (`Mouse2`) and an associated driver class (`Mouse2Driver`), let's create another UML class diagram. The diagram we presented in Figure 6.3 was a pared-down UML class diagram. It did not include all the standard features. This time, in Figure 6.12, we present a UML class diagram that includes all the standard features, plus an extra feature.

Organize.

Figure 6.12's class diagram includes class diagram boxes for both classes—one diagram for the `Mouse2Driver` class and another diagram for the `Mouse2` class. The `Mouse2` class has the same three instance variables as the original `Mouse` class—`age`, `weight`, and `percentGrowthRate`. It also has the same `setPercentGrowthRate` method. But the `getAge` and `getWeight` methods are new, and

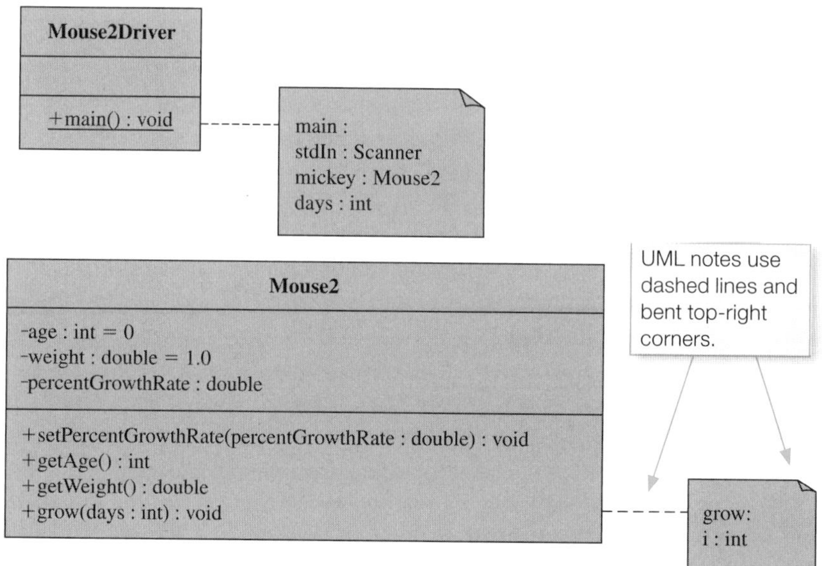

Figure 6.12 UML class diagram for a second-generation Mouse program

the `grow` method is improved. The `getAge` method retrieves a mouse's age. Remember that the `age` variable is `private`, so the only way for the outside world to read a mouse object's `age` is to use a `public` method—the `getAge` method. The `getWeight` method retrieves a mouse's `weight`. The `grow` method simulates a mouse's growth for a specified number of days. Note the `days` parameter. The number of days is passed into the `days` parameter, and that's how the method knows how many days to simulate.

Here are some of the standard UML class diagram features not found in Figure 6.3 that do appear in Figure 6.12:

- To specify member accessibility, prefix all member specifications with a "-" for `private` access or a "+" for `public` access. The instance variables have "-" prefixes, since we want them to be `private`, and the methods have "+" prefixes, since we want them to be `public`.
- To specify initialization, append "= <*value*>" to each variable declaration that includes initialization. For example, note the "= 0" after the `age` instance variable's specification.
- Underline the `main` method in the `MouseDriver` class diagram box, since the `main` method is declared with the `static` modifier. UML standards suggest that you underline all methods and variables that are declared with the `static` modifier. As you learned in Chapter 5, the `static` modifier indicates a class member. You'll learn more about class members in Chapter 7.
- Include a ": <*type*>" suffix with each method. This specifies the type of value that the method returns. All the methods in the `Mouse` class in Figure 6.4 returned `void` (nothing), but in Chapter 5, you saw many Java API class methods with return types like `int` and `double`, and we'll discuss implementation of such methods later in this chapter.

Figure 6.12 also includes an extra UML class diagram feature. It has *notes* for two of its methods—the `main` and `grow` methods. The notes are depicted by the rectangles with the bent top-right corners. Why bent corners? They are supposed to give the impression of a piece of paper with its corner folded, an indication of a hardcopy "note." Including a note in a UML class diagram is purely optional. Usually we won't use them, but this time, we did because we wanted to show how you can include local variables in a UML class diagram.

6.9 Local Variables

A *local variable* is a variable that's declared and used "locally" inside a method. That's different from an instance variable, which is declared at the top of a class, outside all methods. As you perhaps now realize, all the variables we defined in chapters prior to this chapter were local variables. They were all declared within `main` methods, so they were all local variables within the `main` method. We didn't bother to explain the term "local variable" until now because there were no other methods besides `main`, and the idea of a variable being local to `main` wouldn't have made much sense. But the OOP context makes the concept of a local variable more meaningful.

Scope

A local variable has *local scope*—it can be used only from the point at which the variable is declared to the end of the variable's block. A variable's *block* is established by the closest pair of braces that enclose the variable's declaration. Most of the time, you should declare a method's local variables at the top of the method's body. The scope of such variables is then the entire body of the method.

Index variables in a `for` loop are local variables, but they are special. Their scope rule is slightly different from what is described above. As you know from Chapter 4, you should normally declare a `for` loop's index variable within the for loop's header. The scope of such a variable is the for loop's header plus the for loop's body.

Method parameters are usually not considered to be local variables, but they are very similar to local variables in that they are declared and used "locally" inside a method. As with local variables, the scope of a method's parameters is limited to within the body of that method.

Let's round out the discussion of scope by comparing local scope to the scope used by instance variables. While variables with local scope can be accessed only within one particular method, instance variables can be accessed by any instance methods within the instance variable's class. Furthermore, if an instance variable is declared with the public access modifier, it can be accessed from outside of the instance variable's class (with the help of an instantiated object from the instance variable's class).

Mouse2Driver Class

To illustrate local variable principles, we present the Mouse2 program in Figures 6.13 and 6.14. The code includes line numbers to facilitate tracing in an end-of-chapter exercise. The main method in the Mouse2Driver class has three local variables—stdIn, mickey, and days. These appear in the UML class diagram note at the top of Figure 6.12, and they also appear as declarations in the main method in Figure 6.13.

```
/*****************************************************
 * Mouse2Driver.java
 * Dean & Dean
 *
 * This is a driver for the Mouse2 class.
 *****************************************************/

import java.util.Scanner;

public class Mouse2Driver
{
  public static void main(String[] args)
  {
    Scanner stdIn = new Scanner(System.in);
    Mouse2 mickey = new Mouse2();
    int days;

    mickey.setPercentGrowthRate(10);
    System.out.print("Enter number of days to grow: ");
    days = stdIn.nextInt();
    mickey.grow(days);
    System.out.printf("Age = %d, weight = %.3f\n",
      mickey.getAge(), mickey.getWeight());
  } // end main
} // end class Mouse2Driver
```

local variables

Figure 6.13 Mouse2Driver class that drives the Mouse2 class in Figure 6.14

Let's examine Figure 6.13's `Mouse2Driver` class. In the call to `setPercentGrowthRate`, note that we pass in a constant, 10, instead of a variable. Normally, you'll use variables for your arguments, but this example shows that it's also legal to use constants. After setting the percent growth rate, we prompt the user for the number of days of simulated growth, and then we pass the `days` value into the `grow` method. Then we print `mickey`'s `age` and `weight` by embedding `getAge` and `getWeight` method calls within a `printf` statement.

`Mouse2` Class

Now look at the `Mouse2` class in Figure 6.14. Are there any local variables there? The `age`, `weight`, and `percentGrowthRate` variables are instance variables, not local variables, because they're declared outside of all the methods, at the top of the class. Inside the `grow` method, we highlight this fact by prefixing each of these instance variables with a `this` reference. The `grow` method also includes a local variable—the `i` in the `for` loop. Since `i` is declared within the `for` loop header, its scope is limited to the `for` loop block. So you can read and update `i` only within the `for` loop. If you try to access `i` outside the `for` loop, you'll get a compilation error. This `grow` method is similar to the previous Mouse program's `grow` method, but this time we use a `for` loop to simulate multiple days of growth rather than just one day. The `days` parameter determines how many times the loop will repeat.

Previously, we described the default values for instance variables. Now, we'll describe the default values for local variables. Local variables contain *garbage* by default. Garbage means that the variable's value is unknown—it's whatever just happens to be in memory at the time that the variable is created. If a program attempts to access a variable that contains garbage, the compiler generates a compilation error. For example, what would happen if the `=0` initialization were removed from the `for` loop header in the `grow` method in Figure 6.14? In other words, suppose that `for` loop was replaced by this:

```
for (int i; i<days; i++)
{
  this.weight +=
    (0.01 * this.percentGrowthRate * this.weight);
}
```

Because `i` is no longer assigned zero, `i` contains garbage when the `i<days` condition is tested. If you tried to compile code with a statement like this, it wouldn't compile, and the compiler would report:

```
variable i might not have been initialized
```

Local Variable Persistence

OK, let's say you do initialize a local variable. How long will it *persist*? A local variable (or parameter) persists only within its scope and only for the current duration of the method in which it is defined. The next time the method is called, the local variable's value resets to garbage or the value given to it by initialization. The horizontal line drawn in a trace after a method terminates reminds you that method termination converts all the method's local variables into garbage.

6.10 The `return` Statement

If you look back at our original `Mouse` class in Figures 6.4 and 6.10, you'll notice that every method heading has a `void` modifier located at the left of the method name. That means the method does not return any value, and we say "the method has a `void` return type" or more simply "it's a `void` method." But recall from

```java
/*******************************************************************
* Mouse2.java
* Dean & Dean
*
* This class models a mouse for a growth simulation program.
*******************************************************************/

import java.util.Scanner;

public class Mouse2
{
  private int age = 0;                  // age in days
  private double weight = 1.0;          // weight in grams
  private double percentGrowthRate;     // % daily weight gain

  //**************************************************************

  public void setPercentGrowthRate(double percentGrowthRate)
  {
    this.percentGrowthRate = percentGrowthRate;
  } // end setPercentGrowthRate

  //**************************************************************

  public int getAge()
  {
    return this.age;
  }  // end getAge

  //**************************************************************

  public double getWeight()
  {
    return this.weight;
  } // end getWeight

  //**************************************************************

  public void grow(int days)
  {
    for (int i=0; i<days; i++)
    {
      this.weight +=
        (0.01 * this.percentGrowthRate * this.weight);
    }
    this.age += days;
  } // end grow
} // end class Mouse2
```

parameter
parameter
local variable

Figure 6.14 Mouse2 class

Chapter 5 that many of the Java API methods return some kind of value, and in each case, the type of value returned is indicated by an appropriate return type in the method heading located at the left of the method name.

Returning a Value

If you look at the Mouse2 class in Figure 6.14, you'll see that two of the methods have a return type that is different from void. Here is one of those methods:

```
public int getAge()          // return type
{
  return this.age;           // return statement
} // end getAge
```

The return statement in this method allows you to pass a value from the method back to where the method was called. In this case, the getAge method returns age to Mouse2Driver's printf statement in Figure 6.13. Here is that statement again:

```
System.out.printf("Age = %d, weight = %.3f\n",
  mickey.getAge(), mickey.getWeight());
```

method call

In effect, the JVM "assigns" the return value (this.age) to the method call (mickey.getAge()). To perform a mental trace, imagine that the method call is overlaid by the returned value. So if Mickey's age is 2, then 2 is returned, and you can replace the getAge method call by the value 2.

Whenever a method heading's type is different from void, that method must return a value by means of a return statement, and the type of that value must match the type specified in the method heading. For example, the getAge method heading specifies an int return type. The return statement within the getAge method returns this.age. In Figure 6.14, the age instance variable was declared to be an int, and that matches getAge's int return type, so all is well. It's OK to have an expression following the word return; you aren't limited to just having a simple variable. But the expression must evaluate to the method's return type. For example, would it be legal to use this?

```
return this.age + 1;
```

Yes, because this.age + 1 evaluates to an int type, and that matches getAge's return type.

When a method includes conditional branching (with an if statement or a switch statement), it's possible to return from more than one place in the method. In such cases, all returns must match the type specified in the method heading.

Empty return Statement

For methods with a void return type, it's legal to have an *empty* return statement. The empty return statement looks like this:

```
return;
```

The empty return statement does what you'd expect. It terminates the current method and causes control to be passed back to the calling module at the point that immediately follows the method call. Here's a variation of our previous grow method that uses an empty return statement:

```
public void grow(int days)
{
  int endAge = this.age + days;

  while (this.age < endAge)
  {
    if (this.age >= 100)
    {
      return;          ◄──────── empty return statement
    }
    this.weight +=
      .01 * this.percentGrowthRate * this.weight;
    this.age++;
  } // end while
} // end grow
```

In this variation of the grow method, we cut off the aging process at 100 days—after "adolescence"—by checking age inside the loop and returning when age is greater than or equal to 100. Notice the empty return statement. Because nothing is returned, the method heading must specify void for its return type.

It would be illegal to have an empty return statement and a non-empty return statement in the same method. Why? Empty and non-empty return statements have different return types (void for an empty return statement and some other type for a non-empty return statement). There is no way to specify a type in the heading that simultaneously matches two different return types.

The empty return statement is a helpful statement in that it provides an easy way to exit quickly from a method. However, it does not provide unique functionality. Code that uses an empty return statement can always be replaced by code that is devoid of return statements. For example, here's a return-less version of the previous grow method:

```
public void grow(int days)
{
  int endAge = this.age + days;

  if (endAge > 100)
  {
    endAge = 100;
  }
  while (this.age < endAge)
  {
    this.weight +=
      .01 * this.percentGrowthRate * this.weight;
    this.age++;
  }  // end while
} // end grow
```

return Statement Within a Loop

Programmers in industry often are asked to maintain (fix and improve) other people's code. In doing that, they often find themselves having to examine the loops and, more specifically, the loop termination conditions in the program they're working on. Therefore, it's important that loop termination conditions are clear. Normally, loop termination conditions appear in the standard loop-condition section. For `while` loops, that's at the top, for `do` loops, that's at the bottom, and for `for` loops, that's in the header's second component. However, a `return` statement inside a loop results in a loop termination condition that's not in a standard location. For example, in the first `grow` method on the previous page, the `return` statement is inside an `if` statement and the loop termination condition is consequently "hidden" in the `if` statement's condition.

In the interest of maintainability, you should use restraint when considering the use of a `return` statement inside a loop. Based on the context, if inserting `return` statements inside a loop improves clarity, then feel free to insert. However, if it simply makes the coding chores easier and it does not add clarity, then don't insert. So which `grow` implementation is better—the empty `return` version or the `return`-less version? In general, we prefer the `return`-less version for maintainability reasons. However, because the code in both of our adolescent `grow` methods is so simple, it doesn't make much difference here.

6.11 Argument Passing

In the previous section, you saw that when a method finishes, the JVM effectively assigns the return value to the method call. This section describes a similar transfer in the other direction. When a method is called, the JVM effectively assigns the value of each argument in the calling statement to the corresponding parameter in the called method.

Example

Let's examine argument passing by looking at an example—another version of our Mouse program called Mouse3. Here is the code for this new version's driver:

```
public class Mouse3Driver
{
  public static void main(String[] args)
  {
    Mouse3 minnie = new Mouse3();
    int days = 365;

    minnie.grow(days);
    System.out.println("# of days aged = " + days);
  } // end main
} // end class Mouse3Driver
```

The JVM makes a copy of `days`'s value and passes it to the `grow` method.

The `Mouse3Driver` class calls the `grow` method with an argument called `days`, whose value happens to be 365. Then it assigns this value (365) to the parameter called `days` in the `grow` method. The following code shows what happens to the `days` parameter within the `grow` method:

```
public class Mouse3
{
  private int age = 0;                    // age in days
  private double weight = 1.0;            // weight in grams
```

```
  private double percentGrowthRate = 10;  // % daily weight gain
  public void grow(int days)
  {
    this.age += days;
    while (days > 0)
    {
      this.weight +=
        .01 * this.percentGrowthRate * this.weight;
      days--;
    }
  } // end grow
} // end class Mouse3
```

The JVM assigns the passed-in value to the days parameter.

The days parameter decrements down to 0.

Within a method, parameters are treated like local variables. The only difference is that a local variable is initialized inside the method, whereas a parameter is initialized by an argument in the method call. As you can see in the above loop body, the days parameter decrements down to zero. What happens to the days variable in the main method in Mouse3Driver? Because the two days variables are distinct, the days variable in the main method does not change with the days parameter in the grow method. So when Mouse3Driver prints its version of days, it prints the unchanged value of 365 like this:

```
# of days aged = 365.
```

Pass-By-Value

We say that Java uses *pass-by-value* for its argument-passing scheme. As illustrated by Figure 6.15, pass-by-value means that the JVM passes a copy of the argument's value (not the argument itself) to the parameter. Changing the copy does not change the original.

In Mouse3Driver and Mouse3, notice that the calling method's argument is called days and the grow method's parameter also is called days. Is the parameter the same variable as the argument? No! They are separate variables separately encapsulated in separate blocks of code. Because these two variables are in separate blocks of code, there is no conflict, and it's OK to give them the same name. Using the same name is natural because these two variables describe the same kind of thing. When names are in different blocks, you don't have to worry about whether they are the same or not. That's the beauty of encapsulation. Big programs would be horrible nightmares if you were prohibited from using the same name in different blocks of code.

Same Name Versus Different Names for Argument-Parameter Pairs

Most of the time, you'll want to use the same name for an argument/parameter pair. But be aware that using different names is legal and fairly common. When it's more natural and reasonable to use different names for an argument/parameter pair, then use different names. The only requirement is that the argument's type must match the parameter's type. For example, in the Mouse3 program, if num is an int variable, then the following method call successfully passes num's value to the days int parameter:

```
minnnie.grow(num);
```

```
public class Mouse3Driver
{
  public static void main(String[] args)
  {
    ...
    minnie.grow(days);
    ...
  } // end main
} // end class Mouse3Driver
```

```
public class Mouse3
{
  ...
  public void grow(int days)
  {
    ...
  } // end grow
} // end class Mouse3
```

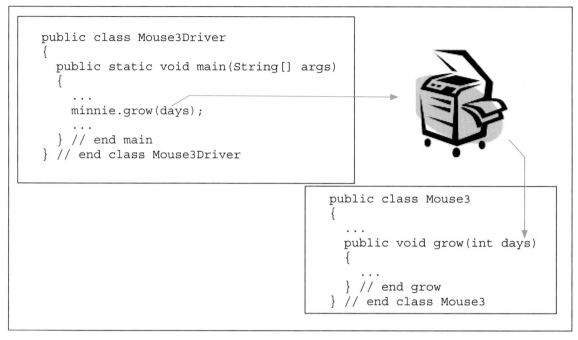

Figure 6.15 Pass-by-value means a copy of the argument's value goes to the corresponding parameter

6.12 Specialized Methods—Accessors, Mutators, and Boolean Methods

Let's now discuss some of the common types of specialized methods. You won't be asked to learn any new syntax; you'll just be asked to apply what you've learned so far.

Accessor Methods

An *accessor* is a method that retrieves part of an object's stored data—typically `private` data. Note the following `getAge` and `getWeight` methods (taken from Figure 6.14's `Mouse2` class). They are accessor methods, as they retrieve the values of the instance variables, `age` and `weight`, respectively.

```
public int getAge()
{
  return this.age;
} // end getAge
public double getWeight()
{
  return this.weight;
} // end getWeight
```

As evidenced by the `getAge` and `getWeight` methods, accessor methods should be named with a "get" prefix. That's why accessor methods are often called *get methods.*

A method should perform only one task. It should be written so that it accomplishes the one thing its name implies. For example, a `getAge` method should simply return its object's `age` instance variable value and do nothing else. We mention this notion because there is sometimes a temptation to provide extra functionality to a method to avoid having to implement that functionality elsewhere. One particularly common *faux pas* (a French term meaning error in etiquette) is to add print statements to a method that doesn't need to print. For example, a novice programmer might implement the `getAge` method like this:

```
public int getAge()
{
  System.out.println("Age = " + this.age);
  return this.age;
} // end getAge
```

Inappropriate print statement

That `getAge` method might work fine for the novice programmer's program, which takes into account the `getAge` method's non-standard print statement. But if another programmer needs to work with the program and call the `getAge` method later, the new programmer would be surprised to find the non-standard print statement. The new programmer would then either have to (1) accommodate the print statement or (2) remove it from the `getAge` method and check for any ripple effects. To avoid that scenario, you should include print statements in a method only if the purpose of the method is to print something.

The exception to the above rule is that it's acceptable and helpful to add print statements temporarily to methods when you're trying to debug a program. For example, if you think there's something wrong with your `getAge` method, you might want to add the above print statement to verify the correctness of the age value just before `getAge` returns it. If you add such debug print statements, don't forget to remove them later, when your program is working.

Debug with temporary print statements.

Mutator Methods

A *mutator* is a method that changes or "mutates" an object's state by changing some or all of that object's stored data—typically `private` data. For example, here is the mutator method for setting or changing a mouse's `percentGrowthRate` instance variable:

```
public void setPercentGrowthRate(double percentGrowthRate)
{
  this.percentGrowthRate = percentGrowthRate;
} // end setPercentGrowthRate
```

As evidenced by the `setPercentGrowthRate` method, mutator methods should be named with a "set" prefix. That's why mutator methods are often called *set methods.*

An accessor allows you to read a `private` instance variable. A mutator allows you to update a `private` instance variable. If you provide a `private` instance variable with both an accessor and a simple mutator like the `setPercentGrowthRate` method above, it effectively converts that `private` instance variable into a `public` instance variable, and it breaks the encapsulation of that variable. There's not much danger with having an accessor alone, but having a simple mutator allows an outsider to enter an unreasonable value that may produce erratic program operation. However, if you include constraint checking and perhaps correcting code in your mutators, they can serve as data *filters* that assign only proper data to your `private` instance variables. For example, the following `setPercentGrowthRate` mutator filters out growth rates that are less than −100%:

Use mutator to filter input.

```
public void setPercentGrowthRate(double percentGrowthRate)
{
  if (percentGrowthRate < -100)
  {
    System.out.println("Attempt to assign an invalid growth rate.");
  }
  else
  {
    this.percentGrowthRate = percentGrowthRate;
  }
} // end setPercentGrowthRate
```

Our examples will occasionally include some mutator error checking to illustrate this filtering function, but to reduce clutter, we'll usually employ the minimal form.

Boolean Methods

A *Boolean method* checks to see whether some condition is true or false. If the condition is true, then `true` is returned. If the condition is false, then `false` is returned. To accommodate the `boolean` returned value, Boolean methods must always specify a `boolean` return type. A Boolean method name should normally start with `is`. For example, here's an `isAdolescent` method that determines whether a `Mouse` object is an adolescent by comparing its `age` value to 100 days (we're making a simplifying assumption that any mouse younger than 100 days is considered to be an adolescent):

```
public boolean isAdolescent()
{
  if (this.age <= 100)
  {
    return true;
  }
  else
  {
    return false;
  }
} // end isAdolescent
```

Here's how this code might be shortened:

```
public boolean isAdolescent()
{
  return this.age <= 100;
} // end isAdolescent
```

To show how the shortened method works, we'll plug in sample values. But first, let's get settled on the goal: Whenever `age` is less than or equal to 100, we want the method to return `true` to indicate adolescence. If `age` is 50, what is returned? `true` (This happens because the `return` statement's `this.age <= 100` expression evaluates to `true`.) If `age` is 102, what is returned? `false` (This happens because the `return` statement's `this.age <= 100` expression evaluates to `false`.) Plug in any number for `age` and you'll see that the shortened function does indeed work properly. In other words, the shortened `isAdolescent` method does indeed return `true` whenever `age` is less than or equal to 100.

Are you bothered by the lack of parentheses around the `return` statement's returned expression? With statements that use a condition (`if` statement, `while` statement, etc.), the condition must be surrounded by parentheses. With the `return` statement's returned expression, the parentheses are optional. You'll see it both ways in industry—sometimes parentheses are included and sometimes they're omitted.
Here's how the `isAdolescent` method could be used in a calling module:

```
Mouse pinky = new Mouse();
. . .
if (pinky.isAdolescent() == false)
{
  System.out.println("The mouse's growth is no longer" +
    " being simulated - too old.");
}
```

Do you know how the above `if` statement can be shortened? Here's a functionally equivalent `if` statement with an improved condition:

```
if (!pinky.isAdolescent())
{
  System.out.println("The mouse's growth is no longer" +
    " being simulated - too old.");
}
```

The goal is to print the warning message if `pinky` is old (not an adolescent). If `isAdolescent` returns `false` (indicating an old Pinky), then the `if` statement's condition is `true` (`!false` evaluates to `true`) and the program prints the warning message. On the other hand, if `isAdolescent` returns `true` (indicating a young Pinky), then the `if` statement's condition is `false` (`!true` evaluates to `false`) and the program skips the warning message.

Although the shortened-version `if` statement might be harder to understand initially, experienced programmers would prefer it. Following that lead, we encourage you to use `!` rather than `== false` for similar situations.

6.13 Problem Solving with Simulation (Optional)

In our previous mouse examples, to keep the focus on OOP concepts rather than mouse growth details, we used a simplistic growth formula. In this section, we show you how to simulate growth in a way that is much closer to the kind of growth that occurs in the real world. Then we show you a simple trick that can be applied to many simulation problems to improve the program's speed and accuracy greatly.

Previously, we modeled growth by assuming that added weight is proportional to weight, like this:

$$addedWeight = fractionGrowthRate \times weight$$

where

$$fractionGrowthRate = .01 \times percentGrowthRate$$

This kind of growth makes weight increase exponentially and continue to curve upward in time, as indicated by Figure 6.16. This is a good approximation for a young plant or animal, where most of the ingested food energy goes into new growth.

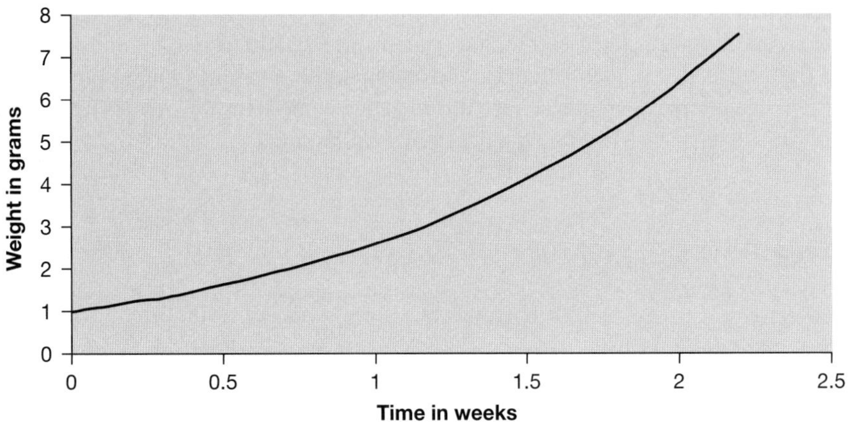

Figure 6.16 Exponential growth

Maturation

But there's a problem with the exponential growth model. Nothing keeps growing forever! After a while, old tissue starts to die, and some of the ingested nutrients must be used to replace the old tissue instead of just adding to it. This slows the growth. As a larger fraction of ingested nutrients go into replacement, the growth curve straightens out, begins to bend the other way, and approaches a maximum. The easiest way to modify the basic exponential growth formula to make it describe maturation is to multiply by another factor to obtain what's called the *logistic equation:*

$$\mathtt{addedWeight} = \mathtt{fractionGrowthRate} \times \mathtt{weight} \times \left(1.0 - \frac{\mathtt{weight}}{\mathtt{maxWeight}}\right)$$

A quick inspection of this improved growth formula shows that as `weight` approaches `maxWeight`, the quantity in parentheses on the right approaches zero, and therefore the added weight on the left approaches zero. At that point, there's no more growth. This provides a reasonable description of an organism reaching maturity.

Computer simulations rely on approximate mathematical models, like the model provided by the above logistic equation. Such simulation models are sometimes good, sometimes not so good, and it's difficult to know how good they are without comparing them to actual live data. But for the current weight gain problem, we have the luxury of being able to compare the simulation model with an exact mathematical model. Here is a closed-form exact mathematical solution that determines the weight of any given time:

$$\mathtt{weight} = \frac{1.0}{\dfrac{1.0}{\mathtt{maxWeight}} + e^{-(\mathtt{fractionGrowthRate} \times \mathtt{time} + g_0)}}$$

This formula contains a growth constant, g_0, which is

$$g_0 = \log_e\left(\frac{\mathtt{minWeight}}{1.0 - \dfrac{\mathtt{minWeight}}{\mathtt{maxWeight}}}\right)$$

You can find g_0 by plugging *minWeight* and *maxWeight* values into the second formula. Then find *weight* by plugging g_0 into the first formula.

Simulation

If you can describe it, you can simulate it.

Usually an exact solution is not available, and the only way to solve a problem is with a simulation. But for this weight gain problem, we have both. Let's look at a program that

```
/*********************************************************************
* Growth.java
* Dean & Dean
*
* This provides different ways to calculate growth.
*********************************************************************/

public class Growth
{
  private double startSize;                   // initial size
  private double endSize;                     // maximum size
  private double fractionGrowthRate;          // per unit time

  //******************************************************************

  public void initialize(double start, double end, double factor)
  {
    this.startSize = start;
    this.endSize = end;
    this.fractionGrowthRate = factor;
  } // end initialize

  //******************************************************************

  public double getSize(double time)
  {
    double g0 = Math.log(startSize / (1.0 - startSize / endSize));

    return 1.0 / (1.0 / endSize +
      Math.exp(-(fractionGrowthRate * time + g0)));
  } // end getSize

  //******************************************************************

  public double getSizeIncrement(double size, double timeStep)
  {
    return fractionGrowthRate *
      size * (1.0 - size / endSize) * timeStep;
  } // end getSizeIncrement
} // end class Growth
```

Figure 6.17 `Growth` class, which implements different ways to evaluate growth

displays time, the exact solution, and the simulated solution together. See the program's `Growth` class in Figure 6.17.

The `Growth` class has three instance variables, `startSize`, `endSize`, and `fractionGrowthRate`, and three methods. The `initialize` method initializes the three instance variables. The `getSize` method uses the closed-form mathematical solution formula provided earlier. It returns the size (e.g., current mouse weight) at the given time. Notice that this method's name starts with "get," so it looks like the name of an accessor method, and it returns a `double` value just as our previous `getWeight` method does. But this class does not have any instance variable named "size." So here's an example of a method that is not really an accessor like the accessors described in Section 6.12, even though its name makes it look like an accessor. The point is this: any method can return a value, not just an accessor method, and any method can have any name that seems appropriate—`getSize` is simply the most appropriate name we could think of for this method that computes and returns a size.

The `getSizeIncrement` method implements one simulation step. It returns the change in size between the current time and the next time. Notice that the `getSize` and `getSizeIncrement` methods do different things. The first one gives the answer directly. The second one gives an incremental value that must be added to a previous answer to get the next answer.

If you are writing your own class and you want to model the growth of one of your class's entities, you could copy and paste the `Growth` class's variables and methods into your class. Alternatively, you could delegate the work to a `Growth` class object just as you delegate work to `Scanner` class objects. To do this, use `new` to instantiate a `Growth` object, initialize it with the growth-related data in your object. Then ask the `Growth` object to solve the growth problem for you by calling its `getSize` or `getSizeIncrement` method. In your program, you could use code like that in the `main` method of the `GrowthDriver` class in Figure 6.18.

This driver class may seem imposing, but it's not difficult. We start by declaring and initializing local variables, and this includes instantiating and initializing a `Growth` object. Then we ask the user to provide a time increment and the total number of time increments. Finally, we use a `for` loop to print time, the exact solution, and the simulated solution for each time step. If you run the program composed of the code in Figures 6.17 and 6.18, you'll get this result:

Sample session:

```
Enter time increment: 1
Enter total time units to simulate: 15
        exact  simulated
time     size     size
 0.0      1.0      1.0
 1.0      2.6      2.0
 2.0      6.4      3.9
 3.0     13.6      7.3
 4.0     23.3     13.3
 5.0     31.7     22.2
 6.0     36.5     32.1
 7.0     38.6     38.4
 8.0     39.5     39.9
 9.0     39.8     40.0
10.0     39.9     40.0
```

```
11.0    40.0    40.0
12.0    40.0    40.0
13.0    40.0    40.0
14.0    40.0    40.0
15.0    40.0    40.0
```

```
/*******************************************************************
* GrowthDriver.java
* Dean & Dean
*
* This compares exact and simulated solutions for growth.
*******************************************************************/

import java.util.Scanner;

public class GrowthDriver
{
  public static void main(String[] args)
  {
    Scanner stdIn = new Scanner(System.in);
    double timeStep;
    double timeMax;
    Growth entity = new Growth();
    double startSize = 1.0;              // weight in grams
    double endSize = 40.0;               // weight in grams
    double fractionGrowthRate = 1.0;     // per unit time
    double size = startSize;

    entity.initialize(startSize, endSize, fractionGrowthRate);
    System.out.print("Enter time increment: ");
    timeStep = stdIn.nextDouble();
    System.out.print("Enter total time units to simulate: ");
    timeMax = stdIn.nextDouble();
    System.out.println("       exact  simulated");
    System.out.println("time    size    size");

    for (double time=0.0; time<=timeMax; time+=timeStep)
    {
      System.out.printf("%4.1f%8.1f%8.1f\n",
        time, entity.getSize(time), size);
      size += entity.getSizeIncrement(size, timeStep);
    } // end for
  } // end main
} // end class GrowthDriver
```

Instantiate Growth object.

Initialize Growth object.

Figure 6.18 GrowthDriver class that demonstrates the Growth class in Figure 6.17

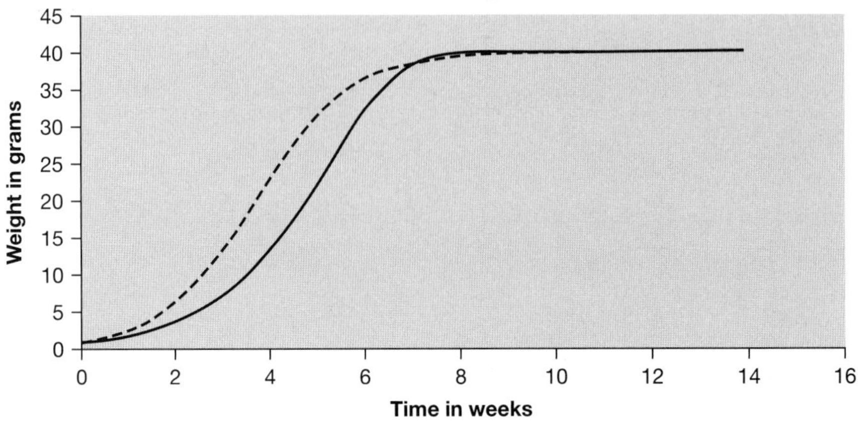

Figure 6.19 Simulated solution with time increment = 1 (solid), compared to the exact solution (dashed)

Figure 6.19 shows what this data looks like in a two-dimensional plot. Alas, the simulated solution doesn't agree very well with the exact solution. It doesn't rise quickly enough, and then it overshoots. The reason for this error is actually quite simple. Each size increment is based on the size at the beginning of the increment. But as time passes, the actual size changes, so for all but the first instant in the increment, the calculation is using old data.

The most straightforward way to fix this accuracy problem is to use a smaller time step. With this simulation algorithm, the error is proportional to the size of the time step. If you cut the time step in half, this cuts the error in half, if you divide the time step by 10, this divides the error by 10, and so on. In the above output, at four weeks, the exact solution says the size is 23.3 grams, but the simulation says it's only 13.3 grams. That's an error of $23.3 - 13.3 = 10$ grams. If we want to reduce this error to less than 1 gram, we need to reduce the time step by a factor of about 10.

If you don't know the exact solution, how do you know your error? Here's a rule of thumb: If you want less than 1% error, make sure the size increment in each time step is always less than about 1% of the average size in that time interval.

This simple algorithm works fine for simple problems. But if you have a tough problem, some things may be sensitive to very small errors, and you may have to take a very large number of very small steps. This might take more time than you can stand. There's also a more insidious problem. Even a double number has limited precision, and when you process many numbers, round-off errors can accumulate. In other words, as you make step sizes smaller, errors initially decrease, but eventually they begin to increase again.

Improved Accuracy and Efficiency Using a Step-with-Midpoint Algorithm[3]

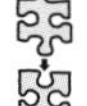

Remove bias. There's a better way to improve accuracy. It's based on a simple principle: Instead of using the condition(s) (e.g., weight) at the beginning of the interval to estimate the change(s) during the interval, use the condition(s) in the middle of the interval to estimate the change(s) during the interval. But how can you know the conditions in the middle of the interval until you get there? Send out a "scouting party"! In other words, make a tentative half-step forward, and evaluate the conditions there. Then go back to the beginning and use the condition(s) at the midpoint to determine what the change(s) will be in a full step forward.

[3]The formal name for this algorithm is "Second-order Runge-Kutta."

At first, this might sound like a hard way to do an easy thing. Why not just cut the step size in half and take two small steps forward? The qualitative answer is: That still leaves a regular bias toward old data. The quantitative answer is: If you use a step-with-midpoint algorithm for your simulation, the size of the error is proportional to the square of the size of the time step. That means that if you reduce the full-step size by a factor of 100, the error goes down by a factor of 10,000. In other words, you can get an extra factor-of-100 accuracy by increasing the computer's work by only a factor of 2.

But what about the work you do? How much harder is it to implement a step-with-midpoint algorithm? Not much. All you have to do is add one simple method. Specifically, to the `Growth` class in Figure 6.17, just add the `getSizeIncrement2` method shown in Figure 6.20.

```
public double getSizeIncrement2(double sizeCopy, double timeStep)
{
  sizeCopy += getSizeIncrement(sizeCopy, 0.5 * timeStep);
  return getSizeIncrement(sizeCopy, timeStep);
} // end getSizeIncrement2
```

No prefix necessary since `getSizeIncrement` and `getSizeIncrement2` are in the same class.

Figure 6.20 A method that implements the step-with-midpoint algorithm
Add this method to the code in Figure 6.17 to improve simulation accuracy and efficiency.

How does this little method work? It simply calls the original `getSizeIncrement` method two times. Notice that the `sizeCopy` parameter in Figure 6.20 is just a copy of the `size` variable in the driver class. The first call to `getSizeIncrement` uses the size at the beginning of the time increment, and it goes only half a time step forward. Then, it uses the returned value to increment `sizeCopy` to the size at the midpoint. The second call to `getSizeIncrement` uses this computed midpoint size and a full time step to determine the change from the beginning to the end of the full time interval.

Within the `getSizeIncrement2` method definition, note the calls to `getSizeIncrement`. There's no reference variable dot prefix at the left of `getSizeIncrement`. Here's why: If you call a method that's in the same class as the current class, then you can call the method directly, with no reference variable dot prefix.

The work required to modify the driver is negligible. All you have to do is change the name of the method called to the name of the new method. In our case, all you have to do is change the last statement in the driver in Figure 6.18 to this:

```
size += entity.getSizeIncrement2(size, timeStep);
```

This appended "2" is the only difference!

Figure 6.21 shows what the improved algorithm produces with a full step size equal to the step size used for Figure 6.19. This takes twice as much computer time as what's in Figure 6.19, but it's clearly much more than twice as good. For example, at four weeks, the error is now only 1.5 grams instead of the previous 10 grams.

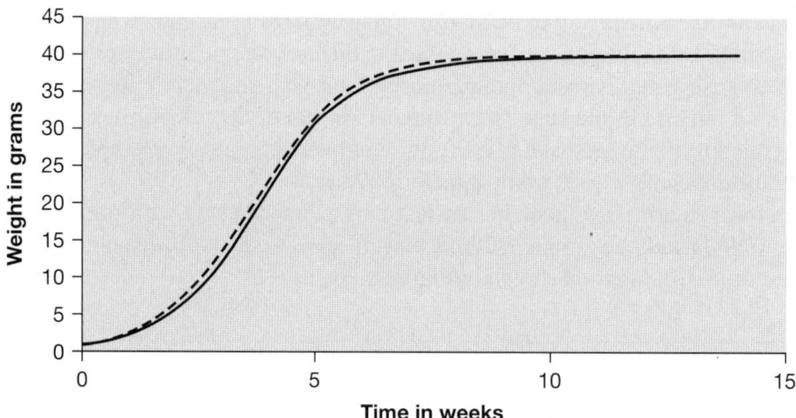

Figure 6.21 Step-with-midpoint simulated solution with time increment = 1 (solid), compared to the exact solution (dashed)

Summary

- An object is a group of related data that identifies the current condition or *state* of the object, plus the methods that describe the *behavior* of that object.
- Objects are *instances* of the classes that define them. A class definition specifies the instance variables which an object of that class contains, and it defines the methods which an object of that class may call. Each object contains its own copy of the instance variables its class defines, and a given instance variable generally has different values in different objects.
- Use the `private` access modifier to specify that a particular variable is encapsulated or hidden. Use the `public` access modifier to make methods accessible to the outside world.
- To make a class as general as possible, drive it from a `main` method in a separate "driver" class. In the driver's `main` method, declare a reference variable of the driven class's type. Then, use Java's keyword `new` to instantiate an object of the driven class, and initialize the reference variable with the object reference returned by `new`.
- Use Java's keyword `this` to refer to the calling object from within one of that object's methods. Use `this` to distinguish an instance variable from a same-named parameter or local variable.
- When you trace an object-oriented program, you need to keep track of which class you're in, which method you're in, which object called that method, parameter and local variable names, and the names of all instance variables in each object.
- A UML class diagram has separate boxes for the class name, a description of the class's variables, and headings for the class's methods. Use a "+" prefix for `public` and a "-" prefix for `private`. Specify variable and method return types and non-default initial values.
- Instance variable default values are zero for numbers, `false` for `boolean` values, and `null` for references. Instance variable values persist for the life of their object. Local variable default values are undefined garbage. Local variables and parameters persist for as long as their method is being executed, and after that, their values are undefined.
- Unless a method's return type is `void`, every path through the method must end with a statement that returns a value of the method's type.

- A method's parameter must have the same type as the method call's argument. What the method gets is a copy of what is in the calling program, so changing a parameter in a method does not change the calling program's value.
- Use `setX` and `getX` methods to modify and retrieve `private` instance variable values. Include filtering in `setX` methods to protect your program from bad input. Use `boolean isX` methods to return `true` or `false`, depending on the value of a particular condition.
- Optionally, improve simulation speed and accuracy by computing the next increment with values determined halfway between that increment's starting and ending points.

Review Questions

§6.2 Object-Oriented Programming Overview

1. A class is an instance of an object. (T / F)
2. How many objects may there be within a single class?

§6.3 First OOP Class

3. A class's instance variables must be declared outside of all _____, and all instance variable declarations should be located at the ________________________.
4. Methods accessible from outside a class are `public`, but instance variables (even those that an outsider may need to change or read) are usually `private`. Why?

§6.4 Driver Class

5. Where does `main` go—in the driver class or in one of the driven classes?
6. When a program has both driver and driven classes, where should most of the program code reside?
7. How do you retrieve a `private` instance variable's value from within a `main` method?
8. A reference variable holds the ___________ of an object.

§6.5 Calling Object, this Reference

9. An instance method might contain a statement like `this.weight = 1.0;` but if that method's class currently has five instantiated objects, there are five different variables called `weight`. How can we determine which one is getting the new value?

§6.6 Instance Variables

10. What are the default values for `int`, `double`, and `boolean` for an object's instance variables?
11. In the Mouse program of Figures 6.4 and 6.5, what is the persistence of `gus`'s `age` variable?

§6.8 UML Class Diagrams

12. After a program is written, a UML class diagram provides a brief outline of each class in the program. It helps other people see what methods are available and what arguments they need. Give some reasons why it might be helpful to have an already created class diagram in front of you while you are implementing the class and writing its methods?.

§6.9 Local Variables

13. Assume the `main` method in `Mouse2Driver` had started more simply with only `Mouse mickey;` What would be the value of `mickey` immediately after this statement?

§6.10 The `return` Statement

14. Usually, the use of multiple `return` statements leads to code that is more understandable. (T / F)

§6.11 Argument Passing

15. How is a method parameter like a local variable, and how do they differ?
16. What is the relationship and difference between a method argument and a method parameter?

§6.12 Specialized Methods—Accessors, Mutators, and Boolean Methods

17. What is the standard prefix for an accessor method?
18. What is the standard prefix for a mutator method?
19. What is the standard prefix for a Boolean method?

§6.13 Problem Solving with Simulation (Optional)

20. Identify two general ways to reduce the size of the error in a simulation. For a given accuracy, which way is more efficient?

Exercises

1. [after §6.2] Suppose you are asked to model plants using an OOP program. For each of the following plant-related entities, specify the most appropriate item to use for its implementation. For each entity, select one of the following: instance variable, object, method, or class.
   a) plant height
   b) sequence of activities that occur when a seed germinates
   c) an indication of whether the plant contains a vascular system
   d) an individual plant
2. [after §6.3] In Java, how do you encapsulate an instance variable?
3. [after §6.4] Describe the relationship between the `main` method and driver and driven classes. Give an example of a class that runs by itself and does not need a separate driver.
4. [after §6.4] Wrapper objects: The wrapper classes discussed in Chapter 5 also provide you with the ability to instantiate objects that are wrapped versions of primitive variables. For example, to create a wrapped version of the `double` number x, you can do this:

```
double x = 55.0;
Double xWrapped = new Double(x)
```

This instantiates an object of type `Double`, which is a wrapped version of the primitive variable, `x`. Then it assigns a reference to that object to the reference variable, `xWrapped`. The `Double` class has a number of prebuilt methods that work with `Double` objects. You can read about these methods in the Java API documentation on the `Double` class. The following program illustrates some of these methods:

```
/*********************************************************
* Wrapper.java
* Dean & Dean
*
* This program exercises some wrapped primitive numbers.
*********************************************************/

public class Wrapper
{
  public static void main(String[] args)
  {
    double x = 44.5;
    double y = 44.5;
```

```
    Double xW = new Double(x);          // the object: wrapped x
    Double yW = new Double(y);          // the object: wrapped y

    System.out.println("object == object? " + (xW == yW));
    System.out.println("value == value? " +
      (xW.doubleValue() == yW.doubleValue()));
    System.out.println(
      "object.equals(object)? " + xW.equals(yW));
    System.out.println("object.compareTo(object)? " +
      xW.compareTo(yW));

    yW = new Double(y + 3.0);
    System.out.println("object.compareTo(largerObject)? " +
      xW.compareTo(yW));

    yW = new Double(Double.NEGATIVE_INFINITY);
    System.out.println("-infinity isInfinite()? " +
      yW.isInfinite());
  } // end main
} // end Wrapper class
```

Compile and run this program, and display the output. Read about the `Double` class in the Java API documentation, and explain why each of the outputs comes out the way it does.

5. [after §6.4] Suppose you have a `Town` class that describes the demographics of small towns. The vital statistics described by this class are `numberOfAdults` and `numberOfChildren`. These vital statistics are encapsulated and not directly accessible from outside the class.

a) Write the following methods for `class Town`:

i. An `initialize` method that establishes initial values of instance variables. Assume that `initialize` gathers all the data it needs by prompting for and inputting values from a user.

ii. A `simulateBirth` method that simulates the birth of one child.

iii. A `printStatistics` method that prints out the current vital statistics.

b) Write a `main` method for a separate driver class that does the following:

i. Creates a town named `newHome`

ii. Calls `initialize` to establish initial values of instance variables for `newHome`.

iii. Simulates the birth of a pair of twins.

iv. Prints out `newHome`'s vital statistics.

6. [after §6.7] Given this PcDesign program:

```
/*******************************************************
* PcDesignDriver.java
* Dean & Dean
*
* This exercises the PcDesign class.
*******************************************************/

public class PcDesignDriver
{
      public static void main(String[] args)
   {
       PcDesign myPc = new PcDesign();
```

```
    myPc.assignRamSize();
    myPc.assignDiskSize();
    myPc.assignProcessor();
    myPc.calculateCost();
    myPc.printSpecification();
  } // end main
} // end class PcDesignDriver
```

```
/*************************************************************
* PcDesign.java
* Dean & Dean
*
* This class collects specifications for a PC.
*************************************************************/

import java.util.Scanner;

public class PcDesign
{
  private long ramSize = (long) 1000000000.0;
  private long diskSize;
  private String processor;
  private double cost;

  //**********************************************************

  public void assignRamSize()
  {
    this.ramSize = (long) 2000000000.0;
  } // end assignRamSize

  //**********************************************************

  public void assignDiskSize()
  {
    Scanner stdIn = new Scanner(System.in);
    long diskSize;
    diskSize = stdIn.nextLong();
  } // end assignDiskSize

  //**********************************************************

  public void assignProcessor()
  {
    Scanner stdIn = new Scanner(System.in);
    this.processor = stdIn.nextLine();
  } // end assignProcessor

  //**********************************************************

```

```
  public void calculateCost()
  {
    this.cost = this.ramSize / 10000000.0 +
      this.diskSize / 100000000.0;
    if (this.processor.equals("Intel"))
    {
      this.cost += 400;
    }
    else
    {
      this.cost += 300;
    }
  } // end calculateCost

  //*******************************************************

  public void printSpecification()
  {
    System.out.println("RAM = " + this.ramSize);
    System.out.println("Hard disk size = " + this.diskSize);
    System.out.println("Processor = " + this.processor);
    System.out.println("Cost = $" + this.cost);
  } // end printSpecification
} // end class PcDesign
```

Use the following trace setup to trace the PC-design program. Note that we have used abbreviations to keep the trace setup's width as small as possible. Don't forget to specify default and initial values, even if they don't affect the final result.

input

```
60000000000
Intel
```

Driver		PcDesign											
	main		aRSize	assignDiskSize		aProc	cCost	printS	obj1				
line#	**myPc**	*line#*	**this**	**this**	**diskSize**	**this**	**this**	**this**	**ramSize**	**dSize**	**proc**	**cost**	**output**

7. [after §6.8] The answer to this question is not in this book—you'll need to look elsewhere. Who are UML's "Three Amigos"?

8. [after §6.8] Construct a UML class diagram for files in a computer directory. The class name should be `File`. Include the following methods: `public String getName()`, `public long length()`, and `public boolean isHidden()`. Also, include the instance variable associated with the first of these methods. Include an indication of whether the member is `public` or `private` and the type of the return value or variable. A `File` class already exists as part of the Java language, and this class also has many other methods, but the API library documentation for this class does not show any instance variables. Does that mean this class has no instance variables?

9. [after §6.9] If an object calls the same method two separate times, in the second execution, the method's local variables begin with the values they had at the end of the previous execution of that method. (T / F)

10. [after §6.9] Trace the Mouse2 program shown in Figures 6.13 and 6.14. Use the following trace setup. Note that we have used abbreviations to keep the trace setup's width as small as possible.

input

2

Mouse2Driver			Mouse											
	main			setPGR		getAge	getWt	grow			obj1			
line#	**mickey**	**days**	*line#*	**this**	**rate**	**this**	**this**	**this**	**days**	**i**	**age**	**wt**	**rate**	**output**

11. [after §6.11] The diagram below shows Mouse2 program methods, with their parameters and local variables indented, and the one instantiated object, with its instance variables indented. Your task is to construct a time line for each method, local variable or parameter, object, and instance variable. Each time line should show that item's persistence (when it starts and ends) relative to the other items. To help you get started, we have provided the time lines for the `main` method and one of its local variables. Provide all the other time lines, and show how they align with each other and those already provided. (Assume that the object and its instance variables come into existence simultaneously.)

```
                                           time →
methods:
  main                       |-----------------------|
    mickey                    |----------------------|
    days
  setPercentGrowthRate
  getAge
  getWeight
  grow
    days
    i
object:
  mickey
    age
    weight
    percentGrowthRate
```

12. [after §6.12] Complete the following `StudentIdDriver` class skeleton by replacing all six occurrences of *<insert-code-here>* with your own code such that the program operates properly. For details, read the comments above or next to the *<insert-code-here>* insertions. Note the `StudentId` class, which is below the `StudentIdDriver` class. The two classes are in separate files.

```
import java.util.Scanner;

public class StudentIdDriver
{
  public static void main(String[] args)
  {
    Scanner stdIn = new Scanner(System.in);
    StudentId student;
```

```
    String name;
    // Instantiate a StudentId object and assign it to student.
    <insert-code-here>

    System.out.print("Enter student name: ");
    name = stdIn.nextLine();

    // Assign name to the student object.
    <insert-code-here>

    System.out.print("Enter student id: ");
    // In a single statement, read an int for the id value,
    // and assign it to the student object.
    <insert-code-here>

    // If invalid id, execute the loop.
    // (Use the isValid method in the while loop condition.)
    while (<insert-code-here>)
    {
      System.out.print("Invalid student id - reenter: ");
      // In a single statement, read an int for the id value
      // and assign it to the student object.
      <insert-code-here>
    }

    System.out.println("\n" + name +
      ", your new e-mail account is: \n" +
      <insert-code-here>                    // Get email account.
  } // end main
} // end class StudentIdDriver

public class StudentId
{
  private String name;
  private int id;

  //*******************************************************

  public void setName(String n)
  {
    this.name = n;
  }

  public String getName()
  {
    return this.name;
  }
```

```
    public void setId(int id)
    {
      this.id = id;
    }

    public int getId()
    {
      return this.id;
    }

    //*******************************************************

    public String getEmailAccount()
    {
      // Include "" in concatenation to convert to strings.
      return "" + this.name.charAt(0) + this.id +
        "@pirate.park.edu";
    }

    //*******************************************************

    public boolean isValid()
    {
      return this.id >= 100000 && this.id <= 999999;
    }
  } // end class StudentId
```

13. [after §6.13] Construct a UML class diagram for the `Growth` class in Figure 6.17, with the `getSizeIncrement2` method of Figure 6.20 included.

Review Question Solutions

1. False. An object is an instance of a class.
2. Any number, including zero.
3. A class's instance variables must be declared outside of all methods, and all instance variable declarations should be located at the top of the class definition.
4. Instance variables are usually `private` to further the goal of encapsulation. That means an object's data is harder to access (and, consequently, harder to mess up). The only way for the data to be accessed from outside of the class is if the data's associated `public` methods are called.
5. The `main` method goes in the driver class.
6. Most of a program's code should be in driven classes.
7. To access a `private` instance variable from within a `main` method, you have to use an instantiated object's reference variable and then call an accessor method. In other words, use this syntax:

   *<reference-variable>***.***<accessor-method-call>*

8. A reference variable holds the <u>memory location</u> of an object.

9. Go back to where the method was called, and look at the reference variable that precedes the method name at that point. That reference variable is the one that the method uses whenever `this` is used.

10. For an object's instance variables, the default values are: `int = 0, double = 0.0, boolean = false`.

11. `gus`'s `age` is an instance variable. Instance variables persist for the duration of a particular object. Because the `gus` object is declared in `main`, `gus` and its instance variables (including `age`) persist for the duration of the `main` method.

12. Some reasons to construct a UML class diagram before writing code are the following:
    a) It provides a complete "to do" list. When you are into the details of writing one method, and wondering whether that method should perform a particular function, the diagram reminds you of what other methods might be able to perform that function.
    b) It provides a complete "parts list," like the parts list of a typical user-assembled "kit." This predefined list helps you avoid accidentally generating different and conflicting names for variables and parameters as you write your code.
    c) It's a working document that can change as work progresses. Changing the UML class diagram helps identify needed alterations to previous work.

13. Immediately after the statement `Mouse mickey;` the value of `mickey` would be garbage.

14. False. Normally, for a method that returns a value, you should have a single return statement at the end of the method. However, it's also legal to have return statements in the middle of a method. That might be appropriate in a very short method, where an internal `return` is immediately obvious. If the method is relatively long, however, a reader might not notice an internal `return`. With a large method, it's better practice to arrange things so that there is only one `return`, located at the end of the method.

15. Parameters and local variables both have method scope and persistence. The code inside the method treats parameters just like it treats local variables. The method initializes the local variables, while the method call initializes the parameters.

16. Arguments and parameters are two different words describing data that passes into a called method. An arguments is the method call's name for the data, and a parameter is the method's name for the same data. A parameter is just a copy of the method call's argument, however, so if the called method changes the value of one if its parameters, this does not alter the value of the method call's argument.

17. The standard prefix for an accessor method is `get`.

18. The standard prefix for a mutator method is `set`.

19. The standard prefix for a Boolean method is `is`.

20. To reduce the error in a simulation, you can reduce step size or switch to a step-with-midpoint algorithm. For a given accuracy, the step-with-midpoint algorithm is more efficient.

CHAPTER 7

Object-Oriented Programming—Additional Details

Objectives

- Improve your understanding of the relationship between a reference variable and an object.
- Learn what happens when you assign a reference.
- Learn how Java recycles memory space.
- Learn how to compare the equality of two different objects.
- Be able to swap the data in two different objects.
- See how a reference parameter can enhance data transfer to and from a called method.
- Learn how to execute a sequence of several method calls in the same statement.
- Learn how to create alternative variations for a method.
- Learn how to combine object creation and initialization in a constructor.
- Learn how to avoid code redundancy by calling a constructor from another constructor.
- Learn how and when to use class variables.
- Learn how to write class methods and when to use them.
- Learn how and when to use class constants.

Outline

7.1 Introduction

In Chapter 6, you learned to write simple object-oriented programming (OOP) programs using simple OOP building blocks. In this chapter, you learn to write more advanced OOP programs using more advanced OOP concepts. In particular, you learn the details of what happens behind the scenes when a program instantiates an object and stores its address in a reference variable. That will help you to appreciate and understand what happens when a program assigns one reference variable to another.

One of the OOP concepts you learn about in this chapter is testing objects for equality. It's common to compare primitives for equality. For example, `if (team1Score == team2Score)` asks whether two variables have the same value. It's also common to compare references for equality. That means we're asking whether two references refer to the same object. Comparing references for equality requires a bit more effort, and in this chapter, you learn what that effort entails. Another concept you learn about is what happens behind the scenes when a program passes a reference as an argument. That's important to know because you'll often need to pass references as arguments.

In addition to presenting more advanced OOP concepts, this chapter also presents more advanced applications of what you already know in regard to OOP. For example, you learn to call several methods in succession, all within one statement. That's called *method-call chaining*, and it can lead to more compact and more elegant code. You also learn about *method overloading*. That's when you have different versions of a method, and each version operates on different types of data. That should sound familiar because you saw it with the `Math` class. Remember the two versions of the `Math.abs` method? One version returns the absolute value of a `double`, and one version returns the absolute value of an `int`.

In the previous chapter, you learned how to instantiate an object in one statement (for example, `Mouse gus = new Mouse();`) and assign a value to the object in a separate statement (for example, `gus.setPercentGrowthRate(10);`). In this chapter, you learn how to combine those two tasks into one statement. To do that, you'll use a special kind of method called a *constructor*. Like methods, constructors can be overloaded by using different types of data for the different constructor versions. But unlike methods, constructors are designed specifically for object creation and initialization.

7.2 Object Creation—A Detailed Analysis

Let's start the chapter with a behind-the-scenes detailed look at what happens when a program instantiates an object and stores its address in a reference variable. Having a clear understanding will help when it comes time to understand other OOP operations, and it will help with some debugging efforts.

Consider the following code fragment:

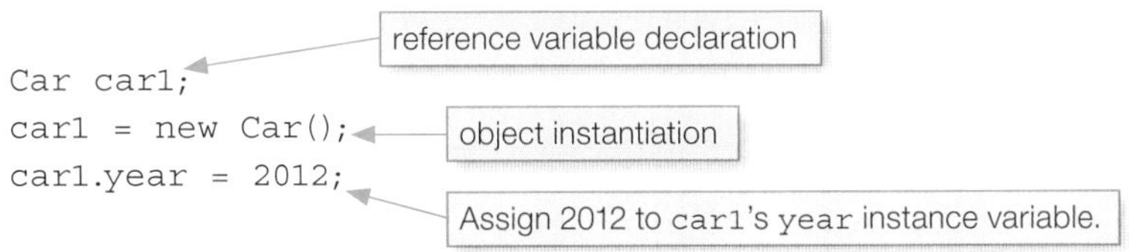

Let's now examine this code in detail one statement at a time.

Statement 1:

The first statement is a variable declaration for the `car1` reference variable. It allocates space in memory for the `car1` reference variable—just the reference variable itself, not an object. Eventually, the `car1` reference variable will hold the address of an object, but since there's no object created for it yet, it doesn't yet hold a legitimate address. What's the default value for a reference variable? It depends. If the reference variable is defined locally within a method (that is, it's a local variable), then it gets garbage initially. If it's defined at the top of a class, above all the method definitions (that is, it's an instance variable), then it gets initialized to `null`. Because Statement 1 doesn't have an access modifier like `public` or `private`, it's reasonable to assume it's a local variable. So `car1` will contain garbage by default, and that's what this picture indicates:

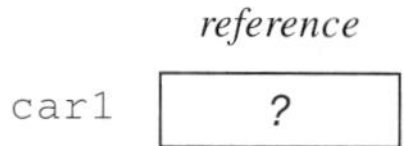

Statement 2:

The second statement's `new` operator allocates space in memory for a new `Car` object, and it initializes members of that new object as specified by the `Car` class definition. The assignment operator assigns the address (memory location) of the allocated space to the `car1` reference variable. Don't forget this operation. Forgetting to instantiate is a common beginner's error.

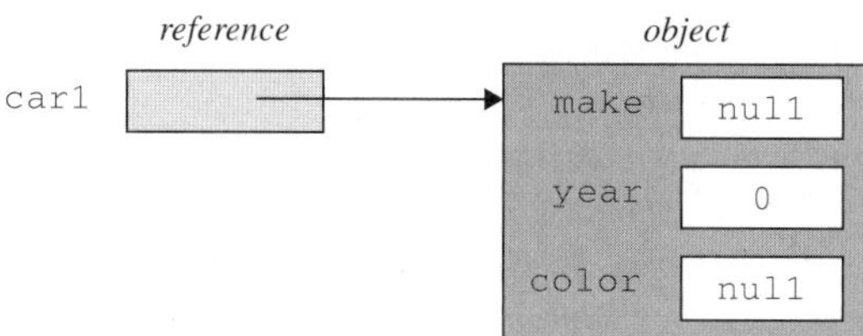

Statement 3:

The third statement uses the `car1` variable's value (the address of a `Car` object) to find a particular `Car` object in memory. Once that `Car` object is found, 2012 is assigned into it. More specifically, 2012 is assigned to the `year` instance variable portion of that `Car` object. Normally, we'd use a method to assign 2012 to `car1`'s `year` instance variable. In the interest of simplification for clarity's sake, we avoided the method call by assuming that `year` is a `public` instance variable.

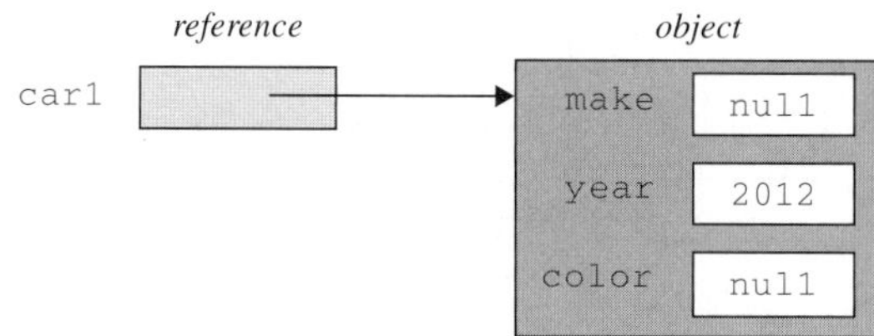

7.3 Assigning a Reference

The result of assigning one reference variable to another is that both reference variables then refer to the same object. Why do they refer to the same object? Because reference variables store addresses, you're actually assigning the right-side reference variable's address into the left-side reference variable. So after the assignment, the two reference variables hold the same address, and that means they refer to the same object. With both reference variables referring to the same object, if the object is updated using one of the reference variables, then the other reference variable will benefit (or suffer) from that change when it attempts to access the object. Sometimes that's just what you want, but if it's not, it can be disconcerting.

An Example

Suppose you want to create two `Car` objects that are the same except for their color. Your plan is to instantiate the first car, use it as a template when creating the second car, and then update the second car's `color` instance variable. Will this code accomplish that?

```
Car johnCar = new Car();
Car stacyCar;
johnCar.setMake("Honda");
johnCar.setYear(2003);
johnCar.setColor("silver");
stacyCar = johnCar;
stacyCar.setColor("peach");
```

This makes `stacyCar` refer to the same object as `johnCar`.

The problem with the above code is that the `stacyCar = johnCar;` statement causes the two references to point to the same single `Car` object. Figure 7.1a illustrates what we're talking about.

Later, we'll see that this *aliasing* (using different names for the same object) can be quite useful, but in this case, it's not what we wanted. In the last statement in the code fragment above, when we use the `setColor` method to change Stacy's car to "peach," we're not specifying the color for a new car. What we're doing is repainting the original car. Figure 7.1a depicts the result. Uh oh . . . John may not be pleased to find his car repainted to peach!

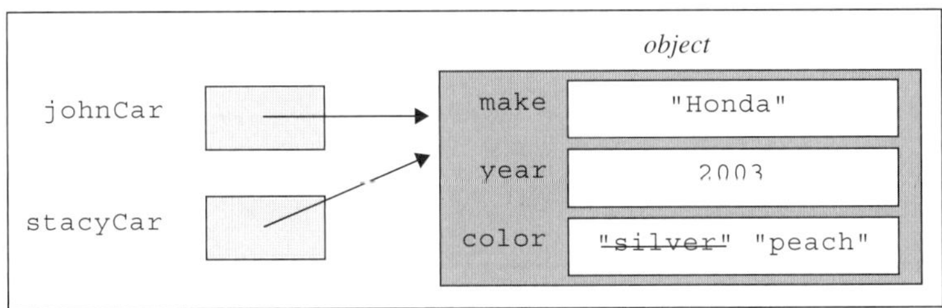

Figure 7.1a Effect of assignment: `stacyCar = johnCar;`
Both reference variables refer to exactly the same object.

If you want to make a copy of a referenced object, you should not assign its reference to another reference. Instead, you should instantiate a new object for the second reference. Then go through each instance variable in the original object, assigning it to its corresponding instance variable in the new object. Figure 7.1b shows what we're talking about, with each dashed line representing an instance variable assignment from the original object to the new object.

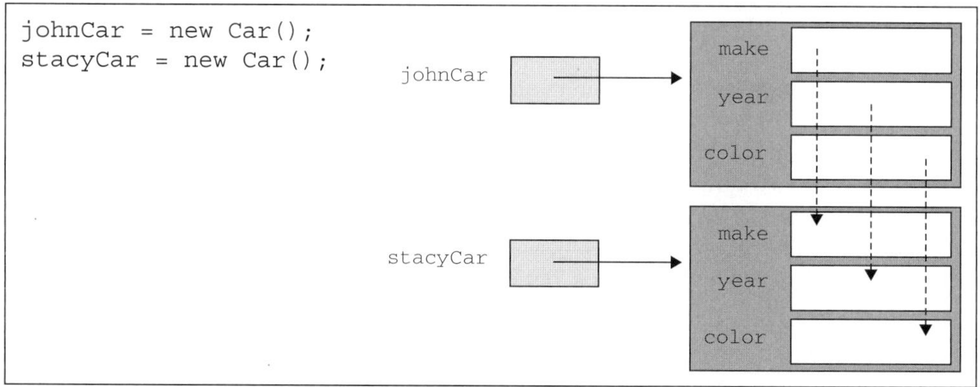

Figure 7.1b Effect of instantiating two separate objects and copying instance variable values from first object into instance variables of second object. Solid arrows are references. Dashed arrows are assignment operations.

To illustrate the strategy outlined in Figure 7.1b, we present the Car program in Figures 7.2 and 7.3. The code includes line numbers to facilitate tracing in an end-of-chapter exercise. Look at the `makeCopy` method in the `Car` class in Figure 7.2. As its name implies, that's the method that's in charge of making a copy of a `Car` object. The `makeCopy` method instantiates a new `Car` object and assigns its reference to a local variable named `car`. Then it copies each of the calling object's instance variable values into `car`'s instance variables. Then it returns `car` to the calling module. By returning `car`, it returns a reference to the newly instantiated `Car` object.

Now look at the driver in Figure 7.3. Note how `main` assigns `makeCopy`'s returned value to `stacyCar`. After `stacyCar` gets the reference to the newly created `Car` object, it calls `setColor` to change the `Car` object's color. Because `stacyCar` and `johnCar` refer to two separate objects, the `stacyCar.setColor("peach")` method call updates only the `stacyCar` object, not the `johnCar` object. Yeah!

Whenever a method finishes, its parameters and locally declared variables are deleted. In our traces, we represent this deletion by drawing a heavy line under all of the terminating method's parameters and local variables. In the `makeCopy` method in Figure 7.2, there is one local variable, the reference variable, `car`. When the `makeCopy` method finishes, the `car` reference variable is deleted. When a reference variable is deleted, the reference it holds is lost, and if that reference is not saved in a separate variable, the program will have no way of finding the object it referred to. In the `makeCopy` method, the `car` reference variable's value does get saved. The assignment is valid because it takes place before `makeCopy`'s local variable gets deleted.

Inaccessible Objects and Garbage Collection

Sometimes you'll want to instantiate a temporary object inside a method, use it for some purpose in that method, and then abandon that object when the method finishes. At other times, you may wish to abandon an object before a method finishes. For example, suppose that in the `main` method in Figure 7.3, after calling `makeCopy` and creating a new `Car` object for `stacyCar`, you want to model John's old car being destroyed in a fire and Stacy volunteering to let him become a co-owner of her new car. You could represent this joint ownership of one car with the statement:

```
johnCar = stacyCar;
```

```
/***************************************************************
* Car.java
* Dean & Dean
*
* This class implements copy functionality for a car.
***************************************************************/

public class Car
{
  private String make;     // car's make
  private int year;        // car's manufacturing year
  private String color;    // car's primary color

  //************************************************************

  public void setMake(String make)
  {
    this.make = make;
  }

  public void setYear(int year)
  {
    this.year = year;
  }

  public void setColor(String color)
  {
    this.color = color;
  }

  //************************************************************

  public Car makeCopy()
  {
    Car car = new Car();   // <-- This instantiates a new object.

    car.make = this.make;
    car.year = this.year;
    car.color = this.color;
    return car;   // <-- This returns a reference to the new object.
  } // end makeCopy

  //************************************************************

  public void display()
  {
    System.out.printf("make= %s\nyear= %s\ncolor= %s\n",
      this.make, this.year, this.color);
  } // end display
} // end class Car
```

Figure 7.2 `Car` class with a `makeCopy` method that returns a reference to a copy of the calling object

```
/****************************************************************
* CarDriver.java
* Dean & Dean
*
* This class demonstrates copying an object.
****************************************************************/

public class CarDriver
{
  public static void main(String[] args)
  {
    Car johnCar = new Car();
    Car stacyCar;

    johnCar.setMake("Honda");
    johnCar.setYear(2003);
    johnCar.setColor("silver");
    stacyCar = johnCar.makeCopy();
    stacyCar.setColor("peach");
    System.out.println("John's car:");
    johnCar.display();
    System.out.println("Stacy's car:");
    stacyCar.display();
  } // end main
} // end class CarDriver
```

This assigns the returned reference to a reference variable in the calling method.

Output:

```
John's car:
make= Honda
year= 2003
color= silver
Stacy's car:
make= Honda
year= 2003
color= peach
```

Figure 7.3 `CarDriver` class that drives the `Car` class in Figure 7.2

Doing this overlays `johnCar`'s previous reference to John's original `Car` object, and that `Car` object becomes inaccessible to the program (abandoned), like `Car` object #1 is in this picture:

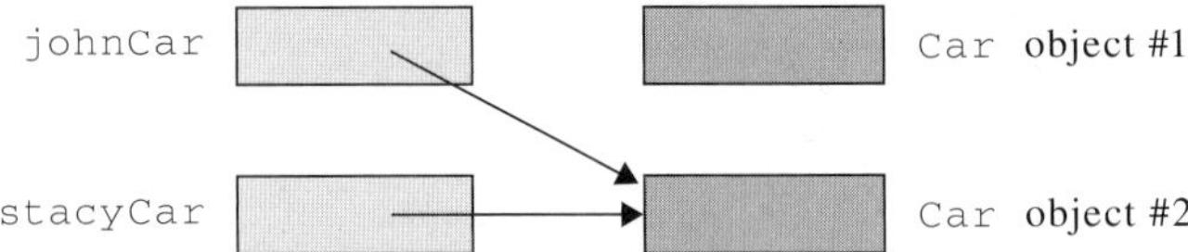

The question is, how does the Java virtual machine (JVM) treat abandoned or inaccessible objects? Inaccessible objects can't participate in the program, so there's no need to keep them around. They become "garbage." In fact, it would be bad to keep them around because they can lead to clogging up the computer's memory. A computer has a finite amount of memory, and each piece of garbage uses up some of that memory. And that means less memory is available for new tasks. If garbage is allowed to accumulate unabated, it would eventually chew up all the *free space* in a computer's memory (free space is the portion of memory that is unused). If there's no free space in memory, there's no space for any new objects, and typically the computer stops working until a reboot kicks the clutter away.

If an inaccessible object persists and uses up space in a computer's memory, that's called a *memory leak*. Memory leaks can occur in computer programs that allocate memory during execution. When a computer language requires the programmer to do something specific to prevent memory leaks, and the programmer forgets to do that, a nasty bug is born—a bug that is very hard to find. In creating the Java language, James Gosling and his Sun co-workers realized this, and they opted to make the language itself deal with the problem. How? By going into the garbage collection business. Not what Dirk and Lenny do when they pick up the trash at your curb every Tuesday, but Java *garbage collection*! Actually, James Gosling didn't invent garbage collection; it's been around since the dawn of garbage. But Java is the first popular programming language to include it as a standard service.[1]

The Java garbage collector is a utility program that searches for inaccessible objects and recycles the space that they occupy by asking the operating system to designate their space in memory as free space. This space might not be used right away, and some computer whiz kid might be able to find some of those old abandoned objects—like wandering through a trash dump, fighting off mean dogs, and looking for furniture—but for practical purposes, you should consider those abandoned objects unrecoverable and gone.

The beauty of Java's automatic garbage collection is that the programmer doesn't have to worry about it—it just happens whenever it's appropriate. And when is it appropriate? Whenever the computer is running low on free space in memory or whenever nothing else is happening, such as when a program is waiting for keyboard input. At that point, the operating system wakes up his buddy, the Java garbage collector, and tells him to go earn his keep.

7.4 Testing Objects for Equality

The previous section illustrated returning a reference from a method. This section illustrates passing a reference to a method to allow the method to read the referenced object's data. One of the most common applications of this occurs in testing two objects for equality. Before looking at this application, it's appropriate to look at the simplest way to evaluate equality.

The == Operator

The == operator works the same for primitive variables and for reference variables. It tests if the values stored in these variables are the same. When applied to reference variables, the == operator returns `true` if and only if the two reference variables refer to the same object; that is, the two reference variables contain the same address and thus are aliases for the same object. For example, what does the following code fragment print?

[1]The concept of garbage collection appeared earlier in the LISP programming language. LISP is a compact language that in principle can do anything, but its primary application is artificial intelligence.

```
Car car1 = new Car();
Car car2 = car1;
if (car1 == car2)
{
  System.out.println("the same");
}
else
{
  System.out.println("different");
}
```

It prints "the same" because car1 and car2 hold the same value—the address of the lone Car object. But if you want to see if two different objects have the same instance-variable values, the == operator is not what you want. For example, what does this code print?

```
Car car1 = new Car();
Car car2 = new Car();

car1.setColor("red");
car2.setColor("red");
if (car1 == car2)
{
  System.out.println("the same");
}
else
{
  System.out.println("different");
}
```

The car1 == car2 expression returns false. Why?

This code prints "different" because car1 == car2 returns false. It doesn't matter that car1 and car2 contain the same data (red). The == operator doesn't look at the object's data; it just looks at whether the two reference variables point to the same object. In this case, car1 and car2 refer to distinct objects, with different storage locations in memory.

The equals Method

If you want to see whether two different objects have the same characteristics, you need to compare the contents of two objects rather than just whether two reference variables point to the same object. To do that, you need an equals method in the object's class definition that compares the two objects' instance variables. Having such an equals method is very common because you often want to test two objects to see whether they have the same characteristics. For Java's API classes, you should use the classes' built-in equals methods. For example, in comparing the contents of two strings, call the String class's equals method. For classes that you implement yourself, adopt the habit of writing your own equals methods.

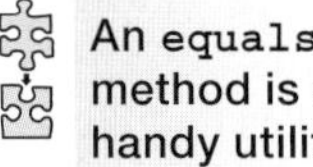

An equals method is a handy utility.

An Example

The following diagram depicts two objects with identical instance variable values. Comparing nathanCar to nickCar with the == operator generates false because the two reference variables point to different objects. However, comparing nathanCar to nickCar with a standard equals method generates true because a standard `equals` method compares instance variable values, and these two objects have identical instance variable values.

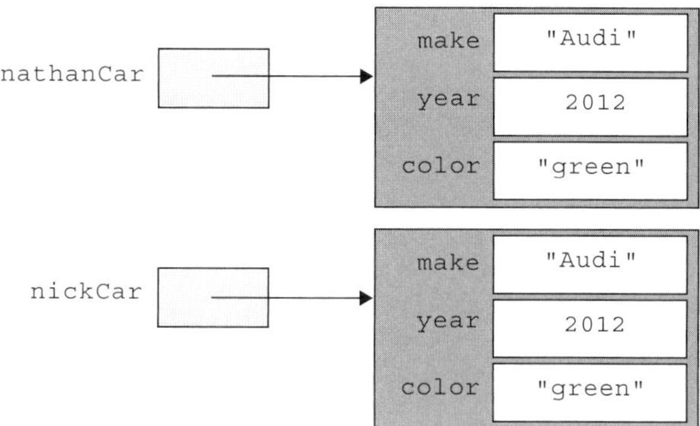

The Car2 program in Figures 7.4 and 7.5 illustrates this example. Figure 7.5's `Car2` class defines an `equals` method, and Figure 7.4's `Car2Driver` class calls the `equals` method while comparing two `Car2` objects. As is common with `equals` method calls, Figure 7.4's `equals` method call is embedded in the condition of an `if` statement. That should make sense when you realize that an `if` statement condition must evaluate to `true` or `false` and an `equals` method does indeed evaluate to `true` or `false`. Typically, an `equals` method evaluates to `true` if the instance variables in two objects contain the same data values, and it evaluates to `false` otherwise. For our Car2 program, the `equals` method evaluates to `true` if `nathanCar` contains the same data (`make`, `year`, and `color`) as `nickCar`. Figure 7.4 shows that `nathanCar` and `nickCar` are assigned the same data. Therefore, the `equals` method returns `true` and the program prints "Cars have identical features."

In the `equals` method call, note how the first `Car2` reference variable, `nathanCar`, appears at the left of the `.equals` and the second `Car2` reference variable, `nickCar`, appears inside the parentheses. Thus, `nathanCar` is the calling object, and `nickCar` is an argument This happens a lot when using two reference variables with a method call—one reference variable will be the calling object and the other one will be the argument.

Let's now examine the `equals` method definition in Figure 7.5. First, note the `equals` method heading. Why is the `return` type `boolean`? Because the `return` type must match the type of the returned value, and `equals` methods always return a Boolean value (either `true` or `false`). Also, note that the type of the `otherCar` parameter is `Car2`. That should make sense when you look back at the `equals` method call in Figure 7.4. It shows that the argument being passed into the `equals` method is `nickCar`, and `nickCar` is a `Car2` reference variable.

OK, now it's time to examine the body of the `equals` method. Notice that there is just one statement—the `return` statement. The return value must be a `boolean`, so the expression after the word `return` must evaluate to either `true` or `false`. This expression is an "anding" together of three `boolean` sub-expressions, each of which evaluates to either `true` or `false`. For the overall expression to be `true`, all three of the sub-expressions must be `true`.

Each sub-expression checks whether a particular instance variable has the same value in the calling object and the passed-in parameter object. For example, to check whether the `year` instance variable has the same value in the calling object and the passed-in parameter object, we do this:

```
this.year == otherCar.year
```

```
/*************************************************************
* Car2Driver.java
* Dean & Dean
*
* This class is a demonstration driver for the Car2 class.
*************************************************************/

public class Car2Driver
{
  public static void main(String[] args)
  {
    Car2 nathanCar = new Car2();
    Car2 nickCar = new Car2();

    nathanCar.setMake("Audi");
    nathanCar.setYear(2012);
    nathanCar.setColor("green");
    nickCar.setMake("Audi");
    nickCar.setYear(2012);
    nickCar.setColor("green");
    if (nathanCar.equals(nickCar))
    {
      System.out.println("Cars have identical features.");
    }
  } // end main
} // end class Car2Driver
```

Note how the equals method call is embedded in an if condition.

Figure 7.4 Car2Driver class that drives the Car2 class in Figure 7.5

In this case, we use the == operator to check for equality. That works fine for the year instance variable because year is an int. But the make and color instance variables are strings, and the == operator is anathema to strings. We must use the equals method for strings! Thus, to check whether the make instance variable has the same value in the calling object and the passed-in parameter object, we do this:

```
this.make.equals(otherCar.make)
```

Hmmm . . . . Does it strike you as odd to use the String class's equals method inside our Car2 class's equals method? That's perfectly OK—the compiler doesn't care if two methods happen to have the same name, as long as they are in different classes. That's part of the beauty of encapsulation!

Can you think of another way to write the body of the Car2 class's equals method? We might have used that boolean expression to the right of the return keyword as the condition of an if statement and then put return true in the if clause and return false in the else clause. But that would have been a harder and longer way to do the same thing—and probably more confusing, too, because it would have required more parentheses. Although Figure 7.5's return statement might appear at first glance to be a Cerberean rat's nest,[2] most veteran programmers would consider it to be rather elegant.

[2]You probably already know what a "rat's nest" is—a tangled mess. But how about "Cerberean"? In Greek mythology, Cerberus is a vicious, three-headed dog creature that guards the entrance to Hades (the world of the dead). We say our return statement might appear to be a Cerberean rat's nest because it's complicated and it has three parts. Which would you rather meet in a dark alley—a vicious, three-headed dog creature or a complicated return statement?

```
/*******************************************************************
* Car2.java
* Dean & Dean
*
* This class implements equals functionality for a car.
*******************************************************************/

public class Car2
{
  private String make;
  private int year;
  private String color;

  //****************************************************************

  public void setMake(String make)
  {
    this.make = make;
  }

  public void setYear(int year)
  {
    this.year = year;
  }

  public void setColor(String color)
  {
    this.color = color;
  }

  //****************************************************************

  // This method tests whether two cars hold the same data.

  public boolean equals(Car2 otherCar)
  {
    return this.make.equals(otherCar.make) &&
           this.year == otherCar.year &&
           this.color.equals(otherCar.color);
  } // end equals
} // end class Car2
```

This compares all instance variables.

Figure 7.5 Car2 class with an equals method

Suppose you want uppercase colors to be considered the same as lowercase colors. In other words, you want a silver 2005 Ford to be considered the same as a Silver 2005 Ford. How should you change the code to handle that? Use `equalsIgnoreCase` instead of `equals` when comparing the color strings:

```
this.color.equalsIgnoreCase(otherCar.color)
```

This shows that you can make your `equals` method return `true` when there is only approximate equality, where you define "approximate" however you wish. We'll discuss the `equals` method in more depth in Chapter 14.

7.5 Passing References as Arguments

By now, you should be fairly comfortable with the concept of passing an argument to a method. We've covered all you need to know about passing primitive types as arguments. But you still need to know a bit more about passing references as arguments. In the example in Figure 7.4, we passed the `nickCar` reference as an argument to the `equals` method. The `equals` method call assigned the `nickCar` reference to its `otherCar` parameter, and then it used the `otherCar` parameter to read the object's data. In that example, we used a passed-in reference to read an object's data. Now let's use a passed-in reference to update an object's data.

Suppose you pass a reference variable to a method, and inside the method, you update the reference variable's instance variables. What happens? Remember that a reference variable holds the address of an object, not the object itself. So in passing a reference variable argument to a method, a copy of the object's address (not a copy of the object itself) is passed to the method and stored in the method's parameter. Because the parameter and the argument hold the same address value, they point to the same object. Thus, if the parameter's instance variables are updated, then the update simultaneously updates the argument's instance variables in the calling module. This is a case where *aliasing* (using two names for the same thing) is really handy.

Person-Swapping Example

Let's see if you understand all of this reference-passing stuff by putting it in the context of a complete program. See the Person program in Figures 7.6 and 7.7. The Person program swaps names for two `Person` objects. As shown in Figure 7.6's `main` method, the `person1` reference variable starts with the name "Jonathan" and the `person2` reference variable starts with the name "Benji." After the `swapPerson` method call, `person1` has the name "Benji," and `person2` has the name "Jonathan." The `swapPerson` method swaps the names by taking advantage of the phenomenon discussed above—if a reference variable is passed to a method, then the parameter and the argument refer to the same object, and an update to one means an update to the other as well. Bottom line: When you pass a reference to a method, you enable the method to modify the referenced object.

General-Purpose Swapping Algorithm

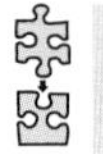

How do you swap two values?

Before digging deeper into the Person program's code, let's come up with a general-purpose swapping algorithm. Having to swap two values is a very common programming requirement, so you should make sure that you fully understand how to do it.

Suppose you're asked to provide an algorithm that swaps the contents of two variables, x and y. To make the goal more concrete, you are given the following algorithm skeleton. Replace <*Insert swap code here.*> with appropriate pseudocode so that the algorithm prints `x = 8, y = 3`.

```
x ← 3
y ← 8
<Insert swap code here.>
print "x = " + x + ", y = " + y
```

```
/**************************************************************
* PersonDriver.java
* Dean & Dean
*
* This class is a demonstration driver for the Person class.
**************************************************************/

public class PersonDriver
{
  public static void main(String[] args)
  {
    Person person1 = new Person();
    Person person2 = new Person();

    person1.setName("Jonathan");
    person2.setName("Benji");
    System.out.println(person1.getName() + ", " +
      person2.getName());

    person1.swapPerson(person2);
    System.out.println(person1.getName() + ", " +
      person2.getName());
  } // end main
} // end class PersonDriver
```

This argument allows the called method to modify the referenced object.

Output:

```
Jonathan, Benji
Benji, Jonathan
```

Figure 7.6 Driver for a program that implements swapping by passing a reference to a method

Would the following code work? Would it swap x and y's contents successfully?

```
y ← x
x ← y
```

The first statement puts x's original value into y. The second statement attempts to put y's original value into x. Unfortunately, the second statement doesn't work because y's original value is gone (overwritten by x in the first statement). If you inserted the above code into the above algorithm, the algorithm would print:

```
x = 3, y = 3
```

That's not what you want! The trick is to save the value of y before you wipe it out with x's value. How do you save it? Use a temporary variable like this:

Swapping requires a temporary variable.

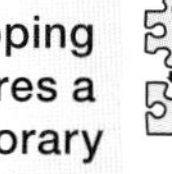

```
temp ← y
y ← x
x ← temp
```

Person-Swapping Example—Continued

Now look at the `Person` class in Figure 7.7. In particular, let's examine how the `swapPerson` method implements the swapping algorithm. The swapped items are the passed-in object's name and the calling object's name. The passed-in object is accessed via the `otherPerson` parameter. Note how we access the passed-in object's name with `otherPerson.name`. And note how we access the calling object's name with `this.name`. And finally, note how we use a `temp` local variable as temporary storage for `otherPerson.name`.

```
/**************************************************************
* Person.java
* Dean & Dean
*
* This stores, retrieves, and swaps a person's name.
**************************************************************/

public class Person
{
  private String name;

  //************************************************************

  public void setName(String name)
  {
    this.name = name;
  }

  public String getName()
  {
    return this.name;
  }

  //************************************************************

  // This method swaps the names for two Person objects.

  public void swapPerson(Person otherPerson)
  {
    String temp;

    temp = otherPerson.name;
    otherPerson.name = this.name;
    this.name = temp;
  } // end swapPerson
} // end class Person
```

swapping algorithm

Figure 7.7 `Person` class, which implements swapping by passing a reference to a method

7.6 Method-Call Chaining

At this point, you should be fairly comfortable with calling a method. Now it's time to go one step further. In this section, you will learn to call several methods in succession, all within one statement. That's called *method-call chaining,* and it can lead to more compact code.

If you look back at Figures 7.3 and 7.4, you'll see several instances where we call several methods one after another, and we use a separate statement for each successive method call, like this code fragment from Figure 7.4:

```
nathanCar.setMake("Audi");
nathanCar.setYear(2012);
```

Wouldn't it be nice to be able to chain the method calls together like this?

```
nathanCar.setMake("Audi").setYear(2012);
```

Method-call chaining is an option, not a requirement. So why use it? Because it can often lead to more elegant code—more compact and easier to understand. But don't overdo it. Using just the right amount of chaining is an art, and moderation is a virtue.

Let's look at method-call chaining in the context of a complete program. See the method-call chain (indicated by a callout) in Figure 7.8's `Car3Driver` class. Because left-to-right precedence applies, the `car.setMake` method executes first. The `setMake` method returns the calling object, which is the `car` object at the left of `car.setMake`. The returned `car` object is then used to call the `setYear` method. In a similar fashion, the `setYear` method returns the calling object, `car`, and it is used to call the `prinIt` method.

```
/***************************************************************
* Car3Driver.java
* Dean & Dean
*
* This drives Car3 to illustrate method-call chaining.
***************************************************************/

public class Car3Driver
{
  public static void main(String[] args)
  {
    Car3 car = new Car3();

    car.setMake("Honda").setYear(2012).printIt();
  } // end main
} // end class Car3Driver
```

Use dots to chain together method calls.

Figure 7.8 Car3 program driver, which illustrates method-call chaining

Method call chaining doesn't work by default. If you want to enable method-call chaining for methods from the same class, you need the following two items in each method definition:

1. The last line in the method body should return the calling object by specifying `return this;`
2. In the method heading, the return type should be the method's class name.

We've implemented those items in the `Car3` class in Figure 7.9. Verify that `setMake` and `setYear` are enabled properly for method-call chaining. Specifically, verify that (1) the last line in each method body is `return this;`, and (2) in each method heading, the return type is the method's class name, `Car3`.

```
/**********************************************************
* Car3.java
* Dean & Dean
*
* This class illustrates methods that can be chained.
**********************************************************/

public class Car3
{
  private String make;
  private int year;

  //*******************************************************

  public Car3 setMake(String make)
  {
    this.make = make;
    return this;
  } // end setMake

  public Car3 setYear(int year)
  {
    this.year = year;
    return this;
  } // end setYear

  //*******************************************************

  public void printIt()
  {
    System.out.println(this.make + ", " + this.year);
  } // end printIt
} // end class Car3
```

The return type is the same as the class name.

Return the calling object.

Figure 7.9 `Car3` class

Whenever you finish a method with a `return this;` statement, you're making it possible to use the same object to call the next method in the chain. However, you can also chain methods called by different types of objects. Just arrange the chain so that the reference type returned by each preceding method matches the class of each following method. So, in general, to make a method chainable, do these two things:

1. In the method heading, specify the return type as the class of a potential following method.

2. Finish the method body with:
   `return` <*reference-to-object-that-will-call-the-following-method*>;

Here is a code fragment from earlier in the book that illustrates chaining of two methods defined in the Java API:

```
ch = stdIn.nextLine().charAt(0);
```

The `stdIn` variable is a reference to an object of the `Scanner` class. It calls `Scanner`'s `nextLine` method, which returns a reference to an object of the `String` class. Then that object calls `String`'s `charAt` method, which returns a character.

7.7 Overloaded Methods

Up until this point, all of the methods we defined for a given class have had unique names. But if you think back to some of the Java API methods presented in Chapter 5, you'll recall that there were several examples where the same name identified more than one method in the same class (For example, there are two `Math.abs` methods-one for a `double` and one for an `int`). This section will show you how to do this in classes you write.

What Are Overloaded Methods?

Overloaded methods are two or more methods in the same class that use the same name to indicate that they perform the same kind of operation in a different context. Since they use the same name, the compiler needs something else besides the name in order to distinguish them. Parameters to the rescue! The parameters establish the context. To make two overloaded methods distinguishable, you define them with different parameters. More specifically, you define them with a different number of parameters or different types of parameters. The combination of a method's name, the number of its parameters, and the types of its parameters is called the method's *signature*. Each distinct method has a distinct signature.

The Java API library provides many examples of overloaded methods. Here are the headings of four overloaded `max` methods from the `Math` class:

```
static double max(double a, double b)
static float max(float a, float b)
static int max(int a, int b)
static long max(long a, long b)
```

Note that each method heading has a distinct signature because each heading is distinguishable in terms of number and types of parameters.

Suppose you're writing software for an alarm system that uses a word, a number, or a combination of a word and a number for the alarm system's password. To activate (or deactivate) the alarm, your program first needs to call a validation method that checks the passed-in word, number, or a combination of word and number. Here are the headings for overloaded methods that could implement that functionality:

```
boolean validate(String word)
boolean validate(int number)
boolean validate(String word, int num)
boolean validate(int num, String word)
```

Here's how you could call the first and second methods:

```
if (alarm.validate("soccer"))
{
  ...
}
else if (alarm.validate(54321))
{
  ...
```

The previous code works fine, but suppose you try to implement an additional `validate` method to handle a different kind of password—one that allows interspersed letters and digits within one string. Here is the proposed method heading:

```
void validate(String code)
```

Would such a method be legal? No. Its signature is the same as the first `validate` method's signature—same method names and same number and types of parameters. It doesn't matter that the parameter names are different—`word` and `code`. It's the parameter types that matter, not the parameter names. It doesn't help that the return types are different—`boolean` and `void`. The return type is not part of the signature. Because the signatures are the same but the return types are different, if you try to include these two method headings in the same class, the compiler will think you're trying to define two different versions of exactly the same method. That will irritate the compiler. Be prepared for it to snarl back at you with a "duplicate definition" compile-time error message.

Benefit of Overloaded Methods

When should you use overloaded methods? When you need to perform essentially the same task with different parameters. For example, the `validate` methods described above perform essentially the same basic task—they determine whether a given value or values form a valid password. But they perform the task on different sets of parameters. Given that situation, overloaded methods are a perfect fit.

Note that the use of overloaded methods is never an absolute requirement. As an alternative, you can use different method names to distinguish different methods. So why are the above `validate` method headings better than the below method headings?

```
boolean validateUsingWord(String word)
boolean validateUsingNumber(int number)
boolean validateUsingWordThenNumber(String word, int num)
boolean validateUsingNumberThenWord(int num, String word)
```

As these examples suggest, using different method names is cumbersome. With only one method name, the name can be simple. As a programmer, wouldn't you prefer to use and remember just one simple name rather than several cumbersome names?

A Complete Example

Look at the class in Figure 7.10. It uses overloaded `setHeight` methods. Both methods assign a `height` parameter to a `height` instance variable. The difference is the technique for assigning the height's units. The first method automatically assigns a hard-coded "cm" (for centimeters) to the `units` instance variable. The second method assigns a user-specified `units` parameter to the `units` instance variable. The second method thus requires two parameters, `height` and `units`, whereas the first method requires only one parameter, `height`. The two methods perform pretty much the same task, with only a slight variation. That's why we want to use the same name and "overload" that name.

Now look at the driver in Figure 7.11 and its two `setHeight` method calls. For each method call, can you tell which of the two overloaded methods is called? Figure 7.11's first method call, `setHeight(72.0, "in")`, calls Figure 7.10's second `setHeight` method because the two arguments in the method call match the two parameters in the second method's heading. Figure 7.11's second method call, `setHeight(180.0)`, calls Figure 7.10's first `setHeight` method because the one argument in the method call matches the one parameter in the first method's heading.

```
/*****************************************************************
* Height.java
* Dean & Dean
*
* This class stores and prints height values.
*****************************************************************/

class Height
{
  double height;  // a person's height
  String units;   // like cm for centimeters

  //**************************************************************

  public void setHeight(double height)
  {
    this.height = height;
    this.units = "cm";
  }

  //**************************************************************

  public void setHeight(double height, String units)
  {
    this.height = height;
    this.units = units;
  }

  //**************************************************************

  public void print()
  {
    System.out.println(this.height + " " + this.units);
  }
} // end class Height
```

Figure 7.10 Height class with overloaded methods

Calling an Overloaded Method from Within an Overloaded Method

Suppose you have overloaded methods and you want one of the overloaded methods to call another one of the overloaded methods. Figure 7.12 provides an example that shows how to do that. Figure 7.12's `setHeight` method is an alternative version of Figure 7.10's one-parameter `setHeight` method. Note how it calls the two-parameter `setHeight` method.

The additional method call makes the program slightly less efficient, but some might consider it more elegant because it eliminates code redundancy. In Figure 7.10, `this.height = height;` appears in both methods, and that's code redundancy—albeit trivial code redundancy.

Why is there no reference variable dot at the left of the `setHeight` method call in the body of the method in Figure 7.12? Because if you're in an instance method, and if you call another method that's

```
/*********************************************************************
* HeightDriver.java
* Dean & Dean
*
* This class is a demonstration driver for the Height class.
*********************************************************************/

public class HeightDriver
{
  public static void main(String[] args)
  {
    Height myHeight = new Height();

    myHeight.setHeight(72.0, "in");
    myHeight.print();
    myHeight.setHeight(180.0);
    myHeight.print();
  } // end main
} // end class HeightDriver
```

Figure 7.11 `HeightDriver` class that drives the `Height` class in Figure 7.10

```
public void setHeight(double height)
{
  setHeight(height, "cm");
}
```

Do not put a reference variable dot prefix here.

Figure 7.12 Example of a method that calls another method in the same class. This helps avoid duplication of code details and possible internal inconsistencies.

in the same class, the reference variable dot prefix is unnecessary. And in this case, the two overloaded `setHeight` methods are instance methods and they are indeed in the same class.

With no reference variable dot prefix in Figure 7.12's `setHeight(height, "cm");` method call, you might be thinking that the method call has no calling object. Actually, there is an implied calling object; it's the same calling object that called the current method. Review quiz: How can you access the current method's calling object? Use the `this` reference. If you want to make the `this` reference explicit, you can add it to Figure 7.12's `setHeight` method call as follows:

```
this.setHeight(height, "cm");
```

We point out this alternative syntax not because we want you to use it, but because we want you to get a clearer picture of calling object details.

Program Evolution

The ability to overload a method name promotes graceful program evolution because it corresponds to how natural language regularly overloads the meanings of words. For example, the first version of your program might define just the one-parameter version of its `setHeight` method. Later, when you decide to enhance your program, it's easier for your existing users if you minimize the new things they have to learn. In this case, you let them either keep using the original method or switch to the improved method. When they want to use the improved method, all they have to remember is the original method name and adding a second argument, for units, to the method call. That's an almost obvious variation, and it's easier to remember than a different method name. It's certainly easier than being forced to learn a new method name for the old task—which would be a necessary cost of upgrading if method overloading were not available.

Keep it simple by reusing good names.

7.8 Constructors

Up to this point, we have used mutators to assign values to the instance variables in newly instantiated objects. That works OK, but it requires having and calling one mutator for each instance variable. As an alternative, you could use a single method to initialize all of an object's instance variables as soon as possible after you create that object. For example, in this chapter's `Car` class in Figure 7.2, instead of defining three mutator methods, you could define a single `initCar` method to initialize `Car` objects. Then you could use it like this:

```
Car lanceCar = new Car();
lanceCar.initCar("Ford", 2011, "lime");
```

This code fragment uses one statement to allocate space for a new object, and it uses another statement to initialize that object's instance variables. Because the instantiation and initialization of an object is so common, wouldn't it be nice if there were a single statement that could handle both of these operations? There is such a statement, and here it is:

```
Car lanceCar = new Car("Ford", 2011, "lime");
```

This unifies the creation of an object and the initialization of its instance variables in just one call. It guarantees that an object's instance variables are initialized as soon as the object is created. The code that follows the word `new` should remind you of a method call. Both that code and a method call consist of a programmer-defined word (`Car` in this case) and then parentheses around a list of items. You can think of that code as a special method call, but it's so special that it has its own name. It's used to construct objects, so it's called a *constructor*.

What Is a Constructor?

A constructor is a method-like entity that's called automatically when an object is instantiated. The above `new Car("Ford", 2011, "lime")` object instantiation calls a constructor named `Car` that has three parameters—a `String`, an `int`, and a `String`. Here's an example of such a constructor:

```
public Car(String m, int y, String c)
{
  this.make = m;
  this.year = y;
  this.color = c;
}
```

As you can see, this constructor simply assigns passed-in parameter values to their corresponding instance variables. After executing a constructor, the JVM returns the address of the newly instantiated and initialized object to where the constructor was called. In the above `Car lanceCar = new Car("Ford", 2011, "lime")` declaration, the address of the instantiated `Car` object gets assigned to the `lanceCar` reference variable.

There are several constructor details you should know before looking at a complete program example. A constructor's name must be the same as the class it's associated with. Thus, a `Car` class's constructor must be named `Car`, with an uppercase "C."

In the heading of a method, you must include a return type, so you might expect the same requirement for the heading of a constructor. Nope. Return types are not used in constructor headings[3] because a constructor call (with `new`) automatically returns a reference to the object it constructs, and the type of this object is always specified by the constructor name itself. Just specify `public` at the left and then write the class name (which is the name of the constructor).

A Complete Example

Let's now look at a complete program example that uses a constructor. See the Car4 program in Figures 7.13 and 7.14. In Figure 7.13, note that we put the constructor above the `getMake` method. In all class definitions, it's good style to put constructors above methods.

```
/***************************************************
* Car4.java
* Dean & Dean
*
* This class stores and retrieves data for a car.
***************************************************/

public class Car4
{
  private String make;    // car's make
  private int year;       // car's manufacturing year
  private String color;   // car's primary color

  //************************************************

  public Car4(String m, int y, String c)
  {
    this.make = m;
    this.year = y;
    this.color = c;
  } // end constructor

  //************************************************

  public String getMake()
  {
    return this.make;
  } // end getMake
} // end class Car4
```

constructor definition

Figure 7.13 `Car4` class, which has a constructor

[3]If you try to define a constructor with a return type specification, the compiler will not recognize it as a constructor and will think it is an ordinary method instead.

```
/*************************************************************
* Car4Driver.java
* Dean & Dean
*
* This class is a demonstration driver for the Car4 class.
*************************************************************/

public class Car4Driver
{
  public static void main(String[] args)
  {
    Car4 lanceCar = new Car4("Ford", 2011, "lime");
    Car4 azadehCar = new Car4("Volt", 2014, "red");

    System.out.println(lanceCar.getMake());
  } // end main
} // end class Car4Driver
```

constructor calls

Output:

```
Ford
```

Figure 7.14 `Car4Driver` class, which drives the `Car4` class in Figure 7.13

Accommodating Java's Fickle Default Constructor

Any time you instantiate an object (with `new`), there must be a matching constructor. That is, the number and types of arguments in your constructor call must match the number and types of parameters in a defined constructor. But until recently, we've instantiated objects without any explicit constructor. So were those examples wrong? No. They all used a zero-parameter freebie *default constructor* that the Java compiler automatically provides if and only if there is no explicitly defined constructor. The Employee program in Figures 7.15a and 7.15b illustrates the use of Java's implicit zero-parameter default constructor.

In Figure 7.15a, note how `main`'s `new Employee()` code calls a zero-parameter constructor. But Figure 7.15b does not define a zero-parameter constructor. No problem. Because there are no other constructors, the Java compiler provides the default zero-parameter constructor, and it matches up with the `new Employee()` zero-argument constructor call.

```
public class EmployeeDriver
{
  public static void main(String[] args)
  {
    Employee emp = new Employee();

    emp.readName();
  } // end main
} // end class EmployeeDriver
```

zero-parameter constructor call

Figure 7.15a Driver for the Employee program

```
import java.util.Scanner;

public class Employee
{
  private String name;

  //*****************************************

  public void readName()
  {
    Scanner stdIn = new Scanner(System.in);

    System.out.print("Name: ");
    this.name = stdIn.nextLine();
  } // end readName
} // end class Employee
```

Figure 7.15b Driven class for the Employee program
This works even though there is no explicitly defined constructor because the Java compiler supplies a matching default zero-parameter constructor.

Note that as soon as you define any kind of constructor for a class, Java's default constructor becomes unavailable. So if your class contains an explicit constructor definition, and if `main` includes a zero-argument constructor call, you must also include an explicit zero-parameter constructor in your class definition.

See the Employee2 program in Figures 7.16a and 7.16b. The driven class in Figure 7.16a compiles successfully, but the driver in Figure 7.16b generates a compilation error. As in Figure 7.15a, the driver code in Figure 7.16b calls a zero-parameter constructor. It worked before, so why doesn't it work this time? This time, the driven class in Figure 7.16a explicitly defines a constructor, so Java does not provide a default zero-parameter constructor. And without that constructor, the compiler complains that there's no matching constructor for the zero-parameter constructor call. How can you fix the Employee2 program to get rid of this error? Add the following zero-parameter `Employee2` constructor to your `Employee2` class:

```
public Employee2()
{ }
```

That's an example of a *dummy constructor*. It's called a dummy constructor because it doesn't do anything other than satisfy the compiler. Note how the braces are on a line by themselves with a blank space between them. That's a style issue. By writing the dummy constructor like that, it makes the empty braces more prominent and clearly shows the intent of the programmer to make the constructor a dummy constructor.

Initializing Instance Constants with a Constructor

As you've seen, the purpose of a constructor is to initialize; that is, assign initial values to an object's attributes. Normally when you think of an object's attributes, you think of its instance variables. But objects can also have instance named constants as attributes (normally called *instance constants*). Constructors are especially important for initializing an object's instance constants. In this subsection, we'll discuss instance constants and how they're initialized within constructors.

In the past, you've seen named constants declared within methods. When a named constant is declared within a method, it's called a *local named constant*, and its scope is limited to that one method. If you want

```
import java.util.Scanner;

public class Employee2
{
  private String name;

  //*************************************

  public Employee2(String n)
  {
    this.name = n;
  } // end constructor

  //*************************************

  public void readName()
  {
    Scanner stdIn = new Scanner(System.in);

    System.out.print("Name: ");
    this.name = stdIn.nextLine();
  } // end readName
} // end class Employee2
```

This one-parameter constructor is the only modification to the Employee class.

Figure 7.16a Driven class for the Employee2 program

```
public class Employee2Driver
{
  public static void main(String[] args)
  {
    Employee2 waitress = new Employee2("Wen-Jung Hsin");
    Employee2 hostess = new Employee2();

    hostess.readName();
  } // end main
} // end class Employee2Driver
```

Zero-parameter constructor call generates a compilation error.

Figure 7.16b Driver for the Employee2 program

an attribute that's constant throughout the life of a particular object, you'll need another kind of named constant, an instance constant. Study Figure 7.17a's Employee3 program. It improves upon our previous employee programs because it uses an instance constant to store an employee's name rather than an instance variable. Here is the instance constant declaration for an employee's name:

```
public final String NAME;
```

The `final` modifier establishes the `NAME` variable as a named constant. It tells the compiler to generate an error if your program attempts to assign a value to the named constant after it's been initialized. We use all uppercase for `NAME` because that is the standard coding convention for named constants.

```
/*************************************************************
* Employee3.java
* Dean & Dean
*
* This gives an employee a permanent name.
*************************************************************/

public class Employee3
{
  public final String NAME;  <---- declaration of instance constant

  //**********************************************************

  public Employee3(String name)
  {
    this.NAME = name;  <---- initialization of instance constant
  } // end constructor
} // end class Employee3
```

Figure 7.17a `Employee3` class uses an instance constant

You should declare an instance constant at the top of its class, above the class's instance variables. Although it's legal to initialize an instance constant as part of its declaration, normally you should not do that. Instead, you should initialize it within a constructor. This allows you to initialize instance constants with different values for different objects. Thus, an instance constant can represent an attribute whose value varies from one object to another, but remains constant throughout the life of any particular object. It represents an inalienable attribute of that object, an attribute that permanently distinguishes that object from all other objects in the same class. With the `Employee3` class using an instance constant for employee names, the `NAME` attribute reflects the fixed nature of people's names.

Because the `final` modifier keeps a named constant from being changed after it's initialized, it's safe to make an instance constant `public`. This makes it especially easy to determine the value of an object's permanent attributes. Just use this syntax:

<reference-variable> . *<instance-constant>*

In Figure 7.17a, here's how the Employee3 program's constructor uses the above syntax to initialize an employee's name (`this` is the reference variable for the constructor's calling object and `NAME` is the instance constant):

```
this.NAME = name;
```

And in Figure 7.17b, here's how the Employee3 program's `main` method uses the above syntax to print the `waitress` employee's name (`waitress` is the reference variable and `NAME` is the instance constant):

```
System.out.println(waitress.NAME);
```

```
/*************************************************************
* Employee3Driver.java
* Dean & Dean
*
* This instantiates an object and prints permanent attribute.
*************************************************************/

public class Employee3Driver
{
  public static void main(String[] args)
  {
    Employee3 waitress = new Employee3("Angie Klein");

    System.out.println(waitress.NAME);
  } // end main
} // end class Employee3Driver
```

direct access to instance constant

Output:

```
Angie Klein
```

Figure 7.17b Driver for the `Employee3` class in Figure 7.17a

Elegance

As described at the beginning of this section, you don't have to implement a constructor to assign initial values to a newly created object. As an alternative, you can instantiate an object by calling the default constructor with no arguments, and then call a method that assigns initial values to the object. For example:

```
Car lexiCar = new Car();
lexiCar.initCar("Tesla", 2014, "zircon blue");
```

That works, but it's much less elegant than using a constructor. By using a constructor, you intimately tie your instance variable initializations with the object you're creating. Also, you simplify things by (1) avoiding a separate method call step and (2) avoiding the need to dream up a separate name for the initialization method (because the constructor's name equals the class name). Bravo, constructors!

7.9 Overloaded Constructors

Overloading a constructor is like overloading a method. Constructor overloading occurs when there are two or more constructors with the same name and different parameters. Like overloaded methods, overloaded constructors have approximately the same semantics, but different syntax. Overloaded constructors are very common (more common than overloaded methods). That's because you'll often want to be able to create objects with different amounts of initialization. Sometimes you'll want to pass in initial values to the constructor. At other times, you'll want to refrain from passing in initial values to the constructor and rely on assigning values later on. To enable both of those scenarios, you need overloaded constructors—one constructor with parameters and one constructor without parameters.

An Example

Suppose you want to implement a `Fraction` class, which stores the numerator and denominator for a given fraction. The `Fraction` class also stores the fraction's quotient, which is produced by dividing the numerator by the denominator. Normally, you'll instantiate the `Fraction` class by passing a numerator argument and a denominator argument to a two-parameter `Fraction` constructor. But for a whole number, you should instantiate a `Fraction` class by passing just one argument (the whole number) to a `Fraction` constructor, rather than passing two arguments. For example, to instantiate a `Fraction` object for the whole number, 3, you should pass in just a 3 to a `Fraction` constructor, rather than a 3 for the numerator and a 1 for the denominator. To handle two-argument `Fraction` instantiations as well as one-argument `Fraction` instantiations, you need overloaded constructors. One way to begin solving a problem is to write a driver that shows how you want the solution to be used. With that in mind, we present a driver in Figure 7.18 that illustrates how the proposed `Fraction` class and its overloaded constructors can be used. The driver's code includes line numbers to facilitate later tracing.

```
/*************************************************************
* FractionDriver.java
* Dean & Dean
*
* This driver class demonstrates the Fraction class.
*************************************************************/

public class FractionDriver
{
  public static void main(String[] args)
  {
    Fraction a = new Fraction(3, 4);
    Fraction b = new Fraction(3);

    a.printIt();
    b.printIt();
  } // end main
} // end class FractionDriver
```

calls to overloaded constructors

Sample session:

```
3 / 4 = 0.75
3 / 1 = 3.0
```

Figure 7.18 `FractionDriver` class, which drives the `Fraction` class in Figure 7.19

Assume that within the `Fraction` class, `numerator` and `denominator` are `int` instance variables and `quotient` is a `double` instance variable. The two-parameter constructor should look something like this:

```
public Fraction(int n, int d)
{
  this.numerator = n;
  this.denominator = d;
  this.quotient = (double) this.numerator / this.denominator;
}
```

Why the (double) cast? Without it, we'd get integer division and truncation of fractional values. The cast converts numerator into a double, the double numerator promotes the denominator instance variable to double, floating-point division occurs, and fractional values are preserved. Our cast to double also provides a more graceful response if the denominator is zero. Integer division by zero causes the program to crash. But floating-point division by zero is acceptable. Instead of crashing, the program prints "Infinity" if the numerator is positive or "-Infinity" if the numerator is negative.

Make it robust.

For a whole number like 3, we could call the above two-parameter constructor with 3 as the first argument and 1 as the second argument. But we want our Fraction class to be friendlier. We want it to have another (overloaded) constructor which has just one parameter. This one-parameter constructor could look like this:

```
public Fraction(int n)
{
  this.numerator = n;
  this.denominator = 1;
  this.quotient = (double) this.numerator;
}
```

Calling a Constructor from Within Another Constructor

The two constructors above contain duplicate code. Duplication makes programs longer. More importantly, it introduces the possibility of inconsistency. Earlier, we used overloaded methods to avoid this kind of danger. Instead of repeating code as in Figure 7.10, in Figure 7.12, we inserted a call to a previously written method that already had the code we wanted. You do the same thing with constructors; that is, you can call a previously written constructor from within another constructor. Constructor calls are different from method calls in that they use the reserved word new, which tells the JVM to allocate space in memory for a new object. Within the original constructor, you could use the new operator to call another constructor. But that would create a separate object from the original object. And most of the time, that's not what you want. Normally, if you call an overloaded constructor, you want to work with the original object, not a new, separate object.

Avoid duplicate code.

To avoid creating a separate object, Java designers came up with special syntax that allows an overloaded constructor to call one of its partner overloaded constructors such that the original object is used. Here is the syntax:

this(<*arguments-for-target-constructor*>);

A this(<*arguments-for-target-constructor*>) constructor call may appear only in a constructor definition, and it must appear as the very first statement in the constructor definition. That means you can't use this syntax to call a constructor from inside a method definition. It also means you can have only one such constructor call in a constructor definition, because only one call statement could be the "very first statement in the constructor definition."

Now look at the Fraction class in Figure 7.19. It has three instance variables—numerator, denominator, and quotient. The quotient instance variable holds the floating-point result of dividing the numerator by the denominator. The second constructor is just like the two-parameter constructor we wrote above. But the first constructor is shorter. Instead of repeating code appearing in the second constructor, it calls the second constructor with the this(...) command.

```
/*************************************************************
* Fraction.java
* Dean & Dean
*
* This class stores and prints fractions.
*************************************************************/

public class Fraction
{
  private int numerator;
  private int denominator;
  private double quotient; // numerator divided by denominator

  //**********************************************************

  public Fraction(int n)
  {
    this(n, 1);
  }

  //**********************************************************

  public Fraction(int n, int d)
  {
    this.numerator = n;
    this.denominator = d;
    this.quotient = (double) this.numerator / this.denominator;
  }

  //**********************************************************

  public void printIt()
  {
    System.out.println(this.numerator + " / " +
      this.denominator + " = " + this.quotient);
  } // end printIt
} // end class Fraction
```

This statement calls the other constructor.

Figure 7.19 `Fraction` class with overloaded constructors

Suppose that during program development, for debugging purposes, you decided to print "In 1-parameter constructor" from within the `Fraction` class's one-parameter constructor. Where would you put that print statement? Because the `this(n, 1)` constructor call must be the first statement in the constructor definition, you would have to put the print statement below the constructor call.

Tracing with Constructors

Figure 7.20 shows a trace of the Fraction program. In the following discussion of it, you'll need to actively refer to not only the trace figure, but also the `FractionDriver` class (Figure 7.18) and the `Fraction` class (Figure 7.19). Note how line 12 in the `FractionDriver` class passes 3 and 4 to the two-parameter

FractionDriver			Fraction											
	main			Fraction		Fraction	printIt	obj1			obj2			
line#	**a**	**b**	*line#*	**n**	**d**	**n**	**this**	**num**	**den**	**quot**	**num**	**den**	**quot**	**output**
12				3	4									
			10					0						
			11						0					
			12							0.00				
			25					3						
			26						4					
			27							0.75				
12	obj1													
13						3								
			10								0			
			11									0		
			12										0.00	
			18	3	1									
			25								3			
			26									1		
			27										3.00	
13		obj2												
15							obj1							
			34											3 / 4 = 0.75
16							obj2							
			34											3 / 1 = 3.00

Figure 7.20 Trace of the Fraction program in Figures 7.18 and 7.19

`Fraction` constructor, which assigns 3 and 4 to the constructor's `n` and `d` parameters. As part of the implied constructor functionality, lines 10–12 in the `Fraction` class initialize `Fraction` instance variables with their default values. Then lines 25–27 overwrite those initialized values. Going back to `FractionDriver`, `new` returns an object reference (obj1) to the reference variable `a`. Then on line 13, the driver passes 3 to the one-parameter constructor. After parameter assignment and instance variable initialization, line 18 in the `Fraction` class passes 3 and 1 to the two-parameter constructor. After the two-parameter constructor overwrites the instance variables, control flows back to the one-parameter constructor, and from there back to `FractionDriver`, where `new` returns an object reference (obj2) to the reference variable `b`. Finally, in lines 15 and 16, the driver prints out the two results.

7.10 Class Variables

So far in this chapter, the focus has been on the "object" part of "object-oriented programming." Thus, when you envision an object-oriented solution, you probably see separate objects, each with its own set of data and behaviors (instance variables and instance methods, respectively). That's a valid picture, but you should be aware that in addition to data and behaviors that are specific to individual objects, you also can have data and behaviors that relate to an entire class. Because they relate to an entire class, such data and behaviors are referred to as *class variables* and *class methods*, respectively.

Let's look at an example. Suppose you are in charge of keeping track of YouTube videos. You need to instantiate a YouTube object for each YouTube video, and within each object, you need to store attributes like the videographer, the video's length, and the video file itself. You should store those attributes in instance variables because they are associated with individual YouTube objects. You also need to store attributes like the number of videos and the most popular video. You should store those attributes in class variables because they relate to the collection of YouTube objects as a whole.

In this section, you'll learn how to declare class variables, when to use them, what their default values are, and what their scope is. In the next section, you'll see examples of using class variables from within class methods.

Class Variable Declaration Syntax

To make a variable a class variable, use the `static` modifier in its declaration. The `static` modifier is why many programmers use the term *static variable* when talking about class variables. Likewise, because class constants and class methods also use the `static` modifier, many programmers use the terms *static constant* and *static method.* We'll stick with the terms *class variable, class constant,* and *class method* because those are the terms Oracle uses.

Here is the syntax for a class variable declaration statement:

<private-or-public> `static` *<type> <variable-name>*

And here is an example:

```
private static int mouseCount; // total number of mouse objects
```

Should class variables be `public` or `private`? The philosophy on this is the same as it is for instance variables. Because you can always write `public` get/set class methods, you don't need `public` class variables any more than you need `public` instance variables. It's best to keep your variables as `private` as possible to maintain control over how they are accessed. Therefore, in addition to making instance variables `private`, you should also make class variables `private`.

Why the Term "static"?

As you know, when the JVM sees the `new` operator in a program, it instantiates an object for the specified class. In so doing, it allocates memory space for all of the object's instance variables. Later, the garbage collector might deallocate (take away) that memory space before the program stops if all references to that space disappear. That sort of memory management, done while the program runs, is called *dynamic allocation*. Class variables are different. The JVM allocates space for a class variable when the program starts, and that class-variable space remains allocated for as long as the program runs. That sort of memory management is called *static allocation*. That's why class variables are called static.

Class Variable Examples

As you know, each use of `new` creates a separate copy of all instance variables for each object. On the other hand, for a particular class, there is only one copy of each class variable, and all objects share that single copy. Thus, you should use class variables to describe properties of a class's objects that need to be shared by all of the objects. For example, consider again the problem of simulating mouse growth. In our previous mouse programs, we kept track of data pertinent to each individual mouse—a mouse's growth rate, a mouse's age, and a mouse's weight. For a more useful simulation program, you'd probably also want to keep track of group data and common environmental data. For example:

`mouseCount` would keep track of the total number of mice.
`averageLifeSpan` would keep track of the average life span for all of the mice.
`simulationDuration` would limit the number of simulation iterations.
`researcher` would identify a person in charge of an experiment on the group of mice.
`noiseOn` would indicate the presence or absence of a stressful noise heard by all the mice.

To see how these class variables would be declared within a Mouse program, see Figure 7.21. If you used instance variables for `mouseCount`, `averageLifeSpan`, and so on, each individual mouse object would have its own copy of that data. So if there were 100 total mice, each of the 100 mice would store the value 100 in its own `mouseCount` variable, the average life span value in its own `averageLifeSpan` variable, and so on. This would mean that every time a new mouse was born or died or aged a year, you would have to update 100 separate copies of `mouseCount`, `averageLifeSpan`, and so on—all with exactly the same information. What a waste of effort! Why not just do it once and let everyone write and read the same common data? If `mouseCount`, `averageLifeSpan`, and so on are class variables, all mouse objects can write to and read from a single record of each of these pieces of information. An outsider can access these class properties by just prefixing the class name to an appropriate class method. It's neither necessary nor desirable to go through a particular instance to get to this group information.

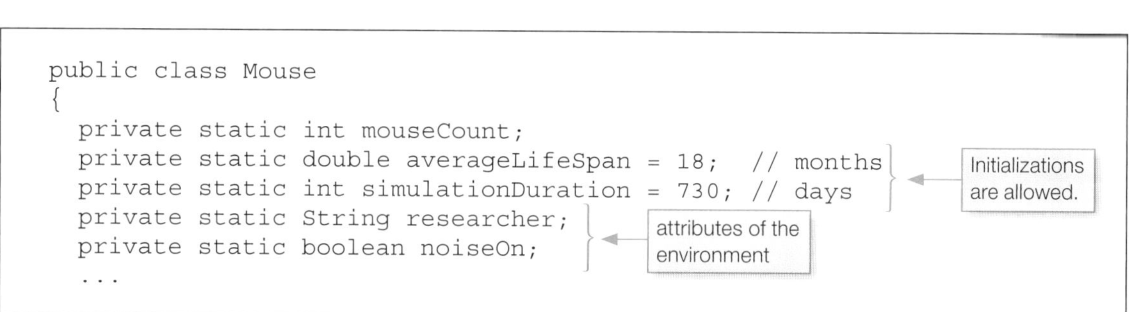

```
public class Mouse
{
  private static int mouseCount;
  private static double averageLifeSpan = 18;  // months
  private static int simulationDuration = 730; // days
  private static String researcher;
  private static boolean noiseOn;
  ...
```


Figure 7.21 Class variable declarations in an enhanced `Mouse` class

Default Values

Class variables use the same default values as instance variables:

Class Variable's Type	Default Value
integer	0
floating point	0.0
`boolean`	`false`
reference	`null`

It follows that the default values for Figure 7.21's class variables are:

```
mouseCount = 0
averageLifeSpan = 0.0
simulationDuration = 0
researcher = null
noiseOn = false
```

Presumably, the program updates `mouseCount` and `averageLifeSpan` as it runs. The default values of `averageLifeSpan` and `simulationDuration` are zero like `mouseCount`, but in Figure 7.21, the defaults don't apply because the declarations include initializations. Even though we expect the program to recompute `averageLifeSpan`, we initialize it to provide documentation of what we think is a reasonable value (18). We also initialize `simulationDuration` (to 730) even though we expect the program to reassign `simulationDuration` with a user-entered value. Presumably, the program prompts the user to enter the number of days to simulate. With appropriate code, the user might be invited to enter −1 to get a "standard" 730-day simulation.

Scope

Let's now compare class variables, instance variables, and local variables in terms of their scopes. You can access a class variable from anywhere within its class. More specifically, that means you can access class variables from instance methods as well as from class methods. That contrasts with instance variables, which you can access only from instance methods. Thus, class variables have broader scope than instance variables. Local variables, on the other hand, have narrower scope than instance variables. They can be accessed only within one particular method. Here is the scope continuum:

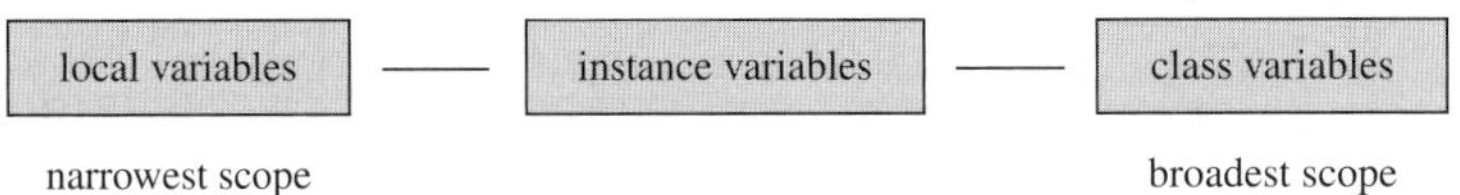

Having narrower scope for local variables might seem like a bad thing because it's less "powerful," but it's actually a good thing. Why? Narrower scope equates to more encapsulation, and as you learned in Chapter 6, encapsulation means you are less vulnerable to inappropriate changes. Class variables, with their broad scope and lack of encapsulation, can be accessed and updated from many different places, and that makes programs hard to understand and debug. Having broader scope is necessary at times, but in general, you should try to avoid broader scope. We encourage you to prefer local variables over instance variables and instance variables over class variables.

7.11 Class Methods

Class methods, like class variables, relate to the class as a whole, and they don't relate to individual objects. As such, if you need to perform a task that involves the class as a whole, then you should implement and use a class method. In Chapter 5, you used class methods defined in the Java API `Math` class; for example, `Math.round` and `Math.sqrt`. Now you'll learn how to write your own class methods. Class methods often access class variables, and in writing your own class methods, you'll get an opportunity to see how to access class variables you've defined.

The PennyJar Program

We'd like to present class method syntax details in the context of a complete program. The program we have in mind is a penny jar program . . .

Jordan (author John's daughter) and Addie (Jordan's BFF) received penny jars for Christmas. They decided to wait until they accumulated a combined total of $40 before breaking open their jars and spending their savings on a huge, ten-pound Hershey bar. They want a program that models their venture. Specifically, they want a program that stores the number of pennies in individual penny jars and also stores the total number of pennies in the collection of all the penny jars. Figure 7.22 shows a Unified Modeling Language (UML) class diagram for a `PennyJar` class that does just that. The class diagram's middle section contains the program's three variables:

- The `goal` variable is the target number of pennies to be saved for all penny jars combined. Because `goal` is an attribute shared by all the penny jars, it's a class variable. In the UML class diagram, you can tell that `goal` is a class variable because it's underlined (as you may recall, UML standards suggest that you underline class members).
- The `allPennies` variable stores the total pennies accumulated so far in all the jars. Because `allPennies` is an attribute of all the penny jars, it's a class variable and therefore, it is underlined in the UML class diagram. Note that the class diagram shows `allPennies` being initialized to 0. Although `allPennies` would have been initialized to 0 by default, the explicit initialization is a form of self-documentation, making it clear that the programmer needs `allPennies` to start at 0 for the program to work properly.
- The `pennies` variable is an instance variable. It keeps track of the number of pennies in one jar.

The class diagram's bottom section contains the program's four methods:

- The `setGoal` method is a mutator method in that it sets/assigns the value of the `goal` variable. Because `setGoal` deals only with classwide data (`goal` is a class variable), `setGoal` is a class method. In the UML class diagram, you can tell that `setGoal` is a class method because it's underlined.
- The `getAllPennies` method retrieves the value of the `allPennies` class variable. Because `getAllPennies` deals only with class-wide data, `getAllPennies` is a class method and is also underlined in the UML class diagram.
- The `getPennies` method is an accessor instance method in that it retrieves the value of the `pennies` instance variable from one particular penny jar.
- The `addPenny` method is another instance method. It simulates adding a penny to a jar.

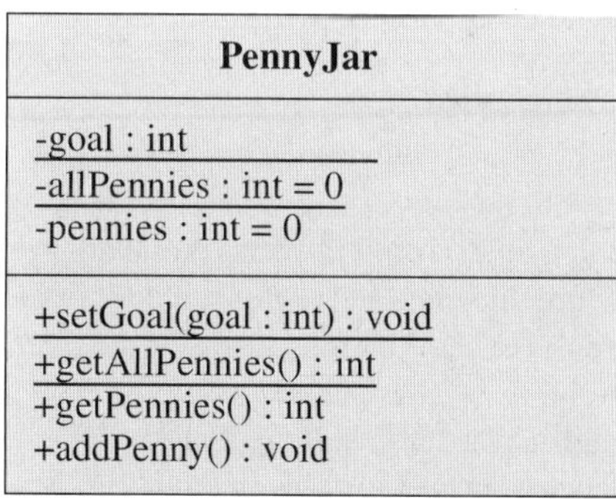

Figure 7.22 Class describing penny jars individually and as a group

How to Define Class Methods

Having introduced the PennyJar program at a high level, it's now time to dig into class method details. To define a class method, use this syntax for the method's heading:

<private-or-public> `static` *<return-type>* *<method-name>* (*<parameters>*)

It's the same syntax that's used for instance methods, except that `static` appears on the left of the return type. Figure 7.23 shows the source code for the PennyJar program's `PennyJar` class. Note that its class methods, `setGoal` and `getAllPennies`, have the `static` modifier in their headings. As with instance methods, most class methods should use the `public` access modifier (we'll discuss `private` methods in the next chapter).

How to Access Class Members

Normally, to access a class member, you should prefix the class member with the class member's class name and then a dot. Prefixing a class member with its class name and then a dot should sound familiar because you've done that with `Math` class members like `Math.round()` and `Math.PI`. Within Figure 7.23's `setGoal` and `getAllPennies` methods, note how the class variables `goal` and `allPennies` are accessed with `PennyJar` dot prefixes—`PennyJar.goal` and `PennyJar.allPennies`. For another example, take a look at the `PennyJarDriver` class in Figure 7.24. In particular, note how its `main` method calls class methods `setGoal` and `getAllPennies` using `PennyJar` dot prefixes:

```
PennyJar.setGoal(4000);
System.out.println(PennyJar.getAllPennies());
```

These examples show class members being accessed from class methods (`setGoal`, `getAllPennies`, and `main` are all class methods because they all use the `static` modifier in their headings). Because class members are supposed to be available to the entire class, they can, of course, be accessed from class methods. But be aware that individual objects can also access class members. Thus, instance methods and constructors (both of which are associated with individual objects) can access class members as well. For example, within the PennyJar program's `addPenny` instance method, note how the `allPennies` and `goal` class variables are accessed:

```
if (PennyJar.allPennies >= PennyJar.goal)
```

Note that it's sometimes legal to omit the class name dot prefix when accessing a class member. In accessing a class member, if the class member is in the same class as the class from which it's accessed, the compiler will allow you to omit the class name dot prefix. For example, the above code fragment accesses `allPennies` and `goal`, which are declared in the `PennyJar` class. Because the code fragment itself comes from the `PennyJar` class, it would be legal to rewrite the code like this:

```
if (allPennies >= goal)
```

Despite the legality of sometimes being able to omit the class name dot prefix when accessing a class member, we recommend that you always include the class name dot prefix because it's a form of self-documentation. It alerts the person reading the code to the fact that the accessed member is special—it deals with classwide information.

Another reason for always including the class name dot prefix for class variables is that it can be used to avoid ambiguity. If you attempt to access a class variable in a place that has a local variable or parameter declared with the same name as the class variable, then to distinguish the class variable from the local variable

```
/*********************************************************************
* PennyJar.java
* Dean & Dean
*
* This class counts pennies stored in a group of penny jars.
*********************************************************************/

public class PennyJar
{
  private static int goal;
  private static int allPennies = 0;
  private int pennies = 0;
}

 //******************************************************************

  public static void setGoal(int goal)
  {
    PennyJar.goal = goal;
  }

 //*******************************************************************

  public static int getAllPennies()
  {
    return PennyJar.allPennies;
  }

  //******************************************************************

  public int getPennies()
  {
    return this.pennies;
  }

  //******************************************************************

  public void addPenny()
  {
    System.out.println("Clink!");
    this.pennies++;
    PennyJar.allPennies++;

    if (PennyJar.allPennies >= PennyJar.goal)
    {
      System.out.println("Time to spend!");
    }
  } // end addPenny
} // end class PennyJar
```

class variables

class method

class method

Figure 7.23 `PennyJar` class, which illustrates both instance members and class members

or parameter with the same name, you must prefix the class variable with its class name and a dot. If you don't do that, then the compiler will bind your access attempt to the local variable or parameter and not to the class variable. For example, the `setGoal` method, copied below for your convenience, uses two `goal` variables—one is the class variable and one is a parameter. To resolve the ambiguity, the class variable must be prefixed with `PennyJar` dot:

```
public static void setGoal(int goal)
{
  PennyJar.goal = goal;
} // end setGoal
```

This identifies a class variable and distinguishes it from a parameter having the same name.

Calling an Instance Method from Within a Class Method

If you're within a class method, you'll get a compilation error if you attempt to directly access an instance member. To access an instance member, you first must have an object, and then you access the object's instance member by prefixing it with the object's reference variable. The reference variable is often referred to as the *calling object*. Does all that sound familiar? The `main` method is a class method, and you've been calling instance methods from `main` for quite a while now. But whenever you did that, you first instantiated an object and assigned the object's reference to a reference variable. Then you called the instance method by prefixing it with the reference variable and a dot. Figure 7.24's `main` method shows what we're talking about:

```
public static void main(String[] args)
{
  PennyJar pennyJar1 = new PennyJar();
  pennyJar1.addPenny();
  ...
```

You need a reference variable dot to call an instance method from within a class method.

If you attempt to access an instance method directly from within a class method without using a reference variable dot prefix, you'll see an error message like this:

```
Non-static <method-name> cannot be referenced from a static context
```

That error message is very common (you've probably seen it many times) because it's easy to forget to prefix instance method calls with a reference variable. When veteran programmers see it, they know what to do; they make sure to prefix the instance method call with a calling object's reference variable. But when beginning programmers see the error message, they often compound the error by trying to "fix" the bug inappropriately. More specifically, when confronted with the non-static method error message, a beginning programmer often will change the offending instance method to a class method, by inserting `static` in the method's heading. (In the PennyJar program, `addPenny` would be changed to a class method). They then get the non-static member error message for any instance variables within the method. They then compound the problem further by changing the method's instance variables to class variables. (In the PennyJar program, `addPenny`'s `pennies` variable would be changed to a class variable). With that change in place, the program compiles successfully and the beginning programmer is happy as a lark, ready to slay the next dragon. Unfortunately, that type of solution leads to a worse problem than a compilation error. It leads to a logic error.

As you know, if a class's member relates to one object rather than to the class as a whole, you should make it an instance member. If you do as described above and "fix" a bug by changing instance members

```
/*************************************************************
* PennyJarDriver.java
* Dean & Dean
*
* This class drives the PennyJar class.
*************************************************************/

public class PennyJarDriver
{
  public static void main(String[] args)
  {
    PennyJar pennyJar1 = new PennyJar();
    PennyJar pennyJar2 = new PennyJar();

    PennyJar.setGoal(4000);
    pennyJar1.addPenny();
    pennyJar1.addPenny();
    pennyJar2.addPenny();
    System.out.println(pennyJar1.getPennies());
    System.out.println(PennyJar.getAllPennies());
  } // end main
} // end class PennyJarDriver
```

Output:

```
Clink!
Clink!
Clink!
2
3
```

Figure 7.24 Driver for the `PennyJar` class in Figure 7.23

to class members, you can get your program to compile and run. And if you have only one object, your program might even produce a valid result. But if you have more than one object, either now or in the future, then with class variables, the objects will share the same data. If you change one object's data, you'll simultaneously change all the objects' data and that would normally be deemed incorrect. For the PennyJar program, all the `PennyJar` objects would share the same `pennies` value, so it would be impossible to keep track of the money in individual penny jars.

When to Use Class Methods

When should you make a method a class method? The general answer is, "when you need to perform a task that involves the class as a whole." But let's get more specific. Here are situations where using class methods is appropriate:

1. If you have a method that uses class variables, calls class methods, or both, then it's a good candidate for being a class method. For example, Figure 7.23's `getAllPennies` is a class method because it

retrieves the value `allPennies`, a class variable. Warning: If in addition to accessing class members, the method accesses instance members, then the method must be an instance method, not a class method.

2. The `main` method is the starting point for all programs and, as such, it gets executed prior to the instantiation of any objects. To accommodate that functionality, you're required to make the `main` method a class method. If your `main` method is rather long and contains well-defined subtasks, you should consider trying to implement those subtasks with their own methods. For an example of that, see the RollDice program in the Interlude "mini-chapter" before Chapter 6. You should use `static` for the subtask methods, thereby making them class methods. By making them class methods, it's easy for `main` to call them (just prefix them with class name dot, rather than having to instantiate an object first and then prefix them with reference variable dot).
3. If you have a general-purpose method that stands on its own, make it a class method. By standing on its own, we mean that the method is not related to a particular object. Such methods are called *utility methods*. You've seen examples of utility methods, like `Math.round` and `Math.sqrt`, in the `Math` class. In the next chapter, you'll learn how to write your own utility methods within the context of a general-purpose utility class.

7.12 Named Constants

Using names instead of hard-coded values makes a program more self-documenting. When a constant value is needed in more than one place in the block of code, establishing the value at one place at the beginning of that block minimizes the chance of inconsistency. In Java, you can define named constants at several levels of scale.

Local Named Constants—A Review from Chapter 3

At the most microscopic level, you can define *local named constants*. Back in Figure 3.5 of Chapter 3, we defined two local named constants, `FREEZING_POINT`, and `CONVERSION_FACTOR`, to self-document the Fahrenheit-to-Celsius conversion formula in a simple program that did nothing more than make a temperature conversion. Usually, we embed this kind of activity in some larger program by putting it in a helper method like this:

```
private double fahrenheitToCelsius(double fahrenheit)
{
  final double FREEZING_POINT = 32.0;
  final double CONVERSION_FACTOR = 5.0 / 9.0;

  return CONVERSION_FACTOR * (fahrenheit - FREEZING_POINT);
} // end fahrenheitToCelsius
```

The local named constants in this method make the code easier to understand.

Instance Constants—A Review from Section 7.8

At the next higher level of scale, sometimes you want a constant that's a permanent property of an object and accessible to all instance methods associated with that object. Those constants are called instance named constants, or, more simply, *instance constants*. Here's an example instance constant declaration that identifies a permanent property of a `Person` object:

```
public final String SOCIAL_SECURITY_NUMBER;
```

An instance constant declaration differs from a local named constant declaration in three ways: (1) An instance constant declaration should appear at the top of the class definition, rather than within a method, (2) An instance constant declaration is preceded by a `public` or `private` access modifier, and (3) Although it's legal to initialize an instance constant in a declaration, it's more common to initialize it in a constructor.

Class Constants

At the next higher level of scale, sometimes you want a constant that's the same for all objects in a class. In other words, you want something that's like a class variable, but it's constant. Those constants are called class named constants, or, more simply, *class constants*. In Chapter 5, you learned about two class constants defined in the Java API `Math` class, `PI` and `E`. Now you'll learn how to write your own class constants. To declare a class constant, use this syntax:

<private-or-public> `static final` *<type>* *<variable-name>* `=` *<initial-value>*`;`

A class constant declaration differs from an instance constant declaration in two ways: (1) A class constant includes the `static` modifier; and (2) A class constant should be initialized as part of its declaration.[4] If you attempt to assign a value to a class constant later on, that generates a compilation error.

As with an instance constant, a class constant declaration should be preceded by a `public` or `private` access modifier. If the constant is needed only within the class (and not outside the class), you should make it `private`. This allows you to modify the constant without upsetting somebody else who previously elected to use your constant in one of their programs. However, if you want the constant to be available to other classes, it's appropriate to make it `public`. It's safe to do that because the `final` modifier makes it immutable (unchangeable).

The following `Human` class contains a `NORMAL_TEMP` named constant. We make it a class constant (with the `static` and `final` modifiers) because all `Human` objects have the same normal temperature of 98.6° Fahrenheit. We make it a `private` class constant because it is needed only within the `Human` class.

```
public class Human
{
  private static final double NORMAL_TEMP = 98.6;
  private double currentTemp;
  ...
  public boolean isHealthy()
  {
    return Math.abs(currentTemp - NORMAL_TEMP) < 1;
  } // end isHealthy
  public void diagnose()
  {
    if ((currentTemp - NORMAL_TEMP) > 5)
    {
      System.out.println("Go to the emergency room now!");
      ...
} // end class Human
```

[4]Although relatively rare, it's legal to declare a class constant as part of a `static` initializer block. For details on *initializer blocks* see http://download.oracle.com/javase/tutorial/java/javaOO/initial.html.

Let's summarize when you should use the three different types of named constants. Use a local named constant if the constant is needed within only one method. Use an instance constant if the constant describes a permanent property of an object. And use a class constant if the constant is a permanent property of the collection of all the objects in the class or of the class in general.

Positions of Declarations

Now for some coding-style issues. We recommend putting all class constant declarations above all instance constant declarations. Putting declarations at the top makes them stand out more, and it's appropriate for class constants to stand out the most because they have the broadest scope. Likewise, we recommend putting all class variable declarations above all instance variable declarations. *Field* is the generic term for a class constant, instance constant, class variable, or instance variable. We recommend that you put all your field declarations at the top, above all your constructor and method declarations. Here is the preferred sequence of declarations within a given class:

class constants
instance constants
class variables
instance variables
constructors
mutator and accessor methods (doesn't matter which type is first)
other methods (doesn't matter which type is first - class methods or instance methods)

7.13 Problem Solving with Multiple Driven Classes

We started this book simply, and we are gradually adding complexity as we go. In Chapters 1 through 5, we showed you programs that contain only one class and one method (the `main` method). In Chapters 6 and 7, we've been showing you programs that contain two classes: (1) a driver class, which contains a single `main` method; and (2) a driven class, which typically contains several methods.

So far, we've used only one driven class to keep things simple, but in the real world, you'll often need more than one driven class. That's because most real-world systems are heterogeneous—they contain mixtures of different types of things. For each different type of thing, it's appropriate to have a different class.

Break up a big problem into isolated simpler problems.

Having more than one driven class allows you to partition a complicated problem into several simpler problems. That lets you focus on one type of thing at a time. When you finish working on one type of thing, you can move onto another type of thing. In this step-by-step fashion, you can gradually build up a large program.

It's no big deal to drive more than one driven class from a single driver. In fact, you saw us do it back in Chapter 5, when statements in a single `main` method called methods from more than one wrapper class, like `Integer` and `Double`. The only thing to remember is when compiling the driver, the compiler must be able to find all the driven classes. If they are prebuilt classes, they must be part of the `java.lang` package or you must import them. If they are classes that you've written yourself, they should be in the same directory as your driver.[5]

[5]It's possible to put your own classes in your own packages in separate directories and import them as you import prebuilt classes. You can learn how to do this in Appendix 4 of this book. However, if all your driven classes are in the same directory as your driver class, it's not necessary to package and import them, and we assume this to be the case for the code that we develop and present in this text.

Summary

- When you declare a reference variable, the JVM allocates space in memory for holding a reference to an object. At that point, there is no memory allocation for the object itself.
- Assigning one reference variable to another does not clone an object. It just makes both reference variables refer to the same object and gives that object an alternate name—an alias.
- To create a separate object, you must use Java's `new` operator. To make a second object be like a first object, copy the first object's instance variable values into the corresponding instance variables in the second object.
- A method can return an assortment of data originating in a method by returning a reference to an internally instantiated object that contains that data.
- Java's garbage collection program searches for inaccessible objects and recycles the space they occupy by asking the operating system to designate their space in memory as free space.
- If you compare two object references with `==`, the result is `true` if and only if the references point to the same object.
- To see whether two different objects contain similar data, you must write an `equals` method that individually compares respective instance variable values.
- To swap two variables' values, you need to store one of the variable's values in a temporary variable.
- If you pass a reference as an argument, and if the reference parameter's instance variables are updated, then the update simultaneously updates the reference argument's instance variables in the calling module.
- If a method returns a reference to an object, you can use what's returned to call another method in the same statement. That's method call chaining.
- To make a program more understandable, you can *overload* a method name by using the same name again in a different method definition that has a different sequence of parameter types. The combination of method name, number of parameters, and parameter types is called a method's *signature*.
- A constructor enables you to initialize instance variables separately for each object. A constructor's name is the same as its class name, and there is no return value specification.
- For a constructor call to work there must be a matching constructor definition, that is, a definition with the same signature.
- If you define a constructor, the default zero-parameter constructor vanishes.
- Use a constructor to initialize instance constants, which represent permanent attributes of individual objects.
- To get help constructing the current object, call an overloaded constructor from within a constructor, by making the first statement in the constructor be: `this(`<*constructor-argument(s)*>`)`.
- Class variables have a `static` modifier. Use class variables for attributes of the collection of all objects in a class. Use instance variables for the attributes of individual objects.
- Remember that class variables have broader scope than instance variables, and instance variables have broader scope than local variables. To improve encapsulation, you should try to use variables with a narrower scope rather than a broader scope.
- A class method can access class members directly, but it cannot access instance members directly. To access an instance member from a class method, you need to use an object reference variable dot prefix.
- An instance method can directly access class members as well as instance members.
- Use class constants for permanent data that is not associated with any particular object. Class constants use the `final` and `static` modifiers.
- Instance constants have a `final` modifier only. Use instance constants for permanent attributes of individual objects.

Review Questions

§7.2 Object Creation—A Detailed Analysis

1. The statement

```
Car car;
```

allocates space in memory for an object. (T / F)

2. What does the `new` operator do?

§7.3 Assigning a Reference

3. Assigning one reference variable to another copies the right-side object's instance variables into the left-side object's instance variables. (T / F)
4. What is a memory leak?

§7.4 Testing Objects for Equality

5. Consider this code fragment:

```
boolean same;
Car carX = new Car();
Car carY = carX;
same = (carX == carY);
```

What is the final value of `same`?

6. What is the return type of an `equals` method?
7. By convention, we use the name `equals` for methods that perform a certain kind of evaluation. What is the difference between the evaluation performed by an `equals` method and the == operator?

§7.5 Passing References as Arguments

8. When you pass a reference to a method, you enable the method to modify the referenced object. (T / F)

§7.6 Method-Call Chaining

9. What two things must be included in a method definition so that it may be called as part of a method-call-chaining statement?

§7.7 Overloaded Methods

10. What is it called when you have two or more methods with the same name in the same class?
11. If you want the current object to call a different method in the same class as the current class, the method call is easy—just call the method directly, with no reference variable dot prefix. (T / F)

§7.8 Constructors

12. What is the return type of a constructor?
13. The name of a constructor must be exactly the same as the name of its class. (T / F)
14. Standard coding conventions suggest that you put constructor definitions after the definitions of all methods. (T / F)

§7.9 Overloaded Constructors

15. If a class's source code contains a single one-parameter constructor, the constructor is overloaded because this one-parameter constructor has the same name as the default zero-parameter constructor. (T / F)
16. Suppose you have a class with two constructors. What are the rules for calling one constructor from the other constructor?

§7.10 Class Variables

17. Normally, you should use the private access modifier for class variables. (T / F)
18. When should you declare a variable to be a class variable, as opposed to an instance variable?
19. What are the default values for class variables?

§7.11 Class Methods

20. What would be wrong with adding the following statement to Figure 7.24's `main` method?

```
PennyJar.addPenny();
```

21. Member access:
    a) It is OK to use `this` in a class method. (T / F)
    b) It is OK to use the class name as a prefix when calling a class method. (T / F)
    c) Within a `main` method, it is OK to omit the class name prefix before the name of another class method being called. (T / F)
22. It is legal to access a class member from an instance method and also from a constructor. (T / F)
23. It is legal to directly access an instance member from a class method. (T / F)

§7.12 Named Constants

24. What keyword converts a variable into a constant?
25. If you want a named constant used by instance methods to have the same value, regardless of which object accesses it, the declaration should include the `static` modifier. (T / F)
26. A class constant should be initialized within a constructor. (T / F)
27. Suppose you have a grading program that instantiates multiple exam objects from an `Exam` class. Provide a declaration for a constant minimum passing score. Assume the minimum passing score for all exams is 59.5.

Exercises

1. [after §7.2] Given a `Car` class with these two instance variables:

```
String make;
int year;
```

Describe all the operations that occur when this statement executes:

```
Car caidenCar = new Car();
```

2. [after §7.3] Trace the Car program shown in Figures 7.2 and 7.3. Use the following trace setup. Note that we have used abbreviations to keep the trace's width as small as possible.

CarDriver			Car																
	main			setMake		setYear		setColor		makeCopy		disp	obj1			obj2			
line#	**jCar**	**sCar**	*line#*	**this**	**make**	**this**	**year**	**this**	**color**	**this**	**car**	**this**	**make**	**year**	**color**	**make**	**year**	**color**	**output**

3. [after §7.3] What is garbage collection?

4. [after §7.5] Suppose that a `Computer` class contains, along with other instance variables, a `hardDrive` string instance variable. Complete the following `swapHardDrive` method, which swaps the calling object's hard drive value with the passed-in parameter's hard drive value.

```
public void swapHardDrive(Computer otherComputer)
{
  <insert code here>
} // end swapHardDrive
```

5. [after §7.5] Normally, we give each object a unique name by assigning its address to only one reference variable. Assigning the value of one reference variable to another reference variable creates two different names for the same thing, which is ambiguous. Identify a situation where this kind of assignment is useful, even though there is name ambiguity.

6. [after §7.6] Given this automobile-specification program:

```
/*************************************************************
* AutoOptionsDriver.java
* Dean & Dean
*
* This exercises the AutoOptions class.
*************************************************************/

import java.util.Scanner;

public class AutoOptionsDriver
{
  public static void main(String[] args)
  {
    Scanner stdIn = new Scanner(System.in);
    String serial;
    AutoOptions auto = new AutoOptions();

    System.out.print("Enter serial number: ");
    serial = stdIn.nextLine();
    auto.specifyEngine(auto.setSerial(serial).
    specifyFrame().specifyBody().isTight());
    auto.specifyTransmission();
    auto.printOptions();
  } // end main
} // end class AutoOptionsDriver
```

```
/************************************************************
* AutoOptions.java
* Dean & Dean
*
* This class records options for "custom" automobiles.
************************************************************/

import java.util.Scanner;
```

```java

public class AutoOptions
{
  private String serial;     // automobile serial number
  private char frame = 'x';  // frame type: A,B
  private String body = "";  // body style: 2Door,4Door
  private int hp = 0;        // engine horsepower: 85, 115, 165

  // transmission: false = manual, true = automatic
  private boolean automatic = false;

  //*************************************************************

  public AutoOptions setSerial(String serial)
  {
    return this;
    this.serial = serial;
  } // end setSerial

  //*************************************************************

  public AutoOptions specifyFrame()
  {
    Scanner stdIn = new Scanner(System.in);

    while (this.frame != 'A' && this.frame != 'B')
    {
      System.out.print("Enter frame (A or B): ");
      this.frame = stdIn.nextLine().charAt(0);
    } // end while
    return this;
  } // end specifyFrame

  //*************************************************************

  public AutoOptions specifyBody()
  {
    Scanner stdIn = new Scanner(System.in);

    while (!this.body.equals("2-door")
      && !this.body.equals("4-door"))
    {
      System.out.print(
        "Enter (2-door or 4-door): ");
      this.body = stdIn.nextLine();
    } // end while
    return this;
  } // end specifyBody

  //*********************************************************
```

```

  public boolean isTight()
  {
    boolean tight = false;

    if (this.frame == 'A' && this.body.equals("4-door"))
    {
      tight = true;
    }
    return tight;
  } // end isTight

  //*******************************************************

  public void specifyEngine(boolean tight)
  {
    Scanner stdIn = new Scanner(System.in);

    if (tight)
    {
      while (this.hp != 85 && this.hp != 115)
      {
        System.out.print("Enter HP (85 or 115): ");
        this.hp = stdIn.nextInt();
      } // end while
    }
    else
    {
      while (this.hp != 85 && this.hp != 115 && this.hp != 165)
      {
        System.out.print("Enter HP (85, 115, 165): ");
        this.hp = stdIn.nextInt();
      } // end while
    } // end if tight else
    stdIn.nextLine(); // flush \r\n after nextInt
  } // end specifyEngine

  //*******************************************************

  public void specifyTransmission()
  {
    Scanner stdIn = new Scanner(System.in);

    System.out.print("Automatic (y/n?): ");
    if (stdIn.nextLine().charAt(0) == 'y')
    {
      this.automatic = true;
    }
  } // end specifyTransmission

  //*******************************************************
```

```
    public void printOptions()
    {
      System.out.printf("serial# %s\n%s frame\n%s\n%-3d HP\n",
        this.serial, this.frame, this.body, this.hp);
      if (automatic)
      {
        System.out.println(" automatic");
      }
      else
      {
        System.out.println("4-speed manual");
      }
    } // end printOptions
  } // end class AutoOptions
```

Use the following trace setup to trace the AutoOptions program. Note that we have used abbreviations to keep the trace setup's width as small as possible.

input

X142R
A
4-door
165
115
Y

AutoOptionsDriver			AutoOptions																
	main			setSerial		spec Frame	spec Body	isTight		specEngine		spec Trans	print Opt	obj1					
line#	ser	auto	*line#*	this	ser	this	this	this	tight	this	tight	this	this	ser	frm	body	hp	auto	output

7. [after §7.6] In the following `Time` and `TimeDriver` class skeletons, replace the italicized <*insert . . .*> lines with your own code such that the program operates properly. More specifically:
   a) In the `Time` class, provide a method definition for the `setHours` method such that `setHours` can be called as part of a method-call chain.
   b) In the `TimeDriver` class, provide a single statement that chains calls to the `setHours`, `setMinutes`, `setSeconds`, and `printTime` methods. Use reasonable values for your method-call arguments. If you pass 8 to `setHours`, 59 to `setMinutes`, and 0 to `setSeconds`, then your method-call-chaining statement should print this:

```
08:59:00
```

```
public class Time
{
  private int hours;
  private int minutes;
  private int seconds;

  //**********************************************************
```

```
  <insert setHours method definition here>

  public Time setMinutes(int minutes)
  {
    this.minutes = minutes;
    return this;
  } // end setMinutes

  public Time setSeconds(int seconds)
  {
    this.seconds = seconds;
    return this;
  } // end setSeconds

  //*************************************************************

  public void printTime()
  {
    System.out.printf("%02d:%02d:%02d\n", hours, minutes, seconds);
  } // end printTime
} // end Time class

public class TimeDriver
{
  public static void main(String[] args)
  {
    Time time = new Time();
    <insert chained-method-calls statement here>
  }
} // end TimeDriver class
```

8. [after §7.7]

a) Modify the two-parameter `setHeight` method in Figure 7.10 to make it test its `units` parameter to see if `units` is equal to one of the following allowable symbols: "m," "cm," "mm," "in," or "ft." If it is equal to one of these, set the instance variables and return `true`. If it is not equal to one of these, return `false`.

b) Does this modification *require* any change to any program that calls the two-argument `setHeight` method? Why or why not?

c) Write a statement that calls the modified method and utilizes the returned information to print an error message "Error: units not recognized" if the units argument is not one of the allowed values.

9. [after §7.8] Provide a standard three-parameter constructor for a class named `JewelryItem`. The class contains three instance variables—`description`, `price`, and `qtyOnHand`. The constructor simply assigns its three parameters to the three instance variables.

10. [after §7.9] Overloaded Constructors:

a) Add a pair of constructors to the `Height` class that implement the initializations provided by the two `setHeight` operations in Figure 7.11. Minimize the total number of statements by having the one-parameter constructor call the one-parameter `setHeight` method and having the two-parameter constructor call the two-parameter `setHeight` method.

b) Provide a complete, rewritten `main` method for the `HeightDriver` class such that the new `main` method uses one of the new constructors from part a) to generate this output:

```
6.0 ft
```

11. [after §7.9] Overloaded Constructors:

Assume that the `Height` class of Figure 7.10 contains only one `setHeight` method—the two-parameter version. Write two constructors for the `Height` class, one with one argument (`double height`), and the other with two arguments (`double height` and `String units`). For the one-argument constructor, use the default of "m" for `units`.

Do not duplicate any internal code. That is, have the one-parameter constructor transfer control to the two-parameter constructor, and have the two-parameter constructor transfer control to the two-parameter `setHeight` method.

12. [after §7.9]: Assume that the following two classes are compiled and run. What is their output?

```
public class SillyClassDriver
{
  public static void main(String[] args)
  {
    SillyClass sc = new SillyClass();
    sc.display();
  }
} // end SillyClassDriver class

public class SillyClass
{
  private int x = 10;

  public SillyClass()
  {
    this(20);
    System.out.println(this.x);
  }

  public SillyClass(int x)
  {
    System.out.println(this.x);
    System.out.println(x);
    this.x = 30;
    x = 40;
  }

  public void display()
  {
    int x = 50;
    display(x);
    System.out.println(x);
  }

  public void display(int x)
  {
    x += 10;
    System.out.println(x);
  }
} // end SillyClass class
```

13. [after §7.10] Given a class with a class variable, all of the class's objects get a separate copy of the class variable. (T / F)

14. [after §7.10] In general, why should you prefer local variables over instance variables and instance variables over class variables?

15. [after §7.10] Given a program that keeps track of book details with the help of a `Book` class, for each of the following program variables, specify whether it should be a local variable, an instance variable, or a class variable.

`bookTitle` (the title of a particular book)
`averagePrice` (the average price of all of the books)
`price` (the price of a particular book)
`i` (an index variable used to loop through all of the books)

16. [after §7.11] If a method accesses a class variable and also an instance variable, the method:

a) must be a local method
b) must be an instance method
c) must be a class method
d) can be either a class method or an instance method—it depends on other factors

17. [after §7.11] If you attempt to directly access an instance method from within a class method, you'll see an error message like this:

`Non-static` *<method-name>* `cannot be referenced from a static context`

Normally, how should you fix the bug?

18. [after §7.11] Consider the following program.

```
public class Test
{
  private int x;
  private static int y;
  public void doIt()
  {
    this.x = 1;
    y = 2;
  }
  public static void tryIt()
  {
    this.x = 3;
    y = 4;
  }
  public static void main(String[] args)
  {
    doIt();
    tryIt();
    Test t = new Test();
    t.doIt();
    Test.doIt();
    Test.tryIt();
  }
} // end Test class
```

a) Mark all of the lines of code that have a compilation error.
b) For each compilation-error line, explain why it is incorrect.

Note:
There are no errors in the variable declarations and there are no errors in the method headings, so don't mark any of those lines as having an error.
For each compilation error, just provide the reason that the error occurs. In particular, do not solve the problem by attempting to show a "fixed" version of the program.

19. [after §7.12] Why is it safe to declare named constants `public`?

20. [after §7.12] Write appropriate declarations for the following constants. In each case, decide whether to include the keyword, `static`, and whether to include initialization in the declaration. Also, make each constant as easily accessible as possible, consistent with protection from inadvertent corruption.
   a) The year of birth of a person.
   b) The format string, `"%-25s%, 13.2f%, 13.2f%(,15.2f\n"`, to use in several `printf` statements in a single method.
   c) The "golden ratio" or width / length of a golden rectangle. It's equal to (sqrt(5) − 1) / 2 = 0.6180339887498949.

Review Question Solutions

1. False. It just allocates memory for a reference variable.
2. The `new` operator allocates memory for an object and returns the address of where that object is stored in memory.
3. False. Assigning a reference variable to another reference variable causes the address in the right side's reference variable to be put into the left side's reference variable. And that makes both reference variables refer to the same object.
4. A memory leak is when an inaccessible object is allowed to persist and use up space in a computer's memory.
5. The final value of `same` is `true`.
6. The return type of an `equals` method is `boolean`.
7. The == operator compares the values of two variables of the same type. If the variables are reference variables, == compares their addresses to see if they refer to the same object. A typical `equals` method compares the values of all the instance variables in the object referred to by its parameter with the values of corresponding instance variables in the object that called it. It returns `true` only if all corresponding instance variables have the same values.
8. True. The reference gives the method access to the reference's object.
9. For a method to be called as part of a method-call-chaining statement, include these things:
   - Within the method body, specify `return` *<reference-variable>*`;`
   - Within the method heading, specify the reference variable's associated class as the return type.
10. If you have two or more methods with the same name in the same class, they're called overloaded methods.
11. True.
12. A constructor does not have a return type and it does not use a `return` statement, but when you call a constructor, `new` returns a reference to the constructed object.
13. True.

14. False. Standard coding conventions suggest that you put constructor definitions before all other method definitions.

15. False. There is only one constructor, because if a class contains a programmer-defined constructor, then the compiler does not provide a default constructor.

16. To call the other constructor, you must insert this statement as the very first statement in the calling constructor's body:

    `this(`*<arguments-for-target-constructor>*`);`

17. True.

18. You should declare a variable to be a class variable, as opposed to an instance variable, if the variable holds data that is associated with the class as a whole. You should use class variables to describe properties of a class's objects that need to be shared by all of the objects.

19. The default values for class variables are the same as they are for instance variables of the same type. Here are the default values:

    integer types get 0
    floating point types get 0.0
    `boolean` types get `false`
    reference types get `null`

20. Because there's no `static` modifier, `addPenny` is an instance method. The `PennyJar.addPenny();` call uses a `PennyJar` dot prefix. It's illegal to use a class name (`PennyJar`) as a prefix for an instance method call. To call an instance method, you need to use a reference variable dot prefix.

21. Member access:
    a) False. You cannot use `this` in a class method.
    b) True. You can always use the class name as a class method prefix.
    c) True, if the `main` method is "merged" into the same class as the other method.
    False, if the other method is in a different class.
    Including the class name prefix allows you to move the `main` method to another class later.

22. True. You can access a class member from an instance method and also from a constructor—just prefix the class member with the class name.

23. False. You can access an instance member from a class method only if you prefix the method name with a reference to a particular object.

24. The keyword `final` converts a variable into a constant.

25. True. Use `static` to make a constant be the same for all objects.

26. False. A class constant should normally be initialized as part of its declaration. If it is assigned a value later on, including within a constructor, it generates a compilation error.

27. Minimum passing score declaration:

```
private static final double MIN_PASSING_SCORE = 59.5;
```

CHAPTER 8

Software Engineering

Objectives

- Develop good coding style.
- Understand preconditions and postconditions.
- Learn how to generate documentation like Java's API documentation.
- Learn how to simplify complicated algorithms by encapsulating subordinate tasks.
- Learn how to adapt a user's point of view.
- Distinguish use of instance variables and local variables.
- Analyze problems by separating concerns.
- Learn when and how to use a top-down design strategy.
- Learn when and how to use a bottom-up design strategy.
- Resolve to use prewritten software whenever feasible.
- Recognize role of prototyping.
- Develop habit of frequent and thorough testing.
- Avoid unnecessary use of the `this` prefix.
- Recognize appropriate uses of class methods.

Outline

8.1 Introduction

In Chapters 6 and 7, we looked mostly at the "science" of Java programming—how to declare objects, define classes, define methods, and so on. In this chapter, we'll be looking more at the "art" and "practice" of Java programming—how to design and develop a program, and how to make it easy to read. The practice of programming is nicely summed up in the term *software engineering,* where software engineering is:[1]

1. The application of a systematic, disciplined, quantifiable approach to the development, operation, and maintenance of software; that is, the application of engineering to software.
2. The study of approaches as in 1.

We start the chapter with an in-depth discussion of coding-style conventions that make code more readable to people who need to know how it works. Then we backtrack a bit and describe techniques that make a block of code understandable to people who do not need to know how it works, only what it does. Next, we show how to divide a large task into a set of smaller tasks by delegating some of the work in a method to other methods. We discuss encapsulation, one of the cornerstones of proper object-oriented programming (OOP) design.

Next, we describe alternative design strategies—top-down, bottom-up, and case-based. As you work with something, your understanding of it improves, and we suggest that you plan to redesign continuously with more sophistication in an evolutionary process called *iterative enhancement.* We emphasize that you'll be happier and your product will be better if you test thoroughly and frequently as you go along.

To facilitate modular testing, we show how you can include a `main` method in each class. Up to now, we've made heavy use of `this` to emphasize that each execution of an instance method is uniquely tied to a particular object, but you can streamline your code by omitting `this` when there is no ambiguity. Sometimes it's useful to create your own utility class with class methods that methods in other classes can call easily. An optional section shows you how to use Java's `Calendar` class for difficult time and date problems. Another optional section shows you how to use simple graphics to construct a handy organizational tool called *CRC cards.*

8.2 Coding-Style Conventions

We'll now present some guidelines for coding style. We've mentioned and illustrated many of these style guidelines previously, so much of this section should be review. We'll provide more guidelines later, as we describe more Java. For a complete list of all the coding-style guidelines used in this book, refer to Appendix 5, "Java Coding-Style Conventions." The coding-style conventions we use are for the most part a simplified subset of the style conventions archived on Oracle's Java Code Conventions website.[2] If you have a style question that is not addressed in Appendix 5, refer to Oracle's website.

[1]Definition taken from Institute of Electrical and Electronics Engineers (IEEE) Standard 610.12.

[2]http://www.oracle.com/technetwork/java/codeconventions-150003.pdf

We realize there are some style issues where there is legitimate disagreement over the best way to do things. Many different standards exist. Oracle attempts to choose the best conventions from among the commonly used conventions. We attempt to do the same. If you're reading this book as part of a course and your teacher disagrees with the book's style conventions or Oracle's style conventions, please follow your teacher's guidelines.

We'll illustrate coding-style conventions by referring to the Student program in Figure 8.1 and Figures 8.2a and 8.2b. This program is a modified version of the Student program at the back of the "Java Coding-Style Conventions" appendix.

Prologue Section

Note the boxed text at the tops of Figures 8.1 and 8.2a. They're called *prologues.* Include a prologue section at the top of each file. The prologue contains these things, in this order:

- line of asterisks
- filename
- programmer name(s)
- blank line with one asterisk
- description
- line of asterisks
- blank line

Enclose the prologue in a `/*. . .*/` comment, and to make the prologue look like a box, insert an asterisk and a space in front of the filename, programmer name, blank line, and description lines.

```
/*********************************************************
* StudentDriver.java
* Dean & Dean
*
* This class acts as a driver for the Student class.
*********************************************************/

public class StudentDriver
{
  public static void main(String[] args)
  {
    Student s1;  // first student
    Student s2;  // second student

    s1 = new Student();
    s1.setFirst("Adeeb");
    s1.setLast("Jarrah");
    s2 = new Student("Heejoo", "Chun");
    s2.printFullName();
  } // end main
} // end class StudentDriver
```

Figure 8.1 `StudentDriver` class

Field Declarations and Initializations

For each class, declare and/or initialize your class's fields at the top of the class's body (a *field* is a generic term for a class constant, an instance constant, a class variable, or an instance variable). Below your field declarations/initializations, you should provide a blank line, a line of asterisks, and another blank line.

Method Descriptions

Note the descriptions above one of the constructors in Figure 8.2a and the methods in Figure 8.2b. Put things in this order above each method:

- blank line
- line of asterisks
- blank line
- description
- blank line

For short obvious methods, it's OK to omit the method description. Between short constructors and between short accessor and mutator methods, it's also OK to omit the line of asterisks.

```
/*******************************************************************
* Student.java
* Dean & Dean
*
* This class handles processing of a student's name.
*******************************************************************/

import java.util.Scanner;

public class Student
{
  private String first = ""; // student's first name
  private String last = "";  // student's last name

  //***************************************************************

  public Student()
  { }

  // This constructor verifies that each passed-in name starts
  // with an uppercase letter and follows with lowercase letters.

  public Student(String first, String last)
  {
    setFirst(first);
    setLast(last);
  }

  //***************************************************************
```

Figure 8.2a Student class—part A

```
  // This method verifies that first starts with an uppercase
  // letter and contains lowercase letters thereafter.

  public void setFirst(String first)
  {
    // [A-Z][a-z]* is a regular expression. See API Pattern class.
    if (first.matches("[A-Z][a-z]*"))
    {
      this.first = first;
    }
    else
    {
      System.out.println(first + " is an invalid name.\n" +
        "Names must start with an uppercase letter and have" +
        " lowercase letters thereafter.");
    }
  } // end setFirst

  //*******************************************************************

  // This method verifies that last starts with an uppercase
  // letter and contains lowercase letters thereafter.

  public void setLast(String last)
  {
    // [A-Z][a-z]* is a regular expression. See API Pattern class.
    if (last.matches("[A-Z][a-z]*"))
    {
      this.last = last;
    }
    else
    {
      System.out.println(last + " is an invalid name.\n" +
        "Names must start with an uppercase letter and have" +
        " lowercase letters thereafter.");
    }
  } // end setLast

  //*******************************************************************

  // Print the student's first and last names.

  public void printFullName()
  {
    System.out.println(this.first + " " + this.last);
  } // end printFullName
} // end class Student
```

Figure 8.2b Student class—part B

Blank Lines

In general, use blank lines to separate logical chunks of code. In Figure 8.1's `StudentDriver` class, note the blank lines:

- between the prologue section and the class definition
- right after a method's local variable declarations

It's not shown in the Student program, but for long methods, it's appropriate to insert blank lines between logically separate chunks of code within the method. Also, when a comment line appears within the body of the code, it's nice to have white space above that comment to make it more visible.

Meaningful Names

Use meaningful names for your classes and variables. For example, `Student` is a good name for the class in Figures 8.2a and 8.2b because the class models a student. Similarly, `setName` would be a good name for a mutator method that sets a student's `first` and `last` name instance variables, and `getLast` would be a good name for an accessor method that returns the last name.

Braces and Indentations

As shown in Figure 8.1 and Figures 8.2a and 8.2b, place opening braces ({) immediately below the first letter of the preceding line. Indent everything that's logically inside the brace. When you're done with a block (that is, when you're ready for the closing brace), "outdent" so the opening and closing braces for a particular block are aligned. By following this indent-outdent scheme, you'll always align opening and closing brace partners in the same column. For example, note how the `Student` class's opening and closing braces are both in the same column.

Our recommendation on where to put the opening brace ({) is different from Oracle's recommendation, which is that the opening brace be at the end of the previous line, like this:

```
public void setName(String first, String last) {
  this.first = first;
  this.last = last;
}
```

This is one of the few places where our recommendation differs from Oracle's recommendation. Many programmers follow the recommendation that we prefer because it provides better visual bracketing of the block of code that the braces define. However, placing the opening brace at the end of the previous line makes the code a little tighter, and if you or your teacher or your boss wants the opening brace at the end of the previous line, you have our blessing to follow that convention.

Be consistent with your indentations. Any indentation width between two and five spaces is acceptable as long as you're consistent throughout your program. We use two spaces in the book because book page widths are less than computer screen widths, and we don't want to run out of room for programs with deep nesting.

Many novice programmers indent improperly. They either don't indent when they should indent, or they indent when they shouldn't indent, or they use inconsistent indentation widths. That leads to programs that are unprofessional-looking and difficult to read. Some novice programmers postpone entering their indents until the end, after they've finished debugging. Big mistake! Use proper indentation as you enter your program. That should be pretty easy since there are really only two rules to remember:

1. Use braces to surround a block of code that is logically inside something else.
2. Indent the code that is inside the braces.

There is one exception to the first rule:

> Code that follows a `switch` statement's `case` clause is considered to be logically inside the `case` clause, but braces are not used.

Variable Declarations

As shown in Figure 8.1's `main` method, place all local variable declarations at the top of the method (even though that's not required by the compiler). Exception: Unless you need a `for` loop iteration variable to persist beyond the end of the `for` loop, declare it in the initialization field of the `for` loop header.

Normally, specify only one variable declaration per line. Exception: If several variables with obvious meanings are intimately related, it's OK to group them on one line.

Include a comment for every variable whose meaning is not obvious. For example, the cryptic local variable declarations in the `main` method in Figure 8.1 definitely need comments, and we also provide comments for the instance-variable declarations in Figure 8.2a. Note how those comments are aligned—their `//`'s are in the same column. In general, if you have comments that appear at the right side of several nearby lines, try to align those comments.

Line Wrap

If you have a statement that is too long to fit on one line, split it at one or more natural breaking points within the statement. For example, note where we break the long print statement in Figure 8.2b's `setFirst` and `setLast` methods. We consider these to be natural breaking points:

- right after the opening parenthesis
- after a concatenation operator
- after a comma that separates parameters
- at whitespace in expressions

After a break point in a long statement, indent the remaining part of the statement on the next line. In Figure 8.2b, note how we indented the continuation lines with the same standard two-space width that we use for all other indentations.

Rather than simply indenting continuation lines with the standard indentation width, some programmers prefer to align continuation lines with a parallel entity on the previous line. For example, in the aforementioned print statement, they would align the continuation line with `first` like this:

```
System.out.println(first +
                   " is an invalid name.\n" +
                   " Names must start with an uppercase" +
                   " letter and have lowercase letters" +
                   " thereafter.");
```

In our opinion, the above code is pushed too far to the right and is unnecessarily chopped up. That's why we prefer to keep it simple and just indent with the normal indentation width.

Braces That Surround One Statement

For a loop statement or an `if` statement that includes only one subordinate, it's legal to omit the braces around the statement. For example, in Figure 8.2b's `setFirst` method, the "if-else" statement could be written like this:

```
if (first.matches("[A-Z][a-z]*"))
  this.first = first;
else
  System.out.println(first + " is an invalid name.\n" +
    "Names must start with an uppercase letter and have" +
    " lowercase letters thereafter.");
```

However, we like to use braces for all loop statements and `if` statements, even if there is only one enclosed statement. Why?

- Braces provide a visual cue for remembering to indent.
- Braces help you avoid a logical mistake if you add code later that's supposed to be within the loop statement or the `if` statement.

The second point can best be understood with an example. Assume that a program contains this code:

```
if (person1.isFriendly())
  System.out.println("Hi there!");
```

Assume that a programmer wants to add a second print statement ("How are you?") for a friendly `person1` object. A careless programmer might do it like this:

```
if (person1.isFriendly())
  System.out.println("Hi there!");
  System.out.println("How are you?");
```

Because the second print statement is not within braces, it is executed regardless of whether `person1` is friendly. And do you want to ask an unfriendly person "How are you?" You might get a scowl for a response.

On the other hand, if the program followed our style guidelines, the original code would look like this:

```
if (person1.isFriendly())
{
  System.out.println("Hi there!");
}
```

Then if a programmer wants to add a second print statement ("How are you?") for a friendly `person1` object, it would be harder to make a mistake. Even a careless programmer would probably code the second print statement correctly like this:

```
if (person1.isFriendly())
{
  System.out.println("Hi there!");
  System.out.println("How are you?");
}
```

In our above discussion, we said that "we like to use braces for all loop statements and `if` statements." More formally stated, we like to use a block for all loop statements and `if` statements. A *block* is a set of statements surrounded by braces.

Comments

As shown in Figure 8.1 and Figures 8.2a and 8.2b, for all but the shortest blocks, include a comment after a closing brace in order to specify the block that is being closed. For example, in Figure 8.2b, note this closing brace line for the `setFirst` method:

```
} // end setFirst
```

Why is that good practice? So someone reading the program can quickly identify the block that is being ended without having to scroll to the top of the block to find out. It's OK to omit closing-curly-brace comments for short blocks of less than about five lines. For short blocks, it's easy to tell what block the closing brace is attached to, and the final comment just adds clutter.

Include comments for code segments that would not be obvious to a typical Java programmer. In Figure 8.2b, notice this comment that appears at the tops of the bodies of the `setFirst` and `setLast` methods:

```
// [A-Z][a-z]* is a regular expression. See API Pattern class.
```

Direct the reader to more info. 

This comment is helpful because the subsequent statement is more obscure than most. A comment should either explain something directly, help the programmer find more information on a topic, or both. A comment like the one above that references an authoritative source is especially important whenever code implements something mysterious—an arbitrary definition like the "regular expression" above, a formula with empirical coefficients, or a mysterious mathematical expression.

Whenever a comment is too long to fit at the right of the line that is being explained, put it on one or more lines by itself above the line that is being explained. The `//` should be indented the same as the described line. If you put a comment on a line by itself, make sure there is sufficient whitespace above it. In the `setFirst` and `setLast` methods of Figure 8.2b, there's sufficient whitespace above the comments because the prior lines happen to be opening braces for their respective method bodies. In other cases, you'll need to insert a full blank line above the comment. It's optional whether you insert a blank line below it.

Do not add individual comments that just restate what the code already tells you. For example, for the first assignment statement in Figure 8.1's `main` method, this comment would be overkill:

```
s1 = new Student(); // instantiate a Student object
```

Developing readable programs is an important skill and a bit of an art form. Having too few comments is bad because it leads to programs that are difficult to understand. But having too many comments is also bad because it leads to cluttered programs that are difficult to wade through. There's a similar balancing act for blank lines. Having too few blank lines is bad because it leads to programs that are difficult to understand; but having too many blank lines is also bad because it leads to programs with too much dead space.

Blank Spaces

As shown in Figure 8.1 and Figures 8.2a and 8.2b, include blank spaces:

- after the single asterisks in the prologue
- before and after all operators (except for the operators inside a `for` loop header)
- between a closing brace and the `//`'s for its associated comment
- after the `//`'s for all comments
- after the `if`, `while`, and `switch` keywords

On the other hand, do not include blank spaces:

- between a method call and its opening parenthesis
- within each of the three components in a `for` loop header

The last point can best be understood with an example. Here is a nicely written `for` loop header:

```
for (int i=0; i<10; i++)
```

Note that there are no spaces surrounding the = operator or the < operator. Why is that good practice? Because the `for` loop header is inherently complex. In order to temper that complexity, we add visual cues to compartmentalize the `for` loop header. More specifically, we consolidate each section (no spaces within each section), and we insert a space after each semicolon to keep the three sections separate.

Grouping Constructors, Mutators, and Accessors

For short, obvious methods, you should omit descriptions. For example, mutators and accessors are short and obvious, so you should omit descriptions for them. Constructors are sometimes short and obvious, but not always. If a constructor simply assigns parameter values to associated instance variables, then it is short and obvious and you should omit a description for it. If, on the other hand, a constructor performs non-obvious input validation on user-entered values prior to assigning them into associated instance variables, then you should include a description for the constructor.

In the interest of grouping similar things together, we recommend omitting the line of asterisks between mutators and accessors and between short obvious constructors. Assuming that a class contains two short, obvious constructors, several mutator and accessor methods, and two short, obvious other methods, here's the framework for such a class:

```
<class-heading>
{
   <instance-variable-declarations>

   //**********************************************************

   <constructor-definition>

   <constructor-definition>

   //**********************************************************

   <mutator-definition>

   <mutator-definition>

   <accessor-definition>

   <accessor-definition>

   //**********************************************************

   <method-definition>

   //**********************************************************

   <method-definition>
}
```

For this case, there are no descriptions for the constructors, the accessors, or the mutators. There is a line of asterisks above the firstmutator, but not above the subsequent mutator and accessors. These omissions make the program more readable by grouping similar things together.

8.3 Documentation for Outsiders

The preceding section described coding strategies that make code easier to understand, so the reader will be better able to know how to edit or improve it, if necessary. This section also describes techniques that make code easier to understand, but this time the goal is for the reader to be able to use the code (not to edit or improve it). As you might expect, there is some overlap, but that's OK. The previous section's style is right for one audience. That was documentation for insiders. This section's style is right for a different audience. This is documentation for outsiders.

Preconditions and Postconditions

Suppose you are thinking about using an existing block of code like a class or a method, but you can't or don't want to dive into the details of how it works. You'll still need to know what it does. Well-written Java code is largely self-documenting. If a program is written with full, descriptive words for its class names and method names, that can help someone using the program discern what the program does.

Comments in a class's prologue might address either what the class does or how it does it. Thus, comments communicate with outsiders as well as insiders. Comments just above a method header might describe what a method does as much as how it does it. So they also communicate with outsiders as well as insiders. Up to now, we've been flexible about the content of those comments. Now, let's move more to the left sides of our brains. Let's think more like lawyers and try to convert some of the comments above our methods into legal *contracts*.

We can make two kinds of contracts. One kind of contract, called a *precondition,* is an assumption. It says the method expects a certain condition or set of conditions to be true before it executes. If the precondition is not satisfied, then the results produced by the method are unpredictable. The other kind of contract, called a *postcondition,* is a guarantee or warranty. It says that the method promises to make certain conditions true by the time it finishes execution. A method's preconditions are what it needs. Its postconditions are what it does.

For many methods, like ordinary accessor and mutator methods, the method and parameter names themselves provide all the necessary precondition and postcondition information. In other cases, an ordinary comment above the method header provides enough additional information to make it clear what the preconditions and postconditions are. Whenever a comment describes the state of a parameter or the state of an instance or class variable that the method relies on, that's an example of providing precondition information. Similarly, whenever a comment describes what a method returns or describes the state of an instance variable or class variable that the method updates, that's an example of providing postcondition information.

If you feel it's appropriate to emphasize a particular precondition assumption or a particular postcondition guarantee, you should consider providing an explicit precondition comment or an explicit postcondition comment. The following example shows an updated version of Figure 7.7's `swapPerson` method, with such explicit precondition and postcondition comments.

```
// This method swaps the names for two Person objects.
// Precondition: The otherPerson parameter must not be null.
// Postcondition: The calling object's name and also the parameter
//                object's name are both updated.

public void swapPerson(Person otherPerson)
{
  ...
```

The precondition above is particularly important because if `otherPerson` is `null`, the program will compile successfully, but it will crash when executed. The postcondition above is also important because it alerts the method user of something fairly unusual. Normally, parameters transport information one way—from the calling module to the called method. But in this case, the `otherPerson` parameter does more. The method updates the value in `otherPerson`'s `name` instance variable, and that updated value is accessible to the calling module.

Starting now, from time to time we will provide explicit precondition and/or postcondition comments to emphasize important conditions in a method. However, to avoid overburdening you, we usually try to avoid situations that require such comments, so you will not see many of them.

Introduction to javadoc

Now let's move further from "how" to "what"—to what we see in Java's application programming interface (API) documentation, where the "how" is completely hidden. Did you ever wonder how they produce that documentation? Do people familiar with Java create web pages from information in their heads, or does a computer program generate those web pages automatically? A computer program generates them automatically. That program is the `javadoc` tool, which (along with `javac` and `java`) is an integral part of Oracle's Java Development Kit (JDK).

Like a word processor, `javadoc` copies certain patterns in Java source code and pastes them onto web pages. The designers of Java API software use `javadoc` to document the code they write. We too can use `javadoc` to document the code we write. To run `javadoc`, at a command prompt, enter this command:

`javadoc -d` *<output-directory>* *<source-files>*

The `-d` *<output-directory>* option[3] ("d" means "destination") causes the output to go to the specified output directory. You can put documentation for more than one source-code file into the same documentation directory. Just put spaces between multiple source file names.

Suppose you want to generate interface documentation on the `Student` class whose source code is presented in Figures 8.2a and 8.2b. Assuming you are currently in the directory that contains the source code, and assuming you want `javadoc`'s output to go to a subdirectory called `docs`, here's what the command would look like:

```
javadoc -d docs Student.java
```

To see the output, open a web browser like Mozilla Firefox, navigate to the `docs` file, and click on `index.html`. Figure 8.3 shows the top part of the interface document that `javadoc` creates—the "Summary" information. This interface document contains an impressive amount of information—but not quite everything we need. For example, it doesn't include the comment in the last line of the prologue that describes the class in general, it doesn't include the comment that describes the two-parameter constructor, and it doesn't include the comments that describe the three methods.

To enable `javadoc` to extract this other information from source code, we need for all interface information to be located immediately above the heading of whatever it is describing. Also, we need for this information to be enclosed in a *javadoc block comment* that begins with a single forward slash followed by two asterisks and ends with a single asterisk followed by a single forward slash, like this:

*/** <extractable-information> */*

[3]For other options and other arguments, enter `javadoc` by itself.

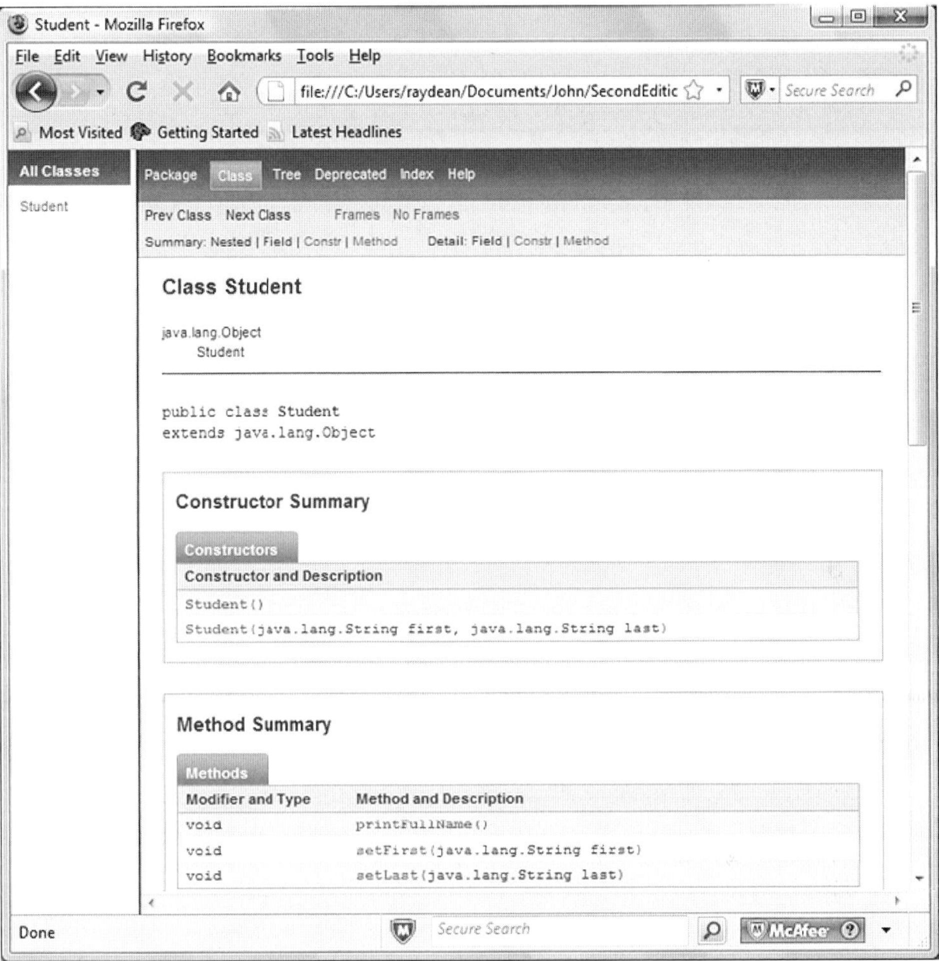

Figure 8.3 Top part of the `javadoc` output for the Student class defined in Figures 8.2a and 8.2b

Since Figure 8.2a has an `import` statement between the general prologue and the class heading, we must move our general comment out of the general prologue and put it into a `javadoc` block comment located just above the class heading. Similarly, we must put individual constructor and method interface information into `javadoc` block comments located just above their respective headings. There is some flexibility. The extractable information in one of these `javadoc` block comments does not need to be on just one line. Also, if you wish, you may put the opening `/**` and the closing `*/` on lines above and below the text, as shown in Figure 8.4.

With these changes implemented in the `Student.java` code, Figure 8.5 shows the top part of what `javadoc` generates. If you compare this with Figure 8.3, you'll see that Figure 8.5 includes the general comment for the whole class and the special comment for the two-parameter constructor. We also changed the rest of the code so that `Student_jd` has `/** ... */` `javadoc` block comments above the method headings too. Therefore, the `javadoc` output also includes special comments for each method. The constructor and method comments also appear in the "Detail" parts of the output display, which is below what you see in Figures 8.3 and 8.5.

```
/*********************************************************************
* Student_jd.java
* Dean & Dean
*********************************************************************/

import java.util.Scanner;

/** This class handles processing of a student's name. */

public class Student_jd
{
  private String first = ""; // student's first name
  private String last = "";  // student's last name

  //******************************************************************

  public Student_jd()
  { }

  /**
  This constructor verifies that each passed-in name starts with
  an uppercase letter and follows with lowercase letters.
  */

  public Student_jd(String first, String last)
  {
    setFirst(first);
    setLast(last);
  }
```

single-line javadoc comment

multiple-line javadoc comment

Figure 8.4 Top part of the `Student` class in Figure 8.2a, modified to accommodate `javadoc`

This is just an introduction. For more details, see Appendix 6. In professional programming, it's important to use `javadoc` to document your code properly. However, as you can see in Appendix 6, this makes code significantly longer. Textbook descriptions accompanying code presentations parallel much of the information in `javadoc` comments and make such comments redundant in the textbook context. Therefore (except for what you see in this section and in Appendix 6), to save book pages and lighten your load, this book's programs do not include the extra comments and apparatus needed for proper `javadoc` documentation.

8.4 Helper Methods

In the first four chapters, we solved essentially every problem we addressed in just one module—the `main` method. As problems get bigger, however, it becomes more and more necessary to partition them into subproblems, each of which has a manageable size. We started doing this in earlier chapters when our `main` method called on some of Java's API methods for help.

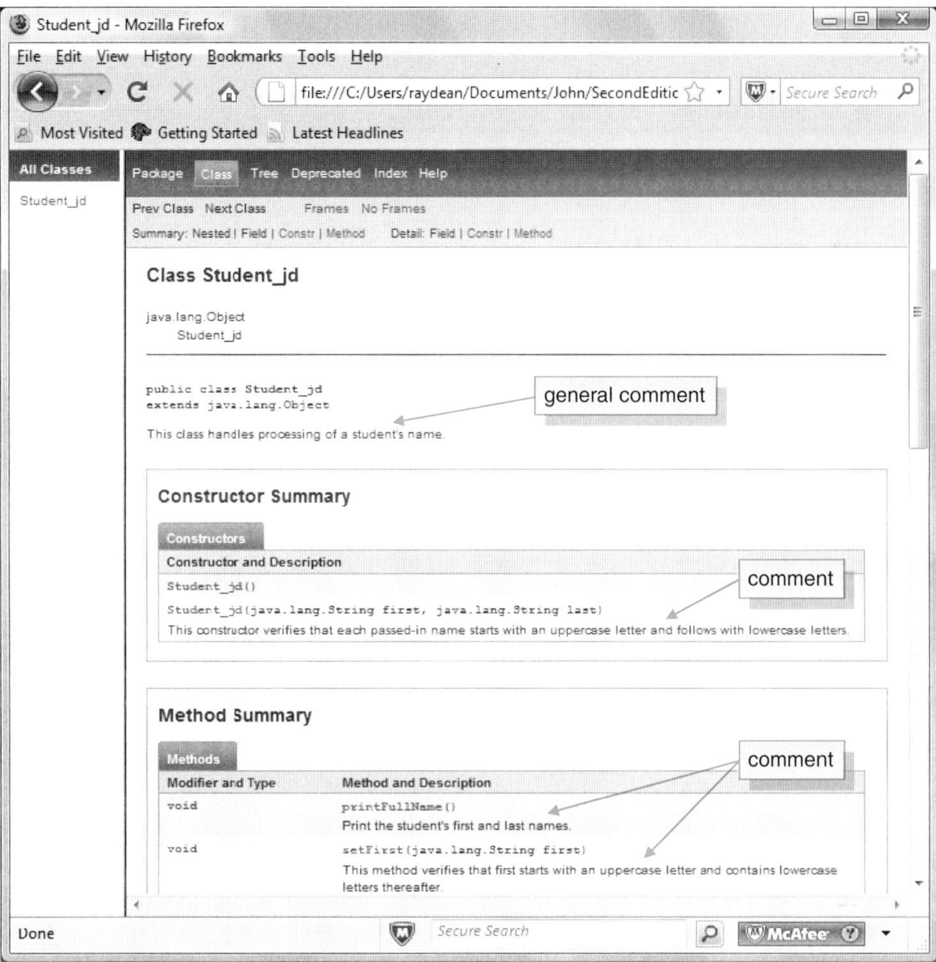

Figure 8.5 Top part of the `javadoc` output for the modified Student class

In a broad sense, any method that is called by another method is a "helper method" because the called method helps the calling method. The calling method is a *client,* and the called method (the broad-sense helper method is a *server.*

You can narrow the definition of helper method by restricting it to a called method that happens to be in the same class as the calling method. In Figure 8.2a, the `Student` constructor calls two methods from the `Student` class-`setFirst` and `setLast`. Presumably, these mutators were written to allow a user to change the instance variables in an object after the object was originally initialized. But once their code is written, why not reuse it? By including calls to these two ordinary methods in the constructor, we avoid duplication of the code in the called methods. Because the `setFirst` and `setLast` mutator methods each include a significant amount of error-checking code that helps the constructor do its job, this organization helps divide the problem into smaller chunks.

You can narrow the definition of helper method even more. Up to this point, all methods we've covered have used the public access modifier. These public methods are part of the class's *interface* because they are responsible for the communication between an object's data and the outside world. Sometimes you'll want to create a method that is not part of the interface; instead, it just supports the operation of other methods within its own class. This special type of method—a method that is in the same class and has a private access modifier—is what we mean when we say *helper method.*

Shirt Program

Suppose you're asked to write a program that handles order entries for sports-uniform shirts. For each shirt order, the program should prompt the user for a shirt's primary color and its trim color. Figure 8.6 shows the driver and a typical sample session.

```
/*****************************************
* ShirtDriver.java
* Dean & Dean
*
* This is a driver for the Shirt class.
*****************************************/

public class ShirtDriver
{
  public static void main(String[] args)
  {
    Shirt shirt = new Shirt();

    System.out.println();
    shirt.display();
  } // end main
} // end ShirtDriver
```

Sample session:

```
Enter person's name: Corneal Conn
Enter shirt's primary color (w, r, y): m
Enter shirt's primary color (w, r, y): r
Enter shirt's trim color (w, r, y): w

Corneal Conn's shirt:
red with white trim
```

Figure 8.6 ShirtDriver class and associated sample session

For each color selection, the program should perform the same input validation. It should verify that the entered color is one of three values—w, r, or y, for white, red, or yellow. That input validation code is non-trivial. It's in charge of:

- prompting the user for a color entry,
- checking whether the entry is valid,
- repeating the prompt if the entry is invalid, and
- converting the single-character color entry to a full-word color value.

These four tasks are a coherent group of activities. Therefore, it's logical to encapsulate them (bundle them together) in a separate module. The fact that each shirt order performs this coherent group of activities two separate times (once for selecting the shirt's primary color and once for selecting the shirt's trim color) provides an additional reason to encapsulate them in a separate module. Thus, instead of repeating the complete code for these four tasks in the constructor each time color selection is needed, you should put this color-selection code in a separate helper method and then call that method whenever color selection is needed. Study the Shirt program and sample session in Figures 8.6, 8.7a, and 8.7b, especially the `public` constructor, `Shirt`, and the `private` helper method, `selectColor`. Note how the constructor calls the `selectColor` method twice. In this particular case, the helper method calls are from a constructor. You can also call a helper method from any ordinary method in the same class.

```
/*********************************************************
* Shirt.java
* Dean & Dean
*
* This class stores and displays color choices for
* a sports-uniform shirt.
*********************************************************/

import java.util.Scanner;

public class Shirt
{
  private String name;      // person's name
  private String primary;   // shirt's primary color
  private String trim;      // shirt's trim color

  //*****************************************************

  public Shirt()
  {
    Scanner stdIn = new Scanner(System.in);

    System.out.print("Enter person's name: ");
    this.name = stdIn.nextLine();

    this.primary = selectColor("primary");
    this.trim = selectColor("trim");
  } // end constructor

  //*****************************************************

  public void display()
  {
    System.out.println(this.name + "'s shirt:\n" +
    this.primary + " with " + this.trim + " trim");
  } // end display
```

No need for a reference variable dot prefix here.

Figure 8.7a `Shirt` class—part A

```
//*******************************************************

Use the private access modifier for a helper method.

// Helper method prompts for and inputs user's selection

private String selectColor(String colorType)
{
  Scanner stdIn = new Scanner(System.in);
  String color; // chosen color, first a letter, then a word

  do
  {
    System.out.print("Enter shirt's " + colorType +
      " color (w, r, y): ");
    color = stdIn.nextLine();
  } while (!color.equals("w") && !color.equals("r") &&
           !color.equals("y"));

  switch (color)
  {
    case "w":
      color = "white";
      break;
    case "r":
      color = "red";
      break;
    case "y":
      color = "yellow";
  } // end switch

  return color;
} // end selectColor
} // end class Shirt
```

Figure 8.7b `Shirt` class—part C

There are two main benefits to using helper methods: (1) By moving some of the details from `public` methods into `private` methods, they enable the `public` methods to be more streamlined. That leads to `public` methods whose basic functionality is more apparent. And that in turn leads to improved program readability. (2) Using helper methods can reduce code redundancy. Why is that? Assume that a particular task (such as color input validation) needs to be performed at several places within a program. With a helper method, the task's code appears only once in the program, and whenever the task needs to be performed, the helper method is called. On the other hand, without helper methods, whenever the task needs to be performed, the task's complete code needs to be repeated each time the task is done.

Note that in Figure 8.7a, we call the `selectColor` method without a reference variable prefix:

```
this.primary = selectColor("primary");
```

Why is there no reference variable dot prefix? If you're in a constructor (or an instance method, for that matter), and you want the current object to call another method that's in the same class, the reference variable dot prefix is unnecessary. Because the constructor and the `selectColor` method are in the same class, the reference variable dot prefix is unnecessary.

8.5 Encapsulation (with Instance Variables and Local Variables)

A program exhibits encapsulation if its data is hidden; that is, if its data is difficult to access from the "outside world." Why is encapsulation a good thing? Because the outside world isn't able to access the encapsulated data directly, it's more difficult for the outside world to mess things up.

Encapsulation Implementation Guidelines

There are two main techniques for implementing encapsulation:

- First, break a big problem into separate classes, where each class defines a set of encapsulated data that describe the current state of an object of that class. Encapsulate this object-state data by using the `private` access modifier for each such data item. As you already know, a class's object-state data items are called *instance variables*.
- Second, break a class's tasks into separate methods, where each method holds a set of additional encapsulated data it needs to do its job. As you already know, a method's data items are called *local variables*.

Declaring instance variables within a class is one form of encapsulation, and declaring local variables within a method is another form of encapsulation. Which is the stronger (more hidden) form of encapsulation? All instance methods have access to all instance variables defined in the same class. On the other hand, only the current method has access to one of its local variables. Therefore, a local variable is more encapsulated than an instance variable. Thus, to promote encapsulation, use local variables, not instance variables, whenever possible.

In writing a method, you'll often find the need for more data than what's provided by the current instance variables. The question then becomes—how should you store that data? In another instance variable? Or locally? Try to resist the urge to add another instance variable. You should use instance variables only for storing fundamental attributes of the class's objects, not for storing additional details. If you can store the data locally, then do so. That furthers the goal of encapsulation. Usually when we think of storing data locally, we think of a local variable declared inside a method's body. Be aware that parameters are another way to store data locally. Remember that a parameter is declared in a method's heading—that tells us it has local scope.

Local Variables Versus Instance Variables in The `Shirt` Class

Now let's see how the above philosophy plays out in the `Shirt` class. The fundamental attributes of a shirt are its name, its primary color and its trim color. That's the basis for our declaration of the three instance variables declared in Figure 8.7a:

```
private String name;       // person's name
private String primary;    // shirt's primary color
private String trim;       // shirt's trim color
```

Now let's look at the other variables we need as we write the class's methods. All of these other variables are somehow associated with the `selectColor` method in Figure 8.7b. We need to transfer data in both directions between the calling `Shirt` constructor and the called `selectColor` method.

First, consider transfer of data into the `selectColor` method. If a shirt's primary color is needed, then `selectColor` should print this prompt message:

```
Enter shirt's primary color (w, r, b):
```

If a shirt's trim color is needed, then `selectColor` should print this prompt message:

```
Enter shirt's trim color (w, r, b):
```

We must transfer data into the `selectColor` method that tells the `selectColor` method which query to print. It would be possible to transfer this data by declaring another instance variable called `colorType`, have the `Shirt` constructor write a value to this instance variable, and then have the `selectorColor` method read the value of this instance variable. But this would be bad practice because it would break the encapsulation within the `selectColor` method and add confusing clutter to our nice clean list of object attributes. The proper way to implement this method-to-method communication is the way we did it, with an argument/parameter transfer.

Second, consider transfer of data out of the `selectColor` method. We also have to transfer data back from the `selectColor` method to the `Shirt` constructor. This data is the string representation of the selected color. There are three good ways to transfer data back to the calling method:

1. If there is only a single return value, you can send it back to the calling module as a `return` value.
2. If there is more than one value to return, you can assemble these values into an object, create that object in the helper method, and return a reference to that locally created "communication object."
3. You can pass into the helper method references to "communication objects" instantiated in the calling module and use code in the helper method to write to those objects.

It's also possible to transfer data back to the calling module by declaring other instance variables, having the helper method write values to them, and having the calling module read from them after the helper method terminates its execution. But this would be bad practice because it would break the encapsulation and add confusing clutter to our nice clean list of object attributes. The proper way to implement this method-to-method communication is the way we did it, with a `return` value. In this case, the `return` value is a reference to a `String` object.

The `Shirt` class has one other variable to consider, the `stdIn` reference to a keyboard communication object. This particular object is used by both the calling constructor and the called helper method, and it is instantiated twice, once in each of those two modules. It is tempting to try to avoid duplicate instantiation by making `stdIn` an instance variable. And it will "work." But we recommend against doing this because `stdIn` is clearly not a fundamental attribute of this class's objects. It's not a variable that describes the state of a shirt! In a later version of the program, you might want to change the method of input from the keyboard to something else, like a data file, described later in Chapter 16. You might even want to use one method of input for the name and a different method of input for the other state variables. Then you'd need to change `stdIn`, and you might want to change it in different ways for different methods. Declaring it local makes future modifications local also, and it's better design practice.

An argument used for not making a variable local is "maybe someday we'll need broader scope." If you have a specific plan that truly requires the broader scope you propose, OK. But if it's just "maybe someday," don't provide broader scope until that "someday" actually comes. Then, at that time, modify your program to increase scope only where it's absolutely necessary.

8.6 Recognizing the User's Point of View

Suppose all you want to do is use somebody else's program, and you don't really care how it's written. In that case, you are a user, and what you want is a user's view of the program. Every time we present a "sample session" or a GUI display, we're showing you what users see.

Sometimes the intended users are other programmers. Then, the most appropriate user's view is documentation designed to be read by programmers. An excellent example of this kind of user's view is the Java API documentation or the output produced by `javadoc`, described in Section 8.3 and Appendix 6. In Chapter 5, you looked at the interfaces of code written by designers of the Java language. We were not showing you how to write that code; we were just showing you how to use it. That view of the Java API was a user's view for users who are themselves programmers.

Sometimes the intended users may not be programmers. They might be artists, accountants, nurses or doctors, detectives, librarians, inventory clerks, or equipment service personnel. Sometimes whole books are devoted to explaining a big program to users who are not programmers. The goal of these books is not to explain how to write the program, or even how it works. It is to explain what the program does and how to use it to solve a problem that is outside the world of programming. Figure 8.8 shows the principal components of program documentation written for users who may not be programmers.

When you write a program that someone else will use, occasionally you should try to put yourself into that other person's "shoes." If you know some users or potential users, visit with them. Ask them what their needs are, and try to determine how they like to think about the subject the program will address. If possible, get them to make a list or draw a picture that portrays the results they would like the program to produce. Ask what kinds of data they would like the program to use as input and what kinds of assumptions they consider to be acceptable.

Try to think about the problem the same way users are inclined and able to think about the problem. Ask them how they have solved the problem or similar problems in the past. When numerical calculations are involved, ask how they have done those calculations. If potential customers are in the market for a computer solution, they will expect greater accuracy and speed from the computer, and they will expect your solution to be better than theirs. But if you know what they have done in the past, you will be more able to give them something they will like, and you will be more prepared for their evaluation of your product.

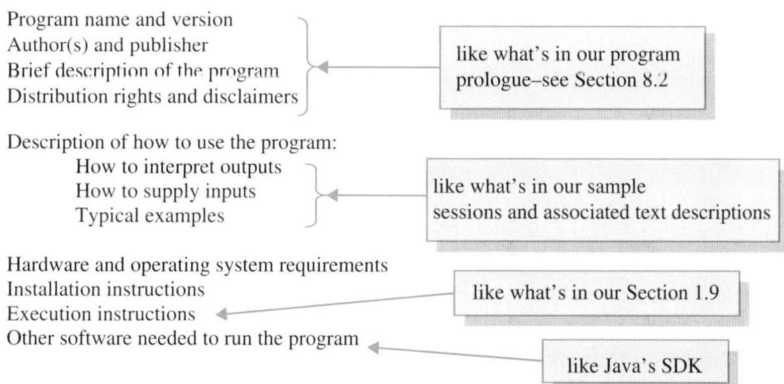

Figure 8.8 Principal components of documentation for users who may not be programmers

As you develop the program, keep thinking about output and input formats—output first, then input. From time to time, test your proposed formats with potential users. When you do this, listen for feedback that goes beyond cosmetics. Are you focusing on the problem users care about? Are your assumptions the same as theirs? Have you left out anything that's important? Learning how to think like a user will improve your effectiveness as a programmer.

8.7 Design Philosophy

In the next several sections, we discuss alternative strategies for solving problems. That's plural "strategies" because there's not just one cookie-cutter strategy that can be used to solve all problems. If there were just one universal strategy, programming would be easy and anyone could do it. But it's not easy. That's why good programmers are in demand and earn a decent wage.

Simplistic Approach to Design

Here's a simplistic recipe for how to design things:

1. Figure out what you want to do.
2. Figure out how to do it.
3. Do it.
4. Test it.

At first, this list seems like obvious common sense. But actually, it works only for very simple problems—problems where everything is easy and you don't need any recipe. What's wrong with this recipe?

First, if a problem is difficult, it's hard to know what its solution will be like. Often we need experience to know even what we want to do. Most clients recognize this and are flexible enough to accept a range of possible solutions. They want to avoid imposing arbitrary specifications that would cause them to miss inexpensive opportunities or incur expensive penalties. With difficult problems, people want to keep their options open.

Second, most problems have several alternate ways in which they can be solved. It takes some experimentation to determine the best way to solve a difficult problem. For very difficult problems, it's impossible to know exactly "how to do it" until we have done it.

Third, when we "do it," we must recognize it will not be perfect. There will be hidden errors. We will discover a better way to do it. The client will discover it would have been better to have asked for something different. And we'll need to do it again.

Fourth, if we defer testing of anything complicated until the end, we are almost sure to fail. The thing might pass its one final "test," but it will probably fail in its ultimate job, because one final test cannot catch all problems.

So, how can you deal with these difficulties?

1. Develop and maintain a sensible compromise between tight specification and flexibility.
2. Partition the problem into distinct concerns.
3. Perform continuous testing at all levels. This helps you identify problems early when they are easy to fix, and it gives you objective assessment of progress. Suppose you're in charge of a large programming project, and you ask your programmers, "How's it coming?" You don't want them just to say, "fine." You want them to show you—by running tests that demonstrate what their current code actually does.

Separation of Concerns

When faced with a large problem, if there is time to reflect, humans like to partition the problem into manageable components. In programming, there's a formal term for the idea of putting code for each task into its own separate place—*separation of concerns.* More formally, we say that separation of concerns is when a program's functional behaviors are implemented in separate modules, with minimal overlap between the functional behaviors.

There are different ways to achieve separation of concerns within a program. At the most granular level, you can break up your program into methods, where each method performs one task. With OOP languages like Java, you can use classes to implement larger concerns, where each class implements its concern by storing related data and providing related methods. With many programming languages, if your program has even still larger concerns, you can keep the concerns separate by grouping a concern's classes together. Java implements that sort of grouping by grouping a concern's classes (where each class is a file) into its own package.

Design Patterns

The strategy of separating a program's concerns is so prevalent that software designers have agreed on separation-of-concern patterns that are particularly useful and popular. Such patterns are examples of *design patterns,* which are general-purpose software solutions to common problems. The software solutions can range in form from high-level descriptions to fully operational source code programs that can be used as starting-point templates. As a programmer, you should take the time to review some of the common design patterns so you know what's out there. That way, when you're confronted with a large, non-trivial problem, before you start implementing it from scratch, you'll have a sense of whether you can borrow from someone else's design. This can help you to avoid reinventing the wheel.

If you'd like to learn about common design patterns, a good place to start is Wikipedia's design pattern page. It describes quite a few design patterns, but we'll focus primarily on just one—the *model-view-controller,* or *MVC* for short. The MVC design pattern has been around a long time—so long that it actually predates the notion of a design pattern. Since it is such a useful design pattern, many integrated development environments include the MVC design pattern as a template that developers can generate with a few button clicks.

Here is how Microsoft describes the three parts (concerns) of the MVC design pattern.[4]

- **Model.** The model manages the behavior and data of the application domain, responds to requests for information about its state (usually from the view), and responds to instructions to change state (usually from the controller).
- **View.** The view manages the display of information.
- **Controller.** The controller interprets the mouse and keyboard inputs from the user, informing the model and/or the view to change as appropriate.

Most of the real work is done by the model. The model stores the program's data (often in a database) and the program's rules (often referred to as *business logic*) that manipulate the data. The view presents the results to the user. Normally, the results are displayed with GUI using windows, text, pictures, etc. The controller is in charge of (1) gathering user input and (2) using that input to tell the model and the view what to do. The model can also tell the view what to do, but typically, it does so indirectly. When the model changes its state, if the view is set up to observe the model's state, the view then updates its presentation automatically. You'll see simple examples of the MVC separation-of-concerns design pattern later in this book, in Section 11.9 and Chapters 17 and 18.

[4]http://msdn.microsoft.com/en-us/library/ff649643.aspx.

Another design pattern is the *Composite* pattern, which organizes objects into compositional trees. We'll describe composition in detail later in this book, in Chapter 13. Another design pattern is the *Observer* pattern, which has objects "broadcast" their state changes to all interested observers.

Testing

It's been said that, on average, experienced programmers make one mistake for every 8 or 10 lines of code.[5] Whew! That's a lot of mistakes. With such a high incidence of mistakes, we hope you're properly convinced about the importance of testing.

Testing has three aspects:

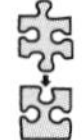

Check the most obvious things first.

- First, subject your program to typical input values. If your program doesn't work with typical input values, you're in real trouble. Co-workers and users may question your competence if your program generates wrong answers for the typical cases.
- Second, subject your program to input values that are at the boundaries of acceptability. These boundary tests often reveal subtle problems that wouldn't show up until later, and such problems might be much harder to fix at that time.
- Third, subject your program to invalid input values. In response to an invalid input value, your program should print a user-friendly message that identifies the problem and prompts the user to try again.

Testing is something that many people envision occurring after a product is finished. That's an unfortunate notion, because a lone test at the end of the production of a complicated product is almost worthless. If the product fails such a test, it may be hard to determine why it failed. If the fix requires many changes, a great deal of work may have been wasted. If the product does not fail a lone final test, you may be lulled into thinking everything is OK even when it's not. Passing a lone final test may actually be worse than failing a lone final test, because passing motivates you to release the product. It's much more costly to fix a problem after a product has been released. (Ray knows about this!) Bottom line—Don't wait until the end to start your testing. Test your program on a regular basis throughout the development process.

Novice programmers sometimes get the idea that it would be "unscientific" to form a pre-conception of what a test result should be before you do the test. That's wrong. It's important that you do have a good idea of what the test result should be before you perform a test. Before you push the "run" button, say out loud what you think the result should be! This improves your chance of recognizing an error.

Testing keeps you on track. In any development program, you should interleave testing and coding so that you get quick feedback. If an experienced programmer makes a mistake in every 8 or 10 lines of code, a new programmer is well advised to perform some kind of test after every 4 or 5 lines of new code! This makes it easy to identify errors, and it reduces your level of stress. The more frequently you test, the more positive feedback you get, and this helps your attitude—it gives you a "warm-fuzzy feeling." Frequent testing makes programming a more pleasant experience.

There is no practical way to verify all the aspects of a complicated system by looking at it only from the outside. Testing should be performed on each component and on combinations of components, at all levels. As you'll see in subsequent discussions, testing typically requires creation of some kind of extra testing code. Sometimes it's a special driver. Sometimes it's a special driven module. Creating such test code may seem like extra work for something that's to be used only in a test environment and not in an actual runtime environment. Yes, it is extra work, but it is well worth the effort. Writing and using test code will save you time in the long run, and it will lead to a better final product.

[5]Of course, we, John and Ray, never make any mistakes. ☺

Using `assert` Statements

Earlier in this chapter, we described preconditions and postconditions. They are examples of a more general construct—an *assertion*. Preconditions and postconditions represent an informal contract with outsiders who may wish to use your code but are not concerned with its details. Java also supports another kind of assertion—an assertion for insiders—that helps programmers debug code during development or modification. This assertion for insiders is a statement that begins with the Java keyword `assert`. It comes in either of two forms:

```
assert <Boolean-condition>;
```

or:

```
assert <Boolean-condition> : <string-describing-an-error>;
```

If its Boolean condition is true, an `assert` statement does nothing. If its condition is false, it prints an output saying an error has occurred.[6] With the second form above, that output includes the descriptive string after the colon. A Java `assert` statement is like an `if` statement that prints an error message when the `if` condition is true. But there are three differences: (1) A true condition in an `assert` statement corresponds to a false condition in the corresponding *`if` (condition) then print* statement. (2) The `assert` statement terminates execution. (3) All `assert` statements are ignored unless you explicitly enable them. You can do that by running Java from a command line with the "enable assert" option (`ea`) specified like this:

```
java -ea <class-name-of-compiled-program>
```

As an alternative, if you're using an integrated development environment to run Java, search for its enable assert configuration option.

It's common practice to use extra print statements to identify errors during program development. Then, after the bugs are gone, you remove those extra print statements to reduce code clutter and improve execution performance. If you use `assert` statements instead, you can leave the `assert` statements in the final code. The clutter they add is minimal, and when you execute without the `-ea` option, there is no performance penalty. If you need to modify the program later, you don't need to reintroduce the original extra print statements to verify the modifications. You can just reuse the still-present `assert` statements by once again executing with the `-ea` option.

To use `assert` statements effectively, look for places in the code where you can ask true/false questions and get consistent and meaningful answers. One logical place is immediately after receiving incoming data. For example, to see if the precondition described above Section 8.3's `swapPerson` method header is satisfied, you might include this as the first statement in `swapPerson`'s body:

```
assert otherPerson != null;
```

Assuming that you run the program with the `-ea` option, when the `assert` statement executes, the JVM checks to see if its condition (`otherPerson != null`) is true or false. If the condition is true, execution continues normally. If the condition is false, the JVM prints an error message and stops the program.

Another logical place is immediately after a series of related operations. For example, look back at the `selectColor` method in Figure 8.7b. The top half of this method has a loop whose continuation condition

[6] If the condition is false, the program throws an `AssertionError` exception and terminates execution. You'll learn about exceptions later in this book, in Chapter 15.

looks at abbreviations for three particular colors. The bottom half of this method has a `switch` statement whose cases are supposed to be those same three colors. Is there consistency between the three colors in the code on the top and the code on the bottom? You can check for consistency in a fairly compact way by inserting these three lines of code at the end of the `switch` statement:

```
    break;
  default:
    assert false : "bad color: " + color.charAt(0);
```

If color specifications are consistent, the `default` case never occurs, and its `assert` statement never executes. If color specifications are inconsistent, execution can sometimes proceed to the `default` case. Then, assuming you run the program with `assert` statements enabled, the `assert` statement executes. Because its condition is always false, the JVM prints an error message showing the mishandled input and stops the program.

8.8 Top-Down Design

The dominant design methodology for large high-performance systems is the top-down design strategy. Top-down design requires the designer to think about the big picture first—that's the "top." After completing the design at the top, the designer works on the design at the next lower level. The design process continues in this iterative manner until the bottom level (the level with the most detail) is reached.

For an OOP project, top-down design means starting with a problem description and working toward a solution using these guidelines:

1. Decide on the classes needed. You should normally include a driver class as one of the classes. To determine the other classes, think of the problem in terms of its component objects. Specify one class for each unique type of object. With large systems that have many classes, pure top-down design defers identification of detailed classes until later because identifying detail classes is itself a detail.
2. For each class, decide on its instance variables, which should be state variables identifying object attributes. The driver class should not have any instance variables.
3. For each class, decide on its `public` methods. The driver class should contain only one `public` method—`main`.
4. For each `public` method, implement in a top-down fashion. Consider each `public` method to be a "top" method. If it is fairly involved and can be broken into subtasks, have it call `private` helper methods to do the subtask work. Finish writing the top methods before starting to write the lower level helper methods. Initially, implement the helper methods as *stubs*. A stub is a dummy method that acts as a placeholder for an actual method. A stub's body typically consists of a print statement that displays something like "In method x, parameters = a, b, c" where x is the name of the method and a, b, and c are values of passed-in arguments. If it's not a `void` method, the stub will also include a default return statement. You'll see examples of `void`-method stubs later in this section.

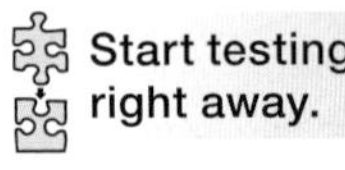
Start testing right away.

5. Test and debug the program. The suggested stub print messages will help you trace the program's actions.
6. Replace stub methods one at a time with fully implemented helper methods. After each replacement, test and debug the program again.

Top-down design is sometimes referred to as *stepwise refinement*. The term stepwise refinement is used because the methodology encourages programmers to implement solutions in an iterative manner where each solution "step" is a refined version of a previous solution step. After implementing top-level tasks, the programmer goes back and refines the solution by implementing the subtasks at the next lower levels.

Benefits of Using Top-Down Design

In top-down design, the designer doesn't worry initially about the details of subtask implementation. The designer focuses on the "big picture" first. Because it focuses on the big picture first, top-down design is good at getting a project going in the right direction. That helps to ensure that the completed program matches the original specifications.

Top-down design is particularly appropriate when a project involves many programmers. Its early emphasis on the big picture forces the project's programmers to agree on common goals. Its strong organizational emphasis promotes coherence and prevents the project from splintering off in different directions. The top-down design methodology facilitates tight managerial control.

Square Program Example: First-Cut Version

Let's now apply the top-down design methodology to a simple example. We'll implement a `Square` class such that each `Square` object can:

- Initialize the square's width.
- Calculate and return its area.
- Draw itself with asterisks using either an asterisks border or a solid pattern of asterisks. Each time the square is drawn, the user is prompted as to whether he or she would like a border format or a solid format, like one of these:

Devise a way to solve the problem.

```
border-format square      solid-format square
     width = 6                 width = 4

   * * * * * *                 * * * *
   *         *                 * * * *
   *         *                 * * * *
   *         *                 * * * *
   *         *
   * * * * * *
```

Using the above top-down design guidelines, the first step is to decide on the classes. In this simple example, it's easy to identify all the classes right at the start—`SquareDriver` and `Square`. The next step is to decide on the instance variables. They should be a minimum definitive set of object properties—state variables. All you need to specify a square is one number. The typical number that people use is the width. So we'll use `width` as our lone instance variable.

But what about the square's area? Area is a property, but it's a simple function of width: area equals width squared. Because we can easily calculate area from width, it would be redundant to include area as another state variable. In principle, we could use `area` as the only state variable, and calculate `width` as the square root of `area` any time we needed `width`. But computing the square root is more difficult than computing the square, and we would frequently end up with a non-integer value for `width`, which would be hard to display in our prescribed asterisk format. So, for our problem, it's a better strategy to use `width` as the lone instance variable.

What about the solidness of the square? This is a conceptual choice. If you want to think of solidness as an inherent property of `Square`-class objects, it's appropriate to create another instance variable like `boolean solid`. On the other hand, if you want to think of solidness as just a temporary display option, solidness should not have state-variable status and it should not be an instance variable. For our example, we've elected to think of solidness as just a temporary display option, so we do not include it as another instance variable.

Identify state variables.

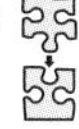

Returning to the top-down design guidelines, we see that the next step is to decide on the `public` methods. The problem description often determines what needs to be `public`. Here's what we need:

- a constructor that sets the square's width
- `getArea`—compute the square's area
- `draw`—display the square with asterisks using either an asterisks border or a solid pattern of asterisks

Let's now step back and look at what we've done so far. See Figure 8.9. It presents a first-cut UML class diagram for our solution's classes, instance variables, and constructor and `public` methods.

Figure 8.9 Square program's UML class diagrams: first-cut version

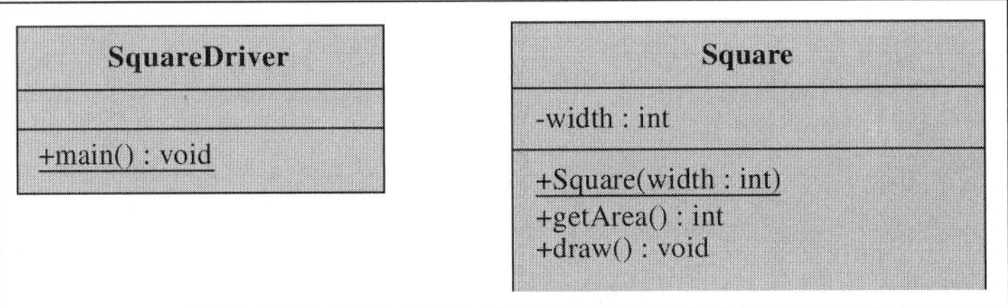

The next step in the top-down design process is to implement the `main` method in the top-level class This implementation appears in Figure 8.10. The code in `main` includes calls to the `Square` constructor and methods identified in Figure 8.9, but it does not yet say anything about how those members of the `Square` class are implemented.

```
/*************************************************************
* SquareDriver.java
* Dean & Dean
*
* This is the driver for the Square class.
*************************************************************/

import java.util.Scanner;

public class SquareDriver
{
  public static void main(String[] args)
  {
    Scanner stdIn = new Scanner(System.in);
    Square square;

    System.out.print("Enter width of desired square: ");
    square = new Square(stdIn.nextInt());
    System.out.println("Area = " + square.getArea());
    square.draw();
  } // end main
} // end class SquareDriver
```

Figure 8.10 `SquareDriver` class

The next step is to implement the public methods in the Square class. This implementation appears in Figure 8.11a. The constructor and getArea methods are straightforward and do not need explanation. But notice that the get in getArea makes this method look like an accessor that simply retrieves an instance variable. Is it OK to create this "false" impression? Yes, it is, because the instance variable is private and therefore hidden from public view. In fact, as noted above, we might actually have used area as the lone

```
/*************************************************************
* Square.java
* Dean & Dean
*
* This class manages squares.
*************************************************************/

import java.util.Scanner;

public class Square
{
  private int width;

  //**********************************************************

  public Square(int width)
  {
    this.width = width;
  }

  //**********************************************************

  public int getArea()
  {
    return this.width * this.width;
  }

  //**********************************************************

  public void draw()
  {
    Scanner stdIn = new Scanner(System.in);

    System.out.print("Print with (b)order or (s)olid? ");
    if (stdIn.nextLine().charAt(0) == 'b')
    {
      drawBorderSquare();
    }
    else
    {
      drawSolidSquare();
    }
  } // end draw
```

Figure 8.11a Square class: first-cut version—part A

instance variable! A user of a class does not have to know exactly how it's implemented. Don't worry about the implementation when you pick a method name. It's the effect that matters, and `getArea` accurately describes the effect of calling that method.

The `draw` method prompts the user to choose either a border format or a solid format for the square's display. It's now becoming apparent that the `draw` method is not trivial. The `drawBorderSquare` and `drawSolidSquare` method calls are examples of subtasks that we should split off into separate helper methods.

Stubs

Top-down design tells us to implement helper methods initially as stubs. For our Square program, that means implementing `drawBorderSquare` and `drawSolidSquare` as stubs. Note the stubs in Figure 8.11b.

```
  //**********************************************************

  private void drawBorderSquare()                     // a STUB
  {
    System.out.println("In drawBorderSquare");
  }

  //**********************************************************

  private void drawSolidSquare()                      // a STUB
  {
    System.out.println("In drawSolidSquare");
  }
} // end class Square
```

Figure 8.11b `Square` class: first-cut version—part B

As you can probably surmise from the examples, a stub doesn't do much. Its main purpose is to satisfy the compiler so that the program is able to compile and run. Its secondary purpose is to provide an output that confirms that the method was called, and (where appropriate) show values passed into that method. When the stubbed Square program runs, it produces either this sample session:

```
Enter width of desired square: 5
Area = 25.0
Print with (b)order or (s)olid? b
In drawBorderSquare
```

or this sample session:

```
Enter width of desired square: 5
Area = 25.0
Print with (b)order or (s)olid? s
In drawSolidSquare
```

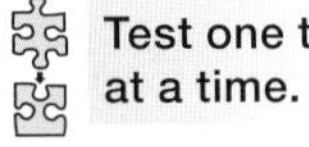

Test one thing at a time.

Using stubs lets programmers test their partially implemented programs to determine whether their behavior is correct down to the stub level. Second, it makes debugging easier. After compiling and running the program successfully with stubs, replace the stubs with actual code one method at a time. As each stub is replaced, test and debug the updated program. If a bug appears, it should be easy to find because you know it's probably in the most recently replaced method.

Square Program Example: Second-Cut Version

The next step in the top-down design process is to replace the helper methods' stub implementations with actual implementations. We have two helper methods to work on—`drawBorderSquare` and `drawSolidSquare`.

Let's start with the `drawBorderSquare` helper method. It prints a horizontal line of asterisks, prints the square's sides, and then prints another horizontal line of asterisks. Here's pseudocode for this algorithm:

```
drawBorderSquare method
   draw horizontal line of asterisks
   draw sides
   draw horizontal line of asterisks
```

All three of `drawBorderSquare`'s draw statements represent non-trivial tasks. Thus, when we translate the `drawBorderSquare` pseudocode into a Java method, we use method calls for each of the draw subtasks:

```
private void drawBorderSquare()
{
  drawHorizontalLine();
  drawSides();
  drawHorizontalLine();
} // end drawBorderSquare
```

Now let's consider the `drawSolidSquare` helper method. It prints a series of horizontal lines of asterisks. Here's pseudocode for its algorithm:

```
drawSolidSquare method
   for (int i=0; i<square's width; i++)
   {
      draw horizontal line of asterisks
   }
```

Again, the draw statement represents a non-trivial task. Thus, when we translate the `drawSolidSquare` pseudocode into a Java method, we use a repeated method call for the draw subtask:

```
private void drawSolidSquare()
{
  for (int i=0; i<this.width; i++)
  {
    drawHorizontalLine();
  }
} // end drawSolidSquare
```

Notice that the `drawBorderSquare` method and the `drawSolidSquare` method both call the same `drawHorizontalLine` helper method. Being able to share the `drawHorizontalLine` method is a nice reward for our diligent use of helper methods, and it provides a good example for this general principle: 

> If two or more methods perform the same subtask, avoid redundant code by having those methods call a shared helper method that performs the subtask.

By writing final code for the `drawBorderSquare` and `drawSolidSquare` methods and writing stub code for the `drawHorizontalLine` and `drawSides` methods, we complete the coding for the Square program's second-cut version. When executed with appropriate print statements in the two stub methods, `drawHorizontalLine` and `drawSides`, the second-cut version produces either this sample session:

```
Enter width of desired square: 5
Area = 25.0
Print with (b)order or (s)olid? b
In drawHorizontalLine
In drawSides
In drawHorizontalLine
```

or this sample session:

```
Enter width of desired square: 5
Area = 25.0
Print with (b)order or (s)olid? s
In drawHorizontalLine
In drawHorizontalLine
In drawHorizontalLine
In drawHorizontalLine
In drawHorizontalLine
```

Square Program Example: Final Version

Keep documentation current.

To facilitate management, it's a good idea to formalize your program's design at various points during the design process. The formalization usually takes the form of UML class diagrams. Having up-to-date UML class diagrams helps to ensure project coherence. At a minimum, current UML class diagrams ensure that all members of a project are using the same classes, instance variables, and method headings. See Figure 8.12.

Figure 8.12 presents a UML class diagram for our complete Square program. It's the same as our earlier UML class diagram except that we've added the helper methods.

The second-cut version of the Square program contains stub implementations for the `draw-HorizontalLine` and `drawSides` methods. Now, we need to replace those stub methods with actual methods. Figures 8.13a and 8.13b contain our final-version `Square` class. The only new items are the

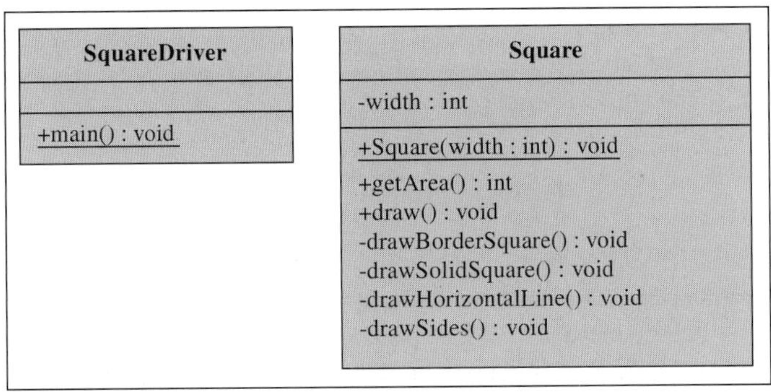

Figure 8.12 Square program's UML class diagram: final version

```
/*********************************************************
* Square.java
* Dean & Dean
*
* This class manages squares.
*********************************************************/

import java.util.Scanner;

public class Square
{
  private int width;

  //*******************************************************

  public Square(int width)
  {
    this.width = width;
  }

  //*******************************************************

  public double getArea()
  {
    return this.width * this.width;
  }

  //*******************************************************

  public void draw()
  {
    Scanner stdIn = new Scanner(System.in);

    System.out.print("Print with (b)order or (s)olid? ");
    if (stdIn.nextLine().charAt(0) == 'b')
    {
      drawBorderSquare();
    }
    else
    {
      drawSolidSquare();
    }
  } // end draw
```

Figure 8.13a `Square` class: final version—part A (an exact copy of Figure 8.11a)

```
  //*****************************************************

  private void drawBorderSquare()
  {
    drawHorizontalLine();
    drawSides();
    drawHorizontalLine();
  } // end drawBorderSquare

  //*****************************************************

  private void drawSolidSquare()
  {
    for (int i=0; i<this.width; i++)
    {
      drawHorizontalLine();
    }
  } // end drawSolidSquare

  //*****************************************************

  private void drawHorizontalLine()
  {
    for (int i=0; i<this.width; i++)
    {
      System.out.print("*");
    }
    System.out.println();
  } // end drawHorizontalLine

  //*****************************************************

  private void drawSides()
  {
    for (int i=1; i<(this.width-1); i++)
    {
      System.out.print("*");
      for (int j=1; j<(this.width-1); j++)
      {
        System.out.print(" ");
      }
      System.out.println("*");
    }
  } // end drawSides
} // end class Square
```

Figure 8.13b `Square` class: final version—part B (a fleshed-out version of Figure 8.11b) `drawHorizontalLine` and `drawSides` methods, which are straightforward. We encourage you to study their implementations on your own in Figure 8.13b.

Top-Down-Design Downside

Almost every human-designed project must necessarily include some form of top-down thinking. However, pure top-down design has some undesirable side effects. One of these is that subordinate modules tend to be overly specialized. A well-known and particularly egregious example of how the top-down way of thinking can lead to excessive specialization is the case of the $660 Pentagon ashtrays. The Pentagon (headquarters of the U.S. Department of Defense) hired a large military contractor to manufacture ashtrays for Pentagon use. Because compatibility is important for many military components, the military generally wants faithful adherence to its specifications, and contractors naturally develop procedures and attitudes that promote conformity. However, sometimes there can be too much of a good thing. The ashtrays conformed perfectly to their specification, but each one had a price of $660. Top-down design went to a ridiculous extreme. Even though some of the top-level specifications may have been unconventional, the contractor probably followed the standard operating procedure and tried to match them perfectly. Hypothetical quote from the contractor's marketing manager: "What was specified did not match anything that was available, so we had to make it by hand in the machine shop."

You may be thinking—Interesting story, but how do the $660 ashtrays relate to programming? The top-down philosophy can lead to inefficient development practices. In the extreme case, that philosophy led to the military contractor expending enormous effort on the design and manufacture of something  as simple as an ashtray. In general, the top-down design philosophy can motivate people to "reinvent the wheel." This tends to increase overall product cost. It also tends to reduce the reliability of the final product. Why? Because with everything being new or reinvented, there's no past history of testing and debugging to rely on.

8.9 Bottom-Up Design

Now, let's look at the logical opposite of top-down design—bottom-up design. Bottom-up design implements specific low-level tasks first. To apply bottom-up design to the Square program, you might implement a `drawSolidSquare` method first. Next, you might implement a `drawBorderSquare` method. After finishing these bottom-level methods, you would implement higher-level methods, which are in charge of more general tasks, like a `draw` method to draw either type of square—a solid square or a border square.

As you implement each program component, you should test it immediately with a custom driver that's tailored to that particular component. You won't need any stubs, since already-tested lower-level methods will be available to be called by whatever higher-level method you are currently testing.

For simple programs like many of those that appear throughout the body of this book, bottom-up design is an appropriate strategy to use because it allows you to focus quickly on the essence of whatever problem is currently most critical, and it allows you to defer presentation details until later. For an example of bottom-up design, look at any program in this book in which we present a driven class before we present a driver for that class. Whenever we do that, we are using a bottom-up presentation, and we are inviting you to think about the program being described from the bottom up.

Bottom-up design also makes it easiest for you to use prewritten software, like that in the Java API and described previously in Chapter 5. The Java API is a particularly good source for prewritten software because its code is (1) optimized for high speed and low memory consumption and (2) highly reliable because it has undergone testing and debugging for years. It's good to use the Java API, but it takes time to learn how to use it. To learn about the Java API, see http://download.oracle.com/javase/7/docs/api/index.html. There, you'll find several ways to look things up. Here are two techniques:

1. Try guessing the name of a class that seems appropriate. Use the scrollbar in the classes frame to search for the guessed class name. There are about 4,000 classes, so finding a particular class requires a well-behaved mouse (we recommend proper diet and exercise to keep your mouse running smoothly). When you find a class name that looks promising, click on it and read about its public constants and methods.
2. Related classes are grouped together in about 166 packages. Use the scrollbar in the packages frame to find a package that looks promising. Click on that package and scroll through its classes. Again, when you find a class name that looks promising, click on it and read about its public constants and methods.

Using prewritten software for your low-level modules reduces development time and project cost. It also improves product quality, because presumably the prewritten parts of your program have already been thoroughly tested and debugged. As in the case of Java API code, you'll often find that prewritten low-level software is quite flexible because it was designed for a broad spectrum of applications. This inherent low-level flexibility will make it easier for you to expand the capabilities of your program when you upgrade it in the future. Using prewritten software can facilitate parallel development. If several different programmers want to use a common subordinate module, they can do it independently. They do not have to coordinate their efforts because that module's design is already established and stable.

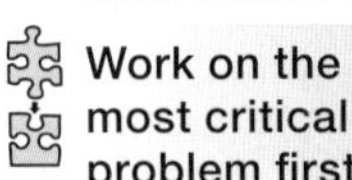

Work on the most critical problem first.

Another benefit of bottom-up design is that it provides the freedom to implement tasks in the most beneficial order. If there's a significant concern as to whether a particular calculation is feasible, it's important to begin working on that calculation as soon as possible. With bottom-up design, there's no need to wait around to address the concern—just attack it immediately. That way, you can determine at the earliest possible time whether the concern will be a showstopper. Likewise, if there is some low-level task that will take a long time to complete, bottom-up design allows you to begin work on it immediately and avoid a potential bottleneck later.

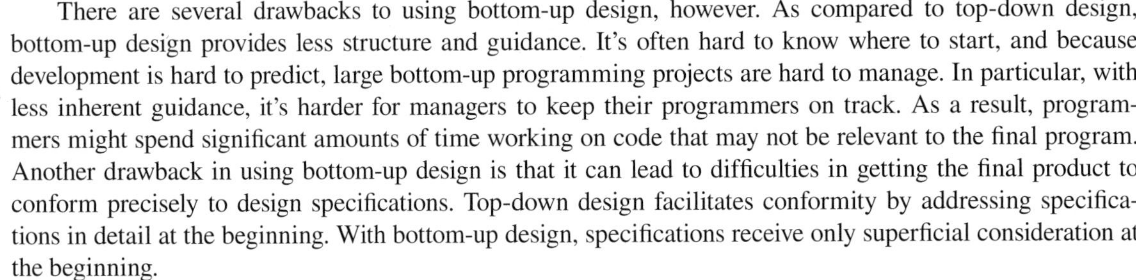

There are several drawbacks to using bottom-up design, however. As compared to top-down design, bottom-up design provides less structure and guidance. It's often hard to know where to start, and because development is hard to predict, large bottom-up programming projects are hard to manage. In particular, with less inherent guidance, it's harder for managers to keep their programmers on track. As a result, programmers might spend significant amounts of time working on code that may not be relevant to the final program. Another drawback in using bottom-up design is that it can lead to difficulties in getting the final product to conform precisely to design specifications. Top-down design facilitates conformity by addressing specifications in detail at the beginning. With bottom-up design, specifications receive only superficial consideration at the beginning.

So, when should you use bottom-up design? When you can use a substantial amount of prewritten and pretested low-level software, the bottom-up design process makes it easy for you to design around that software so that it fits naturally into your complete program. When you can use a substantial amount of prewritten software that is open to your inspection and already designed to fit together (like Java API software[7]), bottom-up design simultaneously promotes high quality and low cost. When low-level details are critical, bottom-up design motivates you to deal with the tough problems first—it gives you the most time to solve them. Thus, bottom-up design can also help you minimize delivery time.

A familiar example of bottom-up software design is the early development of the Microsoft Windows operating system. The original version of Windows was built on top of the already existing and successful DOS

[7]Although we have been encouraging you to think of Java API software as being completely encapsulated, Oracle does not keep the Java API source code secret. It can be downloaded and is available for inspection by Java developers.

operating system.[8] The next major version of Windows was built on top of a novel low-level software core called "NT" (for New Technology). It's important to note that the component source code in these cases was always open to and under the control of system developers, because it was all owned by the same company.[9]

8.10 Case-Based Design

There is another basic way to solve problems and design things. It's what normal people do most of the time in their everyday lives. Instead of going through a formal top-down or bottom-up sequence of steps, you look around for an already-solved problem that's like the problem at hand. Then you figure out how that problem was solved, and you modify that solution to fit your problem. This approach is holistic. It starts with a whole solution and "bends" that whole solution to a different application.

If you have access to source code and the right to copy it or modify it and then redistribute it in a new context, you can modify an existing program or significant parts of existing code. Sometimes the code that you want to borrow is code you wrote yourself for a different application. Such code deserves your consideration, because you'll be intimately familiar with what it does and how it does it. For example, many of this book's projects were designed to show you how to solve a wide range of real-world problems. You can use the algorithms presented in the project assignments to generate Java code that solves particular versions of those problems. Once you have written that code, you'll be completely free to modify it and reuse it in any other context to solve other variations of those problems.

Frequently, the code that you'd like to use will be code that somebody else wrote. Would it be theft or plagiarism to use such code? It might be. If the code is copyrighted, and you don't have permission to use it, you shouldn't try to use it. But you might have permission to use it. Whenever you use code that somebody else wrote, be sure to acknowledge and identify your source.

There is a growing body of what's called "free" software[10] that is debugged and maintained by a select body of experts, and it's available to all people to use and modify for their own purposes, provided they conform to certain reasonable rules: You may charge for products and services that use such software, but you must acknowledge its source, and you may not inhibit others from freely using it. Sometimes this software is low-level code that you can use like Java API software. But sometimes it's a complete program, which you can adapt to a problem you're currently addressing.

8.11 Iterative Enhancement

Often, you have to start working on a problem in order to understand how to solve the problem. That leads to a design process that is often iterative in nature. In the first iteration, you implement a bare-bones solution to the problem. In the next iteration, you add features and implement an improved solution. You continue

[8]The set of commands you can enter into a Windows command-prompt window are essentially DOS commands—they are a software legacy of the IBM PC that came out in the early 1980s.

[9]In principle, it's possible to build software systems out of components that are commercial-off-the-shelf (COTS) programs from different companies. This strategy can be used to avoid "reinventing the wheel" in a big way, and it minimizes new code to the "glue" that provides component interfaces. However, it takes longer to write this glue code than it does to write ordinary code. Moreover, because (in general) the system developer does not have access to component source code and does not have control of component evolution, the development process is relatively risky, and the resulting composite program is relatively brittle. COTS-based system design has a distinctive methodology that is outside the scope of this text.

[10]See http://www.fsf.org. The Free Software Foundation is "dedicated to promoting computer users' rights to use, study, copy, modify, and redistribute computer programs." Two famous examples of this kind of software are the GNU/Linux operating system (GNU stands for "Gnu's Not Unix") and the Apache software that underlies most web servers (http://www.apache.org).

adding features and repeating the design process until you have implemented a solution that does everything you need. This repetitive process is called *iterative enhancement.*

Prototyping—An Optional First Step

A *prototype* is a very "thin" or "bare-bones" implementation or perhaps just a faked "simulation" of a prospective program. Because of a prototype's limited scope, developers can produce prototypes relatively quickly and present them to customers very early in the development process.

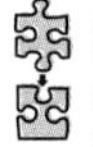
Make sure you are solving the right problem.

A prototype helps users get an early feel for what it will be like to use the program—well before the program is finished. It helps clients provide early feedback that improves the quality of product specification. Thus, prototyping provides a valuable adjunct to the first part of the top-down design process, and it complements early work in a bottom-up design process. Without a prototype, there's always a risk that you'll solve the wrong problem. Even if you solve the problem with great elegance, if it's the wrong problem, the whole effort is a waste of time.

There are two basic ways to generate a prototype. One way is to write a very limited version of the final program in Java. Because a prototype should be relatively simple, you could use whatever design approach seemed easiest. The other way is to use a computer application that provides nice presentations to simulate the final program's user interface for particular "canned" data or a narrow range of user inputs.

Prototyping can be a valuable communication tool, but use it with caution. Suppose that you create a prototype, show it to the client, and the client says: "I like it. Give me a copy so I can start using it tomorrow!" Don't do it! If your prototype is an early iteration of an orderly sequence of planned iterations, fold in what you learn from client reaction, and proceed to the next iteration as originally planned. If your prototype is just a visual presentation pasted together from disparate components, resist the temptation to expand that prototype into a finished product. That's tempting because you might think it would reduce development time. However, adding patches to a cobbled-together mock-up typically produces a messy result that is hard to maintain and upgrade. Eventually, it becomes necessary to rewrite massive amounts of code, and the associated confusion can destroy the program. It's better to think of this kind of prototype as no more than a communication aid that elicits feedback that improves product specification.

Iterating

The first normal design iteration—or the iteration after an optional prototype—should be either a simple adaptation of some already-existing program or a bare-bones implementation developed with either the top-down or bottom-up design strategy. Subsequent iterations may or may not continue to use the same design strategy.

Adjust the design strategy to address the greatest current need with the resources currently available.

How do you decide which strategy to use for each iteration? Select that strategy which best addresses your greatest current need or concern:

- If your greatest current need is to understand what the customer wants, construct a prototype.
- If your greatest concern is on-time delivery, try to use an adaptation of existing software.
- If your greatest current concern is whether some particular functionality can be implemented, use the bottom-up design strategy to implement that functionality as soon as possible.
- If your greatest needs are reliability and low cost, use prewritten software with bottom-up design.
- If your greatest concern is overall performance and managerial control, use the top-down design strategy.

A famous iterated-design example is NASA's man-on-the-moon space program. President John F. Kennedy was thinking top-down when he announced the program. However, the first implementation was a prototype.

Using a modified version of the existing Atlas ICBM rocket, "Project Mercury" shot one man a few hundred miles out into the Atlantic Ocean.

Subsequent iterations of Project Mercury used a bottom-up approach to put astronauts into Earth's orbit. Then, NASA replaced the Atlas booster rocket with the newer and larger Titan ICBM rocket, which carried several people into Earth's orbit in several iterations of "Project Gemini."

NASA's next iteration was a top-down design plan known as "Project Apollo." Project Apollo originally envisioned the use of a gigantic booster rocket called Nova. After working on that for a while, NASA realized that a much smaller booster rocket (called Saturn) would suffice if a smaller moon lander was separated from the mother ship orbiting the moon, and the moon lander's return module was separated from its descent mechanism.

Project Apollo was a top-down design, optimized for NASA's requirements, rather than a bottom-up adaptation of existing military equipment. In the end, the top-down plan involving Nova was scrapped and replaced by a radically different top-down plan. This apparently erratic development sequence is a great example of successful real-world design. The history of successful software is the same. Different design cycles often emphasize different design strategies, and sometimes there are major changes.

Maintenance

After a program has been developed and put into operation, you might think there's no more need to work on it. Not so. In the real world, if a program is useful, programmers are often asked to *maintain* it long after that program is first put into operation. On average, 80% of the work on a successful program is done after the program is first put into operation. Maintenance consists of fixing bugs and making improvements. Maintenance is much easier if good software practices are employed at the beginning and throughout the life of the program. This includes writing the code elegantly in the first place, preserving elegance when you make changes, and providing and keeping complete and well-organized documentation.

Remember that documentation is more than just comments for programmers reading source code. Documentation is also interface information for programmers who want to use already-compiled classes. Section 8.3 in this chapter and Appendix 6 show how to embed interface information in your source code so it can be read by `javadoc` and presented like Oracle's documentation of the Java API. Documentation also includes information for people who are not programmers at all but need to use a finished program. This type of documentation needs to be even more user-oriented than `javadoc`'s output.

If you are responsible for maintaining an existing program, here are some useful thumb rules:

1. Respect your predecessor. Don't change any piece of code you think is wrong until you have spent as much time thinking about it as some other programmer (or you) spent creating it in the first place. There may have been an important reason for doing something in a certain way, even if there is a problem in how it was done, and you want to understand that reason before you make changes.
2. Respect your successor. Whenever you have trouble figuring out what a particular section of code is doing, after you thoroughly understand the problem, fix the code and documentation so that it is easier to figure out next time.
3. Maintain a "standard" bank of test input data (and the corresponding output data), and use it to verify that any changes you have made affect only the problem you are trying to solve and do not have other unwanted effects that ripple through the program.

8.12 Merging the Driver Method into the Driven Class

It's legal to include a `main` method in any class. Figure 8.14 contains a simple Time program that includes its own `main` method.

```
/*******************************************************************
* Time.java
* Dean & Dean
*
* This class stores time in the form of hours, minutes, and
* seconds. It prints the time using military format.
*******************************************************************/

public class Time
{
  private int hours, minutes, seconds;

  //****************************************************************

  public Time(int h, int m, int s)
  {
    this.hours = h;
    this.minutes = m;
    this.seconds = s;
  }

  //****************************************************************

  public void printIt()
  {
    System.out.printf("%02d:%02d:%02d\n",
      this.hours, this.minutes, this.seconds)
  } // end printIt

  //****************************************************************

  public static void main(String[] args)
  {
    Time time = new Time(3, 59, 0);
    time.printIt();
  } // end main
} // end class Time
```

This is a driver for the rest of the code in this class.

Figure 8.14 `Time` class with a built-in `main` driver method

Up until now, we've split each of our OOP programs into separate classes—a driver class and one or more driven classes. It's easiest to grasp the concept of an object if it's associated with one class, while the code that instantiates it is associated with another class. Driven classes and driver classes have distinctive roles. A driven class describes a thing that's being modeled. For example, in our Mouse programs, the `Mouse` class describes a mouse. A driver class contains a `main` method, and it drives the separate `Mouse` class. In our Mouse programs, the `MouseDriver` class instantiates `Mouse` objects and performs actions on those objects. Using two or more classes fosters the habit of putting different types of things in different modules.

Although we'll continue to use separate classes for most of our programs, for short programs that don't do much except demonstrate a concept, we'll sometimes merge `main` into the class that implements the rest of the program. It's a matter of convenience—there's one less file to create and there's slightly less code to enter.

Provide each class with a built-in test method.

In a big program that has one driver class in charge of a large number of driven classes, it's sometimes handy to insert an additional `main` method in some or all of the driven classes. The additional `main` method in a driven class serves as a local tester for the code in that class. Whenever you make a change in the code of a particular class, you can use its local `main` method to test that class directly. It's easy. Just execute the class of interest, and the JVM automatically uses that class's `main` method. Once you've verified the changes you've made locally, you can proceed to execute the driver in a higher-level module to test more or all of the program. You don't have to remove the local `main` methods. You can just leave them there for future local testing or demonstration of the features of each particular class. When you execute the overall program's driver class, the JVM automatically uses the `main` method in that driver class, and it ignores any `main` methods that may happen to be in other classes in the program.

Thus, you can add a `main` method to any class, so the class can be executed directly and act as its own driver. When a multiclass program contains multiple `main` methods (no more than one per class), the particular `main` method that's used is the one in the class that's current when execution starts.

8.13 Accessing Instance Variables Without Using `this`

For a while now, we've used `this` to access the calling object's instance variables from within a method. Here's a formal explanation for when to use `this`:

> Use `this` within an instance method or a constructor to access the calling object's instance variables. The `this` reference distinguishes instance variables from other variables (like local variables and parameters) that happen to have the same name.
>
> However, if there is no name ambiguity, you may omit the `this` prefix when accessing an instance variable.

The code in Figure 8.15 has several places where the `this` prefix is worth mentioning. It's OK to omit `this` in the statement in the `setAge` method because the instance variable name is different from the parameter name. It's not OK to omit `this` in the statement in the `setWeight` method because the identical instance variable and parameter names would create an ambiguity. It is OK to omit `this` in the statement in the `print` method because there is no name ambiguity.

Sometimes an instance method is called by one object and has a parameter that refers to a different object in the same class. `String`'s `equals` method is a familiar example of this situation. Inside such a method, there will be code that needs to refer to two different objects, the calling object and the object referred to by the parameter. The safest and most understandable way to refer to these two objects is to use the `this` prefix to refer to the calling object and the reference-parameter prefix to refer to the other object. However, it's OK to omit the `this` when referring to the calling object, and you'll see this done quite frequently. It makes the code more compact.

8.14 Writing Your Own Utility Class

Up to this point, you've implemented methods that solve problems for a particular class. Suppose you want to implement methods that are more general purpose, so that multiple and unforeseen classes can use them. Those types of methods are called *utility methods*. In the past, you've used utility methods from the `Math`

```
/*******************************************************************
* MouseShortcut.java
* Dean & Dean
*
* This class illustrates uses and omissions of this.
*******************************************************************/

public class MouseShortcut
{
  private int age;              // age in days
  private double weight;        // weight in grams

  //*************************************************************

  public MouseShortcut(int age, double weight)
  {
    setAge(age);
    setWeight(weight);
  } // end constructor

  //*************************************************************

  public void setAge(int a)
  {
    age = a;
  } // end setAge

  //*************************************************************

  public void setWeight(double weight)
  {
    this.weight = weight;
  } // end setWeight

  //*************************************************************

  public void print()
  {
    System.out.println("age = " + age +
      ", weight = " + weight);
  } // end print
} // end class MouseShortcut
```

OK to omit `this` before instance variable, `age` because it's different from parameter, `a`.

Not OK to omit `this` before instance variable, `weight` because it's same as parameter, `weight`.

OK to omit `this` before `age` and `weight` instance variables.

Figure 8.15 `MouseShortcut` class illustrates the use and omission of `this`

class; for example, `Math.round` and `Math.sqrt`. You can also write your own utility methods and put them in your own utility class.

For example, Figure 8.16's `PrintUtilities` class provides print-oriented utility constants and methods. The two constants, `MAX_COL` and `MAX_ROW`, keep track of the maximum column and maximum

```
/*******************************************************************
* PrintUtilities.java
* Dean & Dean
*
* This class contains constants and methods for fancy printing.
*******************************************************************/

public class PrintUtilities
{
  public static final int MAX_COL = 80; // last allowed column
  public static final int MAX_ROW = 50; // last allowed row

  //****************************************************************

  // Print given string horizontally centered.

  public static void printCentered(String s)
  {
    int startingCol; // starting point for string
    startingCol = (MAX_COL / 2) - (s.length() / 2);

    for (int i=0; i<startingCol; i++)
    {
      System.out.print(" ");
    }
    System.out.println(s);
  } // end printCentered

  //****************************************************************

  // Print given string with dashes underneath it.

  public static void printUnderlined(String s)
  {
    System.out.println(s);
    for (int i=0; i<s.length(); i++)
    {
      System.out.print("-");
    }
  } // end printUnderlined
} // end class PrintUtilities
```

Figure 8.16 Example utility class that handles special-needs printing

row for a standard-sized piece of paper. If you have multiple classes that print reports, those constants can help to ensure report-size uniformity. The `printCentered` method prints a given string horizontally centered. The `printUnderlined` method prints a given string with dashes underneath it. These methods belong in a utility class because they perform print routines that might be needed by multiple other classes.

In the `PrintUtilities` class, note that the constants and methods all use the `public` and `static` modifiers. That's normal for utility class members. The `public` and `static` modifiers make it easy for other classes to access `PrintUtilities`' members.

8.15 Problem Solving with the API `Calendar` Class (Optional)

Don't reinvent the wheel.

Although textbooks (including ours) ask you to write little programs that manipulate times and dates, if you get serious about times and dates, you will discover it's a hornet's nest of different number bases, different length months, leap years, daylight savings time, different time zones, and many different formatting conventions. For serious time and date work, you should use Java API prewritten software. Unfortunately, it's not always easy to find the right Java class. For example, if you're working on a program that requires you to keep track of the date and time, you might be inclined to use the `Date` and `Time` classes, but those classes are obsolete. Usually, you should use the `Calendar` class instead. Figure 8.17 contains a CalendarDemo program that exercises some of the methods in the `Calendar` class.

The `Calendar` class is in the `java.util` package. To include it in your program, you could use this `import` statement:

```
import java.util.Calendar;
```

However, since the `Calendar` class is in the same package as the `Scanner` class, which this program also needs, it's easier to make both classes available simultaneously with this one "wildcard" `import` statement:

```
import java.util.*;
```

In the CalendarDemo program's `time` initialization statement, the `time` variable is assigned a reference to an instance of the `Calendar` class. Notice that instead of using `new Calendar()`, we get a `Calendar` instance by calling `Calendar`'s `getInstance` method. If you look up the `getInstance` method in the Java API documentation for the `Calendar` class, you'll see that this method has a `static` modifier, so it's a class method. How do you invoke a class method? Think back to how you invoked methods from the `Math` class in Chapter 5. Instead of using an instance variable before the method name, you use the class name. How does `getInstance` work? You don't need to know, because it's an encapsulated module, but it's reasonable to assume it internally instantiates a `Calendar` object, initializes it with the current time, and then returns a reference to that object. Although this is not the standard way to instantiate new objects, it works. The Java API includes several examples of this indirect type of object construction.

For the rest of the program, you can forget about how the `time` object was created and use it as you would any other object to call instance methods in its own class. The first print statement uses `Calendar`'s `getTime` method to retrieve the time information, and then it prints it all out as shown in the first line of the sample session.

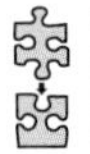

Use ID number in argument to select one of many similar variables.

The next two statements use the object reference with `get` methods to retrieve two particular instance variable values. But wait! There's something wonderfully strange about these two `get` methods. They're not two separate methods like `getDayOfYear` and `getHour` would be. They're both the same method—one method called just plain `get`. Instead of using the method name to identify the instance variable that will be retrieved, the designers of this class decided to use an `int` parameter value to identify that variable. We don't have to know how the method is implemented, because it's encapsulated, but we can use a plausible guess to shed light on what it does. For example, `get`'s parameter could be a `switch` index that steers the control flow to a particular case, where there's code that returns the value of the instance variable that corresponds to that index number.

The problem with using an index number to identify one of many instance variables is that simple integers don't convey much meaning. But you know a solution to this problem. All you have to do is make each such index number a named constant. Then, for the distinguishing method argument, use the named

```
/*******************************************************************
* CalendarDemo.java
* Dean & Dean
*
* This program demonstrates how to use the Calendar class.
*******************************************************************/

import java.util.*;                 // for Scanner and Calendar

public class CalendarDemo
{
  public static void main(String[] args)
  {
    Scanner stdIn = new Scanner(System.in);
    Calendar time = Calendar.getInstance();   // initially now
    int day;                                  // day of year
    int hour;                                 // hour of day

    System.out.println(time.getTime());
    day = time.get(Calendar.DAY_OF_YEAR);
    hour = time.get(Calendar.HOUR_OF_DAY);
    System.out.println("day of year= " + day);
    System.out.println("hour of day= " + hour);

    System.out.print("Enter number of days to add: ");
    day += stdIn.nextInt();
    System.out.print("Enter number of hours to add: ");
    hour += stdIn.nextInt();

    time.set(Calendar.DAY_OF_YEAR, day);
    time.set(Calendar.HOUR_OF_DAY, hour);
    System.out.println(time.getTime());
  } // end main
} // end class CalendarDemo
```

Parameters are int codes that specify the kind of information desired.

Sample session:

```
Wed Oct 23 11:57:31 CDT 2013
day of year= 296
hour of day= 11
Enter number of days to add: 8
Enter number of hours to add: 13
Fri Nov 01 00:57:31 CDT 2013
```

Figure 8.17 Demonstration program for the Calendar class

constant instead of the number. That's how the `Calendar` class implements its generic `get` method. And it's at least as easy for a user to remember one `get` method with different named-constant arguments as it would be to remember different names for `get` methods.

Armed with this concept, you should now be able to see what the rest of the code in our CalendarDemo program is doing. It gets the current day of the year and the current hour of the day. Then it adds a user-input number of days to the current day and a user-input number of hours to the current hour. Then it mutates the object's instance variables for day-of-year and hour by using `Calendar's` generic `set` method. Finally, it prints out the mutated time.

The `Calendar` class nicely illustrates the value of using prewritten software. It really is easier to learn to use that class than it is to write a program that does what it does. Moreover, other people's code sometimes illustrates techniques that may be applicable to code that you write. However, the `Calendar` class also illustrates the kinds of penalties associated with using prewritten software. The biggest penalty is usually the time that you have to spend to locate and figure out what's available. Another penalty is that what you find may not exactly match your immediate needs, and you might have to provide extra code to adapt the prewritten software to your current program. Such penalties motivate many programmers to say, "Oh heck, I'll just write it myself." Sometimes that's the thing to do, but in the long run you'll be ahead if you take time to learn about what others have already developed.

8.16 GUI Track: Problem Solving with CRC Cards (Optional)

When you begin a new design, there's often a period of head-scratching when you're trying to figure out what your classes should be and what they should do. Section 8.8 presented a formal top-down recipe, but sometimes you just need to "muck around" or brainstorm for a while to get your thinking straight.

Explore your options.

Even when you're just mucking around and brainstorming, it's still helpful to write things down. To provide a minimal structure for this informal activity, several years ago computer scientists Kent Beck and Ward Cunningham[11] suggested using old-fashioned 3" × 5" file cards, with a pencil and eraser. Their idea was to allocate one card to each proposed class, with three kinds of information on each card: (1) At the top, put a class name. (2) Below and on the left, make a list of active verb phrases that described what that class will do. (3) Below and on the right, make a list of other classes with which the current class interacts—either actively as a client or passively as a server. The acronym, CRC, helps you remember the kinds of information each card should have. The first C stands for "Class." The R stands for "Responsibility." The last C stands for "Collaboration."

When several different people are participating in a brainstorming session, pencils, erasers, and little white cards might indeed be the best medium to employ. But when you're the only designer, it might be more fun to use little windows on your computer screen. The program presented in Figure 8.18 sets up simulated CRC cards on your computer screen so you can do just that.

This program imports the `javax.swing` package to provide access to three classes in the Java API: `JFrame`, `JTextArea`, and `JSplitPane`. In a `main` method, it repeatedly asks the user for another class until a 'q' entry says it's time to quit. After the user enters each class name, the program instantiates a small `JFrame` window that represents one CRC card. The `JFrame` constructor automatically inserts the text, "Class: <*classname*>" in that window's header and thereby implements the first C in CRC. Then the program instantiates two `JTextArea` "panes," which act like little erasable scratch pads, on which you can write

[11]OOPSLA '89 Conference Proceedings.

```
/*****************************************************************
* CRCCard.java
* Dean & Dean
*
* This program creates a GUI display of CRC cards.
*****************************************************************/

import java.util.Scanner;
import javax.swing.*;  // for JFrame, JTextArea, & JSplitFrame

public class CRCCard
{
  public static void main(String[] args)
  {
    Scanner stdIn = new Scanner(System.in);
    String input;

    System.out.print("Enter class name or 'q' to quit: ");
    input = stdIn.nextLine();
    while (!input.equalsIgnoreCase("q"))
    {
      JFrame frame = new JFrame("Class: " + input);
      JTextArea responsibilities =
        new JTextArea("RESPONSIBILITIES:\n");
      JTextArea collaborators =
        new JTextArea("COLLABORATORS:\n");
      JSplitPane splitPane =
        new JSplitPane(JSplitPane.HORIZONTAL_SPLIT,
        responsibilities, collaborators);

      frame.setSize(500, 300);
      frame.add(splitPane);
      frame.setLocationByPlatform(true);
      frame.setVisible(true);
      frame.toFront();
      splitPane.setDividerLocation(0.67);

      System.out.print("Enter class name or 'q' to quit: ");
      input = stdIn.nextLine();
    } // end while
  } // end main
} // end class CRCCard
```

Create a new window.
Create two containers.
Put containers into split panes.
Put a split pane in the window.

Figure 8.18 Program that puts interactive CRC cards on your computer screen

any text anywhere. The two JTextArea constructor calls automatically write "RESPONSIBILITIES:" and "COLLABORATORS:," respectively, on the first line of each of these two JTextArea panes. Then, the program instantiates a JSplitPane with a HORIZONTAL_SPLIT specification that splits the window into two side-by-side "openings" separated by a moveable vertical partition. The last two JSplitPane parameters paste the individual JTextArea panes into these two openings.

The setSize method call sizes the window to make it about like a 3" × 5" file card. The add method call adds the split pane to the window. The setLocationByPlatform method call tells the computer to offset each additional card so that you can continue to see the titles and borders of previously created cards as they "pile up" on your desktop. The setVisible method makes each new card visible. The toFront method moves it to the front of your screen. The setDividerLocation method positions the JSplitPane divider two-thirds of the way to the right. This provides twice as much space for "responsibilities" text as for "collaboration" text. The specified window dimensions and the location of the split-pane divider are just initial settings, and if you find you need more space, you'll be able to change them interactively on the computer screen at any time while the program is running.

When you run the program in a Windows environment, you'll get a Command Prompt window with a query asking for a class name, like what appears in Figure 8.19.

Move the Command Prompt window to the lower part of the screen. In the Command Prompt window, enter a class name. When you hit the Enter key, an additional window appears in the upper-left corner of the screen. This additional window is your first CRC card. If the Command Prompt window is now underneath the new card, drag the window down and to the right, to get it out of the way. Then, move the cursor to the new CRC card and fill in additional information, like the "run program" entry in the RESPONSIBILITIES pane and the "GarageDoorSystem" entry in the COLLABORATORS pane in Figure 8.20.

Go back to the Command Prompt and enter another class name, and so on. until you have created all the CRC cards that you need. They should pile up automatically in a "stack" that looks something like the four cards in the upper left of the computer screen shown in Figure 8.21.

Now, before you enter a 'q' in the Command Prompt window, you can reduce it to an icon and play around with four CRC cards. You can drag them anywhere on the screen to form logical hierarchies or groupings. On any of the cards, you can change any of the wordings in either of the two panes. If you decide that one of the classes is no good, you can click on its X box to throw it away, reactivate the Command Prompt window, and create more new CRC cards with different class names. When you're through with everything, use Ctrl-PrtScr to print the screen to record your thinking, and then enter 'q' in the Command Prompt window to terminate the program.

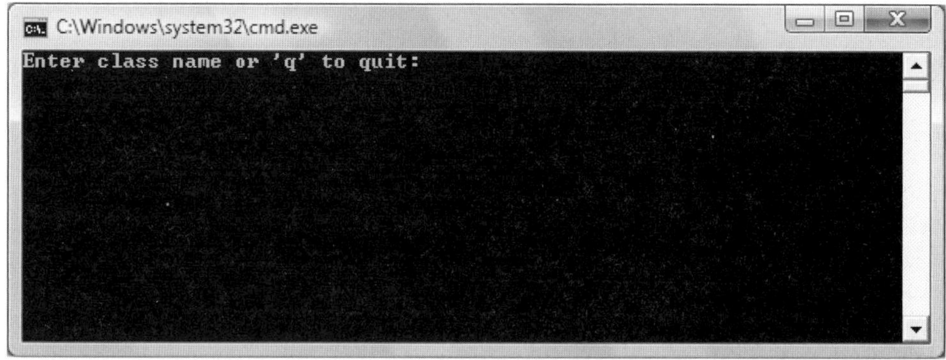

Figure 8.19 Initial Command Prompt display for CRCCards program

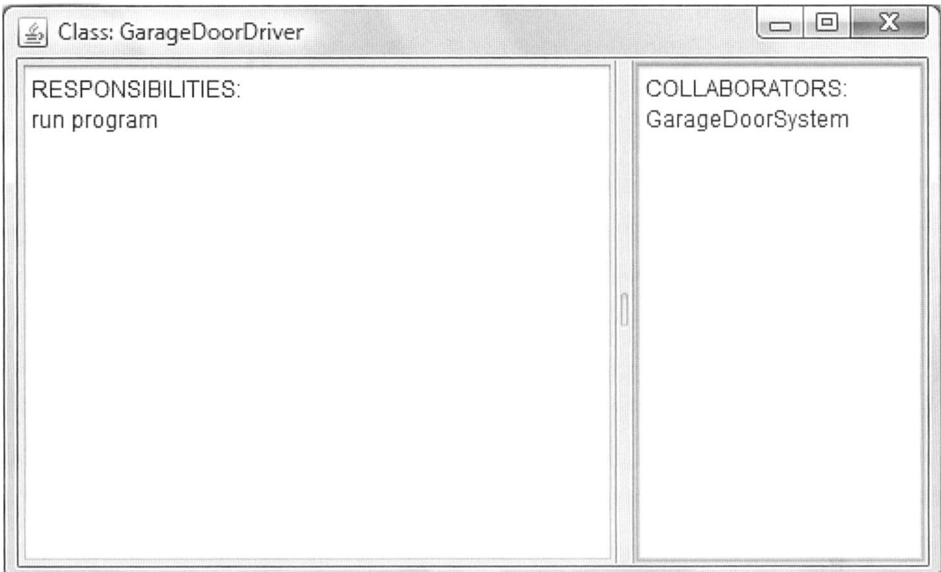

Figure 8.20 First CRC Card after user entries

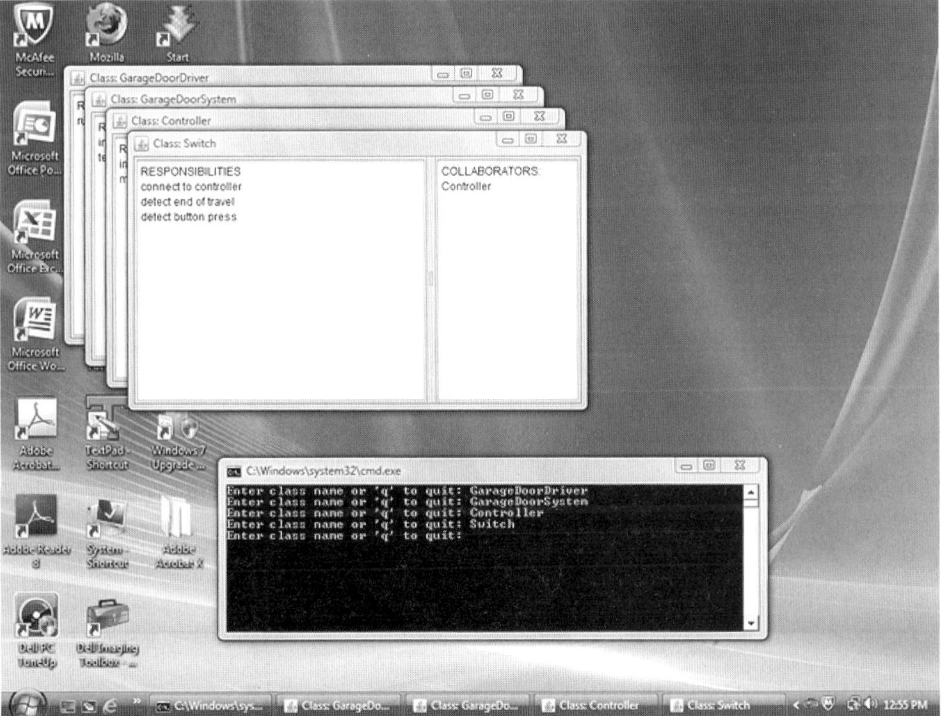

Figure 8.21 What you might see after creating four CRC cards

Summary

- Begin every class with a prologue. Include program name, author(s), and a brief description of what the class does.
- Provide a descriptive comment above or after any code that an experienced Java programmer would not understand.
- Use meaningful names for everything. Do not be cryptic.
- Enclose logical blocks of code in braces. The opening and closing braces should be in the same column as the start of the line that precedes the opening brace.
- Supply a // end *<block-name>* comment after a block's closing brace to improve readability.
- Declare each variable at the beginning of its class or method, or in its `for` loop header. Normally use one line per variable and follow each declaration with an appropriate descriptive comment.
- When a method won't work as might be expected unless certain conditions exist before it's called, warn users with a precondition comment just above the method header.
- When a method modifies variables in ways that might not be expected, warn users with a postcondition comment just above the method header.
- Use `javadoc` with /** *<comment>* */ above constructor and method headers to generate documentation like Java API documentation.
- Practice mentally putting yourself in a user's shoes. Test prospective outputs and inputs with users, not only to tune the cosmetics but also to check basic assumptions and objectives.
- Early in the planning process, separate project concerns into categories like entity, relationship, structure, behavior, and appearance, and try to associate each concern with an appropriate feature of your programming language.
- Use subordinate helper methods to simplify large methods and reduce code redundancy. Make helper methods `private` to minimize clutter in the class interface.
- Use instance variables for object attributes (state information) only. Use local variables and input parameters for calculations within a method and to transfer data into a method. Use return values and/or input reference parameters to transfer data out of a method.
- Plan to test the software you develop frequently and thoroughly as you go along. Include typical, boundary, and unreasonable cases.
- Top-down design is appropriate for large projects that have well-understood objectives. Proceed from general to specific, using stubs to defer implementation of subordinate methods.
- Bottom-up design allows you to give priority to critical details. It fosters reuse of existing software, which reduces development costs and improves system reliability. But this methodology makes large projects hard to manage.
- Expect to go through several design iterations. Use prototyping to help customers get a clearer understanding of what they want, but avoid the trap of trying to convert a clumsy prototype directly into a final product. In each subsequent iteration, select that design strategy which best addresses the greatest current need or concern. A successful program will require ongoing maintenance, and you can make this easier if you preserve and enhance elegance as the program changes and grows.
- To facilitate modular testing, provide a `main` method with every class.
- If there is no name ambiguity, you may omit the `this` prefix when accessing an instance member.
- When you need to perform the same calculations in different applications, put them in `public` class methods in a custom utility class.

Review Questions

§8.2 Coding-Style Conventions

1. One should avoid inserting blank lines between different code sections (because that leads to wasted paper when the program is printed). (T / F)
2. In order, list the seven items that we recommend you include in a file prologue.
3. When adding a comment to a variable declaration, always begin the comment one space after the end of the declaration. (T / F)
4. To get the most on each line of code, always break a long line at the point determined by your text editor or integrated development environment (IDE) (T / F)
5. For an `if` or `while` that has only one statement in its body, braces for the body are optional. The compiler does not require them, but proper style suggests that you should include them. Give at least one reason why it's a good idea to put braces around the body's single statement.
6. What's wrong with the style of a class description that ends like this?

```
      }
    }
}
```

   What might you do to fix it?
7. What should you use to separate large "chunks" of code?
8. For each, write "yes" or "no" to indicate whether it is good style to include a blank space
   - after the single asterisks in the prologue
   - between a method call and its opening parentheses
   - within each of the three components in a `for` loop header
   - after the two semicolons in the `for` loop header
   - between a closing brace and the `//`'s for its associated comment
   - after the `//`'s for all comments
   - after the `if`, `while`, and `switch` keywords

§8.4 Helper Methods

9. Which of the following is a legitimate reason for creating a helper method?
   a) You want the method to be hidden from the outside world.
   b) You have a long and complicated method and would like to partition it into several smaller modules.
   c) Your class contains two or more methods where some of the code is the same in both methods.
   d) All of above.
10. Does a class's interface include the names of `private` methods?

§8.5 Encapsulation (with Instance Variables and Local Variables)

11. In the interest of encapsulation, use local variables instead of instance variables whenever possible. (T / F)
12. If a method modifies a particular instance variable, and if a program calls the same method two separate times, the value of the instance variable at the beginning of the second method call is guaranteed to be the same as the value it had at the end of the first method call. (T / F)

§8.7 Design Philosophy

13. Because some of your preliminary code might change in the course of development, do not waste time testing until everything is done. (T / F)
14. When you are testing a program, it's important to not have any preconceived expectations of what your output should look like. (T / F)

§8.8 Top-Down Design

15. The top-down design methodology is good because:

a) It keeps everyone focused on a common goal. (T / F)
b) It avoids "reinventing the wheel." (T / F)
c) It keeps management informed. (T / F)
d) It minimizes the chances of solving the wrong problem. (T / F)
e) It minimizes overall cost. (T / F)
f) It results in the fewest number of undetected bugs. (T / F)

16. In a top-down design process, which do you decide on first—the classes or the `public` methods?

§8.9 Bottom-Up Design

17. When should you use bottom-up design?

§8.11 Iterative Enhancement

18. If a prototype is successful, what temptation should you resist?

19. Once you select a particular design methodology, keep using that same methodology throughout the entire design process, and do not allow other methodologies to "contaminate" the process originally selected. (T / F)

§8.12 Merging Driver Method into Driven Class

20. You can drive any class from a `main` method within that class, and you can retain that `main` method for future testing of that class even though that class is normally driven from another class in a larger program. (T / F)

§8.14 Writing Your Own Utility Class

21. A utility class's members normally should use the `private` and `static` modifiers. (T / F)

Exercises

1. [after §8.2] Describe the way to declare variables that conforms to good style. Include a description of when and how to include associated comments.

2. [after §8.2] Correct the style of the following class definition.

```
/*Environment.java This class models the world's environment.
It was written by Dean & Dean and it compiles so it must be OK*/
public class Environment{//instance variables
private double sustainableProduction;private double
initialResources;private double currentResources;private
double yieldFactor = 2.0;public void setSustainableProduction
(double production){this.sustainableProduction = production;}
// Set pre-industrial mineral and fossil resources
public void setInitialResources(double resources){this.
initialResources=resources;}
// Initialize remaining mineral and fossil resources
public void setCurrentResources(double resources){this.
currentResources = resources;}
// Fetch remaining mineral and fossil resources
public double getCurrentResources(){return this.
currentResources;}/*Compute annual combination of renewable
and non-renewable environmental production*/public double
```

```
produce(double populationFraction,double extractionExpense){
double extraction;extraction=this.yieldFactor*
extractionExpense*(this.currentResources/this.
initialResources);this.currentResources-= extraction;return
extraction+populationFraction*this.sustainableProduction;}}
```

3. [after §8.4] The following shirt-design program is the same as the Shirt program in Figures 8.6,8.7a, 8.7b, and 8.7c, except for a slight modification in `main`:

```
/*************************************************************
* ShirtDriver.java
* Dean & Dean
*
* This is a driver for the Shirt class.
*************************************************************/

public class ShirtDriver
{
  public static void main(String[] args)
  {
    Shirt shirt1 = new Shirt();
    Shirt shirt2 = new Shirt();

    System.out.println();
    shirt1.display();
    shirt2.display();
  } // end main
} // end ShirtDriver
```

```
/***********************************************************
* Shirt.java
* Dean & Dean
*
* This class stores and displays color choices for
* a sports-uniform shirt.
***********************************************************/

import java.util.Scanner;

public class Shirt
{
  private String name;      // person's name
  private String primary;   // shirt's primary color
  private String trim;      // shirt's trim color

  //*******************************************************

  public Shirt()
  {
    Scanner stdIn = new Scanner(System.in);
```

```
    System.out.print("Enter person's name: ");
    this.name = stdIn.nextLine();

    this.primary = selectColor("primary");
    this.trim = selectColor("trim");
  } // end constructor

  //*******************************************************

  public void display()
  {
    System.out.println(this.name + "'s shirt:\n" +
      this.primary + " with " + this.trim + " trim");
  } // end display

  //*******************************************************

  // Helping method prompts for and inputs user's selection

  private String selectColor(String colorType)
  {
    Scanner stdIn = new Scanner(System.in);
    String color; // chosen color, first a letter, then word

    do
    {
      System.out.print("Enter shirt's " + colorType +
        " color (w, r, y): ");
      color = stdIn.nextLine();
    } while (!color.equals("w") && !color.equals("r") &&
             !color.equals("y"));

    switch (color)
    {
      case "w":
        color = "white";
        break;
      case "r":
        color = "red";
        break;
      case "y":
        color = "yellow";
    } // end switch

    return color;
  } // end selectColor
} // end class Shirt
```

Trace the above program using either the short form or the long form. To help you get started, here's the trace setup, including the input. For the short form, you won't need the line# column.

<u>input</u>

Corneal
r
w
Jill
w
y

ShirtDriver			Shirt												
	main			Shirt	display	selectColor			obj1			obj2			
line#	**sh1**	**sh2**	*line#*	**this**	**this**	**this**	**cType**	**color**	**name**	**prim**	**trim**	**name**	**prim**	**trim**	**output**

4. [after §8.5] This exercise demonstrates using a reference parameter to pass data back to the calling method. Suppose you want a `Car5` class to include a method with this heading:

```
public boolean copyTo(Car5 newCar)
```

This method is supposed to be called by an existing `Car5` object with another `Car5` object as the argument. If any of the calling car's instance variables has not been initialized, the desired method should not try to modify any of the new car's instance variable values, and the method should return `false`. Otherwise, the method should copy all of the calling car's instance variable values into the new car and return `true`. Here's a driver that illustrates the usage:

```
/*************************************************************
* Car5Driver.java
* Dean & Dean
*
* This class is a demonstration driver for the Car5 class.
*************************************************************/

public class Car5Driver
{
  public static void main(String[] args)
  {
    Car5 silviaCar = new Car5();
    Car5 jayCar = new Car5();

    System.out.println(silviaCar.copyTo(jayCar));
    silviaCar = new Car5("Fiat", 2015, "black");
    System.out.println(silviaCar.copyTo(jayCar));
  } // end main
} // end class Car5Driver
```

<u>Output:</u>

```
false
true
```

Write the code for the desired `copyTo` method.

5. [after §8.7] We recommend that you test frequently, even if it means creating special test code that is not used in the final program. Why might it be useful to save such special test code?

6. [after §8.8] Assuming it will be called by the `draw` method in the `Square` class in Figure 8.11a, write a `drawSolidSquare` method that asks the user for the character to print and draws the desired solid square all by itself, without calling any separate `drawHorizontalLine` method.

   Sample session:

```
Enter width of desired square: 5
Area = 25
Print with (b)order or (s)olid? s
Enter character to use: #
#####
#####
#####
#####
#####
```

7. [after §8.8] Assuming that it will be called by the `draw` method in the `Square` class in Figure 8.11a, write a `drawBorderSquare` method that asks the user for two characters to use to draw a bordered square, one character of the border and a different character for the space in the middle. Notice that using the same character for the border and the middle makes this method draw a solid square, and thus this method makes the `drawSolidSquare` method redundant, although this method requires more user interaction.

   Sample session:

```
Enter width of desired square: 5
Area = 25
Print with (b)order or (s)olid? b
Enter character for border: B
Enter character for middle: m
BBBBB
BmmmB
BmmmB
BmmmB
BBBBB
```

8. [after §8.8] Figure 8.2b has two if statement conditions that contain what are called regular expressions. As indicated, these are explained in the Java API `Pattern` class. This exercise is intended to help you get a better feeling for Java's regular expressions and their usage. Use your Java API documentation on the `Pattern` class to get the answers to these questions:
   a) What is the meaning of the regular expression, `"[A-Z][a-z]*"`, which appears in Figure 8.2b?
   b) What is the regular expression for a character string starting with a 'Z' and containing any number of additional characters of any kind except for a space or a tab?
   c) What is the regular expression for a string that represents a U.S. long-distance telephone number (three digits, a dash or space, three digits, a dash or space, and four digits)?

9. [after §8.8] Define "stepwise refinement."

10. [after §8.8] Write stubs for all the constructors and methods in the `Student` class of Figures 8.2a and 8.2b. Each stub should print out the constructor or method name followed by the initial (passed-in) values of all parameters, like this sample output:

```
in Student
in setFirst, first= Adeeb
in setLast, last= Jarrah
in Student, first= Heejoo, last= Chun
in printFullName
```

11. [after §8.9] Write a generic `drawRow` method having this heading:

```
private void drawRow(int startCol, int endCol)
```

`startCol` and `endCol` are the column numbers of the left and right borders, respectively. Then modify the `Square` class's `draw` method in Figure 8.11a to draw either a solid square or a solid triangle whose height and width equal the width of the input width of a square container. What about the area now? Is it a redundant value, or is it a legitimate object attribute? Modify the instance variables and the `getArea` method accordingly. Then drive your modified `Square` class (call it `Square2`) with a `Square2Driver` whose `main` method looks like this:

```
public static void main(String[] args)
{
  Scanner stdIn = new Scanner(System.in);
  Square2 square;

  System.out.print("Enter width of square container: ");
  square = new Square2(stdIn.nextInt());
  square.draw();
  System.out.println("Area = " + square.getArea());
} // end main
```

Sample session:

```
Enter width of square container: 5
Print (s)quare or (t)riangle? t
*
**
***
****
*****
Area = 15
```

12. [after §8.11] Write a prototype of the Square program, using just one class called `SquarePrototype`, with only one method, `main`. Write the minimum amount of code needed to generate the prescribed output for only the simplest case of a solid square. The sample session should look exactly like it would for the final program described in Figures 8.10, 8.13a, and 8.13b, if the user selects the (s)olid option. If the user selects the (b)order option, however, the prototype should respond by printing a "Not Implemented."

13. [after §8.11] When you design something, you should select the design methodology that is best able to address the greatest current design concern. (T / F)

14. [after §8.12] Write a separate driver program that executes the `Time` class shown in Figure 8.14 and sets the time for 17 hours, 30 minutes, and zero seconds. Assume that the `main` method that appears in Figure 8.14 is still there.

15. [after §8.13] Rewrite the `Car` class in Figure 7.2 to eliminate the use of `this`.

16. [after §8.14] Write a utility class called `RandomDistribution`, which contains the following four class methods. You should be able to implement all of these methods with calls to `Math` class methods and/or calls to one of the `uniform` methods within the `RandomDistribution` class.

a) Write a method called `uniform` that generates a `double` random number from a continuous distribution that is uniform between `double` values `min` and `max`.

b) Write another (overloaded) method called `uniform` that generates an `int` random number from a discrete distribution that is uniform between `int` values `min` and `max`, including both of these end points.

c) Write a method called `triangular` that generates an `int` random number from a symmetrical discrete triangular distribution that goes between `int` values `min` and `max`, including both of these end points. (*Hint*: Make two calls to the `int` version of the above `uniform` method.)

d) Write a method called `exponential` that generates a `double` random number from an exponential distribution having an expected time between random arrival events equal to `averageTimeInterval`. Here is the algorithm:

```
return ← averageTimeInterval * logₑ(1.0 - Math.random)
```

17. [after §8.15] The Java API `Calendar` class contains a method called `getTimeInMillisec` which enables you to retrieve the absolute time (in milliseconds) at which any `Calendar` object was created. As indicated in Section 8.15, you can get such an object by calling the `getInstance` class method. You can use this capability to evaluate the runtime of any chunk of code. All you have to do is create a `Calendar` object before the test code starts, create another `Calendar` object right after the code ends, and print out the difference in those two object's times. To demonstrate this capability, write a short program called TestRuntime that asks the user for a desired number of iterations, `num`. Then have it measure the runtime for a loop of `num` iterations that executes the single statement:

```
Math.cos(0.01 * i);
```

The variable `i` is the loop count variable.

Review Question Solutions

1. False. Readability is an important attribute of good computer code. To save printer paper, print on both sides of the page and/or use smaller font.
2. The seven items to include in a file prologue are:
   - line of asterisks
   - filename
   - programmer name(s)
   - blank line with one asterisk
   - description
   - line of asterisks
   - blank line
3. False. That would provide maximum room for each comment, but good programmers make the beginnings of declaration comments line up with each other, and they try to make declaration comments short enough to avoid line wrap.
4. False. Take control, and break a long line at the most logical place(s).
5. Even though it's not necessary, it's a good idea to provide braces with single-statement `if` and `while` statements because
   - Braces provide a visual cue for remembering to indent.
   - Braces help you avoid a logical mistake if you add code later.
6. Unless a block is very short, it may not be immediately obvious which block is being terminated by a particular brace. It's good practice to terminate all but the shortest blocks with a comment, for example,

```
    } // end if
  } // end main
} // end class Whatever
```

7. Separate large chunks of code with blank lines.

8. Yes means include a space, No means do not.
   - Yes, after the single asterisks in the prologue.
   - No, not between a method call and its opening parentheses.
   - No, not within each of the three components in a `for` loop header.
   - Yes, after the two semicolons in the `for` loop header.
   - Yes, between a closing brace and the `//`'s for its associated comment.
   - Yes, after the `//`'s for all comments.
   - Yes, after the `if`, `while`, and `switch` keywords.
9. d) All of the above.
10. No. the interface does not describe `private` members.
11. True. You should generally try to keep things as local as possible, and using local variables instead of instance variables is one way to do this. Instance variables should be reserved for attributes that describe an object's state.
12. False. It's true that an instance variable persists throughout the life of an object, and if the second call of the same method were right after the first call of that method, the final value of the instance variable in the first call would be the same as the initial value in the second call of that method. However, it's possible that some other method could change the value of the instance variable between the two calls of the method in question.
13. False. Test frequently throughout the development process.
14. False. It's important to have a clear idea of what you expect to see before you do a test, so you will have the best chance of recognizing a discrepancy when it occurs.
15. The top down design methodology is the best because:
    a) True.
    b) False. It sometimes forces people to re-invent the wheel.
    c) True
    d) False. If you're worried about solving the wrong problem, use prototyping.
    e) False. To minimize cost, organize the design to reuse existing components.
    f) False. To maximize reliability, organize the design to reuse existing components.
16. In top-down design you decide on the classes before the `public` methods.
17. You should use bottom-up design when your program can utilize a substantial amount of prewritten software or when low-level details are critical and require early attention.
18. If a prototype is successful, it's important to resist the temptation of continuing the development by tinkering with that prototype.
19. False. Many problems need benefits of more than one design methodology. It's a good idea to stick with one methodology through one design cycle (planning, implementation, testing and evaluation), but you might need to switch to a different methodology in the next design iteration.
20. True. The particular `main` method used is the one that's current when execution starts.
21. False. A utility class's members should normally use the `public` and `static` modifiers.

CHAPTER 9

Arrays

Objectives

- Compare an array to other objects.
- Create and initialize arrays.
- Copy values from one array to another.
- Shift data in an array.
- Make histograms.
- Search an array for particular data.
- Sort data.
- Create and use two-dimensional arrays.
- Create and use arrays of objects.
- Learn how to use for-each loops.

Outline

9.1 Introduction

In the past, you've seen that objects typically contain more than one data item, and the different data items each have a different name. Now, we'll look at a special kind of object that holds several items of the same type and uses the same name for all of them. Natural language has ways to give a single name to a population: "pack" of wolves, "herd" of cattle, "pride" of lions, "passel" of possum, "fesnying" of ferrets, and so on. Java has a way to do the same thing.

When you have a collection of items of the same type, and you'd like to use the same name for all of them, you can define them all together as an *array*. Each item in the array is more formally called an array *element*. To distinguish the different elements in the array, you use the array name plus a number that identifies the position of the element within the array. For example, if you stored a collection of song titles in an array named `songs`, you'd distinguish the first song title by saying `songs[0]`, and you'd distinguish the second song title by saying `songs[1]`. As evidenced by this example, array elements start at position 0. An array's position numbers (0, 1, 2, etc.) are more formally called *indexes*. We'll have more to say about array indexes in the next section.

There's an important advantage in using one name for all of a group of similar items and distinguishing them only by a number. It can lead to simpler code. For example, if you need to store 100 song titles, you could declare 100 separate variables. But what a pain it would be to have to write 100 declaration statements and keep track of 100 different variable names. The easier solution is to use an array and declare just one variable—a `songs` array variable.

Readers who want an early introduction to arrays have the option of reading Sections 9.1 through 9.6 after completing Chapter 4. The natural connection between Chapter 4 and this chapter is that Chapter 4 describes loops and arrays rely heavily on loops.

Starting with Section 9.7, we present arrays in an object-oriented context. More specifically, we use constructors, instance methods, and class methods for the programs that illustrate the new array concepts. We discuss techniques for searching an array and sorting an array. We describe different organizational structures for arrays—two-dimensional arrays and arrays of objects. Finally, we describe a special type of `for` loop called a for-each loop, which allows you to loop through the elements in an array without having to declare or use an index variable.

9.2 Array Basics

In this section, we show you how to perform simple operations on an array, such as loading an array with data and printing an array. To illustrate these operations, we'll refer to the `phoneList` array in Figure 9.1. The `phoneList` array holds a list of five contact phone numbers for a cell phone. The first phone number is 8167412000, the second phone number is 2024561111, and so on.

Accessing an Array's Elements

To work with an array, you need to access an array's elements. For example, to print the contents of an array, you need to access the array's first element, print it, access the array's second element, print it, and so on. To access an element within an array, you specify the array's name, followed by square brackets surrounding the element's index. Figure 9.2 shows how to access the individual elements within the `phoneList` array. The first element's index is 0, so you access the first element with `phoneList[0]`. Why is the first element's index 0 instead of 1? The index is a measure of how far you are from the beginning of the array. If you're right at the beginning, the distance from the beginning is 0. So the first element uses 0 for its index value.

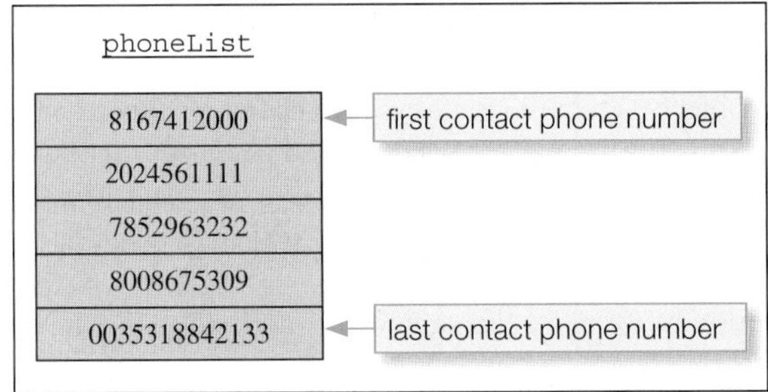

Figure 9.1 Example array—five-element array for holding a list of contact phone numbers

index	phoneList	how to access each element
0	8167412000	`phoneList[0]`
1	2024561111	`phoneList[1]`
2	7852963232	`phoneList[2]`
3	8008675309	`phoneList[3]`
4	0035318842133	`phoneList[4]`

5 elements

Figure 9.2 Accessing elements in a `phoneList` array

Beginning programmers often think that the last index in an array is equal to the number of elements in the array. For example, a beginning programmer might think that the last index in the `phoneList` array equals 5 because the `phoneList` array has 5 elements. Not so. The first index is 0, and the last index is 4. Try to remember this important rule: The last index in an array is equal to one less than the number of elements in the array. If you attempt to access an array element with an index that's greater than the last index or less than zero, you'll get a program crash. So if you specify `phoneList[5]` or `phoneList[-1]`, you'll get a program crash. As part of that crash, the Java virtual machine prints an error message with the word "`ArrayIndexOutOfBoundsException`" in it. `ArrayIndexOutOfBoundsException` is an *exception*. You'll learn about exceptions in Chapter 15, but for now, just think of an exception as a sophisticated type of error that can be used by programmers to determine the source of a bug.

Now that you know how to access an array element, let's put it to use. Here's how you can change the first phone number to 2013434:

```
phoneList[0] = 2013434;
```

And here's how you can print the second phone number:

```
System.out.println(phoneList[1]);
```

Be aware that some people use the term "subscript" rather than "index" because subscripting is the standard English way to represent an element from within a group. In other words, x_0, x_1, x_2, and so on in ordinary writing is the same as `x[0]`, `x[1]`, `x[2]`, and so on in Java.

Example Program

Let's see how arrays are used within the context of a complete program. In Figure 9.3, the ContactList program prompts the user for the number of contact phone numbers that are to be entered, fills up the `phoneList` array with user-entered phone numbers, and prints the created contact list. To fill an array and to print an array's elements, you typically need to step through each element of the array with the help of an index variable that increments from zero to the index of the array's last filled element. Often, the index variable's increment operations are implemented with the help of a `for` loop. For example, the ContactList program uses the following `for` loop header to increment an index variable, `i`:

```
for (int i=0; i<sizeOfList; i++)
```

With each iteration of the `for` loop, `i` goes from 0 to 1 to 2, and so on, and `i` serves as an index for the different elements in the `phoneList` array. Here's how the loop puts a phone number into each element:

```
phoneList[i] = phoneNum;
```

9.3 Array Declaration and Creation

In the previous section, we showed you how to perform simple operations on an array. In so doing, we focused on accessing an array's elements. In this section, we focus on another key concept—declaring and creating arrays.

Array Declaration

An array is a variable and, as such, you must declare it before you can use it. To declare an array, use this syntax:

<element-type>`[]` *<array-variable>*`;`

The *<array-variable>* is the name of the array. The empty square brackets tell us that the variable is defined to be an array. The *<element-type>* indicates the type of each element in the array—`int`, `double`, `char`, `String`, and so on.

Here are some array declaration examples:

```
double[] salaries;
String[] names;
int[] employeeIds;
```

The `salaries` variable is an array whose elements are of type `double`. The `names` variable is an array whose elements are of type `String`. And finally, the `employeeIds` variable is an array whose elements are of type `int`.

Java provides an alternative declaration format for arrays, where the square brackets go after the variable name. Here's what we're talking about:

```
double salaries[];
```

The two formats are identical in terms of functionality. Most folks in industry prefer the first format, and that's what we use, but you should be aware of the alternative format in case you see it in someone else's code.

```
/*********************************************************************
* ContactList.java
* Dean & Dean
*
* This program creates a cell phone contacts phone number
* list and prints the created list.
*********************************************************************/

import java.util.Scanner;

public class ContactList
{
  public static void main(String[] args)
  {
    Scanner stdIn = new Scanner(System.in);
    long[] phoneList;   // list of phone numbers
    int sizeOfList;     // number of phone numbers
    long phoneNum;      // an entered phone number

    System.out.print(
      "How many contact numbers would you like to enter? ");
    sizeOfList = stdIn.nextInt();
    phoneList = new long[sizeOfList];

    for (int i=0; i<sizeOfList; i++)
    {
      System.out.print("Enter phone number: ");
      phoneNum = stdIn.nextLong();
      phoneList[i] = phoneNum;
    } // end for

    System.out.println("\nContacts List:");
    for (int i=0; i<sizeOfList; i++)
    {
      System.out.println((i + 1) + ". " + phoneList[i]);
    } // end for
  } // end main
} // end class ContactList
```

Create an array with a user-specified size.

Fill the array.

Print the array.

Sample session:

```
How many contact numbers would you like to enter? 2
Enter phone number: 8167412000
Enter phone number: 2024561111

Contacts List:
1. 8167412000
2. 2024561111
```

Figure 9.3 ContactList program that shows how to create, fill, and print an array

Array Creation

An array is an object, albeit a special kind of object. As with any object, an array holds a group of data items. As with any object, an array can be created/instantiated using the `new` operator. Here's the syntax for creating an array object with the `new` operator and assigning the array object into an array variable:

```
<array-variable> = new <element-type>[<array-size>];
```

The <*element-type*> indicates the type of each element in the array. The <*array-size*> indicates the number of elements in the array. The following code fragment creates a 10-element array of `long`s:

```
long[] phoneList;
phoneList = new long[10];
```

array creation

These two lines perform three operations: (1) The first line declares the `phoneList` variable, (2) the boxed code creates the array object, and (3) the assignment operator assigns a reference to the array object into the `phoneList` variable.

It's legal to combine an array's declaration, creation, and assignment operations into one statement. The following example does just that. It reduces the previous two-line code fragment to just one line:

```
long[] phoneList = new long[10];
```

Here, we use a constant (10) for the array's size, but you're not required to use a constant. You can use any expression for the array's size. Figure 9.3's ContactList program prompts the user for the size of the array, stores the entered size in a `sizeOfList` variable, and uses `sizeOfList` for the array creation. Here's the array creation code from the ContactList program:

```
phoneList = new long[sizeOfList];
```

Array Element Initialization

Usually, you'll want to declare and create an array in one place and assign values to your array elements in a separate place. For example, the following code fragment declares and creates a `temperatures` array in one statement, and assigns values to the `temperatures` array in a separate statement, inside a loop.

```
double[] temperatures = new double[5];
for (int i=0; i<5; i++)
{
  temperatures[i] = 98.6;
}
```

Declare and create an array.

Assign a value to the ith array element.

On the other hand, sometimes you'll want to declare and create an array and assign values to your array, all in the same statement. That's called an *array initializer*. Here's the syntax:

```
<element-type>[] <array-variable> = {<value1>, <value2>, . . ., <valuen>};
```

The code at the left of the assignment operator declares an array variable using syntax that you've seen before. The code at the right of the assignment operator specifies a comma-separated list of values that are assigned to the array's elements. Note this example:

```
double[] temperatures = {98.6, 98.6, 98.6, 98.6, 98.6};
```

Comparing the above statement to the previous `temperatures` code fragment, you can see that it is the same in terms of functionality but different in terms of structure. Key differences: (1) It's one line, rather than five lines. (2) There's no `new` operator. (3) There's no array-size value. With no array-size value, how do you think the compiler knows the size of the array? The size of the array is dictated by the number of values in the element-values list. In the above example, there are five values in the initializer list, so the compiler creates an array with five elements.

We presented two solutions for assigning values to a temperatures array. Which is better—the five-line code fragment or the one-line array initializer? We prefer the array initializer solution because it's simpler. But remember that you can use the array initializer technique only if you know the assigned values when you first declare the array. For the temperatures example, we do know the assigned values when we first declare the array—we initialize each temperature to 98.6, the normal human body temperature in degrees Fahrenheit. You should limit your use of array initializers to situations where the number of assigned values is reasonably small. For the temperatures example, the number of assigned values is reasonably small—it's five. If you need to keep track of a hundred temperatures, it would be legal to use the array initializer solution, but it would be cumbersome:

```
double[] temperatures =
{
  98.6, 98.6, 98.6, 98.6, 98.6, 98.6, 98.6, 98.6, 98.6, 98.6,
  <repeat above line eight times>
  98.6, 98.6, 98.6, 98.6, 98.6, 98.6, 98.6, 98.6, 98.6, 98.6
}
```

Default Values

You now know how to initialize an array's elements explicitly with an array initializer. But what do an array's elements get by default if you don't use an array initializer? An array is an object, and an array's elements are the instance variables for an array object. As such, an array's elements get default values when the array is created, just as any other instance variables get default values. Here are the default values for array elements:

Array Element's Type	Default Value
integer	0
floating-point	0.0
`boolean`	`false`
reference	`null`

So what are the default values for the elements in the arrays below?

```
double[] rainfall = new double[365];
String[] colors = new String[5];
```

The `rainfall` array gets 0.0 for each of its 365 elements. The `colors` array gets `null` for each of its 5 elements.

9.4 Array `length` Property and Partially Filled Arrays

As illustrated earlier, when working with an array, it's common to step through each element in the array. In doing so, you need to know the size of the array and/or the number of filled elements in the array. In this section, we discuss how to obtain the size of an array and how to keep track of the number of filled elements in an array.

Array `length` Property

Suppose you have a five-element `colors` array that's been initialized like this:

```
String[] colors = {"blue", "gray", "lime", "teal", "yellow"};
```

Here's how to print such an array:

```
for (int i=0; i<5; i++)        // hard-coded array size
{
  System.out.println(colors[i]);
}
```

That works OK, but suppose that you have several other color-related loops in your code, each of them using `i<5`. If you modify your program to accommodate more colors, and change the 5-element array to a 10-element array, you'd have to change all occurrences of `i<5` to `i<10`. To avoid such maintenance work, wouldn't it be nice to replace `i<5` or `i<10` with something generic, like i < array's size? You can do that by using the `color` array's `length` property. Every array object contains a `length` property that stores the number of elements in the array. The `length` property is called a "property," but it's actually just an instance variable with `public` and `final` modifiers. The `public` modifier says that `length` is directly accessible without need of an accessor method. The `final` modifier makes `length` a named constant; so you can't update it. Here's how the `length` property can be used:

```
for (int i=0; i<colors.length; i++)        // number of elements in the array
{
  System.out.println(colors[i]);
}
```

Array `length` Property Versus `String` `length` Method

Remember where else you've seen the word `length` in the Java language? The `String` class provides a `length` method to retrieve the number of characters in a string. Remember that `String`'s `length` is a method, so you must use trailing parentheses when calling it. On the other hand, an array's `length` is a constant, so you don't use trailing parentheses when accessing it. Figure 9.4's ContactList2 program illustrates these concepts. Note that `phoneNum.length()` uses parentheses when checking for the length of the `phoneNum` string as part of input validation. And note that `phoneList.length` does not use parentheses when checking the number of elements in the `phoneList` array to make sure that there's room for another phone number.

If you're like us, you might have a hard time remembering when to use parentheses and when not to. Try using the mnemonic acronym ANSY, which stands for Arrays No, Strings Yes. "Arrays No" means that

```
/*************************************************************
* ContactList2.java
* Dean & Dean
*
* This program creates a contacts phone number list and
* prints the created list. It uses a partially filled array.
*************************************************************/

import java.util.Scanner;

public class ContactList2
{
  public static void main(String[] args)
  {
    Scanner stdIn = new Scanner(System.in);
    String[] phoneList = new String[100]; // phone numbers
    int filledElements = 0;      // number of phone numbers
    String phoneNum;             // an entered phone number

    System.out.print("Enter phone number (or q to quit): ");
    phoneNum = stdIn.nextLine();
    while (!phoneNum.equalsIgnoreCase("q") &&
           filledElements < phoneList.length)
    {
      if (phoneNum.length() < 1 || phoneNum.length() > 16)
      {
        System.out.println("Invalid entry." +
          " Must enter between 1 and 16 characters.");
      }
      else
      {
        phoneList[filledElements] = phoneNum;
        filledElements++;
      }
      System.out.print("Enter phone number (or q to quit): ");
      phoneNum = stdIn.nextLine();
    } // end while

    System.out.println("\nContact List:");
    for (int i=0; i<filledElements; i++)
    {
      System.out.println((i + 1) + ". " + phoneList[i]);
    } // end for
  } // end main
} // end class ContactList2
```

The array `length` property does not use ()'s.

The `String length` method uses ()'s.

Update the number of filled elements.

Use `filledElements` for printing the array.

Figure 9.4 ContactList2 program
This processes a partially filled array, using the array `length` property and the `String length` method.

arrays do not use parentheses when specifying length. "Strings Yes" means that strings do use parentheses when specifying length. If you don't like DFLAs,[1] you can try a more analytical approach to remembering the parentheses rule. Arrays are special-case objects that don't have methods; therefore, an array's `length` must be a constant, not a method. And constants don't use parentheses.

Partially Filled Arrays

In Figure 9.4, note how the ContactList2 program declares the `phoneList` array to have 100 elements. The program repeatedly prompts the user to enter a phone number or enter q to quit. Typically, the user will enter fewer than the maximum 100 phone numbers. That results in the `phoneList` array being partially filled. If you have a partially filled array, as opposed to a completely filled array, you have to keep track of the number of filled elements in the array so you can process the filled elements differently from the unfilled elements. Note how the ContactList2 program uses the `filledElements` variable to keep track of the number of phone numbers in the array. `filledElements` starts at zero and gets incremented each time the program stores a phone number in the array. To print the array, the program uses `filledElements` in the following `for` loop header:

```
for (int i=0; i<filledElements; i++)
```

It's fairly common for programmers to accidentally access unfilled elements in a partially filled array. For example, suppose `ContactList2`'s `for` loop looked like this:

```
for (int i=0; i<phoneList.length; i++)
{
  System.out.println((i + 1) + ". " + phoneList[i]);
} // end for
```

Using `phoneList.length` in the `for` loop header works great for printing a completely filled array, but not so great for printing a partially filled array. In the ContactList2 program, unfilled elements hold `null` (the default value for a string), so the above `for` loop would print `null` for each of the unfilled elements. And that makes for confused and unhappy users. ☹

9.5 Copying an Array

In the previous sections, we focused on array syntax details. In the next several sections, we'll focus less on the syntax and more on the application side of things. In this section, we discuss a general-purpose problem—how to copy from one array to another.

Using Arrays to Hold a Store's Prices

Suppose you use arrays to hold a store's prices, one array for each month's prices. Here's the array for January's prices:

```
double[] pricesJanuary = {1.29, 9.99, 22.50, 4.55, 7.35, 6.49};
```

[1]DFLA = dumb four-letter acronym.

Your intent is to use January's array as a starting point for the other month's arrays. Specifically, you want to copy January's prices into the other months' arrays and modify the other months' prices when necessary. The statement below creates the array for February's prices. Note how `pricesJanuary.length` ensures that February's array is the same length as January's array:

```
double[] pricesFebruary = new double[pricesJanuary.length];
```

Suppose you want the values in February's array to be the same as the values in January's array except for the second entry, which you want to change from 9.99 to 10.99. In other words, you want something like this:

<u>Output:</u>

```
  Jan     Feb
 1.29    1.29
 9.99   10.99
22.50   22.50
 4.55    4.55
 7.35    7.35
 6.49    6.49
```

To minimize re-entry effort and error, it would be nice to have the computer copy the first array's values into the second array and then just alter the one element of the second array that needs changing. Would the following code fragment work?

```
pricesFebruary = pricesJanuary;      ← not a good idea
pricesFebruary[1] = 10.99;
```

An array name is just a reference. It contains the address of a place in memory where the array's data begins. So `pricesFebruary = pricesJanuary;` gets the address of `pricesJanuary`'s data and copies the address into `pricesFebruary`. Then `pricesFebruary` and `pricesJanuary` refer to the same physical data. This picture illustrates the point:

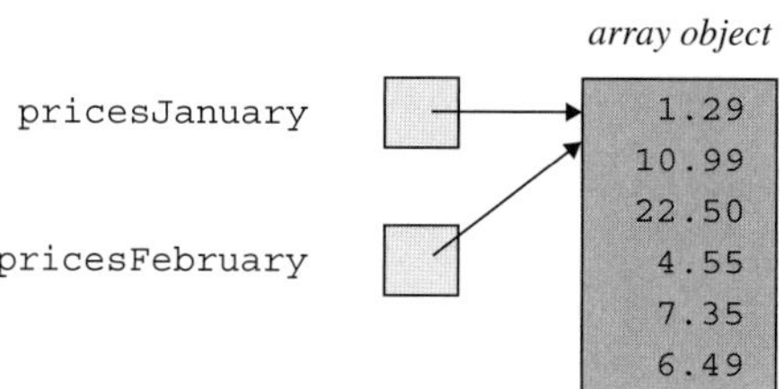

The problem with `pricesFebruary` and `pricesJanuary` referring to the same physical data is that if you change the data for one of the arrays, then you automatically change the data for the other array. For example, the above `pricesFebruary[1] = 10.99;` statement updates not only `pricesFebruary`'s second element, but also `pricesJanuary`'s second element. And that's not what you want.

Usually when you make a copy of an array, you'll want the copy and the original to point to different array objects. To do that, assign array elements one at a time. See Figure 9.5's ArrayCopy program. It uses a `for` loop to assign `pricesJanuary` elements to `pricesFebruary` elements one at a time.

```
/*******************************************************************
* ArrayCopy.java
* Dean & Dean
*
* This copies an array and then alters the copy.
*******************************************************************/

public class ArrayCopy
{
  public static void main(String[] args)
  {
    double[] pricesJanuary =
      {1.29, 9.99, 22.50, 4.55, 7.35, 6.49};
    double[] pricesFebruary = new double[pricesJanuary.length];

    for (int i=0; i<pricesJanuary.length; i++)
    {
      pricesFebruary[i] = pricesJanuary[i];
    }
    pricesFebruary[1] = 10.99;

    System.out.printf("%7s%7s\n", "Jan", "Feb");
    for (int i=0; i<pricesJanuary.length; i++)
    {
      System.out.printf("%7.2f%7.2f\n",
        pricesJanuary[i], pricesFebruary[i]);
    }
  } // end main
} // end class ArrayCopy
```

Figure 9.5 ArrayCopy program that copies an array and then alters the copy

This is what the code in Figure 9.5 produces:

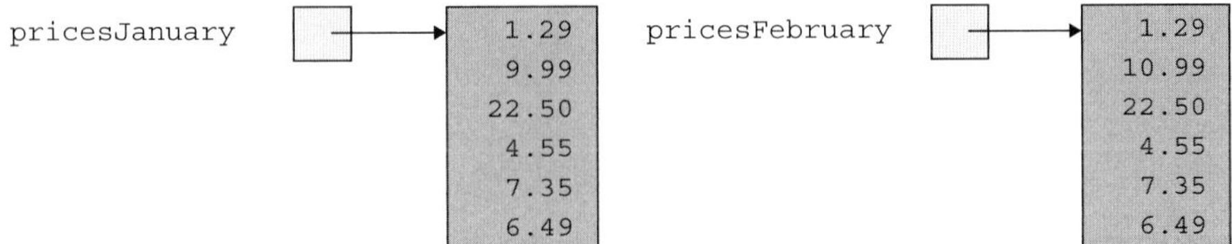

Array Copying Methods Provided by Java's API

Copying data from one array to another is a very common operation, so Java designers provide a special method, `System.arraycopy`, just for that purpose. It allows you to copy any number of elements from any place in one array to any place in another array. Here's how you could use it to copy Figure 9.5's `pricesJanuary` array to the `pricesFebruary` array:

```
System.arraycopy(pricesJanuary, 0, pricesFebruary, 0, 6);
pricesFebruary[1] = 10.99;
```

The first argument is the source array name, that is, the name of the array you're copying from. The second argument is the index of the source array's first element to copy. The third argument is the destination array name, that is, the name of the array you're copying to. The fourth argument is the index of the destination array's first element to replace. The final argument is the total number of elements to copy.

Java 1.6 introduced two additional array-copying methods, `Arays.copyOf` and `Arrays.copyOfRange`. To access these methods, include the following `import` statement:

```
import java.util.Arrays;
```

The `Arrays.copyOf` method starts at index 0 in a specified source array and copies up to a specified number of elements to a returned destination array. The copied elements have the same indices in both arrays. The specified number of elements also establishes the destination array's length. If the destination array's length exceeds the source array's length, the extra elements in the destination array get default values. Here's how you could use it to copy Figure 9.5's `pricesJanuary` array to the `pricesFebruary` array:

```
double[] pricesFebruary = Arrays.copyOf(
  pricesJanuary, pricesJanuary.length);
```

The first argument is the source array's name. The second argument is the destination array's length. Overloaded versions of this method permit elements to be any type of primitive or object, and the destination array's elements automatically become the same type as the source array's elements.

The `Arrays.copyOfRange` method copies elements in a specified range from a specified source array to a returned destination array. Here's how you could use it to copy Figure 9.5's `pricesJanuary` array to the `pricesFebruary` array:

```
double[] pricesFebruary = Arrays.copyOfRange(
  pricesJanuary, 0, 6);
```

As before, the first argument is the source array name. The second argument is the index of the first element to be copied from the source array. The third argument is the index of the last element to be copied from the source array, plus 1. This makes the total number of elements copied equal to the third argument minus the second argument. This difference determines the length of the destination array. If the length of the destination array exceeds the length of the source array, the extra elements in the generated destination array get default values. Overloaded versions of this method permit elements to be any type of primitive or object, and the destination array's elements automatically become the same type as the source array's elements.

9.6 Problem Solving with Array Case Studies

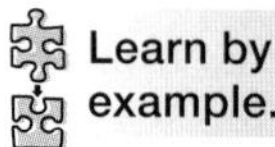

In this section, we present two array-based case studies. For each case study, we present a problem and then examine its solution. The point of these case studies isn't so much that you memorize the details. The point is that you get a feel for how to solve array-oriented problems. Then when you're a programmer in the real world, you'll have a "bag of tricks" that you can draw from. You'll probably have to modify the case-study solutions to make them fit your specific real-world problems, but that's OK. You've got to earn your keep, after all.

Shifting Array-Element Values

Consider the hours array in Figure 9.6. The hours array contains the scheduled work hours for a person for a 31-day period. The first element (hours[0]) contains the scheduled work hours for the person for the current day. The last element (hours[30]) contains the scheduled work hours for the person for the day that's 30 days in the future. At the beginning of each new day, the work hours need to shift to lower-index positions. For example, the hours[1] value needs to shift to the hours[0] element. That should make sense when you realize that when you're going to a new day, you need to make what was the next day's scheduled hours, hours[1], become the current day's scheduled hours, hours[0].

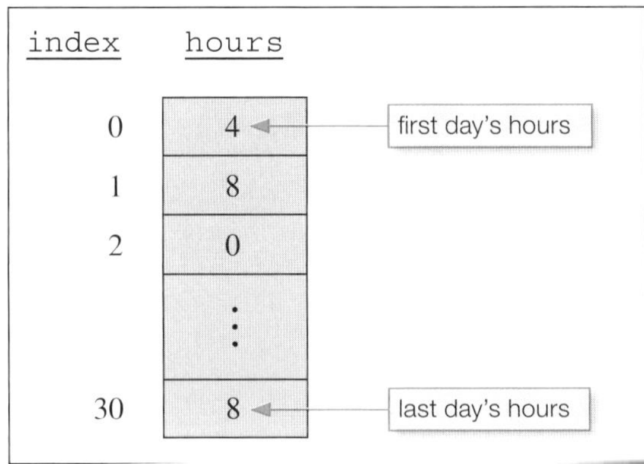

Figure 9.6 Array of scheduled work hours for the next 31 days

Now let's look at Java code that performs this shifting operation. We want to shift each hours element value to its adjacent lower-indexed element. In other words, we want to copy the second element's value into the first element, copy the third element's value into the second element, and so on. Then we want to assign a user-entered value to the last element. Here's the code:

```
for (int d=0; d<hours.length-1; d++)
{
  hours[d] = hours[d+1];
}
System.out.print("Enter last day's scheduled hours: ");
hours[hours.length-1] = stdIn.nextInt();
```

To shift values to lower-index positions, you must start at the low-index end and work toward the high end.

There are several things to note about this code fragment. It's OK to use an expression inside the [ ]'s—we use hours[d+1] to access the element after the hours[d] element. Notice how we shift elements at the low-index end first. What would happen if you started the shifting at the high-index end? You'd overwrite the next element you wanted to move and end up filling the entire array with the value that was originally in the highest element. Not good.

Calculating a Moving Average

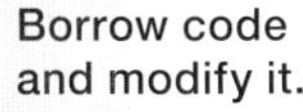

Let's now borrow code from the above example and apply it to another problem. Suppose you need to present a four-day moving average of the Dow Jones Industrial Average (DJIA) at the end of each business day. Assume you already have a four-element array holding the values of

the DJIA at the end of the day on each of the past four days, with four-days-ago's value at index 0, three-days-ago's value at index 1, two-days-ago's value at index 2, and yesterday's value at index 3. For today's four-day moving average, you'll want the sum of the values for the last three days, plus the value for today. This means you'll need to shift everything in the array to lower-index positions and insert today's value at the high-index end. Then you'll need to sum up everything in the array and divide by the length of the array. Presumably, you'll save the shifted array somewhere and then do the same thing again at the end of each day in the future. You could do the shifting and summing in separate loops, but it's easier to do both in the same loop, as shown in Figure 9.7.

```
/*******************************************************************
* MovingAverage.java
* Dean & Dean
*
* This program contains an operation that shifts each array
* element to the next lower element and loads a new input
* into the final element.
*******************************************************************/

import java.util.Scanner;

public class MovingAverage
{
  public static void main(String[] args)
  {
    Scanner stdIn = new Scanner(System.in);
    int[] days = {9400, 9500, 9600, 9700}; // rising market
    double sum;
    int samples;

    System.out.print("Enter number of days to evaluate: ");
    samples = stdIn.nextInt();
    for (int j=0; j<samples; j++)
    {
      // shift down and sum
      sum = 0.0;
      for (int d=0; d<days.length-1; d++)
      {
        days[d] = days[d+1];
        sum += days[d];
      }
      System.out.print("Enter next day's value: ");
      days[days.length-1] = stdIn.nextInt();
      sum += days[days.length-1];
      System.out.printf(
        "Moving average = %5.0f\n", sum / days.length);
    }
  } // end main
} // end class MovingAverage
```

This shifts to lower-index positions.
This accumulates the already shifted values.
This shifts in the latest value.

Figure 9.7 Calculation of a moving average

To allow for different lengths of time, do not hard-code the array length. Instead, always use `<array-name>.length`. Think carefully about each boundary. Notice that the index `[d+1]` onthe right side of the first statement in the inside `for` loop is 1 greater than the count variable value `d`. Remember that the highest index value in an array is always 1 less than the array's length. So the highest value of the count variable should be the array's length minus 2. That's why the loop-continuation condition is `d<days.length-1`. Also, notice that we insert the new final value for the array after the loop terminates, and then we include this final value in the sum before computing the average. Here's an example of what the program does:

Sample session:

```
Enter number of days to evaluate: 4
Enter next day's value: 9800
Moving average = 9650
Enter next day's value: 9800
Moving average = 9725
Enter next day's value: 9700
Moving average = 9750
Enter next day's value: 9600
Moving average = 9725
```

A moving average is smoother than an instantaneous plot, but notice that its values lag behind.

There's a simpler way to do shifting. Do you remember the API `arraycopy` method mentioned in the previous section? You can use it to implement shifts to lower-index positions with this code fragment:

```
System.arraycopy(days, 1, days, 0, days.length-1);
System.out.print("Enter next day's value: ");
days[days.length-1] = stdIn.nextInt();
```

Conceptually, the `arraycopy` method copies everything from element 1 to the last element into a temporary array, and then copies it from this temporary array back into the original array starting at element 0. This eliminates the inner `for` loop in Figure 9.7. Unfortunately, we also used the inner `for` loop to compute the sum needed for the average. But there's a trick you can use, and it makes a program like this more efficient when the array is very large. If you keep track of the sum of all the elements in the array, each time you shift the array element values, you can just correct the sum, rather than completely recomputing it. To correct the sum, subtract the value shifted out and add the value shifted in, like this:

```
sum -= days[0];
System.arraycopy(days, 1, days, 0, days.length-1);
System.out.print("Enter next day's value: ");
days[days.length-1] = stdIn.nextInt();
sum += days[days.length-1];
```

Histograms

In this subsection, we'll use an array as part of a histogram program. But before we present the program, a histogram overview is in order. A *histogram* is a graph that displays quantities for a set of categories. Typically, it indicates category quantities with bars—shorter bars represent smaller quantities, longer bars represent larger quantities. For example, Figure 9.8's histogram shows child mortality rates in different regions

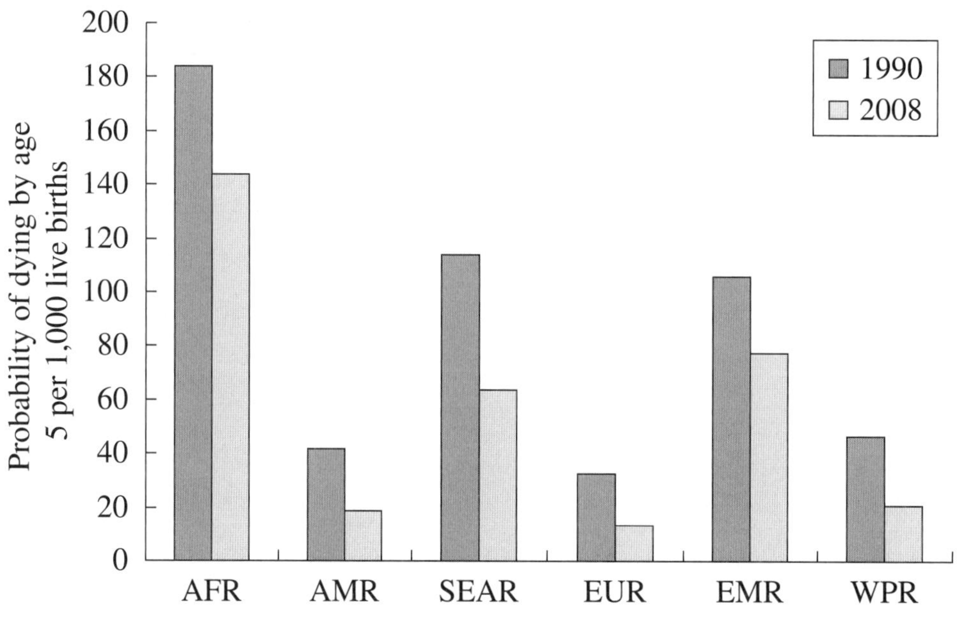

Figure 9.8 Example histogram from World Health Statistics 2010
AFR = African Region, AMR = AmericanRegion, SEAR = Southeast Asia Region, EUR = European Region, EMR = Eastern Mediterranean Region, WPR = Western Pacific Region

of the world at two different times.[2] Histograms are a popular way to present statistical data because they provide a quick and clear representation of the data's distribution.

Suppose you have three coins. When you flip all three, you're curious about the probabilities of getting zero heads, one head, two heads, and three heads. In other words, you're curious about the frequency distribution for the number of heads.

Approximate a mathematical solution with simulation.

You could calculate the frequency distribution mathematically (with the binomial distribution formula), but, instead, you decide to write a program to simulate the coin flips. If you simulate enough coin flips, then the results will approximate the mathematically calculated result.

In your program, you should simulate flipping the three coins a million times. You should print the simulation results in the form of a histogram. For each of the four cases (zero heads, one head, two heads, three heads), print a series of *'s where the number of these asterisks is proportional to the number of times the case occurred. Each series of asterisks represents a histogram bar. That should make more sense by looking at this sample output:

```
Number of times each head count occurred:
 0  124960      **************
 1  375127      *****************************************
 2  375261      *****************************************
 3  124652      **************
```

[2]World Health Organization, World Health Statistics 2010 http://www.who.int/whosis/whostat/EN_WHS10_Full.pdf.

Note the first row of asterisks. That's a horizontal "bar" that pictorially describes the number of times the zero-heads case occurred. The zero at the left is the label for the zero-heads case. The 124960 is the specific number of times the zero-heads case occurred. Or, said another way, 124960 is the *frequency* of the zero-heads case. Note that the zero-heads and three-heads frequencies (124960 and 124652, respectively) are nearly the same, and the one-head and two-heads frequencies (375127 and 375261, respectively) are also nearly the same. Also note that the zero-heads and three-heads frequencies are each approximately one third of the one-head and two-heads frequencies. It's always a good idea to use some kind of independent calculation to predict what a computer's answer should be like. For this simple problem, it's relatively easy to compute an exact answer. Assuming that "T" means "tails" and "H" means "heads," here are all the possible flipping results:

Compare program results with predicted results.

TTT (0 heads)
TTH (1 head)
THT (1 head)
THH (2 heads)
HTT (1 head)
HTH (2 heads)
HHT (2 heads)
HHH (3 heads)

Note that there is only one way to obtain zero heads and only one way to obtain three heads, but there are three ways to obtain one head and three ways to obtain two heads. So the zero-head and three-head frequencies should each be one third of the one-head and two-head frequencies. If you look at the numbers and bar lengths in the above sample output, you'll see that the computer result approximately conforms to this expectation.

See Figure 9.9's CoinFlips program. It does what we want. It simulates flipping three coins a million times, and it prints the simulation results in the form of a histogram. It uses a four-element `frequency` array to keep track of the number of times each head-count value occurs. Each element in the frequency array is called a *bin*. In general, a bin contains the number of occurrences of an event. For the CoinFlips program, the `frequency[0]` element is the first bin, and it holds the number of times none of the three coins lands heads up. The `frequency[1]` element is the second bin, and it holds the number of times one of the three coins lands heads up. After each three-coin-flip simulation iteration, the program adds 1 to the appropriate bin. For example, if a particular iteration generates one head, the program increments the `frequency[1]` bin. And if a particular iteration generates two heads, the program increments the `frequency[2]` bin.

Let's now examine how the CoinFlips program prints the histogram asterisk bars. As specified by the second callout in Figure 9.9, the second large `for` loop prints the histogram. Each iteration of the `for` loop prints the bin label (0, 1, 2, or 3) and then the frequency for that bin. Then it computes the number of asterisks to print by dividing the frequency in the current bin by the total number of repetitions and multiplying by 100. Then it uses an inner `for` loop to display the computed number of asterisks.

9.7 Searching an Array

In order to use an array, you need to access its individual elements. If you know the location of the element you're interested in, then you simply access the element by putting the element's index inside square brackets. But if you don't know the location of the element, you need to search for it. For example, suppose

```
/*********************************************************************
* CoinFlips.java
* Dean & Dean
*
* This generates a histogram of coin flips.
*********************************************************************/

public class CoinFlips
{
  public static void main(String[] args)
  {
    final int NUM_OF_COINS = 3;       // number of coins
    final int NUM_OF_REPS = 1000000;  // repetitions

    // The frequency array holds the number of times
    // a particular number of heads occurred.
    int[] frequency = new int[NUM_OF_COINS + 1];
    int heads;             // heads in current group of flips
    double fractionOfReps; // head count / repetitions
    int numOfAsterisks;    // asterisks in one histogram bar

    for (int rep=0; rep<NUM_OF_REPS; rep++)
    {
      // perform a group of flips
      heads = 0;
      for (int i=0; i<NUM_OF_COINS; i++)
      {
        heads += (int) (Math.random() * 2);
      }
      frequency[heads]++;   // update appropriate bin
    } // end for
    System.out.println(
      "Number of times each head count occurred:");
    for (heads=0; heads<=NUM_OF_COINS; heads++)
    {
      System.out.print(
        " " + heads + "  " + frequency[heads] + " ");
      fractionOfReps = (float) frequency[heads] / NUM_OF_REPS;
      numOfAsterisks = (int) Math.round(fractionOfReps * 100);

      for (int i=0; i<numOfAsterisks; i++)
      {
        System.out.print("*");
      }
      System.out.println();
    } // end for
  } // end main
} // end class CoinFlips
```

This loop fills up the frequency bins. Each iteration simulates one group of three coin flips.

This loop prints the histogram. Each iteration prints one histogram bar.

Figure 9.9 CoinFlips program that generates a histogram for a coin-flips simulation

you're writing a program that keeps track of student enrollments for the courses at your school. The program should be able to add a student, remove a student, view a student's data, and so on. All of those operations require that you first search for the student within a student `ids` array (even the add-a-student operation requires a search, to ensure that the student isn't already in the array). In this section, we present two techniques for searching an array.

Sequential Search

If the array is short (has less than about 20 items), the best way to search it is the simplest way: Step through the array sequentially and compare the value at each array element with the searched-for value. When you find a match, do something and return. Here's a pseudocode description of the sequential-search algorithm:

```
i ← 0
while i < number of filled elements
{
   if list[i] equals the searched-for value
      <do something and stop the loop>
   increment i
}
```

Typically, algorithms are more generic than Java implementations. Part of problem solving is the process of adapting generic algorithms to specific situations. In this case, the "do something" code will be different for different cases. The `findStudent` method in Figure 9.10 illustrates one implementation of the sequential-search algorithm. This particular method might be part of a `Course` class that implements an academic course. The `Course` class stores a course's name, an array of student IDs for the students enrolled in the course, and the number of students in the course. The `findStudent` method searches for a given student ID within the student `ids` array. If it finds the student ID, it returns its index. Otherwise, it returns −1. Note how `findStudent`'s code matches the sequential-search algorithm's logic. In particular, note how `findStudent` implements *<do something and stop the loop>* with a `return i` statement. The `return i` implements "do something" by returning the index of the found student ID. It implements "stop the loop" by returning from the method and terminating the loop simultaneously.

Adapt generic algorithms to specific situations.

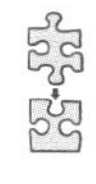

In examining the `findStudent` method, you might be asking yourself "What is the practical use for the returned index?" To do anything with an ID in the `ids` array (add an ID, remove an ID, etc.) you need to know the ID's index. If you don't know the ID's index in advance, the `findStudent` method finds the ID's index for you. Are you still asking yourself "What is the practical use for the returned −1 when the ID is not found?" The −1 can be used by the calling module to check for the case of an invalid student ID.

Figure 9.11's `CourseDriver` class drives Figure 9.10's `Course` class. The `CourseDriver` class is fairly straightforward. It creates an array of student IDs, stores the array in a `Course` object, prompts the user for a particular student ID, and then calls `findStudent` to see whether that particular student is taking the course. To keep things simple, we use an initializer to create the `ids` array. For a more general-purpose driver, you might want to replace the initializer with a loop that repeatedly prompts the user to enter a student ID or −1 to quit. If you choose that option, then you'd need to store the number of filled elements in a `filledElements` variable and pass the `filledElements` variable as the third argument in the `Course` constructor call. This is what the constructor call would look like:

```
Course course = new Course("CS101", ids, filledElements);
```

```
/*********************************************************************
* Course.java
* Dean & Dean
*
* This class represents a particular course in a school.
*********************************************************************/

public class Course
{
  private String courseName;  // name of the course
  private int[] ids;          // IDs for students in the course
  private int filledElements; // number of filled-in elements

  //*******************************************************************

  public Course(String courseName, int[] ids, int filledElements)
  {
    this.courseName = courseName;
    this.ids = ids;
    this.filledElements = filledElements;
  } // end constructor

  //*******************************************************************

  // This method returns index of found ID or -1 if not found.

  public int findStudent(int id)
  {
    for (int i=0; i<filledElements; i++)
    {
      if (ids[i] == id)
      {
        return i;
      }
    } // end for

    return -1;
  } // end findStudent
} // end class Course
```

Figure 9.10 `Course` class, which includes a sequential search method (`findStudent`)

Binary Search

If you have an array with a large number of array elements, like 100,000, a sequential search typically takes quite a long time. If such an array has to be searched many times, it's often worthwhile to use a binary search. *Binary search* gets its name from the way that it bisects a list of values and narrows its search to just half of the bisected list.

```
/*******************************************************************
* CourseDriver.java
* Dean & Dean
*
* This class creates a Course object and searches for a student
* ID within the newly created Course object.
*******************************************************************/

import java.util.Scanner;

public class CourseDriver
{
  public static void main(String[] args)
  {
    Scanner stdIn = new Scanner(System.in);
    int[] ids = {4142, 3001, 6020};
    Course course = new Course("CS101", ids, ids.length);
    int id;        // ID being searched for
    int index;     // index of ID sought or -1 if not found

    System.out.print("Enter 4-digit ID: ");
    id = stdIn.nextInt();
    index = course.findStudent(id);
    if (index >= 0)
    {
      System.out.println("found at index " + index);
    }
    else
    {
      System.out.println("not found");
    }
  } // end main
} // end class CourseDriver
```

<u>Sample session:</u>

```
Enter 4-digit ID: 3001
found at index 1
```

Figure 9.11 Driver for Course program, which illustrates a sequential search

For a binary search to work on an array, the array must be sorted so that everything is in some kind of alphabetical or numerical order. The next section describes one of the many available sorting methods. This initial sorting takes more time than a single sequential search, but you have to do it only once.

After the array has been sorted, you can use a binary search to find values in the array very quickly—  even when the array is extremely long. The worst (slowest) case is when the searched-for value is not in the array. A worst-case sequential search takes an amount of time proportional to the array length. In comparison,

```
// Precondition: array must be sorted lowest to highest
// Postcondition: array not altered

public static int binarySearch(
  int[] array, int filledElements, int value)
{
  int mid;                         // index of middle element
  int midValue;                    // value of middle element
  int low = 0;                     // index of lowest element
  int high = filledElements - 1;   // index of highest element

  while (low <= high)
  {
    mid = (low + high) / 2;        // next midpoint
    midValue = array[mid];         // and the value there
    if (value == midValue)
    {
      return mid;                  // found it!
    }
    else if (value < midValue)
    {
      high = mid - 1;              // next time, use lower half
    }
    else
    {
      low = mid + 1;               // next time, use upper half
    }
  } // end while

  return -1;
} // end binarySearch
```

Figure 9.12 Method that performs a binary search of an array already sorted in ascending order

a worst-case binary search takes an amount of time proportional to the logarithm of the array length. When an array is very long, the difference between linear and logarithmic is huge. For example, suppose the length is 100,000. It works out that $\log_2(100{,}000) \approx 17$. Because 17 is about 6,000 times smaller than 100,000, binary search is approximately 6,000 times faster than sequential search for a 100,000-element array.

Note the `binarySearch` method in Figure 9.12, and, in particular, note its `static` modifier. You can use either an instance method or a class method to implement searching. In the previous subsection, we implemented searching with an instance method. This time, we implement searching with a class method, which is appropriate if you want a method to be used generically. To make it generic (that is, to make it usable by different programs), you should put the method in a separate class and make the method a class method. Because it's a class method, different programs can call the `binarySearch` method easily, using binarySearch's class name, rather than using a calling object. For example, if you put the binarySearch method in a Utilities class, you would call binarySearch like this:

```
Utilities.binarySearch(
  <array-name>, <number-of-filled-elements>, <searched-for-value>);
```

In the binarySearch method call, note the array argument. Being a class method, binarySearch cannot access instance variables. More specifically, it cannot access the searched array as an instance variable. So the searched array must be passed in as an argument. This allows the method to be used for arrays defined outside of the method's class. But as you may recall from Section 7.5, passing a reference (an array name is a reference) allows mutations to propagate back to the caller's code. Not to worry—the postcondition comment at the top of Figure 9.12 guarantees that this particular method will not alter its array argument.

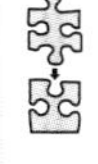

Split up a problem into smaller problems.

The binary search algorithm is based on an ancient strategy—*divide and conquer.* You first identify the middle element in the sorted array. You then figure out whether the searched-for value belongs before or after the middle element. If it belongs before the middle element, you narrow the search range to the lower half of the array (the half with the smaller-indexed elements). If, on the other hand, the searched-for value belongs after the middle element, you narrow the search range to the upper half of the array. You then repeat the process. In other words, within the narrowed-down half of the array, you identify the middle element, figure out whether the searched-for value belongs before or after the middle element, and narrow the search range accordingly. Every time you do this, you cut the problem in half, and this enables you to zero in quickly on the searched-for value—if it's there at all. Splitting the array in half is the "divide" part of "divide and conquer." Finding the searched-for value within one of the halves is the "conquer" part.

Now let's see how the binarySearch method implements the divide-and-conquer algorithm. The method declares mid, low, and high variables that keep track of the indexes for the middle element and the two elements at the ends of the array's search range. For an example, see the left drawing in Figure 9.13.

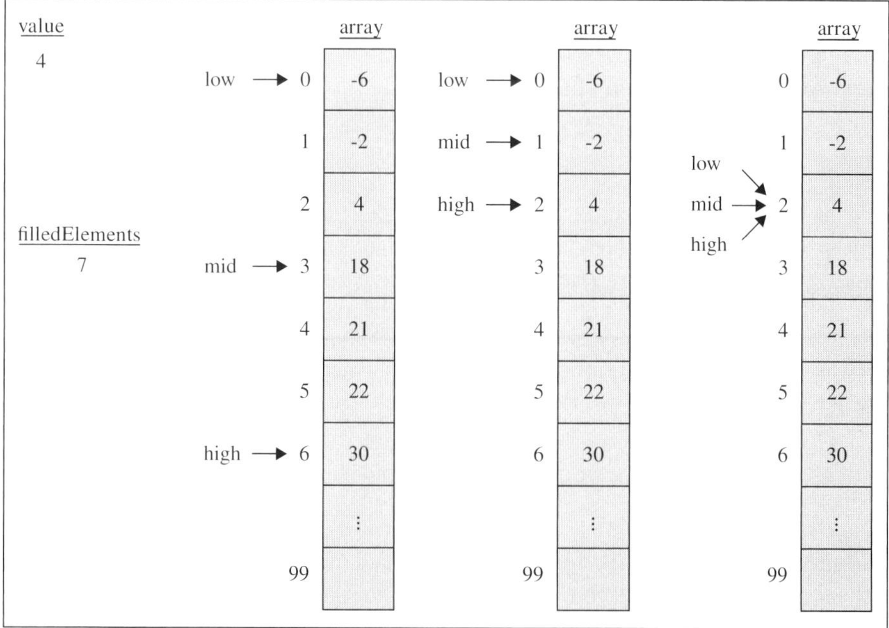

Figure 9.13 Example execution of Figure 9.12's binarySearch method

Using a while loop, the method repeatedly calculates mid (the index of the middle element) and checks whether the mid element's value is the searched-for value. If the mid element's value is the searched-for value, then the method returns the mid index. Otherwise, the method narrows the search range to the low half or the high half of the array. For an example of that narrowing process, see Figure 9.13. The method repeats the loop until either the searched-for value is found or the search range shrinks to the point where low's index is greater than high's index.

9.8 Sorting an Array

Computers are particularly good at storing large quantities of data and accessing that data quickly. As you learned in the previous section, binary search is an effective technique for finding and accessing data quickly. In order to prepare the data for binary search, the data must be sorted. Sorting data is done not only for binary search purposes. Computers also sort data so that it's easier to display in a user-friendly fashion. If you look at the e-mails in your inbox, aren't they normally sorted by date, with the most recent e-mail first? Most e-mail organizers allow you to sort your e-mails using other criteria as well, such as using the "from" person or using the size of the e-mail. In this section, we describe the basics of how sorting is performed. We first present a sorting algorithm, and we then present its implementation in the form of a program that sorts the values in an array.

Selection Sort

There are many different sorting algorithms with varying degrees of complexity and efficiency. Frequently, the best way to solve a problem on a computer is the way a human would naturally solve the problem by hand. To illustrate this idea, we'll show you how to convert one of the common human card-sorting algorithms to a Java sorting program.

If you're sorting cards in a card game, you probably use the *selection sort* algorithm. Assume you're sorting smallest cards first. You search for and select the smallest card and move it to the small-card side of the card group. The small-card side of the card group is where you keep the cards that have been sorted already. You then search for the next smallest card, but in so doing, you look only at cards that are in the unsorted portion of the card group. You move the found card to the second position on the small-card side of the card group. You repeat the search-and-move process until there are no more cards left in the unsorted portion of the card group.

As a first step in implementing the selection sort logic, let's examine a pseudocode solution. Above, we said to "repeat the search-and-move process." Whenever there's a repetition, you should think about using a loop. The following algorithm uses a loop for repeating the search-and-move process. Note how the index i keeps track of where the search starts. The first time through the loop, the search starts at the first element (at index 0). The next time, the search starts at the second position. Each time through the loop, you find the smallest value and move it to the sorted portion of the list (the i also tells you where the smallest value should go).

```
for (i ← 0; i < list's length; i++)
{
   find the smallest value in the list from list[i] to the end of the list
   swap the found value with list[i]
}
```

Notice that the outer loop has another loop inside it. The length of the inner loop shrinks as the outer loop progresses. A picture is worth a thousand words, so we provide Figure 9.14 to show the selection sort algorithm in action. The five pictures show the different stages of a list being sorted using the selection sort algorithm. The white portion of each stage is unsorted. The original stage on the left is all white, indicating that it is entirely unsorted. The shaded portion of each stage is sorted. The stage on the right is all shaded, indicating that it is entirely sorted. The bidirectional arrows show what happens after a smallest value is found. The smallest value (at the bottom of the bidirectional arrow) gets swapped up to the top of the unsorted portion of the list. For example, in going from the first picture to the second picture, the smallest value, −3, gets swapped up to 5's position at the top of the unsorted portion of the list.

Now let's implement a Java version of the selection sort algorithm. You can use either an instance method or a class method. In the previous section, we implemented binary search with a class method. For additional practice, we'll do the same here for selection sort. By implementing selection sort with a class method, you can easily call it from any program that needs to sort a list of numbers—just prefix the method call with class name dot.

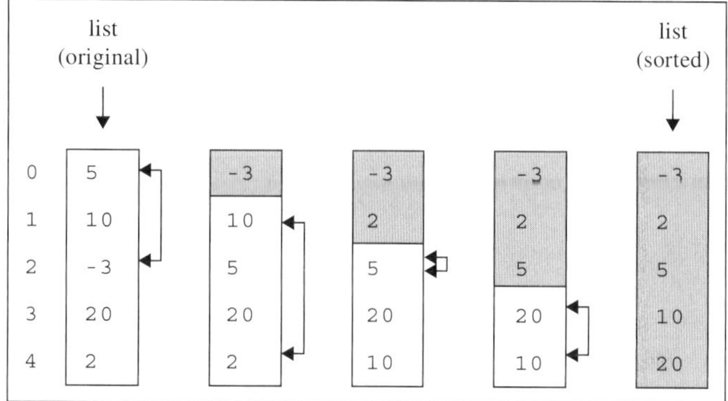

Figure 9.14 Stages in an example execution of the selection sort algorithm

See the `Sort` class in Figure 9.15. Note how the `sort` method body mimics the pseudocode very closely because the sort method uses top-down design. Rather than include the search-for-the-smallest-value code within the `sort` method, the `sort` method calls the `indexOfNextSmallest` helper method. Rather than include the element-swapping code within the `sort` method, the `sort` method calls the `swap` helper method. The only substantive difference between the `sort` method and the sort algorithm is that the `sort` method's `for` loop stops iterating one element before the bottom of the array. That's because there's no need to perform a search when you're at the last element (you already know that the last element is the minimum value for the remainder of the list). We didn't worry about such efficiency details with the algorithm because algorithms are more about basic logic rather than off-by-one details.

Passing Arrays as Arguments

Figure 9.16 contains a driver for Figure 9.15's `Sort` class. Most of the code is fairly straightforward, but please take note of the `studentIds` argument in the `Sort.sort` method call. This is another example of passing an array to a method. An array is an object, and as such, `studentIds` is a reference to an array

```
/*****************************************************************
* Sort.java
* Dean & Dean
*
* This class uses a selection sort to sort a single array.
*****************************************************************/

public class Sort
{
  public static void sort(int[] list)
  {
    int j;                          // index of smallest value

    for (int i=0; i<list.length-1; i++)
    {
      j = indexOfNextSmallest(list, i);
      swap(list, i, j);
    }
  } // end sort

  //***************************************************************

  private static int indexOfNextSmallest(
    int[] list, int startIndex)
  {
    int minIndex = startIndex;  // index of smallest value

    for (int i=startIndex+1; i<list.length; i++)
    {
      if (list[i] < list[minIndex])
      {
        minIndex = i;
      }
    } // end for
    return minIndex;
  } // end indexOfNextSmallest

  //***************************************************************

  private static void swap(int[] list, int i, int j)
  {
    int temp;                       // temporary holder for number

    temp = list[i];
    list[i] = list[j];
    list[j] = temp;
  } // end swap
} // end Sort
```

Figure 9.15 `Sort` class containing a method that sorts an array of integers in ascending order

```
/*************************************************************
* SortDriver.java
* Dean & Dean
*
* This exercises selection sort in class Sort.
*************************************************************/

public class SortDriver
{
  public static void main(String[] args)
  {
    int[] studentIds = {3333, 1234, 2222, 1000};

    Sort.sort(studentIds);
    for (int i=0; i<studentIds.length; i++)
    {
      System.out.print(studentIds[i] + " ");
    }
  } // end main
} // end SortDriver
```

calling the `sort` method

Figure 9.16 Driver that exercises the `sort` method in Figure 9.15

object. As indicated in Section 7.5, a reference argument (in a method call) and its corresponding reference parameter (in a method heading) point to the same object. So if you update the reference parameter's object from within the method, you simultaneously update the reference argument's object in the calling module. Applying that thinking to the Sort program, when you pass the `studentIds` reference to the `sort` method and sort the array there, there's no need to return the updated (sorted) array with a `return` statement. That's because the `studentIds` reference points to the same array object that is sorted within the `sort` method. Thus, we do not include a `return` statement in the `sort` method, and the method works just fine.

Sorting with a Java API Method

Check for efficient API methods.

When an array has more than about 20 elements, it's better to use an algorithm that's more efficient than the relatively simple selection sort algorithm just described. And sure enough, the Java API has a sorting method that uses a more efficient sorting algorithm. It's the `sort` method in the `Arrays` class.

Here's skeleton code for how you might use the `Arrays` class's `sort` method:

```
import java.util.Arrays;
...
  int[] studentIds = {...};
  ...
  Arrays.sort(studentIds);
```

We recommend that you use this API method for heavy-duty sorting. It's an overloaded method, so it also works for arrays of other types of primitive variables. Chapter 11 describes another sorting technique that is almost as efficient as the Java API sorting method, and it's relatively easy to understand. That chapter also describes an alternate implementation of binary search. At this point, if you wish, you can safely jump there.

9.9 Two-Dimensional Arrays

Arrays are good for grouping related data together. Up to this point, we've grouped the data together using standard one-dimensional arrays. If the related data is organized in a table format, consider using a two-dimensional array. This section describes two-dimensional arrays and higher-dimensional arrays.

Two-Dimensional Array Syntax

Two-dimensional arrays use the same basic syntax as one-dimensional arrays except for a second pair of square brackets (`[]`). Each pair of square brackets contains one index. According to standard programming practice, the first index identifies the row and the second index identifies the column position within a row.

For example, here's a two-row by three-column array named `x`:

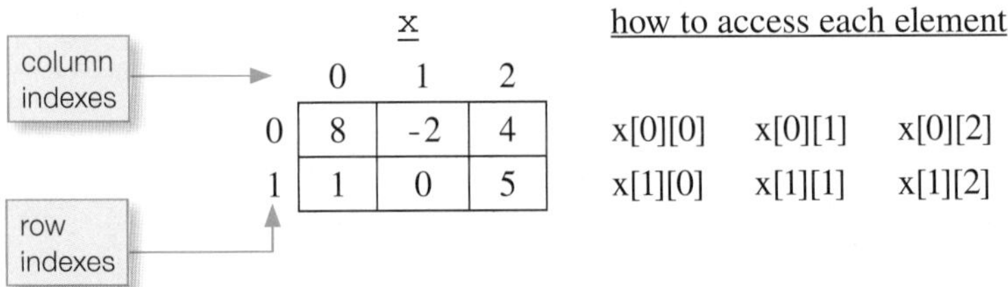

The items at the right, under the "how to access" column heading, show how to access each of the six elements in the array. So to access the value 5, at row index 1 and column index 2, you specify `x[1][2]`.

As with one-dimensional arrays, there are two ways to assign values into a two-dimensional array's elements. You can use an array initializer, where the element assignment is part of the array's declaration. Or you can use standard assignment statements, where the assignment statements are separate from the array's declaration and creation. We'll describe the array initializer technique first. Here's how you can declare the above two-dimensional x array and assign values into its elements, using an array initializer:

initializer for a 2-row by 3-column array

```
int[][] x = {{8, -2, 4}, {1, 0, 5}};
```

Note that the array initializer contains two inner groups, where each inner group represents one row. `{8, -2, 4}` represents the first row. `{1, 0, 5}` represents the second row. Note that elements and groups are separated with commas, and each inner group and the entire set of inner groups are surrounded by braces.

You can use the array initializer technique only if you know the assigned values when you first declare the array. Otherwise, you need to provide array element assignment statements that are separate from the array's declaration and creation. For example, Figure 9.17's code fragment declares and creates the `x` array in one statement, and assigns values to `x`'s elements in a separate statement, inside nested `for` loops.

When working with two-dimensional arrays, it's very common to use nested `for` loops. In Figure 9.17, note the outer `for` loop with index variable `i` and the inner `for` loop with index variable `j`. The outer `for` loop iterates through each row, and the inner `for` loop iterates through each element within a particular row.

```
int[][] x = new int[2][3];
for (int i=0; i<x.length; i++)
{
  for (int j=0; j<x[0].length; j++)
  {
    System.out.print("Enter value for row " + i + ", col " + j + ": ");
    x[i][j] = stdIn.nextInt();
  } // end for j
} // end for i
```

Declare and create a 2-row by 3-column array.

Assign a value to the element at row i, column j.

Figure 9.17 Assigning values to a two-dimensional array using nested `for` loops and the `length` property

Figure 9.17's first line declares `x` to be a 2-row by 3-column array with 6 total elements. So you might expect the first `for` loop's `x.length` property to hold a 6. Not so. Even though it's normal (and useful) to think of `x` as a rectangular box that holds 6 `int` elements, `x` is actually a reference to a 2-element array, and each of the two elements is a reference to its own 3-element array of `int`s. This picture illustrates what we're talking about:

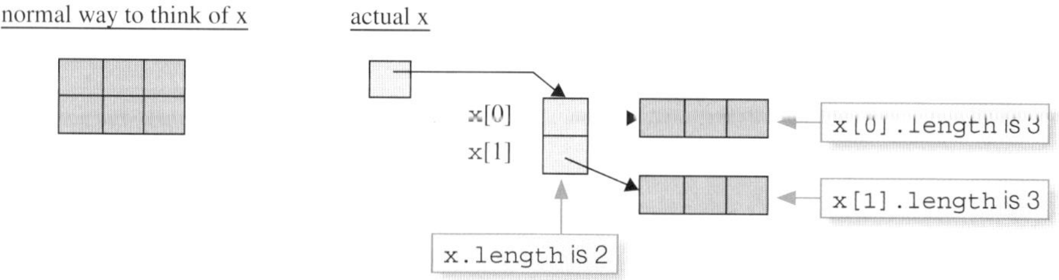

Because `x` is actually a reference to a 2-element array, `x.length` holds the value 2. Or thinking about `x` in the "normal" way (above left picture), `x.length` holds the number of rows in `x`. As you can see above, `x[0]` is a reference to a 3-element array. Thus, `x[0].length` holds the value 3. Or, thinking about `x` in the "normal" way (above left picture), `x[0].length` holds the number of columns in `x`. The point of all this is that the `length` property can be used for iterating through the elements in a two-dimensional array. In Figure 9.17, note how the first loop uses `x.length` to iterate through the rows in `x`, and note how the second loop uses `x[0].length` to iterate through the columns in each row.

Example

Let's put these two-dimensional array concepts into practice by using a two-dimensional array in the context of a complete program. The program, built for a Kansas and Missouri airline company, tells customers when airplanes are expected to arrive at various Kansas and Missouri airports. It uses a two-dimensional array to store flight times between cities, and it displays output like this:

```
       Wch  Top   KC  Col  StL
Wch      0   22   30   42   55
Top     23    0    9   25   37
 KC     31   10    0   11   28
Col     44   27   12    0   12
StL     59   41   30   14    0
```

It takes 25 minutes to fly from Topeka to Columbia.

Different rows correspond to different cities of origin. Different columns correspond to different cities of destination. The labels are abbreviations for city names: "Wch" stands for Wichita, Kansas. "Top" stands for Topeka, Kansas. "KC" stands for Kansas City, Missouri. "Col" stands for Columbia, Missouri. "StL" stands for St. Louis, Missouri. Thus, for example, it takes 25 minutes to fly from Topeka to Columbia. How long does it take to go the other way, from Columbia to Topeka? 27 minutes. Columbia to Topeka takes longer because the trip goes east to west, and airplanes have to contend with headwinds from North America's west-to-east jet stream.

Let's analyze the program by starting with Figure 9.18's `FlightTimesDriver` class. Note how the `main` method declares and creates a `flightTimes` table with a two-dimensional array initializer. And note how the initializer puts each table row on a line by itself. That's not required by the compiler, but it makes for elegant, self-documenting code. It is self-documenting because readers can easily identify each row of table data by looking at a single row of code. After initializing the `flightTimes` table, `main` initializes a one-dimensional array of city names and then calls the `FlightTimes` constructor, the `displayFlightTimesTable` method, and the `promptForFlightTime` method. We'll discuss the constructor and those two methods next.

Figures 9.19a and 9.19b contain the heart of the program—the `FlightTimes` class. In Figure 9.19a, the constructor initializes the `flightTimes` and `cities` instance variable arrays with the data passed to it by the driver's constructor call. Note that it assigns the passed-in `ft` and `c` array references to the instance variables using the = operator. Previously, you learned to use a `for` loop, not the = operator, to make a copy of an array. Why is the = operator acceptable here? Because there's no need to make a second copy of these arrays. After the constructor's first assignment operation, the `flightTimes` instance variable array reference and the `ft` parameter array reference point to the same array object. And that's appropriate. Likewise, after the constructor's second assignment operation, the `cities` instance variable array reference and the `c` parameter array reference point to the same array object.

Figure 9.19a's `promptForFlightTime` method prompts the user for a departure city and a destination city and prints the flight time for that flight. More specifically, it prints a legend of numbers and their associated city names (1 = Wichita, 2 = Topeka, and so on), it prompts the user to enter numbers for the departure and destination cities, and it prints the flight time between the specified cities. Note how user-entered city numbers start with 1 rather than 0 (1 = Wichita). That makes the program more user-friendly because people usually prefer to start counting at one rather than zero. Internally, the program stores city names in an array. Since all arrays start with a 0 index, the program has to translate between user-entered city numbers (which start at 1) and city array indexes (which start at 0). Note how that's done with `+1` and `-1` in the `promptForFlightTime` method.

Figure 9.19b's `displayFlightTimesTable` method displays the flight times table. In doing so, it employs an interesting formatting technique. First, look at the two local named constants, which are separately defined format strings. You have been using literal format strings embedded in strings of text for some time now in the arguments of `printf` method calls. But instead of embedding literal format strings, sometimes it's easier to understand if you declare them separately as named constants. If you go back and count the spaces in the six-column table of flight times, you'll see that each column is exactly 5 spaces wide. So the labels at the top of the columns and the numbers in the columns must both be formatted to use exactly five spaces. Thus, the format string for the labels (`CITY_FMT_STR`) should be `"%5s"`, and the format string for the integer entries (`TIME_FMT_STR`) should be `"%5d"`. Using named constants for format strings allows each format string to be used in many places, and it makes it easy and safe to alter them at any later time—just change the values assigned to the named constants at the beginning of the method.

In the `displayFlightTimesTable` method, note the three `for` loop headers. They all use the `length` property for their termination condition. Since `length` holds 5, the program would run correctly

```
/*********************************************************************
* FlightTimesDriver.java
* Dean & Dean
*
* This manages a table of intercity flight times.
*********************************************************************/

public class FlightTimesDriver
{
  public static void main(String[] args)
  {
    int[][] flightTimes =
    {
      {0, 22, 30, 42, 55},
      {23, 0, 9, 25, 37},
      {31, 10, 0, 11, 28},
      {44, 27, 12, 0, 12},
      {59, 41, 30, 14, 0}
    };
    String[] cities = {"Wch", "Top", "KC", "Col", "StL"};
    FlightTimes ft = new FlightTimes(flightTimes, cities);

    System.out.println("\nFlight times for KansMo Airlines:\n");
    ft.displayFlightTimesTable();
    System.out.println();
    ft.promptForFlightTime();
  } // end main
} // end class FlightTimesDriver
```

Sample session:

```
Flight times for  KansMo Airlines:

        Wch  Top    KC  Col  StL
  Wch     0   22    30   42   55
  Top    23    0     9   25   37
  KC     31   10     0   11   28
  Col    44   27    12    0   12
  StL    59   41    30   14    0

1 = Wch
2 = Top
3 = KC
4 = Col
5 = StL
Enter departure city's number: 5
Enter destination city's number: 1
Flight time = 59 minutes.
```

Figure 9.18 Driver of `FlightTimes` class in Figures 9.19a and 9.19b

```
/*********************************************************************
* FlightTimes.java
* Dean & Dean
*
* This manages a table of intercity flight times.
*********************************************************************/

import java.util.Scanner;

public class FlightTimes
{
  private int[][] flightTimes; // table of flight times
  private String[] cities;     // cities in flightTimes table

  //******************************************************************

  public FlightTimes(int[][] ft, String[] c)
  {
    flightTimes = ft;
    cities = c;
  }

  //******************************************************************

  // Prompt user for cities and print associated flight time.

  public void promptForFlightTime()
  {
    Scanner stdIn = new Scanner(System.in);
    int departure;   // index for departure city
    int destination; // index for destination city

    for (int i=0; i<cities.length; i++)
    {
      System.out.println(i+1 + " = " + cities[i]);
    }
    System.out.print("Enter departure city's number: ");
    departure = stdIn.nextInt() - 1;
    System.out.print("Enter destination city's number: ");
    destination = stdIn.nextInt() - 1;
    System.out.println("Flight time = " +
      flightTimes[departure][destination] + " minutes.");
  } // end promptForFlightTime
```

Print the number-city legend.

Figure 9.19a `FlightTimes` class that displays intercity flight times—part A

```
//**************************************************************

// This method prints a table of all flight times.

public void displayFlightTimesTable()
{
  final String CITY_FMT_STR = "%5s";
  final String TIME_FMT_STR = "%5d";

  System.out.printf(CITY_FMT_STR, ""); // empty top-left corner
  for (int col=0; col<cities.length; col++)
  {
    System.out.printf(CITY_FMT_STR, cities[col]);
  }
  System.out.println();

  for (int row=0; row<flightTimes.length; row++)
  {
    System.out.printf(CITY_FMT_STR, cities[row]);
    for (int col=0; col<flightTimes[0].length; col++)
    {
      System.out.printf(TIME_FMT_STR, flightTimes[row][col]);
    }
    System.out.println();
  } // end for
} // end displayFlightTimesTable
} // end class FlightTimes
```

format strings

Figure 9.19b `FlightTimes` class that displays intercity flight times—part B

if you replaced the length termination conditions with hard-coded 5's. But don't do it. Using the `length` property makes the implementation more scalable. *Scalable* means that it's easy to change the amount of data that the program uses. For example, in the FlightTimes program, using a `cities.length` loop termination condition means that if you change the number of cities in the program, the program will still work properly.

Multidimensional Arrays

Arrays may have more than two dimensions. Arrays with three or more dimensions use the same basic syntax except they have additional `[]`'s. The first pair of brackets corresponds to the largest scale, and each subsequent pair of brackets nests within the previous pair, at progressively smaller levels of scale. For example, suppose the Missouri-Kansas airline company decides to go "green" and expands its fleet with new solar-powered airplanes and airplanes that burn hydrogen generated by wind turbines. The new airplanes have different flight times than the original jet-fuel airplanes. Thus, they need their own flight-times tables. The solution is to create a three-dimensional array where the first dimension specifies the airplane type—0 for the jet-fuel airplanes, 1 for the solar-powered airplanes, and 2 for the hydrogen-powered airplanes. Here's how to declare the new three-dimensional `flightTimes` array instance variable:

```
private int[][][] flightTimes;
```

9.10 Arrays of Objects

You learned in the previous section that a two-dimensional array is actually an array of references where each reference points to an array object. Now let's look at a related scenario. Let's look at an array of references where each reference points to a programmer-defined object. For example, suppose you'd like to store total sales for each sales clerk in a department store. If sales clerk Amanda sells two items for $55.45 and $22.01, then you'd like to store 77.46 as her total-sales value. You can store the sales clerk data in an array, `clerks`, where each element holds a reference to a `SalesClerk` object. Each `SalesClerk` object holds a sales clerk's name and the total sales for that sales clerk. See Figure 9.20 for an illustration of what we're talking about.

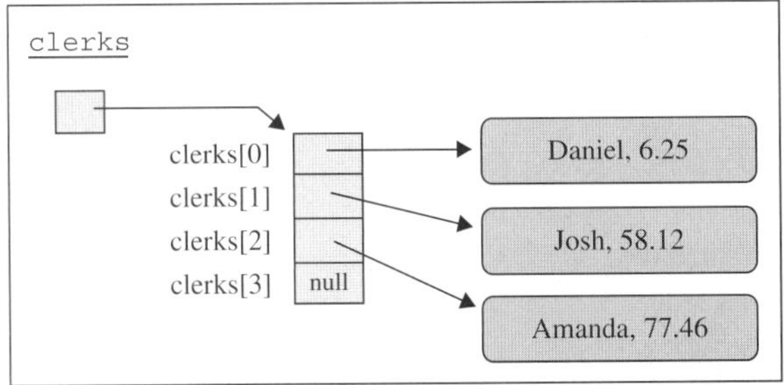

Figure 9.20 Array of objects that stores sales-clerk sales data

The `clerks` array is an array of references. But most folks in industry would refer to it as an array of objects, and that's what we'll do too. An array of objects isn't that much different from an array of primitives. In both cases, you access each array element with square brackets (e.g., `clerks[0]`, `clerks[1]`). But there are some differences you should be aware of, and those differences are the main focus of this section.

Need to Instantiate an Array of Objects <u>and</u> the Objects in That Array

With an array of primitives, you perform one instantiation—you instantiate the array object and that's it. But with an array of objects, you have to instantiate the array object, and you must also instantiate each element object that's stored in the array. It's easy to forget the second step, the instantiation of individual element objects. If you do forget, then the elements contain default values of `null`, as illustrated by `clerks[3]` in Figure 9.20. For the empty part of a partially filled array, `null` is fine, but for the part of an array that's supposed to be filled, you need to overlay `null` with a reference to an object. The following is an example of how to create an array of objects—more specifically, how to create the `clerks` array of objects shown in Figure 9.20. Note the separate instantiations, with the `new` operator, for the `clerks` array and for each `SalesClerk` object.

```
SalesClerk[] clerks = new SalesClerk[4];
clerks[0] = new SalesClerk("Daniel", 6.25);
clerks[1] = new SalesClerk("Josh", 58.12);
clerks[2] = new SalesClerk("Amanda", 77.46);
```

Can't Access Array Data Directly

With an array of primitives, you can access the array's data, the primitives, directly. For example, the following code fragment shows how you can assign and print the first rainfall value in a `rainfall` array. Note how the value is directly accessed with `rainfall[0]`:

```
double[] rainfall = new double[365];
rainfall[0] = .8;
System.out.println(rainfall[0]);
```

On the contrary, with an array of objects, you normally cannot access the array's data directly. The data is stored in variables inside the objects. Because the variables are normally `private`, you normally have to call a constructor or method to access them. For example, the following code fragment shows how you can use a constructor to assign Daniel and 6.25 to the first object in the `clerks` array. It also shows how you can use accessor methods to print the first object's name and sales data:

```
SalesClerk[] clerks = new SalesClerk[4];
clerks[0] = new SalesClerk("Daniel", 6.25);
System.out.println(
  clerks[0].getName() + ", " + clerks[0].getSales());
```

SalesClerks Program

Let's now implement a complete program that adds sales and prints sales for a group of sales clerks in a department store. As described in Section 8.8, we'll first get a big-picture view of things by presenting a UML class diagram. Figure 9.21's class diagram shows two classes. The `SalesClerks` class represents sales data for the entire department store, and the `SalesClerk` class represents total sales for one particular sales clerk.

Start with a UML class diagram to get a big-picture understanding.

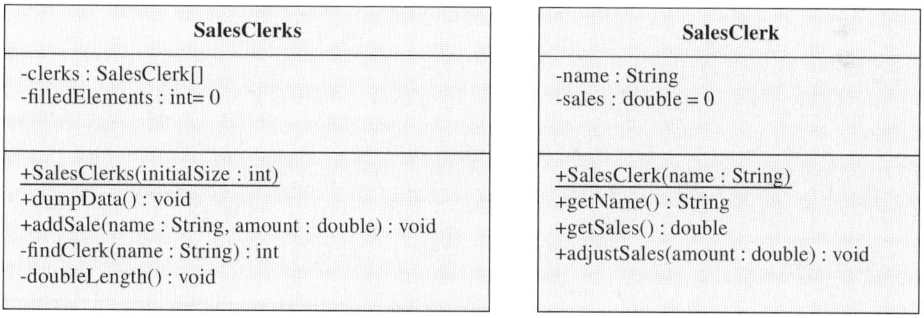

SalesClerks
-clerks : SalesClerk[] -filledElements : int= 0
+SalesClerks(initialSize : int) +dumpData() : void +addSale(name : String, amount : double) : void -findClerk(name : String) : int -doubleLength() : void

SalesClerk
-name : String -sales : double = 0
+SalesClerk(name : String) +getName() : String +getSales() : double +adjustSales(amount : double) : void

Figure 9.21 UML class diagram for the SalesClerks program

The `SalesClerks` class contains two instance variables—`clerks` and `filledElements`. `clerks` is an array of `SalesClerk` objects. `filledElements` stores the number of elements that have been filled so far in the `clerks` array. For a `filledElements` example, see Figure 9.20, where `filledElements` would be 3. The `SalesClerks`'s constructor instantiates the `clerks` array, using the constructor's `initialSize` parameter for the array's size.

The `SalesClerks` class in Figures 9.22a and 9.22b contains four methods—`dumpData`, `addSale`, `findClerk`, and `doubleLength`. The `dumpData` method is the most straightforward of the four. It prints all the data in the `clerks` array. The term *dump* is a computer term that refers to a simple (unformatted) display of a program's data. See the `dumpData` method in Figure 9.22b and verify that it prints the data in the `clerks` array.

```
/*******************************************************************
* SalesClerks.java
* Dean & Dean
*
* This class stores names and sales for sales clerks.
*******************************************************************/

class SalesClerks
{
  private SalesClerk[] clerks;      // contains names and sales
  private int filledElements = 0;   // number of elements filled

  //****************************************************************

  public SalesClerks(int initialSize)
  {
    clerks = new SalesClerk[initialSize];
  } // end SalesClerks constructor

  //****************************************************************

  // Process a sale for the clerk whose name is passed in.
  // Postcondition: If the name is not already in the clerks array,
  // this creates a new object and inserts a reference to it in the
  // next array element, doubling the array length if necessary.

  public void addSale(String name, double amount)
  {
    int clerkIndex = findClerk(name);

    if (clerkIndex == -1)           // add a new clerk
    {
      if (filledElements == clerks.length)
      {
        doubleLength();
      }
      clerkIndex = filledElements;
      clerks[clerkIndex] = new SalesClerk(name);
      filledElements++;
    } // end if

    clerks[clerkIndex].adjustSales(amount);
  } // end addSale
```

Figure 9.22a `SalesClerks` class—part A

```
  //***************************************************************

  // Print all the data - sales clerk names and sales.

  public void dumpData()
  {
    for (int i=0; i<filledElements; i++)
    {
      System.out.printf("%s: %6.2f\n",
        clerks[i].getName(), clerks[i].getSales());
    }
  } // end dumpData

  //***************************************************************

  // Search for the given name. If found, return the index.
  // Otherwise, return -1.

  private int findClerk(String name)
  {
    for (int i=0; i<filledElements; i++)
    {
      if (clerks[i].getName().equals(name))
      {
        return i;
      }
    } // end for
    return -1;
  } // end findClerk

  //***************************************************************

  // Double the length of the array.

  private void doubleLength()
  {
    SalesClerk[] clerks2 = new SalesClerk[2 * clerks.length];
    System.arraycopy(clerks, 0, clerks2, 0, clerks.length);
    clerks = clerks2;
  } // end doubleLength
} // end class SalesClerks
```

Figure 9.22b `SalesClerks` class—part B

The `addSale` method processes a sale for a particular sales clerk. More specifically, the `addSale` method finds the sales clerk specified by its `name` parameter and updates that sales clerk's total sales with the value specified by its `amount` parameter. To find the sales clerk, the `addSale` method calls the `findClerk` helper method. The `findClerk` method performs a sequential search through the `clerks` array, and returns the index of the found sales clerk or −1 if the sales clerk is not found. If the sales clerk is not found, `addSale` adds a new `SalesClerk` object to the `clerks` array in order to store the new sale transaction in it. In adding a new `SalesClerk` object to the `clerks` array, `addSale` checks to make sure that there is available space in the `clerks` array for the new `SalesClerk` object. If the `clerks` array is full (that is, `filledElements` equals `clerks.length`), then `addSale` must do something to provide for more elements. That's where the `doubleLength` helper method comes to the rescue.

```
/*******************************************************************
* SalesClerk.java
* Dean & Dean
*
* This class stores and retrieves a sales clerk's data.
*******************************************************************/

public class SalesClerk
{
  private String name;          // sales clerk's name
  private double sales = 0.0;   // total sales for clerk

  //****************************************************************

  public SalesClerk(String name)
  {
    this.name = name;
  }

  //****************************************************************

  public String getName()
  {
    return name;
  }

  public double getSales()
  {
    return sales;
  }

  //****************************************************************

  // Adjust clerk's total sales by adding the passed-in sale.

  public void adjustSales(double amount)
  {
    sales += amount;
  }
} // end class SalesClerk
```

Figure 9.23 `SalesClerk` class

The `doubleLength` method, as its name suggests, doubles the size of the `clerks` array. To do that, it instantiates a new array, `clerks2`, whose length is twice the length of the original `clerks` arrays. Then it copies all the data from the `clerks` array into the lowest-numbered elements in the `clerks2` array. Finally, it assigns the `clerks2` array to the `clerks` array so the `clerks` array points to the new longer array. See the `addSale`, `findClerk`, and `doubleLength` methods in Figures 9.22a and 9.22b and verify that they do what they're supposed to do.

The `SalesClerk` class, shown on the right side of Figure 9.21, is fairly straightforward. It contains two instance variables, `name` and `sales`, for the sales clerk's name and the sales clerk's total sales. It contains two accessor methods, `getName` and `getSales`. It contains an `adjustSales` method that updates the sales clerk's total sales value by adding the passed-in `amount` to the `sales` instance variable. See the `SalesClerk` class in Figure 9.23 and verify that it does what it's supposed to do.

Now look at the `main` method in the Figure 9.24a's `SalesClerksDriver` class. A declaration instantiates a `SalesClerks` object, passing an initial array-length value of 2 to the `SalesClerks` constructor. Then it repeatedly prompts the user for a sales-clerk name and sales value and calls `addSale` to insert the input

```
/*********************************************************************
* SalesClerksDriver.java
* Dean & Dean
*
* This drives the SalesClerks class.
*********************************************************************/

import java.util.Scanner;

public class SalesClerksDriver
{
  public static void main(String[] args)
  {
    Scanner stdIn = new Scanner(System.in);
    SalesClerks clerks = new SalesClerks(2);
    String name;

    System.out.print("Enter clerk's name (q to quit): ");
    name = stdIn.nextLine();
    while (!name.equals("q"))
    {
      System.out.print("Enter sale amount: ");
      clerks.addSale(name, stdIn.nextDouble());
      stdIn.nextLine();         // flush newline
      System.out.print("Enter clerk's name (q to quit): ");
      name = stdIn.nextLine();
    } // end while
    clerks.dumpData();
  } // end main
} // end SalesClerksDriver
```

Figure 9.24a Driver for the SalesClerks program in Figures 9.22a, 9.22b, and 9.23

data into the SalesClerks object. The looping stops when the user enters a q for the next name. Then main calls dumpData to display the accumulated sales data. Figure 9.24b shows the output.

```
Sample session:
Enter clerk's name (q to quit): Daniel
Enter sale amount: 6.25
Enter clerk's name (q to quit): Josh
Enter sale amount: 58.12
Enter clerk's name (q to quit): Amanda
Enter sale amount: 40
Enter clerk's name (q to quit): Daniel
Enter sale amount: -6.25
Enter clerk's name (q to quit): Josh
Enter sale amount: 12.88
Enter clerk's name (q to quit): q
Daniel:    0.00
Josh:   71.00
Amanda:   40.00
```

Figure 9.24b SalesClerks program output

9.11 For-Each Loops

A *for-each loop* is a modified version of the traditional for loop. It provides a handy way to iterate through all the elements in an array when you don't know or don't care exactly where particular elements are located. Here is the for-each loop syntax:

```
for (<element-type> <element-name> : <array-reference-variable>)
{
  . . .
}
```

Are you wondering why the for-each loop is called a for-each loop, even though there's no "each" in the syntax? It's because most people say "for each" to themselves when reading a for-each loop's header.

Consider, for example, the following code fragment, which prints all the numbers in a given array of prime numbers:

```
int[] primes = {2, 3, 5, 7, 11, 13}

for (int p : primes)
{
  System.out.println(p);
}
```

When we look at this for-each loop, we say to ourselves, "For each p in primes, print p." That verbal expression explains the "for-each" terminology.

Note how the for-each loop header conforms to the previous syntax. The primes reference variable refers to the whole array, and the p variable holds a typical element in that array. It's legal to choose any name for the element. Normally, good Java style dictates use of longer descriptive names for variables, and that rule says we should have used a descriptive word like `prime` for the variable that refers to a typical element in this example. But because the for-each loop header is so self-explanatory, it's accepted practice to abbreviate the for-each element name in a for-each loop, just as it's accepted practice to abbreviate the index in an ordinary `for` loop. Because it's accepted, common practice, we show it here. But go ahead and use a longer word whenever you feel the urge. Descriptive words do improve understanding, and when for-each loops are large, we prefer longer words instead of abbreviations.

As another example, consider this alternative implementation of the `dumpData` method in the `SalesClerks` class in Figure 9.22b. It uses a for-each loop to step through the elements in the `clerks` array:

```
public void dumpData()
{
  for (SalesClerk sc : clerks)
  {
    if (sc != null)
    {
      System.out.printf("%s: %6.2f\n",
        sc.getName(), sc.getSales());
    }
  }
} // end dumpData
```

When you look at the header in this particular for-each loop, say to yourself, "for each `SalesClerk` in `clerks`, . . .".

A for-each loop avoids index initialization, testing, and incrementing. It also simplifies element method calls. In this example, we had to put the `printf` statement inside an `if` statement to avoid an error when the iteration got beyond the `filledElements` limit. At first, it looks like this added complexity offsets the simplifications. But what if one of the existing clerks quits, and we remove that clerk's name from the array? An ordinary `for` loop would generate an error when the iteration hit that `null` cell in the array. So, actually, the `null` test we added to make `dumpData` work with a for-each loop should also be present in an ordinary `for` loop. The next chapter includes an array alternative (an `ArrayList`) that automatically expands and shrinks with element additions and deletions. This makes it possible to use a for-each loop without an explicit `null`-element test.

The for-each loop is great, but you should be aware of several issues when using it: (1) It was introduced in Java 5.0, so it won't work with older compilers. (2) The for-each loop doesn't use an index variable to loop through its elements. That can be a benefit in that it leads to less cluttered code. But it's a drawback if there's a need for an index within the loop. For example, suppose you have an initialized primes array, and you want to display it like this:

```
primes[0] = 2
primes[1] = 3
. . .
primes[5] = 13
```

The numbers inside the square brackets are index values. So if you implemented a solution with a for-each loop, you'd have to add an index variable to your code and increment it each time through the loop. On the other hand, if you implemented a solution with a traditional `for` loop, you'd already have an incrementing index variable built in.

Summary

- Arrays facilitate the representation and manipulation of collections of similar data. You access array elements with *<array-name>*[index], where index is a nonnegative integer, starting at zero.
- You can create and completely initialize an array in one statement, like this:
  <element-type>[] *<array-name>* = {*element0*, *element1*, ...};
- Usually, however, it's more useful to defer element initialization and use new to create an array of uninitialized elements, like this:
  <element-type>[] *<array-name>* = new *<element-type>*[*<array-size>*];
- You can read or write directly to an array element by inserting an appropriate index value in square brackets after the array name at any time after the array has been created.
- Every array automatically includes a public property called length, which you can access directly with the array name. The highest index value is *<array-name>*.length - 1.
- To copy an array, copy each of its elements individually, or use the System.arraycopy method to copy any subset of elements in one array to any location in another array.
- A histogram is an array of elements in which each element's value is the number of occurrences of some event.
- A sequential search is a good way to search for a match in an array whose length is less than about 20, but for long arrays, you should first sort the array with the Arrays.sort method and then use a binary search.
- A two-dimensional array is an array of arrays, declared with two sets of square brackets after the element-type identification. You can instantiate it with an initializer or with new followed by element type and two array-size specifications in square brackets.
- In creating an array of objects, multiple instantiations are required. After instantiating the array, you also need to instantiate the individual element objects within the array.
- Use a for-each loop when you don't know or don't care exactly where the element is.

Review Questions

§9.2 Array Basics

1. It's legal to store ints and also doubles in a single standard array. (T / F)
2. Given an array that's named myArray, you access the first element in the array using myArray[0]. (T / F)

§9.3 Array Declaration and Creation

3. Provide a declaration for an array of strings called names.
4. Consider the heading for any main method:

```
public static void main(String[] args)
```

   What kind of a thing is args?

5. Suppose you create an array with the following statement:

```
int[] choices = new int[4];
```

   What is the default value in a typical element of this array? Is it garbage or something in particular?

§9.4 Array length Property and Partially Filled Arrays

6. The value of an array's length equals the value of the array's largest acceptable index. (T / F)

§9.5 Copying an Array

7. Given

```
String letters = "abcdefghijklmnopqrstuvwxyz";
char[] alphabet = new char[26];
```

Write a `for` loop that initializes `alphabet` with the characters in `letters`.

8. Write a single statement that copies all the elements in

```
char[] arr1 = {'x', 'y', 'z'};
```

to the last three elements of

```
char[] array2 = new char[26];
```

§9.6 Problem Solving with Array Case Studies

9. In Figure 9.7's MovingAverage program, suppose you want to shift in the other direction. How would you write the inner `for` loop header, and how would you write the array assignment statement in the inner `for` loop?
10. What kind of value does a typical histogram "bin" contain?

§9.7 Searching an Array

11. It's possible to search array `ids` for an element equal to `id` with nothing more than this:

```
int i;
for (i=0; i<ids.length && id != ids[i]; i++)
{ }
if (<boolean-expression>)
{
  return i;
}
```

What is the *<boolean-expression>* that indicates that `i` has been found?

§9.8 Sorting an Array

12. We elected to use class methods to implement our sort algorithm. What is an advantage of that?
13. Java's API `sort` method is in what class?

§9.9 Two-Dimensional Arrays

14. We have said that a two-dimensional array is an array of arrays. Consider the following declaration:

```
double[][] myArray = new double[5][8];
```

In the context of the expression, array of arrays, what does `myArray[3]`mean?

§9.10 Arrays of Objects

15. In creating an array of objects, you have to instantiate the array object, and you must also instantiate each element object that's stored in the array. (T / F)

§9.11 For-Each Loop

16. You must use a for-each loop, and not a traditional `for` loop, whenever you need to iterate through a collection of elements. (T / F)

Exercises

1. [after §9.2] The index number of the last element in an array of length 100 is ________.
2. [after §9.3] Declare an array named `scores` that holds `double` values.
3. [after §9.3] Provide a single initialization statement that initializes `myList` to all 1's. `myList` is a 5-element array of `int`'s.
4. [after §9.4] Zoo Animals Program:

   As part of your internship at Parkville's new zoo, you've been asked to write a program that keeps track of the zoo animals. You want to make the program general-purpose so that when you're done, you can sell your program to zoos worldwide and make millions. Thus, you decide to create a generic `Zoo` class.

   Write a `Zoo` class. Your class does not have to do very much—it simply handles the creation and printing of `Zoo` objects. To give you a better idea of the `Zoo` class's functionality, we provide a `main` method:

```
public static void main(String[] args)
{
  Zoo zoo1 = new Zoo();
  String[] animals = {"pig", "possum", "squirrel", "Chihuahua"};
  Zoo zoo2 = new Zoo(animals, "Parkville");
  zoo1.display();
  zoo2.display();
}
```

   When run, the `main` method should print this:

```
The zoo is vacant.
Parkville zoo: pig, possum, squirrel, Chihuahua
```

   Although it's not required, you're encouraged to write a complete program in order to test your `Zoo` class.

5. [after §9.5] Assume that this code fragment compiles and runs. What is its output? Be precise when showing your output.

```
char[] a = new char[3];
char[] b;
for (int i=0; i<a.length; i++)
{
  a[i] = 'a';
}
b = a;
b[2] = 'b';
System.out.println("a[1]=" + a[1] + ", a[2]=" + a[2]);
System.out.println("b[1]=" + b[1] + ", b[2]=" + b[2]);
```

6. [after §9.5] What needs to be added to the following code fragment so that all values except the first two (100000.0 and 110000.0) are copied from `allSalaries` to `workerSalaries`?

```
double[] allSalaries = {100000.0, 110000.0, 25000.0, 18000.0,
  30000.0, 9000.0, 12000.0};
double[] workerSalaries;
```

7. [after §9.5] The following program is supposed to reverse the order of the elements in the `simpsons` array. It compiles and runs, but it doesn't work properly.

```
public class Reverse
{
  public static void main(String[] args)
  {
    String[] simpsons = {"Homer", "Flanders", "Apu"};
    reverse(simpsons);
    System.out.println(
      simpsons[0] + " " + simpsons[1] + " " + simpsons[2]);
  } // end main
  public static void reverse(String[] list)
  {
    String[] temp = new String[list.length];
    for (int i=0; i<list.length; i++)
    {
      temp[i] = list[list.length-i-1];
    }
    list = temp;
  } // end reverse
} // end class Reverse
```

a) What does the program print?

b) Fix the program by providing one or more lines of alternative code for the `list = temp;` line. You are not allowed to change any other code—just provide alternative code for that one line.

8. [after §9.6] Write a program that implements the example described at the beginning of Section 9.6. Your program should shift the array's elements from position x to position x − 1, as described in that section. (Move the value at position 1 to position 0; move the value at position 2 to position 1, and so on).

Start by creating two arrays, `double[] initialHours` and `double[] hours`. In its declaration initialize `initialHours` with the values `{8, 8, 6, 4, 7, 0, 0, 5}`, but don't initialize `hours` when you declare and instantiate it with its 31 elements. Instead, initialize `hours` after its creation by using `System.arraycopy` to copy all the values in `initialHours` into the first elements in hours. Then perform one down-shift operation, and load zero into the (new) highest element.

9. [after §9.7] Write a class method named `allPositive` that receives an array named `arr` of `double` values and returns `true` if all the element values are positive and returns `false` otherwise. Use appropriate access modifiers. Make the method accessible from outside of its class.

10. [after §9.7] Assume that you have already successfully written a class named `Students` that handles student records for the Registrar's office. Assume that the `Students` class:
- Contains a `studentIds` instance variable—an array of `int`s that contains student ID numbers.
- Contains a 1-parameter constructor that initializes the `studentIds` instance variable.
- Contains this `main` method:

```
public static void main(String[] args)
{
  Students s1 = new Students(new int[] {123, 456, 789});
  Students s2 = new Students(new int[] {123, 456, 789, 555});
  Students s3 = new Students(new int[] {123, 456, 789});
  if (s1.equals(s2))
  {
```

```
      System.out.println("s1 == s2");
    }
    if (s1.equals(s3))
    {
      System.out.println("s1 == s3");
    }
  } // end main
```

Write a `public` method named `equals` for your `Students` class that tests whether two `Students` objects are equal. Your `equals` method should be written such that the above `main` method would produce this output:

```
s1 == s3
```

Only provide code for the asked-for `equals` method; do not provide code for the entire `Students` class.

11. [after §9.8] Given the following list array, use the selection sort algorithm to sort the array. Show each step of the selection sort process. Do not provide code; just show pictures of the `list` array after each element swap.

	`list` (original)
0	12
1	2
2	−4
3	0
4	9

	`list` (sorted)
0	−4
1	0
2	2
3	9
4	12

12. [after §9.8] The Insertion Sort algorithm provides an alternative to the selection sort algorithm for sorting small numbers of items (about 20 or less). It's more efficient than selection sort if the array is only slightly out of order. The following code implements the Insertion Sort algorithm:

```
public static void insertionSort(int[] list)
{
  int temp;
  int j;

  for (int i=1; i<list.length; i++)
  {
    temp = list[i];
    for (j=i; j>0 && temp<list[j-1]; j--)
    {
      list[j] = list[j-1];
    }
    list[j] = temp;
  } // end for
} // end insertionSort
```

Note that the scope of the `j` count variable extends beyond the scope of the `for` loop in which it's used. Assume that an array of `int` has been instantiated and the `insertionSort` method has been called with a reference to this array passed in as a parameter. Trace the execution of this method, using the following header and initial entries:

Sort					<arrays>				
	insertionSort				arr1				
line#	**(list)**	**i**	**j**	**temp**	**length**	**0**	**1**	**2**	**3**
					4	3333	1234	2222	1000
10	arr1								

13. [after §9.8] Trace the following code and show the exact output.

```
public class ModifyArray
{
  public static void main(String[] args)
  {
    int sum = 0;
    int[] list = new int[3];

    for (int i=0; i<3; i++)
    {
      list[i] = i + 100;
    }
    modify(list, sum);
    for (int i=0; i<3; i++)
    {
      System.out.print(list[i] + " ");
    }
    System.out.println("\nsum = " + sum);
  }

  public static void modify(int[] list, int sum)
  {
    int temp = list[0];

    list[0] = list[list.length - 1];
    list[list.length - 1] = temp;
    for (int i=0; i<3; i++)
    {
      sum += list[i];
    }
  }
} // end ModifyArray
```

Use the following trace header:

ModifyArray								<arrays>				
	main			modify				arr1				
line#	**i**	**sum**	**list**	**(list)**	**(sum)**	**temp**	**i**	**length**	**0**	**1**	**2**	**output**

14. [after §9.9] Specify a <u>single statement</u> that initializes an array of `int`'s named `myTable` to all 1's. The array should be a two-dimensional array with 2 rows and 3 columns.

15. [after §9.9] As indicated in the text, a two-dimensional array is actually a one-dimensional array of references to other (subordinate) one-dimensional arrays. In the text examples, all the subordinate arrays had the same length. But these subordinate arrays may have different lengths. The following code fragment creates a two-dimensional array whose five rows have lengths ranging from a minimum length of three elements to a maximum length of seven elements:

```
public static void main(String[] args)
{
  int rows = 5;
  int minCols = 3;
  int[][] array2D = new int[rows][0];

  for (int j=0; j<array2D.length; j++)
  {
    array2D[j] = new int[j+minCols];
    System.out.printf("in row %d, number of columns = %d\n",
      j, array2D[j].length);
  }
} // end main
```

Output:

```
in row 0, number of columns = 3
in row 1, number of columns = 4
in row 2, number of columns = 5
in row 3, number of columns = 6
in row 4, number of columns = 7
```

This pattern in which the number of columns increases linearly with the number of rows is typical of many computer-science algorithms. In this particular case, the total number of elements equals $3 + 4 + 5 + 6 + 7 = 25$. Suppose that you wanted a general formula for the total number of elements as a function of arbitrary values for `rows` and `minCols`. Here are two tricks worth remembering:

a) Factor out the minimum number of columns from the rest of the numbers in the series to obtain a series that starts at zero:

```
3 + 4 + 5 + 6 + 7 = (5 * 3) + (0 + 1 + 2 + 3 + 4)
```

b) Add the reversed sequence to the sequence that starts at zero to obtain a sequence of identical values:

```
 0     1     2     3     4
+4    +3    +2    +1    +0
 4     4     4     4     4 = 5 * 4
```

Use these two tricks and the fact that `5 = rows` and `3 = minCols` to write a general formula for the total number of array elements as a function of `rows` and `minCols`.

16. [after §9.9] Write a method named `getMask` that receives a single parameter named `table`, which is a two-dimensional array of `int`'s. The `getMask` method should create and return an array *mask* for the passed-in table array. (The programming term *mask* refers to an array that is built from another array and it contains all 0's and 1's.) For each element in the mask array, if the original array's corresponding element contains a positive number, the mask array's element should contain a 1. And if the original array's corresponding element contains a zero or negative number, the mask array's element should contain a 0. Note this example:

`table` parameter

5	−2	3	1
0	14	0	6
3	6	−1	4

returned array

1	0	1	1
0	1	0	1
1	1	0	1

Note:

- Your method should not change the content of the passed-in table array.
- Your method should work with any sized table, not just the 3-row, 4-column table shown in the example.
- Use appropriate access modifiers. Assume that the method should be accessible from outside its class. In deciding whether the method should be a class method or an instance method, note that the method does not access any instance variables (it only accesses a parameter).

17. [after §9.10] Assume you have the following `City` class:

```
public class City
{
  private String name;
  private double north;           // north latitude in degrees
  private double west;            // west longitude in degrees

  //**********************************************************

  public City(String name, double latitude, double longitude)
  {
    this.name = name;
    this.north = latitude;
    this.west = longitude;
  } // end constructor

  //**********************************************************

  public void display()
  {
    System.out.printf("%12s%6.1f%6.1f\n", name, north, west);
  }
} // end class City
```

Write a code fragment that (1) creates a 4-element array of `City` objects, (2) fills the array with the following data, and (3) uses a loop to display the array's contents like this:

```
New York  41.0  74.0
   Miami  26.0  80.0
 Chicago  42.0  88.0
 Houston  30.0  96.0
```

18. [after §9.11] Suppose you have an array of street addresses that's been initialized and filled as follows:

```
String[] addressList =
{
  "1600 Pennsylvania Avenue",
  "221B Baker Street",
  "8700 N.W. River Park Drive"
};
```

Provide a for-each loop (not a standard `for` loop) that prints the `addressList`'s addresses, one address per line.

Review Question Solutions

1. False. The types of the data elements in a particular array must be the same.
2. True.
3. Declaration for an array of strings called names:

```
String[] names;
```

4. The args parameter in main is an array of strings.
5. The elements of an array are like the instance variables in an object. Array-element default values are not garbage. The default value of an int [] element is 0.
6. False. The largest acceptable index value is one less than the array's length.
7. This code fragment initializes the character array, alphabet:

```
for (int i=0; i<26; i++)
{
  alphabet[i] = letters.charAt(i);
}
```

8. You can copy:

```
arr1[] = {'x', 'y', 'z'}
```

to the end of:

```
arr2[] = new char[26]
```

with the following statement:

```
System.arraycopy(arr1, 0, arr2, 23, 3);
```

9. In the MovingAverage program, to shift in the other direction, the inner for loop header is:

```
for (int d=days.length-1; d>0; d--)
```

The array element assignment statement in this loop is:

```
days[d] = days[d-1];
```

10. A histogram "bin" contains the number of occurrences of an event.
11. The Boolean expression that indicates that i has been found is:

```
(ids.length != 0 && i != ids.length)
```

12. The advantage of using class methods is that the sort method can be used with any passed-in array, not just on a specific instance variable array.
13. Java's API sort method is in the Arrays class.
14. myArray[3] refers to the fourth row, which happens to be an array of eight double values.
15. True.
16. False. You can use a traditional for loop (or a for-each loop) to iterate through a collection of elements.

CHAPTER 10

ArrayLists and an Introduction to the Java Collections Framework

Objectives

- See how the ArrayList class makes arrays more flexible.
- Understand autoboxing.
- Pass anonymous objects to and from methods.
- Appreciate the ease of ArrayList element insertion and deletion.
- Understand the LinkedList structure.
- Compare the performance of alternative software implementations.
- Learn how the Java API groups method interfaces to organize behavioral descriptions.
- Use Java API implementations of lists, queues, stacks, sets, and maps.
- Model a network and simulate flow through that network.

Outline

10.1 Introduction

An object stores a collection of related data, where the data can be of different types. An array stores a collection of related data, where the data must be of the same type. In this chapter, you'll learn about other ways to store collections of related data of the same type.

We start by describing the Java API `ArrayList` class. Like an array, an `ArrayList` stores an ordered list of related data of the same type. But unlike an array, an `ArrayList` grows and shrinks dynamically as you add and remove elements to and from it. That can be a tremendous benefit when you don't know the number of elements ahead of time. `ArrayLists` are able to store objects (actually, references to objects), but they are not able to store primitive values (e.g., `ints` and `doubles`) directly. In this chapter, we discuss how to overcome this limitation by wrapping primitive values in their respective wrapper classes before adding them to an `ArrayList`.

Discussion of `ArrayLists` takes about a third of this chapter. Later chapters use `ArrayLists` extensively, so make sure that you have a good grasp of them before moving on. The `ArrayList` class is part of the Java collections framework, which is a Java API library of classes that store groups of related data. The remainder of this chapter introduces other classes that are in the Java collections framework, as well as additional related topics. Later chapters do not rely on that material, so you'll be OK if you decide to skip over it. But if you skip over it now, you should come back to it later when you have more time. It's useful stuff!

After `ArrayLists`, the chapter introduces another data structure—a *linked list.* Like arrays and `ArrayLists`, linked lists store ordered lists of related data of the same type, but a linked list uses a chain of references to connect its elements. In some programming languages, the strategy of connecting a list's elements with a chain of references can lead to more efficient insertions and deletions in the middle of the list. To see if that holds true with Java, this chapter compares the two Java API classes, `LinkedList` and `ArrayList`, in terms of how fast they perform various operations like accessing, inserting, and removing list elements. You'll see that `ArrayList` is usually significantly faster than `LinkedList`.

Next, the chapter describes the `List` *interface*. We'll have more to say about interfaces in Chapter 14, but for now, just think of an interface as a template for designing classes that share certain qualities. The `ArrayList` and `LinkedList` classes both implement lists, and, as such, they share certain qualities common to all lists. Thus, to help with consistency, they are built using the `List` interface as a template.

`ArrayLists`, linked lists, and the `List` interface are rather general-purpose in nature, while two other data structures, *queues* and *stacks,* have narrower focus. Like the other data structures mentioned so far, queues and stacks store ordered lists of related data of the same type. But unlike the other data structures, queues and stacks are restricted in terms of which elements are manipulated. With a queue, you add elements to the back end and you remove elements from the front end. With a stack, you add elements and remove elements only from one end, which is called the *top*. After describing queue and stack concepts, the chapter presents programs that implement queues and stacks using the Java API `ArrayDeque` class.

Next, the chapter describes quite a few (but far from all) of the classes and interfaces in the Java collections framework and uses a picture to illustrate how they are connected. The last section in this chapter presents a complete program that illustrates how to use some of the more important Java collection framework classes and interfaces. We hope you'll find this program's subject matter—networks—to be engaging, because networks are ubiquitous.

10.2 The ArrayList Class

As you learned in the previous chapter, arrays allow you to work with an ordered list of related data of the same type. Arrays work great for many lists, but if you have a list where the number of elements is hard to predict, they don't work so well. If you don't know the number of elements, you have to either (1) start with an array size that's large enough to accommodate the possibility of a very large number of elements or (2) create a new larger array whenever the array becomes full and you need more room for more elements. The first solution is wasteful of computer memory as it requires allocating space for a large array where most of the elements are unused. The second solution is what we did in the `doubleLength` method in the `SalesClerks` class in Figure 9.22b. It works OK in terms of saving memory, but it requires the programmer to do extra work (writing the code that creates a larger array).

To help with lists where the number of elements is hard to predict, Java language designers came up with the `ArrayList` class. The `ArrayList` class is built using an array, but the array is hidden in the background, so you can't access it directly. With an array in the background, the `ArrayList` class is able to provide the basic functionality that comes with a standard array. With its methods, the `ArrayList` class is able to provide additional functionality that helps when you don't know the number of elements. In this section, we discuss how to create an `ArrayList` and how to use its methods.

How to Create an ArrayList

The `ArrayList` class is defined in the Java API's `java.util` package, so to use the class, you should provide an `import` statement, like this:

```
import java.util.ArrayList;
```

To initialize an `ArrayList` reference variable, use this syntax:

```
ArrayList<element-type> reference-variable = new ArrayList<>();
```

Note the angled brackets around *element-type*. The angled brackets are part of the required syntax. As indicated by the italics, *element-type* and *reference-variable* are descriptions. Normally, we use angled brackets around such descriptions, but we'll refrain from doing so when describing `ArrayList` syntax because description angled brackets might get confused with the `ArrayList`'s required angled brackets. You should replace *element-type* with the type for the `ArrayList`'s elements. You should replace *reference-variable* with an actual reference variable. For example, suppose you've defined a `Student` class, and you want an `ArrayList` of `Student` objects. Here's how to create such an `ArrayList`, named `students`:

Angled brackets are required.

```
ArrayList<Student> students = new ArrayList<>();
```

In the code above, the angled brackets on the right side form the *diamond operator*. The diamond operator is named as such because the two empty angled brackets look like a diamond turned on its side (<>). The diamond operator asks the compiler to perform *type inferencing,* which means the compiler determines the type of the instantiated `ArrayList`'s elements by looking at prior code for clues. In the example above, the "clue" is found in the `ArrayList` declaration at the left, which shows that each element in the `ArrayList` needs to be of type `Student`. You might be asking yourself, "How good is the diamond operator at finding and applying such clues?" The diamond operator's abilities are actually fairly limited. The diamond operator works only for constructor calls, and the type inferencing must be obvious from the program's context.

For clarity, Oracle recommends that you use the diamond operator only if the constructor call is part of a variable declaration initialization statement.

Besides the angled brackets, there are two additional noteworthy items in the above example. First, there is no size specification. That's because ArrayList objects start out with no elements and they automatically expand to accommodate however many elements are added to them. Second, the element type, Student, is a class name. For ArrayLists, you must specify a class name, not a primitive type, for the element type. Specifying a class name means that ArrayLists can hold only references to objects. They cannot hold primitives, like int or double. That's technically true, but there's an easy way to mimic storing primitives in an ArrayList. We'll discuss how to do that in the next section.

As explained above, you can use the diamond operator when instantiating an ArrayList as part of an initialization. On the other hand, if you instantiate a new ArrayList later and assign it into a previously declared ArrayList reference variable, you should instantiate your ArrayList with angled brackets around the ArrayList's element type, like this:

```
ArrayList<String> students;
. . .
students = new ArrayList<String>();
```

In the assignment above, if you use a diamond operator (i.e., use <> instead of <String>), the compiler might generate an error (due to type inferencing failing to work).

Adding Elements to an ArrayList

To convert an instantiated empty ArrayList into something useful, you need to add elements to it. To add an element to the end of an ArrayList, use this syntax:

ArrayList-reference-variable.add(*item*);

The *item* that's added must be the same type as the element type specified in the ArrayList's declaration. Perhaps the simplest type of element object is a string, so let's start with an ArrayList of strings. Suppose you want to write a code fragment that creates this ArrayList object:

	colors
0	"red"
1	"green"
2	"blue"

Try writing the code on your own before proceeding. When you're done, compare your answer to this:

```
import java.util.ArrayList;
. . .
ArrayList<String> colors = new ArrayList<>();
colors.add("red");
colors.add("green");
colors.add("blue");
```

The order in which you add elements determines the elements' positions. Because we added "red" first, it's at index position 0. Because we added "green" next, it's at index position 1. Likewise, "blue" is at index position 2.

API Headings

In describing the ArrayList class, we'll use *API headings* to present the ArrayList class's methods. As you may recall from Chapter 5, API stands for application programming interface, and API headings are the source code headings for the methods and constructors in Java's library of prebuilt Java classes. The API headings tell you how to use the methods and constructors by showing you their parameters and return types. For example, here's the API heading for the Math class's pow method:

```
public static double pow(double num, double power)
```

The above line tells you everything you need to know to use the pow method. To call the pow method, pass in two double arguments: one argument for the base and one argument for the power. The static modifier tells you to preface the call with the class name and then a dot. The double return value tells you to embed the method call in a place that can use a double value. Here's an example that calculates the volume of a sphere:

```
double volume = (4.0 / 3) * Math.PI * Math.pow(radius, 3)
```

How to Access Elements in an ArrayList

With standard arrays, you use square brackets to read and update an element. But with an ArrayList you don't use square brackets. Instead, you use a get method to read an element's value and a set method to update an element's value.

Here's the API heading for the ArrayList's get method:

```
public E get(int index)
```

The index parameter specifies the position of the desired element within the ArrayList calling object. For example, the following method call retrieves the second element in a colors ArrayList:

```
colors.get(1);
```

If the index parameter refers to a nonexistent element, then a runtime error occurs. For example, if colors contains three elements, then this generates a runtime error:

```
colors.get(3);
```

In the get method's API heading, note the E return type:

```
public E get(int index)
```

The E stands for "element." It represents the data type of the ArrayList's elements, whatever that data type happens to be. So if an ArrayList is declared to have String elements, then the get method returns a String value, and if an ArrayList is declared to have Student elements, then the get method returns a Student value. The E in the get method's heading is a generic name for an element type. Using a generic name for a type is an important concept that will come up again with other methods. It's important enough to justify a pedagogical analogy.

Using a generic return type is like saying you're going to the grocery store to get "food." It's better to use a generic term like food rather than a specific term like broccoli. Why? Because you might end up getting Princess Fruit Chews at the store instead of broccoli. By specifying generic food as your "return type," you're free to get Princess Fruit Chews rather than broccoli, as your preschooler sees fit.

Using a generic name for a type is possible with ArrayLists because the ArrayList class is defined to be a *generic class*, by using <E> in its class heading:

```
public class ArrayList<E>
```

You don't need to understand generic class details in order to use ArrayLists, but if you want such details, visit http://download.oracle.com/javase/tutorial/java/generics/index.html.

How to Update an ArrayList Element

Now for the get method's partner, the set method. The set method allows you to assign a value to an ArrayList element. Here is the API heading for ArrayList's set method:

```
public E set(int index, E elem)
```

In the set method's API heading, the index parameter specifies the position of the element you're interested in. If index refers to a nonexistent element, then a runtime error occurs. If index is valid, then set assigns the elem parameter to the specified element. Note that elem is declared with E for its type. As with the set method, the E represents the data type of the ArrayList's elements. So elem is the same type as the type of ArrayList's elements. This example illustrates what we're talking about:

```
String mixedColor;
ArrayList<String> colors = new ArrayList<>();

colors.add("red");
colors.add("green");
colors.add("blue");
mixedColor = colors.get(0) + colors.get(1);
colors.set(2, mixedColor);
```

Note that mixedColor is declared to be a string and colors is declared to be an ArrayList of strings. So in the last statement, when we use mixedColor as the second argument in the set method call, the argument is indeed the same type as the type of color's elements.

Can you determine what the colors ArrayList looks like after the code fragment executes? Draw a picture of the colors ArrayList on your own before proceeding. When you're done, compare your answer to this:

	colors
0	"red"
1	"green"
2	"redgreen"

In the set method's API heading, note the return type, E. Most mutator/set methods simply assign a value and that's it. In addition to assigning a value, the ArrayList's set method also returns a value—the value of the specified element prior to the element being updated. Usually, there's no need to do anything with the original value, so you just call set and the returned value dies. That's what happens in the above code fragment. But if you want to do something with the original value, it's easy to get it because set returns it.

Additional ArrayList Methods

We've now explained the most important methods for the ArrayList class. There are quite a few more methods, and Figure 10.1 provides API headings and brief descriptions of the ones we have just described, as well as several others. As you read through the figure, we hope that you'll find most of the methods to be straightforward. But some items may need clarification. In searching an ArrayList for the first occurrence of a passed-in elem parameter, the indexOf method declares elem's type to be Object. The Object type means the parameter may be any kind of object. Naturally, if the parameter's actual type is different from the type of elements in the ArrayList, then indexOf's search comes up empty and it returns -1 to indicate that elem was not found. By the way, we'll have lots more to say about the Object

`public void add(E elem)`
Appends the specified object to the end of this list.

`public void add(int index, E elem)`
Starting with the specified index position, the add method shifts the original elements at and above the index position to next-higher-indexed positions. It then inserts the elem parameter at the specified index position.

`public boolean contains(Object elem)`
Returns true if the list contains the specified elem parameter.

`public int indexOf(Object elem)`
Searches for the first occurrence of the elem parameter within the list and returns the index position of the found element. If the element is not found, the indexOf method returns -1.

`public E get(int index)`
Returns the object at the specified index.

`public boolean isEmpty()`
Returns true if the list contains no elements.

`public int lastIndexOf(Object elem)`
Searches for the last occurrence of the elem parameter within the list and returns the index position of the found element. If the element is not found, the indexOf method returns -1.

`public E remove(int index)`
Removes and returns the element at the specified index position. To handle the removed element's absence, the remove method shifts all higher-indexed elements by one position to lower-indexed positions.

`public E set(int index, E elem)`
Replaces the element at the specified index with the specified element and returns the replaced element.

`public int size()`
Returns the number of elements currently in the list.

Figure 10.1 API headings and descriptions for several ArrayList methods

type (it's actually an Object class) in Chapter 14. Previously, we covered a one-parameter add method that adds an element at the end of the ArrayList. Figure 10.1's overloaded two-parameter add method adds an element at a specified position within the ArrayList.

Hunger Games Example

To reinforce what you've learned so far, let's take a look at how an ArrayList class is used in a complete working program. See the HungerGames program in Figure 10.2. It creates a list of tributes[1] by instantiating an ArrayList object and calling add to append tributes to the list. It then randomly chooses one of the

```
/*******************************************************************
* HungerGames.java
* Dean & Dean
*
* This class creates an ArrayList of tributes.
* It randomly chooses one tribute and removes him/her.
*******************************************************************/

import java.util.ArrayList;

public class HungerGames
{
  public static void main(String[] args)
  {
    int deceasedIndex; // index of deceased tribute
    String deceased;   // name of deceased tribute
    ArrayList<String> tributes = new ArrayList<>();

    tributes.add("Cato");
    tributes.add("Katniss");
    tributes.add("Peeta");
    tributes.add("Rue");
    tributes.add(1, "Finnick");
    deceasedIndex = (int) (Math.random() * tributes.size());
    deceased = tributes.remove(deceasedIndex);
    System.out.println(deceased + " is no longer in the game.");
    System.out.println("Remaining: " + tributes);
  } // end main
} // end HungerGames
```

Sample output:

```
Peeta is no longer in the game.
Remaining: [Cato, Finnick, Katniss, Rue]
```

Figure 10.2 HungerGames program

[1] *The Hunger Games* (Scholastic Press, 2008), written by Suzanne Collins, describes a dystopian world in which each of 12 districts must provide two "tributes" (a girl and a boy, ages 12–18) to the central government. The tributes are placed in a wilderness arena where they are forced to compete in a game of survival.

tributes and removes that tribute from the list. It prints a message about the removed tribute and the remaining tributes.

Note the format of the tribute list in Figure 10.2's bottom output line—square brackets surrounding a comma-separated list. Can you find the source code that prints that list? If you're looking for square brackets and a loop, forget it—they're not there. So how in the world does the square-bracketed list get printed? In the final `println` statement at the bottom of the program, the `tributes ArrayList` gets concatenated to a string. That causes the JVM to do some work behind the scenes. If you attempt to concatenate an `ArrayList` to a string or print an `ArrayList`, the `ArrayList` returns a comma-separated list of `ArrayList` elements surrounded by square brackets (`[]`). And that's what happens when Figure 10.2's last statement executes.

10.3 Storing Primitives in an `ArrayList`

As mentioned earlier, `ArrayLists` store references. For example, in the Hunger Games, `tributes` is an `ArrayList` of strings, and strings are references. If you need to store primitives in an `ArrayList`, you can't do it directly, but if the primitives are wrapped up in wrapper classes,[2] you can store the resulting wrapped objects in an `ArrayList`. In this section, we show you how to do that.

Stock Average Example

The StockAverage program in Figure 10.3 reads weighted stock values and stores them in an `ArrayList`. In simplified terms, a weighted stock value is the market price of one stock share times a number that scales that price up or down to reflect the importance of the stock's company in the overall marketplace. After the StockAverage program stores the weighted stock values in an `ArrayList`, the program calculates the average of all the entered weighted stock values. Why is an `ArrayList` appropriate for calculating a stock average? An `ArrayList`'s size grows as necessary. That works well for stock averages because there are lots of stock averages (also called *stock indexes*), and they use different numbers of stocks in their calculations. For example, the Dow Jones Industrial Average uses stock values from 30 companies, while the Russell 3000 Index uses stock values from 3,000 companies. Because it uses an `ArrayList`, the StockAverage program works well for both situations.

The StockAverage program stores stock values in an `ArrayList` named `stocks`. The stock values originate from user input in the form of `doubles`, like 25.6, 36.0, and so on. As you know, `ArrayLists` can't store primitives; they can store references only. So the StockAverage program wraps up the `doubles` into `Double` wrapper objects just prior to storing them in the `stocks ArrayList`. As you might imagine, a *wrapper object* is an instance of a wrapper class, and each wrapper object stores one "wrapped up" primitive value. You don't have to worry very much about wrapper objects for `ArrayLists`. For the most part, you can pretend that `ArrayLists` can hold primitives. Case in point: The following line from the StockAverage program appears to add a primitive (`stock`) to the `stocks ArrayList`:

```
stocks.add(stock);
```

What actually happens behind the scenes is that the `stock` primitive gets converted automatically to a wrapper object prior to its being added to the `stocks ArrayList`. Really, there is just one thing you have to worry about when working with primitives in an `ArrayList`. When you create an `ArrayList` object

[2]If you need a refresher on wrapper classes, see Chapter 5.

```
/*************************************************************
* StockAverage.java
* Dean & Dean
*
* This program uses an ArrayList to store user-entered stock
* values. It prints the average stock value.
*************************************************************/

import java.util.Scanner;
import java.util.ArrayList;

public class StockAverage
{
  public static void main(String[] args)
  {
    Scanner stdIn = new Scanner(System.in);
    ArrayList<Double> stocks = new ArrayList<>();
    double stock;                         // a stock value
    double stockSum = 0;                  // sum of stock values

    System.out.print("Enter a stock value (-1 to quit): ");
    stock = stdIn.nextDouble();

    while (stock >= 0)
    {
      stocks.add(stock);
      System.out.print("Enter a stock value (-1 to quit): ");
      stock = stdIn.nextDouble();
    } // end while

    for (int i=0; i<stocks.size(); i++)
    {
      stock = stocks.get(i);
      stockSum += stock;
    }

    if (stocks.size() != 0)
    {
      System.out.printf("\nAverage stock value = $%.2f\n",
        stockSum / stocks.size());
    }
  } // end main
} // end class StockAverage
```

This must be a wrapper class, not a primitive type!

Autoboxing takes place here.

Unboxing takes place here.

Figure 10.3 StockAverage program illustrating an `ArrayList` of `Double` objects

to hold primitive values, the type you specify in the angled brackets must be the wrapped version of the primitive type, that is, Double instead of double, Integer instead of int, and so on. This line from the StockAverage program illustrates what we're talking about:

```
ArrayList<Double> stocks = new ArrayList<>();
```

Autoboxing and Unboxing

In most places, it's legal to use primitive values and wrapper objects interchangeably. The way it works is that the JVM automatically wraps primitive values and unwraps wrapper objects when it's appropriate to do so. For example, if the JVM sees an int value on the right of an assignment statement and an Integer variable at the left, it thinks to itself, hmmm, to make this work, I need to convert the int value to an Integer wrapper object. It then gets out its Styrofoam packing peanuts and duct tape and wraps up the int value into an Integer wrapper object. That process is called *autoboxing*. On the other hand, if the JVM sees an Integer wrapper object on the right of an assignment statement and an int variable at the left, it thinks to itself, hmmm, to make this work, I need to extract the int value from the Integer wrapper object. It then proceeds to tear off the Integer wrapper object's covering, and it gets the int value that's inside. That process is called *unboxing*.

More formally, autoboxing is the process of automatically wrapping a primitive value in an appropriate wrapper class whenever there's an attempt to use a primitive value in a place that expects a reference. Refer to the stocks.add(stock); statement in Figure 10.3. That statement causes autoboxing to occur. The stocks.add method call expects a reference argument. Specifically, it expects the argument to be a reference to a Double wrapper object (since stocks is declared to be an ArrayList of Double references). When the JVM sees a primitive value argument (stock), it automatically wraps the argument in a Double wrapper class.

More formally, unboxing is the process of automatically extracting a primitive value from a wrapper object whenever there's an attempt to use a wrapper object in a place that expects a primitive. Refer to the stock = stocks.get(i); statement in Figure 10.3. That statement causes unboxing to occur. Because stock is a primitive variable, the JVM expects a primitive value to be assigned to it. When the JVM sees a wrapper object on the right of the assignment statement (stocks holds Double wrapper objects and get(i) retrieves the ith such wrapper object), it automatically extracts the primitive value from the wrapper object.

Autoboxing and unboxing take place automatically behind the scenes. That makes the programmer's job easier. Yeah!

Converting Between Primitives and Wrapper Objects the Old-Fashioned Way

In *legacy code* (code written for older compilers), you won't see autoboxing and unboxing. Instead, you'll see explicit conversions. To convert an int value to an Integer object explicitly, you can invoke the Integer constructor with the int value for its argument. To convert a double value into a Double object explicitly, you can invoke the Double constructor with the double value for its argument. To use this technique in Figure 10.3's StockAverage program, you would replace the autoboxing statement (stocks.add(stock);) with this:

```
stocks.add(new Double(stock));
```

To go the other way and explicitly extract an int value from an Integer object, you can have an Integer wrapper object call its intValue method. Likewise, to extract a double value from a

Double object explicitly, you can have a Double wrapper object call its doubleValue method. To use this technique in Figure 10.3's StockAverage program, you would replace the unboxing statement (stock = (stocks.get(i));) with this:

```
stock = stocks.get(i).doubleValue();
```

As mentioned earlier, stocks holds Double wrapper objects, and get(i) retrieves the ith such wrapper object. The doubleValue() method call then extracts the primitive double value from the retrieved wrapper object.

10.4 ArrayList Example Using Anonymous Objects and the For-Each Loop

Anonymous objects and *for-each loops* are programming constructs that are particularly useful when used in conjunction with ArrayLists. In this section, we present for-each loop details and anonymous object details by showing how they're used in the context of an ArrayList program. But before we get to the program, we provide brief introductions for the two new constructs.

Usually, when you create an object, you immediately store the object's reference in a reference variable. That way, you can refer to the object later on by using the reference variable's name. If you create an object and don't immediately assign the object's reference to a reference variable, you've created an anonymous object. It's called anonymous because it doesn't have a name.

A for-each loop is a modified version of the traditional for loop. It can be used whenever there's a need to iterate through all of the elements in a collection of data. An ArrayList is a collection of data, and, as such, for-each loops can be used to iterate through all of the elements in an ArrayList.

A Bear-Store Example

Suppose you want to model a store which sells customized toy bears. You need a Bear class to represent each bear, a BearStore class to represent the store, and a BearStoreDriver class to "drive" the program. Let's start by examining the Bear class in Figure 10.4. The Bear class defines two instance named constants which represent two permanent properties of a particular bear: (1) MAKER, the bear's manufacturer, such as Gund, and (2) TYPE, the bear's type, such as "pooh bear" or "angry campground bear." A constructor initializes these two instance constants, and a display method displays them.

Now let's examine the first part of the BearStore class, shown in Figure 10.5a. The BearStore class has one instance variable, bears, which is declared to be an ArrayList of Bear references. It holds the store's collection of toy bears. The BearStore class's addStdBears method fills the bears ArrayList with a specified number of standard teddy bears. Here's the statement that adds one standard teddy bear to the ArrayList:

```
bears.add(new Bear("Acme", "brown teddy"));
```

The statement instantiates a Bear object and passes the Bear object's reference to the bears.add method call. The statement does not assign the Bear object's reference to a Bear reference variable. Because there's no assignment to a Bear reference variable, that's an example of an anonymous object. As an alternative, the statement could have been written with a Bear reference variable like this:

```
Bear stdBear = new Bear("Acme", "brown teddy");
bears.add(stdBear);
```

```
/*************************************************************
* Bear.java
* Dean & Dean
*
* This class models a toy bear.
*************************************************************/

public class Bear
{
  private final String MAKER; // bear's manufacturer
  private final String TYPE;  // type of bear

  //**********************************************************

  public Bear(String maker, String type)
  {
    MAKER = maker;
    TYPE = type;
  }

  //**********************************************************

  public void display()
  {
    System.out.println(MAKER + " " + TYPE);
  }
} // end Bear class
```

Figure 10.4 Class that represents a toy bear

But why bother with using two statements instead of one? The new bear's reference gets stored in the bears ArrayList and that's where it's processed. There's no need to store it in a second place (e.g., in the stdBear reference variable), so in the interest of code compactness, don't.

Now let's examine the bottom part of the BearStore class, shown in Figure 10.5b. The BearStore class's getUserSpecifiedBear method prompts the user for a customized bear's maker and type and returns the newly created bear. Here's the return statement:

```
return new Bear (maker, type);
```

Note that there's no reference variable for the new bear. Thus, the new bear is considered to be an anonymous object. The return statement returns the new bear to the addUserSpecifiedBears method, where it gets added to the bears ArrayList.

When to Use an Anonymous Object

The bear-store program contains several specific examples of using anonymous objects. In general, you'll see anonymous objects being used in two circumstances:

1. When passing a newly created object into a method or constructor. For example:

```
bears.add(new Bear("Gund", "Teddy"));
```

```
/*********************************************************************
* BearStore.java
* Dean & Dean
*
* This class implements a store that sells toy bears.
*********************************************************************/

import java.util.Scanner;
import java.util.ArrayList;

public class BearStore
{
  ArrayList<Bear> bears = new ArrayList<>();

  //*****************************************************************

  // Fill store with specified number of standard teddy bears.

  public void addStdBears(int num)
  {
    for (int i=0; i<num; i++)
    {
      bears.add(new Bear("Acme", "brown teddy"));
    }
  } // end addStdBears

  //*****************************************************************

  // Fill store with specified number of customized bears.

  public void addUserSpecifiedBears(int num)
  {
    for (int i=0; i<num; i++)
    {
      bears.add(getUserSpecifiedBear());
    }
  } // end addUserSpecifiedBears
```

anonymous object as argument

Returned anonymous object becomes the argument in this method call.

Figure 10.5a Class that implements a toy-bear store—part A

2. When returning a newly created object from a method. For example:

```
return new Bear(maker, type);
```

Embedded Driver

At the bottom of the BearStore class, we've embedded the program's driver, main. It instantiates a BearStore object, adds three standard bears to the bear store, adds two user-specified bears to the bear store, and then displays the store's inventory of bears by calling displayInventory. In displaying the store's inventory, the displayInventory method accesses each bear in the bears ArrayList with the help of a for-each loop. In the next subsection, you'll learn about for-each loop details.

```
  //***************************************************************

  // Prompt user for bear's maker and type and return bear.

  private Bear getUserSpecifiedBear()
  {
    Scanner stdIn = new Scanner(System.in);
    String maker, type;

    System.out.print("Enter bear's maker: ");
    maker = stdIn.nextLine();
    System.out.print("Enter bear's type: ");
    type = stdIn.nextLine();
    return new Bear(maker, type);
  } // end getUserSpecifiedBear        anonymous object as return value

  //***************************************************************

  // Print all the bears in the store.

  public void displayInventory()
  {
    for (Bear bear : bears)
    {
      bear.display();                  for-each loop
    }
  } // end displayInventory

  //***************************************************************

  public static void main(String[] args)
  {
    BearStore store = new BearStore();
    store.addStdBears(3);
    store.addUserSpecifiedBears(2);
    store.displayInventory();
  } // end main
} // end BearStore class
```

Figure 10.5b Class that implements a toy-bear store—part B

For-Each Loop

As mentioned earlier, you can use a for-each loop whenever you need to iterate through all the elements in a collection of data. Here is the for-each loop syntax for an ArrayList:

```
for (<element-type> <element-name> : <ArrayList-reference-variable>)
{
  . . .
}
```

And here is an example for-each loop from Figure 10.5b's displayInventory method:

```
for (Bear bear : bears)
{
  bear.display();
}
```

Note how the for-each loop header matches the preceding syntax: bears is an ArrayList reference variable, bear is the name of an element in the bears ArrayList, and Bear is the type for each element. It's legal to choose any name for the element, and it's common practice to use an abbreviation, but here we chose the descriptive name, bear. With each iteration of the for-each loop, you use the element's name to refer to the current element. For example, bear.display() calls the display method for the current bear element.

Note that, as an alternative, you could implement the displayInventory method using a traditional for loop rather than a for-each loop. Here's an implementation with a traditional for loop:

```
for (int i=0; i<bears.size(); i++)
{
  bears.get(i).display();
}
```

The for-each loop implementation is better because it is simpler. There's no need to declare an index variable, and there's no need to calculate and specify the ArrayList's first and last index values.

10.5 ArrayLists Versus Standard Arrays

There's a lot of overlap in the functionality of an ArrayList and a standard array. So how can you tell which one to use? Your answer will be different for different situations. When deciding on an implementation, consider this table:

Benefits of an ArrayList Over a Standard Array	Benefits of a Standard Array Over an ArrayList
1. It's easy to increase the size of an ArrayList—just call add.	1. A standard array uses [ ]'s to access array elements (which is easier than using get and set methods).
2. It's easy for a programmer to insert or remove an element to or from the interior of an ArrayList—just call add or remove and specify the element's index position.	2. A standard array is more efficient when storing primitive values.

In looking at the table's first ArrayList benefit, easy to increase the size of an ArrayList, think about how much work is required to increase the size of a standard array. For a standard array, the programmer needs to instantiate a larger array and then copy the old array's contents to the new larger array. On the other hand, for an ArrayList, the programmer simply needs to call the add method. Note that behind the scenes, the JVM has to put forth some effort in implementing the add method, but the effort is kept to a minimum. ArrayLists are implemented with the help of an underlying standard array. Usually, the underlying array has a greater number of elements than the ArrayList, so adding another element to the ArrayList is easy—the JVM just borrows an unused element from the underlying array. As a programmer, you don't have to worry about or code those details; the "borrowing" takes place automatically.

The table's second ArrayList benefit, easy for a programmer to insert or remove an element to or from the interior of an ArrayList, is true, but just because it's easy for programmers doesn't mean it's easy for the JVM. Actually, the JVM has to do quite a bit of work when it adds or removes from the interior of an ArrayList. To insert an element, the JVM has to adjust its underlying array by shifting higher indexed elements to make room for the new element. And to remove an element, the JVM has to adjust its underlying array by shifting higher indexed elements to overlay the removed element.

10.6 The LinkedList Class

In the previous chapter and in the first part of this chapter, you learned about arrays and ArrayLists, which store ordered lists of related data of the same type. Now it's time to discuss another structure that stores an ordered list of related data of the same type—a *linked list*. Whereas arrays and ArrayLists store adjacent elements in adjacent memory locations, a linked list uses a chain of references to connect the linked list's elements. Historically, a linked list's elements have been called *nodes*. We'll stick with the term *element* because that's what you're used to, and Oracle uses *element* consistently when referring to items in an array, an ArrayList, and a linked list.

In Java, linked lists are normally implemented using the LinkedList class from the API library. LinkedList objects form elements that are *doubly linked*, which means that each element contains two references—one that points to the element before it, and one that points to the element after it. Study Figure 10.6 to see what we're talking about. The figure shows how elements can be added to and removed from a linked list.

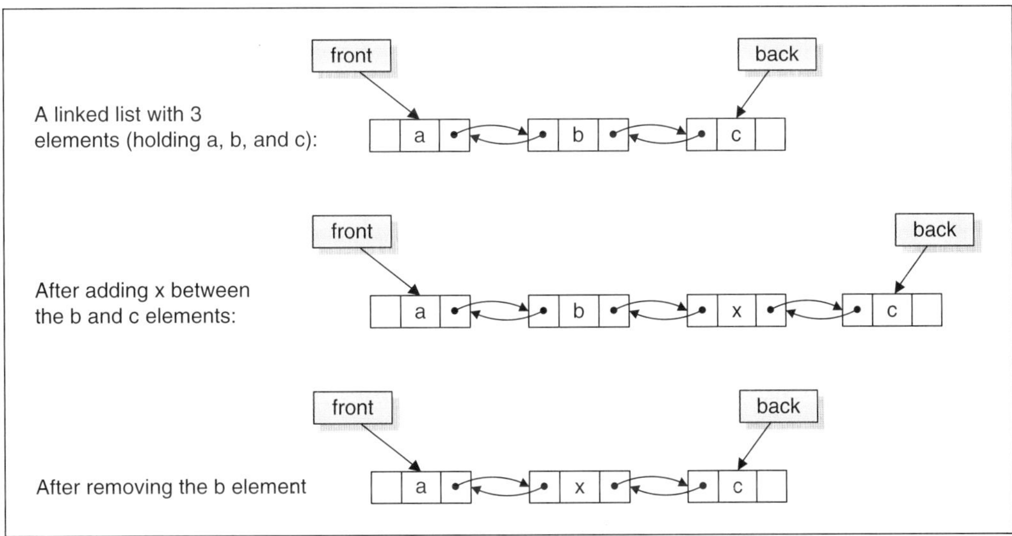

Figure 10.6 Adding and removing elements in a doubly linked list

To create a linked list for storing the names of your favorite Android apps, instantiate the LinkedList class, like this:

```
LinkedList<String> androidApps = new LinkedList<>();
```

The name of the linked list, androidApps, is a reference variable that initially refers to null, which means that the linked list is empty.

A LinkedList implements many of the same methods an ArrayList implements. The add method is an important example. LinkedList's one-parameter add method appends an element to the end of the linked list, just like ArrayList's one-parameter add method appends an element to the high-index end of the ArrayList. The end of a linked list is normally called the list's *back,* and that's what we'll use, but it's sometimes also called the list's *tail.* LinkedList's two-parameter add method, where the first parameter is an index, can insert a new element anywhere in the linked list, just like ArrayList's two-parameter add method. The process of accessing a linked-list element starts from either of the two ends. Thus, it's easiest to find, add, or remove an element that's near either end. Once an element is found in the middle of a linked list, it's relatively easy to remove it or add a new one there, but if a linked list is large, it takes a relatively long time to find an element in the middle. That's much longer than it takes to find an element in the middle of an ArrayList, where the underlying array facilitates jumping quickly to any indexed location.

10.7 The List Interface

ArrayLists and LinkedLists implement many of the same methods—get, set, add, remove, clear, and size, to name a few. Using the same names is convenient because Java programmers have to remember only one set of names, not two. Furthermore, each of the same-named methods uses the same number of parameters and types of parameters. Once again, this makes it easy because Java programmers don't have to remember as much. To encourage and standardize this strategy of having classes use methods with the same signatures (same method names and same parameter types), Java relies on *interfaces.* An interface is a template for designing classes that share certain qualities. The ArrayList and LinkedList classes both implement lists, and, as such, they share certain qualities common to all lists. Thus, to help with consistency, they are implemented using the List interface as a template. To implement a class using an interface, you just append an implements clause to the class's heading. For example, if you look up the ArrayList class in Java's API library, you'll see this:

```
public class ArrayList<E> implements List<E>
```

As indicated in Section 10.2, the E serves as a generic placeholder for the type specified when an ArrayList is declared in the future. For example, if you declare an ArrayList of strings with ArrayList<String> iPhoneApps;, then String will match up with E in ArrayList's class heading. Or, if you declare an ArrayList of Bear objects with ArrayList<Bear> bears;, then Bear will match up with E in ArrayList's class heading.

In the past, you've declared variables with a primitive type at the left or a class name at the left. For example:

```
double distance;
Student student;
```

You can also declare a variable with an interface at the left, like this:

```
List<String> iPhoneApps;
```

If you declare a variable as a double, then you are constrained to assigning numeric values to the variable. Likewise, if you declare a variable with an interface, you are constrained to assigning references to the variable such that the references point to instances of classes that implement the interface. For example, given the iPhoneApps declaration above, you can assign an ArrayList object to iPhoneApps and then, later, assign a LinkedList object to iPhoneApps. Here's the code:

```
iPhoneApps = new ArrayList<String>;
. . .
iPhoneApps = new LinkedList<String>;
```

This code works, but you might be thinking, "Why would I want to assign different type of lists into the same variable?" Suppose you want to write a method that reverses the order of the elements in a passed-in list parameter, and you want your method to be able to handle any type of list that implements the List interface. To do that, you should declare the list parameter with the List interface. Here's the appropriate method heading:

```
public void reverseList(List<String> list)
```

Because ArrayList and LinkedList both implement the List interface, you should be able to pass instances of either of those classes to the reverseList method, and the reverse process should work. Nice!

As you've seen, using an interface for a declaration type can make your code more flexible. On the other hand, using an interface for a declaration type also can constrain your code. It prevents variables from calling methods not specified by the interface. For example, LinkedList specifies several additional methods not specified in List. These additional LinkedList methods enable a LinkedList object to do things an ArrayList object cannot do. But if you declare the variable holding a LinkedList object to be of type List, that object cannot call any of these additional LinkedList methods.

10.8 Problem Solving: How to Compare Method Execution Times

Because the LinkedList class and the ArrayList class both implement the List interface, an instance of either of these classes can call any of the methods specified in the List interface. Therefore, if you use only those methods specified by the List interface, you can freely substitute a LinkedList object for an ArrayList object, and vice versa. Thus, you could go back to any of the previous programs in this chapter, import the java.util package, and replace each occurrence of ArrayList with LinkedList. For example, in the BearStore program in Figure 10.5a, instead of ArrayList<Bear>, you could use LinkedList<Bear>. So if an ArrayList and a LinkedList produce the same results, how do you decide which to use? Use the one that runs faster. But how can you tell that? The best way to determine which of two implementations runs faster is to measure the times it takes them to do the same job. We could just tell you which type of list runs faster (and we'll do that eventually), but the main point of this section is to teach you how to determine the relative performance of methods, so you can apply the technique for other methods of interest.

To determine the execution time of a particular code fragment, surround that fragment with calls to System.nanoTime(), which returns current time in nanoseconds (billionths of a second). Of course, the measured time differences vary with hardware. Even on the same hardware, they vary from one run to another, due to unpredictable background activity. Still, comparative measurement is a useful technique because software performance differences are often quite large and easy to see.

For our first comparison, we'll look at the get and set operations, which are used to retrieve and update a specified list's element value, respectively. We'll determine how fast those operations execute when an ArrayList is used, and then when a LinkedList is used. We'll make those determinations with the help of the ListExecutionTimes program, shown in Figures 10.7a and 10.7b. In the first figure, note how the `list` variable is declared to be an `ArrayList`. Note also the commented-out code that declares the `list` variable as a `LinkedList`. We'll run the program twice, with the comment `//`'s switched the second time.

The ListExecutionTimes program measures the average time to retrieve each element at a random index and then update each element at a different random index. We arbitrarily fix the list length at 1,000, and call the `getIndices` helper method to generate two arrays of indices, `indicesA` and `indicesB`. Each

```
/*******************************************************************
* ListExecutionTimes.java
* Dean & Dean
*
* This measures average time to perform indexed operations.
*******************************************************************/

import java.util.*; // ArrayList, LinkedList, ArrayDeque, Random

public class ListExecutionTimes
{
  public static void main(String[] args)
  {
    String operationType = "average get and set time";
//    String operationType = "average remove and add time";
    int length = 1000;
    int[] indicesA = getIndices(length);  // random sequence
    int[] indicesB = getIndices(length);  // random sequence
    ArrayList<Double> list = new ArrayList<>();
//    LinkedList<Double> list = new LinkedList<>();   <-- alternate implementation
    Double element;
    long time0, time1;

    for (int i=0; i<length; i++)
    {
      list.add(new Double(i));
    }
    time0 = System.nanoTime();
    for (int i=1; i<length; i++)
    {
      element = list.get(indicesA[i]);
      list.set(indicesB[i], element);
//      element = list.remove(indicesA[i]);  }  <-- alternate measured
//      list.add(indicesB[i], element);      }      operations
    }
    time1 = System.nanoTime();
    System.out.println(list.getClass());
    System.out.printf("for length = %d, %s = %,d ns\n",
      length, operationType, (time1 - time0) / length);
  } // end main
```

Figure 10.7a ListExecutionTimes program—part A
For alternate evaluations, substitute commented statements for preceding active statements.

```
  //****************************************************************

  // This returns an array of all integers between zero and
  // length in a random sequence with no duplications

  private static int[] getIndices (int length)
  {
    Random random = new Random();
    ArrayList<Integer> integers = new ArrayList<>();
    int[] indices = new int[length];
    for (int i=0; i<length; i++)
    {
      integers.add(random.nextInt(i+1), new Integer(i));
    }
    for (int i=0; i<length; i++)
    {
      indices[i] = integers.get(i);
    }
    return indices;
  } // end getIndices
} // end class ListExecutionTimes
```

Figure 10.7b ListExecutionTimes program—part B
This helper method generates a random sequence of all integers between zero and `length`.

of these arrays contains all the integers in the range between zero and `length` in a random sequence, and their sequences are different.

In Figure 10.7a, note how calls to `System.nanoTime()` assign current times to `time0` and `time1`. Because these assignments surround the loop that repeatedly calls `get` and `set`, `(time1 - time0)` calculates the total time it takes to perform all the `get` and `set` operations. At the bottom of Figure 10.7a, note the expression `(time1 - time0) / length`. That expression calculates the average time in nanoseconds for each pair of `get` and `set` method calls.

A representative execution time for `get` and `set` operations with an `ArrayList` appears in the first output session in Figure 10.8. To make the comparison with a `LinkedList`, in Figure 10.7a, we replace:

```
ArrayList<Double> list = new ArrayList<>();
```

with:

```
LinkedList<Double> list = new LinkedList<>();
```

Then we recompile and rerun to obtain the result shown in Figure 10.8's second output session. Although the measured times are sensitive to the hardware and background activity of the computer on which the program runs, their relative values should be reasonably accurate. The `LinkedList` implementation needs substantially more time to execute `get` and `set` because instead of jumping directly to the desired element, it must step through the list until the step count equals the index number. For length = 100, the average `get` and `set` time was about 340 nanoseconds (ns) for an `ArrayList` and about 1,586 ns for a `LinkedList`. For length = 10,000, the average `get` and `set` time was about

Output:

```
class java.util.ArrayList
for length = 1000, average get and set time = 174 ns
```

Output:

```
class java.util.LinkedList
for length = 1000, average get and set time = 1,455 ns
```

Figure 10.8 Approximate average times to get and set an indexed element using the programs in Figures 10.7a and 10.7b. The first output uses an ArrayList version of the program. The second output uses a LinkedList version of the program.

122 ns for an ArrayList and about 18,222 ns for a LinkedList. When the list is very long, a LinkedList is much slower.

For our second comparison, we look at two methods that alter the structure of a list—the remove method and the two-parameter add method. Traditionally, people expect a linked list to do better at operations that alter structure. Let's see if that's really the case. To compare the time to move an element from one index to another (by calling remove and add), in Figure 10.7a, we replace:

```
String operationType = "average get and set time";
```

with:

```
String operationType = "average remove and add time";
```

And we replace:

```
element = list.get(indicesA[i]);
list.set(indicesB[i], element);
```

with:

```
element = list.remove(indicesA[i]);
list.add(indicesB[i], element);
```

Recompiling and rerunning with list implemented as an ArrayList generates Figure 10.9's first output session. Recompiling and rerunning with list implemented as a LinkedList generates Figure 10.9's second output session.

In executing the remove method, the JVM performs two operations—it first finds the element and then removes it. With an ArrayList, the find operation is fast (it uses the specified index to go directly to the element), but the remove operation is slow (after removing the element, it must repair its underlying array by shifting all higher elements down by one index number). A LinkedList has the opposite problem. With a LinkedList, the find operation is slow (because it must start its search from one of the ends), but the remove operation is fast (it just changes a couple of references).

Like the remove method, the two-parameter add method is also a two-step operation. First, it finds the element, and then it inserts the element in the list at the specified position. When choosing between an ArrayList and a LinkedList, the performance tradeoffs for add are the same as for remove.

```
Output:
class java.util.ArrayList
for length = 1000, average remove and add time = 1,082 ns

Output:
class java.util.LinkedList
for length = 1000, average remove and add time = 2,543 ns
```

Figure 10.9 Approximate average times to remove the element at one index and add it back at another index. The first output uses an `ArrayList` version of the program. The second output uses a `LinkedList` version of the program.

As Figure 10.9 shows, for most `remove` and `add` operations, an `ArrayList`'s performance is better than a `LinkedList`'s performance. This `ArrayList` advantage extends to very short lengths and very long lengths. For length = 100, the average `remove` and `add` time was about 936 ns for an `ArrayList` and about 3,573 ns for a `LinkedList`. For length = 10,000, the average move time was about 8,503 ns for an `ArrayList` and about 25,855 ns for a `LinkedList`.

A `LinkedList` outperforms an `Arraylist` for structural changes at the low-index end of the list. That's because for an `Arraylist`, to add or remove an element at the low-index end of the list, the JVM must adjust the underlying array by shifting all higher elements up or down by one index position, respectively. This performance advantage for `LinkedLists` over `ArrayLists` is slight.

The relative performance of a `LinkedList` can improve when you add or remove a series of list elements that are near each other (directly connected or connected without too many intervening elements). The nearness is important because, as a programmer, you can use an *iterator* to keep track of the position of the most recently accessed element within a `LinkedList`. With iterators at your disposal, you don't need to start your search for an element at the ends. Instead, you can start your search at the most recently accessed element, and that can save time. To learn about iterators, look up `ListIterator` in Java's API library.

10.9 Queues, Stacks, and the `ArrayDeque` Class

To summarize what you've learned so far about collections of same-typed data: Use an array if you know in advance the number of elements; use an `ArrayList` if you need to add or remove elements anywhere in the list; and use a `LinkedList` in the special case where you need to add or remove a series of list elements that are near to each other. In this section, we present two other special cases—the case where you're allowed to add and remove elements only at the ends and the case where you're allowed to add and remove elements only at one end. Those two cases are handled by *queues* and *stacks,* respectively.

Queues

In society, a queue is a line of waiting people. New people are added to the back of the line as they arrive, and those who have waited the longest are removed from the front of the line as they are served. That strategy is called *first-in first-out (FIFO)* because the people who arrive first are served first. In the world of computer programming, a queue data structure mirrors a line of waiting people. With an ordinary queue,

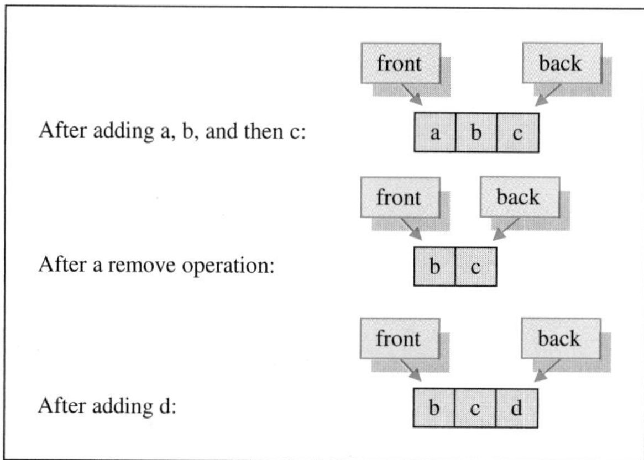

Figure 10.10 Queue operations

you're able to add elements only to the *back* end and remove elements only from the *front* end. Study Figure 10.10 to see what we're talking about.

Although you can use an `ArrayList` or a `LinkedList` to implement a queue, for maximal efficiency, you should use the Java API `ArrayDeque` class. The *deque* (pronounced "deck") in `ArrayDeque` stands for "double-ended queue." With a double-ended queue, you can add or remove from either end.

Behind the scenes, an `ArrayDeque` is implemented with an array, which takes up less room in memory than a linked list, and that's one of the reasons for its improved speed. By its nature, an array is a static entity, so to accommodate a queue's penchant for growing and shrinking dynamically, a little behind-the-scenes trickery is necessary. Trick #1: When an `ArrayDeque` is instantiated, the JVM creates a 16-element empty array. Anytime the array becomes full, and there's an attempt to add another element, the JVM replaces the array with a new array whose capacity is double that of the original array. Trick #2: Anytime there's an attempt to remove an element, there's no need to shift all of the higher-index elements to fill in the vacated front element's spot. Instead, the JVM adjusts the front of the queue, so it points to the next element after the original front element. This frees the original front element and makes it available to hold other elements at the back of the queue later on. Enabling a former front element to hold a future back element is accomplished by using a *circular array.* A circular array is built with a regular array. What makes it "circular" is its ability to treat the high-index element as adjacent to the 0-index element, so the elements form a continuous virtual circle. As part of an add operation, the circular array's back variable will get incremented. If the incrementation causes the back variable's value to become greater than the array's highest index value, then the back variable gets reassigned to 0. The remove operation works the same, but it uses the front variable instead of the back variable.

With the above strategy built into `ArrayDeques`, when you add an element to an `ArrayDeque`, the JVM simply assigns a value to an array element and updates the `ArrayDeque`'s internal back and size properties (except in the case when expanding the underlying array is necessary). Removing an element from an `ArrayDeque` is similarly straightforward.

Now let's dig into the details of how you can use `ArrayDeques` to create and manipulate queues in your own programs. To create an empty queue of people standing in line at a Chipotle Mexican Grill restaurant, instantiate an `ArrayDeque` like this:

```
Queue<String> chipotlesQueue = new ArrayDeque<>();
```

In the code above, note that chipotlesQueue is declared with Queue at its left. Queue is an interface. As explained previously, an interface can be used to constrain an object, so that the object can call only those methods that are included in the interface. The ArrayDeque class includes quite a few methods, and some of those methods are inappropriate for queues. For example, the ArrayDeque class includes a removeLast method, which removes the element at the back of the queue. Because ordinary queues are supposed to remove elements only at the front, not at the back, the Queue interface does not include the removeLast method. Instead, it includes just the remove method, which attempts to remove the element at the front. Be aware that the remove method does not check the queue to see if it's empty before it attempts to remove the front element. So if you call remove from an empty queue, your program will crash. To prevent that, you can call Queue's isEmpty method before calling remove. To see what we're talking about, study the while loop in Figure 10.11's ChipotlesQueue program.

```
/*********************************************************************
* ChipotlesQueue.java
* Dean & Dean
*
* This illustrates the creation and use of an ordinary FIFO queue.
*********************************************************************/

import java.util.*;  // for ArrayDeque and Queue

public class ChipotlesQueue
{
  public static void main(String[] args)
  {
    String servedPerson;  // person removed from the queue's front
    Queue<String> chipotlesQueue = new ArrayDeque<>();

    chipotlesQueue.add("Alexa");
    chipotlesQueue.add("Carolyn");

    while (!chipotlesQueue.isEmpty())
    {
      servedPerson = chipotlesQueue.remove();
      System.out.println("What is your order, " + servedPerson + "?");
    }
  } // end main
} // end class ChipotlesQueue
```

Check for an empty queue before attempting to remove the front element.

Output:

```
What is your order, Alexa?
What is your order, Carolyn?
```

Figure 10.11 ChipotlesQueue program

To add an element to the back of a queue, you can use `Queue`'s `add` method. For example, here's how the ChipotlesQueue program adds "Alexa" to the food ordering line:

```
chipotlesQueue.add("Alexa");
```

Stacks

Like a queue, a *stack* is an ordered list that restricts element access. With a stack, you add and remove elements to and from only one end, and that end is called the *top*. A classic example is a stack of plates. If you're like most people, you arrange your plates in a stack and grab the top plate when you're ready to eat. After you clean and dry your plates, you place them back onto the top of the stack one at a time.[3] The stacking strategy is called *last-in first-out (LIFO)* because the last item placed on the top of the stack is the first item removed from the stack. Study Figure 10.12 to see how elements can be added to and removed from a stack. As indicated in the figure, for a stack, the add operation is called a *push,* and the remove operation is called a *pop.*

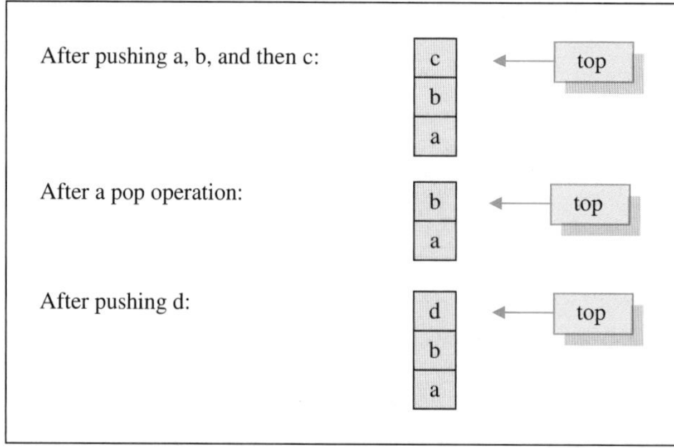

Figure 10.12 Stack operations

Stacks, like queues, can be implemented efficiently with `ArrayDeque`s. For queues, there exists an interface, the `Queue` interface, that can be used to force an `ArrayDeque` to act like a queue. Unfortunately, there is no comparable interface to force an `ArrayDeque` to act like a stack. If you want to create a list that is constrained to stack operations (pushing to and popping from the stack's top), you can (1) use the `Stack` class instead of the `ArrayDeque` class, or (2) write your own stack class with an `ArrayDeque` instance variable and push and pop methods. There are many proponents of these two techniques, and you should feel free to use either of them. However, they both lead to relatively slow solutions (yes, the Java API `Stack` class is slower than the `ArrayDeque` class for stack operations). To avoid the performance hit, we won't worry about providing an airtight stack-operations-only constraint on the `ArrayDeque` class. We'll just use the `ArrayDeque` class as it should be used for a stack and assume that no one edits our source code later on by calling an inappropriate `ArrayDeque` method. (For example, calling `removeLast` would be inappropriate for a stack, because it removes the element at the bottom of the stack.)

In Figures 10.13a, 10.13b, and 10.13c, we show a DrivewayParking program that illustrates how to use `ArrayDeque`s for stacks. The program is designed to help with a low-budget parking service that parks

[3]Although stacking plates is fairly common, alternative practices exist. Tweens with dinner chores have been known to place dishes randomly behind the couch, where they are cleaned later by Sophie the Samoyed.

```
/*******************************************************************
* DrivewayParking.java
* Dean & Dean
*
* This program uses stacks to help with a driveway parking service.
*******************************************************************/

import java.util.*; // ArrayDeque, Scanner

public class DrivewayParking
{
  private ArrayDeque<String> driveway1 = new ArrayDeque<>();
  private ArrayDeque<String> driveway2 = new ArrayDeque<>();

  //****************************************************************

  public void describeDriveways()
  {
    System.out.println("driveway1 " + driveway1);
    System.out.println("driveway2 " + driveway2);
  } // end describeDriveways()

  //****************************************************************

  // This method parks a car in the least full driveway.

  private void parkCar(String licensePlate)
  {
    if (driveway1.size() <= driveway2.size())
    {
      driveway1.push(licensePlate);
    }
    else
    {
      driveway2.push(licensePlate);
    }
  } // end parkCar
```

Park a car by pushing it onto the stack that is least full.

Figure 10.13a DrivewayParking program—part A

college game-day events that parks cars in either of two long, narrow driveways. The program implements the driveways as ArrayDeque-based stacks. Here's the instantiation code:

```
private ArrayDeque<String> driveway1 = new ArrayDeque<>();
private ArrayDeque<String> driveway2 = new ArrayDeque<>();
```

Using stacks is helpful because their LIFO strategy mimics what happens as cars are parked and backed out of the driveways. When a customer arrives, the valet parks the customer's car in the driveway that is less

```
//**************************************************************/

// Return true if and only if licensePlate is found

private boolean getCar(String licensePlate)
{
  String otherPlate;

  if (driveway1.contains(licensePlate))
  {
    otherPlate = driveway1.pop();
    while (!otherPlate.equals(licensePlate))
    {
      driveway2.push(otherPlate);
      otherPlate = driveway1.pop();
    }
    return true;
  }
  else if (driveway2.contains(licensePlate))
  {
    otherPlate = driveway2.pop();
    while (!otherPlate.equals(licensePlate))
    {
      driveway1.push(otherPlate);
      otherPlate = driveway2.pop();
    }
    return true;
  }
  else
  {
    return false;
  }
} // end getCar
```

If the searched for car is found, back out the cars behind it by popping them off of the stack.

Figure 10.13b DrivewayParking program—part B

full and records the car's license plate number by pushing it onto the less full stack. Here's the relevant code from the DrivewayParking program's `parkCar` method:

```
if (driveway1.size() <= driveway2.size())
{
  driveway1.push(licensePlate);
}
else
{
  driveway2.push(licensePlate);
}
```

```
//*******************************************************************

public static void main(String[] args)
{
   Scanner stdIn = new Scanner(System.in);
   char action;
   String licensePlate;
   DrivewayParking attendant = new DrivewayParking();

   do
   {
     attendant.describeDriveways();
     System.out.print("Enter +license to add, " +
       "-license to remove, or q to quit: ");
     licensePlate = stdIn.nextLine();
     action = licensePlate.charAt(0);
     licensePlate = licensePlate.substring(1);
     switch (action)
     {
       case '+':
         attendant.parkCar(licensePlate);
         break;
       case '-':
         if (!attendant.getCar(licensePlate))
         {
           System.out.println("Sorry, couldn't find it.");
         }
      } // end switch
    } while (action != 'q');
  } // end main
} // end class DrivewayParking
```

Figure 10.13c DrivewayParking program—part C

When the customer returns, the valet determines where the car is by calling the `ArrayDeque`'s `contains` method for each driveway stack. The `contains` method returns `true` if the passed-in license plate string is stored in the stack. If the license plate string is found in one of the stacks, cars are popped off of that stack (using the `ArrayDeque`'s `pop` method) and pushed onto the other stack until the customer's car has been popped. The valet then mimics the program's popping and pushing by backing out each car that's behind the customer's car and parking it in the other driveway. To see how that works, mentally trace the DrivewayParking program, using Figure 10.14's sample session as a guide.

Sample session:

```
driveway1 []
driveway2 []
Enter +license to add, -license to remove, or q to quit: +1234
driveway1 [1234]
driveway2 []
Enter +license to add, -license to remove, or q to quit: +2345
driveway1 [1234]
driveway2 [2345]
Enter +license to add, -license to remove, or q to quit: +3456
driveway1 [3456, 1234]
driveway2 [2345]
Enter +license to add, -license to remove, or q to quit: +4567
driveway1 [3456, 1234]
driveway2 [4567, 2345]
Enter +license to add, -license to remove, or q to quit: +5678
driveway1 [5678, 3456, 1234]
driveway2 [4567, 2345]
Enter +license to add, -license to remove, or q to quit: -4321
Sorry, couldn't find it.
driveway1 [5678, 3456, 1234]
driveway2 [4567, 2345]
Enter +license to add, -license to remove, or q to quit: -1234
driveway1 []
driveway2 [3456, 5678, 4567, 2345]
Enter +license to add, -license to remove, or q to quit: q
```

Figure 10.14 DrivewayParking program sample session.
The driveway stacks are displayed with their tops at the left.

10.10 Overview of The Java Collections Framework

The `ArrayList`, `LinkedList`, and `ArrayDeque` classes and the `List`, `Queue`, and `Deque` interfaces are a small part of a large coherent body of Java API software called the *Java collections framework*.[4] In addition to lists and queues, this framework includes interfaces and classes that describe and implement other types of data structures, like *sets* and *maps*.

Figure 10.15 shows part of the top of this framework. All of the interfaces and classes shown here are in the `java.util` package. The boxes in Figure 10.15 with italicized text are interfaces. The boxes with non-italicized text are instantiable classes. The Java collections framework has two interface hierarchies: the collection hierarchy, whose root is the `Collection` interface at the top of the diagram; and the map hierarchy, whose root is the `Map` interface, near the center of the diagram. The upward-pointing dashed arrows[5] from lower interfaces to upper interfaces point to more general interfaces which the lower interfaces

[4]See http://docs.oracle.com/javase/7/docs/technotes/guides/collections/index.html.
[5]These fancy arrows are Unified Modeling Language (UML) symbols. See Appendix 7.

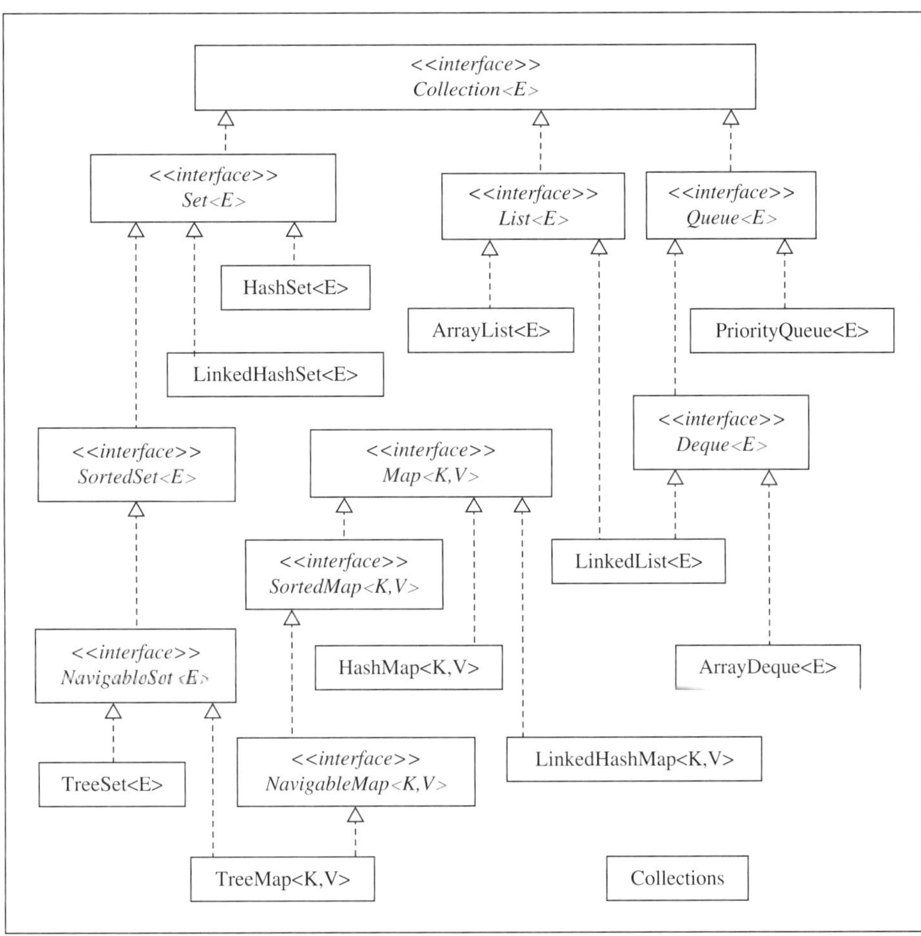

Figure 10.15 Part of the top of the Java collections framework interface hierarchy

extend. *Extending* an interface means specifying additional methods. The upward-pointing dashed arrows from lower classes to upper interfaces point to interfaces which the lower classes implement. Those classes define all the methods specified by the interfaces they implement.

The `Collection` interface specifies methods common to the `Set`, `List`, and `Queue` interfaces, such as the one-parameter `add` method, the `contains` method, the `isEmpty` method, the `set` method, and the `size` method. The `Collection` interface also specifies a `remove(Object elem)` method, which removes the first occurrence of the specified element and returns `true` if the element was found and removed. As indicated in Java API documentation, all Java collections framework classes that implement the `Collection` interface provide a one-parameter constructor that converts any type of collection to any other type of collection. This makes it easy to change a collection's container from one type to another type.

Now let's consider some of the other interfaces in Figure 10.15. You already know much of what the `List` interface specifies, and you know how useful it can be. The `Set` interface does not specify any additional methods beyond those specified in the `Collection` interface. But it adds the stipulation: A `Set` contains no duplicate elements. This is a very useful property.

As mentioned earlier, the `Queue` interface specifies methods that describe waiting-line behavior. Its `add` method adds an element to the back of the queue. Its `remove` method removes an element from the front of the queue. Notice that the `LinkedList` class implements both `List` and `Queue` interfaces, and it also implements additional `Deque` methods, which allow additions to either end or removals from either end.

There's another type of queue, a priority queue, which behaves differently than a regular queue. A priority queue is a queue in which elements are ordered in priority order, regardless of the order in which they are added to the collection. A priority queue's `add` method inserts an element into the queue so that lower-valued elements are always removed before higher-valued elements. A priority queue's `remove` method removes the element with the lowest value. The Java collections framework implements a priority queue's functionality within its `PriorityQueue` class. The `PriorityQueue` class implements the `Queue` interface but not the `Deque` interface.

We put the `Map` interfaces in the middle of Figure 10.15 to emphasize that a map relates an object in a set (on the left side) to another object that could be either another set or a list or a queue (on the right side). A map can perform the same kind of mathematical operation a list can perform. Either of these structures works like a mathematical *function*. With a mathematical function, you supply an independent input value (like an `ArrayList` index), and the function gives back a corresponding output value (like the value at that index). A map generalizes this by also accepting non-integer inputs. For example, the independent variable in a map—the map's *key*—could be the wrapped form of an integer, a character, a string, a floating-point number, or any other type of object. The only constraint is that all keys must be distinct—no duplicates. Thus, a map's keys comprise a set that can be retrieved by calling the `keySet` method.

It's helpful to think of a map as a two-column lookup table. The first column contains items that you know (keys). The second column contains items that you want to find (values). To add a value of type `V` identified by a key of type `K` to a Java `Map`, use this method:

```
public V put(K key, V value)
```

If there is already a value associated with this `key`, the `put` method returns that pre-existing value. To retrieve a reference to the value identified by `key`, use this method:

```
public V get(Object key)
```

And to remove and return the value identified by `key`, use this method:

```
public E remove(K key)
```

In the program in the next section, we'll use a map to relate the integer that serves as an object's identifier to the object it identifies. Whenever it can, that program will work with the identifying integer, but when it needs to access or modify a detail within the object, it will use the map to get the object and look at or change that internal detail.[6]

Note the `Collections` class in the bottom-right corner of Figure 10.15. It's a stand-alone utility class. It has no constructors, and all of its methods are class methods. These methods provide general services to other members of the Java collections framework. Some examples of useful `Collections` methods are `addAll`, `binarySearch`, `fill`, `max`, `min`, `replaceAll`, `reverseOrder`, `rotate`, `shuffle`, `sort`, and `swap`. You can learn more about these in the Java API documentation. The API method headings

[6]A map's key does not need to be just one independent variable. It could be two or more independent variables. And those variables could be different types of objects. To convert a combination of several different objects into a single key, have each of those objects call a universally available `hashCode` method, which returns an `int`. Then add those returned `int`'s to form a single key for the combination. When that single `int` sum goes into one of `Map`'s methods, Java automatically converts it into an `Integer`.

in this class frequently employ strange-looking, generic-type syntax, but if you just do the most obvious thing, there's a good chance it will work and do what you want.

The program in the next section will use four of the classes in Figure 10.15—`TreeSet` (a set), `LinkedHashSet` (another set), `ArrayDeque` (a queue), and `HashMap` (a map). A `TreeSet` sorts its elements into a default or prescribed order. In the upcoming program, `TreeSet` elements will be integers, and by default, the `TreeSet` will keep them ordered from lowest to highest. A `LinkedHashSet` keeps its elements in the order of entry. The upcoming program will use an `ArrayDeque` to implement a FIFO queue. A `TreeMap` has the same order as its `TreeSet` keys. A `LinkedHashMap` has the same order as its `HashSet` keys. A `HashMap`'s order is not predictable, but a `HashMap` is the most efficient type of map. Because order does not matter for a simple lookup, the upcoming program will employ a `HashMap` for its lookup table.

10.11 Collections Example—Information Flow in a Network of Friends

Now let's consider an example that illustrates use of the software in the Java collections framework. Here's the situation. Suppose there is a set of citizens, and each citizen has a random set of friends. Assume all friendships are mutual. That is, all the friends of any particular citizen consider that citizen to be one of their friends, too. Different citizens typically have different friends and different numbers of friends.

After we create citizens and a friendship network, we look to see how information flows through the network by performing the following experiment: We select one particular citizen and give that citizen a message. We ask that citizen to send that message to all of his or her friends and ask those friends to pass it on to their friends. This continues until everyone who can be reached becomes informed.

At the highest level, there are just two basic steps—build the network and distribute a message. So the top-level algorithm is easy. The devil is in the details. The most difficult aspects of building the network are deciding how to store the configuration information and deciding where to put the methods that assemble that information. The most difficult aspect of distributing the message is avoiding infinite repetition.

First, consider how to store configuration information. It's difficult to draw a complete picture of a large random network, and it would be hard to store such a picture if it could be drawn. A better strategy is to store a list of the parts. We could store a list of citizens with associated relationships or we could store a list of relationships with associated citizens. It's more intuitive to store the concrete items (the citizens) and associate the relationships with them. So that's what we'll do.

Because this program makes heavy use of sets, each citizen needs a unique identifier. The obvious way to initialize a citizen's identifier is through a constructor parameter. But what if code calling that constructor happens to use the same ID in two different instantiations? How would we check for that possibility, and how would we respond if it occurred? We'll avoid these problems by encapsulating ID assignment within the class that defines a citizen.

Top-Level Code

Figure 10.16 shows the program's driver. In the `main` method, the first local variable is a `Scanner` object that retrieves keyboard input. The second local variable is an object that represents the community. The third local variable is a map of the community's citizens. The fourth local variable is a set that eventually holds the results of a message-propagation experiment.

The first block of code creates the network. It asks the user to enter two integers separated by whitespace. The first entry is the total number of citizens. The second entry is the total number of distinct

```
/**************************************************************
* CommunityDriver.java
* Dean & Dean
*
* This generates citizens, establishes friend relationships,
* and propagates a message through the network of friends.
**************************************************************/

import java.util.*; // Scanner, Map, and Set

public class CommunityDriver
{
  public static void main(String[] args)
  {
    Scanner stdIn = new Scanner(System.in);
    Community community;
    Map<Integer, Citizen> citizens;
    Set<Integer> informedCitizens;

    // Create the network.
    System.out.print("Enter citizen & relation quantities: ");
    community =
      new Community(stdIn.nextInt(), stdIn.nextInt());
    citizens = community.getCitizens();
    System.out.println("Citizen\tFriends");
    for (Integer id : citizens.keySet())
    {
      // use Citizen's toString method to display citizen info:
      System.out.println(citizens.get(id));
    }

    // Propagate a message through it.
    System.out.print("Enter information source ID: ");
    informedCitizens = community.spreadWord(stdIn.nextInt());
    System.out.println("Citizen\tDelay");
    for (Integer citizenID : informedCitizens)
    {
      System.out.printf("%d\t%d\n",
        citizenID, citizens.get(citizenID).getDelay());
    }
  } // end main
} // end CommunityDriver
```

Figure 10.16 CommunityDriver class

citizen-to-citizen relations (friendships). The next statement instantiates a Community object, whose constructor builds the network. The next statement gets a reference the citizen map. The for loop in this first block of code displays the community's citizens and their friends, with one citizen per row. In each row, the first column contains a citizen's ID, and the second column contains the IDs of all of that citizen's friends. The println statement in this for loop displays citizen information by using the Map called citizens to retrieve a particular citizen object and then asking that object to call its toString method.

The second block of code propagates a message through the network. It asks the user to enter the ID of the citizen who originates the message. The next statement asks the Community method called spreadWord to propagate the message. The remaining statements in this code block describe the message propagation. In each row, the first column contains the ID of a citizen receiving the message, and the second column contains the total number of steps required for the message to get to that citizen—the delay.

Network Composition

Figure 10.17a shows the first part of the Community class. The instance variable, citizens, refers to the map that contains citizen data. This map's keys are individual-citizen IDs, and the corresponding values are the individual-citizen objects corresponding to those IDs. Because the individual-citizen objects eventually include all friend relationships, this map is the repository for all network data. As you saw in the preceding driver code, it's also a handy tool for finding that data.

The Community constructor has parameters for the total number of citizens and the total number of citizen-to-citizen interconnections. It creates a local variable for a random number generator called random. It declares a local variable for a Citizen object. It declares local variables for two citizen ID numbers. The first for loop instantiates each citizen, and using that citizen's internally generated ID, it enters that citizen into the general data repository—the map called citizens. This code assumes that the immutable citizen ID is a public instance variable.

The second for loop creates friend relationships. It randomly picks one citizen. Then it randomly picks another citizen who is not the first citizen and is not one of the first citizen's already-existing friends. Notice how it uses the Java Collection's contains method in the while condition of a do loop to perform expeditiously the tricky logic required to avoid duplication. After identifying a unique new relationship, this method adds each citizen involved in that relationship to the other citizen's set of friends. The pair of addFriend method calls in the last two statements creates bidirectional links. For one-way links, you could omit one of these two method calls.

The getCitizens method at the bottom of Figure 10.17a gives outsiders like drivers access the map called citizens. Once they have this map, they can use a simple citizen ID argument in a citizens.get method call to retrieve the corresponding Citizen object. You saw examples of these operations in the for loops in the driver in Figure 10.16.

Message Propagation

Figure 10.17b contains the method that propagates a message through the network. This method's parameter, sender, initially contains the ID of the message originator. The Set called informedCitizens accumulates IDs of all citizens who have received the message. Because it's a LinkedHashSet, it keeps its contents in their order of entry. The Queue called sendersQueue is a FIFO queue that contains informed citizens who have not yet forwarded the message to their friends.

The setDelay method call records the relative time (delay) when a sender received the message he or she is forwarding. After setting the message originator's delay to zero, spreadWord adds the message originator's ID to informedCitizens and sendersQueue.

```
/*********************************************************************
* Community.java
* Dean & Dean
*
* This describes community structure and behavior.
*********************************************************************/

// Random, Map, HashMap, Set, LinkedHashSet, Queue, ArrayDeque
import java.util.*;

public class Community
{
  private Map<Integer, Citizen> citizens = new HashMap<>();

  //******************************************************************

  // Postcondition: all connections are bidirectional

  public Community(
    int citizenQuantity, int relationQuantity)
  {
    Random random = new Random(0);
    Citizen citizen;                       // any citizen object
    int self, other;                       // ID numbers

    for (int i=0; i<citizenQuantity; i++)
    {
      citizen = new Citizen();
      citizens.put(citizen.ID, citizen); // ID is public
    }
    for (int j=0; j<relationQuantity; j++)
    {
      self = random.nextInt(citizens.size());
      do
      {
        other = random.nextInt(citizens.size());
      } while (other == self ||
          citizens.get(self).getFriends().contains(other));
      citizens.get(self).addFriend(other);
      citizens.get(other).addFriend(self);
    }
  } // end constructor

  //******************************************************************

  public Map<Integer, Citizen> getCitizens()
  {
    return this.citizens;
  } // end getCitizens
```

Figure 10.17a Community class—part A.
This creates citizens and builds the network of friends.

```
  //*******************************************************************

  // Precondition: sender is part of an established network
  // Postcondition: return set includes all connected citizens

  public Set<Integer> spreadWord(int sender)
  {
    Set<Integer> informedCitizens = new LinkedHashSet<>();
    Queue<Integer> sendersQueue = new ArrayDeque<>();

    citizens.get(sender).setDelay(0);         // for originator
    informedCitizens.add(sender);
    sendersQueue.add(sender);
    do
    {
      sender = sendersQueue.remove();
      for (Integer friend : citizens.get(sender).getFriends())
      {
        if (!informedCitizens.contains(friend))
        {
          citizens.get(friend).setDelay(
            citizens.get(sender).getDelay() + 1);
          informedCitizens.add(friend);
          sendersQueue.add(friend);
        }
      } // end for each uninformed friend
    } while (!sendersQueue.isEmpty());
    return informedCitizens;
  } // end spreadWord
} // end class Community
```

Figure 10.17b `Community` class—part B.
This propagates a message through the network.

Then the `spreadWord` method enters a `do` loop. The first statement in this `do` loop extracts the `ID` at the front of the queue and assigns it to `sender`. Initially, this ID extraction from the queue is just a re-assignment of the originator's ID. But in subsequent `do`-loop iterations, `sender` acts like a local variable and takes on different citizens' IDs.

With each new `sender`, `spreadWord` enters a for-each loop that loops through all of that sender's friends. If a particular friend is uninformed, that friend gets a delay equal to the sender's delay plus 1. Then that friend's ID goes into `informedCitizens` and `sendersQueue`. As this process proceeds, more and more friends become informed, and because informed citizens do not satisfy the `if` condition, they do not re-enter the queue. With fewer and fewer new entries, eventually the queue becomes empty, the `do` loop terminates, and `spreadWord` returns the set of all informed citizens.

The Citizen Class

Figure 10.18 shows the Citizen class. The first statement defines a private class variable, nextID, which provides a unique identifier for each new instance. The second statement initializes the ID instance constant with the current value of nextID and then immediately increments nextID in preparation for the next instantiation. Because ID is a constant, we can make it public and avoid the definition and the calling of a getID method.

The friends instance variable is a reference to a set of friend IDs. We give this variable an instance of a TreeSet because we want the friends to be stored in numerical order. If the IDs were names instead of numbers, the variable declaration would be Set<String>, and the corresponding TreeSet would store friends in alphabetical order. If we had instantiated friends as a LinkedHashSet, this would store oldest friends first. The next instance variable, delay, records the relative time when the current citizen receives a message, and that depends on who starts the message.

Notice that there is no explicit constructor and there is no setID method. So it is impossible for an outsider to duplicate citizen identifiers accidentally. Citizen instantiation initializes the current citizen's ID, and it creates a set to hold friend IDs, but it does not populate that friends set. Making friends comes later, in separate successive calls to the addFriend method. The getFriends, setDelay, and getDelay methods are self-explanatory. The toString method displays the current citizen's ID. Then it displays the IDs of that citizen's friends, using the default toString method provided by the Java API TreeSet class.

Typical Output

Figure 10.19 shows results produced by a typical execution. The first two inputs specify a total of 16 citizens and a total of 16 distinct citizen-to-citizen relationships. Under "Citizens Friends" the 16 rows of output describe the network created by the zero-seeded random number generator in the Community constructor. The numbers in the first column are the IDs of the selected citizens. The sets of numbers in the second column are the IDs of friends of the citizens in the first column. In each case, friend IDs are in ascending order because each set of friends is an instance of the TreeSet class. Notice that all friendships are mutual. That is, if citizen B is a friend of citizen A, then citizen A is a friend of citizen B. This makes the total number of friends in the table (32) equal to twice the specified number of (bidirectional) relationships. Also, notice that different citizens have different numbers of friends, and citizen 7 has no friends at all.

The third input arbitrarily specifies a particular message originator: citizen 4. The two columns below that input describe message propagation. The first sender out of the sendersQueue queue (Citizen 4) sends the message to friends 1, 2, 5, and 10, who then enter the queue in that order. The next sender out of the queue (citizen 1) finds 4 is already informed, but she sends to 8, who then enters the queue. The next sender out of the queue (citizen 2) has no uninformed friends. The next sender out of the queue (citizen 5) finds that 2 and 4 already are informed, but he sends to 11. This process continues until the queue becomes empty. Because citizen 7 is nobody's friend (poor guy!), he is never informed.

Because the informedCitizen set returned by the spreadWord method is a LinkedHashSet, the sequence of sender IDs in the output is the same as the sequence of entries into the informed-Citizen set. Notice that this sequence of outputs moves through the network row by row, going left to right across each row before moving down to the next row, until all citizens that can be reached have been reached. If you like fancy terminology, you'll be pleased to learn that this sequence is called a *breadth-first traverse.*

```
/********************************************************************
* Citizen.java
* Dean & Dean
*
* This represents an element in a network of citizens.
********************************************************************/

import java.util.*;                                  // Set & TreeSet

public class Citizen
{
  private static int nextID = 0;                     // for unique IDs
  public final int ID = nextID++;                    // cannot change!
  private Set<Integer> friends = new TreeSet<>();
  private int delay;

  //******************************************************************

  public void addFriend(int friendID)
  {
    this.friends.add(friendID);
  } // end addFriend

  public Set<Integer> getFriends()
  {
    return this.friends;
  } // end getFriends

  public void setDelay(int delay)
  {
    this.delay = delay;
  } // end setDelay

  public int getDelay()
  {
    return this.delay;
  } // end getDelay

  //******************************************************************

  public String toString()
  {
    return String.format("%d\t%s", ID, friends);
  } // end toString
} // end class Citizen
```

Figure 10.18 `Citizen` class
Used by `CommunityDriver` in Figure 10.16 and by `Community` in Figures 10.17a and 10.17b.

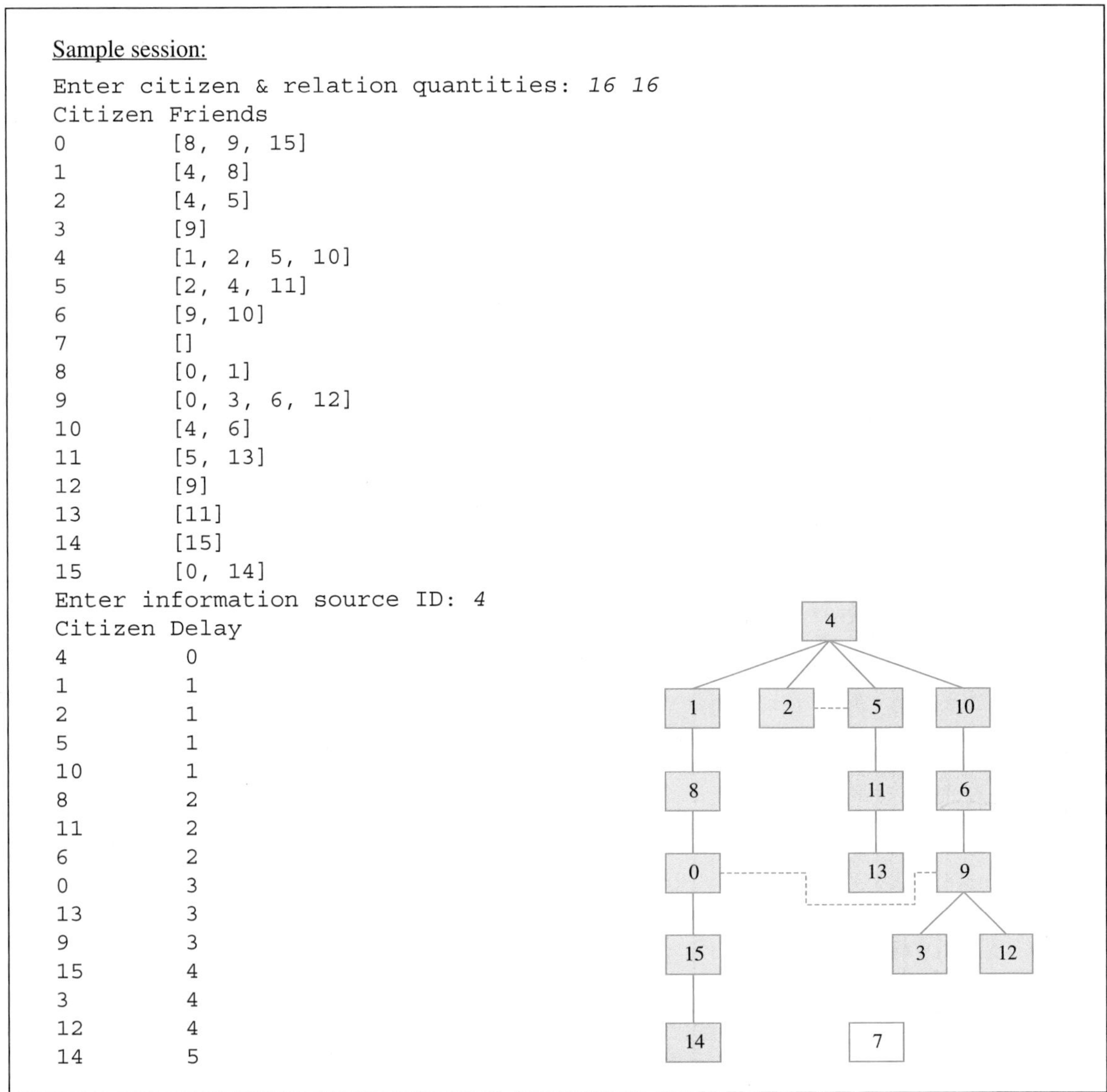

Sample session:

```
Enter citizen & relation quantities: 16 16
Citizen Friends
0       [8, 9, 15]
1       [4, 8]
2       [4, 5]
3       [9]
4       [1, 2, 5, 10]
5       [2, 4, 11]
6       [9, 10]
7       []
8       [0, 1]
9       [0, 3, 6, 12]
10      [4, 6]
11      [5, 13]
12      [9]
13      [11]
14      [15]
15      [0, 14]
Enter information source ID: 4
Citizen Delay
4         0
1         1
2         1
5         1
10        1
8         2
11        2
6         2
0         3
13        3
9         3
15        4
3         4
12        4
14        5
```


Figure 10.19 Typical output produced by the `CommunityDriver` program in Figure 10.16
In the graphical representation at right, solid lines are connections to uninformed friends. Dashed lines are connections to previously informed friends. Friendless Citizen 7 is never informed.

Summary

- If you need to insert or delete several elements within an array, instead of using an array, consider using an `ArrayList`. When you declare an `ArrayList` variable, you must include the element type in angled brackets after the `ArrayList` collection type name. If the declaring statement also instantiates, you may omit the element type in the subsequent instantiation, as in this `ArrayList` of `Car` elements:

```
ArrayList<Car> car = new ArrayList<>();
```

If instantiation comes later, it should include the element type, like this:

```
car = new ArrayList<Car>();
```

- An `ArrayList` automatically adjusts its length to accommodate changes in the number of elements contained. Use `ArrayList`'s `add(E elem)` to add an element to the end of an `ArrayList`, or use `add(int index, E elem)` to insert an element at a particular position. To remove an element from a particular position, use `remove(int index)`. The `size` method returns the total number of currently contained elements.
- An `ArrayList` stores objects only. Autoboxing and unboxing automatically make necessary conversions between primitives and wrapped primitives, so you don't have to worry about that. But if you want an `ArrayList` of primitives like `int`, you must declare it with the wrapped type, like this:

```
ArrayList<Integer> num = new ArrayList<>();
```

- You can pass objects to and from methods anonymously.
- Because a `LinkedList` implements almost all of the methods implemented by an `ArrayList`, it can perform the same functions as an `ArrayList`. But instead of keeping its elements in an array, a linked list chains them together with forward and backward references.
- An interface serves as a template for implementing classes that share common characteristics.
- You can use `System.nanoTime` method calls to measure and compare code performance.
- Chaining makes access operations slower in a `LinkedList` than in an `ArrayList`.
- A queue uses a first in, first-out (FIFO) strategy for adding and removing elements.
- A stack uses a last-in, first-out (LIFO) strategy for adding and removing elements.
- You can use an `ArrayDeque` to implement both queues and stacks efficiently.
- Interfaces like `List`, `Queue`, `Set`, and `Map` organize methods in the Java collections framework into coherent groups.

Review Questions

§10.2 The `ArrayList` Class

1. How is an `ArrayList` more versatile than an array?
2. To avoid runtime errors, you must always specify the size of an `ArrayList` when you declare it. (T / F)
3. What is the return type of the `ArrayList` class's `get` method?
4. If you call the `ArrayList` method, `add(i, x)`, what happens to the element originally at position `i`?

§10.3 Storing Primitives in an `ArrayList`

5. Specifically, under what circumstances does autoboxing take place?
6. Write one statement that appends the `double` value, 56.85, to the end of an existing `ArrayList` called `prices`.
7. Write one statement that displays all of the values in an `ArrayList` of `Doubles` called `prices`. Put the complete list in square brackets and use a comma and a space to separate different values in the list.

§10.4 ArrayList Example Using Anonymous Objects and the For-Each Loop

8. What is an anonymous object?

§10.5 ArrayLists Versus Standard Arrays

9. Given:
    - You have a WeatherDay class that stores weather information for a single day.
    - You like to store WeatherDay objects for a whole year.
    - The primary task of your program is to create distinct sorted collections of WeatherDay objects (sorted by temperature, wind speed, and so on).

   How should you store your WeatherDay objects—in an ArrayList or in a standard array? Provide a rationale for your answer.

§10.6 The LinkedList Class

10. Where does the JVM start the process of trying to access an element in the interior of a linked list?

§10.7 The List Interface

11. Based on what you have learned so far, what is a Java interface?

§10.8 Problem Solving: How to Compare Method Execution Times

12. For a given set of inputs, the execution time of a given program is always the same. (T / F)
13. When we remove an element from the center of an ArrayList, higher-index elements must shift downward. That shifting is not needed when we remove an element from a linked list. Why then does it usually take more time to remove an element from the center of a linked list?

§10.9 Queues, Stacks, and the ArrayDeque Class

14. Because an ArrayDeque is based on an array, when you remove an element from the front of an ArrayDeque, the computer must shift all higher elements down by 1. (T / F)
15. What is the difference between a queue and a stack?

§10.10 Overview of the Java Collections Framework

16. What are the interfaces at the tops of the two interface hierarchies in the Java collections framework?
17. What is a set?

§10.11 Collections Example—Information Flow in a Network of Friends

18. Why don't we use a constructor parameter to establish the ID of a Citizen?

Exercises

1. [after §10.2] What does the ArrayList's remove method do?
2. [after §10.3] Provide a single statement (an initialization statement) that declares an ArrayList named evenNumbers and assigns a newly instantiated ArrayList to it. The instantiated ArrayList should be able to store integers.
3. [after §10.3] Using the evenNumbers ArrayList created in the previous exercise, provide a code fragment that stores the first 10 even numbers in the evenNumbers ArrayList. In other words, put 0 in the first evenNumbers element, put 2 in the second evenNumbers element, and so on . . . , until 18 is put in the tenth evenNumbers element. You must use a standard for loop for your code fragment.

4. [after §10.4] Provide a more elegant (but functionally equivalent) version of this code fragment:

```
ArrayList<Car> cars = new ArrayList<Car>();
Car car1 = new Car("Mustang", 2006, "tiger-striped");
cars.add(car1);
Car car2 = new Car("MiniCooper", 2006, "lime green");
cars.add(car2);
```

5. [after §10.4] Suppose you have an ArrayList of street addresses that's been initialized and filled as follows:

```
ArrayList<String> addressList = new ArrayList<String>();
addressList.add("1600 Pennsylvania Avenue");
addressList.add("221B Baker Street");
. . .
addressList.add("8700 N.W. River Park Drive");
```

Provide a for-each loop (not a standard for loop) that prints the addressList's addresses, one address per line.

6. [after §10.5] Suppose you want to maintain a list of cities described by the City class as defined in Exercise 16 of Chapter 9. And suppose you want to be able to insert or remove cities at any place in the list to maintain a certain order as you add or remove elements. Should you use an array or an ArrayList, and why?

7. [after §10.6] Assume a Customer object includes the instance variables, previous and next. When a Customer object is in a LinkedList, the previous variable refers to the preceding Customer object, and the next variable refers to the following Customer object. Assume the Customer class defines getPrevious, setPrevious, getNext, and setNext methods. Suppose a variable called customer contains a reference to one of the objects in the interior of a LinkedList of Customer objects. Write a code fragment that removes from the linked list the object referred to by customer, by making the preceding object refer to the following object, and vice versa. [Hint: If you use method call chaining, you can do this with no additional variable declarations and two statements.]

8. [after §10.6] Suppose there's a LinkedList of Customer elements as described in Exercise 7. This time, suppose another variable called lineCrasher refers to a new Customer object. Write a code fragment that inserts the lineCrasher object into the linked list just before the customer object. [Hint: If you are careful about the sequence, and you use some chaining, you can do this with no additional variable declarations and just four statements.]

9. [after §10.7] The main method in Figure 10.7's ListExecutionTimes program contains this local variable declaration and initialization:

```
ArrayList<Double> list = new ArrayList<>();
```

After determining ArrayList execution times, we modified the program by replacing the above statement with this:

```
LinkedList<Double> list = new LinkedList<>();
```

Declare the list variable in a way that allows you to assign an ArrayList object to it initially and then later assign a LinkedList object to it. Write the initial declaration and assignment statement, and also write the later assignment statement. Why can you assign either an ArrayList object or a LinkedList object to this same variable?

10. [after §10.8] The times measured by the ListExecutionTimes program includes not only the time required to perform the targeted get and set operations. It also includes time required to perform the for loop operations and the System.nanotime method calls. You can measure these overhead times by running the program with the get and set method calls removed. Modify the ListExecutionTimes class in Figure 10.7a to measure the net average get and set time (with for-loop overhead and nanotime method call overhead removed).

11. [after §10.9] Assume that a stack has been declared and instantiated with the ArrayDeque class. Show the output produced by the following code fragment:

```
stack.push(1);
stack.push(2);
System.out.println(stack.pop());
stack.push(3);
System.out.println(stack.peek());
stack.push(4);

while (!stack.isEmpty())
{
  System.out.println(stack.pop());
}
System.out.println(stack.peek());
```

You should be able to do this by just thinking about it, but feel free to complete and run a program to check your answer.

12. [after §10.10] Assume an ordinary queue has been declared with the Queue interface and instantiated with the ArrayDeque class. Also, assume a priority queue has been declared with the Queue interface and instantiated with the PriorityQueue class. Figure 10.15 showed that the PriorityQueue class also implements the Queue interface. Therefore, the PriorityQueue class also implements add and remove methods. Its remove method does the same thing as ArrayDeque's remove method. But its add method is different. When it adds elements, a priority queue inserts them into the queue so that lower-valued elements are always closer to the front. In other words, in a priority queue, lower-valued elements have priority and "crash" the line. Given this understanding, show the output produced by the following code fragment:

```
ordinaryQueue.add(8);
ordinaryQueue.add(3);
ordinaryQueue.add(12);
System.out.println(ordinaryQueue.remove());
ordinaryQueue.add(5);

while (!ordinaryQueue.isEmpty())
{
  priorityQueue.add(ordinaryQueue.remove());
}
while (!priorityQueue.isEmpty())
{
  System.out.println(priorityQueue.remove());
}
```

You should be able to do this by just thinking about it, but feel free to complete and run a program to check your answer.

13. [after §10.10] A collection like `ArrayList<Bear> bears` could contain two or more references to the same object. Therefore, `bears.size()` would not necessarily return the total number of distinct `Bear` objects in a bear store's inventory. However, all Java collection framework classes have constructors that create the current type of collection from any other type of collection. Given the declaration:

```
int totalBears;
```

Write a single statement that assigns to `totalBears` the total number of distinct `Bear` objects in the `bears` collection.

14. [after §10.10] Enhance the `HungerGames` program in Figure 10.2. Create a `Tribute` class with three instance variables, `name` (a `String`), `gender` (a `char`) , and `district` (an `int`). Provide this new `Tribute` class with a constructor to initialize its instance variables and with `get` methods to access them. Modify the `HungerGames` class by changing `String deceased` to `Tribute deceased`, making the `tributes` list a list of `Tribute` objects, and providing this additional local variable:

```
Map<String, Tribute> tributeMap = new HashMap<>();
```

Populate the `tributes` list with these objects:

```
Tribute("Cato", 'm', 2)
Tribute("Katniss", 'f', 12)
Tribute("Peeta", 'm', 12)
Tribute("Rue", 'f', 11)
Tribute("Finnick", 'm', 4)
```

Then, in a for-each loop that iterates through the `tributes` list, use `Map`'s `put` method to put each tribute into `tributeMap`, using each tribute's name as the key to the object that more completely describes that tribute. Instead of using `Math.random` to pick random indices, use the `Collections` class's `shuffle` method to randomize the sequence of the `tributes` list. Then, in an ordinary `for` loop that iterates through all but the last element in the shuffled `tributes` list, do the following:

- First, use `tributes.get(i)` to get the next object in the `tributes` list.
- Second, use `Map`'s `remove` method to remove this object from the `tributesMap` map.
- Third, display the three attributes of the removed object. After this `for` loop, in a print statement, call `tributesMap.keySet()` to display the keys of all remaining entries in `tributesMap`. Does the operation that removes an entry from the map also remove an element from the list?

15. [after §10.10] In the ListExecutionTimes program in Figure 10.7, the `getIndices` method filled an array by inserting each successive new index at a random position in the existing array. Write code for an alternative `getIndices` method that returns a similar result by using the zero-parameter `add` method to populate the `ArrayList` and the `shuffle` method from the `Collections` class to shuffle it.

16. [after §10.11] Assuming the same information source (citizen 4), what would the second part of the output (Citizen ID and Delay) look like if that `ArrayDeque` were operated as a stack instead of as a queue? Would the same citizens eventually become informed? [Hint: Which lines of code would you change in the `spreadWord` method in order to see what happens?]

Review Question Solutions

1. With an `ArrayList`, you can insert and delete elements anywhere in the sequence, and the list length grows and shrinks dynamically.

2. False. Normally, you specify no size for an `ArrayList` when you declare it.

3. The get method's return type is E, which refers to the type of each element in the ArrayList.
4. The element that is originally at position i shifts to the next higher index position.
5. Autoboxing takes place when a primitive is being used in a place that expects a reference.
6. prices.add(56.85);
7. System.out.println(prices);
8. An anonymous object is an object that's instantiated but it's not stored in a variable.
9. You should store your WeatherDay objects in a standard array.

   Rationale:
   - There's no need for the array to grow or shrink since the size is fixed at 366 (and standard arrays have a fixed size).
   - In sorting, you'll need to access the objects quite often (and access is easier with standard arrays).
10. The process of accessing a linked list element starts from either of the two ends.
11. A Java interface is a template for designing classes that share certain behaviors.
12. False. Executions vary with hardware and within the same hardware from one run to another.
13. Compared to jumping directly to a center element in an ArrayList, it takes a relatively long time to step from one of the ends to a center element in a LinkedList.
14. False. After removing the element at the front of an ArrayDeque, instead of shifting all higher elements down by 1, the computer moves the location of the front up by 1.
15. A queue is a first-in-first-out (FIFO) structure. The end from which elements are removed is opposite the end to which elements are added. A stack is a last-in-first-out (LIFO) structure. The end from which elements are removed (popped) is the same as the end to which elements are added (pushed).
16. The interfaces at the tops of the two interface hierarchies in the Java collections framework are the Collection<E> interface and the Map<K,V> interface.
17. A set is a collection of distinct objects, with no duplications.
18. We don't use a constructor parameter to establish the ID of a Citizen because we want to make it impossible for two distinct citizens to have the same ID. We need distinct citizen IDs because we use citizen IDs as members of the keySet of the Java Map called citizens and as members of the Java Set called informedCitizens.

CHAPTER 11

Recursion

Objectives

- Understand the basic concept of recursion.
- Appreciate the importance of stopping conditions.
- Describe iterative calculations with recurrence relations.
- Use a trace to show exactly what happens in a recursion.
- Convert between iterative and recursive implementations.
- Learn how to evaluate short recursions by hand.
- Use recursion to binary-search an array of ordered data.
- Use recursion to merge-sort an array of unordered data.
- Use recursion to solve the Towers of Hanoi puzzle.
- Partition a GUI problem into Model, View, and Controller concerns.
- Use recursion to model biological trees as fractals.
- Employ Java's `Thread.sleep` method to implement animation.
- Estimate performance and become familiar with Big O notation.

Outline

11.1 Introduction

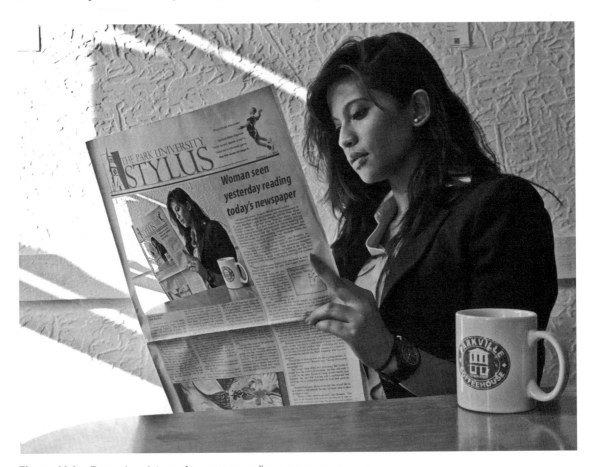

To understand this chapter, you need to know how to write your own multiple-method programs. As such, you need to have read up through the Interlude. For Sections 11.6 and 11.7, you need to know about arrays, so in addition to reading up through the Interlude, you also need to have read Chapter 9 Sections 9.1 through 9.8. For the GUI-track Section 11.9, you need to know about object-oriented programming, so you need to have read up through Chapter 6. Section 11.10 analyzes algorithms from Chapters 9 and 10 as well as from this chapter.

In prior chapters, we provided many examples of methods calling other methods, but we didn't show an example of a method calling itself. If you're not used to it, the idea of a method calling itself probably seems pretty bizarre. Although it's not all that common for a method to call itself, the technique can be used for elegant solutions to problems that otherwise would be very difficult to solve. A method that calls itself is known as a *recursive method,* and a program is said to exhibit *recursion* if it relies on recursive method calls.

In a general sense, recursion is the concept of describing something such that the description relies on a smaller version of the original thing. The smaller version describes itself by relying on a still smaller version of itself, and so on and so on. Typically, recursion appears in pictures and in problem solutions. Figure 11.1 is a recursive picture. In the figure, do you see the newspaper's picture? It's a smaller version of the original

Figure 11.1 Recursive picture of a woman reading a newspaper

picture. If you look closely, you should be able to recognize the newspaper's picture as containing a still-smaller version of the original picture. Near the end of this chapter, there's a section that describes how to draw a recursive picture using Java. But before that section, and even in that section, the chapter's primary focus is not about drawing recursive pictures, but about using recursion to solve problems. Let's begin by thinking about a recursive solution to a non-programming problem . . .

Since the time of Greek antiquity, ships have relied on flags to relay information between them. In wartime, such flag signals have been particularly helpful. Unlike radio signals, which can be intercepted by the enemy from afar, flag signals can be read only by ships close enough to see them.

In battle, an admiral might want to determine the number of casualties his or her fleet has suffered. If the fleet's ships are in a line, with the admiral's ship at one end of that line, the admiral can initiate the process of determining the number of casualties by raising a flag (or set of flags) that signals a request for casualty information. That request propagates to the far end of the line. The last ship returns a flag signal that tells the previous ship the number of casualties that it has suffered. The previous ship then adds that number to its own number of casualties and uses another flag signal to relay this sum back to the previous ship. That continues until the total number of casualties gets back to the admiral's ship.

The naval flag-signaling process described here is recursive in that each ship has the same task—finding the total number of casualties suffered by its own ship plus the number of casualties suffered by the ships farther away. Another reason it's recursive is that each ship must rely on the next ship to gather a subset of the total information that's needed—the casualties suffered by the farther-away ships are a subset of the casualties suffered by all the ships. The flag-signaling process parallels the process of a method making a series of recursive method calls. Each method call has the same task, and each method call relies on the next method call to gather a subset of the final information that's needed.

Because beginning programmers often have a hard time wrapping their heads around recursion, in the first half of this chapter, we stick to the basics—the steps necessary to write a recursive method and a comparison of recursive methods and iterative/loop methods. In the interest of understandability, we use relatively short and simple programs to illustrate those basic concepts. In the second half of the chapter, we let you spread your wings by showing you recursive programs that better illustrate the true power and elegance of recursion. Those programs are more involved, but the effort is worth it. Without recursion, the programs would be very difficult to write. The programs implement algorithms that naturally lend themselves to recursion. Once you understand the recursive nature of the algorithms, the program implementations are fairly straightforward.

11.2 Guidelines for Writing a Recursive Method

In this section, we'll provide you with guidelines for writing recursive methods. Hopefully, they will make sense, but don't worry too much if they don't. They will make more sense in the next section, when you can use the guidelines as the basis for writing an actual recursive method.

If you're given a problem you think might lend itself to a recursive solution, the first step to finding such a solution is to figure out how the problem can be split into successively smaller subproblems. Each subproblem must solve the same type of problem as the original problem, but it must solve its problem using a smaller set of data than the data set for the previous problem. The next step is to identify a condition, called the *stopping condition* or *base case,* which indicates that the current problem is small enough such that it can be solved without having to subdivide the problem further. Can you think of the stopping condition in the naval flag-signaling procedure described earlier? It's when the last ship in the line receives the admiral's request. That last ship solves the problem of finding the number of casualties by returning just its own casualties. It doesn't need to worry about finding the number of casualties in ships farther away because there aren't any.

In implementing a recursive solution, you'll need to define a method with at least one parameter whose value becomes closer to the stopping condition's value with each successive recursive method call. If you don't have such a parameter, the recursive method calls will have no way of reaching the stopping condition. Within your recursive method's body, you'll need to use an `if` statement to check for the stopping condition. The `if` body should contain the solution to the simplest version of the problem. The `else` body should contain one or more calls to the same method with argument values that are closer to the stopping condition's value. When the method is called, the method continues to call itself recursively with argument values that become progressively closer to the stopping condition's value. The calls stop when the stopping condition is satisfied.

When the stopping condition is satisfied, the method solves the simplest version of the problem. Then the method returns the simplest problem's solution to the previous method execution at the next higher level of difficulty. That method execution generates the solution to its version of the problem. This process of returning simpler solutions to previous method executions continues back to the original method execution, which returns the solution to the original problem.

11.3 A Recursive Factorial Method

Calculating a Factorial—Background Details

In this section, we'll use the guidelines described in the previous section and come up with a recursive method that finds the factorial of a given number. Before working on the recursive method, let's review what a factorial is. You can find the factorial of a number by multiplying all the numbers between 1 and the number itself. More formally, here is the formula for the factorial of a number n:

```
n! = n * (n-1) * (n-2) * ... * 2 * 1
```

The exclamation point is the mathematical symbol for factorial, so n! is "n factorial." If n is 5, then

```
5! = 5 * 4 * 3 * 2 * 1 = 120
```

In coming up with a recursive method solution, the first step is to figure out how the factorial problem can be split into successively smaller factorial subproblems. In looking at the formula for the factorial of 5, notice that the right side of the equation is 5 times the factorial of 4. After all, isn't 4 * 3 * 2 * 1 the factorial of 4? So 5 factorial can be written as

```
5! = 5 * 4!
```

Likewise, 4 factorial can be written as

```
4! = 4 * 3 * 2 * 1 = 4 * 3!
```

The pattern tells us that the factorial of any number n can be found by multiplying n times the factorial of n minus 1. Here is the formula:

```
n! = n * (n-1)!
```

This formula is the *recurrence relation* for a factorial. A recurrence relation splits up a problem into successively smaller subproblems. Finding the recurrence relation is the first step in solving a recursion problem.

The next step is to identify the stopping condition. In calculating the factorial of a positive number n, the stopping condition is when n equals 1, because that's as low as you need to go when you're multiplying the numbers together. But you should also handle the case of calculating the factorial of 0. To do that, you should test for n equals 0 and return 1 in that case (because the factorial of 0 is 1). You don't need to find the factorial of a negative number because negative number factorials are undefined.

The Recursive Factorial Method

Look at Figure 11.2 and study its recursive `factorial` method. In the method's heading, note the parameter `n`. It's the number whose factorial is being found. Also, note the statement that contains the recursive method call:

```
nF = n * factorial(n-1);
```

```
import java.util.Scanner;

public class Factorial
{
   public static void main(String[] args)
   {
     Scanner stdIn = new Scanner(System.in);
     int num;

     System.out.print("Enter a non-negative number: ");
     num = stdIn.nextInt();
     if (num >= 0)
     {
       System.out.println(factorial(num));
     }
   } // end main

   //*****************************************************************

   private static int factorial(int n)
   {
     int nF; // n factorial

     if (n == 1 || n == 0)          ← stopping condition
     {
       nF = 1;
     }
     else
     {
       nF = n * factorial(n-1);     ← recursive method call
     }
     return nF;
   } // end factorial
} // end Factorial class
```

Sample session:

```
Enter a non-negative number: 5
120
```

Figure 11.2 Use of recursion to calculate the factorial of an integer

factorial(n-1) is a recursive method call because it's inside a method named factorial, and it calls the factorial method. It calls the factorial method with an argument value of n-1 in order to match the recurrence relation described earlier: n! = n * (n-1)!. Note the n * in the Java code. Even though n * appears at the left of factorial(n-1), the JVM executes the recursive method call before performing the multiplication. Thus, the JVM performs the multiplication on the method call's returned value. It might seem strange to write a statement within a method X that relies on another execution of method X. That reliance requires you to make a leap of faith that the other execution of method X will work properly. Rest assured that it will work properly if the rest of the method is written properly. The leap of faith might be unsettling at first, but if you do it enough, hopefully you will learn to love the "magic of recursion." Remember the first time you rode a bike without training wheels and without someone guiding you? Scary, but after pedaling a while, you realized your momentum would keep you up. Recursion's leap of faith is kind of like that.

As part of a properly written recursive method, you must have a stopping condition. Note the stopping condition in Figure 11.2. The if statement indicates that if n equals 1 or 0, the JVM assigns 1 to nF (where nF stores the factorial of n). That should make sense because 1! equals 1 and 0! also equals 1. The last statement of the factorial method returns nF, so the prior call to factorial receives the calculated factorial of n.

Tracing the Recursive Factorial Method

Figure 11.3 shows a trace of the Factorial program in Figure 11.2, with the user entering a 5 in response to the "Enter a non-negative number:" prompt. Looking at the program, you can see that on line 14, the main method calls the factorial method with an argument of 5. The next four calls to the factorial method are from within the factorial method on line 30 with arguments of 4, 3, 2, and 1, respectively. When the argument 1 is passed to factorial, the stopping condition is satisfied, and nF gets assigned the value 1 on line 26.

<u>input</u>
5

	Factorial										
	factorial		factorial		factorial		factorial		factorial		
line#	n	nF	n	nF	n	nF	n	nF	n	nF	output
14	5	?									
30			4	?							
30					3	?					
30							2	?			
30									1	?	
26										1	
30								2			
30						6					
30				24							
30		120									
14											120

Figure 11.3 Trace of the Factorial program in Figure 11.2

Then the returning process commences. The 1 from the fifth `factorial` call returns to the fourth `factorial` call, which, on line 30, multiplies it by 2 and returns 2. This value returns to the third `factorial` call, which, on line 30, multiplies it by 3 and returns 6. This value returns to the second `factorial` call, which, on line 30, multiplies it by 4 and returns 24. This value returns to the first `factorial` call, which, on line 30, multiplies it by 5 and returns 120. This value returns to the argument of the `println` statement in the `main` method on line 14, and this statement prints out the computed value. In this problem, all the useful work is done in the return sequence, after the stopping condition is reached.

We included the local variable `nF` in the program in Figure 11.2 just to give this trace some substance. Hopefully, it helps you visualize the recursive calling that drills down to the simplest case and the subsequent result accumulation as the nested methods return. In practice, however, experienced programmers would not include the local `nF` variable. Instead, they would probably write the `factorial` method as in Figure 11.4.

```
// Precondition: Method argument is not negative.

public static int factorial(int n)
{
  if (n == 1 || n == 0)
  {
    return 1;
  }
  else
  {
    return n * factorial(n-1);
  }
} // end factorial
```

Figure 11.4 Cleaner implementation of the `factorial` method

Infinite Recursive Loops

In writing a recursive method, beginning programmers sometimes focus exclusively on the method's recursive method call and forget to include a stopping condition. Depending on the nature of the method, that can lead to an infinite recursive loop. An *infinite recursive loop* is when a method calls itself repeatedly without stopping. Each call requires the JVM to save information (such as the state of the program right before the call executes) in a special place in the computer's memory—the *call stack*. To prevent the call stack from eating up a computer's entire memory, the call stack's size is limited. If you run a program with an infinite recursive loop, the call stack eventually will fill up and generate a *stack overflow* runtime error and crash your program. If you'd like to experience the excitement of such a crash, substitute the following code for the `factorial` method in Figure 11.2's program and run the program:

```
public static int factorial(int n)
{
  return n * factorial(n-1);
} // end factorial
```

11.4 Comparison of Recursive and Iterative Solutions

An Iterative Factorial Method

In the prior section, we first presented the factorial of n as

```
n! = n * (n-1) * (n-2) * ... * 2 * 1
```

In looking at the repetitive nature of the above formula (it's a series of multiplications), you might have figured out that you can use a regular loop to calculate the factorial, and there's actually no need to use recursion. We used recursion not because we had to, but because we wanted a simple problem with a simple recursive solution. Recursion can be difficult to grasp initially, and it's best to start learning it with a relatively easy problem, like finding the factorial. Learn to walk before you run.

Anyway, now that you've "learned to walk" with recursion, let's take a step back and show an iterative/loop implementation for a factorial method. Figure 11.5 shows such a method. Instead of having a recursive method call at the bottom of the method, as shown in Figure 11.4, Figure 11.5's method has a loop that multiplies the numbers 1 through n, with one multiplication operation per loop iteration. We could have mimicked the recursive implementation by starting the multiplication at n and going down to 1, but we felt that starting at 1 and going up to n made more sense. In Figure 11.5, can you figure out what happens if the parameter `n` is either 0 or 1? The `for` loop never executes, and the method returns 1. That should make sense when you realize that 0! is 1 and 1! is also 1.

```
public static int factorial(int n)
{
  int fact = 1; // the factorial value so far

  for (int i=2; i<=n; i++)
  {
    fact *= i;
  }
  return fact;
} // end factorial
```

Figure 11.5 Iterative implementation of the `factorial` method

Characteristics of Recursive Solutions

Recursion does not add unique functionality. As exemplified by the factorial example shown above, all recursive programs can be converted to iterative programs that use loops instead of recursive method calls. So why use recursion? Because with some problems, a recursive solution is easier to understand. For example, some mathematical concepts, like the factorial of a number, are defined recursively, and they lend themselves well to programmatic solutions that use recursion. Many puzzles are easiest to solve with recursive thinking, and they also lend themselves well to programmatic solutions that use recursion. Examples are the Towers of Hanoi problem, which we will describe later in this chapter, and maze traversals.

Be aware that there is a downside to recursion. Recursive programs tend to be slow because they generate lots of method calls and method calls have lots of *overhead*. Overhead is work that the computer must do beyond the work described by program code. For each method call, the computer has to (1) save the calling module's local variables, (2) find the method, (3) make copies of call-by-value arguments, (4) pass the arguments, (5) execute the method, (6) find the calling module, and (7) restore the calling module's local variables. Whew! All that work takes time. That's why some recursive implementations can be prohibitively slow. When there a very large number of recursive calls, you should consider rewriting the solution with a loop implementation.

There are different types of recursion. *Mutual recursion* is when two or more methods call each other in a cycle of method calls. For example, if method A calls method B, method B calls method C, and method C calls method A, that's mutual recursion. You should be aware of mutual recursion, but we won't bother to show a mutual recursion program because such programs are not all that common. The most common type of recursion is when a method calls itself. If a method's body includes two (or more) recursive calls to itself and the method executes both calls, then the method is said to exhibit *binary recursion*. You'll see examples of binary recursion programs later in this chapter.

If a method executes just one recursive call to itself, then the method is said to exhibit *linear recursion*. Because it executes just one recursive call to itself, Figure 11.4's `factorial` method exhibits linear recursion. *Tail recursion* is a special case of linear recursion. It's when a recursive method executes its recursive call as its very last operation. You'll see an example of tail recursion in the upcoming subsection, but let's first examine the `factorial` method more closely. It does not exhibit tail recursion. Because its recursive call is positioned at the bottom of its method body, the `factorial` method might appear to exhibit tail recursion. However, it's not just the recursive call's position that matters; it's the recursive call's order of operation that matters. Here is the `factorial` method's recursive call statement:

```
return n * factorial(n-1);
```

In executing the statement, the JVM first calls `factorial` recursively and then multiplies `n` times `factorial`'s returned value. So the last operation is multiplication, not the recursive call.

Of all the different kinds of recursions, tail recursions are the easiest to convert to loop implementations: Just use the changing recursive parameter as the loop index, and use the recursive stopping condition as the loop termination condition. That technique should become more evident as you study the tail recursion example in the next subsection. . . .

Converting from an Iterative Method to a Recursive Method

In the factorial discussion above, you saw an example of converting from a recursive method to an iterative method. Now let's go the other way—converting from an iterative method to a recursive method.

Note the `printReverseMessage` method in Figure 11.6. It receives a string parameter `msg` and prints the string in reverse order. So "Hedgehogs rock!" would print as "!kcor sgohegdeH." It uses a loop to traverse through each of the characters in `msg`, starting with the rightmost character and ending with the leftmost character. Now the challenge is to rewrite that iterative-implementation method as a recursive method.

As mentioned earlier, the first step to writing a recursive method is to figure out how the problem can be split into successively smaller subproblems. After printing a message's rightmost character, what subproblem remains to be solved? Printing the other characters in the message, of course! Solving that subproblem requires working on a substring of the original message—the substring that extends from the message's leftmost character to the message's penultimate character.[1] Before moving on to the next paragraph, try to use that explanation as the basis for writing a recursive `printReverseMessage` method on your own.

[1]Penultimate means "next to the last." Isn't it more fun to say "penultimate" than "next to the last"? Sesquipedalians certainly would agree! ☺

```
/*******************************************************************
* PrintReverseMessageIterative.java
* Dean & Dean
*
* This program prints a given message in reverse order.
*******************************************************************/

import java.util.Scanner;

public class PrintReverseMessageIterative
{
  public static void main(String[] args)
  {
    Scanner stdIn = new Scanner(System.in);
    String msg; // user-entered message

    System.out.print("Enter a message: ");
    msg = stdIn.nextLine();
    printReverseMessage(msg);
  } // end main

  //****************************************************************

  private static void printReverseMessage(String msg)
  {
    int index; // position of character that is to be printed

    index = msg.length() - 1;
    while (index >= 0)
    {
      System.out.print(msg.charAt(index));
      index--;
    }
  } // end printReverseMessage
} // end PrintReverseMessageIterative class
```

Figure 11.6 Iterative implementation of printing a string in reverse order

The following code fragment is a reasonable first attempt at writing a recursive `printReverseMessage` method:

```
private static void printReverseMessage(String msg)
{
  int index = msg.length() - 1;
  System.out.print(msg.charAt(index));
  printReverseMessage(msg.substring(0, index));
}
```

Note how index is assigned the position of the message's last character. That index value is then used to print the message's last character. Note the substring method call. Remember how the substring method call works? It returns a substring of its calling object string (msg in this case), such that the returned string spans from the position of its first argument (0 in this case) to the position just to the left of its second argument (index in this case). Thus, the third line calls printReverseMessage with a substring argument that extends from the message's leftmost character to the message's penultimate character. That is our goal for the recursive call.

Are we done? Is this recursive printReverseMessage method complete? To answer that question, think about what happens when you call it with a msg value of "cow." The JVM prints the w and calls the method recursively with an argument value of "co." The JVM then prints the o and calls the method recursively with an argument value of "c." The JVM then prints the c and calls the method recursively with an argument value of "", the empty string. The JVM then assigns −1 to the index variable and crashes when it attempts to execute msg.charAt(-1).

So what's the problem, and what's the fix? The method needs a stopping condition that stops the recursion when the msg parameter has shrunk all the way down to the empty string. See the corrected recursive printReverseMessage method in Figure 11.7. In particular, note the method's if statement. If msg is non-empty, the JVM prints msg's last character and calls printReverseMessage recursively. On the other hand, if msg contains the empty string, then the JVM does not call printReverseMessage recursively. It just returns. More often than not, recursive methods use if else statements, but in this case, there's no need to do anything special when msg contains the empty string, so no else block is necessary.

```
private static void printReverseMessage(String msg)
{
  int index; // position of last character in msg

  if (!msg.isEmpty())
  {
    index = msg.length() - 1;
    System.out.print(msg.charAt(index));
    printReverseMessage(msg.substring(0, index));
  }
} // end printReverseMessage
```

Figure 11.7 Recursive implementation of the printReverseMessage method

Because printReverseMessage is a recursive method, after the recursion process hits its stopping condition, it returns (same as for all recursive processes). But what does it return? Since it's a void method, it returns nothing! This is like a navy admiral issuing a command for all ships to turn right by 90 degrees. "No discussion. Just do it!" And no information comes back.

The recursive printReverseMessage method executes its recursive call as its very last operation, and, as such, it exhibits tail recursion. With tail recursion, the JVM performs its useful work while proceeding forward with the recursive calls. For the printReverseMessage method, its "useful work" consists of printing a character. On the other hand, the recursive factorial method does not execute its recursive call as its last operation, and the JVM performs its useful work (multiplication) while returning from the recursive calls.

In more complicated problems, you'll see that sometimes some operations occur before recursive calls (and take place during the calling process) and other operations occur after recursive calls (and take place during the return process). In problems like these, the recursive algorithms are usually much easier to understand and implement than their iterative counterparts.

11.5 Recursive Method Evaluation Practice

Perhaps the best way to improve your feeling for recursion is to work through some problems by hand. You did this sort of thing before when you evaluated expressions in Chapter 3, Section 3.16. There, you wrote an expression and then repeated that expression on subsequent lines, with numerical values gradually substituted for variables or functions on lines above them.

Recursive Algorithms Expressed as Functions

When evaluating a recursive method by hand, we'll often rewrite the method as if it were a mathematical function and use a one-letter abbreviation for the method name. For example, consider the Factorial program in Figure 11.2. In tracing the method call `factorial(5)`, we'll rewrite it as `f(5)`, where `f` stands for the factorial function. Using mathematical function notation, here's a compact description of the factorial algorithm:

```
f(n) = { n * f(n-1)        for n > 1
       { 1                 for n <= 1
```

Notice that this algorithmic specification provides both the recurrence relation and the stopping condition together.

To evaluate a recursive method call by hand, start by writing the method with function notation (as shown above). For the first line of your recursion trace, write the recurrence relation with the variables replaced by the initial numbers. Under that, write the recurrence relation for the first subordinate method call with the variables replaced by appropriately altered numbers on both the left and the right sides of the equations. Continue like this until you reach a stopping condition. During the calling sequence, values that might be returned by recursive method calls are not yet known. They correspond to the question marks in Figure 11.3. The stopping condition produces a completely known value. On subsequent rows, rewrite what you previously wrote on the rows above, but in reverse order, replacing right-side unknowns with known values as you go. Eventually, this produces a completely known value for the original recursive call.

Here's how the procedure above would be used to evaluate the `factorial` method being called with `n` equal to 5:

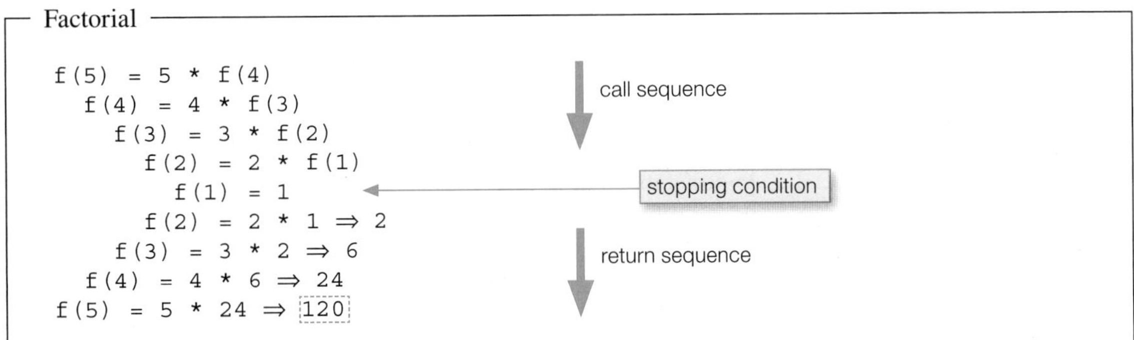

The dashed box at the bottom identifies the desired answer.

For another example, consider a method that returns the balance, b, in a bank account after n equal periodic deposits of amount D, with interest rate R for the time period between deposits. Borrowing from the Java convention, we use lowercase letters for variables and uppercase letters for constants. Here's a compact description of the algorithm, where b stands for the balance function:

```
b(n, D, R) = { D + (1 + R) * b(n-1)        for n >= 1
             { 0                           for n < 1
```

Let's say each deposit is in the amount D = 10, and the interest rate is R = 0.1. Suppose that we want to know the balance when n = 3. The evaluation looks like this:

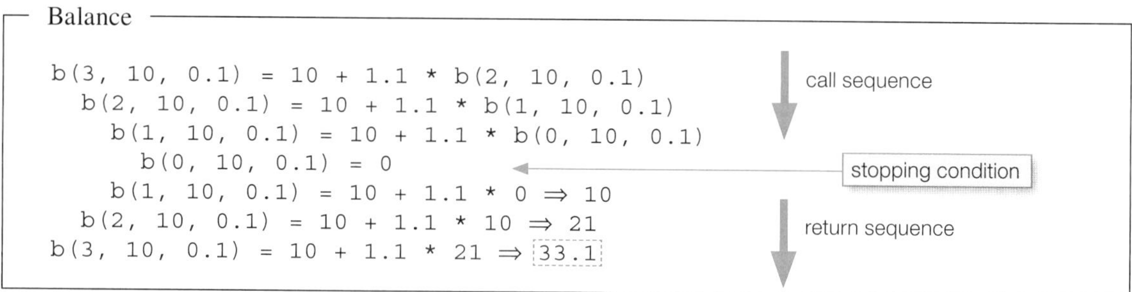

Balance
b(3, 10, 0.1) = 10 + 1.1 * b(2, 10, 0.1)
b(2, 10, 0.1) = 10 + 1.1 * b(1, 10, 0.1)
b(1, 10, 0.1) = 10 + 1.1 * b(0, 10, 0.1)
b(0, 10, 0.1) = 0
b(1, 10, 0.1) = 10 + 1.1 * 0 ⇒ 10
b(2, 10, 0.1) = 10 + 1.1 * 10 ⇒ 21
b(3, 10, 0.1) = 10 + 1.1 * 21 ⇒ 33.1
call sequence
stopping condition
return sequence

More Practice

To improve your recursive method call evaluation skills, you need to practice. In providing you with additional practice problems, we'll cut to the chase and start each problem with a mathematical-function notation. (Every non-void Java method is a mathematical function.) Be aware that in the real world, in evaluating a recursive method call, you'll have to come up with the mathematical-function notation on your own, complete with a recurrence relation and a stopping condition.

Here's a practice problem whose function has two parameters—x and y:

```
         { f(x-3, y-1) + 2        x > 0, x > y
f(x, y) = { f(y, x)               x > 0, x <= y
         { 0                      x <= 0
```

The function is named *f*, where *f* is a generic name for a function. To see how it works, let's evaluate *f*(5, 4):

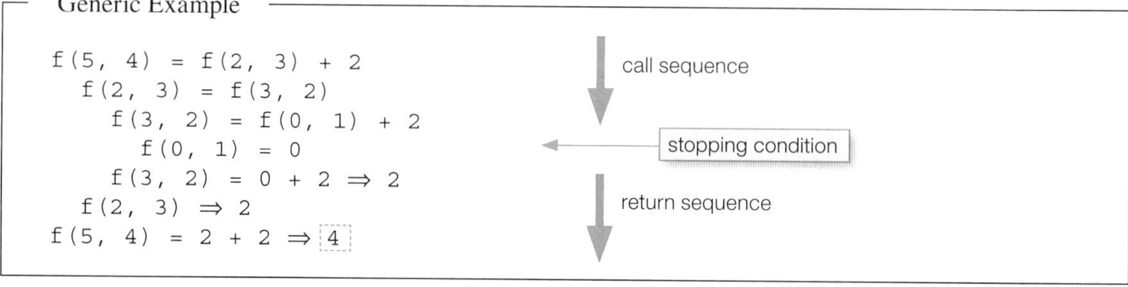

In this example, the evaluation encountered an adequate stopping condition. But sometimes stopping conditions are not adequate. For example, suppose that you have this recurrence relation:

```
           ┌ A * f(x-2, A)          x > 1
f(x, A) = ┤
           └ A                      x = 1
```

Notice how this notation does not distinguish between assignment and equality. The equals sign to the left of the brace represents an assignment. The equals sign on the right represents equality. This kind of recursive specification expects you to make that distinction yourself. And that's OK. But there is still something wrong. What is it? Here's how the evaluation of *f*(4, 3) would go:

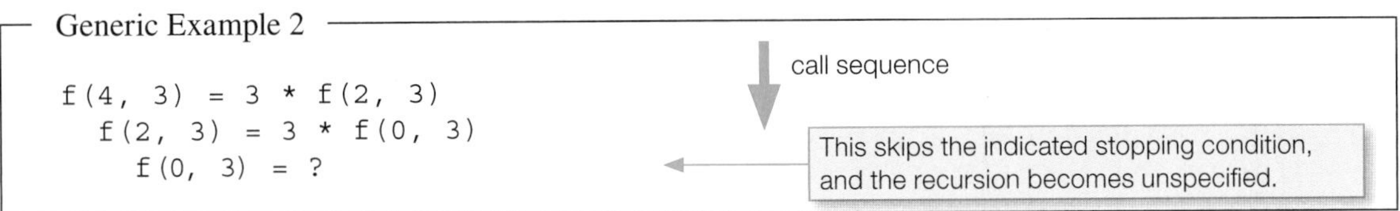
Generic Example 2

```
f(4, 3) = 3 * f(2, 3)
  f(2, 3) = 3 * f(0, 3)
    f(0, 3) = ?
```

call sequence
This skips the indicated stopping condition, and the recursion becomes unspecified.

Because x decrements by two in each iteration, when the starting value for x is even, it misses the indicated stopping condition. This problem is insidious because the indicated stopping condition works for odd starting values. Also, what if the starting x is a floating-point number? In general, inequality provides a better stopping condition than equality.

Here's another example:

```
        ┌ f(x/2)          x > 0
f(x) = ┤
        └ 0               x <= 0
```

The inequality takes care of cases that might skip zero. So is this OK? Suppose you start with x = 4. How many iterations would it take to reach the stopping condition? Let's see, 4, 2, 1, 0.5, 0.25, 0.125, . . . Hey, it never gets there! When you have this kind of problem, you might change that stopping condition to something slightly larger than zero, perhaps something like `x <= 0.000001`.

Here's another example:

```
        ┌ f(x) + 1        x < 3
f(x) = ┤
        └ 4               x >= 3
```

Up until this last example our adequate stopping conditions have been some kind of minimum. The stopping condition does not need to be a minimum. It could also be a maximum. So there's nothing wrong with the fact that this stopping condition is a maximum. But there is still something wrong. What is it? The function gets larger, but the stopping condition does not look at the function. It just looks at x, and x never changes.

Implementing a Recurrence Relation with a Loop

Some recurrence relations lend themselves to loop implementations where you start the loop from the stopping condition and move forward from there. As shown earlier in Figure 11.5, calculating a factorial lends itself well to a loop solution. Another example is a Fibonacci sequence, which a mathematician might describe with a recurrence relation like this:

$$f(n) = \begin{cases} f(n-1) + f(n-2) & n > 1 \\ 1 & n = 1 \\ 0 & n = 0 \end{cases}$$

As with the preceding examples, you could use this to implement a recursive solution and then convert that solution to a loop solution as we did in Section 11.4. But most humans would skip the recursive step. They would use the recursive relation's stopping conditions as initial values and apply its general formula like this:

Fibonacci Sequence

```
f(0) = 0
f(1) = 1
f(2) = f(1) + f(0) = 1 + 0 ⇒ 1
f(3) = f(2) + f(1) = 1 + 1 ⇒ 2
f(4) = f(3) + f(2) = 2 + 1 ⇒ 3
f(5) = f(4) + f(3) = 3 + 2 ⇒ 5
f(6) = f(5) + f(4) = 5 + 3 ⇒ 8
...
```

straightforward evaluation

Using the calculations above as a pattern, here's the loop implementation for finding the Fibonacci value at position n. We're assuming that `fibonacci` is an array that has at least n + 1 elements.

```
fibonacci[0] = 0;
fibonacci[1] = 1;
for (int i=2; i<=n; i++)
{
  fibonacci[i] = fibonacci[i-1] + fibonacci[i-2];
}
```

11.6 Binary Search

Now that you have learned how to work with several forms of recursion, you're ready for some longer recursive programs. We hope that you'll be able to appreciate that their solutions are more elegant than if the programs were implemented with loops. In this section's example, a single recursive call occurs after all the useful work is done. So it's another example of tail recursion.

Here's the problem: Suppose that you want to find the location of a particular value in an array. This is a common database operation. If the array is not sorted, the best you can do is use a sequential search and look at each item individually. If the array is very short, a sequential search is also the fastest way to search because a sequential search is very simple. If the array is long, however, and if it's relatively stable, it's often faster to sort the array and then use a *binary* search.

Back in Section 9.7, in Figure 9.12, we showed how to implement a binary search with a `while` loop. Now we'll show you how to do it with recursion. This will enable you to compare a recursive implementation with an iterative implementation. Figure 11.8 shows our implementation of this algorithm. We have included shaded print statements at appropriate places in the code to show what the code is doing when it executes. After debugging the program, you would want to remove all of these shaded print statements.

```
/*************************************************************
* BinarySearch.java
* Dean & Dean
*
* This uses recursion to find the index of a target value in
* an ascending sorted array. If not found, the result is -1.
*************************************************************/

public class BinarySearch
{
  public static int binarySearch(
    int[] arr, int first, int last, int target)
  {
    int mid;
    int index;

    System.out.printf("first=%d, last=%d\n", first, last);
    if (first == last)              // stopping condition
    {
      if (arr[first] == target)
      {
        index = first;
        System.out.println("found");
      }
      else
      {
        index = -1;
        System.out.println("not found");
      }
    }
    else                            // continue recursion
    {
      mid = (last + first) / 2;
      if (target > arr[mid])
      {
        first = mid + 1;
      }                                  <-- Do some work.
      else
      {
        last = mid;                      Then drill down.
      }
      index = binarySearch(arr, first, last, target);  <--
      System.out.println("returnedValue=" + index);
    }
    return index;
  } // end binarySearch
} // end BinarySearch class
```

Figure 11.8 Implementation of recursive binary search algorithm

Figure 11.9 shows a driver that demonstrates the binary search algorithm implemented in Figure 11.8.

```
/*************************************************************
* BinarySearchDriver.java
* Dean & Dean
*
* This drives the BinarySearch class.
*************************************************************/

public class BinarySearchDriver
{
  public static void main(String[] args)
  {
    int[] array = new int[] {-7, 3, 5, 8, 12, 16, 23, 33, 55};

    System.out.println(BinarySearch.binarySearch(
      array, 0, (array.length - 1), 23));
    System.out.println(BinarySearch.binarySearch(
      array, 0, (array.length - 1), 4));
  } // end main
} // end BinarySearchDriver class
```

Output:

```
first=0, last=8
first=5, last=8
first=5, last=6
first=6, last=6
found
returnedValue=6
returnedValue=6
returnedValue=6
6
first=0, last=8
first=0, last=4
first=0, last=2
first=2, last=2
not found
returnedValue=-1
returnedValue=-1
returnedValue=-1
-1
```

Reduce range and drill down.

Just return the answer.

Figure 11.9 Driver for the BinarySearch class in Figure 11.8

In Figure 11.9's the output section, the shaded areas are outputs generated by the shaded print statements in Figure 11.8. In this recursion, the real work is done while the process is drilling down to the stopping condition. Notice how the`first` and `last` values converge on the match or the place where the match will be if it is there. The answer is generated at the point when the stopping condition is reached. The nested returns just pass this answer back. When you remove the shaded print statements from Figure 11.8, the only output that you will see is the unshaded parts of the output.

In a recursive binary search, as you can see, we simplify the problem by dividing the array into two nearly equally sized arrays. Then we continue dividing until each half contains no more than one element, which is the stopping condition.

Why is a binary search faster than a sequential search? The number of steps required for a sequential search equals <*array*>`.length`, whereas the number of steps required in a binary search equals only $\log_2$(<*array*>`.length`). For example, if there are 1 million items in the array, that's 1 million steps for a sequential search, but only about 20 steps for a binary search. Even if a typical binary-search step is more complicated than a typical sequential-search step, the binary search will still be significantly faster when the array is very long.

As we said in Section 9.7, a precondition for a binary search is that the data searched must be sorted. Section 9.8 provided an implementation for the relatively simple selection sorting algorithm. You could use this to prepare data for either the previous iterative binary search or the present recursive binary search. However, in Section 11.7, we'll show you a more efficient merge sorting algorithm, and that would be a better choice.

11.7 Merge Sort

The binary search discussed in the previous section is a relatively efficient way to find something in a large collection of data. But for a binary search to work, the data must have been sorted already. In Section 9.8 of Chapter 9, we described the relatively simple "selection sort" technique. Selection sort is good for small collections of data, but it's inefficient for large collections of data. Because the binary search you studied in the previous section works well for large collections of data, to prepare for that binary search, you'll want to use a sorting method that is also good for large collections of data. Merge sort is a good option.

The basic strategy in merge sorting is familiar—divide and conquer. In this respect, a merge sort is like a binary search. But this time, instead of making a recursive call for just one of the two halves, it makes recursive calls for both of them. In each of these recursive calls, it divides the current part in half, makes recursive calls for both smaller parts, and so forth, until a part has only one element. That's the stopping condition for the recursive branch that leads to that part.

Return sequences put things back together by merging parts, two at a time. Each return step merges all the elements in one part with all the elements in another part to form a larger part, until eventually, everything is back together in a single whole.

Figure 11.10a shows the recursive `mergeSort` method. The parameter is an unsorted array, called `array`. Local variable declarations create two subordinate arrays, `sub1` and `sub2`. If `array`'s length is an even number, the lengths of `sub1` and `sub2` are both equal to half the length of `array`. If `array`'s length is an odd number, `sub2` is one element longer than `sub1`. The stopping condition stops a recursion whenever `array`'s length becomes one or zero. Negative array lengths are impossible. Whenever the stopping condition occurs, the returned array is the same as the parameter array.

If the parameter array's length is two or more, execution falls to the "else" part of the "if, else" statement. Within the "else" clause, a pair of calls to `System`'s `arrayCopy` method fills the two half-length subordinate arrays, `sub1` and `sub2`, with the values in the lower and upper halves of `array`, respectively. (You learned about

```
/**************************************************************
* MergeSort.java
* Dean & Dean
*
* This performs a recursive merge sort.
**************************************************************/

import java.util.*;

public class MergeSort
{
  public static int[] mergeSort(int[] array)
  {
    int half1 = array.length / 2;
    int half2 = array.length - half1;
    int[] sub1 = new int[half1];
    int[] sub2 = new int[half2];

    if (array.length <= 1)          <-- stopping condition
    {
      return array;
    }
    else
    {
      System.arraycopy(array, 0, sub1, 0, half1);
      System.arraycopy(array, half1, sub2, 0, half2);
      sub1 = mergeSort(sub1);       <-- two recursive
      sub2 = mergeSort(sub2);       <-- method calls
      array = merge(sub1, sub2);
      return array;
    }
  } // end mergeSort
```

Figure 11.10a MergeSort program—part A.
This shows the mergeSort method, which calls itself recursively.

the arrayCopy method in Section 9.5.) Then, the mergeSort method recursively calls itself two times—once to sort the sub1 array and once to sort the sub2 array. After the two recursive calls, the mergeSort method calls the merge helper method. The merge method copies the sub1 and sub2 array elements in sorted order into a new array whose length equals the sum of the lengths of sub1 and sub2. The mergeSort method then returns the new array, which contains the same elements as the original array, but in sorted order.

Take a look at the merge method in Figure 11.10b. It merges its two array parameters, sub1 and sub2, by copying the values from sub1 and sub2 into a new composite array such that the new array's values are sorted in ascending order. This process is made easier by the fact that when merge receives sub1, its elements are already in ascending order. The same is true for sub2. As you can see in Figure 11.10b, the merge method loops through the elements in sub1 and sub2, and for each loop iteration, it copies the smaller of the two elements (from sub1 or sub2) to the new composite array. As that loop iteration comparison

```
//***********************************************************

// precondition: parameters are sorted in ascending order
// postcondition: return is sorted in ascending order

private static int[] merge(int[] sub1, int[] sub2)
{
  int[] array = new int[sub1.length + sub2.length];
  int i1 = 0, i2 = 0;

  for (int i=0; i<array.length; i++)
  {
    // both subgroups have elements
    if (i1 < sub1.length && i2 < sub2.length)
    {
      if (sub1[i1] <= sub2[i2])
      {
        array[i] = sub1[i1];
        i1++;
      }
      else   // sub2[i2] < sub1[i1]
      {
        array[i] = sub2[i2];
        i2++;
      }
    }
    else    // only one subgroup has elements
    {
      if (i1 < sub1.length)
      {
        array[i] = sub1[i1];
        i1++;
      }
      else // i2 < sub2.length
      {
        array[i] = sub2[i2];
        i2++;
      }
    } // end only one subgroup has elements
  } // end for all array elements
  return array;
} // end merge
```

From the two subordinate arrays, copy the smaller element value to the composite array.

After visiting all of the elements in one of the two subordinate arrays, copy the remaining elements from the other one.

Figure 11.10b MergeSort program—part B.
This shows the `merge` method, which merges two sorted parts into one sorted whole.

process continues, the composite array fills up with smaller values first and larger values last. After all of the elements from one of the subordinate arrays have been copied to the composite array, the loop stops and all of the remaining elements from the other subordinate array are copied to the unfilled higher-indexed elements in the composite array. The `merge` method then returns the composite array to `mergeSort`.

Now look at Figure 11.10c. It contains the rest of the MergeSort program—the `main` method and a helper method named `printArray`. The `printArray` method prints the contents of its passed-in `array` parameter. Also, as shown in Figure 11.10c's sample session, it prints a description of the array, using the passed-in `msg` parameter for the description. Having this method available makes it easy to insert temporary diagnostic print statements anywhere in the code while the program is being developed and debugged.

```
  //*************************************************************

  private static void printArray(String msg, int[] array)
  {
    System.out.println(msg);
    for (int i : array)
    {
      System.out.printf("%3d", i);
    }
    System.out.println();
  } // end printArray

  //***********************************************************

  public static void main(String[] args)
  {
    Random random = new Random(0);
    int length = 19;
    int[] array = new int[length];

    for (int i=0; i<length; i++)
    {
      array[i] = random.nextInt(90) + 10;
    }
    printArray("initial array", array);
    printArray("final array", mergeSort(array));
  } // end main
} // end class MergeSort
```

Sample session:

```
initial array
  70 98 59 57 45 93 81 31 79 84 87 27 93 92 45 24 14 25 51
final array
  14 24 25 27 31 45 45 51 57 59 70 79 81 84 87 92 93 93 98
```

Figure 11.10c MergeSort program—part C.
This shows the program's driver and a display helper method.

The MergeSort program's `main` method employs a `Random` object to create a sequence of random numbers that are reproducible from one run to the next. Being able to reproduce random numbers is very helpful during testing and debugging. If you are not already familiar with this capability, you can read more about it in Section 5.8. Speaking of testing, we tested the MergeSort program with arrays of various sizes. You can see an array length value of 19 in Figure 11.10c. We also tested with an array whose length value was an even number and an array whose length value was zero. All the tests gave good results. Likewise, when you write your own programs, you should take the time to thoroughly test your programs with a wide variety of data.

11.8 Towers of Hanoi

In 1883, a French mathematician named Édouard Lucas proposed a puzzle based on an ancient legend. According to one version of this legend, a room in a temple in the city of Hanoi contains 64 golden disks, each with a different diameter and each with a hole in its center. The disks are stacked on three towering posts. At the beginning of the world, all the disks were stacked on just one of the posts, with the largest-diameter disk on the bottom and progressively smaller-diameter disks placed on top of each other. Figure 11.11 shows a simplified version of the problem, with only 4 disks instead of 64.

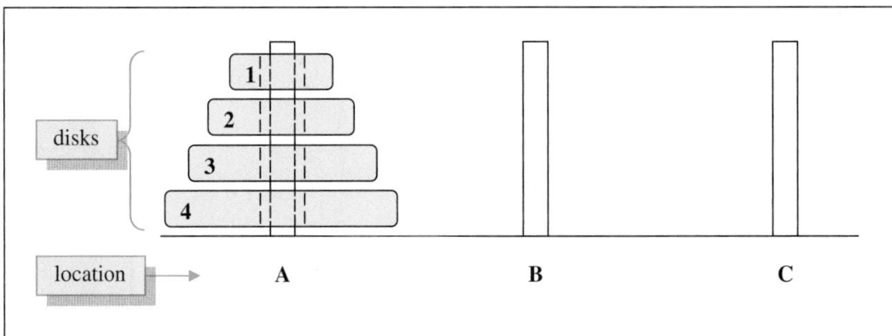

Figure 11.11 Setup for the Towers of Hanoi problem

The temple's monks have a backbreaking, mind-numbing mission. They are supposed to transfer all the disks from post A (the original post) to post C, in accordance with these commandments: (1) Move one disk at a time from one post to another post, (2) Never place any disk on top of a smaller disk. The legend says that when they finish this mission, they will be released from their labor, and the world will end.

Your task is to help the monks obtain the earliest possible release from their labor by giving them a computer program that specifies the optimum transfer sequence. The trick is to find a simple algorithm.

If you attempt to implement your solution with loops, you'll probably create a mess. But if you use recursion, you'll be more likely to succeed. Whenever you want to make a more, you need to identify a source location, `s`, a destination location, `d`, and a temporary location, `t`. For our overall goal, `s` is `A`, `d` is `C`, and `t` is `B`. As you progress toward the final solution, you'll have subordinate goals, with different locations for `s`, `d`, and `t`. Here is the general algorithm. It applies to any subset of disks from disk `n` down to disk 1, where `n` is any number from the maximum number of disks down to 1:

- Move the stack of disks above disk `n` from `s` to `t`.
- Move disk `n` to `d`.
- Move the stack of disks previously above disk `n` from `t` to `d`.

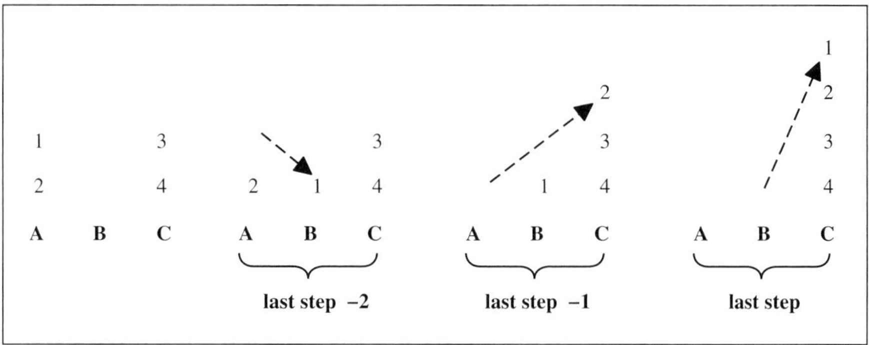

Figure 11.12 Illustration of the Towers of Hanoi algorithm in action

Figure 11.12 shows an example of this algorithm in action. (This particular example happens to be the last few steps in the final solution.) The configuration on the left is a condition that exists shortly before the goal is attained. The configuration on the right is the final condition. The dashed arrows indicate each of the three operations described above for the simplest non-trivial case where there is only one disk above disk 2. The trivial case is the stopping condition. It's when you move the top disk, disk 1, from a source location to a destination location. An example of this appears as the last step in Figure 11.12. This happens to be the final stopping condition, but as you'll see, a program that solves the Towers of Hanoi problem will hit the stopping condition and automatically restart many times during its execution.

Assume that the evaluation process is currently in the left frame of Figure 11.12. The next operation calls the following `move` method with arguments (2, 'A', 'C', 'B'). Within this method, the `else` clause's first subordinate `move` method call uses arguments (1, 'A', 'B', 'C') to implement Figure 11.12's "last step-2." The subsequent `printf` statement implements Figure 11.12 "last step–1." The `else` clause's second subordinate move method call uses arguments (1, 'B', 'C', 'A') to implement Figure 11.12's "last step."

```
private static void move(int n, char s, char d, char t)
{
  if (n == 1)            // recursive stopping condition
  {
    System.out.printf("move %d from %s to %s\n", n, s, d);
  }
  else
  {
    move(n-1, s, t, d); // source to temporary
    System.out.printf("move %d from %s to %s\n", n, s, d);
    move(n-1, t, d, s); // temporary to destination
  }
}
```

The initial call to the recursive method should establish the overall goal, which is to move the entire tower from location `A` to location `C`. To get the largest disk to the new location first, you should start with the maximum possible `n`. The algorithm says you can do this by moving the subordinate set of disks, 1, 2, and 3 from the source location, `A`, to the temporary location, `B`. Then you move disk 4 from the source location, `A`, to the destination location `C`. Then you move the subordinate set of disks 1, 2, and 3 from the temporary location, `B`, to the destination location, `C`, thereby putting them on top of the largest disk, 4. The problem with this process is that the rules don't permit moving more than one disk at a time. So, to move the subordinate set of disks, 1, 2, and 3, you must call the same method recursively to move just disks 1 and 2. To do that, you must call the same method recursively again to move just disk 1.

Of course, the first disk to move is disk 1, but it's hard to know where to put it. Should you move it to location B or to location C? The purpose of the program is to tell you exactly how to proceed. The program is displayed in Figure 11.13. The shaded print statements are not part of the solution, and they should be omitted from a finished product. We inserted them just to help you trace the torturous recursive activity—if you want to. For each method invocation, they print right after the method is called and just before it returns to show you the details of what's happening.

Figure 11.14 displays the output. The shaded lines are lines printed by the shaded print statements in Figure 11.13. As we said, they are just to help you see what happened and are not part of the solution. The solution is given by the unshaded outputs. The most difficult part of tracing a recursive algorithm like this is

```
/*************************************************************
* Towers.java
* Dean & Dean
*
* This uses a recursive algorithm for Towers of Hanoi problem.
*************************************************************/

public class Towers
{
  public static void main(String[] args)
  {
    move(4, 'A', 'C', 'B');
  }

  // Move n disks from source s to destination d using temporary t.
  private static void move(int n, char s, char d, char t)
  {
    System.out.printf(
      "call n=%d, s=%s, d=%s, t=%s\n", n, s, d, t);
    if (n == 1)              // recursive stopping condition
    {
      System.out.printf("move %d from %s to %s\n", n, s, d);
    }
    else
    {
      move(n-1, s, t, d);  // source to temporary
      System.out.printf("move %d from %s to %s\n", n, s, d);
      move(n-1, t, d, s);  // temporary to destination
    }
    System.out.println("return n=" + n);
  }
} // end class Towers
```

two return points

Figure 11.13 Solution to the Towers of Hanoi problem.
Shaded statements are for diagnostic tracing. Remove them for final implementation.

Output:

```
call n=4, s=A, d=C, t=B
call n=3, s=A, d=B, t=C
call n=2, s=A, d=C, t=B
call n=1, s=A, d=B, t=C
move 1 from A to B
return n=1
move 2 from A to C
call n=1, s=B, d=C, t=A
move 1 from B to C
return n=1
return n=2
move 3 from A to B
call n=2, s=C, d=B, t=A
call n=1, s=C, d=A, t=B
move 1 from C to A
return n=1
move 2 from C to B
call n=1, s=A, d=B, t=C
move 1 from A to B
return n=1
return n=2
return n=3
move 4 from A to C
call n=3, s=B, d=C, t=A
call n=2, s=B, d=A, t=C
call n=1, s=B, d=C, t=A
move 1 from B to C
return n=1
move 2 from B to A
call n=1, s=C, d=A, t=B
move 1 from C to A
return n=1
return n=2
move 3 from B to C
call n=2, s=A, d=C, t=B
call n=1, s=A, d=B, t=C
move 1 from A to B
return n=1
move 2 from A to C
call n=1, s=B, d=C, t=A
move 1 from B to C
return n=1
return n=2
return n=3
return n=4
```

Figure 11.14 Output from the Towers program in Figure 11.13.
Unshaded statements are the solution. Shaded ones provide a trace.

keeping track of the place from which a call was made, and therefore where execution resumes after a return. As our note in Figure 11.13 indicates, sometimes the call is made from the "source to temporary" statement, and sometimes the call is made from the "temporary to destination" statement. Fortunately, if you define a recursive algorithm correctly, you can ignore the details of how it plays out during program execution.

We recommend that you cut out four little paper disks of different sizes, number them 1 through 4 from smallest to largest, and build a tower at location "A" on your left. Then move the disks one at a time as the unshaded outputs in Figure 11.14 say. You'll see that the tower does in fact get moved from location "A" to location "C" in precise conformity with the specified rules. This works because the eventual moves are informed by goal information in earlier method calls.

Figure 11.14 shows that the monks would need a minimum of 15 moves to complete the task if there were just four disks. The number 64 is 16 times larger than 4. So if the minimum number of moves increased linearly with the number of disks, the monks would need a minimum of $16 \times 15 = 240$ moves to complete the task and bring about the end of the world. At the time of this book's writing, the world still exists, so perhaps there's something wrong with the calculation? Perhaps the number of moves increases more precipitously as the number of disks increases?

Wikipedia says with 64 disks, the minimum number of moves is $(2^{64} - 1)$, which is 18,446,744,073, 709,551,615 moves. If the monks could make one move per second on a 24/7 basis, they would need a total of 585 billion years to complete the task. Whew! That's a relief. No need to rouse the Four Horsemen of the Apocalypse just yet. But is Wikipedia's formula correct? Can you think of a way to test the validity of that formula? One of the end-of chapter exercises asks you to check that formula with computer simulation and inference.

11.9 GUI Track: Drawing Trees with a Fractal Algorithm (Optional)

Congratulations on making it this far in the recursion chapter! Recursion can be difficult, but as you gain more experience with it, we hope that you'll be able to appreciate its beauty. This section expresses the beauty of recursion not only in terms of problem solving, but also in terms of a picture. In this section, you'll use recursion to draw a grove of trees, where each tree is a recursive picture.

All of the native cells in a living tree contain exactly the same DNA. That identical DNA enables the cells to reproduce themselves and replicate their own attributes in other cells. As a biological tree grows, its cells make decisions about when to start new branches and what angles those branches should take. Because cells throughout the tree have similar attributes, they tend to make the same decisions. Persistent environmental factors also affect those decisions. If one side of a tree gets more sun, it grows faster on that side. A strong prevailing wind influences the direction of growth. Such factors can cause repeated asymmetry.

Computer recursion is a natural analog of the recurrence of similar patterns in an organism. So it's natural for us to use recursion to describe that organism. Our simulated trees will repeat a simple geometrical pattern—a straight section followed by a fork with two branches. The left branch goes 30 degrees to the left and has a length equal to 75% of the length of the straight section. The right branch goes 50 degrees to the right and has a length equal to 67% of the length of the straight section.

An object created by repeating a pattern at different scales is called a *fractal*. A mathematical fractal displays *self-similarity* on all scales. In the natural world—and in computers—self-similarity exists only between upper- and lower-scale limits. In the case of a tree, the upper limit is the size of the trunk and the lower limit is the size of a twig. These upper and lower limits establish recursive starting and stopping conditions.

In addition to focusing on recursion, this section also focuses on two other important subjects: (1) the model-view-controller (MVC) design pattern, and (2) GUI animation. You might recall from Chapter 8 that the MVC pattern enables a program to separate its user view from its internal logic. This section's tree-drawing program implements that pattern by having a separate class for each of the three MVC concerns—model, view, and controller. For animation, the tree-drawing program uses Java's `Thread.sleep` method to insert 50-millisecond delays between successive stages of tree growth. After each delay, the program slightly modifies size attributes and asks the program's window to repaint the scene. This makes the trees grow larger and fill out with more branches as time passes. Very cool!

This section relies on several concepts presented earlier in the book that you might have skipped. You'll need to understand the `drawLine` and `drawString` methods and the `ArrayList` collection class. If you're not familiar with the `Graphics` class's `drawLine` and `drawString` methods, you can learn about them in Section 5.9. If you're not familiar with the `ArrayList` collection class, you can learn about it in Section 10.2 (Chapter 10 was not a prerequisite for the prior sections in this chapter).

Model

Figure 11.15a begins the description of the simulated trees in a class called `Tree`. This class imports the Java API `Graphics` class to help draw the lines that will represent the tree's branches. The `START_X` constant tells where the tree is in the horizontal dimension. This is where the tree first sprouts from the earth. The `START_TIME` constant tells when the tree first sprouts from the earth. The `MAX_TRUNK_LENGTH` constant is the maximum length of the lowest and largest tree member. Because all other members scale down from this length, this constant establishes an upper limit on tree size. The `trunkLength` variable contains the current length of the trunk.

The constructor records the tree's location—its starting position in the x-direction expressed as a number of pixels. It records the time at which the tree starts to grow. It initializes the trunk length. Usually the initial trunk length is a very small value, like 3 pixels, but it also could be any larger value, to represent the planting of a partially grown tree moved from some other place. The first three `get` methods are ordinary accessors.

Figure 11.15b contains methods that describe the tree's structure and cause that structure to change as time passes. First, look at the `drawBranches` method. This describes the tree's structure at any particular time. Considering the apparent complexity of a tree, this is a remarkably simple method. Part of the simplicity comes from the help provided by the `Graphics` parameter, `g`. But most of the simplicity comes from our use of recursion. Each recursive method call specifies one branch. The `x0` and `y0` parameters are the x and y pixel locations of the start of that branch. The `length` parameter is the length of that branch. The `angle` parameter is the angle of that branch in degrees relative to the angle of the branch from which it springs. Angles are measured counterclockwise from zero at horizontal-right (→).

The first statement in the `drawBranches` method converts degrees to radians. The second statement computes the x-position of the other end of the branch. The third statement computes the y-position of the other end of the branch. The computer displays increasing y-pixels in the downward direction. So we use a minus sign to make `y1` appears above `y0` on the screen when `y1` is greater than `y0`.

The `if` condition in the `drawBranches` method makes the method return without doing anything if the length parameter is 2 or less. This is the recursive stopping condition. It corresponds to branches that are twigs. Within the `if` clause, the first statement draws the current branch. The next two statements recursively call the same method to draw the subordinate branches that fork to the left and right, respectively. The left branch's length is 75% of the current branch's length, and the right branch's length is 67% of the current branch's length. So the branches keep getting shorter as the recursion proceeds from the trunk at the bottom of the tree to the twigs at the top and sides of the tree.

```
/*************************************************************
* Tree.java
* Dean & Dean
*
* This defines a recursion that creates a self-similar tree.
*************************************************************/

import java.awt.Graphics;

public class Tree
{
  private final int START_X;     // where the tree sprouts
  private final int START_TIME; // when the tree sprouts
  private final double MAX_TRUNK_LENGTH = 100;
  private double trunkLength;

  //**********************************************************

  public Tree(int location, int startTime, double trunkLength)
  {
    this.START_X = location;
    this.START_TIME = startTime;
    this.trunkLength = trunkLength;
  } // end constructor

  //**********************************************************

  int getStartX()
  {
    return START_X;
  } // end getStartX

  int getStartTime()
  {
    return START_TIME;
  } // end getStartTime

  double getTrunkLength()
  {
    return this.trunkLength;
  } // end getTrunkLength
```

Figure 11.15a `Tree` class—part A.
This class provides the model. It describes the components whose behavior the program simulates.

```
  //***********************************************************

  public void drawBranches(Graphics g,
    int x0, int y0, double length, double angle)
  {
    double radians = angle * Math.PI / 180;
    int x1 = x0 + (int) (length * Math.cos(radians));
    // negate y increment to flip image vertically
    int y1 = y0 - (int) (length * Math.sin(radians));

    if (length > 2)
    {
      g.drawLine(x0, y0, x1, y1);
      drawBranches(g, x1, y1, length * 0.75, angle + 30);
      drawBranches(g, x1, y1, length * 0.67, angle - 50);
    }
  } // end drawBranches

  //***********************************************************

  // This employs the logistic equation to generate bounded
  // growth along an S-curve like that in Figure 6.21.

  public void updateTrunkLength()
  {
    trunkLength = trunkLength + 0.01 * trunkLength *
      (1.0 - trunkLength / MAX_TRUNK_LENGTH);
  } // updateTrunkLength
} // end Tree class
```

Figure 11.15b `Tree` class—part B.
The `drawBranches` method defines the recursion that describes a tree at any one time.

Now look at the `updateTrunkLength` method. The comment at the top explains that we model our tree's growth in time by using a venerable recurrence relation called the *logistic equation*. You can see that the single Java statement in the `updateTrunkLength` method is indeed a recurrence relation if you mentally associate an iteration number, n, with the `trunkLength` variable on the left side and mentally associate the iteration number, n − 1, with each of the three `trunkLength` variables on the right side.

This is an example where we use a recurrence relation simply to generate a next value from a previous value. We are not doing recursion; we're just doing a straightforward evaluation. As shown in Figure 6.21 in Section 6.13, straightforward repeated evaluation of this recurrence relation produces growth that increases exponentially at first, then linearly. Then the growth slows as the size approaches a maximum.

View

Figure 11.16a begins the description of the GUI presentation in a class called `TreePanel`. This class imports the Java API `JPanel` class to hold the display. It imports the Java API `Graphics` class to draw on the display. It imports the Java API `ArrayList` class to hold a collection of tree objects. The

```
/*************************************************************
* TreePanel.java
* Dean & Dean
*
* This displays growing trees.
*************************************************************/

import javax.swing.JPanel;
import java.awt.Graphics;
import java.util.ArrayList;

public class TreePanel extends JPanel
{
  private final int HEIGHT;       // height of frame
  private final int WIDTH;        // width of frame
  private ArrayList<Tree> trees = new ArrayList<>();
  private int time = 0;           // in months

  //**********************************************************

  public TreePanel(int frameHeight, int frameWidth)
  {
    this.HEIGHT = frameHeight;
    this.WIDTH = frameWidth;
  } // end constructor

  //**********************************************************

  public void setTime(int time)
  {
    this.time = time;
  } // setTime

  //**********************************************************

  public void addTree(
    int location, double trunkLength, int plantTime)
  {
    trees.add(new Tree(location, plantTime, trunkLength));
  } // end addTree

  //**********************************************************

  public ArrayList<Tree> getTrees()
  {
    return this.trees;
  } // end getTrees
```

Figure 11.16a `TreePanel` class—part A.
This class manages the view. It keeps track of what will be displayed.

extends JPanel appended to the class header gives our TreePanel class access to already defined methods in the Java API JPanel class. For now, don't worry about how this works. You'll learn more about it later, in Chapter 13.

The HEIGHT and WIDTH constants are the dimensions of the window in which the current panel sits. The trees list identifies the objects to be displayed. The time variable holds the current simulation time.

The constructor initializes the HEIGHT and WIDTH constants with the height and width constants. The first three methods do obvious things. The setTime method allows the driver to update the time variable. The addTree method allows an outsider to insert additional trees into the scene. The getTrees method allows an outsider to see what trees the scene currently contains.

The paintComponent method in Figure 11.16b performs the drawing operations on the display window. The location variable has the horizontal starting position of each tree (that is, where it sits on the ground). The age variable has the simulated age of a tree in years. The super.paintComponent(g) method call initializes the display window by calling the paintComponent method in the JPanel class (which our TreePanel class extends). As indicated previously, you don't need to worry about how this works, because you'll learn about it in Chapter 13.

The g.drawLine method call uses the drawLine method in the Java API Graphics class to draw a horizontal line that represents the surface of the earth. It uses the HEIGHT and WIDTH of the underling window to accommodate that line to the window's size, whatever it might be. The first and second parameters are the x and y positions of one end of that line, and the third and fourth parameters are the x and y positions of the other end of that line. These are in pixels. Because pixels increase downward on a computer screen, and we want y to increase upward, we subtract the y-positions from HEIGHT to put the earth's surface near the bottom of the display.

```
  //*************************************************************

  public void paintComponent(Graphics g)
  {
    int location; // horizontal starting position of a tree
    String age;   // age of a tree in years

    super.paintComponent(g);
    // draw a horizontal line representing surface of the earth:
    g.drawLine(25, HEIGHT - 75, WIDTH - 45, HEIGHT - 75);
    for (Tree tree : trees)
    {
      // draw the current tree:
      location = tree.getStartX();
      tree.drawBranches(
        g, location, HEIGHT - 75, tree.getTrunkLength(), 90);
      // write the age of the current tree:
      age = Integer.toString((time - tree.getStartTime()) / 12);
      g.drawString(age, location - 5, HEIGHT - 50);
    }
  } // end paintComponent
} // end TreePanel class
```

This calls the recursive method in the Tree class.

This converts months to years.

Figure 11.16b TreePanel class—part B.
The paintComponent method draws graphical images on the display window.

The for-each loop iterates through all the objects in the ArrayList called trees. For each tree, it calls the previously described drawBranches method in Figure 11.15b. This call immediately draws the current tree's trunk. Then, because it is a recursive method, it proceeds to call itself recursively until it draws all the rest of the branches in the tree. After that, it subtracts the current time from the current tree's start time to obtain the current tree's age in months, and divides by 12 for the current tree's age in years. Then, at the bottom of the display and under the current tree, it uses the drawString method from the Graphics class to display that tree's current age in years.

It's important to keep in mind that this paintComponent method draws the scene at just one point in time, like a still photograph. For a movie—that is, animation—we need to call TreePanel's paintComponent method many times, with slightly different conditions each time.

Controller

Figure 11.17a begins the description of this program's controller in a class called TreeDriver. This class imports the Java API JFrame class to create a window. It imports the Java API ArrayList class to hold a collection of tree objects.

```
/*************************************************************
* TreeDriver.java
* Dean & Dean
*
* This manages the simulation of growing trees.
*************************************************************/

import javax.swing.JFrame;
import java.util.ArrayList;

public class TreeDriver
{
  private final int WIDTH = 625, HEIGHT = 400;
  private TreePanel panel = new TreePanel(HEIGHT, WIDTH);
  private int time = 0;

  //**********************************************************

  public TreeDriver()
  {
    JFrame frame = new JFrame("Growing Trees");

    frame.setSize(WIDTH, HEIGHT);
    frame.add(panel);
    frame.setVisible(true);
  } // end constructor

  //**********************************************************
```

Figure 11.17a TreeDriver class—part A.
This manages the tree simulation by controlling the Tree model and the TreePanel view classes.

The WIDTH and HEIGHT constants specify the size of the window that will hold our display panel. The panel variable refers to the TreePanel object that will display the simulation. The time variable holds cumulative simulated time in months. The time variable starts at zero, and as the program runs, it adds one every 50 milliseconds.

The constructor instantiates a JFrame window called frame to hold our display panel. The argument passed to the JFrame constructor specifies a title for the title bar at the top of the window. The next three method calls specify the size of the window, add our panel to the window, and make the window visible on the computer screen. You'll see those method calls over and over again later, in Chapter 17 and Chapter 18. They are ubiquitous in Java GUI programs.

The simulate method in Figure 11.17b contains all the controlling operations. The throws Exception appended to the method name deals with a potential problem associated with the Thread.sleep method call near the end of the simulate method. The first declaration assigns the trees variable a reference to the panel's list of trees. The panel and the driver both need access to this trees list, so we might have put it into either the TreePanel class or the TreeDriver class. We chose to put it into the controlled TreePanel class because it's controlled data, and we want the driver to be calling a method in the panel, rather than the panel calling a method in the driver.

The simulate method's second declaration initializes a boolean variable called done. A large while loop, which contains all the method's operations, repeats until done becomes true. The switch statement plants four different trees at four different places at four different times. Then, shortly after the last planting, it makes done equal true. When we look at the final display, we see trees of different ages. The first tree planted is the oldest and largest, and the last tree planted is the youngest and smallest. Because exactly the same algorithm generates all four trees, the final display shows what the largest tree looked like at earlier stages in its development.

The statement immediately after the switch statement calls the repaint() method, which our TreePanel class *inherits* from the Java API JPanel class. You'll learn all about Java inheritance a little later in Chapter 13. But for now, the term *inherits* just means "freely acquires a prewritten method." This repaint method clears the computer screen and then calls our explicitly written paintComponent method, which redraws the trees.

After redrawing the trees, the simulate method increments the driver's time variable and sends the new time to the panel. Then it enters a for-each loop to increase all tree sizes to what they should be at the new time. The last statement in the while loop calls the sleep method in the Thread class, which is in the always-available Java.lang package. This sleep method call could throw an InterruptedException, and that's why we needed to append throws Exception to the simulate method header.

The main method at the bottom of Figure 11.17b instantiates a new driver object and calls the simulate method. Because the simulate method might throw an exception, we also need to append throws Exception to the main method header. This exception handling technique is like the "simple and dirty" technique we suggested using with simple file input in Section 3.24 of Chapter 3. You'll learn how to implement more responsible exception handling later, in Chapter 15.

Final Display

When you execute the program, the first thing you see is a horizontal line across the bottom of the screen, plus a tiny vertical line just above the horizontal line and about 70% of the way over to the right. The long horizontal line represents the surface of the earth. The tiny vertical line represents the shoot of a new tree, just popping out of the earth. This shoot immediately starts growing and sprouting new branches. An

```
  // Start four trees at different times to show what this
  // kind of tree looks like at different stages of its life.

  public void simulate() throws Exception
  {
    ArrayList<Tree> trees = panel.getTrees();
    boolean done = false;

    while(!done)
    {
      switch (time)
      {
        case 0:
          panel.addTree(400, 3, time);
          break;
        case 360:
          panel.addTree(100, 3, time);
          break;
        case 540:
          panel.addTree(300, 3, time);
          break;
        case 630:
          panel.addTree(200, 3, time);
          break;
        case 675:
          done = true;
      } // end switch
      panel.repaint();

      time++;
      panel.setTime(time);
      for (Tree tree : trees)
      {
        tree.updateTrunkLength(); // to correspond to the new time
      }
      Thread.sleep(50);          // throws an InterruptedExeption
    } // end while
  } // end simulate

  //*************************************************************

  public static void main(String[] args) throws Exception
  {
    TreeDriver driver = new TreeDriver();

    driver.simulate();
  } // end main
} // end TreeDriver class
```

Because it extends `JPanel`, our `TreePanel` class acquires this `repaint` method from the `JPanel` class, and `repaint` automatically calls `TreePanel's` `paintComponent` method.

Figure 11.17b `TreeDriver` class—part B.
The `simulate` method implements all control operations. Figure 11.18 shows the final output.

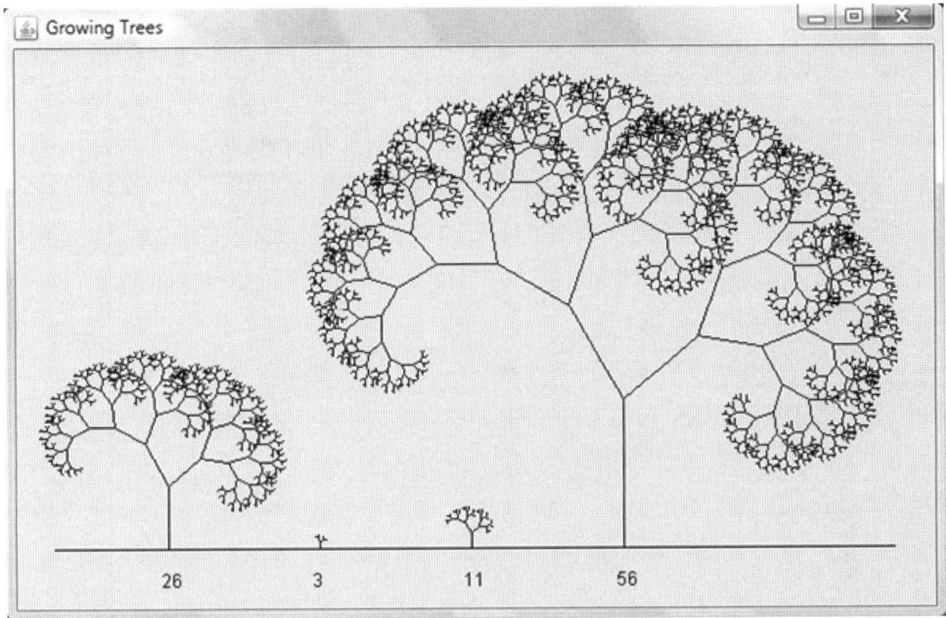

Figure 11.18 Final view of simulated tree growth.
The numbers below the trees indicate their ages in years. Because the same model generated all four trees, you can think of the younger ones as showing earlier stages of the oldest one.

incrementing number just below the tree tells its age in years as it grows. When that tree becomes 30 years old, another tiny shoot appears over to the left, and it immediately starts growing just like the first tree did. Then, 15 years later, a third tree appears and starts growing. Seven and a half years after that, a fourth tree appears and starts growing. Three and three-quarters of a year after that, the simulation stops. Then you see what's in Figure 11.18.

11.10 Performance Analysis

Performance is another word for efficiency. Efficiency is important because if a proposed implementation does not execute in the available time or the available computer memory, it's not useful. Many applications are time-critical. For example, if you speed up the rate at which the trees grow in the previous section's program, eventually that program crashes. Frequently, there are several different ways to solve a problem. The performance estimation techniques reviewed and introduced in this section will help you decide which way is best. 

We quantify the performance of an algorithm by measuring or calculating the execution time or the space (memory) required. High performance means short time and minimal space. Time and space requirements are partially correlated. Usually, when analyzing an algorithm, more attention is paid to time analysis than space analysis, and we follow that pattern in this section. Specifically, we'll focus on estimating the number of computational steps required to complete the algorithm as a function of data. Knowing an algorithm's estimated number of computational steps is a critical part of time complexity analysis (*time complexity analysis* is the formal term for the study of an algorithm's execution speed).

Since Chapter 4, you've seen many examples of loops, and a typical `for` loop header usually tells you the maximum number of iterations. For example, if the loop iterates through an array, the number of iterations equals the array's length. If each iteration takes the same amount of time, the time needed to iterate through the array increases linearly with array length. Sometimes a loop includes another loop nested inside it. Then the total number of iterations is the number of iterations in the outer loop times the number of iterations of the inner loop. The code in Figure 11.19 illustrates this as it implements an algorithm called *insertion sort.* Some people use this instead of selection sort to sort a hand of cards.

```
public static void insertionSort(int[] list)
{
  int itemToInsert;
  int j;

  for (int i=1; i<list.length; i++)
  {
    itemToInsert = list[i];
    for (j=i; j>0 && itemToInsert<list[j-1]; j--)
    {
      list[j] = list[j-1]; // shift up previously sorted items
    }
    list[j] = itemToInsert;
  } // end for
} // end insertionSort
```

Figure 11.19 Insertion sort method

In the insertion sort method, note how the inner `for` loop starts at the outer loop's current index and iterates down through previously sorted items, shifting them upward as the iteration proceeds, until the item to insert is less than a previously sorted item. Then it inserts the new item there. The portion of the array that is already sorted grows as the outer `for` loop progresses.

If the array is already sorted, the condition, `itemToInsert<list[j1]` is always false, and the inner loop never executes. In this best case, the total number of steps is just the number of steps through the outer loop. Because the number of iterations in the outer `for` loop is `list.length 1`, in this best case, the total number of steps is

<minimum steps> = `list.length - 1`

If the array is initially in reverse order, the condition `itemToInsert<list[j-1]` is always true, and the inner loop always shifts up all previously sorted items. In this worst case, the total number of steps is

<maximum steps> = `(list.length - 1) * list.length / 2`

If the array is initially in random order, on average, the inner loop shifts up about half of the previously sorted items. In this average case, the total number of steps is

<average steps> = `(list.length - 1) * list.length / 4`

The analysis above shows that performance depends not only on the nature of the algorithm, but also on the state of the data. The experimental measurements and discussion in Section 10.8 identify another confounding factor. The time needed by computer hardware to perform a particular operation is a function of the number of times it performs that operation. For example, we expect the time to get or set an `ArrayList` element to be independent of list length. In contrast, measured results presented by and associated with Figure 10.8 show that as an `ArrayList`'s length increases from 100 to 1,000 to 10,000, the average `get` and `set` time decreases from 340 ns to 174 ns to 122 ns. Also, we expect that the time to get or set a linked list element to increase linearly with list length. The measured results show that as the length of a `LinkList` increases from 100 to 1,000 to 10,000, the average `get` and `set` time increases from 1,586 ns to 1,917 ns to 18,222 ns. There is a general increase, but it is not exactly linear.

These hard-to-predict performance variations suggest that we should not get too carried away with precision when we do performance analysis. To model performance, we could try to approximate the total time or space required to perform a task with something like this:

<time or space required> ≈ `a * f(n) + b`

In this formula, `n` is the total number of elements, and `f(n)` means "function of `n`." `a` is a scaling factor that indicates the time or space required for each `n`. `b` is a constant that might represent a fixed setup time or a common working space. We might, for example, use `b` to model the unexpectedly high measured `get` and `set` times for the shortest lists described in the previous paragraph. But usually, we elect to simplify the formula by dropping `a` and `b`, and writing:

<time or space required> ≈ $O(n)$

The right side of this equation is what computer scientists call *Big O* notation. The *O* in this term stands for "order of magnitude." A Big O function tells us approximately how time and/or space requirements grow as n (the size of the algorithm's input data) grows. Algorithms with smaller Big O functions tend to run faster and/or use less memory. Algorithms with larger Big O functions tend to run slower and/or use more memory. If the time or space required is approximately the same for all n, we say "the problem is of order one," or more simply, "it's $O(1)$," where "*O*" is the letter, not the number zero.

As `n` increases, if `f(n)` approaches linear dependence on `n`, we say, "The problem is of order n," or more simply, "It's $O(n)$." If `f(n)` approaches parabolic dependence on `n`, we say, "it's $O(n^2)$." If `f(n)` approaches cubic dependence on `n`, we say, "it's $O(n^3)$." And it continues like that. When `f(n)` approaches logarithmic dependence on `n`, and we say, "It's $O(\log n)$." When `f(n)` approaches a dependence like `n * log(n)`, we say, "It's $O(n \log n)$." For some really tough problems, the dependence might increase exponentially, and we say, "It's $O(2^n)$," or perhaps, "It's $O(n^n)$."

Table 11.1 shows these functional dependencies in order of increasing n-dependence. For the logarithms, we use base-2. The #NUM! symbol means the computed number is larger than our computer could hold. Clearly, the n-dependences for $O(1)$ and $O(\log(n))$ are very weak, whereas the n-dependences for $O(2^n)$

n	$O(1)$	$O(\log n)$	$O(n)$	$O(n \log n)$	$O(n^2)$	$O(n^3)$	$O(2^n)$	$O(n^n)$
4	1	2	4	8	16	64	16	256
16	1	4	16	64	256	4,096	65,536	1.8 E19
64	1	6	64	384	4,096	2.6 E5	1.8 E19	3.9 E115
256	1	8	256	2,048	65,536	1.7 E7	1.2 E77	#NUM!
1,024	1	10	1,024	10,240	1.1 E6	1.1 E9	#NUM!	#NUM!

Table 11.1 Big O dependences.
Logarithms are base-2, and #NUM! means the number is larger than our computer could hold.

and $O(n^n)$ are very strong—so strong that computer scientists call exponential dependences *intractable*. Intractable means that for large n, it's virtually impossible to obtain exact answers.

Compared to performing a detailed analysis, it's relatively easy to determine Big O performance. Here are some examples, taken from material in Chapters 9, 10, and 11:

- [§9.7] Sequential search: $O(n)$
- [§9.7] Binary search: $O(\log n)$
- [§9.8] Selection sort: $O(n^2)$
- [§9.9] Two-dimensional array fill: $O(n^2)$
- [§10.2] `List`'s `contains` method: $O(n)$
- [§10.7] `ArrayList`'s `get` and `set` methods: $O(1)$
- [§10.7] `LinkedList`'s `get` and `set` methods: $O(n)$
- [§10.7] `List`'s indexed `remove` and `add` methods: $O(n)$
- [§10.7] `ArrayDeque`'s `offer` and `poll` methods: $O(1)$
- [§10.9] `HashMap`'s `put`, `get`, `contains`, and `remove` methods: $O(1)$
- [§10.9] `TreeSet`'s `add` and `get` methods: $O(\log n)$
- [§10.7] `HashMap`'s `put`, `get` and `contains` methods: $O(1)$

Except for the additional burden of method-call overhead, the performance of a recursive calculation is like that of an iterative counterpart:

- [§11.4] `Factorial`: $O(n)$
- [§11.4] `PrintReverseMessage`: $O(n)$
- [§11.6] Binary search: $O(\log n)$
- [§11.7] Merge sort: $O(n \log n)$
- [§11.8] Towers of Hanoi: $O(2^n)$
- [§11.9] drawBranches: $O(n)$, where n is the number of branches
- [§11.10] Insertion sort: $O(n^2)$, or $O(n)$ if already sorted or nearly sorted

Summary

- A recursive method repeats a given pattern of behavior by calling itself until it reaches a stopping condition.
- We can describe the calculation of a factorial with the recurrence relation, `n! = n * (n-1)!`, and the stopping condition, `if (n == 1 || n == 0)`.
- The trace of a recursion restates the recurrence relation with changing variable values and question marks for unknowns until it reaches a stopping condition. Then it backtracks, substituting previously determined values for question marks until it returns from the first recursive call.
- If a method's body includes just one recursive call to itself, and the call comes at the end of the method's body, it's called tail recursion. It's relatively easy to convert a tail recursion to a loop calculation.
- Binary search is an efficient technique for searching an already ordered array. Binary search lends itself well to recursion, where the binary search is conducted on the two halves of the original array.
- Merge sort is an efficient technique for sorting a large array. Merge sort lends itself well to recursion, where the merge sort is conducted on the two halves of the original array.
- The Towers of Hanoi puzzle starts with different-sized disks stacked at one location with smaller disks above larger disks. The task is to move all the disks to another location, one at a time, never placing any disk on top of a smaller disk, and using only one other location as a temporary repository. There is a recursive solution that solves this difficult problem in a relatively straightforward manner.

- To produce a GUI animation that simulates growing trees, you can partition the tasks into a model class, a view class, and controller class. To model each tree's growth, you can use a recursive fractal algorithm.
- Characterize time performance by estimating the number of computational steps required as a function of data quantity. This often depends on the condition of the data, but for simplicity, we often ignore this and also ignore additive and multiplicative factors with Big O representation.

Review Questions

§11.2 Guidelines for Writing a Recursive Method

1. Fill in the blank: "A recursive method needs to have at least one parameter whose value ___________ with successive methods calls."
2. When a stopping condition is satisfied, a recursive method calls itself. (T / F)

§11.3 A Recursive Factorial Method

3. Write a sufficient stopping condition for the recursive calculation of a factorial.
4. In Figure 11.2's factorial calculation, during the calling sequence (before reaching the stopping condition), what is the value of `nF` when `n` equals 2?

§11.4 Comparison of Recursive and Iterative Solutions

5. Fill in the blank: "If the method's body includes two recursive calls to itself and the method executes both calls, the method is said to exhibit ___________."
6. If it's just as easy to solve a problem with a loop as with recursion, which solution is preferable and why?
7. In the `printReverseMessage` method of Figure 11.7, what variable changes as the recursion progresses, and what is the stopping condition?

§11.5 Recursive Method Evaluation Practice

8. Write the recurrence relation for a factorial using the functional notation, `f(x)`.
9. What's wrong with the stopping condition in this specification?

$$f(x) = \begin{cases} f(x - 2) + x & x > 0 \\ x - 2 & x = 0 \end{cases}$$

§11.6 Binary Search

10. How many recursive method calls does it take to binary search an array with a length of 1,000?
11. Because a recursive binary search uses tail recursion, it's easy to convert it to a loop solution. (T / F)

§11.7 Merge Sort

12. The recursive merge sort algorithm uses binary recursion. (T / F)
13. In a recursive merge sort, what is the recursive stopping condition?
14. In a merge sort, the next element to go into the merged array is the smaller of the smallest elements in the two subordinate arrays. What happens when one of the two subordinate arrays becomes empty?

§11.8 Towers of Hanoi

15. What is the algorithm used to move a stack of disks whose bottom disk is an arbitrary disk n?
16. Describe the arguments used for the first recursive method call in the algorithm in question 15 in terms of the current method's parameters.

§11.9 GUI Track: Drawing a Tree with a Fractal Algorithm (Optional)

17. What is a fractal?
18. In our tree-drawing program, what does the model do?
19. In our tree-drawing program, what does the view do?
20. In our tree-drawing program, what does the controller do?

Exercises

1. [after §11.1] In our ship-signaling analogy, all the data processing occurred as the signal traveled back from the farthest ship to the admiral's ship. With this recursive procedure, only the admiral sees the total number of casualties. The captains of all other ships see only the cumulative casualties from ships that are farther away. Suggest a modification to the recursive procedure that enables every ship's captain to see the total number of casualties.
2. [after §11.3] A problem with factorials is that they rapidly become very large. Sometimes we don't need the factorial itself; we just need the logarithm of the factorial. In that case, rather than computing the factorial and then taking the logarithm, it's better to take the logarithm of each component and then add the components instead of multiplying them. Modify the Factorial program in Figure 11.2 so that it returns the natural logarithm of the factorial of the user-entered number.
3. [after §11.3] Provide code for a method named `sumToInfinity` that returns the sum of numbers from `n` to infinity, where `n` is a `long` parameter. Your method must use an infinite recursive loop.
4. [after §11.3] Many times, you have used `Scanner` methods to read and parse keyboard input. You can also use those same `Scanner` methods to read and parse an ordinary `String`. The text's `printReverseMessage` method in Figure 11.7 used `String`'s `charAt` method to reverse the order of the letters in a string. This exercise asks you to use `Scanner`'s `next` method to reverse the order of the words in a string. The program below should reverse the words in the string, "We Are Many," and return the string, "Many Are We." In the recursive method in this program, supply the code where it says *<code-fragment>*.

```
import java.util.Scanner;

public class ReverseWords
{
  public static void main(String[] args)
  {
    String message = new String("We Are Many");
    Scanner scan = new Scanner(message);
    String reversedMessage = getReverse(scan);

    System.out.println(reversedMessage);
  } // end main

  //************************************************

  public static String getReverse(Scanner scan)
  {
    String nextWord;

    <code-fragment>

  } // end getReverse
} // end class ReverseWords
```

Do not use an array. Use `Scanner's` `next` method to step through the message in the `scan` object, retrieve each `nextWord`, and `return` the return from a recursive call plus that `nextWord`. What changes with each recursive call? What is the stopping condition? What does the `getReverse` method return when the stopping condition is satisfied?

5. [after §11.4] Repeat the solution to the problem in exercise 4, but this time, include an additional method parameter, `String revMsg`, and write the code fragment so that it implements a tail recursion. Again, do not use an array.

6. [after §11.4] Repeat the solution to the problem in exercise 4, but this time, write the code fragment so that it uses iteration instead of recursion. Again, do not use an array.

7. [after §11.4] A palindrome is a string that reads the same in both directions. For example, "mom," "otto," and "abcba" are all palindromes. Write a recursive method named `isPalindrome` that determines whether a string is a palindrome. More specifically, your method should return `true` if the given string parameter (named `str`) is a palindrome and it should return `false` otherwise. Use this method signature:

```
public static boolean isPalindrome(String str, int begin, int end)
```

8. [after §11.5] What's wrong with this method?

```
public double sum(double x)
{
  if (x == 0.5)
  {
    return x;
  }
  else
  {
    return sum(x) + Math.random();
  }
} // end sum
```

9. [after §11.5] Given this recurrence relation and stopping condition:

```
           ⎧ (1 - A) + A * y(k-1, A)        k > 0
y(k, A) =  ⎨
           ⎩ 0                              k <= 0
```

By hand, evaluate `y(5, 0.75)`, using the format presented in Section 11.5.

10. [after §11.5] Given the following method:

```
public static int functionX(int x)
{
  if (x == -4)
  {
    return 4;
  }
  else
  {
    return functionX(x-1) + x;
  }
} // end functionX
```

Determine the value returned by each of the following calls:

a) `functionX(4)`
b) `functionX(0)`
c) `functionX(-4)`
d) `functionX(-8)`

11. [after §11.5] The following recurrence relation is called the *Logistic Equation*:

```
x(k)  = x(k-1)  + gain * x(k-1)  *  (1.0 - x(k-1))
```

Write a program that performs forward evaluation of this recurrence relation, with arbitrary values for `x(0)` and `gain`. Use a `for` loop to iterate through `n` evaluations. After each evaluation, print the value of `x` to six decimal places.

Using `n = 100` and the starting value, `x(0) = 0.2`, describe the qualitative behavior of `x(k)` as `k` increases from zero, for `gain` in each of these three ranges:

```
0.0 < gain < 1.0
1.0 < gain < 2.0
3.0 < gain
```

Change the total number of iterations to `n = 1000`, and describe the behavior in this range:

`2.0 < gain < 2.569945...` (the Feigenbaum Point)

In this range, there are subordinate ranges between thresholds 2.0, 2.4478 . . . , 2.5438 . . . , 2.5643 . . . , 2.5687 . . . , etc., where each of these ranges is smaller than the previous one by a common scaling factor of 4.6692 . . . . This scaling factor, called the *Feigenbaum Constant,* is a universal constant, like pi, and it appears in many natural phenomena. For the clearest picture of what's happening, focus on gains just before subordinate range boundaries, like gains of 2.44, 2.54, 2.56, and 2.568.

For gains in the range between the Feigenbaum Point and 3.0, the dynamics are *chaotic*. When the `gain` equals 3.0, x-values become uniformly distributed in the range, `0 < x < 4/3`, and the behavior seems to be random. But of course, it is not random. It is deterministic, because each value depends on the previous value in accordance with the Logistic Equation's formula.

12. [after §11.8] In the Towers of Hanoi program, declare a new variable called `moveCount` and initialize it to zero. Comment out all current print statements, and just after each commented print statement that says "move . . ." add a statement that increments the `moveCount` variable by 1. Import the Java API `Scanner` class. In the `main` method, prompt for and input an integer that asks the user for desired number of disks, and assign it to a new variable called `disks`. Then substitute `disks` for 4 as the first argument in `main`'s call of the `move` method. After the `move` call in `main`, add a statement that prints the final value of `moveCount` and the value of 2^{disks}. Execute the program for progressively increasing values of the disks, and compare the outputs with Wikipedia's formula. Say how far your computer lets you go toward 64 disks, and if you can't get that far, describe the problem.

13. [after §11.10] Convince yourself that each of the Big O representations at the end of Section 11.10 is a reasonable performance estimate.

Review Question Solutions

1. A recursive method needs to have at least one parameter whose value changes with successive method calls.
2. False. At a stopping condition, a recursive method returns without making another recursive call.
3. A sufficient stopping condition for the recursive calculation of a factorial is `n == 1 || n == 0`. An alternate sufficient stopping condition is `n <= 1`.
4. Before the stopping condition is reached, the value of `nF` is unknown.
5. If the method's body includes two recursive calls to itself, the method is said to exhibit binary recursion.
6. If it's just as easy, a loop is preferable because it takes less overhead.
7. In the `printReverseMessage` method, the variable that changes is `msg`, and the stopping condition is `msg.isEmpty();`
8. The recurrence relation for a factorial using functional notation is: `f(x) = x * f(x-1)`
9. The stopping condition is missed if `x` is less than zero, an odd integer, or not an integer.
10. Because $2^9 = 512$ and $2^{10} = 1024$, it takes 10 recursive method calls to binary-search an array with a length of 1,000.
11. True. Recursive binary search uses tail recursion, and Figure 9.12 in Section 9.7 shows a straightforward loop implementation.
12. True. The recursive merge sort algorithm uses binary recursion.
13. The stopping condition in a recursive merge sort is when the array length is 1.
14. In a merge sort, when one of the two subordinate arrays becomes empty, the merged array gets all remaining elements from the other subordinate array in their already sorted order.
15. The algorithm implemented by the recursive method that moves a stack of disks is:
    a) Move the stack of disks above disk `n` to the temporary location, t.
    b) Move disk `n` to the destination location, d.
    c) Move the stack of disks moved in (a) from the temporary location to the destination location.
16. In the first recursive call [Review Question Solution 15(a)], use n − 1 for the bottom of the pile to move, use s for the source location, use t for the destination location, and use d for the temporary location.
17. A fractal is an object or quantity that displays self-similarity over a range of scales.
18. In our tree-drawing program, model methods specify relative lengths and angles of branches and how trunk size increases in time.
19. In our tree-drawing program, the view specifies screen display at any particular time.
20. In our tree-drawing program, the controller creates and locates new tree instances at particular times and manages the updating of the screen display as time passes.

CHAPTER 12

Type Details and Alternative Coding Mechanisms

Objectives

- Improve your understanding of relationships and differences among primitive data types and your appreciation for their individual limitations.
- Understand how numerical codes identify characters.
- Learn the rules for automatic type conversions and the risks in explicit type casting.
- Understand embedded postfix and prefix increment and decrement operators.
- Understand embedded assignment expressions.
- Learn where and how conditional operator expressions can shorten code.
- See how short-circuit evaluation helps avoid troublesome operations.
- See how empty statement works.
- Learn how to use `break` statements in loops.
- Create and use an `enum`.
- Optionally, use Unicode characters in GUI applications.

Outline

12.1 Introduction

In Chapters 3 and 4, you learned Java language basics. Among other things, you learned about data types, type conversions, and control statements. This chapter describes some additional data types and additional type conversions. It also describes some alternative control statement coding mechanisms.

Chapter 3 introduced you to some of Java's integer and floating-point types of numbers, and Chapter 5 showed you how to find the limits of their ranges. In this chapter, you'll see two more integer types, and for all of the numerical types, you'll learn the amount of storage needed, the precision provided, and how to use range limits. Chapter 3 introduced you to the use of the character type, `char`. In this chapter, you'll see that each character has an underlying numeric value, and you'll learn how to use those values. Chapter 3 introduced you to type conversion with the cast operator. In this chapter, you'll learn more about type conversions. Chapter 3 introduced you to the increment and decrement operators. In this chapter, you'll discover that you can move the positions of these operators (before or after the variable) to control when they act. Chapter 3 introduced you to assignment operators. In this chapter, you'll see how you can embed assignments within expressions to make code more compact.

Chapter 4 introduced you to several kinds of conditional evaluations. In this chapter, you'll learn about the conditional operator that can take on either of two possible values depending on a `boolean` condition. You'll also learn about short-circuit evaluation which can prevent errors by stopping a "dangerous" conditional evaluation in certain situations. In addition, you'll learn more about loops. Specifically, you'll see empty-bodied loops and loops that terminate from within the loop's body. And you'll see alternative coding techniques for `for` loop headers.

The material in this chapter will improve your understanding of several Java nuances and subtleties. This will help you avoid problems in the first place, and it will help you create code that is more efficient and easier to maintain. It will also help you debug code that has problems. It might be your code, or it might be someone else's code. As a real-world programmer, you'll have to work with other people's code, and you'll need to understand what that code is doing.

Much of the material in this chapter could have been inserted at various places earlier in the text. However, it was not necessary for anything we did up until now, and we deferred it until now to keep from encumbering earlier presentations. The assembly of these details into one chapter at this point in the book provides an excellent opportunity for review. As you go through this chapter, integrate this new material into what you learned before and see how it enriches your understanding of those topics.

The last section in this chapter introduces an extended example called GridWorld. The College Board licenses this software to people who want to help new college students obtain advanced placement (AP). It provides a nice example of larger programs, and we will return to it to illustrate new topics as we introduce them.

12.2 Integer Types and Floating-Point Types

This section supplements the numeric data types material you studied in Chapter 3, Section 3.13. 

Integer Types

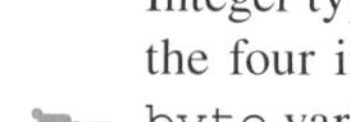

Integer types hold whole numbers (whole numbers are numbers without a decimal point). Figure 12.1 shows the four integer types. The types are ordered in terms of increasing memory storage requirements. Type byte variables require only 8 bits, so they take up the least amount of storage. If you have a program that's taking too much space in memory, you can use smaller types for variables that hold small values. Using smaller types means less storage is needed in memory. Now that memory has become relatively cheap, types byte and short are not used very often.

Type	Storage	Wrapper Class's MIN_VALUE	Wrapper Class's MAX_VALUE
byte	8 bits	−128	127
short	16 bits	−32,768	32767
int	32 bits	−2,147,483,648	2,147,483,647
long	64 bits	≈ $-9*10^{18}$	≈ $9*10^{18}$

Figure 12.1 Properties of Java integer data types

To access an integer's minimum and maximum values, use the MIN_VALUE and MAX_VALUE named constants that come with the integer's wrapper class. As you learned in Chapter 5, Integer and Long are the wrapper classes for the int and long data types. And as you might expect, Byte and Short are the wrapper classes for the byte and short data types. So here's how to print the maximum byte value:

```
System.out.println("Largest byte = " + Byte.MAX_VALUE);
```

The default type for an integer constant is int. But you might have a need for an integer constant that is too big for an int. In that case, you can explicitly force an integer constant to be a long by adding an l or L suffix to the integer constant. For example, suppose that you're writing a solar system program, and you want to store the age of the Earth in a variable named ageOfPlanet. The Earth is 4.54 billion years old, and 4.54 billion is larger than Integer.MAX_VALUE's 2,147,483,647. This generates a compilation error:

```
long ageOfPlanet = 4_540_000_000;
```

But this, with the L suffix, works just fine:

```
long ageOfPlanet = 4_540_000_000L;
```

When you declare a numeric variable, be sure that the type you select is large enough to handle the largest value that your program might put into it. If a value can't fit in the memory space provided, that's called *overflow*. Overflow errors are dramatic, as the ByteOverflowDemo program in Figure 12.2 illustrates.

Integer overflow reverses the sign, so the ByteOverflowDemo program prints negative 128 rather than the correct result, positive 128. In this example, the magnitude of the error is approximately twice as big as the magnitude of the largest allowable value! Overflow also causes sign reversal for types short, int, and long. In such cases, the compiler does not find the problem, and the JVM does not find it either. Java runs the program with no complaints and happily generates a massive error. In the end, it's up to you. Whenever there is any doubt, use a larger type!

```
/*****************************************************************
* ByteOverflowDemo.java
* Dean & Dean
*
* This demonstrates integer overflow.
*****************************************************************/

public class ByteOverflowDemo
{
  public static void main(String[] args)
  {
    byte value = 64;

    System.out.println("Initial byte value = " + value);
    System.out.println("Byte maximum = " + Byte.MAX_VALUE);
    value += value;
    System.out.println("Twice initial byte value = " + value);
  } // end main
} // end ByteOverflowDemo class
```

Output:

```
Initial byte value = 64
Byte maximum = 127
Twice initial byte value = -128
```

A very large error!

Figure 12.2 ByteOverflowDemo program illustrates the overflow problem

Floating-Point Types

As you know, floating-point numbers are real numbers—numbers that allow for non-zero digits to the right of a decimal point. This means you can use floating-point numbers to hold fractional values—values that are smaller than one. Figure 12.3 shows the two floating-point types—`float` and `double`.

Note Figure 12.3's precision column. Precision refers to the approximate number of digits the type can represent accurately. For example, since `float` types have 6 digits of precision, if you attempt to store 1.2345678 in a `float` variable, you would actually store a rounded version—a number like 1.234568. The first 6 digits (1.23456) are precise, but the rest of the number is imprecise. `double` values have 15 digits of precision—quite a bit better than `float` values with their 6 digits of precision. The relatively low precision

Type	Storage	Precision	Wrapper Class's MIN_NORMAL	Wrapper Class's MAX_VALUE
float	32 bits	6 digits	$\approx 1.2 * 10^{-38}$	$\approx 3.4*10^{38}$
double	64 bits	15 digits	$\approx 2.2 * 10^{-308}$	$\approx 1.8*10^{308}$

Figure 12.3 Properties of Java floating-point data types

of a `float` can lead to significant round-off errors when you subtract two numbers that are close in value. If the numbers are close enough, then the difference is a very small number where the rightmost digits are merely approximations. This round-off error is compounded when you have repetitive calculations. Because memory is now relatively inexpensive, you should consider `float` to be an archaic data type, and you should usually avoid it. An exception is when you specify color. Several methods in the Java API `Color` class employ `float` type parameters and/or return values.

Be aware that floating-point numbers do worse than integer numbers when it comes to precision. For example, when comparing the 32-bit `float` type and the 32-bit `int` type, the floating-point type has less precision. `float` numbers have 6 digits of precision, whereas `int` numbers have 9 digits of precision. Likewise, when comparing the 64-bit `double` type and the 64-bit `long` type, the floating-point type has less precision. In particular, `double` numbers have 15 digits of precision, whereas `long` numbers have 19 digits of precision. Why do floating-point numbers lose out on precision? Some of the bits in floating-point numbers are used to specify the exponent that allows these numbers to take on much greater ranges in magnitude than integer numbers can take on. This reduces the bits available to supply precision.

As you learned in Chapter 5, `Float` and `Double` are the wrapper classes for the `float` and `double` data types. To access a floating-point data type's minimum and maximum values, use the `Float` and `Double` classes' `MIN_NORMAL` and `MAX_VALUE` named constants. `MAX_VALUE` is a floating-point data type's largest positive value, and `MIN_NORMAL` is a floating-point data type's smallest full-precision positive value. A floating-point's `MIN_NORMAL` is qualitatively different from an integer's `MIN_VALUE`. Instead of being a large negative value, a floating-point `MIN_NORMAL` is a tiny positive fraction. So what are the limits of negative floating-point numbers? The largest-magnitude negative number a floating-point variable can hold is `-MAX_VALUE`. The smallest-magnitude negative number a floating-point variable can hold safely is `-MIN_NORMAL`, a tiny negative fraction.

Actually, it's possible for a floating-point variable to hold a number whose magnitude is smaller than `MIN_NORMAL`. It can hold a value as small as a floating-point `MIN_VALUE`, which is approximately $1.4 * 10^{-45}$ for `float` and approximately $4.9 * 10^{-324}$ for `double`. But the `MIN_VALUE` of a floating-point number has only one bit of precision, and that could produce a significant error in a computed result—without any explicit indication that an error is present. This is an example of the worst kind of bug because it can go unrecognized for a long time. Therefore, with floating-point numbers, always use `MIN_NORMAL` instead of `MIN_VALUE`.

The default floating-point constant type is `double`. If you declare a variable to be a `float`, you must append an `f` or `F` suffix to all floating-point constants that go into it, like this:

```
float gpa1 = 3.22f;
float gpa2 = 2.75F;
float gpa3 = 4.0;  ←  compilation error, because 4.0 is a double
```

Because of the `f` and `F` suffixes, 3.22f and 2.75F are 32-bit `float` values, so it's legal to assign them into the 32-bit `gpa1` and `gpa2` `float` variables. But 4.0 is a 64-bit `double` value, and attempting to assign it into the 32-bit `gpa3` `float` variable generates a compilation error.

To write a floating-point number in scientific notation, put `e` or `E` before the base-10 exponent value. If the exponent is negative, insert a minus sign between the `e` or `E` and the exponent value. If the exponent is positive, you may use a plus sign after the `e` or `E`, but it's not standard practice. In any event, there must never be any whitespace within the number specification. For example:

```
double x = -3.4e4;  ←  equivalent to −34000.0
double y = 5.6E-4;  ←  equivalent to 0.00056
```

BigInteger and BigDecimal

The largest integer type, long, and the largest decimal type, double, are adequate for most applications. However, every now and then, there's an application that needs a number that won't fit into these types. For example, really good data encryption needs integers that are longer than long. To handle such cases, the java.math package defines special numeric types called BigInteger and BigDecimal. These can represent integer and floating-point numbers of any size.

Like an ArrayList, a BigInteger or BigDecimal automatically expands to accommodate its data. BigInteger constructors allow you to specify the value with a String, with a byte array, or as a random number with a specified number of bits. Here's an example that specifies a BigInteger value with a string:

```
BigInteger bigInt = new BigInteger("25460166057397002366275 71");
```

BigDouble constructors allow you to specify the value with a BigInteger, a long, an int, a double, a String, or a char array. What if the fraction never ends, like 0.333333333 . . .? To handle this, you can specify the number of fractional places directly, in a BigDouble constructor, or indirectly, in a separate object called MathContext. Although it's cumbersome, BigDecimal gives users complete control over rounding behavior—much better control than with double.

This book contains several examples that use double to represent amounts of money. That's fine for informal work. But double is not appropriate for formal accounting because unpredictable rounding can sometimes produce disturbing output. Typically, accountants like everything to be correct down to the last penny. As such, they would not be pleased with a computer program that displays $1999999999.99 for an amount that is supposed to be exactly $2 billion.

BigDecimal's explicit control over rounding enables you to prevent such aberrations. So, in principle, you could use BigDecimal safely for serious accounting. However, BigDecimal is cumbersome and relatively inefficient. For high-quality financial accounting, you should use integers and represent amounts as cents. To convert floating-point input to an internal integer, multiply the input by 100 and then cast to an integer. To convert an internal integer to floating-point, use quotient division (that is, /100) to retrieve the dollars part. Then use modulus division (that is, %100) to retrieve the cents part. If int is not big enough, use long. The long type is big enough for even very large financial amounts (like, for example, the total debt of the United States).

An ordinary double is good enough for most scientific calculations because it's hard to find measuring instruments that have more precision than double. And science without measurement is not science.

12.3 char Type and the ASCII Character Set

This section supplements the char type material that you studied in Chapter 3, Section 3.20. 

Underlying Numeric Values

For most programming languages, including Java, each character has an underlying numeric value. For example, the character 'A' has the underlying decimal value of 65 and the character 'B' has the underlying value of 66. Most programming languages, including Java, get character numeric values from the *American Standard Code for Information Interchange* (*ASCII*, pronounced "askee") character set. All the characters in the ASCII character set are shown in Appendix 1, in Figures A1.1a and A1.1b.

So what's the point of having underlying numeric values for characters? With underlying numeric values, it makes it easier for the JVM to determine the ordering of characters. For example, because 'A' has the value 65 and 'B' has the value 66, the JVM can easily determine that 'A' comes before 'B'. And knowing the order of characters is necessary for string sort operations. For example, suppose a sort method is given the strings "peach", "pineapple", and "apple." The sort method compares the words' first characters 'p', 'p', and 'a', and in doing so, the JVM looks up the characters in the ASCII table. Because 'p' has the value 112 and 'a' has the value 97, "apple" goes first. Then the sort method compares the second characters in "peach" and "pineapple." Since 'e' has the value 101 and 'i' has the value 105, "peach" goes before "pineapple."

You can sometimes use a character's underlying numeric value to your advantage. For example, suppose a program has a `char` variable named `code` whose value determines what happens next. If `code` holds a 'p', the program prints a report, and if `code` holds a 'q', the program quits. Alternatively, if `code` holds a digit character ('0' through '9'), the program can perform a mathematical operation, using `code`'s numeric value as part of that operation. To use `code`'s numeric value, you first need to retrieve it, which requires a little work. Assuming that `digit` has been declared as an `int` variable, here's how to retrieve `code`'s numeric value:

```
digit = code - '0';
```

As the ASCII table shows, the zero character ('0') has an underlying decimal value of 48. If `code` holds the six character ('6'), which has an underlying decimal value of 54, then 6 will be assigned to the `digit` variable, and all is well. This character-to-number code fragment works because the ASCII character set uses successive underlying numeric values (48 through 57) for the digit characters '0' through '9'.

As an alternative to the code fragment above, you could accomplish the same thing with this:

```
digit = code - 48;
```

However, using 48 in your program would be considered inelegant. Why? Because it leads to code that is hard to understand (for someone to understand it, he or she would have to know that 48 is the underlying value of the '0' character). Using '0', on the other hand, would be easier to understand. It's an example of self-documenting code.

Control Characters

Most characters in the ASCII character set represent printable symbols. For example, the 'f' character represents the printable letter *f.* But the first 32 characters and the last character in the ASCII character set are different—they are *control characters.* Control characters perform non-printing operations. For example, the start-of-heading character (ASCII numeric value 1) helps with data being sent from one computer device to another. More specifically, it signals the beginning of transmitted data. When you print a control character, you might be surprised by what appears on the screen. The bell character (ASCII numeric value 7) normally generates a sound and displays nothing, which makes sense, but the start-of-heading character displays something less intuitive. When you print the start-of-heading character, you'll get different results in different environments. For example, in a console window[1] in a Windows environment, a smiley face is displayed. In other environments, a blank square is displayed. Note the following code fragment, with associated output from a console window in a Windows environment:

```
char ch;
for (int code=1; code<=6; code++)
{
  ch = (char) code;
  System.out.print(ch + " ");
}
```

Output:

☺ ☻ ♥ ♦ ♣ ♠

[1]See Chapter 1's "First Program—Hello World" section for a description of how to run a program in a console window.

In the above code fragment, the `(char)` cast operator uses the ASCII table to return the character associated with `code`'s numeric value. So if `code` has the value 1, then `(char) code` returns the start-of-heading character.

Unicode Character Set

The ASCII character set works just fine most of the time, but in some situations, it's insufficient. Sometimes you'll need characters and symbols that are outside the ASCII character set. For example, suppose you want to display a check mark (√) or the pi symbol (π). Those two characters are not part of ASCII. They are part of a newer coding scheme called *Unicode,* which is a superset of ASCII. You can learn about Unicode in the optional section at the end of this chapter (Section 12.14). In that section, we show you how to access the check mark and pi symbols and the many other characters enumerated in the Unicode Standard.

Using the + Operator with chars

Remember how you can use the + operator to concatenate two strings together? You can also use the + operator to concatenate a `char` to a string. Note this example:

```
char first = 'J';
char last = 'D';
System.out.println("Hello, " + first + last + '!');
```

Output:

```
Hello, JD!
```

When the JVM sees a string next to a + sign, it concatenates by first converting the operand on the other side of the + sign to a string. So in the above example, the JVM converts the `first` variable to a string and then concatenates the resulting "J" to the end of "Hello, " to form "Hello, J". The JVM does the same thing with each of the next two characters it sees, `last`'s stored character and '!'. It converts each one to a string and concatenates each one to the string at its left.

Be aware that if you apply the + operator to two characters, the + operator does not perform concatenation; instead, it performs mathematical addition using the characters' underlying ASCII values. Note this example:

```
char first = 'J';
char last = 'D';
System.out.println(first + last + ", What's up?");
```

Output:

```
142, What's up?
```

The intended output is: `JD, What's up?` Why does the code fragment print 142 instead of JD? The JVM evaluates + operators (and most other operators as well) left to right, so in evaluating `println`'s argument, it first evaluates `first + last`. Because both `first` and `last` are `char` variables, the JVM performs mathematical addition using the characters' underlying ASCII values. `first` holds 'J' and J's value is 74. `last` holds 'D' and D's value is 68. So `first + last` evaluates to 142.

There are two ways to fix the above code. You can change the first two lines to string initializations like this:

```
String first = "J";
String last = "D";
```

Or you can insert an empty string at the left of `println`'s argument like this:

```
System.out.println("" + first + last + ", What's up?");
```

12.4 Type Conversions

This section supplements the type casting material you studied in Section 3.19.

Java is a *strongly typed* language, so each variable and each value within a program is defined to have a particular data type. As with all strongly typed languages, you need to be careful when working with more than one data type. In this section, you learn how some, but not all, data types convert to other data types. Java makes some type conversions automatically, and it allows you to force some other type conversions. Either way, be careful. Inappropriate type conversions can cause problems.

To figure out what's allowed in terms of type conversions, learn the ordering scheme in Figure 12.4. Crudely speaking, this picture shows what types can "fit inside" other types. For example, a `byte` value with 8 bits can fit inside a `short` variable that holds 16 bits because an 8-bit entity is "narrower" than a 16-bit entity. We like the terms narrower and wider to describe type sizes, but be aware that those are not formal terms; other people do not use those terms. Notice that the `boolean` type does not appear in this picture. You cannot convert between numeric types and the `boolean` type.

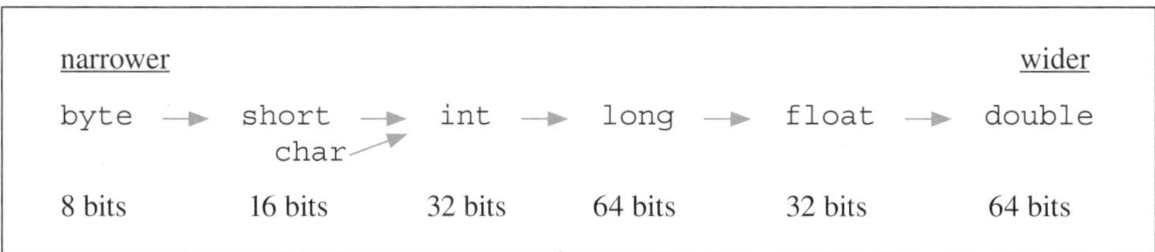

Figure 12.4 Type conversion ordering scheme

Promotion

There are two kinds of type conversion—*promotion* (automatic type conversion) and *type casting* (forced type conversion). You've already seen type casting. We'll revisit it shortly, but let's first discuss promotion.

A promotion is an implicit conversion. It's when an operand's type is automatically converted without having to use a cast operator. It occurs when there's an attempt to use a narrower type in a place that expects a wider type; that is, it occurs when you're going with the flow of the arrows in Figure 12.4. Promotion often occurs in assignment statements. If the expression on the right of an assignment statement evaluates to a type that is narrower than the type of the variable on the left of the assignment statement, then during the assignment the narrower type on the right gets promoted to the wider type on the left. Note these promotion examples:

```
long x = 44;
float y = x;
```

In the first statement, 44 is an `int`. The `int` 44 is narrower than the `long x`, so the JVM promotes 44 to a `long`, and then performs the assignment. In the second assignment statement, `x` is a `long`. The `long x` is narrower than the `float y`, so the JVM promotes `x` to a `float`, and then performs the assignment.

Note these additional promotion examples:

```
double z = 3 + 4.5;
int num = 'f' + 5;
```

mixed expressions

The expressions on the right are *mixed expressions*. A mixed expression is an expression that contains operands of different data types. Within a mixed expression, the narrower operand automatically promotes to the type of the wider operand. In the first statement above, the int 3 is narrower than the double 4.5, so the JVM promotes 3 to a double before adding it to 4.5. In the second statement above, do you know which operand, 'f' or 5, gets promoted to match the other one? 'f' is a char and 5 is an int, and Figure 12.4 shows that char is narrower than int. Thus, the JVM promotes 'f' to an int. More specifically, since f's underlying numeric value is 102 (see Figure A1.1b), the JVM promotes 'f' to 102. Then the JVM adds 102 to 5 and assigns the resulting 107 to num.

Promotions typically occur as part of assignment statements, mixed expressions, and method calls. You've already seen examples with assignment statements and mixed expressions; now let's examine promotions with method calls. As mentioned above, conversions take place any time there's an attempt to use a narrower type in a place that expects a wider type. So if you pass an argument to a method and the method's parameter is defined to be a wider type than the argument's type, the argument's type promotes to match the parameter's type. Figure 12.5's program provides an example of this behavior. Can you determine what promotion takes place within the program? The x argument is a float and it promotes to a double. The 3 argument is an int and it promotes to a double as well.

```
/****************************************************
* MethodPromotion.java
* Dean & Dean
*
* Promote type in method call
****************************************************/

public class MethodPromotion
{
  public static void main(String[] args)
  {
    float x = 4.5f;

    printSquare(x);
    printSquare(3);
  }

  private static void printSquare(double num)
  {
    System.out.println(num * num);
  }
} // end class MethodPromotion
```


Output:

```
20.25
9.0
```

Figure 12.5 Program that demonstrates type promotion in a method call

Type Casting

Type casting is an explicit type conversion. It occurs when you use a cast operator to convert an expression's type. Here's the syntax for using a cast operator:

(*type*) *expression*

It's legal to use a cast operator to convert any numeric type to any other numeric type; that is, the conversion can go in either direction in Figure 12.4's ordering-scheme diagram. For example, the following code fragment casts the `double` `x` to the `int` `y`:

```
double x = 12345.6;
int y = (int) x;
System.out.println("x = " + x + "\ny = " + y);
```

What happens if you omit the `(int)` cast operator? You'd get a compilation error because you'd be directly assigning a `double` into an `int` and that's forbidden (in Figure 12.4's ordering-scheme diagram, there's no arrow going from the `double` type to the `int` type). Why is it illegal to assign a floating-point number directly into an `int`? Because floating-point numbers can have fractions, and `int`s can't handle fractions.

Do you know what the above code fragment prints? `x` remains unchanged (even though `(int)` was applied to it), and `y` gets the whole-number portion of `x` with `x`'s fraction truncated, not rounded. So here's the output:

```
x  = 12345.6
y  = 12345
```

The program in Figure 12.6 further illustrates the use of cast operators. It prompts the user to enter an ASCII value (an integer between 0 and 127). Then it prints the character associated with that ASCII value and also the next character in the ASCII table. In the program, what do the two cast operators do? The first one returns the `char` version of `asciiValue`, an `int` variable. The second one returns the `char` version of `asciiValue + 1`.  The cast operations are needed to print `ch` and `nextCh` as characters, rather than integers. What would happen if you omitted the cast operators? You'd get compile-time errors because you'd be assigning an `int` directly into a `char`, and that's forbidden according to the ordering scheme in Figure 12.4.

Why is it illegal to assign a number directly into a `char`? You'd think it would be safe to assign a small whole number, like a `byte` with 8 bits, into a `char` with 16 bits. It's illegal to assign a number directly into a `char` because numbers can be negative and a `char` can't handle negativity (a `char`'s underlying value is a positive number between 0 and 65535).

12.5 Prefix/Postfix Modes for Increment/Decrement Operators

This section supplements material you studied in the first part of Chapter 3, Section 3.17 (Increment and Decrement Operators), and it uses techniques you studied in Chapter 3, Section 3.18 (Tracing).

The increment operator has two different modes—the *prefix mode* and the *postfix mode*. The prefix mode is when you put the ++ before the variable that is to be incremented. Using the prefix mode causes the variable to be incremented before the variable's value is used. For example:

```
y = ++x        is equivalent to        x = x + 1;
                                       y = x;
```

The postfix mode is when you put the ++ after the variable that is to be incremented. Using the postfix mode causes the variable to be incremented after the variable's value is used. For example:

```
y = x++        is equivalent to        y = x;
                                       x = x + 1;
```

```
/*************************************************************
* PrintCharFromAscii.java
* Dean & Dean
*
* This illustrates manipulation of ASCII code values.
*************************************************************/

import java.util.*;

public class PrintCharFromAscii
{
  public static void main(String[] args)
  {
    Scanner stdIn = new Scanner(System.in);
    int asciiValue; // user entered ASCII value
    char ch;        // the asciiValue's associated character
    char nextCh;    // the character after ch in the ASCII table

    System.out.print("Enter an integer between 0 and 127: ");
    asciiValue = stdIn.nextInt();
    ch = (char) asciiValue;
    nextCh = (char) (asciiValue + 1);
    System.out.println("Entered number: " + asciiValue);
    System.out.println("Associated character: " + ch);
    System.out.println("Next character: " + nextCh);
  } // end main
} // end class PrintCharFromAscii
```

Note the (char) cast operators.

Sample session:

```
Enter an integer between 0 and 127: 67
Entered number: 67
Associated character: C
Next character: D
```

Figure 12.6 Program illustrating the use of cast to convert character codes into characters

To get a better feeling for how this works, trace this code fragment:

```
int x, y;

x = 4;
y = ++x;
System.out.println(x + " " + y);
x = 4;
y = x++;
System.out.println(x + " " + y);
```

Here is the trace:

line#	x	y	output
1	?	?	
3	4		
4	5		
4		5	
5			5 5
6	4		
7		4	
7	5		
8			5 4

Pay attention to the quotes.

Here's a review question to help with your debugging skills. What would the outputs have been if the `println` arguments had been `(x + ' ' + y)`? Instead of specifying the string version of a space, this would have specified the character version of a space, and it would make the computer consider the argument to be a mathematical expression rather than a string concatenation. Because `x` and `y` are integers, it would promote the space character to its underlying numeric value, which is 32 (see Figure A1.1b in Appendix 1). The first print statement would add (5 + 32 + 5) and print 42. The second statement would add (5 + 32 + 4) and print 41.

The decrement operator's prefix and postfix modes work the same as for the increment operator, but they subtract 1 instead of adding 1. To get a feeling for how they work, trace this code fragment:

```
1  int a, b, c;
2
3  a = 8;
4  b = --a;
5  c = b-- + --a;
6  System.out.println(a + " " + b + " " + c);
```

line#	a	b	c	output
1	?	?	?	
3	8			
4	7			
4		7		
5	6			
5			13	
5		6		
6				6 6 13

Let's examine line 5 in more depth:

```
c = b-- + --a;
```

As you might have guessed, in executing this statement, the JVM first decrements a. This should make sense when you look at the operator precedence table in Appendix 1 and confirm that the decrement operator has very high precedence. The JVM also executes b's decrement operator early on, but its execution consists of using b's original value and incrementing b afterwards. The operator precedence table shows that the + operator has higher precedence than the = operator, so the JVM next adds b's original value to a's decremented value. Finally, the JVM assigns the sum to c.

For many people, line 5 is particularly confusing. We showed you this example because you might see this kind of thing in someone else's code, but if you want your code to be understandable, we recommend that you not do this yourself. That is, don't embed ++ or -- expressions within other expressions. Instead of trying to do everything line 5 does in one statement, it would be more understandable to partition line 5 into three separate statements, like this:

```
5a  a--;
5b  c = b + a;
5c  b--;
```

The JVM performs the evaluation in separate steps anyway, so writing it out does not incur any performance penalty. It takes more space on the page, but most people will agree that it's easier to read.

When writing code, how do you decide which mode to use, prefix or postfix? It depends on the rest of your code. Usually, to minimize confusion, you'll put increment and decrement operations on separate lines. Then it doesn't matter which mode you use, but postfix is more common.

12.6 Embedded Assignments

This section supplements material you learned in Chapters 3 and 4. Specifically, it supplements the assignment statements material in Section 3.11 and the while loop material in Section 4.8.

Embedding an Assignment within Another Assignment

Assignments are sometimes embedded as expressions in larger statements. When that happens, remember that (1) an assignment expression evaluates to the assigned value, and (2) assignment operators exhibit right-to-left associativity. To see these concepts in action, consider this code fragment:

```
1  int a, b = 8, c = 5;
2                                same as: a = (b = c);
3  a = b = c;
4  System.out.println(a + " " + b + " " + c);
```

Line 3 shows an assignment expression embedded inside a larger assignment statement. Which of the two assignment operators does the JVM execute first? Because assignment operators exhibit right-to-left associativity, the JVM executes the right assignment operation first. What does the b = c expression evaluate to? It evaluates to 5 because the assigned value, c, is 5. In evaluating line 3, replace the b = c part of the statement with 5 to reduce the statement to:

```
a = 5;
```

Here's what the code fragment's trace looks like:

line#	a	b	c	output
1	?	8	5	
3		5		
3	5			
4				5 5 5

Embedding an Assignment Within a Loop Condition

Except for a pure multiple assignment like `a = b = c;` it's best to avoid embedding multiple assignments as expressions in other statements, because that makes code hard to understand. Nevertheless, it's fairly common to embed a single assignment as an expression in a loop condition. For example, Figure 12.7 contains a program that averages a set of input scores. Note the `(score = stdIn.nextDouble())`

```
/*****************************************************************
* AverageScore.java
* Dean & Dean
*
* This program averages input scores.
*****************************************************************/

import java.util.Scanner;

public class AverageScore
{
  public static void main(String[] args)
  {
    double score;
    double count = 0;
    double totalScore = 0;
    Scanner stdIn = new Scanner(System.in);

    System.out.print("Enter a score (or -1 to quit): ");
    while ((score = stdIn.nextDouble()) != -1)
    {
      count++;
      totalScore += score;
      System.out.print("Enter a score (or -1 to quit): ");
    }
    if (count > 0)
    {
      System.out.println("Average score = " + totalScore / count);
    }
  } // end main
} // end AverageScore class
```

embedded assignment

Figure 12.7 AverageScore program that demonstrates use of embedded assignments

assignment inside the `while` condition. If, for example, the user responds to the prompt by entering 80, `score` gets the value 80, the assignment expression within the parentheses evaluates to 80, and the `while` loop header becomes:

```
while (80 != -1)
```

Because the condition is true, the JVM executes the body of the loop. If the assignment expression were not embedded in the `while` loop condition, it would have to appear twice—once above the loop header and again at the bottom of the loop. Embedding the assignment in the condition improves the loop's structure.

You will sometimes also see embedded assignments in method arguments and array indices. This makes code more compact. Compactness is often a good thing in that it can lead to code that is less cluttered and therefore easier to understand. But don't go too far in trying to make your code compact because compactness can sometimes lead to code that is harder to understand (i.e., it can lead to code that is more *cryptic*). Some programmers get a kick out of making "clever" programs that are as compact as possible. If that's you, try to redirect your efforts to making programs as understandable as possible. You can still use compact code, but do so in a manner that helps, not hinders, understandability.

12.7 Conditional Operator Expressions

This section supplements the material in Chapter 4, Section 4.3 (`if` Statements).

Syntax and Semantics

When you want a logical condition to determine which of two alternate values applies, instead of using the "if, else" form of the `if` statement, you can use a conditional operator expression. The conditional operator is Java's only ternary operator. *Ternary* means three. The conditional relates three operands with the two symbols, `?` and `:`. The `?` goes between the first and second operands, and the `:` goes between the second and third operands.

Here's the syntax:

<condition> `?` *<expression1>* `:` *<expression2>*

If the condition is `true`, the conditional operator expression evaluates to the value of *expression1,* and it ignores *expression2*. If the condition is `false`, the conditional operator expression evaluates to the value of *expression2,* and it ignores *expression1*. Think of *expression1* as the true part of an "if, else" statement. Think of *expression2* as the false part of an "if, else" statement.

For example, consider this expression:

```
(x>y) ? x : y
```

The parentheses around the condition are not required, because > has higher precedence than the `?:` pair, but we recommend using them because they improve readability. What does the JVM do when it sees this expression?

- It compares `x` with `y`.
- If `x` is greater, it evaluates the expression to `x`.
- If `x` is not greater, it evaluates the expression to `y`.

Do you know what general functionality the expression implements? It finds the maximum between two numbers. You can prove this to yourself by plugging in sample numbers. Suppose `x` equals 2 and `y` equals 5. Here's how the expression evaluates to the maximum, 5:

```
(2>5) ? 2 : 5  ⇒
(false) ? 2 : 5  ⇒
5
```

Using the Conditional Operator

A conditional operator expression cannot appear on a line by itself because it is not a complete statement. It is just part of a statement—an expression. The following code fragment includes two examples of embedded conditional operator expressions:

```
int score = 58;
boolean extraCredit = true;

score += (extraCredit ? 2 : 0);
System.out.println(
  "grade = " + ((score>=60) ? "pass" : "fail"));
```

How does it work? Because `extraCredit` is `true`, the first conditional operator evaluates to 2. `score` then increments by 2 from its initial value of 58 to 60. Because `(score>=60)` evaluates to `true`, the second conditional operator evaluates to "pass." The `println` statement then prints:

```
grade = pass
```

In the above code fragment, we like the parentheses the way they are shown, but in the interest of honing your debugging skills, let's examine what happens if you omit each of the pairs of parentheses. As shown in the operator precedence table in Appendix 2, the conditional operator has higher precedence than the += operator. Therefore, it would be legal to omit the parentheses in the += assignment statement. In the `println` statement, the conditional operator has lower precedence than the + operator, so you must keep the parentheses that surround the conditional operator expression. Because the >= operator has higher precedence than the conditional operator, it would be legal to omit the parentheses that surround the `score>=60` condition. Note how we omit spaces in the `score>=60` condition but include spaces around the ? and : that separate the three components of the conditional operator expression. This style improves readability.

You can use the conditional operator to avoid `if` statements. Conditional operator code might look more efficient than `if` statement code because the source code is shorter, but the generated bytecode is typically longer. This is another example of something you might see in someone else's code, but because it's relatively hard to understand, we recommend that you use it with restraint in your own code. For example, the `score += (extraCredit ? 2 : 0);` statement in the above code fragment is rather cryptic. It would be better style to increment the `score` variable like this:

```
if (extraCredit)
{
  score += 2;
}
```

12.8 Expression Evaluation Review

Hand calculation helps you understand.

So far in this chapter, you've learned quite a few type details and operator details. Learning such details will help you debug code that has problems, and it will help you avoid problems in the first place. To make sure that you really understand the details, let's do some expression evaluation practice problems.

Expression Evaluation Practice with Characters and String Concatenation

Note the following three expressions. Try to evaluate them on your own prior to looking at the subsequent answers. While performing the evaluations, remember that if you have two or more operators with the same precedence, use left-to-right associativity (i.e., perform the operation at the left first). So in the first expression, you should perform the + operation in `'1' + '2'` before attempting to perform the second + operation.

1. `'1' + '2' + "3" + '4' + '5'`
2. `1 + 2 + "3" + 4 + 5`
3. `1 + '2'`

Here are the answers:

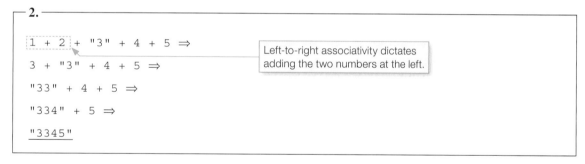

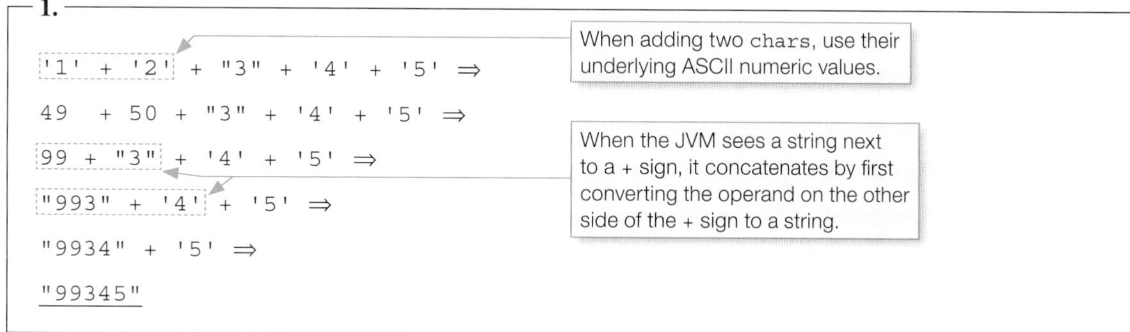

2.

```
1 + 2 + "3" + 4 + 5 ⇒
3 + "3" + 4 + 5 ⇒
"33" + 4 + 5 ⇒
"334" + 5 ⇒
"3345"
```

Left-to-right associativity dictates adding the two numbers at the left.

3.

```
1 + '2' ⇒
1 + 50 ⇒
51
```

Mixed expression—the `char` gets promoted to an `int`, using the underlying ASCII numeric value for '2.'

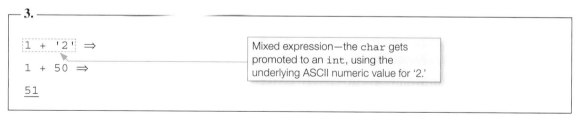

Expression Evaluation Practice with Type Conversions and Various Operators

Assume the following:

```
int a = 5, b = 2;
double c = 3.0;
```

Try to evaluate the following expressions on your own prior to looking at the subsequent answers:

1. `(c + a / b) / 10 * 5`
2. `a + b++`
3. `4 + --c`
4. `c = b = a % 2`

Here are the answers:

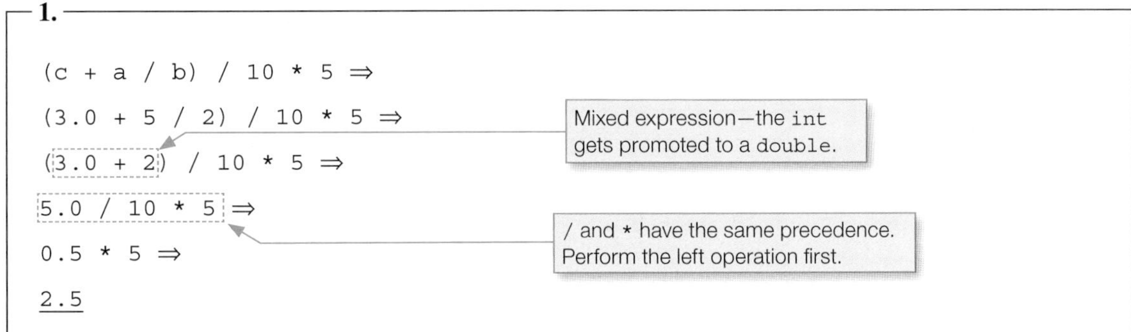

1.
(c + a / b) / 10 * 5 ⇒
(3.0 + 5 / 2) / 10 * 5 ⇒
(3.0 + 2) / 10 * 5 ⇒
5.0 / 10 * 5 ⇒
0.5 * 5 ⇒
2.5
Mixed expression—the int gets promoted to a double.
/ and * have the same precedence. Perform the left operation first.

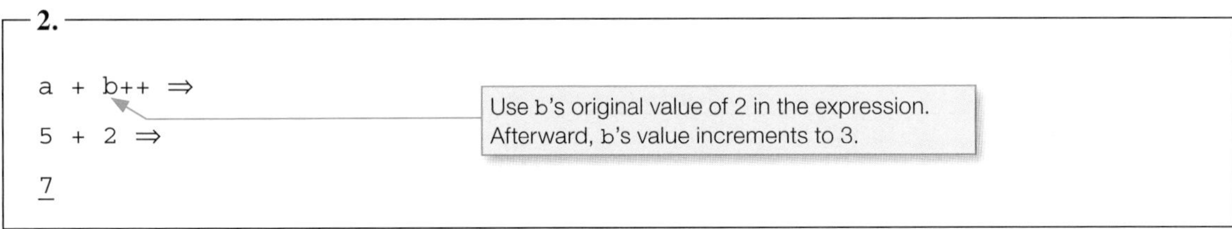

2.
a + b++ ⇒
5 + 2 ⇒
7
Use b's original value of 2 in the expression. Afterward, b's value increments to 3.

3.

```
4 + --c ⇒
4 + 2.0 ⇒
6.0
```

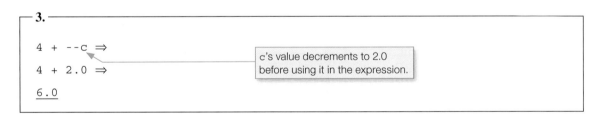

c's value decrements to 2.0 before using it in the expression.

4.

```
c = b = a % 2 ⇒
c = b = 5 % 2 ⇒
c = b = 1 ⇒
c = 1 ⇒
1.0
```

Don't plug in values for variables that are at the left of assignments.

The `b = 1` assignment evaluates to 1.

`c` is a `double`, so the result is a `double`.

More Expression Evaluation Practice

Assume the following:

```
int a = 5, b = 2;
double c = 6.6;
```

Try to evaluate the following expressions on your own prior to looking at the subsequent answers:

1. `(int) c + c`
2. `b = 2.7`
3. `('a' < 'B') && ('a' == 97) ? "yes" : "no"`
4. `(a > 2) && (c = 6.6)`

Here are the answers:

1.

```
(int) c + c ⇒
6 + 6.6 ⇒
12.6
```

`(int) c` evaluates to 6, which is the truncated version of 6.6, but `c` itself doesn't change, so the second `c` remains 6.6.

2.

```
b = 2.7
```

Compilation error. The `double` value won't fit into the narrower `int` variable without a cast operator.

3.

Look up underlying numeric values in ASCII table.

Mixed types, so `char 'a'` converts to `int` 97 before comparison.

```
('a' < 'B') && ('a' == 97) ? "yes" : "no" ⇒
false && true ? "yes" : "no" ⇒
false ? "yes" : "no" ⇒
"no"
```

4.

```
(a > 2) && (c = 6.6) ⇒
(true) && ...
```

`c = 6.6` is an assignment, not an equality condition. Thus, `c = 6.6` evaluates to the `double` value, `6.6`, and a `double` doesn't work with the `&&` operator, so this generates a compilation error. Probably the second operand should be `(c == 6.6)`.

12.9 Short-Circuit Evaluation

This section supplements the `&&` logical operator material you studied in Chapter 4, Section 4.4, and the `||` logical operator material you studied in Chapter 4, Section 4.5.

Consider the program in Figure 12.8. It calculates a basketball player's shooting percentage and prints an associated message. Note the `if` statement's heading, repeated here for your convenience. In particular, note the division operation with `attempted` in the denominator.

```
/*********************************************************************
* ShootingPercentage.java
* Dean & Dean
*
* This program processes a basketball player's shooting percentage.
*********************************************************************/

import java.util.Scanner;

public class ShootingPercentage
{
  public static void main(String[] args)
  {
    int attempted;   // number of shots attempted
    int made;        // number of shots made
    Scanner stdIn = new Scanner(System.in);
    System.out.print("Number of shots attempted: ");
    attempted = stdIn.nextInt();
    System.out.print("Number of shots made: ");
    made = stdIn.nextInt();

    if ((attempted > 0) && ((double) made / attempted) >= .5)
    {
      System.out.printf("Excellent shooting percentage - %.1f%%\n",
        100.0 * made / attempted);
    }
    else
    {
      System.out.println("Practice your shot more.");
    }
  } // end main
} // end class ShootingPercentage
```

If `attempted` is zero, division by zero does not occur.

Use `%%` to print a percent sign.

Sample session:

```
Number of shots attempted: 0
Number of shots made: 0
Practice your shot more.
```

Second sample session:

```
Number of shots attempted: 12
Number of shots made: 7
Excellent shooting percentage - 58.3%
```

Figure 12.8 Program that illustrates short-circuit evaluation

```
if ((attempted > 0) && ((double) made / attempted) >= .5)
```

With division, you should always think about, and try to avoid, division by zero. If `attempted` equals zero, will the JVM attempt to divide by zero? Nope! Short-circuit evaluation saves the day.

Short-circuit evaluation means that the JVM stops evaluating an expression whenever the expression's outcome becomes certain. More specifically, if the left side of an `&&` expression evaluates to `false`, then the expression's outcome is certain (`false && anything` evaluates to `false`) and the right side is skipped. Likewise, if the left side of an `||` expression evaluates to `true`, then the expression's outcome is certain (`true || anything` evaluates to `true`) and the right side is skipped. So in Figure 12.8's `if` statement condition, if `attempted` equals zero, the left side of the `&&` operator evaluates to `false` and the right side is skipped, thus avoiding division by zero.

So what's the benefit of short-circuit evaluation?

1. **Error avoidance:** It can help to prevent problems by enabling you to avoid an illegal operation on the right side of an expression.
2. **Performance:** Because the result is already known, the computer doesn't have to waste time calculating the rest of the expression.

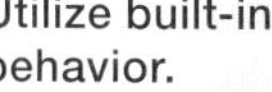
Utilize built-in behavior.

As an aside, note the `%%` in Figure 12.8's `printf` statement. It's a conversion specifier for the `printf` method. Unlike the other conversion specifiers, it is a stand-alone entity; it doesn't have an argument that plugs into it. It simply prints the percent character. Note the printed % at the end of Figure 12.8's second sample session.

12.10 Empty Statement

This section supplements the loop material you studied in Chapter 4.

It's sometimes possible to put all of a loop's functionality inside of its header. For example:

```
for (int i=0; i<1_000_000_000; i++)
{ }
```

The Java compiler requires that you include a statement for the `for` loop's body, even if the statement doesn't do anything. The above empty braces ({ }) form a compound statement[2] and satisfy that requirement. In this section, you learn about an alternative way to satisfy that requirement. You learn about the empty statement.

Using the Empty Statement

The *empty statement* consists of a semicolon by itself. Use the empty statement in places where the compiler requires a statement, but there is no need to do anything. For example, the `for` loop below can be used as a "quick and dirty" way to add a delay to your program:

```
monster.display();
for (int i=0; i<1_000_000_000; i++)
  ;
monster.erase();
```

Coding convention:
Put the empty statement on a line by itself and indent it.

[2]The compound statement, defined in Chapter 4, is a group of zero or more statements surrounded by braces.

Note how the empty statement is appropriate here because all the work is done in the `for` loop header, where `i` counts up to 1 billion. All that counting takes time. Depending on your computer's speed, it might take anywhere from a fraction of a second to 5 seconds.

So why would you want to add a delay to your program? Suppose that you're writing a game program that needs to have a monster appear for only a certain time interval. To implement that functionality, print the monster, execute the delay loop, and then erase the monster.

You might want to use the above code fragment as part of a first-cut attempt at implementing the delay, but don't use it for your final implementation. Why? Because it introduces delay that is dependent on the speed of the computer that runs the program. With varied delay, slow computers would have monsters that linger too long, and fast computers would have monsters that disappear too quickly. In a final implementation, you should use the `Thread` class's `sleep` method to implement the delay. The `sleep` method allows you to specify precisely the amount of delay in milliseconds. Because the `sleep` method might throw an `InterruptedException`, you need to include `throws Exception`, as described in Section 3.24 of Chapter 3, or use `try` and `catch`, which we explain later in Chapter 15.[3]

In the above code fragment, note the coding-convention callout. Can you think of why it's a good idea to put the empty statement on a line by itself? If you put the empty statement on a line by itself and indent it, readers will see it. On the other hand, if you put the empty statement at the end of the previous statement's line, readers probably won't see it. Seeing the code is an important part of making the code understandable. And making code understandable makes it easier to maintain.

Avoid Accidental Misuse of the Empty Statement

It's fairly common for programmers to create unintended empty statements accidentally. Because you enter a semicolon at the end of most lines of Java code, it's easy to get into the habit of hitting the semicolon key at the end of every line of code you write. If you do that at the end of a loop header, it generates an empty statement. Your code might compile and run without a reported error, but it would produce mysterious results. Here is an example:

```
System.out.print("Do you want to play a game (y/n)? ");
while (stdIn.next().equals("y"));
{
  <The code to play the game goes here.>
  System.out.print("Play another game (y/n)? ");
}
```

This semicolon creates an empty statement.

Does the semicolon at the end of the `while` loop header generate a compilation error? No—the semicolon acts as the lone statement (an empty statement) that's inside the `while` loop. The subsequent braces form a compound statement. The compound statement is not part of the `while` loop; it executes after the `while` loop has finished.

So what does the code do? First, suppose the user enters `n`. In the `while` loop header, the JVM compares the entered `n` value to "y." The loop condition is `false`, so the JVM skips the `while` loop's body, the empty statement. The JVM then executes the compound statement and attempts to play a game. That's a logic error: The JVM attempts to play a game even though the user entered `n`.

[3]This adds a delay of 1,000 milliseconds (which equals 1 second):

```
try {Thread.sleep(1000);}
catch (InterruptedException e) { }
```

Now suppose the user enters y. In the `while` loop header, the JVM compares the entered `y` value to "y". The loop condition is `true`, so the JVM executes the `while` loop's body, the empty statement. The JVM then returns to the loop header and executes the `stdIn.next()` method call again. The JVM waits for the user to enter another value. But the user won't know that he or she is supposed to enter anything because there's no prompt. That's a particularly nasty logic error because the program produces no erroneous output and no error message. That means there's no help in determining what is wrong or what to do about it.

You can produce these same types of logic errors by putting semicolons after "if," "else if," or "else" headings. Such semicolons effectively create empty statements, and they're often introduced accidentally during program development or debugging. Be on the alert for empty statements, and whenever you see one, be suspicious and check it out! Better yet, minimize confusion at the end by maximizing care at the beginning.

Haste makes waste.

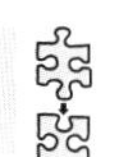

12.11 `break` Statement Within a Loop

This section supplements the loop material you studied in Chapter 4.

In Chapter 4, we introduced you to the use of the `break` statement inside a `switch` statement. It terminates the `switch` statement and transfers control to the next statement after the `switch` statement. In addition, you can use the `break` statement inside a `while`, `do`, or `for` loop. It does the same thing as when it's in a `switch` statement. The `break` terminates the immediately enclosing loop and transfers control to the next statement after the bottom of the loop. We say "immediately enclosing" because you can have a `break` that's nested inside multiple loops. The `break` gets associated with the loop that immediately surrounds it.

The DayTrader program in Figure 12.9 illustrates what's called "day trading." It's a form of gambling in which people buy and sell stock on the stock market every day in hopes of making money off short-term stock movements. This program keeps track of a day trader's stock balance over a three-month period (for day = 1 to 90). The original balance is $1,000. In our simple model, at the beginning of each day, the day trader retains half the initial balance in savings and invests the other half in the stock market. The money returned at the end of the day equals the investment times a random number between 0 and 2. Thus, the money returned ranges anywhere from zero to double the original investment. Each day, the day trader adds the money returned to the balance in savings. If the balance ever goes below $1 or above $5,000, the day trader quits.

Before examining the `break` statement in Figure 12.9, look at the `(day - 1)` argument in the final `printf` statement. This is after the `for` loop, so the scope of `day` needs to be bigger than the scope of the `for` loop. That's why we declared it before the `for` loop with the other local variables. But why did we subtract 1 in the `printf` statement? Because the `day++` operation in the third compartment of the `for` loop header increments `day` one extra time, after the transaction that drives the balance to a terminating value. If we had forgotten to subtract 1 in the `printf` statement, that would be an off-by-one error.

Now look at the DayTrader program's `break` statement. If the balance ever gets outside the $1 to $5,000 range, program control jumps immediately to the next statement below the `for` loop. If you run the program several times, you'll see that sometimes this causes the loop to terminate before `day` reaches 90. You'll get a different result each time you run the program because this program uses `Math.random` to generate a random number in the range between 0.0 and 1.0.

```
/*******************************************************************
* DayTrader.java
* Dean & Dean
*
* This simulates stock market day trading.
*******************************************************************/

public class DayTrader
{
  public static void main(String[] args)
  {
    double balance = 1000.00; // money that's retained
    double moneyInvested;     // money that's invested
    double moneyReturned;     // money that's earned at end of day
    int day;                  // current day, ranges from 1 to 90

    for (day=1; day<=90; day++)
    {
      if (balance < 1.0 || balance > 5000.0)
      {
        break;
      }
      balance = moneyInvested = balance / 2.0;
      moneyReturned = moneyInvested * (Math.random() * 2.0);
      balance += moneyReturned;
    } // end for

    System.out.printf("final balance on day %d: $%4.2f\n",
      (day - 1), balance);
  } // end main
} // end DayTrader
```

Figure 12.9 DayTrader program, which illustrates use of the `break` statement

Be aware that you never really have to use a `break` statement to implement this premature loop-termination capability. For example, you can eliminate the DayTrader program's `if` and `break` statements by changing the `for` loop header to this:

```
for (day=1; day<=90 && !(balance < 1.0 || balance > 5000.0); day++)
```

Don't fall into the trap of using the `break` statement too often. Usually, someone reading your program will look only at the loop header to figure out how the loop terminates. In using a `break` statement, you force the reader to look inside the loop for loop termination conditions. And that makes your program harder to understand. Nonetheless, in certain situations, the `break` statement improves readability rather than hinders it. The DayTrader program's `break` statement is an example where the `break` statement improves readability.

12.12 `for` Loop Header Details

This section supplements the `for` loop material that you studied in Chapter 4, Section 4.10.

Omitting One or More of the `for` Loop Header Components

It's legal, although not all that common, to omit the first and/or third components in the `for` loop header. For example, to print a countdown from a user-entered number, you could use this code:

```
System.out.print("Enter countdown starting number: ");
count = stdIn.nextInt();
for (; count>0; count--)
{
  System.out.print(count + " ");
}
System.out.println("Liftoff!");
```

no initialization component

Actually, it's legal to omit any of the three `for` loop header components, as long as the two semicolons still appear within the parentheses. For example, you can even write a `for` loop header like this:

```
for (;;)
```

When a `for` loop header's condition component (the second component) is omitted, the condition is considered true for every iteration of the loop. With a permanently true condition, such a loop is often an infinite loop and a logic error. But that's not always the case. You can terminate it by using a `break` statement like this:

```
for (;;)
{
  ...
  if (<condition>)
  {
    break;
  }
}
```

You should understand the above example in case you see similar code in someone else's program. But it's rather cryptic, and, as such, you should avoid writing your own code that way.

Multiple Initialization and Update Components

For most `for` loops, one index variable is all that's needed. But every now and then, two or more index variables are needed. To accommodate that need, you can include a list of comma-separated initializations in a `for` loop header. The caveat for the initializations is that their index variables must be the same type. Working in concert with the comma-separated initializations, you can also include a list of comma-separated updates in a `for` loop header. The following code fragment and associated output show what we're talking about. In the `for` loop header, note the two index variables, `up` and `down`, and their comma-separated initialization and update components:

```
System.out.printf("%3s%5s+n", "Up", "Down");
for (int up=1,down=5; up<=5; up++,down--)
{
  System.out.printf("%3d%5d+n", up, down);
}
```

Output:

```
Up  Down
 1     5
 2     4
 3     3
 4     2
 5     1
```

As with many of the techniques presented in this chapter, using multiple initialization and update components in a `for` loop is a bit of an art. It leads to more compact code, which can be a good thing or a bad thing. If the compact code is more understandable, use it. If the compact code is more cryptic, don't use it.

12.13 Enumerated Types

To declare a primitive variable, you use a primitive type, like `int` or `double`. To declare a reference variable, you normally use a class, which serves as a type. Java supports another type, an *enumerated type* (also referred to as an *enum type* or an *enumeration type*), which can be used to restrict a variable to holding one of a fixed set of values defined by the programmer. An enumerated type is sort of like a `boolean` type, in that a `boolean` type restricts a variable to holding just the value `true` or the value `false`. But with an enumerated type, you specify the name of the enumerated type, and you specify the names of the enumerated type's possible values.

Day of the Week

A classic use for an enumerated type is when you need to keep track of the day of the week. To do that, you could declare a `day` variable as an `int`, and use values 0 through 6 to represent Sunday through Saturday. You could get that to work, but the compiler won't stop you from accidentally assigning 7 to `day` (which would presumably lead to a logic error down the road). A better solution uses the reserved word `enum` to define a `Day` enumerated type like this:

```
public enum Day
{
  SUNDAY, MONDAY, TUESDAY, WEDNESDAY, THURSDAY, FRIDAY, SATURDAY
}
```

Note that `SUNDAY`, `MONDAY`, . . . , `SATURDAY` use all uppercase. Because an enumerated type's values are named constants, coding conventions dictate that you use uppercase.

After `Day` is defined, you can use it to declare a `day` variable like this:

```
Day day;
```

And then still later, you can assign one of the enumerated values to `day` like this:

```
day = Day.FRIDAY;
```

This assignment illustrates how you normally need to preface an enumerated value (`FRIDAY` in this example) with its enumerated type name (`Day` in this example).

Letter Grade for a Course

Suppose you want to keep track of the letter grades earned in a series of courses taken by a student. You can do that with the help of a `Grade` enumerated type:

```
private enum Grade {F, D, C, B, A}
```

Note the differences between the `Grade` enumerated type and the prior `Day` enumerated type. `Day` used the `public` access modifier, presumably because there might be a need for multiple classes to declare variables with the `Day` type. On the other hand, `Grade` uses the `private` access modifier, presumably because only one class (the class that defines `Grade`) needs to declare variables with the `Grade` type. Another difference between `Day` and `Grade` is that `Day` uses 4 lines and `Grade` uses only 1 line. That's simply a style thing. If the enumerated constants take up a lot of space, then put them on a separate line or lines. If they don't take up much space, then put them on the same line as the enumerated type's heading.

To see how the `Grade` type can be used in a complete program, study the GradeManagement program in Figures 12.10a and 12.10b. The GradeManagement program reads in a series of course scores and calculates the letter grade for each course and the grade point average (GPA) for all of the courses combined. Within a loop, the `main` method calls `getGrade` to get the letter grade for one course. The `getGrade` method prompts the user to enter a course's overall percentage and uses the percentage value to assign the appropriate letter grade enumerated value into a `grade` variable (which was declared using the `Grade` enumerated type). If the user enters a negative number (to quit), then `grade` sticks with its initial value of `null`. The `getGrade` method returns `grade` to `main`. The `main` method uses a `switch` statement to check `grade`'s value and adds 4, 3, 2, 1, or 0 to the `totalPts` variable, depending on whether `grade` is an A, B, C, D, or F. When the user enters −1 to quit, `main`'s loop terminates, and `main` then determines the GPA by dividing `totalPts` by the number of course percentages entered.

As mentioned earlier, normally you must preface an enumerated type's value with its enumerated type name. So in the GradeManagement program, to assign an A to `grade`, we use `grade = Grade.A;`. However, a `switch` statement's `case` constants are different. If you want to use an enumerated type's value as a `case` constant, you must use the simple form of the enumerated type's value, such as `A`, not the full name, `Grade.A`. To see an example, look for `case A:` in Figure 12.10a.

How to Compare Enumerated Type Values

As you might expect, you can check two enumerated type values for equality or inequality by using the == operator or the != operator. Here's an example that uses the == operator:

```
if (grade == Grade.F)
{
  System.out.println(
    "If this is a required course, you must retake it.");
}
```

Rather than testing for equality, sometimes you might want to test whether a particular enumerated type value is less than another enumerated type value, where "less than" means that an enumerated type value appears at the left of another enumerated type value in the enumerated type's original definition. Look at `Grade`'s enumerated type definition in Figure 12.10a. In the definition, you can see `F`, `D`, `C`, `B`, and `A` specified in that order. So you might expect `Grade.F < Grade.D` to evaluate to `true`. Unfortunately, that doesn't work. To compare two enumerated type values with <, >, <=, or >=, you first have to extract the underlying integers from the two enumerated type values. Remember how each `char` value has an

```
/*************************************************************
* GradeManagement.java
* Dean & Dean
*
* This program reads course percentages and calculates GPA.
*************************************************************/

import java.util.Scanner;

public class GradeManagement
{
  private enum Grade {F, D, C, B, A}      // ordered by value

  //***********************************************************

  public static void main(String[] args)
  {
    int numOfCourses = 0;    // number of course scores entered
    Grade grade;             // holds the grade for a course
    int totalPts = 0;        // total points for all entered courses

    do
    {
      grade = getGrade();

      // null indicates user wants to quit
      if (grade != null)
      {
        numOfCourses++;
        switch (grade)
        {
          case A:
            totalPts += 4;
            break;
          case B:
            totalPts += 3;
            break;
          case C:
            totalPts += 2;
            break;
          case D:
            totalPts += 1;
        } // end switch
      } // end if
    } while (grade != null);
```

Enumerated type definition

For case constants, use the enumerated type's value by itself (no enumerated type name and a dot in front of the A).

Figure 12.10a GradeManagement program—part A

```
    if (numOfCourses == 0)
    {
      System.out.println("No scores were entered.");
    }
    else
    {
      System.out.printf("Overall GPA: %.2f",
        (float) totalPts / numOfCourses);
    }
  } // end main

  //***********************************************************

  // Prompt the user for a course's overall percentage and
  // converts it to a grade. Return null if user wants to quit.

  private static Grade getGrade()
  {
    Scanner stdIn = new Scanner(System.in);
    float percentage;      // overall percentage for one course
    Grade grade = null;    // the course grade

    System.out.print(
      "Enter course overall percentage (-1 to quit): ");
    percentage = stdIn.nextFloat();

    if (percentage >= 90.0)
      grade = Grade.A;
    else if (percentage >= 80.0)
      grade = Grade.B;
    else if (percentage >= 70.0)
      grade = Grade.C;
    else if (percentage >= 60.0)
      grade = Grade.D;
    else if (percentage >= 0.0)
      grade = Grade.F;

    return grade;
  } // end getGrade
} // end class GradeManagement
```

To access an enumerated type's value, preface it with the enumerated type's name and then a dot.

Sample session:

```
Enter course overall percentage (-1 to quit): 77
Enter course overall percentage (-1 to quit): 82
Enter course overall percentage (-1 to quit): 60
Enter course overall percentage (-1 to quit): -1
Overall GPA: 2.00
```

Figure 12.10b GradeManagement program—part B

underlying integer value? Similarly, each enumerated type value has an underlying integer value. The leftmost enumerated type value has a 0 associated with it, the second from the left enumerated type value has a 1 associated with it, and so on. To retrieve the integer value, you need to call the `ordinal` method. Here's an example:

```
if (grade.ordinal() < Grade.C.ordinal())
{
  System.out.println("If this is a prerequisite course for" +
    " a required course, you must retake it.");
}
```

An Enumerated Type as a Class

In the GradeManagement program, note that the `Grade` enumerated type is spelled with an uppercase first letter. That coding convention is an indication that an enumerated type is considered to be a class (albeit a special kind of class). Because an enumerated type is a class, you can define it in its own file, separate from other classes. When you save such a file, you'll need to save it with a `.java` filename extension. And when you compile it, the compiler will generate a `.class` file. These are all earmarks of a class.

If you'd like to define a fixed list of names that can be used for a variable's value and that's it, then there's no great need to think about an enumerated type as a class. On the other hand, if you'd like to define a fixed list of objects that can be used for a variable's value, then you'll need to know more about the enumerated type's classlike qualities. In this subsection, we describe those qualities.

There are three basic steps required to define an enumerated type as a fixed list of objects. It's important to see how to apply the steps to an actual program, so as we explain the steps, please study the step 1, step 2, and step 3 code, as indicated by the callouts in Figure 12.11a.

Step 1: Provide a list of names for the objects, and follow each name with a list of values enclosed in parentheses. Append a semicolon after the last object's parentheses.

Step 2: Provide a list of instance variables that are used to store each object's values. Normally, you should use `public` and `final` for your instance variable declarations, so the instance variables can be accessed outside the enumerated type definition without any risk of their being changed.

Step 3: Provide a `private` constructor that's used to initialize each of the object's instance variables. The compiler requires the constructor to be `private`, so it's impossible to call the enumerated type constructor (and create new enumerated type objects) from outside the enumerated type definition. If no external constructor calls are allowed, where is the constructor called from? From inside the enumerated type's definition! As described in Step 1, an enumerated type definition contains a list of object names with parentheses after each name. Each of those list items is a constructor call. For example, in Figure 12.11a's `City` enumerated type definition, `PARKVILLE (39.2, -94.7)` calls `City`'s constructor and passes 39.2 to the `latitude` parameter and −94.7 to the `longitude` parameter.

If you'd like to manipulate an enumerated type's values, you can use a driver program to retrieve the values and manipulate them from within the driver. On the other hand, if you think the manipulation is general-purpose in nature, such that more than one program might find it useful, then you should consider implementing the manipulation as a `public` method inside the enumerated type definition. For example, see the `getDistance` method in Figure 12.11b. It calculates the distance between two `City` objects—the `City` object that calls `getDistance` and the `City` object that's passed in as an argument to `getDistance`. Lots of programs might need to know the distance between cities, so we chose to put the method within the `City` enumerated type definition, rather than within a driver class.

```
/*************************************************************
* City.java
* Dean & Dean
*
* This enumerated type provides location properties of cities.
*************************************************************/

public enum City
{
  PARKVILLE (39.2, -94.7),        // USA
  HAVANA (23.1, -82.4),           // Cuba
  KINGSTON (18.0, -76.8),         // Jamaica
  NASSAU (25.1, -77.3),           // Bahamas
  SAINT_THOMAS (18.3, -64.9);     // Virgin Islands

  // location of the city in degrees
  public final double latitude;
  public final double longitude;

  private City(double latitude, double longitude)
  {
    this.latitude = latitude;
    this.longitude = longitude;
  }
```

semicolon

Step 1: List of five city objects, each with its own latitude and longitude vaues.

Step 2: Instance variables that store latitude and longitude values for each city object.

Step 3: Constructor that initializes each city object with latitude and longitude values.

Figure 12.11a City enumerated type—part A

Although the getDistance method appears within an enumerated type definition, its syntax is identical to methods defined within a regular class. To illustrate that point, note how the latitude value is accessed with either the calling object (this.latitude) or the parameter object (destination.latitude). That should look familiar because accessing a value within an enumerated type object is just like accessing any other object value—use the dot operator.

The getDistance method's algorithm for finding the *geodesic distance* (the shortest distance between two points on a curved surface) is rather involved. It uses the cities' latitude and longitude values as inputs to the haversine formula. An explanation of the haversine formula is beyond the scope of this book, but if you're curious, Google it. On the other hand, considering the ubiquity of the global positioning system (GPS), everyone should know a little about latitude and longitude, so here goes . . .

Latitude values start at zero degrees (0°) at the equator. As you go north from there, degrees go from 0° to positive 90°. As you go south, degrees go from 0° to negative 90°. The North Pole is at 90°, and the South Pole is at −90°. Longitude values start at 0° at the Prime Meridian, which passes through the Royal Observatory in Greenwich, England. As you go east from there, degrees go in a positive direction from 0° to 180°. As you go west from the Prime Meridian, degrees go in a negative direction from 0° to −180°. The positive and negative 180° longitudes meet at the Anti-Meridian, otherwise known as the International Date Line. The International Date Line is the first place where each new day begins. It primarily passes through open waters in the Pacific Ocean, but it also crosses parts of Russia, Fiji, and Antarctica.

```
  //*************************************************************

  // This method returns the distance in km between two cities.

  public double getDistance(City destination)
  {
    final double R = 6371; // approximate mean radius of Earth in km

    double lat1, lon1; // latitude and longitude of origin city
    double lat2, lon2; // latitude and longitude of destination city
    double a;          // intermediate value used in haversine formula

    // upcoming trig functions work with radians, not degrees
    lat1 = Math.toRadians(this.latitude);
    lon1 = Math.toRadians(this.longitude);
    lat2 = Math.toRadians(destination.latitude);
    lon2 = Math.toRadians(destination.longitude);

    a = Math.pow(Math.sin((lat2 - lat1) / 2), 2) +
        Math.pow(Math.sin((lon2 - lon1) / 2), 2) *
        Math.cos(lat1) * Math.cos(lat2);

    return (2 * Math.atan2(Math.sqrt(a), Math.sqrt(1-a))) * R;
  } // end getDistance
} // end City
```

Accessing City objects' latitude and longitude properties.

Figure 12.11b City enumerated type—part B

Now look at how a driver program can use the City enumerated type. Figure 12.12's CityTravel program determines the distance between Parkville, Missouri, and Kingston, Jamaica—two of the cities in the City enumerated type. The program was designed to help Park University's fledgling Oceanography and Computer Science Department. For lab classes, the department flies its students from its private airfield in Parkville to various island cities in the West Indies.

Retrieving All the Objects in an Enumerated Type

Sometimes you might have a need to retrieve all the objects in an enumerated type. You can do that by calling the enumerated type's values method.[4] It generates an array of the enumerated type's objects. You can then process each of the generated array's objects using a standard loop or a for-each loop. For example, here's how you can retrieve the names of all the City objects from the City enumerated type and print them:

```
for (City city : City.values())
{
  System.out.print(city + " ");
}
```

[4]The compiler automatically creates a values method that matches your enumerated type definition.

```
/*************************************************************
* CityTravel.java
* Dean & Dean
*
* This class prints the distance between two cities.
*************************************************************/

public class CityTravel
{
  public static void main(String[] args)
  {
    final double KM_TO_MILES = 0.62137; // conversion factor
    City origin = City.PARKVILLE;
    City destination = City.KINGSTON;
    double distance = origin.getDistance(destination); // in km

    System.out.printf("%s to %s: %.1f km, or %.1f miles",
      origin, destination, distance, distance * KM_TO_MILES);
  } // end main
} // end CityTravel
```

City.PARKVILLE, calls getDistance.

Output:

```
PARKVILLE to KINGSTON: 2922.1 km, or 1815.7 miles
```

Figure 12.12 CityTravel program

In the CityTravel program, the `main` method uses hardcoded values (`City.PARKVILLE` and `City.KINGSTON`) to specify the origin and destination cities for a particular distance calculation. In an exercise at the end of this chapter, you're asked to improve that code by prompting the user to enter origin and destination cities and using the user-entered values in the `getDistance` method call. If you choose to do the exercise, you'll find you'll need to loop through all of the cities in the `City` enumerated type. You can use the code fragment above as a starting point.

12.14 Hexadecimal, Octal, and Binary Numbers

Normal numbers are expressed as powers of 10, but because computers are binary and 16 is a simple power of 2 ($16 = 2^4$), it's common practice to express computer quantities in base 16 (using powers of 16), rather than base 10 (using powers of 10). Base 10 numbers are called decimal numbers. Base 16 numbers are called *hexadecimal* numbers. The places in decimal numbers are called *digits*. The places in hexadecimal numbers are sometimes called *hexits,* but more often, they're simply called *hexadecimal digits*. Base 10 numbers use the 10 symbols: 0, 1, 2, 3, 4, 5, 6, 7, 8, and 9. Base 16 numbers use the 16 symbols: 0, 1, 2, 3, 4, 5, 6, 7, 8, 9, a, b, c, d, e, and f (uppercase letters A through F are equivalent to a through f). Thus, hexadecimal numbers frequently include one or more of the first six alphabetic characters, as well as one or more of the normal numerical characters.

In Java, you can write any integer in decimal, hexadecimal, or binary form. If you want a number to be interpreted as hexadecimal, you must prefix it with zero followed by `x` or `X`; that is, with `0x` or `0X`. If you want a number to be interpreted as binary, you must prefix it with the character pair zero followed by `b` or `B`; that is, `0b` or `0B`. There is no corresponding prefix for octal. So if you see something like `0x263A`, for example, you can recognize it as a hexadecimal number. Similarly, if you see something like `0b01010111`, you can recognize it as a binary number. For most of us, hexadecimal, octal, and binary numbers are not very intuitive. However, if you can write the number you want to convert (assuming it's hexadecimal, decimal, or binary), it's pretty easy to make conversions. Just use `Integer`'s two-parameter `toString` method:

```
Integer.toString(<starting-number>, <desired-base>);
```

The starting number is the number you want to convert.

For example, if you want to see the decimal equivalent of `0x263A`, use this:

```
System.out.println(Integer.toString(0x263A, 10));
```

This generates an output of `9786`. To go the other way, you could use the formatted print statement

```
System.out.printf("%x\n", 9786);
```

This generates an output of `263a`. Notice that this method's output does not include the `0x` prefix, and it uses lowercase letters for the alphabetic hexadecimal digits.

Similarly, you can use `printf`'s conversion specifiers to print the hexadecimal, decimal, or octal equivalents of hexadecimal, decimal, or binary numbers. Thus, you could print the hexadecimal, decimal, and octal equivalents of the binary number `0b01010111` with these statements:

```
System.out.printf("%x\n", 0b01010111);
System.out.printf("%d\n", 0b01010111);
System.out.printf("%o\n", 0b01010111);
```

These generate the outputs, `57`, `87`, and `127`, respectively.

Conversely, if you want to see the hexadecimal, octal, and binary equivalents of the decimal number 9786, you could use these three print statements:

```
System.out.println(Integer.toString(9786, 16));
System.out.println(Integer.toString(9786, 8));
System.out.println(Integer.toString(9786, 2));
```

These generate the outputs, `263a`, `23072`, and `10011000111010`, respectively.

12.15 GUI Track: Unicode (Optional)

Earlier, you learned that characters get their underlying numeric values from the ASCII character set. That's true for the 128 characters in ASCII, but there are way more than 128 characters in the world. The ASCII character set contains the characters in the Latin alphabet—A through Z—but it does not contain the characters in other alphabets. For example, it does not contain the characters in the Greek, Cyrillic, and Hebrew alphabets. The designers of the Java language wanted Java to be general-purpose, so they wanted to be able to produce text output for many different languages using many different alphabets. To handle the additional characters, the Java designers had to use a bigger character set than the ASCII character set. Thus, they adopted the *Unicode* Standard. The Unicode Standard defines underlying numeric values for a huge set of 65,536 characters.

Why are there 65,536 characters in the Unicode Standard? Because the people who designed the Unicode Standard (the Unicode Consortium) decided that 16 bits would be sufficient to represent all the characters needed in a computer program.[5] And 16 bits can represent 65,536 characters. Here are the binary representations for the first four characters and the last character:

```
0000 0000 0000 0000
0000 0000 0000 0001
0000 0000 0000 0010
0000 0000 0000 0011
. . .
1111 1111 1111 1111
```

Notice that each row is a different permutation of 0's and 1's. If you wrote all such permutations, you'd see 65,536 rows. Thus, with 16 bits, you can represent 65,536 characters. The formula for determining the number of permutations (and consequently the number of rows and the number of characters) is 2 raised to the power of the number of bits. In other words, $2^{16} = 65{,}536$.

You can apply that same reasoning in determining why there are 128 characters in the ASCII character set. Way back in 1963 (when dinosaurs and punched cards roamed the Earth), the people who designed the ASCII character set decided that 7 bits would be sufficient to represent all the characters needed in a computer program. $2^7 = 128$, so 7 bits can represent 128 unique values.

Because the ASCII table was and is such a popular standard with many programming languages, the Unicode designers decided to use the ASCII character set as a subset of the Unicode character set. They inserted the ASCII character set's characters in the first 128 slots of the Unicode character set. That means programmers can find those characters' numeric values by referring to a simple ASCII table; they don't have to wade through the enormous Unicode character set.

Unicode Escape Sequence

Whenever you write an integer, you can write it in either decimal format or hexadecimal format. Likewise, you can specify a character by writing its numeric value in either decimal format or hexadecimal format and then casting it with the `(char)` cast operator. Java also provides another way to specify a character. You can use the *Unicode escape sequence*. The Unicode escape sequence is `\u` followed immediately by the hexadecimal digits of a hexadecimal number. Here's what we're talking about:

'\u####'
This is a single character.

Each # stands for one hexadecimal digit. We elected to show this in single quotes, not double quotes, to emphasize that the six-element escape sequence is just a single character, not a string. It's just like any other escape sequence, however, so you can embed the `\u####` anywhere in a string. The `u` must be lowercase, and there must be exactly four hexadecimal digits.[6]

[5]We're focusing on the original Unicode Standard, which is a subset of the current Unicode Standard. The original Unicode Standard is good enough for almost all Java programming. The original Unicode Standard uses 16 bits for all characters. The current Unicode Standard uses additional bits for additional characters that can't fit in the original Unicode set of 65,536 values. For additional details, see http://www.unicode.org/standard/standard/html

[6]The supplementary Unicode characters have numeric values that require more than four hexadecimal digits. To specify one of these supplementary characters, use a decimal or hexadecimal `int` representation of the character, or prefix the `\u`-representation of the four least-significant hexadecimal digits with an appropriate u-representation in the range, `\uD800` through `\uDFFF`. The prefix, called a *surrogate*, has no independent character association. (For additional information, see documentation on Java's `Character` class and http://www.unicode.org/Public/UNIDATA/Blocks.txt.) There's also another surrogate scheme that represents characters with an 8-bit base value and multiple 8-bit surrogates. This latter scheme is used in communications.

Using Unicode in Java Programs

If you want to print characters using Unicode escape sequences, you can use `System.out.println` in a text-based environment for the first 128 characters, but for the other characters, `System.out.println` in a text-based environment doesn't work consistently. That's because text-based environments recognize just the ASCII portion of the Unicode table; that is, the first 128 characters. To print all the characters in the Unicode table, you need to use graphical user interface (GUI) commands in a GUI environment.

The program in Figure 12.13 provides a GUI window and uses it to illustrate a small sampling of the many characters that are available. The `codes` array contains `int` code values for the Unicode escape sequences for the first characters in blocks of characters that we choose to display. These Unicode escape sequences automatically promote from type `char` to type `int` in the initializing assignment. The array called `descriptions` contains a simple `String` description for each block of characters.

For the window, we use an instance of the Java API `JFrame` class, which is in the `javax.swing` package. We set the window size at 600 pixels wide and 285 pixels high. We include in the window a single `JTextArea` object called `area`, and we enable its line-wrap capability. We use `JTextArea`'s `append` method to add each new string or character to whatever is already there.

Before looping, we display some general font information. The outer `for` loop displays the value of the first code number in one of the chosen blocks of characters and then a description of that block. The inner `for` loop displays the first 73 characters in that block. In the `append` method's argument, notice how we add the loop count, `j` , to the initial Unicode value to get each individual Unicode value as an `int`. Then we cast that `int` into a `char`. Then the concatenated `"  "` converts that `char` into a `String`, which matches the `append` method's parameter type.

Figure 12.14 shows the GUI output this program generates. The characters in the `codes` array in Figure 12.13 are the Unicode escape sequences for the first character in each block of characters shown in Figure 12.14. The hollow squares indicate code numbers that don't have symbols assigned to them or symbols that are not present in the current computer's software library. Notice that both the Greek and Cyrillic blocks include both uppercase and lowercase characters, and they include some additional characters beyond the normal final values of Ω (ω) and Я (я), respectively. These (and other) additional characters are needed for some of the individual languages in the families of languages using these alphabets. Of course, the characters shown in Figure 12.14 are just a tiny sampling of all the characters in Unicode.

Notice that the different characters shown in Figure 12.14 have generally different widths. To get constant-width characters, you'd have to change the font type to something like Courier New. You could do that—and also change the style to bold and size to 10 points—by inserting a statement like this:

```
area.setFont(new Font("Courier New", Font.BOLD, 10));
```

Suppose that you want the Unicode value for ≈. That's the last mathematical operator displayed in Figure 11.14. As indicated by the third `codes` value in the UnicodeDisplay program, the first mathematical operator has a unicode hexadecimal value of 0x2200. The maximum value of the inner `for` loop in Figure 12.13 is 72. The hexadecimal value of 72 is $4 \times 16 + 8 =$ 0x0048. Thus, the Unicode hexadecimal value of the last mathematical operator displayed in Figure 12.14 is 0x2200 + 0x0048 = 0x2248. Sometimes you can use a word processor to help you find the Unicode value of the special symbol that you want. For example, in Microsoft Word, select **Insert / Symbol / Mathematical Operators**, and then select ≈. Then read the Unicode hex value for the selected symbol from the "Character code" field near the bottom of the **Symbol** window. You'll find this also says the Unicode hexadecimal value for the ≈ character is 0x2248.

Look it up. By browsing through the http://unicode.org website, you can find everything Unicode has to offer. If you go there, look for a Code Charts link, and click on it. That should take you to a page that lets you explore the various sub-tables within the huge Unicode table. Try to find the

```
/*******************************************************
* UnicodeDisplay.java
* Dean & Dean
*
* This prints unicode characters.
*******************************************************/

import javax.swing.*;
import java.awt.Font;

public class UnicodeDisplay
{
  public static void main(String[] args)
  {
    int[] codes = {'\u0391',
                   '\u0410',
                   '\u2200',
                   '\u2500',
                   '\u2700'};
    String[] descriptions = {"Greek",
                             "Cyrillic (Russian)",
                             "mathematical operators",
                             "box drawing",
                             "dingbats"};
    JFrame window = new JFrame("Some Unicode Characters");
    JTextArea area = new JTextArea();
    Font font = new Font(Font.DIALOG, Font.PLAIN, 12);

    window.setSize(600,305);     // pixel width, height
    window.setDefaultCloseOperation(JFrame.EXIT_ON_CLOSE);
    window.add(area);
    area.setFont(font);
    area.setLineWrap(true);
    area.append("Font type, style, and size: " +
      font.getFontName() + ", " + font.getSize() + "\n");
    for (int i=0; i<codes.length; i++)
    {
      area.append("0x" + Integer.toString(codes[i], 16) +
        " " + descriptions[i] + ":\n");
      for (int j=0; j<=72; j++)
      {
        area.append((char) (codes[i] + j) + "  ");
      }
      area.append("\n");
    }
    window.setVisible(true);
  } // end main
} // end UnicodeDisplay
```

Figure 12.13 Program that uses GUI to display a sampling of Unicode characters

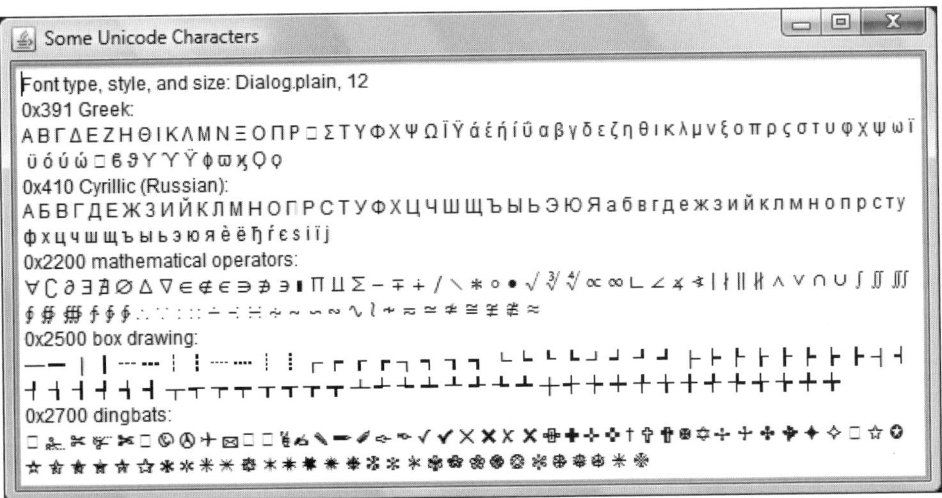

Figure 12.14 Output produced by the program in Figure 12.13

basic Latin link. That takes you to the Basic Latin subtable, which is equivalent to the ASCII table that we present in Appendix 1. This particular Unicode subtable is referred to as Latin because it contains the Latin alphabet—a, b, c, and so on. Visit a few of the other subtables to get an idea of what's available. In every subtable, you'll see a set of characters, and for each character, you'll see its equivalent Unicode value.

There are also several other standards for assigning numbers to characters. Computer applications sometimes include translation tables to make conversions between their own character-coding schemes and Unicode. Be warned, however. The translations don't always work as you might like, and special characters may change in surprising ways when you transfer text with special characters from one application to another.

12.16 Introduction to GridWorld Case Study (Optional)

This section introduces an extended example called GridWorld,[7] which shows how different Java classes work together in the context of a larger program. Starting now, we'll begin using this extended example to help illustrate new topics as we introduce them. Because the classes in this case study are distributed among several different packages, you'll probably want to look at Appendix 4 before proceeding further.

Go to http://www.collegeboard.com/student/testing/ap/compsci_a/case.html to install GridWorld software and documentation on your computer. Then download and unpack the `GridWorldCode.zip` file. This file contains a `framework` directory (the source code for the core classes used by GridWorld programs), a `javadoc` directory (documentation described in Section 8.3 of Chapter 8 and Appendix 6), and a `projects` directory (source code for particular GridWorld program drivers). The `GridWorldCode.zip` file also contains a `gridworld.jar` file, which has bytecode for the core classes used by GridWorld programs. This bytecode is in four packages: `info.gridworld.actor`, `info.gridworld.grid`,

[7]GridWorld software was written by Cay Horstmann for the CollegeBoard Advanced Placement Program (AP). Under a "GNU General Public License" (http://www.gnu.org/copyleft/gpl.html), this program is free for anyone to use, and no one may prevent anyone else from using it, although people may charge for their particular form of delivery. Thus, this software is "free" as in "free speech," but not necessarily free as in "free beer."

info.gridworld.GUI, and info.gridworld.World. To import these packages, your compiler must know where they are. You can tell it where they are by adding the path to the gridworld.jar file in your computer to your computer's *classpaths*. Or you can tell it where they are by including the path to the gridworld.jar file as an option each time that you call the javac compiler. (See the Installation Guide at the above website for details.)

GridWorld's "grid" is a two-dimensional "checkerboard" that holds pieces called "actors." Once you have installed the GridWorld software, try it out. Do this by going to the projects directory and then to the firstProject directory, where you should find a file called BugRunner. This file is the driver for an introductory demonstration. Compile BugRunner and then execute it. This should generate a GUI display with one randomly located rock and one randomly located bug, with the bug pointing up. Clicking a Step button in the lower-left corner of the display makes the bug move forward. As the bug moves, it deposits flowers in its wake, and as time passes, these flowers darken. Figure 12.15 shows a typical example, 16 steps

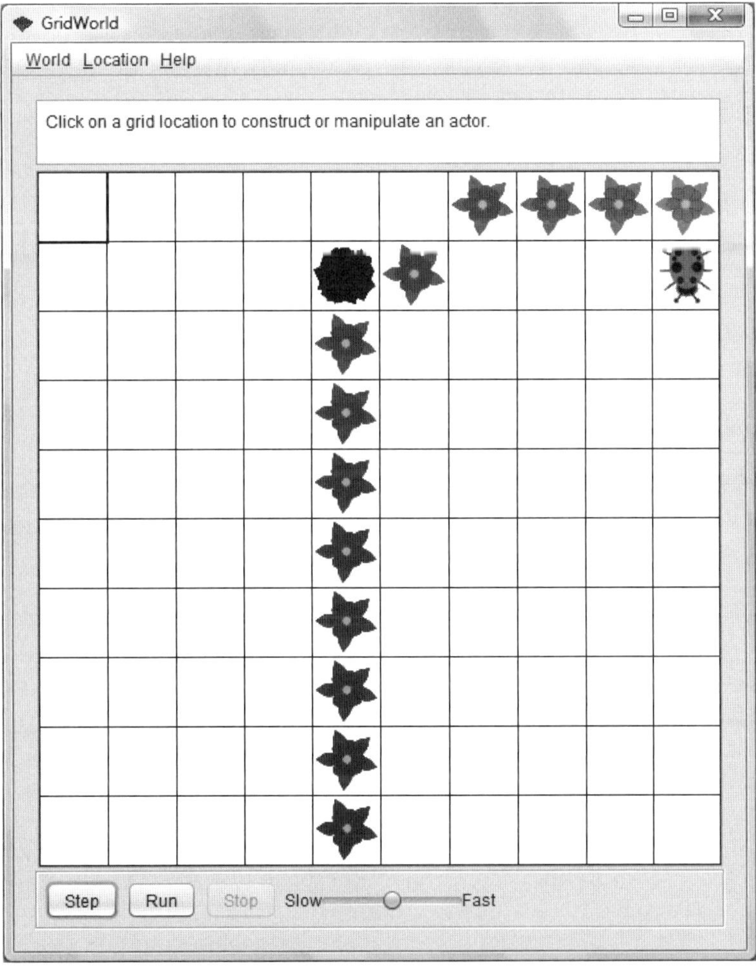

Figure 12.15 Display for GridWorld's firstProject, produced by executing BugRunner and clicking Step 16 times. Starting points are random, so different executions give different results.

after a start with the bug at the bottom of the screen. Whenever the bug encounters an obstacle like a rock or a boundary, instead of moving forward, it turns right by 45°, and it takes two steps to turn 90°.

This program imports code that defines three types of actors—rocks, bugs, and flowers. Program initialization automatically creates and randomly locates one black rock and one red bug. Clicking the Run button makes the bug run forward in repeated steps. You can control how fast the bug runs by adjusting the slow-fast slider. Once you click the Run button, the subdued Stop button brightens. Then you can click that Stop button to make the bug stop running.

Clicking the little icon in the extreme-upper-left corner enables you to move, resize, maximize, minimize, or close the window. Clicking World enables you to alter the grid's dimensions or to quit. By default, the program initially selects the location in the upper-left corner and gives it a darker boundary. Clicking Location enables you to move the selected location, to edit, enter, or delete an actor at the selected location, or to zoom in or zoom out. An option under Help describes the results of various mouse actions and corresponding keyboard shortcuts. Within the grid, clicking an empty location displays a list of constructors. Clicking one of those constructors inserts a rock, bug, or flower into that square. Clicking an occupied location displays a list of the methods you can invoke for the actor currently at that location.

Exercises at the end of the next two chapters will help you investigate existing program structure and behavior and modify program code to alter the behavior.

Summary

- Numerical overflow creates dramatic errors. Whenever there is any doubt about a particular type's ability to hold a value that might be assigned to it, change to a larger type.
- Floating-point numbers have a greater range than integers, but for a given amount of memory, they provide less precision.
- The ASCII character set provides numerical values for the symbols on a standard keyboard.
- Because characters are represented as numbers, `ch1 + ch2` evaluates to the sum of the ASCII values for the `char` variables, `ch1` and `ch2`.
- Type casting allows you to put a numeric value into a numeric variable of a different type, but be careful that you don't get overflow or undesired truncation when you do it.
- When used as a prefix, an increment (++) or decrement (--) operator changes the variable value before that variable participates in other expression operations. When used as a postfix, an increment or decrement operator changes the variable value after it participates in other expression operations.
- If a statement contains multiple assignment operators, the rightmost assignment evaluates first.
- It's sometimes helpful to embed an assignment within a condition, but you should avoid excessive use of embedded increment, decrement, and assignment operations.
- A conditional operator expression provides a compact conditional evaluation. If what's before the `?` is `true`, use what's after the `?`. Otherwise, use what's after the`:`.
- Short-circuit evaluation means that the JVM stops evaluating an expression whenever the expression's outcome becomes certain. Use this feature to avoid illegal operations.
- Use a `break` statement sparingly to terminate loops prematurely.
- An `enum` is a fixed set of distinct constants or a fixed set of objects having distinct identifiers and `final` instance values. It identifies a finite set of alternatives.
- In its extended form, Unicode provides numerical codes for up to a million different characters. You can specify them as decimal or hexadecimal integers or with a Unicode escape sequence. To see the Unicode characters for codes above 127, you must display them in a GUI window.

Review Questions

§12.2 Integer Types and Floating-Point Types

1. For each integer data type, how many bits of storage are used?
2. How would you write the decimal constant, 1.602×10^{-19}, as a double?
3. What is the approximate precision (number of accurate decimal digits) of each of the floating-point types?

§12.3 char Type and the ASCII Character Set

4. How many distinct characters are identified by the basic ASCII character set?
5. What number can you add to an uppercase-letter char variable to convert it to lowercase?

§12.4 Type Conversions

6. Assume the declaration:

```
public final double C = 3.0E10;   // speed of light in cm/sec
```

Write a Java print statement that uses a cast operator to display the value of C in this format:

```
30000000000
```

7. Will this statement be OK, or will it generate a compile-time error? (OK / error)

```
float price = 66;
```

8. Will this statement be OK, or will it generate a compile-time error? (OK / error)

```
boolean done = (boolean) 0;
```

9. Will this statement be OK, or will it generate a compile-time error? (OK / error)

```
float price = 98.1;
```

§12.5 Prefix/Postfix Modes for Increment/Decrement Operators

10. What is the value of z after these statements execute?

```
int z, x = 3;
z = --x;
z += x--;
```

§12.6 Embedded Assignments

11. Write one Java statement that makes w, x, and y all equal to the current value of z.

§12.7 Conditional Operator Expressions

12. Suppose x equals 0.43. Given the following switch statement heading, what does the switch heading's controlling expression evaluate to?

```
switch (x>0.67 ? 'H' : (x>0.33 ? 'M' : 'L'))
```

§12.8 Expression Evaluation Review

13. Assume this:

```
int a = 2;
int b = 6;
float x = 8.0f;
```

Evaluate each of the following expressions, using these guidelines:

- As shown in Section 12.8, put each evaluation step on a separate line and use the ⇒ symbol between steps.
- Evaluate each expression independently of the other expressions; in other words, use the above assumed values for each expression evaluation.
- Expression evaluation problems can be tricky. We encourage you to check your work by running test code on a computer.
- If there would be a compilation error, specify "compilation error."

a) `a + 25 / (x + 2)`
b) `7 + a * --b / 2`
c) `a * --b / 6`
d) `a + b++`
e) `a - (b = 4) % 7`
f) `b = x = 23`

§12.9 Short-Circuit Evaluation

14. Assume `expr1` and `expr2` are expressions that evaluate to `boolean` values. Assume that `expr1` evaluates to `true`. When the computer evaluates each of the following expressions, will it evaluate `expr2`? If yes, just say "yes." If no, explain why, and use the term "short-circuit evaluation" in your explanation.
a) `expr1 || expr2`
b) `expr1 && expr2`

15. Assume this:

```
int a = 2;
boolean flag = true;
```

Evaluate the following expression:

```
a < 3 || flag && !flag
```

§12.10 Empty Statement

16. Assume that the following code fragment is inside of a program that compiles successfully. What does the code fragment print? Hint: This is a trick question. Study the code carefully.

```
int x = 1;
while (x < 4);
{
  System.out.println(x);
  x++;
}
```

§12.11 `break` Statement Within a Loop

17. Usually, you should avoid using `break` except in `switch` statements because using `break` statements forces readers to look for termination conditions inside loop bodies. (T / F)

§12.12 `for` Loop Header Details

18. Assume that the following code fragment is inside of a program that compiles successfully. What does the code fragment print?

```
for (int i=0,j=0; ; i++,j++)
{
  System.out.print(i + j + " ");
}
```

§12.13 Enumerated Types

19. Declare a set of `private` constants called "Kingdom" that identify the five kingdoms of life: Monera, Protoctista, Fungi, Plantae, and Animalia.

§12.15 GUI Track: Unicode (Optional)

20. What is the hexadecimal symbol for the decimal number 13?

21. The Unicode values for characters are the same as the ASCII values in the range `0x00` to `0xFF`. (T / F)

Exercises

1. [after §12.2] If an integer overflows, what type of error is produced—compile-time error, runtime error, or logic error?

2. [after §12.2] Look up `BigInteger` in Java API documentation, and write a program that generates and prints a 128-bit random number.

3. [after §12.3] Write a class method named `getEvenDigit` that prompts the user to enter an even digit and returns the even digit as an `int`. Your method must be robust: Read the input as a string and verify that the input consists of a single even digit. If the user enters invalid input, chastise the user and reprompt him or her to enter an even digit.

   Sample session:

   ```
   Enter an even digit: w
   Invalid entry!
   Enter an even digit: 44
   Invalid entry!
   Enter an even digit: 5
   Invalid entry!
   Enter an even digit: 8 8
   Invalid entry!
   Enter an even digit: 8
   ```

4. [after §12.4] What does this print? `System.out.println('A' + 2);`

5. [after §12.6] Assume `a` and `b` are `boolean` variables. What are their values after this statement executes?

   ```
   a =!((b=4<=5) && (a=4>=5));
   ```

   Hint: First rewrite the statement to make it more readable.

6. [after §12.7] Assume this:

   ```
   int a = 2;
   float x = 8.0f;
   boolean flag = true;
   ```

   To what does the following expression evaluate?

   ```
   (flag) ? (a = (int) (x + .6)) : a
   ```

7. [after §12.8] Assume this:

   ```
   int a = 10;
   int b = 2;
   double x = 6.0;
   ```

Evaluate each of the following expressions. Follow these guidelines:

- As shown in Section 12.8, put each evaluation step on a separate line and use the ⇒ symbol between steps.
- Evaluate each expression independently of the other expressions; in other words, use the above assumed values for each expression evaluation.
- Expression evaluation problems can be tricky. We encourage you to check your work by running test code on a computer.
- If there would be a compilation error, specify "compilation error."

a) `a - 7 / (x - 4)`
b) `8 + a * ++b / 20`
c) `a + b--`
d) `a + (b = 5) % 9`
e) `a = x = -12`

8. [after §12.8] Assume this:

```
String s = "hi";
int num = 3;
char ch = 'm';
```

Evaluate each of the following expressions. Follow these guidelines:

- As shown in Section 12.8, put each evaluation step on a separate line and use the ⇒ symbol between steps.
- Evaluate each expression independently of the other expressions; in other words, use the above assumed values for each expression evaluation.
- Expression evaluation problems can be tricky. We encourage you to check your work by running test code on a computer.
- If there would be a compilation error, specify "compilation error."

a) `s + (num + 4)`
b) `s + num + 4`
c) `s + '!' + "\""`
d) `num + ch`
e) `'8' + 9`

9. [after §12.9] Consider the following code fragment. Line numbers are at the left.

```
1    int a = 2;
2    boolean b = false;
3    boolean c;
4    c = b && ++a == 2;
5    b = a++ == 2;
6    b = !b;
7    System.out.println(a + " " + b + " " + c);
```

Trace the code using this trace setup:

line#	a	b	c	output

10. [after §12.9] Assume:

```
boolean a = false;
boolean b;
double c = 2.5;
```

Determine the output of the following code fragment:

```
b = a && (++c == 3.5);
a = true || (++c == 3.5);
System.out.println(a + " " + b + " " + c);
```

11. [after §12.10] In the Fibonacci sequence, each successive element is the sum of the two previous elements. Starting with 0 and 1, the next element is 0 + 1 = 1. The element after that is 1 + 1 = 2. The element after that is 1 + 2 = 3, and the one after that is 2 + 3 = 5, and so on. Given this declaration:

```
int p, q;
```

Provide a `for` loop that prints this part of the Fibonacci sequence:

```
1 2 3 5 8
```

Your solution should consist of just a `for` loop header and then an empty statement—nothing else. By the way, we recommend that you avoid using code like this for your real programs. This exercise is just for fun (fun for a hacker, anyway ☺).

12. [after §12.10] A common error is to accidentally add a semicolon at the end of a loop header. Run the following `main` method on a computer. What is the output?

```
public static void main(String[] args)
{
  int i;
  int factorial = 1;

  for (i=2; i<=4; i++);
  {
    factorial *= i;
  }
  System.out.println("i = " + i + ", factorial = " + factorial);
} // end main
```

13. [after §12.12] Note the following program. Provide a `for` loop that is functionally equivalent to the given `do` loop.

```
import java.util.Scanner;

public class Test
{
  public static void main(String[] args)
  {
    Scanner stdIn = new Scanner(System.in);
    String entry;

    do
    {
      System.out.println("Enter 'q' to quit: ");
      entry = stdIn.nextLine();
    } while (!entry.equals("q"));
  } // end main
} // end class Test
```

14. [after §12.13] Minimally modify the code in Exercise 16 in Chapter 9 to convert `public class City` to `public enum City2`, such that the new code generates the same output as the original code when the new code is driven by the following `main` method:

```
public static void main(String[] args)
{
  for (City2 city : City2.values())
  {
    city.display();
  }
} // end main
```

15. [after §12.13] Improve the CityTravel program code by prompting the user to enter origin and destination cities and using the user-entered values in the `getDistance` method call. If the user does not spell one of the enumerated cities correctly, print an error message and loop until one of the enumerated cities is entered correctly. To get you started, here is the modified `main` method:

```
public static void main(String[] args)
{
  final double KM_TO_MILES = 0.62137; // conversion factor
  City origin = getCity("origin");
  City destination = getCity("destination");
  double distance = origin.getDistance(destination); // in km

  System.out.printf("%s to %s: %.1f km, or %.1f miles",
    origin, destination, distance, distance * KM_TO_MILES);
} // end main
```

And here is an example of a typical sample session:

```
Enter origin city
(PARKVILLE, HAVANA, KINGSTON, NASSAU, or SAINT_THOMAS): Parkville
Invalid entry. Must use exact spelling.
Enter origin city
(PARKVILLE, HAVANA, KINGSTON, NASSAU, or SAINT_THOMAS): PARKVILLE
Enter destination city
(PARKVILLE, HAVANA, KINGSTON, NASSAU, or SAINT_THOMAS): SAINT_THOMAS
PARKVILLE to SAINT_THOMAS: 3689.9 km, or 2292.8 miles
```

16. [after §12.14] What is the Unicode hexadecimal value for the "∞" (infinity) symbol? Show or explain how you got your answer.

17. [after §12.16] (Case Study) Run GridWorld's BugRunner program. Perform the operations indicated, and answer the four numbered questions.
Move the rock to the location of the bug. Then move the rock to location (6, 0).
   1) What happens when you move a rock to the location of a bug?
   Insert a second rock at location (3, 9). Insert a red bug at (4, 4). Insert a blue bug at (5, 5). Adjust the directions of the two bugs so that they face each other.
   2) After they face each other, what is the red bug's direction and what is the blue bug's direction?
   Click "Step" and see what happens. Then click "Step" again and see what happens. Then Click "Run." Adjust the speed and watch the behavior until you see regular repetition. Then stop the bug from running and delete one of the bugs.
   3) What rules govern bug direction and motion?
   4) What rules govern flower creation, deletion, and coloration?

Review Question Solutions

1. `byte` = 8 bits, `short` = 16 bits, `int` = 32 bits, `long` = 64 bits
2. `1.602E-19` or `1.602e-19`
3. `float` precision ≈ 6 digits, `double` precision ≈ 15 digits
4. The basic ASCII character set describes 128 different characters.
5. To convert uppercase to lowercase, add 32. To go the other way, subtract 32.
6. `System.out.println((long) C); // (int) isn't big enough!`
7. This statement is OK:

```
float price = 66;
```

8. This statement generates a compile-time error because it's illegal to convert between numeric values and `boolean` values:

```
boolean done = (boolean) 0;
```

9. This statement generates a compile-time error because floating-point constants are `double` by default:

```
float price = 98.1;
```

10. `z`'s value is 4. The first decrement uses prefix mode so `x` is first decremented to 2, then 2 is assigned into `z`. The second decrement uses postfix mode so `x` is decremented <u>after</u> its value of 2 is added to `z`.
11. `w = x = y = z;` or any other sequence that has `z` on the right.
12. The `switch` controlling expression evaluates to `'M'`
13. Expression-evaluation practice:

a)
```
a + 25 / (x + 2) ⇒
2 + 25 / (8.0 + 2) ⇒
2 + 25 / 10.0 ⇒
2 + 2.5 ⇒
4.5
```

b)
```
7 + a * --b / 2⇒
7 + 2 * --6 / 2⇒
7 + 2 * 5 / 2⇒
7 + 10 / 2 ⇒
7 + 5 ⇒
12
```

c)
```
a * --b / 6 ⇒
2 * --6 / 6 ⇒
2 * 5 / 6 ⇒
10 / 6 ⇒
1
```

d) `a + b++` ⇒

`2 + 6` (b is updated to 7 after its value is accessed) ⇒

`8`

e) `a - (b = 4) % 7` ⇒

`2 - 4 % 7` ⇒

`2 - 4` ⇒

`-2`

f) `b = x = 23` ⇒

`b = 23.0` ⇒

compilation error (because the `float` 23.0 cannot be assigned to the `int b` without a cast operator)

14. Will it evaluate `expr2`

a) No. Since the left side of the `||` operator is `true`, short-circuit evaluation will cause the right side of the `||` operator (`expr2`) to be ignored (since the result of the entire expression will evaluate to `true` regardless of `expr2`'s value).

b) Yes.

15. Assuming:

```
int a = 2;
boolean flag = true;
```

```
a < 3 || flag && !flag ⇒
2 < 3 || true && !true ⇒
2 < 3 || true && false ⇒
true || true && false ⇒
true      (short-circuit evaluation dictates "true or anything" evaluates to true)
```

16. It prints nothing because, due to the empty statement, the `while` loop header executes repeatedly in an infinite loop.

17. True. Normally, you should avoid using `break` other than in `switch` statements.

18. The code fragment generates an infinite loop because the `for` loop header's missing second component is true by default. The output is:

```
0 2 4 6 ...
```

19.

```
private enum Kingdom
{
  MONERA, PROTOCTISTA, FUNGI, PLANTAE, ANIMALIA;
}
```

20. The hexadecimal symbol for the decimal number 13 is either `d` or `D`.

21. False. They are the same only in the range from `0x00` to `0x7F`.

CHAPTER 13

Aggregation, Composition, and Inheritance

Objectives

- Understand how things are naturally organized in aggregations and compositions.
- Implement aggregation and composition relationships within a program.
- Understand how inheritance can be used to refine an existing class.
- Implement an inheritance hierarchy within a program.
- Learn how to write constructors for derived classes.
- Learn how to override an inherited method.
- Learn how to prevent overriding.
- Learn how to use a class to represent an association.

Outline

13.1 Introduction

Prior to this chapter, the programs you've created have been relatively simple in terms of their object orientation, so you've been able to describe all the objects in a program with just a single class. But for more complex programs, you should consider implementing multiple classes, one for each different type of object within a program. In this chapter, you'll do just that, and you'll focus on the different ways to organize classes in a multiple-class program. First, you'll learn how to organize classes that are parts of a larger containing class. When classes are related like that, where one class is the whole and the other classes are parts of the whole, the classes form an *aggregation.* Then you'll learn how to organize classes where one class, the *base class,* defines common features for a group of objects, and the other classes define specialized features for each of the different types of objects in the group. When classes are related like that, the classes form an inheritance hierarchy. It's called an *inheritance hierarchy* because the specialized classes inherit features from the base class.

In describing inheritance, we present various techniques for working with an inheritance hierarchy's classes. Specifically, we present *method overriding,* which allows you to redefine a method in a specialized class that's already been defined in the base class. We also present the `final` modifier, which allows you to prevent a specialized class from overriding a method defined in the base class.

As a follow-up to the initial presentation of aggregation and inheritance concepts, we describe how the two design strategies can work together. It's sometimes difficult to decide which is the best strategy to use. To give you practice with those decisions, we guide you partway through a program design activity and develop the skeleton for what could be a sophisticated card game program.

As new concepts are introduced, end-of-chapter exercises relate them to the GridWorld case study. An optional section near the end of the chapter uses several of this chapter's concepts together in a GridWorld extension. Another optional section shows you how to improve organization by creating an association class, which defines a set of characteristics that belong to a particular relationship between classes.

By showing you how to organize multiple classes, this chapter provides you with important tools necessary to tackle real-world problems. After all, most real-world programming projects are large and involve multiple types of objects. When you organize objects correctly, it makes programs easier to understand and maintain. And that's good for everyone!

13.2 Composition and Aggregation

There are two primary forms of aggregation. As described above, standard aggregation is when one class is the whole and other classes are parts of the whole. The other form of aggregation also defines one class as the whole and other classes as parts of the whole. But it has an additional constraint that says the whole class is the exclusive owner of the part classes. "Exclusive ownership" means that the part classes cannot be owned by another class while they are being owned by the whole class. This exclusive-ownership form of aggregation is called *composition.* With composition, the whole class is called the *composite,* the part classes are called *components,* and the composite contains the components. Composition is considered to be a strong form of aggregation because the composite-component connections are strong (due to each component having only one owner—the composite).

Composition and Aggregation in the Real World

The concept of composition was not created for computer programming; it's frequently used for complex objects in the real world. Every living creature and most manufactured products are made up of parts. Often, each part is a subsystem that is itself made up of its own set of subparts. Together, the whole system forms a composition hierarchy.

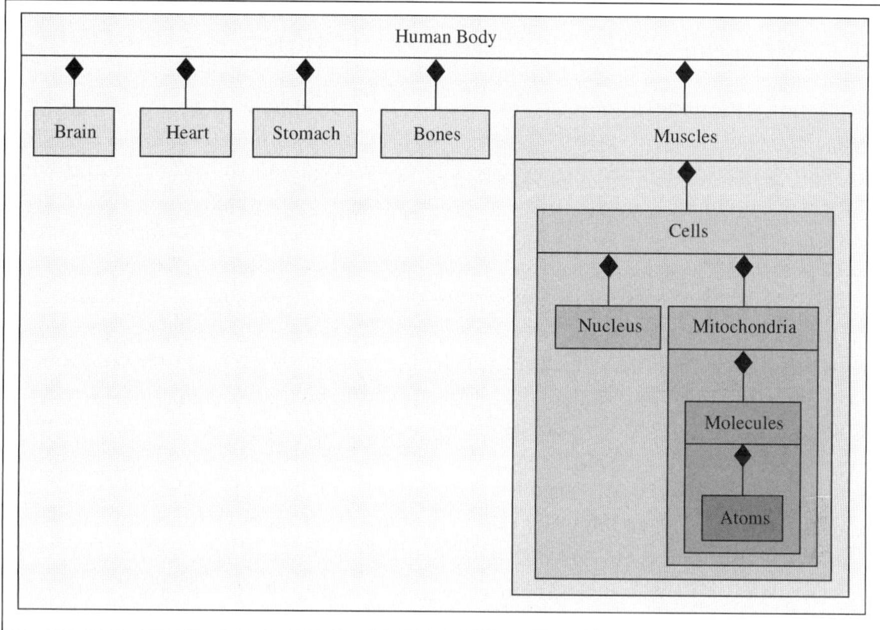

Figure 13.1 Partial representation of a composition hierarchy for a human body

Figure 13.1 shows a composition hierarchy for a human body. At the top of this particular composition hierarchy is a whole body. A human body is composed of several organs—brain, heart, stomach, bones, muscles, and so on. Each of these organs is in turn composed of many cells. Each of these cells is composed of many organelles, like the nucleus (a cell's "brain"), and the mitochondria, (a cell's "stomach"). Each organelle is composed of many molecules. And finally, each organic molecule is composed of many atoms.

In a composition hierarchy (as well as in an aggregation hierarchy), the relationship between a containing class and one of its part classes is known as a *has-a* relationship. For example, each human body has a brain and has a heart. Remember that with a composition relationship, a component part is limited to just one owner at a time. For example, a heart can be in only one body at a time. Although the ownership is exclusive, it's possible for the ownership to change. With a heart transplant, a heart can switch to a new owner, but it still has just one owner at a time.

Note the diamonds in Figure 13.1. In the Unified Modeling Language (UML), solid diamonds denote a composition relationship. They indicate that a whole has exclusive ownership of a part.

Now let's think about an aggregation example where the parts are not exclusively owned by the whole. You can implement a school as an aggregation by creating a whole class for the school and part classes for the different types of people who work and study at the school. The people aren't exclusively owned by the school because a person can be part of more than one aggregation. For example, a person can attend classes at two different schools and be part of two school aggregations. The same person might even be part of a third aggregation of a different type, like a household aggregation.

Composition and Aggregation in a Java Program

Let's look at an example that uses both class relationships—composition (where exclusive ownership is required) and standard aggregation (where exclusive ownership is not required). Suppose you're trying to model a car dealership with a computer program. Because the car dealership is made from several distinct, non-trivial parts, it's a good candidate for being implemented as an aggregation. The "whole" (the top of the aggregation hierarchy) is the dealership. Typically, a business has two kinds of "parts"—people and property. For simplicity, suppose that the only types of people at the car dealership are management and salespeople, and suppose the only type of property is cars. The control that the dealership has over the people is limited. They may also have other relationships, like memberships in families and social clubs. The dealership does not own its employees exclusively. Therefore, the relationship between the dealership and its employees is just aggregation. But the dealership does own its cars exclusively. So that relationship is composition. Note that the dealership can transfer ownership of its cars to its customers. That's OK because composition permits ownership to be transferred. Using a bottom-up design methodology, you should define three classes—`Car`, `Manager`, and `SalesPerson`—for the three types of component objects. Then, you should define a `Dealership` class for the container object.

Before you see the Dealership program's code, let's focus on the big-picture concepts using a UML class diagram. Figure 13.2's UML class diagram shows the Dealership program's four classes and the relationships among them, plus two other classes with which `Manager` or `SalesPerson` might be associated. Because we're now focusing on just the relationships among classes, in each representation of a class, we include just the class name and omit variables and methods. That's OK—UML is very flexible, and such omissions are allowed by the UML standards. UML indicates class relationships with connecting lines that run from one class to another. Formally, each connecting line is called an *association line*.

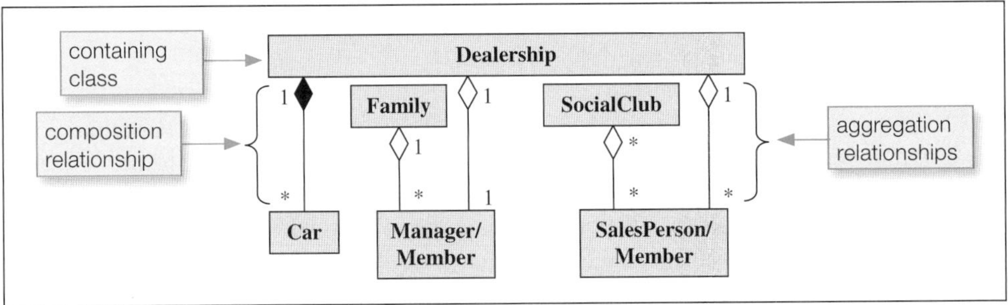

Figure 13.2 Class diagram for Dealership program

In Figure 13.2, note the diamonds on the association lines. Solid diamonds (like the one on the `Dealership-Car` line) indicate composition relationships, and hollow diamonds (like the ones on the `Dealership-Manager` and `Dealership-SalesPerson` lines) indicate aggregation relationships. The diamonds always go next to the containing class, so Figure 13.2's class diagram indicates that `Dealership` is the containing class.

Notice the numbers and asterisks written beside the association lines. These are *multiplicity values* that UML uses to specify the number of objects that participate in associations. The two 1's on the line between `Dealership` and `Manager` indicate a one-to-one association. That means there's one manager for each car dealership. If there were a 2 multiplicity value for the manager class, that would indicate two managers for each car dealership. The combination of 1 and * on other association lines indicates one-to-many associations, where "many" implies an indefinite number. That means you can have lots of cars (or none) and lots of salespeople (or none) for one car dealership.

It's now time to move from the conceptual phase, with emphasis on the dealership's UML class diagram, to the implementation phase, with emphasis on the Dealership program code. Note the `Dealership` class in Figure 13.3, and in particular, note the `manager`, `people`, and `cars` instance variables declared

```
/*************************************************************
* Dealership.java
* Dean & Dean
*
* This represents an auto retail sales organization.
*************************************************************/

import java.util.ArrayList;

public class Dealership
{
  private String company;
  private Manager manager;
  private ArrayList<SalesPerson> people =
    new ArrayList<>();
  private ArrayList<Car> cars = new ArrayList<>();

  //**********************************************************

  public Dealership(String company, Manager manager)
  {
    this.company = company;
    this.manager = manager;
  }

  //**********************************************************

  public void addCar(Car car)
  {
    cars.add(car);
  }

  public void addPerson(SalesPerson person)
  {
    people.add(person);
  }

  //**********************************************************

  public void printStatus()
  {
    System.out.println(company + "\t" + manager.getName());
    for (SalesPerson person : people)
      System.out.println(person.getName());
    for (Car car : cars)
      System.out.println(car.getMake());
  } // end printStatus
} // end Dealership class
```

Containership implemented here.

Figure 13.3 `Dealership` class for the Dealership program

inside the Dealership class. Those instance variable declarations implement the concept of the dealership class containing the other three classes. The general rule is that whenever you have a class that contains another class, declare an instance variable inside the containing class such that the instance variable holds a reference to one or more of the contained class's objects.

Also in the Dealership class, note the use of ArrayLists for the people and cars instance variables. Typically, if you have a class in a UML class diagram with a * multiplicity value, you can use an ArrayList to implement the reference to the asterisked class. ArrayLists are good for implementing * multiplicity values because they can expand to accommodate any number of elements.

Peruse the Car, Manager, and SalesPerson classes in Figures 13.4, 13.5, and 13.6. They simply store and retrieve data. Note the SalesPerson's sales instance variable—it keeps track of the total sales for a salesperson for the current year. There are no methods for accessing or updating the sales instance variable. We omitted those methods to avoid code clutter and to maintain focus on the matter at hand, aggregation and composition. In an actual car dealership program, you'd need to provide those methods.

```
/********************************
* Car.java
* Dean & Dean
*
* This class implements a car.
********************************/

public class Car
{
  private String make;

  //****************************

  public Car(String make)
  {
    this.make = make;
  }

  //****************************

  public String getMake()
  {
    return make;
  }
} // end Car class
```

Figure 13.4 Car class for the Dealership program

See the car dealership program's driver class in Figure 13.7. Most of the code is straightforward. The main method instantiates a Manager object, two SalesPerson objects, and a Dealership object. Then main adds salesPerson and Car objects to the Dealership object. The part of main that merits further attention is the use of local variables for the Manager and SalesPerson objects and the use

```java
/*************************************************************
* Manager.java
* Dean & Dean
*
* This class implements a car dealership sales manager.
*************************************************************/

public class Manager
{
  private String name;

  //*********************************************************

  public Manager(String name)
  {
    this.name = name;
  }

  //*********************************************************

  public String getName()
  {
    return name;
  }
} // end Manager class
```

Figure 13.5 `Manager` class for the Dealership program

```java
/*************************************************
* SalesPerson.java
* Dean & Dean
*
* This class implements a car salesperson.
*************************************************/

public class SalesPerson
{
  private String name;
  private double sales = 0.0; // sales to date

  //**********************************************

  public SalesPerson(String name)
  {
    this.name = name;
  }

  //**********************************************

  public String getName()
  {
    return name;
  }
} // end SalesPerson class
```

Figure 13.6 `SalesPerson` class for the Dealership program

```
/**********************************************************
* DealershipDriver.java
* Dean & Dean
*
* This class demonstrates the car dealership composition.
**********************************************************/

public class DealershipDriver
{
  public static void main(String[] args)
  {
    Manager ryne = new Manager("Ryne Mendez");
    SalesPerson nicole = new SalesPerson("Nicole Betz");
    SalesPerson vince = new SalesPerson("Vince Sola");
    Dealership dealership =
      new Dealership("OK Used Cars", ryne);

    dealership.addPerson(nicole);
    dealership.addPerson(vince);
    dealership.addCar(new Car("GMC"));
    dealership.addCar(new Car("Yugo"));
    dealership.addCar(new Car("Dodge"));
    dealership.printStatus();
  } // end main
} // end DealershipDriver class
```

For aggregations, pass in copies of references.

For compositions, create anonymous objects.

Output:

```
OK Used Cars        Ryne Mendez
Nicole Betz
Vince Sola
GMC
Yugo
Dodge
```

Figure 13.7 Driver for the Dealership program

of anonymous objects for the `Car` objects. Why the discrepancy? Because `Manager` and `SalesPerson` relate to the `Dealership` class with aggregation, and `Car` relates to the `Dealership` class with composition.

Here's the general rule for implementing aggregation relationships. Whenever two classes have an aggregation relationship, you should store the contained class's object in an instance variable in the containing class, and you should also store it in another variable outside the containing class. That way, the object can be added to another aggregation and have two different "owners" (having two different owners is allowed by aggregation). Putting this in the context of the Dealership program, `DealershipDriver` uses local variables when it instantiates the `Manager` and `SalesPerson` objects. That enables the `Manager` and `SalesPerson` objects to exist independently from the dealership, and that mirrors the real world.

Now let's look at the general rule for implementing composition relationships. Whenever two classes have a composition relationship, you should store the contained class's object in an instance variable in the containing class, and you should not store it elsewhere. That way, the object can have only one "owner" (having just one owner is required by composition). Putting this in the context of the Dealership program, `DealershipDriver` creates anonymous objects when it instantiates cars. That gives the dealership exclusive ownership and complete control over the cars, and that mirrors the real world.

13.3 Inheritance Overview

So far in this chapter, we've focused on aggregation and composition hierarchies, where one class is the whole and other classes are parts of the whole. Now we turn to classification hierarchies, which are qualitatively different from composition hierarchies. Whereas a composition hierarchy describes a nesting of things, a classification hierarchy describes an elaboration of concepts. The concept at the top is the most general/generic, and the concepts at the bottom are the most specific.

Classification Hierarchies and Inheritance of Attributes

Before looking at code, let's think about a real-world classification example. Figure 13.8 describes a few of the many possible characteristics of organisms living on the Earth today, with the most general characteristics at the top of the diagram and the most specific characteristics at the bottom of the diagram. Although this chart includes only characteristics of current living organisms, it's helpful to recognize that there was a natural time sequence in the development of these characteristics. The characteristics at the top developed first, and the characteristics at the bottom developed last. The earliest types of life—bacteria—appeared on Earth almost 4 billion years ago as single-celled organisms with no internal partitions. The "is alive" attribute listed in the box at the top—the most general class—endured and became an attribute common to all classes below it.

About 2.3 billion years ago, a nucleus and other components appeared inside cells, creating more sophisticated organisms called Eukaryotes. About 1.3 billion years ago, the first animals appeared. They had more than one cell, and they were vascular (had containers and conveyors like arteries and veins). About 510 million years ago, some of the animals (vertebrates) developed backbones and braincases. About 325 million years ago, the first reptiles appeared. Then, about 245 million years ago, the first mammals appeared. Thus, there is a correspondence between (1) the classification of biological attributes and (2) their sequence of development.

It's useful to identify a similar correspondence in object-oriented programming: (1) A classification hierarchy organizes a program's attributes. (2) A generic-to-specific development sequence organizes our implementation of those attributes. It's good design practice to start with a relatively simple and generic implementation and add specializations and complexity in subsequent design cycles. You'll see some examples of this later on.

Start generic. 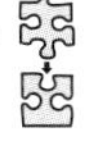

With composition, certain classes contain other classes, but with inheritance, there's no such containership. For example, in Figure 13.8, Animal is above Mammal, but an animal does not contain a mammal. Rather, animal is a generic type, and mammal is a specialized version of animal.

Each descendant type of organism inherits some attributes from its ancestors and adds some new attributes of its own. Ideally, the characteristics associated with each type high in the hierarchy should be just those attributes that are "conserved"—actually inherited by all types descended from that type. Thus, ideally, any type at the bottom of the hierarchy inherits all of the attributes associated with all types above it. For example, mammals have mammary glands and hair. And, because mammals are vertebrates, they

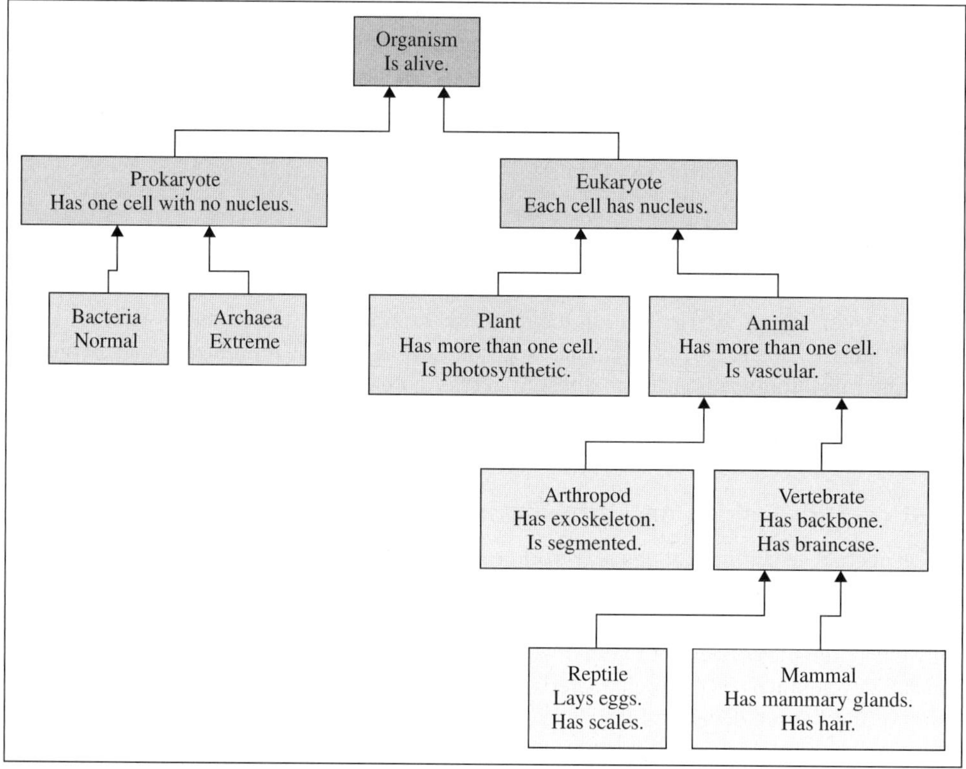

Figure 13.8 Biological example of a classification or inheritance hierarchy

inherit the vertebrate attributes of having a backbone and a braincase. And, because mammals are also animals, they also inherit the animal attributes of having more than one cell and being vascular. And, because mammals are also eukaryotes, they also inherit the eukaryote attribute of having a nucleus in each cell.

The types at the very bottom of a biological inheritance hierarchy do not appear in Figure 13.8, because the complete hierarchy is too big to display in one figure. What's actually at the bottom are species, like Homo sapiens (human beings). In nature, reproduction is possible only among members of the same species. Similarly, in an ideal OOP computer program, the only *realizable* (instantiable) types are the types at the very bottom of inheritance hierarchies. Ideally, all types above instantiable types are abstract types. In the next chapter, you'll learn how you can use Java's abstract keyword to prevent the instantiation of abstract types. Organizing an inheritance hierarchy so that all realizable (instantiable) types appear only at the lowest level (the *leaves* of a hierarchical tree) minimizes duplication, and it minimizes maintenance and enhancement labor.

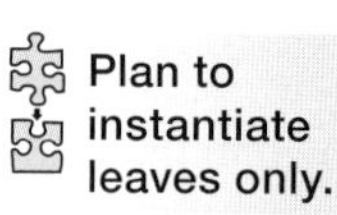
Plan to instantiate leaves only.

UML Class Diagrams for Inheritance Hierarchies

Figure 13.9 shows a UML class diagram for an inheritance hierarchy that keeps track of people associated with a department store. The top class, `Person`, is generic. It contains data and methods that are common to all classes in the hierarchy. Classes below the top class are more specific. For example, the `Customer` and `Employee` classes describe specific types of people in the department store. Because there are two distinct types of store employees, the `Employee` class has two subordinate classes for the two types—the `FullTime` class for full-time employees and the `PartTime` class for part-time employees.

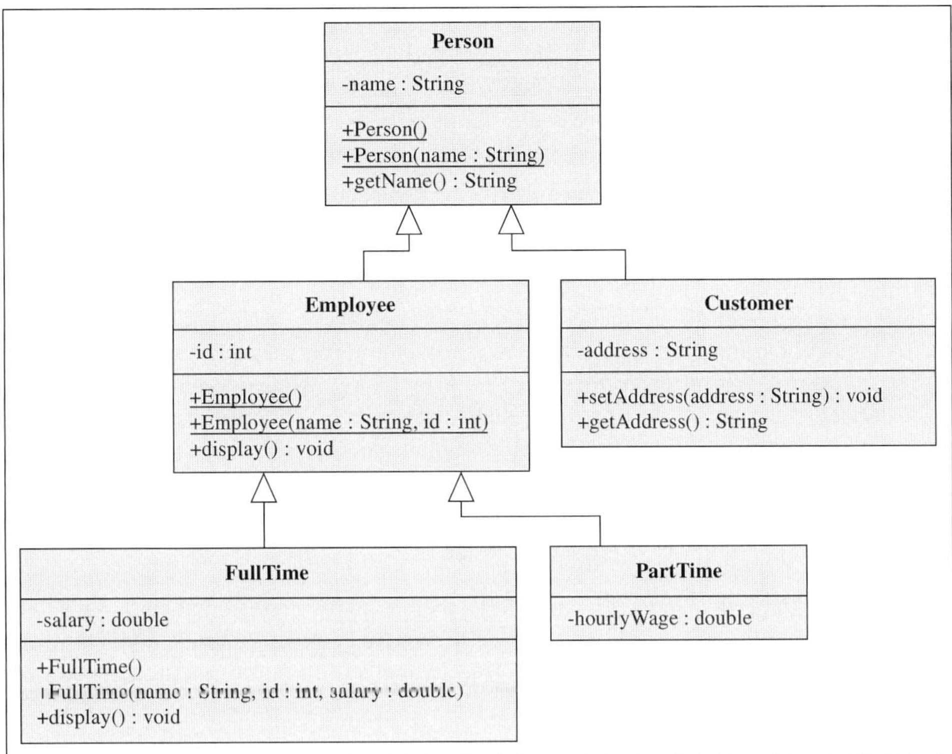

Figure 13.9 UML class diagram for a `Person` class inheritance hierarchy

Within an inheritance hierarchy, lower classes inherit upper classes' members. Thus, the `Employee` and `Customer` classes inherit `name` from the `Person` class. Likewise, the `FullTime` and `PartTime` classes inherit `id` from the `Employee` class. Inheritance travels all the way down the inheritance hierarchy tree, so in addition to inheriting `id` from the `Employee` class, the `FullTime` and `PartTime` classes inherit `name` from the `Person` class.

Within an inheritance hierarchy, classes are linked in pairs. Can you identify the linked pairs in Figure 13.9? The four pairs of linked classes are `Person-Customer`, `Person-Employee`, `Employee-FullTime`, and `Employee-PartTime`. For each pair of linked classes, the more general class is considered to be the *superclass,* and the more specific one is considered to be the *subclass.*

Inheriting a superclass's variables and methods enables a subclass to be a clone of its superclass. But making a subclass that's just a clone would be silly because you could just use the superclass instead. You always want a subclass to be a more specific version of its superclass. That's achieved by establishing additional variables and/or methods inside the subclass's definition. For example, in Figure 13.9, the `Customer` class defines an address instance variable. That means `Customer` objects have a name (inherited from the `Person` class) plus an address. Customer addresses are important in that they enable department stores to mail monthly "Everything-Must-Go Liquidation Sale!" advertisements to their customers.

UML class diagrams usually show superclasses above subclasses. However, that's not always the case. With large projects, you'll have lots of classes and several different types of relationships among the classes. With all that going on, it's sometimes impossible to draw a "clean" class hierarchy picture and preserve the traditional superclass-above-subclass layout. Thus, subclasses sometimes appear at the left, at the right, or even above their superclasses. So how can you tell which is the subclass and which is the superclass? UML class diagrams use a solid line and a hollow arrow for inheritance relationships, with the arrow pointing to the superclass. In Figure 13.9, note how the arrows do indeed point to the superclasses.

Inheritance Terminology

Unfortunately, the terms *superclass* and *subclass* can be misleading. The "super" in superclass seems to imply that superclasses have more capability, and the "sub" in subclass seems to imply that subclasses have less capability. Actually, it's the other way around—subclasses have more capability. Subclasses can do everything that superclasses can do, and more.

For the most part, we'll stick with the terms *superclass* and *subclass* since those are the formal terms used by the designers of the Java language, but be aware that there is alternative terminology. Programmers often use the terms *parent class* or *base class* when referring to a superclass. And they often use the terms *child class* or *derived class* when referring to a subclass. The parent-child relationship between classes is important because it determines inheritance. With a human parent-child relationship, the child normally inherits physical features from the parent.[1] The class parent-child relationship parallels the human parent-child relationship. But with a class parent-child relationship, the child doesn't inherit money; instead, it inherits the variables and methods defined in the superclass.

There are two more inheritance-related terms that you should be aware of. An *ancestor* class refers to any of the classes above a particular class in an inheritance hierarchy. For example, in Figure 13.9's inheritance hierarchy, `Employee` and `Person` are the ancestors of `FullTime`. A *descendant* class refers to any of the classes below a particular class in an inheritance hierarchy. For example, in Figure 13.9's inheritance hierarchy, `Employee`, `Customer`, `FullTime`, and `PartTime` are descendants of `Person`.

Benefits of Inheritance

Long before reading this chapter, you were already convinced of the benefit of modeling your programs with classes, right? (In case you need to be reminded why you love classes so much, it's because classes allow you to encapsulate things.) So you should be able to see the benefit of having a `Customer` class and also an `Employee` class for a department store program. OK, having separate `Customer` and `Employee` classes is good, but why stir up trouble and give them a superclass? If there's no superclass for the `Customer` and `Employee` classes, then the things common to customers and employees would have to be defined in both classes. For example, you'd need a `name` instance variable and a `getName` method in both classes. But redundant code is almost always a bad idea. Why? With redundant code, debugging and upgrading chores are more tedious. After fixing or improving the code in one place, the programmer must remember to fix or improve the code in the other place as well.

In Figure 13.9, notice that classes at different levels in the hierarchy contain different instance variables, and they have different methods (although `Employee` and `FullTime` both have a `display` method, the methods are different; that is, they behave differently). There is no functional duplication, and there is maximal *code reusability*. Code reusability is when you have code that provides functionality for more than one part of a program. Putting common code from two classes into a superclass is one example of code reusability. Code reusability can also take place when you want to add a significant chunk of functionality to an existing class. You might want to implement the functionality by adding code directly to the existing class. But suppose the class works perfectly, and you're scared to touch it for fear of messing it up. Or maybe your know-it-all co-worker wrote the class, and you don't want to risk getting him or her riled up over code modi-

[1]Physical ailments can also be inherited. At age 30, author Ray tore his left Achilles tendon. Sons Stan and John did the same when they turned 30.

fications. No problem. Extend the class (that is, create a subclass) and implement the new functionality in the extended class.

Don't try to teach old dogs new tricks.

You've seen that inheritance gives rise to code reusability, and now you should be properly convinced of the benefits of code reusability. Another benefit of inheritance is that it gives rise to smaller modules (because classes are split into superclasses and subclasses). In general, smaller modules are good because there's less code to wade through when searching for bugs or making upgrades.

13.4 Implementation of a Person/Employee/FullTime Hierarchy

To explain how to implement inheritance, we'll implement the Person/Employee/FullTime hierarchy shown in Figure 13.9. We'll implement the Person and Employee classes in this section, and we'll implement the FullTime class in Section 13.6.

The Person class

Figure 13.10 contains an implementation of the Person class. It will be a superclass, but there's no special code in the Person class indicating that it will be a superclass. The special code comes later, when we define Person's subclasses. That's where we indicate that Person is a superclass for those subclasses.

```
/*************************************************************
* Person.java
* Dean & Dean
*
* This is a base class for an inheritance hierarchy.
*************************************************************/

public class Person
{
  private String name = "";

  //**********************************************************

  public Person()
  { }

  public Person(String name)
  {
    this.name = name;
  }

  //**********************************************************

  public String getName()
  {
    return this.name;
  }
} // end Person class
```

Remember: Once you write your own constructor, the zero-parameter default constructor disappears, and if you want one, you must write it explicitly.

Figure 13.10 Person class, a superclass for the Employee class

The Person class doesn't do much. It just provides a constructor that stores a name and an accessor that retrieves it. However, it contains one item worth examining—the zero-parameter constructor. Normally, when a driver instantiates a Person class, the driver will assign the person's name by passing a name argument to the one-parameter constructor. But suppose you want to test your program with a Person object, and you don't want to hassle with storing a name in the Person object. The zero-parameter constructor allows you to do that. Do you know what name will be given to a Person object created by the zero-parameter constructor? In this case, name is a string instance variable, and the default value for a string instance variable is null. To avoid the ugly null default, note how name is initialized to the empty string.

Quick quiz: Can you achieve the same functionality by omitting the zero-parameter constructor because the compiler automatically provides a default zero-parameter constructor? Nope—remember that once you write any constructor, the compiler no longer provides a default zero-parameter constructor.

The Employee Class

Figure 13.11 contains an implementation of the derived Employee class, which provides an id. Note the extends clause in the Employee class's heading. To enable inheritance, extends *<superclass>* must appear at the right of the subclass's heading. Thus, extends Person appears at the right of the

```
/*******************************************
* Employee.java
* Dean & Dean
*
* This class describes an employee.
*******************************************/

public class Employee extends Person
{
  private int id = 0;

  //****************************************

  public Employee()
  { }

  public Employee(String name, int id)
  {
    super(name);
    this.id = id;
  }

  //****************************************

  public void display()
  {
    System.out.println("name: " + getName());
    System.out.println("id: " + id);
  }
} // end Employee class
```

This means the Employee class is derived from the Person superclass.
This calls the one-parameter Person constructor.
Because name is in a different class and is private, we must use an accessor to get it. Because getName is inherited, we don't need a referencing prefix for it.

Figure 13.11 Employee class, derived from the Person class

Employee class's heading. Note that the Employee class defines just one instance variable, id. Does that mean that an Employee object has no name? No. Employee objects do have names because the Employee class inherits the name instance variable from the Person superclass. Now you'll learn how to access name from within the Employee class.

The Employee class's display method is in charge of printing an employee's information—name and id. Printing the id is easy because id is declared within the Employee class. Printing name requires a bit more work. Since name is a private instance variable in the Person superclass, the Employee class cannot access name directly (that's the same interpretation of private that we've always had). But the Employee class can access name by calling the Person class's public getName accessor method. Here's the relevant code from the display method:

```
System.out.println("name: " + getName());
```

As you might recall, in an instance method, if you call a method that's in the same class as the class you're currently in, the reference variable dot prefix is unnecessary. Likewise, in an instance method, if you call a method that's in the superclass of the class you're currently in, the reference variable dot prefix is unnecessary. Thus, there's no reference variable dot prefix in this call to getName.

While there is complete agreement about the way that a subclass object can access a private instance variable from its superclass (i.e., use a superclass public accessor method), there is disagreement about the terminology used to describe such access. Some textbook authors say that a subclass object inherits private instance variables from its superclass; others say that a subclass object does not inherit private instance variables from its superclass. We fall into the inheritance camp because an instantiated subclass contains private instance variables from the superclass. We know this because they are accessible (via public methods) without having to instantiate a second object for the superclass. In an explanation that would make a politician proud, Oracle first claims that private members are inherited, and then later claims that private members are not inherited (see what we mean by looking at the statement, at http://download.oracle.com/javase/tutorial/java/IandI/subclasses.html). No worries. The terminology doesn't matter, as long as you know how things work in terms of functionality.

13.5 Constructors in a Subclass

Let's now examine Figure 13.11's two-parameter Employee constructor. The goal is to assign the passed-in name and id values to the associated instance variables in the instantiated Employee object. Assigning to the id instance variable is easy because id is declared within the Employee class. But assigning to the name instance variable is harder because name is a private instance variable in the Person superclass. There's no setName mutator method in Person, so how does name get set? Read on. . . .

Using super to Call a Superclass Constructor

Employee objects inherit the name instance variable from Person. It follows that Employee objects should use the Person constructor to initialize their inherited name instance variables. But how can an Employee object call a Person constructor? It's easy—once you know how. To call a superclass constructor, use the reserved word super followed by parentheses and a comma-separated list of arguments that you want to pass to the constructor. For example, here's how Figure 13.11's Employee constructor calls the one-parameter Person constructor:

```
super(name);
```

Calls to `super` are allowed only in one particular place. They're allowed only from within a constructor, and they must be the first line within a constructor. That should sound familiar. In Chapter 7, you learned another usage for the keyword `this`, a usage that is distinct from using `this` dot to specify an instance member. The syntax for this other usage of `this` is:

```
this(<arguments>);
```

This kind of `this` usage calls another (overloaded) constructor from within a constructor in the same class. And recall that you must make such a call on the first line of your constructor.

By the way, would it be legal to have a `this` constructor call and a `super` constructor call within the same constructor? No, because with both constructor calls in the same constructor, that means only one of the constructor calls can be in the first line. The other one would violate the rule that constructor calls must be in the first line.

Default Call to Superclass Constructor

Java developers like to call superclass constructors because doing so promotes software reuse. If you write a subclass constructor and don't include a call to another constructor (with `this` or with `super`), the Java compiler sneaks in and inserts a superclass zero-parameter constructor call by default. Thus, although Figure 13.11's `Employee` zero-parameter constructor has an empty body, the Java compiler automatically inserts `super();` in it. So these two constructors are functionally equivalent:

```
public Employee()
{ }

public Employee()
{
  super();
}
```

The explicit `super();` call makes it clear what's going on. Feel free to include it if you wish, to make your code more self-documenting.

Whenever a constructor is called, the JVM automatically runs up the hierarchical tree to the greatest-grandparent's constructor, and it executes that greatest-grandparent's constructor first. Then it executes the code in the constructor below it, and so on, and finally, it executes the rest of the code in the originally called constructor.[2]

13.6 Method Overriding

From Chapter 7, you know about method overloading. That's when a single class contains two or more methods with the same name but a different sequence of parameter types—approximately the same semantics, but different syntax. Now for a related concept—*method overriding*. That's when a subclass has a method with the same name, the same sequence of parameter types, and the same return type as a method in a superclass. An overriding method has the same syntax but different semantics. The term "overriding" should make sense when you realize that an overriding method overrides/supersedes its associated superclass

[2]This sequence is the same as the sequence that occurs naturally in the embryonic development of a living creature. The characteristics that develop first are the most ancient ones.

method. That means, by default, an object of the subclass uses the subclass's overriding method and not the superclass's overridden method.

The concept of a subclass object using the subclass's method rather than the superclass's method falls in line with this general principle of programming: Local stuff takes precedence over global stuff. Can you think of where else this rule applies? If a local variable and an instance variable have the same name, the local variable takes precedence when you're inside the local variable's method. The same reasoning applies to parameters taking precedence over instance variables when you're inside the parameter's method.

Method Overriding Example

To explain method overriding, we'll continue with the implementation of the Person/Employee/FullTime program. We implemented the `Person` and `Employee` classes in Section 13.4. We implement the `FullTime` class in Figure 13.12. Note `FullTime`'s `display` method. It has the same sequence of

```
/*************************************************************
* FullTime.java
* Dean & Dean
*
* This class describes a full-time employee.
*************************************************************/

public class FullTime extends Employee
{
  private double salary = 0.0;

  //**********************************************************

  public FullTime()
  { }

  public FullTime(String name, int id, double salary)
  {
    super(name, id);
    this.salary = salary;
  }

  //**********************************************************

  @Override
  public void display()
  {
    super.display();
    System.out.printf(
      "salary: $%,.0f\n", salary);
  }
} // end FullTime class
```

Callouts:
- This calls the two-parameter `Employee` constructor. (points to `super(name, id);`)
- This method overrides the `display` method defined in the `Employee` class. (points to `public void display()`)
- This calls the `display` method defined in the `Employee` class. (points to `super.display();`)

Figure 13.12 `FullTime` class, which illustrates method overriding

parameter types as the `display` method in the `Employee` class of Figure 13.11. Because the `FullTime` class `extends` the `Employee` class, the `FullTime` class's `display` method overrides the `Employee` class's `display` method. The Java annotation, `@Override`, asks the compiler to confirm that the method is indeed an overriding method. This is not required, but it provides documentation and helps avoid misunderstanding.

Using `super` to Call an Overridden Method

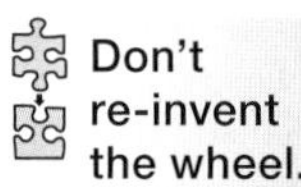
Don't
re-invent
the wheel.

When you override a method, you still can use the overridden method for the services it provides and then provide new code only for additional services. To do this, start your new method with a statement that calls the overridden method by prefixing the overridden method name with `super` dot. During program development, you can tentatively apply `@Override` annotations just above new method headers to see if ancestral methods exist that might already provide some of the services you need. Then, if the compiler does not find an overridden method, it will tell you that, and you can discard your tentative `@Override` annotation.

For example, in Figure 13.12's `FullTime` subclass, note how the `display` method calls the superclass's `display` method with `super.display();`.

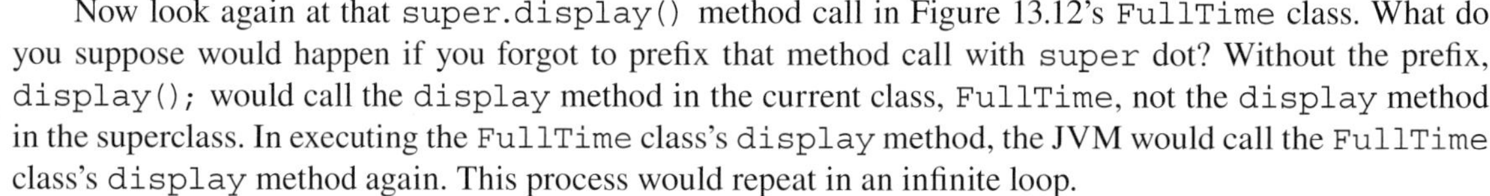

Now look again at that `super.display()` method call in Figure 13.12's `FullTime` class. What do you suppose would happen if you forgot to prefix that method call with `super` dot? Without the prefix, `display();` would call the `display` method in the current class, `FullTime`, not the `display` method in the superclass. In executing the `FullTime` class's `display` method, the JVM would call the `FullTime` class's `display` method again. This process would repeat in an infinite loop.

By the way, you can have a series of overriding methods; that is, you can override an overriding method. But it's illegal to have a series of `super` dot prefixes chained together. In other words, in the `Person/Employee/FullTime` inheritance hierarchy, suppose the `Person` class contains a `display` method that's overridden by the `Employee` and `FullTime` classes. In the `FullTime` class, it would be illegal to call the `Person` class's `display` method like this:

```
super.super.display();   ◄── compilation error
```

To call the `Person` class's `display` method from the `FullTime` class, you'd have to call the `Employee` class's `display` method and rely on the `Employee` class's `display` method to call the `Person` class's `display` method.

Have you noticed that `super` has two different purposes? You can use `super` dot to call an overridden method, and you can also use `super` with parentheses as in `super(name)` to call a superclass's constructor.

Return Types Must Be the Same

An overriding method must have the same return type as the method that it's overriding. If it has a different return type, the compiler generates an error. Said another way, if a subclass and a superclass have methods with the same name, the same sequence of parameter types, and different return types, the compiler generates an error.

This error doesn't occur all that often because if you've got methods with the same names and sequences of parameter types, you'll usually also want the same return types. But you'll see the error crop up every now and then when you're debugging, so be aware of it. By the way, if a subclass and a superclass have methods with the same name and different sequences of parameter types, it doesn't matter if the return types are the same. Why? Because such methods are not in an overriding relationship. They are different methods entirely.

13.7 Using the Person/Employee/FullTime Hierarchy

Now let's reinforce what you've learned about inheritance by looking at what happens when you instantiate an object of the lowest-level derived type and use that object to call overridden methods and inherited methods. Figure 13.13 contains a driver for the FullTime class, and the subsequent output shows what it does. This driver instantiates a fullTimer object from the FullTime class. Then the fullTimer object calls its display method. As shown in Figure 13.12, this display method uses super to call the Employee class's display method, which prints the fullTimer's name and id. Then fullTimer's display method prints the fullTimer's salary.

In the final statement in Figure 13.13, the fullTimer object calls its getName method and prints fullTimer's name. But wait a minute! The FullTime class does not have a getName method, and its superclass, Employee, does not have one either. The code seems to be calling a nonexistent method. What's going on here? What's going on is inheritance—produced by those wonderful little extends clauses. Because there is no explicitly defined getName method in its own FullTime class, the fullTimer object goes up its inheritance hierarchy until it finds a getName method, and then it uses that method. In this case, the first getName method found is in the Person class, so that's the method the fullTimer object inherits and uses. There is no need to use super dot to access the getName method (but using super dot

```
/*************************************************************
* FullTimeDriver.java
* Dean & Dean
*
* The describes a full-time employee.
*************************************************************/

public class FullTimeDriver
{
  public static void main(String[] args)
  {
    FullTime fullTimer = new FullTime("Shreya", 5733, 80000);

    fullTimer.display();
    System.out.println(fullTimer.getName());
  }
} // end FullTimeDriver class
```

Output:

```
name:   Shreya
id:     5733
salary: $80,000
Shreya
```

Figure 13.13 Driver of constructors and methods in an inheritance hierarchy

would work, in case you're curious). If a method is not in the current class, the JVM automatically goes up the inheritance hierarchy and uses the first definition of that method it finds.

Notice that our driver did not instantiate any `Employee` or `Person` objects. It just instantiated an object from a class at the bottom of the inheritance hierarchy only. This is the way a good inheritance hierarchy should be used. Ideally, you should just instantiate objects from classes at the bottom of the hierarchy. Ideally, all the classes above the bottom classes are there to make the bottom classes simple. In real life, we often do use classes above the bottom, but using bottom classes only is the ideal situation.

13.8 The `final` Access Modifier

You've used the `final` access modifier for quite a while now to turn a variable into a named constant. In this section, you'll learn how to use `final` to modify a method and modify a class.

If you use the `final` modifier in a method heading, you'll prevent the method from being overridden with a new definition in a subclass. You might want to do this if you think that your method is perfect and you don't want its original meaning to "drift." You might also want to consider using `final` to help speed things up a bit. Methods that use the `final` modifier should run faster because the compiler can generate more efficient code for them. The code efficiency comes from the compiler not having to prepare for the possibility of inheritance. However, the speed improvement is miniscule for adding `final` to a single method, and you probably won't notice it unless you have a large programming project with lots of subclasses and you use `final` a lot.

If you use the `final` access modifer in a class heading, you prevent the class from having any subclasses. You might want to do this if you have a class that's good and reliable, and you want to preserve its quality and protect it from future "feature creep." By the way, if a class is declared to be a `final` class, there's no point in specifying `final` for any of its methods. A `final` class cannot be extended, so overriding methods cannot exist. The `final` modifier helps prevent invasion by hackers.

Even though it may be difficult to see palpable benefits from the use of `final`, you should use it to improve security. Even if you don't use it for your own programs, you'll need to understand it because you'll see it quite often in the Java API library classes. For example, the `Math` class is defined with the `final` access modifier, so it's illegal to extend the `Math` class and override any of its methods.

13.9 Using Inheritance with Aggregation and Composition

We have described several ways classes can be related—with aggregation, composition, and inheritance. Now let's consider using all three relationships together.

Aggregation, Composition, and Inheritance Compared

Aggregation and composition both implement a has-a relationship. We call aggregation and composition relationships *has-a* relationships because one class, the container class, has a component class inside it. For example, in Section 13.2's Dealership program, a dealership has a sales manager, with non-exclusive ownership rights, and that's why the Dealership program implements the `Dealership-SalesManager` relationship with aggregation. Also, a dealership has an inventory of cars, with exclusive ownership rights, and that's why the Dealership program implements the `Dealership-Car` relationship with composition.

Inheritance implements an is-a relationship. We call an inheritance relationship an *is-a* relationship because one class, a subclass, is a more detailed version of another class. For example, in the Person/Employee/FullTime program, a full-time employee is an employee, and that's why the program implements the `FullTime-Employee` relationship with inheritance. Also, an employee is a person, and that's why the program implements the `Employee-Person` relationship with inheritance as well.

It's important to keep in mind that these are not alternative ways to represent the same relationship. They are ways to represent different relationships. The aggregation and composition relationships are when one class is a whole made up of non-trivial constituent parts defined in other classes. The inheritance relationship is when one class is a more detailed version of another class. More formally, inheritance is when one class, a subclass, inherits variables and methods from another class, a superclass, and then supplements those with additional variables and methods. Because composition and inheritance deal with different aspects of a problem, many programming solutions include a mixture of both paradigms.

Aggregation, Composition, and Inheritance Combined

In the real world, it's fairly common to have aggregation, composition, and inheritance relationships together in the same program. Let's look at an example that uses all three class relationships. Section 13.2's Dealership program uses aggregation and composition, and we repeat it here, with the non-Dealership aggregation relationships omitted for simplicity.

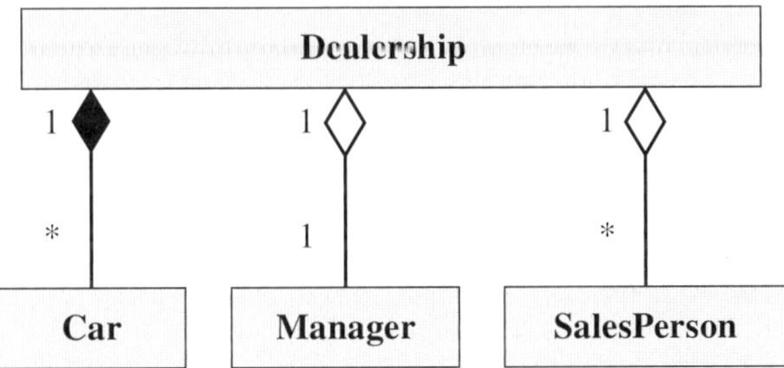

What sort of inheritance relationship could or should be added to the Dealership program? If you look back at Figures 13.5 (the `Manager` class) and 13.6 (the `SalesPerson` class), you'll see that `Manager` and `SalesPerson` both declare the same instance variable, `name`, and they both define the same instance method, `getName`. That's an example of undesirable duplication, and we can use inheritance to eliminate that duplication. Introducing inheritance into that program does not alter the original whole-parts structure. It just introduces a complementary mechanism that eliminates duplication.

Factor out the common code.

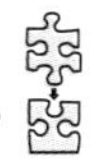

Figure 13.14 shows an improved and expanded UML class diagram for a new Dealership2 program. If you compare this with the previous UML class diagram, you'll see that each class is fleshed out to include instance variables and methods. Figure 13.14's diagram also includes a `Person` class. Our previous `Manager` and `SalesPerson` classes now inherit a variable, two constructors, and a method from this `Person` class. The inheritance reduces the `Manager` and `SalesPerson` classes to the simpler `Manager2` and `SalesPerson2` classes. These simpler classes do not need explicit declaration of `name` and explicit definition of `getName` because they inherit these members from `Person`. Read through the code for the shortened `Manager2` and `SalesPerson2` classes in Figures 13.15 and 13.16.

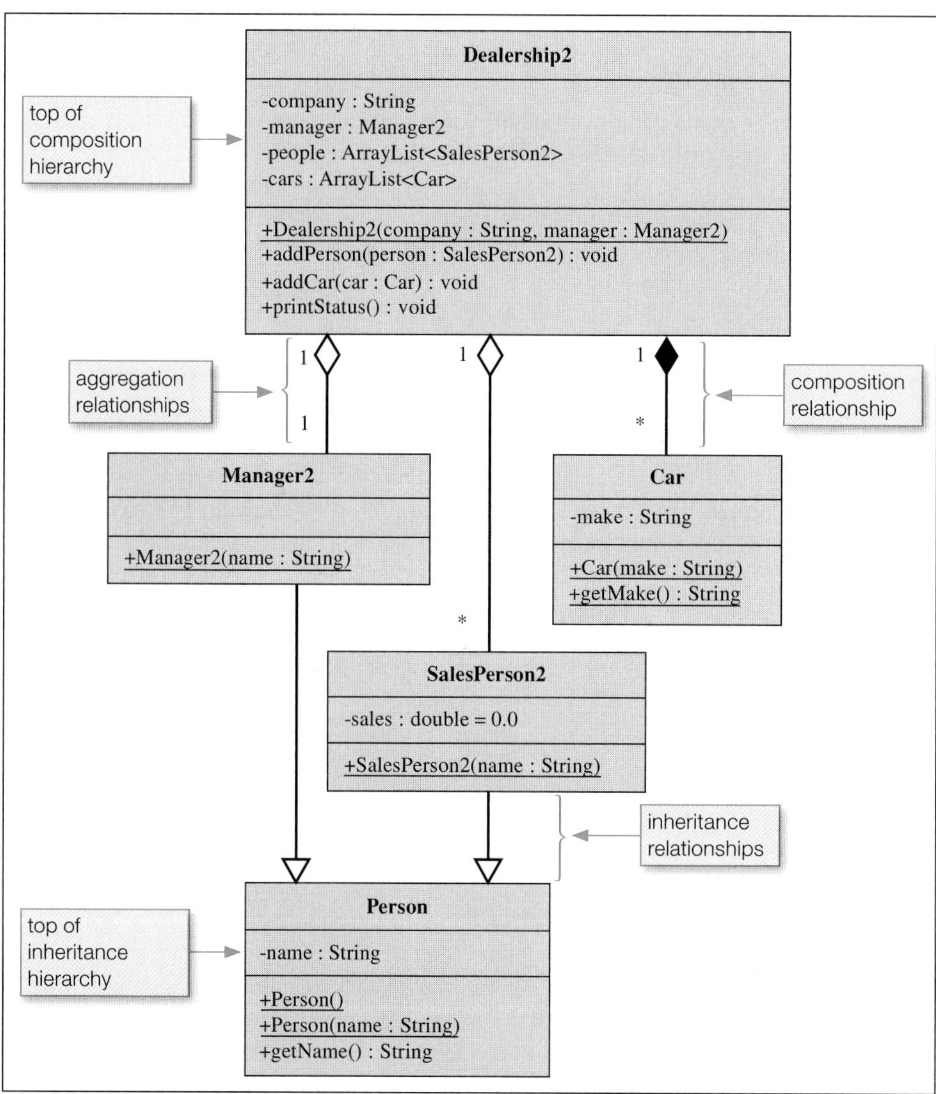

Figure 13.14 Class diagram for the revised car dealership program—Dealership2

The `Car` class is unchanged from the original Dealership program; if you want to see its code, look back at Figure 13.4. The `Dealership2` and `Dealership2Driver` classes are the same as the `Dealership` and `DealershipDriver` classes defined in Figures 13.3 and 13.7, respectively, except that `Dealership` is changed to `Dealership2`, `Manager` is changed to `Manager2`, and `SalesPerson` is changed to `SalesPerson2`.

In Figure 13.14, the addition of the `Person` class makes it look as though we made the Dealership2 program bigger by adding another class. But the additional `Person` class was already defined in another program, the Person/Employee/FullTime program. In borrowing the `Person` class from that program, we

```
/*******************************************************
* Manager2.java
* Dean & Dean
*
* This represents a car dealership manager.
*******************************************************/

public class Manager2 extends Person
{
  public Manager2(String name)
  {
    super(name);
  }
} // end Manager2 class
```

Figure 13.15 `Manager2` class for the Dealership2 program

```
/*********************************************************
* SalesPerson2.java
* Dean & Dean
*
* This represents a car salesperson.
*********************************************************/

public class SalesPerson2 extends Person
{
  private double sales = 0; // sales to date

  //*****************************************************

  public SalesPerson2(String name)
  {
    super(name);
  }
} // end SalesPerson2 class
```

Figure 13.16 `SalesPerson2` class for the Dealership2 program

got something for nothing. The borrowed `Person` class enabled us to shorten two other classes. Being able to borrow classes that have already been written and then inheriting from them in other contexts is an important benefit of OOP. If you look at the prewritten classes in the Java API, you'll see that they do a lot of inheriting from one to another, and in many cases, you have the option of inheriting from them into your own programs as well.

13.10 Design Practice with Card Game Example

In the previous section, you learned how to use different types of class relationships together in a single program. The way you learned was adding inheritance to an existing program. In this section, you'll once again use different types of class relationships, but this time you'll design the program from the ground up. And you'll be doing most of the work, rather than just understanding how it's done by someone else.

Learn by doing.

Your Mission (Should You Choose to Accept It)

Your mission is to design and implement a generic card game program. In carrying out this mission, follow these guidelines:

- Assume it's a game like War or Gin Rummy, where you have a deck of cards and two players.
- Decide on appropriate classes. For each class, draw a UML class diagram and write in the class name.
- Look for composition relationships between classes. For each pair of classes related by composition, draw a compositional association line with appropriate multiplicity values.
- For each class, decide on appropriate instance variables.
- For each class, decide on appropriate `public` methods.
- Look for common instance variables and methods. If two or more classes contain a set of common instance variables and methods, provide a superclass and move the common instance variables and methods to the superclass. The classes originally containing common members now become subclasses of the superclass. For each subclass-superclass pair, draw an association line with an inheritance arrow from the subclass to the superclass to indicate an inheritance relationship.

Now go ahead and use these guidelines to draw a UML class diagram for a generic card game program. Because this is a non-trivial exercise, you may be tempted to look at our solution before trying to come up with a solution on your own. Please resist that temptation! By implementing your own solution, you'll learn more and make yourself aware of potential problems.

Defining the Classes and the Relationships Between Them

Have you finished your class diagram? If so, then you may continue. . . .

In coming up with a class diagram, the first thing to do is to decide on the classes themselves. Unfortunately, that's a bit of an art. The easy classes are the ones that directly correspond to something you can see. In visualizing a card game, can you see two people holding cards and taking additional cards from a deck that sits between them? You should be able to see a deck, two hands, individual cards, and two people. For the deck, use a `Deck` class. For the two hands, use a `Hand` class. For the individual cards, use a `Card` class. You may or may not wish to represent the people. If you're implementing an elaborate card game where players have personalities, use a `Person` class. Otherwise, there's no need for a `Person` class. Let's keep things simple and not implement a `Person` class.

In thinking about the big picture, you should ask yourself, "What is a game?" A game is a composition of several parts, so define `Game` as a whole class and define other classes as the parts of the game. A `Game` is composed of three components/parts—a deck and two hands. Thus, `Deck` and `Hand` are part classes within the `Game` composition class. In Figure 13.17's class diagram, note the association line connecting `Game` to `Deck`. The association line has a solid diamond, which indicates composition, and it has 1-to-1 multiplicity values, which indicate each game has one deck. The `Game` to `Hand` association line also has

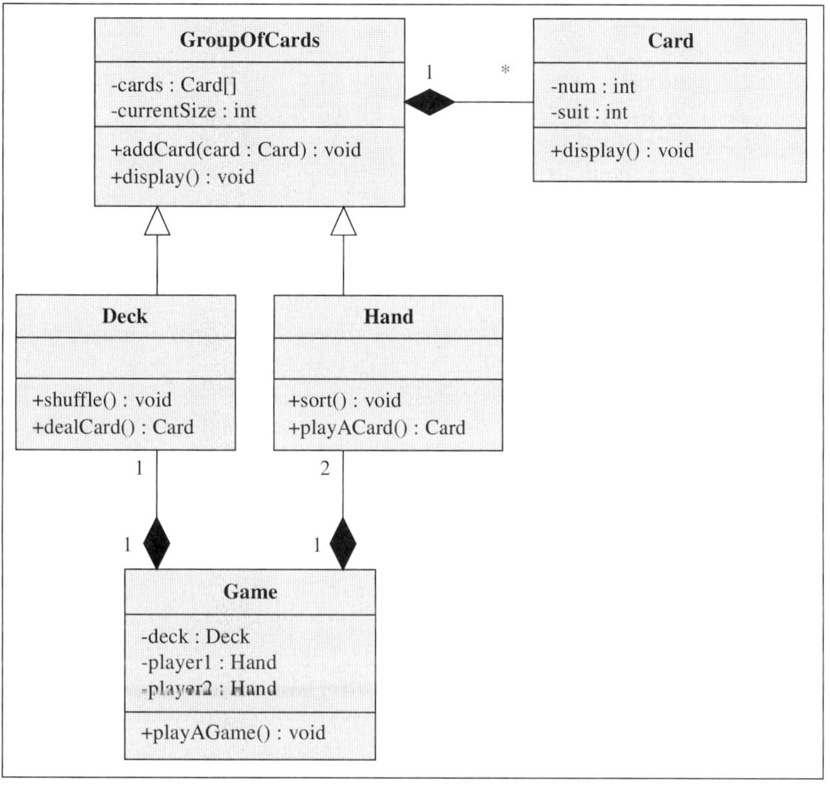

Figure 13.17 Preliminary class diagram for a Card Game program

a solid diamond for composition, but it has 1-to-2 multiplicity values, which indicate each game has two hands.

Coming up with the idea of using a `Game` class is probably more difficult than coming up with the ideas for using `Deck`, `Hand`, and `Card` classes. Why? A game is non-tactile (that is, you can't touch it), so it's hard to see it as a class. Why bother with having a `Game` class? If you omit the `Game` class, you could still implement a card game. Instead of declaring `Deck` and `Hand` objects inside the `Game` class, you could declare them inside the `main` method. But it's more elegant to put them inside a `Game` class. Why? By putting them inside a `Game` class, it furthers the goal of encapsulation. Also, it enables `main` to be streamlined. As you'll see later on, if you have defined a `Game` class, the driver's `main` method just needs to instantiate a `Game` object and then call `playAGame`—that's it. You can't get much more streamlined (and elegant) than that.

For each class in the Card Game program, what are its members (that is, instance variables and methods)? Let's tackle the easy classes first—`Game` and `Card`. The `Game` class needs three instance variables—one for the deck and two for the two hands. It needs a method for playing a game. The `Card` class needs two instance variables—one for a number (two through ace) and one for a suit (clubs through spades). It needs a method to display the card's number and suit values. As a sanity check, verify that Figure 13.17's `Game` and `Card` members match what we've described.

The `Deck` class needs an instance variable for an array of cards such that each card is a `Card` object. The `Deck` class also needs an instance variable to keep track of the current size of the deck. The `Deck` class needs methods for shuffling and dealing. To help with debugging, you should probably also include a method to display all the cards in the deck.

The Hand class needs instance variables for an array of cards and for a current-size value. It needs methods for displaying all the cards, adding a card to the hand, and playing a card from the hand. For most card games, you'd also want a method to sort the hand. Different card games would use different and/or additional Hand methods. We'll keep things simple and not worry about them.

The next step is to try to identify common members and move them to a superclass. The Deck and Hand classes have three common members—a cards array variable, a currentSize variable, and a display method. In moving those members to a superclass, what would be a good name for such a class? It should be something generic that can be used as the superclass for both Deck and Hand. Either GroupOfCards or just plain Cards sounds pretty good. Let's use GroupOfCards. In Figure 13.17's class diagram, note the inheritance association lines connecting Deck and Hand to GroupOfCards.

We've now examined the members in all five classes in the Card Game program, and we've examined the relationships between four of the classes—Game, Deck, Hand, and GroupOfCards. The last piece of the UML-class-diagram puzzle is the relationship between GroupOfCards and Card. Is it an is-a or a has-a relationship? It's not an is-a relationship because it doesn't make sense to say that a group of cards is a card or a card is a group of cards. Instead, it's a has-a relationship because a group of cards has a card (a group of cards usually has more than one card, but that doesn't negate the has-a relationship). In Figure 13.17, note the has-a composition association line connecting GroupOfCards to Card. Figure 13.17 suggests implementing the composition as an array called cards, but it could be an ArrrayList

Note that Figure 13.17's label says "preliminary" class diagram. It's preliminary because for a decent-sized application, it's nearly impossible to get the class diagram 100% right on your first-cut attempt. When you're done coding and testing(!) your prototype program, you should go back and update your class diagram appropriately. The class diagram serves two purposes. Early in the design process, it helps organize ideas and it keeps everybody on the same page. In the post-implementation phase, it serves as documentation that helps interested parties quickly get a handle on the application's organization.

Design is an iterative process.

Inheritance Versus Composition

When deciding on the relationship between two classes, it's usually pretty clear whether to use inheritance or composition. For example, in the Dealership program, a Manager is a Person, so inheritance is used. In the Card Game program, a Game has a Deck, so composition is used.

However, sometimes it's not so clear-cut. For example, you could make the claim that a Deck is a GroupOfCards, and you could also make the claim that a Deck has a GroupOfCards. Where an inheritance is-a relationship and a composition has-a relationship both exist, opinions differ as to which strategy is better. Inheritance implementations are more understandable. Composition implementations are more encapsulated. (If you implement inheritance using the protected modifier for superclass data, the superclass will have direct access to the subclass data, and that breaks the superclass's encapsulation somewhat.) Let's compare the two strategies in the context of the Card Game program.

Figure 13.18's Deck class implements the relationship with inheritance. Figure 13.19's alternative Deck class implements it with composition. We feel that Figure 13.18's inheritance code is more elegant than Figure 13.19's composition code. It has one less line, which is a good thing, but more importantly, it isn't cluttered with references to a groupOfCards variable. In the composition code, you're required to:

```
public class Deck extends GroupOfCards
{
  public static final int TOTAL_CARDS = 52;

  public Deck()
  {
    for (int i=0; i<TOTAL_CARDS; i++)
    {
      addCard(new Card((2 + i%13), i/13));
    }
  } // end constructor
  ...

} // end class Deck
```

This implements inheritance.

With inheritance, there's no need to prefix the method call with an object reference.

Figure 13.18 Inheritance implementation for the `Deck` class

```
public class Deck
{
  public static final int TOTAL_CARDS = 52;
  private GroupOfCards groupOfCards;

  public Deck()
  {
    groupOfCards = new GroupOfCards();

    for (int i=0; i<TOTAL_CARDS; i++)
    {
      groupOfCards.addCard(new Card((2 + i%13), i/13));
    }
  } // end constructor
  ...

} // end class Deck
```

With composition, declaring a `GroupOfCards` variable and instantiating it are required.

With composition, you must prefix the method call with an object reference.

Figure 13.19 Composition implementation for the `Deck` class

(1) declare a `groupOfCards` variable, (2) instantiate the `groupOfCards` variable, and (3) prefix the call to `addCard` with the `groupOfCards` calling object. Isn't the inheritance code nicer, where you don't have to worry about all that? In particular, you can call `addCard` directly (no `groupOfCards` calling object required), and that results in more readable code. By the way, the `addCard` method is defined in the `GroupOfCards` class. With inheritance, the fact that it's in a separate class from `Deck` is transparent. In other words, you call `it` from the `Deck` constructor the same way that you would call any other `Deck` method—without a calling object.

For some class pairs (like Deck and GroupOfCards), it's legal to use either an inheritance or a composition relationship. But it's never OK to use both inheritance and composition for the same feature. What would happen if Deck declared a GroupOfCards local variable and Deck also inherited from a GroupOfCards class? Deck objects would then contain two separate groups of cards and that's wrong!

At this point, you might want to go back to Figure 13.17's preliminary UML class diagram and add more detail. We didn't bother with constants or constructors in Figure 13.17's class diagram. In working with the Deck class skeleton (see Figure 13.18), it's now clear there's a need to (1) add a TOTAL_CARDS constant to the Deck class, (2) add a constructor to the Deck class, and (3) add a constructor to the Card class. For practice, we encourage you to update Figure 13.17's class diagram with these changes in mind. If you don't feel like it, that's OK; our main point here is to make you aware of the iterative nature of the program design process. Try to organize your thoughts as clearly as possible up front, but be prepared to adjust those thoughts later on.

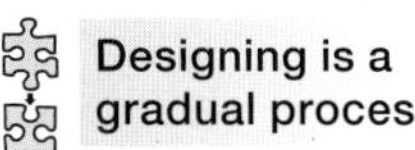
Designing is a gradual process.

Code to Get You Started

Once you've finished with the card game's class diagram, normally the next step would be to implement the classes with Java code. We won't bother to show class implementation details, but we would like to show you how the suggested classes might be driven by a main method. Having followed proper OOP design guidelines, it's easy to produce an elegant main method—see Figure 13.20. Note how short and understandable the main method is. Yeah!

```
public static void main(String[] args)
{
  Scanner stdIn = new Scanner(System.in);
  String again;
  Game game;

  do
  {
    game = new Game();
    game.playAGame();
    System.out.print("Play another game (y/n)?: ");
    again = stdIn.nextLine();
  } while (again.equals("y"));
} // end main
```

Figure 13.20 Card Game program's main method

Another example will illustrate further how the finished classes might be used by the rest of the program. See main's call to playAGame in Figure 13.20. Figure 13.21 shows a partial implementation for the playAGame method. To shuffle the deck, call deck.shuffle(). To deal a card to the first player, call player1.addCard(deck.dealCard()). How's that for straightforward?

We'll leave it to you to finish this program. Two end-of-chapter exercises and a project suggest various elaborations.

```
public void playAGame()
{
  deck.shuffle();

  // Deal all the cards to the two players.
  while (deck.getCurrentSize() > 0)
  {
    player1.addCard(deck.dealCard());
    player2.addCard(deck.dealCard());
  }
  ...

} // end playAGame
```

Figure 13.21 Partial implementation for the `Game` class's `playAGame` method

13.11 GridWorld Case Study Extensions (Optional)

At the end of the previous chapter, in Section 12.16, we introduced the GridWorld case study. The GUI out-put in that section's Figure 12.15 was generated by running this driver:[4]

```
public class BugRunner
{
  public static void main(String[] args)
  {
    ActorWorld world = new ActorWorld();
    world.add(new Bug());
    world.add(new Rock());
    world.show();
  } // end main
} // end class BugRunner
```

In this code, each `add` method call selects a random location in the range of Figure 12.15's two-dimensional grid. Then it asks its parameter (`Bug` or `Rock`) to execute the `putSelfInGrid` method inherited from the `Actor` class. The `putSelfInGrid` method calls the `put` method in the `BoundedGrid` class. This `put` method inserts the `Bug` or `Rock` object into a two-dimensional `occupantArray` at the selected location.

The `show` method call instantiates a `WorldFrame`, which creates the GUI window. `WorldFrame`'s constructor instantiates a `GridPanel` object. An argument passed to the `WorldFrame` constructor gives the `WorldFrame` object a reference back to the `ActorWorld` object that created it. The `WorldFrame` constructor uses this reference to get a reference to the `BoundedGrid`. The `WorldFrame` passes this `BoundedGrid` reference on to the new `GridPanel`. `GridPanel`'s `drawOccupants` method uses this

[4]The `BugRunner` class is in the `GridWorld.jar` file in the `projects/firstProject` directory. The other classes used by this program are in the same `.jar` file in the `framework/info/gridworld` directory in the subdirectories: `actor`, `grid`, `gui`, and `world`.

BoundedGrid reference to call BoundedGrid's getOccupiedLocations method. This method tells GridPanel's paintComponent method where to put Actor images each time it repaints the screen.

Actor Images

The Actor images come from .gif files, where "gif" stands for Graphic Interchange Format.[5] Figure 13.22a shows the images contained in the framework/info/gridworld/actor subdirectory.

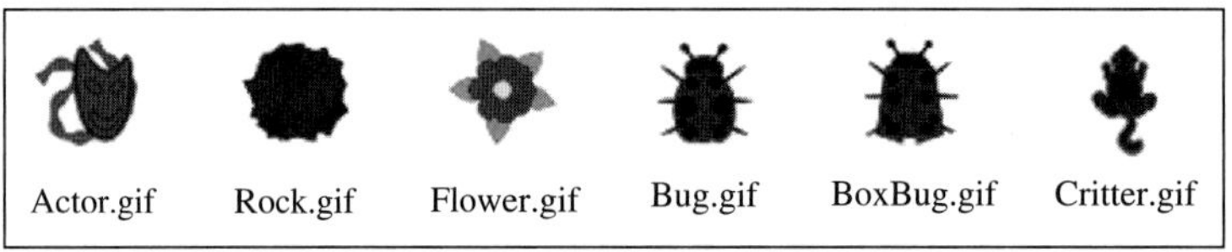

Figure 13.22a GridWorld images in framework/info/gridworld/actor

Figure 13.22b shows the images contained in the projects/critters subdirectory.

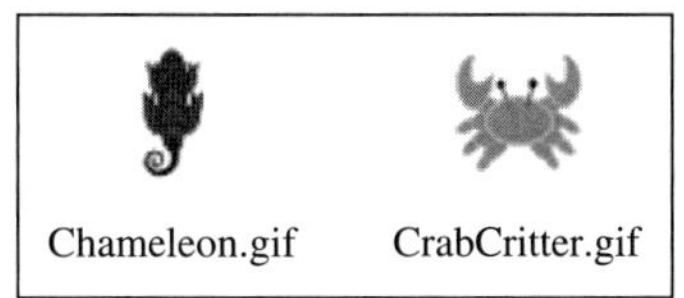

Figure 13.22b GridWorld images in projects/critters

GridWorld's GUI code initially assumes that the appropriate image filename is the corresponding actor's class name followed by .gif. If it can't find a .gif file starting with that name in the same directory as the actor's class, it goes up to that actor's superclass and looks in that superclass's directory for that superclass name followed by .gif, and so forth, up the actor's inheritance hierarchy.

Actor Behavior

When it comes to actor behavior, GridWorld's pivotal method is the act method. In the base class for actors, Actor, the act method doesn't do much. It just reverses the actor's direction, like this:

```
/**
* Reverses the direction of this actor. Override this method in
* subclasses of <code>Actor</code> to define types of actors with
* different behaviors.
*/
public void act()
{
  setDirection(getDirection() + Location.HALF_CIRCLE);
} // end act
```

[5] See Section 5.9 in Chapter 5 for more information about image formats.

But this is not how a Bug in the BugRunner program acts. In that program, each bug moves forward until it hits a grid boundary. Then, if not impeded by rocks, it runs clockwise around the grid's boundary. To get this behavior, the Bug class extends the Actor class and overrides the act method like this:

```
/**
* Moves if it can move, turns otherwise.
*/
@Override
public void act()
{
  if (canMove())
  {
    move();
  }
  else
  {
    turn();
  }
} // end act
```

In GridWorld's projects/boxBug directory, you'll find source code for GridWorld's BoxBug class and its driver, BoxBugRunner. The BoxBug class extends the Bug class. BoxBug's constructor accepts a length value, and its overriding act method causes BoxBug objects to turn 90 degrees clockwise after length steps in any one direction. BoxBugs with length < 9 eventually run around square boundaries, some of which are inside the grid's boundaries. It's not hard to imagine how you might modify BoxBug and extend Bug differently for the CircleBug and SpiralBugclasses.[6]

The Critter Class

GridWorld's framework/info/gridworld/actor directory also contains code for a Critter class, which extends the Actor class. You can drive Critter from CritterRunner, which is in GridWorld's projects/critters directory. Critter's extension of Actor has greater versatility than Bug's extension of Actor. Instead of specifying detailed behavior within its act method, the Critter class delegates detailed specification to five subordinate methods, like this:

```
/**
* A critter acts by getting a list of other actors, processing
* that list, getting locations to move to, selecting one of them,
* and moving to the selected location.
*/
@Override
public void act()
{
  if (getGrid() == null)
  {
    return;
  }
```

[6]These are two of the exercises in GridWorld's "Student Manual," written by Chris Nevison, Barbara Cloud Wells, and Chris Renard.

```
    ArrayList<Actor> actors = getActors();
    processActors(actors);
    ArrayList<Location> moveLocs = getMoveLocations();
    Location loc = selectMoveLocation(moveLocs);
    makeMove(loc);
  } // end act
```

In GridWorld's `Critter` class, each of the five delegate methods is very simple. The `getActors` method returns all immediate neighbors—actors in adjacent squares. The `processActors` method removes all parameter elements that are not of type `Rock` or of type `Critter`—eats neighbors who are edible and different. The `moveLocs` method returns all empty adjacent locations. The `selectMoveLocation` randomly selects one of its parameter elements—a valid empty adjacent location. If there is none, it returns the current location. The `selectMoveLocation` moves to the parameter location—the previously selected location.

You can see `Critter` behavior by executing the `CritterRunner` class in the GridWorld directory, `projects/critters`. Each of the five delegate methods in the `Critter` class does something reasonable. However, if you read the code's comments, you'll see that the designer expects us to `extend` the `Critter` class and override one or more of these five methods. They are *default* methods.

GridWorld Extensions of the `Critter` Class

GridWorld's first extension of the `Critter` class is its `ChameleonCritter` class. This class overrides just two of the five delegate methods, `processActors` and `makeMove`. In its `processActors` method, instead of eating its neighbors, a `ChameleonCritter` changes its color to match the color of a neighbor or one of several neighbors selected at random. In its `makeMove` method, a `ChameleonCritter` turns toward the location it will move to before calling `super.makeMove` to actually make the move. You can see how this works by executing `ChameleonRunner`, which is also in the `projects/critters` directory.

GridWorld's second extension of the `Critter` class is its `CrabCritter` class. This class and its driver, `CrabRunner`, are also in the `projects/critters` directory. `CrabCritter` differs from `Critter` in that it eats only those neighbors who are directly ahead or on either side of the position directly ahead. It differs also in that it can only move sideways, either left or right. If both side locations are available, it randomly chooses one of them. If neither is available, it turns by 90 degrees, randomly either left or right.

`CrabCritter2` Extension of the `Critter` Class

Looking at either BugRunner or CrabRunner output, you'll notice that eventually bugs tend to run clockwise around the grid's boundary. In contrast, CrabRunner's crabs tend to jump back and forth in the middle. Put yourself in the crabs' shells. Suppose you want to catch and eat bugs and flowers more efficiently than a `CrabCritter`. Maybe it would be better to hunt along the boundary and move around that boundary in the counterclockwise direction. Then bugs would be running right into your waiting claws.

These considerations motivate us to create a different breed of crab. Figure 13.23a contains the beginning of an alternate extension of `Critter` called `CrabCritter2`. `CrabCritter2` uses the same imports and has the same constructor as the original `CrabCritter`. The `getActors` method is also essentially the same, although we use a different style. Because we think it's more informative, we use longer variable names, and we change the name of `CrabCritter`'s `getLocationsInDirections` method to `getValidLocations`.

```
/***************************************************************
* CrabCritter2.java
* Dean & Dean
*
* This modifies GridWorld crabs so that they turn half-left.
* GridWorld is distributed under the terms of the GNU General
* Public License as published by the Free Software Foundation.
***************************************************************/

import info.gridworld.actor.*;   // Actor, Critter
import info.gridworld.grid.*;    // Location, Grid
import java.awt.Color;
import java.util.ArrayList;

public class CrabCritter2 extends Critter
{
  public CrabCritter2()
  {
    setColor(Color.RED);
  } // end constructor

  //************************************************************

  @Override
  public ArrayList<Actor> getActors()
  {
    ArrayList<Actor> actors = new ArrayList<>();
    int[] directions =
      {Location.AHEAD, Location.HALF_LEFT, Location.HALF_RIGHT};
    Actor actor;

    for (Location adjacentLoc : getValidLocations(directions))
    {
      actor = getGrid().get(adjacentLoc);
      if (actor != null)
      {
        actors.add(actor);
      }
    }
    return actors;
  } // end getActors
```

Figure 13.23a CrabCritter2—part A

The next part of `CrabCritter2` appears in Figure 13.23b. This contains the other two overriding methods, `getMoveLocations` and `makeMove`. `CrabCritter2`'s `getMoveLocations` method differs from `CrabCritter`'s in that `CrabCritter2` can move only diagonally forward and to the right; that is, +45° clockwise from straight ahead. Consequently, the `directions` array always contains exactly one direction.

```
//*************************************************************

@Override
public ArrayList<Location> getMoveLocations()
{
  ArrayList<Location> locations = new ArrayList<>();
  int[] directions = {Location.HALF_RIGHT};

  for (Location adjacentLoc : getValidLocations(directions))
  {
    if (getGrid().get(adjacentLoc) == null)    // nobody there
    {
      locations.add(adjacentLoc);
    }
  }
  return locations;
} // end getMoveLocations

//*************************************************************

@Override
public void makeMove(Location location)
{
  if (location.equals(getLocation()))
  {
    setDirection(getDirection() + Location.HALF_LEFT);
  }
  else
  {
    super.makeMove(location);
  }
} // end makeMove
```

Figure 13.23b CrabCritter2—part B

If there is a valid location in that direction, the `getMoveLocations` method returns an `ArrayList` with that one location. If there is not a valid location in that direction, the `getMoveLocations` method returns an empty `ArrayList`.

If the `ArrayList` returned by `getMoveLocations` is empty, `Criter's selectMoveLocation` returns the current location. Otherwise, `selectMoveLocation` returns the location in the `ArrayList`. `CrabCritter2's makeMove` method takes this location as its argument. It differs from `CrabCritter's makeMove` method as follows: If a `CrabCritter2` needs to turn, it always turns exactly the same way, `Location.HALF_LEFT` or $-45°$. This makes `CrabCritter2` crabs eventually move around the grid boundary like bugs, but in the opposite direction.

Figure 13.23c shows the final method in our `CrabCritter2` class. It is essentially the same as the `CrabCritter` method called `getLocationsInDirections`, but with a slightly different

```
  //**********************************************************

  private ArrayList<Location> getValidLocations(int[] directions)
  {
    ArrayList<Location> locations = new ArrayList<>();
    Location adjacentLoc;

    for (int d : directions)
    {
      adjacentLoc =
        getLocation().getAdjacentLocation(getDirection() + d);
      if (getGrid().isValid(adjacentLoc))
      {
        locations.add(adjacentLoc);
      }
    }
    return locations;
  } // end getValidLocations
} // end class CrabCritter2
```

Figure 13.23c CrabCritter2—part C

style. Although we created our `CrabCritter2` class to implement significantly different behavior than GridWorld's `CrabCritter` implements, we got most of the algorithm and a substantial amount of the code from GridWorld's implementation. This example shows you another way to reuse existing software.

Now look at the `CrabCritter2Runner` class in Figure 13.24. This is the driver for our CrabCritter2 program. We intentionally made it as similar as possible to GridWorld's `CrabRunner` class, to facilitate comparison. The only difference is that we substituted `CrabCritter2` objects for `CrabCritter` objects.

Do you recall the discussion of GridWorld images at the end of the last chapter? If we do nothing with GridWorld's images, when we run our CrabCritter2 program, the JVM looks for a `CrabCritter2.gif` file in the `CrabCritter2` directory. Failing that, it goes one step up the inheritance hierarchy to the `Critter` class, which is in the `framework/info/gridworld/actor` directory. Figure 13.22a shows that this directory contains a `Critter.gif` file, so the program will use that file's image, which looks something like a squirrel. But we want our image to look like the `CrabCritter.gif` image in Figure 13.22b. So we make a copy of the `CribCritter.gif` image in the `projects/critters` directory and paste it into our `CrabCritter2` directory. Then we change the name of that copied image to `CrabCritter2.gif`.

When we run GridWorld's original CrabCritter program, the three crabs in the middle eventually eat two of the three initial bugs. But one bug survives, continuously eating and replanting flowers around the grid boundary. So the four species in `CrabCritter`'s world—bugs, flowers, crabs, and rocks—all seem to survive. On the other hand, when we run our improved CrabCritter2 program, with its more efficient crabs, eventually these superior crabs consume all the bugs and all the flowers, leaving only crabs and rocks. So we achieved our goal. Hurrah! Was it a worthwhile goal—developing a predator that destroys all its own food? That question lies beyond the scope of computer science.

```
/*********************************************************************
* CrabCritter2Runner.java
* Dean & Dean
*
* This variation of GridWorld's CrabRunner drives
* other GridWorld classes and CrabCritter2.
* GridWorld is distributed under the terms of the GNU General
* Public License as published by the Free Software Foundation.
*********************************************************************/

import info.gridworld.actor.*; // ActorWorld; Bug, Flower, Rock
import info.gridworld.grid.Location;

public class CrabCritter2Runner
{
  public static void main(String[] args)
  {
    ActorWorld world = new ActorWorld();
    world.add(new Location(7, 5), new Rock());
    world.add(new Location(5, 4), new Rock());
    world.add(new Location(5, 7), new Rock());
    world.add(new Location(7, 3), new Rock());
    world.add(new Location(7, 8), new Flower());
    world.add(new Location(2, 2), new Flower());
    world.add(new Location(3, 5), new Flower());
    world.add(new Location(3, 8), new Flower());
    world.add(new Location(6, 5), new Bug());
    world.add(new Location(5, 3), new Bug());
    world.add(new Location(4, 5), new CrabCritter2());
    world.add(new Location(6, 1), new CrabCritter2());
    world.add(new Location(7, 4), new CrabCritter2());
    world.show();
  } // end main
} // end class CrabCritter2Runner
```

Figure 13.24 CrabCritter2Runner—the driver for the `CrabCritter2` class in Figures 13.23a, 13.23b, and 13.23c

13.12 Problem Solving with Association Classes (Optional)

Aggregation, composition, and inheritance implement some of the most common kinds of associations among classes and objects—a has-a association for aggregation and composition, and an is-a association for inheritance. Be aware that there are many other possible kinds of associations, which you can conjure up easily by rattling off a few verb phrases, like "be next to. . . ," "get. . .from. . . ," "set. . .in. . . ," "make. . . with. . . ," run. . .toward. . . ," "sell. . .to. . . ," and so on. Typically, these other kinds of associations are more complicated than is-a or has-a associations. This section describes a powerful way to model other associations.

As you have seen, you can implement simple aggregation and composition associations by giving the container object a reference to each component object. This reference allows container object code to invoke component object methods. But for other kinds of associations, you may need multiple references and additional variables and methods. In other words, you may need a separate class just to describe the association. Such a class is called an *association class*. An association class defines an association object that represents a relationship among other objects. An association object is like an aggregation/composition container, in that it has instance variables that refer to other objects. But it's different in that the objects it refers to also refer to it, and each cannot contain the other. An association object typically receives references to the objects it associates when it is constructed. Whereas an aggregation/composition container contains its component objects, an association object just "knows about" the objects it associates.

Now let's see how this might apply to our previous Dealership program. What we've done so far with that program isn't much to brag about. We created a company with a sales manager, some salespeople, and some cars. What about customers? What about sales? Suppose we add a `Customer` class to our Dealership program. Then suppose some eager salesperson finally makes a sale to that first customer. The next question is, where do we put the information about that sale? In the `Dealership` class? In the `SalesPerson` class (as we seem to be doing in Figure 13.6)? In the `Car` class? In the `Customer` class? Technically, we could put that information in any one of these classes, and then put references to that class in whatever classes need access to that information. We could also scatter the information around among the participating classes in some way. No matter which of these alternatives we picked, however, from some points of view, what we did would seem inappropriate.

A more elegant solution is to encapsulate all of the sale information into one association class and give that class a name that describes the association. That's what we portray in Figure 13.25, which shows an abbreviated class diagram of another version of our previous Dealership program. First, look at the `Customer` class. Because a customer is a person, just like the sales manager and salespeople, we can use inheritance to reduce code and avoid redundancy in the `Customer` class by making the `Customer` class extend the `Person` class. Second, look at the `Sale` class. The `Sale` class appears as just another component in the `Dealership` class diagram. The one-to-many multiplicity suggests that its objects are elements of an `ArrayList`, perhaps named `sales`, which is instantiated in an enhanced version of the 

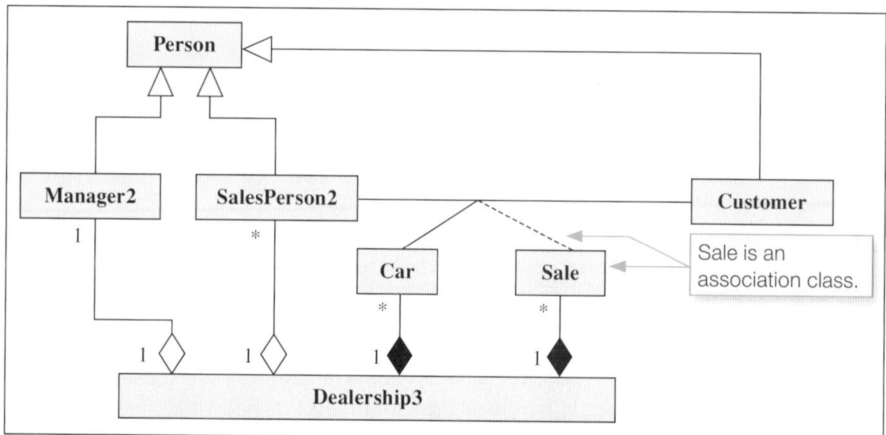

Figure 13.25 Class diagram for another car dealership program with customers and a salesperson-car-customer association

Dealership constructor. As far as the dealership is concerned, a Sale would be just another type of aggregation or composition component, like a SalesPerson2 or a Car.

However, a sale is not a physical entity, like an organism or car is. It's a process—or event—that associates a group of entities. So the Sale class needs to be an association class. What types of objects participate in a Sale association? There's a Car, there's a SalesPerson2, and there's a Customer. Notice how the UML class diagram in Figure 13.25 uses simple solid association lines to interconnect all normal classes that participate in an association. The UML standard suggests ways to decorate these association lines with additional symbols and nomenclature, but just the lines shown convey the message—the idea of an association among objects of the Car, SalesPerson2, and Customer classes. The dashed line that connects the Sale class to the solid association lines graphically identifies the Sale class as an association class describing the related association. The code fragment in Figure 13.26 illustrates the Sale constructor.

```
// This class associates SalesPerson2, Car, and Customer classes

public class Sale
{
  private Car car;                          } <--- references to classes
  private SalesPerson2 salesperson;         }      being associated
  private Customer customer;                }
  private double price;
  ...

  //*****************************************************************

  public Sale(Car car, SalesPerson2 person,   } <--- references passed
    Customer customer, double price)          }      into constructor
  {
    this.car = car;
    this.salesperson = person;
    this.customer = customer;
    this.price = price;
    ...
  } // end constructor
  ...
```

Figure 13.26 Partial implementation of the Sale class shown in Figure 13.25

Caveat—Don't Try to Inherit from an Association Participant

You might be tempted to try to use inheritance to create an association class because you might think that would give you "free access" to at least one of the participants in the association. Don't try to do that. All you'd get would be the ability to make an enhanced clone of one of the objects you want to associate, and you'd have to copy all the details between the clone and the real thing—a waste of effort. Treat an association like an aggregation, with references to the participating objects passed into the association constructor.

Summary

- Object-oriented languages help you organize things and concepts into two basic kinds of hierarchies—a has-a hierarchy for components in an aggregation or composition, and an is-a hierarchy for types in an inheritance.
- An aggregation or composition hierarchy exists when one large object contains several smaller (component) objects.
- For a given whole-part class relationship, if the container contains the only reference to a component, the component association is composition. Otherwise, it's aggregation.
- In an inheritance hierarchy, subclasses inherit all the variables and methods of the superclasses above them, and they typically add more variables and methods to what they inherit.
- To minimize descriptive duplication, organize your ideas so that only the concepts at the very bottom of an inheritance hierarchy (the leaves of the upside-down tree) are specific enough to represent real objects.
- To enable class `B` to inherit all the variables and methods in class `A` and all of class `A`'s ancestors, append `extends A` to the end of class `B`'s heading.
- A constructor should initialize the variables it inherits by immediately calling its superclass's constructor with the statement: `super(`<*arguments*>`);`
- You can override an inherited method by writing a different version of the inherited method in the derived class. Overriding occurs automatically if you use the same method name and the same sequence of parameter types, but if you do this, you must also use the same `return` type.
- You can access an overridden method by prefixing the common method name with `super` and then a dot.
- A `final` access modifier on a method keeps that method from being overridden. A `final` access modifier on a class keeps that class from being extended.
- Programmers frequently use combinations of aggregation, composition, and inheritance to deal with different aspects of an overall programming problem. In a UML class diagram, both relationships are represented by solid lines between related classes, and these lines are called associations. In a composition/aggregation association, there is a solid/hollow diamond at the container end of each association line. In a hierarchical association, there is a hollow arrowhead at the superclass end of the association line.
- Inheritance allows you to reuse code that was written for another context.
- When you have a complicated association among objects, it may help to gather references to those objects together into a common association class.

Review Questions

§13.2 Composition and Aggregation

1. In a UML diagram, what does an asterisk (*) indicate?
2. In a UML diagram, what does a solid diamond indicate?

§13.3 Inheritance Overview

3. Explain how using an inheritance hierarchy can lead to code reusability.
4. What are two synonyms for a superclass?
5. What are two synonyms for a subclass?

§13.4 Implementation of `Person/Employee/FullTime` Hierarchy

6. How do you tell the compiler that a particular class is derived from another class?
7. Based on the UML diagram in Figure 13.9, an instance of the `PartTime` class includes the following instance variables: `name` and `id`. (T / F)

§13.5 Constructors in a Subclass

8. In a subclass's constructor, what do you have to do if you want to begin the constructor with a call to the superclass's zero-parameter constructor?

§13.6 Method Overriding

9. If a superclass and a subclass define methods having the same name and the same sequence of parameter types, and an object of the subclass calls the method without specifying which version, Java generates a runtime error. (T / F).
10. If a subclass method overrides a method in the superclass, is it still possible to call the method in the superclass from the subclass?
11. If a superclass declares a variable to be `private`, can you access it directly from a subclass?

§13.7 Using the `Person/Employee/FullTime` Hierarchy

12. If you wish to call a superclass method, you must always prefix the method name with `super.` (T / F)

§13.8 The `final` Access Modifier

13. A `final` method is called "final" because it's allowed to contain only named constants, not regular variables. (T / F)

§13.9 Using Inheritance with Aggregation and Composition

14. Composition and inheritance are alternative programming techniques for representing what is essentially the same kind of real-world relationship. (T / F).

§13.10 Design Practice with Card Game Example

15. A Deck is a group of cards and a Deck has a group of cards. In our example, it's better to choose the is-a relationship and implement inheritance. In this case, why is inheritance a better choice than composition?

§13.11 Problem Solving with Association Classes (Optional)

16. It's possible to support an association with references, variables, and methods in existing classes. What's the advantage of using an association class instead?

Exercises

1. [after §13.2] (This exercise should be used in combination with Exercises 2 and 3.) Write a definition for a `Point` class. Provide two `double` instance variables, `x` and `y`. Provide a two-parameter constructor that initializes `x` and `y`. Provide a `shiftRight` method that shifts the point in the x direction by the value of the method's `double` parameter, `shiftAmount`. Provide a `shiftUp` method that shifts the point in the y direction by the value of the method's `double` parameter, `shiftAmount`. Make each of these methods return values that enable chaining. Provide accessor methods to retrieve the values of the two instance variables.

2. [after §13.2] (This exercise should be used in conjunction with Exercises 1 and 3.) Write a definition for a `Rectangle` class. Provide two `Point` instance variables, `topLeft` and `bottomRight`, which establish the top-left and bottom-right corners of the rectangle, respectively. Provide a two-parameter constructor that initializes `topLeft` and `bottomRight`. Provide a `shiftRight` method that shifts the rectangle in the

x direction by the value of the method's double parameter, shiftAmount. Provide a shiftUp method that shifts the rectangle in the y direction by the value of the method's double parameter, shiftAmount. Make each of these methods return values that enable chaining. Provide a printCenter method that displays the x and y values of the center of the rectangle.

3. [after §13.2] (This exercise should be used in conjunction with Exercises 1 and 2.) Write a definition for a RectangleDriver class with a main method to do the following: Instantiate a Point named topLeft at x = −3.0 and y = 1.0. Instantiate a Point named bottomRight at x = 3.0 and y = −1.0. Instantiate a Rectangle named rectangle using topLeft and bottomRight as arguments. Call rectangle's printCenter method. Use a single chained statement to shift the rectangle right by one and then up by one. Call rectangle's printCenter method again. The output should be:

```
x = 0.0 y = 0.0
x = 1.0 y = 1.0
```

4. [after §13.2] (Case Study) GridWorld's firstProject file contains the BugRunner driver:

```
public class BugRunner
{
  public static void main(String[] args)
  {
    ActorWorld world = new ActorWorld();
    world.add(new Bug());
    world.add(new Rock());
    world.show();
  } // end main
} // end class BugRunner
```

BugRunner starts by creating an instance of ActorWorld called world. ActorWorld's constructor immediately creates a 10 by 10 BoundedGrid. Next, BugRunner asks the ActorWorld object to add a new Bug at a random location, and then add a new Rock at a different random location. ActorWorld's add method tells each new occupant where it is supposed to go in the grid, and then tells that occupant to put itself into the grid at that location.

BugRunner's last statement calls ActorWorld's show method. This show method instantiates a WorldFrame (a GUI window) with an argument referring back to the ActorWorld object. WorldFrame's constructor instantiates a GridPanel. Then it uses its reference to the ActorWorld object to get a reference to the BoundedGrid, and it passes this BoundedGrid reference on to the new GridPanel. GridPanel's drawOccupants method uses this BoundedGrid reference to call BoundedGrid's getOccupiedLocations method. This allows GridPanel's paintComponent method to put bug and rock images in their proper places each time it repaints the screen.

Draw an abbreviated UML class diagram that contains all classes identified in the above two paragraphs, except for the BugRunner class. Use appropriate composition (black diamond) and aggregation (hollow diamond) links. Include multiplicity symbols where one of the multiplicities might be greater than unity. For the relationship between ActorWorld and WorldFrame, use simple bidirectional association, like this:

Or use bidirectional dependence, like this:

ActorWorld ←- - - uses - - -→ WorldFrame

5. [after §13.3] Suppose that you have three classes—`Shape` (which defines a shape's position in a coordinate system), `Square` (which defines a square's position in a coordinate system plus the square's width), and `Circle` (which defines a circle's position in a coordinate system plus the circle's radius). Assume that the three classes form an appropriate inheritance hierarchy with two inheritance relationships. For each of the two inheritance relationships, specify the superclass and subclass.

6. [after §13.3] Suppose you want to create a computer description of various ways energy is produced. Put your description in the form of a UML diagram with 4 classes (`Electrical`, `EnergyOutput`, `Heat`, and `Mechanical`)and 6 variables (`initialCapitalInvestment`, `typeOfFuelConsumed`, `maxRevolutionsPerMinute`, `maxTemperature`, `powerOutput`, and `volts`). Draw an appropriate UML class diagram with class names, variable names, and inheritance arrows. You may omit type specifications and methods.

7. [after §13.4] Ellipse program:
Java's API classes make extensive use of inheritance. For example, Oracle's Java API documentation shows that the `java.awt.geom.Ellipse2D` package has a class named `Double` that has these instance variables:[7]

`double`	`height` The overall height of the `Ellipse2D`.
`double`	`width` The overall width of this `Ellipse2D`.
`double`	`x` The x-coordinate of the upper-left corner of this `Ellipse2D`.
`double`	`y` The y-coordinate of the upper-left corner of this `Ellipse2D`.

And it has these constructors:

`Double()` Constructs a new `Ellipse2D`, initialized to location (0, 0) and size (0, 0).
`Double(double x, double y, double w, double h)` Constructs and initializes an `Ellipse2D` from the specified coordinates.

It has accessors for the instance variables, and an initializing method, but that's about all. Fortunately, this class `extends` a class called `Ellipse2D`, which has several other useful methods, including:

`boolean`	`contains(double x, double y)` Tests if a specified point is inside the boundary of this `Ellipse2D`.
`boolean`	`contains(double x, double y, double w, double h)` Tests if the interior of this `Ellipse2D` entirely contains the specified rectangular area.
`boolean`	`intersects(double x, double y, double w, double h)` Tests if the interior of this `Ellipse2D` intersects the interior of a specified rectangular area.

[7]These boxed descriptions were copied from Oracle's Java API website.

Write a short program in a class called `EllipseDriver`:

Import `java.awt.geom.Ellipse2D.Double`, and write a `main` method that calls the 4-parameter `Double` constructor to instantiate an ellipse like that shown in the picture below.[8] Then, in `println` statements, call the superclass's 2-parameter `contains` method to show whether the points $x=3.5$, $y=2.5$ and $x=4.0$, $y=3.0$ are contained within the specified ellipse.

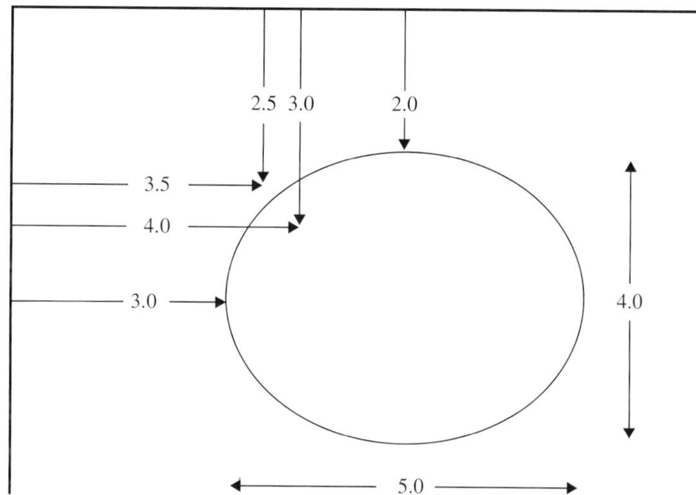

Output:

```
contains x=3.5, y=2.5? false
contains x=4.0, y=3.0? true
```

8. [after §13.4] (Case Study) Using GridWorld documentation, draw a UML class diagram like the diagram in Figure 13.9 for the `Actor` class and its `Bug`, `Rock`, and `Flower` descendants. Include UML specifications of all class and instance variables and constants, as well as all methods in these classes.

9. [after §13.5] Define a class named `Circle` that is derived from this API superclass:

```
java.awt.geom.Ellipse2D.Double
```

See Exercise 6 for a brief description of this `Double` superclass. Your subclass should declare the two `private` instance variables, `xCtr` and `yCtr`, initialized to `0.0`. These variables are the x- and y-coordinates of the circle's center. Your class should include a zero-parameter constructor, and it should also include a three-parameter constructor whose parameters are the x- and y-distances to the circle's center and the circle's diameter. This three-parameter constructor should not only initialize the new instance variables but also use the four-parameter constructor of the superclass to initialize the four instance variables in the superclass. Your class should also provide the following accessor methods: `getXCtr`, `getYCtr`, and `getRadius`, where the radius is half the height of the superclass shape. Verify the code you write by compiling it and running it with representative values for x-center, y-center, and diameter.

[8] The program does not actually draw the ellipse, but the `...Ellipse2D.Double` class has a mathematical understanding of it.

10. [after §13.6] Suppose you have two classes related by inheritance that both contain a zero-parameter method named doIt. Here's the subclass's version of doIt:

```
public void doIt()
{
  System.out.println("In subclass's doIt method.");
  doIt();
} // end doIt
```

The doIt(); call is an attempt to call the superclass's version of doIt.

a) Describe the problem that occurs when another method calls the above doIt method and the doIt method executes.

b) How should you fix the problem?

11. [after §13.5] (Case Study) This chapter's exercise 4 presented code containing the statement

```
ActorWorld world = new ActorWorld();
```

After that, we said, "ActorWorld's constructor immediately creates a 10 by 10 BoundedGrid." But if you look at GridWorld's source code for the ActorWorld class, you'll see that the zero-parameter constructor is empty:

```
/**
 * Constructs an actor world with a default grid.
 */
public ActorWorld()
{
}
```

How does this empty constructor create "a 10 by 10 BoundedGrid"?

12. [after §13.6] (Case Study) Find and describe the uses of explicit super calls in the GridWorld classes mentioned in this chapter's Exercise 4: ActorWorld, BoundedGrid, Bug, Rock, WorldFrame, and GridPanel. In each class that makes a super call, identify the class, the method or constructor that includes the super call, the constructor or method called by super, and the (extended) class in which that called method resides.

13. [after §13.6] (Case Study) Create a RandomBug class that extends the Bug class in GridWorld. In this extended class, override the move method inherited from Bug. Do this by copying the code from Bug's move method and then replacing the code fragment that creates flowers and deposits them in the bug's wake with a setDirection call that randomly selects one of the directions, 0, 45, 90, 135, 180, 225, 270, or 315. Get a random value the same way that GridWorld gets a random value for a bounded grid in the World class's getRandomEmptyLocation method. But instead of using

```
generator.nextInt(emptyLocs.size())
```

use

```
45 * generator.nextInt(8)
```

Also create a RandomBugRunner class to exercise your new RandomBug class. In this class's main method, instantiate an ActorWorld object called world. Then have world add ten RandomBug objects, and then have world call its show method.

Executing RandomBugRunner should generate a display similar to that in Figure 11.15, except this time there will be ten randomly located bugs and no rock. Push the Step button a few times to see how the bugs move. Then click World > UnboundedGrid, adjust the sliders on the sides to move the cluster of bugs to the center of the larger display, click Run, and watch the bugs gradually spread apart as time passes.

14. [after §13.8] What does it mean when you use the `final` modifier for a method?

15. [after §13.8] What does it mean when you use the `final` modifier for a class?

16. [after §13.9] Fill in the blanks:

If thing A "has a" thing B and "has a" thing C, there is a(n) ______ association, and A's class definition will contain declarations for ______ variables. If A "is a" special form of B, there is a(n) ______ association, and the right side of A's class heading will contain the words ______ .

17. [after §13.9] Identification of type of association:

Given the following list of word pairs, for each word pair, identify the association between the two words. More specifically, identify whether the two words are related by composition or inheritance. To get you started, we've provided the answers to the first two word pairs. Bicycle and front wheel are related by composition because a bicycle "has a" front wheel. Bicycle and mountain bike are related by inheritance because a mountain bike "is a" bicycle.

		inheritance or composition?
bicycle	front wheel	composition
bicycle	mountain bike	inheritance
structural member	beam	____
building	floor	____
company	fixed assets	____
employee	salesperson	____
forest	tree	____
bird	robin	____
class	method	____
neurosis	paranoia	____

18. [after §13.10] Shuffling:

Suppose you are developing the Card Game program suggested by Figure 13.17 in the text. The following partial UML class diagram shows where you are in the developmental process:

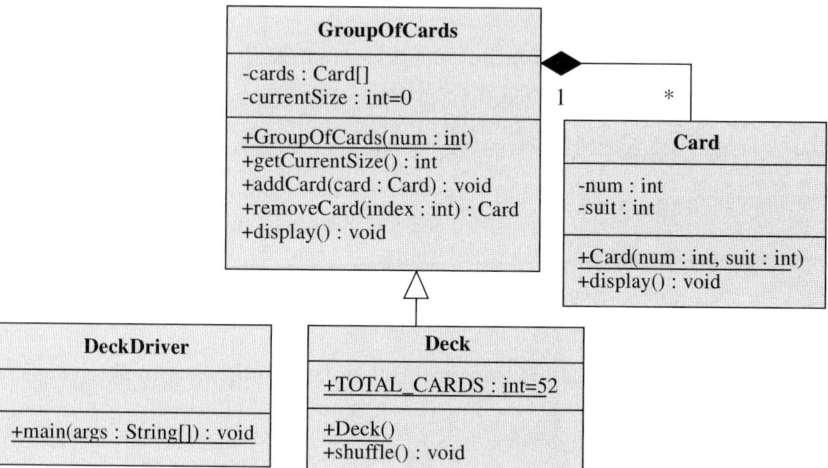

Assume that you have written methods for a `Card` class and a `GroupOfCards` class. Also, assume that the `addCard` method increments `currentSize` after adding the input card to the end of the currently filled part of the `cards` array. Finally, assume that the `removeCard` method retrieves a reference to the card at `index` in the `cards` array, decrements the `currentSize` of the cards array, shifts all array elements above `index` down by one place, and returns the reference to the card originally at `index`.

To shuffle the deck, use a `for` loop that starts with `unshuffled = getCurrentSize()` and steps down to one. In each iteration, use `Math.random` to pick an index in the unshuffled range, remove the card at that index, and then add it to the high end of the array. Include all that functionality in a `Deck.shuffle` method.

Extra credit:
Write Java code that tests your `shuffle` method. To do that, you'll need to implement all of the classes and methods in the above UML class diagram. Your `main` method should instantiate a `deck`, display it, shuffle it, and display it again.

19. [after §13.10] First part of the `Game` class for the game of Hearts:

The `Game` class introduced in the text contained a deck and exactly two players. Improve that `Game` class by including an array of type `Trick[]` and a `numberOfTricks` variable that keeps track of the number of tricks played so far. Include a `final` instance variable called `PLAYERS`, which will be initialized in a constructor whose parameter value is the number of players participating in the game. Replace the individual `player1` and `player2` instance variables by instances in an array of type `Hand[]`. Include two `boolean` instance variables, `hearts` and `queenOfSpades`, whose values switch from `false` to `true` whenever the first heart or the queen of spades is played.

Write Java code that defines that part of the `Game` class that includes the class heading, the instance variable declarations, and the one-parameter custom constructor whose parameter is the number of players. This constructor should instantiate a `Hand` array with a length equal to the number of players. It should instantiate individual `Hand` objects for each player, using a two-parameter `Hand` constructor. The first parameter is the player number, starting with 0. The second parameter is the maximum number of cards that the player will receive, which depends on the total number of cards in the deck and the number of players. The game constructor should also instantiate a `Trick` array, but not populate it with any individual tricks.

Review Question Solutions

1. A * on a UML diagram means the multiplicity can be "any number."
2. In a UML diagram, a solid diamond is placed on an association line next to a containing class in a composition association. It indicates that the containing class exclusively contains the class that's at the other end of the association line.
3. Putting common code from two classes into a superclass is one example of code reusability. Code reusability can also take place when you want to add a significant chunk of functionality to an existing class, and you implement the solution with a new subclass.
4. Two synonyms for a superclass—parent class, base class.
5. Two synonyms for a subclass—child class, derived class.
6. To tell the compiler that a class is derived from another class, you write `extends` *<other-class>* at the end of your new class's heading line.
7. True. An instance of a subclass includes that class's instance variables and its ancestors' instance variables.
8. Nothing. This happens automatically. You can, however, preempt this by writing `super();` as the first line in your derived constructor.

9. False. There is no problem. The JVM selects the method in the subclass.
10. Yes. In the call statement, preface the common method name with `super.`
11. No. If a superclass's instance variable is `private`, you cannot access it directly from a subclass. You can access it by calling an accessor method (assuming the accessor is `public`). In calling the superclass's method, there's no need to prefix the method call with a reference dot.
12. False. The `super.` prefix is necessary only when you want to call a superclass method that has been overridden.
13. False. A `final` method is allowed to contain regular variables. It's called "final" because it's illegal to create an overriding version of the method in a subclass.
14. False. Composition and inheritance are completely different class relationships. Composition is when a class is comprised of non-trivial constituent parts and the parts are defined to be classes. Inheritance is when one class is a more detailed version of another class. More formally, inheritance is when one class, a subclass, inherits variables and methods from another class, a superclass.
15. Inheritance is better because with this example, there's a second class that is also a group of cards. Because there are two classes that share some of the same properties, you should put those common properties in a shared superclass, `GroupOfCards`. Doing this promotes software reuse and avoids code redundancy.
16. You can make a complicated association easier to recognize and understand by organizing the references to all association participants and other association information and methods in a single class that represents the association only.

CHAPTER 14

Inheritance and Polymorphism

Objectives

- Understand the role of the `Object` class.
- Learn why you need to redefine the `equals` and `toString` methods.
- Learn how polymorphism and dynamic binding improve program versatility.
- Understand what the compiler checks and what the JVM does when a reference variable is associated with a method name.
- Understand the constraints affecting assignment of an object of one class to a reference variable of another class.
- See how to use an array of ancestor reference variables to implement polymorphism among descendant methods.
- See how an `abstract` method declaration in an `abstract` superclass eliminates the need for a dummy method definition in the superclass.
- See how you can use an interface to specify common method headings, store common constants, and implement multiple polymorphisms.
- Learn where to use `protected` member access.
- Optionally, learn how to draw a three-dimensional object.

Outline

14.1 Introduction

This is the second of two chapters on inheritance. The previous chapter applied a broad brush stroke to fundamental inheritance concepts. In this chapter, we narrow the focus and describe several inheritance-related topics in depth. We start with the `Object` class, which is the (Java API) superclass of all other classes. We then discuss one of the cornerstones of object-oriented programming (OOP)—polymorphism. *Polymorphism* is the ability for a particular method call to perform different operations at different times. It occurs when you have a reference variable that refers to different types of objects during the course of a program's execution. When the reference variable calls the polymorphic method, the reference variable's object type determines which method is called at that time. Pretty cool, eh? Polymorphism provides programs with a great deal of power and versatility.

After introducing polymorphism, we describe its partner, dynamic binding. *Dynamic binding* is the mechanism used by Java to implement polymorphism. We then provide alternative implementations of polymorphism, using `abstract` classes to make code cleaner.

Chapter 10 introduced you to several examples of interfaces in the prewritten Java collections framework. In this chapter, we'll show you how to define your own multiple-signature interfaces and use them to manage code development, store universal constants, and make polymorphism more versatile.

Then we'll describe the `protected` modifier, which simplifies access to inherited code. Finally, in an optional section, we present a three-dimensional graphics problem that illustrates polymorphism with the Java API.

The material in this chapter is relatively difficult, but once you get it, you'll truly understand what OOP is about, and you'll know how to craft elegantly structured programs.

14.2 The `Object Class` and Automatic Type Promotion

The `Object` class is the ancestor of all other classes. It is the primordial ancestor—the root of the inheritance hierarchy. Any class that explicitly extends a superclass uses `extends` in its definition. Whenever anyone creates a new class that does not explicitly extend some other class, the compiler automatically makes it extend the `Object` class. Therefore, all classes eventually descend from the `Object` class. The `Object` class doesn't have many methods, but the ones it has are significant because they are always inherited by all other classes. In the next two sections, you'll learn about the `Object` class's two most important methods, `equals` and `toString`. Because any class that you write automatically includes these two methods, you need to be aware of what happens when these methods are called.

Before diving into the details of these two methods, however, we want to make you aware of a Java process that's very similar to the numerical type promotion that you studied in Chapters 3 and 12. There, you saw that in the course of making an assignment or copying an argument into a parameter, the JVM performs a promotion when the two involved types conform to a certain numerical hierarchy. For example, when an `int` value is assigned into a `double` variable, the JVM automatically promotes the `int` value to a `double` value.

An analogous promotion can also occur with reference types. When an assignment or argument passing operation involves different reference types, the JVM automatically promotes the source reference type to the target reference type if the target reference type is above the source reference type in the inheritance hierarchy. In particular, since the `Object` class is an ancestor of every other class, when the need arises, Java automatically promotes any class type to the `Object` type. The next section describes a situation that stimulates this kind of type promotion.

14.3 The equals Method

Syntax

The Object class's equals method—which is inherited automatically by all other classes—has this public interface:

```
public boolean equals(Object obj)
```

Because all classes automatically inherit this method, unless a similarly defined method takes precedence, any object, objectA, can invoke this method to compare itself with any other object, objectB, with a method call like this:

```
objectA.equals(objectB)
```

This method call returns a boolean value of either true or false. Notice that we did not specify the type of either objectA or objectB. In general, they can be instantiations of any class, and they do not need to be objects of the same class. The only constraint is that objectA must be a non-null reference. For example, if Cat and Dog classes exist, this code works correctly:

```
Cat cat = new Cat();
Dog dog = new Dog();

System.out.println(cat.equals(dog));
```

<u>Output:</u>

```
false
```

The equals method that is called here is the equals method that the Cat class automatically inherits from the Object class. The parameter in this inherited method is of type Object, as specified in the method's public interface above. But the dog argument we pass to this method is not of type Object. It is of type Dog. So what's happening? When we pass the dog reference into the inherited equals method, the reference type automatically promotes from type Dog to type Object. Then the inherited equals method performs an internal test to see if the passed in dog is the same as the calling cat. Of course it is not, so the output is false, as you can see.

Semantics

Notice that we just said, "performs an internal test." Now let's focus on that mysterious "internal test." How can you tell if two objects are the same or "equal"? When you say "objectA equals objectB," you could mean this:

1. objectA is just an alias for objectB, and both objectA and objectB refer to exactly the same object.

Or you could mean this:

2. objectA and objectB are two separate objects that have the same attributes.

The equals method that all classes inherit from the Object class implements the narrowest possible meaning of the word "equals." That is, this method returns true if and only if objectA and objectB refer to exactly the same object (definition 1). This behavior of equals is exactly the same as the behavior of the == operator when it tests the equality of two reference variables. That operator also returns true if and only if both references refer to exactly the same object.

Suppose you have a Car class with three instance variables, make, year, and color, and you have a constructor that initializes these instance variables with corresponding argument values. Suppose this Car class does not define an equals method itself, and the only equals method it inherits is the one it inherits automatically from the Object class. The following code illustrates that the equals method inherited from the Object class does exactly the same thing as the == operator does:

```
Car car1 = new Car("Honda", 2014, "red");
Car car2 = car1;
Car car3 = new Car("Honda", 2014, "red");

System.out.println(car2 == car1);
System.out.println(car2.equals(car1));
System.out.println(car3 == car1);
System.out.println(car3.equals(car1));
```

different names for the same object

different objects with the same attributes

Output:

```
true
true
false
false
```

This narrow sense of the word "equals" is not always what you want. For example, suppose that your spouse decides to buy a new car, goes to a particular auto dealer, and orders a red 2014 Honda, as suggested by the above car1 instantiation. When you see the brochures that your spouse brings home, you're impressed and decide you would like a new car for yourself too. You'd like it to be just like your spouse's car except for the color, which you want to be blue. So you go to the same dealer and say, "I want the same car my spouse just ordered, but I want the color to be blue." A month later, the dealer calls both you and your spouse at your separate places of work and says to each of you separately, "Your car is ready. Please come in to pick it up at 5:30 PM today." You both show up as requested, and the dealer takes you outside and proudly exclaims, "Here it is. How do you like it?" You say "Great, it's just what I wanted!" Then your spouse says, "But where is my car?" And the dealer replies, "But I thought you were to be joint owners of the same car, and your spouse told me to change the color of that car to blue." Oops, somebody made a mistake. . . .

The mistake occurred in the communication between you and the dealer when you said, "the same car." You meant the second meaning above: objectA and objectB are two separate objects which have the same attributes. But the dealer heard the first meaning above: objectA is just another name for objectB, and both objectA and objectB refer to exactly the same object.

Defining Your Own equals Method

Now let's see how you can implement the second meaning. To do so, include in your class an explicit version of an equals method that tests for equal attributes. Then, when your program runs, and an instance of your class calls the equals method, your equals method takes precedence over Object's equals method, and the JVM utilizes the equals method you defined. The equals method in Figure 14.1's Car class tests for equal attributes by comparing the values of all three instance variables, which are the object's attributes. It returns true only if all three instance variables have the same values, and it returns false otherwise. Notice that this equals method includes two subordinate equals method calls—one made by the make instance variable and the other made by the color instance variable. As explained in Chapter 3, these calls to String's equals method check to see if two different strings have the same character sequence.

```
/*************************************************
* Car.java
* Dean & Dean
*
* This defines and compares cars.
*************************************************/

public class Car
{
  private String make;  // car's make
  private int year;     // car's listed year
  private String color; // car's color

  //**********************************************

  public Car(String make, int year, String color)
  {
    this.make = make;
    this.year = year;
    this.color = color;
  } // end Car constructor

  //**********************************************

  public boolean equals(Car otherCar)
  {
    return otherCar != null &&
           make.equals(otherCar.make) &&
           year == otherCar.year &&
           color.equals(otherCar.color);
  } // end equals
} // end class Car
```

This "overrides" the Object class's equals method.

Figure 14.1 Car class, which defines equals to mean having the same instance variable values

In the equals method's return expression, notice the otherCar != null subexpression. If this evaluates to false (indicating that otherCar is null), Java's short-circuit evaluation keeps the computer from trying to use a null reference to access the other car's make and color reference variables. Such short-circuit evaluation prevents runtime errors. You should always strive to make your code robust. In this case, that means you should consider the possibility of someone passing in a null value for otherCar. If null gets passed in and there's no test for null, the JVM generates a runtime error when it sees otherCar.make. This is a fairly common error—attempting to access a member from a null reference variable—and you can avoid it easily. Just test for null prior to accessing the member. For our equals method, if otherCar is null, then the otherCar != null subexpression is false, and the return statement returns false. Returning false is appropriate because a null otherCar is clearly not the same as the calling object Car.

Get in the habit of writing equals methods for most of your programmer-defined classes. Writing equals methods is usually straightforward because they tend to look the same. Feel free to use the Car class's equals method as a template.

Remember that any reference variable can call the equals method, even if the reference variable's class doesn't define an equals method. You know what happens in that case, right? When the JVM realizes that there's no local equals method, it looks for the equals method in an ancestor class. If it doesn't find an equals method prior to reaching the Object class at the top of the tree, it uses the Object class's equals method. This default operation often appears as a bug. To fix the bug, make sure that your classes implement their own equals methods. 

As the callout in Figure 14.1 suggests, it's OK to think informally that the Car class's equals method "overrides" the Object class's equals method. But technically speaking, it is not an overriding method. If you precede your "overriding" equals method with an @Override tag, the compiler will complain and tell you it is not an overriding method. Instead, it's a method that has the same name as a superclass method, but a different signature because the parameter's type is different. As shown earlier, the heading of the Object class's equals method looks like this:

```
public boolean equals(Object obj)
```

On the other hand, the heading of our Car class's equals method looks like this:

```
public boolean equals(Car otherCar)
```

You can see that the parameter types are different—obj is an Object, while otherCar is a Car.

There are many situations where actually overriding a method is useful. As illustrated in the Employee and FullTime classes in Figures 13.11 and 13.12, an overriding method can use super to obtain the services of the overridden method and then provide additional statements for additional services. But there is no good reason to actually override Object's equals method because that method works just fine for equals method calls whose arguments match the generic Object parameter.

equals Methods in API Classes

Note that equals methods are built into many API classes.[1] For example, the String class and the wrapper classes implement equals methods. As you'd expect, these equals methods test whether two references point to data that is identical (not whether two references point to the same object).

You've seen the String class's equals method before, so the following example should be fairly straightforward. It illustrates the difference between the == operator and the String class's equals method. What does this code fragment print?

```
String s1 = "hello";
String s2 = "he";

s2 += "llo";
if (s1 == s2)
{
  System.out.println("same object");
}
if (s1.equals(s2))
{
  System.out.println("same contents");
}
```

[1]To get an idea of how common equals methods are, go to Oracle's Java API website (http://download.oracle.com/javase/7/docs/api/) and search for occurrences of equals.

This code fragment prints "same contents." Let's make sure you understand why. The == operator returns `true` only if the two reference variables being compared refer to the same object. In the first `if` statement, `s1 == s2` returns `false` because `s1` and `s2` do not refer to the same object. In the second `if` statement, `s1.equals(s2)` returns `true` because the characters in the two compared strings are the same.

Actually, there's another twist to the `String` class. To minimize storage requirements, the Java compiler makes string references refer to the same `String` object whenever an assignment refers to a duplicate string literal. That's called *string pooling*. For example, suppose the preceding code included a third declaration that looked like this:

```
String s3 = "hello";
```

Then, if the first `if` condition were `(s1 == s3)`, the output would say "same object" because `s1` and s3 would refer to the same "hello" string object.

14.4 The `toString` Method

The `Object` Class's `toString` Method

Let's now consider another important method that all classes inherit from the `Object` class. The `Object` class's `toString` method returns a string that's a concatenation of the calling object's full class name, an @ sign, and a sequence of digits and letters. For example, consider this code fragment:

```
Object obj = new Object();
Car car = new Car();

System.out.println(obj.toString());
System.out.println(car.toString());
```

When executed, the code fragment produces this:

```
java.lang.Object@601BB1
Car@1BA34F2
```

full class name

These digits and letters form a hashcode.

Note how `obj.toString()` generates `java.lang.Object` for the full class name. The full class name consists of the class name prefixed by the class's package. The `Object` class is in the `java.lang` package, so its full class name is `java.lang.Object`. Note how `car.toString()` generates `Car` for the full class name. Because the `Car` class is not part of a package, its full class name is simply `Car`.

Note how `obj.toString()` generates 601BB1 for its *hashcode* value. A hashcode helps the JVM find data that might be anywhere in a large block of memory. In Java, hashcode values, like 601BB1, are written as hexadecimal numbers. We described the hexadecimal number system in Section 12.14. What follows is a review of that description.

Hexadecimal Numbers

Hexadecimal numbers use digits that can have one of 16 values—0, 1, 2, 3, 4, 5, 6, 7, 8, 9, A, B, C, D, E, and F (the lowercase letters a through f are also acceptable). The A through F values represent the numbers 10 through 15. With 16 unique digits, hexadecimal numbers form what is known as a base-16 number system. In the familiar base-10 number system for decimal numbers, suppose you're counting up and you get to the largest digit, 9. To form the next number, 10, you need two digits—a 1 at the left and a 0 at the right; the result is 10. Likewise, suppose that you're counting up with hexadecimal numbers and you get to the largest digit, F for 15. To form the next number, 16, you need two digits—a 1 at the left and a zero at the right; the result is 10. In other words, 10 is how you write 16 in hexadecimal. For additional help with hexadecimal counting, see Section 12.14 or Appendix 1. In Appendix 1, you'll see a sequence of hexadecimal numbers and their associated decimal numbers, in the context of the Unicode/ASCII character set. Appendix 8 explains how to make conversions back and forth between decimal and any other number system.

You know the hexadecimal number A is equivalent to the decimal number 10. What about the 601BB1 value generated by the previous code fragment—what is its equivalent decimal number? Appendix 8 explains the mathematics of converting any hexadecimal number to its decimal equivalent, but we'll present a shortcut. In Windows Vista, for example, select Start / Programs / Accessories / Calculator / View / Scientific. In the calculator window, click the Hex button, enter 601BB1, and then click the Dec button. This changes the display to 6298545, which is the decimal equivalent of 601BB1. Thus, in the previous code fragment, when `obj.toString()` returns a string with 601BB1 at the right of the @ sign, it means the `obj` object's memory address is one of just a few memory addresses in the 6,298,545th row of an object *hash table*. The JVM can jump immediately to that row and, within that row, quickly perform a short search for the correct address.

Overriding the toString Method

Retrieving the class name, an @ sign, and a hashcode is usually worthless, so you'll almost always want to avoid calling the `Object` class's `toString` method and instead call an overriding `toString` method. The reason we're discussing the `Object` class's `toString` method is because it's easy to call it accidentally, and when that happens, we want you to understand what's going on.

Because the `Object` class defines a `toString` method, every class has a `toString` method, even if it does not define one or inherit one through some other class it explicitly `extends`. Many Java API classes define overriding `toString` methods. For example, the `String` class's `toString` method trivially returns the string that's stored in the `String` object. As described in Chapter 10, the `ArrayList` class's `toString` method (inherited from the `AbstractCollection` class) returns a square-bracketed comma-delimited list of strings that represent the individual array elements. The `Date` class's `toString` method returns a `Date` object's month, day, year, hour, and second values as a single concatenated string. In general, `toString` methods should return a string that describes the calling object's contents.

Because retrieving the contents of an object is such a common need, you should get in the habit of providing an explicit `toString` method for most of your programmer-defined classes. Typically, your `toString` methods should simply concatenate the calling object's stored data and return the resulting string. Your `toString` methods should not print the concatenated string value; they should just return it. We're mentioning this point because novice programmers have a tendency to put print statements in their `toString` methods, and that's wrong. A method should do only what it's supposed to do and nothing more. The `toString` method is supposed to return a string value, and that's it!

For example, look at the `toString` method in the Car2 program in Figure 14.2. It returns a string that describes the calling object's contents.

Implicit toString Method Calls

In the Car2 program, the `main` method has no explicit `toString` method call. So how does this program illustrate use of the `toString` method? Whenever a reference appears alone inside a print statement (`System.out.print` or `system.out.println`), the JVM automatically calls the referenced

```
/*****************************************************
* Car2.java
* Dean & Dean
*
* This instantiates a car and displays its properties.
*****************************************************/

public class Car2
{
  private String make;  // car's make
  private int year;     // car's listed year
  private String color; // car's color

  //**************************************************

  public Car2(String make, int year, String color)
  {
    this.make = make;
    this.year = year;
    this.color = color;
  } // end Car2 constructor

  //**************************************************

  @Override
  public String toString()
  {
    return "make = " + make + ", year = " + year +
      ", color = " + color;
  } // end toString

  //**************************************************

  public static void main(String[] args)
  {
    Car2 car = new Car2("Honda", 1998, "silver");
    System.out.println(car);
  } // end main
} // end class Car2
```

This overrides the `Object` class's `toString` method.

Figure 14.2 Car2 program that illustrates overriding the `toString` method

object's `toString` method. In Figure 14.2, this statement generates a call to the `toString` method in the `Car2` class:

```
System.out.println(car);
```

Let's look at another example that uses the `toString` method. See the Counter program in Figure 14.3. Once again, there's a `toString` method and no explicit call to it. So how does it get called? When you

```
/***************************************************
* Counter.java
* Dean & Dean
*
* This creates a counter and displays its count value.
***************************************************/

public class Counter
{
  private int count;

  //************************************************

  public Counter(int count)
  {
    this.count = count;
  } // end constructor

  //************************************************

  @Override
  public String toString()
  {
    return Integer.toString(count);
  } // end toString

  //************************************************

  public static void main(String[] args)
  {
    Counter counter = new Counter(100);
    String message = "Current count = " + counter;
    System.out.println(message);
  } // end main
} // end class Counter
```

This overrides the Object class's toString method.

Figure 14.3 Counter program that illustrates implicitly calling the `toString` method

concatenate a reference variable and a string (with the + operator), the JVM automatically calls the reference's `toString` method. Thus, in Figure 14.3, this statement's `counter` reference generates a call to the `Counter` class's `toString` method:

```
String message = "Current count = " + counter;
```

Note that you'll often see the `toString` method explicitly called with the standard call syntax even when it's not necessary. For example, in the Counter program's `main` method, we might have used this alternative implementation for the `message` assignment statement:

```
String message = "Current count = " + counter.toString();
```

Some programmers would claim that this alternative implementation is better because the code is more self-documenting. Some programmers would claim that the original implementation is better because the code is more compact. We don't have a preference as to which implementation is better—either way is fine.

Counter Program's toString Method—A Detailed Analysis

Let's revisit the toString method in Figure 14.3's Counter program. Because the Counter class contains only one piece of data, count, there's no need for concatenation code as part of the toString implementation. Just return count's value and that's it. So this might have been your first-cut implementation for toString:

```
public int toString()
{
  return count;
}
```

But this produces a compile-time error. Do you know why? An overriding method must have the same return type as the method it's overriding. Because the Counter class's toString method is an overriding implementation of the Object class's toString method, the two methods must have the same return type. Because the Object class's return type is a String, the above int return type generates an error. With that in mind, this might have been your second-cut implementation for toString:

```
public String toString()
{
  return count;
}
```

But this also produces an error. Why? Incompatible types. The returned value, count, is an int, and the return type is defined to be a String. The solution is to convert count explicitly to a String before returning it, like this:

```
public String toString()
{
  return Integer.toString(count);
}
```

Do you understand the Integer.toString code? In Chapter 5, you learned that all primitive types have a corresponding wrapper class. Integer is one such class—it wraps up the int primitive. The Integer class's toString method returns a string representation of its passed-in int argument. So if count is 23, then Integer.toString(count) returns the string "23."

Quick quiz: Is the Integer class's toString method a class method or an instance method? Look at the method call's prefix. The method call, Integer.toString, uses a class name for the prefix. When a method call uses a class name for a prefix instead of a reference variable, you know the method is a class method. Thus, Integer's toString is a class method.

Note that all the wrapper classes have toString methods. They all do the same thing—they return a string representation of their passed-in argument. Here are some examples:

```
Double.toString(123.45)      : evaluates to string "123.45"
Character.toString('G')      : evaluates to string "G"
```

String's valueOf Method

There's another way to convert primitive variables to strings. Use the String class's valueOf method. This takes a primitive value and returns a string. Like the wrapper toString methods described above, it's a class method, so you must use its class name, String, as a prefix. Thus, instead of the previous method calls, you could use these method calls:

```
String.valueOf(123.45)  : evaluates to string "123.45"
String.valueOf('G')     : evaluates to string "G"
```

The valueOf method works with different data types because it's an overloaded method, and the JVM automatically selects that particular method whose parameter type matches the type of the data provided. Because we frequently customize toString methods, String's valueOf method is likely to produce more consistent standard conversions.

In addition to converting primitives to strings, the valueOf method can also be used to convert an array of vowel characters to a string. This code fragment prints the string "aeiou":

```
char[] vowels = {'a', 'e', 'i', 'o', 'u'};
System.out.print(String.valueOf(vowels));
```

14.5 Polymorphism and Dynamic Binding

Polymorphism Overview

If you ask an object-oriented programming (OOP) aficionado to name the three most important characteristics of OOP, he or she will probably answer "encapsulation, inheritance, and polymorphism." The previous chapter discussed encapsulation and inheritance. Now it's time to discuss polymorphism. The word *polymorphism* means "having many forms." It comes from the Greek root words *poly* (many) and *morph* (form). In chemistry and mineralogy, polymorphism is when a substance can crystallize in two or more alternative forms. In zoology, polymorphism is when a species has two or more different forms, like the different castes of bees spawned by the same queen to perform different functions in a beehive. In computer science, polymorphism is when different types of objects respond differently to the same method call.

Here's how it works. You declare a general type of reference variable that is able to refer to objects of different types. What is the most general type of reference variable? It's an Object reference variable, declared, for example, like this:

```
Object obj;
```

Once you have declared a reference variable of type Object, you can use it to refer to any type of object. For example, suppose you define a class named Dog, as in Figure 14.4, and another class named Cat, as in Figure 14.5. Each of the two derived classes contains a toString method that overrides the toString method in the Object class. Notice that the two toString methods shown override Object's toString method in different ways. One returns what a dog says, "Woof! Woof!," and the other returns what a cat says, "Meow! Meow!"

The different toString method definitions in the Dog and Cat classes enable the toString method to be polymorphic. If you call toString with a reference to a Dog object, it responds the way a dog would respond, but if you call toString with a reference to a Cat object, it responds the way a cat would respond. The driver in Figure 14.6 demonstrates this effect. Notice how the obj reference variable can contain a reference to either a Dog object or a Cat object, and that object determines which toString method is called.

```
/****************************************
* Dog.java
* Dean & Dean
*
* This class implements a dog.
****************************************/

public class Dog
{
  @Override
  public String toString()
  {
    return "Woof! Woof!";
  }
} // end Dog class
```

Figure 14.4 `Dog` class for the Pets program driven by code in Figure 14.6

```
/***************************************
* Cat.java
* Dean & Dean
*
* This class implements a cat.
***************************************/

public class Cat
{
  @Override
  public String toString()
  {
    return "Meow! Meow!";
  }
} // end Cat class
```

Figure 14.5 `Cat` class for the Pets program driven by code in Figure 14.6

Why does the program print "Woof! Woof!" twice? There are two print statements. The first one explicitly calls a `toString` method. The second one uses an implicit call to a `toString` method—when a reference variable appears alone in a `String` context, the compiler automatically appends `.toString()` to the bare reference variable. So the last two statements in the `Pets` class are equivalent.

Dynamic Binding

The terms *polymorphism* and *dynamic binding* are intimately related, but they're not the same. Polymorphism is a form of behavior. Dynamic binding is the mechanism for that behavior—how it's implemented. Specifically, polymorphism is when different types of objects respond differently to the same method call. Dynamic binding is what the JVM does to match a polymorphic method call with a particular method.

Just before the JVM executes a method call, it determines the type of the method call's actual calling object. If the actual calling object is from class `X`, the JVM *binds* class `X`'s method to the method call. If the actual calling object is from class `Y`, the JVM binds class `Y`'s method to the method call. After the JVM binds the appropriate method to the method call, the JVM executes the bound method. For example, note the `obj.toString` method call in the following statement near the bottom of Figure 14.6:

```
System.out.println(obj.toString());
```

Depending on which type of object is referred to by `obj`, the JVM binds either `Dog`'s `toString` method or `Cat`'s `toString` method to the `obj.toString` method call. After binding takes place, the JVM executes the bound method and prints either "Woof! Woof!" or "Meow! Meow!"

Dynamic binding is "dynamic" because the JVM does the binding operation while the program is running. Binding takes place at the latest possible moment, right before the method is executed. That's why dynamic binding is often called *late binding*. Some programming languages bind method calls at compile time rather than at runtime. That is called *static binding*. Java's designers decided to go with dynamic binding rather than static binding because dynamic binding facilitates polymorphism.

Compilation Details

The Pets program illustrated polymorphic behavior by calling `Dog` and `Cat` versions of the `toString` method. Could we have done the same thing with `Dog` and `Cat` versions of a `display` method? In other words, if `Dog` implemented a display method that prints "I'm a dog," would the following code work?

```
Object obj = new Dog();
obj.display();
```

```
/*************************************************************
* Pets.java
* Dean & Dean
*
* This illustrates simple polymorphism.
*************************************************************/

import java.util.Scanner;

public class Pets
{
  public static void main(String[] args)
  {
    Scanner stdIn = new Scanner(System.in);
    Object obj;

    System.out.print("Which type of pet do you prefer?\n" +
      "Enter d for dogs or c for cats: ");
    if (stdIn.next().equals("d"))
    {
      obj = new Dog();
    }
    else
    {
      obj = new Cat();
    }
    System.out.println(obj.toString());
    System.out.println(obj);
  } // end main
} // end Pets class
```

The `obj` reference variable can contain a reference to either a `Dog` object or a `Cat` object.

That object determines which version of the `toString` method is called here.

These two statements are equivalent.

Sample session:

```
Which type of pet do you prefer?
Enter d for dogs or c for cats: d
Woof! Woof!
Woof! Woof!
```

Figure 14.6 Driver for the Pets program that includes classes in Figures 14.4 and 14.5

According to our dynamic binding discussion, the code would work just fine. The JVM would see a `Dog` object in the `obj` reference variable and bind the `Dog`'s `display` method to the `obj.display` method call. But it doesn't matter that the code works fine in terms of dynamic binding. The code won't compile successfully because the compiler senses there might be a problem.

When the compiler sees a method call, *<reference-variable>*.*<method-name>*`()`, it checks to see if the reference variable's class contains a method definition for the called method. Note the `obj.toString` and `obj.display` method calls in the following examples. In the left example, the compiler checks to see if `obj`'s class, `Object`, contains a `toString` method. Since it does, the code compiles successfully. In the right example, the compiler checks to see if `obj`'s class, `Object`, contains a `display` method. Since the `Object` class does not contain a `display` method, the code produces a compilation error.

```
Object obj = new Dog();
System.out.println(obj.toString());
```
legal

```
Object obj = new Dog();
obj.display();
```
compilation error

Does this mean that polymorphism works only for the methods defined in the `Object` class? Fortunately, that's not the case. Later in this chapter, you'll learn how to make polymorphism work for any method.

The `instanceof` Operator

As you've seen, whenever a generic reference calls a polymorphic method, the JVM uses the type of the referenced object to decide which method to call. Figure 14.7 shows how you can do a similar thing explicitly

```
/*************************************************************
* Pets2.java
* Dean & Dean
*
* This illustrates use of the instanceof operator.
*************************************************************/

import java.util.Scanner;

public class Pets2
{
  public static void main(String[] args)
  {
    Scanner stdIn = new Scanner(System.in);
    Object obj;

    System.out.print("Which type of pet do you prefer?\n" +
      "Enter d for dogs or c for cats: ");
    if (stdIn.next().equals("d"))
    {
      obj = new Dog();
    }
    else
    {
      obj = new Cat();
    }
    if (obj instanceof Dog)
    {
      System.out.println("Wag tail");
    }
  } // end main
} // end Pets2 class
```

This condition evaluates to `true` if the object referred to is an instance of the `Dog` class or a class descended from the `Dog` class.

Sample session:

```
Which type of pet do you prefer?
Enter d for dogs or c for cats: d
Wag tail
```

Figure 14.7 Demonstration of the `instanceof` operator

in your code. In particular, suppose you want to see if a referenced object is an instance of some particular class. You can do this with a special operator called the `instanceof` operator (notice that the "o" in `instanceof` is lowercase). Using the Pets example again, suppose you want to print "Wags tail" if `obj`'s object is an instance of class `Dog` or any class descended from class `Dog`. You can do that with the `if` statement at the bottom of the `main` method in Figure 14.7. Thus, the `instanceof` operator provides a simple and direct way to sort out the various object types that might be referred to a by a generic reference variable.

14.6 Assignments When the Two Sides' Classes are Different

Let's now look at something that's quite common with polymorphic programs—assigning an object to a reference where the object's class and the reference's class are different. In the following code fragment, assume that `Student` is a subclass of `Person`. What does this code fragment do?

```
Person p = new Student();
Student s = new Person();
```

This generates a compile-time error.

The first line assigns a `Student` object (actually a reference to a `Student` object) to a `Person` reference variable. It's assigning a subclass object to a superclass reference variable. That's a legal assignment because a `Student` "is a" `Person`. It's going up the inheritance hierarchy—the direction in which automatic type promotion occurs. The second line tries to assign a `Person` object to a `Student` reference variable. It's trying to assign a superclass object to a subclass reference variable. That's illegal because a `Person` is not necessarily a `Student`. The second line generates a compile-time error.

The "is a" mnemonic can help you remember the rule, but if you're a Curious George,[2] you probably want more. If you want to understand the true rationale behind the rule, here goes. It's OK to assign a descendant-class object into an ancestor-class reference variable because all the compiler cares about is whether the assigned-in descendant-class object has all the members that any object of the reference variable's class should have. And if you assign a descendant-class object to an ancestor-class reference variable, it does. Why? Because descendant-class objects always inherit all ancestor-class members!

As with primitives, if there is compatibility, you can go the other way by using a cast. In other words, you can use a cast to force an object referred to by a more generic reference variable into a more specific type—a type that's below it in the same inheritance hierarchy. For example, if `p` is a `Person` reference variable, and `Student` inherits from `Person`, the compiler will accept this:

```
Student s = (Student) p;
```

Although the compiler will accept this statement, that does not necessarily mean the program will run successfully. For successful execution, when dynamic binding occurs, the object actually referred to by the `p` reference variable must be at least as specific as a `Student`. That is, the referenced object must be either an instance of the `Student` class or an instance of a descendant of the `Student` class. Why? Because after the assignment of the reference to a `Student` reference variable, the object will be expected to have all of the members that a `Student` has, which is generally more than all the members a `Person` has.

[2]Curious George is the main character in a series of books written by Margret and H. A. Rey. George is a curious monkey. Author John's grade-schooler, Caiden, is a Curious-George wannabe.

14.7 Polymorphism with Arrays

So far, you've seen polymorphism in the context of code fragments and a simple Pets program. Those examples served their purpose—they illustrated the basics. But they didn't illustrate the real usefulness of polymorphism. The real usefulness of polymorphism comes when you have an array or `ArrayList` of generic reference variables and assign different types of objects to different elements. That allows you to step through the array or `ArrayList` and call a polymorphic method for each element. At runtime, the JVM uses dynamic binding to pick out the particular method that applies to each type of object found.

Polymorphism in an Explicit Inheritance Hierarchy

The Pets program used polymorphic `toString` methods for the `Dog` and `Cat` classes. The compiler accepted the `Object` reference variable with the `toString` method calls because the `Object` class defines its own `toString` method. Recall that polymorphism did not work for `Dog` and `Cat` `display` methods because the `Object` classs does not define its own `display` method. Suppose the method you want to make polymorphic is not defined in the `Object` classs. How can you have polymorphism and still satisfy the compiler? Actually, there are several related ways. One way is to create a superclass for the classes that define the different versions of the polymorphic method, and define the method within the superclass. Then use that superclass name when declaring the polymorphic reference variable(s). Another way to satisfy the compiler is to *declare the method* (specify the method heading only) in an `abstract` ancestor class and then use that ancestor class name for the reference variable type. Still another way to satisfy the compiler is to `implement` an *interface* that declares the method and then use that interface name for the reference variable type. We'll illustrate the first way in this section and the other two ways in subsequent sections.

Payroll Example

To illustrate polymorphism in an explicit inheritance hierarchy, we'll develop a payroll program that uses dynamic binding to select the appropriate method for calculating an employee's pay. Employees that happen to be salaried get dynamically bound to a `Salaried` class's `getPay` method. Employees that happen to be hourly get dynamically bound to an `Hourly` class's `getPay` method.

Let's start with the UML class diagram in Figure 14.8. It describes the Payroll program's class structure. `Employee` is a superclass and `Salaried` and `Hourly` are subclasses. The fourth class, `Payroll`, is the program driver. Its `main` method drives the `Salaried` and `Hourly` classes by instantiating them and then calling their methods. What is the association between `Payroll` and the other classes—inheritance or composition/aggregation? The UML class diagram's solid (black) diamonds indicate composition associations between the `Payroll` container and the `Salaried` and `Hourly` components. That should make sense when you realize that the `Payroll` class "has a" heterogeneous array of `Salaried` and `Hourly` objects. Assuming the `Payroll` class has exclusive control over these objects, its association with them is a composition, and the diamonds should be solid.

Before we get bogged down in the code for what is a fairly substantial program, let's look at how the program will work for some hypothetical employees. Simon is a salaried employee, with an annual salary of $48,000. The program pays salaried employees twice a month, so Simon gets paid $2,000 on the 15th and 30th of every month (the program makes the simplifying assumption that months have 30 days). Anna and Donovan are hourly employees, who work 8 hours a day, 5 days a week (Monday through Friday). With a pay rate of $25 per hour, Anna earns $1000 per week. With a pay rate of $20 per hour, Donovan

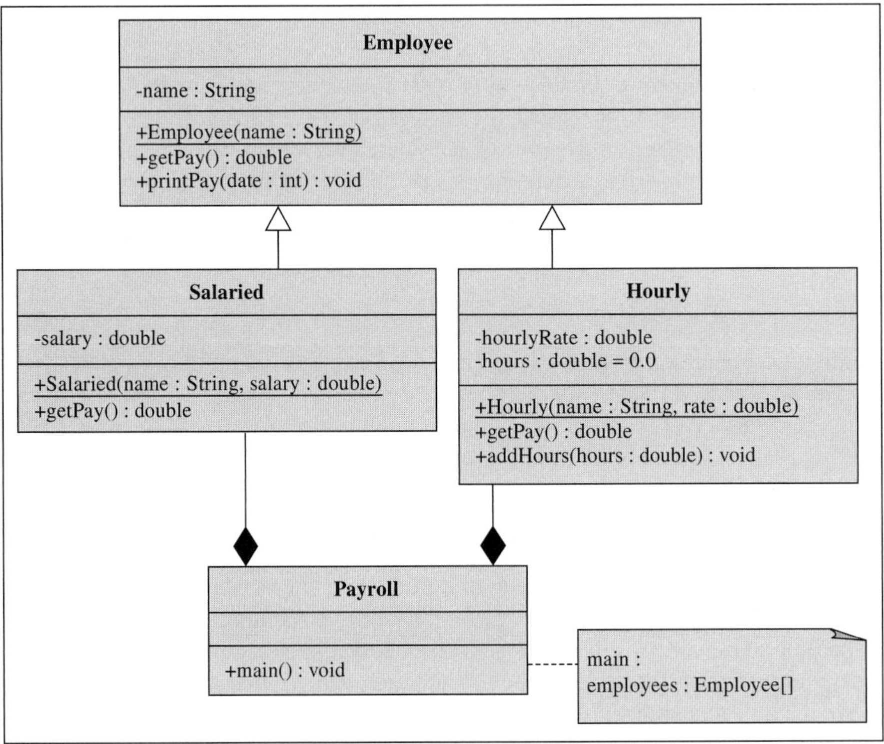

Figure 14.8 Class diagram for the Payroll program

earns $800 per week. If the first pay period spans Tuesday through Friday, with Friday occurring on the 4th, then this is what the program will print for a month's worth of paychecks:

Output:

```
 4      Anna:   800.00
 4   Donovan:   640.00
11      Anna:  1000.00
11   Donovan:   800.00
15     Simon:  2000.00
18      Anna:  1000.00
18   Donovan:   800.00
25      Anna:  1000.00
25   Donovan:   800.00
30     Simon:  2000.00
```

Let's begin implementation with the `main` method in the driver in Figure 14.9. Note `main`'s local variable, `employees`. It's declared to be a 100-element array of `Employee` objects. That's what it's declared as, but that's not exactly what it holds. As you can see from the assignment statements, the first three `employees` elements are an `Hourly`, a `Salaried`, and another `Hourly`. This is a heterogeneous array. All the elements in the array are instances of classes derived from the array's class, and none of them is an instance of the `Employee` class itself. Even though there may be no instances of the array's class in the array,

```
/*************************************************************
* Payroll.java
* Dean & Dean
*
* This class hires and pays employees.
*************************************************************/

public class Payroll
{
  public static void main(String[] args)
  {
    Employee[] employees = new Employee[100];
    Hourly hourly;
    employees[0] = new Hourly("Anna", 25.0);
    employees[1] = new Salaried("Simon", 48000);
    employees[2] = new Hourly("Donovan", 20.0);

    // This arbitrarily assumes that the payroll's month
    // starts on a Tuesday (day = 2), and it contains 30 days.
    for (int date=1,day=2; date<=30; date++,day++,day%=7)
    {
      for (int i=0;
        i<employees.length && employees[i] != null; i++)
      {
        if (day > 0 && day < 6
          && employees[i] instanceof Hourly)
        {
          hourly = (Hourly) employees[i];
          hourly.addHours(8);
        }
        if ((day == 5 && employees[i] instanceof Hourly) ||
          (date%15 == 0 && employees[i] instanceof Salaried))
        {
          employees[i].printPay(date);
        }
      } // end for i
    } // end for date
  } // end main
} // end class Payroll
```

This selects appropriate elements.

This casts elements into their native class.

This selects the appropriate time to print each different type.

Figure 14.9 Driver for the simple Payroll program

the array's type is the right type to use because it is able to accommodate instances of all classes descended from the array's class.

Continuing with the `main` method, the outer `for` loop steps through 30 days, keeping track of two variables. Notice how the first compartment in the `for` loop header declares more than one variable of the specified type. The `date` variable represents the date of the month. It determines when salaried employees are paid. For simplicity, this program assumes 30 days per month. If you want to learn how to get the actual number of days in each month, go to Oracle's Java API website and read up on the `Calendar` class.[3] The `day` variable represents the day of the week. It determines when hourly employees are paid. Assuming that day 1 is a Monday, because the initial value of `day` is 2, the program's month starts on a Tuesday. Notice how the third compartment of the `for` loop header executes more than one operation. It increments both `date` and `day` and then uses `day%=7` to make the `day` variable roll over to 0 whenever it reaches 7.

The inner `for` loop steps through the heterogeneous array of employees. The for loop header's second component employs a compound continuation condition. The `i<employees.length` condition alone would allow looping through all 100 elements of the `employees` array. What's the point of the `for` loop header's `employees[i] != null` condition? The program instantiates only three objects for this array, and 97 elements still contain the default value of `null`. If the program tries to call a method with a `null` reference, it crashes. More specifically, it generates a `NullPointerException` error the first time it tries to use the `null` reference. The `employees[i] != null` condition avoids that by stopping the loop when it gets to the first `null` element.

Inside the inner `for` loop, the first `if` statement accumulates hours for hourly workers. It checks to see if `day` is a weekday (not 0 or 6). It also checks to see if the object referenced by the current array element is an instance of the `Hourly` class. This enables the program to accumulate hours only during working days of the week and only for hourly workers. Once we know that the actual object is an instance of the `Hourly` class, it's safe to cast the generic reference into an `Hourly` reference. So we proceed to cast that reference into type `Hourly`, assign it to an `Hourly` reference variable, then use that specific type of reference variable to call the `addHours` method. Why did we jump through those hoops? Suppose we tried to call the `addHours` method with our generic reference in a statement like this:

```
employees[i].addHours(8);
```

The compiler would generate the following error message:

```
cannot find symbol
symbol : method addHours(int)
```

Because there is no `addHours` method in the `Employee` class, but there is one in the `Hourly` class, we must cast the array element explicitly into an `Hourly` type of reference and use that reference to call the method that we need.

Now look at the second `if` statement in the inner `for` loop. Instead of accumulating hours, its purpose is generating output for a payroll report. This `if` statement executes if either of two or'd conditions is `true`. The first condition is `true` if it's Friday (day = 5) and if the calling object is an instance of the `Hourly` class. The second condition is `true` if it's the middle of the month and if the calling object is not an instance of the `Hourly` class. If either of these conditions is satisfied, the raw array element calls the method like this:

```
employees[i].printPay(date);
```

This strategy wouldn't have worked with the `addHours` method in the first `if` statement, but it does work with the `printPay` method in the second `if` statement. Why? Look at the UML specification of the `Employee` class in Figure 14.8. This time, the method being called, `printPay`, is supposed to be defined in the array's class.

Now let's work on implementation of that `printPay` method in the `Employee` class. In Figure 14.10, note how `printPay` prints the date and the employee's name and then calls `getPay`. The `getPay` method is supposed to calculate an employee's pay. But the `Employee` class's `getPay` method simply returns 0.0.

[3] Here's an example of how the last day in the current month can be found:

```
int lastDayInCurrentMonth =
  Calendar.getInstance().getActualMaximum(Calendar.DAY_OF_MONTH);
```

What's up with that? Are employees really paid nothing? Certainly not! The `Employee` class's `getPay` method is simply a dummy method that's never executed. The "real" `getPay` methods (i.e., the ones that are executed) are the overriding definitions in the `Salaried` and `Hourly` subclasses. These overriding definitions make the `getPay` method polymorphic! How does the JVM know to use those methods and not the dummy method in the `Employee` class? When it performs dynamic binding, the JVM looks at the method's calling object. For the `getPay` case, the calling object is an instance of either the `Salaried` class or the `Hourly` class. Can you see why?

Go back to Figure 14.9's `main` method and note the assignment of an `Hourly` object into `employees[0]`. When `employees[i].printPay()` gets called with `i` equal to 0, the calling object is an `Hourly` object. Within the `printPay` method, when `getPay` is called, the calling object is still an `Hourly` object. Therefore,

```
/*********************************************
* Employee.java
* Dean & Dean
*
* This is a generic description of an employee.
*********************************************/

public class Employee
{
  private String name;

  //*******************************************

  public Employee(String name)
  {
    this.name = name;
  }

  //*******************************************

  public void printPay(int date)
  {
    System.out.printf("%2d %10s: %8.2f\n",
      date, name, getPay());
  } // end printPay

  //*******************************************

  // This dummy method satisfies the compiler.

  public double getPay()
  {
    System.out.println("error! in dummy");
    return 0.0;
  } // end getPay
} // end class Employee
```

Figure 14.10 Employee class

the JVM uses the `Hourly` class's `getPay` method. And that's what we want—the `employees[0]` object is an `Hourly`, so it uses the `Hourly` class's `getPay` method, which resets `hours` after each payment. The same argument can be applied to the `employees[1]` object. Because it's a `Salaried` object, it uses the `Salaried` class's `getPay` method, which is independent of hours. Thanks to polymorphism and dynamic binding, life is good.

The really cool thing about polymorphism and dynamic binding is being able to program generically. In the `main` method, we can call `printPay` for all the objects in the array and not worry about whether the object is `Hourly` or `Salaried`. We just assume that `printPay` works appropriately for each employee. This ability to program generically enables programmers to think about the big picture without getting bogged down in details.

In the `Employee` class, were you bothered by the dummy `getPay` method? Were you thinking, "Why include a `getPay` method in the `Employee` class even though it's never executed?" It's needed because if there were no `getPay` method in the `Employee` class, the compiler would generate an error. Why? Because when the compiler sees a method call with no dot prefix, it checks to make sure that the method can be found within the current class. The `getPay()` method call (within the `printPay` method) has no explicit calling object, so the compiler requires the `Employee` class to have a `getPay` method.

Now it's time to implement the "real" `getPay` methods. See Figures 14.11 and 14.12. The methods in these two classes are both simple, but they are different. To keep the JVM from selecting the dummy `getPay` method in the base class during dynamic binding, all derived classes should override that method.

```
/*********************************************
* Salaried.java
* Dean & Dean
*
* This class implements a salaried employee.
*********************************************/

public class Salaried extends Employee
{
  private double salary;      // per year

  //******************************************

  public Salaried(String name, double salary)
  {
    super(name);
    this.salary = salary;
  } // end constructor

  //******************************************

  @Override
  public double getPay()
  {
    return this.salary / 24;  // per half month
  } // end getPay
} // end class Salaried
```

Figure 14.11 `Salaried` class

```
/*************************************************************
* Hourly.java
* Dean & Dean
*
* This class implements an employee paid by the hour.
*************************************************************/

public class Hourly extends Employee
{
  private double hourlyRate;
  private double hours = 0.0;

  //**********************************************************

  public Hourly(String name, double rate)
  {
    super(name);
    hourlyRate = rate;
  } // end constructor

  //**********************************************************

  // Postcondition: This resets hours to zero.

  @Override
  public double getPay()
  {
    double pay = hourlyRate * hours;
    hours = 0.0;
    return pay;
  } // end getPay

  //**********************************************************

  public void addHours(double hours)
  {
    this.hours += hours;
  } // end addHours
} // end class Hourly
```

Figure 14.12 The `Hourly` class

14.8 abstract Methods and Classes

The dummy `getPay` method in Figure 14.10 is an example of a *kludge* (pronounced "klooj"). Kludgy code is ugly, inelegant code that provides a workaround for a problem. Usually, inelegant code is hard to understand. And hard-to-understand code is hard to maintain. So try to avoid kludges. Sometimes that's not possible, but in this case, we can indeed avoid the dummy-method kludge. Here's how. . . .

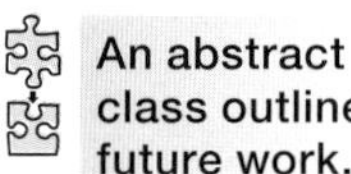
An abstract class outlines future work.

If you find yourself writing a dummy method that will be overridden by methods defined in all instantiable descendant classes, stop and reconsider. There's a better way. Use an *abstract class* to tell the compiler what you're trying to do ahead of time. In the abstract class, provide an abstract method declaration for each method that is merely a "dummy" method for overriding method(s) in subclass(es). To declare a method, just write the method heading with the additional modifier `abstract`, and terminate this modified method heading with a semicolon. For example, note the `abstract` modifier in the `Employee2` class heading in Figure 14.13.

An `abstract` declaration doesn't contain enough information to define the method. It just specifies its outside-world interface and says that definition(s) will exist somewhere else. Where? In all instantiable descendant classes! Using an `abstract` method avoids the inelegant dummy method definition, and it's a better way to implement polymorphism.

The `abstract` modifier is well named. Something is abstract if it is general in nature, not detailed. An `abstract` method declaration is general in nature. It doesn't provide method details. It just serves notice that the method exists and that it must be fleshed out by "real" method definitions in all instantiable descendant classes. Have we followed this rule for our program? In other words, do we have definitions of the `getPay` method in all the `Employee2` descendant classes? Yes, the `Salaried` and `Hourly` classes

```
/*****************************************************
* Employee2.java
* Dean & Dean
*
* This abstract class describes employees.
*****************************************************/

public abstract class Employee2
{
  private String name;
  public abstract double getPay();

  //*************************************************

  public Employee2(String name)
  {
    this.name = name;
  }

  //*************************************************

  public void printPay(int date)
  {
    System.out.printf("%2d %10s: %8.2f\n",
      date, name, getPay());
  } // end printPay
} // end class Employee2
```

If there's an `abstract` method, the class is `abstract`, too.

This `abstract` method declaration replaces the dummy method definition.

Figure 14.13 `Employee2` class, using the `abstract` modifier to replace a dummy method definition with a simpler method declaration

in Figures 14.11 and 14.12 already contain the required getPay method definitions. However, we need to revise the Salaried, Hourly, and Payroll classes by making these replacements:

```
Employee → Employee2
Salaried → Salaried2
Hourly → Hourly2
Payroll → Payroll2
```

Then the Salaried2, Hourly2, and Payroll2 classes will start out looking like this:

```
public class Salaried2 extends Employee2
{
  ...

public class Hourly2 extends Employee2
{
  ...

public class Payroll2
{
  public static void main(String[] args)
  {
    Employee2[] employees = new Employee2[100];
    ...
```

Here's another thing to note when declaring an abstract method. Because an abstract method declaration does not provide a definition for that method, the class definition is incomplete. Because the class definition is incomplete, it can't be used to construct objects. The compiler recognizes this and complains if you don't recognize it in your code. To satisfy the compiler, you must add an abstract modifier to the class heading whenever you have a class that contains one or more abstract methods. For example, note the abstract modifier in the Employee2 class heading in Figure 14.13.

Adding an abstract modifier to a class heading makes it impossible to instantiate an object from that class. If a program attempts to instantiate an abstract class, the compiler generates an error. For example, because Employee2 is an abstract class, we'd get a compilation error if we had a main method like this:

```
public static void main(String[] args)
{
  Employee2 emp = new Employee2("Benji");
}
```

Because Employee2 is abstract, this generates a compilation error.

Sometimes you don't want a subclass to define a method that was declared to be abstract in its superclass. Instead you want to defer the method definition to the next generation. It's easy to do this. In the subclass, just ignore that method and declare the subclass abstract also (because at least that method is still undefined). You can defer method definitions like this as far as you want, provided you ultimately define them all in any non-abstract descendant class you use to instantiate objects.

We have said that if any method in a class is abstract, that class must be abstract. But this does not mean all methods in an abstract class must be abstract. It's frequently useful to include one or more non-abstract method definitions in an abstract class. Thus, classes descended from an abstract class can inherit non-abstract methods from that class and are not required to redefine those non-abstract methods.

Do you recall the admonition in the margin in the previous chapter's Section 13.3? It said, "Plan to instantiate leaves only." You can enforce this plan by declaring all intended non-leaf classes to be `abstract`. Although they cannot be instantiated, these `abstract` classes can reduce programming work and improve consistency by defining common, reusable non-abstract methods.

Illegal to Use `private` or `final` with `abstract`

An `abstract` method declaration cannot be `private`, and the definitions of the method that appear in descendant classes cannot be `private` either. Why? An `abstract` method declaration must mirror the headings for all its associated overriding methods. So if an `abstract` method declaration were allowed to be `private`, then its associated overriding methods would have to be `private` too. But `private` doesn't make sense for methods that override `abstract` methods. After all, the point of overriding an `abstract` method is to enable a superclass to call the subclass's overriding method. For example, in Figure 14.13, the overriding method `getPay` is called from `printPay`, which is in the `Employee2` superclass. That sort of call would be impossible if the subclass method were `private`. Because `private` doesn't make sense for methods that override `abstract` methods, Java's designers decided that `private` is not allowed for `abstract` method declarations and for methods that override `abstract` methods.

An `abstract` class or method cannot be `final`. The `final` modifier keeps a class from being extended and keeps a method from being overridden. But an `abstract` class is supposed to be extended and an `abstract` method is supposed to be overridden, so it's illegal to use `final` with `abstract`.

14.9 Interfaces

Java interfaces can do lots of different things, and one of those things is help implement polymorphism. But before we get into that, we'd like to mention a couple of other uses of a Java interface.

Using Interfaces to Standardize Inter-Class Communication

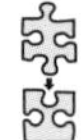

Establish communication protocols early.

The most obvious use of a Java interface is what its name implies—to specify the headings for a set of methods that a class must implement. A Java interface is a contract between a program designer and program implementers that standardizes communication among different classes. This use of interfaces is essential to the success of large programming projects.

Suppose, for example, that you are designing an accounting system, and you're currently focusing on "asset" accounts, which keep track of the value of things the company owns or has rights to. Typical asset accounts are Cash, Accounts Receivable, Inventory, Furniture, Manufacturing Equipment, Vehicles, Buildings, and Land. These things are different from each other, so it would not be natural for classes representing them to be in a single inheritance hierarchy. Some of these accounts (Furniture, Manufacturing Equipment, Vehicles, and Buildings) describe long-term or "fixed" assets whose values depreciate gradually over time. Each year, an accountant prepares a set of financial statements, like the Balance Sheet and a Profit and Loss Statement. This preparation requires access to the objects representing the depreciating assets to get information like original cost, date of acquisition, and depreciation rate.

To facilitate this access, references to these objects could be in a common array or `ArrayList`. Then a program could step through that array or `ArrayList` and call identically named polymorphic `get` methods to retrieve values of the `originalCost`, `acquisitionDate`, and `depreciationRate` instance variables in objects that represent depreciating assets. Suppose different programmers are writing the classes for different accounts. The best way to assure that all programmers are "reading from the same page" is to require that all classes that access a certain set of data implement the same Java interface. In our accounting system example, the interface for the `get` methods that access `originalCost`, `acquisitionDate`, and `depreciationRate` instance variables might be called the `AssetAging` interface. The `AssetAging` interface would contain declarations/headings for its methods, but not definitions.

If a particular class includes a definition of all methods declared in some interface (like `AssetAging`), you can tell the world (and the Java compiler) that that class provides such definitions by appending an `implements` clause to its class heading, like this:

```
public <class-name> implements <interface-name>
{
  . . .
```

For multiple interfaces, separate their names with commas, like this:

```
public <class-name> implements <interface-name1>, <interface-name2>, . . .
{
  . . .
```

For inheritance and an interface, do it like this:

```
public <class-name> extends <parent-class-name> implements <interface-name>
{
  . . .
```

A given class can extend only one superclass, but it can `implement` any number of interfaces. A Java interface is like a "pure" `abstract` class. It's pure in that it never defines any methods. It's less versatile than an `abstract` class, however. It can't declare any `static` methods, it can't declare any variables, and it can't declare any instance constants. In other words, it provides only `public static final` named constants and only `public abstract` method declarations. Here's the syntax for an interface definition:

```
interface <interface-name>
{
  <type> <CONSTANT_NAME> = <value>;
   . . .
  <return-type> <method-name>(<type> <parameter-name> . . .);
   . . .
}
```

You begin the definition of an interface with the keyword, `interface`, just like you begin the definition of a class with the keyword, `class`. You define named constants in an interface in the same way that you define named constants in a class, and you declare methods in an interface by appending a semicolon to the method headings. Note that the keyword `public` does not appear anywhere in our syntax template. You may include the `public` modifier, but it's not necessary, and it's standard practice to omit it because it simply does not make sense for an interface or any of its components to be anything but `public`. Also, note that the keyword `abstract` does not appear anywhere in our syntax template. You may include an `abstract` modifier in any method declaration, and you may include the `abstract` modifier in the interface heading, but again, it's not necessary, and it's standard practice to omit it because it simply does not make sense to have an interface that is not completely `abstract`. Also, note that the keyword `static` does not appear anywhere in our syntax template. It's understood that any constant is `static`, and it's understood that any method is not `static`. Finally, note that the keyword `final` does not appear in our syntax template. It's understood that all constants are `final`, and because no methods are defined, it's understood that any method declaration is not `final`.

Using an Interface to Store Universal Constants

In addition to telling the world that your class defines a certain minimum set of methods, implementing an interface also gives your class free access to all the named constants which that interface defines. Putting common named constants into an interface and then giving multiple classes access to those named constants by having them implement that interface is a handy way to provide easy access to a large set of common physical constants and/or empirical factors or constants. You avoid duplicate definitions of those constants, and you don't have to use a class-name dot prefix to access those constants. In principle, you could use an inheritance hierarchy to provide direct access to common constants, but that would be bad practice because it would waste your one inheritance opportunity on nothing more than a bunch of constants. If you use an interface to do this, you're still free to use inheritance and/or additional interfaces for other purposes.

Using Interfaces to Implement Additional Polymorphisms

Now suppose that you have already created an inheritance hierarchy, and you are already using it to implement some particular polymorphism, as we did in our Payroll program. Then suppose that you want to add another polymorphism that doesn't fit the structure of the original inheritance hierarchy. For example, you might want a method to be polymorphic among only some of the classes in the original hierarchy, and/or you might want a polymorphism to include classes that are outside that hierarchy. A Java class cannot participate in more than one inheritance—it can extend only one other class. Thus, you cannot use `abstract` classes to support polymorphisms that span more than one inheritance hierarchy. But as the previous accounting system example suggests, you can span more than one inheritance hierarchy with a Java interface. And one of the principal reasons to use Java interfaces is to implement multiple polymorphisms.

Make it polymorphic without distorting inheritance.

To illustrate this, we'll enhance the previous Payroll program by adding two classes of commissioned employees.[4] One of those classes gets a "straight" commission. The other class gets a salary plus a commission. In both cases, the commission is based on a common fixed percentage of sales. Figure 14.14 contains

```
/*******************************************************
* Commission.java
* Dean & Dean
*
* This inteface specifies a common attribute
* and declares common behavior of commissioned employees.
*******************************************************/

interface Commission
{
  double COMMISSION_RATE = 0.10;

  void addSales(double sales);
} // end interface Commission
```

Figure 14.14 `Commission` interface, for an enhanced version of the Payroll program

[4]See Appendix 7 for a complete UML diagram of the enhanced Payroll program developed in this subsection.

the code for an interface that defines this fixed percentage as a named constant and declares a method that must be defined in all classes that implement the interface.

Figure 14.15 shows the code for a `Commissioned` class, which describes a class of employees who work on a straight commission. The `Commissioned` class extends Figure 14.13's `Employee2` class. `Employee2` is an `abstract` class, and as such, the `Commissioned` subclass must define all `Employee2`'s `abstract` methods. The only `abstract` method in the `Employee2` class is the `getPay` class, so the `Comissioned` class must define the `getPay` method, and yes, it does. This increases the total number of polymorphic `getPay` methods to three. In the `commissioned` class's heading, note the clause, `implements interface Commission`. This provides direct access to the `COMMISSION_RATE` named constant, which the `Commissioned` class's `getPay` method uses to do its job. When it implements the `Commission` interface, the `Commissioned` class also takes on an obligation. It must define all the methods declared in that interface. The only method declared in the `Commission` interface is the `addSales` method, and yes, the `Commissioned` class defines this method, too.

```
/*************************************************************
* Commissioned.java
* Dean & Dean
*
* This class represents employees on straight commission.
*************************************************************/

public class Commissioned extends Employee2 implements Commission
{
  private double sales = 0.0;

  //**********************************************************

  public Commissioned(String name)
  {
    super(name);
    this.sales = sales;
  } // end constructor

  //**********************************************************

  public void addSales(double sales)
  {
    this.sales += sales;
  } // end addSales

  //**********************************************************

  // Postcondition: This resets sales to zero.

  @Override
  public double getPay()
  {
    double pay = COMMISSION_RATE * sales;

    sales = 0.0;
    return pay;
  } // end getPay
} // end class Commissioned
```

The interface requires this method definition.
Inheritance from an abstract class requires this method definition.
The interface supplies this constant value.

Figure 14.15 `Commissioned` class, for straight-commission employees in enhanced Payroll program

```
/*************************************************************
* SalariedAndCommissioned.java
* Dean & Dean
*
* This class represents salaried and commissioned employees.
*************************************************************/

public class SalariedAndCommissioned
  extends Salaried2 implements Commission
{
  private double sales;

  //**********************************************************

  public SalariedAndCommissioned(String name, double salary)
  {
    super(name, salary);
  } // end constructor

  //**********************************************************

  public void addSales(double sales)
  {
    this.sales += sales;
  } // end addSales

  //**********************************************************

  // Postcondition: This resets sales to zero.

  @Override
  public double getPay()
  {
    double pay =
      super.getPay() + COMMISSION_RATE * sales;

    sales = 0.0;    // reset for next pay period
    return pay;
  } // end getPay
} // end class SalariedAndCommissioned
```

The interface requires this method definition.

The interface supplies this constant value.

This method overrides the method defined in the parent class.

Figure 14.16 SalariedAndCommissioned class, for employees in enhanced Payroll program

Figure 14.16 shows the code for a SalariedAndCommissioned class. This class extends the Salaried2 class. The Salaried2 class is like the Salaried class in Figure 14.11, except for one difference: whereas the Salaried class extends Employee, the Salaried2 class extends Employee2. The SalariedAndCommissioned class describes a class of employees that earn a salary and a commission. The Salaried2's class defines a getPay method, so the compiler does not insist that the SalariedAndCommissioned class also define a getPay method, but logically, we need to override the Salaried2 getPay method. Notice how the overriding method uses the super prefix to call the method that it overrides. This additional getPay method definition increases the total number of polymorphic getPay methods to four.

The SalariedAndCommissioned class also implements the Commission interface. This provides direct access to the COMMISSION_RATE named constant, which the getPay method uses to do its job. Because it implements the `Commission` interface, the `SalariedAndCommissioned` class must define all methods declared in that interface, and yes, it does define the `addSales` method.

To execute those additional classes, we need a `Payroll3` class like that shown in Figure 14.17. The `Payroll3` class adds two more objects (Glen and Carol) to the array. Then it uses those objects to call the `addSales` methods in the new classes. To make these method calls, we cast the array elements into

```
/*************************************************************
* Payroll3.java
* Dean & Dean
*
* This class hires and pays four different types of employees.
*************************************************************/

public class Payroll3
{
  public static void main(String[] args)
  {
    Employee2[] employees = new Employee2[100];
    Hourly2 hourly;
    employees[0] = new Hourly2("Anna", 25.0);
    employees[1] = new Salaried2("Simon", 48000);
    employees[2] = new Hourly2("Donovan", 20.0);
    employees[3] = new Commissioned("Glen");
    employees[4] = new SalariedAndCommissioned("Carol", 24000);

    ((Commission) employees[3]).addSales(15000);
    ((Commission) employees[4]).addSales(15000);

    // This arbitrarily assumes that the payroll's month
    // starts on a Tuesday (day = 2), and it contains 30 days.
    for (int date=1,day=2; date<=30; date++,day++,day%=7)
    {
      for (int i=0; i<employees.length && employees[i] != null; i++)
      {
        if (day > 0 && day < 6
          && employees[i] instanceof Hourly2)
        {
          hourly = (Hourly2) employees[i];
          hourly.addHours(8);
        }
        if ((day == 5 && employees[i] instanceof Hourly2) ||
          (date%15 == 0 &&
            (employees[i] instanceof Salaried2 ||
            employees[i] instanceof Commissioned)))
        {
          employees[i].printPay(date);
        }
      } // end for i
    } // end for date
  } // end main
} // end class Payroll3
```

Figure 14.17 Driver for the third version of the Payroll program

the interface type. The compiler requires a cast because the `addSales` method does not appear in the `Employee2` class. Note that we need an extra set of parentheses surrounding the `(Commission)` cast operator and the calling object. We could have used more specific casts like this:

```
((Commissioned) employees[3]).addSales(15000);
((SalariedAndCommissioned) employees[4]).addSales(15000);
```

But it's more elegant to cast into the more generic `Commission` interface type and let the JVM select among the polymorphic alternatives as it does its dynamic binding. Using either type of casting, here's what the `Payroll3` driver generates:

Output:

```
 4        Anna:    800.00
 4     Donovan:    640.00
11        Anna:   1000.00
11     Donovan:    800.00
15       Simon:   2000.00
15        Glen:   1500.00
15       Carol:   2500.00
18        Anna:   1000.00
18     Donovan:    800.00
25        Anna:   1000.00
25     Donovan:    800.00
30       Simon:   2000.00
30        Glen:      0.00
30       Carol:   1000.00
```

In our coded examples, notice the similarity between the use of an interface name and the use of a class name! It's not possible to instantiate an interface because it's inherently `abstract`, but you can use it like any ordinary class to specify type. For example, you can declare an array of elements whose type is an interface name, you can populate that array with instances of classes that implement that interface, and then you can pull objects out of that array and cast them into any type (class or interface) that those objects conform to. The Payroll4 driver in Figure 14.18 and the subsequent output illustrate these possibilities.

The trick is to think about what the compiler needs and what the JVM does. For example, you can create an array of interface references because the elements in the array are just references, not instantiated objects. The compiler lets you populate that array with references to objects from classes that implement that interface because it knows those objects can call any method the interface declares. In a method call, the compiler lets you cast a reference into the type of any class that declares or defines any version of that method because it knows the JVM can find at least one method to bind. At runtime, the JVM selects the most appropriate method to bind.

14.10 The `protected` Access Modifier

So far, we've discussed only two modes of accessibility for a class's members—`public` and `private`; `public` members can be accessed from anywhere; `private` members can be accessed only from inside the members' class. There is another access modifier that is a limited form of the `public` access modifier—the

```
/***************************************************************
* Payroll4.java
* Dean & Dean
*
* This class hires and pays employees some kind of commission.
***************************************************************/

public class Payroll4
{
  public static void main(String[] args)
  {
    Commission[] people = new Commission[100];

    people[0] = new Commissioned("Glen");
    people[1] = new SalariedAndCommissioned("Carol", 24000);

    people[0].addSales(15000);
    people[1].addSales(15000);
    for (int i=0; i<people.length && people[i] != null; i++)
    {
      ((Employee2) people[i]).printPay(15);
    }
  } // end main
} // end class Payroll4
```

Although you can't instantiate an interface itself, you can declare interface references.

The compiler accepts this cast because Employee2 defines a printPay method, but the JVM binds the objects to methods in classes descended from Employee2.

Output:

```
15    Glen:   1500.00
15   Carol:   2500.00
```

Figure 14.18 Demonstration of class-like properties of an interface

protected access modifier. It specifies an accessibility that's between public and private. Members that are protected can be accessed only from within the same package[5] or from within the member's inheritance hierarchy. As you might recall, an inheritance hierarchy is the class in which the member resides, plus all its descendant classes.

When should you use the protected modifier? The general rule is that you should use it when you want easy access to a member, but you don't want to advertise it to the general public. In other words, you want it to have more exposure than a private member, but less exposure than a public member.[6] Hmmm . . . that's still kind of vague. Let's elaborate with an example.

[5] If you want to learn more about packages and how to group your classes into a programmer-defined package, see Appendix 4.

[6] Because a protected member can be accessed from any class descended from the class that defines the protected member, anyone could extend the class that defines the protected member and thereby gain direct access to it. In other words, the protected modifier doesn't actually provide much protection. If you're an outsider, stay away from someone else's protected members. Consider them to be non-standard products that are not guaranteed.

The Payroll Program with a `protected` Method

Suppose you want to enhance the Payroll program to include calculation of FICA taxes (FICA stands for Federal Insurance Contribution Act, and it funds the U.S. Social Security program). This tax calculation is best done in a separate method. Where should that method go? The only time this calculation will be done is when employees are paid. So, logically, it's a helper method called by the `getpay` method.

Where is the `getPay` method? It's a polymorphic method that has an overriding redefinition in all classes that directly and indirectly extend the `Employee` class—`Commissioned`, `Salaried`, `Hourly`, and `SalariedAndCommissioned`. But hey! This set of classes, together with the `Employee` class itself, make up the `Employee` hierarchical tree. So instead of repeating the definition of the FICA calculation in all classes that have a `getPay` method, it's more logical and more efficient to put this common FICA calculation in the tree's root class, `Employee`, and make it `protected`.

To avoid trampling on previous versions of the program, we use new class names in our new FICA enhanced Payroll program. See Figure 14.19. It shows the program's UML diagram with the new class names—`Payroll5`, `Employee3`, `Commissioned2`, `Salaried3`, `Hourly3`, and `SalariedAndCommissioned2`.

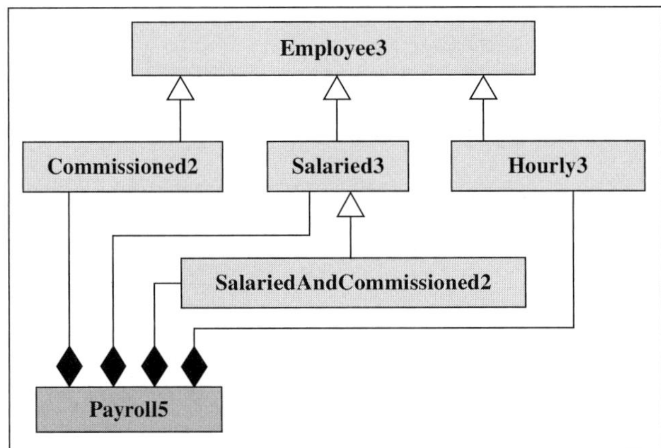

Figure 14.19 Abbreviated class diagram for an enhanced Payroll program

Figure 14.20 shows the definition of `Employee3`, which includes this additional common helper method, `getFICA`. `Employee3` also includes some named constants used in the FICA calculation. The details of this calculation are not relevant to the present discussion, so to save space, we are implementing it in a fairly cryptic form using the conditional operator. What this little `getFICA` method does is a reasonable representation of what actually happens to people's paychecks. So if you're curious, you might want to expand the cryptic code into a more readable form. (An end-of-chapter exercise asks you to do this.)

Each of the polymorphic `getPay` methods includes a call to this new `getFICA` method. The code for this call is essentially the same in each of the `getPay` methods, so we'll show it just once, in the `Salaried3` class in Figure 14.21.

For the most part, the `SalariedAndCommissioned2` class that extends the `Salaried3` class is like what's shown in Figure 14.16, with appropriate changes in the version numbers at the ends of the class names. However, in the `getPay` method, we cannot use `super.getPay()` to access `salary` in the `Salaried3` class, because the FICA tax makes the value returned by `Salaried3`'s `getPay` method different from the value of `salary`.

```
/***************************************************************
* Employee3.java
* Dean & Dean
*
* This abstract class describes employees, and it includes
* Social-Security tax calculation.
***************************************************************/

public abstract class Employee3
{
  public abstract double getPay();
  private String name;
  private final static double FICA_TAX_RATE = 0.08; // fraction
  private final static double FICA_MAX = 90000;     // dollars
  private double ytdIncome;         // total year-to-date income

  //************************************************************

  public Employee3(String name)
  {
    this.name = name;
  }

  //************************************************************

  public void printPay(int date)
  {
    System.out.printf("%2d %10s: %8.2f\n",
      date, name, getPay());
  } // end printPay

  //************************************************************

  // Postcondition: ytdIncome is increased by pay.

  protected double getFICA(double pay)
  {
    double increment, tax;

    ytdIncome += pay;
    increment = FICA_MAX - ytdIncome;
    tax = FICA_TAX_RATE *
      (pay < increment ? pay : (increment > 0 ? increment : 0));
    return tax;
  } // end getFICA
} // end class Employee3
```

This limits accessibility to classes in the subtree or in the same package.

Figure 14.20 Employee3 class, which includes the protected getFICA method

```
/*******************************************************
* Salaried3.java
* Dean & Dean
*
* This class represents salaried employees.
*******************************************************/

public class Salaried3 extends Employee3
{
  protected double salary;

  //*****************************************************

  public Salaried3(String name, double salary)
  {
    super(name);
    this.salary = salary;
  } // end constructor

  //*****************************************************

  @Override
  public double getPay()
  {
    double pay = salary;

    pay -= getFICA(pay);
    return pay;
  } // end getPay
} // end class Salaried3
```

This allows direct access from descendant classes.

This calls `protected` method at top of subtree.

Figure 14.21 Enhanced version of `Salaried` class that includes tax deduction

With the FICA tax, there must be another way to access `salary`. Although `Salaried3` could include a `getSalary` accessor method, the code would be simpler if `salary` were `public`. But would you want everybody's salary to be `public`? Probably not. The most appropriate thing to do here is to elevate the accessibility of the `salary` variable in the `Salaried3` class from `private` to `protected`. This gives descendant classes direct access to the `salary` variable, but it does not expose it as much as a `public` modifier would. Because the `SalariedAndCommissioned2` class `extends Salaried3`, if there is a `protected` modifier on the `salary` variable in `Salaried3`, you can define the `getPay` method in the `SalariedAndCommissioned2` class like this:

```
public double getPay()
{
  double pay = salary + COMMISSION_RATE * sales;

  pay -= getFICA(pay);
  sales = 0.0; // reset for next pay period
  return pay;
} // end getPay
```

`protected` in `Salaried3`

`protected` in `Employee3`

So there you have it. Polymorphism enables you to put heterogeneous objects into generic arrays whose type is either a class the objects' classes descend from or an interface the objects' classes implement. Then you can cast array elements into subclass or interface types, so the array elements can make method calls that are specific to their subclass or interface. The JVM finds the method that best matches the calling object and executes that method. The `protected` modifier allows direct access to variables and methods from anywhere in the `protected` member's subtree.

14.11 GUI Track: Three-Dimensional Graphics (Optional)

Now that you know how inheritance and interface-based polymorphism work, you should be able to follow some of the subtleties that make graphical painting work. This section provides a typical illustration of polymorphism in Java API usage.

The Java API provides several classes which together enable you to draw and color many two-dimensional shapes. In addition, the `Graphics2D` class includes two methods (`draw3DRect` and `fill3DRect`) that enable you to portray a simple three-dimensional shape. They draw a rectangle with shading that makes it look like the rectangle is either raised slightly above the page or depressed slightly below the page. In Chapters 17 and 18, we'll show you how you can use these two methods to simulate an unpressed or a pressed button. But that's about the extent of the help you can get from the Java API in the creation of what appear to be three-dimensional images.

Portraying a general three-dimensional image in Java requires consideration of geometry and trigonometric calculations. In this section, we'll give you a taste of this by portraying an arbitrarily oriented solid cylinder. Figure 14.22 shows a driver for a class that displays such an object. In the declarations section, the `JFrame` constructor instantiates a window called `frame`. The subsequent method calls, `setSize` and `setDefaultCloseOperation`, establish that window's size in pixels and what should happen when the user clicks the X box in its upper-right corner.

Now look at the two user prompts. This program uses a spherical coordinate system. In this kind of coordinate system, elevation is an angle that's like latitude. A zero elevation input says the cylinder should lie on a side, with its axis pointing at the equator. A plus or minus 90° elevation input says the cylinder axis should stand on an end, with its axis pointing at either the North Pole or the South Pole. Azimuth is an angle that's like east longitude. With elevation at zero, a zero azimuth input says the cylinder axis should point right at the viewer. A positive azimuth input says it should point to the right, and a negative azimuth input says it should point to the left. The `Cylinder` constructor call instantiates the `Cylinder` object, and the subsequent `add` method call puts that object in the window. The `setVisible` method call makes the window's contents visible. Figure 14.23 shows the resulting display for the input −15° elevation and +60° azimuth angles.

The class that defines this shape and describes how to paint it appears in Figures 14.24a, 14.24b, and 14.24c. In Figure 14.24a, notice that our `Cylinder` class extends the `JPanel` class, which is imported from the Java API. The instance variables include the basic attributes of the object—its height (`cylH`) and diameter (`cylD`). These are pixel values, which are inherently integers, but we declare them to be `double`. Why do we use `double` instead of `int`? Three-dimensional graphics typically involves a considerable amount of calculation. Declaring variables to be `double` forces automatic promotion of any `int` factors that might appear in expressions. This keeps track of fractional information and provides the best possible visual display.

The other instance variables describe attributes of the displayed image—its orientation angles and its illumination extremes. For simplicity, the program uses "white" illumination, which contains identical values

```
/***************************************************************
* CylinderDriver.java
* Dean & Dean
*
* This drives the Cylinder class.
***************************************************************/

import java.util.Scanner;
import javax.swing.*;    // for JFrame and JPanel

public class CylinderDriver
{
  public static void main(String[] args)
  {
    Scanner stdIn = new Scanner(System.in);
    JFrame frame = new JFrame("Three-Dimensional Cylinder");
    Cylinder cylinder;
    double elev;      // cylinder axis elevation angle in degrees
    double azimuth;   // cylinder axis azimuth angle in degrees

    frame.setSize(600, 600);
    frame.setDefaultCloseOperation(JFrame.EXIT_ON_CLOSE);
    System.out.print("Enter axis elevation (-90 to +90): ");
    elev = stdIn.nextDouble();
    System.out.print("Enter axis azimuth (-90 to +90): ");
    azimuth = stdIn.nextDouble();
    cylinder = new Cylinder(elev, azimuth);
    frame.add(cylinder);
    frame.setVisible(true);
  } // end main
} // end CylinderDriver class
```

Sample session:

```
Enter axis elevation (-90 to +90): -15
Enter axis azimuth (-90 to +90): 60
```

Figure 14.22 Driver for `Cylinder` class in Figures 14.24a, 14.24b, and 14.24c

for red, green, and blue components. The `c1` and `c2` variables represent the intensity of these components for shaded and directly illuminated surfaces, respectively. The program declares these variables to be of type `float` because that's the type of the parameters in the Java API `Color` constructor. Zero corresponds to pitch black, and unity corresponds to pure white, so the specified `c1` and `c2` values correspond to two different shades of gray—the darkest and lightest shades seen on the curved sides of the cylinder in Figure 14.23.

Note the `Cylinder` constructor in Figure 14.24a. This transforms the input elevation and azimuth angles from degrees to radians, and it also constrains the magnitudes of the input angles to less than 90°. This avoids spurious results. It lets the user see only one end of the cylinder, but because the other end is the same, the program still portrays everything of interest.

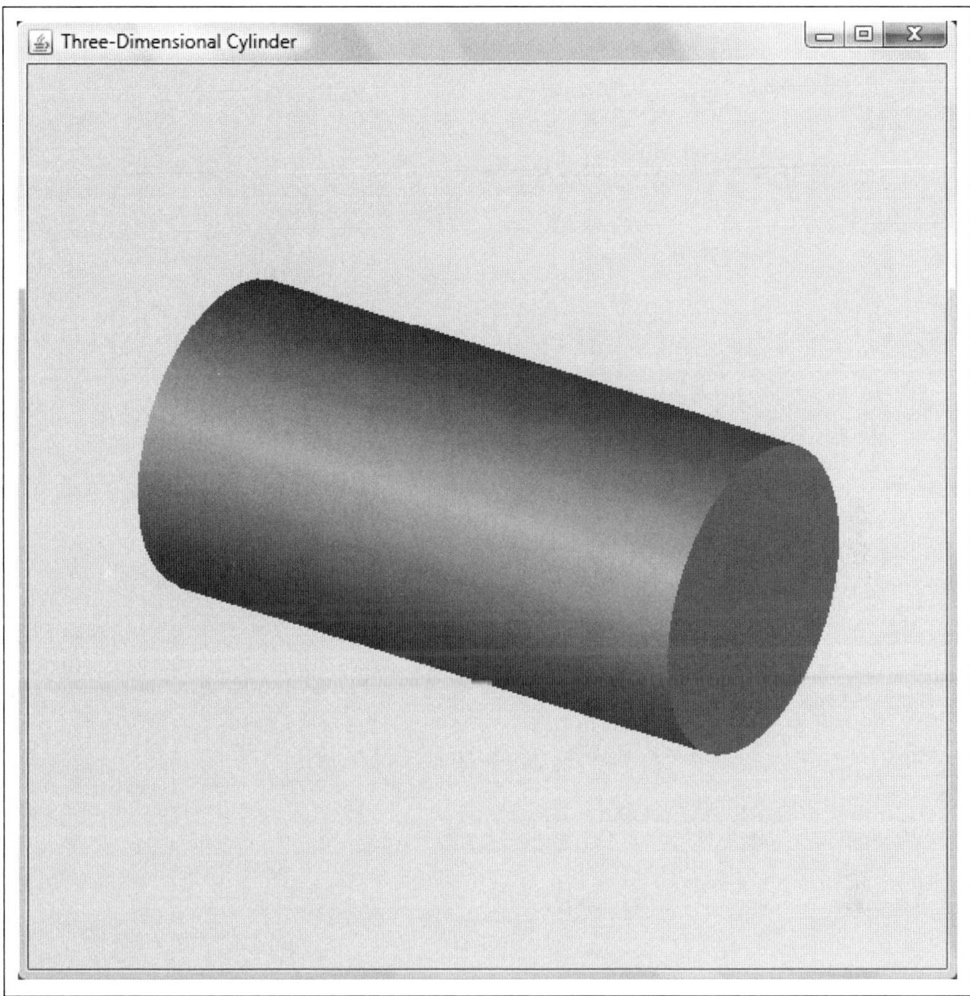

Figure 14.23 Three-dimensional portrayal of a solid cylinder

Now look at Figure 14.24b. This contains the first part of a large `paintComponent` method. The JVM automatically calls the `paintComponent` method when the program first runs and whenever a user does something to alter the contents of the program's window. (For example, if a user maximizes a window, the JVM calls the program's `paintComponent` method.) The `paintComponent` method defined here overrides a `paintComponent` method defined in the `JComponent` class, which is the superclass of the `JPanel` class and therefore an ancestor of the `Cylinder` class. To ensure that graphical components get painted properly, if you ever implement an overriding `paintComponent` method, that you should always call the `paintComponent` method the superclass defines or inherits as the first line in your overriding `paintComponent` method. Here's the relevant code from Figure 14.24b:

```
super.paintComponent(g);
```

```
/*************************************************************
* Cylinder.java
* Dean & Dean
*
* This displays a cylinder illuminated from viewing direction.
*************************************************************/

import javax.swing.JPanel;
import java.awt.*;            // for Graphics, Graphics2D, Color
import java.awt.Rectangle;
import java.awt.geom.*;       // for Ellipse2D and GeneralPath
import java.awt.GradientPaint;

public class Cylinder extends JPanel
{
  private double cylElev;     // cylinder axis elevation radians
  private double cylAzm;      // cylinder axis azimuth radians
  private double cylH = 400;  // cylinder height in pixels
  private double cylD = 200;  // cylinder diameter in pixels
  private float c1 = 0.3f;    // minimum illumination brightness
  private float c2 = 0.7f;    // maximum illumination brightness

  //*********************************************************

  public Cylinder(double elev, double azimuth)
  {
    cylElev = Math.toRadians(elev);
    if (Math.abs(cylElev) >= Math.PI / 2.0)
    {
      cylElev = Math.signum(cylElev) * Math.PI / 2.0001;
    }
    cylAzm = Math.toRadians(azimuth);
    if (Math.abs(cylAzm) >= Math.PI / 2.0)
    {
      cylAzm = Math.signum(cylAzm) * Math.PI / 2.0001;
    }
  }
```

inheritance from the API JPanel class

Figure 14.24a Cylinder class—part A

By definition, when a method overrides another method, the two methods must have the exact same signature (same name and same sequence of parameter types). Because the PaintComponent method in JComponent declares a Graphics object parameter, the PaintComponent method in Cylinder also declares a Graphics object parameter, named g. Even though g is declared as a Graphics object, the JVM actually passes a Graphics2D argument to the g parameter. That's good because the Cylinder class relies on the g parameter to perform sophisticated graphics operations found only in the Graphics2D class. The Graphics2D class is a subclass of Graphics. As you may recall from earlier in the chapter, it's always legal to pass a descendant-class argument into a ancestor-class -parameter. Also, you may recall that to call descendant-class methods with the ancestor-class parameter, first you need to assign the ancestor-class parameter into a descendant-class variable. Here's the relevant code from Figure 14.24b:

```
Graphics2D g2d = (Graphics2D) g;
```

```
//*************************************************************

@Override
public void paintComponent(Graphics g)
{
  super.paintComponent(g);
  final double MIDX = 0.5 * getWidth();
  final double MIDY = 0.5 * getHeight();
  Graphics2D g2d = (Graphics2D) g;
  double imageRotAngle;                // image rotation angle
  GeneralPath shape;                   // curved cylinder side
  float c;                             // current color level

  // Apparent tipping of cylinder
  double tipCosine = Math.cos(cylAzm) * Math.cos(cylElev);
  double tipSine = Math.sqrt(1.0 - tipCosine * tipCosine);
  double frontEndAngle =
    Math.acos(tipCosine) * 2.0 / Math.PI;

  // Minor diameter of end ovals & apparent cylinder height
  double minorD = cylD * tipCosine;
  double apparentH = cylH * tipSine;

  // Shapes of curved sides and oval ends
  Rectangle rectangle = new Rectangle(
    (int) Math.round(MIDX - cylD / 2),
    (int) Math.round(MIDY - apparentH / 2),
    (int) Math.round(cylD),
    (int) Math.round(apparentH));
  Ellipse2D.Double frontEllipse = new Ellipse2D.Double(
    (int) Math.round(MIDX - cylD / 2),
    (int) Math.round(MIDY - apparentH / 2 - minorD / 2),
    (int) Math.round(cylD),
    (int) Math.round(minorD));
  Ellipse2D.Double backEllipse = new Ellipse2D.Double(
    (int) Math.round(MIDX - cylD / 2),
    (int) Math.round(MIDY + apparentH / 2 - minorD / 2),
    (int) Math.round(cylD),
    (int) Math.round(minorD));

  // Color for sides of cylinder
  GradientPaint gradientPaint = new GradientPaint(
    (float) (MIDX - cylD / 2), 0.0f, new Color(c1, c1, c1),
    (float) (MIDX), 0.0f, new Color(c2, c2, c2), true);
```

The passed-in object's class must be in the `Graphics2D` subtree.

Figure 14.24b `Cylinder` class—part B

```
      // Rotate image from vertical around center
      if (cylElev == 0.0)
      {
        imageRotAngle =
          Math.signum(Math.sin(cylAzm)) * Math.PI / 2.0;
      }
      else
      {
        imageRotAngle =
          Math.atan(Math.sin(cylAzm) / Math.tan(cylElev));
        if (Math.tan(cylElev) < 0)
        {
          imageRotAngle += Math.PI;
        }
      }
      g2d.rotate(imageRotAngle, MIDX, MIDY);

      // Define and paint curved sides of cylinder
      shape = new GeneralPath(rectangle);
      shape.append(backEllipse, false);
      g2d.setPaint(gradientPaint);
      g2d.fill(shape);

      // Paint visible end of cylinder
      c = c2 - (float) ((c2 - c1) * frontEndAngle);
      g2d.setColor(new Color(c, c, c));
      g2d.fill(frontEllipse);
    } // end paintComponent
  } // end Cylinder class
```

Parameter type is `interface Shape`, which `Rectangle` implements.

Parameter type is `interface Paint`, which `GradientPaint` implements.

Figure 14.24c `Cylinder` class—part C

Notice that everything in Figure 14.24b is some kind of declaration. As before, the program uses `double` for the `MIDX` and `MIDY` pixel positions to preserve fractional information. For use later, it casts the incoming `Graphics` parameter into a more specific `Graphics2D` reference. After declaring the working variables, `imageRotAngle`, `shape`, and `c`, it begins a sequence of initialized declarations. These declarations actually implement most of the method's calculations. Using declarations to implement sequential steps in a calculation provides self-documentation. It makes a long tedious calculation easier to understand by giving intermediate variables understandable names.

You can use your understanding of geometry and trigonometry to verify the calculation of the apparent angle, apparent cylinder height, and minor diameter of the elliptical ends of a cylinder that's oriented as the input specifies. The declarations under "// Shapes of curved sides and oval ends" use these values to define a rectangle and two ellipses. The `rectangle` and `backEllipse` objects will help configure the cylinder's sides in a `GeneralPath` object called `shape`, and the `frontEllipse` object will enable the program to paint the cylinder's visible end.

The last declaration in Figure 14.24b instantiates a `GradientPaint` object. This establishes a color gradation that makes the sides of the cylinder appear to be round. In the range between x = (`MIDX` –`clyD` / 2) and x = (MIDX), this creates 16 narrow vertical stripes with color varying linearly from an intensity given by c1 on the left to an intensity given by c2 on the right. The true argument makes the method ramp the shading back down again in the range between x = (MIDX) and x = (MIDX + clyD / 2).

Notice that the rectangle and both ellipses are defined in a vertical orientation, but the cylinder in Figure 14.23 is not vertical! It is oriented at an angle sloping down and to the right. This requires a rotation. Now look at Figure 14.24c. The subordinate statement in the else part of the if statement computes the amount of rotation. The call to Graphics2D's rotate method after the else clause tells the computer how to perform this rotation around the window's midpoint. But the rotation doesn't actually occur until after the object is painted.

The next step is instantiating a GeneralPath object and assigning it to the reference variable called shape. Initially, this shape is nothing more than our previously defined rectangle. Then shape's append method adds the backEllipse component, so shape now includes everything in either the rectangle or the back-side ellipse.

Before it actually paints, the program must specify the coloring scheme with a setPaint method call. The setPaint parameter must be a reference to an object that implements the Paint interface. If you look up the API documentation for the GradientPaint class, you'll see that, yes, it does implement the Paint interface. Because our gradientPaint object is an instance of the GradientPaint class, and because the GradientPaint class implements the Paint interface, the program can call setPaint with a reference to the gradientPaint object as the argument. At this point, the program is ready to paint the sides of the cylinder with the fill(shape) method call. This tells the computer to perform the previously specified method of painting in the previously specified shape, and then rotate the result as previously specified.

The last three statements in the program paint the visible end of the cylinder. The first statement calculates intensity based on the angle of the visible end—using a lighter shade when we're looking directly at the end and a darker shade when we see it at a grazing angle. The setColor method call changes the paint mode from the previously established gradient painting to a flat gray having the just-computed intensity. The final fill(frontEllipse) method call tells the computer to paint the front ellipse shape and then rotate the result as previously specified.

Figure 14.24c contains several examples where API input parameter types are interfaces rather than classes. The GeneralPath constructor's parameter is of type Shape, where Shape is an interface. The object called rectangle conforms to this because its class, Rectangle, implements the Shape interface. The first parameter in GeneralPath's append method and the parameter in Grahpics2D's fill method are also of type Shape, The objects called backEllipse, shape, and frontEllipse all conform because their classes, Ellipse2D and GeneralPath, both implement the Shape interface, too. Graphics2D's setPaint method receives a parameter of type Paint, where Paint is another interface. The object called gradientPaint conforms to this because its class, GradientPaint, implements the Paint interface. These are all polymorphic references!

Summary

- The Object class is the ancestor of all other classes.
- To avoid using the Object class's equals method, for each of your classes, you should define an equals method that compares instance variable values.
- To avoid the Object class's mysterious response to a toString method call, for each of your classes, you should define a toString method that outputs a string concatenation of instance variable values.

- At compile time, the compiler confirms that a reference variable's class is able to handle each of the reference variable's method calls in some way. At runtime, the JVM looks at the particular type of the object referred to by the reference variable to determine which one of several alternative polymorphic methods should actually be called, and it binds the object to that method.
- The `instanceof` operator enables you to determine explicitly whether the object referred to by a reference variable is an instance of a particular class or descended from that class.
- You can always assign an object to a more generic reference variable, because the object's methods include methods inherited from the reference variable's class.
- You can safely cast a more generic reference into a more specific type only if you know the actual object referred to will be as specific as or more specific than the cast.
- You can implement polymorphism in an array of heterogeneous objects by declaring the array elements to be instances of a common inheritance ancestor. To satisfy the compiler, you can write a dummy method in that ancestor class and override it in all classes instantiated in the array. Or you can declare the method in an `abstract` ancestor class and then define overriding methods in all descendant classes instantiated in the array. Or you can declare the method in an interface and implement that interface in all classes instantiated in the array.
- A class can extend one inherited superclass and/or implement any number of interfaces.
- An interface provides simple access to common constants.
- The `protected` access modifier provides direct access to members of classes in the same package or in the inheritance subtree whose root is the class in which the protected member is declared.
- With the aid of explicit trigonometric calculations, you can use Java API classes to draw what appear to be three-dimensional objects.

Review Questions

§14.2 The `Object` Class and Automatic Type Promotion

1. If you want a class you define to inherit methods from the `Object` class, you must append the suffix `extends Object` to your class's heading. (T / F)

§14.3 The `equals` Method

2. When used to compare reference variables, the == operator works the same as the `Object` class's `equals` method. (T / F)
3. What does the `equals` method defined in the `String` class compare?

§14.4 The `toString` Method

4. What is returned by the `Object` class's `toString` method?
5. What's wrong with replacing the `println` statement in Figure 14.2's `main` method with these two statements?

```
String description = car.toString();
System.out.println(description);
```

6. The return type of an overriding method must be the same as the return type of the overridden method. (T / F)

§14.5 Polymorphism and Dynamic Binding

7. In Java, polymorphic method calls are bound to method definitions at compile time (not runtime). (T / F)

§14.6 Assignments When the Two Sides' Classes are Different

8. Assume one reference variable's class is descended from another reference variable's class. To be able to assign one reference variable to the other one (without using a cast operator), the left-side variable's class must be a(n) ____________ of the right-side reference variable's class.

§14.7 Polymorphism with Arrays

9. A given array may contain different types of elements. (T / F)

§14.8 `abstract` Methods and Classes

10. What are the syntax features of an `abstract` method?
11. Any class that contains an `abstract` method must be declared to be an `abstract` class. (T / F)
12. You cannot instantiate an abstract class. (T / F)

§14.9 Interfaces

13. You can use an interface to provide direct access to a common set of constants from many different classes. (T / F)
14. You can declare reference variables to have an interface type and use them just as you would use reference variables declared to be the type of a class in an inheritance hierarchy. (T / F)

§14.10 The `protected` Access Modifier

15. Describe the access provided by the `protected` modifier.
16. It's illegal to use `private` for any method that overrides an abstract method. (T / F)

Exercises

1. [after §14.3] Write a `sameColorAs` method for the `Car` class in Figure 14.1. It should return `true` if the compared cars' colors are the same, regardless of their other attributes.

2. [after §14.4] Write the output produced by the program in Figure 14.2.

3. [after §14.4] What does the following program output? For each dog's output, describe how the output is generated (be specific).

```
public class Animal
{
  public static void main(String[] args)
  {
    Animal sparky = new Dog();
    Animal lassie = new Animal();

    System.out.println(
      "sparky = " + sparky + "\tlassie = " + lassie);
  } // end main
} // end Animal

public class Dog extends Animal
{
  public String toString()
  {
    return "bark, bark";
  }
} // end class Dog
```

4. [after §14.4] What happens if you add an object of a class that does not define a `toString` method to an `ArrayList`, and then you try to print the `ArrayList`? (Assume that the object's class is a programmer-defined class that does not have an `extends` phrase in its heading.)

5. [after §14.5] Why is dynamic binding often called *late binding*?

6. [after §14.6] Given: `Animal` = superclass, `Dog` = subclass.
Identify all compilation errors in this code fragment. Provide a compilation error message if you like, but it's not required.

```
Animal animal;
Dog fido, sparky = new Dog();
animal = sparky;
fido = new Animal();
```

7. [after §14.6] Suppose you have an object called `thing`, but you are not sure what type it is, and you would like to have your program print out that type. The `Object` class (and therefore any class!) has another method, `getClass`, that returns a special object of type `Class` that contains information about the class of the object calling the `getClass` method. The `Class` class has a method called `getName` which returns the name of the class described by its calling object. Write a statement that prints the name of `thing`'s class.

8. [after §14.7] Given: `Animal` = superclass, `Dog` = subclass, `Cat` = subclass.
In the following code fragment, the bottom two lines generate compile-time errors. Provide corrected versions of those two lines. Preserve the spirit of the original code. For example, the bottom line should assign the second `animals` element into the `fluffy` variable.

```
Animal[] animals = new Animal[20];
animals[0] = new Dog();
animals[1] = new Cat();
Dog lassie = animals[0];
Cat fluffy = animals[1];
```

9. [after §14.7] (Case Study) Looking in a previously discussed GridWorld class, identify an example of a two-dimensional array that can hold objects of any type. Provide the class name and the variable declaration. List three types of objects that have gone into this array in previous executions of GridWorld programs.

10. [after §14.8] Each `abstract` method in a superclass must be overridden by a corresponding method in every non-`abstract` class descended from it. (T / F).

11. [after §14.8] Given the Pets3 program below, write an `abstract Animal2` class that contains just one item—an `abstract` declaration for a `speak` method. Write `Dog2` and `Cat2` classes that extend `Animal2`, so that when you run the Pets3 program and input either a 'c' or a 'd,' the program prints either "Meow! Meow!" or "Woof! Woof."

```
import java.util.Scanner;

public class Pets3
{
  public static void main(String[] args)
  {
    Scanner stdIn = new Scanner(System.in);
    Animal2 animal;

    System.out.print("Which type of pet do you prefer?\n" +
      "Enter c for cats or d for dogs: ");
    if (stdIn.nextLine().charAt(0) == 'c')
```

```
        {
          animal = new Cat2();
        }
        else
        {
          animal = new Dog2();
        }
        animal.speak();
      } // end main
    } // end Pets3 class
```

12. [after §14.9] (Case Study) Identify a GridWorld `abstract` class extended by a non-abstract class in previous executions of the `GridWorld` software. List the interfaces (method headings) of each of this `abstract` class's `abstract` methods and each of its non-abstract methods.

13. [after §14.9] Rewrite the `Commission` interface shown in Figure 14.14 to explicitly show the `abstract`, `public`, `static`, and `final` modifiers in every place where they apply. (Your elaborated interface definition should compile.)

14. [after §14.9] Change the `Pets3` class in Exercise 11, above, as follows: Replace all instances of `Pets3` by `Pets4`, and replace `Animal2` by `Animal3`. Then, write an `Animal3` interface and `Dog3` and `Cat3` classes that implement `Animal3`, so that when you run the Pets4 program and input either a 'c' or a 'd,' the program prints either "Meow! Meow!" or "Woof! Woof."

15. [after §14.9] (Case Study) Identify a GridWorld `interface` that is `implemented` by a class in previous executions of the `GridWorld` software. List the method headings of each method that this `interface` specifies.

16. [after §14.9] (Case Study) Describe GridWorld's `Location` class. What Java API interface does it implement? Identify its instance variables, summarize its class constants, and list its method headers.

17. [after §14.10] Expand the cryptic code in the `getFICA` method of the `Employee3` class in Figure 14.20 into "if else" statements so that the algorithm is easier to understand.

Review Question Solutions

1. False. Every class is a descendant of the `Object` class, so specifying `extends Object` is not necessary. In fact, it is undesirable because it prevents the extension of some other class.

2. True.

3. The `equals` method defined in the `String` class compares a string's characters.

4. The `Object` class's `toString` method returns a string concatenation of these three text components:
   - full classname
   - @ character
   - a hexadecimal hashcode value

5. Nothing. It's just a matter of style—whether you want more compactness or more self-documentation.

6. True.

7. False. At runtime. The JVM determines which method is called.

8. To be able to assign one reference variable to the other one (without using a cast operator), the left-side variable's class must be a(n) superclass/ancestor of the right-side reference variable's class.

9. True, if each element's type is either the type defined in the array declaration or a descendant of that type (or conforms to the interface that defines the array's type—see Section 14.9).

10. The syntax features of an `abstract` method are:
- The method heading contains the `abstract` modifier.
- There is a semicolon at the end of the heading.
- There is no method body.

11. True.

12. True.

13. True

14. True.

15. It is legal to access a `protected` member:
- from within the same class as the `protected` member
- from within a class descended from the `protected` member
- from within the same `package`

16. True. An `abstract` method must be `public` or `protected` (it cannot be `private`). An overriding method must be no more restrictive than its overridden method. Therefore, if a method overrides an `abstract` method, it cannot be `private`.

CHAPTER 15

Exception Handling

Objectives

- Understand what an exception is.
- Use `try` and `catch` blocks for numeric input validation.
- Understand how `catch` blocks catch an exception.
- Explain the difference between checked and unchecked exceptions.
- Look up exception details on Oracle's Java API website.
- Catch exceptions with the generic `Exception` class.
- Use the `getClass` and `getMessage` methods
- Catch exceptions with multiple `catch` blocks and multiple-exception `catch` parameters.
- Understand exception messages.
- Propagate exceptions back to the calling module with the help of a `throws` clause.
- Learn how to close resources explicitly or close them implicitly in a try-with-resources block.

Outline

15.1 Introduction

As you know, programs sometimes generate errors. Compile-time errors deal with incorrect syntax, like forgetting parentheses around an `if` statement condition. Runtime errors deal with code that behaves inappropriately, like trying to divide by zero. In previous chapters, we fixed compile-time errors by correcting the erroneous syntax, and we fixed runtime errors by making code more robust. In this chapter, we deal with errors using a different technique—exception handling. We'll describe exceptions more formally later on, but for now, think of an exception as an error, or simply something that goes wrong with a program. Exception handling is an elegant way to deal with such problems.

We start this chapter by looking at a common problem—making sure that users enter a valid number when they are asked for a numeric input. You'll learn how to implement such input validation using `try` and `catch` blocks, two of the key exception handling constructs. There are different types of exceptions, and you'll learn how to deal with the different types appropriately. In the chapter's final section, you'll use exception handling as part of a GUI line-plot program.

To understand this chapter, you need to be familiar with object-oriented programming, arrays, and inheritance basics. As such, you need to have read up through Chapter 13. This chapter does not depend on material covered in Chapter 14.

Different readers may want to read different parts of this chapter (Exception Handling) and the next chapter (Files, Buffers, Channels, and Paths). If you plan to read the next chapter, then you'll need to read this chapter in its entirety because the topic addressed in the next chapter, file manipulation, relies heavily on exception handling. On the other hand, if you plan to skip the next chapter and go directly to Chapters 17 and 18 (GUI programming) or to Chapters S17 and S18 on the website (FX GUI programming), then you'll need to read only the first part of this chapter, Sections 15.1 through 15.7.

15.2 Overview of Exceptions and Exception Messages

As defined by Oracle,[1] an *exception* is an event that disrupts the normal flow of instructions during the execution of a program. *Exception handling* is a technique for handling such exceptions gracefully.

The first exceptions we'll look at deal with invalid user input. Have you ever crashed a program (made it terminate ungracefully) due to invalid input? If a program calls the `Scanner` class's `nextInt` method and a user enters a non-integer, the JVM generates an exception, displays a big, ugly error message, and terminates the program. Here's a sample session that illustrates what we're talking about:

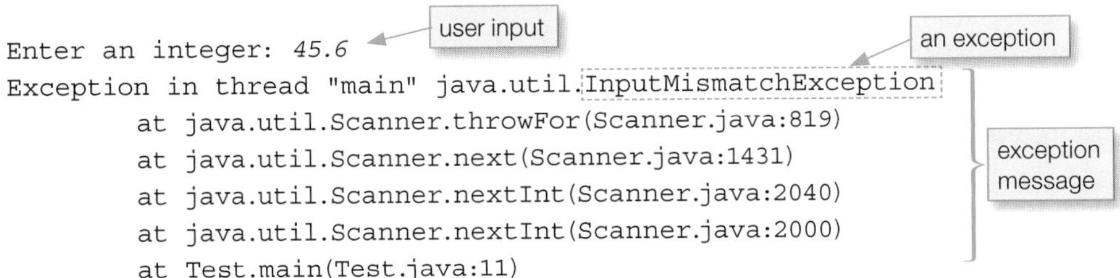

Enter an integer: 45.6
Exception in thread "main" java.util.InputMismatchException
at java.util.Scanner.throwFor(Scanner.java:819)
at java.util.Scanner.next(Scanner.java:1431)
at java.util.Scanner.nextInt(Scanner.java:2040)
at java.util.Scanner.nextInt(Scanner.java:2000)
at Test.main(Test.java:11)
user input
an exception
exception message

[1]http://download.oracle.com/javase/tutorial/essential/exceptions/index.html.

Note the `InputMismatchException` above. That's the type of exception that's generated when a user enters a non-integer in response to a `nextInt` method call. Note the *exception message*. Exception messages can be annoying, but they serve a useful purpose. They provide information about what's gone wrong. Toward the end of this chapter, we cover exception message details. But first, a more important issue—how to avoid getting ugly exception messages in the first place. Let us begin.

15.3 Using try and catch Blocks to Handle "Dangerous" Method Calls

Some method calls, like `nextInt`, are dangerous in that they can lead to exceptions, and exceptions can lead to program crashes. By the way, "dangerous" is not a standard exception handling term, but we'll use it because it helps with explanations. In this section, we describe how to use `try` and `catch` blocks to fend off exception messages and program crashes. Use a `try` block to "try" out one or more dangerous method calls. If there's a problem with the dangerous method call(s), the JVM jumps to a `catch` block and the JVM executes the `catch` block's enclosed statements. Drawing an analogy, a `try` block is like a circus trapeze act. A trapeze act contains one or more dangerous stunts, like a triple flip or a triple twist. The dangerous stunts are like dangerous method calls. If something goes wrong with one of the stunts and an acrobat falls, there's a net to catch the acrobat. Likewise, if something goes wrong with one of the dangerous method calls, control passes to a `catch` block. If nothing goes wrong with the trapeze stunts, the net isn't used at all. Likewise, if nothing goes wrong with the dangerous method calls, the `catch` block isn't used at all.

Syntax and Semantics

Here's the syntax for `try` and `catch blocks`:

```
try
{
   <statement(s)>
}
catch (<exception-class> <parameter>)
{
   <error-handling-code>
}
```

Normally, one or more of these statements will be a "dangerous" API method call or constructor call.

The exception class should match the type of exception that the `try` block might throw.

As shown above, a `try` block and its associated `catch` block (or multiple `catch` blocks, which we'll address later) must be contiguous. You can put other statements before the `try` block or after the (last) `catch` block, but not between them. Note the parameter in the `catch` block's heading. We'll explain `catch` block parameters in the context of the following example program.

See Figure 15.1's LuckyNumber program. Note how the `try` and `catch` blocks follow the syntax pattern shown above. Within the `try` block, the `nextInt` method call tries to convert a user entry to an integer. For the conversion to work, the user entry must contain only digits and an optional preceding minus sign. If the user entry conforms to that format, the JVM assigns the user entry to the `num` variable, skips the `catch` block, and continues with the code below the `catch` block. If the user entry does not conform to that format, an exception occurs. If an exception occurs, the JVM immediately exits from the `try` block and instantiates an *exception object*—an object that contains information about the exception event.

```
/*************************************************************
* LuckyNumber.java
* Dean & Dean
*
* This program reads the user's lucky number as an int.
*************************************************************/

import java.util.Scanner;
import java.util.InputMismatchException;

public class LuckyNumber
{
  public static void main(String[] args)
  {
    Scanner stdIn = new Scanner(System.in);
    int num; // lucky number
    try
    {
      System.out.print("Enter your lucky number (an integer): ");
      num = stdIn.nextInt();
    }
    catch (InputMismatchException e)
    {
      System.out.println(
        "Invalid entry. You'll be given a random lucky number.");
      num = (int) (Math.random() * 10) + 1;   // between 1-10
    }
    System.out.println("Your lucky number is " + num + ".");
  } // end main
} // end LuckyNumber class
```

Import `InputMismatchException` for use below.

The `e` parameter receives an `InputMismatchException` object.

Sample session 1:

```
Enter your lucky number (an integer): 27
Your lucky number is 27.
```

Sample session 2:

```
Enter your lucky number (an integer): 33.42
Invalid entry. You'll be given a random lucky number.
Your lucky number is 8.
```

Figure 15.1 LuckyNumber program, which uses `try` and `catch` blocks for numeric user entry

In this example, the JVM instantiates an `InputMismatchException` object. The JVM then passes the `InputMismatchException` object to the `catch` block heading's `e` parameter. Because `e` is declared to be an `InputMismatchException` and `InputMismatchException` is not part of the core Java language, at the top of the program, we need to include:

```
import java.util.InputMismatchException;
```

After passing the exception object to the catch block, the JVM executes the catch block's body. In this example, the catch block prints an "Invalid entry . . ." message and assigns a random number to the num variable. Then execution continues with the code below the catch block.

Throwing an Exception

When the JVM instantiates an exception object, we say that the JVM *throws an exception*. We'd prefer to say "throws an exception object" rather than "throws an exception" because the thing that's being thrown is an exception object. But most programmers don't worry about the difference between an exception, which is an event, and an exception object. No big deal. We'll go with the flow and use the standard terminology—throwing an exception.

When the JVM throws an exception, the JVM looks for a matching catch block. If it finds a matching catch block, it executes it. If it does not find a matching catch block, the JVM prints the exception object's exception message and terminates the program. What is a "matching catch block"? A catch block is "matching" if the catch heading's parameter type is the same as the type of the thrown exception.[2] For example, in the LuckyNumber program, the InputMismatchException parameter matches the InputMismatchException object thrown by the nextInt method call. Therefore, the InputMismatchException parameter's catch block is a matching catch block if and when the nextInt method call throws an InputMismatchException.

An exception object contains information about the error, including the error's type and a list of the method calls that led to the error. We'll use some of the exception object's information later, but for now, all we need the exception object for is its ability to match up with the proper catch block.

15.4 Line Plot Example

Now let's see how try and catch are used in the context of a more complicated program. We start by presenting a program without try and catch blocks. Then we analyze the program and determine how it can be improved by adding try and catch blocks.

First-Cut LinePlot Program

The program in Figure 15.2 plots a line by reading in coordinate positions for a series of points. The best way to get a handle on what the LinePlot program does is to show a sample session. Below, the user chooses to plot a line that goes from the origin (the default starting point) to point (3,1) to point (5,2):

Sample session:

```
Enter x & y coordinates (q to quit): 3 1
New segment = (0,0) - (3,1)
Enter x & y coordinates (q to quit): 5 2
New segment = (3,1) - (5,2)
Enter x & y coordinates (q to quit): q
```

[2]Actually, as you'll see in Section 15.9, a catch block is also considered to be matching if the catch heading's parameter type is a superclass of the thrown exception's class.

```
/**************************************************************
* LinePlot.java
* Dean & Dean
*
* This program plots a line as a series of user-specified
* line segments.
**************************************************************/

import java.util.Scanner;

public class LinePlot
{
  private int oldX = 0;  // oldX and oldY save previous point
  private int oldY = 0;  // starting point is the origin (0,0)

  //************************************************************

  // This method prints the description of a line segment from the
  // previous point to the current point.

  public void plotSegment(int x, int y)
  {
    System.out.println("New segment = (" + oldX + "," + oldY +
      ")-(" + x + "," + y + ")");
    oldX = x;
    oldY = y;
  } // end plotSegment

  //************************************************************

  public static void main(String[] args)
  {
    Scanner stdIn = new Scanner(System.in);
    LinePlot line = new LinePlot();
    String xStr, yStr;   // coordinates for point in String form
    int x, y;            // coordinates for point

    System.out.print("Enter x & y coordinates (q to quit): ");
    xStr = stdIn.next();
    while (!xStr.equalsIgnoreCase("q"))
    {
      yStr = stdIn.next();
      x = Integer.parseInt(xStr);
      y = Integer.parseInt(yStr);
      line.plotSegment(x, y);
      System.out.print("Enter x & y coordinates (q to quit): ");
      xStr = stdIn.next();
    } // end while
  } // end main
} // end class LinePlot
```

These could generate runtime errors.

Figure 15.2 LinePlot program that plots a line—first draft

As you can see, the program's display is very primitive—it uses text to represent each line segment. In a real line-plotting program, you'd use Java's `drawPolyLine` method to display the line. That's what we do in the GUI section at the end of this chapter. But for now, we'll keep it simple and use a text-based display rather than a GUI-based display. That way, we can maintain focus on this chapter's primary topic, exception handling.

Using "q" as a Sentinel Value

In the past, when you entered numbers inside a loop, you often terminated the loop with a numeric sentinel value. This program employs a more elegant solution because it allows a non-numeric "q" as the sentinel value. How can you read in numbers and the string "q" with the same input statement? Use strings for both types of input—for the "q" and also for the numbers. For each number input, the program converts the number string to a number by calling the `Integer` class's `parseInt` method.

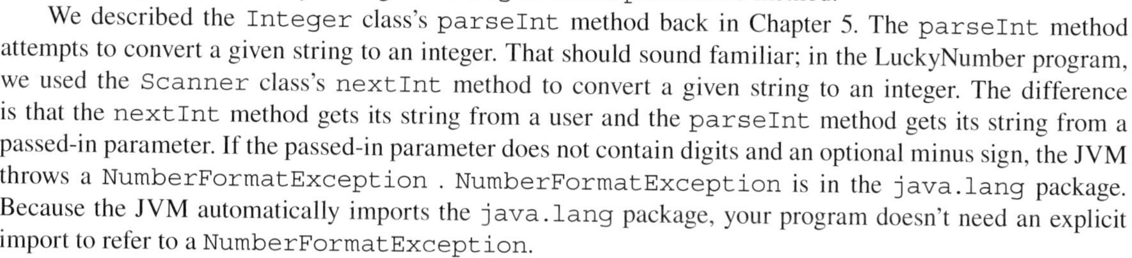

We described the `Integer` class's `parseInt` method back in Chapter 5. The `parseInt` method attempts to convert a given string to an integer. That should sound familiar; in the LuckyNumber program, we used the `Scanner` class's `nextInt` method to convert a given string to an integer. The difference is that the `nextInt` method gets its string from a user and the `parseInt` method gets its string from a passed-in parameter. If the passed-in parameter does not contain digits and an optional minus sign, the JVM throws a `NumberFormatException` . `NumberFormatException` is in the `java.lang` package. Because the JVM automatically imports the `java.lang` package, your program doesn't need an explicit import to refer to a `NumberFormatException`.

Input Validation

Note how the LinePlot program calls `stdIn.next` to read x coordinate and y coordinate values into `xStr` and `yStr`, respectively. Then the program attempts to convert `xStr` and `yStr` to integers by calling `Integer.parseInt`. The conversions work fine as long as `xStr` and `yStr` contain digits and an optional minus sign. But what happens if the user enters a non-integer for `xStr` or `yStr`? With invalid input, the program crashes, like this:

<u>Sample session:</u>

```
Enter x & y coordinates (q to quit): 3 1.25
Exception in thread "main" java.lang.NumberFormatException: For input string: "1.25"
. . .
```

To deal with this possibility, let's rewrite the `while` loop in the `main` method of Figure 15.2 so that it includes input validation using a `try-catch` mechanism. The first step is to identify the dangerous code. Can you find the dangerous code? The two `parseInt` method calls are dangerous in that they might throw a `NumberFormatException`. So let's put those two statements into a `try` block and add a matching `catch` block, as shown in Figure 15.3.

Look for potential problems.

Do you see any logic errors in Figure 15.3's `while` loop? What happens if there's invalid input? A `NumberFormatException` object is thrown and caught, and then an error message is printed. Then `line.plotSegment` executes. But you wouldn't want to print the line segment if the input values were messed up. To avoid that possibility, move the `line.plotSegment(x, y);` line to the last line in the `try` block. This way, it gets executed only if the two `parseInt` method calls work properly. Figure 15.4 shows the final version of the LinePlot program's `while` loop.

```
while (!xStr.equalsIgnoreCase("q"))
{
  yStr = stdIn.next();
  try
  {
    x = Integer.parseInt(xStr);
    y = Integer.parseInt(yStr);
  }
  catch (NumberFormatException nfe)
  {
    System.out.println("Invalid entry: " + xStr + " " + yStr
      + "\nMust enter integer space integer.");
  }

  line.plotSegment(x, y);
  System.out.print("Enter x & y coordinates (q to quit): ");
  xStr = stdIn.next();
} // end while
```

These statements should be inside a `try` block.

Figure 15.3 First attempt at improving the LinePlot program's `while` loop

```
while (!xStr.equalsIgnoreCase("q"))
{
  yStr = stdIn.next();
  try
  {
    x = Integer.parseInt(xStr);
    y = Integer.parseInt(yStr);
    line.plotSegment(x, y);
  }
  catch (NumberFormatException nfe)
  {
    System.out.println("Invalid entry: " + xStr + " " + yStr
      + "\nMust enter integer space integer.");
  }
  System.out.print("Enter x & y coordinates (q to quit): ");
  xStr = stdIn.next();
} // end while
```

This statement should be inside the `try` block, not after the `try-catch` structure.

Figure 15.4 Final version of the LinePlot program's `while` loop

15.5 try Block Details

Now that you know the basic idea behind `try` blocks, it's time to flesh out some subtle `try` block details.

try Block Size

Deciding on the size of your `try` blocks is a bit of an art. Sometimes it's better to use small `try` blocks, and sometimes it's better to use larger `try` blocks. It's legal to surround an entire method body with a `try` block, but that's usually counterproductive because then dangerous code is harder to identify. In general, you should make your `try` blocks small enough so that your dangerous code is easily identified.

On the other hand, if you need to execute several related dangerous statements in succession, you should consider surrounding the statements with one inclusive `try` block rather than surrounding each statement with its own small `try` block. Multiple small `try` blocks can lead to cluttered code. One inclusive `try` block can lead to improved readability. The improved LinePlot program includes both `parseInt` statements in a single `try` block because they are conceptually related and physically close together. That improves readability.

Assume That try Block Statements Are Skipped

If an exception is thrown, the JVM immediately jumps out of the current `try` block. The immediacy of the jump means that if there are statements in the `try` block after the exception-throwing statement, those statements get skipped. The compiler is a pessimist. It knows that statements inside a `try` block might possibly be skipped, and it assumes the worst; that is, it assumes that all statements inside a `try` block get skipped. Consequently, if there's a `try` block that contains an assignment to x, the compiler assumes that the assignment is skipped. If there's no assignment to x outside of the `try` block and x's value is needed outside of the `try` block, you'll get this compile-time error:

```
variable x might not have been initialized
```

If you get that error, usually you can fix it by initializing the variable prior to the try block. Let's look at an example. . . .

Your goal is to implement a `getIntFromUser` method that performs robust input for an `int` value. Your method should prompt the user for an integer, read the entered value as a string, and then convert the string to an `int`. If the conversion fails, your method should reprompt the user for an integer. If the user eventually enters a valid integer value, `getIntFromUser` should return it to the calling module.

Figure 15.5 is a first-cut attempt at implementing the `getIntFromUser` method. It does a good job with the logic, but it contains compile-time errors that are due to the initializations inside the `try` block. We'll fix the `try` block's errors soon enough, but let's first explain the `try` block's logic.

The `try` block contains these three lines:

```
valid = false;
x = Integer.parseInt(xStr);
valid = true;
```

Note how the three-line code fragment assigns `valid` to `false` and then turns around and assigns it back to `true`. Strange, eh? Actually, it's a fairly common strategy to assume one thing, try it out, and then change the assumption if it's proven wrong. And that's what's happening here. This code starts by assuming that the user entry is invalid. It calls `parseInt` to test whether it's actually valid; that is, it checks to see if the user entry is an integer. If it is valid, the next statement executes, and `valid` gets set to `true`. But what happens if the `parseInt`

Assume one thing, then change as required.

```
public static int getIntFromUser()
{
  Scanner stdIn = new Scanner(System.in);
  String xStr;    // user entry
  boolean valid; // is user entry a valid integer?
  int x;          // integer form of user entry

  System.out.print("Enter an integer: ");
  xStr = stdIn.next();

  do
  {
    try
    {
      valid = false;
      x = Integer.parseInt(xStr);
      valid = true;
    }
    catch (NumberFormatException nfe)
    {
      System.out.print("Invalid entry. Enter an integer: ");
      xStr = stdIn.next();
    }
  } while (!valid);

  return x;
} // end getIntFromUser
```

compile-time error: `valid` might not have been initialized

compile-time error: `x` might not have been initialized

Figure 15.5 A method that illustrates the problem with initializing inside a `try` block

conversion fails? The `valid` variable never gets set to `true` because an exception is thrown and the JVM immediately jumps out of the `try` block. So this code seems reasonable. Unfortunately, "seems reasonable" isn't good enough this time.

Can you figure out the compile-time errors? If not, don't feel bad; we didn't see them until after the compiler helped us. As shown by the callouts in Figure 15.5, the compiler complains that the `valid` and `x` variables might not have been initialized. Why all the fuss? Can't the compiler see that `valid` and `x` are assigned values in the `try` block? Yes, the compiler can see the assignments, but remember that the compiler is a pessimist. It assumes that all statements inside a `try` block are skipped. Even though we know that the `valid = false;` statement is in no actual danger of being skipped (it's a simple assignment, and it's the first line in the `try` block), the compiler still assumes that it gets skipped.

What's the solution? (1) Move the `valid = false;` assignment up to `valid`'s declaration line. (2) Initialize `x` to `0` as part of `x`'s declaration line. Figure 15.6 contains the corrected implementation.

15.6 Two Categories of Exceptions—Checked and Unchecked

Exceptions fall into two categories—*checked* and *unchecked*. Checked exceptions must be checked with a `try-catch` mechanism. Unchecked exceptions can optionally be checked with a `try-catch` mechanism, but it's not a requirement.

```
public static int getIntFromUser()
{
  Scanner stdIn = new Scanner(System.in);
  String xStr;              // user entry
  boolean valid = false; // is user entry a valid integer?
  int x = 0;                // integer form of user entry

  System.out.print("Enter an integer: ");
  xStr = stdIn.next();

  do
  {
    try
    {
      x = Integer.parseInt(xStr);
      valid = true;
    }
    catch (NumberFormatException nfe)
    {
      System.out.print("Invalid entry. Enter an integer: ");
      xStr = stdIn.next();
    }
  } while (!valid);

  return x;
} // end getIntFromUser
```

These initializations before the `try` block meet the compiler's demands.

Figure 15.6 Corrected version of the `getIntFromUser` method in Figure 15.5

Identifying an Exception's Category

How can you tell whether a particular exception is classified as checked or unchecked? An exception is an object, and as such, it is associated with a particular class. To find out if a particular exception is checked or unchecked, look up its associated class on Oracle's Java API website.[3] Once you find the class, look at its ancestors. If you find that it's a descendant of the `RuntimeException` class, then it's an unchecked exception. Otherwise, it's a checked exception.

For example, if you look up `NumberFormatException` on Oracle's Java API website, you'll see this:

```
java.lang.Object
   java.lang.Throwable
      java.lang.Exception
         java.lang.RuntimeException
            java.lang.IllegalArgumentException
               java.lang.NumberFormatException
```

If you see this class in the hierarchy, the exception is unchecked.

This shows that the `NumberFormatException` class is a descendant of the `RuntimeException` class, so the `NumberFormatException` class is an unchecked exception.

[3]http://download.oracle.com/javase/7/docs/api/index.html.

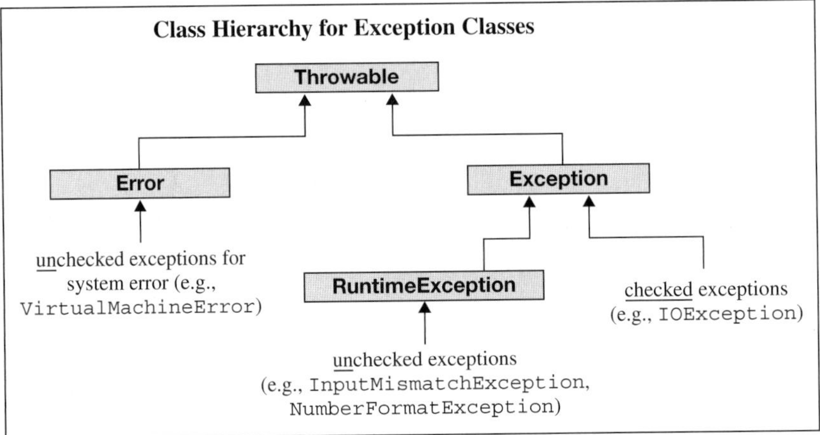

Figure 15.7 Exception class hierarchy

Figure 15.7 shows the class hierarchy for all exceptions. It reiterates the point that unchecked exceptions are descendants of the `RuntimeException` class. It also shows that some unchecked exceptions are descendants of the `Error` class. In the interest of keeping things simple, we didn't mention the `Error` class previously. You probably won't encounter its exceptions unless you do a lot of recursive programming, as described in Chapter 11. In that case you probably will encounter `StackOverflowException`, which derives from `VirtualMachineError`, which derives from `Error`.

Programmer-Defined Exception Classes

It's possible for programmers to define their own exception classes. Such programmer-defined exception classes must be derived from the `Exception` class or from a subclass of the `Exception` class. Generally speaking, you should limit yourself to predefined exception classes, because programmer-defined exception classes tend to fragment error-handling activities, and that makes programs harder to understand.

15.7 Unchecked Exceptions

As you learned in the previous section, unchecked exceptions need not be checked with a `try-catch` mechanism. However, at runtime, if the JVM throws an unchecked exception and there's no `catch` block to catch it, the program will crash.

Strategies for Handling Unchecked Exceptions

If your program contains code that might throw an unchecked exception, there are two alternate strategies for dealing with it:

1. Use a `try-catch` structure.

or

2. Don't attempt to catch the exception, but write the code carefully so as to avoid the possibility of the exception being thrown.

In the `getIntFromUser` method in Figure 15.6, we employed the first strategy—we used a `try-catch` structure to handle the dangerous `parseInt` method call. Normally, you should use a `try-catch` structure for parse method calls (parseInt, parseLong, parseDouble, and so on) because that leads to cleaner solutions. In the next example, the preferred strategy isn't so clear cut. We'll use both strategies and compare the results.

StudentList Example

Figure 15.8 presents a StudentList class that manages a list of student names. The class stores student names in an ArrayList named students. The class contains a constructor for initializing the students list, a display method for printing the students list, and a removeStudent method that removes a specified student from the students list. We'll focus on the removeStudent method.

```
/*****************************************************
* StudentList.java
* Dean & Dean
*
* This class manages an ArrayList of students.
*****************************************************/

import java.util.ArrayList;

public class StudentList
{
  ArrayList<String> students = new ArrayList<>();

  //**************************************************

  public StudentList(String[] names)
  {
    for (int i=0; i<names.length; i++)
    {
      students.add(names[i]);
    }
  } // end constructor

  //**************************************************

  public void display()
  {
    for (int i=0; i<students.size(); i++)
    {
      System.out.print(students.get(i) + " ");
    }
    System.out.println();
  } // end display

  //**************************************************

  public void removeStudent(int index)
  {
    students.remove(index);  <- This is a dangerous method call.
  } // end removeStudent
} // end StudentList
```

Figure 15.8 First draft of StudentList class, which maintains a list of students

The `students.remove` method call is dangerous because it might throw an unchecked exception,

`IndexOutOfBoundsException`. If its `index` argument holds the index of one of the `students`' elements, then that element is removed from the `students ArrayList`. But if its `index` argument holds an invalid index, an `IndexOutOfBoundsException` is thrown. This occurs, for example, if we use Figure 15.9's `StudentListDriver` class as the driver. Note how the `StudentListDriver` class uses an index value of 6, even though there are only four students in the student list. The `StudentListDriver` and `StudentList` classes compile just fine, but when run, the `students.remove` method call throws an exception and the JVM terminates the program and prints the error message shown at the bottom of Figure 15.9.

```
/*************************************************************
* StudentListDriver.java
* Dean & Dean
*
* This is the driver for the StudentList class.
*************************************************************/

public class StudentListDriver
{
  public static void main(String[] args)
  {
    String[] names = {"Caleb", "Izumi", "Mary", "Usha"};
    StudentList studentList = new StudentList(names);

    studentList.display();
    studentList.removeStudent(6);
    studentList.display();
  } // end main
} // end StudentListDriver
```

This argument value generates a runtime error.

Output:

```
Caleb Izumi Mary Usha
Exception in thread "main" java.lang.IndexOutOfBoundsException: Index: 6,
Size: 4
        at java.util.ArrayList.RangeCheck(ArrayList.java:547)
        at java.util.ArrayList.remove(ArrayList.java:390)
        at StudentList.removeStudent(StudentList.java:43)
        at StudentListDriver.main(StudentListDriver.java:17)
```

Figure 15.9 Driver of `StudentList` class

Improve the `removeStudent` Method

Let's now make the `removeStudent` method more robust by gracefully handling the case where it's called with an invalid index. Figures 15.10a and 15.10b show two different robust implementations for the `removeStudent` method. The first implementation uses a `try-catch` mechanism and the second implementation uses careful code. These are the two strategies mentioned earlier for handling unchecked exceptions.

```
public void removeStudent(int index)
{
  try
  {
    students.remove(index);
  }
  catch (IndexOutOfBoundsException e)
  {
    System.out.println("Can't remove student because " +
      index + " is an invalid index position.");
  }
} // end removeStudent
```

Figure 15.10a Using a `try-catch` structure for the `removeStudent` method

```
public void removeStudent(int index)
{
  if (index >= 0 && index < students.size())
  {
    students.remove(index);
  }
  else
  {
    System.out.println("Can't remove student because " +
      index + " is an invalid index position.");
  }
} // end removeStudent
```

Figure 15.10b Using a careful-code strategy for the `removeStudent` method

Which solution is better—a `try-catch` mechanism or careful code? The solutions are about the same in terms of readability. With things being equal in terms of readability, go with the careful-code implementation because it's more efficient. Exception handling code is less efficient because it requires the JVM to instantiate an exception object and find a matching `catch` block.

15.8 Checked Exceptions

Let's now look at checked exceptions. If a code fragment has the potential of throwing a checked exception, the compiler forces you to associate that code fragment with a `try-catch` mechanism. If there is no associated `try-catch` mechanism, the compiler generates an error. With unchecked exceptions, you have a choice of how to handle them—a `try-catch` mechanism or careful code. With checked exceptions, there's no choice—you must use a `try-catch` mechanism.

CreateNewFile Program

For good examples of checked exceptions, we need files. The next chapter explains file details, but you don't need to know file details now. You can understand how to deal with file-code exceptions before you understand file-code details. The draft of a CreateNewFile program in Figure 15.11 attempts to create an empty file with a user-specified name. To do this, the program needs to use the `Path`, `Paths` and `Files` classes in the API's `java.nio.file` package. So it imports that package.

This program is supposed to prompt the user for the name of a file to create. If a file with that name already exists, it is supposed to print "Sorry, that file already exists." If a file with that name does not

```
/***************************************************************
* CreateNewFile.java
* Dean & Dean
*
* This attempts to create a new file.
***************************************************************/

import java.util.Scanner;
import java.nio.file.*;            // Path, Paths, Files

public class CreateNewFile
{
  public static void main(String[] args)
  {
    Scanner stdIn = new Scanner(System.in);
    String filename;
    Path path;                      // file location

    System.out.print("Enter name of file to create: ");
    filename = stdIn.nextLine();
    path = Paths.get(filename);
    if (Files.exists(path))
    {
      System.out.println("Sorry, that file already exists.");
    }
    else
    {
      Files.createFile(path);
      System.out.println(filename + " created.");
    }
  } // end main
} // end CreateNewFile class
```

COMPILATION ERROR!
unreported `IOException`

Figure 15.11 Draft of CreateNewFile program, which is supposed to create a new file

already exist, it is supposed to create a new file with that name. Unfortunately, when we try to compile the code in Figure 15.11, the complier complains:

```
<path-to-source-code-file>: 28: error: unreported exception IOException;
must be caught or declared to be thrown
  Files.createFile(path);
                  ^
```

This tells us that the `createFile` method throws a checked exception, specifically, the `IOException`. The "quick and dirty" way to satisfy the compiler is to append `throws IOException` to the method header, as suggested back in Chapter 3, in Section 3.24.

A more responsible way to satisfy the compiler is to provide another import:

```
import java.io.IOException;
```

Then surround the `createFile` method call with a `try` block like this:

```
else
{
  try
  {
    Files.createFile(path);
  }
  catch (IOException ioe)
  {
    System.out.println("File I/O error");
  }
  System.out.println(filename + " created.");
}
```

This variable name is arbitrary. For example, it could be just e or x.

This results in a program that compiles successfully and runs. But is it a good program? Novice programmers often solve problems by trying something out without thoroughly thinking it through, and if it leads to reasonable results, they quickly move on. Try to resist that urge.

Don't move on until you are sure of your solution.

Although the above code compiles and runs, it doesn't behave appropriately when an `IOException` is thrown. Can you identify the inappropriate behavior? If an `IOException` is thrown, the `catch` block prints the message:

```
"File I/O error"
```

But then, even though the file was not created, the program also prints the message:

```
filename + " created."
```

Remember: Just because a program runs, that doesn't mean it's correct. Also, just because you've fixed one problem, that doesn't mean you've fixed all problems. Read on . . .

Using API Documentation When Writing Exception Handling Code

When you want to use a method or constructor from one of the API classes and you're not sure about it, you can look it up in the API documentation. On its API documentation page, look for a "throws" section. This identifies specific types of exceptions that constructor or method might throw. For an explanation of a specific exception, click on that exception's link. This takes you to the API documentation for that exception's class. On the exception class's API page, scroll down and read the class's description. Then scroll back up and look at the class's class hierarchy. As indicated earlier, if `RuntimeException` is an ancestor, the exception is an unchecked exception. Otherwise, it's a checked exception.

If you apply this API-lookup strategy to the CreateNewFile program, you'll learn the following about the exceptions it might throw:

- The `Paths.get(filename)` method call throws an `InvalidPathException` if the `String` argument cannot be converted to a valid `Path`. The `InvalidPathException` class derives from the `RuntimeException` class, so it's an unchecked exception.

- The `Files.exists(path)` method call throws a `SecurityException`. This derives from the `RuntimeException` class, so it's an unchecked exception.

- The `Files.createFile(path)` method throws an `UnsupportedOperationsException` in addition to a `SecurityException`. Because these both derive from the `RuntimException` class, they are unchecked exceptions. It also throws a `FileAlreadyExistsException` and an `IOException`. Since these do not derive from the `RuntimeException` class, they are checked exceptions, and they must be in a `try` block. Since the `FileAlreadyExistsException` derives from the `IOException` class, a `catch` block that catches an `IOException` will catch them both. The `FileAlreadyExistsException` is the checked exception we thought we handled but really didn't.

The correct way to handle the two checked exceptions is to move the "created" printout up into the `try` block, after the statement that creates the new file, like this:

```
else
{
  try
  {
    Files.createFile(path);
    System.out.println(filename + " created.");
  }
  catch (IOException ioe)
  {
    System.out.println("File I/O error");
  }
}
```

This statement is now in a better location.

Now the program prints the "created" message only if the file is actually created. Yeah!

15.9 Generic catch Block with Exception Class

Previous programs have used the try-catch mechanism to catch only one type of exception. Because the FileAlreadyExistsException derives from the IOException, the catch block in the CreateNewFile program is able to catch either of these two exceptions, However, the CreateNewFile program used an if statement to avoid the first of these two exceptions. So it, too, used the try-catch mechanism to catch only one type of exception. 

Alternatively, we can use exception handling alone to catch more than one exception and distinguish them one from another. There are two ways to do this: (1) Provide one generic catch block that catches every type of exception that might be thrown, and in that block provide code that identifies the exception actually thrown. (2) Provide a sequence of catch blocks that catch different exceptions. This section describes the generic-catch-block technique. The next section describes the sequence-of-catch-blocks technique. In practice, as we develop a program, we typically start with the generic-catch-block technique and then perhaps move to the sequence-of-catch-blocks technique.

Generic catch Block

To provide a generic catch block, define a catch block with an Exception type parameter. Then, inside the catch block, call the getClass method (inherited by Exception from Object), and also call the getMessage method (inherited by Exception from Throwable). The getClass method call identifies the particular type of exception that was thrown. The getMessage method call explains why that exception was thrown. Just one of these two method calls might leave you wondering, but the combination of both of them usually tells you exactly what you need to know.[4]

Here is what the generic catch block looks like:

```
catch (Exception e)
{
  System.out.println(e.getClass());
  System.out.println(e.getMessage());
}
```

Because the Exception class is in the always available java.lang package, you do not need to import it. If a catch block uses an Exception parameter, it matches all thrown exceptions. Why? Because when an exception is thrown, it looks for a catch parameter that's either identical to the thrown exception or a superclass of the thrown exception. The Exception class is the superclass of all checked exceptions. Therefore, all checked exceptions consider an Exception catch parameter to be a match.

Figure 15.7 showed that the Exception class also includes unchecked exceptions. So in addition to catching and describing all checked exceptions, except for rare Error exceptions, the generic catch block above catches and describes all unchecked exceptions thrown from the corresponding try block.

ReadFromFile Example

Figure 15.12a's ReadFromFile program opens a user-specified file and prints all lines of text in that file. The Scanner constructor that instantiates the stdIn object does not throw any exceptions. The System.out.print method does not throw any exceptions. So we do not include them in the try block.

[4]For more information, you can call the printStackTrace method. This generates voluminous error output like that described later in Section 15.11.

```
/*************************************************************
* ReadFromFile.java
* Dean & Dean
*
* This opens an existing text file and prints its lines.
*************************************************************/

import java.util.Scanner;
import java.nio.file.Paths;

public class ReadFromFile
{
  public static void main(String[] args)
  {
    Scanner stdIn = new Scanner(System.in);
    Scanner fileIn;       // file handler
    String filename;      // user-specified file name
    String line;          // line of text

    System.out.print("Enter a filename: ");
    try
    {
      filename = stdIn.nextLine();
      fileIn = new Scanner(Paths.get(filename));
      while (fileIn.hasNext())
      {
        line = fileIn.nextLine();
        System.out.println(line);
      }
    } // end try
    catch (Exception e)
    {
      System.out.println(e.getClass());
      System.out.println(e.getMessage());
    }
  } // end main
} // end ReadFromFile class
```

The `Scanner` constructor throws a checked exception.

Figure 15.12a ReadFromFile program—a simple file-reader

The `stdIn.nextLine` method call throws an unchecked `NoSuchElementException` and an unchecked `IllegalStateException`. As indicated earlier, the `Paths.get(filename)` method call throws an unchecked `InvalidPathException`. Because it takes extra code to check for these problems, and our generic `catch` clearly identifies them if they occur, we (optionally) include these method calls in our enlarged `try` block. The `Scanner` constructor that instantiates the `fileIn` object throws a checked `IOException`, so we (necessarily) include it in the `try` block.

The while loop iterates through all lines of text in the file. The fileIn.hasNext method throws an unchecked IllegalStateException, and the fileIn.nextLine method throws an unchecked IllegalStateException and an unchecked NoSuchElementException. Because these exceptions are unchecked, one might think we could remove the code that throws them from the try block—perhaps putting the while loop after the catch block. But if we attempt that, the compiler complains:

```
error: variable fileIn might not have been initialized
  while (fileIn.hasNext())
```

This is another example of the principle we discussed back in Section 15.5. The compiler is a pessimist. It knows that statements inside a try block might be skipped, and it assumes the worst. In the Section 15.5 example, we were able to move the required initialization out of the try block and put it before the try block. But this time, we can't do that, because the initialization throws a checked exception. Therefore, we must include the while loop in the try block along with that initialization.

Figure 15.12b contains outputs from three separate executions of the ReadFromFile program in Figure 15.12a. In the first sample session, the user employs an illegal character in the filename specification. The e.getClass method call says this input generates an InvalidPathException. Java API documentation and the first bullet in the last subsection of Section 15.8 say this exception must have come from the Paths.get method call.

Sample session #1:
```
Enter a filename: Einstein.*
class java.nio.file.InvalidPathException
Illegal char <*> at index 9: Einstein.*
```

Sample session #2:
```
Enter a filename: Einstein
class java.nio.file.NoSuchFileException
Einstein
```

Sample session #3:
```
Enter a filename: Einstein.txt
A scientific theory should be as simple as possible,
but not simpler.
```

Figure 15.12b Outputs from the ReadFromFile program in Figure 15.12a

In the second sample session, the user specifies a nonexistent file. The e.getClass method says that this input generates a NoSuchFileException. This Class information is right "on target" and quite helpful, but it's not one of the exceptions that API documentation associates with our program's constructor and method calls. What's happening here? NoSuchFileException derives from the IOException that API documentation says the Scanner(Path source) constructor throws. So the actual output is more specific than the API documentation. The Java compiler is getting smarter as it gets older!

In the third sample session, the user correctly specifies the file, and the program correctly displays its contents.

15.10 Multiple `catch` Blocks and Multiple Exceptions Per Block

When a `try` block might throw more than one type of exception, instead of providing one generic `catch` block, you can provide a sequence of `catch` blocks. Multiple `catch` blocks help you provide different responses for different types of exceptions. A sequence of `catch` blocks is like the "if, else if, else if, . . ." form of the `if` statement. In any particular situation, the first `catch` with a matching parameter is the one that executes. If the final `catch` block is generic (and can catch any type of exception), it is like the "else" part of an "if, else if, . . . else" statement.

When you are developing a new program, it's convenient to start with a generic `catch` block and insert specific `catch` blacks ahead of it as you discover the need to respond differently to different kinds of exceptions.

The ReadFromFile Program Revisited

The ReadFromFile program in Figure 15.12a showed how you can use the compiler's knowledge and experiment with a simple generic `catch` block to discover the types of exceptions that might be thrown by various forms of bad input. Once you have discovered all the types of exceptions bad user entries might generate, you can put these particular exceptions into an additional `catch` block inserted before the generic `catch` block. Then, you can modify the code so that if any of the bad entries occurs, the program prints a short error message, repeats the original prompt, and takes new input—in an iteration that continues until the input is satisfactory. The ReadFromFile2 program in Figure 15.13a includes these modifications.

In the ReadFromFile2 program, we saved four lines of code by eliminating the `filename` and `line` variables. Then we added the variable, `boolean makeEntry`, which is initially `true`. This new variable allows entry into the outer `while` loop, which includes all other code. When the file opens successfully, execution proceeds through to the `makeEntry = false;` statement. This allows the current iteration to continue, but it prevents another iteration.

Notice how the new `catch` block's heading includes two exceptions types. The single "or" symbol, |, which separates them says either of them will be caught by the same `catch` block. Thus, this one block catches all plausible user-entry errors and prints a message that applies to any of them. Assuming user entries like those in Figure 15.12b, Figure 15.13b shows the output generated in just one execution of the ReadFromFile2 program in Figure 15.13a.

`catch` Block Ordering—The Order Matters

Whenever you use more than one `catch` block, and the class of one `catch` block's exception is derived from the class of another `catch` block's exception, you must arrange the `catch` blocks so that the more general exceptions come later.

For example, if you look up `FileNotFoundException` on the Java API website, you'll see this hierarchy:

```
java.lang.Object
   java.lang.Throwable
      java.lang.Exception
         java.io.IOException
            java.io.FileNotFoundException
```

⚠ If you choose to have a `catch` block with a `FileNotFoundException` and a `catch` block with an `IOException` in the same `catch`-block sequence, you must put the `IOException` block after the

```
/*************************************************************
* ReadFromFile2.java
* Dean & Dean
*
* This opens an existing text file and prints its lines.
*************************************************************/

import java.util.Scanner;
import java.nio.file.*;    // Paths, specific exceptions

public class ReadFromFile2
{
  public static void main(String[] args)
  {
    Scanner stdIn = new Scanner(System.in);
    Scanner fileIn;        // file handler
    boolean makeEntry = true;

    while (makeEntry)
    {
      System.out.print("Enter a filename: ");
      try
      {
        fileIn = new Scanner(Paths.get(stdIn.nextLine()));
        makeEntry = false; // because user entry is now OK
        while (fileIn.hasNext())
        {
          System.out.println(fileIn.nextLine());
        }
      } // end try
      catch (InvalidPathException | NoSuchFileException e)
      {
        System.out.println("Filename invalid or not found.");
      } // end catching exceptions user can handle
      catch (Exception e)
      {
        System.out.println(e.getClass());
        System.out.println(e.getMessage());
      }
    } // end while makeEntry
  } // end main
} // end ReadFromFile2 class
```


Figure 15.13a ReadFromFile2 program—an improved file-reader

```
Sample session:
Enter a filename: Einstein.*
Filename invalid or not found.
Enter a filename: Einstein
Filename invalid or not found.
Enter a filename: Einstein.txt
A scientific theory should be as simple as possible,
but not simpler.
```

Figure 15.13b Output produced by the ReadFromFile2 program in Figure 15.13a

`FileNotFoundException` block. If you put a `catch` block for `IOException` first, it would match `FileNotFoundException` too, and the other `catch` block would always be skipped. As long as you understand these principles, there's no need to memorize the hierarchical relationships among all types of exceptions because the compiler will tell you in a compile-time error if you try to arrange multiple `catch` blocks in the wrong order.

The compiler does not allow you to put an exception and an ancestor of that exception in the same `catch` block. For example, it will complain if you try to put a `FileNotFoundDxception` and an `IOException` together in the same multiple-exception `catch` header. However, the compiler did not complain when we put an `InvalidPathException` and a `NoSuchFileException` together in the same `catch`-block parameter list in Figure 15.13a. So what can you infer about the relationship between these last two exceptions? Since the compiler accepts them both in the same `catch` header, neither of these two classes is a subclass of the other.

Removing Redundant Generic `catch` and Handling Unchecked Exceptions

Earlier, we suggested starting program development with a generic `catch` block and inserting specific `catch` blocks ahead of it as the compiler identifies exceptions that need specific responses. As development proceeds, newly introduced specific `catch` blocks will catch more and more checked exceptions. If specific `catch` blocks catch all checked exceptions, the generic `catch` block is redundant. If the generic `catch` block becomes redundant, you should simplify the program by removing it.

The above procedure deals with checked exceptions, but it does not deal with unchecked exceptions. Any time after your program compiles, you can start looking for unchecked exceptions by executing it with different types of input. If your program still has a generic `catch` block, that `catch` block will catch an unchecked `RuntimeException` as well as all checked exceptions. If you have removed the generic `catch` block, the JVM will still identify any unchecked exception that might occur. In either event, you should try to deal with possible unchecked exceptions by using "careful code"; that is, rewriting your code so that situations which might generate unchecked exceptions never occur. Code that deals with possible unchecked exceptions may be in a `try` block, but it should not involve the relatively inefficient `catch` mechanism.

15.11 Understanding Exception Messages

Unless you're incredibly careful, you've probably written programs that have generated runtime error messages. But prior to this chapter, you weren't properly prepared to understand those error messages thoroughly. Now you are. In this section, we describe exception messages by showing exception message details in the context of a complete program.

NumberList Program

The program in Figures 15.14a and 15.14b reads in a list of numbers and calculates the mean. The program compiles and runs successfully most of the time, but it's not very robust. There are three types of entries that make the program crash. We'll describe those three entry types, but before you read about them, first try to determine them on your own.

Identify possible input errors.

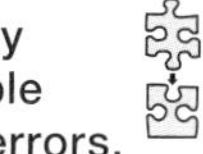

```
/*************************************************
* NumberListDriver.java
* Dean & Dean
*
* This is the driver for the NumberList class.
*************************************************/

public class NumberListDriver
{
  public static void main(String[] args)
  {
    NumberList list = new NumberList();
    list.readNumbers();
    System.out.println("Mean = " + list.getMean());
  } // end main
} // end class NumberListDriver
```

Figure 15.14a NumberList program driver that drives the class in Figure 15.14b

User Enters a Non-Integer

In the `readNumbers` method, note the `parseInt` call. If the user enters a `q`, the `while` loop terminates and `parseInt` is not called. But if the user enters something other than q, `parseInt` is called. If `parseInt` is called with a non-integer argument, then `parseInt` throws a `NumberFormatException`. And because there's no `try-catch` structure the JVM prints a detailed error message and then terminates the program. For example, if the user enters `hi`, the JVM prints a detailed error message and terminates the program, like this:[5]

Sample session:

```
Enter a whole number (q to quit): hi
Exception in thread "main" java.lang.NumberFormatException:
For input string: "hi"
    at java.lang.NumberFormatException.forInputString(
      NumberFormatException.java:65)
    at java.lang.Integer.parseInt(Integer.java:492)
    at java.lang.Integer.parseInt(Integer.java:527)
    at NumberList.readNumbers(NumberList.java:28)
    at NumberListDriver.main(NumberListDriver.java:13)
```

thrown exception

call stack trace

[5]The formatting of the error message may be slightly different, but the information will be similar. In a `catch` block, you could generate a message like this intentionally by calling `Exception`'s `printStackTrace` method.

```
/*************************************************************
* NumberList.java
* Dean & Dean
*
* This inputs numbers and calculates their mean value.
*************************************************************/

import java.util.Scanner;

public class NumberList
{
  private int[] numList = new int[100]; // array of numbers
  private int size = 0;                 // number of numbers

  //**********************************************************

  public void readNumbers()
  {
    Scanner stdIn = new Scanner(System.in);
    String xStr;   // user-entered number (String form)
    int x;         // user-entered number

    System.out.print("Enter a whole number (q to quit): ");
    xStr = stdIn.next();

    while (!xStr.equalsIgnoreCase("q"))
    {
      x = Integer.parseInt(xStr);
      numList[size] = x;
      size++;
      System.out.print("Enter a whole number (q to quit): ");
      xStr = stdIn.next();
    } // end while
  } // end readNumbers

  //**********************************************************

  public double getMean()
  {
    int sum = 0;

    for (int i=0; i<size; i++)
    {
      sum += numList[i];
    }
    return sum / size;
  } // end getMean
} // end class NumberList
```

Figure 15.14b `NumberList` class that computes the mean of input numbers

Let's analyze the error message. First, the JVM prints the exception that was thrown. In this case, it is `NumberFormatException`. Then it prints a *call stack trace*. A call stack trace is a listing of the methods that were called prior to the crash, in reverse order. What methods were called? First `main`, then `readNumbers`, then `parseInt`. Note the numbers at the right side of the call stack trace. They are the line numbers in the source code for where the methods are called. For example, the 13 in the bottom line says that `main`'s 13th line is a call to the `readNumbers` method.

User Immediately Enters `q` to Quit

At the bottom of the `getMean` method, note the division operation. Whenever you perform integer division, you should always be sure to avoid division by zero. In the NumberList program, it's not avoided. The `size` instance variable is initialized to zero, and if the user immediately enters `q` to quit, `size` stays at zero and `getMean` performs division by zero. Integer division by zero throws an `ArithmeticException`. Because there's no `try-catch` mechanism, the JVM prints a detailed error message and terminates the program, like this:

<u>Sample session:</u>

```
Enter a whole number (q to quit): q
Exception in thread "main"
    java.lang.ArithmeticException: / by zero
    at NumberList.getMean(NumberList.java:47)
    at NumberListDriver.main(NumberListDriver.java:14)
```

Note that if you perform floating-point division with a denominator of zero, there is no exception. If the numerator is a positive number, division by 0.0 returns the value `Infinity`. If the numerator is a negative number, division by 0.0 returns the value `-Infinity`. If the numerator is also 0.0, division by 0.0 returns the value `NaN` (`NaN` is a named constant in the `Double` class, and it stands for "not a number").

User Enters More Than 100 Numbers

In the NumberList program's instance-variable declarations, note that `numList` is a 100 element array. In the `readNumbers` method, note how this statement assigns user-entered numbers into the `numList` array:

```
numList[size] = x;
```

If the user enters 101 numbers, then the `size` variable increments to 100. That's bigger than the maximum index (99) in the instantiated array. If you access an array element with an index that's greater than the maximum index or less than zero, the operation throws an `ArrayIndexOutOfBoundsException`. Because there are no `try` and `catch` blocks, the JVM prints a detailed error message and then terminates the program, like this:

<u>Sample session:</u>

```
...
Enter a whole number (q to quit): 32
Enter a whole number (q to quit): 49
Enter a whole number (q to quit): 51
Exception in thread "main"
  java.lang.ArrayIndexOutOfBoundsException: 100
  at NumberList.readNumbers(NumberList.java:29)
  at NumberListDriver.main(NumberListDriver.java:13)
```

We've now finished our description of the NumberList program's three runtime errors. Normally, when you see such errors, you should fix your code so as to avoid the runtime errors in the future. So for the Number-List program, you should add fixes for the three runtime errors. One of the chapter exercises asks you to do just that.

15.12 Using `throws` *<exception-type>* to Postpone the `catch`

In all the examples so far, we've handled thrown exceptions locally; that is, we've put the `try` and `catch` blocks in the method that contains the dangerous statement. But sometimes that's not feasible.

Moving `try` and `catch` Blocks Back to the Calling Method

When it's not feasible to use local `try` and `catch` blocks, you can move the `try` and `catch blocks` out of the dangerous statement's method and back to the calling method. If you do that and the dangerous statement throws an exception, the JVM immediately jumps out of the dangerous statement's method and passes the exception back to the `try` and `catch` blocks in the calling method.[6]

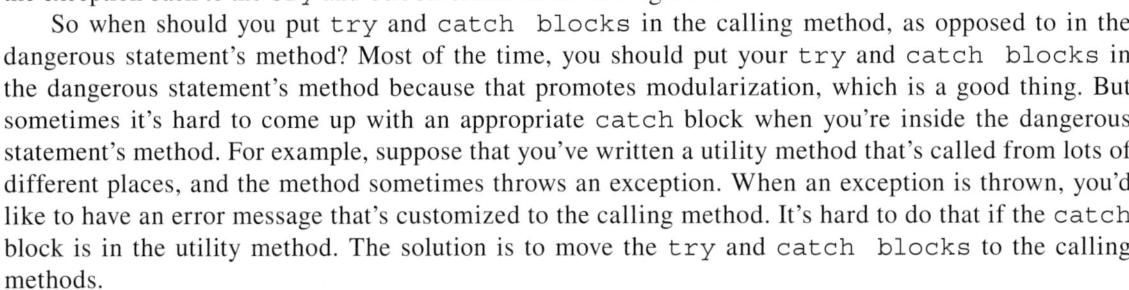

So when should you put `try` and `catch blocks` in the calling method, as opposed to in the dangerous statement's method? Most of the time, you should put your `try` and `catch blocks` in the dangerous statement's method because that promotes modularization, which is a good thing. But sometimes it's hard to come up with an appropriate `catch` block when you're inside the dangerous statement's method. For example, suppose that you've written a utility method that's called from lots of different places, and the method sometimes throws an exception. When an exception is thrown, you'd like to have an error message that's customized to the calling method. It's hard to do that if the `catch` block is in the utility method. The solution is to move the `try` and `catch blocks` to the calling methods.

Consider another example. Suppose you've written a method with a non-`void` return type that sometimes throws an exception. With a non-`void` return type, the compiler expects the method to return a value. But when an exception is thrown, you normally don't want to return a value because there's no appropriate value to return. So how can you have a non-`void` method and not return a value? Move the `try` and `catch` blocks to the calling method. Then when an exception is thrown, the JVM returns to the calling method without returning a value. The calling method's `try` and `catch` blocks handle the thrown exception, most likely with an error message. Let's see how this works in a Java program.

StudentList Program Revisited

Figure 15.15 contains a modified version of Figure 15.8's `StudentList` class. The main difference is that the `removeStudent` method now returns the name of the student it removes. This enables the calling method to do something with the removed element.

In the `removeStudent` method, note the `return` statement. The `students.remove` method call attempts to remove the element at the position indicated by `index`. If `index` is less than zero or greater than the index of the last element, then the JVM throws an `IndexOutOfBoundsException`. In our previous `StudentList` class, we handled the exception locally, within the `removeStudent`

[6]Actually, the jump to the calling method is not immediate if there's a `finally` block below the `try` block(s). In that case, the JVM jumps to the `finally` block prior to jumping to the calling method. We describe the `finally` block in the next section.

```
/*************************************************************
* StudentList2.java
* Dean & Dean
*
* This program manages an ArrayList of students.
*************************************************************/

import java.util.ArrayList;

public class StudentList2
{
  private ArrayList<String> students = new ArrayList<>();

  //***********************************************************

  public StudentList2(String[] names)
  {
    for (int i=0; i<names.length; i++)
    {
      students.add(names[i]);
    }
  } // end constructor

  //***********************************************************

  public void display()
  {
    for (int i=0; i<students.size(); i++)
    {
      System.out.print(students.get(i) + " ");
    }
    System.out.println();
  } // end display

  //***********************************************************

  public String removeStudent(int index)
    throws IndexOutOfBoundsException
  {
    return students.remove(index);
  } // end removeStudent
} // end StudentList2
```

Throw the error-handling job to the calling method.

Figure 15.15 StudentList2 class, which is driven by the class in Figure 15.16

method. This time, because we're returning a value, it's more convenient to transfer the exception handling work back to the calling method. We do that by putting `try` and `catch blocks` in the calling method and by putting a `throws` clause in the `removeStudent` method's heading. Here's the heading:

```
public String removeStudent(int index)
  throws IndexOutOfBoundsException
```

Adding the `throws` clause reminds the compiler that the method might throw an unhandled exception. The `throws` clause is required if the unhandled exception is a checked exception, and it's just recommended if the unhandled exception is an unchecked exception. Because the `IndexOutOfBoundsException` is an unchecked exception, it's legal to omit the above `throws` clause. But it's good style to include it because it provides valuable self-documentation. If a programmer later wants to use the `removeStudent` method, the `throws` clause warns the programmer to provide a "remote" `try-catch` mechanism to handle the `IndexOutOfBoundsException` when calling `removeStudent`.

To see how to implement this "remote" `try-catch` mechanism, look at the `StudentList2Driver` class in Figure 15.16. It displays a list of students, asks the user which student should be removed, and attempts to remove that student. If the `removeStudent` method call throws an exception, it's handled by the `catch` block in `StudentList2Driver`, which asks the user again which student should be removed.

Do you remember how we used `throws Exception` back in Section 3.24? We characterized that use as "quick and dirty" because it swept the compiler's complaint "under the rug." Are we doing that again here? No. This time, the calling method takes responsibility and deals with whatever exception might be thrown. As long as some caller eventually deals with thrown exceptions, it's possible to postpone the `catch` more than once by rethrowing to a previous caller. However, to keep debugging manageable, you should avoid deep hierarchies of postponed `catches` because they make it more difficult to determine where a failure occurred. In particular, you should not postpone the `catch` if you make a recursive call (Chapter 11) within a `try` block. The next section addresses another possible problem—idle resource accumulation—which can create problems regardless of whether the `catch` is postponed.

15.13 Automatic Cleanup Using Try-With-Resources

In implementing an exception handler, you'll sometimes need to provide "cleanup code" that executes regardless of whether an exception is thrown. Although in principle that cleanup code could be anything, in practice, it's almost always just the closing of a `Closeable` resource like a file handler. After writing to a file, you must *close* that file somehow to complete the writing process. Closing files also releases system resources and improves system performance.

Unfortunately, if some operation between file opening and file closing throws an exception, the close operation might be skipped. If a method throws exceptions back to a caller, the StudentList2 program in Figures 15.15 and 15.16 shows that normally you don't need a `try-catch` mechanism in the method throwing the exception. But if that method needs to close a file, you can put the file opening and file processing code inside a `try` block, and then follow that (otherwise unnecessary) `try` block with a separate `finally` block, which looks something like this:

```
/*************************************************************
* StudentList2Driver.java
* Dean & Dean
*
* This drives the StudentList2 class.
*************************************************************/

import java.util.Scanner;

public class StudentList2Driver
{
  public static void main(String[] args)
  {
    Scanner stdIn = new Scanner(System.in);
    String[] names = {"Caleb", "Izumi", "Mary", "Usha"};
    StudentList2 studentList = new StudentList2(names);
    int index;
    boolean reenter;

    studentList.display();

    do
    {
      System.out.print("Enter index of student to remove: ");
      index = stdIn.nextInt();
      try
      {
        System.out.println(
          "removed " + studentList.removeStudent(index));
        reenter = false;
      }
      catch (IndexOutOfBoundsException e)
      {
        System.out.print("Invalid entry. ");
        reenter = true;
      }
    } while (reenter);

    studentList.display();
  } // end main
} // end StudentList2Driver
```

If there is no error, this method returns name of student removed.

If exception is thrown in `removeStudent` method, this `catch` block catches it.

Sample session:

```
Caleb Izumi Mary Usha
Enter index of student to remove: 6
Invalid entry. Enter index of student to remove: 1
removed Izumi
Caleb Mary Usha
```

Figure 15.16 Driver for the `StudentList2` class

```
finally
{
  if (fileOut != null)
  {
    fileOut.close();
  }
} // end finally
```

The `finally` block gets executed after the `try` block, regardless of whether the `try` block's code throws an exception. Thankfully, starting with Java 1.7, there's a better way to close a file.

All classes that `implement Closeable` also `implement AutoCloseable`. When an object is `AutoCloseable`, instead of closing it with an explicit `close` statement, we can ask the JVM to close it for us. We request this service by opening the `AutoCloseable` object in a *try-with-resources* header. This alternative approach simplifies code and makes programs more robust because Java automatically closes the resource (typically a file handler) regardless of whether all the code in the try-with-resources header and the following `try` block executes successfully or throws an exception.

In the try-with-resources approach, the creation of the `AutoCloseable` resource (both declaration and initialization) moves into a header in a set of parentheses after the keyword, `try`. When the try-catch mechanism is all in the same method, this avoids an explicit `close` statement. When the try-with-resources is in a method that throws its exceptions, this avoids a `finally` block. With semicolon separators between them, we can put any number of `AutoCloseable` resource-creation statements in the parentheses after the `try`, and the JVM closes all of them automatically at the end of that `try` block, regardless of whether exceptions are thrown.

The WriteToFile program in Figure 15.17 shows the case in which the try-with-resources is in a `write` method that `throws` an `IOException`. This is like `StudentList2`'s `removeStudent` method that `throws` an `IndexOutOfBoundsException`. It's different in that `WriteToFile`'s `write` method creates a `PrintWriter` file handler, which it must eventually close. Notice, however, that `WriteToFile`'s `write` method contains no explicit `close` method call, and it contains no `finally` block. It avoids these by opening the file in a try-with-resources header.

The WriteToFile program in Figure 15.17 postpones the `catch`. However, try-with-resources also works when the `catch` block(s) are in the same method, immediately after the `try`. One of the exercises asks you to implement this alternative version of the WriteToFile program.

While the try-with-resources mechanism replaces the most common application of a `finally` block (releasing opened resources), the try-with-resources mechanism does not prevent you from also using a `finally` block for other cleanup activities. In addition, be aware that you can nest a `try-catch` mechanism inside a larger `try` block or inside a larger try-with-resources block. Similarly, you can nest a try-with-resources block and its associated `catch` block(s) inside a larger `try` block or inside a larger try-with-resources block. In other words, so long as every statement in a try-with-resources header opens something that is `AutoCloseable`, you can use a try-with-resources block just like an ordinary `try` block.

15.14 GUI Track and Problem Solving: Line Plot Example Revisited (Optional)

Problem Description

Earlier in the chapter, we implemented a LinePlot program that plotted a line defined by a sequence of user-specified points. The line plot's display was less than ideal. It "displayed" the line as a text description of

```
/*************************************************************
* WriteToFile.java
* Dean & Dean
*
* This writes to a file using try-with-resources and a postponed catch.
*************************************************************/

import java.io.*;             // PrintWriter, IOException

public class WriteToFile
{
  public int write(String filename, String text)
    throws IOException
  {
    try (PrintWriter fileOut = new PrintWriter(filename))
    {
      fileOut.println(text);
      return text.length(); // if exception is not thrown
    } // end try and close fileOut automatically
  } // end writeToFile

  //*********************************************************

  public static void main(String[] args)
  {
    String filename = "Feynman.txt";
    String text = "It is fundamentally impossible to make "
      + "a precise prediction\n of exactly what will happen "
      + "in a given experiment.";
    int length = 0;
    WriteToFile writer = new WriteToFile();

    try
    {
      length = writer.write(filename, text);
      System.out.println("written string length = " + length);
    }
    catch (Exception e)
    {
      System.out.println(e.getClass());
      System.out.println(e.getMessage());
    }
  } // end main
} // end class WriteToFile
```

Opening a file in a try-with-resources header avoids an explicit `close`, which might also require an otherwise unnecessary `finally` block.

Sample session:

```
written string length = 111
```

Figure 15.17 WriteToFile program, which uses try-with-resources and automatic closing

line segments. For example, this is what the program produces for a five-segment line that goes from point (0,0) to (1,3) to (2,1) to (3,2) to (4,2) to (5,1):

(0,0)–(1,3), (1,3)–(2,1), (2,1)–(3,2), (3,2)–(4,2), (4,2)–(5,1)

Figure 15.18 shows how LinePlotGUI, an improved version of the LinePlot program, displays the above (0,0) to (1,3) to . . . to (5,1) line. In the interest of simplicity, there are no interval hash marks on the x and y axes. As you can perhaps guess, the shown x axis has six implied hash marks for the values 0, 1, 2, 3, 4, and 5. And the shown y axis has four implied hash marks for the values 0, 1, 2, and 3.

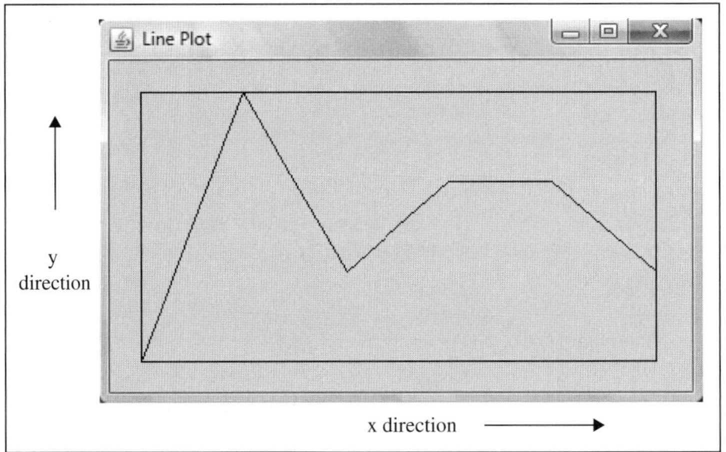

Figure 15.18 Sample output for the LinePlotGUI program

Prewritten Software and the drawPolyLine Method

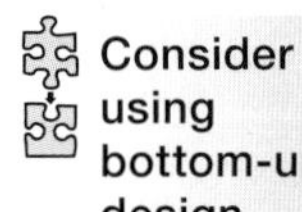

Consider using bottom-up design.

In solving problems, it's fairly common to take a top-down approach: Write the driver method first (`main`), then write the `public` methods that provide an interface to the outside world, then write the `private` helper methods. That approach frequently works fine, but sometimes it can lead to reinventing the wheel. If you used a pure top-down approach to implement the LinePlotGUI program, you'd probably implement the line plot as a sequence of `drawLine` method calls, one for each segment of the line (for a discussion of the `drawLine` method, see Chapter 5). Using `drawLine` would work, but it would require a loop and probably some debugging effort. The better approach is to dust off your Java API tome (http://docs.oracle.com/javase/7/docs/api/index.html) and search for a pre-built line-drawing method.

Lo and behold, there's a line drawing method that does exactly what you want. The `drawPolyline` method draws a line by connecting a sequence of points. Java API documentation shows that the `Graphic` class's `drawPolyline` method has three parameters—`xPixels`, `yPixels`, and `numOfPoints`. The `numOfPoints` parameter holds the number of points in the line. The `xPixels` parameter holds an array of the horizontal pixel positions for each of the points. The `yPixels` parameter holds an array of the vertical pixel positions for each of the points. For example, `xPixels[0]` and `yPixels[0]` hold pixel positions for the first point, `xPixels[1]` and `yPixels[1]` hold pixel positions for the second point, and so on. Figure 15.19's `drawPolyline` method call displays a line that connects four points in the shape of an N.

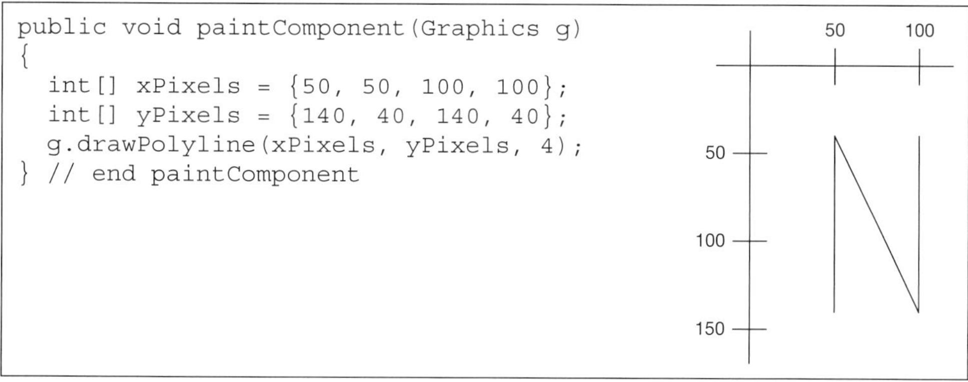

public void paintComponent(Graphics g)
{
int[] xPixels = {50, 50, 100, 100};
int[] yPixels = {140, 40, 140, 40};
g.drawPolyline(xPixels, yPixels, 4);
} // end paintComponent
50
100
50
100
150

Figure 15.19 An example `drawPolyline` method call that displays a line in the shape of an N

Algorithm Development

Once you learn about the `drawPolyline` method, the LinePlotGUI program's basic algorithm becomes clear: Fill up the `xPixels` and `yPixels` arrays with pixel values for a sequence of points. Then use those arrays to call `drawPolyline`.

Before fleshing out that basic algorithm with more details, you need to be aware of an important assumption. The plotted line's points are evenly spaced along the x axis. More specifically, the points occur at positions x = 0, x = 1, x = 2, and so on. So when the program prompts the user for a point, only a y value is needed, not an x value (because x's value is already known: x = 0, x = 1, x = 2, and so on).

Here's a high-level description of the algorithm:

Organize your thoughts as an algorithm.

1. Prompt the user for the number of points and the maximum y value (`maxY`).
2. For each point, prompt the user for a y coordinate value and store the value in `yCoords`.
3. Determine the number of horizontal pixels between adjacent points (`pixelInterval`).
4. Fill up the `xPixels` array:

   `xPixels[i] ← i * pixelInterval`

5. Fill up the `yPixels` array by scaling the values in the `yCoords` array:

   `yPixels[i] ← (yCoords[i]/maxY) *` height in pixels of plotted line's border

6. Call `drawRect` to display a border for the plotted line.
7. Call `drawPolyline` to display the plotted line.

Program Structure

Now that you have a high-level description of the algorithm, you might be tempted to translate it into source code immediately. First, you should determine where the different parts of the program should go. As is customary for a non-trivial program like this one, you should implement the solution with two classes—a `LinePlotGUI` class for driving the program and a `LinePlotPanel` class for drawing the pictures. Clearly, the `drawRect` and `drawPolyline` method calls should go in the `LinePlotPanel` class because they involve drawing. But what about the code that prompts the user for y coordinate values, and what about the code that calculates the `xPixels` and `yPixels` values? It's important to have the code in the right class to ensure that each class has a clearly defined role. That helps with program development, readability, and maintainability. The `LinePlotGUI` class drives the program and input plays a big role in that effort. Therefore, you should put the user-prompt code in the `LinePlotGUI` class. The `LinePlotPanel` class draws the pictures, and calculations play a big role in that effort. Therefore, you should put the drawing-calculations code in the `LinePlotPanel` class.

Modularity

Now it's time to look at the LinePlotGUI program's source code in Figures 15.20a, 15.20b, and 15.20c. Note the modular nature of the `LinePlotGUI` class. The `JFrame` calls (`setSize`, `setTitle`, `setDefaultCloseOperation`) are in their own module, the `LinePlotGUI` constructor. The user prompt code is in its own module, the `readYCoordinates` method. The `readYCoordinates` method prompts the user for the number of points. It also prompts the user for the maximum value on the y axis. The two inputs both need to undergo the same type of input validation. To avoid redundant code, the input validation code is in a common helping method, `getIntFromUser`. The `readYCoordinates` method also prompts the user for the y value of each point, which can be anywhere in the range between zero and the maximum y value.

The `LinePlotPanel` class in Figures 15.21a and 15.21b descends from `JPanel`, so `LinePlotPanel` is a `JPanel`. Because `JPanel` descends from `JComponent`, `LinePlotPanel` is also a `JComponent`. When a GUI program starts, and whenever a user does something to alter its window, the JVM automatically calls `JComponent's` `paintComponent` method or an overriding `paintComponent` method defined in a class descended from `JComponent`. In Figure 15.21b you can see that our `LinePlotPanel` redefines the `paintComponent` method. So `LinePlotPanel's` `paintComponent` method manages the picture drawing.

We might have combined the drawing calculations and the graphics code by putting them together in the `paintComponent` method, but as shown in Figures 15.21a and 15.21b, the drawing-calculations code is in its own module, the `LinePlotPanel` constructor. That's a good idea for several reasons. It furthers the goal of modularization in that separate tasks are performed in separate modules. It improves program speed. The `LinePlotPanel` constructor executes only once, when the `LinePlotPanel` object is instantiated in `main`. The `paintComponent` method executes every time the user does something to alter the program's window. There's no need to redo the drawing calculations every time that happens, so moving the drawing calculations to the `LinePlotPanel` constructor works well.

For a line plot with a small number of points, it would be no big deal if we'd made the mistake of putting all the drawing calculations code in the `paintComponent` method. But if there are many points, the window-resizing slowdown might be noticeable. Slow is acceptable for some things, like initially loading a program, but not for GUI things, like resizing a window. Users are an impatient bunch. Have you ever thumped your mouse in anger because you were in a hurry, but your computer wasn't?

Scalability

A program is *scalable* if it's able to support larger or smaller amounts of data and more or fewer users. The support for more data and more users might require changes to the software, but the changes should be cost-effective incremental add-ons, not massive rewrites.

A professional-grade line-plotting program should be able to handle large quantities of data and different types of data. Our LinePlotGUI program may not achieve that level of scalability (it handles only one type of data, and all the data must fit in one window), but it's not too bad, considering that we needed to keep it reasonably short to fit in an introductory textbook. The LinePlotGUI program is scalable in that its scope

```
/*************************************************************
* LinePlotGUI.java
* Dean & Dean
*
* This program plots a line as a sequence of connected,
* user-specified points.
*************************************************************/

import javax.swing.*;  // for JFrame, JOptionPane

public class LinePlotGUI extends JFrame
{
  private static final int FRAME_WIDTH = 400;
  private static final int FRAME_HEIGHT = 250;
  private static final int MARGIN = 20; // space between frame
                                        // and line plot

  int numOfPoints;      // points go from N=0 to N=numOfPoints-1
  int maxY;             // y coordinate values go from y=0 to y=maxY
  double[] yCoords;     // y coordinate values for all the points

  //**********************************************************

  public LinePlotGUI()
  {
    setSize(FRAME_WIDTH, FRAME_HEIGHT);
    setTitle("Line Plot");
    setDefaultCloseOperation(JFrame.EXIT_ON_CLOSE);
  } // end LinePlotGUI

  //**********************************************************

  int getMargin()
  {
    return MARGIN;
  }

  int getMaxY()
  {
    return maxY;
  }

  double[] getYCoords()
  {
    return yCoords;
  }
```

Figure 15.20a LinePlotGUI class—part A

```
//***************************************************************

// This method prompts the user for y coordinates for points
// at positions x=0, x=1, etc.

public void readYCoordinates()
{
  String yStr;  // user's entry for a point's y coordinate
  numOfPoints = getIntFromUser("Enter number of points: ");
  maxY = getIntFromUser("Enter maximum point value: ");
  yCoords = new double[numOfPoints];

  for (int i=0; i<numOfPoints; i++)
  {
    yStr = JOptionPane.showInputDialog(
      "At x = " + i + ", what is y's value?\n" +
      "Enter an integer between 0 and " +
      maxY + " inclusive:");
    try
    {
      yCoords[i] = Integer.parseInt(yStr);
      if (yCoords[i] < 0 || yCoords[i] > maxY)
      {
        JOptionPane.showMessageDialog(null,
          "Invalid entry. Value must be between 0 and " + maxY);
        i--;
      }
    }
    catch (NumberFormatException e)
    {
      JOptionPane.showMessageDialog(null,
        "Invalid entry. Must enter an integer.");
      i--;
    }
  } // end for
} // end readYCoordinates
```

These initializations use a helper method.

Figure 15.20b `LinePlotGUI` class—part B

is constrained primarily by user input and a few named constants, not by hard-to-change coding constructs. Examples: The user specifies the number of points and the maximum y value for each point; named constants specify the window size and the window margin.

Robustness and the `getInsets` Method

In the `LinePlotPanel` constructor in Figure 15.21a, note how `frame` (the program's output window) calls the `getInsets` method. The `getInsets` method returns a window's `Insets` object. The `Insets` object stores the thicknesses of the window's four borders. For example, `frame.getInsets().left`

```
  //*************************************************************

  // This method prompts the user for an integer, performs input
  // validation, and returns the entered integer.

  private static int getIntFromUser(String prompt)    ◄── helper method
  {
    String entry;           // user entry
    boolean valid = false; // is user entry a valid integer?
    int entryInt = 0;       // integer form of user entry

    entry = JOptionPane.showInputDialog(prompt);
    do
    {
      try
      {
        entryInt = Integer.parseInt(entry);
        valid = true;
      }
      catch (NumberFormatException e)
      {
        entry = JOptionPane.showInputDialog(
          "Invalid entry. Enter an integer:");
      }
    } while (!valid);

    return entryInt;
  } // end getIntFromUser

  //*************************************************************

  public static void main(String[] args)
  {
    LinePlotGUI linePlotGUI = new LinePlotGUI();
    linePlotGUI.readYCoordinates();
    LinePlotPanel linePlotPanel = new LinePlotPanel(linePlotGUI);
    linePlotGUI.add(linePlotPanel);
    linePlotGUI.setVisible(true);
  } // end main
} // end class LinePlotGUI
```

Figure 15.20c `LinePlotGUI` class—part C

```
/*************************************************************
* LinePlotPanel.java
* Dean & Dean
*
* This class displays a line as a sequence of connected points.
*************************************************************/

import javax.swing.*; // for JPanel
import java.awt.*;    // for Graphics

public class LinePlotPanel extends JPanel
{
  private int[] xPixels; // holds x value for each plotted point
  private int[] yPixels; // holds y value for each plotted point

  // Line plot is surrounded by a rectangle with these specs:
  private int topLeftX, topLeftY;
  private int rectWidth, rectHeight;

  //**********************************************************

  // Calculate dimensions for the line-plot rectangle, using the
  // passed-in frame, which contains the frame's dimensions and
  // coordinate values. Fill in xPixels and yPixels arrays.

  public LinePlotPanel(LinePlotGUI frame)
  {
    int numOfPoints = frame.getYCoords().length;
    int pixelInterval;   // distance between adjacent points

    topLeftX = topLeftY = frame.getMargin();

    // getInsets works only if setVisible is called first
    frame.setVisible(true);
    rectWidth =
      frame.getWidth() - (2 * topLeftX +
      frame.getInsets().left + frame.getInsets().right);
    rectHeight =
      frame.getHeight() - (2 * topLeftY +
      frame.getInsets().top + frame.getInsets().bottom);
```

Figure 15.21a `LinePlotPanel` class—part A

returns the width (in pixels) of the `frame` window's left border and `frame.getInsets().top` returns the height (in pixels) of the `frame` window's top border. The top border includes the height of the title bar.

If you don't want to bother with the `getInsets` method, you might be tempted to use hard-coded guesses for the border sizes. Don't do it. Different Java platforms (e.g., Windows, UNIX, and Macintosh

```
    // Calculate integer pixel interval between adjacent points
    pixelInterval = rectWidth / (numOfPoints - 1);

    // Make rectangle's actual width = multiple of pixelInterval
    rectWidth = (numOfPoints - 1) * pixelInterval;

    xPixels = new int[numOfPoints];
    yPixels = new int[numOfPoints];

    for (int i=0; i<numOfPoints; i++)
    {
      xPixels[i] = topLeftX + (i * pixelInterval);
      yPixels[i] = topLeftY + rectHeight - (int) Math.round(
        (frame.getYCoords()[i] / frame.getMaxY()) * rectHeight);
    }
  } // end LinePlotPanel constructor

  //*************************************************************

  // This class displays line as sequence of connected points.

  @Override
  public void paintComponent(Graphics g)
  {
    super.paintComponent(g);
    g.drawRect(topLeftX, topLeftY, rectWidth, rectHeight);
    g.drawPolyline(xPixels, yPixels, xPixels.length);
  } // end paintComponent
} // end class LinePlotPanel
```

Figure 15.21b `LinePlotPanel` class—part B

platforms) have different window border sizes. So even if you guess right for your current Java platform, your guesses won't necessarily work for alternative Java platforms. Moral of the story: Be robust and use `getInsets`. Don't use hard-coded guesses.

Summary

- An exception is an event that occurs during the execution of a program that disrupts the program's normal flow of instructions.
- Exception handling is a technique for handling exceptions gracefully.
- Use a `try` block to "try out" one or more dangerous method calls. If there's a problem with the dangerous method calls, the JVM throws an exception and looks for a "matching" `catch` block.
- A `catch` block is matching if the `catch` heading's parameter type is the same as or an ancestor of the type of the thrown exception.

- If an exception is thrown, the JVM immediately jumps out of the current `try` block. That means that if there are statements in the `try` block after the exception-throwing statement, those statements get skipped.
- Checked exceptions must be checked with a `try-catch` mechanism.
- Unchecked exceptions may optionally be checked with a `try-catch` mechanism, but it's not a requirement.
- Unchecked exceptions are descendants of the `RuntimeException` class.
- To implement a simple, general-purpose exception handler, define a `catch` block with an `Exception` type parameter, and inside the `catch` block, call the `Exception` class's `getClass` and `getMessage` methods.
- To define an exception handler with more specificity, define a sequence of `catch` blocks. Use multiple exception `catch` parameters for common responses. Arrange the `catch` blocks with the more general exception classes later, and end with the general-purpose exception handler.
- If a program crashes, the JVM prints a call stack trace. A call stack trace is a listing of the methods called prior to the crash, in reverse order.
- When objects like a file handler are `AutoCloseable`, declare and initialize them (open the files) in a try-with-resources header. A single header may contain multiple opening statements, separated by semicolons. Then the JVM automatically closes all opened objects without an explicit `close` call, regardless of whether operations in the try-with-resources header or the associated `try` block throw exceptions.
- To propagate an exception back to the calling module, append throws *<exception-type>* to the called method header. Then the called method does not need `try` and `catch` blocks, but it may include a try-with-resources block and/or a `finally` block for cleanup.

Review Questions

§15.3 Using `try` and `catch` Blocks to Handle "Dangerous" Method Calls

1. If your program contains an API method call, you should put it inside a `try` block. To be fully compliant with proper coding practice, you should apply this rule for all your API method calls. (T / F)
2. A `try` block and its associated `catch` block(s) must be contiguous. (T / F)

§15.5 `try` Block Details

3. Usually, you should try to aggregate related dangerous statements in the same `try` block to minimize clutter. (T / F)
4. Where should you put safe statements that use the results of dangerous operations?
5. If an exception is thrown, the JVM jumps to a matching `catch` block, and after executing the `catch` block, it returns to the `try` block at the point where the exception was thrown. (T / F)
6. In checking for compile-time errors, the compiler takes into account that all statements inside a `try` block might get skipped. (T / F)

§15.6 Two Categories of Exceptions—Checked and Unchecked

7. If an exception is derived from the `RuntimeException` class it is a(n) ________________ exception.
8. Checked exceptions are exceptions that are in or derived from the ________________ class, but not in or derived from the ________________ class.

§15.7 Unchecked Exceptions

9. In the following list, indicate whether each option is a viable option for an unchecked exception that you know your program might throw:
   a) Ignore it.
   b) Rewrite the code so that the exception never occurs.
   c) Put it in a `try` block, and catch it in a following `catch` block.

§15.8 Checked Exceptions

10. When a statement might throw a checked exception, you can keep the compiler from complaining if you put that statement in a `try` block and follow the `try` block with a `catch` block whose parameter type is the same as the exception type. (T / F)
11. You can determine whether a particular statement contains a checked exception and the type of that exception by attempting to compile with no `try-catch` mechanism. (T / F)

§15.9 Generic `catch` Block with `Exception` Class

12. Is it OK to include code that can throw both unchecked and checked exceptions in the same `try` block?
13. What type of exception matches all checked exceptions and all unchecked exceptions except those derived from the `Error` class?
14. What does the `getMessage` method return?

§15.10 Multiple `catch` Blocks and Multiple Exceptions Per Block

15. The compiler automatically checks for out-of-order `catch` blocks. (T / F)
16. Write the header for a `catch` block that catches either an `InvalidPathException` or a `NoSuchFileException`.

§15.11 Understanding Exception Messages

17. What are the two types of information displayed by the JVM when it encounters a runtime error that terminates execution?

§15.12 Using `throws <exception-type>` to Postpone the `catch`

18. Suppose you want to postpone catching of a `NumberFormatException`. What should you append to the heading of a method to alert the compiler and a potential user that something in the method might throw that type of exception?
19. Given a non-`void` method that contains no `try` and `catch blocks`. If the method throws an exception, we know that the JVM transfers the thrown exception back to the calling method. But does the JVM return a value (with a `return` statement) to the calling module?

§15.13 Automatic Cleanup Using Try-With-Resources

20. The JVM will close your file automatically if you instantiate its file handler in a try-with-resources header, like this (T / F):

```
public void writeToFile(String filename, String text)
  throws IOException
{
  PrintWriter fileOut;
  try (fileOut = new PrintWriter(new File(filename)))
  {
    fileOut.println(text);
  }
} // end writeToFile
```

Exercises

1. [after §15.3] Given the program below, what is the output if the user enters "one" in response to the prompt?

```
/*************************************************************
* FantasyFootball.java
* Dean & Dean
*
* This prints out names of football players.
*************************************************************/

import java.util.Scanner;
import java.util.ArrayList;

public class FantasyFootball
{
  public static void main(String[] args)
  {
    Scanner stdIn = new Scanner(System.in);
    ArrayList<String> players = new ArrayList<>();
    String indexStr;
    int index = 0;
    players.add("Aaron Rodgers");
    players.add("Ray Rice");
    players.add("Andrew Luck");
    System.out.print("Enter a number between 1 and 3: ");
    indexStr = stdIn.nextLine();
    try
    {
      index = Integer.parseInt(indexStr);
      System.out.println("Entered index OK.");
    }
    catch (NumberFormatException e)
    {
      System.out.println("Entered index wasn't an integer");
    }
    try
    {
      System.out.println(players.get(index - 1));
    }
    catch (IndexOutOfBoundsException e)
    {
      System.out.println(
      "Can't access players[" + (index - 1) + "]");
    }
   System.out.println("done");
  } // end main
} // end class FantasyFootball
```

2. Given the above program, what is the output if the user enters 1 in response to the prompt?

3. [after §15.4] Add a `try-catch` structure to the following program to make it compile and execute correctly, even when the divisor is zero. Note that division by zero throws an `ArithmeticException`.

```
import java.util.Scanner;

public class Division
{
  public static void main(String[] args)
  {
    Scanner stdIn = new Scanner(System.in);
    int n, d, q;
    System.out.print("Enter numerator: ");
    n = stdIn.nextInt();
    System.out.print("Enter divisor: ");
    d = stdIn.nextInt();
    q = n / d;
    System.out.println(q);
  } // end main
} // end Division class
```

To help you out, we've provided the `catch` block, below:

```
catch (ArithmeticException e)
{
  System.out.println("Error, but keep going anyway.");
}
```

There's no need to check for correct input; you may assume that the user enters two properly formatted `int` values for input.

4. [after §15.5] Program Improvement:

The following program performs division and does not throw an exception when you input a zero for the divisor. It also does not detect input number format exceptions. Minimize the total lines of code required to meet the requirements.

```
/*********************************************************
* Division2.java
* Dean & Dean
*
* This attempts to prevent division by zero.
*********************************************************/

import java.util.Scanner;

public class Division2
{
  public static void main(String[] args)
  {
    Scanner stdIn = new Scanner(System.in);
    double dividend;
    int divisor;
    System.out.print("Enter dividend: ");
    dividend = stdIn.nextDouble();
```

```
      System.out.print("Enter divisor: ");
      divisor = stdIn.nextInt();
      System.out.println(dividend / divisor);
    } // end main
  } // end Division2 class
```

a) First, rewrite the program so that it still employs a `double` dividend and `int` divisor, but if the user input for the divisor is zero, it refuses to perform the division operation, and keeps asking for the divisor until the user supplies something other than zero.

b) Next, rewrite the program of part a so that if the user inputs an improper format for either the dividend or the divisor, the entire input query repeats until both formats are OK. Hint: Put `try` and `catch blocks` in a loop that executes `while (OK == false)`, and set `OK = true` after all of the critical operations in the `try` block have succeeded. Note: If the scanned format is bad, you'll get infinite looping unless you re-instantiate `stdIn` in each iteration, or use a two-step operation for each input (input a string and then parse it).

5. [after §15.7] What happens if an unchecked exception is thrown and never caught?

6. [after §15.10] Multiple `catch` Blocks:

Suppose the code in a `try` block might throw any of the following exceptions:

a) `Exception`
b) `IllegalArgumentException`
c) `IOException`
d) `NumberFormatException`
e) `RuntimeException`

Identify an acceptable sequence for multiple `catch` blocks of these types.

7. [after §15.11] Correcting Problems:

Fix the problems in the NumberList program without making any changes to the `NumberListDriver` class.

If a user immediately enters "q" to quit, print "NaN" by making a small program correction that utilizes `double`'s `NaN` value, and avoid using the `try-catch` mechanism to catch the `int` arithmetic exception.

<u>Sample session:</u>

```
Enter a whole number (q to quit): q
Mean = NaN
```

If the entry is not a "q," and if it is not a legal integer, catch the exception, and in the `catch` block use the `getClass` method inherited from the `Object` class to print the name of the exception class followed by the error message with the statement:

```
System.out.println(e.getClass() + " " + e.getMessage());
```

Avoid the possibility of an `ArrayIndexOutOfBoundsException` by adding to the `while` condition `size < numList.length`, and perform the query and entry at the end of the `while` loop only if `size < numList.length`.

8. [after §15.12] TestExceptions:

What does this program output? Because this program converts between string and numeric values, use quotes to denote string values.

```
/*************************************************************
* TestExceptions.java
* Dean & Dean
*
* This looks up the value at a calculated index.
*************************************************************/
```

```
public class TestExceptions
{
  private double[] value =
    new double[] {1.0, 0.97, 0.87, 0.7, 0.47, 0.17};
  private int num;

  //*******************************************************

  public double eval(String n1, String n2)
    throws IndexOutOfBoundsException
  {
    try
    {
      num = Integer.parseInt(n1) / Integer.parseInt(n2);
    }
    catch (NumberFormatException nfe)
    {
      num++;
      System.out.println("in first catch");
    }
    catch (ArithmeticException ae)
    {
      num++;
      System.out.println("in second catch");
    }
    return value[num];
  }
  //*******************************************************
  public static void main(String[] args)
  {
    TestExceptions te = new TestExceptions();
    try
    {
      System.out.println(te.eval("5.0", "4"));
      System.out.println(te.eval("5", "0"));
      System.out.println(te.eval("22", "5"));
      System.out.println(te.eval("33", "5"));
    }
    catch (Exception e)
    {
      System.out.println("in main's catch");
    }
    System.out.println("Bye");
  } // end main
} // end TestExceptions class
```

9. [after §15.13] Simple try-with-resources:

Write a program called WriteToFileEx that writes and displays what the WriteToFile program in Figure 15.17 writes and displays. But do not use a `separate write method`. Put the try-with-resources block together with the `catch` block in the `main` method.

Review Question Solutions

1. False. Many API method calls are safe, and there's no need to put those method calls inside a `try` block.
2. True. You cannot put any statements between associated `try` and `catch` blocks.
3. True.
4. Put safe statements that use the results of dangerous operations inside the `try` block and after those dangerous operations.
5. False. After executing the `catch` block, the JVM continues downward; it does not jump back to the `try` block. Consequently, `try`-block statements get skipped if they follow an exception-throwing statement.
6. True.
7. If an exception is derived from the `RuntimeException` class, it is an unchecked exception.
8. Checked exceptions are exceptions that are in or derived from the `Exception class`, but not in or derived from the `RuntimeException` class.
9. Viable options for an unchecked exception that you know might be thrown:
   a) Not viable! You don't want your program to crash at runtime.
   b) Viable.
   c) Viable.
10. True.
11. True. If the statement contains a checked exception, the compiler will say so and identify the exception type.
12. Yes.
13. The `Exception` exception.
14. The `Exception` class's `getMessage` method returns a text description of the thrown exception.
15. True. The compiler complains if an earlier more generic `catch` block preempts a later more specific `catch` block.
16. `catch (InvalidPathException | NoSuchFileException e)`
17. The two types of information displayed by the JVM when it encounters a runtime error are:
   a) Identification of the particular exception thrown.
   b) A call-stack trace, which is a reverse-order listing of the methods called just prior to the crash, along with the line numbers where the error occurred in each method.
18. You must append `throws NumberFormatException` to the end of the method heading.
19. No. When an exception is thrown back to the calling method, the JVM does not return a value (with a `return` statement) to the calling module.
20. False. The try-with-resources header must also include the file handler's declaration.

CHAPTER 16

Files, Buffers, Channels, and Paths

Objectives

- Copy simple text from and to small local files.
- Learn the rudiments of HTML.
- Copy HTML from a remote website.
- Learn how to copy Java objects to and from Java object files.
- Be able to use different character sets and different file-opening options.
- Write and read large text files efficiently through text buffers.
- Copy primitive data to or from any place in a primitive buffer.
- Copy bytes randomly to or from any place in a file through a Java `Channel`.
- Establish a non-volatile connection to file data with a memory map.
- Learn how to manipulate paths.
- Optionally, implement a GUI file chooser.

Outline

16.1 Introduction

Except for the simple ReadFromFile and WriteToFile programs in the previous chapter, up until now, program input has come from the keyboard and program output has gone to the computer screen. That type of input/output (I/O) is temporary. When you enter input from the keyboard, the input is not saved. If you want to run the program again, you must enter the input again. Likewise, when you send output to the computer screen, it's not saved. If you want to analyze the output, you must analyze it on the screen. A day later, if you want to look at it again, you must run the program again.

For permanent or reusable I/O, you can store input or output data in a file. A *file* is a group of data that is typically stored in a contiguous block on a non-volatile storage device (such as a hard disk). Data files are fundamentally the same as the `.java` and `.class` program files that you have been using all along to hold your Java programs. But instead of holding programs, data files hold data that programs read from for input or write to for output. For your data files, use an extension that suggests how the data is formatted and the type of program that understands that format. For example, `.txt` identifies a text file that a simple word-processing program can understand.

This chapter begins with Hypertext Markup Language (HTML) writing and reading. The HTML file-generator example reviews the simpler exception-handling techniques introduced in the previous chapter in the context of HTML text. This example opens two local files in a single try-with-resources header. The website-reader example in Section 16.3 opens a connection with a remote website. This shows how Web access relates to file handling. Text files are easy to understand, and you can create or view text files with almost any text editor.

Section 16.4 shows how you can use software already in the Java language to perform the intricate structural conversion between a program object and the stream of bytes that flow to and from a file. Java's built-in software greatly simplifies the custom code needed to write objects to a file and read objects from a file. You'll learn how to write more than one distinct object to the same file and how to write modifications of previously written objects. And you'll learn how to read back any or all of those objects. Assuming the code that reads is consistent with the code that writes, and you know exactly how things are organized, object files are easy to use.

Section 16.5 will show you how to specify particular character sets to support different languages and how to control the way opened files may be manipulated. Section 16.6 shows how to transfer large blocks of text with maximum efficiency.

Section 16.7 shows how to convert each of the different types of primitive data or arrays of any of these types to the stream of bytes that flow to and from a file. It also shows how to access or mutate primitive data at random locations in one of those streams of bytes. Although it's more tedious than working with text or using Java's automatic processing of large objects, working directly with primitive data, and especially bytes, tends to improve a program's efficiency.

Section 16.8 extends these techniques to a larger scale and shows how to connect the streams of bytes to files. Memory mapping establishes a "permanent" link between your program and a particular file's data—a link that survives normal file closing.

Section 16.9 explains how you can specify and manipulate paths to directories and files; it explains how to copy or move files from one directory to another; and it shows how to display what's in a directory. Section 16.10 shows how to search the entire directory structure for files that have specified features in their names. And section 16.11, which is optional, shows how to display a directory's contents with a graphical user interface.

This chapter focuses on transferring data to and from a local file, which directly or indirectly involves conversion to or from a stream of bytes. But converting program data to or from a stream of bytes has broader applications. If you want to transfer data to or from other computers through local networks or through the Internet, you also need to make the conversion between your program's data types and streams of bytes. This chapter shows you how to make those conversions. So while you are learning about data storage in files, you are also learning about data transfer through communication channels.

16.2 Simple Text-File Example: HTML File Generator

Readers who want to use file I/O early have the option of reading this section after completing Section 3.23 in Chapter 3. If you jump from Chapter 3 to here, you should be aware that some of the material in this section won't make sense. But if you treat the file I/O code in the program in Figures 16.1a and 16.1b as a recipe, it will show you how to read from a file anything you can read from the keyboard, and it will show you how to write to a file anything you can print to the computer screen. This section consolidates what we explained in Chapter 15 about simple text file I/O with an example that reads from a text file and writes to a text file and opens both files in the same try-with-resources header. The program translates a user-specified text file into web page format. Then it writes the translation to a newly generated HTML file.

In Figure 16.1a, the program starts by reading a user-specified filename into the `filenameIn` variable. Then it composes the output file name. The name of the output file should be the same as the name of the input file except for the extension, which should be `.html`. To compose the name of the output file, the `String` method `lastIndexOf` finds the index of the last dot in `filenameIn`. If there is no dot, the `lastIndexOf` method returns a value of `-1`, and the program simply appends `.html` to the original filename. If there is a dot, `String`'s `substring` method returns the part of the string up through the character immediately before the dot, and the program adds `.html` to that. This process replaces the original filename's extension with a `.html` extension, and it assigns the result to `filenameOut`.

In the try-with-resources header, the program uses `filenameIn` to open the input file and create a `Scanner` object called `fileIn` to handle file-reading operations. Also in the try-with-resources header, it uses `filenameOut` to open the output file and create a `PrintWriter` object called `fileOut` to handle file-writing operations. Both `fileIn` and `fileOut` close automatically when the `try` block ends. Within the `try` block, the code starts by checking for an empty input file. If it's empty, it prints a warning message. Otherwise, it proceeds to transform the input file's data into the output file's format.

To help you understand this transformation, we need to digress and provide a brief overview of HTML (the computer language used to create Internet web pages). This book is not about HTML, but it's worth learning a little about HTML because Internet web pages are what got the Java language going.

HTML overview:

- At the top of every web page, you should include a doctype declaration, which tells the web browser the language the page is written in. For most web pages, the language will be *html,* and the doctype declaration will be `<!DOCTYPE html>`.
- HTML *tags* are surrounded by angled brackets, and they describe the purpose of their associated text.
- `<html>` and `</html>` tags surround the entire web page.
- The content between `<head>` and `</head>` tags is the heading for an HTML page. The heading contains information that describes the HTML page. This information is used by the browser and by search engines, but it is not visible on the HTML page.
- `<title>` and `</title>` tags surround the text that appears in a web page's title bar. Internet search engines use the `<title>` content to find web pages.
- The content between `<body>` and `</body>` tags is the body for the HTML page. The body contains the text that's displayed on the HTML page.

```
/*************************************************************
* HTMLGenerator.java
* Dean & Dean
*
* This program copies the contents of a user-specified
* file and pastes it into a newly generated HTML file.
*************************************************************/

import java.util.Scanner;
import java.io.PrintWriter;
import java.nio.file.Paths;

public class HTMLGenerator
{
  public static void main(String[] args)
  {
    Scanner stdIn = new Scanner(System.in);
    String filenameIn;         // original file's name
    int dotIndex;              // position of dot in filename
    String filenameOut;        // HTML file's name
    String line;               // a line from the input file

    System.out.print("Enter file's name: ");
    filenameIn = stdIn.nextLine();

    // Compose the new filename
    dotIndex = filenameIn.lastIndexOf(".");
    if (dotIndex == -1)        // no dot found
    {
      filenameOut = filenameIn + ".html";
    }
    else                       // dot found
    {
      filenameOut =
        filenameIn.substring(0, dotIndex) + ".html";
    }

    try (
      Scanner fileIn = new Scanner(Paths.get(filenameIn));
      PrintWriter fileOut = new PrintWriter(filenameOut))
    {
      // First line used for title and header elements
      line = fileIn.nextLine();
      if (line == null)
      {
        System.out.println(filenameIn + " is empty.");
      }
```

This opens an input file and an output file.

Figure 16.1a HTMLGenerator program—part A

```
      else
      {
        // Write the top of the HTML page.
        fileOut.println("<!DOCTYPE html>");
        fileOut.println("<html>");
        fileOut.println("<head>");
        fileOut.println("<title>" + line + "</title>");
        fileOut.println("</head>");
        fileOut.println("<body>");
        fileOut.println("<h1>" + line + "</h1>");

        while (fileIn.hasNextLine())
        {
          line = fileIn.nextLine();

          // Blank lines generate p tags.
          if (line.isEmpty())
          {
            fileOut.println("<p>");
          }
          else
          {
            fileOut.println(line);
          }
        } // end while

        // Write ending HTML code.
        fileOut.println("</body>");
        fileOut.println("</html>");
      } // end else
    } // end try and close fileOut and fileIn automatically

    catch (Exception e)
    {
      System.out.println(e.getClass());
      System.out.println(e.getMessage());
    } // end catch
  } // end main
} // end class HTMLGenerator
```

Figure 16.1b HTMLGenerator program—part B

- <h1> and </h1> tags surround the text that appears as a heading within a web page. Web browsers use large fonts to display text that's surrounded by <h1> tags.
- <p> tags indicate the beginning of a new paragraph. Web browsers generate a blank line for each <p> tag, and this helps set paragraphs apart.

Now let's analyze Figure 16.1b, the second half of the HTMLGenerator program. In the `else` clause, the code writes <html> and <head> tags to the output file. It writes the input file's first line to the output file, surrounded by <title> and </title> tags. Then it ends the web page's head section by writing </head> to the output file, and it begins the web page's body section by writing <body> to the output file. Then it reuses the input file's first line and writes it to the output file, surrounded by <h1> and </h1> tags. Then it loops through the subsequent lines in the input file. For each blank line, it writes a <p> tag to the output file, indicating a new paragraph. It writes each line that's not blank to the output file as is. The last two statements in the `else` clause complete the output file's HTML code. The two files close automatically when the `try` block ends. The generic `catch` block catches and describes any type of exception that might be thrown.

To see how the HTMLGenerator program works when applied to an actual input file, study the input file and resulting output file in Figure 16.2. If you'd like to verify that the HTMLGenerator program generates a working web page, create Figure 16.2's `historyChannel.txt` file, and run the HTMLGenerator program with `historyChannel.txt` as input. That should generate the `historyChannel.html` file in Figure 16.2. Open a browser window, and within that browser window, open the `historyChannel.html` file. For example, open a Windows Internet Explorer browser and perform a **File/Open** command. Voila—you should see the `historyChannel.html` file displayed as a web page!

HTML tags without slashes (like <html>, <head>, and <title> tags) are called *start tags*. HTML tags with slashes (like </title>, </head>, and </html> tags) are called *end tags*. For most tags, it is

Example input file, `historyChannel.txt`:

```
When Chihuahuas Ruled the World

Around 8000 B.C., the great Chihuahua Dynasty ruled the world.
What happened to this ancient civilization?

Join us for an extraordinary journey into the history
of these noble beasts.
```

Resulting output file, `historyChannel.html`:

```
<!DOCTYPE html>
<html>
<head>
<title>When Chihuahuas Ruled the World</title>
</head>
<body>
<h1>When Chihuahuas Ruled the World</h1>
<p>
Around 8000 B.C., the great Chihuahua Dynasty ruled the world.
What happened to this ancient civilization?
<p>
Join us for an extraordinary journey into the history
of these noble beasts.
</body>
</html>
```

This works, but to conform to strict HTML standards, web pages should have a </p> at the end of each paragraph.

Figure 16.2 Example input file for the HTMLGenerator program and its resulting output file

required that you have an accompanying end tag for every start tag. You can see that that rule is followed for the `<html>`, `<head>`, and `<title>` start tags in Figure 16.2, but it's not followed for the `<p>` start tags. Although not required by the HTML standards, having a `</p>` end tag for every `<p>` start tag can help with a web page's understandability and maintainability. An end-of-chapter exercise asks you to improve the HTMLGenerator program so that it generates `</p>` end tags at the end of every paragraph. Figure 16.2's callout shows where the `</p>` end tags should go in the generated `historyChannel.html` file.

16.3 A Website Reader

From accessing a file, it's a relatively small step to accessing a remote website. In either case, while data is in transit, it travels as a stream of bytes. In the previous section, when you read text from a file, the stream of bytes was hidden in the background. But when you read text from a remote website, this stream of bytes appears explicitly as an instance of the `InputStream` class.

Figure 16.3a displays HTML code that was retrieved from Park University's home page. Note that the website's DOCTYPE line specifies HTML 4.01. That means the web page stored at http://www.park.edu uses version 4.01 of HTML code. Many web pages, like this one, use old versions of HTML. In case you're curious, the more recent version of HTML, called HTML5, has quite a few features not found in the earlier versions. Some of those new features are:

- Allow web applications to work without an Internet connection.
- Make web pages flashier with more special effects.
- Allow objects to move and react to cursor movements.
- Run games from a web browser.
- Create interactive 3-D effects.
- Play audio without a plug-in.

```
Enter a full URL address: http://www.park.edu
Enter number of lines: 10

<!DOCTYPE html PUBLIC "-//W3C//DTD HTML 4.01 Transitional//EN"
"http://www.w3.org/TR/html4/loose.dtd">

<html lang="en">
<head>
<link title="Park University" rel="search" type="application/opens
earchdescription+xml" href="http://www.park.edu/searchpark.xml">
<title>Park University - Online, Undergraduate, and Graduate
Degrees</title>
<meta http-equiv="content-type" content="text/html;charset=ISO-8859-1">
<meta name="Title" content="Park University: Online, Bachelor's
and Master's Degrees">
<meta name="alexaVerifyID" content="WW8pCEhGJqjJ118X9XjQZGhMbXE" />
<meta name="Description" content="Park University offers accredited
online, undergraduate, and graduate degrees at campuses and military
bases nationwide.">
```

Figure 16.3a Typical output of the WebPageReader program in Figure 16.3b

```
/*************************************************************
* WebPageReader.java
* Dean & Dean
*
* This reads a web page.
*************************************************************/

import java.util.Scanner;
import java.net.*;            // URL, URLConnection
import java.io.InputStream;

public class WebPageReader
{
  public static void main(String[] args)
  {
    Scanner stdIn = new Scanner(System.in);
    Scanner webIn;
    URL url;
    URLConnection connection;
    InputStream inStream;     // stream of bytes
    int i = 0, maxI;          // line number and max line number

    try
    {
      System.out.print("Enter a full URL address: ");
      url = new URL(stdIn.nextLine());
      connection = url.openConnection();
      inStream = connection.getInputStream();
      webIn = new Scanner(inStream);
      System.out.print("Enter number of lines: ");
      maxI = stdIn.nextInt();
      while (i < maxI && webIn.hasNext())
      {
        System.out.println(webIn.nextLine());
        i++;
      }
    } // end try

    catch (Exception e)
    {
      System.out.println(e.getClass());
      System.out.println(e.getMessage());
    }
  } // end main
} // end WebPageReader
```

Figure 16.3b WebPageReader program—a simple website reader

Figure 16.3b contains a program that reads what you see in Figure 16.3a. This is the initial activity of a typical web browser. In addition to `Scanner`, the program imports packages that give access to the `URL`, `URLConnection`, and `InputStream` classes. After making declarations, the program enters a `try` block. The `try` block is required because the URL constructor can throw a `MalformedURLException`, the `openConnection` method call can throw an `IOException`, and the `getInputStream` method call can throw an `IOException`. After successfully completing its dangerous operations, the program asks for a maximum number of lines to print. Then it enters a `while` loop that prints up to this number of lines.

16.4 Object File I/O

When file data is text, you can write to a file with a simple word processor (like Windows Notepad or UNIX vi) and read from that file with a Java program, or vice versa. When a file is not text, it's more difficult, but when the programs that write and read a particular file are both Java programs, you have an advantage. You can use software that's built into the Java language to perform the structural conversion between program objects and streams of bytes. This section explains how to use that built-in software.

Enabling Java Objects to be Stored in a File

The Java language has a built-in mechanism that *serializes* each object's data as it goes into a file and *unserializes* that data as it comes out of a file and goes back into object form. Whenever a program writes serialized data into a file, it also writes the recipe it used to serialize that data. That recipe includes the type of the object, the type of each data item, and the sequence in which the data items are stored. When another program reads serialized data from a file, it also reads the recipe to learn how to reconstruct the object from the serialized data. To enable a class to use Java's built-in serializing mechanism, you must append the following clause to that class's heading:

```
implements Serializable
```

This makes it look as though your class is implementing an interface. But this interface doesn't define any named constants, and it doesn't require the class to implement any particular methods. It just identifies the class's objects as objects needing serialization services. For example, look at the `TestObject` class in Figure 16.4. Notice that this class implements the `Serializable` interface.

If a class is `Serializable`, all classes derived from it are automatically `Serializable`, too. Suppose your `Serializable` class has instance variables that refer to other objects. Those objects' classes also must be `Serializable`. This must be true through all levels in a composition hierarchy. Does that sound like a pain? It's not, really. Just be sure to include `implements Serializable` in the definition of all classes that define objects that you'd like to store as objects. The alternative would be a pain, though. If you couldn't store a whole object, you'd have to provide explicit code to write and read each primitive data item in the container object, and in all component objects in that container object, down the composition tree to all primitive leaves.

If you define or extend a class that implements `Serializable`, to verify that the class read into is compatible with the class written from, the compiler automatically gives that class a distinctive version number. You can override this automatically generated version number with an explicit version number by including a declaration like this:

```
private static final long serialVersionUID = 0L;
```

```
/*************************************************************
* TestObject.java
* Dean & Dean
*
* A typical heterogeneous object.
*************************************************************/

import java.io.Serializable;

public class TestObject implements Serializable
{
  private int id;
  private String text;
  public double number; // This variable's access is public!

  //**********************************************************

  public TestObject(int id, String text, double number)
  {
    this.id = id;
    this.text = text;
    this.number = number;
  } // end constructor

  //**********************************************************

  public void display()
  {
    System.out.print(this.id + "\t");
    System.out.print(this.text + "\t");
    System.out.println(this.number);
  } // end display
} // end TestObject class
```

To be writable to and readable from a file, an object must be an instance of a class that implements this interface.

Figure 16.4 Typical definition of a `Serializable` object

Then, each time you modify the class, you can increment the version number—to 1L, 2L, 3L, etc. If you use the `-Xlint` compiler option to show all warnings, the compiler issues a warning if you do not give a `Serializable` class an explicit version number. For all situations we consider in this book, you can safely refrain from providing an explicit version number and safely ignore the corresponding compiler warning that might occur.

Notice that one of the instance variables in the `TestObject` class in Figure 16.4 is `public`. In general, making an instance variable `public` is not good practice, but we do it here to make it easier for you to understand certain modifications to a later program that uses this particular class.

Writing a Serializable Object to a File

Figure 16.5 contains a program that writes instances of `TestObject` to a user-specified file. The try-with-resources header opens the user-specified file by instantiating a `FileOutputStream` and then an `ObjectOutputStream`. Inside the `try` block, the program instantiates and writes two distinct objects. Upon reaching the end of the `try` block, the JVM automatically closes the file. The generic `catch` block describes any exception that might be thrown.

```
/*******************************************************************
* WriteObject.java
* Dean & Dean
*
* This writes two distinct objects to an object file.
*******************************************************************/

import java.util.Scanner;
import java.io.*;     // ObjectOutputStream, FileOutputStream

public class WriteObject
{
  public static void main(String[] args)
  {
    Scanner stdIn = new Scanner(System.in);
    TestObject testObject;

    System.out.print("Enter filename: ");
    try (ObjectOutputStream fileOut = new ObjectOutputStream(
      new FileOutputStream(stdIn.nextLine())))
    {
      testObject = new TestObject(1, "first", 1.0);
      fileOut.writeObject(testObject);
      testObject = new TestObject(2, "second", 2.0);
      fileOut.writeObject(testObject);
    } // end try and close fileOut automatically
    catch (Exception e)
    {
      System.out.println(e.getClass());
      System.out.println(e.getMessage());
    } // end catch
  } // end main
} // end WriteObject class
```

Open the file.

Write two objects.

Sample session:

```
Enter filename: objectFile
```

Figure 16.5 WriteObject program, which writes `Serializable` objects to a file

Reading a Serializable Object from a File

Figure 16.6's ReadObject program reads data for all objects of the `TestObject` class from a user-specified file. The code down through the try-with-resources header should look familiar—because it is similar to the code down through the try-with-resources header in the preceding WriteObject program.

Within its `try` block, the ReadObject program contains an infinite `while` loop, which keeps reading objects until it throws an `EOFException` (end-of-file exception). When reading each object, we must cast the returned value explicitly to our specific object type. The first `catch` block does nothing except catch the `EOFException` and terminate the infinite `while` loop. We're forced to use this odd loop-termination trick because the `ObjectInputStream` class does not have anything comparable to the

```
/*************************************************************
* ReadObject.java
* Dean & Dean
*
* This reads all objects in an object file.
*************************************************************/

import java.util.Scanner;
// for ObjectInputStream, FileInputStream, and EOFException
import java.io.*;

public class ReadObject
{
  public static void main(String[] args)
  {
    Scanner stdIn = new Scanner(System.in);
    TestObject testObject;

    System.out.print("Enter filename: ");
    try (ObjectInputStream fileIn = new ObjectInputStream(
      new FileInputStream(stdIn.nextLine())))
    {
      while (true)
      {
        testObject = (TestObject) fileIn.readObject();
        testObject.display();
      }
    } // end try and close fileIn automatically
    catch (EOFException e)
    {} // end-of-file exception terminates infinite while loop
    catch (Exception e)
    {
      System.out.println(e.getClass());
      System.out.println(e.getMessage());
    }
  } // end main
} // end ReadObject class
```

Open the file.
Must cast to specific object type.

Sample session:

```
Enter filename: objectFile
1       first   1.0
2       second  2.0
```

Figure 16.6 `ReadObject` class that reads all `Serializable` objects from a file

`hasNext` method you'd normally expect to see in the parentheses after `while`. But this trick works just fine, because try-with-resources automatically performs the necessary file closing. The subsequent generic `catch` block describes any other exceptions that might be thrown in the event of a real problem.

The sample session cleanly displays the two previously written objects without fussing about the `EOFException`, which terminated the `while` loop.

Writing an Updated Version of a Previously Written Object

If you ask ObjectOutputStream's writeObject method to write exactly the same object again while the file is still open, the serializing software recognizes the repetition and writes just a reference to the previously written object. This is like what happens when you instantiate a new String that is exactly the same as a previously instantiated String. This is a nice space-saving feature, but it can be a problem if you're simulating the behavior of a particular object, and you want a file to accumulate a record of that object's changing state as the simulation progresses. To see this problem, in the WriteObject program in Figure 16.5, replace this statement:

```
testObject = new TestObject(2, "second", 2.0);
```

with this:

```
testObject.number *= 1.1;
```

Then, if you execute the revised WriteObject program and re-execute the ReadObject program, the ReadObject program will generate this output:

```
Enter name of file to read: objectFile
1       first    1.0
1       first    1.0
```

The second record of the object's state is just a copy of the first record. It doesn't reflect the change in the value of the number variable. To make Java store the latest state of an object instead of just a reference to the original state, you need to invoke ObjectOutputStream's reset method sometime before you write the updated version of the object. To see how this works, in the WriteObject program in Figure 16.5, this time, replace

```
testObject = new TestObject(2, "second", 2.0);
```

with this:

```
fileOut.reset();
testObject.number *= 1.1;
```

This allows writing an updated version of a previously written object.

Then execute the revised WriteObject program and the ReadObject program, and you'll get the result you want:

```
Enter name of file to read: ObjectFile
1       first    1.0
1       first    1.1
```

16.5 Character Sets and File-Access Options

This section describes ways to read and write text files with alternate character sets and ways to specify file-access options.

Character Sets

In the ReadFromFile programs in Figures 15.12a and 15.13a, the WriteToFile program in Figure 15.17, and the HTMLGenerator program in Figures 16.1a and 16.1b, all text representation employed a certain set of characters—the current computer's *default character set*. The output in Figure 16.3b says this website uses the ISO-8859-1 character set. You can determine the name of the default character set your computer uses by importing

```
java.nio.charset.Charset
```

and calling

```
Charset.defaultCharset()
```

Java guarantees support for certain standard character sets, like the 128 characters in US-ASCII (see Figures A1.1a and A1.1b in Appendix 1), the 256 characters in ISO-8859-1, and several 16-bit Unicode formats. You can read about these in Java API documentation. You can determine the names of all the particular character sets your own computer can identify, read, and write by executing this code fragment:

```
for (String s : Charset.availableCharsets().keySet())
{
  System.out.println(s);
}
```

In our ReadFromFile, WriteToFile, and HTMLGenerator programs, we could have specified one of these other character sets by using slightly different constructors. For example, to modify the ReadFromFile program to read data that was written in the particular character set having the name "US-ASCII" instead of using the one-parameter `Scanner` constructor, use this two-parameter `Scanner` constructor:

```
fileIn = new Scanner(Paths.get(filename), "US-ASCII");
```

Suppose you write "Hello, World!" to a text file with UTF-16 (a 16-bit representation) and then read it back with US-ASCII (an 8-bit representation). What would you expect to see? Here's what we get:

```
?? H e l l o ,   W o r l d !
```

When we write with US-ASCII and read with UTF-16, we get this:

```
????????
```

When we write and read with consistent character sets, either both with US-ASCII or both with UTF-16, we get what you would expect:

```
Hello, World!
```

As you might expect and these results suggest, a UTF-16 file is approximately twice as big as a US-ASCII file.

File-Access Options

Up until now, the types of file-access options we have employed have been the default options built into the particular file-handling constructors we used. For example, whenever we opened a file for writing, the computer assumed the file did not exist and should be created from scratch. Or if the file did exist, the computer assumed we wanted to discard all its previous data and treat it as a brand new file. But sometimes we would rather add or append new data to data already in an existing file. To do this, we open the file with an `APPEND` option. This is one of the constants in `Enum StandardOpenOption`, found in the `java.nio.file` package.

You can display the names of all of Java's standard open options by executing this code:

```
for (StandardOpenOption opt : StandardOpenOption.values())
{
  System.out.println(opt);
}
```

Upcoming examples will include `StandardOpenOptions` identified by these `String` designations: `APPEND`, `CREATE`, `READ`, `TRUNCATE_EXISTING`, and `WRITE`.

16.6 Buffered Text File I/O

The previous examples of text file reading in the ReadFromFile programs in Figures 15.12a and 15.13a, of text file writing in the WriteToFile program in Figure 15.17, and of text file I/O in Figures 16.1a and 16.1b were fine for small and local text files. But the file-handling techniques used in those examples are not suitable for large files or files that might be transferred from one computer to another.

For large and/or remote file I/O, you should always employ an intermediate buffer. A *buffer* is a sequential storage structure that acts like a first-in first-out (FIFO) queue, or "waiting line." As time passes, a queue's length changes dynamically with independent variations in arrival rate and service rate. Because a buffer resides in high-speed memory, a program can transfer data into or out of a buffer very quickly. On the other hand, files reside in persistent storage, which is always slower than the computer's main memory, and it is occasionally located a considerable distance from the computer itself—perhaps even on the other side of the world. The computer accesses files in intermittent bursts whose timing depends on extraneous factors like disk position, other traffic on the bus or communication link, and distance.[1] A buffer decouples program transfers from file transfers, so that transfers between the program and the buffer do not need to be synchronized with transfers between the buffer and the file. With large files, a buffer provides a substantial performance advantage.

Using Java's `BufferedWriter`

The BufferedWriteToFile program in Figure 16.7a shows how to write text to a file through a buffer. It also illustrates specification of character set and open options. The `java.io` package includes the `BufferedWriter` class. The `java.nio.file` package includes the `Paths` class, whose class `get` method returns a `Path` object. That package also contains the `Files` class, whose `newBufferedWriter` method returns a `BufferedWriter` object configured for a particular character set and particular open options. Java API documentation says the `newBufferedWriter` method conforms to this interface:

```
public static BufferedWriter newBufferedWriter(
  Path path, Charset cs, OpenOption... options)
  throws IOException
```

The character-set and open-options parameters do not appear in the `BufferedWriter` constructors in the `java.io` package, and we'd need lots of extra code to include them if the `Files.newBufferedWriter` method were not available.

The `OpenOption...` notation is called a *varargs*. It means that we may supply any number of arguments of the specified type, including none. A varargs works for any type of object, but in this particular case, the specified type is `OpenOption`. Only the last parameter may be a varargs parameter. The argument passed to a varargs parameter may be either an array or a sequence of objects separated by commas. In the Java API, varargs are frequently `enum`s, and sure enough, classes that implement the `OpenOption` interface, like `StandardOpenOption`, are `enum`s. (For a discussion of `enum`, see Section 12.13.)

After prompting for and reading a filename, the program prompts for and reads a `String` designating the second of two open options. In a try-with-resources header, it asks `Files` to return an appropriately configured instance of `BufferedWriter`. Our `BufferedWriteToFile` class uses the local computer's default character set, but the callout indicates that we could specify another character set with something like `Charset.forName("US-ASCII")`. `StandardOpenOption`'s `valueOf` method converts the `String` designation for the second open option into its internal value.

[1]It takes more than one-tenth of a second for light or a radio signal to travel to the other side of the world and back—hundreds of millions of computer cycles.

```
/*************************************************************
* BufferendWriteToFile.java
* Dean & Dean
*
* This writes a string through a buffer to a text file.
*************************************************************/

import java.util.Scanner;
import java.io.BufferedWriter;
import java.nio.file.*; // Paths, Files, StandardOpenOption
import java.nio.charset.Charset;

public class BufferedWriteToFile
{
  public static void main(String[] args)
  {
    Scanner stdIn = new Scanner(System.in);
    String fileName, openOption;

    System.out.print("Enter filename: ");
    fileName = stdIn.nextLine();
    System.out.print("Enter TRUNCATE_EXISTING or APPEND: ");
    openOption = stdIn.nextLine();

    try (BufferedWriter fileOut = Files.newBufferedWriter(
      Paths.get(fileName),
      Charset.defaultCharset(),
      StandardOpenOption.CREATE,
      StandardOpenOption.valueOf(openOption)))
    {
      System.out.println("Enter a line of text:");
      fileOut.write(stdIn.nextLine() + "\n");
    } // end try
    catch (Exception e)
    {
      System.out.println(e.getClass());
      System.out.println(e.getMessage());
    }
  } // end main
} // end BufferedWriteToFile class
```

For a special character set, substitute: `Chaset.forName` *("<set-name>")*

Figure 16.7a `BufferedWriteToFile` class
This demonstrates `Files.newBufferedWriter` with default character set and standard open options. The callout shows entry for a special character set.

Within the `try` block, the program asks the user to enter a line of text. Then it writes that line to the file. The explicit `"\n"` in the `write` method's argument restores the line termination that `Scanner`'s `nextLine` method automatically removes. For simplicity, this program reads and writes just one line, but it would be straightforward to include a loop that reads and writes any number of lines.

The `BufferedWriter` instance closes automatically at the end of the `try` block. We use a generic `catch` block because the `newBufferedWriter` method call can throw an `InvalidPathException`. This is an unchecked exception, which does not descend from the checked `IOException` we must catch. Our generic display is easier to read than the default stack trace generated by the JVM if the path is invalid.

Figure 16.7b shows two sample-session outputs. In the first sample session, the user specifies the second open option as "TRUNCATE_EXISTING" to delete all existing text from any existing file whose name happens to match the specified filename. Then the user enters what will be the first line of text in the specified file. In the second sample session, the user specifies the second open option as "APPEND" to append the subsequently entered line of text.

Sample session 1:
```
Enter filename: Ecclesiastes
Enter TRUNCATE_EXISTING or APPEND: TRUNCATE_EXISTING
Enter a line of text:
Do not be over-virtuous
```

Sample session 2:
```
Enter filename: Ecclesiastes
Enter TRUNCATE_EXISTING or APPEND: APPEND
Enter a line of text:
nor play too much the sage,
```

Figure 16.7b Outputs for two executions of the BufferedWriteToFile program in Figure 16.7a
`APPEND` and `TRUNCATE_EXISTING` identify particular `StandardOpenOptions`.

Using Java's `BufferedReader`

The BufferedReadFromFile program in Figure 16.8 shows how to read text from a file through a buffer. `Files`' `newBufferedReader` method also includes specification of the character set. It does not accept any open-option specification—it just assumes that we want the `READ` option only. The imports and local variables are the same as those in the BufferedWriteToFile program in Figure 16.7a. The user interaction is like that in the BufferedWriteToFile program, except it's simpler because there is no open-option specification. In the try-with-resources header, the creation of a `BufferedReader` is like the creation of a `BufferedWriter` was in Figure 16.7a, except it's simpler because there is no open-option specification.

Within the `try` block, the program employs a `while` loop to print all lines in the specified file. `BufferedReader's ready` method provides the same loop condition as `Scanner's hasNextLine` method. The `BufferedReader` instance closes automatically at the end of the `try` block. As indicated earlier, we use a generic `catch` block because the `newBufferedReader` method call throws an unchecked `InvalidPathException`, which does not descend from the checked `IOException`, and our generic display is easier to read than the default stack trace generated by the JVM.

The output at the bottom of Figure 16.8 displays the text written to the file in the two sample sessions in Figure 16.7b.

Web Reading

You can also use Java's `BufferedReader` to read from the Web. To do this, create a `URL` object. Then create a `URLConnection`. Then create an `InputStream`. Then use that `InputStream` as the argument

```
/*************************************************************
* BufferedReadFromFile.java
* Dean & Dean
*
* This reads strings through a buffer from a text file.
*************************************************************/

import java.util.Scanner;
import java.io.BufferedReader;
import java.nio.file.*;    // Paths, Files
import java.nio.charset.Charset;

public class BufferedReadFromFile
{
  public static void main(String[] args)
  {
    Scanner stdIn = new Scanner(System.in);
    String fileName;

    System.out.print("Enter filename: ");
    fileName = stdIn.nextLine();
    try (BufferedReader fileIn = Files.newBufferedReader(
      Paths.get(fileName),
      Charset.defaultCharset()))
    {
      while (fileIn.ready())
      {
        System.out.println(fileIn.readLine());
      }
    } // end try
    catch (Exception e)
    {
      System.out.println(e.getClass());
      System.out.println(e.getMessage());
    }
  } // end main
} // end BufferedReadFromFile class
```

For a special character set, substitute: `Charset.forName` ("*<set-name>*")

<u>Sample session:</u>

```
Enter filename: Ecclesiastes
Do not be over-virtuous
nor play too much the sage;
```

Figure 16.8 `BufferedReadFromFile` class, which demonstrates `BufferedReader` with a default character set
The callout shows an entry for a special character set.

in an InputStreamReader constructor, and finally, use that InputStreamReader as the argument in a BufferedReader constructor. Here is the sequence of operations:

```
URL url = new URL(webAddress);
URLConnection connection = url.openConnection();
InputStream in = connection.getInputStream();
BufferedReader reader =
  new BufferedReader(new InputStreamReader(in));
```

The URL and URLConnection are in the java.net package. The InputStream, InputStreamReader, and BufferedReader classes are in the java.io package.

If you need a character set different from your computer's default character set, you can specify the character set with a second argument supplied to an alternate InputStreamReader constructor. An exercise at the end of this chapter asks you to read a web page with a BufferedReader.

16.7 Primitive Buffers with Random Access

We can also buffer primitive data, like byte, char, short, int, float, long, and double. To facilitate this, the java.nio package provides the base class, Buffer, which descends directly from Object, plus Buffer's direct descendants: ByteBuffer, CharBuffer, DoubleBuffer, FloatBuffer, IntBuffer, LongBuffer, and ShortBuffer.

These classes make it easy to read and write each kind of primitive variable. Because computer systems universally transmit and store data as bytes, the ByteBuffer class plays a central role. It provides single-variable put and get methods that make conversions between other primitive types and the byte type. The ByteBuffer class also provides methods like asIntBuffer and asDoubleBuffer, which provide alternate access to chunks of a ByteBuffer's data. With help from these methods, the data in the array underlying a ByteBuffer can represent any primitive type. Beyond that, it can represent any combination of primitive types. Still more, it can represent any combination of primitive types and Object types. We won't try to mix primitives and objects from arbitrary classes, but a later example will put an int, a String, and a double into the same ByteBuffer.

Buffer Methods—Inherited by All Buffer Descendants

All particular primitive buffer types inherit common methods from the abstract parent class, Buffer. Figure 16.9 presents several of these Buffer methods as interfaces and short descriptions. Many of these methods would normally be void, but they return a Buffer instead. Because these returned Buffers are this references to the calling object, these Buffer methods support method-call chaining.

The capacity, limit, position, and mark methods get and set corresponding buffer indices. These are like bookmarks that identify particular buffer locations where you may want to go or to which you may want to return. They are in units of the current primitive type, like the length of or index into an array of that type. They satisfy these constraints:

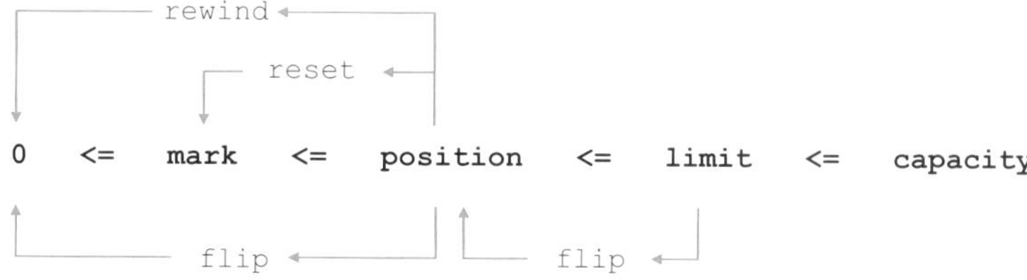

`public int capacity()`
Returns this buffer's capacity as the maximum number of this buffer's elements.

`public Buffer clear()`
Clears the buffer.

`public Buffer flip()`
Sets upper limit to the current position. Then deletes any mark and sets the current position to zero.

`public boolean hasRemaining()`
Tells whether there are any elements between the current position and the upper limit.

`public int limit()`
Returns this buffer's upper limit as the maximum number of this buffer's elements.

`public Buffer limit(int newLimit)`
Sets this buffer's upper limit as the maximum number of this buffer's elements.

`public Buffer mark()`
Sets a "mark" at this buffer's current position.

`public int position()`
Returns the index number of this buffer's next element.

`public Buffer position(int newPosition)`
Sets this buffer's current position to the index of the next element to be read or written.

`public int remaining()`
Returns the number of elements between the current position and the upper limit.

`public Buffer reset()`
Resets this buffer's position to the (required) current "mark" position.

`public Buffer rewind()`
Deletes any mark and resets this buffer's position to zero.

Figure 16.9 Several `Buffer` methods—those that return `Buffer` can be chained

By default, there is no `mark`, the initial `position` is zero, and the initial `limit` equals `capacity`.

As a program reads or writes elements, `position` automatically increments to the start of the next element. A `mark` method call makes the current `position` be the destination of any subsequent `reset`. A `rewind` method call deletes any mark and makes the next `position` zero. A `flip` method call sets the limit to the current `position`, deletes any `mark`, and makes the next `position` zero.

As we have said, a buffer is like a queue or waiting line. Assume that a quantity of elements does not exceed the buffer's `capacity`, and suppose you want the buffer to transport that quantity to or from a file or through the Internet. In a loop, you could put these elements into the buffer. Then, in another loop, you could get these elements from the buffer and pass them on. To facilitate this operation, insert a `flip` method call after the `put` loop, and execute the `get` loop `while(position<limit)`.

Your ability to set `position` to any index between zero and `limit` gives you *random read and write access* to any element(s) in the range between `position(0)` and `position(limit-1)`.

Elementary `ByteBuffer` Methods

The `ByteBuffer` class defines many additional methods. The class method

```
public static ByteBuffer allocate(int capacity)
```

acts like a ByteBuffer constructor. It creates and returns an empty ByteBuffer whose maximum number of bytes equals the capacity parameter. The initial limit also equals this capacity. Any method call that tries to put into the buffer more bytes than limit throws a BufferOverflowException.

There are get methods for all primitive variables: get for byte, getChar for char, getDouble for double, getFloat for float, getInt for int, getLong for long, and getShort for short. There are also put methods for all primitive variables: put for byte, putChar for char, putDouble for double, putFloat for float, putInt for int, putLong for long, and putShort for short.

These methods are overloaded. Zero-parameter get methods like getChar() are *relative*. They read and return the primitive at the calling buffer's current position and then increment that position to the start of the next element. The position increment is two bytes for getChar, four bytes for getInt, etc. One-parameter get methods like getChar(int index) are *absolute*. The index parameter is a primitive index. It specifies the location of the desired primitive not in bytes, but in number of primitives before the desired primitive. These absolute get methods do not change the buffer's current position.

One-parameter put methods like putChar(char value) are relative. They write the primitive, starting at the calling buffer's current position. Then they increase that position to just after the inserted primitive. Two-parameter put methods like putChar(int index, char value) are absolute. The index parameter tells where the value parameter goes. These absolute put methods do not change the buffer's current position.

The relative method

```
public ByteBuffer put(ByteBuffer source)
```

copies all remaining source bytes to the calling buffer, starting at the calling buffer's current position. As indicated in Figure 16.9, *remaining* means all bytes between current position (inclusive) and current limit (not inclusive). If the calling buffer is not big enough, the method throws a BufferOverflow Exception. The next program will use this method just to copy all data from one buffer to another. But because the two buffers' limit and position values may be different, you can use this method to copy an arbitrary subset of the data in the first buffer to an arbitrary position in the second buffer.

Figure 16.10 contains a simple program that uses ByteBuffer's relative putInt method to put a single int value at the beginning of a ByteBuffer. Then it uses ByteBuffer's absolute putDouble method to put a single double value into that same ByteBuffer, with seven blank spaces between the end of the four-byte int and the start of the eight-byte double. The first two lines of output show how the relative putInt method call adds four to buffer1's position and how the absolute putDouble method call does not change that position. Even though a double needs eight bytes, this program shows that it can start anywhere in the buffer, not just at a multiple of eight.

After populating buffer1, the program copies all its data to buffer2, using the buffer-to-buffer put method. But before doing that, it must call buffer1's rewind method. Here's why: The sample session shows that after it finishes populating buffer1, buffer1's position is 4. If the program did not have a buffer1.rewind() statement before the buffer2.put(buffer1) statement, buffer2 would not receive buffer1's first four bytes. The buffer1.rewind() statement resets buffer1's position to zero, and this enables buffer2 to receive all buffer1's data. After the buffer-to-buffer put, buffer2's position is 19, which is its limit. The buffer2.flip() statement moves that position back to zero again, preparing it for subsequent get method calls.

Array Methods

We can also copy arrays of primitives into or out of a ByteBuffer. The class method

```
public static ByteBuffer wrap(byte[] byteArray)
```

```
/*************************************************************
* ByteBufferAccess.java
* Dean & Dean
*
* This puts different primitive elements into one byte buffer.
*************************************************************/

import java.nio.ByteBuffer;

public class ByteBufferAccess
{
  public static void main(String[] args)
  {
    int bufLength = 4 + 7 + 8; // int + empty spaces + double
    ByteBuffer buffer1 = ByteBuffer.allocate(bufLength);
    ByteBuffer buffer2 = ByteBuffer.allocate(bufLength);

    // populate output buffer
    buffer1.putInt(2);
    System.out.println("afterIntPos= " + buffer1.position());
    buffer1.putDouble(11, 2.0);
    System.out.println("afterDblPos= " + buffer1.position());
    // Transfer everything to input buffer
    buffer1.rewind();
    buffer2.put(buffer1);
    // display transferred data
    buffer2.flip();
    System.out.println(buffer2.getInt());
    System.out.println(buffer2.getDouble(11));
  } // end main
} // end ByteBufferAccess class
```

<u>**Sample session:**</u>

```
afterIntPos= 4
afterDblPos= 4
2
2.0
```

Figure 16.10 ByteBufferAccess program has different types of elements in the same buffer

acts like another `ByteBuffer` constructor. It creates, populates, and returns a `ByteBuffer` filled with the parameter's elements. Alternatively, you can use the previously described `allocate` method to create the buffer and then populate it with separate method call(s). This alternative requires more code, but it is more versatile.

To understand how a `ByteBuffer` can help organize arrays, you need to become familiar with some additional `ByteBuffer` methods. The `ByteBuffer` class has additional overloaded `get` and `put` methods for arrays of bytes. The one-parameter `get(byte[] destination)` method is *relative*. Starting at the calling buffer's current `position`, this method copies all remaining buffer bytes into a previously defined `destination` array, and it increases the calling buffer's `position` to its `limit`.

The three-parameter get(byte[] destination, int offset, int length) is *absolute*. It copies length bytes, starting at the buffer's current position, into the destination array, starting at offset array index. This absolute get method does not change the buffer's current position.

The one-parameter put(byte[] source) is relative. This method copies all bytes in the source array into the calling buffer, starting at the calling buffer's current position. It increases the calling buffer's position by the total number of bytes written. The three-parameter put(byte[] source, int offset, int length) is absolute. It copies length bytes from the source array, starting at the offset array index, into the calling buffer, starting at the buffer's current position. This absolute put method does not change the buffer's current position.

Another ByteBuffer method, public byte[] array(), returns a byte array that is a *view* or an alias of the array that underlies the calling buffer. Changes to the returned array appear immediately as changes in the calling buffer, and vice versa.

The ByteBuffer class provides the additional view methods asCharBuffer, asDoubleBuffer, asFloatBuffer, asIntBuffer, asLongBuffer, and asShortBuffer. These additional view methods give other types of primitives access to a ByteBuffer's underlying array. The returned primitive-buffer view starts at the ByteBuffer's current position. This returned view must be assigned to a declared variable, like DoubleBuffer doubleView, but it is not a separate object. Therefore, it doesn't need to be created with a separate allocate.

Other classes in the java.nio package—CharBuffer, DoubleBuffer, FloatBuff, IntBuffer, LongBuffer, and ShortBuffer—define get methods like get(char[] destination) and get(char[] destination, int offset, int length). Like ByteBuffer's corresponding get methods, these copy a range of primitives from the calling buffer to a previously defined destination array. These other classes (CharBuffer, etc.) also define put methods like put(char[] source)and put(char[] source, int offset, int length). Like ByteBuffer's corresponding put methods, these methods copy source array values into the calling buffer.

With these methods, it's relatively easy to copy an array of primitives that are not bytes to or from a ByteBuffer. For example, given double[] doubles, and ByteBuffer buffer, we can copy the doubles array into the byte buffer with a statement like this:

```
buffer.asDoubleBuffer().put(doubles)
```

Or we can copy double values from the byte buffer into the doubles array with a statement like this:

```
buffer.asDoubleBuffer().get(doubles)
```

A powerful feature of the combination of a ByteBuffer and the DoubleBuffer obtained from ByteBuffer's asDoubleBuffer method is the independence of the two buffers' position variables. Whenever you use the DoubleBuffer's put or get, it may or may not (depending on the DoubleBuffer method) change that DoubleBuffer's position. But it never changes the corresponding ByteBuffer's position, and vice versa. The zero position in the ByteBuffer is the initial byte in the underlying array. But the zero position in the DoubleBuffer is the current position in the ByteBuffer. This refers to the place in the underlying array where eight-byte double values begin. Because the two buffer positions are independent, you can change ByteBuffer's position without affecting DoubleBuffer's position, and vice versa. Each unit change in DoubleBuffer's position corresponds to eight bytes in the underlying array and therefore a change of eight in ByteBuffer's position.

The ByteBufferArrayAccess program in Figure 16.11 shows how to put and get arrays of primitive elements—an array of ints and an array of doubles. It also shows how to put and get a String.

```
/*************************************************************
* ByteBufferArrayAccess.java
* Dean & Dean
*
* This buffers: byte array, string, and double array.
*************************************************************/

import java.util.Arrays;
import java.nio.ByteBuffer;

public class ByteBufferArrayAccess
{
  public static void main(String[] args)
  {
    int[] ints = new int[]{1, 1, 2, 3, 5, 8};
    String str =
      "The purpose of computing is insight, not numbers.";
    double[] doubles = new double[]{1.0, 2.0, 1.5, 1.67, 1.6};
    byte[] strBytes = str.getBytes();
    ByteBuffer buffer = ByteBuffer.allocate(
      4 * ints.length + strBytes.length + 8 * doubles.length);

    // put to buffer
    buffer.asIntBuffer().put(ints);
    buffer.position(4 * ints.length);
    buffer.put(strBytes).asDoubleBuffer().put(doubles);
    // fill working arrays with zeros and rewind buffer
    Arrays.fill(ints, 0);
    Arrays.fill(strBytes, (byte) 0);
    Arrays.fill(doubles, 0.0);
    str = "";
    buffer.rewind();
    // get from buffer
    buffer.asIntBuffer().get(ints);
    buffer.position(4 * ints.length);
    buffer.get(strBytes).asDoubleBuffer().get(doubles);
    str = new String(strBytes);
    // display transferred data
    System.out.println(Arrays.toString(ints));
    System.out.println(str);
    System.out.println(Arrays.toString(doubles));
  } // end main
} // end ByteBufferArrayAccess class
```

Sample session:

```
[1, 1, 2, 3, 5, 8]
The purpose of computing is insight, not numbers.
[1.0, 2.0, 1.5, 1.67, 1.6]
```

Figure 16.11 ByteBufferArrayAccess program has different types of arrays in the same buffer

To put a String into a ByteBuffer, we first convert it to an array of bytes, and then we put that array of bytes into the ByteBuffer just as we would put an array of primitive bytes into it. To get a String from a ByteBuffer, we first get into a byte array the part of the ByteBuffer that represents the String, and then we pass that byte array to a String constructor.

The first three declarations specify the data that will go into the buffer—an array of ints, a String,[2] and an array of doubles. The next declaration converts the String to an array of bytes. The last declaration creates the buffer with just enough capacity to hold the int array, the String, and the double array. By the way, when working with buffers, it's important to have independent knowledge of how big the whole buffer needs to be, the type(s) of data it will hold, and if more than one type, the amounts of each type. Without this independent information, you don't know how much capacity to allocate, you don't know what methods to use to put and get the data, and you don't know where each type of data starts.

After creating local variables, the program populates the buffer. First, buffer obtains an IntBuffer view. Then, in a chained method call, it uses that returned IntBuffer view to put the array of ints into the buffer. Because putting through the IntBuffer view does not alter buffer's position, we need the subsequent explicit position statement to advance buffer's position to where the next data goes. The next statement puts the strbytes array into the buffer. Then, in a chained method call, buffer obtains a DoubleBuffer view that starts just after the last byte of strbytes. Then, in another chained method call, the returned DoubleBuffer view puts the array of doubles into the buffer.

The asIntBuffer and asDoubleBuffer method calls return objects of the classes, IntBuffer and DoubleBuffer, respectively. So the methods that these returned objects call in chained operations are IntBuffer and DoubleBuffer methods. But it is not necessary to import the IntBuffer and DoubleBuffer classes because these class names never appear explicitly in the program.

Next, the program employs some class methods from the Java API Arrays class in the java.util package to zero out all the previously filled working arrays and the String variable, str. This guarantees that our final results are not just leftover initial values. Then buffer1 rewinds itself to prepare for subsequent get operations.

The statements that get data from the buffer are similar to the statements that put data into the buffer, except they use get methods rather than put methods. The String constructor reverses the getBytes method call in the declarations.

The last three statements display the retrieved data, with help from toString methods defined in Java's Arrays class. The output confirms that the values retrieved from the buffer are exactly the same as the values previously put into the buffer.

Character Sets Revisited

For simplicity, the getBytes method call that converts the specified String to an array of bytes in the declarations, and the later String constructor with the byte[] argument both use the default character set. To work with a different character set, we could use the getBytes method and the String constructor that accept character-set specification. Specifically, we could include this additional import:

```
import java.nio.charset.Charset;
```

Then, to convert the specified String to a byte array, we could replace the strBytes declaration with this:

```
byte[] strBytes = str.getBytes(Charset.forName("US-ASCII"));
```

[2]This particular string is "Hamming's Motto." Richard W. Hamming was a founder and president of the Association of Computing Machinery.

And later, to convert the byte array back into a String, we could use this:

```
str = new String(strBytes, Charset.forName("US-ASCII"));
```

Of course, the choice of "US-ASCII" here is arbitrary. In practice, if the default character set is not OK, you might use one of the other character sets.

16.8 Channel I/O and Memory-Mapped Files

ByteBuffer methods organize in sequences of primitives or arrays of primitives. FileChannel methods organize in sequences of buffers or arrays of buffers. A channel is a large-scale view of what's in a file. To put heterogeneous data into a file, first put primitives into buffers, and then put those buffers into channels. Conversely, to get heterogeneous data from a file, first extract the buffers from the channels, and then extract the primitives from the buffers.

Channel position and size are measured in bytes just like buffer position and limit. If you don't alter the channel's position explicitly, position automatically increments as you write bytes. Like an ArrayList a channel's size expands automatically to accept more data. If you specify a starting place, you could go back and overwrite any sequence of bytes. Then you could use channel's size method to restore its position to just after the last contained byte and continue from there. After writing from any combination of buffers, you can read with a different combination of buffers. But it's not like object I/O. To retrieve information properly, you must remember exactly where everything is and exactly what type it is.

To work with file channels, include these imports:

```
import java.nio.channels.FileChannel;
import java.nio.file.*; // Path, Paths, Files, StandardOpenOption
```

To specify the file you want, create a new Path, using something like this:

```
System.out.print("Enter filename: ");
Path path = Paths.get(stdIn.nextLine());
```

To open the file for either reading or writing (over any pre-existing data), in a try-with-resources header, create a channel like this:

```
try(FileChannel channel = FileChannel.open(
  path, StandardOpenOption.CREATE,
  StandardOpenOption.WRITE, StandardOpenOption.READ))
{
  // ...
} // end try
```

Because the file opens in a try-with resources header, it closes automatically at the end of the try block.

FileChannel Methods

To write data to a file through a channel, first write the data to one or more buffers with ByteBuffer write methods, as described in the previous section. Then use one of FileChannel's four write methods in Figure 16.12 to copy the data from the buffer(s) to the channel, and thus to the file. The channel expands automatically as required to accept whatever is written to it. The first three write methods in Figure 16.12 start writing at the channel's current position, or at the end of the channel if the channel-opening statement

`public static FileChannel open(Path path, OpenOption... options)`
Opens or creates a file and returns a channel to access it.

`public long position()`
Returns this channel's current position in the file.

`public FileChannel position(long newPosition)`
Sets this channel's current position in the file.

`public long size()`
Returns the current size of this channel's file.

`public int write(ByteBuffer source)`
Writes the remaining `bytes` from the `source` buffer into this channel, starting at the current channel position or its end. Returns the number of bytes written.

`public long write(ByteBuffer[] sources)`
Writes the remaining `bytes` from `sources` buffers into this channel, starting at the current channel position or its end. Returns the number of bytes written.

`public long write(ByteBuffer[] sources, int offset, int length)`
Writes the remaining bytes from `length` buffers, starting at `sources[offset]`, into this channel, starting at the channel's current position or its end. Returns the number of bytes written.

`public int write(ByteBuffer source, long start)`
Writes the remaining `bytes` from the `source` buffer into this channel, starting at `start` in the channel. Does not change the channel's current position. Returns the number of bytes written.

`public int read(ByteBuffer destination)`
Reads the remaining bytes from this channel into the remaining positions in `destination` buffer. Returns the number of bytes read or `-1` if buffer not filled.

`public long read(ByteBuffer[] destinations)`
Reads the remaining bytes from this channel into the remaining positions in the buffers in the `destinations` array. Returns the number of bytes read or `-1` if buffers are not filled.

`public long read(ByteBuffer[] destinations, int offset, int length)`
Reads the remaining bytes from this channel into the remaining positions in `length` buffers, starting at `destinations[offset]`. Returns the number of bytes read or `-1` if buffers are not filled.

`public int read(ByteBuffer destination, long start)`
Reads bytes after `start` in this channel into the remaining positions in the `destination` buffer. Returns the number of bytes read or `-1` if buffer is not filled. Does not change channel `position`.

```
public MappedByteBuffer map(
  FileChannel.MapMode mode, long position, long size)
```
Creates a persistent view of `size` bytes of this channel's file, starting at this channel's current `position`, with modes `READ_ONLY`, `READ_WRITE` (mutable view), or `PRIVATE` (copy).

Figure 16.12 Selected `FileChannel` methods

included `StandardOpenOption.APPEND`. With any of these three methods, channel `position` automatically increments as new data arrives.

The first `write` method copies data from a single source buffer, starting at that buffer's current `position`. The second `write` method copies data from a sequence of source buffers, starting at each one's current `position`. The third `write` method copies data from only those buffers in a specified subarray of an array of buffers. The fourth `write` method is like the first `write` method, except it starts writing at a user-specified byte index in the channel, and the channel's internal `position` does not change.

To read data from a file through a channel, use one of `FileChannel`'s four `read` methods in Figure 16.12. All four of these `read` methods keep reading from the channel until the remaining spaces in the target buffers are filled. The remaining spaces are those from the current `position` to the `limit`. For homogeneous buffers, size each buffer to match a coherent range of bytes in the channel, so that each buffer's `capacity` determines the number of bytes copied. For a heterogeneous buffer, set `position` and `limit` so that the difference between them determines the number of bytes copied.

The first three `read` methods start reading at the current channel `position`, and this `position` automatically increments as data is copied. The first `read` method puts data into a single destination buffer, starting at that buffer's current `position`. The second `read` method puts data into a sequence of destination buffers, starting at each one's current `position`. The third `read` method puts data into only those buffers in a specified subarray of an array of buffers. The fourth `read` method is like the first `read` method, except it starts reading at the user-specified byte-index, `start`, and the channel's internal `position` does not change. After copying data from the channel to buffer(s), you can copy that data from the buffer(s) using `ByteBuffer`'s appropriate `read` method.

The last method in Figure 16.12, `map`, is particularly interesting. You invoke this method while the channel is open, and the buffer it creates continues to exist and be accessible after the channel closes. For very large files, using a map improves I/O performance. When you need repeated access to the same file in different contexts, it simplifies your program.

File Channel I/O and Memory Map Example

The ChanneledFileAccess program in Figures 16.13a and 16.13b shows how all this fits together. This program organizes the data in a file like the data in a table—a spreadsheet table or in a relational database table. This program's table has three columns. The first column contains `int`s. Because an `int` is 4 bytes, this column is four bytes wide. The second column contains `String`s. It is arbitrarily 12 bytes wide. The third column contains `double`s. Because a double is 8 bytes, this column is 8 bytes wide. Thus, each row in the table has a 4-byte `int` field, followed by a 12-byte `String` field, followed by an 8-byte `double` field, for a total of 24 bytes. All data in one row is one *record*. We use a 24-byte `ByteBuffer` to copy records into and out of the same file channel with the simplest `FileChannel` methods in Figure 16.12:

```
write(ByteBuffer source)
read(ByteBuffer destination)
```

This program requires a substantial number of imports. It defines a `TEXT` constant to establish the text field length and a `RECORD` constant to establish total record length and the `capacity` of a buffer that holds exactly one record. Later, the program uses a multiple of the `RECORD` constant to allocate capacity for another buffer that can hold several records.

The program's `writeRecord` method writes individual records to a specified channel. This method's parameters are the channel and values for each of the fields in one record. Notice that this method's header includes `throws IOException`. The first declaration creates an array of bytes whose length equals the length of our record's text field, and it initializes this array with the value of the method's `string` parameter. The `Arrays.copyOfRange` method discards bytes beyond 12 if `string` is too long for the text field. The second declaration creates a buffer whose capacity is `RECORD` bytes. The first statement after the declarations chains three method calls. The first method call `puts` the 4-byte ID number into the ID field. This advances the buffer's internal position to 4. The second method call `puts` the 12-byte string into the

```
/*************************************************************
* ChanneledFileAccess.java
* Dean & Dean
*
* This channel buffered data to and from a file's table.
*************************************************************/

import java.nio.channels.FileChannel;
import java.io.IOException;
import java.util.*;     // Arrays, Scanner
import java.nio.*;      // ByteBuffer, MappedByteBuffer
import java.nio.file.*; // Path, Paths, StandardOpenOption

public class ChanneledFileAccess
{
  public final static int TEXT = 12;
  public final static int RECORD = 4 + TEXT + 8;

  //**********************************************************

  // This adds one buffered record to a file channel.

  public void writeRecord(FileChannel channel,
    int id, String string, double value) throws IOException
  {
    byte[] strBytes =
      Arrays.copyOfRange(string.getBytes(), 0, TEXT);
    ByteBuffer buffer = ByteBuffer.allocate(RECORD);

    buffer.putInt(id).put(strBytes).putDouble(value);
    buffer.rewind();
    channel.write(buffer);
  } // end writeRecord

  //**********************************************************

  // This reads a specified record from a file channel.

  public void readRecord(FileChannel channel,
    int recordIndex) throws IOException
  {
    ByteBuffer buffer = ByteBuffer.allocate(RECORD);

    channel.read(buffer, recordIndex * RECORD);
    buffer.rewind();
    displayRecord(buffer);
  } // end readRecord
```

Figure 16.13a ChanneledFileAccess program—part A

```
  //*********************************************************

  private static void displayRecord(ByteBuffer buffer)
  {
    int id;
    byte[] strBytes = new byte[TEXT];
    double value;

    id = buffer.getInt();
    buffer.get(strBytes);
    value = buffer.getDouble();
    System.out.printf("%4d %10s %6.1f\n",
      id, new String(strBytes), value);
  } // end displayRecord

  //*********************************************************

  public static void main(String[] args)
  {
    Scanner stdIn = new Scanner(System.in);
    ChanneledFileAccess cio = new ChanneledFileAccess();
    ByteBuffer mappedBuffer = ByteBuffer.allocate(3 * RECORD);

    System.out.print("Enter filename: ");
    Path path = Paths.get(stdIn.nextLine());
    try(FileChannel channel = FileChannel.open(
      path, StandardOpenOption.CREATE,
      StandardOpenOption.WRITE, StandardOpenOption.READ))
    {
      cio.writeRecord(channel, 1, "first", 1.0);
      cio.writeRecord(channel, 2, "second", 2.0);
      cio.writeRecord(channel, 3, "third", 3.0);
      System.out.print("Enter file's record index (0,1,2): ");
      cio.readRecord(channel, stdIn.nextInt());
      mappedBuffer = channel.map(
        FileChannel.MapMode.READ_WRITE, 0, channel.size());
    }
    catch(IOException e)
    {
      System.out.println(e.getClass());
      System.out.println(e.getMessage());
    }
    // Now, channel is gone, but mappedBuffer still exists.
    System.out.print("Enter map's record index (0,1,2): ");
    mappedBuffer.position(stdIn.nextInt() * RECORD);
    displayRecord(mappedBuffer);
  } // end main
} // end class ChanneledFileAccess
```

Figure 16.13b ChanneledFileAccess program—part B

text field. This advances the buffer's internal position to 12. The third method call puts the 8-byte double into the value field. The next statement rewinds the buffer. The last statement writes the buffer's bytes into the channel. This channel.write method call is what throws the IOException.

The program's readRecord method reads the record identified by a specified recordIndex. This method's header also includes throws IOException. The declaration creates a buffer whose capacity is RECORD bytes. The first statement after the declaration reads bytes from the channel into the buffer. It starts reading at the channel position given by recordIndex * RECORD, and the total number of bytes it reads equals the buffer's size, which is RECORD. This channel.read method call is what throws the IOException. The next statement rewinds the buffer. The last statement calls the program's displayRecord method to display the record that was read.

In Figure 16.13b, we made the program's displayRecord method a class method so that we can call it directly from main, as well as from readRecord. The displayRecord method prints one record's data by reading exactly RECORD bytes from the buffer parameter, starting at that parameter's current position. This method declares three variables for the three fields in a record. The first statement after the declarations gets the 4-byte integer, id. This advances the buffer's position to the start of the current record's text field. The next statement gets enough of the subsequent buffer's bytes to just fill the 12-byte strBytes array. This advances the buffer's position to the start of the current record's value field. The next statement fills value with an 8-byte double. The last statement prints the specified record's data.

The main method's declarations create a Scanner for keyboard input and a ChanneledFileAccess object to access the program's writeRecord and readRecord methods. They also create a ByteBuffer, which will contain a mapped view of the file's contents that continues to exist after the channel closes.

After the declarations, main asks the user for a filename and uses it to create a Path. Then main asks FileChannel to open a new file channel for both WRITE and READ access. Because this opening method call throws a checked exception, and because we would like later file closing to be automatic, we put this opening method call in a try-with-resources header. Inside the associated try block, main calls writeRecord three times to write three distinct records to the channel. Then it asks the user to specify the index of a particular record in the file and calls the readRecord method to read and display the selected record's data.

While the channel is still open, main calls FileChannel's map method to create a mappedBuffer object that provides a persistent view of the file's data. The map method call throws a checked exception, so it, too, must be inside the try block. Because the preceding calls to our writeRecord and readRecord methods are necessarily also in the try block, the associated catch block also can catch the exceptions these methods throw.

Finally, after the catch block, when the channel is closed, main asks the user to specify the index of a particular record in the mappedBuffer object. Then it calls the displayRecord method to print the data selected from mappedBuffer. The success of this last operation demonstrates the persistence of a mappedBuffer.

Figure 16.13c shows output generated by the ChanneledFileAccess program for particular record index specifications. The first output comes from readRecord's call of the displayRecord method while the channel is still open. In this case, the argument passed to displayRecord is a 24-byte buffer containing only the selected record. The selected record starts at byte 48 in the file, but the 24-byte buffer containing that record enters displayRecord with buffer position equal to zero. The second output comes from main's call of the displayRecord method after the channel closes. In this case, the selected record starts at byte 24 in the file, and the 72-byte mapped buffer enters displayRecord with buffer position equal to 24. These record index selections are arbitrary. Each selection could have been any of the three indices, 0, 1, or 2, and if they had been the same, the displayed data would have been the same.

Sample session:
```
Enter filename: Records
Enter file's record index (0,1,2): 2
   3  third           3.0
Enter map's record index (0,1,2): 1
   2  second          2.0
```

Figure 16.13c ChanneledFileAccess Program output

This relatively simple ChanneledFileAccess program demonstrates Java's ability to organize and randomly access file data and to provide persistent views of file data. These capabilities appear even with just the simplest of the `FileChannel`'s `read` and `write` methods. If you go back to Figure 16.12 and look at what `FileChannel`'s other `read` and `write` methods can do, you'll see that Java provides a great variety of file organization and access alternatives.

16.9 Path, Whole-File, and Directory Operations

Up until now, this chapter has dealt with details—the details of writing data to a file and reading data from a file. Now, we'll back away from details and consider operations on whole files. In this section, we'll look at alternate ways to describe the path to a file. We'll look at ways to copy or move a file from one directory to another. And we'll look at ways to display directory contents. Later sections will show how to search the directory tree for particular file names and how to display directory contents with GUI.

Defining and Manipulating Paths

To perform any of these operations, we must tell the computer where each file of interest is. We do that by creating a path that leads to the file. Then we can use the path to specify the file. As you have seen, the simplest way to create a `Path` is to use the `Files.get` method in the `java.nio.file` package in the Java API:

```
Path path = Files.get("<path-to-directory-or-file>");
```

The *<path-to-directory-or-file>* string can have any of several different forms. It can be an *absolute path*. An absolute path starts at the root of the directory hierarchy (`C:/` in Windows or just `/` in Unix). It progresses down the directory hierarchy with each subdirectory followed by a forward slash. If it's a path to a directory, that directory's name is the name just before the last forward slash. If it's a path to a file, that file's name is the name just after the last forward slash. It can be a *relative path*, which starts at the current directory.[3] The relative path to the current directory is a single dot (.). The relative path to another file named "sisterSally" in the current directory is just the name of that other file (`sisterSally`). Assuming the current directory contains a subdirectory called, "sisterSonia," the relative path to a file named "nieceNedra" in that subdirectory is

```
sisterSonia/nieceNedra
```

[3]By "current directory," we mean the default directory that the JVM chooses for the current program.

The relative path to the directory above the current directory—the parent directory—is a pair of dots (`..`). The relative path the directory above that—the grandparent directory—is two pairs of dots (`../..`). And so forth. Assuming the parent directory contains another directory called "auntAgnes," the relative path to a file named "cousinCora" in that other directory is

```
../auntAgnes/cousinCora
```

Suppose you have a path created by using just the name of a file in the current directory. That's a relative path. Let's call it `pathR`. You can obtain the corresponding absolute path, `pathA`, like this:

```
Path pathA = pathR.toAbsolutePath();
```

Given two absolute paths, `pathA1` and `pathA2`, you can obtain the relative path from the end of `pathA1` to the end of `pathA2` like this:

```
Path path1_to_path2 = pathA1.relativize(pathA2);
```

Given an absolute (or a relative) `Path` called `path`, to find the total number of elements after the root (or current directory), use

```
path.getNameCount();
```

To identify the `subpath` from the element after `int start` through `int end`, use

```
path.subpath(start, end);
```

To obtain the combination of `path1` followed by `path2`, use

```
Path pathComb = path1.resolve(path2);
```

Creating, Moving, Copying, and Deleting Files

The file designated by a path may not exist when a path is created. The `Files.exists(path)`, method call returns `true` if a file already exists at the location specified by the path, and if you have permission to access that file. To create a new directory, use `Path.createDirectory(path)`. To create a new regular file, you can use `Files.createFile(path)` as described in Section 15.8, but usually you'll just open a new file for writing, as illustrated in Chapter 15 and several times in this chapter. Use `Files.isDirectory(path)` to see if `path` leads to a directory. Use `Files.isRegularFile` to see if `path` leads to a regular file.

To move or copy an existing directory or file from `path1` to `path2`, in a `try` block, use

```
Files.move(path1, path2, <option(s)>);
```

or

```
Files.copy(path1, path2, <option(s)>);
```

There may be zero or more options. If there are no options, and `path2` refers to an existing file, either a `move` or a `copy` method call throws a `FileAlreadyExistsException`. To replace an existing file of the same name, include the option

```
StandardCopyOption.REPLACE_EXISTING
```

Java API documentation describes other possible options.

To delete a file referred to by `path`, in a `try` block, use

```
Files.delete(path);
```

If path refers to a nonexistent file, this call throws a NoSuchFileException. If path refers to a directory that is not empty, this call throws a DirectoryNotEmptyException. Before you can delete a directory, you must first delete or remove all the files it contains.

Describing a Directory's Contents

Figure 16.14 shows how you might display the contents of a directory. The first statement in the main method creates a relative path to the current directory. The statement in the try-with-resources header uses this path to create a DirectoryStream, which contains information about what's in the directory at the

```
/*************************************************************
* DirectoryDescription.java
* Dean & Dean
*
* This describes files in the current directory.
*************************************************************/

import java.nio.file.*; // Path, Paths, DirectoryStream, Files

public class DirectoryDescription
{
  public static void main(String[] args)
  {
    Path pathToDirectory = Paths.get(".");

    try (DirectoryStream<Path> paths =
      Files.newDirectoryStream(pathToDirectory))
    {
      for (Path path : paths)
      {
        System.out.printf("%-30s%6d bytes\n",
        path.getFileName(), Files.size(path));
      }
    }
    catch (Exception e)
    {
      System.out.println(e.getClass());
      System.out.println(e.getMessage());
    }
  } // end main
} // end DirectoryDescription class
```

Sample session:

```
DirectoryDescription.class     1793 bytes
DirectoryDescription.java       772 bytes
FileSizesGUI.class             1813 bytes
FileSizesGUI.java              1975 bytes
```

Figure 16.14 DirectoryDescription program

end of the specified path. The `printf` statement in the `try` block displays filename and file size (length in bytes), as shown in the sample session below the code. Size is just one of several file attributes. Java API documentation describes other file attributes and explains how to access and modify them.

16.10 Walking a Directory Tree

There's another way to explore a file's environment. That's by using the interfaces, classes, and `enum`s in the `java.nio.file` package. To illustrate this other approach, this section presents a program that searches a user-specified subtree of your computer's file directory for files whose names conform to a user-specified text pattern. The program's output shows exactly where the files it finds are located.

Figure 16.15a shows the driver of this `FindFiles` program. The first declared variable is for the root of the user-specified subtree. It must be an absolute path, starting with a forward slash and ending with the name of the root of the desired subtree, like this: /.../<*sub-tree root*>. The second declared variable is for keyboard input. The third declared variable is for the user-specified pattern that determines which files will be identified. This text string should match all desired filenames, with a ? substituted for each character that could be anything and/or a * substituted for each substring that could be anything. For example, the entry, `*.java`, tells the program to find all Java source-code files. The last declared variable, `visitor`, is a special object that does most of the program's work.

After the program prompts for and reads the two user specifications, it creates the `visitor` object. The `"glob:"` part of the argument passed to the `FileVisitor` constructor tells the computer to treat ? and * as wildcards whenever it finds them in the `pattern` part of the constructor's argument.[4] Then the program calls the `walkFileTree` method, which throws a checked exception. Figure 16.15b shows output produced by a typical execution. Whenever the program finds a matching file, it displays that file's name and the number of bytes in that file. Names without reported bytes are directory names.

Now, let's look at heart of the program—the `FileVisitor` class in Figures 16.16a and 16.16b. Figure 16.16a contains the imports, declarations, and constructor. You can see that the Java API code that does most of the work comes from various places in the `java.nio.file` package. And you can see that our `FileVisitor` class is actually just an extension of Java's `SimpleFileVisitor` class. Our class's first instance variable is a `PathMatcher`, which compares the user's specified pattern with the name of each file it encounters. Our class's second instance variable, `tab`, controls the output's line indentation. The constructor generates a `PathMatcher` with the user-specified pattern.

Figure 16.15b shows that the program visits files in the specified subtree in what mathematicians call a *depth-first traverse*. That's a formal term for what Java designers call "walking the tree." The program actually visits all files in the subtree, but it generates printouts only for directories and for regular files whose names match the user's specified pattern. What you see in Figure 16.15b indicates that the program does something in three situations: (1) when it encounters a new directory going down the hierarchy, (2) when it encounters a regular file, and (3) when it returns to a directory above after visiting all its contained files and subdirectories.

Figure 16.16b contains three methods that define what this program does in each of these three situations. Its methods override three corresponding methods in the extended `SimpleFileVisitor` class to produce the particular results that we want. The `@Override` notation just above each method's heading

[4]The patterns specified by "glob" syntax also have other options. See Java API documentation of the `getPathMatcher` method in the `java.nio.file.FileSystem` class.

```
/*************************************************************
* FindFiles.java
* Dean & Dean
*
* This searches the directory tree for files matching a pattern.
*************************************************************/

import java.nio.file.*;                    // Path, Paths, Files
import java.util.Scanner;
import java.io.IOException;

public class FindFiles
{
  //**********************************************************

  public static void main(String[] args)
  {
    Path startDir;
    Scanner stdIn = new Scanner(System.in);
    String pattern;     // ? is wild char; * is wild substring
    FileVisitor visitor;

    System.out.println(
      "Enter absolute path to starting directory:");
    startDir = Paths.get(stdIn.nextLine());
    System.out.print("Enter filename search pattern: ");
    pattern = stdIn.nextLine();

    visitor = new FileVisitor("glob:" + pattern);
    try
    {
      Files.walkFileTree(startDir, visitor);
    }
    catch (IOException e)
    {
      System.out.println(e.getClass());
      System.out.println(e.getMessage() );
    }
  } // end main
} // end FindFiles class
```

Figure 16.15a FindFiles program driver

helps the compiler confirm that the method does indeed correspond to a method in the inherited class. As it walks the tree, when the program encounters any new directory going down the hierarchy, it calls the `preVisitDirectory` method. After producing appropriate indentation, that method prints that directory's name, and then it increments `tab` to increase subsequent indentation. When the program encounters a file whose filename matches the user's specification, it calls the `visitFile` method. After producing appropriate indentation, that method prints that file's name and size in bytes. When the program returns to

```
Sample session:
Enter absolute path to starting directory:
/Users/raydean/Documents/John/ITPJSourceCode/pgmsInChapBodies
Enter filename search pattern: *2.java
pgmsInChapBodies
    chap01
    chap03
    chap04
    chap05
    chap06
        Mouse2.java                 1264 bytes
        MouseDriver2.java            687 bytes
    chap07
        Car2.java                    950 bytes
        Employee2.java               448 bytes
    chap08
    chap09
    chap10
        SpeedDialList2.java         1459 bytes
    chap11
    chap12
        Dealership2.java            1240 bytes
        DealershipDriver2.java       817 bytes
        Manager2.java                297 bytes
        SalesPerson2.java            413 bytes
    chap13
        Car2.java                    984 bytes
        Employee2.java               588 bytes
        Hourly2.java                 832 bytes
        Pets2.java                   713 bytes
        Salaried2.java               621 bytes
    chap14
        GetIntFromUser2.java         978 bytes
        PrintLineFromFile2.java     1210 bytes
        StudentList2.java           1056 bytes
    chap15
        WriteObject2.java            999 bytes
        WriteTextFile2.java          804 bytes
    chap16
        FactorialButton2.java       2980 bytes
    chap17
```

Figure 16.15b Typical FindFiles program output

a directory after visiting all its files and subdirectories, it calls the `postVisitDirectory` method. All it does is decrement `tab` to decrease subsequent indentation.

Suppose all you want to see is the directory structure, not any of the files in that structure. For this result, just run the program with nothing but a simple carriage return (Enter) for the pattern specification.

```
/****************************************************************
* FileVisitor.java
* Dean & Dean
*
* This displays a "glob" filtered file in a file system tree.
****************************************************************/

// for SimpleFileVisitor, Path, PathMatcher, FileSystem,
//     FileSystems, FileVisitResult, and Files:
import java.nio.file.*;
import java.nio.file.attribute.BasicFileAttributes;
import java.io.IOException;

public class FileVisitor extends SimpleFileVisitor<Path>
{
  private PathMatcher matcher;
  private int tab = 0;

  //************************************************************

  public FileVisitor(String syntaxAndPattern)
  {
    FileSystem system = FileSystems.getDefault();

    this.matcher = system.getPathMatcher(syntaxAndPattern);
  } // end constructor

  //************************************************************
```

Figure 16.16a `FileVisitor` class—part A

Suppose that what you want to see is a listing of matching files only, with each file identified as the total path from the directory's root to that file. For this result, do not override the `preVisitDirectory` and `postVisitDirectory` methods. That is, delete or comment out these two methods in Figure 16.16b. Delete the `tab` variable and replace the `for` loop in the `visitFile` method with this statement:

```
name = name.toAbsolutePath();
```

Suppose you want to truncate or terminate the search. In Figure 16.16b, with appropriate code adjustments, use the following code:

```
return FileVisitResult.SKIP_SIBLINGS
```

or

```
return FileVisitResult.SKIP_SUBTREE
```

or

```
return FileVisitResult.TERMINATE
```

One of the end-of-chapter exercises asks you to use the `SKIP_SUBTREE` return option in a modification of our FindFiles program that displays files and file sizes in one directory only.

```
  @Override
  public FileVisitResult preVisitDirectory(Path path,
    BasicFileAttributes attributes) throws IOException
  {
    for (int i=0; i<tab; i++)
    {
      System.out.print(" ");
    }
    System.out.println(path.getFileName());    // directory
    tab++;
    return FileVisitResult.CONTINUE;
  } // end preVisitDirectory

  //*********************************************************

  @Override
  public FileVisitResult visitFile(Path path,
    BasicFileAttributes attributes) throws IOException
  {
    Path name = path.getFileName();

    if (name !=null && matcher.matches(name))
    {
      for (int i=0; i<tab; i++)
      {
        System.out.print(" ");
      }

      System.out.printf("%-25s%6d bytes\n",
        name, Files.size(path));                // ordinary file
    }
    return FileVisitResult.CONTINUE;
  } // end visitFile

  //*********************************************************

  @Override
  public FileVisitResult postVisitDirectory(Path path,
    IOException exc)
  {
    tab--;
    return FileVisitResult.CONTINUE;
  } // end postVisitDirectory
} // end FileVisitor class
```

Figure 16.16b `FileVisitor` class—part B

16.11 GUI Track: The JFileChooser Class (Optional)

In Section 16.9's FileSizes program, we displayed the filenames and file sizes of all the files in the current directory. In Section's 16.10's FindFiles program, we visited other directories. Now suppose you would like to display filenames and file sizes for any directory you select. This section presents a program that does that using a GUI format. The program gets the user's desired directory with the help of the Java API `JFileChooser` dialog box.

A file-chooser dialog box allows the user to select a file or a directory from a graphical, interactive directory structure. File choosers are everywhere in modern software. For example, a word processor employs a file chooser whenever the user selects **Open** from the **File** menu. Figure 16.17 shows how a user selects a file with the help of a `JFileChooser` dialog box.

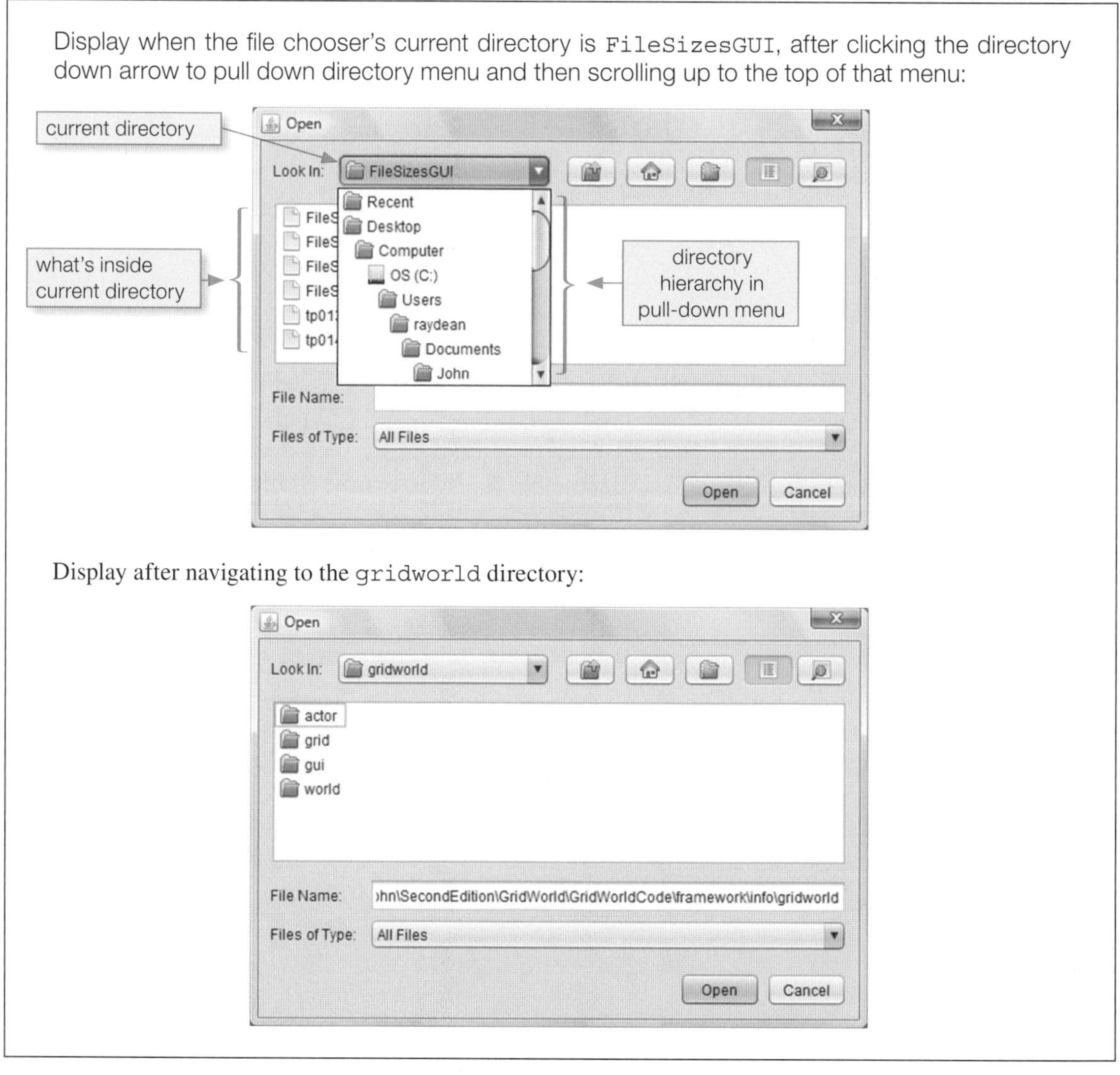

Figure 16.17 Selecting a file with a `JFileChooser` component

JFileChooser Usage

The JFileChooser class is defined in the javax.swing package, so you must import that package to access this class. To create a JFileChooser dialog box, call the JFileChooser constructor like this:

```
JFileChooser chooser = new JFileChooser(<current-directory>);
```

The *current-directory* argument specifies the name of the directory that initially appears at the top of the file-chooser dialog box. That's the file chooser's current directory. This statement shows how we created Figure 16.17's file chooser:

```
JFileChooser chooser = new JFileChooser(".");
```

As mentioned previously, the "." argument represents the JVM's current directory. Be aware that the file chooser's current directory and the JVM's current directory aren't always the same. If you called the JFileChooser constructor with a "C:/spreadsheets" argument, the file chooser's current directory would be C:/spreadsheets, but the JVM's current directory would be unaffected.

The JFileChooser class has many methods. In this section's example, we'll use just three of them—the setFileSelectionMode, showOpenDialog, and getSelectedFile methods. Here are their API headings and descriptions:

`public void setFileSelectionMode(int mode)`
Specifies the type of file the user can choose—a file, a directory, or either one.

`public int showOpenDialog(null)`
Displays a file-chooser dialog box. Returns a named constant, which indicates whether the user selected Open or Cancel.

`public File getSelectedFile()`
Returns the selected file or directory.

When you call setFileSelectionMode, you pass in a mode argument to specify the type of file that the user can choose. If the mode is JFileChooser.FILES_ONLY, the user may choose only a file. If the mode is JFileChooser.DIRECTORIES_ONLY, the user may choose only a directory. If the mode is JFileChooser.FILES_AND_DIRECTORIES, the user may choose either a file or a directory.

You call showOpenDialog to display a file-chooser dialog box. After the dialog box displays, if the user clicks the file chooser's Open button, the showOpenDialog method call returns the JFileChooser.APPROVE_OPTION named constant. If the user clicks the file chooser's Cancel button, the showOpenDialog method call returns the JFileChooser.CANCEL_OPTION named constant.

After a user selects a file with the JFileChooser dialog box, the program calls getSelectedFile to retrieve the selected file or directory. At that point, the program will probably want to do something with the file or directory. But before it does, it should use File's isFile or isDirectory method to determine the selected file's type.

JOptionPane Usage

You may need to use some of the methods of the JOptionPane class as well. This class is also in the javax.swing package, so if you imported this package for the JFileChooser class, you'll have access to the JOptionPane class too. The JOptionPane class provides many useful class methods, which you can access directly with the class name. We'll look at just two of them—the showConfirmDialog method and the showMessageDialog method. Here are the API headings and descriptions:

```
public static int showConfirmDialog(Component parentComponent,
  Object message, String title, int optionType
```

This brings up a dialog box in which the number of choices is determined by optionType.

```
public static void showMessageDialog(Component parentComponent,
  Object message, String title, int messageType)
```

This brings up a dialog box that displays a message.

When you call showConfirmDialog, you can use a reference to another frame within which you want the box displayed, or you can use just a null to position it relative to the whole screen. In addition, you may supply a text message, and you must supply a text title. For the optionType parameter, supply one of these two arguments:

JOptionPane.YES_NO_OPTION or JOptionPane.YES_NO_CANCEL_OPTION

The value returned is the option selected by the user, one of these two constants:

JOptionPane.YES_OPTION or JOptionPane.NO_OPTION

When you call showMessageDialog, you may use a reference to another frame within which you want the box displayed, or you may use just a null to position it relative to the whole screen. In addition, you may supply a text message, and you must supply a text title. For the messageType parameter, supply one of the following arguments:

```
JOptionPane.ERROR_MESSAGE
JOptionPane.INFORMATION_MESSAGE
JOptionPane.WARNING_MESSAGE
JOptionPane.QUESTION_MESSAGE
JOptionPane.PLAIN_MESSAGE
```

FileSizesGUI Program

Now we're ready to incorporate these ideas in that improved FileSizes program that we mentioned earlier. Our FileSizesGUI program uses a JFileChooser dialog box to retrieve a user-specified file or directory. If the user selects a file, the program displays the file's name and size. If the user selects a directory, the program displays the filenames and sizes of all of the files in the directory. To get a better idea of how the program will operate, see the sample session in Figure 16.18.

The program itself is in Figures 16.19a and 16.19b. In Figures 16.19a and 16.19b, look for each of the following operations:

- A JFileChooser constructor call
- A call to JOptionPane's showConfirmDialog method, and use of the value returned
- A call to JFileChooser's setFileSelectionMode method
- A call to JFileChooser's showOpenDialog method, and use of the value returned
- A call to JFileChooser's getSelectedFile method, and use of the value returned
- A call to JOptionPane's showMessageDialog method

After calling getSelectedFile, the program needs to determine the type of the user's selection—file or directory. The isFile and isDirectory calls take care of that. Within the directory-processing

When you run the program, a `JOptionPane.showConfirmDialog` method call displays this:

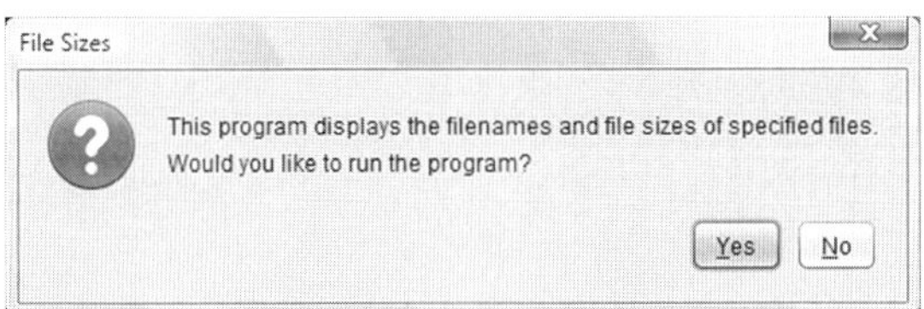

Clicking Yes causes a call to `JFileChooser`'s `showOpenDialog` method to display another window, and after you navigate to the gridworld directory and select grid, it looks like this:

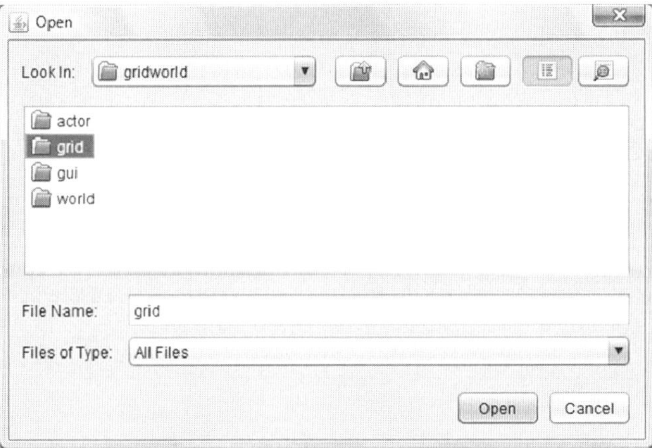

Clicking Open causes a call to `JOptionPane`'s `showMessageDialog` method to display this:

And clicking OK terminates the program.

Figure 16.18 Sample session for the FileSizeGUI program

```
/***************************************************************
* FileSizesGUI.java
* Dean & Dean
*
* This displays your computer files in GUI format.
***************************************************************/

import java.io.File;
import javax.swing.*;    // for JFileChooser and JOptionPane;

public class FileSizesGUI
{
  public static void main(String[] args)
  {
    File fileDir;         // user-specified file or directory
    int response;         // user's response to GUI prompts
    File[] files;         // array of files in specified directory
    String output = "";  // list of filenames and sizes
    JFileChooser chooser = new JFileChooser(".");

    response = JOptionPane.showConfirmDialog(null,
      "This program displays the filenames and file sizes of" +
      " specified files.\nWould you like to run the program?",
      "File Sizes", JOptionPane.YES_NO_OPTION);
```

Figure 16.19a FileSizesGUI program—part A

code, note how `fileDir`, a `File` object, calls `listFiles`. The `listFiles` method returns the files and directories that are in the `fileDir` directory. The returned files and directories are stored as `File` objects in an array named `files`. After filling the `files` array, the program loops through each of its `File` elements. For each element, it prints filename and file size by calling `getName` and `length`, respectively.

In the FileSizesGUI program, note the `String.format` method calls. The `String.format` method works the same as the `printf` method except instead of printing a formatted value, it returns a formatted value. We use the `String.format` method calls in an attempt to display values with uniform widths. Specifically, we want to display the filenames and file sizes with uniform widths, so the file-size values display in an aligned fashion. However, the bottom dialog box in Figure 16.18 shows that the file-size values are not aligned. The problem is that with GUI output, different characters print with different widths. For example, you can see that the 17-character string, "`AbstractGrid.java`" is shorter than the 16-character string, "`BoundedGrid.java`". That's because the characters in "`Abstract`" are narrower than the characters in "`Bounded`". The "`Grid.java`" item is even more compressed because it has the most spaces, and spaces are extremely narrow in this display's font. Having different characters print with different widths is stylish most of the time, but in our FileSizesGUI program, it's annoying. To fix this problem, you could embed a `JTextArea` component into the `JOptionPane` dialog box and set the `JTextArea` component's font to a monospaced font. (With monospaced font, every character prints with the same width.) You'll learn about the `JTextArea` component in Chapter 18.

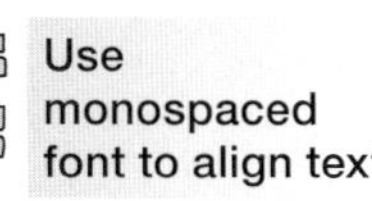
Use
monospaced
font to align text

```
      if (response == JOptionPane.YES_OPTION)
      {
        chooser.setFileSelectionMode(
          JFileChooser.FILES_AND_DIRECTORIES);
        response = chooser.showOpenDialog(null);
        if (response == JFileChooser.APPROVE_OPTION)
        {
          fileDir = chooser.getSelectedFile();
          if (fileDir.isFile())
          {
            output += String.format("%-25s%12s%n",
              fileDir.getName(), fileDir.length() + " bytes");
          }
          else if (fileDir.isDirectory())
          {
            files = fileDir.listFiles();
            for (int i=0; i<files.length; i++)
            {
              output += String.format("%-25s%12s%n",
                files[i].getName(), files[i].length() + " bytes");
            } // end for
          } // end else
          else
          {
            output = "Invalid entry. Not a file or directory.";
          }
          JOptionPane.showMessageDialog(null, output,
            "File Sizes", JOptionPane.INFORMATION_MESSAGE);
        } // end if
      } // end if
    } // end main
  } // end FileSizesGUI class
```

Figure 16.19b FileSizesGUI program—part B

Summary

- Most of the file-transfer classes you'll need are in the `java.io`, `java.nio`, `java.nio.files`, and `java.nio.channels` packages.
- To read text from a simple local file, supply a `String` filename as the argument in a `Paths.get` method call to obtain a `Path` description of the file's location. Then use this `Path` as the argument in a new `Scanner`, and use familiar `Scanner` methods to read the file's data.
- To write text to a simple local file, instantiate a `PrintWriter` object with a `String` filename as its argument. Then use `PrintWriter` methods like `println` to write data to the file.
- You can use simple text file I/O to translate plain-text information into HTML format for a web page.
- You can use a `Scanner` with an `InputStream` from a `URLConnection` to read the HTML text version of what's on a remote website.

- You can store whole objects in a file as a stream of `bytes`, provided those objects and all their component objects implement the `Serializable` interface.
- To write an object to a file, construct a `FileOutputStream` with a `String` filename as the argument. Then with that stream as the argument, construct in a new `ObjectOutputStream`. Then use `ObjectOutputStream`'s `writeObject` method to write the object's data to the file.
- To read an object from a file, construct a `FileInputStream` with a `String` filename as the argument. Then with that stream as the argument, construct a new `ObjectInputStream`. Then use `ObjectInputStream`'s `readObject` method to read the object's data from the file.
- To specify a character set, import `java.nio.charset.Charset`, and use:
  <character-set> `= Charset.defaultCharset()`, or something like this:
  <character-set> `= Charset. forName("US-ASCII")`
- With large text files, write through a buffer by opening the file with
  `new PrintWriter(Files.newBufferedWriter(`
  `Paths.get(`*<filename>*`),` *<character-set>*`,` *<open-option(s)>*`))`
  Then write the text with methods like `println`.
- With large text files, read through a buffer by opening the file with
  `Files.newBufferedReader(`
  `Paths.get(`*<filename>*`),` *<character-set>*`))`
  Then read the text with a method like `readLine`.
- `ByteBuffer`'s `put` and `get` methods copy any single primitive or an array of bytes to or from the next available buffer `position` or an `index` number of primitive elements from the start of the buffer.
- To copy an array of *<primitive>* to or from a `ByteBuffer`'s underlying array, assign the return from `ByteBuffer`'s `as`*<Primitive>*`Buffer` to a corresponding *<Primitive>*`Buffer`. Then use *<Primitive>*`Buffer`'s `put` and `get` array methods to access the data in `ByteBuffer`'s underlying array.
- `FileChannel` methods organize heterogeneous data into sequences of `ByteBuffers` or arrays of `ByteBuffers`. To open `FileChannel channel` for either input or output, use:
  `channel = FileChannel.open(Path path, OpenOption... options)`
  where `...` means there can be any number of options, including none. Java API options are `enums` like `java.nio.file.StandardOpenOption.READ`
- The return from `FileChannel`'s method,
  `map(FileChannel.MapMode mode, long position, long size)`
  is a non-volatile connection to file data, which persists beyond `FileChannel` closure.
- The Java API supplies many useful ways to manipulate paths and describe the files in a directory.
- With an extension of Java's `SimpleFileVisitor`, you can search a directory tree for files that match a "glob" pattern, which includes `?` and `*` wildcards. To do this, use the class method call, `Files.walkFileTree(startDirectory, fileVisitor)`.
- Optionally, with Java's `JFileChooser` class, you can enable a user of one of your programs to find any file in his or her computer by interacting with a familiar graphical user interface.

Review Questions

§16.2 Simple Text-File Example—HTML File Generator

1. Assuming the object that manages output will be called `writer`, write a statement that opens a text file called `dogs.html` for output by `println` statements.
2. Where do the `<h1>` and `</h1>` tags go in an HTML file?

§16.4 Object File I/O

3. To be file-writable and file-readable, an object must be an instance of a class that `implements` __________________.
4. Write a statement that opens a file for input of objects and assigns a reference to the connection to `objectIn`. Assume the file is in the current directory, with the name `automobiles.data`.
5. Assuming you have made the declaration, `TestObject testObject;` you can assign the value returned by `ObjectInputStream`'s `readObject` method directly to `testObject`. (T / F)

§16.5 Character Sets and File-Access Options

6. Given the filename "TradingPartners," write a try-with-resources header that opens a file for writing text with a `PrintWriter` called `fileOut` and stores that text using the UTF-16 character set.
7. Given the filename "Suppliers," write a try-with-resources header that opens a text file for reading with a `Scanner called fileIn` and interprets the file's data using the UTF-16 character set.
8. What Java API package contains `Enum StandardOpenOption`?

§16.6 Buffered Text File I/O

9. In a Java interface, what does the `...` notation (an ellipsis) in the parameter specification
   `OpenOption... options`
   say about the arguments you should supply to the constructor or method?
10. When a text file is large, for good performance, what type of file handler should you use?

§16.7 Primitive Buffers with Random Access

11. What does a `flip()` method call do?
12. What is the difference between the `ByteBuffer` methods, `get()` and `get(int index)`?
13. `ByteBuffer`'s `putDouble(double value)` method inserts the 8-byte `value` into the buffer starting at the current `position` and increments `position` to just after that inserted value. (T / F)
14. Given the array, `int[] integers` and `ByteBuffer buffer`, write a statement that copies `integers[3]` into `buffer`, starting at `buffer`'s current `position`.

§16.8 Channel I/O and Memory-Mapped Files

15. Assuming `import java.nio.file.*;` and an existing file called "CanadianFriends," write a Java statement that declares and opens a read-only `FileChannel` to that file.
16. Given a `FileChannel` called `channel` and a `ByteBuffer` called `buffer` whose `capacity` and `limit` equal `channel.size()`, write a statement that maps the entire file into the buffer for read-only access.

§16.9 The File Class

17. Write code for a try-with-resources header that creates a stream containing information about what's in the parent of the current directory.

§16.10 Walking a Directory Tree

18. What is a "glob"?

Exercises

1. [after §16.2] Edit the HTMLGenerator.java program given in the text so that </p> end tags are inserted properly in the generated HTML file. The </p> end tags should be inserted at the bottom of each paragraph.

Note:

- Do not allow a </p> end tag to be generated when there's no accompanying <p> start tag (start and end tags must always be partnered).
- Your program should be robust (i.e., handle the weird cases). In particular, handle the case where there's only a title and no paragraphs at all.

2. [after §16.4] Modify the code in Chapter 15's `StudentList` class in Figure 15.8 so that objects from that class can be written to a file. For this exercise, you may simply ignore the "dangerous method call" in the `removeStudent` method. Using the `WriteObject` class in Figure 16.5 as a guide, encode a `StudentListFileWriter` class that writes three `StudentList` objects to a file called `StudentLists`, as follows: Construct the first object with the same four names as those used in the `StudentListDriver` in Figure 15.9, and write it to the file. Construct the second object with those same four names plus the additional name "Anna," and write it to the file. For the third object, use the remove method to remove "Izumi" from the second object, and write this modification of the second object to the file so that it appears as a distinct third object in the file. Finally, using the `ReadObject` class in Figure 16.6 as a guide, encode a `StudentListFileReader` class that reads and displays all objects in the file. Run this `StudentListFileReader` to confirm that the file holds three objects with different name lists.

3. [after §16.5] Describe the difference between `StandardOpenOption.CREATE` and `StandardOpenOption.CREATE_NEW`.

4. [after §16.6] Modify the BufferedWriteToFile program in Figure 16.7a so that you can write with `PrintWriter`'s `print`, `println`, and `printf` methods. Instead of defining `fileOut` as a `BufferedWriter`, define it as a `PrintWriter`, using the `PrintWriter` constructor that takes a `BufferedWriter` argument.

5. [after §16.6] Modify the BufferedReadFromFile program in Figure 16.8 so that you can read with `Scanner`'s `next`, `nextLine`, `nextInt`, `nextDouble`, etc. Instead of defining `fileIn` as a `BufferedReader`, define it as a `Scanner`, using the `Scanner` constructor that takes a `BufferedReader` argument.

6. [after §16.6] Consider the following code:

```
/*************************************************************
* WebPageReader.java
* Dean & Dean
*
* This reads a web page through a buffer.
*************************************************************/
import java.util.Scanner;

public class WebPageReader
{
  BufferedReader reader;
  public WebPageReader(String webAddress)
  {
    URL url = new URL(webAddress);
    URLConnection connection = url.openConnection();
    InputStream in = connection.getInputStream();
    reader = new BufferedReader(new InputStreamReader(in));
  } // end constructor
```

```
  //*************************************************************

  public String readLine()
  {
    return reader.readLine();
  } // end readLine

  //*************************************************************

  public static void main(String[] args)
  {
    Scanner stdIn = new Scanner(System.in);
    String url, line;
    System.out.print("Enter a full URL address: ");
    url = stdIn.nextLine();
    WebPageReader wpr = new WebPageReader(url);
    while ((line = wpr.readLine()) != null)
    {
      System.out.println(line);
    }
  } // end main
} // end WebPageReader
```

Use this program as a starting point and make these adjustments to it:

- Add `import` statements as required.
- For each method call and constructor call:
  If it throws an unchecked exception, ignore it.
  If it throws a checked exception:
    Specify the specific exception in a `throws` clause.
    Within a generic catch block in `main`, catch any exception and print its class and message using `getClass` and `getMessage` method calls.

Be aware that your program might be correct, but it might not be able to access web pages successfully. To access web pages, your computer needs to have Internet access capabilities. If your firewall asks if it's OK for Java to access the Internet, click "yes" and continue.

Here are three sample sessions:

First sample session:

```
Enter a full URL address: htp://www.park.edu
class java.net.MalformedURLException
unknown protocol: htp
```

Second sample session:

```
Enter a full URL address: http:/www.park.edu
class java.lang.IllegalArgumentException
protocol = http host = null
```

Third sample session:

```
Enter a full URL address: http://www.park.edu
<first-ten-lines-like-Figure-16.3>
```

7. [after §16.7] The following program uses a `DoubleBuffer` to implement a random-access queue. The input data could represent 24 temperatures measured at each hour of a typical day. The program reads the first 12 data items into the buffer. Then it prints the temperature at index 9. Then it copies the first 6 buffer items to an output array. Then it reads the remaining input data items into the buffer. Finally, it copies the remaining buffer items to the output array. Insert and append comments that explain each operation, and in particular, explain how the buffer's mark and limit variables relate to the head and tail of the queue.

```
/**************************************************************
* Temperatures.java
* Dean & Dean
*
* This uses a buffer to implement a random-access data queue.
* mark is ________________ limit is _____________________
**************************************************************/

import java.nio.DoubleBuffer;

  public class Temperatures
  {
    public static void main(String[] args)
    {
      double[] dataIn = new double[] {1.3, 0.4, -0.6,
        -1.2, -1.8, -2.1, -2.3, -1.3, -0.8, 0.2, 1.3, 2.3, 3.5,
        4.8, 5.7, 6.5, 5.1, 4.2, 3.0, 1.9, 0.6, 0.1, -0.1, -0.6};
      double[] dataOut = new double[24];
      DoubleBuffer buffer = DoubleBuffer.allocate(24);

      buffer.mark();

      buffer.put(dataIn, 0, 12);
      buffer.limit(buffer.position());
      System.out.println(buffer.get(9));
      buffer.reset();

      buffer.get(dataOut, 0, 6);
      buffer.mark();
      buffer.position(buffer.limit());
      buffer.limit(buffer.capacity());

      buffer.put(dataIn, buffer.position(), buffer.remaining());
      buffer.reset();

      buffer.get(dataOut, buffer.position(), buffer.remaining());
    } // end main
  } // end class Temperatures
```

8. [after §16.7] The first sequence in Figure 16.11's sample session is a Fibonacci sequence, and the second sequence is ratios of consecutive pairs of the Fibonacci sequence. The second sequence converges on the "golden ratio," a beautiful proportion found in ancient architecture and used to find roots of nonlinear equations. (a) Write iterative code that extends the Fibonacci sequence and computes the golden ratio to an

accuracy of eight decimal places. (b) Write a formula for the sequence and use it to derive an exact algebraic expression for the golden ratio, or its reciprocal, the "golden mean."

9. [after §16.8] Write a text file through a `FileChannel`. Within the program, create the filename "Zapata" and create a string of text, "It is better to die on your feet than to live on your knees." Convert the filename to a `Path`, and convert the text `String` to a `Byte` array. Use `ByteBuffer`'s `wrap` method to wrap the `Byte` array into a buffer. In a try-with-resources header open a `FileChannel` for writing, and call the channel's `write` method to write the buffer to the file. Confirm that your program works by reading the new "Zapata" file with a simple text editor.

10. [after §16.9] Use a program like Microsoft Notepad or UNIX vi to create a text file, "Zapata," which contains the text string given in Exercise 8, distributed over two lines. Write a program that creates a subdirectory called MexicanHistory and moves the Zapata file from the current directory to that new subdirectory. In your program, use one declaration to create the path to Zapata and another declaration to create a path to MexicanHistory. In a third declaration, use the `resolve` method to create a destination path. Create the new directory only if that directory does not exist already. Allow the move to fail if it already happened in a previous execution, but report the failure with `catch`-block statements.

11. [after §16.10] Modify this chapter's DirectoryDescription program to print name and size information for only those files whose "glob" pattern matches a user-specified string. To do this, you'll need to prompt for and read the desired "glob" pattern. In addition to `Scanner`, you'll need to use the Java API `FileSystem` and `PathMatcher` classes to create a `matcher` object like that in the `FileVisitor` constructor in Figure 16.16a. Then, you'll need to put the for-each loop's `printf` statement inside an `if` statement whose condition is something like `matcher.matches(name)`, where `name = path.getFileName()`.

12. [after §16.10] Modify this chapter's FindFiles program to restrict its search to just one directory. Do this by conditionally returning `SKIP_SUBTREE` at an appropriate place in the `FileVisitor` class.

Review Question Solutions

1. `writer = new PrintWriter("dogs.html");`
2. The `<h1>` and `</h2>` tags enclose the visible web page header.
3. To be file-writable and file-readable, an object must be an instance of a class that `implements Serializable`.
4.
```
ObjectInputStream objectIn = new ObjectInputStream(
  new FileInputStream("automobiles.data"));
```
5. False. The `readObject` method returns an `Object` type. You must explicitly cast the returned value to the type of your reference variable before assigning it to that variable.
6.
```
try (PrintWriter fileOut =
  new PrintWriter("TradingPartners", "UTF-16"))
```
7.
```
try (Scanner fileIn =
  new Scanner(Paths.get("Suppliers", "UTF-16"))
```
8. `Enum StandardOpenOption` is in the `java.nio.file` package.
9. The `...` notation (an ellipsis) says you may enter zero or more arguments of the type specified just before the ellipsis.
10. When a file is large, for good performance, use a buffered file handler. For text files, use a `BufferedReader` or a `BufferedWriter`.

11. A `flip()` method call makes a buffer's `limit` equal its current `position`, and then it makes the current `position` equal to zero.

12. `ByteBuffer`'s `get()` method is relative. It returns the `byte` at the buffer's current `position` and increments that `position`. `ByteBuffer`'s `get(int index)` method is absolute. It returns the `byte` at `index` and does not alter the buffer's current `position`.

13. True. `ByteBuffer`'s `putDouble(double value)` method inserts `value` into the buffer starting at the current `position` and increments `position` to just after that inserted value.

14. To copy `integers[3]` to `buffer`'s current `position`, use:

```
buffer.asIntBuffer().put(integers, 3, 1);
```

15. To open a read-only `FileChannel` to an existing file called "CanadianFriends," use this:

```
FileChannel channel = new FileChannel.open(
  Paths.get("CanadianFriends"), StandardOpenOption.READ);
```

16. To map all of `channel` into `buffer` for read-only access, use this:

```
buffer = channel.map(FileChannel.MapMode.READ, 0, channel.size());
```

17. This creates a stream containing information about the parent directory:

```
DirectoryStream<Path> paths =
  Files.newDirectoryStream(Paths.get(".."))
```

It goes in a try-with-resources header.

18. A "glob" is a pattern used to see if a particular string matches a specified form. In a glob, each `?` identifies a position that may contain any character, and a `*` identifies a contiguous sequence of positions that may contain any characters.

CHAPTER 17

GUI Programming Basics

Objectives

- Understand the event-driven programming paradigm. In particular, understand what it means to fire an event, and understand the terms *listener* and *event handler.*
- Use the `JFrame` class to implement window functionality.
- Create and use `JLabel`, `JTextField`, and `JButton` components.
- Implement a listener for the `JTextField` and `JButton` components.
- Understand what an interface is and implement the `ActionListener` interface.
- Understand what an inner class is and implement a listener as an inner class.
- Know the difference between an anonymous inner class and a standard inner class.
- Create and use `JOptionPane` dialog boxes.
- Be able to distinguish between multiple events.
- Describe the primary GUI packages.
- Describe the difference between lightweight and heavyweight components.
- Learn how to use mouse listeners with images.

Outline

17.1 Introduction

Hopefully, you've been on the edge of your seat in reading the prior chapters. If not, be prepared to be on the edge of your seat now. It's time for the really good stuff: *graphical user interface* (*GUI*) programming.

You've probably heard the term *GUI,* and you probably know that it's pronounced "gooey." But do GUI's three words, graphical user interface, make sense? "Graphical" refers to pictures, "user" refers to a person, and "interface" refers to communication. Thus, GUI programming employs pictures—like windows, labels, text boxes, buttons, and so on—to communicate with users. For example, Figure 17.1 shows a window with two labels, a text box, and a button. We'll describe windows, labels, text boxes, and buttons in detail later on.

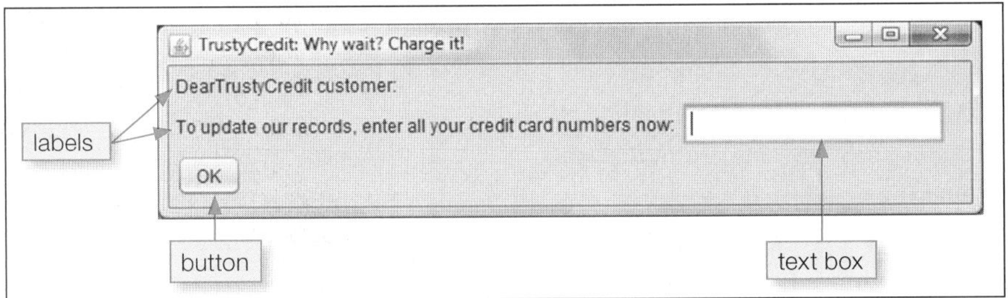

Figure 17.1 Example window that uses two labels, a text box, and a button

In the old days, program interfaces consisted of just text. Programs would prompt the user with a text question, and users would respond with a text answer. That's what we've been using for all of our programs so far. Text input/output (I/O) works well in many situations, but you can't get around the fact that some people consider text display to be boring. Many of today's users expect programs to be livelier. They expect windows, buttons, colors, and so on for input and output. They expect GUI.

Although companies still write many text-based programs for internal use, they normally write GUI-based programs for programs that are to be used externally. It's important for external programs to be GUI based because external programs go to customers, and customers typically won't buy programs unless they are GUI based. So if you want to write programs that people will buy, you'd better learn GUI programming.

We start this chapter with an overview of basic GUI concepts and terminology. We then move on to a bare-bones program where we introduce basic GUI syntax. We next cover listeners, inner classes, and several rudimentary *GUI components,* which are objects that sit inside a window, including `JLabel`, `JTextField`, and `JButton`. We describe the `JOptionPane` class (for generating a dialog box) and the `Color` class (for generating a color). Finally, we present mouse listeners and images. Section 17.18 shows how to use a mouse to drag an image around in a window.

You may have noticed optional GUI-track sections at the end of about half of the prior chapters. The GUI material in this chapter and the next is different from the GUI material in the earlier chapters, and it does not depend on the earlier chapters' GUI material. So if you skipped the earlier GUI material, no worries.

To understand this chapter, you need to be familiar with object-oriented programming, arrays, inheritance, and exception handling. As such, you need to have read up through Chapter 15. This chapter does not depend on material covered in Chapter 16.

17.2 Event-Driven Programming Basics

GUI programs usually use *event-driven programming* techniques. The basic idea behind event-driven programming is that the program waits for events to occur and the program responds to events if and when they occur.

Terminology

So what is an event? An *event* is a message that tells the program that something has happened. For example, if the user clicks a button, then an event is generated, and it tells the program that a particular button was clicked. More formally, when the user clicks a button, we say that the button object *fires an event.* Note these additional event examples:

User Action	What Happens
Pressing the Enter key while the cursor is inside a text box.	The text box object fires an event, and it tells the program that the Enter key was pressed within the text box.
Clicking a menu item.	The menu item object fires an event, and it tells the program that the menu item was selected.
Closing a window (clicking on the window's top-right-corner "X" button).	The window object fires an event, and it tells the program that the window's close button was clicked.

If an event is fired, and you want your program to handle the fired event, then you need to create a *listener* for the event. For example, if you want your program to do something when the user clicks a particular button, you need to create a listener for the button. For now, think of a listener as an ear. If an event is fired and there's no ear listening to it, then the fired event is never "heard" and there's no response to it. On the other hand, if there is an ear listening to a fired event, then the ear "hears" the event and the program then responds to the fired event. The way the program responds is by executing a chunk of code known as an *event handler.* See Figure 17.2. It depicts a button being pressed (see the mouse pointer), an event being fired (see the sound waves), a listener hearing the event (see the ear), and an event handler being executed (see the arrow going down the event-handler code). This system of using listeners for event handling is known as the *event-delegation model*—event handling is "delegated" to a particular listener.

The Event-Driven Programming Framework

Based on the above description, event-driven programming may feel like an altogether new type of programming. Particularly the part about firing an event and listening for a fired event. Many people are fine with the idea of event-driven programming being a new type of programming. But the truth of the matter is that it's

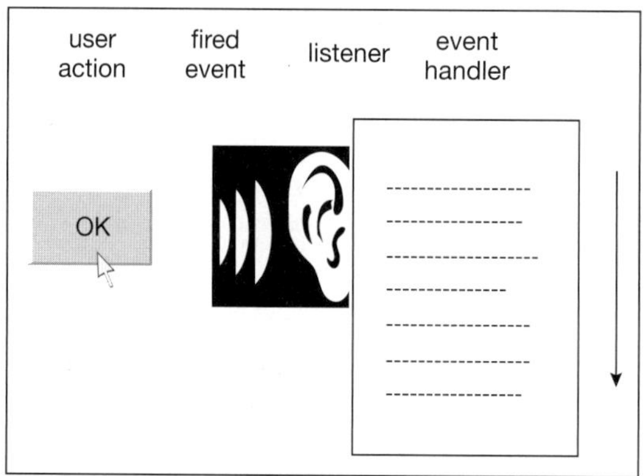

Figure 17.2 What happens when a button is pressed

really just object-oriented programming with window dressing. Make that lots of window dressing. Oracle provides an extensive collection of GUI classes that, together, form a framework on which to build GUI applications. That framework is comprised of classes, methods, inheritance, and so on. In other words, it's comprised of OOP components. As a programmer, you don't have to understand all the details of how the framework works; you just have to understand it well enough to use it. For example, you have to know how to plug in your event handlers properly. Figure 17.3 provides a high-level, graphic illustration of what we're talking about.

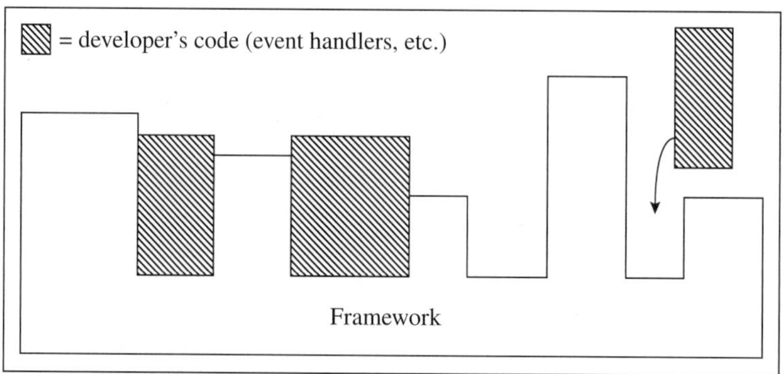

Figure 17.3 Event-driven programming framework

Why did the Java language designers bother to provide the event-driven programming framework? It satisfies the goal of getting maximum benefit from minimum code. With the help of the framework, Java programmers can get a GUI program up and running with a relatively small amount of effort. Initially, the effort might not seem so small, but when you consider all that the GUI program does (automatic event firing, listening for fired events, and so on), you'll find that your return on investment is quite good.

17.3 A Simple Window Program

OK. Enough talk about concepts. Time to roll up your sleeves and get your hands dirty with some code. To get a feel for the big picture, let's start with a simple GUI program and discuss the GUI commands at a high level. Later, we'll cover the GUI commands in greater detail.

In Figure 17.4, we present a SimpleWindow program that displays a line of text inside a window. Note the two `import` statements at the top of the program. They import the `javax.swing` and `java.awt`

```
/*******************************************************
* SimpleWindow.java
* Dean & Dean
*
* This program displays a label in a window.
*******************************************************/

import javax.swing.*; // for JFrame, JLabel
import java.awt.*;    // for FlowLayout

public class SimpleWindow extends JFrame
{
  private static final int WIDTH = 250;
  private static final int HEIGHT = 100;

  //****************************************************

  public SimpleWindow()
  {
    setTitle("Simple Window");
    setSize(WIDTH, HEIGHT);
    setLayout(new FlowLayout());
    setDefaultCloseOperation(EXIT_ON_CLOSE);
    createContents();
    setVisible(true);
  } // end SimpleWindow constructor

  //****************************************************

  private void createContents()
  {
    JLabel label = new JLabel("Hi! I'm Larry the label!");
    add(label);
  } // end createContents

  //****************************************************

  public static void main(String[] args)
  {
    new SimpleWindow();
  } // end main
} // end class SimpleWindow
```

This adds the label to the window.

This instantiates an anonymous window object.

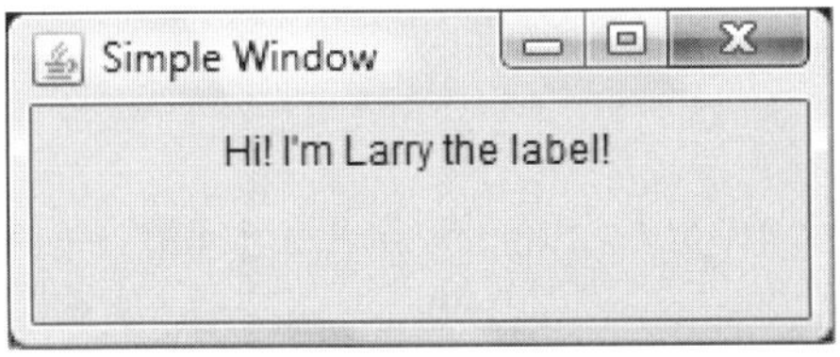

Simple Window

Hi! I'm Larry the label!

Figure 17.4 SimpleWindow program and its output

packages. In writing GUI programs, you'll use many of Java's prebuilt GUI classes from Oracle's API library. To use the prebuilt GUI classes, you'll need to import them into your GUI programs. You could import the classes individually, but there's a better way. Recall that a package is a collection of prebuilt classes. Because most of the critical prebuilt GUI classes come from the `javax.swing` and `java.awt` packages, import those two packages and you'll import most of the critical prebuilt GUI classes. Get used to importing those two packages in every one of your GUI programs. Recall that to import a package, you need to use an asterisk, like this:

```
import javax.swing.*;
```

The `*` is a wildcard, and it allows you to import all the classes within a particular package.

In `SimpleWindow`'s class heading, note the `extends JFrame` clause. The `JFrame` class is part of the GUI framework mentioned previously. The `JFrame` class provides standard Windows features such as a title bar, a minimize button, and so on. Below the class heading, note the `WIDTH` and `HEIGHT` named constants. They're used by the `setSize` method call to specify the dimensions of the window.

Let's now examine the `main` method. GUI programs typically create a window with GUI components, and then they just sit around waiting for the user to do something like click a button, select a menu option, and so on. Thus, `main` is very short—it just instantiates the window and that's it. In this simple example, we don't even bother to assign the instantiated window object to a reference variable. Review: What do you call an object that isn't stored in a reference variable? An anonymous object.

In performing the anonymous-object instantiation, `main` calls the `SimpleWindow` constructor. The `SimpleWindow` constructor (1) calls `setTitle` to assign the window's title, (2) calls `setSize` to assign the window's size, (3) calls `setLayout` to assign the window's layout scheme, and (4) calls `setDefaultCloseOperation` to enable the *close-window button* (the "X" in the top-right corner) to work properly.

In the interest of modularization, the `SimpleWindow` constructor then calls a helper method, `createContents`, to create the components that go inside the window. The `createContents` method contains only two lines. With only two lines, there's really no need for a helper method, but we want you to form good habits. For this trivial example, there's only one component and there's no event handler for the component. Thus, two lines are all that's needed. But normal GUI programs have multiple components and multiple event handlers. For that, quite a few lines are needed. If you stick those lines in the constructor, you'd have a long constructor. Better to break things up and stick them in a helper method.

The `createContents` method instantiates a `JLabel` component and then calls the `add` method to add the `JLabel` component to the window. A `JLabel` component is the simplest type of GUI component. It's a piece of text that the user can read but cannot change.

After executing `createContents`, the JVM returns to the `SimpleWindow` constructor. The `SimpleWindow` constructor then calls `setVisible` to make the window visible.

17.4 JFrame Class

In the previous section, we introduced you to the `JFrame` class. In this section, we describe the `JFrame` class in more depth. More specifically, we cover its characteristics and its methods.

JFrame Basics

These days, most purchasable software is windows-based. When you load such software, you'll see a window and that window will have a title bar, a border, a minimize button, a close-window button, the ability to resize the window, and so on. You could implement all those features from scratch in your own classes, but why "reinvent the wheel"? The `JFrame` class implements the standard windows features that you've come to know and love. To get all that cool windows functionality for free, just implement your classes by extending the JFrame class. What a deal!

The JFrame class should be the superclass for most of your GUI application windows, so a programmer-defined window will normally have extends JFrame in its class heading. For the extends JFrame to work, you must import the JFrame class or import JFrame's package, javax.swing. As explained above, it's common to import the javax.swing package for all GUI programs.

The JFrame class is called a *container* because it contains components (like labels, buttons, menus, and so on). It inherits the ability to contain components from its superclass, the Container class.

JFrame Methods

By extending the JFrame class, you automatically get the standard windows functionality mentioned above. In addition, you inherit a host of windows-related methods. In the SimpleWindow program, we use these inherited methods—setTitle, setSize, setLayout, setDefaultCloseOperation, add, and setVisible. The setLayout and setDefaultCloseOperation methods come directly from the JFrame class. The other methods come from ancestors of the JFrame class—setTitle from the Frame class, add from the Container class, setSize and setVisible from the Component class.

The setTitle method displays a specified string in the current window's title bar. If setTitle is not called, then the window's title bar is empty.

The setSize method assigns the width and height of the current window. See Figure 17.4 and note how the SimpleWindow program assigns the width to 300 and the height to 200. The width and height values are specified in terms of *pixels*. A pixel is a computer monitor's smallest displayable unit, and it displays as a dot on the screen. If you call setSize with a width of 300 and a height of 200, then your window will consist of 200 rows where each row contains 300 pixels. Each pixel displays with a certain color. The pixels form a picture by having different colors for the different pixels. For example, the window depicted in Figure 17.4 might contain blue pixels on the perimeter (for the window's border), and black pixels in the center (for the window's message).

To give you perspective on how big a 300-by-200 pixel window is, you need to know the dimensions, in pixels, of an entire computer screen. The dimensions of a computer screen are referred to as the screen's *resolution*. Resolution settings are adjustable. Two common resolution settings are 800-by-600 and 1024-by-768. The 800-by-600 setting displays 600 rows where each row contains 800 pixels.

If you forget to call the setSize method, your window will be really small. It will display only the beginning of the title and the three standard window-adjustment buttons—minimize, maximize, and close-window. It won't display the window's contents unless you manually resize the window. Here's what the SimpleWindow program displays if you omit the setSize method call:

The setLayout method assigns a specified *layout manager* to the current window. The layout manager is prebuilt Java API software that determines the positioning of components. In the SimpleWindow program's setLayout call, we specify the FlowLayout manager, and the FlowLayout manager causes components to be positioned in the top-center position. The FlowLayout class is defined in the java.awt package, so don't forget to import that package. In the next chapter, we describe the FlowLayout manager and other layout managers in more detail. We're using the FlowLayout manager (as opposed to other layout managers) in this chapter because the FlowLayout manager is the easiest to use, and we're trying to keep things simple for now.

By default, a program's close-window button (the X in the top-right corner) doesn't work very nicely. When the user clicks it, the window closes, but the program still runs in the background. To remedy this situation, call `setDefaultCloseOperation(EXIT_ON_CLOSE)`. Then when the user clicks the close-window button, the window closes and the program terminates. Having a closed program run in the background is usually unnoticeable, and that's why many programmers have a hard time remembering to call `setDefaultCloseOperation(EXIT_ON_CLOSE)`. Nonetheless, you should try to remember to call it. If you forget to call it, and a user's computer has limited memory and there are many programs running in the background, the computer's performance will degrade.

The `add` method adds a specified component to the current window. Once the component is added, it stays with the window for the life of the program. We mention this so that you're comfortable using a local variable declaration for a component. In the following example, although `label` is defined locally within `createContents`, the instantiated `JLabel` component stays with the window after `createContents` finishes:

```
private void createContents()
{
  JLabel label = new JLabel("Hi! I'm Larry the label!");
  add(label);
} // end createContents
```

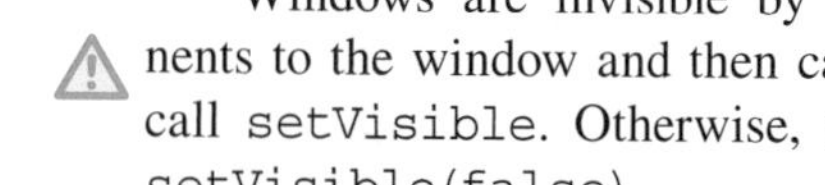

Windows are invisible by default. To make a window and its contents visible, add the components to the window and then call `setVisible(true)`. Do it in that order—add components first, then call `setVisible`. Otherwise, the added components won't display. To make a window invisible, call `setVisible(false)`.

The `JFrame` class contains many additional methods, too many to mention here. If you've got some time on your hands, we encourage you to find out what's available by looking up the `JFrame` class on Oracle's Java API website— http://download.oracle.com/javase/7/docs/api/

17.5 Java Components

Now let's consider the objects that sit inside a window—the components. Here are some examples of Java components:

- `JLabel, JTextField, JButton`
- `JTextArea, JCheckBox, JRadioButton, JComboBox`
- `JMenuBar, JMenu, JMenuItem`

These aren't all of the Java components, just some of the more commonly used ones. We'll describe the first three components in this chapter and the other components in the next chapter.

All of these component classes are in the `javax.swing` package, so you must import that package to use them. But remember that you're already importing the `javax.swing` package to access the `JFrame` class. There's no need to import it twice.

Component classes typically are derived from the `JComponent` class, which supports many useful inheritable features. Along with many other methods, the `JComponent` class contains methods that handle these component features:

- foreground and background colors
- text font

- border appearance
- tool tips
- focus

For detailed information on these features, look up the JComponent class on Oracle's Java API website.

17.6 JLabel Component

User Interface

The JLabel component doesn't do much. It simply displays a specified single line of text. It's considered to be a read-only component because the user can read it, but the user cannot interact with it.

Normally, the JLabel component displays a single line of text, not multiple lines. If you want to display multiple lines, use the JTextArea component, which is covered in the next chapter.

Implementation

To create a JLabel object, call the JLabel constructor like this:

```
JLabel <JLabel-reference> = new JLabel(<label-text>);
```

optional

The *label-text* is the text that appears in the JLabel component. If the label-text argument contains a newline character, \n, it's ignored (remember that the JLabel component displays only a single line of text). If the label-text argument is omitted, then the JLabel component displays nothing. Why instantiate an empty label? So you can fill it in later on with text that's dependent on some condition.

To add a JLabel object to your JFrame window, use this syntax:

```
add(<JLabel-reference>);
```

JLabel-reference comes from the above initialization statement.

The JLabel class needs the javax.swing package, but that should be available already because it's needed for the JFrame class.

Methods

The JLabel class, like all the GUI component classes, has quite a few methods. We'll just mention two of them—the getText and setText accessor and mutator methods. Here are their API headings and descriptions:

```
public String getText()
```

Returns the label's text.

```
public void setText(String text)
```

Assigns the label's text. Note that the programmer can update the label's text even though the user cannot.

17.7 `JTextField` Component

User Interface

The *JTextField* component displays a rectangle and allows the user to enter text into the rectangle. Here's an example:

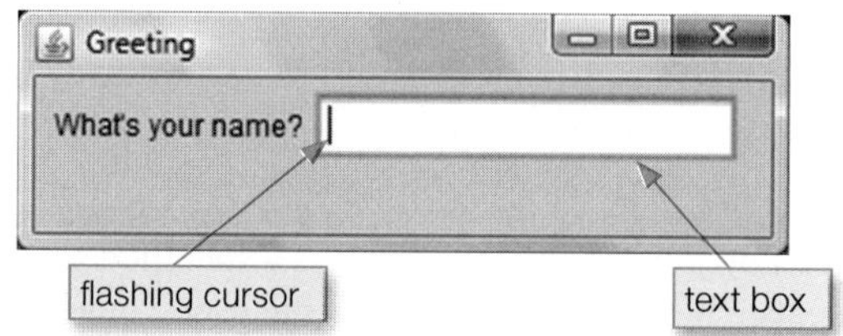

Implementation

To create a `JTextField` object, call the `JTextField` constructor like this:

```
JTextField <JTextField-reference> = new JTextField(<default-text>, <width>);
```

optional

The *default-text* is the text that appears in the text box by default. The *width* is the number of characters that can display in the text box at one time. If the user enters more characters than can display at one time, then the leftmost characters scroll off the display. If the default-text argument is omitted, then the empty string is used as the default. If the width argument is omitted, then the box's width is slightly greater than the width of the default text.

To add a `JTextField` object to your `JFrame` window, use this syntax:

```
add(<JTextField-reference>);
```

The `JTextField` class needs the `javax.swing` package, but that should be available already because it's needed for the `JFrame` class.

Methods

The `JTextField` class has quite a few methods. Here are API headings and descriptions for some of the more useful ones:

`public String getText()`
Returns the text box's contents.

`public void setText(String text)`
Assigns the text box's contents.

`public void setEditable(boolean flag)`
Makes the text box editable or non-editable.

`public void setVisible(boolean flag)`
Makes the text box visible or invisible.

```
public void addActionListener(ActionListener listener)
```

Adds a listener to the text box.

Text boxes are editable by default, which means users can type inside them. If you want to prevent users from editing a text box, call `setEditable` with an argument value of `false`. Calling `setEditable(false)` prevents users from updating a text box, but it does not prevent programmers from updating a text box. Programmers can call the `setText` method regardless of whether the text box is editable or non-editable.

Components are visible by default, but there are some instances where you might want to call `setVisible(false)` and make a component disappear. After you calculate a result, you might want just the result to appear without the clutter of other components. When a component is made to disappear, its space is automatically reclaimed by the window so other components can use it.

When a `JTextField` component calls `addActionListener`, the JVM attaches a listener object to the text box, and that enables the program to respond to the user pressing Enter within the text box. We'll cover listeners in more detail soon enough, but first we're going to step through an example program that puts into practice what you've learned so far. . . .

17.8 Greeting Program

In Figures 17.5a and 17.5b, we present a Greeting program that displays a personalized greeting. It reads the user's name from a text box (a `JTextField` component) and displays the entered name in a label (a `JLabel` component).

Most of the code in the Greeting program should look familiar because it closely parallels the code in the SimpleWindow program. For example, notice the short `main` method with the anonymous object instantiation. Also notice how the constructor contains calls to `setTitle`, `setSize`, `setLayout`, `setDefaultCloseOperation`, and `setVisible`. Finally, note the `createContents` helper method that creates the components and adds them to the window. Now let's focus on what's new about the Greeting program—a text box and an event handler.

The `Greeting` program uses a text box called `nameBox` to store the user's name. Note how the `createContents` method instantiates `nameBox` with a width of 15. Note how the `createContents` method calls the `add` method to add `nameBox` to the window. That code is straightforward. But something that's not so straightforward is `nameBox`'s declaration. It's declared as an instance variable at the top of the class. Why an instance variable instead of a `createContents` local variable? Aren't local variables preferred? Yes, but in this case, we need to access `nameBox` not only in `createContents`, but also in the `actionPerformed` event handler (which we'll get to next). It's possible to use a local variable within `createContents` and still access it from the event handler, but that's a bit of a pain.[1] For now, we'll keep things simple and declare the `nameBox` as an instance variable. We'll do the same with the `greeting` label, because we need to access it in `createContents` and also in the `actionPerformed` event handler.

The Greeting program's `actionPerformed` event handler specifies what happens when the user presses **Enter** within the text box. Note that the `actionPerformed` method is inside our `Listener` class. We cover listeners and event-handler mechanics in the next section.

[1]If you declare a variable locally within `createContents`, you can retrieve it from an event handler by calling `getSource`. The `getSource` method is covered in Section 17.14.

```
/*************************************************************
* Greeting.java
* Dean & Dean
*
* This program demonstrates text boxes and labels.
* When the user presses Enter after typing something into the
* text box, the text box value displays in the label below.
*************************************************************/

import javax.swing.*;     // for JFrame, JLabel, JTextField
import java.awt.*;        // for FlowLayout
import java.awt.event.*; // for ActionListener, ActionEvent

public class Greeting extends JFrame
{
  private static final int WIDTH = 325;
  private static final int HEIGHT = 100;
  private JTextField nameBox; // holds user's name
  private JLabel greeting;    // personalized greeting

  //**********************************************************

  public Greeting()
  {
    setTitle("Greeting");
    setSize(WIDTH, HEIGHT);
    setLayout(new FlowLayout());
    setDefaultCloseOperation(EXIT_ON_CLOSE);
    createContents();
    setVisible(true);
  } // end constructor

  //**********************************************************

  // Create components and add them to window.
  private void createContents()
  {
    JLabel namePrompt = new JLabel("What's your name?");
    nameBox = new JTextField(15);
    greeting = new JLabel();
    add(namePrompt);
    add(nameBox);
    add(greeting);
    nameBox.addActionListener(new Listener());
  } // end createContents
```

4. Import this package for event handling.

3. Register a listener.

Figure 17.5a Greeting program—part A

```
  //***************************************************************

  // Inner class for event handling.

  private class Listener implements ActionListener
  {
    public void actionPerformed(ActionEvent e)
    {
      String message; // the personalized greeting
      message = "Glad to meet you, " + nameBox.getText()+ "!";
      nameBox.setText("");
      greeting.setText(message);
    } // end actionPerformed
  } // end class Listener

  //***************************************************************

  public static void main(String[] args)
  {
    new Greeting();
  } // end main
} // end class Greeting
```

1. listener class heading

2. event handler

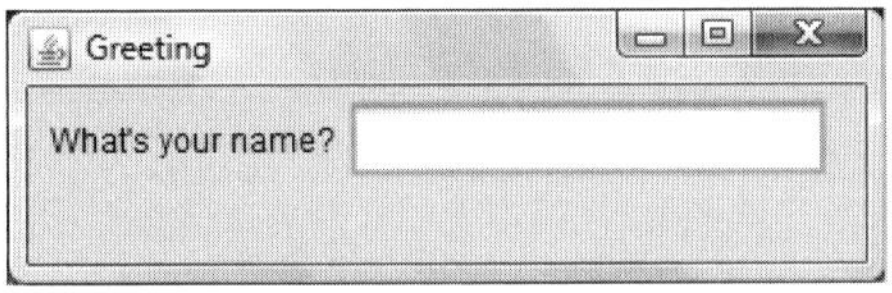

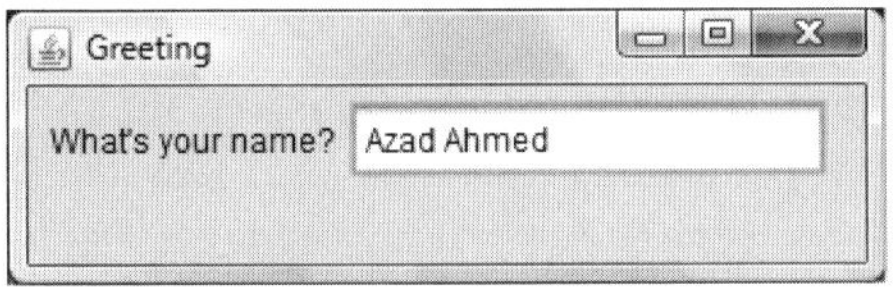

After pressing **Enter** in the text box:

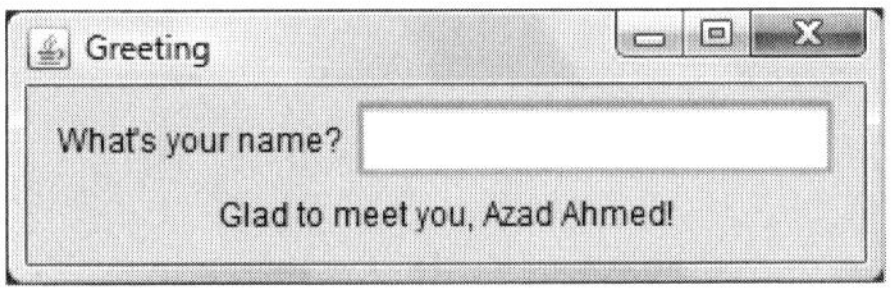

Figure 17.5b Greeting program—part B, and its associated output

17.9 Component Listeners

When the user interacts with a component (e.g., when the user clicks a button or presses **Enter** while in a text box), the component fires an event. If the component has a listener attached to it, the fired event is "heard" by the listener. Consequently, the listener handles the event by executing its `actionPerformed` method. In this section, you'll learn how to make all that work by creating a listener and an associated `actionPerformed` method.

How to Implement a Listener

Below, we show the steps needed to implement a listener for a text box. These steps correspond to the numbered callouts in Figures 17.5a and 17.5b:

1. Define a class with an `implements ActionListener` clause appended to the right of the class's heading. To see an example, look at callout 1 in Figure 17.5b. The `implements ActionListener` clause means that the class is an implementation of the `ActionListener` interface. We discuss interfaces in the next subsection.

2. Include an `actionPerformed` event handler method in your listener's class. Here's a skeleton of an `actionPerformed` method inside a listener class:

```
private class Listener implements ActionListener
{
  public void actionPerformed(ActionEvent e)
  {
    <do-something>
  }
}
```

   Even if your `actionPerformed` method doesn't use the `ActionEvent` parameter (e, above), you still must include that parameter in the method heading to make your method conform to the requirements of a listener.

   To see an example of a complete `actionPerformed` method, look at callout 2 in Figure 17.5b. It refers to a listener class that's named `Listener`. `Listener` is not a reserved word—it's just a good descriptive name we picked for the listener class in the Greeting program.

3. *Register* your listener class. More specifically, that means adding your listener class to a text box component by calling the `addActionListener` method. Here's the syntax:

```
<text-box-component>.addActionListener(new <listener-class>());
```

   To see an example, look at callout 3 in Figure 17.5a.

   The point of the registration process is so your text box can find a listener when an *enter event* is fired. An enter event is fired whenever the user presses **Enter** from within the text box.

   Registering a listener is like registering your car. When you register your car, nothing much happens at that point. But later, when some event occurs, your car registration comes into play. What event would cause your car registration to be used? If you get caught speeding, the police can use your registration number as part of a traffic citation. If you get into a wreck, your insurance company can use your registration number to raise your insurance rates.

4. Import the `java.awt.event` package. Event handling requires the use of the `ActionListener` interface and the `ActionEvent` class. Those entities are in the `java.awt.event` package, so that package must be imported for event handling to work. To see the `import` statements within a complete program, look at callout 4 in Figure 17.5a.

The `ActionListener` Interface

In the Greeting program, we specified `implements ActionListener` in the listener's class heading. `ActionListener` is an *interface*. You might recall interfaces from Chapter 14. An interface is somewhat like a class in that it contains variables and methods. But unlike a class, an interface's variables must be constants (implemented with the `final` modifier), its methods must be empty (implemented as method headings), and an interface cannot be instantiated. If a programmer uses an interface to derive a new class, the compiler requires the new class to implement methods for all of the interface's method headings.

So what's the point of having an interface with all empty methods? The answer is that it can be used as a template or pattern when creating a class that falls into a certain category. More specifically, what's the point of the `ActionListener` interface? Because all action-event listeners must implement it, it means that all action-event listeners will be similar and therefore understandable. It means that all action-event listeners will implement the `ActionListener`'s one method, the `actionPerformed` method. And in implementing that method, they'll be forced to use this prescribed heading:

```
public void actionPerformed(ActionEvent e)
```

By using the prescribed heading, it ensures that fired action events will be received properly by the listener.

17.10 Inner Classes

Here's a reprint of the Greeting program, in skeleton form:

```
public class Greeting extends JFrame
{
  ...
  private class Listener implements ActionListener
  {
    public void actionPerformed(ActionEvent e)
    {
      String message; // the personalized greeting
      message = "Glad to meet you, " + nameBox.getText();
      nameBox.setText("");
      greeting.setText(message);
    } // end actionPerformed
  } // end class Listener
  ...
} // end class Greeting
```

Do you notice anything odd about the position of the `Listener` class in the `Greeting` program? See how the `Listener` class is indented and how its closing brace is before the `Greeting` class's closing brace? The `Listener` class is inside the `Greeting` class!

If a class is limited in its scope such that it is needed by only one other class, you should define the class as an *inner class* (a class inside another class). Because a listener is usually limited to listening to just one class, listeners are usually implemented as inner classes.

It's not required by the compiler, but inner classes should normally be `private`. Why? Because the main point of using an inner class is to further the goal of encapsulation and using `private` means the outside world won't be able to access the inner class. Note the `private` modifier in the above `Listener` class heading.

Besides furthering the goal of encapsulation, there's another reason to use an inner class as opposed to a top-level class (*top-level class* is the formal term for a regular class—a class not defined inside another class). An inner class can access its enclosing class's instance variables directly. Because listeners normally need to access their enclosing class's instance variables, this is an important benefit.

17.11 Anonymous Inner Classes

Take a look at the GreetingAnonymous program in Figures 17.6a and 17.6b. It's virtually identical to the previous Greeting program. Can you identify the difference between the GreetingAnonymous program and the Greeting program?

```
/*************************************************************
* GreetingAnonymous.java
* Dean & Dean
*
* This program demonstrates an anonymous inner class.
*************************************************************/

import javax.swing.*;     // for JFrame, JLabel, JTextField
import java.awt.*;        // for FlowLayout
import java.awt.event.*; // for ActionListener, ActionEvent

public class GreetingAnonymous extends JFrame
{
  private static final int WIDTH = 325;
  private static final int HEIGHT = 100;
  private JTextField nameBox; // holds user's name
  private JLabel greeting;    // personalized greeting

  //**********************************************************

  public GreetingAnonymous()
  {
    setTitle("Greeting Anonymous");
    setSize(WIDTH, HEIGHT);
    setLayout(new FlowLayout());
    setDefaultCloseOperation(EXIT_ON_CLOSE);
    createContents();
    setVisible(true);
  } // end constructor
```

Figure 17.6a GreetingAnonymous program, which has an anonymous inner class—part A

```
  //**********************************************************

  // Create components and add them to window.

  private void createContents()
  {
    JLabel namePrompt = new JLabel("What's your name?");
    nameBox = new JTextField(15);
    greeting = new JLabel();
    add(namePrompt);
    add(nameBox);
    add(greeting);
    nameBox.addActionListener(

      // anonymous inner class for event handling
      new ActionListener()
      {
        public void actionPerformed(ActionEvent e)
        {
          String message; // the personalized greeting
          message = "Glad to meet you, " + nameBox.getText();
          nameBox.setText("");
          greeting.setText(message);
        } // end actionPerformed
      } // end anonymous inner class
    ); // end addActionListener call
  } // end createContents

  //**********************************************************

  public static void main(String[] args)
  {
    new GreetingAnonymous();
  } // end main
} // end class GreetingAnonymous
```

Figure 17.6b GreetingAnonymous program, which has an anonymous inner class—part B

In the Greeting program, we implemented a listener class named `Listener`, using this code:

```
private class Listener implements ActionListener
{
```

That code is omitted in the GreetingAnonymous program—there's no class named `Listener`. But we still need a listener object so that the text box's enter event is detected and acted upon. This time, instead of declaring a listener class with a name (e.g., `Listener`), we implement a listener class anonymously (without a name).

We've discussed anonymous objects previously. That's where you instantiate an object without storing its reference in a variable. In our previous Greeting program, we instantiated an anonymous `Listener` object with this line:

```
nameBox.addActionListener(new Listener());
```

The point of using an anonymous object is to avoid cluttering the code with a variable name when an object needs to be used only one time. The same idea can be applied to classes. The point of using an *anonymous inner class* is to avoid cluttering up the code with a class name when a class needs to be used only one time. For example, if a particular listener class listens to just one object, then the listener class needs to be used only one time as part of an addActionListener method call. Therefore, to unclutter your code, you may want to use an anonymous inner class for the listener.

Using an anonymous inner class is not a compiler requirement. It's an elegance issue. In industry, you'll find some people who say anonymous inner classes are elegant and you'll find other people who say anonymous inner classes are confusing. Do as you see fit. Better yet, do as your teacher sees fit.

Below, we show the syntax for an anonymous inner class. Naturally, there's no class name. But there is an interface name. So anonymous inner classes aren't built from scratch; they're built with the help of an interface.[2] Note the new operator. Formally speaking, the new operator isn't part of the anonymous inner class. But practically speaking, because there's no point in having an anonymous inner class without instantiating it, you can think of the new operator as being part of the anonymous inner class syntax.

```
new <interface-name> ()
{
  <class-body>
}
```

Here's an example of an anonymous inner class, taken from the GreetingAnonymous program:

```
nameBox.addActionListener(
  new ActionListener()
  {                              ActionListener is an interface.
    public void actionPerformed(ActionEvent e)
    {
      ...
    } // end actionPerformed
  } // end inner-class constructor
);
```

For comparison purposes, here's an example of a named (non-anonymous) inner class. It's taken from the Greeting program:

```
private void createContents()
{
  ...
  nameBox.addActionListener(new Listener());
} // end createContents

private class Listener implements ActionListener
{
  public void actionPerformed(ActionEvent e)
  {
    ...
  } // end actionPerformed
} // end class Listener
```

[2]As an alternative, it's legal to define an anonymous class with a superclass instead of an interface. The details of doing this are beyond the scope of this textbook.

There are only two syntactic differences between the two code fragments—the addActionListener call and the listener class heading. There are no semantic differences between the two code fragments, so the Greeting program and the GreetingAnonymous program behave the same.

17.12 JButton Component

It's now time to learn another GUI component—a button component.

User Interface

If you press a button on an electronic device, something usually happens. For example, if you press the power button on a TV's remote control, the TV turns on or off. Likewise, if you press/click a GUI *button* component, something usually happens. For example, in Figure 17.1's TrustyCredit window, if you click the OK button, the entered credit card numbers get processed by the TrustyCredit company.

Implementation

To create a button component, call the *JButton* constructor like this:

```
JButton helloButton = new JButton("Press me");
```

button label's text

When this button is displayed, it says "Press me" in the center of the button. The label argument is optional. If it's omitted, the label gets the empty string by default and the button displays with a blank face (no writing or icons on it).

After you have created the helloButton, add it to your window, like this:

```
add(helloButton);
```

To make the button useful, you'll need to implement a listener. As with the text box listeners, button listeners must implement the ActionListener interface. The ActionListener interface dictates that you must have an actionPerformed event handler method. The code skeleton looks like this:

```
private class Listener implements ActionListener
{
  public void actionPerformed(ActionEvent e)
  {
    <do-something>
  }
}
```

We're using private instead of public for the listener class because a listener normally is implemented as an inner class, and inner classes are normally private. We're using a named inner class instead of an anonymous inner class because named inner classes are slightly more flexible. They allow you to create a listener that's used on more than one component. We'll provide an example in an upcoming program.

To register the above listener with our helloButton component, do this:

```
helloButton.addActionListener(new Listener());
```

The JButton class needs the javax.swing package, but that should be available already because it's needed for the JFrame class. The ActionListener interface and the ActionEvent class need the java.awt.event package, so import that package.

Methods

Here are API headings and descriptions for some of the more useful `JButton` methods:

`public String getText()`

Returns the button's label.

`public void setText(String text)`

Assigns the button's label.

`public void setVisible(boolean flag)`

Makes the button visible or invisible.

`public void addActionListener(ActionListener listener)`

Adds a listener to the button. The listener "listens" for the button being clicked.

FactorialButton Program

It's time to put all this `JButton` syntax into practice by showing you how it's used within a complete program. We've written a FactorialButton program that uses a `JButton` component to calculate the factorial for a user-entered number.[3] To give you a better idea of how the program operates, see the sample session in Figure 17.7.

Figures 17.8a and 17.8b contain the FactorialButton program listing. Most of the code should already make sense because the program's structure parallels the structure in our previous GUI programs. We'll skip the more familiar code and focus on the more difficult code.

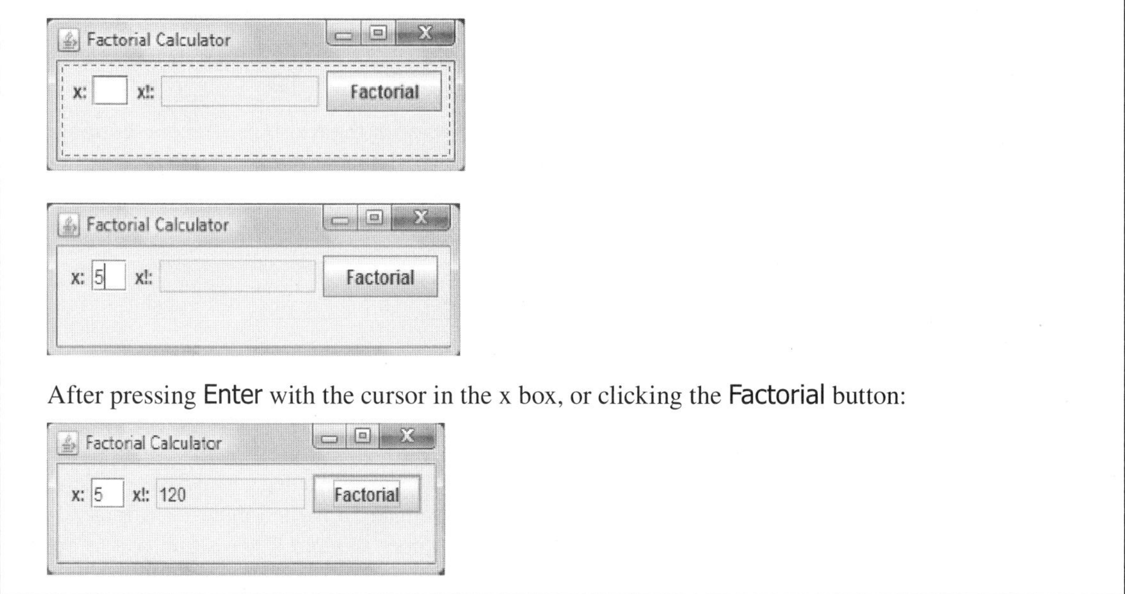

Figure 17.7 Sample session for the FactorialButton program

[3]The factorial of a number is the product of all positive integers less than or equal to that number. The factorial of n is written as n! Example: The factorial of 4 is written as 4!, and 4! is equal to 24 because 1 times 2 times 3 times 4 equals 24.

```
/*************************************************************
* FactorialButton.java
* Dean & Dean
*
* When user clicks button or presses Enter with cursor in input
* text box, entered number's factorial displays in output text box.
*************************************************************/

import javax.swing.*;
import java.awt.*;
import java.awt.event.*;

public class FactorialButton extends JFrame
{
  private static final int WIDTH = 300;
  private static final int HEIGHT = 100;
  private JTextField xBox;  // holds user entry
  private JTextField xfBox; // holds generated factorial

  //**********************************************************

  public FactorialButton()
  {
    setTitle("Factorial Calculator");
    setSize(WIDTH, HEIGHT);
    setLayout(new FlowLayout());
    setDefaultCloseOperation(EXIT_ON_CLOSE);
    createContents();
    setVisible(true);
  } // end FactorialButton constructor

  //**********************************************************

  private void createContents()
  {
    JLabel xLabel = new JLabel("x:");
    JLabel xfLabel = new JLabel("x!:");
    JButton btn = new JButton("Factorial");
    Listener listener = new Listener();

    xBox = new JTextField(2);
    xfBox = new JTextField(10);
    xfBox.setEditable(false);
    add(xLabel);
    add(xBox);
    add(xfLabel);
    add(xfBox);
    add(btn);
    xBox.addActionListener(listener);
    btn.addActionListener(listener);
  } // end createContents
```

Here, we register the same listener with two different components.

Figure 17.8a FactorialButton program—part A

```
  //***************************************************

  // Inner class for event handling.

  private class Listener implements ActionListener
  {
    public void actionPerformed(ActionEvent e)
    {
      int x;          // numeric value for user-entered x
      int xf;         // x factorial

      try
      {
        x = Integer.parseInt(xBox.getText());
      }
      catch (NumberFormatException nfe)
      {
        x = -1;   // indicates an invalid x
      }

      if (x < 0)
      {
        xfBox.setText("undefined");
      }
      else
      {
        if (x == 0 || x == 1)
        {
          xf = 1;
        }
        else
        {
          xf = 1;
          for (int i=2; i<=x; i++)
          {
            xf *= i;
          }
        } // end else

        xfBox.setText(Integer.toString(xf));
      } // end else
    } // end actionPerformed
  } // end class Listener

  //***************************************************

  public static void main(String[] args)
  {
    new FactorialButton();
  } // end main
} // end class FactorialButton
```

Convert the user-entered number from a string to a number.

Figure 17.8b FactorialButton program—part B

We declare most of our GUI variables locally within createContents, but we declare the two text box components as instance variables at the top of the program. Why the difference? As discussed earlier, normally you should declare components as local variables to help with encapsulation. But if a component is needed in createContents and also in an event handler, it's fine to declare it as an instance variable where it can be shared more easily. In the FactorialButton program, we declare the two text boxes as instance variables because we need to use them in createContents and also in the actionPerformed event handler.

Note this line from the createContents method:

```
xfBox.setEditable(false);
```

This causes the factorial text box, xfBox, to be non-editable (i.e., the user won't be able to update the text box). That should make sense because xfBox holds the factorial, and it's up to the program (not the user) to generate the factorial. Note in Figure 17.7 that the factorial text box is grayed out. You get that visual cue free of charge whenever you call setEditable(false) from a text box component. Cool!

Again from the createContents method:

```
Listener listener = new Listener();
...
xBox.addActionListener(listener);
btn.addActionListener(listener);
```

Note that we're registering the same listener with two different components. By doing this, we give the user two ways to trigger a response. The user can press Enter when the cursor is in the input text box (xBox) or the user can click on the button (btn). Either way causes the listener to react. Whenever you register the same listener with two different components, you need to have a name for the listener. That's why we use a named inner class for this program (an anonymous inner class wouldn't work).

Figure 17.8b's actionPerformed method is chock full of interesting code. The first thing you see is perhaps the most important—the Integer.parseInt method call. If you ever need to read numbers or display numbers in a GUI program, you have to use string versions of the numbers. Thus, to read a number from the input text box, we first read it in as a string, and then we convert the string to a number. To accomplish this, we read the string using xBox.getText(), and we convert it to a number using Integer.parseInt.

Ideally, you should always check user input to make sure it's valid. In the actionPerformed method, we check for two types of invalid input—a non-integer input and a negative number input. Those inputs are invalid because the factorial is mathematically undefined for those cases. The negative number case is easier, so we'll start with it. Note this code in the middle of the ActionPerformed method:

```
if (x < 0)
{
  xfBox.setText("undefined");
}
```

x is the user's entry after it's been converted to an integer. If x is negative, the program displays undefined in the xfBox component.

Now for the non-integer input case. Note this code near the top of the `ActionPerformed` method:

```
try
{
  x = Integer.parseInt(xBox.getText());
}
catch (NumberFormatException nfe)
{
  x = -1; // indicates an invalid x
}
```

The `Integer.parseInt` method attempts to convert `xBox`'s user-entered value to an integer. If `xBox`'s user-entered value is a non-integer, then `parseInt` throws a `NumberFormatException`. To handle that possibility, we put the `Integer.parseInt` method call inside a `try` block, and we include an associated `catch` block. If `parseInt` throws an exception, we want to display `undefined` in the `xfBox` component. To do that, we could call `xfBox.setText("undefined")` in the `catch` block, but then we'd have redundant code—`xfBox.setText("undefined")` in the catch block and also in the subsequent if statement. To avoid code redundancy and its inherent maintenance problems, we assign −1 to `x` in the `catch` block. That causes the subsequent `if` statement to be `true` and that in turn causes `xfBox.setText("undefined")` to be called.

After validating the input, the `actionPerformed` method calculates the factorial. It first takes care of the special case when `x` equals 0 or 1. It then takes care of the `x` ≥ 2 case by using a `for` loop. Study the code. It works fine, but do you see a way to make it more compact? You can omit the block of code that starts with `if (x == 0 || x == 1)` because that case is handled by the `else` block. More specifically, you can delete the six lines above the second `xf = 1;` line.

Write compact code.

17.13 Dialog Boxes and the JOptionPane Class

A *dialog box*—often referred to simply as a *dialog*—is a specialized type of window. The primary difference between a dialog box and a standard window is that a dialog box is more constrained in terms of what it can do. While a standard window usually remains on the user's screen for quite a while (often for the duration of the program) and performs many tasks, a dialog box remains on the screen only long enough to perform one specific task. While a standard window is highly customizable, a dialog box typically is locked into one particular format.

User Interface

There are three types of `JOptionPane` dialogs—a *message dialog,* an *input dialog,* and a *confirmation dialog*. Each type performs one specific task. The message dialog displays output. The input dialog displays a question and an input field. The confirmation dialog displays a yes/no question and yes/no/cancel button options. See what the different types look like in Figure 17.9. The optional GUI section at the end of Chapter 3 used the first two of these types, the message dialog and the input dialog.

Implementation

To create a message dialog box, call the `showMessageDialog` method like this:

```
JOptionPane.showMessageDialog(<container>, <message>);
```

Because `showMessageDialog` is a class method in the `JOptionPane` class, you need to prefix the `showMessageDialog` call with "`JOptionPane` dot". Remember—call instance methods using "*<reference-variable>* dot" syntax, and call class methods using "*<class-name>* dot" syntax.

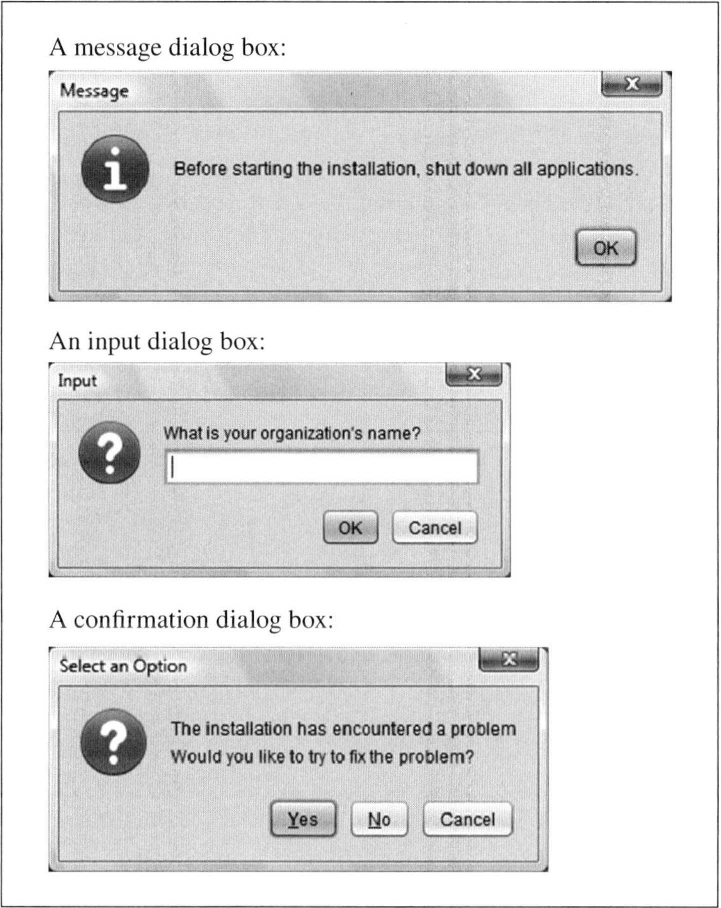

Figure 17.9 Three types of `JOptionPane` dialog boxes

To create an input dialog box, call the `showInputDialog` method like this:

```
String input = JOptionPane.showInputDialog(<container>, <message>);
```

The returned `String` value, `input`, is the value the user enters in the text field. To create a confirmation dialog box, call the `showConfirmDialog` method like this:

```
int selection = JOptionPane.showConfirmDialog(<container>, <message>);
```

The returned `int` value, `selection`, is one of `JOptionPane`'s `static` constants. If the user clicks Yes, the returned value is `JOptionPane.YES_OPTION`. If the user clicks No, the returned value is `JOptionPane.NO_OPTION`. If the user clicks Cancel, the returned value is `JOptionPane.CANCEL_OPTION`. You can use these `JOptionPane` constants as `case` values in a `switch` statement.

As indicated in Java API documentation, the `JOptionPane` class also provides other methods and options. Sometimes, as in the GUI program at the end of Chapter 3, we just want the dialog box to pop out all by itself—not within any container. For that, we make the *<container>* argument `null`. At other times, we want the dialog to appear inside a larger window. For that, we make the *<container>* argument be a `JFrame` object. In all cases, the *<message>* argument is the dialog box's text.

Note the showMessageDialog call in Figure 17.10's HelloWithAFrame program. This call uses helloFrame for showMessageDialog's container argument. What type of container is it? As the code indicates, helloFrame is an instance of the HelloWithAFrame class, and the HelloWithAFrame class extends the JFrame container. Therefore, by inheritance, helloFrame is a JFrame container. And consequently, the dialog box displays in the center of the program's JFrame container. Verify this by looking at Figure 17.10's output.

Suppose you don't want to bother with centering the dialog within a particular container. In that case, use null for showMessageDialog's container argument. That causes the dialog to display in the center of the screen. For example, this code fragment generates a screen-centered dialog:

```
JOptionPane.showMessageDialog(
  null, "Before starting the installation,\n" +
  "shut down all applications.");
```

```
import javax.swing.*;
public class HelloWithAFrame extends JFrame
{
  public HelloWithAFrame()
  {
    setTitle("Hello");
    setSize(400, 200);
    setDefaultCloseOperation(EXIT_ON_CLOSE);
    setVisible(true);
  } // end HelloWithAFrame constructor

  //***********************************************************

  public static void main(String[] args)
  {
    HelloWithAFrame helloFrame = new HelloWithAFrame();
    JOptionPane.showMessageDialog(helloFrame, "Hello, world!");
  } // end main
} // end class HelloWithAFrame
```

Hello

Message

Hello, world!

OK

Figure 17.10 HelloWithAFrame program and its output

By the way, it's very common to use `null` for `showMessageDialog`'s container argument, probably more common than using a non-`null` value.

The `JOptionPane` class needs the `javax.swing` package. If you've imported the `javax.swing` package for the `JFrame` class already, there's no need to import it again.

Method Details

In Figure 17.10, note the message dialog's title-bar message—it's "Message." Kinda boring, eh? To liven things up, add a third argument to the `showMessageDialog` call that specifies the dialog's title. Also in Figure 17.10, note the message dialog's icon—it's an **i** inside a circle. That's the default icon. To specify an icon explicitly, add a fourth argument to the `showMessageDialog` call that specifies one of the named constants in Figure 17.11.

JOptionPane Named Constants (for Specifying a Dialog Box's Icon)	Icon	When to Use
INFORMATION_MESSAGE		For a dialog box that provides informational text.
WARNING_ MESSAGE		For a dialog box that warns the user about a problem.
ERROR_MESSAGE		For a dialog box that warns the user about an error. Normally, an error is considered to be more serious than a warning.
QUESTION_MESSAGE		For a dialog box that asks the user a question. Normally, the question-mark icon is used with a confirm dialog box or an input dialog box. But it is legal to use it with a message dialog box as well.
PLAIN_MESSAGE	no icon	For a plain-looking dialog box. The dialog box contains a message, but no icon.

Figure 17.11 Icon options within a `JOptionPane` dialog

Here's how to call the four-parameter version of `showMessageDialog`:

```
JOptionPane.showMessageDialog(
  <null-or-container>, <message>, <title>, <icon_constant>);
```

Here's an example four-argument `showMessageDialog` call and the resulting dialog:

```
JOptionPane.showMessageDialog(null, "A virus has been detected.", "Warning",
  JOptionPane.WARNING_MESSAGE);
```

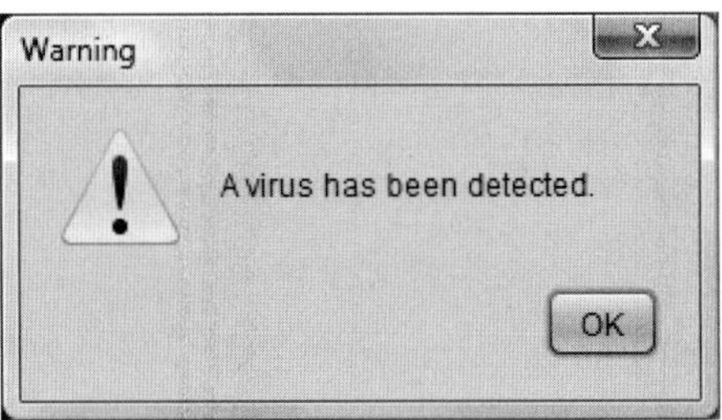

17.14 Distinguishing Between Multiple Events

Now that you understand the basic building blocks of GUI programming (`JFrame` and `JOptionPane` windows; `JLabel`, `JTextField`, and `JButton` components), you're prepared to consider more complex situations that GUI programmers encounter. In this section, you'll learn how to use a single listener to distinguish between two different component events.

The `getSource` Method

Suppose that you register a listener with two components. When the listener hears an event, you'll probably want to determine which component fired the event. That way, you can customize your event handling: Do one thing if component X fired the event, and do another thing if component Y fired the event.

From within a listener, how can you determine the source of an event? In other words, how can you identify the component that fired an event? Call `getSource`, of course! More specifically, within the `actionPerformed` method, use the `actionPerformed` method's `ActionEvent` parameter to call `getSource`. The `getSource` method returns a reference to the component whose event was fired. To see which component that was, use == to compare the returned value with the components in question. For example, in the below code fragment, we compare the returned value to a button component named `okButton`.

```
public void actionPerformed(ActionEvent e)
{
  if (e.getSource() == okButton)
  {
    ...
```

Improved FactorialButton Program

Remember the FactorialButton program from Figure 17.8? It calculated the factorial of a user-entered number. The calculations were triggered by the user clicking the factorial button or the user pressing **Enter** in the input text box. With our simple first-cut FactorialButton program, we didn't bother to distinguish between the button-click event and the text-box-enter event. Let's now improve the program by having the different events trigger different results. The button click will still display the factorial, but the text box enter will display this dialog-box message:

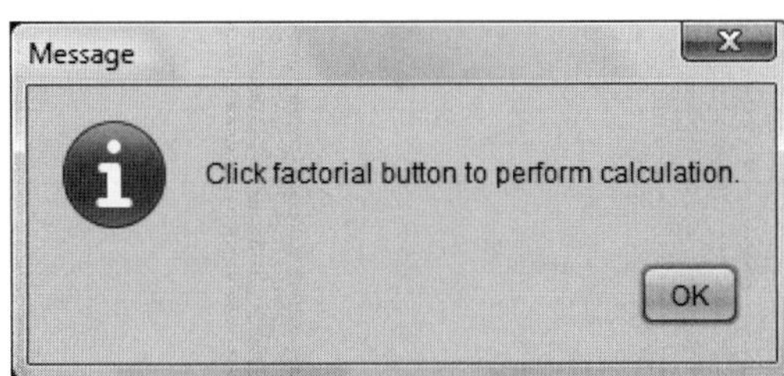

See Figure 17.12. It shows the `Listener` class for our new and improved FactorialButton program. We're showing only the `Listener` class because the rest of the program hasn't changed. If you want to see the rest of the program, refer back to Figure 17.8. In our new `Listener` class, note how we call `getSource` and compare its returned value to `xBox`. `xBox` is the text box component that holds the

```
private class Listener implements ActionListener
{
  public void actionPerformed(ActionEvent e)
  {
    int x;   // numeric value for user entered x
    int xf;  // x factorial

    if (e.getSource() == xBox)
    {
      JOptionPane.showMessageDialog(null,
        "Click factorial button to perform calculation.");
    }

    else // the button must have been clicked
    {
      try
      {
        x = Integer.parseInt(xBox.getText());
      }
      catch (NumberFormatException nfe)
      {
        x = -1;   // indicates an invalid x
      }

      if (x < 0)
      {
        xfBox.setText("undefined");
      }
      else
      {
        if (x == 0 || x == 1)
        {
          xf = 1;
        }
        else
        {
          xf = 1;
          for (int i=2; i<=x; i++)
          {
            xf *= i;
          }
        } // end else

        xfBox.setText(Integer.toString(xf));
      } // end else
    } // end else button was clicked
  } // end actionPerformed
} // end class Listener
```

This is the input text box.

Figure 17.12 Modified `Listener` class for the FactorialButton program

user's entry for x. If getSource returns xBox, we call showMessageDialog and display the above dialog message.

17.15 Using getActionCommand to Distinguish Between Multiple Events

In this section, we continue our discussion of distinguishing between multiple events. But instead of calling getSource, this time we call getActionCommand.

getSource Is Somewhat Limited

In Figure 17.12's Listener class, we call getSource to identify the component whose event was fired. That works fine most of the time, but not always. Note the following cases where calling getSource is inadequate:

1. If the event-firing components are in a different class from the listener class.

   A listener class's getSource method can retrieve the component responsible for the fired event, but there is no way to identify the type of the returned component because that requires comparing the returned component with the original components (using ==). If the original components are in a different class and private, using them in the listener class generates a compile-time error.

2. If there's a need to have a *modal* component.

   A modal component is a component with more than one state or status. For example, suppose there's a button whose label toggles between "Show Details" and "Hide Details." The two labels correspond to two different modes of operation—in one mode details are shown, and in another mode details are hidden. If a modal button is clicked, getSource can retrieve the button, but it cannot retrieve the button's mode. In the show details/hide details example, getSource cannot directly determine whether the button's mode is show details or hide details.

getActionCommand to the Rescue

If you need to identify an event from within a listener and getSource is inadequate, you have another alternative—the getActionCommand. The getActionCommand method returns the "action command" associated with the component whose event was fired. Typically, the action command is the component's label. For example, the default action command for a button is the button's label.

Let's revisit the case where a button's label toggles between "Show Details" and "Hide Details." In the following code fragment, assume that instructions is a label component, detailedInstructions and briefInstructions are string local variables, and btn is the "Show Details/Hide Details" button. Note how getActionCommand determines the button's mode by retrieving the button's label:

```
public void actionPerformed(ActionEvent e)
{
  if (e.getActionCommand().equals("Show Details"))
  {
    instructions.setText(detailedInstructions);
    btn.setText("HideDetails");
  }
```

```
    else
    {
      instructions.setText(briefInstructions);
      btn.setText("ShowDetails");
    }
  } // end actionPerformed
```

17.16 Color

So far in this chapter, all of our components have been simple in terms of color—black text on white background or black text on light-gray background. It's time to add some color. You should get used to adding color to most of your GUI applications. After all, color can enhance a user's experience with a program by providing visual cues and visual appeal. Remember, color is fun!

Color Methods

Most GUI components are composed of two colors. The *foreground color* is the color of the text, and the *background color* is the color of the area behind the text. Let's jump right into an example that shows you how to set the colors. This code fragment creates a dark-green button with white text:

```
JButton btn = new JButton("Click Me");
btn.setBackground(Color.GREEN.darker());
btn.setForeground(Color.WHITE);
```

And here's what the dark-green button with white text looks like:

Click Me

The `setBackground` and `setForeground` methods are mutator methods. Here are the API headings and descriptions for their associated accessor methods:

`public Color getBackground()`

Returns the component's background color.

`public Color getForeground()`

Returns the component's foreground color.

Here's an example that uses the `getBackground` and `getForeground` methods with a text box:

```
JTextField nameBox = new JTextField();
Color originalBackground = nameBox.getBackground();
Color originalForeground = nameBox.getForeground();
```

Why might you want to save a text box's original colors? As a visual cue, you might want to change a text box's colors when the user enters something invalid. And when the user fixes the entry, you'd change back to the original colors. In order to do that, you need to retrieve and save the original colors when the window is first loaded.

You've now seen color examples with a button and a text box. Color works the same way for most other components. An exception is the `JLabel` component. Its background is transparent by default, so if you apply color to it, you won't see the color. To change a label's background color, you first have to make it opaque by having the label reference variable call `setOpaque(true)`. After that, if you call `setBackground(`*<color>*`)`, you'll see the specified color.

`Color` Named Constants

Let's now talk about color values. You can specify color values with named constants or with instantiated `Color` objects. We'll start with named constants.

The `Color` class defines this set of named constants:

```
Color.BLACK        Color.GREEN         Color.RED
Color.BLUE         Color.LIGHT_GRAY    Color.WHITE
Color.CYAN         Color.MAGENTA       Color.YELLOW
Color.DARK_GRAY    Color.ORANGE
Color.GRAY         Color.PINK
```

These named constants are class members, so we access them using *<class name>* dot syntax, where the class name is `Color`. For darker or brighter shades, chain `darker()` or `brighter()` method calls.

The `Color` class is in the `java.awt` package, so don't forget to import that package when working with colors.

`Color` Objects

To obtain a color that is not in the `Color` class's list of named constant colors, instantiate a `Color` object with a specified mixture of red, green, and blue. Here's the `Color` constructor call syntax:

```
new Color(<red 0–255>, <green 0–255>, <blue 0–255>)
```

Each of the three `Color` constructor arguments is an `int` value between 0 and 255. The `int` value represents an amount of color, with 0 indicating no color and 255 indicating the maximum amount of color. For example, this line sets a button's background color to a dark magenta:

```
button.setBackground(new Color(128, 0, 128));
```

The instantiated `Color` object uses half the maximum for red (128), no green (0), and half the maximum for blue (128). For the brightest magenta, increase the red and blue values from 128 to 255.

White light is the combination of all colors,[4] so `new Color(255, 255, 255)` produces white. Black is the absence of all colors, so `new Color(0, 0, 0)` produces black.

This technique of creating a color by mixing specified amounts of red, green, and blue is used by many programming languages. The red, green, blue 3-tuple is commonly referred to as an *RGB value*. When coming up with RGB values for your programs, it's perfectly acceptable to use trial and error, but to save time, you may want to visit an RGB color table online. For example—http://web.njit.edu/~kevin/rgb.txt.html.

`JFrame` Background Color

Setting the background color for a `JFrame` window is slightly trickier than setting it for a component. First, you have to get the `JFrame`'s *content pane*, and then you have to apply the background color to it. As shown below, the content pane is the inner part of the `JFrame`.

[4] In 1666, Isaac Newton discovered that white light is composed of all the colors of the color spectrum. He showed that when white light is passed through a triangular prism, it separates into different colors. And when the resulting colors are passed through a second triangular prism, they are brought back together to form the original white light.

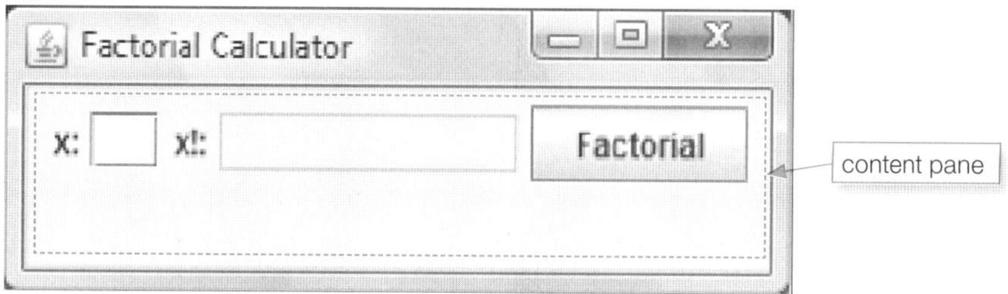

While the `JFrame` class handles perimeter features such as window dimensions, the title bar, and the close-out button, the content pane handles interior features such as components, layout, and background color. So when you add components, set the layout, and set the background color, you do it to the content pane, not the `JFrame`. These three statements illustrate what we're talking about:

```
getContentPane().add(btn);
getContentPane().setLayout(new FlowLayout());
getContentPane().setBackground(Color.YELLOW);
```

In versions of Java prior to Java 5.0, `JFrame`'s `getContentPane` method was required for all three tasks—adding a component, setting the layout, and setting the window's background color. With the advent of Java 5.0, the Java language designers made things easier for the first two tasks. Now, if you want to add a component or set the layout, as an option, you may omit the call to `getContentPane`. In other words, this works:

```
add(btn);
setLayout(new FlowLayout());
```

The reason this code works is that with the current version of Java, `JFrame`'s `add` and `setLayout` methods automatically get the content pane behind the scenes. And the retrieved content pane is used for the ensuing `add` and `setLayout` operations. So which is better—`getContentPane().add(btn)` or just `add(btn)`? They are functionally equivalent, but the second one is generally preferred because it's less cluttered. Ditto for the `setLayout` method call.

For setting the window's background color, the current version of Java still requires that you call `getContentPane` before calling `setBackground`. If you call `setBackground` without calling `getContentPane`, it sets the `JFrame`'s background color, not the content pane's background color. And because the content pane sits on top of the `JFrame`, the `JFrame`'s color is covered up and not seen.

Now you know that setting a window's background color requires `getContentPane`. Similarly, getting a window's background color requires `getContentPane`. For example:

```
Color saveColor = getContentPane().getBackground();
```

ColorChooser Program

Let's put what you've learned about color into practice by using it within a complete program. In our ColorChooser program, we implement light gray and green buttons that set the window's background color to gray or dark green, respectively. See Figure 17.13 to get an idea of what we're talking about.

See the ColorChooser program listing in Figures 17.14a, 17.14b, and 17.14c. Most of the code should already make sense because its structure mirrors the structure in our previous GUI programs. We'll focus on the new code—the color code.

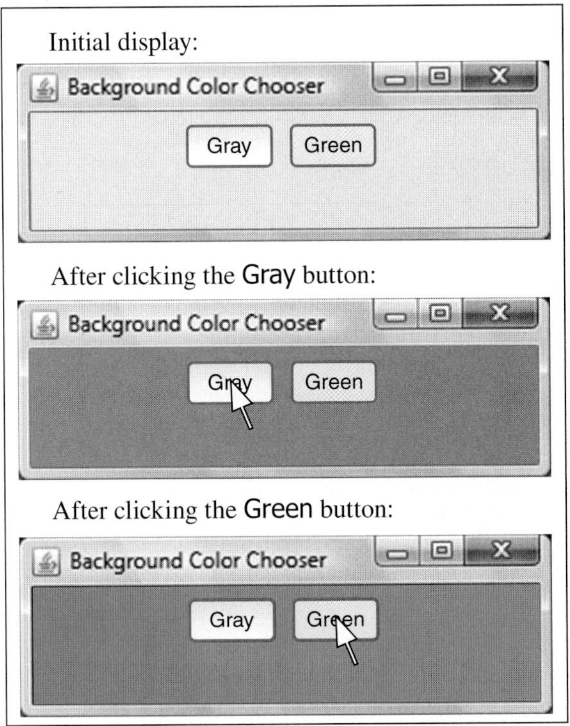

Figure 17.13 Sample session for the ColorChooser program

```
/***********************************************************
* ColorChooser.java
* Dean & Dean
*
* This program's buttons allow the user to set the window's
* background color to gray or green.
***********************************************************/

import javax.swing.*;      // for JFrame & JButton
import java.awt.*;         // for FlowLayout, Color, & Container
import java.awt.event.*;   // for ActionListener & ActionEvent

public class ColorChooser extends JFrame
{
  private static final int WIDTH = 330;
  private static final int HEIGHT = 100;

  private JButton grayButton;  // changes background to gray
  private JButton greenButton; // changes background to green
```

Figure 17.14a ColorChooser program—part A

```
//***************************************************************

public ColorChooser()
{
  setTitle("Background Color Chooser");
  setSize(WIDTH, HEIGHT);
  setLayout(new FlowLayout());
  setDefaultCloseOperation(EXIT_ON_CLOSE);
  createContents();
  setVisible(true);
} // end ColorChooser constructor

//***************************************************************

private void createContents()
{
  grayButton = new JButton("Gray");
  grayButton.setBackground(Color.LIGHT_GRAY);
  grayButton.addActionListener(new ButtonListener());
  add(grayButton);

  greenButton = new JButton("Green");
  greenButton.setBackground(Color.GREEN);
  greenButton.addActionListener(new ButtonListener());
  add(greenButton);
} // end createContents

//*****************************************************************

// Inner class for event handling.

private class ButtonListener implements ActionListener
{
  public void actionPerformed(ActionEvent e)
  {
    Container contentPane = getContentPane();
    if (e.getSource() == grayButton)
    {
      // Change the window background color to gray.
      contentPane.setBackground(Color.GRAY);
    }
    else
    {
      // Change the window background color to dark green.
      contentPane.setBackground(Color.GREEN.darker());
    }
  } // end actionPerformed
} // end class ButtonListener
```

This sets the Gray button's color.

This sets the Green button's color.

These lines change the window's background color.

Figure 17.14b ColorChooser program—part B

```
  //*************************************************************

  public static void main(String[] args)
  {
    new ColorChooser();
  }
} // end class ColorChooser
```

Figure 17.14c ColorChooser program—part C

Note the difference between the `setBackground` calls in the `createContents` method and the `setBackground` calls in the `actionPerformed` method. In `createContents`, we're dealing with the gray and green button components, so it is not necessary to call `getContentPane` prior to calling `setBackground`. In `actionPerformed`, we're dealing with the `JFrame` window, so it is necessary to call `getContentPane` prior to calling `setBackground`.

Finally, note that instead of using a `color` constant, in the `createContents` method we could generate the green for the button's background by creating a new `color` object, like this:

```
greenButton.setBackground(new Color(0, 255, 0));
```

And in the `actionPerformed` method we could generate the darker green color for the window's background by creating a new `Color` object, like this:

```
ContentPane.setBackground(new Color(0, 192, 0));
```

17.17 How GUI Classes Are Grouped Together

Throughout this chapter, you've used Java's prebuilt GUI classes from Oracle's Java API library. For example, you used the `JFrame` class for creating a window, the `JButton` class for creating a button, and the `Color` class for creating a color. In this section, we describe how Java API prebuilt GUI classes are grouped and organized.

Subpackages

The Java API is a huge class library that adds functionality to the core Java language. To simplify things, the classes are organized into a hierarchy of packages where each package contains a group of classes. To avoid having too many classes in one package, packages are often split into *subpackages*. A subpackage is a group of classes from within a larger group of classes. For example, rather than putting all the GUI classes (and there are lots of them!) within the `java.awt` package, Java language designers split off the GUI event-handling classes and put them in their own subpackage, `java.awt.event`. To import all the classes in the `java.awt` package and the `java.awt.event` subpackage, do this:

```
import java.awt.*;
import java.awt.event.*;
```

Because the `java.awt.event` subpackage contains `java.awt` in its name, is it OK to omit the `java.awt.event import` statement and do just the following?

```
import java.awt.*;
```

This imports classes in the `java.awt` package only.

No—you must import `java.awt` and `java.awt.event` separately. Think of the `java.awt` package and the `java.awt.event` subpackage as completely separate entities. The fact that they share the common name "java.awt" is irrelevant as far as the compiler is concerned. The compiler treats them as separate packages. So then why the shared name? The shared name helps programmers remember that the classes in `java.awt.event` are conceptually related to the classes in `java.awt`.

We've been referring to `java.awt.event` as a "subpackage". It's just as common to refer to it as a "package". We'll use both terms because both are valid.

The AWT and Swing Libraries

In Sun's first Java compiler, all GUI classes were bundled into one library known as the Abstract Windowing Toolkit (AWT). The AWT's GUI commands generate GUI components that look different on different platforms. In other words, if your program instantiates an AWT button component, the button will have a Macintosh look and feel if the program is run on a Macintosh computer, but a Windows look and feel if the program is run on a Windows computer.[5] That leads to portability issues. Your programs are still portable in the sense that they'll run on different platforms. But they'll run differently on different platforms. If you have a persnickety customer who demands one precise appearance on all platforms, then AWT components probably won't be satisfactory.

One of Java's strongest selling points was (and is) its portability, so soon after Java's initial release, the Java language designers proceeded to develop a set of more portable GUI components. They put their new, more-portable components in a brand new library named Swing. To make the relationship clear between the new Swing components and the AWT components, they used the same component names except that they prefaced the new Swing components with a "J." For example, the AWT has a `Button` component, so Swing has a `JButton` component.

The AWT GUI components are known as *heavyweight components,* while the Swing GUI components are known as *lightweight components.* The AWT components are said to be *heavyweight* because they are built by the native environment that runs the bytecode. (The term *native environment* refers to the low-level instructions inherent to a particular computer platform, such as Windows or Macintosh.) For each AWT component, the bytecode asks the native environment to create the component, and it's up to the native environment to find native code for the requested component. That's the reason AWT components have a different look and feel when run on different computer platforms. On the other hand, Swing components are said to be *lightweight* because they're built with Java code. Being built with Java code means they're "light" enough to move to different platforms and have the same look and feel everywhere.

The Swing library includes more than just GUI component classes. It adds lots of functionality to the AWT, but it does not replace the AWT entirely. Today, Java GUI application programmers use both libraries—the AWT and Swing. The primary AWT packages are `java.awt` and `java.awt.event`. The primary Swing package is `javax.swing`. The "x" in `javax` stands for "extension" because the `javax` packages (`javax.swing` is one of several `javax` packages) are considered to be a major extension to the core Java platform.

17.18 Mouse Listeners and Images (Optional)

The Java API provides several different types of listeners. Earlier in this chapter, you learned about the most common listener—the `ActionListener`. You should use the `ActionListener` for events where the user does something to a component, such as clicking a button or pressing Enter within a text box. In this

[5] *Look and feel* is a standard GUI term, and it refers to the appearance of something and the way in which the user interacts with it.

section, you'll learn about mouse listeners. As the name implies, you should use mouse listeners for events where the user does something with the mouse. Also in this section, you'll learn about images (pictures). You'll learn how to display an image and drag an image with your mouse.

Mouse Listeners

In creating a mouse listener, you use the same basic steps that you use for the `ActionListener`—you define a listener class, you define an event handler method(s) within the listener class, and you register your listener class with a component. Although the same basic steps are used, mouse listeners are slightly more complicated than the `ActionListener`. There are several different types of mouse listeners, and each type of mouse listener handles multiple types of mouse events.

We describe two mouse listener types, and they are defined by their two interfaces—`MouseListener` and `MouseMotionListener`. Figure 17.15 shows the API headings and descriptions for the methods

`MouseListener` Interface Event Handlers

`public void mouseClicked(MouseEvent event)`

Called when the user presses and releases the mouse button while the mouse cursor is stationary on a `MouseListener`-registered component.

`public void mouseEntered(MouseEvent event)`

Called when the mouse cursor enters the bounds of a `MouseListener`-registered component.

`public void mouseExited(MouseEvent event)`

Called when the mouse cursor exits from the bounds of a `MouseListener`-registered component.

`public void mousePressed(MouseEvent event)`

Called when the user presses the mouse button while the mouse cursor is on a `MouseListener`-registered component.

`public void mouseReleased(MouseEvent event)`

Called when the user releases the mouse button, but only if the prior mouse press was on a `MouseListener`-registered component.

`MouseMotionListener` Interface Event Handlers

`public void mouseDragged(MouseEvent event)`

Called when the user holds the mouse button down while moving the mouse cursor, but only if the initial mouse press was on a `MouseMotionListener`-registered component.

`public void mouseMoved(MouseEvent event)`

Called when the user moves the mouse while the mouse cursor is on a `MouseMotionListener`-registered component.

Figure 17.15 API headings and descriptions for the methods in the `MouseListener` and `MouseMotionListener` interfaces

defined by the two interfaces. Read through the API headings and descriptions to get an idea of what's possible in terms of mouse event handling.

As a programmer, you don't have to worry about calling the mouse event handler methods. They're called automatically when their associated mouse events occur. For example, if the user presses the mouse button while the mouse cursor is on a `MouseListener`-registered component, the JVM automatically calls the `mousePressed` event handler.

In the upcoming program, the goal is to enable a user to drag an image across a window using the mouse. To do that, you need to detect the mouse being pressed and moved (i.e., dragged) while the mouse cursor is on the image. And to do that, you need to register a mouse listener. But you can register a mouse listener only with a component, not with an image. So what's the solution? You're already familiar with some components—`JLabel`, `JTextField`, and `JButton`. Those classes are component classes because they are descendants of the `Component` class. There's another component class that's a bit different. It doesn't feel like a component in the normal sense of the word, but it's a Java component nonetheless (because it's a descendant of the `Component` class), and it works great for handling mouse events. So what is the mystery component? `JPanel`!

Think of a `JPanel` object as a generic storage area for other objects. More formally, the `JPanel` class is a descendant of the `Container` class, and as such, it's a container and you can add objects to it. In the next chapter, you'll add Swing components (`JLabel`, `JTextField`, and so on) to `JPanel` containers. In the upcoming program example, you add an image object to a `JPanel` container. By surrounding the image with a `JPanel` container, you provide a platform that mouse listeners can attach to. In the upcoming program example, the `JPanel` listeners allow you to detect mouse events on the image object.

The DragSmiley Program

See Figure 17.16. It contains a driver class and a sample session for a DragSmiley program. As indicated in the sample session, the program initially displays a smiley face in the top-left corner of the program's window. If the user presses the mouse button, the smiley image changes to a scared image (presumably because the smiley is apprehensive of what the user might do to it). When the user releases the mouse button, the scared image changes back to the smiley image. If the mouse cursor resides on the image and the user drags the mouse, the image follows the mouse cursor.

Study Figure 17.16's `DragSmiley` constructor. In it, the following two statements instantiate a `JPanel` container named `smileyPanel` and add the `JPanel` container to `DragSmiley`'s window.

```
smileyPanel = new SmileyPanel();
add(smileyPanel);
```

See the `SmileyPanel` class in Figures 17.17a, 17.17b, and 17.17c. The `SmileyPanel` class is where the bulk of the program's logic is. We'll describe the `SmileyPanel` class by first focusing on the listeners. Note how the `SmileyPanel` constructor creates the mouse listeners and adds them to the `JPanel` container. Note the mouse listener class headings, repeated here for your convenience:

```
private class ClickListener extends MouseAdapter
private class DragListener extends MouseMotionAdapter
```

The `extends` clauses indicate inheritance from the `MouseAdapter` and `MouseMotionAdapter` classes. For each event handling interface with more than one method, the Java API provides an associated class that already implements the interface's methods for you. Those classes are called

```
/*******************************************************
* DragSmiley.java
* Dean & Dean
*
* This program displays a smiley face image.
* When the user presses the mouse, the image changes to a
* scared image. The user can drag the image.
*******************************************************/

import javax.swing.JFrame;

public class DragSmiley extends JFrame
{
  private static final int WIDTH = 250;
  private static final int HEIGHT = 250;
  private SmileyPanel smileyPanel;      // drawing panel

  //****************************************************

  public DragSmiley()
  {
    setTitle("Drag Smiley");
    setSize(WIDTH, HEIGHT);
    setDefaultCloseOperation(EXIT_ON_CLOSE);
    smileyPanel = new SmileyPanel();
    add(smileyPanel);
    setVisible(true);
  } // end DragSmiley constructor

  //**************************************

  public static void main(String[] args)
  {
    new DragSmiley();
  }
} // end class DragSmiley
```

Initial display: While dragging smiley: After releasing mouse button:

Drag Smiley Drag Smiley Drag Smiley

Figure 17.16 Driver class and sample output for the DragSmiley program

```
/*************************************************************
* SmileyPanel.java
* Dean & Dean
*
* This class contains a smiley image and listeners
* that enable image dragging and image swapping.
*************************************************************/

import javax.swing.*;         // for JPanel and ImageIcon
import java.awt.*;            // for Point and Graphics
// for MouseAdapter, MouseEvent, and MouseMotionAdapter:
import java.awt.event.*;

public class SmileyPanel extends JPanel
{
  private final ImageIcon SMILEY = new ImageIcon("smiley.gif");
  private final ImageIcon SCARED = new ImageIcon("scared.gif");
  private final int WIDTH = SMILEY.getIconWidth();
  private final int HEIGHT = SMILEY.getIconHeight();

  private Point imageCorner; // image's top-left corner location
  private Point prevPt;      // mouse location for previous event
  private ImageIcon image;   // toggles between smiley and scared
  private boolean grabbed;   // mouse has a hold on the icon

  //**********************************************************

  public SmileyPanel()
  {
    image = SMILEY;
    imageCorner = new Point(0, 0); // image starts at top left
    ClickListener clickListener = new ClickListener();
    DragListener dragListener = new DragListener();
    this.addMouseListener(clickListener);
    this.addMouseMotionListener(dragListener);
  } // end SmileyPanel constructor

  //**********************************************************

  // Draw the window, including the updated image.

  @Override
  public void paintComponent(Graphics g)
  {
    super.paintComponent(g);
    image.paintIcon(this, g,
      (int) imageCorner.getX(), (int) imageCorner.getY());
  } // end paintComponent
```

Add mouse listeners to the JPanel container.

Call paintIcon to display the image.

Figure 17.17a DragSmiley program's SmileyPanel class—part A

```
//*************************************************************

private class ClickListener extends MouseAdapter
{
  // When mouse pressed, change to scared image.

  @Override
  public void mousePressed(MouseEvent e)
  {
    image = SCARED;
    repaint();
    prevPt = e.getPoint(); // save current position

    // Make sure mouse was pressed within the image.
    if (prevPt.getX() >= imageCorner.getX() &&
        prevPt.getX() <= imageCorner.getX() + WIDTH &&
        prevPt.getY() >= imageCorner.getY() &&
        prevPt.getY() <= imageCorner.getY() + HEIGHT)
    {
      grabbed = true;
    }
  } // end mousePressed

  // When mouse released, return to smiley image.

  @Override
  public void mouseReleased(MouseEvent e)
  {
    image = SMILEY;
    repaint();
    grabbed = false;
  } // end mouseReleased
} // end class ClickListener
```

Figure 17.17b DragSmiley program's `SmileyPanel` class—part B

adapter classes. The `MouseAdapter` class implements the `MouseListener` interface's methods. Likewise, the `MouseMotionAdapter` class implements the `MouseMotionListener` interface's methods. Adapter classes don't do much. They simply implement their associated interface's methods as dummy methods, like this:

```
public void mousePressed(MouseEvent event)
{ }
```

To implement a listener that detects the mouse being pressed, you extend the `MouseAdapter` class and provide an overriding `mousePressed` method. For an example, see Figure 17.17b. As an alternative, you can implement a listener using an interface rather than an adapter. But remember that an interface is a contract, and when you implement an interface, you're required to provide methods for all the interface's methods. So if you wanted to replace the `SmileyPanel` class's adapters with interfaces, you'd have to provide

```
  //**********************************************************

  private class DragListener extends MouseMotionAdapter
  {
    // Enable an image to be dragged by a mouse.

    @Override
    public void mouseDragged(MouseEvent e)
    {
      Point currentPt = e.getPoint(); // current position

      // Make sure mouse was pressed within the image.
      if (grabbed)
      {
        imageCorner.translate(
          (int) (currentPt.getX() - prevPt.getX()),
          (int) (currentPt.getY() - prevPt.getY()));
        prevPt = currentPt; // save current position
        repaint();
      }
    } // end mouseDragged
  } // end class DragListener
} // end class SmileyPanel
```

Figure 17.17c DragSmiley program's `SmileyPanel` class—part C

dummy methods for methods you won't use. On the other hand, when you extend adapter classes, you just override those methods you will use.

Displaying an Image

It's now time to see how the `SmileyPanel` class draws its images. At the top of the class, the `SMILEY` and `SCARED` named constants are initialized as follows:

```
final private ImageIcon SMILEY = new ImageIcon("smiley.gif");
final private ImageIcon SCARED = new ImageIcon("scared.gif");
```

The `ImageIcon` constructor creates an image object from its passed-in filename parameter. So in the above code fragment, two image objects are created from the `smiley.gif` and `scared.gif` files, respectively.[6]

In the `SmileyPanel` constructor, the `mousePressed` event handler, and the `mouseReleased` event handler, note how `SMILEY` and `SCARED` get assigned into the `image` instance variable. Those assignments are what cause the image to change when the user presses the mouse button and releases it.

The `JPanel` class has a `paintComponent` method that's in charge of drawing Swing components (e.g., text boxes and buttons) within the `JPanel` container. But it doesn't handle drawing lines, shapes, or

[6] *gif* stands for Graphics Interchange Format. It's used for an exact representation of a simple drawn image. The alternative Joint Photographic Experts Group (JPEG) format, which uses the `.jpg` extension, uses data compression to save storage space, but it loses information in the process.

images. To draw those things, you need to provide an overriding `paintComponent` method with calls to graphics methods. For example, here is `SmileyPanel`'s overriding `paintComponent` method:

```
public void paintComponent(Graphics g)
{
  super.paintComponent(g);
  image.paintIcon(this, g, (int) imageCorner.getX(), (int) imageCorner.getY());
} // end paintComponent
```

Note the `paintComponent` method's `g` parameter. It's a `Graphics` object, and it's used to call graphics methods within the `paintComponent` method. For example, the `image.paintIcon` method call draws `image` (a smiley face or a scared face), and it requires a `Graphics` object, `g`, for its second argument. In calling the `paintIcon` method, you provide three arguments in addition to the `Graphics` argument: (1) an *image observer,* which listens for the completion of the image being loaded, (2) the x coordinate of the image's top-left corner, and (3) the y coordinate of the image's top-left corner.

The first argument is an image observer component that manages repainting as the image file downloads initially or as an animated image file changes. It's common for an image to take longer to display than the rest of the window because image files tend to be relatively large. If you want to do something special when the image finishes loading, then add an overriding `imageUpdate` event handler method to the image observer's class. In the DragSmiley program, we don't implement an `imageUpdate` event handler. In the above code fragment, we use `this` for the image observer component, where `this` refers to the `SmileyPanel` calling object. `SmileyPanel` is a subclass of `JPanel`, and `JPanel` implements the automatic repaint functionality required by an image observer component.

In the above code fragment, note the `super.paintComponent(g)` method call. You should always include that call as the first statement within an overriding `paintComponent` method. Without it, the background for `paintComponent`'s associated object might be displayed improperly.

Notice that there's no explicit call to the DragSmiley program's `paintComponent` method. You should never call the `paintComponent` method directly. Instead, you should call the `repaint` method and let the `repaint` method call the `paintComponent` method for you. The `repaint` method waits until the program's window is properly prepared to handle the `paintComponent` method. Note in the `SmileyPanel` class how `repaint` is called at the bottom of the three event handlers. That's where there's a need to redraw the image. By the way, in addition to calling `paintComponent` whenever `repaint` is called, the JVM calls `paintComponent` automatically when the program starts up and whenever a user does something to alter the program's window (e.g., when the user resizes the window or moves another window off the window).

Summary

- The `JFrame` class should be used as the superclass for most of your GUI application windows.
- The `JFrame` class implements all the standard window features such as a border, a title bar, a minimize button, a close-window button (the "X"), the ability to resize the window, and so on.
- `JLabel` is a read-only component; the user simply reads the label's message.
- The `JTextField` component allows the user to enter text into a text box.
- When the user interacts with a component (e.g., when the user clicks a button or presses enter while in a text box), the component fires an event.
- If a component has a listener attached to it, the fired event is "heard" by the listener and consequently handled by the listener.

- A listener handles an event by executing its `actionPerformed` event-handler method.
- Listeners often are implemented with the `ActionListener` interface. An interface is a class-like entity whose methods are all empty. If a programmer uses an interface to derive a new class, the compiler requires the new class to implement methods for all the interface's methods.
- If a class is limited in its scope such that it is needed by only one other class, then you should define the class as an inner class (a class inside another class).
- An anonymous inner class is an inner class without a name.
- To display a simple window with a message, call `JOptionPane`'s `showMessageDialog` method.
- To identify the component whose event was fired, use the `actionPerformed` method's `ActionEvent` parameter to call `getSource` or `getActionCommand`.
- To adjust a GUI component's text color, call `setForeground`. To adjust the color behind the text, call `setBackground`.
- To adjust a window's background color, call the content pane's `setBackground` method.
- To detect and handle mouse events, use the `MouseAdapter` and `MouseMotionAdapter` classes, which implement the `MouseListener` and `MouseMotionListener` interfaces, respectively.

Review Questions

§17.2 Event-Driven Programming Basics

1. What is a listener?
2. What is an event handler?

§17.3 A Simple Window Program

3. Write a statement that adds functionality to a program's close-window button such that when the close-window button is clicked, it causes the program to terminate.

§17.4 `JFrame` Class

4. What is the name of the superclass for classes that contain components?

§17.5 Java Components

5. What package are `JButton` and many other J-prefixed components defined in?

§17.6 `JLabel` Component

6. Provide an initialization statement that declares a `JLabel` reference variable named `hello` and assigns "Hello World" to the reference variable.

§17.7 `JTextField` Component

7. Provide an initialization statement that instantiates a 10-character-wide text box object. As part of the initialization, assign the text box object to a reference variable named `input`.

§17.9 Component Listeners

8. Write a statement that registers a listener reference variable named `responder` with a component named `component`.
9. If you want a class to handle an event, what clause must be added to the right side of the class's heading?
10. What is the heading of the one method specified by the `ActionListener` interface?

§17.10 Inner Classes

11. If a class is limited in scope such that it is only needed internally within another class, you should define the class to be an ________________.

§17.11 Anonymous Inner Classes

12. If you want to implement an event handler with an anonymous inner class, what argument do you give to the `addActionListener` method to register the listener?

§17.12 `JButton` Component

13. In the `createContents` method of the FactorialButton program in Figure 17.8a, what type of object calls the add methods?
14. In the FactorialButton program in Figures 17.8a and 17.8b, what component fires the event that the listener handles?

§17.13 Dialog Boxes and the `JOptionPane` Class

15. What package contains the `JOptionPane` class?
16. Write a statement that displays a dialog in the center of the screen. The dialog should display "This is only a test." in the message area, "TEST" in the title area, and no icon.

§17.14 Distinguishing Between Multiple Events

17. Suppose that you have several components registered with the same listener, and the components and listener are defined within the same class. Within the listener, what `ActionEvent` method should you call to determine which component fires an event?

§17.15 Using `getActionCommand` to Distinguish Between Multiple Events

18. Assume there's a listener that's been registered for several different buttons. Assume the listener uses an `actionPerformed` method with an `ActionEvent` parameter named `action`. Assume the user clicks one of the registered buttons. Provide a statement that retrieves the text label from the clicked button and assigns the retrieved label to a `String` variable named `buttonLabel`.

§17.16 Color

19. Write a statement that sets the text color to blue for a `JButton` object named `button1`.
20. How do you get a reference to the container that surrounds all of the components in a `JFrame` object?

§17.17 How GUI Classes Are Grouped Together

21. If your program needs the `java.awt.event` subpackage, you can implicitly import it by importing the `java.awt` package. (T / F)

Exercises

1. [after §17.2] Give three examples of how a user might cause an event to be fired.
2. [after §17.3] For each of the following, what Java API package must you import?
   a) `JFrame` and `JLabel`
   b) `FlowLayout`
3. [after §17.4] For our previous GUI programs, we've done setup work (setting the title, adding components, and so on) within a constructor. That's generally preferred, but it's not a compiler requirement. For practice purposes, write a minimal, but fully functional, program that displays this:

Your program should not include a constructor. It should include only one method—a `main` method with only five statements (or four statements if you find a shortcut for setting the frame's title).

4. [after §17.6] Provide a complete program that displays this `Hello World` message:

Note these label characteristics: (1) a raised bevel border, (2) italics, (3) a large font size (30 points), (4) a tool tip that says "Life is Great!" Use this program skeleton as a starting point:

```
import javax.swing.*;
import java.awt.*;

//*************************************************************

public class BigHello extends JFrame
{
  public BigHello()
  {
    JLabel label = <instantiation> ;
    setSize(200, <height> );
    setLayout(new FlowLayout());
    add(label);

    <3-statement code fragment>

    setVisible(true);
  } // end constructor

  //*************************************************************

  public static void main(String[] args)
  {
    BigHello hello = new BigHello();
  } // end main
} // end BigHello class
```

To figure out how to do this, in Java's API, look up the `setFont`, `setBorder`, and `setToolTipText` methods that `JLabel` inherits from `JComponent`. For the `setFont` argument, use `Component`'s `getFont` to get the default font, and then alter it by using `Font`'s two-parameter `deriveFont` method in which the first parameter specifies an italic font style and the second parameter specifies a 30-point size. Use `JComponent`'s `setBorder` method, and for its `Border` argument, use the appropriate class method from the `BorderFactory` class.

5. [after §17.7] The width parameter in the `JTextField` constructor specifies the width of the text box in pixels. (T / F)

6. [after §17.7] What can you do to prevent users from updating a `JTextField` component?

7. [after §17.9] Write the heading for the method you must define in a class that implements an `ActionListener`.

8. [after §17.9] The `ActionListener` interface and the `ActionEvent` class are in what Java API package?

9. [after §17.9] An interface is a class-like thing whose methods are all empty. If an interface is applied to a class, then the interface acts like a template that the class must conform to. (T / F)

10. [after §17.10] An inner class can access its enclosing class's instance variables directly. (T / F)

11. [after §17.12] It's appropriate to use an anonymous inner class if you are going to use the class only once. In the Factorial program in Figures 17.8a and 17.8b, we use the `listener` object twice, so that `listener` object needed to have a name. However, we used that object's class only once, to instantiate that one object. Therefore, that object's class did not need to have a name, and we could have used an anonymous class to create our `listener` object. For this exercise, modify the Factorial program to use an anonymous `ActionListener` class instead of the named `Listener` class. [Hint: The program is already set up to facilitate this change—it's mostly cut-and-paste.]

12. [after §17.13] Do you have to create a `JFrame` window to use a `JOptionPane` dialog box?

13. [after §17.13] To answer this question, you may need to look up `JOptionPane`'s `showInputDialog` and `showConfirmDialog` methods on Oracle's Java API website. What does this program do?

```
import javax.swing.JOptionPane;
public class UncertainHello
{
  public static void main(String[] args)
  {
    String name;
    int response;
    do
    {
      name = JOptionPane.showInputDialog("What's your name? ");
      response = JOptionPane.showConfirmDialog(null, "Are you sure?");
      if (response == JOptionPane.NO_OPTION)
      {
        name = "there";
        break;
      }
    } while (response == JOptionPane.CANCEL_OPTION);

    System.out.println("Hello " + name);
  } // end main
} // end class UncertainHello
```

14. [after §17.14] By calling `setEnabled(false)`, you can disable a button and give it a muted appearance and make its listener unresponsive to clicks on it. Modify Figure 17.12's program so that the factorial button is initially disabled. Enable it only after the user enters a character in the `xBox` text box. To enable it, create a *key listener* for the `xBox` text box, and have the key listener's `keyTyped` event handler call `setEnabled(true)`. Use the following key listener code skeleton:

```
private class KeyListener extends KeyAdapter
{
  public void keyTyped(KeyEvent e)
  {
    ...
  }
} // end class KeyListener
```

Note `extends KeyAdapter` in the above class heading. An *adapter* class implements an interface by providing an empty-bodied method for each method in the interface. In this case, the `KeyAdapter` API class implements the `KeyListener` API interface.

15. [after §17.16] To set a `JFrame`'s background color, what method should you call before calling `setBackground`?
16. [after §17.17] What do the letters in "awt" stand for?

Review Question Solutions

1. A listener is an object that waits for events to occur.
2. An event handler is a method that responds to an event.
3. `setDefaultCloseOperation(EXIT_ON_CLOSE);`
4. The superclass for objects that contain other objects is the `Container` class.
5. Many J-prefixed components are defined in the `javax.swing` package.
6. `JLabel hello = new JLabel("Hello World!");`
7. `JTextField input = new JTextField(10);`
8. `component.addActionListener(responder);`
9. For a class to handle an event, add this to the right side of the class's heading:
   `implements ActionListener`
10. The heading of the method specified by the `ActionListener` interface is:
    `public void actionPerformed(ActionEvent e)`
11. If a class is limited in scope such that it is only needed internally within another class, you should define the class to be an <u>inner class</u>.
12. The argument to give to the `addActionListener` method to register an anonymous listener class is

```
new ActionListener()
{
  <implementation-of-ActionListener-interface>
}
```

13. The object that calls the `add` methods is a `JFrame` object.
14. It's ambiguous. It could be either `xBox` or `btn`.
15. The package that contains the `JOptionPane` class is the `javax.swing` package.
16. This code generates the asked-for dialog box:

```
JOptionPane.showMessageDialog(null,
  "This is only a test.", "TEST", JOptionPane.PLAIN_MESSAGE);
```

17. To identify the firing component, call the `getSource` method.
18. `buttonLabel = action.getActionCommand();`
19. `button1.setForeground(Color.BLUE);`
20. Call `JFrame`'s `getContentPane` method.
21. False. The `java.awt` and `java.awt.event` packages contain separate classes. To import classes from `java.awt.event`, you must import that package explicitly, like this:

```
import java.awt.event.*;
```

CHAPTER 18

GUI Programming—Component Layout, Additional GUI Components

Objectives

- Know GUI design basics.
- Know the benefits of using layout managers.
- Understand `FlowLayout` manager details.
- Understand `BoxLayout` manager details.
- Understand `BorderLayout` manager details.
- Be able to use the `SwingConstants` interface.
- Understand `GridLayout` manager details.
- Use embedded layout managers and `JPanel`s for windows that have a substantial number of components.
- Implement `JTextArea` components for text that spans more than one line.
- Implement a `JCheckBox` component for yes/no user input.
- Implement `JRadioButton` and `JComboBox` components when the user needs to choose a value from among a list of predefined values.
- Become familiar with additional Swing components such as menus, scroll panes, and sliders.

Outline

18.1 Introduction

This is the second chapter in our two-chapter sequence on GUI programming. In the previous chapter, you learned GUI basics. You learned about windows, components, and listeners. Almost all GUI programs need those things. In this chapter, you'll learn how to make your GUI programs more functional and more visually appealing. You'll improve the functionality by implementing some additional GUI components—`JTextArea`, `JCheckBox`, `JRadioButton`, and `JComboBox`. You'll improve the visual appeal by applying various layout techniques to your windows' components. More specifically, you'll learn how to apply these layout managers—`FlowLayout`, `BoxLayout`, `BorderLayout`, and `GridLayout`. And you'll learn how to apply different layout managers to different areas of your windows.

For an example of what you'll be learning, see Figure 18.1. Note the combo box, radio button, and check box components. Also, note how the radio buttons are grouped in the center, the check buttons are grouped at the right, and the Next and Cancel buttons are grouped at the bottom center. In this chapter, you'll learn how to make such groupings, and you'll learn how to position them appropriately.

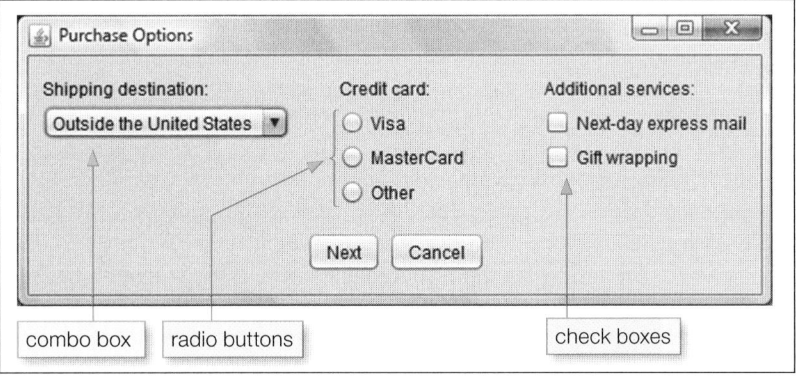

Figure 18.1 Example window that uses radio buttons, check boxes, and a combo box

18.2 GUI Design and Layout Managers

With text-based programs, it's relatively easy to tell users what to do. As a programmer, you just provide text instructions, and the user enters input when prompted to do so. With GUI programs, it's more difficult to tell users what to do. As a programmer, you display a window with various components, set up listeners, and then wait for the user to do something. It's important that your display be easy to understand; otherwise, your users won't know what to do. To make your display easy to understand, follow these guidelines:

- Choose the right components.
- Be consistent.
- Position components appropriately.

GUI Design Basics

In Figure 18.1, note the small circles next to Visa, MasterCard, and Other. Those circles are radio button components (we describe radio buttons in Section 18.14). Using radio buttons for these credit card options is an example of choosing the right component. Radio buttons provide implicit instructions to the user about how to proceed. Most users recognize small circles as radio buttons, and when they see them, they know to click one of them with the mouse.

In Figure 18.1, note the Next and Cancel buttons at the bottom center of the window. Assume that the window is one of several windows in a purchasing application. Assume that other windows in the application also display Next and Cancel buttons in the bottom-center position. Placing Next and Cancel buttons in the same position is an example of being consistent. Consistency is important because users are more comfortable with things they've seen before. As another example, be consistent with color schemes. In a given application, if you choose red for a warning message, use red for all your warning messages.

In Figure 18.1, note how the three radio button components (Visa, MasterCard, and Other) and the "Credit card:" label component are positioned together as a group. More specifically, they're aligned in a vertical column and they're physically close together. That's an example of positioning components appropriately. Positioning them together as a group provides a visual cue that they're logically related. As another example of appropriate positioning, note that there are sizable gaps separating the left, center, and right component groups. Finally, note how the "Shipping destination:", "Credit card:", and "Additional services:" labels are aligned in the same row. That alignment, the aforementioned gaps, and the aforementioned component groupings all lead to a more appealing and understandable display.

Layout Managers

As you now know, positioning components appropriately is an important part of GUI design. In the old days, positioning components was a tedious, manual process. Programmers would spend hours calculating the space needed for each component and the pixel coordinate positions for each component. Today, programmers are freed from that tedium by having layout managers do those calculations for them. As you may recall from the previous chapter, a *layout manager* is an object that controls the positioning of components within a container. In general, the layout manager's goal is to arrange components neatly. Usually, the neatness goal equates to making sure components are aligned and making sure components are appropriately spaced within the layout manager's container. For example, in Figure 18.1, layout managers are responsible for aligning the left components, aligning the middle components, aligning the right components, and spacing the three component groups across the width of the window.

If a user adjusts a window's size, the JVM consults with the layout manager, and the layout manager then recalculates the pixel coordinate positions for each component. All this takes place automatically, without any intervention on the programmer's part. How convenient! Hail to the layout manager!

There are different types of layout managers, and they have different strategies for positioning components within a container. See the table in Figure 18.2. It describes several layout managers from Oracle's API library.

Layout Manager	Description
BorderLayout	Splits container into five regions—north, south, east, west, and center. Allows one component per region.
BoxLayout	Allows components to be arranged in either a single column or a single row.
FlowLayout	Allows components to be added left to right, flowing to the next row as necessary.
GridLayout	Splits container into a rectangular grid of equal-sized cells. Allows one component per grid cell.
GridBagLayout	A more flexible and complex version of GridLayout. Allows grid cells to vary in size.

Figure 18.2 Several of the more popular layout managers

In the previous chapter, we used the simplest type of layout manager—the FlowLayout manager. The FlowLayout manager is useful for some situations, but we'll often need alternative layout managers for other situations. In this chapter, we'll describe the FlowLayout manager in more detail, and we'll also describe the BorderLayout, BoxLayout, and GridLayout managers. Those are the four most popular layout managers, so you should know them well.

Assigning a Layout Manager

To assign a particular layout manager to a JFrame window from within a class that extends JFrame, call the setLayout method as follows:

```
setLayout (new <layout-manager-class>(<arguments>));
```

In this code template, replace *<layout-manager-class>* by a layout manager class (like FlowLayout) and replace *<arguments>* by zero or more arguments. The arguments will be different for the different layout managers, and we'll get to those details later. If setLayout is not called, then the BorderLayout manager is used, because that's the default layout manager for a JFrame window.

The BorderLayout, FlowLayout, GridLayout, and GridbagLayout manager classes are in the java.awt package. The BoxLayout manager class is in the javax.swing package. Import accordingly.

18.3 FlowLayout Manager

In the previous chapter, we wanted to present GUI basics without getting bogged down in layout manager details. So we chose a simple layout manager, FlowLayout, that didn't require much explanation. We just used it and didn't dwell on particulars. Now it's time to explain the particulars, so you can take advantage of its functionality more fully.

Layout Mechanism

The FlowLayout class implements a simple one-compartment layout scheme that allows multiple components to be inserted into the compartment. When a component is added to the compartment, it is placed to the right of any components that were previously added to the compartment. If there is not enough room to add a component to the right of the previously added components, the new component is placed on the next line (i.e., it "flows" to the next line). Note the following example.

Assume that you've implemented a program that prompts the user to enter his or her name and prints a personalized greeting after the user presses enter. We'll show you a sample session that starts with a wide window and a short name. Here's what the program displays after the user types Tom:

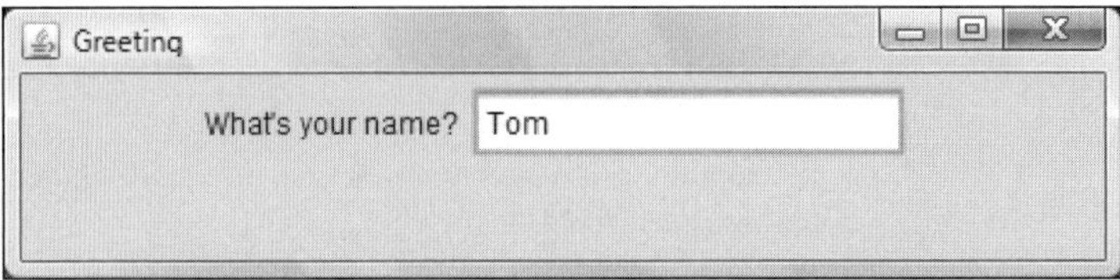

And here's what the program displays after the user presses enter:

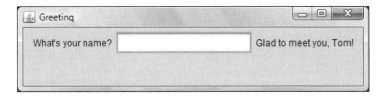

If the user enters a longer name, like Fidelis Kiungua, the greeting label can't fit on the first line, so it wraps to the next line:

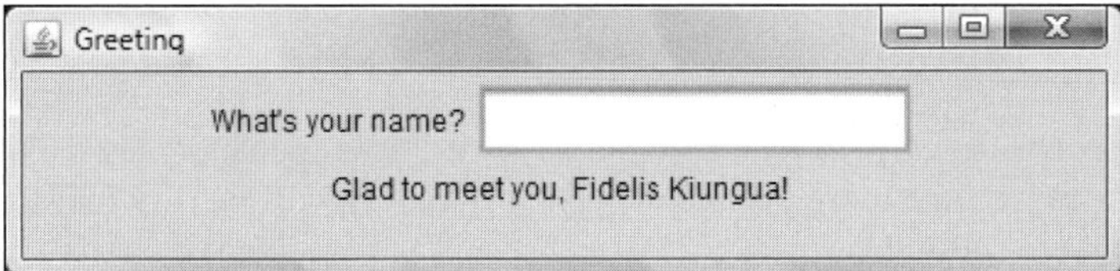

If the user manually resizes the window to make it narrower, the text box can no longer fit on the first line, so it wraps to the next line:

Alignment

By default, the `FlowLayout` manager positions its components using center alignment. For example, in the above window, note how the "What's your name?" label is centered between the left and right borders. If you'd like to change the `FlowLayout` manager's alignment, insert one of the `FlowLayout` alignment constants (`FlowLayout.LEFT`, `FlowLayout.CENTER`, `FlowLayout.RIGHT`) in the `FlowLayout` constructor call. For example, here's how to specify left alignment:

```
setLayout(new FlowLayout(FlowLayout.LEFT));
```

Here's what our Greeting program displays when left alignment is used:

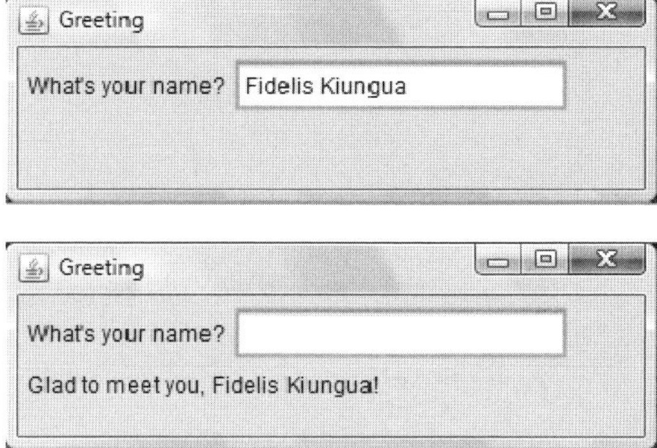

Layout Changes

Normally, `setLayout` is called just once in a program—when the program initially lays out its components. But if there's a need to adjust the layout scheme dynamically, call `setLayout` again. For example, if you want the user to be able to adjust text alignment, add **Align Left**, **Align Center**, and **Align Right** buttons. Add a listener to each button. In each listener, call `setLayout`. This would be the listener for the **Align Left** button:

```
private class Listener implements ActionListener
{
  public void actionPerformed(ActionEvent e)
  {
    setLayout(new FlowLayout(FlowLayout.LEFT));
    validate();
  } // end actionPerformed
} // end class Listener
```

Note the `validate` method in this code fragment. It causes the layout manager to regenerate the component layout. If you attempt to change your window's layout in some way (after the window is initially displayed with `setVisible(true)`) you should call `validate` to make sure the change takes effect. These method calls attempt to change the layout:

- `setLayout`—Change the window's layout manager.
- `add`—Add a component to the window.
- `setSize`—Change the window's size.
- `remove`—Remove a component from the window.

If your window is already visible and you call one of those methods, don't forget to call `validate` afterward. If you have a series of such calls, there's no need to have separate `validate` method calls. Putting one `validate` method call at the end works fine. 

18.4 BoxLayout Manager

The `FlowLayout` manager is popular because it's easy to use. Just add components, and the container accumulates them like words on a page of text. It starts at the top, filling each line from left to right, and then moves automatically to the next line. But suppose you want to arrange items vertically in a column. In that case, you should use a BoxLayout manager. A BoxLayout manager can arrange items in either one row or one column. If the container is wide enough, you can use a FlowLayout manager to arrange items in a row. So a BoxLayout manager is most useful when you want to arrange items in a column.

Example—a Dance Recital Poster

Suppose you want to implement a program that prints a poster with details about an upcoming dance recital. You want your program to be flexible, so it can be reused later for other dance recital events. Thus, rather than hardcoding the event details in print statements, the program prompts the user for event details, stores them, and then prints them in a pleasing format as a poster. The poster should show the following five items, one above the other:

- the name of the dance recital performance
- a promotional image, like a photograph of the featured performer
- the dance recital's date
- the dance recital's time
- the dance recital's location

Figure 18.3 shows an example of what we're talking about.

Figure 18.4 contains the dance recital program's driver, which first instantiates a DanceRecital object, where DanceRecital is the class that stores a dance recital's event details (name of the performance, image, date, time, and venue). For each of those details, the driver prompts the user to enter the relevant information and then stores it in the DanceRecital object. The driver then calls the displayPoster method, which displays the dance recital's details using the BoxLayout manager for its layout scheme.

Study the DanceRecital class in Figures 18.5a and 18.5b. The first figure contains standard mutator methods, which assign dance recital details to the DanceRecital object. The second figure contains the displayPoster method, which generates a window designed to look like a dance recital poster. The programs in the previous chapter also generated windows, but they did so in a different manner. Those programs generated their windows immediately, when they called the constructor for the program's class as the first statement in main. For example, the SimpleWindow program called new SimpleWindow() as the first statement in main, and the SimpleWindow constructor was in charge of configuring the instantiated window. On the other hand, the DanceRecital program generates its window by calling the JFrame constructor from within its displayPoster method. Here's the relevant code, where the constructor's argument, "Dance Recital," specifies the window's title:

```
JFrame frame = new JFrame("Dance Recital");
```

So why did we choose this different strategy for the DanceRecital program? For each GUI program in the last chapter, the window served as the container for input components (text boxes and buttons), and displaying the input components was essential for setting up the rest of the program's execution. But for the DanceRecital program, the input is text based, so there's no need to display a window right away. Actually, there might not be a need to display a window at all. In the future, a programmer might want to improve the program by adding a method that generates promotional emails for the dance recital.[1] Presumably, the programmer would rewrite the driver to allow the user to choose between

[1]Generating emails from a Java program is beyond the scope of this book, but the starting point is installing Oracle's JavaMail package. To learn how to use the JavaMail package, go to http://www.oracle.com/technetwork/java/javamail.

Sample session:

```
Name of the performance: Swan Lake
Image file: dancer.jpg
Dance recital's date: February 14, 2014
Dance recital's time: 7:30 pm
Dance recital's venue: Park University, Alumni Hall
```

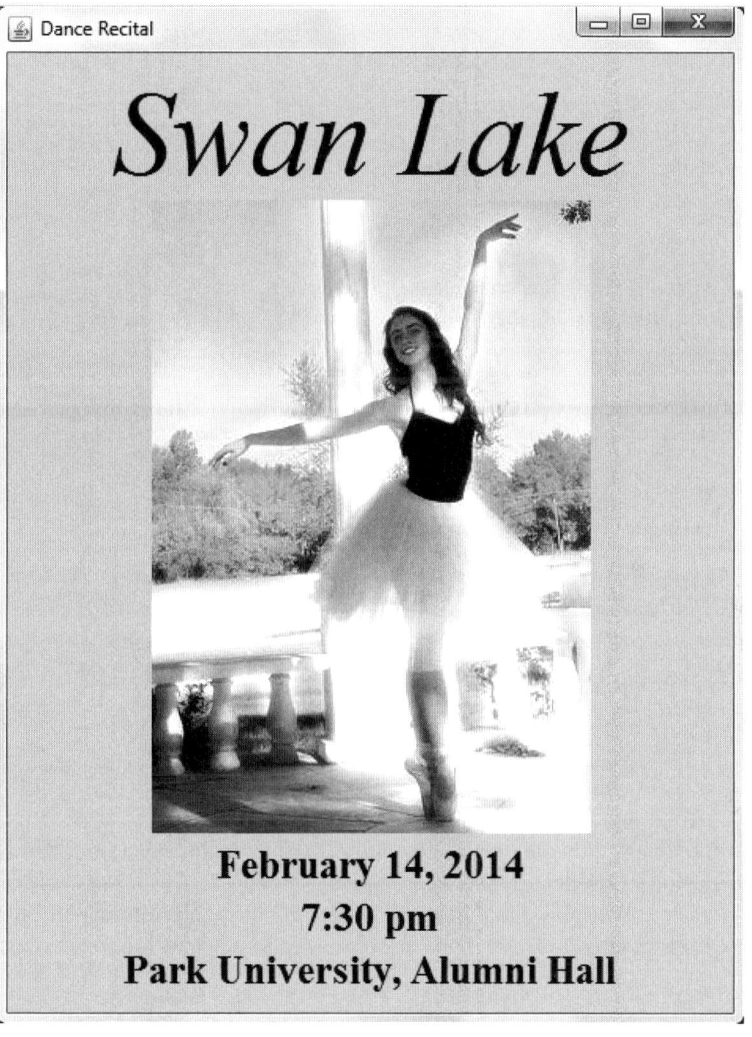

Figure 18.3 Representative output for the DanceRecital program
The pictured ballerina, Jessica Sapenaro, is a dance teacher and a Park University computer science student.

```
/*************************************************************
* DanceRecitalDriver.java
* Dean & Dean
*
* This is the driver for the DanceRecital class.
*************************************************************/

import java.util.Scanner;

public class DanceRecitalDriver
{
  public static void main(String[] args)
  {
    Scanner stdIn = new Scanner(System.in);
    DanceRecital danceRecital = new DanceRecital();

    System.out.print("Name of the performance: ");
    danceRecital.setPerformance(stdIn.nextLine());
    System.out.print("Image file: ");
    danceRecital.setImage(stdIn.nextLine());
    System.out.print("Dance recital's date: ");
    danceRecital.setDate(stdIn.nextLine());
    System.out.print("Dance recital's time: ");
    danceRecital.setTime(stdIn.nextLine());
    System.out.print("Dance recital's venue: ");
    danceRecital.setVenue(stdIn.nextLine());
    danceRecital.displayPoster();
  } // end main
} // end class DanceRecitalDriver
```

Figure 18.4 Driver of the `DanceRecital` class in Figures 18.5a and 18.5b

displaying a poster window and generating promotional emails. By instantiating the window within the `displayPoster` method, we allow for that possibility.

We still haven't gotten to the program's *raison d'être* (a French expression meaning "reason for being"), which is to illustrate how to use the `BoxLayout` manager. The `displayPoster` method uses the `BoxLayout` manager to display the dance recital details in a column format. To carry out that task, `displayPoster` assigns a `BoxLayout` manager to the window by calling `setLayout` like this:

```
frame.setLayout(
  new BoxLayout(frame.getContentPane(),BoxLayout.Y_AXIS));
```

Note the embedded `BoxLayout` constructor call. The constructor call's first argument specifies the window's content pane (so the `BoxLayout` manager knows what to attach itself to). The second argument specifies the window's orientation—vertical or horizontal. To arrange all the components vertically, one on top of the other, use `BoxLayout.Y_AXIS`, as shown in the code above. To arrange all the components horizontally, side by side, use `BoxLayout.X_AXIS`.

```
/*************************************************************
* DanceRecital.java
* Dean & Dean
*
* This program stores and prints dance recital information.
*************************************************************/

import javax.swing.*;  // JFrame, JLabel, BoxLayout, ImageIcon
import java.awt.*;     // Font, Component

public class DanceRecital
{
  private String performance;  // name of the performance
  private String image;        // promotional image
  private String date;         // dance recital's date
  private String time;         // dance recital's time
  private String venue;        // dance recital's location

  //**********************************************************

  public void setPerformance(String performance)
  {
    this.performance = performance;
  } // end setPerformance

  public void setImage(String imageFile)
  {
    this.image = imageFile;
  } // end setImage

  public void setDate(String date)
  {
    this.date = date;
  } // end setDate

  public void setTime(String time)
  {
    this.time = time;
  } // end setTime

  public void setVenue(String venue)
  {
    this.venue = venue;
  } // end setVenue
```

Figure 18.5a DanceRecital class—part A

```
  //***********************************************************

  // This method prints a poster for a dance recital.

  public void displayPoster()
  {
    JFrame frame = new JFrame("Dance Recital"); // the window
    JLabel pictureLabel; // container for dancer picture

    frame.setSize(480, 640); // pixel width, height
    frame.setDefaultCloseOperation(JFrame.EXIT_ON_CLOSE);
    frame.setLayout(
      new BoxLayout(frame.getContentPane(), BoxLayout.Y_AXIS));
    addLabel(frame, this.performance, Font.ITALIC, 75);

    pictureLabel = new JLabel(new ImageIcon(image));
    pictureLabel.setAlignmentX(Component.CENTER_ALIGNMENT);
    frame.add(pictureLabel);

    addLabel(frame, this.date, Font.BOLD, 25);
    addLabel(frame, this.time, Font.BOLD, 25);
    addLabel(frame, this.venue, Font.BOLD, 25);
    frame.setVisible(true);
  } // end displayPoster

  //***********************************************************

  // This method instantiates a label and adds it to the window.

  private void addLabel(
    JFrame frame, String labelText, int style, int size)
  {
    JLabel label = new JLabel(labelText);
    label.setAlignmentX(Component.CENTER_ALIGNMENT);
    label.setFont(new Font("Serif", style, size));
    frame.add(label);
  } // end addLabel
} // end class DanceRecital
```

Assign a vertical layout to the window.

Create a JLabel component with an embedded image.

Figure 18.5b DanceRecital class—part B

After creating a vertical layout for the window, displayPoster adds all its components to the window. Because adding each of the poster's performance name, date, time, and venue components to the window requires much of the same code, we use a helper method, addLabel, to implement that code. As you can see in Figure 18.5b, the addLabel method creates a JLabel object, specifies center alignment, assigns the font, and finally adds the component to the window. The dancer image is a bit different than the other components. Actually, the ImageIcon is not a component at all, in that it's not a subclass of the Component class. As such, the compiler won't allow you to add it directly to the window. The trick is to first embed the ImageIcon in a JLabel object and then add the resulting JLabel object to the window. Study the displayPoster method to see how that's done.

With a FlowLayout manager, if you add more components than can fit in one row, the extra components wrap around (flow) to the next row. With a BoxLayout manager (vertical or horizontal), such wraparounds do not take place. With a BoxLayout manager window, if you add components to a full window, the new elements won't be visible. However, the components are part of the window, and the user can expose them by increasing the window's size (by maximizing or dragging a corner).

18.5 BorderLayout Manager

In this section, we discuss the BorderLayout manager, which allows you to establish the position of particular components in both horizontal and vertical dimensions.

BorderLayout Regions

The BorderLayout manager is particularly useful for windows that need components near their edges. It's common to put a title near the top edge of a window. It's common to put a menu near the left edge of a window. It's common to put buttons near the bottom edge of a window. The BorderLayout manager accommodates all those situations by splitting up its container into five *regions,* or compartments. Four of the regions are near the edges, and one is in the center. You access the four edge regions with geographical names—north, south, east, and west. Note the regions' positions in Figure 18.6.

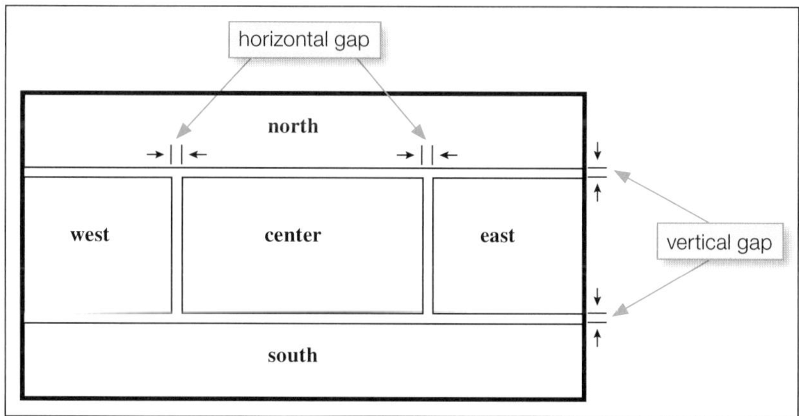

Figure 18.6 BorderLayout regions

Assume that you're inside a container class. To assign a BorderLayout manager to the container, call the setLayout method like this:

```
setLayout (new BorderLayout (<horizontal-gap>, <vertical-gap>));
```

The *horizontal-gap* argument specifies the number of pixels of blank space that separate the west, center, and east regions. Figure 18.6 illustrates this. The *vertical-gap* argument specifies the number of pixels of blank space that separate the north region from the other regions and the south region from the other regions. Once again, Figure 18.6 illustrates this. If you omit the gap arguments, the gap values are zero by default. In other words, if you call the `BorderLayout` constructor with no arguments, there will be no gaps between the regions.

The sizes of the five regions are determined at runtime, and they're based on the contents of each region. Thus, if the west region contains a long label, the layout manager attempts to widen the west region. Likewise, if the west region contains a short label, the layout manager attempts to narrow the west region.

If an outer region is empty, it collapses so that it does not take up any space. But what exactly happens during the collapse? Each outer region controls only one dividing line, so only one dividing line moves for each collapsed region. Figure 18.7 shows you that the west region's dividing line is the boundary between west and center, the north region's dividing line is the boundary between north and below, and so on. So if the north

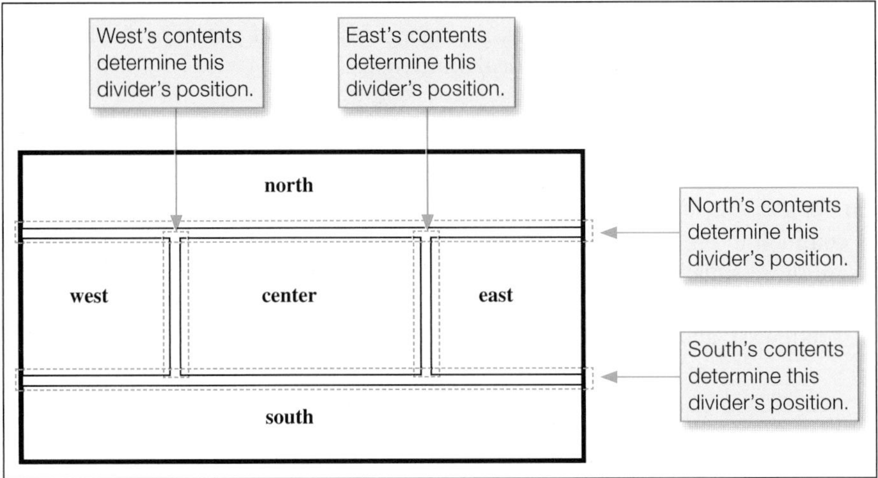

Figure 18.7 `BorderLayout` regions

region is empty, the north dividing line moves all the way up to the top border, and the west, center, and east regions all expand upward. What happens if the east and south regions are both empty? The east region being empty causes the east dividing line to move all the way to the right border. The south region being empty causes the south dividing line to move all the way down to the bottom border. Here's the resulting layout:

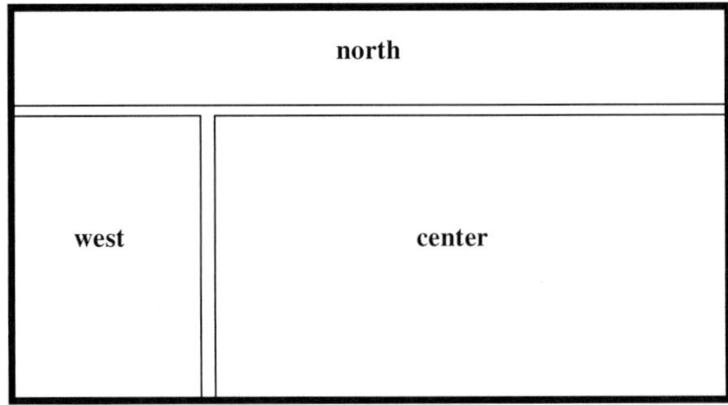

What happens if the center region is empty? The center region doesn't control any of the dividing lines, so nothing happens.

Adding Components

Suppose you have a container class that uses a `BorderLayout` manager. To add a component to one of the container's `BorderLayout` regions, call the container's `add` method like this:

```
add (<component>, <region>);
```

Replace *<component>* by a component like a `JLabel` or a `JButton`, and replace *<region>* by one of these named constants: `BorderLayout.NORTH`, `BorderLayout.SOUTH`, `BorderLayout.WEST`, `BorderLayout.EAST`, or `BorderLayout.CENTER`. For example, here's how to add a Tunisia button to the north region:

```
add(new JButton("Tunisia"), BorderLayout.NORTH);
```

If you call the `add` method with no region argument, the center region is used by default. Thus, to add a Central African Republic button to the center region, you can use either of these two statements:

```
add(new JButton("Central African Republic"), BorderLayout.CENTER);
add(new JButton("Central African Republic"));
```

Which statement is better? We prefer the first statement because it makes the code easier to understand. More formally, we say that the first statement is *self-documenting*.

With a `FlowLayout`, you can add as many components as you like. With a `BorderLayout`, you can add only five components total, one for each of the five regions. If you add a component to a region that already has a component, then the new component replaces the old component. Thus, in executing the following lines, the Somalia button replaces the Djibouti button:

```
add(new JButton("Djibouti"), BorderLayout.EAST);
add(new JButton("Somalia ", BorderLayout.EAST));
```

If you need to add more than one component to a region, it's easy to make the mistake of calling `add` twice for the same region. After all, there's no compile-time error to warn you of your misdeed. But what you really need to do is add a `JPanel` component. We'll discuss the `JPanel` component later in the chapter. It allows you to store multiple components in a place where only one component is allowed.

AfricanCountries Program with Buttons

OK, now let's put this `BorderLayout` material into practice by using it within a complete program. In our AfricanCountries program, we add African-country buttons to the five regions of a `BorderLayout` window. See the program's output window at the bottom of Figure 18.8. The five rectangles you see are the five regions, but they're also the five buttons. The buttons are the same size as the regions because, with a `BorderLayout` manager, components automatically expand to fill their entire region. Note how the outer four regions' sizes conform nicely to their contents. In other words, the west region is wide enough to show "Western Sahara," the south region is tall enough to show "South Africa," and so on. In contrast, note how the center region is unable to display its full "Central African Republic" content. This is because the outer regions control the dividing lines. The center region gets whatever room is left over.

Skim through the AfricanCountries program listing in Figure 18.8. Most of the code is straightforward. But this statement is rather quirky:

```
add(new JButton("<html>South<br>Africa</html>"), BorderLayout.SOUTH);
```

```
/*************************************************************
* AfricanCountries
* Dean & Dean
*
* This program shows component layout for BorderLayout manager.
*************************************************************/

import javax.swing.*;  // for JFrame and JButton
import java.awt.BorderLayout;

public class AfricanCountries extends JFrame
{
  private static final int WIDTH = 310;
  private static final int HEIGHT = 170;

  public AfricanCountries()
  {
    setTitle("African Countries");
    setSize(WIDTH, HEIGHT);
    setDefaultCloseOperation(EXIT_ON_CLOSE);
    setLayout(new BorderLayout());
    add(new JButton("Tunisia"), BorderLayout.NORTH);
    add(new JButton("<html>South<br>Africa</html>"),
      BorderLayout.SOUTH);
    add(new JButton("Western Sahara"), BorderLayout.WEST);
    add(new JButton("Central African Republic"),
      BorderLayout.CENTER);
    add(new JButton("Somalia"), BorderLayout.EAST);
    setVisible(true);
  } // end AfricanCountries constructor

  //**********************************************************

  public static void main(String[] args)
  {
    new AfricanCountries();
  } // end main
} // end class AfricanCountries
```

Figure 18.8 AfricanCountries program and its output

Let's review briefly those angled-bracket commands that you see here—<html>, <br>, and </html>. As you may recall from the HTMLGenerator program in Chapter 16, the angled bracket elements are called tags. The `<html>` tag indicates the start of a HTML file, the `<br>` tag indicates a line break (i.e., a new line), and the `</html>` tag indicates the end of an HTML file. Normally, you insert tags into an HTML file. But here we're inserting them into component text in order to produce a new line. When used in `JLabel` and `JButton` text, the `<html>` and `</html>` tags tell the Java compiler that the enclosed text (the text between the `<html>` and `</html>` tags) should be interpreted as HTML text. And the `<br>` tag tells the Java compiler to insert a newline character in the text.[2]

We'd like to mention one additional item in the AfricanCountries program. The `setLayout` method call can be omitted. As we said previously, the `BorderLayout` is the default layout manager for `JFrame` windows. Therefore, if you omit the `setLayout` method call, the program works just fine. But we prefer to keep the `setLayout` call because it makes the program easier to understand.

AfricanCountries Program with Labels

You might have noticed the dividing lines in Figure 18.8's output window. Those come from the buttons' borders, not from the `BorderLayout` manager. If we used label components instead of button components, you would see no dividing lines. Likewise in Figure 18.8, the margins around the words come from the button components. If we used label components instead of button components, you would see no margins around the words. Below, we show what the AfricanCountries program displays when the button components are replaced with label components. Be aware that the dashed lines don't appear on the actual window. We've drawn them in to show you the region boundaries.

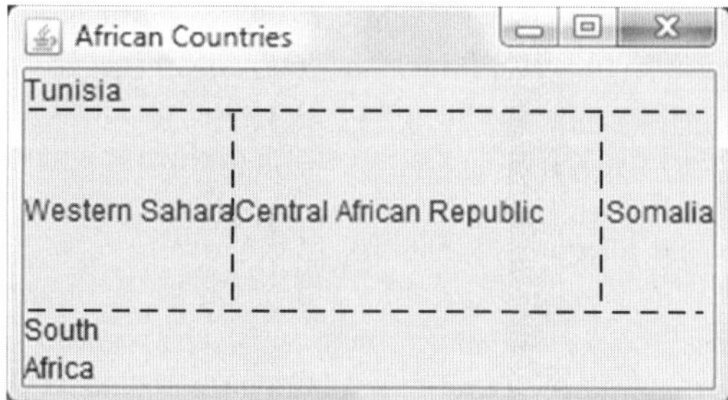

The regions are much the same as before, except that the west and east regions are narrower. That's because there are no margins around the words. Narrower west and east regions means there's more room for the center region. Thus, the center region displays its entire "Central African Republic" text.

Note that the African-country labels are left aligned. That's the default for a label in a `BorderLayout` region. If you want a different alignment than the default, instantiate the label with an alignment constant like this:

```
new JLabel(<label's-text>, <alignment-constant>)
```

[2]It may have occurred to you to insert the newline character, `\n`, into the component's text. Unfortunately, that doesn't work for `JButton` and `JLabel` components. However, it does work for the `JTextArea` component. We'll describe the `JTextArea` component later in the chapter.

Replace <*alignment-constant*> by one of the following named constants: `SwingConstants.LEFT`, `SwingConstants.CENTER`, or `SwingConstants.RIGHT`. Here's an example that adds a center-aligned label to a `BorderLayout` north region:

```
add(new JLabel("Tunisia", SwingConstants.CENTER), BorderLayout.NORTH);
```

If we apply that line of code to our AfricanCountries program and we apply similar center-alignment code to our center and south regions, the program displays this:

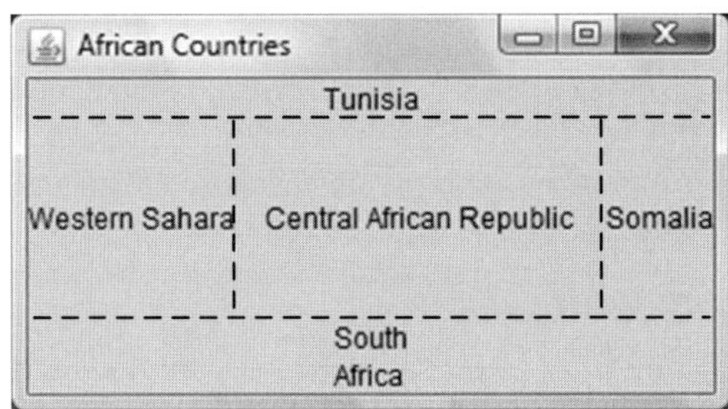

Once again, the dashed lines don't appear on the actual window. We've drawn them to show you the region boundaries. There's no point in applying center alignment to the west and east labels. For these labels, the alignment is irrelevant because the west and east labels have no room to move. As evidenced by the dashed lines, they're already aligned with both their left and right boundaries.

Now back to the alignment constants—`SwingConstants.LEFT`, `SwingConstants.CENTER`, and `SwingConstants.RIGHT`. You might think that `SwingConstants` is a class because its first letter is capitalized. If it were a class, then it would describe an object. But it doesn't describe an object, and it's not a class. Actually, `SwingConstants` is an interface, defined in the `javax.swing` package. The Java API provides the `SwingConstants` interface as a repository for various GUI-related named constants. To access a named constant in the `SwingConstants` interface, preface the named constant with the interface name. For example, to access the `LEFT` alignment constant, use `SwingConstants.LEFT`. If you want additional details about interfaces, see Chapter 14, Section 14.9.

It's easy to get confused between label alignment for a `BorderLayout` container and label alignment for a `FlowLayout` container. With a `BorderLayout` container, if you want to specify a label's alignment, you need to specify a `SwingConstants` value as part of the label's instantiation. If you do that with a `FlowLayout` container, the code will compile, but it won't affect the label's alignment. With a `FlowLayout` container, individual component alignment is irrelevant. What matters is the container's alignment. If the container uses left alignment, then all its components are left aligned; if the container uses center alignment, then all its components are center aligned; and so on. To set the container's alignment, insert one of the `FlowLayout` alignment constants (`FlowLayout.LEFT`, `FlowLayout.CENTER`, `FlowLayout.RIGHT`) in the `FlowLayout` constructor call. Here's how to specify left alignment for all the components in a `FlowLayout` container:

```
setLayout(new FlowLayout(FlowLayout.LEFT));
```

18.6 `GridLayout` Manager

The `BorderLayout` manager's partitioning scheme (north, south, east, west, center) works well for many situations, but not for all situations. Often, you'll need to display information using a table format; that is, you'll need to display information that's organized by rows and columns. The `BorderLayout` manager doesn't work well for table formats, but the `GridLayout` manager works great!

GridLayout Cells

The GridLayout manager lays out a container's components in a rectangular grid. The grid is divided into equal-sized cells. Each cell can hold only one component.

Assume that you're inside a container class. To assign a GridLayout manager to the container, call the setLayout method like this:

```
setLayout(new GridLayout(<number-of-rows>, <number-of-columns>,
  <horizontal-gap>, <vertical-gap>));
```

The *<number-of-rows>* and *<number-of-columns>* arguments specify the number of rows and number of columns, respectively, in the rectangular grid. The *<horizontal-gap>* argument specifies the number of pixels of blank space that appear between each column in the grid. The *<vertical-gap>* argument specifies the number of pixels of blank space that appear between each row in the grid. If you omit the gap arguments, the gap values are zero by default. In other words, if you call the GridLayout constructor with only two arguments, there will be no gaps between the cells.

Adding Components

Assume that you're inside a GridLayout container class. To add a component to one of the container's cells, call the add method like this:

```
add(<component>);
```

Note the simplicity of the add method call. In particular, note that there's no mention of the cell that the component plugs into. So how does the GridLayout manager know which cell to plug the component into? The GridLayout manager positions components within the container using left-to-right, top-to-bottom order. The first added component goes in the top-left-corner cell, the next added component goes in the cell to the right of the first component, and so on.

The code fragment below generates a two-row, three-column table with six buttons. The code fragment specifies gaps of 5 pixels between the rows and columns.

```
setLayout(new GridLayout(2, 3, 5, 5));
add(new JButton("1"));
add(new JButton("2"));
add(new JButton("3"));
add(new JButton("4"));
add(new JButton("5"));
add(new JButton("6"));
```

Assume that the above code fragment is part of a complete, working program. Here's what the program displays:

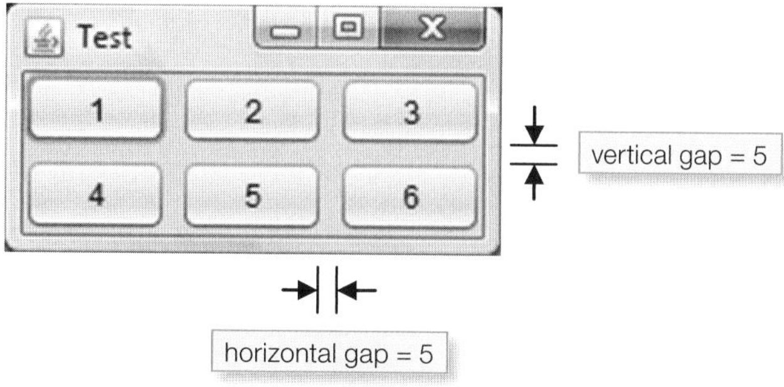

The six rectangles you see are the six cells, but they're also the six buttons. The buttons are the same size as the cells because, with a GridLayout manager, components expand to fill their cells. That should sound familiar; BorderLayout components do the same thing.

Specifying Number of Rows and Number of Columns

When creating a GridLayout manager, you call the GridLayout constructor with a number-of-rows argument and a number-of-columns argument. Those two arguments require some explanation. To help with the explanation, consider three different cases.

Case one:

If you know the number of rows and columns in your table and the table will be completely filled in (i.e., there are no empty cells), call the GridLayout constructor with the actual number of rows and the actual number of columns. That's what we did in our previous example. We knew we wanted a two-row by three-column table with six buttons, so we specified 2 for the rows argument and 3 for the columns argument.

Case two:

Sometimes you might want a row-oriented display. In other words, you want a certain number of rows displayed, and you don't care about or aren't sure about the number of columns. If that's the case, call the GridLayout constructor with the actual number of rows for the rows argument and 0 for the columns argument. A 0 for the columns argument indicates that you're leaving it up to the GridLayout manager to determine the number of columns.

The code fragment below generates a two-row GridLayout with five buttons. Because the setLayout call does not specify gap values, the GridLayout displays no gaps between the buttons.

```
setLayout(new GridLayout(2, 0));
add(new JButton("1"));
add(new JButton("2"));
add(new JButton("3"));
add(new JButton("4"));
add(new JButton("5"));
```

Assume the above code fragment is part of a complete, working program. Here's what the program displays:

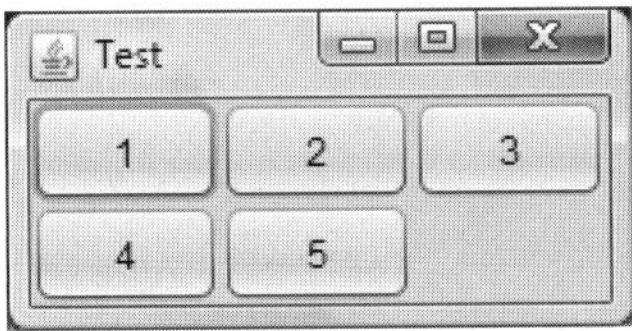

Case three:

Sometimes you might want a column-oriented display. In other words, you want a certain number of columns displayed, and you don't care about or aren't sure about the number of rows. If this is the case, call the GridLayout constructor with the actual number of columns for the columns argument and 0 for the rows argument. A 0 for the rows argument indicates that you're leaving it up to the GridLayout manager to determine the number of rows.

The code fragment below generates a four-column `GridLayout` with five buttons:

```
setLayout(new GridLayout(0, 4));
add(new JButton("1"));
add(new JButton("2"));
add(new JButton("3"));
add(new JButton("4"));
add(new JButton("5"));
```

Assume that the above code fragment is part of a complete, working program. Here's what the program displays:

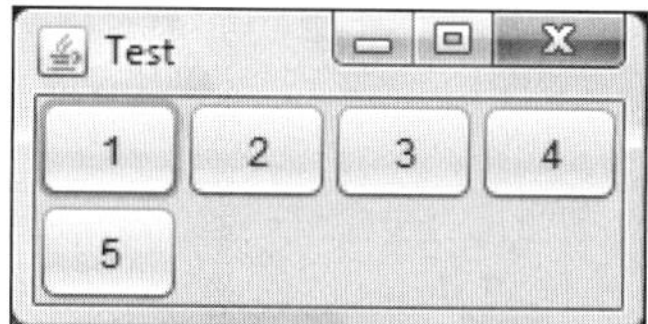

Now for a couple of things to watch for. As you know, there's special significance when you call the `GridLayout` constructor with rows = 0 or columns = 0. It puts the `GridLayout` manager in charge of choosing the number of rows or the number of columns. But it only works if you have one 0-value argument, not two. If you call the `GridLayout` constructor with two 0-value arguments, you'll get a run-time error.

What about the opposite case—when you call the `GridLayout` constructor with two non-0 values for the rows and columns arguments. That's fine, as long as your table is completely filled. If it's not completely filled, you might get unexpected results. For example, the above four-column window is not completely filled. Suppose you accidentally specify a value for the rows argument:

```
setLayout(new GridLayout(2, 4));
```

Here's what the program displays:

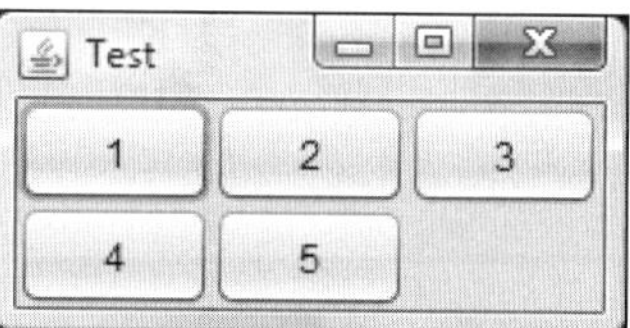

Now that's strange! There are three columns, even though we specified four. Moral of the story: Call the `GridLayout` constructor with two non-0 values only if your table is completely filled.[3]

[3]Here's the inside skinny. If you call the `GridLayout` constructor with two non-0 values for the rows and columns arguments, the columns argument is ignored and the `GridLayout` manager determines the number of columns on its own. For the case where you have two non-0 values and the table is completely filled, the `GridLayout` manager still determines the number of columns on its own. But the determined number of columns works out perfectly (that is, the determined number of columns matches the specified number of columns).

18.7 Tic-Tac-Toe Example

In this section, we present a simple tic-tac-toe program. We've chosen tic-tac-toe because we wanted to illustrate `GridLayout` details. And tic-tac-toe, with its three-row by three-column board, provides the perfect opportunity for that.

User Interface

The program initially displays a three-row, three-column grid of blank buttons. Two users, player X and player O, take turns clicking blank buttons. Player X goes first. When player X clicks a button, the button's label changes from blank to X. When player O clicks a button, the button's label changes from blank to O. Player X wins by getting three X's in a row, 3 X's in a column, or 3 X's in a diagonal. Player O wins in the same manner, except that O's are looked at instead of X's. To get a better handle on all this, see the sample session in Figure 18.9.

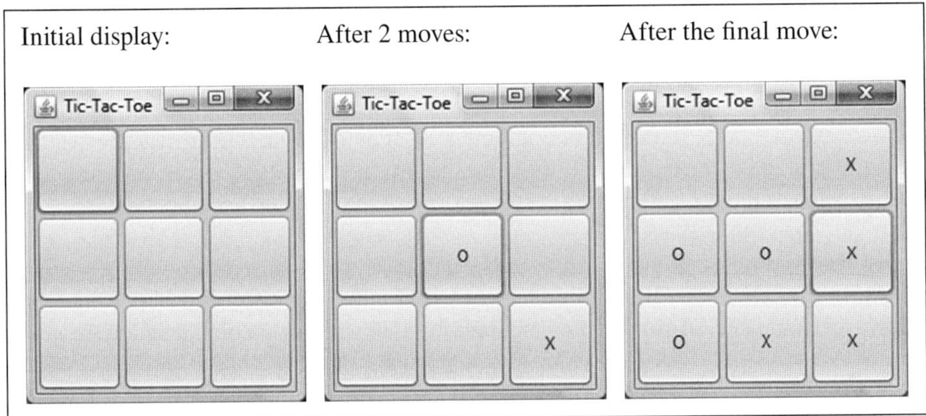

Figure 18.9 Sample session for the TicTacToe program

Program Details

See the TicTacToe program listing in Figures 18.10a and 18.10b. Most of the code should make because already because its structure parallels the structure in our previous GUI programs. We'll skip the familiar code and focus on the more difficult code.

Note the `setLayout` method call in Figure 18.10a. It contains a `GridLayout` constructor call that specifies three rows and three columns. The constructor call does not include horizontal-gap and vertical-gap arguments, so the tic-tac-toe buttons display with no gaps between them.

Now let's take a look at the `Listener` class in Figure 18.10b. In particular, note the statement where we get the clicked button and save it in a local variable:

```
JButton btn = (JButton) e.getSource();
```

⚠ The `(JButton)` cast operator is used because if there were no cast operator, the compiler would generate an error. Why? Because the compiler would see an `Object` at the right being assigned to a `JButton` at the left (it sees an `Object` at the right because `getSource` is defined with an `Object` return type). In this case, because `getSource` really returns a `JButton`, it's legal to cast its returned value to `JButton`, and that satisfies the compiler and eliminates the error.

```
/***************************************************************
* TicTacToe.java
* Dean & Dean
*
* This program implements a simplified version of tic-tac-toe.
* As a user clicks blank buttons, the buttons' labels change
* to X and O in alternating sequence.
***************************************************************/

import javax.swing.*;     // for JFrame and JButton
import java.awt.GridLayout;
import java.awt.event.*;  // for ActionListener and ActionEvent

public class TicTacToe extends JFrame
{
  private static final int WIDTH = 200;
  private static final int HEIGHT = 220;
  private boolean xTurn = true; // Is it X's turn?

  //*************************************************************

  public TicTacToe()
  {
    setTitle("Tic-Tac-Toe");
    setSize(WIDTH, HEIGHT);
    setDefaultCloseOperation(EXIT_ON_CLOSE);
    createContents();
    setVisible(true);
  } // end TicTacToe constructor

  //*************************************************************

  // Create components and add to window.

  private void createContents()
  {
    JButton button; // re-instantiate this button to fill board
    Listener listener = new Listener();

    setLayout(new GridLayout(3, 3));
    for (int i=0; i<3; i++)
    {
      for (int j=0; j<3; j++)
      {
        button = new JButton();
        button.addActionListener(listener);
        add(button);
      } // end for j
    } // end for i
  } // end createContents
```

Figure 18.10a TicTacToe program—part A

```
  //***************************************************************

  // If user clicks a button, change its label to "X" or "O".

  private class Listener implements ActionListener
  {
    public void actionPerformed(ActionEvent e)
    {
      JButton btn = (JButton) e.getSource();
      if (btn.getText().isEmpty())
      {
        btn.setText(xTurn ? "X" : "O");
        xTurn = !xTurn;
      }
    } // end actionPerformed
  } // end class Listener

  //***************************************************************

  public static void main(String[] args)
  {
    new TicTacToe();
  }
} // end class TicTacToe
```

Figure 18.10b TicTacToe program—part B

Let's examine the `Listener` class's `if` statement:

```
if (btn.getText().isEmpty())
{
  btn.setText(xTurn ? "X" : "O");
  xTurn = !xTurn;
}
```

We first check to ensure that the button is a blank button. We then reassign the button's label by using a conditional operator. If `xTurn` holds `true`, then X is assigned to the button label. Otherwise, O is assigned to the button label. We then change the value of `xTurn` by assigning its negated value into it. More specifically, if `xTurn` is `false`, we assign `true` to `xTurn`. And if `xTurn` is `true`, we assign `false` to `xTurn`.

18.8 Problem Solving: Winning at Tic-Tac-Toe (Optional)

As you might have noticed, the previous section's TicTacToe program doesn't check for a winning move. As a problem-solving exercise, let's now discuss how to add that functionality. Rather than provide you with a Java solution, we'll provide you with the thought process for coming up with a solution. We'll codify the thought process using pseudocode. One of this chapter's projects asks you to finish the job by implementing a complete Java program solution.

Iterate to enhance.

To check for a win (i.e., to check for three in a row, three in a column, or three in a diagonal), the listener needs to access multiple buttons. As it stands now, the `TicTacToe` listener can access only one button—the button that was clicked. It gets that button by calling `getSource`. So how should you change the program so the listener can access multiple buttons?

To access multiple buttons, you need to declare multiple buttons. You could declare nine separate buttons, but the more elegant solution is to declare a three-row, three-column, two-dimensional array of buttons. The next question is, where should you declare the array? Do you declare it as a local variable inside the listener or as an instance variable at the top of the program? In general, local variables are preferred, but in this case, a local variable won't work. Local variables don't persist. You need to be able to update a button from within the listener and have that update be remembered the next time the listener is called. Thus, you need to declare the buttons array as an instance variable.

You need to check for a win only when the user clicks a button. So add check-for-a-win code to the `actionPerformed` method inside the button's listener. In adding the code, use top-down design. In other words, don't worry about the low-level details; just assume they work. Here's the updated `actionPerformed` method. The added code is in pseudocode:

```
public void actionPerformed(ActionEvent e)
{
  JButton btn = (JButton) e.getSource();
  if (btn.getText().isEmpty())
  {
    btn.setText(xTurn ? "X" : "O");
    if there's a win
    {
      print winning player
      prepare for new game
    }
    else
    {
      xTurn = !xTurn;
    }
  }
} // end actionPerformed
```

pseudocode

The pseudocode contains three tasks—checking for a win, printing the winner, and preparing for a new game. Checking for a win requires the most thought, so we'll postpone that task for now. Let's discuss the other two tasks first.

Printing the winner should be straightforward. Just call `JOptionPane.showMessageDialog` with a congratulatory message. The message should include the player's name, X or O, which can be obtained by re-using the conditional operator code, `xTurn ? "X" : "O"`.

Preparing for a new game should be straightforward as well. Just assign the empty string to the board's button labels and assign `true` to the `xTurn` variable (X always goes first).

Feel free to implement the print-winning-player and prepare-for-new-game tasks as embedded code inside the `if` statement or as separate helper methods. Either way is fine. But the checking-for-a-win task definitely should be implemented as a separate helper method. Why? Note how cleanly `win` is called in the above pseudocode. You can retain that clean look in the final Java code only if you implement the checking-for-a-win task as a method, not as embedded code.

In implementing the win method, you need to check the two-dimensional buttons array for three in a row, three in a column, or three in a diagonal. Normally, when you access a group of elements in an array, you should use a `for` loop. So you might want to use a `for` loop to access the elements in the first row, use another `for` loop to access the elements in the second row, and so on. But that would require eight `for` loops:

```
for loop to handle first row
for loop to handle second row
...
for loop to handle second diagonal
```

Yikes! That's a lot of `for` loops! Is there a better way? How about taking the opposite approach and using no `for` loops? Use one big `if` statement like this:

```
if (btns[0][0] equals X and btns[0][1] equals X and btns[0][2] equals X) or
   (btns[1][0] equals X and btns[1][1] equals X and btns[1][2] equals X) or
    ...
   (btns[0][2] equals X and btns[1][1] equals X and btns[2][0] equals X)
{
   return true
}
else
{
   return false
}
```

That works fine, but if you're bothered by the length of the `if` condition (eight lines long), you might want to try the following. Use one `for` loop for all the rows, one `for` loop for all the columns, and one `if` statement for the two diagonals:

```
for (i←0; i<3; i++)
{
   if (btns[i][0] equals X and btns[i][1] equals X and btns[i][2] equals X)
   {
      return true
   }
}
for (j←0; j<3; j++)
{
   if (btns[0][j] equals X and btns[1][j] equals X and btns[2][j] equals X)
   {
      return true
   }
}
if (btns[0][0] equals X and btns[1][1] equals X and btns[2][2] equals X) or
   (btns[0][2] equals X and btns[1][1] equals X and btns[2][0] equals X)
{
   return true
}
return false
```

Of the three solutions, we prefer the last one because we feel its code is the most understandable.

To make the tic-tac-toe program more "real world," you'd probably want to provide additional functionality. In particular, you'd want to check for a "cat's game," which is when the board is filled and no one has won. You're asked to implement that functionality in one of the chapter's projects.

18.9 Embedded Layout Managers

Suppose that you'd like to implement this math-calculator window:

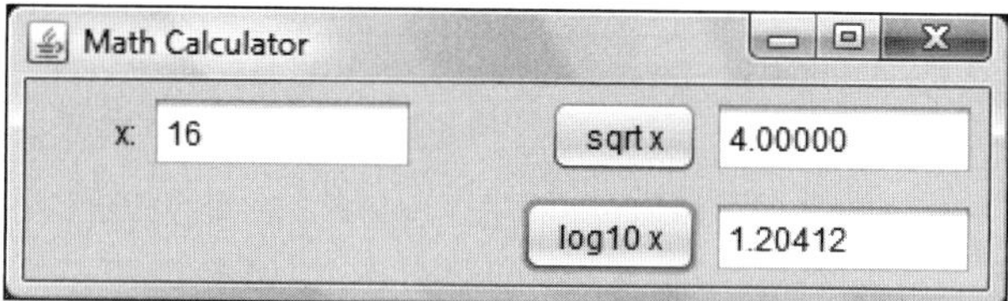

What type of layout scheme should you use? Coming up with a good layout scheme often requires creativity. We'll walk you through the creative process for this math-calculator example.

Trying Out the Different Layout Managers

The math calculator window appears to have two rows and four columns. So is a two-row by four-column `Gridlayout` scheme appropriate? The `GridLayout` manager is usually adequate for positioning components in an organized, tabular fashion, but it's limited by one factor—each of its cells must be the same size. If we use a two-row by four-column `GridLayout` scheme for the math-calculator window, then we'll have eight same-sized cells. That's fine for most of the cells, but not for the top-left cell. The top-left cell would hold the x: label. With such a small label, we would want a relatively small cell for it. But with a `GridLayout` scheme, a "relatively small cell" is not an option.

Because the `GridLayout` manager is less than ideal, you might want to think about the `FlowLayout` manager. That could sort of work if you use right-aligned components. But then you'd be at the mercy of the user not resizing the window. If the user widens the window, then the log10 x button would flow up to the top line, and you don't want that. So the `FlowLayout` manager is also less than ideal. The `BorderLayout` manager isn't even close. So what's the solution?

Using an Embedded Layout Scheme

Delegate.

In coming up with layouts for more complex windows, the key is often to embed layout managers inside other layout managers. Let's first tackle the outer layout manager. For the math calculator window, we want the input at the left and the output at the right. Those two entities are approximately the same width, so it makes sense to consider using a two-column `GridLayout` for them. The left column would contain the input components—the x label and the input text box. The right column would contain the output components—the square root's button and output text box and the logarithm's button and output text box. We'd like to organize the output components so that the square root's items are above the logarithm's items. That means using two rows for our `GridLayout`. See the two-row by two-column `GridLayout` scheme in Figure 18.11.

As you know, `GridLayout` managers only allow one component per cell. But Figure 18.11 shows two components in the top-left cell and two components in the top-right cell. To implement that organization

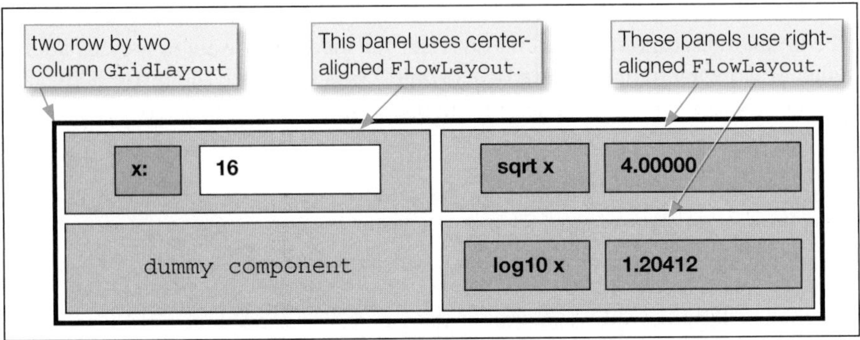

Figure 18.11 `GridLayout` with embedded `FlowLayout` panels in three of the cells

scheme, you'll need to group each of the two-component pairs into their own separate containers. And to achieve the proper layout, you'll need to apply layout managers to each of those containers. The top-left cell's container uses a center-aligned `FlowLayout` manager. The right cells' containers use right-aligned `FlowLayout` managers. Voilà—layout managers inside a layout manager. Pretty cool, eh?

When you have a non-trivial window, it's very common to have embedded layout managers. And when that happens, it can take a considerable amount of tweaking to get your windows to look right. Despite the tweaking, using embedded layout managers is still a lot easier than having to position components manually with pixel values like in the old days. The next section provides details on the containers for the embedded layout managers.

18.10 `JPanel` class

Before continuing with the implementation of the math-calculator program, we need to discuss the `JPanel` class. A `JPanel` container object is a generic storage area for components. If you have a complicated window with many components, you might want to compartmentalize the components by putting groups of components in `JPanel` containers. `JPanel` containers are particularly useful with `GridLayout` and `BorderLayout` windows because each compartment in those layouts can store only one component. If you need a compartment to store more than one component, let that one component be a `JPanel` container, and put multiple components into the `JPanel` container.

Implementation

As you may recall, GUI classes that begin with *J* come from the `javax.swing` package. So that's where the `JPanel` container class comes from, and you need to import the `javax.swing` package in order to use `Jpanel`.

To instantiate a `JPanel` container, use this syntax:

```
JPanel <JPanel-reference> = new JPanel(<layout-manager?>);
```

The *layout-manager* argument is optional. If it's omitted, the default is a center-aligned `FlowLayout` manager.

So the `JPanel` container's default layout manager is `FlowLayout`. Quick quiz: Do you remember the `JFrame` container's default layout manager? It's `BorderLayout`. That should make sense when you realize that `JFrame` containers are designed to handle the window as a whole. For the window as a whole, the default BorderLayout scheme works well because its report-oriented regions (north for a header, south for a footer, center for a main body) match the needs of many program windows. On the other hand, JPanel containers are designed for compartments within a window. For such compartments, the default FlowLayout scheme works well because its free-form flow matches the needs for many compartments.

Adding Components to a JPanel

After instantiating a JPanel, you'll want to add components to it. Adding components to a JPanel is the same as adding components to a JFrame. Call the add method. As you know, the add method works differently for the different layout managers. If your JPanel uses either a FlowLayout manager or a GridLayout manager, call the add method like this:

```
<JPanel-reference>.add(<component>);
```

If your JPanel uses a BorderLayout manager, you should add a second argument to specify the component's region:

```
<JPanel-reference>.add(<component>, <BorderLayout-region>);
```

Adding JPanel to a Window

After adding components to a JPanel, you'll need to add the JPanel to a window. If your window uses a FlowLayout manager or a GridLayout manager, call the add method like this:

```
add(<JPanel-reference>);
```

If your window uses a BorderLayout manager, you'll want to add a second argument to specify the component's region:

```
add(<JPanel-reference>, <BorderLayout-region>);
```

In the next section, we return to the math-calculator program. That will give us an opportunity to see how JPanel works in the context of a complete program.

18.11 MathCalculator Program

See the MathCalculator program listing in Figures 18.12a, 18.12b, and 18.12c. You should peruse the entire program on your own, but we'll focus primarily on the panel-related code.

From the MathCalculator program's createContents method, here's the code that creates the top-left cell's panel:

```
xPanel = new JPanel(new FlowLayout(FlowLayout.CENTER));
xPanel.add(xLabel);
xPanel.add(xBox);
```

The first statement instantiates the JPanel container. Because the JPanel constructor uses a center-aligned FlowLayout by default, you can write the first statement like this and get the same result:

```
xPanel = new JPanel();
```

But we prefer the original statement because it's self-documenting. The second and third statements add the x: label and the input text box to the panel.

```
/*************************************************************
* MathCalculator.java
* Dean & Dean
*
* This program uses embedded layout managers to display
* the square root and logarithm of a user-entered number.
*************************************************************/

// for JFrame, JButton, JTextField, JLabel, and JPanel
import javax.swing.*;
import java.awt.*;         // for GridLayout and FlowLayout
import java.awt.event.*;  // for ActionListener and ActionEvent

public class MathCalculator extends JFrame
{
  private static final int WIDTH = 380;
  private static final int HEIGHT = 110;

  private JTextField xBox;     // user's input value
  private JTextField xSqrtBox; // generated square root
  private JTextField xLogBox;  // generated logarithm

  //**********************************************************

  public MathCalculator()
  {
    setTitle("Math Calculator");
    setSize(WIDTH, HEIGHT);
    setDefaultCloseOperation(EXIT_ON_CLOSE);
    createContents();
    setVisible(true);
  } // end MathCalculator constructor

  //**********************************************************

  // Create components and add to window.

  private void createContents()
  {
    JPanel xPanel;     // holds x label and its text box
    JPanel xSqrtPanel; // holds "sqrt x" label and its text box
    JPanel xLogPanel;  // holds "log x" label and its text box
    JLabel xLabel;
    JButton xSqrtButton;
    JButton xLogButton;
    Listener listener;

    setLayout(new GridLayout(2, 2));
```

Figure 18.12a MathCalculator program—part A

```
    // Create the x panel:
    xLabel = new JLabel("x:");
    xBox = new JTextField(8);
    xPanel = new JPanel(new FlowLayout(FlowLayout.CENTER));
    xPanel.add(xLabel);
    xPanel.add(xBox);

    // Create the square-root panel:
    xSqrtButton = new JButton("sqrt x");
    xSqrtBox = new JTextField(8);
    xSqrtBox.setEditable(false);
    xSqrtPanel = new JPanel(new FlowLayout(FlowLayout.RIGHT));
    xSqrtPanel.add(xSqrtButton);
    xSqrtPanel.add(xSqrtBox);

    // Create the logarithm panel:
    xLogButton = new JButton("log10 x");
    xLogBox = new JTextField(8);
    xLogBox.setEditable(false);
    xLogPanel = new JPanel(new FlowLayout(FlowLayout.RIGHT));
    xLogPanel.add(xLogButton);
    xLogPanel.add(xLogBox);

    // Add panels to the window:
    add(xPanel);
    add(xSqrtPanel);
    add(new JLabel()); // dummy component
    add(xLogPanel);

    listener = new Listener();
    xSqrtButton.addActionListener(listener);
    xLogButton.addActionListener(listener);
  } // end createContents

  //*************************************************************

  // Inner class for math calculations.

  private class Listener implements ActionListener
  {
    public void actionPerformed(ActionEvent e)
    {
      double x;      // numeric value for user entered x
      double result; // calculated value
```

Add dummy component so the bottom-left cell gets filled in.

Figure 18.12b MathCalculator program—part B

```
      try
      {
        x = Double.parseDouble(xBox.getText());
      }
      catch (NumberFormatException nfe)
      {
        x = -1;   // indicates an invalid x
      }

      if (e.getActionCommand().equals("sqrt x"))
      {
        if (x < 0)
        {
          xSqrtBox.setText("undefined");
        }
        else
        {
          result = Math.sqrt(x);
          xSqrtBox.setText(String.format("%7.5f", result));
        }
      } // end if

      else // calculate logarithm
      {
        if (x < 0)
        {
          xLogBox.setText("undefined");
        }
        else
        {
          result = Math.log10(x);
          xLogBox.setText(String.format("%7.5f", result));
        }
      } // end else
    } // end actionPerformed
  } // end class Listener

  //************************************************************

  public static void main(String[] args)
  {
    new MathCalculator();
  } // end main
} // end class MathCalculator
```

Figure 18.12c MathCalculator program—part C

Farther down in the createContents method, here's the code that adds the panels to the window:

```
add(xPanel);
add(xSqrtPanel);
add(new JLabel()); // dummy component
add(xLogPanel);
```

The first, second, and fourth statements add the three panels to the window's top-left, top-right, and bottom-right cells, respectively. The third statement adds a dummy component (a blank label) to the bottom-left cell. The dummy component is necessary because without it, the xLogPanel would go into the bottom-left cell, and that's not what you want.

There's one additional item worth mentioning in this program. Note the String.format method call in Figure 18.12c's actionPerformed method. The String.format method works the same as the printf method, except that instead of printing a formatted value, it returns a formatted value. In the actionPerformed method, we call String.format to retrieve a formatted version of the calculated logarithm value. Specifically, the %7.5f conversion specifier returns a floating-point value with 5 decimal places and 7 total characters.

18.12 JTextArea Component

In the previous chapter, we introduced you to a few GUI components—JLabel, JTextField, JButton, and JOptionPane—that provide basic input/output functionality. Now we'll introduce you to a few more GUI components—JTextArea, JCheckBox, JRadioButton, and JComboBox—that provide more advanced input/output functionality. We'll start with the JTextArea component.

User Interface

The JLabel component works great for displaying a single line of text. You can use a JLabel component to display multiple lines of text, but achieving multiple lines requires cluttering up your code with HTML tags—<html> ...
...</html>. The preferred technique for displaying multiple lines of text is to use a JTextArea component. The large white area in Figure 18.13 is a JTextArea component. By the way, the small shaded area at the bottom of the figure is a JCheckBox component. We'll describe JCheckBox components in the next section.

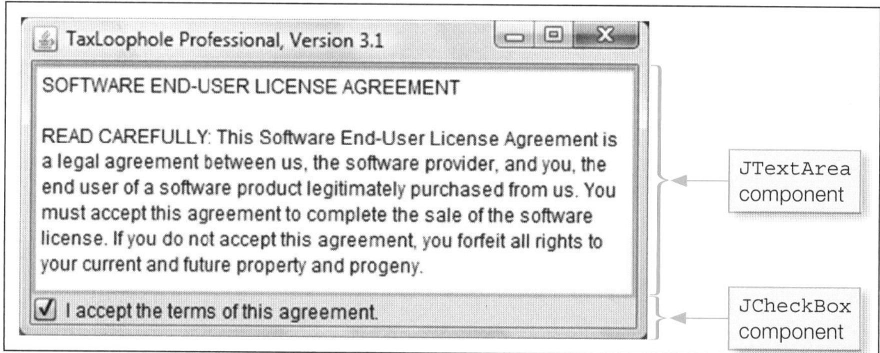

Figure 18.13 A window with a JTextArea component and a JCheckBox component

Implementation

To create a `JTextArea` component, call the `JTextArea` constructor like this:

```
JTextArea <JTextArea-reference> = new JTextArea(<display-text>);
```

The *display-text* is the text that appears in the `JTextArea` component. If the display-text argument is omitted, then the `JTextArea` component displays no initial text.

Methods

The `JTextArea` class, like all the GUI component classes, has quite a few methods. Here are the API headings and descriptions for the more popular `JTextArea` methods:

`public String getText()`

Returns the text area's text.

`public void setText(String text)`

Assigns the text area's text.

`public void setEditable(boolean flag)`

Makes the text box editable or non-editable.

`public void setLineWrap(boolean flag)`

Turns line wrap on or off.

`public void setWrapStyleWord(boolean flag)`

Specifies whether word boundaries are used for line wrapping.

`JTextArea` components are editable by default, which means users are allowed to type inside them. If you want to prevent users from editing a `JTextArea` component, call `setEditable` with an argument value of `false`. Doing so prevents users from updating the text area, but it does not prevent programmers from updating the text area. Programmers can call the `setText` method regardless of whether the text area is editable or non-editable.

`JTextArea` components have line wrap turned off by default. Normally, you'll want to turn line wrap on by calling `setLineWrap(true)`. That way, long lines automatically wrap to the next row, instead of disappearing when they reach the text area's right boundary.

For `JTextArea` components with line wrap turned on, the default is to perform line wrap at the point where the text meets the text area's right boundary, regardless of whether that point is in the middle of a word. Normally, you'll want to avoid that draconian[4] default behavior and have line wrap occur only at word boundaries. To change to a word-boundary line-wrap policy, call `setWrapStyleWord(true)`.

License-Agreement Example

Look back at the license-agreement `JTextArea` component in Figure 18.13. Figure 18.14 contains the code associated with that component. Let's now examine Figure 18.14's code. Notice the `\n\n` in the `JTextArea` constructor call. As you might recall, `\n`'s are ignored inside `JLabel` text. But they work fine inside `JTextArea` text. Notice the three calls - `setEditable(false)`, `setLineWrap(true)`, and `setWrapStyleWord(true)`. Those calls are common for `JTextArea` components.

[4]A draconian policy is a policy that is harsh or severe. The word *draconian* comes from Draco, a 7th-century B.C. government official from Athens who was in charge of codifying local law. Draco's laws were exceedingly severe. For example, even minor offenses were punishable by the death penalty.

```
private void createContents()
{
  JTextArea license;
  JCheckBox confirmBox;

  setLayout(new BorderLayout());
  license = new JTextArea(
    "SOFTWARE END-USER LICENSE AGREEMENT\n\n" +
    "READ CAREFULLY: This Software End-User License Agreement" +
    " is a legal agreement between us, the software provider," +
    " and you, the end user of a software product legitimately" +
    " purchased from us. You must accept this agreement to" +
    " complete the sale of the software license. If you do not" +
    " accept this agreement, you forfeit all rights to your" +
    " current and future property and progeny.");
  license.setEditable(false);
  license.setLineWrap(true);
  license.setWrapStyleWord(true);
  confirmBox = new JCheckBox(
    "I accept the terms of this agreement.", true);

  add(license, BorderLayout.CENTER);
  add(confirmBox, BorderLayout.SOUTH);
} // end createContents
```

Figure 18.14 The code that created Figure 18.13's display

In Figure 18.13, notice how the blocks of text in the `JTextArea` and the `JCheckBox` components crowd up against their boundaries. The image in Figure 18.15 is 40 pixels higher and 20 pixels wider than the image in Figure 18.13. This enlargement enables the image in Figure 18.15 to include a 10-pixel wide empty border between each component's contents and its respective boundary.

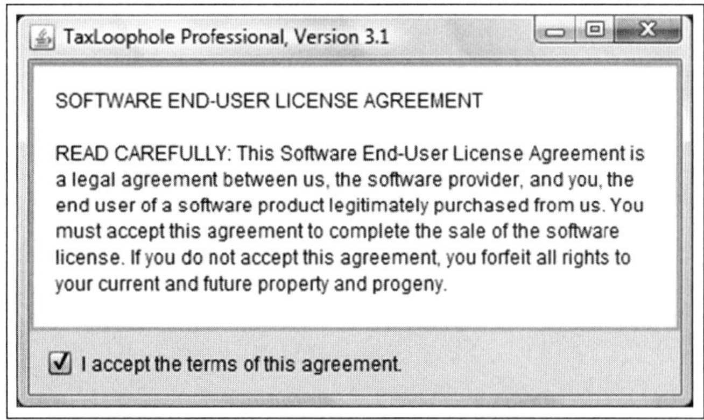

Figure 18.15 Insertion of empty borders around contents of the components in Figure 18.13's window

The program containing the code for Figure 18.14 imports the `javax.swing` and `java.awt` packages. Suppose it also imports `javax.swing.border.EmptyBorder`. Then, the following statements can insert the space you see in Figure 18.15 between each component's boundary and its contents:

```
license.setBorder(new EmptyBorder(10, 10, 10, 10));
confirmBox.setBorder(new EmptyBorder(10, 10, 10, 10));
```

18.13 JCheckBox Component

User Interface

Look at the *check box* component at the bottom of Figure 18.15. Use a check box component if you want to present an option. A check box component displays a small square with a label to its right. When the square is blank, the check box is unselected. When the square contains a check mark, the check box is selected. Users click on the check box in order to toggle between selected and unselected.

Implementation

To create a check box component, call the `JCheckBox` constructor like this:

```
JCheckBox <JCheckBox-reference> = new JCheckBox(<label>, <selected>);
```

The *label* argument specifies the text that appears at the right of the check box's square. If the label argument is omitted, then no text appears at the right of the check box's square. The *selected* argument specifies whether the check box is selected initially—`true` means selected, `false` means unselected. If the selected argument is omitted, then the check box is initially unselected.

Here's how the check box was created in the license-agreement window:

```
confirmBox = new JCheckBox("I accept the terms of this agreement.", true);
```

Methods

Here are the API headings and descriptions for the more popular `JCheckBox` methods:

`public boolean isSelected()`

Returns `true` if the check box is selected and `false` otherwise.

`public void setVisible(boolean flag)`

Makes the check box visible or invisible.

`public void setSelected(boolean flag)`

Makes the check box selected or unselected.

`public void setEnabled(boolean flag)`

Makes the check box enabled or disabled.

`public void addActionListener(ActionListener listener)`

Adds a listener to the check box.

The `isSelected` and `setVisible` methods are straightforward, but the other three methods need further explanation. Let's start with `setSelected`. Why might you want to call `setSelected` and adjust

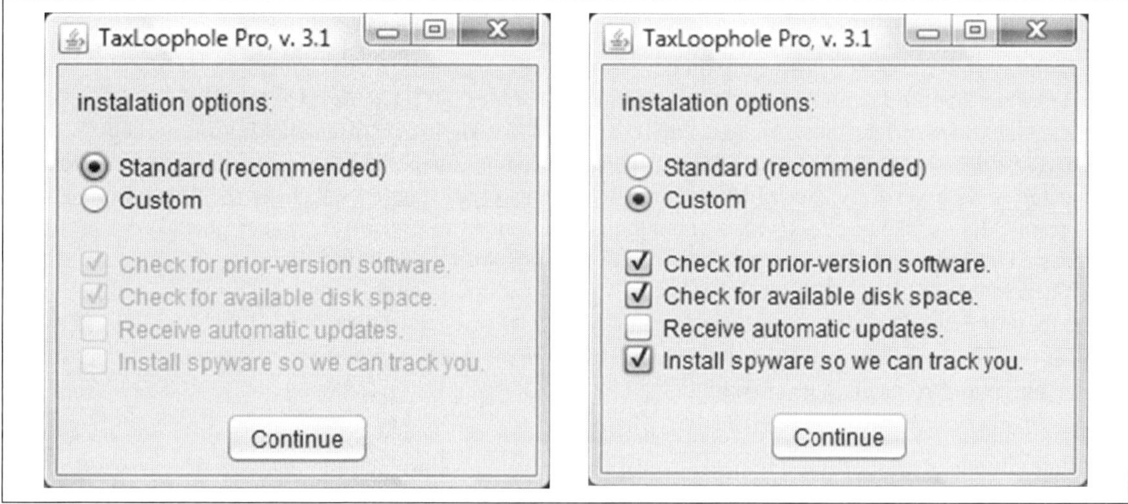

Figure 18.16 Example that the illustrates `JCheckBox`'s `setSelected` and `setEnabled` methods

the selection status of a check box? Because you might want one user input to affect another user input. For example, in Figure 18.16, the user's selection of standard versus custom[5] should affect the check box selections. More specifically, if the user selects the Standard option, the check box selections should go to their "standard" settings. As you can see in Figure 18.16's left window, the standard settings for the check boxes are the top two selected and the bottom two unselected. To have your program select the top two check boxes, those two check boxes should call `setSelected(true)`. To have your program unselect the bottom two check boxes, those two check boxes should call `setSelected(false)`.

To have your program disable a check box, the checkbox should call `setEnabled(false)`. Why might you want to call `setEnabled(false)` and disable a check box? Because you might want to keep the user from modifying that box's value. For example, if the user selects the Standard option as shown in Figure 18.16's left window, the check box selections should be set to their standard settings (as explained above), and then each check box should call `setEnabled(false)`. That way, the user cannot make changes to the standard-configuration check box values. In Figure 18.16's left window, note that the check boxes are gray. We say that they're *grayed out*. That's the standard GUI way of telling the user that something is disabled.

Check Box Listeners

With a `JButton` component, you'll almost always want an associated listener. But with a `JCheckBox` component, you may or may not want an associated listener. If you have a check box with no listener, then the check box simply serves as an input entity. If that's the case, then the check box's value (checked or unchecked) typically would get read and processed when the user clicks a button. On the other hand, if you want something to happen immediately, right when the user selects a check box, then add a listener to the check box component. Suppose you have a Green Background check box. If you want the window's background color to change to green right when the user clicks the check box, add a listener to the check box.

[5]The Standard and Custom circles at the top of Figure 18.16 are called radio buttons. We'll describe `JRadioButton` components in the next section.

The syntax for adding a listener to a JCheckBox component is the same as the syntax for adding a listener to a JButton component. Provide a listener that implements the ActionListener interface and then add the listener to the JCheckBox component by calling addActionListener.

Be aware that the Java API provides an alternative listener interface for the JCheckBox component—the ItemListener interface. An ActionListener listens for the user clicking on a check box. An ItemListener listens for a *state change*; that is, it listens for a check box changing from selected to unselected or vice versa. A check box state change is triggered when a user clicks the check box or when a program calls setSelected with a value that's different from the current value. Because we prefer the ActionListener interface in most situations, we'll stick with it when implementing JCheckBox listeners. When we get to the JRadioButton and JComboBox components in the next sections, we'll continue to use the ActionListener interface, not the ItemListener interface.

Installation-Options Example

It's now time to put these check box concepts into practice by showing you some code. Look back at the installation-options windows in Figure 18.16. In Figure 18.17, we provide the listener code associated with those windows. Let's walk through the code. In the if statement's condition, we check to see whether the standard option was selected. If that's the case, we disable the check boxes by calling setEnabled(false) for each check box. We then assign the check boxes to their standard settings by calling setSelected(true)

```
private class Listener implements ActionListener
{
  public void actionPerformed(ActionEvent e)
  {
    if (e.getSource() == standard) // standard option chosen
    {
      prior.setEnabled(false);
      diskSpace.setEnabled(false);
      updates.setEnabled(false);
      spyware.setEnabled(false);
      prior.setSelected(true);
      diskSpace.setSelected(true);
      updates.setSelected(false);
      spyware.setSelected(false);
    }
    else                              // custom option chosen
    {
      prior.setEnabled(true);
      diskSpace.setEnabled(true);
      updates.setEnabled(true);
      spyware.setEnabled(true);
    }
  } // end actionPerformed
} // end Listener
```

Figure 18.17 Listener code for Figure 18.16's installation-options windows

or `setSelected(false)` for each check box. In the `else` block, we handle the custom option being selected. We enable the check boxes by having each box call `setEnabled(true)`. This enables the user to control whether the box is selected or not.

18.14 JRadioButton Component

User Interface

Look at the circles in the windows in Figure 18.16. They're called *radio buttons.* A `JRadioButton` component displays a small circle with a label to its right. When the circle is blank, the radio button is unselected. When the circle contains a large dot, the radio button is selected.

According to the description so far, radio buttons sound a lot like check boxes. They display a shape and a label, and they keep track of whether something is on or off. The key difference between radio buttons and check boxes is that radio buttons almost always come in groups. And within a radio button group, only one radio button can be selected at a time. If a user clicks an unselected radio button, the clicked button becomes selected, and the previously selected button in the group becomes unselected. If a user clicks a selected radio button, no change occurs (i.e., the clicked button remains selected). In contrast, if a user clicks a selected check box, the check box changes its state from selected to unselected.

Implementation

To create a `JRadioButton` component, call the `JRadioButton` constructor like this:

```
JRadioButton <JRadioButton-reference> =
  new JRadioButton(<label>, <selected>);
```

The *label* argument specifies the text that appears at the right of the radio button's circle. If the label argument is omitted, then no text appears at the right of the radio button's circle. The *selected* argument specifies whether the radio button is initially selected—`true` means selected, `false` means unselected. If the selected argument is omitted, then the radio button is initially unselected.

This example shows how we created the `standard` and `custom` radio buttons in the installation-options program:

```
standard = new JRadioButton("Standard (recommended)", true);
custom = new JRadioButton("Custom");
```

To enable the functionality of a radio button group (that is, only one button selected at a time), create a `ButtonGroup` object and add individual radio button components to it. Here's how:

```
ButtonGroup <ButtonGroup-reference> = new ButtonGroup();
<ButtonGroup-reference>.add(<first-button-in-group>);
. . .
<ButtonGroup-reference>.add(<last-button-in-group>);
```

The following example shows how we created the radio button group for the `standard` and `custom` radio buttons in the installation-options program:

```
ButtonGroup rbGroup = new ButtonGroup();
rbGroup.add(standard);
rbGroup.add(custom);
```

After adding radio buttons to a radio button group, you still have to add them to a container. Radio buttons work the same as other components in terms of adding them to a container. Call the container's `add` method like this:

```
add(<first-button-in-group>);
. . .
add(<last-button-in-group>);
```

That's a lot of adding. You need to add each radio button twice—once to a radio button group and once to a container. If you like shortcuts, you might be thinking, Why does Java make you add the individual radio buttons to the container? Why not add the button group to the container and rely on that to take care of getting the buttons in the container? Adding the buttons individually to the container gives you freedom in positioning the buttons. If you wanted to, you could even put them in different panels.

Because the `JRadioButton` class begins with a *J,* you can correctly assume that it's defined in the `javax.swing` package. But what about the `ButtonGroup` class? Even though it doesn't begin with a *J,* it's also defined in the `javax.swing` package.

Methods

Here are the API headings and descriptions for the more popular `JRadioButton` methods:

`public boolean isSelected()`

Returns `true` if the radio button is selected and `false` otherwise.

`public void setSelected(boolean flag)`

Makes the radio button selected if the argument is true. Does nothing if the argument is `false`.

`public void setEnabled(boolean flag)`

Makes the radio button enabled or disabled. If enabled, it responds to mouse clicks.

`public void addActionListener(ActionListener listener)`

Adds a listener to the radio button.

We described these same methods in the `JCheckBox` section. Only one of them needs further attention—the `setSelected` method. To understand how `setSelected` works, you first need to understand fully how a user interacts with a radio button group. To select a radio button, a user clicks it. That causes the radio button to become selected and all other radio buttons in the group to become unselected. To programmatically select a radio button, you have the radio button call `setSelected(true)`. That causes the radio button to become selected and all other radio buttons in the group to become unselected. As mentioned above, there is no way for a user to unselect a button. Likewise, there is no way for a program to unselect a button. That's why calling `setSelected(false)` doesn't do anything. It compiles and runs, but it doesn't cause any buttons to change their selected status.

18.15 JComboBox Component

User Interface

A *combo box* allows a user to select an item from a list of items. Combo boxes are sometimes called *drop-down lists* because if a user clicks a combo box's down arrow, a list of selection items drops down from the original display. Then, if a user clicks a selection from the drop-down list, the list disappears, and only the selected item remains displayed. To get a better idea of what we're talking about, see the select-a-day combo box in Figure 18.18.

Combo boxes and radio button groups are similar in that they both allow the user to select one item from a list of items. But a combo box takes up less space on the window. So if you have a long list of items to choose from, and you want to save space, use a combo box rather than a group of radio buttons.

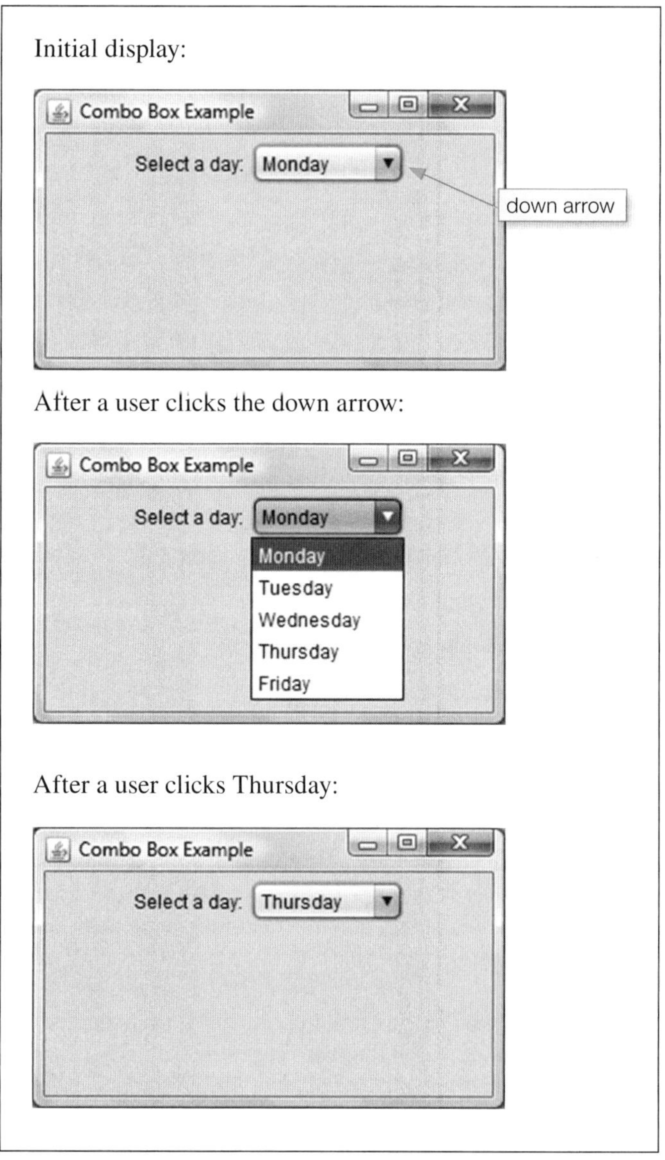

Figure 18.18 Select-a-day combo box example

Implementation

Creating a combo box component is a two-step process. First, instantiate an array of options. Then, use the array as part of a `JComboBox` instantiation. Here's the syntax for a typical `JComboBox` instantiation:

`JComboBox` *<JComboBox-reference>* `= new JComboBox(`*<array-of-options>*`);`

The following code fragment shows how we created the combo box in Figure 18.18:

```
String[] days =
  {"Monday", "Tuesday", "Wednesday", "Thursday", "Friday"};
daysBox = new JComboBox(days);
```

When a combo box first displays, the first item in its array is selected. So in the above example, Monday is selected when the combo box first displays.

Methods

Here are the API headings and descriptions for the more popular `JComboBox` methods:

`public void setVisible(boolean flag)`
Makes the combo box visible or invisible.

`public void setEditable(boolean flag)`
Makes the combo box's top portion editable or non-editable.

`public Object getSelectedItem()`
Returns the item that is currently selected.

`public void setSelectedItem(Object item)`
Changes the currently selected item to the item that's passed in.

`public int getSelectedIndex()`
Returns the index of the item that is currently selected.

`public void setSelectedIndex(int index)`
Changes the currently selected item to the item at the given index position.

`public void addActionListener(ActionListener listener)`
Adds a listener to the combo box.

The `setVisible` and `addActionListener` methods should look familiar by now. The other methods are new and require further explanation. Let's start with `setEditable`. If a combo box calls `setEditable(true)`, the combo box's top portion becomes editable. That means that a user can enter text into it the same as if it were a text box component. Additionally, the user still can use the drop-down portion of the combo box the same as always. If the user enters text into the top portion of an editable combo box, the drop-down options don't change, but the selected value does change to the entered value, as evidenced by the returned value from `getSelectedItem`. Combo boxes are named "combo" for "combination" because they are capable of implementing a mixture of components—part drop-down list, part text box. But most programmers don't bother with the combo box's text-box capability. They usually stick with the default behavior, where the top portion of the combo box is not editable.

The `getSelectedItem` method returns the currently selected item. For example, here's how you can retrieve the currently selected `daysBox` item and store it in a `favoriteDay` variable:

```
String favoriteDay = (String) daysBox.getSelectedItem();
```

What's the point of the `(String)` cast operator? The `getSelectedItem` method is defined to have a return type of `Object`. Therefore, if there were no cast operator, the compiler would see an `Object` at the right being assigned into a `String` at the left, and that would generate a compile-time error. But there is a cast operator, so the compiler sees a `String` at the right being assigned into a `String` at the left. And that makes the compiler happy.

If you'd like to select an option from a combo box programmatically, call `setSelectedItem` and pass in the item that you want to select. For example, to select `Friday` from the `daysBox` component, do this:

```
daysBox.setSelectedItem("Friday");
```

Normally, you'll call `setSelectedItem` with an argument that matches one of the combo box's items. But that's not always the case. If you'd like to clear a combo box so that no options are selected, call `setSelectedItem(null)`. If you call `setSelectedItem` with a different item (not `null` and not a combo box item), then nothing happens. Well, actually, nothing happens if it's a standard combo box. But if you first make it an editable combo box by calling `setEditable(true)`, then the item you pass to `setSelectedItem` gets put into the editable top portion of the combo box.

There are two ways to access items in a combo box. Use either item names or item indexes. The `getSelectedItem` and `setSelectedItem` methods use item names. The `getSelectedIndex` and `setSelectedIndex` methods use item indexes. For example, note how this code fragment calls `setSelectedIndex` with an index value of 2:

```
String[] days =
  {"Monday", "Tuesday", "Wednesday", "Thursday", "Friday"};
daysBox = new JComboBox(days);
daysBox.setSelectedIndex(2);
```

Because combo boxes store their items in arrays, combo box item indexes are 0-based. Therefore, in this code fragment, Monday is 0, Tuesday is 1, and Wednesday is 2. Thus, `daysBox.setSelectedIndex(2)` changes the selected item to Wednesday.

Indexing helps you process data.

Now for a short brain-teaser. Given the above code fragment, how can you change the currently selected day to the next day? Do arithmetic with the index, like this:

```
daysBox.setSelectedIndex(daysBox.getSelectedIndex() + 1);
```

18.16 Job Application Example

In this section, we put into practice what you've learned in the previous three sections. We present a complete program that uses check boxes, radio buttons, and a combo box. The program implements a job application form. If the user enters values that are indicative of a good employee, the program displays an encouraging message ("Thank you for your application submission. We'll contact you after we process your information."). Study the sample session in Figure 18.19 to get a better idea of what we're talking about.

See the JobApplication program listing in Figures 18.20a, 18.20b, and 18.20c. You should peruse the entire program on your own, particularly the listener code, but we'll focus only on the most difficult part—the layout design.

1. Initial display:

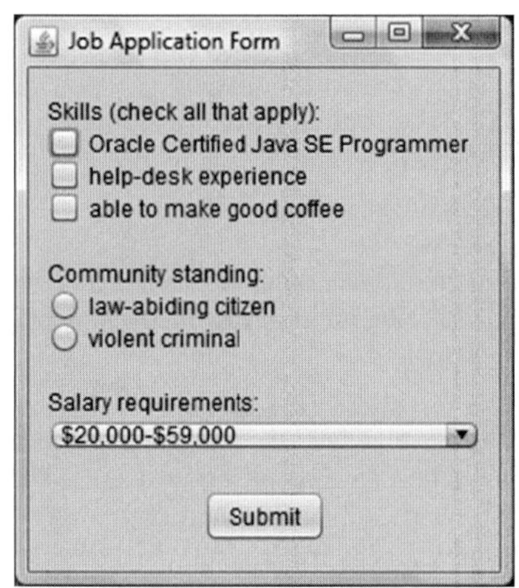

2. After the user enters good values:

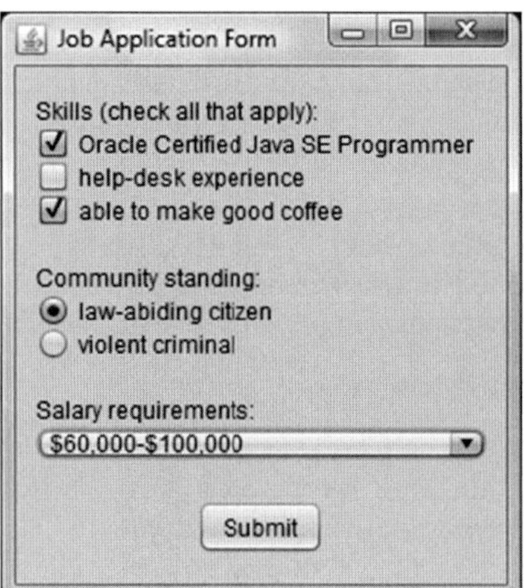

3. After the user clicks **Submit**:

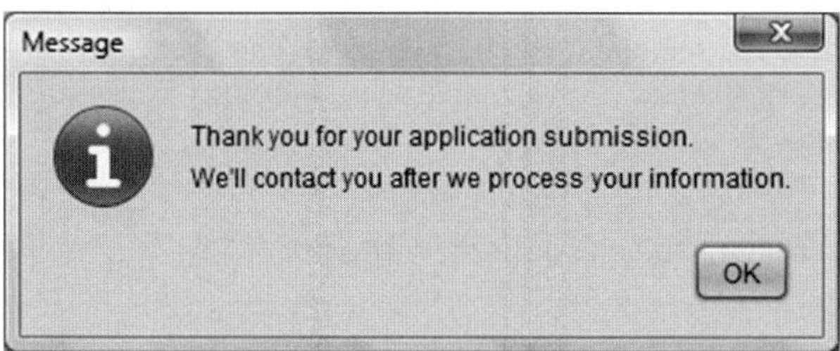

4. After the user enters not-so-good values:

5. After the user clicks **Submit**:

Figure 18.19 Sample session for the JobApplication program

```
/*************************************************************
* JobApplication.java
* Dean & Dean
*
* This program implements job application questions
* with check boxes, radio buttons, and a combo box.
*************************************************************/

// for JFrame, JCheckBox, JRadioButton, JComboBox, JButton,
// ButtonGroup, JPanel, JLabel, & JOptionPane:
import javax.swing.*;
import java.awt.*;       // BorderLayout, GridLayout, FlowLayout
import java.awt.event.*; // ActionListener, ActionEvent
import javax.swing.border.EmptyBorder;

public class JobApplication extends JFrame
{
  private static final int WIDTH = 250;
  private static final int HEIGHT = 300;

  private JCheckBox java;       // Java certified?
  private JCheckBox helpDesk;   // help-desk experience?
  private JCheckBox coffee;     // good coffee maker?
  private JRadioButton goodCitizen, criminal;
  private JComboBox salary;
  private String[] salaryOptions =
    {"$20,000-$59,000", "$60,000-$100,000", "above $100,000"};
  private JButton submit;       // submit the application

  //**********************************************************

  public JobApplication()
  {
    setTitle("Job Application Form");
    setSize(WIDTH, HEIGHT);
    setDefaultCloseOperation(EXIT_ON_CLOSE);
    createContents();
    setVisible(true);
  } // end JobApplication constructor

  //**********************************************************

  // Create components and add to window.
```

Figure 18.20a JobApplication program—part A

We spent quite a bit of time on the JobApplication's layout in order to get things to look right. Initially, we thought a simple one-column GridLayout manager would work. We added one component per cell, and we added three filler components (empty JLabels) to create gaps between the four different input areas. We thought that plan would yield the layout shown in Figure 18.21's left picture. Unfortunately, when we entered the code, the actual program yielded the layout shown in Figure 18.21's right picture. There are three problems with the actual layout—the Submit button is too wide, the top two gaps are missing, and the components are touching the top and left boundaries. We'll now discuss how to fix those problems.

If at first you don't succeed, try again.

```
private void createContents()
{
  ButtonGroup radioGroup;

  // Note:
  // The most straightforward implementation is to use a
  // GridLayout manager for the JFrame and add all components
  // to its cells. That doesn't work well because:
  // 1) Can't apply a margin to JFrame.
  // 2) The button panel is taller than the other components.

  // Need windowPanel for south-panel separation & outer margin
  JPanel windowPanel = new JPanel(new BorderLayout(0, 10));
  windowPanel.setBorder(new EmptyBorder(10, 10, 10, 10));

  // centerPanel holds all components except button
  JPanel centerPanel = new JPanel(new GridLayout(11, 1));

  // Need a panel for button so it can be center aligned
  JPanel southPanel = new JPanel(new FlowLayout());

  java = new JCheckBox("Oracle Certified Java SE Programmer");
  helpDesk = new JCheckBox("help-desk experience");
  coffee = new JCheckBox("able to make good coffee");
  goodCitizen = new JRadioButton("law-abiding citizen");
  criminal = new JRadioButton("violent criminal");
  radioGroup = new ButtonGroup();
  radioGroup.add(goodCitizen);
  radioGroup.add(criminal);
  salary = new JComboBox(salaryOptions);
  submit = new JButton("Submit");
  submit.addActionListener(new ButtonListener());

  centerPanel.add(new JLabel("Skills (check all that apply):"));
  centerPanel.add(java);
  centerPanel.add(helpDesk);
  centerPanel.add(coffee);
  centerPanel.add(new JLabel());  // filler
  centerPanel.add(new JLabel("Community standing:"));
  centerPanel.add(goodCitizen);
  centerPanel.add(criminal);
  centerPanel.add(new JLabel());  // filler
  centerPanel.add(new JLabel("Salary requirements:"));
  centerPanel.add(salary);

  windowPanel.add(centerPanel, BorderLayout.CENTER);
  southPanel.add(submit);
  windowPanel.add(southPanel, BorderLayout.SOUTH);
  add(windowPanel);
} // end createContents
```

Figure 18.20b JobApplication program—part B

```
  //***************************************************************

  // Read entered values and display an appropriate message.

  private class ButtonListener implements ActionListener
  {
    public void actionPerformed(ActionEvent e)
    {
      if (
        (java.isSelected() || helpDesk.isSelected()
                           || coffee.isSelected()) &&

        (goodCitizen.isSelected()) &&
        (!salary.getSelectedItem().equals("above $100,000")))
      {
        JOptionPane.showMessageDialog(null,
          "Thank you for your application submission.\n" +
          "We'll contact you after we process your information.");
      }
      else
      {
        JOptionPane.showMessageDialog(null,
          "Sorry, no jobs at this time.");
      }
    } // end actionPerformed
  } // end class ButtonListener

  //***************************************************************

  public static void main(String[] args)
  {
    new JobApplication();
  }
} // end class JobApplication
```

Figure 18.20c JobApplication program—part C

Problem 1: Submit Button is Too Wide

As you may recall from earlier in the chapter, buttons expand if they're added directly to a `GridLayout` cell. That explains the wide Submit button. You can fix this cosmetic problem by embedding a `FlowLayout` panel into the Submit button's area, and then adding the Submit button to the `FlowLayout` panel. With a `FlowLayout` manager, buttons don't expand; they keep their natural size.

Embed another manager. 

Problem 2: Top Two Gaps are Missing

In our first cut of the program, we used this code to implement the filler components:

```
JLabel filler = new JLabel();
...
add(filler);
```

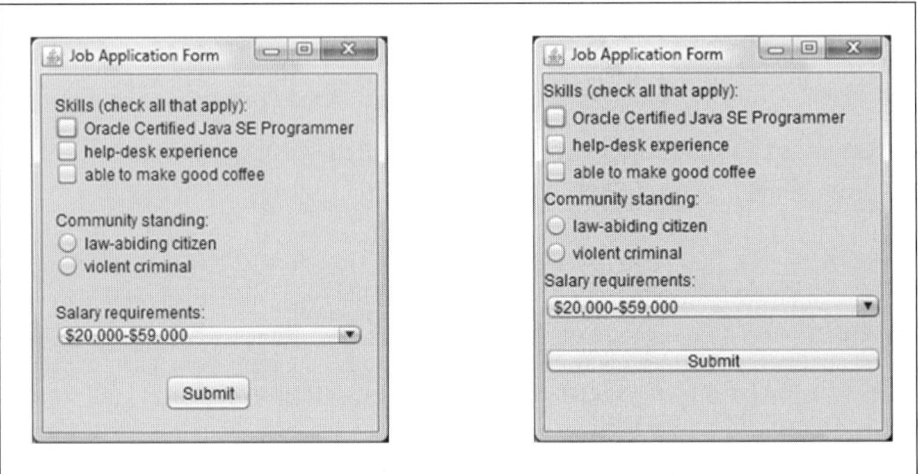

Figure 18.21 Intended versus actual layouts with the 13-row by 1-column `GridLayout` scheme

```
...
add(filler);
...
add(filler);
```

We instantiated only one label and reused it three times. You like to reuse, right? Well, the layout manager doesn't. The layout manager sees only one object, so it makes only one cell. It does not make cells for the first two `add(filler)` calls; it only makes a cell for the last `add(filler)` call. You can fix the problem by using three anonymous `JLabel` objects like this:

Use separate objects.

```
add(new JLabel());
...
add(new JLabel());
...
add(new JLabel());
```

Problem 3: Components Are Too Close to the Top and Left Boundaries

By default, containers have no margins. So if a container has left-aligned components, those components will be very close to the container's boundaries. That explains the boundary ugliness in Figure 18.21's right picture. Just as we did for Figure 18.15, you can add a margin by calling `setBorder` like this:

```
<container>.setBorder(new EmptyBorder(<top>, <left>, <bottom>, <right>));
```

In calling `setBorder`, you'll need to pass a border object as an argument. There are several different types of border classes. You should use the `EmptyBorder` class because an empty border produces a margin, which is what you want. In calling the `EmptyBorder` constructor, you'll need to pass in pixel values for the widths of the border's top, left, bottom, and right sides. For example, this constructor call passes in 10-pixel values for all four of the border's sides:

```
windowPanel.setBorder(new EmptyBorder(10, 10, 10, 10));
```

Be aware that the `EmptyBorder` class is in the `javax.swing.border` package. So import that package if you want to create an empty border.

You might think that the `setBorder` method works for all containers. Not so. It works for the `JPanel` container, but not the `JFrame` container. Therefore, you need to add a `JPanel` container to the `JobApplication` `JFrame` window and call `setBorder` from the `JPanel` container.

Use a panel. 

What type of layout manager is appropriate for the new `JPanel` container? If you use a `GridLayout` manager, that works OK, but not great. With a `GridLayout`, all rows are the same height. In Figure 18.19, note how the **Submit** button is slightly taller than the other components. The **Submit** button's added height provides a visual cue for the button's importance. To accommodate the button being taller than the other components, use a `BorderLayout` manager. Add the button panel to the south region and add all the other components to the center region. Actually, because the center region allows for only one component, you need to add the components to a `GridLayout` panel and then add the `GridLayout` panel to the center region.

Document Difficult Code

The JobApplication program's layout design is rather complicated and somewhat non-intuitive. If you ever write complicated and non-intuitive code, you should document it with detailed comments. If you don't, then someone (maybe you) might waste time later in trying to figure it out. See all the comments for the panel declarations in Figure 18.20b. Those comments help to clarify the layout-design code.

18.17 More Swing Components

In this chapter and the previous chapter, you've learned quite a bit about the Swing library. Enough to get up and running for most basic GUI needs. If you decide you want to know more, refer to Oracle's Java API website. In particular, refer to this web page within Oracle's Java API website:

http://download.oracle.com/javase/tutorial/uiswing/components/componentlist.html

It contains links to web pages for each of the standard Swing components. Each Swing component web page provides detailed information about the component, with picture examples. By perusing the Swing components web pages now, you'll know what's available.

Menus and Scroll Panes

As a first attempt at learning Swing components on your own, we recommend that you look up the `JMenuBar`, `JMenu`, and `JMenuItem` classes on the Java API website. Those classes allow you to add a *menu bar* and *menus* to the top of a window. Also, look up the `JScrollPane` class. It allows you to create a scrollable container. See Figure 18.22. It shows a window with a menu bar and a scroll bar. The menu bar contains two menus—one allows the user to adjust the brightness of the window's background color and one allows the user to adjust the font size of the window's text. The scroll bar is part of what is known

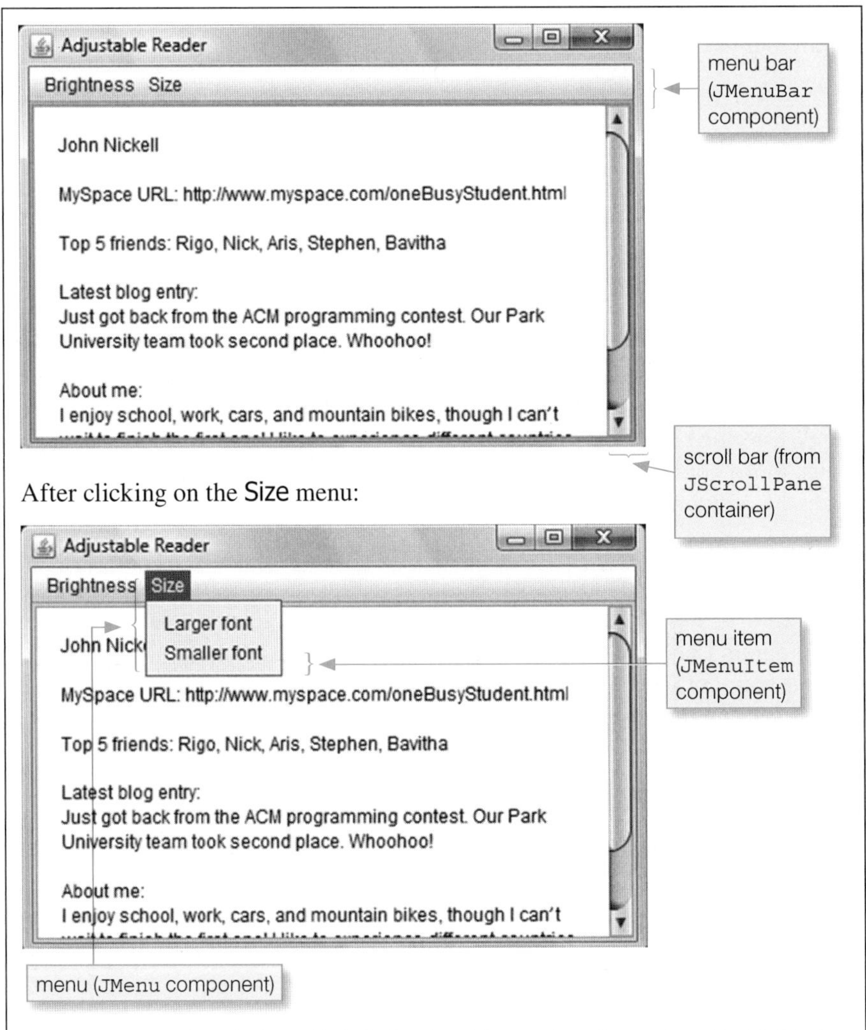

Figure 18.22 A reader program window that uses a menu bar and a scroll bar to adjust the view

as a *scroll pane*. The scroll bar allows the user to scroll up and down and view the contents of the entire window.

If you'd like to see Figure 18.22's program in its entirety, view the `ReaderMenu.java` file on the book's website. Figure 18.23 shows a portion of that program—the portion that creates the menu bar, menus, and menu items. And the following statement shows the portion of the program that creates the scroll pane. More specifically, the following statement creates a scroll pane for a text area component and then adds the scroll pane to the window.

```
add(new JScrollPane(textArea));
```

```
private JMenuBar mBar;                  // the menu bar
private JMenu menu1, menu2;             // the two menus
private JMenuItem mi1, mi2, mi3, mi4; // the four menu items
:
menu1 = new JMenu("Brightness");
menu2 = new JMenu("Size");

mi1 = new JMenuItem("Lighter background");
mi2 = new JMenuItem("Darker background");
mi3 = new JMenuItem("Larger font");
mi4 = new JMenuItem("Smaller font");

mi1.addActionListener(new BrightnessListener());
mi2.addActionListener(new BrightnessListener());
mi3.addActionListener(new SizeListener());
mi4.addActionListener(new SizeListener());

menu1.add(mi1);
menu1.add(mi2);
menu2.add(mi3);
menu2.add(mi4);

mBar = new JMenuBar();
mBar.add(menu1);
mBar.add(menu2);
setJMenuBar(mBar);
```

Figure 18.23 Code that creates menu bar, menus, and menu items for the window in Figure 18.22

Sliders

For another learn-on-your-own example, we recommend that you look up the `JSlider` class on the Java API website. The `JSlider` class allows you to add a *slider* component to a window. A slider allows the user to select a value from a range of values. To select a value, the user drags a "thumb" along a bar of values. See Figure 18.24. It mimics a lunar eclipse by covering a white circle (the moon) with a gray circle (the Earth's shadow). When the user drags the slider's thumb right, the shadow moves right. When the user drags the slider's thumb left, the shadow moves left. The slider uses an event handler to adjust the shadow's position.

If you'd like to see Figure 18.24's program in its entirety, view the `EclipseSlider.java` file on the book's website. The following code from the lunar eclipse program shows how to instantiate a slider, set properties, and add a listener:

```
slider = new JSlider(SwingConstants.HORIZONTAL, 0, 100, 0);
slider.setMajorTickSpacing(5);
slider.setPaintTicks(true);
slider.addChangeListener(new Listener());
```

This adds the slider to the current `JFrame`:

```
add(slider, BorderLayout.SOUTH);
```

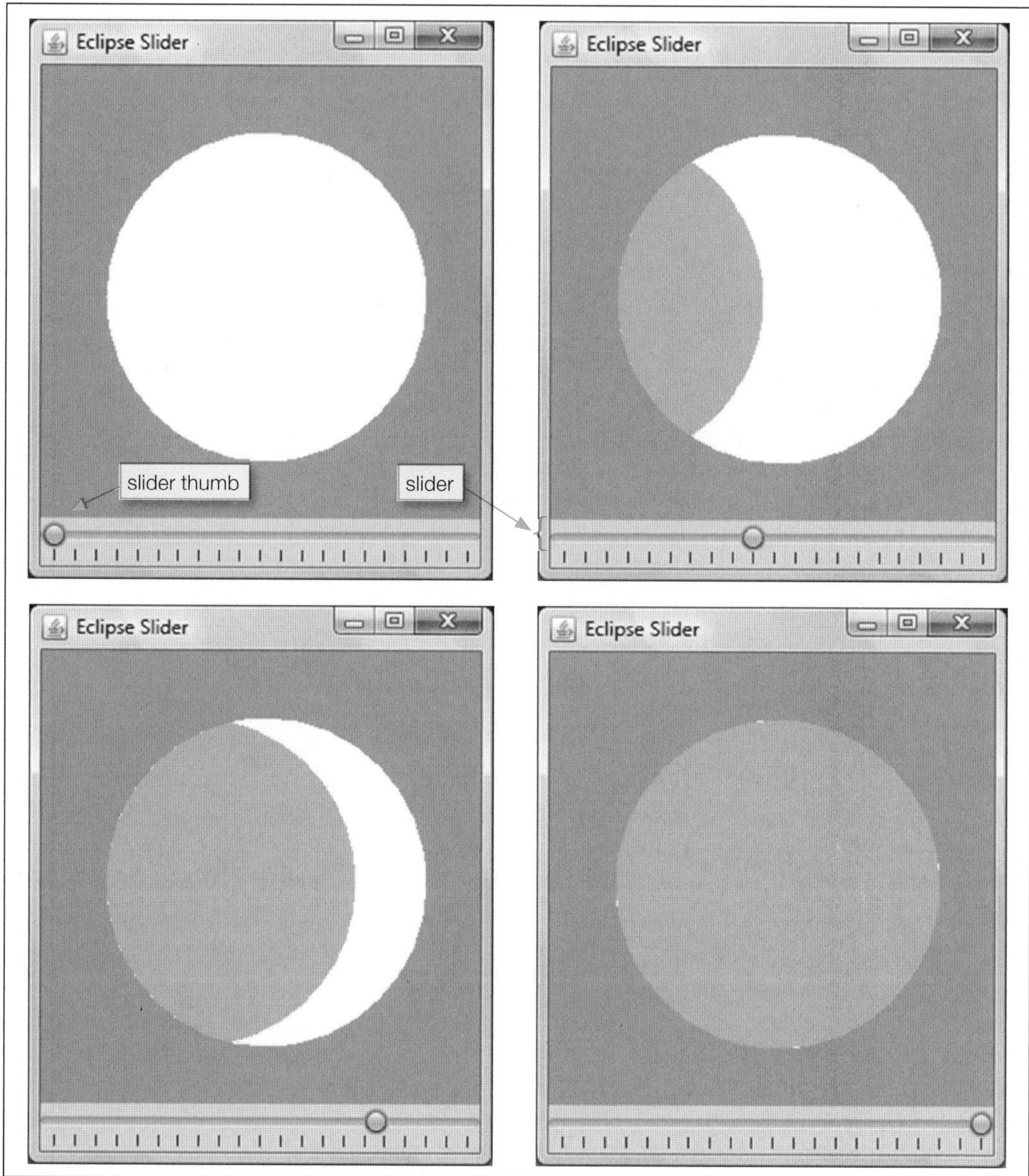

Figure 18.24 Four displays of a window that uses a slider to mimic an eclipse

Summary

- Layout managers automate the positioning of components within containers.
- The `FlowLayout` class implements a simple one-compartment layout scheme that allows multiple components to be appended to the right end of a row that wraps onto subsequent rows.

- The `BoxLayout` manager implements a simple one-row or one-column layout scheme.
- The `BorderLayout` manager provides five regions/compartments—north, south, east, west, and center—in which to insert components.
- The `SwingConstants` interface stores a set of GUI-related constants that are commonly used by many different GUI programs.
- The `GridLayout` manager lays out a container's components in a rectangular grid of equal-sized cells. Each cell can hold only one component.
- If you have a complicated window with many components, you might want to compartmentalize them by storing groups of components in `JPanel` containers.
- To display multiple lines of text, use a `JTextArea` component.
- A `JCheckBox` component displays a small square with an identifying label. Users click the check box in order to toggle it between selected and unselected.
- A `JRadioButton` component displays a small circle with a label to its right. If an unselected button is clicked, the clicked button becomes selected, and the previously selected button in the group becomes unselected.
- A `JComboBox` component allows the user to select an item from a list of items. `JComboBox` components are called "combo boxes" because they are a combination of a text box (normally, they look just like a text box) and a list (when the down arrow is clicked, they look like a list).

Review Questions

§18.2 GUI Design and Layout Managers

1. Layout managers adapt automatically to changes in the size of a container or one of its components. (T / F)
2. Which package contains layout managers?

§18.3 `FlowLayout` Manager

3. How does the `FlowLayout` manager arrange components?
4. Write a single statement that gives the current container a flow layout with right alignment.

§18.4 `BoxLayout` Manager

5. Write a statement that gives a `Frame` object called `frame` a `BoxLayout` manager that aligns elements in a column.

§18.5 `BorderLayout` Manager

6. What are the five regions established by the `BorderLayout` manager?
7. The sizes of the five regions in a border layout are determined at runtime based on the contents of the four outer regions. (T / F)
8. By default, how many components can you put in any one region of a border layout?
9. Write a single statement that adds a new `JLabel` with the text "Stop" to the center region of a `BorderLayout` manager. The label should be centered within the center region.

§18.6 `GridLayout` Manager

10. When you instantiate a `GridLayout` manager, you should always specify both the number of rows and the number of columns. (T / F)
11. In a grid layout, all cells are the same size. (T / F)

§18.7 Tic-Tac-Toe Example

12. What happens to the `xTurn` variable in the Tic-Tac-Toe program if you click the same cell twice?

§18.10 `JPanel` Class

13. Why are `JPanel` containers particularly useful with `GridLayout` and `BorderLayout` windows (as opposed to `FlowLayout` windows)?

§18.11 `MathCalculator` Program

14. In the MathCalculator program's `createContents` method, indicate the purpose of the statement: `add(new JLabel());`

§18.12 `JTextArea` Component

15. `JTextArea` components are editable by default. (T / F).
16. `JTextArea` components employ line wrap by default. (T / F).

§18.13 `JCheckBox` Component

17. What happens if you click a check box that's already selected?
18. Provide a statement that creates a check box named `attendance`. The check box should be preselected, and it should have an "I will attend" label.

§18.14 `JRadioButton` Component

19. What happens if you click a radio button that is already selected?
20. What happens if you click an initially unselected radio button that is a member of a `RadioGroup`?

§18.15 `JComboBox` Component

21. How are combo boxes and radio button groups similar?
22. What two methods can be called to determine the current selection for a combo box?

§18.16 Job Application Example

23. The JobApplication program contains the following code fragment. What happens to the program if the code fragment is omitted?

```
radioGroup = new ButtonGroup();
radioGroup.add(goodCitizen);
radioGroup.add(criminal);
```

24. Provide a statement that adds a 20-pixel blank margin to a `JPanel` container named `panel`.

§18.17 More Swing Components

25. Provide a `JSlider` constructor call where the minimum value is 0, the maximum value is 50, and the initial value is 10. Hint: Look up the answer on Oracle's Java API website.

Exercises

1. [after §18.2] What is the default layout manager for a `JFrame` window?

2. [after §18.3] With a `FlowLayout` manager, a button component expands so that it completely fills the size of the region in which it is placed. (T / F)

3. [after §18.5] Provide a complete program that is a modification of Chapter 17's Greeting program. Your new program should use a `BorderLayout` manager (instead of a `FlowLayout` manager), and it should generate the following display after a name has been entered. Make the frame size 300 pixels wide and 80 pixels high.

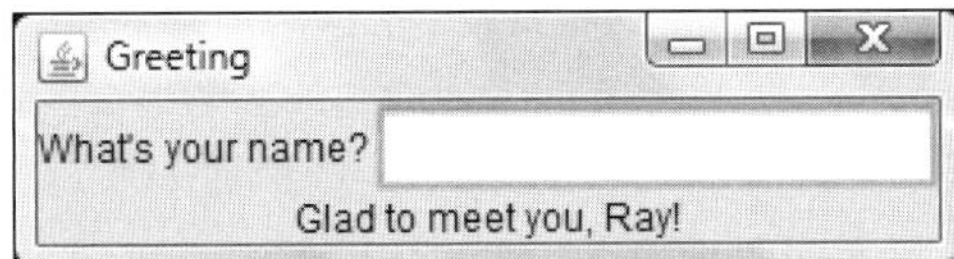

4. [after §18.5] With a `BorderLayout`, what happens if the east region is empty? Said another way, which region(s), if any, expand(s) if the east region is empty?

5. [after §18.5] Assume you have this program:

```
import javax.swing.*;
import java.awt.*;
public class BorderLayoutExercise extends JFrame
{
  public BorderLayoutExercise()
  {
    setTitle("Border Layout Exercise");
    setSize(300, 200);
    setDefaultCloseOperation(EXIT_ON_CLOSE);
    setLayout(new BorderLayout());
    add(new JLabel("Lisa the label"), BorderLayout.NORTH);
    add(new JLabel("LaToya the label"), BorderLayout.CENTER);
    add(new JLabel("Lemmy the label"), BorderLayout.SOUTH);
    setVisible(true);
  } // end BorderLayoutExercise constructor

  //*****************************************************

  public static void main(String[] args)
  {
    new BorderLayoutExercise();
  }
} // end class BorderLayoutExercise
```

a) Specify the changes you would make to the above code to produce this output:

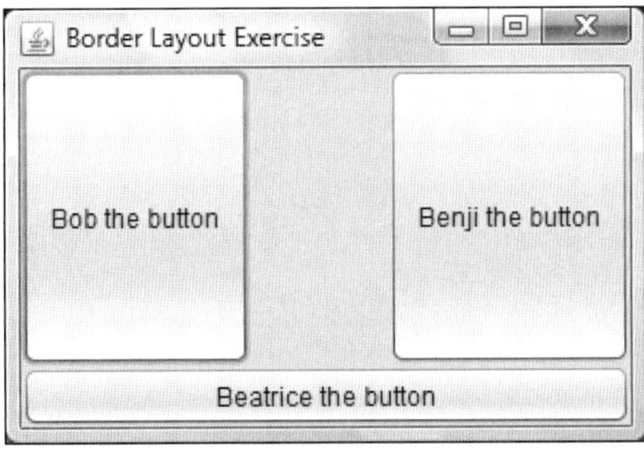

b) Specify the changes you would make to the above code to produce this output:

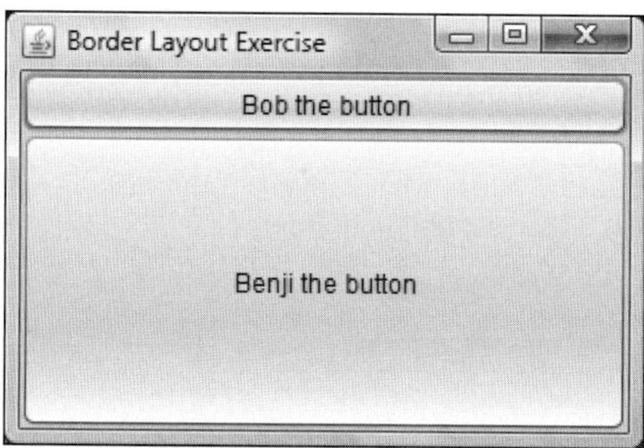

6. [after §18.6] If a `JButton` component is directly added to a `GridLayout` cell, it expands so that it completely fills the size of its cell. (T / F)

7. [after §18.6] Given the following code fragment, draw a picture that illustrates the buttons' positions within the program's window.

```
setLayout(new GridLayout(0, 3));
add(new JButton("1"));
add(new JButton("2"));
add(new JButton("3"));
add(new JButton("4"));
add(new JButton("5"));
add(new JButton("6"));
add(new JButton("7"));
```

8. [after §18.10] What kind of container should you put into an individual grid layout cell or an individual border layout region to allow that cell or region to contain more than one component?

9. [after §18.12] Suppose you're given a window with two `JTextArea` components, `msg1` and `msg2`, and a `JButton` component. When clicked, the button swaps the contents of the two text areas. Provide the code that performs the swap operation. More specifically, provide the code that goes inside the following `actionPerformed` method:

```
private class Listener implements ActionListener
{
  public void actionPerformed(ActionEvent e)
  {
    ...
  }
}
```

10. [after §18.13] Provide a statement that creates a check box named `bold`. The check box should be unselected, and it should have a "boldface type" label.

11. [after §18.13] How can your code determine whether a check box is selected or not?

12. [after §18.14] Provide a `createContents` method for a program that displays this window:

The male and female radio buttons should behave in the normal fashion—when one is selected, the other is unselected. Note that the male button is selected when the window initially displays. Your `createContents` method must work in conjunction with this program skeleton:

```
import javax.swing.*;
import java.awt.*;

public class MaleFemaleRadioButtons extends JFrame
{
  private JRadioButton male;
  private JRadioButton female;

  public MaleFemaleRadioButtons()
  {
    setTitle("Male-Female Radio Buttons");
    setSize(300, 100);
    setDefaultCloseOperation(EXIT_ON_CLOSE);
    createContents();
    setVisible(true);
  } // end MaleFemaleRadioButtons constructor

  <The createContents method goes here.>

  public static void main(String[] args)
  {
    new MaleFemaleRadioButtons();
  }
} // end class MaleFemaleRadioButtons
```

13. [after §18.15] The `JCheckBox`, `JRadioButton`, and `JComboBox` components are defined in what package?

14. [after §18.15] Provide a `createContents` method for a program that initially displays this window:

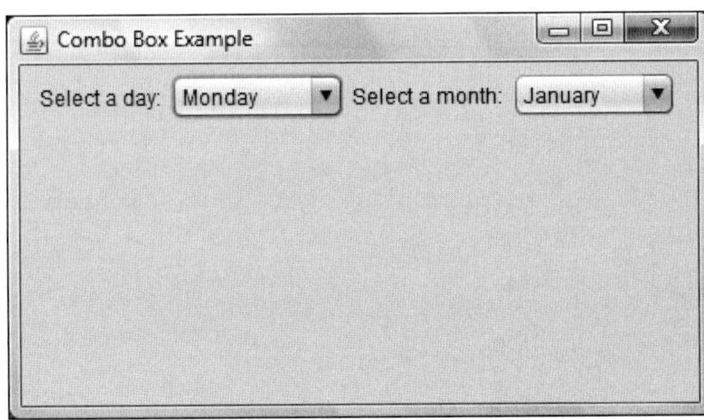

When the user clicks the left combo box, this displays:

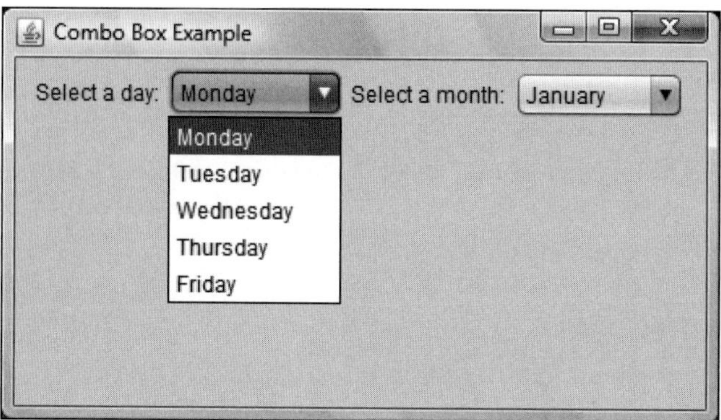

When the user clicks the right combo box, this displays:

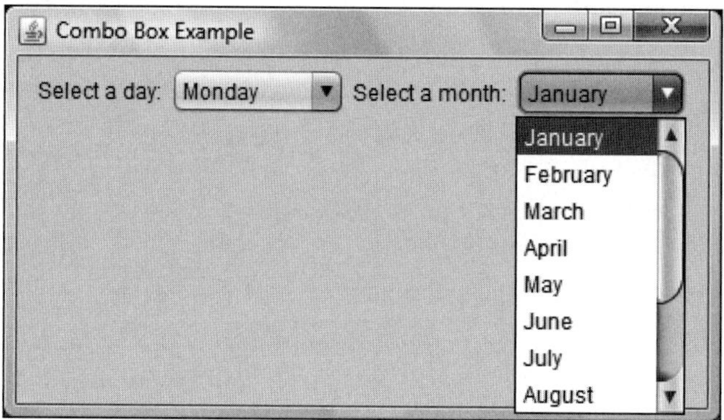

Your `createContents` method must work in conjunction with this program skeleton:

```
import javax.swing.*;
import java.awt.*;
public class ComboBoxExample extends JFrame
{
  private JComboBox daysBox;
  private JComboBox monthsBox;
  private String[] days =
    {"Monday", "Tuesday", "Wednesday", "Thursday", "Friday"};
  private String[] months =
    {"January", "February", "March", "April", "May", "June",
     "July", "August", "September", "October", "November",
     "December"};

  public ComboBoxExample()
  {
    setTitle("Combo Box Example");
    setSize(400, 225);
    setDefaultCloseOperation(EXIT_ON_CLOSE);
```

```
    createContents();
    setVisible(true);
  } // end ComboBoxExample constructor

  <The createContents method goes here.>

  public static void main(String[] args)
  {
    new ComboBoxExample();
  }
} // end class ComboBoxExample
```

Review Question Solutions

1. True.
2. Layout managers are in the `java.awt` package.
3. The `FlowLayout` manager places components left to right in a row until it runs out of space, and then it goes to the next row and does the same thing, and so on.
4. `setLayout(new FlowLayout(FlowLayout.RIGHT));`
5. `frame.setLayout`
   `(new BoxLayout(frame.getContentPane(), BoxLayout.Y_AXIS));`
6. The five regions of a border layout are North at the top, South at the bottom, and West, Center, and East in a row between them.
7. True.
8. 0 or 1.
9. `add(new JLabel("Stop", SwingConstants.CENTER), BorderLayout.CENTER);`
   or
   `add(new JLabel("Stop", SwingConstants.CENTER));`
10. False. Specify both values, for rows and columns, only if you know the number of rows and columns in your table and the table is completely filled in (i.e., there are no empty cells). Otherwise, specify just one dimension that you're sure of and specify zero for the other dimension.
11. True.
12. Nothing. It does not change value.
13. `JPanel` containers are particularly useful with `GridLayout` and `BorderLayout` windows because each compartment in those layouts can store only one component. If you need a compartment to store more than one component, let that one component be a `JPanel` container and put multiple components into the `JPanel` container.
14. The `add(new JLabel());` statement adds a dummy component (a blank label) to the bottom-left cell. The dummy component is necessary because, without it, the `xLogPanel` would go into the bottom-left cell, and that's inappropriate.
15. True. `JTextArea` components are editable by default.
16. False. `JTextArea` components do not employ line wrap by default.
17. If you click a check box that's already selected, the check box becomes unselected.

18. The following code creates a check box named `attendance`. The check box is preselected, and it has an "I will attend" label.

```
JCheckBox attendance = new JCheckBox("I will attend", true);
```

19. Nothing. It stays selected.
20. The clicked button becomes selected, and all other buttons in the group become unselected.
21. Combo boxes and radio button groups are similar in that they both allow the user to select one item from a list of items.
22. To determine the current selection for a combo box, call either `getSelectedItem` or `getSelectedIndex`.
23. If the `radioGroup` code is omitted from the JobApplication program, the program still compiles and runs, but the radio buttons operate independently. In other words, clicking one radio button will not cause the other one to be unselected.
24. This statement adds a 20-pixel blank margin to a `JPanel` container named `panel`:

```
panel.setBorder(new EmptyBorder(20, 20, 20, 20));
```

25. `JSlider` constructor call:

```
new JSlider(0, 50, 10);
```

APPENDIX 1

ASCII Character Set

Java uses the Unicode Standard to define all the characters that can be processed by a Java program. The Unicode Standard is huge—it allows up to 1,112,064 characters. For an overview of Unicode, see Section 12.15 in Chapter 12. The first 128 characters of the Unicode character set are particularly important because they include the characters in the Latin alphabet—A through Z—plus other common characters, such as digits and punctuation marks. Those 128 characters used to be the only characters available to most programming languages, and they formed the American Standard Code for Information Interchange (ASCII) character set. See Figures A1.1a and A1.1b, which show the ASCII characters and their associated decimal and hexadecimal values.

In the 1960s, when the ASCII character set was born, the characters' positions were chosen carefully in order to satisfy the most people and build support for a common standard. For example, the #, $, and % characters were chosen to be next to each other because they appear next to each other on most typewriters. In

Dec	Hex	Character
0	0	null
1	1	start of heading
2	2	start of text
3	3	end of text
4	4	end of transmission
5	5	enquiry
6	6	acknowledge
7	7	audible bell
8	8	backspace
9	9	horizontal tab (\t)
10	A	line feed (\n)
11	B	vertical tab
12	C	form feed
13	D	carriage return (\r)
14	E	shift out
15	F	shift in

Dec	Hex	Character
16	10	data link escape
17	11	device control 1
18	12	device control 2
19	13	device control 3
20	14	device control 4
21	15	negative acknowledge
22	16	synchronous idle
23	17	end transmission block
24	18	cancel
25	19	end of medium
26	1A	substitute
27	1B	escape
28	1C	file separator
29	1D	group separator
30	1E	record separator
31	1F	unit separator

Figure A1.1a First 32 characters in the ASCII character set

Dec	Hex	Character
32	20	space
33	21	!
34	22	"
35	23	#
36	24	$
37	25	%
38	26	&
39	27	'
40	28	(
41	29	)
42	2A	*
43	2B	+
44	2C	,
45	2D	-
46	2E	.
47	2F	/
48	30	0
49	31	1
50	32	2
51	33	3
52	34	4
53	35	5
54	36	6
55	37	7
56	38	8
57	39	9
58	3A	:
59	3B	;
60	3C	<
61	3D	=
62	3E	>
63	3F	?

Dec	Hex	Character
64	40	@
65	41	A
66	42	B
67	43	C
68	44	D
69	45	E
70	46	F
71	47	G
72	48	H
73	49	I
74	4A	J
75	4B	K
76	4C	L
77	4D	M
78	4E	N
79	4F	O
80	50	P
81	51	Q
82	52	R
83	53	S
84	54	T
85	55	U
86	56	V
87	57	W
88	58	X
89	59	Y
90	5A	Z
91	5B	[
92	5C	\
93	5D	]
94	5E	^
95	5F	_

<table>
<tr><th>Dec</th><th>Hex</th><th>Character</th></tr>
<tr><td>96</td><td>60</td><td>`</td></tr>
<tr><td>97</td><td>61</td><td>a</td></tr>
<tr><td>98</td><td>62</td><td>b</td></tr>
<tr><td>99</td><td>63</td><td>c</td></tr>
<tr><td>100</td><td>64</td><td>d</td></tr>
<tr><td>101</td><td>65</td><td>e</td></tr>
<tr><td>102</td><td>66</td><td>f</td></tr>
<tr><td>103</td><td>67</td><td>g</td></tr>
<tr><td>104</td><td>68</td><td>h</td></tr>
<tr><td>105</td><td>69</td><td>i</td></tr>
<tr><td>106</td><td>6A</td><td>j</td></tr>
<tr><td>107</td><td>6B</td><td>k</td></tr>
<tr><td>108</td><td>6C</td><td>l</td></tr>
<tr><td>109</td><td>6D</td><td>m</td></tr>
<tr><td>110</td><td>6E</td><td>n</td></tr>
<tr><td>111</td><td>6F</td><td>o</td></tr>
<tr><td>112</td><td>70</td><td>p</td></tr>
<tr><td>113</td><td>71</td><td>q</td></tr>
<tr><td>114</td><td>72</td><td>r</td></tr>
<tr><td>115</td><td>73</td><td>s</td></tr>
<tr><td>116</td><td>74</td><td>t</td></tr>
<tr><td>117</td><td>75</td><td>u</td></tr>
<tr><td>118</td><td>76</td><td>v</td></tr>
<tr><td>119</td><td>77</td><td>w</td></tr>
<tr><td>120</td><td>78</td><td>x</td></tr>
<tr><td>121</td><td>79</td><td>y</td></tr>
<tr><td>122</td><td>7A</td><td>z</td></tr>
<tr><td>123</td><td>7B</td><td>{</td></tr>
<tr><td>124</td><td>7C</td><td>|</td></tr>
<tr><td>125</td><td>7D</td><td>}</td></tr>
<tr><td>126</td><td>7E</td><td>~</td></tr>
<tr><td>127</td><td>7F</td><td>delete</td></tr>
</table>

Figure A1.1b Remaining characters in the ASCII character set

Figure A1.1b, note that #, $, and % are at positions 35, 36, and 37, respectively. Also, note in Figure A1.1b that the first uppercase letter, A, is at hexadecimal position 41, and the first lowercase letter, a, is at hexadecimal position 61. To change case, you can simply add or subtract hexadecimal 20. The people who created the ASCII character set were thinking in hex!

APPENDIX 2

Operator Precedence

The table in Figures A2.1a and A2.1b shows operator precedence. The operator groups at the top of the table have higher precedence than the operator groups at the bottom of the table. All operators within a particular precedence group have equal precedence. If an expression has two or more same-precedence operators, then within that expression, those operators execute from left to right or right to left, as indicated in the group heading.

1. **grouping, access, and postfix-mode operators (left to right):**

`(`*<expression>*`)`	expressions
`(`*<list>*`)`	arguments or parameters
`[`*<expression>*`]`	indices
<type-or-member>`.`*<type-or-member>*	member access
`x++`	increment, postfix mode
`x--`	decrement, postfix mode

2. **unary operators (right to left):**

`++x`	increment, prefix mode
`--x`	decrement, prefix mode
`+x`	plus
`-x`	minus
`!x`	logical inversion
`~`	bit inversion
`new` *<classname>*	object instantiation
`(`*<type>*`) x`	cast

3. **multiplication and division operators (left to right):**

`x * y`	multiplication
`x / y`	division
`x % y`	remainder

4. **addition, subtraction, and concatenation operators (left to right):**

`x + y`	addition
`x - y`	subtraction
`s1 + s2`	string concatenation

Figure A2.1a Operator precedence—part A

The unconditional operators in groups 8, 9, and 10 evaluate all operands, even though the final condition may be determined before all operands have been evaluated. The conditional operators in groups 11, 12, and 13 stop the evaluation process as soon as the final condition is determined. Groups 8, 9, 10, 11, and 12 describe expressions with two different types of operations, depending on the types of operators. If the

5. bit shift operators (left to right):

`x << n`	left shift (shift `x`'s bits left by `n` positions, add 0s at right)
`x >> n`	signed right shift (shift `x`'s bits right by `n` positions, add 0s or 1s at left to match the original leftmost bit)
`x >>> n`	unsigned right shift (shift `x`'s bits right by `n` positions, add 0s at left)

6. relational operators (left to right):

`x < y`	less than
`x <= y`	less than or equal to
`x >= y`	greater than or equal to
`x > y`	greater than
<object> `instanceof` *<class>*	conforms to

7. equality operators (left to right):

`x == y`	equal
`x != y`	not equal

8. unconditional `boolean` or bitwise AND operator (left to right):

`x & y`

9. unconditional `boolean` or bitwise EXCLUSIVE OR operator (left to right):

`x ^ y`

10. unconditional `boolean` or bitwise OR operator (left to right):

`x | y`

11. conditional `boolean` AND operator (left to right):

`x && y`

12. conditional `boolean` OR operator (left to right):

`x || y`

13. ternary conditional operator (right to left):

`x ? y : z`	if `x` is `true`, `y`, else `z`

14. assignment (right to left):

`y = x`	`y` ← `x`
`y += x`	`y` ← `y + x`
`y -= x`	`y` ← `y - x`
`y *= x`	`y` ← `y * x`
`y /= x`	`y` ← `y / x`
`y %= x`	`y` ← `y % x`
`y <<= n`	`y` ← `y << n`
`y >>= n`	`y` ← `y >> n`
`y >>>= n`	`y` ← `y >>> n`
`y &= x`	`y` ← `y & x`
`y ^= x`	`y` ← `y ^ x`
`y |= x`	`y` ← `y | x`

Figure A2.1b Operator precedence—part B

operators are `boolean`, the operations are logical, with a `true` or `false` result for each case. If the operators are integers or characters, each bit evaluates separately.

APPENDIX 3

Java Reserved Words

The table in Figures A3.1a, A3.1b, A3.1c, and A3.1d presents Java *reserved words*. Java reserved words are words you cannot use for the name of anything you define because they already have special meanings. Most of these words are *keywords*—they play particular roles in a Java program. An asterisk indicates that the word in question is not used in the body of this text.

`abstract`—not realizable. This is a modifier for classes and methods and an implied modifier for interfaces. An `abstract` method is not defined. An `abstract` class contains one or more `abstract` methods. All of an interface's methods are abstract. You cannot instantiate an interface or `abstract` class.

`assert`—claim something is true. Anywhere in a program, you can insert statements saying `assert` *<boolean-expression>*`;` Then, if you run the program with the option `enableassertions`, the JVM throws an `AssertionError` exception when it encounters an `assert` that evaluates to `false`.

`boolean`—a logical value. This primitive data type evaluates to either `true` or `false`.

`break`—jump out of. This command causes execution in a `switch` statement or loop to jump forward to the first statement after the end of that `switch` statement or loop.

`byte`—8 bits. This is the smallest primitive integer data type. It is the type stored in binary files.

`case`—a particular alternative. The `byte`, `char`, `short`, or `int` value immediately following the `case` keyword identifies one of the `switch` alternatives.

`catch`—capture. A `catch` block contains code that is executed when code in a preceding `try` block `throws` an exception, which is a special object that describes an error.

`char`—a character. This is a primitive data type that contains the integer code number for a text character or any other symbol defined in the Unicode Standard.

`class`—a complex type. This block of Java code defines the attributes and behavior of a particular type of object. Thus, it defines a data type that is more complex than a primitive data type.

`const`*—a constant. This archaic term is superceded by `final`.

Figure A3.1a Reserved words—part A

continue*—skip to end. This command causes execution in a loop to skip over the remaining statements in the loop's code and go directly to the loop's continuation condition.

default—otherwise. This is usually the last case in a `switch` statement. It represents all other cases (cases not identified in previous `case` blocks).

do—execute. This is the first keyword in a do-while loop. The continuation condition appears in parentheses after the `while` keyword at the end of the loop.

double—twice as much. This primitive floating-point data type requires twice as much storage, 8 bytes, as the older floating-point data type, `float`, which requires only 4 bytes.

else—otherwise. This keyword may be used in a compound `if` statement as the header (or part of the header) of a block of code that executes if the previous `if` condition is not satisfied.

enum—enumeration. This special type of `class` defines a set of named constants, which are implicitly `static` and `final`.

extends—derives from. This class heading extension specifies that the class being defined will inherit all members of the class named after the `extends` keyword.

false—no. This is one of the two possible `boolean` values.

final—last form or value. This modifier keeps classes and methods from being redefined, and it says that a named value is a constant.

finally—last operation. This may be used after `try` and `catch` blocks to specify operations that need to be performed after a `catch` processes an exception.

float—floating point. This is an older floating-point data type. It requires 4 bytes.

for—the most versatile type of loop. This keyword introduces a loop whose header specifies and controls the range of iteration.

goto*—jump to. This deprecated command specifies an unconditional branch. Don't use it.

if—conditional execution. This keyword initiates execution of a block of code if an associated condition is satisfied.

implements—defines. This class heading extension specifies that the class being defined will define all methods declared by the `interface` named after the `implements` keyword.

import—bring in. This tells the compiler to make subsequently identified classes available for use in the current program.

Figure A3.1b Reserved words—part B

inner—internal. When followed by the keyword `class`, this specifies that the class defined in the subsequent code block be nested inside the current class.

instanceof—conforms to. This `boolean` operator tests whether the object on the left is an instance of the class on the right or whether the object on the left as an instance of a subclass of the class on the right.

int—integer. This is the standard integer data type. It requires 4 bytes.

interface—what an outsider sees. A Java interface declares a set of methods but does not define them. A class that `implements` an `interface` must define all the methods declared in that `interface`. An `interface` can also define `static` constants. Another kind of interface just conveys a particular message to the compiler.

long—long integer. This is the longest integer data type. It requires 8 bytes.

native—indigenous. Native code is code that has been compiled into the (low-level) language of the local processor. Sometimes called *machine code*.

new—fresh instance of. This Java command calls a class constructor to create a new object at runtime.

null—nothing. This is the value in a reference variable that does not refer to anything.

package—an associated group. In Java, this is a container for a group of related classes that a programmer can `import`.

private—locally controlled. This modifier of methods and variables makes them accessible only from within the class in which they are declared.

protected—kept from public exposure. This is a modifier for methods and variables that makes them accessible only from within the class in which they are declared, descendants of that class, or other classes in the same `package`.

public—accessible to everyone. This modifier of classes, methods, and variables makes them accessible from anywhere. A Java `interface` is implicitly `public`.

return—go and perhaps send back to. This command causes program control to leave the current method and go back to the point that immediately follows the point from which the current method was called. A value or reference may be sent back too.

short—small integer. This integer data type requires only 2 bytes.

static—always present. This modifier for methods and variables gives them class scope and continuous existence.

Figure A3.1c Reserved words—part C

strictfp*—strict floating point. This modifier for a class or method restricts floating-point precision to the Java specification and keeps calculations from using extra bits of precision that the local processor might provide.

super—parent or progenitor. This is a reference to a constructor or method that would be inherited by the object's class if it were not overridden by a new definition in that class.

switch—select an alternative. This causes program control to jump forward to the code following the `case` that matches the condition supplied immediately after the `switch` keyword.

synchronized*—This modifier for methods prevents simultaneous execution of a particular method by different threads. It avoids corruption of shared data in a multithreading operation.

this—the current object's. The `this` dot reference distinguishes an instance variable from a local variable or parameter, or it says the object calling another method is the same as the object that called the method in which the calling code resides, or it yields initiation of object construction to another (overloaded) constructor in the same class.

throw*—generate an exception. This command followed by the name of an exception type causes an exception to be `thrown`. It enables a program to throw an exception explicitly.

throws—might throw an exception. This keyword followed by the name of a particular type of exception may be appended to a method heading to transfer the `catch` responsibility to the method that called the current method.

transient*—may be abandoned. This variable modifier tells Java serializing software that the value in the modified variable should not be saved to an object file.

true—yes. This is one of the two `boolean` values.

try—attempt. A `try` block contains code that might throw an exception, plus code that would be skipped if an exception were thrown.

void—nothing. This describes the type of a method that does not return anything.

volatile*—erratic. This keyword keeps the compiler from trying to optimize a variable that might be asynchronously altered.

while—so long as. This keyword plus a `boolean` condition heads a `while` loop, or it terminates a do-while loop.

Figure A3.1d Reserved words—part D

APPENDIX 4

Packages

As you may recall, a package is a group of related classes. In this appendix, we describe the packages that organize Oracle's library of Java application programming interface (API) classes. We then show you how to create your own packages for programmer-defined classes. Finally, we introduce you to some nifty advanced options.

Java API Packages

When you download a version of Java from Oracle, you get the API package hierarchies as part of the Java Development Kit (JDK). Installation of that "kit" automatically makes the Java API packages part of your Java environment.

Java API classes are organized in package hierarchies. Figure A4.1 shows part of these Java API package hierarchies. Notice that this shows two hierarchies, one with the `java` package at its root, and another

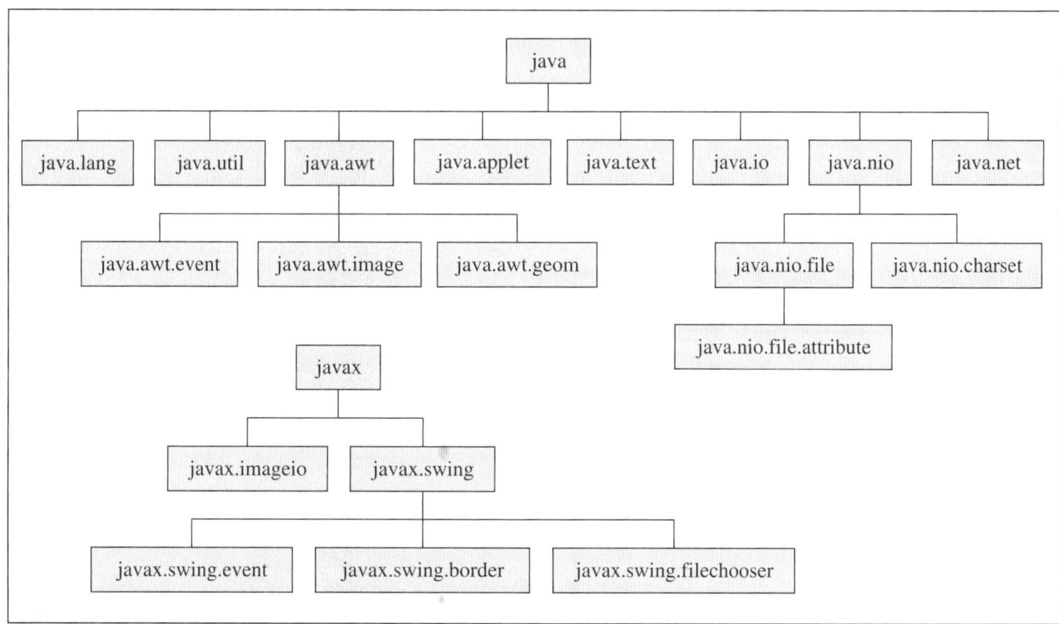

Figure A4.1 Abbreviated Java API package hierarchies

with the `javax` package at its root. This picture includes all of the API packages we have imported at some point in this book, but the packages shown in Figure A4.1 are only a small fraction of all the packages in the API package hierarchies.

This hierarchical organization helps people locate particular classes they need to use. It's OK for several different classes to have the same class name if they are all in different packages. So encapsulating small groups of classes into individual packages allows a given class name to be reused in different contexts. Also, each package protects the `protected` members of the classes it includes from access from outside that package.

To make the classes in a particular package available to a program you are writing, you import that package, as in these statements, which import the `java.util` and `java.awt` packages:

```
import java.util.*;
import java.awt.*;
```

Figure A4.1 includes several packages three levels down from the top. For example, in the `java` tree, under `java.awt`, consider the `java.awt.event` package. When we import the `java.awt` package in the statement above, this provides access to all classes in the `java.awt` package itself, but it does not also provide access to packages under it. In other words, it does not also import the `java.awt.event` package. If you also need access to classes in the `java.awt.event` package, you also must import that package explicitly by adding this third `import` statement:

```
import java.awt.event.*;
```

Custom Packages

The Java language permits you to create your own packages for organizing programmer-defined classes into package hierarchies. This involves the following steps:

First, design a package structure that makes sense for the program you are creating. Then create a directory structure that corresponds exactly to that package structure. (Later, we'll show how to create this directory structure automatically, but the manual process we're describing now is easier to understand.) Figure A4.2a shows part of a package structure that could be used for this book's examples. Figure A4.2b shows the corresponding directory structure. Note the "IPWJ" at the top of both figures. IPWJ is an acronym for our book's title, *Introduction to Programming with Java.*

Whenever you compile a class you want to be in a package, insert this line at the top of the class, above all statements, even the `import` statements:

```
package <package-path>
```

The *package path* is a subdirectory path, except it uses a dot (.) instead of a slash (/ or \). The first name in the package path should be the name of the root of the package hierarchy and the name of the highest directory in the part of the directory structure that corresponds to it. The last name in the package path should be the name of the directory that will contain the class being defined. So, for example, if you are defining a `Car` class and you intend for the `Car.class` bytecode file to be in the `IPWJ.chapter13.things` package shown in Figure A4.2a, the first statement in your `Car.java` file should be:

```
package IPWJ.chapter13.things;
```

When you compile your `Car.java` source code, by default the generated `Car.class` bytecode goes into the current directory, and the `package` statement above does not by itself change that. Thus, if you choose to write your source code in the `...IPWJ/chapter14/things` directory, the `Car.class` file goes

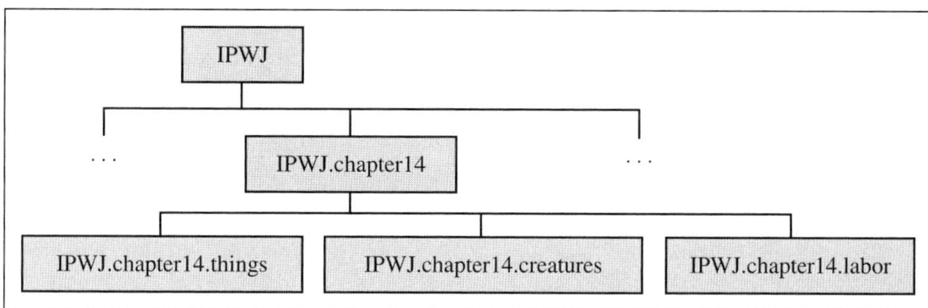

Figure A4.2a A typical programmer-defined package structure

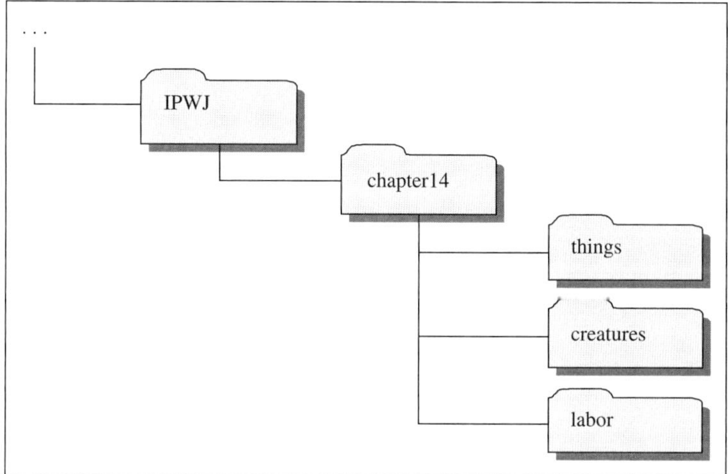

Figure A4.2b Directory structure that corresponds to the package structure in Figure A4.2a

immediately into this directory also. If you want this directory to include both source code and bytecode, you're done. If you want source code and bytecode to be in separate directories, you'll probably decide to write your source code in a separate source-code directory and then move the generated bytecode to the directory that matches its specified package. (Later, we'll show how you can ask the compiler to move it for you.)

For the Java compiler to import a class that's in a separate package, that class must be accessible through a *class path* that has been established previously in your operating system's environment. There may be more than one class path. On a Windows machine, you can see all registered `CLASSPATH`'s by opening a command-prompt window and entering the command `set`. (In UNIX, the command is `env`.) After `CLASSPATH`, you'll see a list of several class path specifications, with semicolons between them. (In UNIX, the separators are colons.) Typically, the first class path in the list is a single dot. That means "current directory." Suppose `myJava` is a root directory in the `C:` drive, and suppose the `IPWJ` directory shown in Figure A4.2b is in the `myJava` directory. To make classes in the `IPWJ` package hierarchy accessible to the Java compiler, your computer's `CLASSPATH` must include the following path:

```
C:/myJava.
```

Thus, the full pathname of the `Car.class` file in the `things` directory shown in Figure A4.2b would be:

```
C:/myJava/IPWJ/chapter14/things/Car.class
```

In Windows, the appropriate way to add a class path to the operating system's environment is to go to the **Control Panel** icon and click **System**. Then under the **Advanced** tab, click **Environment Variables**. . . . Then select **CLASSPATH** and click **Edit**. . . . Add a semicolon to the end of the list, and then enter your desired new class path, like this:

```
C:/myJava
```

In UNIX, use the `setenv` command, like this:

```
setenv classpath .:/myJava
```

Some Advanced Options

Optionally, you can ask the Java compiler to put the compiled `.class` file into your desired destination directory automatically. To do this, invoke the compiler from a command prompt with the `-d` option, like this:

```
javac -d <class-path> <source-code>
```

For our `Car` example, this would be:

```
javac -d E:/myjava Car.java
```

The full pathname of the directory that gets the compiled code is *<class-path>/<package-path>*. If the destination directory exists already, the generated `.class` file goes into that directory. If it does not exist, the compiler automatically creates the required directory and then inserts the generated `.class` file into it. Thus, if you plan to use this option, you do not need to create the directory structure in Figure A4.2b explicitly. You can let the compiler do it for you as you go along. Typical IDEs also provide ways to do this.

If you are developing a Java application for use by others, you'll probably want to organize your application's classes into a package structure and store the `.class` files in a corresponding directory structure, as previously described. When you have finished developing your application, it's straightforward to use a program like WinZip to compress all your application's files into a single file, perhaps called `IPWJ.zip`. After downloading this `.zip` file, your customer can insert this `.zip` file anywhere in his or her directory structure and then establish a class path to that `.zip` file, which includes the name of the `.zip` file itself as the final element in the class path. For example, if the `IPWJ.zip` file is in your customer's `C:/myJava` directory, your customer's `CLASSPATH` should include this class path:

```
C:/myJava/IPWJ.zip
```

This enables your customer's Java compiler to access all your package classes while they are still in their compressed form. Notice that the class path to a `.zip` file should include the `.zip` file itself, but if you unzip the file, the class path should go only to the containing directory.

Of course, this also works the other way. If you acquire a Java application developed by someone else, it will probably have its classes prepackaged and compressed into a `.zip` file (or some other type of *archive*). In that case, you may be able to put that compressed `.zip` file wherever you want in your computer's directory structure, and then just add a class path to that `.zip` file, with the `.zip` filename as the final element in that class path.

APPENDIX 5

Java Coding-Style Conventions

This appendix describes Java coding-style conventions. Most of these guidelines are widely accepted. However, alternative guidelines do exist in certain areas. The coding conventions presented in this document are for the most part a simplified subset of the archived coding conventions presented on the Java Code Conventions web page:

http://www.oracle.com/technetwork/java/codeconventions-150003.pdf

If you have a style question that is not addressed in this document, refer to Oracle's website.

While reading the following sections, refer to the example program in the last section. You can mimic the style in that example.

Prologue

1. Put this prologue section at the top of the file:

```
/*********************************************
* <filename>
* <programmer's-name>
*
* <file-description>
*********************************************/
```

2. Include a blank line below the prologue section.

Section Delimiting

1. After state variable definitions and between constructor and method definitions, enter a line of stars, like this:

```
//*********************************************************
```

   Leave a blank line above and below this line of stars.

2. Within a large constructor or method, insert blank lines between logical sections of code. For example, unless the loops are small and intimately related, enter a blank line between the end of one loop and the beginning of another loop.

Embedded Comments

1. Provide comments for code that might be confusing to someone reading your program for the first time. Assume the reader understands Java syntax.

2. Do not comment code that is obvious. For example, this comment is unnecessary and therefore exhibits poor style:

   This comment just adds clutter.

```
for (int i=0; i<10; i++)    // for loop header
```

3. Write your programs with clear, self-documenting code in order to reduce the need for comments. For example, use mnemonic (descriptive) identifier names.

4. Always include a single space between the `//` and the comment text.

5. The comment's length determines its format.

   - If the comment will occupy more than one line, use complete lines, like this:

```
// This is a block comment. Use it for comments that
// occupy more than one line. Note the alignment for /'s
// and words.
```

   - If a comment is to reside on a line by itself, position it above the line of code it describes. Indent the `//` the same as the described line of code. Include a blank line above the comment line. Here's an example:

```
// Display error if invalid file name.
if (fileName == null || filename.getName().equals(""))
{
  JOptionPane.showMessageDialog(this, fileErrorMsg);
}
```

   - Many comments are small enough to fit in space to the right of the code they describe. Whenever possible, all such comments should start in the same column, as far to the right as possible. The following example demonstrates proper positioning for short comments:

```
double testScores = new double[80]; // test scores for one class
int student;

...
while (testScores[student] >= 0)     // negative score quits
{
  testScores[student] = score;
  ...
```

6. Provide an end comment for each closing brace that is a significant number of lines (five or more?) down from its matching opening brace. For example, note the `// end for row` and `// end getSum` comments in the following method definition:

```
public double getSum(double table[][], int rows, int cols)
{
  double sum = 0.0;

  for (int row=0; row<rows; row++)
  {
    for (int col=0; col<cols; col++)
    {
      sum += table[row][col];
    } // end for col
  } // end for row

  return sum;
} // end getSum
```

Variable Declarations

1. Normally, you should declare only one variable per line. For example:

```
double avgScore;       // average score on the test
int numOfStudents;     // number of students in the class
```

Exception:
If several variables are intimately related, it is acceptable to declare them together on one line. For example:

```
int x, y, z;           // coordinates for a point
```

2. Normally, you should include a comment for each variable declaration line.

Exception:
Don't include a comment for names that are obvious (i.e., `studentId`) or standard (i.e., `i` for a `for` loop index variable, `ch` for a character variable).

Braces That Surround One Statement

1. For `if`, `else`, `for`, `while`, and `do` constructs that execute only one statement, it's good practice to treat that one statement as though it were a compound statement and enclose it in braces, like this:

```
for (int i=0; i<scores.length; i++)
{
  sumOfSquares += scores[i] * scores[i];
}
```

2. Exception:
If it would be illogical to add another statement to the construct at a later time, you may omit the curly braces when the omission improves readability. For example, an experienced programmer might write something like this:

```
for (; num>=2; num--)
  factorial *= num;
```

Placement of Braces

1. Place opening and closing braces on lines by themselves, such that the braces are aligned with the line above the opening brace. For `do` loops, put the `while` condition on the same line as the closing brace.

2. Examples:

```
public class Counter
{
   <field-and-method-declarations>
}

if (...)
{
   <statements>
}
else if (...)
{
   <statements>
}
else
{
   <statements>
}

for/while (...)
{
   <statements>
}

do
{
   <statements>
} while (...);

switch (...)
{
   case ... :
      <statements>
      break;
   case ... :
      <statements>
      break;
      ...
   default:
      <statements>
}
```

```
int doIt()
{
  <statements>
}
```

3. Brace alignment is a contentious issue. Oracle's Java Code Conventions website recommends putting the opening curly brace at the end of the previous line. This is one place where our conventions diverge from Oracle's conventions. We recommend that you put the opening curly brace on its own line because that helps make compound statements stand out.

4. For empty-bodied constructors, place the opening and closing braces on the same line and separate them with a space, like this:

```
public Counter()
{ }
```

The `else if` Construct

1. If the body of an `else` is just another `if`, form an `else if` construct (put the `else` and the `if` on the same line). See the above brace placement section for an example of a proper `else if` construct.

Alignment and Indentation

1. Align all code that is logically at the same level. See the above brace placement section for examples of proper alignment.

2. Indent all code that is logically inside other code. That is, for nested logic, use nested indentation. For example:

```
for (...)
{
  while (...)
  {
    <statements>
  }
}
```

3. You may use an indentation width of two to five spaces. Once you choose an indentation width, you should stick with it. Use the same indentation width throughout your program.

4. When a statement is too long to fit on one line, write it on multiple lines such that the continuation lines are indented appropriately. If the long statement is followed by a single statement that is logically inside of the long statement, use braces to enclose the single statement. Use either of the following techniques to indent continuation lines:

   - Indent to a column position such that similar entities are aligned. In the example below, the entities that are aligned are the three method calls:

```
while (bucklingTest(expectedLoad, testWidth, height) &&
       stressTest(expectedLoad, testWidth) &&
       slendernessTest(testWidth, height))
{
  numOfSafeColumns++;
}
```

- Indent the same number of spaces as all other indents. For example:

```
while (bucklingTest(expectedLoad, testWidth, height) &&
  stressTest(expectedLoad, testWidth) &&
  slendernessTest(testWidth, height))
{
  numOfSafeColumns++;
}
```

Multiple Statements on One Line

1. Normally, each statement should be put on a separate line.

   Exception:
   If statements are intimately related and very short, it is acceptable (but not required) to put them together on one line. For example:

   ```
   a++; b++; c++;
   ```

2. For assignment statements that are intimately related and use the same assigned value, it is acceptable (but not required) to combine them into one assignment statement. For example:

   ```
   x = y = z = 0;
   ```

Spaces within a Line of Code

1. Never put a space at the left of a semicolon.

2. Parentheses:

   - Never enter a space on the inside of enclosing parentheses.
   - If the entity to the left of a left parenthesis is an operator or a construct keyword (`if`, `switch`, etc.), then precede the parenthesis with a space.
   - If the entity to the left of a left parenthesis is a method name, then do not precede the parenthesis with a space.

   For example:

   ```
   if ((a == 10) && (b == 10))
   {
     printIt(x);
   }
   ```

3. Operators:

   - Normally, an operator should be surrounded by spaces. For example:

   ```
   if (x >= 3 && x <= 7)
   {
     y = (a + b) / 2;
   }
   ```

- Special cases:
  - Complex expressions:
    - —Within an inner component of a complex expression, do not surround the inner component's operators with spaces.
    - —Two common occurrences of complex expressions are conditional expressions and `for` loop headers. See the examples below.
  - Dot operator—no spaces at its left or right.
  - Unary operators—no space between unary operator and its associated operand.

For example:

```
if (zeroMinimum)
{
  x = (x<0 ? 0 : x);
}

while (list1.row != list2.row)
{
  <statements>
}

for (int i=0,j=0; i<=bigI; i++,j++)
{
  <statements>
}
```

Shortcut Operators

1. Use increment and decrement operators instead of their equivalent longer forms. For example:

Do not use	Use this
`x = x + 1`	`x++` or `++x` (depending on the context)
`x = x - 1`	`x--` or `--x` (depending on the context)

2. Use compound assignments instead of their equivalent longer forms. For example:

Do not use	Use this
`x = x + 5`	`x += 5`
`x = x * (3 + y)`	`x *= 3 + y`

Naming Conventions

1. Use meaningful names for your identifiers.

2. For named constants, use all uppercase letters. If there are multiple words, use underscores to separate the words. For example:

```
public static final int SECONDS_IN_DAY = 86400;
private final int ARRAY_SIZE;
```

3. For class names (and their associated constructors), use uppercase for the first letter and lowercase for all other letters. If there are multiple words in the class name, use uppercase for the first letter of all words. For example:

```
public class InnerCircle
{
  public InnerCircle(radius)
  {
    <constructor-body>
  }
```

4. For all identifiers other than constants and constructors, use all lowercase letters. If there are multiple words in the identifier, use uppercase for the first letter of all words that follow the first word. For example:

```
double avgScore;     // average score on the test
int numOfStudents;   // number of students in the class
```

Method and Constructor Organization

1. Normally, each method definition should be preceded by these items:

   - a blank line
   - a line of *'s
   - a blank line
   - a description of the purpose of the method
   - a blank line
   - parameter descriptions (for non-obvious parameters)
   - a blank line

   Ideally, all method parameters should use descriptive enough names so that the purpose of each parameter is inherently obvious. However, if this is not the case, then include a list of parameters and their descriptions in a method prologue above the method heading. For example, in a tic-tac-toe program, a method that handles a player's move would be relatively complicated and would require a method prologue like this:

```
//*****************************************************

// This method prompts the user to enter a move, validates the
// entry, and then assigns that move to the board. It also checks
// whether that move is a winning move.
//
// Parameters: board - the tic-tac-toe board/array
//             player - holds the current player ('X' or 'O')

public void handleMove(char[][] board, char player)
{
```

   Assuming you describe instance and class variables when you declare them, you should not provide prologues for "trivial" accessors, mutators, and constructors that just read or write instance and class variables. On the other hand, if a mutator performs validation on a parameter prior to assigning it to its associated instance variable, then it is not trivial, and you should include a prologue with it. The same reasoning applies to a constructor. A simple-assignment constructor should not have a prologue. A validation constructor should have a prologue.

2. In the interest of grouping similar things together, you should omit asterisk lines between trivial constructors, and you should omit asterisk lines between mutators and accessors.
Assume that a class contains two trivial constructors, several mutator and accessor methods, and two other simple methods. Here's the framework for such a class:

```
<class-heading>
{
  <instance-variable-declarations>

  //***********************************************************

  <trivial-constructor-definition>

  <trivial-constructor-definition>

  //***********************************************************

  <mutator-definition>

  <mutator-definition>

  <accessor-definition>

  <accessor-definition>

  //***********************************************************

  <simple-method-definition>

  //***********************************************************

  <simple-method-definition>
}
```

In the above framework, note that there are no descriptions for trivial constructors, accessors, or mutators, or for simple methods. Note also that there is a line of asterisks above the first mutator, but not above the subsequent mutator and accessors. Those omissions help to make a program more readable by grouping similar things together. Also, note that there are no comments above each of the two simple methods at the bottom of the class, but there are lines of asterisks.

3. Place local variable declarations immediately below the method heading. Do not place local variable declarations within the executable code.
Exception: Declare a `for` loop index variable within its `for` loop header.

Class Organization

1. Each of your classes may contain the following items (in the following order):
   a) class prologue section
   b) `import` statements
   c) constant class variables
   d) non-constant class variables
   e) instance variables
   f) abstract methods
   g) constructors
   h) instance methods
   i) class methods
2. Normally, you should place a `main` method and any of its helper methods in its own separate driver class. But it's sometimes appropriate to include a short `main` method within the class that it drives as an embedded testing tool. Put such a method at the end of the class definition.

Sample Java Program

For example, look at the coding style in the Student program in Figures A5.1a, A5.1b, and A5.2.

```
/*******************************************************************
* Student.java
* Dean & Dean
*
* This class handles the processing of a student's name.
*******************************************************************/

import java.util.Scanner;

public class Student
{
  private String first = ""; // student's first name
  private String last = "";  // student's last name

  //*****************************************************************

  public Student()
  { }

  // This constructor verifies that each passed-in name starts
  // with an uppercase letter and follows with lowercase letters.

  public Student(String first, String last)
  {
    setFirst(first);
    setLast(last);
  }
```

Figure A5.1a `Student` class, used to illustrate coding conventions—part A

```
//***************************************************************

// This method verifies that first starts with an uppercase
// letter and contains lowercase letters thereafter.

public void setFirst(String first)
{
  // [A-Z][a-z]* is a regular expression. See API Pattern class.
  if (first.matches("[A-Z][a-z]*"))
  {
    this.first = first;
  }
  else
  {
    System.out.println(first + " is an invalid name.\n" +
      "Names must start with an uppercase letter and have" +
      " lowercase letters thereafter.");
  }
} // end setFirst

//***************************************************************

// This method verifies that last starts with an uppercase
// letter and contains lowercase letters thereafter.

public void setLast(String last)
{
  // [A-Z][a-z]* is a regular expression. See API Pattern class.
  if (last.matches("[A-Z][a-z]*"))
  {
    this.last = last;
  }
  else
  {
    System.out.println(last + " is an invalid name.\n" +
      "Names must start with an uppercase letter and have" +
      " lowercase letters thereafter.");
  }
} // end setLast

//***************************************************************

// Print the student's first and last names.

public void printFullName()
{
  System.out.println(first + " " + last);
} // end printFullName
} // end class Student
```

Figure A5.1b `Student` class, used to illustrate coding conventions—part B

```
/*************************************************
* StudentDriver.java
* Dean & Dean
*
* This class acts as a driver for the Student class.
*************************************************/

public class StudentDriver
{
  public static void main(String[] args)
  {
    Student s1;  // first student
    Student s2;  // second student

    s1 = new Student();
    s1.setFirst("Adeeb");
    s1.setLast("Jarrah");
    s2 = new Student("Heejoo", "Chun");
    s2.printFullName();
  } // end main
} // end class StudentDriver
```

Figure A5.2 `StudentDriver` class, used with the `Student` class in Figures A5.1a and A5.1b

APPENDIX 6

Javadoc with Tags

Chapter 8's Section 8.2 and Appendix 5 describe a programming style that is optimized for code presentation in a textbook and for students writing relatively simple programs. Most of the suggestions there carry over to professional programming practice. However, in professional programming, you should provide interface documentation for your classes that looks like the documentation Oracle provides for its Java application programming interface (API) classes.

As described in Section 8.3, you can run the `javadoc` executable that comes with your Java Development Kit (JDK) and generate this documentation automatically. To run javadoc, at a command prompt, enter this command:

```
javadoc  d <output-directory> <source-files>
```

The `-d` *<output-directory>* option[1] causes the output to go to another directory. If you omit this `-d` option symbol, by default, the output goes to the current directory, but that's not a good idea because `javadoc` creates many files that would clutter up the current directory. You can put documentation for more than one class in the same directory by putting spaces between the source-file names.

As indicated in Section 8.3, for a comment to appear on a `javadoc` web page, it must be located either immediately above a class heading (after `import` statements) or immediately above a method heading. In addition, it must be within a special (`javadoc`) comment block that starts with `/**` and ends with `*/`. The opening and closing symbols may be on the same line as the enclosed comment, like this:

```
import java.util.Scanner;

/** This class handles processing of a student's name. */

public class Student_jd
{
```

Or the opening and closing symbols may be on separate lines, like this:

```
/**
Precondition: Each passed-in name must start with an uppercase
letter and all subsequent letters must be lowercase.
*/
public Student_jd(String first, String last)
{
```

[1]For other options and arguments, enter `javadoc -help`.

In both cases, it's good style to leave a blank line above the `javadoc` comment. When the closing `*/` is on the same line as the commenting text, there should also be a blank line after the comment. When the closing `*/` is on a separate line, the following blank line is frequently omitted.

Within a `/** ... */` `javadoc` comment block, `javadoc` also recognizes several special tags, which enable it to extract other kinds of information. For a complete description, see:

http://www.oracle.com/technetwork/java/javase/documentation/index-jsp-135444.html

Figure A6.1 contains an abbreviated list of `javadoc` tags.

Description of a constructor or method parameter:
`@param` *<parameter-name>* *<explanation>*

Description of a return value:
`@return` *<explanation>*

Description of an exception that might be thrown:
`@throws` *<exception-type>* *<explanation>*

Hyperlink reference to another documented item:
`@see` *<package-name>*.*<class-name>*
`@see` *<package-name>*.*<class-name>*#*<variable-name>*
`@see` *<package-name>*.*<class-name>*#*<method-name>*(*<type1>*, ...)

Figure A6.1 Abbreviated list of `javadoc` tags

The most important tags are the `@param` and the `@return` tags. Figure A6.2 shows a class like that defined in Figure 14.11, but with its comments modified for `javadoc`. The functionality of this class is exactly the same as that defined in Figure 14.11. But this version enables several `javadoc` features. Notice how the general class description has been moved from the prologue into a separate `javadoc` comment block immediately above the class heading. In the `javadoc` comment block above the constructor there are two tagged parameter descriptions. In the `javadoc` comment block above the method there is a tagged `return` value description. The `@Override` is not a `javadoc` tag. It's a compiler tag.

Suppose the current directory contains source code for the `Employee` class copied from Figure 14.10, and it also contains the source code for the `Salaried_jd` class shown in Figure A6.2. Then suppose that we open a command-prompt window and enter the following command:

```
javadoc -d docs Employee.java Salaried_jd.java
```

This creates interface documentation for both classes and outputs that combined documentation to the `docs` subdirectory. Figure A6.3a shows what you'll see if you open a web browser, navigate to the `docs` directory, click on `index.html`, and select `Salaried_jd` in the left panel under "All Classes."

In Figure A6.3a's right panel, near the top, you can see the documentation of `Salaried_jd`'s inheritance from `Employee`. In the `Salaried_jd` documentation, `Employee` is colored and underlined in several places. These are links, and if you click on any of them, the display switches immediately to the `Employee` class's documentation. In Figure A6.2, our general comment had two sentences, and both of these sentences appear in the general comment in Figure A6.3a. Notice that the constructor and method summary blocks do not contain any comments. The `@param` and `@return` tags do not produce any

```
/*************************************************************
* Salaried_jd.java
* Dean & Dean
*************************************************************/

/**
This class implements a salaried employee.
It has same functionality as the Salaried class in Chapter 14.
*/

public class Salaried_jd extends Employee
{
  private double salary;

  //**********************************************************

  /**
  @param name    person's name
  @param salary  annual salary in dollars
  */

  public Salaried_jd(String name, double salary)
  {
    super(name);
    this.salary = salary;
  } // end constructor

  //**********************************************************

  /** @return    half month's pay in dollars  */

  @Override
  public double getPay()
  {
    return this.salary / 24;
  } // end getPay
} // end class Salaried_jd
```


Figure A6.2 Modification of Figure 14.11's `Salaried` class with `javadoc` comments

summary-block output. If we had included text in the `javadoc` comment block above the constructor or method heading in Figure A6.2, only the first sentence of that text (the "summary" sentence) would appear in the corresponding summary block in Figure A6.3a.

Now suppose you use the scroll bar on the right to scroll down. This displays what you see in Figure A6.3b. Notice that the "Detail" blocks do display the tagged parameter and return information supplied in `javadoc` comment blocks above the constructor and method headings in Figure A6.2.

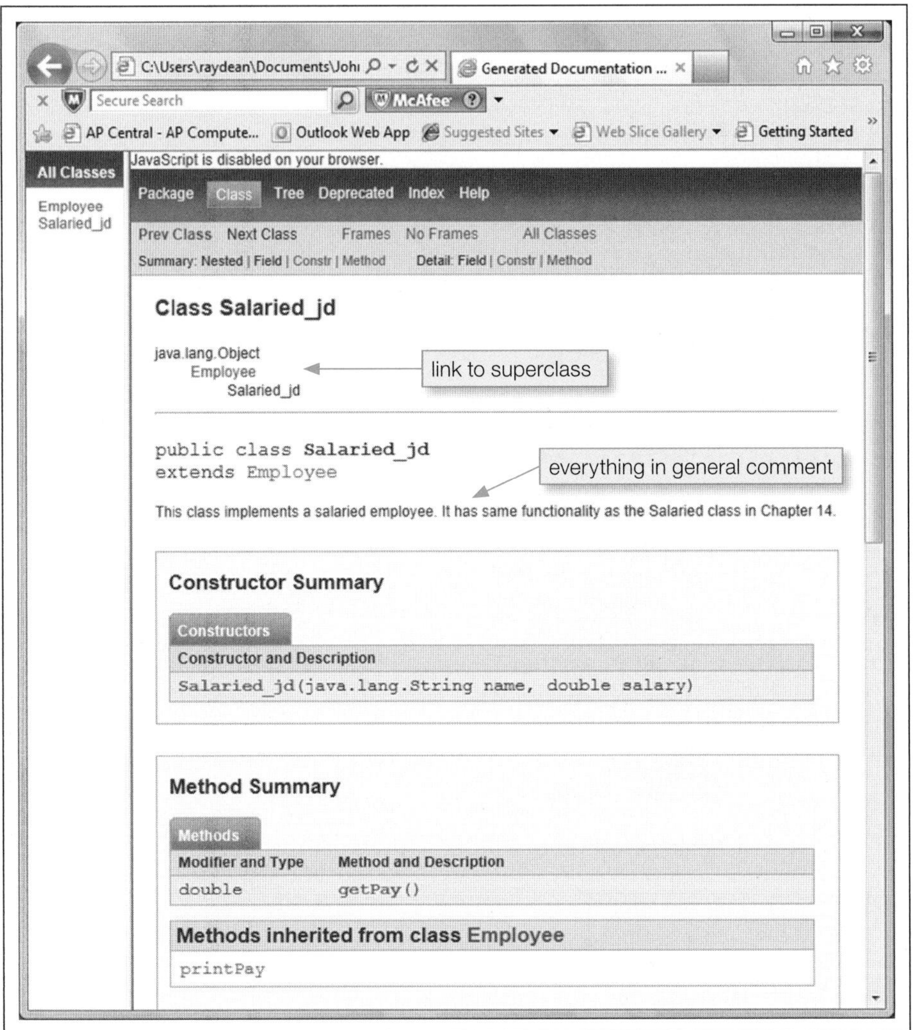

Figure A6.3a `javadoc` output for the `javadoc`-commented `Salaried` class—part A

If you had included text in a `javadoc` comment block preceding the constructor or method heading in Figure A6.2, all of this text would appear in the corresponding "Detail" block in Figure A6.3b. Finally, notice that `javadoc` also tells us that the `getPay` method defined in `Salaried_id` overrides a `getPay` method defined in `Employee`.

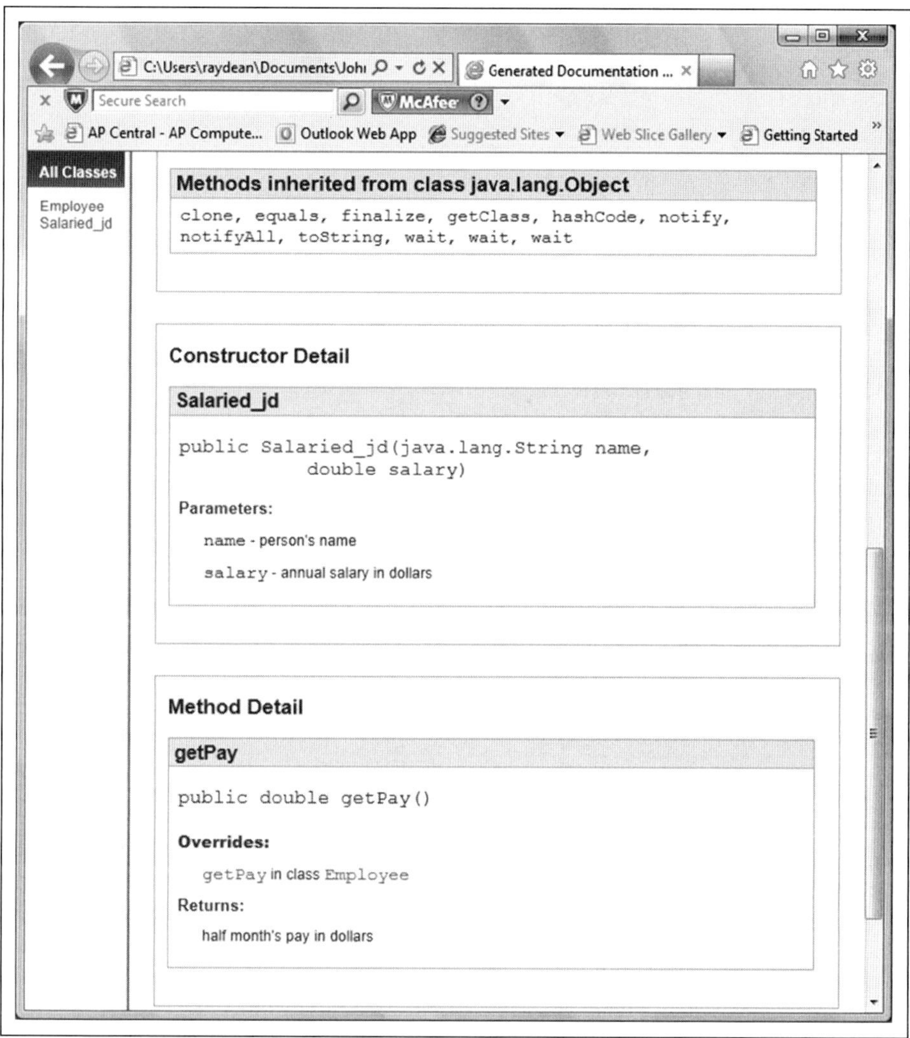

Figure A6.3b `javadoc` output for the `javadoc`-commented `Salaried` class—part B

APPENDIX 7

UML Diagrams

The Unified Modeling Language (UML) is a descriptive language that helps program designers organize the subject matter of a prospective object-oriented program, and it provides high-level documentation of both structure and behavior. It's independent of any particular programming language, and it doesn't compile into an executable program. It's just an organizational tool. It was developed by the "Three Amigos"—Grady Booch, James Rumbaugh, and Ivar Jacobson, at Rational Software Corp., which is now part of IBM. Currently, it is maintained by the nonprofit Object Management Group (OMG) consortium.

UML specifies many different kinds of visualizing diagrams.[1] In this appendix, we'll focus on just two of them—*activity diagrams* (which depict behavior) and *class diagrams* (which depict structure). When UML describes behavior, arrows point to what happens next. When UML describes structure, arrows point to what provides support, and this is opposite to the direction of "information flow." So in the following discussion, be prepared for a switch in arrow directionality as we move from activity diagrams to class diagrams.

UML Activity Diagrams

Activity diagrams are UML's version of the flowcharts we introduced in Chapter 2. They portray an algorithm's flow of control. Figure A7.1 shows an example of a UML activity diagram for the Happy Birthday algorithm presented as a flowchart in Figure 2.9. The solid black circle is an *initial state,* and the black dot in a white circle is a *final state.* The oval boxes represent *action states* or *activities.* They contain informal descriptions of coherent actions. The arrows are *transitions.* The labels in square brackets next to some of the transitions are `boolean` conditions called *guards*—a particular transition occurs if and only if the adjacent guard value is `true`. The actions or activities shown in Figure A7.1 represent low-level or primitive operations.

At a higher level of scale, the activity described in a single oval could represent a whole set of actions. For example, you could use a single activity symbol to represent the whole looping operation shown in Figure A7.1 like this:

print "Happy birthday!"
100 times

[1]The full UML specification is 1,000 pages long. For a simple 20-page introduction to the UML specification, see ftp://ftp.software.ibm.com/software/rational/web/whitepapers/2003/intro_rdn.pdf. For a more complete description, see Sinan Si Alhir, *UML in a Nutshell* (O'Reilly, 1998). Also see Ivar Jacobson, Grady Booch, and James Rumbaugh, *The Unified Software Development Process* (Addison-Welsey, 2005).

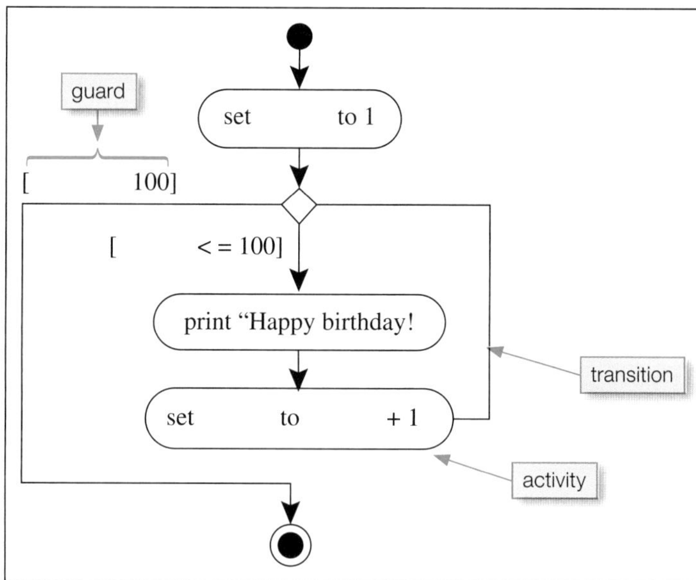

Figure A7.1 UML activity diagram for the Happy Birthday algorithm in Figure 2.9

Or you could use a single activity symbol to represent all the actions performed by a complete method. An activity symbol is not supposed to represent code itself. It's supposed to represent the code's "activity." Thus, it's appropriate to repeat an activity symbol that represents a complete method when you call that method more than once.

When there is more than one class, and perhaps several objects, UML suggests that you organize the activities into columns, such that all the activities for any one class or object are in a single column dedicated to that class or object. UML calls these separate columns *swimlanes.* Vertical dashed lines separate adjacent swimlanes. At the top of the diagram, over appropriate swimlane(s), write the class name for the lane or lanes below. Precede each class name with a colon and put it into a separate rectangular box. When you mean to instantiate an object, write that object's name followed by a colon and its class name. Underline it and put it into a separate rectangular box located just after the activity that creates it.

Figure A7.2 shows the UML activity diagram for the Mouse2 program defined in Figures 6.13 and 6.14. Notice how each activity (oval) is aligned under its own class and (if applicable) its own object. Activities for the lowest-level objects typically represent complete methods. Activities for higher-level objects typically represent code fragments. Solid black arrows represent control flow. They always go from one activity to another activity. Notice how the control flow moves continuously downward.

Dashed black arrows represent data flow associated with each activity. They go from an activity to an object, or from an object to an activity, but never from one activity to another activity. These dashed lines are often omitted to reduce clutter, but you can see how they help to show what the activities do. For example, notice how the dashed line from the "mickey : Mouse2" object to the "print mickey's attributes" activity helps explain what happens and allows us to suppress the two "get" method calls embedded in the print statement:

```
System.out.printf("Age = %d, weight = %.3f\n",
  mickey.getAge(), mickey.getWeight());
```

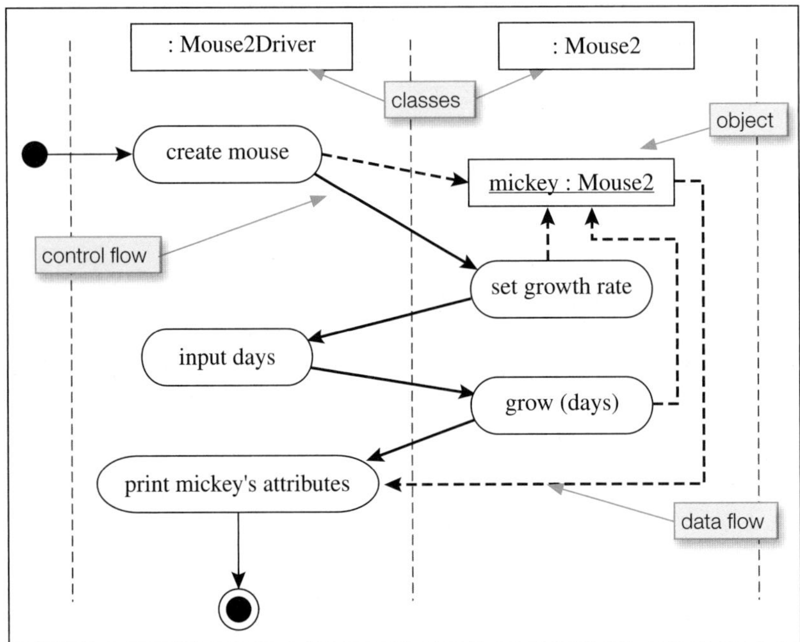

Figure A7.2 UML activity diagram for the Mouse2 program in Figures 6.13 and 6.14.
Ovals are activities. Rectangles are classes or objects—objects are underlined. Dashed gray vertical lines separate adjacent *swimlanes*, with one lane for each class or object. Solid black arrows represent control flow. Dashed black arrows represent data flow.

The introduction of constructors in Chapter 7 makes it possible to include the "set growth rate" activity within the "create mouse" activity. This would replace the top two swimlane-crossing transitions with a single transition from the "create mouse" activity to the "input days" activity in the same left-side swimlane. Minimizing swimlane crossings is a good design goal.

UML Class and Object Diagrams

Starting in Chapter 6, we gradually introduced you to various features of UML class diagrams. UML object diagrams are similar, except the title (object name followed by a colon followed by class name) is underlined—as in the UML activity diagram in Figure A7.2. An object block does not include a methods compartment, and only those variables of current interest should be listed in the attribute compartment. Object diagrams are context-dependent snapshots, with attribute values being current values rather than initial values. Class diagrams have more general application, and from now on, we'll restrict our attention to them.

We'll use a comprehensive example to summarize most of the features of UML class diagrams presented throughout the main part of the book. The example we'll use is the Payroll3 program described in Section 14.9. Figure A7.3 portrays a first-cut class diagram in which each class is represented by a simple, one-compartment rectangle that contains nothing more than the class name. The solid lines drawn between related classes are simple association lines. A simple (unadorned) association line implies bidirectional knowledge—the class at each end knows about the class at the other end. Thus a simple line says dependencies are mutual, but it says nothing else about the nature of the relationship between connected classes.

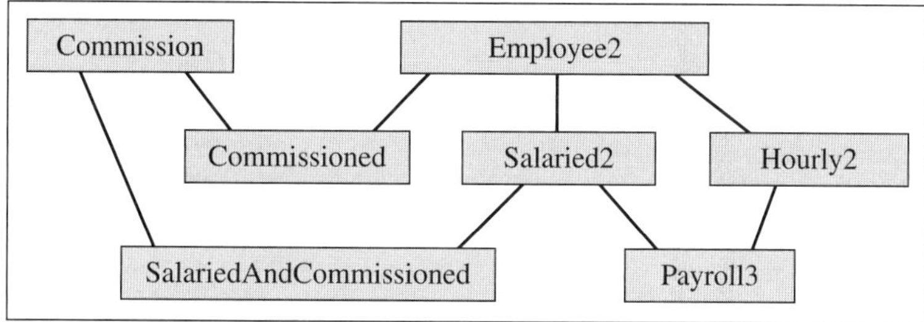

Figure A7.3 First-cut UML class diagram for the Payroll3 program in Section 14.9

As you progress in your design thinking, you'll flesh out class descriptions, perhaps deciding to make some of the classes `abstract` or converting them to interfaces. In addition, you'll modify many of the association lines by adding special symbols that describe particular types of relationships. In addition, you might add barbed arrowheads to convert associations from bidirectional to unidirectional and make dependencies go just one way. Unidirectional dependencies are preferable to bidirectional dependencies because they simplify software management—software changes to one class are less likely to require changes to other classes.

Figure A7.4 contains a fleshed-out and modified version of the first-cut UML class diagram in Figure A7.3. Notice that we italicize the `Commission` and `Employee2` class names. That means they have at least one `abstract` method and cannot be instantiated. We also italicize all the `abstract` methods they contain. Next, look at the hollow arrowheads, which indicate inheritance. Inheritance arrowheads on solid lines indicate extension of a class. Inheritance arrowheads on dashed lines indicate implementation of an interface. The arrowheads point in the direction of generalization—toward the more general entity. The more specific entities know about the more general entities and depend on them. Because of this dependency, changes to ancestor classes or interfaces can force changes to descendant or implementing classes. On the other hand, since an ancestral class or interface does not know about its descendants, changes in descendants or implementations never force changes in ancestors or interfaces. Inheritance is automatically a unidirectional association.

Now look at the composition indicators.[2] We chose to show them as (solid diamond) compositions rather than (hollow diamond) aggregations because the class that instantiates the components (`Payroll3`) inserts anonymous components into its containing array. All the composition lines have multiplicities. These indicate that there is always exactly one payroll and there could be any number of employees of any of the four types. Because `Hourly2`, `Salaried2`, `Commissioned`, and `SalariedAndCommissioned` all descend from the `Employee2` class, we can put instances of all four of these classes into a common `Employees2` array, as we do in the `Payroll3` class definition in Figure 14.17.

Finally, look at the barbed arrowheads we have added to the composition association lines. As we said, all association lines are bidirectional by default, and one design objective is to convert bidirectional associations into unidirectional associations. The barbed arrowheads on the non-diamond ends of the four composition lines do that. They say the composition's components have no knowledge of their container. That's appropriate in this case because this container is just a driver, and many drivers are ephemeral—here today and gone tomorrow.

[2]In Figure A7.4, notice how the association line between `Payroll3` and `Commissioned` arcs over the association line between `SalariedAndCommissioned` and `Salaried2`. This UML detail helps distinguish a cross-over from a junction.

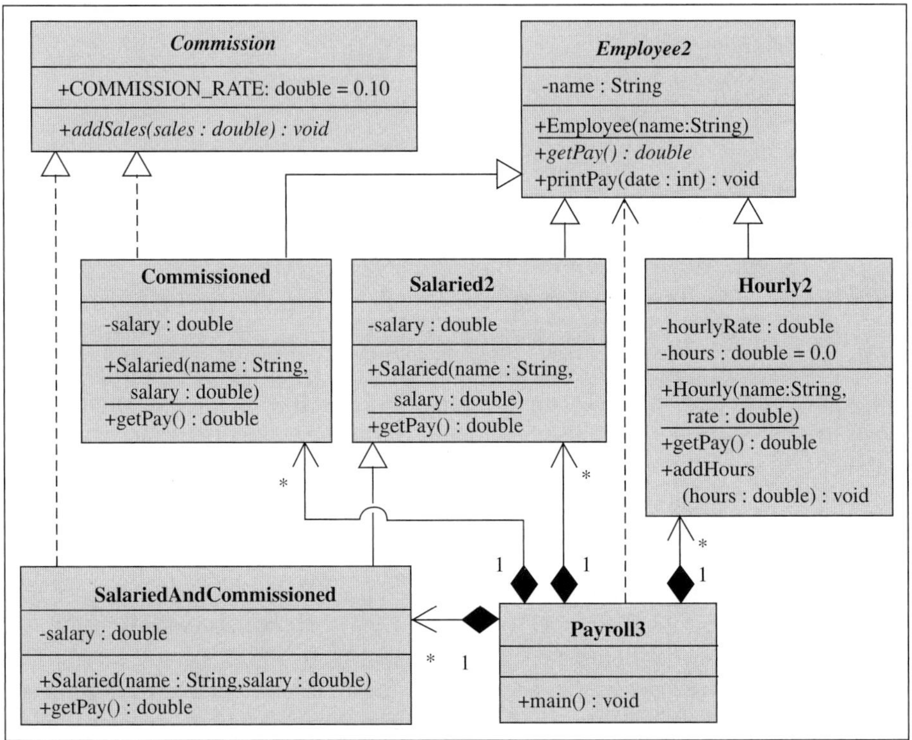

Figure A7.4 UML class diagram for the Payroll3 program
This shows inheritance from classes and implementation of an interface. It also shows composition. Because every association line in this figure has some kind of arrowhead, all its associations are unidirectional. The dashed association between `Payroll3` and `Employee2` is a simple dependence. That means the `Employee2` type appears in the declaration of a parameter or local variable somewhere in `Payroll3`'s code.

Figure A7.4 also includes a dashed association line with a barbed arrow that points to the `abstract` class, `Employee2`. This acknowledges that `main`'s local variable, `employees`, depends on the class, `Employee2`, because the type of its elements is `Employees2`. The barbed arrow at the `Employees2` end of this dashed association line indicates that the association is unidirectional. `Payroll3` knows about `Employees2`, but `Employees2` does not know about `Payroll3`. Thus, changes to `Employees2` might require changes to `Payroll3`, but changes to `Payroll3` never require changes to `Employees2`. UML uses dashed association lines for parameter and local variable dependencies, and it uses solid association lines for instance and class variable dependencies.

As described in Section 13.12, UML also uses dashed association lines to connect an association class to an association between or among other classes. Figure 13.25 shows a solid association line connecting the three classes: `SalesPerson2`, `Customer`, and `Car`. Although we did not discuss this detail in Chapter 13, the fact that this association line is solid and has no barbed arrowheads at its ends suggests that each of these three classes has instance variables that refer to particular instances of the other two classes.

The association class called `Sales` makes such additional references unnecessary because the `Sales` class can hold all these references itself, in one place. Thus, this extra association class reduces the number of reference variables. More importantly, it eliminates the need to alter the definition of the `SalesPerson2`

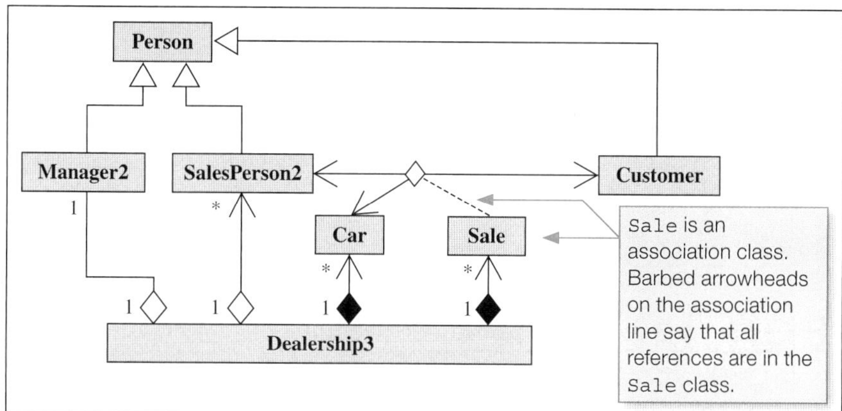

Figure A7.5 Improved version of the class diagram in Figure 13.25.
An arrowhead on an association line means the adjacent class does not have references to the other classes in that association.

and `Car` classes when we add a `Customer` class and a `Sale` association to the program. To reflect the fact that the `SalesPerson2`, `Car`, and `Customer` classes do not need any references to instances of other classes in the common association, we put barbed arrowheads on the three ends of the association line that connects them. This changes Figure 13.25 to what appears in Figure A7.5.[3]

Notice that Figure A7.5 also includes a composition association between `Dealership3` and `Sale`. The barbed arrowheads at the `Sale` and `Car` ends of their respective composition lines and at the `SalesPerson2` end of its aggregation line say that `Dealership3` depends on these other classes. In other words, `Dealership3` has references to instances of the `Sale`, `Car`, and `SalesPerson2` classes, but not vice versa. In contrast, the aggregation association between `Dealership3` and `Manager2` does not have any arrowheads. This says each has a reference to the other.

[3]Notice the small diamond at the intersection of the `Sale` association lines. This UML detail helps distinguish a junction from a cross-over.

APPENDIX 8

Number Systems and Conversions Between Them

Probably because we have ten fingers, ordinary humans use a base-ten number system called *decimal*. In the decimal system, the smallest component is a *digit*. Each digit has one of ten distinct values, 0, 1, 2, 3, 4, 5, 6, 7, 8, and 9. Computer programmers also use two other number systems, *binary* and *hexadecimal*.

In Chapter 1 Section 1.2, we provided a brief introduction to the base-two number system, called *binary*. In the binary system, the smallest component is a *bit*. A bit has one of only two distinct values, 0 and 1. In Chapter 12 Section 12.2, we used bits to describe the amount of memory required to hold different types of Java numbers.

In Section 12.14, we used binary numbers to describe the base-sixteen number system, called *hexadecimal*. In the hexadecimal system, the smallest component is a *hexit*. A hexit has one of sixteen distinct values, 0, 1, 2, 3, 4, 5, 6, 7, 8, 9, A, B, C, D, E, and F. In hexadecimal, A, B, C, D, E, and F correspond to decimal 10, 11, 12, 13, 14, and 15, respectively. Instead of the uppercase letters A, B, C, D, E, and F, Java allows us to substitute the lowercase letters a, b, c, d, e, and f. Computers typically use hexadecimal for memory addressing, and as described in Section 12.15 and in Appendix 1, they also use hexadecimal to organize and represent characters.

Sometimes people use a base-eight number system, which is called *octal*. In principle, the base can be any positive integer.

To represent a number greater than its base, a number system uses multiple components. It puts each component in a different position, and it assigns each position a distinctive weight. In any one number system, the weights are distinct integer powers of that system's base, with the powers in descending order.[1] For example, when we write a decimal number like 1609.34, what we actually mean is

$$1*10^{+3} + 6*10^{+2} + 0*10^{+1} + 9*10^{0} \quad + 3*10^{-1} + 4*10^{-2}$$

decimal point

The decimal point separates the integer part of the number on the left from the fraction part on the right, and the decimal point's location determines the weight in each component position.

It's easy to generalize this decimal example. For the more general case, let's rewrite this number using N_{+3}, N_{+2}, N_{+1}, N_0, N_{-1}, N_{-2}, and N_{-3} for the number's components, like 1, 6, 0, 9, 3, 4, and a trailing 0. The mathematical term for the base is *radix,* so we use R for the base value.

[1]When the highest powers come first, it's called a *big endian*. Because mathematical operations like addition, subtraction, and multiplication start with the least-significant elements, computer hardware sometimes arranges bits with the lowest powers first. That's called *little endian*.

Expression 1:

$$N_{+3} * R^{+3} + N_{+2} * R^{+2} + N_{+1} * R^{+1} + N_0 * R^0 \quad + N_{-1} * R^{-1} + N_{-2} * R^{-2} + N_{-3} * R^{-3}$$

radix point

"In" Conversion—from Another Base to Decimal

Now, suppose we have a memory location in hexadecimal, like A34C. What is that in decimal? Because it is hexadecimal, R is 16. Because it has no radix point, it is an integer. Recognizing that hexadecimal A is decimal 10 and hexadecimal C is decimal 12, N_{+3} is 10, N_{+2} is 3, N_{+1} is 4, and N_0 is 12. So with A34C in Expression 1, the evaluation looks like this:

```
10 * (16)^+3 + 3 * (16)^+2 + 4 * (16)^+1 + 12 * (16)^0 ⇒
10 * 4096 + 3 * 256 + 4 * 16 + 12 * 1 ⇒
40960 + 768 + 64 + 12 ⇒ 41804
```

Therefore, $A34C_{hexadecimal}$ is the same as $41804_{decimal}$.

This evaluation requires three additions and six multiplications (assuming we drop the unnecessary final multiplication by unity). Expression 1 is a polynomial. When we use this algorithm to evaluate a polynomial, the number of multiplications increases as the square of the number of terms. If we care about efficiency, either in a computer or in a hand calculation, wherever we evaluate a polynomial, we should not do what we have just done. There is a better way—namely, to group things using parentheses, like this:

Expression 2:

$$((N_{+3}*R + N_{+2})*R + N_{+1})*R + N_0 \quad + ((N_{-3}/R + N_{-2})/R + N_{-1})/R$$

radix point

Notice how similar the fractional part of Expression 2 is to the integer part. If you know the integer part and want the fractional part, just copy the integer part without N_0, and then replace each plus subscript with a minus subscript and replace each multiplication with a division. Expression 2 tells how to convert integers and fractions from another base to decimal with best efficiency. You'll see shortly that this same expression also tells how to convert integers and fractions from decimal to another base. For conversions "in" from another base, calculations move from the outer components N_{+3} and N_{-3} "in" toward the radix point. For conversions "out" to another base, you'll see shortly that calculations move from the radix point "out" to the outer components, N_{+3} and N_{-3}. Whenever you write this expression, remember that R is always the other base. It's never ten.

Now let's repeat our previous calculation of the decimal equivalent of a hexadecimal integer, using Expression 2 instead of Expression 1. With the same hexadecimal integer, A34C, in Expression 2, the evaluation looks like this:

```
((10 * 16 + 3) * 16 + 4) * 16 + 12 ⇒
(163 * 16 + 4) * 16 + 12 ⇒ 41804
```

Of course, we get the same result as before. But this time, the number of multiplications increases linearly with the number of terms, so there are only three multiplications instead of six. Using a typical hand calculator, we just enter the numbers and operators in the order in which they appear, and update the evaluation at each closing parenthesis.

Converting the fractional part of a fractional hexadecimal number is also easier with Expression 2 than with Expression 1. Suppose we have the hexadecimal fraction, 0.5B7. What is that in decimal? Because it is hexadecimal, R is 16. Because it does have a radix point, it is a fraction. Recognizing that hexadecimal B is decimal 11, N_{-3} is 7, N_{-2} is 11, and N_{-1} is 5. So with 0.5B7 in expression 2, the evaluation looks like this:

```
((7 / 16 + 11) / 16 + 5) / 16 ⇒
(11.4375 / 16 + 5) / 16 ⇒
5.71484375 / 16 ⇒ 0.357177734
```

Therefore, $0.5B7_{hexadecimal}$ is the same as $0.357177734_{decimal}$.

"Out" Conversion—from Decimal to Another Base

Expression 2 also makes it easy to remember how to perform conversion in the other direction—from a decimal number to the same number in another base. To see how this works, let's start with an integer or the integer part of a floating-point number. This is what is to the left of the radix point in Expression 2:

Integer Part of Expression 2: — Think of this as a dividend.

$$((N_{+3}*R + N_{+2})*R + N_{+1})*R + N_0$$

Think of this expression as the dividend in a division problem. If we divide this dividend by the base, R, the quotient is

$$(N_{+3}*R + N_{+2})*R + N_{+1}$$

And the remainder is N_0. This remainder is the least-significant integer component in the other base.

Now, consider this quotient to be a dividend and divide it by the base, R. This time, the remainder is N_{+1}. Do this again, and the remainder is N_{+2}. Continue until the quotient is zero. The remainders generated by this algorithm are the other base's integer components, from least significant to most significant.

If you're doing it by hand, you need to store intermediate results, and if you're like us, even if your hand calculator has storage, you'll want to use pencil and paper to keep track of your progress. Suppose you want to reverse our previous conversion that led to $41804_{decimal}$. The picture below illustrates what you might write on the paper as you proceed. The picture shows a sequence of divisions by the new base value, 16, starting at the bottom and moving up. The result is the reverse of the remainder sequence. When you finish, the converted number appears vertically on the right, read from top to bottom.

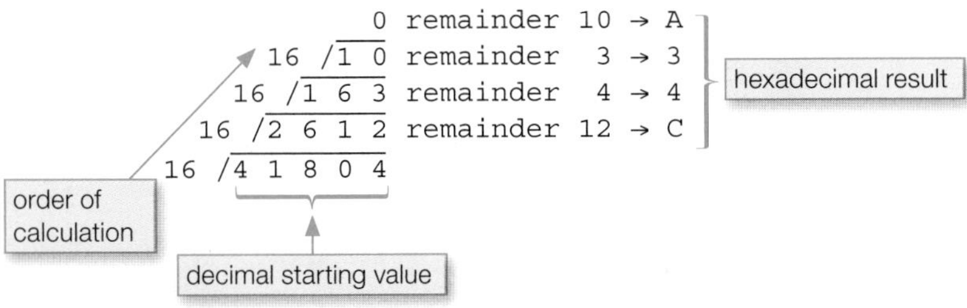

Now consider the fractional part of a floating-point number by looking at what's to the right of the radix point in Expression 2:

Fractional Part of Expression 2: — Think of this as a multiplicand.

$$((N_{-3}/R + N_{-2})/R + N_{-1})/R$$

Think of this expression as the multiplicand in a multiplication problem. Multiply this multiplicand by the base, R, to obtain the floating-point quotient:

$$(N_{-3}/R + N_{-2})/R + N_{-1}$$

The last term, N_{-1}, is the integer part of this floating-point quotient. It is the most significant component of the original fraction. Now look at the fractional part of this floating-point quotient:

$$(N_{-3}/R + N_{-2})/R$$

Consider this to be another multiplicand. Multiply it by the base, R, to obtain another fraction followed by the integer N_{-2}, which is the second most significant component of the original fraction. Continue until the fractional part equals zero or you have as much precision as you want.

To illustrate this process, we'll start with a rounded version of the decimal fraction we generated before, `0.357178`, and make the conversion in the other direction, from decimal to hexadecimal. With a simple hand calculator, this conversion is easier than the integer conversion. It doesn't require any storage—just use pencil and paper to write down the resulting components as they appear. The picture below shows how the calculation proceeds.

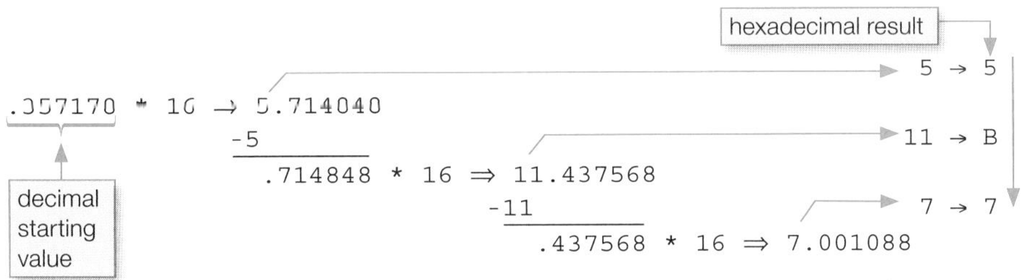

Index

SYMBOLS

A

B

E

F

H

I

N

O